Principal Functional Groups

Family Name	Example	Functional Group*	Text Reference
Amides	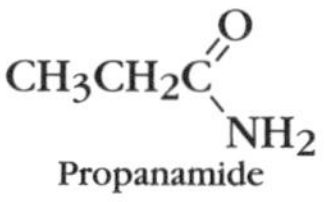 CH_3CH_2C with $=O$ and NH_2 (Propanamide)	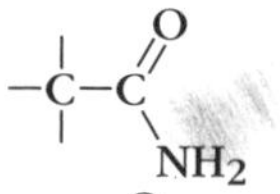 functional group structure	Chapter 16
N-Substituted Amides	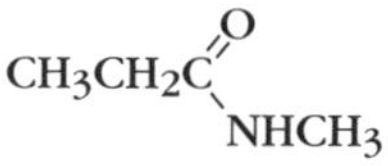 CH_3CH_2C with $=O$ and $NHCH_3$ (N-Methylpropanamide)	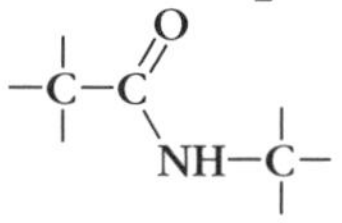 functional group structure	Chapter 16
N,N-Disubstituted Amides	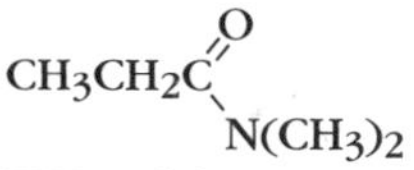 CH_3CH_2C with $=O$ and $N(CH_3)_2$ (N,N-Dimethylpropanamide)	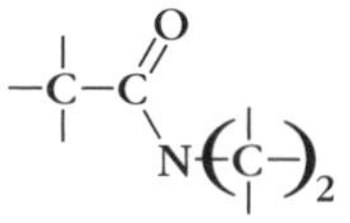 functional group structure	Chapter 16
Nitriles	$CH_3CH_2C \equiv N$ (Propanenitrile)	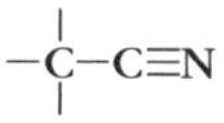 $-C-C \equiv N$	Chapter 16
Conjugated Polyenes	$CH_2 = CH - CH = CH_2$ (1,3-Butadiene)	 $C=C(C=C)_n$	Chapter 19
Arenes	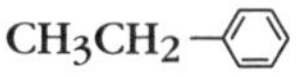 $CH_3CH_2 -$ (Ethylbenzene)	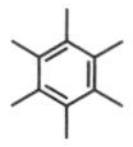 benzene ring structure	Chapters 20 and 21
1° Amines	$CH_3CH_2CH_2NH_2$ (1-Propanamine)	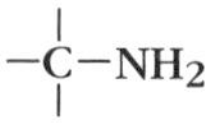 $-C-NH_2$	Chapter 22
2° Amines	$CH_3CH_2CH_2NHCH_3$ (N-Methyl 1-propanamine)	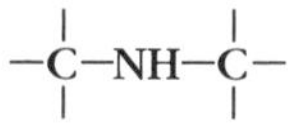 $-C-NH-C-$	Chapter 22
3° Amines	$CH_3CH_2CH_2N(CH_3)_2$ (N,N-Dimethyl 1-propanamine)	 $(-C-)_3N$	Chapter 22
Phenols	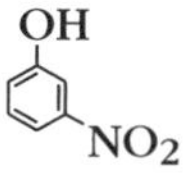 OH, NO_2 (m-Nitrophenol)	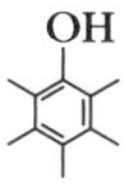 OH on benzene ring	Chapter 23
Thiols	$CH_3CH_2CH_2SH$ (1-Propanethiol)	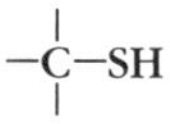 $-C-SH$	Section 11.19
Sulfides	$CH_3CH_2SCH_2CH_3$ (Diethyl sulfide)	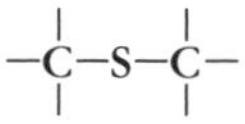 $-C-S-C-$	Section 13.15

* A bond line indicates that the carbon atom is bonded to carbon or hydrogen atoms in the rest of the molecule.

ORGANIC CHEMISTRY

ORGANIC CHEMISTRY

George H. Schmid

University of Toronto

with 630 illustrations
(including 230 spectra)

 Mosby

St. Louis Baltimore Boston Carlsbad Chicago Naples New York Philadelphia Portland
London Madrid Mexico City Singapore Sydney Tokyo Toronto Wiesbaden

Publisher: James M. Smith
Executive Editor: Lloyd W. Black
Developmental Editor: John S. Murdzek
Project Manager: John Rogers
Production Editor: Lavon Wirch Peters
Designer: Frank Loose
Design Coordinator and Cover Design: Renée Duenow
Cover Art: ArtScribe, Inc.
Art Developer: John S. Murdzek
Illustrations: Network Graphics, Kurt Griffin, ArtScribe, Inc.
Manufacturing Manager: Theresa Fuchs

Printed in the United States of America
Composition by Black Dot Graphics
Printing/binding by Rand McNally

Mosby–Year Book, Inc.
11830 Westline Industrial Drive
St. Louis, Missouri 63146

Library of Congress Cataloging in Publication Data

Schmid, George H.
   Organic chemistry / George H. Schmid.
     p.   cm.
   Includes index.
   ISBN 0-8016-7490-5
   1. Chemistry, Organic. I. Title.
QD251.2.S354 1995
547—dc20   95-24791
                              CIP

95 96 97 98 99 / 9 8 7 6 5 4 3 2 1

To Bernadette
For her patience and understanding

PREFACE

This text is intended for students majoring in the sciences. It is organized in the functional group framework with reaction mechanisms introduced at appropriate places. The purpose of writing this text is to bridge the gap between organic chemistry and biochemistry by teaching the basic mechanistic principles of organic chemistry and showing how these principles apply to certain biochemical reactions.

Before comparisons can be made between organic and biochemical reactions, however, certain principles of organic chemistry must be established. One of the most important is stereochemistry. Principles of conformational analysis are introduced in Chapter 4; chirality is introduced in Chapter 6. The introduction of these principles early in the text allows stereochemistry to be included in discussions of the reactions of many of the functional groups. Knowing both the details of the reaction and its stereochemistry facilitates the presentation of its mechanism.

In this text, the stereochemistry of each of the following reactions is discussed in considerable detail: electrophilic addition reactions of alkenes (Chapters 7 and 8) and alkynes (Chapter 9), reduction of aldehydes and ketones (Chapter 11), nucleophilic substitution and elimination reactions (Chapter 12), ring opening reactions of epoxides (Chapter 13), and nucleophilic addition reactions of carbonyl containing compounds (Chapters 14, 15, 16, and 17). Thus the stereochemistry of the reactions of the major functional groups is a recurring theme throughout the text.

The biological applications of organic mechanisms are not isolated from the text in special topic boxes but rather are presented in sections close to the discussion of the organic reaction mechanism. In this way comparisons between reactions in the laboratory and the analogous reactions in living systems are readily made. For example, the discussion of the mechanism of the formation of alcohols by reduction of aldehydes and ketones (Section 11.10) is followed closely by a discussion of the mechanism of the enzyme catalyzed reduction of carbonyl groups to form optically active alcohols (Section 11.12). Other examples include the reaction of chiral methyl as a biological example of the S_N2 mechanism (Section 12.3) and acetyl transfer reactions of acetyl coenzyme A (Section 16.10) as an example of the addition elimination mechanism (Section 16.6) of acyl transfer reactions in living systems.

The use of spectroscopy to determine the structures of organic compounds is started in Chapter 5 with the introduction of IR and broadband decoupled ^{13}C NMR spectroscopy. Limiting the introduction of spectroscopy to these two relatively simple techniques allows an early introduction to the methodology of determining the structure of simple organic compounds from their IR and

^{13}C NMR spectra. Exercises in Chapters 5, 7, 8, and 9 provide students with the opportunity to enhance their skill in the use of these two spectral techniques.

Once students have become familiar with the use of IR and ^{13}C NMR spectroscopy, ^{1}H NMR spectroscopy is introduced in Chapter 10. Additional exercises that require the use of IR, ^{13}C NMR, and ^{1}H NMR spectroscopy to determine the structure of organic compounds are placed at the ends of chapters throughout the text. In this way, the uses of spectroscopy in structure determination are reviewed with the study of each new functional group.

The early introduction of spectroscopy has certain pedagogical advantages. The introduction of IR spectroscopy in Chapter 5 reinforces the concept of functional groups first introduced in Chapter 2. In addition, students gain a great sense of accomplishment early in their study of organic chemistry when they can successfully determine the structure of an organic compound (even a relatively simple one) from its IR and ^{13}C NMR spectra.

The arrow symbolism that is used in organic chemistry to demonstrate electron movement is used extensively in this text. Many students, however, find such "electron pushing" artificial and divorced from any physical reality. To aid students in making the connection between the arrow symbolism and electron movement in the mechanisms of organic reactions, space-filling three-dimensional electron density contour drawings obtained by semiempirical computational methods have been included in the mechanisms of several reactions. The mechanism of the addition of HCl to 2,3-dimethyl-2-butene (Section 7.9), for example, is represented in the usual way (structural formula and curved arrows) and by artists' renditions of the electron density contour diagrams of the reactants, intermediates, transition states, and products obtained by the use of the Spartan computational program. Other examples are found in Chapters 12, 13, and 14.

Study Aids

The following study aids are provided to help students learn the material in the text.

Colored screens are used to highlight parts of molecules and to follow the course of many reactions.

Sample Problems provide worked-out solutions to typical exercises. For example, a number of Sample Problems in Chapters 5 and 10 show how to use the molecular formula of a compound and its spectral data to determine its structural formula.

Exercises are carefully placed throughout to reinforce the concepts discussed in the text immediately preceding the exercise.

Additional Exercises are included at the end of each chapter to help students evaluate their knowledge of the material presented in each chapter.

Summaries are placed at the end of each chapter to provide a quick and convenient overview of the important concepts in the chapter. Many chapters have **Reaction Summaries** that summarize and provide references to the important reactions discussed in the chapter. Where appropriate, chapters have **Mechanism Summaries** that summarize and provide references to the mechanisms discussed in the chapter.

A **Glossary** of key terms that defines the important terms is included at the end of the text.

An **Answers to Selected Exercises** section at the end of the text provides answers to selected Exercises. Detailed solutions to both Exercises and Additional Exercises are provided in the Study Guide and Solutions Manual.

Acknowledgements

It is a pleasure to thank the many people who assisted me in the creation of this text.

Jim Smith, Vice-President and Publisher at Mosby, whose enthusiasm, interest, and continued support in this project has never waned.

John Murdzek, Developmental Editor, who guided the long and difficult journey from manuscript to final text and who was always available with excellent advice on matters ranging from pedagogy to art design.

John Smith and the artists of Network Graphics, Kurt Griffin, and Carolyn Duffy and Greg Holt of ArtScribe, Inc., who transformed my crude drawings into the magnificent illustrations in the final text.

Errol Lewars (Trent University), who carried out the Spartan semiempirical calculations.

Katherine Aiken, Copy Editor and Proofreader, and Catherine Vale, Proofreader, who were dedicated to producing a perfect text.

John Rogers (Project Manager), Lavon Peters (Production Editor), Renée Duenow (Design Coordinator), and Theresa Fuchs (Manufacturing Manager), who put all the pieces together and produced this attractive text.

Thanks are also due to Helen Hudlin and Judy Hauck for their assistance in the early stages of this project and to Dan Mathers and Aviva Kats for technical assistance.

I wish to thank James L. Coke (University of North Carolina, Chapel Hill), Steven A. Hardinger (California State University, Fullerton), Joseph M. Prokipcak (University of Guelph), James Schreck (University of North Colorado), and Keith E. Taylor (University of Windsor), for their thoughtful and constructive reviews of the manuscript, which were particularly helpful. I am also grateful to Steve Hardinger for the excellent job of proofreading the galley pages.

I also want to thank the other reviewers who contributed valuable comments and suggestions to this text:

Margaret-Ann Armour
University of Alberta

Robert G. Carlson
University of Kansas

James Chapman
Rockhurst College

William P. Dailey
University of Pennsylvania

James A. Deyrup
University of Florida, Gainesville

Morris Fishman
New York University

Jeremiah P. Freeman
University of Notre Dame

Adrian V. George
University of Sydney

David Goldsmith
Emory University

Steven M. Graham
Oklahoma State University

Robert B. Grossman
University of Kentucky

Gamini U. Gunawardena
University of Utah

Ernie Harrison
Pennsylvania State University—York

David G. Hewitt
Monash University

Mark J. Kurth
University of California, Davis

Olivier R. Martin
Binghamton University

Eugene A. Mash
University of Arizona

Daniel H. O'Brien
Texas A&M University

Albert R. Matlin
Oberlin College

Louis D. Quinn
University of Massachusetts, Amherst

Michael G. McGrath
College of the Holy Cross

Roger W. Read
University of New South Wales

Nancy Mills
Trinity University

John P. Richard
State University of New York at Buffalo

Paul M. Nave
Arkansas State University

Carl H. Schiesser
University of Melbourne

Instructional Supplements

Study Guide and Solutions Manual by Stanislaw Skonieczny and George H. Schmid. This carefully prepared supplement contains answers to all in-text Exercises and end-of-chapter Additional Exercises and explains in detail how the answers are obtained. The Study Guide and Solutions Manual was written for all students but should be particularly invaluable for students having difficulty with problem solving. Also included in each chapter is a summary of the important concepts. The appendix contains the following three summaries: methods of preparing functional groups, reactions of functional groups, and uses of important reagents.

Molecules 3D! Molecular Model Building Software. Faster, more powerful, and easier to use than plastic model kits, this new software allows students to build and rotate in the *x-, y-,* and *z*-planes virtually any molecule (up to 400 atoms). With versions for Windows and Macintosh, Molecules 3D! lets students construct organic molecules, biomolecules, polymers, inorganics, and everything in between—with just the click of a mouse. Students can select from seven distinct molecular geometries. Bond lengths, angles, and torsions are easily determined. Powerful algorithms ensure correct conformations so students know their models are accurate. Finished molecules can be saved, printed, or exported to other programs.

ViewStudy™ CD-ROM. The ViewStudy™ CD-ROM is a unique, easy-to-use teaching and learning tool available in both Windows and Macintosh formats. Artwork from the text is contained on the disc, including each piece's full caption. Images are arranged by chapter, by specifically created topic (the concept-based Study Views), and by figure number indexed to the text. Images and captions can be quickly perused, the images can be enlarged for projection, and images and captions can be printed out for use as "study cards." Images can be arranged to create custom-designed slide shows. Images can also be exported for use in word processing programs and computerized testing systems. This innovative disc is free to qualified adopters of the text and is available for sale to students at a very low cost.

Transparency Acetates. Over 100 full-color transparency acetates have been chosen from the artwork in the text to enhance lectures and reinforce key concepts.

CONTENTS

DETAILED CONTENTS

3   Energy Changes in Chemical Reactions, 87

4   Conformations of Alkanes and Cycloalkanes, 128

5   Introduction to Spectroscopy, 171

6   Stereochemistry, 225

7 Alkenes: Structure, Acid Additions, and Preparation, 276

8 More Addition Reactions of Alkenes, 325

12 Nucleophilic Substitution and Elimination Reactions, 504

15 Carboxylic Acids, Their Salts, and Their Esters, 650

16 Acyl Trasfer Reactions: Interconversion of Carboxylic Acid Derivatives, 697

17 Enols and Enolate Anions, 752

18 Free Radical Reactions, 799

19   π Electron Delocalizaton in Acyclic Compounds and Intermediates, 838

20   Aromaticity: π Electron Delocalization in Cyclic Compounds, 885

21   Chemistry of Benzene and Its Derivatives, 921

22   Amines, 961

23   Halobenzenes, Phenols, and Quinones, 1007

24   Chemistry of Difunctional Compounds, 1046

25 Carbohydrates, 1092

ORGANIC CHEMISTRY

CHAPTER

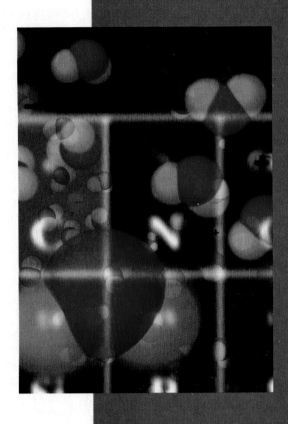

1

RETROSPECT and PRELUDE

1.1   Why Study Organic Chemistry?

"Why do I have to take organic chemistry?" is a question that has been asked by countless students over the years. A typical answer is that you need it to get into some specialized program such as medical or dental school. This rather practical response evades the real question, "Why do people who set entrance requirements to many professional schools consider organic chemistry so important?" The answer to this question is that they recognize that an understanding of organic chemistry is essential to an understanding of the chemistry of life. This answer raises other questions, however, such as "What is organic chemistry and how did the realization of its importance come about?"

Since the dawn of civilization people have been using materials obtained from nature. It was not until the 19th century, however, that scientists began the systematic identification, characterization, and classification of compounds obtained from living organisms. These compounds were classified as organic compounds because they were derived from living organisms. This was the beginning of organic chemistry. Compounds isolated from nonliving sources were classified as inorganic compounds.

Organic compounds have one thing in common: they all contain the element carbon. Chemists soon discovered that, contrary to their original assumption, organic compounds did not have to come from living organisms. They could be prepared in the laboratory instead and sometimes even from inorganic compounds. Over the years, therefore, the original distinction between compounds obtained from living and nonliving sources was lost. Despite this fact chemists have found it convenient to retain the name *organic* to designate carbon-containing compounds. Today we define *organic chemistry* as the *chemistry of carbon-containing compounds.*

Organic chemistry is used in the chemical industry to prepare many important compounds that were originally available only from nature. The pain reliever, aspirin, for example, was originally obtained as an extract from the bark of the willow tree so it was available in only small quantities. Now hundreds of tons of aspirin are produced industrially every year. Organic chemistry is used to prepare organic compounds *not* found in nature too, such as synthetic fibers, plastics, paints, blood substitutes, and insecticides.

Many organic compounds carry out important functions in living organisms. Examples include the large and complicated molecules of DNA that contain all our genetic information, the proteins that make up the muscles and skin of our bodies, the enzymes that catalyze the chemical reactions in our bodies, and most of the foods that provide us with the energy we need. Applying the principles of organic chemistry to explain how these compounds carry out their biological functions is a major area of research in organic chemistry.

Organic compounds also cause some serious environmental problems. Insecticides such as DDT, methoxychlor, and lindane, although initially useful in controlling disease-bearing insects, are no longer used for this purpose because of their harmful effects on other species. Chlorofluorocarbons (CFCs) used as propellents in aerosols are gradually being replaced because they persist in the atmosphere and threaten to destroy the stratospheric ozone layer that protects us from harmful ultraviolet radiation. It is important to learn how and why these compounds react in order to prevent a reoccurrence of these kinds of environmental problems.

These few examples illustrate the impact that organic compounds have on our lives and why it is important to study and learn the principles of organic chemistry. Before we begin our study of organic chemistry, however, let's review some basic principles of atomic structure and chemical bonding.

1.2 Atoms

Atoms are made up of a nucleus surrounded by one or more electrons. The nucleus contains one or more protons, each with a positive charge, and uncharged neutrons. The number of protons in an atom defines its **atomic number** while the total number of protons and neutrons defines its **mass number** or **nucleon number.** The overall charge on an atom is zero and each electron has a single negative charge, so the number of electrons in an atom is equal to the number of protons in its nucleus. The masses of protons and neutrons are about the same. The mass of a proton is about 1800 times the mass of an electron. Thus most of the mass of an atom is contained in its nucleus. Atomic nuclei are small compared to the sizes of atoms, however, so the electrons occupy most of the volume of an atom.

Particle	Mass
Electron	9.1094×10^{-31} kg
Proton	1.6726×10^{-27} kg
Neutron	1.6749×10^{-27} kg

Electrons and atomic nuclei are held together by the attractive force between them. This attraction is an electrical force that acts between charged particles and is described by **Coulomb's law:**

$$F = \frac{k\,(Q_1\,Q_2)}{r^2}$$

where Q_1 and Q_2 are the magnitudes of the charges, r is the distance between them, and k is a constant called the *Coulomb constant*. When two charges have opposite signs, the force is attractive. When the two charges have the same sign, the force is repulsive.

The arrangement of the elements into a periodic table is based on the periodic variations of their properties. This periodic behavior is due to differences in the number of outermost or **valence electrons** of the atoms of different elements. The number of valence electrons in an atom is responsible for its ability to combine with other atoms. This combining power for other atoms is called the **valence** of an atom.

We can conveniently indicate the number of valence electrons in each atom by **Lewis electron-dot symbols.** These symbols, introduced by the American chemist Gilbert Lewis, represent valence electrons by dots around the symbol of the element. Table 1.1 lists the number of valence electrons, the valence, and the Lewis electron-dot symbol for the elements of the second period. The valence of each of the elements of Groups I to IV of the periodic table is the number of valence electrons. For Groups V to VIII, however, the valence is eight minus the number of valence electrons.

While each element is characterized by the number of protons in its nucleus, the number of neutrons may vary. Atoms whose nuclei contain the same number of protons but different numbers of neutrons are called **isotopes.** Some of the isotopes that we will encounter in this book are deuterium, ^{13}C, ^{14}C, and ^{18}O.

Table 1.1	Valence Electrons, Valences, and Lewis Electron-Dot Symbols							
Group	I	II	III	IV	V	VI	VII	VIII
Number of valence electrons	1	2	3	4	5	6	7	8
Valence	1	2	3	4	3	2	1	0
Lewis electron-dot symbol	Li	·Be	·B·	·C·	·N·	:O·	:F:	:Ne:

Deuterium is the isotope of hydrogen that contains a neutron in addition to a proton in its nucleus and consequently has a mass number of 2 (written ^{2}H or D). ^{13}C is a nonradioactive isotope of carbon that contains seven neutrons in its nucleus and has a mass number of 13. ^{14}C is a radioactive isotope of carbon that has eight neutrons in its nucleus. ^{18}O is a nonradioactive isotope of oxygen that contains ten neutrons in its nucleus.

While Lewis electron-dot symbols indicate the number of valence electrons in an atom, they don't tell us anything about where the electrons are located in space about the nucleus. For this information we must turn to atomic orbitals.

1.3 Hydrogen Atomic Orbitals

Electrons have properties of both particles and waves. Describing electrons in atoms as particles led in 1913 to the Bohr model of the atom. This model has since been rejected because it correctly describes only the properties of the hydrogen atom and not the properties of any other atom. Describing electrons in atoms as waves is the basis of the modern theory of the electron structure of atoms. Before we describe this theory, however, let's briefly review the properties of waves.

There are two kinds of waves, standing waves and traveling waves. The waves that carry sound from a loudspeaker or the waves that form the wake of a boat are examples of traveling waves. Standing waves, on the other hand, vibrate in a fixed place because they are secured at both ends. Plucking a guitar string causes it to vibrate and form standing waves, as shown in Figure 1.1. Notice that only certain vibrations are allowed because only integral or half-integral wavelengths are possible with a string that is attached at both ends.

Upward displacements of the waves in Figure 1.1 are assigned plus signs, while downward displacements are given minus signs. Places where the displacements, or amplitudes, of the waves are zero are called **nodes.** The lowest energy vibration of a guitar string, the fundamental vibration, has no nodes. The next-

Figure 1.1
Standing waves caused by the plucking of a guitar string. Plus and minus signs show the relative phases of the wave and a dot represents a node (a point of zero displacement). The points of attachment of the string do not define a node.

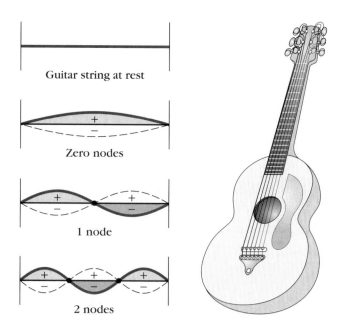

Guitar string at rest

Zero nodes

1 node

2 nodes

highest energy vibration, called the first overtone, has a single node. The next-highest energy vibration, called the second overtone, has two nodes and so on. The higher the energy of the wave, the greater the number of nodes.

The shapes of the waves illustrated in Figure 1.1 can be described by mathematical equations called **wave functions.** Wave functions are not limited to describing the shape of the vibrations of mechanical systems like strings, drums, or the air column in an organ pipe. They can also be used to describe the motion (vibration) of an electron about the nucleus of an atom. The mathematical equation that describes the motion of the electron in a hydrogen atom as a wave was first formulated in 1926 by the Austrian scientist Erwin Schrödinger and is called the **Schrödinger wave equation.** The mathematical solutions to the Schrödinger wave equation are electron wave functions that describe the amplitude of the wave as a function of the three coordinates needed to describe motion in three dimensions. These electron wave functions, given the symbol ϕ, are called **atomic orbitals.**

Just as there are only certain vibrations possible for a guitar string, there are only certain values of the electron wave functions of a hydrogen atom. We will concern ourselves only with the atomic orbitals that are the most important in organic chemistry, namely the 1s, 2s, and 2p atomic orbitals. The 1s atomic orbital has the lowest possible energy, while the 2s and 2p atomic orbitals are of slightly higher energy.

All s orbitals are spherical in shape. Figure 1.2 shows two ways in which a 1s atomic orbital can be depicted. The graph of ϕ vs r in Figure 1.2, A, shows how ϕ varies with distance in any direction from the nucleus while Figure 1.2, B, shows how ϕ varies in three-dimensional space. In order to help you visualize the orbitals, we have chosen a small but constant value of ϕ to represent the size and shape of atomic orbitals. This boundary surface is usually chosen so that it encloses 90% to 95% of the value of ϕ. The boundary surface that represents a 1s atomic orbital is a sphere. That is, the 1s orbital, which has no nodes, is spherical. This means that ϕ varies in exactly the same way in any direction from the nucleus. Notice in Figure 1.2, A, moreover, that the electron density is greatest right at the nucleus and falls off as the distance from the nucleus increases. The 2s orbital is also spherical but has one node, as shown in Figure 1.3.

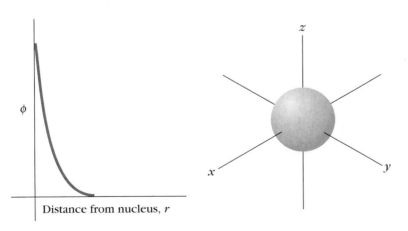

A

B

Figure 1.2
The 1s atomic orbital of the hydrogen atom. **A,** The variation of ϕ with distance (r) in any direction from the nucleus. **B,** Spherical shape of the 1s atomic orbital.

ϕ

Distance from nucleus, r

Figure 1.3
The 2s atomic orbital of the hydrogen atom. **A,** The variation of ϕ with distance (r) in any direction from the nucleus. **B,** Boundary surface of the 2s orbital. Notice that the interior node is not visible in this representation of the 2s atomic orbital.

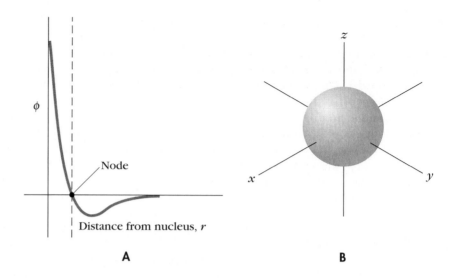

A hydrogen atom has three $2p$ orbitals whose shapes are different from s orbitals. The value of ϕ for each of the $2p$ orbitals is a maximum along one of the three cartesian coordinates. Thus the value of ϕ is a maximum along the x-axis for the orbital we designate as the $2p_x$ orbital. Similarly the values of ϕ for the $2p_y$ and $2p_z$ orbitals are at a maximum along the y- and z-axes, respectively. The boundary surfaces for the $2p_x$, $2p_y$, and $2p_z$ atomic orbitals are shown in Figure 1.4.

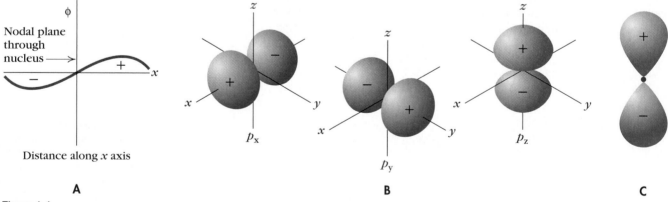

Figure 1.4
The 2p atomic orbitals of the hydrogen atom. **A,** The variation of ϕ along the x-axis for the $2p_x$ orbital. **B,** Boundary surface of the $2p_x$, $2p_y$, and $2p_z$ atomic orbitals resembles a dumbbell in shape. **C,** How $2p$ atomic orbitals will be depicted in this text.

While the Schrödinger wave equation for the hydrogen atom can be solved to give exact mathematical wave functions, it cannot be solved exactly for other atoms or molecules. However, approximate solutions are available that are sufficiently accurate to use them to determine the electron configurations of other atoms.

1.4 Electron Configurations of Atoms

The **electron configuration** of an atom is a description of how its electrons are distributed among its various orbitals. The electron configurations of the elements of the periodic table are obtained by using the Aufbau principle. We build up the most stable or ground state electron configuration of an atom by using the following three principles:

> ■ Aufbau *is a German word meaning "building up."*

1. **Electrons occupy the lowest energy orbitals available.** The energies of the orbitals of the first 36 elements increase in the following order:

$$1s < 2s < 2p < 3s < 3p < 4s < 3d$$

$$\boxed{\text{INCREASING ENERGY}}\!\!\!>$$

2. **The Pauli exclusion principle.** Only two electrons may exist in the same orbital, and they must have opposite spins.
3. **Hund's rule.** When orbitals of identical energy are available, electrons first occupy these orbitals singly.

The most stable electron configuration of an element (ground state configuration) is obtained by adding electrons one at a time to atomic orbitals, starting with the one of lowest energy (the $1s$ orbital). Thus the electron configuration of the ground state of the hydrogen atom is described by a single electron in its $1s$ orbital. The electron configuration of the helium atom contains two electrons of opposite spin in its $1s$ orbital, while the electron configuration of lithium contains two electrons of opposite spin in the $1s$ orbital and a third electron in the $2s$ orbital. The electron configurations of the elements of the first and second periods of the periodic table are shown in Table 1.2. Notice that the third and fourth valence electrons of the carbon atom occupy separate $2p$ orbitals because of Hund's rule. As a result, two different $2p$ orbitals contain one electron each in the

Table 1.2	Electron Configurations of the Elements of the First and Second Periods of the Periodic Table		
Element	Electron Configuration	Number of Valence Electrons	Valence
H	$1s^1$	1	1
He	$1s^2$	2	0
Li	$1s^2\,2s^1$	1	1
Be	$1s^2\,2s^2$	2	2
B	$1s^2\,2s^2\,2p_x^{1*}$	3	3
C	$1s^2\,2s^2\,2p_x^1\,2p_y^1$	4	4
N	$1s^2\,2s^2\,2p_x^1\,2p_y^1\,2p_z^1$	5	3
O	$1s^2\,2s^2\,2p_x^2\,2p_y^1\,2p_z^1$	6	2
F	$1s^2\,2s^2\,2p_x^2\,2p_y^2\,2p_z^1$	7	1
Ne	$1s^2\,2s^2\,2p_x^2\,2p_y^2\,2p_z^2$	8	0

*The $2p_x$, $2p_y$, and $2p_z$ orbitals are of equal energy. We have arbitrarily placed the first electron in the $2p_x$ orbital. It could have been placed as well in either the $2p_y$ or the $2p_z$ orbital.

electron configuration of carbon. For nitrogen, one electron is located in each of the three $2p$ orbitals. For oxygen, however, there are four electrons for three $2p$ orbitals, so two of the electrons must be paired.

Atoms combine in definite ways to form molecules. One of the aims of chemistry is to use atomic orbitals to construct the bonds in molecules. Before we can learn how to do this we must know which atoms are bonded together in a molecule and also the shape of the molecule. Lewis structures are a convenient shorthand that provides some of this information.

1.5   Lewis Structures

Gilbert N. Lewis proposed a method of representing the structure of compounds in 1915 that is still useful today. The representations are called **Lewis structures.** They are based on Lewis' theory of bonding: an atom will form one or more bonds with other atoms by sharing electrons until it has an electron configuration that resembles that of the nearest noble gas.

We write Lewis structures of molecules by combining the Lewis electron-dot symbols of the atoms making up the molecule in such a way that all the atoms achieve a noble gas electron configuration by sharing electrons with their neighboring atoms. The best way to illustrate how this is done is by writing the Lewis structure of a few molecules. We begin with the hydrogen molecule because it is a simple example. A hydrogen atom has a single valence electron. If we join two hydrogen atoms together, each hydrogen atom attains a helium electron configuration by sharing the two electrons. This forms the single bond in a hydrogen molecule.

$$H\cdot \ + \ \cdot H \longrightarrow H:H$$

Next, consider methane (CH_4). Its Lewis structure is obtained by placing the hydrogen atoms around the carbon atom so that the carbon atom has eight electrons, an octet. Four of the eight electrons come from carbon and the other four come from the four hydrogen atoms, one from each hydrogen. The eight electrons around carbon give it an electron configuration like neon. The two electrons shared between carbon and hydrogen give each hydrogen atom an electron configuration like helium.

While Lewis structures can be written with two dots to represent a pair of bonding electrons, it is more convenient to write a structure in which they are replaced by a dash (—). Both of the following are correct Lewis structures of methane.

$$
\begin{array}{ccc}
& H & & H \\
& \ddot{} & & | \\
H : & C : H & \text{or} & H - C - H \\
& \ddot{} & & | \\
& H & & H
\end{array}
$$

Methane

The structure of ethane (C_2H_6) is slightly more complicated than methane. We write its structure by distributing its fourteen valence electrons so that each carbon atom shares eight electrons and each hydrogen atom shares two electrons.

$$
\begin{array}{ccc}
H \ \ H & & H \ \ H \\
\ddot{} \ \ \ddot{} & & | \ \ \ | \\
H : \ddot{C} : \ddot{C} : H & \text{or} & H - C - C - H \\
\ddot{} \ \ \ddot{} & & | \ \ \ | \\
H \ \ H & & H \ \ H
\end{array}
$$

Ethane

So far we have written the structures of compounds that have just one pair of electrons between any two atoms. The sharing of a single pair of electrons between two atoms forms a **single bond.** Other molecules are known that have adjacent atoms that share two or three pairs of electrons. Sharing two pairs of electrons forms a **double bond,** while sharing three pairs of electrons forms a **triple bond.**

Ethylene (C_2H_4) is an example of a compound that contains a double bond. The only way its Lewis structure can be written so that both carbon atoms have an octet is by allowing them to share two pairs of electrons. The two pairs of electrons of a double bond are usually represented by two dashes ($=$).

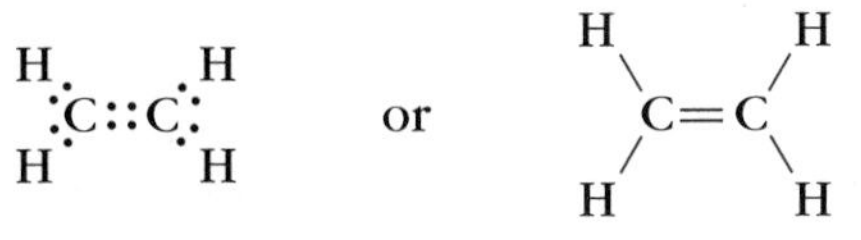

Ethylene

Acetylene (C_2H_2) is an example of a compound that contains a triple bond. The two carbon atoms must share three pairs of electrons to give each of them an octet. A triple bond is usually represented by three dashes ($\equiv$).

$$H\!:\!C\!:\!:\!:\!C\!:\!H \quad \text{or} \quad H-C\equiv C-H$$

Acetylene

Many compounds have structures that contain nonbonding or lone pairs of valence electrons. These are electrons that are not shared between atoms. Oxygen atoms, nitrogen atoms, and the halogen atoms (F, Cl, Br, and I) have unshared pairs of electrons in most of their compounds. The following are examples of compounds that have at least one lone pair of electrons on one of their atoms.

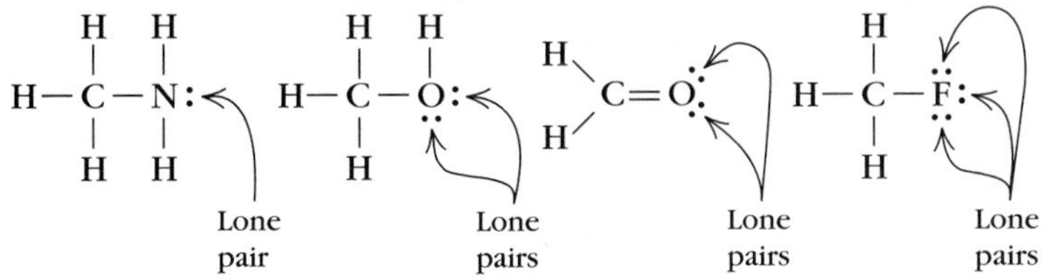

A single lone pair of electrons is located on the nitrogen atom of methylamine, two lone pairs on the oxygen atoms of methanol and formaldehyde, and three lone pairs on the fluorine atom of fluoromethane.

Sample Problem 1.1

What is the Lewis structure of carbon dioxide (CO_2)?

Solution:

Carbon has the Lewis symbol $:\!C\!:$, so it needs four more electrons to complete its octet. The Lewis symbol for oxygen is $\ddot{O}\!:$, so it needs two electrons to complete its octet. If the carbon atom shares its electrons with two electrons from each oxygen atom, then all three atoms will attain octets. This forms two carbon-oxygen double bonds as follows:

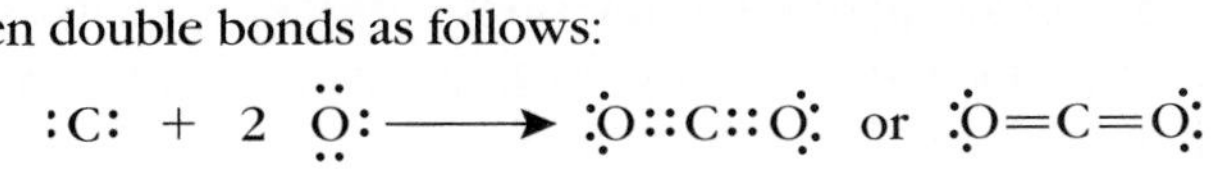

$$:\!C\!: \; + \; 2 \; \ddot{\underset{..}{O}}\!: \longrightarrow \; \ddot{\underset{..}{O}}\!:\!:\!C\!:\!:\!\ddot{\underset{..}{O}} \; \text{ or } \; \ddot{\underset{..}{O}}\!=\!C\!=\!\ddot{\underset{..}{O}}$$

| Exercise 1.1 | Write Lewis structures for each of the following compounds. Be sure to include all lone pairs of electrons. |

(a) Hydrogen fluoride, HF
(b) Iodine monochloride, ICl
(c) Water, H_2O
(d) Chloromethane, CH_3Cl
(e) Hydrogen cyanide, HCN

(f) Hydrazine, H_2NNH_2
(g) Acetonitrile, CH_3CN
(h) Tetrafluoroethylene, C_2F_4
(i) Formaldimine, H_2CNH

| Exercise 1.2 | Why is the following an incorrect Lewis structure for CO_2? |

$$:\overset{..}{\underset{..}{O}}:C:\overset{..}{\underset{..}{O}}:$$

Sometimes lone pairs of electrons are involved in bond formation. When this happens, writing Lewis structures becomes a bit more complicated.

1.6   Formal Charges

The structures of most organic molecules can be accurately represented by Lewis structures. Sometimes, however, application of the octet rule requires that we place **formal charges** on specific atoms of the molecule. The formal charge of an atom in a molecule or ion is calculated by arbitrarily assigning all lone pair electrons and one half of its bonding electrons to that atom. If the number of electrons assigned to that atom in this way is greater than the number of valence electrons of the same neutral isolated atom, the atom is given a negative formal charge. If the number is less, the atom is given a positive formal charge.

We can illustrate this idea by calculating the formal charges on the carbon and oxygen atoms that make up carbon monoxide, CO. First we need to write a Lewis structure that correctly places an octet of electrons around both the carbon and oxygen atoms.

$$:O:::C:$$

Next, we assign one electron of each bonding pair to each atom. Thus the oxygen atom in CO is assigned a total of five valence electrons (half of the six electrons of the triple bond plus its two lone pair electrons). An isolated oxygen atom has six valence electrons. Thus the oxygen in the structure $:O:::C:$ has formally lost an electron, so it is assigned a formal charge of $+1$. Similarly, the carbon atom is assigned five valence electrons (three from the triple bond plus the two lone pair electrons). An isolated carbon atom has four valence electrons. Thus the carbon atom in the structure has formally gained an electron, so it is assigned a -1 formal charge. To indicate the formal charges, we write the Lewis structure of CO as follows:

$$:\overset{+}{O}:::\overset{-}{C}:$$

Notice that while the carbon and oxygen atoms have formal charges, CO is a neutral molecule because the sum of the formal charges is zero. This is a general rule: the total charge on a molecule or ion is the sum of its formal charges.

We can summarize this method of calculating formal charges by means of the following equation:

$$\begin{bmatrix} \text{Formal charge} \\ \text{on an atom in} \\ \text{a Lewis structure} \end{bmatrix} = \begin{bmatrix} \text{Number of valence} \\ \text{electrons in the} \\ \text{isolated atom} \end{bmatrix} - \begin{bmatrix} \text{Number of valence} \\ \text{electrons assigned in} \\ \text{the Lewis structure} \end{bmatrix}$$

It should be noted that we have counted only the valence electrons when determining formal charges. It is unnecessary to count the inner electrons because they are not involved in bond formation, so neglecting them does not affect the calculation of formal charge on an atom. For example, the total number of electrons about the oxygen atom in CO is seven: two inner $1s$ electrons and five valence electrons. An isolated oxygen atom has a total of eight electrons. Thus the oxygen atom in CO has one fewer electron than an isolated oxygen atom, which means it is assigned a formal charge of $+1$. The carbon atom of CO also has a total of seven electrons. An isolated carbon atom has a total of only six electrons, however, so the carbon atom in CO is assigned a formal charge of -1. Thus the results are identical whether or not the inner electrons are counted. Because it is more convenient, formal charges are usually calculated by considering only the valence electrons of the Lewis structure.

Sample Problem 1.2 provides a few more examples of using this method to determine formal charges on the atoms in Lewis structures.

| **Sample Problem 1.2** | What are the formal charges on nitrogen, oxygen, and carbon in ammonia (NH_3), hydronium ion (H_3O^+), and methyl cation (CH_3^+), respectively? |

Solution:

Step 1 Begin by writing the Lewis structure of ammonia.

$$\begin{array}{c} H \\ | \\ H-\overset{\displaystyle}{\underset{\displaystyle \cdot\cdot}{N}}-H \end{array}$$

Step 2 Count the number of valence electrons assigned to the nitrogen atom in the Lewis structure of NH_3.

Number of valence electrons from N—H bonds $= 3$
Number of valence electrons from lone pairs $= 2$
Total $= 5$

Step 3 Count the number of valence electrons in an isolated nitrogen atom. The number is 5.

Step 4 Calculate the formal charge on nitrogen:

$$\text{Formal charge} = \begin{bmatrix} \text{Number of valence} \\ \text{electrons in the} \\ \text{isolated atom} \end{bmatrix} - \begin{bmatrix} \text{Number of valence} \\ \text{electrons assigned in} \\ \text{the Lewis structure} \end{bmatrix}$$

Formal charge $= 5 - 5 = 0$

For the hydronium ion (H_3O^+)

Step 1 $$\begin{bmatrix} \begin{array}{c} H \\ | \\ H-\overset{\displaystyle}{\underset{\displaystyle \cdot\cdot}{O}}-H \end{array} \end{bmatrix}^+$$

Step 2 Number of valence electrons on the oxygen atom in the Lewis structure:

Number from O—H bonds $= 3$
Number from lone pairs $= 2$
Total $= 5$

Step 3 Number of valence electrons on an isolated oxygen atom is 6.

Step 4 Formal charge = 6 - 5 = +1

For the methyl cation (CH₃⁺)

Step 1

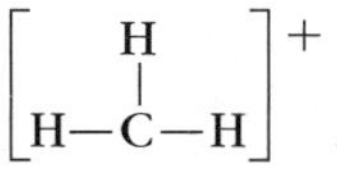

$$\left[\begin{array}{c} H \\ | \\ H-C-H \end{array} \right]^{+}$$

Step 2 Number of valence electrons on the carbon atom in the Lewis structure:

$$\begin{aligned} \text{Number from C—H bonds} &= 3 \\ \text{Number from lone pairs} &= 0 \\ \text{Total} &= 3 \end{aligned}$$

Step 3 Number of valence electrons on an isolated carbon atom is 4.

Step 4 Formal charge = 4 - 3 = +1

Formal charges have no physical significance. The assumption that bonding pairs of electrons are shared equally is incorrect, as we will learn in Section 1.13, so that formal charges do not represent actual charges.

| **Exercise 1.3** | Calculate the formal charges on each atom and the net charge on each of the following ions or molecules. |

(a) Nitric acid

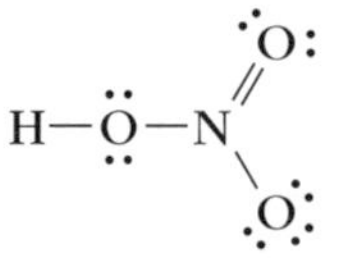

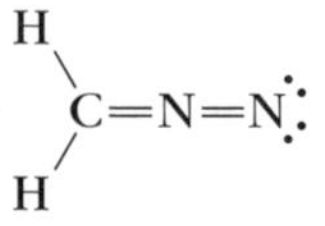

(b) Methoxide ion

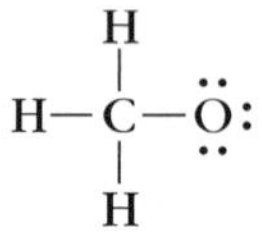

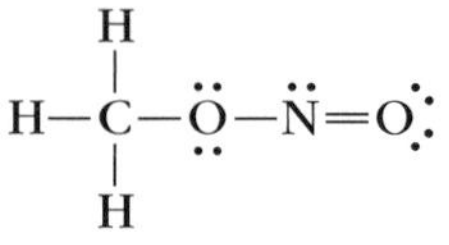

(c) Borohydride ion

(d) Diazomethane

(e) Methyl nitrite

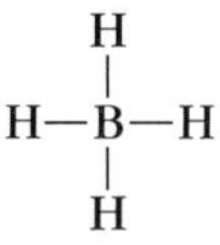

| **Exercise 1.4** | Thionyl chloride, a reagent that we will encounter later in this text, has the structure shown below. Verify by calculation that thionyl chloride must have formal charges on sulfur and oxygen. |

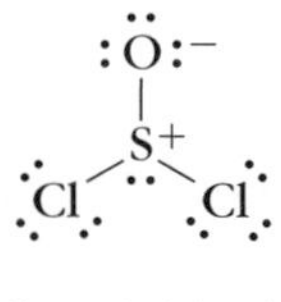

Thionyl chloride

Lewis structures are good representations of how atoms are joined together in a molecule. They do not show the three-dimensional arrangement of the atoms in molecules, however, nor do they explain why atoms combine to form molecules. In the next section, we learn how to use Lewis structures to predict the three-dimensional shapes of molecules.

1.7 Predicting Molecular Structure by Minimizing Electron-Pair Repulsions

The three-dimensional arrangement of atoms in a molecule, called its **molecular structure** or **molecular geometry,** plays an important role in determining the molecule's chemical properties. Many accurate experimental methods are now available for determining molecular structure. These methods are useful when precise information about molecular structure is required. These methods can be tedious or time-consuming to perform, however, so it is useful to have a method of quickly predicting the approximate geometry of a molecule. A simple model that allows us to do this is called the **valence-shell electron-pair repulsion (VSEPR) model.** The main postulate of VSEPR is that the positions of atoms or groups about a central atom are determined principally by minimizing electron-pair repulsion. Thus bonding and nonbonding pairs of electrons around a given atom will be located as far apart as possible.

To learn how this model works, let's consider the molecule BeH_2, which has the following Lewis structure:

$$H—Be—H$$

There are two pairs of electrons about the central beryllium atom. Placing the two bonding electron pairs on opposite sides of the beryllium atom at 180° from each other allows them to be as far away from each other as possible. This linear arrangement provides the maximum possible separation of the two electron pairs and is in accord with the experimentally determined geometry for BeH_2.

$$H—Be—H$$
$$\overset{\nwarrow\quad\nearrow}{180°}$$

Next, consider BF_3, in which a central boron atom is bonded to three fluorine atoms. The Lewis structure of BF_3 is the following:

$$:\ddot{F}:$$
$$|$$
$$:\ddot{F}—B—\ddot{F}:$$

Repulsion between three electron pairs can be minimized by placing them in a plane separated by angles of 120°:

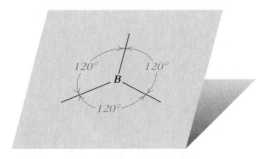

This geometry, which is called trigonal planar, places the electron pair of each B—F bond as far from the other two B—F bonds as possible. Thus the molecular structure of BF_3 is predicted to be the following:

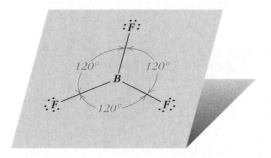

The experimentally determined geometry of BF_3 is trigonal planar.

Now, consider methane, in which a central carbon atom is bonded to four hydrogen atoms:

$$\begin{array}{c} H \\ | \\ H-C-H \\ | \\ H \end{array}$$

Repulsion between four pairs of electrons is minimized by adopting a tetrahedral arrangement. We can visualize this arrangement by placing the carbon atom at the center of a tetrahedron and pointing each of its four pairs of electrons toward a corner of the tetrahedron. The bond angles around the carbon atom are 109.5° in this arrangement.

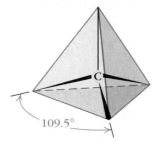

■ *A regular tetrahedron has four identical faces in the shape of an equilateral triangle and four identical corners.*

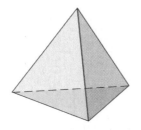

In methane each of the four electron pairs is shared between the carbon atom and a hydrogen atom. Placing each hydrogen atom at a corner of the tetrahedron gives a structure of methane that is in agreement with its actual structure.

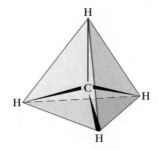

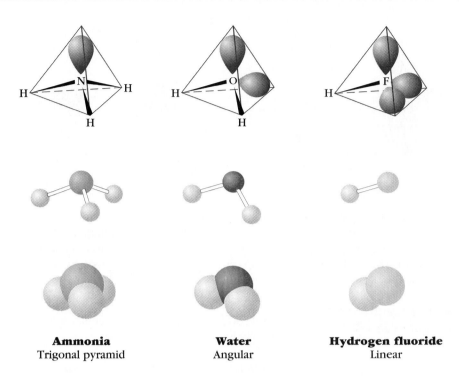

Figure 1.5
Structures of NH_3, H_2O, and HF. In each of these molecules there is a tetrahedral arrangement of the four valence electron pairs around the central atom. Their structures, however, are described only by the location of their atoms. Thus the structure of ammonia is a trigonal pyramid with its three hydrogen atoms at the base and the nitrogen atom at the apex. Water is a bent molecule, and HF has to be linear because it is a diatomic molecule.

Ammonia
Trigonal pyramid

Water
Angular

Hydrogen fluoride
Linear

Four electron pairs about a central atom always adopt a tetrahedral arrangement. Whether a molecule is itself tetrahedral depends on whether all of its four electron pairs form bonds. Ammonia (NH_3), water (H_2O), and hydrogen fluoride (HF), for example, all have four valence electron pairs around their central atoms, but these four electron pairs are bonding *and* nonbonding pairs. While the four electron pairs have a tetrahedral arrangement, the structures of these molecules differ, as shown in Figure 1.5.

How does the VSEPR model account for the geometry about multiple bonds? The electron pairs of a double or triple bond must be in the space between the nuclei of the two atoms in order to form a multiple bond. In other words, the electron pairs in a multiple bond act as one region of electron density. Because of this, the VSEPR model counts a multiple bond as one effective pair of electrons. Consider HCN, whose Lewis structure is the following:

$$H-C\equiv N:$$

The triple bond and the electron pair of the single bond of the central carbon atom are each regarded in the model as a pair of electrons. As a result, they adopt a linear arrangement in order to be as far away from each other as possible. In doing so, they make HCN a linear molecule.

How do we use the VSEPR model to predict the structure of a molecule like ethane, which has two carbon atoms? The answer is to count the number of valence electron pairs around each carbon atom. Because there are four pairs of valence electrons around each carbon atom, ethane adopts a tetrahedral arrangement about each carbon atom, as shown in Figure 1.6.

In summary, the VSEPR model correctly predicts the molecular structures of most molecules formed by nonmetallic elements. It is simple to apply if you adhere to the following rules:

1. Write the Lewis structure of the molecule.

Figure 1.6
The molecular structure of ethane as determined by VSEPR model. The geometry around each carbon atom is tetrahedral.

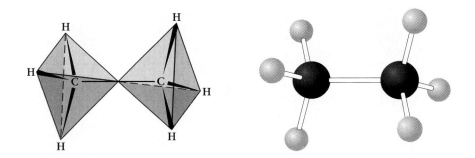

2. Count the valence electron pairs around a central atom. In counting valence electron pairs, each multiple bond counts as a single pair.

3. Two valence electron pairs about the central atom lead to a linear arrangement; three valence electron pairs lead to a trigonal planar arrangement; and four valence electron pairs lead to a tetrahedral arrangement.

Exercise 1.5	Use the VSEPR model to predict the molecular structure of each of the following compounds.

    (a) CCl_4 **(c)** NH_4^+ **(e)** OF_2 **(g)** C_2H_4
    (b) $BeCl_2$ **(d)** H_3O^+ **(f)** CH_3OH **(h)** C_2H_2

Exercise 1.6	Estimate the bond angles around both carbon atoms and around the oxygen atom of dimethyl ether (CH_3OCH_3).

Lewis structures can be used together with the VSEPR model to conveniently predict molecular structures. Lewis structures do not adequately explain why bonds are formed, however.

1.8   Bond Formation

■ *The units for energy, kilo-calories/mole and kilo-joules/mole, are abbreviated kcal/mol and kJ/mol, respectively. The conversion factor between kcal and kJ is 4.18 kJ/kcal. Energy is given in kcal/mol in this text followed by its equivalent in kJ/mol in parentheses—e.g., 104 kcal/mol (435 kJ/mol).*

Chemical bonds form because the resulting molecules are more stable than their isolated atoms. For example, the combination of two hydrogen atoms to form a hydrogen molecule releases 104 kilocalories/mole (435 kilojoules/mole) of energy. In other words, a hydrogen molecule is more stable than two isolated hydrogen atoms by 104 kcal/mol (435 kJ/mol).

$$H\cdot + H\cdot \longrightarrow H_2 + 104 \text{ kcal/mol (435 kJ/mol)}$$

The same Coulombic forces that are responsible for the stability of atoms are also responsible for the formation of bonds between two atoms. In the case of bond formation, both attractive and repulsive Coulombic forces are important. The energy diagram in Figure 1.7 illustrates the importance of these opposing Coulombic forces in formation of a bond between two hydrogen atoms.

The energy of the two hydrogen atoms in Figure 1.7 is plotted as a function of the distance between them, *r;* this is the internuclear distance. The energy of the two atoms at an infinite distance apart is defined as zero because there is no inter-

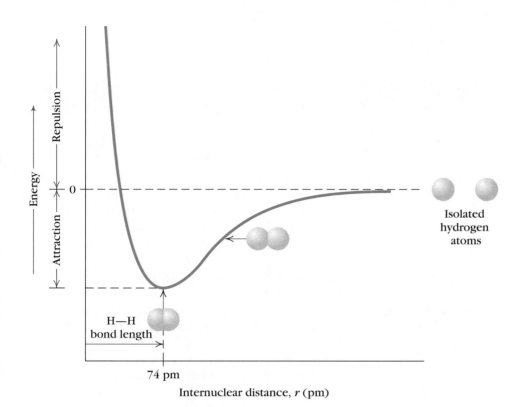

Figure 1.7
Energy profile for two H atoms as a function of their internuclear distance r. As the atoms approach, their energy decreases until a minimum is reached at 74 pm, the bond length of H_2.

action between them. As the atoms move closer together, the energy of the two hydrogen atoms decreases because each negatively charged electron feels the attraction of both positively charged nuclei. The energy continues to decrease as the atoms move closer together until it reaches a minimum value, which corresponds to the most stable configuration of the two nuclei and the two electrons. The equilibrium internuclear distance at this minimum value of energy is called the **bond length.** The difference in energy between the minimum value and zero, the energy of two isolated hydrogen atoms, is a measure of the bond strength. If the atoms are pushed any closer together than their equilibrium bond length, there is a rapid rise in energy because of the strong repulsion between the positively charged nuclei.

The energy diagram in Figure 1.7 demonstrates that two atoms form a bond to attain a state of lowest possible energy. For two hydrogen atoms, the internuclear distance where energy is a minimum is 74 picometers (pm), the bond length of the H_2 molecule.

The tendency of systems in nature to achieve the lowest possible energy explains why atoms join together to form chemical bonds in molecules, but it does not describe the behavior of electrons in chemical bonds. For this, we need to consider modern theories of chemical bonding.

■ *1 picometer = 1 pm = 10^{-12} meter. 100 picometers = 1 angström = 1 Å.*

1.9   Chemical Bonds: The Localized Valence Bond Orbital Model

One modern theory of chemical bonding is the localized valence bond orbital model. This model considers only the outer or valence electrons of the atoms

Figure 1.8
Formation of the bond in the hydrogen molecule by the overlap of the 1s atomic orbitals of each hydrogen atom.

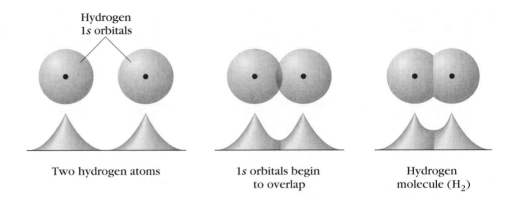

Hydrogen
1s orbitals

Two hydrogen atoms   1s orbitals begin   Hydrogen
to overlap   molecule (H_2)

involved and visualizes molecules as being held together by bonds localized between neighboring atoms. These localized orbitals are constructed by superimposing the wave functions of the valence electrons of the two atoms. Superimposing the wave functions is a process called **orbital overlap,** and it results in a new wave function with electron density between the nuclei. The new orbital replaces the atomic orbitals as the description of the electrons in a molecule. To illustrate the principles of this model, let's use it to describe the bond in a hydrogen molecule.

The bond in a hydrogen molecule is formed by the overlap of the 1s atomic orbitals of two hydrogen atoms, as shown in Figure 1.8. The result is a new orbital with its electron density localized between the two nuclei and containing two electrons. The shape of localized valence bond orbitals, like atomic orbitals, is represented by a surface that encompasses 95% of the orbital.

Exercise 1.7	Use the localized valence bond orbital model to describe bond formation between the following pairs of atoms.

(a) A hydrogen atom and a fluorine atom bond to form HF.
(b) Two fluorine atoms bond to form F_2.

Overlap of hydrogen 1s atomic orbitals provides a satisfactory picture of the bond in a hydrogen molecule. However, overlap of the valence 2s or 2p atomic orbitals does not provide a satisfactory picture of the bonds formed by elements in the second period of the periodic table. For these elements, we must modify the 2s and 2p atomic orbitals.

1.10   Hybrid Atomic Orbitals

The ground state electron configuration of beryllium is $1s^2 \, 2s^2$, as shown in Section 1.4. In order to form BeH_2, we must use both the 2s and a 2p orbital of beryllium to overlap with the 1s orbitals of hydrogen. This kind of overlap, however, cannot form BeH_2 with a linear geometry.

Exercise 1.8	If we construct the two Be—H bonds in BeH_2 by overlap of the filled 2s orbital of beryllium and an empty 1s orbital of a hydrogen atom to form one bond, and overlap of an empty 2p orbital of beryllium and a filled 1s orbital of the second hydrogen atom to form the second bond, what H—Be—H bond angle would result?

The correct geometry of BeH_2 can be obtained if we use **hybrid atomic orbitals.** Hybrid orbitals are obtained by mathematically combining the *valence orbitals of an atom.* In the case of beryllium, we combine the 2s orbital with any one of the 2p orbitals. Mathematical combination of an atom's atomic orbitals to form a special set of directional orbitals is called **hybridization.** Combination can be carried out in two ways, either by adding the two wave functions or subtracting them, as shown in Figure 1.9.

The hybridization of two atomic orbitals forms two hybrid atomic orbitals. Hybrid orbitals are named for the atomic orbitals from which they were constructed. The hybrids in Figure 1.9, for instance, are called *sp* **hybrid orbitals** because they were made by combining one 2s orbital and one 2p orbital.

Hybrid orbitals are directional because they have a lobe of high electron density pointing in one specific direction. The two *sp* hybrid orbitals of beryllium point directly away from each other so they form a bond angle of 180°. Each *sp* orbital contains a single electron and the remaining unhybridized 2p atomic orbitals are empty, as shown in Figure 1.10.

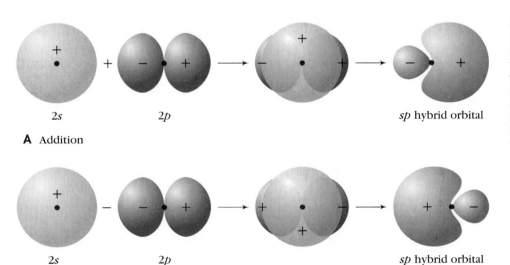

| 2s | 2p | *sp* hybrid orbital |

A Addition

| 2s | 2p | *sp* hybrid orbital |

B Subtraction

C How *sp* hybrid orbitals will be depicted in this text

Figure 1.9
Hybridization (mathematical combination) of the 2s orbital with one of the 2p orbitals of beryllium to form two *sp* hybrid atomic orbitals. Hybridization can be accomplished either by addition (**A**) or subtraction (**B**) of the two wave functions involved. **C,** How *sp* hybrid orbitals will be depicted in this text.

Figure 1.10
The atomic orbitals of an *sp* hybrid atom. **A,** The two *sp* hybrid atomic orbitals point away from each other. **B,** All the atomic orbitals of an *sp* hybrid atom. The *p* atomic orbitals are perpendicular to each other and to the linear *sp* hybrid orbitals. The back lobes of the *sp* hybrid atomic orbitals are omitted for simplicity and the orbitals are drawn with narrowed lobes for easier visualization.

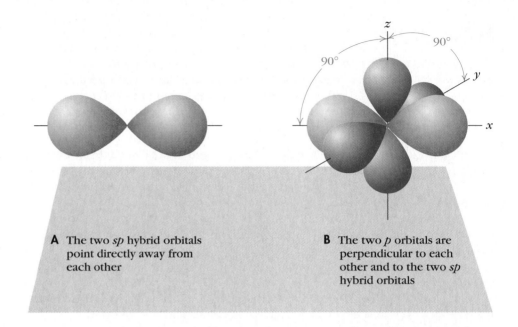

A The two *sp* hybrid orbitals point directly away from each other

B The two *p* orbitals are perpendicular to each other and to the two *sp* hybrid orbitals

Two of the solutions to the Schrödinger wave equation for the beryllium atom are the wave functions that describe its 2*s* and 2*p* orbitals. Mathematically it has been shown that the two *sp* hybrid orbitals formed by hybridization of the 2*s* and 2*p* orbitals of a beryllium atom are also solutions to its Schrödinger wave equation. The ground state electron configuration of a beryllium atom, therefore, can be equally well described as $1s^2$, $2sp$, $2sp$. This means that we can use these hybrid orbitals to construct localized valence bond orbitals for compounds containing a beryllium atom. Thus the electron structure of BeH_2 can be obtained by the overlap of each *sp* hybrid orbital of a beryllium atom with the 1*s* atomic orbital of hydrogen, as shown in Figure 1.11. This generates two new localized orbitals, the Be—H bonds, with a H—Be—H bond angle of 180°. Each localized orbital contains two electrons.

Figure 1.11
Localized valence bond orbital description of the bonds in BeH_2. The Be—H bonds form from the overlap of two *sp* hybrid orbitals on Be with the 1*s* atomic orbitals on each H atom. The two *sp* hybrid orbitals on Be point directly away from each other so the H—Be—H bond angle is 180°.

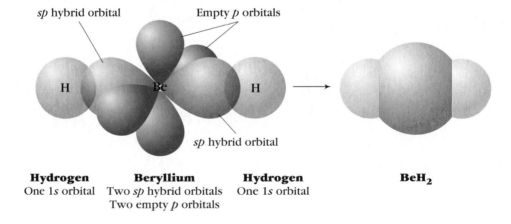

Hydrogen
One 1*s* orbital

Beryllium
Two *sp* hybrid orbitals
Two empty *p* orbitals

Hydrogen
One 1*s* orbital

BeH₂

The *sp* hybrid orbitals that so accurately describe the bonding of BeH_2 do not work for all molecules. One example is boron trifluoride, BF_3, which has trigonal

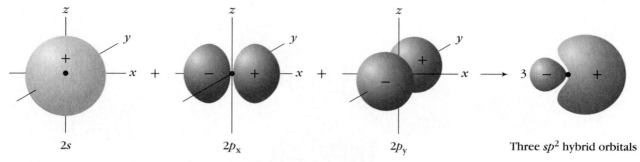

A Formation of three sp^2 hybrid orbitals from one $2s$ and two $2p$ orbitals

B How sp^2 hybrid orbitals will be depicted in this text

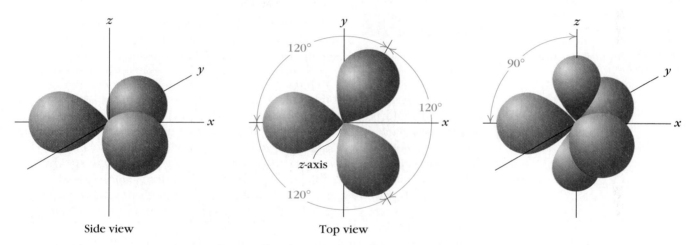

C Side and top views of the three sp^2 hybrid orbitals from **A**   **D** All the orbitals of an sp^2 hybrid atom

Figure 1.12
A, Combination of $2s$ and two $2p$ atomic orbitals (arbitrarily taken here as $2p_x$ and $2p_y$) forms three sp^2 hybrid atomic orbitals. **B,** How sp^2 hybrid orbitals will be depicted in this text. **C,** Side and top view of the three sp^2 hybrid orbitals from **A**. **D,** All the orbitals of an sp^2 hybrid atom. The three hybrid atomic orbitals are separated by 120° and the unhybridized p orbital is perpendicular to the plane of the sp^2 orbitals.

planar geometry with F—B—F bond angles of 120°. The valence atomic orbitals of boron in the ground state ($2s^2$, $2p$) do not have the correct geometry to account for this structure but neither do a set of sp hybrid orbitals. The trigonal planar geometry of the boron atom is best represented by using a different set of hybrid orbitals.

Mathematically combining the $2s$ orbital and two of the $2p$ orbitals gives three new orbitals called **sp^2 hybrid orbitals,** as shown in Figure 1.12, *A*. Each sp^2 hybrid orbital looks very much like an sp orbital because both have high electron density in a single direction. However, sp and sp^2 hybrid orbitals have different orientations in space. The three sp^2 hybrid orbitals lie in a plane and point to the three corners of an equilateral triangle. The other $2p$ orbital, not used to construct the hybrid orbitals, remains unchanged and is oriented perpendicular to the plane containing the three hybrid orbitals. A representation of the orbitals of an sp^2 hybrid atom is shown in Figure 1.12, and Figure 1.13 shows how these orbitals on boron can be combined with $2p$ atomic orbitals of fluorine to form BF_3.

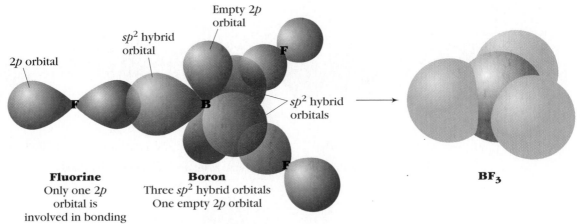

Fluorine
Only one 2p
orbital is
involved in bonding

Boron
Three sp^2 hybrid orbitals
One empty 2p orbital

BF$_3$

Figure 1.13
Localized valence bond orbital representation of the bonds in boron trifluoride. The B—F bonds form from the overlap of the three sp^2 hybrid orbitals on the B atom with the 2p atomic orbital on each F atom. For clarity, only the 2p orbitals of the fluorine atoms involved in bond formation are shown.

Exercise 1.9	Use the localized valence bond orbital model to describe bond formation in borane (BH_3), which is a molecule that exists only at low pressures.

Neither sp nor sp^2 hybrid orbitals can be used to describe the structure of a tetrahedral molecule such as methane, CH_4. Instead, the four equivalent C—H bonds in methane can be correctly represented with a set of four hybrid orbitals obtained by combining a 2s atomic orbital on a carbon atom with all three of its 2p atomic orbitals. Each of these four orbitals, which are called **sp^3 hybrid orbitals,** points toward a different corner of a tetrahedron. These hybrid atomic orbitals are shown in Figure 1.14. Bond formation in methane can be viewed as the overlap of each hydrogen 1s orbital with one sp^3 hybrid of the carbon atom, as shown in Figure 1.15.

In summary, hybrid atomic orbitals are constructed by mathematically combining atomic orbitals on a *single atom.* The combination of atomic orbitals is chosen to form hybrid orbitals that agree with the experimentally observed geometry. Thus, the bonding in a linear molecule, such as BeH_2, can be described as localized valence bond orbitals constructed from two sp hybrid orbitals on a central beryllium atom; the localized orbitals that describe the bonding in a trigonal planar molecule, such as BF_3, are constructed from three sp^2 hybrid orbitals on a central boron atom; and the bonding in a tetrahedral molecule such as methane can be described as localized orbitals constructed from a set of four equivalent sp^3 hybrid orbitals on a central carbon atom.

Exercise 1.10	Use the localized valence bond orbital model to describe the bonds in tetrafluoromethane, CF_4.

Because we are mostly interested in carbon-containing compounds in organic chemistry, in Section 1.11 we apply the use of hybrid orbitals to the localized valence bond orbital description of molecules that contain carbon-carbon single, double, and triple bonds.

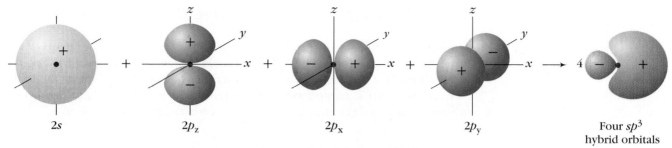

2s   2p$_z$ 2p$_x$ 2p$_y$ Four sp^3
hybrid orbitals

A Formation of four sp^3 hybrid orbitals from one 2s and three 2p orbitals

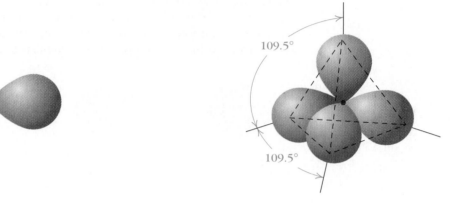

B How sp^3 hybrid orbitals will
be depicted in this text

C The four sp^3 hybrid orbitals formed
in **A** form a tetrahedron

Figure 1.14
A, Combination of 2s and three 2p
atomic orbitals forms four sp^3 hybrid
atomic orbitals. **B,** How sp^3 hybrid
orbitals will be depicted in this text.
C, All the orbitals of an sp^3 hybrid
atom. Each hybrid orbital points
toward a different corner of a tetrahedron. The hybrid orbitals are separated by an angle of 109.5°.

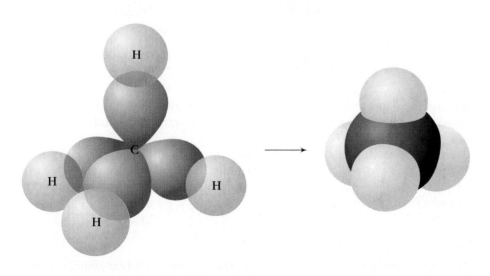

Carbon
Four sp^3 hybrid
orbitals

Hydrogen
Each has one
1s orbital

Methane (CH$_4$)

Figure 1.15
Localized valence bond orbital representation of the bonds in methane.
The 1s atomic orbitals on the four
hydrogen atoms overlap with the
four sp^3 hybrid orbitals of the carbon
atom.

1.11 Bonding in Representative Organic Compounds

A special characteristic of carbon is that it can form stable single, double, and triple bonds with other atoms. These bonds are adequately described by the localized valence bond orbital model. However, to correctly represent the structures of the organic compounds that contain these bonds, we must use hybrid orbitals of carbon. Let's start by describing a carbon-carbon single bond.

A   Carbon-Carbon Single Bonds

The simplest organic compound that contains a carbon-carbon single bond is ethane, C_2H_6. The tetrahedral geometry about the carbon atoms of ethane is correctly represented by using sp^3 hybrid carbon atoms to form localized orbitals. Thus a carbon-carbon single bond can be represented by the end-on overlap of two sp^3 hybrid orbitals (one from each carbon atom):

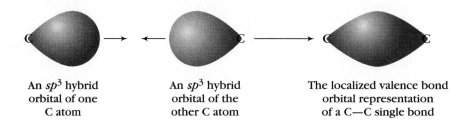

An sp^3 hybrid orbital of one C atom

An sp^3 hybrid orbital of the other C atom

The localized valence bond orbital representation of a C—C single bond

The bond formed by end-on overlap of two sp^3 hybrid orbitals is an example of a **sigma bond** (σ bond). Sigma bonds are cylindrically symmetrical about the axis joining the two nuclei; this axis is called the *bond axis*.

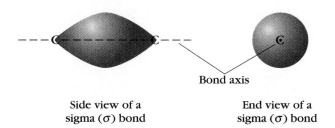

Bond axis

Side view of a sigma (σ) bond

End view of a sigma (σ) bond

Carbon-carbon single bonds are not the only examples of σ bonds. Carbon-hydrogen bonds are also σ bonds because they too are cylindrically symmetrical about the C—H bond axis.

We complete the localized orbital picture of ethane by forming six C—H bonds by the overlap of each of the remaining six sp^3 hybrid orbitals, three from each carbon atom, with the 1s orbital of each of the six hydrogen atoms. This process is illustrated in Figure 1.16.

B   Carbon-Carbon Double Bonds

The simplest organic compound that contains a carbon-carbon double bond is ethylene, C_2H_4. Its geometry is different from that of ethane. Experimental evidence shows that all six atoms of ethylene lie in a plane with the bond angles shown on p. 26.

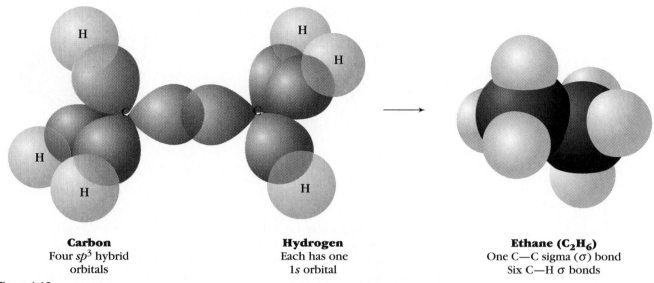

Carbon
Four sp^3 hybrid orbitals

Hydrogen
Each has one $1s$ orbital

Ethane (C_2H_6)
One C—C sigma (σ) bond
Six C—H σ bonds

Figure 1.16
Localized valence bond orbital representation of the bonds in ethane. All C—C and C—H bonds are σ bonds. The C—C single bond forms from the overlap of two sp^3 hybrid orbitals. The C—H bonds form from the overlap of carbon sp^3 hybrid orbitals with the hydrogen $1s$ atomic orbitals.

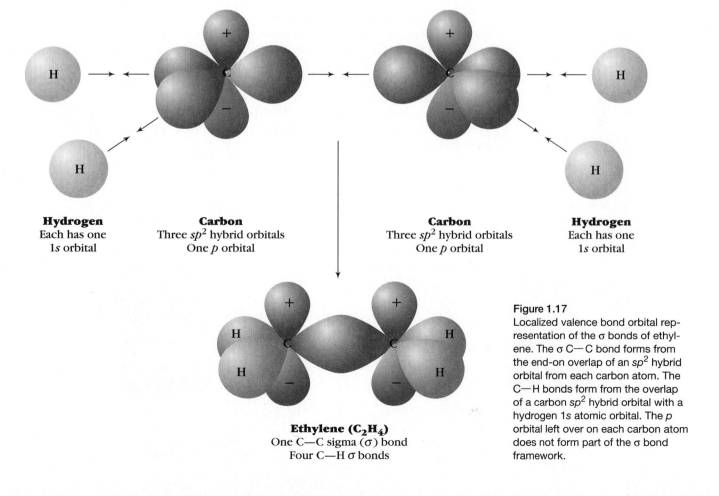

Hydrogen
Each has one $1s$ orbital

Carbon
Three sp^2 hybrid orbitals
One p orbital

Carbon
Three sp^2 hybrid orbitals
One p orbital

Hydrogen
Each has one $1s$ orbital

Ethylene (C_2H_4)
One C—C sigma (σ) bond
Four C—H σ bonds

Figure 1.17
Localized valence bond orbital representation of the σ bonds of ethylene. The σ C—C bond forms from the end-on overlap of an sp^2 hybrid orbital from each carbon atom. The C—H bonds form from the overlap of a carbon sp^2 hybrid orbital with a hydrogen $1s$ atomic orbital. The p orbital left over on each carbon atom does not form part of the σ bond framework.

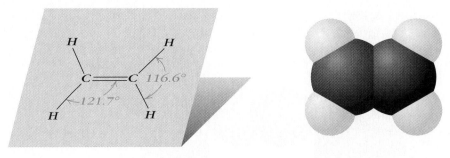

Geometry of ethylene (C_2H_4)

The experimentally observed geometry of ethylene can be constructed from localized valence bond orbitals if both carbon atoms are sp^2 hybridized. One carbon-carbon bond is formed by the end-on overlap of an sp^2 hybrid orbital from each carbon atom. Each of the four C—H bonds is formed by the overlap of one of the remaining sp^2 hybrid orbitals from each carbon atom with a $1s$ orbital from each hydrogen, as shown in Figure 1.17.

The C—H bonds and the C—C bond formed by overlap of two sp^2 orbitals are σ bonds because they have the required cylindrical symmetry about their bond axes. These bonds all lie in a plane and define the sigma-bond framework of ethylene.

The second carbon-carbon bond is formed by sideways overlap of the remaining p orbital on each carbon atom, as shown in Figure 1.18. This kind of sideways overlap results in the formation of a different type of localized orbital called a **pi (π) bond.** Notice that these p orbitals overlap above and below the plane of the sigma-bond framework. As a result, half the π bond lies above the molecular plane and the other half lies below.

Notice that σ and π bonds have different shapes. A σ bond is cylindrically symmetrical about its bond axis, while a π bond extends above and below the molecular plane.

Figure 1.18
Localized valence bond representation of the π bond of ethylene. The π bond in ethylene forms from the sideways overlap of the p orbitals on sp^2 hybrid carbon atoms. Notice that half of the π bond lies above the molecular plane and half below. For clarity, the σ bonds are represented by lines.

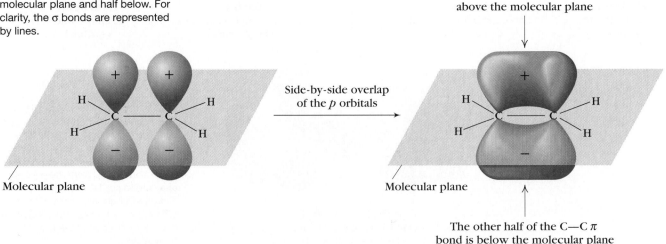

   Thus the localized valence bond orbital model represents a C=C double bond as one σ bond and one π bond with H—C—H and H—C—C bond angles of 120°. While this model does not predict the exact bond angles in ethylene, it is still useful as an aid to understanding the chemistry of compounds containing a C=C double bond. These compounds, which are called *alkenes,* are described in detail in Chapters 7 and 8.

C   Carbon-Carbon Triple Bonds

Acetylene, C_2H_2, is the simplest organic compound that contains a C≡C triple bond. A molecule of acetylene is linear, so its C—C—H bond angle is 180°.

   We can construct localized orbitals that agree with this geometry if both carbon atoms are *sp* hybridized. The C—C σ bond is formed by the end-on overlap of an *sp* hybrid orbital from each carbon atom. Each of the two C—H bonds is formed by the overlap of the remaining *sp* hybrid orbital on each carbon atom with a 1*s* orbital on a hydrogen atom, as shown in Figure 1.19.

   Each carbon atom has two unhybridized *p* orbitals left. Sidewise overlap of these *p* orbitals with identical orbitals on the other carbon atom forms two π localized molecular orbitals, as shown in Figure 1.20. Thus a C≡C triple bond in the localized valence bond orbital model consists of one C—C σ bond and two C—C π bonds.

   Localized valence bond orbitals describe the electron structure of molecules in much more detail than the simple dashes used in Lewis structures to represent one or more bonds between two atoms. Both Lewis structures and localized orbital models localize the two electrons of a bond between the two bonding atoms. Electrons of a bond are not always localized between the two bonding atoms, however.

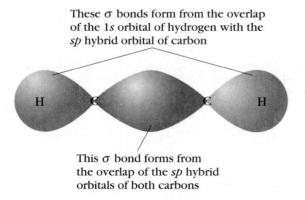

These σ bonds form from the overlap of the 1*s* orbital of hydrogen with the *sp* hybrid orbital of carbon

This σ bond forms from the overlap of the *sp* hybrid orbitals of both carbons

Figure 1.19
Localized valence bond orbital representation of the σ bonds of acetylene. The C—H σ bonds are formed by the overlap of an *sp* hybrid orbital on each carbon atom with a 1*s* orbital on each hydrogen atom. The C—C σ bond is formed by overlap of an *sp* hybrid orbital on each carbon atom.

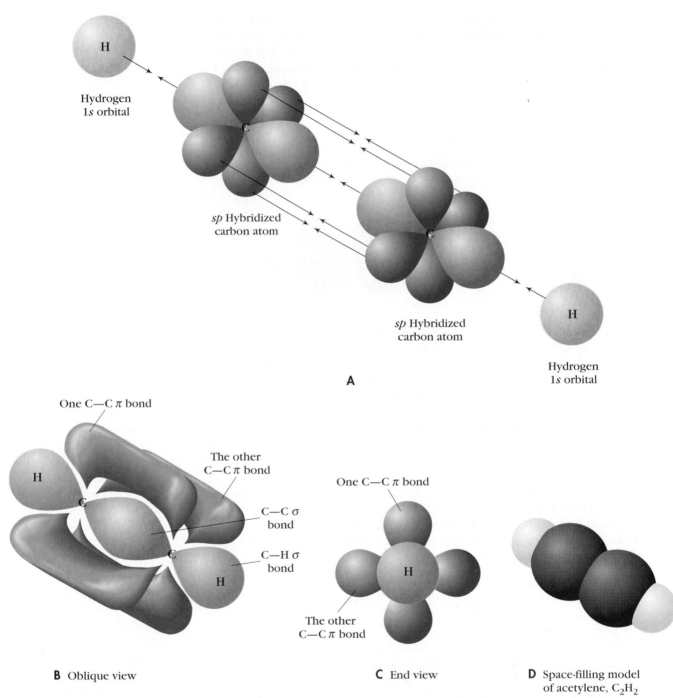

B Oblique view

C End view

D Space-filling model
of acetylene, C_2H_2

Figure 1.20
Localized valence bond orbital representation of the bonds of acetylene. **A,** Atomic orbitals from
each C and H atoms that overlap. **B,** Overlap of $2p_z$ orbital from each carbon atom forms one π
bond and overlap of $2p_y$ orbitals forms second π bond. **C,** End view of the σ and π bonds along
the C—C bond axis. **D,** Cylindrical shape of the total electron density of the triple bond of acetylene.

1.12   The Structures of Some Molecules Are Best Represented as Resonance Hybrids

More than one correct Lewis structure can be written for certain molecules and ions. For example, the Lewis structure of the formate anion, HCO_2^-, can be written in either of the following two ways:

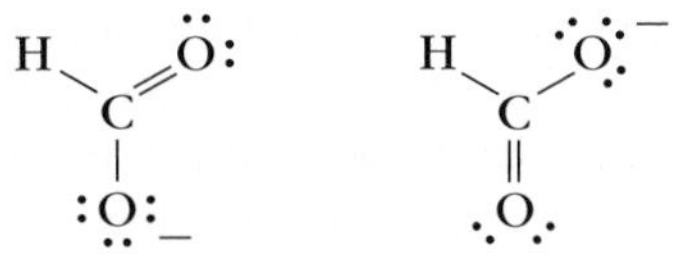

If either of these structures correctly represents the bonding in the formate anion, there should be two kinds of C—O bonds; there should be a longer single C—O bond and a shorter C=O double bond. Experiments have shown, however, that the two C—O bonds in the formate anion are the same length and that the bond length is somewhere between that of a C—O single and a C=O double bond. Thus these structures, while valid Lewis structures, do not correctly represent the bonding in the formate anion.

A molecule or ion for which more than one Lewis structure can be written but which is accurately described by none of them is said to be a **resonance hybrid** of all of them. For HCO_2^-, this means that its real structure is a resonance hybrid of the two Lewis structures shown previously. These two Lewis structures are called **resonance** or **contributing structures** because they *contribute* their characteristics to the real structure of HCO_2^-. We represent the structure of a molecule that is a resonance hybrid by writing its contributing Lewis structures and placing double-headed arrows between them. Thus the structure of the formate anion, which is a resonance hybrid of two Lewis structures, is represented in the following way:

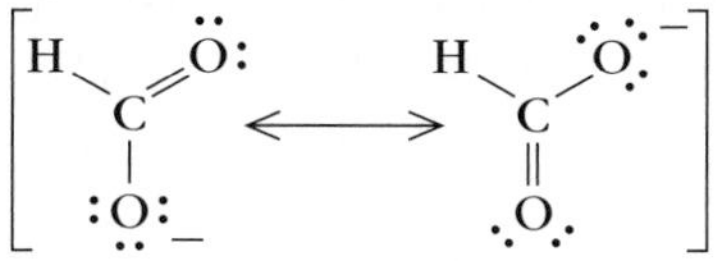

The double-headed arrow $\longleftrightarrow$ between two or more correct Lewis structures of a molecule or ion has special significance in chemistry. It indicates that the Lewis structure at the end of each arrowhead contributes to the resonance hybrid of the real molecule or ion (HCO_2^- in this case). This does *not* mean that the real structure of HCO_2^- is represented some of the time by one Lewis structure and the rest of the time by the other. The real molecule has only one structure, which we can visualize as being made up of a contribution of two fictitious Lewis structures. The Lewis structures are fictitious because neither alone correctly represents the electron structure of HCO_2^-.

Individual Lewis structures localize electron pairs between two atoms. In reality, however, valence electrons in molecules or ions are not always localized. Sometimes they are delocalized; sometimes, that is, they are spread over more than two atoms. Some of the valence electrons in HCO_2^- are delocalized. This is why the electron structure of HCO_2^- must be represented as a resonance hybrid of two contributing Lewis structures.

In an attempt to depict electron delocalization, chemists sometimes write resonance hybrids with dotted lines to represent delocalization of electrons over several atoms. Thus the electron structure of HCO_2^- is sometimes represented by the following structure:

$$H-C \overset{O}{\underset{O}{\diagup}} -$$

The dotted lines represent a bond between the carbon and oxygen atoms that is neither a single nor a double bond; it is intermediate between them. While this representation of electron delocalization in molecules is sometimes used in chemistry, we will not use it in this text. We will use resonance hybrids instead to describe electron delocalization in organic molecules.

Let's return to the Lewis structure representation of resonance hybrids. How do we know what Lewis structures contribute to a resonance hybrid? The answer to this question is given by the following series of rules.

1. All resonance structures must be correct Lewis structures for the molecule or ion. Thus the following structure does *not* contribute to the resonance hybrid of HCO_2^-:

$$H-C^-\overset{\ddot{O}:}{\underset{\ddot{O}:}{}}$$

This is not a correct Lewis structure because there are more than eight valence electrons about the carbon atom.

2. The position of the nuclei must be the same in all resonance structures. Only the location of electrons can be shifted from one contributing structure to another. Nonbonding electron pairs and electrons in double bonds are the ones that are usually shifted.

3. The number of paired or unpaired electrons must be the same in all resonance structures. The following structure does not contribute to the resonance hybrid of HCO_2^- because it has two unpaired electrons while the electrons in the other two resonance structures, shown previously, are paired.

$$\overset{H}{\underset{:\ddot{O}:^-}{\diagdown}}C\cdot\overset{\cdot\ddot{O}:}{\diagup}$$

4. All contributing structures do not contribute equally to a resonance hybrid. Each structure contributes in proportion to its stability. The most stable form contributes the most.

The following are additional examples of molecules or ions whose electron structures are best represented as resonance hybrids of the contributing structures shown.

$$NO_2^- \qquad \left[\quad :N\overset{\ddot{O}:}{\underset{\ddot{O}:^-}{\diagup}} \quad \longleftrightarrow \quad :N\overset{\cdot\ddot{O}:^-}{\underset{O:}{\diagup}} \quad \right]$$

$$CH_3NO_2 \qquad \left[\quad H-\overset{H}{\underset{H}{\overset{|}{\underset{|}{C}}}}-N^+\overset{\ddot{O}:}{\underset{\ddot{O}:^-}{\diagup}} \quad \longleftrightarrow \quad H-\overset{H}{\underset{H}{\overset{|}{\underset{|}{C}}}}-N^+\overset{\cdot\ddot{O}:^-}{\underset{O:}{\diagup}} \quad \right]$$

CO_3^{2-}

$$\left[\quad \ddot{\overset{..}{O}} \underset{\underset{\ddot{\overset{..}{O}} :^-}{|}}{\overset{}{C}} \overset{..}{\ddot{O}}:^- \quad \longleftrightarrow \quad ^-:\overset{..}{\ddot{O}} \underset{\underset{.\ddot{O}.}{\|}}{\overset{}{C}} \overset{..}{\ddot{O}}:^- \quad \longleftrightarrow \quad :\overset{..}{\ddot{O}} \underset{\underset{:\ddot{O}:^-}{|}}{\overset{}{C}} \overset{..}{\ddot{O}}:^- \quad \right]$$

Exercise 1.11   Write the contributing structures to the resonance hybrid of each of the following ions or molecules.

(a) $CH_3CO_2^-$ **(b)** NO_3^- **(c)** $CH_2{=}CHCH_2^+$

■ *Be careful not to confuse the two symbols $\longleftrightarrow$ and $\overset{\longrightarrow}{\longleftarrow}$. Two arrows pointing in opposite directions $\overset{\longrightarrow}{\longleftarrow}$ is the symbol for an equilibrium between reactants and products; this symbol joins chemical species that really exist. The double-headed arrow $\longleftrightarrow$, on the other hand, joins fictitious structures that contribute to the electron structure of a real molecule.*

In summary, the electron structures of certain molecules and ions are better represented by a resonance hybrid made up of two or more contributing structures. This concept is introduced in order to account for the delocalization of electrons in certain molecules and ions. In most cases, the delocalized electrons are nonbonding electrons and electrons in multiple bonds. A double-headed arrow between contributing structures is the symbol that we use to represent the resonance hybrid structure of a molecule or ion.

The linear combination of atomic orbitals model of bonding, which is described in the next section, provides a more accurate representation of electron delocalization in molecules.

1.13 The LCAO Model of Chemical Bonding

The **linear combination of atomic orbitals** model, which is abbreviated **LCAO,** mathematically combines the atomic orbitals (the wave functions) of the atoms to form **molecular orbitals.** Molecular orbitals are orbitals that can encompass more than just two atoms. We can illustrate this model with a few examples, beginning with the hydrogen molecule.

The bond in a hydrogen molecule can be described by mathematically combining the $1s$ orbitals (the $1s$ wave functions) of each hydrogen atom. The wave functions are combined by adding them *and* by subtracting them, as shown in Figure 1.21. The result of this mathematical operation is two molecular orbitals, one of which is called a *bonding molecular orbital* and the other, which is called an *antibonding molecular orbital.*

The bonding molecular orbital is lower in energy than the antibonding molecular orbital because the bonding orbital has electron density concentrated in the region between the two positively charged nuclei. The bonding molecular orbital in the hydrogen molecule is designated σ_{1s} because it is a σ orbital (cylindrically symmetrical about the axis joining the two nuclei), and it is formed by the combination of $1s$ atomic orbitals. The energy difference between the σ_{1s} molecular orbital and the $1s$ atomic orbitals is related to the strength of the H—H bond in a hydrogen molecule.

The antibonding molecular orbital is higher in energy because there is little electron density between the nuclei. As a result, repulsion between the bare nuclei tends to force the nuclei apart. The antibonding molecular orbital in the hydrogen molecule is designated σ^*_{1s} in order to distinguish it from the bonding molecular orbital. The energy of the σ^*_{1s} molecular orbital is higher than the energy of the $1s$ atomic orbitals by the same amount that the energy of the σ_{1s} molecular orbital is lower than the energy of the $1s$ atomic orbitals.

The number of molecular orbitals that results from the LCAO model is always

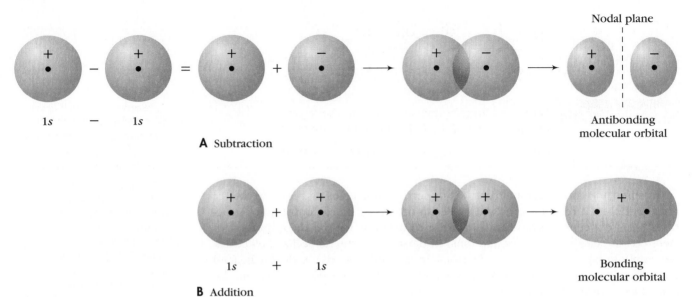

A Subtraction

B Addition

Figure 1.21
Two molecular orbitals of hydrogen obtained by the combination of two 1s atomic orbitals, one from each hydrogen atom. **A,** Subtraction of the two wave functions forms an antibonding molecular orbital. **B,** Addition of the two wave functions forms a bonding molecular orbital.

equal to the number of atomic orbitals mathematically combined. This is a general principle of the mathematics involved in this model. As we have seen in the case of the hydrogen molecule, mathematical combination of two atomic orbitals results in two molecular orbitals. Combination of three atomic orbitals in other molecules or ions forms three molecular orbitals, and so on.

The Pauli exclusion principle applies to molecular orbitals as well as to atomic orbitals, so each molecular orbital can accommodate a maximum of two electrons of opposite spin. In filling molecular orbitals we use the same Aufbau principle that we used to fill atomic orbitals. Thus the two electrons that form the H—H bond in the hydrogen molecule occupy the low-energy σ_{1s}-bonding molecular orbital of the hydrogen molecule, as shown in Figure 1.22.

Bond formation in a hydrogen molecule is relatively easy to describe using the LCAO model because we need consider the mathematical combination of only two electrons in two 1s orbitals. Bond formation in other molecules is more complicated because more electrons and more atomic orbitals are involved in forming molecular orbitals. An example of a more complicated system is ethylene, C_2H_4.

Considering only the valence electrons, the six atoms of ethylene have a total of twelve filled or partially filled atomic orbitals (four from each carbon atom and one from each hydrogen atom). To combine all these orbitals using the LCAO model is mathematically very complicated and provides us with twelve molecular orbitals, far more than we need. Fortunately, understanding the chemical reactions of alkenes makes it possible to greatly simplify the LCAO model. Reactions of alkenes usually occur at the π bond, so the σ bond framework is rarely involved. As a result, we can use localized valence bond orbitals to describe the σ bond framework and use the LCAO model to construct the molecular orbitals of the C=C double bond from the 2p atomic orbitals of the two sp^2 hybrid carbon atoms.

Mathematical combination of the two 2p orbitals of the sp^2 hybrid carbon atoms, just like the 1s orbitals of two hydrogen atoms, forms two molecular orbitals. One of these molecular orbitals is bonding and the other is antibonding, as shown in Figure 1.23. Both are π molecular orbitals. The bonding π molecular

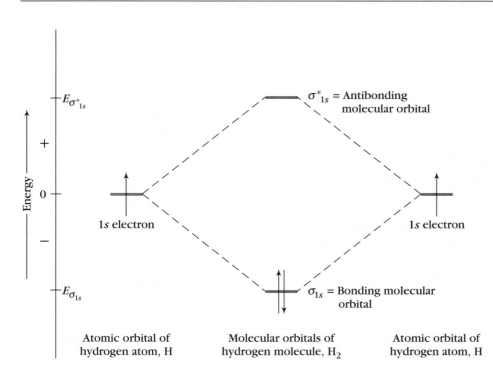

Figure 1.22
Atomic and molecular orbital energies of hydrogen atoms and the hydrogen molecule. The energies of the two molecular orbitals σ_{1s} ($E_{\sigma_{1s}}$) and σ^*_{1s} ($E_{\sigma^*_{1s}}$) are equal in magnitude but differ in sign. The sign of $E_{\sigma^*_{1s}}$ is positive, indicating a repulsive force, while the sign of $E_{\sigma_{1s}}$ is negative, indicating an attractive force.

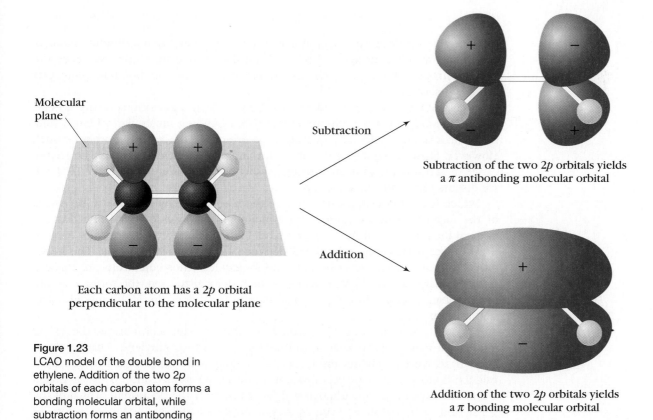

Each carbon atom has a 2p orbital perpendicular to the molecular plane

Subtraction of the two 2p orbitals yields a π antibonding molecular orbital

Addition of the two 2p orbitals yields a π bonding molecular orbital

Figure 1.23
LCAO model of the double bond in ethylene. Addition of the two 2p orbitals of each carbon atom forms a bonding molecular orbital, while subtraction forms an antibonding orbital.

Figure 1.24
Atomic orbital energies for the $2p$ atomic orbitals of sp^2 hybrid carbon atoms and energies of the two molecular orbitals obtained by adding and subtracting them. The energies of the two molecular orbitals, π_{2p} ($E_{\pi_{2p}}$) and π^*_{2p} ($E_{\pi^*_{2p}}$), are equal in magnitude but differ in sign. The sign of $E_{\pi^*_{2p}}$ is positive, indicating a repulsive force, while the sign of $E_{\pi_{2p}}$ is negative, indicating an attractive force.

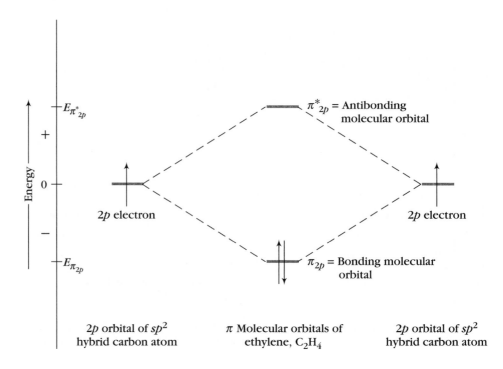

orbital, which is designated π_{2p}, is lower in energy than the antibonding molecular orbital, which is designated π^*_{2p}. Using the Aufbau principle, we place the two electrons (one from each $2p$ atomic orbital) in the bonding molecular orbital, as shown in Figure 1.24.

The LCAO and the localized valence bond orbital descriptions of the π bond in ethylene are similar yet different. The LCAO bonding molecular orbital is similar to the π bond obtained by overlap of $2p$ orbitals in the localized orbital model. However, LCAO describes an antibonding molecular orbital as well. The significance of antibonding molecular orbitals will become evident in Chapter 19 when we discuss ultraviolet spectroscopy.

Which is the better representation of the π bond in ethylene, the LCAO model or the localized valence bond orbital model? Both provide reasonable descriptions of the π bond, but the localized model is adequate for our purposes because the π bond in ethylene is localized between the two carbon atoms. The LCAO model is best used to describe ions and molecules whose electrons are delocalized. We learned how to represent one such delocalized ion, the formate anion, as a resonance hybrid in the previous section. How does the LCAO model represent this ion?

To construct the molecular orbitals of the formate anion using the LCAO model, we make the same assumptions that we did for ethylene. That is, the σ bond framework is chemically inert and is adequately represented by localized valence bond orbitals so we need apply the LCAO model only to the π framework. Thus we need to mathematically combine only the $2p$ atomic orbital of the sp^2 hybrid carbon atom and the $2p$ atomic orbital of each oxygen atom, a total of three orbitals.

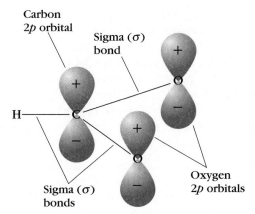

These three atomic orbitals combine to form the three molecular orbitals shown in Figure 1.25.

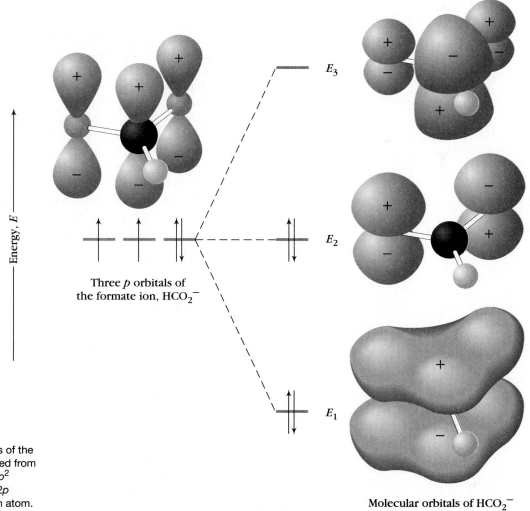

Figure 1.25
The three π molecular orbitals of the formate anion. They are formed from the 2p atomic orbital of the sp^2 hybrid carbon atom and the 2p atomic orbital of each oxygen atom.

Molecular orbitals of HCO_2^-

The molecular orbital of lowest energy (E_1) in Figure 1.25 surrounds not just two atoms but three (both oxygen atoms and the carbon atom). The next-highest energy molecular orbital (E_2) has one node, which passes through the carbon atom. This means that most of its electron density is found on the two oxygen atoms. The highest energy molecular orbital (E_3) has two nodes and is not involved in the bonding. Thus the LCAO picture of the formate anion consists of an inner molecular orbital whose two electrons encompass the carbon and oxygen atoms. The electron pair of the next-highest energy molecular orbital is shared between both oxygen atoms.

Exercise 1.12	Show how to use orbital overlap to construct the localized valence bond orbitals of the σ bond framework of the formate anion.

Exercise 1.13	Use the LCAO model to construct the two molecular orbitals of He_2 and place the valence electrons of the two helium atoms in their correct molecular orbitals. Based on this diagram, explain why He_2 is not a stable molecule.

The molecular orbitals obtained using the LCAO model and the valence bond resonance Lewis structures that contribute to the resonance hybrid are equivalent ways of representing the delocalized π electron structure of the formate anion. Both indicate that the negative charge is shared equally between both oxygen atoms. While giving the same result in this case, each has advantages and disadvantages. The molecular orbitals clearly illustrate the delocalization of the π electrons but are difficult to draw. Resonance hybrids are easier to write but do a poorer job of representing electron delocalization. In this text, we will use mainly the resonance hybrid representations of molecules in which electron delocalization occurs. The major exception will be encountered in Chapter 20, where we use the LCAO model to arrive at a definition of aromaticity.

So far, our concern has been with how we picture the electron distribution in molecules. We have not said anything about whether the electrons are distributed evenly in a molecule or not. In the next section we begin to discuss the charge distribution in molecules.

1.14   Polar Covalent Bonds

Electron distribution in most chemical bonds is unsymmetrical. Sometimes this results in an unsymmetrical charge distribution in a molecule. In this and following sections we will examine what causes an unsymmetrical charge distribution in a chemical bond and what effect this has on certain properties of molecules.

Electron distribution in a chemical bond depends on the relative electron-attracting abilities of the two atoms that form the bond. The intrinsic ability of an element to attract electrons is called its **electronegativity.** Linus Pauling established a semiquantitative scale in which each element is assigned an electronegativity value. On this scale, which is shown in Figure 1.26, a larger number signifies a greater affinity for electrons. Values of electronegativity range from 4.0 for fluorine, the most electronegative element, to values below 1 for the alkali met-

															H 2.1	He

Li 1.0	Be 1.5											B 2.0	C 2.5	N 3.0	O 3.5	F 4.0	Ne
Na 0.9	Mg 1.2											Al 1.5	Si 1.8	P 2.1	S 2.5	Cl 3.0	Ar
K 0.8	Ca 1.0	Sc 1.3	Ti 1.5	V 1.6	Cr 1.6	Mn 1.5	Fe 1.8	Co 1.9	Ni 1.9	Cu 1.9	Zn 1.6	Ga 1.6	Ge 1.8	As 2.0	Se 2.4	Br 2.8	Kr
Rb 0.8	Sr 1.0	Y 1.2	Zr 1.4	Nb 1.6	Mo 1.8	Tc 1.9	Ru 2.2	Rh 2.2	Pd 2.2	Ag 1.9	Cd 1.7	In 1.7	Sn 1.8	Sb 1.9	Te 2.1	I 2.5	Xe
Cs 0.7	Ba 0.9	Lu 1.2	Hf 1.3	Ta 1.5	W 1.7	Re 1.9	Os 2.2	Ir 2.2	Pt 2.2	Au 2.4	Hg 1.9	Tl 1.8	Pb 1.9	Bi 1.9	Po 2.0	At 2.2	Rn

Electronegativity (EN) < 2.0   EN > 2.4
EN 2.0-2.4   EN unknown

Figure 1.26
Pauling values for the average electronegativities of the elements.

als, the most electropositive elements. Electronegativity generally increases going from left to right across a period and decreases going down a group in the periodic table.

Electron distribution is symmetrical in the bonds of homonuclear diatomic molecules such as H_2 and Cl_2. The two bonding electrons in these compounds are distributed evenly because the electronegativities of the constituent atoms are identical. These kinds of bonds are called **nonpolar covalent bonds.**

When two atoms of very different electronegativities interact, electron transfer occurs from the more electropositive atom to the more electronegative one to produce a positively charged ion (a cation) and a negatively charged ion (an anion). Electropositive sodium (Na), for example, readily transfers an electron to electronegative chlorine (Cl) to form a sodium cation (Na^+) and a chloride anion (Cl^-). The attractive force between these two ions of opposite charges is called an **ionic bond.**

Nonpolar covalent bonds and ionic bonds represent the two extremes of electron distribution in a bond. The electron distribution in the majority of bonds is somewhere in between. That is, the bonding electrons in most bonds are attracted slightly more by one of the atoms than the other. A bond formed by this kind of unequal sharing of electrons is called a **polar covalent bond.** Thus the bonds in all compounds fit somewhere on a continuum of bond types. At one extreme is a nonpolar covalent bond; at the other extreme is an ionic bond; and in between are the polar covalent bonds.

INCREASING IONIC CHARACTER ⟩

Type of bond	Nonpolar covalent	Polar covalent	Ionic
Electronegativity difference between bonding atoms	Zero	Intermediate	Large

The pair of electrons in a polar covalent bond has a greater chance of being found in the region of the more electronegative element. Thus a center of partial negative charge is created on the more electronegative element and a center of partial positive charge is created on the less electronegative element. With the exception of hydrogen, most of the atoms that bond to carbon are more electronegative than carbon, so the carbon atom of most bonds in organic compounds is a center of partial positive charge.

For the carbon-hydrogen bond, the electronegativity of hydrogen (2.1) is slightly less than that of carbon (2.5), so there is a slight tendency for the electrons of the C—H bond to be attracted toward the carbon atom. This electronegativity difference is so small, however, especially compared to the bonds of other atoms to carbon, that we can consider the C—H bond to be essentially a nonpolar covalent bond.

We indicate a partial charge on an atom in a molecule by the symbol delta, δ. The symbol $\delta+$ indicates a partial positive charge on an atom, while the symbol $\delta-$ indicates a partial negative charge. These symbols are often added to a bond to indicate that an electron pair is shared unequally in a polar covalent bond, as shown in the case of the carbon-chlorine bond in chloromethane.

$$
\begin{array}{c}
\text{H} \\
| \\
\text{H}-\text{C}\overset{\delta+}{-}\text{Cl}^{\delta-} \\
| \\
\text{H}
\end{array}
$$

Sometimes carbon atoms have a partial negative charge. This usually happens when carbon is bonded to an element to the left of it in the periodic table. These elements are less electronegative than carbon so they attract electrons less strongly. In the bond between carbon and one of these elements, the carbon has a partial negative charge and the other element has a partial positive charge. In an organometallic compound such as diethyl magnesium, for example, the carbon atoms bonded to magnesium have partial negative charges.

$$
\begin{array}{c}
\text{H} \quad \text{H} \qquad\qquad \text{H} \quad \text{H} \\
| \qquad | \quad {}^{\delta-}\;\;{}^{\delta+}\;\;{}^{\delta-}| \qquad | \\
\text{H}-\text{C}-\text{C}-\text{Mg}-\text{C}-\text{C}-\text{H} \\
| \qquad | \qquad\qquad | \qquad | \\
\text{H} \quad \text{H} \qquad\qquad \text{H} \quad \text{H}
\end{array}
$$

Exercise 1.14	Which of the following molecules have polar covalent bonds? In those molecules that do, use $\delta+$ to indicate which atoms have partial positive charges and $\delta-$ to indicate which have partial negative charges.

(a) H_2 **(b)** HCl   **(c)** ICl   **(d)** I_2 **(e)** BrCl

Polar covalent bonds are common in many molecules. In the next section we learn that many of these molecules have dipole moments and are polar.

1.15   Bond Polarity and Molecular Dipole Moments

Experimentally it is known that when **nonpolar molecules** are placed in an electric field they orient themselves randomly, as shown in Figure 1.27, *A*. **Polar molecules,** on the other hand, tend to orient themselves so that the negative part of the molecule points toward the positive plate and the positive part points toward the negative plate, as shown in Figure 1.27, *B*.

Polar molecules behave in this way because they have a region of negative charge and a region of positive charge. Why are the charges separated in a polar molecule? We can answer this question by combining the concepts of polar covalent bonds and molecular geometry. We begin with a simple example, HCl.

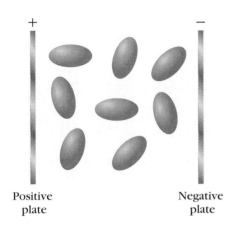

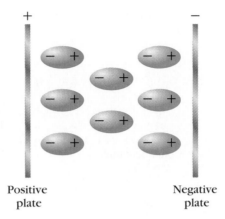

Positive plate   Negative plate   Positive plate   Negative plate

A Nonpolar molecules   **B** Polar molecules

Figure 1.27
The effect of an electric field on the orientation of nonpolar molecules **(A)** and polar molecules **(B).** Nonpolar molecules orient themselves randomly, whereas polar molecules orient themselves so that their negative ends point toward the positive plate and their positive ends point toward the negative plate.

The uneven electron distribution in the polar covalent bond of gaseous HCl results in a center of partial positive charge on the hydrogen atom and a center of partial negative charge on the chlorine atom.

$$\overset{\delta+}{H}\!\!-\!\!\overset{\delta-}{\ddot{\underset{\cdot\cdot}{Cl}}}:$$

Two separated, equal, and opposite charges constitute a **dipole.** A dipole is expressed quantitatively by its **dipole moment,** μ, which is usually expressed in debye units, D. The dipole moment of a molecule can be measured experimentally. The dipole moment of HCl, for example, is 1.1 D.

The diatomic HCl molecule contains only a single bond so its dipole moment is both the dipole moment of the H—Cl bond, called a **bond dipole moment,** and the dipole moment of the whole HCl molecule, called a **molecular dipole moment.** Bond dipole moments and molecular dipole moments are the same only in diatomic molecules. In polyatomic molecules we must distinguish between the two because the value of the molecular dipole moment is the vector sum of the individual bond dipole moments. The vector sum represents both the direction and magnitude of each individual bond dipole moment. Thus both the shape of the molecule and the electronegativity difference between atoms in the individual bonds contribute to the molecular dipole moment. Several examples of the contributions of bond dipole moments to the molecular dipole moments of molecules are shown in Figure 1.28.

Three compounds in Figure 1.28, carbon dioxide, boron trifluoride, and tetrachloromethane, have polar covalent bonds yet do not have molecular dipole moments. These molecules are symmetrical (carbon dioxide is linear, BF_3 is trigonal, and tetrachloromethane is tetrahedral) so the vector sum of the individual bond dipole moments is zero. That is, the bond dipole moments cancel each other.

Exercise 1.15	Write the three-dimensional structure for each of the following molecules. Indicate the expected bond dipole moments and decide which of the molecules have a molecular dipole moment.

(a) CH_3F **(b)** CF_4 **(c)** H_2CO **(d)** H_2S **(e)** NCl_3

Figure 1.28
The molecular dipole moment is the vector sum of the individual bond dipole moments.

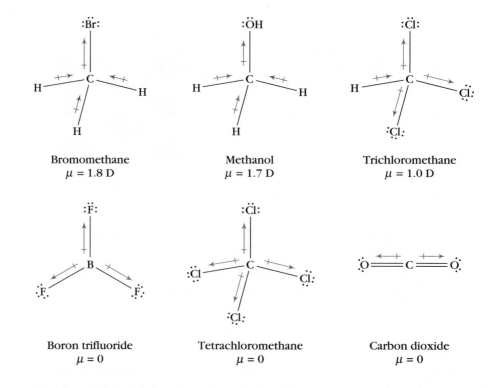

Bromomethane
$\mu = 1.8$ D

Methanol
$\mu = 1.7$ D

Trichloromethane
$\mu = 1.0$ D

Boron trifluoride
$\mu = 0$

Tetrachloromethane
$\mu = 0$

Carbon dioxide
$\mu = 0$

One important consequence of uneven electron distribution in molecules is that polar molecules are attracted to one another.

1.16   Intermolecular Attractions and Repulsions

Why are some compounds solid at room temperature, while others are liquids and still others are gases? Why do two liquid compounds with the same molecular weight have different boiling points? Part of the answer to these questions is the strength of the interactions between molecules. These intermolecular interactions are electrostatic in nature so they include attractions and repulsions. Thus intermolecular attraction is simply the attraction between a center of positive charge on one molecule or ion and a center of negative charge on another molecule or ion. Repulsions, on the other hand, occur between centers of like charge in adjacent molecules.

The strength of these attractions and repulsions determines the physical state of a substance at a particular temperature. It is generally true that substances are solids where attractive forces are the strongest. Inorganic salts such as sodium chloride are solids with high melting points because the positive sodium ions and the negative chloride ions strongly attract each other. In contrast, substances tend to be gases where attractive forces are very weak. Simple molecules like H_2, N_2, and O_2 are gases with very low melting and boiling points because their molecules only weakly attract each other. Attractive forces in liquids are intermediate between the strong attractions in solids and the weak attractions in gases.

The specific interactions that are responsible for the physical state of simple molecules and salts are easy to describe because the centers of negative and positive charges are easy to identify. While it is more difficult to describe and evaluate the attractions and repulsions in complex molecules, chemists have identified four major kinds of intermolecular attractions in these molecules. These interactions, which are called *ion-dipole, dipole-dipole, hydrogen bonds,* and *London forces,* deserve to be examined in some detail.

A   Ion-Dipole Interactions

We learned in Section 1.15 that many molecules are polar because they have a dipole moment. The charge separation in polar molecules attracts ions. Thus a positive ion is attracted to the negative end and repelled by the positive end of a polar molecule. Similarly a negative ion is attracted to the positive end and repelled by the negative end of a polar molecule.

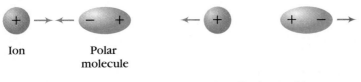

Ion   Polar
molecule

Ion-dipole attraction   Ion-dipole repulsion

Such interaction occurs when solid sodium chloride dissolves in water. Its ions enter the solution where each ion is surrounded by water molecules. Positively charged sodium ions are attracted to the negative oxygen atom of water molecules and repelled by the positive hydrogen atoms. As a result, sodium ions orient themselves next to water molecules as follows:

$$Na^+$$
$$\overset{..}{\underset{..}{O}}\; 2\delta^-$$
$$\delta^+ H \qquad H\, \delta^+$$

There is an overall attraction between a sodium ion and a water molecule because the attraction between the sodium ion and the oxygen atom of water is greater than the repulsion between the sodium ion and the hydrogen atoms.

Negatively charged chloride ions are attracted to the positive hydrogen atoms of water molecules and repelled by the negative oxygen atoms. As a result, chloride ions orient themselves next to water molecules as follows:

$$2\delta^-$$
$$\overset{..}{\underset{..}{O}}$$
$$\delta^+ H \qquad H\, \delta^+$$
$$Cl^-$$

As a consequence of these attractions, an aqueous solution of sodium chloride consists of a uniform distribution of sodium and chloride ions surrounded by water molecules, as shown in Figure 1.29.

Any positive or negative ion is attracted to a polar molecule in the same manner. However, this attraction is usually weaker than the ionic bond formed between two ions of opposite charge. Attraction between ions and polar molecules is important in solutions and is called **ion solvation.** We will have more to say about solvation in Chapter 3.

Figure 1.29
Molecular picture of an aqueous solution of sodium chloride. The water molecules surrounding the ions are oriented so that their oxygen atoms are near sodium ions and their hydrogen atoms are near chloride ions.

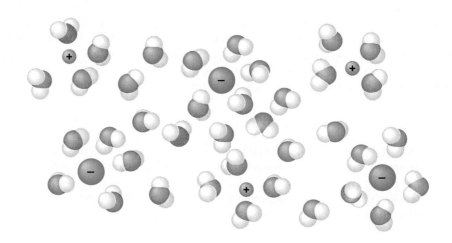

B   Dipole-Dipole Interactions

Polar molecules can attract or repel other polar molecules. If the dipoles of two polar molecules are oriented so that the negative end of one dipole is next to the positive end of the other dipole, then the dipoles will attract each other. If the two ends of like charge are next to each other, however, the dipoles will repel each other.

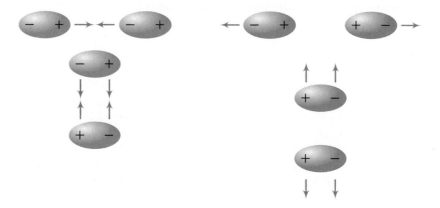

Dipole-dipole attraction   Dipole-dipole repulsion

The attraction of unlike charges lowers the energy of a system, so a collection of polar molecules will arrange itself so that the centers of opposite charge attract each other. Keep in mind, however, that the charges on polar molecules are usually small, so dipole-dipole interactions are weaker than ion-dipole or ion-ion interactions.

Although dipole-dipole interactions are relatively weak, they can be important in the liquid and solid states where molecules are close together. Consider, for example, the difference between the boiling points of iodine monochloride, ICl (bp = 97 °C) and bromine, Br_2 (bp = 58 °C). This difference of almost 40° is not due to a difference in their molecular masses because they are almost the same: 162.3 amu for ICl and 159.8 amu for Br_2. Instead the major difference is that there are no dipole-dipole interactions between nonpolar Br_2 molecules while there are between ICl molecules. The higher boiling point of ICl means that more energy must be added to ICl than to Br_2 to convert it to a gas. This additional energy is needed to overcome the dipole-dipole attractions present in ICl but not in Br_2.

■ *The boiling point of a compound is the temperature at which the intermolecular forces in the liquid state are overcome and the substance becomes a gas.*

C   Hydrogen Bonds

Hydrogen bonds are not bonds at all. They are a particularly strong type of dipole-dipole attraction. Bonds between hydrogen atoms and nitrogen, oxygen, and fluorine atoms are strongly polarized. The hydrogen atom of these bonds is strongly attracted by a nonbonding electron pair on oxygen and nitrogen atoms in other molecules. The most important hydrogen bonds in organic chemistry are between the following combinations of atoms, in which the hydrogen bond is depicted by a dotted line:

$$O—H\cdots\cdots:O \qquad N—H\cdots\cdots:O$$
$$O—H\cdots\cdots:N \qquad N—H\cdots\cdots:N$$

The hydrogen bonds in water and ammonia are shown in Figure 1.30.

The energies of most hydrogen bonds are about 6.0 kcal/mol (25 kJ/mol). This is much weaker than strong covalent and ionic bonds, whose energies range from 25 to over 120 kcal/mol (100 to 500 kJ/mol). Thus the strength of hydrogen bonds is intermediate between other intermolecular attractions (ion-dipole and dipole-dipole) and strong covalent and ionic bonds. A hydrogen bond also differs from other intermolecular attractions because it is localized between a hydrogen atom and a specific pair of electrons. Other intermolecular forces act between molecules as a whole rather than specific atoms on adjacent molecules.

The effect of hydrogen bonds on physical properties is dramatically demonstrated by examining the melting and boiling points of methane (CH_4), ammonia (NH_3), water (H_2O), and hydrogen fluoride (HF), as shown in Table 1.3.

The tremendous differences in the melting and boiling points of the compounds shown in Table 1.3 are not due to molecular mass differences, because their molecular masses differ only slightly. The differences result, instead, from the varying ability of the molecules of these compounds to form hydrogen bonds.

Figure 1.30
Hydrogen bonding in water **(A)** and ammonia **(B)**. Hydrogen bonding puts each oxygen atom in water and each nitrogen atom in ammonia at the center of a distorted tetrahedron of hydrogen atoms.

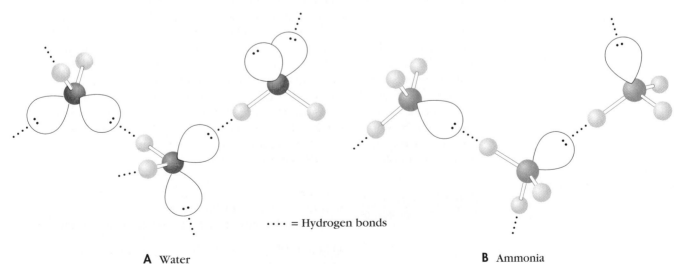

···· = Hydrogen bonds

A Water   **B** Ammonia

Table 1.3	Melting and Boiling Points of Methane, Ammonia, Water, and Hydrogen Fluoride.			
Molecule	CH_4	NH_3	H_2O	HF
Molecular mass	16.0	17.0	18.0	20.0
Melting point (°C)	–182	–77	0	–83
Boiling point (°C)	–164	–33	100	19

Methane, for example, cannot form hydrogen bonds so its boiling and melting points are the lowest. The other three can form hydrogen bonds. The hydrogen bonds in water must be particularly strong, however, because water has the highest melting and boiling points among the three. In other words, it must take more energy to break the hydrogen bonds of water than of ammonia or hydrogen fluoride. From this we conclude that hydrogen bonds between hydrogen and oxygen atoms are stronger than those between hydrogen and nitrogen or fluorine atoms.

D   London Forces

So far we have considered only intermolecular interactions between polar molecules or between polar molecules and ions. But intermolecular attractions must also exist between nonpolar molecules, such as the diatomic gases H_2, O_2, and N_2, because they condense to liquids when cooled to sufficiently low temperatures. An explanation of these forces was given in 1926 by Fritz London, who proposed that nonpolar molecules and atoms may have instantaneous molecular dipole moments caused by momentary unsymmetrical electron distribution in a molecule.

Electrons in molecules are in constant motion. Viewed over a long time their average distribution in a nonpolar molecule is such that the center of negative charge of the electrons coincides with the center of positive charge of the nuclei. If we were to take an instantaneous picture of the electron distribution, however, the centers of positive and negative charge would not generally coincide. As a result the nonpolar molecule has an instantaneous dipole moment that changes constantly in size and direction with the movement of electrons in the molecule. This changing dipole alternately attracts and repels the electrons of adjacent molecules, causing the formation of an ever-changing instantaneous dipole in the second molecule. Thus the dipole in the second molecule fluctuates in phase with the dipole in the first molecule, and these two instantaneous dipoles in adjacent molecules cause an attraction between the two nonpolar molecules.

These attractive forces, which are usually called *London forces,* are the reason why diatomic gases condense. Symmetrical diatomic molecules such as O_2 (bp = –183.0 °C), N_2 (bp = –195.8 °C), and H_2 (bp = –252.8 °C) have very low boiling points, so we conclude that London forces are very weak compared to hydrogen bonds, ion-dipole, or dipole-dipole interactions.

Exercise 1.16	Identify the most important intermolecular interaction between molecules of each of the following substances.

    (a) Liquid HF   **(b)** He   **(c)** IBr

The fact that the examples used in this section are not organic compounds does not mean that these intermolecular interactions are not important in organic molecules. On the contrary, as we introduce each class of organic compound in future chapters, it will become apparent that their physical properties are determined by one or more of these attractive forces. Which force predominates depends on the structure of the compound.

1.17   Summary

Organic chemistry is the study of carbon-containing compounds. Organic compounds are used to manufacture many of the products we consume and are essential to the life of all living organisms. An understanding of organic chemistry, therefore, is essential not only for people wishing to enter the health professions but also for any well-educated person.

Atoms are made up of a positively charged nucleus surrounded by one or more negatively charged electrons. **Lewis electron-dot symbols** are a convenient way of representing atoms and their valence electrons. A more detailed picture of the electron structure of atoms is provided by the **Schrödinger wave equation,** which describes the electrons in atoms as waves. The solutions to this wave equation are called **atomic orbitals.** Atomic orbitals are regions in space where there is a high chance of finding an electron. Different atomic orbitals have different energies and shapes. All *s* orbitals, for example, are spherical in shape, while *p* orbitals are shaped like dumbbells.

Lewis structures provide a convenient method of indicating how atoms in a molecule are connected. Plus (+) and minus (–) signs are used to indicate the **formal charge** on specific atoms in Lewis structures. Formal charges are a way of keeping track of the valence electrons of atoms in molecules.

$$\begin{array}{c}\text{Formal charge} \\ \text{on an atom in} \\ \text{a Lewis structure}\end{array} = \begin{bmatrix}\text{Number of valence} \\ \text{electrons in the} \\ \text{isolated atom}\end{bmatrix} - \begin{bmatrix}\text{Number of valence} \\ \text{electrons assigned in} \\ \text{the Lewis structure}\end{bmatrix}$$

Molecular structure or **molecular geometry,** the three-dimensional arrangement of atoms in a molecule, is an important property of a molecule. According to the **valence-shell electron-pair repulsion (VSEPR)** model, atoms or groups of atoms about a central atom arrange themselves to minimize electron pair repulsion. In this way, two valence electron pairs about an atom adopt a linear arrangement; three valence electron pairs adopt a trigonal planar arrangement; and four adopt a tetrahedral arrangement.

Chemical bonds form because the resulting molecules are more stable than their isolated atoms. Bonds in molecules can be described by **localized valence bond orbitals** formed by **orbital overlap.** Orbital overlap is the addition of the wave functions of the valence electrons of the two atoms forming the bond. To form bonds in organic molecules, carbon atoms use **hybrid atomic orbitals.** Hybrid atomic orbitals form by the combination of atomic orbitals on an atom. The hybrid orbitals of carbon atoms are chosen to agree with the known structure of the molecule. When forming the bonds of a tetrahedral carbon atom, an sp^3 hybrid carbon atom is used, which has four equivalent sp^3 **hybrid atomic orbitals** pointing to the corners of a regular tetrahedron. The bonds of trigonal carbon atoms, such as those in molecules that contain a double bond, are formed by an sp^2 hybrid carbon atom. An sp^2 hybrid carbon atom has three equivalent sp^2 **hybrid atomic orbitals** (all lying in a plane) and one unhybridized *p* orbital. A linear bond arrangement about a carbon atom, such as that in a triple bond, is formed by *sp* hybrid carbon atoms. An *sp* hybrid carbon atom has two equivalent **sp hybrid atomic orbitals** and two unhybridized *p* orbitals.

The structure of certain molecules cannot be adequately represented by a single Lewis structure because they have delocalized electron structures. Such molecules can be represented, however, as a **resonance hybrid** of two or more contributing Lewis structures. Delocalization of electrons in molecules can also be represented by the **linear combination of atomic orbitals (LCAO)** model. The LCAO model mathematically combines the atomic orbitals (wave functions) of the atoms to form **molecular orbitals.** Molecular orbitals encompass the entire molecule.

The **electronegativity** of an atom is its intrinsic ability to attract electrons. The type of bond between two atoms depends on the difference in their electronegativities. **Ionic bonds** form between atoms with large electronegativity differences, while **nonpolar covalent bonds** form between atoms of identical electronegativity. **Polar covalent bonds** are intermediate between these extremes of bond polarity. Electronegativity differences in polar covalent bonds are not large, so the bonding electrons are attracted only slightly more to the more electronegative atom.

The uneven electron distribution in a polar covalent bond results in a **bond dipole moment.** The vector sum of these bond dipole moments in a molecule determines its **molecular dipole moment. Polar molecules** have molecular dipole moments, but not all molecules with polar covalent bonds are necessarily polar. A symmetrical molecule in which the vector sum of its bond dipole moments is zero, for example, has no molecular dipole moment, so it is a **nonpolar molecule.** Examples of nonpolar compounds with polar covalent bonds are CCl_4, a tetrahedral molecule; BF_3, a trigonal-shaped molecule; and CO_2, a linear molecule.

Intermolecular interactions are electrostatic in nature and are responsible for many physical properties of molecules. Four kinds of intermolecular attractions are **ion-dipole, dipole-dipole, hydrogen bonds,** and **London forces.** Hydrogen bonds, which are really strong dipole-dipole attractions not bonds, are the strongest of the intermolecular forces. Hydrogen bonds, however, are not as strong as covalent or ionic bonds.

Additional Exercises

1.17   Define and give an example for each of the following terms.

(a)	Isotopes	**(j)**	Localized valence bond orbital	**(s)**	Resonance hybrid
(b)	Valence electrons	**(k)**	Ionic bond	**(t)**	sp^3 hybrid carbon atom
(c)	Lewis structures	**(l)**	Bonding molecular orbital	**(u)**	Hydrogen bond
(d)	Sigma (σ) bond	**(m)**	Polar covalent bond	**(v)**	Dipole-dipole interaction
(e)	Pi (π) bond	**(n)**	Nonpolar covalent bond	**(w)**	sp^2 hybrid carbon atom
(f)	Triple bond	**(o)**	Bond dipole moment	**(x)**	Ion-dipole interaction
(g)	Formal charge	**(p)**	Molecular dipole moment	**(y)**	sp hybrid carbon atom
(h)	Nonbonding electrons	**(q)**	Polar molecule	**(z)**	Antibonding molecular orbital
(i)	Atomic orbital	**(r)**	Resonance structures		

1.18   Give the number of valence electrons in an atom of each of the following elements.

(a)	Chlorine	**(b)**	Nitrogen	**(c)**	Phosphorus	**(d)**	Oxygen	**(e)**	Silicon

1.19   Give the name of the element corresponding to each of the following ground state electron configurations.

     (a) $1s^2\,2s^2\,2p^3$ **(b)** $1s^2\,2s^2\,2p^5$ **(c)** $1s^2\,2s^2\,2p^2$

1.20   Write the ground state electron configuration for each of the following elements.

     (a)   Lithium   **(b)**   Sulfur   **(c)**   Phosphorus   **(d)**   Chlorine   **(e)**   Silicon

1.21   Write Lewis structures for each of the following compounds. Be sure to include any nonbonding electrons that may be present.

     (a) CH_3NHCH_3 **(c)** $CH_2{=}CH_2$ **(e)** CH_3F **(g)** $CH_3C{\equiv}N$

     (b) CH_3OCH_3 **(d)** $H_2C{=}O$ **(f)** $CH_3{-}\overset{\underset{\displaystyle CH_3}{|}}{\underset{\underset{\displaystyle CH_3}{|}}{N^{\pm}}}{-}CH_3$ **(h)** CCl_4

1.22   Write Lewis structures for the following molecules or ions.

     (a) CH_3NH_2 **(b)** $NH_4{}^+$ **(c)** H_2CO **(d)** SiF_4 **(e)** $HC_2{}^-$ **(f)** CN^-

1.23   Calculate the formal charge on each atom and the net charge on each of the following ions or compounds.

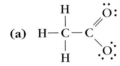 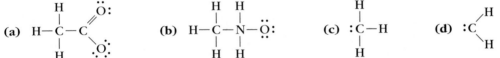

     (a)   **(b)**   **(c)**   **(d)**

1.24   Write the Lewis structure of the perchlorate ion, in which chlorine is bonded to four oxygen atoms, and the sulfate ion, in which sulfur is bonded to four oxygen atoms. Determine the formal charge on each atom and the net charge on both of these ions.

1.25   Predict the molecular geometry about the atoms indicated in each of the following compounds.

     (a)   Carbon atom of $CH_3{}^+$ **(e)**   Boron atom of $(CH_3O)_3B$
     (b)   Carbon atom of $CH_3{}^-$ **(f)**   Nitrogen atom of $(CH_3)_4N^+$
     (c)   Phosphorus atom of PCl_3 **(g)**   Iodine atom of ICl_3
     (d)   Oxygen atom of $(CH_3)_3O^+$ **(h)**   Nitrogen atom of CH_3NH_2

1.26   Nitryl chloride, O_2NCl, is a planar molecule. Write its Lewis structure, indicate the formal charge on each atom, and predict the bond angles about the nitrogen atom.

1.27   Identify the hybrid that can be used to describe the bonds formed by carbon in each of the following compounds or ions.

     (a) CCl_4 **(b)** $C{\equiv}N^-$ **(c)** $CH_3{}^+$ **(d)** $:CH_2$

1.28   Cyanamide, H_2NCN, is an important industrial chemical used to prepare fertilizers and melamine plastics.

     (a)   Write the Lewis structure of cyanamide.
     (b)   Predict the geometry about the carbon atom.
     (c)   Construct the localized valence bond orbital representation of cyanamide.
     (d)   How many σ bonds and how many π bonds are in cyanamide?

1.29   The $2s$ and $2p$ valence orbitals of an oxygen atom can also be combined to form hybrid atomic orbitals. Which hybrid oxygen atom would you choose, sp, sp^2, or sp^3, to construct the localized valence bond orbitals of water?

1.30   The $2s$ and $2p$ valence orbitals of a nitrogen atom can also be combined to form sp, sp^2, and sp^3 hybrid atomic orbitals. Which hybrid nitrogen atom would you use to construct the localized valence bond orbitals of each of the following molecules?

     (a) $:NH_3$ **(b)** $:N{\equiv}N:$ **(c)** $\overset{\displaystyle H}{\underset{\displaystyle H}{{>}}}C{=}\overset{\displaystyle H}{N\cdot}$

1.31   Write the Lewis structure of $F_2C{=}CH_2$. What are the predicted bond angles about each carbon atom? Using the correct hybrid carbon atom to approximate the needed geometry, draw the localized valence bond orbital representation of $F_2C{=}CH_2$.

1.32   Draw an orbital overlap model that shows the localized valence bond orbitals of formaldehyde.

$$\underset{H}{\overset{H}{>}}C=\ddot{O}:$$

1.33   Indicate the hybridization of each carbon atom in propene and acetonitrile and predict the value of the bond angles around each carbon atom.

Propene   Acetonitrile

1.34   Write the two resonance structures that contribute to the resonance hybrid of the linear molecule N_2O.

1.35   One contributing structure is written for each of the following molecules whose structure is represented as a resonance hybrid. Write the remaining contributing structures.

(a)   :$\ddot{O}=\overset{+}{O}-\ddot{O}:^-$ One additional contributing structure

(b)   :$\ddot{O}=\overset{+}{S}-\ddot{O}:^-$ One additional contributing structure

(c) $^-:\ddot{O}-\overset{O}{\underset{:\ddot{O}:^-}{Cr}}\overset{+}{-}\ddot{O}:^-$ Three additional contributing structures

1.36   The two structures shown below contain the same number of atoms and electrons. Are they both resonance structures of a resonance hybrid? Explain your answer.

$$H-\ddot{O}-C\equiv N: \qquad H-\ddot{N}=C=\ddot{O}:$$

1.37   What is the difference in electron distribution and energy between bonding molecular orbitals and antibonding molecular orbitals?

1.38   Use the LCAO model to construct the molecular orbitals of each of the following molecules, then predict which ones would be stable.

(a) H_2^+ (b) Li_2 (c) H_2^- (d) He_2^+ (e) Be_2

1.39   Show the direction of the bond dipole moment for each of the following bonds.

(a) Si—F   (b) C—N   (c) C—Li   (d) C—O   (e) N—F   (f) C—F

1.40   Classify the bonds in each of the following substances as ionic, nonpolar covalent, or polar covalent.

(a) NaF   (b) CCl_4 (c) MgF_2 (d) IBr   (e) I_2

1.41   Which of the following are polar molecules? Explain your answers.

(a) BCl_3 (b) NH_3 (c) CH_3OH (d) SO_2 (e) SiF_4 (f) CH_3OCH_3

1.42   Explain why CF_4 does not have a molecular dipole moment but CF_2Cl_2 does.

1.43   You have been given a sample of a compound that is either *cis-* or *trans*-1,2-dichloroethene. Would it be possible to determine the identity of the sample by experimentally determining its dipole moment? Explain.

cis-1,2-Dichloroethene   *trans*-1,2-Dichloroethene

1.44 Which of the following molecules form hydrogen bonds in the liquid state? Place a dotted line between two adjacent molecules to indicate the intermolecular hydrogen bond.

(a) CH_3OH **(b)** CH_3F **(c)** CH_3CH_3 **(d)** $CH_3C\equiv N$ **(e)** CH_3NH_2 **(f)** $HCCl_3$

1.45 Use intermolecular interactions to account for the differences in boiling points of the following compounds:

Compound	Structural Formula	Boiling Point (bp), °C
Water	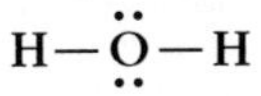 H—Ö—H	100
Ethyl alcohol	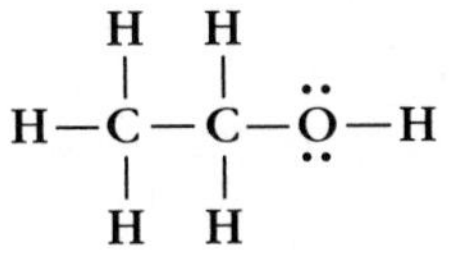 H—C—C—Ö—H	78
Ethane	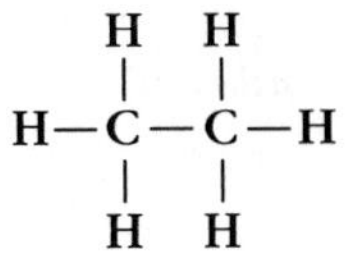 H—C—C—H	−89
Dimethyl ether	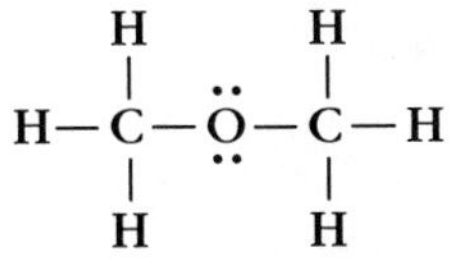 H—C—Ö—C—H	−25

CHAPTER

ALKANES, CYCLOALKANES, and THEIR DERIVATIVES

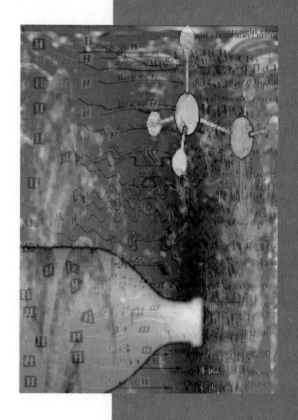

2.1   Introduction

Organic chemistry, like any branch of work or study, has its own terms and definitions that must be learned in order to understand the subject. We introduce a number of terms, definitions, and concepts in this chapter that serve as a foundation for all organic chemistry. We begin by defining alkanes.

Alkanes are a family of compounds that contain only C—H and C—C single bonds. Methane, CH_4, and ethane, C_2H_6, which were introduced in Chapter 1, are the simplest members of this family. Like methane and ethane, the carbon atoms of all alkanes have tetrahedral geometry. Thus we can correctly represent the geometry and electron structure of alkanes by using sp^3 hybrid carbon atoms to form localized orbitals, as discussed in Section 1.11A.

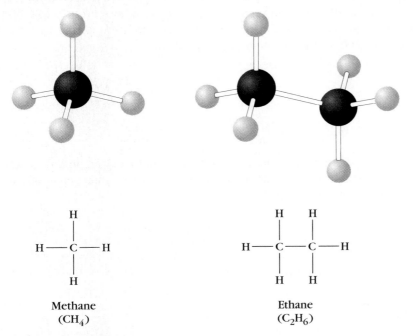

Methane
(CH_4)

Ethane
(C_2H_6)

Alkanes whose carbon atoms are joined together in a straight chain are called **straight chain** or **unbranched alkanes.** No carbon atom in a straight chain alkane is bonded to more than two other carbon atoms. Alkanes that contain at least one carbon atom bonded to more than two other carbon atoms are called **branched chain alkanes.** The **branch point** is the carbon atom that is bonded to more than two other carbon atoms. The following are examples of straight and branched chain alkanes:

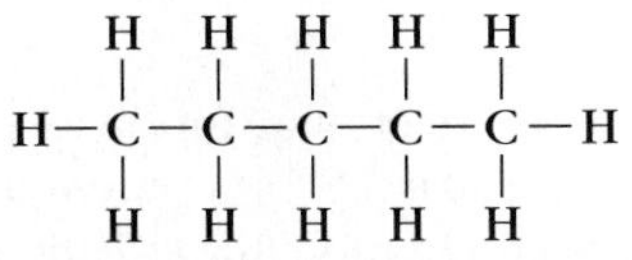

Branch point

Straight chain or unbranched alkane

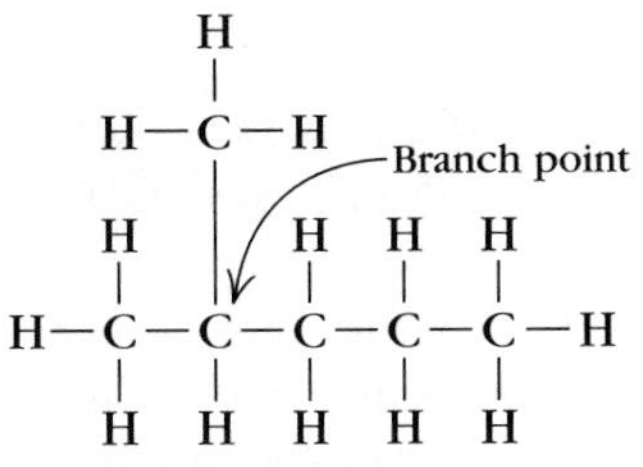

Branched chain alkane

A compound is said to be **acyclic** if its structure does not contain a ring. Both branched and straight chain alkanes are examples of **acyclic compounds.**

The atoms of many organic compounds are joined together to form rings. Compounds containing rings are called **cyclic compounds.** Alkanes that contain rings of carbon atoms are called **cycloalkanes.** The following are examples of cycloalkanes:

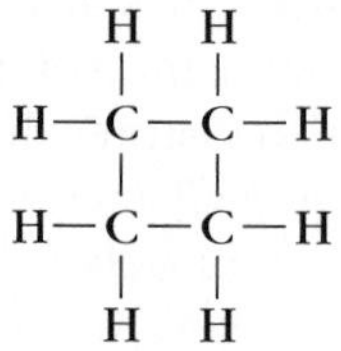

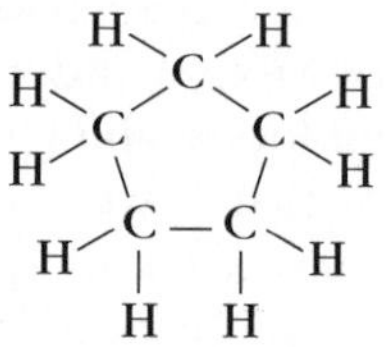

Compounds that contain only the elements carbon and hydrogen are called **hydrocarbons.** Alkanes are a special subclass of hydrocarbons. They are usually called **saturated hydrocarbons** because they contain the maximum number of hydrogen atoms per carbon. Hydrocarbons that contain a carbon-carbon double bond are called **alkenes** and those that contain a carbon-carbon triple bond are called **alkynes.** Ethylene and acetylene, which were introduced in Chapter 1, are examples of each of these families of compounds.

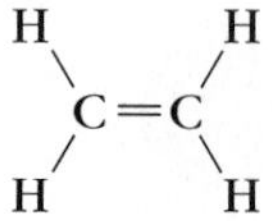

Ethylene
(an alkene)

Acetylene
(an alkyne)

Benzene is the most common example of a family of compounds called **aromatic hydrocarbons.** All members of this family contain an aromatic ring, which is usually a benzene ring.

Benzene
(an aromatic hydrocarbon)

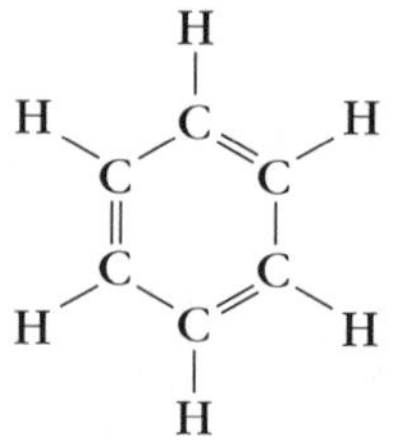

Alkenes, alkynes, and aromatic hydrocarbons are sometimes referred to as **unsaturated hydrocarbons** because they don't contain the maximum number of hydrogen atoms per carbon atom.

Alkanes, alkenes, alkynes, and aromatic hydrocarbons differ in their chemical properties, so we shall study each in separate chapters. We study alkanes in this chapter, alkenes in Chapters 7 and 8, alkynes in Chapter 9, and aromatic compounds in Chapters 20 and 21.

2.2   How to Write Condensed Structural Formulas of Alkanes and Cycloalkanes

We learned in Section 1.5 how to write Lewis or structural formulas of simple molecules. Although such structural formulas are useful, they occupy a lot of space and are time-consuming to write. As a result, chemists have devised several ways of condensing them. One way to condense these structures is to use **condensed structural formulas.** In condensed structural formulas carbon-hydrogen and carbon-carbon bonds are implied instead of shown explicitly. For example, the condensed structural formula of ethane is written as follows:

$$\begin{array}{c} H \quad H \\ | \quad | \\ H-C-C-H \\ | \quad | \\ H \quad H \end{array} \quad \equiv \quad CH_3CH_3$$

Lewis structural formula   Condensed
structural formula

Each carbon atom in ethane is bonded to three hydrogen atoms, so this is simply represented by writing CH_3. The three C—H bonds aren't shown, nor is the C—C bond. Instead we simply place the two CH_3 groups next to each other. In a similar way, a carbon atom bonded to two hydrogen atoms is written as CH_2, while a carbon atom bonded to a single hydrogen atom is written as CH, as illustrated in the following examples:

$$\begin{array}{c} H \quad H \quad H \\ | \quad | \quad | \\ H-C-C-C-H \\ | \quad | \quad | \\ H \quad H \quad H \end{array} \quad \equiv \quad CH_3CH_2CH_3$$

Condensed
structural formulas

$$\begin{array}{c} H \\ | \\ H-C-H \\ \\ H \quad | \quad H \quad H \\ | \quad | \quad | \quad | \\ H-C-C-C-C-H \\ | \quad | \quad | \quad | \\ H \quad H \quad H \quad H \end{array} \quad \equiv \quad \begin{array}{c} CH_3 \\ | \\ CH_3CHCH_2CH_3 \end{array}$$

2-Methylbutane   Condensed
structural formula

Sometimes a C—C bond is included in the condensed structural formula for clarity. An example is the vertical C—C bond in the condensed formula of 2-methylbutane.

Another simplification is to place identical CH_2 units of a chain or identical groups bonded to the same atom in parentheses and indicate the number of them with a subscript. The following examples illustrate this method:

$$\begin{array}{c} H \quad H \quad H \quad H \quad H \\ | \quad | \quad | \quad | \quad | \\ H-C-C-C-C-C-H \\ | \quad | \quad | \quad | \quad | \\ H \quad H \quad H \quad H \quad H \end{array} \quad \equiv \quad CH_3CH_2CH_2CH_2CH_3 \quad \equiv \quad CH_3(CH_2)_3CH_3$$

$$\begin{array}{c} H \\ | \\ H-C-H \\ \\ H \quad | \quad H \quad H \\ | \quad | \quad | \quad | \\ H-C-C-C-C-H \\ | \quad | \quad | \quad | \\ H \quad | \quad H \quad H \\ | \\ H-C-H \\ | \\ H \end{array} \quad \equiv \quad \begin{array}{c} CH_3 \\ | \\ CH_3CCH_2CH_3 \\ | \\ CH_3 \end{array} \quad \equiv \quad (CH_3)_3CCH_2CH_3$$

Finally the most drastic simplification is to write structural formulas in which neither carbon nor hydrogen atoms are shown explicitly. The following are examples of these structures, which are called **line** or **skeletal structures**:

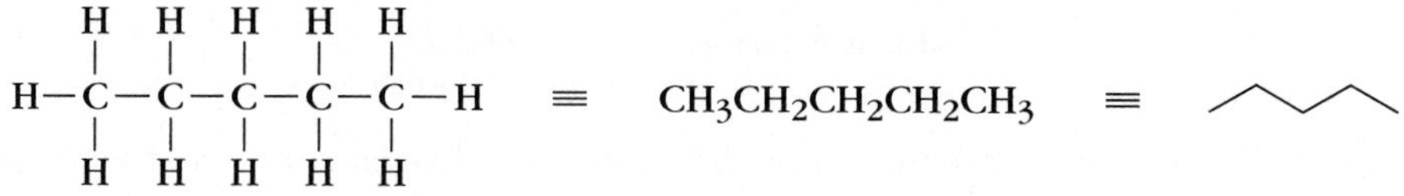

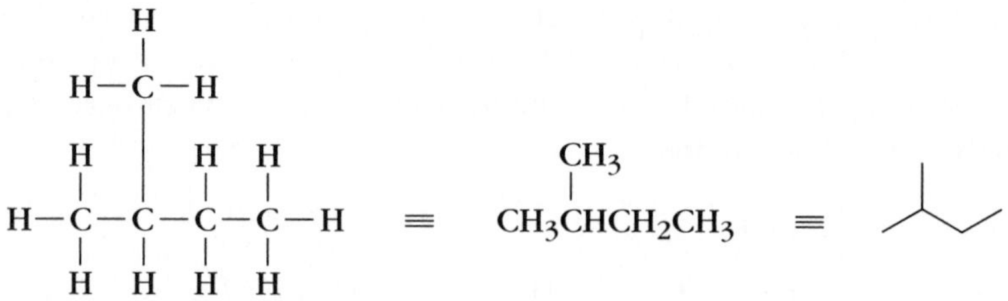

In line structures, a carbon atom is assumed to be at the end of each line and at the intersection of any two, three, or four lines. The correct number of hydrogen atoms is assumed to be bonded to each carbon atom to complete its valence of four.

Line or skeletal structures are particularly useful for writing the structures of cycloalkanes. For example:

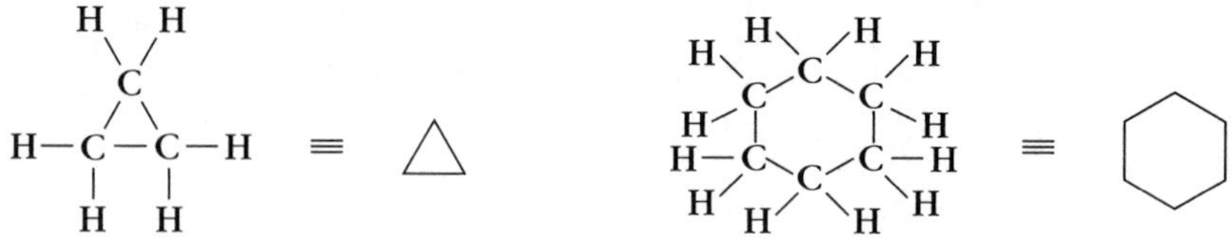

| **Exercise 2.1** | Convert each of the following into structural formulas. |

(a) $CH_3(CH_2)_6CH(CH_3)_2$

(b) $(CH_3)_3C(CH_2)_2\overset{\displaystyle CH_2CH_3}{\overset{|}{C}}HCH_3$

(c)

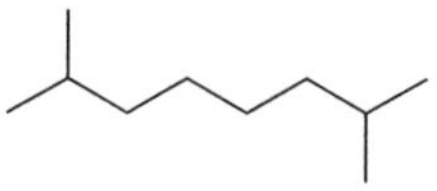

(d)

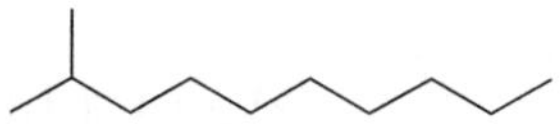

(e) $CH_3CH_2\overset{\displaystyle CH_2CH(CH_3)_2}{\overset{|}{C}}H(CH_2)_2C(CH_3)_3$

(f)

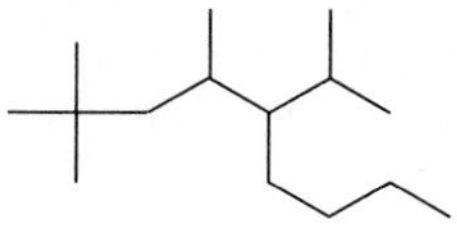

(g)

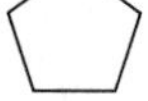

(h)

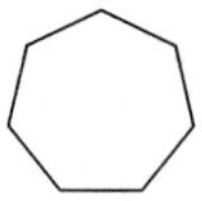

(i)

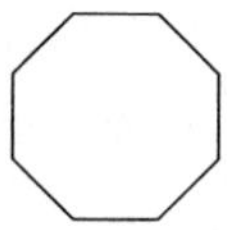

A word of caution—condensed structural formulas do not represent the arrangement of the atoms of a molecule in space. For example, the tetrahedral arrangement of atoms and groups of atoms about an sp^3 hybrid carbon atom is not represented by any condensed structural formula. In practical terms this means that the atoms of a structural formula can be arranged on a page in any way as long as the atoms are correctly joined together. While it is convenient to write the longest chain of a molecule horizontally, other equally correct representations are possible. For example, all four of the following condensed structural formulas correctly represent the structure of butane, C_4H_{10}, an alkane that contains four carbon atoms:

$$
\begin{array}{llll}
CH_3 & & CH_3 & \\
| & CH_2 & | & CH_3 \\
CH_2CH_2CH_3 \quad H_3C \diagdown CH_2 & CH_2 & | & CH_2CH_2 \\
 & H_3C \diagup & CH_2 & | \\
 & & CH_3 & CH_3
\end{array}
$$

As we convert the molecular formula of an organic compound into its structural formula, we soon find that we can write more than one structural formula for most molecular formulas.

2.3 Constitutional Isomers

The variety of structures of molecules in organic chemistry arises because of the different numbers of atoms or types of atoms that combine to form organic molecules. However, variations in structure can also result from a difference in the way the same atoms are connected. For example, we can write two structural formulas for the molecular formula, C_2H_6O.

$$
\begin{array}{cc}
CH_3CH_2\overset{..}{\underset{..}{O}}H & CH_3\overset{..}{\underset{..}{O}}CH_3 \\
C_2H_6O & C_2H_6O \\
\text{Ethanol} & \text{Dimethyl ether} \\
\text{(bp 78 °C)} & \text{(bp −24 °C)}
\end{array}
$$

These two structural formulas differ in the way their atoms are connected. Ethanol contains one carbon-carbon single bond and one carbon-oxygen single bond, while dimethyl ether contains two carbon-oxygen single bonds. Thus these structures represent two different compounds. One, dimethyl ether, is a gas (bp −24 °C), while the other, ethanol, is a liquid (bp 78 °C) at room temperature.

Two or more different compounds that have the same molecular formula, such as ethanol and dimethyl ether, are called **isomers.** Ethanol and dimethyl ether, furthermore, are special kinds of isomers called **constitutional isomers** because they differ in the order in which their atoms are connected.

Alkanes that contain three or fewer carbon atoms have no isomers because the atoms can be arranged in only one way:

$$
\begin{array}{ccc}
CH_4 & CH_3CH_3 & CH_3CH_2CH_3 \\
\text{Methane} & \text{Ethane} & \text{Propane}
\end{array}
$$

Only one compound has ever been isolated for each of these molecular formulas.

The carbon atoms of the four-carbon alkane, C_4H_{10}, can be arranged in two ways:

Table 2.1

Molecular Formulas and the Number of Their Constitutional Isomers

Molecular Formula	Number of Constitutional Isomers
C_5H_{12}	3
C_6H_{14}	5
C_7H_{16}	9
C_9H_{20}	35
$C_{11}H_{24}$	159
$C_{15}H_{32}$	4347
$C_{30}H_{62}$	4,111,846,763

$$CH_3CH_2CH_2CH_3$$
Butane
(bp $-0.5\ °C$)

$$CH_3\overset{\overset{\displaystyle CH_3}{|}}{CH}CH_3$$
Methylpropane
(bp $-12\ °C$)

Butane and methylpropane are constitutional isomers just like ethanol and dimethyl ether.

The number of structural isomers that can exist increases rapidly as the number of carbon atoms in an alkane increases, as shown in Table 2.1

Exercise 2.2 Write the structural formulas of the three constitutional isomers of molecular formula C_5H_{12}, the five constitutional isomers of molecular formula C_6H_{14}, and the nine constitutional isomers of molecular formula C_7H_{16}.

The two constitutional isomers of molecular formula C_4H_{10} have distinctive names. How did they get these names? To answer this question, we must learn how to name organic compounds.

2.4 How Do We Name Organic Compounds?

Until about the end of the 19th century, compounds were routinely named by their discoverers. Often the name reflected the source of the compound. The name *formic acid,* for example, was given to a substance obtained from a certain species of ant (from the Latin name for ants, *formicae*), while the crystalline substance isolated from urine was given the name *urea.* Other names were chosen more arbitrarily. The tranquillizing agent, barbituric acid, for example, was named by its discoverer for a friend named *Barbara.* The problem with these names was that the structures of most of the substances were unknown so it was impossible to relate the names of the substances to their structures.

As the structures of more and more compounds became known, it became possible to devise a systematic method that would give a unique name to a compound based on its structure. Such a system has been developed over the years by the International Union of Pure and Applied Chemistry (IUPAC). The IUPAC system is a set of rules by which any organic compound can be named from its structure in a logical and consistent manner. The reverse is also true; from the IUPAC name of a compound, its structural formula can be written.

The goal of assigning a unique and systematic name to every organic compound is laudable but impractical. One reason is that long before the IUPAC system was devised, many compounds in common use already had names. Some of these names, called trivial or **common names,** are still used today.

There are other reasons for common names. The structures of compounds isolated from natural sources have to be determined. A name must be given to the compound while this work is going on. Because its structure is not known, a common name is given that is usually based on the source of the compound. Even when the structure of the compound is finally determined, its IUPAC name may be so complex that the common name is retained for everyday use. Despite these minor problems, the IUPAC system is extensively used, and learning a few fundamental IUPAC rules of naming compounds is essential to the study of organic chemistry.

Let's start by learning how to name alkanes.

2.5 Naming Alkanes

According to the IUPAC system, the structure of any acyclic compound can be regarded as being formed from a straight chain alkane by *replacing one or more hydrogen atoms of the chain by other atoms or groups.* This means that straight chain alkanes are the parent structures for all acyclic compounds. 2-Methylbutane, for example, a five-carbon acyclic alkane is considered to be formed from butane,

$$CH_3$$
$$|$$
$$CH_3CHCH_2CH_3$$
2-Methylbutane

the four-carbon straight chain alkane, by replacing one specific hydrogen atom with a CH_3 group.

Replacing
this hydrogen atom
with a CH_3 group forms

H CH_3
| |
$CH_3CHCH_2CH_3$ $\longrightarrow$ $CH_3CHCH_2CH_3$

Butane   2-Methylbutane

2-Methylbutane then is a derivative of butane. In other words, the structure of butane has been modified in some way to get this structure. Atoms or groups of atoms that replace one or more hydrogen atoms of a straight chain alkane (like the CH_3 group in 2-methylbutane) are called **substituents.**

The IUPAC names of acyclic compounds are constructed in a similar way. A compound's IUPAC name consists of the name of the parent structure plus the name or names of any substituents. Therefore to name an alkane by the IUPAC method, we must learn both the names of the parent structures, the straight chain alkanes, and any substituents. We begin with the straight chain alkanes.

A   Straight Chain Alkanes

The names of straight chain alkanes containing 1 to 24 carbon atoms are given in Table 2.2 on p. 58. The structures of the compounds in Table 2.2 are arranged so that each one differs from its neighbors by only one CH_2 group. A CH_2 group is called a **methylene group.** A series of compounds such as the alkanes in Table 2.2 is called a **homologous series,** and the compounds in the series are called **homologs.**

Straight chain alkanes consist of a chain of CH_2 groups with an extra hydrogen atom at the ends of the chain. Thus we can represent a typical straight chain alkane in the following way:

$$H-(CH_2)_n-H$$

A straight chain alkane that contains n carbon atoms must also contain $(2n + 2)$ hydrogens. Thus its general formula is $C_nH_{2n + 2}$, where n is the number of carbon atoms in the alkane.

We are already familiar with the names of the first four members of this series: methane, ethane, propane, and butane. Other members are simply named by adding the ending *-ane* to the Greek name for the number of carbon atoms in the compound. Thus pentane is derived by adding the ending *-ane* to the Greek prefix *pent,* meaning five. The names of all alkanes end in *-ane,* which is the IUPAC ending denoting an alkane.

Table 2.2	Names of Straight Chain Alkanes Containing 1 to 24 Carbon Atoms	
Number of Carbon Atoms	Name	Structural Formula
1	Methane	CH_4
2	Ethane	CH_3CH_3
3	Propane	$CH_3CH_2CH_3$
4	Butane	$CH_3(CH_2)_2CH_3$
5	Pentane	$CH_3(CH_2)_3CH_3$
6	Hexane	$CH_3(CH_2)_4CH_3$
7	Heptane	$CH_3(CH_2)_5CH_3$
8	Octane	$CH_3(CH_2)_6CH_3$
9	Nonane	$CH_3(CH_2)_7CH_3$
10	Decane	$CH_3(CH_2)_8CH_3$
11	Undecane	$CH_3(CH_2)_9CH_3$
12	Dodecane	$CH_3(CH_2)_{10}CH_3$
13	Tridecane	$CH_3(CH_2)_{11}CH_3$
14	Tetradecane	$CH_3(CH_2)_{12}CH_3$
15	Pentadecane	$CH_3(CH_2)_{13}CH_3$
16	Hexadecane	$CH_3(CH_2)_{14}CH_3$
17	Heptadecane	$CH_3(CH_2)_{15}CH_3$
18	Octadecane	$CH_3(CH_2)_{16}CH_3$
19	Nonadecane	$CH_3(CH_2)_{17}CH_3$
20	Icosane	$CH_3(CH_2)_{18}CH_3$
21	Henicosane	$CH_3(CH_2)_{19}CH_3$
22	Docosane	$CH_3(CH_2)_{20}CH_3$
23	Tricosane	$CH_3(CH_2)_{21}CH_3$
24	Tetracosane	$CH_3(CH_2)_{22}CH_3$

Now that we know the names of the parent compounds, let's learn how to name simple carbon-containing substituents.

B   Naming Unbranched Alkyl Groups

An **alkyl group** is the part of the structure that remains after removing one hydrogen atom from an alkane. For example, a methyl group (CH_3—) is the alkyl group that is formed by the removal of one hydrogen atom from methane. Removal of a hydrogen atom from the end carbon atom of any straight chain alkane forms an unbranched alkyl group. Unbranched alkyl groups are named by replacing the *-ane* endings of the names of the straight chain alkanes given in Table 2.2 by an *-yl* ending.

$CH_3CH_2CH_2$— $CH_3(CH_2)_4CH_2$— $CH_3(CH_2)_{10}CH_2$—

Propyl group   Hexyl group   Dodecyl group

A bond line in these structures indicates that the alkyl group is bonded at that carbon atom to another unspecified group or atom.

Exercise 2.3	Write the structures of the alkyl groups formed by the removal of one hydrogen atom from the end carbon atom of each of the following alkanes.

(a) Butane   **(b)** Octane   **(c)** Undecane   **(d)** Icosane   **(e)** Tridecane

We're now ready to use the IUPAC rules to construct the names of branched alkanes from the names of alkyl groups and the parent alkane.

C   Naming Branched Alkanes

The names of all but the most complex branched chain alkanes can be obtained by combining the names of alkyl groups and the parent alkane name by means of the following four steps.

Step 1   Determine the parent name by finding the longest continuous chain of carbon atoms in the molecule.

The parent name is the name of the alkane corresponding to the longest continuous chain of carbon atoms. Keep in mind that the longest continuous chain of carbon atoms may not always be written horizontally!

$CH_3CH_2CHCH_3$
$\quad\quad\quad\;|$
$\quad\quad CH_2CH_2CH_3$

Named as a substituted hexane (6 carbon atoms) not butane (4 carbon atoms) or propane (3 carbon atoms)

$\quad\quad CH_2CH_2CH_3$
$\quad\quad\quad\;|$
$CH_3CH_2CHCHCH_2CH_3$
$\quad\quad\quad\quad\;|$
$\quad\quad\quad CH_2CH_2CH_2CH_3$

Named as a substituted nonane

If the molecule contains two chains of equal length, choose the one with the greater number of substituents as the parent chain.

$\quad\; CH_3$
$\quad\;\;|$
$CH_3CHCHCH_2CH_2CH_2CH_3$
$\quad\quad\;|$
$\quad\; CH_2CH_3$

Named as a disubstituted heptane

not

$\quad\; CH_3$
$\quad\;\;|$
$CH_3CHCHCH_2CH_2CH_2CH_3$
$\quad\quad|$
$\quad CH_2CH_3$

Named as a monosubstituted heptane

Step 2   Number the carbon atoms in the parent chain starting at the end nearest the first branch point.

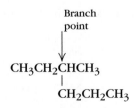

Branch point

$CH_3CH_2CHCH_3$
$\quad\quad\quad|$
$\quad\quad CH_2CH_2CH_3$

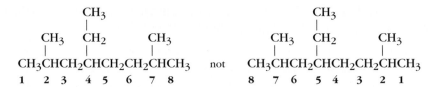

$$
\begin{array}{c}
\quad\; 1 \quad 2 \quad 3 \\
\text{Numbered as} \quad CH_3CH_2CHCH_3 \\
\qquad\qquad\qquad\quad | \\
\qquad\qquad\quad CH_2CH_2CH_3 \\
\qquad\qquad\qquad 4 \quad 5 \quad 6
\end{array}
\qquad \text{not} \qquad
\begin{array}{c}
6 \quad 5 \quad 4 \\
CH_3CH_2CHCH_3 \\
\qquad\qquad | \\
\quad CH_2CH_2CH_3 \\
\qquad 3 \quad 2 \quad 1
\end{array}
$$

If the branch point is the same number of carbon atoms from each end, then begin numbering at the end nearer the second branch point.

$$
\begin{array}{c}
\qquad\qquad\; CH_3 \\
\qquad\qquad\;\; | \\
CH_3 \quad CH_2 \qquad CH_3 \\
| \qquad\; | \qquad\quad | \\
CH_3CHCH_2CHCH_2CH_2CHCH_3 \\
1 \quad 2 \;\; 3 \quad 4 \;\; 5 \;\; 6 \;\; 7 \;\; 8
\end{array}
\quad \text{not} \quad
\begin{array}{c}
\qquad\qquad\; CH_3 \\
\qquad\qquad\;\; | \\
CH_3 \quad CH_2 \qquad CH_3 \\
| \qquad\; | \qquad\quad | \\
CH_3CHCH_2CHCH_2CH_2CHCH_3 \\
8 \quad 7 \;\; 6 \quad 5 \;\; 4 \;\; 3 \;\; 2 \;\; 1
\end{array}
$$

Step 3　Name each substituent and precede this name with the number of the carbon atom on the parent chain to which it is bonded.

$$
\begin{array}{c}
1 \quad 2 \quad 3 \\
CH_3CH_2CHCH_3 \\
\qquad\qquad | \\
\quad CH_2CH_2CH_3 \\
\qquad 4 \quad 5 \quad 6
\end{array}
$$

Parent name: hexane
Substituent: methyl group ($-CH_3$)
Location: attached to C3; 3-methyl

If two substituents are attached to the same carbon atom of the parent chain, give them both the same number.

$$
\begin{array}{c}
\qquad\qquad CH_3 \\
1 \quad 2 \;\; 3| 4 \quad 5 \quad 6 \\
CH_3CH_2CCH_2CH_2CH_3 \\
\qquad\qquad | \\
\qquad\;\; CH_2CH_3
\end{array}
$$

Parent name: hexane
Substituents: methyl group ($-CH_3$)
　　　　　　ethyl group ($-CH_2CH_3$)
Location: methyl group attached to C3; 3-methyl
　　　　 ethyl group attached to C3; 3-ethyl

Step 4　Write the name of the compound as a single word.

The name and number of the substituents are prefixes. Hyphens are used to separate the various prefixes and commas are used to separate numbers.

$$
\begin{array}{c}
1 \quad 2 \quad 3 \\
CH_3CH_2CHCH_3 \\
\qquad\qquad | \\
\quad CH_2CH_2CH_3 \\
\qquad 4 \quad 5 \quad 6
\end{array}
$$

Parent name: hexane
Substituent: methyl group ($-CH_3$)
Location: attached to C3; 3-methyl
NAME: 3-methylhexane

　Two or more identical substituents present in a molecule are indicated by the use of the following prefixes.

Number of identical substituents	2	3	4	5	6	7	8
Prefix	di-	tri-	tetra-	penta-	hexa-	hepta-	octa-

Each of the identical substituents must have a number.

$$
\begin{array}{c}
\qquad CH_3 \\
\qquad | \;\; 3 \;\; 4 \\
CH_3CHCHCH_3 \\
1 \quad 2 \;\; | \\
\qquad\quad CH_3
\end{array}
$$

Parent name: butane
Substituents: two methyl groups ($-CH_3$)
Locations: one attached to C2; 2-methyl
　　　　　 other attached to C3; 3-methyl
NAME: 2,3-dimethylbutane

When two or more substituents are present, their names are listed alphabetically. Ethyl, for example, is placed before methyl. When deciding the alphabetical order, ignore the prefixes di-, tri-, tetra-, and so forth.

$$CH_3$$
$$\overset{1\quad 2\quad 3|\ 4\quad 5\quad 6}{CH_3CH_2CCH_2CH_2CH_3}$$
$$|$$
$$CH_2CH_3$$

Parent name: hexane
Substituents: methyl group ($—CH_3$)
             ethyl group ($—CH_2CH_3$)
Location: methyl group attached to C3; 3-methyl
        ethyl group attached to C3; 3-ethyl
NAME: 3-ethyl-3-methylhexane

The IUPAC name of a compound such as 3-methylhexane is made up of four parts. The first part is the name of the parent chain, which indicates the number of carbon atoms in the chain. For 3-methylhexane, the parent chain is hexane, so it contains six carbons. The second part is the ending or suffix that classifies the compound. The ending *-ane* indicates that this compound is an alkane. The third part is the prefix or prefixes that tell us what kind of substituents are attached to the parent chain. For 3-methylhexane, the substituent is a methyl group. Finally one or more numbers indicate where the substituents are located. For 3-methylhexane, the number 3 indicates that the methyl group is located on carbon atom 3 of the hexane chain.

Name of substituent ⎤   ⎡ Number of carbon atoms
                              in parent chain

3–methyl<u>hex</u>ane

Location of substituent ⎤   ⎡ Suffix indicates class of
                             compound (an alkane)

Sample Problem 2.1   Give an IUPAC name to the following compound.

$$CH_3CH_2CHCH_2CH_2CH_2CH_3$$
$$|$$
$$CH_3CH_2CHCH_3$$

Solution:

Step 1 The longest continuous chain contains eight carbon atoms. Therefore the parent name is octane.

Step 2 Numbering the carbon chain from left to right gives the lowest number to the first branch.

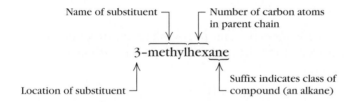

Step 3 There are two substituents:
    one ethyl group attached to C4; 4-ethyl
    one methyl group attached to C3; 3-methyl

Step 4 The IUPAC name is 4-ethyl-3-methyloctane.

| Sample Problem 2.2 | Write the structure of 2,4-dimethyl-4-propyldecane. |

Solution:

Step 1 Look at the parent name to determine how many carbon atoms are in the parent chain and write the compound's carbon skeleton. The parent name is decane so the longest chain contains 10 carbon atoms.

$$C-C-C-C-C-C-C-C-C-C$$

Step 2 Identify the substituents and their locations, then place them on the proper carbon atoms. There are two methyl groups, one on C2 and the other on C4. There is also one propyl group on C4.

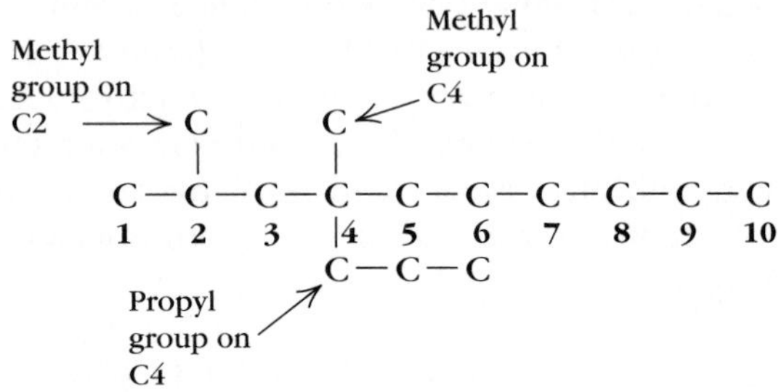

Methyl
group on
C2 $\longrightarrow$ C

Methyl
group on
C4

$$C-C-C-C-C-C-C-C-C-C$$
$$1 \quad 2 \quad 3 \quad |4 \quad 5 \quad 6 \quad 7 \quad 8 \quad 9 \quad 10$$
$$C-C-C$$

Propyl
group on
C4

Step 3 Add the hydrogen atoms to complete the structure. Make sure enough hydrogen atoms are added so that each carbon atom forms a total of four bonds.

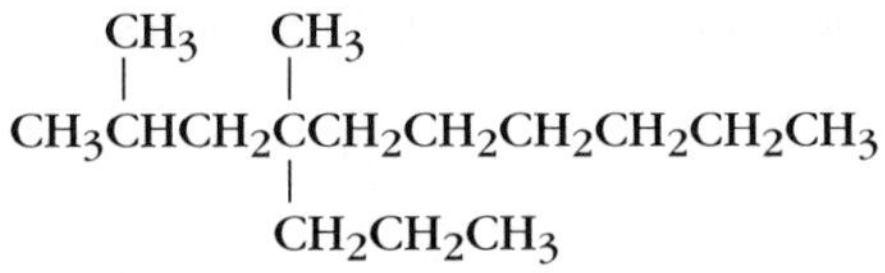

$$\begin{array}{cc} CH_3 & CH_3 \\ | & | \\ CH_3CHCH_2CCH_2CH_2CH_2CH_2CH_2CH_3 \\ & | \\ & CH_2CH_2CH_3 \end{array}$$

| Sample Problem 2.3 | Give an IUPAC name to the following compound. |

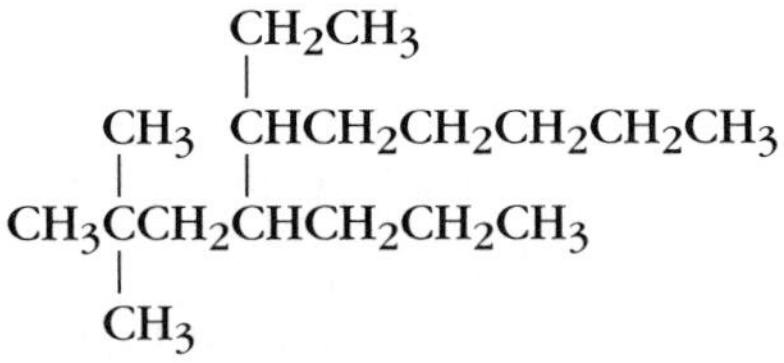

$$\begin{array}{c} CH_2CH_3 \\ | \\ CH_3 \quad CHCH_2CH_2CH_2CH_2CH_3 \\ | \quad \quad | \\ CH_3CCH_2CHCH_2CH_2CH_3 \\ | \\ CH_3 \end{array}$$

Solution:

Step 1 The parent name is decane.

Step 2 Numbering the carbon chain from left to right gives the lowest number to the first branch.

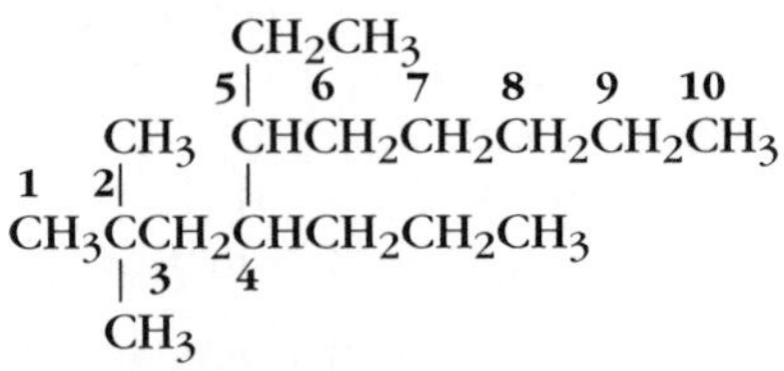

$$\begin{array}{c} CH_2CH_3 \\ 5| \quad 6 \quad 7 \quad 8 \quad 9 \quad 10 \\ CH_3 \quad CHCH_2CH_2CH_2CH_2CH_3 \\ 1 \quad 2| \quad \quad | \\ CH_3CCH_2CHCH_2CH_2CH_3 \\ |3 \quad 4 \\ CH_3 \end{array}$$

Step 3 There are four substituents:
one ethyl group attached to C5; 5-ethyl

two methyl groups attached to C2; 2,2-dimethyl
one propyl group attached to C4; 4-propyl

Step 4 The IUPAC name is 5-ethyl-2,2-dimethyl-4-propyldecane.

Notice that the di- in dimethyl does not affect the alphabetical order in which the substituents are listed; that is, ethyl is listed before methyl.

Exercise 2.4   Give the correct IUPAC name for each of the following compounds.

(a) CH$_3$CHCH$_2$CH$_3$
      |
    CH$_2$CH$_3$

(d)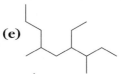

          CH$_3$
          |
(b) CH$_3$CH$_2$CCH$_2$CHCH$_2$CH$_3$
         | |
        CH$_3$ CH$_2$CH$_2$CH$_3$

(e)

(c) CH$_3$CH(CH$_2$)$_{10}$CH$_3$
      |
    CH$_3$

(f)

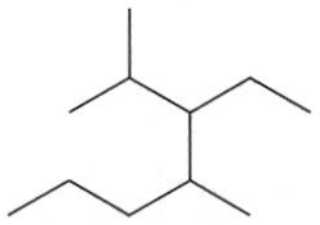

Exercise 2.5   Write the structure corresponding to each of the following IUPAC names.

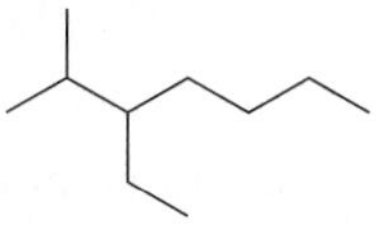

(a) 2-Methylpentane
(b) 3-Ethylheptane
(c) 3,3-Diethylpentane

(d) 6-Ethyl-2,2-dimethyloctane
(e) 2,3,7-Trimethyl-5-propyldodecane

Exercise 2.6   Why is the IUPAC name 1-ethylhexane incorrect?

Although the use of the IUPAC system of naming organic compounds is encouraged, some simple alkanes have common names that are used quite frequently.

D   Common Names

The constitutional isomers of alkanes containing four and five carbon atoms were given common names long ago. We must learn these common names because they are used so widely. The two isomeric alkanes of molecular formula C$_4$H$_{10}$ are called *n*-butane and isobutane.

                                        CH$_3$
                                        |
     CH$_3$CH$_2$CH$_2$CH$_3$ CH$_3$CHCH$_3$

         n-Butane   Isobutane

The common names of the three isomers of C_5H_{12} are *n*-pentane, isopentane, and neopentane.

$$CH_3CH_2CH_2CH_2CH_3 \qquad \underset{\overset{|}{\underset{}{}}}{CH_3CHCH_2CH_3} \qquad \underset{\overset{|}{CH_3}}{\overset{\overset{CH_3}{|}}{CH_3CCH_3}}$$

$$\overset{CH_3}{\underset{}{}}$$

n-Pentane   Isopentane 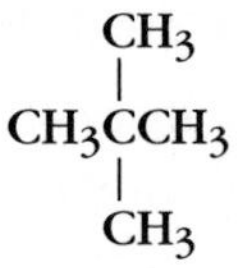   Neopentane

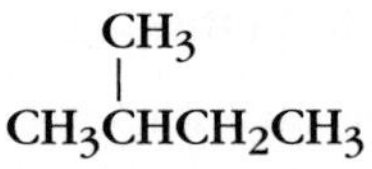

The prefix *n*- is also used in the common names of larger alkanes to indicate a straight chain compound. For example:

$$CH_3(CH_2)_6CH_3 \qquad\qquad CH_3(CH_2)_{10}CH_3$$

n-Octane   *n*-Dodecane

The prefix *iso*- is used in common names to indicate that one end of the alkane chain has the following structure:

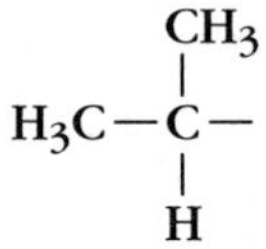

$$\underset{\overset{|}{H}}{\overset{\overset{CH_3}{|}}{H_3C-C-}}$$

For example:

$$\underset{\overset{|}{H}}{\overset{\overset{CH_3}{|}}{H_3C-C-CH_2CH_2CH_3}} \qquad\qquad \underset{\overset{|}{H}}{\overset{\overset{CH_3}{|}}{H_3C-C-CH_2CH_2CH_2CH_3}}$$

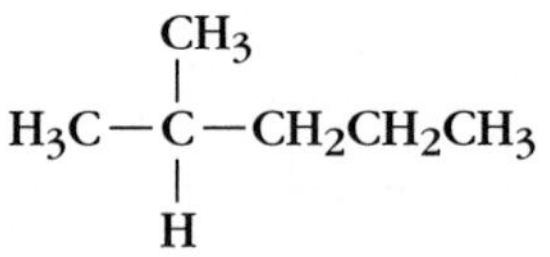

Isohexane 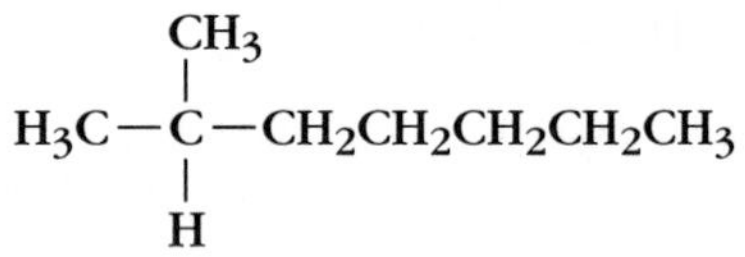   Isooctane

■ *H_3C-, CH_3-, and $-CH_3$ are equivalent ways of writing the structure of a methyl group that we will use in this text.*

The prefix *neo*- is used in common names to indicate that one end of the alkane chain has the following structure:

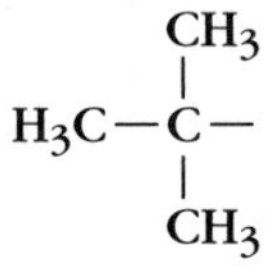

$$\underset{\overset{|}{CH_3}}{\overset{\overset{CH_3}{|}}{H_3C-C-}}$$

For example:

$$\underset{\overset{|}{CH_3}}{\overset{\overset{CH_3}{|}}{H_3C-C-CH_2CH_3}} \qquad\qquad \underset{\overset{|}{CH_3}}{\overset{\overset{CH_3}{|}}{H_3C-C-CH_2CH_2CH_2CH_3}}$$

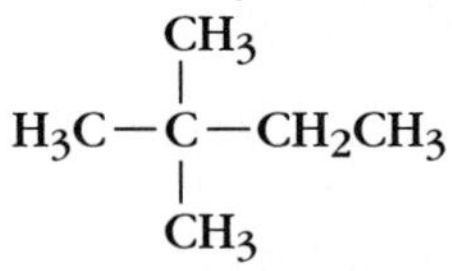

Neohexane 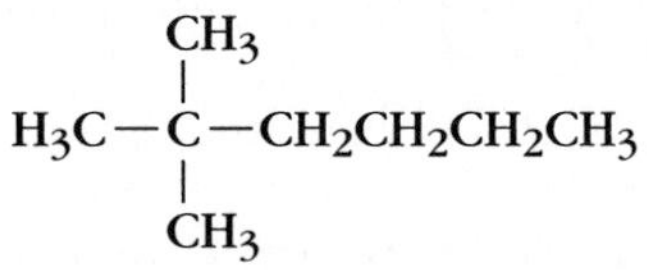   Neooctane

Rather than use common names for more complex alkanes, chemists use the systematic IUPAC method. However, to name more complicated branched alkanes by the IUPAC rules, we must learn how to name branched alkyl groups.

E   Naming Branched Alkyl Groups

If we remove one hydrogen atom from an internal carbon atom of a straight chain alkane instead of one of its end carbon atoms, we form a **branched alkyl group.** For example, we can form two isomeric alkyl groups from propane in the following way:

$$CH_3CHCH_2 - \boxed{H} \leftarrow \quad \text{Removing this H atom forms this alkyl group} \rightarrow CH_3CH_2CH_2 -$$

$$| \qquad \boxed{H}$$

Straight chain
alkyl group

Removing this H atom forms this alkyl group $\longrightarrow CH_3CHCH_3$
$$|$$

Branched
alkyl group

Branched alkyl groups are named by using the same basic four steps used to name a branched alkane. Let's illustrate this method by deriving the name for the following group:

$$CH_3$$
$$|$$
Parent chain $- CHCH_3$
$$1 \quad 2$$

The longest continuous chain of carbon atoms in the branched alkyl group that begins at the point of attachment to the parent chain contains two carbon atoms. Thus the alkyl group is a substituted ethyl group. We begin numbering at the point of attachment and assign numbers to each of the substituents. In this case, the substituent is a methyl group attached to carbon atom 1. Thus this branched alkyl group is called 1-methylethyl. To avoid confusion, the name of branched alkyl groups is enclosed in parentheses when writing the name of the complete alkane.

$$CH_3$$
$$|$$
1–Methylethyl alkyl group
$$CH_3CH_2CH_2CH_2CH_2CHCHCH_3$$
$$|$$
$$CH_2CH_2CH_3$$

4-(1-methylethyl)nonane

The IUPAC names of some of the more frequently encountered alkyl groups are summarized in Table 2.3 on p. 67.

Some of the simpler branched alkyl groups were given names before the IUPAC rules were devised. These common names are also given in Table 2.3. The prefixes *sec-*, *tert-*, and *n-* in these common names are abbreviations; *sec-* stands for secondary, *tert-* for tertiary, and *n-* for normal. The prefix *iso-* is used to designate a special structural feature of alkyl groups. Isoalkyl groups have a single one-carbon branch on the next-to-last carbon atom farthest from the point of attachment to the parent chain. For example:

$$CH_3$$
$$\diagdown$$
$$CH -$$
$$\diagup$$
$$CH_3$$

Isopropyl
group

$$CH_3$$
$$\diagdown$$
$$CHCH_2CH_2 -$$
$$\diagup$$
$$CH_3$$

Isopentyl
group

$$CH_3$$
$$\diagdown$$
$$CH(CH_2)_2CH_2 -$$
$$\diagup$$
$$CH_3$$

Isohexyl
group

■ *Often the prefix* tert- *is further abbreviated to* t-.

These common names have become so well established in the chemical literature that IUPAC rules make allowances for them. Unfortunately there is little relationship between the structures of these groups and their common names, so there's really no choice but to memorize them. The following examples show how the names of these groups are used.

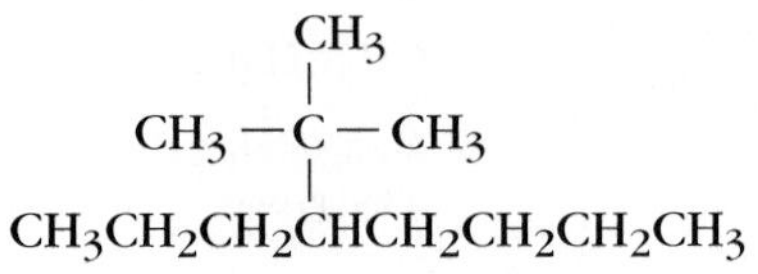

$$CH_3$$
$$|$$
$$CH_3-C-CH_3$$
$$|$$
$$CH_3CH_2CH_2CHCH_2CH_2CH_2CH_3$$

4-(1,1-Dimethylethyl)octane
or
4-*tert*-butyloctane

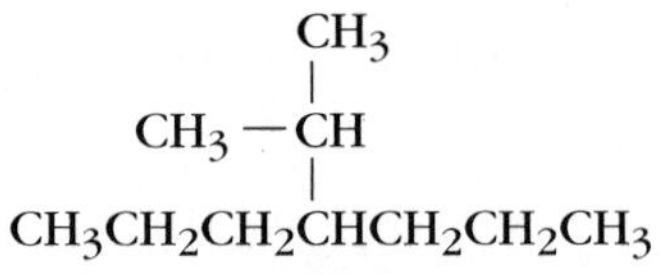

$$CH_3$$
$$|$$
$$CH_3-CH$$
$$|$$
$$CH_3CH_2CH_2CHCH_2CH_2CH_3$$

4-(1-Methylethyl)heptane
or
4-isopropylheptane

When using the common names of these branched alkyl groups, the prefixes *iso*- and *neo*- are considered as part of the name when alphabetizing the alkyl groups. However, the hyphenated prefixes *sec*- and *tert*- are not. Thus *sec*-butyl and *tert*-butyl (letter *b*) are listed alphabetically before isobutyl (letter *i*) or neopentyl (letter *n*).

Exercise 2.7   Give the IUPAC name for each of the following compounds.

(a)

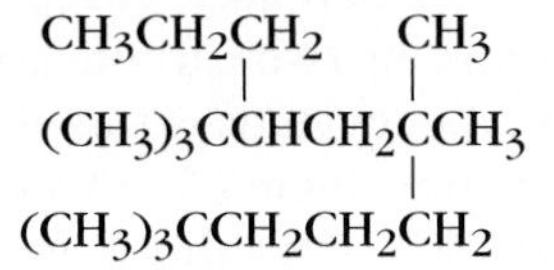

$$CH_3CH_2CH_2 \quad CH_3$$
$$| \qquad |$$
$$(CH_3)_3CCHCH_2CCH_3$$
$$|$$
$$(CH_3)_3CCH_2CH_2CH_2$$

(d)

$$CH_2(CH_2)_2CH_3$$
$$|$$
$$(CH_3)_2CHCHCH_2CH_2CHCH_3$$
$$|$$
$$CH_2CHCH_2CH_2CH_2CH_3$$
$$|$$
$$CH_2CH(CH_3)_2$$

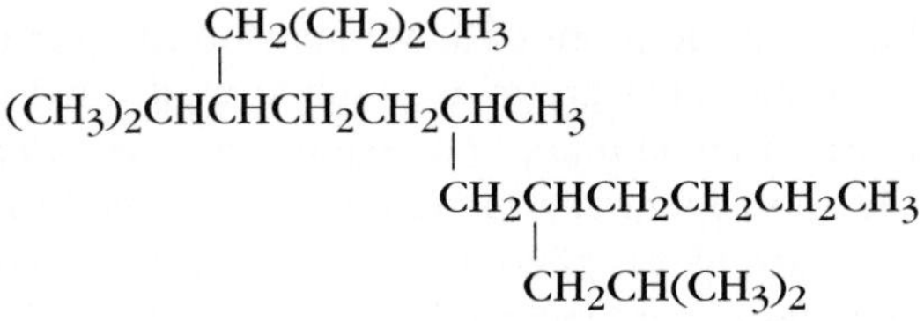

(b)

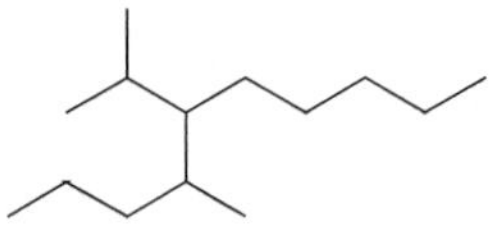

(e)

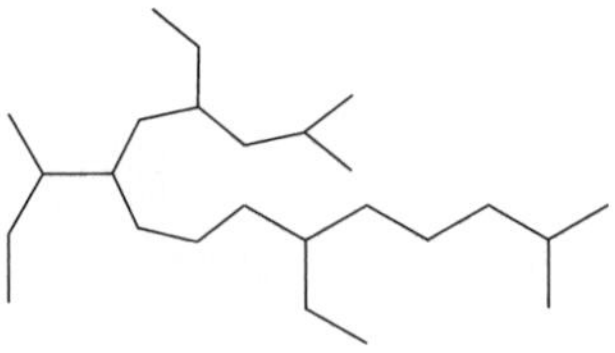

(c)

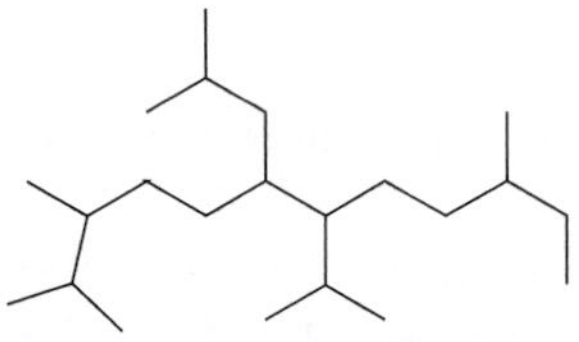

Exercise 2.8   Write the structure for each of the following IUPAC names.

(a) 2-Methyl-3-(1-methylethyl)hexane
(b) 5-(1,1,2-Trimethylpropyl)nonane
(c) 4-*tert*-Butyl-2,2,3,6-tetramethyloctane
(d) 3,3-Diisopropyl-2,4-dimethylpentane
(e) 3-Ethyl-5-(1-ethyl-1-methylpropyl)nonane

Exercise 2.9   According to the IUPAC rules, 1-ethylpropyl is correct yet 1-ethylpropane is wrong. Explain.

So far we have learned how to name only acyclic alkanes. We now turn our attention to cycloalkanes.

Table 2.3	IUPAC and Common Names of Frequently Encountered Alkyl Groups	

Structure	IUPAC Name	Common Name
CH_3-	Methyl	Methyl
CH_3CH_2-	Ethyl	Ethyl
$CH_3CH_2CH_2-$	Propyl	*n*-Propyl
$\overset{\overset{\displaystyle CH_3}{\mid}}{CH_3CH}- \equiv (CH_3)_2CH-$	1-Methylethyl	Isopropyl
$CH_3CH_2CH_2CH_2-$	Butyl	*n*-Butyl
$\overset{\overset{\displaystyle CH_3}{\mid}}{CH_3CH_2CH}-$	1-Methylpropyl	*sec*-Butyl
$\overset{\overset{\displaystyle CH_3}{\mid}}{CH_3CHCH_2}-$	2-Methylpropyl	Isobutyl
$\overset{\overset{\displaystyle CH_3}{\mid}}{\underset{\underset{\displaystyle CH_3}{\mid}}{CH_3C}}- \equiv (CH_3)_3C-$	1,1-Dimethylethyl	*tert*-butyl
$CH_3CH_2CH_2CH_2CH_2-$	Pentyl	*n*-Pentyl
$\overset{\overset{\displaystyle CH_3}{\mid}}{\underset{\underset{\displaystyle CH_3}{\mid}}{CH_3CCH_2}}- \equiv (CH_3)_3CCH_2-$	2,2-Dimethylpropyl	Neopentyl

2.6   Naming Cycloalkanes

Cycloalkanes that contain only one ring are named by attaching the prefix *cyclo-* to the name of the straight chain alkane containing the same number of carbon atoms:

Cyclopropane   Cyclobutane   Cyclopentane   Cyclohexane   Cycloheptane

The unsubstituted cycloalkanes are rings containing only methylene groups (CH_2). Because each ring contains exactly twice as many hydrogen atoms as carbon atoms, the general molecular formula of cycloalkanes is C_nH_{2n}. *The general formula of alkanes is C_nH_{2n+2},* so a cycloalkane and an alkane with the same number of carbon atoms are not isomeric. The cycloalkane contains two fewer hydrogens than the alkane.

Substituted cycloalkanes are named by using the name of the unsubstituted ring as the parent name with the alkyl group named as the substituent. If only one substituent is present, no numbering is needed.

CH₃   CH₂CH₃   CH₂CH(CH₃)₂

Methylcyclooctane   Ethylcyclopentane   Isobutylcyclohexane
                                                   (2-Methylpropyl)cyclohexane

If two or more substituents are present, the ring carbon atoms are numbered. If the substituents are identical, the numbering begins with one of the substituted ring carbons and continues in the direction that gives the lowest possible number to the remaining substituents.

CH₃   CH₃

    6 1 2   not   2 1 6
    5   3   3   5
       4   4
           CH₃   CH₃

1,3-Dimethylcyclohexane   1,5-Dimethylcyclohexane

If the two substituents are different, the lowest number is given to the first substituent listed alphabetically.

CH₃   CH₃CH₂   CH₂CH₃

    4 3 2
    5   1
       6   CH(CH₃)₂
           CH₂CH₃
1-Ethyl-3-methylcyclohexane   1,1-Diethyl-3-isopropylcyclopentane
          not
3-Ethyl-1-methylcyclohexane

The cycloalkane is not necessarily the parent structure in all cyclic compounds. If the acyclic portion of the compound contains more carbon atoms than the ring or if more than one ring is attached to the acyclic chain, the ring is named as a cycloalkyl substituent on the acyclic portion.

        CH₂(CH₂)₅CH₃

1-Cyclopentylheptane   1,5-Dicyclohexylpentane

Exercise 2.10   Give IUPAC names for each of the following compounds.

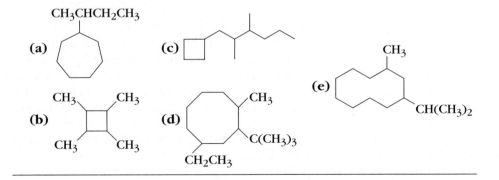

CH₃CHCH₂CH₃

(a)

(c)

(b)   CH₃   CH₃
                                    CH₃
            CH₃   CH₃   **(d)**
                                    C(CH₃)₃
                        CH₂CH₃

                                    CH₃
                            (e)
                                    CH(CH₃)₂

Exercise 2.11	Write the structures that correspond to each of the following IUPAC names.

(a) 1,3-Diethylcyclobutane   (c) 1-Chloro-3-isopropylcyclopentane

(b) 1-Isobutyl-2,4-dimethylcycloheptane   (d) *tert*-Butylcyclononane

2.7 Classifications of Carbon and Hydrogen Atoms

A carbon atom is classified according to the number of other carbon atoms bonded to it. There are four possibilities, designated primary (1°), secondary (2°), tertiary (3°), and quaternary (4°).

A **primary (1°) carbon atom** is bonded to only one other carbon atom, a **secondary (2°) carbon atom** is bonded to two carbon atoms, a **tertiary (3°) carbon atom** to three, and a **quaternary (4°) carbon atom** to four. Examples of these four kinds of carbon atoms are found in 2,2,4-trimethylpentane.

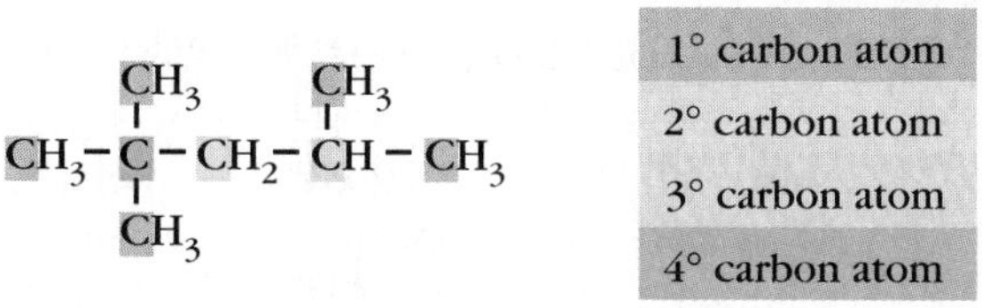

$$CH_3 - \overset{\overset{\displaystyle CH_3}{|}}{\underset{\underset{\displaystyle CH_3}{|}}{C}} - CH_2 - \overset{\overset{\displaystyle CH_3}{|}}{CH} - CH_3$$

1° carbon atom
2° carbon atom
3° carbon atom
4° carbon atom

The classification of hydrogen atoms is based on the type of carbon atom to which they are bonded. A **primary (1°) hydrogen atom** is attached to a primary carbon atom; a **secondary (2°) hydrogen atom** is bonded to a secondary carbon atom; and a **tertiary (3°) hydrogen atom** is bonded to a tertiary carbon atom. 2-Methylbutane has examples of all three kinds of hydrogen atoms.

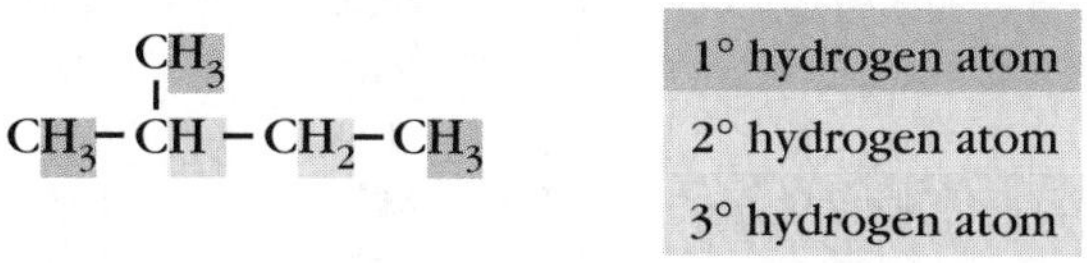

$$CH_3 - \overset{\overset{\displaystyle CH_3}{|}}{CH} - CH_2 - CH_3$$

1° hydrogen atom
2° hydrogen atom
3° hydrogen atom

Exercise 2.12	Identify the types of carbons in each of the following structures.

$$(a)\ CH_3 \overset{\overset{\displaystyle CH_3}{|}}{\underset{\underset{\displaystyle CH_3}{|}}{C}} CH_3$$

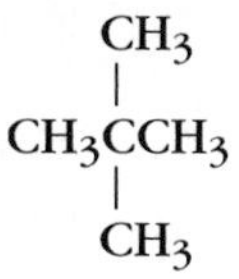

(b) [cyclopentane structure with CH₃ groups]

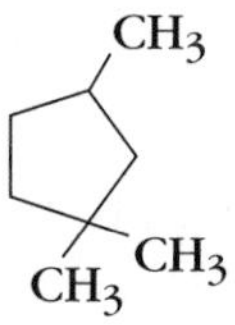

(c) [cyclohexane with C(CH₃)₃ group]

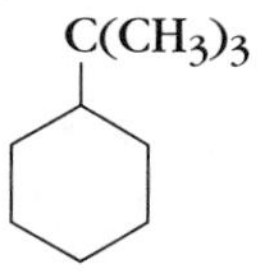

We've learned to identify various kinds of hydrocarbons, and we've learned how to name alkanes in a very systematic way. In the following sections, we learn something about their physical and chemical properties and where we obtain these compounds.

2.8 Physical Properties

Physical properties of substances are properties that can be observed without change to the identity of the substance. Solubilities, densities, melting points, and boiling points are the most frequently cited physical properties of organic compounds.

Alkanes and cycloalkanes are nonpolar compounds; therefore, they dissolve in nonpolar or weakly polar organic solvents. They are insoluble in polar solvents such as water so they are said to be **hydrophobic,** or "water hating." The hydrophobic property of alkanes and cycloalkanes makes them good preservatives for metals because they keep water from reaching the metal.

The densities, melting points, and boiling points of a number of straight chain alkanes, branched chain alkanes, and cycloalkanes are listed in Table 2.4. Notice that alkanes and cycloalkanes containing 1 to 4 carbon atoms are gases at room temperature, while those containing 5 to 17 carbon atoms are liquids, and those with 18 or more carbon atoms are solids.

Table 2.4	Physical Properties of Selected Alkanes and Cycloalkanes			
Number of Carbon Atoms	Name	Melting point (°C)	Boiling point (°C)	Density (g/ml)
Straight Chain Alkanes				
1	Methane	−182.5	−164	0.554
2	Ethane	−183.3	−88.6	0.546
3	Propane	−189	−42.1	0.500
4	Butane	−138	−0.5	0.578
5	Pentane	−129	36	0.626
6	Hexane	−95	69	0.660
7	Heptane	−91	98	0.684
8	Octane	−57	126	0.703
9	Nonane	−51	151	0.718
10	Decane	−30	174	0.730
20	Icosane	37	343	0.789
25	Pentacosane	53	402	0.801
30	Triacontane	66	450	0.810
Branched Chain Alkanes				
4	Isobutane	−159	−11	0.579
5	Isopentane	−160	27.8	0.620
5	Neopentane	−16.5	9.5	0.613
6	Isohexane	−154	60	0.654
8	Isooctane	−107	99	0.692
Cycloalkanes				
3	Cyclopropane	−127	−33	0.720
4	Cyclobutane	−50	13	0.750
5	Cyclopentane	−94	49	0.741
6	Cyclohexane	7	81	0.780
7	Cycloheptane	−12	118	0.81
8	Cyclooctane	14	148	0.83

The densities of the alkanes and cycloalkanes listed in Table 2.4 are all between 0.50 and 0.83 g/ml. This makes them all less dense than water (1.0 g/ml). Because alkanes are insoluble in water, a mixture of water and alkanes quickly separates into two layers with the alkane on top. This is why an oil slick floats on top of water.

Table 2.4 also shows that the boiling points of constitutional isomers depend on the structures of the isomers; the more branched the isomer, the lower its boiling point. For example, the boiling point of neopentane, which has two branches, is 9.5 °C, whereas the boiling point of isopentane, which has one branch, is 27.8 °C, and pentane, which has no branches, is 36 °C. Similar relationships exist between isobutane and butane, isohexane and hexane, and isooctane and octane.

These trends can be understood by considering the London forces acting between alkane molecules. As molecular size increases, London forces also increase and contribute to an increase in boiling point. Changes in intermolecular forces are also responsible for the effect of branching on boiling points. Branched chain alkanes are more spherelike and have less surface area than their straight chain isomers. As a result branched chain alkanes have weaker intermolecular forces than their straight chain isomers. These weaker forces are overcome at a lower temperature so the branched isomer has a lower boiling point.

■ *London forces are discussed in Section 1.16, D.*

| Exercise 2.13 | Make molecular models of pentane, 2-methylbutane (isopentane), and 2,2-dimethylpropane (neopentane). Which isomer is most nearly spherical? |

2.9 Chemical Properties

Alkanes were originally called **paraffins** because chemists found them to be unreactive compared to other classes of organic compounds. Alkanes are not highly reactive, but they are not inert either. They react with several halogens and, like most organic compounds, burn in air.

The reaction of alkanes with halogens is called a **halogenation reaction.** Alkanes react spontaneously with fluorine and react with chlorine and bromine when heated or irradiated with light. No reaction occurs with iodine. The products are **haloalkanes,** compounds in which one or more hydrogen atoms of an alkane are replaced by halogen atoms. The reaction of methane and chlorine when heated or irradiated with light, for example, generally forms a mixture of chlorinated products whose composition depends on the concentrations of the two reactants and the time allowed for the reaction. This mixture occurs by a sequential substitution of hydrogen atoms by chlorine atoms.

■ *The word **paraffin** is derived from the two Latin words par(um) and affin (is), and means slight affinity.*

■ *The halogenation of alkanes is discussed in more detail in Chapter 18.*

■ *For clarity, nonbonded electrons will no longer be placed around halogen atoms.*

$$CH_4$$
$$Cl_2 \downarrow \text{Heat or light}$$
$$CH_3Cl \ + \ HCl$$
$$\xrightarrow[\text{Heat or light}]{Cl_2} CH_2Cl_2 \ + \ HCl$$
$$\xrightarrow[\text{Heat or light}]{Cl_2} CHCl_3 \ + \ HCl$$
$$\xrightarrow[\text{Heat or light}]{Cl_2} CCl_4 \ + \ HCl$$

When alkanes are mixed with oxygen in proper proportions and then ignited, they react to produce carbon dioxide, water, and large amounts of heat. This reaction is called **combustion** and is used to heat homes and power cars all over the world. Methane, for instance, is the principal constituent of natural gas.

$$CH_4 + 2O_2 \longrightarrow CO_2 + 2H_2O \quad \Delta H° = -213 \text{ kcal/mol} \quad (-890 \text{ kJ/mol})$$

Chemists also use combustion to determine the molecular formulas of organic compounds.

2.10 Determination of Molecular Formulas by Quantitative Combustion

The relative amounts of carbon, hydrogen, and oxygen in an organic compound can be determined by a quantitative combustion method that has changed little over the last 100 years. Figure 2.1 shows a schematic view of an apparatus for combustion analysis. In this method, a small sample (5 to 10 mg) of an organic

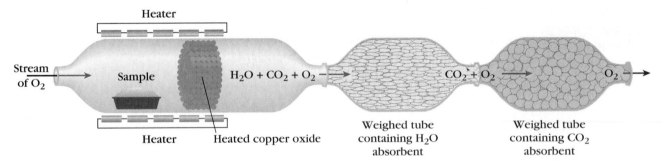

Figure 2.1
Combustion analysis apparatus for determining the percentage of carbon and hydrogen in a sample of an organic compound. The sample is vaporized in a slow stream of oxygen gas and passed over hot copper oxide, which burns the sample completely to H_2O and CO_2. The increase in weight of the absorbent tubes gives the weights of H_2O and CO_2 formed by combustion of the compound.

compound is carefully weighed, then vaporized in a slow stream of oxygen gas. The mixture is passed through heated copper oxide, which burns the sample quantitatively to carbon dioxide and water. The water produced is absorbed by a tube containing a drying agent, and the carbon dioxide is absorbed by another tube that contains a strong base (usually sodium hydroxide on asbestos). The increase in weight of each tube gives the weight of water and carbon dioxide formed on combustion of the sample.

The amounts of hydrogen and carbon in the sample are calculated from the weights of water and carbon dioxide produced, respectively. The amount of oxygen present in the sample cannot be determined directly because the sample is burned in the presence of excess oxygen. For compounds containing only carbon, hydrogen, and oxygen, however, the percent of oxygen present is the difference between 100% and the sum of the percents of carbon and hydrogen.

Sample Problem 2.4 demonstrates how to use combustion analysis to determine the formula of a compound.

| Sample Problem 2.4 | An 8.53 mg sample of a liquid organic compound forms 5.08 mg of H_2O and 12.50 mg of CO_2 on combustion. Find the empirical and molecular formulas of the compound. |

Solution:

Of the weight of CO_2 formed, only the fraction $C/CO_2 = 12.01/44.01$ is carbon, and only the fraction $2H/H_2O = 2.02/18.02$ of the weight of water is hydrogen. Thus the actual weight of carbon and hydrogen formed is:

$$\text{wt C} = (12.50 \text{ mg}) \frac{(12.01)}{(44.01)} = 3.41 \text{ mg}$$

$$\text{wt H} = (5.08 \text{ mg}) \frac{(2.02)}{(18.02)} = 0.569 \text{ mg}$$

The percent composition of carbon and hydrogen in the sample is:

$$\% \text{ C} = \frac{\text{Wt C}}{\text{Wt of sample}} (100\%) = \frac{3.41 \text{ mg}}{8.53 \text{ mg}} (100\%) = 40.0\%$$

$$\% \text{ H} = \frac{\text{wt H}}{\text{Wt of sample}} (100\%) = \frac{0.569 \text{ mg}}{8.53 \text{ mg}} (100\%) = 6.67\%$$

The percent of hydrogen and carbon is 46.67%, which is less than 100%, so the remainder, 53.33%, is the percent of oxygen in the sample.

To calculate the molecular formula of our sample, we must first calculate its **empirical formula.** The empirical formula is the simplest formula that gives the relative numbers of different atoms in a molecule. We do this as follows:

We assume a sample size of 100 g because the percent values immediately give the number of grams of each element in the sample. Thus there are 40.0 g of C; 6.67 g of H; and 53.33 g of O in 100 g of our sample. The number of moles of each element is obtained by dividing the number of grams of each element by its atomic weight.

$$\text{C:} \quad \frac{40.0 \text{ g}}{12.01 \text{ g/mol}} = 3.33 \text{ mol}$$

$$\text{H:} \quad \frac{6.67 \text{ g}}{1.01 \text{ g/mol}} = 6.60 \text{ mol}$$

$$\text{O:} \quad \frac{53.33 \text{ g}}{16.0 \text{ g/mol}} = 3.33 \text{ mol}$$

We convert these numbers to their smallest whole numbers by dividing each by 3.33, the smallest number of moles:

$$\text{C:} \frac{3.33}{3.33} = 1 \qquad \text{H:} \frac{6.60}{3.33} = 1.98 \approx 2 \qquad \text{O:} \frac{3.33}{3.33} = 1$$

The final result is a ratio of 1 C atom to 2 H atoms to 1 O atom. This gives CH_2O as the empirical formula of our sample. The molecular formula can be any multiple of this empirical formula because they all have the same ratio of elements. Thus CH_2O, $C_2H_4O_2$, $C_3H_6O_3$, and so forth are all possible molecular formulas for our sample.

To convert the empirical formula of our sample to its molecular formula, we need to know its molecular weight. Historically, molecular weights were determined by freezing point depression or boiling point elevation. These methods have been replaced in recent years by mass spectrometry. Using one of these methods, we determine that the molecular weight of our sample is 60. The weight of the empirical formula CH_2O is 30, so molecules of our sample must contain twice as many atoms as the empirical formula. Thus the molecular formula of our sample is $C_2H_4O_2$.

■ *We will study mass spectrometry in Chapter 18.*

While the molecular formula tells us the total number of atoms of each element in a molecule, it doesn't tell us how these atoms are joined together. In Chapters 5 and 10, we will learn techniques that allow us to determine the complete structures of molecules.

Exercise 2.14	A 9.76 mg sample of a gas whose molecular weight is found to be 16 forms 26.53 mg of CO_2 and 21.56 mg of H_2O on combustion. Calculate its molecular formula.

2.11   Sources and Uses of Alkanes

Petroleum, or crude oil, and coal are the two largest reservoirs of organic compounds readily available for our use. They are fossil fuels formed over many thousands of years from prehistoric plants. The principal constituents of petroleum are straight chain alkanes, some branched alkanes, and certain aromatic compounds. In addition, there are oxygen-, nitrogen-, and sulfur-containing organic compounds, although their quantities vary depending on the source of the crude oil. Alkanes are obtained for commercial use by refining petroleum. Crude oil is refined by fractional distillation into mixtures of alkanes with useful boiling point ranges. The major mixtures obtained by distillation, which are called *fractions,* the alkanes they contain, and some of their uses are listed in Table 2.5.

Although alkanes are readily available from petroleum, they are less abundant in living systems. One place where alkanes are found in nature, however, is the waxy coating of the leaves of many plants. This waxy coating is believed to prevent the plant from losing water. One alkane, hentriacontane, $CH_3(CH_2)_{29}CH_3$, is a component of the leaf wax of tobacco plants and is also present in beeswax.

Organic compounds are classified into families according to their structures. We have already learned that the structures of alkanes consist of only C—H and C—C single bonds. The structural features that allow us to classify organic compounds are called *functional groups.*

Table 2.5	Commercial Products Obtained by Fractional Distillation of Crude Oil		
Fraction	Boiling Point Range (°C)	Number of Carbon Atoms	Uses
Natural gas	<20	1–4	Heating fuel
Petroleum ether	30–100	5–7	Degreasing solvents
Gasoline	30–200	5–12	Motor fuel
Kerosene	175–275	12–15	Jet fuel
Diesel	190–330	15–20	Diesel fuel
Fuel oil	230–360	18–22	Heating fuel
Lubricating oil	>350	20–30	Lubricants
Asphalt	Nonvolatile	>40	Roofing and road surface materials

2.12   Functional Groups

Chemists have learned from countless experiments that there is a direct relationship between the structure of organic compounds and their chemical reactivity.
This relationship is used to classify compounds into families. The alkanes, for
instance, have similar structures that contain only C—H and C—C single bonds,
and they are generally less reactive than other organic compounds. We don't
have to learn the reactions of thousands of different alkanes, therefore, because
their reactions are all basically the same. Thus a fundamental principle of organic
chemistry is that if we know the structure of a compound, then we know the
family it belongs to, and, if we know its family, then we can roughly predict its
chemical reactivity.

The group of atoms and their associated bonds that define the structure and
chemical properties of a particular family of compounds is called a **functional
group.** For example, alkenes are a family of compounds whose characteristic
functional group is the carbon-carbon double bond. The electron structure of a
carbon-carbon double bond remains much the same in all alkenes, so the chemical reactivity of all members of the family also remains much the same. Thus ethylene, the simplest alkene, and 2-cholestene, a steroid that contains a double
bond, both readily react with bromine.

$$CH_2 = CH_2 \; + \; Br_2 \; \longrightarrow \; \overset{\displaystyle Br}{\underset{\displaystyle Br}{CH_2CH_2}}$$

Ethylene

2-Cholestene

Once we learn the reactions of one kind of alkene, we have learned the reactions
of many others.

In our discussions of the chemical reactions of each family, we will focus our
attention mostly on the functional group and will be less concerned with the
structure of the rest of the molecule. This allows us to adopt several symbols to
represent the nonfunctional group part of organic compounds.

2.13   Using R and Ar to Represent Parts of Organic Structures

As we learn the characteristic reactions of the various families of organic compounds, it is often convenient to use a general notation for that part of an organic
compound that does not undergo reaction. In this text we will use the letter *R* as
a general symbol to indicate any alkyl group. For example, if we wish to emphasize a particular general reaction of the carbon-carbon double bond in alkenes,
we can write a general formula for an alkene by adding one or more of the letter
R to the carbon atoms of a carbon-carbon double bond.

$RCH=CH_2$ represents $CH_3CH=CH_2$ and $(CH_3)_3CCH=CH_2$ and many other alkenes

$$\begin{matrix} R \\ \diagdown \\ \quad C=CH_2 \\ \diagup \\ R' \end{matrix} \quad \text{represents} \quad \begin{matrix} CH_3 \\ \diagdown \\ \quad C=CH_2 \\ \diagup \\ CH_3CH_2 \end{matrix} \quad \text{and} \quad \begin{matrix} (CH_3)_2CH \\ \diagdown \\ \quad C=CH_2 \\ \diagup \\ CH_3CH_2 \end{matrix} \quad \text{and many others}$$

$$\begin{matrix} R \\ \diagdown \\ \quad C=CH_2 \\ \diagup \\ R' \end{matrix} \quad \text{does not represent} \quad \begin{matrix} CH_3 \\ \diagdown \\ \quad C=CH_2, \\ \diagup \\ CH_3 \end{matrix} \quad \begin{matrix} \text{because R and R' represent} \\ \text{different alkyl groups} \end{matrix}$$

Using this symbol, we can write the following general equation for the reaction of most alkenes with bromine:

$$RCH=CH_2 \ + \ Br_2 \ \longrightarrow \ \underset{\underset{Br}{|}}{\overset{\overset{Br}{|}}{RCH-CH_2}}$$

We learned earlier in this chapter that alkyl groups are substituents derived from alkanes. Similarly, **aryl groups** are substituents derived from aromatic compounds. We will use the symbol *Ar* to represent a general aryl group. For example:

ArCl represents   and many others

This text is organized for the most part around the common functional groups. Although we will study each of the major functional groups in detail in subsequent chapters, it is important at this early stage of the study of organic chemistry to be able to recognize and name some of the more common examples of the most important functional groups.

2.14   Representative Examples of Common Functional Groups

A   Aromatic Compounds

■ *We will study in detail the chemistry of aromatic compounds in Chapters 20 and 21.*

Benzene is the most common example of a family called *aromatic compounds.* The structure of benzene is a resonance hybrid of the following contributing structures:

■ *Resonance hybrids are discussed in Section 1.12.*

This representation means that the π bond electrons of benzene are not localized as three carbon-carbon double bonds but rather are delocalized about the entire ring. Because of this electron delocalization, the reactions of benzene are much different from those of typical alkenes. In fact, benzene is relatively unreactive. Only alkanes as a class are less reactive than benzene.

The structure of benzene is usually represented by writing only one of its two resonance structures. Even though only one structure is written, it is understood that it represents a resonance hybrid.

When a benzene ring is a substituent on a carbon chain, it is called a **phenyl group** and is represented in any one of the following ways:

$$\text{(benzene ring)} \qquad \text{or} \qquad C_6H_5 - \qquad \text{or} \qquad Ph - \qquad \text{or} \qquad \phi -$$

B   Haloalkanes or Alkyl Halides

Haloalkanes are compounds in which a hydrogen atom of an alkane is replaced by a halogen atom (fluorine, chlorine, bromine, or iodine). Thus the functional group of this family is a halogen atom bonded to an sp^3 hybrid carbon atom.

$$-\overset{|}{\underset{|}{C}} - X \qquad \left. \right\} \quad \text{Functional group of a haloalkane}$$

sp^3 Hybrid carbon atom

$$X = F, Cl, Br, \text{ or } I$$

The following are examples of haloalkanes:

$$CH_3I \qquad\qquad CH_3CH_2Cl \qquad\qquad (CH_3)_2CHBr$$

Haloalkanes are also called **alkyl halides.**

Haloalkanes are named by the same IUPAC rules used to name branched alkanes. The halogen atom, rather than the alkyl group, is the substituent. When halogen atoms are substituents, they are given the following names:

Substituent	Prefix Name
Fl—	fluoro-
Cl—	chloro-
Br—	bromo-
I—	iodo-

The following examples illustrate the way haloalkanes are named according to the IUPAC rules:

$$CH_3CH_2CH_2CH_2Cl \qquad (CH_3)_2CHCHCH_2CH_3 \qquad ClCH_2CCH_3$$

with Br substituent above on the pentane; Cl above and I below on the propane.

1-Chlorobutane	3-Bromo-2-methylpentane	1,2-Dichloro-2-iodopropane

Common names are often used for the simpler haloalkanes. These common names consist of two parts. The first part is the name of the alkyl group bonded to the halogen atom and the second is the word *chloride, bromide,* and so forth. For example:

Methyl group	Isopropyl group	*n*-Propyl group	Methylene group
CH_3I	$(CH_3)_2CHCl$	$CH_3CH_2CH_2Br$	CH_2Cl_2
Methyl iodide	Isopropyl chloride	*n*-Propyl bromide	Methylene chloride

Chloroform and carbon tetrachloride are two common names widely used for the following two compounds:

$$CHCl_3 \qquad\qquad\qquad CCl_4$$

Chloroform   Carbon tetrachloride

Exercise 2.15   Give the IUPAC name for each of the following haloalkanes.

(a) 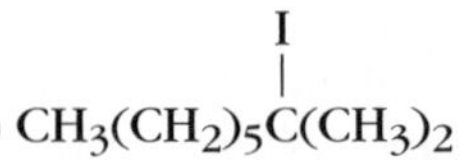 $CH_3(CH_2)_5\overset{\displaystyle I}{\underset{\displaystyle \;}{C}}(CH_3)_2$
    (b) 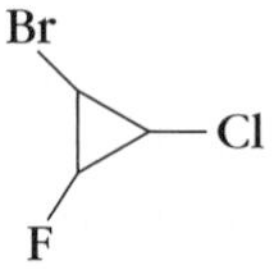 [cyclopropane with Br, F, and Cl substituents]
    (c) [cyclohexane ring with Br and CH(CH₃)₂ substituents]

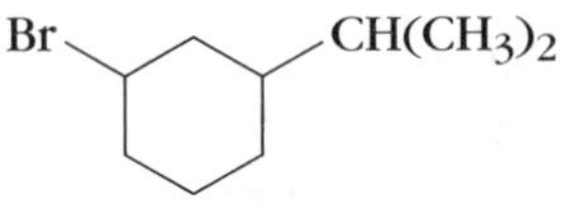

Exercise 2.16   Write the structures of each of the following haloalkanes.

(a) *n*-Butyl bromide
(b) 2,3,7-Trichloro-4-(1,1-dimethylethyl) octane
(c) *tert*-Butyl bromide

(d) Hexachloroethane
(e) Isobutyl chloride

C   Alcohols

The functional group of the alcohol family is the hydroxy (—OH) group bonded to an sp^3 hybrid carbon atom.

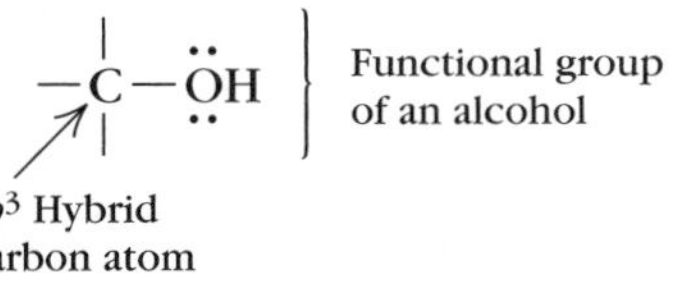

sp^3 Hybrid carbon atom   Functional group of an alcohol

■ *The modification of IUPAC rules needed to name alcohols will be discussed in Chapter 11.*

Most people are familiar with at least some of the simpler members of the alcohol family. Methyl alcohol, CH_3OH, (IUPAC name methanol) is the simplest member. It was originally obtained by heating wood at high temperature in the absence of air. Because of this method of preparation, it was known as wood alcohol. Methanol causes blindness when ingested in small amounts and death in larger amounts. It is an important industrial chemical that is used both as a solvent and as a starting material for plastics. Ethyl alcohol, CH_3CH_2OH, (IUPAC name ethanol) is one of the products of fermentation of grains, fruits, and sugars and is the intoxicating ingredient in alcoholic beverages. Ethanol is used industrially as a solvent and as a starting material for the production of other chemicals.

Common names are often used for many of the simpler alcohols. Like the common names of haloalkanes, these common names are obtained by naming the alkyl group attached to the —OH group followed by the word *alcohol:*

$(CH_3)_2CH\overset{..}{\underset{..}{O}}H$ $CH_3CH_2CH_2\overset{..}{\underset{..}{O}}H$ $(CH_3)_3C\overset{..}{\underset{..}{O}}H$ $CH_3\overset{\displaystyle :\overset{..}{O}H}{\underset{\displaystyle \;}{C}}HCH_2CH_3$

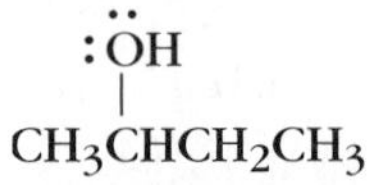

Isopropyl alcohol   *n*-Propyl alcohol   *tert*-Butyl alcohol   *sec*-Butyl alcohol

Exercise 2.17 Write the structures of each of the following alcohols.

(a) Neopentyl alcohol **(b)** Isobutyl alcohol **(c)** *n*-Hexyl alcohol

---

D **Ethers**

The functional group of ethers is an oxygen atom bonded to two carbon atoms. Both carbon atoms can be part of alkyl groups, aryl groups, or any combination of the two.

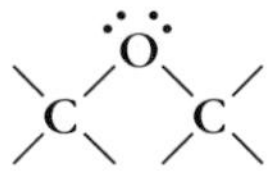

Functional group of an ether

The best known ether may be diethyl ether because it has been used as an anesthetic and is frequently used as a solvent in the laboratory. Another useful solvent is the cyclic ether tetrahydrofuran (THF). Anisole, which smells like anise, is an ether that contains an aryl group; it is used as a food flavoring.

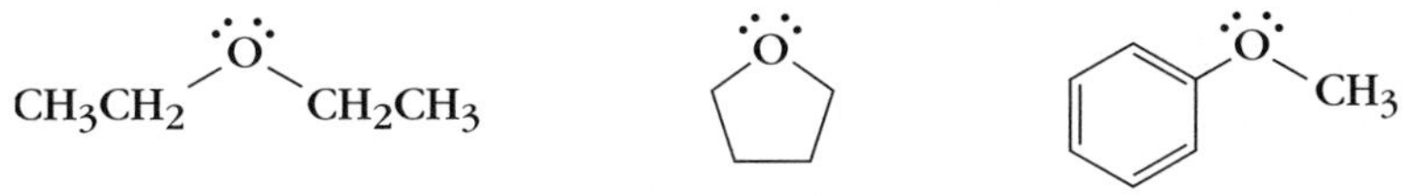

CH_3CH_2 — O — CH_2CH_3

Diethyl ether

Tetrahydrofuran
(THF)

O — CH_3

Anisole

E **Amines**

The functional group of amines is a nitrogen atom bonded to one, two, or three organic groups. The groups can be any combination of alkyl or aryl groups. Amines are classified as primary (1°), secondary (2°), or tertiary (3°) according to the number of carbon atoms that are attached to the nitrogen atom.

Functional group
of a 1° amine

Functional group
of a 2° amine

Functional group
of a 3° amine

The following are examples of each type of amine:

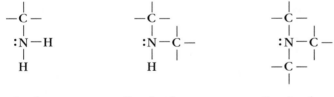

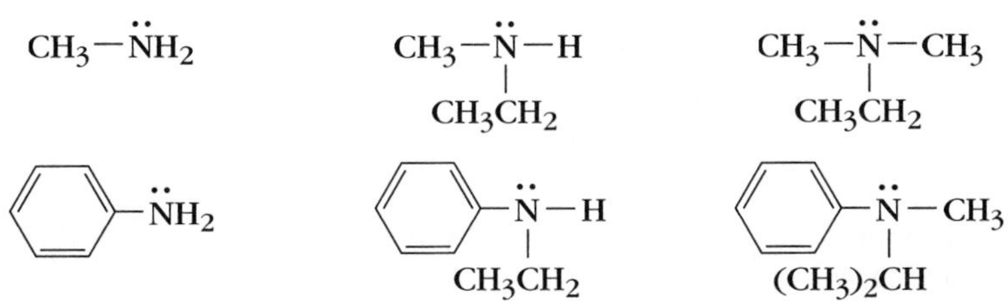

CH_3 — NH_2

CH_3 — N — H
        |
     CH_3CH_2

CH_3 — N — CH_3
        |
     CH_3CH_2

NH_2

N — H
   |
CH_3CH_2

N — CH_3
   |
$(CH_3)_2CH$

Primary (1°) amines

Secondary (2°) amines

Tertiary (3°) amines

F **Aldehydes and Ketones**

Both aldehydes and ketones are families of compounds that contain a carbon-oxygen double bond. The carbon-oxygen double bond is called a **carbonyl group.**

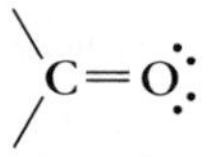

A carbonyl group

The carbon atom of the carbonyl group in aldehydes is bonded to at least one hydrogen atom, while in ketones it is bonded to two organic groups. The organic groups can be either alkyl or aryl in both aldehydes and ketones.

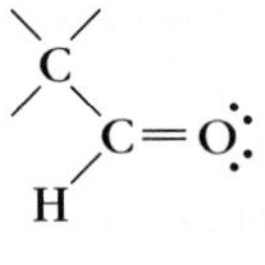

Functional group
of an aldehyde

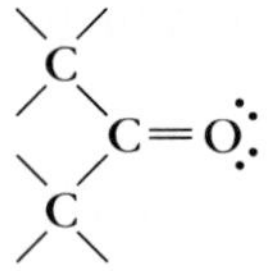

Functional group
of a ketone

The following are examples of some readily available aldehydes and ketones and their common names:

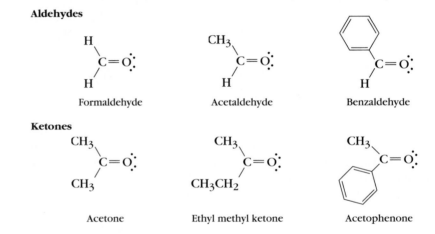

Aldehydes

Formaldehyde

Acetaldehyde

Benzaldehyde

Ketones

Acetone

Ethyl methyl ketone

Acetophenone

G   Carboxylic Acids

The functional group of carboxylic acids is called a carboxyl group. It contains a carbonyl group whose carbon atom is bonded to a hydroxy group.

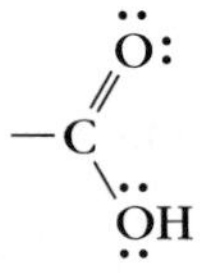

$$-C \overset{\overset{\displaystyle ..}{O:}}{\underset{\overset{\displaystyle}{OH}}{\big\|}} \qquad \text{also written as} \qquad -COOH$$

Carboxyl group, the
functional group of
carboxylic acids

The following are a few examples of simple carboxylic acids found in nature and their common names:

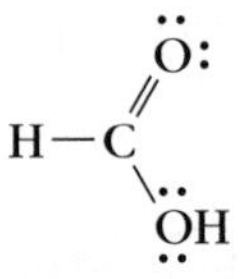

Formic acid

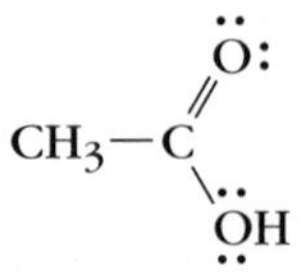

Acetic acid

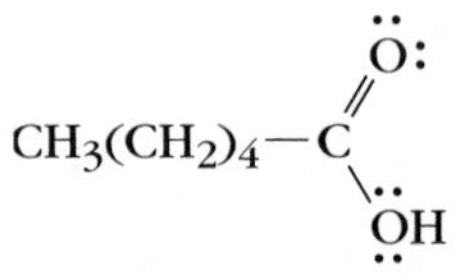

Caproic acid

Formic acid was first isolated from ants, and it is partly responsible for the irritating sting of red ants. Acetic acid is the major organic ingredient of vinegar (vinegar has an acetic acid content of 4% to 5%), while caproic acid is responsible for the strong odor of goats and of dirty socks.

H   Carboxylic Acid Esters

Carboxylic acid esters are another family of compounds that contain a carbonyl group as part of their functional group.

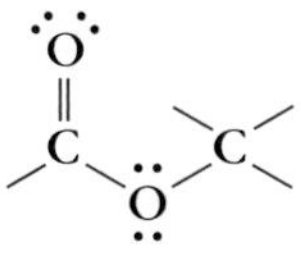

also written as $-COOC-$

The following are some examples of simple esters and their common names:

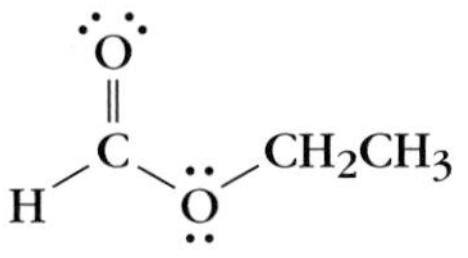

or $HCOOCH_2CH_3$

Ethyl formate

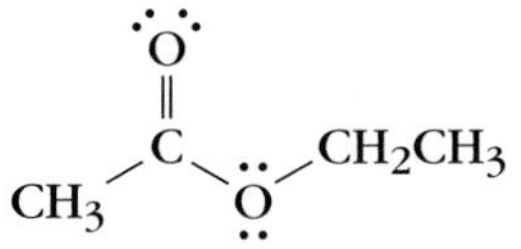

or $CH_3COOCH_2CH_3$

Ethyl acetate

I   Amides

Amides are compounds that contain the following functional group:

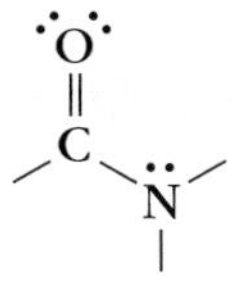

The nitrogen atom can form two bonds with any combination of hydrogen atoms, alkyl groups, and aryl groups, as shown in the following examples:

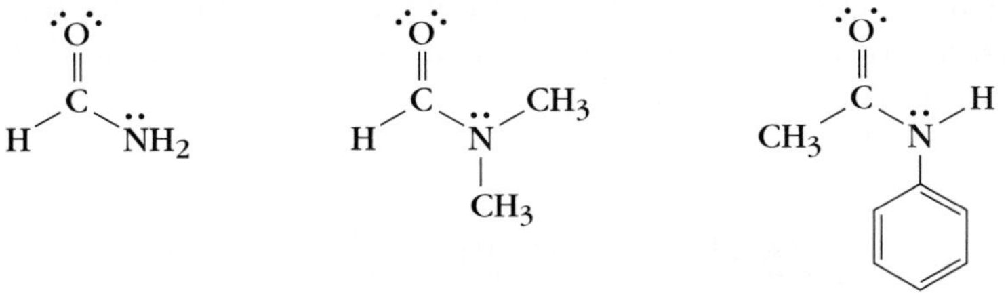

Formamide   N,N-Dimethylformamide   N-Phenylacetamide

The prefixes N and N,N indicate that the substituents are attached to the nitrogen atom.

The amide functional group is also called a peptide linkage because it links together amino acids in proteins (proteins are also called *polypeptides*).

■ *Proteins are discussed in Chapter 26.*

Exercise 2.18   Identify the functional group in each of the following compounds.

(a) CH_2=$CHCH_2CH_3$

(d) $CH_3CH_2\overset{\overset{\displaystyle ..\ddot{O}..}{\|}}{C}CH_2CH_3$

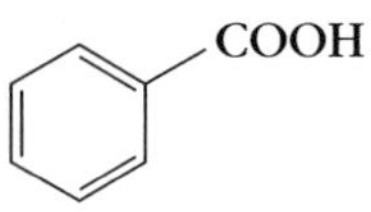

(g) CH_3C≡CCH_2CH_3

(b) ⬡—COOH

(e) $CH_3(CH_2)_4\ddot{N}H_2$

(h) (structure with Cl)

(c) $CH_3\ddot{N}HCH_2CH_3$

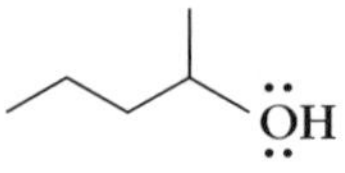

(f) (structure with $\ddot{O}H$)

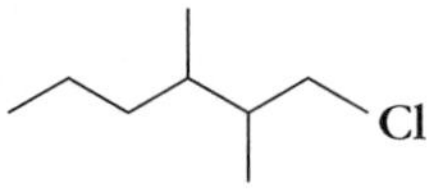

(i) 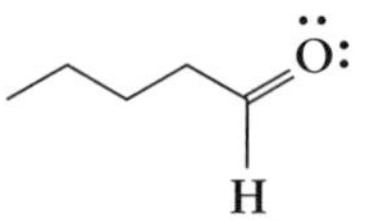 (structure with $\ddot{O}:$ and H)

Exercise 2.19   Write structural formulas for compounds of chemical formula C_3H_6O that have:

(a) A ketone functional group

(b) An aldehyde functional group

Exercise 2.20   Write structural formulas for compounds of chemical formula C_3H_8O that have:

(a) An alcohol functional group

(b) An ether functional group

The functional group determines the chemistry of a particular compound. But what about the rest of the molecule?

2.15   Chemical Reactions Occur at or Near Functional Groups

Knowing the structural formula of an organic compound makes it possible to identify its functional group or groups. Understanding the chemistry of its functional groups makes it possible, in turn, to predict the chemical and physical properties of the compound. Predicting properties can be simplified somewhat if we recognize that there are three parts to the structural formulas of all organic compounds, as illustrated in Figure 2.2.

Figure 2.2
The three parts of a structural formula are the functional group (or groups), the hydrogen atoms adjacent to the functional group(s), and the remaining alkanelike portion.

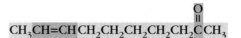

$$CH_3CH=CHCH_2CH_2CH_2CH_2CH_2\overset{\overset{\displaystyle O}{\|}}{C}CH_3$$

Functional group

C and H atoms adjacent to functional groups

Alkanelike region between functional groups

The first and most important part of the structure is the functional group or groups. There are two functional groups in the structure in Figure 2.2, a carbon-carbon double bond and a ketone group. Thus the compound represented by this structure will undergo reactions typical of ketones and alkenes. But what about the rest of the structure?

The carbon and hydrogen atoms adjacent to the functional groups are the second most important part of the structure. These hydrogen atoms are affected by the presence of the functional group, so they are much more reactive than the hydrogen atoms bonded to carbon atoms in alkanes. We will also learn these reactions when we study the chemistry of the various functional groups in subsequent chapters. The final part of the structure resembles an alkane because it consists of only C—H and C—C single bonds not adjacent to any functional group. Consequently, this part of the structure is chemically unreactive like the alkanes. In other words, *the alkanelike portion of a structure serves as a carrier of the functional group or groups and the adjacent CH_2 or CH_3 groups but is rarely involved in chemical reactions.*

In conclusion, no matter how complicated the structure of a particular compound, we will be able to predict fairly accurately its chemical reactions if we can identify its functional groups, if we know the reactions of those functional groups, and if we know the reactions of the hydrogen atoms adjacent to the functional groups.

2.16 Summary

Alkanes are compounds that contain only C—H and C—C single bonds; their general formula is C_nH_{2n+2}. **Cycloalkanes** are alkanes whose carbon atoms form a ring and have a general molecular formula C_nH_{2n}. Alkanes and cycloalkanes with the same number of carbon atoms are not isomers.

Alkanes are relatively unreactive; their major reactions are halogenation and combustion. Combustion is the source of much of the energy used in our world.

Alkanes and cycloalkanes are named systematically by a series of IUPAC rules. The IUPAC name of a compound consists of the name of the parent structure with one or more prefixes. These prefixes are both words and numbers that tell us the kind and number of substituents attached to the parent structure and where they are attached.

A fundamental principle of organic chemistry is that if we know the structure of a compound, we know the family it belongs to. This, in turn, makes it possible to roughly predict its chemical reactivity.

The structural features that make it possible to classify compounds by their structure and reactivity are called **functional groups.** Functional groups are the atoms and their associated bonds that define the structure and chemical properties of a particular family of compounds.

There are three parts to the structural formulas of most organic compounds: the functional group or groups, the CH_2 group or groups adjacent to the functional group, and the rest of the structure that resembles an alkane in that it contains only C—H and C—C single bonds not adjacent to any functional group. Chemical reactions occur at the functional group(s) and adjacent CH_2 groups. The alkanelike portion serves as a carrier of the functional group or groups and the adjacent CH_2 or CH_3 groups but is rarely involved in chemical reactions.

...ditional Exercises

2.21   Define and give an example of each of the following terms.

(a) Alkyl group
(b) Unsaturated hydrocarbon
(c) Molecular formula
(d) Branched chain alkane
(e) Acyclic compound
(f) Ether
(g) Ketone
(h) Cycloalkane
(i) Hydrocarbon

(j) Functional group
(k) Empirical formula
(l) Amide
(m) 2° Hydrogen atom
(n) Hydrophobic
(o) Line structure
(p) Amine
(q) Aldehyde
(r) Saturated hydrocarbon

(s) Constitutional isomers
(t) 1° Carbon atom
(u) Aromatic compound
(v) Haloalkane
(w) Carboxylic acid
(x) Alcohol
(y) Ester
(z) Straight chain alkane

2.22   In each of the following sets of three compounds, decide whether the structural formulas represent the same compound. Verify your answers by writing the IUPAC name for each compound.

(a) $(CH_3)_2CHCH_2CH_3$

CH_3
$CH_3CHCH_2CH_3$

(b) $ClCH_2CH(CH_3)_2$ $(CH_3)_2CHCH_2Cl$ CH_3CHCH_2Cl
                                                         CH_3

(c)

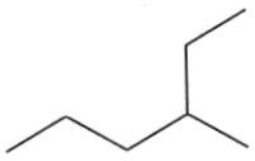

$CH_3CH_2CHCH_2CH_2CH_3$ $CH_3(CH_2)_2CHCH_3$
            CH_3 CH_2CH_3

2.23   Give the IUPAC name for each of the following compounds:

(a) $(CH_3)_2CCHCH_3$ with F above and F below

(b) CCl_3CCl_3

(c) 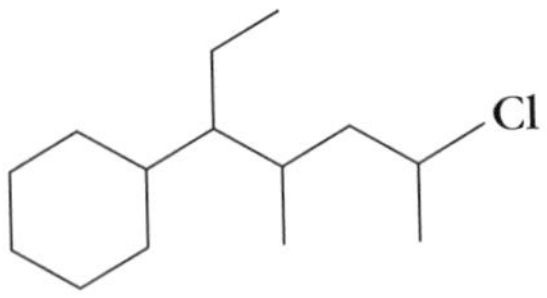 Cl

(d)

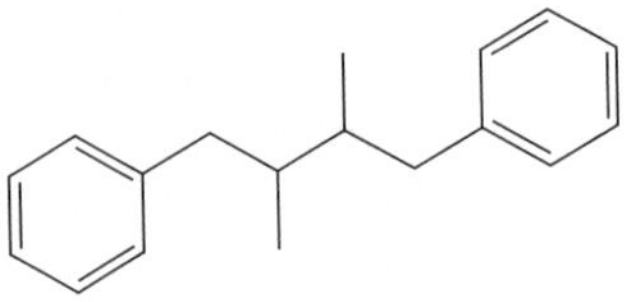

(e)

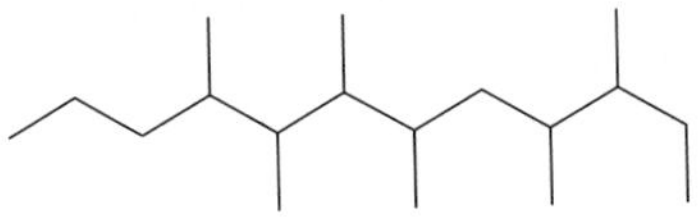

(f) 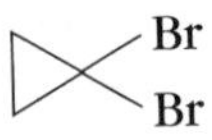 Br Br

(g)

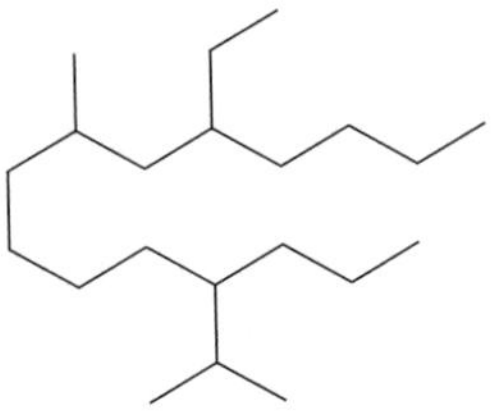

(h) 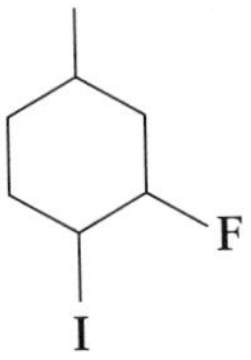 F I

(i) $C(CH_3)_3$

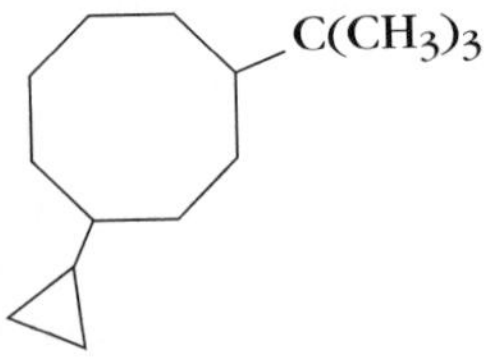

2.24   Write the structure that corresponds to each of the following IUPAC names.

(a) 4-Ethyl-5-methylnonane
(b) 2,4,6-Trifluorodecane
(c) 1-Bromo-2,4-dimethylhexane
(d) 2,2,3,3-Tetramethylbutane
(e) 4,4-Diethyl-2-methylheptane

(f) 1-Chloro-4-ethyl-3,3-dimethyl-4-propyldecane
(g) 1,1-Dibromocyclopentane
(h) 1,2-Dichloro-3-(1-methylethyl)cyclobutane
(i) 1-Chloro-2,2-dimethyl-5-(3,3-dimethylbutyl)dodecane

2.25 The following are incorrect IUPAC names. Determine what is incorrect about the name and give the correct IUPAC name.

(a) 2-*tert*-Butylpentane   (c) 2-Chloro-2-butylpentane   (e) 3-Cyclohexylbutane
(b) 4-Methylpentane   (d) 1,5-Dimethylpentane   (f) 2-Dimethylpentane

2.26 Write the structure that corresponds to each of the following common names.

(a) Isononane   (d) Neopentyl chloride   (f) Isopentyl bromide
(b) *n*-Hexyl chloride   (e) *n*-Butyl alcohol   (g) Cyclohexyl alcohol
(c) *sec*-Butyl bromide

2.27 Write the structure that corresponds to each of the following IUPAC names.

(a) 1,2-Diethylcycloheptane   (f) 1,1,2,2-Tetraphenylethane
(b) 1-Chloro-3-pentylcyclohexane   (g) 1,2-Dicyclohexyl-1-iodopropane
(c) 1-Bromo-1-phenylcyclobutane   (h) Phenylcyclohexane
(d) 2,3-Dicyclopropylhexane   (i) 1,1-Difluorocyclopentane
(e) 2-Chloro-3-cyclopentyl-5,5-dimethyl-4-(1-methylpropyl)octane   (j) 1,2-Dichloro-3-methylcyclobutane

2.28 Designate the carbon atoms in each of the following compounds as 1°, 2°, 3°, or 4°.

(a) $CH_3CCH_2CH(CH_3)_2$ with CH_3 above and CH_2CH_3 below   (b)   (c)   (d)

2.29 Designate the hydrogen atoms in each of the following compounds as 1°, 2°, or 3°.

(a) $\triangleright$—$CH(CH_3)_2$ (b)   (c)   (d)

2.30 Write the structure and give the IUPAC name of an alkane that has:

(a) Primary and quaternary carbon atoms but no secondary or tertiary carbon atoms
(b) Three primary, two secondary, and one tertiary carbon atom but no quaternary carbon atoms
(c) Only six primary hydrogen atoms and two secondary hydrogen atoms
(d) Nine carbon atoms with 12 primary hydrogen atoms and eight secondary hydrogen atoms

2.31 Write the structure and give the IUPAC name for all the isomers of molecular formula C_5H_{10} that contain a cyclopropane ring.

2.32 Predict which compound will have the lowest boiling point in each of the following sets of compounds. Explain your answer.

(a) Octane; propane; dodecane
(b) Heptane; 2-methylhexane; 3,3-dimethylpentane
(c) CH_3OH; CH_3CH_3; cyclooctane

2.33 There are eight different alkyl groups that contain five carbon atoms. Write their structures and give their IUPAC names.

2.34 Write a suitable structure for each of the following:

(a) Two alcohols of molecular formula C_3H_8O
(b) Three ethers of molecular formula $C_4H_{10}O$
(c) Two esters of molecular formula $C_4H_8O_2$
(d) Four aldehydes of molecular formula $C_5H_{10}O$
(e) Five cyclic ketones of molecular formula C_5H_8O

2.35 There are four isomeric amines of molecular formula C_3H_9N. Write their structures and classify each as 1°, 2°, or 3°.

2.36 Bromoform contains 94.85% bromine, 0.40% hydrogen, and 4.75% carbon and has a molecular weight of 253. What is the molecular formula of bromoform?

2.37   A 10.53 mg sample of a liquid compound with an odor very much like pineapple forms 9.78 mg of H_2O and 23.92 mg of CO_2 on combustion. Its molecular weight is found to be 116. What is the molecular formula for this compound?

2.38   Write structural formulas for the following:

(a)   Two isomers of formula $C_3H_6O_2$; one contains a carboxylic acid group, the other is an ester
(b)   Two isomeric ketones of formula $C_5H_{10}O$
(c)   Two isomers of formula C_4H_8; one is an alkene, the other a cyclopropane derivative

2.39   Circle and name the functional group or groups in the following compounds.

(a) 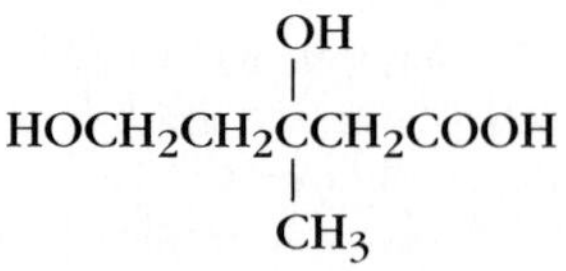 $HOCH_2CH_2\overset{\overset{\displaystyle OH}{|}}{\underset{\underset{\displaystyle CH_3}{|}}{C}}CH_2COOH$

Mevalonic acid

(d)

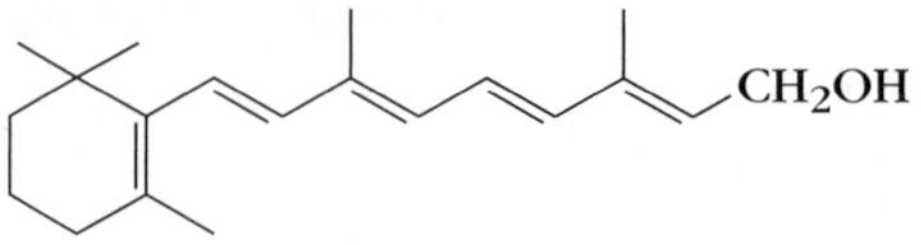

Vitamin A

(b)

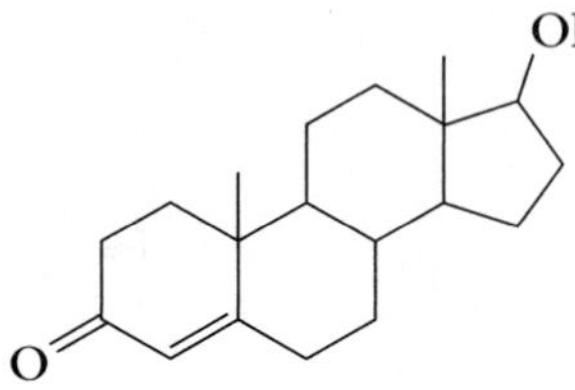

Testosterone

(e)

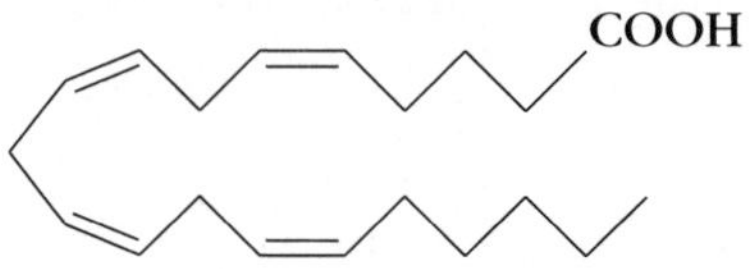

Arachidonic acid

(c)

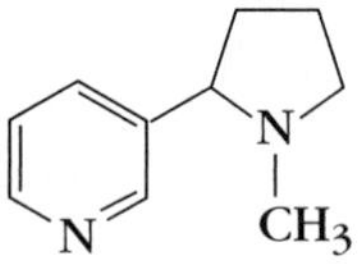

Nicotine

(f)

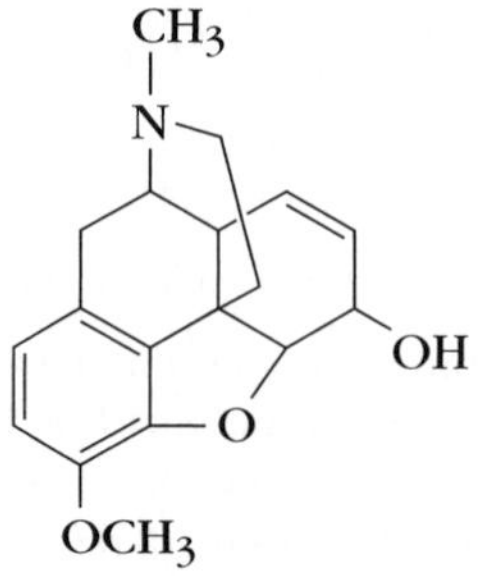

Codeine

CHAPTER

ENERGY CHANGES in CHEMICAL REACTIONS

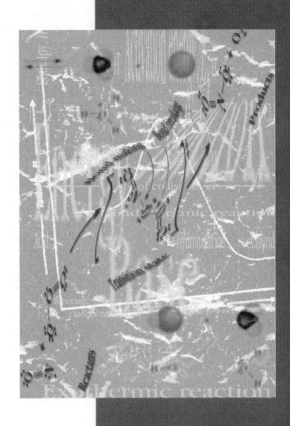

3.1   Introduction

Acid-base reactions are one of the most important types of reactions in chemistry. We begin this chapter with a review of the basic principles of acids and bases and then show that these principles apply equally well to certain families of organic compounds. After learning how to predict the extent of acid-base reactions, we turn our attention to the larger problem of predicting whether or not a reaction will occur spontaneously.

We will learn that the major factor in determining the spontaneity of a reaction is its tendency to proceed to products of lowest energy. Heat is usually produced in these reactions, but this is not a sufficient criterion for a spontaneous reaction. Another factor that contributes to the spontaneity of a reaction is the tendency to proceed in the direction of greater molecular disorder. We will learn how to combine these two tendencies into an energy term that not only tells us if a reaction will occur spontaneously but also allows us to calculate its equilibrium constant.

This energy term does not include time. Consequently we don't know how long it takes for a spontaneous reaction to occur. We will learn that the major factor that determines the speed or rate of a chemical reaction is an energy barrier that must be overcome for the reaction to occur.

We begin this chapter with a review of Brønsted-Lowry acids and bases.

3.2   Brønsted-Lowry Acids and Bases

In 1923 the Danish chemist J. Brønsted and the English chemist T. Lowry proposed a definition of acids and bases that is still widely used. According to this definition a **Brønsted-Lowry acid is any compound or ion that can donate a proton (a hydrogen ion, H^+) to another molecule or ion. A Brønsted-Lowry base is any compound or ion that can accept a proton.** These definitions are illustrated by the reaction that occurs when gaseous HCl is dissolved in water.

H—Cl	$\overset{..}{\underset{..}{O}} \overset{\diagup H}{\diagdown H}$	$H—\overset{+}{\underset{..}{O}} \overset{\diagup H}{\diagdown H}$	Cl$^-$
Acid	Base	Hydronium ion	Chloride ion

In this reaction, HCl donates a proton to water to form a hydronium ion and a chloride ion. Thus HCl is an acid and water is a base.

When an acid such as HCl donates a proton, the remaining species, in this case a chloride ion Cl$^-$, retains the electron pair of the original H—Cl bond. Because a chloride ion can accept a proton in the reverse reaction, it is also a base. It is called the **conjugate base** of the acid HCl. The two species HCl and Cl$^-$ are a **conjugate acid-base pair.** The acid HCl is the **conjugate acid** of Cl$^-$. Similarly water and a hydronium ion are a conjugate acid-base pair; water is the conjugate base of the hydronium ion, and the hydronium ion is the conjugate acid of water. These relationships are illustrated as follows:

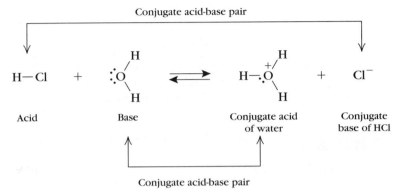

There is a simple relationship between the structure of an acid and its conjugate base that may help you remember what these terms mean. The structure of the conjugate base is that of the acid minus its acidic proton. For example, the conjugate base of hydrogen cyanide, HCN, is obtained by removing the acidic proton to form CN^-. Similarly the conjugate acid of a base is obtained by adding a proton to the structure of the base. Thus HF is the conjugate acid of the base F^-.

Exercise 3.1 Write the structure of the conjugate acid of each of the following bases:

(a) OH^- **(b)** Br^- **(c)** NO_3^- **(d)** HCO_3^- **(e)** NH_3 **(f)** SO_4^{2-}

Exercise 3.2 Write the structure of the conjugate base of each of the following acids:

(a) $HClO_4$ **(b)** HI **(c)** H_2O **(d)** HCO_3^- **(e)** NH_4^+ **(f)** H_3PO_4

Exercise 3.3 Identify the two conjugate acid-base pairs in each of the following reactions:

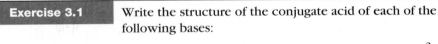

(a) $HBr + H_2O \rightleftharpoons H_3O^+ + Br^-$
(b) $H_2SO_4 + H_2O \rightleftharpoons H_3O^+ + HSO_4^-$
(c) $NH_2^- + H_2O \rightleftharpoons NH_3 + OH^-$

The ability of an acid to donate a proton to a particular base depends on its acid strength.

3.3 Relative Strengths of Acids and Bases

The strength of any acid, HA, in water is defined as its ability to donate a proton to water. In dilute aqueous solution, this strength is expressed quantitatively by the equilibrium constant of the following reaction:

$$HA + H_2O \xrightleftharpoons{K} H_3O^+ + A^-$$

$$K = \frac{[H_3O^+][A^-]}{[HA][H_2O]}$$

In dilute aqueous solutions, the concentration of water is essentially constant at 55.5 M. Therefore we can incorporate the concentration of water [H_2O] and K into a new constant K_a called the **acid dissociation constant.**

Table 3.1	Acid Dissociation Constants of Various Acids at 25 °C				
	Acid	K_a	pK_a	Conjugate Base	pK_b
STRONGEST ACID	HI	$\approx 10^{10}$	≈ -10	I$^-$	24 **WEAKEST**
	H_2SO_4	$\approx 10^9$	≈ -9	HSO_4^-	23 **BASE**
	HBr	$\approx 10^9$	≈ -9	Br$^-$	23
	HCl	$\approx 10^7$	≈ -7	Cl$^-$	21
	H_3O^+	55.5	-1.7	H_2O	15.7
	HF	6.8×10^{-4}	3.2	F$^-$	10.8
	RCOOH	10^{-4}-10^{-5}	4-5	RCOO$^-$	9-10
	CH_3COOH	1.7×10^{-5}	4.76	CH_3COO^-	9.24
	H_2CO_3	4.3×10^{-7}	6.4	HCO_3^-	7.6
	H_2S	8.9×10^{-8}	7.1	HS$^-$	6.9
	HCN	7.9×10^{-10}	9.1	CN$^-$	4.9
	NH_4^+	4.9×10^{-10}	9.2	NH_3	4.8
	RSH	$\approx 10^{-10}$	≈ 10	RS$^-$	4
	R_3NH^+ $R_2NH_2^+$ RNH_3^+	10^{-10}-10^{-11}	10-11	R_3N R_2NH RNH_2	3-4
	CH_3OH	6.3×10^{-16}	15.2	CH_3O^-	-1.2
	H_2O	2×10^{-16}	15.7	OH$^-$	-1.7
	RCH_2OH R_2CHOH R_3COH	10^{-16}-10^{-17}	16-17	RCH_2O^- R_2CHO^- R_3CO^-	-2-(-3)
	H_2	$\approx 10^{-35}$	35	H$^-$	-21
	NH_3	$\approx 10^{-38}$	38	NH_2^-	-24
WEAKEST ACID	CH_3CH_3	$\approx 10^{-50}$	50	$CH_3CH_2^-$	-36 **STRONGEST BASE**

$$K_a = K\,[H_2O] = \frac{[H_3O^+][A^-]}{[HA]}$$

The acid dissociation constants of a number of acids are listed in Table 3.1.

The K_a values of the acids listed in Table 3.1 range from about 10^{10} for strong inorganic acids to about 10^{-50} for the weak organic acid ethane. A large value of K_a for an acid means that it has completely transferred its proton to water at equilibrium. The K_a of HCl is $\approx 10^7$, for example, so the following reaction is essentially complete at equilibrium:

$$HCl \quad + \quad H_2O \quad \longrightarrow \quad H_3O^+ \quad + \quad Cl^-$$

Hydrochloric acid   Water   Hydronium ion   Chloride ion

A small value of K_a for an acid means that it has incompletely transferred its proton to water at equilibrium. The K_a of HCN is 7.9×10^{-10}, for example, so no proton transfer to water has occurred at equilibrium.

$$HCN \quad + \quad H_2O \quad \longleftarrow \quad CN^- \quad + \quad H_3O^+$$

Hydrocyanic acid   Water   Cyanide ion   Hydronium ion

The enormous range of acidities listed in Table 3.1 can be conveniently compressed by using a logarithmic scale. In this way acid strength is expressed as pK_a, the negative logarithm of the dissociation constant:

$$pK_a = -\log K_a$$

Strong acids have negative values of pK_a. An acid whose acidity constant, K_a, is 10^5, for instance, has a pK_a of -5:

$$pK_a = -\log 10^5 = -5$$

Weak acids have positive values of pK_a; a K_a of 10^{-5} gives a pK_a of $+5$. Thus the more negative the value of the pK_a, the larger the value of K_a and the stronger the acid. Conversely, the weaker an acid, the more positive its value of pK_a and the smaller its value of K_a.

Exercise 3.4   Convert each of the following K_a values into pK_a.

(a) 10^5 **(b)** 10^{-6} **(c)** 1.8×10^{-16} **(d)** 1.5×10^8

Exercise 3.5   Convert each of the following pK_a values into K_a values.

(a) 10   **(b)** -10 **(c)** 3.5   **(d)** -1.5 **(e)** 0.5

Exercise 3.6   In each of the following pairs of compounds, indicate which is the stronger acid.

(a) Methanesulfonic acid ($pK_a = -1.8$) or formic acid ($pK_a = 3.7$)
(b) Phenol ($pK_a = 10.0$) or acetic acid ($pK_a = 4.7$)
(c) Glucose ($pK_a = 12.3$) or ethanol ($pK_a = 15.9$)

An acid is strong because its conjugate base is weak. That is, a strong acid readily gives up its proton because its conjugate base has little tendency to accept a proton. This establishes a qualitative relationship between the strength of an acid and the strength of its conjugate base; *a strong acid has a weak conjugate base.* Quantitatively this means that *the weaker the base the more negative is the* pK_a *of its conjugate acid.* For example, the conjugate acid of chloride ion (Cl$^-$) is the strong acid HCl whose pK_a is about -7. The conjugate acid of cyanide ion (CN$^-$) is the weak acid HCN whose pK_a is $+9.1$. Because the pK_a of HCl is more negative than the pK_a of HCN, Cl$^-$ is a weaker base than CN$^-$. Thus the pK_a value of an acid tells us not only its strength but also the strength of its conjugate base.

Exercise 3.7	For each of the following pairs of bases, use the data in Table 3.1 to decide which one is the stronger base.

   (a) F$^-$ or HCO$_3^-$ **(b)** H$_2$O or NH$_3$ **(c)** CN$^-$ or HSO$_4^-$

The hydronium ion (H$_3$O$^+$) *is the strongest acid that can exist in water.* Adding any acid, such as HCl (p$K_a \approx -7$), that is stronger than the hydronium ion (p$K_a = -1.7$) to water results in the complete transfer of protons from the acid to water to form hydronium ions.

Similarly *the hydroxide ion* (OH$^-$) *is the strongest base that can exist in water.* Any base stronger than hydroxide ion removes a proton from water to form hydroxide ions. That is, any base whose conjugate acid is weaker than water (p$K_a = 15.7$) deprotonates water to form hydroxide ions. For example adding amide ions, NH$_2^-$, (p$K_a \approx 38$ for NH$_3$, the conjugate acid of NH$_2^-$), to water results in the complete transfer of a proton from water to the amide ions to form hydroxide ions and ammonia.

$$NH_2^- \ + \ H_2O \ \longrightarrow \ OH^- \ + \ NH_3$$

If acids stronger than the hydronium ion and bases stronger than the hydroxide ion cannot exist in water, how can we determine their strengths in water? The answer is that we determine their relative strengths in nonaqueous solvents and then extrapolate these values to aqueous solutions. Consequently the pK_a values more negative than -1.7 and more positive than 15.7 listed in Table 3.1 are determined by extrapolation.

Extrapolated values are fairly reliable for acids whose pK_a values are slightly more negative than -1.7 and slightly more positive than 15.7. The worst extrapolated values are those of very strong and very weak acids. Thus the pK_a values <-3 and >25 listed in Table 3.1 may be incorrect by several pK_a units.

So far we have concentrated mostly on the behavior of inorganic acids and bases. Many organic compounds, however, are also Brønsted-Lowry acids and bases.

3.4   Organic Compounds as Brønsted-Lowry Acids and Bases

■ *Carboxylic acids were introduced in Section 2.14.*

The functional group or groups of an organic compound determine if it is a Brønsted-Lowry acid or base. Carboxylic acids are acids, for example, because they donate a proton to water to form a carboxylate anion and a hydronium ion.

| Acetic acid A carboxylic acid (acid) | Water (base) | Acetate ion (conjugate base of acetic acid) | Hydronium ion (conjugate acid of water) |

Carboxylic acids are an acidic family of organic compounds. They are much weaker than inorganic acids, however, as shown by their relative values of pK_a. The pK_a of inorganic acids range from -7 to -10 while the pK_a of carboxylic acids range from 4 to 5.

Carboxylate anions, such as the acetate anion, are the conjugate bases of carboxylic acids. Because most carboxylic acids are weaker acids than inorganic acids, carboxylate anions are stronger bases than Cl^- or HSO_4^-. Carboxylate anions, however, are not as strong as hydroxide ions because water is a much weaker acid than most carboxylic acids.

Compounds that contain an amine functional group are organic bases. Primary, secondary, or tertiary amines are all bases because they readily accept a proton from a hydronium ion to form an alkylammonium ion and water.

■ *If you don't remember what 1°, 2°, or 3° amines are, reread Section 2.14.*

| 1° Amine (base) | Hydronium ion (acid) | Alkylammonium ion (conjugate acid of 1° amine) | Water (conjugate base of H_3O^+) |

The acid-base reaction of amines is the organic analog of the reaction of ammonia with a strong acid.

| Ammonia (base) | Hydronium ion (acid) | Ammonium ion (conjugate acid of NH_3) | Water (conjugate base of H_3O^+) |

In both of these reactions, a proton is transferred from the oxygen atom of the hydronium ion to a nitrogen atom. The new N—H covalent bond is formed by the sharing of the original lone pair of electrons from the nitrogen atom of the amine with the hydrogen atom. Amines are bases, therefore, because they have an unshared pair of electrons.

Based on the pK_a of their conjugate acids, the alkylammonium ions ($pK_a \approx$ 10–11 from Table 3.1), 1°, 2°, and 3° alkyl amines are stronger bases than either chloride ion or carboxylate anions but they are weaker bases than hydroxide ion.

Alcohols are another family of compounds that are bases. The oxygen atom of an alcohol has unshared electron pairs and accepts a proton from an acid to form a protonated alcohol.

$$CH_3-\overset{H}{\underset{\cdot\cdot}{\overset{|}{O}}}: \quad + \quad H-\overset{+}{\underset{|}{\overset{H}{\underset{\cdot}{O}}}} \quad \rightleftharpoons \quad CH_3-\overset{+}{\underset{|}{\overset{H}{O}}} \quad + \quad \overset{H}{\underset{|}{\overset{\cdot\cdot}{O}}}$$

Methanol   Hydronium ion   Protonated methanol   Water
(base)   (acid)   (conjugate acid of   (conjugate
                                            methanol)   base of H_3O^+)

The base strength of alcohols is about the same as water. In other words, alcohols are relatively weak bases that are much weaker than hydroxide ion or carboxylate ions but stronger than Cl^-.

    Alcohols are also weak acids. An alcohol can donate a proton to water to form an alkoxide ion and a hydronium ion.

$$CH_3-\overset{H}{\underset{\cdot\cdot}{\overset{|}{O}}}: \quad + \quad \overset{H}{\underset{|}{\overset{\cdot\cdot}{O}}} \quad \rightleftharpoons \quad CH_3-\overset{\cdot\cdot}{\underset{\cdot\cdot}{O}}:^- \quad + \quad H-\overset{+}{\overset{H}{\underset{|}{O}}}$$

Methanol   Water   Methoxide ion   Hydronium ion
(acid)   (base)   An alkoxide ion   (conjugate acid
                             (conjugate base of   of water)
                             methanol)

■ *Compounds such as alcohols and water that can accept and donate a proton are said to be **amphoteric**.*

The pK_a values of alcohols range from 16 to 17, which is very similar to that of water whose $pK_a = 15.7$. As a result, alcohols and water are of comparable acid strength.

    Ethers, aldehydes, and ketones are other compounds that have oxygen-containing functional groups and react as bases. They are weaker bases than water or alcohols, so their conjugate acids are very strong acids. In fact, the conjugate acids of ethers, aldehydes, and ketones are all stronger acids than the hydronium ion. The following equations illustrate the reactions of ethers, aldehydes, and ketones as bases.

$$\overset{CH_2CH_3}{\underset{CH_2CH_3}{\overset{|}{\underset{|}{\overset{\cdot\cdot}{O}}}}} \quad + \quad H-\overset{+}{\overset{H}{\underset{|}{O}}} \quad \rightleftharpoons \quad H-\overset{+}{\underset{CH_2CH_3}{\overset{CH_2CH_3}{O}}} \quad + \quad \overset{H}{\underset{|}{\overset{\cdot\cdot}{O}}}$$

Diethyl ether   Hydronium ion   Protonated ether   Water
(base)   (acid)   (conjugate acid of   (conjugate base
                                diethyl ether)   of H_3O^+)

$$\overset{\overset{\cdot\cdot}{O}:}{\underset{H}{\overset{\|}{CH_3C}}} \quad + \quad H-\overset{+}{\overset{H}{\underset{|}{O}}} \quad \rightleftharpoons \quad \overset{\overset{H}{\underset{+}{\overset{|}{O}}}}{\underset{H}{\overset{\|}{CH_3C}}} \quad + \quad \overset{H}{\underset{|}{\overset{\cdot\cdot}{O}}}$$

Acetaldehyde   Hydronium ion   Protonated   Water
An aldehyde   (acid)   acetaldehyde   (conjugate base
(base)   (conjugate acid   of H_3O^+)
                                      of acetaldehyde)

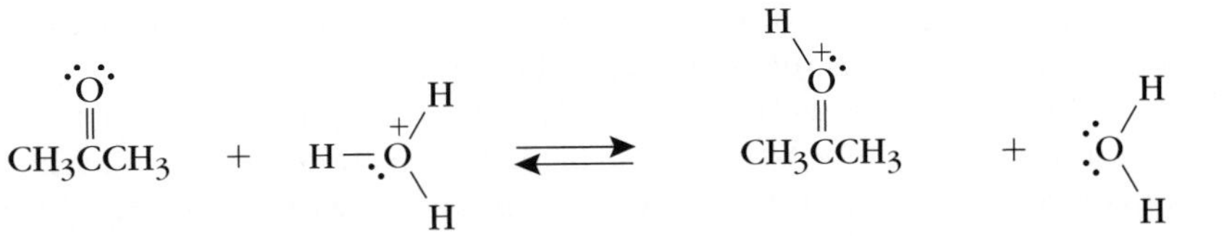

Acetone   Hydronium ion   Protonated   Water
A ketone   (acid)   acetone   (conjugate base
(base)   (conjugate acid   of H_3O^+)
                                        of acetone)

Table 3.2	pK_a Values of Various Families of Organic Compounds and Their Conjugate Acids Relative to Water[*]

	Family of Compounds	Acid or Conjugate Acid	pK_a	Base or Conjugate Base	
STRONGEST ACID	Protonated aldehydes	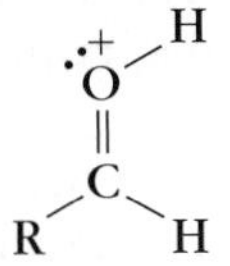	≈ -10	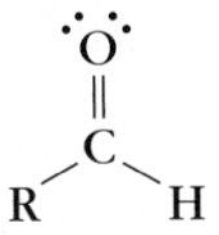	WEAKEST BASE
	Hydrogen chloride	HCl	≈ -7	Cl^-	
	Protonated ketones	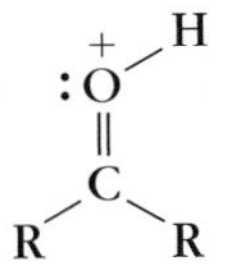	≈ -7	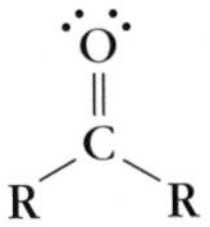	
	Protonated esters	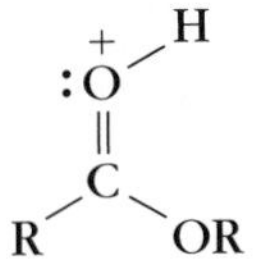	≈ -6	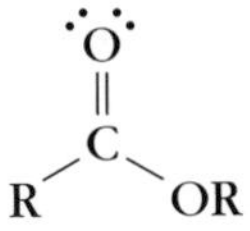	
	Protonated ethers	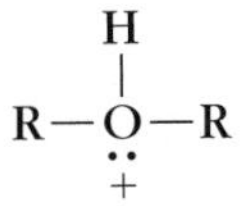 $R{-}\overset{H}{\underset{+}{O}}{-}R$	≈ -4	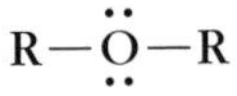 $R{-}\ddot{O}{-}R$	
	Protonated alcohols	$\begin{Bmatrix} RCH_2OH_2{}^+ \\ R_2CHOH_2{}^+ \\ R_3COH_2{}^+ \end{Bmatrix}$	≈ -2	$\begin{Bmatrix} RCH_2OH \\ R_2CHOH \\ R_3COH \end{Bmatrix}$	
	Hydronium ion	H_3O^+	-1.7	H_2O	
	Carboxylic acids	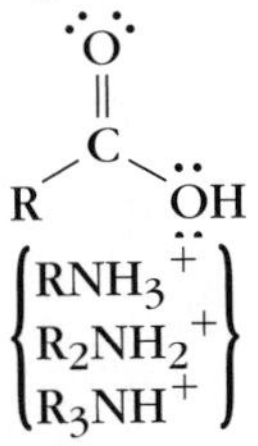	4-5	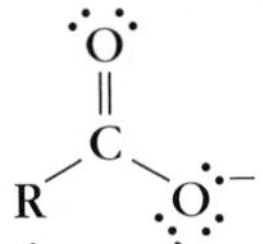	
	Alkyl ammonium ions	$\begin{Bmatrix} RNH_3{}^+ \\ R_2NH_2{}^+ \\ R_3NH^+ \end{Bmatrix}$	10-11	$\begin{Bmatrix} RNH_2 \\ R_2NH \\ R_3N \end{Bmatrix}$	
	Water	H_2O	15.7	OH^-	
WEAKEST ACID	Alcohols	$\begin{Bmatrix} RCH_2OH \\ R_2CHOH \\ R_3COH \end{Bmatrix}$	16-17	$\begin{Bmatrix} RCH_2O^- \\ R_2CHO^- \\ R_3CO^- \end{Bmatrix}$	STRONGEST BASE

[*]Acidities of HCl, hydronium ion, and water are included in this table for comparison.

The estimated pK_a values of several families of organic compounds and their conjugate acids or bases are summarized in Table 3.2.

Most of the organic bases listed in Table 3.2 have an unshared pair of electrons on a nitrogen or an oxygen atom. This is a general feature of organic compounds that are bases. Thus a nitrogen or oxygen atom with an unshared pair of electrons is usually the site of protonation of a neutral organic compound. This is the organic equivalent of the protonation of water to form a hydronium ion or the protonation of ammonia to form an ammonium ion.

While there are similarities between the protonation of water and ammonia and organic compounds containing nitrogen and oxygen atoms, there is one major difference. Frequently the organic compound undergoes further reactions after protonation. These reactions will be covered in detail in future chapters.

Before we leave the subject of organic acids and bases, we must say something about the acidity of alkanes. The estimated pK_a value of about 50 for ethane given in Table 3.1 makes alkanes the weakest organic acids. As a general rule, hydrogen atoms bonded to carbon atoms are not very acidic. In acetic acid, for instance, the most acidic hydrogen is the one bonded to oxygen—not any of the ones bonded to carbon.

We will encounter in later chapters a number of compounds that have specific hydrogens that are exceptions to this general rule. Special notice will be taken of these compounds, and the reasons for the enhanced acidity of these hydrogens will be explained at that time.

Exercise 3.8   Write the structure of the product of protonation of each of the following compounds.

(a) [structure: cyclohexane ring with ÖH group]

(b) [structure: cyclopentane ring with Ö]

(c) $(CH_3)_3N:$

(d) [structure: cyclopentane ring with =Ö]

(e) CH_3CH_2 [C=O, with H]

(f) H_3C [C=O, with Ö–cyclopentyl]

Exercise 3.9   Write the structure of the conjugate base of each of the following acids.

(a) [structure: H_3C–C(=Ö$^+$–H)–CH_3]

(b) $CH_3CH_2OCH_2CH_3$ [with H on O$^+$]

(c) [structure: pyrrolidine ring with N$^+$ bearing two H]

(d) [structure: benzene ring with COOH]

(e) $(CH_3)_3COH_2^+$

(f) [structure: cyclopentyl–C(=Ö$^+$–H)–H]

Knowing the relative acidities of organic compounds makes it possible to predict whether or not an acid-base reaction will take place.

3.5   Using pK_a Values to Predict the Extent of Acid-Base Reactions

In Section 3.4, we learned that the pK_a of an acid is a measure (or an estimate) of its ability to donate a proton to water. Frequently, however, we want to know the ability of a particular acid to donate a proton to a base other than water. For example, is acetic acid a strong enough acid to protonate the oxygen atom of ethanol?

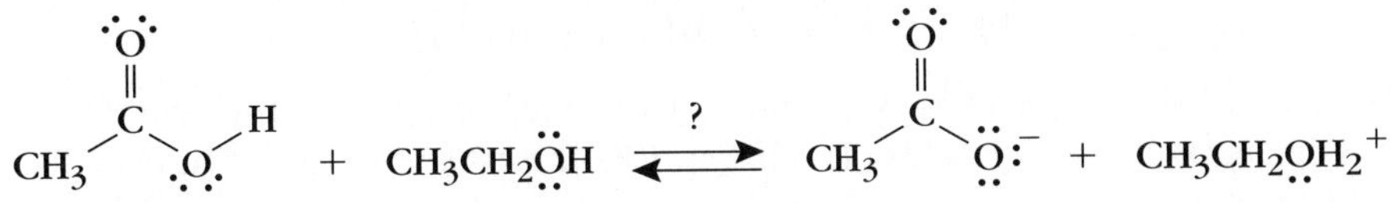

| Acetic acid (acid) | Ethanol (base) | Acetate ion | Protonated ethanol |

We can answer this question by viewing this equilibrium reaction as a competition between two bases, ethanol and acetate ion, for protons. The stronger base wins. We know from the data in Table 3.2 that acetate ion is a stronger base than ethanol, so acetate ion should compete more effectively than ethanol for protons. As a result, the equilibrium lies to the left, and acetic acid is not a strong enough acid to effectively protonate ethanol.

This specific example illustrates a general rule. *The equilibrium of an acid-base reaction lies in the direction of the weaker acid and weaker base.* In the case of acetic acid and ethanol, ethanol is a weaker base than acetate ion and acetic acid is a weaker acid than protonated ethanol. Thus the equilibrium lies toward acetic acid and ethanol.

Another way of reaching the same conclusion is to look at the relative positions of the acids and their conjugate bases in compilations of pK_a values like those in Tables 3.1 and 3.2. An acid will donate a proton to the conjugate base of any acid weaker than itself. A weaker acid has a pK_a value that is more positive. For example, H_2S (pK_a = 7.1) will transfer a proton to the conjugate base of all acids that appear below it in Table 3.1.

Exercise 3.10   Using the data in Table 3.1, decide which of the following bases will remove a proton from H_2CO_3.

(a) CH_3O^- (b) CH_3COO^- (c) H_2O (d) RNH_2 (e) HSO_4^-

Exercise 3.11   Using the data in Table 3.1, decide which of the following acids will transfer a proton to NH_3.

(a) CH_3OH (b) CH_3COOH (c) H_2O (d) H_2SO_4 (e) HF

Not only can we predict whether or not proton transfer will occur in an acid-base reaction, but we can also calculate the value of the reaction's equilibrium constant. For example, we can calculate the equilibrium constant of the reaction of acetic acid and ethanol by using the pK_a values of acetic acid and protonated ethanol in the following way. We begin by writing the equilibrium expression for this reaction:

$$CH_3COOH \ + \ CH_3CH_2OH \ \underset{}{\overset{K_{eq}}{\rightleftharpoons}} \ CH_3COO^- \ + \ CH_3CH_2OH_2^+$$

| Acetic acid (acid) | Ethanol (base) | Acetate ion | Protonated ethanol |

$$K_{eq} = \frac{[CH_3COO^-][CH_3CH_2OH_2^+]}{[CH_3COOH][CH_3CH_2OH]}$$

The equilibrium constant, K_{eq}, can be expressed in terms of the acid dissociation constants of the two acids in solution, CH_3COOH and $CH_3CH_2OH_2^+$.

For acetic acid

$$CH_3COOH \ + \ H_2O \ \rightleftharpoons \ CH_3COO^- \ + \ H_3O^+$$

$$(K_a)_{CH_3COOH} = \frac{[CH_3COO^-][H_3O^+]}{[CH_3COOH]}$$

For protonated ethanol

$$CH_3CH_2OH_2^+ \ + \ H_2O \ \rightleftharpoons \ CH_3CH_2OH \ + \ H_3O^+$$

$$(K_a)_{CH_3CH_2OH_2^+} = \frac{[H_3O^+][CH_3CH_2OH]}{[CH_3CH_2OH_2^+]}$$

$$\frac{(K_a)_{CH_3COOH}}{(K_a)_{CH_3CH_2OH_2^+}} = \left[\frac{[CH_3COO^-][H_3O^+]}{[CH_3COOH]}\right]\left[\frac{[CH_3CH_2OH_2^+]}{[H_3O^+][CH_3CH_2OH]}\right]$$

Therefore

$$\frac{(K_a)_{CH_3COOH}}{(K_a)_{CH_3CH_2OH_2^+}} = K_{eq}$$

or

$$pK_{eq} = (pK_a)_{CH_3COOH} - (pK_a)_{CH_3CH_2OH_2^+}$$

Using the value for the pK_a of $CH_3CH_2OH_2^+$ ($pK_a \approx -2$) listed in Table 3.2 and the value of the pK_a of acetic acid ($pK_a \approx 5$), we obtain a value of $K_{eq} \approx 10^{-7}$ ($pK_{eq} \approx 7$) for the reaction of acetic acid and ethanol. Clearly acetic acid is too weak an acid to transfer a proton to ethanol.

Sample Problem 3.1

Based on the relative acid and base strengths of the products and reactants, determine whether the following equilibrium reaction lies to the right or the left.

$$RSH \ + \ R-\overset{\overset{\displaystyle \ddot{O}:}{\|}}{\underset{\underset{\displaystyle \ddot{O}:^-}{}}{C}} \ \rightleftharpoons \ RS^- \ + \ R-\overset{\overset{\displaystyle \ddot{O}:}{\|}}{\underset{\underset{\displaystyle :\ddot{O}H}{}}{C}}$$

Solution:

Step 1 Identify the two pairs of conjugate acids and bases in the reaction. RCOOH and RCOO⁻ are one pair and RSH and RS⁻ are the other.

Step 2 Identify the two acids and the two bases. RCOOH is one acid and RSH is the other; the two bases are RS⁻ and RCOO⁻.

Step 3 Find the pK_a values of RSH and RCOOH in Table 3.1. The pK_a of RSH is ≈10 and that of RCOOH is ≈5. RCOOH is the stronger acid. Consequently, its conjugate base, RCOO⁻, is the weaker base.

Step 4 The reactants RCOO⁻ and RSH are the weaker base and weaker acid, respectively, so the equilibrium lies to the left.

| **Exercise 3.12** | Using the pK_a values in Tables 3.1 and 3.2, determine whether the following equilibrium reactions lie to the right or the left. |

(a) $RCOOH \ + \ R_2NH \ \rightleftharpoons \ RCOO^- \ + \ R_2NH_2^+$

(b) $H_3O^+ \ + \ R\ddot{O}R \ \rightleftharpoons \ H_2O \ + \ \overset{+}{\underset{\underset{H}{|}}{R\ddot{O}R}}$

| **Exercise 3.13** | Using the pK_a values in Tables 3.1 and 3.2, calculate K_{eq} for each of the reactions in Exercise 3.12. |

The second answer to Exercise 3.13 demonstrates that the extent of proton transfer from a hydronium ion to many organic compounds is very small in water. Such compounds are simply too weak to undergo acid-base reactions with either the hydronium ion or the hydroxide ion in water. It's possible to form stronger acids and bases, however, in solvents that contain no water.

3.6   Acid-Base Reactions in Nonaqueous Solvents

Acid-base reactions of most large organic compounds are difficult to carry out in aqueous solutions for several reasons. First, most organic compounds are not very soluble in water. This problem can be overcome by carrying out the reaction in a solvent in which the organic compound is soluble. We will have more to say about solvents in Section 3.15.

A second problem is the strength of acids in aqueous solution. Recall that the hydronium ion is the strongest acid that can exist in water. Because the conjugate acids of many organic compounds are stronger than H_3O^+, these compounds cannot be protonated with aqueous acid solutions. Instead the conjugate acid gives up a proton to water to form more hydronium ions (H_3O^+). We need an acid stronger than H_3O^+ to protonate these compounds. We can obtain stronger acid solutions by switching from water as the solvent to a nonaqueous solvent that is a weaker base than water; that is, we must switch to a solvent whose conjugate acid is stronger than H_3O^+.

Commercially available 98% to 100% sulfuric acid is the most common strongly acidic solution used in organic chemistry. The acid in concentrated sulfuric acid is $H_3SO_4^+$ and the base is HSO_4^- formed by the following self-ionization reaction:

$$2\ H_2SO_4 \rightleftharpoons H_3SO_4^+ + HSO_4^-$$

Conjugate acid of H_2SO_4 Conjugate base of H_2SO_4

This reaction is similar to the self-ionization of water.

$$2\ H_2O \rightleftharpoons H_3O^+ + OH^-$$

Most organic compounds are converted to their conjugate acids when dissolved in concentrated sulfuric acid. For example:

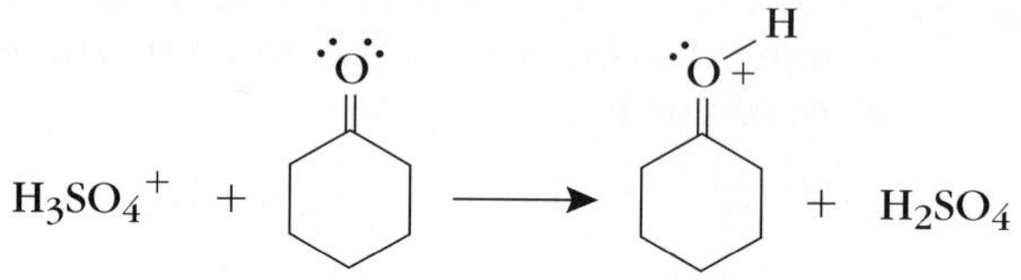

H₃SO₄⁺ + (Acid)   (Base) → (Weaker acid) + H₂SO₄ (Weaker base)

Another way of preparing strongly acidic solutions is to dissolve an acid (such as gaseous HCl) in an ether (such as diethyl ether or the cyclic ether tetrahydrofuran, THF). For example, when gaseous HCl is bubbled into THF, a proton is transferred from HCl to THF as follows:

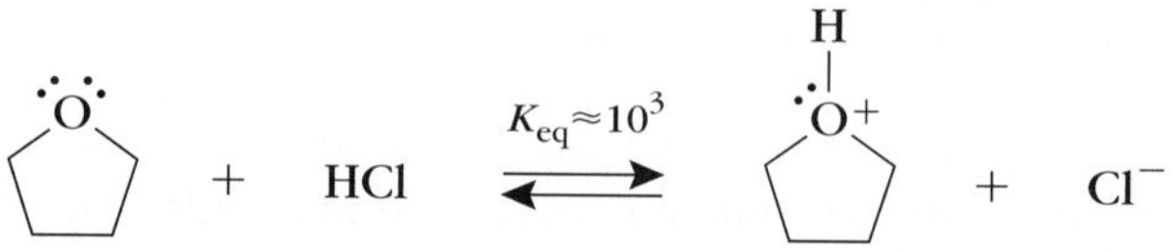

Tetrahydrofuran (THF) (base)   Acid $K_{eq} \approx 10^3$ Weaker acid in solution (conjugate acid of THF)   Weaker base in solution (conjugate base of HCl)

The acid in this solution is protonated THF, which is a stronger acid than H_3O^+.

Other acidic solutions can be made by adding gaseous HCl or the strong acid p-toluenesulfonic acid ($pK_a \approx -7$), a derivative of sulfuric acid, to hydrocarbons such as hexane or benzene. Hydrocarbons are such weak bases that they do not react with either HCl or p-toluenesulfonic acid. As a result, the acid in these solutions is HCl or p-toluenesulfonic acid. For example, p-toluenesulfonic acid reacts with methyl benzoate in hexane solvent as follows:

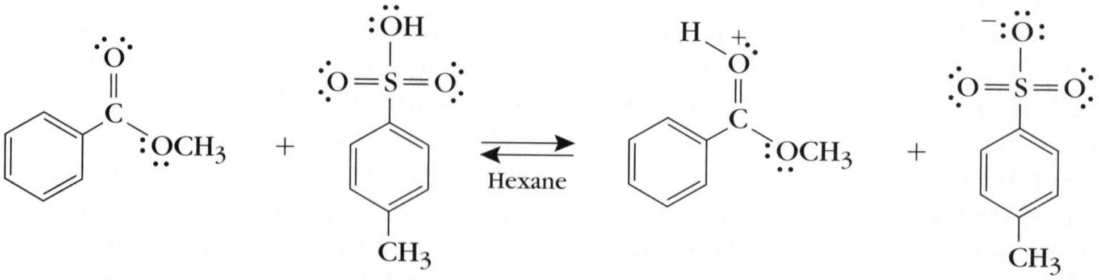

Methyl benzoate An ester (base) p-Toluenesulfonic acid (acid)   Hexane   Conjugate acid of methyl benzoate   Conjugate base of p-Toluenesulfonic acid

Another problem with aqueous solutions is that the strongest base in water is the hydroxide ion. This base is not strong enough to remove protons from many organic compounds. Thus to obtain stronger basic solutions we need a solvent that is a weaker acid; that is, we need a solvent whose conjugate base is a stronger base than hydroxide ion. Again, strongly basic solutions can be made in nonaqueous solvents.

A more strongly basic solution than sodium hydroxide in water can be prepared by adding sodium amide, $NaNH_2$, to liquid ammonia. In liquid ammonia the base is the amide ion, NH_2^-, and the acid is the ammonium ion, NH_4^+, which forms by the following self-ionization reaction of ammonia:

$$2\ NH_{3(l)} \rightleftharpoons NH_2^- + NH_4^+$$

Liquid
ammonia   Conjugate   Conjugate
                               base of   acid of
                               NH_3 NH_3

The amide ion is a strong enough base in liquid ammonia to remove a hydrogen atom from acetylene, $HC{\equiv}CH$. An aqueous solution of hydroxide ion, on the other hand, is not a strong enough base even though the hydrogens of acetylene are more acidic ($pK_a \approx 25$) than those of alkanes ($pK_a \approx 50$). The equilibrium constants of the following equations demonstrate these differences.

$$HC{\equiv}CH + NH_2^- \xrightarrow[NH_{3(l)}]{K \approx 10^{13}} HC{\equiv}C^- + NH_3$$

$$HC{\equiv}CH + OH^- \xleftarrow[H_2O]{K \approx 10^{-7}} HC{\equiv}C^- + H_2O$$

Other strongly basic solutions can be prepared in other solvents that are weaker acids than water. Common examples are pentane, hexane, diethyl ether, and THF. Thus a mixture of solid sodium amide in hexane or diethyl ether is also a source of amide ions. Hydride ion, the conjugate base of H_2 ($pK_a \approx 35$), is another strong base. It is usually prepared and used as a mixture of sodium hydride (NaH) in hexane or diethyl ether.

The strongest bases available are the alkyllithiums. Butyllithium, for example, is prepared in hexane by the following reaction:

$$CH_3CH_2CH_2CH_2Br + 2Li \xrightarrow{\text{Hexane}} CH_3CH_2CH_2CH_2Li + LiBr$$

1-Bromobutane   Butyllithium

Butyllithium and other alkyllithium compounds react as if they contain the alkanide ion, R^-. This anion is a very strong base because it is the conjugate base of the alkanes, which are the weakest organic acids. In fact, they are the strongest bases that we will use in this text.

Why are we so concerned with forming conjugate bases or conjugate acids of organic compounds? Frequently these organic acids or bases must be prepared as the first step in a desired chemical reaction. We postpone any further discussion of this point until we begin to discuss each functional group in turn. For now, it's enough to realize that we have at our disposal a number of strongly basic solutions that are capable of removing acidic hydrogens from most organic com-

■ *Notice that liquid ammonia is liquefied ammonia gas, not the aqueous ammonia solution that is found in most general chemistry laboratories. We distinguish between the two by using the formula $NH_{3(l)}$ to represent liquid ammonia and $NH_{3(aq)}$ to represent aqueous ammonia.*

pounds and other very acidic solutions that can protonate most organic compounds.

Exercise 3.14 Write the equation for the acid-base reaction that occurs between the following pairs of reagents. Label the weaker acid and weaker base in each reaction.

(a) Gaseous HCl is added to methanol (CH_3OH).
(b) Solid sodium bicarbonate ($NaHCO_3$) is added to concentrated sulfuric acid.
(c) Gaseous methylamine (CH_3NH_2) is bubbled into a solution of *p*-toluenesulfonic acid in hexane.
(d) Solid sodium hydroxide is dissolved in $(CH_3)_3COH$ ($pK_a \approx 17$).
(e) Gaseous NH_3 is bubbled into a solution of butyllithium in hexane.
(f) Solid sodium methoxide ($NaOCH_3$) is added to water.
(g) Solid sodium amide ($NaNH_2$) is added to ethanol (CH_3CH_2OH).

Exercise 3.15 Estimate the value of the equilibrium constant for each reaction in Exercise 3.14.

The Brønsted-Lowry definition of acids and bases is limited to compounds that donate or accept protons. G.N. Lewis proposed a more general definition of acids and bases.

3.7 Lewis Acids and Bases

The theory of acids and bases was expanded in 1923 when G.N. Lewis proposed that all molecules and ions that react by accepting a pair of electrons should be called acids. These acids are called **Lewis acids** to distinguish them from Brønsted-Lowry acids. Molecules and ions that donate a pair of electrons in a reaction are called **Lewis bases.** Thus a Lewis acid is an electron-pair acceptor, and a Lewis base is an electron-pair donor.

The following equations are examples of reactions between Lewis acids and Lewis bases.

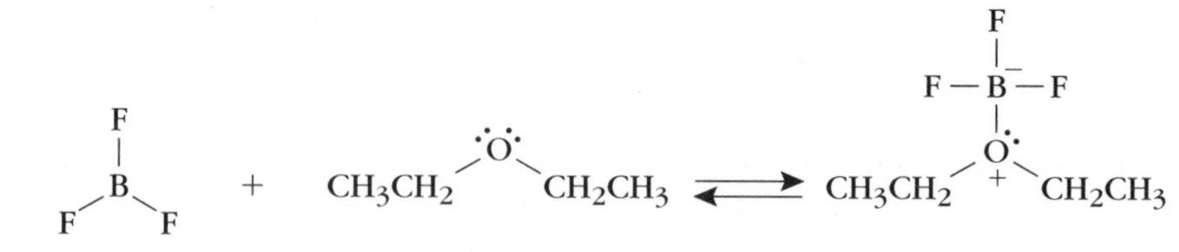

Boron trifluoride
Lewis acid
(electron acceptor)

Diethyl ether
Lewis base
(electron donor)

$$H^+ + CH_3CH_2\ddot{O}H \rightleftharpoons CH_3CH_2\ddot{O}H_2^+$$

Proton
Lewis acid
(electron acceptor)

Ethanol
Lewis base
(electron donor)

Compounds or ions that are bases according to the Brønsted-Lowry definition are also bases according to the Lewis definition. Diethyl ether and ethanol, which are shown in the examples previously, act as Brønsted-Lowry bases by accepting a proton. At the same time, they can share a pair of electrons with Lewis acids so they act as Lewis bases too.

All Brønsted-Lowry acids are also Lewis acids, but the reverse is not true. Boron trifluoride (BF_3), for instance, is a Lewis acid because the boron atom can accept a pair of electrons in its valence shell. It does not fit the Brønsted-Lowry definition of an acid, however, because it has no proton to donate. Any electron-deficient atom acts as a Lewis acid by accepting a pair of electrons from a Lewis base. Compounds of many metals, such as Al, Fe, and Zn, act as Lewis acids in reactions with organic compounds, as shown in the following examples.

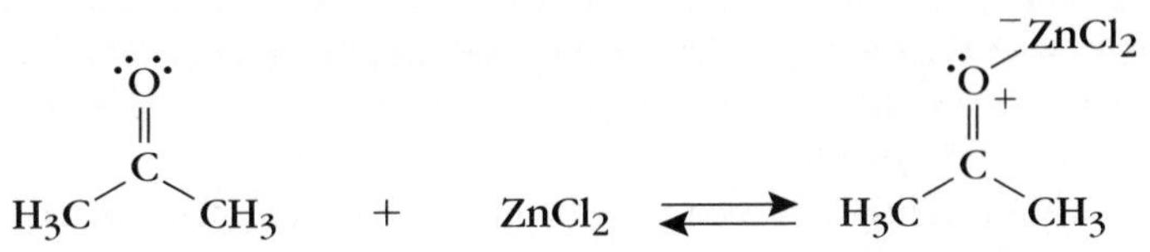

Lewis base Lewis acid

$$FeBr_3 \quad + \quad Br_2 \rightleftharpoons FeBr_4^- \; Br^+$$

Lewis acid Lewis base

Alkanes are neither acids nor bases according to the Lewis definition because all carbon atoms in alkanes have a complete octet and every hydrogen has two electrons. As a result alkanes can neither easily accept nor donate electrons. This is one reason alkanes are one of the least reactive classes of organic compounds.

Exercise 3.16 Identify the Lewis acid and Lewis base in each of the following reactions.

(a) $NH_3 + H^+ \rightleftharpoons NH_4^+$

(b) $BF_3 + F^- \rightleftharpoons BF_4^-$

(c) $AlCl_3 + Cl^- \rightleftharpoons AlCl_4^-$

(d)

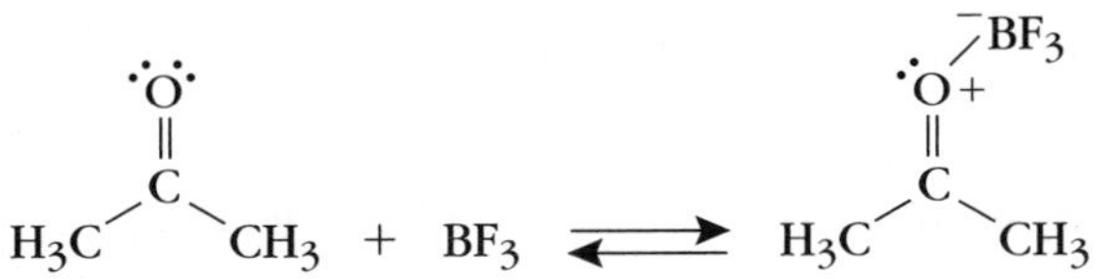

(e) $NH_3 + BCl_3 \rightleftharpoons \overset{+}{H_3N}-\overset{-}{BCl_3}$

Organic chemists have developed a symbolic device, a curved arrow, to indicate the movement of electrons in chemical reactions.

3.8 Curved Arrows Represent the Direction of Electron Pair Movement in Chemical Reactions

The reaction of ethanol with hydrogen chloride illustrates how curved arrows are used to show the movement of an electron pair from an electron donor to an electron acceptor.

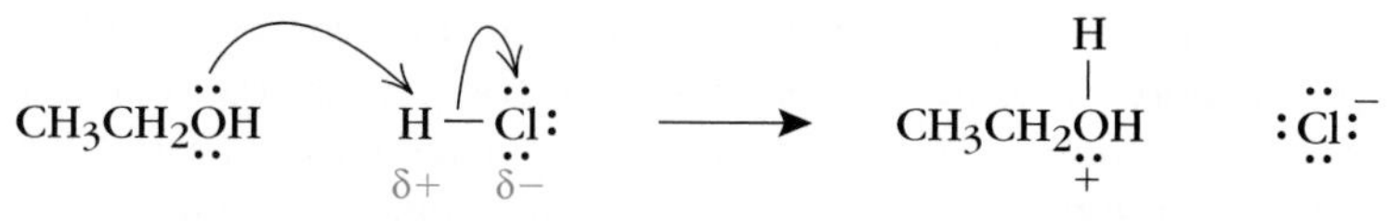

$$CH_3CH_2\overset{..}{\underset{..}{O}}H \qquad H-\overset{..}{\underset{..}{C}l:\\ \delta+\delta- \qquad\longrightarrow\qquad CH_3CH_2\overset{\overset{H}{|}}{\underset{+}{O}}H \qquad :\overset{..}{\underset{..}{C}l:}^-$$

Ethanol   Electron acceptor
Electron donor

In this reaction, ethanol is the electron donor (Lewis base) and the proton in HCl is the electron acceptor (Lewis acid). The nonbonding electron pairs are shown on both reagents. A curved arrow is drawn from the electron source, one of the nonbonding electron pairs on the oxygen atom of ethanol, to the electron deficient atom of the Lewis acid. The direction of the arrow indicates the direction of electron pair movement. The tail of the arrow is placed next to the electron pair that is going to move. The head of the arrow is placed next to the electron deficient atom where the electron pair will finally end up. *The curved arrow represents the movement of electron pairs, not atoms, during the transformation from reactants to products.*

Moving a nonbonding electron pair from the oxygen atom of ethanol to the hydrogen atom of HCl forms a new bond between these two atoms. This new bond gives the hydrogen atom two bonds, one to oxygen and another to chlorine. Structures in which a hydrogen atom forms two bonds are not permitted, so we draw a second curved arrow to represent the simultaneous release of the H—Cl bonding electron pair to the chlorine atom. Thus the chlorine atom detaches itself to become a chloride ion.

This curved arrow convention is how organic chemists indicate the electron movement in a reaction. In the ethanol-HCl example, the arrows show us that it is a nonbonding electron pair on the oxygen that becomes the bonding pair between oxygen and hydrogen in the product and the bonding pair between hydrogen and chlorine becomes localized on the chloride ion.

Keep in mind, however, that although curved arrows are a convenient device to keep track of electron movement in a chemical reaction, they do not necessarily represent the way electrons actually move in the reaction.

Sample Problem 3.2   Use curved arrows to show electron pair movement in the following reaction.

$$CH_3CH_2SH_2{}^+ \;+\; H_2O \;\rightleftharpoons\; CH_3CH_2SH \;+\; H_3O^+$$

Solution:

Step 1 Rewrite the equation so that all nonbonding electrons are shown explicitly.

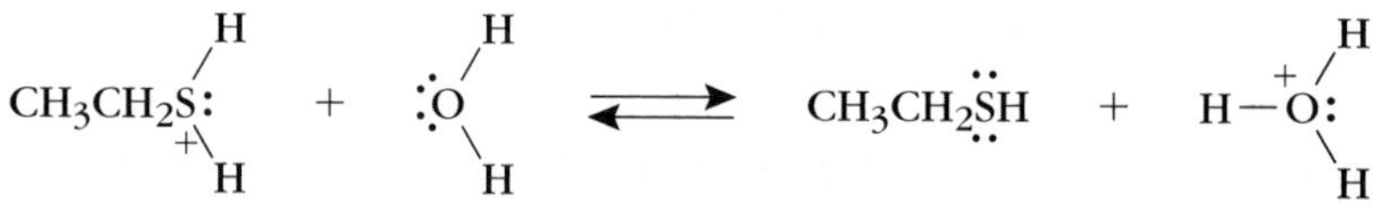

$$CH_3CH_2\overset{\overset{H}{/}}{\underset{+\backslash}{\overset{..}{S}:}}\quad +\quad :\overset{\overset{H}{/}}{\underset{\backslash}{O}}\quad \rightleftharpoons\quad CH_3CH_2\overset{..}{\underset{..}{S}}H\quad +\quad H-\overset{\overset{+}{\overset{H}{/}}}{\underset{\backslash}{\overset{..}{O}:}}$$
$$HHH$$

Step 2 Identify the electron donor and the electron acceptor. The electron pair for the new bond between oxygen and hydrogen in the product comes from one of the lone pairs on the oxygen atom of water. Therefore water is the electron donor and $CH_3CH_2SH_2{}^+$ is the electron acceptor.

Step 3 Draw an arrow from the source of electrons, one nonbonding pair on the oxygen atom of water, to their destination, one of the hydrogen atoms on sulfur in $CH_3CH_2SH_2{}^+$.

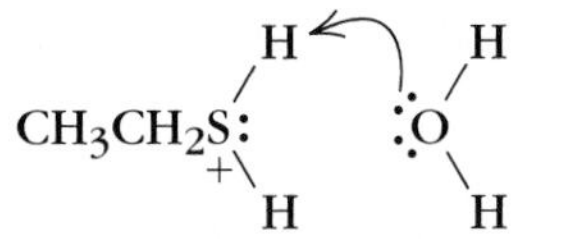

$$CH_3CH_2\overset{+}{\underset{|}{S}}: \qquad \overset{H}{\underset{H}{\overset{\frown}{O}}}$$

Step 4 Add a second arrow to indicate the movement of the electron pair displaced by the first electron movement.

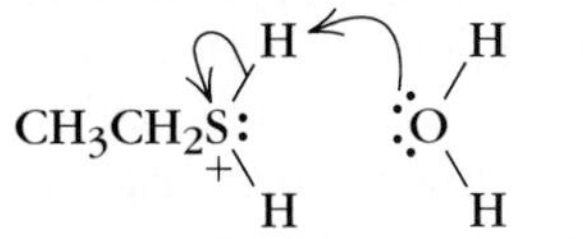

$$CH_3CH_2\overset{+}{\underset{|}{S}}: \qquad \overset{H}{\underset{H}{\overset{\frown}{O}}}$$

The arrows show that a nonbonding electron pair from oxygen forms a new O—H bond in the hydronium ion. The bonding electron pair between hydrogen and sulfur becomes one of the two nonbonding electron pairs on the sulfur atom of CH_3CH_2SH in the product.

Exercise 3.17	Use curved arrows to show electron pair movement in each of the following reactions.

(a) $NH_4^+ \ + \ OH^- \ \rightleftarrows \ NH_3 \ + \ H_2O$

(b) $(CH_3)_3C^+ \ + \ Cl^- \ \rightleftarrows \ (CH_3)_3CCl$

(c)

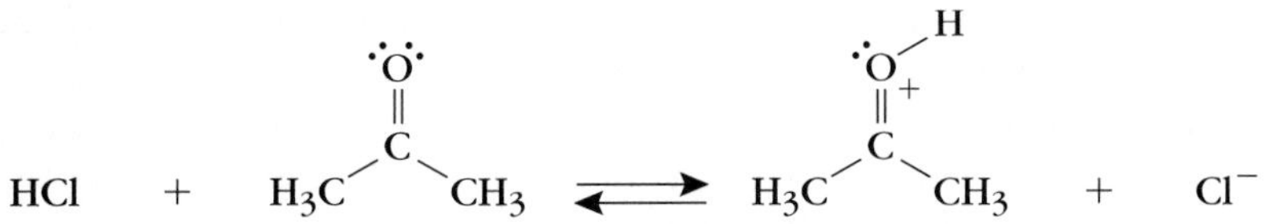

$$HCl \quad + \quad \underset{H_3C}{\overset{\overset{\displaystyle \ddot{O}}{\|}}{C}}\overset{}{\underset{}{CH_3}} \quad \rightleftarrows \quad \underset{H_3C}{\overset{\overset{\displaystyle \overset{H}{\underset{+}{\ddot{O}}}}{\|}}{C}}\overset{}{\underset{}{CH_3}} \quad + \quad Cl^-$$

(d) $\underset{H}{\overset{H}{\underset{|}{\overset{|}{H-B-H}}}} \ + \ \overset{H}{\underset{H}{\ddot{O}}} \ \rightleftarrows \ \underset{H}{\overset{H}{\underset{|}{\overset{|}{H-B}}}} \ + \ H_2 \ + \ :\ddot{O}H^-$

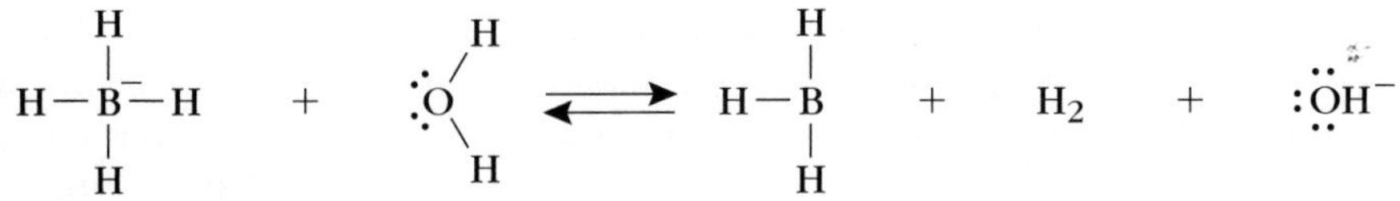

Exercise 3.18	Curved arrows are used incorrectly in each of the following reactions. Explain what is wrong and rewrite the equations with the curved arrows drawn correctly.

(a) $\underset{H}{\overset{H}{\underset{|}{\overset{+|}{H_3C-N-H}}}} \quad + \quad :\ddot{C}l: \quad \longrightarrow \quad \underset{H}{\overset{H}{\underset{|}{\overset{|}{H_3C-N}}}}\overset{H}{\underset{\cdot\ddot{C}l:}{}}$

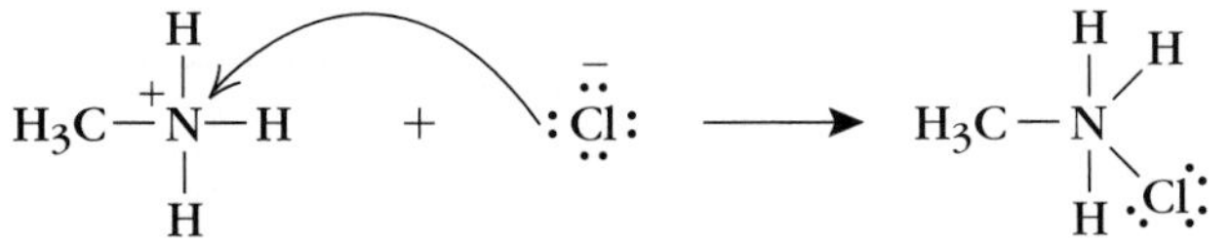

(b) $H_3N: \quad H-OSO_3H \quad \longrightarrow \quad H_3\overset{+}{N}-H \quad + \quad {}^-OSO_3H$

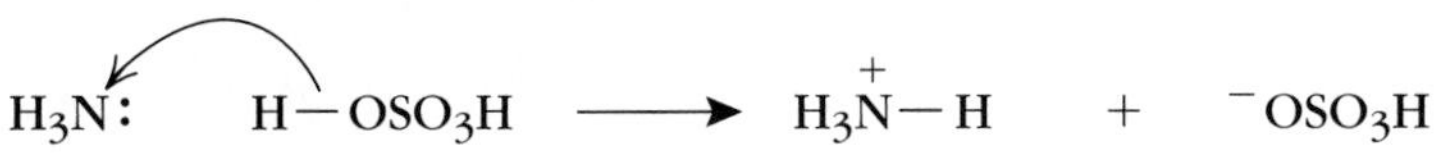

(c) $\underset{F}{\overset{F}{\underset{|}{\overset{|}{F-B}}}}\overset{H}{\underset{H}{\ddot{O}}} \quad \longrightarrow \quad \underset{F}{\overset{F}{\underset{|}{\overset{|}{F-B}}}}\overset{H}{\underset{H}{\overset{-\ddot{O}+}{}}}$

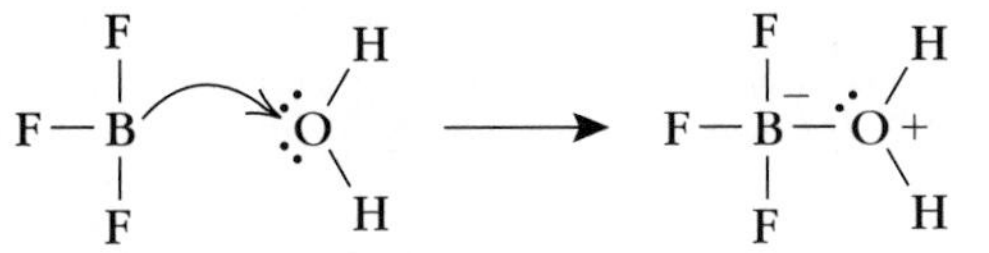

While we've learned to predict the extent of acid-base reactions from the relative pK_a values of the acids in solution, we still haven't discussed what determines the pK_a of acids. To understand what determines pK_as, we must examine the energy changes that occur in chemical reactions.

3.9   Free Energy and the Direction of Chemical Reactions

We learned in Section 3.5 how to estimate equilibrium constants of acid-base reactions from the pK_a values of acids. As a result, we discovered that the extent of proton transfer ranges widely. In some reactions, proton transfer is almost complete; in others, there is no proton transfer at all; and in still others partial proton transfer occurs. What determines the extent of proton transfer? Or, more generally, what determines the position of equilibrium in any chemical reaction?

We can try to understand the factors that affect the equilibria of chemical reactions by examining processes that are spontaneous. By spontaneous, we mean processes that occur by themselves, given enough time, without outside intervention.

Spontaneous processes involving macroscopic objects occur with a decrease in potential energy. For example, water runs spontaneously downhill from a state of higher potential energy to one of lower potential; water never runs spontaneously uphill. Similarly, a rock spontaneously rolls down a mountain but it never rolls back up to the top. From these examples, we might conclude that chemical reactions should also proceed spontaneously to an equilibrium state of lowest energy. Many reactions do, but the situation is more complicated.

Consider a process in which there is no energy change at all. Suppose that we place a container of ammonia gas in a room. The container and the room are at the same temperature and pressure. If the container of ammonia is opened, the pungent odor of ammonia is soon evident throughout the room. If the temperature of the container and its surroundings remains constant and if there is no energy change in the system, then why does the smell of ammonia permeate the room? The answer has to do with the molecular nature of matter. Ammonia molecules move rapidly and randomly throughout the space available to them. By opening the container, we allow the molecules of ammonia to move into the room until they are evenly spread throughout the container and the room. Thus the spontaneity of this process is associated with the random motion of large numbers of molecules, which leads to a tendency for all molecules to become more dispersed in space.

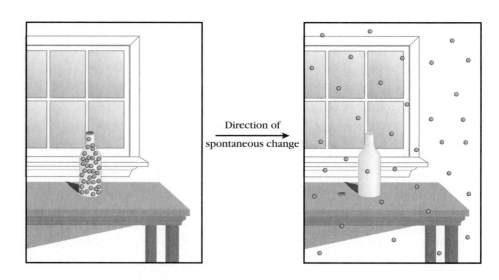

By observing a great many spontaneous processes, we conclude that there is a *tendency for systems to proceed to a state in which there is a greater disorder on the molecular level as well as a tendency to proceed to a state of lower potential energy.* The best compromise between these two tendencies at any temperature is the equilibrium state. These ideas are described quantitatively by thermodynamics, the branch of science concerned with the energy changes that accompany chemical and physical processes.

The thermodynamic property that measures the amount of molecular disorder is called **entropy** and is denoted *S*. The change of entropy for any process is given the symbol ΔS, which is defined as follows:

$$\Delta S = S_{\text{final state}} - S_{\text{initial state}}$$

The state referred to in this definition is made up of two parts. One part, called the **system,** is whatever we are investigating. A system has clearly defined boundaries, so a chemical reaction in a flask is an example of a system. The rest of the state is all the rest of the universe outside the boundaries of the system, and it is called the **surroundings.** *In any spontaneous process the total entropy of a system and its surroundings increases. Thus the amount of molecular disorder increases in any spontaneous process.*

$$\Delta S_{\text{(system)}} + \Delta S_{\text{(surroundings)}} > 0$$

While this equation is a criterion for a spontaneous change, it is not very useful for chemists because they do not usually carry out reactions in isolated systems. We can obtain a better criterion but first we must learn how to express quantitatively the changes in energy in a chemical reaction.

Most of the chemical reactions of interest to organic chemists occur at constant pressure and in solution. Under these conditions, the heat gained or lost can be regarded as the change of internal energy in the reaction. The loss or gain of heat in this manner is called the **enthalpy change** for the reaction and is given the symbol ΔH.

$$\Delta H = \begin{bmatrix} \text{Sum of the enthalpies} \\ \text{of products} \end{bmatrix} - \begin{bmatrix} \text{Sum of the enthalpies} \\ \text{of reactants} \end{bmatrix}$$

A chemical reaction in which heat is evolved has a negative value of ΔH and is called an **exothermic reaction.** In the combustion of methane, for example, 213 kcal (890 kJ) of heat are released for every mole of methane consumed at constant pressure so $\Delta H = -213$ kcal/mol (-890 kJ/mol) for this reaction.

$$CH_4 + 2O_2 \longrightarrow CO_2 + 2H_2O \quad \Delta H = -213 \text{ kcal/mol} (-890 \text{ kJ/mol})$$

Heat is given off because the sum of the enthalpies of the reactants is more than the sum of the enthalpies of the products, as shown in Figure 3.1, *A*.

A chemical reaction in which heat is absorbed has a positive value of ΔH and is called an **endothermic reaction.** The gas phase reaction of CH_4 and I_2 to form CH_2I_2 and HI at constant pressure is an example of an endothermic reaction because it requires 16.6 kcal (69.4 kJ) of heat for each mole of CH_4:

$$CH_4 + 2I_2 \longrightarrow CH_2I_2 + 2HI \quad \Delta H = +16.6 \text{ kcal/mol} (+69.4 \text{ kJ/mol})$$

Heat is required for this reaction because the sum of the enthalpies of the products is more than the sum of the enthalpies of the reactants, as shown in Figure 3.1, *B*.

■ *Changes in variables are conventionally designated by the Greek capital letter Δ (delta). Each such change is defined as the final value of the variable minus the initial value of the variable.*

Figure 3.1
Enthalpy changes in exothermic **(A)** and endothermic **(B)** reactions.

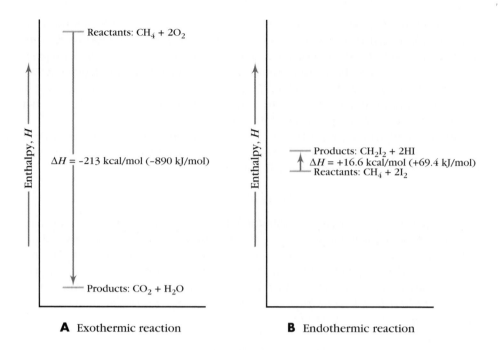

A Exothermic reaction   **B** Endothermic reaction

The **Gibbs free energy, G,** relates enthalpy, H, entropy, S, and temperature, T in kelvin, according to the following equation.

$$G = H - TS$$

This function is named in honor of J. Willard Gibbs, an American mathematician and physicist who developed many areas of thermodynamics.

At constant temperature the change in Gibbs free energy for a chemical reaction is

$$\Delta G = \Delta H - T\Delta S$$

This is an important equation because it allows us to predict whether or not a particular reaction will occur spontaneously at constant temperature and pressure. The key to determining the spontaneity of a reaction is the sign of ΔG. *When ΔG is negative (<0), the reaction will take place spontaneously at constant temperature and pressure.*

We can understand why ΔG must be negative for a spontaneous process by examining the signs of both ΔH and ΔS for a spontaneous process. The value of ΔH is negative ($\Delta H < 0$) for any reaction that proceeds toward a state of lower energy. The value of ΔS, on the other hand, is positive ($\Delta S > 0$) for a spontaneous reaction, so the term $-T\Delta S$ is negative. Thus, in a spontaneous reaction, both a decrease in enthalpy (a tendency toward lower energy) and an increase in total entropy (a tendency toward greater molecular disorder) contribute negative terms to ΔG, so the Gibbs free energy must decrease during a spontaneous reaction.

Conversely, a reaction with a positive value of ΔG is not spontaneous *in the direction written*. The reverse reaction has a negative value of ΔG, however, so the reverse reaction will proceed spontaneously instead. Finally if $\Delta G = 0$, the reaction has no tendency to proceed in either direction because the reaction is at equilibrium.

In summary:
- If $\Delta G < 0$ at constant T and P, reaction is spontaneous and it is called an **exergonic** reaction.
- If $\Delta G > 0$ at constant T and P, reaction is not spontaneous and it is called an **endergonic** reaction. While the forward reaction is not spontaneous, the reverse reaction is spontaneous.
- If $\Delta G = 0$ at constant T and P, reaction is at equilibrium so no net reaction occurs.

The value of the free energy change in a chemical reaction allows us not only to determine if a reaction will occur spontaneously but also to calculate the value of its equilibrium constant.

3.10 Standard Free Energy Changes and Equilibrium Constants Are Related

The quantitative relationship between the standard free energy change of a reaction and its equilibrium constant is given by the following:

$$\Delta G° = -2.30 \, RT \, [\log K_{eq}]$$

where T is temperature in kelvin, R is the gas constant, and $\Delta G°$ is the standard Gibbs free energy. The standard Gibbs free energy is defined as the free energy change in the reaction when all reactants and products are in their standard states (1 atm pressure and 25 °C for all elements and compounds and 1 M concentration for all solutions).

The value of R must be chosen so that the term RT and $\Delta G°$ have the same units. Since we usually express $\Delta G°$ in kilocalories per mole (kcal/mol), R must be 1.987×10^{-3} kcal/mol K. If $\Delta G°$ is given in kilojoules per mole (kJ/mol), R must be 8.315×10^{-3} kJ/mol K. Sample Problem 3.3 illustrates how to calculate the equilibrium constant of a reaction from its standard free energy.

| Sample Problem 3.3 | Calculate the equilibrium constant for the following reaction whose standard free energy change is $\Delta G° = -7.0$ kcal/mol (-29.2 kJ/mol). |

$$CO \; + \; 2H_2 \; \rightleftharpoons \; CH_3OH$$

Solution:

Begin by writing the equation that relates K_{eq} to $\Delta G°$, then substitute the values of $\Delta G°$, R, and T into the equation:

$$\Delta G° = -2.30 \, RT \, [\log K_{eq}]$$

$$-7.0 \, \frac{kcal}{mol} = -2.30 \, (1.987 \times 10^{-3} \, \frac{kcal}{mol \, K})(298.2 \, K) \log K_{eq}$$

$$\log K_{eq} = +5.139$$

$$K_{eq} = 10^{5.139} = 1.38 \times 10^5$$

The value of the equilibrium constant for this reaction is very large, so the position of the equilibrium must be far to the right. Therefore, at equilibrium, there must be a high concentration of CH_3OH and extremely low concentrations of CO and H_2. At equilibrium, in other words, CO and H_2 have been converted almost completely to CH_3OH.

The relationship between the sign of $\Delta G°$ of a reaction and the value of its equilibrium constant is summarized in Table 3.3.

Table 3.3	Qualitative Relationship Between the Standard Free Energy Change, $\Delta G°$, of a Reaction and its Equilibrium Constant, K_{eq}	
	$\Delta G°$	K_{eq}
Equilibrium	$\Delta G° = 0$	$K_{eq} = 1$
Not spontaneous	$\Delta G° > 0$	$K_{eq} < 1$
Spontaneous	$\Delta G° < 0$	$K_{eq} > 1$

Exercise 3.19   Given $\Delta G°$, calculate the equilibrium constant for each of the following reactions.

(a) $CH_2{=}CH_2 + Br_2 \rightleftharpoons BrCH_2CH_2Br$ $\Delta G° = -21.2$ kcal/mol
(b) $CH_3OH + HBr \rightleftharpoons CH_3Br + H_2O$ $\Delta G° = -41.8$ kJ/mol

Exercise 3.20   Given K_{eq}, calculate $\Delta G°$ for each of the following reactions.

(a) $CH_3COOH + H_2O \rightleftharpoons CH_3COO^- + H_3O^+$ $K_{eq} = 1.8 \times 10^{-5}$
(b) $HF + CN^- \rightleftharpoons F^- + HCN$ $K_{eq} = 1.3 \times 10^6$

We have learned that chemical reactions proceed in the direction of lower energy and greater molecular disorder. Up to now, however, we have not worried about how long it takes for these processes to occur. What are the factors that influence the speeds of chemical reactions?

3.11   Rates of Chemical Reactions

The free energy change in a reaction is the thermodynamic criterion that determines if a reaction will be spontaneous. But the free energy change doesn't tell us how long it will take for a reaction to go to completion. A spontaneous reaction may be fast or slow. For example, the spontaneous reaction, $\Delta G° = -19.1$ kcal/mol (-80 kJ/mol), of a strong base (NaOH) and a strong acid (HCl) in aqueous solution occurs as fast as one reagent is added to the other. In contrast the spontaneous reaction of methane (natural gas) and oxygen, $\Delta G° = -191.6$ kcal/mol (-801 kJ/mol) is so slow in the absence of a spark that we can leave oxygen and methane in a closed container for decades without any reaction occurring.

In order to understand why some spontaneous reactions occur rapidly while others do not, we must understand the factors that influence the rate. In the following sections we will examine how the concentration of the reactants, the solvent, the temperature, and the presence of catalysts affect the rates of reactions. Before we begin, however, we need a precise definition of the rate of a reaction.

The words *rate, velocity,* and *speed* are often used interchangeably to describe a distance travelled in a certain time.

$$\text{Speed of object} = \frac{\text{Distance travelled}}{\text{Time taken}} = \frac{\text{Change in its position}}{\text{Time for change}}$$

The rate of a chemical reaction is similarly defined as the change in concentration of a reactant (or product) in a given time interval. The concentration of reactants decreases during the reaction, so $\Delta[\text{Reactant}]/\Delta t$ is negative; conversely the concentration of products increases with time, so $\Delta[\text{Product}]/\Delta t$ is positive.

$$\text{Reaction rate} = -\frac{\Delta[\text{Reactant}]}{\Delta t} = -\frac{\text{Change in reactant concentration}}{\text{Time needed for change}}$$

$$\text{Reaction rate} = +\frac{\Delta[\text{Product}]}{\Delta t} = +\frac{\text{Change in product concentration}}{\text{Time needed for change}}$$

One factor that affects the rate of reaction is the concentration of reagents.

3.12   The Rate Law

The only way to find out how the concentration of reagents affects the rate of a reaction is to carry out a series of experiments in which the concentration of each reagent is varied independently. Consider, for example, the gas phase reaction of nitrogen monoxide (NO) and ozone (O_3), which is believed to be one of the reactions responsible for the depletion of the ozone layer of the upper atmosphere.

$$NO + O_3 \longrightarrow NO_2 + O_2$$

We find experimentally that if the concentration of NO is doubled while the concentration of O_3 is held constant, the rate of reaction doubles. If the concentration of O_3 is doubled while the concentration of NO is held constant, the rate also doubles. If the concentration of both NO and O_3 is doubled, the rate increases by a factor of four. From these kinds of observations, we conclude that the rate of reaction is proportional to the product of the concentrations of NO and O_3. This relationship is expressed quantitatively as a **rate law** for the reaction. A rate law is an equation that relates the rate of a reaction to the concentrations of its reactants raised to various powers. The rate law of the reaction of NO and O_3 is given by the following:

$$\text{Rate of reaction} = k[\text{NO}][\text{O}_3]$$

The quantity k is called the **rate constant** of the reaction. It is a proportionality constant in the relationship between rate and concentration. It has a fixed value for a particular reaction at any given temperature, but its value varies with temperature.

| **Exercise 3.21** | Predict the change in the rate of reaction of NO and O_3 if the initial concentrations are changed as follows: |

(a) The concentration of NO is decreased by half while the concentration of O_3 remains constant.

(b) The concentration of O_3 is decreased by half while the concentration of NO remains constant.

(c) The concentrations of both NO and O_3 are decreased by half.

Both reactant concentrations in the rate law of the reaction of NO and O_3 are raised to a power of one. These exponents represent the **reaction order** of NO and O_3 in the experimentally determined rate law. Thus the reaction of NO and O_3 to form NO_2 and O_2 is first order with respect to NO and first order with respect to O_3. The overall order of a reaction is the sum of the orders of each of the species in the rate law. In the reaction of NO and O_3, the overall order is two. Thus this reaction is second order overall.

Not all reactions are second order overall. The decomposition of dinitrogen pentoxide, for instance,

$$2N_2O_5 \longrightarrow 4NO_2 + O_2$$

has the following rate law:

$$\text{Rate} = k[N_2O_5]$$

This reaction is first order with respect to N_2O_5 and first order overall. Notice that there is no connection between the order with respect to N_2O_5 in the rate law and the coefficient of N_2O_5 in the balanced equation.

The reaction of NO and Cl_2 has the following rate law:

$$2NO + Cl_2 \longrightarrow 2NOCl$$
$$\text{Rate} = k[NO]^2[Cl_2]$$

This reaction is third order overall because it is first order in Cl_2 and second order in NO.

Sometimes there is a coincidental relationship between the orders of the reactants and their coefficients in the balanced chemical equation. The reaction of NO and Cl_2 is an example. The coefficient of NO in the balanced chemical equation and its order are both two. Similarly, the coefficient of Cl_2 and its order are both one. This is a coincidence. Generally there is no correspondence between the coefficients of the reactants and their orders in the rate law.

| **Exercise 3.22** | The experimentally determined rate law is given for each of the following reactions. Give the overall order and the order with respect to each species in each rate law. |

(a) △ $\longrightarrow$ $CH_2{=}CHCH_3$ Rate $= k[\text{cyclopropane}]$

      Cyclopropane

(b) $(CH_3)_3CCl + H_2O \longrightarrow (CH_3)_3COH + HCl$ Rate $= k[(CH_3)_3CCl]$

(c) $CH_2{=}CH_2 + Br_2 \longrightarrow BrCH_2CH_2Br$ Rate $= k[CH_2{=}CH_2][Br_2]^2$

(d) $CH_3Cl + NaOH \longrightarrow CH_3OH + NaCl$ Rate $= k[CH_3Cl][NaOH]$

We can experimentally determine the rate law of a reaction, but what does this mean at the molecular level? The Transition State Theory is one way to describe in molecular terms the effect of changes in concentration and temperature on chemical reactions.

3.13   From Rate to Mechanism

The reaction resulting from the collision of two molecules is described in molecular terms by the **Transition State Theory.** According to this theory, two molecules react if they can overcome a free energy barrier called the **free energy of activation** ($\Delta G^{\ddagger}$). Like ΔG°, $\Delta G^{\ddagger}$ consists of an enthalpy component, called the enthalpy of activation ($\Delta H^{\ddagger}$) and an entropy component, called the entropy of activation ($\Delta S^{\ddagger}$), which are related by the following equation:

$$\Delta G^{\ddagger} = \Delta H^{\ddagger} - T\Delta S^{\ddagger}$$

The enthalpy of activation is related to the change in internal energy needed to react, and the entropy of activation is associated with the change in orientation of the reactants needed for the reaction to occur. Thus two molecules react if they collide with sufficient energy and the proper orientation. Let's examine these two requirements in more detail.

Consider first the requirement that molecules must have the proper orientation in order to react. The gas phase reaction of NO and O_3 introduced in Section 3.12 is believed to occur in a single step in which an O_3 molecule collides with an NO molecule to form the products. Such a reaction, in which only a single molecular event takes place between reactants and products, is called an **elementary reaction.** We can write this reaction in the following way.

| Reactants | Transition state | Products |

The structure in the preceding equation that is enclosed in brackets with the symbol ‡ superscripted outside the brackets is called a **transition state.** The transition state structure represents the geometry of the collision and the bond breaking and bond making at an energy maximum in the path from reactants to products. The dotted lines in the structure of the transition state indicate bonds that are made and broken in the transition state. Because of its high energy, a transition state is only a fleeting arrangement of atoms. As a result, we cannot isolate or directly observe a transition state.

The two molecular collisions depicted in Figure 3.2 show why reactions occur only when the reactant molecules are properly oriented. In Figure 3.2, *A,* the NO molecule approaches with its N atom toward the O_3 molecule at an angle expected for the formation of bonds in the NO_2 molecule. This orientation leads to reaction. In Figure 3.2, *B,* the NO molecule approaches with its O atom toward the O_3 molecule. This orientation does not allow formation of a bond between N and O atoms so no reaction occurs. Instead the NO and O_3 molecules simply collide and move away from each other. In most reactions, especially those involving

■ *Throughout this text, transition state structures are enclosed in brackets with the symbol ‡ superscripted outside the brackets.*

Figure 3.2
The importance of molecular orientation in the reaction of NO and O_3. **A,** Collision between an NO and an O_3 molecule in this orientation leads to reaction. **B,** Collision between an NO and an O_3 molecule in this orientation does *not* lead to reaction because there is no way to form a new N—O bond.

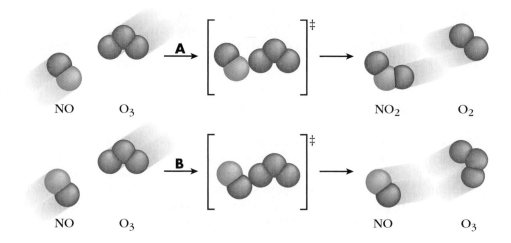

large and complicated molecules, only a very small fraction of the collisions have the required orientation for reaction.

In addition to correct spatial orientation, the collisions of reactant molecules need to have sufficient energy. Energy is needed to overcome intermolecular repulsions between the approaching molecules and to begin bond breaking at the onset of the reaction.

The free energy change that occurs as the reactants approach each other, collide, and form products is plotted on a free energy–reaction path diagram and is shown in Figure 3.3 for the reaction of NO and O_3.

The free energy curve starts at the left with the free energy of the reactants, NO and O_3. As the reaction progresses (going from left to right in the diagram), the reactants come together and the free energy increases until it reaches a maximum at the transition state. The energy difference between the transition state and the reactants is the free energy of activation ($\Delta G^{\ddagger}$) of the forward reaction. $\Delta G^{\ddagger}$ is positive and is the barrier that must be overcome for the reaction to occur. $\Delta G^{\ddagger}$ is positive because both $\Delta H^{\ddagger}$ and $\Delta S^{\ddagger}$ are positive in the activation process. $\Delta H^{\ddagger}$ is positive because energy must be added to bring the molecules together and begin to break bonds, and $\Delta S^{\ddagger}$ is positive because the system becomes more ordered since the reactants must adopt a specific orientation to react.

The reverse reaction also has a free energy of activation, which is the barrier that must be overcome for the reaction to occur in the reverse direction. Also represented in the free energy–reaction path diagram shown in Figure 3.3 is the free energy change of the reaction. While most reactions have a positive free energy of activation for either the forward or reverse reaction, the overall reaction may be either exergonic or endergonic. The reaction of NO and O_3 shown in Figure 3.3 is exergonic ($\Delta G < 0$) because the free energy of the products, NO_2 and O_2, is less than the free energy of the reactants.

The diagram in Figure 3.3 graphs the free energy change along a specific path between the reactants and products. This path describes in detail how the reactants are converted to products and is called the **mechanism** of the reaction.

The mechanism of a reaction cannot be observed directly. Instead it is proposed to explain experimental observations. In future chapters, we will learn a great many experimental facts about various functional groups. For a number of

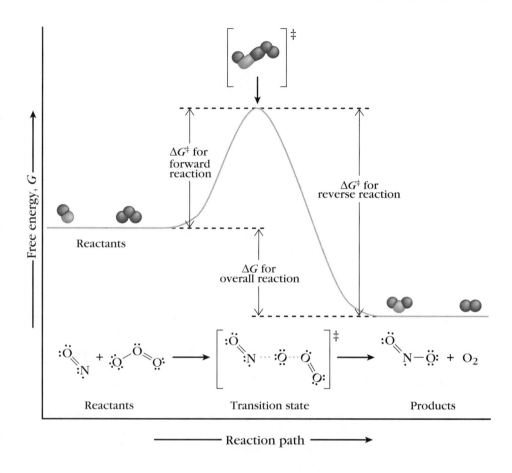

functional groups, we will learn how these experimental facts have been used to propose reaction mechanisms.

3.14 Temperature and Rates of Reaction

The rates of most chemical reactions increase as temperature increases. Generally it is found that the rate increases by a factor of two to four for every 10 K increase in temperature.

The relationship between the rate constant of a reaction and the temperature is expressed by the **Arrhenius equation,** which is named after the Swedish chemist Svante Arrhenius who first proposed it. The Arrhenius equation is

$$\log K = \log A - \frac{E_a}{2.303\ RT}$$

where E_a is the activation energy, R is the gas constant (1.987×10^{-3} kcal/mol K or 8.315×10^{-3} kJ/mol K), and T is the absolute temperature. The symbol A is called the frequency factor. The quantities in the Arrhenius equation have been related by Transition State Theory to $\Delta H^{\ddagger}$ and $\Delta S^{\ddagger}$. The equation that relates activation energy (E_a) and the enthalpy of activation ($\Delta H^{\ddagger}$) is

$$E_a = \Delta H^{\ddagger} + RT$$

Figure 3.4
Distribution of kinetic energies of collision. The shaded areas show that the relative fractions of molecules with kinetic energies of collision greater than the activation energy increase with increasing temperature.

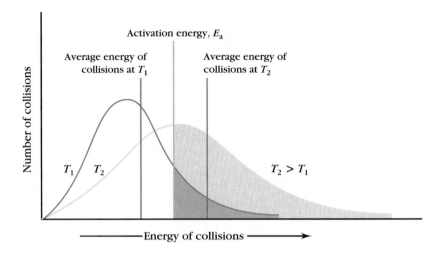

The frequency factor (A) is identified as the frequency of collisions of reactant molecules with proper orientation and, consequently, is related to $\Delta S^{\ddagger}$. The frequency factor is relatively independent of temperature over small temperature ranges ($\Delta T \approx 30$ K).

The energy needed for a reaction to occur is provided by the kinetic energy of the reactant molecules. To understand the molecular explanation of the Arrhenius equation, we must examine the effect of temperature on the distribution of the kinetic energies of collisions between molecules in the gas phase. Figure 3.4 shows the distribution of energies of collisions at two temperatures, T_1 and T_2, where T_2 is the higher temperature ($T_2 > T_1$). The violet line in Figure 3.4 indicates the activation energy for the reaction of the colliding molecules. At the lower temperature, the activation energy is greater than the average energy of the colliding molecules, so only a small fraction of the collisions have the required energy to reach the transition state. As a result, the reaction at T_1 is slow. At the higher temperature, a larger fraction of the molecules have the energy needed to overcome the activation energy, so the reaction is faster at T_2. The effect of temperature, therefore, is to increase the fraction of collisions that have sufficient energy for reaction.

The same principles that determine the spontaneity of a chemical reaction also apply to the formation of solutions.

3.15 Solvents and Solutions

A solution is a homogeneous mixture of two or more substances in which the one in greater concentration is called the *solvent* and the others are called *solutes*. Solutions play important roles in our lives. Most of the reactions in the world, including those in living organisms, occur in solution. While solutions may exist in any of the three states of matter, our interest is mostly in homogeneous solutions made by dissolving solids or liquids in liquids.

Experience has led chemists to the conclusion that in general "like dissolves like." This means that structurally similar substances are soluble, so they form homogeneous solutions. Dissimilar substances are insoluble, and, therefore, they do not form solutions. Nonpolar compounds, such as hydrocarbons, therefore, are soluble in other nonpolar compounds. Alcohols, like methanol and ethanol, are similar to water in structure because they all have —OH groups that can form hydrogen bonds; consequently, methanol and ethanol form homogeneous solutions with water. However, the structures of hexane and methanol are very different; methanol is polar and hydrogen bonding while hexane is nonpolar and non-hydrogen bonding. As a result, methanol and hexane are insoluble, and mixing them doesn't form a homogeneous solution.

Whether the structures of two compounds are similar or not is an important factor in determining if they will form a homogeneous solution. Consequently we can use the structures of compounds as a rough guide to predict whether or not they will form a solution on mixing.

One broad structural feature used to classify organic solvents, for example, is whether their molecules contain —OH groups. Solvents whose molecules contain —OH groups are classified as **protic solvents.** Water, alcohols, and carboxylic acids are all protic solvents. Compounds that can form hydrogen bonds tend to dissolve in protic solvents. Solvents whose molecules do not contain —OH groups are called **aprotic solvents.** Chloroform, hexane, acetone, acetonitrile, and benzene are aprotic solvents.

Solvents are classified further based on their polarity. The dielectric constant of a solvent is often taken as a measure of solvent polarity. The **dielectric constant** of a liquid is defined as the ratio of work needed to separate two oppositely charged particles a given distance in a vacuum to the work required to separate them the same distance when they are in the liquid. Thus the larger the dielectric constant of a solvent, the more soluble is an ionic substance such as NaCl.

Table 3.4 lists the dielectric constants (symbolized ϵ) for a number of liquids. Their values range from about 2 to greater than 100. Water ($\epsilon = 78$) has one of the larger values, while hexane ($\epsilon = 1.9$) has one of the smaller values. We arbitrarily define **polar solvents** as liquids that have a dielectric constant greater than 15. Thus water, ethanol ($\epsilon = 24.5$), methanol ($\epsilon = 32.7$), and formic acid ($\epsilon = 58.5$) are all polar solvents. They are also protic solvents, so they are often referred to as *polar protic solvents.* Ionic and highly polar substances tend to dissolve in polar protic solvents.

Acetone ($\epsilon = 20.7$) and acetonitrile ($\epsilon = 37.5$) are also polar solvents, but they are aprotic solvents so they are usually referred to as *polar aprotic solvents.* Polar substances also tend to dissolve in polar aprotic solvents.

Liquids that have dielectric constants below 15 and are not hydrogen bond donors are called **nonpolar or apolar solvents.** Examples include chloroform ($\epsilon = 4.8$), hexane ($\epsilon = 1.9$), and benzene ($\epsilon = 2.3$). Only nonpolar substances dissolve to any great extent in nonpolar solvents.

Solvents can be classified further based on their ability to donate a pair of electrons. Solvents whose molecules can donate a pair of electrons are called **donor** or **Lewis base solvents** because they act as Lewis bases. Diethyl ether, ethanol, and tetrahydrofuran are donor solvents. Lewis acids, such as $AlCl_3$ or BF_3, tend to be soluble in electron donor solvents.

These three classification schemes are summarized in Table 3.4 for most common solvents used in organic chemistry. As the data in Table 3.4 show, solvents are not restricted to one class. Acetic acid, for example, is apolar but protic while tetrahydrofuran is an apolar, aprotic, and electron donating solvent.

To understand why a solute dissolves in a solvent, we must examine the free energy change of solvation. *Solvation* is a term that refers to the total of the solute-solvent interactions. The free energy change of solvation is the free energy change that occurs when a solute and a solvent are mixed to form a solution.

Table 3.4   **Classification of Some Common Organic Solvents**

Solvent	Structure	ϵ^*	Protic	Aprotic	Polar	Nonpolar	Electron donor
Formamide	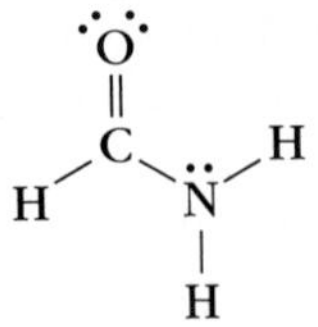	111		√	√		√
Water	H_2O	78	√		√		√
Formic acid	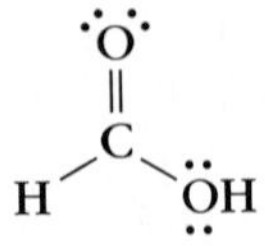	58.5	√		√		√
Dimethyl sulfoxide	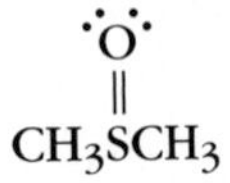 CH_3SCH_3	46.6		√	√		√
Acetonitrile	$CH_3C \equiv N:$	37.5		√	√		√
N, N-dimethylformamide	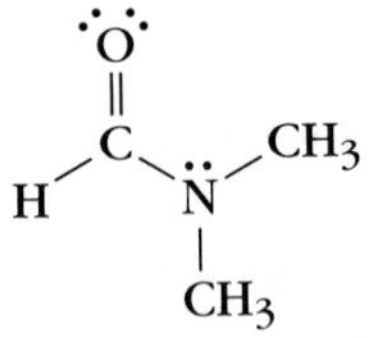	36.7		√	√		√
Methanol	CH_3OH	32.7	√		√		√
Ethanol	CH_3CH_2OH	24.5	√		√		√
Acetone	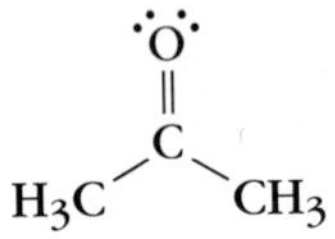	20.7		√	√		√
Methylene chloride	CH_2Cl_2	9.1		√		√	
Tetrahydrofuran	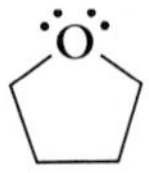	7.6		√		√	√
Acetic acid	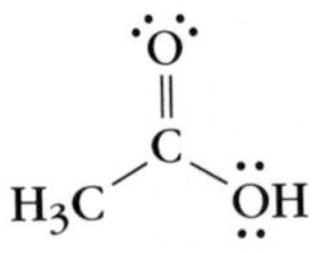	6.2	√			√	√
Chloroform	$CHCl_3$	4.8		√		√	
Diethyl ether	$C_2H_5OC_2H_5$	4.3		√		√	√
Benzene		2.3		√		√	
Carbon tetrachloride	CCl_4	2.2		√		√	
Hexane	$CH_3(CH_2)_4CH_3$	1.9		√		√	

*ϵ = dielectric constant.

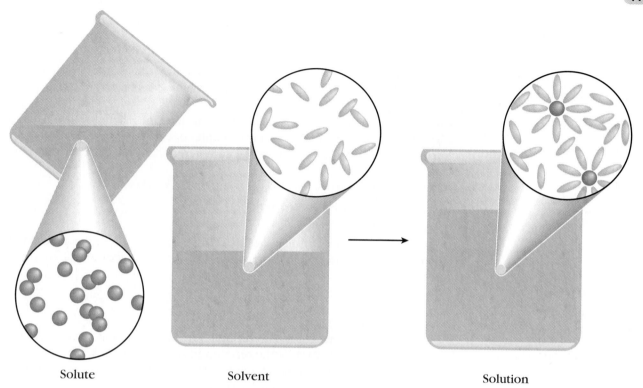

Solute   Solvent   Solution

$$\Delta G_{\text{solvation}} = G_{\text{solution}} - (G_{\text{solute}} + G_{\text{solvent}})$$

A solute dissolves in a solvent to form a solution when $\Delta G_{\text{solvation}} < 0$, whereas the solute does not dissolve when $\Delta G_{\text{solvation}} > 0$. When a solution is formed, $\Delta S_{\text{solvation}} > 0$ because molecular disorder increases, so entropy change always favors solution formation. When $\Delta H_{\text{solvation}} < 0$, solution formation occurs. For $\Delta H_{\text{solvation}} < 0$, the enthalpy of the solution must be smaller than the sum of the enthalpy of the solute and enthalpy of the solvent.

Solvation then depends on the attractive forces between the solute molecules and the solvent molecules. On a molecular level, this means that the stronger these attractive forces are in the solution compared to the intermolecular attractions in the solute and solvent, the greater the tendency to form a solution.

In practical terms this means that the attractive forces between polar solvents and ionic or dipolar solutes result in the formation of solutions. Hydrogen bonding favors solution formation when either solute or solvent molecules contain hydroxy groups (—OH) and the other an unshared pair of electrons, as in groups such as —OH, —OR, —C=O, and —NR$_2$.

■ *The intermolecular attractive forces responsible for formation of solution are the ones that we learned about in Section 1.16, namely ion-dipole, dipole-dipole, hydrogen bonding, and London forces.*

Exercise 3.23 Classify each of the following solvents according to the classification given in Table 3.4.

(a) $CH_3\ddot{O}CH_2CH_2\ddot{O}CH_3$

(b) [benzene ring with CH_3 substituent]

(c) $H_3C-\overset{\displaystyle \overset{..}{\underset{..}{O}}:}{\underset{\displaystyle :\!\underset{..}{O}-CH_2CH_3}{C}}$

(d) CH_3NO_2

Similar interactions are responsible for the effect of solvents on rates of reaction.

3.16   Effect of Solvents on Rates of Reactions

Changing solvents can have a great effect on rates of reactions. The reaction of triethylamine and iodomethane, for example, occurs 390 times faster in acetonitrile ($CH_3C{\equiv}N$) as solvent than when carbon tetrachloride (CCl_4) is the solvent.

$$(CH_3CH_2)_3N \ + \ CH_3I \ \longrightarrow \ (CH_3CH_2)_3\overset{+}{N}CH_3 \ + \ I^-$$

Triethylamine   Iodomethane

Solvents affect rates of reactions by relative stabilization of reactants and transition states. The effect of changing solvents on the hypothetical spontaneous reaction

$$X \ + \ Y \ \longrightarrow \ \text{Products}$$

is illustrated in Figure 3.5.

The free energy–reaction path diagram for the reaction in solvent A is shown in Figure 3.5, *A*, while that for the reaction in solvent B is shown in Figure 3.5, *B*. The rates of the two reactions are determined by their free energies of activation. According to the diagram, the rate of the reaction of X and Y is faster in solvent B than in solvent A because $(\Delta G^{\ddagger})_B < (\Delta G^{\ddagger})_A$. The free energy of activation in solvent B is less because the reactants are less stable in solvent B. That is, the reactants have a higher free energy in solvent B than in solvent A; the reactants are destabilized by solvent B compared to solvent A. The free energy of the transition state is hardly affected by changing from solvent A to solvent B, so the effect of

Figure 3.5
Free energy–reaction path diagrams for the hypothetical reaction of X and Y in solvent A **(A)** and solvent B **(B)**. Reaction occurs faster in solvent B than in solvent A because $(\Delta G^{\ddagger})_B < (\Delta G^{\ddagger})_A$. $(\Delta G^{\ddagger})_B < (\Delta G^{\ddagger})_A$ because solvent B destabilizes the reactants compared to solvent A.

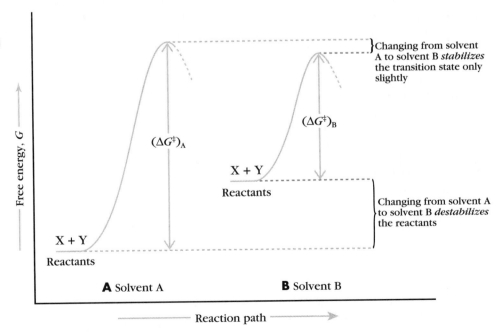

changing solvent is due to changes in relative stabilities of the reactants in the two solvents.

In this example we have chosen the case where changing solvents destabilizes the reactants. Other cases are possible in which a change of solvent causes the transition state to be destabilized. We will encounter both types of solvent effects in future chapters.

Exercise 3.24	Changing the solvent of the hypothetical reaction of X and Y discussed previously from A to C causes the following free energy changes in the reactants and transition state. In solvent C, the free energy of the transition state is lower than in solvent A, while the free energy of the reactants is unaffected by a change from solvent A to C. In which solvent does the reaction occur faster?

While no satisfactory theory has been developed for the effect of solvent on the rate of reactions at the molecular level, we can offer certain general conclusions.

Transition states of reactions between nonpolar molecules are often also nonpolar. We might expect, therefore, that any solvent that stabilizes the transition state would also stabilize the reactants to about the same extent, with the result that there would be little or no rate change with changing solvent.

Rates of reactions between nonpolar molecules that generate a polar transition state are more affected by a change of solvent. In such reactions, the transition state can be stabilized by an increase in the polarity of the solvent, which has the effect of decreasing the free energy of activation and so increasing the rate.

Reactions in which polar reactants form less polar transition states would be expected to show a decrease in rate in changing to a more polar solvent. In these cases, the reactants are stabilized in the more polar solvent but the transition states are not. This results in an increase in the free energy of activation in the more polar solvent with a consequent decrease in rate. In future chapters, we will encounter specific examples of these kinds of solvent effects.

3.17   Catalysts

We learned in Section 3.14 that one way to make a reaction go faster is to raise the temperature of the reactants. This doesn't change the activation energy of the reaction, but it does increase the number of reactant molecules with enough energy to surmount the activation barrier. Another way to increase the rate of a reaction is to change its mechanism in a way that lowers the activation energy. We can do this for some reactions by adding a **catalyst.** A catalyst is a substance that increases the rate of a chemical reaction without being used up in the reaction. Catalysts, therefore, do not appear in the overall balanced chemical equation.

Catalysis can be classified as either homogeneous or heterogeneous. In **homogeneous catalysis** the catalyst is present in the same phase as the reactants. An example of homogeneous catalysis is the effect of a strong acid such as sulfuric acid on the reaction of acetic acid and ethanol.

$$H_3C-\overset{\overset{\displaystyle ..\,\ddot{O}:}{\|}}{C} \quad + \quad CH_3CH_2\ddot{O}H \quad \underset{\longleftarrow}{\overset{H^+}{\longrightarrow}} \quad H_3C-\overset{\overset{\displaystyle ..\,\ddot{O}:}{\|}}{C} \quad + \quad H_2O$$
$$\qquad\quad :\!\underset{..}{O}\!-\!H \qquad\qquad\qquad\qquad\qquad\qquad\qquad :\!\underset{..}{O}\!-\!CH_2CH_3$$

Acetic acid   Ethanol   Ethyl acetate   Water

Without the catalyst, the equilibrium is reached only very slowly. The sulfuric acid is soluble in the reaction mixture, and it causes the equilibrium to be reached rapidly.

Notice that in writing the balanced equation of a catalyzed chemical reaction, the symbol or name of the catalyst is placed above the arrow or arrows between reactants and products. In the reaction of acetic acid (CH_3COOH) and ethanol (CH_3CH_2OH), a proton, (not H_2SO_4) is placed above the two arrows because any strong acid will catalyze the equilibrium reaction of acetic acid and ethanol.

In **heterogeneous catalysis,** the catalyst is present as a distinct phase. Frequently the catalyst is a solid, such as a metal or metal oxide, that is not soluble in solution or in the gas phase. An important example of a heterogeneous catalyst is the metal-catalyzed addition of hydrogen to ethylene to form ethane.

$$CH_2\!=\!CH_2 \quad + \quad H_2 \quad \overset{Pd}{\longrightarrow} \quad CH_3CH_3$$

Ethylene   Ethane

This reaction does not occur in the absence of a catalyst. In the presence of finely divided Ni or Pd, however, the reaction occurs rapidly at room temperature.

A catalyst speeds up a chemical reaction by lowering its energy of activation. It usually does this by altering the path (mechanism) of the reaction. However, a catalyst does not change the energy of the reactants and products, so it does not change the equilibrium constant of the reaction.

Enzyme-catalyzed reactions are a particularly striking example of how catalysts work. **Enzymes** are giant protein molecules that function as catalysts for biological reactions. Only a small specific region of an enzyme, called its **active site,** is responsible for its catalytic activity.

Enzymes catalyze reactions so effectively that rate increases of a factor of a million are common. In fact, most reactions in living organisms do not occur without enzymes. The reaction of water and carbon dioxide to form carbonic acid, for example, occurs ten million times (10,000,000) faster in the presence of the enzyme carbonic anhydrase than in the absence of the enzyme.

$$CO_2 \quad + \quad H_2O \quad \rightleftarrows \quad H_2CO_3$$

One way that enzymes operate as catalysts is illustrated in Figure 3.6. In step 1, the enzyme brings together two molecules at the active site, where it binds them to itself by means of the usual intermolecular attractions discussed in Chapter 1. In step 2, the two molecules are brought sufficiently close together and with the proper orientation for a reaction to occur. After the reaction has occurred, the molecule of product is released from the active site in step 3 and the enzyme is ready to begin again.

In summary, molecules react when the energies of their collisions are greater than the activation energy of the reaction and they collide with the proper orientation. If these two conditions are met, reaction will occur. Catalysts speed up

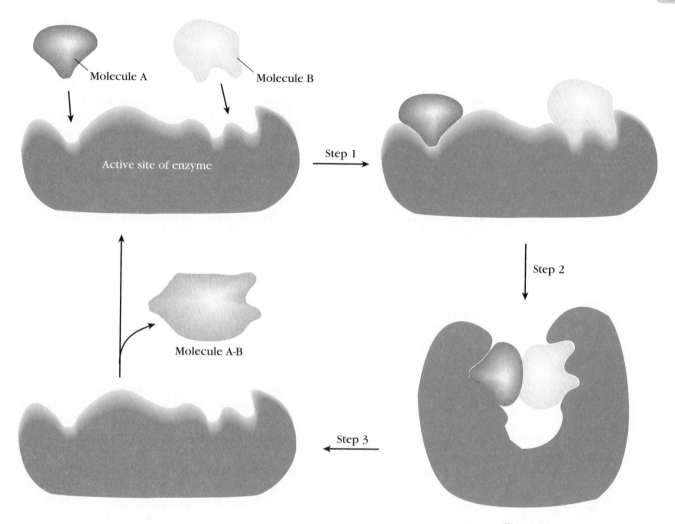

Active site of enzyme

Molecule A

Molecule B

Step 1

Step 2

Molecule A-B

Step 3

Figure 3.6
Pictorial representation of the action of an enzyme catalyst. In step 1, the reactant molecules are held at the active site of the enzyme by attractive forces such as hydrogen bonds or dipole-dipole attractions. In step 2, the reactants are brought close enough together with the proper orientation for reaction to occur. After the reaction occurs, the product leaves the active site in step 3 and the enzyme returns to the form it had before step 1. While enzymes participate in the reaction, they are not permanently altered and can repeat the catalytic cycle again and again.

reactions by providing a path for the reaction that has a lower activation energy than the path available in the absence of a catalyst. Sometimes, as in the case of enzymes, the catalyst positions the reactants in the proper orientation for reaction.

3.18   Summary

Brønsted-Lowry acids are compounds or ions that can donate a proton to another molecule or ion. **Brønsted-Lowry bases** are compounds or ions that can accept a proton. The loss of a proton from an acid forms its **conjugate base.** The addition of a proton to a base forms its **conjugate acid.** An acid-base reaction consists of two conjugate acid-base pairs.

The ability of an acid to transfer a proton to water is a standard measure of acid strength. The quantitative measure of the strength of an acid is its **acid dissociation constant, K_a.** For acids stronger than the hydronium ion or weaker than water, values of K_a (or $pK_a = -\log K_a$) must be obtained by indirect methods. The use of water as the common base allows us to construct a unified scale of strengths of acids and their conjugate bases. We can compare the strengths of acids and bases on the same scale and, consequently, predict if products or reactants are favored at equilibrium. An acid-base equilibrium is established to favor the side with the weaker acid and the weaker base.

Lewis acids are electron acceptors and **Lewis bases** are electron donors. The movement of electron pairs between Lewis acids and bases is represented by a curved arrow. The tail of the arrow is placed next to the electron pair that is going to move and the arrowhead is placed next to the electron deficient atom on the Lewis acid where the electron pair will end up.

In spontaneous reactions, there is a tendency to proceed to a state in which there is greater molecular disorder as well as a tendency to proceed to a state of lower energy. These tendencies are summarized in a function called the **Gibbs free energy.** The change in Gibbs free energy of a reaction is a criterion of its spontaneity. For a spontaneous reaction, $\Delta G < 0$; for a nonspontaneous reaction $\Delta G > 0$; for a reaction at equilibrium, $\Delta G = 0$.

The relationship between the concentration of reactants and the rate of a chemical reaction is expressed as a **rate law.** The **Transition State Theory** proposes that a reaction occurs when molecules collide with the proper orientation and enough energy to overcome an energy barrier called the **free energy of activation.** The top of this energy barrier is the point of highest energy along the path from reactants to products and is called the **transition state.** For a reaction to occur, molecules must collide with enough energy to reach the transition state *and* have the correct orientation.

The tendency toward a state of lower energy and of greater molecular disorder also determines whether two substances will form a solution. The solution forms when the attractive forces between solvent and solute molecules are stronger than those between solute-solute molecules and solvent-solvent molecules.

Solvents are classified according to their structures. Solvents whose molecules contain —OH groups, such as water, methanol, and formic acid, are called **protic solvents. Aprotic solvents** cannot form hydrogen bonds because they lack —OH groups. **Polar solvents** are defined as liquids that have a dielectric constant greater that 15. **Apolar or nonpolar solvents** have dielectric constants less than 15. Polar and apolar solvents can be either protic or aprotic.

Changing solvents affects the rates of reactions by preferential solvation of either the reactants or the transition states. This increases or decreases the free energy of activation that results in an increase or decrease in the rate.

Catalysts are substances that increase the rate of chemical reactions but are not consumed in the reaction. Catalysts can be heterogeneous or homogeneous. A **heterogeneous catalyst** is not soluble in the reaction mixture, while a **homogeneous catalyst** is soluble in the reaction mixture. A catalyst functions by providing an alternate path from the reactants to the products that has a lower free energy of activation than the uncatalyzed reaction.

Additional Exercises

3.25 Define each of the following terms.

(a) Rate law
(b) Brønsted-Lowry base
(c) Activation energy
(d) Conjugate acid-base pair
(e) Lewis acid
(f) K_a

(g) Brønsted-Lowry acid
(h) Endergonic reaction
(i) Catalyst
(j) Aprotic solvent
(k) Lewis base

(l) Spontaneous reaction
(m) Exergonic reaction
(n) pK_a
(o) Transition state
(p) Protic solvent

3.26 In the following equation, identify each species as an acid or a base and indicate each conjugate acid-base pair.

$$CH_3COOH + HCO_3^- \longrightarrow CH_3COO^- + H_2O + CO_2$$

3.27 Write the structure of the conjugate base of each of the following acids.

(a) $(CH_3)_3NH^+$

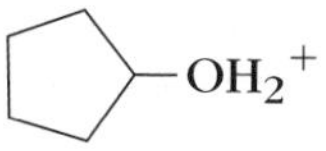

(c)
$$\overset{H}{\underset{+}{\overset{|}{CH_3OCH_3}}}$$

(e) $CH_3CH{=}CHCOOH$

(b) ⬠$-OH_2^+$

(d) CH_4

(f) $(CH_3)_2NH$

3.28 Write the structure of the conjugate acid of each of the following bases.

(a)

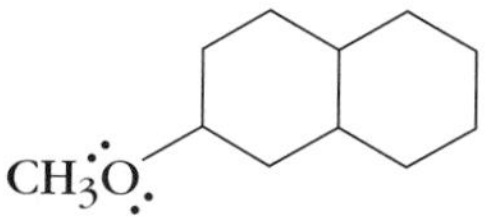

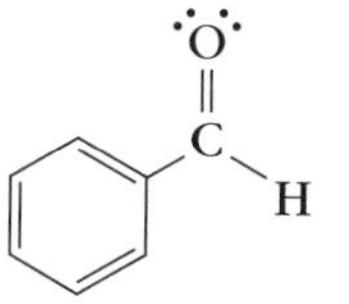

(c)

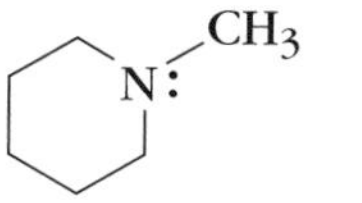

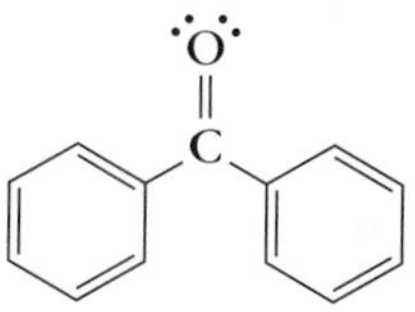

(e)

(b)

(d)

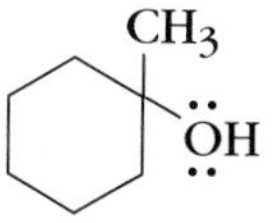

(f)

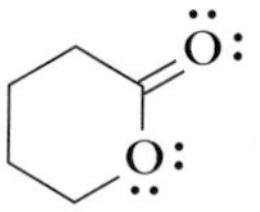

3.29 Arrange each of the following sets of compounds or ions in order of increasing acid strength in water.

(a) CH_3COOH, CH_3CH_2OH, H_2O
(b) Cl^-, CN^-, H_2O
(c) CH_3OH, CH_3COO^-, Br^-

(d) H_3O^+, CH_3COOH, $CH_3NH_3^+$
(e) CH_3NH_2, CH_3COO^-, H_2O

3.30 Using the data in Tables 3.1 and 3.2, predict which of the following acids is strong enough to react with an aqueous solution of trimethylamine $(CH_3)_3N$ to form its conjugate acid.

(a) H_3O^+
(b) HCO_3^-
(c) CH_3COOH

3.31 Using the data in Tables 3.1 and 3.2, estimate the equilibrium constant for each of the following reactions.

(a) $Na^+NH_2^- + CH_3(CH_2)_2CH_3 \rightleftharpoons CH_3(CH_2)_2CH_2^-Na^+ + NH_3$
(b) $NaH + (CH_3)_3COH \rightleftharpoons (CH_3)_3CO^-Na^+ + H_2$
(c) $CH_3(CH_2)_2CH_2^-Li^+ + H_2O \rightleftharpoons LiOH + CH_3(CH_2)_2CH_3$
(d) Aqueous concentrated $HCl + CH_3CH_2OH \rightleftharpoons CH_3CH_2OH_2^+ + Cl^-$

(e) Concentrated H_2SO_4 +

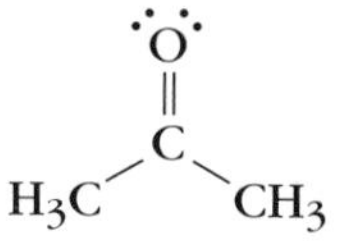

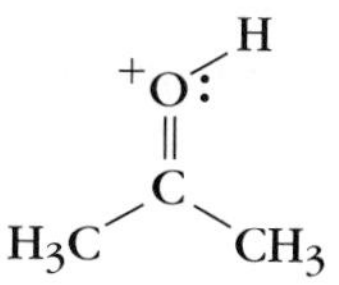

3.32 Calculate an approximate value of $\Delta G°$ for each of the reactions in Exercise 3.31.

3.33 An aqueous solution of HCl is added to an aqueous solution of acetone. **(a)** Will HCl donate a proton to acetone according to the following equation? **(b)** Verify your answer in **(a)** by calculating the value of K_{eq} for the reaction.

$$HCl \ + \ CH_3\overset{\overset{\displaystyle \cdot\cdot\overset{\cdot\cdot}{O}}{\|}}{C}CH_3 \ \rightleftarrows \ CH_3\overset{\overset{\displaystyle \overset{+}{:}OH}{\|}}{C}CH_3 \ + \ Cl^-$$

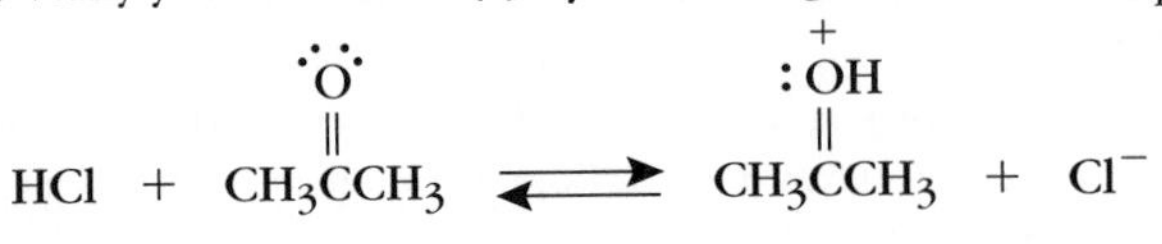

Acetone

3.34 If the reaction in Exercise 3.33 is carried out in hexane as solvent rather than water, will proton transfer from HCl to acetone occur? Again verify your answer by calculating the value of K_{eq} for the reaction between HCl and acetone in hexane.

3.35 The pK_a of acetylene (HC≡CH) is about 25. Will the following reaction occur if acetylene is bubbled into a methanol solution of sodium methoxide (NaOCH$_3$)?

$$HC{\equiv}CH \ + \ CH_3O^- \ \longrightarrow \ HC{\equiv}C^- \ + \ CH_3OH$$

Acetylene   Methoxide ion
(from NaOCH$_3$)

3.36 You want to prepare a solution of sodium methoxide (NaOCH$_3$) by reacting a base with methanol. Which base would form methoxide ion in higher concentration, sodium hydroxide (NaOH) or sodium hydride (NaH)?

3.37 Calculate the equilibrium constant of the acid-base reaction between HCl and ethanol when water is the solvent and when tetrahydrofuran is the solvent. Which solvent would you use to obtain the largest amount of protonated ethanol?

3.38 Identify the Lewis acid and the Lewis base in each of the following reactions.

(a)

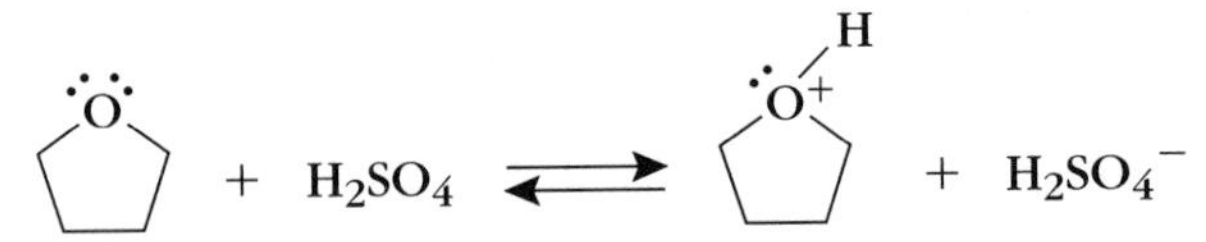

$$\overset{\cdot\cdot}{\underset{\cdot\cdot}{O}} + H_2SO_4 \rightleftarrows \overset{\overset{H}{\diagup}}{\overset{\cdot\cdot}{\underset{}{O^+}}} + H_2SO_4^-$$

(b)

$$CH_3\overset{\overset{\displaystyle \cdot\cdot\overset{\cdot\cdot}{O}}{\|}}{C}\overset{}{\underset{\cdot\overset{\cdot\cdot}{O}\cdot}{}}CH_3 + BF_3 \rightleftarrows CH_3\overset{\overset{\displaystyle \overset{\overset{\textstyle ^-BF_3}{\diagup}}{\cdot\cdot\overset{}{O^+}}}{\|}}{C}\overset{}{\underset{\cdot\overset{\cdot\cdot}{O}\cdot}{}}CH_3$$

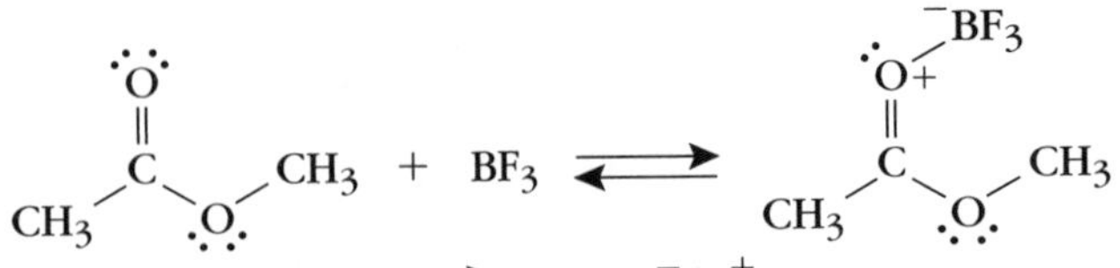

(c) $Br_2 + FeBr_3 \rightleftarrows FeBr_4^- Br^+$

3.39 Use curved arrows to follow electron pair movement in each of the reactions in Exercise 3.38.

3.40 Use curved arrows to follow electron pair movement in each of the following reactions.

(a) $NH_4^+ \ + \ NH_2^- \ \rightleftarrows \ 2\,NH_3$

(b)

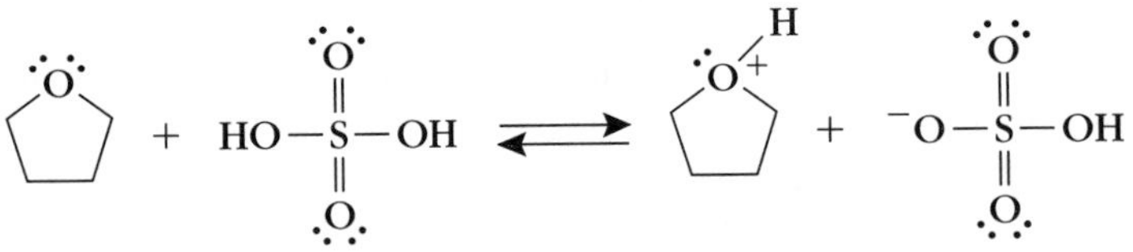

$$\overset{\cdot\cdot}{\underset{\cdot\cdot}{O}} + HO{-}\overset{\overset{\displaystyle \cdot\cdot\overset{\cdot\cdot}{O}}{\|}}{\underset{\underset{\displaystyle \cdot\overset{\cdot\cdot}{O}\cdot}{\|}}{S}}{-}OH \rightleftarrows \overset{\overset{H}{\diagup}}{\overset{\cdot\cdot}{\underset{}{O^+}}} + \ ^-O{-}\overset{\overset{\displaystyle \cdot\cdot\overset{\cdot\cdot}{O}}{\|}}{\underset{\underset{\displaystyle \cdot\overset{\cdot\cdot}{O}\cdot}{\|}}{S}}{-}OH$$

(c)

$$\overset{\overset{H}{\diagup}}{\overset{\cdot\cdot}{\underset{}{O^+}}} + H_2O \rightleftarrows H_3O^+ \ + \ \overset{\cdot\cdot}{\underset{\cdot\cdot}{O}}$$

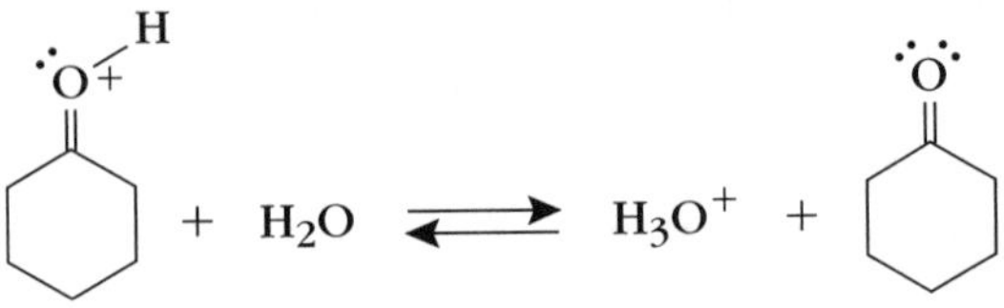

(d) $NH_2^- \ + \ HC{\equiv}CH \ \rightleftarrows \ NH_3 \ + \ ^-C{\equiv}CH$

3.41 An organic cation such as the *tert*-butyl cation can react as both a Lewis acid and a Brønsted-Lowry acid.

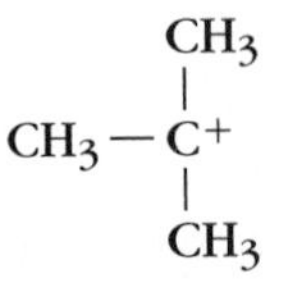

$$CH_3{-}\overset{\overset{\displaystyle CH_3}{|}}{\underset{\underset{\displaystyle CH_3}{|}}{C^+}}$$

tert-Butyl cation

Write the equations for the reaction of a *tert*-butyl cation with water in which the cation acts as **(a)** a Lewis acid and **(b)** a Brønsted-Lowry acid.

3.42   On the diagram, label the location of each of the following:

(a)   The reactants for the forward reaction

(b)   The transition state

(c)   The enthalpy of activation, $\Delta H^{\ddagger}$, for the forward reaction

(d)   The enthalpy of activation, $\Delta H^{\ddagger}$, for the reverse reaction

(e)   The enthalpy change for the reaction, ΔH

(f)   The products of the forward reaction

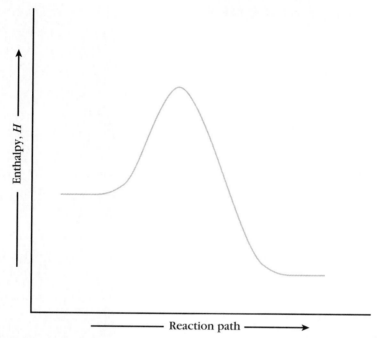

3.43   Dissolving 1-isopropylcyclohexene in aqueous strong acid at 25 °C causes it to isomerize to isopropylidenecyclohexane according to the following equation:

H
|
$H_3C-C-CH_3$ $\xrightleftharpoons[K_{eq}]{H^+}$ H_3C CH_3

1-Isopropylcyclohexene   Isopropylidenecyclohexane

At equilibrium, the product mixture contains 70% 1-isopropylcyclohexene and 30% isopropylidenecyclohexane. Calculate:

(a) the equilibrium constant, K_{eq}, for this isomerization reaction

(b) the standard free energy change, $\Delta G°$, in this isomerization reaction

C H A P T E R

CONFORMATIONS of ALKANES and CYCLOALKANES

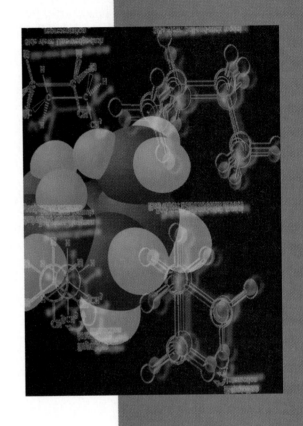

4.1   Introduction

We learned in Section 1.5 that Lewis structures are a convenient way to specify which atoms are joined together in a molecule. Based on the Lewis structure of a molecule, we might believe that its atoms occupy fixed positions. On the contrary, atoms in molecules are constantly in motion relative to each other. Consider, for example, the carbon-hydrogen bonds in a typical alkane. These bonds undergo various stretching and bending motions in a manner similar to two balls attached by a spring.

Rotation about carbon-carbon single bonds is also common in organic molecules. In this chapter, we develop the fundamental principles of **conformational analysis,** which is the energy changes that a molecule undergoes as groups rotate about a single bond. The principles are introduced with acyclic compounds, then they are applied to cycloalkanes. While rotation is restricted in small cycloalkanes, it still plays a significant role in the structure of cyclohexane and its derivatives.

Before we can describe rotation about carbon-carbon single bonds, we must learn how to represent the three-dimensional structure of a molecule on a page.

4.2   Writing Three-Dimensional Structures of Molecules

Representing a three-dimensional object on a two-dimensional surface (like a page) is a problem confronted by artists and graphic designers all the time. They have developed techniques such as shading and sizing that allow the viewer to recognize the three dimensions of an object. We have used these techniques in ball-and-stick representations of molecules in previous chapters. The following representations of methane and ethane, for example, have been used previously.

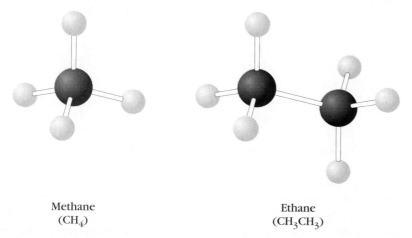

Methane
(CH_4)

Ethane
(CH_3CH_3)

These kinds of representations of molecules in three dimensions require a certain artistic ability as well as time to draw them, so they are rarely used by chemists. Instead, a number of simpler ways to represent three-dimensional structures of molecules have been developed over the years.

One method is to use symbols for bonds that indicate to the viewer where the atoms are located relative to the plane of the page. We do this by using one of three symbols, a line ——— , a dashed wedge .... , or a solid wedge ———◀ . The meanings of these symbols are as follows:

1. **X——Y:** Both atoms X and Y and the bond between them are in the plane of the page.
2. **X⋯⋯Z:** Atom X is in the plane of the page, but atom Z is below the plane; the bond extends below the plane.
3. **X——◣V:** Atom X is in the plane of the page, but atom V is above the plane; the bond extends above the plane.

We can use these line-dash-wedge structures to represent the three-dimensional orientation of atoms in methane and ethane as follows:

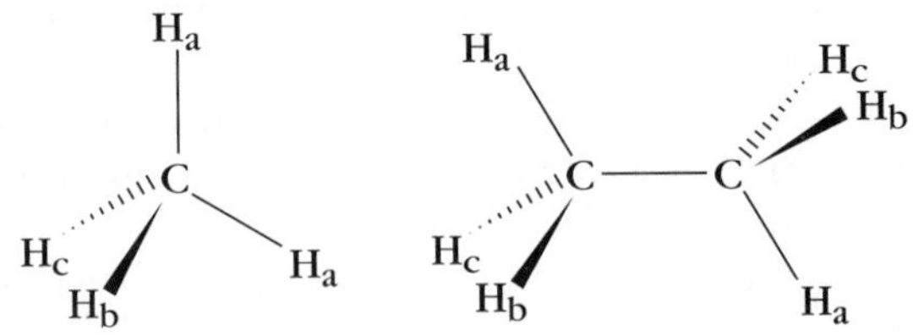

All carbon atoms, hydrogen atoms labelled H_a, and the bonds joining them are in the plane of the page. Bonds joining the carbon atoms to H_b project toward the viewer, while the bonds from carbon atoms to H_c project away from the viewer. Thus hydrogen atoms H_b all lie above the plane and hydrogen atoms H_c all lie below the plane of the page.

Another way to visualize the three-dimensional relationships of atoms bonded to the carbon atoms of a carbon-carbon single bond is by a sawhorse structure. In a sawhorse structure, the two carbon atoms and the single bond joining them are viewed from an oblique angle. Ethane, for example, is represented by the following sawhorse structure.

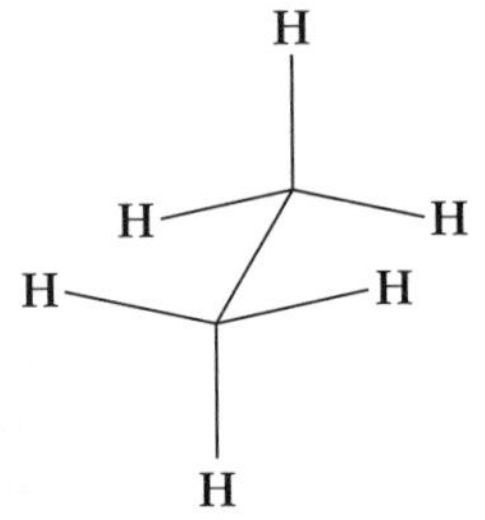

Sawhorse structure of ethane

In order to use sawhorse structures properly, it must be understood that a carbon atom is located at each intersection of four lines.

Rather than attempt to show a structure in three dimensions, organic chemists sometimes use a projection of the structure onto a page. One such projection is called a **Newman projection formula.** A Newman formula is a projection onto the page of a view along the single bond connecting two carbon atoms. The front carbon atom is represented by a dot and the rear carbon atom by a circle. The three remaining bonds are drawn symmetrically around each carbon atom, as shown in Figure 4.1.

The advantage of Newman projection formulas is that the spatial relationship between atoms on adjacent carbon atoms can be easily visualized. We can specify this relationship by means of a **dihedral** or **torsional angle,** (θ), which is defined as the angle between C—H bonds on adjacent carbon atoms.

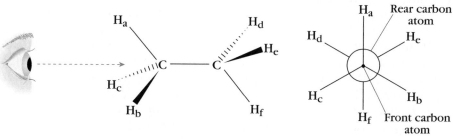

θ = Dihedral or torsional angle

Notice that the H—C—H bond angles of ethane are no longer 109.5° in its Newman projection. The reason for this is that Newman projection formulas are a projection of the molecule onto the page and not a three-dimensional representation. Thus the 109.5° H—C—H bond angles of ethane appear to be 120° when the structure of ethane is flattened onto the page! The reader is urged to make a molecular model to verify this point.

In the next section, we use these methods to visualize the rotation about the carbon-carbon bond of ethane.

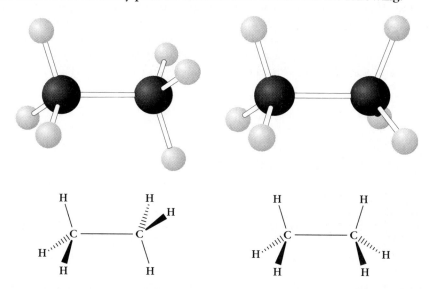

Line-dash-wedge representation   Newman projection formula

Figure 4.1
Projection of the view along the carbon-carbon bond of ethane onto the page gives its Newman projection formula.

4.3 Conformations of Ethane

When we start to write the structure of ethane in three dimensions, we find that we can write many different structures, all of which have the required H—C—H bond angles of 109.5°. For example, the ball-and-stick and line-dash-wedge representations of two of many possible structures of ethane are the following:

Either one of these structures can be converted to the other by simply rotating one methyl group about the carbon-carbon single bond relative to the other methyl group. Different arrangements of atoms in a molecule that are converted into another by rotation about a single bond are called **conformations** or **conformers.**

Exercise 4.1	Make a molecular model of ethane and arrange its atoms in the same way as the line-dash-wedge structure shown at the bottom of p. 131, on the left. Draw its Newman projection formula. Now arrange the atoms in the same way as the line-dash-wedge structure shown at the bottom of p. 131, on the right. Draw its Newman projection formula.

We can use the dihedral angle to describe two limiting conformations of ethane. In one, called the **eclipsed conformation,** the dihedral angle is 0°. In the other, called the **staggered conformation,** the dihedral angle is 60°. The Newman projections of these conformations of ethane are shown in Figure 4.2.

There are an infinite number of other conformations of ethane, each differing by only tiny increments in their dihedral angles. Because there are many different conformations of ethane, their potential energies are not all the same. The graph in Figure 4.3 shows how the potential energy of ethane changes as the dihedral angle goes from 0° to 360°. In this figure, a dihedral angle of 0° is defined by an eclipsed conformation. Also two hydrogen atoms, one on each carbon atom, are highlighted so that the rotation can be followed more easily.

Let's follow the rotation starting from an eclipsed conformation ($\theta = 0°$). As we hold one carbon atom fixed and rotate the other about the C—C bond, the dihedral angle increases and we pass through a staggered conformation ($\theta = 60°$) and then another eclipsed conformation ($\theta = 120°$). We alternate through a staggered conformation (at $\theta = 180°$), an eclipsed conformation (at $\theta = 240°$), and another staggered conformation (at $\theta = 300°$) before rotation brings the molecule back to its original eclipsed conformation ($\theta = 360° = 0°$). You'll find that the use of a molecular model of ethane will help you to visualize this rotation.

The graph in Figure 4.3 shows that all the staggered conformations of ethane are 3.0 kcal/mol (12 kJ/mol) lower in energy than all of the eclipsed conformations. This makes the staggered the most stable conformation of ethane. Thus a staggered conformation must acquire 3.0 kcal/mol (12 kJ/mol) of energy in order to rotate to an eclipsed conformation. The energy needed to overcome this barrier to rotation is called **torsional energy.**

Figure 4.2
Newman projection formulas of the eclipsed and staggered conformations of ethane. Notice that the Newman projection of the eclipsed conformation is not accurately drawn. The dihedral angle should be zero, but to clearly show the eclipsed hydrogens they are drawn with a small but not zero dihedral angle.

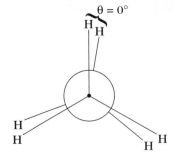

Eclipsed conformation of ethane

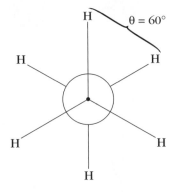

Staggered conformation of ethane

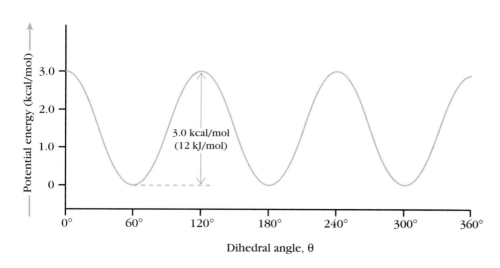

Figure 4.3
Change in potential energy with rotation about the carbon-carbon bond in ethane. The potential energy is highest in the eclipsed conformation and lowest in the staggered conformation.

Why is the staggered conformation of ethane more stable than the eclipsed conformation? The most important factor seems to be bonding electron pair repulsion. Thus electron pair repulsion between electrons in the C—H bonds on neighboring carbon atoms is the greatest when the two bonds are aligned (in the eclipsed conformation) and at a minimum when they are farthest apart (in the staggered conformation). This destabilizes the eclipsed conformation relative to the staggered conformation. Destabilization of a molecule caused by the eclipsing of bonds on adjacent atoms is called **torsional strain.**

What does all this have to do with the structure of ethane? This analysis tells us that gaseous ethane molecules are not static. Instead of being frozen into a particular conformation, they easily undergo internal rotations because molecules of ethane in staggered conformations need to acquire only 3.0 kcal/mol (12 kJ/mol) of energy to overcome the barrier to rotation to another staggered conformation. Unless the temperature is very low (−200 °C), most ethane molecules acquire this energy from normal thermal motion. As a result, most ethane molecules spend the majority of their time in a staggered conformation, passing only fleet-

■ *We used electron pair repulsion in Section 1.7 to successfully predict molecular geometry by the VSEPR model.*

ingly through an eclipsed conformation to another staggered conformation. Thus the internal rotation in an ethane molecule can be pictured as a succession of abrupt rotations from one staggered conformation to another.

We can extend the principles to propane, which is a slightly more complicated system.

4.4   Conformations of Propane

The next higher homolog of the alkane family is propane. Its three-dimensional structure and Newman projection along one single bond are shown in Figure 4.4.

Figure 4.4
Three-dimensional representation along one carbon-carbon bond of a staggered conformation of propane and its Newman projection formula.

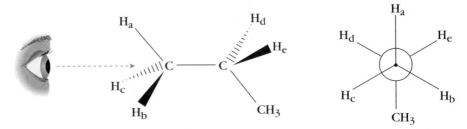

The graph in Figure 4.5 shows how the potential energy of propane changes with rotation about one carbon-carbon bond. The torsional energy is slightly higher in propane, 3.4 kcal/mol (14 kJ/mol), than in ethane, 3.0 kcal/mol (12 kJ/mol). This slight increase arises because one eclipsed repulsion between two C—H bonds in ethane is replaced by an eclipsed repulsion between one C—H bond and one C—C bond in propane.

Figure 4.5
Change in potential energy with rotation about one of the single carbon-carbon bonds in propane. The eclipsed conformation is 3.4 kcal/mol (14 kJ/mol) higher in energy than the staggered conformation.

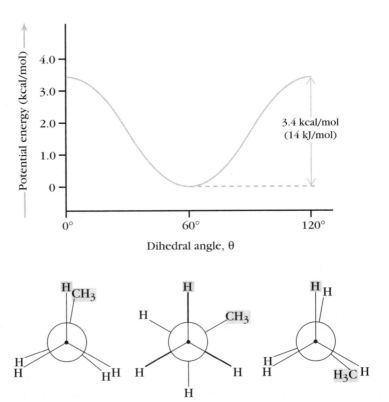

From the differences in torsional energy between ethane and propane, we can assign a value to the repulsion between two eclipsed C—H bonds and the repulsion between eclipsed C—H and C—C bonds. Because the torsional energy of ethane is composed of three equal C—H bond eclipsing interactions, we can assign a value of 1 kcal/mol (4 kJ/mol) to each C—H interaction. If we assume that this value is the same in propane, we can say that each eclipsed interaction between two C—H bonds contributes 1 kcal/mol (4 kJ/mol) to its torsional energy. There are two of these interactions in propane for a total of 2 kcal/mol (8 kJ/mol). The remainder of the torsional energy, 1.4 kcal/mol (6 kJ/mol), must be due to repulsion between eclipsed C—H and C—C bonds.

Exercise 4.2 The plot of potential energy versus dihedral angle in Figure 4.5 shows rotations only from 0° to 120°. Complete the diagram by plotting the potential energy for rotations from 120° to 360°. Write the structure of each staggered and eclipsed conformation.

4.5 Conformations of Butane and Higher Alkanes

The conformations formed by rotation about the bond between carbon atoms 2 and 3 of butane, and their potential energies, are shown in Figure 4.6.

Exercise 4.3 Write the sawhorse and the line-dash-wedge representations of each of the Newman projections shown in Figure 4.6.

The graph in Figure 4.6 shows that there are two staggered conformations of butane that differ in energy. The one of lower energy is called the **anti conformation** and the one of higher energy is called the **gauche conformation.** The anti conformation is 0.77 kcal/mol (3.2 kJ/mol) lower in energy because its dihedral angle of 180° places the two methyl groups as far away from each other as possible. In the gauche conformation, the two methyl groups are separated by an angle of only 60°.

In the anti conformation, therefore, the two methyl groups are so far apart that they do not interact. In the gauche conformation, on the other hand, the two methyl groups are close enough to intrude on each other's space. A repulsive interaction that occurs when two groups are forced to be closer to each other than their atomic radii permit is called **steric hindrance, steric strain,** or **van der Waals strain.** All three of these names refer to the same repulsive interaction, which is caused by forcing two atoms or groups so close together that they begin to occupy the same space. Steric strain, which causes the gauche conformation of butane to be less stable than the anti conformation, is shown in Figure 4.7 (p. 137).

The graph in Figure 4.6 also shows that there are two eclipsed conformations that differ in energy. The one whose dihedral angle is 120° is 3.6 kcal/mol (15 kJ/mol) higher in energy than the anti conformation. This increase can be accounted for by its torsional energy. In this eclipsed conformation there are repulsions between two eclipsed C—H and C—C bonds, each of which con-

Figure 4.6
Change in potential energy with rotation about the bond between carbon atoms 2 and 3 of butane. The dihedral angle is here defined as the angle between the two methyl groups.

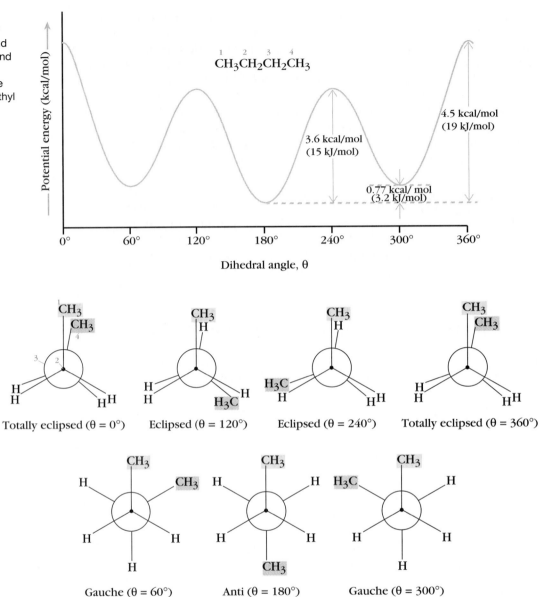

tributes 1.4 kcal/mol (6 kJ/mol), and one between the eclipsed C—H bonds, which contributes 1.0 kcal/mol (4 kJ/mol), for a total of 3.8 kcal/mol (16 kJ/mol) of torsional energy. This is very close to the observed value.

The conformation of highest energy is the eclipsed conformation in which the dihedral angle is 0°. In this conformation the two methyl groups are as close together as they can get. The energy of this conformation is about 4.5 kcal/mol (19 kJ/mol) higher than the anti conformation of butane. The potential energy of this eclipsed conformation is so high because of both torsional and steric strain since the methyl groups in this conformation are forced even closer together than in the gauche conformation.

Using the same method as before, we can estimate the repulsion between eclipsing C—C bonds. Assuming that the repulsive energy between eclipsing

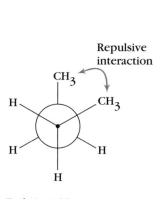

Repulsive
interaction

End view: Newman
projection formula

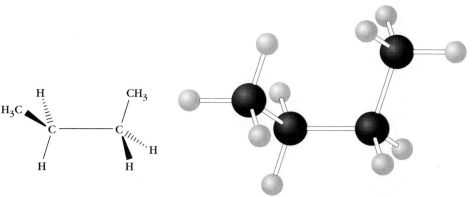

End view: Ball-and-stick model

Repulsive
interaction

Figure 4.7
Steric strain in the gauche conformation of butane is caused by repulsion between the two hydrogen
atoms indicated.

Side view: Line-dash-wedge
representation

Side view: Ball-and-stick model

C—H bonds is still about 1.0 kcal/mol (4 kJ/mol), we calculate a value of 2.5
kcal/mol (11 kJ/mol) for the repulsion energy between eclipsing C—C bonds.
The energies of these various interactions are summarized in Table 4.1.

The same principles that we have developed for butane apply equally well to
pentane, hexane, and the other straight chain higher alkanes. As a result, these
alkanes are most stable when all their carbon-carbon bonds are in anti conformations, as shown by the ball-and-stick model of octane in Figure 4.8.

Table 4.1	Contributions of Various Interactions to the Energy of Alkane Conformations	
Interaction	Energy	
	(kcal/mol)	(kJ/mol)
Two eclipsed C—H bonds	1.0	4
Eclipsed C—H and C—C bonds	1.4	6
Two eclipsed C—C bonds	2.5	11
Gauche butane	0.77	3.2

Figure 4.8
Ball-and-stick model of the most stable conformation of octane. All of the carbon-carbon bonds are arranged in the anti conformation.

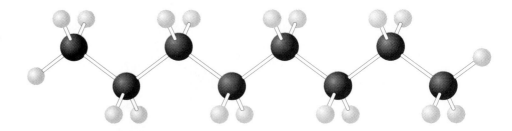

Exercise 4.4   The following is a Newman projection formula of one of the staggered conformations of methylpropane (isobutane) obtained by looking along the single bond between carbon atoms 1 and 2.

$$
\begin{array}{c}
\text{H} \\
\text{H}_3\text{C} \quad\quad\quad \text{CH}_3 \\
\text{H} \quad\quad\quad \text{H} \\
\text{H}
\end{array}
$$

(a) By rotating the front carbon atom relative to the rear one, write the Newman projection formulas of all its staggered and eclipsed conformations.
(b) How many equivalent staggered conformations are there?
(c) How many equivalent eclipsed conformations are there?
(d) If the repulsive energy between eclipsing C—H and C—C bonds is 1.4 kcal/mol (6 kJ/mol), calculate how much more stable the staggered conformation is than the eclipsed conformation.
(e) Construct a potential energy versus dihedral angle diagram for methylpropane. Start with a dihedral angle of 0° (defined as one of the eclipsed conformations) and rotate the single bond by 360°.

Exercise 4.5   Repeat Exercise 4.4 for dimethylpropane.

The same interactions between atoms and groups that are present in straight chain alkanes also contribute to the potential energy of cycloalkanes. In certain cycloalkanes, however, there is another factor that becomes apparent as we examine the relative stabilities of cycloalkanes.

4.6   Relative Stabilities of Cycloalkanes

We learned in Chapter 2 that alkanes are generally unreactive compounds. Therefore it is surprising to find that cyclopropane and to a lesser degree cyclobutane react with a variety of reagents to form ring-opened products. For example, both cyclopropane and cyclobutane react with hydrogen gas in the presence of a nickel catalyst to form propane and butane:

$$\triangle \quad + \quad H_2 \quad \xrightarrow[120\ °C]{\text{Ni catalyst}} \quad CH_3CH_2CH_3$$

Cyclopropane   Propane

$$\square \quad + \quad H_2 \quad \xrightarrow[200\ °C]{\text{Ni catalyst}} \quad CH_3CH_2CH_2CH_3$$

Cyclobutane   Butane

Alkanes and cycloalkanes, such as cyclopentane and cyclohexane, on the other hand, do *not* react with hydrogen under these conditions.

$$\pentagon \quad + \quad H_2 \quad \xrightarrow[200\ °C]{\text{Ni catalyst}} \quad \text{No reaction}$$

Cyclopentane

$$\hexagon \quad + \quad H_2 \quad \xrightarrow[200\ °C]{\text{Ni catalyst}} \quad \text{No reaction}$$

Cyclohexane

Why do cyclopropane and cyclobutane react differently than alkanes and other cycloalkanes? When chemists find that similar kinds of compounds react differently, one of the first things they do is examine the relative energies of the compounds. One of the ways to do this is to compare their heats of combustion.

When the energy from the complete combustion of organic compounds is not used to do work, it is released as heat. The amount of heat given off when an organic compound is burned with an excess of oxygen can be measured very accurately by means of a calorimeter. When combustion is carried out under standard conditions of temperature and pressure, the heat released is called the **standard heat of combustion** or the **standard enthalpy of combustion ($\Delta H°$).**

We can illustrate the use of standard enthalpies of combustion by determining the stabilities of two constitutional isomers, butane and methylpropane. The two combustion reactions and their standard enthalpies of combustion are as follows:

$$CH_3CH_2CH_2CH_3 \quad + \quad 13/2\ O_2 \quad \longrightarrow \quad 4CO_2 \quad + \quad 5H_2O \quad \Delta H° = -686.8 \text{ kcal/mol } (-2873.6 \text{ kJ/mol})$$

   Butane

$$\underset{\text{Methylpropane}}{CH_3\overset{\displaystyle \overset{CH_3}{|}}{C}HCH_3} \quad + \quad 13/2\ O_2 \quad \longrightarrow \quad 4CO_2 \quad + \quad 5H_2O \quad \Delta H° = -684.8 \text{ kcal/mol } (-2865.2 \text{ kJ/mol})$$

In order to compare the enthalpies of these two isomeric C_4H_{10} alkanes, we need some standard of reference. The enthalpy content of CO_2 and water is the standard because one mole of either one of the isomeric alkanes forms the *same* number of moles of CO_2 and water. Thus the difference in their heats of combustion is a measure of their relative stability, as shown in Figure 4.9.

Figure 4.9
One mole of methylpropane and one mole of butane burn in excess oxygen to give the same number of moles of CO_2 and H_2O. The enthalpy of methylpropane, however, is 2.0 kcal/mol (8.4 kJ/mol) less than the enthalpy of combustion of butane. Thus methylpropane is 2.0 kcal/mol (8.4 kJ/mol) more stable than butane.

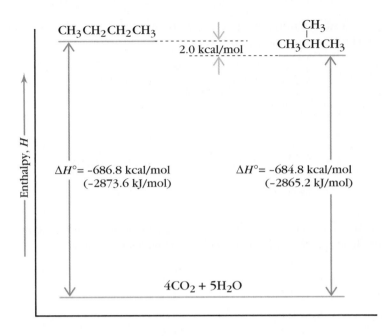

The combustion of butane releases more heat than methylpropane, yet both form the same amounts of the same products. Therefore butane contains more energy and methylpropane is the more stable isomer.

The situation is slightly different for the cycloalkanes because combustion of each cycloalkane forms different numbers of moles of CO_2 and water. One mole of cyclohexane forms six moles each of CO_2 and water, for instance, while one mole of cyclopropane forms only three moles of each. To overcome this problem, we compare the heats of combustion per CH_2 unit. Thus the total heat of combustion of a cycloalkane is divided by the number of CH_2 units in the ring. These values are listed in Table 4.2.

From the data in Table 4.2, we see that cyclohexane has the lowest heat of combustion per CH_2 group (-157.4 kcal/mol or -658.6 kJ/mol) of all of the cycloalkanes. This value is the same as that of unbranched alkanes. As a result, the other cycloalkanes are higher in energy than cyclohexane and unbranched alkanes. This increased energy relative to cyclohexane or unbranched alkanes is called the **ring strain energy** of a cycloalkane. Values of strain energy per CH_2 group and the total ring strain energy for cycloalkanes containing up to 10 carbon atoms and the method of calculating these values are given in Table 4.2.

From these data, we conclude that cyclopropane has the largest total ring strain energy, followed by cyclobutane. The other cycloalkanes all have a certain amount of ring strain energy, but none has as much as cyclopropane or cyclobutane.

The increased ring strain of cyclopropane and cyclobutane explains why they react with H_2 to form propane and butane, respectively. Highly strained cycloalkanes tend to undergo reactions that open their rings to form less strained acyclic compounds. This tendency is greater for cyclopropane than for cyclobutane because cyclopropane has the higher ring strain energy. Cyclohexane, on the other hand, has little tendency to undergo ring opening reactions because it has no ring strain.

Why is the cyclohexane ring strain-free, while both cyclopropane and cyclobutane are highly strained?

Table 4.2				Heats of Combustion and Ring Strain of Cycloalkanes							
Name	Number of CH$_2$ Units		$\Delta H°$		$\Delta H°$ per CH$_2$ unit*		Strain Energy per CH$_2$ Unit†		Total Ring Strain Energy‡		
		kcal/mol	(kJ/mol)	kcal/mol	(kJ/mol)	kcal/mol	(kJ/mol)	kcal/mol	(kJ/mol)		
Cyclopropane	3	– 499.8	(–2091)	–166.6	(–697.0)	9.2	(38.4)	27.6	(115)		
Cyclobutane	4	– 655.8	(–2744)	–164.0	(–686.0)	6.6	(27.4)	26.4	(110)		
Cyclopentane	5	– 793.5	(–3320)	–158.7	(–664.0)	1.3	(5.4)	6.5	(27)		
Cyclohexane	6	– 944.5	(–3952)	–157.4	(–658.6)	0	(0)	0	(0)		
Cycloheptane	7	–1108	(–4637)	–158.3	(–662.4)	0.9	(3.8)	6.3	(27)		
Cyclooctane	8	–1269	(–5309)	–158.6	(–663.6)	1.2	(5.0)	9.6	(40)		
Cyclononane	9	–1429	(–5981)	–158.8	(–664.6)	1.4	(6.0)	12.6	(54)		
Cyclodecane	10	–1586	(–6636)	–158.6	(–663.6)	1.2	(5.0)	12.0	(50)		
Unbranched alkanes				–157.4	(–658.6)	0	(0)	0	(0)		

*$\Delta H°$ per mol divided by number of CH$_2$ groups.
†Difference between the value of $\Delta H°$ per CH$_2$ group for that cycloalkane and $\Delta H°$ per CH$_2$ for unbranched alkanes.
‡Strain energy per CH$_2$ group multiplied by the number of CH$_2$ groups.

4.7 Origins of Strain in Cycloalkanes

There are three factors that contribute to ring strain. The first is **angle strain,** which is the energy needed to distort normal carbon tetrahedral bond angles from 109.5° to the angles in a particular cycloalkane. The second is **torsional strain,** which was introduced in Section 4.3 and defined as the destabilization of a molecule caused by the eclipsing of bonds on adjacent atoms. Finally there is **steric strain,** which was introduced in Section 4.5 and defined as the repulsive interaction between two groups that are forced to be closer to each other than their atomic radii permit. This factor is not important in small-ring cycloalkanes (C$_3$ to C$_6$) but becomes more important in medium-ring cycloalkanes (C$_7$ to C$_{12}$).

We begin our study of the ring strain in cycloalkanes by looking at cyclopropane.

A   Cyclopropane

The three carbon atoms of cyclopropane define the plane of the molecule as well as describe an equilateral triangle with C—C—C bond angles of 60°. As a result, the angle strain in cyclopropane is particularly severe because the normal tetrahedral carbon bond angles of 109.5° must be compressed by 49.5° to form the C—C—C bond angles in a cyclopropane ring.

The carbon-carbon bonds in cyclopropane are weaker than those in acyclic alkanes because the small C—C—C bond angles in cyclopropane prevent effective orbital overlap. A sigma bond is strongest when the overlap of atomic orbitals occurs along the internuclear bond axis. Bond strength is reduced if the overlap of atomic orbitals is not along the bond axis.

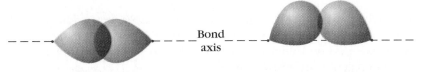

Good overlap = Strong bond   Poor overlap = Weaker bond

The atomic orbitals on adjacent carbon atoms of cyclopropane cannot point directly at each other because of the small C—C—C bond angles. Instead, they overlap at an angle that results in the formation of weaker bent bonds.

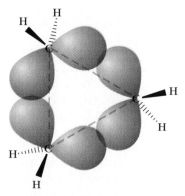

Bent C—C bonds in cyclopropane

Torsional strain also contributes to the total ring strain of cyclopropane because all the bonds are eclipsed. We can see this with a Newman projection along one of the carbon-carbon bonds of cyclopropane, as shown in Figure 4.10. The conformations of the cyclopropane bonds resemble those in the totally eclipsed conformation of butane (Figure 4.6, p. 136).

Figure 4.10
Newman projection along one carbon-carbon bond of cyclopropane. This projection shows how adjacent carbon-hydrogen bonds are totally eclipsed.

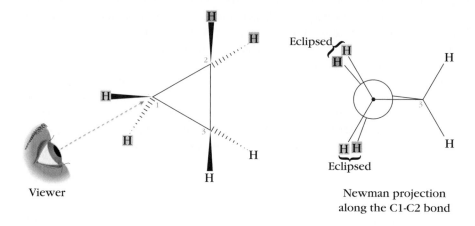

In Section 4.4, we learned that two eclipsed C—H bonds in ethane raise its energy by 1.0 kcal/mol (4 kJ/mol). In cyclopropane, there are six pairs of eclipsed C—H bonds that should contribute about 6.0 kcal/mol (24 kJ/mol) to the energy of the molecule. This is a relatively small portion of the 27.5 kcal/mol (115 kJ/mol) total strain energy of cyclopropane. So most of the ring strain in cyclopropane must be due to angle strain, not to torsional strain.

B   Cyclobutane

The four-carbon cyclobutane ring is not planar. It assumes a folded or slightly bent conformation instead, as shown in Figure 4.11. If cyclobutane were planar, all four C—C—C bond angles would be 90°. Because the ring is bent, the C—C—C bond angles are 88° instead.

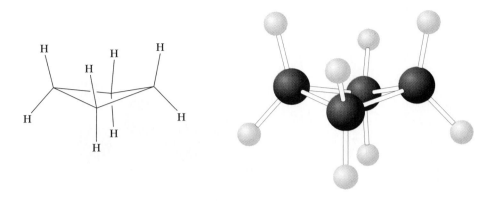

Figure 4.11
The conformation of cyclobutane. The four-carbon cyclobutane ring is slightly folded or bent so that all four C—C—C bond angles are 88° not 90°.

If cyclobutane were planar, all eight C—H bonds would be eclipsed (analogous to cyclopropane). This conformation would contribute about 8.0 kcal/mol (32 kJ/mol) of torsional strain to the molecule, so cyclobutane folds slightly to overcome it. While its angle strain is increased slightly by decreasing the bond angle from the 90° of a planar structure to 88°, this increased strain is compensated for by a large decrease in torsional strain.

| **Exercise 4.6** | Use molecular models to make a model of the bent conformation of cyclobutane, then draw the Newman projection along one C—C bond. If your models permit, make a planar cyclobutane and draw the Newman projection along one of its C—C bonds. |

C   Cyclopentane

If cyclopentane were planar, it would have C—C—C bond angles of 108°. These angles would be very close to the 109° of normal tetrahedral bond angles. As a result, a planar cyclopentane would have very little angle strain. However, all the C—H bonds in a planar cyclopentane would be eclipsed as in cyclopropane and planar cyclobutane. In order to overcome some of this torsional strain, cyclopentane assumes the slightly bent or envelope conformation shown in Figure 4.12.

The five-membered ring of cyclopentane is sufficiently flexible that it is not confined to the one bent conformation shown in Figure 4.12. Thermal motion causes slight twisting of its carbon-carbon bonds. As a result, each of the five methylene groups in turn moves in and out of the plane of the other four atoms. For example, a slight twist causes carbon atom 2 in Figure 4.12 to be bent upwards out of the plane while carbon atom 1 returns to the plane. This twisting makes it seem as if the atom bent out of the plane moves around the ring as the molecule undulates.

| **Exercise 4.7** | Assume that a planar cyclopentane has no angle strain or steric strain. What would be the ring strain in planar cyclopentane due to torsional strain if each eclipsed pair of C—H bonds contributes 1.0 kcal/mol (4 kJ/mol)? Is this greater than or less than the value given in Table 4.2? How much ring strain is relieved by adopting a bent conformation? |

Figure 4.12
The bent or envelope conformation of cyclopentane. In this representation, carbon atoms 2, 3, 4, and 5 all lie in a plane and carbon atom 1 is bent upwards. As a result, most of the carbon-hydrogen bonds are nearly staggered with respect to neighboring carbon-hydrogen bonds.

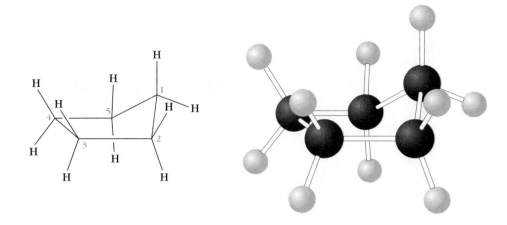

Cyclopropane, in summary, can do nothing about its angle and torsional strains because of its planar geometry. Both cyclobutane and cyclopentane, on the other hand, can minimize their ring strains by adopting nonplanar conformations that relieve torsional strain.

Now we examine cyclohexane to learn how it minimizes its angle and torsional strains so that it has no ring strain.

4.8   Conformations of Cyclohexane

Angle and torsional strain are at a minimum when the six carbon atoms of cyclohexane are arranged in the **chair conformation** shown in Figure 4.13. The chair conformation is nonplanar with C—C—C bond angles of 111.5°. These bond angles are close to the normal tetrahedral bond angle. The Newman projection formula in Figure 4.14 shows, moreover, that all bonds are in a staggered conformation. As a result, there is no torsional strain in the six-membered ring.

Figure 4.13
Ball-and-stick **(A)** and line drawing **(B)** representations of the chair conformation of cyclohexane.

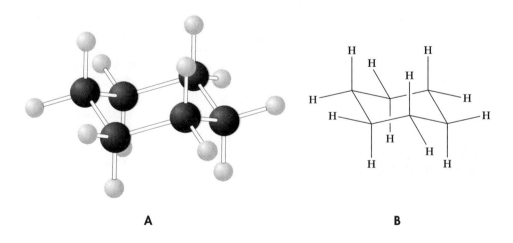

A   B

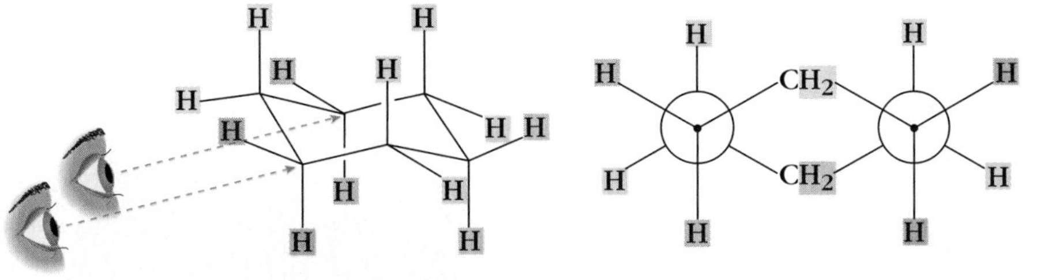

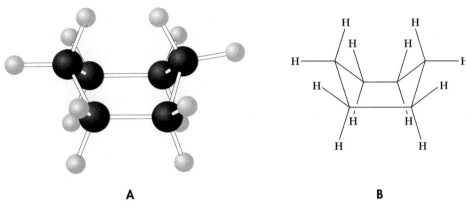

Newman projection formula

Figure 4.14
Newman projection obtained by viewing the chair conformation of cyclohexane along the bonds indicated.

■ *You should verify the Newman projection in Figure 4.14 by examining a molecular model of the chair conformation of cyclohexane. No matter which C—C bonds are chosen for viewing, all C—H bonds on adjacent carbon atoms are in staggered conformations.*

Another nonplanar conformation of cyclohexane is the **boat conformation** shown in Figure 4.15. The boat conformation is also free of angle strain because all of its C—C—C bond angles are approximately tetrahedral. The *boat conformation is not free of torsional strain,* however, because several adjacent C—H bonds are in an eclipsed conformation. We can see this from the Newman projection obtained by viewing the boat form along the C—C bonds indicated in Figure 4.16.

A   **B**

Figure 4.15
Ball-and-stick **(A)** and line drawing **(B)** representations of the boat conformation of cyclohexane.

In the boat conformation, there is also a repulsion between one hydrogen atom on carbon atom 1 and another on carbon atom 4. The interaction between these so-called flagpole hydrogen atoms is shown in Figure 4.17. The interaction illustrated in Figure 4.17 arises because the two flagpole hydrogen atoms are forced closer to each other than their atomic radii permit. This is the first example of steric strain (Section 4.5) in a cycloalkane. Because the boat conformation has torsional *and* steric strain, both of which are absent in the chair conformation, the boat conformation is less stable than the chair.

The boat form of cyclohexane can relieve some of its torsional and steric strain by twisting slightly, as shown in Figure 4.18. In this new conformation, which is called the **twist boat conformation,** the flagpole interaction is reduced because the hydrogen atoms move away from each other and the

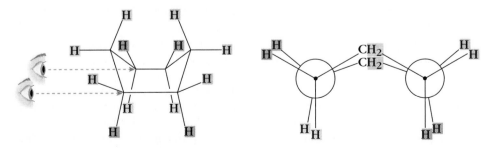

Newman projection formula

Figure 4.16
Newman projection formula obtained by viewing the boat conformation of cyclohexane along the bonds indicated. The boat conformation has torsional strain because adjacent carbon-hydrogen bonds must assume an eclipsed conformation.

Figure 4.17
Line drawing **(A)** and space-filling model **(B)** of the flagpole interaction between the C1 and C4 hydrogen atoms of the boat conformation of cyclohexane. These two hydrogen atoms are located across the cyclohexane ring but the boat conformation brings them into close proximity.

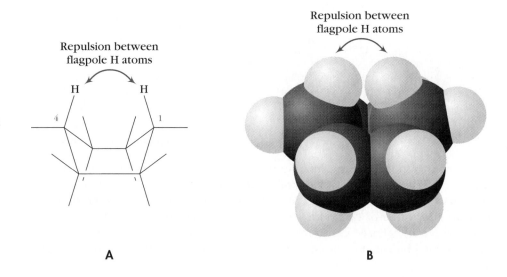

eclipsed C—H bonds are slightly less eclipsed. Because torsional and steric strain are less in the twist conformation than in the boat conformation, the twist conformation is more stable than the boat. However, the twist conformation is still higher in energy than the chair conformation.

We learned in previous sections that alkanes are easily converted from one conformation to another. Similarly the conformations of cyclohexane are easily interconverted too.

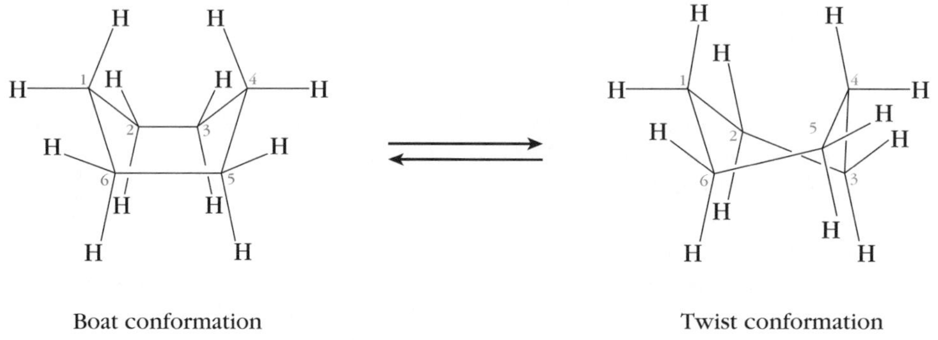

Figure 4.18
Twisting the boat conformation of cyclohexane forms the twist boat conformation. The twist boat conformation has less torsional and steric strain than the boat conformation. The twist is accomplished by a slight rotation about one carbon-carbon bond of the boat conformation. The reader should verify this with the aid of molecular models.

Boat conformation   Twist conformation

4.9   Interconversion of the Conformations of Cyclohexane

We learned in Section 4.5 that the various conformations of butane are easily and rapidly interconverted at room temperature by rotation about its central C—C bond. While rotation in a cyclohexane ring is more restricted than in butane, it can still occur and the result is interconversion of the various conformations of cyclohexane.

The path and the potential energy changes that occur when the chair and boat conformations of cyclohexane are interconverted are shown in Figure 4.19. The rotation involved in this process is illustrated in Figure 4.20. Carbon atoms 2 and 4 are rotated in the direction indicated relative to carbon atoms 3 and 5. This

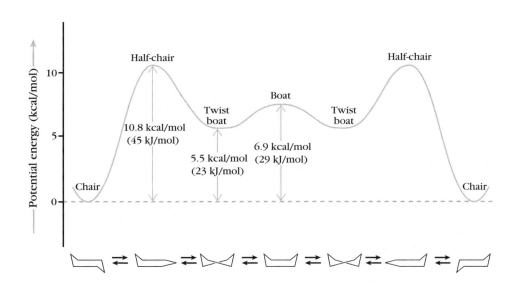

Figure 4.19
Graph of the potential energy change when one chair conformation of cyclohexane is converted to the other chair conformation.

converts some of the staggered bonds in the chair conformation into the eclipsed bonds of the boat conformation.

Along the path from the chair to the boat conformation, the highest energy point is the half-chair conformation. The half-chair conformation has five of its six carbon atoms in a plane and it is 10.8 kcal/mol (45 kJ/mol) higher in energy than the chair conformation. The boat conformation, 6.9 kcal/mol (29 kJ/mol), and the twist conformation, 5.5 kcal/mol (23 kJ/mol), are both higher in energy than the chair conformation.

As in alkanes, the energy barrier is sufficiently low between the various conformations of cyclohexane that it is impossible to isolate any conformation at room temperature. The thermal energies of cyclohexane molecules at room temperature are great enough that the various conformations interconvert many

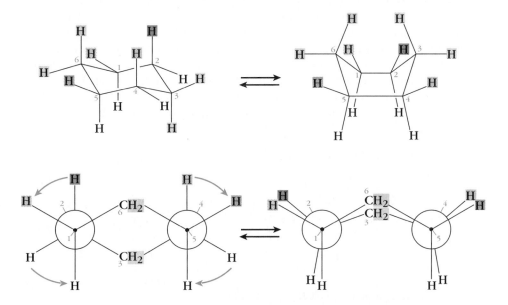

Chair conformation   Boat conformation

Figure 4.20
The boat conformation of cyclohexane can be formed by rotating carbon atoms 2 and 4 of the chair conformation in the directions indicated by the arrows.

times each second. Because of its greater stability, the chair conformation is favored at equilibrium. In fact, the chair is favored over the twist conformation at equilibrium by 10^5. Because most of the molecules in a sample of cyclohexane at room temperature are in chair conformation, we will confine our discussions to this conformation.

Let's look a little more closely at the hydrogen atoms of the chair conformation.

4.10   Axial and Equatorial Bonds

The spatial orientations of the twelve C—H bonds of the chair conformation of cyclohexane are not all the same. This difference can be easily visualized by placing a molecular model of cyclohexane on a table. Six C—H bonds extend roughly parallel to the table top in a kind of equatorial belt and are called **equatorial bonds.** The other six C—H bonds are perpendicular to the table top. These bonds are called **axial bonds.** Three of them are on each side of the general plane of the ring. Thus each carbon atom of cyclohexane has one axial and one equatorial C—H bond.

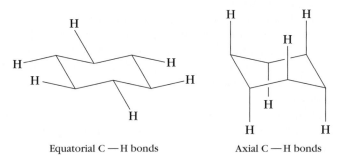

Equatorial C—H bonds   Axial C—H bonds

Another way to visualize the relative locations of the twelve C—H bonds of the chair conformation is with a Newman projection looking along the C1—C2 and C5—C4 bonds. This representation, which appears in Figure 4.21, shows clearly that the axial bonds on carbon atoms 1, 3, and 5 are on the opposite side of the ring to those on carbons 2, 4, and 6. It also shows four of the six equatorial bonds that are located near the plane of the ring. The two equatorial C—H bonds on carbon atoms 6 and 3 have been omitted from Figure 4.21 for clarity.

Students studying organic chemistry need to know how to draw the chair conformation of cyclohexane and its axial and equatorial bonds. Take a few minutes now to learn how to do this by following the instructions in Box 4.1.

Figure 4.21
Newman projection of the chair conformation of cyclohexane showing the equatorial and axial carbon-hydrogen bonds. H_a designates the axial hydrogen atoms, while H_e designates the equatorial ones. The equatorial hydrogen atoms of carbon atoms 6 and 3 point toward and away from the viewer, respectively, so they have been omitted for clarity.

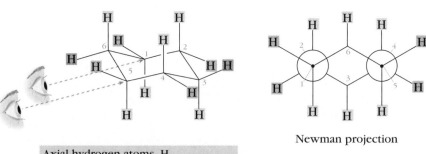

Axial hydrogen atoms, H_a
Equatorial hydrogen atoms, H_e
H_e omitted from Newman projection

Newman projection

Box 4.1   Drawing the Chair Conformation of Cyclohexane and its Axial and Equatorial Bonds

First, make a molecular model of the chair conformation of cyclohexane and use it to follow along with these instructions.

Drawing the Ring

1. Start by drawing two parallel lines offset at a slight angle. The ends of these lines are the locations of carbon atoms 2, 3, 5, and 6:

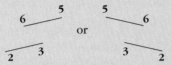

2. From carbon atoms 2 and 5 draw parallel lines in opposite directions. The ends of these lines are the locations of carbon atoms 1 and 4:

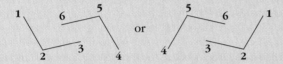

3. Complete the ring by drawing lines between carbon atoms 1 and 6 and 3 and 4:

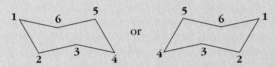

Notice that the chair conformation is drawn as three sets of parallel lines of different slopes:

Adding Axial and Equatorial Bonds

1. Draw axial bonds vertical to the ring; alternate up and down so you end up with three up and three down:

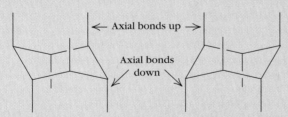

2. Draw equatorial bonds on carbon atoms 1 and 4 so they are parallel to the bonds between C2—C3 and C5—C6:

Continued.

Box 4.1—cont'd   **Drawing the Chair Conformation of Cyclohexane and its Axial and Equatorial Bonds**

3. Draw equatorial bonds on carbon atoms 2 and 5 so they are parallel to the bonds between C6—C1 and C3—C4:

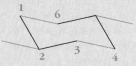 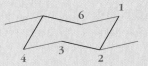

4. Draw equatorial bonds on carbon atoms 3 and 6 so they are parallel to the bonds between C1—C2 and C4—C5.

5. Drawing all six equatorial bonds of the chair conformation results in the following structures:

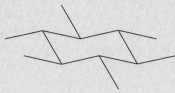 

6. Drawing the axial and equatorial bonds of the chair conformation results in the following structures:

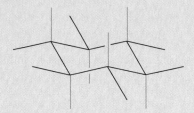 

We learned in the previous section that the chair conformations of cyclohexane are continually interconverting. This process is called **ring inversion** or **ring flipping** and is illustrated in Figure 4.22. The important result of ring flipping is that axial and equatorial C—H bonds are interconverted. The energy barrier to this process is low enough, 10.7 kcal/mol (45 kJ/mol), to make ring flipping rapid at room temperature.

Molecular models can be used to demonstrate the interchange of axial and equatorial bonds by ring flipping. Make a model of the chair form of cyclohexane and place balls of one color in the axial positions and balls of a different color in the equatorial positions. Hold the middle four carbon atoms and flip the two ends

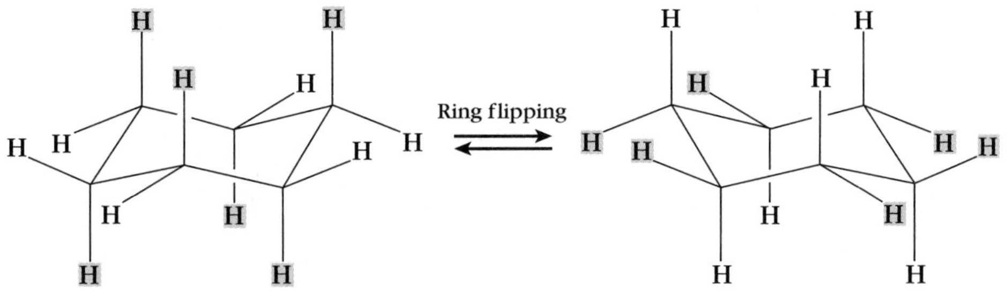

Figure 4.22
Chair to chair interconversion of cyclohexane. Notice that the axial carbon-hydrogen bonds in one conformation are converted to equatorial carbon-hydrogen bonds and vice versa by this process.

in opposite directions to convert one chair conformation into the other. Notice that the colored balls in the axial and equatorial positions interchange when you do this.

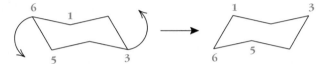

While the energies of all chair conformations of cyclohexane are the same, this is not true of monosubstituted cyclohexanes.

4.11   Conformations of Monosubstituted Cycloalkanes

Substituents in the chair conformation of cyclohexane can occupy either an axial or an equatorial position. The two conformations of methylcyclohexane, for example, are shown in Figure 4.23. In one, called the *axial conformer*, the methyl group is located in the axial position; in the other, called the *equatorial conformer*, the methyl group is in the equatorial position. Ring flipping converts one to the other, so these two conformations are in equilibrium at room temperature.

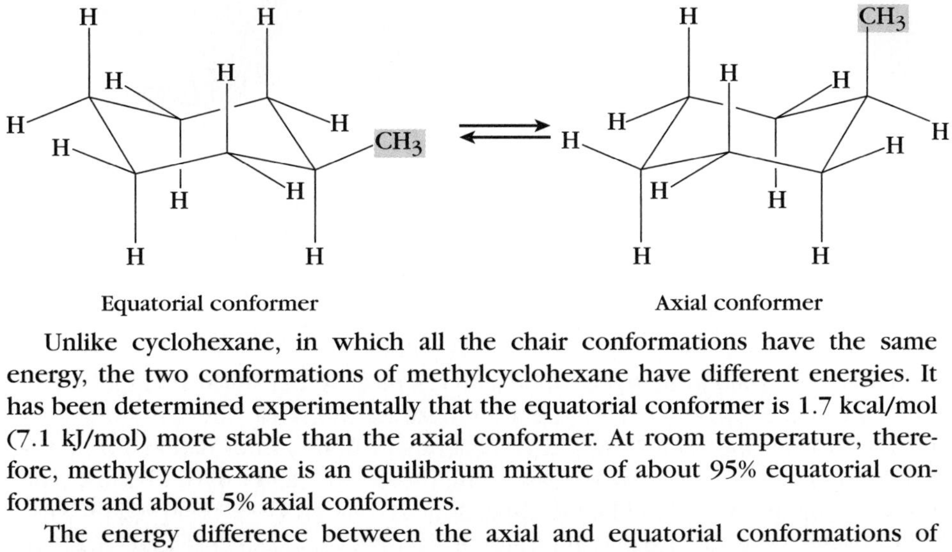

Figure 4.23
Methylcyclohexane exists as an equilibrium mixture of two chair conformations. In one, the methyl group is axial; in the other, it is equatorial.

Equatorial conformer   Axial conformer

Unlike cyclohexane, in which all the chair conformations have the same energy, the two conformations of methylcyclohexane have different energies. It has been determined experimentally that the equatorial conformer is 1.7 kcal/mol (7.1 kJ/mol) more stable than the axial conformer. At room temperature, therefore, methylcyclohexane is an equilibrium mixture of about 95% equatorial conformers and about 5% axial conformers.

The energy difference between the axial and equatorial conformations of methylcyclohexane is due to steric strain present in the axial conformer. When the methyl group on C1 is axial, it is too close to the axial hydrogens on both C3 and C5. The resulting steric strain is called a **1,3-diaxial interaction** and is shown in Figure 4.24.

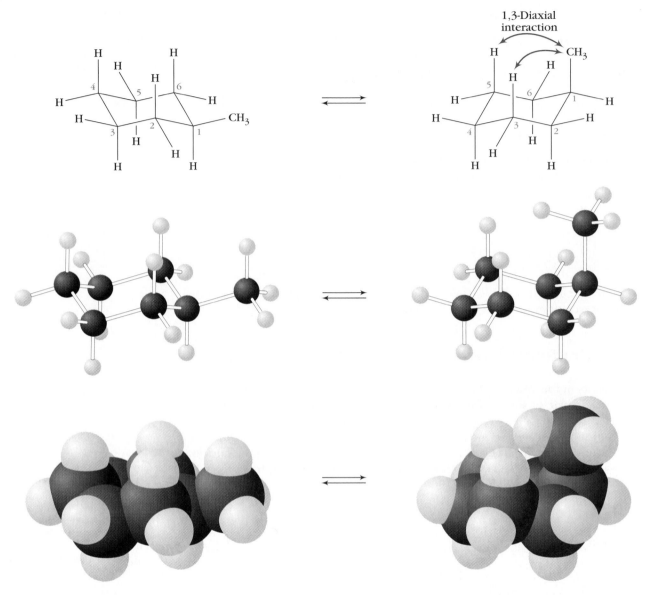

A No 1,3-diaxial interaction between equatorial methyl group and H atoms

B 1,3-Diaxial interaction between axial methyl group and axial H atoms on C3 and C5

Figure 4.24
A, No 1,3-diaxial interaction between the equatorial methyl group on C1 and the axial hydrogens on C3 and C5. **B,** 1,3-Diaxial interaction between the axial methyl group on C1 and the axial hydrogens on C3 and C5.

The steric strain in methylcyclohexane is similar to the steric strain in butane. The Newman projections in Figure 4.25 illustrate the steric strain in butane and methylcyclohexane. The Newman projections of both conformations of methylcyclohexane are obtained by looking along the C1—C2 and C5—C4 bonds, while those of butane are obtained by looking along its central C—C bond (C2—C3).

Consider first the axial conformer of methylcyclohexane (Figure 4.25, *A*). Looking along the C1—C2 bond, we see that the interaction between the axial H on C3 and the axial methyl group on C1 is just like that in the gauche conformation of butane.

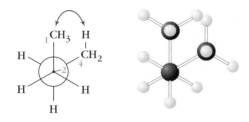

Gauche conformation of butane
Steric strain = 0.77 kcal/mol (3.2 kJ/mol)

Anti conformation of butane
Steric strain = 0

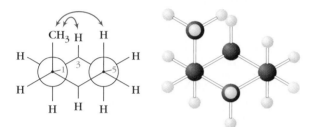

Axial conformation of methylcyclohexane
Steric strain = 1.54 kcal/mol (6.4 kJ/mol)

A

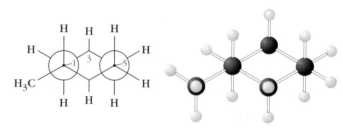

Equatorial conformation of methylcyclohexane
Steric strain = 0

B

Figure 4.25
Similarities in the steric strain in gauche conformation of butane and the axial conformation of methylcyclohexane (**A**) and anti conformation of butane and the equatorial conformation of methylcyclohexane (**B**).

Make a molecular model of the axial conformation of methylcyclohexane and look along its C1—C6 bond. You will see that an identical interaction exists between the axial methyl group on C1 and the axial H on C5. These 1,3-diaxial interactions are absent in the equatorial conformer of methylcyclohexane shown in Figure 4.25, *B.*

Recall from Section 4.5 that steric interference like these 1,3-diaxial interactions contributes approximately 0.77 kcal/mol (3.2 kJ/mol) to the energy of the gauche conformation of butane. Because there are two similar interactions in the axial conformer of methylcyclohexane, we would expect it to be 1.54 kcal/mol (6.4 kJ/mol) less stable than the equatorial conformer that lacks such steric interactions. The experimental value is 1.7 kcal/mol (7.1 kJ/mol), as previously mentioned.

Monosubstituted cycloalkanes other than methylcyclohexane exist as equilibrium mixtures of axial and equatorial conformers too. Generally, there is less steric interaction when a substituent is in the equatorial position than in the axial position, so the equatorial conformer is usually more stable than the axial conformer. The difference in energy between the two conformers depends on the size of the substituent. The bulkier the substituent, the more likely is it to occupy the equatorial position. In other words, the bulkier the substituent, the larger the 1,3-diaxial interaction, and the greater the energy difference between the axial and equatorial conformers. The differences in energies between the axial and equatorial conformers of several monosubstituted cyclohexanes are listed in Table 4.3.

Table 4.3

Energy Difference Between the Axial and Equatorial Conformers of Several Monosubstituted Cyclohexanes at 25 °C

Substituent	Energy Difference*	
	kcal/mol	(kJ/mol)
—CN	0.2	0.8
—Cl	0.6	2.4
—Br	0.7	2.8
—OH	1.0	4.2
—COOH	1.4	5.9
—CH₃	1.7	7.1
—CH₂CH₃	1.9	8.0
—CH(CH₃)₂	2.2	9.2
—C(CH₃)₃	5.4	22.6

*Energy of axial conformer minus energy of equatorial conformer.

As we can see in Table 4.3, the bulk of the alkyl substituents and their steric strain, which is expressed as the energy difference between the axial and equatorial conformers, increases steadily along the series $-CH_3$ < $-CH_2CH_3$ < $-CH(CH_3)_2$ < $-C(CH_3)_3$. The 1,3-diaxial interaction of the *tert*-butyl group is so large that *tert*-butylcyclohexane exists almost entirely as the equatorial conformer.

Exercise 4.8   From the data in Table 4.3, decide whether each of the following compounds exists predominantly in the equatorial or axial conformation.

(a) CH₂CH₃ structure

(b) OH structure

(c) COOH structure

(d) CN structure

When a cyclohexane ring has more than one substituent, we must consider their locations relative to each other as well as their preference for the axial or equatorial positions.

4.12   *cis*- and *trans*-Isomers of Disubstituted Cycloalkanes

Two substituents on different carbon atoms of a cycloalkane ring can be located on opposite sides or the same side of the ring. Two substituents on opposite sides of the ring are said to be *trans*- to each other, while substituents on the same side are said to be *cis*- to each other. 1,2-Dimethylcyclopropane, for instance, can exist as either *trans*- or *cis*-.

trans-1,2-Dimethylcyclopropane   *cis*-1,2-Dimethylcyclopropane

cis- and *trans*-1,2-dimethylcyclopropane are called **stereoisomers** because they are isomers that differ only in the arrangement of their atoms in space. Stereoisomers can't be interconverted without breaking bonds. In order to convert *cis*-1,2-dimethylcyclopropane into its *trans*-isomer, for example, we must switch the positions of a hydrogen atom and a methyl group on the same carbon atom. To do this we would have to break a C—H and a C—C bond. Because breaking bonds requires a great deal of energy (somewhere between 90–110 kcal/mol [375–460 kJ/mol] for C—H and C—C bonds), each of the two stereoisomers is a stable compound that can be isolated, stored in a container, and kept indefinitely at room temperature. Thus *cis*- and *trans*-1,2-dimethylcyclopropane are different compounds with different physical properties.

From their heats of combustion, we know that *trans-*1,2-dimethylcyclopropane is 1.2 kcal/mol (5 kJ/mol) more stable than its *cis-*isomer. The *cis-*isomer is higher in energy because of steric hindrance between the two methyl groups on the same side of the ring. This steric interaction is absent in the *trans-*isomer.

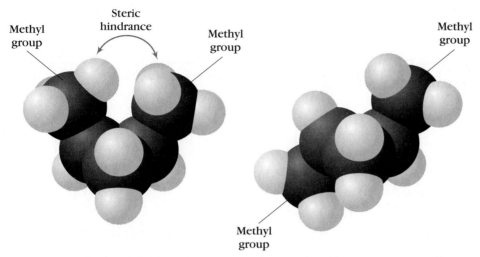

*cis-*1,2-Dimethylcyclopropane   *trans-*1,2-Dimethylcyclopropane

cis-trans Stereoisomers also exist in other cycloalkanes such as cyclobutane and cyclopentane. It is more convenient to represent cycloalkanes by means of a planar ring than it is to draw their three-dimensional representations. The stereochemistry of the substituents on a planar ring is indicated by wedges and dashes. Planar ring line-dash-wedge representations of *cis-* and *trans-*1,2-dimethylcyclobutane, for example, are shown in Figure 4.26.

*cis-*1,2-Dimethylcyclobutane   *trans-*1,2-Dimethylcyclobutane

Figure 4.26
Line-dash-wedge representations of the stereoisomers of 1,2-dimethylcyclobutane. For the sake of convenience and speed, a planar cyclobutane ring is used. Keep in mind, however, that the cyclobutane ring is not planar but is slightly puckered.

Line-dash-wedge representations are not exactly correct because, except for cyclopropane, cycloalkanes are not planar, as we learned in Section 4.7. However, these representations are sufficiently accurate for our purposes and they make it much easier to see the relative positions of the substituents on the rings of cycloalkanes. The following are additional examples of *cis-trans* stereoisomers containing cyclobutane and cyclopentane rings.

*cis-*1-Chloro-2-methylcyclobutane   *trans-*1-Chloro-2-methylcyclobutane

cis-1,2-Dichlorocyclopentane 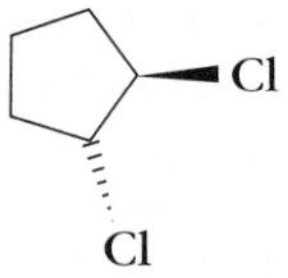   *trans*-1,2-Dichlorocyclopentane

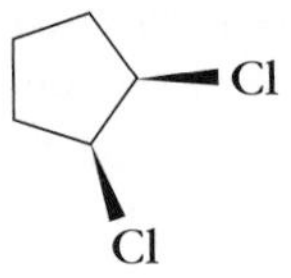

Sample Problem 4.1   Name the following compound.

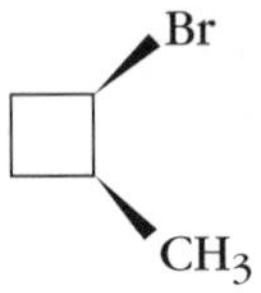

Solution:

Step 1 The parent name is cyclobutane.

Step 2 There are two substituents, a methyl group and a bromine atom.

Step 3 The two substituents are located on the same side of the cyclobutane ring so they are *cis-* to each other.

Step 4 The name is *cis*-1-bromo-2-methylcyclobutane.

Exercise 4.9   Give a name to each of the following compounds.

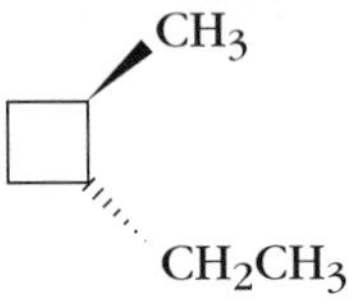

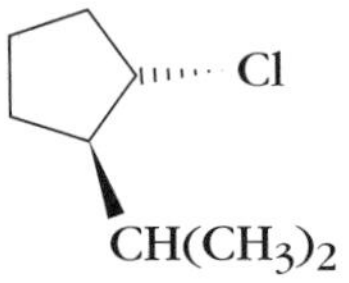

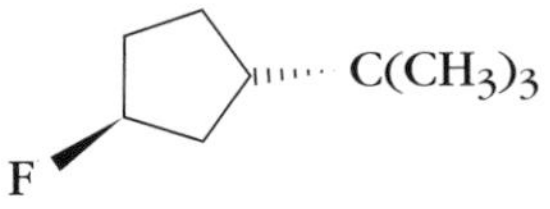

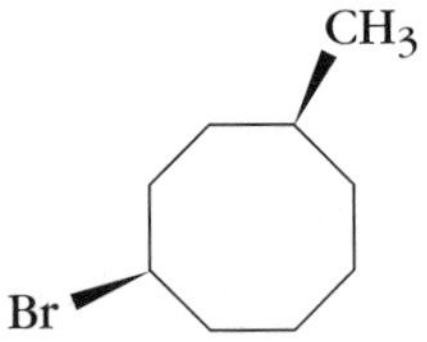

Exercise 4.10   Write the structure for the *cis-* and *trans*-stereoisomers of:

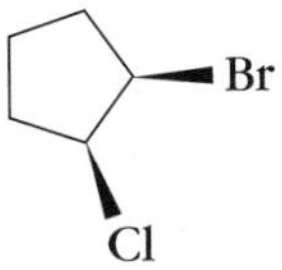

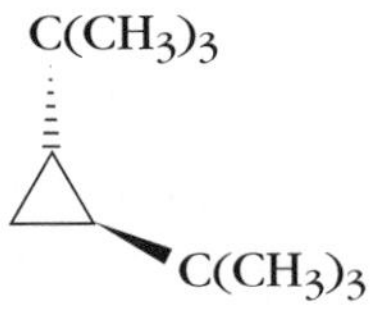

(a) 1,2-Dichlorocyclopropane   **(e)** 1,3-Diethylcyclopentane
(b) 1-Isopropyl-2-methylcyclobutane   **(f)** 1,3-Dibromocycloheptane
(c) 1,3-Dimethylcyclobutane   **(g)** 1,2-Dimethylcyclohexane
(d) 1-Bromo-2-methylcyclopentane   **(h)** 1,3-Dimethylcyclohexane

The substituent in monosubstituted cyclohexanes prefers the equatorial position. The situation is more complex, however, in disubstituted cyclohexanes.

4.13 Conformations of Disubstituted Cyclohexanes

There are two things that we would like to know about disubstituted cyclohexanes. First, which is the more stable, the *cis-* or the *trans*-isomer? Second, what is the preferred conformation of both the *cis-* and the *trans*-isomers? We can obtain this information by analyzing the steric interactions of the two substituents in chair conformations. Before doing this, however, we must look at how the *cis-* and *trans*-stereoisomers of disubstituted cyclohexanes are represented.

We begin by representing the *cis-* and *trans*-stereoisomers of disubstituted cyclohexane with a planar ring. Then we convert these planar structures into the chair conformations of disubstituted cyclohexanes as shown for *cis*-1,2-dimethyl-cyclohexane in Figure 4.27. The planar cyclohexane ring is viewed in Figure 4.27 from one edge. If you imagine firmly holding carbon atoms 2, 3, 5, and 6 and pushing down on carbon atom 1 and up on carbon atom 4, a chair conformation of cyclohexane is formed. In the chair conformation of the *cis*-isomer, one methyl is in the axial position while the other is in the equatorial position.

Figure 4.27
Converting the planar structure of *cis*-1,2-dimethylcyclohexane into its chair conformations.

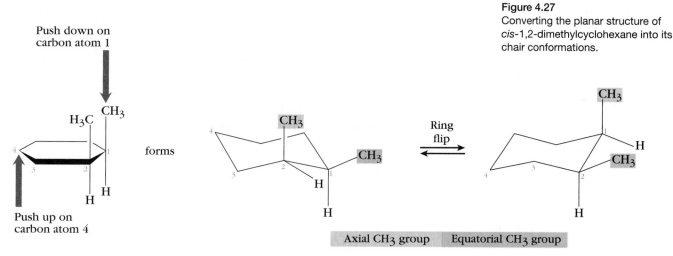

Axial CH3 group   Equatorial CH3 group

Cyclohexane rings flip easily at room temperature, so *cis*-1,2-dimethylcyclo-hexane exists in the two chair conformations shown in Figure 4.27. Both chair conformations have one methyl group axial and the other equatorial.

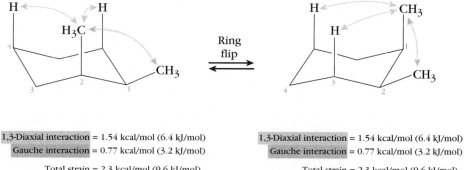

1,3-Diaxial interaction = 1.54 kcal/mol (6.4 kJ/mol)
Gauche interaction = 0.77 kcal/mol (3.2 kJ/mol)

Total strain = 2.3 kcal/mol (9.6 kJ/mol)

1,3-Diaxial interaction = 1.54 kcal/mol (6.4 kJ/mol)
Gauche interaction = 0.77 kcal/mol (3.2 kJ/mol)

Total strain = 2.3 kcal/mol (9.6 kJ/mol)

The two chair conformations of *cis*-1,2-dimethylcyclohexane are exactly equal in energy because their steric interactions are identical. Each conformation has a 1,3-diaxial interaction and a gauche butane interaction between the two methyl groups. Both of these interactions give a total strain energy of 2.3 kcal/mol (9.6 kJ/mol).

Let's now look at *trans*-1,2-dimethylcyclohexane. Begin again with a planar ring, then push up C4 and push down C1 to form the chair conformation. This chair conformation is called the *diequatorial conformation* because both methyl groups are equatorial. Ring flipping forms the other chair conformation. It is called the *diaxial conformation* because both methyl groups are axial.

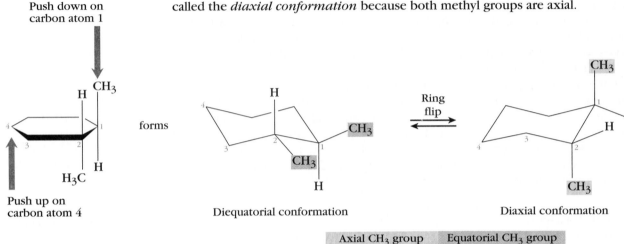

Push down on
carbon atom 1

Push up on
carbon atom 4

forms

Diequatorial conformation

Ring
flip

Diaxial conformation

Axial CH₃ group   Equatorial CH₃ group

The strain energies of the two chair conformations of *trans*-1,2-dimethylcyclo-hexane are different. The diequatorial conformation has only a gauche butane interaction between the two methyl groups and no 1,3-diaxial interactions. As a result, the total strain energy is 0.77 kcal/mol (3.2 kJ/mol) for the diequatorial conformation. The diaxial conformation has two 1,3-diaxial interactions and no gauche butane interactions.

Ring
flip

1,3-Diaxial interaction = 0
Gauche interaction = 0.77 kcal/mol (3.2 kJ/mol)

Total strain = 0.77 kcal/mol (3.2 kJ/mol)

1,3-Diaxial interaction = 2 X 1.54 kcal/mol (2 X 6.4 kJ/mol)
Gauche interaction = 0

Total strain = 3.1 kcal/mol (12.8 kJ/mol)

Each 1,3-diaxial interaction contributes 1.54 kcal/mol (6.4 kJ/mol), for a total strain energy of 3.1 kcal/mol (12.8 kJ/mol) for the diaxial conformation. Consequently, the diequatorial conformation is estimated to be 2.3 kcal/mol (9.7 kJ/mol) more stable than the diaxial conformation, which means in turn that *trans*-1,2-dimethylcyclohexane exists almost entirely (>99%) in the diequatorial conformation at room temperature.

This analysis also allows us to estimate the relative energies of the stereoisomers of 1,2-dimethylcyclohexane. We have estimated a strain energy of 0.77 kcal/mol (3.2 kJ/mol) for the diequatorial *trans*-conformation and a total strain energy of 2.3 kcal/mol (9.7 kJ/mol) for the equivalent conformations of the *cis*-isomer. The difference in strain energy between the two is an estimate of the stability of the *trans*- relative to the *cis*-isomer. Thus we calculate that the diequatorial conformation of *trans*-1,2-dimethylcyclohexane is 1.53 kcal/mol (6.4 kJ/mol) more stable than either conformation of *cis*-1,2-dimethylcyclohexane. This is in agreement with the experimental value of 1.48 kcal/mol (6.2 kJ/mol) obtained from heats of combustion data.

We can repeat this conformational analysis for *cis*- and *trans*-1,3-dimethylcyclohexane. The more stable chair conformation of *cis*-1,3-dimethylcyclohexane is the one that has both methyl groups in equatorial positions. This diequatorial conformation is estimated to be 3.7 kcal/mol (15.5 kJ/mol) more stable than the diaxial conformation. Consequently, *cis*-1,3-dimethylcyclohexane exists exclusively in the diequatorial conformation at room temperature.

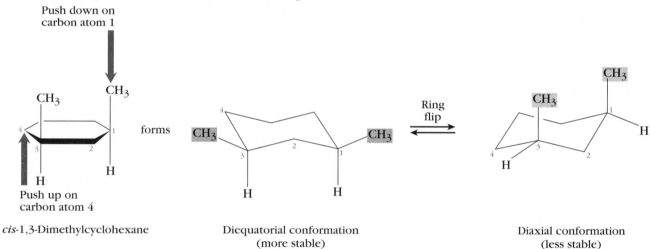

cis-1,3-Dimethylcyclohexane   Diequatorial conformation   Diaxial conformation
                                              (more stable)   (less stable)

Axial CH$_3$ group   Equatorial CH$_3$ group

The two chair conformations of *trans*-1,3-dimethylcyclohexane both have one axial and one equatorial methyl group so they are of equal energy.

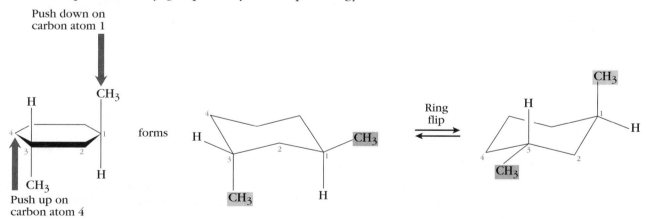

trans-1,3-Dimethylcyclohexane   Axial CH$_3$ group   Equatorial CH$_3$ group

The preferred diequatorial conformation of *cis*-1,3-dimethylcyclohexane has no strain energy due to 1,3-diaxial or gauche butane interactions. The axial methyl group of the *trans*-isomer, on the other hand, has one 1,3-diaxial interaction that contributes 1.54 kcal/mol (6.4 kJ/mol) of strain energy. Thus we calculate that the diequatorial conformation of the *cis*-isomer is 1.54 kcal/mol

(6.4 kJ/mol) more stable than the *trans*-isomer. Again, this is in good agreement with the experimental value of 1.51 kcal/mol (6.3 kJ/mol) obtained from heats of combustion data.

The most stable conformation of *trans*-1,4-dimethylcyclohexane has both methyl groups equatorial.

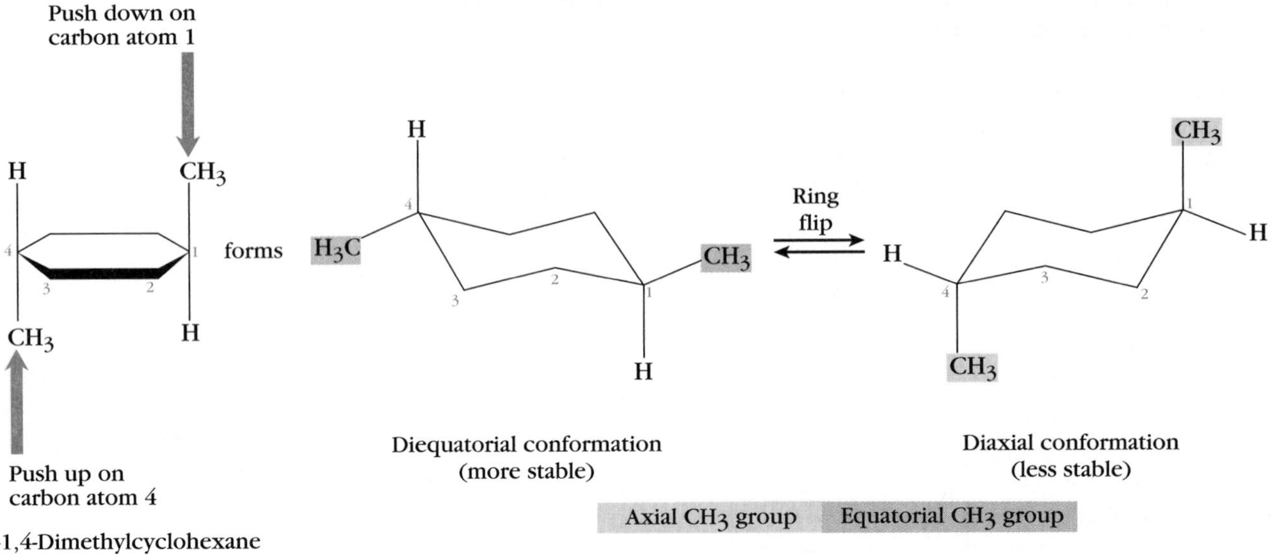

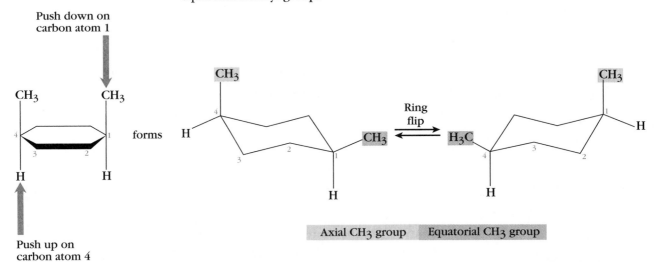

Push down on
carbon atom 1

Push up on
carbon atom 4

trans-1,4-Dimethylcyclohexane

forms

Ring
flip

Diequatorial conformation
(more stable)

Diaxial conformation
(less stable)

Axial CH₃ group   Equatorial CH₃ group

Both chair conformations of *cis*-1,4-dimethylcyclohexane have one axial and one equatorial methyl group.

Push down on
carbon atom 1

Push up on
carbon atom 4

cis-1,4-Dimethylcyclohexane

forms

Ring
flip

Axial CH₃ group   Equatorial CH₃ group

From heats of combustion data, we know that the diequatorial conformation of the *trans*-isomer is 1.6 kcal/mol (6.7 kJ/mol) more stable than the *cis*-isomer.

Exercise 4.11 Using the values in Table 4.1, carry out a conformational analysis of *cis*- and *trans*-1,4-dimethylcyclohexane and calculate the energy difference between these two stereoisomers.

The results of the conformational analysis for the isomers of dimethylcyclohexane are summarized in Table 4.4.

Table 4.4	Most Stable Conformations and Relative Stabilities of Stereoisomers of Dimethylcyclohexane	
Dimethyl-cyclohexane	Orientation of Methyl Groups in Most Stable Conformation	More Stable Stereoisomer
cis-1,2-	Axial-equatorial	
trans-1,2-	Equatorial-equatorial	trans-
cis-1,3-	Equatorial-equatorial	cis-
trans-1,3-	Axial-equatorial	
cis-1,4-	Axial-equatorial	
trans-1,4-	Equatorial-equatorial	trans-

The data in Table 4.4 show that the relative stabilities of the *cis-trans* isomers of dimethylcyclohexane depend on the number of methyl groups that are axial and/or equatorial. No matter whether the methyl groups are 1,2-, 1,3-, or 1,4-, the conformation in which the two methyl groups are both equatorial is more stable than when one methyl is equatorial and one is axial. The least stable conformation is the one with two axial methyl groups.

It is important to remember that in disubstituted cyclohexanes, ring flipping interconverts only the conformations of one particular stereoisomer. *Ring flipping does not interconvert stereoisomers.* For example, ring flipping will never convert one conformation of a *trans*-1,2-dimethylcyclohexane into a conformation of *cis*-1,2-dimethylcyclohexane. Bond breaking must occur to convert a *cis*-isomer to a *trans*-isomer.

In many disubstituted cyclohexanes the substituents are different. How can we determine the most stable conformation in these cases? We know, for instance, that the energy difference between the axial and equatorial positions is roughly related to the bulk of the substituent. The larger the group, the greater its tendency to occupy the equatorial position. Therefore if both substituents can't be equatorial, the most stable conformation is generally the one with the bulkier group equatorial and the other axial.

| Sample Problem 4.2 | Write the structures of the two chair conformations of *cis*-1-bromo-4-*tert*-butylcyclohexane and indicate which is the more stable. |

Solution:

Each of the two chair conformations of *cis*-1,4-disubstituted cyclohexanes has one substituent axial and the other equatorial. Thus the two chair conformations are the following:

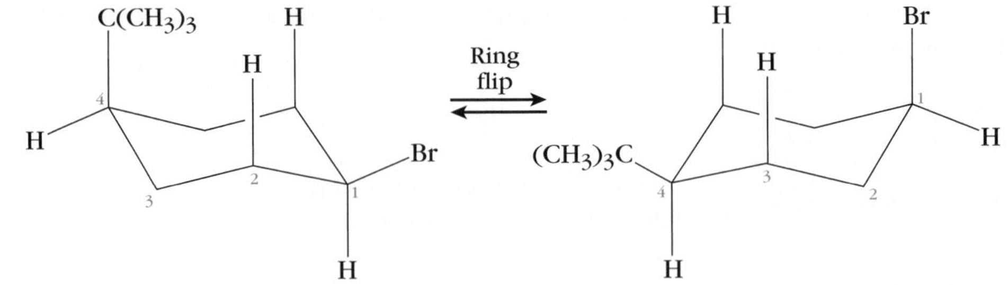

The less stable conformation has the bulkier group in the axial position

The more stable conformation has the bulkier group in the equatorial position

cis-1-Bromo-4-*tert*-butylcyclohexane

The bulky *tert*-butyl group has a greater preference for the equatorial position, so the most stable conformation has an equatorial *tert*-butyl group and an axial bromine atom.

Exercise 4.12 Write the structure of the two chair conformations of each of the following substituted cyclohexanes and indicate which is the more stable.

(a) *trans*-1-Ethyl-4-isopropylcyclohexane
(b) *trans*-1-Ethyl-3-methylcyclohexane
(c) *cis*-1-Bromo-2-isopropylcyclohexane

Exercise 4.13 Identify each of the following compounds as either the *cis*- or *trans*-stereoisomer.

(a)

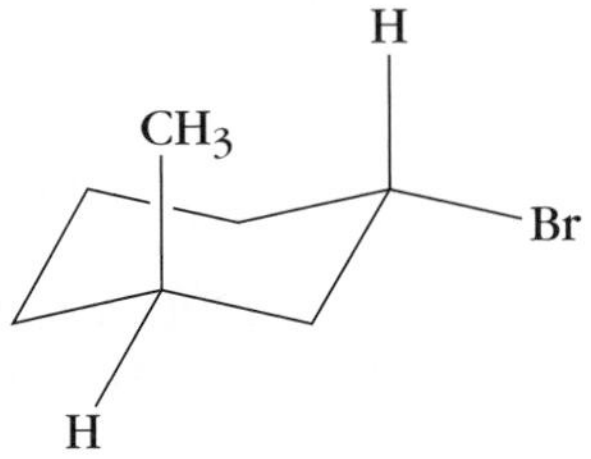

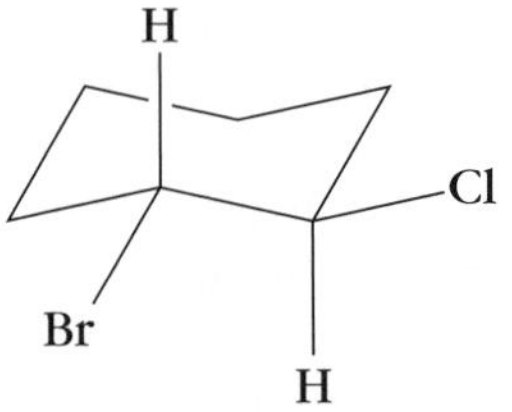

(b)

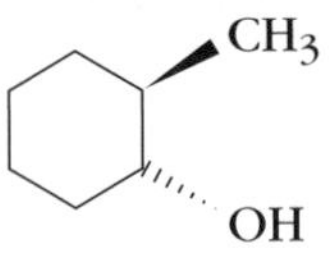

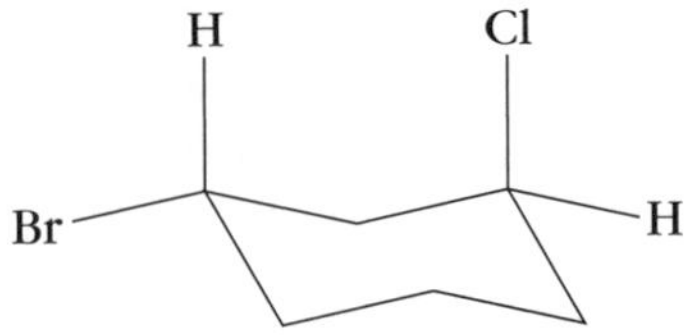

(c)

(d)

(e)

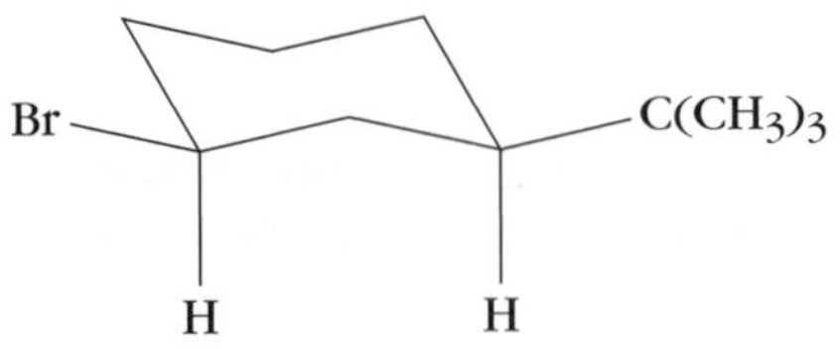

(f)

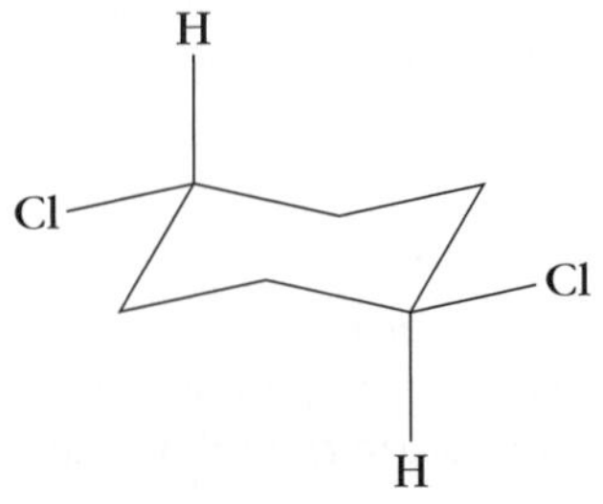

The principles of conformational analysis derived from the cyclohexane ring system can be applied equally well to larger and more complex ring systems.

4.14 Fused Polycyclic Molecules

Polycyclic fused ring systems contain rings of atoms that share two adjacent carbon atoms and the bond between them. The two shared atoms are called **bridgehead carbon atoms.** The most common fused rings are cyclohexane rings and the simplest example is decalin, which contains two fused cyclohexane rings.

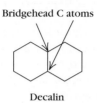

Bridgehead C atoms

Decalin

Two stereoisomers of decalin are known, *cis*-decalin and *trans*-decalin. The cyclohexane rings of each are in the chair conformation. The bridgehead hydrogen atoms of *trans*-decalin are both in axial positions, as shown in Figure 4.28. As a result the two rings are joined together in such a way that each ring is like two equatorial substituents in the other ring. This makes *trans*-decalin analogous to *trans*-1,2-dimethylcyclohexane.

In *cis*-decalin, on the other hand, one bridgehead hydrogen atom is in the axial position while the other is in the equatorial position, as shown in Figure 4.28. As a result, the two rings are joined together as one equatorial and one axial substituent. This makes *cis*-decalin analogous to *cis*-1,2-dimethylcyclohexane.

At room temperature *trans*-decalin is found to be 2.6 kcal/mol (10.9 kJ/mol) more stable than the *cis*-isomer. This difference in stability is due to more 1,3-diaxial interactions in the *cis*-isomer.

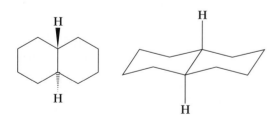

trans-Decalin

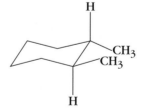

trans-1,2-Dimethylcyclohexane

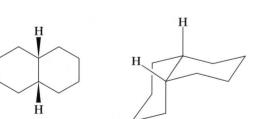

cis-Decalin

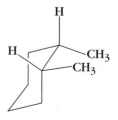

cis-1,2-Dimethylcyclohexane

A   **B**

Figure 4.28
Representation of *cis*- and *trans*-decalin. **A,** The line-dash-wedge representations are drawn with planar cyclohexane rings. The stereochemistry at the ring junction is emphasized in this view. **B,** The line representations emphasize that both bridgehead hydrogen atoms occupy axial positions in *trans*-decalin (analogous to *trans*-1,2-dimethylcyclohexane), whereas one hydrogen atom is axial and one equatorial in *cis*-decalin (analogous to *cis*-1,2-dimethylcyclohexane).

Exercise 4.14	Build molecular models of *cis*- and *trans*-decalin in the conformations represented in Figure 4.28.

(a) Carry out a chair-chair interconversion on both rings of *cis*-decalin.

(b) Can you carry out a chair-chair interconversion on both rings of *trans*-decalin?

(c) Identify all the 1,3-diaxial and gauche butane interactions in both *cis*- and *trans*-decalin.

(d) Using the data in Table 4.1, estimate the difference in stability of *cis*- and *trans*-decalin.

Many compounds containing polycyclic fused ring systems are found in nature. One example is the biologically important class of compounds called steroids. The two stereoisomers 5-α-cholestane and 5-β-cholestane are examples of the two types of ring systems that are found in natural steroids. Both have four fused rings, three six-membered and one five-membered, as shown in Figure 4.29. They differ in the fusion between the rings A and B. The ring junction is *trans*- for 5-α-cholestane but *cis*- for 5-β-cholestane. The same principles that we learned from simple cyclohexanes can be applied equally well to these more complex fused ring systems.

The ring system shown in Figure 4.29 is the basic skeleton for the majority of the steroids such as cholesterol and cortisone, as well as the steroidal hormones testosterone and progesterone.

4.15 Summary

Rotation can occur about carbon-carbon single bonds, so alkanes exist as a large number of rapidly interconverting **conformations** or **conformers.** These conformations can be shown in three dimensions by using either a line-dash-wedge

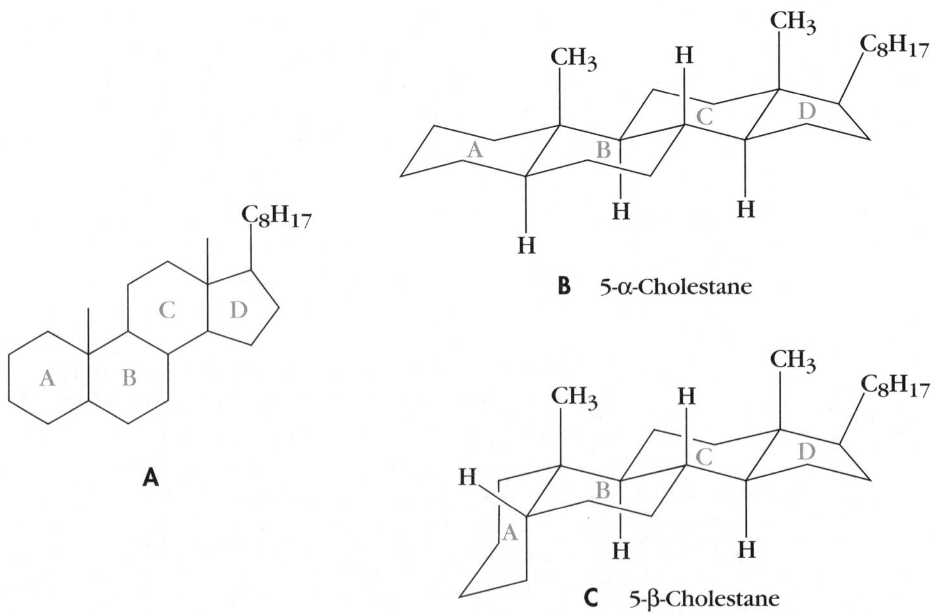

Figure 4.29
A, The structure of the basic ring in steroids using planar cycloalkane rings. **B,** Line drawing of 5-α-cholestane showing the *trans*- junction of rings *A* and *B*. **C,** Line drawing of 5-β-cholestane showing *cis*-junction of rings *A* and *B*.

representation or sawhorse drawings. Projecting the view along the carbon-carbon single bond onto a plane forms **Newman projection formulas.**

The **staggered** conformation of alkanes is more stable than the **eclipsed** conformation because of **torsional strain.** The most stable staggered conformation of butane and other straight chain alkanes is the **anti** conformation, in which the two alkyl groups are as far away as possible.

Based on heats of combustion data, cyclopropane is found to have the most **ring strain** of the cycloalkanes, followed by cyclobutane. Cyclohexane is strain free, and the other cycloalkanes have varying degrees of ring strain.

Three factors that contribute to total ring strain are **angle strain,** the energy needed to distort the normal carbon tetrahedral bond angle to the angles needed to form a particular cycloalkane; **torsional strain,** the destabilization due to eclipsing of bonds on adjacent atoms; and **steric strain,** a repulsive interaction that arises when two groups are forced closer to each other than their atomic radii permit.

The cyclohexane ring is strain free because it can adopt a **chair conformation** in which all bond angles are close to 109° and all C—H bonds on adjacent carbon atoms are in a staggered conformation. There are two kinds of C—H bonds in the chair conformation of cyclohexane, six **axial** (a) and six **equatorial** (e).

Each carbon atom of the chair conformation of cyclohexane has one equatorial and one axial C—H bond.

The chair conformation of cyclohexane is conformationally mobile, so axial and equatorial bonds can be interconverted by means of ring flipping. While all chair conformations of cyclohexane are equal in energy, the most stable conformation of monosubstituted cyclohexanes is generally the one in which the substituent is in an equatorial position. This preference for equatorial positions is due to 1,3-diaxial interactions. Generally, the bulkier the substituent, the greater its 1,3-diaxial interaction, and the greater its preference for the equatorial position.

Disubstituted cycloalkanes exist as **stereoisomers,** which are isomers that differ only in the arrangements of their atoms in space. Stereoisomers of cycloalkanes can't be interconverted by ring flipping. Conformational analysis of disubstituted cyclohexanes shows that the *trans*-isomer is the more stable stereoisomer for 1,2- and 1,4-disubstituted cyclohexanes, while the *cis*-isomer is the more stable one for 1,3-disubstituted cyclohexanes. If both substituents can't occupy equatorial positions, the bulkier of the two usually occupies the equatorial position.

The principles of conformational analysis apply equally well to molecules that contain fused cyclohexane rings. Fused cyclohexane rings are two or more cyclohexane rings that share two adjacent carbon atoms and the bond between them. Many compounds that contain fused cyclohexane rings are biologically important.

Additional Exercises

4.15   Define or give an example of each of the following terms.

(a)	Eclipsed conformation	**(j)**	Newman projection formula
(b)	Torsional energy	**(k)**	Gauche conformation
(c)	Steric strain	**(l)**	Staggered conformation
(d)	Standard heat of combustion	**(m)**	Torsional strain
(e)	Ring strain	**(n)**	Anti conformation
(f)	Angle strain	**(o)**	Boat conformation of cyclohexane
(g)	Chair conformation of cyclohexane	**(p)**	Axial bond
(h)	Equatorial bond	**(q)**	1,3-Diaxial interaction
(i)	Fused polycyclic molecule	**(r)**	Stereoisomers

4.16   For 1,2-diiodoethane, represent the anti, the gauche, and two different eclipsed conformations by:

(a)   Newman projection formulas   **(c)**   Line-dash-wedge representations
(b)   Sawhorse structures

4.17   For 2-methylbutane, represent all three staggered and all three eclipsed conformations by:

(a)   Newman projection formulas along its C2—C3 bond   **(c)**   Line-dash-wedge representations
(b)   Sawhorse structures

4.18   Repeat Exercise 4.17 for 2,3-dimethylbutane.

4.19   Using the data in Table 4.1, draw a quantitative diagram of potential energy versus dihedral angle for:

(a)   2-Methylbutane   **(b)**   2,3-Dimethylbutane.

In both cases, define the dihedral angle as 0° for the eclipsed conformation of highest energy.

4.20   Cyclobutane also undergoes ring flipping, which causes its C—H bonds to interchange positions as follows:

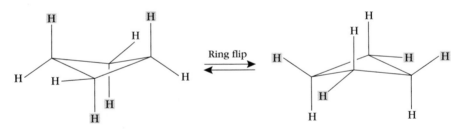

(a)   Write the structures of the two conformations of *cis*-1,3-dimethylcyclobutane.
(b)   Which would you expect to be the more stable? Explain.

4.21   On the basis of their standard heats of combustion, which is the more stable of the following pairs of isomers?

	Compound	Heat of Combustion			Compound	Heat of Combustion	
		kcal/mol	(kJ/mol)			kcal/mol	(kJ/mol)
(a)	$CH_3CH{=}CH_2$	−461.8	(−1932.2)	or	Cyclopropane	−469.6	(−1964.8)
(b)	Pentane	−784.2	(−3281.1)	or	2-Methylbutane	−782.4	(−3273.6)
(c)	Pentane	−784.2	(−3281.1)	or	2,3-Dimethylpropane	−779.0	(−3259.3)

4.22   In each of the following structures, indicate whether the substituent is in an axial or an equatorial position.

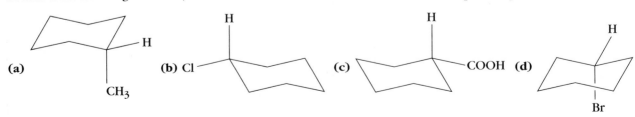

4.23 Convert each of the following structures into its two chair conformations. In each case indicate which one should be the more stable conformation.

(a)

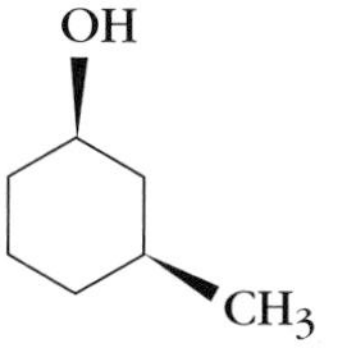

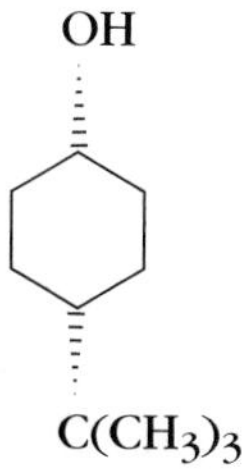

(b)

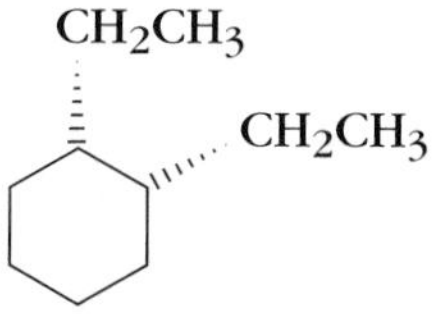

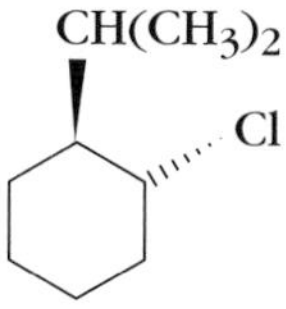

(c)

(d)

(e)

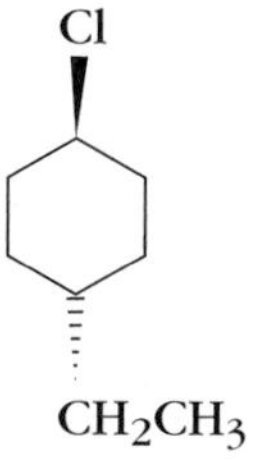

(f)

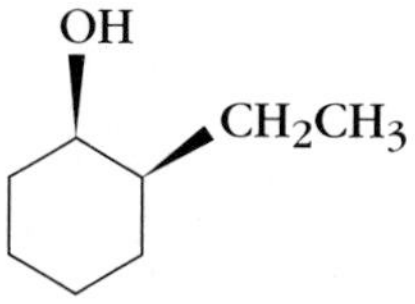

4.24 Write the structure of the preferred conformation of:

(a) 1-Isopropyl-1-methylcyclohexane
(b) cis-1-Bromo-2-isopropylcyclohexane
(c) trans-1-Methyl-3-isopropylcyclohexane
(d) cis-1-Chloro-4-isopropylcyclohexane

4.25 Identify each of the following compounds as either the cis- or the trans-isomer:

(a)

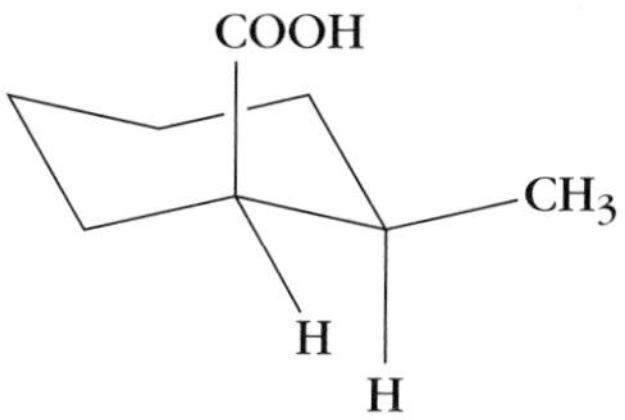

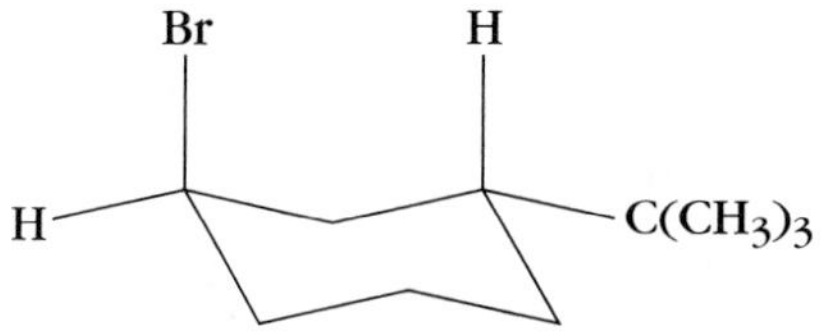

(b)

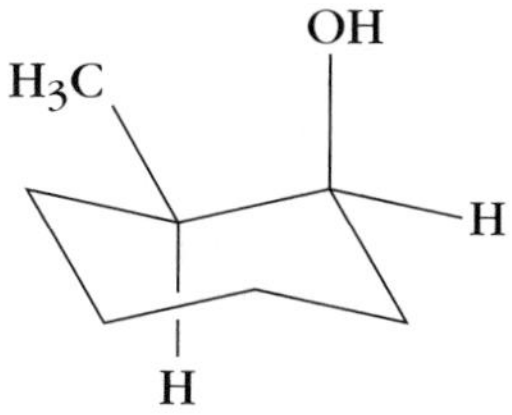

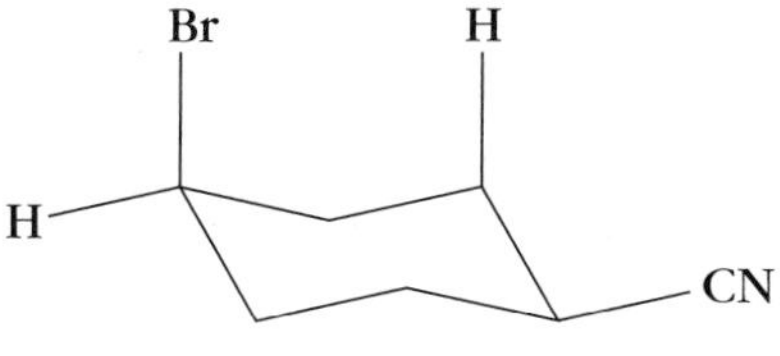

(c)

(d)

(e)

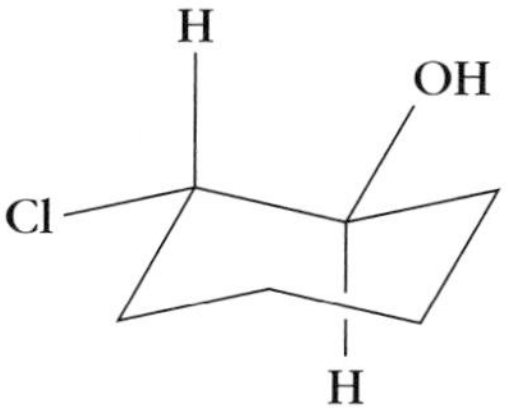

(f)

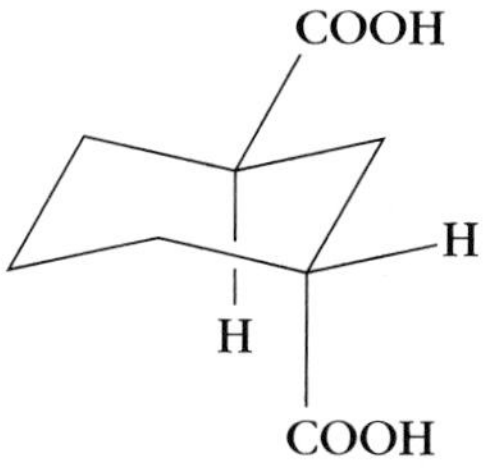

4.26 Write the two chair conformations of all trans-1,2,3,4,5,6-hexachlorocyclohexane.

4.27 Decide if the two structures in the following pairs of compounds represent: (a) conformers, (b) stereoisomers, or (c) constitutional isomers.

(a)

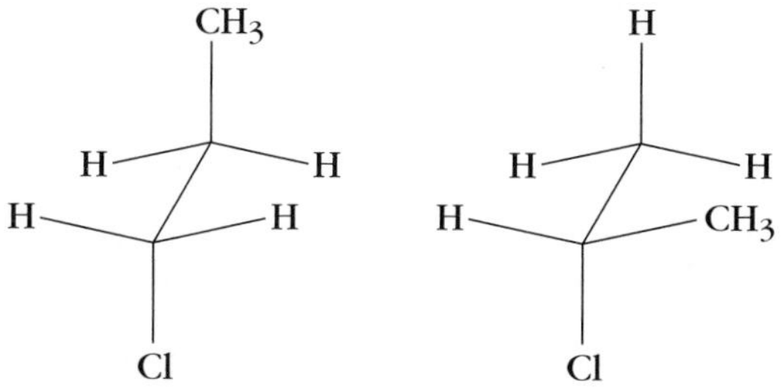

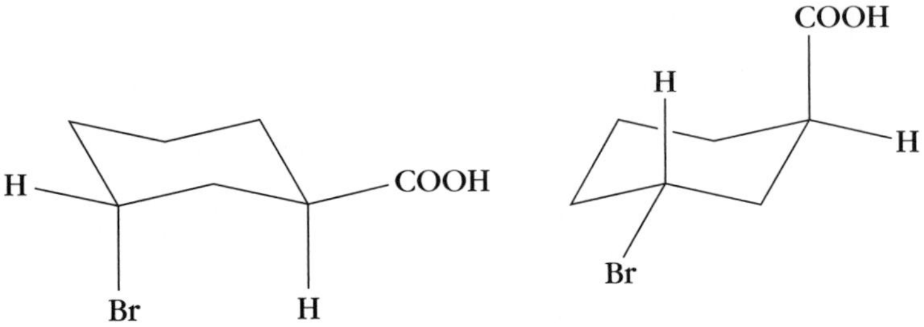

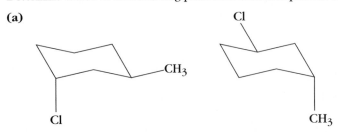

4.28   Write the following structures of 1,2-dichlorocyclohexane:

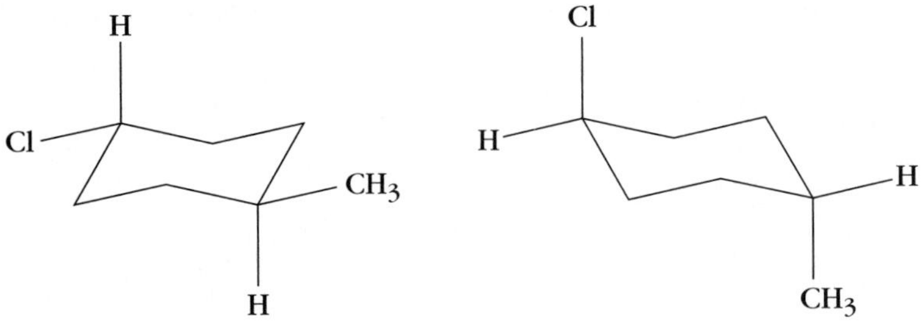

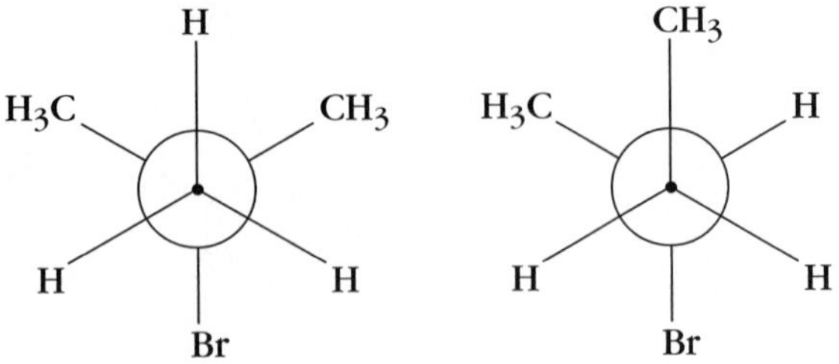

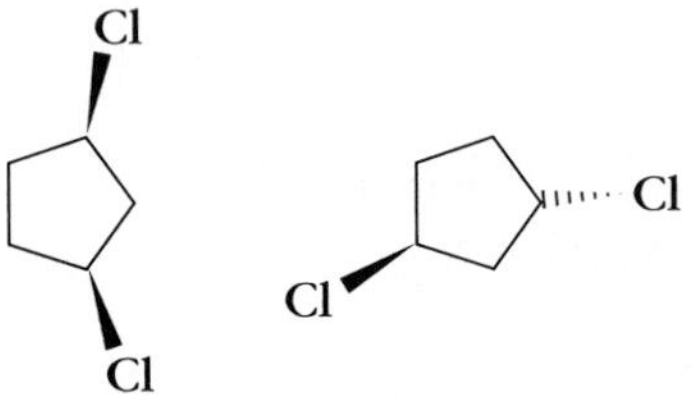

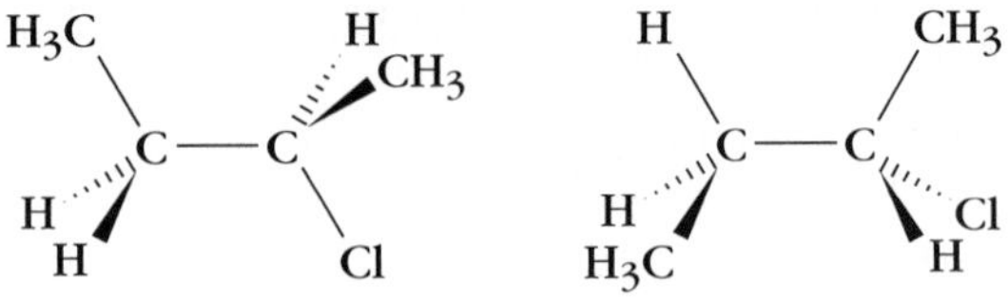

    (a)   Two chair conformations that can be interconverted by ring flipping

    (b)   Two chair conformations that can't be interconverted by ring flipping

4.29   Determine which of the following pairs of structures represent conformers or stereoisomers.

(b)

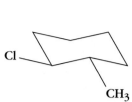

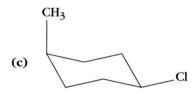

(c)

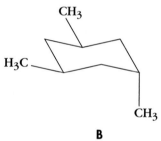

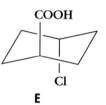

4.30 Arrange the following eight structures into four pairs of conformational isomers.

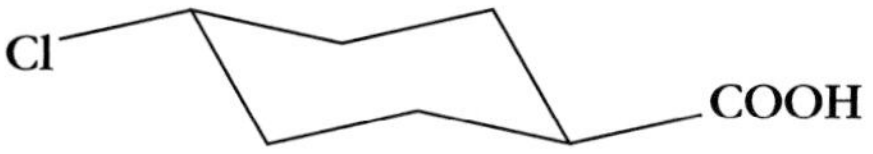

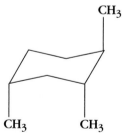

A

E

B

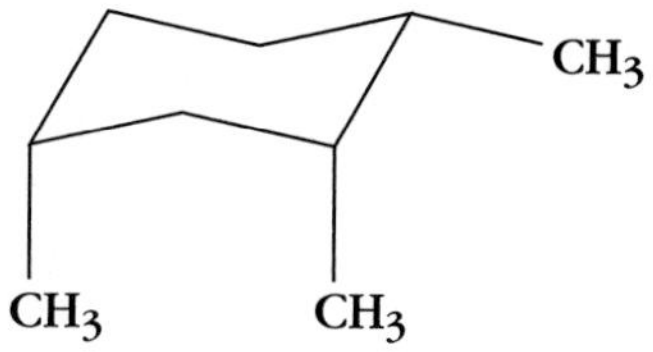

F

C

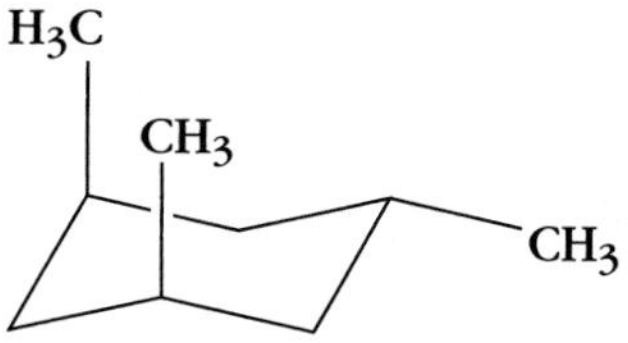

G

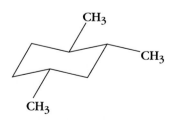

D

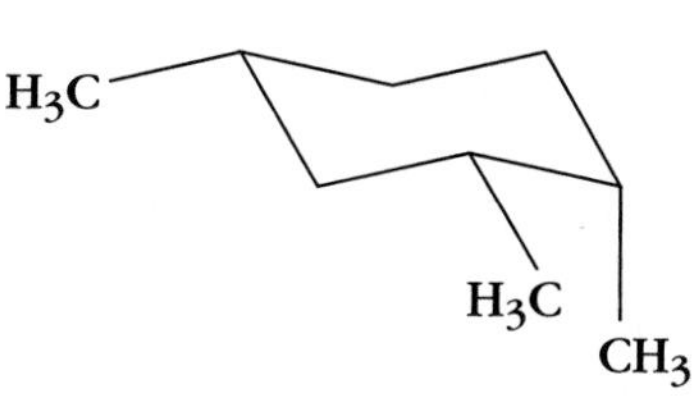

H

4.31 *trans*-Decalin is 2.7 kcal/mol (11.3 kJ/mol) more stable than *cis*-decalin. Based on this information and the data in Table 4.1, estimate the relative stabilities of *cis*- and *trans*-9-methyldecalin.

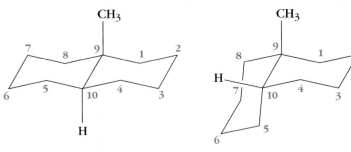

trans-9-Methyldecalin   *cis*-9-Methyldecalin

CHAPTER 5

INTRODUCTION to SPECTROSCOPY

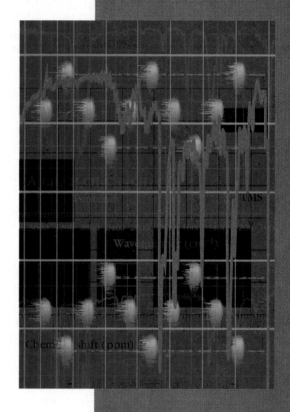

Horse racing fans across the country were shocked when on May 4, 1968, Dancer's Image, the first horse across the finish line at the 94th running of the Kentucky Derby, was disqualified because chemists reported that his postrace urine test showed traces of the pain-killing drug phenylbutazone, commonly known as Butazolidin.

Phenylbutazone has been used to treat rheumatoid arthritis and related disorders in humans. It is also used in equine medicine to reduce inflammation and pain. While it may be given to injured thoroughbreds in order to allow them to continue training, using it to allow a thoroughbred to compete at race tracks that permit parimutuel wagering is strictly forbidden.

The detection of phenylbutazone in the postrace test of Dancer's Image's urine is an example of one of the most important things that chemists do, namely to determine the structure of one or more compounds in a complex mixture by the use of **spectroscopy,** the study of the interactions of matter with electromagnetic radiation. How chemists use spectroscopy is introduced in this chapter.

Before we can use spectroscopy to determine chemical structures, we must first have a pure sample of the compound. Thus this chapter begins by explaining how chemists isolate and purify compounds. After a review of electromagnetic radiation, we learn how to use two spectroscopic techniques, infrared and nuclear magnetic resonance spectroscopy, to determine the structures of compounds.

5.1   Detection, Isolation, and Purification of the Components of a Mixture of Compounds

Knowing the structure of an organic compound is important because we can use that structure to predict its chemical reactions. Before we can determine its structure, however, we must have a pure sample of the compound. Among the techniques that can be used to isolate and purify organic compounds, the two simplest are crystallization and distillation.

Crystallization is a method of purifying a mixture of solids based on their differences in solubility in a suitable solvent. The crude solid is dissolved in a min-

Figure 5.1
A simple distillation apparatus.

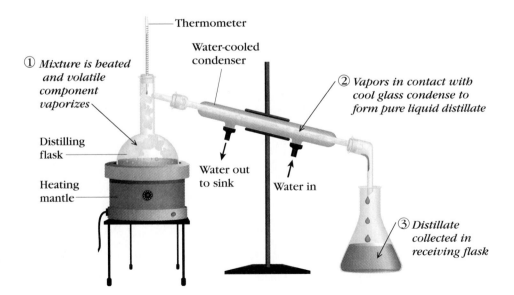

imum amount of a hot solvent, then the resulting solution is allowed to cool slowly and undisturbed. As the solution cools, crystals of the least soluble solid slowly form and precipitate, leaving the more soluble solids dissolved in solution. The crystals are isolated from the solution by filtration. Often crystallization must be repeated several times before pure crystals are obtained. Crystallization is most effective when the mixture in question contains one component in large excess (> 90%).

Distillation is a method of purifying volatile liquids based on their differences in boiling points. The apparatus used for distillation is shown in Figure 5.1. The crude liquid is placed in the distillation flask and heated to boiling. Vapors of the lowest-boiling liquid rise up the column to the condenser where they are condensed and collected in a receiving flask. Increasing the heat causes the next-lowest-boiling liquid to boil; its vapors follow the same path and are collected in a different receiver flask. Thus the liquids distill in order of their boiling points with the lowest-boiling liquid distilling first.

The ability to separate two liquids by distillation depends on the nature of the distillation column. In practice, liquids that differ by about 10 °C in boiling point can be separated in the laboratory with a simple apparatus such as the one shown in Figure 5.1.

Distillation is widely used in industry. Crude oil, for example, is separated into fractions of different boiling point ranges by distillation, as we discussed in Section 2.11. Alcoholic beverages such as scotch, whiskey, and bourbon are called distilled spirits because they are distilled several times during their production.

If distillation or crystallization can't provide the necessary purification of a mixture, the next step is to try **chromatography.** A common type of chromatography is called **column** or **liquid chromatography,** which uses an apparatus like the one shown in Figure 5.2. The column is packed with a solid material called the **stationary phase.** The stationary phase is usually silica gel (hydrated SiO_2) or alumina (Al_2O_3).

The mixture to be separated is dissolved in a suitable solvent. This solution is then placed at the top of the stationary phase in the glass tube. More solvent,

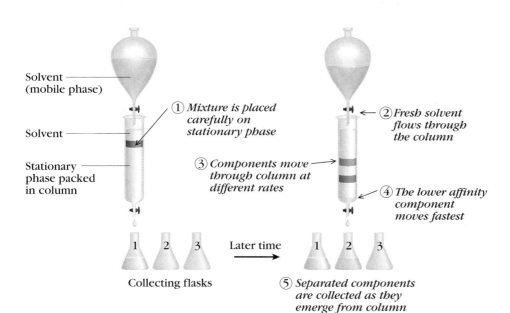

Solvent (mobile phase)

Solvent

Stationary phase packed in column

① *Mixture is placed carefully on stationary phase*

② *Fresh solvent flows through the column*

③ *Components move through column at different rates*

④ *The lower affinity component moves fastest*

Collecting flasks

Later time

⑤ *Separated components are collected as they emerge from column*

Figure 5.2
A simple chromatography column consists of a vertical glass tube open at the top with a stopcock at the other end. The glass tube is packed with a solid stationary phase such as silica gel (hydrated SiO_2) or alumina (Al_2O_3).

called the **mobile phase,** is continually added to the top of the stationary phase so that the mixture slowly passes through the column. The rate at which the mobile phase flows through the column is regulated by the stopcock. Best results are obtained by allowing the mobile phase to emerge (elute) dropwise from the bottom of the column.

Column chromatography separates mixtures because the intermolecular attractions between the various compounds of the mixture and the stationary phase differ. Thus the amount of time it takes for a compound to pass through the column strongly depends on the attractions between it and the stationary phase. The strongest attractions are generally between polar- and hydrogen-bonded molecules and the stationary phase. These molecules are held more tightly so they move down the column more slowly than nonpolar compounds.

It is fairly simple to separate compounds that differ in polarity by column chromatography. For example, a mixture of 2-butanol and 2-chlorobutane is easily separated by this technique. The relatively nonpolar haloalkane passes down the column more quickly than the more polar alcohol.

It is more difficult to separate compounds whose attractions to the stationary phase are very similar, such as compounds with the same functional group. In order to do this, several improvements have been made to the apparatus shown in Figure 5.2. First, very small spherical particles of uniform size are used as the stationary phase because they provide a larger surface area for better adsorption. They can be packed closer together so they provide a large surface area in a small column space. Second, high-pressure pumps are used to force the mobile phase through the tightly packed column. Finally, special detectors are used to detect the appearance of material as it elutes from the column. The technique that uses these improvements is called **high performance liquid chromatography (HPLC).** Figure 5.3 shows the results of HPLC analysis of a mixture of fatty acids obtained by the hydrolysis of a natural fat. Notice that the fatty acids differ only in the length of their carbon chains.

Vapor phase or **gas chromatography** is another technique frequently used to analyze the composition of mixtures. Gas chromatography is different from column chromatography or HPLC because it uses a gas for the mobile phase, rather than a liquid. The gas used as the mobile phase, which is usually molecular nitrogen or helium, is called the *carrier gas.* Gas chromatography requires an instrument called a *gas chromatograph,* which is shown schematically in Figure 5.4.

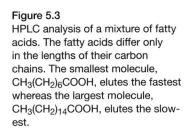

OH
|
$CH_3CHCH_2CH_3$

2-Butanol

Cl
|
$CH_3CHCH_2CH_3$

2-Chlorobutane

Figure 5.3
HPLC analysis of a mixture of fatty acids. The fatty acids differ only in the lengths of their carbon chains. The smallest molecule, $CH_3(CH_2)_6COOH$, elutes the fastest whereas the largest molecule, $CH_3(CH_2)_{14}COOH$, elutes the slowest.

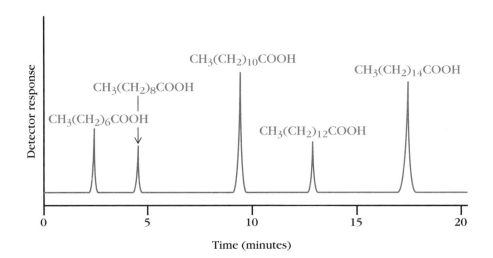

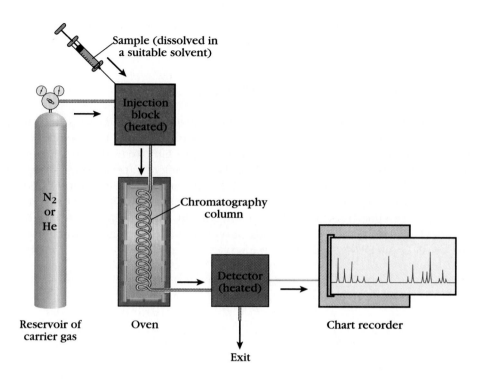

Figure 5.4
Schematic of a gas chromatograph. The important parts of a gas chromatograph are the reservoir of carrier gas, the sample injection block, the oven (containing the chromatography column), the detector, and the chart recorder.

Experimentally, the mixture of compounds is dissolved in a suitable solvent and the resulting solution is injected by syringe into the heated injection block of the gas chromatograph. The sample is instantly vaporized and swept through the heated chromatography column by the carrier gas. Special detectors detect the appearance of the components as they elute from the column and indicate their presence as a peak on a chart recorder.

Gas chromatography is an extremely powerful method of separating compounds of similar structure. Figure 5.5 shows, for example, the results of an analysis of a mixture of hydrocarbons containing four carbon atoms.

Once a pure sample of a compound is obtained, we can proceed to determine its structure. However, most of the modern techniques used to determine struc-

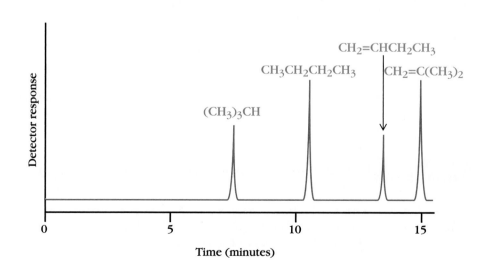

Figure 5.5
Gas chromatography analysis of a mixture of C_4 hydrocarbons. Even though these compounds have similar structures and molecular weights, gas chromatography can separate all four of them.

tures depend on the interaction of electromagnetic radiation and matter, so we begin by briefly reviewing electromagnetic radiation.

5.2 Electromagnetic Radiation

Visible light is the form of electromagnetic radiation most familiar to us. Other types of electromagnetic radiation that have found common use include x-rays (used in medicine and dentistry), ultraviolet (UV) light (emitted by sunlamps), infrared (IR) radiation (emitted by heat lamps), microwave radiation (used in microwave ovens), and radio waves (used to carry radio and television signals).

One model scientists have for electromagnetic radiation is a wave travelling at the speed of light. This description allows us to characterize electromagnetic radiation according to its **wavelength, λ,** which is the distance between successive crests or successive troughs of the wave.

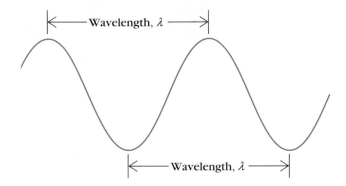

Different types of electromagnetic radiation have different wavelengths. The wavelengths, for instance, of x-rays are about 10^{11} times shorter than the wavelengths of AM radio waves.

Another way to describe a wave is by its **frequency, ν,** which is the number of wave crests or troughs that pass a given point in one second. Frequency has the unit s^{-1}, which is given the special name **hertz (Hz).**

$$1 \text{ Hz} = 1 \text{ s}^{-1}$$

For example, a frequency of 25 Hz means that 25 wave crests or complete cycles (1 wavelength) pass a given point in one second. Thus 25 Hz is equal to 25 cycles per second.

Wavelength and frequency are inversely related by the equation:

$$\lambda = c/\nu$$

where c is the speed of light (2.99×10^8 m/s), λ is the wavelength in meters, and ν is the frequency in hertz.

Exercise 5.1   What is the wavelength equivalent in meters of each of the following?

(a) Red light of 4.41×10^{14} Hz
(b) A radio station that broadcasts at 102.7 megahertz (MHz, 1 MHz = 10^6 Hz)
(c) 60 cycle-per-second alternating electric current

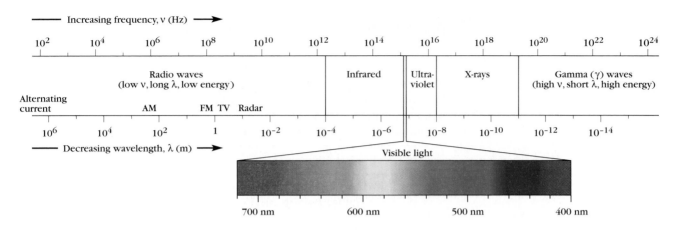

Figure 5.6
The electromagnetic spectrum. It ranges from short wavelength (high frequency) gamma waves to long wavelength (low frequency) alternating current.

The complete range of electromagnetic waves is called the **electromagnetic spectrum** and is shown in Figure 5.6. Wavelengths range from the very short, 10^{-12} m (1 picometer = 1 pm), to the very long, 10^6 m (1 megameter = 1 Mm). At the short wavelength end of the spectrum are the gamma rays and x-rays. At the long wavelength end of the spectrum are the radio waves. Our eyes are sensitive only to visible light, which occupies a very narrow portion of the spectrum, $\lambda = 4 \times 10^{-7}$ to $\lambda = 7.5 \times 10^{-7}$ m.

The relationship between the frequency of electromagnetic radiation and its energy is expressed by the following equation:

$$\text{Energy} = h\nu = hc/\lambda$$

where ν is the frequency of the electromagnetic radiation, λ is its wavelength, c is the speed of light, and h is called Planck's constant. The value of Planck's constant is 1.58×10^{-37} kcal s/mol (6.63×10^{-37} kJ s/mol). Using this equation, we can calculate the energy of electromagnetic radiation of a particular frequency or wavelength. Sample Problem 5.1 demonstrates how.

Sample Problem 5.1 | Calculate the energy in kcal/mol corresponding to yellow light of frequency 5.50×10^{14} Hz.

Solution:

$$\begin{aligned}
\text{Energy} &= h\nu \\
&= (1.58 \times 10^{-37} \text{ kcal s/mol})(5.50 \times 10^{14} \text{ s}^{-1}) \\
&= 8.69 \times 10^{-23} \text{ kcal/mol}
\end{aligned}$$

The energy of electromagnetic radiation is directly proportional to its frequency but inversely proportional to its wavelength. As a result, high-energy electromagnetic radiation has high frequencies and short wavelengths, whereas low-energy radiation has low frequencies and long wavelengths, as shown in Figure 5.6.

Exercise 5.2	Calculate the energy in kcal/mol that corresponds to each of the following.

(a) Infrared radiation with $\lambda = 2.3 \times 10^{-3}$ cm
(b) X-rays with $\nu = 10^{17}$ Hz
(c) Ultraviolet radiation with $\nu = 6.5 \times 10^{15}$ Hz
(d) Visible radiation with $\lambda = 450$ nm
(e) A shortwave radio wave with $\lambda = 1$ m

Exercise 5.3	Arrange each group of three types of radiation in order of increasing energy.

(a) Radio waves, x-rays, and visible light
(b) Infrared light, ultraviolet light, and visible light
(c) Radiation with $\nu = 10^5$ Hz, radiation with $\nu = 10^{26}$ Hz, radiation with $\nu = 10^{14}$ Hz
(d) Radiation with $\lambda = 10$ m, radiation with $\lambda = 10$ nm, radiation with $\lambda = 10^6$ m
(e) Radiation with $\lambda = 10^{-3}$ m, radiation with $\nu = 10^{26}$ Hz, radiation with $\lambda = 10^{-9}$ m

An organic compound placed in a beam of electromagnetic radiation may absorb some of the radiation. Absorption of electromagnetic radiation in the infrared region is particularly important in organic chemistry.

5.3   Absorption of Infrared Radiation

Absorption of electromagnetic radiation by molecules is measured experimentally by means of instruments called *spectrometers*. An instrument that measures the radiation absorbed in the infrared (IR) region of the electromagnetic spectrum is called an **infrared (IR) spectrometer.** An infrared spectrometer is shown schematically in Figure 5.7.

Figure 5.7
Schematic diagram of an infrared spectrometer. The important parts of a spectrometer are the infrared light source, the sample and reference cells, the chopper, the monochromator, the detector, and the chart recorder.

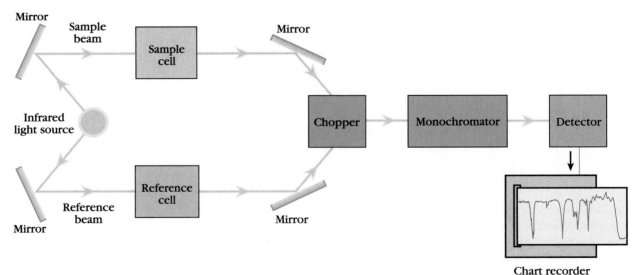

Chart recorder

The IR light is split by mirrors into two equal beams. One beam of light passes through a tube or cell containing the sample, usually in a solvent, whose spectrum is to be measured. The other, called the *reference beam,* travels through the reference tube or cell that contains only the solvent. The function of such a double beam operation is to measure the difference in intensities between the two beams at each wavelength.

After passing through the two cells, both beams are reflected to a rotating mirror called a *chopper.* As the chopper rotates, it causes the sample and reference beams to be reflected alternately to the monochromator. The monochromator separates each beam into its component wavelengths. The detector measures the difference in intensities between the reference and sample segments of the beam at each wavelength. This difference in intensities is passed along to the recorder, which records the **infrared spectrum** (IR spectrum) of the compound. An IR spectrum represents the amount of IR radiation absorbed by the sample as a function of the wavelength.

Modern IR spectrometers use a Michelson interferometer rather than a monochromator to separate the IR radiation into its component wavelengths. An interferometer can be thought of as an instrument that encodes the initial wavelengths into a special form that the detector can observe. This encoding produces an interferogram. When a compound is placed in the beam (either before or after the interferometer), it absorbs particular wavelengths so that their intensities are reduced in the interferogram. The data from the interferogram is fed into a computer where it undergoes Fourier transformation to produce the IR spectrum. Fourier transformation is a mathematical means of sorting out the individual wavelengths again for the final presentation of the IR spectrum. IR spectrometers that use this technique are called **Fourier Transform Infrared** (abbreviated **FTIR**) spectrometers. All IR spectra in this text have been recorded by a FTIR spectrometer.

IR spectra can be determined for solids, liquids, or gases. IR spectra of solids are usually measured as a dispersion in either anhydrous potassium bromide or in mineral oil. IR spectra of pure liquid samples are often obtained by placing the sample between two small optically clear flat disks of sodium or potassium chloride, which do not absorb in the IR region. IR spectra can also be obtained for samples in solution. The only requirement is that the solvent have few absorptions in the IR region. In practice, this means that chloroalkanes, such as carbon tetrachloride (CCl_4) and methylene chloride (CH_2Cl_2), are the most widely used solvents for IR spectra.

Let's now take a look at an IR spectrum.

5.4   The Infrared Spectrum

An IR spectrum is a plot of the amount of IR radiation absorbed by a compound as a function of the wavelength of IR radiation passed through the compound. See, for example, the infrared spectrum of pentane in Figure 5.8.

The left vertical scale of the spectrum is the **percent transmittance,** which is defined as the percent of light entering the sample cell that is transmitted to the detector. One hundred percent transmittance means that the sample does not absorb any light, while zero transmittance means that all the light is absorbed. Because 100% transmittance is located at the top of the spectrum, absorption bands in the IR spectrum project downward.

The **intensity** of an absorption band is related to its percent transmittance.

Figure 5.8
Infrared spectrum of pentane.

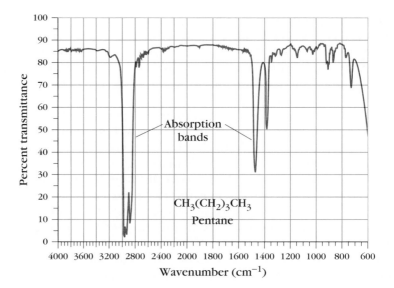

The smaller the percent of transmittance, the more intense the absorption band. The absorption bands farthest to the left in Figure 5.8 are intense because their transmittance is only 5% to 10%.

The wavelengths of the light passed through the sample are recorded on the horizontal scale. Instead of recording wavelengths in meters, a unit called wavenumbers (cm^{-1}) is used. The wavenumber (cm^{-1}) of light is the number of its wavelengths in a centimeter and is related to its wavelength (in centimeters) as follows:

$$\text{Wavenumber } (\nu) = 1/\lambda \text{ (in cm)}$$

An IR spectrum records absorptions from 4000 cm^{-1} to about 600 cm^{-1}. Another horizontal scale is sometimes found at the top of IR spectra. This scale is in units of micrometers (μm). Chemists prefer wavenumbers as the units for reporting the positions of absorption bands because wavenumbers (ν), unlike wavelengths (λ), are directly proportional to energy. Thus the higher the wavenumber, the higher its energy:

$$\text{Energy} = hc/\lambda = hc\nu$$

In Figure 5.8, as in all IR spectra, the larger wavenumbers are located toward the left, so the energy of the absorbed light also increases toward the left. In Figure 5.8, for instance, the absorption band at approximately 1500 cm^{-1} is at higher energy than the one at about 1400 cm^{-1}, but both of the absorptions are at lower energy than the ones near 3000 cm^{-1}.

The infrared spectrum of pentane (Figure 5.8) contains many absorption bands. How can we relate these bands to the structure of pentane?

5.5 Functional Groups Have Characteristic Infrared Absorption Bands

The IR spectrum of any organic compound contains many absorption bands. To understand how we can use some of these bands to determine structure, we must examine the IR spectra of a number of compounds that contain the same functional group. Consider the alcohol functional group, for instance. The IR spectra of four alcohols are shown in Figure 5.9.

■ One micrometer is 10^{-6} m.

■ In Section 2.14, C, we learned that alcohols are compounds that contain a hydroxy group bonded to an sp³ hybrid carbon atom.

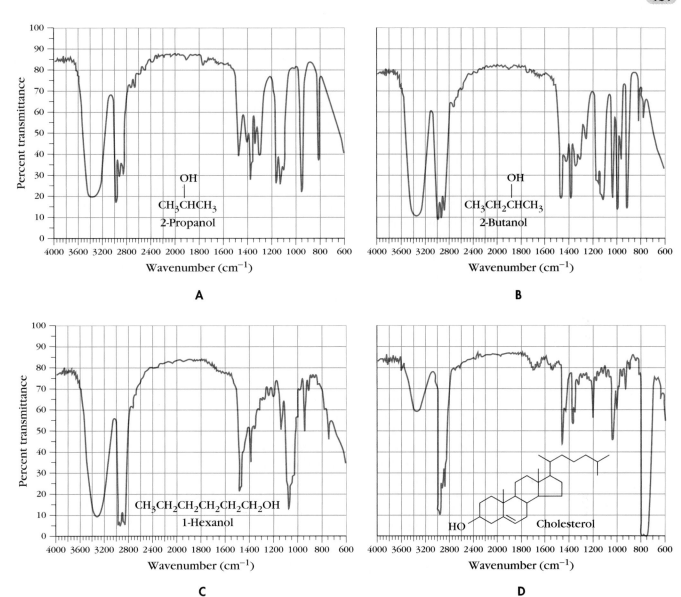

Figure 5.9
IR spectra of 2-propanol, 2-butanol, 1-hexanol, and cholesterol.

The first thing we notice is that the four IR spectra in Figure 5.9 are similar but different. The sizes, shapes, and positions of the absorption bands are similar but not exactly the same. First, they all contain a number of absorption bands in the 3000 to 2800 cm^{-1} region of the spectrum just like pentane. Second, the alcohols all have a large, broad absorption band centered near 3300 cm^{-1} that is absent in the spectrum of pentane. If we were to examine the IR spectra of several hundred alcohols, instead of just four, we would find the same thing. All alcohols have an absorption band centered near 3300 cm^{-1} that is absent in the spectra of alkanes.

On the basis of the comparisons of the IR spectra of thousands of compounds containing various functional groups, chemists have concluded that *most functional groups have characteristic infrared absorption bands whose location varies little in the infrared spectra of all compounds that contain that functional group.*

For example, the absorption bands that appear in the 3000 to 2800 cm^{-1} region of the IR spectra of alkanes are due to the presence of C—H bonds. Because most alcohols also have C—H bonds, we would expect to find a similar absorption band in this region of the IR spectra of alcohols. The alcohols whose IR spectra are shown in Figure 5.9 have C—H bonds, and their IR spectra show absorption bands in the 3000 to 2800 cm^{-1} region. *The large majority of organic compounds contain C—H bonds, so their IR spectra contain one or more absorption bands in the 3000 to 2800 cm^{-1} region.*

In alcohols, the absorption band centered near 3300 cm^{-1} is due to the O—H bond of the hydroxy group (—OH). Alkanes do not have this functional group, so we would not expect to see absorption bands near 3300 cm^{-1} in the IR spectra of alkanes. In Figure 5.8, for example, there is no absorption band in the 3400 to 3200 cm^{-1} region of the IR spectrum of pentane.

Chemists use IR spectra of compounds in this way to determine their functional group or groups. We look for absorption bands in particular regions of the IR spectrum that are characteristic of the various functional groups. The presence of such bands indicates that the compound contains a particular functional group. Conversely, the absence of an absorption band characteristic of a particular functional group indicates that the compound does not contain that group.

Chemists sometimes use IR spectra to identify a particular compound. This is possible because, as we noted earlier, the IR spectra of compounds with the same functional group are similar but *not identical.* In Figure 5.9, for example, the IR spectra of the alcohols are particularly different in the region from 1400 to 625 cm^{-1}. This region usually contains a very complex series of absorption bands, and correlations between any of them and a specific functional group are difficult to make. It turns out, however, that this complex pattern of absorption bands is unique for every compound. Thus these absorption patterns serve as spectroscopic fingerprints and the 1400 to 625 cm^{-1} region of an IR spectrum is called the **fingerprint region.**

It was the fingerprint region that provided definite evidence of the presence of phenylbutazone in the urine of Dancer's Image. The fingerprint region of the IR spectrum of the substance obtained from the urine sample was identical to that in the IR spectrum of an authentic sample of phenylbutazone. This proved that phenylbutazone was in the urine sample.

Chemists use IR spectra most frequently to identify the type of functional group in a compound. This requires a knowledge of the characteristic absorption bands of most functional groups. This information is summarized in the following sections.

A   Hydrocarbons

Hydrocarbons, as we learned in Chapter 2, are compounds that contain only carbon and hydrogen atoms. At this time, we will examine the characteristic IR absorption bands of only three families of hydrocarbons, the alkanes, the alkenes, and the alkynes. Recall that alkanes contain only C—H bonds and C—C single bonds, alkenes have a carbon-carbon double bond (C=C), and alkynes have a carbon-carbon triple bond (C≡C).

$$CH_3CH_2CH_2CH_2CH_3 \qquad CH_3CH=CHCH_2CH_3 \qquad HC≡CCH_2CH_2CH_3$$

| Pentane | 2-Pentene | 1-Pentyne |
| (an alkane) | (an alkene) | (an alkyne) |

The IR spectra of alkanes contain absorption bands in only two regions, as shown by the IR spectrum of pentane in Figure 5.8. One group of absorption bands is found between 2960 and 2850 cm^{-1}, and the other is found between 1500 and 1350 cm^{-1}. Since almost all organic compounds contain C—H bonds, these absorption bands are present in the IR spectra of almost all organic compounds.

In addition to the alkane C—H absorption bands, the IR spectra of alkenes contain two absorption bands that are characteristic of a carbon-carbon double bond. Both appear in the IR spectrum of 1-pentene shown in Figure 5.10. One is observed at 3080 cm^{-1} and is due to the presence of sp^2 hybrid carbon-hydrogen bonds. The other is observed at 1645 cm^{-1} and is due to the carbon-carbon double bond.

The IR spectra of alkynes contain a characteristic absorption band at 2260 to 2100 cm^{-1} due to the carbon-carbon triple bond (C≡C). This band is found at 2120 cm^{-1} in the spectrum of 1-hexyne shown in Figure 5.10, *B*. In addition, terminal alkynes, which have a hydrogen atom bonded to a carbon atom of the triple bond, have another absorption band at 3350 to 3260 cm^{-1} due to the sp hybrid carbon-hydrogen bond. This absorption appears at 3300 cm^{-1} in the IR

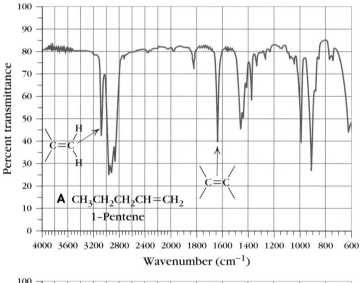

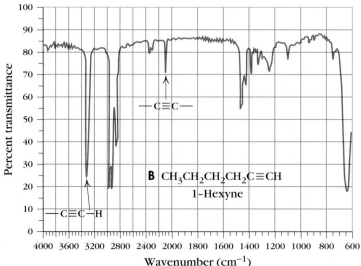

Figure 5.10
Infrared spectra of 1-pentene and 1-hexyne.

spectrum of 1-hexyne in Figure 5.10, *B*. This absorption band is not observed in alkynes that do not have a hydrogen atom bonded to the carbon atom of a triple bond. The IR spectrum of 2-pentyne ($CH_3C{\equiv}CCH_2CH_3$), for instance, does not have this absorption.

B   Haloalkanes

Haloalkanes are compounds that contain a halogen atom bonded to an sp^3 hybrid carbon atom. Absorption bands due to the carbon-halogen bond appear in the fingerprint region from 1350 to 670 cm^{-1} of the IR spectra of haloalkanes. Because other absorptions interfere in the fingerprint region, these bands cannot be used to verify the presence of halogens in an organic compound. Additional data from other techniques, such as nuclear magnetic resonance (NMR) spectroscopy and mass spectrometry, are needed to verify the presence of halogens in an organic compound.

C   Alcohols

We learned earlier (see Figure 5.9) that the characteristic feature of the IR spectra of alcohols is a broad, strong absorption band centered near 3300 cm^{-1} caused by the O—H bond.

D   Amines

Figure 5.11
Comparison of the 4000 to 2400 cm^{-1} region of the IR spectra of a primary alcohol, a primary amine, a secondary amine, and a tertiary amine. Primary amines have two sharp absorption bands, secondary amines have one sharp absorption band, and tertiary amines have no absorption band in the 3200 to 3500 cm^{-1} region.

The functional group of amines is a nitrogen atom bonded to one, two, or three organic groups.

$$CH_3CH_2NH_2 \qquad (CH_3CH_2)_2NH \qquad (CH_3CH_2)_3N$$

Ethylamine
(An example of
a primary amine)

Diethylamine
(An example of
a secondary amine)

Triethylamine
(An example of
a tertiary amine)

A $CH_3CH_2CH_2CH_2CH_2OH$ **B** $CH_3CH_2CH_2CH_2NH_2$ **C** $(CH_3CH_2)_2NH$ **D** $(CH_3CH_2)_2NCH_3$

1-Pentanol
A primary alcohol

N-Butylamine
A primary amine

Diethylamine
A secondary amine

N, *N*-Diethylmethylamine
A tertiary amine

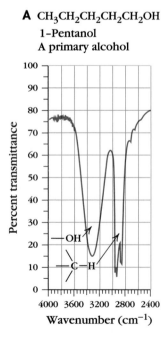
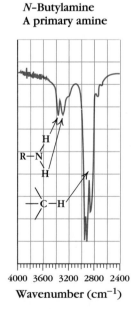
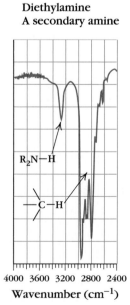
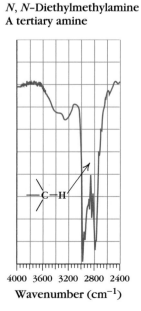

The IR spectra of amines that contain an N—H bond show characteristic absorption bands in the 3300 to 3500 cm^{-1} range. While alcohols also have a characteristic absorption band in this region, the amine absorption is much smaller and much sharper. Primary amines have two sharp absorption bands, and secondary amines have a single sharp absorption band in this region. Tertiary amines do not have N—H bonds, so they do not have absorption bands in this region. These differences are illustrated by the spectra in Figure 5.11.

E   Ethers

Ethers are compounds that contain an oxygen atom bonded to two carbon atoms. Diethyl ether and tetrahydrofuran (THF) are examples of ethers.

$$CH_3CH_2OCH_2CH_3$$

Diethyl ether   Tetrahydrofuran (THF)

The IR spectra of simple ethers have an intense absorption band in the region 1050 to 1250 cm^{-1} due to the C—O bonds. This absorption band is found at 1110 cm^{-1} in the spectrum of diethyl ether in Figure 5.12. Alcohols (ROH) and esters (RCOOR) also have C—O bonds, so they too have an absorption band in the region 1050 to 1250 cm^{-1}. The absorption band due to the C—O bond falls in the fingerprint region, so in complex molecules it is often difficult to distinguish it from other strong absorption bands in the region.

F Carbonyl-Containing Compounds

A carbon-oxygen double bond (C=O) is called a carbonyl group. We will consider only the IR absorption bands of the carbonyl group in aldehydes, ketones, and carboxylic acids at this time. The following compounds are representative of these three functional groups.

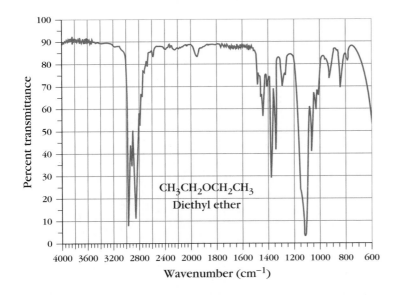

Figure 5.12
Infrared spectrum of diethyl ether.

$$CH_3CH_2CH_2CH_2C \overset{O}{\underset{H}{\big\backslash}} \qquad CH_3\overset{\overset{O}{\|}}{C}CH_2CH_2CH_3 \qquad CH_3CH_2CH_2CH_2C \overset{O}{\underset{OH}{\big\backslash}}$$

Pentanal	2-Pentanone	Pentanoic acid
(an aldehyde)	(a ketone)	(a carboxylic acid)

One of the most characteristic absorption bands in an IR spectrum is the one due to the presence of the carbonyl group. This strong band is found at wavenumbers from 1750 to 1680 cm^{-1}.

The spectrum of pentanal, a typical aliphatic aldehyde, is shown in Figure 5.13, *A*. The spectrum contains an intense absorption band at 1725 cm^{-1} due to the carbonyl group. Aldehydes have a hydrogen bonded to the carbon of the carbonyl group, and the presence of this C—H bond gives rise to absorptions at 2810 and 2715 cm^{-1}. Both of these bands are sharp but have a low intensity, and the band at around 2800 cm^{-1} is sometimes hidden by the absorptions of the aliphatic C—H bonds.

Figure 5.13
Infrared spectra of pentanal **(A)** and 2-pentanone **(B)**.

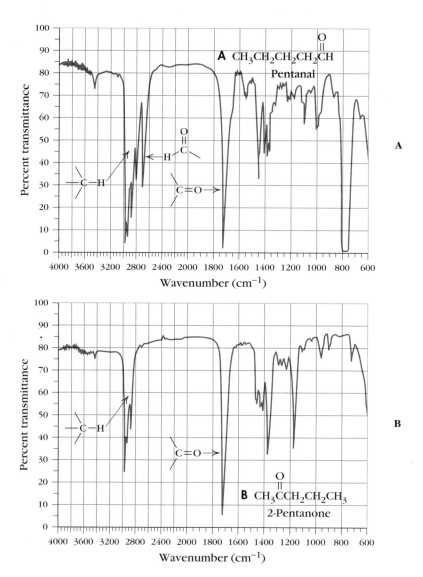

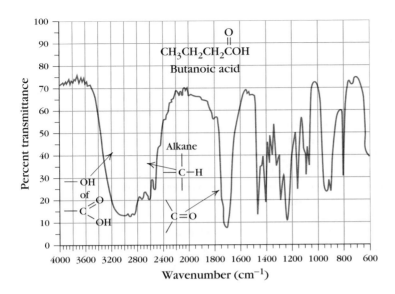

Figure 5.14
Infrared spectrum of butanoic acid. The absorption band due to the carbonyl group is at 1700 cm^{-1} while the band due to the —OH group overlaps with aliphatic C—H absorption bands to form a large absorption band that extends from 3500 to 2500 cm^{-1}.

In general, the absorption band of the carbonyl group of ketones is found in the region 1715 to 1700 cm^{-1}. In the spectrum of 2-pentanone shown in Figure 5.13, *B,* for example, the absorption band of the carbonyl group is found at 1715 cm^{-1}. Notice that the absorption bands due to the C—H bond on the carbonyl carbon atom present in aldehydes is missing from the spectrum of the ketone.

| Exercise 5.4 | Mineral oil is a mixture of a number of high molecular weight alkanes. Why can the IR spectra of ketones be obtained as a dispersion in mineral oil when we know that alkanes absorb in the IR region? |

The IR spectra of carboxylic acids have two characteristic absorption bands, as shown by the spectrum of butanoic acid in Figure 5.14. The first is the absorption band due to the carbonyl group at about 1700 cm^{-1}. The second and most distinctive feature is the broad absorption band due to the O—H bond of the carboxylic acid group, which begins at about 3400 cm^{-1} and extends to about 2400 cm^{-1}. The breadth of this absorption overlaps the aliphatic C—H absorption bands.

G   Summary of Characteristic IR Absorption Bands

The use of IR spectroscopy to identify the functional group in an organic compound seems to involve the memorization of a lot of numbers. In certain respects this is true. There are hundreds of characteristic absorption bands for different kinds of functional groups, and the detailed analysis of an IR spectrum is a complicated task. Rather than memorizing the wavenumbers, it's enough at this time to recognize characteristic regions of an IR spectrum where the absorption bands of specific functional groups are found. This information is often enough to identify the functional group in a compound from its IR spectrum. The various regions of IR absorption bands characteristic of several important functional groups are listed in Table 5.1. These regions are also set out in a correlation chart in Figure 5.15.

Table 5.1	Summary of the Characteristic IR Absorption Bands of Important Functional Groups*		
Range of Absorption Band Positions (cm^{-1})	Bond of the Functional Group Responsible for the Absorption Band	Remarks	
3400-3200	Alcohol O—H	Usually broad and intense	
3500-3300	Amine N—H	Sharp and low intensity	
3350-3260	Alkyne $\equiv$C—H	Sharp and medium intensity	
3080-3020	Alkene $=$C—H	Sharp and medium intensity	
3400-2400	Carboxylic acid —OH	Overlaps with alkane C—H absorption bands	
2820-2800 and 2720-2700	H—C of H—C$=$O	Two sharp bands of medium intensity	
2250-2100	Alkyne —C$\equiv$C	Usually sharp of medium to low intensity	
2260-2200	Nitrile —C$\equiv$N	Usually sharp of medium to low intensity	
1750-1730	Ester C$=$O	Intense band	
1730-1720	Aldehyde C$=$O	Intense band	
1720-1680	Carboxylic acid C$=$O	Usually broad and intense	
1715-1700	Ketone C$=$O	Intense band	
1670-1645	Alkene C$=$C	Usually sharp and medium to low intensity	
1250-1050	Ether C—O—C	Intense band	
1300-1050	Ester C—O—C	Usually broad and intense	

*Most organic compounds have absorption bands in the region 3000-2800 cm^{-1} due to alkane C—H bonds in addition to the characteristic IR absorption bands of the functional groups that they contain.

Figure 5.15
Correlation chart showing the location of characteristic IR absorption bands of several important functional groups.

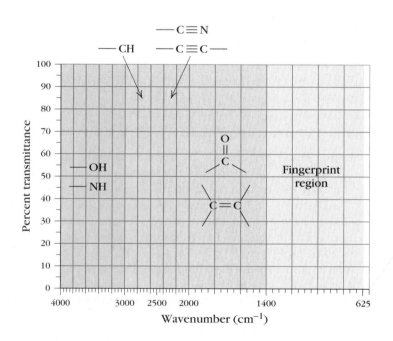

Exercise 5.5   Identify the functional group or groups in each of the following structures and the regions of the IR spectrum where each functional group will absorb.

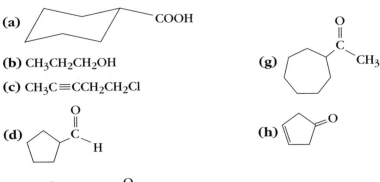

(a)

COOH

(b) CH₃CH₂CH₂OH

(c) CH₃C≡CCH₂CH₂Cl

(d)

(e)

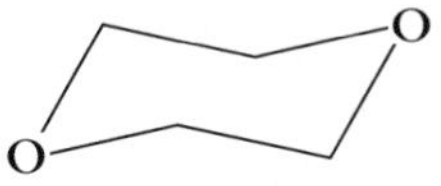

(f) CH₂=CHCH₂CH=CH₂

(g)

(h)

(i) 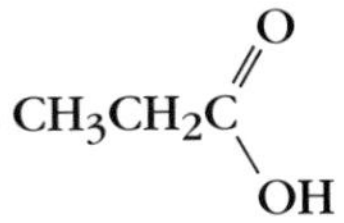 CH₃CH₂C

5.6   Interpreting Infrared Spectra

Chemists rarely determine the complete structure of a compound solely from its IR spectrum. Instead, an IR spectrum is generally used to determine the kind of functional group the compound contains. We can illustrate how this is done by analyzing the following IR spectrum to determine the functional group present.

Sample Problem 5.2   What functional group is present in the compound that gives the following IR spectrum?

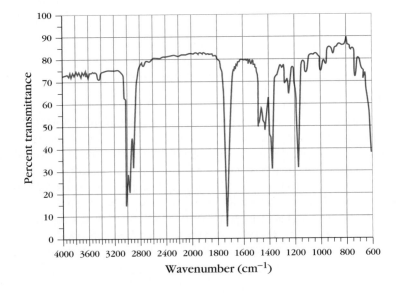

Solution:

In addition to the C—H absorption bands around 3000 cm^{-1} there is a carbonyl absorption at 1715 cm^{-1}. This means that the compound could be an aldehyde, ketone, or carboxylic acid. It can't be a carboxylic acid, however, because the broad distinctive absorption due to the —OH group of the carboxyl group (—COOH) in the region from 3400 to 2400 cm^{-1} is missing. It can't be an aldehyde, either, because absorptions due to the C—H bond on the carbonyl carbon atom of an aldehyde at about 2800 cm^{-1} and 2700 cm^{-1} are also missing. There is no absorption due to a carbon-carbon double bond near 1650 cm^{-1} and no sp^2 hybrid C—H absorption above 3000 cm^{-1} so it can't be an alkene. There is no absorption band at about 2100 cm^{-1} due to a carbon-carbon triple bond so it probably isn't an alkyne. We conclude, therefore, that the compound is probably a simple ketone.

Sample Problem 5.3   What functional group is present in the compound that gives the following IR spectrum?

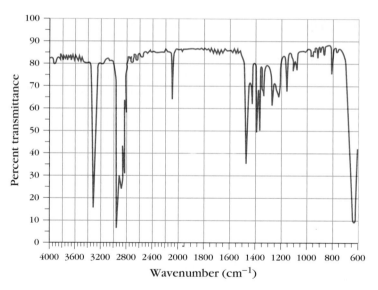

Solution:

The usual alkane C—H absorption bands appear around 3000 cm^{-1}. There is also a sharp absorption band at 2100 cm^{-1}, indicating the presence of a carbon-carbon triple bond. The absorption band at 3300 cm^{-1} is too sharp to be due to the —OH group of an alcohol and too intense to be due to the —N—H bond of a secondary amine. It indicates instead that the alkyne has a hydrogen atom bonded to a carbon atom of the triple bond ($\equiv$C—H). There are no absorption bands in the region 1600 to 1750 cm^{-1} so it can't be a carbonyl-containing compound or one containing a carbon-carbon double bond. We conclude that the compound is probably an alkyne of the structure RC$\equiv$C—H.

| Sample Problem 5.4 | What functional group is present in the compound that gives the following IR spectrum? |

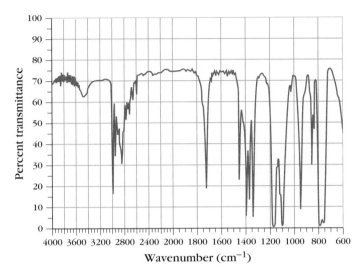

Solution:

The presence of an absorption band at 1730 cm^{-1} means that the compound contains a carbonyl group. The absence of a broad absorption band from 3400 to 2400 cm^{-1} means that the compound is not a carboxylic acid. The absorption bands at 2700 and 2800 cm^{-1} indicate the presence of a hydrogen atom bonded to the carbonyl carbon atom of an aldehyde. Therefore we conclude that the compound is probably an aldehyde.

Exercise 5.6   The IR spectra of four compounds are shown below. Determine the functional group in each of the compounds.

Spectrum **A**

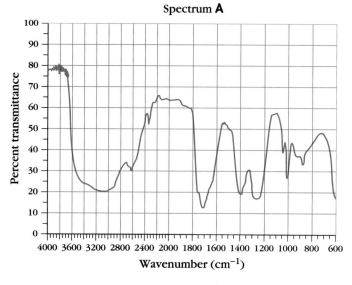

Spectrum **C**

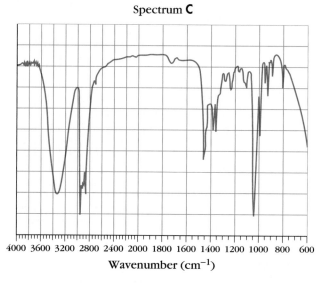

Spectrum **B**

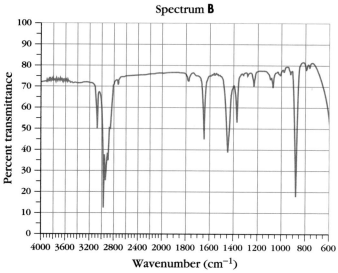

Spectrum **D**

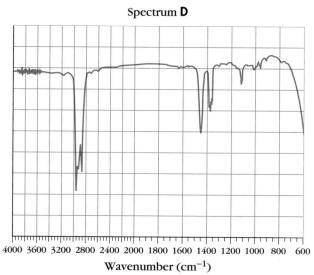

Sample Problem 5.5   What differences in the IR spectra of the following pair of compounds would allow you to distinguish between them?

$$CH_3(CH_2)_4CH_3 \text{ and } CH_2{=}CHCH_2CH_2CH_2CH_3$$

Solution:

One compound is an alkane; the other is an alkene. Alkenes have an absorption band in the 1680 to 1620 cm^{-1} region due to their double bonds. Alkenes with a double bond at the end of the chain also have two bands slightly above 3000 cm^{-1} due to their sp^2 hybrid C—H bonds. These latter bands are sometimes obscured by overlap with the absorption bands of sp^3 C—H bonds. Consequently the presence of bands in these regions distinguishes an alkene from an alkane because bands in these regions are absent in the alkane.

Exercise 5.7   What differences in the IR spectra of the following pairs of compounds would allow you to distinguish between them?

(a) 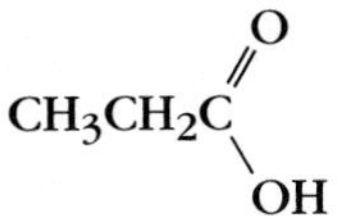 $CH_3CH_2C\overset{O}{\underset{OH}{\Vert}}$ and $CH_3CH_2C\overset{O}{\underset{H}{\Vert}}$

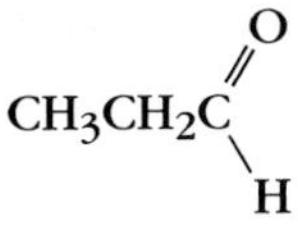

(b) 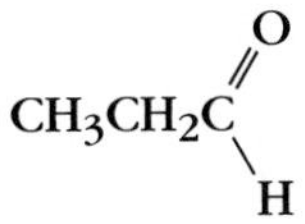 $CH_3CH_2C\overset{O}{\underset{H}{\Vert}}$ and $CH_3CH_2CH_2OH$

(c) $CH_3CH{=}CHCH_2CH_3$ and $CH_2{=}CHCH_2CH_2CH_3$

| Exercise 5.8 | A single IR spectrum is given for each of the following pairs of compounds. Match the compound with its infrared spectrum. |

(a) $CH_2{=}CH(CH_2)_3CH_3$ or $HC{\equiv}C(CH_2)_3CH_3$

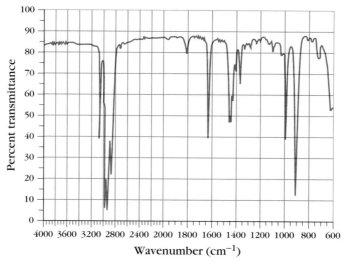

(b) $(CH_3)_2CHCH_2CH_2OH$ or $(CH_3)_2CHCH_2COOH$

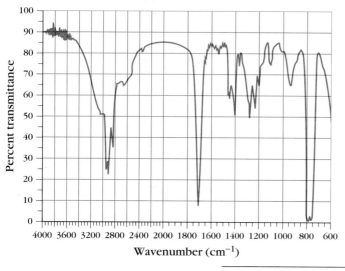

(c) $CH_3CH_2CH{=}CH_2$ or $CH_3CH_2CH_2CH_2OH$

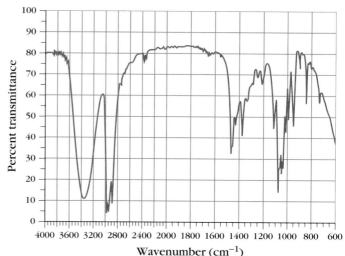

We've been introduced to the practical side of infrared spectroscopy by learning what to look for in a spectrum to determine the functional group or groups in an organic compound. Now let's learn why organic compounds absorb infrared radiation.

5.7   Absorption of Infrared Radiation and Molecular Vibrations

Within a molecule, the atoms have a certain freedom to move about. Rotations about single bonds are one of these motions, as we discussed in Chapter 4. Bond vibrations are another motion of atoms in molecules. During bond vibrations, atoms joined by a bond vibrate back and forth about an average bond length. Bond vibrations are of particular interest to organic chemists because such vibrations absorb infrared radiation. Let's examine the bond vibrations of a diatomic molecule, which is the simplest example of such vibrations.

Vibrational motion, just like the motion of electrons in atoms or molecules, is quantized. These quantized vibrational states for a diatomic molecule are usually represented on an energy-internuclear distance diagram such as the one shown in Figure 5.16. Figure 5.16 resembles Figure 1.7, which showed the energy changes as two hydrogen atoms combine to form a hydrogen molecule. In Figure 5.16, however, we have added vibrational quantum states.

At room temperature most of the molecules are in their lowest vibrational state. A model sometimes used to illustrate these vibrations is the stretching and compressing of a spring (the bond) connected to two balls (the atoms). This vibration has both an amplitude and a frequency, as shown in Figure 5.17.

Absorption of infrared radiation occurs only when its energy is exactly equal to the difference between one vibrational state (usually the lowest excited state) and an excited vibrational state. In Figure 5.16, the energy needed to excite the bond from its lowest vibrational state to the one of next higher energy is designated ΔE. This excitation can also be represented by our vibrating spring model, as shown in Figure 5.18.

The frequency of the vibration of a bond of a diatomic molecule increases with increasing bond strength and decreases with increasing mass of the two atoms. As a result, bonds between different atoms have different vibrational frequencies and absorb infrared radiation of differing energy.

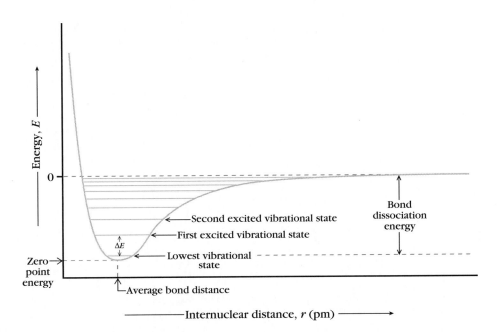

Figure 5.16
Energy–internuclear distance diagram showing the vibrational levels of a diatomic molecule.

Figure 5.17
The amplitude and frequency of the vibration of two balls attached to a spring. This model is meant to represent the bond vibrations of a diatomic molecule.

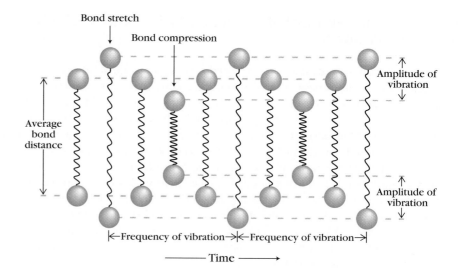

Figure 5.18
Infrared radiation of frequency ν has exactly the energy needed to excite the bond from its lowest vibrational state to an excited vibrational state. Notice that the absorption of infrared energy changes the amplitude of the vibration but not its frequency.

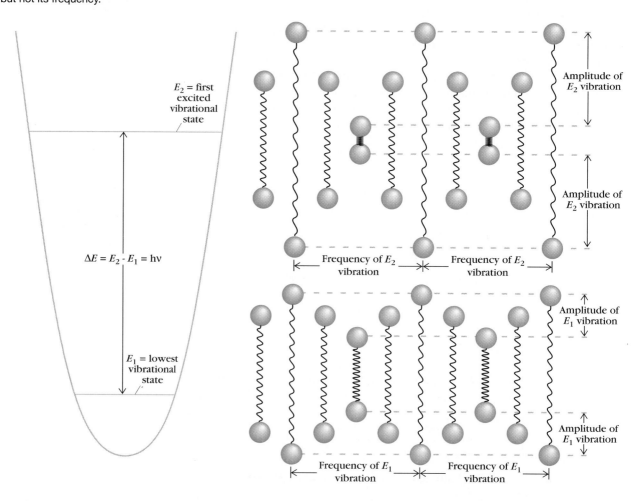

Molecules more complex than a diatomic molecule have many more possible vibrations. A nonlinear molecule with n atoms has $3n-6$ possible fundamental vibrations. Thus pentane, C_5H_{12}, has 45 possible infrared absorption bands. Not all of these bands are present in the IR spectrum of pentane as we can see from Figure 5.8. Some vibrations occur at the same wavelength, while others are not seen because the vibrations occur outside of the range recorded on the spectrum.

The $3n-6$ molecular vibrations of polyatomic molecules are of two types, stretching and bending. The vibrations illustrated in Figures 5.17 and 5.18 for diatomic molecules are stretching vibrations. Several examples of stretching and bending vibrations for atoms bonded to a tetrahedral carbon atom are shown in Figure 5.19.

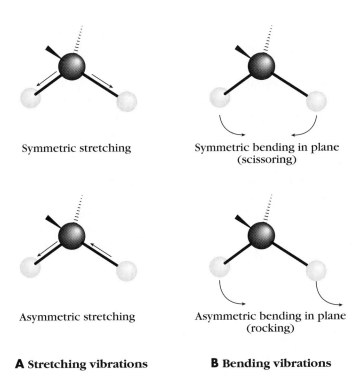

Symmetric stretching

Symmetric bending in plane
(scissoring)

Asymmetric stretching

Asymmetric bending in plane
(rocking)

A Stretching vibrations   **B** Bending vibrations

Figure 5.19
Examples of the molecular stretching and bending vibrations possible for atoms bonded to a tetrahedral carbon atom.

Infrared absorption bands due to the stretching vibrations of functional groups are the ones that are most characteristic and predictable. The positions of the absorption bands given in Table 5.1, for example, are due to excitation of stretching vibrations of the functional group. The absorption bands due to bending vibrations generally appear in the fingerprint region of the IR spectrum, so they are less useful for identifying functional groups.

Many molecules also absorb radiation of other wavelengths.

5.8 Absorption of Other Electromagnetic Radiation by Molecules

In Section 5.7, we learned that the range of energies associated with infrared radiation corresponds to the energies necessary to excite various molecular vibrations. Molecules can also absorb radiation of other wavelengths. The requirement for absorption of any radiation is always the same. Absorption of radiation occurs when its energy is exactly equal to the energy needed to excite a molecule from one energy state to another of higher energy.

There are three kinds of molecular excitations that are of particular interest to chemists because of the information they give about molecular structure. The first is absorption of infrared radiation, which was discussed in Sections 5.3 to 5.7. It is of interest to chemists because it tells us the kind of functional group or groups that are present in the molecule.

A second type is the excitation of the π electrons of alkenes and aromatic compounds. The amount of energy needed to excite π electrons depends on the compound, but it can be as much as 40 to 140 kcal/mol (167 to 585 kJ/mol). This is more energy than infrared radiation can provide, but it is not too much for ultraviolet and visible light. Thus compounds that absorb ultraviolet and visible light usually contain π bonds. We will discuss ultraviolet and visible spectroscopy in Chapter 19.

A third type of molecular excitation, called *nuclear magnetic resonance (NMR),* requires a change in the alignment of certain nuclei with respect to an external magnetic field. This process requires only 1.0×10^{-6} to 1.0×10^{-7} kcal/mol (4.18×10^{-6} to 4.18×10^{-7} kJ/mol), which is much less than the energy needed to excite π electrons or vibrate molecular bonds. The amount of energy needed in NMR is so small, in fact, that it is available from radio waves. In Section 5.9, we discuss how chemists use NMR to gather information about the carbon skeletons of molecules.

In summary, chemists use three spectroscopic techniques, IR, ultraviolet/visible, and NMR, to determine the structures of organic compounds. Each type of spectroscopy gives specific information about the structure of a particular compound, as summarized in Table 5.2.

In Chapter 10, we will learn how NMR can also provide information about the hydrogen atoms in organic compounds by means of a technique called ^1H (hydrogen) NMR spectroscopy.

Table 5.2	Spectroscopic Techniques and the Structural Information They Provide	
Type of Spectroscopy	Physical Process	Structural Information Obtained
Ultraviolet and visible	Excitation of π electrons	Presence of π bonds
Infrared	Excitation between vibrational states	Type of functional group
Nuclear Magnetic Resonance (NMR)	Reorientation of spin of nuclei in magnetic field	Carbon skeleton and number of hydrogen atoms on each carbon atom

5.9   Nuclear Magnetic Resonance

Nuclear magnetic resonance (NMR) is the most valuable technique for determining the structures of organic compounds because atomic nuclei in different environments produce signals in different positions on an NMR spectrum. NMR analysis is limited, however, to those atomic nuclei with an odd atomic number or an odd mass number with a nuclear spin, such as ^1H, ^2H, ^{13}C, ^{15}N, ^{19}F, and ^{31}P. Hydrogen nuclei (^1H, protons) and one isotope of carbon (^{13}C) are the two elements detected by NMR spectroscopy that provide chemists with the most information about the structure of organic compounds. The most abundant isotope of carbon, ^{12}C, does not have a spin so it is not detected by NMR spectroscopy. While this may seem to be a disadvantage for determining structures, we will learn later in this chapter that this is a very definite advantage.

By varying the experimental conditions, NMR analysis of a sample can be carried out so that we detect signals from only the atoms of one element. In hydrogen NMR spectroscopy, for example, we record signals from hydrogen nuclei (^1H, protons) in different structural environments, while in carbon NMR spectroscopy we record signals from ^{13}C nuclei in different structural environments.

We begin our introduction to NMR spectroscopy in this chapter by learning how a ^{13}C NMR spectrum is obtained, what it looks like, and how it is used to obtain structural information about a compound. Then we learn how to use the molecular formula, the ^{13}C NMR spectrum, and the IR spectrum of a compound to determine its structure. Finally we examine the physical process responsible for NMR spectroscopy.

An NMR analysis is performed by means of an NMR spectrometer, which is shown schematically in Figure 5.20. An NMR spectrum is obtained by dissolving the sample in a suitable solvent, placing the solution in the magnetic field of the spectrometer, and then subjecting it to radio frequency (rf) radiation. When a nucleus absorbs energy the detector senses this energy change and transmits the information to a recorder that plots the results on a chart.

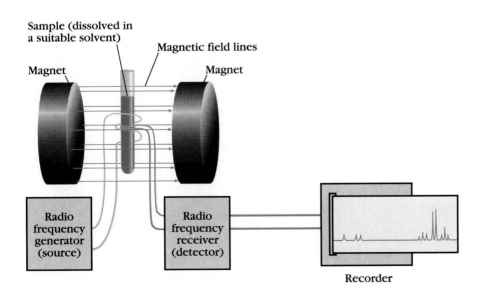

Sample (dissolved in a suitable solvent)

Magnetic field lines

Magnet

Magnet

Radio frequency generator (source)

Radio frequency receiver (detector)

Recorder

Figure 5.20
Schematic diagram of an NMR spectrometer. The principal components consist of a magnet (either an electromagnet or a superconducting magnet operating at liquid helium temperatures), a source of radiation in the radio frequency range of the electromagnetic spectrum, a detector to measure the absorption of radio frequency radiation by the sample, and a recorder.

Figure 5.21
An NMR spectrum. The horizontal axis is the applied magnetic field strength. The strength of the field is lowest to the left (downfield) and highest to the right (upfield).

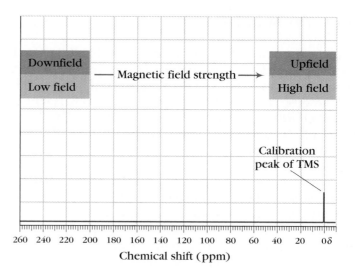

Chemical shift (ppm)

Certain NMR instruments operate by maintaining a constant magnetic field and varying the rf radiation, while others vary the magnetic field at a constant rf value. Both types of instruments are available commercially, and the NMR spectrum produced from either type is the same.

NMR spectra are recorded on a chart such as the one shown in Figure 5.21. The horizontal axis is the applied field strength, which increases from left to right. Thus the left side of the spectrum is the low-field end, also referred to as downfield, and the right side is the high-field end, also referred to as upfield.

The reference point used to define the positions of signals in both carbon and hydrogen NMR spectra is the signal obtained from tetramethylsilane, $(CH_3)_4Si$, which is abbreviated TMS. TMS is used because it gives rise to a single signal that occurs upfield (farther right on the spectrum) of most other NMR signals found in organic compounds. Experimentally, a small amount of TMS is added to the sample solution so that the standard reference peak is produced as the spectrum is recorded.

Absorption positions are recorded on the spectrum in frequency units (Hz) downfield (to the left) of the TMS peak. The TMS peak is defined as zero Hz and is shown in Figure 5.21. Frequency units are more convenient to use than gauss, the usual units of magnetic field strength, but the two units are directly related. Hertz and gauss will be discussed in greater detail in Section 5.11.

The position of an absorption in the spectrum relative to the position of the absorption of TMS is called its **chemical shift.** By convention, chemical shifts are not expressed in Hz but on an arbitrary scale called the **delta scale (δ scale).** The δ scale expresses chemical shifts as a dimensionless fraction of the total applied field. The chemical shift of a particular absorption peak in either a carbon or hydrogen NMR spectrum is calculated by the following equation:

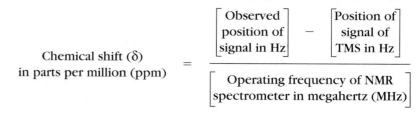

$$\text{Chemical shift }(\delta)\text{ in parts per million (ppm)} = \frac{\begin{bmatrix}\text{Observed}\\ \text{position of}\\ \text{signal in Hz}\end{bmatrix} - \begin{bmatrix}\text{Position of}\\ \text{signal of}\\ \text{TMS in Hz}\end{bmatrix}}{\begin{bmatrix}\text{Operating frequency of NMR}\\ \text{spectrometer in megahertz (MHz)}\end{bmatrix}}$$

Because values of δ for a particular carbon or proton are a dimensionless fraction of the total applied field, they are the same regardless of the spectrometer used to obtain the spectrum. Thus values of chemical shifts of the carbon nuclei of a particular compound are the same even when the ^{13}C NMR spectra are recorded on ^{13}C NMR spectrometers that have different operating magnetic fields and/or frequencies.

Sample Problem 5.6   An NMR signal is observed at 930 Hz downfield from the TMS signal. If the operating frequency of the NMR spectrometer is 15.1 MHz, what is the chemical shift of the signal on the delta scale?

Solution:

The observed position of the signal in Hz is 930 Hz. The position of the signal of TMS in Hz is by definition 0. The NMR spectrometer operates at 15.1 MHz. We substitute these values in the definition of the delta scale given above to obtain the answer as follows:

$$\delta = \frac{930 \text{ Hz} - 0 \text{ Hz}}{15.1 \text{ MHz}} = 61.5$$

Exercise 5.9   The following signals were recorded on an NMR spectrometer operating at 90 MHz. Their chemical shifts are expressed in hertz downfield from the TMS signal. Convert these chemical shifts to the delta (δ) scale.

(a) 1550 Hz   **(b)** 2150 Hz   **(c)** 2790 Hz

Next, we learn how to use ^{13}C NMR spectra to obtain information about the structures of organic compounds.

5.10 ^{13}C NMR Spectroscopy

The only isotope of carbon that has a nuclear spin is ^{13}C. However, ^{13}C represents only 1.1% of the carbon nuclei in naturally occurring carbon. This low abundance of ^{13}C originally caused many technical problems in obtaining ^{13}C NMR spectra. Since 1970, these problems have been overcome by the use of improved electronic and computer techniques so that ^{13}C NMR spectroscopy is now routinely used by organic chemists.

The ^{13}C NMR spectrum obtained from a ^{13}C NMR spectrometer operating in the normal mode consists of a series of sharp signals. Each signal represents a different ^{13}C nucleus in the molecule, as shown by the spectrum of acetaldehyde in Figure 5.22. In addition to the calibration signal of TMS, there are two signals in this spectrum that correspond to the two different carbon atoms of acetaldehyde.

The chemical shifts of ^{13}C nuclei in organic compounds range from δ 0 to δ 260. The chemical shift of a particular carbon atom depends on the identities of its neighboring atoms and the type of bonds it makes with them. In other words,

Figure 5.22
^{13}C NMR spectrum of acetaldehyde. The peak at zero on the δ scale is due to TMS.

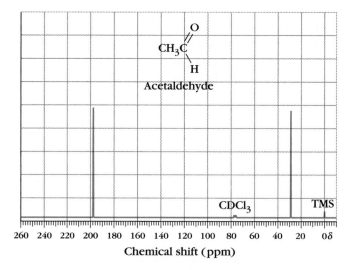

■ *A group of small signals also appears at δ 78.0 in the ^{13}C spectrum in Figure 5.22. These signals are due to deuteriochloroform (CDCl₃), which is added to the sample to facilitate the recording of the ^{13}C spectrum. These signals appear in all ^{13}C spectra reproduced in this text.*

the chemical shift of a carbon atom depends on its structural environment. Based on data obtained by examination of ^{13}C NMR spectra of many organic compounds, chemists have established chemical shift ranges of carbon atoms in various functional groups. These ranges are summarized in the correlation chart shown in Figure 5.23.

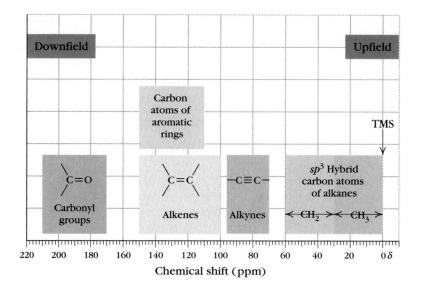

Figure 5.23
Correlation chart showing a carbon atom in various functional groups and its approximate ^{13}C NMR chemical shift value.

Based on the data in Figure 5.23, sp^3 hybrid carbon atoms of alkanes tend to absorb in the δ 0 to 60 range, while sp^2 hybrid carbon atoms tend to absorb more downfield in the δ 100 to 210 range. The chemical shifts of carbonyl carbon atoms are particularly distinct in ^{13}C NMR spectra and are found in the extreme downfield end of the spectrum (δ 170 to 210). In Figure 5.22, therefore, we assign the signal at δ 198.5 to the carbonyl carbon atom of acetaldehyde and the other at δ 29.3 to the carbon atom of the methyl group.

Carbon atoms of carbon-carbon double bonds and of aromatic rings are found in the δ 100 to 160 range, as seen in the ^{13}C NMR spectra of 1-pentene and 4-chlorophenol shown in Figure 5.24.

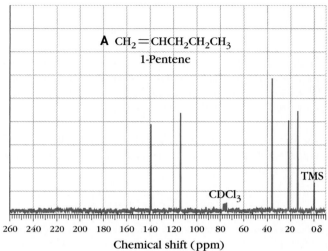

Figure 5.24
13C NMR spectra of 1-pentene **(A)**
and 4-chlorophenol **(B)**.

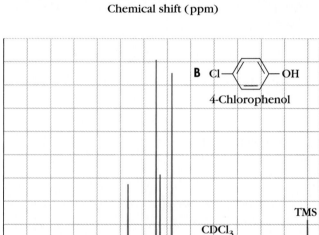

4-Chlorophenol has six carbon atoms, but its ^{13}C NMR spectrum shows only four carbon signals because ring carbon atoms 2 and 2′ and carbon atoms 3 and 3′ are in the same structural environments.

OH

4-Chlorophenol

They are equivalent, so their chemical shifts are identical. As a result, four signals are seen instead of six.

Box 5.1 discusses in more detail how to distinguish between equivalent and nonequivalent carbon atoms in a molecule. The information in the box will help you to properly interpret ^{13}C NMR spectra.

Box 5.1   How to Distinguish Between Equivalent and Nonequivalent Carbon Atoms in a Molecule

One way to determine if two or more atoms are in equivalent environments in a structure is to imagine that each is separately replaced by a different atom or group. If the products of such a replacement are identical, then the atoms are equivalent. If the products are different, then the atoms are nonequivalent.

Consider the carbon atoms in butane, for example. We will replace each carbon atom by a silicon atom. This is convenient because silicon is in the same group as carbon and forms four bonds.

$$\overset{1}{C}H_3\overset{2}{C}H_2\overset{3}{C}H_2\overset{4}{C}H_3$$

To determine if carbon atoms 1 and 4 are equivalent, we write one structure in which carbon atom 1 is replaced by a silicon atom and another in which carbon atom 4 is replaced by a silicon atom.

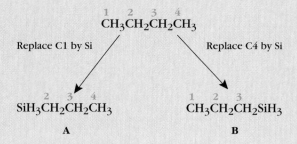

Rotating structure A by 180° gives structure C, which is identical to structure B.

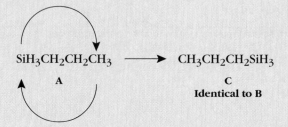

Therefore carbons 1 and 4 are equivalent. In a similar way, we can show that carbon atoms 2 and 3 are equivalent.

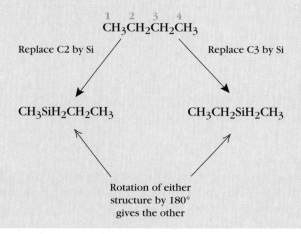

| Box 5.1—cont'd | How to Distinguish Between Equivalent and Nonequivalent Carbon Atoms in a Molecule |

Carbon atoms 1 and 2, however, are nonequivalent:

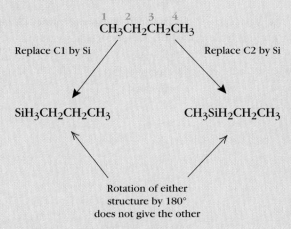

A word of caution—structural formulas indicate only how atoms in a molecule are joined together; they give no indication of the three-dimensional structure of a molecule. In some cases it is necessary to examine the three-dimensional structure of a molecule to determine if atoms are equivalent. For now, looking at the structural formula is sufficient. Beginning in Chapter 6, we will learn how the three-dimensional structure of molecules affects the equivalence of atoms.

| Exercise 5.10 | Determine the number of nonequivalent carbon atoms in each of the following compounds. |

(a) 2-Chloropropane   **(c)** 3-Methylhexane   **(e)** Dimethylpropane
(b) 1-Chloropropane   **(d)** Methylpropane

We can use the information in a ^{13}C NMR spectrum in two ways. First, the chemical shifts of the various signals give us information about the kinds of carbon atoms in the molecule. Second, the number of signals tells us how many nonequivalent carbon atoms are present in the compound. Keep in mind, however, that the number of signals is not always equal to the total number of carbon atoms in the molecule because several of the carbon atoms may be equivalent. Sample Problem 5.7 illustrates how to use the number of signals to distinguish between isomers.

Sample Problem 5.7	Three isomers of molecular formula C_5H_{12} are known. Can we distinguish them from their ^{13}C NMR spectra?

Solution:

First write the structural formulas of the three isomers. Second, determine the number of nonequivalent carbon atoms for each isomer.

Structural Formula	Number of Nonequivalent Carbons	Number of Signals in ^{13}C NMR Spectrum
$CH_3CH_2CH_2CH_2CH_3$	3	3

$CH_3CHCH_2CH_3$
   |
   CH_3
	4	4

CH_3
   |
CH_3CCH_3
   |
   CH_3
	2	2

All the carbon atoms are sp^3 hybrids so their chemical shifts are all found in the region δ 0 to 60. The major difference between the isomers is the number of nonequivalent carbon atoms. Because each nonequivalent carbon atom gives rise to a signal in the ^{13}C NMR spectrum, simply counting the number of signals serves to distinguish between the various isomers. Thus the number of signals as well as their chemical shifts can provide us with information about the structure of a particular compound.

Exercise 5.11	Would it be possible to use ^{13}C NMR spectroscopy to distinguish between the two isomers 1-chloropropane and 2-chloropropane? Explain.

Neither IR nor ^{13}C NMR spectroscopy should be regarded as a substitute for the other. Rather, the two are complementary. IR spectroscopy tells us the kind or kinds of functional group or groups in a compound, while ^{13}C NMR spectroscopy tells us the number and kinds of nonequivalent carbon atoms in the molecule.

Additional structural information about a compound can be obtained with these two techniques when we also have the molecular formula of the compound. From its molecular formula, we can calculate the degree of unsaturation of a molecule.

5.11   Calculating the Degree of Unsaturation from Molecular Formulas

In the process of using spectroscopy to determine the structure of a compound, we can often eliminate potential structures by calculating the **degree of unsaturation** from its molecular formula. Degree of unsaturation is defined as the number of pairs of hydrogen atoms that must be added to the molecular formula of the compound of unknown structure to obtain the molecular formula of the alkane with the same number of carbon atoms.

Consider, for example, a hydrocarbon of unknown structure whose molecular formula is C_4H_8. We know that this compound is not an alkane because the molecular formula of butane, the alkane with four carbon atoms, is C_4H_{10}. A pair of hydrogen atoms needs to be added to the formula C_4H_8 to obtain C_4H_{10}. Therefore C_4H_8 has one degree of unsaturation.

■ *Keep in mind that the general molecular formula of alkanes is C_nH_{2n+2}, where n is any integer.*

$$\text{Molecular formula of butane} = C_4H_{10}$$
$$\text{Molecular formula of unknown compound} = C_4H_8$$
$$\text{Difference} = H_2 \text{ (1 pair H atoms)}$$
$$\text{Degree of unsaturation} = 1$$

What does the degree of unsaturation tell us about the structure of the molecule? We can answer this question by comparing the general molecular formulas of alkanes, cycloalkanes, alkenes, alkynes, and dienes (compounds that contain two double bonds), as shown in Table 5.3.

The information in Table 5.3 tells us that the structure of a compound with one degree of unsaturation can contain either a double bond or a ring. A compound with two degrees of unsaturation can have a triple bond, two double bonds, a ring with a double bond, or two rings that share a side in its structure.

■ *Compounds with two rings that share a side are called bicyclic compounds.*

The hydrocarbon of molecular formula C_4H_8 has one degree of unsaturation, so it has either a double bond or a ring in its structure. Thus this compound could have any of the following structures:

CH_3		$CH_2{=}CHCH_2CH_3$	$CH_3CH{=}CHCH_3$	$(CH_3)_2C{=}CH_2$
C_4H_8	C_4H_8	C_4H_8	C_4H_8	C_4H_8

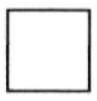

Table 5.3   **Degrees of Unsaturation of Alkanes, Cycloalkanes, Alkenes, Alkynes, Dienes, Cycloalkenes, and Bicyclic Hydrocarbons**

Family	General Molecular Formula	Examples	Degree of Unsaturation
Alkanes	C_nH_{2n+2}	$CH_3CH_2CH_3$ (C_3H_8)	0
Cycloalkanes	C_nH_{2n}	 (C_3H_6)	1
Alkenes	C_nH_{2n}	$CH_2{=}CHCH_3$ (C_3H_6)	1
Alkyne	C_nH_{2n-2}	$HC{\equiv}CCH_3$ (C_3H_4)	2
Diene	C_nH_{2n-2}	$CH_2{=}C{=}CH_2$ (C_3H_4)	2
Cycloalkene	C_nH_{2n-2}	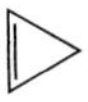 (C_3H_4)	2
Bicyclic Hydrocarbon	C_nH_{2n-2}	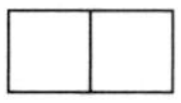 (C_6H_{10})	2

While the number of structures of the unknown compound of molecular formula C_4H_8 is limited to five, we have not yet solved the problem of determining its structure.

Exercise 5.12   Can you use ^{13}C NMR spectroscopy to distinguish between the five isomeric compounds of molecular formula C_4H_8 given previously?

Exercise 5.13   Calculate the degrees of unsaturation for each of the following molecular formulas.

(a) C_6H_{12} (b) C_6H_6 (c) C_8H_{14} (d) $C_{14}H_{28}$ (e) $C_{20}H_{34}$

We can also calculate degrees of unsaturation for organic compounds other than hydrocarbons. To determine the degree of unsaturation for organic compounds containing halogen atoms, for example, we simply replace the halogen atom in the molecular formula by hydrogen and repeat the procedure given previously. Thus the degree of unsaturation for the compound of molecular formula C_3H_3Cl is calculated as follows:

$$\begin{aligned}
\text{Molecular formula of alkane} &= C_3H_8 \\
\text{Molecular formula of chloroalkane} &= C_3H_3Cl \\
\text{Replace Cl by H to give} &= C_3H_4 \\
\text{Difference} &= H_4 \text{ (2 pair H atoms)} \\
\text{Degree of unsaturation} &= 2
\end{aligned}$$

According to Table 5.3, this compound contains two double bonds, one ring and one double bond, one triple bond, or two rings that share a side.

This method works because halogens, like hydrogen, are univalent, so replacing a halogen atom in any organic compound by a hydrogen atom does not change the compound's degree of unsaturation. For example:

$CH_2{=}CHCH_2CH_2Cl$ Replace Cl atom by H atom to give $CH_2{=}CHCH_2CH_3$

C_4H_7Cl C_4H_8
1 degree of   1 degree of
unsaturation   unsaturation

To determine the degree of unsaturation for organic compounds containing oxygen atoms, simply ignore the oxygen atom in the molecular formula and repeat the procedure as before. For example, the degree of unsaturation for the compound of molecular formula C_4H_8O is calculated as follows:

$$\begin{aligned}
\text{Molecular formula of alkane} &= C_4H_{10} \\
\text{Molecular formula of compound} &= C_4H_8O \\
\text{Remove O atom to give} &\quad C_4H_8 \\
\text{Difference} &= H_2 \text{ (1 pair H atoms)} \\
\text{Degree of unsaturation} &= 1
\end{aligned}$$

According to Table 5.3, this compound contains either a double bond or a ring. The following are several possible structures.

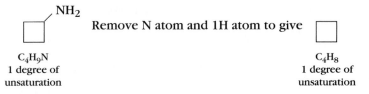

Notice that the double bond can be either a carbon-carbon or a carbon-oxygen double bond.

This method works because oxygen is divalent, so removing it doesn't affect the degree of unsaturation.

$$CH_2=CHCH_2CH_2OH \quad \text{Remove O atom to form} \quad CH_2=CHCH_2CH_3$$

$$\begin{matrix} C_4H_8O \\ \text{1 degree of} \\ \text{unsaturation} \end{matrix} \qquad\qquad\qquad \begin{matrix} C_4H_8 \\ \text{1 degree of} \\ \text{unsaturation} \end{matrix}$$

To determine the degree of unsaturation for organic compounds containing nitrogen atoms, subtract one hydrogen atom for each nitrogen atom, then ignore the nitrogen atom and repeat the calculation as before. For example, the compound of molecular formula C_4H_9N is considered to be the same as the hydrocarbon C_4H_8, so they both have one degree of unsaturation.

This method works because nitrogen is trivalent and a nitrogen-containing organic compound has one more hydrogen atom than its corresponding hydrocarbon.

Remove N atom and 1H atom to give

$$\begin{matrix} C_4H_9N \\ \text{1 degree of} \\ \text{unsaturation} \end{matrix} \qquad\qquad\qquad\qquad \begin{matrix} C_4H_8 \\ \text{1 degree of} \\ \text{unsaturation} \end{matrix}$$

Exercise 5.14   Calculate the degree of unsaturation for the compound of molecular formula C_4H_6 and give four examples that contain the following structural features: one with a triple bond; one with two double bonds; one with a ring and a double bond; and one with two rings that share a side.

Exercise 5.15   For each of the following compounds, calculate the degree of unsaturation and write three possible structures in agreement with the molecular formula.

(a) C_3H_4ClBr **(c)** C_4H_5OCl **(e)** C_5H_8NOBr
(b) $C_6H_{10}O$ **(d)** C_5H_9N **(f)** $C_3H_6O_2$

Knowing the degree of unsaturation of a compound allows us to write possible structures for it. To distinguish between them, we use IR and ^{13}C NMR spectroscopy.

5.12   Using IR and ^{13}C NMR Spectroscopy in Structure Determinations

The best way to illustrate how to use IR and ^{13}C NMR spectroscopy to determine the structure of an organic compound is to work out a number of examples.

Sample Problem 5.8   Determine the structure of the compound of molecular formula C_6H_{12} whose IR and ^{13}C NMR spectra are as follows:

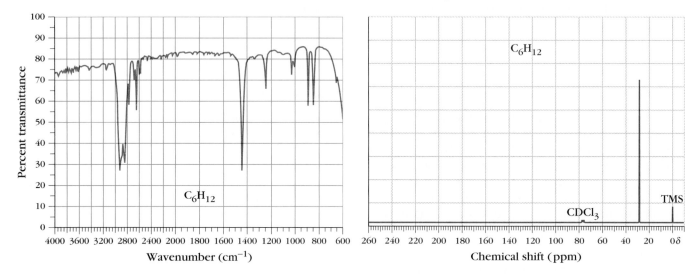

Solution:

Begin by determining the degree of unsaturation using the method described in the previous section. For a hydrocarbon of molecular formula C_6H_{12}, the degree of unsaturation is one.

We conclude from this degree of unsaturation that the compound contains either a double bond or a ring. Because the compound is a hydrocarbon, the double bond must be a carbon-carbon double bond. Therefore the compound contains either a ring of carbon atoms or a carbon-carbon double bond:

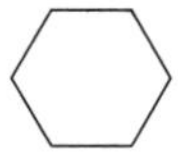

$$CH_2 = CHCH_2CH_2CH_2CH_3$$

Cycloalkane   An alkene

At this point it's not necessary to write the structure of all possible isomeric compounds because we're still trying to determine which functional group is in the molecule.

Next, look at the spectra for evidence of the presence or absence of these functional groups. The absence of a carbon-carbon double bond absorption band in the IR spectrum (near 1650 cm^{-1}) suggests that the compound is a cycloalkane.

Now we must write possible isomeric structures of molecular formula C_6H_{12} that contain a single ring of carbon atoms in order to determine the identity of the cycloalkane.

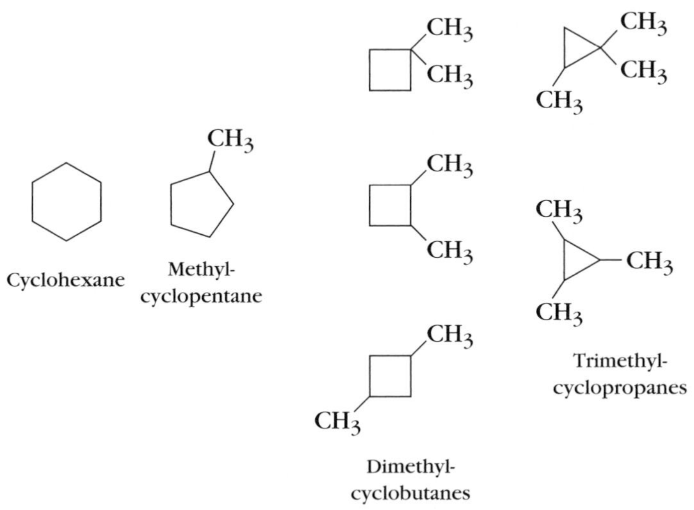

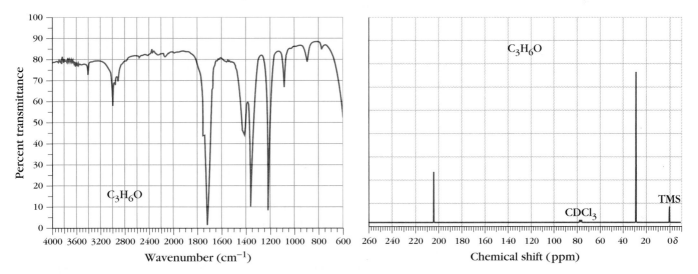

Cyclohexane

Methyl-
cyclopentane

Trimethyl-
cyclopropanes

Dimethyl-
cyclobutanes

We return to the spectra to find further evidence to distinguish between the isomers. The ^{13}C NMR spectrum contains only one signal, so all six carbon atoms in the compound are equivalent. The only structure in which all six carbon atoms are equivalent is cyclohexane. Thus the compound must be cyclohexane.

Sample Problem 5.9   Determine the structure of the compound of molecular formula C_3H_6O whose IR and ^{13}C NMR spectra are as follows:

Solution:

For a compound of molecular formula C_3H_6O, the degree of unsaturation is one. According to Table 5.3, this compound contains either a double bond or a ring. Because the molecule contains oxygen, the double bond could be either a carbon-carbon double bond or a carbon-oxygen double bond. If the compound contains a ring, the oxygen atom could be in the ring. The following are examples of molecules that contain these structural features:

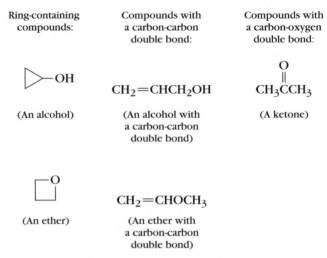

Ring-containing compounds:

▷—OH

(An alcohol)

☐—O

(An ether)

Compounds with a carbon-carbon double bond:

$CH_2{=}CHCH_2OH$

(An alcohol with a carbon-carbon double bond)

$CH_2{=}CHOCH_3$

(An ether with a carbon-carbon double bond)

Compounds with a carbon-oxygen double bond:

$$\overset{O}{\overset{\|}{CH_3CCH_3}}$$

(A ketone)

The presence of an absorption band in the IR spectrum at 1710 cm^{-1} as well as the signal at δ 205 in the ^{13}C NMR spectrum indicate that the compound contains a carbonyl group (C=O). Also, the absence of absorption bands due to a hydroxy group (—OH) and a carbon-carbon double bond rules out structures with these functional groups. Thus only compounds containing a carbonyl group agree with the spectral data.

The following are the two isomeric compounds of molecular formula C_3H_6O that contain a carbonyl group:

$$\overset{O}{\overset{\|}{CH_3CCH_3}} \qquad \overset{O}{\overset{/\!/}{CH_3CH_2C}}\!\!\diagdown_H$$

A ketone   An aldehyde

The ^{13}C NMR spectrum contains only two signals, and the IR spectrum lacks absorption bands in the 2800 to 2700 cm^{-1} region where an H atom bonded to the carbonyl carbon atom of the aldehyde would absorb, indicating that the compound is the ketone.

The sequence of steps used in Sample Problems 5.8 and 5.9 can be summarized as follows:
- Begin with the molecular formula.
- Determine the degree of unsaturation.
- Use Table 5.3 to propose functional groups and structures or partial structures that agree with the calculated degree of unsaturation.
- Write possible structural formulas that contain these functional groups.
- Look at the available spectra for evidence of the presence or absence of these functional groups.
- Determine which functional group or groups are present.
- Write structures of as many isomers as possible that contain the functional group(s).

- Consult the spectra again for data that distinguish between isomers.
- Determine which structure is consistent with *all* spectral data.

This sequence of steps should serve as a general guide. Sometimes it will be possible to eliminate steps, while in other cases it may be necessary to repeat steps. It all depends on the complexity of the problem and your experience in solving such problems. As you gain more experience, you may tailor the scheme to your unique way of solving problems.

Exercise 5.16	Based on their molecular formulas, IR spectra, and ^{13}C NMR spectra, determine the structures of the following three compounds.

(a) C_2H_6O

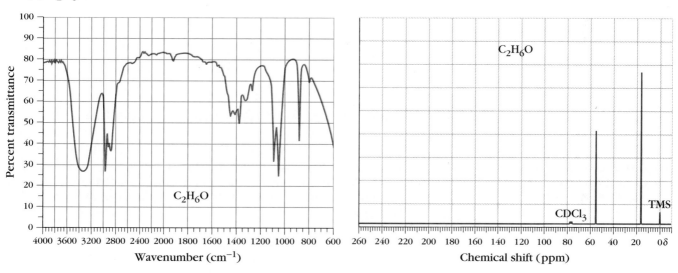

(b) $C_5H_{10}O$

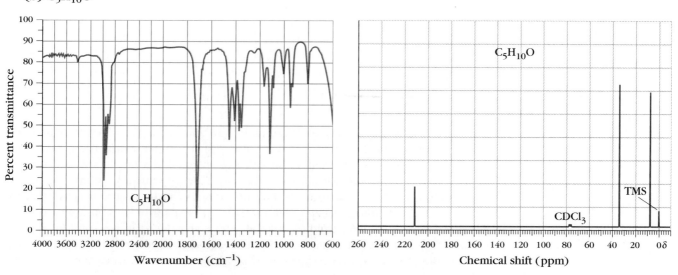

(c) $C_5H_{10}O_2$

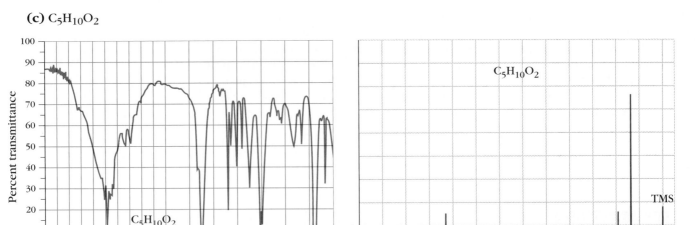

5.13   Theory of Nuclear Magnetic Resonance

■ *A magnetic moment is a vector quantity so it has both magnitude and direction.*

Nuclei of atoms with odd atomic numbers or an odd mass number behave as if they are spinning on an axis. These nuclei are positively charged and any spinning charge, either positive or negative, generates a magnetic field that coincides with its axis. Spinning nuclei, therefore, behave like tiny bar magnets with a **magnetic moment.**

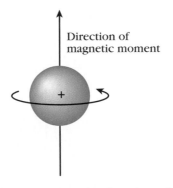

Nuclear magnetic moments are randomly oriented in the absence of an external magnetic field. When placed in an external magnetic field (designated H_0), however, only certain orientations of these magnetic moments are allowed. The allowed orientations correspond to their nuclear spin quantum numbers. For ^1H, ^{13}C (but not ^{12}C), ^{19}F, and ^{31}P, the nuclear spin can have only two orientations associated with the quantum spin numbers of $+\frac{1}{2}$ and $-\frac{1}{2}$. Spin number $+\frac{1}{2}$ is also designated α spin and spin number $-\frac{1}{2}$ is also designated β spin. In an external magnetic field, the magnetic moments are aligned either with the field (corresponding to an α spin) or against the field (corresponding to a β spin).

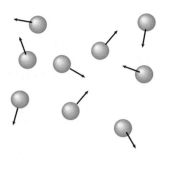

No magnetic field: Magnetic moments of nuclei are randomly oriented

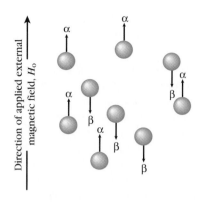

Applied external magnetic field: Nuclei with α spin are oriented with the field; nuclei with β spin are oriented against the field

A magnet tends to align itself with an external magnetic field. A compass needle, for example, points north due to the Earth's magnetic field. A magnet aligned against an external magnetic field is in a higher energy state than one aligned with the field. For example, energy must be added to force a compass needle to point south. For nuclei in an external field, those with α spin are lower in energy than those with β spin.

If nuclei in spin states α and β are now irradiated with electromagnetic radiation of energy exactly equal to the energy difference between the two spin states, energy will be absorbed and nuclei in the α spin state will be converted to the higher energy β spin state. This process is known as spin flipping. When spin flipping occurs, the nuclei are said to be in resonance with the exciting radiation, hence the origin of the word *resonance* in the name nuclear magnetic resonance.

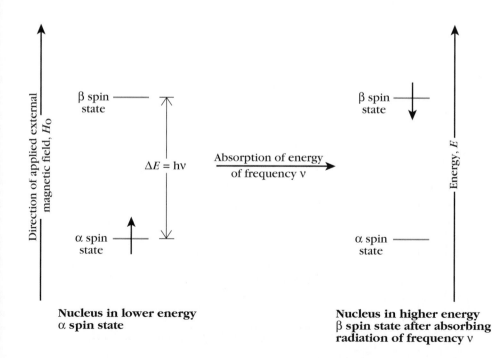

Nucleus in lower energy α spin state

Nucleus in higher energy β spin state after absorbing radiation of frequency ν

This is the basis of NMR. To summarize, placing nuclei that have a magnetic moment, such as ^{13}C or ^{1}H, in an external magnetic field creates two spin states of different energies. The phenomenon that we observe in NMR is the excitation of nuclei from the lower energy spin state to the higher energy spin state. The excitation (spin flipping) occurs when radiation of the proper frequency is passed through the sample.

From our explanation so far, we might expect that all ^{13}C nuclei would absorb excitation radiation at exactly the same frequency. This is not correct, however, because there is a difference in chemical shifts for different types of ^{13}C nuclei. To understand why this is so, we must examine the relationship between the applied magnetic field and the energy difference between the α and β spin states of a particular nucleus. This relationship is given by the following equation:

$$\Delta E = \gamma \frac{h}{2\pi} H$$

where H is the magnetic field strength *at the nucleus,* h is Planck's constant, and γ is the gyromagnetic ratio. The gyromagnetic ratio is a constant that depends on the magnetic moment of the nucleus and is different for each nucleus. This equation tells us that the energy difference between the two spin states in a magnetic field is proportional to the magnetic field felt at the nucleus, as shown in Figure 5.25.

When there is no magnetic field at the nucleus, there is no difference in energy between the α and β spin states and the magnetic moments of the nuclei are randomly oriented in space. In Figure 5.25, this corresponds to the point where the lines for the α and β spin states intersect at the left of the graph. If the Earth lost its magnetic field, for example, compass needles would point randomly in all directions. As the magnetic field felt at the nucleus increases, the energy necessary to change the spin state of the nuclei also increases. Thus if the magnetic field of the Earth were to suddenly increase, the energy needed to force a compass needle to point south would also increase. Thus the magnetic field felt at the nucleus determines the energy between the two spin states and also deter-

Figure 5.25
The energy difference ΔE between the α and β spin states depends directly on the magnetic field felt at the nucleus.

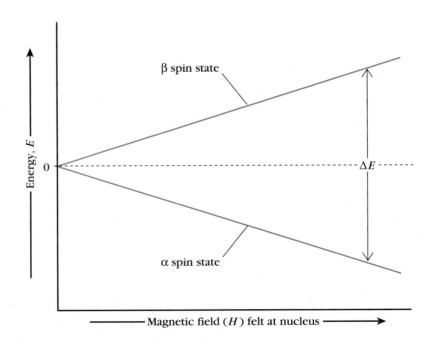

mines the frequency of excitation energy necessary to cause spin flipping to occur. This is the basis of the chemical shift.

5.14   The Chemical Shift

Up to now we have considered the effect of a magnetic field on the energy difference between the two spin states of a bare nucleus. This is certainly not representative of the situation in real molecules, however, because all nuclei in molecules are surrounded by electrons. When a molecule is placed in a magnetic field, its electrons also generate a small induced magnetic field, H_i, which opposes the external applied magnetic field, H_O. The result is that the magnetic field felt at the nucleus of atoms in molecules is a bit smaller than the applied magnetic field. We say, therefore, that the nuclei of a molecule are **shielded** from the full effect of the applied magnetic field.

Strength of magnetic field felt at nucleus

Strength of induced magnetic field caused by circulation of neighboring electrons

$$H = H_O - H_i$$

Strength of external applied magnetic field

In order for a nucleus surrounded by electrons to undergo spin flipping, the applied magnetic field must be increased slightly to overcome the effect of the induced magnetic field caused by neighboring electrons. For example, a bare ^{13}C nucleus at a frequency of 15.1 MHz undergoes spin flipping when the magnetic field at its nucleus equals 14,092.0 gauss. However, a ^{13}C nucleus in a molecule that is shielded by 0.5 gauss due to neighboring electrons will not undergo spin flipping at an applied magnetic field of 14,092.0 gauss because the magnetic field felt at its nucleus is only 14,091.5 gauss. Therefore we must increase the applied magnetic field to 14,092.5 gauss in order for a magnetic field of 14,092.0 gauss to be felt at the nucleus. Spin flipping will occur at this applied field.

The electron environment of a nucleus depends on the types of atoms to which it is bonded and the kinds of bonds it forms with these atoms. Types of atoms and kinds of bonds vary from nucleus to nucleus, so the electron environments of the various ^{13}C nuclei in molecules differ as well. This means, in turn, that their magnetic environments also differ. Consequently, they will undergo spin flipping at different applied external magnetic fields. This is the basis of chemical shift differences in NMR spectroscopy.

As we discuss the various functional groups in future chapters, we will examine how their presence in a molecule affects the chemical shifts of neighboring ^{13}C and ^1H nuclei and how this information can be used to deduce even more details about the structure of organic compounds.

■ *The various ^1H nuclei in molecules have different magnetic environments, too. This and other aspects of ^1H NMR spectroscopy are discussed in Chapter 10.*

5.15   Summary

Separating and purifying mixtures of organic compounds is an important part of the work of chemists. **Crystallization** is a method of purifying solids based on their differences in solubility in a solvent. **Distillation** purifies volatile liquids based on their boiling point differences.

If crystallization and distillation are unsuccessful, some type of **chromatography** can be used. In chromatography, the mixture is dissolved in an appropriate solvent, and the resulting solution is placed at one end of a column containing an inert **stationary phase.** The mixture is eluted by means of a **mobile phase,** which can be either a liquid or a gas. Mixtures can be separated by chromatography because the intermolecular attractions between the stationary phase and the various components of the mixture differ. As a result, the components elute from the column at different rates.

Once a pure sample of a compound is obtained, we can use spectroscopy to determine its structure. Spectroscopy is the study of the interaction of electromagnetic radiation with matter. Absorption of electromagnetic radiation by molecules is measured experimentally by an instrument called a **spectrometer.** Most organic compounds absorb electromagnetic radiation in the infrared region of the spectrum. The record of the absorption of light in the infrared region by a compound is called its **infrared (IR) spectrum.** The position of the absorption bands in an IR spectrum provides information about the functional group or groups in a molecule. *Most functional groups have characteristic infrared absorption bands. The location of these bands varies little in the infrared spectra of all compounds that contain that functional group.* The absorption bands correspond to the amount of energy needed to increase the amplitude of various **molecular vibrations,** such as bond stretching and bending.

Placing nuclei such as ^{13}C and ^1H in a strong external magnetic field causes their **magnetic moment** to be oriented either with the external field (α spin) or against it (β spin). If nuclei in these two spin states are now irradiated with electromagnetic radiation of energy exactly equal to the energy difference between the two spin states, energy absorption occurs to convert nuclei in the lower energy α spin state to the higher energy β spin state. This process is called spin flipping. The absorption of energy is detected, amplified, and recorded as a **nuclear magnetic resonance (NMR) spectrum.**

A ^{13}C NMR spectrum gives a single signal for each nonequivalent carbon atom in the molecule. The position of each of these signals in the spectrum relative to the signal of tetramethylsilane, TMS [$(CH_3)_4Si$], is called its **chemical shift.** Chemical shifts are measured by the **delta (δ) scale.** TMS is the reference point of this scale, and its delta value is zero.

The signals in ^{13}C NMR spectra can be used in two ways. The chemical shifts of the signals give us information about the kinds of carbon atoms in a molecule because ^{13}C nuclei with similar magnetic environments appear at about the same chemical shift. Thus the chemical shifts of sp^2 hybrid carbon atoms are found in a region that is different from the region where the signals of sp^3 hybrid carbon atoms are found. Finally, the number of signals in a ^{13}C NMR spectrum tells us how many nonequivalent carbon atoms are in the molecule.

Additional Exercises

5.17 Define each of the following terms.

(a) Functional group	**(f)** TMS	**(k)** Fingerprint region
(b) Spectroscopy	**(g)** Electromagnetic spectrum	**(l)** Chemical shift
(c) Distillation	**(h)** Wavelength	**(m)** ^{13}C NMR spectrum
(d) IR spectrum	**(i)** Chromatography	**(n)** Frequency
(e) Degree of unsaturation	**(j)** Wavenumber	**(o)** Crystallization

5.18 Convert each of the following wavelengths into wavenumbers (cm^{-1}).

(a) 3.05 μm	**(c)** 4.71 μm	**(e)** 6.06 μm
(b) 3.35 μm	**(d)** 5.81 μm	

5.19 Each of the wavelengths given in Exercise 5.18 represents a characteristic absorption band of a functional group. Identify the functional group that corresponds to each absorption and write its structure.

5.20 The five IR spectra on p. 220 that are numbered 1 to 5 correspond to five of the following compounds. Match the compound to its correct spectrum and explain what absorption bands in the spectrum led to your conclusion.

(a) 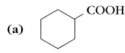 COOH

(d) $(CH_3)_2CHC\!\!\overset{\displaystyle O}{\underset{\displaystyle H}{\big\|}}$

(f) $CH_3CH{=}CH(CH_2)_4CH_3$

(b) $(CH_3)_3CC{\equiv}CH$

(e) $[(CH_3)_2CH]_2NH$

(g) $CH_3CH_2\overset{\displaystyle CH_3}{\underset{\displaystyle CH_3}{\overset{\displaystyle |}{\underset{\displaystyle |}{C}}}}{-}OH$

(c)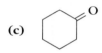

5.21 One of the two IR spectra that are designated A and B corresponds to *tert*-butyl alcohol; the other corresponds to *tert*-butylamine. Match each compound to the correct spectrum and explain what absorption bands in the spectrum led to your conclusion.

$(CH_3)_3COH$ $\qquad\qquad$ $(CH_3)_3CNH_2$

tert-Butyl alcohol $\qquad\qquad$ *tert*-Butylamine

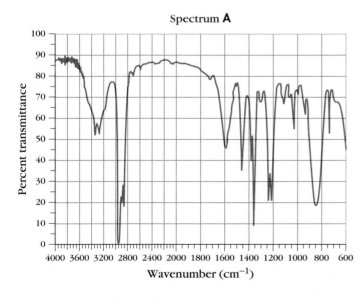

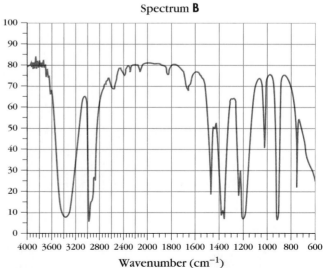

Spectrum 1

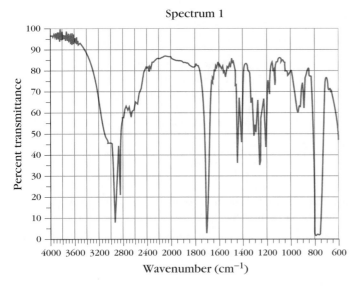

Spectrum 2

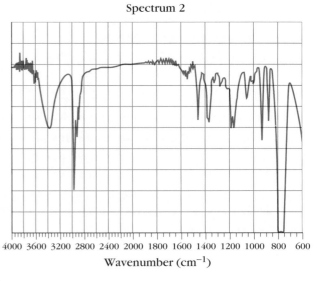

Spectrum 3

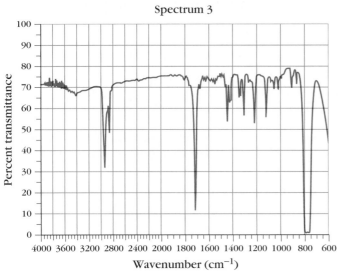

Spectrum 4

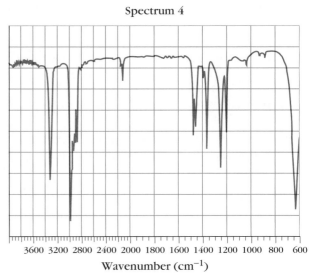

Spectrum 5

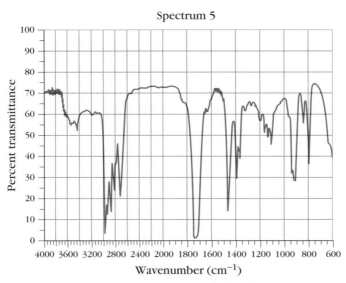

5.22   The structure and IR spectrum of a compound containing several functional groups are given below. Identify each functional group and locate its characteristic absorption band in the IR spectrum.

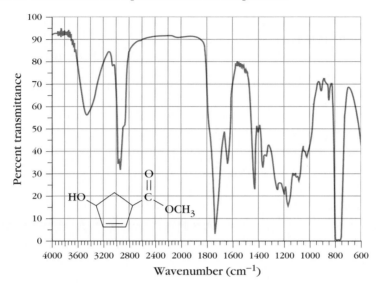

5.23   Calculate the degrees of unsaturation for each of the following molecular formulas.

(a) $C_{10}H_{18}$ (c) C_3H_9N (e) $C_8H_{12}O_3$
(b) $C_{14}H_{22}O$ (d) C_5H_5N (f) C_6H_6NO

5.24   How many degrees of unsaturation does each of the following compounds have?

(a) $CH_3CCH_2CH_3$ (with =O above C)

(b)

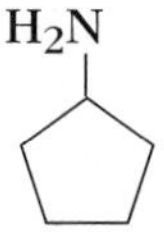

(c) (cyclohexanone)

(d) 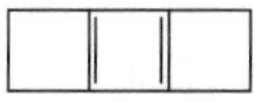 (box figure)

5.25   In each of the following structures identify the equivalent carbon atoms and give the total number of nonequivalent carbon atoms.

(a) 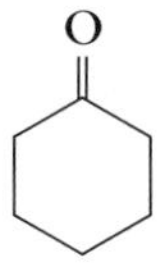 (cyclohexanone)

(b) $CH_3CHCH_2CCH_3$ (with OH and CH$_3$ and CH$_3$ substituents)

(c) $(CH_3)_2CHCH(CH_3)_2$

5.26   A single ^{13}C NMR spectrum is given for each of the following pairs of isomers. Match the ^{13}C NMR spectrum with the correct isomer.

(a)    or $(CH_3)_2C{=}C(CH_3)_2$

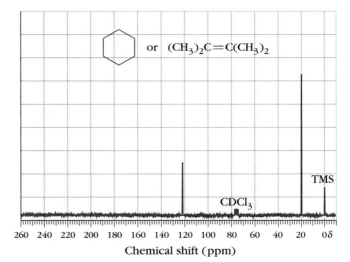

(b) $(CH_3)_3COH$ or $(CH_3)_2CHCH_2OH$

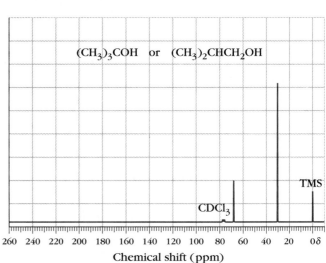

(c) 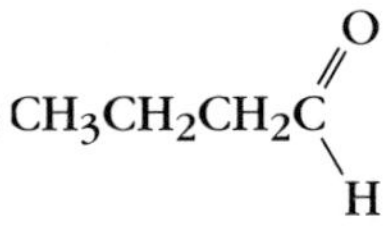 $CH_3CH_2CH_2C\overset{O}{\underset{H}{\big\backslash}}$ or

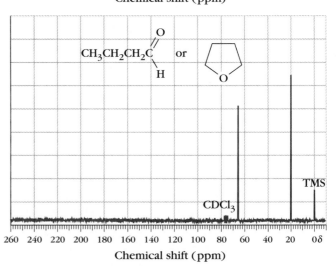

5.27 From their molecular formulas, IR spectra, and ^{13}C NMR spectra, determine the structures of the following compounds on pp. 223 and 224.

(a) $C_3H_6Br_2$

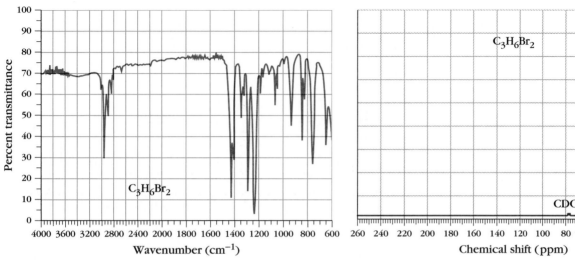

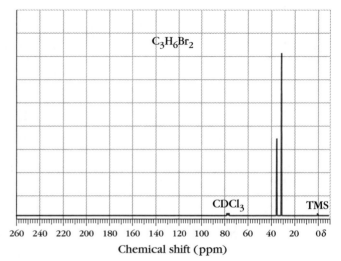

(b) C_5H_8O

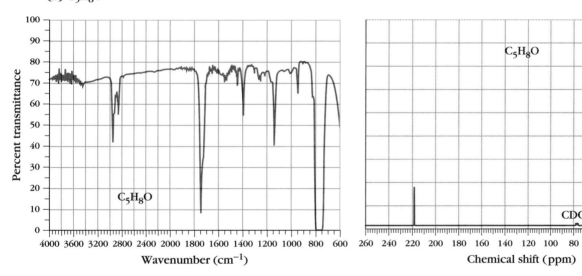

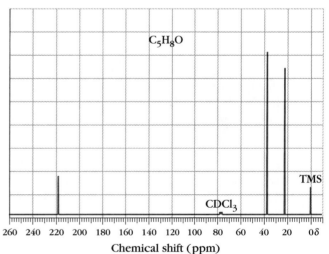

(c) $C_5H_{11}N$

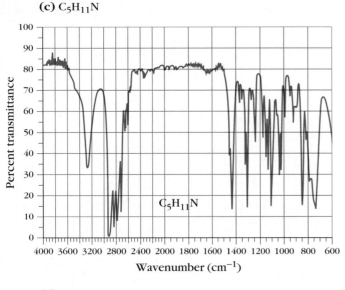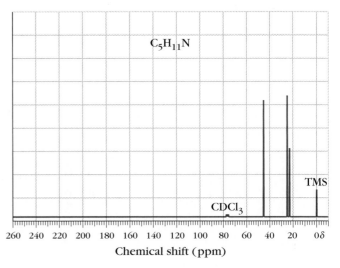

(d) $C_4H_8O_2$

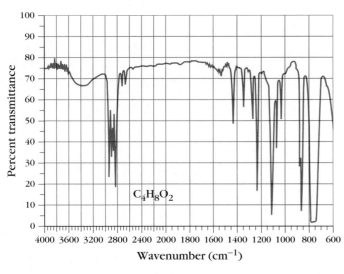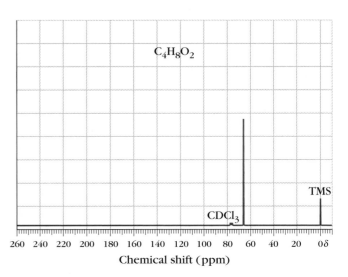

(e) $C_5H_{10}O$

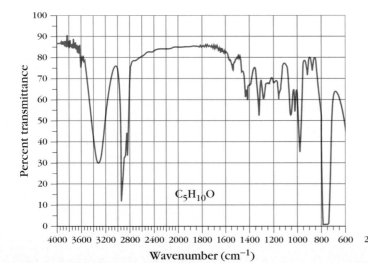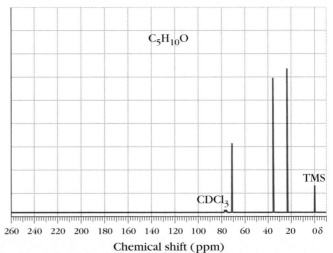

STEREOCHEMISTRY

The idea that the chemical reactions of a compound can be predicted from its structural formula has been repeatedly stressed in previous chapters. However, a number of compounds were discovered in the 19th century that seemed to contradict this idea. One of them is the amino acid 2-amino-4-methylpentanoic acid, which is usually known by its common name *leucine*. The chemical formula of leucine is $C_6H_{13}NO_2$, and it has the following structural formula:

$$\underset{\substack{| \\ CH_3}}{\overset{\substack{CH_3 \\ \backslash}}{}}CHCH_2\underset{\substack{| \\}}{\overset{\substack{NH_2 \\ |}}{}}CHCOOH$$

Leucine
(2-Amino-4-methylpentanoic acid)

Two isomers of leucine are known. While both are known to have the same structural formula, one tastes bitter, while the other tastes sweet. How can two compounds with the same structural formula be so different? To answer this question, we must learn the principles of **stereochemistry,** which is chemistry in three dimensions.

We are all familiar with the three-dimensional arrangement of objects in our world. Less familiar to many people is the three-dimensional aspect of molecules. Yet the spatial arrangement of atoms in molecules plays an important role in determining many of their physical and chemical properties. Understanding the principles of stereochemistry leads to a clearer understanding of the chemical reactions in organic chemistry and biochemistry.

In this chapter, we introduce the basic principles and terms of stereochemistry. We begin by examining objects and their mirror images.

6.1 Enantiomers and Molecular Chirality

All objects have mirror images. Some objects are identical to and can be superposed on their mirror images. If an object is superposable on its mirror image, then we can place the object on top of its mirror image and all parts of both images will coincide. The cup in Figure 6.1, for example, is superposable on its mirror image.

Figure 6.1
A, A cup and its mirror image.
B, The cup and its mirror image are superposable.

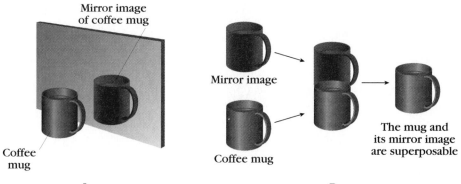

A   B

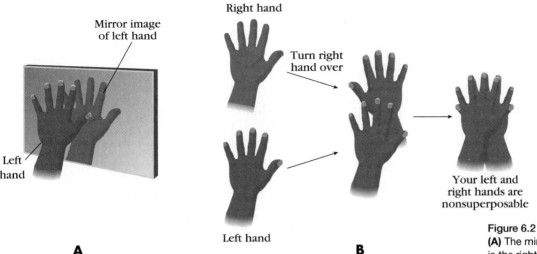

Figure 6.2
(A) The mirror image of the left hand is the right hand, yet **(B)** the left and right hands are nonsuperposable.

Other objects, such as a person's hands, are nonsuperposable on their mirror images. The right hand is the mirror image of the left hand, yet the two are non-superposable, as shown in Figure 6.2.

Any object nonsuperposable on its mirror image is said to be **chiral.** The word *chiral* comes from the Greek word *cheir,* meaning hand, and is used to refer to objects that have the same relationship as your left and right hands. The opposite of chiral is **achiral.** Any object superposable on its mirror image is defined as achiral.

Many familiar objects are chiral. Nuts and bolts have left- or right-handed threads. Certain musical instruments such as bassoons and clarinets are chiral, as are shoes and gloves. The chirality of any object is often apparent only when we look to see whether or not the object and its mirror image are superposable.

■ *Chiral is pronounced* **ki**-ral; *it rhymes with spiral.*

| **Exercise 6.1** | Which of the following common objects are chiral? |

(a) A rake **(d)** A screw
(b) A baseball glove **(e)** A screwdriver
(c) A baseball bat **(f)** A sweatshirt with *TORONTO* on the front

Certain molecules such as 2-chlorobutane are chiral. We can better under-stand why 2-chlorobutane is chiral if we examine three-dimensional models of the molecule. In Figure 6.3, for instance, the two three-dimensional structures of 2-chlorobutane differ only in the way the atoms and groups of atoms are arranged in space about C2.

The two structures in Figure 6.3 are mirror images of each other. If the struc-ture on the left is placed before a mirror, the structure on the right is seen in the mirror and vice versa. Making molecular models of these two structures will help you to see this relationship.

Not only are the two structures in Figure 6.3 mirror images but they are also nonsuperposable, as shown in Figure 6.4. Again, use your molecular models to verify this relationship. Because the two structures in Figure 6.3 are nonsuperpos-

When you make a molecular model, use a single ball to repre-sent each atom or group bonded to carbon atom 2 of 2-chlorobu-tane. For example, a blue ball can represent the ethyl group, a red one the methyl group, a green one the hydrogen atom, and a yellow one the chlorine atom. This is a quick and easy way of making molecular models of these two structures.

Figure 6.3
Ball and stick **(A)** and line-dash-wedge **(B)** representations of 2-chlorobutane. Remember that a solid line represents a bond between two atoms in the plane of the page, a solid wedge represents a bond between an atom in the plane and another in front of the plane, and a dashed wedge represents a bond between an atom in the plane and another behind the plane.

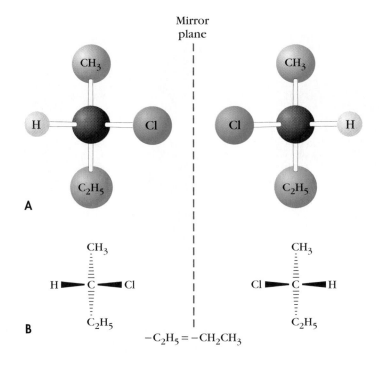

Figure 6.4
The structures of 2-chlorobutane are nonsuperposable mirror images. **A,** When the —CH$_3$ and —C$_2$H$_5$ groups match, the H and Cl atoms do not. **B,** If we rotate the mirror image by 180°, we can match the H and Cl atoms but then the —CH$_3$ and —C$_2$H$_5$ groups don't match.

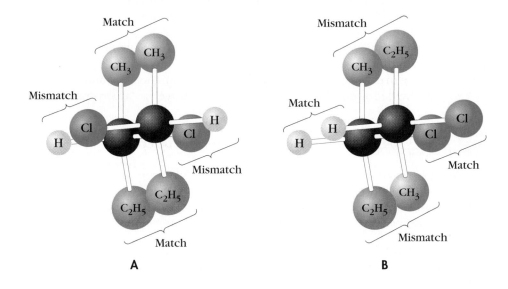

Figure 6.5
Carbon atom 2 is called a stereo-center because it is bonded to four different substituents. Such atoms are often identified by placing an asterisk (*) next to them.

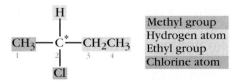

able mirror images of each other, the two molecules they represent are called **enantiomers.**

How can we recognize molecules that exist as a pair of enantiomers? The only sure way to recognize a chiral molecule is to determine whether it can be superposed on its mirror image. This test of superposability requires a well-developed ability to visualize objects in three dimensions. Most people find it necessary to build three-dimensional models of the molecule and its mirror image and then try to superpose them. But there are other ways to recognize compounds that can exist as enantiomers.

One way is to look for a carbon atom in the molecule that is bonded to four different groups or atoms. For example, C2 of the chiral compound 2-chlorobutane has four different substituents bonded to it; it has a chlorine atom, a hydrogen atom, a methyl group, and an ethyl group, as shown in Figure 6.5. A carbon atom bonded to four different groups or atoms is called a **stereocenter.** A molecule that contains a single stereocenter always exists as a pair of enantiomers.

An important property of enantiomers is that interchanging any two substituents of the stereocenter forms the enantiomer of the original structure. Figure 6.6 shows how to do this for 2-chlorobutane. Interchanging the positions of the chlorine and hydrogen atoms in either structure forms its enantiomer. Again you should verify this with your molecular models.

■ *In Section 6.9, we encounter some molecules that contain more than one tetrahedral carbon atom bonded to four different substituents that can also exist as enantiomers.*

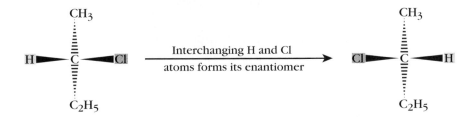

Figure 6.6
Switching the places of any two atoms or groups of atoms bonded to the stereocenter in either structure forms its enantiomer. For 2-chloro-butane, for example, switching the places of the H and Cl atoms forms the corresponding enantiomer.

The broader definition of a stereocenter is an atom bearing several groups of such a nature that an interchange of two groups will produce a stereoisomer. Therefore C2 of 2-chlorobutane is a stereocenter because it is bonded to four different substituents and interchange of any two will form the enantiomer of the original structure.

Interchanging groups or atoms as we have done in this chapter is very simple to imagine and to do on paper or with molecular models. Keep in mind, however, that covalent bonds must be broken in order to carry out these transformations on actual compounds and breaking bonds requires a great deal of energy. As a result, enantiomers (like those of 2-chlorobutane) do not interconvert spontaneously at room temperature. In practical terms, this means that enantiomers can be separated from one another and stored indefinitely without interconverting.

Some molecules, such as 2-chloropropane, are superposable on their mirror images. In Figure 6.7, for example, the representation of 2-chloropropane and its mirror image are superposable because rotating the mirror image 180° generates the original representation. 2-Chloropropane is an achiral molecule and C2 is not a stereocenter.

Figure 6.7
A, 2-Chloropropane and its mirror image. **B,** When the mirror image is rotated by 180°, the two are superposable. Both structures, the original and its mirror image, represent two orientations in space of the same molecule.

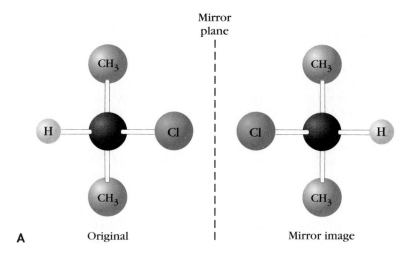

A   Original   Mirror image

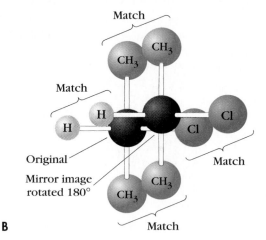

B

Exercise 6.2   Which of the following compounds are chiral? Place an asterisk next to the stereocenter.

(a) Chloroethane   **(d)** 3-Bromohexane
(b) 1-Bromo-1-chloroethane   **(e)** 2-Chloro-1,5-diiodohexane
(c) 3-Chloropentane   **(f)** 3,4,5-Trichloro-3,5-diethylheptane

---

Exercise 6.3   Write three-dimensional structures of the enantiomers of the compounds in Exercise 6.2 that have stereocenters.

Another way of determining whether a molecule is chiral is to look for an element of symmetry, such as a plane of symmetry. A **plane of symmetry** is an imaginary plane that cuts a molecule (or object) into equal halves that are mirror images of each other. The cup shown in Figure 6.8, *A,* has a plane of symmetry. The left half of the cup is the mirror image of the right half. The left hand shown in Figure 6.8, *B,* does not have a plane of symmetry; its left half is not the mirror image of the right half.

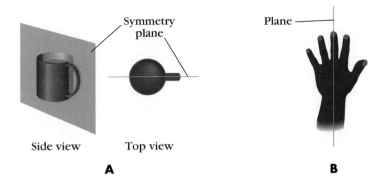

Side view   Top view

A   **B**

Figure 6.8
Symmetry planes that cut objects in half. **A,** The cup has a plane of symmetry because the left half is the mirror image of the right half. **B,** The plane through the left hand is not a plane of symmetry because the left half is not the mirror image of the right.

Molecules that have a plane of symmetry are superposable on their mirror images, so they are achiral. An example is chloroiodomethane shown in Figure 6.9, *A.* The structure is written so that the page is the plane of symmetry that bisects the spherically symmetrical iodine, chlorine, and carbon atoms. As a result, one half of the structure is the mirror image of the other and chloroiodomethane is achiral. Replacing one hydrogen of chloroiodomethane with another atom or group destroys its symmetry, as shown in Figure 6.9, *B.* Thus 1-chloro-1-iodoethane has no plane of symmetry and is a chiral compound.

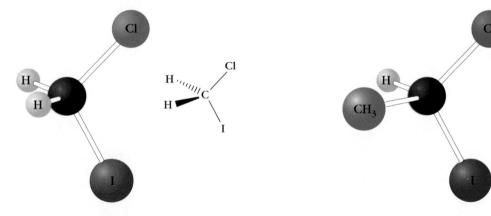

A Chloroiodomethane:
The page is a plane of symmetry
so the molecule is achiral

B 1-Chloro-1-iodoethane:
There is no plane of symmetry
so the molecule is chiral

Figure 6.9
A, Chloroiodomethane has a plane of symmetry, making one side of the molecule the mirror image of the other side. Therefore chloroiodomethane is achiral. **B,** 1-Chloro-1-iodoethane has no plane of symmetry, so it is chiral.

Exercise 6.4	The cup in Figure 6.8 has a plane of symmetry. Where would you place the word MOM or DAD on the cup so that **(a)** the cup no longer has a plane of symmetry or **(b)** the plane of symmetry is retained?

The presence of a single stereocenter is simply another way of saying that a compound lacks a plane of symmetry. This is evident from the chiral compound 1-chloro-1-iodoethane. It lacks a plane of symmetry (Figure 6.9), but it contains a stereocenter (C1). In this case, therefore, either method leads to the same conclusion.

Chiral compounds are very common both as substances obtained from nature and as the products of chemical reactions. Detecting stereocenters or planes of symmetry in complex molecules takes practice because these features are not always readily apparent. This is particularly the case in cyclic compounds. For example, are the two following cyclic compounds chiral?

Chlorocyclohexane
(achiral)

2-Chlorocyclohexanone
(chiral)

To determine the chirality of chlorocyclohexane, let's try to find a stereocenter. Carbon atoms 2, 3, 4, 5, and 6 are not stereocenters because each one is bonded to two hydrogen atoms. The remaining carbon atom, C1, is bonded to a chlorine atom and a hydrogen atom. The remaining two substituents are part of the ring. To decide if they are identical or different, we compare the path around the ring in both directions to find the first point of difference. If there is no difference, the two groups are identical; if there is a difference the two groups are different. In chlorocyclohexane, the path from C2 to C6 is equivalent to the path C6 to C2. That is, both paths are equivalent because they both contain a chain of five CH_2 groups. Thus C1 is not a stereocenter and chlorocyclohexane is achiral.

An alternative way of arriving at the same conclusion is to realize that chlorocyclohexane has a plane of symmetry passing through the chlorine atom, the H atom on C1, C1, the two H atoms on C4, and C4.

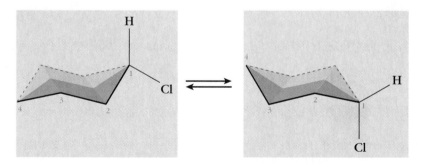

The situation is different for 2-chlorocyclohexanone because it has no plane of symmetry and C2 is a stereocenter. The four substituents bonded to C2 are a chlorine atom, a H atom, a C=O group, and a —CH_2 group. Thus 2-chlorocyclohexanone is chiral. Make molecular models of chlorocyclohexane and 2-chlorocyclohexanone to verify these conclusions.

There is one final point about recognizing chiral compounds. Frequently, the mirror image of compounds such as substituted cyclohexanes and acyclic hydrocarbons can be written so that they appear to have an enantiomeric relationship. For example, we can write the structure of the gauche conformation of butane and its mirror image as shown in Figure 6.10. While these two structures appear to be nonsuperposable mirror images, the two are interconverted simply by rotation about the central C—C bond. Rotation occurs rapidly at room temperature so the two gauche conformations of butane are not enantiomers. Therefore a

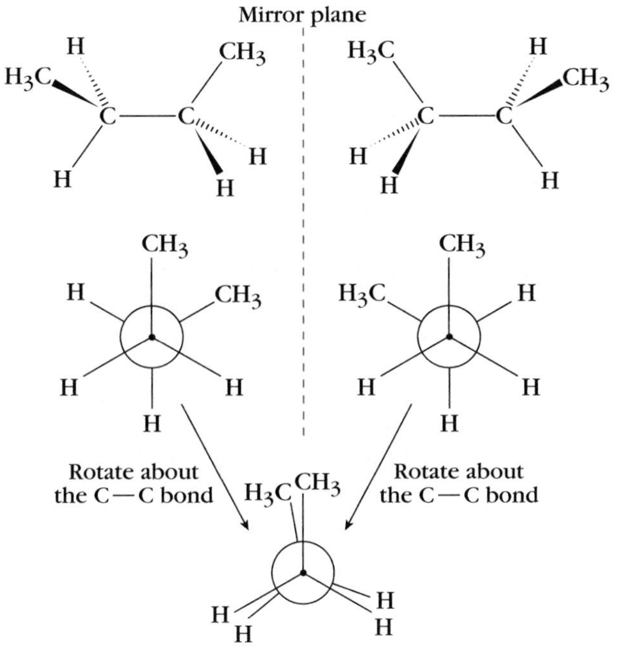

Figure 6.10
The gauche conformation of butane and its mirror image. The two are interconverted simply by rotation about the C—C bond, so butane is achiral.

molecule cannot exist as a pair of enantiomers if it can be converted into its mirror image by a conformational change such as rotation about a single bond or by ring flipping.

■ *The conformations of butane are described in Section 4.5.*

Exercise 6.5   Which of the following compounds are chiral?

(a) 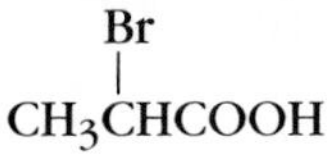 CH$_3$CHCOOH   (Br on CH)

(b) 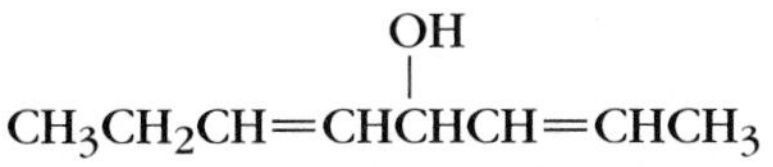 CH$_3$CH$_2$CH=CHCHCH=CHCH$_3$ (OH on middle CH)

(c) 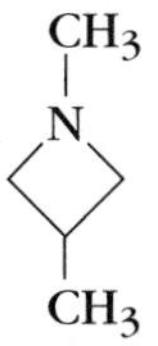 (azetidine ring with N—CH$_3$ and CH$_3$)

(d) 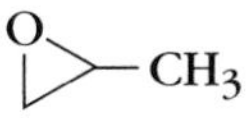 (epoxide)—CH$_3$

(e) 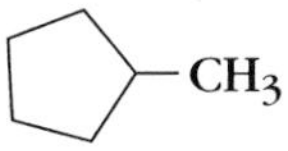 (cyclopentane)—CH$_3$

We have learned to identify chiral compounds from their structural formulas. Now let's learn how to name them.

6.2   Naming Enantiomers

We have learned to convey pictorially the stereochemistry of a molecule in a number of ways. But drawing pictures or structures is cumbersome and time-consuming. We need a method to indicate the particular arrangement of atoms and groups in space characteristic of a given stereocenter (its **configuration**). Furthermore, we must be able to incorporate the configuration of the stereocenter of a compound into its name.

A set of rules to do this was devised by three chemists, R.S. Cahn (England), C.K. Ingold (England), and V. Prelog (Switzerland). This set of rules is called the **Cahn-Ingold-Prelog Rules,** and it designates the configuration of the stereocenter of one enantiomer (*R*), from the Latin *rectus,* meaning right handed, and the other (*S*), from the Latin *sinister,* meaning left handed. This designation is then added to the IUPAC name of the compound. According to this system, one enantiomer of 2-chlorobutane is named (*S*)-2-chlorobutane, while the other is named (*R*)-2-chlorobutane. The (*R*) and (*S*) configurations of a stereocenter are assigned as the following discussion indicates.

Each substituent attached to the stereocenter is assigned a **priority.** Priorities are assigned by applying the following four rules:

Rule 1. Priorities are first assigned on the basis of atomic number. *The higher the atomic number of the atom bonded directly to the stereocenter, the higher its priority.* For example:

Element	Br	>	Cl	>	F	>	O	>	N	>	C	>	H
Atomic Number	35		17		9		8		7		6		1

⟸ INCREASING PRIORITY

Rule 2. In the case of isotopes, the isotope of highest atomic mass has the higher priority. For the three isotopes of hydrogen, tritium (T, atomic mass 3), deuterium (D, atomic mass 2), and hydrogen (H, atomic mass 1), the order of priority is:

Isotope	T	>	D	>	H
Atomic Mass	3		2		1

⟸ INCREASING PRIORITY

Rule 3. If priority cannot be assigned by applying Rules 1 and 2 because two atoms of the same atomic number are bonded to the stereocenter, compare atomic numbers of the second atoms in each group. Continue comparing atomic numbers as necessary until a difference is found. A priority assignment is made at this first point of difference on the basis of atomic number. For example, the priorities of the methyl and ethyl groups bonded to the stereocenter of 2-chlorobutane are determined as follows.

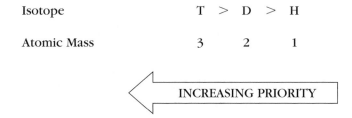

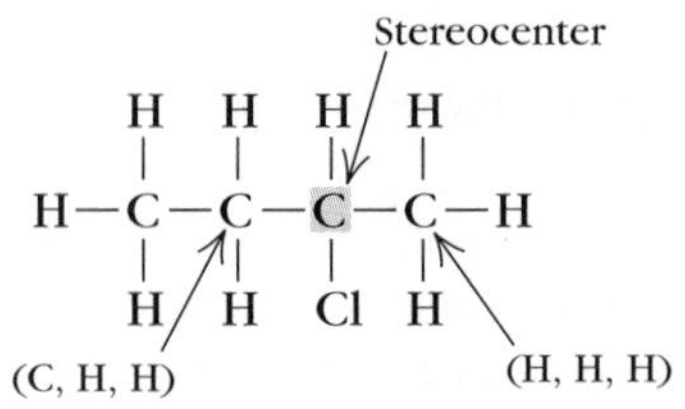

For both ethyl and methyl, the atom bonded to the stereocenter is a carbon, so we must compare the second atoms in each group. In the methyl group, these are three hydrogen atoms (H, H, H). In the ethyl group, the second atoms are one carbon and two hydrogen atoms (C, H, H). This is the first point of difference between the two groups. Because carbon (atomic number 6) has a higher priority than hydrogen (atomic number 1), we assign the higher priority to the ethyl group. The complete order of priority of the four groups bonded to the stereocenter of 2-chlorobutane is:

$$-\text{Cl (highest priority)} > -\text{C}_2\text{H}_5 > -\text{CH}_3 > -\text{H (lowest priority)}$$

■ *An ethyl group (CH_3CH_2—) is frequently abbreviated C_2H_5—.*

Once the priorities of the four substituents bonded to the stereocenter have been assigned, *the molecule (or the molecular model) is rotated so that the substituent of lowest priority, given the number 4, is directed away from us.* We then look at the remaining three substituents and trace a path from the substituent of highest priority, number 1, to the one of second-highest priority, number 2, and finally to the one of third-highest priority, number 3. If the direction of this path is clockwise, the enantiomer is assigned the (*R*) configuration. If the direction is counterclockwise, the enantiomer is assigned the (*S*) configuration.

The application of these rules is illustrated in Sample Problems 6.1 and 6.2.

Sample Problem 6.1   Give the IUPAC name for the following enantiomer of bromochlorofluoromethane:

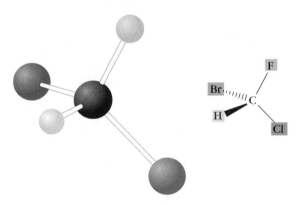

Solution:

Step 1 Assign priorities to all atoms bonded to the stereocenter. The four atoms in order of decreasing atomic number are —Br > —Cl > —F > —H. The order of priority of the atoms is:

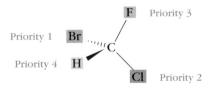

Step 2 Rotate the structure so that the group of lowest priority, the hydrogen atom, is pointed away from the viewer.

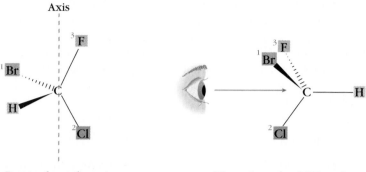

Rotate about the axis
until you can view along
the C-H bond

View along the C-H bond

When viewed along the carbon-hydrogen bond, the remaining three substituents look like this:

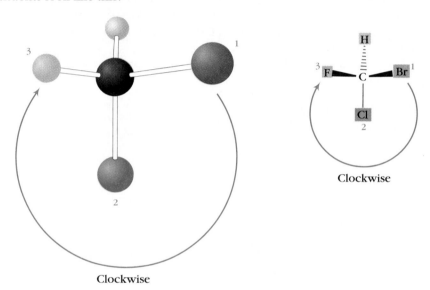

Clockwise

Clockwise

The path from the substituent of priority 1 (Br) to 2 (Cl) to 3 (F) is clockwise; therefore the stereocenter is designated (*R*). The correct name of this enantiomer is (*R*)-bromochlorofluoromethane.

Exercise 6.6   Exchange the positions of the fluorine and chlorine atoms in the three-dimensional representation of bromochlorofluoromethane given in Sample Problem 6.1 and determine whether the stereocenter is (*R*) or (*S*).

Sample Problem 6.2   The structural formula of 2-chloro-1-propanol is:

$$\begin{array}{c} \text{Cl} \\ | \\ \text{CH}_3\text{CHCH}_2\text{OH} \end{array}$$

Draw the three-dimensional representations for each enantiomer and give (*R*) and (*S*) designations to each.

Solution:

The following is one way of representing the three-dimensional structures of the two enantiomers:

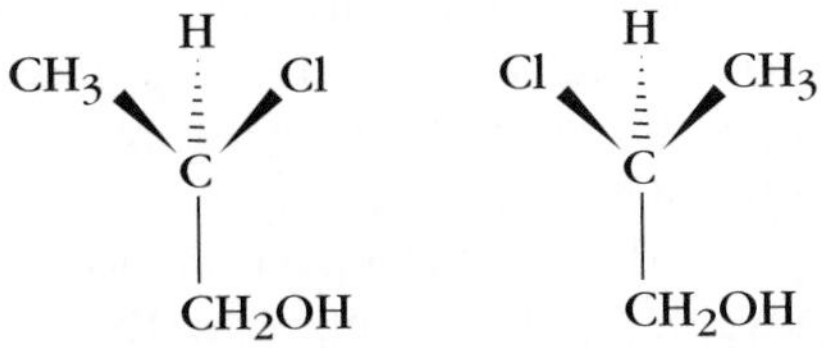

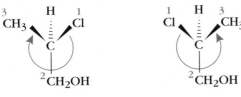

The chlorine has the highest priority and the hydrogen has the lowest. The methyl group has three hydrogens bonded to its carbon, while the $-CH_2OH$ group has two hydrogens and one oxygen atom. Oxygen has a higher priority than hydrogen so the $-CH_2OH$ group has higher priority than the $-CH_3$ group. The order of priority is:

$$-Cl > -CH_2OH > -CH_3 > -H$$

The structures shown previously are already correctly oriented with the substituent of lowest priority pointing away from us. Therefore:

<table>
<tr>
<td align="center">

$\overset{3}{CH_3}$ H $\overset{1}{Cl}$
C
$\overset{2}{CH_2OH}$

</td>
<td align="center">

$\overset{1}{Cl}$ H $\overset{3}{CH_3}$
C
$\overset{2}{CH_2OH}$

</td>
</tr>
<tr>
<td align="center">

Clockwise path from
1 to 2 to 3; therefore
this is the (*R*) configuration

</td>
<td align="center">

Counterclockwise
path from 1 to 2 to 3;
therefore this is the (*S*)
configuration

</td>
</tr>
</table>

The structures of the two enantiomers in this problem were drawn specifically so that the hydrogen atom, the atom of lowest priority, is directed away from the viewer. This will not always be the case. Usually, the stereochemical representation of a molecule must be rotated to place the substituent of lowest priority away from the viewer. It is therefore important to develop the ability to mentally visualize the rotation of a structure in space. Without such an ability, assigning the configuration of stereocenters is extremely difficult. Comparing the written structures with molecular models is invaluable in developing this ability. In Sample Problem 6.3, we demonstrate one example of the kinds of movement of structures in space that must be mastered.

| Sample Problem 6.3 | Are the following structures of 2-chlorobutane, *A* and *B*, superposable or do they represent a pair of enantiomers? |

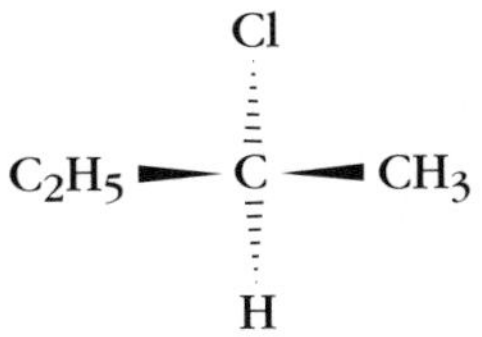

<table>
<tr>
<td align="center">

Cl
|
C_2H_5 — C — CH_3
|
H

A

</td>
<td align="center">

    Cl
C_2H_5 C
CH_3
      H

B

</td>
</tr>
</table>

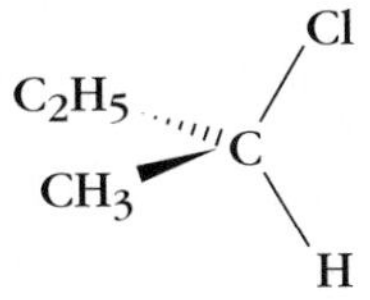

Solution:

To determine the relationship between these two structures, we must rotate one so that they are both represented in the same way. One method is to rotate structure *B* so that its groups are arranged like those in structure *A*. Imagine a vertical axis in the plane of the page through the stereocenter of structure *B*. Rotating *B* 90° around this axis places the Cl and H behind the plane of the page in the same position as the Cl and H atoms in structure *A*. It also places the —C_2H_5 and —CH_3 groups in the same position, so structure *B* is the same as structure *A*. They both represent the same molecule.

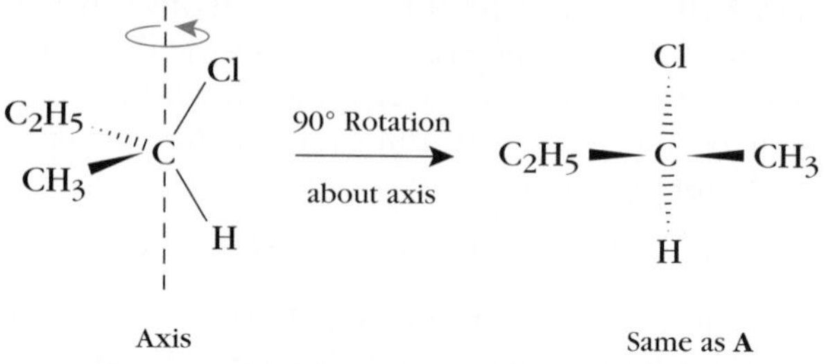

There are many ways of mentally rotating structures *A* and *B* to compare them. One way is shown here. You may have another way that is easier for you. Use whichever method is the simplest. But check your conclusions with molecular models until you become confident of your ability to mentally rotate structures written on a page.

| Exercise 6.7 | Decide whether the following pairs of structures are superposable or are enantiomers. |

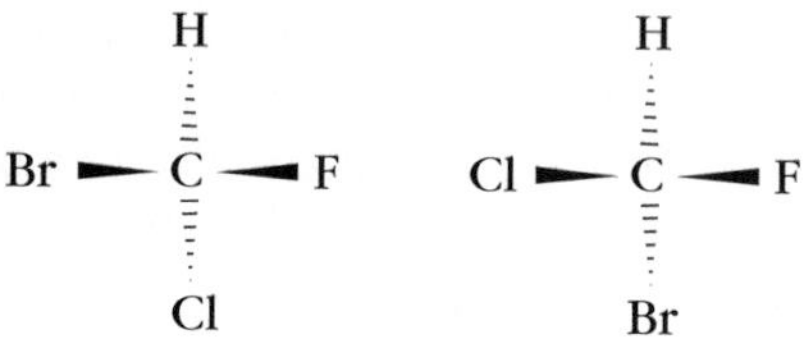

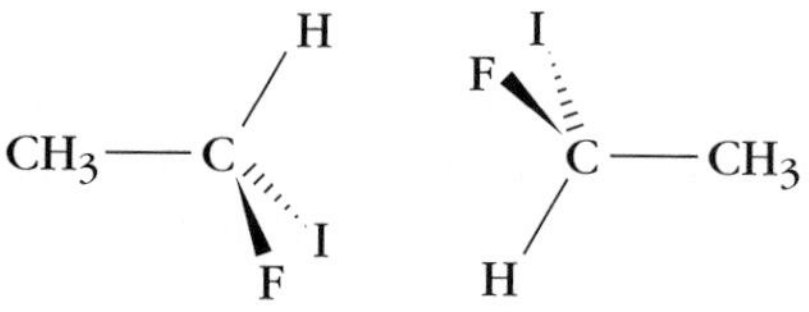

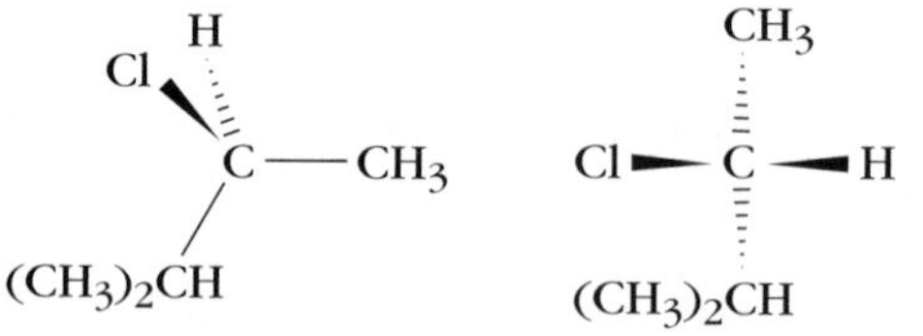

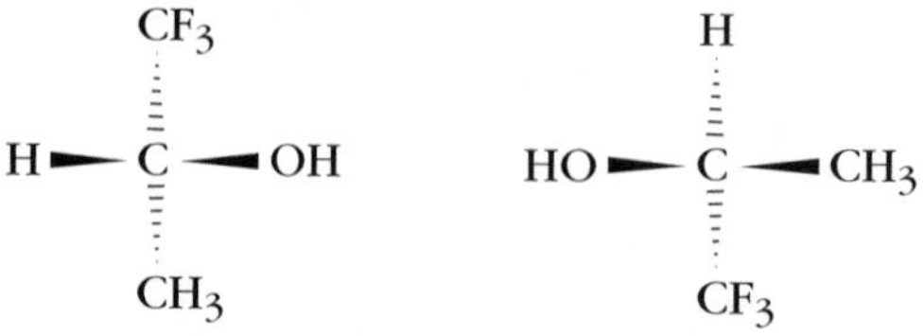

Exercise 6.8 Arrange the following sets of atoms and groups in order of increasing priority.

(a) $-I$, $-H$, $-CH_3$, $-Cl$
(b) $-C(CH_3)_3$, $-CH_3$, $-CH_2CH_3$, $-CH(CH_3)_2$
(c) $-CH_3$, $-CH_2OH$, $-OH$, $-CH_2Cl$
(d) $-CH_3$, $-CH_2D$, $-CH_2CH_3$, $-H$

Exercise 6.9 Give the correct IUPAC name to each of the following enantiomers:

(a) 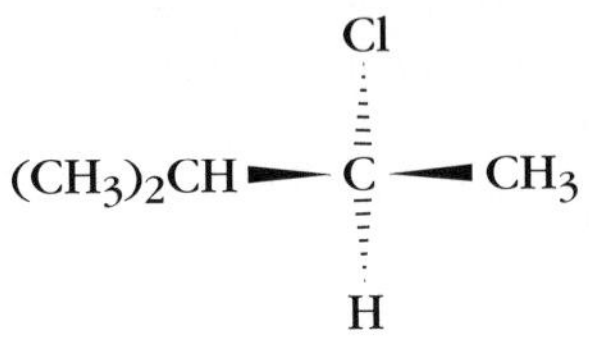 (CH$_3$)$_2$CH—C—CH$_3$ (with Cl up, H down)

(c) Cl—C—CH$_2$Cl (with H up, CH$_3$ down)

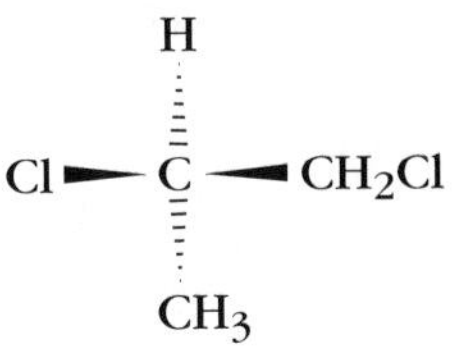

(b)

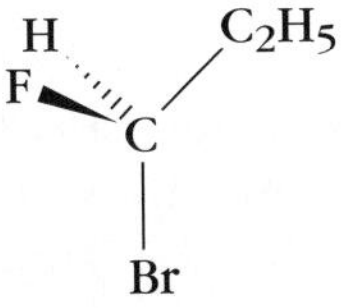

(d)

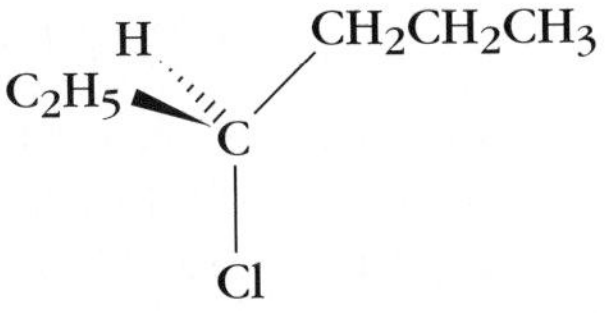

Exercise 6.10 Write three-dimensional structural formulas for each of the following names:

(a) (*S*)-1-Chloro-1-fluoroethane
(b) (*R*)-3-Iodo-2-methylpentane
(c) (*R*)-3-Chloro-2,2,3-trimethylpentane
(d) (*S*)-1-Chloro-1-deuterioethane (CH$_3$CHDCl)

The Cahn-Ingold-Prelog Rules given so far allow us to assign priorities to all substituents that contain single bonds. An additional rule is necessary to determine the priority of substituents that contain multiple bonds.

Rule 4. For the purpose of assigning priorities, a double bond is considered as two single bonds for both atoms of the double bond.

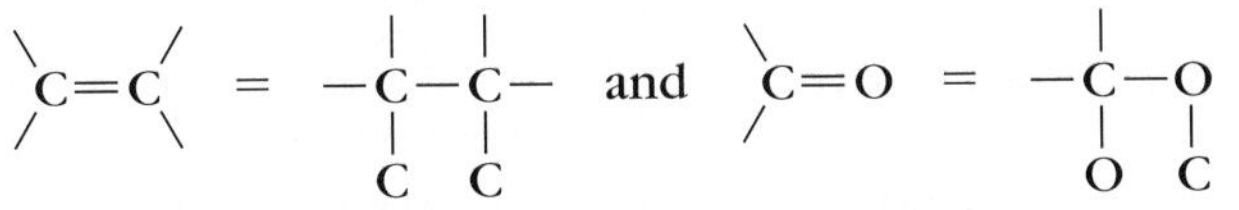

Similarly triple bonds are regarded as three single bonds for both atoms.

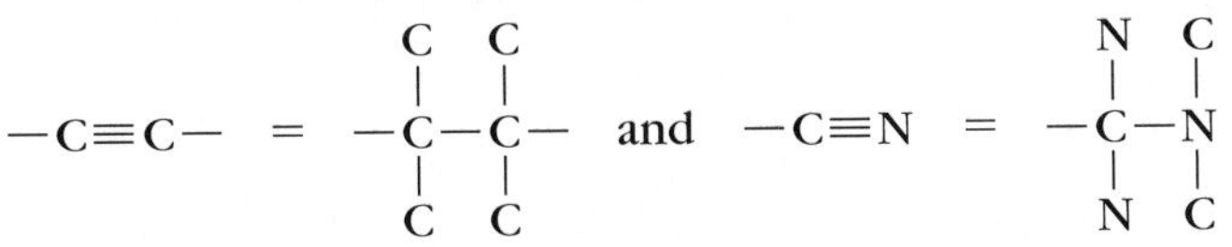

The use of this additional rule is illustrated in Sample Problems 6.4 and 6.5.

Sample Problem 6.4

Assign the configuration of the stereocenter in the following compound.

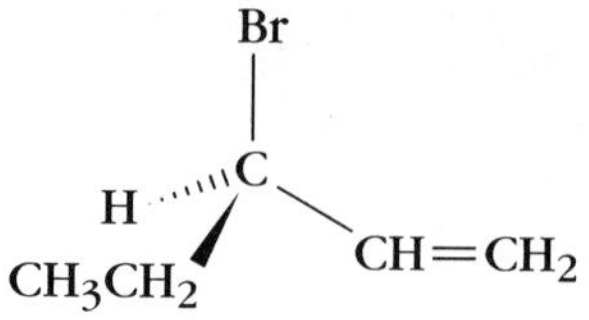

Solution:

The substituent of highest priority is Br, while H is the one of lowest priority. The carbon of the ethyl group bonded to the stereocenter contains two hydrogen atoms and a carbon.

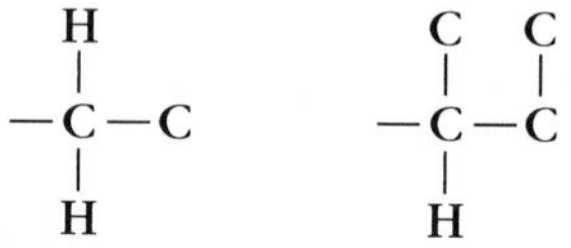

The carbon atom of the other group bonded to the stereocenter, $CH=CH_2$, is considered to have one hydrogen atom and two carbon atoms. Thus the order of priority is $-Br > CH=CH_2 > -C_2H_5 > -H$, and the configuration is assigned as follows:

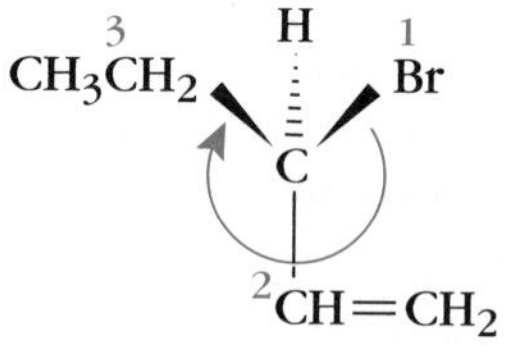

Direction of decreasing priority
is clockwise so the configuration is (R)

Exercise 6.11   Assign priorities to each of the following sets of substituents.

(a) $-OH$, $-CH_3$, $-CH(CH_3)_2$, $-CH_2CH_2Cl$

(b)

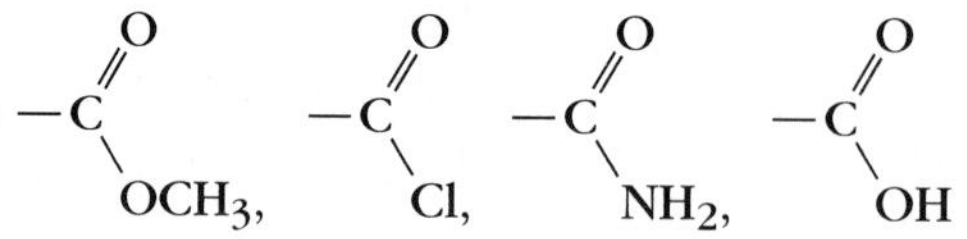

(c) $-C\equiv N$, $-C(CH_3)=CH_2$, $-CH_2NH_2$, $-CH_2Cl$

Sample Problem 6.5

Designate the stereocenter in the following compound as (R) or (S):

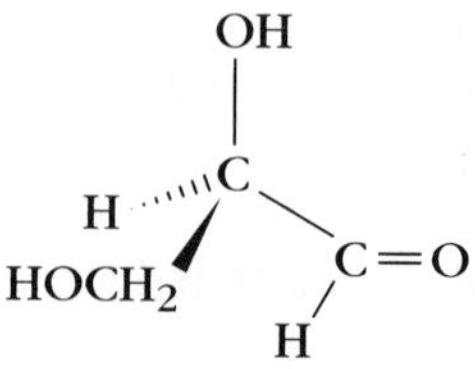

Solution:

The substituent of highest priority is the —OH group, while H is the one of lowest priority. The carbon of the —CH_2OH group has two hydrogens and an oxygen atom bonded to it, while the carbon of the —CHO group is considered to have two oxygens and one hydrogen bonded to it:

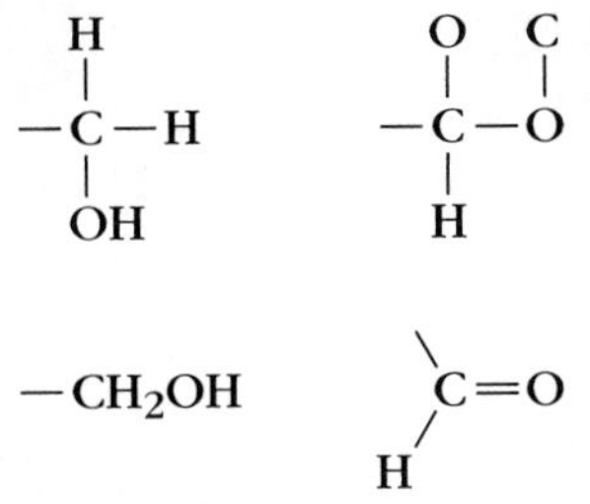

$$
\begin{array}{cc}
\overset{\textstyle H}{\underset{\textstyle OH}{-C-H}} & \overset{\textstyle O \quad C}{\underset{\textstyle H}{-C-O}}
\end{array}
$$

$$
-CH_2OH \qquad \overset{\textstyle \diagdown}{\underset{\textstyle H}{\diagup}}C=O
$$

Thus the order of priority is —OH > —CHO > —CH_2OH > —H, and the configuration is assigned as follows:

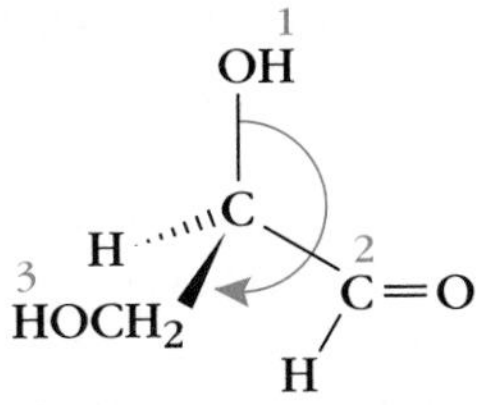

Direction of decreasing priority
is clockwise so the configuration is (R)

Exercise 6.12   Designate the stereocenter in each of the following compounds as (R) or (S).

(a)

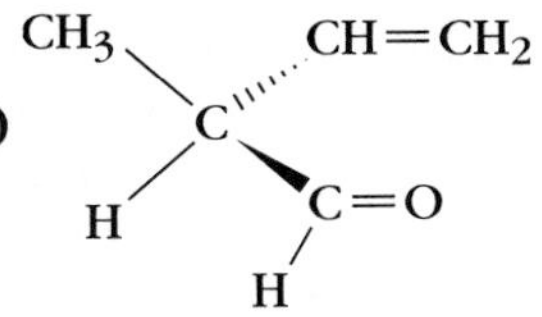

(c) CH_3 ━ C ◄ C≡CH with $CH(CH_3)_2$ above and $CH=CH_2$ below

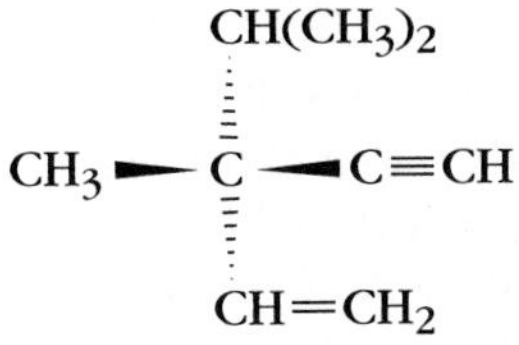

(b)

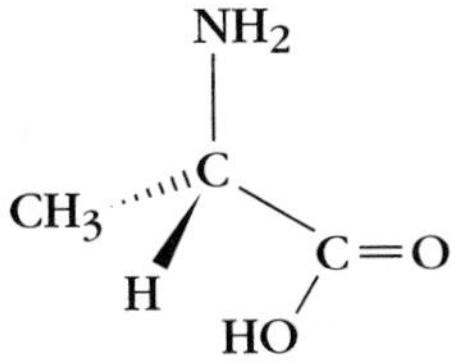

(d)

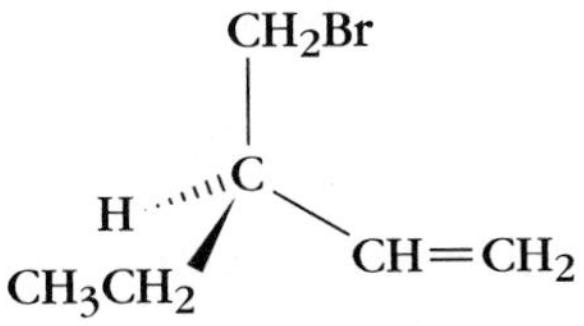

We've learned how to identify enantiomers and how to name them based on their structural formulas. We will now learn how to detect enantiomers in the laboratory.

6.3   Optical Activity

Enantiomers are different compounds because they're nonsuperposable mirror images that can't be interconverted without breaking chemical bonds. While we

Figure 6.11
Schematic representation of **(A)** end view of ordinary light showing that vibrations occur in all perpendicular directions simultaneously and **(B)** end view showing vibrations of plane-polarized light occur entirely in a single plane.

A Ordinary light   **B** Plane-polarized light

can recognize the difference between enantiomers on the molecular level, what techniques do we use in the laboratory to see these differences? One easily observed difference is the behavior of enantiomers toward plane-polarized light.

What is plane-polarized light? In Chapter 5, an oscillating or vibrating wave model of light was introduced to explain the interaction of light and matter. These vibrations occur in a plane perpendicular to the direction of light. Thus the vibrations of the whole beam of light occur in all the perpendicular directions simultaneously so that if we were to look at the beam end-on, we would see the vibrations in various directions, as shown in Figure 6.11, *A*. When ordinary light is passed through a polarizer, the polarizer interacts with the light beam so that waves of the light that emerges oscillate in only one plane, as shown in Figure 6.11, *B*. Such light is called *plane-polarized light*.

We can illustrate the different behavior of enantiomers toward plane-polarized light by using quartz crystals as an example. Quartz crystals exist in two forms that are nonsuperposable mirror images of each other. These two forms of quartz crystals have the same relationship as enantiomers of a chiral compound.

When a beam of plane-polarized light passes through these two forms of quartz crystals, the beam of light is affected differently. One form of quartz rotates the plane of polarized light in one direction, while the other form rotates the plane in the opposite direction. The crystal form that rotates the plane of polarized light to the right when facing the source of light is called **dextrorotatory** (from Latin *dextro,* meaning *right*). The other crystal form rotates the plane of polarized light to the left, so it is called **levorotatory** (from Latin *levo,* meaning *left*). When substances rotate the plane of polarized light they are said to be **optically active.**

Enantiomers are optically active substances too. Thus when a beam of plane-polarized light is passed through a solution containing only one enantiomer, the light is rotated in one direction. Passing the same beam through a solution containing only the other enantiomer causes the beam to rotate in the opposite direction. By convention, a minus sign ($-$) is placed before the name of an optically active compound that is levorotatory and a plus sign ($+$) is placed before the name of an optically active compound that is dextrorotatory. For example, ($-$)-cholesterol is levorotatory and ($+$)-sucrose is dextrorotatory.

The effect of plane-polarized light on optically active compounds is observed and measured with an instrument called a **polarimeter.** A schematic representation of a polarimeter is shown in Figure 6.12. A polarimeter consists of a light source (usually a sodium lamp), a polarizer, a sample tube in which the optically active substance (usually in solution) is held in the light beam, and an analyzer (usually another polarizer) that measures the extent (in degrees) by which the beam of plane-polarized light has been rotated by the optically active substance.

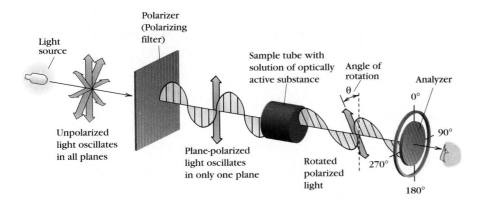

Figure 6.12
Schematic representation of a polarimeter.

When the substance in the sample tube is optically inactive, the plane-polarized light emerges from the tube with its plane unchanged. If a solution of an optically active compound is placed in the sample tube, it will rotate the plane-polarized light. That is, the plane of the light that emerges from the tube is not parallel to the plane of the light that entered the tube. Also, the intensity of the light that reaches the eyepiece is reduced. The original intensity of the light can be restored by rotating the analyzer left or right until its axis is parallel to the plane of the light emerging from the sample tube. The angle of rotation of the analyzer (expressed in degrees) is designated α and is called the *observed rotation.*

The observed rotation of an optically active compound depends on a number of factors that are all taken into account in a quantity called the *specific rotation* of an optically active compound.

6.4 Specific Rotation

The amount of rotation that plane-polarized light undergoes as it passes through a solution depends on the number of optically active molecules that it encounters. The number of molecules in its path depends on the length of the sample tube and the concentration of the solution. Suppose, for example, that we have two sample tubes, one twice as long as the other, and both contain the same concentration of the same optically active compound. The observed rotation of the longer tube will be twice that of the shorter one because the longer tube contains twice the number of optically active molecules as the shorter tube. Similarly, a solution twice as concentrated produces twice the observed rotation. Other factors that influence the rotation are the temperature of the sample and the wavelength of the polarized light. In order to take these factors into account and to standardize observed rotations, chemists have defined a quantity called **specific rotation,** which is given the symbol $[\alpha]$. It is calculated from the observed rotation by the following equation:

$$[\alpha] = \frac{[\text{Observed rotation in degrees}]}{[\text{Path length, l, in dm}]\,[\text{Concentration, }C\text{, in g/ml}]} = \frac{\alpha}{1 \times C}$$

Because specific rotation also depends on the temperature, the wavelength of

Table 6.1	Specific Rotations of Some Organic Compounds	
Compound	Solvent	$[\alpha]_D$ (Degrees)
(+)-Camphor	Ethyl alcohol	+43.8
(−)-Cholesterol	Chloroform	−39.5
(+)-Tartaric acid	Water	+12.7
(+)-Sucrose	Water	+66.5
β-(+)-Glucose	Water	+52.7
β-(−)-Fructose	Water	−92.4
(−)-Mandelic acid	Water	−158
(+)-Mandelic acid	Water	+158

light, and the solvent, these quantities are usually reported as well. Thus the specific rotation of a compound would be reported as follows:

$$[\alpha]_D^{25} = -5.75° \ (H_2O)$$

The subscript D means that the observed rotation was measured while using the light of wavelength 589 nm emitted from a sodium lamp (the so-called D-line of sodium). The superscript 25 means that a temperature of 25 °C was maintained during the measurement. The solvent used is indicated by placing its chemical formula in parentheses after the value of the specific rotation. The specific rotation, −5.75°, means that a sample dissolved in water containing 1 g/mL of the optically active substance in a 1-dm tube produces a rotation of 5.75 degrees in a counterclockwise direction. When specific rotations are expressed in this standard way, they are a physical constant of an optically active compound just like its boiling point or melting point. The specific rotations of some organic compounds are listed in Table 6.1.

Exercise 6.13	The observed rotation, taken in a 10-cm sample tube, of a solution of 2.350 grams of (+)-nicotine in 100 ml of ethyl alcohol is +3.835 at 20 °C. What is the specific rotation of (+)-nicotine?

We have focused our attention on the optical rotation of pure enantiomers. What happens to the optical rotation of a solution that contains both enantiomers?

6.5 Equimolar Mixtures of Enantiomers Are Optically Inactive

Many chemical reactions form products that contain a stereocenter, yet most of these products are optically inactive. How can this be? The answer is that the product is an equimolar mixture of the two enantiomers. We can explain this result by examining a chemical reaction that forms 2-chlorobutane.

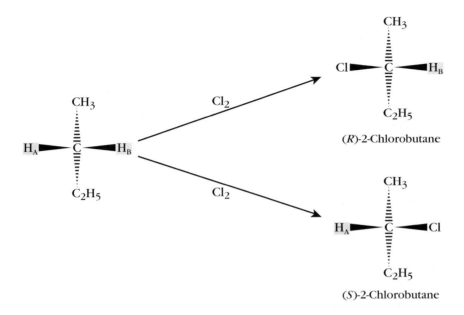

Figure 6.13
The formation of optically inactive 2-chlorobutane by the chemical reaction of chlorine and butane. Substitution of H_A or H_B by chlorine is equally likely so equal quantities of the two enantiomers are formed.

(R)-2-Chlorobutane

(S)-2-Chlorobutane

Recall from Chapter 2 that chlorine reacts with alkanes to form chloroalkanes. Chlorine reacts with butane to form two monochlorinated products, 1-chlorobutane and 2-chlorobutane. The substitution of a hydrogen on C2 of butane by a chlorine atom in this chemical reaction forms 2-chlorobutane. There is no reason why one hydrogen atom of C2 should be substituted in preference to the other, so half the molecules of 2-chlorobutane are formed by replacing one hydrogen, H_A, and the other half are formed by replacing the other hydrogen, H_B, as shown in Figure 6.13. The result of this chemical reaction is that equal quantities of the two enantiomers, (R)-2-chlorobutane and (S)-2-chlorobutane, are formed. A 50:50 mixture of two enantiomers is called a **racemic mixture** or **racemate.**

A solution of a racemic mixture is optically inactive because it contains equal concentrations of the two enantiomers. The molecules of one enantiomer rotate the plane of plane-polarized light to the right, but the molecules of the other enantiomer rotate the plane by an equal amount to the left. The net result is that the plane of the polarized light leaves the sample tube unchanged.

A sample of a compound that is a racemic mixture is designated by the symbol (±) at the beginning of the name of the compound. For example, the specific rotation, $[\alpha]_D$, of (+)-leucine is +10.8°, while $[\alpha]_D$ of (−)-leucine is −10.8°. An equal mixture of (−)-leucine and (+)-leucine has a specific rotation of 0° and it is designated (±)-leucine.

A sample of a pure enantiomer has a specific rotation that is an intrinsic property of that stereoisomer. An equal mixture of enantiomers is optically inactive, $[\alpha]_D = 0$. These are the two extreme cases: a pure enantiomer and a mixture of equal amounts of two enantiomers. What about mixtures that contain unequal amounts of two enantiomers? These kinds of mixtures are discussed next.

6.6 Enantiomeric Excess and Optical Purity

A sample of a compound that contains a single enantiomer is said to be **enantiomerically pure.** An example of an enantiomerically pure sample is one that contains only (+)-leucine or only (−)-leucine. A sample that contains (+)-leucine and less than an equimolar amount of (−)-leucine has an **enantiomeric excess** of (+)-leucine. Enantiomeric excess is defined as the fractional excess of one enantiomer over the other.

$$\begin{matrix}\text{Percent} \\ \text{Enantiomeric} \\ \text{Excess}\end{matrix} = 100\%\left(\frac{\begin{matrix}\text{Moles of} \\ \text{enantiomer in excess}\end{matrix} - \begin{matrix}\text{Moles of} \\ \text{other enantiomer}\end{matrix}}{\text{Total moles of both enantiomers}}\right)$$

The use of this equation is illustrated in Sample Problem 6.6.

Sample Problem 6.6   Calculate the percent enantiomeric excess of a solution that contains one mole of a dextrorotatory enantiomer and a half mole of its levorotatory isomer.

Solution:

There are more moles of the dextrorotatory isomer in solution, and therefore the dextrorotatory isomer is present in enantiomeric excess. The percent of enantiomeric excess in the solution, which is usually abbreviated as %ee, is calculated as follows:

$$\%ee = (100\%)\left(\frac{1\text{ mole} - 0.5\text{ mole}}{1.5\text{ mole}}\right) = (100\%)\left(\frac{0.5\text{ mole}}{1.5\text{ mole}}\right) = 33\%$$

We can also use this equation to calculate the relative amounts of each enantiomer present when we know the percent enantiomeric excess of a sample. This is illustrated in Sample Problem 6.7.

Sample Problem 6.7   If a sample of 2-chlorobutane has an enantiomeric excess of 70% (−)-2-chlorobutane, what percent of each enantiomer is present in the sample?

Solution:

Imagine that our sample contains a certain number of molecules, such as 100. Seventy of those molecules (70%) are (−)-2-chlorobutane. The remaining 30 molecules (100 − 70) are an equal mixture of 15 molecules of (+)-2-chlorobutane and 15 molecules of (−)-2-chlorobutane. The total number of (−)-2-chlorobutane molecules is 70 + 15 = 85. Thus 85% of our sample is (−)-2-chlorobutane and 15% is (+)-2-chlorobutane. We can check this conclusion by calculating the %ee and seeing if it agrees with the value given originally.

$$\%ee = \left(\frac{85 - 15}{100}\right)(100\%) = 70\%$$

Percent enantiomeric excess is equal to and is another way of expressing **optical purity.** The percent optical purity (%op) of a mixture is defined as the ratio of its specific rotation to the specific rotation of a pure enantiomer.

$$\%\text{op} = (100\%)\left(\frac{\text{Observed specific rotation of mixture}}{\text{Observed specific rotation of pure enantiomer}}\right)$$

Optical purity is often easier to calculate because the specific rotation of a mixture is easily obtained in the laboratory, and the specific rotation of a pure enantiomer is frequently available in the chemical literature. Sample Problem 6.8 illustrates how to calculate the percent optical purity of an optically active sample of 2-butanol.

Sample Problem 6.8

The observed specific rotation of a sample of 2-butanol is +7.53°. The specific rotation of an enantiomerically pure sample of (+)-2-butanol is +13.52°. What is the percent optical purity of this sample?

Solution:

$$\%\text{op} = (100\%)\left(\frac{\text{Observed specific rotation of mixture}}{\text{Observed specific rotation of pure (+)-2-butanol}}\right)$$

$$\%\text{op} = (100\%)\left(\frac{7.53°}{13.52°}\right) = 55.7\%$$

The specific rotation of the sample of 2-butanol described in Sample Problem 6.8 is dextrorotatory because there is an excess of the dextrorotatory enantiomer in the sample. We can show that this is true by calculating the percent of each enantiomer in the sample by the method shown in Sample Problem 6.7. Thus a sample whose optical purity is 55.7% contains a mixture containing 22.2% (−)-2-butanol and 77.8% (+)-2-butanol.

Exercise 6.14

A solution is made by dissolving 28.35 g of pure cane sugar in 100 ml of water. Its rotation is determined in a 10-cm sample tube and found to be +18.88°. What is the specific rotation of cane sugar?

Exercise 6.15

An aqueous cane sugar solution is placed in a 10-cm sample tube, and its observed rotation is found to be +5.38°. The specific rotation of pure cane sugar in water is +66.6°. What is the concentration of cane sugar in the solution?

Exercise 6.16

Predict the observed rotation if the solution in Exercise 6.15 is **(a)** diluted by adding 100 mL of water and **(b)** the rotation of the diluted solution is observed in a 20-cm sample tube.

Exercise 6.17   What is the molar composition of a sample that contains a 90% enantiomeric excess of (−)-2-chlorobutane?

The specific rotation is a physical property that allows us to distinguish between a compound and its enantiomer. Certain chemical reactions can do this, too.

6.7 Reaction with an Optically Active Reagent Is Another Way to Distinguish Between Enantiomers

The only physical property that distinguishes between a pair of enantiomers is the direction in which they rotate the plane of polarized light. However, enantiomers can also have different chemical properties. For example, when a racemic mixture of malic acid is injected into a rabbit, only the levorotatory isomer is recovered from the rabbit's urine. It does not react in the rabbit's body and is excreted, whereas the dextrorotatory isomer reacts completely.

$$
\begin{array}{c}
COOH \\
| \\
CHOH \\
| \\
CH_2COOH
\end{array}
$$

Malic acid

Epinephrine

Epinephrine is one of a number of hormones secreted in response to low blood glucose level. It increases the amount of glucose released into the blood by the liver and decreases the utilization of glucose by muscle. Epinephrine isolated from animal kidneys is the levorotatory form, and it is thirty times more active in releasing glucose into the blood than the isomeric dextrorotatory form synthesized in the laboratory.

Not all chemical reactions of enantiomers are different. Glyceraldehyde, for example, reacts with mercury (II) oxide (HgO) to form glyceric acid.

Glyceraldehyde $\xrightarrow{\text{HgO}}$ Glyceric acid

There is no detectable difference in the reaction of the two enantiomers of glyceraldehyde.

Why do enantiomers react exactly the same sometimes but quite differently other times? The answer is that *enantiomers react differently only when they react with one stereoisomer of a chiral reagent.* When enantiomers react with achiral reagents, their chemical reactivities are identical. This fact is of supreme importance to the chemistry of all living systems.

The locations where chemical reactions occur in living systems are chiral. Enzymes, for example, which are the catalysts of biochemical reactions, are very elaborate chiral protein molecules. Only one stereoisomer of a particular enzyme

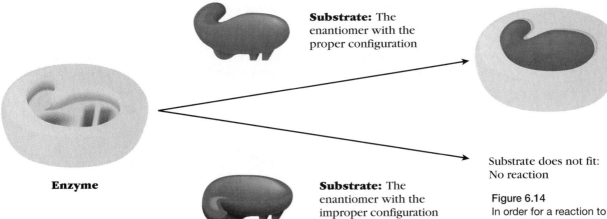

Substrate: The enantiomer with the proper configuration

Enzyme

Substrate: The enantiomer with the improper configuration

Substrate does not fit: No reaction

Figure 6.14
In order for a reaction to occur, the molecule must fit into the chiral reactive site of an enzyme. Only the enantiomer with the correct configuration can fit, so only the correct enantiomer will react. The other enantiomer is unreactive.

is present in a living system, and it has the ability to discriminate between stereoisomers. A simplified view of how enzymes do this is shown in Figure 6.14.

The reactions of enantiomers with reagents are analogous to putting on socks and shoes. Socks are achiral because one sock is the superposable mirror image of the other. Either sock will fit your left foot. Either sock will also fit your right foot. Both socks are the same to your feet. In chemical terms, either member of a pair of enantiomers (one foot) reacts the same with an achiral reagent (a sock). A shoe, however, is chiral. The left shoe is the nonsuperposable mirror image of the right shoe. The left shoe fits (reacts with) the left foot but not the right one (no reaction). Thus your feet (like the enzyme) can discriminate between a pair of enantiomers (left and right shoes).

The ability to discriminate between stereoisomers means that enzymes selectively pick the isomer of correct configuration for a particular reaction. We can now understand why one enantiomer of leucine tastes bitter and the other tastes sweet. Because of their different configurations, the two enantiomers react with different sites on the tongue; one site is responsible for a sweet taste, the other for a bitter taste.

Because of the chirality of living systems, many of the substances obtained from plants and animals are optically active and frequently consist of a single enantiomer. Historically, most optically active compounds available to chemists have been obtained from living organisms. A few examples are listed in Table 6.1 (p. 244).

We can experimentally distinguish between two enantiomers on the basis of the sign of their specific rotations. Also, we may be able to distinguish them on the basis of their reactivity with chiral reagents. But neither of these experiments tells us the configuration of the stereocenters of the enantiomers. How do we relate the configuration of an enantiomer to its specific rotation?

6.8   Absolute and Relative Configurations

Scientists of the late 19th and early 20th centuries were well acquainted with naturally occurring optically active compounds. They were able to isolate such compounds, determine their structural formulas, and measure their specific rotations.

They were frustrated, however, in their attempts to determine the absolute configuration of the stereocenters in these compounds. The **absolute configuration** of a compound is the actual orientation in space of the four substituents bonded to a stereocenter. The problem can be illustrated by examining glyceraldehyde, a compound that contains a single stereocenter and can exist as two enantiomers.

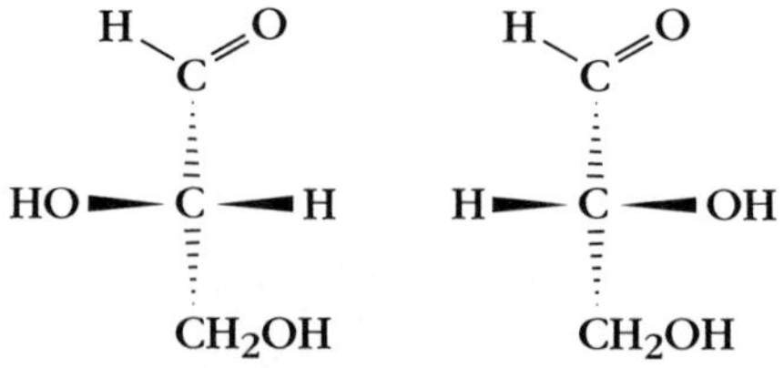

(S)-Glyceraldehyde   (R)-Glyceraldehyde

One of these enantiomers is levorotatory, while the other is dextrorotatory. How do we know whether (R)- or (S)-glyceraldehyde is the levorotatory isomer? The answer is that we can't know by simply looking at the structures. To determine the absolute configuration of an optically active compound requires an experimental method that actually locates the positions of the atoms. Such a technique was not available until 1951.

In 1951, J.M. Bijvoet reported the first successful determination of the actual arrangement in space of the atoms of (+)-tartaric acid by means of a special x-ray diffraction technique. It was only after Bijvoet's work that chemists were able to assign absolute configurations to optically active compounds. Unfortunately, the technique developed by Bijvoet is difficult, time-consuming, and cannot be applied to all compounds. Nevertheless, chemists were able to assign absolute configurations to many compounds without using the x-ray technique of Bijvoet. How is this possible? The answer is that once the absolute configuration of one compound is known, the configurations of others can be related to it by means of chemical reactions.

The following reaction provides a simple example of how configurations of compounds can be related by chemical reactions.

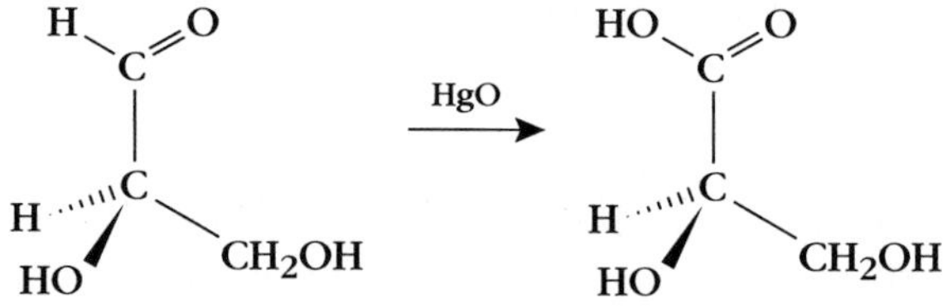

(+)-Glyceraldehyde   (−)-Glyceric acid

The reaction of (+)-glyceraldehyde with mercury (II) oxide forms (−)-glyceric acid. This reaction occurs *without breaking any of the bonds to the stereocenter in (+)-glyceraldehyde.* As a result, the spatial arrangement of the substituents in (+)-glyceraldehyde is the same as that of the substituents in (−)-glyceric acid. Therefore the relative configuration of the two compounds must be the same. If we know the absolute configuration of one of these compounds, we immediately know the absolute configuration of the other. Thus *the relative configurations of two optically active compounds can be determined by converting one into the other by chemical reactions that do not involve breaking of bonds to the stereocenter.*

How do we know that (+)-glyceraldehyde has the configuration shown previously? We know because its configuration was related to (+)-tartaric acid by a series of chemical reactions, none of which broke a bond to the stereocenter. Because we know the absolute configuration of (+)-tartaric acid from Bijvoet's work, we can assign the (R) absolute configuration to (+)-glyceraldehyde. This means that (−)-glyceric acid also has the (R) configuration. Thus we have obtained the absolute configuration of (−)-glyceric acid without doing an x-ray structure determination.

We now know both the specific rotation and the absolute configuration of the enantiomers of glyceraldehyde and glyceric acid. This information is frequently included in the names of optically active compounds. For example, the name (R)-(+)-glyceraldehyde tells us that the dextrorotatory enantiomer of glyceraldehyde has the (R) configuration. Using this method, we can write the complete stereochemical description of the reaction of glyceraldehyde with mercury (II) oxide as follows:

$$(R)\text{-}(+)\text{-Glyceraldehyde} \xrightarrow{\text{HgO}} (R)\text{-}(-)\text{-Glyceric acid}$$

The reaction of glyceraldehyde with HgO illustrates an important point. While the configurations of the stereocenters of both reactant and product have not changed in this reaction, the sign of the specific rotation has changed between reactant and product. This means that *there is no correlation between the sign of the specific rotation of a compound and the absolute configuration of its stereocenter.*

Exercise 6.18   (S)-(−)-2-Methyl-1-butanol reacts with KMnO$_4$ to form (+)-2-methylbutanoic acid. What is the absolute configuration of the stereocenter of (+)-2-methylbutanoic acid? Explain your answer.

$$\overset{\overset{\textstyle CH_3}{|}}{C_2H_5CHCH_2OH} \xrightarrow{\text{KMnO}_4} \overset{\overset{\textstyle CH_3}{|}}{C_2H_5CHCOOH}$$

(S)-(−)-2-Methyl-1-butanol   (+)-2-Methylbutanoic acid

Compounds with a single stereocenter are fairly simple to deal with. Compounds with more than one stereocenter are more complex.

6.9   Molecules with More than One Stereocenter

We've learned that molecules with one stereocenter exist in only two enantiomeric configurations, (R) and (S). As the number of stereocenters in a molecule increases, the total number of possible stereoisomers also increases. In general, *a molecule with n stereocenters has a maximum of 2^n stereoisomers.* Thus a molecule that contains two stereocenters has a maximum of four stereoisomers ($2^2 = 4$).

The amino acid 2-amino-3-methylpentanoic acid, which is better known as isoleucine, has two stereocenters.

$$\overset{5\quad\;4\quad\;3\quad\overset{\textstyle NH_2}{|}}{CH_3CH_2CHCHCOOH}$$
$$\underset{\underset{\textstyle CH_3}{|}}{}_{2\;\;1}$$

Isoleucine
(2-Amino-3-methylpentanoic acid)

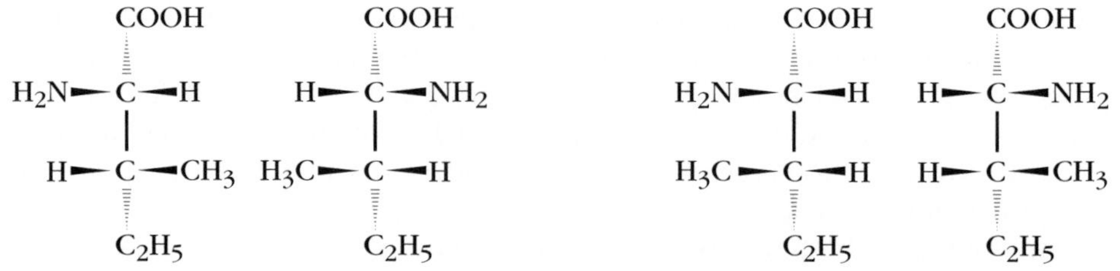

Figure 6.15
Three-dimensional line-dash-wedge representations of the four stereo-isomers of isoleucine. In these representations, the two stereocenters are placed vertically in the plane of the page. The other groups are placed either above or below the plane of the page. By convention, carbon atom 1 is placed at the top.

$$
\begin{array}{cccc}
\text{COOH} & \text{COOH} & \text{COOH} & \text{COOH} \\
| & | & | & | \\
H_2N\!-\!C\!-\!H & H\!-\!C\!-\!NH_2 & H_2N\!-\!C\!-\!H & H\!-\!C\!-\!NH_2 \\
| & | & | & | \\
H\!-\!C\!-\!CH_3 & H_3C\!-\!C\!-\!H & H_3C\!-\!C\!-\!H & H\!-\!C\!-\!CH_3 \\
| & | & | & | \\
C_2H_5 & C_2H_5 & C_2H_5 & C_2H_5 \\
(2S,3R) & (2R,3S) & (2S,3S) & (2R,3R) \\
\mathbf{A} & \mathbf{B} & \mathbf{C} & \mathbf{D}
\end{array}
$$

While there are a number of ways to represent the three-dimensional structures of molecules with two stereocenters, chemists have found that the representations shown in Figure 6.15 are particularly useful. The four structures in Figure 6.15 represent the eclipsed conformation of the molecule (Section 4.3). These are not the most stable conformations of the molecules, yet we write them in this way because it is easier to recognize the planes of symmetry. We will see the need for this in Section 6.10.

In molecules containing more than one stereocenter, the configuration of each stereocenter is designated (R) or (S) using the Cahn-Ingold-Prelog Rules introduced in Section 6.2. These designations are given in Figure 6.15 for the four stereoisomers of isoleucine. We analyze each center separately and use numbers to show the designation for each stereocenter. For example, carbon atoms 2 and 3 of structure 6.15, A, are assigned (S) and (R) as follows.

The four atoms or groups of atoms bonded to C2 are a hydrogen atom, a —COOH group, an —NH$_2$ group, and the alkyl group —CH(CH$_3$)CH$_2$CH$_3$. Their order of priority according to the Cahn-Ingold-Prelog Rules is:

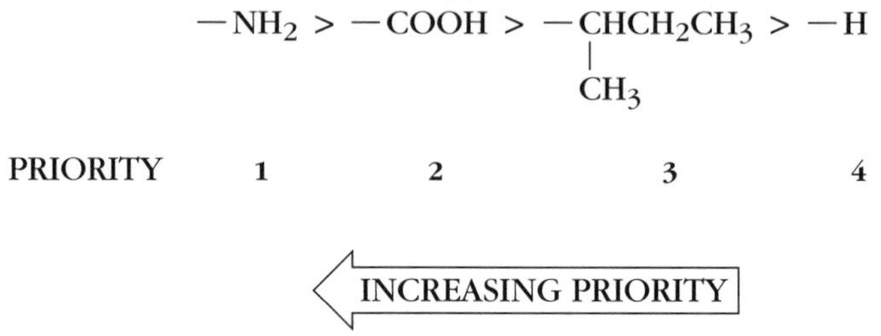

$$
-NH_2 \;>\; -COOH \;>\; \underset{\underset{CH_3}{|}}{-CHCH_2CH_3} \;>\; -H
$$

PRIORITY   1   2   3   4

◁ INCREASING PRIORITY

When the molecule is rotated so that the hydrogen atom on C2 is pointed away from the viewer, the orientation of the groups around C2 is as follows:

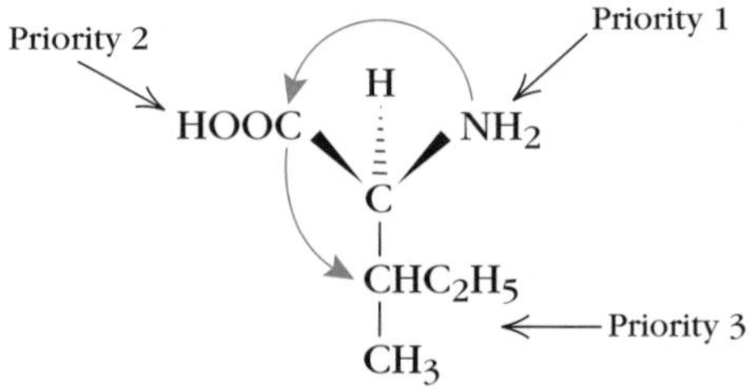

Counterclockwise
(2S)

The direction from the group of highest priority to the one of lowest priority is counterclockwise, so C2 has the (S) configuration.

Repeating the procedure with C3 shows that C3 has the (*R*) configuration.

Clockwise
(3*R*)

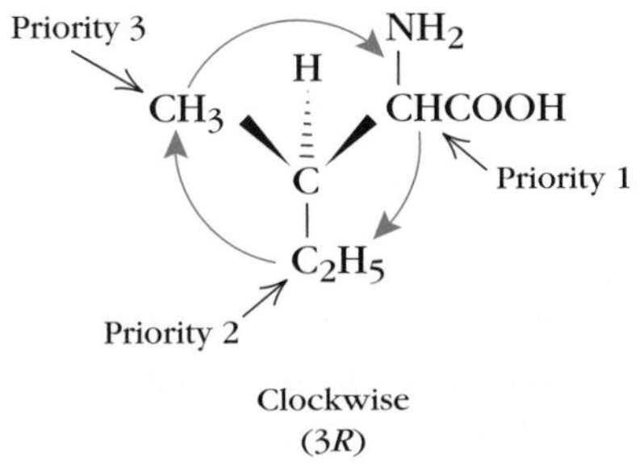

The four structures in Figure 6.15 are two pairs of enantiomers. The (2*S*,3*R*) (structure 6.15, *A*) and (2*R*,3*S*) (structure 6.15, *B*) stereoisomers are one pair of enantiomers because they are nonsuperposable mirror images of each other. The second pair of enantiomers are the (2*S*,3*S*) (structure 6.15, *C*) and (2*R*,3*R*) (structure 6.15, *D*) stereoisomers. Notice that the configuration of *each* stereocenter of one enantiomer is opposite that of its mirror image. For example, the configuration is (*S*) for C2 and (*R*) for C3 of structure 6.15, *A*. Its mirror image, structure 6.15, *B*, has the opposite configuration at both C2 (*R*) and C3 (*S*).

A different relationship exists for structures 6.15, *A*, and 6.15, *C*. Carbon atom 2 of *both* structures has the (*S*) configuration, but in structure 6.15, *A*, C3 has the (*R*) configuration while C3 in structure 6.15, *C*, has the (*S*) configuration. Stereoisomers 6.15, *A*, and 6.15, *C*, clearly are not mirror images. Stereoisomers that are not mirror images of each other are called **diastereomers.**

Note carefully the fundamental difference between enantiomers and diastereomers. Enantiomers have the opposite configuration at all stereocenters. Diastereomers have the same configuration at one or more stereocenters but the opposite configuration at the remaining stereocenters. The relationship between the four stereoisomers of isoleucine is as follows:

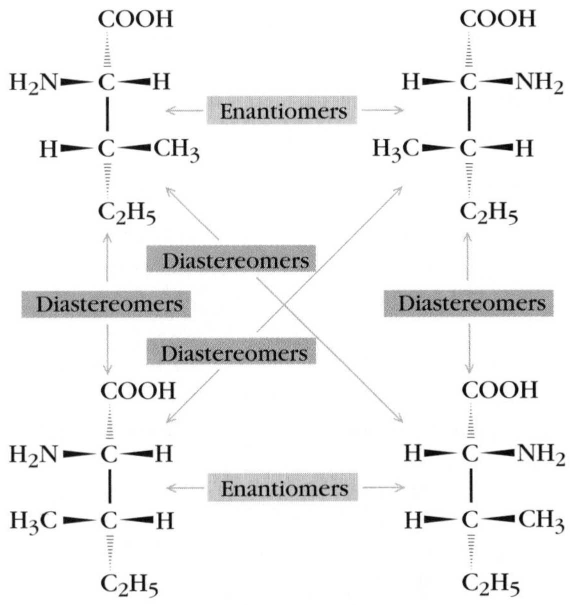

Out of the four possible stereoisomers of isoleucine, only the one with the (2S,3R) configuration is formed in nature. The ability to form only one of several stereoisomers is a characteristic of enzyme-catalyzed reactions, as we discussed in Section 6.7.

Exercise 6.19   Write the structures of all stereoisomers of the following compounds using the line-dash-wedge representation of three-dimensional structures. Identify the stereoisomers that are enantiomers and those that are diastereomers.

        Br   OH

        |

(a) 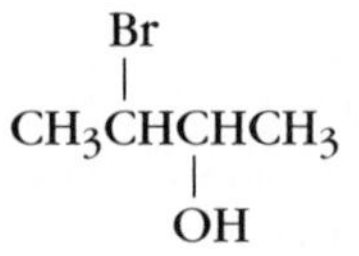 $CH_3CHCHCH_3$ **(b)** 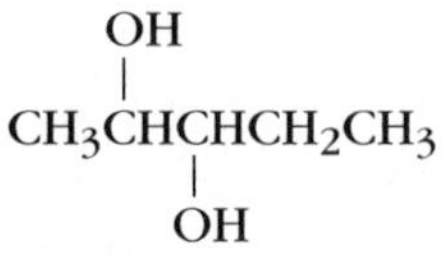 $CH_3CHCHCH_2CH_3$

              | |

             OH   OH

The 2^n rule gives us the maximum number of stereoisomers for a compound with n stereocenters. However, some compounds with n stereocenters have fewer than the maximum number of stereoisomers.

6.10   *meso* Compounds

Tartaric acid, like isoleucine, has two stereocenters.

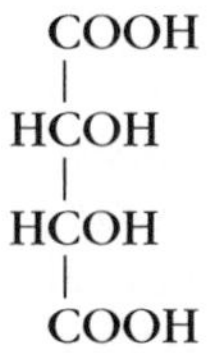

$$COOH$$
$$|$$
$$HCOH$$
$$|$$
$$HCOH$$
$$|$$
$$COOH$$

Three-dimensional representations of the four stereoisomers of tartaric acid are shown in Figure 6.16.

The (2R,3R) (structure 6.16, A) and the (2S,3S) compounds (structure 6.16, B) are a pair of enantiomers. The other two structures, 6.16, C, and 6.16, D, are mirror images of each other, but they are *not* a pair of enantiomers because they are superposable. In fact, structures 6.16, C, and 6.16, D, represent different ori-

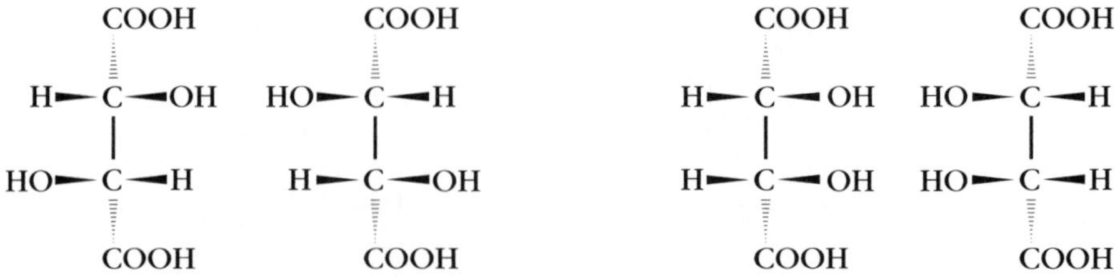

Figure 6.16
Stereoisomers of tartaric acid.

entations in space of the same molecule. We can see this by simply rotating one structure by 180° in the plane of the page.

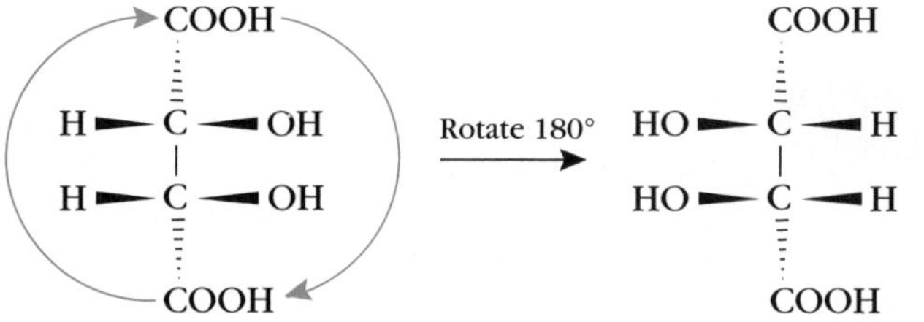

$$
\begin{array}{ccc}
\text{COOH} & & \text{COOH} \\
\text{H} \longrightarrow \text{C} \longleftarrow \text{OH} & \text{Rotate 180°} & \text{HO} \longrightarrow \text{C} \longleftarrow \text{H} \\
\text{H} \longrightarrow \text{C} \longleftarrow \text{OH} & \longrightarrow & \text{HO} \longrightarrow \text{C} \longleftarrow \text{H} \\
\text{COOH} & & \text{COOH} \\
(2R, 3S) & & (2S, 3R) \\
6.16, C & & 6.16, D
\end{array}
$$

Because the molecule represented by structure 6.16, *C* (or 6.16, *D*) is superposable on its mirror image, it is achiral even though it contains stereocenters. An achiral molecule that has stereocenters is called a ***meso* compound** or a ***meso* form.**

Tartaric acid exists, therefore, as three instead of the maximum four stereoisomers. The three stereoisomers are a pair of enantiomers, (2R,3R)-tartaric acid and (2S,3S)-tartaric acid, and a *meso* compound, (2R,3S)-tartaric acid. The *meso* compound and the enantiomers are diastereomers.

There are a number of ways to recognize whether a compound with stereocenters is a *meso* compound. First, a *meso* compound is superposable on its mirror image. Thus the test of superposability identifies a *meso* compound, and it is the method we have used to identify the *meso* form of tartaric acid.

Another method is to look at the four atoms or group of atoms bonded to the two stereocenters. If the four groups or atoms bonded to one stereocenter are identical to those bonded to the other, the compound will have a *meso* form. In 2,3-dibromobutane, for example,

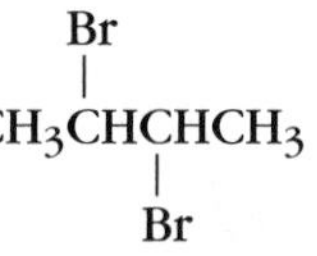

$$
\begin{array}{c}
\text{Br} \\
| \\
\text{CH}_3\text{CHCHCH}_3 \\
| \\
\text{Br}
\end{array}
$$

carbon atoms 2 and 3 are stereocenters. A methyl group, a bromine atom, a hydrogen atom, and a —CH(Br)CH$_3$ group are bonded to both carbon atoms 2 and 3. Compounds with this structural feature exist as three stereoisomers, a pair of enantiomers, and a *meso* form.

The presence of four groups bonded to one stereocenter that are identical to the four on the other stereocenter in a compound creates a plane of symmetry in one of the compound's stereoisomers. This, then, is a third method of identifying

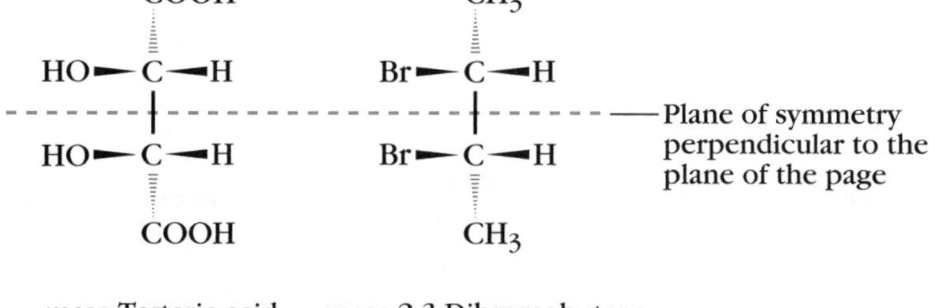

$$
\begin{array}{cc}
\text{COOH} & \text{CH}_3 \\
\text{HO} \longrightarrow \text{C} \longleftarrow \text{H} & \text{Br} \longrightarrow \text{C} \longleftarrow \text{H} \\
\text{- - - - - - - - - - - - - - - - - -} & \\
\text{HO} \longrightarrow \text{C} \longleftarrow \text{H} & \text{Br} \longrightarrow \text{C} \longleftarrow \text{H} \\
\text{COOH} & \text{CH}_3 \\
\end{array}
$$

——Plane of symmetry perpendicular to the plane of the page

meso-Tartaric acid   *meso*-2,3-Dibromobutane

Figure 6.17
The planes of symmetry of *meso*-tartaric acid and *meso*-2,3-dibromobutane. The plane divides the molecule into halves that are mirror images of each other.

meso forms of compounds that contain more than one stereocenter. The planes of symmetry of the *meso* forms of tartaric acid and 2,3-dibromobutane are shown in Figure 6.17.

Exercise 6.20	For each of the following compounds, write three-dimensional structures for all stereoisomers. Identify the stereoisomers that are enantiomers, diastereomers, and *meso* compounds.

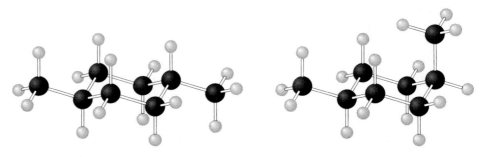

(a) HOCH$_2$CHCHCH$_2$OH   **(b)** CH$_3$CHCHCHCH$_3$ **(c)** CH$_3$CHCHCOOH

So far, we've looked only at the possible stereoisomers of acyclic compounds. How does the presence of a ring affect the possible number and type of stereoisomers?

6.11 Stereoisomers of Disubstituted Cyclohexanes

In Section 4.12 we discussed two stereoisomers of disubstituted cyclohexanes, *cis*- and *trans*-disubstituted cyclohexanes. *Cis*- and *trans*-stereoisomers of cycloalkanes are frequently called geometrical isomers even though they are, by definition, also diastereomers.

■ *Use of molecular models will help in understanding the following discussion.*

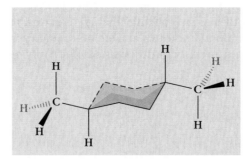

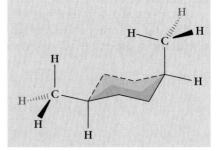

Figure 6.18
Both *cis*- and *trans*-1,4-dimethyl-cyclohexane have a plane of symmetry.

trans-1,4-Dimethylcyclohexane   *cis*-1,4-Dimethylcyclohexane

We conclude our study of the stereochemistry of cyclohexanes by examining the effect of conformation on the chirality of disubstituted cyclohexanes, beginning with 1,4-disubstituted cyclohexanes.

A   1,4-Disubstituted Cyclohexanes

Structural formulas of 1,4-disubstituted cyclohexanes, containing either two identical substituents or two different substituents, do not contain a stereocenter. Also, both the *cis-* and *trans*-isomers have a plane of symmetry passing through carbon atoms 1 and 4 and the two substituents. Therefore *cis-* and *trans*-1,4-disubstituted cyclohexanes are achiral as long as the substituents themselves are not chiral. The plane of symmetry is evident in either the flat or the three-dimensional chair representation of *cis-* and *trans*-1,4-dimethylcyclohexane, as shown in Figure 6.18.

B   1,3-Disubstituted Cyclohexanes

1,3-Disubstituted cyclohexanes have two stereocenters and therefore can exist as a maximum of four stereoisomers. When the two substituents are identical, how-

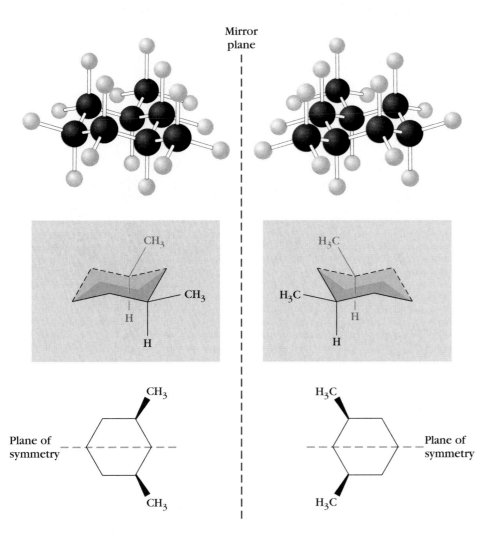

Figure 6.19
cis-1,3-Dimethylcyclohexane is superposable on its mirror image and has a plane of symmetry.

ever, as in the case of 1,3-dimethylcyclohexane, the *cis*-isomer is a *meso* compound because it has a plane of symmetry, as shown in Figure 6.19.

trans-1,3-Dimethylcyclohexane has no plane of symmetry so it exists as a pair of enantiomers.

Recall from Chapter 4 that the two chair conformations of *trans*-1,3-dimethylcyclohexane are of equal energy and both have one axial and one equatorial methyl group. Neither of these chair conformations is superposable on its mirror image. Thus ring flipping can't make a *trans*-1,3-disubstituted cycloalkane superposable on its mirror image, as shown in Figure 6.20.

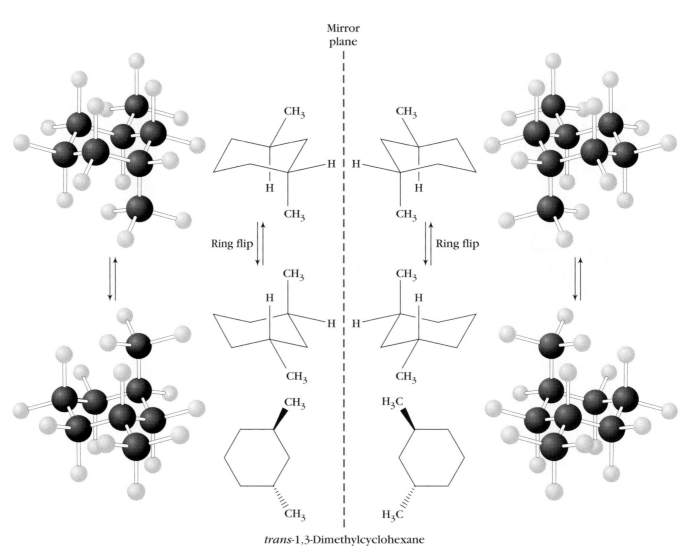

trans-1,3-Dimethylcyclohexane

Figure 6.20
Ring flipping can't make *trans*-1,3-dimethylcyclohexane superposable on its mirror image.

C   1,2-Disubstituted Cyclohexanes

1,2-Disubstituted cyclohexanes also have two stereocenters and four possible stereoisomers. When the two substituents are identical, as in 1,2-dimethylcyclohexane, the *trans*-isomer does not have a plane of symmetry so it exists as a pair of enantiomers:

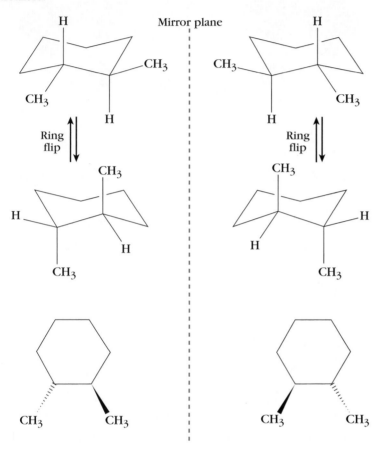

trans-1,2–Dimethylcyclohexane

The mirror images of *trans*-1,2-dimethylcyclohexane are nonsuperposable. Changing the conformation by flipping the ring of either mirror image still does not make them superposable. Thus no matter what their conformation, the mirror images of *trans*-1,2-dimethylcyclohexane are nonsuperposable.

The same cannot be said of *cis*-1,2-dimethylcyclohexane. Structures **6.1** and **6.2** in Figure 6.21 represent *cis*-1,2-dimethylcyclohexane and its mirror image. Notice, however, that a chair-chair interconversion of structure **6.1** by ring flip gives the same structure as **6.2.** Thus the two mirror images of *cis*-1,2-dimethylcyclohexane are constantly interconverting because they are simply different conformations of the same molecule. *cis*-1,2-Dimethylcyclohexane, therefore, is a *meso* form because it's superposable with its mirror image because of the conformational equilibrium of the various chair forms.

Figure 6.21
The mirror images of *cis*-1,2-dimethylcyclohexane are different conformations of the same molecule.

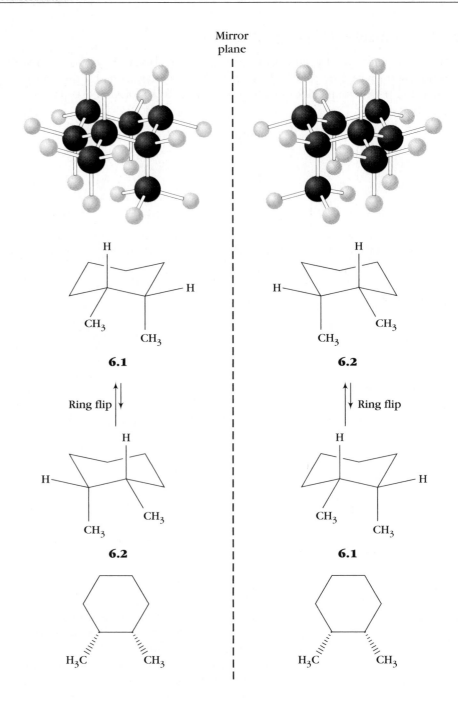

In general, it is easier to predict the presence or absence of chirality in ring compounds by using a flat polygon for the ring and dashed lines and wedges to represent the stereochemistry of the substituents. As we have seen in the case of dimethylcyclohexane, the correct conclusion is reached by the use of these structures even though the reason for chirality or lack of chirality may be more complex than realized from the simple flat structures. The chirality of identical disubstituted cyclohexanes is summarized in Table 6.2.

Table 6.2	Chirality of Identical Disubstituted Cyclohexanes	
Position of Substituent	Diastereomer	Chirality
1,4	*cis-*	Achiral
1,4	*trans-*	Achiral
1,3	*cis-*	*meso*
1,3	*trans-*	Enantiomers
1,2	*cis-*	*meso*
1,2	*trans-*	Enantiomers

Exercise 6.21 Write the structures for all the stereoisomers of both 1,2-dichlorocyclopropane and 1-bromo-2-chlorocyclopropane. Identify the structures that are pairs of enantiomers and those that are achiral.

Exercise 6.22 Construct a table similar to Table 6.2 that summarizes the chirality of all stereoisomers of bromochlorocyclohexane.

Because of the different spatial arrangements of their atoms, stereoisomers may also have different physical and spectral properties.

6.12   Physical and Spectral Properties of Stereoisomers

Each of the various types of stereoisomers of a chiral compound that we have identified in previous sections can be isolated as pure compounds. Do these different stereoisomers have different physical and spectral properties? We can answer this question by looking at the physical properties of the stereoisomers and the racemic mixtures of mandelic and tartaric acids given in Table 6.3.

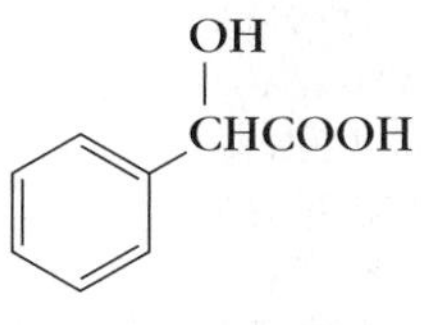

OH
|
CHCOOH

Mandelic acid

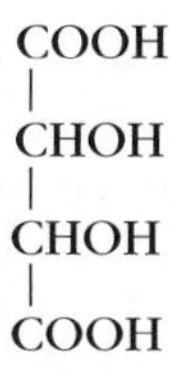

COOH
|
CHOH
|
CHOH
|
COOH

Tartaric acid

The data in Table 6.3 show that enantiomers have the same physical properties. For example, the two enantiomers of mandelic acid, (+)- and (−)-mandelic acid, have the same melting points, densities, heats of combustion, and solubilities in water. The only differences between the two are the signs of their specific rotations and their reactions with optically active reagents.

| Table 6.3 | | Some Physical Properties of the Stereoisomers and Racemic Mixtures of Mandelic and Tartaric Acids | | | |

Stereoisomer	Melting Point (°C)	$[\alpha]_D$ (Degrees) in Water	Density (g cm^{-3})	Heat of Combustion (kcal/mol)	Solubility in Water (g/100 ml) at 20 °C
(−)-Mandelic acid	133-135	−158	1.341	−886.3	8.64
(+)-Mandelic acid	133-135	+158	1.341	−886.3	8.64
(±)-Mandelic acid	131-132	0	1.300	−885.9	16.0
(−)-Tartaric acid	171-174	− 12.7	1.7598	−274.4	139
(+)-Tartaric acid	171-174	+ 12.7	1.7598	−274.4	139
(±)-Tartaric acid	206	0	1.7880	−272.4	20.6
meso-Tartaric acid	146-148	0	1.6660	−275.0	120*

*15 °C.

The *meso* form of tartaric acid, on the other hand, is diastereomeric with either enantiomer of tartaric acid so it is a different compound. As a result, its physical properties differ from those of the enantiomers.

While the racemic mixture is a mixture of enantiomers, it acts as though it were a pure compound different from either enantiomer. Thus the physical properties of a racemic mixture are different from those of the *meso* form (if one exists for a chiral compound) and the pure enantiomers. In Table 6.3, for example, the melting point of racemic tartaric acid is 206 °C, whereas the melting point of the pure enantiomer is 171 to 174 °C and that of the *meso* form is 146 to 148 °C.

While enantiomers have identical IR, ^{13}C NMR, and ^{1}H NMR spectra, their spectra differ from those of either a racemic mixture or the *meso* form. The spectra of these stereoisomers differ mostly in the fingerprint region, as we can see by examining the IR spectra of racemic, *meso-*, and (+)-tartaric acids shown in Figure 6.22.

Chemists use these differences in physical properties to separate a racemic mixture into its pure components.

6.13   Resolution of Enantiomers

In future chapters, we will see that organic chemists frequently use optically active compounds to investigate how certain organic reactions occur. How do chemists obtain optically active compounds? The first optically active compounds were obtained from natural sources. Usually, however, the type of compound that we need for a particular experiment is not available from natural sources. In that

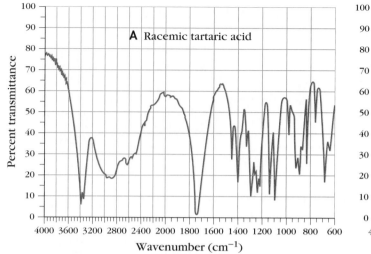

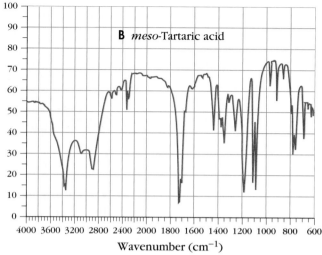

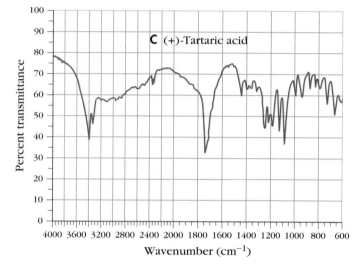

Figure 6.22
A comparison of the IR spectra of racemic (**A**), *meso-* (**B**), and (+)-tartaric acids (**C**). The spectra are similar but different, especially in the fingerprint region.

case, we must start with a racemic mixture of the desired compound and separate its enantiomers. The separation of a racemic mixture into its enantiomers is called **resolution.**

The techniques used to separate a mixture of compounds, which were discussed in Section 5.1, depend on differences in the physical properties of the compounds in the mixture. These techniques cannot be used to resolve a racemic mixture, however, because the physical properties of the enantiomers are identical (except for the sign of their specific rotation).

Diastereomers, unlike enantiomers, have different physical properties. We can use this difference to resolve a racemic mixture. The idea is to form a pair of diastereomers by the reaction of the racemic mixture with a single pure enantiomer, called the *resolving agent.* The two diastereomers can be separated because they have different physical properties. Once the diastereomers are separated and purified, the pure enantiomer can then be regenerated by chemical reaction. We can summarize the general steps involved in the resolution of a racemic mixture as follows:

Step 1: Reaction of racemic mixture with a chiral resolving agent forms a mixture of two diastereomers.

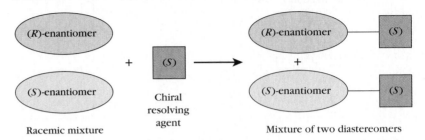

Step 2: The differences in physical properties of diastereomers allows the two diastereomers to be separated.

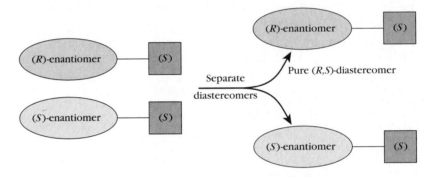

Step 3: Once the pure (*R,S*)-diastereomer is isolated, a reaction is chosen that cleaves it to yield the pure (*R*)-enantiomer .

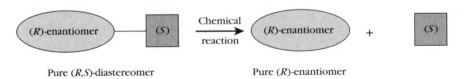

A racemic mixture of 1-phenylethylamine (amphetamine) can be resolved using this method. To carry out this resolution, a readily available single enantiomer of a chiral compound that can react with amphetamine is needed. One such compound is (*S*)-(−)-malic acid, which is available in large quantities from natural sources.

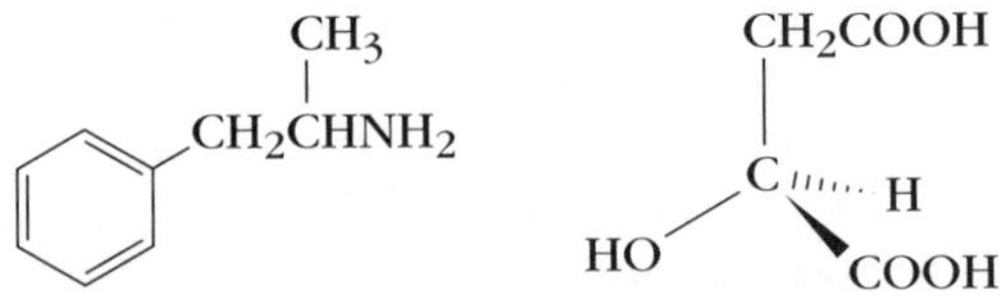

$$CH_3$$
$$CH_2CHNH_2$$

1-Phenylethylamine
(Amphetamine)

$$CH_2COOH$$
$$C\ \text{\small\textit{Iıııı..}}\ H$$
$$HO \qquad\qquad COOH$$

(*S*)-(−)-Malic acid

Malic acid, a carboxylic acid, and 1-phenylethylamine, an amine, undergo an acid-base reaction of the type discussed in Section 3.4 to form a salt with the following structure:

$$CH_3$$
$$CH_2\overset{*}{\underset{*}{C}}H\overset{+}{N}H_3 \quad {}^-OOCCH_2\overset{*}{C}HCOOH$$
$$OH$$

This salt has two stereocenters and can exist as four stereoisomers. However, by using a single enantiomer of malic acid to form the salt, we decrease the number of stereoisomers from four to two. The two stereoisomers are not enantiomers because they are not mirror images of each other. Rather, they are diastereomers. The diastereomers can be separated, usually by repeated crystallization, and purified. Reaction of each pure diastereomer with aqueous base forms one pure enantiomer of amphetamine. Thus the reaction of (R)-1-phenylethylammonium (S)-malate with an aqueous solution of sodium hydroxide yields sodium (S)-malate and pure (R)-1-phenylethylamine. The two are easily separated because sodium (S)-malate is soluble in the aqueous solution, while the amine is not. When the same reaction is carried out with the other diastereomer, (S)-1-phenylethylammonium (S)-malate, (S)-1-phenylethylamine (the other enantiomer) is obtained. The reactions involved in this resolution of a racemic mixture of 1-phenylethylamine are shown in Figure 6.23.

Figure 6.23
The product of the reaction of (S)-(−)-malic acid and racemic 1-phenylethylamine (amphetamine) is a mixture of two diastereomers. The two diastereomers have different physical properties so they can be separated by conventional means, such as recrystallization. Once separated, reaction of one diastereomer with aqueous NaOH forms one of the pure enantiomers.

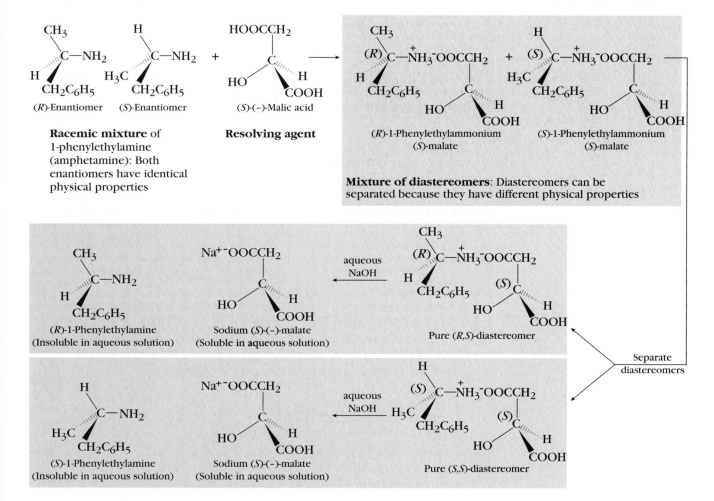

Box 6.1   **Chiral Drugs**

Although most pharmaceutical drugs are chiral, they are sold as racemic mixtures because of the difficulty pharmaceutical chemists have of devising large-scale methods of resolving enantiomers that will not substantially increase the cost of manufacture. Usually this is not a health problem since the stereoisomer that is not biologically active passes through the body without effect.

One tragic exception is the drug thalidomide prescribed as an antidepressant in the early 1960s. Despite strong warnings that it not be used by pregnant women or even women who might become pregnant, it was prescribed in Canada and Europe, but never in the United States, for the treatment of morning sickness. Thalidomide was sold as a racemic mixture even though only the (R)-enantiomer has the desired antidepressant properties. The (S)-enantiomer was found to be mutagenic and antiabortive and, therefore, caused not only deformed fetuses but prevented their natural expulsion. The use of thalidomide by pregnant women led to the birth of many very seriously deformed children with underdeveloped arms and legs. Clearly (S)-thalidomide doesn't pass harmlessly through the body but reacts in ways that harm a developing human fetus.

(R)-Thalidomide   (S)-Thalidomide

Concern that racemic mixtures of drugs could have undesirable side effects caused by the presence of the "other" enantiomer has led organic chemists to develop reactions that form a single stereoisomer. Several examples of such reactions are discussed in future chapters.

A variation of this method of resolving racemic mixtures is to use a chromatography column (Section 5.1) containing particles whose surfaces are coated with one enantiomer of a chiral compound. The structure of the enantiomer is chosen so that it interacts weakly with the racemic mixture. As the solution of the racemic mixture passes through the column, the two enantiomers have different intermolecular attractions with the chiral column packing so they move down the column at different rates. As a result, separation of the two enantiomers occurs. This is another example of how one stereoisomer of a chiral compound can discriminate between two enantiomers.

The two methods of resolving racemic mixtures are variants of the same concept. Reaction between a single stereoisomer of a chiral compound and a racemic mixture always forms diastereomers. Up to now, we have used this concept to resolve a racemic mixture. In future chapters, however, we will see that this concept also applies to many reactions in living systems.

6.14   Stereochemical Nonequivalence of Atoms

Butane has two kinds of hydrogen atoms; there are those of the two methylene groups and those of the two methyl groups. While butane doesn't have a stereocenter, we can create one by replacing one methylene hydrogen atom by an achiral atom or group other than methyl, ethyl, or H. For example, if we replace one methylene hydrogen atom of butane by a deuterium atom, we create a stereocenter at C2.

$$CH_3CH_2CH_2CH_3 \xrightarrow[\text{H atom by D}]{\substack{\text{Replace 1} \\ \text{methylene}}} CH_3CH_2\underset{\underset{H}{|}}{\overset{\overset{D}{|}}{C}}CH_3$$

No stereocenter   Stereocenter

The atom at which substitution must take place in order to create a stereocenter is called a **prochiral atom.** Carbon atom 2 in butane, therefore, is a prochiral carbon atom. Compounds that contain prochiral atoms are called **prochiral compounds.**

A stereocenter is formed at C2 because the two methylene hydrogen atoms on C2 are stereochemically nonequivalent. We can show this by replacing each hydrogen atom in turn by a deuterium atom, as shown in Figure 6.24.

The two hydrogen atoms on C2 of butane are called **enantiotopic hydrogens** because the replacement of first one, then the other, can give rise to enantiomers. In contrast, the methyl hydrogen atoms of butane are not enantiotopic. Substitution of any of the three methyl hydrogen atoms by a deuterium atom doesn't make C1 a stereocenter because it's still bonded to two hydrogen atoms.

$$CH_3CH_2CH_2CH_3 \xrightarrow[\text{atom by D}]{\substack{\text{Replace 1} \\ \text{methyl H}}} CH_3CH_2CH_2CH_2D$$

No stereocenter   Still no stereocenter

It's frequently necessary to differentiate between the enantiotopic atoms or groups in a compound. They can be designated as pro-*R* or pro-*S* in the following way. One of the two enantiotopic atoms or groups is arbitrarily given priority

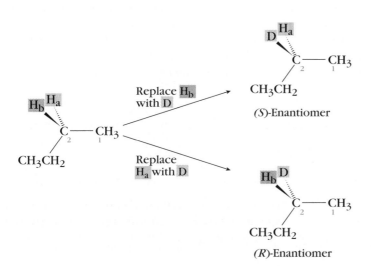

Figure 6.24
Replacement of hydrogen atom labelled H$_a$ by a deuterium atom forms (*R*)-2-deuteriobutane, while replacement of H$_b$ forms (*S*)-2-deuteriobutane.

over the other. The configuration of the prochiral center is then determined according to the Cahn-Ingold-Prelog Rules. If the configuration is (*R*), the atom or group given the higher priority is designated as pro-*R*. The other is designated pro-*S*. When the atoms are hydrogens, pro-*S* is abbreviated as H_S and pro-*R* is abbreviated as H_R. This method is applied to butane in Figure 6.25. The designation of one enantiotopic atom or group as pro-*S* isn't changed by reversing their arbitrary order of priority.

Figure 6.25
Designating the two methylene hydrogen atoms of butane as H_R or H_S.

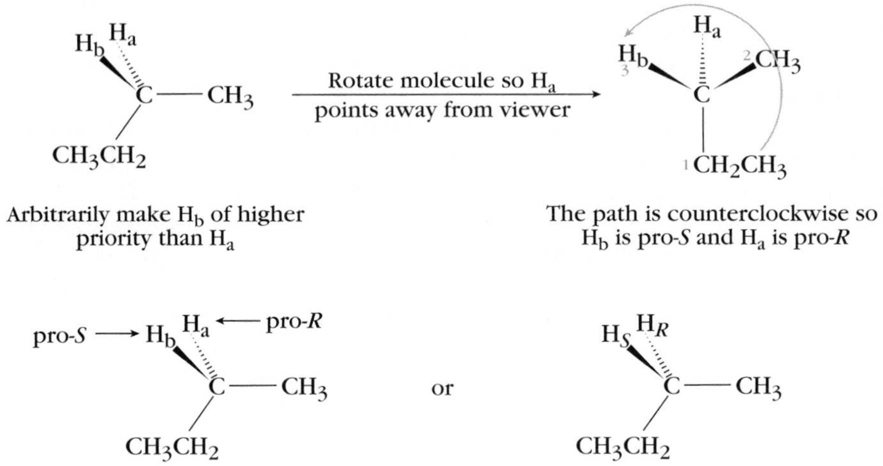

Arbitrarily make H_b of higher priority than H_a

Rotate molecule so H_a points away from viewer

The path is counterclockwise so H_b is pro-*S* and H_a is pro-*R*

pro-*S* ⟶ H_b H_a ⟵ pro-*R*

or

H_S H_R

Exercise 6.23   Using the stereochemical representation of butane in Figure 6.25, show that H_a is H_R even when the priority of H_a is made higher than H_b.

Enantiotopic hydrogens or groups are identical in all respects except when they are placed in a chiral environment. They are placed in a chiral environment when they react with a chiral reagent such as an enzyme. The enzyme-catalyzed phosphorylation of glycerol with ATP, for instance, forms exclusively (*R*)-(−)-glycerol-1-phosphate. The enzyme that catalyzes this reaction is called glycerokinase.

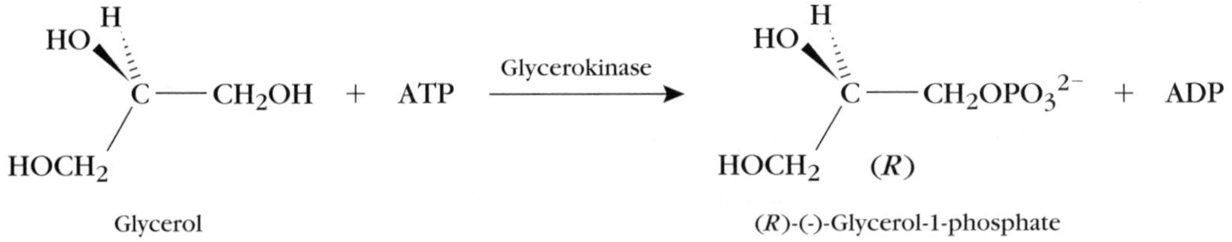

Glycerol   (*R*)-(-)-Glycerol-1-phosphate

Glycerokinase converts glycerol, an achiral compound, into a single enantiomer of glycerol-1-phosphate because the enzyme can distinguish between the two —CH_2OH groups of glycerol. To glycerokinase, therefore, the two —CH_2OH groups are not equivalent.

| Sample Problem 6.9 | Identify the enantiotopic hydrogen atoms in CH_3CH_2OH. |

Solution:

The methyl hydrogens are not enantiotopic because the same compound is formed by the successive replacement of the three hydrogen atoms by deuterium or another atom or achiral group.

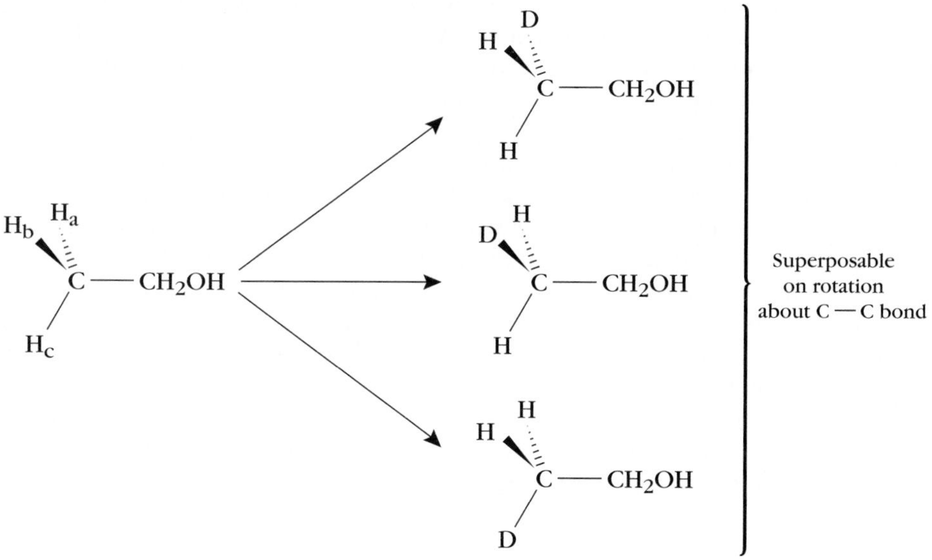

The methylene hydrogens are enantiotopic because their successive replacement by deuterium forms enantiomers.

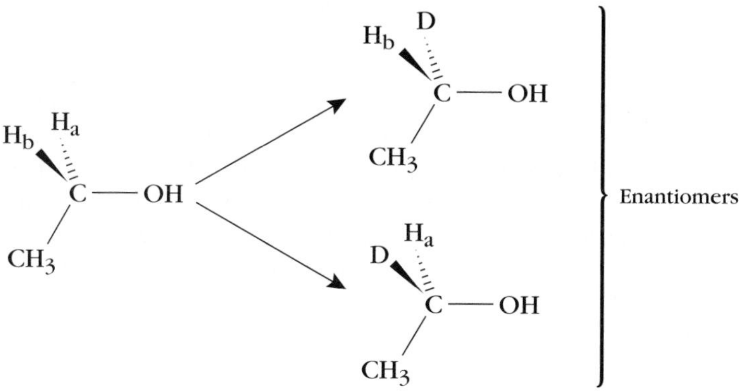

| Exercise 6.24 | Identify the enantiotopic hydrogen atoms in each of the following compounds. |

(a) $ClCH_2CH_2CH_2Cl$ **(b)** CH_3CH_2COOH **(c)**

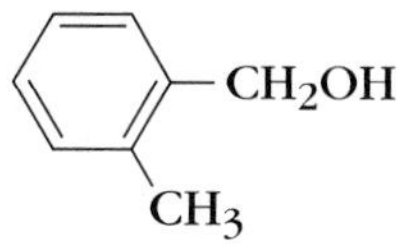

Exercise 6.25   Designate the enantiotopic hydrogens of the compounds in Exercise 6.24 as pro-*S* or pro-*R*.

6.15   Fischer Projection Formulas

In the last few chapters, we have made extensive use of three-dimensional structures to represent various stereoisomers. We have used three-dimensional structures because they clearly represent the spatial arrangement of atoms and groups in molecules and they can be moved about on the page without restriction as long as we don't break any bonds.

Chemists sometimes represent the structures of stereoisomers by two-dimensional projection formulas called **Fischer projection formulas.** These formulas were developed by the German chemist Emil Fischer to represent the stereochemistry about the many stereocenters of carbohydrate molecules. Fischer projection formulas are useful for representing these compounds because they are easy to draw and they save space. They must be used with great care, however, because they can lead to incorrect stereochemical conclusions if they are used incorrectly.

Let's use Fischer projection formulas to represent (*S*)-(−)-lactic acid, a compound that is produced by active skeletal muscles in the body. Lactic acid is a simple compound that contains a single stereocenter.

$$\text{HO} \cdots \underset{\displaystyle H}{\overset{\displaystyle COOH}{\diagup \!\! C \diagdown}} CH_3$$

(*S*)-(−)-Lactic acid

In a Fischer projection, the stereocenter in lactic acid is represented by two crossed lines. By convention, a carbon atom is understood to be located where the two lines cross and carbon atom 1 is placed at the top. The horizontal lines represent bonds that extend above the plane of the page, while vertical lines represent bonds that extend below the plane. This convention can be seen by comparing the following Fischer projection with a three-dimensional representation of (*S*)-lactic acid.

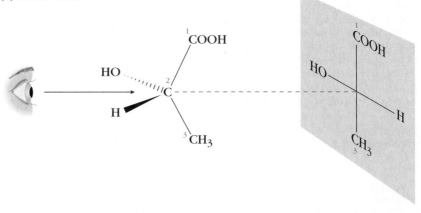

Line-dash-wedge representation
of (*S*)-lactic acid

Fischer projection formula
of (*S*)-lactic acid

A Fischer projection may be rotated by 180° in the plane of the page, but it may not be flipped over or rotated by any other angle. These manipulations result in a Fischer projection that does not represent the original three-dimensional structure so they may lead to incorrect stereochemical results.

Exercise 6.26   The Fischer projection formula of (S)-lactic acid has been manipulated as indicated to give a new Fischer projection formula. Is the configuration of the stereocenter in the two new Fischer projection formulas still (S)?

(a)

$$
\begin{array}{c}
\text{COOH} \\
\text{HO} \!-\!\!\!\mid\!\!\!-\! \text{H} \\
\text{CH}_3
\end{array}
\xrightarrow[\text{formula over}]{\substack{\text{Flip Fischer}\\\text{projection}}}
\begin{array}{c}
\text{COOH} \\
\text{H} \!-\!\!\!\mid\!\!\!-\! \text{OH} \\
\text{CH}_3
\end{array}
$$

(S)-lactic acid

(b)

$$
\begin{array}{c}
\text{COOH} \\
\text{HO} \!-\!\!\!\mid\!\!\!-\! \text{H} \\
\text{CH}_3
\end{array}
\xrightarrow[\text{formula by 90°}]{\substack{\text{Rotate Fischer}\\\text{projection}}}
\begin{array}{c}
\text{OH} \\
\text{CH}_3 \!-\!\!\!\mid\!\!\!-\! \text{COOH} \\
\text{H}
\end{array}
$$

(S)-lactic acid

While Fischer projection formulas are a convenient shorthand way of representing stereoisomers, the limitations placed on their movement means that they must be used carefully. We recommend that you convert Fischer projection formulas to three-dimensional structures such as line-dash-wedge representations or make molecular models when performing any manipulations of a Fischer projection on the page. In future chapters, we will use three-dimensional formulas rather than Fischer projection formulas until we discuss the chemistry of carbohydrates in Chapter 25.

6.16   Summary

An object that is nonsuperposable on its mirror image is **chiral. Chirality** is the property of an object that makes it nonsuperposable on its mirror image. Objects that are superposable on their mirror images are **achiral. Enantiomers** are molecules that are nonsuperposable on their mirror images. A molecule can exist as a pair of enantiomers if it contains a single tetrahedral carbon atom, called a **stereocenter,** bonded to four different substituents. Interchanging two substituents on a stereocenter forms an enantiomer of the original structure.

A **plane of symmetry** is an imaginary plane that cuts a molecule in half, making one half the mirror image of the other. Chiral molecules lack a plane of symmetry. A molecule with a plane of symmetry is superposable on its mirror image and is achiral.

Compounds that rotate plane-polarized light are said to be **optically active.** A pair of enantiomers have identical physical properties and differ in only two respects. First, they rotate the plane of polarized light in equal and opposite directions, and second, they react differently with optically active compounds.

A **racemic mixture** is a mixture of equal molar quantities of two enantiomers. A racemic mixture is optically inactive. A mixture of enantiomers is optically active only if one enantiomer is present in greater amounts than the other. The fractional excess of one enantiomer in a mixture over the other is called its **enantiomeric excess.**

Configuration is defined as the three-dimensional arrangement of substituents around a stereocenter and is designated (R) or (S) according to the Cahn-Ingold-Prelog Rules. There is no relationship between the optical rotation of a compound and its assigned (R) or (S) configuration.

The **absolute configuration** of a stereoisomer is its actual three-dimensional structure. Two compounds have the same **relative configuration** if one is obtained from the other by one or more chemical reactions in which no bonds to the stereocenter are broken. If the relative configuration of a series of compounds is known, it is necessary to determine the absolute configuration of only one compound in the series to know the absolute configuration of the entire series.

Molecules that contain n stereocenters have a maximum of 2^n stereoisomers. Stereoisomers that have the opposite configuration at *all* stereocenters are enantiomers. **Diastereomers,** on the other hand, have at least one stereocenter with the same configuration, while the others have opposite configurations. Diastereomers, therefore, are not mirror images. *meso* **Compounds** contain stereocenters but are achiral because they have a plane of symmetry.

Enantiomers have the same physical properties except for the direction that they rotate plane-polarized light. The physical and spectral properties of enantiomers, diastereomers, and *meso* compounds are all different. These differences can be used to separate a racemic mixture into its enantiomers in a process called **resolution of a racemic mixture.** The idea behind resolution is to react a racemic mixture with an optically active compound to form a pair of diastereomers. Because the diastereomers differ in their physical properties, they can be separated. The individual diastereomers can then be converted into the pure enantiomers by means of a specific chemical reaction.

Additional Exercises

6.27 Define and provide examples of each of the following terms.

(a) Stereocenter	**(g)** Plane of symmetry	**(m)** Specific rotation
(b) Enantiomer	**(h)** Diastereomer	**(n)** Relative configuration
(c) Racemic mixture	**(i)** *meso* Compound	**(o)** Enantiotopic hydrogens
(d) Chiral molecule	**(j)** Superposable	**(p)** Achiral molecule
(e) Prochiral center	**(k)** Optically active	
(f) Enantiomeric excess	**(l)** Absolute configuration	

6.28 Imagine that each hydrogen of 2-methylbutane is replaced by a single chlorine atom to form a number of isomers of the formula $C_5H_{11}Cl$. Write the structural formula of each isomer. Draw the three-dimensional structures of any stereoisomers, assign the (R) and (S) configurations of the stereocenter, and give a name to each stereoisomer.

6.29 Write the structure of a chiral alkane that contains the fewest possible number of carbon atoms.

6.30 Place an asterisk next to the stereocenter, if any, in the following structures.

(a) $(CH_3)_2CHCH_2CH_3$ **(b)** $ClCH_2\overset{\overset{\displaystyle Cl}{|}}{C}HCH_3$ **(c)** $CH_2{=}CHCHCH_2CH_2CH_3$ with CH_3 below

6.31   The following three-dimensional structures represent one particular stereoisomer. Which would you expect to be optically active?

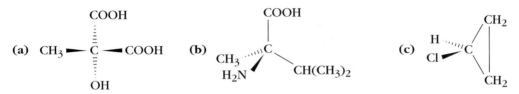

(a) $CH_3{-}\overset{\overset{\displaystyle COOH}{|}}{\underset{\underset{\displaystyle OH}{|}}{C}}{-}COOH$ **(b)** structure with COOH, CH₃, H₂N, CH(CH₃)₂   **(c)** structure with CH₂, H, Cl, CH₂

6.32   Give an IUPAC name to each of the following compounds.

(a) $ClCH_2{-}\overset{\overset{\displaystyle H}{|}}{\underset{\underset{\displaystyle Br}{|}}{C}}{-}CH_2I$ **(b)** structure with (CH₃)₃C, H, CH₃, C₂H₅   **(c)** structure with BrCH₂CH₂, CHCl₂, CH₃, H

6.33   Assign the (R) or (S) configuration to the stereocenter in each of the following compounds.

(a) $HO{-}\overset{\overset{\displaystyle CH_3}{|}}{\underset{\underset{\displaystyle CH_2OH}{|}}{C}}{-}\overset{\overset{\displaystyle O}{\|}}{\underset{\underset{\displaystyle NH_2}{|}}{C}}$ **(b)** structure with O, C, CH₃O, NH₂, COOH, H   **(c)** $CH_2{=}CH{-}\overset{\overset{\displaystyle CH_3}{|}}{\underset{\underset{\displaystyle CH_2CH_2CH_3}{|}}{C}}{-}C(CH_3)_3$

6.34   Draw the three-dimensional structures that correspond to each of the following names.

(a) (S)-1-Chloro-2-fluoropropane   **(c)** (R)-2-Nitrobutane ($-\overset{\overset{\displaystyle O}{\overset{+}{\|}}}{\underset{\underset{\displaystyle O^-}{}}{N}}$ = nitro group)

(b) (R)-2,3-Dimethylpentane   **(d)** (S)-2-Fluoro-1-nitropropane

6.35   The optical rotation of a compound is determined to be $+17.3°$ in water. Determine the value of the observed rotation if **(a)** a 5-cm sample tube is used instead of the original 10-cm tube, and **(b)** water is added so that the concentration of the solution is half its original value (sample tube is still 10 cm).

6.36   Pure (+)-amphetamine has a specific rotation $[\alpha]_D$ of $+40.1°$. What proportion of dextrorotatory and levorotatory amphetamine must be mixed to give a mixture with an optical rotation of $-10.3°$?

6.37   Mercury(II)oxide, HgO, reacts with a carbon-carbon double bond to form a carboxylic acid, carbon dioxide, and water according to the following equation:

$$RCH{=}CH_2 \xrightarrow{\text{HgO}} RC\overset{\overset{\displaystyle O}{\|}}{\underset{\underset{\displaystyle OH}{}}{}} + H_2O + CO_2$$

What is the configuration of the stereocenter of the product of oxidation of the following compound with HgO?

$$(R)\text{-}(+)\text{-}CH_3\underset{\underset{\displaystyle CH_2CH_3}{|}}{C}HCH{=}CH_2$$

6.38 Explain why mercury(II)oxide oxidation of the following optically active compound forms an optically inactive product.

$$(S)\text{-}(-)\text{-}CH_3CH_2CHCH{=}H_2$$
$$|$$
$$COOH$$

6.39 Are the following pairs of structures superposable or are they enantiomers?

(a)

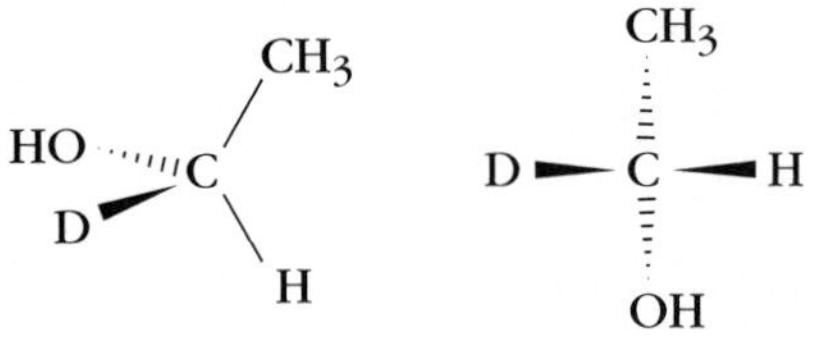

(b)

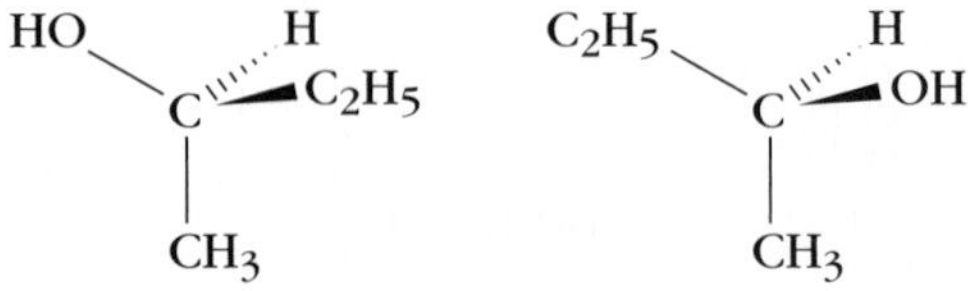

(c)

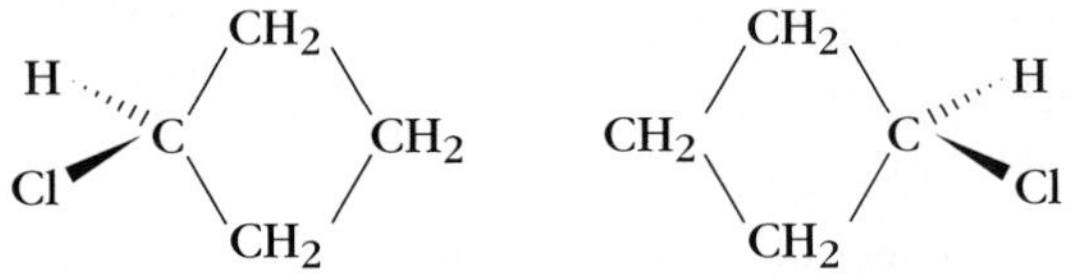

(d)

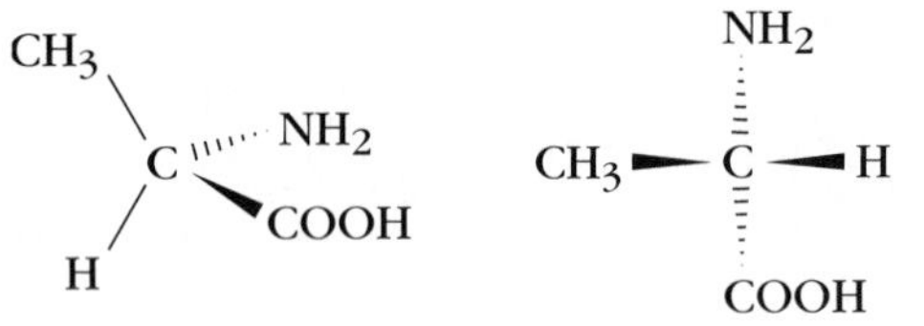

(e)

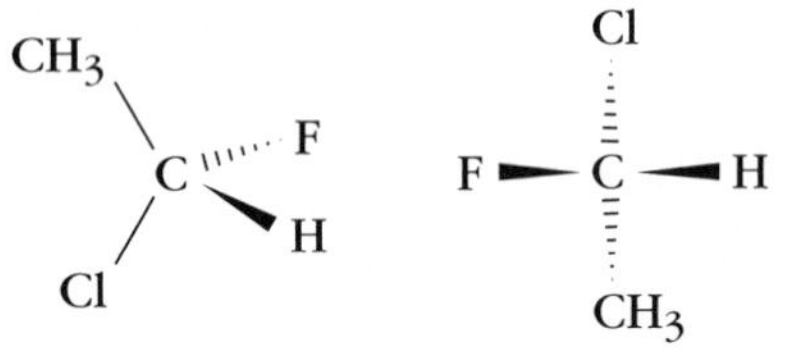

(f)

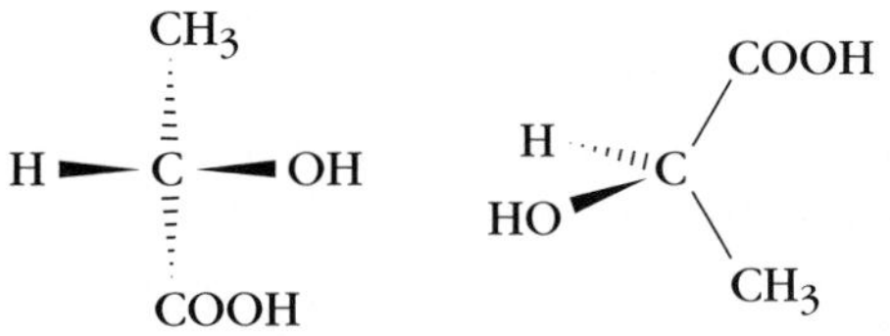

(g)

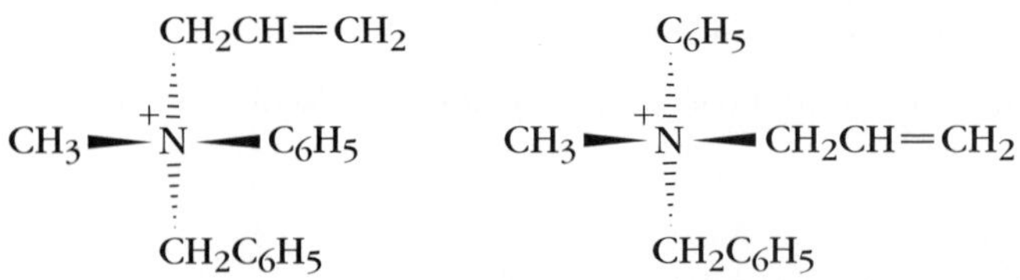

6.40 (−)-Menthol is a compound isolated from peppermint oil and widely used as a flavoring ingredient.

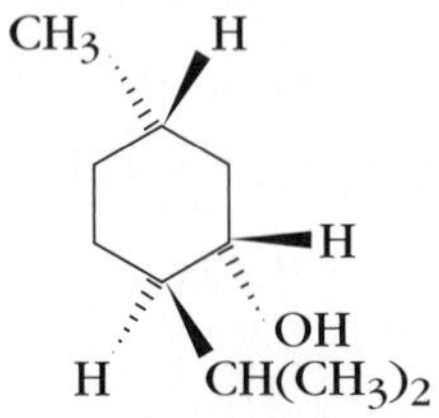

(−)-Menthol

Identify the three stereocenters of (−)-menthol. How many stereoisomers of (−)-menthol are there?

6.41 Write three-dimensional representations for each of the following compounds.

(a) (S)-2-Fluorobutane (d) (S)-3-Chloro-3-methylhexane
(b) (R)-2-Bromohexane (e) (2S,3S)-2,3-Dibromobutane
(c) (R)-Cysteine, HSCH₂CH(NH₂)COOH (f) (2R,4R)-2-Chloro-4-methylhexane

6.42 Assign (R) or (S) configurations to each stereocenter in the following compounds.

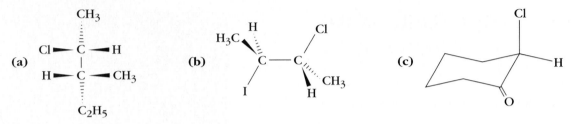

6.43 Write the three-dimensional structure of one compound that corresponds to the following:

(a) An optically active chloroalkane of molecular formula $C_5H_{11}Cl$
(b) A cyclic *meso* compound of molecular formula $C_{10}H_{20}$
(c) A racemic mixture of a cyclic compound of molecular formula $C_{10}H_{20}$

6.44 Consider the following structures **A** to **E**.

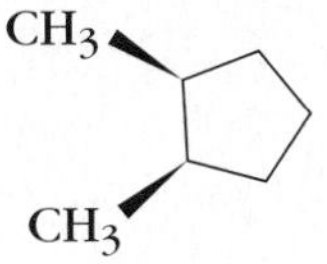

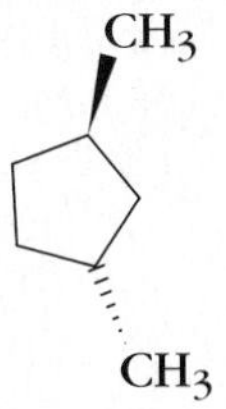

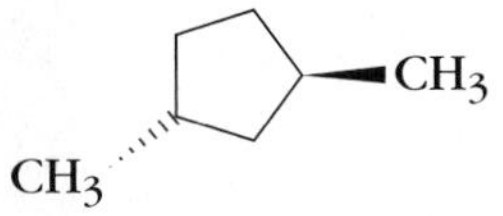

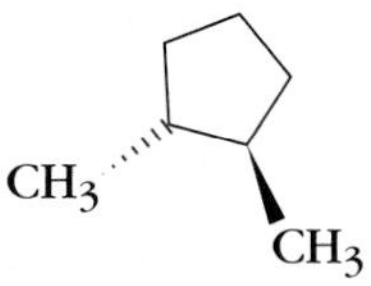

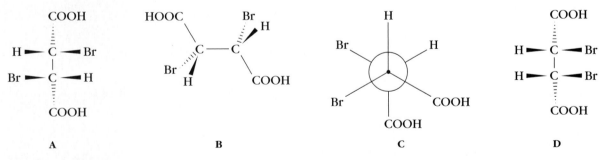

Identify:
(a) Any two identical structures (d) A pair of enantiomers
(b) Any two constitutional isomers (e) A *meso* compound
(c) Any two diastereomers

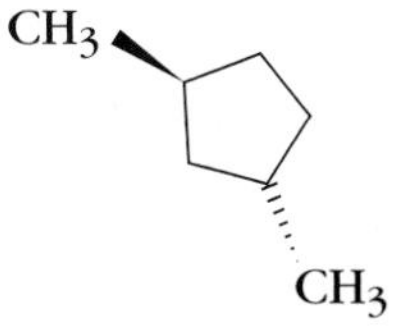

6.45 Consider the following structures **A** to **D**.

Identify the following pairs of structures as enantiomers, diastereomers, or the same compound.

(a) A and B (b) A and C (c) B and C (d) B and D

6.46 How many stereoisomers are there of 2,3,4-trichloropentane? Draw a three-dimensional structure for each stereoisomer and assign the (R) or (S) configuration to each stereocenter.

6.47 If you replaced the chlorine atom on carbon atom 3 of 2,3,4-trichloropentane by a bromine atom, would this change the number of stereoisomers? What happens to the number of stereoisomers if you replace the chlorine on carbon atom 2 by a bromine atom? Verify your answers by drawing three-dimensional structures for each stereoisomer.

ALKENES: STRUCTURE, ACID ADDITIONS, and PREPARATION

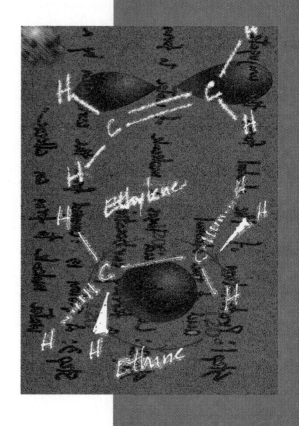

Hydrocarbons that contain a carbon-carbon double bond as their only functional group are called **alkenes** (Section 2.12). Another name for alkenes is **olefins,** from olefiant gas, an old name for ethylene.

Alkenes are widely used commercially. The simplest alkene, ethylene, is an important industrial starting material that is obtained from petroleum. It is used to prepare a wide variety of commercial products, including antifreeze, plastic bottles, tubing, and molded plastic objects. Ethylene is also found in living systems. Some plants generate small quantities of ethylene during the ripening of their fruit. Fruit shippers can pick fruit while it is still green and then ripen it by treating it with ethylene gas at its destination.

Certain flowers and plants contain alkenes that are used as food additives because of their pleasant smell or taste. The essence of a plant's fragrance is called its *essential oil.* The oily drops are obtained by heating the leaves or petals in steam. Blending perfumes and flavoring foods are major uses of essential oils. γ-Terpinene, for example, is an alkene found in the essential oils of coriander seed, which is used as a food flavoring.

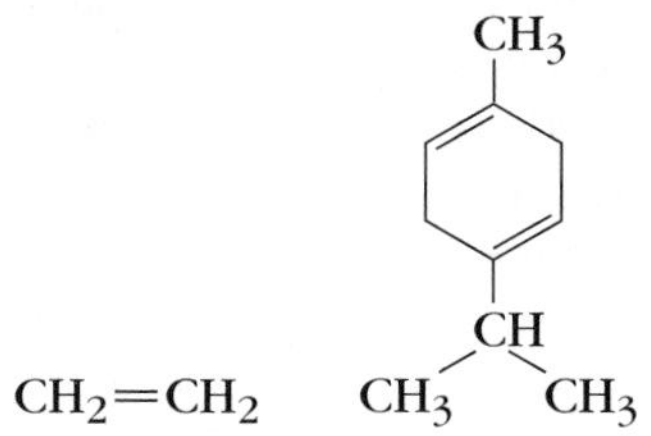

CH$_2$=CH$_2$
Ethylene

γ-Terpinene

The carbon-carbon double bond has a rich and interesting chemistry. In this chapter and Chapter 8, we explore the structure and chemical properties of alkenes. Let's begin by reviewing the structure of the carbon-carbon double bond.

7.1   Structure and Bonding in Alkenes

The general molecular formula of alkenes is C_nH_{2n}, as we learned in Section 5.11. Alkenes are isomeric with cycloalkanes, and both have one degree of unsaturation. Because alkenes have fewer hydrogen atoms than alkanes with the same number of carbon atoms, alkenes are frequently referred to as unsaturated hydrocarbons.

Ethylene is a planar molecule, as we learned in Section 1.11, B, whose carbon atoms have trigonal geometry. This is in contrast to the tetrahedral geometry about the carbon atoms of ethane. As a result, the bond angles about the carbon atoms and the carbon-carbon bond lengths in ethylene and ethane are different, as shown in Figure 7.1.

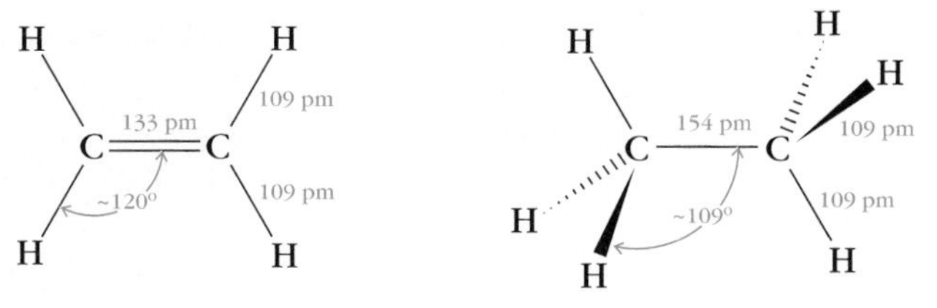

Ethylene

Ethane

Figure 7.1
Observed bond lengths and bond angles in ethane and ethylene. The carbon atoms in ethylene have trigonal geometry, while the carbon atoms in ethane have tetrahedral geometry.

The bond angles about the carbon atoms of ethylene are approximately 120°, while in ethane the bond angles are about 109°. The carbon-carbon double bond in ethylene is much shorter (133 pm) than the carbon-carbon single bond in ethane (154 pm). The carbon-hydrogen bond lengths are about the same in both compounds (109 pm).

The carbon-carbon bond strengths in ethane and ethylene are also different. Experimentally it is found that it takes 145 kcal/mol (606 kJ/mol) to break the carbon-carbon double bond in ethylene compared to 88 kcal/mol (368 kJ/mol) for the carbon-carbon single bond in ethane. Thus a carbon-carbon double bond is stronger than a carbon-carbon single bond, but it is not twice as strong.

The localized valence bond orbital model of the electron structure of ethylene was presented in Section 1.11, B. To review, the correct geometry about a carbon-carbon double bond is obtained with sp^2 hybrid carbon atoms. The planar framework of σ bonds is constructed as shown in Figure 7.2, A. The carbon-carbon σ bond is formed by the end-on overlap of one of the three sp^2 hybrid orbitals on each carbon atom. The C—H bonds are formed by the overlap of the two remaining sp^2 hybrid orbitals on each carbon atom with a 1s orbital of hydrogen. Sideways overlap of the remaining p atomic orbitals of the two sp^2 hybrid carbon atoms forms the π bond, as shown in Figure 7.2, B. Thus the double bond of ethylene (or any alkene) is made up of one σ bond and one π bond.

Recall from Chapter 4 that it takes very little energy to cause rotation about a carbon-carbon σ bond in alkanes such as butane. Rotation, however, does not take place easily about carbon-carbon double bonds. Consequently, it is possible to isolate stereoisomers of alkenes such as the two dichloroethylenes shown in Figure 7.3.

If rotation about the double bond occurred as easily as the rotation about a carbon-carbon single bond, the two dichloroethylenes in Figure 7.3 should quickly interconvert. They do not, however. Instead, these two compounds have been separated and each is stable and can be distinguished by its spectroscopic properties. So unless an alkene is heated to high temperatures (250 °C to 350 °C) or subjected to strong light, rotation doesn't occur about a carbon-carbon double bond.

The bonding model illustrated in Figure 7.2 is consistent with the absence of rotation about the double bond. In order for rotation to occur, one carbon atom of the double bond must be rotated by 90° relative to the other one. This rotation breaks the π bond and requires energy, as shown in Figure 7.4. The energy

Figure 7.2
A, The planar framework of σ bonds of ethylene. The C—C σ bond is formed by end-on overlap of an sp^2 hybrid orbital on each carbon atom. Each carbon atom also forms σ bonds to two hydrogen atoms by sp^2-1s orbital overlap. **B,** Sideways overlap of p atomic orbital on each carbon atom forms the π bond.

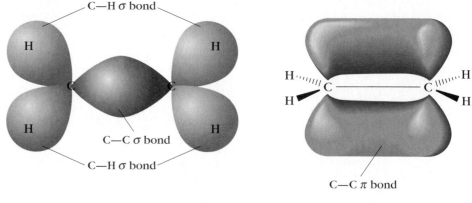

A   **B**

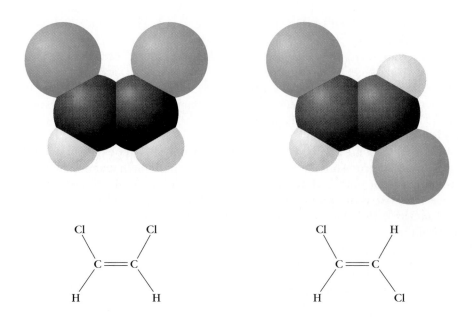

Figure 7.3
Two stereoisomers of dichloroethylene. The two are neither superposable nor mirror images, so they are diastereomers.

needed for this process is about 65 kcal/mol (272 kJ/mol) compared to only 4.5 kcal/mol (19 kJ/mol) for rotation about the central carbon-carbon bond of butane (Section 4.5).

The rigidity of the double bond is responsible for the existence of stereoisomers of alkenes. The two dichloroethylenes shown in Figure 7.3 are examples of these kinds of stereoisomers because they have the same sequence of bonded atoms but differ in the orientation of the atoms in space. Because these stereoisomers are neither superposable nor mirror images of each other, they are diastereomers. The name **geometric isomers** is sometimes given to these stereoisomers.

Many of the stereoisomers of alkanes can be easily converted by rotation about a C—C single bond. As we learned in Section 4.3, these stereoisomers are called conformational isomers. The stereoisomers of alkenes, on the other hand,

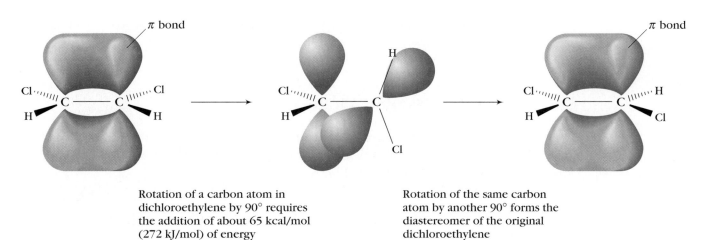

Rotation of a carbon atom in dichloroethylene by 90° requires the addition of about 65 kcal/mol (272 kJ/mol) of energy

Rotation of the same carbon atom by another 90° forms the diastereomer of the original dichloroethylene

Figure 7.4
Rotation about a carbon-carbon double bond requires the breaking of its π bond.

are not easily converted so they, like enantiomers and diastereomers, are **configurational isomers.**

Recall from Section 6.1 that a tetrahedral carbon atom bonded to four different groups or atoms is a stereocenter. The two trigonal carbon atoms in the stereoisomers of dichloroethylene (Figure 7.3) are also stereocenters, because if we interchange the two substituents on a carbon atom of either stereoisomer, we obtain the other. This, then, is an example of a trigonal carbon atom that is a stereocenter.

The double bond in all alkenes is rigid, yet not all alkenes exist as diastereomers or geometric isomers. How can we determine which alkenes can exist as diastereomers? If both carbon atoms of the double bond are stereocenters, the alkene can exist as diastereomers. Thus *if one or both carbon atoms of the double bond are bonded to identical atoms or groups, diastereomers cannot exist.* For example:

|One carbon atom of the double bond is bonded to identical Cl atoms, so these two compounds are identical | Neither carbon atom of the double bond is bonded to identical atoms, so these two compounds are diastereomers |

The pair of compounds on the left are superposable, while the pair on the right are not. It will be easier to see the difference between these pairs of compounds if you make molecular models of them.

Exercise 7.1   Identify each of the following alkenes that can exist as a pair of diastereomers. Write the structures of the pair of diastereomers in each case.

(a) $CH_3CH{=}CHCH_3$ **(c)** $(CH_3)_2C{=}CHCl$
(b) $BrFC{=}CHCH_3$ **(d)** $F_2C{=}CCl_2$

Because some alkenes can exist as diastereomers, any system of naming alkenes must be able to give these isomers a unique name. We'll see how this is done as we learn to name alkenes.

7.2   Naming Alkenes

Certain alkenes were known and given names long before the introduction of the IUPAC system of naming compounds. A few of these compounds are still called by their common or nonIUPAC names:

$$CH_2{=}CH_2 \qquad CH_3CH{=}CH_2 \qquad CH_3\overset{\overset{\displaystyle CH_3}{|}}{C}{=}CH_2 \qquad CH_2{=}CH\overset{\overset{\displaystyle CH_3}{|}}{C}{=}CH_2$$

Ethylene   Propylene   Isobutylene   Isoprene

The systematic IUPAC name of an alkene is obtained in four steps similar to those used to name alkanes.

Step 1. Choose the longest continuous chain of carbon atoms that contains *all the carbon-carbon double bonds* as the parent chain.

Step 2. Substitute the ending *-ene* for the ending *-ane* in the name of the corresponding alkane to obtain the parent name. If the molecule contains more than one double bond, the ending becomes *-diene* for two double bonds, *-triene* for three, and so forth. The position of each double bond should be indicated by a number. The letter *a* is placed before the ending -diene, -triene, etc., as an aid to pronunciation. For example: but*a*diene, pent*a*diene, etc.

Step 3. Number the chain to give the lowest number to the first carbon of the double bond.

Step 4. Write the complete name of the compound with the correct number for all substituents, which are listed in alphabetical order.

Sample Problem 7.1 shows how to apply these rules.

Sample Problem 7.1   Give the IUPAC name for the following compound.

$$\underset{\displaystyle CH_3CH_2\overset{\textstyle \overset{\textstyle CH_3}{|}}{C}HCH=CH_2}{}$$

Solution:

Step 1 The longest continuous chain containing the double bond is five carbon atoms long.

Step 2 The name of the straight chain alkane with five carbon atoms is *pentane*. There is one double bond in the molecule so the *-ane* ending of pentane is changed to *-ene*, making *pentene* the parent name of this alkene.

Step 3 Numbering the chain from right to left gives the lowest number, number 1, to the first carbon of the double bond.

Step 4 The methyl group is the only substituent, and it is located on carbon atom 3. The complete name of this alkene is *3-methyl-1-pentene*.

Exercise 7.2   Give the IUPAC name for each of the following compounds.

(a) $(CH_3)_2C=CHCHCHCH_3$ with Br substituent and CH_2CH_3 substituent

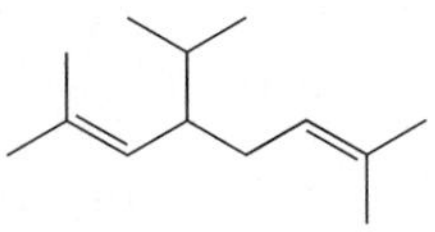

(b)

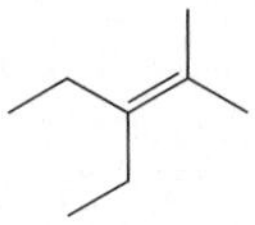

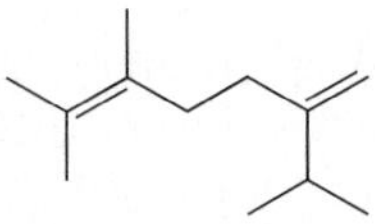

(c) $Cl_2CHCH=CH_2$

(d)

(e)

Exercise 7.3   Write the structure corresponding to each of the following IUPAC names.

(a) 2,3-Dimethyl-2-pentene   (c) 2,3,4,5-Tetramethyl-1,3,5-hexatriene
(b) 2,5-Dibromo-3-hexene   (d) 3,7,7-Trimethyl-4-octene

Simple cyclic alkenes with no substituents are named by changing the *-ane* ending of the name of the corresponding cycloalkane to the *-ene* ending. For example:

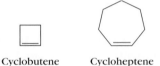

Cyclobutene   Cycloheptene

Numbers are used to indicate the position of the substituents on a cycloalkene ring. The direction of numbering the ring carbon atoms is chosen so that the carbon atoms of the double bond receive numbers 1 and 2 *and* the substituents receive the lowest numbers possible. The naming of substituted cycloalkenes is illustrated in Sample Problem 7.2.

Sample Problem 7.2   Give the IUPAC name of the following compound.

CH₃   CH₃

CH₃CH₂

Solution:

Step 1 The six-membered ring is the parent structure, so this compound is a derivative of cyclohexane.

Step 2 The *-ane* ending of cyclohexane is changed to *-ene,* making *cyclohexene* the name of the parent structure.

Step 3 There are two ways of numbering the ring carbon atoms that give the numbers 1 and 2 to the two carbon atoms of the double bond:

CH₃ ₃ CH₃   CH₃ ₆ CH₃
   4   2   5   1
CH₃CH₂ ⁵   ₆ 1   CH₃CH₂ ⁴ ₃ 2

The numbering on the left figure is preferred because it gives the substituents the lowest numbers possible.

Step 4 The complete name is *5-ethyl-3,3-dimethylcyclohexene.* The number 1, which indicates the position of the double bond, is understood and is not included in the name.

Exercise 7.4   Give the IUPAC name for each of the following structures.

(a) 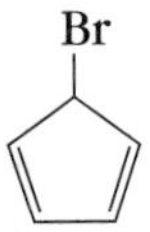   (b) 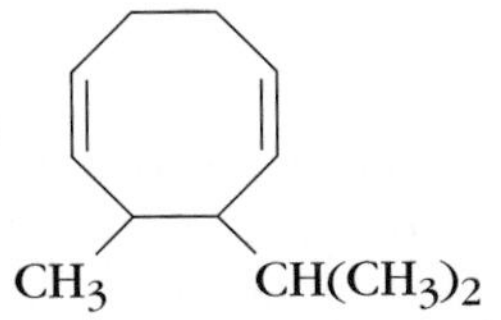   (c)

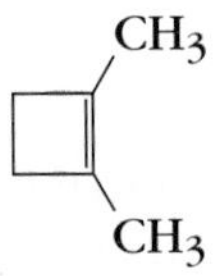

Br

CH₃   CH(CH₃)₂

CH₃

CH₃

Exercise 7.5   Write the structure that corresponds to each of the following IUPAC names.

(a) 1-Chloro-2-methylcyclopentene
(b) 2-Ethyl-4-methylcycloheptene
(c) 3,6-Diisopropylcyclohexene

Some substituents also contain double bonds. The following are a few examples that have common names that are widely used and should be learned:

$$CH_2{=}CH{-} \qquad CH_2{=}CHCH_2{-} \qquad \underset{\displaystyle CH_2{=}\overset{\textstyle CH_3}{\overset{|}{C}}{-}}{}$$

Vinyl group   Allyl group   Isopropenyl group

Like alkyl groups, these group names are used to construct common names of certain compounds, as illustrated by the following examples:

$$CH_2{=}CHCl \qquad CH_2{=}CHCH_2OH \qquad \overset{\textstyle CH_3}{\overset{|}{CH_2{=}CBr}}$$

Vinyl chloride   Allyl alcohol   Isopropenyl bromide

7.3   Naming Stereoisomers of Alkenes

Stereoisomers of alkenes are named in two ways. One way is to use the terms *cis-* and *trans-*. We have used these terms before to indicate the relative locations of two substituents on a cycloalkane ring (Section 4.12). These terms are used for alkenes in a similar way to indicate the relative positions of the two substituents bonded to each carbon atom of the double bond. In the *cis-* isomer, two identical groups are located on the same side of the double bond. In the *trans-* isomer, they are located on opposite sides. For example:

cis-1,2–Dichloroethene   *trans*-1,2–Dichloroethene

Exercise 7.6   Four isomeric butenes (C_4H_8) are known. Write the structure and name of each isomer.

Configurational isomers of alkenes can also be named by using the Cahn-Ingold-Prelog Rules (Section 6.2). The priorities of the two groups bonded to each carbon atom of the double bond are determined using the rules given in Section 6.2. The alkene with the two groups or atoms of higher priority on the same side of the double bond is called Z (for *zusammen,* the German word for *together*). The other configurational isomer is called E (for *entgegen,* the German word for *opposite*). The letter E or Z, enclosed in parentheses, is added to the beginning of the name of the alkene. Sample Problem 7.3 illustrates how to apply the Cahn-Ingold-Prelog Rules to configurational isomers of alkenes.

Sample Problem 7.3   Assign the E or Z configuration to the double bond in the following compound and give its complete name.

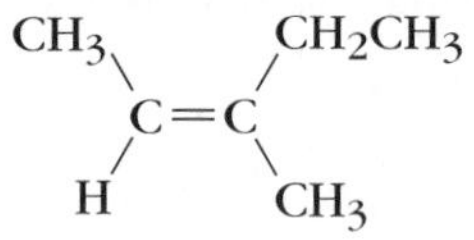

$$CH_3 \diagdown \qquad \diagup CH_2CH_3$$
$$C = C$$
$$H \diagup \qquad \diagdown CH_3$$

Solution:

Step 1 Identify the two atoms or groups bonded to each carbon atom of the double bond. Carbon atom 2 has a H atom and a —CH_3 group. Carbon atom 3 has two groups, a CH_3CH_2— group and a —CH_3 group.

Step 2 Determine the priority of the two atoms or groups of atoms on each carbon. On carbon atom 2, the —CH_3 group has a higher priority than the H atom, while on carbon 3, the CH_3CH_2— group has the higher priority.

Step 3 Determine if the atoms or groups of higher priority on each carbon atom are on the same side or opposite sides of the double bond.

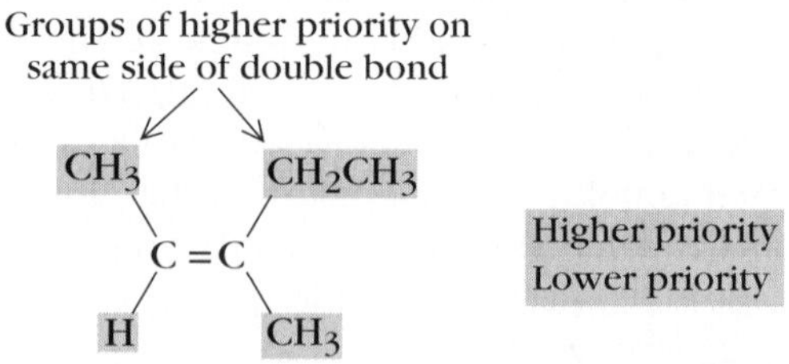

Groups of higher priority on
same side of double bond

$$CH_3 \qquad CH_2CH_3$$
$$C = C$$
$$H \qquad CH_3$$

Higher priority
Lower priority

Because the two groups of higher priority are on the same side of the double bond, we assign the Z configuration to the double bond.

The complete name is (Z)-3-methyl-2-pentene.

Exercise 7.7	Which atom or group of atoms in each set has the higher priority?

(a) I or Cl

(b) CH₃CH₂ or CH₃

(c) (CH₃)₂CH or Cl

(d) OH or Cl

(e) OH or CH₃

(f) CH₃ or CH₂OH

Exercise 7.8	Give the IUPAC name for each of the following structures using the *E* or *Z* designation to assign the configuration of the double bond.

(a)

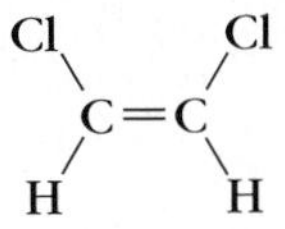

$$\underset{H}{\overset{Cl}{}}C=C\underset{H}{\overset{Cl}{}}$$

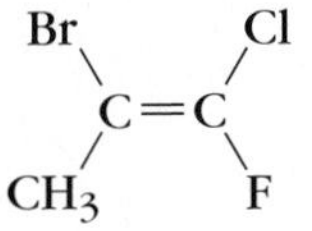

(b)

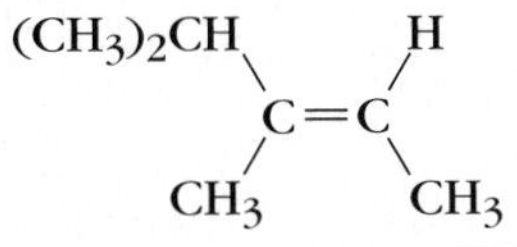

$$\underset{CH_3}{\overset{(CH_3)_2CH}{}}C=C\underset{CH_3}{\overset{H}{}}$$

(c)

$$\underset{CH_3}{\overset{Br}{}}C=C\underset{F}{\overset{Cl}{}}$$

(d)

$$\underset{ClCH_2}{\overset{CH_3}{}}C=C\underset{H}{\overset{CH_2CH_3}{}}$$

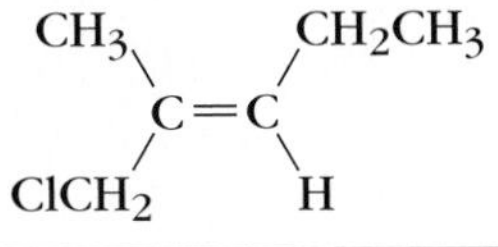

Exercise 7.9	Write the structure corresponding to each of the following names.

(a) (*E*)-4,4-Dimethyl-2-pentene

(b) (*Z*)-3-Ethyl-2-methyl-3-hexene

(c) (*E*)-4-Chloro-2-methyl-3-hexene

(d) (*Z*)-4-Bromo-1-chloro-2-butene

(e) (*E*)-4-Chloro-1-fluoro-2,4-dimethyl-2-hexene

The same method can be used to designate the configuration about the double bonds of dienes, trienes, and polyenes. Once the configuration of each double bond has been determined, its configuration and the number that indicates the position of the double bond on the parent chain are placed in parentheses before the name.

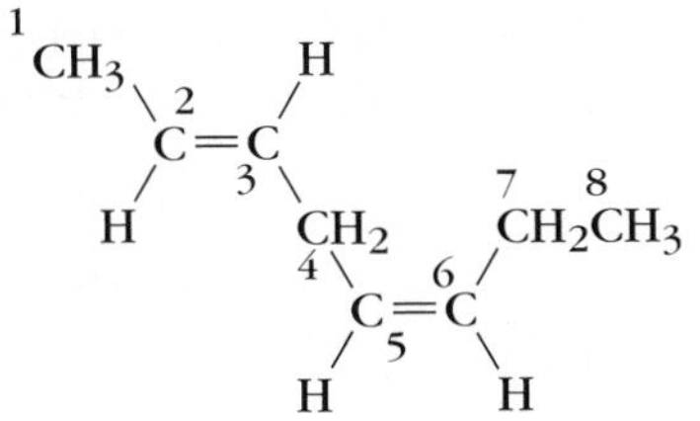

$$\overset{1}{CH_3}\diagdown\overset{2}{\underset{H}{\diagup}}C=C\overset{H}{\underset{\overset{4}{CH_2}}{\diagdown}}\overset{7}{\underset{\overset{6}{\underset{5}{C}}=C}{\diagup}}\overset{8}{\underset{H}{\overset{CH_2CH_3}{\diagup}}}$$

(2*E*,5*Z*)-2,5-Octadiene

Exercise 7.10	Give the IUPAC name of the following compound.

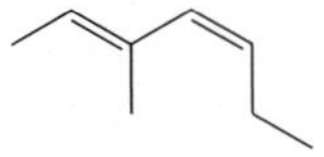

Box 7.1   **Geometric Isomers and Biological Activity**

Many animals and insects use chemicals to communicate with others of the same or different species. These chemicals, called **pheromones,** are secreted by the insect or animal and can mark a trail, issue a warning, or attract a mate. Compounds used as a sex attractant are usually secreted by the female of the species to attract a male. The following compounds have been identified as sex attractants for the insects specified.

Cabbage looper moth

Codling moth

Silkworm moth

European corn borer moth

The sex attractant for the silkworm moth is detectable by the male silkworm moth in extremely low concentrations; in air 10^{-13} moles/L is sufficient. It also very specifically attracts the male silkworm moth. It has no effect on other species. Changing the shape of the molecule reduces its effectiveness dramatically. Simply changing the configuration of one double bond from Z to E decreases the effectiveness of the silkworm moth sex attractant by a factor of 10^{12}!

Pheromones have great potential for insect control. The scent of a sex attractant can be used to attract a destructive insect to a trap, where it can be caught and destroyed. The value of this method is that it is specific for one type of insect; consequently it does not cause generalized harm to the environment.

Exercise 7.11   Write the structures and name all the configurational isomers of the following compound.

$$\underset{\qquad\qquad\qquad CH_3CH_2CH=CHCH=CHCHCH_3}{\overset{\displaystyle CH(CH_3)_2}{\overset{\displaystyle |}{}}}$$

7.4   Physical Properties

Table 7.1 compares the physical properties of representative alkenes and alkanes. The data in Table 7.1 show that the physical properties of alkenes and alkanes are similar. For example, alkenes and alkanes are both less dense than water, and compounds of both families that contain four or fewer carbon atoms are gases at room temperature. Their boiling points and densities increase, moreover, as the number of carbon atoms in their chains increases. Both alkanes and alkenes are

Table 7.1		A Comparison of Selected Physical Properties of Alkenes and Alkanes					
$n*$	Name of Alkene	Boiling Point (°C)	Density (g/ml)	Name of Alkane	Boiling Point (°C)	Density (g/ml)
2	Ethene	−103.7	—	Ethane	−88.6	—
3	Propene	−47.8	0.514	Propane	−42.1	0.500
4	1-Butene	−6.3	0.595	Butane	−0.5	0.578
5	1-Pentene	30.0	0.641	Pentane	36	0.626
6	1-Hexene	63.5	0.674	Hexane	69	0.660
7	1-Heptene	93.6	0.697	Heptane	98	0.684
8	1-Octene	101.7	0.715	Octane	126	0.703
9	1-Nonene	150.8	0.718	Nonane	151	0.718
10	1-Decene	170.6	0.741	Decane	174	0.730
12	1-Dodecene	213	0.758	Dodecane	216	0.749
20	1-Icosene	341	0.788	Icosane	343	0.789

*Number of carbon atoms.

insoluble in water but very soluble in nonpolar solvents such as ether, benzene, and carbon tetrachloride.

The similarities of the physical properties of alkanes and alkenes are due to the similarities of their structures. Both families of compounds contain only non-polar C—C and C—H bonds. They are all nonpolar compounds so they are insoluble in water but soluble in nonpolar solvents. The presence of a double bond in an alkene does not significantly alter its polarity compared to an alkane. Consequently alkanes and alkenes with the same number of carbon atoms have similar physical properties.

7.5   Relative Stabilities of Alkenes

The relative stabilities of alkenes can be determined from their heats of combustion in a manner similar to that used to determine the stabilities of cycloalkanes in Section 4.6. Let's use this method to determine the relative stabilities of the isomeric butenes.

The isomeric butenes all undergo combustion according to the following equation.

$$C_4H_8 \ + \ 6O_2 \ \longrightarrow \ 4CO_2 \ + \ 4H_2O$$

The heat given off in this reaction is a measure of their relative stabilities. Thus the stability of the butene decreases as the amount of heat released on combustion increases. The values of their experimental heats of combustion are compared in Figure 7.5. Based on the data in Figure 7.5, we conclude that the relative order of stabilities of the butenes is

1-Butene   <   *cis*-2-Butene   <   *trans*-2-Butene   <   Methylpropene

INCREASING STABILITY

Figure 7.5
Heats of combustion of isomeric butenes. The most stable isomer is methylpropene; the least stable isomer is 1-butene.

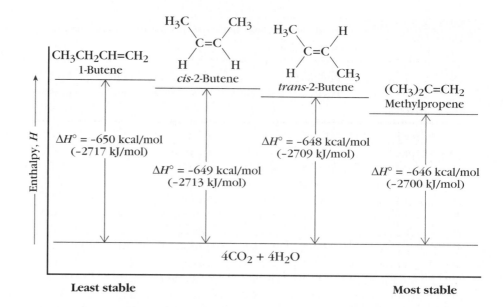

These results for the isomeric butenes can be generalized and expanded by examining similar heats of combustion data that have been obtained from other isomeric alkenes. These data lead to the conclusion that the *relative stabilities of alkenes increase as the degree of substitution of the double bond increases.* The degree of substitution is defined as the total number of alkyl groups that are bonded to the carbon atoms of the double bond. Figure 7.6 shows the general effect of degree of substitution on the relative stabilities of isomeric hexenes.

Another general trend in alkene stabilities shown by the data in Figure 7.5 is that a *trans- alkene is more stable than its cis- isomer.* For example, *trans*-2-butene is 1.0 kcal/mol (4.2 kJ/mol) more stable than *cis*-2-butene. The higher energy of *cis*- alkenes compared to their *trans*- isomers is attributed to the crowding of the two alkyl substituents on the same side of the double bond, as shown in Figure 7.7. This is another example of steric strain or steric hindrance, which we saw in Section 4.5 for the gauche conformation of butane and in Section 4.11 for the axial conformation of methylcyclohexane.

The *trans*- isomers of small ring cyclic alkenes are not more stable than their *cis*- isomers. Cycloalkenes that contain five or fewer carbon atoms exist only in their *cis*- forms.

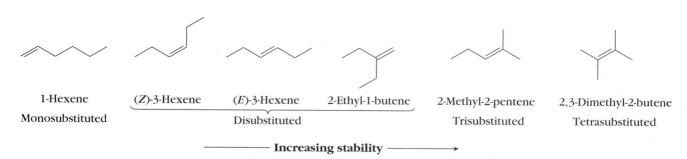

1-Hexene	(Z)-3-Hexene	(E)-3-Hexene	2-Ethyl-1-butene	2-Methyl-2-pentene	2,3-Dimethyl-2-butene
Monosubstituted		Disubstituted		Trisubstituted	Tetrasubstituted

—————— **Increasing stability** ——————→

Figure 7.6
Relative stabilities of isomeric hexenes. Generally, alkenes with more highly substituted double bonds are more stable than isomers with less substituted double bonds.

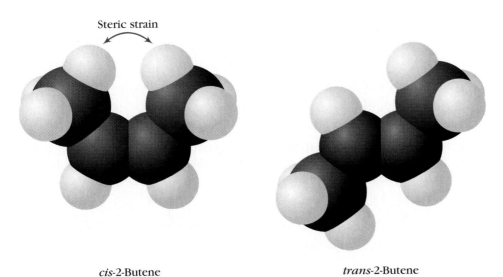

Steric strain

cis-2-Butene   *trans*-2-Butene

Figure 7.7
cis- And *trans*-2-butene. Steric strain between methyl groups on the same side of the double bond makes the *cis*- isomer less stable than the *trans*- isomer.

Cyclopropene and cyclobutene have even more ring strain than the corresponding cycloalkanes. This ring strain arises mostly from angle strain due to the compression of the 120° bond angle of the two sp^2 hybrid carbon atoms to either 60° for cyclopropene or 90° for cyclobutene.

Cyclopropene   Cyclobutene   Cyclopentene

trans-Cycloalkenes with a ring of five or fewer carbon atoms have never been isolated. The reason becomes clear if we try to make *trans*-cyclohexene. If we place two carbon atoms *trans*- to each other on the double bond, we must add two more carbon atoms to connect these two carbon atoms to form a six-membered ring. A great deal of strain is introduced into the molecule when we try to close the *trans*-cyclohexene ring. Try this with your molecular models, but be careful not to strain them so that they break.

To form a *trans*-cycloalkene, we must connect the two highlighted carbon atoms with a chain of carbon atoms

When $n > 3$, the *trans*-cycloalkene is stable at room temperature; when $n = 1$, 2, or 3, it is not

trans-Cyclohexene and *trans*-cycloheptene are too strained to be isolated. However, there is spectroscopic evidence that *trans*-cyclohexene and *trans*-cycloheptene are formed briefly in certain chemical reactions. An eight-membered ring is the smallest ring that can accommodate a *trans*- double bond. Although *trans*-cyclooctene has been prepared, it is highly strained and is less stable by 9.1 kcal/mol (38.1 kJ/mol) than its *cis*- isomer.

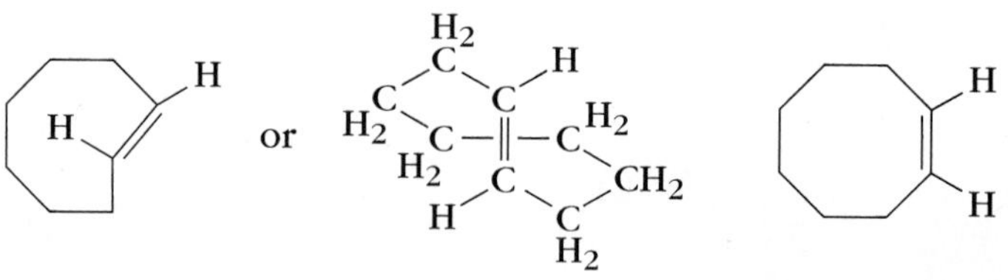

trans-Cyclooctene   *cis*-Cyclooctene

Rings that contain more than 10 carbon atoms can easily accommodate a *trans*- double bond. For these cycloalkenes, there is little energy difference between the *cis*- and *trans*- isomers.

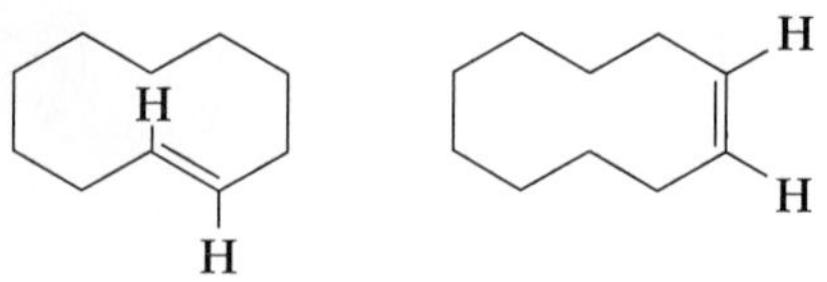

trans-Cyclodecene   *cis*-Cyclodecene

Exercise 7.12   Arrange the following sets of alkenes in order of increasing stability.

(a) 2-Pentene; 2-methyl-2-butene; 1-pentene
(b) (*E*)-3-Methyl-3-hexene; 2-methyl-1-hexene; (*Z*)-3-methyl-3-hexene
(c) 1-Heptene; 2,3-dimethyl-2-pentene; 2-methyl-1-hexene

7.6   Spectroscopic Properties of Alkenes

We learned in Section 5.5 that alkenes have two characteristic absorption bands in their IR spectra. One, due to the carbon-carbon double bond stretching vibration, is found in the region 1670 to 1645 cm^{-1}. The other is observed in the region 3080 to 3020 cm^{-1} and is due to the sp^2 hybrid carbon-hydrogen bond stretching vibration.

The absorption due to the double bond is most intense for monosubstituted alkenes. As more alkyl groups are added to the double bond, its intensity diminishes. For tri-, tetra-, and relatively symmetrically substituted alkenes, this band is often of such low intensity that it can't be observed. In Figure 7.8, for example, is the spectrum of 2,3-dimethyl-2-butene, a symmetrical alkene, which shows no absorption due to the double bond in the region 1670 to 1645 cm^{-1}.

Why doesn't 2,3-dimethyl-2-butene show an absorption due to its double bond? In order to absorb IR energy, a vibration must change the dipole moment of the molecule. The double bond in 2,3-dimethyl-2-butene has no dipole moment because it is symmetrically substituted. There is still no dipole moment when the bond is stretched or compressed; because the vibration of its double bond produces no change in the dipole moment of the molecule, no absorption of IR energy occurs.

Generally, the stretching and compressing of bonds with dipole moments causes an absorption in the IR spectrum. Furthermore, the larger the change in

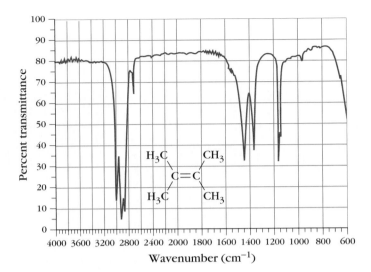

Figure 7.8
The IR spectrum of 2,3-dimethyl-2-butene. Because the molecule is symmetrical, its C=C absorption band in the region 1670 to 1645 cm^{-1} is absent.

dipole moment, the more intense the absorption. Large changes in dipole moment are associated with the stretching of carbon-oxygen double bonds and the O—H bond of alcohols, so the absorption bands due to these vibrations are intense.

Recall from Section 5.10 that the ^{13}C NMR chemical shifts of the sp^2 hybrid carbon atoms of a carbon-carbon double bond are located at δ 100-150. This is far downfield from the ^{13}C NMR chemical shifts of sp^3 hybrid carbon atoms of alkanes. As a result, carbon atoms of a carbon-carbon double bond are readily distinguished from the sp^3 carbon atoms in alkanes.

The ^{13}C NMR chemical shift of an sp^2 hybrid carbon is influenced by the number of alkyl groups bonded to it, as illustrated by the following examples.

$$CH_2=CH_2 \qquad \overset{\delta 19.4}{\overset{\curvearrowright}{CH_3}}\overset{\delta 115.9}{\overset{\downarrow}{CH}}=CH_2 \qquad \overset{\delta 24.2}{\overset{\curvearrowright}{(CH_3)_2}}\overset{\delta 113.2}{\overset{\downarrow}{C}}=CH_2$$

$$\underset{\delta 123.5}{\overset{\curvearrowleft}{}} \qquad \qquad \underset{\delta 133.4}{\overset{\curvearrowleft}{}} \qquad \qquad \underset{\delta 141.8}{\overset{\curvearrowleft}{}}$$

As a general rule, the range of values of the ^{13}C chemical shifts of carbon atoms of the double bond with zero, one, or two alkyl substituents is found to be the following:

$$\begin{aligned} \delta \quad =CH_2 \quad &< \quad 120 \text{ ppm} \\ \delta \quad =CHR \quad &> \quad 120 \text{ ppm} \\ \delta \quad =CR_2 \quad &> \quad 135 \text{ ppm} \end{aligned}$$

Exercise 7.13	Write the structure of the compound of molecular formula C_5H_8, whose IR and ^{13}C NMR spectra are given on p. 292.

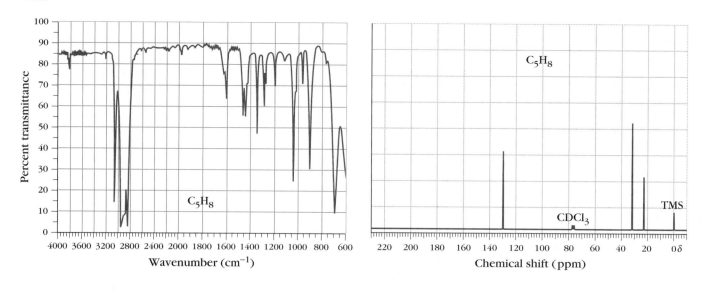

We've learned about the structures of alkenes, their stabilities, and their physical and spectroscopic properties. Now let's focus our attention on their chemical properties.

7.7   Addition to the Double Bond Is the Predominant Reaction of Alkenes

Many reagents readily react with the double bond of alkenes. Bromine, for example, reacts with ethylene to form 1,2-dibromoethane.

$$H_2C=CH_2 \ + \ Br-Br \ \longrightarrow \ \begin{array}{c} H_2C-CH_2 \\ | \quad\ | \\ Br \ \ Br \end{array}$$

        Ethylene   Bromine   1,2-Dibromoethane

The reaction of bromine and ethylene is an example of a class of reactions called **addition reactions.** They are called addition reactions because the product is formed by the addition of a reagent (in this case Br_2) to the double bond. In addition reactions, the π bond of the double bond is broken and is replaced by two σ bonds. In this example, the two σ bonds are two C—Br bonds.

Addition reactions of alkenes are generally thermodynamically favored processes. For example, the addition of HCl, H_2O, Cl_2, or H_2 to ethylene in the gas phase at 25 °C is an exergonic process.

$$CH_2=CH_2 \ + \ HCl \ \longrightarrow \ H-CH_2CH_2-Cl \qquad \Delta G° \ = \ -2.8 \text{ kcal/mol } (-11.7 \text{ kJ/mol})$$

$$CH_2=CH_2 \ + \ H_2O \ \longrightarrow \ H-CH_2CH_2-OH \qquad \Delta G° \ = \ -1.83 \text{ kcal/mol } (-7.65 \text{ kJ/mol})$$

$$CH_2=CH_2 \ + \ Cl_2 \ \longrightarrow \ Cl-CH_2CH_2-Cl \qquad \Delta G° \ = \ -34.0 \text{ kcal/mol } (-142.1 \text{ kJ/mol})$$

$$CH_2=CH_2 \ + \ H_2 \ \longrightarrow \ H-CH_2CH_2-H \qquad \Delta G° \ = \ -24.2 \text{ kcal/mol } (-101.2 \text{ kJ/mol})$$

Just because a reaction is thermodynamically favored, however, does not necessarily mean that the reaction will occur readily, as you will recall from Section

Box 7.2   **Writing Chemical Equations for Organic Reactions**

A chemical equation is a shorthand notation that tells us what happens in a chemical reaction. Organic chemists simplify the writing of the equations of organic reactions by the use of a number of conventions. Let's examine these conventions because we will use them throughout this text.

A balanced chemical equation is the best and most complete way of describing a chemical reaction. Several examples of balanced equations representing organic reactions are given in Section 7.7.

Organic chemists frequently write an organic chemical equation by placing one reactant, usually the organic compound, to the left of the arrow and the second reactant above the arrow. Catalyst, solvent, temperature, and any special reaction conditions are listed below the arrow.

$$\underset{\substack{\text{Organic}\\ \text{reactant}}}{} \xrightarrow[\substack{\text{Catalyst, solvent,}\\ \text{and temperature}}]{\text{Second reactant}} \underset{\substack{\text{Organic}\\ \text{product}}}{}$$

The second reactant is frequently, but not always, an inorganic reagent.

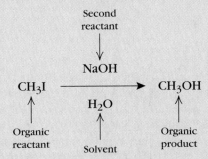

The emphasis in these types of chemical equations is on the transformation of the organic compound. Inorganic products are usually not included in the equation. Iodide ion, for example, which is also a product of the above reaction, is not included in the equation.

Organic chemists often abbreviate a series of reactions by numbering the reactions consecutively above and below the arrow. For example, the following sequence of two reactions converts ethylene to ethanol, as we will discuss in Chapter 8.

Reaction 1

$$3CH_2{=}CH_2 \;+\; BH_3 \;\xrightarrow[\text{THF}]{}\; (CH_3CH_2)_3B$$

$$\underset{\text{Ethylene}}{} \qquad\qquad \underset{\text{Borane}}{} \qquad\qquad \underset{\text{Triethylborane}}{}$$

Reaction 2

$$(CH_3CH_2)_3B \;\xrightarrow[\text{H}_2\text{O}]{\text{H}_2\text{O}_2/\text{NaOH}}\; 3CH_3CH_2OH$$

$$\underset{\text{Triethylborane}}{} \qquad\qquad\qquad \underset{\text{Ethanol}}{}$$

Continued.

Box 7.2—cont'd   **Writing Chemical Equations for Organic Reactions**

In reaction 1, ethylene reacts with borane in the solvent tetrahydrofuran (THF) to form triethylborane. In reaction 2, triethylborane forms ethanol when treated with an aqueous solution of hydrogen peroxide and sodium hydroxide. These two reactions can be condensed into the following single chemical equation.

Reagent and
solvent used
in first reaction

$$CH_2=CH_2 \xrightarrow[\text{(2) } H_2O_2/NaOH/H_2O]{\text{(1) } BH_3/THF} CH_3CH_2OH$$

Reagents and solvent
used in second reaction

Often both reactions are carried out consecutively in the same reaction vessel.

When the reactions of organic compounds are represented in these ways, minor products are not shown, and in most cases the equation is not balanced. These chemical equations are used because they save space and time.

3.13. In order for a reaction to occur, the reagents must come together with the proper orientation and an energy greater than the free energy of activation. Many addition reactions meet these criteria; that is, they are thermodynamically favored and they have an energetically accessible pathway (or mechanism). While others are thermodynamically favored, they need a catalyst to provide an alternate pathway for the reaction to occur. In this chapter and Chapter 8, we will learn about both catalyzed and uncatalyzed additions to carbon-carbon double bonds.

Let's begin by examining additions of Brønsted-Lowry acids.

Figure 7.9
Addition of strong acids to cyclohexene.

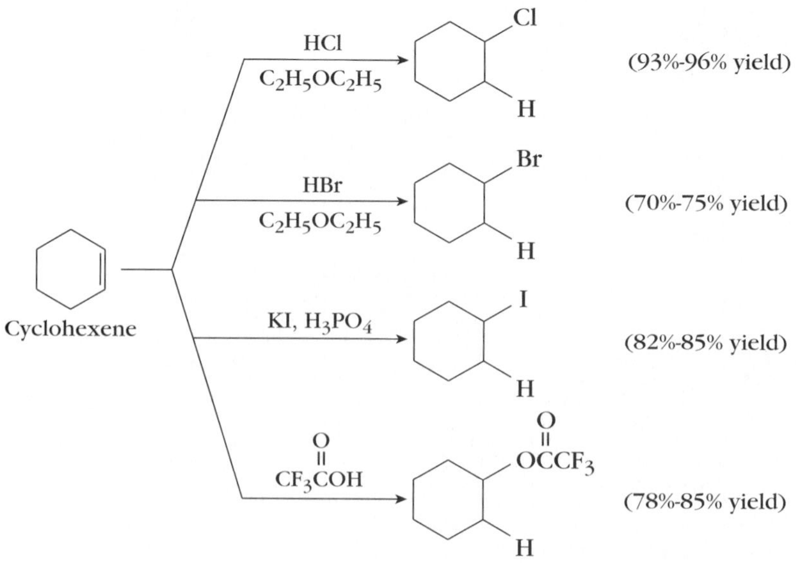

7.8 Additions of Brønsted-Lowry Acids to Alkenes

The addition of strong acids is a typical reaction of alkenes. Equations for the additions of the strong acids HCl, HBr, HI, and trifluoroacetic acid (CF_3COOH) to cyclohexene are shown in Figure 7.9.

The addition of HCl or HBr is usually carried out by passing the dry gaseous acid directly into the liquid alkene. Sometimes a solvent such as diethyl ether ($C_2H_5OC_2H_5$) or dichloromethane (CH_2Cl_2) is used. HI, on the other hand, is not added directly. Instead, it is made from a mixture of KI and 95% orthophosphoric acid (H_3PO_4) in the reaction mixture. Trifluoroacetic acid is a liquid and is both the reactant and the solvent in its reactions with alkenes.

Weaker acids add to alkenes only with the aid of an acid catalyst. For example, water, alcohols, and acetic acid (CH_3COOH) react with cyclohexene in the presence of a small amount of sulfuric acid, as shown in Figure 7.10.

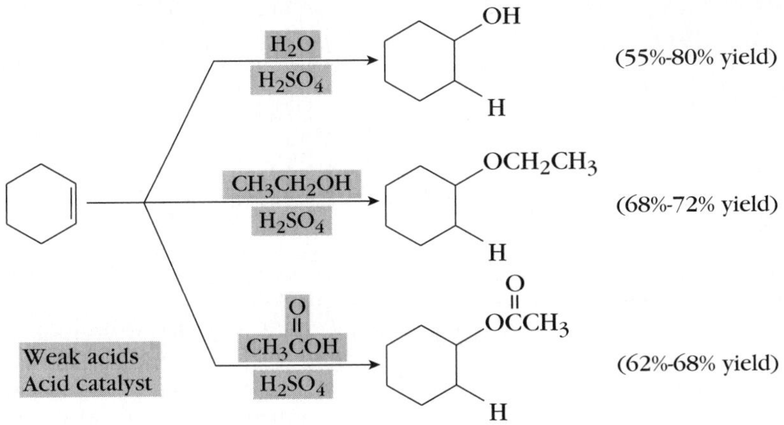

Figure 7.10
Addition of weak acids to alkenes requires an acid catalyst.

Exercise 7.14   Write the equations for the reactions of ethylene with each of the following reagents.

(a) HCl

(b) CF_3COOH

(c) KI and H_3PO_4

(d) CH_3OH with a few drops of H_2SO_4

(e) H_2O with a few drops of H_2SO_4

Exercise 7.15   Write the equations for the reactions of 1,2-dimethylcyclopentene with each of the reagents given in Exercise 7.14.

From these and other experimental data, chemists have proposed a mechanism of acid additions to alkenes.

7.9   The Mechanism of Acid Additions to Alkenes

We learned in Section 3.13 that a mechanism of a chemical reaction is a detailed description of how the reactants are converted to products. A mechanism of a reaction is written as a series of elementary reactions; reactions, as we also learned in Section 3.13, in which only a single molecular event takes place between reactants and products.

The accepted two-step mechanism for the addition to alkenes is shown for the reaction of HCl and 2,3-dimethyl-2-butene in Figure 7.11.

Figure 7.11
The two-step mechanism for the addition of HCl to 2,3-dimethyl-2-butene. The usual way of writing mechanisms of organic reactions is to use curved arrows to show the electron movement between reactants and products in each step of the mechanism. Another way of picturing the electron movement is by using three-dimensional electron density contour drawings (90% to 95% probability contours). The drawing above a structural formula is its electron density contour, which was created by using semiempirical computational methods. Both the electron density contour drawings and the structural formulas with curved arrows portray the electron movement in each step of the reaction mechanism.

Step 1: Addition of a proton to the double bond forms a carbocation

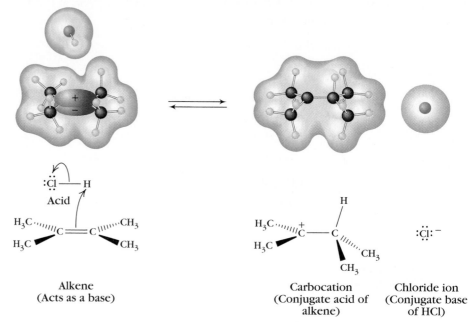

Step 2: Reaction of carbocation with a chloride ion forms the product

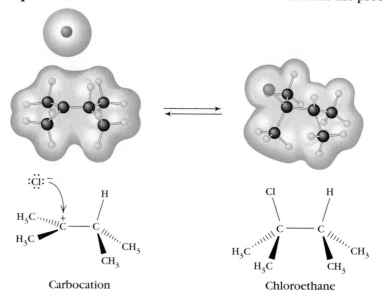

Step 1 of the mechanism is an acid-base reaction in which the alkene acts as the base. In this step, the two electrons of the π bond are donated to the proton, with the result that a σ bond is formed between the proton and one of the alkene carbon atoms. As the bond between the carbon and the proton is forming, the two electrons in the HCl bond are transferred to Cl to form Cl^-. This electron movement is shown by the curved arrows in Figure 7.11.

The movement of the π electrons to form a σ bond leaves only six electrons on the other carbon atom of the double bond. A positive ion that contains a carbon atom with only six electrons in its valence shell is called a **carbocation.** A carbocation is electron deficient, and it has a positive charge. The carbocation formed in Step 1 accepts an electron pair from a chloride ion in Step 2 to form the product.

We learned in Section 3.7 that Lewis acids are electron pair acceptors. An equivalent way to describe Lewis acids is to say that they are **electrophiles.** Electrophiles are electron-loving or electron-seeking reagents. According to this definition a proton is an electrophile. Because an electrophile (a proton) reacts with the electrons of the double bond in Step 1 of the mechanism, it is called an electrophilic addition mechanism. Reactions that occur by this kind of mechanism are called **electrophilic addition reactions.** Electrophilic addition reactions are a general class of organic reactions because many electrophiles (electron-deficient compounds) other than protons add to double bonds. Additional examples of electrophilic addition reactions of alkenes are presented in Chapter 8.

Lewis bases are electron pair donors. Because they donate a pair of electrons to a center of positive charge, they are also called **nucleophiles.** Nucleophiles are nucleus-loving or nucleus-seeking reagents. In Step 2 of the electrophilic addition mechanism, for example, the chloride ion is a nucleophile because it donates a pair of electrons to the carbocation.

■ *The suffix -phile means "loving" so electrophile means an electron-loving reagent.*

Exercise 7.16	Is the alkene in Step 1 of the mechanism shown in Figure 7.11 a nucleophile or an electrophile? What about the carbocation in Step 2?

Exercise 7.17	Write a two-step mechanism for the addition of trifluoroacetic acid (CF_3COOH) to ethylene.

In Section 3.13, we described how the free energy change in an elementary reaction can be shown by means of a free energy-reaction path diagram. A similar diagram for the addition of HCl to 2,3-dimethyl-2-butene is shown in Figure 7.12. The diagram in Figure 7.12 is a combination of the free energy-reaction path diagrams for each step of the mechanism. The two steps are joined so that the product of the first step, the carbocation, is the reactant for the second step.

Figure 7.12 contains the free energy levels of the reactants, products, and carbocation, as well as the path from reactants to product via the carbocation. The path passes through two transition states labelled T.S.#1 and T.S.#2. The standard free energy change for the overall reaction is labelled ΔG°, while the free energy of activation to reach T.S.#1 from the reactants is $\Delta G_1^\ddagger$ and the free energy of activation needed to reach T.S.#2 from the carbocation is $\Delta G_2^\ddagger$.

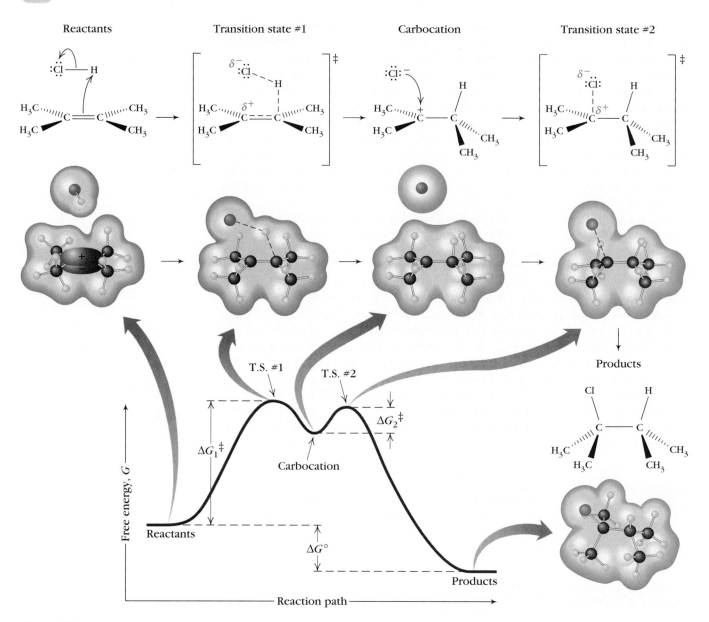

Figure 7.12
The free energy-reaction path diagram for the reaction of 2,3-dimethyl-2-butene and HCl.

$\Delta G_1^\ddagger$ is much greater than $\Delta G_2^\ddagger$ in Figure 7.12. The physical meaning of this difference in free energies of activation is that the carbocation is formed more slowly from reactants than it reacts with chloride ion, so it has a very short lifetime. This explains why the carbocation is never observed in acid additions to alkenes. Any chemical structure whose energy is greater than either reactants or products and has a short lifetime is called a **reactive intermediate.** The carbocation in the mechanism of acid additions to alkenes is a reactive intermediate.

Because $\Delta G_1^\ddagger \gg \Delta G_2^\ddagger$, Step 1 is much slower than Step 2 of the mechanism. The slowest step in a mechanism is called the **rate-limiting step** or **rate-determining step**. No reaction can occur faster than its rate-determining step.

The concept of a rate-determining step can be illustrated by means of the following analogy. A small company builds bicycles. It takes two wheels, a frame, a set of handlebars, and a chain mechanism to build a single bicycle. These bicycles are built on an assembly line where a different person performs each of the steps needed to build a bicycle. First, it takes 5 minutes to add the wheels to the frame. Next, the handlebars are added in a process that takes 2 minutes. Finally, it takes 10 minutes to add the complicated chain mechanism. The slowest or rate-determining step in this assembly line is adding the chain mechanism. No matter how fast the wheels or handlebars are added to the frame, the speed at which a complete bicycle can be built cannot be faster than the speed of adding the chain mechanism.

For the addition of HCl to 2,3-dimethyl-2-butene, Step 1 in Figure 7.11 is the rate-determining step because the overall rate of the reaction cannot be faster than the rate of Step 1. *For acid additions to all alkenes, therefore, the formation of the carbocation (Step 1) is the slow step that determines the rate of the reaction.* Anything that we do to affect the rate-determining step affects the overall rate of the reaction. The stability of the carbocation is a major factor that affects the rate-determining step.

7.10   Structure and Stabilities of Carbocations

Carbocations are classified as primary, secondary, and tertiary depending on whether one, two, or three alkyl groups are bonded to the electron-deficient carbon atom. The names of simple carbocations are obtained by adding the word cation to the alkyl name representing their structure.

$$CH_3^+ \qquad \overset{+}{CH_3CH_2} \qquad \overset{+}{CH_3CHCH_3} \qquad \overset{\overset{\displaystyle CH_3}{\displaystyle |}}{\underset{+}{CH_3CCH_3}}$$

Methyl cation   Ethyl cation   Isopropyl cation   *tert*-Butyl cation
                     Primary (1°)   Secondary (2°)   Tertiary (3°)
                     carbocation   carbocation   carbocation

A carbocation has an electron-deficient carbon atom bonded to three atoms or groups of atoms. Atoms that have only three pairs of electrons in their valence shell have a trigonal planar shape, as we discussed in Section 1.7. Thus the three atoms or groups of atoms bonded to the electron-deficient carbon atom of the carbocation are all in a plane with bond angles of about 120°, as shown in Figure 7.13.

To construct the orbital representation of a carbocation with this geometry, we must represent the electron-deficient carbon atom as an sp^2 hybrid. Its three

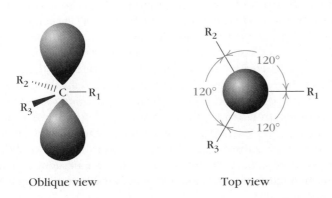

Oblique view   Top view

Figure 7.13
The planar geometry of a carbocation. The R—C—R bond angles are all about 120°.

Figure 7.14
A localized valence bond representation of the electron structure of a tertiary carbocation.

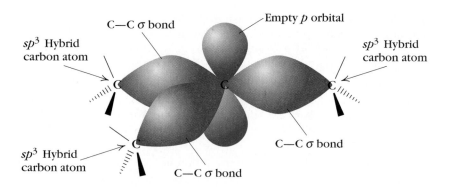

sp^2 hybrid orbitals lie in a plane with the lobes of the empty $2p$ orbital above and below the plane, as shown in Figure 7.14.

The relative stabilities of many alkyl cations have been determined in the gas phase by several sophisticated techniques that we will not discuss. The stabilization energies of several alkyl carbocations relative to the methyl cation are summarized in Table 7.2. The more negative the stabilization energy, the more stable is the carbocation relative to the methyl cation.

Based on the data in Table 7.2, we conclude that the order of stability of these alkyl cations is:

$$\overset{+}{C}H_3 \quad < \quad CH_3CH_2CH_2\overset{+}{C}H_2 \quad < \quad CH_3CH_2\overset{+}{C}HCH_3 \quad < \quad (CH_3)_3\overset{+}{C}$$

INCREASING STABILITY ⟹

From these, as well as other experimental data, chemists have concluded that the general order of alkyl cation stability is:

$\overset{+}{C}H_3$	<	Primary alkyl cation	<	Secondary (2°) alkyl cation	<	Tertiary (3°) alkyl cation

INCREASING STABILITY ⟹

Why do the stabilities of alkyl cations depend on the number of alkyl groups bonded to their electron-deficient carbon atom? The number of alkyl groups is important because the relative stability of a carbocation depends chiefly on how well it accommodates the charge on its electron-deficient carbon atom. We know from the laws of electrostatics that dispersing the charge of a charged system increases its stability. Therefore any change in its structure that tends to disperse or spread out the positive charge of the electron-deficient carbon atom over the rest of the ion will stabilize a carbocation. Conversely, anything that tends to concentrate charge on its electron-deficient carbon atom will destabilize a carbocation.

How do substituents concentrate or disperse the charge on an electron-deficient carbon atom? Substituents disperse charge by their abilities to either withdraw electrons or donate electrons. Consider a substituent Y that replaces a hydrogen atom bonded to the electron-deficient carbon atom of a methyl cation.

| Table 7.2 | Stabilization Energies of Several Alkyl Cations Relative to the Methyl Cation | | |

Cation		Stabilization Energy	
Name	Structure	kcal/mol	kJ/mol
Methyl	$\overset{+}{C}H_3$	0	0
n-Butyl	$CH_3CH_2CH_2\overset{+}{C}H_2$	−47	−197
sec-Butyl	$CH_3CH_2\overset{+}{C}HCH_3$	−68	−284
tert-Butyl	$(CH_3)_3\overset{+}{C}$	−84	−351

$$
\begin{array}{c}
\text{H} \\
| \\
\text{H—C+} \\
| \\
\text{H}
\end{array}
\qquad
\begin{array}{c}
\text{Replacing one H by} \\
\text{substituent Y gives}
\end{array}
\qquad
\begin{array}{c}
\text{Y} \\
| \\
\text{H—C+} \\
| \\
\text{H}
\end{array}
$$

Methyl cation   Substituted methyl cation

Compared to a hydrogen atom, the substituent Y may be electron donating or electron withdrawing. An **electron-withdrawing group** or **substituent** removes electron density and increases the positive charge on the electron-deficient carbon atom compared to a hydrogen atom. This makes the substituted carbocation less stable than a methyl cation. An **electron-donating group** or **substituent** disperses electron density so it reduces the positive charge on the electron-deficient carbon atom compared to a hydrogen atom. This makes the substituted carbocation more stable than a methyl cation. As a result of this charge dispersal, the electron-donating substituent itself acquires a slight positive charge.

Are alkyl groups electron-withdrawing or electron-donating substituents? We can decide by looking at their effect on the stabilities of carbocations. We've seen, for instance, that the stability of a carbocation increases as the number of alkyl groups bonded to its electron-deficient carbon atom increases. As a result, *alkyl groups are electron-donating substituents.*

Alkyl groups are thought to be electron-donating substituents because of their **inductive effect;** this term is used to describe the polarization of the electrons in a bond by a nearby electropositive or electronegative atom. In a carbocation, the electron-deficient carbon atom is an electropositive center. This causes a shift in the electron density of the σ bond toward the positive carbon atom. This shift of electron density causes a partial positive charge on the adjacent carbon atoms. In this way the positive charge of the electron-deficient carbon atom is spread to the carbon atoms of the adjacent alkyl groups.

We can indicate this electron-donating inductive effect of alkyl groups by using an arrow in place of the usual line representation of a bond. The arrowhead shows the direction of the shift in electron density of the bond caused by the alkyl group. Thus the stabilizing effect of the methyl groups of the *tert*-butyl cation is represented as follows:

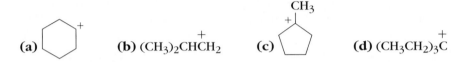

$$CH_3$$

Electrons of σ bond
are drawn toward the
positive charge

Exercise 7.18   Classify the following as primary (1°), secondary (2°),
or tertiary (3°) carbocations.

(a) ⬡⁺   **(b)** $(CH_3)_2CHCH_2^+$ **(c)** ⬠ with CH_3 and + on ring   **(d)** $(CH_3CH_2)_3\overset{+}{C}$

Exercise 7.19   List the following carbocations in order of increasing
stability.

(a) $(CH_3CH_2)_2\overset{+}{CH}$ **(b)** $CH_3(CH_2)_3\overset{+}{CH_2}$ **(c)** $\overset{+}{CH_3}$ **(d)** $CH_3CH_2\overset{+}{C}(CH_3)_2$

The formation and the relative stabilities of carbocations help explain why
only one of two possible products is observed in the additions of acids to unsymmetrical alkenes.

7.11   Regiochemistry and Stereochemistry of Acid Additions to Alkenes

The alkenes used so far to illustrate their reactions with acids are **symmetrical
alkenes.** Symmetrical alkenes have the same two substituents on each of the two
carbon atoms of the double bond. Other alkenes such as propene, $CH_3CH=CH_2$,
are **unsymmetrical alkenes** because the two substituents on each carbon atom
of the double bond are different. It's important to distinguish between these two
kinds of alkenes because the products of addition reactions depend on whether
the structure of the starting alkene is symmetrical or unsymmetrical.

The addition of HCl (or any other acid) to a symmetrical alkene such as ethylene forms only a single product because the two carbons of the double bond of
ethylene are equivalent. The addition to an unsymmetrical alkene such as
propene shows, on the other hand, that more than one constitutional isomer is
possible because the two carbon atoms of the double bond of propene are nonequivalent. Thus it should be possible to form two constitutional isomers, 1-
chloropropane and 2-chloropropane. This doesn't occur, however. Instead only
one product, 2-chloropropane, is formed.

$$CH_3CH=CH_2 \ + \ H{-}Cl \longrightarrow \underset{\substack{\text{2-Chloropropane}\\\text{(Sole product)}}}{CH_3\underset{\displaystyle |}{\underset{\displaystyle Cl}{C}H}{-}\underset{\displaystyle |}{\underset{\displaystyle H}{C}H_2}} \qquad \underset{\substack{\text{1-Chloropropane}\\\text{(Not formed)}}}{CH_3\underset{\displaystyle |}{\underset{\displaystyle H}{C}H}{-}\underset{\displaystyle |}{\underset{\displaystyle Cl}{C}H_2}}$$

From the results of many addition reactions of acids to unsymmetrical alkenes, the Russian chemist V. Markownikoff summarized his observations in a simple statement called the **Markownikoff Rule:**

> **In the addition of protic acids to alkenes, the proton adds to the carbon atom of the double bond that has the greater number of hydrogen atoms.**

The following are examples of addition reactions that follow the Markownikoff Rule.

One hydrogen atom
bonded to this carbon atom

$$CH_3CH_2\underset{H}{\overset{}{C}}=\underset{H}{\overset{H}{C}} + HCl \xrightarrow{(C_2H_5)_2O} CH_3CH_2\underset{H}{\overset{Cl}{C}}HCH_2$$

Two hydrogen atoms
bonded to this carbon atom

(92% yield)

No hydrogen atoms
bonded to this carbon atom

$$CH_3CH_2\underset{CH_3}{\overset{}{C}}=\underset{H}{\overset{CH_3}{C}} + HBr \xrightarrow{(C_2H_5)_2O} CH_3CH_2\underset{CH_3}{\overset{Br\ H}{C}}-CHCH_3$$

One hydrogen atom
bonded to this carbon atom

(86% yield)

Exercise 7.20 Write the equation and the structure of the product of addition of HCl to each of the following alkenes:

(a) 1-Hexene **(c)** 2-Methyl-2-butene
(b) 2-Methyl-1-butene **(d)** 1-Methylcyclohexene

Reactions that can potentially form two or more constitutional isomers but actually form only one are called **regioselective reactions.** The additions of Brønsted-Lowry acids to unsymmetrical alkenes are regioselective reactions. More generally, reactions that form products according to the Markownikoff Rule are regioselective reactions.

The stereochemistry of the addition of acids to alkenes is illustrated by the addition of HCl to (Z)-3-methyl-3-hexene, which forms 3-chloro-3-methylhexane.

■ *Regioselective is pronounced* ree-gee-oh-selective.

$$CH_3CH_2\underset{CH_3}{\overset{}{C}}=\underset{H}{\overset{CH_2CH_3}{C}} + HCl \xrightarrow{C_2H_5OC_2H_5} CH_3CH_2\underset{CH_3}{\overset{Cl}{C}}CH_2CH_2CH_3$$

(Z)-3-Methyl-3-hexene

3-Chloro-3-methylhexane
(97% yield)

Stereocenter

This is the expected product of Markownikoff addition. While 3-chloro-3-methylhexane has a stereocenter, the product formed is found to be optically inactive.

This means that the addition of HCl to (Z)-3-methyl-3-hexene forms a product that is a racemic mixture (Section 6.5). Thus there is no stereochemical discrimination in this reaction. Equal amounts of both possible stereoisomers, (R)- and (S)-3-chloro-3-methylhexane, are formed. Reactions that form all possible stereoisomers are called **nonstereospecific reactions.** The additions of Brønsted-Lowry acids to alkenes, therefore, are examples of nonstereospecific reactions.

This reaction illustrates an important general rule in organic chemistry. Whenever *the reactants of a chemical reaction are optically inactive or achiral, the product is always optically inactive because it is either a racemic mixture or achiral.*

Let's now return to the proposed mechanism of additions of acids to alkenes given in Section 7.9 to see if it is in accord with these new experimental observations. Let's start with the stereochemistry of the reaction.

7.12 Mechanistic Explanation of the Stereochemistry of Acid Additions

The mechanism for the addition of HCl to (Z)- or (E)-3-methyl-3-hexene is analogous to that of the addition of HCl to 2,3-dimethyl-2-butene, discussed in Section 7.9. Step 1 is formation of a planar carbocation, while the product is formed by the reaction of the carbocation with a chloride ion in Step 2. Reaction of a planar carbocation with a nucleophile such as chloride ion can occur equally well from either side of its plane, as shown in Figure 7.15. Reaction from one side of the plane forms one stereoisomer, while reaction from the other side forms the other stereoisomer. Thus equal amounts of (R)- and (S)-3-chloro-3-methylhexane (a racemic mixture) are formed by reaction of chloride ion with the carbocation

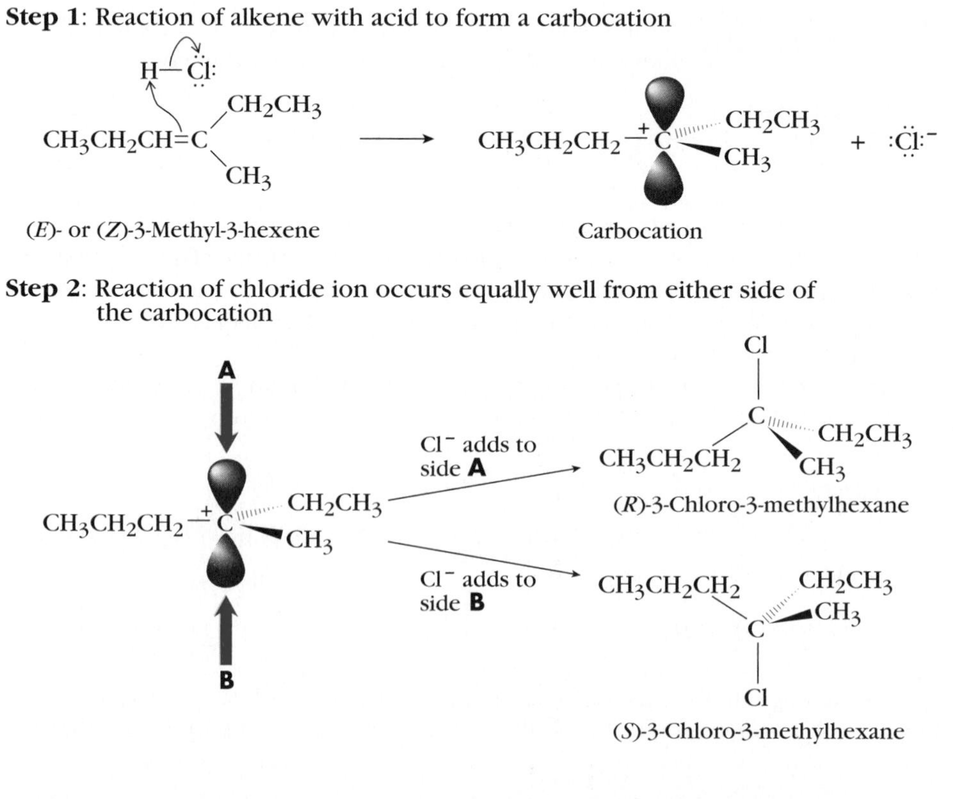

Figure 7.15
Mechanism of addition of HCl to (Z)- or (E)-3-methyl-3-hexene. Step 1 is the formation of a planar carbocation by addition of a proton to 3-methyl-3-hexene. In Step 2, addition of the chloride ion to the carbocation occurs equally well from either side of the carbocation's plane. As a result, a racemic mixture is formed in this reaction.

Step 1: Reaction of alkene with acid to form a carbocation

Step 2: Reaction of chloride ion occurs equally well from either side of the carbocation

formed by addition of HCl to either (*E*)- or (*Z*)-3-methyl-3-hexene. Notice, moreover, that the same carbocation can be formed by the addition of HCl to *either* (*Z*)- or (*E*)-3-methyl-3-hexene. As a result, a racemic mixture of 3-chloro-3-methylhexane is formed in both reactions.

Exercise 7.21	Use your molecular models to construct the two carbocations formed by addition of a proton to (*E*)- and (*Z*)-3-methyl-3-hexene. Are they superposable?

7.13   Mechanistic Explanation of the Markownikoff Rule

The mechanism for acid additions to alkenes is in accord with the Markownikoff Rule. Consider, for example, the mechanism for the addition of HCl to propene. Step 1 is the addition of a proton to form a carbocation. However, in the case of propene, two carbocations can be formed. One is a secondary and the other is a primary carbocation.

$$CH_3CH=CH_2 \;+\; H{-}Cl \longrightarrow \underset{\text{Secondary carbocation}}{CH_3\overset{+}{CH}{-}\underset{|}{CH_2}{-}H} \xrightarrow{\;Cl^-\;} \underset{\substack{\text{2-Chloropropane}\\\text{Observed product}}}{CH_3\underset{\underset{Cl}{|}}{CH}{-}\underset{\underset{H}{|}}{CH_2}}$$

$$\xcancel{\longrightarrow}\; \underset{\text{Primary carbocation}}{CH_3\underset{\underset{H}{|}}{CH}{-}\overset{+}{CH_2}}$$

We know that the only product formed in this reaction is 2-chloropropane. It is formed by reaction of chloride ion with the secondary carbocation in Step 2 of the mechanism. This means that the more stable secondary carbocation is formed faster than the primary one in Step 1. Why should the more stable carbocation be formed faster than the less stable one? After all, we learned in Section 3.13 that the thermodynamic stability of a compound and the rate at which it is formed are not directly related.

In 1955, the American chemist G.S. Hammond proposed that for any elementary reaction (Section 3.13) the *structure and energy of its transition state resemble the structure and energy of whichever species is the higher in energy, the reactants or the products.* This relationship between the energy of a reactive intermediate and the energy of the transition state leading to its formation is called the **Hammond Postulate.**

Let's apply this postulate to Step 1, the rate-determining step, of the mechanism for acid additions to alkenes. The reactants are the alkene and the acid, while the product is the carbocation. The carbocation is higher in energy than the alkene or the acid. According to the Hammond Postulate, the structure and energy of the transition state leading to the formation of a carbocation in Step 1 will be similar to the structure and energy of the carbocation.

The Hammond Postulate means that we can use the relative stabilities of carbocations as a guide to the relative stabilities of the transition states. Thus the more stable a carbocation, the lower in energy the transition state for its formation, and the faster it is formed. In this way, we relate the rate of formation of reactive intermediates to their stability.

Let's now return to the addition of HCl to propene and examine the free energy changes that occur in Step 1 of the mechanism by means of the free energy-reaction path diagram shown in Figure 7.16. Of the two paths shown in Figure 7.16, one leads to a primary carbocation via transition state T.S.#1, while the other leads to a secondary carbocation via transition state T.S.#2. The free energy of activation is designated $\Delta G_1^{\ddagger}$ and $\Delta G_2^{\ddagger}$ for the path to the primary and secondary carbocations, respectively.

Recall from Section 3.13 that the free energy of activation determines the rate of a chemical reaction. The larger the value of $\Delta G^{\ddagger}$, the slower the reaction. According to the Hammond Postulate, the structure and energy of the transition state leading to the formation of a carbocation will be similar to the structure and energy of that carbocation. Thus the structure and energy of T.S.#1 resemble a primary carbocation, while the structure and energy of T.S.#2 resemble a secondary carbocation. Because the energy of a primary carbocation is greater than the energy of a secondary carbocation, it follows that T.S.#1 must be higher in energy than T.S.#2. This means that $\Delta G_1^{\ddagger}$ is larger than $\Delta G_2^{\ddagger}$. Therefore the secondary carbocation should be formed faster than the primary one. In fact, the difference in free energies of activation ($\Delta G_1^{\ddagger} - \Delta G_2^{\ddagger}$) is so large that the secondary carbocation is the only carbocation ever formed.

The use of the Hammond Postulate to relate the relative stabilities of the carbocations involved as intermediates in the mechanism of acid additions to alkenes and the relative energies of the transition states leading to them allows us to formulate a modern definition of the Markownikoff rule:

An electrophile reacts with the double bond of an alkene to form the more stable carbocation.

Exercise 7.22   Using the mechanism of acid additions to alkenes and the Hammond Postulate, explain why the addition of HCl to methylpropene forms only 2-chloro-2-methylpropane as product.

Figure 7.16
Free energy-reaction path diagram for the formation of a primary and a secondary carbocation by addition of a proton to propene.

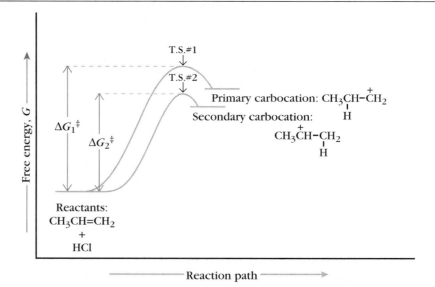

Additions of acids to alkenes do not always form a single product. The addition of HCl to 2-pentene, for example, forms a mixture of 2-chloropentane and 3-chloropentane. The mechanism of acid additions to alkenes can account for this result. The two carbocations formed in Step 1 of the mechanism are both secondary carbocations:

Step 1: Addition of a proton to 2-pentene forms two carbocations of approximately equal energy.

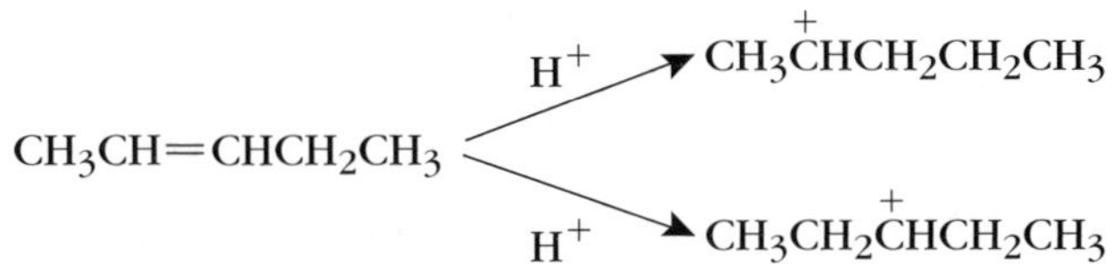

$$CH_3CH=CHCH_2CH_3 \begin{array}{c} \xrightarrow[H^+]{} CH_3\overset{+}{C}HCH_2CH_2CH_3 \\ \\ \xrightarrow[H^+]{} CH_3CH_2\overset{+}{C}HCH_2CH_3 \end{array}$$

The energies of these two carbocations are about the same, so the two free energies of activation for their formation should also be about the same. The result is that both are formed and both go on in Step 2 to react with chloride ion to form the two observed products.

Step 2: Chloride ion reacts with both carbocations to form a mixture of 2-chloropentane and 3-chloropentane.

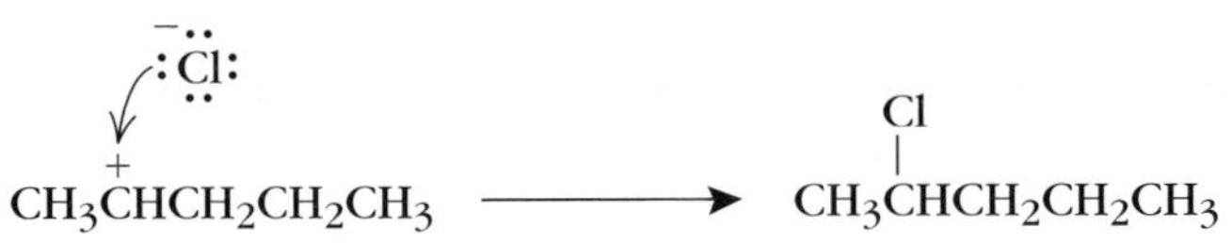

$$CH_3\overset{+}{C}HCH_2CH_2CH_3 \xrightarrow{:\ddot{C}l:^-} \underset{\text{2-Chloropentane}}{CH_3\underset{|}{\overset{Cl}{C}}HCH_2CH_2CH_3}$$

$$CH_3CH_2\overset{+}{C}HCH_2CH_3 \xrightarrow{:\ddot{C}l:^-} \underset{\text{3-Chloropentane}}{CH_3CH_2\underset{|}{\overset{Cl}{C}}HCH_2CH_3}$$

So far we have examined the mechanism of additions of strong acids to alkenes. However, the mechanism is essentially the same for the reactions of alkenes and weak acids, which require an acid catalyst. We can illustrate this by examining the acid-catalyzed hydration of alkenes.

7.14   Acid-Catalyzed Hydration of Alkenes

The acid-catalyzed hydration of alkenes forms alcohols as a product. For example, the reaction of (Z)-3-methyl-3-hexene with a dilute aqueous solution of sulfuric acid forms 3-methyl-3-hexanol.

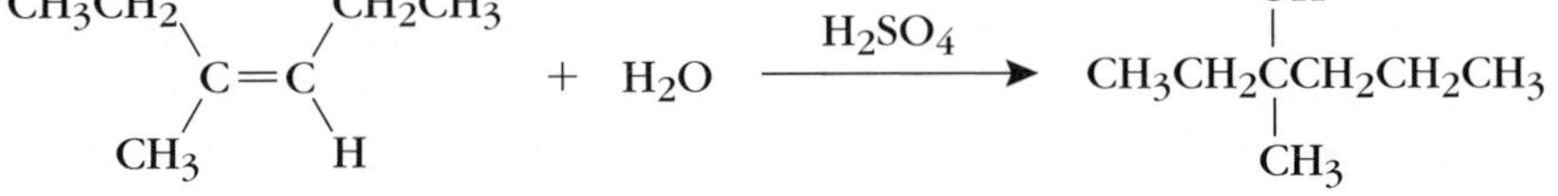

$$\underset{\text{(Z)-3-Methyl-3-hexene}}{\underset{CH_3}{\overset{CH_3CH_2}{}}\underset{H}{\overset{CH_2CH_3}{C=C}}} + H_2O \xrightarrow{H_2SO_4} \underset{\substack{\text{3-Methyl-3-hexanol} \\ \text{(75\% yield)}}}{CH_3CH_2\underset{CH_3}{\overset{OH}{\underset{|}{\overset{|}{C}}}}CH_2CH_2CH_3}$$

As in the addition of HCl, the same product is obtained from either the (Z)- or the (E)- isomer, both products are optically inactive, and they have the Markown-

Step 1: Proton is transferred from a hydronium ion to the alkene to form a carbocation

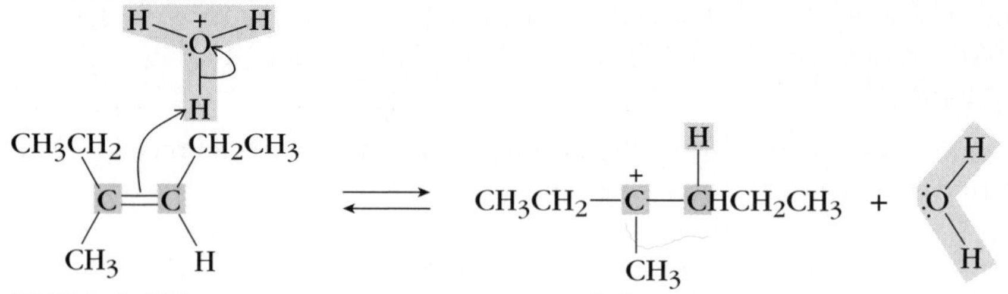

(Z)-3-Methyl-3-hexene   Carbocation

Step 2: Carbocation reacts with water to form the conjugate acid of the alcohol

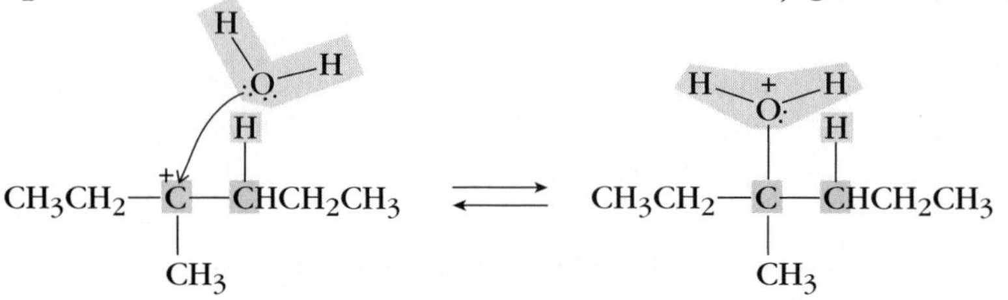

Conjugate acid of the alcohol

Step 3: Proton is transferred from conjugate acid of the alcohol to water

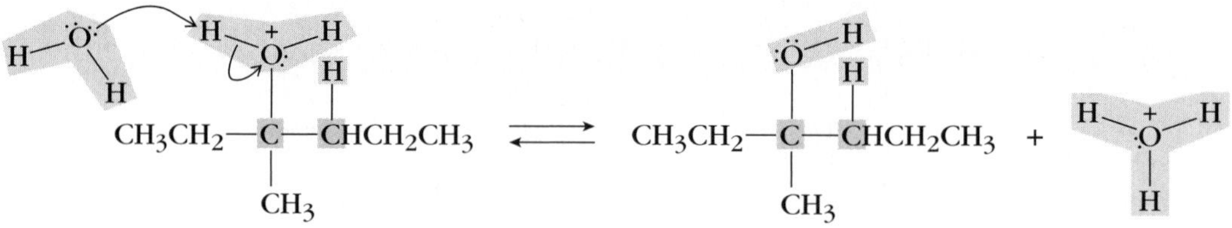

3-Methyl-3-hexanol

Figure 7.17
Mechanism for the acid-catalyzed hydration of (Z)-3-methyl-3-hexene. Step 1 is the reaction of a hydronium ion with the alkene to form a carbocation. The carbocation reacts in Step 2 with water, which acts as a nucleophile. Step 3 is an acid-base reaction in which a proton is transferred from the conjugate acid of the alcohol to water.

ikoff orientation. Consequently we can write a mechanism for this acid-catalyzed hydration reaction that is similar to the one for the addition of HCl. The mechanism for the hydration reaction is shown in Figure 7.17.

A third step is needed in the hydration mechanism because the product of the first two steps is not an alcohol but the conjugate acid of an alcohol (Section 3.4). Step 3 is an acid-base reaction between the conjugate acid of an alcohol (an acid) and water (a base). The equilibrium constant of Step 3 is close to one because, as we discussed in Section 3.4, the acid strengths of a hydronium ion and a protonated alcohol are about the same. In the reaction mixture, however, the concentration of water is so much greater than the concentration of the alcohol that the equilibrium is shifted to the right. As a result, the alcohol is formed as product and the hydronium ion catalyst is regenerated.

Exercise 7.23	Using the mechanism of acid-catalyzed hydration of alkenes as a model, write a three-step mechanism for the acid-catalyzed addition of methanol (CH_3OH) to cyclohexene.

7.15   More About the Stabilities of Carbocations

The rates of hydration of alkenes have been extensively studied, and the results of these experiments demonstrate how substituents affect the rates of electrophilic addition reactions. These results in turn show how substituents affect the stabilities of carbocations. These studies compare the rates of hydration of a series of alkenes whose structures are systematically varied. An example is a study of the rates of hydration of the following alkenes.

$$CH_3CH{=}CH_2 \qquad CH_3\overset{\overset{\displaystyle CH_3}{|}}{C}{=}CH_2 \qquad CH_3\overset{\overset{\displaystyle Br}{|}}{C}{=}CH_2 \qquad CH_3\overset{\overset{\displaystyle OCH_3}{|}}{C}{=}CH_2$$

Propene   Methylpropene   2-Bromopropene   2-Methoxypropene

Methylpropene, 2-bromopropene, and 2-methoxypropene are all derived by replacing the hydrogen atom bonded to carbon atom 2 of propene with a methyl group, a bromine atom, and a methoxy group ($-OCH_3$), respectively. The rates of hydration of these four alkenes are determined under the same experimental conditions. Then, setting the rate of hydration of propene equal to 1, the relative rates of the other alkenes are found to be as follows:

$$CH_3\overset{\overset{\displaystyle OCH_3}{|}}{C}{=}CH_2 \qquad CH_3\overset{\overset{\displaystyle CH_3}{|}}{C}{=}CH_2 \qquad CH_3CH{=}CH_2 \qquad CH_3\overset{\overset{\displaystyle Br}{|}}{C}{=}CH_2$$

2-Methoxypropene   Methylpropene   Propene   2-Bromopropene
1.7×10^{10} 7500   1   0.22

We see from these data that replacing the hydrogen atom of propene with either a methyl group or a methoxy group greatly increases the rate of hydration. Replacing the hydrogen atom with a bromine atom, on the other hand, decreases the rate slightly. We can understand these results by examining the structures of the carbocations formed in Step 1, the rate-determining step, of the mechanism of acid-catalyzed hydration. Consider first the two carbocations formed by hydration of propene and methyl propene.

$$CH_3CH{=}CH_2 \; + \; H_3O^+ \;\rightleftharpoons\; CH_3\overset{+}{C}HCH_3$$

Propene   Isopropyl cation
                                 (a 2° carbocation)

$$CH_3\overset{\overset{\displaystyle CH_3}{|}}{C}{=}CH_2 \; + \; H_3O^+ \;\rightleftharpoons\; CH_3\overset{\overset{\displaystyle CH_3}{|}}{\underset{+}{C}}CH_3$$

Methylpropene   *tert*-Butyl cation
                                      (a 3° carbocation)

Protonation of propene forms a secondary carbocation, while protonation of methylpropene forms a tertiary one. Recall from Section 7.10 that a tertiary carbocation is more stable than a secondary one. Furthermore, by applying the Hammond Postulate (Section 7.13), we know that a tertiary carbocation is formed faster than a secondary one. As a result, hydration of methylpropene is faster than propene because protonation of methylpropene forms the more stable carbocation.

Let's now look at the carbocations formed by protonation of 2-methoxypropene and 2-bromopropene. Because replacing a hydrogen atom by a methoxy group leads to an even faster rate of hydration, a carbocation with a methoxy group bonded to the electron-deficient carbon atom must be even more stable than a tertiary carbocation. A carbocation with a bromine atom bonded to the electron-deficient carbon atom, on the other hand, must be less stable than a secondary carbocation because the rate of hydration of 2-bromopropene is slower than that of propene.

In Section 7.10, we learned that some substituents donate electrons to an electron-deficient carbon atom, while other substituents withdraw electrons from it. Electron-donating substituents stabilize a positive charge on an adjacent carbon atom, while electron-withdrawing substituents destabilize it. Thus we conclude that alkyl and methoxy groups are electron-donating substituents, while bromine (and the other halogen atoms) are electron-withdrawing.

In summary, we can determine whether a substituent donates electrons to an electron-deficient carbon atom (an electron-donating substituent) compared to hydrogen or withdraws electrons (an electron-withdrawing substituent) from its effect on the rate of hydration of alkenes. An electron-donating substituent stabilizes the carbocation intermediate; this in turn increases the rate of hydration and other electrophilic addition reactions of alkenes. Conversely, electron-withdrawing groups slow the rate of hydration because they destabilize the carbocation intermediate. The effect of a number of substituents on the stability of an adjacent electron-deficient carbon atom is summarized in Table 7.3.

How do the substituents listed in Table 7.3 donate or withdraw electrons from an adjacent electron-deficient center? One way, already discussed in Section 7.10, is by an inductive effect. Alkyl groups, for example, donate electrons by an inductive effect. This disperses or delocalizes the charge and stabilizes a carbocation. In contrast, adjacent electronegative elements such as the halogens withdraw electrons by an inductive effect. This further concentrates the charge on the posi-

Table 7.3		Effect of Substituents Compared to Hydrogen on the Stability of an Adjacent Electron-Deficient Carbon Atom	
Substituent (Y)		Electronic Effect	Effect on Stability of Substituted Carbocation
Name	Structure	Compared to Hydrogen	YCH_2^+ Compared to CH_3^+
Methoxy	$-OCH_3$	Strong electron donor	Strongly stabilizing
Amino	$-NH_2$	Strong electron donor	Strongly stabilizing
Alkyl	$-R$	Electron donor	Stabilizing
Aryl	$-Ar$	Electron donor	Stabilizing
Halogen	$-Cl, -Br, -I$	Electron withdrawing	Destabilizing
Trialkylammonium	$-NR_3^+$	Strong electron withdrawing	Strongly destabilizing
Nitro	$-NO_2$	Strong electron withdrawing	Strongly destabilizing

tive carbon atom and destabilizes the cation. Groups that have positive charges adjacent to the electron-deficient center, such as trialkylammonium ions ($-NR_3^+$), also withdraw electrons by an inductive effect and destabilize carbocations.

Another way that some electron-donating substituents stabilize carbocations is by a resonance effect. The methoxy group, for example, stabilizes a carbocation by sharing a pair of its electrons with the electron-deficient carbon atom. The electrons that it shares are one of its oxygen atom's lone pairs. Sharing occurs by overlap of the orbital containing this lone pair of electrons with the empty p orbital of the electron-deficient carbon atom. We can represent this electron delocalization from the oxygen atom to the carbon atom by the localized valence bond orbital model (Section 1.9) in the following way:

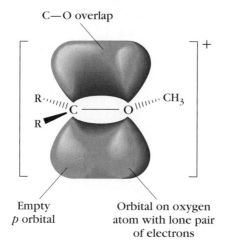

C—O overlap

Empty
p orbital

Orbital on oxygen
atom with lone pair
of electrons

This electron delocalization can also be represented as a resonance hybrid (Section 1.12) of the following two structures:

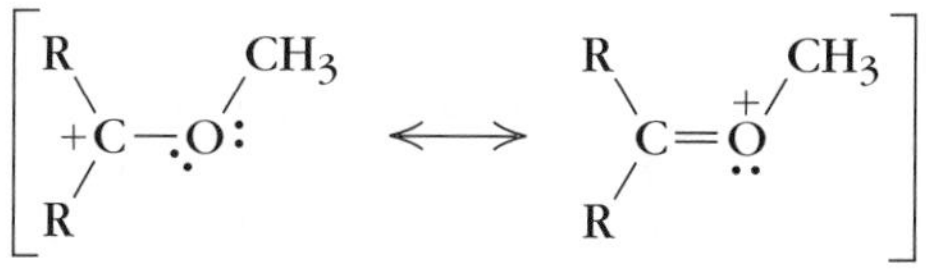

Electron-donating resonance effects are stabilizing effects for carbocations and amount to the release of a pair of electrons to a positively charged center.

Some substituents affect the stability of an adjacent electron-deficient center by both inductive and resonance effects. A methoxy group, for example, stabilizes an adjacent carbocation by its electron-donating resonance effect but destabilizes the cation by its electron-withdrawing inductive effect. The resonance effect is the stronger of the two, however, because a methoxy group stabilizes an adjacent electron-deficient center.

Exercise 7.24   Write the other contributing structure of the resonance hybrid of the following carbocation stabilized by an amino group.

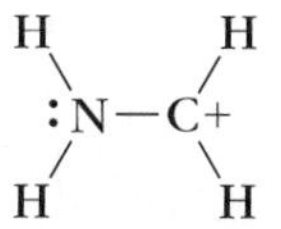

In the laboratory, acid catalysts must be used for the hydration of alkenes. These conditions are too drastic for the hydration of alkenes in living systems, however, so hydration occurs by means of an enzyme catalyst.

7.16 Enzyme-Catalyzed Hydration of Double Bonds

One of the remarkable aspects of chemical reactions in living systems is the ability of enzymes to discriminate between stereoisomers. An example is the enzyme-catalyzed hydration of *trans*-butenedioate, which is usually referred to by its common name of fumarate. This reaction takes place in humans and animals in the Citric Acid Cycle. The Citric Acid Cycle is a series of chemical reactions that make up the final pathway for the oxidation of food to CO_2, H_2O, and energy.

The enzyme that catalyzes the hydration of fumarate is called fumarase. The product is **optically active** and is called (*S*)-malate. Fumarase also catalyzes the reverse reaction, the dehydration of (*S*)-malate to form fumarate.

Fumarate
(*trans*-butenedioate)

(*S*)-Malate

Maleate
(*cis*-Butenedioate)

Fumarase is very discriminating. It catalyzes the hydration of fumarate but not the hydration of maleate, the *cis*-isomer of fumarate. It catalyzes the dehydration of (*S*)-malate but not the dehydration of (*R*)-malate. This is completely different from the acid-catalyzed hydration of a double bond in which the same optically inactive product is formed from either an (*E*)- or a (*Z*)-alkene. To understand why there is such a difference between acid-catalyzed and enzyme-catalyzed hydrations, we must first examine the general features of enzyme-catalyzed reaction mechanisms.

In the first step of an enzyme-catalyzed mechanism, the enzyme (E) binds with the substrate (S) to form an enzyme-substrate complex (ES). A **substrate** is an organic compound that binds to and forms a complex with the enzyme. Fumarate, therefore, is a substrate for the enzyme fumarase. After the chemical reactions have occurred in the enzyme-substrate complex, the complex breaks up to give the product and the regenerated enzyme in the second step. The enzyme is ready to catalyze the reaction of another molecule of substrate.

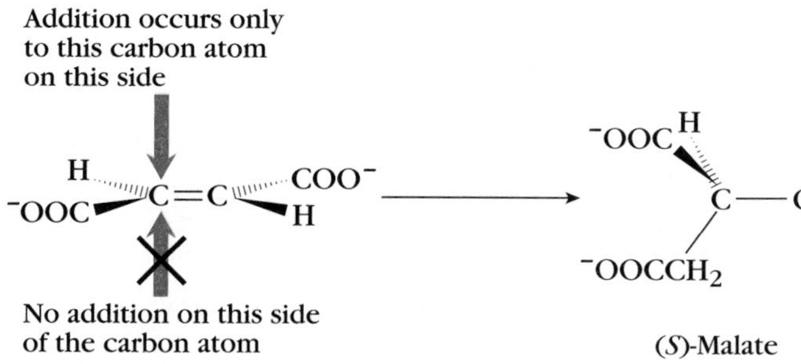

Addition occurs only
to this carbon atom
on this side

$$H \cdots C = C \cdots COO^-$$

No addition on this side
of the carbon atom

$^-OOC \cdots C \cdots H$

$C \longrightarrow OH$

$^-OOCCH_2$

(S)-Malate

Figure 7.18
Enzyme-catalyzed hydration of
fumarate occurs only to one side of
the double bond to form only one
enantiomer, (S)-malate.

From the experimental facts given above, we can deduce the following structural features of the enzyme-substrate complex formed in the mechanism of the enzyme-catalyzed hydration. Because only fumarate reacts with the enzyme, the sites that bind the substrate are located so that only fumarate forms a complex with the enzyme and maleate does not. Only one enantiomer of the product is formed, (S)-malate, which means that addition of the —OH group occurs to only one side of the double bond, as shown in Figure 7.18.

The key difference between acid-catalyzed and enzyme-catalyzed hydrations of alkenes is their stereochemistry. Acid-catalyzed hydration forms optically inactive products, while the product of fumarase-catalyzed hydration of fumarate is optically active. Enzymes (such as fumarase) must be chiral molecules, therefore, because the only way that an achiral molecule such as fumarate can form an optically active product is by reaction with a chiral reagent. In future chapters, we will encounter other examples of optically active products formed by enzyme-catalyzed reactions.

7.17 Carbocation Rearrangements

Additions of acids to alkenes sometimes give unexpected results. The addition of HCl to 3,3-dimethyl-1-butene, for example, forms the expected product of Markownikoff addition, 2-chloro-3,3-dimethylbutane, as well as a second product, 2-chloro-2,3-dimethylbutane. Both products are formed in about equal amounts.

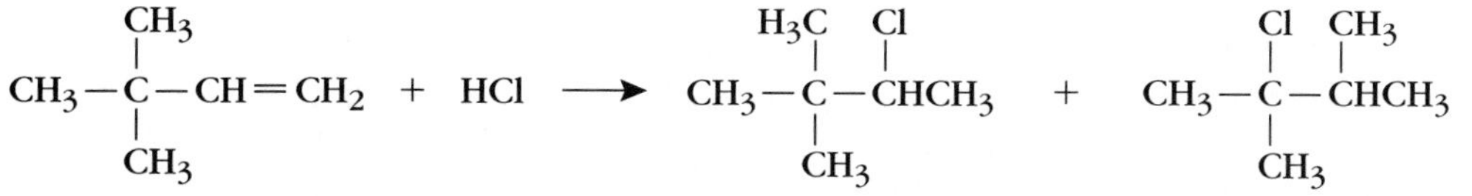

$$CH_3 - \overset{\overset{\displaystyle CH_3}{|}}{\underset{\underset{\displaystyle CH_3}{|}}{C}} - CH = CH_2 \ + \ HCl \ \longrightarrow \ CH_3 - \overset{\overset{\displaystyle H_3C}{|}}{\underset{\underset{\displaystyle CH_3}{|}}{C}} - \overset{\overset{\displaystyle Cl}{|}}{CHCH_3} \ + \ CH_3 - \overset{\overset{\displaystyle Cl}{|}}{\underset{\underset{\displaystyle CH_3}{|}}{C}} - \overset{\overset{\displaystyle CH_3}{|}}{CHCH_3}$$

3,3-Dimethyl-1-butene   2-Chloro-3,3-dimethylbutane   2-Chloro-2,3-dimethylbutane
                             (Unrearranged product)   (Rearranged product)

The unexpected product has an entirely different structure from the starting alkene. During the formation of this product, there has been a **rearrangement** of the carbon skeleton of the alkene. The products, therefore, are constitutional isomers.

Many other examples are known of acid additions that form products whose structures are the results of rearrangement of the original alkene. Because the mechanisms of these reactions involve carbocations, we are led to the conclusion that these carbocations can rearrange. Why should a carbocation want to rearrange its structure? To answer this question, let's look at the mechanism for the addition of HCl to 3,3-dimethyl-1-butene shown in Figure 7.19.

Figure 7.19

Mechanism of addition of HCl to 3,3-dimethyl-1-butene.

Step 1: Reaction with alkene and acid to form a carbocation

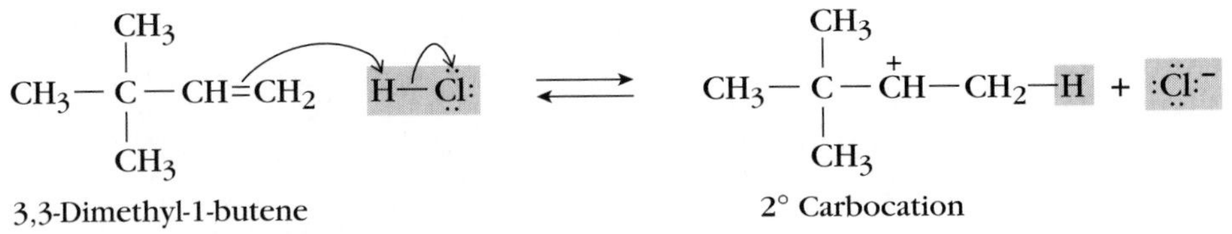

3,3-Dimethyl-1-butene   2° Carbocation

Step 2: Secondary carbocation reacts with chloride ion to form unrearranged product

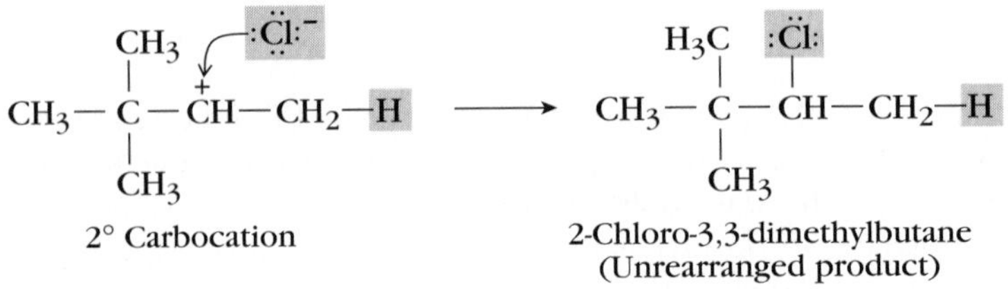

2° Carbocation   2-Chloro-3,3-dimethylbutane (Unrearranged product)

Rearrangement of secondary carbocation to tertiary carbocation

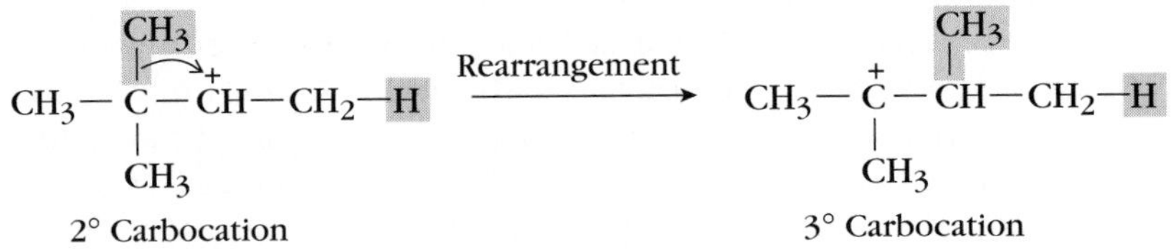

2° Carbocation   3° Carbocation

Step 3: Reaction of tertiary carbocation with chloride ion to form rearranged product

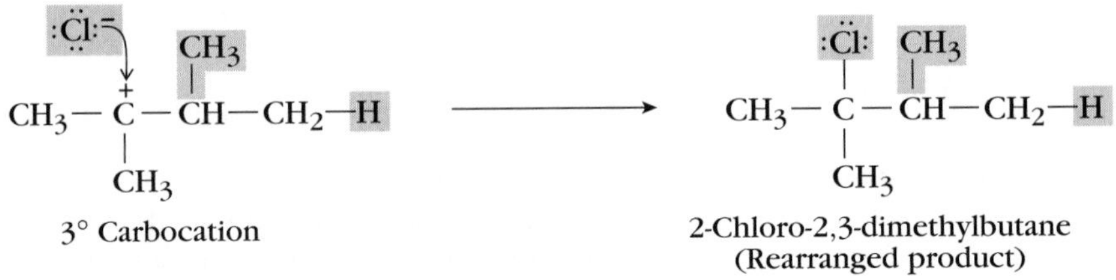

3° Carbocation   2-Chloro-2,3-dimethylbutane (Rearranged product)

Step 1 of the mechanism is the donation of a pair of electrons from the π bond of the alkene to the electrophile to form a secondary carbocation. This carbocation can undergo two reactions in Step 2. It can react with chloride ion to form 2-chloro-3,3-dimethylbutane, and it can rearrange to a tertiary carbocation. *Rearrangement occurs because a secondary carbocation is converted into a more stable tertiary carbocation.* It can be pictured as the movement of a *methyl group with its bonding electrons* from an adjacent carbon atom to the carbon atom bearing the positive charge. The carbon atom that loses the methyl group becomes electron deficient and acquires a positive charge. This rearranged carbocation reacts with chloride ion to form the rearranged product in Step 3.

Carbocations can also rearrange by the movement of a hydrogen and its pair of electrons. This kind of rearrangement occurs in the mechanism of the addition of HCl to 3-methyl-1-butene, as shown in Figure 7.20.

These two rearrangements are similar in that an atom $(:H^-)$ or a group of atoms $(:CH_3^-)$ moves to a positively charged carbon atom, taking its bonding

Step 1: Protonation of alkene forms secondary carbocation

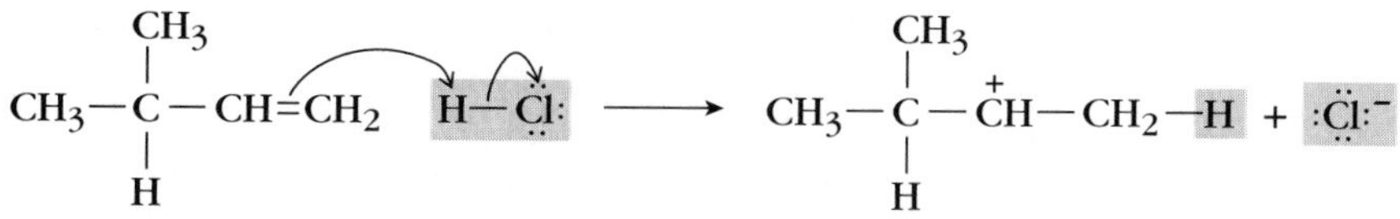

3-Methyl-1-butene   2° Carbocation

Figure 7.20
The mechanism of the addition of
HCl to 3-methyl-1-butene.

Step 2: Secondary carbocation reacts with chloride ion to form
    unrearranged product

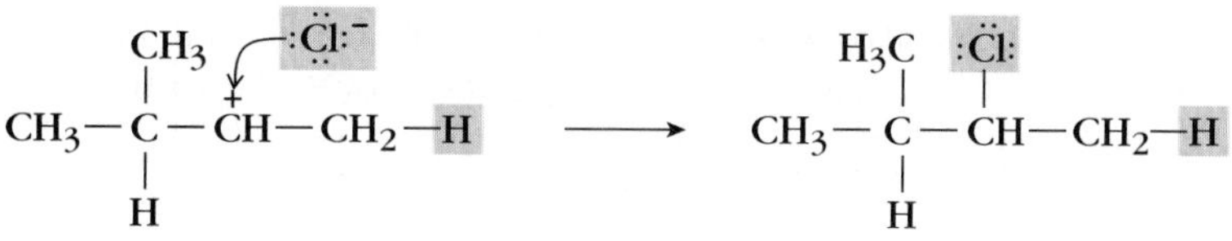

2° Carbocation   2-Chloro-3-methylbutane
                              (Unrearranged product)

Rearrangement of secondary carbocation to tertiary carbocation by
hydride shift

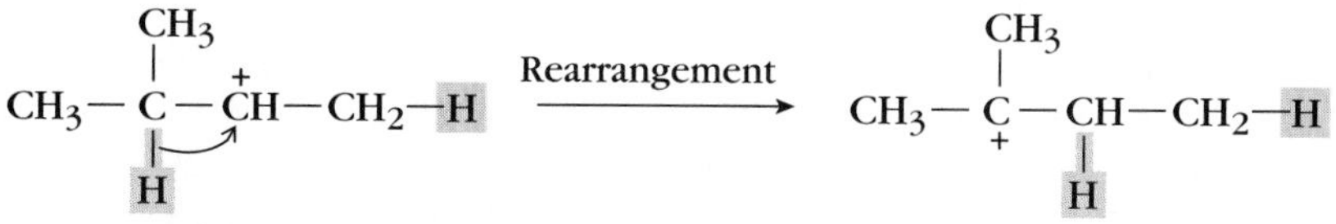

2° Carbocation   3° Carbocation

Step 3: Rearranged carbocation reacts with chloride ion

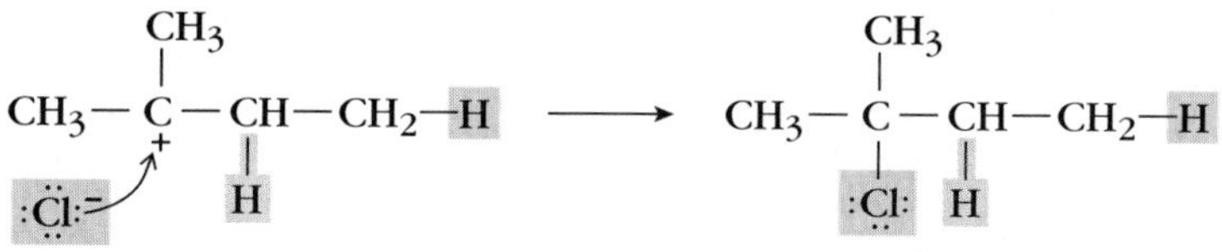

3° Carbocation   2-Chloro-2-methylbutane
                              (Rearranged product)

electron pair with it. The movement of a hydrogen with its two bonding electrons is called a **hydride shift,** while an analogous shift of an alkyl group is
called an **alkyl shift.** Rearrangements in which a group moves from one atom to
the very next one are called **1,2-shifts.** *A 1,2-shift of a hydrogen atom or an
alkyl group takes place if the rearrangement forms a more stable carbocation.*
These kinds of rearrangements are a common reaction of carbocations.

Exercise 7.25	Explain why the initial carbocation formed in the mech-anism of the addition of HCl to 3-methyl-1-butene rearranges by means of a hydride shift rather than a methyl shift.

| **Exercise 7.26** | Each of the following carbocations undergoes rearrangement to form a more stable carbocation. Write the structure of the rearranged carbocation. |

(a) $(CH_3)_3\overset{+}{C}CH_2$ **(b)** $(CH_3)_2CH\overset{+}{C}HCH_3$ **(c)**

We have learned that carbocations are formed in the mechanism of acid additions to alkenes. Carbocations are also formed in the mechanisms of other reactions, especially certain reactions used to prepare alkenes.

7.18   Preparation of Alkenes

Hydration of alkenes to form alcohols is a reversible reaction.

$$(CH_3)_2C{=}CH_2 \;+\; 2H_2O \;\rightleftharpoons\; (CH_3)_3C{-}OH$$

Methylpropene   *tert*-Butyl alcohol

The forward reaction is the hydration of methylpropene, while the reverse reaction is the acid-catalyzed dehydration (loss of water) of an alcohol. When alcohols are dehydrated, they lose a hydrogen atom and a hydroxy group from adjacent carbon atoms to form the π bond of an alkene. Dehydration of alcohols is an example of an **elimination reaction** because a molecule of water is eliminated from the alcohol to form an alkene.

$$-\overset{|}{\underset{H}{C}}-\overset{|}{\underset{OH}{C}}- \;\underset{}{\overset{H_2SO_4}{\rightleftharpoons}}\; \overset{\backslash}{\underset{/}{C}}{=}\overset{/}{\underset{\backslash}{C}} \;+\; HOH$$

Alcohol   Alkene   Water

In many cases the equilibrium constant for this reaction is not large. By choosing the proper conditions, the reaction can be made to go in either direction. Dehydration, therefore, can be made to occur by removal of the alkene by distillation, or hydration can be made the dominant reaction by the use of a large excess of water.

The mechanism of acid-catalyzed dehydration of alcohols shown in Figure 7.21 is the exact reverse of the mechanism of acid-catalyzed hydration of alkenes shown in Figure 7.17. Step 1 of the dehydration mechanism is the protonation of the hydroxy group of the alcohol. As we know, the equilibrium constant of this reaction is about one, so a strong acid must be used to shift this equilibrium to the right to form enough of the conjugate acid of the alcohol for the reaction to occur. In Step 2, the conjugate acid of the alcohol loses a water molecule to form the *tert*-butyl cation. In Step 3, the carbocation loses one of its nine equivalent hydrogen atoms to a water molecule to form an alkene.

There are a number of other methods of preparing alkenes that we will learn in future chapters. Table 7.4, p. 318, lists these reactions and the sections in which they are discussed in detail.

Step 1: Protonation of the alcohol forms the conjugate acid of the alcohol

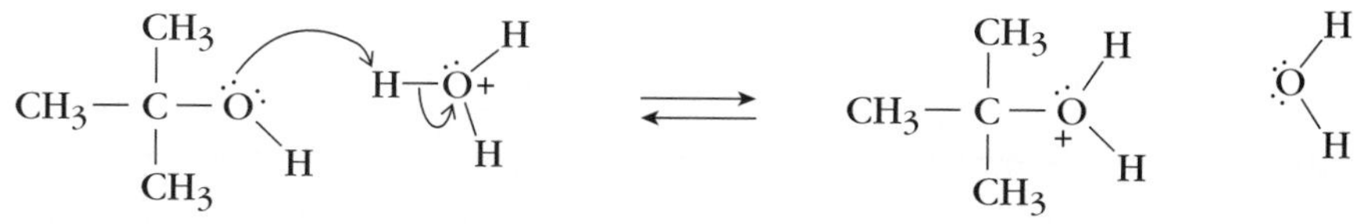

tert-Butyl alcohol Conjugate acid of alcohol

Figure 7.21
The mechanism of the dehydration of *tert*-butyl alcohol. This mechanism is the exact reverse of the mechanism of hydration of (Z)-3-methyl-3-hexene (Figure 7.17).

Step 2: Conjugate acid loses a molecule of water to form a tertiary carbocation

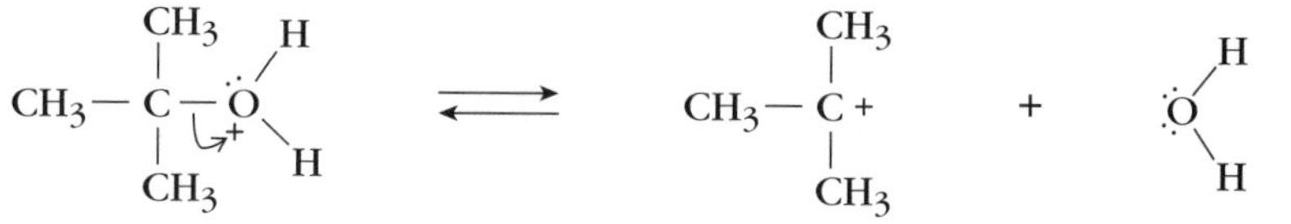

tert-Butyl cation Water

Step 3: A proton is removed from the carbocation by a water molecule to form an alkene

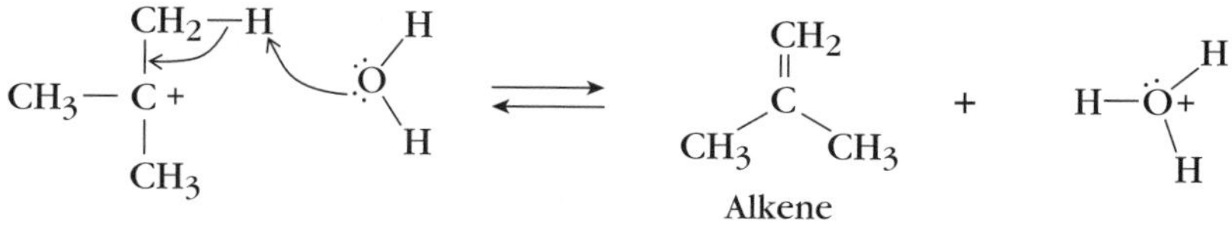

Alkene

7.19 Summary

Alkenes are compounds that contain one or more carbon-carbon double bonds. A double bond consists of a σ bond made by end-on overlap of two sp^2 hybrid orbitals and a π bond formed by sideways overlap of p orbitals. The overall bond strength of a carbon-carbon double bond is greater than the C—C single bond in an alkane. The strength of the π part of the double bond is estimated to be about 65 kcal/mol (272 kJ/mol).

Because rotation about a double bond is restricted, certain alkenes exist as a pair of stereoisomers. The geometry of a double bond can be specified in two ways. One way is to use the terms *cis*- and *trans*-. In the *cis*-isomer, the two similar groups are located on the same side of the double bond. In the *trans*-isomer, they are located on opposite sides. The second way is to use the Cahn-Ingold-Prelog Rules (Section 6.2). The priorities of the two groups bonded to each carbon of the double bond must be determined. The alkene with the two atoms or groups of atoms of higher priority on the same side of the double bond is called (Z)-, while the other configurational isomer is called (E)-.

The relative stabilities of alkenes increase as the degree of substitution of the double bond increases:

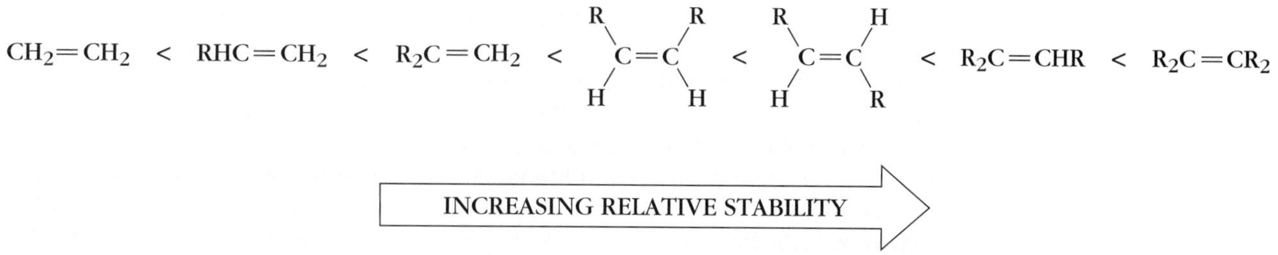

$$CH_2=CH_2 \ < \ RHC=CH_2 \ < \ R_2C=CH_2 \ < \ \overset{R}{\underset{H}{C}}=\overset{R}{\underset{H}{C}} \ < \ \overset{R}{\underset{H}{C}}=\overset{H}{\underset{R}{C}} \ < \ R_2C=CHR \ < \ R_2C=CR_2$$

INCREASING RELATIVE STABILITY

Table 7.4	Summary of Laboratory Methods of Preparing Alkenes	
	Reaction	Section Reference

Dehydration of Alcohols　　　　　　　　　　　　　　　　　　7.18

$$\underset{\overset{|}{\underset{}{OH}}}{R_2CCHR_2} \xrightarrow{H_2SO_4} R_2C{=}CR_2$$

Dehydrohalogenation of Haloalkanes　　　　　　　　　　　12.9

$$\underset{\overset{|}{\underset{}{X}}}{R_2CCHR_2} \xrightarrow{Base} R_2C{=}CR_2$$

Wittig Reaction　　　　　　　　　　　　　　　　　　　　14.16

$$R_2C{=}O \;+\; (C_6H_5)_3P{=}CHR' \longrightarrow R_2C{=}CHR'$$

Partial Hydrogenation of Alkynes　　　　　　　　　　　　9.8

Lindlar's Catalyst

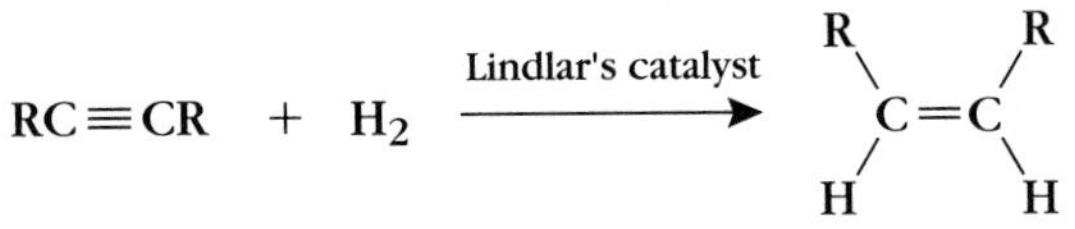

$$RC{\equiv}CR \;+\; H_2 \xrightarrow{Lindlar's\ catalyst} \underset{cis\text{-Alkene}}{\overset{\displaystyle R\quad R}{\underset{\displaystyle H\quad H}{C{=}C}}}$$

Na/NH₃

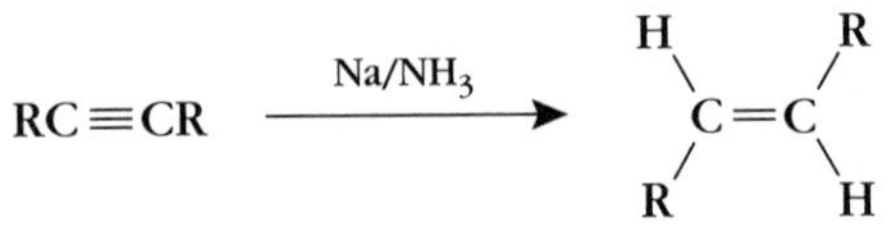

$$RC{\equiv}CR \xrightarrow{Na/NH_3} \underset{trans\text{-Alkene}}{\overset{\displaystyle H\quad R}{\underset{\displaystyle R\quad H}{C{=}C}}}$$

Hydroboration-Protonolysis

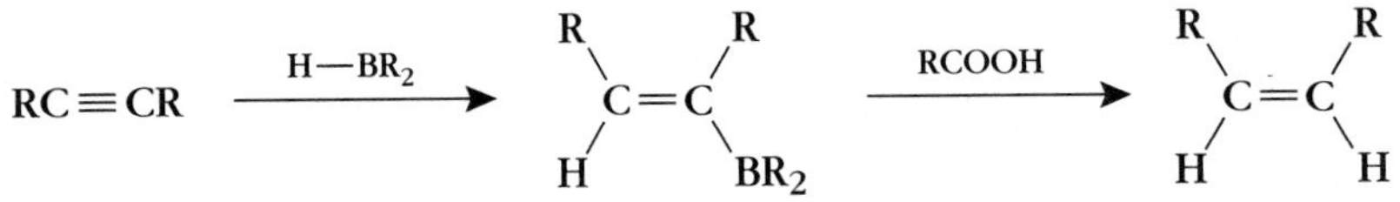

$$RC{\equiv}CR \xrightarrow{H-BR_2} \overset{\displaystyle R\quad R}{\underset{\displaystyle H\quad BR_2}{C{=}C}} \xrightarrow{RCOOH} \underset{cis\text{-Alkene}}{\overset{\displaystyle R\quad R}{\underset{\displaystyle H\quad H}{C{=}C}}}$$

The principal reaction of alkenes is addition to the double bond. When this reaction occurs, the π bond is broken and two new single bonds are formed. Addition reactions of alkenes are generally a thermodynamically favored process. Additions of strong acids such as HCl, HBr, HI, and trifluoroacetic acid (CF_3COOH) occur easily. Weaker acids, however, such as water, alcohols, and acetic acid (CH_3COOH) add to alkenes only with the aid of an acid catalyst.

A planar carbocation intermediate is formed in the first step of the accepted two-step mechanism of acid additions. A **carbocation** is a positively charged ion that contains a carbon atom with only six electrons in its valence shell. A carbocation intermediate reacts with nucleophiles in the second step of the mechanism to form the product.

Reagents that are electron pair acceptors, such as a proton, are classified as **electrophiles.** Because the two electrons of the π bond are donated to a proton in Step 1, this mechanism is called an **electrophilic addition mechanism.** Reactions in which the two electrons of the π bond are donated to an electrophile are called **electrophilic addition reactions.**

Nucleophiles donate a pair of electrons to a center of positive charge. In Step 2 of the electrophilic addition mechanism of the addition of HCl, therefore, the chloride ion is a nucleophile because it donates a pair of electrons to the electron-deficient carbon atom of the carbocation.

The relative stabilities of carbocations depend on their structures. Any change in structure that tends to disperse or spread out the positive charge of the electron-deficient carbon atom over the rest of the ion will stabilize a carbocation. Conversely, anything that tends to concentrate the charge on its electron-deficient carbon atom will destabilize a carbocation. Thus replacing a hydrogen atom on a methyl cation by an **electron-withdrawing group** or **substituent** removes electron density and increases the positive charge on the electron-deficient carbon atom. This makes the substituted carbocation less stable than a methyl cation. Replacing a hydrogen atom by an **electron-donating group** or **substituent** disperses and reduces the positive charge on the electron-deficient carbon atom. This makes the substituted carbocation more stable than a methyl cation.

Alkyl groups are electron donating so the stability of alkyl cations increases as the number of alkyl groups bonded to the electron-deficient carbon atom increases:

Methyl cation	<	Primary carbocation	<	Secondary carbocation	<	Tertiary carbocation
$CH_3{}^+$		$RCH_2{}^+$		R_2CH^+		R_3C^+

INCREASING STABILITY $\Longrightarrow$

When acids (HX) add to unsymmetrical alkenes, only a single product is usually formed. The **Markownikoff Rule** predicts that the hydrogen of the acid will add to the carbon atom of the double bond that has the greater number of hydrogen atoms and the group X will add to the carbon atom having the least number of hydrogen atoms.

Reactions that can potentially form two or more constitutional isomers but actually form only one are called **regioselective reactions.** The additions of Brønsted-Lowry acids to unsymmetrical alkenes are regioselective reactions.

The **Hammond Postulate** states that for any elementary reaction, the structure and energy of its transition state resemble the structure and energy of whichever species (the reactants or the products) is higher in energy. We can use this postulate to relate the relative stabilities of the carbocations involved as intermediates in the mechanism of acid additions to alkenes and the relative energies of the transition states leading to them. Thus the more stable a carbocation, the lower in energy the transition state for its formation, and the faster it is formed. This allows us to formulate a modern definition of the Markownikoff Rule: When an acid (HX) adds to an alkene, the more stable carbocation is formed.

If the product of additions of acids to alkenes has a stereocenter, the product is found to be optically inactive. This result is in accord with the mechanism of acid additions to alkenes. In Step 2 of the mechanism, reaction of a nucleophile with the planar carbocation occurs equally well from either side of its plane. Thus equal quantities of each enantiomer are formed, resulting in an optically inactive racemic mixture.

In contrast to acid-catalyzed hydration, the enzyme fumarase catalyzes the hydration of fumarate, but not its *cis*-isomer maleate, to form an optically active product, (*S*)-malate. The discriminating behavior of fumarase can be explained by a mechanism in which an **enzyme-substrate complex** initially forms. Reaction occurs in this complex, which then dissociates to form the product and regenerate the enzyme. Because of physical constraints in the enzyme-substrate complex, the —OH group can be added to only one side of the alkene so that only one of the two possible enantiomers is formed.

One general method of preparing alkenes is by elimination reactions. The dehydration of an alcohol is an elimination reaction. Dehydration forms alkenes and is the reverse of the hydration of an alkene. An equilibrium exists between the hydration of an alkene to form an alcohol and its dehydration. In many cases, the equilibrium constant is not large so the reaction can be made to occur in either direction. Hydration can be made to predominate by using large quantities of water; dehydration can be forced to completion by removing the alkene by distillation as it forms. Because dehydration is the reverse of hydration, their mechanisms are identical. Thus carbocations are involved as intermediates in both acid additions to alkenes *and* the dehydration of some alcohols.

Additions of acids to alkenes sometimes form rearranged products. These products arise because the carbocation intermediate rearranges to a more stable carbocation. Thus carbocations undergo three reactions: loss of a proton to form an alkene, reaction with nucleophiles such as halide ions, and rearrangement to form a more stable carbocation.

Additional Exercises

7.27   Define each of the following terms either in words or by means of a chemical equation.

    (a) Electron-withdrawing substituent   (d) Electrophilic addition reaction   (g) Markownikoff Rule
    (b) Electrophile   (e) Nucleophile   (h) Electron-donating substituent
    (c) Carbocation   (f) Rate-determining step   (i) Enzyme-substrate complex

7.28   Give the IUPAC name of each of the following alkenes.

(a) $CH_2=CHCH(CH_3)_2$

(b) $(CH_3)_2C=CBr_2$

(c) $CH_2=CCH_2CH(CH_3)_2$
              |
              CH_3

(d)

(e)

(f)

(g)

(h)

(i)

7.29 Write the structure corresponding to each of the following IUPAC names.

(a) 4-Chloro-3-methyl-1-butene (e) 4-Isopropyl-1-nonene (h) (Z)-1-Chloro-1,2-dibromoethene
(b) 4-Propyl-4-octene (f) 1-Bromopropene (i) (E)-3-Methyl-2-heptene
(c) 1,2-Pentadiene (g) 1,1,1-Trichloro-3-methyl-2-butene (j) *trans*-2,2,5,5-Tetramethyl-3-hexene
(d) 2-Chloro-3-ethyl-2-pentene

7.30 The following names are incorrect. Write the structure and give the correct IUPAC name.

(a) 3-Butene (b) 5-Methyl-4-octene (c) 3-Isopropyl-2-butene

7.31 Assign the configuration (E or Z) to each of the carbon-carbon double bonds in the following alkenes.

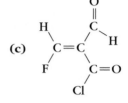

7.32 Write the structure of the product of the reaction of cyclopentene with each of the following reagents.

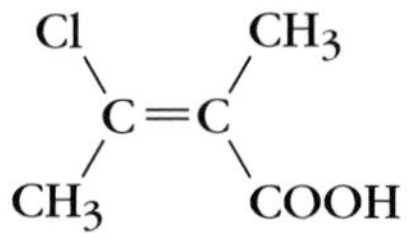

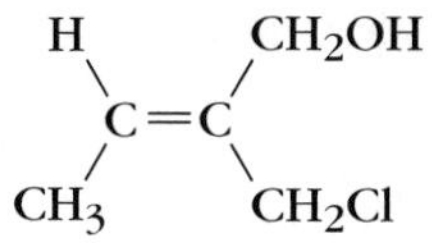

(a) KI, H_3PO_4
(b) HCl in diethyl ether ($C_2H_5OC_2H_5$) solvent
(c) Dilute aqueous acid solution
(d) Ethanol (CH_3CH_2OH) with a catalytic amount of H_2SO_4
(e) CF_3COOH
(f) HBr in diethyl ether ($C_2H_5OC_2H_5$) solvent

7.33 Write the structure of the product of the reaction of 1-hexene with each of the reagents listed in Exercise 7.32.

7.34 In each of the following reactions, write the structural formula of the missing starting material or reagents.

(a) 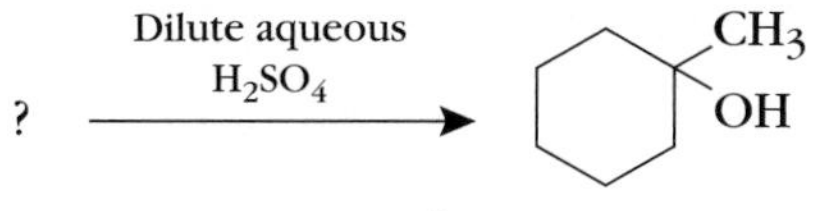 ? $\xrightarrow[\text{H}_2\text{SO}_4]{\text{Dilute aqueous}}$ [cyclohexane ring with CH₃ and OH]

(b) 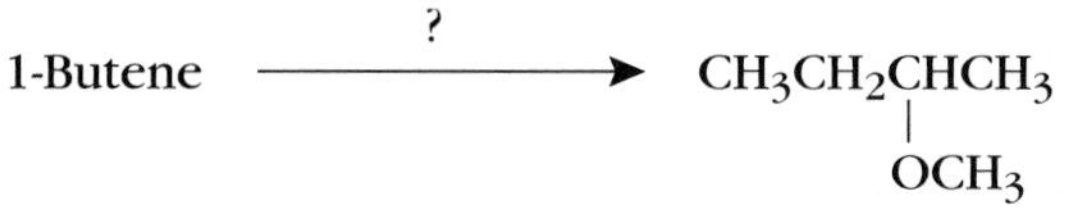 1-Butene $\xrightarrow{?}$ $CH_3CH_2CHCH_3$ with OCH_3

(c) 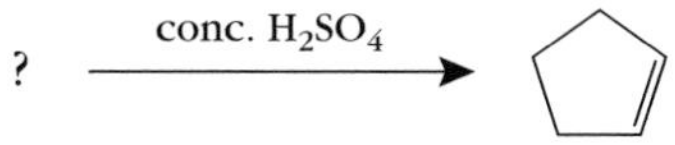 ? $\xrightarrow{\text{conc. H}_2\text{SO}_4}$ [cyclopentene]

(d) 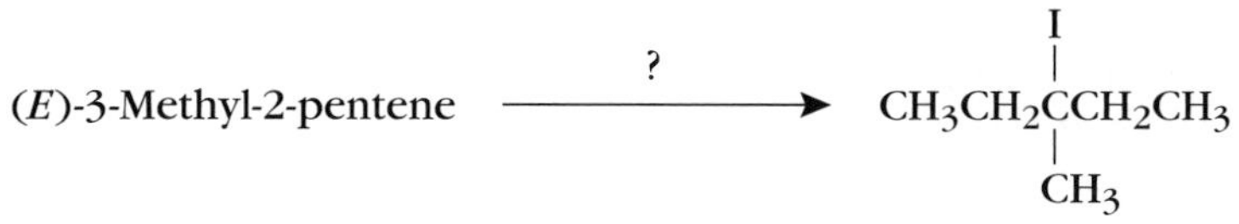 (E)-3-Methyl-2-pentene $\xrightarrow{?}$ $CH_3CH_2CCH_2CH_3$ with I and CH_3

7.35 Write the structures and names of the product or products expected from the addition of HCl in diethyl ether to each of the following compounds.

(a) Methylpropene

(b) 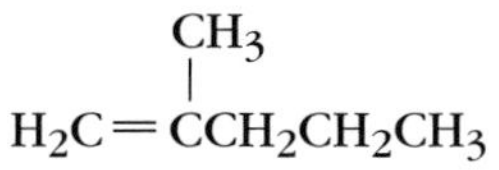 $H_2C=CCH_2CH_2CH_3$ with CH_3

(c) 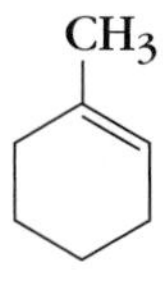 [cyclohexene with CH₃]

(d) 3-Methyl-1-butene

(e) $H_2C=CHBr$

(f) 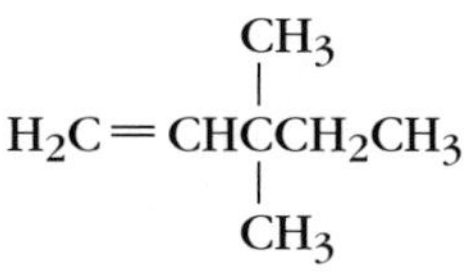 $H_2C=CHCCH_2CH_3$ with CH_3 and CH_3

(g) 2-Methyl-2-butene

7.36 In each of the following pairs of alkenes, which one undergoes acid-catalyzed hydration faster?

(a) Methylpropene or propene

(b) $H_2C{=}CCH_3$ or $H_2C{=}CHCl$
         |
         CH_3

(c) 2-Methyl-1-butene or $H_2C{=}CHCH_2CH_2CH_3$

(d) Propene or 3,3,3-trifluoropropene

(e) $CH_3CH{=}CCH_3$ or 2-butene
              |
              OCH_3

7.37 Heating neopentyl alcohol, $(CH_3)_3CCH_2OH$, in concentrated sulfuric acid converts it into an 85:15 mixture of two alkenes. Write the structure of the two alkenes and propose a mechanism for their formation. Which alkene would you expect to be the major product? Explain.

7.38 Based on what you know about the mechanism of the acid-catalyzed hydration of alkenes, propose a mechanism for the following reaction. Be sure to indicate the stereochemistry of the product.

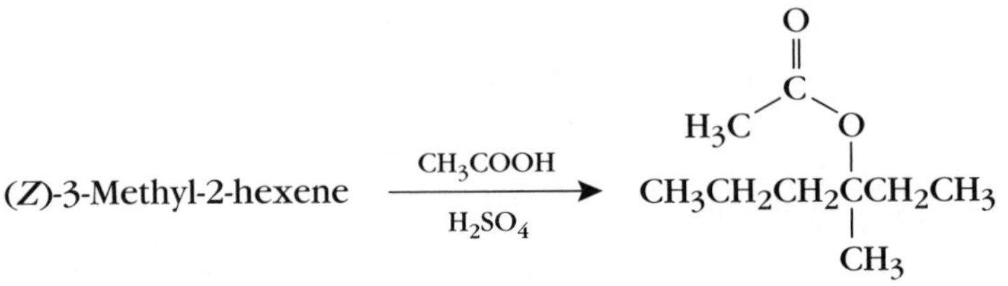

$$(Z)\text{-3-Methyl-2-hexene} \xrightarrow[\text{H}_2\text{SO}_4]{\text{CH}_3\text{COOH}} $$

$$\begin{array}{c} \quad\quad\;\; O \\ \quad\quad\;\; \| \\ \quad\quad\;\; C \\ H_3C \diagup \; \diagdown O \\ \qquad\qquad | \\ CH_3CH_2CH_2CCH_2CH_3 \\ \qquad\qquad | \\ \qquad\qquad CH_3 \end{array}$$

7.39 HCl slowly adds to 1-bromocyclohexene to form two stereoisomers. Write the mechanism of this reaction and write the structure of the two products, clearly showing their stereochemistry.

7.40 Propose a mechanism for each of the following reactions.

(a)

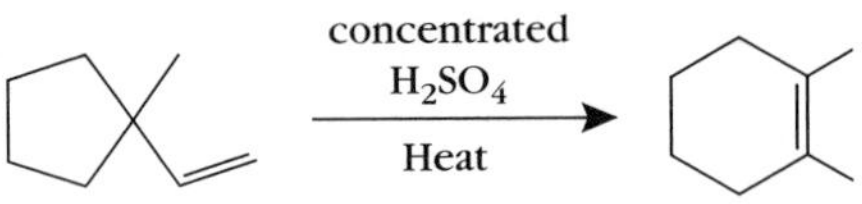

$$\xrightarrow[\text{Heat}]{\substack{\text{concentrated} \\ \text{H}_2\text{SO}_4}}$$

(b)

Formic acid
(a strong acid)

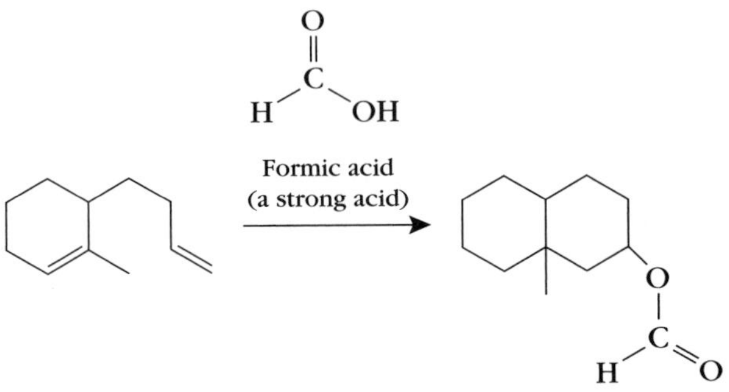

7.41 A student wants to prepare 2-chloro-2-methylpentane by adding HCl gas to a solution of 2-methyl-1-pentene in diethyl ether. The cylinder of HCl gas is empty, however, so the student decides to use a dilute aqueous solution of HCl instead. Two products are isolated. One is the expected 2-chloro-2-methylpentane. The IR spectrum of the other product is given below. What is the structure of the second product, and how was it formed?

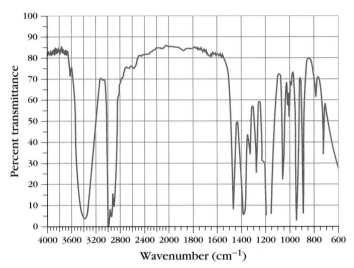

7.42 Explain how IR and ^{13}C NMR spectroscopy can be used to distinguish between the following pairs of compounds. Indicate precisely the peaks that serve to identify each compound.

(a) $CH_3CH=CHOCH_2CH_3$ and ⬠—OH

(b) $CH_3CH_2CH_2C\overset{\displaystyle O}{\underset{\displaystyle OH}{\big\|}}$ and $CH_3CH=CHCH_2OH$

(c) $CH_2=CHCCH_3$ and $CH_2=CHCH_2C\overset{\displaystyle O}{\underset{\displaystyle OH}{\big\|}}$

7.43   IR and ^{13}C NMR spectra are given for one of each of the following pairs of compounds. Match the spectra with the correct compound.

(a)   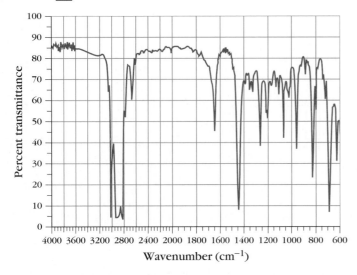   and CH_2=$CHCH_2CH_2CH_2CH$=CH_2

(b) CH_2=$CHOCH_3$ and CH_2=$CHCH_2OH$

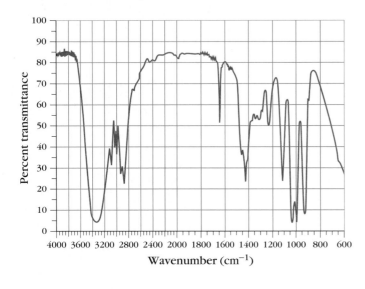

7.44   Explain why the ^{13}C spectrum of a sample taken from a bottle labeled 1-bromopropene has 6 peaks.

7.45   Determine the structure of each of the following compounds from the data given.

     (a)   Molecular formula: $C_2H_2Cl_2$; IR: 1610 cm^{-1}; ^{13}C NMR: δ 128.4, δ 115.6.
     (b)   Molecular formula: $C_4H_6Cl_2$; IR: small broad peak at 1630 cm^{-1}; ^{13}C NMR: δ 130.4, δ 38.9.
     (c)   Molecular formula: C_6H_{10}; IR: 1650 cm^{-1}; ^{13}C NMR: δ 143.4, δ 113.0, δ 20.6.
     (d)   Molecular formula: C_3H_3N; IR: 2220 cm^{-1}, 1605 cm^{-1}; ^{13}C NMR: δ 137.3, δ 117.0, δ 107.7.

CHAPTER

MORE ADDITION REACTIONS
of ALKENES

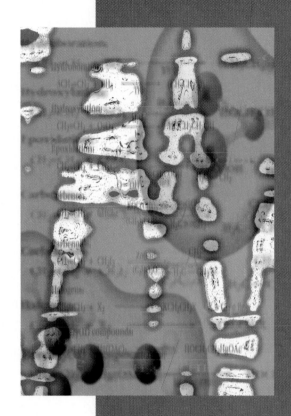

The reaction of Brønsted-Lowry acids with alkenes, discussed in Chapter 7, is only one example of electrophilic additions to alkenes. Additional examples that we will study in this chapter are shown in Figure 8.1. Each will be studied in some detail, beginning with hydroboration.

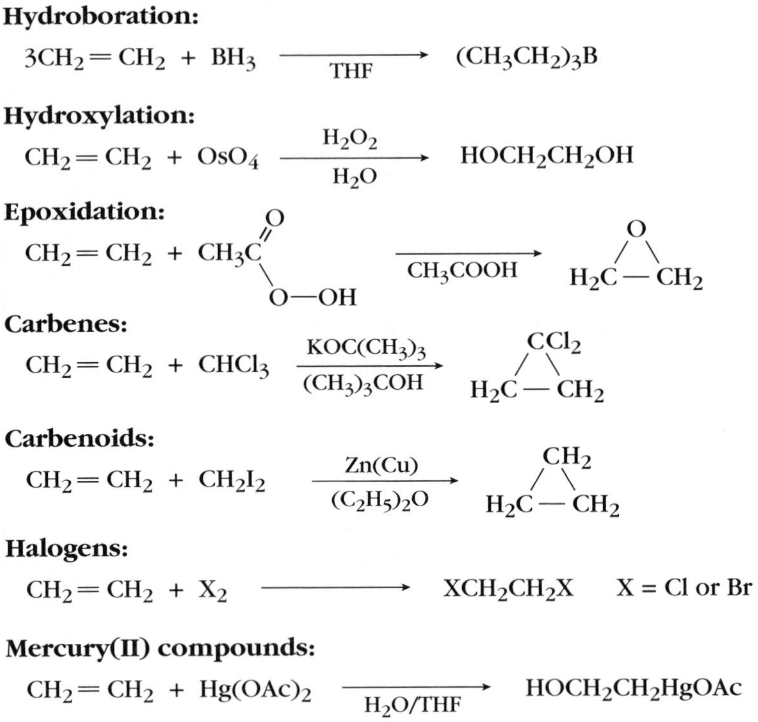

Figure 8.1
Examples of important electrophilic additions to alkenes studied in this chapter.

Hydroboration:

$$3CH_2=CH_2 + BH_3 \xrightarrow{THF} (CH_3CH_2)_3B$$

Hydroxylation:

$$CH_2=CH_2 + OsO_4 \xrightarrow[H_2O]{H_2O_2} HOCH_2CH_2OH$$

Epoxidation:

$$CH_2=CH_2 + CH_3C\overset{O}{\underset{O-OH}{}} \xrightarrow{CH_3COOH} H_2C\overset{O}{-}CH_2$$

Carbenes:

$$CH_2=CH_2 + CHCl_3 \xrightarrow[(CH_3)_3COH]{KOC(CH_3)_3} H_2C\overset{CCl_2}{-}CH_2$$

Carbenoids:

$$CH_2=CH_2 + CH_2I_2 \xrightarrow[(C_2H_5)_2O]{Zn(Cu)} H_2C\overset{CH_2}{-}CH_2$$

Halogens:

$$CH_2=CH_2 + X_2 \longrightarrow XCH_2CH_2X \quad X = Cl \text{ or } Br$$

Mercury(II) compounds:

$$CH_2=CH_2 + Hg(OAc)_2 \xrightarrow{H_2O/THF} HOCH_2CH_2HgOAc$$

8.1   Hydroboration of Alkenes Forms Organoboranes

Hydroboration is the addition of borane (BH_3) to an alkene in an ether solvent, such as tetrahydrofuran (THF), to form an organoborane:

$$3CH_2=CH_2 + BH_3 \xrightarrow{THF} [CH_3-CH_2]_3B$$
Ethylene   Borane   Triethylborane (an organoborane)

Borane is an electrophile because it has only six valence electrons about its boron atom. It completes its octet by reacting with the two electrons of the π bond of the double bond. Each equivalent of borane reacts with three equivalents of alkene. The first addition forms a monoalkylborane. Because the boron atom of the monoalkylborane contains only six valence electrons, it is also an electrophile that reacts with a second molecule of alkene to form a dialkylborane. The dialkylborane then reacts with a third molecule of alkene to form a trialkylborane.

■ *Borane does not exist as BH_3 under normal conditions but is in equilibrium with the dimer, diborane (B_2H_6), which is the more stable form.*

■ *Hydroboration was discovered in 1956 by the American chemist H.C. Brown.*

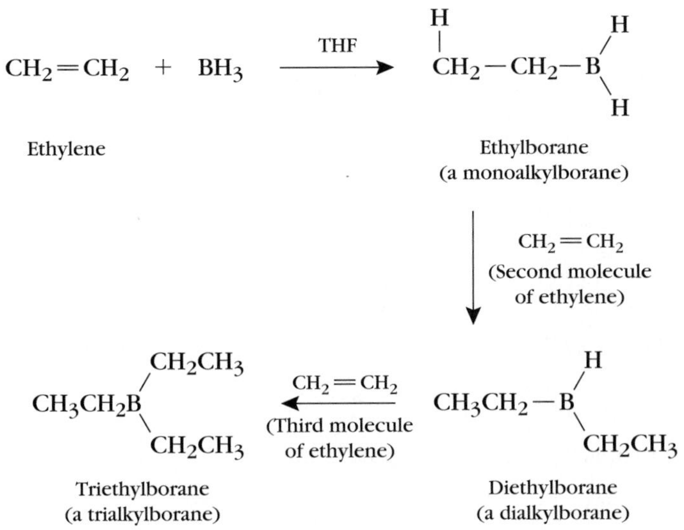

$$CH_2=CH_2 \; + \; BH_3 \xrightarrow{\text{THF}} \overset{\overset{\displaystyle H}{|}}{CH_2}-CH_2-B\overset{\overset{\displaystyle H}{\diagdown}}{\diagdown H}$$

Ethylene   Ethylborane
                                                         (a monoalkylborane)

$$\downarrow \begin{matrix} CH_2=CH_2 \\ \text{(Second molecule} \\ \text{of ethylene)} \end{matrix}$$

$$CH_3CH_2B\overset{\diagup CH_2CH_3}{\diagdown CH_2CH_3} \xleftarrow[\substack{\text{(Third molecule} \\ \text{of ethylene)}}]{CH_2=CH_2} CH_3CH_2-B\overset{\diagup H}{\diagdown CH_2CH_3}$$

Triethylborane   Diethylborane
(a trialkylborane)   (a dialkylborane)

Because boron is the electrophilic atom of borane, it adds mainly but not exclusively to the **less substituted** carbon atom of the double bond. In the hydroboration of 1-hexene shown in Figure 8.2, for example, the major product is formed by addition of the boron atom of BH_3 to Cl of 1-hexene.

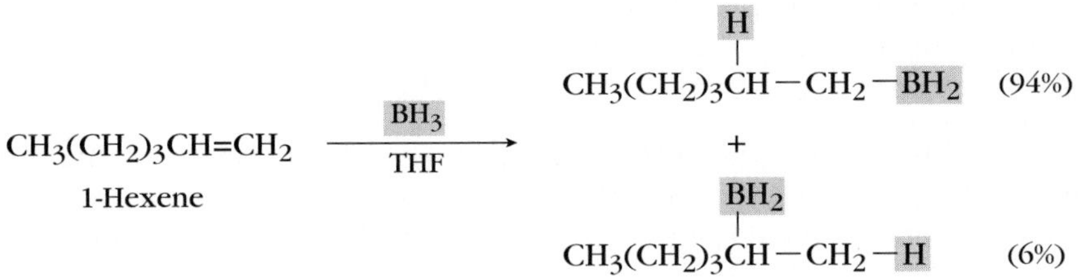

$$CH_3(CH_2)_3CH=CH_2 \xrightarrow[\text{THF}]{BH_3}$$

$$CH_3(CH_2)_3\overset{\overset{\displaystyle H}{|}}{CH}-CH_2-\boxed{BH_2} \quad (94\%)$$

$$+$$

$$CH_3(CH_2)_3\overset{\overset{\displaystyle \boxed{BH_2}}{|}}{CH}-CH_2-\boxed{H} \quad (6\%)$$

Figure 8.2
First formed products in the reaction of borane and 1-hexene. For simplicity, the reactions to form the di- and trialkylboranes are not shown. The di- and trialkylboranes are formed with the same regiochemistry as the monoalkylborane.

1-Hexene

Better control of the regiochemistry can be accomplished by the use of other hydroborating reagents. One particularly useful reagent is 9-borabicyclo[3.3.1]nonane (9-BBN), which is prepared by carefully adding one equivalent of borane to one equivalent of 1,5-cyclooctadiene.

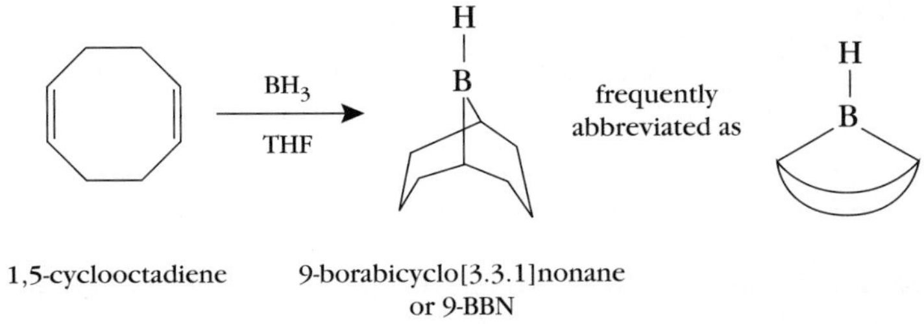

$$\xrightarrow[\text{THF}]{BH_3}$$

frequently abbreviated as

1,5-cyclooctadiene   9-borabicyclo[3.3.1]nonane
                              or 9-BBN

The boron atom of 9-BBN adds to the less sterically hindered carbon atom of the double bond of alkenes. That is, the boron atom adds to the carbon atom that contains the fewest groups or the smallest groups, as shown by the examples in Figure 8.3.

Figure 8.3
9-BBN adds to C1 of 1-hexene because the H atom is smaller than the $CH_3(CH_2)_3$— group, although it adds to C2 of 4-methyl-2-pentene because the CH_3— group is smaller than the $(CH_3)_2CH$— group.

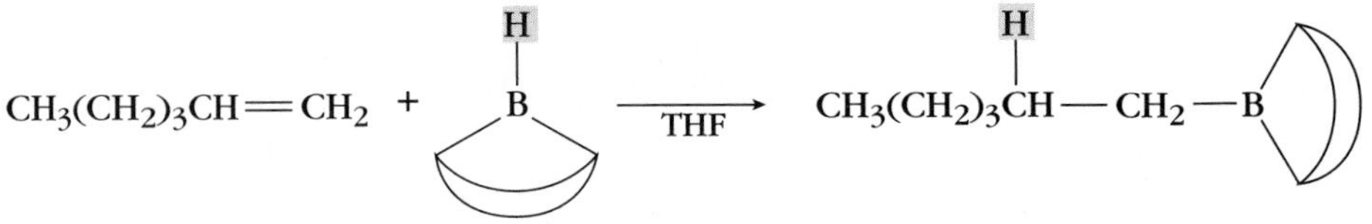

1-Hexene   9-BBN   (99.9% yield)

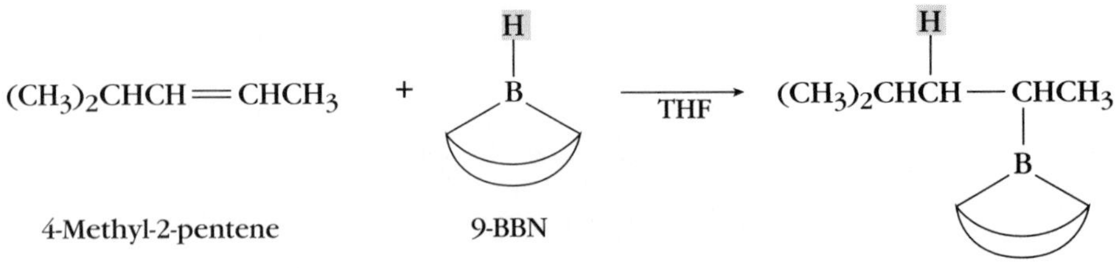

4-Methyl-2-pentene   9-BBN   (99.8% yield)

Exercise 8.1   Write equations for the reaction of methylpropene with:

(a) A dilute aqueous solution of H_2SO_4
(b) A tetrahydrofuran (THF) solution of borane (BH_3)
(c) A tetrahydrofuran solution of 9-BBN

Addition of borane or 9-BBN occurs without rearrangement of the carbon skeleton of the alkene. For example, unlike addition of acids (Section 7.17), no rearranged products are formed by addition of 9-BBN to 3,3-dimethyl-1-butene.

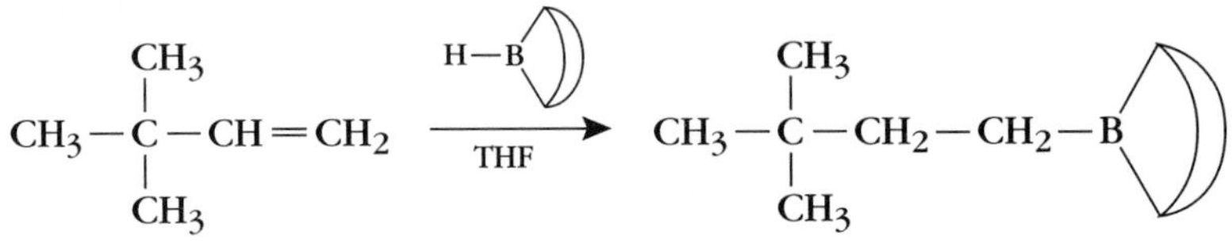

(99% yield)

Exercise 8.2   Write equations for the hydroboration of each of the following alkenes using 9-BBN as the hydroborating reagent.

(a) 1-Butene   (b) Methylpropene   (c) 3-Methyl-1-butene

An important feature of hydroboration of alkenes is its stereochemistry.

8.2   Stereochemistry of Hydroboration of Alkenes

The addition of 9-BBN to 1-methylcyclopentene shown in Figure 8.4 yields a product in which the boron atom is *trans-* to the methyl group. This means that

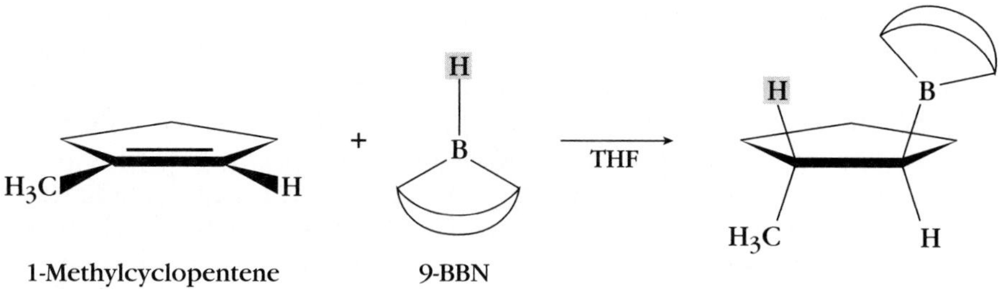

Figure 8.4
The reaction of 9-BBN and 1-methyl-cyclopentene forms a product in which the boron and hydrogen atoms of 9-BBN add to the same face of the double bond. This reaction is a typical example of the stereochemistry obtained from the hydroboration of alkenes.

1-Methylcyclopentene   9-BBN

the boron and hydrogen atoms of 9-BBN are added to the *same* face of the double bond. Addition of two atoms or groups to the same face or side of the plane of a double bond is called **syn addition.** The *hydroboration of most alkenes occurs by syn addition.*

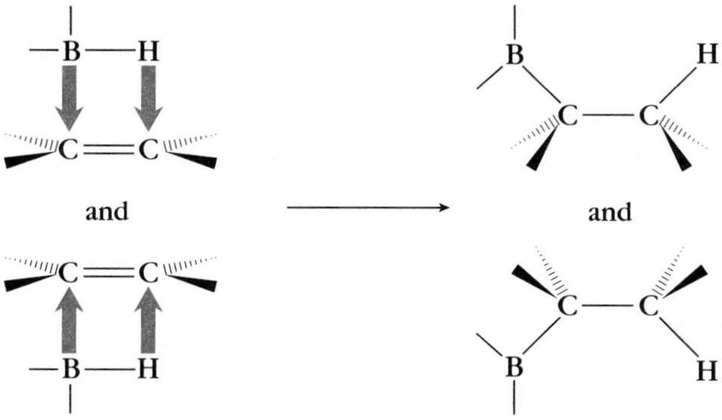

and

and

In general, syn addition occurs equally well to either side of a carbon-carbon double bond. As a result, the products of hydroboration of alkenes with either borane or 9-BBN are either achiral or a racemic mixture, as shown in Figure 8.5.

Figure 8.5
The products of the syn addition of 9-BBN to 1-methylcyclopentene are a racemic mixture because addition occurs equally well to either side of the double bond.

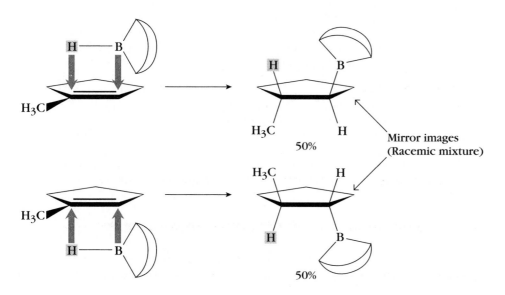

50%

Mirror images
(Racemic mixture)

50%

Hydroboration reactions are additional examples of a principle that we discussed in Section 7.11. *When the reactants of a chemical reaction are optically inactive or achiral, the product is always optically inactive because it is either a racemic mixture or achiral.* Keep this principle in mind as you examine the chemical reactions introduced in this text. Many of the reactions discussed in future chapters also form products that are racemic mixtures. Sometimes, the structures of only one enantiomer of the racemic mixture product will be written. This is done for simplicity and does not mean that an optically active product is formed!

| **Sample Problem 8.1** | Write an equation for the hydroboration of (*Z*)-2-butene using 9-BBN as the hydroborating reagent. Indicate the stereochemistry of the product where appropriate. |

Solution:

Addition occurs equally well from either side of the double bond to form a racemic mixture.

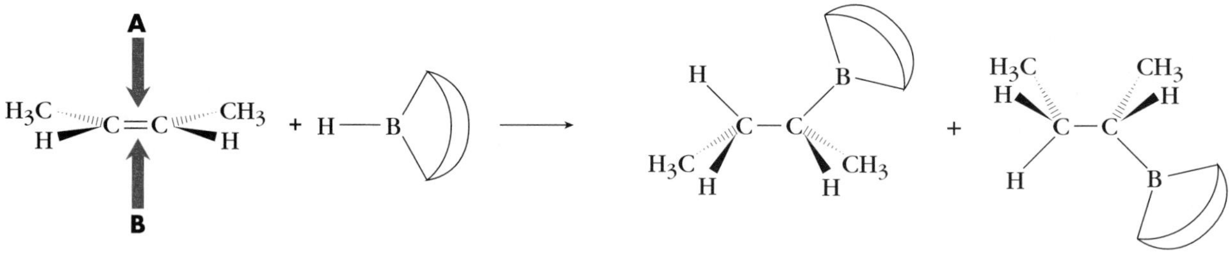

Product formed by addition from side **A**   Product formed by addition from side **B**

Racemic mixture

| **Exercise 8.3** | Write an equation for the hydroboration of each of the following alkenes using 9-BBN as the hydroborating reagent. Indicate the stereochemistry of the product where appropriate. |

(a) 2-Methyl-1-butene   **(d)** (*E*)-3-Methyl-2-pentene
(b) (*E*)-4-Methyl-2-hexene   **(e)** 1,2-Dimethylcyclohexene
(c) 1-Methylcyclohexene   **(f)** (*Z*)-4,4-Dimethyl-2-pentene

---

Based on its regiochemistry, stereochemistry, and lack of rearranged products, we can now propose a mechanism for the hydroboration of alkenes.

8.3   Mechanism of Hydroboration of Alkenes

Based on the experimental results presented in the previous sections, we can reach the following conclusions about the mechanism for the hydroboration of

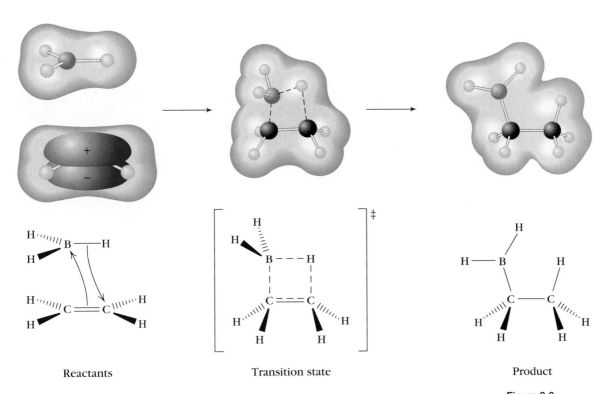

Reactants   Transition state   Product

Figure 8.6
Mechanism proposed for the hydroboration of alkenes. The dashed
lines indicate bonds being made or
broken.

alkenes. Because the products are formed by syn addition, formation of the new
C—B and C—H bonds must occur from the same side of the double bond.
Because no rearranged products are found, the formation of these two bonds
must occur at about the same time. As a result, neither carbon atom of the double bond must acquire much positive charge during the addition; thus no carbocation intermediate is involved in the mechanism. A two-step mechanism consistent with these conclusions is shown in Figure 8.6.

The addition occurs by the interaction of the filled π bond of the alkene with
the empty *p* orbital of the boron atom while C—H bond formation occurs. This
mechanism is viewed as taking place by a **four-center transition state** because
four atoms (two carbons, one boron, and one hydrogen) undergo the changes in
bonding shown in Figure 8.6 at about the same time.

While this mechanism is in accord with the stereochemistry of addition, what
about its regiochemistry? Recall from Section 8.1 that replacing two hydrogen
atoms of borane by the large and bulky organic group in 9-BBN increased the percent of product formed by addition to C1 of 1-hexene. This result suggests that
the most important factor that affects the regiochemistry is steric hindrance
between the groups bonded to boron and the alkyl groups bonded to the carbon
atoms of the double bond. We can see this by examining two possible approaches
of a hydroborating reagent and an unsymmetrical alkene, as shown in Figure 8.7.
Approach of the boron atom to the less substituted carbon atom of the double
bond is favored because there is less steric hindrance that way. This pathway leads
to the transition state with less steric hindrance, which in turn is lower in energy.

Organoboranes are particularly valuable compounds because the reactions
they undergo allow chemists to prepare or synthesize compounds that contain
several other functional groups.

Figure 8.7
Two approaches of a hydroborating reagent and an unsymmetrical alkene. Approach of the boron atom to the less substituted carbon atom is favored because there is less steric hindrance.

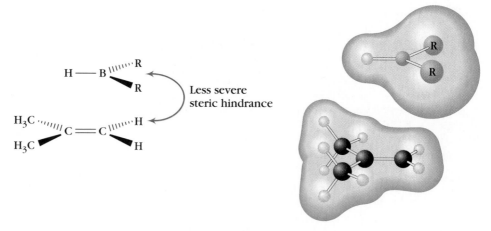

A Lower energy approach: Boron adds to less substituted carbon atom

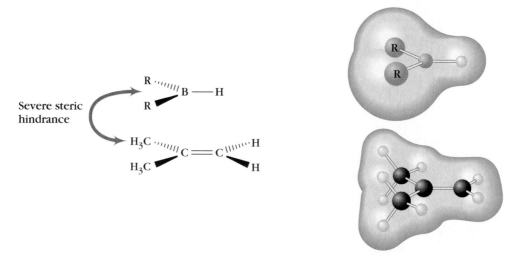

B Higher energy approach: Boron adds to more substituted carbon atom

8.4   Reactions of Organoboranes

Organoboranes undergo many reactions in which the boron atom is replaced by other atoms or functional groups. Three examples of these reactions are summarized in Figure 8.8.

Protonolysis is the reaction of an organoborane with propanoic acid (CH_3CH_2COOH). In this reaction, the boron atom is replaced by a hydrogen atom to form an alkane. Protonolysis is the second reaction of a two-reaction sequence that allows us to convert an alkene into an alkane; the entire sequence of reactions is called **hydroboration-protonolysis.**

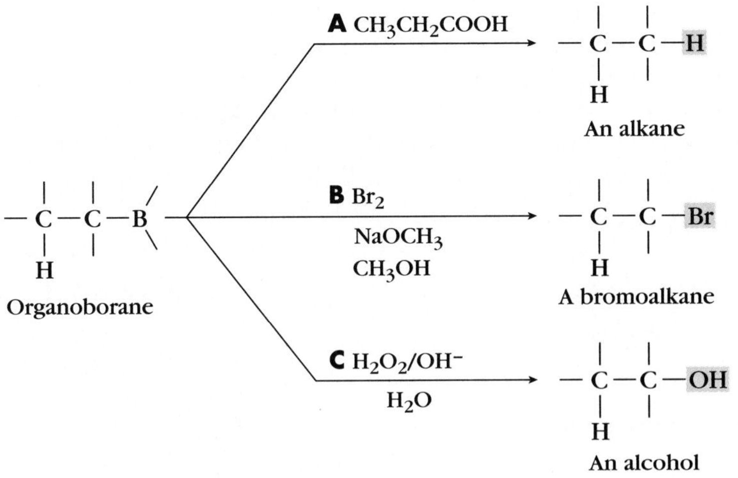

Figure 8.8
Reactions in which the boron atom of an organoborane is replaced by **A,** a hydrogen atom; **B,** a bromine atom; and **C,** a hydroxy group.

Reaction 1: Hydroboration

$$3CH_3CH_2CH_2CH_2CH=CH_2 \xrightarrow[\text{THF}]{\text{BH}_3} [CH_3(CH_2)_4CH_2]_3B$$

1-Hexene   Trihexylborane

Reaction 2: Protonolysis

$$[CH_3(CH_2)_4CH_2]_3B \xrightarrow{\text{CH}_3\text{CH}_2\text{COOH}} 3CH_3(CH_2)_4CH_3$$

Hexane

Bromination of organoboranes replaces the boron atom by a bromine atom to form a bromoalkane. Again we can use this reaction as the second reaction in a two-reaction sequence to convert an alkene into a bromoalkane. This sequence of reactions is called **hydroboration-bromination.** Notice that this sequence of two reactions results in the anti-Markownikoff addition of HBr to the carbon-carbon double bond of an alkene.

Reaction 1: Hydroboration

$$3CH_3CH_2CH_2CH_2\underset{\underset{CH_3}{|}}{C}=CH_2 \xrightarrow[\text{THF}]{\text{BH}_3} (CH_3CH_2CH_2CH_2\underset{\underset{CH_3}{|}}{CH}CH_2)_3B$$

2-Methyl-1-hexene

Reaction 2: Bromination

$$(CH_3CH_2CH_2CH_2\underset{\underset{CH_3}{|}}{CH}CH_2)_3B \xrightarrow[\substack{\text{NaOCH}_3 \\ \text{CH}_3\text{OH}}]{\text{Br}_2} 3CH_3CH_2CH_2CH_2\underset{\underset{CH_3}{|}}{CH}CH_2Br$$

1-Bromo-2-methylhexane
(Product of anti-Markownikoff
addition of HBr to 2-methyl-1-hexene)

The reaction of organoboranes with an alkaline solution of hydrogen peroxide forms alcohols. In this reaction, the boron atom is replaced by a hydroxy group. This sequence of two reactions that converts an alkene to an alcohol is called the **hydroboration-oxidation** of alkenes. Hydroboration-oxidation results in the anti-Markownikoff addition of water to the carbon-carbon double bond of an alkene.

Reaction 1: Hydroboration

$$3CH_3CH{=}CH_2 \quad + \quad BH_3 \quad \xrightarrow[\text{THF}]{} \quad (CH_3CH_2CH_2)_3B$$

Propene

Reaction 2: Oxidation

$$(CH_3CH_2CH_2)_3B \quad \xrightarrow[\text{H}_2\text{O}]{\text{H}_2\text{O}_2/\text{OH}^-} \quad 3CH_3CH_2CH_2OH$$

1-Propanol
(Product of anti-Markownikoff
addition of water to propene)

Protonolysis, bromination, and oxidation of organoboranes occur so that the carbon atom originally bonded to boron in the organoborane maintains its configuration in the product. For example, *trans*-2-methylcyclopentanol is the product of the oxidation of the organoborane formed by the addition of 9-BBN to 1-methylcyclopentene.

Reaction 1: Hydroboration

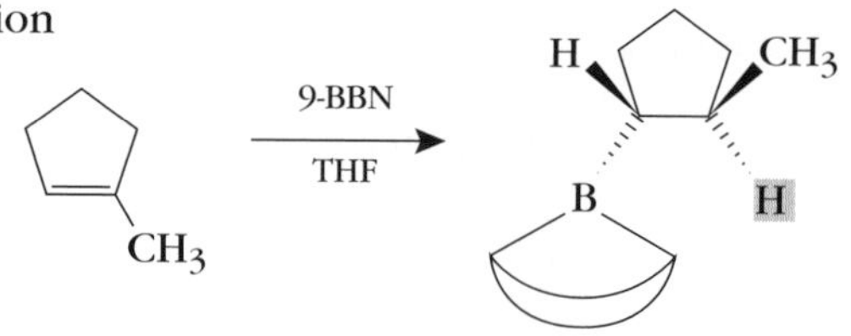

1-Methylcyclopentene

Reaction 2: Oxidation

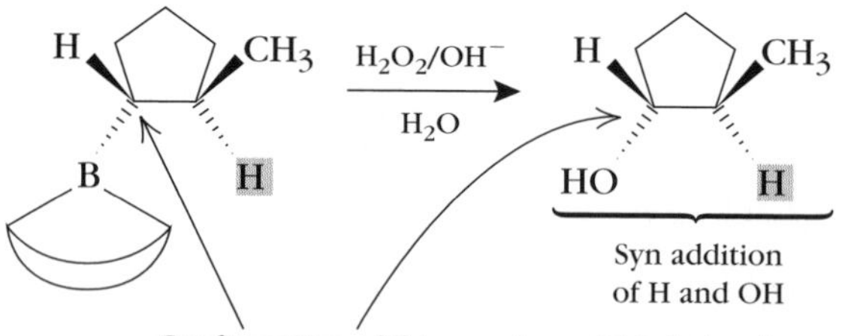

Configuration of this
C atom is the same in
reactant and product

Syn addition
of H and OH

trans-2-Methylcyclopentanol
(82-87% yield)

Thus the overall result of *hydroboration-oxidation, hydroboration-protonolysis, and hydroboration-bromination of alkenes is syn addition of water, hydrogen, and HBr to the double bond.* Moreover, hydroboration-oxidation and hydroboration-bromination form products that are the result of anti-Markownikoff addition to the alkenes.

Another reason the reactions of organoboranes are valuable methods of preparing alcohols, alkanes, and bromoalkanes is that rearranged products are not formed. The hydroboration-oxidation of 3,3-dimethyl-1-butene, for example, forms the unrearranged product 3,3-dimethyl-1-butanol.

Reaction 1: Hydroboration

$$3(CH_3)_3CCH=CH_2 \xrightarrow[THF]{BH_3} [(CH_3)_3CCH_2CH_2]_3B$$

3,3-Dimethyl-1-butene

Reaction 2: Oxidation

$$[(CH_3)_3CCH_2CH_2]_3B \xrightarrow[H_2O]{H_2O_2/OH^-} 3(CH_3)_3CCH_2CH_2OH$$

3,3-Dimethyl-1-butanol
(76-80% yield)
(Unrearranged product of
anti-Markownikoff addition
of water to 3,3-dimethyl-1-butene)

The organoboranes formed in the first reaction of each of these two-reaction sequences are not usually isolated. Once their formation is completed, the second reaction is carried out in the same flask by the addition of the second reagent.

Of the three reactions of organoboranes presented so far, the oxidation to

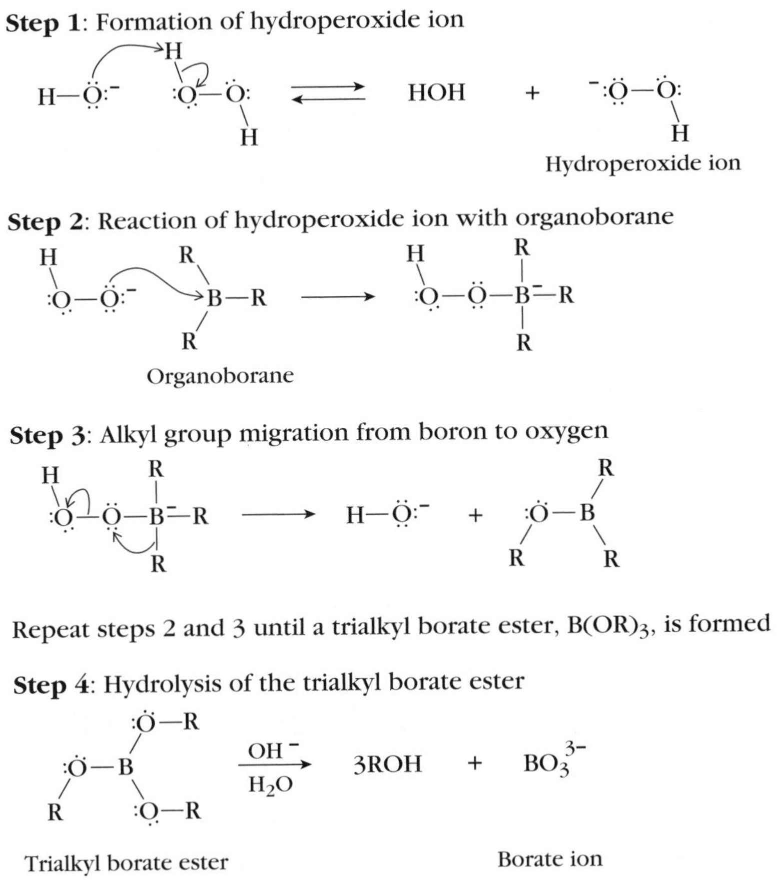

Step 1: Formation of hydroperoxide ion

Hydroperoxide ion

Step 2: Reaction of hydroperoxide ion with organoborane

Organoborane

Step 3: Alkyl group migration from boron to oxygen

Repeat steps 2 and 3 until a trialkyl borate ester, $B(OR)_3$, is formed

Step 4: Hydrolysis of the trialkyl borate ester

Trialkyl borate ester   Borate ion

Figure 8.9
Mechanism of the oxidation of organoboranes to alcohols by aqueous alkaline hydrogen peroxide.

alcohols has been the most studied, and a proposed mechanism is shown in Figure 8.9, p. 335.

Step 1 of the mechanism is the formation of the hydroperoxide ion by an acid-base reaction between hydrogen peroxide and hydroxide ion. Hydrogen peroxide is a stronger acid ($pK_a = 11.6$) than water ($pK_a = 15.7$), so hydroperoxide ion and water are present in large concentrations in aqueous solution at equilibrium. Step 2 is also an acid-base reaction. In this case, the organoborane is a Lewis acid that reacts with a Lewis base, the hydroperoxide ion. In Step 3, the alkyl group migrates with its electron pair from the boron atom to the oxygen atom, expelling a hydroxide ion in the process. Steps 2 and 3 are repeated twice more until a trialkyl borate ester is formed. In Step 4, the trialkyl borate ester is hydrolyzed to three moles of alcohol and a borate ion.

In summary, alcohols, alkanes, and bromoalkanes can be conveniently prepared from alkenes by a sequence of two reactions. The first is hydroboration of an alkene to form an organoborane; the second is reaction with the appropriate reagent: propanoic acid to form an alkane, bromine to form a bromoalkane, and alkaline hydrogen peroxide to form an alcohol.

The first step, hydroboration, is highly regioselective with the boron atom adding to the less sterically hindered carbon atom of the double bond. Hydroboration is also a stereospecific reaction in which the products are formed by syn addition of the boron and hydrogen atoms to the double bond. The second step replaces the boron atom with an H atom, an —OH group, or a Br atom. The overall result of the two-step sequence is syn addition to the double bond of an alkene to form achiral or racemic mixtures of products with anti-Markownikoff regiochemistry. Rearrangement of the carbon skeleton so typical of acid-catalyzed addition reactions of alkenes does not occur in these sequences of reactions.

Exercise 8.4   Write the structure of the product of the reaction of the tri-(2-methylcyclohexyl)borane with each of the following reagents. Indicate the stereochemistry of the product wherever appropriate.

R$_3$B where R is

2-Methylcyclohexyl

(a) CH_3CH_2COOH **(c)** $Br_2/NaOCH_3$ in CH_3OH
(b) $H_2O_2/NaOH/H_2O$ **(d)** CH_3CH_2COOD

8.5   Hydroxylation of Alkenes

Alkenes react with osmium tetraoxide (OsO_4) to form vicinal glycols.

Alkene   Vicinal glycol

Glycols or **diols** are compounds that contain two —OH groups. In a vicinal glycol, the two —OH groups are on adjacent carbons. The reaction in which an —OH group is added to each of the two carbon atoms of the double bond is called **hydroxylation.**

Hydroxylation of alkenes by osmium tetraoxide/H_2O_2 forms vicinal glycols by syn addition. The hydroxylation of cyclohexene, for example, forms *cis*-1,2-cyclohexanediol, a *cis*- glycol:

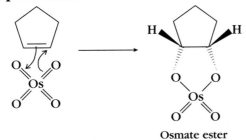

cis-1,2-Cyclohexanediol
(40% yield)
a *cis*-glycol

A two-step mechanism, shown in Figure 8.10, has been proposed to account for the observed stereochemistry of the product. In Step 1 of the mechanism, a cyclic intermediate, called an osmate ester, is formed by the addition of osmium tetraoxide to the alkene. Two oxygen atoms of osmium tetraoxide add to the same side of the double bond because addition to opposite sides of the double bond would require a highly strained *trans*- five-membered ring. Addition to the same side of the double bond results in syn addition. In Step 2, the osmate ester is hydrolyzed to the *cis*-glycol by reaction with aqueous hydrogen peroxide.

Osmium tetraoxide is both toxic and expensive so it is used in only catalytic amounts with an excess of an oxidizing agent such as hydrogen peroxide (H_2O_2). The small amount of OsO_4 reacts with some of the alkene to form the osmate ester, which is converted by aqueous hydrogen peroxide to the glycol and

Step 1: Formation of the osmate ester

Osmate ester

Step 2: Hydrolysis of the osmate ester and regeneration of OsO_4

Osmate ester

Figure 8.10
A two-step mechanism for the hydroxylation of cyclopentene by OsO_4.

reduced osmium. Hydrogen peroxide then reoxidizes the osmium to OsO_4, which then continues the hydroxylation of more alkene. In this way, the expense and the toxic effects of OsO_4 are minimized.

Exercise 8.5	Write the structure of the product of the reaction of OsO_4/H_2O_2 with each of the following alkenes. Indicate the stereochemistry of the product where appropriate.

    (a) 1-Hexene   **(b)** Cycloheptene   **(c)** (*Z*)-3-Hexene

Exercise 8.6	Hydroxylation of (*E*)-2-butene and (*Z*)-2-butene with $OsO_4/H_2O_2/H_2O$ forms different glycols. Write the structures of the products of the two reactions and explain why they are different.

Suppose we wish to prepare a particular glycol by the hydroxylation of an alkene. How do we decide what alkene to use? A solution is provided in Sample Problem 8.2.

Sample Problem 8.2	Given the following glycol, write the equation for its preparation by hydroxylation of the appropriate alkene.

$$CH_3CHCH_2-OH$$
$$|$$
$$OH$$

Solution:

The carbon atoms bonded to the two —OH groups of the diol were the carbon atoms of the double bond of the original alkene. The structure of the alkene is obtained by removing both —OH groups and placing a double bond between the two carbon atoms.

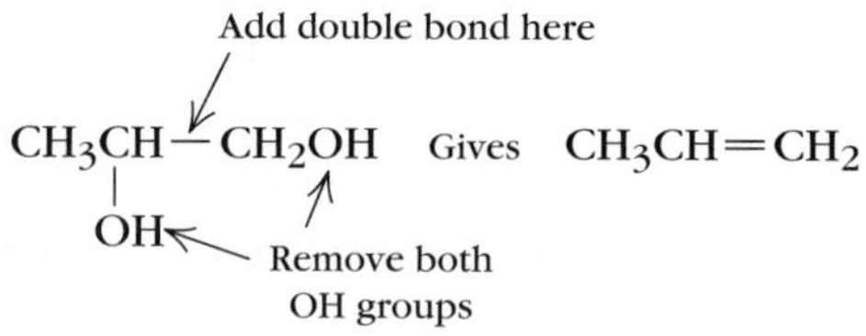

Add double bond here

$$CH_3CH{-}CH_2OH \quad \text{Gives} \quad CH_3CH{=}CH_2$$
$$|$$
$$OH$$

Remove both
OH groups

The reaction of propene with OsO_4 to form the desired glycol is:

$$CH_3CH{=}CH_2 \ + \ OsO_4 \ \xrightarrow[\text{H}_2\text{O}]{\text{H}_2\text{O}_2} \ CH_3CHCH_2OH$$
$$|$$
$$OH$$

Exercise 8.7   For each of the following glycols, write the equation for its preparation by hydroxylation of the appropriate alkene.

(a)

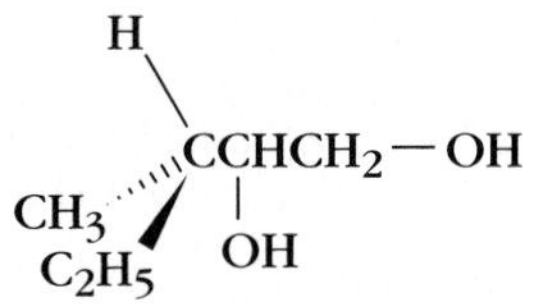

$$H$$

$$\overset{CH_3}{\underset{C_2H_5}{\text{C}}}CHCH_2-OH$$

$$OH$$

(b)

$$CH_3$$
$$HO$$
$$HO$$
$$CH_3$$

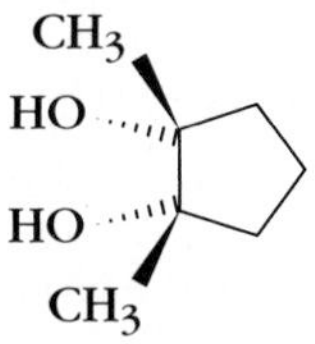

Another addition reaction of alkenes that forms carbon-oxygen bonds is epoxidation.

8.6   Epoxidation of Alkenes

Alkenes react with peroxycarboxylic acids to form compounds with an oxygen-containing three-membered ring:

$$\text{C}=\text{C} \quad + \quad R-\overset{O}{\underset{O-OH}{C}} \quad \longrightarrow \quad \text{C}-\text{C} \quad + \quad R-\overset{O}{\underset{O-H}{C}}$$

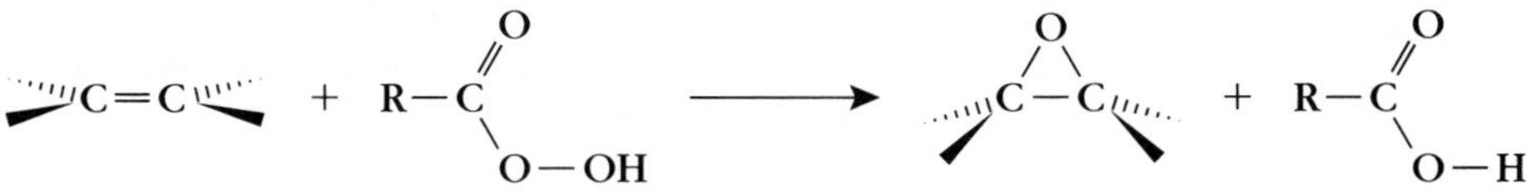

Alkene   Peroxycarboxylic acid   Epoxide   Carboxylic acid

Compounds with an oxygen-containing three-membered ring are called **epoxides** or **oxiranes** and the reaction that forms them is called **epoxidation.**

Epoxidation reactions are usually carried out in organic solvents such as diethyl ether, acetic acid, or chlorinated hydrocarbons. Two commonly used peroxycarboxylic acids are peroxyacetic acid and peroxybenzoic acid.

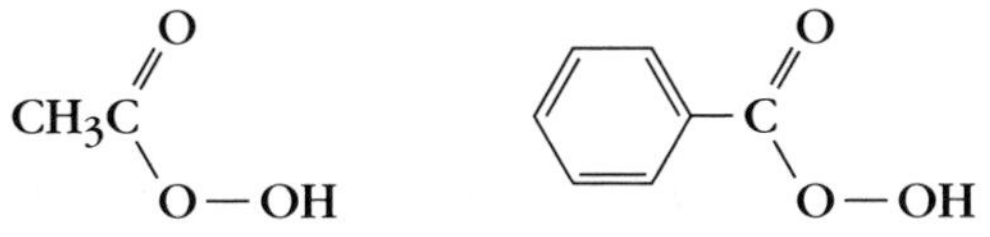

$$CH_3C\overset{O}{\underset{O-OH}{}}$$

Peroxyacetic acid

$$C\overset{O}{\underset{O-OH}{}}$$

Peroxybenzoic acid

Epoxidation of alkenes is a stereospecific reaction because stereoisomeric alkenes react to form different epoxide stereoisomers. (*Z*)-2-Butene, for example, reacts with peroxyacetic acid in acetic acid to form *meso cis*-2,3-dimethyloxirane, while epoxidation of (*E*)-2-butene forms racemic *trans*-2,3-dimethyloxirane.

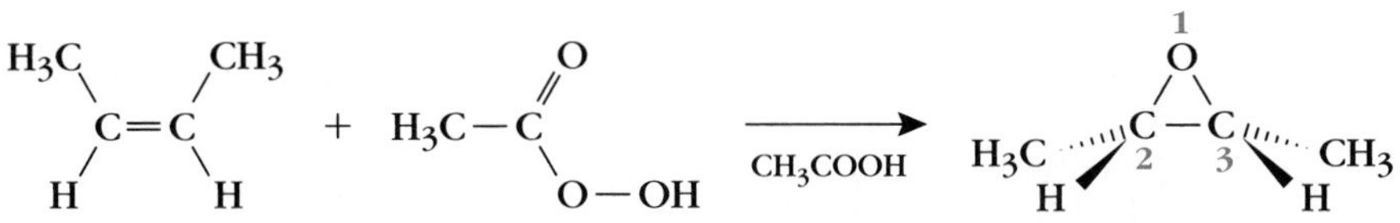

$$\underset{H}{\overset{H_3C}{}}\text{C}=\text{C}\underset{H}{\overset{CH_3}{}} \quad + \quad H_3C-\overset{O}{\underset{O-OH}{C}} \quad \xrightarrow{CH_3COOH} \quad H_3C\underset{H}{\overset{1}{\underset{2}{C}}}\overset{O}{\underset{3}{C}}\underset{H}{\overset{}{CH_3}}$$

(*Z*)-2-Butene   Peroxyacetic acid   *cis*-2,3-Dimethyloxirane
(*meso*)
(75% yield)

$$\underset{\substack{(E)\text{-2-Butene}}}{\overset{\displaystyle H \qquad CH_3}{\underset{\displaystyle H_3C \qquad H}{C=C}}} \quad + \quad \underset{\substack{\text{Peroxyacetic acid}}}{\overset{\displaystyle O}{H_3C-C}}\overset{\displaystyle }{\underset{\displaystyle O-OH}{}} \quad \xrightarrow{\ CH_3COOH\ } \quad \underset{\substack{\textit{trans}\text{-2,3-Dimethyloxirane} \\ \text{(racemic mixture)} \\ \text{(85\% yield)}}}{\overset{\displaystyle O}{\underset{\displaystyle H_3C \qquad H}{H\ \ C-C\ \ CH_3}}}$$

From these results, we see that epoxidation is characterized by syn addition of an oxygen atom to the double bond. Substituents that are *cis-* to each other in the alkene are also *cis-* in the oxirane; those that are *trans-* in the alkene are also *trans-* in the oxirane.

The data in Table 8.1 show that increasing the number of methyl groups on a double bond increases the rate of epoxidation. We can rationalize these data by saying that electron-donating methyl groups increase the electron density of the π bond of the double bond. This in turn increases the rate at which the double bond reacts with peroxyacetic acid. Thus peroxycarboxylic acids act as electrophiles in epoxidation reactions.

Table 8.1	Effect on the Rate of Epoxidation of Increasing the Number of Methyl Groups on a Double Bond	
Alkene	Structure	Relative Rate
Ethylene	$CH_2{=}CH_2$	1
Propene	$CH_3CH{=}CH_2$	22
Methylpropene	$(CH_3)_2C{=}CH_2$	484
2-Methyl-2-butene	$(CH_3)_2C{=}CHCH_3$	5082

We can take advantage of the data in Table 8.1 to selectively epoxidize one of two differently substituted double bonds in a diene. The more highly substituted double bond will react more quickly with the one equivalent of peroxycarboxylic acid used in the reaction.

$$\underset{\substack{\text{More highly substituted} \\ \text{double bond (more reactive)}}}{\overset{\displaystyle CH_2{=}CHCH_2CH_2 \quad CH_3}{\underset{\displaystyle H_3C \qquad CH_3}{C=C}}} \quad \overset{\substack{\text{1 equivalent} \\ O \\ CH_3C \\ O-OH}}{\xrightarrow{\qquad CH_3COOH \qquad}} \quad \overset{\displaystyle CH_2{=}CHCH_2CH_2 \quad CH_3}{\underset{\displaystyle H_3C \quad O \quad CH_3}{C-C}}$$

The mechanism of epoxidation is illustrated in more detail in Section 13.10, p. 570.

The mechanism of epoxidation is believed to occur in a single step (a concerted mechanism) in which the oxygen atom farthest from the carbonyl group is transferred to the double bond of the alkene.

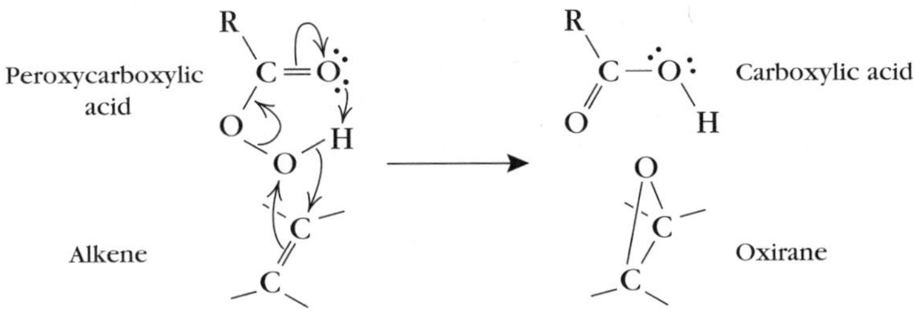
Peroxycarboxylic acid
Alkene
Carboxylic acid
Oxirane

Exercise 8.8 Write the structure of the major product in the epoxidation of each of the following alkenes with one equivalent of peroxyacetic acid in acetic acid. Show the stereochemistry of the product where appropriate.

(a) 1-Hexene
(b) (*E*)-2-Pentene
(c) Cyclohexene
(d) (*Z*)-2-Pentene
(e)

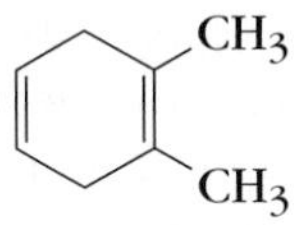
CH3
CH3

Alkenes also undergo addition reactions that form compounds with cyclopropane rings.

8.7 Additions of Carbenes and Carbenoids

Substituted cyclopropanes are easily and conveniently prepared by the reaction of alkenes and chloroform in a strongly basic solution. For example, when chloroform ($CHCl_3$) is added to a *t*-butyl alcohol solution of methylpropene and potassium *t*-butoxide, the following addition reaction occurs to form 1,1-dichloro-2,2-dimethylcyclopropane.

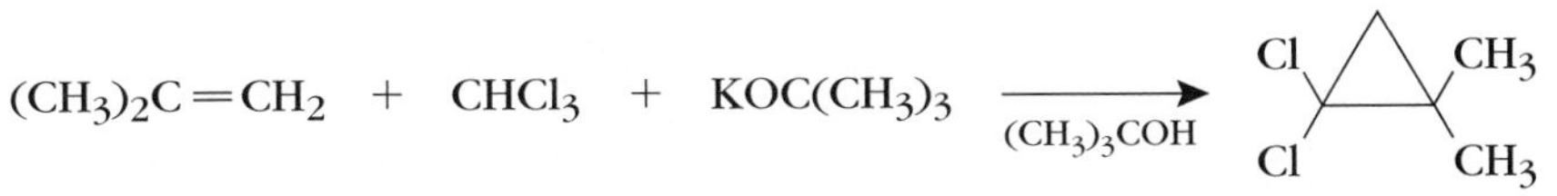
(CH3)2C=CH2 + CHCl3 + KOC(CH3)3
(CH3)3COH
Cl
CH3
Cl
CH3

Methylpropene — Chloroform — Potassium *t*-butoxide — 1,1-Dichloro-2,2-dimethylcyclopropane (60-65% yield)

In this addition reaction, the carbon atoms of the double bond of the original alkene become two of the three carbon atoms of the cyclopropane ring. The third carbon atom comes from chloroform.

As in the epoxidation of alkenes, the product of this reaction retains the stereochemistry of the original alkene. Under the same reaction conditions, for example, *trans*-2-pentene forms only *trans*-1,1-dichloro-2-ethyl-3-methylcyclopropane.

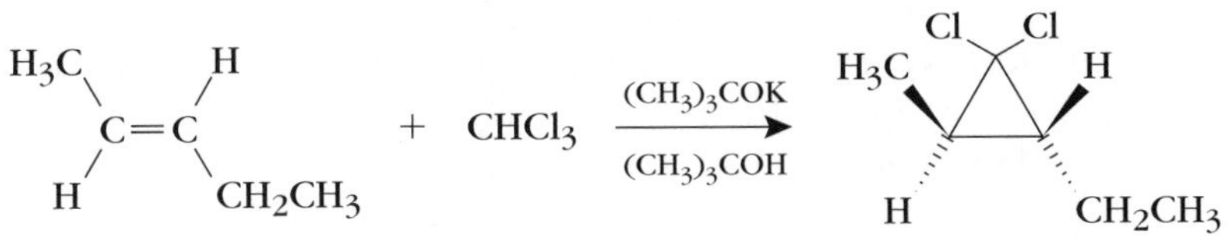
H3C
H
C=C
H
CH2CH3
+ CHCl3
(CH3)3COK
(CH3)3COH
Cl
Cl
H3C
H
H
CH2CH3

trans-2-Pentene — *trans*-1,1-Dichloro-2 ethyl-3-methylcyclopropane (58-63% yield)

None of the *cis*-isomer of the cyclopropane is formed. Similarly *cis*-2-pentene reacts to form only *cis*-1,1-dichloro-2-ethyl-3-methylcyclopropane.

The reagent that is generated and adds to the double bond of the alkene in this reaction is a **carbene.** A carbene is a reactive intermediate that has the following Lewis structure:

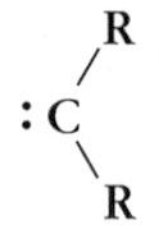

$$: C \overset{R}{\underset{R}{<}}$$

The carbon atom of a carbene is bonded to only two atoms or groups of atoms and has only six valence electrons. While carbenes are neutral, they still behave as electrophiles because they seek electrons to complete their octet.

Carbenes are formed by the reaction of chloroform with a strong base. Chloroform (as well as bromoform and iodoform) is quite acidic for an organic compound ($pK_a \approx 24$), due to the strong electron-withdrawing inductive effects of the three halogen atoms. The reaction of chloroform with a base such as an alkoxide ion in a weakly acidic solvent such as an alcohol ($pK_a \approx 18$) forms a small equilibrium concentration of trichloromethyl anion.

CHBr₃	CHI₃
Bromoform	Iodoform

| Alkoxide ion $pK_a \approx 24$ | | Trichloromethyl anion $pK_a \approx 18$ |

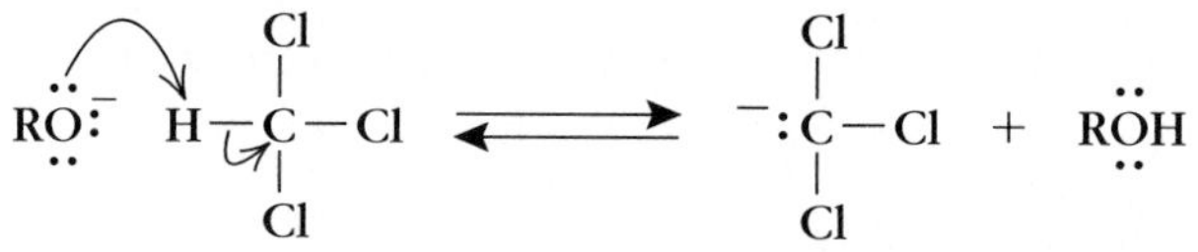

This anion loses a chloride ion to form a dichlorocarbene.

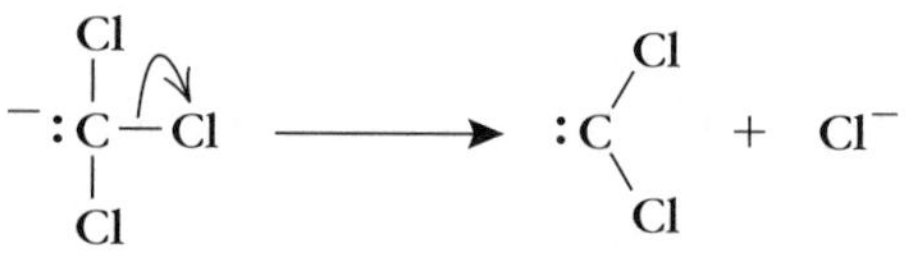

Dichlorocarbene

The net result of these two reactions is the removal or elimination of HCl from the carbon atom of chloroform.

In the presence of an alkene, dichlorocarbene adds to the double bond to form a dichlorocyclopropane derivative.

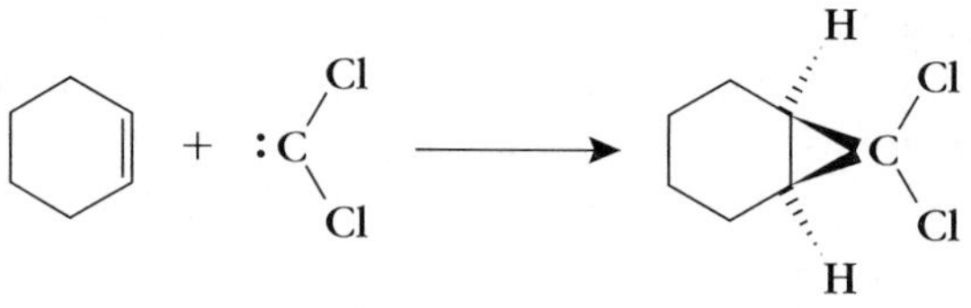

(65% yield)

Because the product retains the *cis*- or *trans*-stereochemistry of the alkene, the addition reaction must occur by a single-step mechanism in which both new bonds to the carbon atoms of the double bond are formed at about the same time.

Another reaction that converts alkenes to cyclopropanes is the **Simmons-Smith** reaction, named in honor of the two American chemists, H.E. Simmons and R.D. Smith, who discovered it. Simmons and Smith found that a cyclopropane derivative is formed when a diethyl ether solution of an alkene reacts with a mixture of methylene iodide (CH_2I_2) and zinc dust that has been activated by alloying it with a small amount of copper.

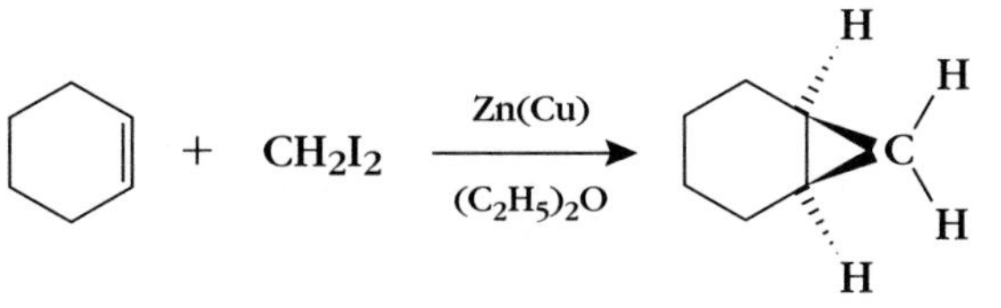

(62-67% yield)

The mixture of methyl iodide and activated zinc dust is called the Simmons-Smith reagent. The active species in the mixture is believed to resemble iodomethylzinc iodide (ICH_2ZnI).

$$CH_2I_2 \ + \ Zn(Cu) \ \xrightarrow[(C_2H_5)_2O]{} \ ICH_2ZnI$$

Simmons-Smith reagent
(a carbenoid)

This reagent is called a **carbenoid** because it reacts in some ways like a carbene, even though it does not contain a divalent carbon atom.

Exercise 8.9   Write the structure of the product formed by the reaction of each of the following alkenes with chloroform and potassium *t*-butoxide in *t*-butyl alcohol.

(a) Cyclopentene   **(b)** (*Z*)-3-Methyl-2-pentene   **(c)** (*E*)-3-Hexene

---

Exercise 8.10   Write the structure of the product formed by the reaction of each of the following alkenes with CH_2I_2 and Zn(Cu) in diethyl ether.

(a) Methylcyclohexene   **(c)** (*E*)-3,4-Dimethyl-3-heptene
(b) 2,3-Dimethyl-2-butene

In summary, substituted cyclopropanes can be prepared by reacting alkenes with a carbene prepared by the base-promoted elimination of HCl from chloroform or by the Simmons-Smith reaction.

8.8   Additions of Halogen

Halogens add to the carbon-carbon double bond of alkenes to form vicinal dihalides.

Alkene   X = Cl, Br   Vicinal dihalide

Table 8.2	Effect on the Rate of Bromination of Increasing the Number of Methyl Groups on a Double Bond	
Alkene	Structure	Relative Rate of Br_2 Addition (in methanol)
Ethylene	$CH_2{=}CH_2$	1
Propene	$CH_3CH{=}CH_2$	61
Methylpropene	$(CH_3)_2C{=}CH_2$	5410
2-Methyl-2-butene	$(CH_3)_2C{=}CHCH_3$	1.32×10^5
2,3-Dimethyl-2-butene	$(CH_3)_2C{=}C(CH_3)_2$	1.82×10^6

For practical purposes, the addition of halogens to alkenes is a general reaction only for chlorine and bromine. Most organic compounds undergo explosive reactions with fluorine so special apparatus and techniques are needed to use it. The addition of iodine, on the other hand, forms an equilibrium mixture in which the alkene and iodine are generally favored at room temperature.

The data in Table 8.2 show that the rates of addition of bromine to an alkene increase as the number of methyl groups on the double bond increases. These data suggest that the methyl groups increase the rate of bromination by increasing the electron density of the π bond portion of the carbon-carbon double bond. Thus chlorine and bromine act as electrophiles in their addition reactions with alkenes, analogous to peroxycarboxylic acids in epoxidation reactions.

The addition of either chlorine or bromine to cyclopentene forms only *trans*-1,2-dihalocyclopentane. None of the *cis*-isomer is formed.

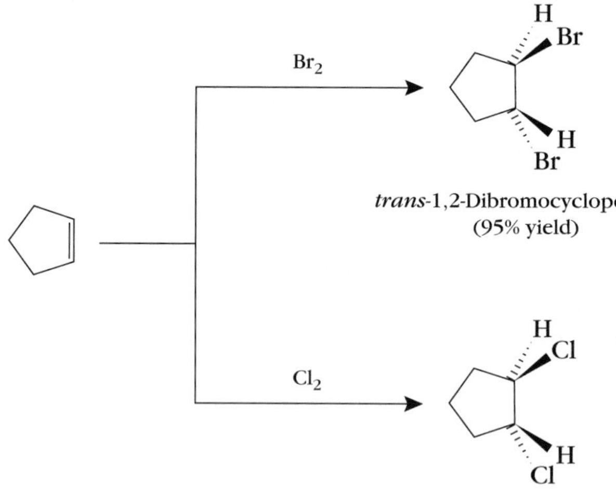

trans-1,2-Dibromocyclopentane
(95% yield)

trans-1,2-Dichlorocyclopentane
(90% yield)

The formation of products with the two halogen atoms *trans*- to each other means that the two halogen atoms must have added to different sides of the plane of the double bond. This is called **anti addition** of the two halogen atoms.

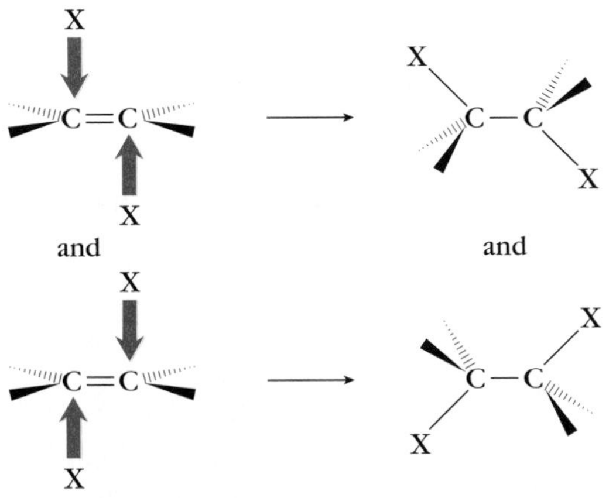

Anti addition of halogen to a carbon-carbon double bond

The product of anti addition of halogens to alkenes is optically inactive. For example, bromination of cyclopentene forms a racemic mixture of *trans*-1, 2-dibromocyclopentane.

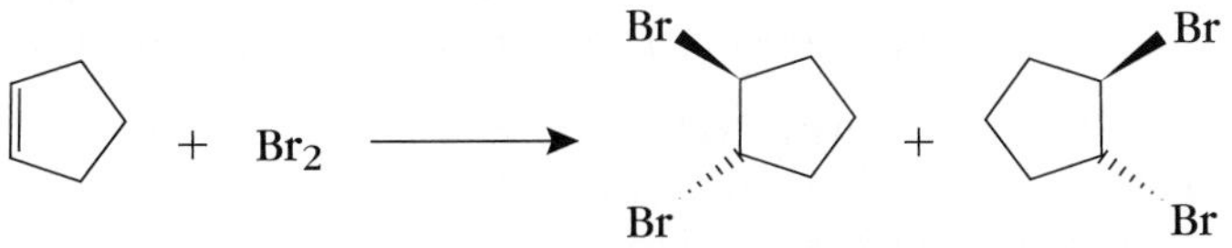

Racemic *trans*-1,2-dibromocyclopentane

The anti addition of bromine and chlorine to alkenes is stereospecific. A reaction is **stereospecific** if stereoisomeric reactants form stereoisomerically different products under the same reaction conditions. Bromination of (*Z*)-2-butene, for example, is stereospecific because it forms a racemic mixture of (2*R*,3*R*)- and (2*S*,3*S*)-2,3-dibromobutane, while bromination of its stereoisomer, (*E*)-2-butene, forms the *meso* diastereomer.

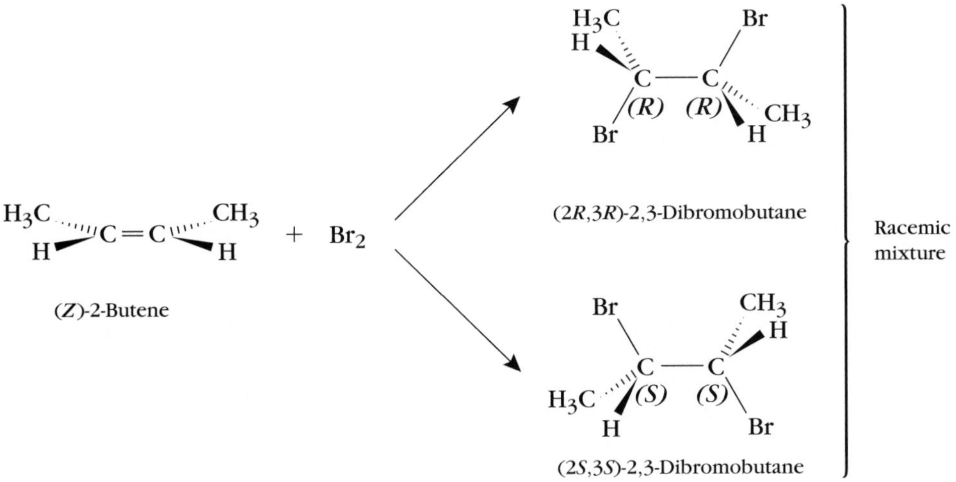

(*Z*)-2-Butene

(2*R*,3*R*)-2,3-Dibromobutane

(2*S*,3*S*)-2,3-Dibromobutane

Racemic mixture

$$H_3C \cdots C = C \cdots H \atop H \qquad CH_3 \qquad + \quad Br_2$$

(E)-2-Butene

$$\underset{Br}{\overset{H_3C}{\underset{\scriptstyle (R)}{\overset{\scriptstyle H}{}}}} C - C \underset{CH_3}{\overset{Br}{\underset{\scriptstyle (S)}{}}}$$

meso-2,3-Dibromobutane

$$\underset{H}{\overset{Br}{}} C - C \underset{Br}{\overset{H}{}} CH_3 \atop H_3C \cdots (S) \quad (R)$$

meso-2,3-Dibromobutane

Identical

Exercise 8.11   Write the equations for the additions of bromine and chlorine to each of the following alkenes. Show the stereochemistry of the product where appropriate.

(a) $CH_2{=}CHCH_2CH_3$ **(c)** Cyclohexene
(b) Methylpropene   **(d)** 1-Methylcyclohexene

The proposed mechanism for the anti addition of bromine to alkenes is shown in Figure 8.11. In Step 1 of the mechanism, the nucleophilic π bond of the alkene

Figure 8.11
Two-step mechanism for the bromination of an alkene. In the first step, the alkene acts as a nucleophile to displace a bromide ion from bromine to form a bromonium ion.

Step 1: Formation of bromonium ion

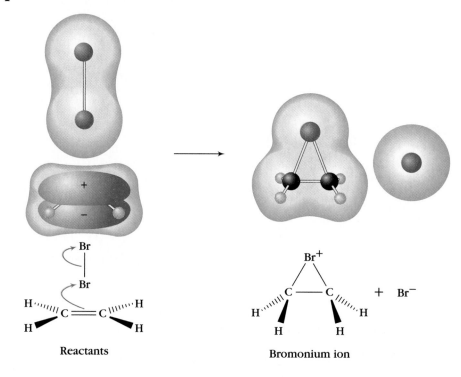

Reactants   Bromonium ion

Step 2: Reaction of bromonium ion with bromide ion to form a vicinal dibromide

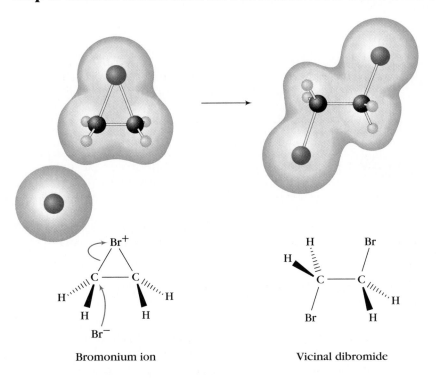

Bromonium ion   Vicinal dibromide

Figure 8.11, cont'd
In the second step, reaction of a bromide ion can occur only from the side opposite the bromine atom of the bromonium ion. As a result, the bromonium ion is opened stereospecifically.

reacts with one end of the electrophilic bromine molecule to displace the other bromine atom as a bromide ion. The product of this reaction is a **bromonium ion intermediate.**

A bromonium ion intermediate contains a three-membered ring composed of a bromine atom bridged to both carbon atoms of the original double bond. The structure of this ion is rigid, and it can undergo reaction with nucleophiles only from the side opposite the bromine atom. As a result, a bromide ion reacts with the bromonium ion in Step 2 to form vicinal dibromide by anti addition. Because the bromide ion can react with either carbon atom of the bromonium ion, an optically inactive product is formed.

■ *The intermediate formed in the addition of chlorine to alkenes is called a chloronium ion.*

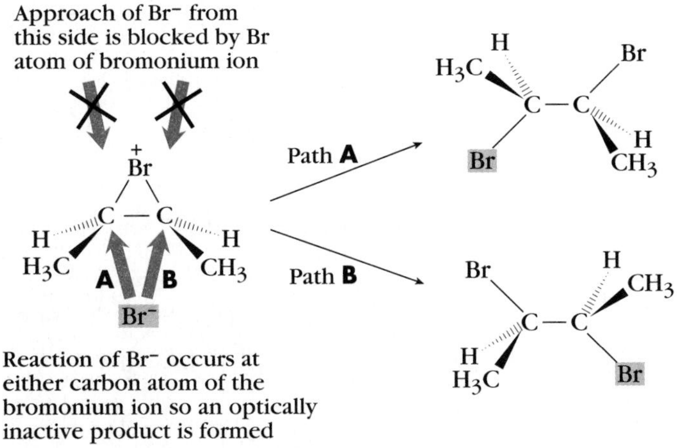

Approach of Br⁻ from
this side is blocked by Br
atom of bromonium ion

Path **A**

Path **B**

Reaction of Br⁻ occurs at
either carbon atom of the
bromonium ion so an optically
inactive product is formed

| **Exercise 8.12** | Write the mechanism for the reaction of chlorine and ethylene. |

| **Exercise 8.13** | Write the mechanism for the reaction of chlorine and (Z)-3-hexene. Is the product *meso* or a racemic mixture? |

| **Exercise 8.14** | Repeat Exercise 8.13 using (E)-3-hexene as the alkene. Is the product *meso* or a racemic mixture? |

Products other than vicinal dihalides can be formed when halogens are added to alkenes in the presence of other nucleophiles or a nucleophilic solvent.

8.9 Halogenation in the Presence of Other Nucleophiles

Additions of chlorine and bromine to alkenes are usually carried out in inert solvents such as carbon tetrachloride or chloroform. Under these conditions, the halide ion is the only nucleophile in the reaction mixture that can react with a

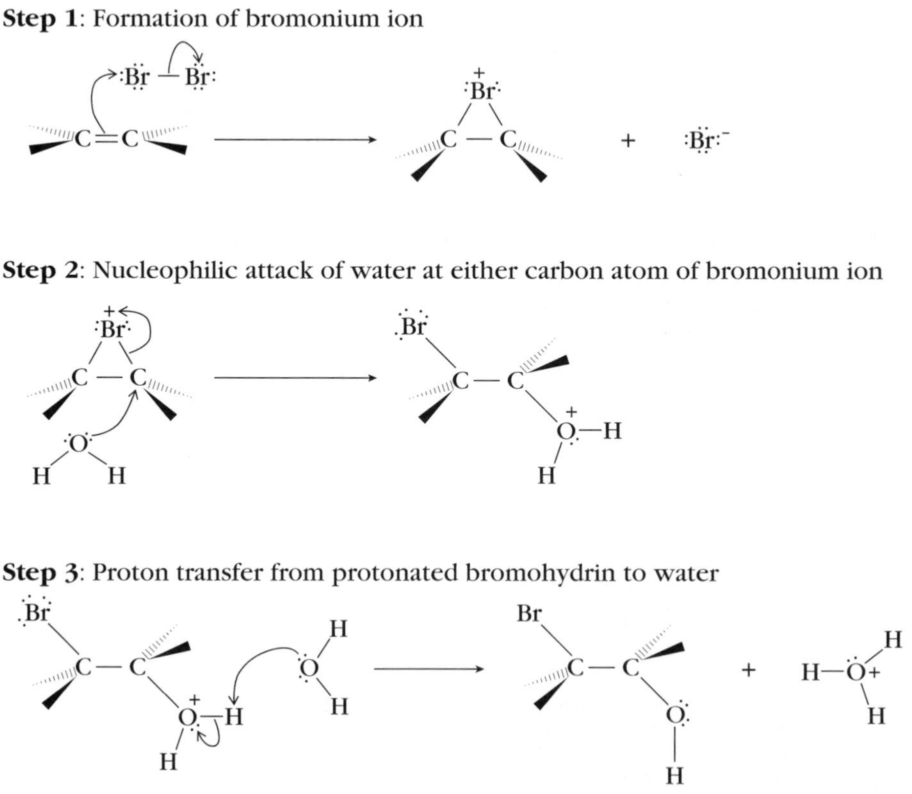

Figure 8.12
Proposed mechanism for the formation of bromohydrins by addition of bromine to an alkene in water.

Step 1: Formation of bromonium ion

Step 2: Nucleophilic attack of water at either carbon atom of bromonium ion

Step 3: Proton transfer from protonated bromohydrin to water bromonium or chloronium ion. As a result, only a vicinal dihalide is formed as product.

When halogenation reactions are carried out in the presence of a nucleophilic solvent such as water, however, the product is a vicinal halo alcohol, commonly called a **halohydrin.** The halohydrin is called a **chlorohydrin** when chlorine is the halogen and a **bromohydrin** when bromine is the halogen atom. The reaction of bromine with cyclopentene in water as solvent, for example, forms racemic *trans*-2-bromocyclopentanol.

Cyclopentene   Racemic *trans*-2-bromocyclopentanol
                                      (a bromohydrin)
                                       (62-67% yield)

The net result of this reaction is the anti addition of BrOH to the double bond.

The proposed mechanism for this reaction is shown in Figure 8.12. Step 1 is the formation of a bromonium ion. Water is the most plentiful nucleophile in the reaction mixture so it reacts with the bromonium ion in Step 2 to form the protonated bromohydrin, which transfers a proton to water in Step 3 to form the product.

If an alcohol is used as a solvent instead of water in the bromination reaction, the product is a vicinal halo ether. An example is the addition of bromine to cyclopentene in methanol as solvent.

Cyclopentene   *trans*-1-Bromo-2-methoxycyclopentane
                                   (74-78% yield)

Halogenation of an unsymmetrical alkene in the presence of other nucleophiles is a regioselective reaction (Section 7.11). For example, only 1-bromo-2-methyl-2-propanol is formed in the reaction of methylpropene with an aqueous solution of bromine. None of the other isomer, 2-bromo-2-methyl-1-propanol, is formed.

Methylpropene   1-Bromo-2-methyl-   2-Bromo-2-methyl-
                                   2-propanol   1-propanol
                                   (77% yield)

In the presence of other nucleophiles, the electrophilic halogen atom is always bonded to the less substituted carbon atom of the original double bond and the nucleophile is bonded to the more highly substituted carbon atom. This orientation is in accord with the Markownikoff Rule of addition of acids to alkenes (Section 7.11). However, the mechanisms of these reactions involve halonium ion intermediates, not intermediate carbocations as in the Markownikoff addition of acids.

Why should reaction with the nucleophile occur preferentially at the more highly substituted carbon atom of an unsymmetrical halonium ion intermediate? To answer this question, we must examine more closely the structure of halonium ions. We will use a bromonium ion as an example.

The bromonium ion formed by addition to a symmetrical alkene can be represented by the localized valence bond orbital model as a resonance hybrid (Section 1.12) of the following three contributing structures:

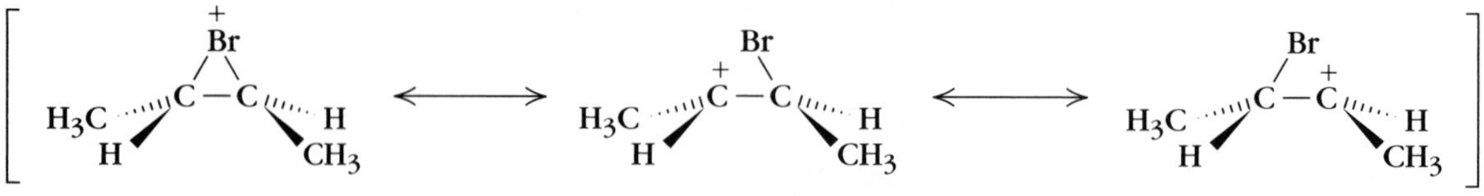

The two acyclic contributing structures are equivalent so they contribute equally to the overall structure of the bromonium ion. As a result, the electron deficiency at both carbon atoms of a symmetrical bromonium ion is the same.

The bromonium ion formed by addition to an unsymmetrical alkene can be represented by the following three contributing structures:

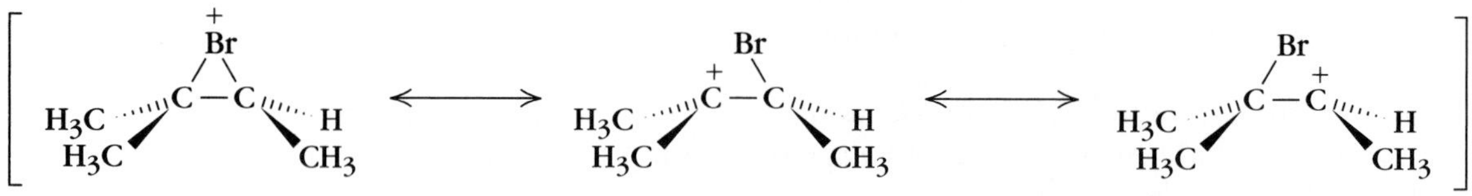

In this case, the two acyclic structures are different. The middle structure is a tertiary carbocation, while the one on the right is a secondary carbocation. The tertiary carbocation is more stable so it contributes more than the other acyclic structure to the resonance hybrid of the bromonium ion. As a result, the C—Br bond to the more substituted carbon atom is the weakest so it is easily broken by the nucleophile.

Exercise 8.15   Write the equation and the structure of the product of addition of an aqueous chlorine solution to each of the following alkenes.

(a) Cyclohexene   **(b)** $(CH_3)_2C{=}CH_2$ **(c)** 1-Methylcyclohexene

Exercise 8.16   Write the mechanism for the addition of chlorine to methylpropene in methanol (CH_3OH) as solvent.

8.10   Addition of Mercury(II) Compounds

The mercury atom of certain mercury(II) compounds is electrophilic because it is positively charged, so these compounds add to the double bond of alkenes. One such compound that contains an electrophilic mercury atom is mercury(II) acetate, a white water-soluble solid.

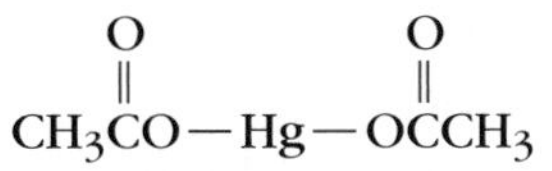

$$CH_3CO-Hg-OCCH_3$$

Mercury(II) acetate

■ Mercury(II) acetate is usually abbreviated as AcO-Hg-OAc or Hg(OAc)$_2$, where AcO- and -OAc are abbreviations for:

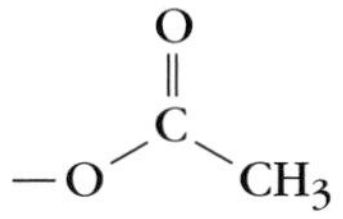

The reaction of mercury(II) acetate with alkenes is called **oxymercuration.** This addition reaction is regiospecific. The product contains a —HgOAc group bonded to the less substituted carbon atom of the double bond and a hydroxy group bonded to the more substituted carbon atom.

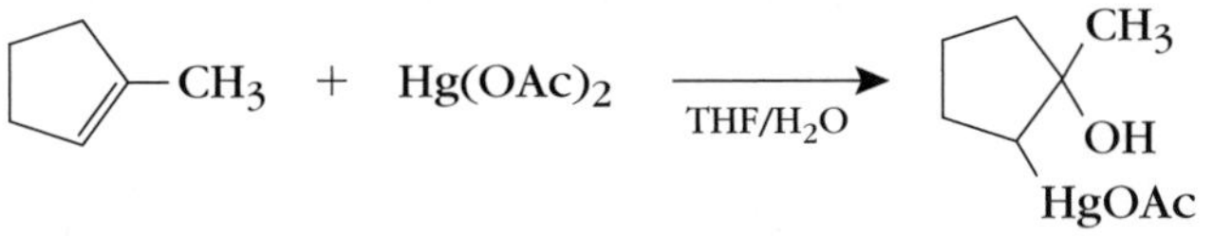

Like organoboranes, the products of oxymercuration are valuable because they can be used to synthesize compounds containing other functional groups. In this case, alcohols are easily prepared by reacting the product of oxymercuration with sodium borohydride (NaBH$_4$) in an aqueous basic solution.

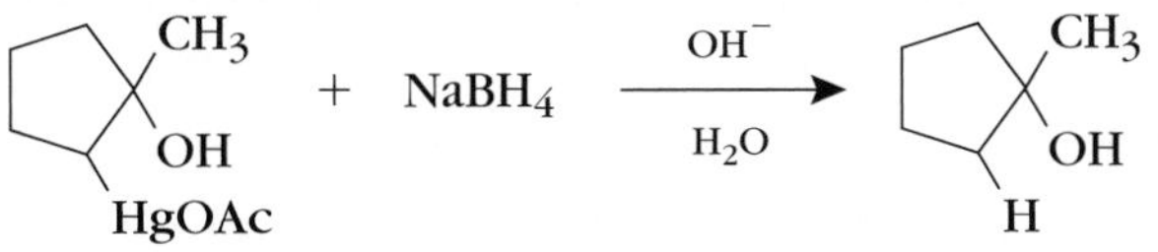

In this reaction, which is called **demercuration,** a hydrogen atom from NaBH$_4$ replaces the mercury atom. This sequence of two reactions, oxymercuration of alkenes followed by demercuration of the product, is an excellent method of preparing alcohols. The important point is that *the net result of the oxymercuration-demercuration sequence of reactions is the Markownikoff addition of water to the double bond of the alkene.*

The stereochemistry of oxymercuration is variable. Oxymercuration of some alkenes such as the 2-butenes occurs by anti addition, while oxymercuration of other alkenes occurs by syn addition or by both syn and anti additions. The relationship between alkene structure and the stereochemistry of oxymercuration is not well understood, so the mechanism of oxymercuration is still the subject of investigation. Despite lack of mechanistic information about the oxymercuration step, oxymercuration-demercuration is still an extremely useful way to prepare alcohols in yields as high as 85% to 95%.

Oxymercuration-demercuration and acid-catalyzed hydration are both methods of forming alcohols by the addition of water to the double bond of an alkene according to the Markownikoff Rule. Thus either method can be used to prepare racemic 2-butanol from 1-butene.

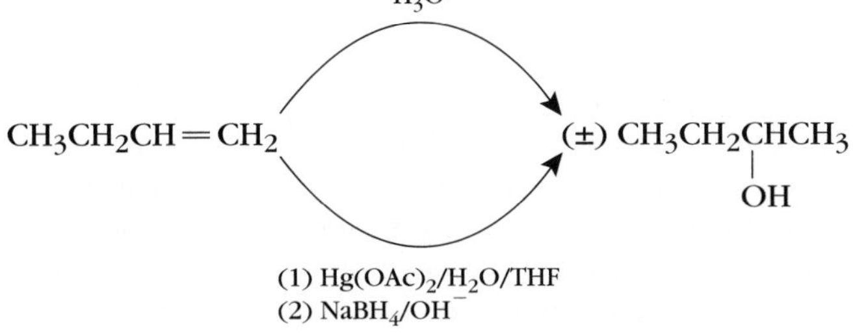

$CH_3CH_2CH=CH_2$ $\xrightarrow{H_3O^+}$ $(\pm)\ CH_3CH_2CHCH_3$ OH

(1) Hg(OAc)$_2$/H$_2$O/THF
(2) NaBH$_4$/OH$^-$

Of the two methods, oxymercuration-demercuration is preferred for several reasons. First, oxymercuration-demercuration is not an equilibrium reaction, so yields of the alcohol tend to be higher than in acid-catalyzed hydration. Second, rearrangement products so easily formed in acid-catalyzed hydration are not formed by oxymercuration-demercuration.

Exercise 8.17 Write the equations for the reaction of each of the following alkenes with first mercury(II) acetate in water/THF and then NaBH$_4$ in aqueous basic solution.

(a) 1-Hexene   (b) 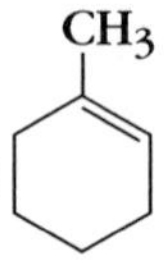 [structure with CH$_3$]   (c) 2-Methyl-1-pentene

---

Exercise 8.18 Write the equation that shows how to prepare each of the following alcohols by oxymercuration-demercuration of the appropriate alkene.

(a) 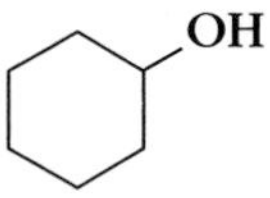 [cyclohexanol structure with OH]   (b) (C$_2$H$_5$)$_3$COH   (c) CH$_3$CH$_2$CH$_2$CHCH$_3$ with OH

Exercise 8.19 Two alcohols are formed by the acid-catalyzed hydration of 3,3-dimethyl-1-butene. One alcohol has the same structure as the alcohol formed by oxymercuration-demercuration of the same alkene, while the other alcohol has a different structure. What is the structure of each alcohol and why are they different?

8.11 Catalytic Hydrogenation of Alkenes

Hydrogen gas reacts with an alkene in the presence of a suitable catalyst to form an alkane.

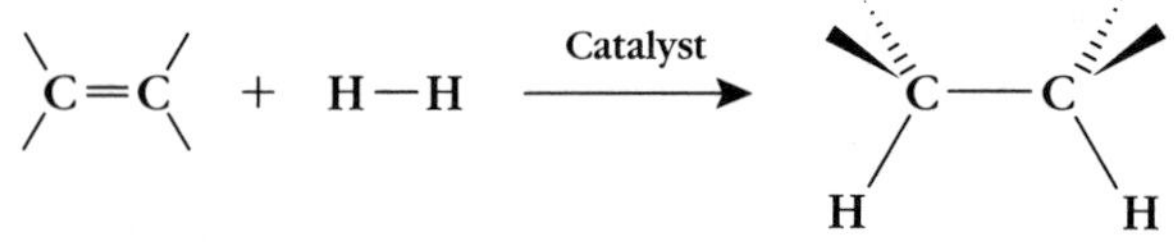

$$\overset{\displaystyle}{C=C} + H-H \xrightarrow{\text{Catalyst}} \overset{\displaystyle}{C-C}$$

Alkene   Alkane

This reaction is called the **catalytic hydrogenation** of alkenes. The reaction of hydrogen gas and ethylene is an exergonic process, as we learned in Section 7.7, but the uncatalyzed reaction occurs extremely slowly. The rate of hydrogenation is greatly increased by carrying out the reaction in the presence of certain metals. Platinum, added to the reaction mixture as PtO$_2$, is a hydrogenation catalyst often used, but metals such as palladium, nickel, and rhodium are also widely used. It has been found experimentally that best results are obtained when the metal is finely divided (to increase its surface area) and absorbed on an inert material such as charcoal.

The solvents used in hydrogenation reactions are typically ethanol, hexane, or acetic acid. These solvents are chosen for their ability to dissolve the alkene. The metal catalysts are not soluble in these solvents so the hydrogenation is a heterogeneous reaction. That is, the reaction occurs at the surface of the metal and not in solution. This reaction is different from the other additions to alkenes that we've studied because catalytic hydrogenation is not an electrophilic addition reaction of alkenes.

Experimental results show that the catalytic hydrogenation of alkenes occurs by syn addition. For example, catalytic hydrogenation of 1-ethyl-2-methylcyclohexene forms *cis*-1-ethyl-2-methylcyclohexane.

1-Ethyl-2-methylcyclohexene

cis-1-Ethyl-2-methylcyclohexane
(racemic mixture)
(73-80% yield)

The detailed mechanism of catalytic hydrogenation of alkenes is complex because it involves formation of bonds between atoms of the catalyst and hydrogen atoms and the carbon atoms of the double bond. The major features of the mechanism, suitable for our purposes, are given in the simple schematic representation shown in Figure 8.13.

While hydrogenation of alkenes usually forms a racemic mixture or an achiral product, hydrogenation of chiral alkenes may form only a single diastereomer if

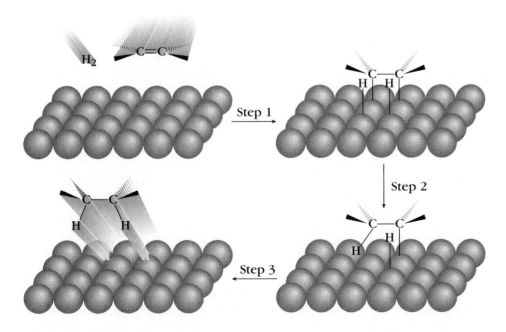

Step 1

Step 2

Step 3

Figure 8.13
Schematic representation of the syn addition of hydrogen in the catalytic hydrogenation of alkenes. In Step 1, both molecular hydrogen and the alkene form bonds with metal atoms of the catalyst. In Step 2, one carbon-hydrogen bond is formed. Formation of the second carbon-hydrogen bond and diffusion of the saturated product from the catalyst surface occur in Step 3.

addition occurs preferentially to one face of the double bond. The Pt-catalyzed hydrogenation of α-pinene, for example, forms only a single stereoisomer.

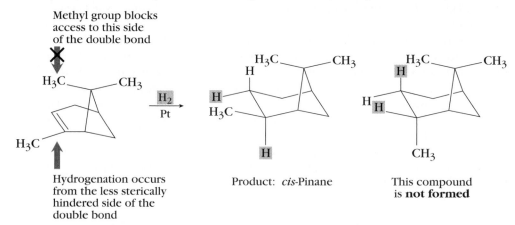

One of the methyl groups bonded to the four-membered ring of α-pinene is located above the double bond so it sterically blocks hydrogenation from that side. As a result hydrogenation occurs exclusively from the other, less sterically hindered, side of the double bond to form only *cis*-pinane.

Exercise 8.20   Write the structure of the product of catalytic hydrogenation of each of the following alkenes. Indicate the stereochemistry where appropriate.

(a) CH$_3$CH$_2$CH=C(CH$_3$)$_2$ **(c)** 3,3-Dimethyl-1-butene

(b) 3-Methylcyclopentene   **(d)**

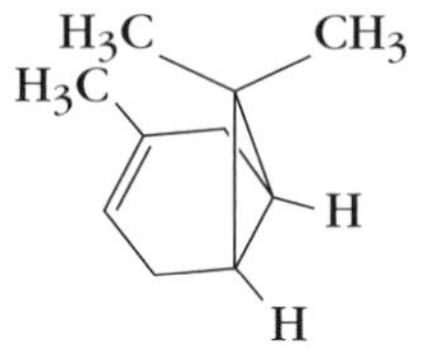

Exercise 8.21   Catalytic hydrogenation of (*Z*)-4,5-dimethyl-4-octene produces a single *meso* compound, whereas catalytic hydrogenation of its (*E*)-isomer forms a racemic mixture. Explain this difference in product composition and write the structures and IUPAC names of each product of these two reactions.

Hydrogenation of alkenes is not only a convenient way to prepare alkanes, but it can also be used to determine relative stabilities of unsaturated compounds.

8.12   Heats of Hydrogenation

The catalytic hydrogenation of alkenes is an exothermic as well as an exergonic reaction. The enthalpy change ($\Delta H°$) for this reaction is called the **heat of hydrogenation** of an alkene. Heats of hydrogenation can be used to determine the relative stabilities of isomeric alkenes. Thus heats of hydrogenation complement the use of heats of combustion (Section 7.5) in determining the relative stabilities of isomeric organic compounds.

The idea is to compare the heats of hydrogenation of isomeric alkenes *that form the same alkane*. Then the difference in their heats of hydrogenation is a measure of their difference in stability. This method can be illustrated using three isomeric butenes.

■ *Recall from Chapter 3 that in an exergonic reaction ΔG < 0, while ΔH < 0 in an exothermic reaction.*

$$\begin{array}{c} CH_3 \quad CH_3 \\ \diagdown \quad \diagup \\ C=C \\ \diagup \quad \diagdown \\ H \quad H \end{array} \quad + \quad H_2 \quad \xrightarrow[C_2H_5OH]{Pt} \quad CH_3CH_2CH_2CH_3 \quad \Delta H^\circ = -28.6 \text{ kcal/mol} \ (-119.7 \text{ kJ/mol})$$

(Z)-2-Butene

$$\begin{array}{c} H \quad CH_3 \\ \diagdown \quad \diagup \\ C=C \\ \diagup \quad \diagdown \\ CH_3 \quad H \end{array} \quad + \quad H_2 \quad \xrightarrow[C_2H_5OH]{Pt} \quad CH_3CH_2CH_2CH_3 \quad \Delta H^\circ = -27.6 \text{ kcal/mol} \ (-115.5 \text{ kJ/mol})$$

(E)-2-Butene

$$CH_3CH_2CH=CH_2 \quad + \quad H_2 \quad \xrightarrow[C_2H_5OH]{Pt} \quad CH_3CH_2CH_2CH_3 \quad \Delta H^\circ = -30.3 \text{ kcal/mol} \ (-126.8 \text{ kJ/mol})$$

1-Butene

In each reaction, one mole of hydrogen is consumed per mole of alkene and the same product is formed. Nevertheless, ΔH° is different in each reaction. The differences in ΔH° are a measure of the relative stabilities of the three alkenes. Thus (Z)-2-butene is more stable than 1-butene by 1.7 kcal/mol (7.1 kJ/mol), while (E)-2-butene is more stable than (Z)-2-butene by 1.0 kcal/mol (4.2 kJ/mol). The relationship between the heats of hydrogenation of the isomeric butenes and their relative stabilities is shown in Figure 8.14. Notice that the more stable the alkene, the smaller its heat of hydrogenation. The order of stability

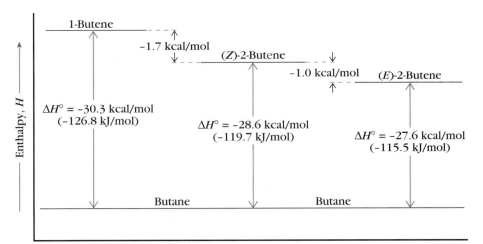

Figure 8.14
Relative stabilities of 1-butene, (Z)-2-butene, and (E)-2-butene, as measured by their heats of hydrogenation.

based on these heats of hydrogenation is the same as that determined from the heats of combustion (Section 7.5).

$$1\text{-Butene} < (Z)\text{-}2\text{-Butene} < (E)\text{-}2\text{-Butene}$$

INCREASING STABILITY

Comparisons of the heats of hydrogenation of other alkenes show that their order of relative stabilities is exactly the same as that determined by heats of combustion.

$$CH_2{=}CH_2 \ < \ RCH{=}CH_2 \ < \ R_2C{=}CH_2 \ < \ \underset{H}{\overset{R}{}}C{=}C\underset{H}{\overset{R}{}} \ < \ \underset{H}{\overset{R}{}}C{=}C\underset{R}{\overset{H}{}} \ < \ RCH{=}CR_2 \ < \ R_2C{=}CR_2$$

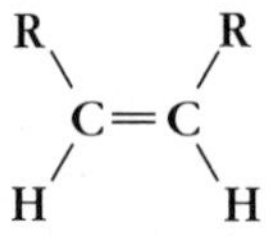

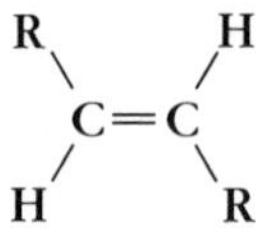

INCREASING RELATIVE STABILITY

Exercise 8.22   Catalytic hydrogenation of the five alkenes A-E forms the same alkane. Their heats of hydrogenation are as follows:

Alkene	Heat of Hydrogenation (kcal/mol)	(kJ/mol)
A	−30.8	−128.9
B	−29.5	−123.4
C	−26.6	−111.3
D	−27.0	−112.9
E	−32.8	−137.2

List the alkenes in order of increasing stability.

Exercise 8.23   Consider the following isomeric alkenes:

(1) 2-Methyl-2-butene   (2) 3-Methyl-1-butene   (3) 2-Methyl-1-butene

(a) Write the equation for the catalytic hydrogenation of each alkene.
(b) Arrange the alkenes in order of increasing relative stability.

8.13 Summary of Alkene Addition Reactions

1. Addition of Hydrogen Halides   SECTION 7.8

Racemic products with Markownikoff regiochemistry are formed; H always adds to less
substituted carbon atom of double bond.
Proposed mechanism involves a carbocation
intermediate.

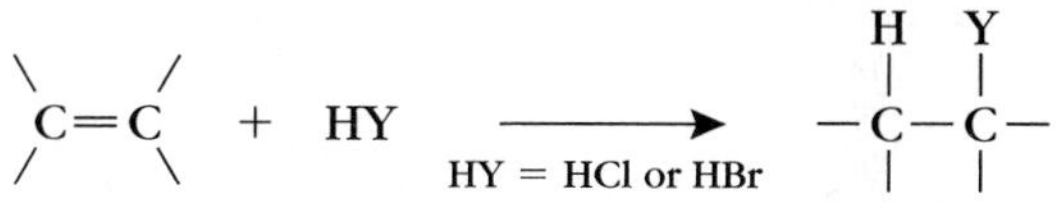

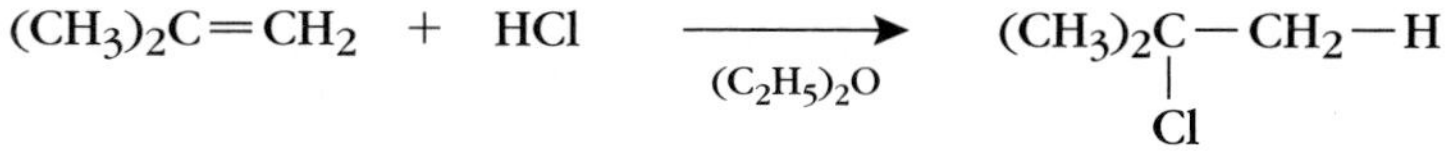

$$\text{C}=\text{C} \quad + \quad \text{HY} \xrightarrow[\text{HY = HCl or HBr}]{} -\overset{|}{\underset{|}{\text{C}}}-\overset{|}{\underset{|}{\text{C}}}- \quad \begin{array}{c}\text{H} \quad \text{Y}\end{array}$$

$$(CH_3)_2C=CH_2 \quad + \quad HCl \xrightarrow[(C_2H_5)_2O]{} (CH_3)_2C-CH_2-H \\ \qquad\qquad\qquad\qquad\qquad\qquad\qquad\qquad\qquad \overset{|}{Cl}$$

2. Hydroxylation   SECTION 8.5

Products formed by syn addition. Proposed mechanism
involves a cyclic osmate ester intermediate.

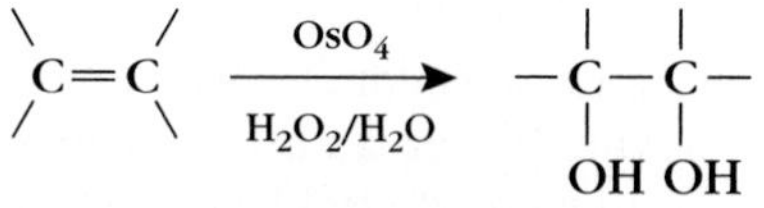

$$\text{C}=\text{C} \xrightarrow[\text{H}_2\text{O}_2/\text{H}_2\text{O}]{\text{OsO}_4} -\overset{|}{\underset{|}{\text{C}}}-\overset{|}{\underset{|}{\text{C}}}- \\ \qquad\qquad\qquad\qquad \text{OH} \quad \text{OH}$$

Vicinal glycol

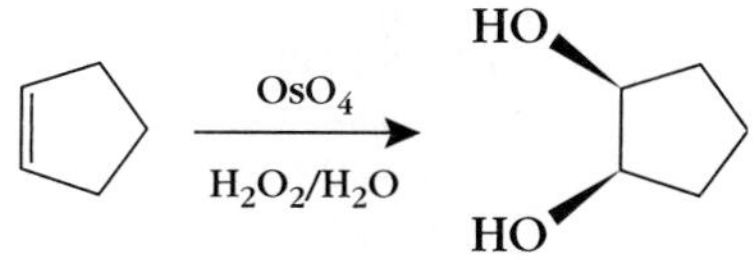

3. Epoxidation   SECTION 8.6

Syn addition of oxygen atom to the double bond by
means of a concerted mechanism.

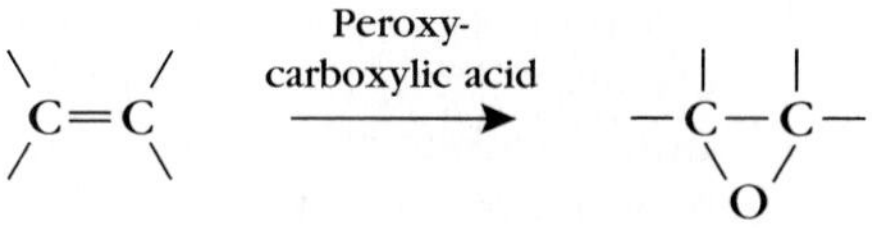

$$\text{C}=\text{C} \xrightarrow{\text{Peroxy-carboxylic acid}} -\overset{|}{\text{C}}-\overset{|}{\text{C}}- \\ \qquad\qquad\qquad\qquad \underset{O}{\diagdown\diagup}$$

Epoxide

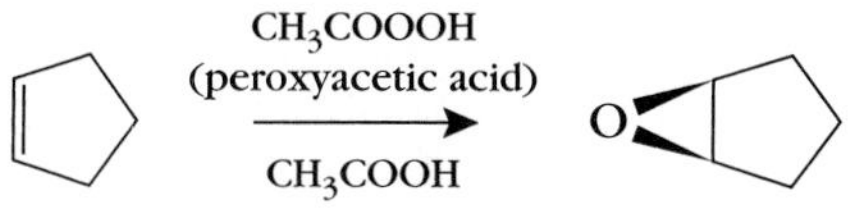

4. Additions of Carbenes   SECTION 8.7

Product is formed by syn addition of intermediate
dichlorocarbene to the double bond. Dichlorocarbene
is formed by base-promoted elimination of HCl from
chloroform.

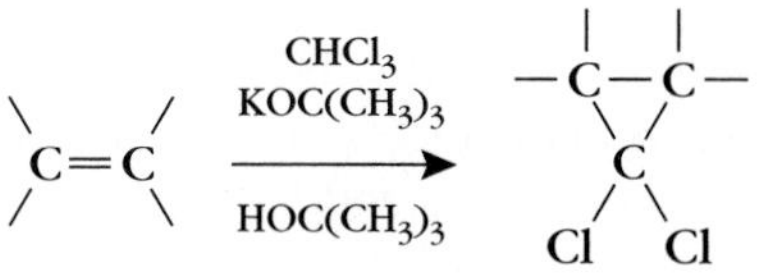

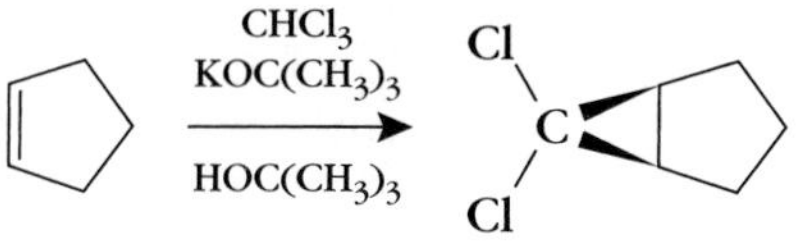

$$\text{C}=\text{C} \xrightarrow[\text{HOC(CH}_3)_3]{\overset{\text{CHCl}_3}{\text{KOC(CH}_3)_3}} \begin{array}{c} -\overset{|}{\text{C}}-\overset{|}{\text{C}}- \\ \diagup\text{C}\diagdown \\ \text{Cl} \quad \text{Cl} \end{array}$$

5. **Additions of Carbenoids**   SECTION 8.7

Reaction believed to involve intermediate formation of
ICH_2ZnI, which reacts like a carbene toward alkenes.

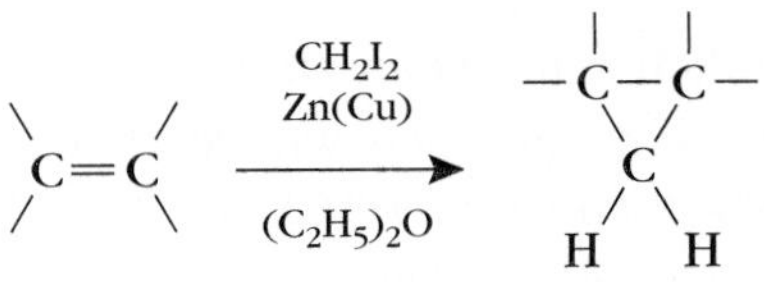

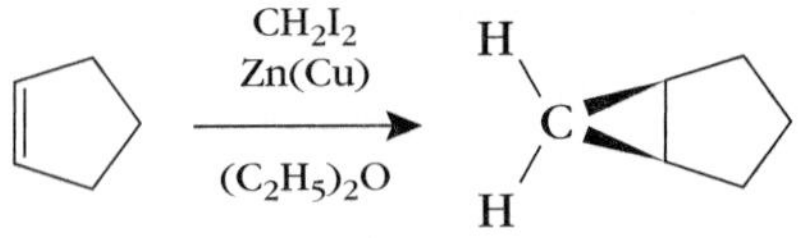

6. **Halogenation**   SECTION 8.8

Halogenation of aliphatic alkenes and cycloalkenes
occurs by anti addition. Proposed mechanism involves
a three-membered halonium ion.

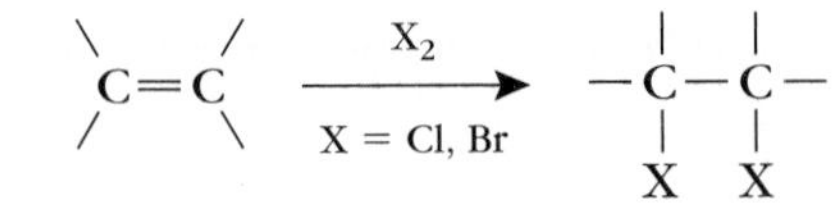

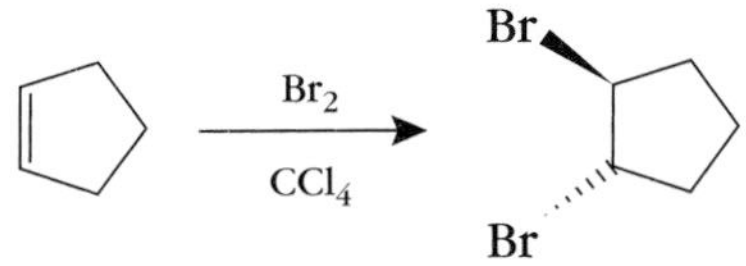

7. **Halohydrin Formation**   SECTION 8.9

Product is formed by anti addition and according to the
Markownikoff Rule; the —OH group bonds to the more
substituted carbon atom of the double bond. Proposed
mechanism involves a halonium ion intermediate.

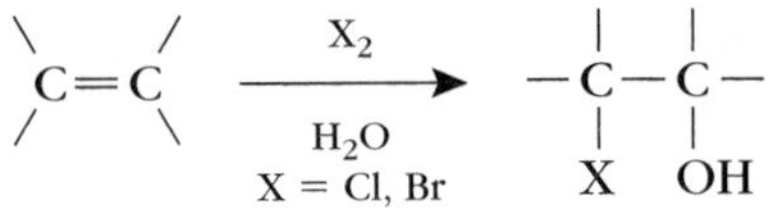

Halohydrin

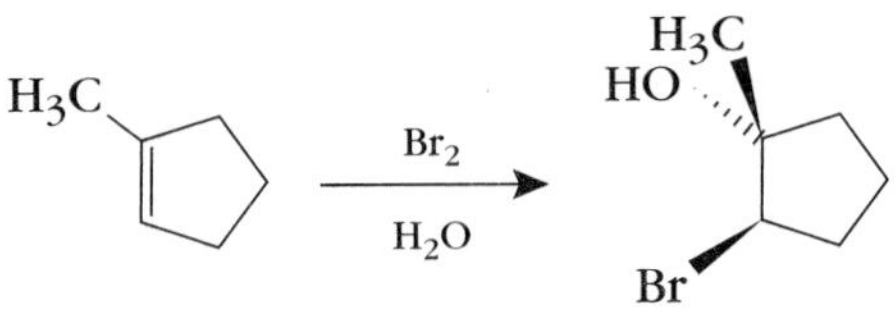

8. **Catalytic Hydrogenation**   SECTION 8.11

Reaction occurs on the surface of the catalyst to form
products of syn addition.

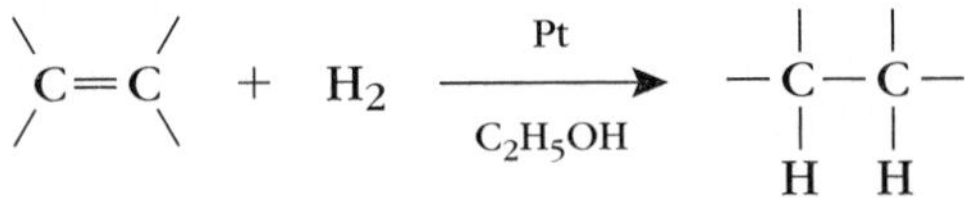

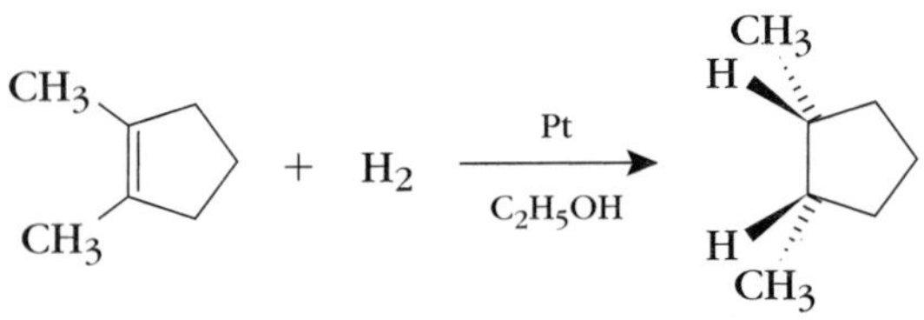

9. **Hydration**

   a. **Acid-catalyzed hydration**   SECTION 7.14

Racemic mixture or achiral products formed by Markownikoff addition. Proposed mechanism involves carbocation intermediate.

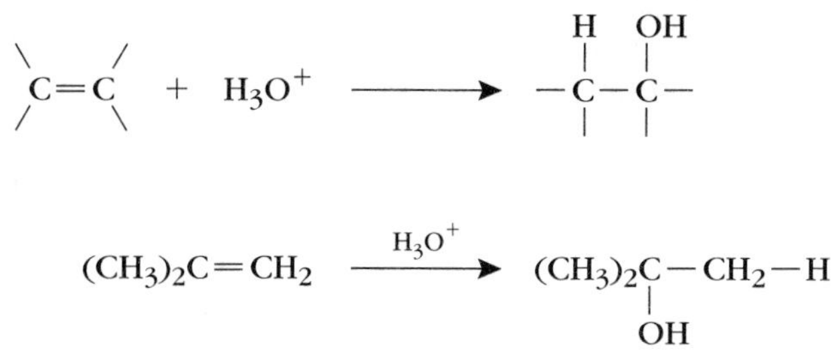

$$\ce{\underset{/}{\overset{\backslash}{C}}=\underset{\backslash}{\overset{/}{C}} + H3O+ ->} \quad \ce{-\overset{\overset{\displaystyle H}{|}}{C}-\overset{\overset{\displaystyle OH}{|}}{C}-}$$

$$\ce{(CH3)2C=CH2 ->[H3O+] (CH3)2\underset{\underset{\displaystyle OH}{|}}{C}-CH2-H}$$

   b. **Hydroboration-oxidation**   SECTIONS 8.1–8.4

Reaction occurs by first forming an organoborane by syn addition of borane. Proposed mechanism involves a four-centered transition state. Oxidation of organoborane occurs so that the carbon atom originally bonded to boron maintains its configuration in the alcohol product. The net result of this sequence of two reactions is the syn and anti Markownikoff addition of water to an alkene.

$$\ce{\underset{/}{\overset{\backslash}{C}}=\underset{\backslash}{\overset{/}{C}} ->[1) BH3][2) H2O2/OH^-]} \quad \ce{-\overset{|}{\underset{\underset{\displaystyle H}{|}}{C}}-\overset{|}{\underset{\underset{\displaystyle OH}{|}}{C}}-}$$

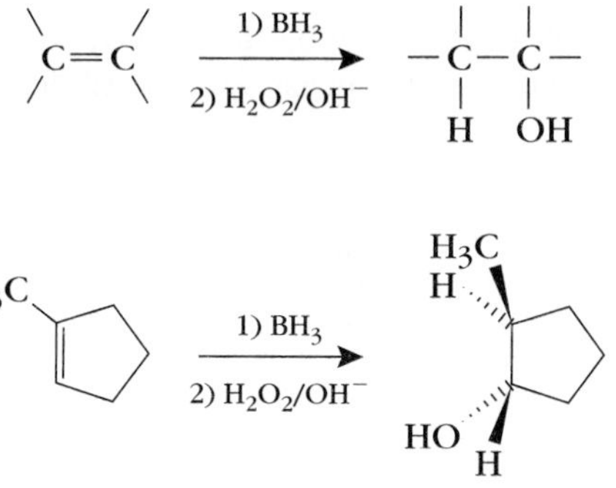

   c. **Oxymercuration-demercuration**   SECTION 8.10

Reaction occurs by first forming an organomercury compound, which is reduced by NaBH₄ to form an alcohol. The net result of this sequence of two reactions is the Markownikoff addition of water to an alkene.

$$\ce{\underset{/}{\overset{\backslash}{C}}=\underset{\backslash}{\overset{/}{C}} ->[1) Hg(OAc)2][2) NaBH4/OH^-]} \quad \ce{-\overset{\overset{\displaystyle H}{|}}{C}-\overset{\overset{\displaystyle OH}{|}}{C}-}$$

$$\ce{(CH3)2C=CH2 ->[1) Hg(OAc)2][2) NaBH4/OH^-] (CH3)2\underset{\underset{\displaystyle OH}{|}}{C}-CH2-H}$$

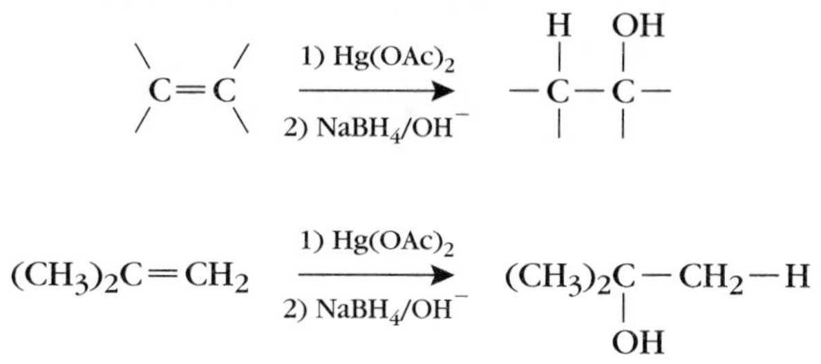

   d. **Fumarase-catalyzed hydration**   SECTION 7.16

Reaction forms an optically active product. The proposed mechanism involves the formation of an enzyme-substrate complex in which the hydroxy group is added to only one side of the double bond to form a single enantiomer.

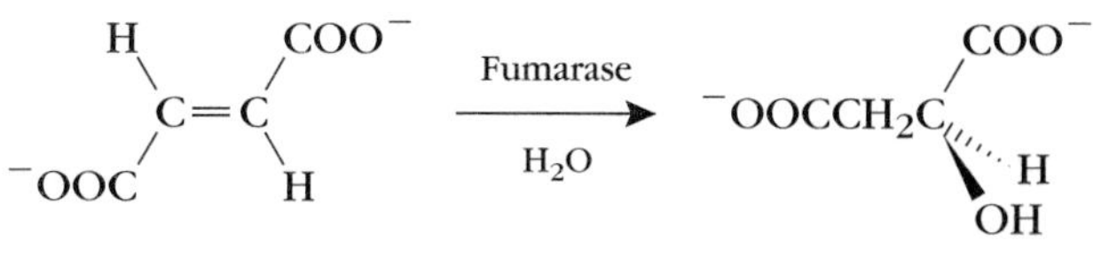

8.14 Summary of Organoborane Reactions

1. Protonolysis   SECTION 8.4

Alkanes are formed by replacing the boron atom of an organoborane by a hydrogen atom.

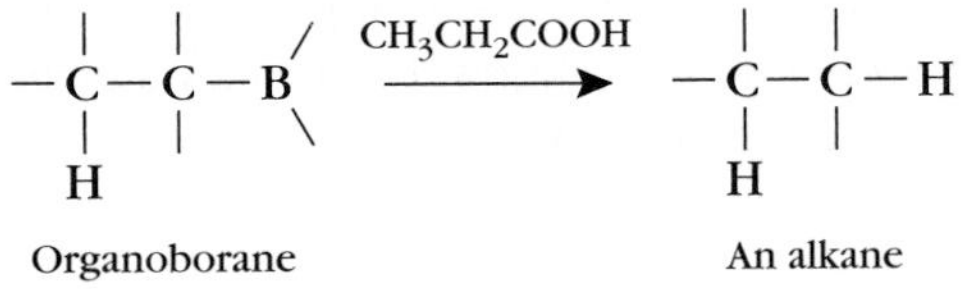

$$-\overset{|}{\underset{|}{C}}-\overset{|}{\underset{H}{C}}-B\diagup \xrightarrow{CH_3CH_2COOH} -\overset{|}{\underset{|}{C}}-\overset{|}{\underset{H}{C}}-H$$

Organoborane   An alkane

$$[CH_3(CH_2)_4CH_2]_3B \xrightarrow{CH_3CH_2COOH} 3CH_3(CH_2)_4CH_3$$

2. Bromination   SECTION 8.4

Bromoalkanes are formed by replacing the boron atom of the organoborane with a bromine atom.

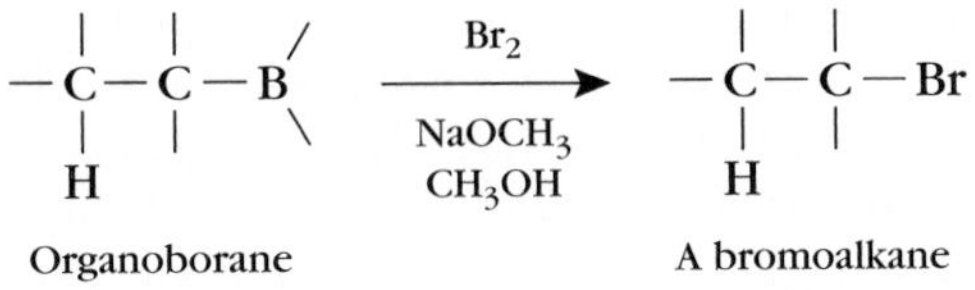

$$-\overset{|}{\underset{|}{C}}-\overset{|}{\underset{H}{C}}-B\diagup \xrightarrow[\substack{NaOCH_3 \\ CH_3OH}]{Br_2} -\overset{|}{\underset{|}{C}}-\overset{|}{\underset{H}{C}}-Br$$

Organoborane   A bromoalkane

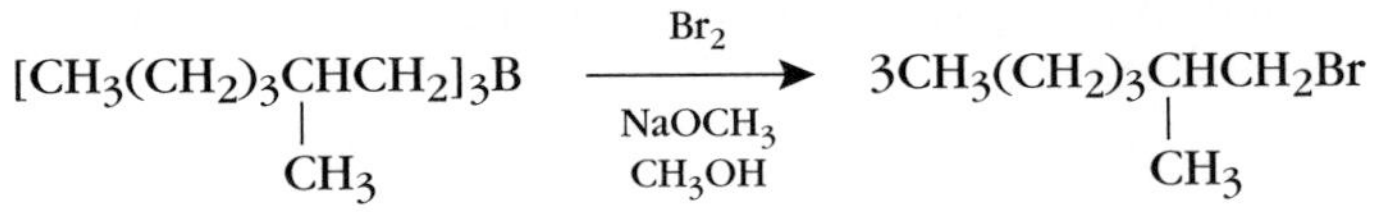

$$[CH_3(CH_2)_3\underset{\underset{CH_3}{|}}{C}HCH_2]_3B \xrightarrow[\substack{NaOCH_3 \\ CH_3OH}]{Br_2} 3CH_3(CH_2)_3\underset{\underset{CH_3}{|}}{C}HCH_2Br$$

3. Oxidation   SECTION 8.4

Oxidation of organoboranes by hydrogen peroxide forms alcohols. The stereochemistry of the carbon atom bonded to the boron atom in the organoborane is retained in the alcohol.

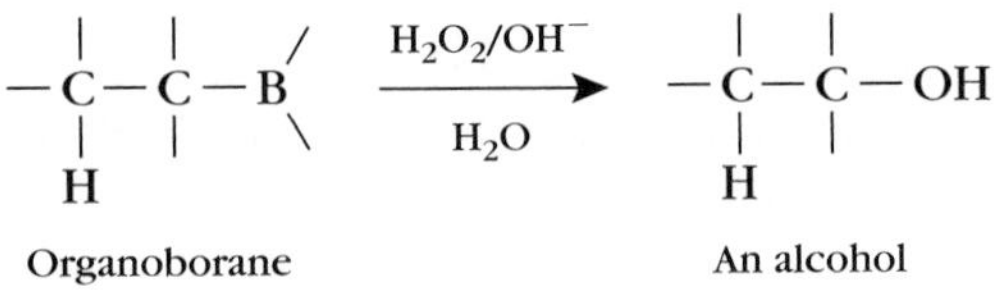

$$-\overset{|}{\underset{|}{C}}-\overset{|}{\underset{H}{C}}-B\diagup \xrightarrow[\substack{H_2O}]{H_2O_2/OH^-} -\overset{|}{\underset{|}{C}}-\overset{|}{\underset{H}{C}}-OH$$

Organoborane   An alcohol

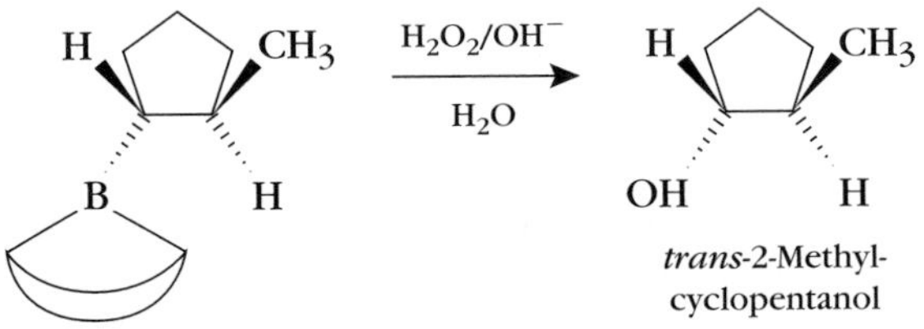

trans-2-Methyl-cyclopentanol

Additional Exercises

8.24   Define each of the following terms either in words, by a structure, or by means of a chemical equation.

(a) Anti addition
(b) Acid-catalyzed hydration
(c) Hydroboration
(d) Carbene addition
(e) Hydroxylation
(f) Stereospecific addition
(g) Syn addition

(h) Epoxidation
(i) Halogenation
(j) Carbenoid
(k) Catalytic hydrogenation
(l) Oxymercuration-demercuration
(m) Halohydrin

(n) Vicinal glycol
(o) Hydroboration-oxidation
(p) Heats of hydrogenation
(q) Hydroboration-protonolysis
(r) Organoborane
(s) Hydroboration-bromination

8.25 Write the structure of the product of the reaction of each of the following reagents with 4-isopropylmethylenecyclohexane.

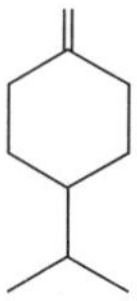

4-Isopropylmethylenecyclohexane

(a) Solution of 9-BBN in THF followed by reaction with an aqueous alkaline solution of hydrogen peroxide
(b) Solution of Br_2 in CCl_4
(c) Aqueous solution of Br_2
(d) Hydrogen gas dissolved in hexane in the presence of finely divided Pd metal
(e) THF solution of mercury(II) acetate followed by reaction with an alkaline solution of $NaBH_4$
(f) Diethyl ether solution of diiodomethane in the presence of solid activated zinc dust
(g) Solution of 9-BBN in THF followed by reaction with a solution of Br_2 and $NaOCH_3$ in methanol
(h) Solution of H_2O_2 in THF containing a catalytic amount of OsO_4

8.26 Write the equations of the reactions that form each of the following compounds from propene.

(a) Propane

(b) 2-Bromopropane

(c) 1,2-Dichloropropane

(d) 1-Bromopropane

(e) 1-Chloro-2-propanol, $ClCH_2\overset{\displaystyle OH}{\underset{\displaystyle |}{C}}HCH_3$

(f) 2-Propanol, $CH_3\overset{\displaystyle OH}{\underset{\displaystyle |}{C}}HCH_3$

(g) 1,2-Propandiol, $HOCH_2\overset{\displaystyle OH}{\underset{\displaystyle |}{C}}HCH_3$

(h) 2-Methyloxirane

8.27 Indicate the reagents needed to convert 1-ethylcyclohexene into each of the following compounds.

(a)

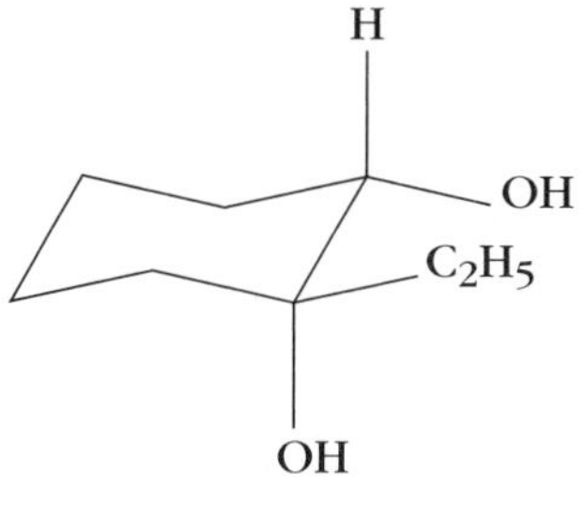

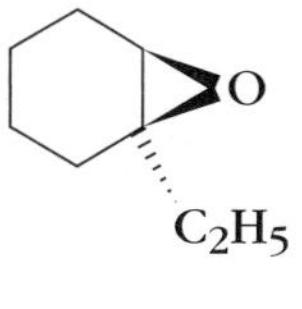

(b)

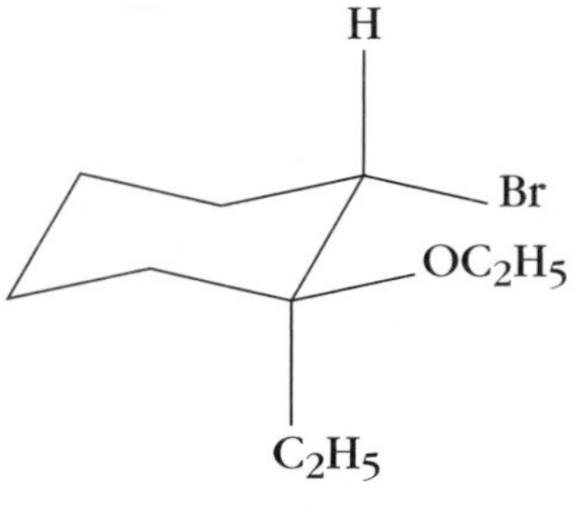

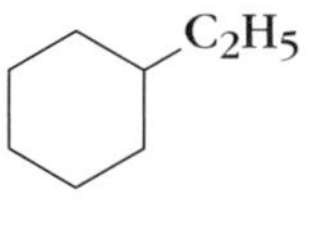

(c)

(d)

(e)

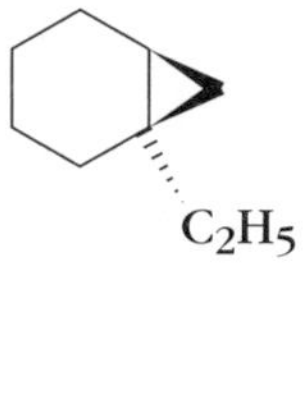

(f)

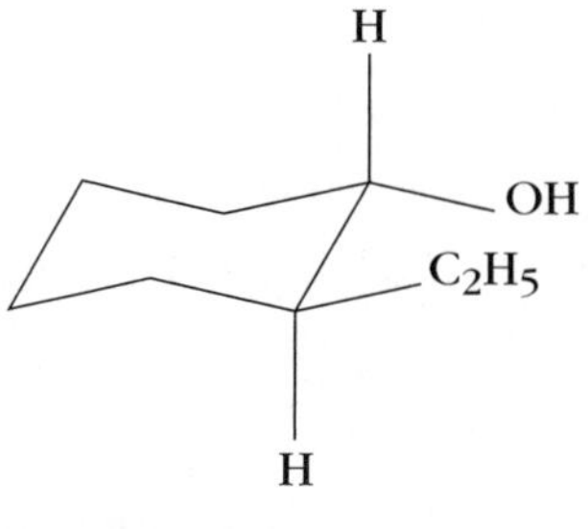

8.28   Write the structure of the product of each of the following reactions. Indicate stereochemistry where appropriate.

(a)

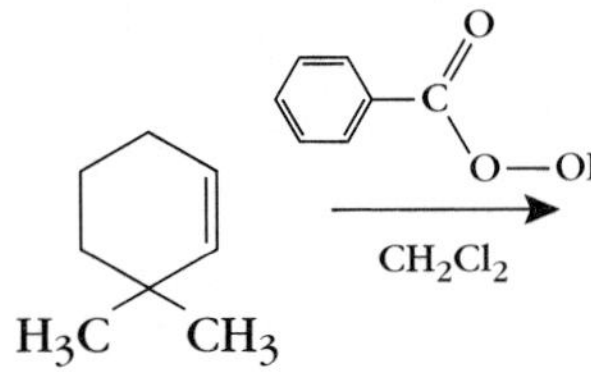

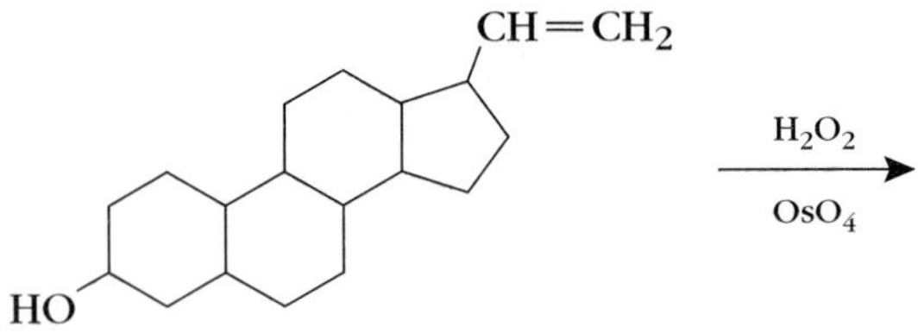

(b)

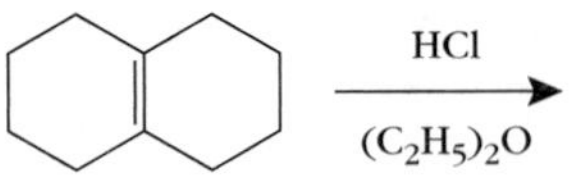

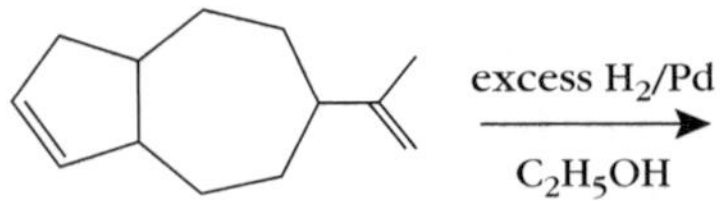

(c)

(d)

(e)

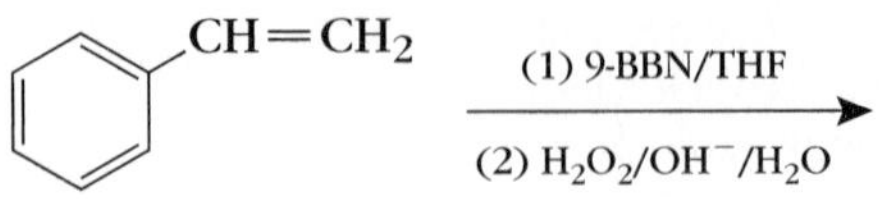

(f)   *trans*-Cyclooctene

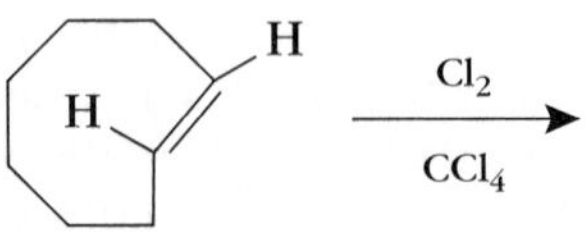

8.29   Write the structure of an alkene that forms the same product when reacted with $Hg(OAc)_2$ followed by $NaBH_4$, or with BH_3 followed by H_2O_2.

8.30   How would you use ^{13}C NMR spectroscopy to distinguish between the following two isomeric alkenes?

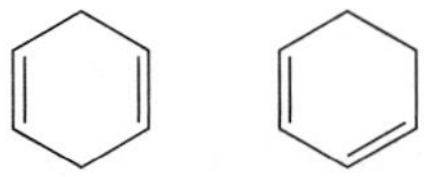

8.31   The reaction of 4-methylcyclopentene with peroxyacetic acid in acetic acid forms two isomeric epoxides. Write the structure of each epoxide and explain how they are formed.

8.32   Complete each of the following reactions by replacing the question mark with the correct reactant, product, or reagents.

(a) $(CH_3)_2C\!=\!CHCH_3$ $\xrightarrow{\ ?\ }$ $(CH_3)_2CH\!-\!\underset{\underset{OH}{|}}{C}HCH_3$

(b)   ? $\xrightarrow{\ ?\ }$

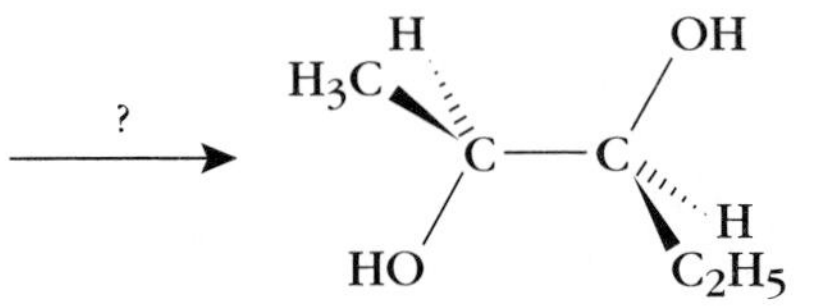

(c)

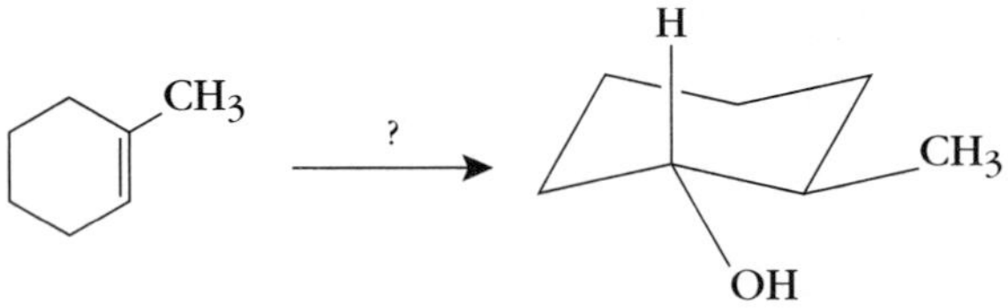

(d)

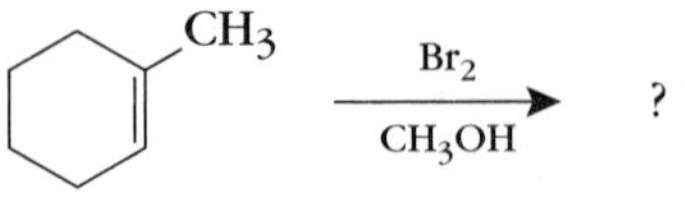

(e)   ? $\xrightarrow{\ ?\ }$ *meso*-2,3-dichlorobutane

(f)   ? $\xrightarrow{\ ?\ }$

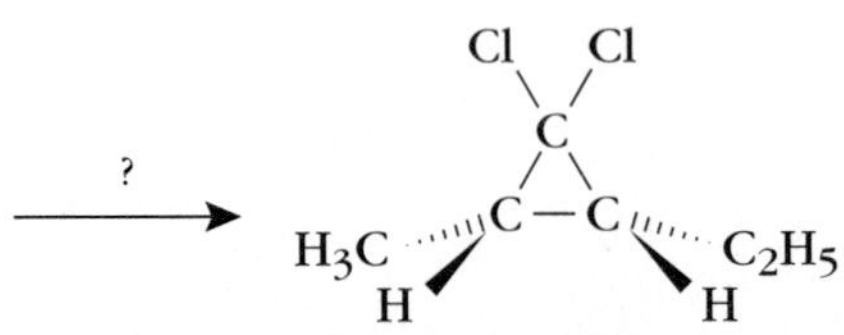

8.33 An alkene (A) with molecular formula C_5H_{10} reacts with a THF solution of 9-BBN to form an organoborane. Reaction of the organoborane with an alkaline solution of H_2O_2 forms compound B, an alcohol. The same alkene undergoes oxymercuration-demercuration to form alcohol C, which is isomeric with alcohol B. Alcohol C can also be prepared by acid-catalyzed hydration of alkene A. Catalytic hydrogenation of alkene A forms an alkane, compound D, whose ^{13}C NMR spectrum is given below. From these data, determine the structures of compounds A, B, C, and D.

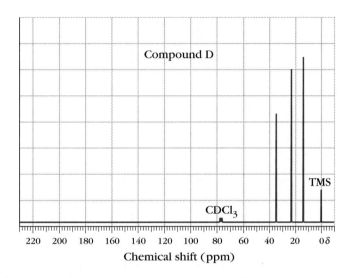

8.34 In acetic acid (CH_3COOH), *cis*-2-pentene reacts with chlorine to form racemic mixtures of the following compounds. Based on your knowledge of the mechanism of halohydrin formation, propose a mechanism that accounts for the formation of these products.

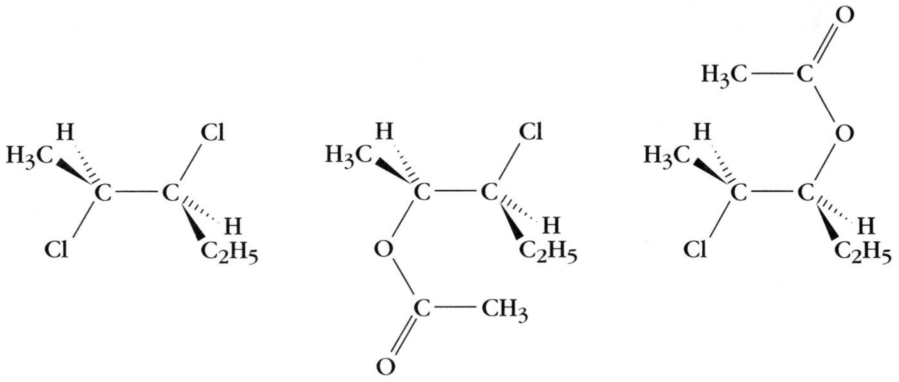

8.35 Catalytic hydrogenation of (*R*)-2,3-dimethyl-1-hexene forms a single optically active compound. Write its structure and explain why the product is optically active. What is the configuration of the stereocenter of the product?

8.36 When deuterium gas (D_2) is used instead of hydrogen gas in catalytic hydrogenation, two deuterium atoms are added to the double bond of the alkene. Write the structure of the major product expected by the reaction of D_2/PtO_2 with each of the following compounds. Indicate the stereochemistry of the product where appropriate.

(a) Cyclohexene
(b) *cis*-3-Hexene
(c) *trans*-3-Hexene (d) 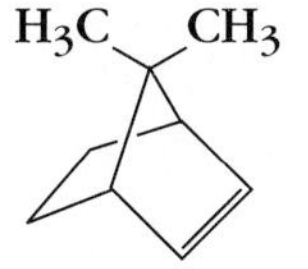
(e)

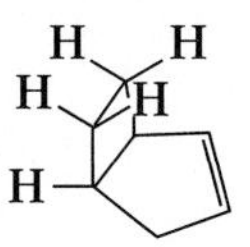

8.37 Based on your knowledge of the mechanism of additions of hydrogen halides to alkenes, propose a mechanism for the following reaction:

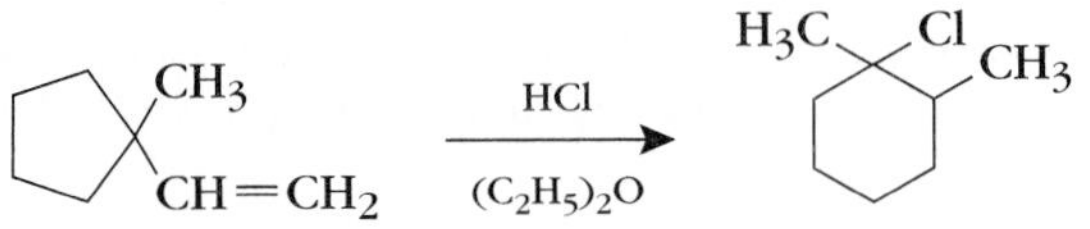

8.38 BrCl adds to fumaric acid and maleic acid to give the following products.

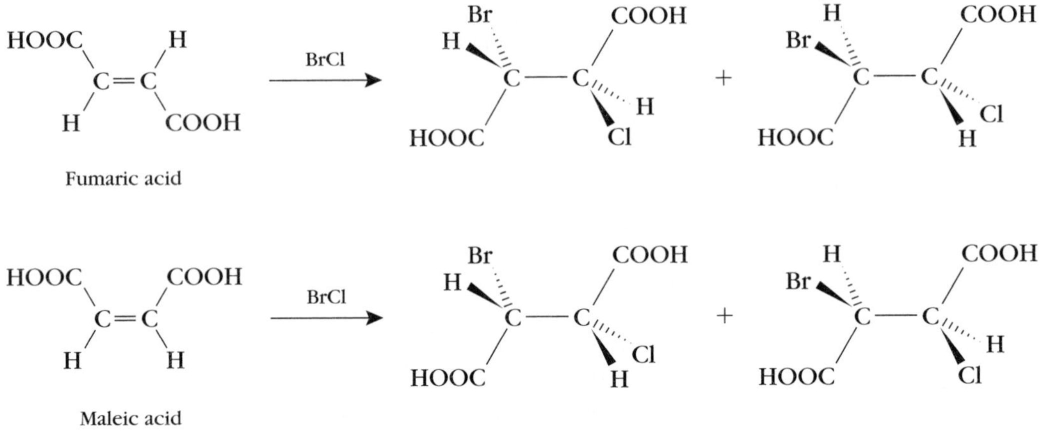

Fumaric acid

Maleic acid

These results are typical of the addition of BrCl to alkenes. What is the stereochemistry of this addition reaction? Based on your knowledge of the mechanism for the bromination of alkenes, write a two-step mechanism for the addition of BrCl that accounts for the stereochemistry of the addition to both fumaric and maleic acids.

8.39 BrCl adds to methylpropene to form only 1-bromo-2-chloro-2-methylpropane. Is this result in accord with the mechanism you proposed in Exercise 8.38?

8.40 Hydrogenation of alkenes can also be accomplished by the use of homogeneous catalysts. An example is Wilkinson's catalyst, which is a complex of rhodium, $RhCl[P(C_6H_5)_3]_3$. Hydrogenation of (E)-3,4-dimethyl-3-hexene in the presence of Wilkinson's catalyst forms a racemic mixture of 3,4-dimethylhexane, while hydrogenation of the (Z)-isomer forms the *meso* compound. What is the stereochemistry of the catalytic hydrogenation of alkenes using Wilkinson's catalyst?

8.41 Based on your knowledge of the mechanism for halogenation, propose a mechanism for the following reaction.

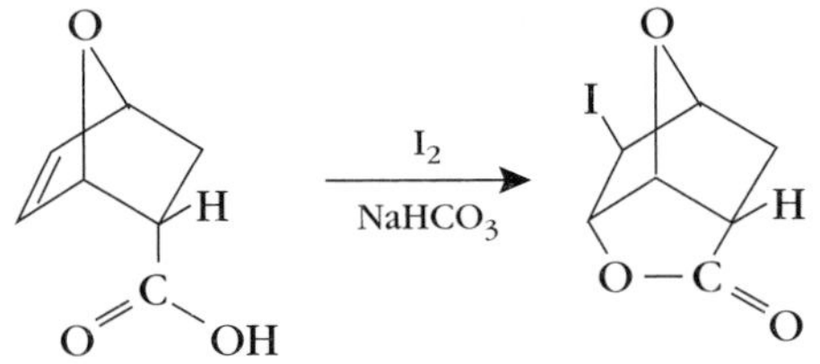

8.42 The difference in the regioselectivity of hydroboration of (E)-4-methyl-2-pentene with borane and 9-BBN is given in the following table:

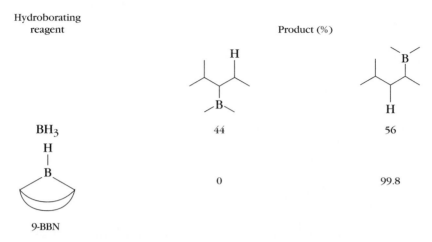

Hydroborating reagent	Product (%)	
BH₃	44	56
9-BBN	0	99.8

Using the accepted mechanism for hydroboration, explain why hydroboration using 9-BBN forms only a single product, while both isomers are formed in the reaction with borane.

8.43 Explain how IR and ^{13}C NMR spectroscopy can be used to distinguish between the following pairs of compounds. Indicate precisely the peak or peaks that serve to distinguish between the two.

(a) $\overset{\text{Cl}}{\underset{|}{\text{CH}_3\text{CHCH}_3}}$ and $\text{CH}_3\text{CH}_2\text{CH}_2\text{Cl}$

(b) [cyclohexene oxide] O and [cyclohexanone] O

(c) $\overset{\text{OH}}{\underset{|}{\text{CH}_3\text{CHCH}_2\text{Br}}}$ and $\overset{\text{Br}}{\underset{|}{\text{CH}_3\text{CHCH}_2\text{Br}}}$

8.44 Determine the structure of each of the following compounds from the data given.

(a) Molecular formula: $C_2H_4Br_2$; ^{13}C NMR: δ 30.5.
(b) Molecular formula: $C_6H_{14}O_2$; IR: 3300 cm^{-1}; ^{13}C NMR: δ 24.8, δ 75.1.

8.45   Determine the structure of each of the following compounds from its molecular formula and IR and ^{13}C NMR spectra.

   (a) C_2H_5ClO **(b)** C_4H_8O **(c)** $C_6H_{12}O_2$

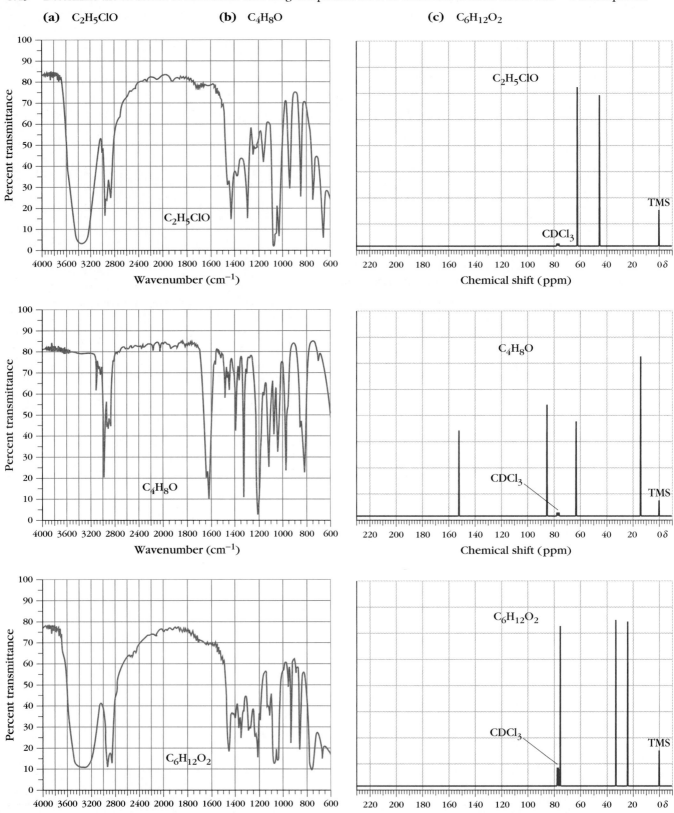

CHAPTER

ALKYNES

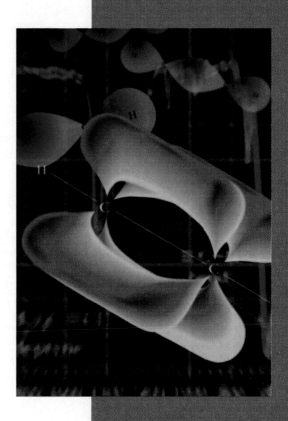

So far we have studied the chemistry of two families of hydrocarbons, alkanes and alkenes. Alkanes are compounds that contain only carbon-carbon and carbon-hydrogen single bonds. The single bonds in alkanes are nonpolar, so they are particularly unreactive. In fact, alkanes are the most unreactive of all the classes of organic compounds. As a result, the alkane portion of most molecules serves simply as a carrier of functional groups and is rarely involved in chemical reactions.

Alkenes also contain single bonds, but their most important structural feature is a carbon-carbon double bond. Because of this double bond, alkenes are much more reactive than alkanes. Alkenes, for instance, undergo electrophilic addition reactions due to the availability of electrons in the π portion of the double bond.

In this chapter we study the chemistry of alkynes, which are hydrocarbons that contain a carbon-carbon triple bond. Alkynes, like alkenes, are unsaturated hydrocarbons so they undergo many of the same reactions. The reason for this similarity becomes evident as we compare the structure and bonding in alkynes and alkenes.

9.1   Structure and Bonding in Alkynes

The general molecular formula of alkynes is C_nH_{2n-2}. One triple bond, therefore, introduces two degrees of unsaturation into a molecule. The geometry about each carbon atom of the triple bond is linear, as shown in Figure 9.1.

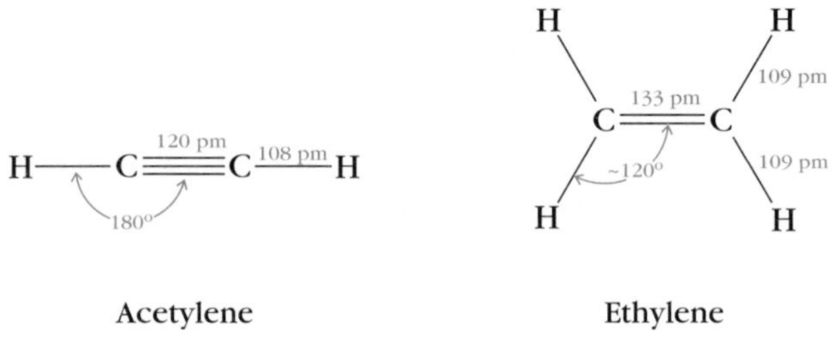

Figure 9.1
Observed bond lengths and bond angles in acetylene and ethylene.

Acetylene   Ethylene

The H—C—C bond angles in acetylene are 180° compared to approximately 120° in ethylene. The triple bond in acetylene is much shorter (120 pm) than the double bond in ethylene (133 pm) or the single bond in ethane (154 pm). Another difference is their carbon-carbon bond strengths. Experimentally, it is found that a carbon-carbon triple bond is stronger than either a carbon-carbon double or single bond. For instance, it takes about 200 kcal/mol (836 kJ/mol) to break the triple bond in acetylene compared to 145 kcal/mol (606 kJ/mol) to break the double bond in ethylene and 88 kcal/mol (368 kJ/mol) to break the single bond in ethane.

The localized valence bond orbital model of the electron structure of acetylene was presented in Section 1.11, C. To review, the linear geometry of a carbon-carbon triple bond in acetylene is obtained by using sp hybrid carbon atoms. The carbon-carbon σ bond is formed by end-on overlap of the sp hybrid orbitals of the two carbon atoms, as shown in Figure 9.2, A. Each of the two C—H bonds is formed by the overlap of the remaining sp hybrid orbital on each carbon atom with a $1s$ orbital of a hydrogen atom. Sideways overlap of the parallel p orbitals on each carbon atom forms a π bond. Because each carbon atom has two p orbitals, sideways overlap forms two π bonds, as shown in Figure 9.2, B. Thus the triple bond of acetylene (or any alkyne) is made up of a σ bond and two π bonds.

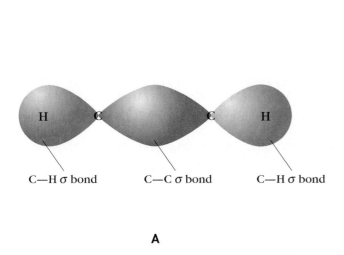

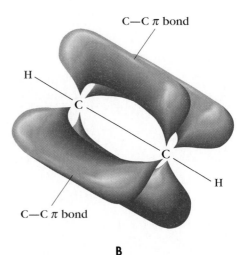

Figure 9.2
A, The σ-bond framework of acetylene. The C—C σ bond is formed by overlap of two *sp* hybrid orbitals (one from each C atom). Each C—H σ bond is formed by overlap of an *sp* hybrid orbital of a C atom with a 1*s* orbital of a H atom. **B,** Sideways overlap of the two sets of parallel *p* atomic orbitals forms the two π bonds.

Many of the chemical reactions of alkenes and alkynes are similar because both have π bonds that contain electrons that are readily available for reaction with electrophilic reagents. Before we discuss reactions, however, we must learn how to name alkynes.

9.2   Naming Alkynes

The common name acetylene is not only given to the simplest alkyne, but it is also used as a common class name for all alkynes. Specific alkynes are usually named according to the IUPAC rules.

The IUPAC names of alkynes are obtained by a simple modification of the general IUPAC rules for naming alkanes discussed in Chapter 2. The longest continuous carbon chain that contains the triple bond determines the parent name. The ending *-ane* of the name of the corresponding alkane is replaced by the ending *-yne* and the triple bond is given the lowest possible number. For example:

$$CH_3CH_2CH_2C≡CH \qquad \text{1-Pentyne}$$

$$CH_3CH_2C≡CCH_2CH_3 \qquad \text{3-Hexyne}$$

$$CH_3CH_2C≡CCHCH_2CH_3 \qquad \text{5-Methyl-3-heptyne}$$
$$\underset{\displaystyle CH_3}{|}$$

The ending *-adiyne* is given to a compound with two triple bonds, *-atriyne* to one with three triple bonds, and so forth.

$$\overset{\displaystyle CH_2Cl}{\underset{\displaystyle CH_3}{\overset{|}{CH_3CC≡CCH_2CH_2C≡CH}}} \qquad \text{8-chloro-7,7-dimethyl-1,5-octadiyne}$$

Alkynes with the general structure RC≡CH are called **terminal alkynes.** Alkynes where the triple bond is at least one carbon atom from the end of the carbon chain are called **internal alkynes.**

$$CH_3CH_2C{\equiv}CH \qquad CH_3C{\equiv}CCH_3$$

1-Butyne   2-Butyne
A terminal alkyne   An internal alkyne

In straight chain compounds that contain both double and triple bonds, the chain is numbered from the end nearest the first unsaturated bond. As a result, sometimes the triple bond gets the lower number, while other times the double bond gets the lower number. If there is a choice of numbering, the double bond gets the lower number. When the name is written, the ending for the double bond is always given first so these compounds are called *-enynes*, not *-ynenes*.

$$HC{\equiv}CCH{=}CH_2 \qquad \text{Butenyne}$$

$$CH_3CH_2CH{=}CHC{\equiv}CH \qquad \text{3-Hexen-1-yne}$$

$$HC{\equiv}CCH_2CH_2CH{=}CH_2 \qquad \text{1-Hexen-5-yne}$$

Alkynyl groups are substituents that contain a triple bond. Ethynyl and 2-propynyl are two examples of such substituents.

$$HC{\equiv}C{-} \qquad \overset{1}{-}CH_2\overset{2}{C}{\equiv}CH$$

Ethynyl group   2-Propynyl group

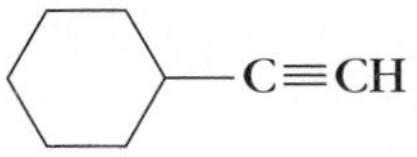

Ethynylcyclohexane 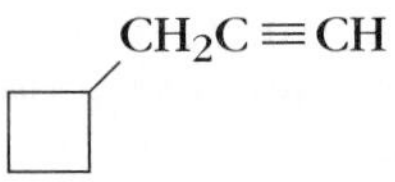   2-Propynylcyclobutane

---

Exercise 9.1   Give the IUPAC name for each of the following compounds.

(a)
$$\begin{array}{c} CH_3 \\ | \\ CH_3CH_2CC{\equiv}CH \\ | \\ CH_3 \end{array}$$

(d)
$$CH_3CH_2CHC{\equiv}CCH_2CHCH_2CH_3$$ with cyclopentyl and $CH_2CH(CH_3)_2$ substituents

(b) (structure with Br)

(e) (cyclopentane with $C{\equiv}CH$)

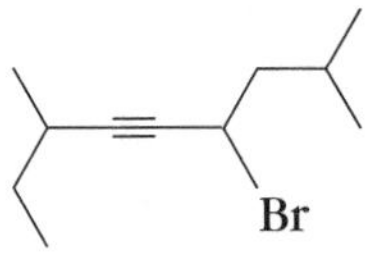

(c) $(CH_3)_3CCH_2C{\equiv}C(CH_2)_4CH_3$

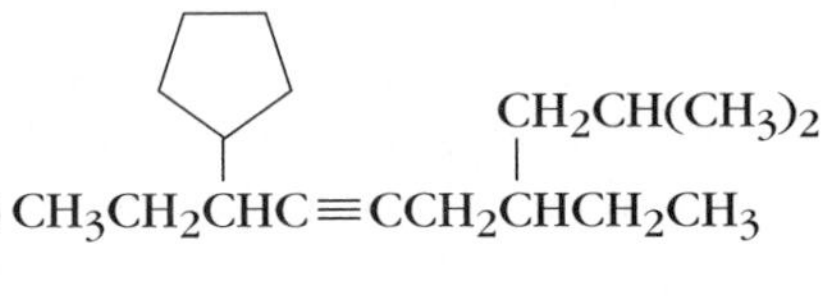

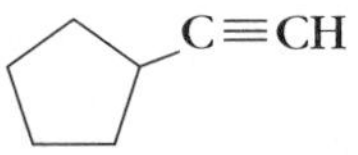

(f)
$$CH_3CH_2C{\equiv}CCH_2 \diagdown \quad CH_2CH_3$$
$$C{=}C$$
$$H \qquad H$$

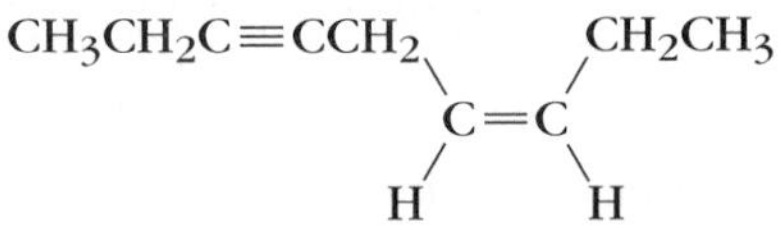

Exercise 9.2   Write the structure of each of the following compounds.

(a) 3-Heptyne
(b) 2-Chloro-5-methyl-3-heptyne
(c) 4-Bromo-2-hexyne
(d) 1,3-Hexadiyne

(e) 3-*sec*-Butyl-1-heptyne
(f) 3,4-Dimethylcyclodecyne
(g) 1,3-Octadien-5-yne
(h) 1-Ethyl-2-(2-propynyl)-cyclopentene

Table 9.1	Physical Properties of Selected Alkanes, Alkenes, and Alkynes			
Name	Structure	Melting point (°C)	Boiling point (°C)	Density (g/mL) (20 °C)
Propane	$CH_3CH_2CH_3$	−189	− 47.8	0.514
Propene	$CH_2=CHCH_3$	−185	− 42.1	0.500
Propyne	$HC\equiv CCH_3$	−101.5	− 23	0.670
Hexane	$CH_3(CH_2)_4CH_3$	− 95	69	0.660
1-Hexene	$CH_2=CH(CH_2)_3CH_3$	−140	63	0.674
1-Hexyne	$HC\equiv C(CH_2)_3CH_3$	−124	71	0.733
Decane	$CH_3(CH_2)_8CH_3$	− 30	174	0.741
1-Decene	$CH_2=CH(CH_2)_7CH_3$	− 66	170.6	0.730
1-Decyne	$HC\equiv C(CH_2)_7CH_3$	− 36	182	0.770

9.3   Physical Properties of Alkynes

The data in Table 9.1 show that the melting points, boiling points, and densities of representative alkynes, alkenes, and alkanes are similar. Notice, for example, that all are less dense than water and compounds of all three families that contain four or fewer carbon atoms are gases at room temperature. Notice as well that their boiling points and densities increase as the number of carbon atoms in their chains increases.

The similarities in physical properties arise because all three families contain nonpolar C—C and C—H bonds. The presence of π bonds in alkenes and alkynes makes them slightly more polar than alkanes but not enough to affect their physical properties. They are all nonpolar compounds so they are insoluble in water but very soluble in nonpolar solvents.

9.4   Physiologically Active Alkynes

Compounds that contain carbon-carbon triple bonds are also found in nature. Most have been isolated from plants, and they usually contain other functional groups besides one or more carbon-carbon triple bonds.

Many naturally occurring alkynes are physiologically active. There are several polyacetylenes, isolated from plant sources, for example, that humans use as food or for flavoring. These include norcapillen found in tarragon and falcarinone found in caraway, common sunflowers, and Jerusalem artichoke.

It's not known whether the triple bond confers any part of the distinctive flavors of these foods.

Other polyacetylenes are poisonous. Ichthyotherol is the active ingredient in the plant extract used by natives in the Lower Amazon Basin to poison their arrowheads. Ichthyotherol causes convulsions in mammals.

Ichthyotherol   Capillin

Other polyacetylenes can be beneficial. The alkynyl ketone known as capillin, for example, is active against fungi of the skin.

While little is known about the plant physiology of polyacetylenes, there are indications that these compounds may be linked with special hormonal functions in plant metabolism. In several instances, it has been observed that the concentration of certain acetylenes varies drastically during the vegetation period.

Certain acetylenes also act as hormones in humans. 17-Ethynylestradiol, for example, is used as a birth control agent. It is used in birth control pills instead of naturally occurring hormones because it is more effective at preventing pregnancy and consequently can be used in smaller doses.

17-Ethynylestradiol

9.5   Spectroscopic Properties of Alkynes

Recall from Section 5.5 that terminal alkynes show two characteristic IR absorption bands. The one due to stretching of the triple bond is found at 2250 to 2100 cm^{-1}, while the one due to the stretching of the terminal C—H bond is found at 3350 to 3260 cm^{-1}. A third intense absorption band due to the bending of the terminal C—H bond can sometimes be found at 600 to 700 cm^{-1}. All three absorptions are clearly visible in the spectrum of 1-heptyne shown in Figure 9.3.

The absorption band due to stretching of the triple bond in internal alkynes is found at 2200 to 2260 cm^{-1}. This band is often weak because there is little change in dipole when the triple bond is stretched. Recall from Section 7.6 that the absorption band due to stretching of the double bond in the IR spectrum of symmetrically substituted alkenes is weak for the same reason.

The ^{13}C NMR chemical shifts of sp hybrid carbon atoms of the triple bond are usually found in the narrow range of δ 70 to 95. These chemical shifts are very different from those of similarly substituted sp^2 and sp^3 hybrid carbon atoms of alkenes and alkanes. ^{13}C NMR spectroscopy, therefore, can be used to detect the presence of a carbon-carbon triple bond in a molecule.

The ^{13}C NMR chemical shifts of sp hybrid carbon atoms, like sp^2 hybrid carbon atoms, depend on the number of alkyl groups bonded to the carbon atom.

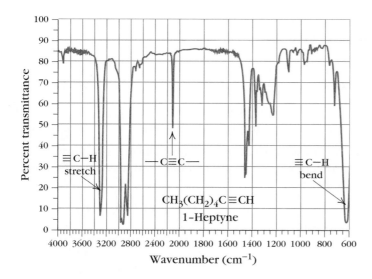

Figure 9.3
Infrared spectrum of 1-heptyne. The three absorptions that are characteristic of terminal alkynes are the terminal C—H stretch at about 3300 cm^{-1}, the C≡C stretch at about 2100 cm^{-1}, and bending of the terminal C—H bond at about 650 cm^{-1}.

The chemical shift of an *sp* carbon atom bonded to hydrogen is usually found in the region δ 65 to 70, whereas one bonded to an alkyl group is found at about δ 75 to 85, as shown by the following examples:

$$
\underset{\delta\,75\text{-}85}{\overset{\delta\,65\text{-}70}{RC\equiv CH}}
\qquad
\underset{\delta\,85.5}{\overset{\delta\,67.1}{HC\equiv CCH_2CH_3}}
\qquad
\underset{\delta\,80.4}{H_3CC\equiv CCH_3}
$$

9.6 Additions of Brønsted-Lowry Acids and Halogens to Alkynes

Both alkynes and alkenes have π bonds that contain electrons that are readily available for reactions with electrophilic reagents. As a result, most of the addition reactions of alkynes are similar to those of alkenes. Alkynes, however, have two π bonds so they can react with two equivalents of electrophilic reagent. Addition of the second equivalent of electrophile tends to be more difficult to accomplish so that the reaction can usually be stopped after the addition of only one equivalent.

A   Additions of Brønsted-Lowry Acids

The addition of one equivalent of a Brønsted-Lowry acid, such as trifluoroacetic acid, to internal alkynes, such as 3-hexyne, usually forms equal mixtures of the corresponding (*E*)- and (*Z*)-alkenes.

$$
C_2H_5C\equiv CC_2H_5 \xrightarrow{\;\;CF_3C\underset{OH}{\overset{O}{\parallel}}\;\;}
$$

3-Hexyne

(*Z*)-3-Trifluoroacetyl-
3-hexene

(*E*)-3-Trifluoroacetyl-
3-hexene

Addition of a second equivalent forms a geminal disubstituted alkane with regioselectivity according to the Markownikoff Rule.

$$CH_3C\equiv CCH_3 \xrightarrow[\text{of HBr}]{\text{1st eq}} CH_3CH\!=\!CCH_3 \xrightarrow[\text{of HBr}]{\text{2nd eq}} CH_3CH_2CCH_3$$

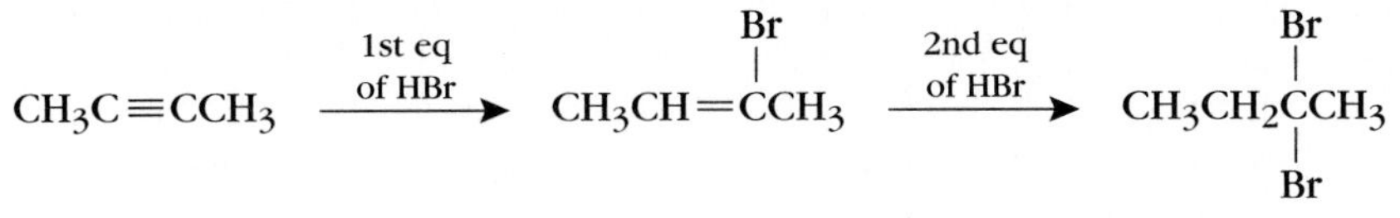

2-Butyne   (E) or (Z)-2-Bromo-2-butene   2,2-Dibromobutane

Additions of hydrogen halides to terminal alkynes also form products according to the Markownikoff Rule so the hydrogen atom adds to the least substituted carbon atom of the triple bond.

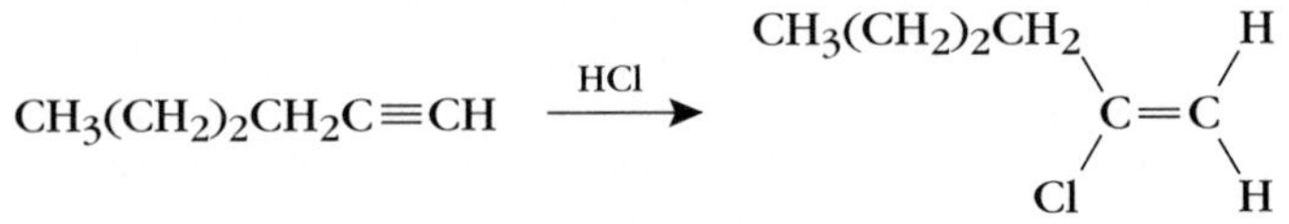

$$CH_3(CH_2)_2CH_2C\equiv CH \xrightarrow{\text{HCl}}$$

1-Hexyne   2-Chloro-1-hexene

A proposed mechanism for the reaction of Brønsted-Lowry acids with alkynes consistent with the stereochemistry and regiochemistry of the addition is shown in Figure 9.4.

Figure 9.4
A two-step mechanism for the addition of trifluoroacetic acid (CF₃COOH) to propyne. A vinyl carbocation is formed in Step 1. Reaction of the vinyl carbocation in Step 2 with trifluoroacetate ion forms the product.

Step 1: Protonation of the triple bond forms a vinyl carbocation

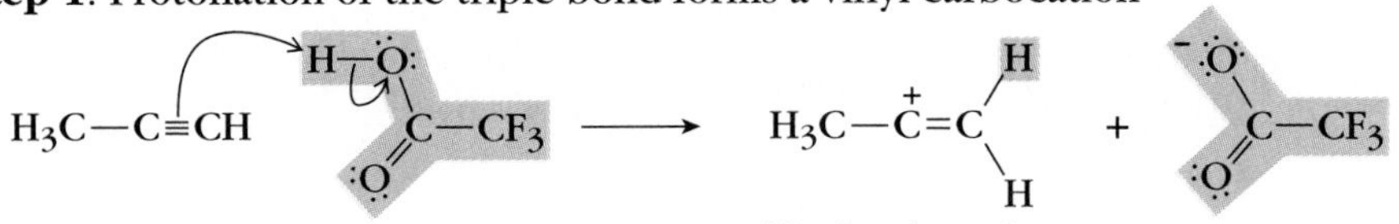

Propyne   Vinyl carbocation

Step 2: Reaction of the vinyl carbocation with trifluoroacetate ion forms the product

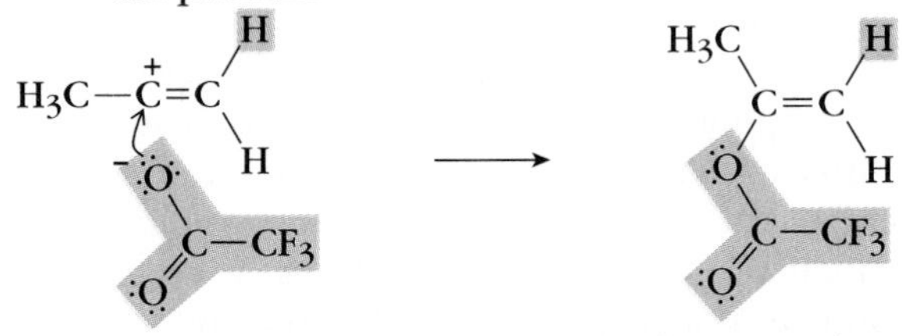

2-Trifluoroacetylpropene

The mechanisms of additions of Brønsted-Lowry acids to alkenes and alkynes are similar in that rate-determining protonation in Step 1 leads to the formation of a carbocation. The major difference between the two mechanisms is the nature of the carbocation. Protonation of alkynes forms an intermediate vinyl carbocation, whereas protonation of alkenes forms an alkyl carbocation.

Protonation of a terminal alkyne can form, at least in theory, two isomeric vinyl carbocations. Protonation of propyne, for example, might yield the following primary and secondary vinyl carbocations:

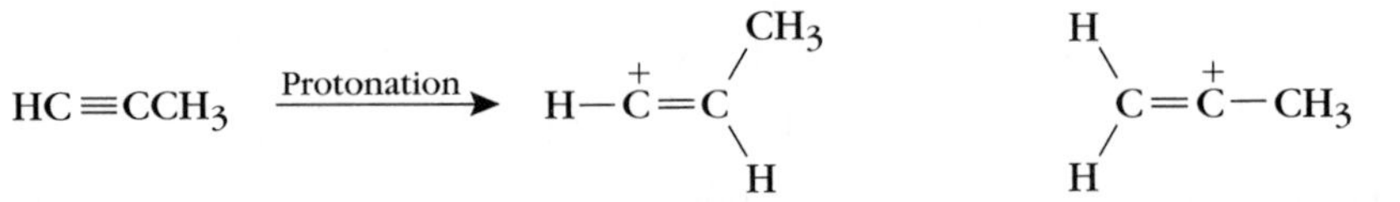

$$HC\equiv CCH_3 \xrightarrow{\text{Protonation}}$$

A primary vinyl carbocation   A secondary vinyl carbocation

Vinyl carbocations have the same order of stabilities as their similarly substituted alkyl carbocations. Thus a secondary vinyl carbocation is more stable than a primary one. Additions of Brønsted-Lowry acids to terminal alkynes form products according to the Markownikoff Rule, therefore, because the transition state for protonation of a terminal alkyne to form a secondary vinyl carbocation is lower in energy than the transition state to form a primary one. In other words, the lowest energy pathway is the one that leads to formation of the product with Markownikoff orientation.

B   Halogenation

Bromine and chlorine react with alkynes just as they do with alkenes. Addition of one equivalent of halogen forms a dihaloalkene, and the products are often mixtures of (*E*)- and (*Z*)-isomers.

$$CH_3CH_2C\equiv CCH_2CH_3 \xrightarrow{\ Cl_2\ }$$

$$\underset{Cl}{\overset{CH_3CH_2}{>}}C=C\underset{CH_2CH_3}{\overset{Cl}{<}} \quad + \quad \underset{CH_3CH_2}{\overset{Cl}{>}}C=C\underset{CH_2CH_3}{\overset{Cl}{<}}$$

3-Hexyne   (*E*)-3,4-Dichloro-3-hexene   (*Z*)-3,4-Dichloro-3-hexene

A tetrahaloalkane is formed when two equivalents of halogen are added to an alkyne.

$$HC\equiv CCH_2CH_3 \xrightarrow{\ Br_2\ } \underset{Br}{\overset{Br}{HC=CCH_2CH_3}} \xrightarrow{\ Br_2\ } \underset{Br\ \ Br}{\overset{Br\ \ Br}{HC-CCH_2CH_3}}$$

1-Butyne   (*E*)- and (*Z*)-1,2-Dibromo-1-butene   1,1,2,2,-Tetrabromobutane

---

Exercise 9.3   Write the structure of the product of the reaction of 1-hexyne with each of the following reagents.

(a) One equivalent of HCl   (c) One equivalent of Cl_2
(b) Two equivalents of HBr   (d) Two equivalents of Br_2

Exercise 9.4   Write the structure of the product of the reaction of 4-octyne with each of the reagents listed in Exercise 9.3.

9.7   Hydration of Alkynes

Recall that addition of water to the double bond of alkenes can be accomplished by acid-catalyzed hydration (Section 7.14) or hydroboration-oxidation (Section 8.4). These two methods add water to the triple bond, too, but the reactions are not entirely analogous.

A   Acid-Catalyzed Hydration

Alkynes, like alkenes, undergo acid-catalyzed addition of water according to the Markownikoff Rule. At this point, however, the similarity between the two reactions ends. The initial product of hydration of alkynes is a vinyl alcohol, which is a compound with an —OH group bonded to a carbon atom of a double bond. The preferred term for these compounds is **enols** (-*ene* for the double bond + -*ol* for the alcohol). Enols are in equilibrium with ketones and are rapidly converted to them. As a result the final product of hydration of an alkyne is an aldehyde or ketone.

$$C_2H_5C \equiv CH \xrightarrow{H_3O^+} \underset{\text{An enol}}{\underset{H}{\overset{HO}{C_2H_5C = CH}}} \rightleftharpoons \underset{\text{Ketone}}{\overset{O}{C_2H_5C - CH_3}}$$

An enol and its corresponding ketone are constitutional isomers. Constitutional isomers that are readily interconvertible through a rapid equilibration are called **tautomers.** The process of interconverting tautomers is called **tautomerism.** The most common kind of tautomerism involves structures that differ in the location of a hydrogen atom such as keto-enol tautomerism:

■ *We will study tautomerism in more detail in Chapter 17.*

$$\underset{\text{Enol tautomer}}{\overset{\boxed{H}}{\underset{}{O}}\overset{}{\diagdown}{C = C}} \rightleftharpoons \underset{\text{Keto tautomer}}{\overset{O}{\diagdown}{C - C}\overset{\boxed{H}}{\diagup}}$$

The following keto-enol equilibrium constants of simple ketones and aldehydes are very small so the keto tautomer is favored over the enol tautomer.

$$K = \frac{[\text{Enol}]}{[\text{Keto}]}$$

$$\underset{}{\overset{O}{CH_3CCH_3}} \rightleftharpoons \underset{}{\overset{OH}{CH_3C = CH_2}} \qquad 4.7 \times 10^{-9}$$

$$\overset{O}{\bigcirc} \rightleftharpoons \overset{OH}{\bigcirc} \qquad 4.1 \times 10^{-7}$$

$$\underset{H}{\overset{O}{CH_3C}} \rightleftharpoons \underset{H}{\overset{OH}{CH_2 = C}} \qquad 5.9 \times 10^{-7}$$

We can understand the position of a keto-enol equilibrium by comparing the relative acidities of the two tautomers. Recall from Chapter 3 that hydrogen atoms bonded to oxygen atoms tend to be more acidic than hydrogen atoms bonded to carbon atoms. The keto form is favored at equilibrium because it is a weaker acid than the enol form.

Not only does the equilibrium constant favor the keto tautomer but the conversion of the enol to the keto tautomer occurs very rapidly in acid solutions. As a result, the enols of simple ketones and aldehydes are never observed and acid-catalyzed hydration of alkynes forms ketones or aldehydes as products. Furthermore, the number of products formed depends on the structure of the alkyne. Symmetrical alkynes form a single ketone, while unsymmetrical alkynes form a mixture of two isomeric ketones. Terminal alkynes form a single methyl ketone and acetylene is the only alkyne that forms an aldehyde. These results are summarized in Figure 9.5.

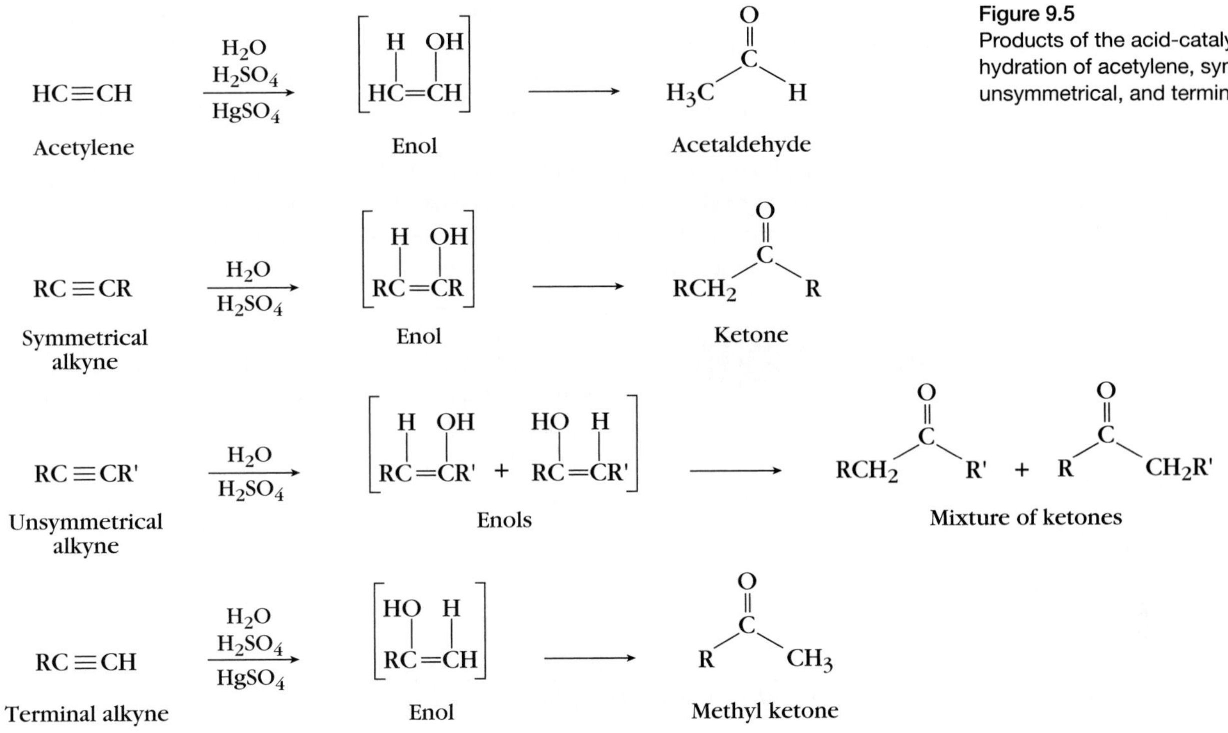

$$HC\equiv CH \xrightarrow[\text{HgSO}_4]{\substack{H_2O \\ H_2SO_4}} \left[\begin{array}{c} H \quad OH \\ | \quad | \\ HC=CH \end{array} \right] \longrightarrow \underset{\text{Acetaldehyde}}{H_3C-\overset{\overset{\displaystyle O}{\|}}{C}-H}$$

Acetylene   Enol

Figure 9.5
Products of the acid-catalyzed hydration of acetylene, symmetrical, unsymmetrical, and terminal alkynes.

$$RC\equiv CR \xrightarrow[H_2SO_4]{H_2O} \left[\begin{array}{c} H \quad OH \\ | \quad | \\ RC=CR \end{array} \right] \longrightarrow \underset{\text{Ketone}}{RCH_2-\overset{\overset{\displaystyle O}{\|}}{C}-R}$$

Symmetrical alkyne   Enol

$$RC\equiv CR' \xrightarrow[H_2SO_4]{H_2O} \left[\begin{array}{c} H \quad OH \qquad HO \quad H \\ | \quad | \qquad\quad | \quad | \\ RC=CR' \;+\; RC=CR' \end{array} \right] \longrightarrow RCH_2-\overset{\overset{\displaystyle O}{\|}}{C}-R' \;+\; R-\overset{\overset{\displaystyle O}{\|}}{C}-CH_2R'$$

Unsymmetrical alkyne   Enols   Mixture of ketones

$$RC\equiv CH \xrightarrow[\text{HgSO}_4]{\substack{H_2O \\ H_2SO_4}} \left[\begin{array}{c} HO \quad H \\ | \quad | \\ RC=CH \end{array} \right] \longrightarrow \underset{\text{Methyl ketone}}{R-\overset{\overset{\displaystyle O}{\|}}{C}-CH_3}$$

Terminal alkyne   Enol

Exercise 9.5   Write the equation for the acid-catalyzed hydration of each of the following alkynes.

(a) 2-Butyne   (c) $CH_3CH_2CH_2C\equiv CCH_2CH_2CH_3$
(b) 1-Hexyne   (d) 2,5-Dimethyl-3-hexyne

Exercise 9.6   For each of the following enols, write the structure of its keto form.

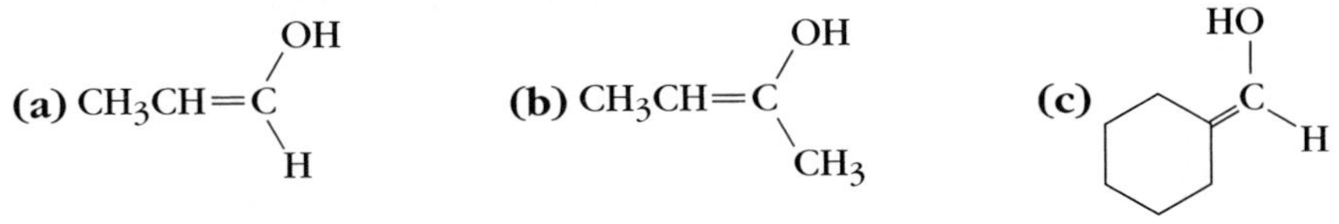

(a) $CH_3CH=\overset{\displaystyle OH}{\underset{\displaystyle H}{C}}$ (b) $CH_3CH=\overset{\displaystyle OH}{\underset{\displaystyle CH_3}{C}}$ (c)

Exercise 9.7   Write the structures of the alkynes whose acid-catalyzed hydration would produce each of the following ketones as the only product.

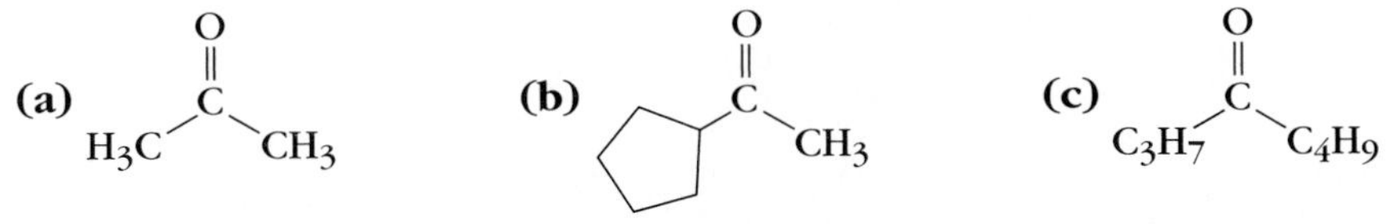

(a) $H_3C-\overset{\overset{\displaystyle O}{\|}}{C}-CH_3$ (b)   (c) $C_3H_7-\overset{\overset{\displaystyle O}{\|}}{C}-C_4H_9$

The mechanism proposed for the acid-catalyzed hydration of alkynes is shown in Figure 9.6. It is similar to the mechanism for the acid-catalyzed hydration of alkenes. The regiochemistry of the reaction, for example, can be explained by protonation of the carbon-carbon triple bond to form the more stable secondary vinyl carbocation.

Internal alkynes undergo acid-catalyzed hydration readily at room temperature. Acid-catalyzed hydration of acetylene and terminal alkynes, however, is slow. Their rates of hydration can be increased by adding mercury(II) sulfate ($HgSO_4$) to the reaction mixture as a catalyst.

Exercise 9.8 Acid-catalyzed hydration of 2-hexyne forms a mixture of two isomeric ketones. Write a mechanism that explains the formation of these two products.

Figure 9.6
Mechanism of the acid-catalyzed hydration of a terminal alkyne.

Step 1: Protonation of the carbon-carbon triple bond forms the more stable secondary vinyl carbocation

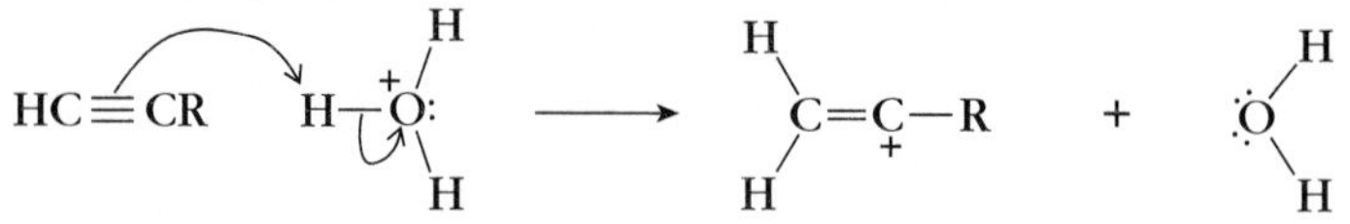

A terminal alkyne

Secondary vinyl carbocation

Step 2: A water molecule reacts with the vinyl carbocation

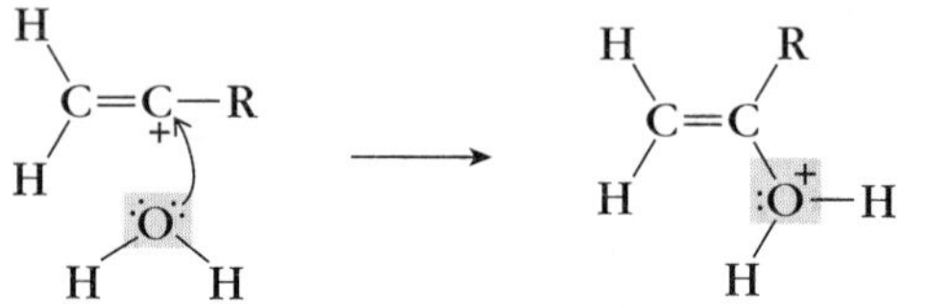

Step 3: Loss of proton to solvent water molecule forms an enol

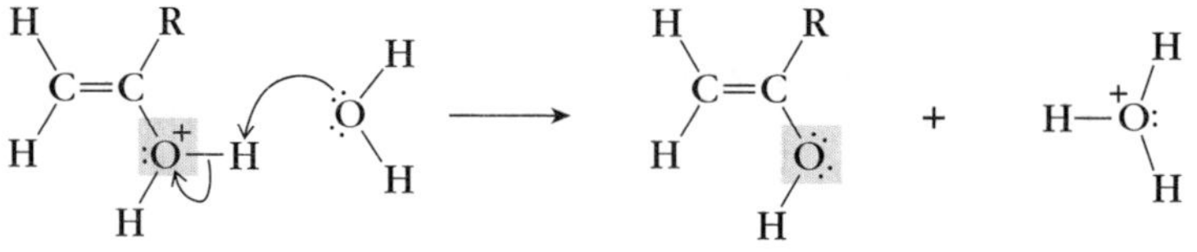

Enol

Step 4: Enol tautomerizes to keto product

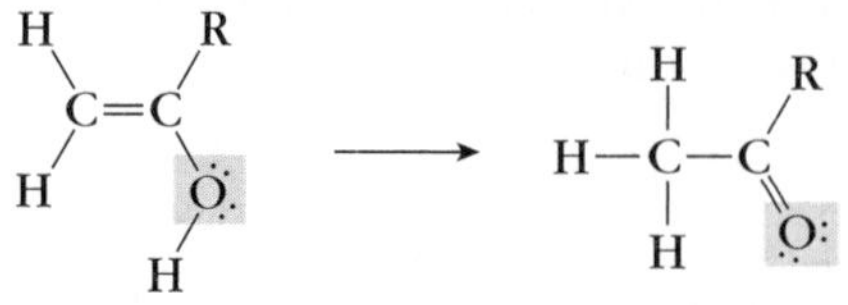

Enol

Keto

B Hydration of Alkynes by Hydroboration-Oxidation

Borane also reacts rapidly with alkynes to form a vinyl organoborane by syn addition. Oxidation of these vinyl organoboranes by alkaline hydrogen peroxide forms an enol, which tautomerizes to a ketone.

$$3\ CH_3C\equiv CCH_3 \xrightarrow[\text{THF}]{BH_3} \underset{\substack{CH_3 \quad\quad CH_3}}{\overset{\substack{H \quad\quad BR_2}}{C=C}} \xrightarrow[\substack{H_2O \\ OH^-}]{H_2O_2} 3\ \underset{\substack{CH_3 \quad\quad CH_3}}{\overset{\substack{H \quad\quad OH}}{C=C}}$$

2-Butyne   A vinyl organoborane   Enol

$$R=\ \underset{\substack{CH_3 \quad\quad CH_3}}{\overset{\substack{H}}{C=C}}$$

$$3\ CH_3CH_2\overset{O}{\underset{CH_3}{\overset{\|}{C}}}$$

2-Butanone
(65% yield)

Hydroboration-oxidation of symmetrical alkynes forms a single ketone as product, while unsymmetrical alkynes form a mixture of two isomeric ketones.

In theory, the hydroboration of alkynes with boranes is a straightforward method of preparing ketones from alkynes. In practice, however, the vinyl organoborane formed in the first step of addition has a double bond that can react with a second equivalent of borane, thus leading to complex mixtures of products. This is a particularly serious problem in the reaction of borane with terminal alkynes where the reaction cannot be stopped after the first addition. It is possible to control the hydroboration of internal alkynes by carefully adjusting the concentrations of reagents, but a better way of preventing a second addition is by the use of a bulky sterically-hindered hydroborating reagent, such as 9-BBN.

■ *Recall from Section 8.1 that 9-BBN, 9-borabicyclo[3.3.1]nonane, was particularly useful for the hydroboration of alkenes.*

Frequently
abbreviated as

9-Borabicyclo[3.3.1]nonane
or 9-BBN

Reaction of an alkyne with 9-BBN occurs with the usual syn addition of boron and hydrogen to the triple bond. The boron atom adds to the less substituted terminal carbon atom of the triple bond. Addition of a second molecule of 9-BBN to the intermediate vinyl organoborane is slow because the large groups bonded to boron physically block approach of the second molecule of 9-BBN. Consequently, only a single addition of 9-BBN occurs. Reaction of this vinyl organoborane with alkaline hydrogen peroxide forms an enol, which quickly isomerizes to an aldehyde.

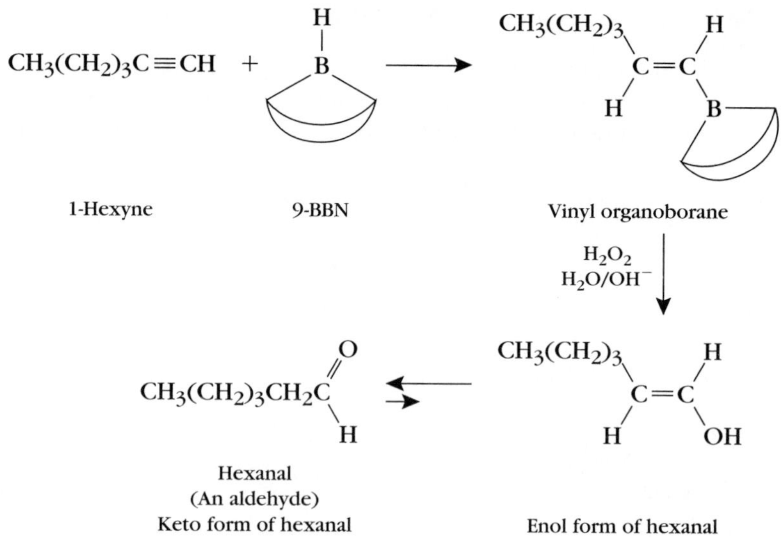

CH3(CH2)3C≡CH
+
H
B
CH3(CH2)3
H
C=C
H
B
1-Hexyne
9-BBN
Vinyl organoborane
H2O2
H2O/OH−
O
CH3(CH2)3CH2C
H
CH3(CH2)3
H
C=C
H
OH
Hexanal
(An aldehyde)
Keto form of hexanal
Enol form of hexanal

The use of bulky hydroborating reagents such as 9-BBN makes it possible to prepare good yields of aldehydes from terminal alkynes and ketones from symmetrical internal alkynes. In the case of symmetrical internal alkynes, both hydroboration-oxidation and acid-catalyzed hydration form the same ketone product.

B
H
C=C
R
R
9-BBN
H2O2
H2O/OH−
RC≡CR
O
RC—CH2R
H3O+

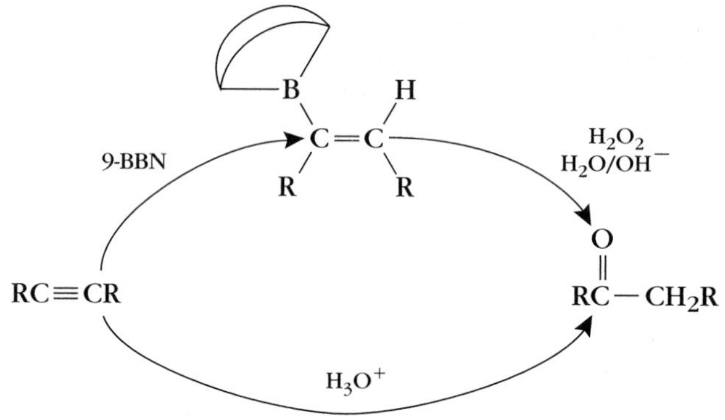

Hydroboration-oxidation and acid-catalyzed hydration of unsymmetrical internal alkynes, on the other hand, are not useful reactions to prepare ketones because a mixture of two isomeric ketones is formed.

B
H
C=C
R
R'
+
H
B
C=C
R
R'
9-BBN
H2O2
H2O/OH−
RC≡CR'
O
RC—CH2R'
+
O
RCH2—CR'
H3O+

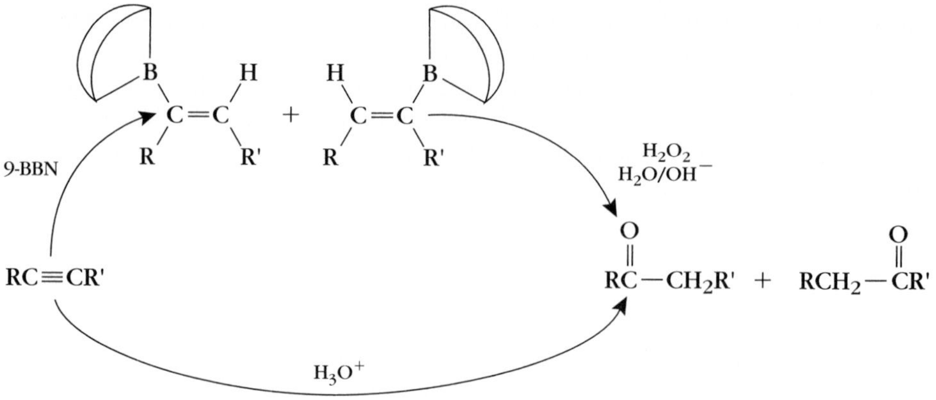

For terminal alkynes, hydroboration-oxidation forms aldehydes, whereas acid-catalyzed hydration forms methyl ketones.

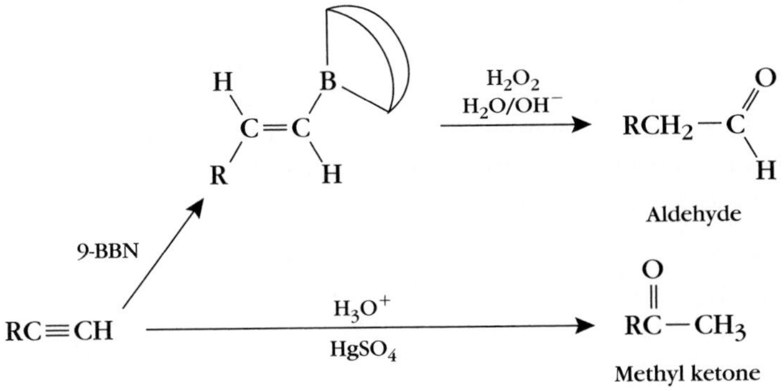

Exercise 9.9   Write the equation for the reaction of each of the following alkynes with (a) 9-BBN followed by aqueous alkaline hydrogen peroxide and (b) aqueous solution of H_2SO_4.

(a) 1-Pentyne   **(b)** 4-Octyne   **(c)**

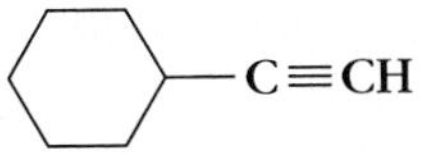

Exercise 9.10   Write the structures of the two ketones formed by the reaction of 2-pentyne and 9-BBN followed by aqueous alkaline hydrogen peroxide.

Alkynes can be reduced to alkanes, but the reaction is more complicated than the reduction of alkenes.

9.8   Hydrogenation of Alkynes

Alkynes, like alkenes, undergo catalytic hydrogenation to form alkanes. The same catalysts, PtO_2, Pd, and Ni, are used for hydrogenation of either alkenes or alkynes.

$$CH_3CH_2C{\equiv}CCH_3 \;+\; 2H_2 \;\xrightarrow{\;PtO_2,\; Pd,\; or\; Ni\;}\; CH_3CH_2CH_2CH_2CH_3$$

2-Pentyne   Pentane
                                        (95% yield)

Catalytic hydrogenation of alkynes takes place in two steps. The first formed product is an alkene, which then adds another equivalent of hydrogen to form an alkane.

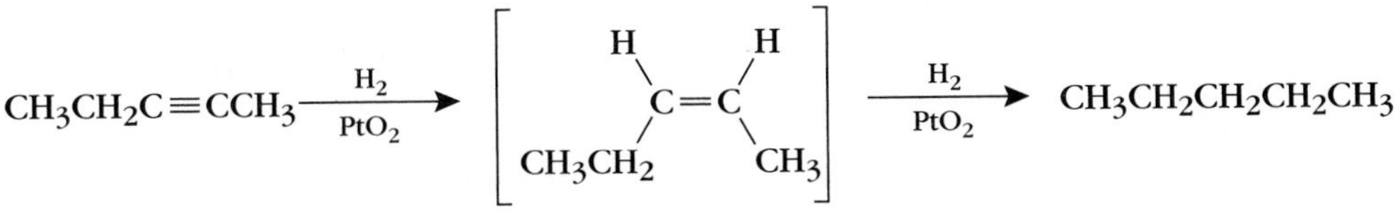

2-Pentyne   (Z)-2-Pentene   Pentane
                          (Not isolated)

It's impossible to stop catalytic hydrogenation of alkynes at the alkene stage when the reaction is carried out in the presence of the usual metal catalysts. The alkene can be made the final product, however, if a special catalyst known as **Lindlar's catalyst** is used instead. Lindlar's catalyst is a mixture of finely divided palladium metal and $CaCO_3$ that has been washed with $Pb(OAc)_2$. This mixture catalyzes the hydrogenation of alkynes to alkenes but *not* the hydrogenation of alkenes to alkanes. Thus hydrogenation stops at the alkene stage. Lindlar's catalyst, like those used in the hydrogenation of alkenes, adds hydrogen syn to the triple bond to form the (*Z*)- alkene as a product.

$$CH_3CH_2C \equiv CCH_2CH_3 \xrightarrow[\text{Lindlar's catalyst}]{H_2}$$

3-Hexyne

(*Z*)-3-Hexene
(97% yield)

The hydroboration-protonolysis sequence of reactions, which we used to add hydrogen to alkenes in Section 8.4, can be used to add hydrogen to alkynes to form (*Z*)-alkenes. Vinyl organoboranes are the product of the syn addition of 9-BBN to alkynes as we learned in Section 9.7. Reaction of these vinyl boranes with propanoic acid replaces the boron atom by a hydrogen atom with retention of configuration at the vinyl carbon atom, thus producing (*Z*)-alkenes from alkynes in high yields.

$$C_3H_7C \equiv CC_3H_7 \ + \ \text{9-BBN} \longrightarrow \xrightarrow{CH_3CH_2COOH}$$

4-Octyne   9-BBN

(*Z*)-4-octene
(98% yield)

Hydroboration-protonolysis of terminal alkynes also occurs smoothly in high yields to form the 1-alkene.

$$C_6H_{13}C \equiv CH \ + \ \text{9-BBN} \longrightarrow \xrightarrow{CH_3CH_2COOH}$$

1-Octyne   9-BBN

1-Octene
(96% yield)

Reactions can be carried out at −78 °C by placing the reaction flask in a bath containing a mixture of dry ice (solid carbon dioxide) and acetone.

A third way to prepare alkenes from alkynes is the reaction of internal alkynes with lithium or sodium metal in liquid ammonia. Ammonia is a gas at room temperature (bp −33 °C) but it can be kept as a liquid by carrying out the reaction at low temperature, such as −78 °C. In contrast to catalytic hydrogenation and hydroboration-protonolysis, the reaction of alkynes with lithium or sodium in liquid ammonia forms an (*E*)-alkene.

$$CH_3(CH_2)_2C \equiv CCH_2CH_3 \xrightarrow[-78\ ^\circ C]{Li/NH_3}$$

3-Heptyne

(*E*)-3-heptene
(92% yield)

Terminal alkynes do not form alkenes when treated with lithium or sodium metal in liquid ammonia. Instead terminal alkynes react with sodium metal to form a sodium acetylide and hydrogen gas as the only products.

$$RC\equiv CH \quad + \quad Na \quad \xrightarrow[NH_{3(l)}]{} \quad RC\equiv C^-Na^+ \quad + \quad \tfrac{1}{2}H_2$$

| Terminal alkyne | Sodium metal | Liquid ammonia | Sodium acetylide | Hydrogen gas |

The proposed mechanism for the reaction of an internal alkyne with sodium metal in liquid ammonia to form an (*E*)-alkene is outlined in Figure 9.7. Three different kinds of intermediates are formed in this mechanism. The first (formed in Step 1) is a vinyl radical anion. A radical anion has both a negative charge on one carbon atom and an unpaired electron on another. The radical anion is a strong base, so in Step 2 of the mechanism it removes a proton from an ammonia molecule to form a vinyl radical. A radical is a species that has an unpaired electron on a carbon atom. The vinyl radical formed in Step 2 is more stable in the *trans*- configuration. In Step 3, the *trans*-vinyl radical accepts an electron from a sodium atom to form the final intermediate, a vinyl anion. Anions where the negative charge is on a carbon atom are called **carbanions**. Carbanions are strong bases, so in Step 4 the vinyl anion removes a proton from an ammonia molecule to form an (*E*)-alkene.

Step 1: An electron is transferred from a sodium atom to the alkyne to form a vinyl radical anion

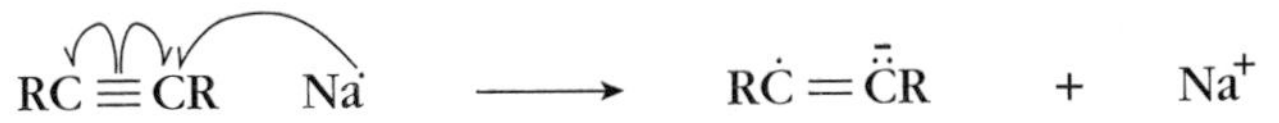

Vinyl radical anion

Figure 9.7
Proposed mechanism for the reduction of an alkyne to an (*E*)- alkene by sodium in liquid ammonia. A single barbed curved arrow, ⌒, designates the movement of a single electron.

Step 2: The vinyl radical anion removes a proton from the liquid ammonia solvent to form a vinyl radical with a preferred *trans* geometry

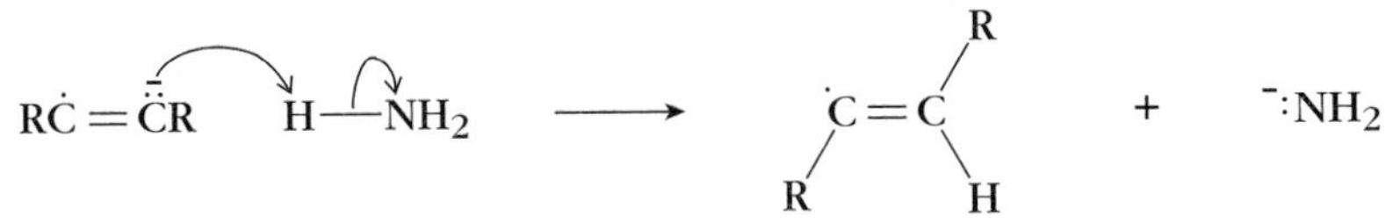

trans Vinyl radical

Step 3: An electron is transferred from a sodium atom to the vinyl radical to form a vinyl anion

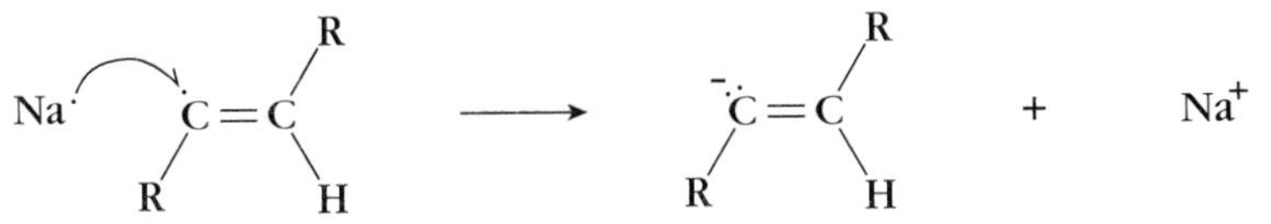

Vinyl anion

Step 4: The vinyl anion removes a proton from the liquid ammonia solvent to form an (*E*)-alkene

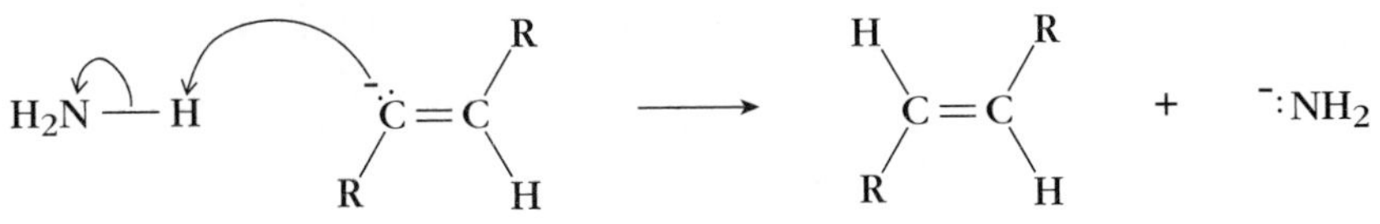

(*E*)-Alkene

The mechanism for the reduction of alkynes by sodium in liquid ammonia shown in Figure 9.7 may seem rather complicated, but it is really the stepwise addition of two electrons and two protons to an alkyne, which is the same thing as adding a hydrogen molecule, H_2. Thus catalytic hydrogenation, hydroboration-protonolysis, and sodium in liquid ammonia all add two protons and a pair of electrons to an alkyne to form an alkene. The only differences among the three are the reaction conditions, the mechanism of addition, and the stereochemistry of the product.

Exercise 9.11   Write the equation for the reaction of 3-nonyne with each of the following reagents.

(a) $H_2/PtO_2/C_2H_5OH$
(b) 9-BBN followed by CH_3CH_2COOH
(c) Na in liquid ammonia, $NH_{3(l)}$

Exercise 9.12   What alkyne and what reaction conditions would you choose to prepare each of the following alkenes?

(a) (*E*)-2-Hexene   **(b)** (*Z*)-5-Decene   **(c)**

$$\begin{array}{ccc} D & & D \\ \diagdown & & \diagup \\ & C{=}C & \\ \diagup & & \diagdown \\ C_2H_5 & & C_2H_5 \end{array}$$

Exercise 9.13   A chemist believes that a mixture of finely divided Ni and Pd on carbon will catalyze the hydrogenation of alkynes to (*E*)-alkenes. The catalyst is prepared and used to hydrogenate 4-octyne. The product is isolated and its ^{13}C NMR spectrum is recorded. Based on the following ^{13}C NMR spectrum, is the product of this reaction (*E*)-4-octene?

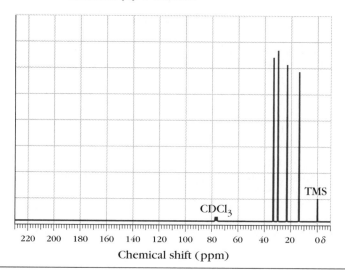

Chemical shift (ppm)

Catalytic hydrogenation can also be used to determine the relative stabilities of alkynes.

9.9   Relative Stabilities of Alkynes

When we compared heats of hydrogenation in Section 8.12, we discovered that the relative stabilities of isomeric alkenes increased as the number of alkyl groups bonded to the carbon atoms of the double bond increased. Using the same technique, we can determine the relative stabilities of alkynes. For example, hydrogenation of both isomeric butynes forms butane, but the hydrogenation of 1-butyne releases 4.5 kcal/mol (18.8 kJ/mol) more heat.

$$CH_3CH_2C \equiv CH \xrightarrow[\text{Pt}]{2H_2} CH_3CH_2CH_2CH_3 \quad \Delta H° \ = \ -69.6 \text{ kcal/mol } (-290.9 \text{ kJ/mol})$$

1-Butyne   Butane

$$CH_3C \equiv CCH_3 \xrightarrow[\text{Pt}]{2H_2} CH_3CH_2CH_2CH_3 \quad \Delta H° \ = \ -65.1 \text{ kcal/mol } (-272.1 \text{ kJ/mol})$$

2-Butyne   Butane

From these data, we conclude that 2-butyne, an internal alkyne, is more stable than 1-butyne, its terminal isomer, by 4.5 kcal/mol (18.8 kJ/mol). It has been found for other alkynes as well that internal alkynes are more stable than their terminal isomers. Just like isomeric alkenes, the more alkyl groups bonded to the carbon atoms of the triple bond, the more stable the alkyne. This is what we would expect on the basis of the electron-donating ability of alkyl groups (Section 7.10), which act to stabilize alkynes and alkenes, thus lowering their heats of hydrogenation.

Heats of hydrogenation of alkynes are slightly greater than twice the heats of hydrogenation of similarly substituted alkenes.

$$CH_3(CH_2)_3C \equiv CH \xrightarrow{H_2} CH_3(CH_2)_3CH = CH_2 \quad \Delta H° \ = \ -39.2 \text{ kcal/mol } (-163.8 \text{ kJ/mol})$$

1-Hexyne   1-Hexene

$$CH_3(CH_2)_3CH = CH_2 \xrightarrow{H_2} CH_3(CH_2)_4CH_3 \quad \Delta H° \ = \ -30.1 \text{ kcal/mol } (-125.8 \text{ kJ/mol})$$

1-Hexyne   Hexane

From these data, we conclude that not only is the first hydrogenation of an alkyne more exothermic than the second, but alkynes have a higher energy content than either alkenes or alkanes. The high energy content of alkynes has been put to practical use in the acetylene torches used in welding where a flame temperature above 2500 °C is needed.

9.10   Acidity of Terminal Alkynes

So far, we have stressed the similarities in the chemistry of alkenes and alkynes. The data in Table 9.2 show, however, that terminal alkynes are much stronger acids than either alkenes or alkanes.

Table 9.2		pK$_a$ Values of Acetylene, Ethylene, and Ethane		
	Acid	Approximate pK$_a$ Relative to H$_2$O	Conjugate Base	
Strongest acid	HC≡CH	25	HC≡C⁻	Weakest base
	H$_2$C=CH$_2$	44	H$_2$C=CH⁻	
Weakest acid	CH$_3$CH$_3$	50	CH$_3$CH$_2$⁻	Strongest base

Because the pK$_a$ values vary so widely (from 25 for acetylene to 50 for ethane), the strengths and stabilities of the conjugate bases vary as well. These differences may be explained by examining the percent of s character in the hybrid atomic orbitals occupied by the lone pair of electrons in each of the conjugate bases.

The lone pair of electrons of the conjugate base of an alkane is located in an sp^3 hybrid atomic orbital. Recall from Section 1.10 that an sp^3 hybrid atomic orbital is made up by combining one s and three p atomic orbitals. Thus an sp^3 hybrid atomic orbital contains 1/4 or 25% s character. The lone pair of electrons in the conjugate base of an alkene is located in an sp^2 hybrid orbital, which has 33% s character (1/3 s and 2/3 p). In the conjugate base of acetylene, the lone pair of electrons is located in an sp hybrid orbital, which has 50% s character (1/2 s and 1/2 p).

Conjugate base of ethane   Conjugate base of ethylene   Conjugate base of acetylene

On average, electrons in s orbitals are held closer to the nucleus of an atom than electrons in p orbitals. As a result, s electrons have increased electrostatic attraction, have lower energy, and have greater stability than electrons in p atomic orbitals (Section 1.4). The more s character in a hybrid atomic orbital that contains a pair of electrons, the less basic is that pair of electrons. This means that the basicity of the conjugate bases of acetylene, ethylene, and ethane increases as the amount of s character in the hybrid orbital that contains the pair of electrons decreases. The hybrid orbital of the conjugate base of acetylene has the highest percent of s character (50%), so it is the weakest base. The conjugate base of ethylene (33% s character) is stronger than HC≡C⁻, but weaker than the conjugate base of ethane (25% s character).

INCREASING BASE STRENGTH

The stronger the conjugate base, the weaker its acid. The order of acid strength of acetylene, ethylene, and ethane, therefore, is the following:

$$
\underset{\substack{H \quad H \\ | \quad | \\ H-C-C-H \\ | \quad | \\ H \quad H}}{} \quad < \quad \underset{\substack{H \\ \diagdown \\ C=C \\ \diagup \quad \diagdown \\ H \quad H}}{\overset{H}{\diagup}} \quad < \quad H-C\equiv C-H
$$

INCREASING ACID STRENGTH ⟩

While terminal alkynes are stronger acids than other hydrocarbons, they are still very weak acids. Their pK_a (≈ 25) makes terminal alkynes much weaker acids than water ($pK_a = 15.7$) or alcohols ($pK_a \approx 18$). Consequently hydroxide ion and alkoxide ion are not strong enough bases to convert terminal alkynes to their conjugate bases, which are called **acetylide ions.** Another way of saying the same thing is that acetylide ions are strong enough bases to remove protons completely from water, alcohols, or carboxylic acids. Thus the position of the following equilibrium reaction lies far to the left:

Terminal alkyne   Acetylide ion

$$
R-C\equiv C-H \ + \ ^-OH \ \rightleftharpoons \ R-C\equiv C^- \ + \ HOH
$$

Weaker acid $pK_a \approx 25$ Weaker base   Stronger base   Stronger acid $pK_a = 15.7$

Because water, alcohols, and carboxylic acids react readily with acetylide ions, they are not suitable solvents for the formation of these anions. Moreover, acetylide ions are readily converted to their conjugate acids (terminal alkynes) by reaction with any compound that contains hydroxy groups.

Acetylide ions can be formed if we use a strong enough base. The requirement for such a base, as discussed in Section 3.5, is that its conjugate acid must have a pK_a value more positive than about 25 (the pK_a of terminal alkynes). One such base is the amide ion (NH_2^-), the conjugate base of ammonia ($pK_a = 35$). Amide ion is prepared by the Fe^{3+} catalyzed reaction of sodium metal with liquid ammonia at $-78\,°C$.

$$
\underset{\substack{H \\ | \\ H-N-H \\ \cdot\cdot}}{} + \ Na \ \xrightarrow[\substack{NH_{3(1)}-78\,°C}]{Fe^{3+}\ catalyst} \ Na^+ \ ^-:\underset{\substack{| \\ \cdot\cdot}}{\overset{H}{N}}-H \ + \ \tfrac{1}{2}H_2
$$

Liquid ammonia (used as solvent and reactant)   Sodium metal   Sodium amide

Solutions of sodium acetylides can be prepared by adding acetylene or terminal alkynes to a solution of sodium amide in liquid ammonia as solvent.

$$
H-C\equiv C-H \ + \ NaNH_2 \ \longrightarrow \ H-C\equiv C^-Na^+ \ + \ NH_3
$$

Acetylene

Stronger acid $pK_a = 25$ Stronger base   Weaker base   Weaker acid $pK_a = 35$

$$R-C\equiv C-H \;+\; NaNH_2 \;\longrightarrow\; R-C\equiv C^-Na^+ \;+\; NH_3$$

Terminal alkyne

$pK_a \approx 25$ $pK_a = 35$

Exercise 9.14   If you placed a terminal alkyne such as 1-hexyne in a solution of sodium hydride in hexane, would you expect to form the acetylide ion? Explain. Why can hexane be used as the solvent?

Exercise 9.15   A chemist adds water by mistake to a solution of $CH_3(CH_2)_3C\equiv C^-Na^+$ in liquid ammonia. A vigorous reaction occurs. What products are formed?

Acetylide ions are used to prepare larger and more complicated alkynes by combining smaller structural units.

9.11   Preparation of Alkynes by Alkylation of Acetylide Ions

We learned in Section 7.9 that compounds that donate a pair of electrons to a center of positive charge are called nucleophiles. According to this definition, the acetylide ions formed from acetylene or terminal alkynes are nucleophiles. Nucleophiles typically react with methyl or primary haloalkanes as shown by the reaction of sodium acetylide with 1-bromobutane in liquid ammonia as solvent.

■ *The reaction of nucleophiles with primary haloalkanes is discussed in more detail in Chapter 12.*

$$H-C\equiv C^-Na^+ \;+\; CH_3(CH_2)_2CH_2Br \;\xrightarrow{NH_{3(l)}}\; H-C\equiv C(CH_2)_3CH_3 \;+\; NaBr$$

Sodium acetylide   1-Bromobutane   1-Hexyne

                        (Primary haloalkane)   (75% yield)

The reaction of sodium acetylide with 1-bromobutane is called an **alkylation reaction** because the product is formed by joining an alkyl group to an existing molecule. In this specific case, 1-hexyne is formed by joining a butyl group (from 1-bromobutane) to one of the carbon atoms of acetylene (from sodium acetylide) by means of a new carbon-carbon single bond. Alkylation is a very useful reaction because it allows us to make larger molecules by joining together parts of two smaller ones.

The two-reaction sequence of forming an acetylide ion then alkylating it is an important and general method of preparing alkynes. Terminal alkynes are conveniently prepared by starting with acetylene. The first reaction is formation of sodium acetylide in liquid ammonia, followed by alkylation with a methyl or primary haloalkane in the second reaction. This is the method used previously to prepare 1-hexyne.

Internal alkynes can be prepared by starting with a terminal alkyne.

$$H-C\equiv CCH_2CH_3 \;+\; NaNH_2 \;\xrightarrow{NH_{3(l)}}\; Na^+ \; ^-C\equiv CCH_2CH_3 \;+\; NH_3$$

1-Butyne

$$Na^{+\ -}C \equiv CCH_2CH_3 + ICH_2CH_2CH_2CH_3 \xrightarrow{NH_{3(l)}} CH_3CH_2CH_2CH_2C \equiv CCH_2CH_3 + NaI$$

1-Iodobutane   3-Octyne
(70% yield)

Internal alkynes can also be prepared by the dialkylation of acetylene.

First alkylation:

$$H-C \equiv C-H \ + \ NaNH_2 \xrightarrow{NH_{3(l)}} H-C \equiv C^- Na^+ \ + \ NH_3$$

Acetylene

$$H-C \equiv C^- Na^+ \ + \ CH_3CH_2I \xrightarrow{NH_{3(l)}} H-C \equiv CCH_2CH_3 \ + \ NaI$$

Iodoethane

Second alkylation:

$$H-C \equiv CCH_2CH_3 \ + \ NaNH_2 \xrightarrow{NH_{3(l)}} Na^{+\ -}C \equiv CCH_2CH_3 \ + \ NH_3$$

1-Butyne

$$Na^{+\ -}C \equiv CCH_2CH_3 \ + \ ICH_3 \xrightarrow{NH_{3(l)}} CH_3C \equiv CCH_2CH_3 \ + \ NaI$$

Iodomethane   2-Pentyne
(65% Overall yield)

Successful alkylation of acetylides occurs only when methyl or primary haloalkanes are used. Alkylation of acetylides cannot be carried out with secondary and tertiary haloalkanes for reasons that we will discuss in Chapter 12.

Exercise 9.16   Write the structure of the product in each of the following reactions.

(a) $CH_3C \equiv C^- Na^+ \ + \ $ 1-bromopropane $\longrightarrow$

(b) Excess $\ C_2H_5C \equiv C^- Na^+ \ + \ $ 1,4-dibromobutane $\longrightarrow$

Exercise 9.17   Write two chemical equations for the preparation of 3-heptyne by the alkylation of an acetylide ion. The haloalkane and the acetylide ion must be different in each equation.

Up to now, we've been concerned with individual reactions of the various functional groups. Let's now learn how to combine them into an appropriate multistep sequence of reactions to prepare new and more complicated compounds.

9.12   Planning an Organic Synthesis

The preparation (or synthesis) of organic compounds is a major function of organic chemistry. The purpose of organic synthesis, however, is not only to produce important compounds in the chemical and pharmaceutical industries but also to teach the principles and concepts of organic chemistry. Successfully synthesizing a compound requires both a *knowledge of the reactions of the various functional groups* and *the ability to properly fit them together into a multistep sequence that leads to the desired product in high yield.*

The secret to planning the synthesis of any compound is to work backward. Thus we begin with the compound to be prepared and work our way backward to the starting compounds. Sample Problem 9.1 illustrates how to use this method.

Sample Problem 9.1   Propose a synthesis of 2-pentyne starting with acetylene, any alkyl halide, any inorganic chemicals, and any solvents needed.

Solution:

We start the synthesis by identifying the functional group or groups in the compound we want to synthesize. In 2-pentyne the functional group is a carbon-carbon triple bond.

Next, we begin the process of thinking backwards by asking a series of questions starting with "How are internal alkynes prepared?" The answer to this question is by the alkylation of acetylide ions (Section 9.11). The general equation for this reaction is as follows:

$$RC \equiv C^- \ + \ R'X \ \longrightarrow \ RC \equiv CR'$$

We must transform this general equation into a specific equation for the synthesis of 2-pentyne from a haloalkane and an acetylide ion. "What haloalkane and acetylide ion should we combine to form 2-pentyne?" We obtain the answer by replacing R with C_2H_5 and R′ with CH_3. In this way we obtain the specific equation for the formation of 2-pentyne.

$$\underset{\text{Acetylide ion}}{C_2H_5C \equiv C^- Na^+} \ + \ \underset{\text{Haloalkane}}{CH_3I} \ \longrightarrow \ \underset{\text{2-Pentyne}}{C_2H_5C \equiv CCH_3} \ + \ NaI$$

We've shown how to prepare 2-pentyne, but we're not finished yet. The original instructions for this synthesis say to start with acetylene. Neither of the reagents used to prepare 2-pentyne by the reaction shown above are acetylene. Thus we must continue working backwards. At this point, we repeat the process by asking, "How are acetylide ions prepared?" The answer is by the reaction of a terminal alkyne with sodium amide in liquid ammonia (Section 9.10).

$$C_2H_5C \equiv CH \ + \ NaNH_2 \ \xrightarrow[\text{NH}_{3(l)}]{} \ C_2H_5C \equiv C^- Na^+$$

Acetylene is still not a reactant so we continue by asking "How is 1-butyne formed?" Terminal alkynes are prepared by alkylation of sodium acetylide.

$$HC \equiv C^- Na^+ \ + \ C_2H_5I \ \longrightarrow \ C_2H_5C \equiv CH \ + \ NaI$$

"How is $HC\equiv C^-Na^+$ formed?" It is formed by the reaction of acetylene with sodium amide in liquid ammonia:

$$HC\equiv CH \;+\; NaNH_2 \xrightarrow[\;NH_{3(l)}\;]{} HC\equiv C^-Na^+$$

The synthesis is finished, therefore, because we have worked our way back to acetylene, the designated starting material. Let's now write the complete synthetic scheme, starting with acetylene, and building up to 2-pentyne.

$$HC\equiv CH \xrightarrow{NaNH_2/NH_{3(l)}} HC\equiv C^-Na^+ \xrightarrow{C_2H_5I} C_2H_5C\equiv CH$$

$$\Big\downarrow NaNH_2/NH_{3(l)}$$

$$C_2H_5C\equiv CCH_3 \xleftarrow{CH_3I} C_2H_5C\equiv C^-Na^+$$

At this point, it's a good idea to check that we have followed the original instructions for the synthesis. They specified that we must begin with acetylene and that we can use any haloalkanes, any inorganic reagents, and any solvents.

Our multistep synthesis satisfies these criteria because acetylene is the starting material, two haloalkanes (iodomethane and iodoethane) are used, and the other reagents are either inorganic chemicals or solvents. Thus we have followed the instructions and synthesized 2-pentyne in four steps from acetylene.

Designing a synthetic scheme by starting with the desired compound and working backwards is called a **retrosynthesis.** The steps in a retrosynthesis can be summarized as follows:

Step 1. Identify the functional group in the compound to be synthesized.

Step 2. Answer the question "How is that functional group prepared?" by writing the general equations for all of the chemical reactions that you know that form that particular functional group as a product.

Step 3. Choose the general chemical reaction in Step 2 that is the best method to prepare the product; that is, choose the most convenient one or the one that gives the better yield of product.

Step 4. Write the chemical equation with the specific compounds that undergo the reaction to form the desired product.

Step 5. Repeat this cycle of steps until the reactants are the permitted starting materials.

The ability to place organic reactions in a multistep sequence that will form a particular compound is a skill that must be developed through practice. After a bit of practice, you'll be able to do a retrosynthesis without writing all the possible reactions as suggested in Step 2. It turns out that there are usually only a few high-yield reactions that convert one functional group to another; you will become familiar with these methods as you work more and more synthetic problems.

| Sample Problem 9.2 | Propose a synthesis of (Z)-3-hexene starting with any organic compound containing no more than four carbon atoms and any inorganic reagents and solvents needed. |

Solution:

First, identify the functional group or groups in the compound we want to synthesize. The functional group is a carbon-carbon double-bond with the (Z)- configuration.

Second, ask yourself, "How are alkenes prepared?" So far we have learned only three ways to prepare alkenes:

      (1) Catalytic hydrogenation of alkynes
      (2) Hydroboration-protonolysis of alkynes
      (3) Lithium or sodium in liquid ammonia reduction of alkynes

Because all three methods use an alkyne, you must ask, "What alkyne or alkynes would we use in each of these reactions to prepare 3-hexene?" The carbon skeleton of the alkyne is not changed in any of these three reactions, so the answer is 3-hexyne. The equations of the reactions of 3-hexyne with each of these reagents are as follows:

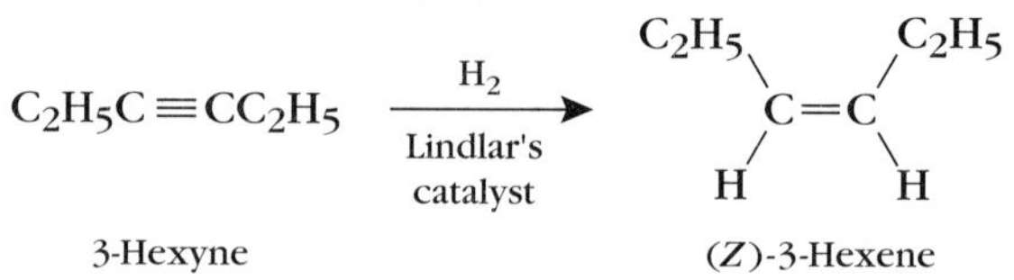

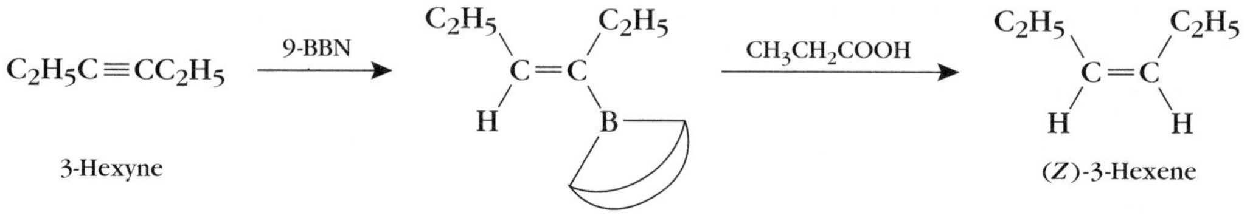

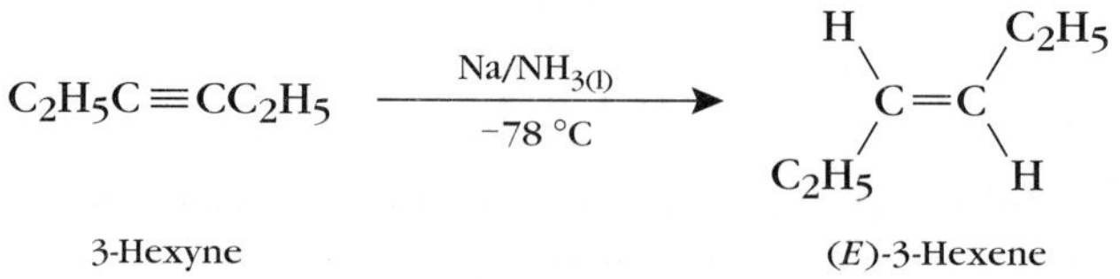

The reaction of sodium in liquid ammonia is not an acceptable method because the wrong isomer, the (E)- rather than the (Z)-, is formed. This reaction, therefore, cannot be part of the synthesis of (Z)-3-hexene.

Both the catalytic hydrogenation and the hydroboration-protonolysis of 3-hexyne form the desired product. So both methods are acceptable. Which is the better method? In this case, both methods give good yields so either method is satisfactory, at least in theory. In the laboratory, the choice would depend on very practical matters such as whether or not the laboratory is equipped with the necessary equipment or whether the chemicals for one method are readily available. For a paper synthesis such as this one, however, we don't have to worry about the availability of chemicals or equipment.

Now we need to synthesize 3-hexyne. "How are internal alkynes prepared?" The alkylation of acetylide ions is the only method of preparing alkynes that we know. The general equation for this reaction is as follows:

$$RC \equiv C^- + R'X \longrightarrow RC \equiv CR'$$

"What haloalkane and acetylide ion would we react to form 3-hexyne?" If we substitute C_2H_5 for both R and R' in the general formula $RC \equiv CR'$, we obtain the formula for 3-hexyne. Making the same replacement in the reactants, we obtain the specific reaction for the preparation of 3-hexyne.

$$C_2H_5C \equiv C^- + C_2H_5X \longrightarrow C_2H_5C \equiv CC_2H_5$$

The last step is to prepare the acetylide ion by reaction of 1-butyne and sodium amide in liquid ammonia. We can stop here because our instructions were to begin with any organic compound of no more than four carbon atoms. Both 1-butyne and haloethane contain no more than four carbon atoms so we have accomplished our synthesis.

Let's now write the complete synthetic scheme starting with the simplest compound and building up to the desired product.

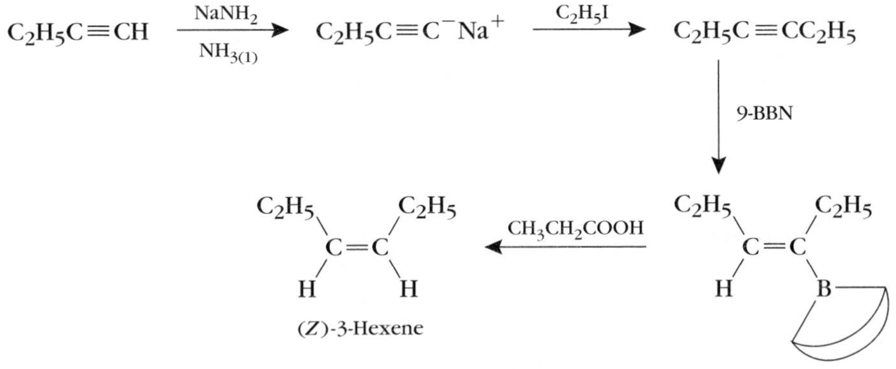

(Z)-3-Hexene

In this way, we have prepared (Z)-3-hexene, a molecule that contains six carbon atoms, in four steps from two smaller molecules, one which contains two carbon atoms and the other which contains four carbon atoms.

Exercise 9.18   Starting with acetylene, any haloalkane, any inorganic reagent, and any solvent, propose a synthesis for each of the following compounds.

(a) $CH_3CH_2\overset{\overset{\displaystyle Br}{|}}{C}=CH_2$ (b) $CH_3\overset{\overset{\displaystyle O}{||}}{C}CH_2CH_3$ (c) $CH_3CH_2CH_2CH_2OH$

In addition to the general rules of synthesis listed above, the yield of a chemical reaction must be considered before it is chosen for a synthesis. This is particularly true in a synthetic scheme that contains several steps. Consider, for example, a hypothetical reaction sequence in which compound A reacts to form compound B which in turn reacts to form compound C. The effect of the yield of each individual step on the overall yield of C based on A as the starting material is shown as follows:

$$A \longrightarrow B \longrightarrow C$$

Yield of each reaction				Total yield of C
90%	1 mole	0.9 mole	0.81 mole	81%
70%	1 mole	0.7 mole	0.49 mole	49%

If we start with one mole of A and the reaction of A to form B occurs in 90% yield, we obtain 0.90 mole of B. Reaction of B to form C in 90% yield gives a total yield of 81% (0.90 mole × 0.90 = 0.81 mole) of C based on the one mole of A that we started with. If the yield of each step is reduced to 70%, the final yield of C based on A as the starting material is only 49%. Thus a 20% decrease in the yield of both steps corresponds to a 32% decrease in the overall yield!

Not only is the yield of each individual step important, but the total number of steps is also important. Consider an alternate hypothetical synthesis of C from A involving X and Y as follows:

$$A \longrightarrow X \longrightarrow Y \longrightarrow C$$

Yield of each reaction					Total yield of C
90%	1 mole	0.9 mole	0.81 mole	0.73 mole	73%
70%	1 mole	0.7 mole	0.49 mole	0.34 mole	34%

Once again we begin with one mole of A. In one case the yield of the three reactions is 90% each, but because there is one additional step the overall yield of C is only 73%. This is a decrease of 8% from the sequence containing only two steps. In the other case, where each step occurs with a 70% yield, the overall yield of C is only 34% (down from 49%).

These two hypothetical syntheses illustrate a fundamental rule of planning an organic synthesis: *the most efficient synthetic scheme is the one with the fewest number of reactions and the one where each reaction occurs in the highest possible yield.*

Exercise 9.19   What is the yield of (Z)-2-butene obtained from propyne in the following series of reactions?

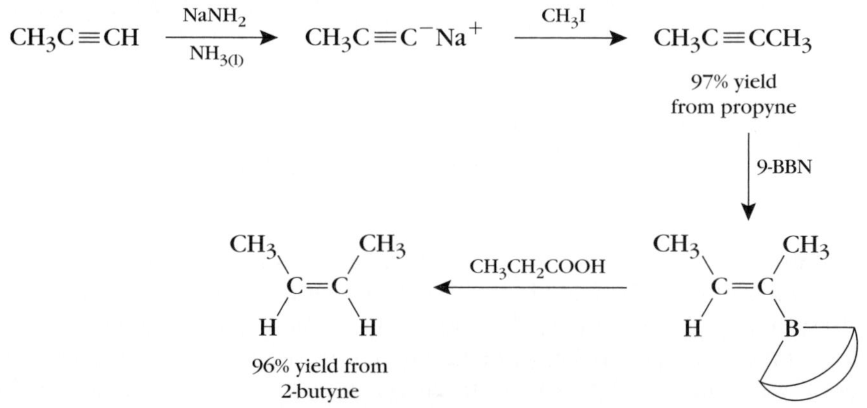

A final word of warning: reactions that form more than one product are usually best avoided in planning a synthesis. For example, the mercury(II)-catalyzed hydration of 2-pentyne forms two products in about equal amounts.

$$C_2H_5C \equiv CCH_3 \xrightarrow[HgSO_4]{H_2SO_4/H_2O} C_2H_5\overset{\displaystyle O}{\overset{\|}{C}}-CH_2CH_3 \ + \ C_2H_5CH_2-\overset{\displaystyle O}{\overset{\|}{C}}CH_3$$

2-Pentyne   3-Pentanone   2-Pentanone

This is not an acceptable synthesis of either 3-pentanone or 2-pentanone. First of all, the product is a mixture of two compounds that have to be separated in order to obtain either compound in pure form. Separation and purification of a mixture of products is a time-consuming and sometimes difficult process. Second, even if the products can be separated, the yield will be low (probably around 40% to 50%). If at all possible, reactions that form mixtures of products should be avoided in planning any synthesis.

9.13 Summary

Alkynes are hydrocarbons that contain one or more carbon-carbon triple bonds. The carbon atoms of the triple bond are *sp* hybrid carbon atoms. The triple bond consists of one σ bond, which is formed by end-on overlap of *sp* hybrid orbitals, and two π bonds, which are formed by sideways overlap of *p* orbitals.

Reactions of alkynes are similar to those of alkenes. Both undergo **electrophilic addition reactions** to form products according to the Markownikoff Rule. Either one or two equivalents of HCl and HBr can add to alkynes. **Vinyl halides** are the products of the addition of one equivalent, while **geminal dihalides** are the products of addition of two equivalents.

Water can be added to an alkyne either by acid-catalyzed hydration or hydroboration-oxidation. Acid-catalyzed hydration of alkynes adds water to an alkyne according to the Markownikoff Rule to form an **enol** that immediately **tautomerizes** to a ketone. Acid-catalyzed hydration of acetylene and terminal alkynes usually requires $HgSO_4$ as a catalyst. This reaction is a good way to prepare methyl ketones from terminal alkynes or ketones from symmetrical internal alkynes.

The hydroboration of an alkyne with 9-BBN followed by oxidation with alkaline hydrogen peroxide also adds water to form an intermediate enol that tautomerizes to an aldehyde or ketone. Hydroboration-oxidation is a good way to form aldehydes from terminal alkynes and ketones from symmetrical internal alkynes.

Depending on the reaction conditions, alkynes can be reduced to either alkanes or alkenes. Reaction of alkynes with hydrogen gas in the presence of a metal catalyst (catalytic hydrogenation) forms an alkane, while catalytic hydrogenation using **Lindlar's catalyst** forms (*Z*)-alkenes. Hydroboration of either terminal or internal alkynes with 9-BBN followed by reaction with propanoic acid forms an alkene. In the case of internal alkynes, the (*Z*)-isomer is formed. Reduction of internal alkynes by lithium or sodium in liquid ammonia forms (*E*)-alkenes.

The major difference in reactivity between alkenes and alkynes is their difference in acidity. The alkyne hydrogen atom of a terminal alkyne is much more acidic than either a vinyl hydrogen atom of an alkene or the hydrogen atoms of alkanes. Consequently, reaction of terminal alkynes with sodium amide in liquid ammonia forms an **acetylide ion.** Neither alkenes nor alkanes undergo such a reaction.

An **alkylation reaction** is the reaction of an acetylide ion with primary haloalkanes. These reactions are particularly important in organic chemistry because they form carbon-carbon single bonds, thus making it possible to synthesize large molecules from smaller ones.

Alkylation of acetylide ions obtained from a terminal alkyne provides one way to synthesize a wide variety of compounds with different functional groups, as illustrated in the following chart.

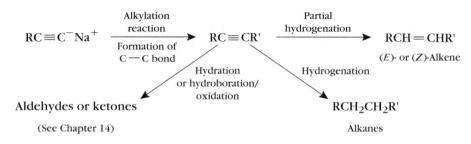

9.14 Summary of Reactions of Alkynes

A Addition Reactions

1. Addition of Brønsted-Lowry Acids

   a. One equivalent of HY   SECTION 9.6, A

Addition of one equivalent of HY to internal alkynes forms a mixture of (*E*)- and (*Z*)-products. Proposed mechanism occurs via a vinyl carbocation intermediate.

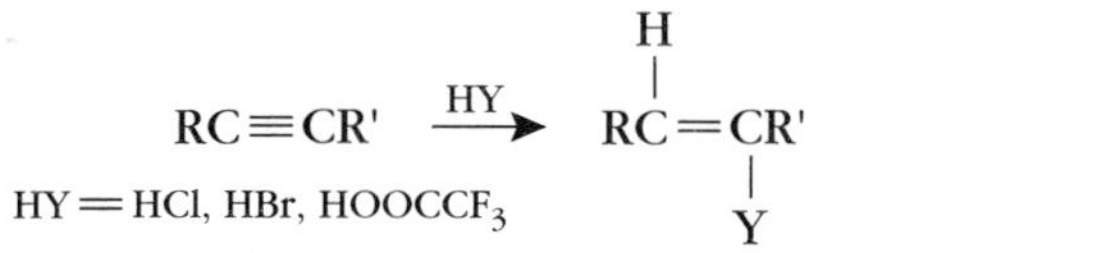

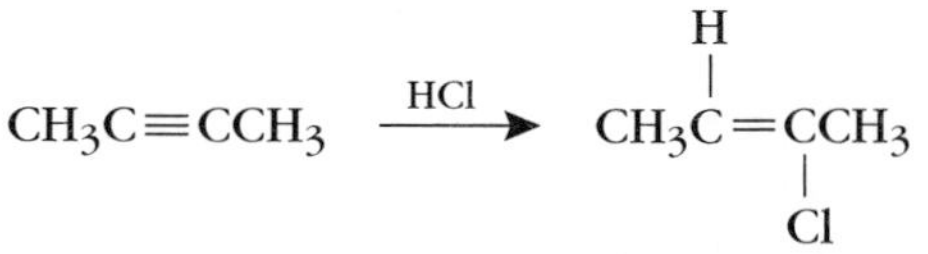

$$RC\equiv CR' \xrightarrow{HY} \underset{\underset{Y}{|}}{RC}=CR' \;\; \overset{H}{|}$$

HY = HCl, HBr, HOOCCF$_3$

$$CH_3C\equiv CCH_3 \xrightarrow{HCl} \underset{\underset{Cl}{|}}{CH_3C}=CCH_3 \;\; \overset{H}{|}$$

Mixture of (*E*)- and (*Z*)-isomers

Addition of one equivalent of HY to terminal alkynes forms products according to the Markownikoff Rule; H always adds to less substituted carbon atom of triple bond.

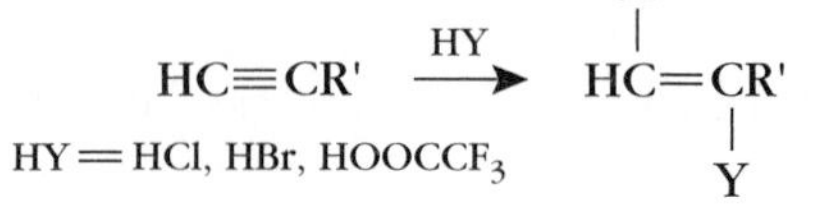

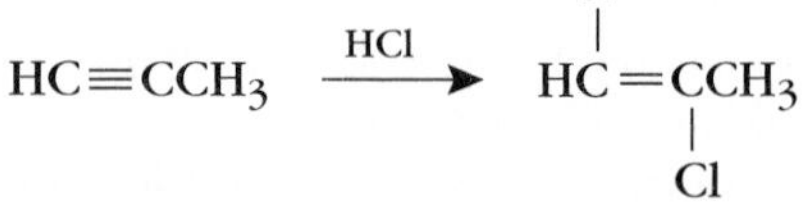

$$HC\equiv CR' \xrightarrow{HY} \underset{\underset{Y}{|}}{HC}=CR' \;\; \overset{H}{|}$$

HY = HCl, HBr, HOOCCF$_3$

$$HC\equiv CCH_3 \xrightarrow{HCl} \underset{\underset{Cl}{|}}{HC}=CCH_3 \;\; \overset{H}{|}$$

b. Two equivalents of HY

Addition of two equivalents of HY forms geminal products.

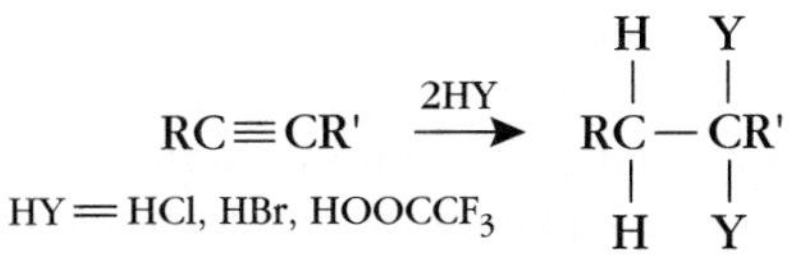

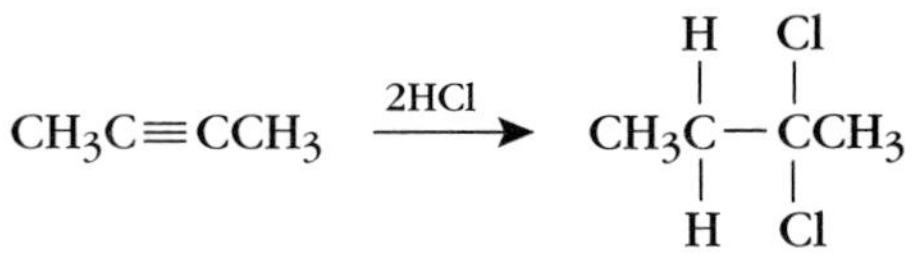

$$RC \equiv CR' \xrightarrow{2HY} \underset{\underset{H}{|} \; \underset{Y}{|}}{\overset{\overset{H}{|} \; \overset{Y}{|}}{RC - CR'}}$$

HY = HCl, HBr, HOOCCF$_3$

$$CH_3C \equiv CCH_3 \xrightarrow{2HCl} \underset{\underset{H}{|} \; \underset{Cl}{|}}{\overset{\overset{H}{|} \; \overset{Cl}{|}}{CH_3C - CCH_3}}$$

2. Addition of Halogens

a. One equivalent of X$_2$ SECTION 9.6, B

Addition of one equivalent of X$_2$ forms a mixture of (E)- and (Z)-isomers.

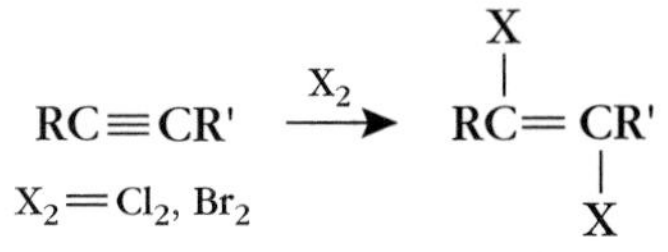

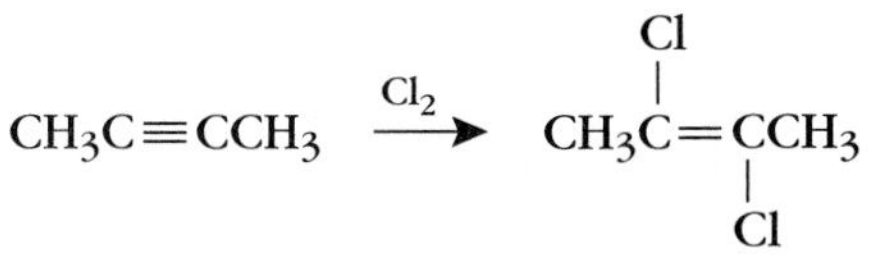

$$RC \equiv CR' \xrightarrow{X_2} \underset{\underset{X}{|}}{\overset{\overset{X}{|}}{RC = CR'}}$$

X$_2$ = Cl$_2$, Br$_2$

$$CH_3C \equiv CCH_3 \xrightarrow{Cl_2} \underset{\underset{Cl}{|}}{\overset{\overset{Cl}{|}}{CH_3C = CCH_3}}$$

Mixture of (E)- and (Z)-isomers

b. Two equivalents of X$_2$

Addition of two equivalents of X$_2$ forms a tetrahaloalkane.

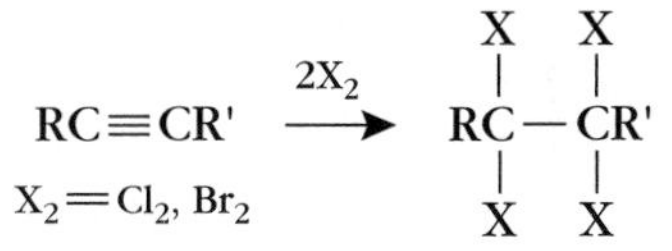

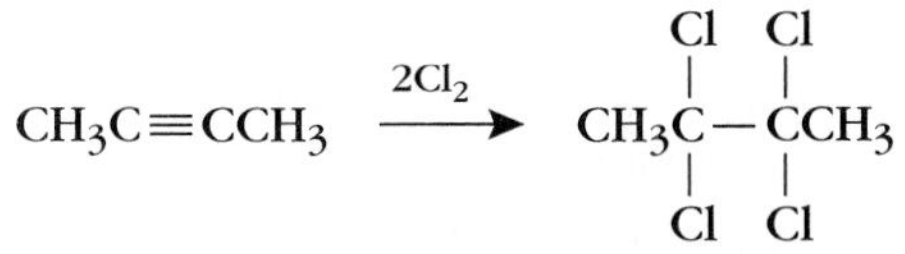

$$RC \equiv CR' \xrightarrow{2X_2} \underset{\underset{X}{|} \; \underset{X}{|}}{\overset{\overset{X}{|} \; \overset{X}{|}}{RC - CR'}}$$

X$_2$ = Cl$_2$, Br$_2$

$$CH_3C \equiv CCH_3 \xrightarrow{2Cl_2} \underset{\underset{Cl}{|} \; \underset{Cl}{|}}{\overset{\overset{Cl}{|} \; \overset{Cl}{|}}{CH_3C - CCH_3}}$$

3. Hydration of Alkynes

a. Acid-catalyzed hydration   SECTION 9.7, A

Addition of water to the triple bond occurs according to the Markownikoff Rule to form an enol, which quickly isomerizes to a ketone. The product of acid-catalyzed hydration of a terminal alkyne is a methyl ketone.

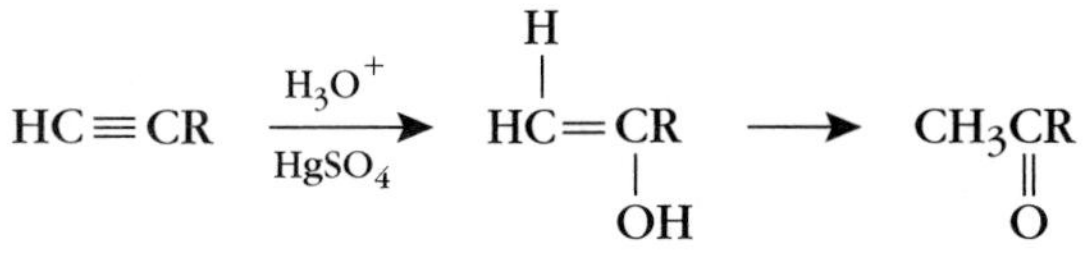

$$HC \equiv CR \xrightarrow[HgSO_4]{H_3O^+} \underset{\underset{OH}{|}}{\overset{\overset{H}{|}}{HC = CR}} \longrightarrow \underset{\underset{O}{\|}}{CH_3CR}$$

Enol   Methyl ketone

$$HC \equiv C(CH_2)_3CH_3 \xrightarrow[HgSO_4]{H_3O^+} \underset{\underset{O}{\|}}{CH_3C(CH_2)_3CH_3}$$

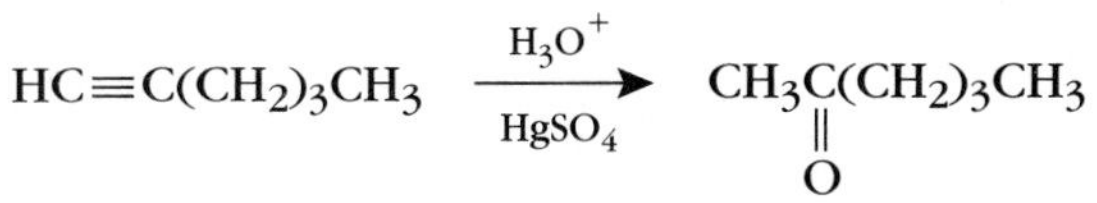

b. Hydroboration-oxidation   SECTION 9.7, B

Hydroboration occurs by syn addition to the triple bond to form an organoborane. Oxidation of the organoborane forms a carbonyl-containing compound. Terminal alkynes form aldehydes, while internal alkynes form ketones.

$$RC{\equiv}CH \xrightarrow{\text{H}-\text{BR}_2} \underset{\overset{|}{R}}{\overset{\overset{H}{|}}{C}}{=}\underset{\overset{|}{H}}{\overset{\overset{BR_2}{|}}{C}}$$

$$\underset{\overset{|}{R}}{\overset{\overset{H}{|}}{C}}{=}\underset{\overset{|}{H}}{\overset{\overset{BR_2}{|}}{C}} \xrightarrow[\text{OH}^-]{\text{H}_2\text{O}_2} \underset{\overset{|}{R}}{\overset{\overset{H}{|}}{C}}{=}\underset{\overset{|}{H}}{\overset{\overset{OH}{|}}{C}} \longrightarrow RCH_2C\overset{O}{\underset{H}{\big\|}}$$

Aldehyde

$$CH_3CH_2C{\equiv}CH \xrightarrow[\text{(2) H}_2\text{O}_2/\text{OH}^-]{\text{(1) H}-\text{BR}_2} CH_3(CH_2)_2C\overset{O}{\underset{H}{\big\|}}$$

$$CH_3C{\equiv}CCH_3 \xrightarrow[\text{(2) H}_2\text{O}_2/\text{OH}^-]{\text{(1) H}-\text{BR}_2} CH_3CH_2\overset{O}{\overset{\|}{C}}CH_3$$

4. Hydrogenation of Alkynes   SECTION 9.8

a. Complete hydrogenation

Catalytic hydrogenation with metal catalysts generally adds two equivalents of hydrogen to the triple bond to form an alkane.

$$RC{\equiv}CR' \xrightarrow[\substack{\text{Metal} \\ \text{catalyst}}]{2H_2} \underset{\overset{|}{H}}{\overset{\overset{H}{|}}{R}}C{-}\underset{\overset{|}{H}}{\overset{\overset{H}{|}}{C}}R'$$

$$CH_3C{\equiv}CCH_3 \xrightarrow[\text{Pd}]{2H_2} \underset{\overset{|}{H}}{\overset{\overset{H}{|}}{CH_3}}C{-}\underset{\overset{|}{H}}{\overset{\overset{H}{|}}{C}}CH_3$$

b. Partial hydrogenation

(i) Lindlar's catalyst

Use of a less reactive catalyst such as Lindlar's catalyst partially hydrogenates the triple bond to form the (Z)-isomer of an alkene.

$$RC{\equiv}CR' \xrightarrow[\substack{\text{Lindlar's} \\ \text{catalyst}}]{H_2} \underset{R}{\overset{H}{\diagdown}}C{=}C\underset{R'}{\overset{H}{\diagup}}$$

$$CH_3C{\equiv}CCH_3 \xrightarrow[\substack{\text{Lindlar's} \\ \text{catalyst}}]{H_2} \underset{H_3C}{\overset{H}{\diagdown}}C{=}C\underset{CH_3}{\overset{H}{\diagup}}$$

(ii) Hydroboration-protonolysis

Reaction of the product of hydroboration of an alkyne with CH_3CH_2COOH forms an alkene. This sequence of reactions converts an internal alkyne into the (*Z*)-isomer of an alkene.

$$RC \equiv CR' \xrightarrow{H-BR_2} \underset{R}{\overset{H}{\diagdown}} C = C \overset{BR_2}{\underset{R'}{\diagup}}$$

$$\underset{R}{\overset{H}{\diagdown}} C = C \overset{BR_2}{\underset{R'}{\diagup}} \xrightarrow{CH_3CH_2COOH} \underset{R}{\overset{H}{\diagdown}} C = C \overset{H}{\underset{R'}{\diagup}}$$

$$CH_3C \equiv CCH_3 \xrightarrow[\text{(2) } CH_3CH_2COOH]{\text{(1) } H-BR_2} \underset{H_3C}{\overset{H}{\diagdown}} C = C \overset{H}{\underset{CH_3}{\diagup}}$$

(*Z*)-2-Butene

(iii) Lithium or sodium in liquid ammonia

Internal alkynes are partially hydrogenated to (*E*)-alkenes

$$RC \equiv CR' \xrightarrow[NH_{3(l)}]{Na} \underset{R}{\overset{H}{\diagdown}} C = C \overset{R'}{\underset{H}{\diagup}}$$

$$CH_3C \equiv CCH_3 \xrightarrow[NH_{3(l)}]{Na} \underset{H_3C}{\overset{H}{\diagdown}} C = C \overset{CH_3}{\underset{H}{\diagup}}$$

(*E*)-2-Butene

B Special Reactions of Alkynes

1. Formation of Acetylide Ions   SECTION 9.10

The hydrogen atom bonded to a carbon atom of the triple bond in a terminal alkyne can be removed by any base whose conjugate acid has a $pK_a > 25$.

$$(CH_3)_2CHC \equiv CH \xrightarrow[NH_{3(l)}]{NaNH_3} (CH_3)_2CHC \equiv C^- Na^+$$

2. Reactions of Terminal Alkynes with Sodium Metal

Sodium metal reacts with terminal alkynes to form an acetylide ion and hydrogen gas.

$$2\ RC \equiv CH \xrightarrow{2Na} 2\ RC \equiv C^- Na^+ \ + \ H_2$$

$$2\ (CH_3)_2CHC \equiv CH \xrightarrow{2Na} 2\ (CH_3)_2CHC \equiv C^- Na^+ \ + H_2$$

C Alkylation of Acetylide Ions   Section 9.11

Reaction of an acetylide ion with a methyl or primary (1°) haloalkane is a good way to prepare an internal alkyne.

$$\underset{\substack{\text{Acetylide} \\ \text{ion}}}{RC \equiv C^- Na^+} \ + \ \underset{\substack{R' = H \text{ or} \\ \text{alkyl group}}}{R'CH_2X} \longrightarrow RC \equiv CCH_2R'$$

$$(CH_3)_2CHC \equiv C^- Na^+ \ + \ CH_3I \longrightarrow (CH_3)_2CHC \equiv CCH_3$$

Additional Exercises

9.20   Define each of the following terms either in words, by a structure, or by means of a chemical equation.

   (a)   Alkyne   **(e)**   Hydroboration-oxidation of an alkyne
   (b)   Enol   **(f)**   Acetylide ion
   (c)   Tautomerism   **(g)**   Alkylation of acetylide ion
   (d)   Vinyl cation   **(h)**   Hydroboration-protonolysis of an alkyne

9.21   Write structures and give the IUPAC names of all alkynes of molecular formula C_6H_{10}. Indicate which of the isomers will react with $NaNH_2$ in liquid ammonia. Indicate which of the isomers exists as a pair of enantiomers.

9.22   Give the IUPAC name of each of the following alkynes.

(a)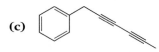

(d)

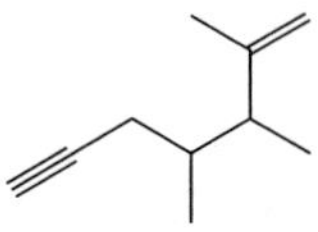

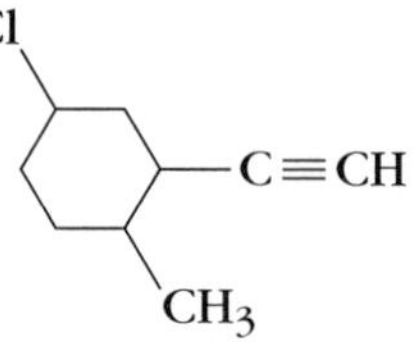

(b)

(e)

(c)

(f)

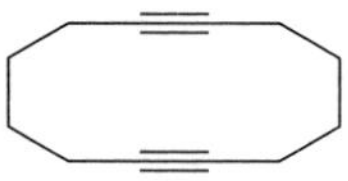

9.23   Write the structure that corresponds to each of the following names.

   (a)   *cis*-1-Ethynyl-4-methylcyclohexane   **(d)**   (*R*)-4-Bromo-2-hexyne
   (b)   5-*tert*-Butyl-2-methyl-3-nonyne   **(e)**   3-Chloro-5-methyl-1,6,8-decatriyne
   (c)   3,4-Dichlorocyclodecyne

9.24   Write the structure of the product formed in the reaction of 1-pentyne with each of the following reagents.

   (a)   One equivalent of HCl   **(g)**   Two equivalents of HCl
   (b)   Excess hydrogen gas/PtO_2 **(h)** $NaNH_2$ in liquid ammonia
   (c) $H_2SO_4/H_2O/HgSO_4$ **(i)**   The product of reaction **(h)** and CH_3I
   (d)   9-BBN followed by H_2O_2/OH^- **(j)**   Two equivalents of Br_2
   (e)   9-BBN followed by CH_3CH_2COOH **(k)**   Excess hydrogen/Lindlar's catalyst
   (f)   One equivalent of Br_2 **(l)** CF_3COOH

9.25   Write the structure of the product formed (if any) in the reaction of 2-pentyne with each of the following reagents.

   (a)   One equivalent of HCl   **(h)** $NaNH_2$ in liquid ammonia
   (b)   Excess hydrogen gas/PtO_2 **(i)**   The product of reaction **(h)** and CH_3I
   (c) $H_2SO_4/H_2O/HgSO_4$ **(j)**   Two equivalents of Cl_2
   (d)   9-BBN followed by H_2O_2/OH^- **(k)**   Sodium metal in liquid ammonia
   (e)   9-BBN followed by CH_3CH_2COOH **(l)** CF_3COOH
   (f)   One equivalent of Br_2 **(m)**   Excess hydrogen/Lindlar's catalyst
   (g)   Two equivalents of HCl

9.26   In each of the following reactions, write the structural formula of the missing starting material or reagent.

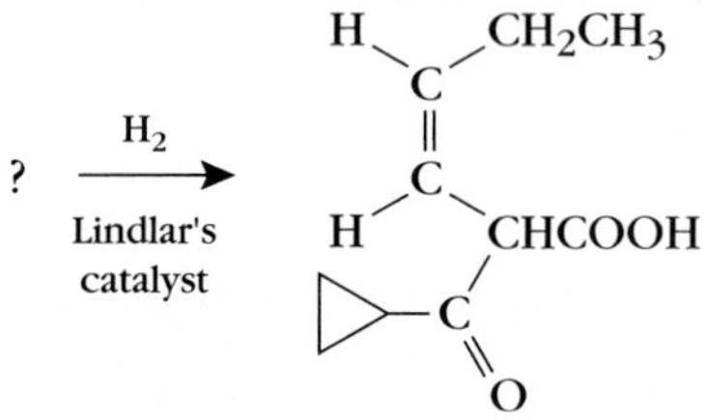

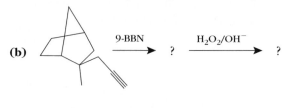

(c) $CH_2=CHCH_2CHCH_2CH_2Br$ $\xrightarrow{CH_3C\equiv C^-Na^+}$ **(d)** 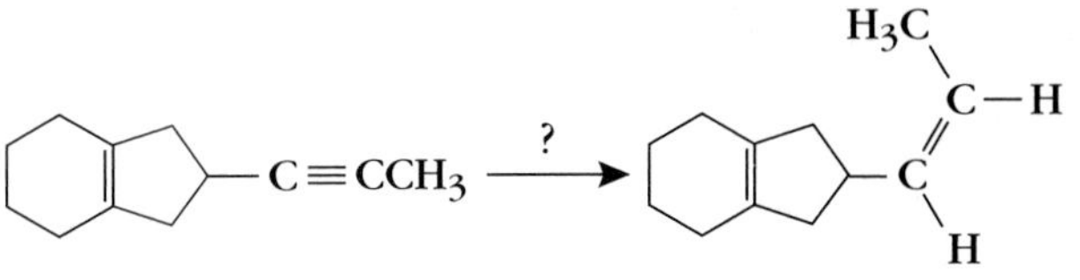 [bicyclic ring]$-C\equiv CCH_3$ $\xrightarrow{?}$ [bicyclic ring with substituent]
  |
  CH_3

9.27 Starting with 1-butyne, any haloalkane, any inorganic reagent, and any solvent, propose a synthesis for each of the following compounds.

(a) 1,2-Dibromobutane

(b) $CH_3CH_2CHCH_2Br$
    |
    OH

(c) 2-Chlorobutane

(d) 1-Butene

(e) $CH_3CH_2CH_2C\overset{\displaystyle O}{\underset{\displaystyle H}{\diagup}}$

(f) $CH_3\overset{\displaystyle O}{\overset{\|}{C}}CH_2CH_3$

(g) 3-Heptyne

9.28 Write the reaction or reactions that convert 1-hexyne into each of the following compounds.

(a) Hexane

(b) 2-Chloro-1-hexene

(c) $CH_3CH_2CH_2CH_2\overset{\displaystyle O}{\overset{\|}{C}}CH_3$

(d) 2-Heptyne

(e) $CH_3CH_2CH_2CH_2CH_2C\overset{\displaystyle O}{\underset{\displaystyle H}{\diagup}}$

9.29 Starting with acetylene, any haloalkane, any inorganic reagents, and any solvent, propose a synthesis for each of the following compounds.

(a) 2-Chlorooctane

(b) *meso*-2,3-Dibromobutane

(c) $CH_3(CH_2)_2CH_2C\overset{\displaystyle O}{\underset{\displaystyle H}{\diagup}}$

(d) $CH_3CH_2CH_2CH_2CH_2OH$

(e) 1-Bromohexane

(f) Racemic $CH_3CHCHCH_2CH_3$
        | |
        OH OH

(g) $CH_3CH_2\overset{\displaystyle O}{\overset{\|}{C}}CH_2CH_2CH_3$

(h) $CH_3\overset{\displaystyle O}{\overset{\|}{C}}CH_2CH_2CH_2CH_3$

(i) $CH_3CH_2CH_2CH_2CH_2C\equiv CD$

(j) $\underset{D}{\overset{H_3C}{\diagdown}}C=C\underset{D}{\overset{CH_2CH(CH_3)_2}{\diagup}}$

(k) $\underset{D}{\overset{H_3C}{\diagdown}}C=C\underset{CH_2(CH_2)_6CH_3}{\overset{D}{\diagup}}$

9.30 Muscalure is the sex pheromone of the common house fly. It can be synthesized according to the following series of reactions. Write the structures of the intermediate compounds A, B, and C, as well as the structure of muscalure.

$$CH_3(CH_2)_{12}C\equiv CH \; + \; NaNH_2 \; \xrightarrow{NH_{3(l)}} \; A\;(C_{15}H_{27}Na)$$

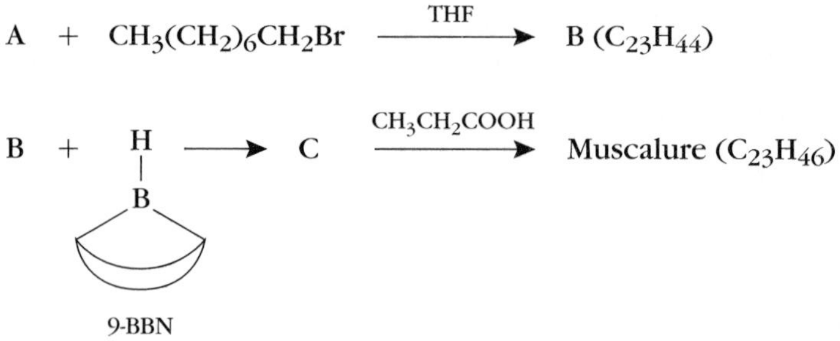

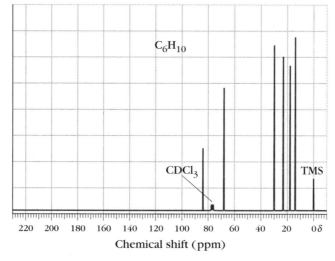

9.31 What absorption bands in the IR spectrum of a compound would you look for to determine whether that compound is an alkyne or an alkene? If you determine that the compound is an alkyne, what absorption bands would you use to establish that it is an internal or a terminal alkyne?

9.32 A compound isolated as a side product in a reaction has the molecular formula C_6H_{10}. On the basis of its ^{13}C NMR spectrum, which is given below, is this product an alkyne or a diene? What is the structure of the compound?

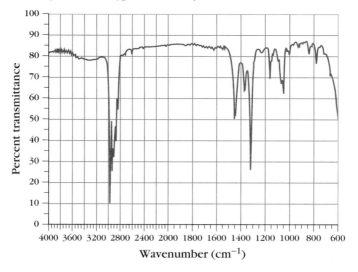

9.33 Before carrying out a reaction, chemists frequently use spectroscopy to check the purity of the starting materials. A chemist wants to use 3-hexyne in a reaction. She takes a sample from the bottle and obtains the IR spectrum shown below. This IR spectrum lacks the absorption bands typical of an alkyne. Should the chemist be concerned? Explain.

9.34 Explain how IR and ^{13}C NMR spectroscopy can be used to distinguish between the following pairs of compounds. Indicate precisely the peak or peaks that serve to distinguish between the two.

(a) CH_2=$CHCH_2CH_2CH$=CH_2 and $(CH_3)_2CHCH_2C$≡CH

(b) CH_3C≡CCH_3 and CH_2=$CHCH$=CH_2

(c) HC≡$CCH_2\overset{\overset{\displaystyle O}{\|}}{C}CH_3$ and HC≡CCH=$CHCH_2OH$

9.35 Determine the structure of each of the following compounds from its molecular formula and IR and ^{13}C NMR spectra.

(a) C_3H_4O

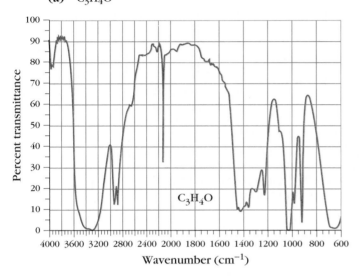

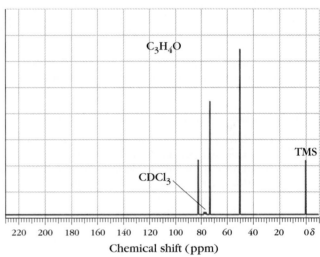

(b) C_5H_8

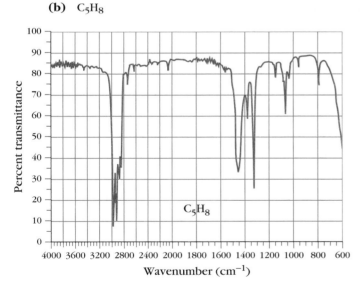

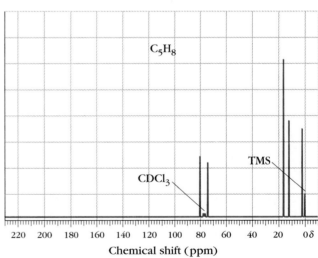

(c) C_8H_6

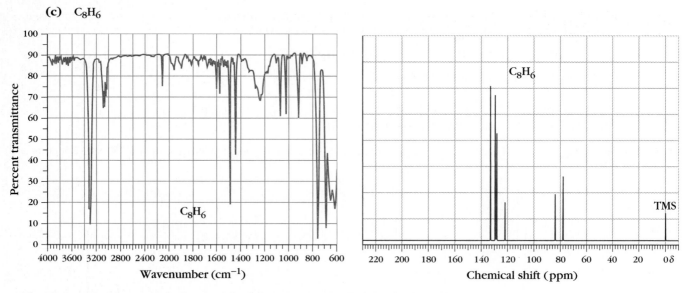

9.36   Determine the structure of each of the following compounds from the data given.

    (a)   Molecular formula: C_5H_8; IR: 2120 cm^{-1}, 3300 cm^{-1}; ^{13}C NMR: δ 13.4, δ 20.4, δ 22.0, δ 68.2, δ 84.5.

    (b)   Molecular formula: C_5H_7Cl; IR: 2100 cm^{-1}, 3290 cm^{-1}; ^{13}C NMR: δ 34.5, δ 56.9, δ 71.8, δ 86.5.

    (c)   Molecular formula: $C_3H_2O_2$; IR: 2100 cm^{-1}, 3500-2600 cm^{-1}; ^{13}C NMR: δ 73.9, δ 77.7, δ 157.3.

    (d)   Molecular formula: C_4H_7N; IR: 2240 cm^{-1}; ^{13}C NMR: δ 13.3, δ 19.0, δ 19.2, δ 119.8.

CHAPTER 10

NUCLEAR MAGNETIC RESONANCE SPECTROSCOPY

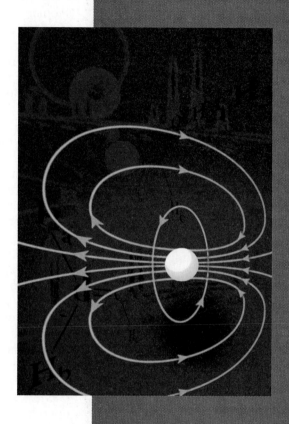

In Chapter 5, we learned how to use ^{13}C NMR and infrared spectroscopy to determine the structure of an organic compound. ^{13}C NMR spectroscopy is used to determine the number and kinds of nonequivalent carbon atoms in a molecule, and infrared spectroscopy is used to determine what functional groups are present. Taken together these techniques provide valuable information about the complete structure of an unknown compound. In this chapter, we introduce another useful spectroscopic technique called ^1H nuclear magnetic resonance (^1H NMR), also frequently referred to as proton magnetic resonance (PMR). ^1H NMR spectroscopy is complementary to ^{13}C NMR and IR spectroscopy. It allows us to determine the number and kinds of nonequivalent hydrogen atoms in a molecule. We begin this chapter by reviewing and extending the elementary principles of NMR introduced in Section 5.9.

10.1   The Nuclear Magnetic Experiment Revisited

Certain nuclei behave as if they are spinning about an axis. When charged spheres spin, they generate magnetic moments that make them behave (to a first approximation) like small bar magnets. Nuclei with odd atomic numbers and odd mass numbers (such as ^1_1H, $^{19}_9F$, and $^{31}_{15}P$) and even atomic numbers and odd mass numbers (such as $^{13}_6C$) all have these net magnetic properties.

Nuclei with even atomic numbers and even mass numbers (such as $^{12}_6C$, $^{16}_8O$, and $^{32}_{16}S$) have no net magnetic moment because the particles in the nucleus are paired. One half spin in one direction while the other half spin in the opposite direction, with the result that there is no net nuclear spin and consequently no magnetic moment.

The two nuclei with magnetic properties that are of particular interest to organic chemists are ^1H and ^{13}C. In the absence of any external perturbation, the magnetic moments of these nuclei are randomly oriented. If these nuclei are placed in an external magnetic field, H_O, however, they orient themselves in one of two ways. One is with the field (the lower energy orientation), and the other is opposite the magnetic field (the higher energy orientation). The nuclei aligned with H_O are said to have spin α, while those aligned against H_O are said to have spin β. The energy difference between the α and β spin states is very small, 0.006 cal (.025 J). At a magnetic field of 14,000 gauss, therefore, there are 1,000,007 nuclei in the lower energy state (spin α) for every 1,000,000 nuclei in the higher (spin β).

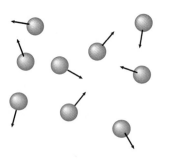 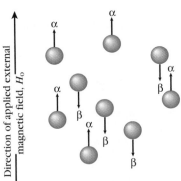

No magnetic field: Magnetic moments of nuclei are randomly oriented

Applied external magnetic field: Nuclei with α spin are oriented with the field; nuclei with β spin are oriented against the field

Two different energy spin states, α and β, create the necessary conditions for NMR spectroscopy (Section 5.8). These nuclei, still in an external magnetic field H_O, are irradiated with electromagnetic radiation of a frequency that corresponds to the energy difference between spins α and β. This causes some of the nuclei in spin state α to absorb enough energy to "flip" to spin state β. NMR spectroscopy is possible because we can observe and measure the absorption of energy.

The energy difference between spin states α and β depends on the identity of the nuclei and the strength of the magnetic field. The stronger the magnetic field, the greater the energy difference between the two spin states. This relationship is expressed as follows:

$$\nu = \gamma H / 2\pi \qquad (10.1)$$

where ν is the frequency of the electromagnetic radiation (in Hz), H is the magnetic field strength (in gauss) felt at the nucleus, and γ is the gyromagnetic ratio. The gyromagnetic ratio is a nuclear constant that is characteristic of particular nuclei. Notice that the magnetic field strength at the nucleus and energy difference between the two spin states are linearly related, as shown in Figure 10.1.

The first commercial NMR spectrometers were made in the 1960s and were continuous-wave instruments. With these spectrometers, an NMR experiment can be carried out in one of two ways. We can place the sample in a constant magnetic field, pass electromagnetic radiation of steadily changing frequency through it, then record the frequency at which radiation is absorbed. For technical reasons, however, it is more convenient to keep constant the frequency of the electromagnetic radiation and vary the strength of the magnetic field. Absorption of energy occurs when the strength of the magnetic field exactly matches the energy necessary to flip the nuclei from spin state α to spin state β.

More modern NMR spectrometers use a technique called **Fourier transform NMR (FT NMR) spectroscopy.** In a Fourier transform spectrometer, the sample in a magnetic field is irradiated with a brief intense pulse of radio frequency radia-

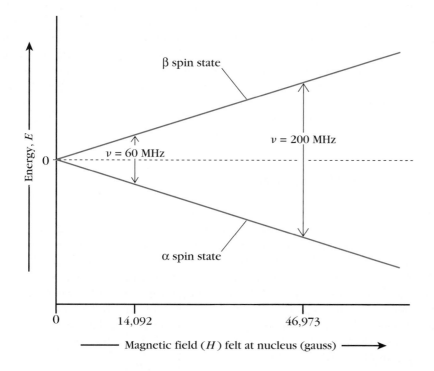

Figure 10.1
The stronger the magnetic field (in gauss) felt at the nucleus, the higher the frequency (in Hz) needed to flip the nucleus. A frequency of 60 MHz is needed to flip a nucleus that feels a magnetic field of 14,092 gauss, but a frequency of 200 MHz is needed if the nucleus feels a magnetic field of 46,973 gauss.

tion. The range of frequencies of the radio frequency radiation is broad enough that the nuclei under study, either ^{1}H or ^{13}C, flip to spin state β. As some of the nuclei return to their α spin states, the frequencies of their energy loss are measured simultaneously and stored in a computer memory. An FT NMR spectrometer generates a single ^{1}H or ^{13}C NMR spectrum in a few seconds. As a result, it can scan the sample repeatedly in a short period of time. Each of these scans is stored in the computer memory, and they are computer averaged. Computer averaging is a mathematical procedure that averages the random background noise to zero and increases the intensities of the real signals. Then a program in the computer carries out a Fourier transform analysis of the data and produces an NMR spectrum.

The principal advantage of FT NMR spectroscopy is the great increase in sensitivity (ratio of signal to random background noise). A continuous wave spectrometer takes about 100 times longer than an FT NMR spectrometer to obtain a single spectrum. An FT NMR spectrometer can accumulate many scans during the time that it takes a continuous-wave spectrometer to scan the spectrum once. This results in an increase in sensitivity proportional to the square root of the number of scans. This increase in sensitivity brought about by the introduction of FT NMR spectroscopy has led to the routine use of ^{13}C NMR spectroscopy as an important technique in the determination of structures of organic compounds. Most of the NMR spectra shown in this text were obtained by FT NMR. The ^{1}H NMR spectra were recorded on a 200 MHz FT NMR spectrometer, while ^{13}C NMR spectra were obtained on a 50 MHz FT NMR spectrometer, unless otherwise indicated.

The magnet in an NMR spectrometer must produce a magnetic field of between 14,100 and 117,500 gauss in order to bring a ^{1}H nucleus into resonance. This corresponds to electromagnetic radiation of a frequency of between 60 and 500 MHz. Commercial ^{1}H NMR spectrometers are now available that operate at one of the following frequencies: 100 MHz, 200 MHz, 300 MHz, or 500 MHz.

Nuclei of ^{13}C and ^{1}H have different gyromagnetic ratios so they resonate at different magnetic fields. The practical consequence of this fact is that ^{13}C NMR and ^{1}H NMR spectra cannot be recorded on a spectrometer that operates at only a single frequency and single magnetic field. Originally it was necessary to have two different spectrometers to measure ^{13}C and ^{1}H NMR spectra. Recent advances in NMR instrumentation and manufacture have resulted in the development of instruments that can be quickly and easily modified to record both ^{13}C and ^{1}H NMR spectra.

The results of a nuclear magnetic resonance experiment are recorded on an NMR spectrum. Let's review its important features.

10.2   The NMR Spectrum

Typical ^{1}H and ^{13}C NMR spectra are shown in Figure 10.2.

Several terms used to describe the relative locations of absorptions in an NMR spectrum were introduced in Chapter 5 but merit review at this time. The magnetic field applied to the sample increases from left to right in both ^{1}H and ^{13}C NMR spectra. Signals on the left of the spectrum are said to be deshielded and are located downfield (low field), while those on the right are said to be shielded and are located upfield (high field). While it is possible to record magnetic field strength along the top of an NMR spectrum in hertz (cps—cycles per second), most spectra do not include this information. Only a few NMR spectra in this text are calibrated in hertz along the top.

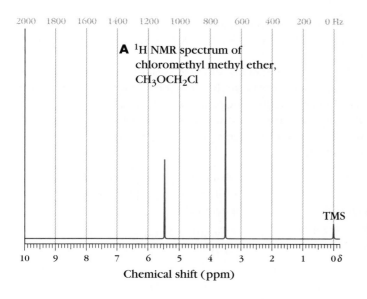

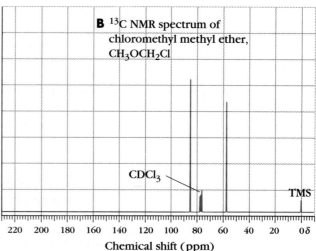

Figure 10.2
The ^{1}H NMR spectrum **(A)** and the ^{13}C NMR spectrum **(B)** of chloromethyl methyl ether.

An NMR spectrometer does not measure and record the actual value of the magnetic field at which absorption occurs because it is impractical to do so. The frequencies of signals are measured with reference to an internal standard instead. The reference compound used in both ^{1}H and ^{13}C NMR spectroscopy is tetramethylsilane (TMS).

$$CH_3 - \underset{\underset{CH_3}{|}}{\overset{\overset{CH_3}{|}}{Si}} - CH_3$$

Tetramethylsilane
(TMS)

A small amount of TMS is added to the solution of the compound whose spectrum is to be measured, and the sharp signal from the twelve equivalent hydrogen atoms is assigned a value of zero hertz in ^{1}H NMR; the sharp signal from the four equivalent carbon atoms is also assigned a value of zero hertz in ^{13}C NMR. TMS was chosen as a reference standard because its signals in ^{13}C and ^{1}H NMR are at a higher field than almost any other signal, so they do not obscure the signals of most hydrogen or carbon atoms. Furthermore, TMS doesn't react with most organic compounds.

The positions of signals in the NMR spectrum of a compound are called their **chemical shifts.** They may be cited relative to the reference signal in two ways. One is to measure the distance in hertz between the reference signal and a particular signal. For example, the two signals in the ^{1}H NMR spectrum of chloromethyl methyl ether in Figure 10.2 are located at 704 and 1096 hertz downfield from the TMS signal.

There is one problem with this method, namely the chemical shift of a particular resonance (in hertz) varies with the strength of the external magnetic field. Since the strengths of the magnetic field and the electromagnetic radiation are directly proportional (Equation 10.1), doubling the strength of the electromagnetic radiation will double the distance (in hertz) between the signal and the reference signal. This is a problem because the NMR spectrometers commonly used in laboratories around the world often operate at different fixed frequencies (60 MHz, 100 MHz, etc.). As a result, chemical shifts given in hertz will vary from one instrument to another.

The second way to report the chemical shifts of peaks in an NMR spectrum uses an arbitrary scale, called the **delta scale (δ scale).** The advantage of this scale is that it reports chemical shifts independent of the field strength of the instrument. Chemical shifts on the delta scale are obtained by dividing the distance between that signal and the reference signal (in hertz) by the frequency of the spectrometer in megahertz.

$$\begin{array}{c} \text{Chemical shift } (\delta) \\ \text{in parts per million (ppm)} \end{array} = \dfrac{\begin{bmatrix} \text{Observed} \\ \text{position of} \\ \text{signal in Hz} \end{bmatrix} - \begin{bmatrix} \text{Position of} \\ \text{signal of} \\ \text{TMS in Hz} \end{bmatrix}}{\begin{bmatrix} \text{Operating frequency of NMR} \\ \text{spectrometer in megahertz (MHz)} \end{bmatrix}}$$

For example, the chemical shift on the delta scale of a signal found 400 Hz downfield from TMS on a spectrum obtained on a 200 MHz NMR spectrometer is:

$$\delta = \frac{400 \text{ Hz}}{200 \text{ MHz}} = 2.00$$

Chemical shifts on the delta scale are usually given two decimal places for ^{1}H and one decimal place for ^{13}C. *TMS is the standard reference for both ^{1}H and ^{13}C NMR, so the chemical shift of both its hydrogen and carbon atoms is defined as δ 0.00.* As we move downfield from TMS the values of chemical shifts on the delta scale increase as nuclei become progressively deshielded. By convention, the delta (δ) scale is printed at the bottom of ^{1}H and ^{13}C spectra.

Because the δ scale doesn't vary from spectrometer to spectrometer, it can be used to conveniently tabulate NMR data. In Figure 10.2, for example, the chemical shifts of the two hydrogen atoms of chloromethyl methyl ether are δ 3.52 and δ 5.48.

Exercise 10.1   Arrange each of the following groups of chemical shifts in order of increasing shielding.

(a) δ 5.34; δ 9.56; δ 2.74   **(c)** 750 Hz; 150 Hz; 450 Hz
(b) δ 38.5; δ 105.4; δ 205.7

Exercise 10.2   Fill in the missing values in the following table.

Frequency of Spectrometer (MHz)	Distance of Signal Downfield from TMS (Hz)	Chemical Shift of Signal (δ)
200	453	
60		2.60
	1054	3.50
400	1000	
200	500	

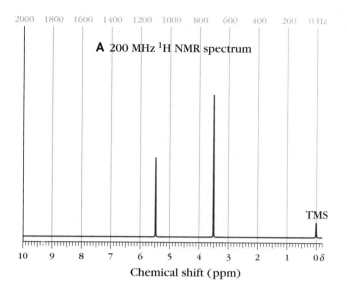

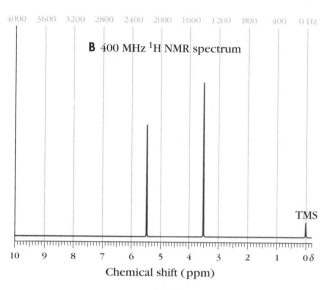

Figure 10.3
^{1}H NMR spectra of chloromethyl methyl ether at **(A)** 200 MHz and **(B)** 400 MHz. The chemical shifts of the two signals on the delta scale are the same whether the spectrum is recorded at 200 MHz or 400 MHz. In contrast, the chemical shifts on the frequency scale are different on the spectrum recorded at 200 MHz than on the one recorded at 400 MHz.

The important point to remember is that *the chemical shift of a particular signal given on the delta scale is a constant regardless of the operating frequency of the NMR spectrometer.* This fact is illustrated by the two ^{1}H NMR spectra of chloromethyl methyl ether shown in Figure 10.3. One was recorded on a 200 MHz NMR spectrometer and the other on a 400 MHz spectrometer. The chemical shifts of the two signals expressed in hertz are 704 Hz and 1096 Hz at 200 MHz, and 1408 Hz and 2192 Hz at 400 MHz, while the chemical shifts on the delta scale for both hydrogen atoms are δ 3.52 and 5.48 whether at 200 or 400 MHz.

The chemical shift ranges for most ^{1}H and ^{13}C NMR spectra are noticeably different. Almost all ^{1}H NMR signals are found between δ 0 and 10, while most ^{13}C NMR signals are found between δ 0 and 220. The wide range of chemical shifts means that it is less likely that two nonequivalent signals will accidently overlap in ^{13}C NMR. In order to decrease the chance of accidental overlap in ^{1}H NMR, chemists try to use instruments of high field strength, such as 500 MHz, rather than ones of low field strength, such as 60 MHz. The example shown in Figure 10.3 illustrates how signals of nonequivalent nuclei are more widely separated at high field strength. The chemical shifts of the two hydrogen atoms of chloromethyl methyl ether are separated by 392 Hz in the spectrum recorded at 200 MHz, while they are separated by 784 Hz at 400 MHz.

Exercise 10.3 Most 60 MHz NMR spectrometers cannot resolve two signals that differ by δ 0.02. How many hertz (Hz) would separate these two signals if the NMR spectrum were obtained using a 400 MHz spectrometer?

Why does the ^{1}H NMR spectrum of chloromethyl methyl ether show only two signals when the molecule contains five hydrogen atoms? We answer this question next.

10.3   The Number of NMR Signals

We learned in Chapter 5 that nuclei in the same magnetic environment have the same chemical shift in an NMR spectrum. Nuclei in the same magnetic environment are said to be **magnetically equivalent.** Nuclei in different magnetic environments are said to be **magnetically nonequivalent,** and they have different chemical shifts.

Magnetically equivalent nuclei in NMR spectroscopy are generally the same as chemically equivalent nuclei. Simple visual inspection is usually enough to determine how many chemically distinct nuclei are present in a molecule. For example, in a molecule of dimethoxymethane, $CH_3OCH_2OCH_3$, the six hydrogen atoms of the methyl groups are magnetically and chemically equivalent to each other. The two hydrogen atoms of the methylene group are also magnetically and chemically equivalent to each other. However, the six hydrogen atoms of the methyl groups are not equivalent to the two methylene hydrogen atoms. There are two sets of equivalent hydrogen atoms in dimethoxymethane, and consequently its ^1H NMR spectrum shows only two signals. There are two signals in the ^{13}C NMR spectrum of dimethoxymethane because the two sets of carbon atoms are magnetically and chemically nonequivalent.

If any doubt exists, the replacement method, which was introduced in Chapter 5, can be used to determine whether two nuclei are equivalent or not. To apply this method to carbon atoms, we must separately replace two carbon atoms by silicon. The two carbon atoms are equivalent if the products of this replacement are identical; they are nonequivalent if the products are different. The same method can be used to determine whether two hydrogen atoms are equivalent or not. In this case, the two hydrogen atoms are separately replaced by deuterium (D) atoms.

$H_3COCH_2OCH_3$	$H_3C-\overset{\overset{\displaystyle CH_3}{\displaystyle |}}{\underset{\underset{\displaystyle CH_3}{\displaystyle |}}{C}}-Br$	$\overset{H_3C}{\underset{H_3C}{>}}C=C\overset{CH_3}{\underset{CH_3}{<}}$
Dimethoxymethane	*tert*-Butyl bromide	2,3-Dimethyl-2-butene

Equivalent hydrogen atoms

Exercise 10.4   Identify each group of equivalent hydrogen atoms in the following compounds.

(a) $CH_3CH_2SCH_2CH_3$ (b) $CH_3-\overset{\overset{\displaystyle CH_3}{\displaystyle |}}{\underset{\underset{\displaystyle CH_2Cl}{\displaystyle |}}{C}}-H$ (c) $CH_3CH_2CH_3$

Exercise 10.5   For each of the following compounds, identify the chemically nonequivalent hydrogen atoms.

(a) (*E*)-2-Butene   (d) 1,1-Dichlorocyclopropane
(b) (*Z*)-2-Butene   (e) *cis*-1,2-Dichlorocyclopropane
(c) 1,2-Dichloropropane   (f) *trans*-1,2-Dichlorocyclopropane

Another thing that we can learn from the ^{1}H NMR spectrum of a compound is the relative numbers of equivalent nuclei.

10.4 Counting the Relative Numbers of Equivalent Hydrogens in an NMR Spectrum

The area under a peak in an ^{1}H NMR spectrum is directly related to the relative abundance of the hydrogen atoms causing that signal. The larger the signal area, the greater the number of hydrogen atoms responsible for that signal. Thus the relative numbers of different kinds of hydrogen atoms in a molecule can be determined by comparing their relative peak areas. All ^{1}H NMR spectrometers have the ability to electronically measure the area under a signal or peak. This process is called **integration.**

The integration of an NMR spectrum is recorded on the spectrum in a stepwise manner. An example is the stepwise line drawn on the spectrum of chloromethyl methyl ether shown in Figure 10.4. The relative heights of these steps, which can easily be measured with a ruler, correspond to the relative numbers of equivalent hydrogen atoms responsible for these various peaks. In Figure 10.4, for example, the measured ratio of the integrated area of the two peaks is 4.7:3.2 (1.47:1 = 2.94:2 ≈ 3:2). A ratio of 3:2 is consistent with a molecule that contains three equivalent —CH$_3$ hydrogen atoms and two equivalent —CH$_2$ hydrogen atoms.

In the example shown in Figure 10.4, the measured ratio 3:2 corresponds to the actual number of distinct hydrogen atoms in the molecule. This is not always the case. For example, the integrated ratio of the areas of the two peaks in the spectrum of 1,2-dimethoxyethane shown in Figure 10.5 is also 3:2. A molecule of 1,2-dimethoxyethane contains six equivalent —CH$_3$ hydrogen atoms and four equivalent —CH$_2$ hydrogen atoms. The ratio of the actual number of hydrogen atoms in the molecule is 6:4, which is a multiple of 3:2. Thus the actual number of nonequivalent hydrogen atoms in a molecule may be a multiple of the ratio measured from the integrated spectrum.

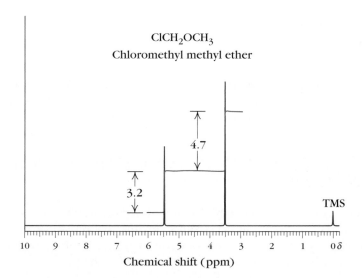

Figure 10.4
The integrated ^{1}H NMR spectrum of chloromethyl methyl ether. The ratio of the integrated area measured by a ruler is 4.7:3.2, which corresponds to a ratio of approximately 3:2 (2.94:2).

Figure 10.5
The integrated ^{1}H NMR spectrum of 1,2-dimethoxyethane. The measured ratio of the integrated area is 1.48:1, which corresponds to a ratio of 2.96:2 ≈ 3:2. The actual ratio of non-equivalent hydrogen atoms is 6:4, which is a multiple of 3:2.

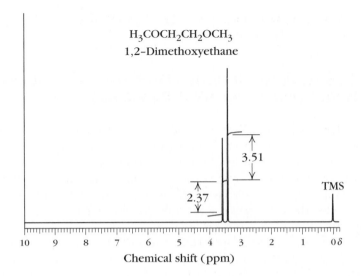

It is more difficult to obtain correct integration of peak areas in a ^{13}C NMR spectrum because of the differences in the rates at which various ^{13}C nuclei relax. In this process, called relaxation, an excited nucleus in a β spin state loses energy to the surroundings and returns to an α spin state. The energy released by the nucleus is converted into a small amount of heat within the sample. In practical terms this means that ^{13}C NMR spectra are usually not integrated.

Exercise 10.6 For each of the following compounds, give the ratio of the actual number of chemically different hydrogen atoms in the molecule and the ratio that would be obtained for these hydrogens from the integrated ^{1}H NMR spectrum of the compound.

(a) $CH_3CCH_2CCH_3$ (with two O double-bonded to the C's)

(b) 1,4-bis(CH_2CH_3) benzene

(c) $(CH_3)_3CCH_2I$

In order to use the unique pattern of signals in the NMR spectrum of a compound to deduce its structure, we must first learn how the chemical shifts of carbon and hydrogen atoms depend on molecular structure.

10.5 ^{1}H and ^{13}C NMR Chemical Shifts and Molecular Structure

Recall from Chapter 5 that the hybridization of carbon atoms is one structural feature that affects ^{13}C NMR chemical shifts. In this section, we learn how other structural features, such as the electronegativity of neighboring atoms or groups of atoms, affect the chemical shifts of hydrogen and carbon.

The hydrogen chemical shifts of various structural units important in organic chemistry are listed in Table 10.1.

Many of the values listed in Table 10.1 are for hydrogens that are part of or adjacent to functional groups that we have not yet discussed. While a detailed analysis of these chemical shifts is postponed until these functional groups are introduced in future chapters, we can correlate certain chemical shifts with the structural environments of hydrogens in classes of compounds that have already been discussed.

All the hydrogen atoms of alkanes (Chapter 2) are found in the relatively high field region between δ 0.8 and 1.7. Notice in Table 10.1, however, that primary hydrogen atoms are more shielded than secondary hydrogen atoms, which are more shielded than tertiary hydrogen atoms. Thus chemical shift values (δ) for alkane hydrogen atoms increase in the order $1° < 2° < 3°$.

The chemical shifts of hydrogens on carbon atoms bonded to electronegative elements such as oxygen, nitrogen, and the halogens are shifted downfield to the

Table 10.1	Typical Hydrogen Chemical Shift Ranges	
Type of Hydrogen	Structure	Range of Chemical Shifts (δ)
1° Alkyl	RCH_3	0.8-1.1
2° Alkyl	R_2CH_2	1.2-1.4
3° Alkyl	R_3CH	1.4-1.7
1° Allylic	$R_2C{=}C\begin{smallmatrix}R\\CH_3\end{smallmatrix}$	1.6-1.9
Vinylic	$R_2C{=}CH_2$	4.5-5.0
Vinylic	$R_2C{=}C\begin{smallmatrix}H\\R\end{smallmatrix}$	5.0-6.5
Acetylenic	$RC{\equiv}CH$	2.5-3.0
Benzylic	$ArCH_3$	2.2-2.7
1° Alkyl chloride	RCH_2Cl	3.6-3.8
1° Alkyl bromide	RCH_2Br	3.4-3.6
1° Alkyl iodide	RCH_2I	3.1-3.3
Ether	$ROCH_2R$	3.3-4.0
Aromatic	Ar-H	6.0-9.0
Methyl ketone	$R{-}\underset{\underset{\displaystyle CH_3}{}}{\overset{\overset{\displaystyle O}{\|}}{C}}$	2.1-2.6
Aldehyde	$R{-}\underset{\underset{\displaystyle H}{}}{\overset{\overset{\displaystyle O}{\|}}{C}}$	9.7-10.0

region between δ 2.5 and 4.5. Adjacent electronegative atoms affect chemical shifts of hydrogen in the following ways:

- The more electronegative the nearby group or atom, the more it deshields the adjacent hydrogen atoms. The more the hydrogen atoms are deshielded, the further downfield are their chemical shifts. The following data demonstrate this effect.

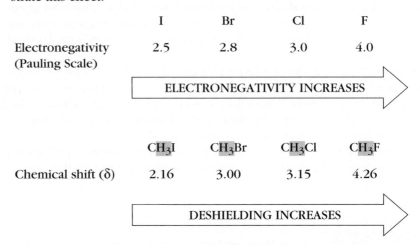

	I	Br	Cl	F
Electronegativity (Pauling Scale)	2.5	2.8	3.0	4.0

ELECTRONEGATIVITY INCREASES

	CH_3I	CH_3Br	CH_3Cl	CH_3F
Chemical shift (δ)	2.16	3.00	3.15	4.26

DESHIELDING INCREASES

- Several electronegative groups or atoms bonded to a single carbon atom have a cumulative effect on the chemical shift of the adjacent hydrogen atom. For example:

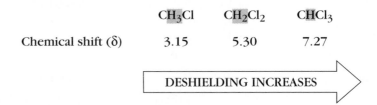

	CH_3Cl	CH_2Cl_2	$CHCl_3$
Chemical shift (δ)	3.15	5.30	7.27

DESHIELDING INCREASES

- The deshielding effect on hydrogen atoms diminishes rapidly with increasing distance from the carbon atom bearing the electronegative group or atom, as shown by the following example.

	$CH_3CH_2CH_2CH_2Cl$	$CH_3CH_2CH_2CH_3$
Chemical shift δ	0.95	0.90

Notice that the chemical shift of the hydrogen atoms on C4 of 1-chlorobutane is about the same as the methyl hydrogen atoms of butane. This is clear evidence that the hydrogen atoms of C4 of 1-chlorobutane no longer experience the deshielding effect of the chlorine atom.

The presence of a carbon-carbon double bond deshields most hydrogen atoms in its vicinity. The extent of deshielding depends on the location of the hydrogen atom relative to the double bond. Thus chemical shifts of primary allylic hydrogen atoms are found slightly downfield in the region between δ 1.6 and 1.9, while the chemical shifts of hydrogen atoms bonded directly to the carbon atoms of the double bond (vinylic hydrogen atoms) are found downfield between δ 4.5 and 6.5.

$1°$ Allylic hydrogen atoms (δ 1.6–1.9)
Vinylic hydrogen atoms (δ 4.5–6.5)

The triple bond of alkynes also deshields most hydrogen atoms in its vicinity but not as much as a carbon-carbon double bond. Thus acetylenic hydrogen atoms of terminal alkynes are found between δ 2.5 and 3.0.

Exercise 10.7	Only one signal appears in the ¹H NMR spectrum of each of the following compounds. At approximately what chemical shift would you expect to find that signal?

(a) CH_3OCH_3
(b) $ClCH_2CH_2Cl$
(c) $(CH_3)_2C{=}C(CH_3)_2$

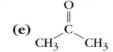

Many of the factors that influence the chemical shifts of hydrogen atoms also influence the chemical shifts of carbon atoms.

B   Typical Chemical Shifts in ¹³C NMR Spectra

A few ¹³C NMR chemical shifts were given in Section 5.10. A more complete list is given in Table 10.2.

While the factors that influence the chemical shifts of carbon atoms are more complicated than those for hydrogen atoms, the electronegativity of the group or atom bonded to a carbon atom is still an important factor. As in the case of ¹H NMR spectroscopy, the more electronegative the group bonded to a carbon atom, the further downfield its ¹³C NMR chemical shift, irrespective of the hybridization of the carbon atom. The following data demonstrate this effect.

Table 10.2	**Typical ¹³C NMR Chemical Shift Ranges**	

Type of Carbon	Structure	Range of Chemical Shifts (δ)
$1°$ Alkyl	RCH_3	1-30
$2°$ Alkyl	R_2CH_2	20-44
$3°$ Alkyl	R_3CH	32-60
$4°$ Alkyl	R_4C	38-70
Benzylic	$ArCH_3$	20-65
$1°$ Alkyl chloride	RCH_2Cl	35-80
$1°$ Alkyl bromide	RCH_2Br	25-70
$1°$ Alkyl iodide	RCH_2I	0-40

ELECTRONEGATIVITY INCREASES ⟹

	Br	Cl	O	F
Electronegativity (Pauling Scale)	2.8	3.0	3.5	4.0
sp^3 hybrid carbon	CH_3Br	CH_3Cl	CH_3OH	CH_3F
Chemical shift (δ)	9.6	25.6	49.9	71.6

sp^2 hybrid carbon	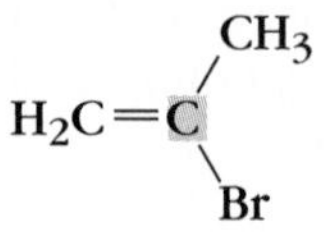 $H_2C=C\overset{CH_3}{\underset{Br}{}}$	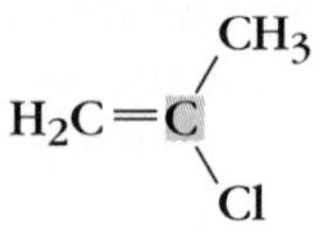 $H_2C=C\overset{CH_3}{\underset{Cl}{}}$
Chemical shift (δ)	114.3	124.9

DESHIELDING INCREASES ⟹

As in the case of ^{1}H NMR spectroscopy, increasing the number of electronegative atoms or groups bonded to a single carbon atom has a cumulative effect on the chemical shift of the carbon atom.

^{13}C	CH_3Cl	CH_2Cl_2	$CHCl_3$	CCl_4
Chemical shift (δ)	25.6	54.4	77.7	96.7

INCREASING DESHIELDING OF CARBON ⟹

Exercise 10.8	At approximately what chemical shifts would you expect to find the signals in the ^{13}C NMR spectrum of each of the compounds in Exercise 10.6?

In summary, the NMR chemical shifts of carbon and hydrogen atoms are affected similarly by neighboring atoms or groups of atoms.

10.6   Using ^{1}H NMR to Determine Structures of Compounds

So far we have learned that the ^{1}H NMR spectrum of a compound provides the following information about its structure:

- The number of signals in the spectrum reveals the number of nonequivalent hydrogen atoms in the structure.
- The chemical shifts of the signals give information about the chemical environment of hydrogen atoms in the structure.
- Integration of the area under each of the peaks reveals the relative numbers of each kind of nonequivalent hydrogen atoms.

The use of this information to determine the structure of a compound is illustrated in Sample Problem 10.1.

| Sample Problem 10.1 | Determine the structure of the compound $C_5H_{11}Cl$ from its ^{1}H NMR spectrum. |

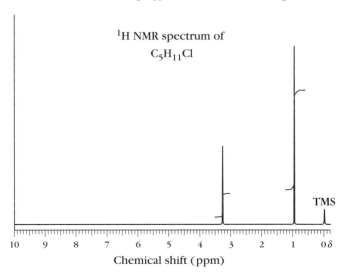

^{1}H NMR spectrum of
$C_5H_{11}Cl$

TMS

Chemical shift (ppm)

Solution:

The molecular formula, $C_5H_{11}Cl$, indicates that there are no elements of unsaturation in the compound. There are two signals in the spectrum, indicating that the molecule contains two types of hydrogen atoms and their integrated ratio is 4.5:1. A molecule can't contain half a hydrogen atom so the ratio of the actual number of hydrogen atoms must be a multiple of the integrated ratio. The multiple 9:2 indicates that there are 9 hydrogen atoms of one kind and 2 of another, which accounts for the 11 hydrogen atoms in the compound.

A peak at chemical shift δ 1.01 indicates that the hydrogen atoms responsible for this signal are probably in methyl groups bonded to a carbon atom. These 9 hydrogen atoms are chemically equivalent. One useful thing to learn about ^{1}H NMR spectra is that a methyl group accounts for 3 equivalent hydrogen atoms and a *tert*-butyl group accounts for 9. Thus we can write the following partial structure of the compound.

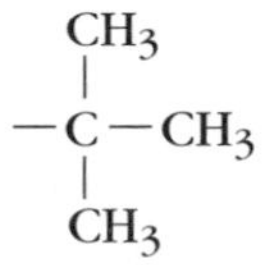

$$\begin{array}{c} CH_3 \\ | \\ -C-CH_3 \\ | \\ CH_3 \end{array}$$

The remaining 2 hydrogens are equivalent, indicating that they are a methylene group. However, the chemical shift of the remaining 2 hydrogen atoms is downfield from the usual chemical shift of methylene hydrogen atoms so we conclude that they are near an electronegative atom. Chlorine is the only electronegative atom in the molecule so the methylene hydrogen atoms must be near it. The following is another partial structure.

$$Cl-CH_2-$$

Attaching the *tert*-butyl group to the methylene group gives the complete structure.

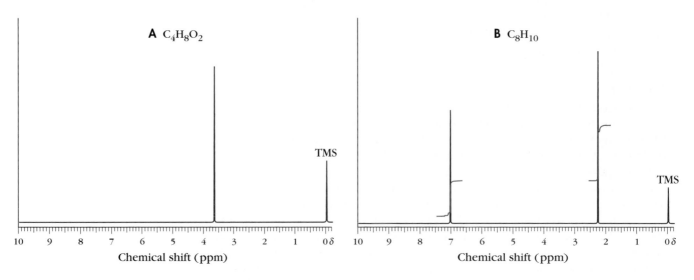

$$Cl-CH_2-\underset{\underset{CH_3}{|}}{\overset{\overset{CH_3}{|}}{C}}-CH_3 \longleftarrow \delta\,1.01$$

δ 3.32

| Exercise 10.9 | Determine the structure of each of the following compounds from its chemical formula and ^{1}H NMR spectrum. |

(a) $C_4H_8O_2$ **(b)** C_8H_{10} **(c)** $C_6H_{12}O$

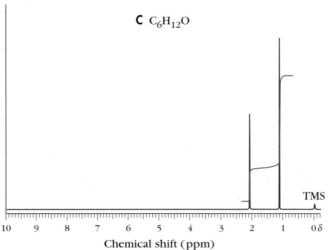

10.7   Spin-Spin Coupling in ^{1}H NMR Spectra

The ^{1}H NMR spectra presented so far have consisted of only a series of single peaks. These kinds of spectra are obtained only for compounds whose nonequivalent hydrogen atoms are separated from each other by at least one atom that contains no hydrogen atoms. Atoms that can insulate nonequivalent hydrogen atoms from each other in this way include oxygen, a quaternary carbon, a carbonyl carbon, sulfur, and nitrogen. The nonequivalent methyl and methylene hydrogen atoms in chloromethyl methyl ether, for instance, are separated by an oxygen atom. The result is an ^{1}H NMR spectrum consisting of two single peaks or singlets, one for each kind of nonequivalent hydrogen atom. In the absence of this kind of structural feature, single absorption peaks are split into multiplets by magnetic interactions between neighboring nonequivalent hydrogen atoms. This magnetic interaction is called **spin-spin coupling** or **spin-spin splitting.**

Spin-spin coupling between chemically equivalent hydrogen atoms is not observed in ^{1}H NMR spectroscopy. This is why only a singlet is observed for the equivalent methyl hydrogens in the ^{1}H NMR spectrum of chloromethyl methyl ether. Coupling occurs between hydrogen nuclei and nonequivalent neighboring hydrogen nuclei. The strongest spin-spin coupling occurs between vicinal hydrogen atoms and is called **vicinal coupling.** Vicinal hydrogen atoms are separated by three bonds.

$$\begin{array}{ccc} H & & H \\ \diagdown 1 & 3 \diagup & \\ C & \!\!\!- 2 -\!\!\! & C \\ \diagup \diagdown & & \diagup \diagdown \end{array}$$

Spin-spin coupling between hydrogen atoms separated by two bonds is called **geminal coupling** and is observed only if the two hydrogen atoms are nonequivalent.

$$\begin{array}{c} H \diagdown 1 \quad 2 \diagup H \\ C \\ \diagup \quad \diagdown \end{array}$$

While spin-spin coupling between hydrogen atoms separated by four bonds is weaker, it can sometimes be observed.

The ^{1}H NMR spectrum of 1,1,2-trichloroethane in Figure 10.6 shows a typical splitting pattern caused by spin-spin coupling between neighboring nonequivalent hydrogen nuclei. Notice that there are two distinct groups of peaks in the spectrum, corresponding to the two kinds of nonequivalent hydrogen atoms present in 1,1,2-trichloroethane. The chemical shift of the hydrogen atom on C1 would be expected to be further downfield than that of the hydrogen atoms on C2 because two electronegative chlorine atoms deshield more than one (Section 10.5). This is confirmed by the integration, which shows that the ratio of the area of the downfield to the upfield group is 1:2.

The two hydrogen atoms that give rise to the upfield resonance are chemically equivalent, so no coupling between them is observed. Both these hydrogen atoms, however, are coupled to the downfield hydrogen. As a general rule, n equivalent neighboring hydrogen atoms split the signal of the observed hydrogen atom into $n + 1$ peaks. This is called the $n + 1$ rule. Thus the upfield signal is split by the Cl_2CH hydrogen atom into two peaks called a doublet. The two peaks of the doublet have the same area, giving a relative ratio of 1:1 in their intensities. Spectra whose splitting patterns conform to the n + 1 rule are called **first-order spectra.**

Figure 10.6
The ^{1}H NMR spectrum of 1,1,2-trichloroethane.

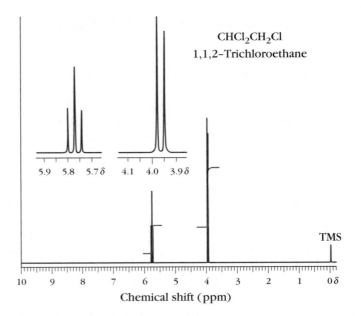

Coupling is a reciprocal interaction so the Cl_2CH hydrogen atom is also coupled to the $ClCH_2$ hydrogen atoms. Consistent with the $n + 1$ rule, the downfield signal is split into three peaks called a triplet. The middle peak in a triplet has twice the area of either outer peak so the relative ratios of the three peaks are 1:2:1.

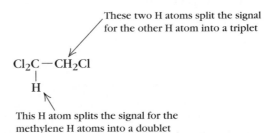

Spin-spin coupling may seem like a complicating factor to the interpretation of ^{1}H NMR spectra. In fact, it often makes structural determination easier because of the additional information it provides. Spin-spin coupling reveals how many hydrogen atoms are vicinal to one or more hydrogen atoms responsible for a particular signal. As a result, certain patterns of peaks are characteristic of several common structural features of organic compounds.

One of the most common of these patterns is due to the hydrogen atoms of an ethyl group, shown in the ^{1}H NMR spectrum of bromoethane in Figure 10.7. The signals of the hydrogen atoms of the ethyl group in compounds of the type CH_3CH_2X appear as a triplet-quartet pattern. The signal of the methylene hydrogen atoms is split into a quartet by coupling with the three methyl hydrogen atoms. The relative ratios of the peak areas in this quartet are 1:3:3:1 and are characteristic of this pattern.

The methyl signal is split into a triplet by coupling with the two methylene hydrogen atoms. The relative ratios of the peak areas of this triplet are 1:2:1 and are characteristic of this pattern.

These three H atoms split the signal
of the methylene H atoms into a quartet

$$Br-CH_2-CH_3$$

These two H atoms split the signal of
the methyl H atoms into a triplet

The ¹H NMR spectrum of bromoethane shown in Figure 10.7 has the expected spin-spin coupling pattern of an ethyl group, an upfield triplet, and a downfield quartet, but the relative intensities of the patterns are skewed. In both the triplet and the quartet the outer lines are shorter than the inner lines, although they should be identical. Departure from the theoretical relative ratios in this fashion is called leaning. Leaning is quite useful for the analysis of complex spectra because the multiplet leans in the direction of the signal of the nucleus responsible for the splitting. In Figure 10.7, for example, the triplet and quartet lean toward each other.

The coupling of the hydrogen atoms of an isopropyl group produces another characteristic pattern as shown for 2-chloropropane in Figure 10.8. The signal of the six equivalent hydrogen atoms is split into a doublet by the single hydrogen atom on C2. The signal of the hydrogen atom on C2 is split into a septet by the six equivalent methyl hydrogen atoms. This doublet-septet pattern of peaks is characteristic of the presence of an isopropyl group in a molecule. All seven peaks of the single hydrogen atom of an isopropyl group are not always visible. The outermost lines of the septet are often weak and do not show up on the spectrum. While there should be seven peaks, often only five are visible in the spectrum.

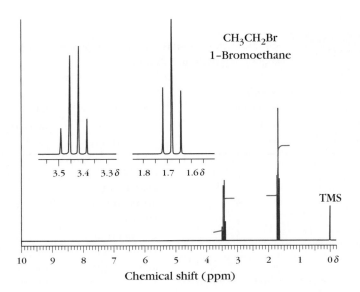

CH_3CH_2Br
1–Bromoethane

Figure 10.7
¹H NMR spectrum of bromoethane showing the characteristic triplet-quartet pattern of an ethyl group.

Chemical shift (ppm)

Figure 10.8
^{1}H NMR spectrum of 2-chloro-
propane showing the characteristic
doublet-septet pattern of an isopropyl group.

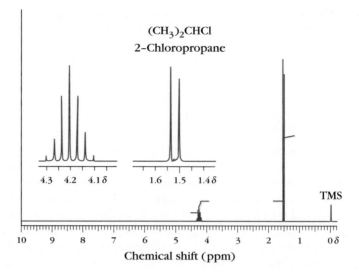

The splitting patterns and ratios of relative peak areas expected for coupling to various numbers of hydrogen atoms are summarized in Table 10.3.

How do we determine the chemical shift of a hydrogen atom that is split into a multiplet by neighboring nonequivalent hydrogen atoms? The answer is simply to take the center of the multiplet. For a doublet, the chemical shift is halfway between the two peaks; for a triplet, the center peak is the chemical shift of the hydrogen.

The effects of spin-spin coupling on ^{1}H NMR signals can be summarized as follows:

- The signal of a hydrogen atom separated from n equivalent neighboring hydrogens by three bonds is split into a multiplet of $n + 1$ peaks.
- Coupling between equivalent hydrogen atoms does not appear in ^{1}H NMR spectra. The equivalent hydrogens may be on the same carbon or on different carbon atoms.

Because of the reciprocal nature of coupling, coupling constants give additional information about which hydrogen atoms are coupled.

Table 10.3	Splitting Patterns that Result from Presence of n Equivalent Neighboring Hydrogen Atoms		
Number of Equivalent Neighboring Hydrogens (n)	Number of Peaks Observed in Multiplet ($n + 1$)	Name of Pattern and Abbreviation	Ratios of Relative Peak Areas
0	1	Singlet (s)	1
1	2	Doublet (d)	1:1
2	3	Triplet (t)	1:2:1
3	4	Quartet (q)	1:3:3:1
4	5	Quintet (quint)	1:4:6:4:1
5	6	Sextet (sex)	1:5:10:10:5:1
6	7	Septet (sept)	1:6:15:20:15:6:1

Exercise 10.10   Two ^{1}H NMR spectra and the structures of three compounds are given below. Match the structure to the correct spectrum.

(a) Cl_2CHCH_3 **(b)** $Cl_3CCH_2CCl_3$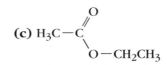

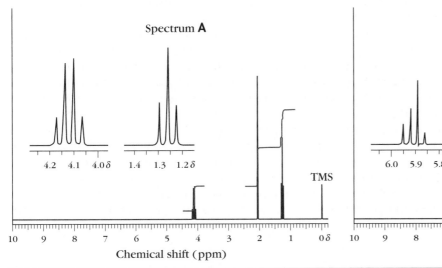

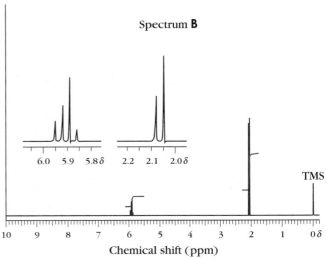

Exercise 10.11   Two ^{1}H NMR spectra are given below. One is the spectrum of 1-bromopropane; the other is the spectrum of 2-bromopropane. Match the structure to the correct spectrum.

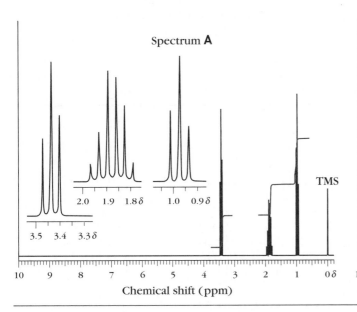

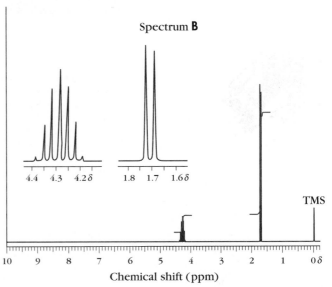

10.8   Coupling Constants

The distance between adjacent peaks in a multiplet is called the **coupling constant.** Coupling constants are measured in hertz and are given the symbol J. The coupling constant between two nonequivalent hydrogen atoms, designated H_a and H_b, is represented by J_{ab}. Typical values of hydrogen coupling constants are given in Table 10.4. The value of a particular coupling constant is independent of the magnitude of the applied field for reasons that we will learn shortly. Coupling constants, therefore, are the same whether measured on a spectrometer operating at 60 MHz, 200 MHz, or 400 MHz.

The coupling constant for the doublet and triplet in the ^1H NMR spectrum of 1,1,2-trichloroethane are the same because coupling is a reciprocal process. Thus the $ClCH_2$ hydrogen atoms split the signal for the Cl_2CH hydrogen atom into a triplet with a coupling constant of 7 Hz. The Cl_2CH hydrogen atom, in turn, splits the signal of the $ClCH_2$ hydrogen atoms into a doublet with the same coupling constant, 7 Hz.

The reciprocal nature of coupling is frequently used to determine which hydrogen atoms are coupled in complex spectra. NMR spectra often contain many multiplets so it can be difficult to identify which hydrogen atoms are coupled. If two multiplets have exactly the same coupling constant, the hydrogen atoms responsible for these multiplets must be on adjacent carbon atoms.

Under certain conditions, spin-spin coupling can also be observed in ^{13}C NMR spectra.

10.9   Spin-Spin Coupling in ^{13}C NMR Spectra

In the ^{13}C NMR spectra shown so far, all peaks are singlets. Why is there no splitting due to either $^{13}C-^{13}C$ or $^{13}C-H$ coupling? Splitting due to $^{13}C-^{13}C$ coupling is rarely observed because the natural abundance of ^{13}C is so low (1.108%)

Table 10.4		Typical Values of Hydrogen Coupling Constants			
Type of Coupling	Structure	Range of Coupling Constants (Hz)	Type of Coupling	Structure	Range of Coupling Constants (Hz)
Vicinal		6-8	Geminal		0-22
trans-		11-18	cis-		6-14
Vinyl geminal		0-3			

that there is only a small probability that the same molecule will contain two ^{13}C atoms. If a molecule should contain two ^{13}C atoms, the probability that they are sufficiently close to couple is smaller still.

The absence of ^{13}C—H coupling is due to a deliberate decision made to simplify ^{13}C spectra. Strong spin-spin coupling occurs between a ^{13}C nucleus and not only the nuclei of hydrogen atoms bonded to it but also the nuclei of hydrogen atoms separated by two, three, or more bonds. The result of such extensive coupling is a spectrum too complicated for easy interpretation.

When the spectrum is measured while simultaneously irradiating the sample with a second strong electromagnetic radiation in the ^{1}H radio frequency range, no coupling is observed. This technique, called **broadband decoupling,** removes all ^{13}C—H coupling and produces a spectrum that contains only single peaks. The ^{13}C NMR spectra presented so far are all decoupled. A decoupled spectrum is especially useful for counting the number of nonequivalent carbon atoms in a molecule.

Another technique called off-resonance decoupling removes all ^{13}C—H coupling except for those hydrogen nuclei directly bonded to carbon. As a result, spectra obtained by this technique are often called **proton-coupled ^{13}C NMR spectra.** In a proton-coupled spectrum, the signal of the carbon atom of a methyl group is split into a quartet. The signal of the carbon atom of a methylene group is split into a triplet, and the signal of a methine carbon atom is split into a doublet. Quaternary carbon atoms have no hydrogen atoms bonded to them, so their signals appear as a singlet. Figure 10.9 shows the proton-coupled ^{13}C NMR spectrum of chloromethyl methyl ether.

In the determination of complex structures, two ^{13}C NMR spectra are often obtained. First, the broadband decoupled spectrum is obtained to determine the number of nonequivalent carbon atoms and their chemical shifts. Then a proton-coupled spectrum is obtained to determine the number of hydrogen atoms bonded to each carbon atom.

Let's learn how to use the data from ^{13}C and ^{1}H NMR spectra to determine the structures of complex organic compounds.

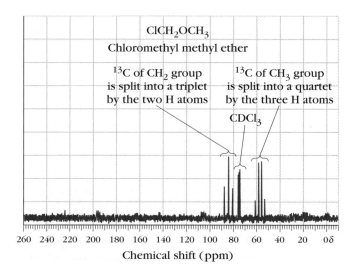

Figure 10.9
The proton-coupled ^{13}C NMR spectrum of chloromethyl methyl ether. The signal of the methyl carbon atom is split into a quartet, while the signal of the methylene carbon atom is split into a triplet.

10.10 Using NMR Spectroscopy to Determine Structures of Compounds

The combination of ^{13}C and ^{1}H NMR spectroscopy is a powerful technique used by chemists to determine the structures of organic compounds. Let's review the information that we can obtain from nuclear magnetic resonance spectroscopy.

The broadband decoupled ^{13}C NMR spectrum reveals the number of non-equivalent carbon atoms and their chemical shifts. The proton-coupled ^{13}C NMR spectrum reveals the number of hydrogen atoms bonded to each carbon atom.

The ^{1}H NMR spectrum of a compound provides us with the following four pieces of information:

- The number of different signals or multiplets reveals how many different kinds of hydrogen atoms are present in the molecule.
- The integration of these different signals reveals the relative number of each type of hydrogen atom present in the molecule.
- The chemical shift of the signals reveals something about the chemical environment of each type of hydrogen atom.
- The spin-spin coupling pattern of the different signals reveals the number and kind of neighboring hydrogen atoms.

The use of this information to deduce the structure of a compound is illustrated in Sample Problem 10.2.

Sample Problem 10.2   Determine the structure of the compound of molecular formula $C_6H_{12}Cl_2O_2$ whose broadband decoupled ^{13}C, proton-coupled ^{13}C, and ^{1}H NMR spectra are shown on p. 429.

Solution:

In order to determine the structure of a compound from its NMR spectra, we must obtain as much information as possible from the data given in a systematic manner. One way to do this is by the method outlined below. This method is not unique and is intended only as a guide. With practice, most people develop their own way of solving these kinds of problems.

First we use the molecular formula to determine the degree of unsaturation in the compound (Section 5.11). The compound of molecular formula $C_6H_{12}Cl_2O_2$ is saturated, and therefore it contains no rings or multiple bonds.

Examination of the broadband decoupled ^{13}C NMR spectrum indicates that the compound contains four nonequivalent carbon atoms. This is confirmed by the ^{1}H NMR spectrum, which shows four groups of hydrogen atoms.

The splitting patterns of the two groups of hydrogen atoms at high field suggest the presence of an ethyl group. This is confirmed by the coupled ^{13}C NMR spectrum, which reveals that the signal at δ 15.2 is split into a quartet and the one at δ 63.8 is split into a triplet. The $^{1}H—^{1}H$ coupling constants of the triplet and quartet are identical, consistent with an ethyl group.

The relative areas of the four groups of hydrogen atoms are 1:1:4:6, which accounts for all twelve hydrogen atoms. The relative areas suggest that there is not one but two identical ethyl groups in the molecule. The chemical shift of the methylene hydrogen atoms indicates that the ethyl groups are bonded to an electronegative element. Based on the molecular formula, oxygen atoms are the logical choice so we can write the following partial structure:

■ *NMR spectra occupy a great deal of space, so NMR spectra in books are frequently recorded in an abbreviated form. In the form used for ^{1}H NMR spectra in this text, the chemical shift of each peak is followed by its splitting pattern; then its relative area in parentheses; and if split, its coupling constant if known. For example, the spectrum of bromoethane shown in Figure 10.7 is abbreviated as follows:*
δ 1.68 t (3H) J = 7 Hz;
δ 3.44 q (2H) J = 7 Hz.

$$C \begin{matrix} \diagup OCH_2CH_3 \\ \diagdown OCH_2CH_3 \end{matrix}$$

This accounts for five of the six carbon atoms and ten of the twelve hydrogen atoms.

The chemical shifts of the remaining two hydrogen atoms are different, which indicates that they are on different carbon atoms. This is confirmed from the coupled ^{13}C spectrum, which reveals that the two downfield carbon atoms are both split into doublets, as well as the ^1H—^1H coupling constants of the two doublets, which are identical. The complete structure is the following:

$$Cl-\overset{\overset{\displaystyle H}{|}}{\underset{\underset{\displaystyle Cl}{|}}{C}}-\overset{\overset{\displaystyle H}{|}}{\underset{\underset{\displaystyle OCH_2CH_3}{|}}{C}}-OCH_2CH_3$$

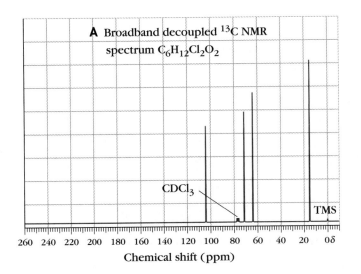

A Broadband decoupled ^{13}C NMR spectrum $C_6H_{12}Cl_2O_2$

CDCl$_3$

TMS

260 240 220 200 180 160 140 120 100 80 60 40 20 0δ
Chemical shift (ppm)

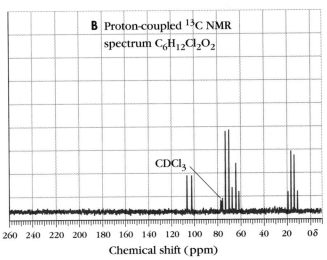

B Proton-coupled ^{13}C NMR spectrum $C_6H_{12}Cl_2O_2$

CDCl$_3$

260 240 220 200 180 160 140 120 100 80 60 40 20 0δ
Chemical shift (ppm)

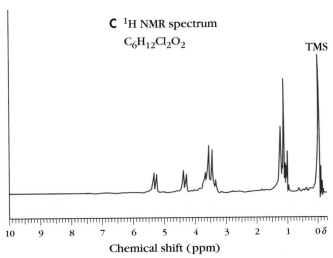

C ^1H NMR spectrum $C_6H_{12}Cl_2O_2$

TMS

10 9 8 7 6 5 4 3 2 1 0δ
Chemical shift (ppm)

Exercise 10.12 Several isomeric structures of the solution to Sample Problem 10.2 were not considered. Are the spectra given in Sample Problem 10.2 consistent with either of the following isomeric structures?

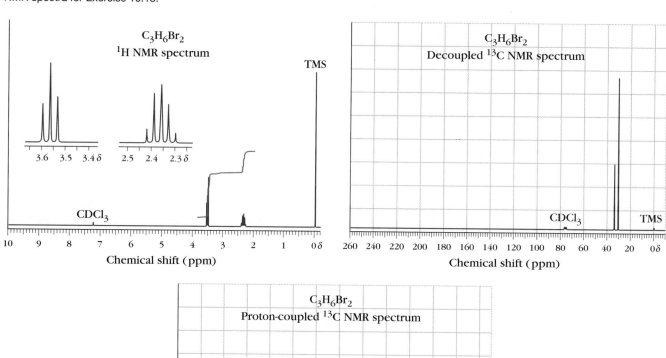

Exercise 10.13 Determine the structure of the compound of molecular formula $C_3H_6Br_2$ from its ^1H, decoupled ^{13}C, and coupled ^{13}C spectra given in Figure 10.10.

Figure 10.10
NMR spectra for Exercise 10.13.

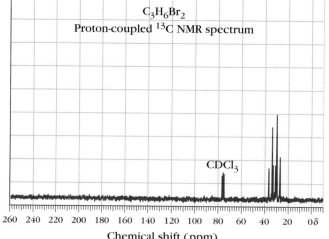

Exercise 10.14   Draw the ^{1}H NMR spectrum you would expect for each of the following compounds.

(a) $(CH_3)_2CCH_2Br$
with Br above the second C

(b) $ClCH_2CH_2C$ with $=O$ and CH_3

(c) $CH_3CH_2OCH(CH_3)_2$

(d)
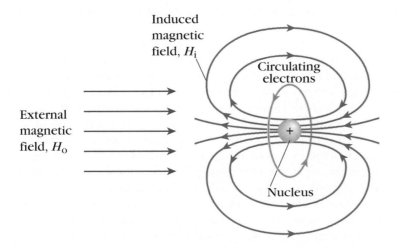

We have seen how chemical shifts and spin-spin couplings in NMR spectra can help us determine the structures of compounds. In the next several sections, we discover the physical basis of chemical shifts and why neighboring atoms couple.

10.11 The Physical Basis of Chemical Shifts

A signal in the NMR spectrum, as explained in Sections 5.14 and 10.1, represents absorption of the energy necessary to cause a nucleus to flip from one spin state to another. This explanation focuses on the nucleus and its behavior in a magnetic field, while ignoring any outside electronic effects. Atoms in a molecule are not bare nuclei, however, because they are surrounded by bonding and nonbonding electrons. The density of these electrons and their circulation in the molecule have an important effect on the chemical shift of nuclei in a molecule.

When a nucleus, surrounded by bonding electrons, is placed in an external magnetic field, H_0, the electrons move around the nucleus and generate a small local magnetic field, as shown schematically in Figure 10.11. The small magnetic field generated by circulating electrons is called an induced magnetic field, H_i, and it opposes the external magnetic field felt at the nucleus. As a result, the magnetic field felt at the nucleus is slightly less than the external magnetic field, H_0. The induced magnetic field shields the nucleus from the full force of the external magnetic field so the electrons are said to shield the nucleus.

Induced
magnetic
field, H_i

Circulating
electrons

External
magnetic
field, H_0

Nucleus

Figure 10.11
An external magnetic field, H_0, generates a small induced magnetic field H_i at the nucleus.

Figure 10.12
Electrons around a nucleus cause it to be shielded from the external magnetic field H_0. As a result, it will absorb at higher external magnetic field strength than a bare nucleus.

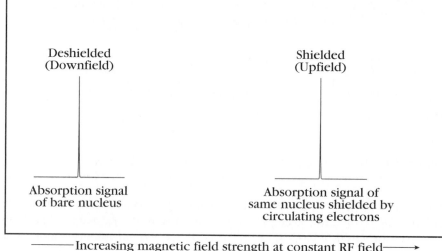

Deshielded
(Downfield)

Shielded
(Upfield)

Absorption signal
of bare nucleus

Absorption signal of
same nucleus shielded by
circulating electrons

Increasing magnetic field strength at constant RF field

A nucleus shielded by electrons will not absorb at the same external magnetic field strength as a bare nucleus. A shielded nucleus will absorb at higher external field strength because the external field must be increased to compensate for the small induced magnetic field that opposes it (Section 5.14). The chemical shift difference between a bare nucleus and one shielded by circulating electrons is shown schematically in Figure 10.12.

The electron density surrounding the nucleus determines the extent to which it is shielded. The greater the electron density, the greater the shielding. Conversely, removing electron density from a nucleus deshields it. The distribution of electrons in chemical bonds determines the electron densities at hydrogen and carbon atoms in an organic molecule. Thus nuclei in electron-poor environments are deshielded, and the greater the deshielding, the larger the chemical shift on the delta scale. Nuclei in electron-rich environments are affected in the opposite way.

We learned in Section 10.5 that the presence of a carbon-carbon double bond deshields most hydrogen atoms in its vicinity. The extent of deshielding depends on the location of the hydrogen atom relative to the double bond. In particular vinylic hydrogen atoms are found further downfield than allylic hydrogen atoms.

Why are vinylic hydrogens so deshielded by the double bond of alkenes? To help answer this question, consider an alkene oriented with respect to the magnetic field as shown in Figure 10.13. The applied magnetic field induces the π electrons to circulate in a closed loop above and below the plane of the double bond. The motion of the π electrons generates a local magnetic field, H_i, that opposes the applied magnetic field in the region of the nuclei of the carbon atoms of the double bond but enhances it in the region of the vinylic hydrogen atoms. The vinylic hydrogens, therefore, are deshielded where the induced field enhances the applied field causing the deshielding of vinylic hydrogens.

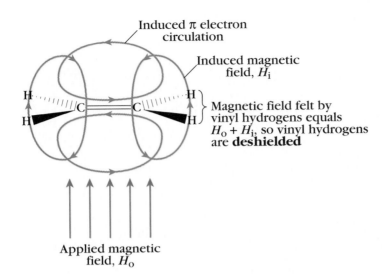

Induced π electron
circulation

Induced magnetic
field, H_i

Magnetic field felt by
vinyl hydrogens equals
$H_o + H_i$, so vinyl hydrogens
are **deshielded**

Applied magnetic
field, H_o

Figure 10.13
The applied magnetic field H_o
causes the π electrons of the double
bond to circulate and generate an
induced magnetic field, H_i. H_i reinforces the applied field in the region
of vinylic hydrogens, causing them
to be deshielded.

10.12 The Origin of Spin-Spin Coupling

Spin-spin coupling patterns occur because nuclei behave like tiny magnets in an applied magnetic field. They align either with the field (α spin) or against the field (β spin). The energy difference between these two spin states is so small that at room temperature a little more than half the nuclei have α spin and the rest have β spin. A hydrogen atom in a molecule, therefore, has two types of neighboring hydrogen atoms: those with α spin and those with β spin.

How does this difference in the spin states of neighboring hydrogen atoms affect the absorption signal of a hydrogen? To answer this question, let us examine the simplest case, which is two hydrogen atoms bonded to adjacent carbon atoms. We designate one hydrogen atom H_a and the other H_b.

$$\begin{array}{c} \quad\quad\quad H_b \\ \backslash| \quad / \\ C - C \\ / \quad |\backslash \\ H_a \end{array}$$

In the absence of H_b, H_a feels only the applied external magnetic field, H_o, at its nucleus. In the presence of H_b with α spin (aligned with the external magnetic field), the nucleus of H_a feels both H_o and an additional strengthening magnetic field caused by the spin of H_b. Consequently the H_a resonance occurs at a lower field strength than would be required in the absence of H_b. If H_a is located in the presence of H_b with β spin (aligned against the external magnetic field), the magnetic field at the nucleus of H_a is less than H_o due to the spin of H_b. In this case, the H_a resonance occurs at a higher field than would be expected in the absence of H_b.

Notice that the effect of the two spin states of a hydrogen nucleus on a neighboring nucleus is due to the magnetic spin of the hydrogen nucleus. The magnitude of this magnetic spin is independent of the external magnetic fields. Thus the splitting (the coupling constant) is the same for spectra measured at 60 MHz, 200 MHz, or 400 MHz.

The two spin states of H_b are nearly equally populated (Section 10.1), so their contributions to the external magnetic field have the same magnitude but oppo-

■ *Recall that the coupling constant between two nonequivalent hydrogen atoms, H_a and H_b, is represented by J_{ab}.*

Figure 10.14
Spin-spin coupling between H_a and H_b, illustrated by a tree diagram. The coupling constant J_{ab} is the same for both doublets.

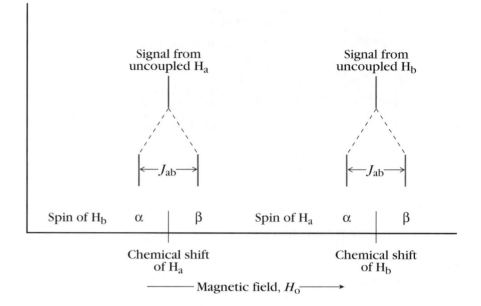

site directions; one increases the field, the other decreases the field. Consequently, the single absorption expected for H_a in the absence of H_b is said to be split into a doublet; the upfield shift is exactly equal to the downfield shift. Integration of the entire doublet gives a value of one hydrogen relative to the other hydrogen atoms in the compound.

The signal for H_b is split into a doublet in the same way by the two spin states of H_a. Thus H_a splits H_b and vice versa with identical coupling constants. The spin-spin coupling between H_a and H_b is illustrated schematically by a tree diagram shown in Figure 10.14.

We have seen how an NMR signal is split into a doublet by one nearby hydrogen atom. The spin-spin coupling pattern of hydrogen H_a with two or more nearby hydrogens can be explained in a similar manner. Consider first the presence of two hydrogens adjacent to H_a. These hydrogens are chemically equivalent and have the same chemical shift, but we must be able to distinguish between them in order to explain their individual effect on the signal of H_a. Therefore we designate one of them H_b and the other $H_{b'}$:

$$
\begin{array}{c}
\backslash | \qquad / H_b \\
C - C \\
/ \quad | \backslash \\
H_a \qquad H_{b'}
\end{array}
$$

The nuclei of hydrogens H_b and $H_{b'}$ have three different ways in which their spins can be oriented with respect to the external magnetic field. These are shown in Figure 10.15.

Both states in which one nucleus has α spin and the other β spin are of equal energy. As a result, the nucleus of H_a feels one of three different magnetic fields at any given moment. One is stronger than, one is equal to, and one is weaker than the external field. The probability that H_a feels any one of these magnetic fields is 1:2:1. Consequently, the signal for H_a is split into three peaks, a triplet, with relative intensities 1:2:1. The chemical shift of H_a is the center peak of the triplet.

H_b	H_b'	Effect on magnetic field felt by H_a	Relative intensity
α	α	Stronger than external field	1
α	β	Same as external field	2
β	α		
β	β	Weaker than external field	1

Figure 10.15
Three spin alignments with the external field of two hydrogen nuclei and their effect on the magnetic field felt by the nucleus of a hydrogen atom on an adjacent carbon atom.

The nuclear spin of H_a splits the signal for H_b and H_b' into a doublet. The spin-spin coupling between H_a and two chemically equivalent hydrogens H_b and H_b' is shown schematically by a tree diagram in Figure 10.16.

Exercise 10.15   The nuclei of three chemically equivalent hydrogen atoms have eight possible spin orientations. These represent four different interactions with an external magnetic field. What are these spin combinations, and how are they responsible for the observed splitting of a signal of an adjacent hydrogen atom into a quartet of relative intensities 1:3:3:1? Draw a tree diagram to illustrate the couplings.

---

^{1}H NMR spectra become more complicated when coupling occurs between a hydrogen atom and nonequivalent neighbors.

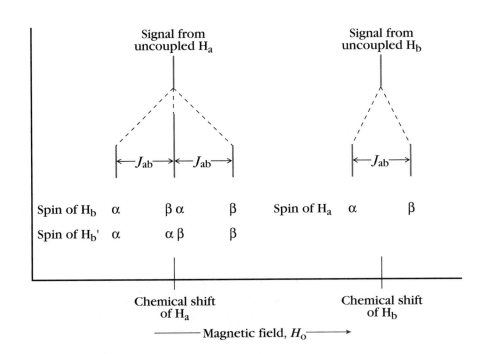

Figure 10.16
Spin-spin coupling between H_a and two chemically equivalent hydrogens H_b and H_b', illustrated by a tree diagram. The coupling constant J_{ab} is the same for the doublet and the triplet. Notice that $J_{ab} = J_{ab'}$ because H_b and H_b' are magnetically equivalent.

10.13 Coupling to Nonequivalent Neighbors

Spin-spin coupling can become more complex when coupling occurs to more than one type of adjacent hydrogen atom with different coupling constants. The hydrogen atoms of the vinyl group of ethyl vinyl ether are an example.

$$H_a\!\!\diagdown\qquad OCH_2CH_3$$
$$C\!\!=\!\!C$$
$$H_b\diagup\qquad \diagdown H_c$$

Ethyl vinyl ether

The ^{1}H NMR spectrum of ethyl vinyl ether is shown in Figure 10.17.

The spectrum in Figure 10.17 contains the typical pattern expected for an ethyl group. The other peaks further downfield are the result of coupling between the three nonequivalent hydrogen atoms of the vinyl group.

Consider first the hydrogen atom labelled H_c whose chemical shift is δ 6.44, deshielded by both the oxygen atom and the double bond. The signal for H_c is split into a doublet by coupling with H_a. Each of these peaks is further split into a doublet by coupling with H_b. Thus the signal of H_c is split into four signals called a doublet of doublets, as shown by the tree diagram in Figure 10.18.

The signal for H_a is also split by two nonequivalent hydrogen atoms with different coupling constants. It is split into a doublet by H_c, and each of the doublets is further split by H_b.

Exercise 10.16 | Explain how H_b is split into a doublet of doublets by coupling with H_a and H_c.

Figure 10.17
^{1}H NMR spectrum of ethyl vinyl ether.

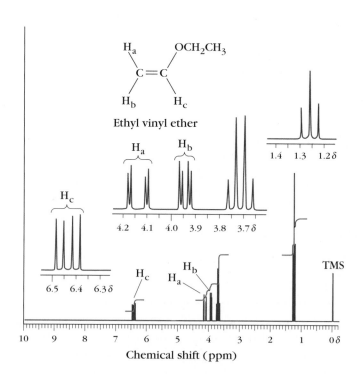

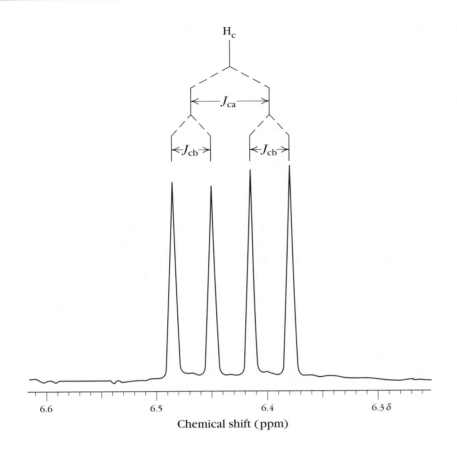

Figure 10.18
The signal of H_c is split into a doublet of doublets by coupling with two nonequivalent hydrogen atoms that have different coupling constants.

Exercise 10.17   In the tree diagram shown in Figure 10.18, the signal for H_c is first split by coupling with H_a, then split by coupling with H_b. Show that the same splitting pattern will result if the signal for H_c is first split by H_b then H_a.

Exercise 10.18   If $J_{ac} = J_{bc}$ for ethyl vinyl ether, use a tree diagram to show that the resulting splitting of H_c is a triplet with relative ratio of areas 1:2:1.

Exercise 10.19   Draw the ^{1}H NMR spectrum of vinyl bromide. The chemical shifts of the three hydrogen atoms have the following values:

$$\underset{\delta\,6.45}{\text{H}}\,\text{C}=\text{C}\,\overset{\text{H} \leftarrow \delta\,5.85}{\underset{\text{H} \leftarrow \delta\,5.95}{}}$$

Br, H ← δ 5.85
δ 6.45 → H, H ← δ 5.95

Vinyl bromide

10.14 More Complex Spin-Spin Coupling Patterns

Not all ^{1}H NMR spectra can be analyzed by applying the $n + 1$ rule. Most compounds that contain an alkyl chain of more than four carbon atoms or a cycloalkyl group exhibit a large number of overlapping peaks in the region δ 0 to 3 in their ^{1}H NMR spectra. The 60 MHz spectrum of 1-bromobutane shown in Figure 10.19, for example, is clearly not first order. The signal of the methyl group that appears near δ 1.0 has a pattern vaguely resembling a triplet, but there are additional splittings not predicted by the $n + 1$ rule. The signals of the hydrogen atoms on C2 and C3 appear as a broad undefined multiplet between δ 1.1 and 2.2 ppm. Finally the signal of the hydrogen atoms on C1 appears as the expected triplet at δ 3.10.

Why is the ^{1}H NMR spectrum of 1-bromobutane so complicated? The answer is that in order to observe a first order spectrum, the chemical shift difference between coupled hydrogen atoms must be much greater than their coupling constants. This means that for a first order spectrum

$$\Delta\delta_{AB} \gg J_{AB}$$

where $\Delta\delta_{AB}$ is the difference between the chemical shifts of two hydrogen atoms A and B in Hz and J_{AB} is their coupling constant also in Hz.

If $\Delta\delta_{AB}$ is more than ten times greater than J_{AB}, a first order spectrum is generally observed. Coupling constants between hydrogen atoms in alkyl chains have values of between 6 and 8 Hz (Table 10.4). Consequently, when chemical shift differences between coupled hydrogen atoms are more than 60 Hz, a first order spectrum is observed. Conversely, when the chemical shift difference is much less than 60 Hz, the spectrum observed is not first-order.

■ Recall that spectra whose spin-spin coupling patterns conform to the $n + 1$ rule are called first order spectra.

Figure 10.19
The 60 MHz ^{1}H NMR spectrum of 1-bromobutane. The signals for the hydrogen atoms on C1, C2, and C3 are not simple patterns expected from the $n + 1$ rule. This is not a first order spectrum.

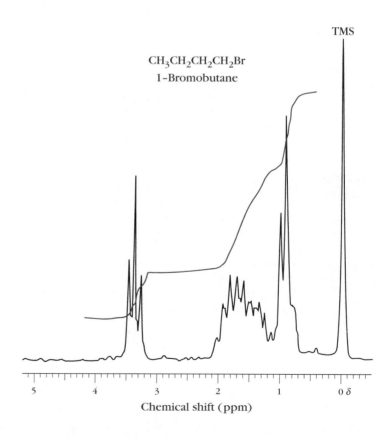

CH$_3$CH$_2$CH$_2$CH$_2$Br
1-Bromobutane

TMS

Chemical shift (ppm)

In practice it is difficult to obtain the structure of a compound if its ^{1}H NMR spectrum is not first order. Typically, however, we can obtain information about parts of the structure. For example, if we did not know the structure of 1-bromobutane, we could conclude based on the splitting patterns and the integration in Figure 10.19 that there is probably a methyl group and a CH_2X group in the structure. We couldn't conclude much more, however, and 1-bromobutane is a relatively simple molecule.

The problem of spectra that are not first order is not entirely intractable. A spectrum is not first order because $\Delta\delta_{AB} \approx J_{AB}$. If we could somehow increase the difference between the resonance frequencies of the two coupled hydrogen atoms while keeping their coupling constants the same, we might be able to make $\Delta\delta_{AB} \gg J_{AB}$. Recall that the resonance frequency of a hydrogen nucleus is proportional to the external field strength (and vice versa) but coupling constants are field independent. Thus our chances of obtaining a first order spectrum increase as the operating field strength of the NMR spectrometer increases. This is illustrated in Figure 10.20, where the spectrum of 1-bromobutane obtained on a 400 MHz NMR instrument is shown.

Exercise 10.20	If $\Delta\delta_{AB}$ has a value of 60 Hz at 60 MHz, what is its value when measured on an NMR spectrometer operating at the following frequencies?

(a) 100 MHz **(b)** 300 MHz **(c)** 500 MHz

In deducing the structures of compounds from their NMR spectra, we have ignored the fact that molecules continually undergo conformational changes. How do rapid molecular movements affect NMR spectra?

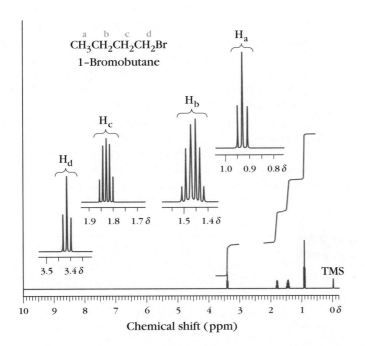

Figure 10.20
^{1}H NMR spectrum of 1-bromobutane obtained at 400 MHz. Increasing the field strength from 60 MHz to 400 MHz results in a spectrum that is first order.

10.15 NMR Spectra and the NMR Time Scale

We learned in Chapter 4 that alkanes and substituted alkanes exist in a number of conformations and that the most stable is the staggered conformation. For chloroethane, the staggered conformations are shown in Figure 10.21. Let us focus our attention on the methyl hydrogen atoms, which are labelled H_1, H_2, and H_3 in Figure 10.21.

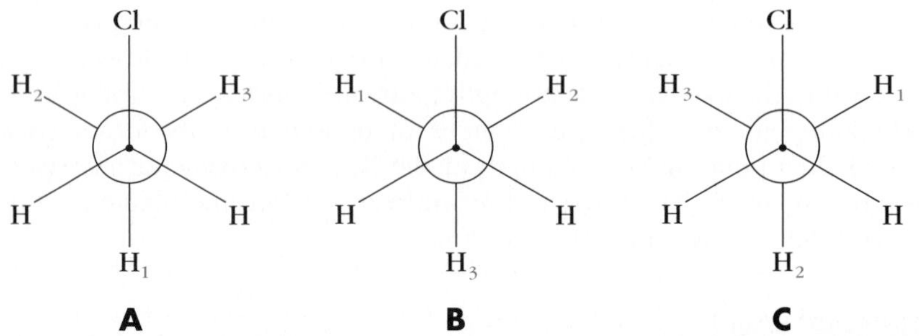

Figure 10.21
The Newman projections of the staggered conformations of chloroethane.

A   **B**   **C**

In projection formula *A*, the methyl hydrogen H_1 is located anti to the chlorine while H_2 and H_3 are gauche to the chlorine atom. If chloroethane were locked in this conformation, we would expect H_1 to have a different magnetic environment and consequently a different chemical shift than H_2 and H_3. In actual fact, we observe only one methyl signal, split into the expected triplet, in the ^1H NMR spectrum of chloroethane.

Only one signal is observed because the rate of interconversion of these three staggered conformations of chloroethane is so fast that the NMR spectrometer detects only a single average signal for all three methyl hydrogen atoms. The rate of rotation is said to be fast on the NMR time scale. An NMR instrument, like the human eye or a camera, has only a limited ability to record a rapid sequence of events. Anything that exceeds that ability appears blurred or averaged. For chloroethane, this means that the signals for the methyl hydrogen atoms are averaged into a single signal by fast rotation about the carbon-carbon single bond.

In theory, by cooling a sample of chloroethane, we should be able to slow the rotation so that the NMR instrument would detect signals from both the anti and gauche methyl hydrogens. In practice, however, this is very difficult to do because the energy needed to convert one conformation into another is only a few kcal/mol. The temperatures needed to prevent interconversion between staggered conformations of chloroethane would be so low that most solvents used in NMR spectroscopy would freeze before they could be attained, making it impossible to record the spectrum.

| Exercise 10.21 | The methyl carbon of methylcyclohexane is found at δ 23.1 ppm in its ^{13}C NMR spectrum at room temperature. When the ^{13}C NMR spectrum of methylcyclohexane is recorded at −90 °C, two methyl signals appear. Explain this observation. |

10.16 Magnetic Resonance Imaging

^{1}H NMR spectroscopy has been applied to diagnostic medicine by a technique called **magnetic resonance imaging** or **MRI.** In the human body, MRI detects the hydrogen atoms of water and other substances such as lipids and proteins. Figure 10.22 shows an example of the use of MRI to detect a brain tumor. The soft tissue of the brain in Figure 10.22 appears in the image as various shades of grey, while the bones of the skull appear as dark areas. Bones appear darker in the image because bones do not contain much water or other substances that contain hydrogen atoms.

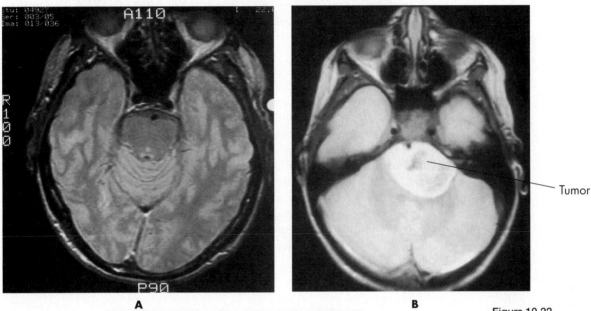

A   Tumor   B

Figure 10.22
An MRI cross section, just above the eyes, of (**A**) a normal brain and (**B**) a brain with a tumor at the base of the brain stem. The tumor is a lighter shade of gray. The dark areas extending from the tumor indicate hemorrhaging. For clinical use, color provides no advantage over black and white images. *(Courtesy Dr. Kucharczyk, University of Toronto, Toronto, Ontario.)*

MRI has several advantages for medical diagnosis over another medical imaging technique called x-ray computerized tomography or x-ray CT. MRI uses radio frequency radiation, which is lower in energy and less damaging to the body than x-rays (Figure 5.6, p. 177), MRI requires no injection of imaging or contrasting agents into the body as does x-ray CT, and MRI images have a higher intrinsic contrast than x-ray CT images.

Whole body NMR scanners such as the one pictured in Figure 10.23 are used today to obtain MRI images. They produce MRI images by using radio frequency pulses (Section 10.1) to excite the hydrogen atoms of the water in the tissue under observation. The radio frequency pulse time is varied while a rotating magnetic field gradient is superimposed onto the main magnetic field. This makes the resonance frequency a function of the spatial origin of the signal. Fourier transformation of the raw data produces a series of projections from which various images, cross sections or three-dimensional, can be obtained by using back projection techniques.

Recall that ^{1}H NMR spectroscopy provides chemical shift differences of hydrogen atoms in different magnetic environments in a molecule. MRI, on the other hand, doesn't measure chemical shift differences of the hydrogen atoms. MRI instead measures a difference in the relaxation times of hydrogen atoms in different environments. The relaxation time for a nucleus is the rate at which the spin

Figure 10.23
An MRI instrument. *(From Coletta VP: College Physics,* St. Louis, 1995, *Mosby.)*

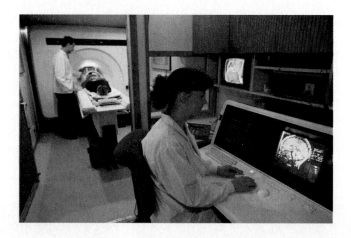

states of the nucleus, following excitation, return to equilibrium. Relaxation times of hydrogen atoms in liquids range from seconds to milliseconds.

Although the mechanism of relaxation of water in tissue is not completely understood, its rate of relaxation is related to the extent of binding of water to the surface of biological molecules. Increased binding slows molecular motions, which decreases relaxation times. The brain, for example, consists mostly of grey and white matter adjacent to fluid-filled cavities. Water is more tightly bound in white matter than in grey matter, so water molecules in white matter reorient more slowly than those in grey matter. Water in white matter consequently has a shorter relaxation time than water in grey matter. Relaxation of the hydrogen atoms in water in most diseased tissue is prolonged compared to water in white or grey tissue, which is the basis for the image contrast between normal and diseased tissue.

Figure 10.24
A cross section of a functional MRI of a human brain with a tumor. The red area is the region of the brain activated when the person performed a sequence of finger taps with the right hand. Notice how close this region is to the tumor. *(Courtesy Dr. D. Mikulis, Toronto Hospital, Toronto, Ontario.)*

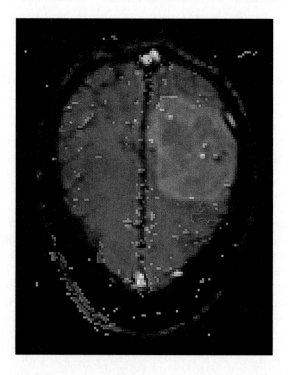

It takes many minutes to obtain an image of a patient with most MRI instruments in use today. Recent advances in NMR instrumentation make it possible to collect the entire raw data set for an MRI image in 50 msec or less. This rapid measurement time allows the use of MRI to study blood flow in the brain by a technique called **functional MRI.** Blood flows to the specific region of the brain involved in responding to some form of external stimulus. Functional MRI detects this blood flow because magnetic resonance imaging is extraordinarily sensitive to blood flow.

Figure 10.24 shows a functional MRI of the brain of a patient with a tumor. This image was obtained by comparing the data accumulated while the patient repeated a specific sequence of finger taps with the right hand and while the patient remained at rest. The image reveals the region of the brain responsible for the sequence of finger taps, which is shown in red in Figure 10.24.

Nuclear magnetic resonance imaging has great potential in medicine. Research using functional MRI is underway to map the functions of the brain, which will help psychiatrists identify and treat many mental illnesses. If the rate of conversion of raw data into the final image can be increased, real time imaging will become a reality. Then it may be possible to perform surgical procedures guided by MRI.

10.17 Summary

Nuclei of ^{13}C and ^{1}H act as tiny bar magnets. When placed in a strong magnetic field, these nuclei orient themselves either with (α-spin) or against (β-spin) the applied magnetic field. When these nuclei are irradiated in a magnetic field with radio frequency radiation, enough energy is absorbed to "spin flip" some of the nuclei from α spin to β spin. This absorption of energy can be recorded as a **nuclear magnetic resonance (NMR) spectrum.**

An NMR spectrum is recorded by means of an instrument called an **NMR spectrometer.** Hydrogen and carbon atoms in different chemical environments absorb energy at slightly different applied magnetic fields. As a result, an NMR spectrum consists of a number of different signals corresponding to each different kind of ^{1}H and ^{13}C in the molecule.

The positions of signals of the NMR spectrum are called their **chemical shifts.** Chemical shifts are measured on the **delta scale (δ scale)** using **tetramethylsilane (TMS)** as an internal standard. Values of delta are independent of the frequency at which the spectrometer operates. TMS is arbitrarily given a delta value of zero. In general, the more electronegative the atom or group adjacent to a particular nucleus (either ^{1}H or ^{13}C), the further downfield its chemical shift.

The area under each signal can be integrated to give the relative number of nuclei responsible for each signal.

Signals of specific hydrogen atoms are often split into multiplets by spin-spin coupling with adjacent nonequivalent hydrogen atoms. The signal of a hydrogen atom with equivalent neighboring hydrogen atoms is split into $n + 1$ signals. The distance between the $n + 1$ signals in hertz is called the **coupling constant**, which is designated J.

An NMR spectrum provides us with the following information about the structure of a compound:

- The number of different signals or multiplets reveals how many kinds of chemically different carbon or hydrogen atoms are present in the molecule.
- The value of the integration reveals the relative number of chemically different hydrogen atoms that are present in the molecule.

• The chemical shift of the signals provides information about the chemical environment of each type of hydrogen or carbon atom.
• The spin-spin coupling pattern of the different signals reveals the number and kind of neighboring hydrogen atoms.

From these data, the structures of most simple compounds can be deduced.

Additional Exercises

10.22 For each of the following structures, identify the chemically nonequivalent hydrogen atoms.

(a) $CH_3CH_2CH_2CH_2CH_3$
(b) $CH_3CH{=}CH_2$
(c) $CH_3CH_2OCH_3$
(d) *cis*-BrCH=CHBr
(e) *trans*-BrCH=CHBr
(f) $ClCH_2CH_2CH(CH_3)_2$

(g)

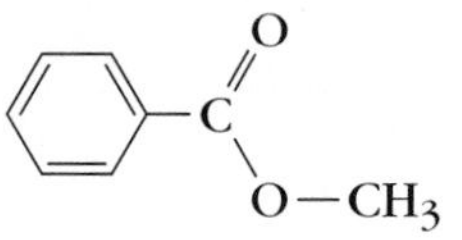

(h)

(i)

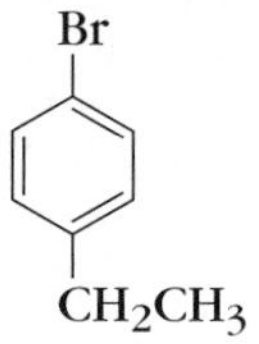

10.23 How many signals would you expect in the broadband decoupled ^{13}C NMR spectrum of each compound in Exercise 10.22?

10.24 In each of the following pairs of compounds, indicate which of the highlighted hydrogen atoms will be found farther downfield.

(a) $CH_3\ CH_2Br$ $CH_3\ CH_2Cl$

(b) 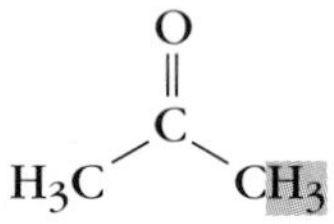 $CH_3CH_2CH_3$

(c) $CH_2{=}CH_2$ $ClCH{=}CH_2$

(d) 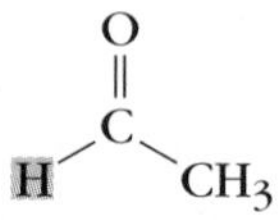 $CH_2{=}CH_2$

(e)

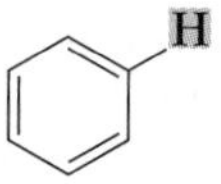

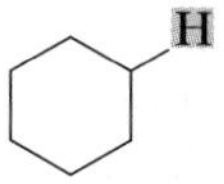

10.25 Predict the chemical shifts of the highlighted hydrogen atoms in the compounds given in Exercise 10.24.

10.26 Frequently a decision on the structure of a compound can be made from the results of a single spectroscopic technique. Which single spectroscopic technique would you use to distinguish each of the following pairs of compounds? Explain clearly what feature in the spectrum allows the compound to be distinguished.

(a) CH_3CH_2OH and $CH_3CH_2OCH_2CH_3$

(b) 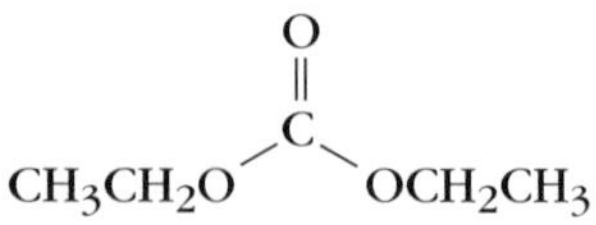 and

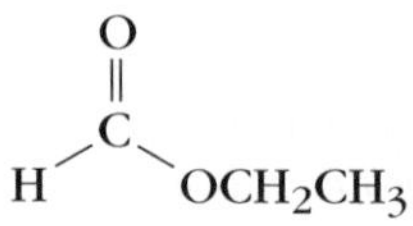

(c) 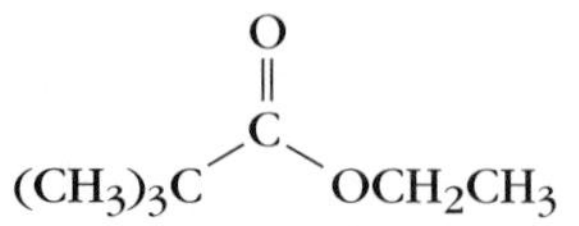 and

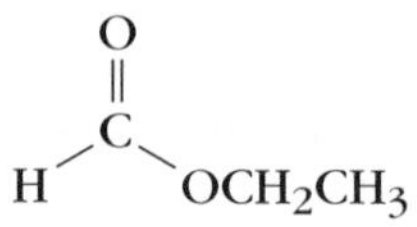

(d) $CH_3CH_2C{\equiv}CH$ and $CH_3CH{=}CHCH_3$

(e) 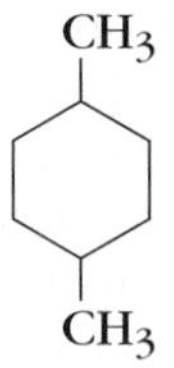 and

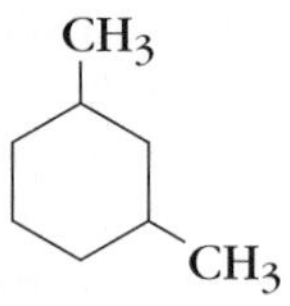

(f) 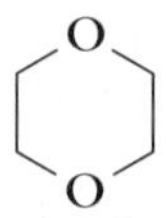 and

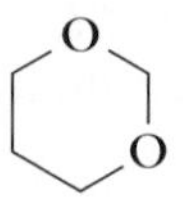

10.27 Five broadband decoupled ^{13}C NMR spectra are given on p. 445. Match each of the following structures with its correct spectrum.

(a) $(CH_3)_2CHCH_2OH$ **(b)** 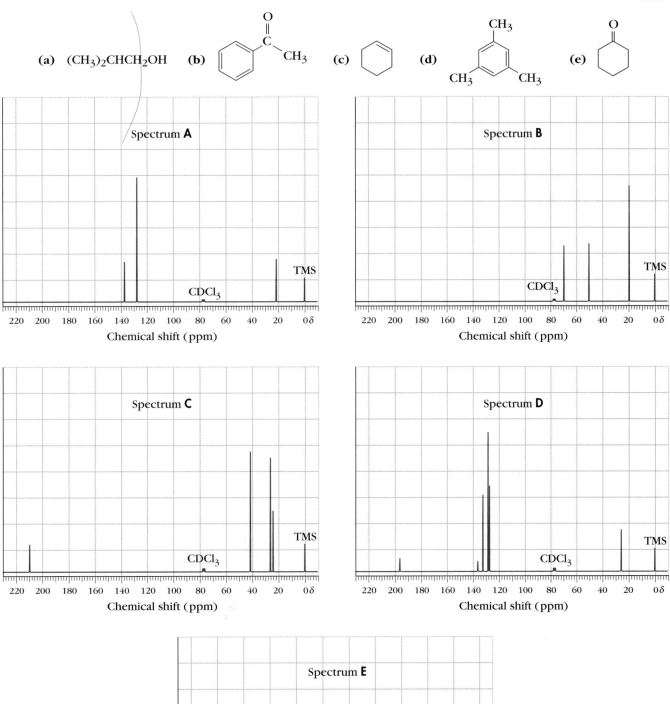 **(c)** **(d)** **(e)**

Spectrum **A**

CDCl₃

TMS

Spectrum **B**

CDCl₃

TMS

Spectrum **C**

CDCl₃

TMS

Spectrum **D**

CDCl₃

TMS

Spectrum **E**

CDCl₃

TMS

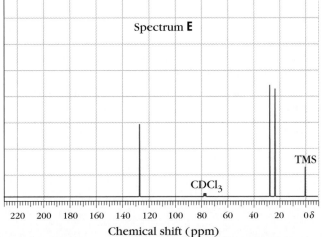

10.28 Sketch the splitting pattern for the highlighted hydrogen atom in each of the following compounds.

(a) Cl_2CHCH_2I

(b) 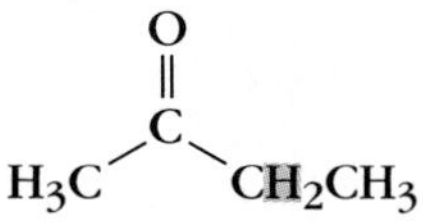 H_3C—C(=O)—CH_2CH_3

(c) $BrCH_2CH_2Cl$

(d) 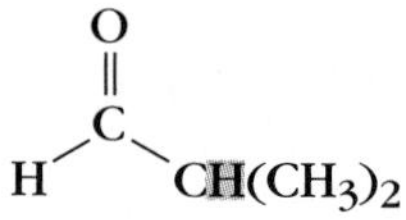 H—C(=O)—$CH(CH_3)_2$

(e) 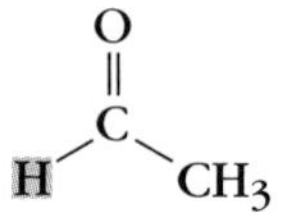 H—C(=O)—CH_3

10.29 Explain what features of their ^1H NMR spectra would allow you to distinguish between each of the following pairs of compounds.

(a) 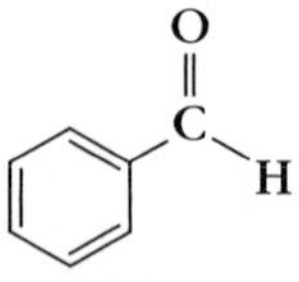 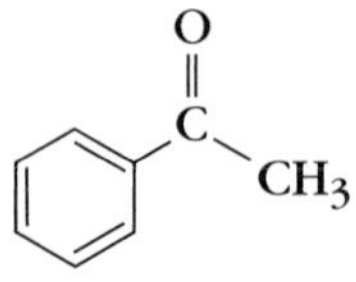 C_6H_5—C(=O)—H C_6H_5—C(=O)—CH_3

(b) $CH_3C{\equiv}CCH_3$ $CH_3CH_2C{\equiv}CH$

(c) 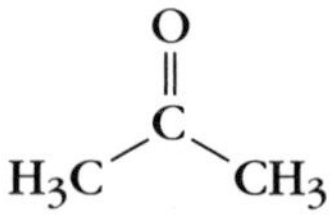 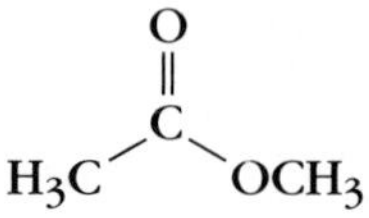 H_3C—C(=O)—CH_3 H_3C—C(=O)—OCH_3

(d) $Br_2C{=}CH_2$ cis-$BrCH{=}CHBr$

10.30 The ^1H NMR spectrum of a compound of molecular formula C_4H_8O is shown in the following figure.

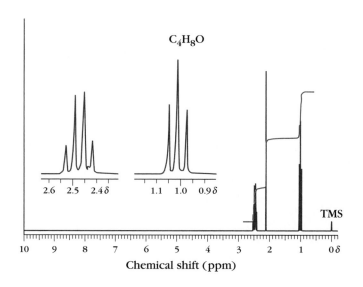

(a) Propose a structure for this compound.
(b) Indicate which hydrogen atoms give rise to which peaks in the spectrum.
(c) Use a tree diagram to explain the coupling patterns.

10.31 Sketch the ^1H NMR spectrum of each of the following compounds. Include the approximate chemical shifts, splitting patterns, and appropriate peak areas.

(a) $(CH_3)_3CCH_2CH_3$
(b) $CH_3OCH(CH_3)_2$
(c) 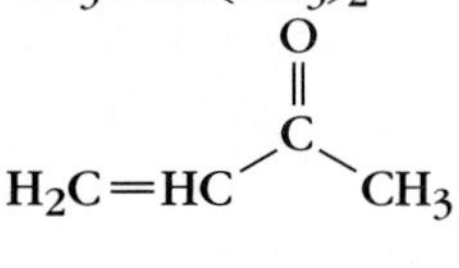 $H_2C{=}HC$—C(=O)—CH_3

10.32 Determine the structure of the following compounds from the data given.

(a) $C_9H_{18}O$ IR 1670 cm^{-1}. ^{1}H NMR δ 0.95 s
(b) C_4H_7NO IR 2240 cm^{-1}. ^{1}H NMR δ 2.63 t (2H); δ 3.30 s (3H); δ 3.55 t (2H)
(c) C_5H_{12} ^{13}C NMR δ 27.4; δ 31.4
(d) $C_5H_{10}O$ IR 1720 cm^{-1}. ^{13}C NMR δ 8.0; δ 35.5; and δ 210.7
(e) C_4H_7N IR 2240 cm^{-1}. ^{13}C NMR δ 19.0; δ 19.8; δ 123.7. ^{1}H NMR δ 1.33 d (6H); δ 2.72 sept (1H)

10.33 The ^{1}H NMR spectrum of a compound of molecular formula C_4H_9Br is shown in the following figure.

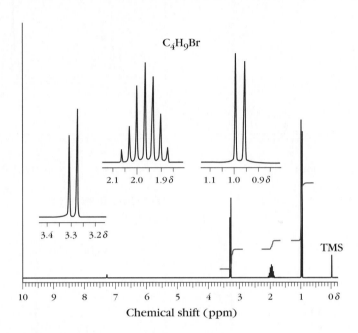

(a) Propose a structure for this compound.
(b) Indicate which hydrogen atoms give rise to which peaks in the spectrum.
(c) Use a tree diagram to explain the coupling patterns.

10.34 Write the structure of each of the compounds whose molecular formula, ^{1}H NMR spectra, and other spectral data are given in the following figures.

(a) C_6H_8 ^{13}C NMR δ 26.0; δ 124.5

(b) $C_{10}H_{18}O$ IR 1720 cm^{-1}

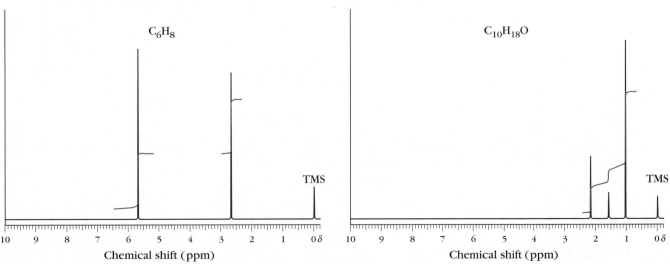

(c) $C_7H_{14}O$ IR 1695 cm^{-1}

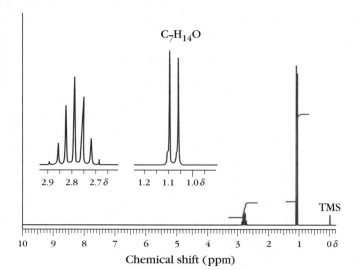

(e) $C_5H_{11}Cl$

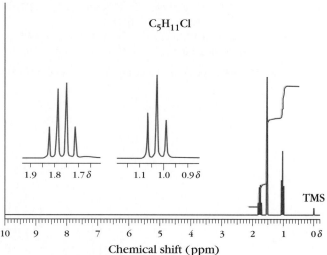

(d) C_6H_{10} IR 3250 cm^{-1}; 2150 cm^{-1}

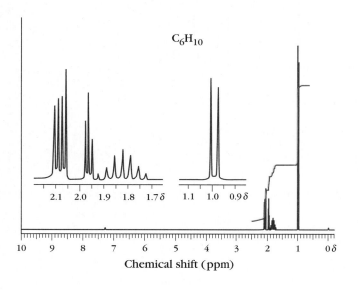

(f) C_5H_8 IR 2220 cm^{-1}

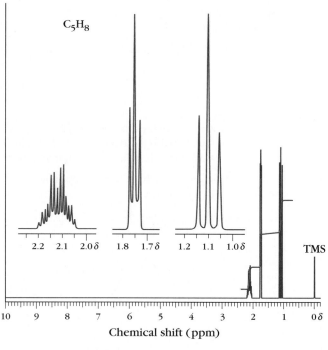

10.35 The splitting pattern of the signal due to the hydrogen H_b in the following compound depends on the relative values of J_{ab} and J_{bc}. Using a tree diagram, draw the resulting splitting pattern of H_b in each of the following cases.

$$H_3C-CH_2-CH_2X$$

$$\uparrow \qquad \uparrow \qquad \uparrow$$

$$H_a \qquad H_b \qquad H_c$$

(a) $J_{ab} = J_{bc} = 7$ Hz   **(b)** $J_{ab} = 7$ Hz; $J_{bc} = 14$ Hz   **(c)** $J_{ab} = 14$ Hz; $J_{bc} = 8$ Hz

ALCOHOLS

Alcohols are a large family of compounds whose functional group is a hydroxy group ($-$OH) bonded to an sp^3 hybrid carbon atom (Section 2.14, C).

$$\text{HO}\longrightarrow\text{C}$$

Hydroxy group

sp^3 Hybrid carbon atom

Ethyl alcohol is the most common example of this family of compounds. Our ancestors thousands of years ago knew how to obtain ethyl alcohol by the natural fermentation of sugars and starch. Fermentation is a complex series of chemical reactions by which the enzymes of yeast decompose aqueous solutions of sugars and starches to produce ethanol and CO_2. Sugars and starches come from a variety of natural sources such as sugar cane and various grains.

$$(C_6H_{10}O_5)_n \;+\; nH_2O \xrightarrow{\text{Enzymes in yeast}} 2n\,CO_2 \;+\; 2n\,CH_3CH_2OH$$

Sugars and starches obtained from grains and fruits

Ethyl alcohol

Until the middle of the 20th century, fermentation of starch was the major source of ethyl alcohol. Then the use of petroleum as the major source of hydrocarbons led to the method of producing ethyl alcohol by the acid-catalyzed hydration of ethylene (Section 7.14).

$$CH_2{=}CH_2 \;+\; H_2O \xrightarrow{H_2SO_4} CH_3CH_2OH$$

Methyl alcohol is another widely used alcohol. It was originally obtained by heating wood in the absence of air so it was called wood alcohol. Catalytic reduction of carbon monoxide by hydrogen gas is the modern method of producing methanol.

$$2H_2 \;+\; CO \xrightarrow[\text{Heat}]{\text{Catalyst}} CH_3OH$$

Methanol

Methanol is used principally as a raw material for the production of other commercially important compounds such as formaldehyde ($H_2C{=}O$). Methanol is also used as an antifreeze in windshield washer solutions and as an additive to jet fuels.

Many complex compounds containing one or more hydroxy groups are present in living organisms. **Cholesterol** is a steroid that contains a hydroxy group. Cholesterol has received a great deal of attention because of the suspected link between its concentration in human blood and certain types of heart diseases.

Cholesterol

In this chapter, we'll study the chemistry of alcohols. Let's begin by learning how alcohols are classified.

11.1 Classification of Alcohols

In Section 2.7, an sp^3 hybrid carbon atom was classified as primary, secondary, or tertiary depending on the number of other carbon atoms bonded to it. This designation is also the basis of the classification of alcohols. The hydroxy group of a **primary alcohol** is bonded to a primary sp^3 hybrid carbon atom, the hydroxy group of a **secondary alcohol** is bonded to a secondary sp^3 hybrid carbon atom, and the hydroxy group of a **tertiary alcohol** is bonded to a tertiary sp^3 hybrid carbon atom. This classification of alcohols is summarized in Figure 11.1.

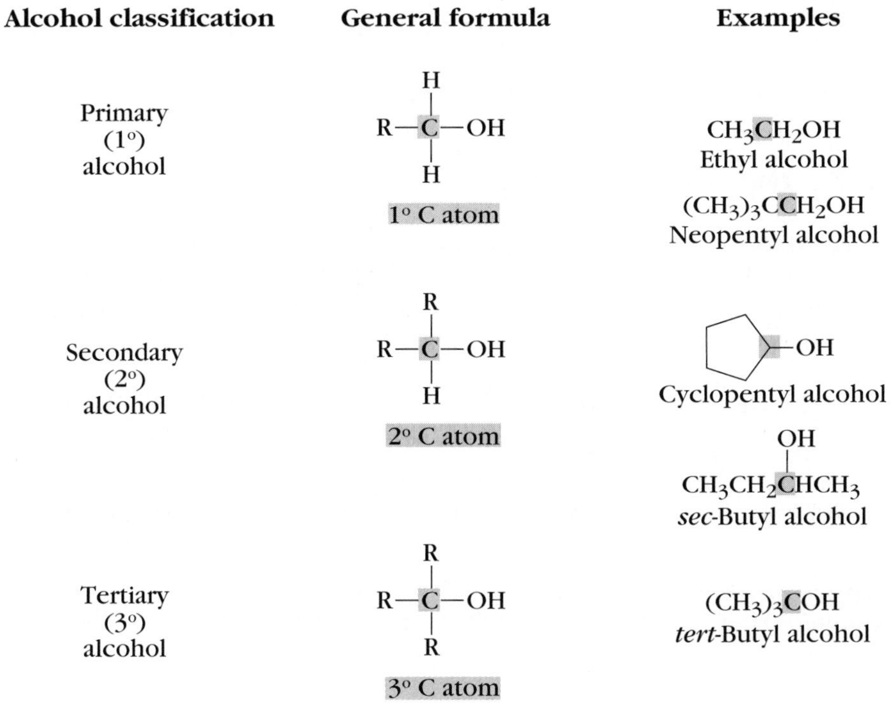

Alcohol classification	General formula	Examples
Primary (1°) alcohol	R—C(H)(H)—OH 1° C atom	CH_3CH_2OH Ethyl alcohol $(CH_3)_3CCH_2OH$ Neopentyl alcohol
Secondary (2°) alcohol	R—C(R)(H)—OH 2° C atom	Cyclopentyl alcohol $CH_3CH_2CHCH_3$ (with OH) sec-Butyl alcohol
Tertiary (3°) alcohol	R—C(R)(R)—OH 3° C atom	$(CH_3)_3COH$ tert-Butyl alcohol

Figure 11.1
Classification of alcohols as primary (1°), secondary (2°), or tertiary (3°).

We must be careful to distinguish between alcohols and those compounds that have a hydroxy group bonded to an sp^2 carbon atom such as phenol and enols. Enols react differently than alcohols. Enols easily tautomerize to their keto form as we learned in Chapter 9 (Section 9.7). Alcohols don't tautomerize. Phenol is a compound in which the hydroxy group is bonded to an aromatic ring. As we'll learn in Chapter 23, most of the reactions of phenol and its derivatives are different from those of alcohols.

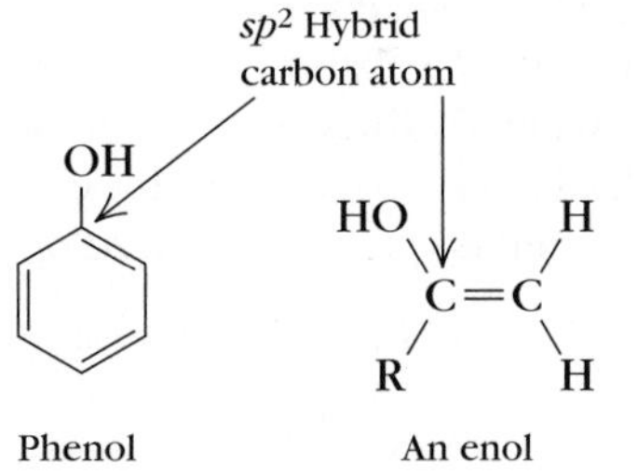

sp^2 Hybrid carbon atom

Phenol

An enol

Exercise 11.1	Identify each of the following as a primary, a secondary, or a tertiary alcohol.

$\overset{\displaystyle OH}{\underset{\displaystyle |}{}}$

(a) $(CH_3)_2CHCHCH_3$ **(d)** $(CH_3CH_2)_3COH$

$\overset{\displaystyle CH_3}{\underset{\displaystyle |}{}}$ $\overset{\displaystyle CH_3}{\underset{\displaystyle |}{}}$

(b) $CH_3CH_2CHCH_2OH$ **(e)** $CH_2{=}CHCH_2CHCH_2OH$

$\overset{\displaystyle CH_3}{\underset{\displaystyle |}{}}$

(c) HO— [cyclohexane ring structure]   **(f)** [cyclopropyl]—$\overset{\displaystyle CH_3}{\underset{\displaystyle |}{}}$CHOH

                CH₃

11.2   Naming Alcohols

The **common names** of alcohols, as we discussed in Section 2.14, C, consist of two words. The first word is the name of the alkyl group (Section 2.5) bonded to the —OH group, and the second word is alcohol. The following are examples of common names of several simple alcohols.

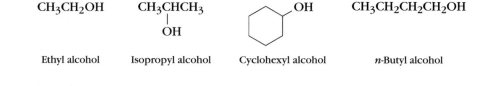

CH_3CH_2OH $\overset{\displaystyle CH_3CHCH_3}{\underset{\displaystyle \overset{|}{OH}}{}}$ [cyclohexane]—OH $CH_3CH_2CH_2CH_2OH$

Ethyl alcohol   Isopropyl alcohol   Cyclohexyl alcohol   *n*-Butyl alcohol

Exercise 11.2	Give a common name to each of the following alcohols.

(a) CH_3OH **(c)** [cyclobutane]—OH

(b) [pentane chain]—OH   **(d)** $(CH_3)_2CHCH_2OH$

Alcohols with more complex structures are best named by the IUPAC rules. The following additions to the IUPAC rules, first introduced in Section 2.5, are needed to name alcohols.

1. Select as the parent structure the longest continuous chain of carbon atoms that contains the —OH group.
2. Obtain the parent name by substituting the ending *-ol* for the ending *-e* of the corresponding alkane. If the compound contains more than one —OH, the endings are *-diol* for two, *-triol* for three, and so forth. In these cases, the *-e* of the alkane name is retained for ease of pronunciation.

$$HOCH_2CH_2OH$$

1,2-ethanediol

3. Number the parent chain to give the lowest number to the carbon atom bonded to the —OH group. The position of the —OH group is given the lowest number in molecules that contain carbon-carbon double bonds, alkyl groups, or halogen atoms. The order of preference of these groups is the following:

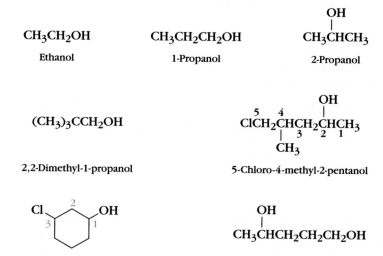

X = F, Cl, Br, I

The following compounds are named according to the IUPAC rules:

CH₃CH₂OH

Ethanol

CH₃CH₂CH₂OH

1-Propanol

OH
|
CH₃CHCH₃

2-Propanol

(CH₃)₃CCH₂OH

2,2-Dimethyl-1-propanol

OH
|
ClCH₂CHCH₂CHCH₃
5   4 | 2   1
            3
          CH₃

5-Chloro-4-methyl-2-pentanol

Cl ⟨ ⟩ OH
3   1
    2

3-Chlorocyclohexanol

OH
|
CH₃CHCH₂CH₂CH₂OH

1,4-Pentanediol

Exercise 11.3   Give the IUPAC name to each of the compounds in Exercise 11.2.

Exercise 11.4   Write the structure for each of the following compounds.

(a) 2,2-Dimethyl-1-butanol
(b) 3-Methyl-1-pentanol
(c) 1-Methylcyclohexanol
(d) cis-3-Chlorocyclohexanol

(e) 1,3-Propanediol
(f) 3-Cyclohexenol
(g) (E)-2-Buten-1,4-diol
(h) (2S, 3R)-3-Methyl-4-penten-2-ol

Exercise 11.5   Give the IUPAC name corresponding to each of the following structures.

(a) $ClCH_2CH_2OH$

(b) $(CH_3)_2CH(CH_2)_2CH_2OH$

(c) 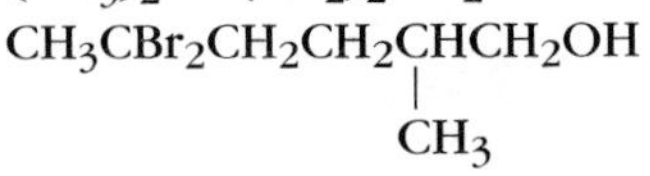 $CH_3CBr_2CH_2CH_2CHCH_2OH$
$\quad\qquad\qquad\qquad\qquad\quad |$
$\quad\qquad\qquad\qquad\qquad\quad CH_3$

(d)

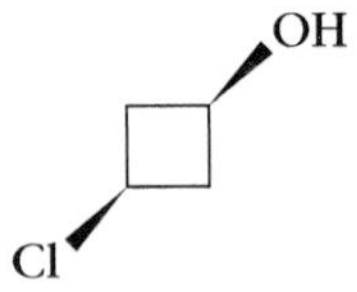

(e)

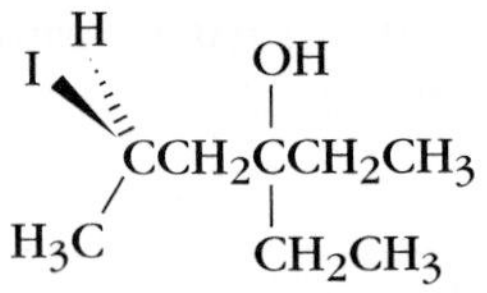

(f) 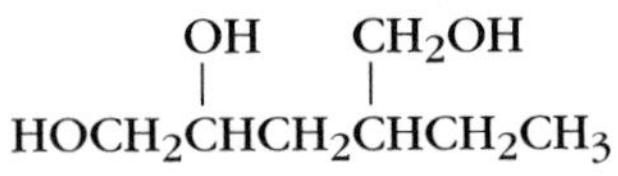 $HOCH_2CHCH_2CHCH_2CH_3$

11.3   Physical Properties

The variations in some of the physical properties of straight chain alcohols with increasing molecular weight are best understood if we consider alcohols as organic derivatives of water. That is, an alcohol can be visualized as being formed by replacing one hydrogen atom of a water molecule by an alkyl group.

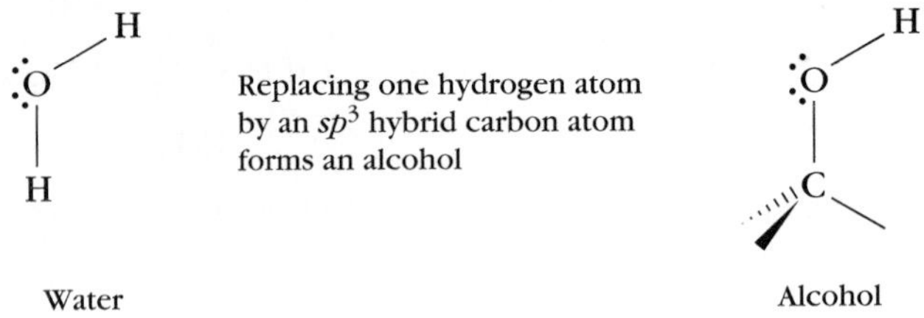

Replacing one hydrogen atom by an sp^3 hybrid carbon atom forms an alcohol

Water   Alcohol

According to this view, an alcohol consists of two parts. One part, the nonpolar alkane-like alkyl group, is insoluble in water and is called the **hydrophobic** (water-hating) or **lipophilic** (fat-loving) part. The term hydrophobic emphasizes the insolubility in water of a substance, while lipophilic emphasizes its solubility in nonpolar solvents. The two terms are used interchangeably. The second part,

Table 11.1		Solubilities of Alcohols in Water	
Name	Carbon Atoms	Formula	Solubility (g/100 ml) (20 °C)
Methanol	1	CH_3OH	∞*
1-Propanol	3	$CH_3CH_2CH_2OH$	∞*
1-Butanol	4	$CH_3CH_2CH_2CH_2OH$	7.9
1-Hexanol	6	$CH_3(CH_2)_4CH_2OH$	0.6
1-Octanol	8	$CH_3(CH_2)_6CH_2OH$	0.05
1-Nonanol	9	$CH_3(CH_2)_7CH_2OH$	Insoluble
1-Dodecanol	12	$CH_3(CH_2)_{10}CH_2OH$	Insoluble

*∞ Means completely soluble in water in all proportions.

Table 11.2	Comparison of Boiling Points of Alcohols and Alkanes of Similar Molecular Weight		
Name	Structure	Molecular Weight	Boiling point (°C)
Methanol	CH_3OH	32	64.5
Ethane	CH_3CH_3	30	−88.5
1-Propanol	$CH_3CH_2CH_2OH$	60	97
Butane	$CH_3(CH_2)_2CH_3$	58	−0.5
1-Dodecanol	$CH_3(CH_2)_{10}CH_2OH$	186	255
Tridecane	$CH_3(CH_2)_{11}CH_3$	184	234

the polar water-like hydroxy group, is soluble in water and is called the **hydrophilic** (water-loving) or **lipophobic** (fat-hating) part.

The solubilities of several straight chain alcohols in water are given in Table 11.1. The data in Table 11.1 show that the solubility in water of straight chain alcohols containing a single hydroxy group decreases as the number of carbon atoms increases. Alcohols containing more than nine carbon atoms resemble alkanes in their water solubilities rather than low molecular weight alcohols.

In Table 11.2, the boiling points of alcohols and alkanes of similar molecular weight are compared. The data in Table 11.2 show that the boiling points of methanol and 1-propanol are much higher than the boiling points of the alkanes of about the same molecular weight and shape. This difference becomes smaller, however, when the alkyl part of the alcohol becomes very large. The boiling point of 1-dodecanol, for example, is not much higher than the boiling point of tridecane.

Differences in physical properties between low molecular weight and high molecular weight alcohols are due to differences in their intermolecular forces. The important structural feature of low molecular weight alcohols is their hydroxy group. Just as in water, hydrogen bonding in these alcohols is very important to their physical properties. It is the ability of these alcohols to form hydrogen bonds that is responsible for their higher boiling points and water solubilities compared to alkanes of the same molecular weight.

As the alkyl portion of the alcohol increases in size with increasing molecular weight, the alkyl portion of the molecule becomes the dominant feature, and structurally these alcohols resemble alkanes more than water. As a result, hydrophobic attractions are more important than any hydrogen bonding, with the result that their boiling points and water solubilities resemble alkanes.

■ *Intermolecular forces are discussed in Section 1.16.*

11.4 Spectroscopic Properties of Alcohols

The IR spectra of alcohols have a characteristic strong absorption band in the region from 3400 to 3200 cm^{-1} due to the O—H stretching (Section 5.5). The exact position and appearance of this band depends on the degree of hydrogen bonding in the sample. The infrared spectrum of pure 1-butanol in Figure 11.2, for example, shows a broad and intense absorption band at 3350 cm^{-1}. There is a great deal of hydrogen bonding in a pure liquid alcohol, so the peak is broad

Figure 11.2
The solid red line represents the IR spectrum of pure liquid 1-butanol. The band at 3350 cm^{-1} due to O—H stretching is broad and intense because of hydrogen bonding. The dotted red line inserted from 3800 to 3000 cm^{-1} represents the IR spectrum in that region of a dilute solution of 1-butanol in carbon tetrachloride. A sharp band due to free O—H stretching appears at 3650 cm^{-1}, in addition to the hydrogen bonded O—H band at 3350 cm^{-1}.

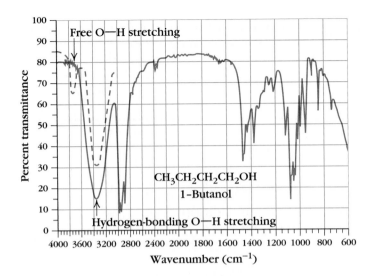

and intense. The dotted line inserted from 3800 to 3000 cm^{-1} represents the IR spectrum of a dilute solution of 1-butanol (about 1%) in carbon tetrachloride. There is less hydrogen bonding in a dilute solution of the alcohol, so a sharp band appears at about 3650 cm^{-1} (due to the free OH stretching), in addition to the broad and intense band at 3350 cm^{-1}.

Compounds with C—O bonds (alcohols and ethers) generally have a strong absorption band in the range 1000 to 1200 cm^{-1}. Since other functional groups also have absorption bands in this fingerprint region, this band is less characteristic of alcohols than the absorption band in the region from 3640 to 3200 cm^{-1}.

In the ^{13}C NMR spectra of alcohols, the signal of a carbon atom bonded to a hydroxy group is found downfield from the corresponding unsubstituted carbon atom because of the electron-withdrawing oxygen atom of the hydroxy group. The carbon atom bonded to the hydroxy group of cyclohexanol, for example, is downfield (δ 70.0) compared to the unsubstituted carbon atoms in the molecule.

OH
δ 36.0 →   ← δ 70.0
                        ↘ δ 25.0
δ 26.4

■ *Vicinal spin-spin coupling of hydrogen atoms is discussed in Section 10.7.*

The chemical shifts of hydrogen atoms bonded to oxygen in the ^{1}H NMR spectra of alcohols are temperature and concentration dependent. The signal of the hydrogen atom of the —OH group, if present, is usually found in the δ 0.5 to 4.5 region of the spectrum. Vicinal spin-spin coupling of hydrogen atoms bonded to oxygen atoms with neighboring hydrogens on carbon atoms is not usually observed because traces of acidic impurities in the sample catalyze an exchange between the hydroxy hydrogen atoms and the proton of the acidic impurity.

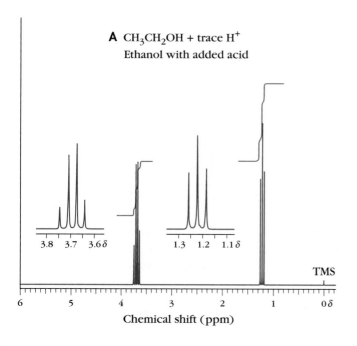

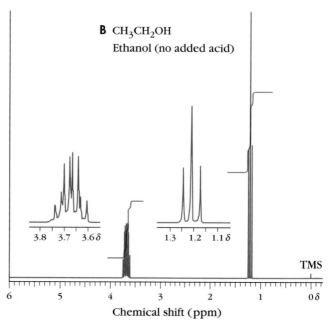

Figure 11.3
A, The δ 6.0 to 0.0 region of the ¹H NMR spectrum of ethanol with added HCl shows the typical splitting pattern of an ethyl group, because the hydrogen atom of the —OH group and the methylene hydrogen atoms are decoupled by proton exchange. **B,** Each of the four signals of the methylene group in the ¹H NMR spectrum of purified ethanol is split into a doublet by coupling with the hydrogen atom of the —OH group.

This process is rapid on the NMR time scale, so the effects of spin-spin coupling are removed and no coupling is observed.

The rate of hydrogen exchange can be substantially reduced by carefully purifying the alcohol and the solvent to remove acidic impurities that catalyze hydrogen exchange. This is illustrated by the two ¹H NMR spectra of ethanol shown in Figure 11.3. Only the typical splitting pattern of an ethyl group (Section 10.7) appears in Figure 11.3, *A*, the ¹H NMR spectrum of a sample of ethanol with added HCl. No coupling is observed, because rapid exchange causes the methylene hydrogen atoms to see an equal number of hydroxy hydrogen atoms with spin α and spin β, which means that the average hydroxy hydrogen atoms have a nuclear spin of zero. In contrast, the rate of hydrogen exchange is slowed in a purified sample of ethanol, whose ¹H NMR spectrum is shown in Figure 11.3, *B*, so each of the four signals of the methylene group is split into a doublet by coupling with the hydrogen atom of the —OH group.

The peak due to a hydroxy hydrogen atom can usually be identified by exchanging it with deuterium. This is done by shaking the sample with deuterium oxide, D_2O. Any exchangeable hydrogen atoms are quickly replaced by deuterium atoms.

$$RCH_2OH \ + \ D_2O \ \rightleftharpoons \ RCH_2OD \ + \ DOH$$

Signals due to deuterium are not observed in ¹H NMR spectra, so exchange of the hydroxy hydrogen atom by deuterium leads to the disappearance of its ¹H NMR signal.

Exercise 11.6	Deduce the structure of the alcohols whose molecular formulas and ¹H NMR spectra are given on p. 458.

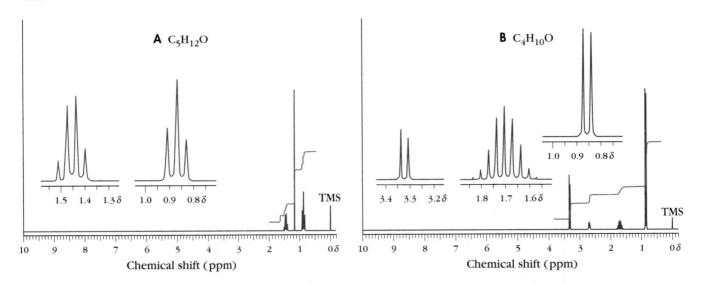

11.5   Hydration of Alkenes Forms Alcohols: A Review

Several electrophilic additions to alkenes that produce alcohols were discussed in Chapters 7 and 8 and are summarized and compared in Table 11.3.

Oxymercuration-demercuration and hydroboration-oxidation are complementary because one method forms an alcohol with Markownikoff regiochemistry, while the other method forms an alcohol with anti-Markownikoff regiochemistry.

1-Methylcyclopentanol
Markownikoff product
(91% yield)

trans-2-Methylcyclopentanol
anti-Markownikoff product
(Racemic mixture, 86% yield)

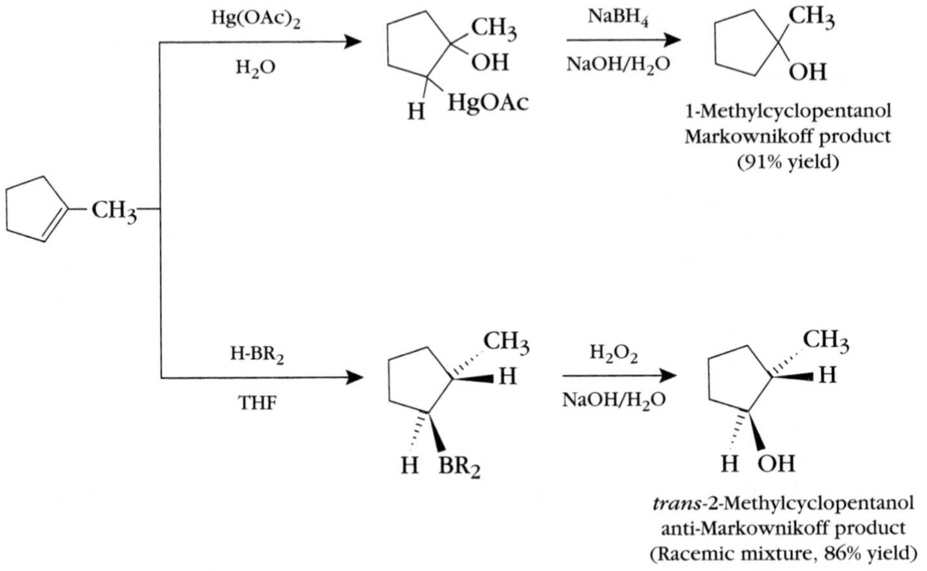

| Exercise 11.7 | Write the equations for the reaction of 2-methyl-2-hexene with each of the following reagents. |

(a) A THF solution of 9-BBN followed by H_2O_2 and $NaOH/H_2O$

(b) An aqueous solution of $Hg(OAc)_2$ followed by an alkaline solution of $NaBH_4$

■ *The structure of 9-BBN is given in Section 8.1.*

Table 11.3	Comparison of Three Methods of Forming Alcohols from Alkenes

Method and General Equation	Stereochemistry of Product	Orientation of Product	Comments	Reference
Acid-catalyzed hydration: 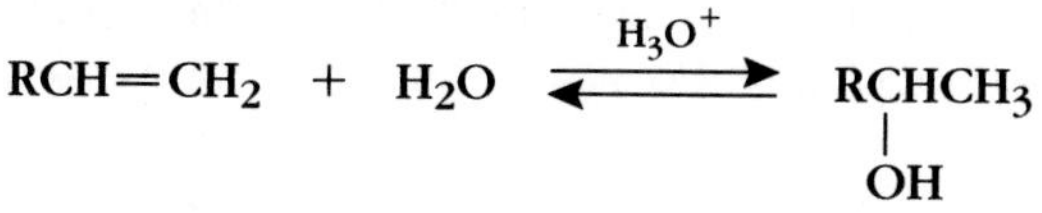 $$RCH{=}CH_2 \ + \ H_2O \ \underset{}{\overset{H_3O^+}{\rightleftharpoons}} \ \underset{OH}{RCHCH_3}$$	Nonstereospecific	Addition of water according to Markownikoff Rule	Reversible reaction; carbocation is an intermediate in the reaction so there is a possibility of rearrangement	Section 7.14
Oxymercuration-demercuration: 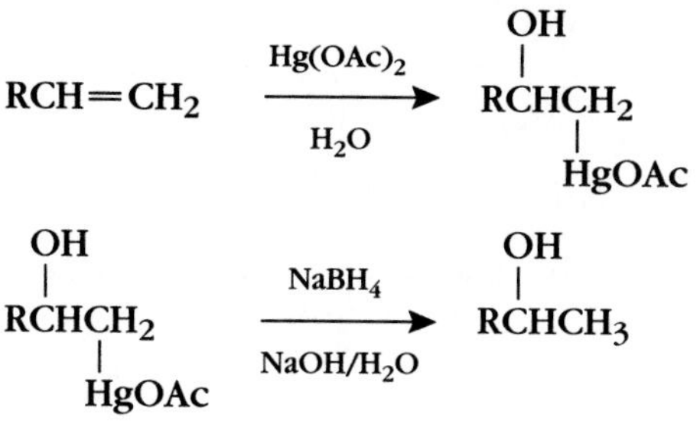 $$RCH{=}CH_2 \ \underset{H_2O}{\overset{Hg(OAc)_2}{\longrightarrow}} \ \underset{HgOAc}{\overset{OH}{RCHCH_2}}$$ $$\underset{HgOAc}{\overset{OH}{RCHCH_2}} \ \underset{NaOH/H_2O}{\overset{NaBH_4}{\longrightarrow}} \ \underset{}{\overset{OH}{RCHCH_3}}$$	Nonstereospecific	Addition of water according to Markownikoff Rule	No rearrangement	Section 8.10
Hydroboration-oxidation: 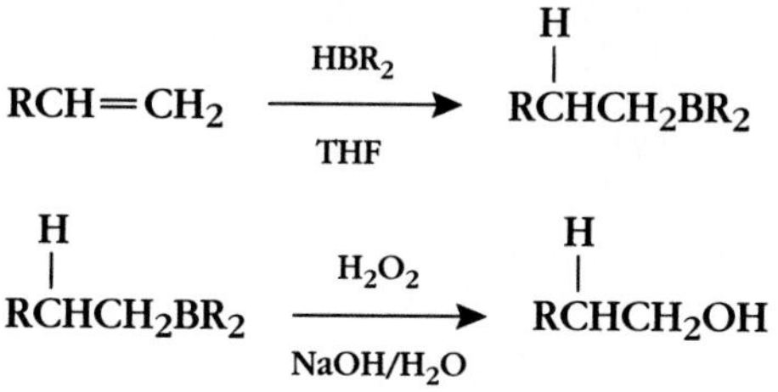 $$RCH{=}CH_2 \ \underset{THF}{\overset{HBR_2}{\longrightarrow}} \ \overset{H}{RCHCH_2BR_2}$$ $$\overset{H}{RCHCH_2BR_2} \ \underset{NaOH/H_2O}{\overset{H_2O_2}{\longrightarrow}} \ \overset{H}{RCHCH_2OH}$$	Products form by *syn* addition of water	Alcohol formed by anti-Markownikoff addition of water	No rearrangement	Section 8.1 to 8.4

Two useful syntheses of alcohols involve reactions with the carbonyl groups of aldehydes and ketones. Before we can discuss these reactions, we must examine the structure of the carbonyl group.

11.6   An Introduction to the Carbonyl Group

A number of the functional groups introduced in Section 2.14 contain a **carbonyl group.** A carbonyl group consists of a carbon atom linked to an oxygen atom by a double bond.

$$\diagdown C = O$$
$$\diagup$$

A carbonyl group

The hydration of alkynes (Section 9.7) forms aldehydes and ketones, which are two classes of compounds that contain a carbonyl group.

$$\begin{array}{cc} O & O \\ \parallel & \parallel \\ C & C \\ R \quad H & R \quad R \\ \text{Aldehyde} & \text{Ketone} \end{array}$$

- *The aldehyde functional group* $R-C\overset{\displaystyle O}{\underset{\displaystyle H}{\diagup\diagup}}$ *is often abbreviated as RCHO.*

A carbonyl group and the atoms bonded to it lie in a plane, as shown in Figure 11.4

Figure 11.4
The plane of a carbonyl group.

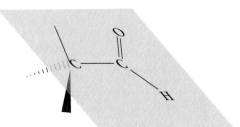

Like a carbon-carbon double bond, a carbonyl group undergoes addition reactions. Additions to a carbonyl group differ from additions to a carbon-carbon double bond, however, because of the different electronegativities of the two atoms of the carbonyl group. Oxygen is more electronegative than carbon, so the carbonyl oxygen is electron rich and the carbonyl carbon is electron deficient. As a result, both the sigma and pi bonds of a carbonyl group are polar covalent bonds (Section 1.14) and a carbonyl group is strongly polarized in the following way:

$$\diagdown C = O$$
$$\diagup {}_{\delta^+ \quad \delta^-}$$

Addition of an unsymmetrical reagent occurs so that its nucleophilic (basic) part adds to the carbonyl carbon, while its electrophilic (acidic) part adds to the carbonyl oxygen. Notice that this orientation of addition is simply an example of the attraction of unlike charges. *No matter what the reaction mechanism, addition to a carbonyl group follows this orientation.*

Nucleophilic part of reagent adds here ——→ Electrophilic part of reagent adds here

$$\underset{\delta^+\ \ \delta^-}{C=O} \longrightarrow \underset{\underset{X}{|}\ \ \underset{Y}{}}{-C-O}$$

Product of addition

$$\underset{\delta^-\ \ \delta^+}{X-Y}$$

The reactions of a carbonyl group are very different from those of a carbon-carbon double bond. While the predominant reaction of alkenes is electrophilic addition (Section 7.7), a predominant reaction of carbonyl-containing compounds is nucleophilic addition to the carbonyl carbon. Two of these nucleophilic addition reactions are used to prepare alcohols. We discuss first the reaction of the carbonyl groups of aldehydes and ketones with organometallic reagents.

11.7   Synthesis of Alcohols by the Use of Grignard and Organolithium Reagents

Compounds that contain a carbon-metal bond such as Grignard and organolithium reagents are called **organometallic compounds.** Grignard reagents contain a carbon-magnesium bond while lithium reagents contain a carbon-lithium bond.

Grignard reagents, named after their discoverer the French chemist Victor Grignard, are prepared by the reaction of a halogen-containing compound with magnesium metal turnings in diethyl ether or THF as solvent.

General Reaction:

$$RX \ + \ Mg \ \xrightarrow[\text{or THF}]{(C_2H_5)_2O} \ RMgX \qquad \begin{matrix} X = Cl, Br, \text{ or I} \\[4pt] R = \text{alkyl or aryl} \end{matrix}$$

Grignard reagent

Examples:

$$CH_3I \ + \ Mg \ \xrightarrow{(C_2H_5)_2O} \ CH_3MgI$$

Iodomethane   Methylmagnesium iodide

Bromobenzene + Mg $\xrightarrow{\text{THF}}$ Phenylmagnesium bromide

Organolithium reagents are prepared by a similar reaction. Instead of magnesium metal, lithium metal is used to react with a halogen-containing compound.

General Reaction:

$$RX \ + \ 2Li \ \xrightarrow[\text{or THF}]{(C_2H_5)_2O} \ RLi \ + \ LiX \qquad \begin{matrix} X = Cl, Br, \text{ or I} \\[4pt] R = \text{alkyl or aryl} \end{matrix}$$

Organo-lithium reagent

■ *Recall that THF is the abbreviation for tetrahydrofuran,*

Examples:

$$CH_3CH_2CH_2CH_2Br + 2Li \xrightarrow[(C_2H_5)_2O]{} CH_3CH_2CH_2CH_2Li + LiBr$$

1-Bromobutane   Butyllithium

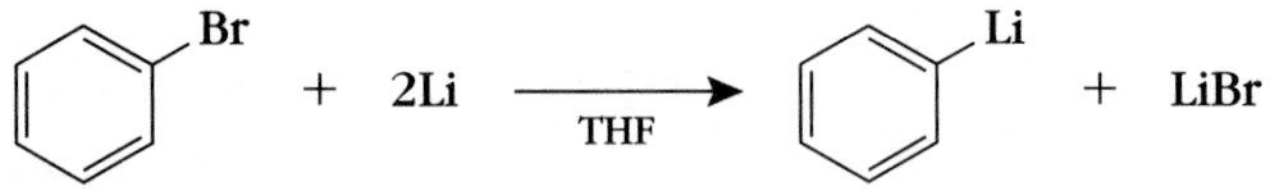

Bromobenzene   Phenyllithium

Lithium and magnesium are less electronegative elements than carbon, with the result that the C—Li and C—Mg bonds are polarized such that the carbon atom is a center of partial negative charge (a nucleophilic center) and the metal atom is a center of partial positive charge.

The important feature of these compounds is their nucleophilic carbon atom that reacts as if it were a carbanion, despite the fact that the C—Mg and C—Li bonds are more polar covalent than ionic. The nucleophilic carbon atom of both Grignard and organolithium reagents reacts with the electropositive carbonyl carbon atom of carbonyl-containing compounds, as shown for the Grignard reagent in Figure 11.5.

The reaction occurs in two steps. The first is the addition of the Grignard reagent to the carbonyl carbon atom to form the magnesium salt of the conjugate base of an alcohol (an alkoxide ion). Reaction of the magnesium salt with aqueous acid in the second step forms an alcohol.

Exercise 11.8   Write the reaction of an organolithium reagent, RLi, with a carbonyl group followed by reaction with aqueous acid.

The reaction of Grignard and organolithium reagents with the carbonyl group of aldehydes and ketones is a useful synthesis of alcohols because a new carbon-

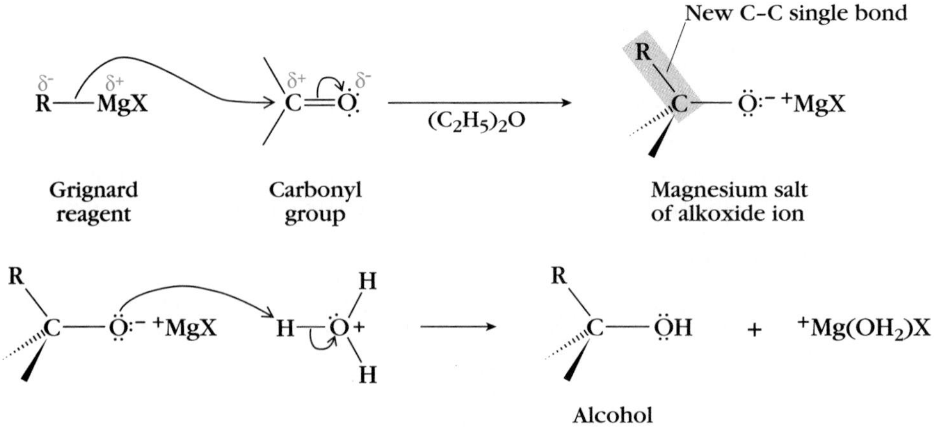

Figure 11.5
Use of a Grignard reagent in the synthesis of alcohols.

carbon single bond is formed between the electrophilic carbon atom of the carbonyl group and the nucleophilic carbon atom of the organometallic compound. As a result, a large and complex alcohol can be formed by combining several smaller molecules. Primary (1°), secondary (2°), and tertiary (3°) alcohols can be prepared by this method because the structure of the carbonyl-containing compound as well as the Grignard and organolithium reagent can be varied.

The reaction of formaldehyde, $H_2C{=}O$, for example, with Grignard and organolithium reagents forms primary alcohols.

$$CH_3CH_2CH_2CH_2Br \xrightarrow[(C_2H_5)_2O]{Mg} CH_3CH_2CH_2CH_2MgBr$$

Butylmagnesium bromide

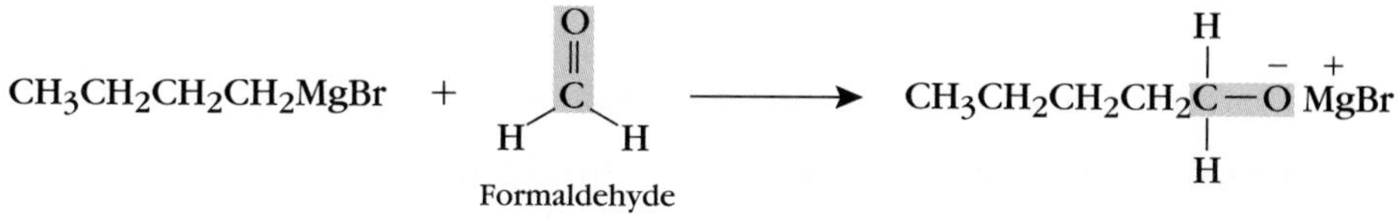

Formaldehyde

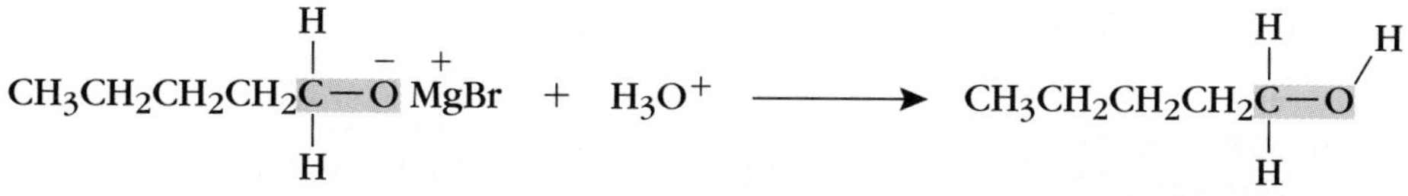

1-Pentanol
(72% yield)

Grignard and organolithium reagents react with aldehydes (other than formaldehyde) to form secondary alcohols.

$$(CH_3)_2CHCH_2Br \xrightarrow[(C_2H_5)_2O]{Mg} (CH_3)_2CHCH_2MgBr$$

Isobutylmagnesium bromide

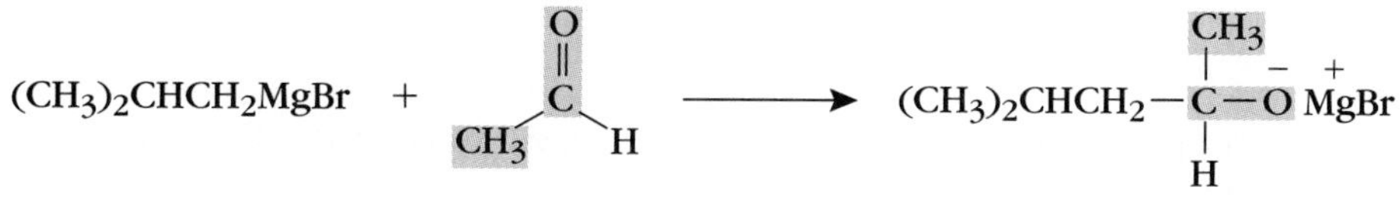

Acetaldehyde

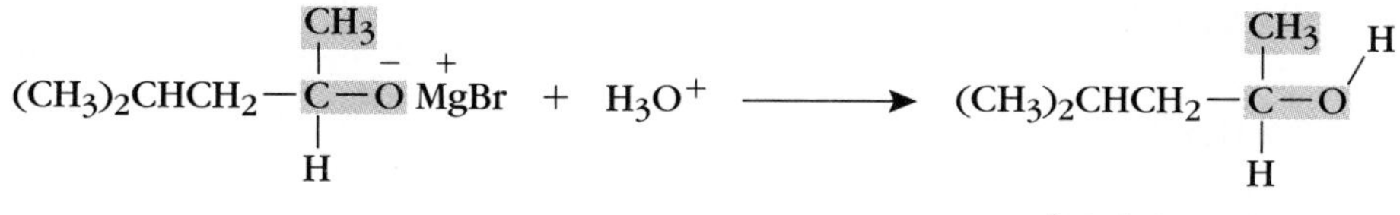

4-Methyl-2-pentanol
(65% yield)

Ketones react with Grignard reagents and organolithium reagents to form tertiary alcohols.

$$CH_3CH_2CH_2Br \xrightarrow[(C_2H_5)_2O]{Mg} CH_3CH_2CH_2MgBr$$

Propylmagnesium bromide

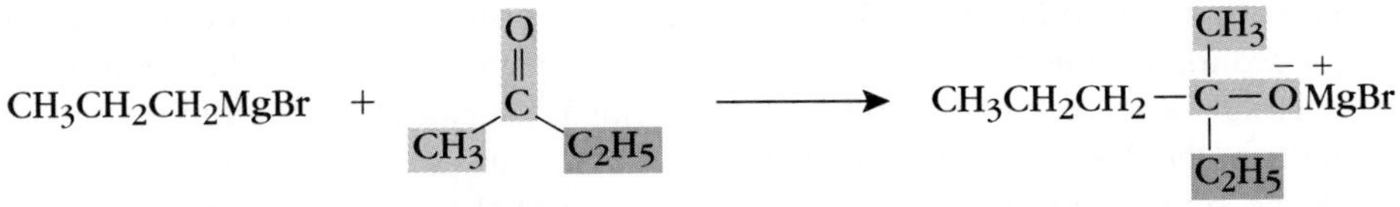

$$CH_3CH_2CH_2MgBr \; + \; \underset{\substack{CH_3 \quad C_2H_5}}{\overset{O}{\underset{\|}{C}}} \longrightarrow CH_3CH_2CH_2 - \overset{CH_3}{\underset{C_2H_5}{\overset{|}{\underset{|}{C}}}} - \overset{-}{O}\overset{+}{MgBr}$$

2-Butanone
(A ketone)

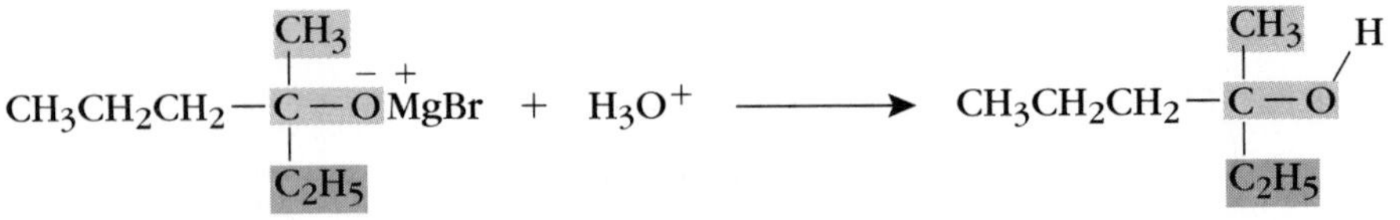

$$CH_3CH_2CH_2 - \overset{CH_3}{\underset{C_2H_5}{\overset{|}{\underset{|}{C}}}} - \overset{-}{O}\overset{+}{MgBr} \; + \; H_3O^+ \longrightarrow CH_3CH_2CH_2 - \overset{CH_3}{\underset{C_2H_5}{\overset{|}{\underset{|}{C}}}} - \overset{H}{\overset{|}{O}}$$

3-Methyl-3-hexanol
(68% yield)

In summary, Grignard and organolithium reagents form:

1. **Primary alcohols** by reaction with **formaldehyde.**
2. **Secondary alcohols** by reaction with **aldehydes** other than formaldehyde.
3. **Tertiary alcohols** by reaction with **ketones.**

Exercise 11.9   Write the equations for each of the following.

(a) The formation of an organolithium reagent by the reaction of lithium metal and 2-bromobutane.
(b) The reaction of the organolithium reagent prepared in **(a)** with formaldehyde, $H_2C=O$.
(c) The hydrolysis of the product in **(b)** to form an alcohol.
(d) The reaction of the organolithium reagent prepared in **(a)** with

$$CH_3CH_2\overset{O}{\overset{\diagup\!\!\!\diagup}{\underset{\diagdown}{C}}}_H$$ followed by hydrolysis to form the product.

(e) Repeat **(d)** but add the organolithium reagent to $CH_3CH_2\overset{O}{\overset{\|}{C}}CH_2CH_3$.

Exercise 11.10   Write the product for each of the following reactions.

(a) $(CH_3)_2CHI \xrightarrow[\text{(C}_2\text{H}_5)_2\text{O}]{\text{Mg}}$?

(b) 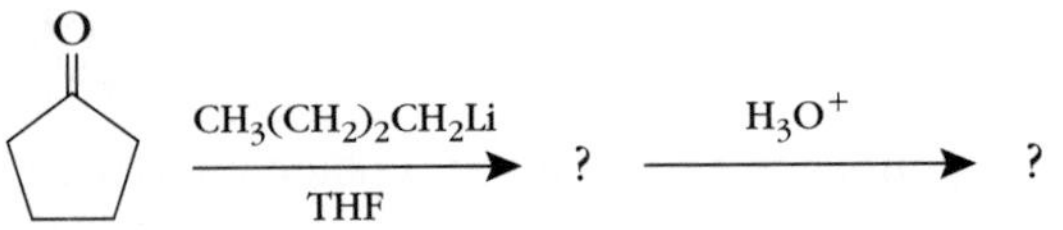 $\xrightarrow[\text{THF}]{\text{CH}_3(\text{CH}_2)_2\text{CH}_2\text{Li}}$? $\xrightarrow{\text{H}_3\text{O}^+}$?

(c) 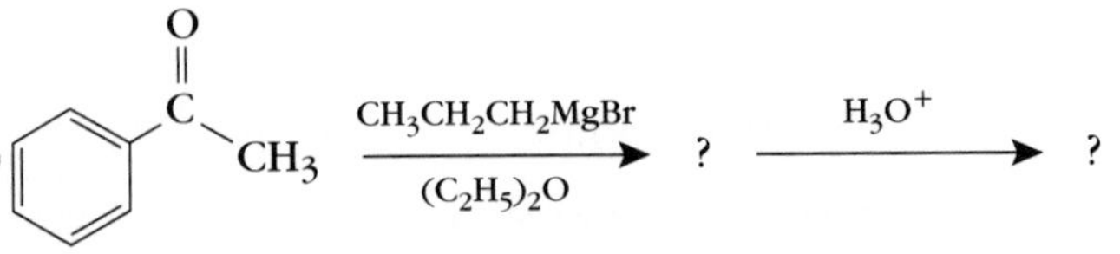 $\xrightarrow[\text{(C}_2\text{H}_5)_2\text{O}]{\text{CH}_3\text{CH}_2\text{CH}_2\text{MgBr}}$? $\xrightarrow{\text{H}_3\text{O}^+}$?

(d) 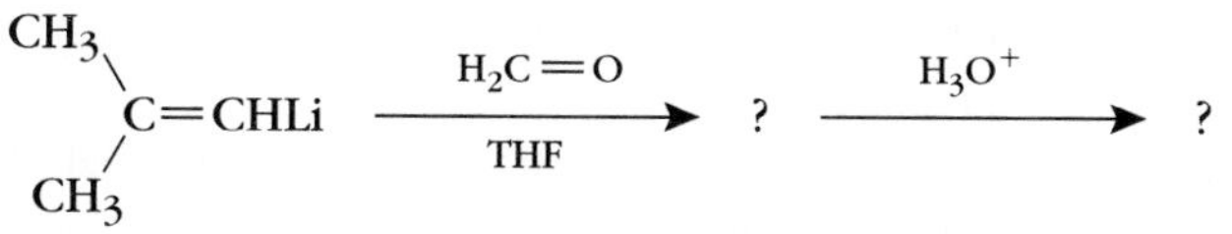 $\underset{CH_3}{\overset{CH_3}{\diagdown}}C=CHLi$ $\xrightarrow[\text{THF}]{H_2C=O}$? $\xrightarrow{H_3O^+}$?

(e) 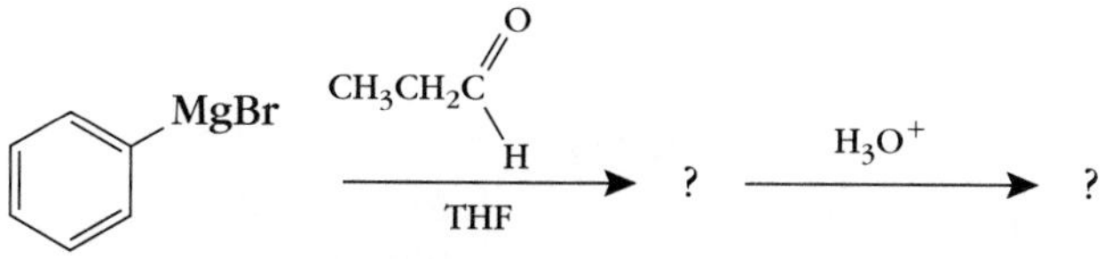   benzene—MgBr $\xrightarrow[\text{THF}]{CH_3CH_2\overset{O}{\overset{\|}{C}}\diagdown H}$? $\xrightarrow{H_3O^+}$?

The utility of Grignard and organolithium reagents in the synthesis of a wide variety of alcohols has been emphasized in this section. How do we decide which Grignard or organolithium reagent and which carbonyl-containing compound to use to synthesize a particular alcohol?

11.8   Planning a Grignard Synthesis of an Alcohol

We learned in Section 9.12 that an effective way to plan a synthesis is to start with the desired product and work backwards until we arrive at the starting materials. The same method can be used to plan a Grignard synthesis of an alcohol. Again let us illustrate this method with a Sample Problem.

Sample Problem 11.1

Propose a synthesis of 2-butanol starting with any organic compound containing two or fewer carbon atoms, any inorganic reagent, and any solvent.

Solution:

We know that the functional group is an alcohol. So the first step is to classify the alcohol as 1°, 2°, or 3°. 2-Butanol is a 2° alcohol. How are 2° alcohols formed in a Grignard (or organolithium) reaction? The answer is by the reaction of a Grignard or organolithium reagent with an aldehyde.

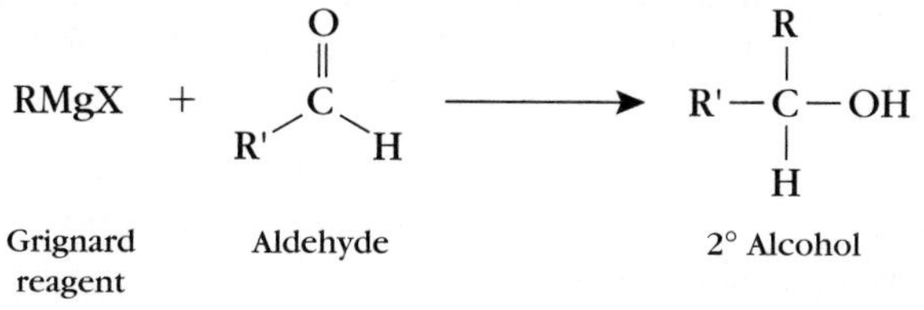

$$RMgX\ +\ \underset{R'}{\overset{O}{\overset{\|}{\underset{\diagdown}{C}}}}\diagup H\ \longrightarrow\ R'-\underset{H}{\overset{R}{\underset{|}{\overset{|}{C}}}}-OH$$

Grignard   Aldehyde   2° Alcohol
reagent

The important thing to notice in this reaction is that the carbon atom bearing the —OH group in the product is the carbonyl carbon atom in the starting aldehyde. Also the hydrogen atom bonded to the carbonyl carbon atom of the starting aldehyde ends up bonded to the carbon atom bearing the —OH group.

$$RMgX\ +\ \underset{R'}{\overset{O}{\overset{\|}{\underset{\diagdown}{C}}}}\diagup H\ \longrightarrow\ R'-\underset{H}{\overset{R}{\underset{|}{\overset{|}{C}}}}-OH$$

Grignard
reagent

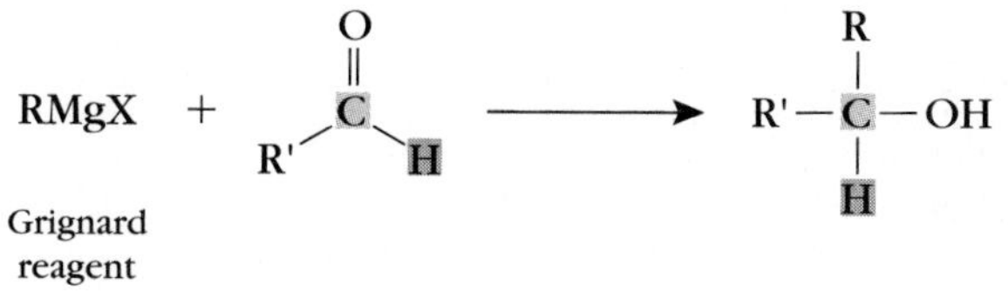

One of the two groups R or R′ must come from the aldehyde, while the other comes from the Grignard reagent.

With this in mind, we can dissect the product and discover what aldehyde and Grignard reagent to use to form the product.

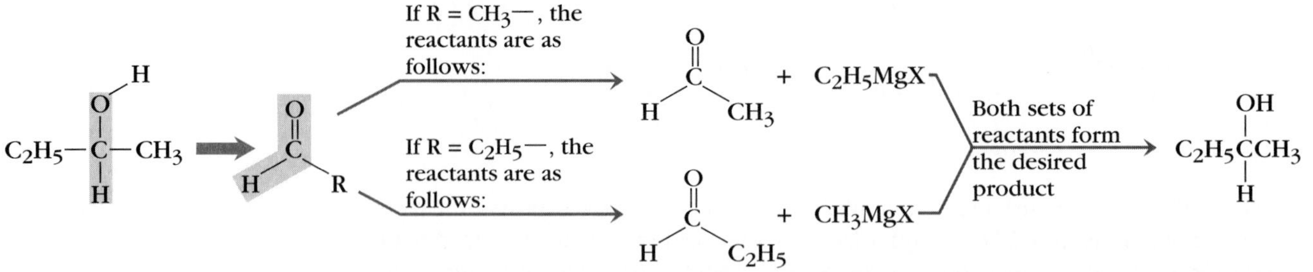

2-Butanol can be synthesized by reaction of either of these two sets of reactants. Many alcohols can be synthesized in two ways in the reactions of Grignard or organolithium reagents with aldehydes or ketones. So which is the preferred synthesis of 2-butanol? The answer depends on a number of factors. For our purposes, the most important factor is the starting materials that we are allowed. In this case, we are limited to organic compounds containing two or fewer carbon atoms as starting materials. If we choose the reaction of methylmagnesium halide (CH_3MgX) and CH_3CH_2CHO for our synthetic plan, we must synthesize CH_3CH_2CHO, something that we will not learn to do until Section 11.18. If we choose the reaction of ethylmagnesium halide (CH_3CH_2MgX) and acetaldehyde (CH_3CHO), both of which contain two carbon atoms, we have achieved our synthesis:

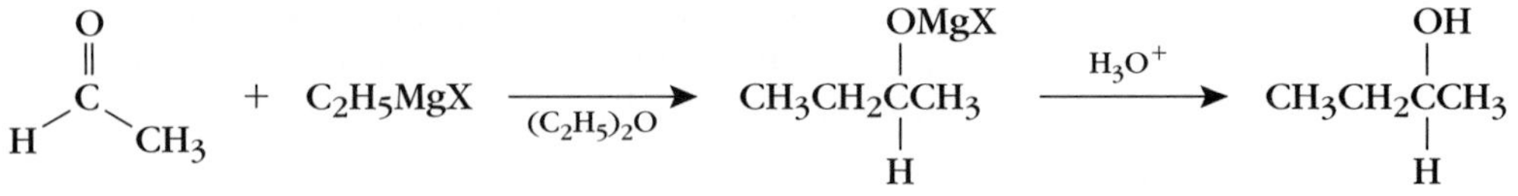

Exercise 11.11　Starting with any Grignard reagent, any carbonyl-containing compound, any inorganic reagent, and any solvent, plan a synthesis of each of the following alcohols.

(a) 2-Hexanol
(b) 1-Hexanol
(c) 1-Methylcyclopentanol

(d) 3-Ethyl-3-pentanol
(e) 1-Cyclohexyl-1-butanol

Exercise 11.12　Starting with any organolithium reagent, any carbonyl-containing compound, any inorganic reagent, and any solvent, plan a synthesis of each of the following alcohols.

(a) 2-Methyl-2-pentanol
(b) Isobutyl alcohol
(c) 3-Buten-2-ol

(d)

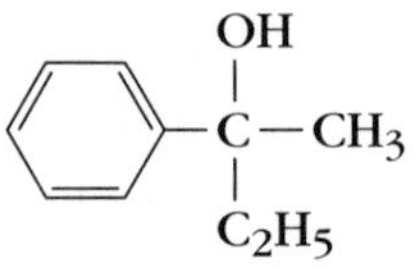

Exercise 11.13	Starting with any organic compound containing four or fewer carbon atoms, any inorganic reagent, and any solvent, plan a synthesis of each of the following alcohols.

(a) 1-Isopropylcyclobutanol   **(b)** 2-Methyl-4-heptanol

The reduction of aldehydes and ketones is a second way to synthesize alcohols. Before we discuss this synthetic method, we need to review oxidation and reduction reactions.

11.9 Oxidation and Reduction of Organic Compounds

Proton transfer reactions between acids and bases and oxidation-reduction reactions are two very important reactions in chemistry. We learned about the relative acidities of many organic compounds in Section 3.4, and we've seen the importance of proton transfer in the mechanisms of several electrophilic addition reactions. Now let's examine the oxidation and reduction of organic compounds.

Oxidation is defined in inorganic chemistry as a loss of electrons, while reduction is a gain of electrons by an atom or ion.

$$Li \longrightarrow Li^+ + e^-$$

Lithium   Lithium
metal   cation

Oxidation of lithium metal forms lithium cation

$$Br_2 + 2e^- \longrightarrow 2Br^-$$

Bromine   Bromide
ion

Reduction of bromine forms bromide ion

These definitions are based on the full transfer of an electron between ions or atoms. Since most oxidation reactions of organic compounds don't involve such a transfer of electrons, we must adopt slightly different definitions of oxidation and reduction.

For practical purposes, we define oxidation of carbon atoms in organic compounds as reactions that remove hydrogen atoms from carbon atoms and/or add oxygen, nitrogen, or halogen atoms to carbon atoms.

$$RCH{=}CHR \xrightarrow{OsO_4} \overset{\displaystyle OH \quad OH}{\underset{}{RCH{-}CHR}}$$

Oxidation: Addition of O atoms to C atoms of C=C

$$RCH_3 \xrightarrow[CCl_4]{Cl_2} RCH_2Cl$$

Oxidation: Removal of H atom from C atom and addition of Cl atom to C atom

$$RCH_2OH \xrightarrow[\text{reagent}]{\text{Collin's}} RC\overset{O}{\underset{H}{\diagdown}}$$

Oxidation: Removal of H atoms from C atom

We define reduction of carbon atoms of organic compounds as reactions that add hydrogen atoms to carbon atoms and/or remove oxygen, nitrogen, or halogen atoms from carbon atoms.

$$RCH{=}CHR \xrightarrow[\text{PtO}_2]{\text{H}_2} RCH_2CH_2R$$

Reduction: Addition of H atoms
to C atoms of C=C

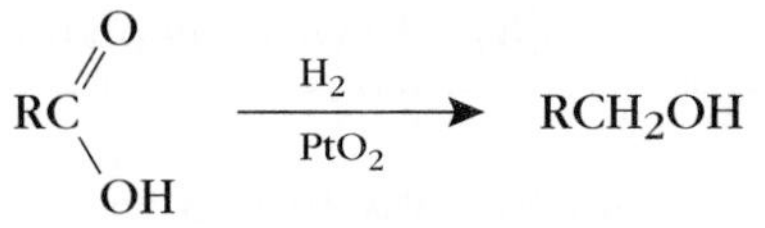

$$RC\overset{\displaystyle O}{\underset{\displaystyle OH}{\big\|}} \xrightarrow[\text{PtO}_2]{\text{H}_2} RCH_2OH$$

Reduction: Addition of H atoms and
removal of O atoms from C atom

Not all reactions involve oxidation or reduction. A reaction that adds one hydrogen atom and one oxygen, nitrogen, or halogen atom is neither an oxidation nor a reduction; neither is a reaction that loses one hydrogen atom and one oxygen, nitrogen, or halogen atom. An example is the hydration of alkenes, in which both a hydrogen atom and an oxygen atom are added to the carbon atoms of the double bond.

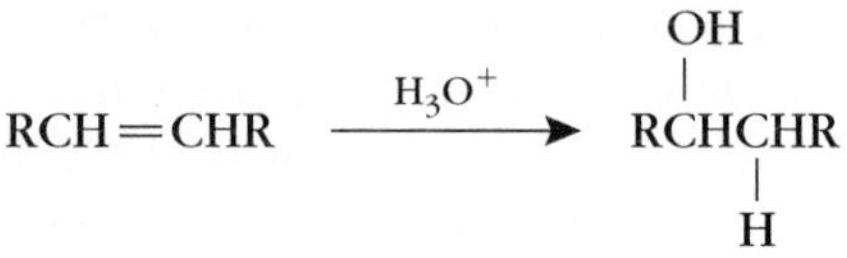

$$RCH{=}CHR \xrightarrow{\text{H}_3\text{O}^+} \underset{\displaystyle H}{\overset{\displaystyle OH}{RCHCHR}}$$

Neither oxidation nor reduction because both
H and O atoms are added to the double bond

A number of functional groups are listed in Figure 11.6 in order of increasing oxidation state. An organic compound is oxidized if a reaction converts its functional group into one of a higher oxidation state. For example, oxidation of a secondary alcohol forms a ketone. An organic compound is reduced if a reaction converts its functional group into one of lower oxidation state. Catalytic hydrogenation, for example, reduces alkenes to alkanes.

Exercise 11.14   In each of the following reactions, indicate whether the starting organic compound is oxidized, reduced, or no change in oxidation state occurs.

(a) 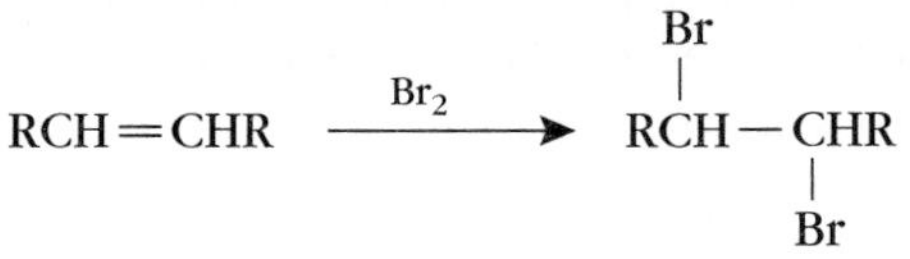 $RCH{=}CHR \xrightarrow{\text{Br}_2} \underset{\displaystyle Br}{\overset{\displaystyle Br}{RCH{-}CHR}}$

(b) 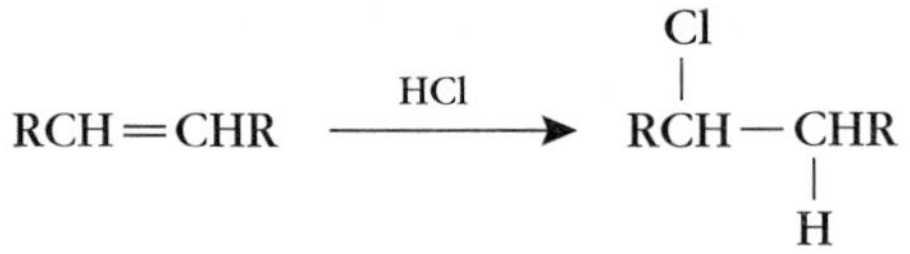 $RCH{=}CHR \xrightarrow{\text{HCl}} \underset{\displaystyle H}{\overset{\displaystyle Cl}{RCH{-}CHR}}$

(c) 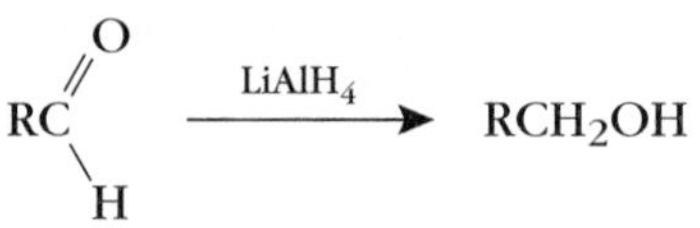 $RC\overset{\displaystyle O}{\underset{\displaystyle H}{\big\|}} \xrightarrow{\text{LiAlH}_4} RCH_2OH$

(d)  $RCH{=}CHR \xrightarrow{\text{RCO}_3\text{H}} \overset{\displaystyle O}{RHC{-}CHR}$

(e) $RCH_2OH \xrightarrow{\text{HCl}} RCH_2Cl$

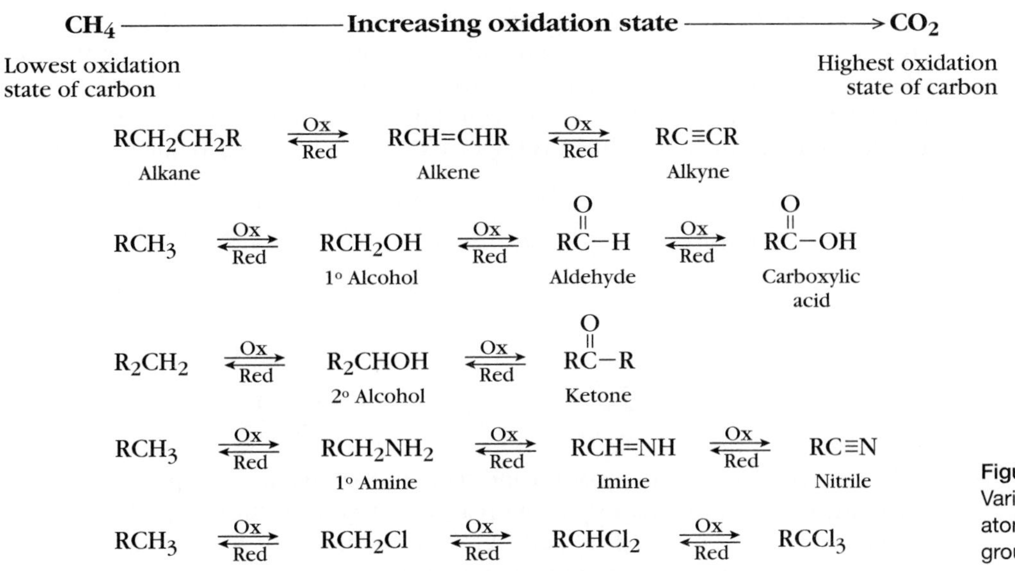

$$CH_4 \longrightarrow \text{Increasing oxidation state} \longrightarrow CO_2$$

Lowest oxidation
state of carbon

Highest oxidation
state of carbon

$$RCH_2CH_2R \underset{\text{Red}}{\overset{\text{Ox}}{\rightleftarrows}} RCH{=}CHR \underset{\text{Red}}{\overset{\text{Ox}}{\rightleftarrows}} RC{\equiv}CR$$

Alkane   Alkene   Alkyne

$$RCH_3 \underset{\text{Red}}{\overset{\text{Ox}}{\rightleftarrows}} RCH_2OH \underset{\text{Red}}{\overset{\text{Ox}}{\rightleftarrows}} \overset{\overset{\displaystyle O}{\|}}{RC}{-}H \underset{\text{Red}}{\overset{\text{Ox}}{\rightleftarrows}} \overset{\overset{\displaystyle O}{\|}}{RC}{-}OH$$

       1° Alcohol   Aldehyde   Carboxylic acid

$$R_2CH_2 \underset{\text{Red}}{\overset{\text{Ox}}{\rightleftarrows}} R_2CHOH \underset{\text{Red}}{\overset{\text{Ox}}{\rightleftarrows}} \overset{\overset{\displaystyle O}{\|}}{RC}{-}R$$

       2° Alcohol   Ketone

$$RCH_3 \underset{\text{Red}}{\overset{\text{Ox}}{\rightleftarrows}} RCH_2NH_2 \underset{\text{Red}}{\overset{\text{Ox}}{\rightleftarrows}} RCH{=}NH \underset{\text{Red}}{\overset{\text{Ox}}{\rightleftarrows}} RC{\equiv}N$$

       1° Amine   Imine   Nitrile

$$RCH_3 \underset{\text{Red}}{\overset{\text{Ox}}{\rightleftarrows}} RCH_2Cl \underset{\text{Red}}{\overset{\text{Ox}}{\rightleftarrows}} RCHCl_2 \underset{\text{Red}}{\overset{\text{Ox}}{\rightleftarrows}} RCCl_3$$

Figure 11.6
Various oxidation states of carbon
atoms in a variety of functional
groups.

Table 11.4	Common Oxidizing Reagents Used in Organic Chemistry		

Name	Chemical Formula	Name	Chemical Formula
Oxygen	O_2	Mercury(II) acetate	$Hg(OAc)_2$
Hydrogen peroxide	H_2O_2	Nitric acid	HNO_3
Peroxycarboxylic acids	$\overset{\overset{\displaystyle O}{\|}}{RC}\diagdown OOH$	Permanganate ion	MnO_4^-
N-Bromosuccinimide	 (structure: NBr succinimide)	Chromic anhydride	CrO_3
Selenium dioxide	SeO_2	Dichromate ion	$Cr_2O_7^{2-}$
Halogens	X_2 X = F, Cl, Br, I	Osmium tetraoxide	OsO_4

Table 11.5	Common Reducing Reagents Used in Organic Chemistry

Reducing Reagent	Examples
Catalytic hydrogenation	H_2 + Metal (Pt, Pd, or Ni)
Metal hydrides	$LiAlH_4$, $NaBH_4$
Borohydrides	BH_3, R_2BH
Metals	Li, Na, K, Zn, Hg

So far we have focused our attention on the oxidation and reduction of organic compounds. However, oxidation of an organic compound cannot occur without a simultaneous reduction of some other reactant. In the reaction of ethanol and chromic anhydride (CrO_3) to form acetic acid, for example, ethanol is oxidized to acetic acid, while CrO_3 (the oxidizing reagent) is reduced to Cr^{3+}.

$$CH_3CH_2OH \longrightarrow H_3CC\begin{smallmatrix}O\\OH\end{smallmatrix}$$

Oxidation of ethanol forms acetic acid

Ethanol   Acetic acid

$$CrO_3 \longrightarrow Cr^{3+}$$

Reduction of CrO_3 forms Cr^{3+}

The reagents used for oxidation and reduction of organic compounds are usually inorganic oxidizing and reducing reagents. A general summary of useful oxidizing reagents is collected in Table 11.4, while reducing reagents are listed in Table 11.5. Remember to use the correct reagent when writing the equation for oxidation or reduction reactions of organic compounds. To oxidize an organic compound, an oxidizing reagent from Table 11.4 is needed. The reduction of an organic compound requires one of the reducing reagents from Table 11.5.

Exercise 11.15   Decide whether an oxidizing or reducing agent is needed to carry out each of the following transformations.

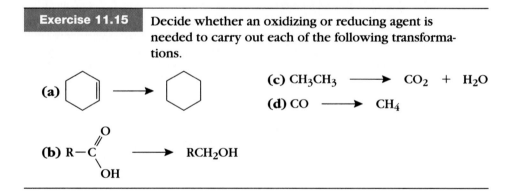

(a)

(c) $CH_3CH_3 \longrightarrow CO_2 + H_2O$

(d) $CO \longrightarrow CH_4$

(b) $R-C\begin{smallmatrix}O\\OH\end{smallmatrix} \longrightarrow RCH_2OH$

11.10 Synthesis of Alcohols by Reduction of Aldehydes and Ketones

An alcohol can be formed by the reduction of an aldehyde or ketone, as we discussed in Section 11.9. Primary alcohols are formed by reduction of aldehydes.

$$\underset{\text{Aldehyde}}{R-\overset{O}{\underset{}{\overset{\|}{C}}}-H} \xrightarrow{\text{Reduction}} \underset{\text{1° Alcohol}}{R-\overset{H}{\underset{H}{\overset{|}{\underset{|}{C}}}-H}}$$

R = alkyl or aryl group

Secondary alcohols are formed by reduction of ketones.

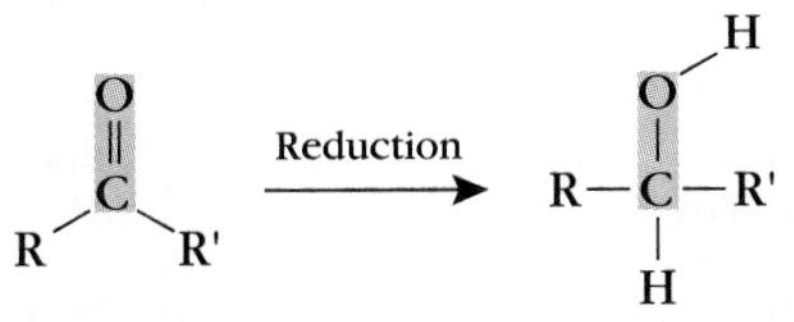

Ketone

2° Alcohol

R and R' can be
alkyl or aryl groups

Many different reducing agents can be used to reduce aldehydes and ketones. Lithium aluminum hydride (LiAlH$_4$) is particularly effective in the laboratory. Reductions are usually carried out by adding a diethyl ether or THF solution of the aldehyde or ketone to a diethyl ether or THF solution of LiAlH$_4$. The reaction is rapid, and on completion of the reaction, the alcohol is present as a mixture of lithium and aluminum salts. Addition of dilute acid solution to the reaction mixture converts these salts to the alcohol. The following examples illustrate the synthesis of primary and secondary alcohols by the LiAlH$_4$ reduction of aldehydes and ketones.

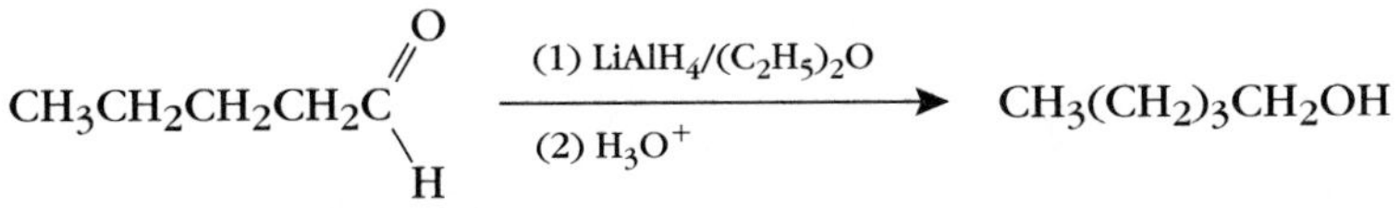

$$CH_3CH_2CH_2CH_2C\overset{O}{\underset{H}{\diagup}} \xrightarrow[\text{(2) } H_3O^+]{\text{(1) LiAlH}_4/(C_2H_5)_2O} CH_3(CH_2)_3CH_2OH$$

Pentanal
(an aldehyde)

1-Pentanol
(82%–86% yield)

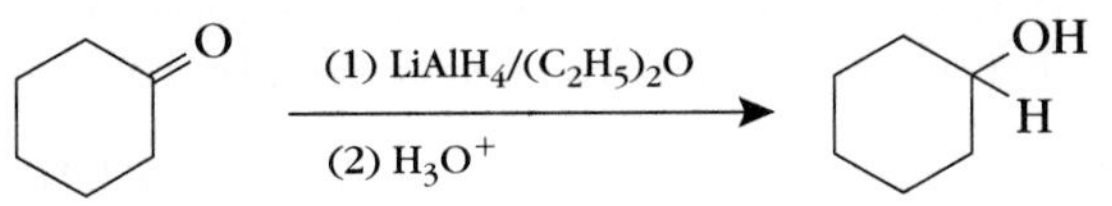

Cyclohexanone
(a ketone)

Cyclohexanol
(90%–95% yield)

Lithium aluminum hydride reacts vigorously with water to release hydrogen gas.

$$LiAlH_4 + 4 H_2O \longrightarrow LiOH + Al(OH)_3 + 4 H_2$$

LiAlH$_4$ reduction reactions therefore must be carried out under anhydrous conditions to prevent hydrolysis of LiAlH$_4$.

Sodium borohydride (NaBH$_4$), which also reduces aldehydes and ketones to alcohols, is much less reactive than LiAlH$_4$. It can be safely handled and weighed in the open atmosphere, and the reduction can be carried out in water and alcohols as solvents.

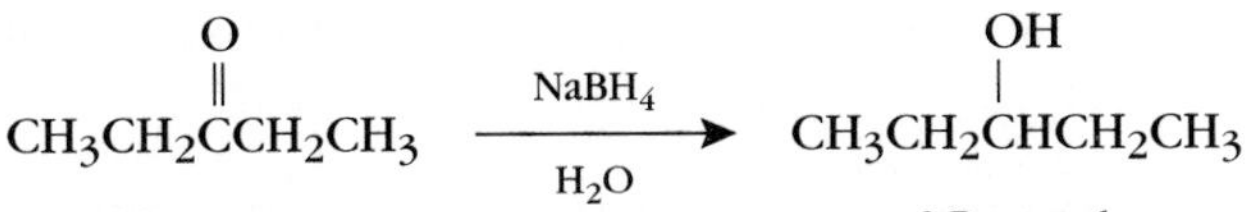

$$CH_3CH_2\overset{O}{\overset{\|}{C}}CH_2CH_3 \xrightarrow[\text{H}_2\text{O}]{\text{NaBH}_4} CH_3CH_2\overset{OH}{\overset{|}{C}H}CH_2CH_3$$

3-Pentanone
(a ketone)

3-Pentanol
(85% yield)

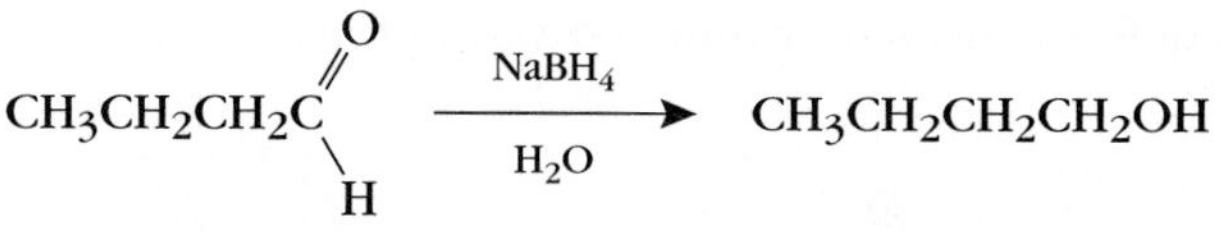

$$CH_3CH_2CH_2C\overset{\displaystyle O}{\underset{\displaystyle H}{\big|}} \xrightarrow[\text{H}_2\text{O}]{\text{NaBH}_4} CH_3CH_2CH_2CH_2OH$$

Butanal   1-Butanol
(an aldehyde)   (87% yield)

Both $LiAlH_4$ and $NaBH_4$ reduce a carbonyl group by transferring a hydride ion (H^-) to the carbonyl carbon atom to form an alkoxide ion. The alcohol is formed by reaction of the alkoxide ion (as part of a salt) with aqueous acid.

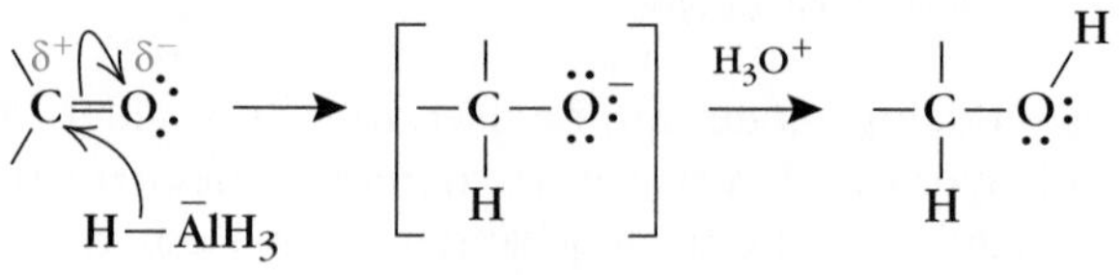

Alkoxide ion   Alcohol
(present as
mixture of Li
and Al salts)

The carbonyl group of aldehydes and ketones, like a carbon-carbon double bond, can be reduced to alcohols by catalytic hydrogenation (Section 8.11).

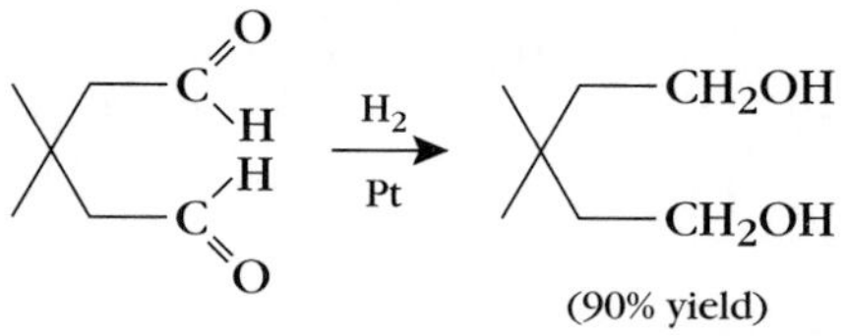

(90% yield)

Catalytic hydrogenation of ketones and aldehydes that also contain a carbon-carbon double bond results in reduction of both double bonds. In contrast, carbon-carbon double bonds are generally not reduced by $LiAlH_4$ or $NaBH_4$.

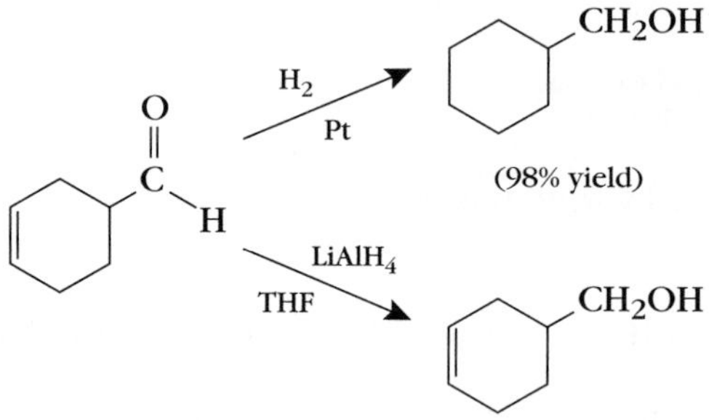

Exercise 11.16   Give the products of the reaction of each of the following compounds with lithium aluminum hydride followed by hydrolysis with an aqueous acid solution.

(a) 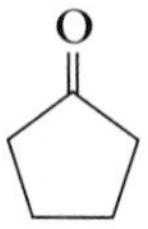   (b) $(CH_3)_2CHCH_2\overset{\displaystyle O}{\overset{\|}{C}}CH_3$ (c) $CH_2{=}CHCH_2CH_2C\overset{\displaystyle O}{\underset{\displaystyle H}{\big\langle}}$

Exercise 11.17   Give the product of each of the following reactions.

(a) 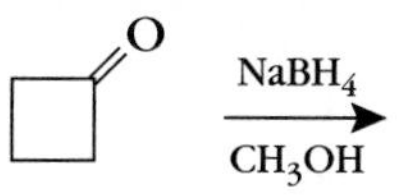 $\xrightarrow[\text{CH}_3\text{OH}]{\text{NaBH}_4}$

(c) 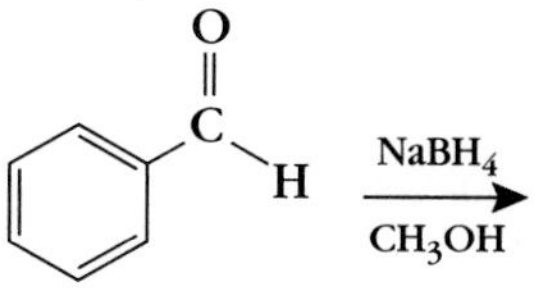 $\xrightarrow[\text{CH}_3\text{OH}]{\text{NaBH}_4}$

(b) 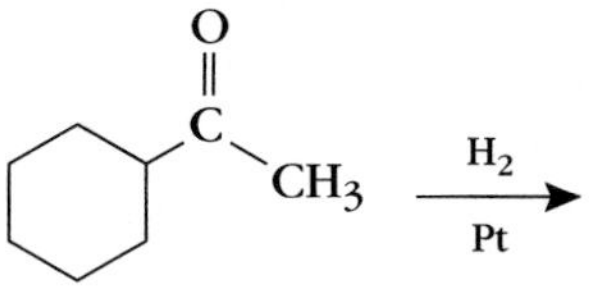 $\xrightarrow[\text{Pt}]{\text{H}_2}$

(d) 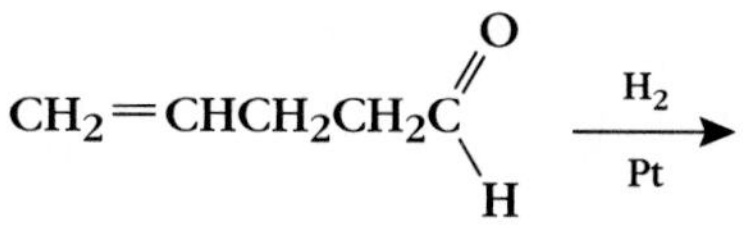 $\text{CH}_2\!=\!\text{CHCH}_2\text{CH}_2\overset{\displaystyle O}{\overset{\|}{\text{C}}}\!\!-\!\text{H}$ $\xrightarrow[\text{Pt}]{\text{H}_2}$

Exercise 11.18   Explain why one of the following alcohols *cannot* be prepared by the reduction of a carbonyl-containing compound.

(a) 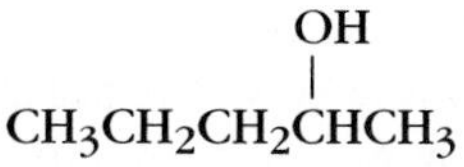 $\underset{\underset{\text{OH}}{|}}{\text{CH}_3\text{CH}_2\text{CH}_2\text{CHCH}_3}$

(b) 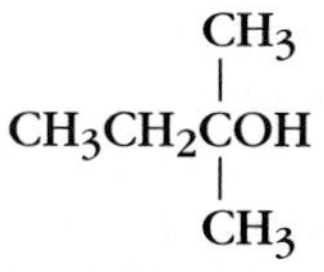 $\underset{\underset{\text{CH}_3}{|}}{\overset{\overset{\text{CH}_3}{|}}{\text{CH}_3\text{CH}_2\text{COH}}}$

(c) $(\text{CH}_3)_3\text{CCH}_2\text{CH}_2\text{OH}$

11.11 Distinguishing the Sides of Carbonyl Groups of Prochiral Aldehydes and Ketones

The LiAlH$_4$ reduction of 2-butanone forms 2-butanol, which has a stereocenter.

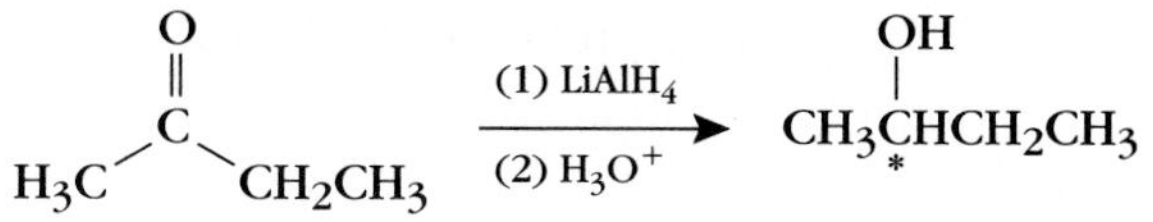

$$\text{H}_3\text{C}\overset{\displaystyle O}{\overset{\|}{\text{C}}}\text{CH}_2\text{CH}_3 \xrightarrow[\text{(2) H}_3\text{O}^+]{\text{(1) LiAlH}_4} \underset{*}{\overset{\overset{\text{OH}}{|}}{\text{CH}_3\text{CHCH}_2\text{CH}_3}}$$

2-Butanone   2-Butanol

Compounds such as 2-butanone that react to form products containing one or more stereocenters are called prochiral (Section 6.14).

Exercise 11.19   Which of the following aldehydes or ketones is prochiral?

(a) 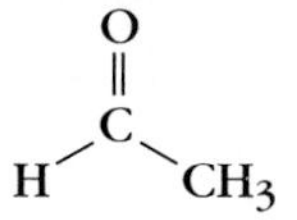 $\text{H}\overset{\displaystyle O}{\overset{\|}{\text{C}}}\text{CH}_3$

(c) $\text{H}\overset{\displaystyle O}{\overset{\|}{\text{C}}}\text{H}$

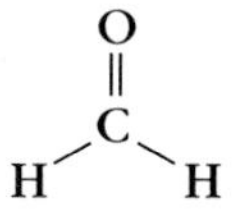

(b) 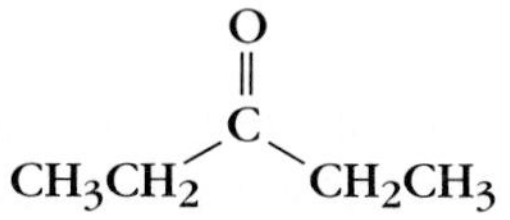 $\text{CH}_3\text{CH}_2\overset{\displaystyle O}{\overset{\|}{\text{C}}}\text{CH}_2\text{CH}_3$

(d)

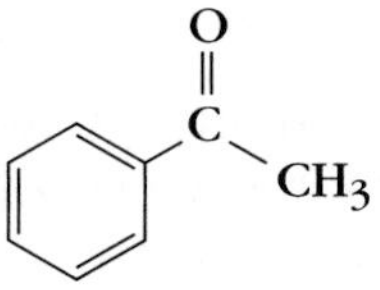

The 2-butanol formed by the reduction of 2-butanone is a racemic mixture because addition of hydride ion occurs equally well to either side of the carbonyl group to form equal amounts of the two enantiomers, as shown in Figure 11.7.

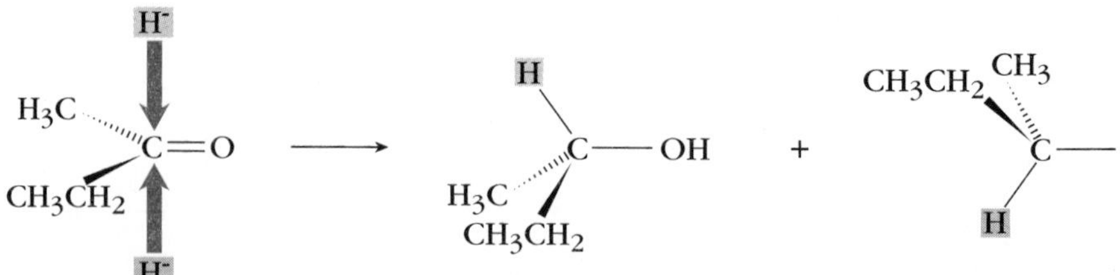

Figure 11.7
Addition of hydride ion to one side of the carbonyl group of 2-butanone forms (R)-2-butanol, while addition to the other forms (S)-2-butanol. The product is racemic since addition occurs equally well to either side.

(S)-2-Butanol   *(R)-2-Butanol*

Enantiomers

■ *The term* Re *is pronounced "ray," and the term* Si *is pronounced "see."*

The two sides or faces of a carbonyl group in a prochiral aldehyde or ketone are different because addition to one side forms the (S)-enantiomer and addition to the other forms the (R)-enantiomer. We can distinguish between these different sides by using the Cahn-Ingold-Prelog Rules (Section 6.2) in two dimensions. The order of priority of the groups bonded to the carbonyl carbon atom is determined using the Cahn-Ingold-Prelog Rules. Then, we view the carbonyl group from one side of its plane. If the direction from highest to lowest priority of the groups is clockwise, that side is called the Re face or side. If the direction of the groups from highest to lowest priority is counterclockwise, that side is called the Si side. The application of this method of distinguishing the sides of a carbonyl group is illustrated in Sample Problem 11.2.

Sample Problem 11.2   Designate the side of the carbonyl group of 2-butanone shown below.

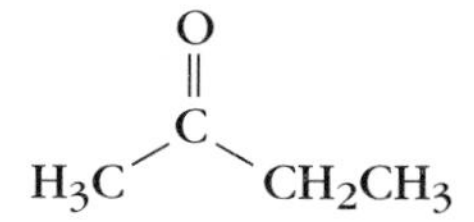

One side of the carbonyl
group of 2-butanone

Solution:

First, the priorities of the atoms and groups of atoms bonded to the carbonyl carbon atom are determined.

$$CH_3 < CH_3CH_2 < O$$

◁ DECREASING PRIORITY

The order of decreasing priority of the atoms and groups bonded to the carbonyl carbon atom is clockwise (*rectus*), so this side of the carbonyl group is called the Re side or face.

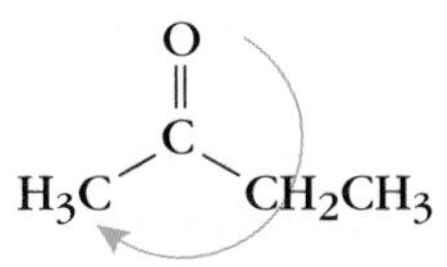

Order of decreasing
priority is clockwise

The opposite side is therefore the Si side.

Exercise 11.20 Designate as Re or Si the face of the carbonyl group in each of the following compounds:

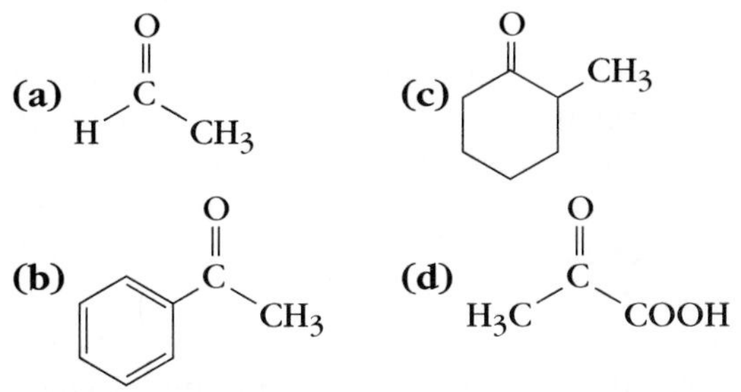

(a)

(b)

(c)

(d)

Metal hydride (LiAlH$_4$ or NaBH$_4$) reduction of prochiral ketones and aldehydes almost always results in the formation of alcohols that are racemic mixtures. If we could somehow add hydride ion to one side of the carbonyl group in preference to the other, an optically active alcohol consisting of exclusively one enantiomer would be formed. Such preference for one side of a carbonyl group occurs routinely in enzyme-catalyzed reductions of carbonyl groups.

11.12 Enzyme-Catalyzed Reduction of Carbonyl Groups

Enzymes catalyze almost all biochemical reactions that comprise life. Although enzymes are subject to the same chemical principles that govern all catalysts, they differ from ordinary chemical catalysts in their high degree of specificity toward their reactants (substrates). Two aspects of this substrate specificity that make enzymes useful in organic synthesis are their stereospecificity and structural specificity.

Stereospecificity refers to the enzyme's specificity both in binding chiral substrates and in catalyzing their reactions. The fumarase-catalyzed hydration of fumarate (Section 7.16) is one example of enzyme stereospecificity. **Structural (or geometrical) specificity** refers to the ability of an enzyme to distinguish between substrates of similar structure. A few enzymes are absolutely specific for only one substrate. Most enzymes, however, catalyze the reactions of a small number of structurally related compounds.

The stereospecificity, easy availability, and the mild conditions under which they catalyze reactions have led to the growing use of enzymes as catalysts in the synthesis of certain organic compounds. The most valuable enzymes for organic synthesis are those that catalyze a broad range of structurally similar substrates yet retain their stereospecific function.

L-Lactate dehydrogenase is an example of an enzyme that has found synthetic application because it is stereospecific yet has a fairly wide structural specificity. It catalyzes the reduction of the keto group in more than 20 α-keto acids to (*S*)-2-hydroxy acids.

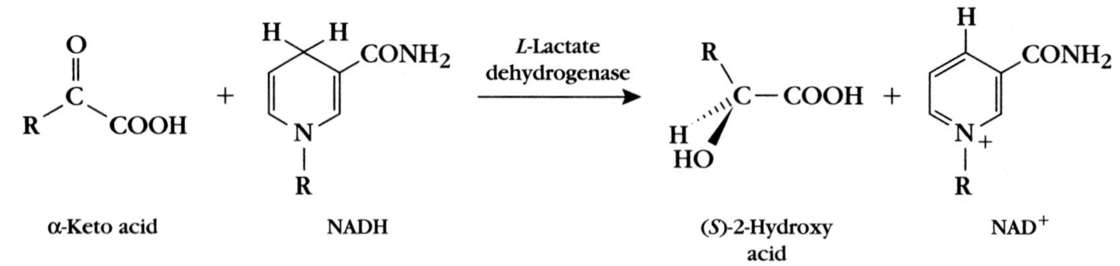

NADH is the reduced form and NAD^+ is the oxidized form of nicotinamide adenine dinucleotide in this reaction. NADH and NAD^+ are one of several pairs of reducing and oxidizing agents in living organisms. NADPH and $NADP^+$ are another pair of reducing and oxidizing agents, which are structurally related to NADH and NAD^+. NADPH is the reduced form and $NADP^+$ is the oxidized form of nicotinamide adenine dinucleotide phosphate.

NADH and NAD^+ are **coenzymes** in the reduction of α-keto acids. Coenzymes are complex organic molecules that are coordinated with the enzyme and are essential for the enzyme to carry out its catalytic function. NADH is a reducing agent that is oxidized to NAD^+ as it reduces a compound. NAD^+ is an oxidizing agent that is reduced to NADH as it oxidizes a compound.

$$NADH \rightleftharpoons NAD^+$$

Reducing agent: Reduced form of nicotinamide adenine dinucleotide

Oxidizing agent: Oxidized form of nicotinamide adenine dinucleotide

The structures of NADH and NADPH are shown in Figure 11.8. While this structure is large and complicated, only the nicotinamide part is involved in oxidation-reduction reactions, so we'll concentrate only on this part of the mole-

Figure 11.8
Structures of reduced forms of nicotinamide adenine dinucleotide, NADH, and nicotinamide adenine dinucleotide phosphate, NADPH.

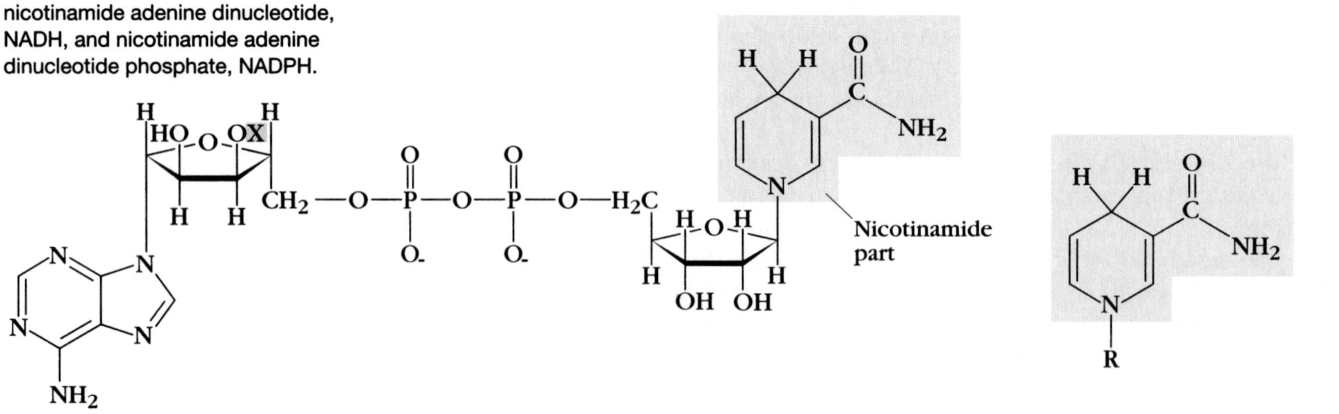

X = H: Reduced form of nicotinamide adenine dinucleotide, NADH

X = PO_3^{2-}: Reduced form of nicotinamide adenine dinucleotide phosphate, NADPH

Abbreviated structure of NADH and NADPH

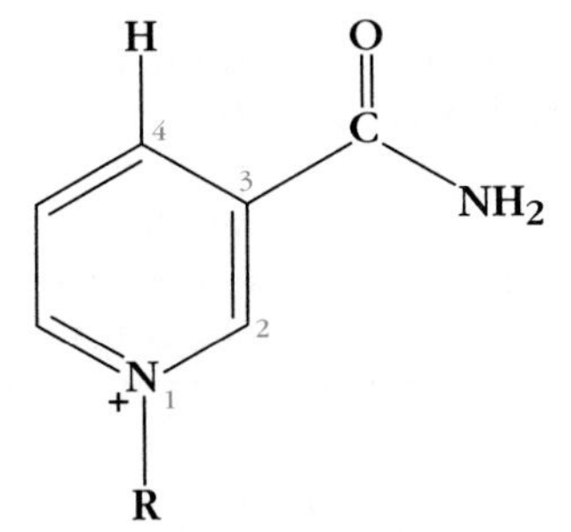

H

O
‖
C
NH₂

NAD⁺ and NADP⁺

Figure 11.9
Abbreviated structures of the oxidized forms of nicotinamide adenine dinucleotide, NAD⁺, and nicotinamide adenine dinucleotide phosphate, NADP⁺. R refers to the remainder of the structure shown in Figure 11.8.

cule and abbreviate by "R" the remainder of the structure, as shown in Figure 11.8. While the remainder of the molecule is not involved in the oxidation-reduction function of the molecule, it does play an important role by anchoring the coenzyme in its proper place on the enzyme surface.

The abbreviated structures of NAD⁺ and NADP⁺ are shown in Figure 11.9. They differ from NADH and NADPH in that they lack a hydrogen atom and two electrons on C4. All atoms of the six-member nitrogen-containing ring of NAD⁺ and NADP⁺, the nicotinamide ring, lie in a plane.

NADH is the biological equivalent of metal hydrides such as LiAlH₄ or NaBH₄ because it also reduces the carbonyl group by transferring a hydride ion to the carbonyl carbon atom. NADH transfers a hydride ion from C4 of the ring to the Re face of the α-keto acid to form an (S)-2-hydroxy acid, as shown in Figure 11.10.

The reduction is stereospecific, as shown by the data in Table 11.6. The products are formed in yields greater than 90% and are enantiomerically pure within the limits of experimental determination.

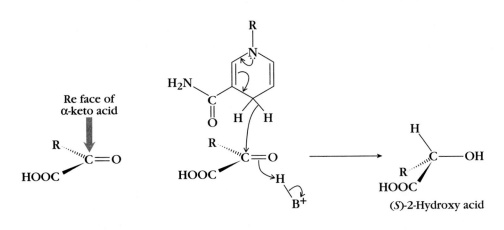

Figure 11.10
NADH transfers a hydride ion to the Re face of the carbonyl group of an α-keto acid to form an (S)-2-hydroxy acid. B represents any base in the vicinity that is part of the enzyme, such as a hydroxy group or an amine group, RNH₂.

| Exercise 11.21 | Write the structure of the product of the following reaction. |

$$CH_3CH_2\overset{\overset{\displaystyle O}{\|}}{C}COOH \xrightarrow[\textit{L-Lactate dehydrogenase}]{NADH}$$

Table 11.6 *L*-Lactate Dehydrogenase-Catalyzed Reduction of Several α-Keto Acids (RCOCOOH)

R	Yield (%)	Absolute Configuration of Alcohol	Enantiomeric Excess (ee) (%)
CH_3CH_2	99	*S*	>99
$CH_3(CH_2)_2$	97	*S*	>99
$C_6H_5CH_2$	94	*S*	>99

11.13 Mechanism of Stereospecific Enzyme-Catalyzed Interconversion of Acetaldehyde and Ethanol

A mechanism involving direct hydride ion transfer from NADH to the carbonyl carbon atom was proposed for the reduction of α-keto acids by NADH in Section 11.12. The evidence for this mechanism was obtained by a series of experiments carried out on the alcohol dehydrogenase-catalyzed interconversion of acetaldehyde and ethanol.

Acetaldehyde NADH Ethanol NAD^+

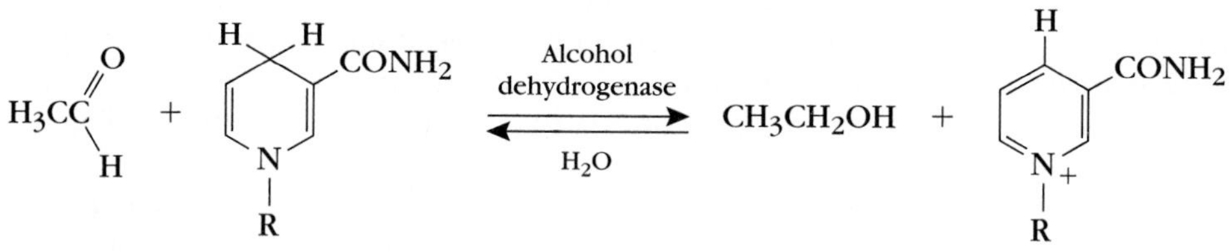

The following five experiments provide details of the mechanism of this biological oxidation-reduction reaction:

1. When acetaldehyde is reduced in deuterium oxide (D_2O) instead of water as solvent, the ethanol produced contains no deuterium atoms.

Acetaldehyde NADH Ethanol NAD^+

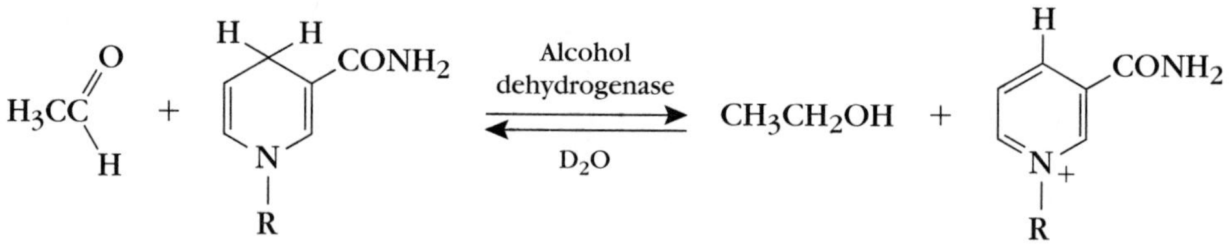

This result suggests that the transfer of hydride ion occurs directly from NADH to acetaldehyde and not via the assistance of any water or D_2O molecule.

2. When 1,1-dideuterioethanol is oxidized in H_2O, the NADD formed contains a single deuterium atom per molecule.

NAD^+ NADD

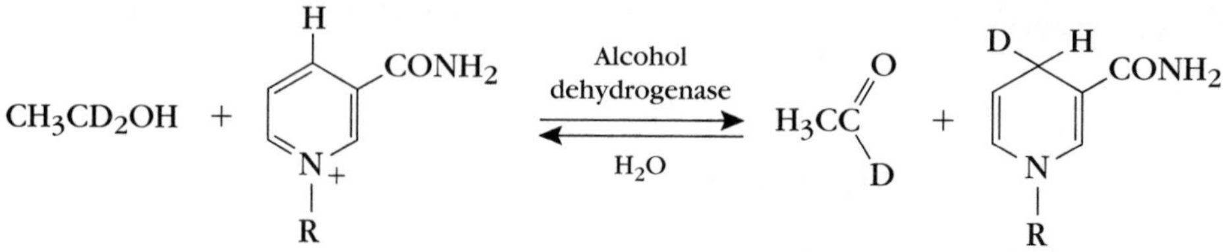

This is further evidence that the hydride ion is transferred directly from NADH to acetaldehyde.

3. NADD, the deuterated reduced form of NADH, was isolated from the oxidation of 1,1-dideuterioethanol and shown to have the (R) configuration at C4.

(R) configuration
at C4

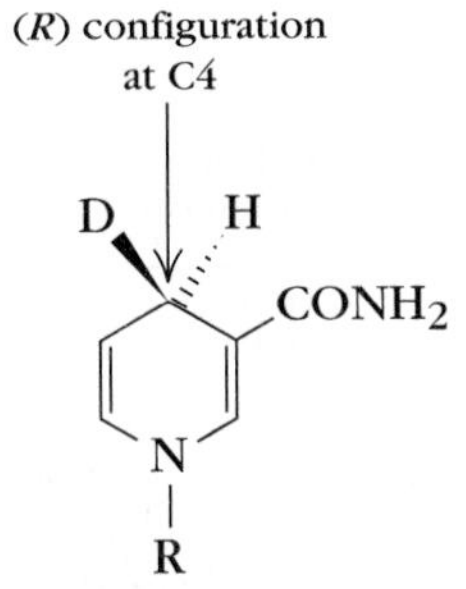

Since only the stereoisomer with the (R) configuration at C4 is formed, the deuteride ion is transferred from 1,1-dideuterioethanol to the Re face of the nicotinamide ring of NAD$^+$.

4. Only (R)-1-deuterioethanol is formed by the reduction of acetaldehyde by NADD, the deuterated reduced form of NADH, which has the (R) configuration at C4.

Acetaldehyde   NADD   NAD$^+$

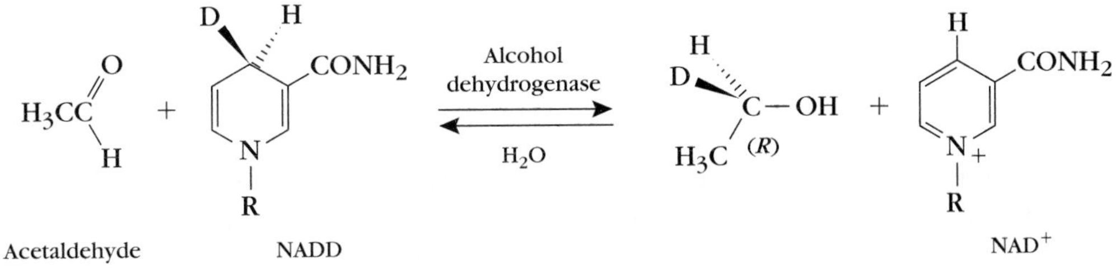

Formation of only one enantiomer of 1-deuterioethanol means that the deuteride ion is transferred only to the Re face of the carbonyl group of acetaldehyde.

5. No NADD, only NADH, is formed, when (S)-1-deuterioethanol is oxidized by NAD$^+$.

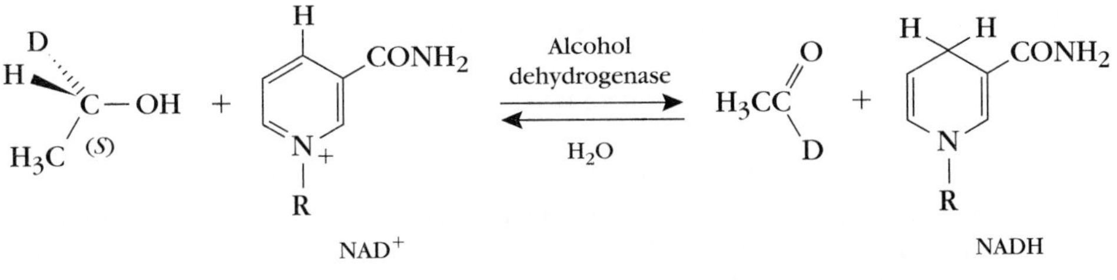

NAD$^+$ NADH

Formation of only NADH means that the enzyme can distinguish between the pro-R and pro-S hydrogen atoms of ethanol.

These results show that there is direct hydride ion transfer from NADH to acetaldehyde (Experiments 1 and 2) in the alcohol dehydrogenase–catalyzed interconversion of ethanol and acetaldehyde. Also, the enzyme distinguishes between the Si and Re faces of both the nicotinamide ring of NAD$^+$ (Experiment 3) and the carbonyl group of acetaldehyde (Experiment 4). It can also distinguish between the pro-S and pro-R hydrogen atoms of ethanol (Experiment 5). These results allow us to picture this stereospecific reaction, as shown in Figure 11.11.

■ *pro-S and pro-R hydrogen atoms are discussed in Section 6.14.*

Figure 11.11
Pictorial representation of the stereospecific transfer of hydride ion to and from the carbonyl group of acetaldehyde. B represents any base in the vicinity that is part of the enzyme, such as a hydroxy group or an amine group, RNH$_2$. H$_R$ is the pro-*R* hydrogen atom, and H$_S$ is the pro-*S* hydrogen atom.

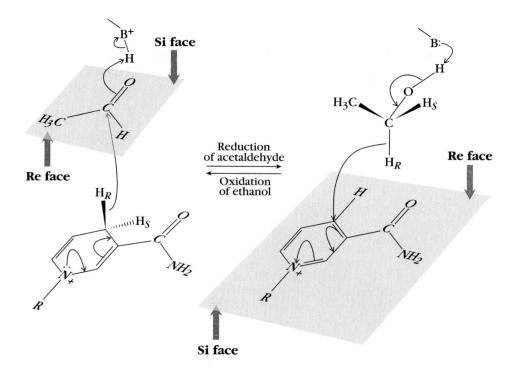

In the reduction of acetaldehyde, the pro-*R* hydrogen of carbon atom 4 of the ring of NADH is transferred to the Re face of the carbonyl group of acetaldehyde. In the reverse reaction, oxidation, the pro-*R* hydrogen of ethanol is transferred to the Re face of the nicotinamide ring.

Such a stereospecific reaction can occur only if the enzyme brings the reagents together in the proper orientation. Let's discuss what is involved in positioning reagents on an enzyme so that such stereospecific reactions can occur.

11.14 Origins of Enzyme Specificity

We learned in Section 7.16 that enzyme-catalyzed reactions occur in an enzyme-substrate complex formed by interaction of an enzyme with a substrate. The substrate complexes with only a small region of the large enzyme molecule, which is called the active site. An active site is usually a pocket or cleft on the surface of the enzyme molecule whose three-dimensional shape is complementary to the substrate. It is the shape of the active site that determines the geometrical specificity of an enzyme.

Also, the parts of the enzyme around the active site form a three-dimensional arrangement of functional groups that interact specifically with the substrate by means of intermolecular attractions, such as the electrostatic attractions and hydrogen bonds described in Section 1.16, and hydrophobic interactions. The precise locations of these groups and their interactions determine the stereospecificity of an enzyme. These features of the active site are illustrated in Figure 11.12.

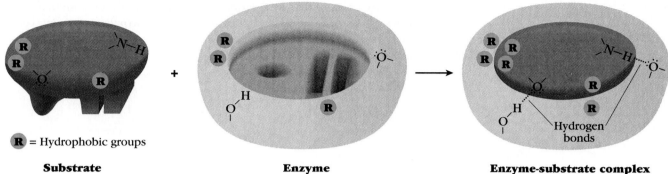

R = Hydrophobic groups

Substrate   **Enzyme**   **Enzyme-substrate complex**

Figure 11.12
Formation of an enzyme-substrate complex illustrating the complementary relationship between the enzyme and the substrate. Hydrophobic groups are represented by an R in a circle, and hydrogen bonds are represented by dotted lines.

Molecules that differ from the substrate in shape or arrangement of functional groups cannot form the enzyme-substrate complex that leads to the formation of products, as illustrated in Figure 11.13. In this way enzymes exhibit specificities towards substrates.

So far we've discussed several methods to synthesize alcohols. Let's now discuss some of their reactions.

11.15 Conversion of Alcohols into Alkoxide Ions

Alkoxide ions, RO^-, are the conjugate bases of alcohols, as we learned in Section 3.4. Alkoxide ions are commonly used in organic chemistry as nucleophiles in nucleophilic substitution reactions (Chapter 12) and as bases in the base-promoted elimination reactions of haloalkanes (Chapter 12).

Alkoxide ions are prepared by the reaction of strong bases with an alcohol. Any base whose conjugate acid has a pK_a value more positive than about 25 can be used to quantitatively prepare an alkoxide ion from an alcohol. Sodium amide ($NaNH_2$) is a strong base commonly used for this purpose.

$$CH_3CH_2OH + NaNH_2 \longrightarrow CH_3CH_2O^-Na^+ + NH_3$$

Ethanol	Sodium amide	Sodium ethoxide	Ammonia
$pK_a \approx 16$			$pK_a \approx 38$

Figure 11.13
A substrate that lacks the proper shape or arrangement of functional groups cannot form an enzyme-substrate complex.

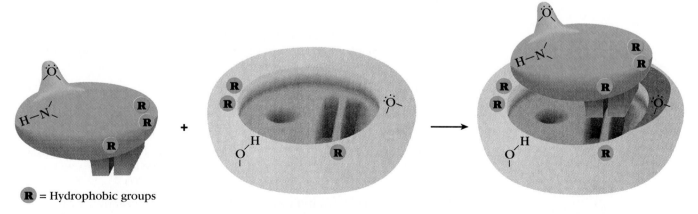

R = Hydrophobic groups

Substrate   **Enzyme**   **An enzyme-substrate complex does not form**

Other strong bases that react with alcohols to form alkoxide ions include Grignard and organolithium reagents and acetylide ions. Hydroxide ion, however, is too weak a base ($[pK_a]_{H_2O} = 15.7$) to quantitatively convert an alcohol into its alkoxide ion.

Alcohols, like water, react with the alkali metals (Li, Na, and K) to form the corresponding alkoxide ion and hydrogen gas.

$$H_2O \ + \ Na \ \longrightarrow \ HO^-Na^+ \ + \ 1/2\ H_2$$

Water   Sodium   Sodium   Hydrogen
            metal   hydroxide   gas

$$CH_3CH_2OH \ + \ Na \ \longrightarrow \ CH_3CH_2O^-Na^+ \ + \ 1/2\ H_2$$

Ethanol   Sodium   Sodium ethoxide   Hydrogen
           metal   gas

Sodium metal reacts less vigorously with alcohols than with water. As the size of the alkyl portion of the alcohol increases, the vigor of the reaction with sodium metal continually decreases. The more reactive potassium metal is used to prepare alkoxides from alcohols containing four or more carbon atoms.

Exercise 11.22   Write the equation for the reaction of 2-propanol with each of the following reagents.

(a) $CH_3MgI/C_2H_5OC_2H_5$ **(c)** Aqueous NaOH
(b) $CH_3C\equiv C^-\ Na^+/C_2H_5OC_2H_5$ **(d)** $CH_3CH_2CH_2Li/hexane$

11.16 Formation of Esters from Alcohols

Alcohols react with acids to form esters and water.

$$\text{Alcohol} \ + \ \text{Acid} \ \rightleftarrows \ \text{Ester} \ + \ \text{Water}$$

The acid can be a carboxylic acid or one of a number of other acids such as arenesulfonic acids, sulfuric acid, or phosphoric acid.

Alcohols react with carboxylic acids in the presence of a strong acid (H_2SO_4) catalyst to form esters of carboxylic acids. This reaction is called **esterification,** and it is discussed in detail in Chapter 15. The acid-catalyzed reaction of acetic acid (a carboxylic acid) with ethanol to form ethyl acetate (an ester) and water is an example of an esterification reaction.

Acetic acid   Ethanol   Ethyl acetate   Water
(a carboxylic acid)   (a carboxylic
                                         acid ester)

Alkyl arenesulfonates are the esters of alcohols and arenesulfonic acids. Alkyl arenesulfonates are prepared in the laboratory by the reaction of an alcohol with an arenesulfonyl chloride. The reaction of *p*-toluenesulfonyl chloride and methanol, for instance, forms methyl *p*-toluenesulfonate. Pyridine, a base, is added to react with the HCl also formed in the reaction.

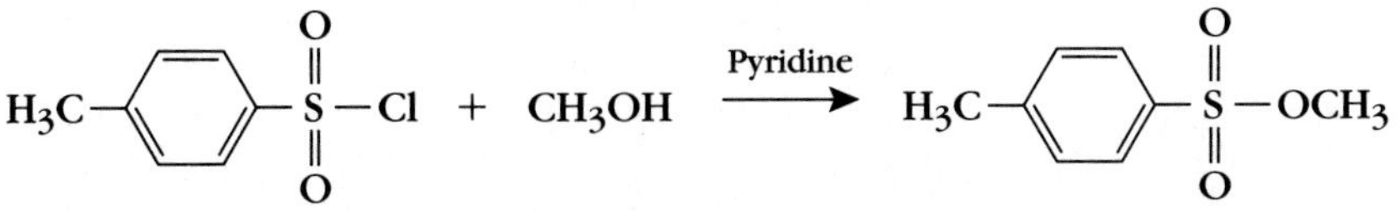

p-Toluenesulfonyl chloride   Methanol   Methyl *p*-toluenesulfonate
                                                                        (methyl tosylate)

The carbon-oxygen bond of the alcohol is not broken in its reaction with an arenesulfonyl chloride. If the carbon atom bonded to the hydroxy group is a stereocenter, its configuration is retained in the alkyl arenesulfonate.

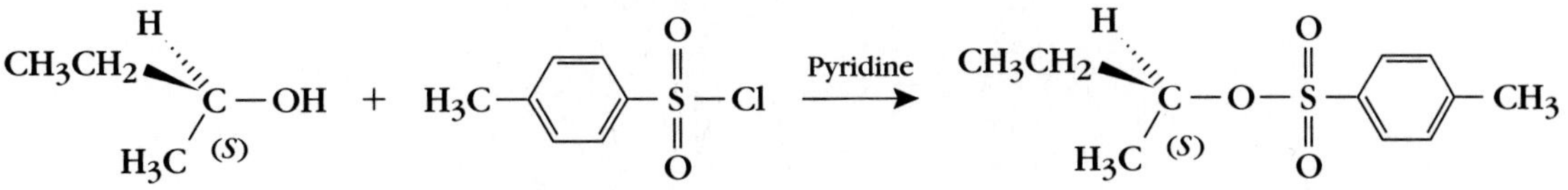

(*S*)-2-Butanol   *p*-Toluenesulfonyl chloride   (*S*)-2-Butyl *p*-toluenesulfonate

The name *p*-toluenesulfonyl is usually shortened to tosyl, with the result that the name of alkyl *p*-toluenesulfonate esters becomes alkyl tosylates.

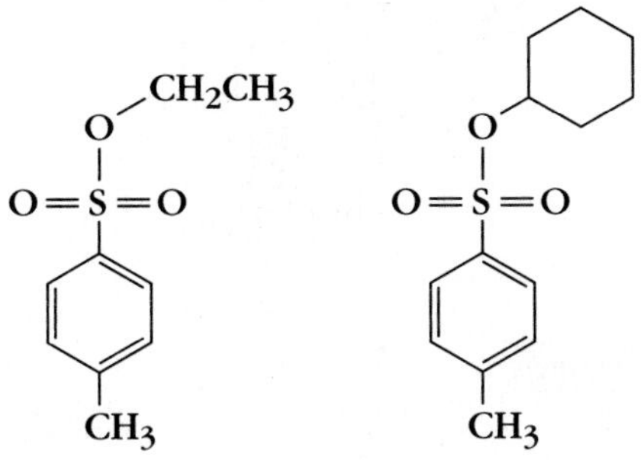

Ethyl tosylate   Cyclohexyl tosylate

Alcohols react with sulfuric acid to form either monoalkyl or dialkyl sulfate esters.

Methanol   Sulfuric   Methyl sulfate
                   acid

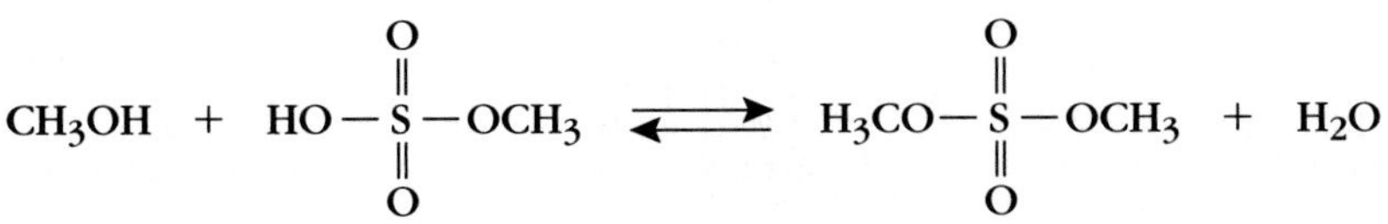

Dimethyl sulfate

Notice that there are no sulfur-carbon bonds in sulfate esters, unlike sulfonate esters, which contain one sulfur-carbon bond.

Alcohols and phosphoric acid can react to form mono-, di-, and trialkyl phosphate esters.

$$
\begin{array}{cccc}
\underset{\text{O}}{\overset{\text{O}}{\parallel}} & \underset{\text{O}}{\overset{\text{O}}{\parallel}} & \underset{\text{O}}{\overset{\text{O}}{\parallel}} & \underset{\text{O}}{\overset{\text{O}}{\parallel}} \\
\text{HO}-\text{P}-\text{OH} & \text{HO}-\text{P}-\text{OR} & \text{RO}-\text{P}-\text{OR} & \text{RO}-\text{P}-\text{OR} \\
\underset{\text{OH}}{|} & \underset{\text{OH}}{|} & \underset{\text{OH}}{|} & \underset{\text{OR}}{|}
\end{array}
$$

| Phosphoric acid | Monoalkyl phosphate | Dialkyl phosphate | Trialkyl phosphate |

Mono- and dialkyl phosphate esters are an important part of many molecules involved in the reactions of living organisms. An example is the phosphate linkage in the polynucleotide chain of nucleic acids shown in Figure 11.14.

Figure 11.14
A small segment of the polynucleotide chain of nucleic acids.

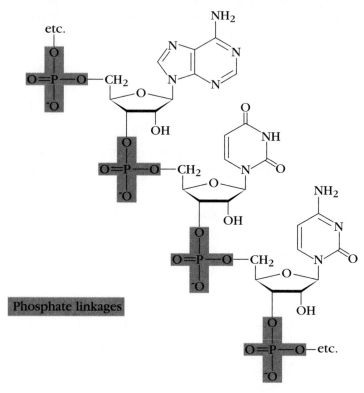

Phosphate linkages

Esters of diphosphoric acid, $H_4P_2O_7$, (also called pyrophosphoric acid) and triphosphoric acid, $H_5P_3O_{10}$, are also known in living systems. ADP is an example of a monoalkyl diphosphate ester, while ATP is an example of a monoalkyl triphosphate ester (see Chapter 27). NADH and NAD$^+$ are two examples of compounds of biological importance that contain a dialkyl diphosphate group.

■ *The structure of NADH is given in Figure 11.8 (p. 476) and the structure of NAD$^+$ is given in Figure 11.9 (p. 477).*

$$
\begin{array}{ccc}
\underset{\text{OH}\quad\text{OH}}{\overset{\text{O}\quad\text{O}}{\text{HO}-\text{P}-\text{O}-\text{P}-\text{OH}}} &
\underset{\text{OH}\quad\text{OH}}{\overset{\text{O}\quad\text{O}}{\text{RO}-\text{P}-\text{O}-\text{P}-\text{OH}}} &
\underset{\text{OH}\quad\text{OH}}{\overset{\text{O}\quad\text{O}}{\text{RO}-\text{P}-\text{O}-\text{P}-\text{OR}}}
\end{array}
$$

| Diphosphoric acid (pyrophosphoric acid) | Monoalkyl diphosphate ester | Dialkyl diphosphate ester |

$$
\begin{array}{cc}
\underset{\text{OH}\quad\text{OH}\quad\text{OH}}{\overset{\text{O}\quad\text{O}\quad\text{O}}{\text{HO}-\text{P}-\text{O}-\text{P}-\text{O}-\text{P}-\text{OH}}} &
\underset{\text{OH}\quad\text{OH}\quad\text{OH}}{\overset{\text{O}\quad\text{O}\quad\text{O}}{\text{RO}-\text{P}-\text{O}-\text{P}-\text{O}-\text{P}-\text{OH}}}
\end{array}
$$

| Triphosphoric acid | Monoalkyl triphosphoric ester |

11.17 Conversion of Alcohols into Haloalkanes

Alcohols react with hydrogen halides (HCl, HBr, or HI) to form haloalkanes.

General Reaction

$$ROH \ + \ HX \ \longrightarrow \ RX \ + \ H_2O$$

Alcohol   Hydrogen   Haloalkane
           halide

$$X = Cl, Br, or I$$

Example

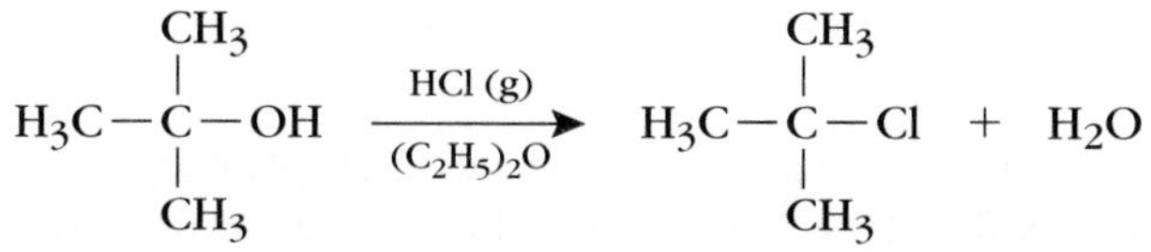

$$
\underset{\substack{\text{2-Methyl-2-propanol}\\(\textit{tert}\text{-butyl alcohol})}}{H_3C-\overset{\displaystyle CH_3}{\underset{\displaystyle CH_3}{\overset{\vert}{\underset{\vert}{C}}}}-OH} \ \xrightarrow[\text{(C}_2\text{H}_5)_2\text{O}]{\text{HCl (g)}} \ \underset{\substack{\text{2-Chloro-2-methylpropane}\\(\textit{tert}\text{-butyl chloride})\\(90\%\text{-}95\%\text{ yield})}}{H_3C-\overset{\displaystyle CH_3}{\underset{\displaystyle CH_3}{\overset{\vert}{\underset{\vert}{C}}}}-Cl} \ + \ H_2O
$$

The hydroxy group of the alcohol is replaced by a halogen atom in this reaction, which can be used to prepare 1°, 2°, and 3° haloalkanes. The mechanism of this reaction is discussed in Chapter 12.

Other reagents that are used to replace a hydroxy group by a halogen atom are thionyl chloride and phosphorus tribromide. Phosphorus tribromide (PBr$_3$) converts alcohols to bromoalkanes, while thionyl chloride (SOCl$_2$) converts them to chloroalkanes.

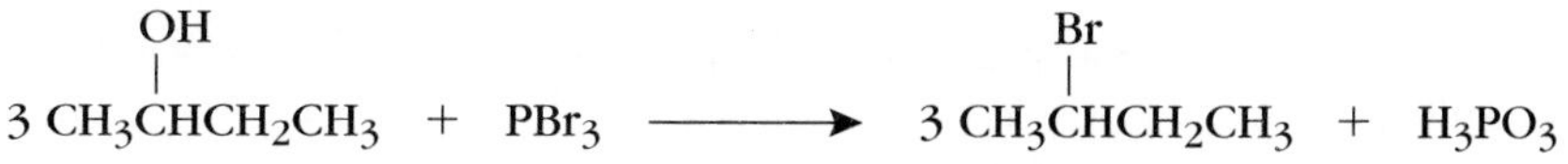

$$
\underset{\text{2-Butanol}}{3 \ CH_3\overset{\overset{\displaystyle OH}{\vert}}{CH}CH_2CH_3} \ + \ \underset{\substack{\text{Phosphorus}\\\text{tribromide}}}{PBr_3} \ \longrightarrow \ \underset{\substack{\text{2-Bromobutane}\\(85\%\text{ yield})}}{3 \ CH_3\overset{\overset{\displaystyle Br}{\vert}}{CH}CH_2CH_3} \ + \ \underset{\substack{\text{Phosphorus}\\\text{acid}}}{H_3PO_3}
$$

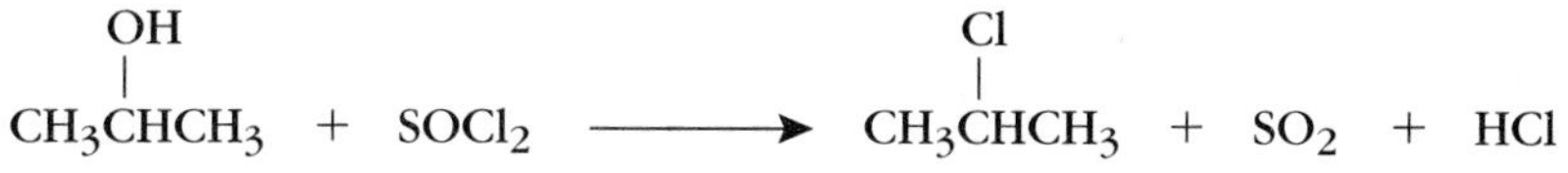

$$
\underset{\substack{\text{2-Propanol}\\(\text{isopropyl}\\\text{alcohol})}}{CH_3\overset{\overset{\displaystyle OH}{\vert}}{CH}CH_3} \ + \ \underset{\substack{\text{Thionyl}\\\text{chloride}}}{SOCl_2} \ \longrightarrow \ \underset{\substack{\text{2-Chloropropane}\\(\text{isopropyl chloride})\\(95\%\text{ yield})}}{CH_3\overset{\overset{\displaystyle Cl}{\vert}}{CH}CH_3} \ + \ SO_2 \ + \ HCl
$$

Exercise 11.23   Write the structure of the product of each of the following reactions.

(a) $CH_3CH_2CH_2CH_2OH \ + \ HBr \ \xrightarrow[\text{H}_2\text{O}]{\text{Heat}}$

(c) ⬠$-CH_2OH \ + \ PBr_3 \longrightarrow$

(b) ⬡$-OH \ + \ SOCl_2 \longrightarrow$

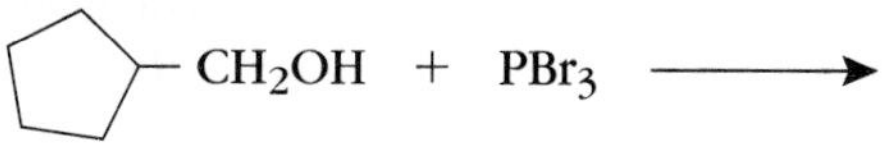

(d) ⬡$-\overset{\displaystyle CH_3}{\underset{\displaystyle CH_3}{\overset{\vert}{\underset{\vert}{C}}}}-OH \ + \ HCl \longrightarrow$

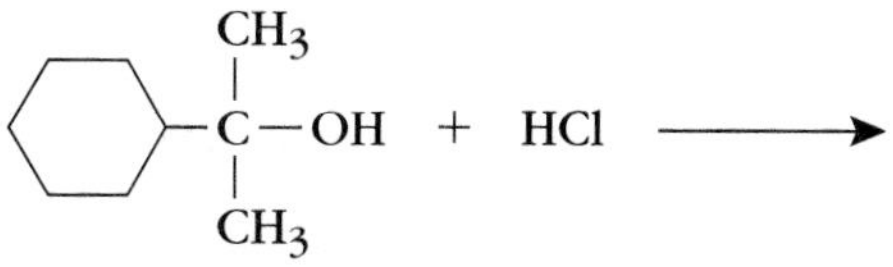

11.18 Oxidation of Alcohols

We discussed in Section 11.9 that aldehydes and ketones react with reducing agents to form primary and secondary alcohols, respectively. The reverse reaction, the oxidation of primary and secondary alcohols to aldehydes and ketones, can be accomplished by reacting these alcohols with oxidizing agents.

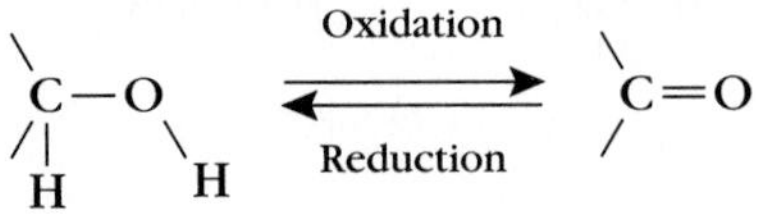

The reagents that are commonly used to oxidize primary and secondary alcohols are shown in Table 11.7. These oxidizing agents don't normally oxidize tertiary alcohols.

All of the reagents listed in Table 11.7 oxidize a secondary alcohol to a ketone. Cyclohexanol, for example, is oxidized to cyclohexanone by aqueous acidic sodium dichromate.

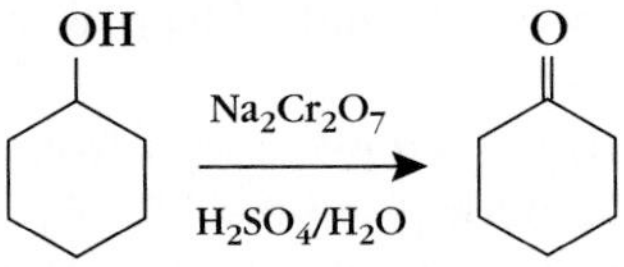

Primary alcohols are oxidized first to aldehydes that are easily oxidized further to carboxylic acids.

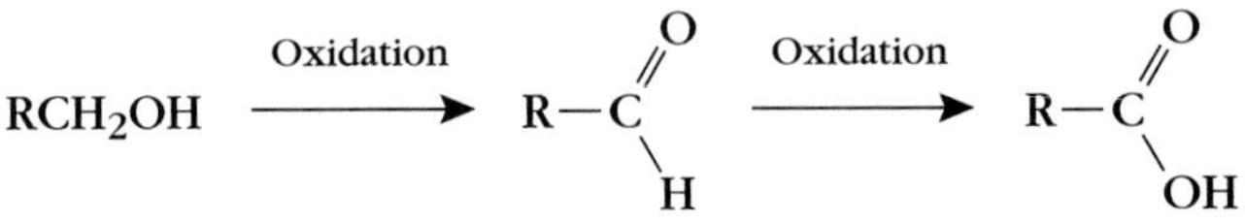

Table 11.7	Reagents Commonly Used to Oxidize Alcohols
Reagent	Chemical Formula
Aqueous acidic sodium dichromate	$Na_2Cr_2O_7/H_2SO_4/H_2O$
Aqueous acidic chromium trioxide	$CrO_3/H_2SO_4/H_2O$
Aqueous acidic potassium permanganate	$KMnO_4/H_2SO_4/H_2O$
Pyridinium chlorochromate (PCC) in anhydrous solvent	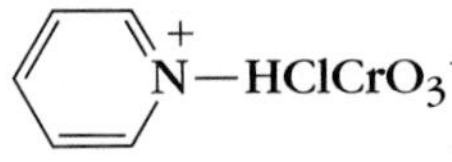 $\overset{+}{N}-HClCrO_3^-$
Collins' reagent	$\left(N: \right)_2 CrO_3$

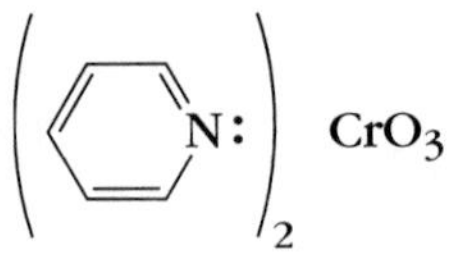

Oxidation of primary alcohols can be stopped at the aldehyde stage only by using special reagents like pyridinium chlorochromate (PCC) or Collins' reagent in anhydrous methylene chloride.

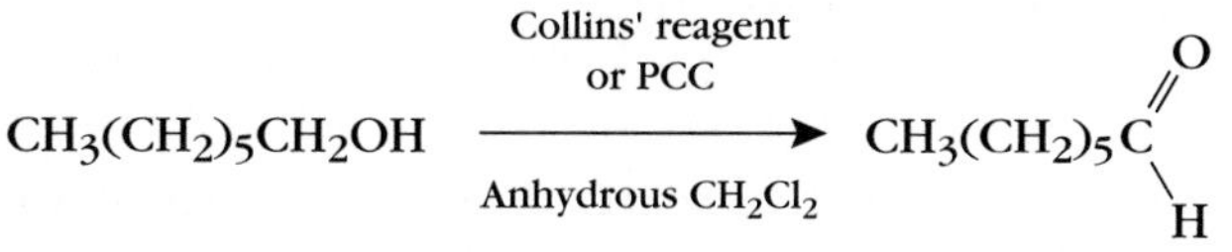

CH$_3$(CH$_2$)$_5$CH$_2$OH $\xrightarrow[\text{Anhydrous CH}_2\text{Cl}_2]{\text{Collins' reagent or PCC}}$ CH$_3$(CH$_2$)$_5$C⟨O, H⟩

1-Heptanol

Heptanal
(74%-82% yield)

The other oxidizing agents listed in Table 11.7 oxidize primary alcohols to carboxylic acids. Aldehydes are involved as intermediates in these oxidation reactions but can't be isolated because they are easily oxidized to carboxylic acids.

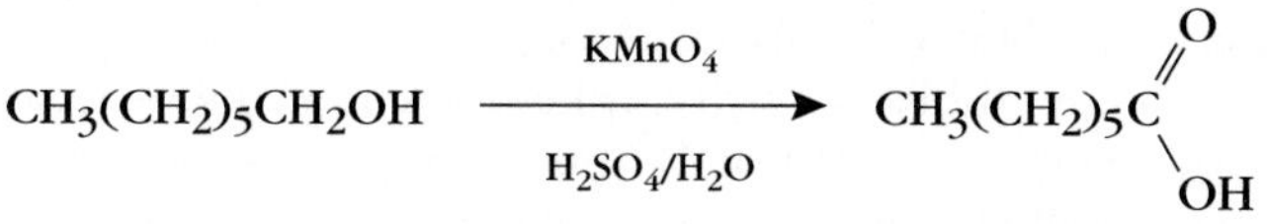

CH$_3$(CH$_2$)$_5$CH$_2$OH $\xrightarrow[\text{H}_2\text{SO}_4/\text{H}_2\text{O}]{\text{KMnO}_4}$ CH$_3$(CH$_2$)$_5$C⟨O, OH⟩

1-Heptanol

Heptanoic acid
(93% yield)

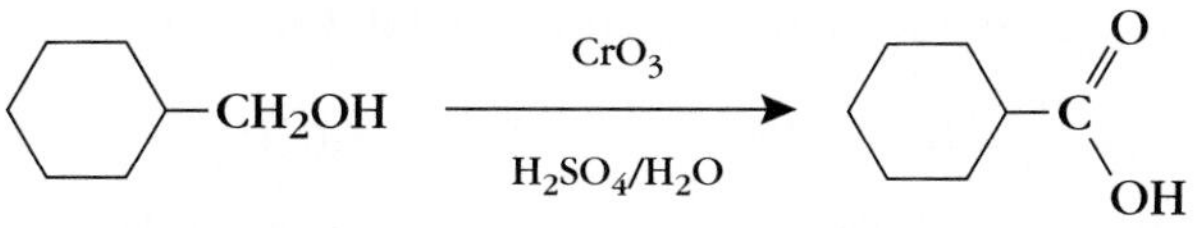

⬡—CH$_2$OH $\xrightarrow[\text{H}_2\text{SO}_4/\text{H}_2\text{O}]{\text{CrO}_3}$ ⬡—C⟨O, OH⟩

Cyclohexylmethanol

Cyclohexanecarboxylic acid
(88% yield)

Exercise 11.24   What oxidizing agent would you use to carry out each of the following reactions?

(a) 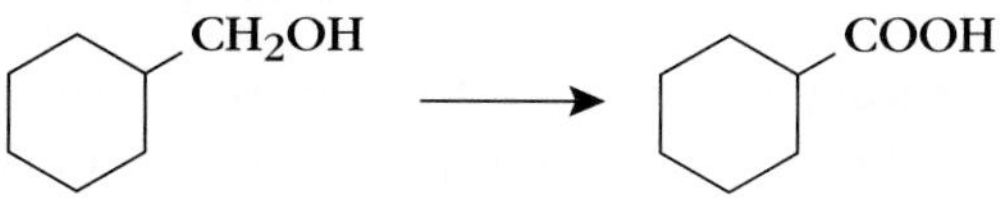 ⬡—CH$_2$OH   ⟶   ⬡—COOH

(b) 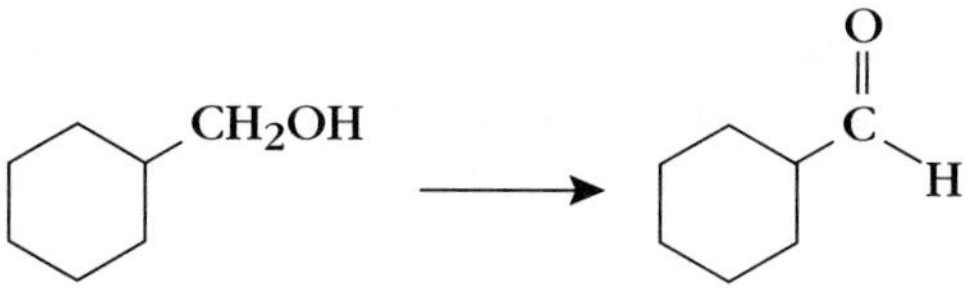 ⬡—CH$_2$OH   ⟶   ⬡—CHO

(c)  ⬠—OH   ⟶   ⬠=O

11.19 Thiols

Sulfur is directly below oxygen in the periodic table; consequently, sulfur is the element most like oxygen. As a result, many oxygen-containing organic compounds have sulfur analogs. Thiols (RSH), for example, are the sulfur analogs of alcohols (ROH).

Thiols are named by the same system used for alcohols, with the suffix *-thiol* added to the name of the alkane parent name instead of *-ol.* The final *-e* of the alkane parent name is retained for ease of pronunciation. When the —SH group is named as a substituent, it is called a *mercapto* group.

SH
(structure)

$(CH_3)_2CHCH_2CH_2SH$ $HSCH_2CH_2OH$

Cyclopentanethiol   3-Methylbutanethiol   2-Mercaptoethanol

The most characteristic physical property of low molecular weight thiols is their disagreeable odor. 3-Methylbutanethiol and *cis-* and *trans-*2-butene-1-thiol, for example, are the major constituents of skunk scent. The offensive odor of thiols decreases as the number of carbon atoms increases because both their relative sulfur content and their volatility decrease.

Hydrogen bonding in thiols is much weaker than in alcohols because the S—H bond is less polar than the O—H bond. This difference is the reason that thiols have much lower boiling points than alcohols of about the same molecular weight. Methanethiol (MW = 48), for example, is a gas at room temperature (bp = 6 °C) while ethanol (MW = 46) is a liquid (bp = 78 °C).

Thiols can be prepared by the reaction of a haloalkane with excess hydrosulfide ion, a reaction whose mechanism is discussed in Chapter 12.

$$CH_3(CH_2)_3CH_2Br \ + \ Na^{+\ -}SH \longrightarrow CH_3(CH_2)_3CH_2SH \ + \ NaBr$$

1-Bromopentane   Sodium   1-Pentanethiol
               hydrosulfide

Yields are usually poor unless an excess of hydrosulfide ion is used because the product thiol can react further with the haloalkane to form a symmetrical sulfide (RSR).

$$CH_3(CH_2)_3CH_2SH \ + \ CH_3(CH_2)_3CH_2Br \longrightarrow CH_3(CH_2)_3CH_2SCH_2(CH_2)_3CH_3 \ + \ HBr$$

Symmetrical sulfide

The problem of forming the unwanted symmetrical sulfide as a byproduct can be overcome by using thiourea, $(NH_2)_2C{=}S$, in the preparation of thiols from haloalkanes. This method is a two-reaction sequence in which an alkylisothiouronium salt is formed in the first reaction. The mechanism of this reaction is discussed in Chapter 12. Hydrolysis of the alkylisothiouronium salt in base is the second reaction, which forms the desired thiol.

(reaction scheme)

Thiourea   1-Bromopentane   S-Pentylisothiouronium bromide

(reaction scheme with conditions: 1) NaOH/H_2O 2) H_3O^+)

S-Pentylisothiouronium bromide   1-Pentanethiol   Urea
                                    (85% yield)

Both reactions are usually carried out in the same reaction vessel without isolation of the alkylisothiouronium salt.

Alkanethiols ($pK_a \approx 10$) are stronger acids than alcohols ($pK_a \approx 17$). As a result, alkanethiols are quantitatively converted to their conjugate bases, called alkanethiolate ions, by hydroxide ion.

$$RS\!-\!H \qquad \overset{-}{:}OH \longrightarrow RS\overset{-}{:} \qquad HOH$$

Alkanethiol	Hydroxide ion		Alkanethiolate ion	Water
Stronger acid ($pK_a \approx 10$)	Stronger base		Weaker base	Weaker acid ($pK_a \approx 15.7$)

A major difference between similar oxygen- and sulfur-containing compounds is their products of oxidation. Oxidation of alcohols, for example, results in oxidation of the carbon atom bonded to the hydroxy group to form aldehydes or ketones (Section 11.18). In contrast, oxidation of thiols results in oxidation of the sulfur atom to form disulfides (RSSR).

$$2RSH \underset{\text{Reduction}}{\overset{\text{Oxidation}}{\rightleftarrows}} RSSR$$

A thiol   A disulfide

Oxidation of thiols is a reversible reaction, so disulfides are easily reduced to thiols. The reversible oxidation-reduction reactions of thiols and disulfides are very important in biochemistry. The amino acid cysteine, for example, is oxidized to cystine. Cystine in turn can be reduced to cysteine.

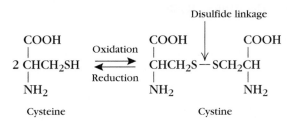

As we discuss in Chapter 26, disulfide linkages are important in determining the overall shape of protein molecules.

11.20 Summary

Alcohols are compounds that contain a hydroxy group bonded to an sp^3 hybrid carbon atom. The hydroxy group of a **1° alcohol** is bonded to a primary carbon atom, the hydroxy group of a **2° alcohol** is bonded to a secondary carbon atom, and the hydroxy group of a **3° alcohol** is bonded to a tertiary carbon atom.

Three common methods of preparing alcohols are (1) hydration of alkenes, which was discussed in Chapters 7 and 8, (2) reactions of Grignard or organolithium compounds with carbonyl groups, and (3) reduction of carbonyl groups.

Magnesium metal reacts with halogen-containing compounds in anhydrous diethyl ether or THF to form a **Grignard reagent.** Replacing magnesium metal by lithium metal in this reaction results in the formation of an **organolithium**

reagent. Additions of Grignard and organolithium reagents to ketones form tertiary alcohols, while additions to aldehydes form secondary alcohols. Primary alcohols are formed by adding either Grignard or organolithium reagents to formaldehyde, $H_2C=O$.

Carbonyl groups of aldehydes and ketones are reduced to alcohols by **sodium borohydride,** $NaBH_4$, or **lithium aluminum hydride,** $LiAlH_4$. Reduction of aldehydes forms primary alcohols, while reduction of ketones forms secondary alcohols. Neither $LiAlH_4$ nor $NaBH_4$ usually distinguishes between the two faces of a carbonyl group. As a result, reduction of a prochiral aldehyde or ketone forms a product that is a racemic mixture.

Certain enzymes along with the coenzyme NADH also reduce carbonyl groups to alcohols. NADH is the reduced form of **nicotinamide adenine dinucleotide,** and it is the biological equivalent of $LiAlH_4$ and $NaBH_4$. The enzyme-catalyzed reduction can distinguish between the two faces of a carbonyl group, with the result that the product is a single stereoisomer.

Alcohols react with strong bases such as sodium hydride, sodium amide, Grignard, and organolithium reagents to form **alkoxide ions.** Alcohols and acids react to form **esters** and water. Examples include the esters of carboxylic acids, arenesulfonic acids, sulfuric acid, and phosphoric acids. Esters of phosphorus acids are particularly important in living organisms.

Oxidation of alcohols forms carbonyl-containing compounds. This reaction of alcohols is the reverse of one method of preparing them: reduction of aldehydes and ketones. Depending on the oxidizing agent, primary alcohols are oxidized to either aldehydes or carboxylic acids, while secondary alcohols are oxidized to ketones. Oxidation of tertiary alcohols does not usually occur.

Thiols, RSH, are the sulfur analogs of alcohols. Thiols are usually prepared by the reaction of a haloalkane with thiourea. While oxidation of alcohols occurs at the carbon atom bonded to the hydroxyl group, oxidation of thiols occurs at the sulfur atom to form **disulfides, RSSR.**

11.21 Summary of Methods of Synthesizing Alcohols

1. Tertiary Alcohols

a. Oxymercuration-Demercuration   SECTION 8.10

Markownikoff addition of water to a carbon-carbon double bond without rearrangement. Achiral alkenes form optically inactive products (racemic mixtures or achiral alcohols).

(i) Unsymmetrical disubstituted alkene

3° Alcohol

(ii) Trisubstituted alkene

3° Alcohol

(iii) Symmetrical tetrasubstituted alkene

$$(CH_3)_2C = C(CH_3)_2 \xrightarrow[\text{(2) NaBH}_4\text{/NaOH}]{\text{(1) Hg(OAc)}_2\text{/H}_2\text{O}} (CH_3)_2\underset{\underset{HO}{|}}{C} - \underset{\underset{H}{|}}{C}(CH_3)_2$$

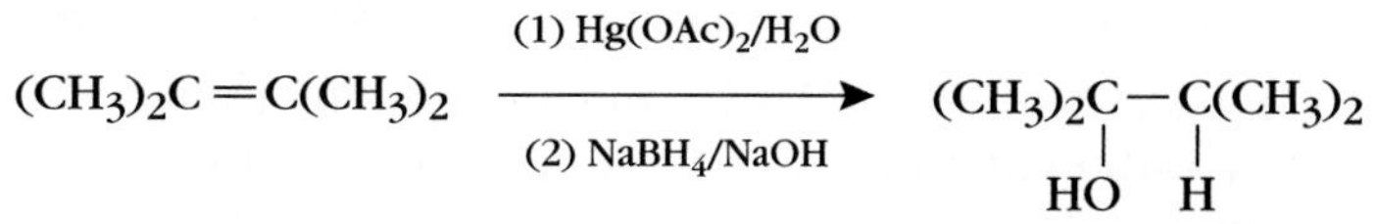

3° Alcohol

b. Hydroboration-Oxidation of Symmetrical Tetrasubstituted Alkenes   SECTIONS 8.1–8.4

Syn addition of water to a carbon-carbon double bond.

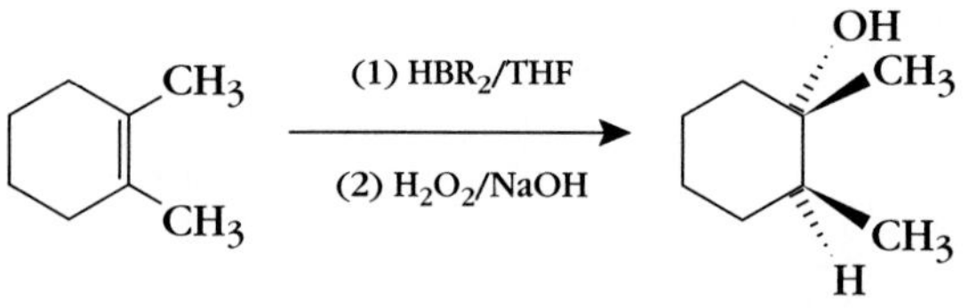

3° Alcohol
(Racemic mixture)

c. Reaction of Organometallic Compounds and Ketones   SECTION 11.7

Addition of either Grignard or organolithium reagents to the carbonyl carbon atom forms optically inactive or achiral alcohols.

(i) Grignard reagent

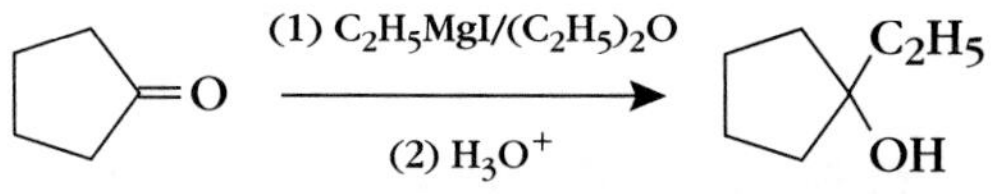

3° Alcohol

(ii) Organolithium reagent

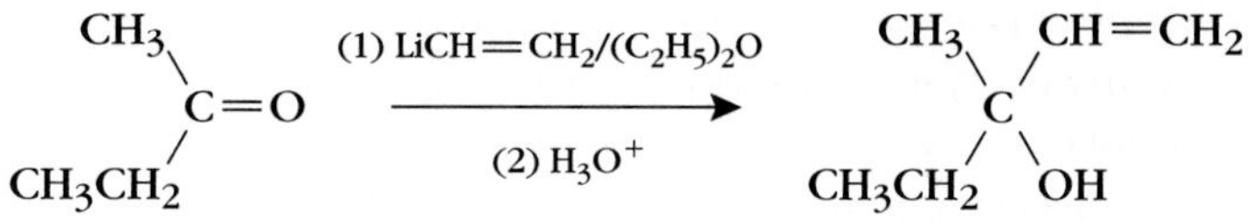

3° Alcohol

2. Secondary Alcohols

a. Oxymercuration-Demercuration   SECTION 8.10

Markownikoff addition of water to carbon-carbon double bond without rearrangement.

(i) Terminal alkenes

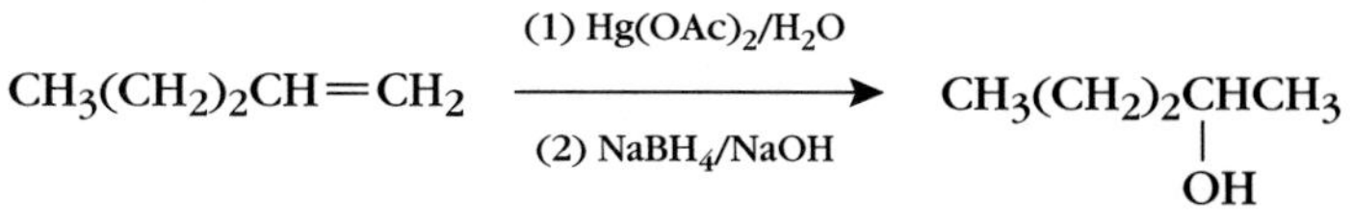

$$CH_3(CH_2)_2CH = CH_2 \xrightarrow[\text{(2) NaBH}_4\text{/NaOH}]{\text{(1) Hg(OAc)}_2\text{/H}_2\text{O}} CH_3(CH_2)_2\underset{\underset{OH}{|}}{C}HCH_3$$

2° Alcohol

(ii) Symmetrical disubstituted alkenes

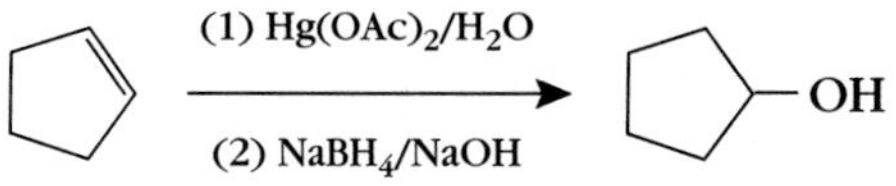

2° Alcohol

b. Hydroboration-Oxidation SECTIONS 8.1–8.4

Syn addition and anti-Markownikoff addition of water to a carbon-carbon double bond without rearrangement.

(i) Symmetrical disubstituted alkenes

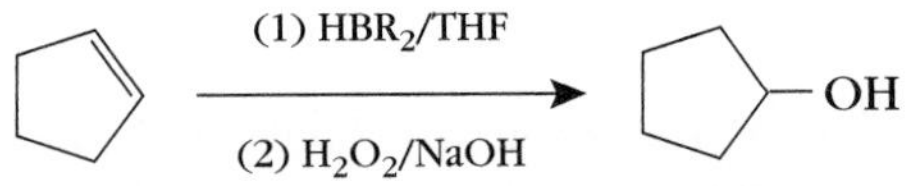

$2°$ Alcohol

(ii) Trisubstituted alkenes

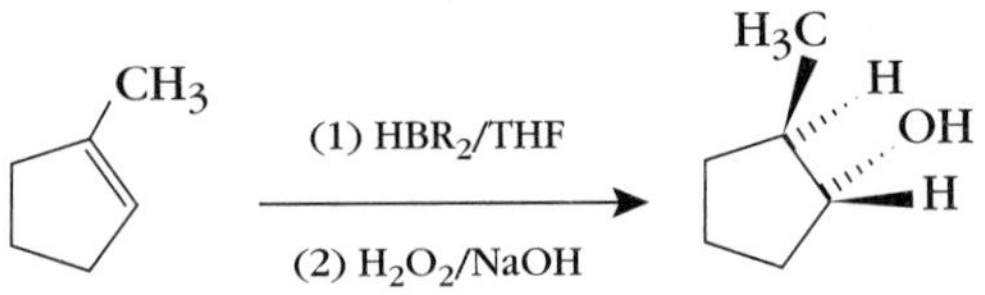

$2°$ Alcohol

c. Reduction of a Ketone SECTION 11.10

Reduction of a ketone forms an optically inactive or achiral alcohol.

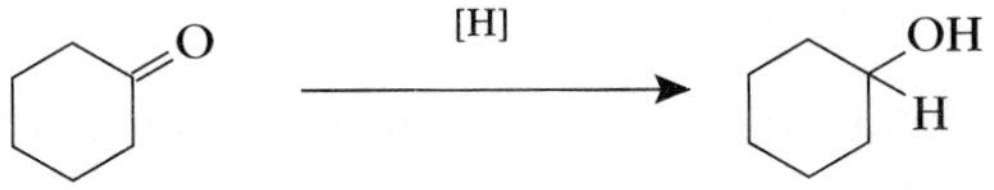

$2°$ Alcohol

$[H] = Pt/H_2$, $NaBH_4$, or $LiAlH_4$

d. Reaction of Organometallic Compounds and Aldehydes Section 11.7

Addition of either Grignard or organolithium reagents to the carbonyl carbon atom forms optically inactive or achiral alcohols.

(i) Grignard reagent

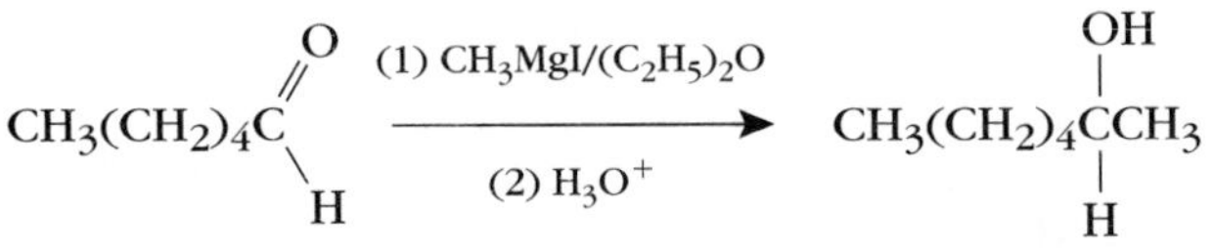

$2°$ Alcohol

(ii) Organolithium reagent

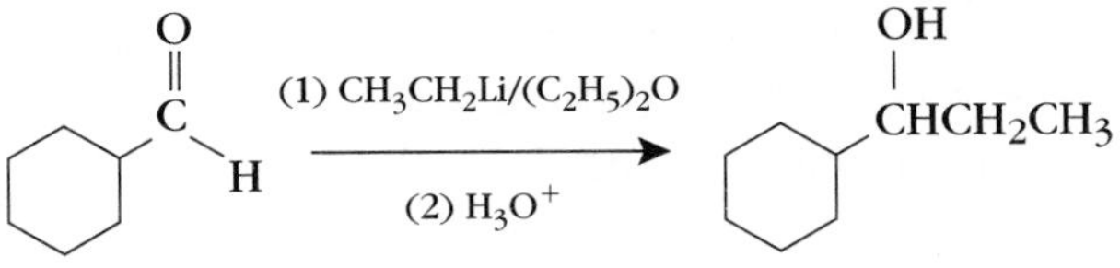

$2°$ Alcohol

3. **Primary Alcohols**
   a. **Hydroboration-Oxidation of Terminal Alkenes**   SECTIONS 8.1–8.4

Anti-Markownikoff addition of water to carbon-carbon double bond.

$$(CH_3)_2C{=}CH_2 \xrightarrow[\text{(2) } H_2O_2/NaOH]{\text{(1) } HBR_2/THF} (CH_3)_2CHCH_2OH$$

1° Alcohol

   b. **Reduction of Aldehydes**   SECTION 11.10

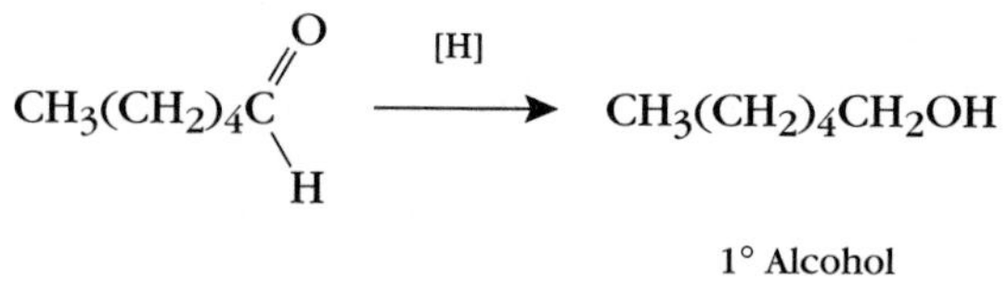

$$CH_3(CH_2)_4\overset{\displaystyle O}{\underset{\displaystyle H}{C}} \xrightarrow{[H]} CH_3(CH_2)_4CH_2OH$$

1° Alcohol

$$[H] = H_2/Pt, \ NaBH_4, \ or \ LiAlH_4$$

   c. **Reaction of Organometallic Compounds and Formaldehyde**   SECTION 11.7
      (i) **Grignard reagent**

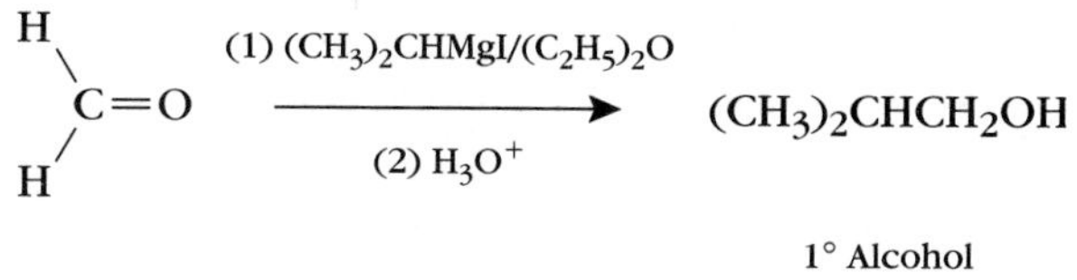

$$\underset{H}{\overset{H}{C}}{=}O \xrightarrow[\text{(2) } H_3O^+]{\text{(1) } (CH_3)_2CHMgI/(C_2H_5)_2O} (CH_3)_2CHCH_2OH$$

1° Alcohol

      (ii) **Organolithium reagent**

$$\underset{H}{\overset{H}{C}}{=}O \xrightarrow[\text{(2)}H_3O^+]{\text{(1) } CH_3(CH_2)_2CH_2Li/(C_2H_5)_2O} CH_3(CH_2)_2CH_2CH_2OH$$

1° Alcohol

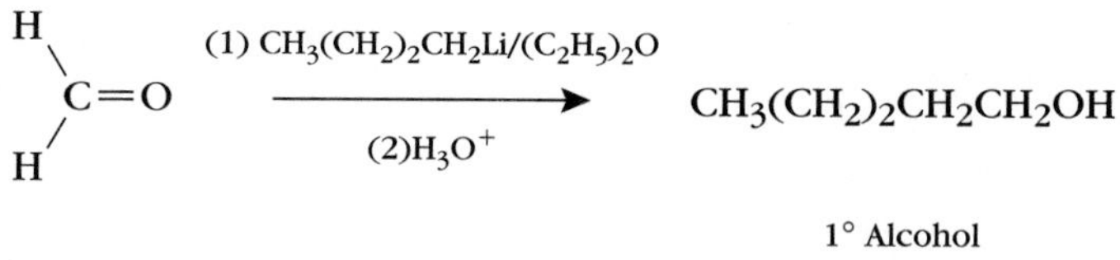

11.22 Summary of Reactions of Alcohols

1. **Reaction with Strong Bases Forms Alkoxide Ions**   SECTION 11.15

pK_a of conjugate acid of base must be more positive than about 25.

$$RO{-}H \quad :Base^- \longrightarrow RO^- + H{-}Base$$

$:Base = NH_2^-, \ H^-$, Grignard or organolithium reagent

$$CH_3CH_2OH + NaNH_2 \longrightarrow CH_3CH_2O^-Na^+ + NH_3$$

Sodium ethoxide

2. **Reaction with Alkali Metals forms Alkoxide Ions**   SECTION 11.15

$$2ROH + 2M \longrightarrow 2RO^-M^+ + H_2$$

$M = Li, \ Na, \ or \ K$

$$2(CH_3)_3COH + 2K \longrightarrow 2(CH_3)_3CO^-K^+ + H_2$$

Potassium *tert*-butoxide

3. Esters from Alcohols

SECTION 11.16

a. Carboxylic Acid Esters

Acid-catalyzed reaction of an alcohol and a carboxylic acid forms a carboxylic acid ester. Also discussed in Chapter 15.

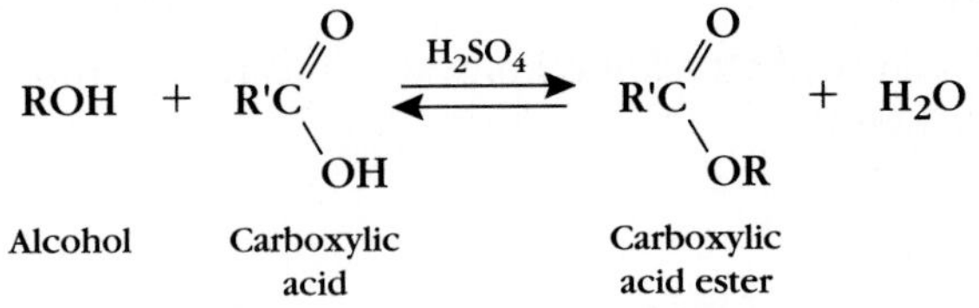

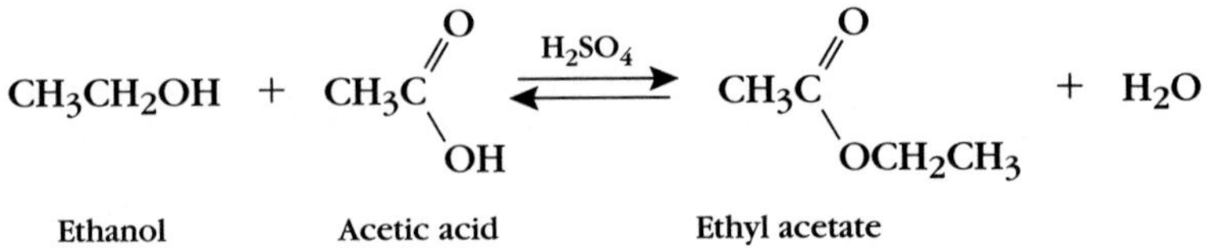

$$ROH + R'C\overset{O}{\underset{OH}{\big\backslash}} \underset{\longleftarrow}{\overset{H_2SO_4}{\longrightarrow}} R'C\overset{O}{\underset{OR}{\big\backslash}} + H_2O$$

Alcohol   Carboxylic   Carboxylic
           acid   acid ester

$$CH_3CH_2OH + CH_3C\overset{O}{\underset{OH}{\big\backslash}} \underset{\longleftarrow}{\overset{H_2SO_4}{\longrightarrow}} CH_3C\overset{O}{\underset{OCH_2CH_3}{\big\backslash}} + H_2O$$

Ethanol   Acetic acid   Ethyl acetate

b. Arenesulfonate Esters

$$ROH + ArSO_2Cl \overset{Pyridine}{\longrightarrow} ArSO_2OR$$

Alcohol   Arenesulfonyl   Arenesulfonate
           chloride   ester

$$CH_3CH_2OH + \text{(p-toluenesulfonyl chloride)} \overset{Pyridine}{\longrightarrow} \text{(ethyl p-toluenesulfonate)}$$

Ethanol   p-Toluenesulfonyl   Ethyl p-Toluenesulfonate
           chloride   (Ethyl tosylate)
           (Tosyl chloride)

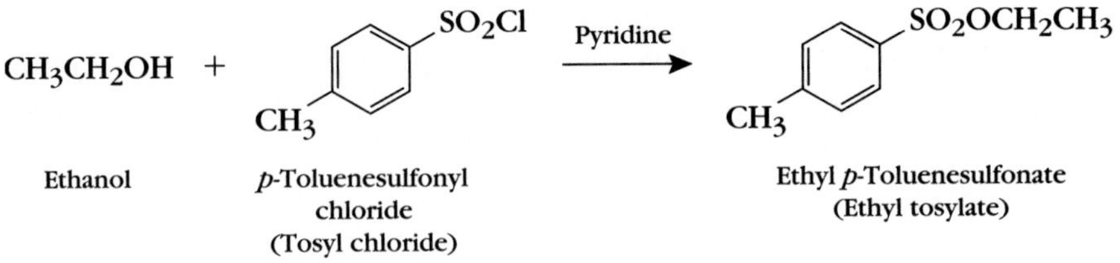

c. Sulfate Esters

$$HO-\overset{O}{\underset{O}{\overset{\|}{\underset{\|}{S}}}}-OH \underset{\longleftarrow}{\overset{ROH}{\longrightarrow}} RO-\overset{O}{\underset{O}{\overset{\|}{\underset{\|}{S}}}}-OH \underset{\longleftarrow}{\overset{ROH}{\longrightarrow}} RO-\overset{O}{\underset{O}{\overset{\|}{\underset{\|}{S}}}}-OR$$

Sulfuric   Alkyl   Dialkyl
acid   sulfate   sulfate

$$HO-\overset{O}{\underset{O}{\overset{\|}{\underset{\|}{S}}}}-OH \overset{CH_3OH}{\longrightarrow} CH_3O-\overset{O}{\underset{O}{\overset{\|}{\underset{\|}{S}}}}-OH \overset{CH_3OH}{\longrightarrow} CH_3O-\overset{O}{\underset{O}{\overset{\|}{\underset{\|}{S}}}}-OCH_3$$

Methyl   Dimethyl
sulfate   sulfate

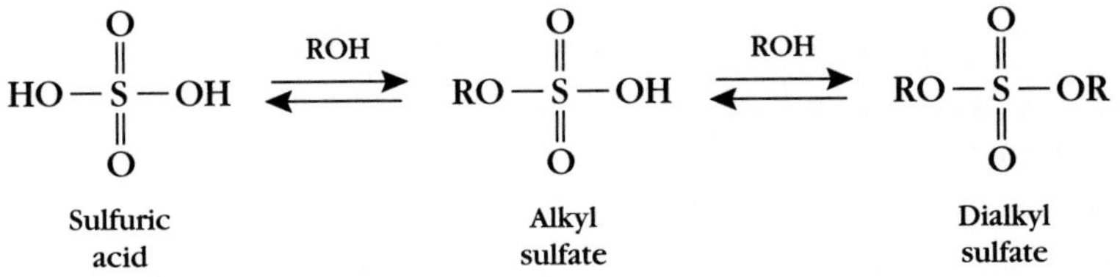

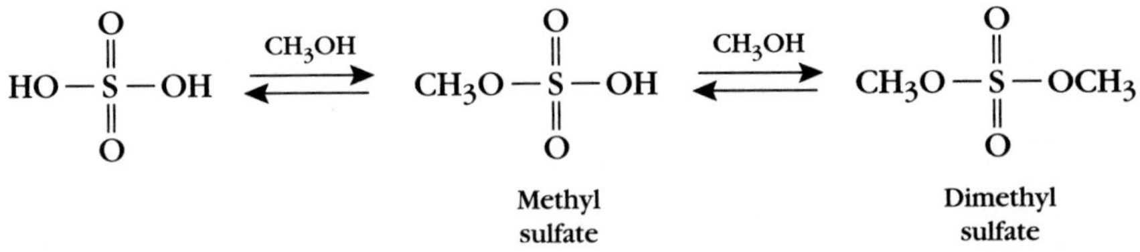

4. Haloalkane from Alcohols

SECTION 11.17

a. By Reaction with Thionyl Chloride

$$ROH \ + \ SOCl_2 \ \longrightarrow \ RCl \ + \ SO_2 \ + \ HCl$$

Alcohol   Thionyl   Chloro-
               chloride   alkane

$$(CH_3)_2CHOH \ + \ SOCl_2 \ \longrightarrow \ (CH_3)_2CHCl \ + \ SO_2 \ + \ HCl$$

2-Propanol   2-Chloropropane

b. By Reaction with PBr₃

$$3ROH \ + \ PBr_3 \ \longrightarrow \ 3RBr \ + \ P(OH)_3$$

Alcohol   Phosphorus   Bromo-   Phosphorus
          tribromide   alkane   acid

$$3(CH_3)_2CHCH_2OH \ + \ PBr_3 \ \longrightarrow \ 3(CH_3)_2CHCH_2Br \ + \ H_3PO_3$$

2-Methyl-1-propanol   1-Bromo-2-methylpropane

c. By Reaction with Hydrogen Halides

$$ROH \ + \ HX \ \longrightarrow \ RX \ + \ H_2O$$

Alcohol   Hydrogen   Haloalkane
         halide

HX = HI, HBr, or HCl

$$(CH_3)_2CHOH \ + \ HCl \ \longrightarrow \ (CH_3)_2CHCl \ + \ H_2O$$

2-Propanol   2-Chloropropane

5. Oxidation of Alcohols

SECTION 11.18

a. Aldehydes from Primary Alcohols

Oxidation of primary alcohols can be stopped at the aldehyde stage only by using special reagents like PCC or Collins' reagent.

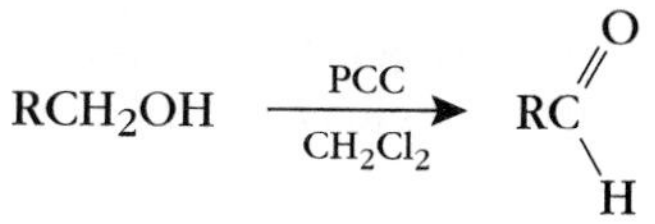

1° Alcohol   Aldehyde

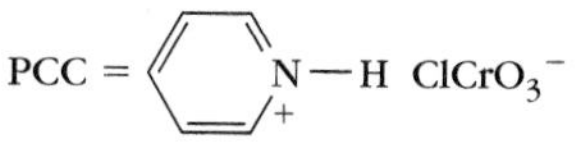

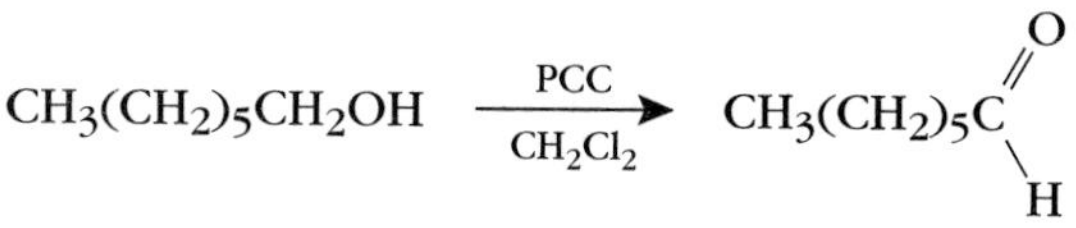

$$PCC = \text{pyridinium} \ N-H \ ClCrO_3^-$$

1-Heptanol   Heptanal

b. Carboxylic Acids from Primary Alcohols

Aldehydes are intermediates in the oxidation of primary alcohols to carboxylic acids.

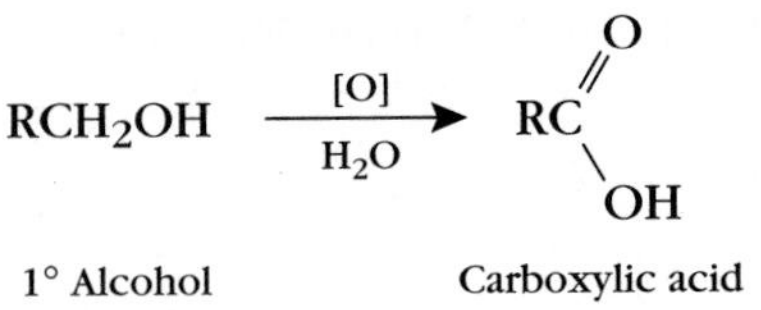

$$RCH_2OH \xrightarrow[H_2O]{[O]} RC{\overset{\displaystyle O}{\underset{OH}{\big\|}}}$$

1° Alcohol   Carboxylic acid

[O] = $KMnO_4/H_2SO_4$, CrO_3/H_2SO_4, or $Na_2Cr_2O_7/H_2SO_4$

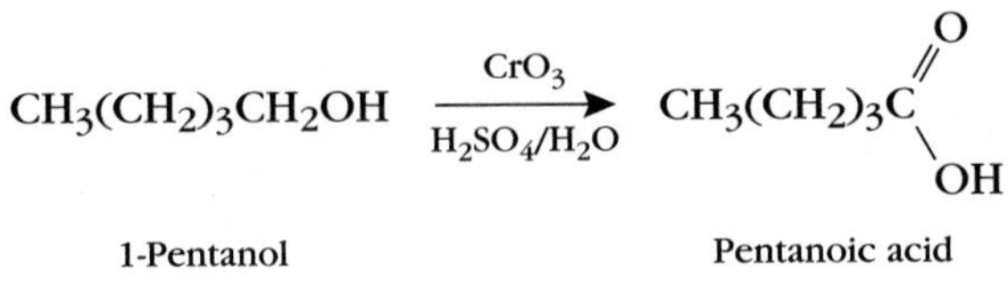

$$CH_3(CH_2)_3CH_2OH \xrightarrow[H_2SO_4/H_2O]{CrO_3} CH_3(CH_2)_3C{\overset{\displaystyle O}{\underset{OH}{\big\|}}}$$

1-Pentanol   Pentanoic acid

c. Ketones from Secondary Alcohols

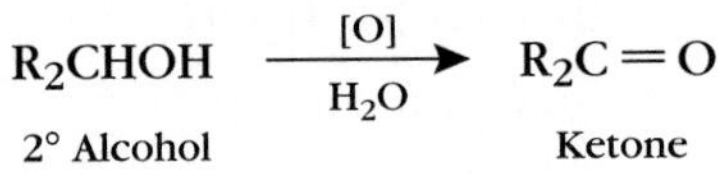

$$R_2CHOH \xrightarrow[H_2O]{[O]} R_2C{=}O$$

2° Alcohol   Ketone

[O] = $KMnO_4/H_2SO_4$, CrO_3/H_2SO_4, $Na_2Cr_2O_7/H_2SO_4$, PCC/CH_2Cl_2, or Collins' reagent

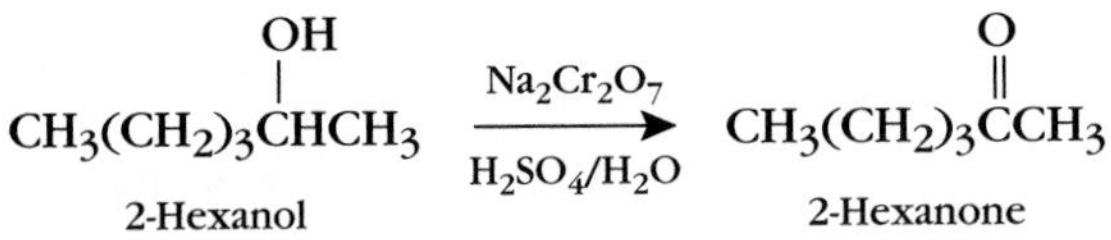

$$CH_3(CH_2)_3\overset{\displaystyle OH}{\underset{\big|}{C}}HCH_3 \xrightarrow[H_2SO_4/H_2O]{Na_2Cr_2O_7} CH_3(CH_2)_3\overset{\displaystyle O}{\underset{\big\|}{C}}CH_3$$

2-Hexanol   2-Hexanone

6. Enzyme-Catalyzed Interconversion of Ethanol and Acetaldehyde   SECTION 11.13

The mechanism of oxidation occurs by the direct transfer of the pro-*R* hydrogen atom of ethanol to the Re face of the nicotinamide ring of NAD^+. In reduction, the pro-*R* hydrogen atom of C4 of the NADH ring is transferred to the Re face of the carbonyl group of acetaldehyde.

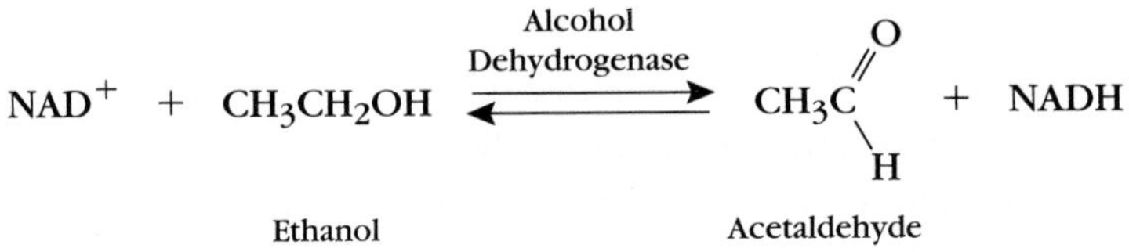

$$NAD^+ + CH_3CH_2OH \underset{}{\overset{\text{Alcohol Dehydrogenase}}{\rightleftharpoons}} CH_3C{\overset{\displaystyle O}{\underset{H}{\big\|}}} + NADH$$

Ethanol   Acetaldehyde

Additional Exercises

11.25 Define each of the following terms either in words, by a structure, or by means of a chemical equation.

(a) Carbonyl group
(b) Grignard reagent
(c) 1° Alcohol
(d) Oxidation
(e) Thiol
(f) Organolithium reagent
(g) Re face of a carbonyl group
(h) 2° Alcohol
(i) Si face of a carbonyl group
(j) Reduction
(k) 3° Alcohol
(l) Disulfide

11.26 Give the IUPAC name of each of the following structural formulas.

(a)
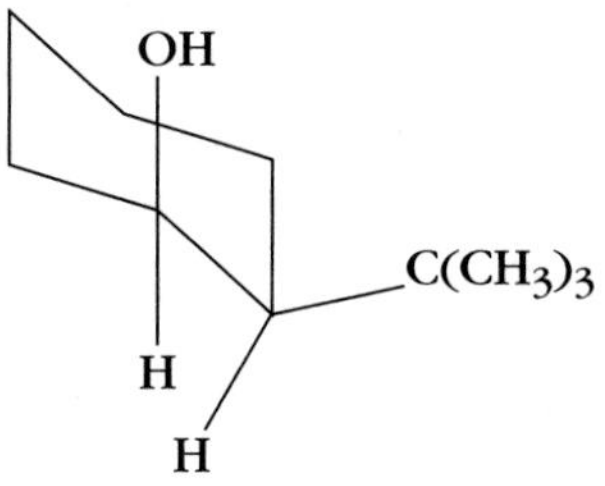

(b)
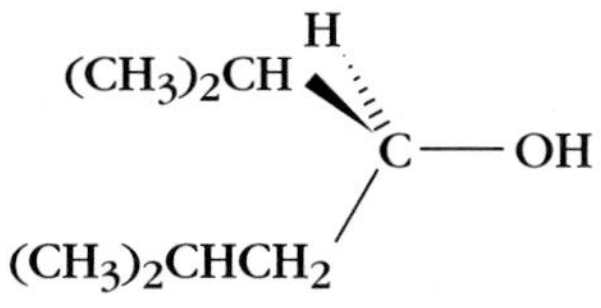

(c)
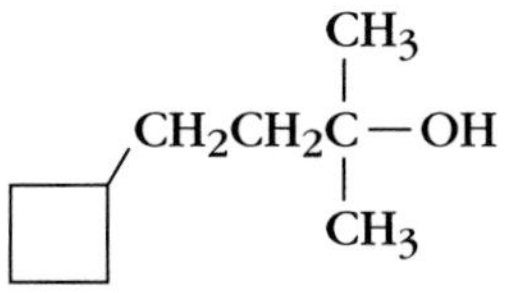

(d)
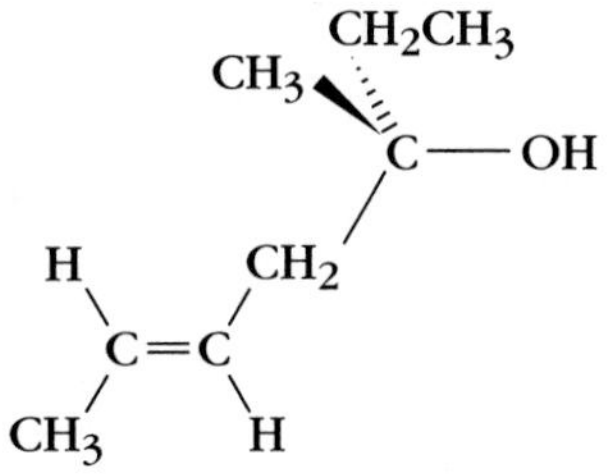

(e)
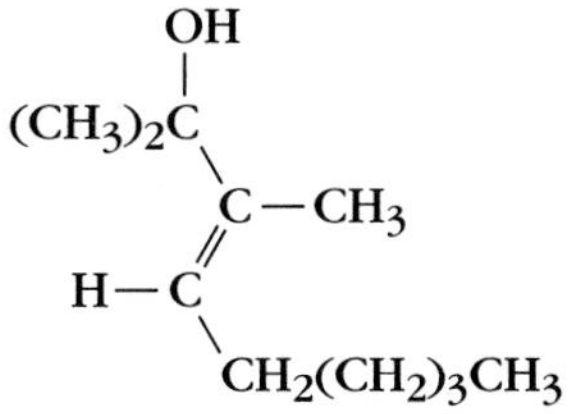

(f)
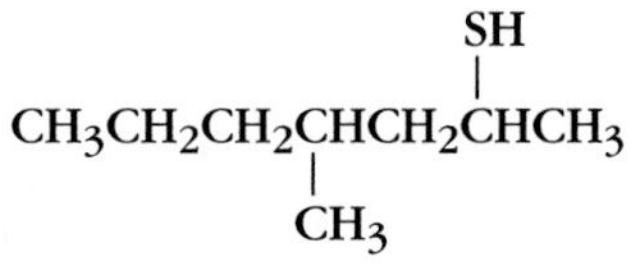

11.27 Write the structural formula corresponding to each of the following names.

(a) 4,4-Diethylcyclohexanol
(b) 2-Bromo-2-chloro-3-isopropyl-5-methyl-1-octanol
(c) *cis*-4-(2-Bromoethyl)cyclohexanol
(d) (1*R*,2*S*)-1,2-Cyclopentanediol
(e) (2*S*,4*S*)-1,2,4-Pentanetriol
(f) 1,2-Ethanedithiol

11.28 Write the structure of the major product in the reaction of methylpropene with

(a) Borane (BH_3) followed by reaction with a basic solution of hydrogen peroxide.
(b) Aqueous $Hg(OAc)_2$ followed by $NaBH_4$.

11.29 Write the structure of the major product of the reaction of cyclopentanone with each of the following reagents.

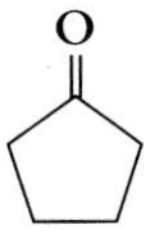

Cyclopentanone

(a) $NaBH_4/CH_3OH$
(b) $(CH_3)_3C^-Li^+/(C_2H_5)_2O$, followed by reaction with aqueous acid
(c) $(CH_3)_2CHCH_2MgBr/(C_2H_5)_2O$, followed by reaction with aqueous acid

11.30 Write the equation for the synthesis of 2-hexanol by

(a) A reduction reaction
(b) Using a Grignard reagent
(c) Using an organolithium reagent

11.31 Including all stereoisomers, there are 11 isomeric alcohols of molecular formula $C_5H_{11}OH$.

    (a)   Write the structural formula for each.

    (b)   Give each its correct IUPAC name.

    (c)   Designate each as a 1°, 2°, or 3° alcohol.

    (d)   Identify each pair of enantiomers.

    (e)   Decide which of the possible syntheses of alcohols could be used to prepare each isomer and write the complete equation.

11.32 Starting with iodoethane, any other organic compound, and any other reagent, write the equation for the synthesis of each of the following compounds.

    (a)

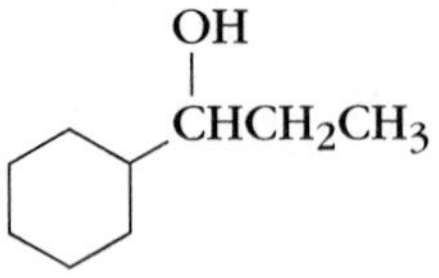

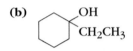

    (c)   1-Propanol

    (d)   3-Methyl-3-hexanol

    (e)   1-Deuterioethane

    (b)

11.33 Reaction of a carbonyl-containing compound, $C_7H_{14}O$, with $NaBH_4$ followed by reaction with an aqueous acid solution forms a product whose ^1H NMR spectrum is shown below. Write the structures of both the product and the starting carbonyl-containing compound.

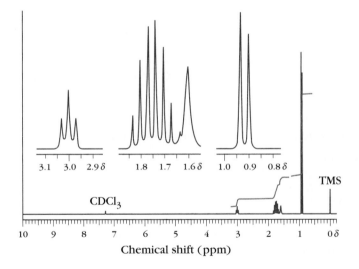

11.34 Determine the structure of each of the following alcohols from its molecular formula and ^1H NMR spectrum shown below.

   (a) C_3H_8O **(b)** $C_9H_{10}O$ **(c)** $C_5H_{10}O$

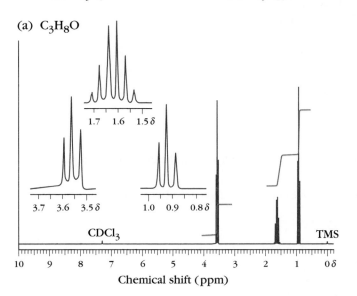

(a) C_3H_8O

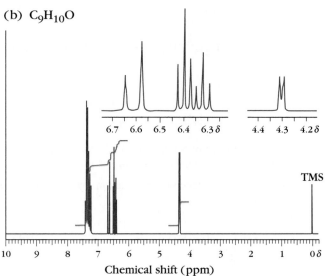

(b) $C_9H_{10}O$

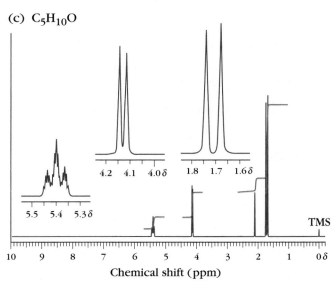

(c) $C_5H_{10}O$

11.35 How many signals are present in the ^{13}C NMR spectrum of each of the following alcohols?

   (a) 3-Hexanol

   (b) 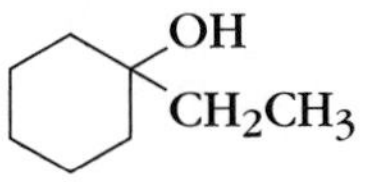 cyclohexane ring with OH and CH_2CH_3

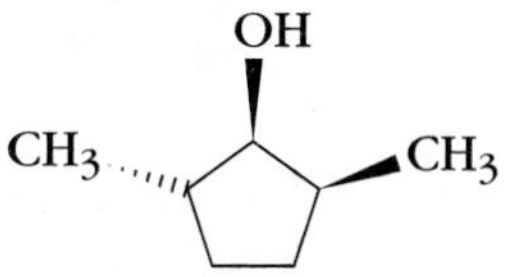

   (c) 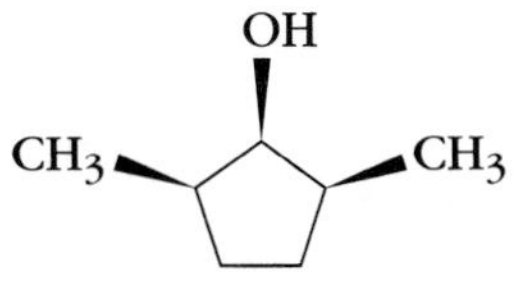 cyclopentane with OH and two CH_3 (cis)

   (d) cyclopentane with OH and two CH_3 (trans)

   (e) 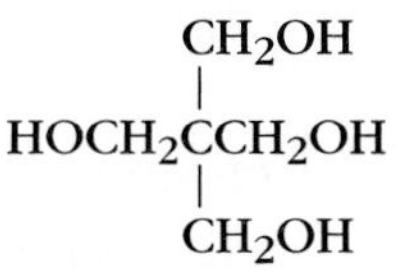 $HOCH_2CCH_2OH$ with CH_2OH up and CH_2OH down

11.36 Write the equation (or equations if more than one reaction is necessary) for the preparation of each of the following compounds, starting with the compound given and any other reagents necessary.

(a) 1-Methylcycloheptene $\longrightarrow$ 1-Methylcycloheptanol

(b) 1-Methylcycloheptene $\longrightarrow$ *trans*-2-Methylcycloheptanol

(c)
$$CH_3(CH_2)_3\overset{\overset{\displaystyle O}{\|}}{C}CH_3 \longrightarrow \text{2-Bromohexane}$$

(d)

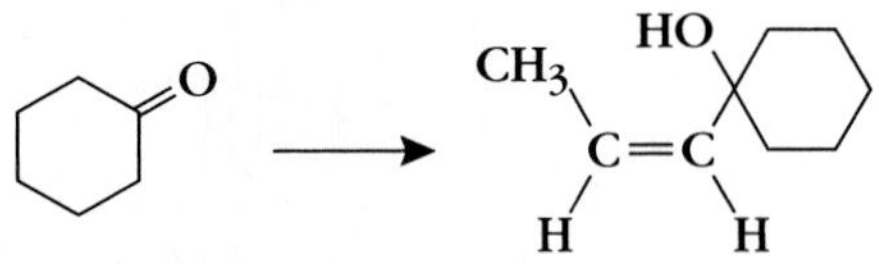

11.37 In the presence of NADH, the enzyme yeast alcohol dehydrogenase reduces aldehydes and ketones to optically active alcohols that have the (*S*) configuration.

(a) Write the structure of the alcohol formed by the reduction of each of the following carbonyl compounds by this enzyme.

(i) $CH_3(CH_2)_5\overset{\overset{\displaystyle O}{\|}}{C}CH_3$

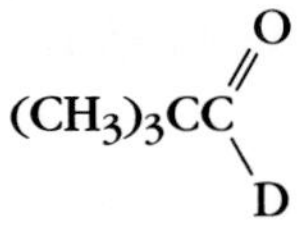

(iii) $CH_3\overset{\overset{\displaystyle O}{\|}}{C}CH_2CH_2OH$

(ii) $(CH_3)_3CC\overset{\diagup O}{\diagdown D}$

(iv) $CH_3\overset{\overset{\displaystyle O}{\|}}{C}-\overset{\overset{\displaystyle O}{\|}}{C}CH_3$

(b) To what face of the carbonyl group does the enzyme transfer a hydride ion?

11.38 Explain the following facts. A diol is formed when two moles of Grignard reagent are used per mole of ketone. When only one mole of Grignard reagent is used, the starting material is recovered.

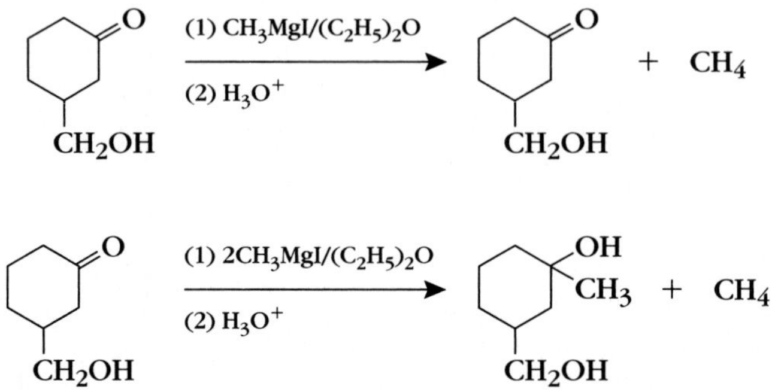

11.39 Write the equation for the reaction of 1-pentanol with each of the following reagents:

(a) NaH/pentane
(b) Hot concentrated aqueous HBr
(c) $SOCl_2$
(d) *p*-Toluenesulfonyl chloride

(e) $Na_2Cr_2O_7$/aqueous H_2SO_4
(f) PCC/dichloromethane
(g) PBr_3

11.40 Write the equation for the reaction of 2-pentanol with each of the following reagents:

(a) $NaNH_2/NH_3$
(b) $CH_3MgI/(C_2H_5)_2O$
(c) *p*-Toluenesulfonyl chloride
(d) CrO_3/aqueous H_2SO_4

(e) $KMnO_4$/aqueous H_2SO_4
(f) Na metal
(g) NaI/aqueous H_2SO_4

11.41 Write the equation for the reaction of 2-methyl-2-propanol with each of the following reagents:

(a) CH_3Li
(b) $KMnO_4$/aqueous H_2SO_4

(c) Aqueous concentrated HCl
(d) PCC/dichloromethane

11.42 Starting with 1-butanol, any solvent, and any organic or inorganic reagent, write the equation (or equations if more than one step is needed) for the synthesis of each of the following compounds:

(a) 1-Chlorobutane

(b) 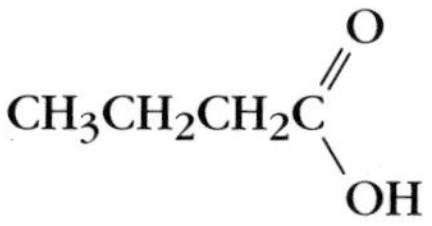 $CH_3CH_2CH_2C\overset{O}{\underset{OH}{\diagup\diagdown}}$

(c) 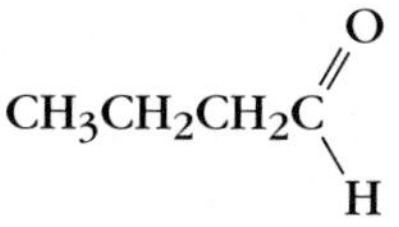 $CH_3CH_2CH_2C\overset{O}{\underset{H}{\diagup\diagdown}}$

(d) 1-Hexyne
(e) 4-Octanol
(f) Octane

11.43 Compounds labelled with isotopes such as deuterium (D) and ^{13}C at specific positions are valuable in determining the mechanisms of organic reactions and following the progress of the reactions in living organisms. Starting with $^{13}CH_3OH$ and D_2O as sources of ^{13}C and D, respectively, and any other organic or inorganic reagents needed, write equations for the synthesis of the following labelled compounds.

(a) 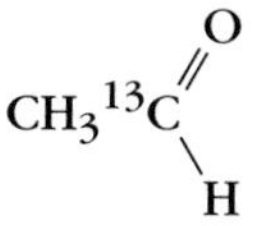 $CH_3{}^{13}C\overset{O}{\underset{H}{\diagup\diagdown}}$

(b) 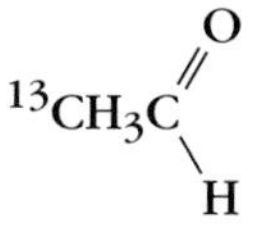 $^{13}CH_3C\overset{O}{\underset{H}{\diagup\diagdown}}$

(c) $^{13}CH_3\overset{O}{\overset{\|}{C}}CH_3$

(d) $^{13}CH_3CH_2D$

(e) $D^{13}CH_2CH_3$

(f) $(CH_3)_2CH^{13}CH_2OH$

11.44 The sex attractant for the Douglas fir tussock moth has been synthesized by the following series of reactions. Give the structures of all the intermediates and the sex attractant.

1-Heptyne $\xrightarrow[\text{NH}_3]{\text{NaNH}_2}$ $C_7H_{11}Na$ (Compound A)

Compound A + $Cl(CH_2)_3Br$ $\longrightarrow$ $C_{10}H_{17}Cl$ (Compound B)

Compound B + Mg $\longrightarrow$ (Compound C)

Compound C + $C_{10}H_{21}C\overset{O}{\underset{H}{\diagup\diagdown}}$ $\longrightarrow$ $\xrightarrow[\text{H}_2\text{O}]{\text{H}_2\text{SO}_4}$ $C_{21}H_{40}O$ (Compound D)

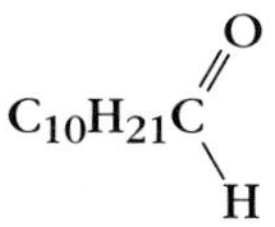

Compound D $\xrightarrow[\substack{\text{Lindlar's}\\\text{catalyst}}]{\text{H}_2}$ $C_{21}H_{42}O$ (Compound E)

Compound E $\xrightarrow{\text{CrO}_3}$ $C_{21}H_{40}O$ (Desired compound)

11.45 Explain why oxidation of either (R)- or (S)-1-deuterioethanol by PCC in CH_2Cl_2 forms a mixture of CH_3CDO and CH_3CHO, while alcohol dehydrogenase–catalyzed oxidation of (S)-1-deuterioethanol forms *only* CH_3CDO.

11.46 Write the structures of compounds F to L.

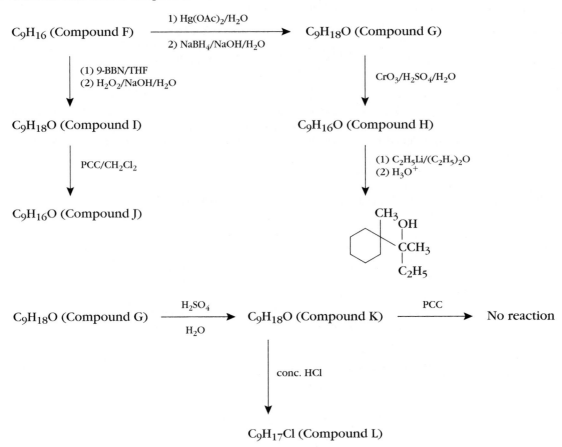

C_9H_{16} (Compound F) $\xrightarrow[\text{2) } NaBH_4/NaOH/H_2O]{\text{1) } Hg(OAc)_2/H_2O}$ $C_9H_{18}O$ (Compound G)

(1) 9-BBN/THF
(2) $H_2O_2/NaOH/H_2O$

$CrO_3/H_2SO_4/H_2O$

$C_9H_{18}O$ (Compound I)

$C_9H_{16}O$ (Compound H)

PCC/CH_2Cl_2

(1) $C_2H_5Li/(C_2H_5)_2O$
(2) H_3O^+

$C_9H_{16}O$ (Compound J)

$C_9H_{18}O$ (Compound G) $\xrightarrow[H_2O]{H_2SO_4}$ $C_9H_{18}O$ (Compound K) $\xrightarrow{PCC}$ No reaction

conc. HCl

$C_9H_{17}Cl$ (Compound L)

11.47 Match each of the structures on the left with its correct ^1H NMR data given on the right. Recall that the ^1H NMR data are given in the following way: the chemical shift; the spin-spin splitting pattern where s = singlet, d = doublet, t = triplet, q = quartet, and m = multiplet; and the relative areas of the peaks (given in parentheses).

(a) $\underset{\overset{|}{Br}}{CH_3CHCHCH_3}$
     $\overset{\underset{|}{Br}}{}$

(1) δ 2.00 t (1H); δ 3.44 t (1H)

(b) $BrCH_2CH_2CH_2CH_2Br$

(2) δ 1.88 s (3H); δ 3.87 s (1H)

(c) $\underset{\overset{|}{Br}}{(CH_3)_2CCH_2Br}$

(3) δ 1.76 d (3H); δ 2.28 q (2H)
     δ 3.56 t (2H); δ 4.30 m (1H)

(d) $\overset{\overset{Br}{|}}{CH_3CHCH_2CH_2Br}$

(4) δ 1.78 d (3H); δ 4.20 q (1H)

11.48 Determine the structure of each of the following compounds from its molecular formula, ^{1}H NMR spectrum, and spectral data given in the following figures.

    (a) $C_4H_{10}O_2$ **(b)** C_4H_6O **(c)** C_6H_{12}

(a) $C_4H_{10}O_2$ IR 3350 cm^{-1}. ^{13}C NMR δ 16.9; δ 70.9

(b) C_4H_6O IR 3295 cm^{-1}; 2120 cm^{-1}. δ 3.5; δ 51.1; δ 77.6; δ 81.8

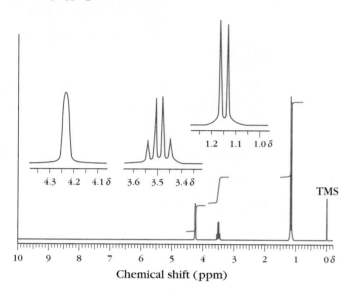

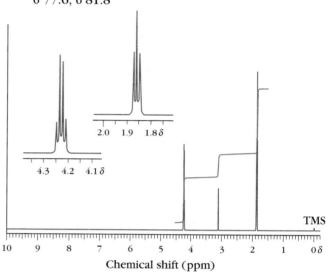

(c) C_6H_{12} IR 1640 cm^{-1}. ^{13}C NMR δ 29.2; δ 33.6; δ 108.8; δ 149.8

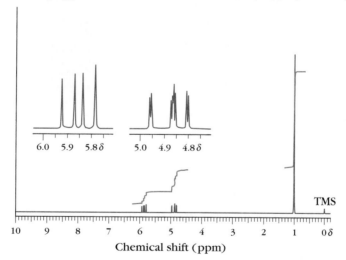

CHAPTER

12

NUCLEOPHILIC SUBSTITUTION and ELIMINATION REACTIONS

Our study of organic chemistry has been organized so far in terms of the physical and chemical properties of different functional groups such as alkenes, alkynes, and alcohols. We depart from this organization in this chapter by studying two reaction types rather than a functional group. The two types are nucleophilic substitution and elimination reactions. We make this change because compounds with many different functional groups undergo these types of reactions. Thus it's more instructive to consider these reactions as unifying themes that encompass several different functional groups. Let's begin by learning about nucleophilic substitution reactions.

12.1   Nucleophilic Substitution Reactions

Iodomethane reacts with hydroxide ion (from an aqueous solution of sodium hydroxide) to form methanol and iodide ion.

$$CH_3-I \;+\; OH^- \longrightarrow CH_3-OH \;+\; I^-$$

Iodomethane   Methanol

The iodine atom of iodomethane is substituted by a hydroxy group, so this reaction is called a substitution reaction. The substitution occurs at an sp^3 hybrid carbon atom by a hydroxide ion, which is a nucleophile. The substitution of an atom or group of atoms bonded to a carbon atom by a nucleophile is called a **nucleophilic substitution reaction.** The reaction of iodomethane and hydroxide ion, shown previously, is an example of a nucleophilic substitution reaction.

The reagents in all nucleophilic substitution reactions that occur at an sp^3 hybrid carbon atom have three features in common.

■ *The definition of a nucleophile is given in Section 7.9.*

Substitution occurs at an sp^3 hybrid carbon atom, which is the reaction site. This carbon atom is bonded to a **leaving group,** which refers to any atom or group that can be substituted for by a nucleophile. Halogen atoms are common leaving groups. Other examples include arene sulfonates, $ArSO_3-$; ammonium ions, $-NR_3^+$; and alkyl sulfonium ions, $-SR_2^+$. The nucleophile is the reagent that substitutes for the leaving group. While many nucleophiles are negatively charged, such as OH^- and CN^-, others are neutral molecules such as NH_3, H_2O, and alcohols. All nucleophiles have at least one pair of electrons to form a new bond with the sp^3 hybrid carbon atom. A number of examples of nucleophilic substitution reactions are shown in Figure 12.1.

| **Exercise 12.1** | In each of the nucleophilic substitution reactions in Figure 12.1, identify the nucleophile, the reaction site, and the leaving group. |

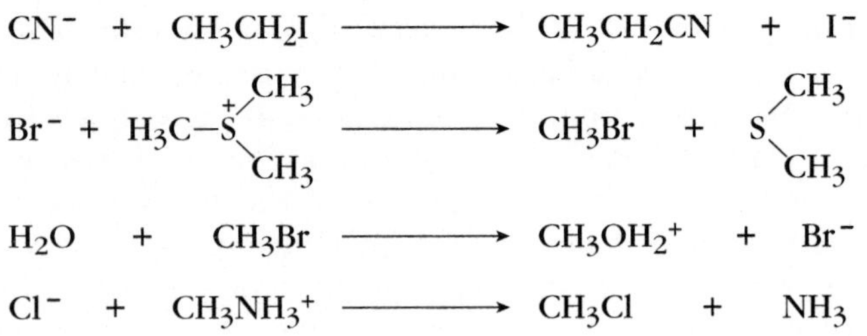

Figure 12.1
Examples of nucleophilic substitution reactions.

$$CN^- + CH_3CH_2I \longrightarrow CH_3CH_2CN + I^-$$

$$Br^- + H_3C\overset{+}{\underset{CH_3}{\overset{CH_3}{-S}}} \longrightarrow CH_3Br + S\underset{CH_3}{\overset{CH_3}{\diagup}}$$

$$H_2O + CH_3Br \longrightarrow CH_3OH_2^+ + Br^-$$

$$Cl^- + CH_3NH_3^+ \longrightarrow CH_3Cl + NH_3$$

Nucleophilic substitution reactions are used extensively in organic synthesis. The reaction of iodomethane and sodium hydroxide, for example, converts a haloalkane into an alcohol by replacing one functional group with another. Other examples of the use of nucleophilic substitution reactions in organic synthesis are given in Section 12.18.

Nucleophilic substitution reactions occur by two mechanisms. Let's begin by discussing the S_N2 mechanism.

12.2 The S_N2 Mechanism

The S_N2 mechanism is based on the rate law and stereochemistry of certain nucleophilic substitution reactions. The rate of the reaction of iodomethane and sodium hydroxide, for example, is directly proportional to the concentration of both iodomethane and hydroxide ion (from sodium hydroxide), which means that the rate law is second order overall: first order in sodium hydroxide and first order in iodomethane.

$$CH_3-I + NaOH \longrightarrow CH_3-OH + NaI$$

Iodomethane   Methanol

$$Rate = k[CH_3I][NaOH]$$

The reaction of 2-bromooctane with sodium hydroxide also follows the same rate law.

$$\underset{\text{2-Bromooctane}}{CH_3(CH_2)_5\overset{\overset{\displaystyle Br}{|}}{C}HCH_3} + NaOH \longrightarrow \underset{\text{2-Octanol}}{CH_3(CH_2)_5\overset{\overset{\displaystyle OH}{|}}{C}HCH_3} + NaBr$$

If optically active 2-bromooctane is used, the product is an optically active alcohol whose configuration is opposite to that of the starting 2-bromooctane.

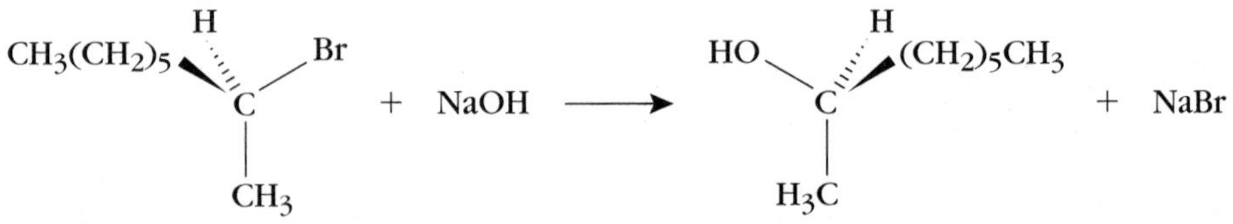

(S)-$(+)$-2-Bromooctane (R)-$(-)$-2-Octanol

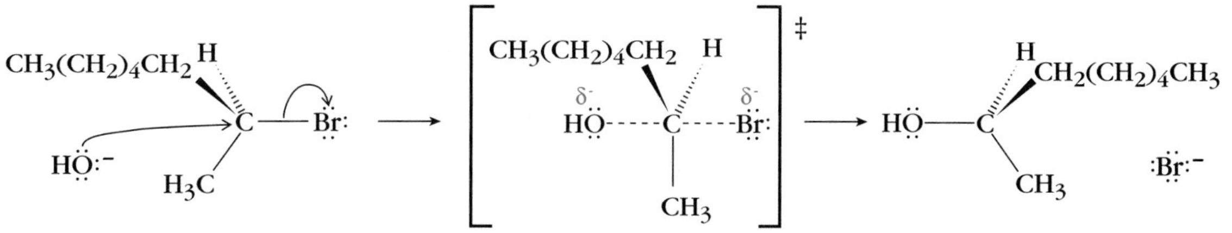

(S)-(+)-2-Bromooctane   Transition state   (R)-(+)-2-Octanol

Figure 12.2
The S$_N$2 mechanism. In this single-step mechanism, the nucleophile approaches the reaction center from the side opposite the bond to the leaving group. This backside attack results in an inversion of the configuration of the reaction site of the starting material. This inversion is similar to the inversion of an umbrella caught in a strong wind.

From this and many other examples, we conclude that nucleophilic substitution reactions whose rate laws are second order overall occur with inversion of configuration at the reaction site.

The English chemists E.D. Hughes and C.K. Ingold proposed a mechanism, called an **S$_N$2 mechanism,** for nucleophilic substitution reactions that follow a second order rate law with inversion of configuration at the reaction site. This mechanism is illustrated for the reaction of hydroxide ion and (S)-(+)-2-bromooctane in Figure 12.2. In this single-step or concerted mechanism, the nucleophile approaches the reaction site from the side opposite the leaving group. As the nucleophile approaches, its unshared electron pair begins bond formation to the carbon atom. As bond formation between the nucleophile and the carbon atom increases, bond breaking between the carbon atom and the leaving group also increases. At the transition state, both the nucleophile and the leaving group are partially bonded to the carbon atom; the oxygen atom of the hydroxide ion, the carbon atom, and the bromine atom all lie in a straight line. Once past the transition state, the leaving group is expelled with its pair of electrons and bond formation is completed between the nucleophile and the carbon atom of the reaction site.

■ *The approach of a nucleophile from the side opposite the leaving group is also called a backside approach.*

The designation S$_N$2 emphasizes the important features of this mechanism. The letter *S* stands for substitution, the letter *N* stands for nucleophilic, and *2* stands for bimolecular, meaning that two molecules or ions are present in the rate determining transition state.

The free energy–reaction path diagram for the reaction of chloromethane and iodide ion is shown in Figure 12.3. The free energy of the products in Figure 12.3 is slightly lower than the free energy of reactants because the reaction is known to be exergonic. A reaction that occurs by an S$_N$2 mechanism is an elementary reaction because it has a single transition state between reactants and products. The transition state is the point of highest free energy along the reaction path because it contains a carbon atom surrounded by five atoms or groups of atoms, two of which are partially bonded to carbon.

Nucleophilic substitution reactions that occur by an S$_N$2 mechanism are also known in living systems.

12.3   A Biological Example of a Reaction That Occurs by an S$_N$2 Mechanism

One of the many essential reactions in living organisms is the nucleophilic substitution reaction at the carbon atom of the methyl group of *S*-adenosylmethionine.

S-Adenosylmethionine

While *S*-adenosylmethionine is a complicated molecule, we can focus on its reaction site by abbreviating two of the groups bonded to the sulfur atom.

Figure 12.3
Free energy–reaction path diagram for the reaction of chloromethane and iodide ion, which occurs by an S_N2 mechanism.

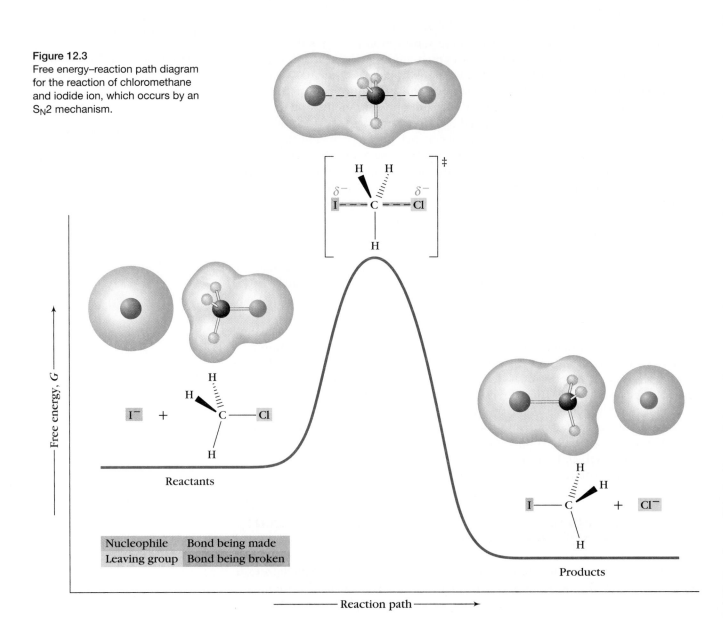

Nucleophile	Bond being made
Leaving group	Bond being broken

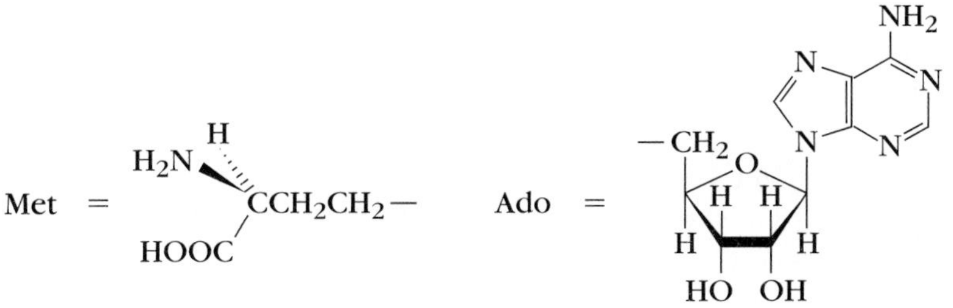

Using these abbreviations the structure of *S*-adenosylmethionine becomes

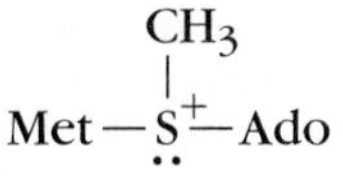

$$\text{Met} - \overset{\overset{\textstyle CH_3}{|}}{\underset{\textstyle ..}{S}} \overset{\textstyle +}{-} \text{Ado}$$

S-Adenosylmethionine is an important biological methylating agent. It transfers the methyl group on its sulfur atom to oxygen and nitrogen atoms of other molecules in a reaction that is sometimes called a methylation reaction. We see that a methylation reaction is the same as a nucleophilic substitution reaction when we write the reaction of a nucleophile with *S*-adenosylmethionine.

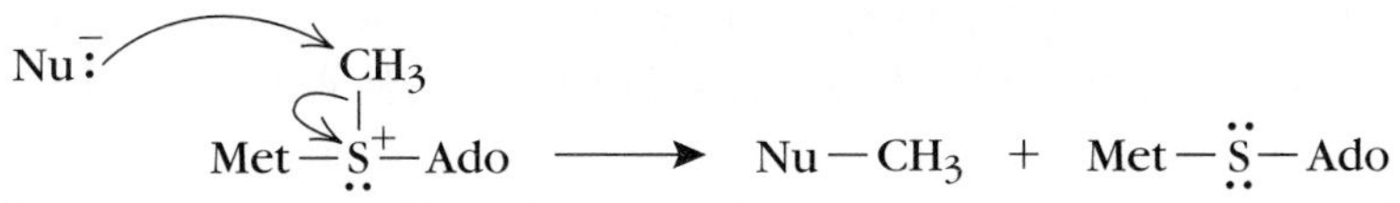

The Met-S-Ado group is the leaving group in this reaction, and an oxygen or nitrogen atom is usually the nucleophile. The enzyme *O*-methyltransferase, for instance, catalyzes the transfer of the methyl group from *S*-adenosylmethionine to the oxygen atom of protocatechuic acid.

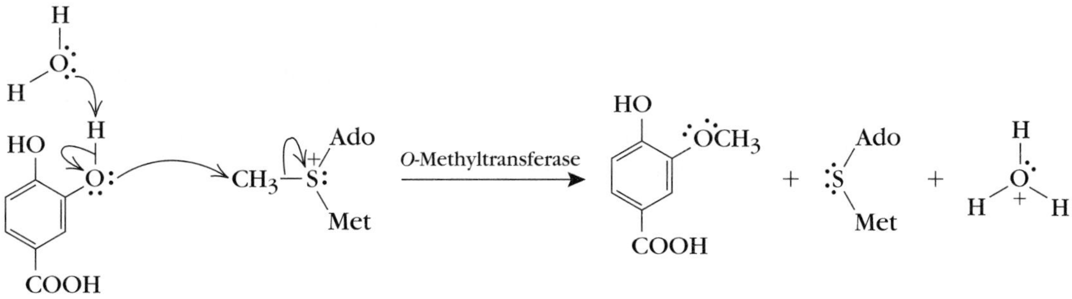

Protocatechuic
acid

Evidence that the methyl group comes from *S*-adenosylmethionine has been obtained from *in vivo* experiments in which $^{14}CH_3$-labeled *S*-adenosylmethionine was administered to living plants and the $^{14}CH_3$-labeled methylated products were isolated and identified.

The final evidence that this methyl transfer reaction is a nucleophilic substitution reaction that occurs by an S$_N$2 mechanism was provided by H.G. Floss in 1980. Floss prepared optically active *S*-adenosylmethionine by placing a deuterium (D) atom, a tritium (T) atom, and a hydrogen atom on the carbon atom of the reaction site. The *O*-methyltransferase–catalyzed methylation of protocatechuic acid using chiral *S*-adenosylmethionine occurs with inversion of configuration.

■ In vivo experiments *means experiments carried out within a living organism.*

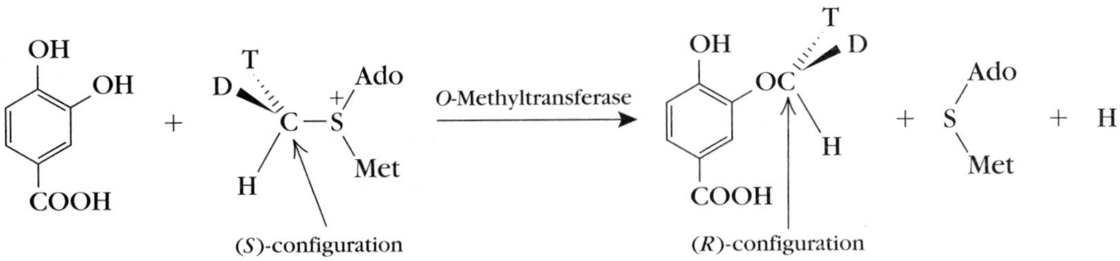

(S)-configuration   (R)-configuration

The optical activity of S-adenosylmethionine is so small that it cannot be used to follow the stereochemical course of this reaction. The configurations of the methyl carbon atoms in the product and reactant were related by a sequence of stereochemically unambiguous reactions to a compound whose absolute configuration is known.

This specific methylation reaction is only one of many examples of nucleophilic substitution reactions whose stereochemistry has been determined by the use of chiral S-adenosylmethionine. The results of this work indicate that the S_N2 mechanism is common in biological reactions.

12.4 Factors That Influence the Rates of Substitution Reactions That Occur by an S_N2 Mechanism

In this section, we will discuss four variables that influence the rate of nucleophilic substitution reactions: (1) the number of alkyl groups bonded to the reaction site, that is, the structure of the alkyl group, (2) the nature of the nucleophile, (3) the nature of the leaving group, and (4) the effect of solvents. Let's begin with the effect of alkyl group structure.

A   The Effect of Alkyl Group Structure

Relative rates are used to determine how a change in alkyl group structure affects the rate of reaction that occurs by an S_N2 mechanism. The effect of replacing a hydrogen atom of bromomethane by a methyl group, for example, can be determined by comparing the rates of the two following reactions.

$$CH_3Br \ + \ I^- \ \xrightarrow[\text{Acetone}]{25\,°C} \ CH_3I \ + \ Br^-$$

$$CH_3CH_2Br \ + \ I^- \ \xrightarrow[\text{Acetone}]{25\,°C} \ CH_3CH_2I \ + \ Br^-$$

In both reactions, the reaction temperature, the solvent (acetone), the leaving group (Br^-), and the nucleophile (I^-) are identical. The only differences are the structures of the bromoalkane and the product, which are bromomethane and iodomethane in one reaction and bromoethane and iodoethane in the other. As a result, we measure *only* the effect of changing the structure of the alkyl group from methyl to ethyl on the rate of reaction. This change is expressed as a relative rate defined as a ratio of the second order rates of reaction of bromomethane and bromoethane.

$$\text{Relative rate} = \frac{S_N2 \text{ rate of } CH_3Br}{S_N2 \text{ rate of } CH_3CH_2Br} = \frac{k_1[I^-][CH_3Br]}{k_2[I^-][CH_3CH_2Br]} = \frac{k_1[CH_3Br]}{k_2[CH_3CH_2Br]}$$

Table 12.1   Relative Rates of Reaction of Representative Bromoalkanes with Iodide Ion in Acetone at 25 °C

Bromoalkane	Name	Relative Rate	Time Needed for One Half of Starting Bromoalkane to React		
			Minutes	Hours	Days
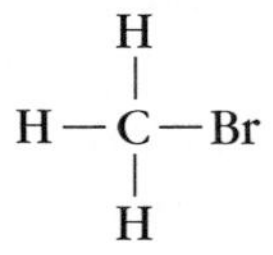 H—C(H)(H)—Br	Bromomethane	145	1.65	–	–
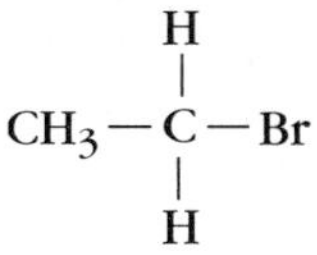 CH_3—C(H)(H)—Br	Bromoethane (1°)	1.0	240	4	–
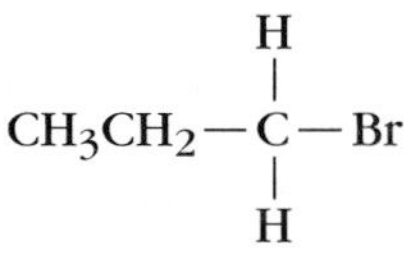 CH_3CH_2—C(H)(H)—Br	1-Bromopropane (1°)	0.82	300	5	–
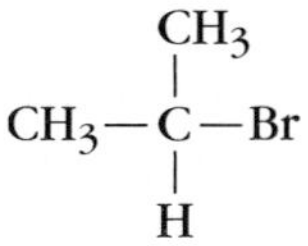 CH_3—C(CH_3)(H)—Br	2-Bromopropane (2°)	7.8×10^{-3}	3.1×10^4	513	21
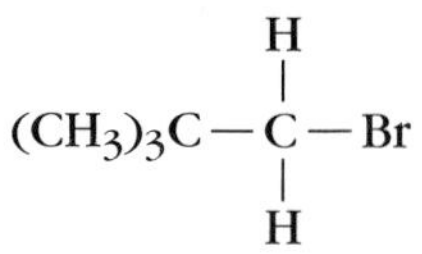 $(CH_3)_3C$—C(H)(H)—Br	1-Bromo-2,2-dimethylpropane (Neopentyl bromide)	1.3×10^{-5}	1.8×10^7	3.1×10^5	1.3×10^4 (35 years)
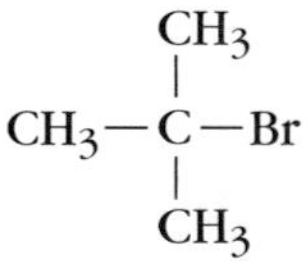 CH_3—C(CH_3)(CH_3)—Br	2-Bromo-2-methylpropane	Negligible	–	–	–

The effect of alkyl group structure on the relative rates of nucleophilic substitution reactions that occur by an S_N2 mechanism is summarized in Table 12.1.

The effect of alkyl structure on relative rates is dramatically illustrated by expressing the data in terms of hours or minutes for 50% of the bromoalkane to react, as shown in the last three columns of Table 12.1. It takes 240 minutes for 50% of bromoethane to react with iodide ion in acetone at 25 °C. On the other hand, it takes 35 years for 50% of 1-bromo-2,2-dimethylpropane (neopentyl bromide) to react under the same conditions! Clearly the structure of the alkyl group of a bromoalkane has a great effect on its rate of reaction with iodide ion.

While the data in Table 12.1 were obtained from a specific reaction, they are characteristic of the general effect of the structure of alkyl groups on the rate of reactions that occur by an S_N2 mechanism.

$$CH_3-X \ > \ RCH_2-X \ > \ R_2CH-X \ >> \ (CH_3)_3CCH_2-X \ >>> \ R_3C-X$$

Methyl   1°   2°   Neopentyl   3°

DECREASING RATE OF REACTION BY AN S_N2 MECHANISM ⟶

How do we explain these results? To answer this question, let's return to our description of the S_N2 mechanism. As shown in Figure 12.2, a nucleophile approaches the reaction site from the side opposite the leaving group. Alkyl substituents bonded to the reaction site sterically hinder this backside approach of the nucleophile, as shown by the reaction of iodide ion with bromoethane and bromomethane in Figure 12.4. Increasing steric hindrance to approach of the nucleophile increases the free energy of activation, which results in a slower rate of reaction. Consequently the rate of reaction with iodide ion is slower with bromoethane than bromomethane.

An examination of the backside approach of the iodide ion to the reaction site of other bromoalkanes, such as 2-bromopropane, 2-bromo-2-methylpropane, and 1-bromo-2,2-dimethylpropane, reveals a progressive increase in steric hindrance, which is illustrated pictorially in Figure 12.5. Replacing all three hydrogen atoms of bromomethane with methyl groups blocks backside approach so effectively that nucleophilic substitution reactions of *tert*-butyl bromide do not occur by an S_N2 mechanism. Notice that there is severe steric hindrance to the backside approach of the nucleophile to 1-bromo-2,2-dimethylpropane, *even though it is a primary bromoalkane.*

Thus increasing steric hindrance to backside approach of a nucleophile and decreasing rate of reaction are directly related in the S_N2 mechanism.

Figure 12.4
Steric hindrance to approach of iodide ion is greater for bromoethane than bromomethane. This increase in steric hindrance causes the free energy of activation for the reaction of bromoethane to be larger than that of bromomethane, with the result that bromoethane reacts slower with iodide ion.

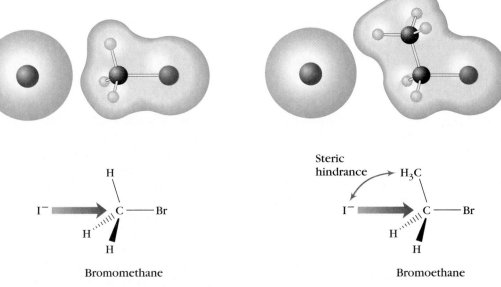

Bromomethane

A

Bromoethane

B

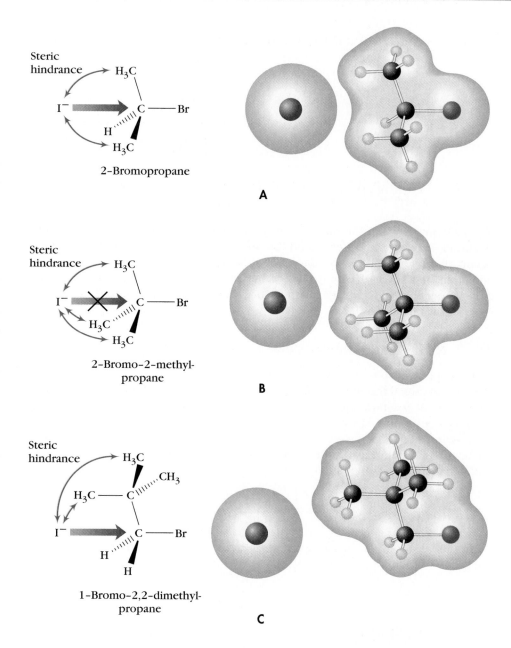

Figure 12.5
Steric hindrance for backside approach of iodide ion to 2-bromo-propane, 2-bromo-2-methylpropane, and 1-bromo-2, 2-dimethylpropane.

Steric hindrance

2-Bromopropane

A

Steric hindrance

2-Bromo-2-methyl-propane

B

Steric hindrance

1-Bromo-2,2-dimethyl-propane

C

Exercise 12.2 Predict which one of each of the following pairs of compounds will react faster with sodium iodide in acetone.

(a) $(CH_3CH_2)_3CCH_2OSO_2$—⟨benzene ring⟩—CH_3 $CH_3CH_2CH_2CH_2OSO_2$—⟨benzene ring⟩—CH_3

(b) $(CH_3)_2CHCl$ $CH_3CH_2CH_2Cl$

(c) ⟨cyclohexane⟩—Br   ⟨cyclohexane with CH_3 and Br⟩

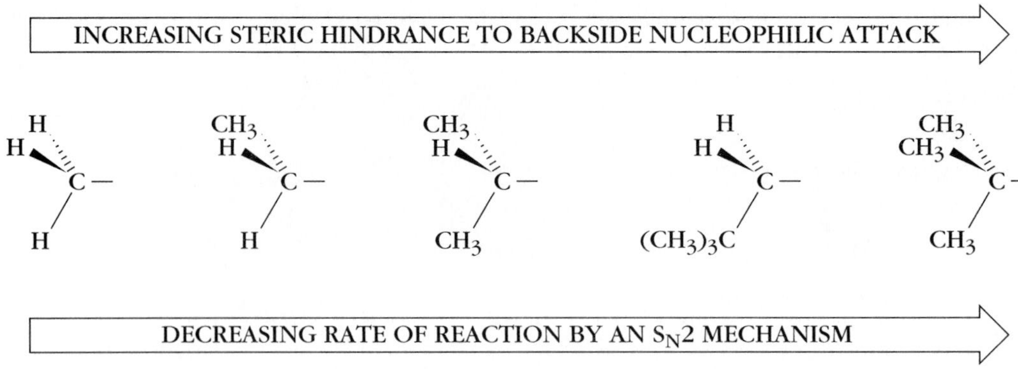

INCREASING STERIC HINDRANCE TO BACKSIDE NUCLEOPHILIC ATTACK ⟩

DECREASING RATE OF REACTION BY AN S_N2 MECHANISM ⟩

In summary, when the reaction center is a methyl carbon atom, nucleophilic substitution reactions occur readily by an S_N2 mechanism. An S_N2 mechanism occurs more slowly at a secondary carbon atom, very slowly at a neopentyl carbon atom, and not at all at a tertiary carbon atom. Nucleophilic substitution reactions do occur at tertiary carbon atoms, but they occur by another mechanism, as we will learn in Section 12.6.

B   Nature of the Nucleophile

A nucleophile displaces a leaving group from the reaction center in the S_N2 mechanism. Since this step is the rate-determining step, it follows that the rate of a substitution reaction will vary as the nucleophile is changed. **Nucleophilicity** is a measure of how fast a nucleophile displaces a leaving group from a particular reaction center. A quantitative measure of relative nucleophilicity has been obtained by measuring the rates of reaction of a variety of nucleophiles with iodomethane in methanol at 25 °C.

$$Nu:^- \quad CH_3-I \xrightarrow[25\ °C]{CH_3OH} CH_3-Nu \ + \ I^-$$

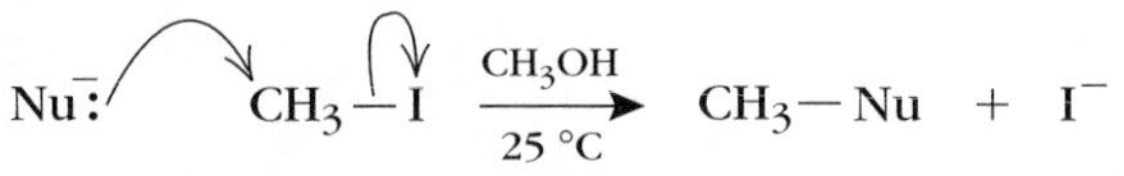

In this reaction, the temperature, the reactive site (a methyl group), the leaving group (iodide ion), and the solvent (methanol) are all kept constant and only

Table 12.2		Relative Reactivity of Nucleophiles in Their Reaction With Iodomethane in Methanol at 25 °C		
	Nucleophile	Relative Rate	Nucleophile	Relative Rate
Good nucleophiles	$C_6H_5Se^-$	2.33×10^6	$C_6H_5O^-$	24
	$C_6H_5S^-$	357,000	C_6H_5SH	21
	I^-	1,130	$(CH_3)_2S$	15
	CN^-	213	NH_3	14
	$(CH_3)_2Se$	89	Cl^-	1.0
	CH_3O^-	83	CH_3COO^-	0.9
	Br^-	27	F^-	.017	Poor
	N_3^-	26	CH_3OH	.000043	nucleophiles

the nucleophile is changed. In this way, the effect of changing only the nucleophile on the relative rate of this reaction can be measured. The results are given in Table 12.2, where the rate of reaction with chloride ion is arbitrarily set at 1.

One of the nucleophiles listed in Table 12.2 is the solvent, methanol. The reaction of iodomethane with methanol, in which methanol is both the solvent and the nucleophile, is one example of a class of nucleophilic substitution reactions called **solvolysis reactions.** Part of the solvent is incorporated in the product of solvolysis reactions, as shown by the reaction of iodomethane with methanol to form dimethyl ether.

$$CH_3I \quad + \quad CH_3OH \quad \longrightarrow \quad CH_3OCH_3 \quad + \quad HI$$

Iodomethane   Methanol   Dimethyl ether
(acts as both solvent
and nucleophile)

| Exercise 12.3 | The reaction of iodomethane and methanol occurs by an S$_N$2 mechanism. However, dimethyl ether is not the product formed in this one-step mechanism. What is the product? What must happen to it in order to form the observed product? |

The data in Table 12.2 show that the best nucleophile reacts with iodomethane in methanol at 25 °C 54 billion times faster than methanol, the poorest nucleophile! The data in Table 12.2 were obtained in methanol as solvent, so these relative nucleophilicities apply only to that solvent. The relative reactivities of nucleophiles in other solvents may be slightly different from the order listed in Table 12.2. Nevertheless, we can still use the data in Table 12.2 to note a few general trends.

- **Nucleophilicity increases with increasing negative charge.** A conjugate base is a stronger nucleophile than its acid.

	Acid	Conjugate base

Relative Rate   21   357,000

Relative Rate   0.000043   83

INCREASING NUCLEOPHILICITY ⟩

- **Nucleophilicity usually increases going down a column of the periodic table.** The nucleophilicity of either anions or neutral compounds whose nucleophilic atoms are in the same column of the periodic table generally increases going down the column.

	$:\!\ddot{F}\!:^{-}$	$:\!\ddot{C}\!l\!:^{-}$	$:\!\ddot{B}\!r\!:^{-}$	$:\!\ddot{I}\!:^{-}$
Relative Rate	0.017	1.0	27	1,130

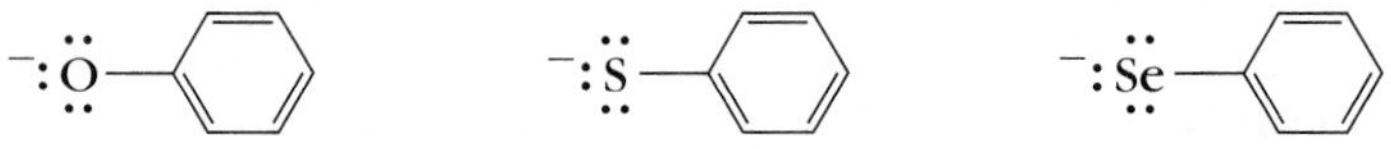

	$^{-}\!:\!\ddot{O}\!-\!\text{C}_6\text{H}_5$	$^{-}\!:\!\ddot{S}\!-\!\text{C}_6\text{H}_5$	$^{-}\!:\!\ddot{S}\!e\!-\!\text{C}_6\text{H}_5$
Relative Rate	24	357,000	2,333,000

	$\ddot{S}(CH_3)_2$	$\ddot{S}e(CH_3)_2$
Relative Rate	15	89

INCREASING NUCLEOPHILICITY ⟩

• **Nucleophilicity and basicity are generally related as long as the nucleophilic atom is the same.**

↑ INCREASING NUCLEOPHILICITY

Nucleophile	Relative Rate of Reaction With Iodomethane	Conjugate Acid	pK_a of Conjugate Acid
$CH_3\ddot{O}\!:^{-}$	83	$CH_3\ddot{O}H$	15.7
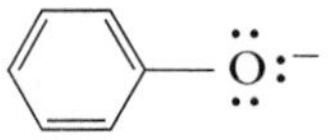 $\text{C}_6\text{H}_5\!-\!\ddot{O}\!:^{-}$	24	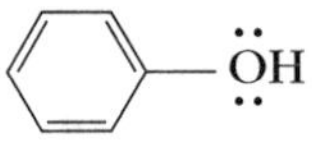 $\text{C}_6\text{H}_5\!-\!\ddot{O}H$	9.89
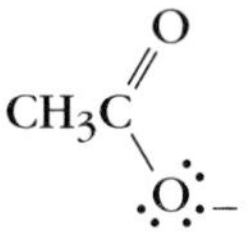 $CH_3C\!\!\begin{smallmatrix}O\\ \\ \ddot{O}\!:^{-}\end{smallmatrix}$	0.9	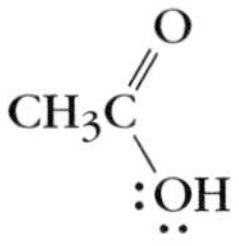 $CH_3C\!\!\begin{smallmatrix}O\\ \\ \ddot{O}H\end{smallmatrix}$	4.75

The nucleophilic atom in each case is a negatively charged oxygen atom, and in this limited series, the stronger the base, the faster the substitution reaction.

C   Nature of the Leaving Group

The effect of changing the leaving group on the rate of reactions that occur by an S_N2 mechanism is given in Table 12.3 for the following reaction:

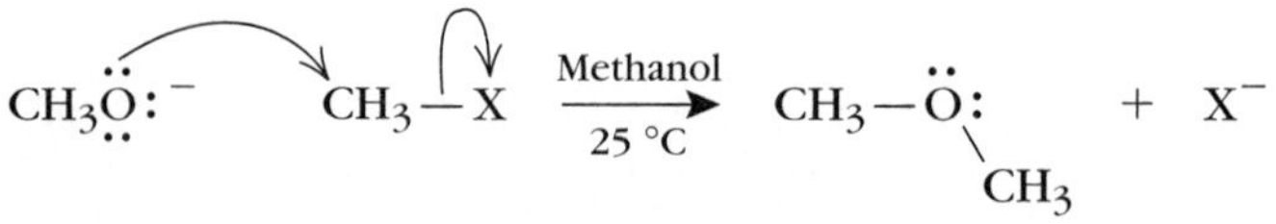

$$CH_3\ddot{O}\!:^{-} \quad CH_3\!-\!X \xrightarrow[\text{25 °C}]{\text{Methanol}} CH_3\!-\!\ddot{O}\!\!\diagdown\!\!_{CH_3} \quad + \quad X^{-}$$

Again all factors are kept constant except for the leaving groups, which are designated by X.

The data in Table 12.3 show that in general the less basic the leaving group, the more easily it leaves in a reaction that occurs by an S_N2 mechanism. Iodomethane, for example, reacts faster than methyl acetate with methoxide ion

Table 12.3	Effect of Leaving Group on the Relative Rates of Reaction of CH_3X With Methoxide Ion in Methanol at 25 °C		
Name of Leaving Group	Structure of Leaving Group	Relative Rate	pK_a of Conjugate Acid
Benzenesulfonate	$\langle\bigcirc\rangle$—SO_3^-	5.8	−6.5
Iodide	$:\ddot{I}:^-$	1.9	≈ −10
Bromide	$:\ddot{Br}:^-$	1.0	≈ −9
Chloride	$:\ddot{Cl}:^-$	0.016	≈ −7
Fluoride	$:\ddot{F}:^-$	10^{-4}	+3.2
Acetate	$CH_3\overset{O}{\underset{.\ddot{O}:^-}{C}}$	≈0	+4.8
Hydroxide	$^-:\ddot{O}H$	≈0	+15.7
Methoxide	$CH_3\ddot{O}:^-$	≈0	+15.2
Amide	$H_2\ddot{N}^-$	≈0	≈ +38

because iodide is a weaker base (pK_a of HI ≈ −10) than acetate ion (pK_a of CH_3COOH = 4.8). Consequently iodide ion is said to be a better leaving group than an acetate group. The one exception to this general correlation is the benzenesulfonate group, which is a better leaving group than predicted by its basicity.

A rough trend exists between the basicity of an atom or group and its leaving group ability because in the transition state of the S$_N$2 mechanism the leaving group begins to acquire a negative charge. The less basic atom or group of atoms should be the more easily displaced because it can more easily accommodate a negative charge.

A word of caution is necessary. The pK_a values of several of the compounds in Table 12.3 are only estimates (Section 3.3). Consequently, we can't use these data to make quantitative predictions of the exact relative rates of leaving groups in an S$_N$2 mechanism. We can conclude, however, that Cl, Br, I, and arenesulfonates are good leaving groups, while fluoride, acetate, hydroxide, alkoxide, and amide are so poor that they *are never* leaving groups in nucleophilic substitution reactions.

While the hydroxide group is a poor leaving group, it can be transformed into a good leaving group in a number of ways that we will learn in Section 12.9.

Exercise 12.4	Predict the products of the substitution reaction of one mole of $ClCH_2(CH_2)_4CH_2Br$ with the following:

(a) One mole of $C_2H_5C{\equiv}C^-Na^+$ **(b)** Two moles of $C_2H_5C{\equiv}C^-Na^+$

Exercise 12.5	List the following compounds in order of decreasing rate of nucleophilic substitution reaction with sodium methoxide in methanol.

(a) CH_3C (with =O and OCH$_3$)   (b) CH_3Br (c) CH_3OSO_2—〈 〉—CH_3 (d) CH_3I

D   Role of Solvents

A change of solvent often has a strong effect on the rate of a reaction. Changing the solvent from methanol to *N,N*-dimethylformamide (DMF), for instance, increases the rate of the reaction of 1-bromobutane and azide ion by a factor of 2700.

$$CH_3CH_2CH_2CH_2Br \;+\; N_3^{-} \longrightarrow CH_3CH_2CH_2CH_2N_3 \;+\; Br^{-}$$

1-Bromobutane   Azide ion

Solvent	Relative Rate
CH_3OH Methanol	1.0
N,N-Dimethylformamide (H–C(=O)–N(CH$_3$)$_2$)	2,700

■ Intermolecular interactions are discussed in Section 1.16.

Changing solvents affects the rates of reactions because of the different abilities of solvents to solvate reagents and transition states. **Solvation** refers to specific interactions between solvent molecules and dissolved reagents and/or transition states. These interactions are hydrogen bonding, dipole-dipole, and ion-dipole interactions.

To understand the effect of changing the solvent from methanol to *N,N*-dimethylformamide on the reaction of 1-bromobutane and azide ion, we must examine how this change affects the stabilities of both the starting materials and the rate-determining transition state. In the polar protic solvent methanol, hydrogen bonding between methanol molecules and azide ions strongly solvates and stabilizes the azide ions. In contrast, anions like azide ion and the halide ions are not solvated by polar aprotic solvents like *N,N*-dimethylformamide. Because methanol stabilizes one of the reactants and the transition state is stabilized to about the same extent by both solvents, the free energy of activation of the reaction in *N,N*-dimethylformamide is smaller than that in methanol, which results in an acceleration of the rate of reaction in *N,N*-dimethylformamide compared to methanol. The free energy-reaction path diagrams in the two solvents are shown in Figure 12.6.

On the other hand, changing the solvent from methanol, one polar protic sol-

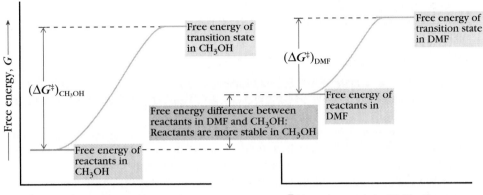

A Reaction path in CH_3OH solvent   **B** Reaction path in DMF solvent

Figure 12.6
Free energy–reaction path diagrams for the reaction of 1-bromobutane and azide ion in the solvents methanol (**A**) and N,N-dimethylformamide (DMF) (**B**). Both solvents stabilize the transition state to about the same extent. Because DMF is a poor solvent for anions, the free energy of the reactants is higher in DMF than in methanol. As a result the free energy of activation in DMF is smaller and the rate is faster than in methanol.

vent, to water, another polar protic solvent, increases the rate of reaction of 1-bromobutane and azide ion by only a factor of seven. The solvation of the reactants and transition states is about the same in water and methanol because they are both polar protic solvents. Consequently there is little difference in their free energies of activation and the two rates are very similar.

In summary, when the nucleophile in a nucleophilic substitution reaction that occurs by an S_N2 mechanism is an anion, its rate of reaction will be accelerated by changing from a polar protic to a polar aprotic solvent. Changing from one polar protic solvent to another has only a small effect on the rate.

Exercise 12.6   Solvent A solvates the transition state of a reaction that occurs by an S_N2 mechanism more than solvent B does. Both solvents solvate the reactants to about the same extent. Would the reaction occur faster in solvent A or B? Draw two free energy-reaction path diagrams, one for solvent A, the other for B, to explain your answer.

E   Summary of Factors That Affect the S_N2 Mechanism

The rates of reactions that occur by an S_N2 mechanism are influenced by the following factors:

• Increasing the branching near the reactive site causes an increase in steric hindrance and a corresponding decrease in the rate according to the following order:

$$CH_3X > CH_3CH_2X > (CH_3)_2CHX \gg (CH_3)_3CCH_2X \ggg (CH_3)_3CX$$

• Rates increase with increasing nucleophilicity. In general, nucleophilicity increases in going down a column of the periodic table and with increasing negative charge; that is, when comparing an acid-base pair, the conjugate base is a stronger nucleophile.

• Decreasing the base strength of the leaving group generally increases the rate.

• For reactions in which the nucleophile is an anion, changing from a polar protic solvent to a polar aprotic one greatly increases the rate of reaction.

It is erroneous to think that nucleophilic substitution reactions do not occur at tertiary carbon atoms. They do, but these reactions occur by a different mechanism.

12.5 Rate Law and Stereochemistry of Nucleophilic Substitution Reactions at Tertiary Carbon Atoms

We have learned that nucleophilic substitution reactions occur by an S_N2 mechanism when the reaction site is a methyl or secondary carbon atom. The S_N2 mechanism is very sensitive to steric hindrance, and as a result, when the reactive site is a tertiary carbon atom, reaction does not occur by this mechanism. Yet if the proper conditions are chosen, a tertiary haloalkane will readily undergo nucleophilic substitution reactions. For example, 2-bromo-2-methylpropane reacts with water in acetone as a solvent to form 2-methyl-2-propanol.

2-Bromo-2-methylpropane   2-Methyl-2-propanol
(*tert*-butyl bromide)   (*tert*-butyl alcohol)

This reaction is only one example of a number of nucleophilic substitution reactions that occur with neutral or nonbasic nucleophiles at tertiary carbon atoms. Under these conditions, the characteristics of these nucleophilic substitution reactions differ from those that react by an S_N2 mechanism. The first difference is the rate law.

Nucleophilic substitution reactions that occur by an S_N2 mechanism follow a second order rate law: first order in both nucleophile and substrate (Section 12.2). In contrast, the reaction of 2-bromo-2-methylpropane with water in acetone depends only on the concentration of 2-bromo-2-methylpropane and is independent of the concentration of water. In other words, the reaction follows a **first order rate law;** the reaction is first order only in 2-bromo-2-methylpropane.

$$\text{Rate} = k\left[(CH_3)_3CBr\right]$$

A second difference is the stereochemistry of the two reactions. Inversion of configuration at the reaction site is a consequence of reactions that occur by an S_N2 mechanism. In contrast, the products of the nucleophilic substitution reactions with optically active tertiary and some optically active secondary alkyl derivatives are a mixture of enantiomers. For example, the reaction of (*S*)-3-bromo-3-methylhexane with water in acetone as solvent forms both enantiomers of 3-methyl-3-hexanol as product.

(*S*)-3-Bromo-   (*S*)-3-Methyl-   (*R*)-3-Methyl-
3-methylhexane   3-hexanol   3-hexanol

Mixture of enantiomers

Since the S_N2 mechanism does not explain the experimental results of nucleophilic substitution reactions at tertiary carbon atoms, a new mechanism is needed to explain these observations.

12.6 The S_N1 Mechanism

Nucleophilic substitution reactions at sp^3 hybrid carbon atoms occur by two mechanisms. The reaction occurs by an S_N2 mechanism when the reaction site is a methyl, a primary, or a secondary carbon atom, as we discussed earlier. The S_N1 mechanism was proposed to explain the experimental observations obtained for nucleophilic substitution reactions at tertiary carbon atoms. This two-step mechanism is shown in Figure 12.7 for the reaction of (S)-3-bromo-3-methylhexane with a nucleophile.

Step 1 of the S_N1 mechanism is the spontaneous dissociation or ionization of 3-bromo-3-methylhexane to a tertiary carbocation intermediate. This step is slow and rate-determining. Recall that carbocation intermediates are also formed in acid-catalyzed hydrations of alkenes and additions of Brønsted-Lowry acids to alkenes, discussed in Chapter 7. Just like the carbocations formed in addition reactions to alkenes, the carbocation formed in the S_N1 mechanism reacts rapidly with a nucleophile to form the product in Step 2 of the mechanism. Recall from Section 7.17 that an initially formed carbocation will rearrange, if possible, to a

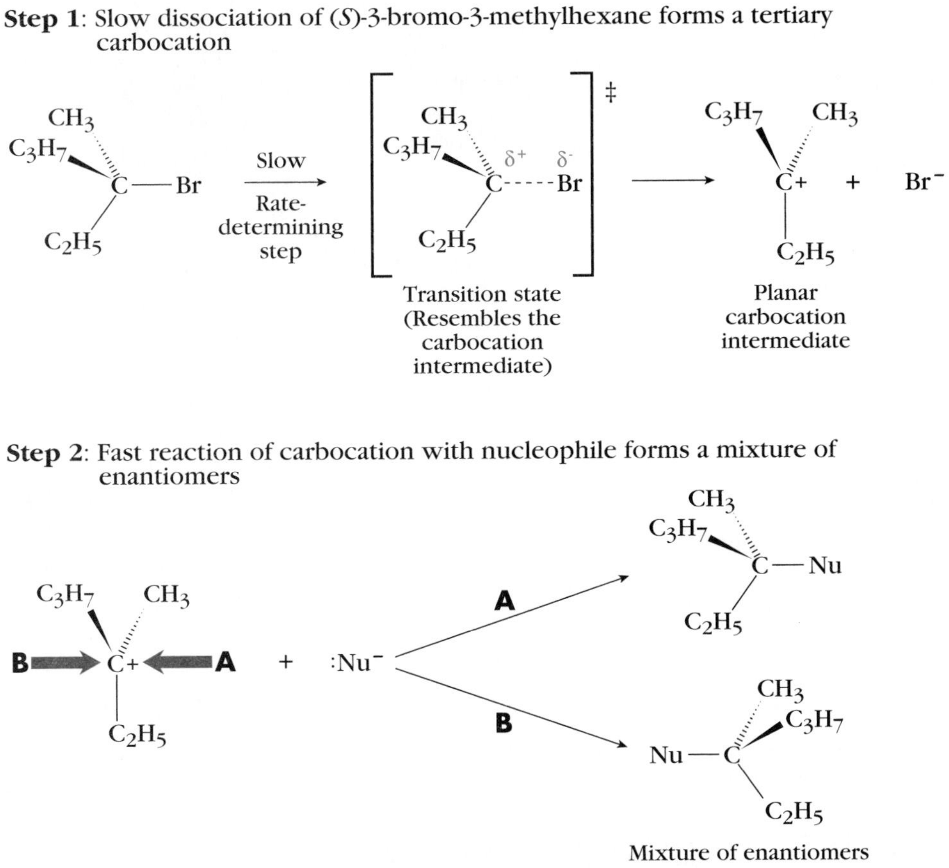

Step 1: Slow dissociation of (S)-3-bromo-3-methylhexane forms a tertiary carbocation

Transition state
(Resembles the carbocation intermediate)

Planar carbocation intermediate

Figure 12.7
The two-step S_N1 mechanism. Step 1 is slow and is the rate-determining step. In Step 1, the carbon-bromine bond is broken to form a tertiary carbocation. Step 2 is the reaction of the carbocation with a nucleophile. The nucleophile reacts with either side of the planar carbocation to form a mixture of enantiomers.

Step 2: Fast reaction of carbocation with nucleophile forms a mixture of enantiomers

Mixture of enantiomers

more stable carbocation. Such rearrangements sometimes take place in nucleophilic substitution reactions that occur by S_N1 mechanisms.

The structure of the carbocation intermediate in this mechanism explains why the product is a mixture of both enantiomers. As discussed in Section 7.10, a carbocation is planar and approach of the nucleophile occurs from either face, with the result that a mixture of the two possible enantiomers is formed.

According to the S_N1 mechanism, the rate of the reaction is determined by the dissociation of the haloalkane into a carbocation, an event that gives this mechanism its name, S_N1. The letters S and N, as before, stand for substitution and nucleophilic but 1 stands for unimolecular; it is unimolecular because the rate-determining transition state contains only one reactant, the bromoalkane.

Figure 12.8
Free energy–reaction path diagram for the reaction of *tert*-butyl chloride and iodide ion, which occurs by an S_N1 mechanism. Step 1 is rate-determining and its free energy of activation, $\Delta G^{\ddagger}$, determines the rate of reaction.

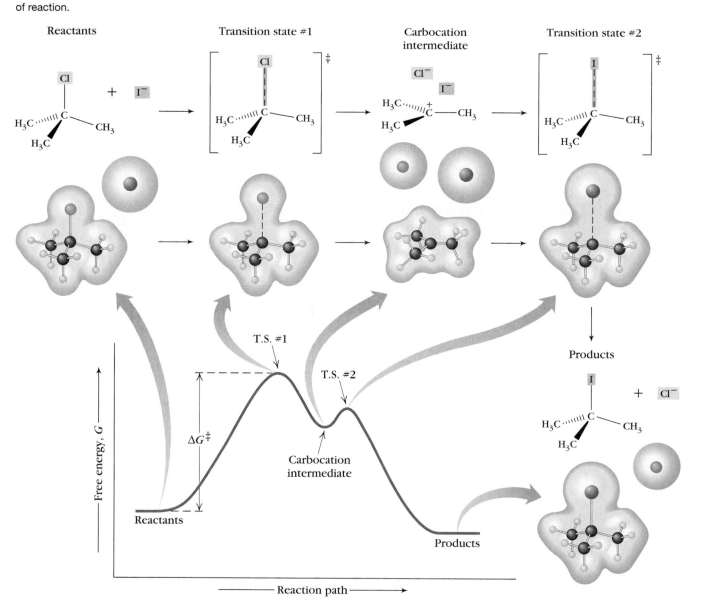

Nucleophile   Leaving group   Bond being made   Bond being broken

The rate of the reaction is independent of the nucleophile concentration because the nucleophile plays no role in the slow, rate-determining first step of the mechanism. The rate depends only on the concentration of 3-bromo-3-methylhexane, and therefore the concentration of nucleophile does not appear in the rate law.

The free energy–reaction path diagram for an S_N1 mechanism is shown in Figure 12.8. The diagram in Figure 12.8 is a combination of the free energy–reaction path diagrams of Step 1 and Step 2 of the S_N1 mechanism. The diagrams for Steps 1 and 2 are joined so that the product of Step 1, the carbocation, is the reactant of Step 2. The reaction path passes through two transition states. Step 1 is the ionization step, and it is strongly endergonic with a large free energy of activation that determines the rate of reaction. Step 2 is the reaction of the carbocation with a nucleophile, and it is strongly exergonic with a small free energy of activation. Consequently, a nucleophile reacts with the carbocation almost as soon as the cation is formed.

12.7 Factors That Influence the Rates of Substitution Reactions That Occur by an S_N1 Mechanism

In this section we will examine the effect of changing the structure of the alkyl group, and changing the nucleophile and solvent, on the rates of reactions that occur by an S_N1 mechanism.

The effect of alkyl group structure on the first order rate of reaction of several bromoalkanes with water is shown in Table 12.4. These data show that reaction at a tertiary reaction site is faster than the rate of the corresponding reaction at a primary or secondary carbon atom. The rates of reactions that occur by an S_N1 mechanism are in the same order as carbocation stability.

Table 12.4	Relative Rate of the Reaction of Some Bromoalkanes With Water in Acetone as Solvent

$$RBr + H_2O \longrightarrow ROH + HBr$$

Bromoalkane	Type	Relative Rate
CH_3Br	Methyl	≈ 0
CH_3CH_2Br	1°	1.00
$(CH_3)_2CHBr$	2°	1.39
$(CH_3)_3CBr$	3°	3,530

INCREASING RATE OF REACTION

3° >> 2° > 1° >> Methyl

INCREASING CARBOCATION STABILITY

The more stable the carbocation formed by a compound, the faster its reaction by an S_N1 mechanism. This explains the order of reactivity, $3° \gg 2° > 1° \gg$ methyl. Thus we conclude that the more likely a compound is to form a carbocation, the more likely it is to undergo nucleophilic substitution reactions by an S_N1 mechanism.

Changing the nucleophile has a different effect on the rates of reactions that occur by an S_N1 mechanism than on those that react by means of an S_N2 mechanism. Recall from Section 12.4, B, that changing the nucleophile in an S_N2 mechanism strongly affects the rate of a reaction. In contrast, the rates of reactions that occur by an S_N1 mechanism are not greatly affected because the nucleophile does not take part in the rate-determining step of the mechanism.

Exercise 12.7   Write the mechanism for the following solvolysis reaction. Be sure to include the stereochemistry of the product.

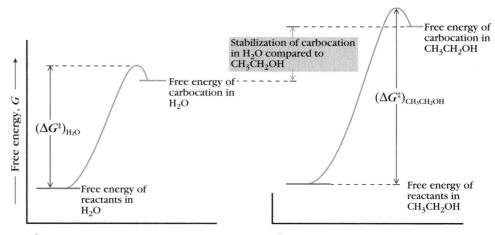

Generally reactions that occur by an S_N1 mechanism occur faster in more polar solvents. The rate of reaction of 2-chloro-2-methylpropane, for example, is about 3 million times faster in water, a more polar solvent, than in ethanol. Polar solvents are particularly good at solvating and stabilizing charged species. As a result, polar solvents stabilize the carbocation intermediate and the transition state leading to it more than they do the neutral reactant molecules. The more polar the solvent, the more it stabilizes the rate-determining transition state compared to the reactants, which causes a decrease in the free energy of activation and an acceleration in the rate. The major effect of changing solvent polarity on the rate of a reaction that occurs by an S_N1 mechanism, therefore, is to change the free energy of the rate-determining transition state, as illustrated in Figure 12.9.

■ *Polarity of solvents is discussed in Section 3.15.*

Figure 12.9
The free energy–reaction path diagrams for the reaction of *tert*-butyl chloride in **(A)** water and **(B)** ethanol. Solvation of the carbocation and the transition state leading to its formation is greater in water than in ethanol. As a result, the free energy of activation is lower in water, leading to a faster rate of reaction.

With so many variables, what determines whether a particular compound undergoes nucleophilic substitution reactions by an S$_N$1 or S$_N$2 mechanism?

12.8   What Determines Whether a Compound Reacts by an S$_N$1 or S$_N$2 Mechanism?

We now know the basic differences between S$_N$1 and S$_N$2 mechanisms. But why does a particular compound react by an S$_N$1 mechanism rather than an S$_N$2 mechanism or vice versa? The major factor that influences the choice of mechanism is the structure of the alkyl group. The alkyl group influences the mechanism by affecting the free energy of activation of the two mechanisms.

Steric hindrance to backside approach of a nucleophile is particularly severe when the reaction site is a tertiary carbon atom. This causes the free energy of activation for the reaction by an S$_N$2 mechanism at a tertiary reaction site to be much higher than the free energy of activation for formation of a relatively stable tertiary carbocation. *Consequently nucleophilic substitution reactions at a tertiary reactive site occur **exclusively** by an S$_N$1 mechanism.* This is illustrated by the free energy–reaction path diagram in Figure 12.10.

When the reaction site is a methyl carbon atom, the situation is reversed. The lack of serious steric hindrance at a methyl carbon atom results in a lower free energy of activation for reaction by an S$_N$2 mechanism than for formation of the relatively high energy methyl cation by an S$_N$1 mechanism. *Consequently nucleophilic substitution reactions at a methyl or primary carbon atom occur **exclusively** by an S$_N$2 mechanism.* The relative free energies of activation are illustrated in Figure 12.11.

When the reaction site is a secondary carbon atom, attempting to compare the relative free energies of activation of the two mechanisms becomes more difficult. There are two major factors working against each other. On the one hand, as we go from a primary to a secondary reaction site, steric hindrance to the S$_N$2

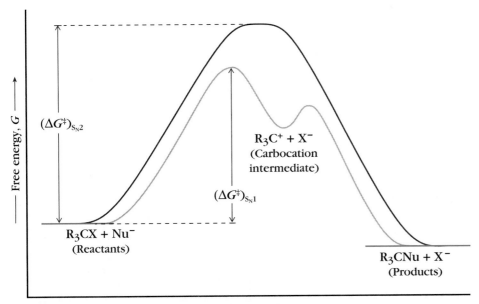

Figure 12.10
A comparison of the free energy–reaction path diagrams for nucleophilic substitution reactions that occur at a tertiary reactive site by S$_N$1 and S$_N$2 mechanisms. Reaction occurs by an S$_N$1 mechanism because it has the lower free energy of activation.

Figure 12.11
A comparison of the free energy–reaction path diagrams for nucleophilic substitution reactions that occur at a methyl carbon atom by S_N1 and S_N2 mechanisms. The reaction occurs by an S_N2 mechanism because it has a much lower free energy of activation. In fact, there is no evidence that a methyl cation is ever formed in a nucleophilic substitution reaction.

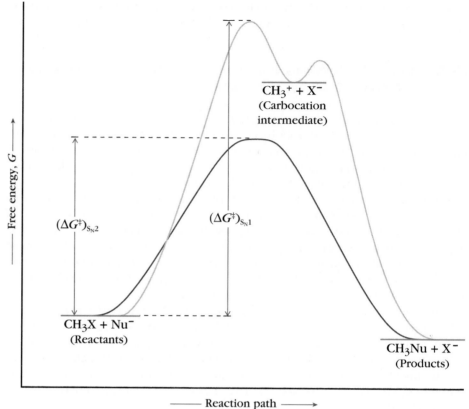

Exercise 12.8   Give a mechanistic explanation for the following observed facts:

(a)

Rate = k [NaCN] [Alkyl chloride]

(b)

Formic acid

Rate = k [Alkyl tosylate]

mechanism increases, resulting in an increase in the activation energy. This accounts for the decrease in the rate of reaction of secondary compounds relative to primary ones in nucleophilic substitution reactions that occur by an S_N2 mechanism (Table 12.1). On the other hand, the activation energy for formation of a secondary carbocation by an S_N1 mechanism is less than that for formation of a primary carbocation (Section 7.10). This accounts for the small increase in the first order rate of reaction of secondary compounds relative to primary in nucleophilic substitution reactions that occur by an S_N1 mechanism (Table 12.4).

In general, for secondary compounds, substitution by an S_N1 mechanism is favored in a reaction in which the leaving group is very good, the nucleophile is particularly weak, and the solvent is very polar. These factors favor formation of a carbocation over backside approach of the nucleophile. Substitution by an S_N2 mechanism is favored in a reaction in which the secondary carbon is bonded to a moderately good leaving group, the nucleophile is very good, and the reaction is carried out in a polar aprotic solvent. These factors favor backside approach over ionization to a carbocation. Consequently no general conclusion about the mechanism, S_N1 or S_N2, can be made for substitution reactions at secondary carbon atoms because each case must be evaluated separately.

Let's apply these conclusions to a reaction that we have already studied, the formation of haloalkanes from alcohols (Section 11.17).

12.9   The Mechanism of the Reactions of Alcohols and Hydrogen Halides

Alcohols react with hydrogen halides to form haloalkanes, as we discussed in Section 11.17. 1-Butanol, for example, reacts with hydrogen bromide to form 1-bromobutane.

$$CH_3CH_2CH_2CH_2OH \ + \ HBr \ \longrightarrow \ CH_3CH_2CH_2CH_2Br \ + \ HOH$$

  1-Butanol   1-Bromobutane

The mechanism of this nucleophilic substitution reaction, however, is a little different from the S_N1 and S_N2 mechanisms discussed in previous sections because it involves an additional step. This additional step is protonation of the hydroxy group, which converts the hydroxy group into a good leaving group as shown in Figure 12.12.

Step 1: Protonation of the oxygen atom of the hydroxy group occurs in an acid-base reaction between the alcohol and HBr

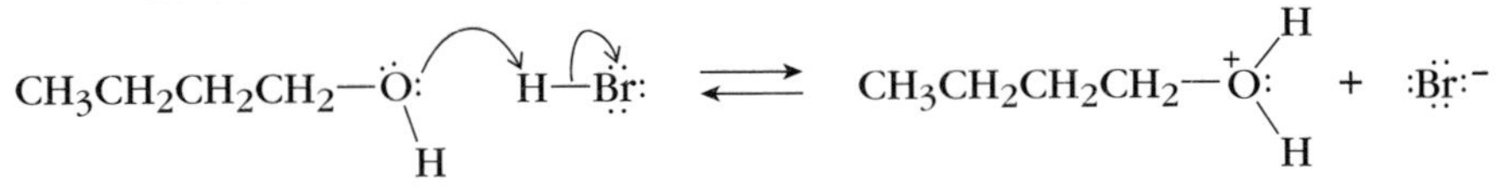

  1-Butanol   Conjugate acid of 1-butanol

Step 2: Bromide ion displaces a water molecule from the reaction site by an S_N2 mechanism

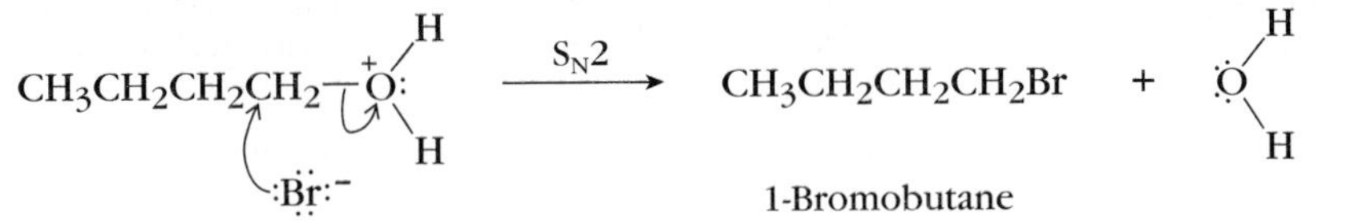

                                    1-Bromobutane

Figure 12.12
The mechanism of the reaction of 1-butanol and HBr. Step 1 converts the hydroxy group into a good leaving group, while Step 2 is an S_N2 mechanism in which the bromide ion displaces a stable water molecule.

This reaction occurs because the acid protonates the hydroxy group, thereby converting it into an $-OH_2^+$ group. The $-OH_2^+$ group is a good leaving group because it leaves as a stable water molecule. The second step is either an S_N1 or S_N2 mechanism depending on the structure of the alcohol. In the absence of acid, no reaction occurs between bromide ion and alcohols.

Exercise 12.9   Write the mechanism of the reaction of *tert*-butyl alcohol and HCl to form *tert*-butyl chloride and water.

Elimination reactions also occur in many nucleophilic substitution reactions.

12.10 Elimination Reactions

A reaction in which a π bond is formed by the loss of atoms or groups of atoms from adjacent atoms is called an **elimination reaction.** In this chapter, we will be concerned mostly with **dehydrohalogenation reactions,** which are the elimination of a hydrogen and a halogen atom from adjacent carbon atoms to form a carbon-carbon double bond. Other elimination reactions are also known in which atoms or groups other than halogen are the leaving group.

Leaving group

L = Halogen, arenesulfonate, $-\overset{+}{S}R_2$, or $-\overset{+}{N}R_3$
If L = Cl, Br, or I, the elimination is called a dehydrohalogenation reaction
HOS = Protic polar solvent

The carbon atom bonded to the leaving group is called the α carbon atom and any adjacent carbon atoms are called β carbon atoms. Since this elimination occurs by loss of atoms on adjacent carbon atoms, it is often called a **1,2-elimination** or a **β-elimination.** Both designations are used by chemists.

Elimination reactions occur to a greater or lesser degree along with substitution reactions. The reaction of 2-bromo-2-methylpropane and ethanol, for example, forms products of both substitution, 2-ethoxy-2-methylpropane, and elimination, methylpropene.

2-Bromo-2-methylpropane (*tert*-butyl bromide) | 2-Ethoxy-2-methylpropane (product of substitution formed in 73% yield) | Methylpropene (product of elimination formed in 27% yield)

Elimination reactions occur by two mechanisms. One is unimolecular and the other is bimolecular.

12.11 Unimolecular Elimination: The E1 Mechanism

Step 1 of both the S_N1 mechanism of nucleophilic substitution reactions and the E1 mechanism of elimination reactions is a slow, rate-determining ionization of the haloalkane to form a carbocation. Step 1 of the mechanism of the reaction of ethanol and 2-bromo-2-methylpropane, for example, is formation of a *tert*-butyl cation. The substitution product is formed by reaction of the *tert*-butyl cation with a nucleophile, and the elimination product is formed by transfer of a proton from the *tert*-butyl cation to a base, as shown in Figure 12.13.

The mechanism given in Figure 12.13 that describes the formation of the elimination product is called an **E1 mechanism.** The *E* stands for elimination, and *1* indicates that the rate-determining step is unimolecular as in the S_N1 mechanism.

Any hydrogen atom bonded to a β carbon atom can be lost in Step 2 of the E1 mechanism. In the case of the *tert*-butyl cation, there are nine equivalent hydrogen atoms in this position and the same product is formed no matter which one of the nine hydrogens is removed by the base. In most carbocations, however, all the hydrogen atoms bonded to the β carbon atoms are not equivalent. As a result, more than one alkene can be formed in the elimination reaction. In such cases,

Step 1: Slow dissociation of 2-bromo-2-methylpropane forms a carbocation, which is a common step in both the E1 and S_N1 mechanisms

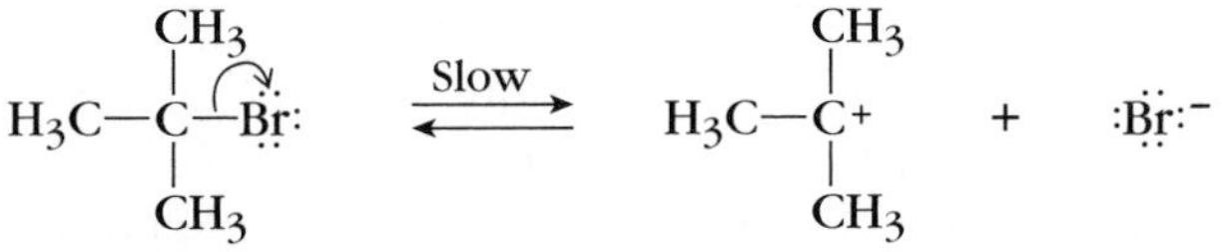

2-Bromo-2-methylpropane
(*tert*-Butyl bromide)

Step 2 of E1 mechanism: A base removes a hydrogen atom from a β carbon atom to form a carbon-carbon double bond

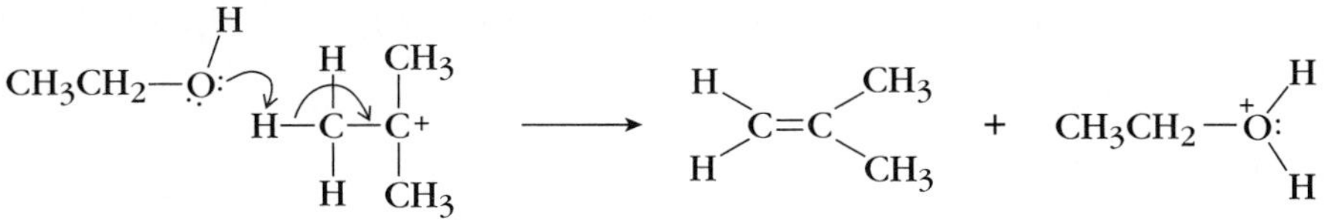

Methylpropene

Step 2 of S_N1 mechanism: Nucleophile reacts with carbocation to form the substitution product

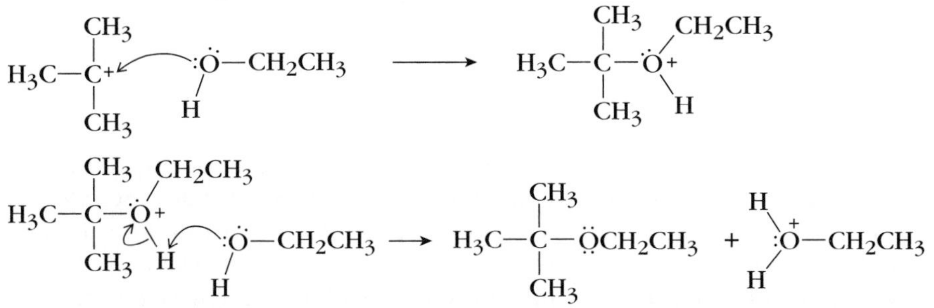

2-Ethoxy-2-methylpropane

Figure 12.13
Comparison of E1 and S_N1 mechanisms. The slow and rate-determining formation of a carbocation is the same in Step 1 in both mechanisms. Removal by a base of a β hydrogen atom from the carbocation to form an alkene is Step 2 of the E1 mechanism, while addition of a nucleophile to the carbocation is Step 2 of the S_N1 mechanism.

the more stable alkene is usually formed in larger amounts. Consequently elimination by an E1 mechanism forms the more highly substituted alkene in larger amounts.

$$CH_3CH_2\underset{\underset{CH_3}{|}}{\overset{\overset{Br}{|}}{C}}CH_3 \quad \xrightarrow[55\,°C]{CH_3CH_2OH} \quad CH_3CH_2\underset{\underset{CH_3}{|}}{\overset{\overset{OCH_2CH_3}{|}}{C}}CH_3 \quad + \quad \underset{H}{\overset{H_3C}{>}}C=C\underset{CH_3}{\overset{CH_3}{<}} \quad + \quad \underset{H}{\overset{H}{>}}C=C\underset{CH_2CH_3}{\overset{CH_3}{<}}$$

2-Bromo-2-methylbutane   2-Ethoxy-2-methylbutane   2-Methyl-2-butene   2-Methyl-1-butene
                                   (65% yield)   (30% yield)   (5% yield)

■ *Recall that the more substituted the alkene, the greater its stability. You may wish to review the stability of alkenes in Section 7.5.*

Exercise 12.10   Write a stepwise mechanism to explain how all the products are formed in the reaction of 2-bromo-2-methylbutane in ethanol shown previously.

While the E1 and S_N1 mechanisms have a common rate-determining first step, the elimination reaction can frequently be made the predominant reaction by using a stronger base.

12.12 Bimolecular Elimination: The E2 Mechanism

The amount of elimination product in a reaction that forms both substitution and elimination products can be increased by increasing the basicity of the nucleophile. Recall that the reaction of 2-bromo-2-methylpropane with ethanol forms a mixture of products in which the substitution product, 2-ethoxy-2-methylpropane, is formed in greater amounts than the elimination product, methylpropene (Section 12.11). When sodium ethoxide, the conjugate base of ethanol, is added to this reaction mixture, the only product formed is methylpropene.

$$CH_3-\underset{\underset{CH_3}{|}}{\overset{\overset{CH_3}{|}}{C}}-Br \quad \xrightarrow[55\,°C]{CH_3CH_2OH} \quad CH_3-\underset{\underset{CH_3}{|}}{\overset{\overset{CH_3}{|}}{C}}-OCH_2CH_3 \quad + \quad H_2C=C\underset{CH_3}{\overset{CH_3}{<}}$$

2-Bromo-2-methylpropane   2-Ethoxy-2-methylpropane   Methylpropene
(*tert*-butyl bromide)   (73% yield)   (27% yield)

$$CH_3-\underset{\underset{CH_3}{|}}{\overset{\overset{CH_3}{|}}{C}}-Br \quad \xrightarrow[\underset{50\text{-}60\,°C}{CH_3CH_2OH}]{CH_3CH_2O^-\ Na^+} \quad H_2C=C\underset{CH_3}{\overset{CH_3}{<}}$$

Methylpropene
(100% yield)

Not only does the product composition change, but reaction occurs more quickly in the presence of ethoxide ion. In addition to a change in rate, the rate law of the two reactions is different. The reaction in the absence of ethoxide ion occurs by an $S_N1/E1$ mechanism, whose rate law is first order in haloalkane. The rate law of the elimination reaction in the presence of ethoxide ion is overall second order, first order in both 2-bromo-2-methylpropane and sodium ethoxide.

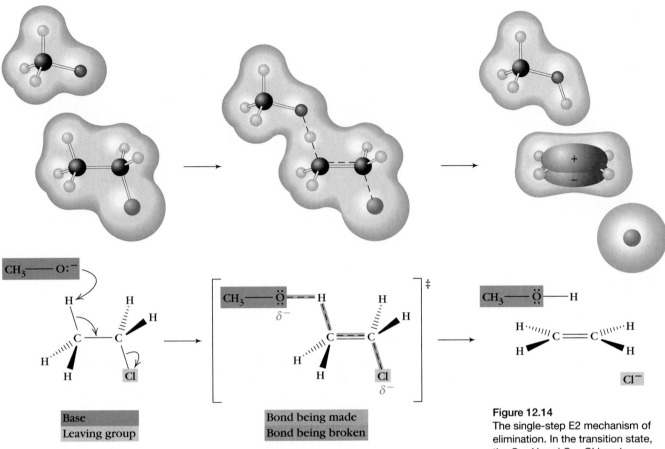

$$\text{Rate} = k[(CH_3)_3CBr]\ [C_2H_5ONa]$$

Figure 12.14
The single-step E2 mechanism of elimination. In the transition state, the C—H and C—Cl bonds are being broken, a bond between the base and a hydrogen atom is being formed, and the π bond of the carbon-carbon double bond is being formed.

The single-step mechanism proposed for this elimination reaction is called an **E2 mechanism** and is shown in Figure 12.14 for the reaction of chloroethane and methoxide ion. The *E* in the designation E2 mechanism stands for elimination, and *2* stands for bimolecular because the transition state contains molecules of both haloalkane and base.

In an E2 mechanism, three things occur simultaneously. First the electrons of the base begin to abstract a β hydrogen, with the result that the β carbon-hydrogen bond begins to break. At about the same time, a new carbon-carbon double bond begins to form as the leaving group begins to leave with its pair of electrons. Thus in the transition state, two bonds, the B—H bond and the π part of the carbon-carbon double bond, are being formed and two, the C—H and the C—L bonds, are being broken.

In order for an elimination reaction to occur by an E2 mechanism, the C—H and C—L bonds must be properly oriented.

12.13 Stereochemistry of the E2 Mechanism

1-Bromo-1,2-diphenylpropane has two stereocenters and consequently exists as four stereoisomers.

■ *Recall that phenyl is the name of* ⟨phenyl structure⟩ *(C_6H_5).*

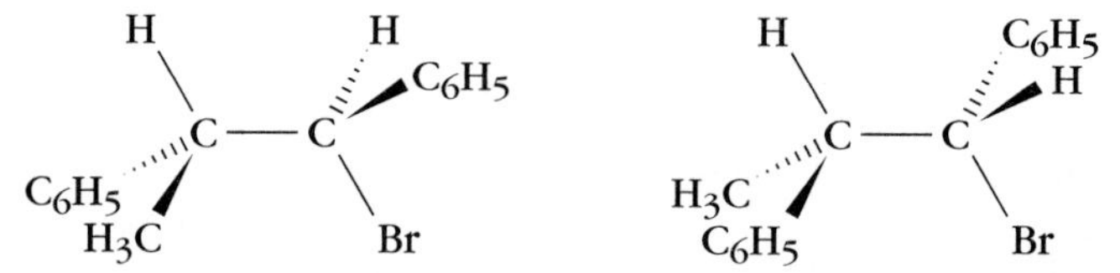

(1S,2R)-1-Bromo-
1,2-diphenylpropane

(1R,2S)-1-Bromo-
1,2-diphenylpropane

ENANTIOMERS

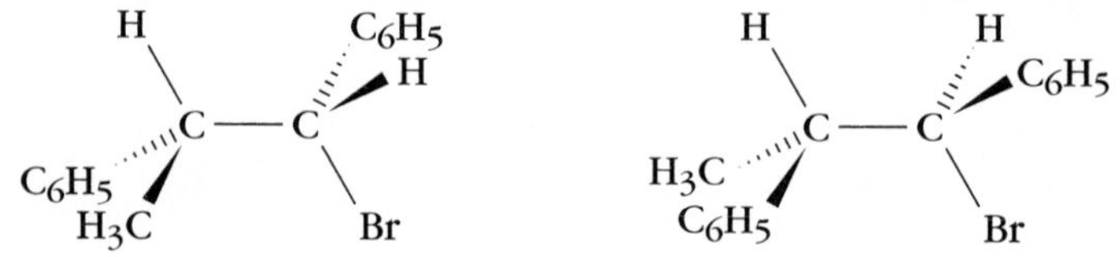

(1R,2R)-1-Bromo-
1,2-diphenylpropane

(1S,2S)-1-Bromo-
1,2-diphenylpropane

ENANTIOMERS

The elimination reaction of all four of these stereoisomers with sodium methoxide in methanol occurs by means of an E2 mechanism. The product of elimination, however, depends on the starting compound. The elimination reaction of (1S,2R)-1-bromo-1,2-diphenylpropane and its enantiomer forms exclusively (E)-1,2-diphenyl-1-propene. None of the (Z)- isomer is formed.

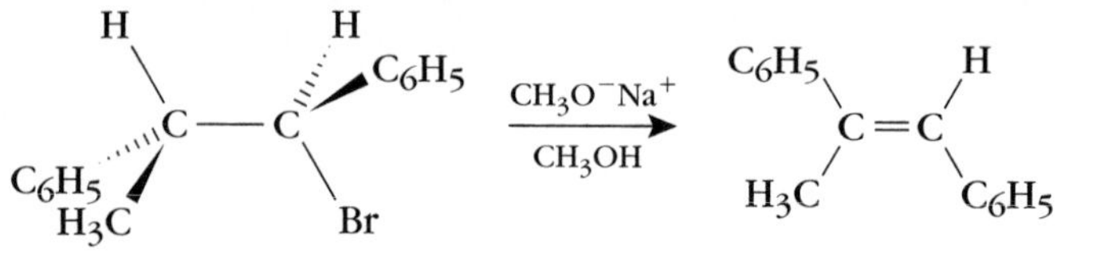

(1S,2R)-1-Bromo-
1,2-diphenylpropane
and its enantiomer

(E)-1,2-Diphenyl-1-propene
No (Z)- isomer formed

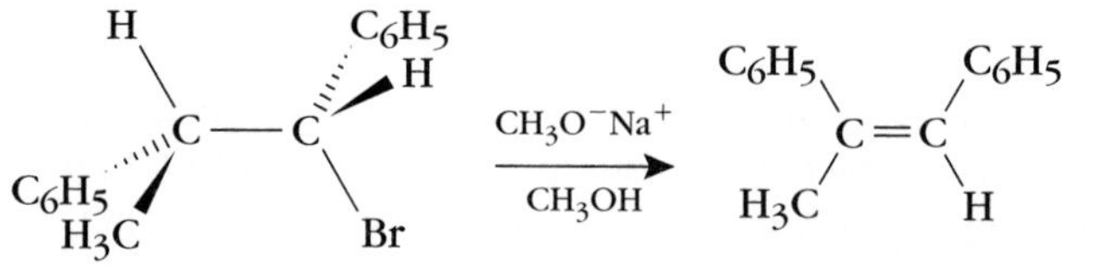

(1R,2R)-1-Bromo-
1,2-diphenylpropane
and its enantiomer

(Z)-1,2-Diphenyl-1-propene
No (E)- isomer formed

Under the same conditions, (1R,2R)-1-bromo-1,2-diphenylpropane and its enantiomer form only (Z)-1,2-diphenyl-1-propene. None of the (E)- isomer is formed.

Such stereospecific elimination reactions mean that a concerted E2 mechanism must occur only when the bonds being made and broken all lie in a plane. These bonds are planar when the starting compound is in a staggered conformation with the β hydrogen and the leaving group anti to each other, as shown in Figure 12.15. Elimination reactions in which the β hydrogen and the leaving group are in a staggered conformation and are anti to each other are called **anti**

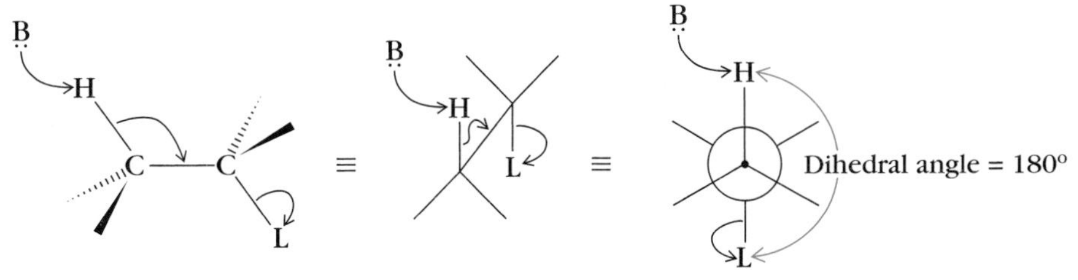

Figure 12.15
Three representations of the preferred conformation for anti elimination by an E2 mechanism. The groups are in a staggered conformation with the β hydrogen and the leaving group anti to each other (a dihedral angle of 180°).

elimination reactions. The anti elimination of HBr from the stereoisomers of 1-bromo-1,2-diphenylpropane by an E2 mechanism is represented as follows:

We have encountered the term anti before when we discussed the stereochemistry of electrophilic additions to alkenes. Anti elimination is simply the reverse of anti addition to a carbon-carbon double bond.

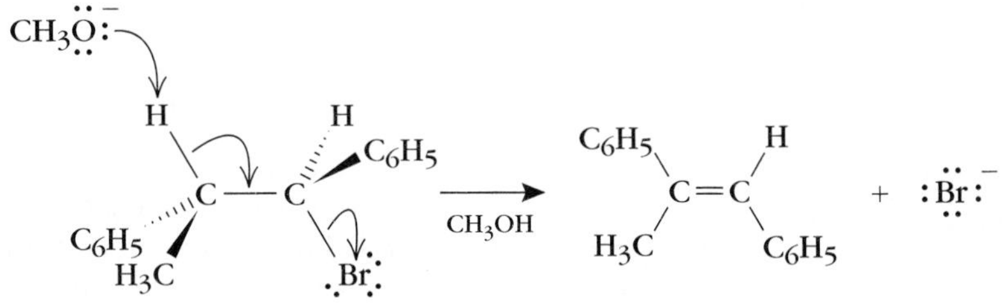

(1*S*,2*R*)-1-Bromo-1,2-diphenylpropane

(*E*)-1,2-Diphenyl-1-propene

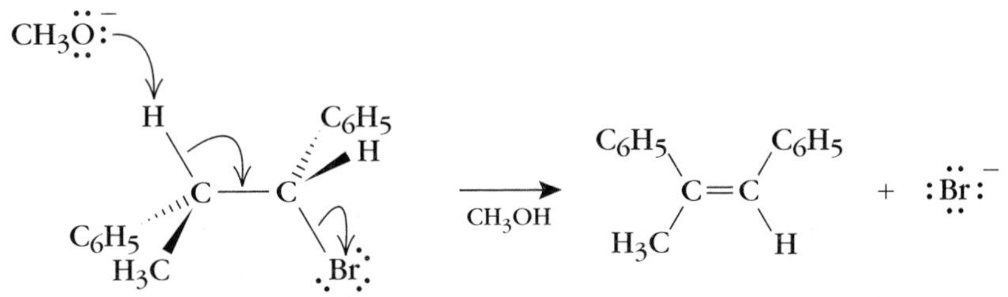

(1*R*,2*R*)-1-Bromo-1,2-diphenylpropane

(*Z*)-1,2-Diphenyl-1-propene

The E2 mechanism requires a precise geometry for elimination because the orbitals of the carbon atoms in the original C—H and C—L bonds in the starting material must begin to overlap in the transition state to form the π bond in the product. For this to occur, the C—H and C—L bonds of the starting compound must be coplanar. The C—H and C—L bonds can be anti coplanar or syn coplanar, as shown in Figure 12.16.

An E2 mechanism usually occurs by anti rather than syn elimination because in the transition state for syn elimination, the C—H and C—L bonds are eclipsed, rather than staggered as in the transition state for anti elimination. As a result, the base and the leaving group tend to repel each other in the transition state for syn elimination, as shown in Figure 12.17. This repulsive interaction is absent in the transition state for anti elimination, so the transition state for syn elimination is higher in energy than the one for anti elimination.

The last feature of the E2 mechanism that we must examine is the product composition when isomeric alkenes can be formed.

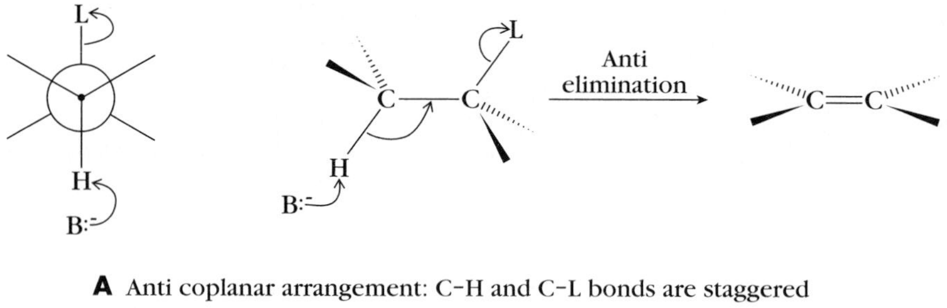

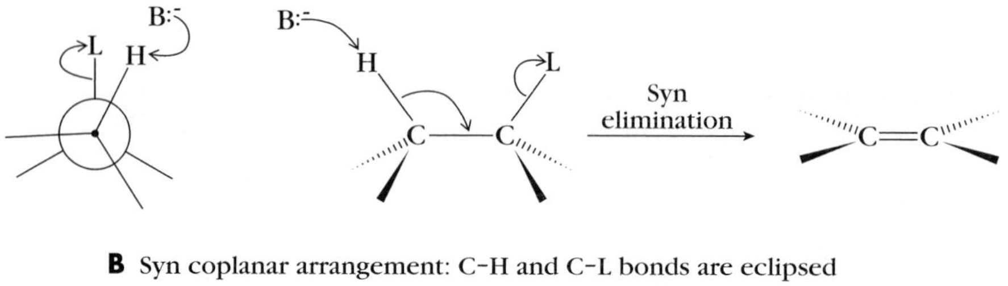

Figure 12.16
Two coplanar arrangements of the C—H and C—L bonds. In **A,** the molecule is in a staggered conformation and the C—H and C—L bonds are anti coplanar. This leads to anti elimination. In **B,** the molecule is in an eclipsed conformation and the C—H and C—L bonds are syn coplanar, which leads to syn elimination.

A Anti coplanar arrangement: C–H and C–L bonds are staggered

B Syn coplanar arrangement: C–H and C–L bonds are eclipsed

Exercise 12.11	The reaction of *trans*-2-methyl-bromocyclohexane with potassium *tert*-butoxide in *tert*-butyl alcohol forms only 3-methylcyclohexene in good yield. Under the same experimental conditions *cis*-2-methyl-bromocyclohexane forms a mixture of 1-methylcyclohexene and 3-methylcyclohexene. Explain why these two stereoisomers form different products.

Figure 12.17
The C—H and the C—L (in this case L = halogen atom) bonds are eclipsed in the transition state for syn elimination, which causes a repulsive interaction between the base and the leaving group. This interaction is missing in the transition state for anti elimination so the transition state for syn elimination is higher in energy.

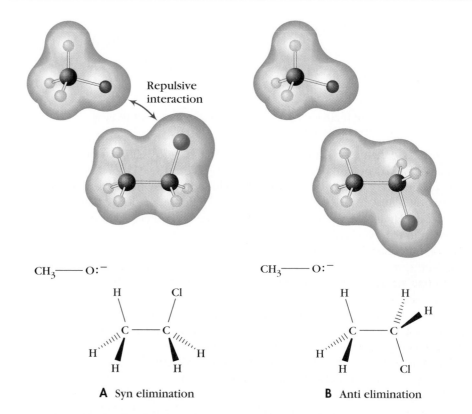

A Syn elimination

B Anti elimination

12.14 Product Composition of Elimination Reactions That Occur by an E2 Mechanism

Elimination by an E1 mechanism forms the more stable alkene as product when a molecule contains more than one type of β hydrogen. The same occurs for eliminations that occur by an E2 mechanism. The dehydrohalogenation of 2-bromo-2-methylbutane with ethoxide ion, for instance, forms the more stable alkene as the major product.

| 2-Bromo-2-methylbutane | 2-Methyl-1-butene (30% yield) | 2-Methyl-2-butene (70% yield) |
| | Less stable alkene | More stable alkene |

Generally the use of hydroxide, methoxide, or ethoxide ions as bases in elimination reactions forms the more stable alkene as the major product. Elimination reactions that form the more highly substituted alkene as the major product are said to follow the **Saytzeff Rule,** named for the 19th century Russian chemist A. N. Saytzeff who first formulated it.

Exceptions to the Saytzeff Rule occur when dehydrohalogenation reactions are carried out using bulky bases such as potassium *tert*-butoxide, $(CH_3)_3CO^-K^+$, in *tert*-butyl alcohol. In this case, the less-substituted alkene is the major product.

| 2-Bromo-2-methylbutane | 2-Methyl-1-butene (74% yield) | 2-Methyl-2-butene (26% yield) |

The reasons for this change in product composition with change in base are complicated but seem to be related to the increased strength of the base (pK_a of *tert*-butyl alcohol = 19, while pK_a of methanol = 16) and the steric bulk of *tert*-butoxide ion. *tert*-Butoxide ion is larger than ethoxide or hydroxide ions and has trouble approaching sufficiently close to react with the internal secondary hydrogen atoms of 2-bromo-2-methylbutane. Instead, it reacts with one of the more exposed primary hydrogen atoms of the methyl group to form the less-substituted alkene. An elimination reaction that forms the least-substituted alkene as the major product is said to follow the **Hofmann Rule.**

12.15 Summary of the E2 Mechanism

The characteristic features of elimination reactions that occur by an E2 mechanism are as follows:
- They follow a second order rate law—first order in base and first order in substrate.
- Elimination usually occurs when the leaving group, the two carbon atoms, and the β hydrogen are in an anti coplanar arrangement.
- The major product of elimination is the most stable alkene except in the case where a bulky base is used.

12.16 Mechanism of Acid-Catalyzed Dehydration of Alcohols

So far our discussion of elimination reactions has focused on base-promoted elimination of hydrogen halides from haloalkanes. Acid-catalyzed dehydration of alcohols, which was discussed briefly in Section 7.18, is another example of an elimination reaction. Dehydration of alcohols occurs by heating the alcohol in concentrated sulfuric or phosphoric acid.

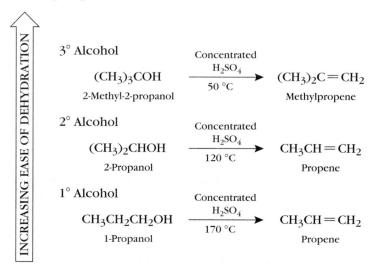

■ *The mechanism of the acid-catalyzed dehydration of 2-methyl-2-propanol is shown in Figure 7.21 (p. 317).*

Notice that tertiary alcohols are more easily dehydrated than secondary alcohols, which in turn are dehydrated more easily than primary alcohols.

Acid-catalyzed dehydration of secondary and tertiary alcohols occurs by an E1 mechanism. Step 1 is protonation of the oxygen atom of the hydroxy group of the alcohol. Step 2 is loss of a water molecule to form a carbocation that loses a proton to form an alkene. Any weak base in solution such as water or HSO_4^- can abstract a proton from the carbocation in Step 3.

The reason tertiary alcohols react fastest in acid-catalyzed dehydration is now clear. In reactions that occur by an E1 mechanism, tertiary compounds will always react faster than secondary or primary compounds because tertiary compounds form a more stable tertiary carbocation intermediate. Thus the order of ease of acid-catalyzed dehydration of alcohols is the same as the order of carbocation stability.

$$RCH_2OH \quad << \quad R_2CHOH \quad < \quad R_3COH$$

1° Alcohol   2° Alcohol   3° Alcohol

INCREASING EASE OF DEHYDRATION ⇒

$$RCH_2^+ \quad << \quad R_2CH^+ \quad < \quad R_3C^+$$

1° Carbocation   2° Carbocation   3° Carbocation

INCREASING CARBOCATION STABILITY ⇒

Recall that carbocations will rearrange, if possible, to a more stable carbocation by a hydride or alkyl shift (Section 7.17). Similar rearrangements are observed in dehydration of certain alcohols. The acid-catalyzed dehydration of 3,3-dimethyl-2-butanol, for example, forms a mixture of 2,3-dimethyl-1-butene and 2,3-dimethyl-2-butene because the first formed intermediate carbocation undergoes rearrangement to a more stable tertiary carbocation.

$$(CH_3)_3CCHCH_3 \xrightarrow[120\ °C]{\substack{conc. \\ H_2SO_4}} \underset{H}{\overset{H}{\diagdown}}C=C\underset{CH(CH_3)_2}{\overset{CH_3}{\diagup}} + \underset{H_3C}{\overset{H_3C}{\diagdown}}C=C\underset{CH_3}{\overset{CH_3}{\diagup}}$$

3,3-Dimethyl-2-butanol   2,3-Dimethyl-1-butene   2,3-Dimethyl-2-butene

Exercise 12.12 Write the mechanism for the acid-catalyzed dehydration of 3,3-dimethyl-2-butanol. Which of the two products, 2,3-dimethyl-1-butene or 2,3-dimethyl-2-butene, will be formed in the greatest amount? Explain your answer.

Acid-catalyzed dehydration of primary alcohols larger than 1-propanol forms mixtures of alkenes because the terminal alkene, which is the first product formed, undergoes acid-catalyzed isomerization to the more stable internal alkene.

$$CH_3(CH_2)_2CH_2OH \xrightarrow[170\ °C]{\substack{conc. \\ H_2SO_4}} CH_3CH_2CH=CH_2$$

1-Butanol   1-Butene

$$CH_3CH_2CH=CH_2 + H_3O^+ \longrightarrow CH_3CH_2\overset{+}{C}HCH_3 + H_2O$$

sec-Butyl cation

$$H_2O + CH_3CH_2\overset{+}{C}HCH_3 \longrightarrow CH_3CH=CHCH_3 + CH_3CH_2CH=CH_2 + H_3O^+$$

cis- and *trans*-2-Butene   1-Butene
(Major product)   (Minor product)

Exercise 12.13 Write the mechanism for the acid-catalyzed dehydration of 1-butanol to form a mixture of 1-butene and *cis*- and *trans*-2-butene.

We now know the major features of the four mechanisms involved in nucleophilic substitution and elimination reactions. We now must discuss what factors determine the type of reaction, elimination or substitution, and the type of mechanism, unimolecular or bimolecular, that a particular compound will undergo.

12.17 Summary of Factors That Influence the Mechanisms of Substitution and Elimination Reactions

In the preceding sections, we discussed the complexities of the reaction of nucleophiles with an alkyl group that contains a suitable leaving group. How do these factors influence a particular reaction? Is the major product formed by a nucleophilic substitution reaction or an elimination reaction? Does the reaction occur by a unimolecular mechanism, S_N1 or E1, or a bimolecular mechanism, S_N2 or E2? While we cannot make hard and fast rules to answer these questions, we can reach certain general conclusions based on the structure of the alkyl group that contains the leaving group. Let us begin with primary alkyl derivatives.

A   Primary Alkyl Derivatives

Straight chain aliphatic primary alkyl derivatives always react by a bimolecular mechanism, either S_N2 or E2, because their dissociation would form an extremely unstable primary carbocation. If a good nucleophile is used, such as I^-, CN^-, or CH_3S^-, nucleophilic substitution is the predominant product, usually to the exclusion of the elimination product. Even in the case of strong bases such as ethoxide and methoxide ions, the major product is formed by nucleophilic substitution. Only in the case of highly hindered bases, such as *t*-butoxide ion, is the rate of the S_N2 mechanism slowed down sufficiently by steric hindrance to allow elimination by an E2 mechanism to occur. Substitution reactions are exceedingly slow with weak nucleophiles such as methanol or water. The following are examples that illustrate these generalizations:

$$CH_3CH_2CH_2Br \xrightarrow[\text{Acetone}]{I^-} CH_3CH_2CH_2I \ + \ Br^-$$

$$CH_3CH_2CH_2Br \xrightarrow[\text{CH}_3\text{CH}_2\text{OH}]{CH_3CH_2O^- Na^+} CH_3CH=CH_2 \ + \ CH_3CH_2CH_2OCH_2CH_3$$
$$\text{(9\% yield)} \qquad\qquad \text{(91\% yield)}$$

$$CH_3CH_2CH_2Br \xrightarrow[\text{(CH}_3)_3\text{COH}]{(CH_3)_3CO^- K^+} CH_3CH=CH_2 \ + \ CH_3CH_2CH_2OC(CH_3)_3$$
$$\text{(85\% yield)} \qquad\qquad \text{(15\% yield)}$$

Branched primary alkyl derivatives react with good nucleophiles to form predominately substitution products but react with strong bases to form mainly elimination products. Exceptions to this general rule are the 2,2-dialkyl derivatives that cannot undergo β-elimination reactions and undergo substitution reactions extremely slowly.

$$(CH_3)_2CHCH_2Br \ + \ I^- \xrightarrow[\text{Acetone}]{} (CH_3)_2CHCH_2I \ + \ Br^-$$

$$(CH_3)_2CHCH_2Br \xrightarrow[\substack{\text{CH}_3\text{CH}_2\text{OH} \\ (55\,°C)}]{CH_3CH_2O^- Na^+} (CH_3)_2C=CH_2 \ + \ (CH_3)_2CHCH_2OCH_2CH_3$$
$$\text{(60\% yield)} \qquad\qquad \text{(40\% yield)}$$

$$(CH_3)_2CHCH_2Br \xrightarrow[\text{(CH}_3)_3\text{COH}]{\text{(CH}_3)_3\text{CO}^-\text{K}^+} (CH_3)_2C{=}CH_2 \ + \ (CH_3)_2CHCH_2OC(CH_3)_3$$

(92% yield)   (8% yield)

B   Secondary Alkyl Derivatives

Depending on the conditions, reactions of secondary alkyl derivatives occur by either a unimolecular or a bimolecular mechanism to form products of both elimination and substitution. Secondary alkyl derivatives react with strong bases such as ethoxide ion or hydroxide ion to form a mixture of elimination products, which occur by an E2 mechanism, and substitution products, which occur by an S_N2 mechanism. The proportion of elimination product can be increased by using either a stronger base such as amide ion (NH_2^-) or a more bulky base such as *tert*-butoxide ion.

Br
|
CH_3CHCH_3 $\xrightarrow[\text{CH}_3\text{CH}_2\text{OH}]{\text{Na}^{+-}\text{OCH}_2\text{CH}_3}$ $CH_3CH{=}CH_2$ + CH_3CHCH_3 (with OCH$_2$CH$_3$)

80%   20%
Formed by E2   Formed by S_N2
mechanism   mechanism

Br
|
CH_3CHCH_3 $\xrightarrow[\text{(CH}_3)_3\text{COH}]{\text{K}^{+-}\text{OC(CH}_3)_3}$ $CH_3CH{=}CH_2$

Formed by E2 mechanism

Good nucleophiles such as RS^-, CN^-, and I^- or weakly basic nucleophiles such as acetate ion (CH_3COO^-) favor substitution reactions by an S_N2 mechanism. Substitution reactions can also be favored by choosing a polar aprotic solvent.

Br
|
CH_3CHCH_3 $\xrightarrow[\text{CH}_3\text{COOH}]{\text{CH}_3\text{C}({=}\text{O})\text{O}^-\text{Na}^+}$ CH_3CHCH_3 (with $O{-}C({=}O)CH_3$) + Na^+Br^-

Secondary alkyl derivatives that contain a good leaving group react in a nonnucleophilic polar protic solvent to form a mixture of elimination and substitution products by an S_N1/E1 mechanism.

Br
|
CH_3CHCH_3 $\xrightarrow[\text{heat}]{\text{CH}_3\text{CH}_2\text{OH}}$ $CH_3CH{=}CH_2$ + CH_3CHCH_3 (with OCH$_2$CH$_3$)

15%   85%
Formed by E1   Formed by S_N1
mechanism   mechanism

C   Tertiary Alkyl Derivatives

Tertiary alkyl derivatives undergo substitution reactions only by an S_N1 mechanism; steric hindrance at the tertiary carbon atom prevents substitution by an S_N2

mechanism. In competition with substitution is elimination by an E1 mechanism that forms alkenes as a side product.

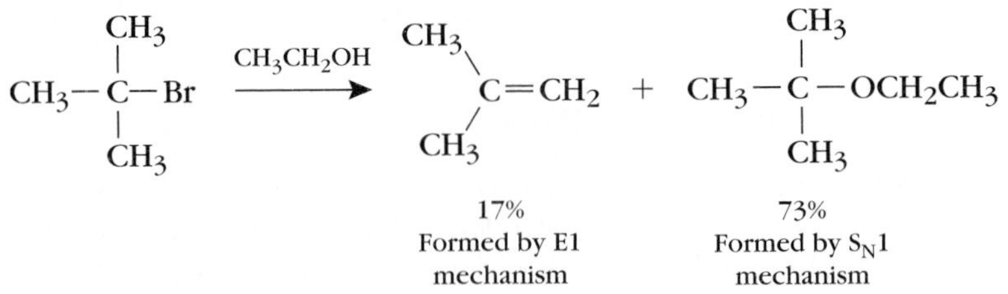

$$CH_3-\underset{\underset{CH_3}{|}}{\overset{\overset{CH_3}{|}}{C}}-Br \xrightarrow{CH_3CH_2OH} \underset{CH_3}{\overset{CH_3}{>}}C=CH_2 \;+\; CH_3-\underset{\underset{CH_3}{|}}{\overset{\overset{CH_3}{|}}{C}}-OCH_2CH_3$$

17%
Formed by E1
mechanism

73%
Formed by S_N1
mechanism

Elimination by an E2 mechanism is the predominant reaction of tertiary alkyl derivatives with bases.

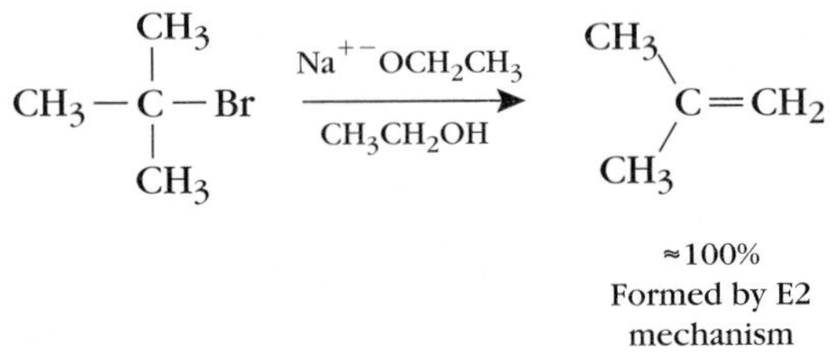

$$CH_3-\underset{\underset{CH_3}{|}}{\overset{\overset{CH_3}{|}}{C}}-Br \xrightarrow[CH_3CH_2OH]{Na^{+-}OCH_2CH_3} \underset{CH_3}{\overset{CH_3}{>}}C=CH_2$$

≈100%
Formed by E2
mechanism

Exercise 12.14   Write the structure of the products in each of the following reactions. If more than one product is formed, indicate the one formed in highest yield.

I. The reaction of 1-bromobutane with
(a) Sodium iodide in acetone
(b) Methanol
(c) Sodium methoxide in methanol
(d) Potassium *tert*-butoxide in *tert*-butyl alcohol
II. The reaction of 2-bromobutane with
(a) Ethanol
(b) Sodium methanethiolate ($NaSCH_3$) in ethanol
(c) Sodium cyanide in *N,N*-dimethylformamide
(d) Sodium methoxide in methanol
(e) Potassium *tert*-butoxide in *tert*-butyl alcohol
III. The reaction of 2-methyl-2-bromobutane with
(a) Aqueous acetone
(b) Ethanol
(c) Sodium methoxide in methanol
(d) Potassium *tert*-butoxide in *tert*-butyl alcohol

Let's now apply these general conclusions to the problem of how to use nucleophilic substitution reactions to synthesize compounds in the best yield possible.

12.18 Using Nucleophilic Substitution Reactions in Synthesis

Nucleophilic substitution reactions are useful in synthesis because we can transform haloalkanes into compounds that contain other functional groups. The alkylation of acetylide anions, which we discussed in Section 9.11, is an example of a reaction in which a haloalkane is converted into an alkyne.

$$H—C≡C^-Na^+ \quad + \quad CH_3(CH_2)_2CH_2Br \quad \xrightarrow[NH_{3\,(l)}]{} \quad H—C≡C(CH_2)_3CH_3 \quad + \quad NaBr$$

Sodium
acetylide

1-Bromobutane

1-Hexyne
(75% yield)

This reaction occurs by an S_N2 mechanism, so we can now understand why best results in alkylation of acetylide anions are obtained only with primary haloalkanes. An acetylide anion is a strong base as well as a nucleophile, so when a secondary or tertiary haloalkane is used, elimination by an E2 mechanism competes with alkylation by an S_N2 mechanism.

The synthesis of thiols by the reaction of haloalkanes with either excess hydrosulfide ion or thiourea, discussed in Section 11.19, are other examples of reactions used in synthesis that occur by an S_N2 mechanism. A number of other reactions that convert haloalkanes to other families of organic compounds are given in Figure 12.18.

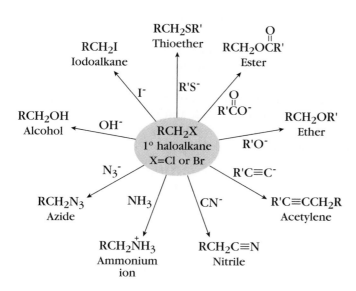

Figure 12.18
Nucleophilic substitution reactions that convert a primary haloalkane into compounds with other functional groups.

12.19 Summary

A **nucleophilic substitution reaction** is the substitution of an atom or group of atoms, called the leaving group, on an sp^3 hybrid carbon atom by a nucleophile. An **elimination reaction** forms a π bond by the elimination of atoms or groups of atoms from adjacent carbon atoms. Such elimination reactions are also called **β-eliminations** or **1,2-eliminations**.

Nucleophilic substitution reactions occur by two mechanisms, S_N2 and S_N1. The rate law of the **S_N2 mechanism** is overall second order; it is first order in each reagent. In the S_N2 mechanism, the entering nucleophile attacks the electrophilic carbon atom from directly opposite the leaving group, resulting in an inversion of configuration at the reaction site. Because of this backside approach, the rate of substitution reactions that occur by an S_N2 mechanism is greatly inhibited by increasing steric bulk of either reagent. Consequently only methyl, primary, and, under some conditions, secondary alkyl derivatives undergo nucleophilic substitution reactions by an S_N2 mechanism. The factors that favor substitution by an S_N2 mechanism over elimination for primary and secondary alkyl groups are (1) a polar aprotic solvent, (2) a good leaving group, and (3) a good but weakly basic nucleophile.

Step 1 of an **S_N1 mechanism** is the slow dissociation of the alkyl derivative into a carbocation, followed by a second step in which the carbocation intermediate reacts with a nucleophile. Reactions that occur by an S_N1 mechanism follow a rate law that is first order in only the alkyl derivative. Optically active compounds that react by an S_N1 mechanism form products that are a mixture of enantiomers. Tertiary alkyl derivatives and any other compound that can form a relatively stable carbocation react by an S_N1 mechanism.

Elimination reactions of alkyl derivatives occur by either an E1 or E2 mechanism. Ionization to a carbocation intermediate is a first step common to both the **E1 mechanism** and the S_N1 mechanism. As a result, tertiary derivatives and any other compound that can form a relatively stable carbocation usually react by E1 and S_N1 mechanisms. As a result of this similarity in the rate-determining step, polar solvents and good leaving groups favor both E1 and S_N1 mechanisms. E1 and S_N1 mechanisms differ in their second steps. Elimination of a proton is the second step of the E1 mechanism, while reaction with a nucleophile is the second step of the S_N1 mechanism.

The **E2 mechanism** involves the simultaneous abstraction of a β hydrogen by a base with the formation of a new carbon-carbon double bond as the leaving group leaves with its pair of electrons. The mechanism occurs in a single step by means of an **anti coplanar** transition state in which the β hydrogen, the two carbon atoms of the new double bond, and the leaving group are all in the same plane. The elimination products are usually a mixture of alkenes in which the more substituted alkenes predominate.

12.20 Summary of Mechanisms of Substitution and Elimination Reactions

1. **Nucleophilic Substitution Reactions**

   a. **S_N2 mechanism**　　　　　　　SECTION 12.2

   A concerted mechanism in which the nucleophile approaches the reaction site from the backside with a resulting inversion of configuration at the reaction site. Methyl and primary haloalkanes and their derivatives react by this mechanism.

Backside approach　　Reaction site　　Transition state

L = Leaving group　　　Nu = Nucleophile

b. S_N1 mechanism   SECTION 12.6

A two-step mechanism. Step 1 is a slow ionization to a carbocation, which reacts with a nucleophile in Step 2. Tertiary haloalkanes and their derivatives react by this mechanism.

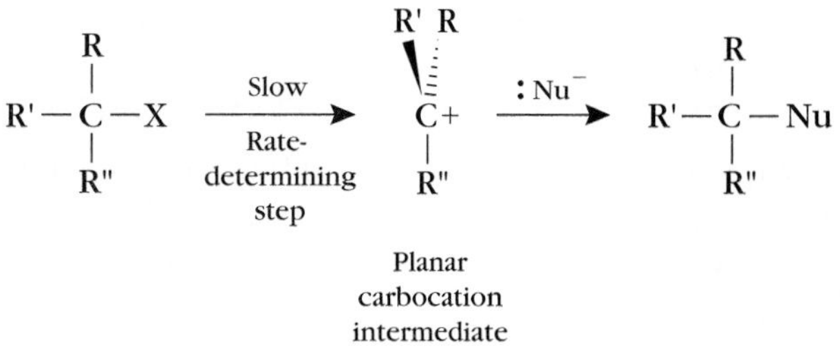

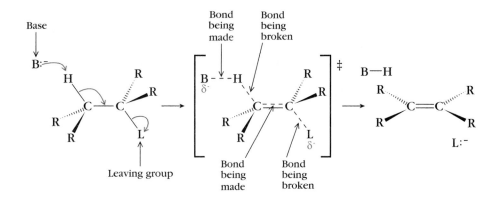

Planar carbocation intermediate

2. Elimination Reactions

a. E1 mechanism   SECTION 12.11

A two-step mechanism. Step 1 is formation of a carbocation intermediate from which a base removes a proton to form the more highly substituted alkene in Step 2.

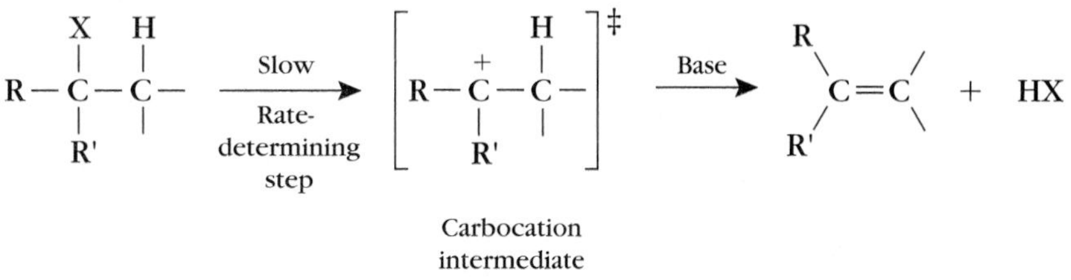

Carbocation intermediate

b. E2 mechanism   SECTIONS 12.12–12.15

A concerted mechanism that requires anti coplanar arrangements of C—H and C—L bonds.

Base

Bond being made   Bond being broken

Leaving group

Bond being made   Bond being broken

Additional Exercises

12.15 Define each of the following terms in words or by means of an example:

(a) Nucleophile	**(e)** Leaving group	**(i)** Hofmann Rule
(b) S_N1 mechanism	**(f)** E2 mechanism	**(j)** E1 mechanism
(c) Saytzeff Rule	**(g)** S_N2 mechanism	**(k)** Carbocation
(d) Nucleophilic substitution reaction	**(h)** Elimination reaction	**(l)** Second order rate law

12.16 In each of the following pairs of ions or compounds, decide which one is the more nucleophilic in methanol as solvent.

(a) $(CH_3)_3P$; $(CH_3)_3N$	**(c)** CH_3OH; CH_3Cl
(b) NH_2^-; NH_3	**(d)** CH_3COO^-; CH_3COOH

12.17 Write the structures of the major products formed by the reaction of 1-bromobutane with each of the following reagents:

(a) $NaNH_2/NH_{3(l)}$	**(d)** Aqueous NaOH	**(g)** $Na^+ \ ^-C{\equiv}CCH_3/NH_{3(l)}$
(b) NaI/acetone	**(e)** $NaCN/CH_3OH$	**(h)** $CH_3NH_2/(C_2H_5)_2O$
(c) H_2O	**(f)** CH_3COONa (sodium acetate)$/CH_3COOH$	

12.18 What reagents, organic or inorganic, would you use to convert 1-bromobutane to each of the following compounds?

(a) $CH_3(CH_2)_2CH_2OCH_3$	**(c)** 1-Butene	**(e)** 2-Methyl-2-hexanol
(b) 1-Butanol	**(d)** Pentanenitrile ($CH_3CH_2CH_2CH_2CN$)	

12.19 Write the structure of the product in each of the following reactions. Indicate the stereochemistry of the product where appropriate.

(a) $(CH_3)_2CHBr \ + \ NaSCH_3 \ \xrightarrow{\text{Acetone}}$

(b) $(CH_3)_3COH \ \xrightarrow{\text{Concentrated HCl}}$

(c) $(CH_3)_2\overset{\overset{\displaystyle Cl}{|}}{C}CH_2CH_2CH_3 \ \xrightarrow[\text{(CH}_3)_3\text{COH}]{(CH_3)_3CO^-K^+}$

(d)

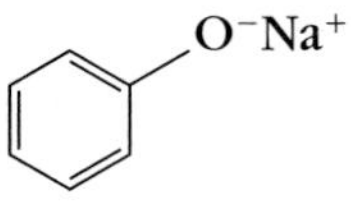

$+ \ CH_3\overset{\overset{\displaystyle Br}{|}}{C}HCH_2CH_3 \ \xrightarrow{CH_3CH_2OH}$

(e)

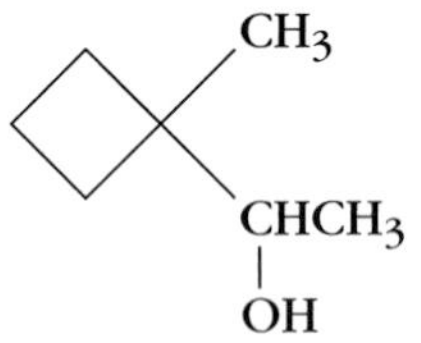

$\xrightarrow[130\,°C]{\text{Concentrated H}_2\text{SO}_4}$

(f)

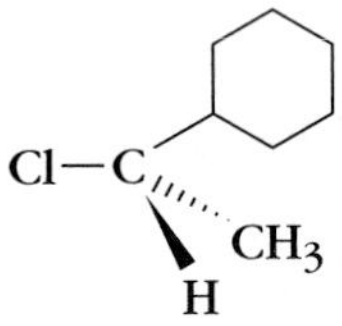

$\xrightarrow{CH_3COOH}$

(g) $(CH_3)_2CHCH_2\overset{\overset{\displaystyle OH}{|}}{C}HCH_3 \ \xrightarrow[130\,°C]{\text{Concentrated H}_2\text{SO}_4}$

(h) *meso*

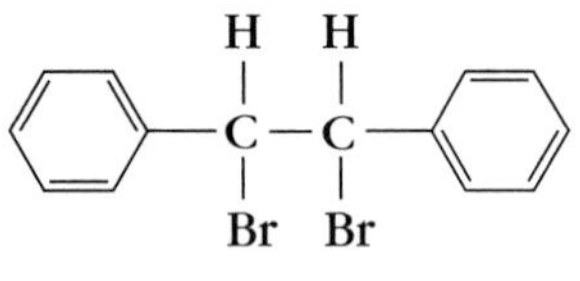

$\xrightarrow[CH_3CH_2OH]{CH_3CH_2O^- Na^+}$

(i) $CH_3\overset{\overset{\displaystyle Br}{|}}{C}HCH_2CH_2CH_3 \ \xrightarrow[CH_3CH_2OH,\ \text{heat}]{KOH}$

12.20 Many cyclic compounds are prepared by intramolecular (within the same molecule) nucleophilic substitution reactions. The following are several examples. For each, write a stepwise mechanism that describes how each of the cyclic products is formed from the acyclic starting material.

(a) $BrCH_2CH_2CH_2CH_2Br \ \xrightarrow[CH_3OH]{NaOH}$

$+ \ 2NaBr$

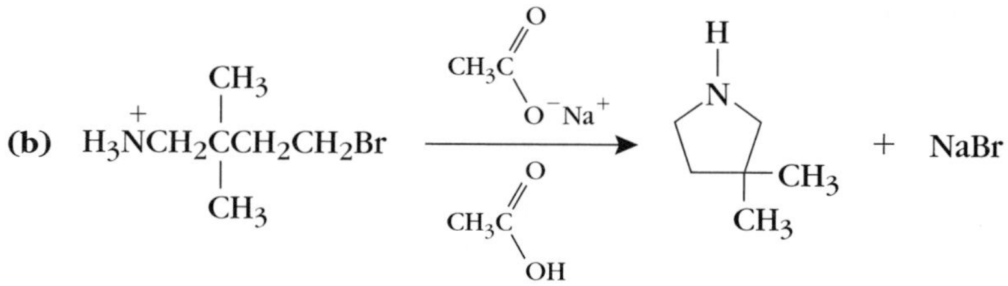

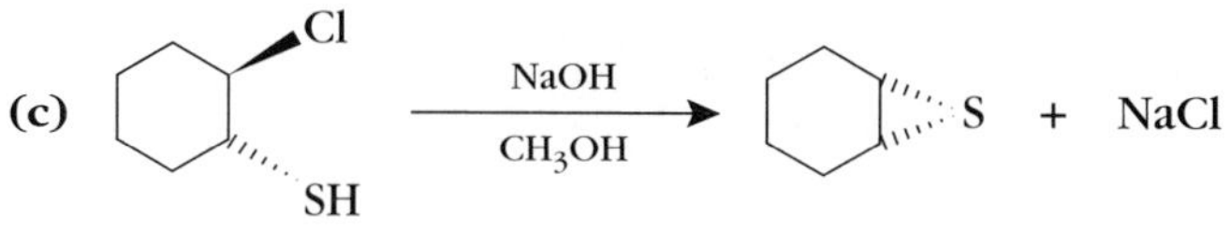

(b)

(c)

12.21 (R)-1-Deuterioethanol retains its optical activity for long periods of time when dissolved in pure water. However, addition of a small amount of sulfuric acid causes a rapid loss of its optical activity. Explain.

12.22 The solvolysis reaction of 2-bromo-2-methylbutane in methanol at room temperature forms the following products in the yields indicated:

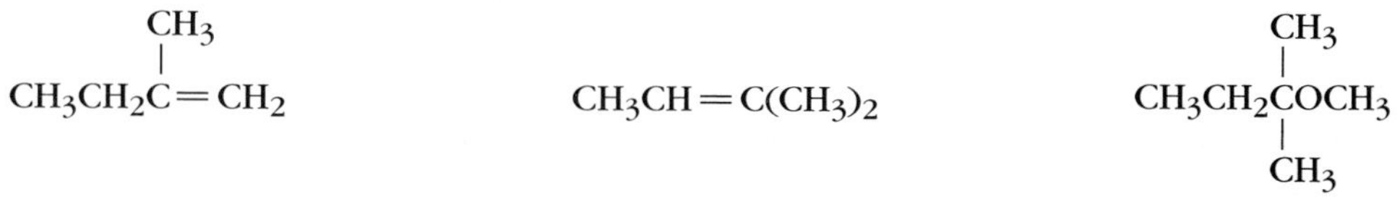

$$CH_3CH_2C=CH_2$$

$$CH_3CH=C(CH_3)_2$$

$$CH_3CH_2COCH_3$$

7%

30%

63%

How would the relative yields of the elimination and substitution products change if the reaction conditions were changed as follows:

(a) Sodium methoxide is added to the methanol solvent.

(b) Methanol is replaced by 90% water/10% methanol.

(c) Methanol is replaced by potassium *tert*-butoxide in *tert*-butyl alcohol.

12.23 The heating of ethanol and concentrated sulfuric acid to 130 °C forms the following three products:

$$CH_2=CH_2 \qquad CH_3CH_2OCH_2CH_3 \qquad CH_3CH_2OSO_3H$$

Write a multistep mechanism that explains the formation of these three products.

12.24 The reaction of optically active 3-methyl-3-hexanol shown below results in the formation of an optically inactive bromoalkane. Write the structure of the bromoalkane and propose a mechanism that accounts for this observation.

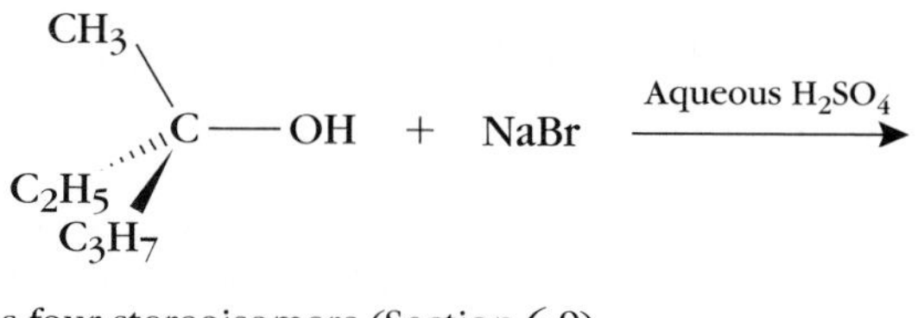

12.25 3-Phenyl-2-butyl tosylate exists as four stereoisomers (Section 6.9).

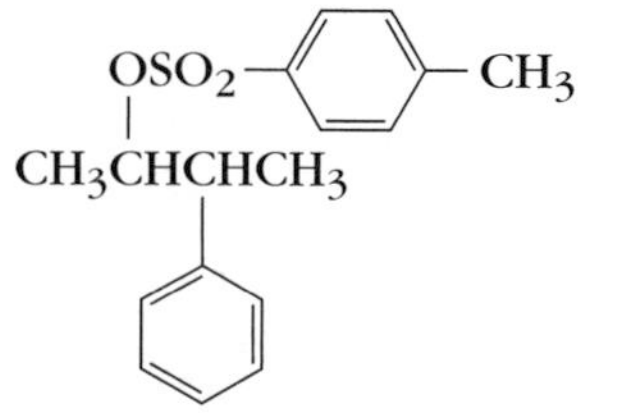

3-Phenyl-2-butyl tosylate

(a) Write three-dimensional structures for each of the four stereoisomers.

(b) For each stereoisomer, designate the configuration of each stereocenter as (R) or (S).

(c) Each of the stereoisomers reacts with sodium methoxide in methanol to form 2-phenyl-2-butene by an E2 mechanism. Write the structure of the alkene that you would expect from each stereoisomer.

12.26 The free energy-reaction path diagram for the elimination reaction of *t*-butyl bromide in ethanol with sodium ethoxide is shown below. Label the location of the starting material, product, and intermediate. Indicate the meaning of the free energies labelled *a, b,* and *c.*

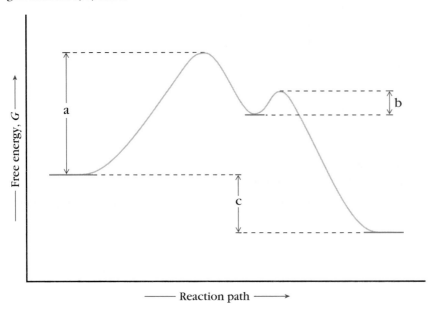

12.27 Predict the effect of the changes given below on the rate of the following reaction.

$$CH_3I \ + \ Na^{+\ -}OCH_3 \ \xrightarrow{CH_3OH} \ CH_3OCH_3 \ + \ Na^+I^-$$

  (a) Change the nucleophile from $NaOCH_3$ to $NaSeCH_3$.
  (b) Change the leaving group from I to F.
  (c) Change the leaving group from I to OTs.
  (d) Change CH_3I to $(CH_3)_2CHI$.
  (e) Change the solvent from methanol to DMF.

12.28 Write a mechanism to explain the following experimental result.

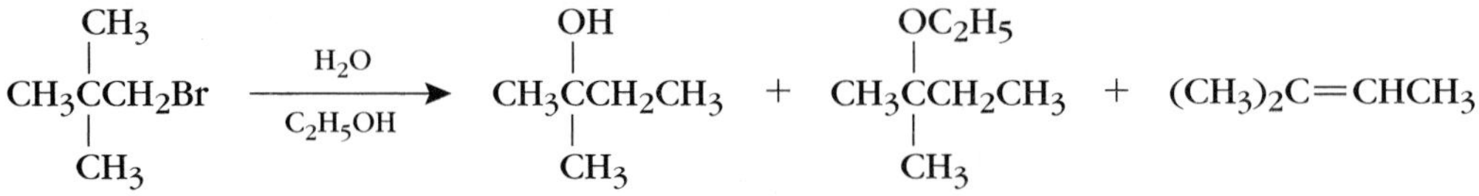

$$\underset{\underset{CH_3}{|}}{\overset{\overset{CH_3}{|}}{CH_3CCH_2Br}} \ \xrightarrow[C_2H_5OH]{H_2O} \ \underset{\underset{CH_3}{|}}{\overset{\overset{OH}{|}}{CH_3CCH_2CH_3}} \ + \ \underset{\underset{CH_3}{|}}{\overset{\overset{OC_2H_5}{|}}{CH_3CCH_2CH_3}} \ + \ (CH_3)_2C{=}CHCH_3$$

12.29 Optically active (2S,3S) 3-methoxy-2-butyl tosylate undergoes a nucleophilic substitution reaction with sodium methoxide in methanol to form an optically inactive product. Write the three-dimensional structure of the product and explain why it is optically inactive.

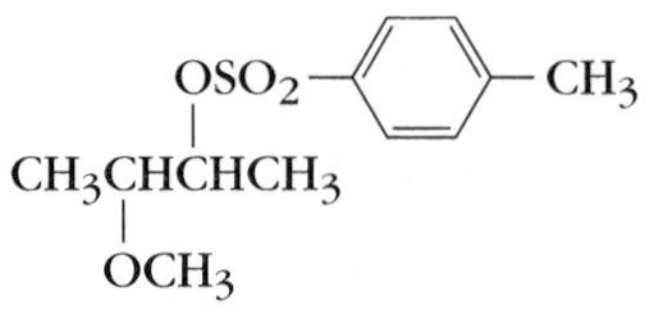

$$\underset{\underset{OCH_3}{|}}{\overset{\overset{OSO_2-\bigcirc-CH_3}{|}}{CH_3CHCHCH_3}}$$

3-Methoxy-2-butyl tosylate

12.30 In each of the following equations only the starting material and the products are given. Write the equations for the necessary steps to accomplish the synthesis. Be sure to show all reagents needed and the structure of any intermediate products. If there is more than one way to carry out the synthesis, write all the methods and decide which would be the best synthesis.

(a) $(CH_3)_2CHOH \longrightarrow (CH_3)_2CHSCH_2CH_3$

(b) $CH_3CH=CH_2 \longrightarrow CH_3\underset{\underset{CN}{|}}{C}HCH_3$

(c) $(CH_3)_2CHCH_2CH_2OH \longrightarrow (CH_3)_2CHCH_2CH_2N(CH_3)_2$

(d) (S)-1-Deuterioethanol $\longrightarrow$ $CH_3-C\overset{\overset{H}{}}{\underset{N_3}{}}D$

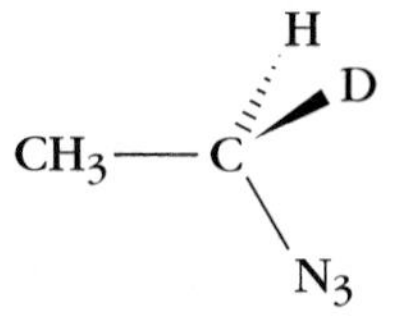

(e)

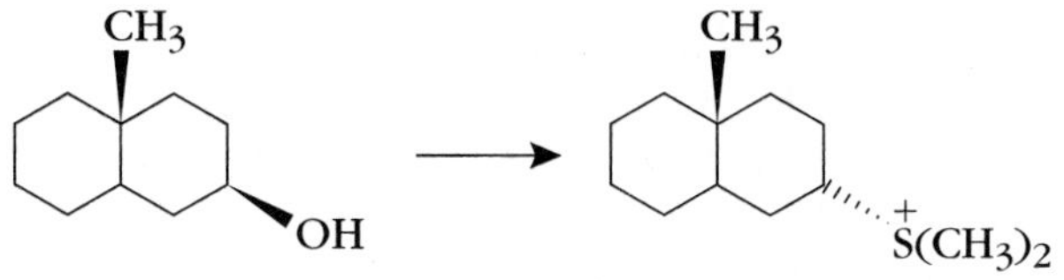

(f)

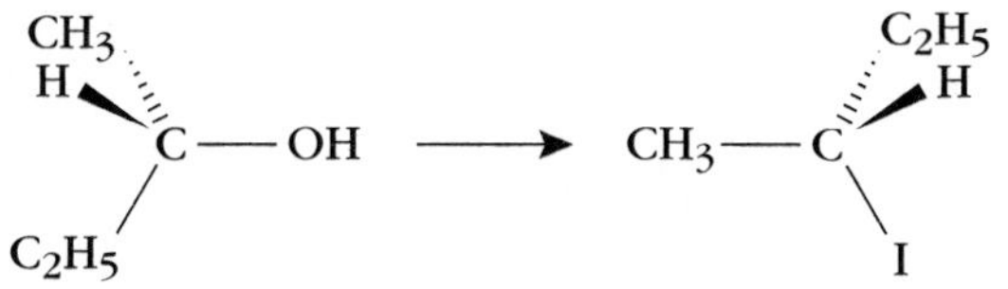

12.31 The elimination products of menthyl chloride depend on the reaction conditions as shown by the following data. Explain this experimental observation in terms of the mechanisms of elimination.

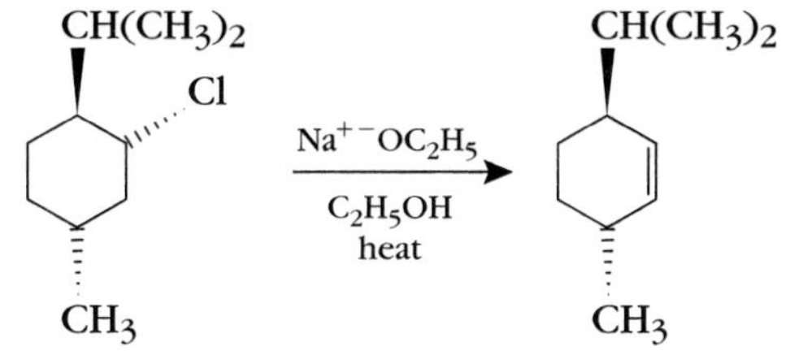

Menthyl chloride

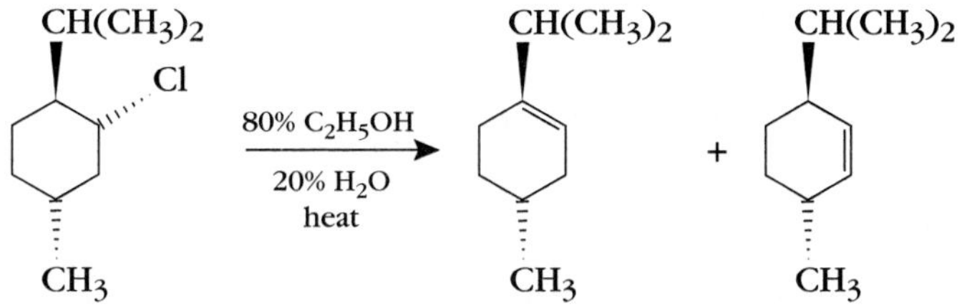

12.32 Explain why the solvolysis reaction of compound A in acetic acid occurs approximately 10,000,000 times more slowly than the corresponding reaction of compound B.

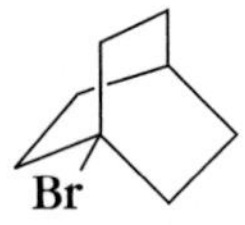

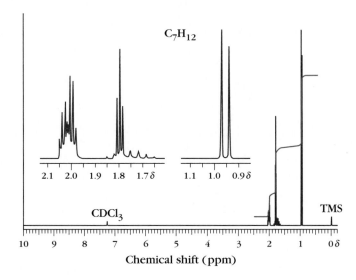

$$CH_2CH_3$$
$$|$$
$$CH_3CH_2CCH_2CH_3$$
$$|$$
$$Br$$

Compound A 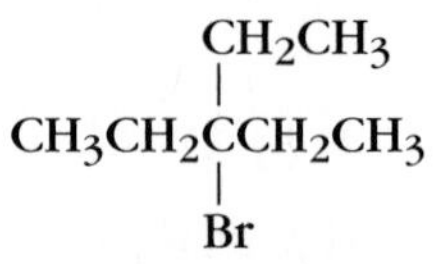   Compound B

12.33 A sodium acetylide and a bromoalkane react to form the compound whose ^1H NMR spectrum is shown in the following figure. Determine the structure of the product and write the equation for its formation.

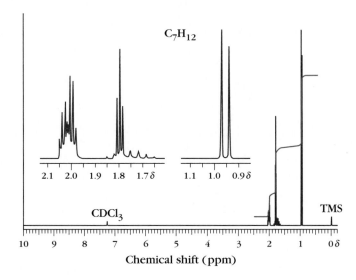

12.34 The compound $C_6H_{14}O$, whose ^1H NMR spectrum is shown in the following figure reacts with concentrated H_2SO_4 to form alkene, C_6H_{12}, whose ^1H NMR spectrum consists of a slnglet at δ 1.70. Write the structure of the starting material and the product and give the mechanism for the reaction.

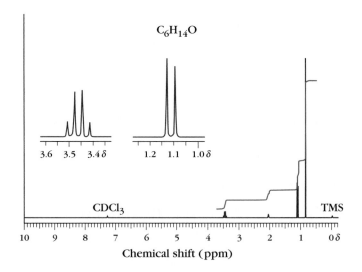

CHAPTER

13

ETHERS and EPOXIDES

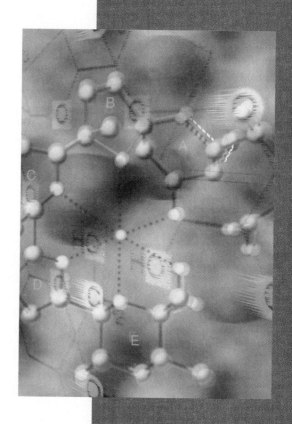

Ethers are a class of compounds with an oxygen atom bonded to two carbon atoms that are part of alkyl, aryl, or vinyl groups. The oxygen atom of an ether can be part of a straight chain or a ring, as shown by the following examples.

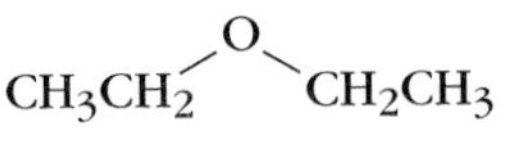

$$CH_3CH_2 \overset{O}{\diagup\diagdown} CH_2CH_3$$

Diethyl ether

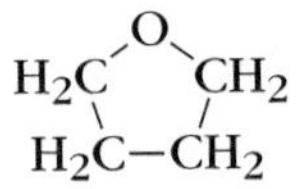

$$H_2C \overset{O}{\diagup\diagdown} CH_2$$
$$H_2C - CH_2$$

Tetrahydrofuran
(THF)

$$CH_2{=}CH \overset{O}{\diagup\diagdown} CH_2CH_3$$

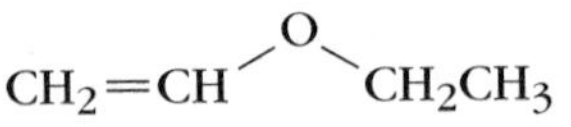

Ethyl vinyl ether

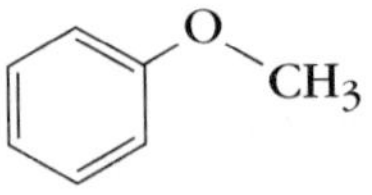

Methyl phenyl ether
(Anisole)

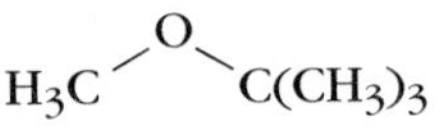

$$H_3C \overset{O}{\diagup\diagdown} C(CH_3)_3$$

tert-Butyl methyl ether

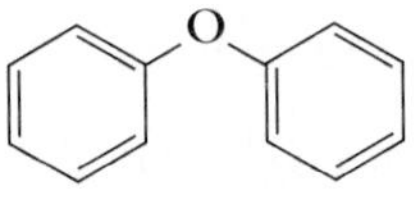

Diphenyl ether

Both THF and diethyl ether have been used in previous chapters as solvents for organic reactions. Diethyl ether is produced commercially in large quantities and is used as an industrial solvent and has been used medicinally as an anesthetic. *tert*-Butyl methyl ether is used as a gasoline additive to prevent automobile engines from knocking (premature ignition of the fuel). Anisole is used commercially to impart a licorice-like flavor to foods.

13.1   Naming Ethers

IUPAC rules allow two ways of naming ethers. Relatively simple ethers are named by specifying the two groups bonded to the oxygen atom and adding the word ether.

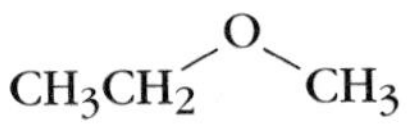

$$CH_3CH_2 \overset{O}{\diagup\diagdown} CH_3$$

Ethyl methyl ether

$$CH_3CH_2 \overset{O}{\diagup\diagdown} CH_2Cl$$

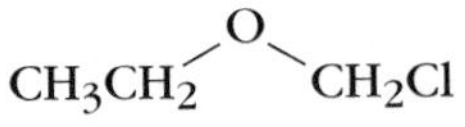

Chloromethyl ethyl ether

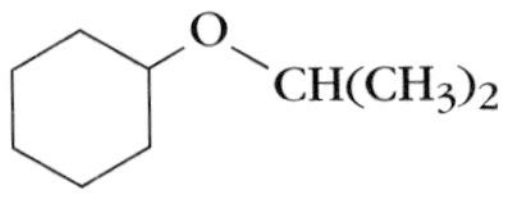

Cyclohexyl isopropyl ether

The names of the groups bonded to the oxygen atom are arranged in alphabetical order, and they, along with the word *ether,* are all separated by a space.

Ethers are named systematically by choosing the more complex group as the parent chain or structure. The oxygen atom and the other group are then treated as a substituent. Substituents of general structural formula —OR or —OAr are named by replacing the "yl" of the name of the group bonded to the oxygen atom by "oxy." For example:

$$CH_3{-}O$$

Methyl group – yl = meth + oxy = Methoxy group

We can illustrate this method by developing the IUPAC name of the following ether:

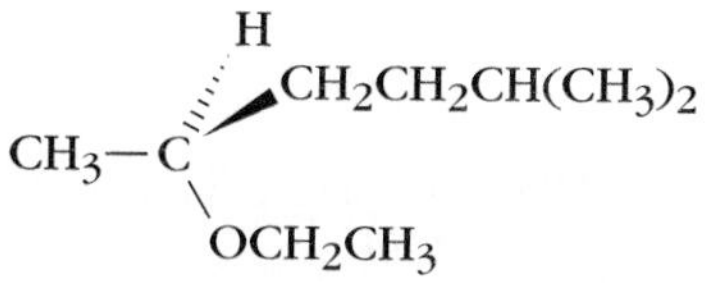

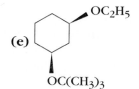

The six-carbon atom chain is the parent chain because it is more complex than the two-carbon atom chain. It contains two substituents, an ethoxy group on C2 and a methyl group on C5. Carbon atom 2 is a stereocenter with the (S) configuration. The IUPAC name of this ether is (S)-2-ethoxy-5-methylhexane.

Exercise 13.1   Give the IUPAC name to each of the following ethers.

(a) $CH_3CH_2OCH(CH_3)_2$
(b) $CH_2{=}CHOCH_3$
(c) $(CH_3)_2CHOCH(CH_3)_2$
(d) $CH_3OCH_2CH_2OCH_3$

(e)

Exercise 13.2   Write the structure of the compounds corresponding to each of the following names.

(a) Cyclobutyl methyl ether
(b) 2-Chloro-1-cyclohexoxy-2-methyloctane
(c) (R)-2-*tert*-Butoxybutane

(d) (Z)-4-Ethoxy-2-butene
(e) 1,2,3-Triethoxypropane

The systematic names of cyclic ethers are based on their ring size. The following are the parent names of cyclic ethers containing three to seven atoms.

Oxirane   Oxetane   Oxolane   Oxane   Oxepane

If the carbon atoms of these rings contain substituents, a numbering scheme is needed to systematically name the compounds. The oxygen atom is always number 1, and the substituents are numbered as shown in the following examples.

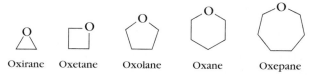

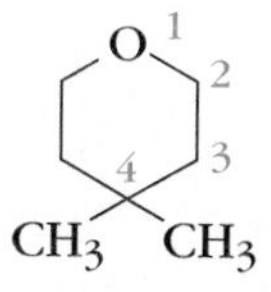

4,4-Dimethyloxane 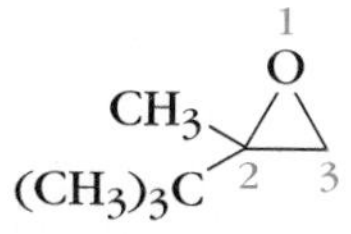   2-*tert*-Butyl-2-methyloxirane

Exercise 13.3   Give the systematic name to each of the following compounds.

(a)   (b)   (c)

Exercise 13.4   Write the structure corresponding to each of the following names.

(a) 2,2-Dimethyloxirane   (c) 2,3,4,5-Tetrachlorooxolane
(b) *cis*-2,3-Dimethyloxirane   (d) 3-Cyclohexyloxepane

Ethers with three-membered rings that contain the oxygen atom (oxiranes) are commonly known as **epoxides.** Specific simple epoxides are named by adding the word *oxide* to the name of the alkene from which it is formed by epoxidation (see Section 8.6). For example:

$$CH_2{=}CH_2 \xrightarrow[\text{CH}_3\text{COOH}]{\substack{\text{CH}_3\text{COOOH}\\ \text{(Peroxyacetic acid)}}} H_2C{-}CH_2$$

Ethylene   Ethylene oxide

Cyclohexene   Cyclohexene oxide

More complicated epoxides are named by treating the oxygen atom of the ring as a substituent, giving it the name epoxy and indicating by two numbers where the oxygen is bonded.

1,2-Epoxy-2-methylhexane   *trans*-5-Chloro-2,3-epoxyhexane

The common parent names of five- and six-membered ring ethers are derived from the names of the corresponding unsaturated cyclic ethers. The saturated five- and six-membered cyclic ethers contain four more hydrogen atoms than the corresponding unsaturated ethers, so the prefix tetrahydro is added to the name of the saturated ether.

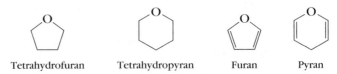

Tetrahydrofuran   Tetrahydropyran   Furan   Pyran

Exercise 13.5   Give a name to each of the following compounds.

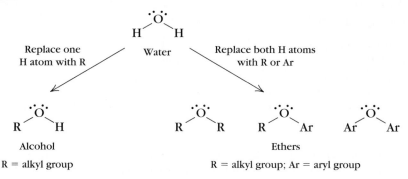

(a)

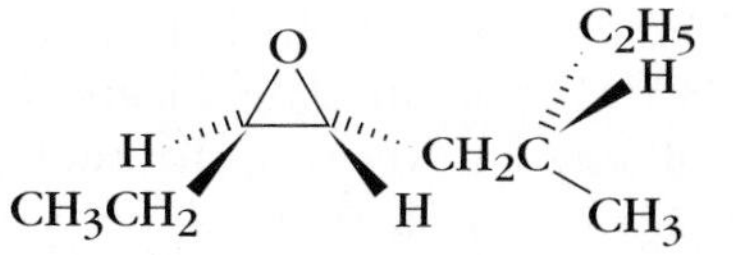

(b) 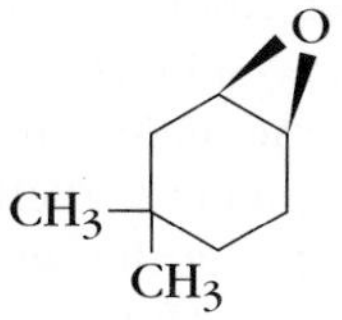 CH₃—

(c)

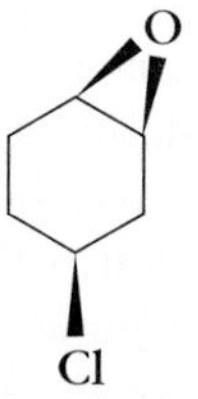

Cl

Exercise 13.6   Write the structure of the compound that corresponds to each of the following names.

(a) 2-Methyltetrahydrofuran   (c) *trans*-2,3-Epoxy-5-ethoxyhexane
(b) 3-Chlorofuran   (d) *trans*-3-Methylepoxycyclopentane

13.2   Structure of Ethers

The structures of ethers, like alcohols, can be looked at as derivatives of water. Replacing one hydrogen atom of a water molecule with an alkyl group makes an alcohol, while replacing both hydrogens with alkyl, aryl, or vinyl groups makes an ether.

$\overset{\cdot\cdot}{\underset{H \quad H}{\text{O}}}$

Replace one   Water   Replace both H atoms
H atom with R   with R or Ar

$\overset{\cdot\cdot}{\underset{R \quad H}{\text{O}}}$ $\overset{\cdot\cdot}{\underset{R \quad R}{\text{O}}}$ $\overset{\cdot\cdot}{\underset{R \quad Ar}{\text{O}}}$ $\overset{\cdot\cdot}{\underset{Ar \quad Ar}{\text{O}}}$

Alcohol   Ethers

R = alkyl group   R = alkyl group; Ar = aryl group

The C—O—C bond angle in ethers is very close to tetrahedral (109.5°), similar to the H—O—H bond angle in water. The C—O—C bond angle in dimethyl ether, for example, is 112°.

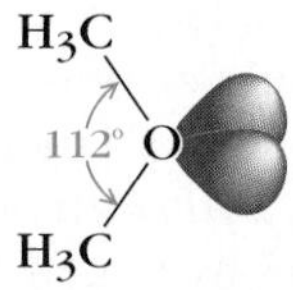

H₃C

112° O

H₃C

Dimethyl ether

Ethers lack the O—H bond present in water and alcohols. As a result, the physical properties of ethers do not resemble those of water and alcohols.

13.3   Physical Properties of Ethers

The boiling points of ethers are very similar to those of alkanes and chloroalkanes of similar molecular weight, but they are very much lower than the boiling points of alcohols of the same molecular weight. The data in Table 13.1 show, for example, that the boiling points of diethyl ether (35 °C), pentane (36 °C), and 1-chloropropane (47 °C) are about the same but are all much lower than the boiling point of 1-butanol (118 °C). The data in Table 13.1 also show that the difference between the boiling points of ethers and alcohols decreases as their molecular weights increase. Notice, for instance, that the boiling points of dibutyl ether and 1-octanol are 142 °C and 195 °C, respectively.

Table 13.1	Comparison of Boiling Points of Ethers, Alcohols, Alkanes, and Chloroalkanes of Similar Structure and Molecular Weights	
Compound	Molecular Weight	Boiling Point (°C)
CH_3OCH_3	46	−25
$CH_3CH_2CH_3$	44	−42
CH_3Cl	50	−24
CH_3CH_2OH	46	78
$CH_3CH_2OCH_2CH_3$	74	35
$CH_3(CH_2)_3CH_3$	72	36
$CH_3CH_2CH_2Cl$	78	47
$CH_3(CH_2)_2CH_2OH$	74	118
$CH_3(CH_2)_3O(CH_2)_3CH_3$	130	142
$CH_3(CH_2)_7CH_3$	128	151
$CH_3(CH_2)_5CH_2Cl$	134	160
$CH_3(CH_2)_6CH_2OH$	130	195

The solubilities of ethers in water, on the other hand, are more like alcohols than alkanes or chloroalkanes. The solubilities of diethyl ether (7.5 g/100 mL) and 1-butanol (7.9 g/100 mL) in water are about the same, for example, while 1-chloropropane and pentane are both insoluble in water.

The best way to understand these properties is by looking at alcohols and ethers as organic derivatives of water. Alcohols and water both have —OH groups, so both can form intermolecular hydrogen bonds. The boiling points of both water and low molecular weight alcohols are unusually high because energy is needed to break the hydrogen bonds. Ethers, alkanes, and chloroalkanes lack —OH groups, so they cannot form hydrogen bonds in the liquid. No extra energy is needed to break intermolecular hydrogen bonds, so boiling points of these compounds are much lower than those of alcohols of similar molecular weight.

Ethers, alcohols, and water all have an electronegative oxygen atom with two unshared pairs of electrons. This electronegative oxygen atom causes water and low molecular weight alcohols and ethers to be polar molecules with the following dipole moments.

	H_2O	CH_3CH_2OH	CH_3OCH_3	$CH_3CH_2CH_3$
Dipole moment (debyes)	1.9	1.7	1.3	0.1

The polarity of ethers is one reason why they are such good polar aprotic solvents.

Although ethers cannot form hydrogen bonds with other ether molecules, one of the unshared pairs of electrons on the ether oxygen atom can act as a hydrogen bond acceptor with water or an alcohol.

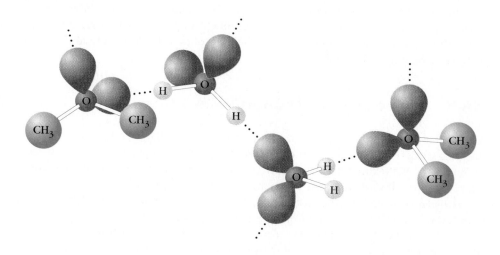

■ *Hydrogen bonding between water and dimethyl ether molecules. The dotted lines represent hydrogen bonds.*

Thus low molecular weight ethers are soluble in water because they fit into the hydrogen-bonded structure of water as alcohols do. Alkanes and chloroalkanes, on the other hand, lack the unshared pairs of electrons necessary to act as hydrogen bond acceptors, so they are essentially insoluble in water.

13.4   Conformations of Tetrahydropyrans

Tetrahydropyrans, which are six-membered saturated rings that contain a single oxygen atom, occur very widely in nature. They are most abundant in the pyranose sugars (Chapter 25). In order to better understand properties of these complicated carbohydrates, chemists have studied the simpler tetrahydropyrans in great detail. These investigations demonstrate that there are many similarities in the conformational preferences of cyclohexane and tetrahydropyran.

■ *The conformations of cyclohexane are discussed in Section 4.9.*

Microwave studies show that tetrahydropyran preferentially exists in the chair conformation just like cyclohexane. Oxygen-carbon single bonds are shorter than carbon-carbon single bonds, however, and the C—O—C bond angle is slightly smaller too, so the chair conformation of tetrahydropyran must be modified to accomodate these slight differences. Consequently, the chair conformation of tetrahydropyran is slightly more puckered than that of cyclohexane.

There are two types of hydrogens in tetrahydropyran just as there are in cyclohexane. Five hydrogen atoms are in the axial and five hydrogen atoms are in the equatorial positions in tetrahydropyran. The axial and the equatorial hydrogen atoms can be interconverted by a ring flip similar to that in cyclohexane. The free energy of activation for ring flipping is similar for tetrahydropyran, 10.3 kcal/mol (43 kJ/mol) and cyclohexane 10.8 kcal/mol (45 kJ/mol).

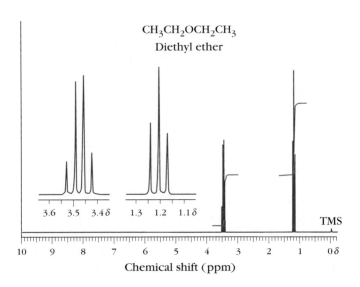

Axial hydrogen   **Equatorial hydrogen**

Substituents generally prefer to occupy the equatorial positions of tetrahydropyran because there is less steric hindrance compared to the axial positions. Thus the conformational preferences of substituents in cyclohexane and tetrahydropyran are generally similar.

13.5 Spectroscopic Properties of Ethers

The infrared spectra of ethers show a characteristic band due to the carbon-oxygen single bond stretch in the 1050 to 1250 cm^{-1} range, as discussed in Section 5.5, E. Unfortunately, many other absorptions occur in this range, which makes it difficult to distinguish ethers by infrared spectroscopy.

The electronegative ether oxygen deshields hydrogen atoms on carbon atoms bonded to the ether oxygen atom. The chemical shifts of hydrogen atoms in the H—C—O—C group, for instance, are usually found in the δ 3.0 to 4.0 range, as shown in the ^{1}H NMR spectrum of diethyl ether in Figure 13.1.

The ^{1}H NMR spectra of epoxides are often complex. The ^{1}H NMR spectrum of 1,2-epoxybutane, for instance, consists of five groups of signals, as shown in Figure 13.2. The methyl hydrogen atoms of 1,2-epoxybutane are found in Figure 13.2 as a triplet at δ 1.0. The methylene hydrogen atoms are found as a multiplet at about δ 1.6, and the three hydrogen atoms bonded to the carbon atoms of the epoxide ring are found as multiplets at δ 2.45, δ 2.65, and δ 2.85.

Figure 13.1
The ^{1}H NMR spectrum of diethyl ether.

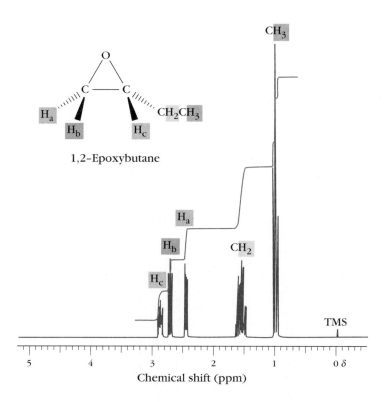

Figure 13.2
The ^{1}H NMR spectrum of 1,2-epoxybutane.

The ^{1}H NMR spectrum of 1,2-epoxybutane is complex because the three hydrogen atoms bonded to the carbon atoms of the epoxide ring are nonequivalent, and coupling occurs between them. The coupling is shown by the tree diagram in Figure 13.3.

The electronegative oxygen atom also deshields carbon atoms bonded to it, so their chemical shifts are usually found in the δ 50 to 80 range, as shown in the ^{13}C NMR spectrum of diethyl ether in Figure 13.4.

■ *The representation of coupling between nonequivalent hydrogen atoms by a tree diagram is discussed in Section 10.12.*

13.6 Ethers Are Lewis Bases

The unshared pairs of electrons on the oxygen atom make it possible for ethers to form Lewis acid-base complexes with many compounds and cations. Boron trifluoride (BF_3), a toxic gas that is used as a Lewis acid catalyst in a wide variety of reactions, forms complexes with ethers. Bubbling gaseous BF_3 into diethyl ether forms a stable solution of the complex that is convenient to use and store. The name of this complex of diethyl ether and BF_3 is boron trifluoride ethyl etherate.

Boron trifluoride	Diethyl ether	Boron trifluoride ethyl etherate

Recall from Chapter 11 that THF and diethyl ether are commonly used as solvents in the formation of Grignard reagents. One or more molecules of ether are

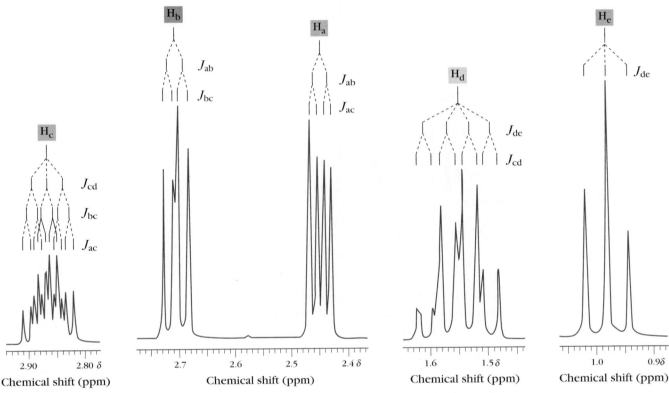

1,2–Epoxybutane

Figure 13.3
The tree diagram for the coupling of the five kinds of hydrogen atoms of 1,2-epoxybutane. Hydrogen atom H_a is split into a doublet by coupling with H_b. Each signal of the doublet is further split into a doublet by coupling with H_c. In a similar manner, H_b is split into a doublet of doublets by coupling with H_a and H_c. Notice that J_{cis} (J_{bc}) > J_{trans} (J_{ac}). H_c is split into a triplet by coupling with the methylene hydrogen atoms (H_d). Each signal of the triplet is split into a doublet by coupling with H_b. Each signal of the doublets is further split into a doublet by coupling with H_a. H_d is split into a quartet by coupling with the methyl hydrogen atoms (H_e). Each signal of the quartet is further split into a doublet by coupling with H_c. Finally, the methyl hydrogen atoms are split into the usual triplet of an ethyl group.

believed to form a complex with the magnesium ion, which stabilizes the Grignard reagent and keeps it in solution.

$$CH_3CH_2 \quad\quad R \quad\quad CH_2CH_3$$
$$\underset{CH_3CH_2}{\overset{\quad}{}}\ddot{O}\cdots\overset{+}{Mg}\cdots\ddot{O}\underset{CH_2CH_3}{\overset{\quad}{}}$$
$$Cl^-$$

Ethers also form Lewis acid-base complexes with cations.

$$M^+ \ + \ \ddot{:}\overset{R}{\underset{R}{O}} \quad\longrightarrow\quad M\cdots\overset{+}{\ddot{O}}\overset{R}{\underset{R}{}}$$

Cation   Ether   Ether-cation
(Lewis acid) (Lewis base)   complex

A number of macrocyclic polyethers have been prepared recently that contain four or more ether functional groups in a ring of twelve or more atoms. They are

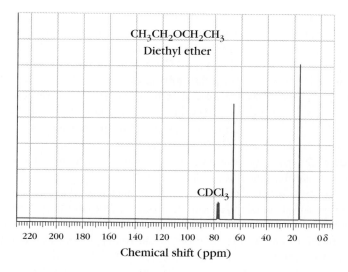

Figure 13.4
The ^{13}C NMR spectrum of diethyl ether.

given the name **crown ethers** because their three-dimensional shape resembles a crown.

The IUPAC names of these crown ethers are very complex, so a shorthand method of naming them is commonly used. The word *crown* is preceded by a number that indicates the total number of atoms in the ring and followed by another number that indicates the number of oxygen atoms in the ring. Thus 12-crown-4 ether is a twelve-atom ring that contains four oxygen atoms.

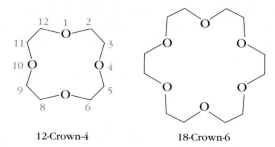

12-Crown-4   18-Crown-6

A crown ether such as 18-crown-6 is shaped much like a doughnut with a hole in the middle. Lining the inside of the hole are the oxygen atoms, while the hydrogen atoms of the —CH_2 groups cover the outside. A cation fits into the hole of the crown, where it is held by electrostatic interactions with the unshared electron pairs of the oxygen atoms. Only cations that fit comfortably in the hole in the crown form complexes with a particular crown ether. A potassium ion, for example, fits exactly in the hole of 18-crown-6. As a result, 18-crown-6 ether forms a strong complex with the potassium ions of potassium salts.

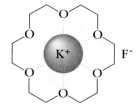

18-Crown-6 potassium fluoride complex

Complexes of a crown ether and a cation have both hydrophilic and hydrophobic regions. The outside of the complex, made up of carbon-hydrogen bonds, is hydrophobic; the inside of the complex, made up of the charged cation and the oxygen atoms, is hydrophilic. The hydrophobic exterior region makes these complexes soluble in nonpolar solvents. As a result, they can be used to carry out reactions in nonpolar solvents with inorganic compounds that ordinarily require aqueous solutions. Potassium permanganate, for example, is usually used to oxidize alcohols in aqueous acidic solutions because it is insoluble in nonpolar solvents such as benzene (Section 11.18). Adding 18-crown-6 to benzene, however, dissolves potassium permanganate to form a valuable reagent for the oxidation of alcohols in a nonpolar solvent.

A wide variety of crown ethers of various sizes and shapes have been made and their ability to form complexes with cations is being extensively studied. The purpose of this research is not only to learn how to prepare better reagents for organic synthesis, but also to learn more about the types of interactions that bind the cation to the organic Lewis base. The results of this research are applicable not only to macrocyclic polyethers synthesized in the laboratory, but also to the naturally occurring compounds that bind cations.

13.7   Naturally Occurring Ethers Bind Cations and Increase Their Mobility Across Membranes

The flow of molecules and ions across the membrane that separates the inside of a cell from its environment is carefully regulated by specific transport systems. The flow of sodium and potassium ions into and out of a cell is particularly important because the ratio of their concentrations must be maintained within a narrow range to permit enzyme-catalyzed systems to function normally.

Some microorganisms synthesize compounds, called **ionophores,** that increase the permeability of cell membranes to sodium, potassium, and other ions. Ionophores form complexes with these cations, and these complexes are

Figure 13.5
A, Structure of monensin with the oxygen atoms that complex the sodium ion indicated by orange screens. **B**, Ball-and-stick model of monensin-sodium ion complex showing the octahedral geometry of the oxygen atoms about the sodium ion. For clarity, hydrogen atoms are omitted.

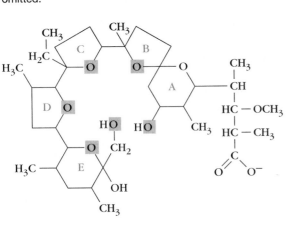

A

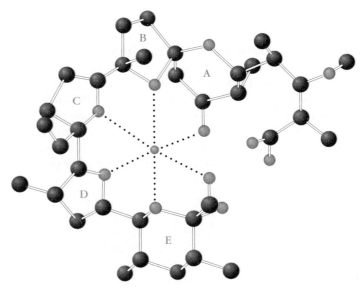

B

able to traverse cell membranes and bypass the normal ion transport system. Ionophores disrupt the careful balance of ions in a cell and interfere with the normal operation of the cell. This often leads to the death of the cell.

A number of antibiotics work by transporting sodium and potassium ions through cell membranes. An example is the complex formed between monensin and sodium ions. The structure of this complex is shown in Figure 13.5. The monensin-sodium ion complex, shown in Figure 13.5, has both hydrophobic and hydrophilic regions. The alkanelike periphery is the hydrophobic portion of this complex, analogous to the crown ether complexes. The hydrophilic portion is the interior of the complex, where the unshared electron pairs of oxygen atoms, from either ether groups or hydroxy groups, bind the cation by means of electrostatic interactions. Ionophores work, therefore, by wrapping a cation in a hydrophobic blanket, which effectively hides the hydrophilic part of the complex. This makes the complex soluble in the fatty (hydrophobic) cell membrane and allows it to easily pass through the membrane.

13.8 Preparation of Ethers

Acyclic and cyclic ethers are prepared by reactions already introduced, such as nucleophilic substitution reactions, or modifications of reactions already introduced, such as a modified oxymercuration-demercuration reaction. In addition to these laboratory methods, ethers used in large quantities as solvents are prepared industrially by a modification of the acid-catalyzed reactions of alcohols first introduced in Section 7.18. We begin by examining the industrial preparation of ethers.

A   Industrial Preparation

Diethyl ether and other symmetrical acyclic ethers are prepared industrially by the sulfuric acid–catalyzed intermolecular dehydration of primary (1°) alcohols.

$$2\ CH_3CH_2OH \xrightarrow[140\ °C]{H_2SO_4} CH_3CH_2OCH_2CH_3\ +\ H_2O$$

These same reagents were used previously to form an alkene in an acid-catalyzed dehydration reaction. Why is an ether formed in one case and an alkene in another? The answer is that the product of the reaction depends on the experimental conditions. In order to form mostly ether, the temperature of the reaction must be kept relatively low. For example, diethyl ether is the major product when excess ethanol and concentrated sulfuric acid are heated to 140 °C. If the temperature is raised to 180 °C, however, then ethylene is the major product. In general, if the alcohol is 2° or 3° or the temperature is too high, the balance between substitution and elimination shifts in favor of elimination.

| Exercise 13.7 | The mechanism for the formation of diethyl ether by the sulfuric acid–catalyzed intermolecular dehydration of ethanol has many features in common with the acid-catalyzed unimolecular dehydration of ethanol to form ethylene. Compare and contrast the two mechanisms. |

| **Exercise 13.8** | You have just been hired for the summer by a chemical company that decides to begin producing 1,4-dioxane. Your boss, knowing that you have taken a course in organic chemistry, asks you to propose a synthesis of 1,4-dioxane using the sulfuric acid–catalyzed intermolecular dehydration of an alcohol. Which alcohol would you use as starting material? Based on the mechanism of the reaction, what products other than 1,4-dioxane might be formed? |

1,4-Dioxane

The sulfuric acid–catalyzed dehydration of alcohols is a useful industrial synthesis of a limited number of symmetrical ethers. In the laboratory, however, alkoxymercuration-demercuration of alkenes is a much better method of preparing ethers because a wide range of acyclic ethers can be prepared in which the alkyl groups are primary, secondary, and tertiary.

B   Alkoxymercuration-Demercuration of Alkenes

The oxymercuration-demercuration method of preparing alcohols was introduced in Chapter 8 (Section 8.10). The first step, the oxymercuration, is the reaction of an alkene with water in the presence of mercury (II) acetate.

$$\underset{\text{Alkene}}{\overset{\diagdown}{\diagup}}C=C\underset{}{\overset{\diagup}{\diagdown}} \quad + \quad \underset{\text{Mercury (II) acetate}}{Hg(OAc)_2} \quad \xrightarrow{H_2O/THF} \quad HO-\overset{|}{\underset{|}{C}}-\overset{|}{\underset{|}{C}}-HgOAc$$

This reaction can be modified to serve as a synthesis of ethers by simply employing an alcohol as solvent rather than a mixture of water and THF. The product of this reaction then is an alkoxy mercury (II) acetate.

$$\underset{\text{Alkene}}{\overset{\diagdown}{\diagup}}C=C\underset{}{\overset{\diagup}{\diagdown}} \quad + \quad \underset{\text{Mercury (II) acetate}}{Hg(OAc)_2} \quad \xrightarrow{ROH} \quad \underset{\text{Alkoxy mercury (II) acetate}}{RO-\overset{|}{\underset{|}{C}}-\overset{|}{\underset{|}{C}}-HgOAc}$$

Demercuration of this product is accomplished by reaction with an aqueous solution of NaOH and NaBH$_4$. This two-step sequence of reactions is called **alkoxymercuration-demercuration.** The net result of this reaction is the addition of alcohol to the alkene double bond according to the Markownikoff Rule.

ALKOXYMERCURATION

$$CH_3(CH_2)_3CH{=}CH_2 \quad + \quad Hg(OAc)_2 \quad \xrightarrow{CH_3OH} \quad CH_3(CH_2)_3\overset{\displaystyle OCH_3}{\overset{|}{CH}}-CH_2-HgOAc$$

1-Hexene

DEMERCURATION

$$CH_3(CH_2)_3\overset{\overset{\displaystyle OCH_3}{|}}{CH}-CH_2-HgOAc \xrightarrow[\text{NaOH/H}_2\text{O}]{\text{NaBH}_4} CH_3(CH_2)_3\overset{\overset{\displaystyle OCH_3}{|}}{CH}-CH_2-H$$

2-Methoxyhexane
(80% yield)

Just like oxymercuration-demercuration, alkoxymercuration-demercuration occurs without rearrangement.

$$(CH_3)_3C-CH=CH_2 \xrightarrow[\text{(2) NaBH}_4/\text{NaOH/H}_2\text{O}]{\text{(1) Hg(OAc)}_2/\text{C}_2\text{H}_5\text{OH}} (CH_3)_3C-\overset{\overset{\displaystyle}{|}}{\underset{\underset{\displaystyle OC_2H_5}{|}}{CH}}-CH_3$$

3,3-Dimethyl-1-butene

2-Ethoxy-3,3-dimethylbutane
(78% yield)

As a result, alkoxymercuration-demercuration can be used to synthesize ethers of a wide variety of structural types.

In devising a synthesis using the alkoxymercuration-demercuration method, we must decide which alkene and alcohol to react with mercury (II) acetate in order to form the desired product. Sample Problem 13.1 illustrates how this decision is made.

Sample Problem 13.1   Propose a synthesis of the following ether using the alkoxymercuration-demercuration method.

$$CH_3CH_2CH_2\overset{\overset{\displaystyle OCH_2CH_2CH_3}{|}}{CH}CH_3$$

Solution:

First, identify the two alkyl groups of the ether. The two alkyl groups are:

$$CH_3CH_2CH_2\overset{\overset{\displaystyle |}{}}{CH}CH_3 \quad \text{and} \quad -CH_2CH_2CH_3$$

Next, decide which alkyl group will come from the alkene and which from the alcohol. There are three possible combinations of alkenes and alcohols.

Combination of reagents	Alkene	Alcohol
1	$CH_3CH_2CH=CHCH_3$	$CH_3CH_2CH_2OH$
2	$CH_3CH_2CH_2CH=CH_2$	$CH_3CH_2CH_2OH$
3	$CH_3CH=CH_2$	$CH_3CH_2CH_2\overset{\overset{\displaystyle OH}{|}}{CH}CH_3$

The reaction of 1-propene and 2-pentanol (combination 3) is unacceptable because it gives the wrong product.

$$CH_3CH=CH_2 \xrightarrow[\text{(2) NaBH}_4/\text{NaOH/H}_2\text{O}]{\text{(1) CH}_3\text{CH}_2\text{CH}_2\overset{\overset{\displaystyle OH}{|}}{\text{CH}}\text{CH}_3/\text{Hg(OAc)}_2} CH_3CH_2CH_2\overset{\overset{\displaystyle OCH(CH_3)_2}{|}}{CH}CH_3$$

Not the desired product

The reaction of 2-pentene and 1-propanol (combination 1) gives the desired product, but it is an unacceptable synthesis because the product is a mixture of two constitutional isomers in comparable yield.

$$
\begin{array}{c}
\text{OCH}_2\text{CH}_2\text{CH}_3 \\
| \\
\text{CH}_3\text{CH}_2\text{CH}_2-\overset{}{\text{CHCH}_3}
\end{array}
$$

The desired product

$$\text{CH}_3\text{CH}_2\text{CH}=\text{CHCH}_3 \xrightarrow[\text{(2) NaBH}_4/\text{NaOH}/\text{H}_2\text{O}]{\text{(1) Hg(OAc)}_2/\text{CH}_3\text{CH}_2\text{CH}_2\text{OH}}$$

2-Pentene

$+$

$$
\begin{array}{c}
\text{CH}_3\text{CH}_2\text{CH}-\text{CH}_2\text{CH}_3 \\
| \\
\text{OCH}_2\text{CH}_2\text{CH}_3
\end{array}
$$

Not the desired product

Combination 2 forms the desired product (and only the desired product), so it is the best synthesis of the three.

$$\text{CH}_3\text{CH}_2\text{CH}_2\text{CH}=\text{CH}_2 \xrightarrow[\text{(2) NaBH}_4/\text{NaOH}/\text{H}_2\text{O}]{\text{(1) Hg(OAc)}_2/\text{CH}_3\text{CH}_2\text{CH}_2\text{OH}} \begin{array}{c} \text{OCH}_2\text{CH}_2\text{CH}_3 \\ | \\ \text{CH}_3\text{CH}_2\text{CH}_2\text{CHCH}_3 \end{array}$$

Exercise 13.9   Propose a synthesis of each of the following ethers using the alkoxymercuration-demercuration method.

(a) Cyclopentyl ethyl ether   **(c)** 2-Ethoxy-2-methylbutane
(b) 1-Butoxy-1-methylcyclohexane   **(d)** 3-Propoxyhexane

Exercise 13.10   Ethanol is available in most laboratories as either absolute ethanol (100% ethanol) or a mixture of 95% ethanol/5% water. A student, in an attempt to prepare 2-ethoxyhexane by the alkoxymercuration-demercuration reaction, uses 95% ethanol instead of absolute ethanol. How will this mistake affect the synthesis of the product?

C   The Williamson Ether Synthesis

The Williamson ether synthesis is the reaction of alkoxide ions with methyl or primary haloalkanes or alkyl tosylates to form ethers.

Cyclohexanol   Methoxycyclohexane
                                         (92% yield)

The Williamson ether synthesis is a practical example of a nucleophilic substitution reaction that occurs by an S_N2 mechanism. Recall from Chapter 12 that secondary and tertiary alkyl derivatives react with alkoxide ions to form substantial quantities of elimination products by either an E1 or E2 mechanism. In practical terms this means that secondary and tertiary alkyl groups are less satisfactory as reaction sites than primary alkyl groups, as illustrated in Sample Problem 13.2.

| Sample Problem 13.2 | Propose a synthesis of the following ether using the Williamson ether synthesis. |

$$CH_3CH_2CH_2 \overset{O}{\diagdown} CH(CH_3)_2$$

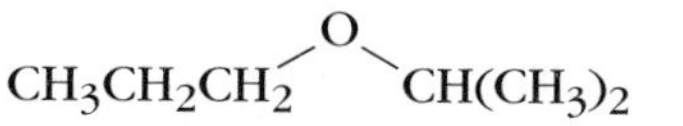

Solution:

One alkyl group of the ether must be derived from the alkoxide ion and the other from the haloalkane or alkyl tosylate. There are two possibilities, just as in most cases of the Williamson ether synthesis.

$$\underset{\text{Alkoxide ion}}{CH_3CH_2CH_2O^-} + \underset{\substack{\text{Secondary} \\ \text{alkyl compound}}}{\overset{\overset{\displaystyle X}{|}}{CH_3CHCH_3}}$$

or $X = Cl, Br, I, \text{ or } OTs$

$$\underset{\text{Alkoxide ion}}{(CH_3)_2CHO^-} + \underset{\substack{\text{Primary} \\ \text{alkyl compound}}}{CH_3CH_2CH_2X}$$

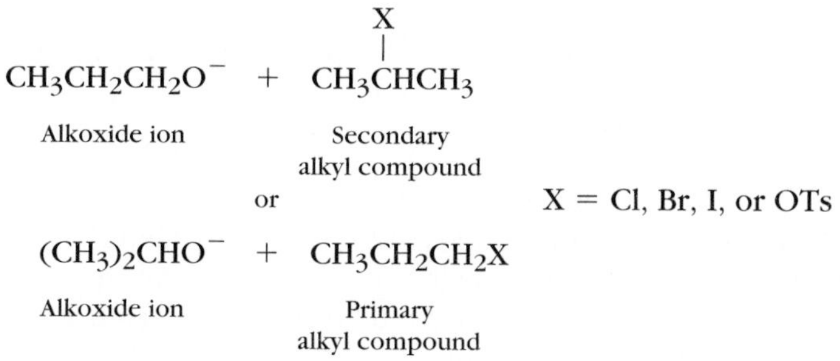

Alkoxide ions form more substitution products when they react with primary alkyl compounds than they do with secondary alkyl compounds. The reaction of $(CH_3)_2CHO^-$ with $CH_3CH_2CH_2X$ gives a higher yield of ether, which makes it the synthetic method of choice.

$$(CH_3)_2CHO^- + CH_3CH_2CH_2Br \longrightarrow CH_3CH_2CH_2 \overset{O}{\diagdown} CH(CH_3)_2$$

If one alkyl group of the ether is highly branched, it is generally best to obtain that group from the alkoxide ion, as shown in the following example.

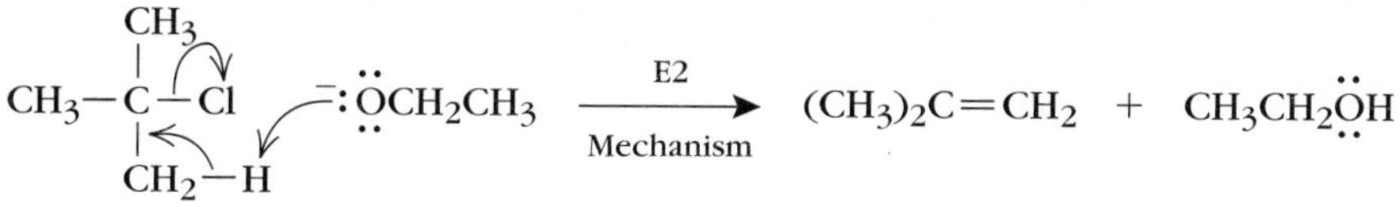

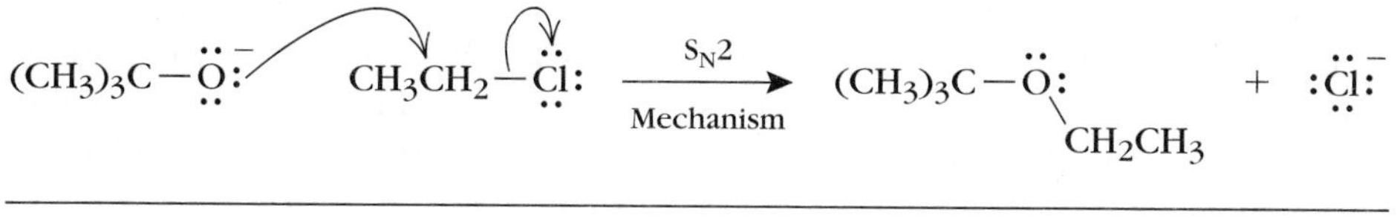

$$CH_3 - \overset{\overset{\displaystyle CH_3}{|}}{\underset{\underset{\displaystyle CH_2 - H}{|}}{C}} - Cl \quad :\overset{..}{\underset{..}{O}}CH_2CH_3 \xrightarrow[\text{Mechanism}]{E2} (CH_3)_2C{=}CH_2 + CH_3CH_2\overset{..}{\underset{..}{O}}H$$

$$(CH_3)_3C - \overset{..}{\underset{..}{O}}{:}^- \quad CH_3CH_2 {-} \overset{..}{\underset{..}{C}}l{:} \xrightarrow[\text{Mechanism}]{S_N2} (CH_3)_3C - \overset{..}{\underset{..}{O}}{:} \underset{CH_2CH_3}{} \quad + \quad :\overset{..}{\underset{..}{C}}l{:}^-$$

The Williamson ether synthesis is a particularly good method of preparing phenyl ethers, which have the general formula:

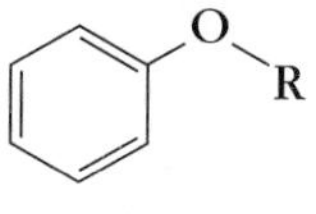

The only restriction to this method is that the aromatic ring must be introduced as part of a phenoxide ion rather than a halogen-containing aromatic compound.

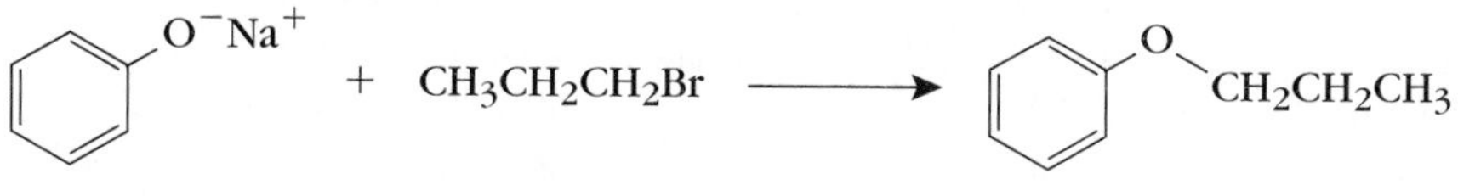

| Sodium phenoxide | 1-Bromopropane | Phenyl propyl ether |

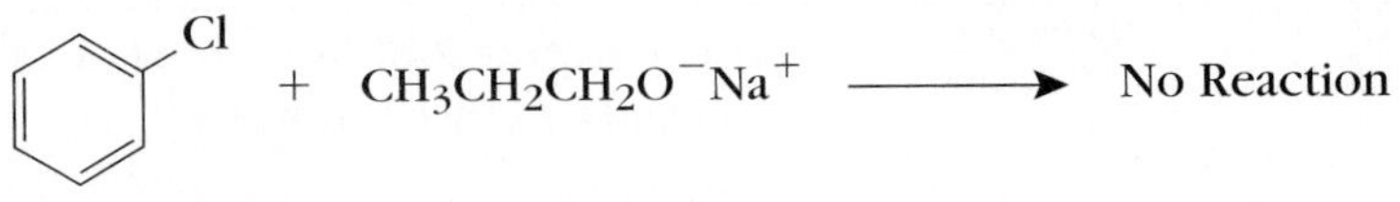

Chlorobenzene   Sodium propoxide

The reasons for this restriction are discussed in Chapter 23.

    The Williamson ether synthesis can also be used to prepare cyclic ethers. An example is the preparation of tetrahydrofuran from the reaction of 4-chloro-1-butanol and aqueous sodium hydroxide.

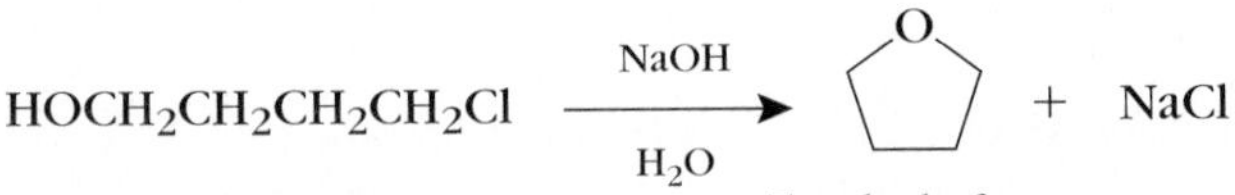

4-Chloro-1-butanol   Tetrahydrofuran
(70% yield)

The nucleophile is an alkoxide ion formed by the removal of the hydrogen atom of the hydroxy group of the alcohol by hydroxide ion. In the second step, the alkoxide ion reacts intramolecularly to displace the chloride ion in what amounts to an internal S_N2 mechanism.

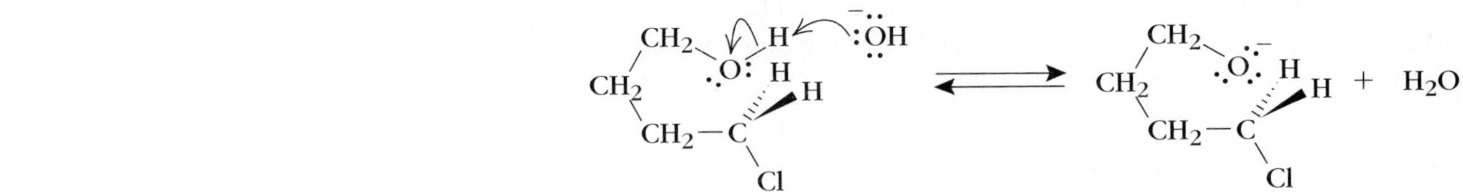

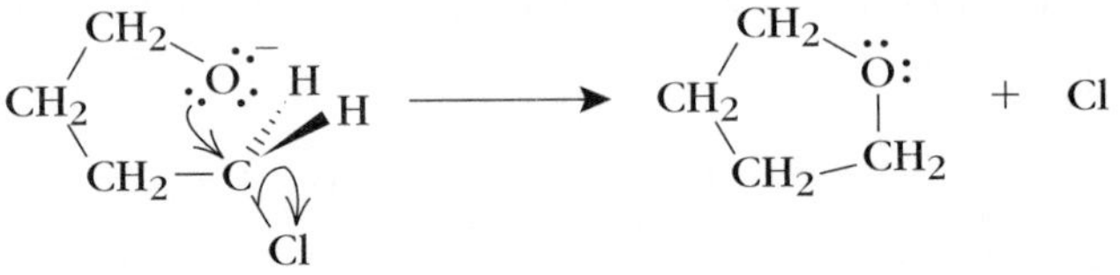

    There are two noteworthy features of this reaction. First, an intramolecular reaction is favored over an intermolecular one. None of the product of intermolecular nucleophilic substitution of chloride ion by hydroxide ion, 1,4-butanediol, is found because the intramolecular reaction is faster than the intermolecular one.

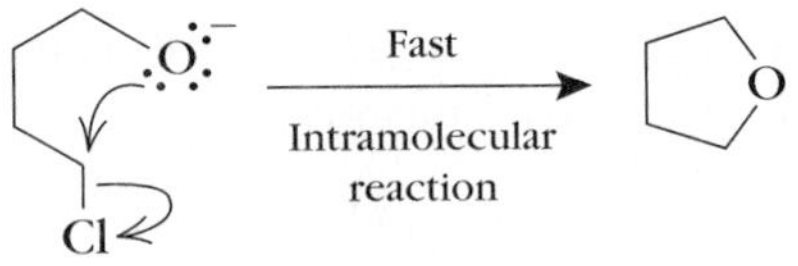

Tetrahydrofuran

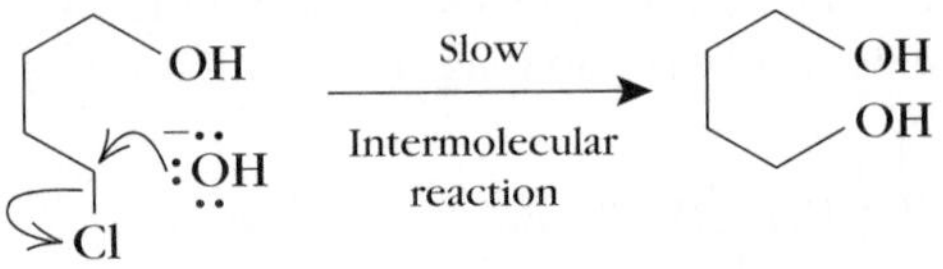

1,4-Butanediol

Second, sodium hydroxide is the base that generates the alkoxide ion. Stronger bases than hydroxide ion are usually needed to completely convert an alcohol into its conjugate base, as discussed in Chapter 11 (Section 11.15). Hydroxide ion works in this case, however, because the alcohol does not need to be converted completely into its conjugate base. As alkoxide ion cyclizes in the second step, more alkoxide is generated to restore the equilibrium in the first step. This process continues until all the alcohol is converted into alkoxide ion and the reaction is complete.

Exercise 13.11 Propose a synthesis of each of the following ethers by means of the Williamson ether synthesis.

(a) *N*-Butyl methyl ether
(b) 1,2-Dimethoxyethane
(c) 3-Isopropoxy-1-propene

(e)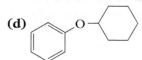

(d)

(f) Tetrahydropyran

Exercise 13.12 A student tries to synthesize methoxyethane by heating a mixture of iodomethane, ethanol, and aqueous sodium hydroxide. The student, puzzled by the low yield of ether obtained, asks for your advice. Why did the student get a low yield of ether?

13.9 Reactions of Ethers

Ethers are widely used as polar aprotic solvents because they do not react with most reagents used in organic reactions. Ethers, except for oxiranes, as we will learn in Section 13.12, do not react with halogens, weak acids, bases, or nucleophiles. Ethers do react, however, with strong acids. Heating a solution of a dialkyl ether in concentrated aqueous HI or HBr cleaves the ether bond and initially forms a haloalkane and an alcohol. The alcohol can react further in excess HI or HBr to form a second mole of haloalkane.

■ *The reaction of alcohols and hydrogen halides to form haloalkanes is discussed in Sections 11.17 and 12.9.*

$$\underset{R}{\overset{R}{\diagdown}}O \xrightarrow{\text{HX}} ROH + RX$$

$$\xrightarrow[\text{HX}]{} RX + H_2O$$

$$\underset{R}{\overset{Ar}{\diagdown}}O \xrightarrow{\text{HX}} ArOH + RX$$

Ar = aromatic ring
HX = HCl < HBr < HI

INCREASING REACTIVITY

Alkyl phenyl ethers react with HI and HBr to form a phenol and a haloalkane. The phenol *does not* react with strong acids to form halobenzenes for reasons that are discussed in Chapter 23. The following specific examples illustrate the acid-induced cleavage of ethers:

$$CH_3CH_2OCH_2CH_3 \xrightarrow[\text{Heat}]{\substack{\text{Excess} \\ \text{conc. HI}}} 2\ CH_3CH_2I$$

Diethyl ether   Iodoethane

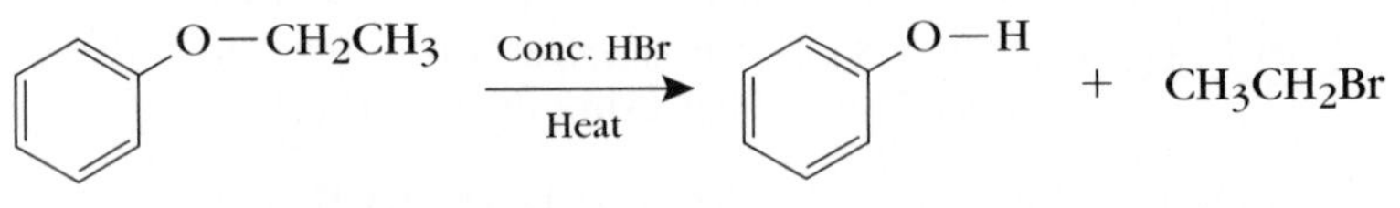

Ethyl phenyl ether   Phenol   Bromoethane

The mechanism of the acid-induced cleavage of ether is shown in Figure 13.6. Step 1 is an acid-base reaction that forms the conjugate acid of the ether. This reaction converts a poor leaving group, the alkoxy group, into a good leaving group, an alcohol. Step 2 is a nucleophilic substitution reaction in which the halide ion is the nucleophile and the alcohol is the leaving group.

Ethers are cleaved by HI and HBr for two reasons. First, both are strong enough acids to protonate the oxygen atom of the ether, and second, their conjugate bases, bromide and iodide ions, are good nucleophiles in the subsequent nucleophilic substitution reaction. Since HI is the stronger acid and iodide ion is the better nucleophile, HI cleaves ethers more rapidly than HBr or HCl do.

The nucleophilic substitution reaction in step 2 of the mechanism in Figure 13.6 occurs by either an S_N1 or S_N2 mechanism, depending on the structure of the ether. If the carbon atom bonded to the oxygen atom of an ether is either primary or secondary, step 2 occurs by an S_N2 mechanism. If the two carbon atoms are tertiary, then step 2 occurs by an S_N1 mechanism because the alkyl group is capable of forming a stable carbocation.

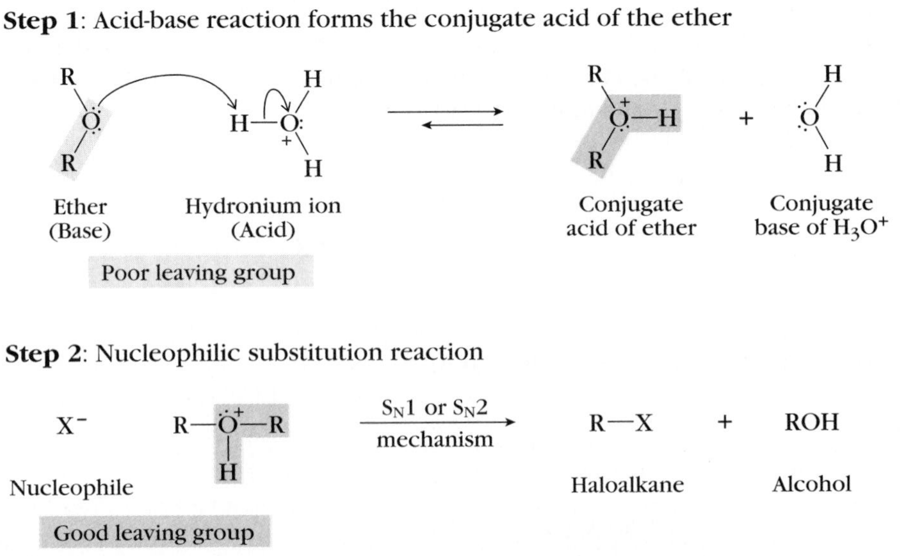

Figure 13.6
The mechanism of the acid-induced cleavage of an ether. The key to the mechanism is step 1 in which a poor leaving group, the alkoxy group, is protonated thereby converting it into a good leaving group. Step 2 occurs by either an S_N1 or S_N2 mechanism depending on the structure of the carbon atoms bonded to the ether oxygen atom. Remember that HI and HBr are strong enough acids to protonate water, so a concentrated aqueous solution of HBr contains hydronium ions and bromide ions but no free HBr.

Step 1: Acid-base reaction forms the conjugate acid of the ether

Ether   Hydronium ion   Conjugate   Conjugate
(Base)   (Acid)   acid of ether   base of H_3O^+

Poor leaving group

Step 2: Nucleophilic substitution reaction

Nucleophile   Haloalkane   Alcohol

Good leaving group

Exercise 13.13	Write the structure of the products formed in the reaction of each of the following ethers with excess concentrated aqueous HI at 130 °C.

(a) Tetrahydropyran   **(b)** Butoxycyclopentane   **(c)** Phenyl *tert*-butyl ether

Exercise 13.14	The reaction of 2-methoxy-2-methylpropane with the strong acid trifluoroacetic acid (CF_3COOH) forms methanol and methylpropene. Write a mechanism that accounts for these products.

Exercise 13.15	Explain why diethyl ether reacts with concentrated aqueous HCl very slowly at room temperature, while di-*tert*-butyl ether is rapidly cleaved under the same conditions.

13.10 Preparation of Epoxides

The principal method of preparing epoxides (oxiranes) is by epoxidation. Recall from Chapter 8 (Section 8.6) that epoxidation forms epoxides by the reaction of alkenes with peroxyacids. Reaction of cyclohexene with peroxyacetic acid, for instance, forms epoxycyclohexane.

Cyclohexene   Peroxyacetic acid   Epoxycyclohexane   Acetic acid
                                                              (85% yield)

Epoxidation is a mild reaction (it occurs at room temperature) that can be used to convert a wide variety of alkenes into epoxides. It is particularly useful because the stereochemistry of the alkene is retained in the epoxide. Epoxidation is a concerted mechanism in which the oxygen atom farthest from the carbonyl group is transferred to one face of the double bond of the alkene, as at the top of p. 570.

Optically active oxiranes can be formed if epoxidation occurs preferential to one of the faces of a prochiral alkene. This occurs in enzyme-catalyzed epoxidation reactions, as we will learn in Section 13.11.

Exercise 13.16	Epoxidation of *E*-2-pentene forms an optically inactive oxirane. Show that this product is actually an equal mixture of enantiomers.

Another way to prepare epoxides is by an intramolecular Williamson ether synthesis starting with a halohydrin. Recall that halohydrins are conveniently pre-

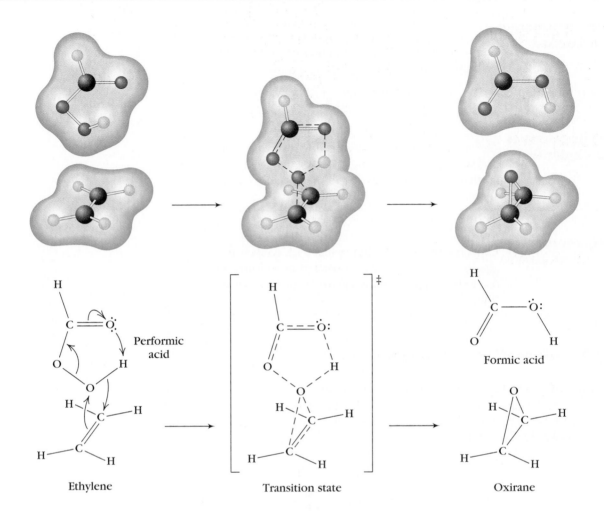

| Ethylene | Transition state | Oxirane |

pared by halogenation of alkenes in aqueous solution (Section 8.9). When treated with base, a halohydrin forms an alkoxide ion that reacts intramolecularly to displace the halogen atom on the adjacent carbon atom. This is essentially an intramolecular nucleophilic substitution reaction that occurs by an S_N2 mechanism.

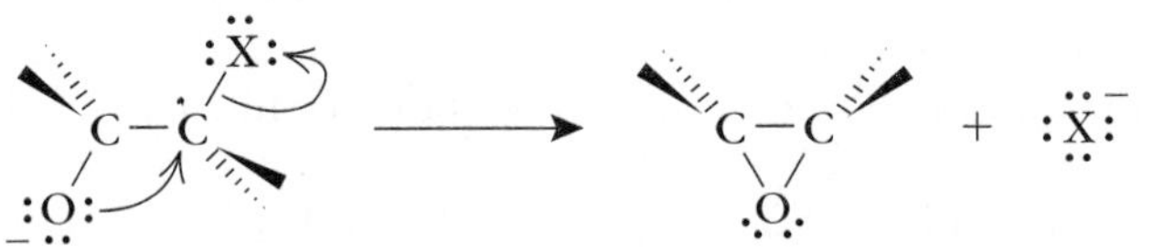

In order for the alkoxide ion to react by an intramolecular S_N2 mechanism, the alkoxide oxygen atom and the leaving group must be anti to each other. In

---

Exercise 13.17   Would you expect that the same epoxide would be formed by each of the following reactions? **(a)** Epoxidation of Z-2-butene with peroxyacetic acid and **(b)** chlorination of Z-2-butene in aqueous solution followed by reaction with aqueous sodium hydroxide.

Box 13.1   Hazards of Using Diethyl Ether in the Laboratory

Diethyl ether is particularly hazardous in the laboratory because it is extremely flammable and highly volatile and can form explosive mixtures in air. As a result, open flames should never be used as heat sources when diethyl ether and other low molecular weight dialkyl ethers are used in the laboratory.

Another hazard of dialkyl ethers is their tendency to slowly form hydroperoxides and peroxides by air oxidation at room temperature. Diethyl ether, for example, is oxidized by atmospheric oxygen to form 1-ethoxyethylhydroperoxide and diethyl peroxide.

$$CH_3CH_2OCH_2CH_3 \xrightarrow{\text{Atmospheric oxygen}} \overset{\displaystyle OOH}{\underset{\displaystyle |}{CH_3CHOCH_2CH_3}} + CH_3CH_2O\!-\!OCH_2CH_3$$

Diethyl ether   1-Ethoxyethyl-   Diethyl peroxide
                       hydroperoxide

Hydroperoxides and peroxides are explosive substances. They are unstable and decompose explosively because they contain a relatively weak oxygen-oxygen bond. The oxygen-oxygen bond energy is only about 33 kcal/mol (138 kJ/mol) compared to about 88 kcal/mol (368 kJ/mol) for a carbon-carbon single bond. Hydroperoxides and peroxides are a hazard because they tend to accumulate in the containers of ethers stored in the laboratory. Purification of dialkyl ethers by distillation causes the concentration of these peroxides to increase since their boiling points are higher than the ether. Consequently, distilling dialkyl ethers is particularly dangerous unless precautions are taken to destroy the peroxides before distillation by treating the ether with a solution of ferrous ions to reduce the peroxides.

The hazards of using diethyl ether in the laboratory cannot be overemphasized. It is volatile, flammable, and forms explosive mixtures with atmospheric oxygen. Great care must be taken when using diethyl ether in the laboratory.

this way the alkoxide attacks the backside of the reactive site, as shown in the following example:

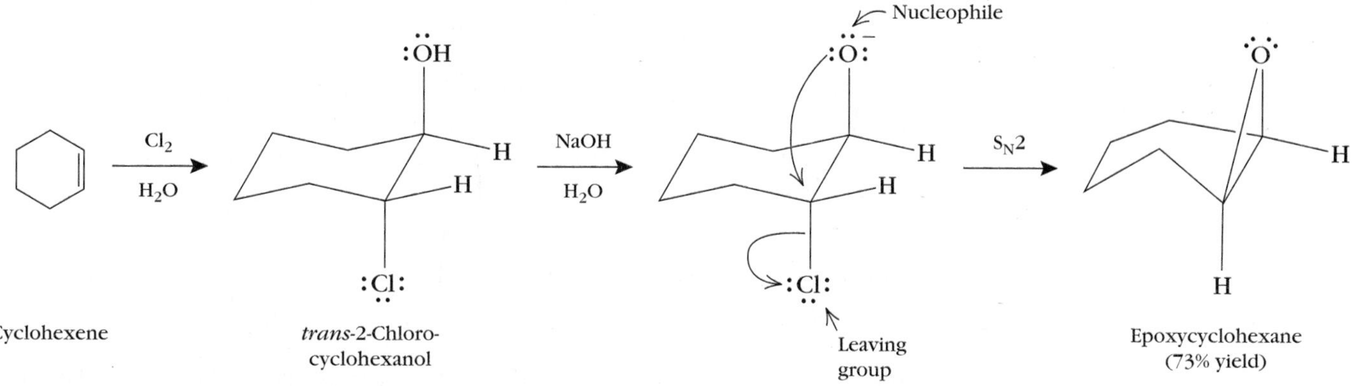

Cyclohexene   *trans*-2-Chloro-   Leaving   Epoxycyclohexane
                     cyclohexanol   group   (73% yield)

Halohydrins in which the nucleophile and the leaving group cannot achieve this anti conformation do not react to form epoxides.

| **Exercise 13.18** | Write equations for the reaction or reactions required to accomplish each of the following transformations. |

(a) 1,2-Epoxypentane from 2-chloro-1-pentanol
(b) 2,2-Diethyloxirane from 2-ethyl-1-butene
(c) 2-Methyl tetrahydropyran from 6-chloro-1-hexene
(d) Ethyloxirane from 1-butanol

Epoxides occur in living systems too. They are synthesized by enzyme-catalyzed reactions that are very similar to those used in the laboratory.

13.11 Enzyme-Catalyzed Preparation of Optically Active Epoxides

Optically inactive epoxides form when prochiral alkenes react with peroxycarboxylic acids because epoxidation takes place equally well on both faces of the alkene.

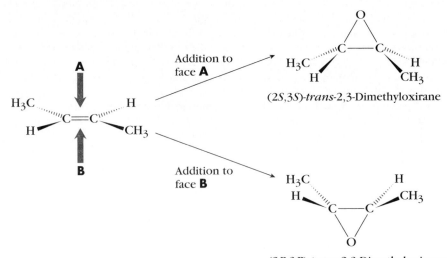

Recall from Section 11.12 that certain enzymes can distinguish between the faces of the carbonyl group of prochiral aldehydes and ketones, so these enzymes can form optically active products.

A number of enzymes can differentiate between the faces of the double bond of simple prochiral alkenes. Enzymes present in a number of fermentation processes, for example, form optically active 1,2-epoxyoctane in 70% enantiomeric excess from 1-octene.

$$CH_2=CH(CH_2)_5CH_3 \xrightarrow{\text{Fermentation reaction}} \overset{O}{\triangle}\cdots H \ (CH_2)_5CH_3$$

This enzyme-catalyzed reaction is the biological equivalent of the epoxidation of alkenes by peroxycarboxylic acids. The major difference is that the enzyme-catalyzed reaction provides a route to optically active epoxides.

Now that we have discussed some of the ways in which epoxides are formed, we can turn our attention to their reactivity.

13.12 Ring Opening Is the Principal Reaction of Epoxides

Epoxides are important synthetic intermediates because they undergo a number of ring-opening reactions in which one of the C—O bonds is broken. The reaction of oxirane (ethylene oxide) with an aqueous solution of HBr, for example, forms 2-bromoethanol.

$$\underset{\substack{\text{Oxirane} \\ \text{(ethylene oxide)}}}{\text{(epoxide structure)}} \xrightarrow[\substack{H_2O \\ 25\,°C}]{HBr} \underset{\substack{\text{2-Bromoethanol} \\ \text{(90\% yield)}}}{BrCH_2CH_2OH}$$

At first glance this reaction may seem analogous to the cleavage of ethers by acids, which was described in Section 13.9. There is a major difference, however. The cleavage of epoxides occurs readily with dilute acid under very mild conditions, whereas other ethers require concentrated solutions of strong acids and high temperatures.

Ring-opening reactions of epoxides also occur with bases. For example, oxirane reacts with ethylmagnesium iodide (a Grignard reagent) in diethyl ether to form 1-butanol.

$$\underset{\substack{\text{Oxirane} \\ \text{(ethylene oxide)}}}{\text{(epoxide structure)}} \xrightarrow[\substack{(CH_3CH_2)_2O}]{\substack{CH_3CH_2MgI \\ \text{(Ethylmagnesium} \\ \text{iodide)}}} CH_3CH_2CH_2CH_2OMgI \xrightarrow{H_3O^+} \underset{\substack{\text{1-Butanol} \\ \text{(74\% yield)}}}{CH_3CH_2CH_2CH_2OH}$$

This example illustrates the difference in reactivity between epoxides and acyclic ethers because both oxirane and diethyl ether are present in the reaction mixture. Diethyl ether is inert to a Grignard reagent so it can be used as the solvent in the reaction, while oxirane reacts rapidly with the Grignard reagent to form a primary alcohol as product.

Exercise 13.19   Write the structure of the product formed by the reaction of oxirane (ethylene oxide) with each of the following reagents.

(a) Aqueous HCl
(b) Sodium cyanide (NaCN) in ethanol
(c) Sodium hydroxide in water
(d) Ethylmagnesium bromide in diethyl ether, followed by addition of dilute sulfuric acid
(e) Butyllithium ($CH_3CH_2CH_2CH_2Li$) in heptane, followed by addition of dilute sulfuric acid

---

Epoxides react so readily because their three-membered rings open easily. Epoxides, like cyclopropane (Chapter 4, Section 4.6), are very strained molecules because their bond angles are about 60°, which is much smaller than the normal tetrahedral bond angle (109.5°) of acyclic compounds. Compressing these bond angles from 109.5° to 60° causes angle strain, which is relieved when the ring of an epoxide is opened.

The two carbon-oxygen bonds in a symmetrical epoxide are equivalent, so only one product is formed when the ring is opened. The two carbon-oxygen bonds in an unsymmetrical epoxide, on the other hand, are not equivalent, so two products can be formed when the ring is opened. Which of these two products predominates? It depends on whether the reaction is carried out under acidic or basic conditions.

A   Regiochemistry and Stereochemistry of Nucleophilic Ring Opening

Under basic conditions, nucleophiles react with unsymmetrical epoxides by breaking the bond between the oxygen atom and the less substituted, less sterically hindered ring carbon atom. For example, 2,2-dimethyloxirane reacts with methoxide ion in methanol by breaking the bond between the oxygen atom and the primary ring carbon atom to form 1-methoxy-2-methyl-2-propanol.

2,2-Dimethyloxirane

$CH_3O^- Na^+$
CH_3OH

1-Methoxy-2-methyl-2-propanol
(80% yield)

This reaction illustrates that *under basic conditions, nucleophiles react with epoxides at the less sterically hindered ring carbon atom.*

The configuration of the carbon atom at which substitution occurs is inverted. The reaction of methoxide ion with (2R,3R)-*trans*-2,3-Dimethyloxirane, for example, yields (2R,3S)-3-methoxy-2-butanol.

(2R,3R)-*trans*-2,3-Dimethyloxirane

$CH_3O^- Na^+$
CH_3OH

(2R,3S)-3-Methoxy-2-butanol
(70% yield)

Reaction of methoxide ion occurs at either equivalent ring carbon with inversion of its configuration. Notice, however, that the configuration of the other ring carbon atom is retained because it is not involved in the reaction.

Reaction site with (R) configuration
Nucleophile Leaving group (R) configuration

Configuration inverted to (S)
Configuration retained as (R)

These observations are consistent with a nucleophilic substitution reaction in which methoxide ion (the nucleophile) reacts at a ring carbon atom (the reaction site) with displacement of an alkoxide ion (the leaving group). The reaction occurs at the less substituted carbon atom with inversion of configuration, which suggests an S_N2 mechanism. Thus the principles developed in Chapter 12 for nucleophilic substitution reactions that occur by an S_N2 mechanism can be applied to the mechanism of this nucleophilic ring opening of epoxides as well. Bond formation between the nucleophile and the ring carbon accompanies carbon-oxygen bond breaking. In the transition state, most of the ring strain of the three-membered ring is relieved as it begins to open. Finally the initially formed alkoxide ion rapidly abstracts a proton from the solvent to form a β-substituted alcohol as the isolated product.

| **Exercise 13.20** | Write the structure of the product of the reaction of (2R,3R)-*trans*-2,3-dimethyloxirane with each of the following reagents. |

(a) Aqueous ammonia solution
(b) LiAlH₄/(C₂H₅)₂O followed by H₃O⁺

(c) CH₃C≡C⁻ Na⁺/NH₃(l) followed by H₃O⁺

| **Exercise 13.21** | Which one of the following compounds—*A*, *B*, or *C*— correctly represents the product of the reaction of (2S,3S)-*trans*-2,3-dimethyloxirane with an aqueous NaOH solution? |

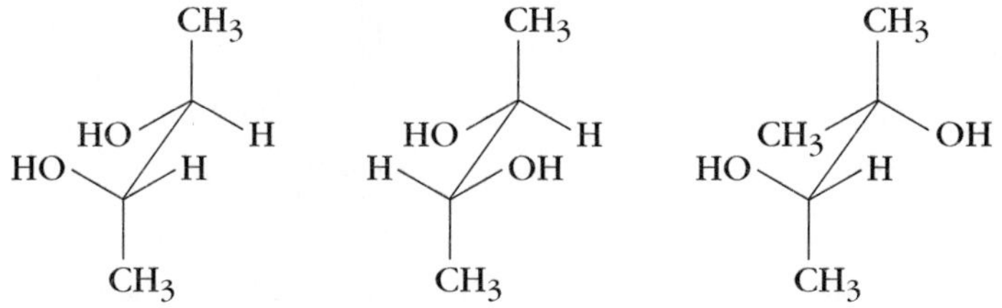

A   B   C

B   Regiochemistry and Stereochemistry of Acid-Catalyzed Ring Opening

The bond to the more substituted carbon atom of the ring is broken in acid-catalyzed ring openings of unsymmetrical epoxides. For example, the acid-catalyzed reaction of 2,2,3-trimethyloxirane with methanol forms 3-methoxy-3-methyl-2-butanol as the major product.

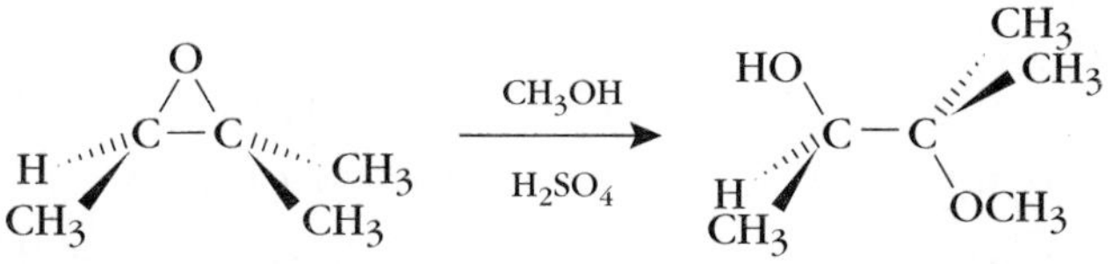

2,2,3-Trimethyloxirane

3-Methoxy-3-methyl-2-butanol
(76% yield)

This reaction illustrates that *under acidic conditions, the major product is formed by reaction of nucleophiles at the more sterically hindered ring carbon atom of the epoxide.*

While the regiochemistry is the opposite of that observed in nucleophilic ring openings of epoxides under basic conditions, the stereochemistry of the reaction is still the same. *The configuration at the reaction site is inverted in both acid- and base-catalyzed ring openings of epoxides.* Thus the acid-catalyzed hydrolysis of epoxycyclohexane forms *trans*-1,2-cyclohexanediol and not its *cis*- isomer.

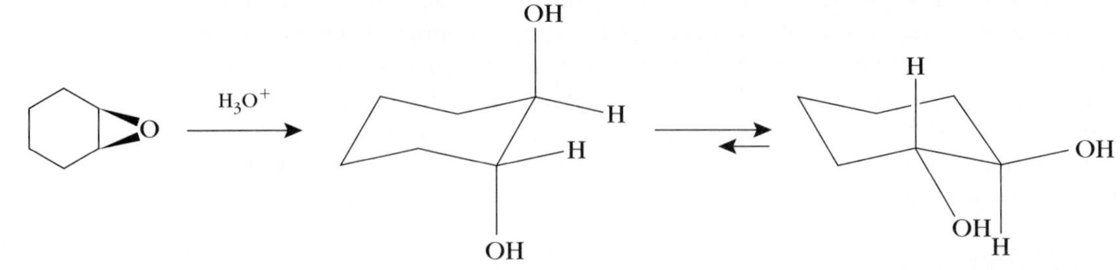

Epoxycyclohexane   *trans*-1,2-Cyclohexanediol

The proposed mechanism for the acid-catalyzed ring opening of epoxides is shown in Figure 13.7 for the reaction of methanol and 2,2-dimethyloxirane. Step 1 is an acid-base reaction that forms the conjugate acid of the epoxide by protonation of the oxygen atom of the epoxide. Step 2 is the reaction of the nucleophile at the more substituted carbon atom of the three-membered ring of the

Figure 13.7
Mechanism of acid-catalyzed ring opening of an epoxide. Step 2 is S$_N$2-like because methanol is the nucleophile, the more substituted carbon atom is the reaction site, and the protonated ether oxygen is the leaving group.

Step 1: Acid-base reaction forms the conjugate acid of the epoxide

Conjugate acid
of epoxide

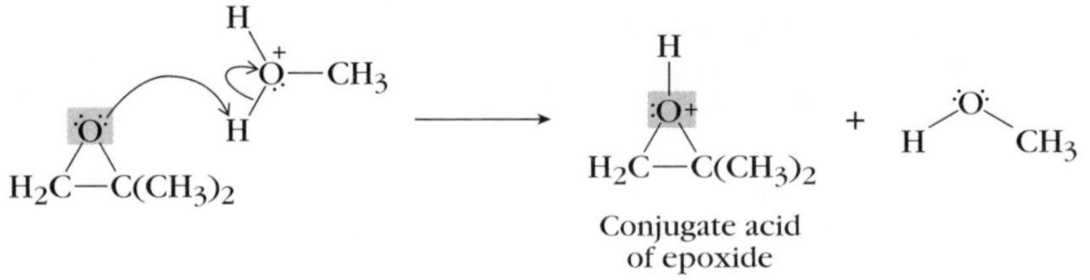

Step 2: Nucleophile (CH$_3$OH) reacts with the most substituted carbon atom of the epoxide ring

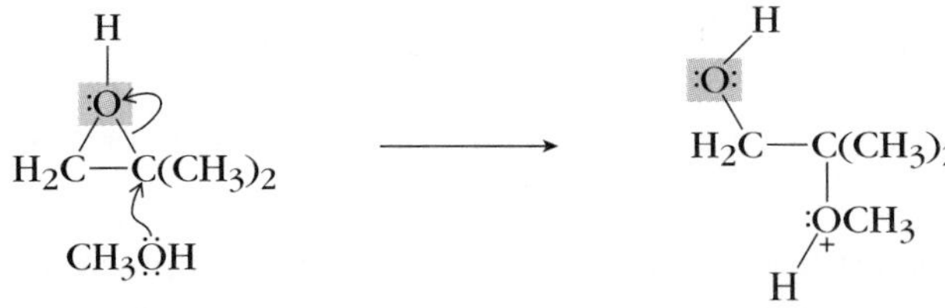

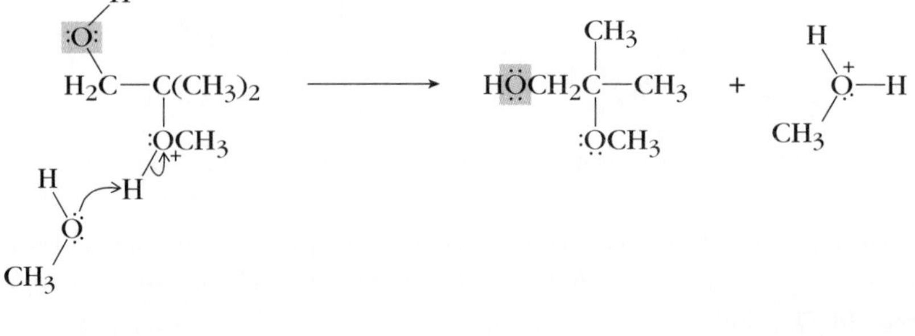

Step 3: Transfer of proton from the conjugate acid of the product to CH$_3$OH

protonated epoxide. Step 3 of the mechanism is loss of a proton to a solvent molecule to form the product, 2-methoxy-2-methyl-1-propanol.

The reaction of the nucleophile with the more substituted carbon atom of the conjugate acid of the epoxide is a surprising feature of the mechanism shown in Figure 13.7 because nucleophiles usually react faster with the less substituted carbon atom in S_N2-like mechanisms. Why does the nucleophile react with the more substituted carbon atom? The carbon atom that reacts with the nucleophile depends on the electron structure of the conjugate acid of the epoxide, which can be represented by the localized valence bond model (Section 1.12) as a resonance hybrid of the following three contributing structures:

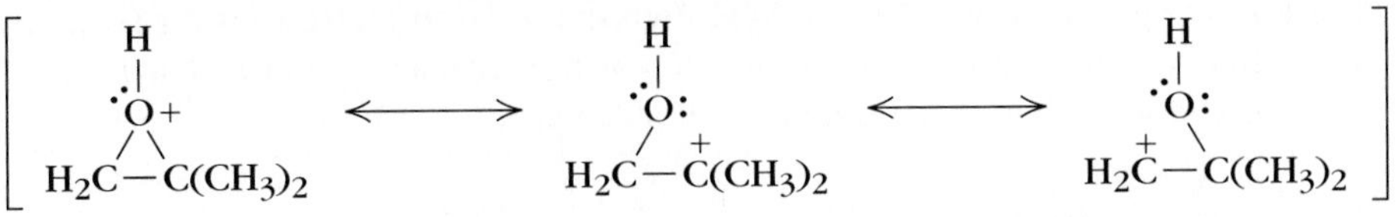

The middle resonance structure, which is a tertiary carbocation, is the most stable, so it contributes much more to the resonance hybrid than the structure on the right, which is a primary carbocation. As a result, the C—O bond to the more substituted carbon atom is the weakest, so it is easily broken by the nucleophile.

In summary, nucleophiles react with epoxides at the less hindered ring carbon atom under basic conditions and with the more substituted ring carbon atom of the protonated epoxide under acidic conditions. In both acid- or base-catalyzed ring openings, the configuration of the reaction site is inverted.

Exercise 13.22 Write the structure of the product of the reaction of 2,2-dimethyloxirane with each of the following reagents.

(a) Aqueous HBr   (b) Ethanol with a trace of H_2SO_4 (c) Aqueous HCl

---

13.13 Synthetic Uses of Epoxides

Epoxides are valuable synthetic intermediates because their reactions are facile and stereospecific, and their regiochemistry can be controlled by choosing acidic or basic conditions. In this section, we will examine several general synthetic reactions in which epoxides play an important role.

One particularly useful reaction that involves an epoxide intermediate is the **anti hydroxylation of alkenes.** This two-reaction sequence combines two stereospecific reactions. The first reaction is syn epoxidation of the double bond while the second is the hydrolysis of the epoxide formed in the first reaction. The anti hydroxylation of cyclohexene to *trans*-1,2-cyclohexanediol illustrates this sequence of reactions.

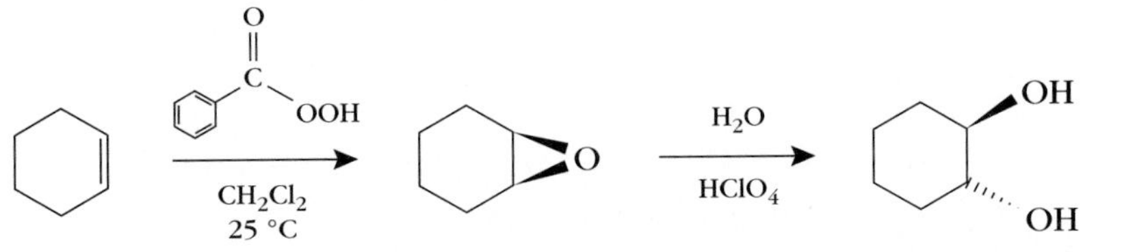

Cyclohexene   Epoxycyclohexane   *trans*-1,2-Cyclohexanediol
                                                (racemic mixture 86% yield)

Anti hydroxylation is another method of preparing diols from alkenes. Recall from Section 8.5 that direct hydroxylation of alkenes with OsO_4 forms diols. There is a major difference between these two methods, however. The direct hydroxylation of alkenes occurs by syn addition to form *cis-* diols, while anti hydroxylation is formally an anti addition that forms *trans-* diols. These two methods, therefore, make it possible to prepare either a *cis-* or *trans*-diol using these two complementary reactions.

Epoxides can also be used to add carbon atoms to preexisting chains. Epoxides react with Grignard or organolithium reagents, for instance, to form alcohols that contain two more carbon atoms than the original organometallic compound. In the following example, 1-bromo-3-methylbutane is converted into a Grignard reagent that reacts with ethylene oxide to form a primary alcohol that has two more carbon atoms than the original bromoalkane.

$$(CH_3)_2CHCH_2CH_2Br \xrightarrow[\text{(C}_2\text{H}_5)_2\text{O}]{\text{Mg}} (CH_3)_2CHCH_2CH_2MgBr$$

1-Bromo-3-methylbutane
(4-carbon parent chain)

$$\begin{array}{c} O \\ / \backslash \\ H_2C-CH_2 \end{array}$$

$$(CH_3)_2CHCH_2CH_2\underline{CH_2CH_2OH} \xleftarrow{H_3O^+} (CH_3)_2CHCH_2CH_2\underline{CH_2CH_2}OMgBr$$

5-Methyl-1-hexanol
(6-carbon parent chain)
(70% yield)

When this reaction is extended to unsymmetrical epoxides, the magnesium halide in solution acts as a Lewis acid and catalyzes the epoxide ring opening to form a carbocation.

The carbocation may subsequently rearrange, leading to a mixture of products. A mixture of products can be avoided by using an organolithium reagent instead. Organolithium reagents react with the epoxide at the less sterically hindered ring carbon atom, and the lithium ion does not catalyze epoxide ring opening. Consequently, unsymmetrical alkenes can be converted into secondary or tertiary alcohols in two steps: first, form an epoxide; second, react it with an organolithium reagent.

$$CH_3CH_2CH_2CH\overset{O}{\overset{/\backslash}{-}}CH_2 \xrightarrow[\text{(2) H}_3\text{O}^+]{\text{(1) C}_6\text{H}_5\text{Li/pentane}} CH_3CH_2CH_2\overset{OH}{\underset{|}{CH}}-CH_2$$

(70% yield)

$$(CH_3)_2C\overset{O}{\overset{/\backslash}{-}}CH_2 \xrightarrow[\text{(2) H}_3\text{O}^+]{\text{(1) CH}_3\text{(CH}_2)_2\text{CH}_2\text{Li/pentane}} (CH_3)_2\overset{OH}{\underset{|}{C}}-CH_2-(CH_2)_3CH_3$$

(75% yield)

Exercise 13.23   Write the equation for the reaction in which an epoxide is used to prepare each of the following compounds.

(a) $C_2H_5C \equiv CCH_2CH_2OH$

(b) $(CH_3)_2CCH_2OH$
$\quad\quad\quad\quad |$
$\quad\quad\quad\quad Cl$

(c) $CH_3(CH_2)_7CH_2OH$

(d)

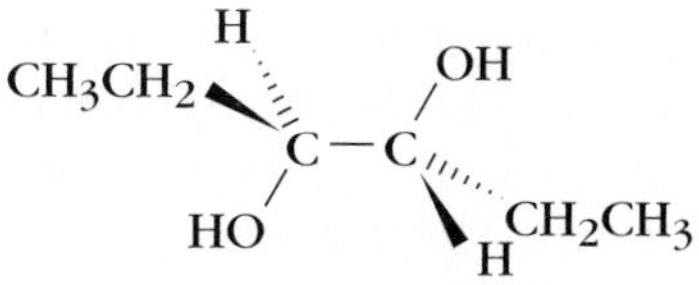

Nucleophilic ring opening of an epoxide is a key reaction in the formation of steroid hormones from squalene.

13.14 2,3-Epoxysqualene Is an Intermediate in the Biological Synthesis of Steroids

Most steroids in humans and animals are obtained from cholesterol. Cholesterol, in turn, is obtained from squalene, an acyclic polyene that contains 30 carbon atoms. The first step in the biosynthesis of steroids is the formation of 2,3-epoxysqualene by the epoxidation of squalene. This reaction is catalyzed by a microsomal enzyme that uses oxygen as the epoxidizing agent and needs the biological reducing agent NADPH. This enzyme transfers only one of the oxygen atoms of O_2 to squalene so it is called monooxygenase. Notice the selectivity of this enzyme-catalyzed epoxidation. Of the six double bonds in squalene, only one is epoxidized.

The epoxide ring of 2,3-epoxysqualene undergoes an enzyme-catalyzed nucleophilic ring-opening reaction that sets in motion a complicated series of reactions that first forms lanosterol, a constituent of lanolin, the waxy fat from wool. Lanosterol is then converted into cholesterol. The mechanistic details of these reactions are postponed until Chapter 19. It is enough to realize that epoxides are valuable intermediates not only in the laboratory but also in complicated biosynthetic reaction schemes.

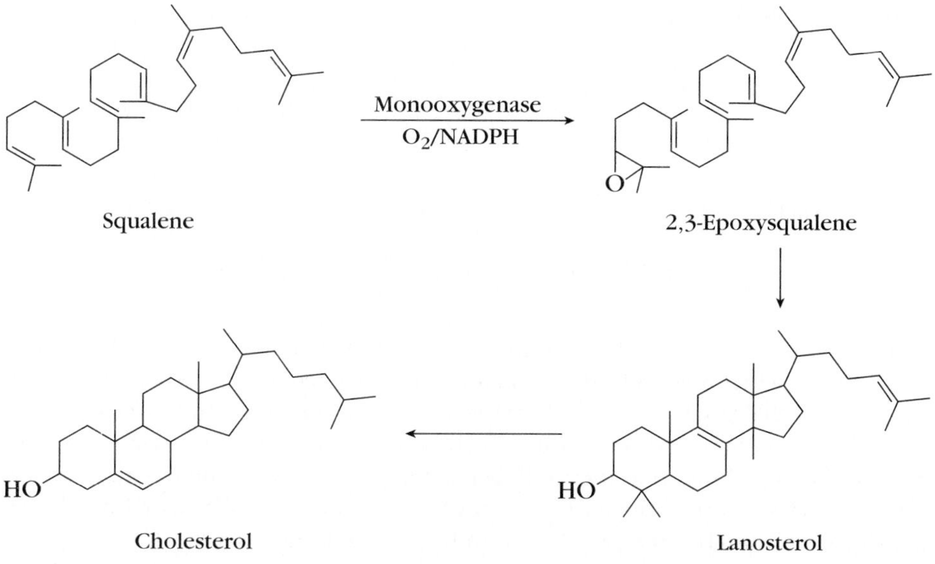

Squalene

2,3-Epoxysqualene

Cholesterol

Lanosterol

13.15 Sulfides

Sulfides are a class of compounds that are the sulfur analogs of ethers in which a sulfur atom, instead of an oxygen atom, is bonded to two carbon atoms. The two carbon atoms bonded to sulfur can be any alkyl, aryl, or vinyl group, and the sulfur atom can be part of a straight chain or ring.

The same rules used to name ethers apply to sulfides. The word *sulfide* is used in place of *ether* for simple compounds, and *alkylthio* is used in place of *alkoxy* for more complex compounds.

$$CH_3CH_2\overset{S}{\diagdown}CH_2CH_3 \qquad \overset{S-CH_3}{\diagup} \qquad CH_2{=}CHCH_2CH_2{-}S\diagdown_{CH_3}$$

Diethyl sulfide   Methyl phenyl sulfide   4-Methylthio-1-butene

Sulfides are prepared by the reaction of primary or secondary haloalkanes with a thiolate anion (RS^-).

$$RS^-Na^+ \quad + \quad R'X \quad \longrightarrow \quad RSR' \quad + \quad Na^+X^-$$

Sodium thiolate   Primary or secondary   Sulfide   Sodium halide
                     haloalkane or tosylate

This nucleophilic substitution reaction occurs by an S_N2 mechanism analogous to the Williamson synthesis of ethers (Section 13.8, C). Thiolate anions are very good nucleophiles and good yields of sulfides are formed by this method.

$$CH_3S^-Na^+ \quad + \quad ClCH_2CH_2OH \quad \xrightarrow{\;CH_3CH_2OH\;} \quad CH_3SCH_2CH_2OH$$

Sodium   2-Chloroethanol   2-Methylthioethanol
methanethiolate   (75% to 80% yield)

Ethers and sulfides differ in their behavior toward oxidizing agents. Ethers undergo oxidation at the carbon atom bonded to the oxygen atom (see Box 13.1). Sulfides, on the other hand, are oxidized at sulfur. Oxidation of a sulfide with one equivalent of hydrogen peroxide (H_2O_2) or sodium metaperiodate ($NaIO_4$), for example, forms the corresponding sulfoxide (R_2SO). A sulfoxide can be further oxidized to a sulfone (R_2SO_2) by a peroxycarboxylic acid, such as peroxyacetic acid (CH_3COOOH):

$$R-\overset{\cdot\cdot}{\underset{\cdot\cdot}{S}}-R' \quad \xrightarrow[\text{of } H_2O_2 \text{ or } HIO_4]{\text{1 equivalent}} \quad R-\overset{:\overset{..}{O}\overset{-}{\cdot}}{\underset{\overset{+}{\cdot}}{S}}-R' \quad \xrightarrow{CH_3COOOH} \quad R-\overset{:\overset{..}{O}\overset{-}{\cdot}}{\underset{:\overset{..}{O}:^-}{\overset{2+}{S}}}-R'$$

Sulfide   Sulfoxide   Sulfone

$$\boxed{\text{INCREASING OXIDATION STATE OF SULFUR} \Rightarrow}$$

The sulfur atom of sulfides, like the oxygen atom of ethers, has two lone pairs of electrons, which makes both ethers and sulfides nucleophiles. Recall from Section 12.4, B that nucleophilicity increases in going down a column of the periodic table, so compounds of sulfur, which is below oxygen in the same column, are more nucleophilic than their oxygen analogs. As a result, sulfides react much faster than ethers with primary haloalkanes. The reaction of sulfides with primary haloalkanes occurs by an S_N2 mechanism to form sulfonium salts.

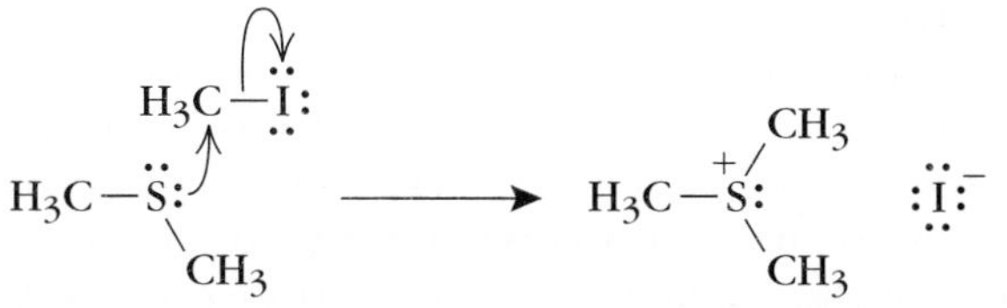

Dimethyl sulfide   Trimethylsulfonium iodide

Sulfonium salts react with nucleophiles by an S_N2 mechanism.

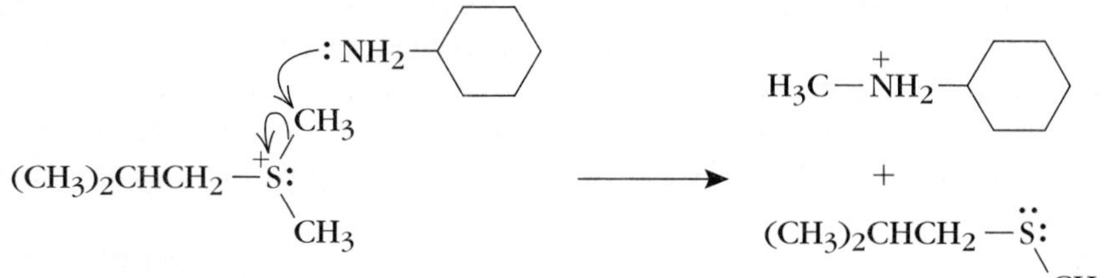

Isobutyldimethylsulfonium ion

The nucleophile reacts at the less sterically hindered carbon atom to transfer an alkyl group (in this case a methyl group) from the sulfonium ion to the nucleophile. Nature makes extensive use of this alkylation reaction, as we discussed in Section 12.3. The methylating agent in many biological reactions is *S*-adenosylmethionine, which is a sulfonium salt. *S*-Adenosylmethionine is the biological equivalent of iodomethane.

Exercise 13.24   Write the structure of the organic product formed in each of the following reactions.

(a) $+$ CH_3I $\longrightarrow$

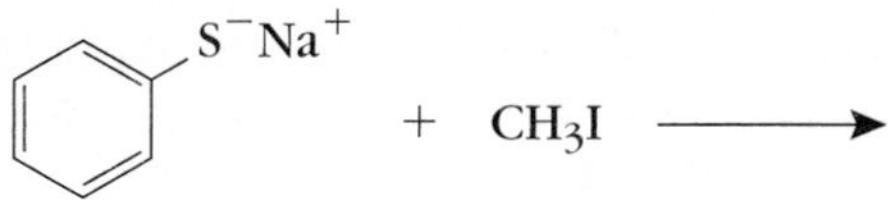

(b) $(CH_3)_2S$ $+$ $ClCH_2-$ $\longrightarrow$

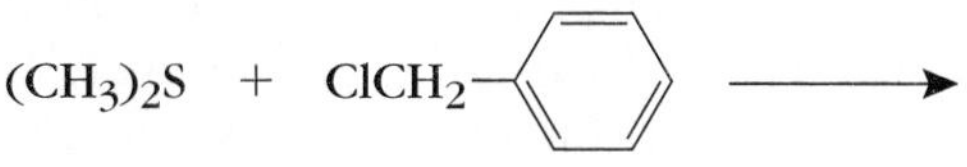

(c) $\xrightarrow[H_2O_2]{1 \text{ equivalent}}$

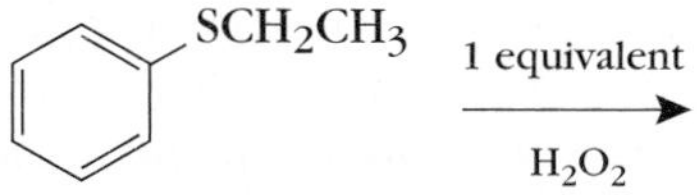

(d) $(CH_3)_2\overset{+}{S}CH_2C(CH_3)_3$ $+$ $:NH_3$ $\longrightarrow$
      I^-

13.16 Summary of Concepts and Terms

Ethers are compounds that contain an oxygen atom bonded to two carbon atoms. The two carbon atoms may be part of alkyl, aryl, or vinyl groups. If the two groups are identical, the ether is said to be **symmetrical**; if they are different, it is said to be **unsymmetrical**. Ethers may also be **cyclic** if the oxygen atom is in a ring or **acyclic** if it is not.

    Tetrahydropyrans are saturated six-membered rings that contain a single oxygen atom in the ring. They occur very widely in nature. Like cyclohexane, tetrahydropyrans are most stable in the chair conformation. Substituents generally prefer to occupy the equatorial positions of tetrahydropyrans.

Ethers can act as Lewis bases because they have two unshared pairs of electrons on their oxygen atoms. As a result, ethers form Lewis acid-base complexes with many ions and compounds. **Crown ethers** are compounds that contain four or more ether oxygen atoms in a ring of twelve or more atoms. Crown ethers form complexes with cations and can be used to dissolve many inorganic salts in nonpolar solvents. Crown ethers have a hydrophobic exterior and a hydrophilic interior where the charged cation resides. The hydrophobic exterior is what makes these complexes soluble in nonpolar solvents. In this way, many inorganic salts can be dissolved in nonpolar solvents. Naturally occurring compounds called **ionophores** resemble crown ethers because they too have hydrophobic exteriors and hydrophilic interiors. The complexes formed by ionophores with a number of cations are able to transport these cations across cell membranes.

Ethers are prepared in the laboratory by either an **alkoxymercuration-demercuration** sequence of reactions or the **Williamson ether synthesis.** The alkoxymercuration-demercuration reactions first form an intermediate alkoxy-organomercury (II) compound, which is then reduced by $NaBH_4$. The result is addition of an alcohol to the double bond of the alkene with Markownikoff orientation. The Williamson ether synthesis is a nucleophilic substitution reaction of an alkoxide ion with a haloalkane. This reaction occurs by an S_N2 mechanism, so it is limited to the preparation of ethers in which one alkyl group is primary.

Ethers are inert to most reagents. The major exceptions are HI and HBr, which react with ethers to form cleavage products. The cleavage reaction occurs by an S_N2 mechanism when the alkyl groups bonded to oxygen are primary or secondary and by an S_N1 mechanism if the alkyl groups bonded to the oxygen are tertiary.

Ethers such as diethyl ether must be handled with care in the laboratory because they are flammable and form explosive hydroperoxides and peroxides when exposed to atmospheric oxygen for extended periods of time.

Compounds with a three-membered oxygen-containing ring are called **epoxides.** Epoxides are most easily prepared by the reaction of alkenes with a peroxycarboxylic acid. The three-membered epoxide ring opens readily under acidic or basic conditions because opening is accompanied by relief of ring strain. Epoxide ring opening under basic conditions occurs by an S_N2 mechanism at the least sterically hindered ring carbon atom. Ring opening under acidic conditions occurs preferentially at the most substituted ring carbon atom. In both reactions, the configuration of the reactive site is inverted.

Epoxides are valuable intermediates in the synthesis of organic compounds, not only in the laboratory, but also in the synthesis of steroids in humans and other animals.

Sulfides are the sulfur analogs of ethers. Sulfides are oxidized at the sulfur atom to form **sulfoxides,** which can be further oxidized to **sulfones.** Sulfides react with primary haloalkanes by an S_N2 mechanism to form sulfonium ions.

13.17 Summary of Reactions

1. **Preparation of Ethers**

   a. **Alkoxymercuration-demercuration of alkenes** SECTIONS 8.10 AND 13.8

   Markownikoff addition of ROH to carbon-carbon double bond without rearrangement.

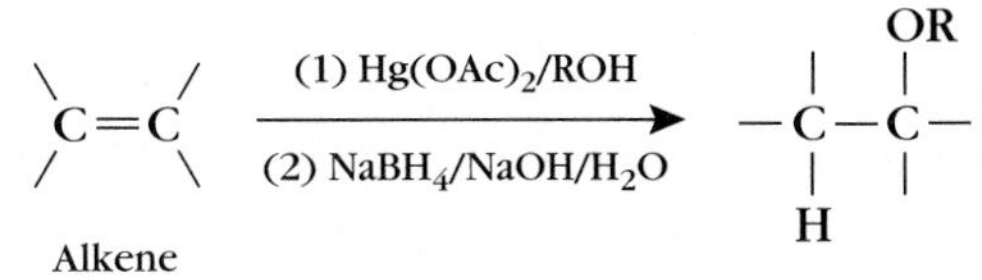

$$\underset{\text{Alkene}}{\overset{\diagdown}{\underset{\diagup}{C}}=\overset{\diagup}{\underset{\diagdown}{C}}} \xrightarrow[\text{(2) NaBH}_4\text{/NaOH/H}_2\text{O}]{\text{(1) Hg(OAc)}_2\text{/ROH}} \quad -\overset{\overset{\text{OR}}{|}}{\underset{\overset{|}{H}}{C}}-\overset{|}{\underset{|}{C}}-$$

$$\underset{\text{1-Pentene}}{CH_3(CH_2)_2CH=CH_2} \xrightarrow[\text{(2) NaBH}_4\text{/NaOH/H}_2\text{O}]{\text{(1) Hg(OAc)}_2\text{/CH}_3\text{OH}} \underset{\text{2-Methoxypentane}}{CH_3(CH_2)_2\overset{\overset{\text{OCH}_3}{|}}{C}HCH_3}$$

   b. **Williamson ether synthesis**

   Nucleophilic substitution reaction that occurs by an S_N2 mechanism, so best results are obtained by using methyl, 1° haloalkanes, or alkyl tosylates.

$$\underset{\substack{\text{Alkoxide} \\ \text{ion}}}{R'O^-} + \underset{\text{1° Haloalkane}}{RCH_2-X} \longrightarrow R'O-CH_2R + X^-$$

$$\underset{\substack{\text{2-Propanoxide} \\ \text{ion}}}{(CH_3)_2CHO^-} + \underset{\text{Iodoethane}}{CH_3CH_2I} \longrightarrow \underset{\text{2-Ethoxypropane}}{CH_3CH_2OCH(CH_3)_2}$$

2. **Reactions of Ethers with Hydrogen Halides** SECTION 13.9

   With excess HX, the initially formed alcohol reacts further to form a haloalkane; ArOH doesn't react further with HX.

$$R-O-R' \xrightarrow[\text{H}_2\text{O, Heat}]{\text{HX}} ROH + R'X$$

$$Ar-O-R \xrightarrow[\text{H}_2\text{O, Heat}]{\text{HX}} ArOH + RX$$

$$HX = HCl < HBr < HI$$

   INCREASING REACTIVITY ⟩

$$\underset{\text{Diisopropyl ether}}{(CH_3)_2CHOCH(CH_3)_2} \xrightarrow[\text{H}_2\text{O, 135 °C}]{\text{Excess HBr}} \underset{\text{2-Bromopropane}}{2\ (CH_3)_2CHBr}$$

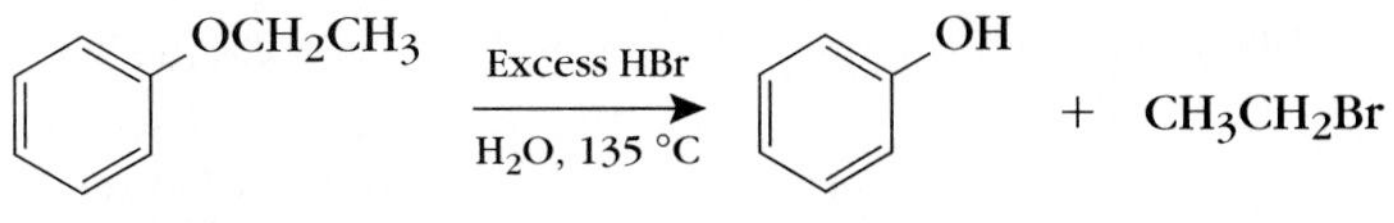

Ethyl phenyl ether →(Excess HBr, H₂O, 135 °C) Phenol + Bromoethane

3. **Preparation of Epoxides** SECTIONS 8.6 AND 13.10

   Syn addition of oxygen atom to carbon-carbon double bond.

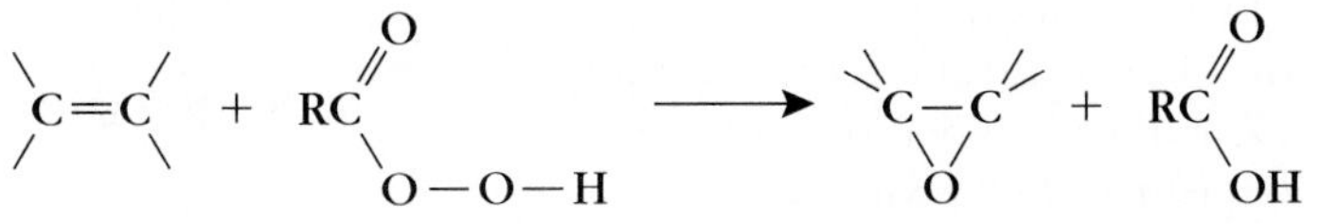

Alkene + Peroxycarboxylic acid ⟶ Epoxide + Carboxylic acid

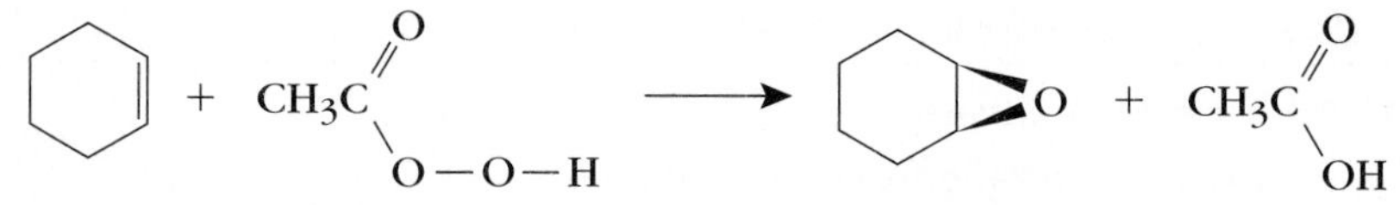

| Cyclohexane | Peroxyacetic acid | | Epoxy-cyclohexane | Acetic acid |

4. Reactions of Epoxides   SECTION 13.12

a. Ring openings in basic solutions

Nucleophile adds to the least sterically hindered carbon atom of the epoxide ring. Inversion of configuration occurs at the reaction site.

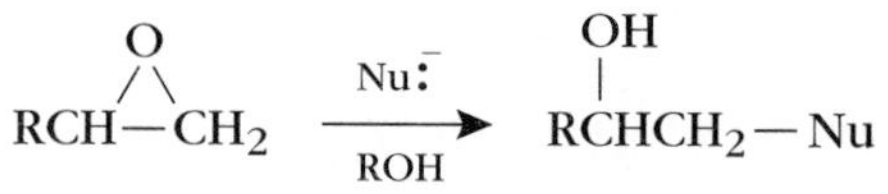

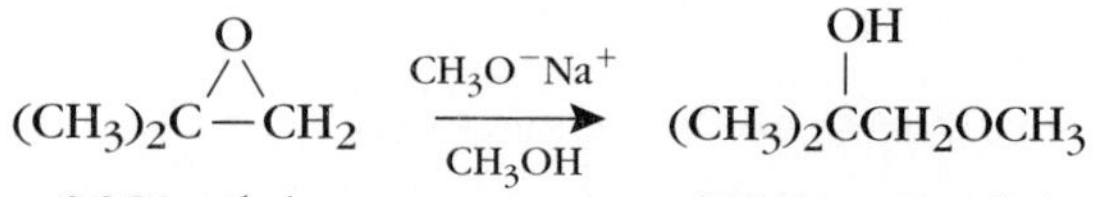

2,2-Dimethyl-oxirane   1-Methoxy-2-methyl-2-propanol

b. Acid-catalyzed ring openings

Nucleophile reacts at the most sterically hindered carbon atom of the conjugate acid of the epoxide ring. Inversion of configuration occurs at the reaction site.

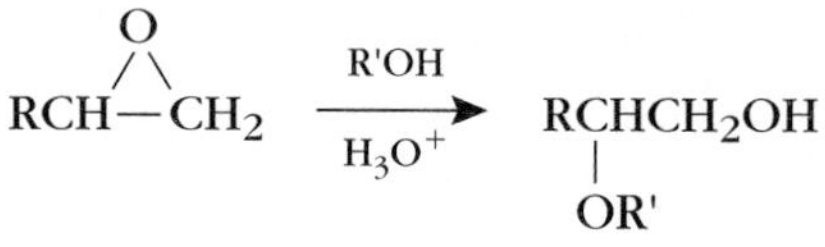

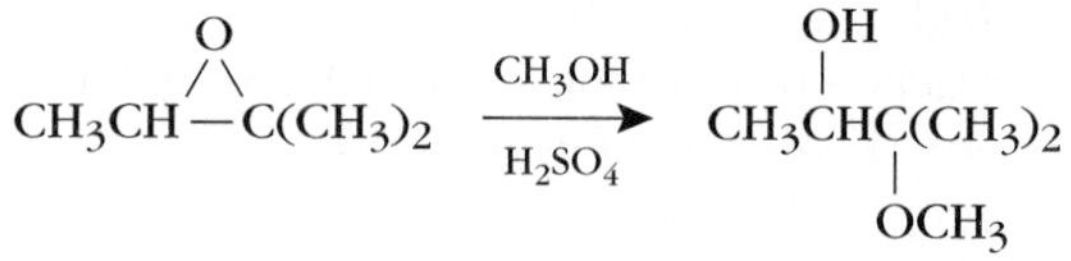

3-Methoxy-3-methyl-2-butanol

5. Preparation of Sulfides   SECTION 13.15

Nucleophilic substitution reactions that occur by an S_N2 mechanism. Thiolate anions are better nucleophiles than alkoxide ions, so good yields of sulfides are formed by use of methyl, 1° and 2° haloalkanes, or alkyl tosylates.

$$RS^-Na^+ \ + \ R'X \longrightarrow RSR' \ + \ NaX$$

| Thiolate anion | 1° or 2° Haloalkane or tosylate | Sulfide | Sodium halide |

$$CH_3S^-Na^+ \ + \ CH_3(CH_2)_2CH_2Br \longrightarrow CH_3(CH_2)_2CH_2SCH_3$$

Sodium methane-thiolate   1-Bromobutane   1-Methyl-thiobutane

6. Oxidation of Sulfides   SECTION 13.15

The sulfur atom of sulfides is oxidized in contrast to ethers where the carbon atom adjacent to the oxygen atom is oxidized.

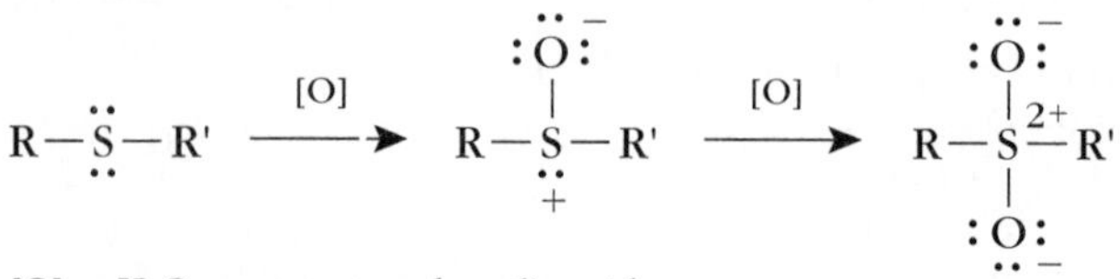

[O] = H_2O_2 or peroxycarboxylic acid

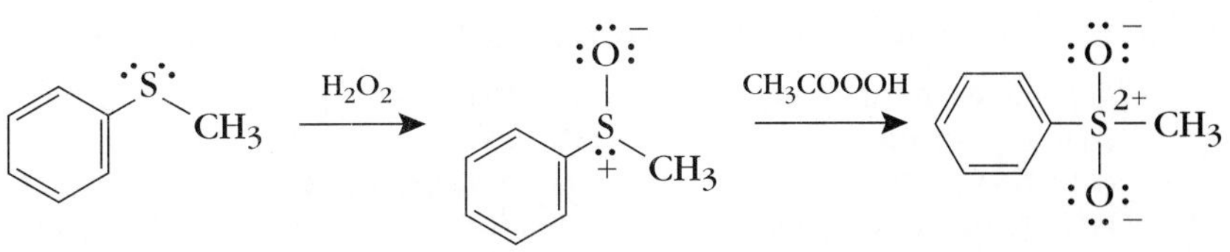

| Methyl phenyl sulfide | | Methyl phenyl sulfoxide | | Methyl phenyl sulfone |

7. Alkylation of Sulfides   SECTION 13.15

Nucleophilic substitution reactions that occur by S_N2 mechanism.

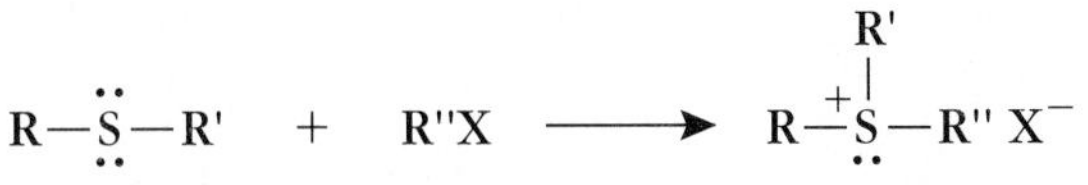

| Sulfide | Methyl, 1° or 2° haloalkanes, or alkyl tosylates | | Sulfonium salt |

| Dimethyl sulfide | Iodomethane | | Trimethyl sulfonium iodide |

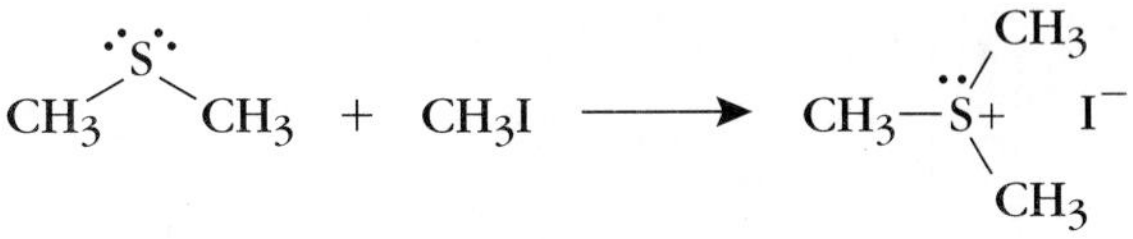

Additional Exercises

13.25 Write the structure that corresponds to each of the following names.

   (a) 3,4-Epoxyheptane
   (b) 1-Ethoxy-2-methylpentane
   (c) *trans*-2-Methyl-3-ethyloxirane
   (d) 1,2-Dimethoxyethane
   (e) *cis*-2,3-Dimethyltetrahydrofuran
   (f) Cyclopropoxycyclopentane

13.26 Give the IUPAC name to each of the following compounds.

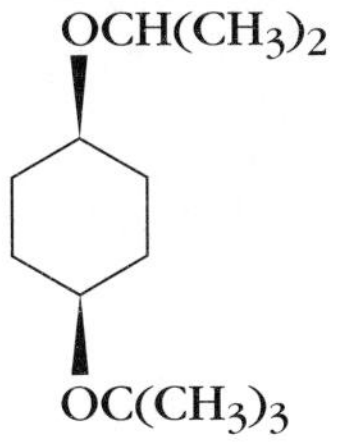

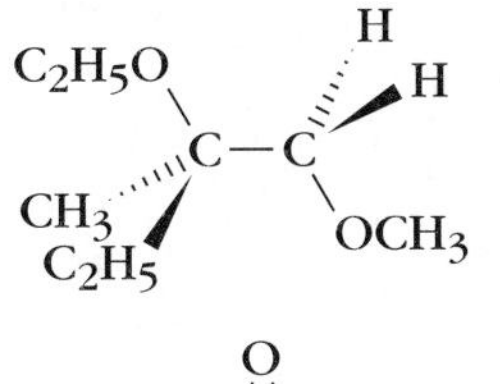

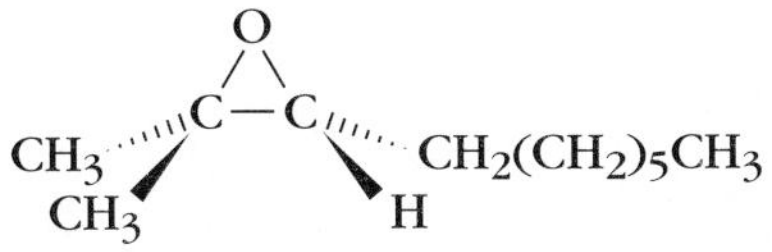

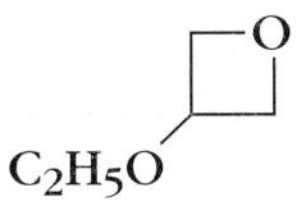

13.27 Write the structures of the products formed by the reaction of oxirane (ethylene oxide) with each of the following reagents.

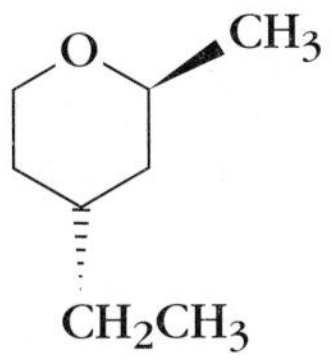

   (a) Boiling water
   (b) Liquid ammonia
   (c) Sodium cyanide (NaCN)/C_2H_5OH
   (d) NaN_3/C_2H_5OH
   (e) $CH_3C{\equiv}C^-Na^+$/$NH_{3(l)}$ followed by addition of dilute H_2SO_4
   (f) CH_3CH_2MgI/$(C_2H_5)_2O$ followed by addition of dilute H_2SO_4
   (g) $CH_3S^-Na^+$/C_2H_5OH followed by addition of dilute H_2SO_4

13.28 Write the structures of the products formed by the reaction of 2,2-dimethyloxirane with each of the following reagents.

   (a) $LiAlH_4/(C_2H_5)_2O$ **(e)** H_3O^+

   (b) $C_2H_5MgBr/(C_2H_5)_2O$ **(f)** $HC\equiv C^-Na^+/NH_{3(l)}$

   (c) $NaNH_2/NH_{3(l)}$ **(g)** $NaSH/C_2H_5OH$

   (d) C_2H_5SNa/C_2H_5OH

13.29 Write the structure of the major product in each of the following reactions. Include the stereochemistry where appropriate.

(a)

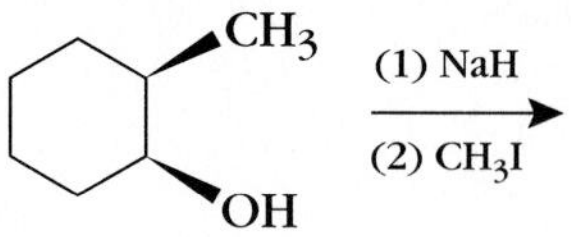

(b) 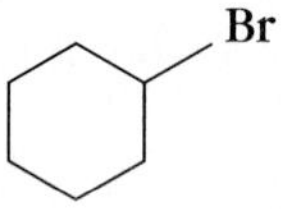   + $(CH_3)_2CHS^-Na^+$ ⟶

(c) 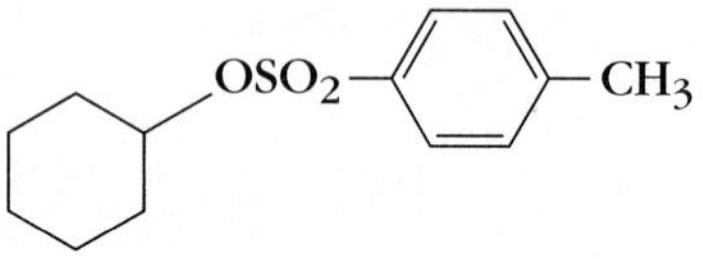   + $(CH_3)_3CO^-K^+$ ⟶

(d) 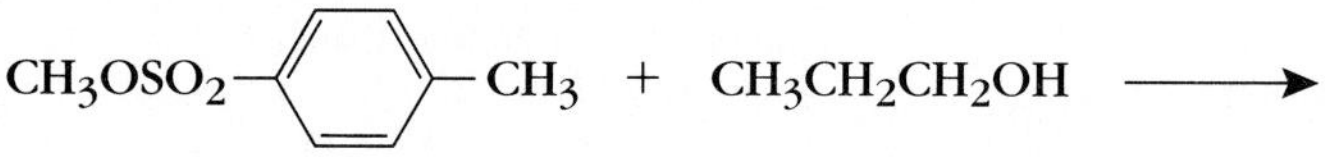 CH_3OSO_2—⟨ ⟩—CH_3 + $CH_3CH_2CH_2OH$ ⟶

(e) CH_3SCH_3 $\xrightarrow{\text{2 equivalents of } H_2O_2}$

(f) $[(CH_3)_3CS(CH_3)_2]^+I^-$ + CH_3CH_2OH ⟶

13.30 Write the structure, including the correct stereochemistry, of the principal product in each of the following reactions.

(a) 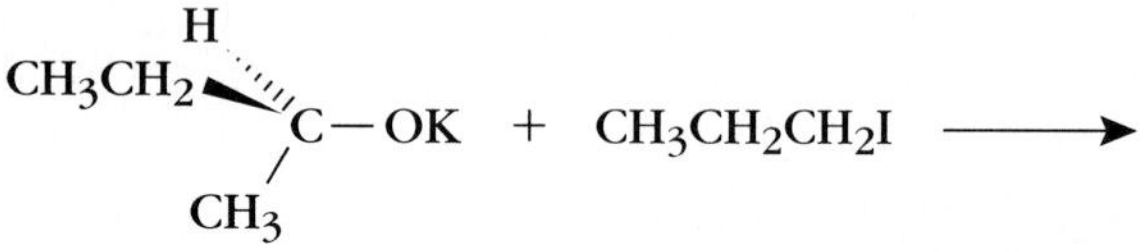 + $CH_3CH_2CH_2I$ ⟶   **(d)** 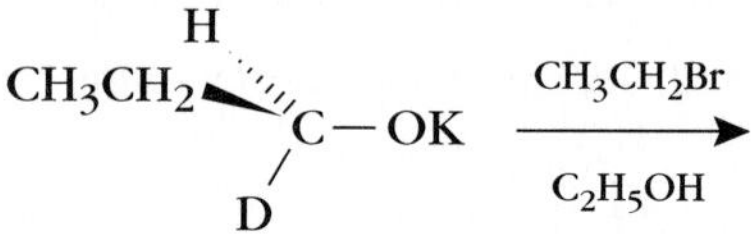 $\xrightarrow[C_2H_5OH]{CH_3CH_2Br}$

(b) 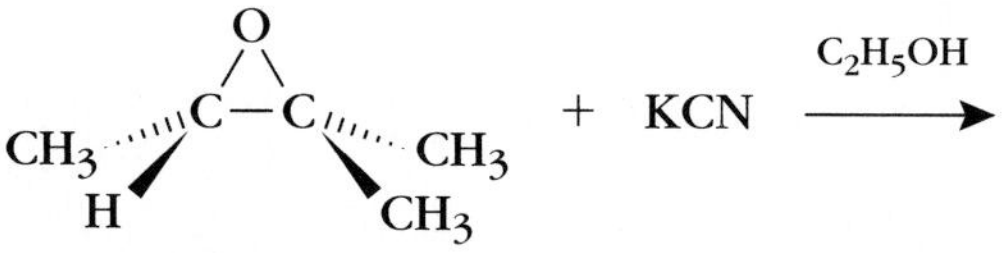 + KCN $\xrightarrow{C_2H_5OH}$ **(e)** 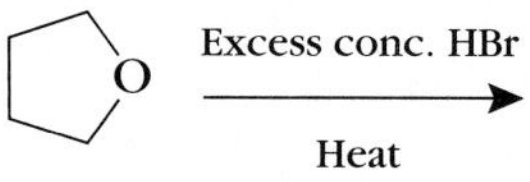 $\xrightarrow[\text{Heat}]{\text{Excess conc. HBr}}$

(c) 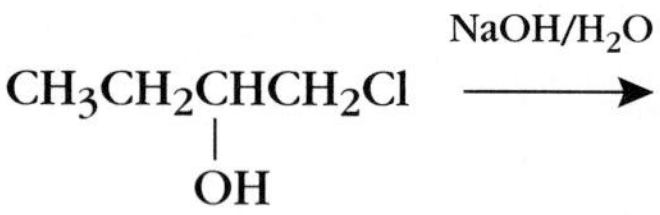 $CH_3CH_2CHCH_2Cl$ $\xrightarrow{NaOH/H_2O}$ **(f)** 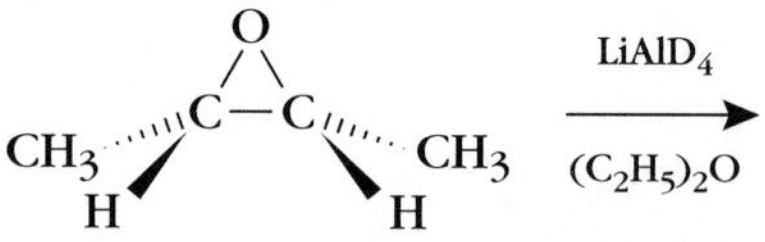 $\xrightarrow[(C_2H_5)_2O]{LiAlD_4}$
         |
       OH

13.31 Write the reactions (more than one may be needed) to convert cyclohexene into each of the following compounds.

(a) 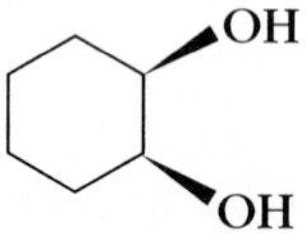   **(c)** 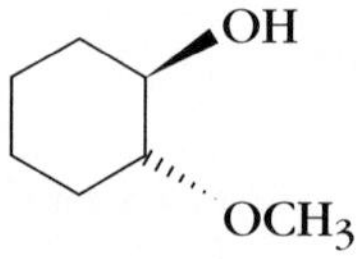   **(e)**

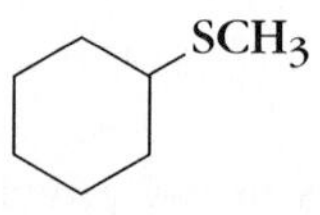

(b) 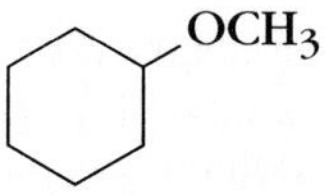   **(d)** 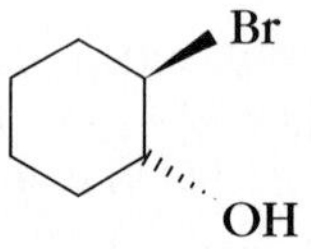   **(f)**

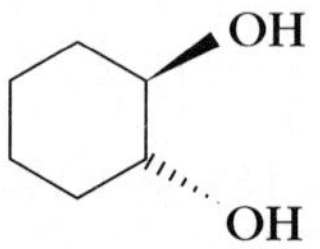

13.32 Write the equations for two different methods of preparing *trans*-2,3-dimethyloxirane from (*E*)-2-butene.

13.33 Starting with 2,2-dimethyloxirane and any other reagents, propose a synthesis for each of the following compounds. In some cases more than one step may be necessary.

(a) 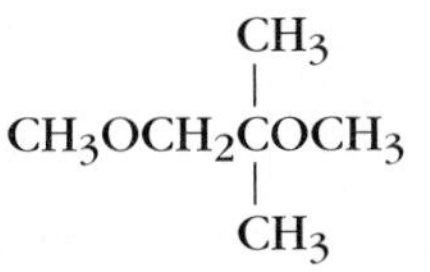 $CH_3OCH_2\underset{\underset{CH_3}{|}}{\overset{\overset{CH_3}{|}}{C}}OCH_3$

(d) $(CH_3)_2\underset{\overset{|}{OH}}{C}CH_2OCH_3$

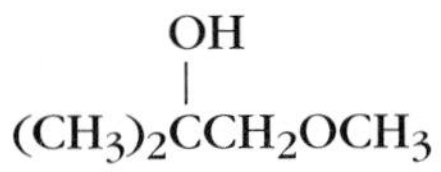

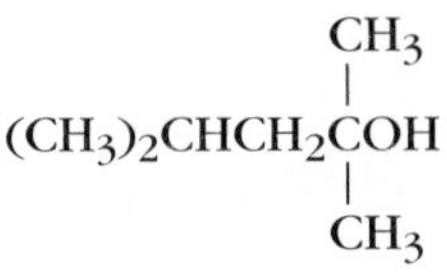

(b) $(CH_3)_2CHCH_2\underset{\underset{CH_3}{|}}{\overset{\overset{CH_3}{|}}{C}}OH$

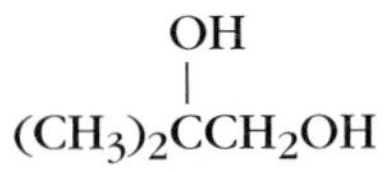

(e) $(CH_3)_2\underset{\overset{|}{OH}}{C}CH_2OH$

(c) $CH{=}C(CH_3)_2$ attached to a benzene ring

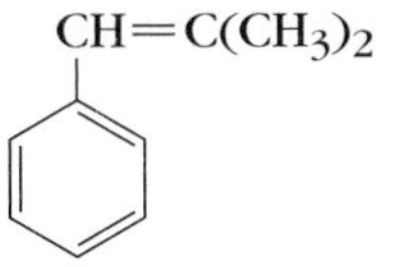

(f) $CH_2{=}C(CH_3)_2$

13.34 Tetrahydrofuran is formed when 1,4-dibromobutane is heated in an aqueous sodium hydroxide solution. Propose a mechanism to account for this result.

13.35 Heating (*S*)-2-methoxybutane with one equivalent of anhydrous HBr forms bromomethane and optically active 2-butanol. Propose a mechanism to explain this result. What is the configuration of the alcohol?

13.36 Propose a synthesis for each of the following compounds. You may start with any organic compound containing four or fewer carbon atoms, any aromatic compound, and use any solvent or inorganic reagent.

(a) 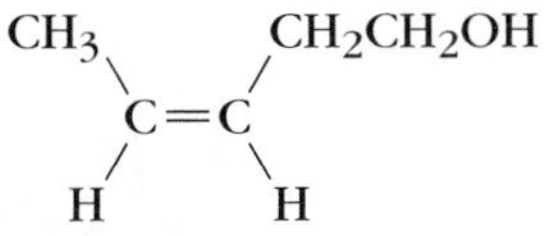 $\underset{H}{\overset{CH_3}{C}}{=}\underset{H}{\overset{CH_2CH_2OH}{C}}$

(d) 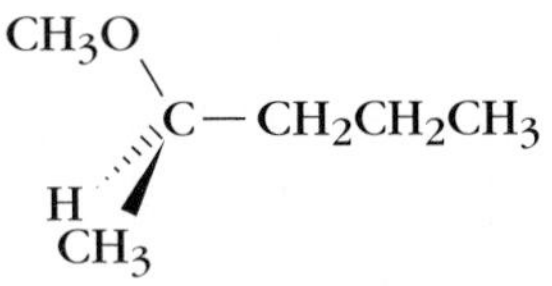 $\underset{\underset{CH_3}{H}}{\overset{CH_3O}{C}}{-}CH_2CH_2CH_3$

(b) $(CH_3)_3CCH_2CH_2OCH_3$

(c) 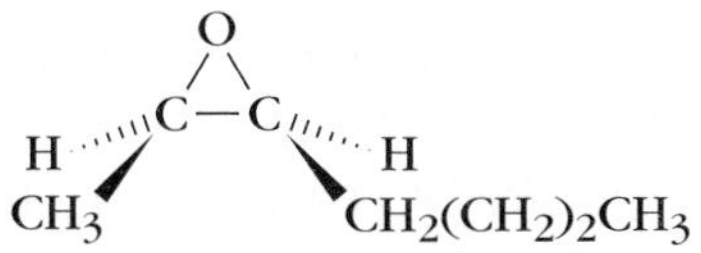 $\underset{CH_3}{\overset{H}{\underset{}{C}}}\overset{\overset{O}{\diagup\diagdown}}{-}\underset{CH_2(CH_2)_2CH_3}{\overset{H}{C}}$

(e) benzene ring with $OCH_2CH{=}CH_2$

13.37 A student finds a bottle labelled 2-bromocyclohexanol on the laboratory shelf and carries out a reaction with some of it with aqueous sodium hydroxide in an attempt to prepare epoxycyclohexane. The isolated product is a mixture of 40% epoxycyclohexane and 60% of another compound. Repeating the reaction with a fresh sample of the same 2-bromocyclohexanol does not improve the yield of epoxycyclohexane. The puzzled student comes to you for advice. Explain why the yield of epoxycyclohexane is so low and give the structure of the other compound found in the product mixture.

13.38 Optically active 2-ethoxyoctane, $[\alpha]_D = -15.6°$, is formed when bromoethane is added to the product of the reaction of (*R*)-2-octanol with NaH. On the other hand, reaction of (*R*)-2-bromooctane with sodium ethoxide in ethanol also forms optically active 2-ethoxyoctane with the same specific rotation but opposite in sign, $[\alpha]_D = +15.6°$. Explain why the specific rotation of the two products differs in sign, even though the starting materials for both reactions have the (*R*) configuration. What is the configuration of the stereocenter in the two ethers?

13.39 Propose a mechanism for the following reaction.

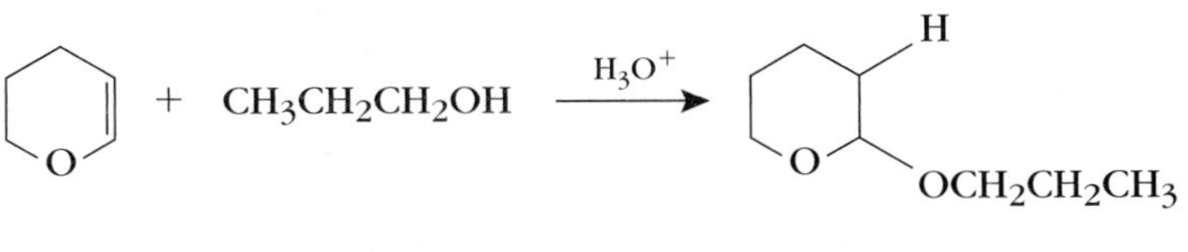

$$\text{dihydropyran} + CH_3CH_2CH_2OH \xrightarrow{H_3O^+} \text{tetrahydropyran with } OCH_2CH_2CH_3$$

13.40 What spectroscopic technique would you use to distinguish between the following pairs of compounds? Explain exactly what difference you would expect to find in the spectra of the two compounds.

(a) $CH_3OCH_2CH_2OCH_3$ and $CH_3OCH_2CH_2OH$ (c) $CH_3CH_2OCH_2CH_3$ and $ClCH_2CH_2OCH_2CH_2Cl$

(b) $CH_3CH_2OCH_2CH_3$ and

13.41 Write the structural formula of the following two compounds whose molecular formula and ^1H NMR spectra are as follows:

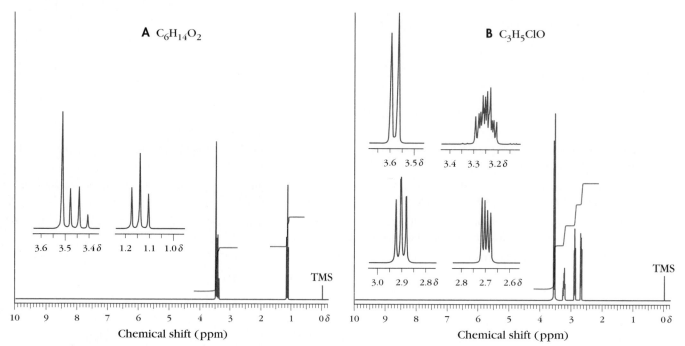

Chemical shift (ppm)   Chemical shift (ppm)

13.42 Write the structural formula of each of the following compounds based on the spectral information provided.

(a) $C_2H_4Cl_2O$ IR: 1200 cm^{-1}. ^1H NMR: δ 1.27 d (3H); δ 3.07 q (1H).

(b) $C_3H_8O_2$ IR: 3350 cm^{-1}; 1200 cm^{-1}. ^{13}C NMR: δ 74.0; δ 61.6; δ 58.8. ^1H NMR: δ 3.40 s (3H); δ 3.51 t (2H); δ 3.72 m (2H); δ 3.10 m (1H).

(c) C_4H_8O ^1H NMR: δ 1.27 d (3H); δ 3.07 q (1H); ^{13}C NMR: δ 52.5; δ 12.9.

(d) $C_4H_8O_2$ ^1H NMR: δ 1.78 m (1H); δ 3.91 t (2H); δ 4.84 s (1H) ^{13}C NMR: δ 94.2; δ 61.6; δ 26.5.

CHAPTER

14

ALDEHYDES and KETONES

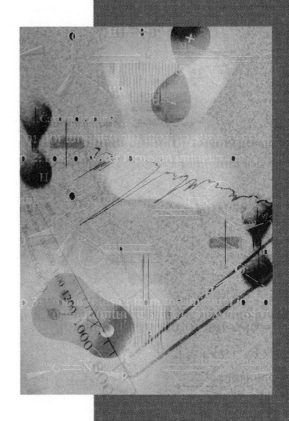

Aldehydes and ketones are two important classes of compounds that contain a carbonyl group, C=O, as their functional group (Section 2.14, F). While we've encountered aldehydes and ketones in previous chapters, let's briefly review their structures.

Recall that the simplest aldehyde, formaldehyde, has two hydrogen atoms bonded to the carbonyl carbon. In all other aldehydes, the carbonyl carbon atom is bonded to one hydrogen atom and either an alkyl or an aryl group. In ketones, the carbonyl carbon atom is bonded to two carbon atoms, which can be part of any combination of aryl and alkyl groups.

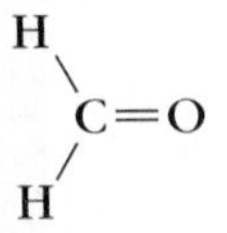

$$\begin{array}{ccc} \underset{\underset{H}{|}}{\overset{\overset{H}{|}}{C}}=O & \underset{\underset{H}{|}}{\overset{\overset{R}{|}}{C}}=O & \underset{\underset{R'}{|}}{\overset{\overset{R}{|}}{C}}=O \end{array}$$

Formaldehyde   General formula   General formula
                      of an aldehyde   of a ketone
                      R = alkyl or aryl group   R = alkyl or aryl group
                                                R' = alkyl or aryl group

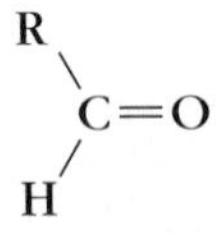

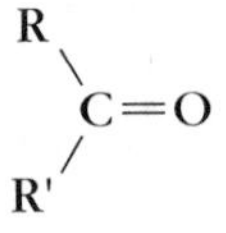

Organic chemists write the structure of the functional group of aldehydes in a number of equivalent ways. One way, shown above, clearly indicates the presence of the double bond between the oxygen and carbon atoms. Line drawings can also be used, and finally, the functional group can be abbreviated as RCHO. The following illustrates the different ways of representing the structures of aldehydes:

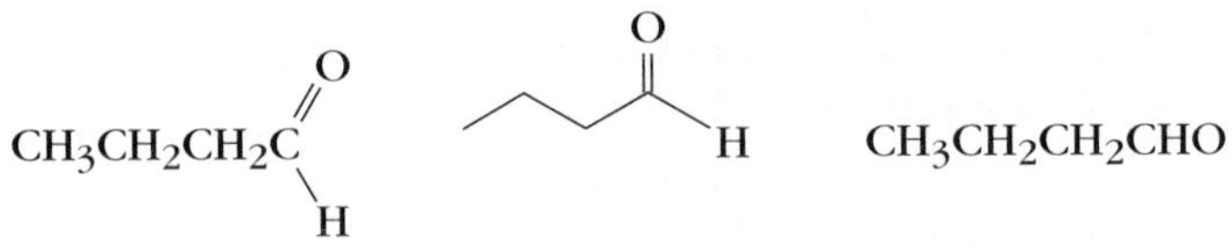

$$CH_3CH_2CH_2\overset{\overset{O}{\|}}{\underset{\underset{H}{}}{C}} \qquad \qquad CH_3CH_2CH_2CHO$$

Three ways of writing the structure of butanal

Aldehydes and ketones are widely used commercially. Formaldehyde, HCHO, which is used in the preparation of adhesive resins, is synthesized industrially by passing a mixture of methanol and air over a metal oxide catalyst at high temperature. The resulting gases are dissolved in water and concentrated by distillation to form an aqueous solution called formalin, which contains 37% formaldehyde.

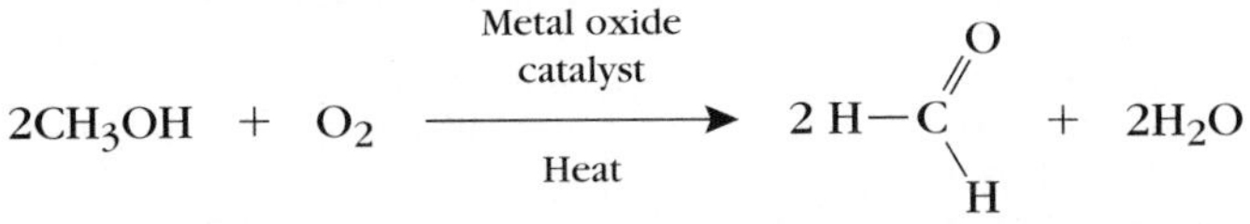

$$2CH_3OH + O_2 \xrightarrow[\text{Heat}]{\substack{\text{Metal oxide} \\ \text{catalyst}}} 2\;H{-}\overset{\overset{O}{\|}}{\underset{\underset{H}{}}{C}} + 2H_2O$$

Methanol   Formaldehyde

Acetone, $(CH_3)_2C{=}O$, is by far the most industrially important ketone. Several billion pounds of acetone are prepared every year for use as a solvent and in the production of other chemicals. It is prepared commercially by high temperature metal-catalyzed oxidation of 2-propanol.

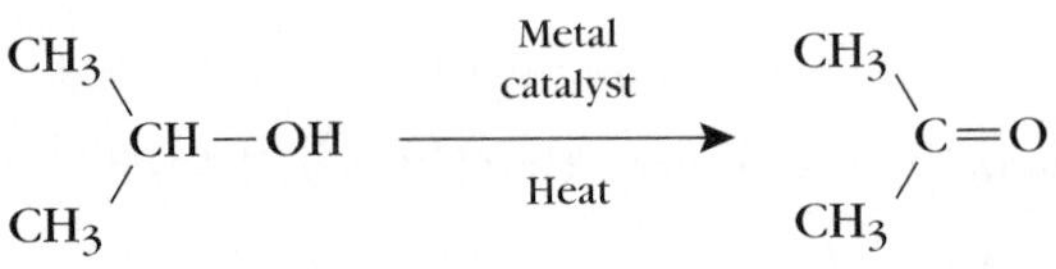

$$\underset{\underset{CH_3}{|}}{\overset{\overset{CH_3}{|}}{CH}}{-}OH \xrightarrow[\text{Heat}]{\substack{\text{Metal} \\ \text{catalyst}}} \underset{\underset{CH_3}{|}}{\overset{\overset{CH_3}{|}}{C}}=O$$

2-Propanol   Acetone

Acetone is not usually produced in the human body. People suffering from diabetes mellitus, however, will, if untreated, exhibit a pathological condition called ketosis, a symptom of which is exhalation of acetone. People with diabetes mellitus cannot utilize glucose properly. As a result, their bodies increase the chemical breakdown of proteins and fatty acids, forming acetone in the process. Because of its volatility, acetone is easily eliminated from the body when the person exhales.

Many aldehydes and ketones are important to the normal function of the human body. For example, the aldehyde, 11-*cis*-retinal, plays an important role in our ability to see, while the steroid hormone progesterone, a ketone, is secreted in women at the time of ovulation.

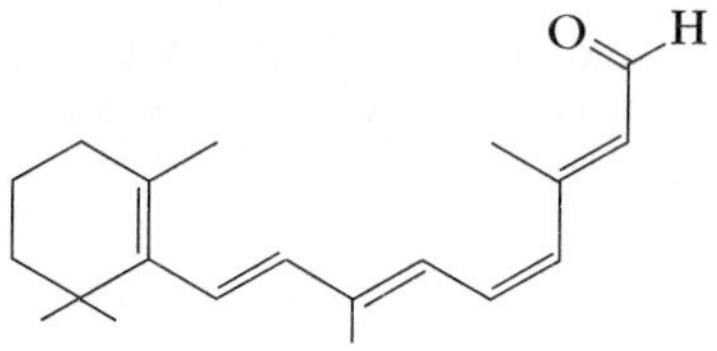

11-*cis*-Retinal 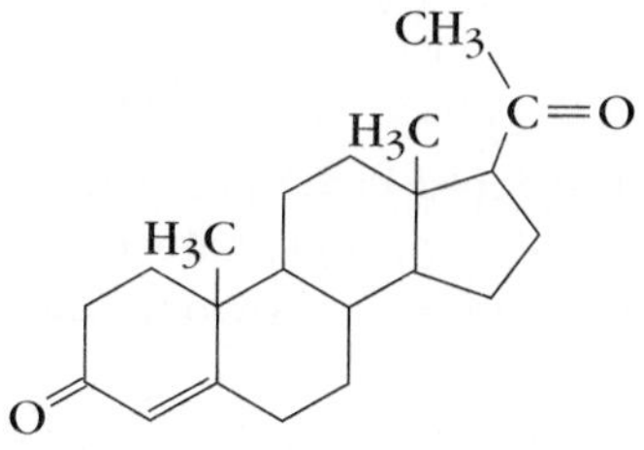   Progesterone

As well as the usual sections that serve to acquaint us with aldehydes and ketones, this chapter has a central mechanistic theme, the nucleophilic additions to the carbonyl groups of aldehydes and ketones. The principles developed in this chapter will have broad applicability to the mechanisms of the reactions of various derivatives of carboxylic acids that we will discuss in Chapters 15, 16, and 17.

We begin our study of aldehydes and ketones by learning how to name them.

14.1 Naming Aldehydes and Ketones

Aldehydes are named according to the IUPAC system by identifying the longest continuous chain of carbon atoms that contains the —CHO group. This establishes the parent alkane name. The terminal -*e* of the parent alkane name is then replaced by -*al* to signify the presence of an aldehyde group. The carbon atom of the —CHO group is always assigned number 1. The suffix -*dial* is used to indicate the presence of two —CHO groups, -*trial* for three, and so forth. As with *diols,* the final -*e* of the name of the parent alkane chain is retained when there is more than one —CHO group in the chain. The following examples illustrate these IUPAC rules.

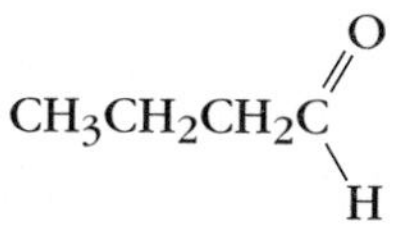

Butanal 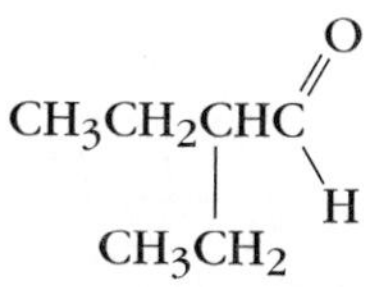   2-Ethylbutanal

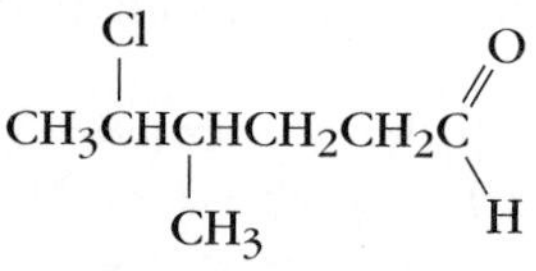

5-Chloro-4-methylhexanal 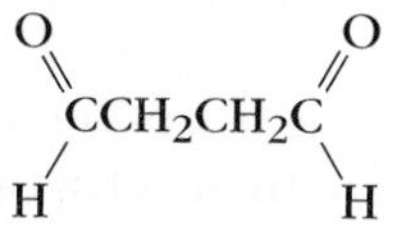   Butanedial

Notice that the longest continuous chain of carbon atoms in 2-ethylbutanal is a pentane. This is not chosen as the parent chain, however, because it does not contain the —CHO group.

Certain aldehydes, particularly those with the aldehyde group attached to a ring, are more easily named by considering the —CHO group as a substituent. In these cases the suffix *-carbaldehyde* is added to the name of the ring. For example:

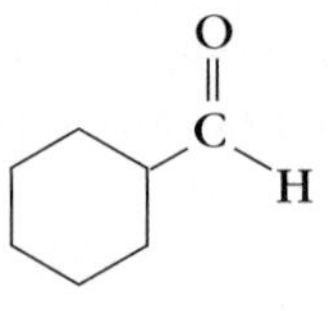

Cyclohexanecarbaldehyde

A —CHO group attached to a ring is assigned the number 1 regardless of the presence of alkyl groups, halogen atoms, hydroxy groups, or carbon-carbon double or triple bonds, as shown in the following example.

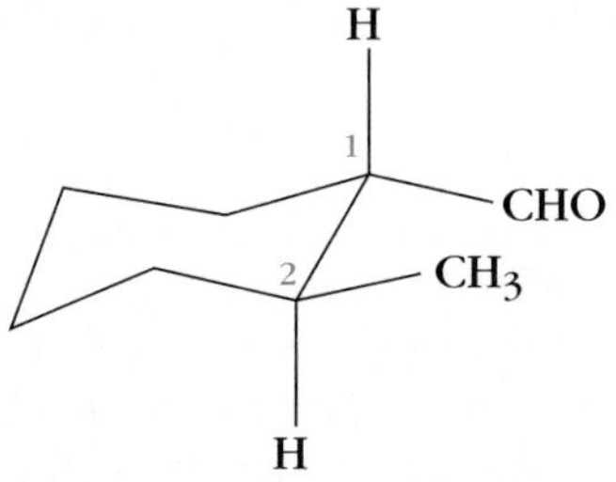

trans-2-Methylcyclohexanecarbaldehyde

Common names of many low molecular weight aldehydes are often used. A few examples along with their IUPAC names are given in Table 14.1.

Table 14.1	Common and IUPAC Names of Several Low Molecular Weight Aldehydes	
Structure	Common Name	IUPAC Name
HCHO	Formaldehyde	Methanal
CH_3CHO	Acetaldehyde	Ethanal
CH_3CH_2CHO	Propionaldehyde	Propanal
$CH_3CH_2CH_2CHO$	Butyraldehyde	Butanal
$(CH_3)_2CHCHO$	Isobutyraldehyde	2-Methylpropanal
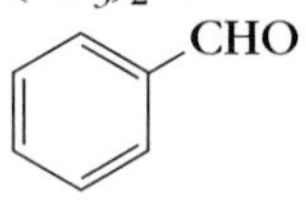 ⬡CHO	Benzaldehyde	Benzenecarbaldehyde

We shall use the common names *formaldehyde, acetaldehyde,* and *benzaldehyde* in this text because they are used far more frequently than their systematic names.

Low molecular weight substituted aldehydes are sometimes named by designating the positions of their substituents with Greek letters. The position next to the carbonyl group is designated α in these common names.

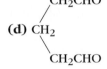

γ   β   α
$CH_3-CH_2-CH_2-C$ CH_3CH_2CHCHO
4   3   2   1

α-Chlorobutyraldehye

Notice the difference between the way the chain is numbered in the common names and IUPAC system. In the IUPAC system, the carbonyl carbon atom is designated number 1, whereas in the common names the carbon atom *next* to the carbonyl group is designated α. Thus the α-carbon atom is C2, the β-carbon atom is C3, and so on.

Exercise 14.1   Give the IUPAC name to each of the following aldehydes.

(a) $CH_3CH_2(CH_2)_3CHO$
(b) $(CH_3)_2CHCH_2CHO$
(c) 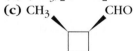 CH_3 CHO

(d)
CH_2CHO
CH_2
CH_2CHO

Exercise 14.2   Write the structure of the compound corresponding to each of the following names.

(a) 3-Methylpentanal
(b) *cis*-3-Chlorocyclohexanecarbaldehyde
(c) Ethanedial
(d) 3-Bromo-2-isopropylpentanedial
(e) β-Iodobutyraldehyde
(f) (*Z*)-Butenedial

Ketones are named according to the IUPAC system by identifying the longest continuous chain of carbon atoms that contains the carbonyl group. This establishes the parent alkane name. The terminal *-e* of the parent alkane name is then replaced by *-one* to identify the compound as a ketone. The carbonyl carbon atom is assigned the lowest possible number in the chain, regardless of the presence of alkyl groups, halogen atoms, hydroxy groups, or carbon-carbon double or triple bonds.

The suffix *-dione* is used to indicate a ketone that contains two carbonyl groups, *-trione* for three, and so forth. As with aldehydes, the final *-e* of the name of the parent alkane chain of the ketone is retained when there is more than one carbonyl group. The following examples illustrate these IUPAC rules.

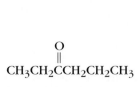

$CH_3CH_2CCH_2CH_2CH_3$

3-Hexanone

4-Chlorocyclohexanone

$CH_3CH_2CH=CHCCH_3$

3-Hexen-2-one

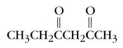

$CH_3CH_2CCH_2CCH_3$

2,4-Hexanedione

Notice that the carbonyl carbon atom in a ring is assigned the number 1.

The common name of propanone, *acetone,* is so widely used that we will refer to it only by that name in this book.

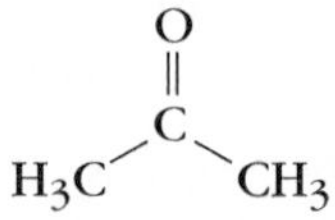

$$H_3C \overset{\overset{\textstyle O}{\|}}{\underset{}{C}} CH_3$$

Acetone
(Propanone)

Sometimes ketones are named by designating *both* the groups bonded to the carbonyl carbon atom followed by the word ketone. The groups are listed alphabetically, as shown in the following examples.

Ethyl methyl ketone 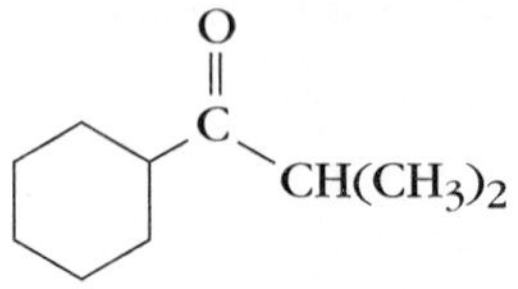   Cyclohexyl isopropyl ketone 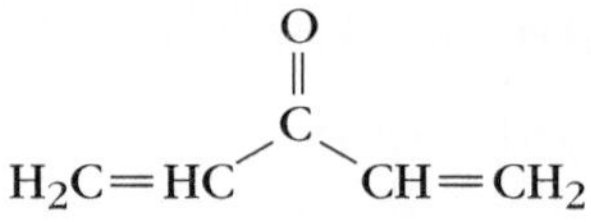   Divinyl ketone

---

Exercise 14.3   Give an IUPAC name to each of the following ketones.

(a)

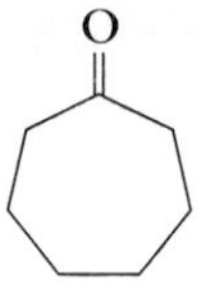

(b)

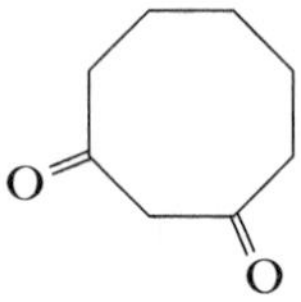

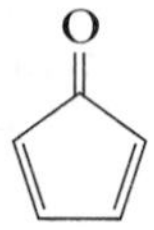

(c) 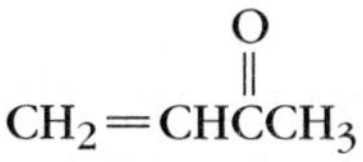 $CH_2{=}CHCCH_3$

(d) $(CH_3)_2CHCCH_2CH_3$

(e)

(f) 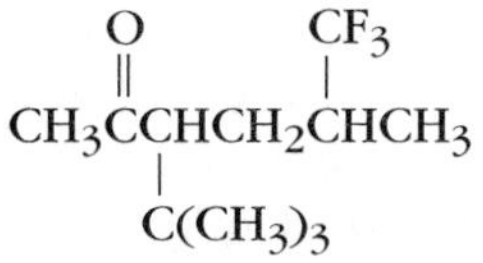 $CH_3CCHCH_2CHCH_3$ with CF_3 and $C(CH_3)_3$ substituents

---

Exercise 14.4   Write the structure of the compound corresponding to each of the following names.

(a) 2-Pentanone   **(d)** 2,5-Octanedione
(b) 4-Hepten-3-one   **(e)** Cyclopentyl ethyl ketone
(c) 3-Methylcyclopentanone   **(f)** 4-Hydroxy-2-pentanone

14.2   Structure of the Carbonyl Group

The carbonyl group in aldehydes and ketones and the atoms attached to it lie in a plane, as we discussed in Section 11.6. Formaldehyde, for example, is a planar molecule whose bond angles around the carbonyl carbon atom are close to 120°. The average carbon-oxygen double bond length is 121 pm, which is much shorter than the typical carbon-carbon double bond length of 134 pm found in simple alkenes. The following are the bond angles of some simple aldehydes and ketones.

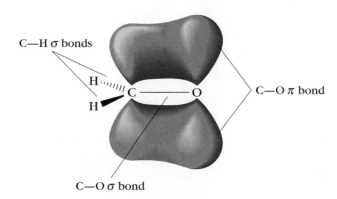

| Formaldehyde | Acetaldehyde | Acetone |

The localized valence bond orbital model of the electron structure of formaldehyde is shown in Figure 14.1. The correct geometry is obtained by using an sp^2 hybrid carbon atom for the carbon atom of the carbonyl group. The σ bonds to the two hydrogen atoms and a single σ bond to oxygen are formed by end-on overlap of the three sp^2 hybrid orbitals of the carbonyl carbon atom with the orbitals of the two hydrogen atoms and the oxygen atom. Sideways overlap of the remaining p atomic orbital of the sp^2 hybrid carbon with a p orbital of oxygen forms the π bond.

C—H σ bonds

C—O π bond

C—O σ bond

Figure 14.1
The σ bonds of formaldehyde are formed by overlap of the three sp^2 hybrid orbitals of the carbonyl carbon atom with atomic orbitals from two hydrogen atoms and the oxygen atom. The π bond is formed by sideways overlap of a p orbital from oxygen and the remaining p orbital of the sp^2 hybrid carbon atom.

The electron structure of the carbonyl group can also be described as a resonance hybrid of the following two resonance structures.

Oxygen is more electronegative than carbon, so the resonance structure containing a negative charge on the oxygen atom and a positive charge on the carbonyl carbon atom is an important contributor to the resonance hybrid of the carbonyl group. The resonance hybrid description of the carbonyl group indicates that the electron distribution in the carbon-oxygen double bond is not symmetrical. It tells us that the oxygen atom has an excess of electron density while the carbon atom is electron deficient. This electron distribution in the carbonyl group is consistent with the chemistry of the carbonyl group of aldehydes and ketones, as we have already discussed in Section 11.7.

The π electron structure of the carbonyl group can be described by the simplified LCAO method introduced in Section 1.13. Recall that in this method, the σ

■ *The construction of the π bonds of the carbonyl group is similar to the LCAO method used to construct the π bonds of the carbon-carbon double bond in alkenes (see Section 1.13).*

bonds are described by localized valence bond orbitals and the LCAO method is used only to describe the π bonds. Combining the *p* atomic orbital of the *sp*² hybrid carbonyl carbon atom and the *p* atomic orbital of the *sp*² carbonyl oxygen atom results in the formation of two molecular orbitals, a π bonding, and a π anti-bonding molecular orbital, which are shown in Figure 14.2.

It's difficult to see the unsymmetrical electron distribution in the π bond of formaldehyde from the molecular orbitals shown in Figure 14.2, *B*. Rather than looking at just the π bond, the total charge distribution of formaldehyde in the LCAO model can be obtained by calculating the total charge density on each atom. One way to represent such charge distribution is shown for formaldehyde in Figure 14.2, *C*. The surface of the total charge distribution is colored red where there is an excess of negative charge and dark blue where there is an excess of positive charge. The results of the total charge distribution of formaldehyde, therefore, show that the oxygen atom is electron rich and the carbonyl carbon and the two hydrogen atoms are electron deficient, which makes formaldehyde a polar molecule.

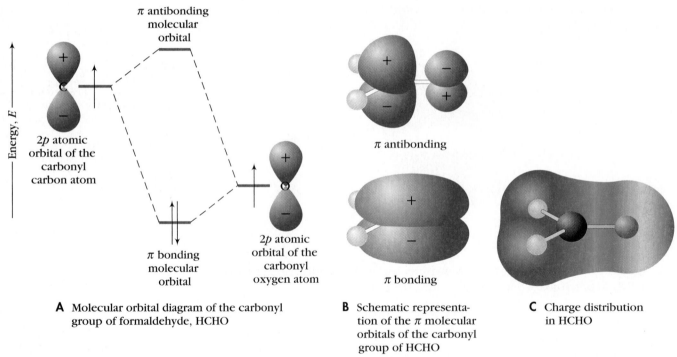

A Molecular orbital diagram of the carbonyl group of formaldehyde, HCHO

B Schematic representation of the π molecular orbitals of the carbonyl group of HCHO

C Charge distribution in HCHO

Figure 14.2
A, Relative energies of the two π molecular orbitals of the carbonyl group of formaldehyde. **B,** Schematic three-dimensional representation of the two π molecular orbitals of the carbonyl group of formaldehyde. **C,** Charge distribution in formaldehyde. Red on the surface indicates excess negative charge, and blue indicates excess positive charge.

Because there is a partial separation of charges, the carbonyl group has a bond dipole moment. This bond dipole, in turn, makes aldehydes and ketones polar molecules, which affects their chemical and physical properties.

14.3   Physical Properties

The boiling points of low molecular weight aldehydes and ketones are higher than those of alkanes, alkenes, chloroalkanes, and ethers of comparable molecular weights, as shown by the data in Table 14.2.

Table 14.2	Comparison of the Boiling Points of Aldehydes and Ketones With Other Types of Compounds of Similar Molecular Weight	

Compound	Molecular Weight	Boiling Point (°C)
Butane	58	0
1-Butene	56	−4
Acetone	58	56
Propanal	58	49
1-Propanol	60	97
Methoxyethane	60	−25
Chloromethane	50	−24
Nonanal	142	192
2-Nonanone	142	195
1-Nonanol	144	212
Decane	142	174

The boiling points of aldehydes and ketones are higher than alkanes and other relatively nonpolar molecules because of the attraction between their molecular dipoles. This difference is largest with low molecular weight aldehydes and ketones. As the polar carbonyl group becomes a smaller and smaller part of the molecule, the boiling points of aldehydes and ketones tend to come closer to the boiling points of hydrocarbons of about the same molecular weight. For example, there is a 56 °C difference in boiling point between acetone and butane, yet only a 21 °C difference between 2-nonanone and decane.

Notice, however, that the boiling point of 1-propanol, an alcohol, is about 50 °C higher than the boiling point of either acetone or propanal. Low molecular weight alcohols have higher boiling points because they form intermolecular hydrogen bonds, which are stronger than the dipole-dipole interactions between aldehyde or ketone molecules. As the molecular weight of the compounds increases, however, this difference tends to disappear.

Aldehydes and ketones are not hydrogen bond donors, but the oxygen atoms are hydrogen bond acceptors, so low molecular weight aldehydes and ketones are quite soluble in water.

As in the case of other functional groups, the water solubility of aldehydes and ketones rapidly decreases as the hydrophobic part of the molecule increases along a homologous series. For example, while formaldehyde, acetaldehyde, and acetone are miscible with water, aldehydes and ketones containing more than eight carbon atoms are essentially insoluble in water.

Exercise 14.5	Arrange the following compounds in order of increasing boiling points.

(a) 1-Butanol   (c) 2-Butanone
(b) Diethyl ether   (d) Butane

■ *Representation of acetone dissolved in water. The dotted lines indicate hydrogen bonds.*

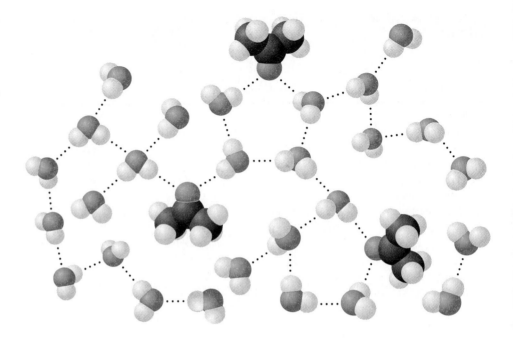

14.4   Spectroscopic Properties of Aldehydes and Ketones

The carbonyl group is a functional group that is very easy to detect by IR spectroscopy, as we discussed in Section 5.5, F. Recall that IR spectra of aldehydes and ketones show a strong C=O bond stretching absorption in the region of 1700 to 1730 cm^{-1}. The absorptions of the carbonyl groups of most acyclic saturated ketones and cyclohexanones, for example, are typically found in the region 1700 to 1715 cm^{-1}. The strong C=O bond stretching absorption of acyclic saturated aldehydes is typically found in the region 1720 to 1730 cm^{-1}, as illustrated by the spectrum of butanal in Figure 14.3.

Figure 14.3
Infrared spectrum of butanal. The carbonyl C=O stretching frequency appears as a strong absorption at 1720 cm^{-1}. The carbonyl C—H stretching appears as two peaks at 2800 cm^{-1} and 2700 cm^{-1}.

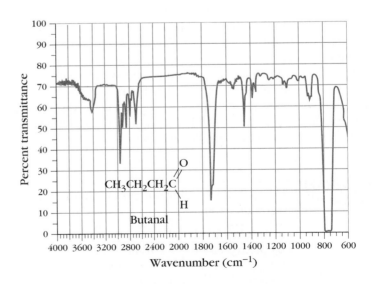

The IR spectra of aldehydes differ from those of ketones in an important way. The stretching of the carbonyl carbon-hydrogen bond gives rise to two absorptions at about 2800 and 2700 cm^{-1}. The IR spectrum of butanal in Figure 14.3 shows these two absorption bands at 2800 cm^{-1} and 2700 cm^{-1}. Both are sharp but weak, and the band near 2800 cm^{-1} can be hidden sometimes by the absorptions of the aliphatic C—H bonds. These two absorption bands are absent in the IR spectra of ketones because ketones lack a hydrogen atom bonded to their carbonyl carbon atom.

| **Exercise 14.6** | Laboratory workers often need to determine when a reaction has reached completion. How would you use IR spectroscopy to determine if all the starting material, 2-hexanone, has been reduced by $LiAlH_4$ to 2-hexanol? |

| **Exercise 14.7** | How would you use IR spectroscopy to distinguish between each of the following pairs of compounds? |

(a) 2-Butanone and butanal
(b) 3-Hexanone and acetic acid (CH_3COOH)
(c) 3-Pentanone and 1-pentanol

Aldehydes are also easily identified by ^{1}H NMR spectroscopy. The chemical shift of the —CHO hydrogen is found in the region δ 9–10, where few other peaks appear. In the ^{1}H NMR spectrum of propanal, shown in Figure 14.4, the hydrogen atom bonded to the carbonyl carbon atom is a triplet centered at δ 9.82 because it is coupled to the two hydrogen atoms on the α-carbon atom.

Hydrogen atoms on the α-carbon atom in both aldehydes and ketones are slightly deshielded, with chemical shifts in the region δ 2.0 to 2.6. Methyl ketones, for example, can often be identified by the presence of a sharp methyl

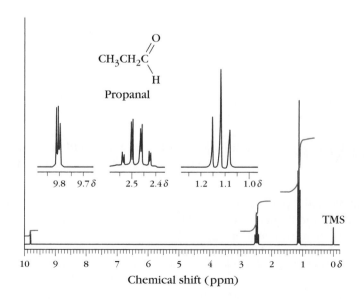

Figure 14.4
The ^{1}H NMR spectrum of propanal. The hydrogen atom bonded to the carbonyl carbon atom is a triplet centered at δ 9.82 because it is coupled to the two hydrogen atoms on the α-carbon atom. Each signal of the methylene quartet is also split into a doublet by coupling to the hydrogen atom bonded to the carbonyl carbon atom.

singlet near δ 2.1. The following examples illustrate the chemical shifts of hydrogen atoms near the carbonyl group of simple aldehydes and ketones.

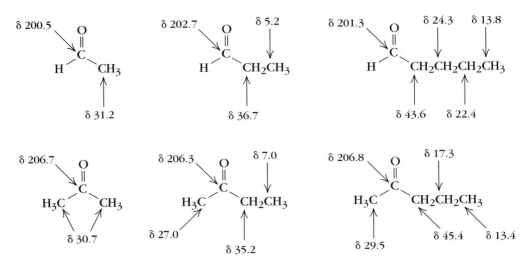

Aldehydes and ketones can also be readily identified by the chemical shift of the carbonyl carbon in their ^{13}C NMR spectra. In Section 5.10, we discussed that the peaks due to the sp^2 hybrid carbonyl carbon atoms are found in the low field region (approximately δ 200) of their ^{13}C NMR spectra due, in part, to the electronegativity of the carbonyl oxygen atom, which deshields the carbonyl carbon atom. Carbon atoms adjacent to the carbonyl carbon are also deshielded compared to those further away. The following examples of the ^{13}C NMR chemical shifts of some aldehydes and ketones illustrate these observations.

Exercise 14.8   Explain how NMR spectroscopy could be used to distinguish between each of the following pairs of compounds.

(a) Pentane and 3-pentanone
(b) 1-Butene and 2-butanone
(c) Cyclopentanone and pentanal
(d) Cyclohexanone and 2-cyclohexenone

Exercise 14.9 The ^{1}H NMR spectra of three isomeric compounds (Compounds I, II, and III) of molecular formula $C_5H_{10}O$ are shown in Figure 14.5. The IR spectra of all three compounds have an absorption band in the region of 1700 to 1730 cm^{-1}. Assign a structure to each of the compounds and indicate which hydrogen atoms give rise to which peaks in each ^{1}H NMR spectrum.

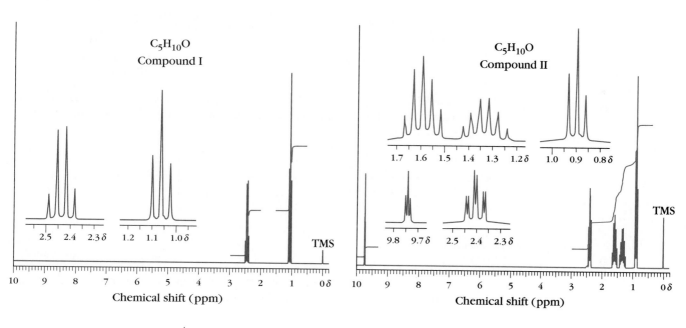

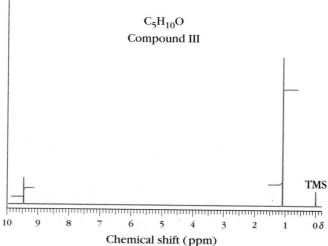

Figure 14.5
NMR spectra for Exercise 14.9.

14.5   Synthesis of Aldehydes

Two methods of preparing aldehydes have been discussed in previous chapters in connection with the chemistry of other functional groups. One method is the oxidation of primary alcohols and the other is hydroboration-oxidation of terminal alkynes.

Because there are so many ways of preparing alcohols, the oxidation of alcohols is one of the most important synthetic methods of preparing aldehydes. Recall that oxidation of primary alcohols by pyridinium chlorochromate (PCC) in dichloromethane at room temperature forms aldehydes in excellent yields (Section 11.18). Other oxidizing agents, such as sodium dichromate in aqueous acid, are to be avoided because they oxidize aldehydes further to carboxylic acids.

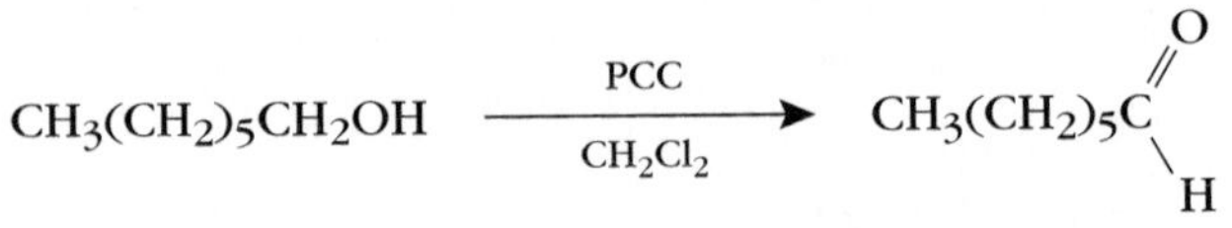

$$CH_3(CH_2)_5CH_2OH \xrightarrow[CH_2Cl_2]{PCC} CH_3(CH_2)_5C\overset{\displaystyle O}{\diagdown}H$$

1-Heptanol

Heptanal
(78% yield)

Aldehydes can also be prepared from alkynes. Hydroboration-oxidation of terminal alkynes using a bulky hydroborating reagent such as 9-BBN forms aldehydes (Section 9.7, B).

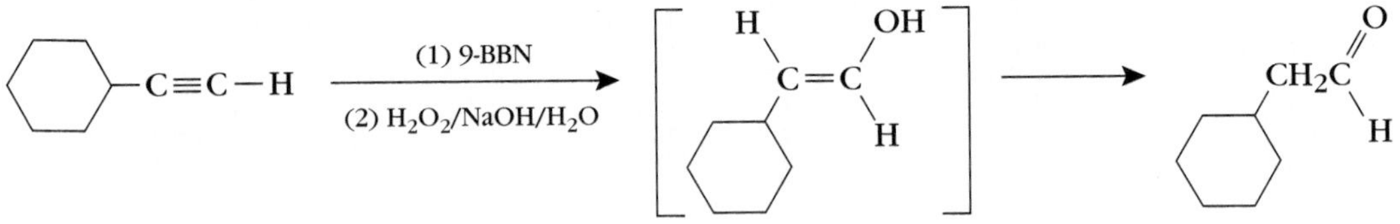

Ethynylcyclohexane

Enol

Cyclohexylethanal
(68% yield)

We briefly mention here a third method of preparing aldehydes by the partial reduction of certain carboxylic acid derivatives, where Y = Cl or OR'.

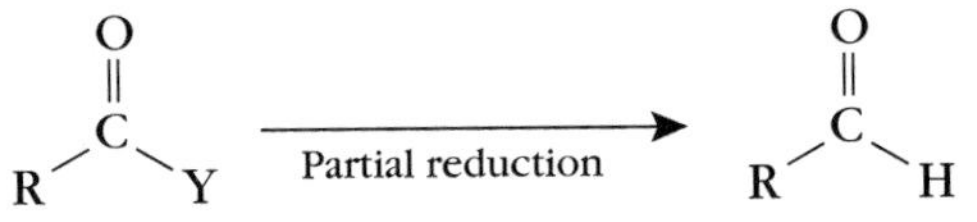

$$R\overset{\displaystyle O}{\underset{\displaystyle Y}{\diagup}}\xrightarrow{\text{Partial reduction}} R\overset{\displaystyle O}{\underset{\displaystyle H}{\diagup}}$$

The partial reduction of a carboxylic acid chloride, $R\overset{O}{\diagup}Cl$, by lithium aluminum tri(*t*-butoxy)hydride, Li[O$-$C(CH$_3$)$_3$]$_3$AlH or Li(*t*-OC$_4$H$_9$)$_3$AlH, for example, forms aldehydes.

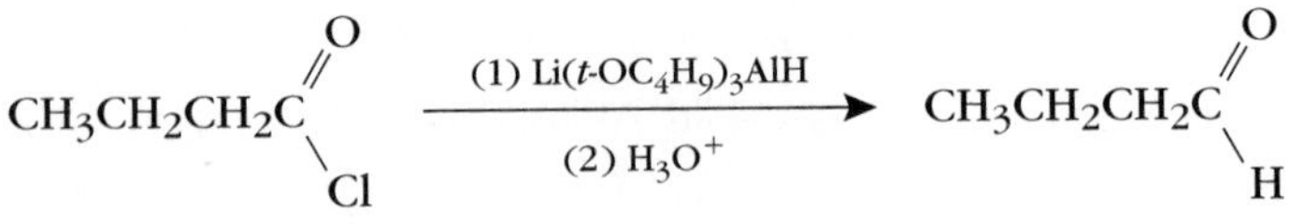

$$CH_3CH_2CH_2C\overset{\displaystyle O}{\underset{\displaystyle Cl}{\diagup}}\xrightarrow[(2)\ H_3O^+]{(1)\ Li(t\text{-}OC_4H_9)_3AlH} CH_3CH_2CH_2C\overset{\displaystyle O}{\underset{\displaystyle H}{\diagup}}$$

Butanoyl chloride

Butanal
(70% yield)

We will return to this reaction for a more detailed discussion in Section 16.14, A.

Specific aldehydes can also be prepared by the oxidative cleavage of alkenes. This method consists of two reactions. In the first reaction, a mixture of 3% to 4% ozone in dry oxygen, prepared in the laboratory by passing a stream of dry oxygen through an electric discharge, is bubbled into a solution of a symmetrically disubstituted alkene to form an **ozonide.** Reaction of the ozonide with a mild reducing agent, such as zinc in acetic acid or dimethyl sulfide [$(CH_3)_2S$], forms an aldehyde.

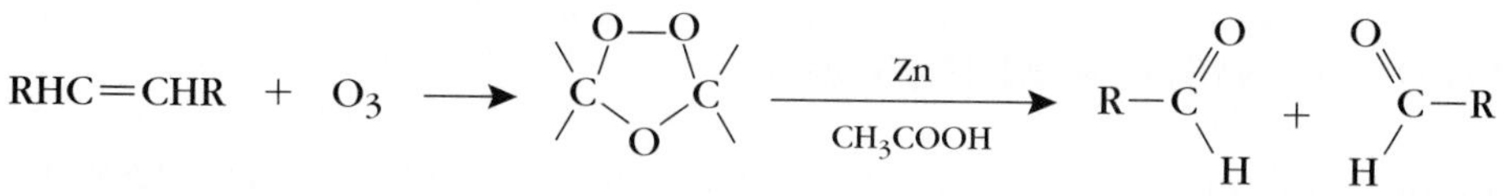

RHC=CHR + O_3 $\longrightarrow$

Symmetrical
disubstituted
alkene
(*E* or *Z* isomer)

Ozonide

$\xrightarrow[CH_3COOH]{Zn}$

Aldehyde

$CH_3CH_2CH=CHCH_2CH_3$ $\xrightarrow[(2)\ Zn/CH_3COOH]{(1)\ O_3}$ 2 CH_3CH_2C

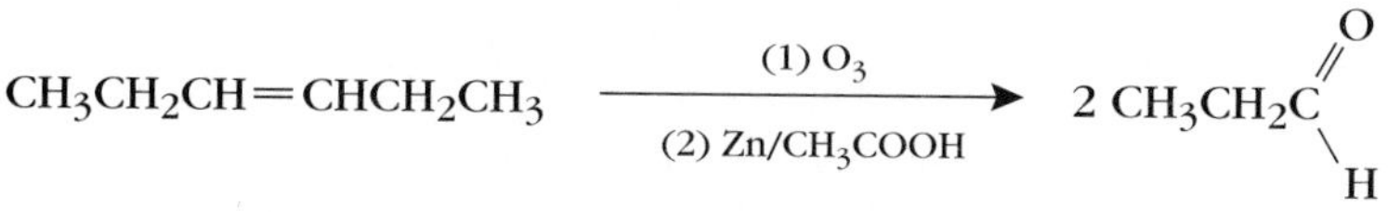

(*E*) or (*Z*)-3-Hexene

Propanal
(85% yield)

Oxidative cleavage of unsymmetrical alkenes is not usually a good synthetic method because it forms a mixture of products.

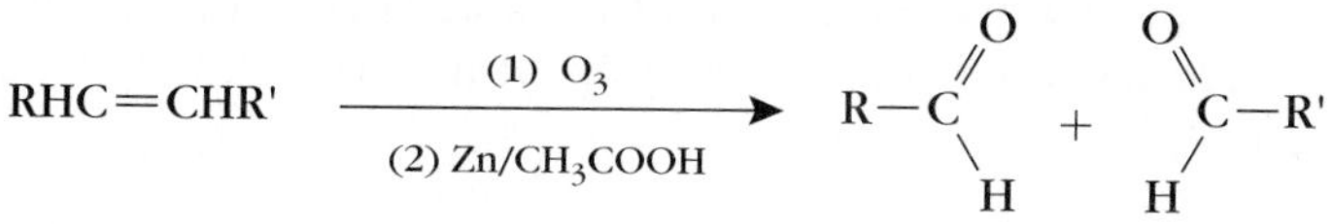

RHC=CHR' $\xrightarrow[(2)\ Zn/CH_3COOH]{(1)\ O_3}$ R—C + C—R'

(*E* or *Z* isomer)

Two different aldehydes

As a result, highest yields of aldehydes are obtained by oxidative cleavage of symmetrically disubstituted alkenes. Oxidative cleavage of cycloalkenes, for example, is a good method of preparing dialdehydes.

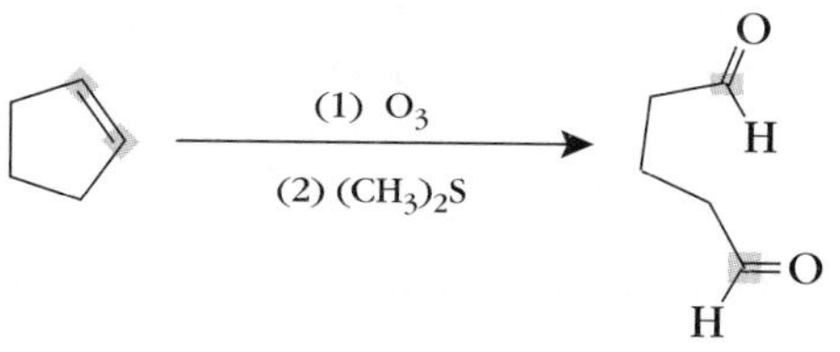

$\xrightarrow[(2)\ (CH_3)_2S]{(1)\ O_3}$

Cyclopentene

Pentanedial
(68% yield)

Exercise 14.10   Write the structure of the product of each of the following reactions.

(a) $(CH_3)_3CCH_2C{\equiv}CH$ $\xrightarrow[(2)\ H_2O_2/NaOH/H_2O]{(1)\ 9\text{-}BBN}$

(b) ⬡—CH_2OH $\xrightarrow[CH_2Cl_2]{PCC}$

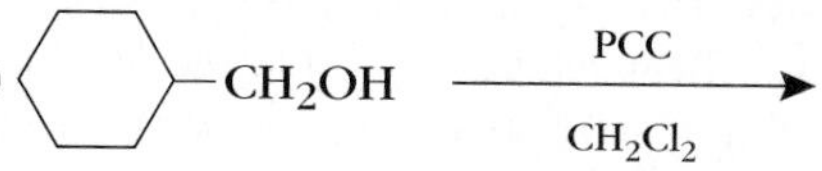

(c) $(CH_3)_2CHC$ $\xrightarrow[(2)\ H_3O^+]{(1)\ Lithium\ aluminum\ tri\ (t\text{-}butoxy)\ hydride}$

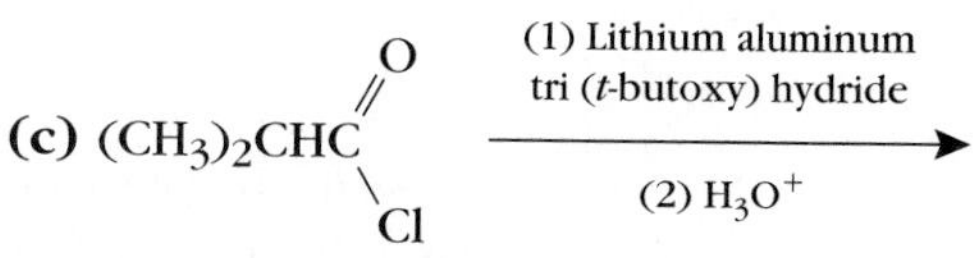

| **Exercise 14.11** | How would you prepare hexanal from each of the following compounds? |

(a) 1-Hexanol   **(b)** 1-Hexyne   **(c)** 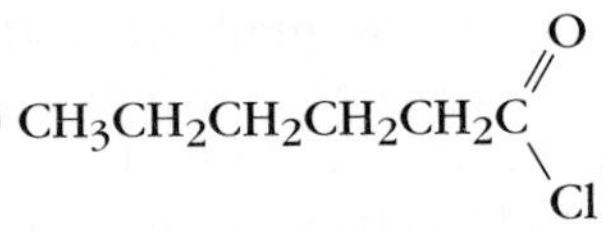 $CH_3CH_2CH_2CH_2CH_2C\overset{\displaystyle O}{\underset{\displaystyle Cl}{\Vert}}$

14.6   Synthesis of Ketones

Several of the methods of preparing ketones, like those of preparing aldehydes, have been discussed in previous chapters. Recall that oxidation of secondary alcohols forms ketones (Section 11.18). Sodium dichromate in aqueous acid, CrO_3 in aqueous acid, or PCC in methylene chloride, for example, effectively oxidize secondary alcohols to ketones.

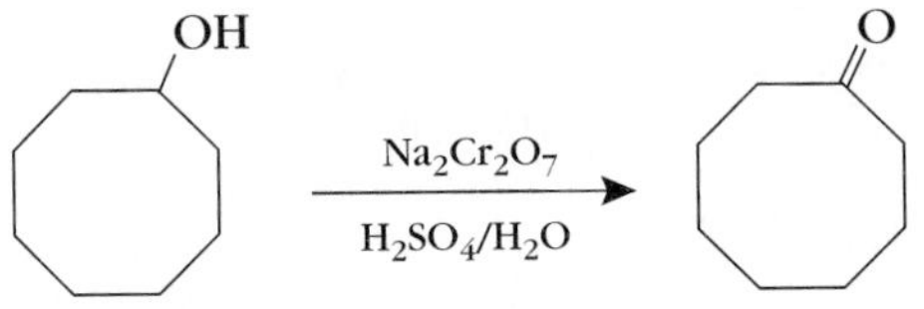

Cyclooctanol → Cyclooctanone (92% - 96% yield)

Certain ketones can also be prepared from alkynes. Methyl ketones can be prepared by acid-catalyzed hydration of terminal alkynes, with $HgSO_4$ as a catalyst (Section 9.7, A).

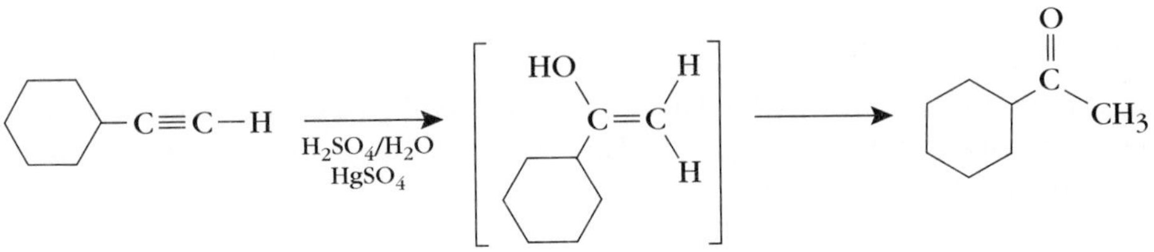

Ethynylcyclohexane   Enol   Cyclohexyl methyl ketone (85% yield)

Hydroboration-oxidation of symmetrical alkynes also forms ketones (Section 9.7, B).

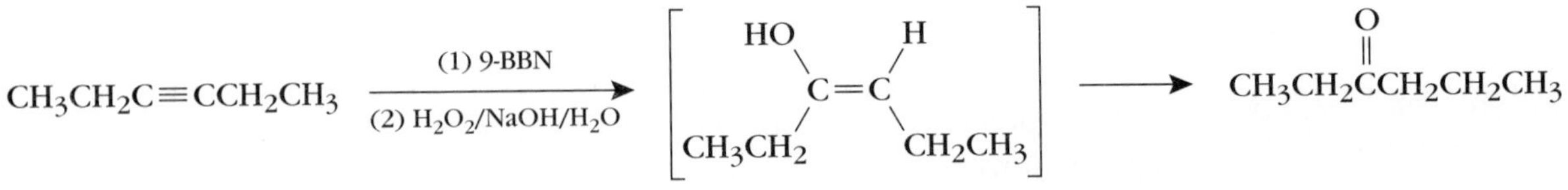

3-Hexyne   Enol   3-Hexanone (85% yield)

As in the preparation of aldehydes (Section 14.5), certain ketones can be prepared by the oxidative cleavage of alkenes. Best results are obtained by the oxidation of symmetrically tetrasubstituted alkenes because only a single ketone is formed as the product.

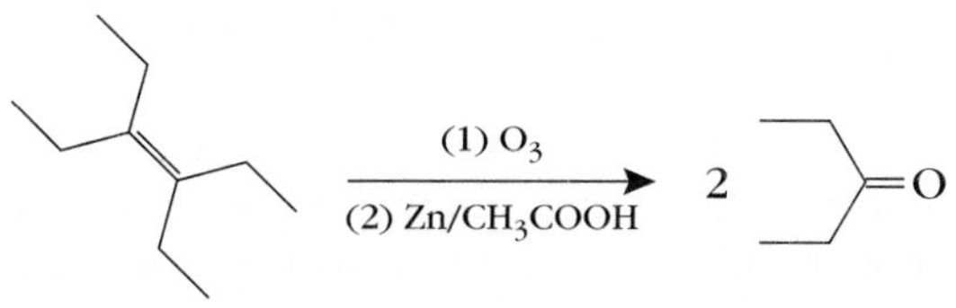

3,4-Diethyl-3-hexene   3-Pentanone

We briefly mention two other methods of preparing ketones that we will discuss in more detail in future chapters. Aryl ketones are often prepared by Friedel-Crafts acylation, which is the reaction of a carboxylic acid chloride with an aromatic ring in the presence of $AlCl_3$ catalyst (Section 21.9).

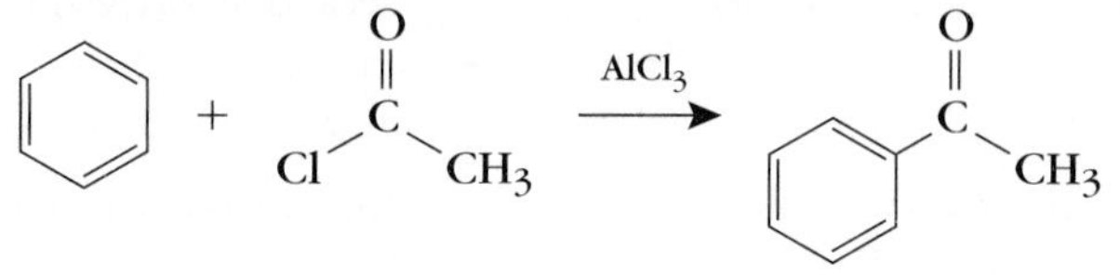

Benzene   Acetyl chloride   Acetophenone (92% yield)

Ketones can also be prepared by the reaction of carboxylic acid chlorides and an organocopper reagent, a **lithium dialkylcuprate** (R_2CuLi) (Section 16.14, B).

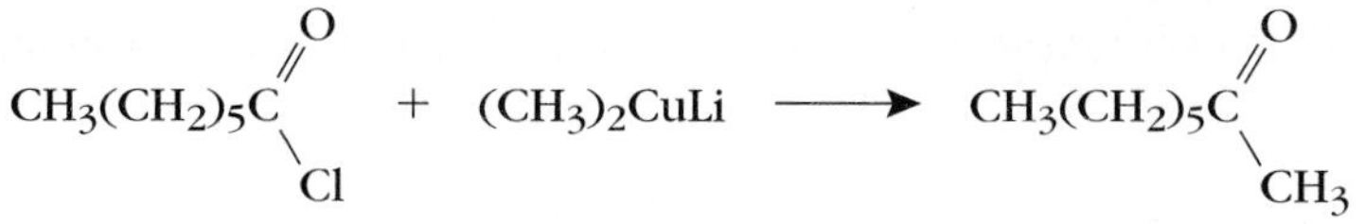

Heptanoyl chloride   Lithium dimethylcuprate   2-Octanone (85% yield)

Exercise 14.12   Show how you would carry out each of the following transformations. More than one reaction may be necessary.

(a)

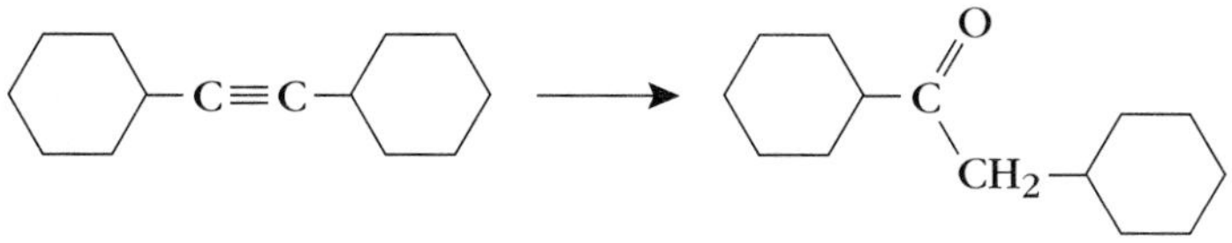

(b)

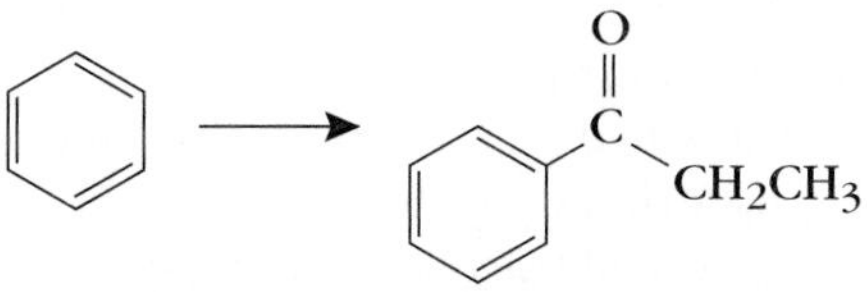

(c)

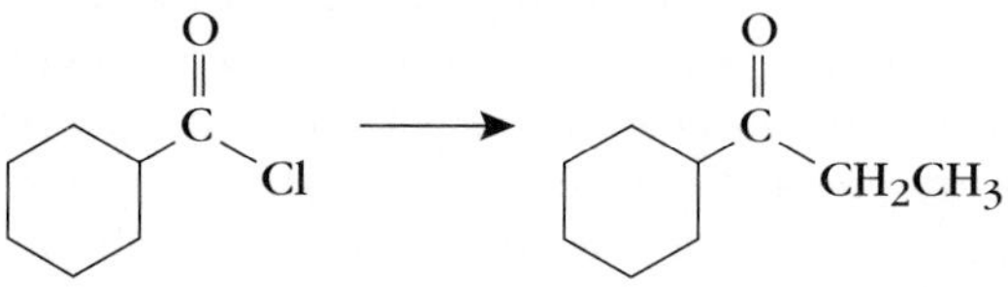

(d) $(CH_3)_2C{=}CHCH_3 \longrightarrow (CH_3)_2CHC$

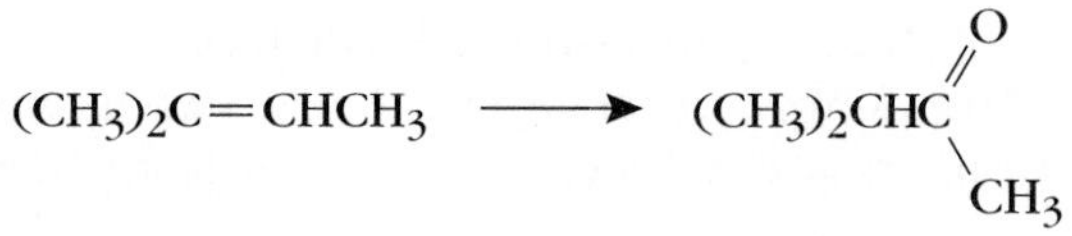

14.7   Basicity of Aldehydes and Ketones

As we discussed in Sections 11.6 and 14.2, the electronegativity difference between carbon and oxygen of the carbonyl group results in a partial positive charge on the carbonyl carbon atom and a partial negative charge on the carbonyl oxygen atom. As a result, the oxygen atom is nucleophilic and the site of reactions with acids and electrophiles, while the carbonyl carbon atom is electrophilic and the site of reactions with bases and nucleophiles.

Electrophilic carbon atom   Nucleophilic oxygen atom

$$\delta+ \quad \delta-$$
$$C=O$$

Bases and nucleophiles react here   Acids and electrophiles react here

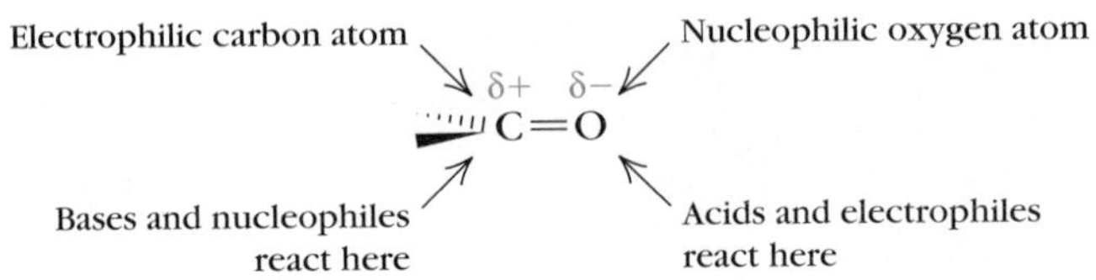

Aldehydes and ketones react with protons at the carbonyl oxygen atom to form their conjugate acid. The electron structure of the conjugate acid of acetaldehyde can be represented as a resonance hybrid of two resonance structures.

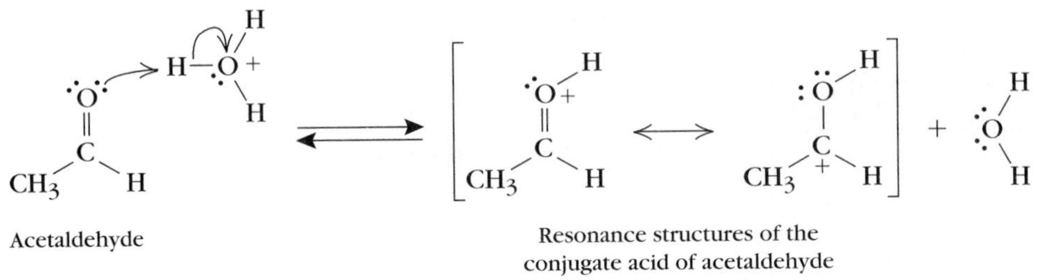

Acetaldehyde   Resonance structures of the conjugate acid of acetaldehyde

One of these structures, the one on the right, has a positive charge on the carbonyl carbon atom. As a result, protonation increases the charge on the carbonyl carbon atom compared to the original aldehyde or ketone. This is important because protonation of the carbonyl group of aldehydes and ketones enhances their reaction with weak nucleophiles, as we will learn in Section 14.10.

Table 14.3 compares the pK_a values of the conjugate acids of aldehydes and ketones with a number of other acids. Based on these values, we conclude that the conjugate acids of aldehydes and ketones are very strong acids (as strong or stronger than HCl), so aldehydes and ketones are very weak bases in solution. Aldehydes are generally weaker bases than ketones, and both aldehydes and ketones are weaker bases than ethers or alcohols.

Despite being relatively weak bases, protonation of aldehydes and ketones, just like protonation of ethers and alcohols, is an important first step in the mechanism of many of their reactions in acid solution, as we shall learn in Section 14.10.

Table 14.3	pK_a Values of Selected Organic Acids*		
Acid		Base	pK_a

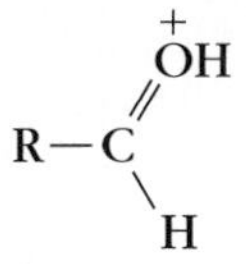

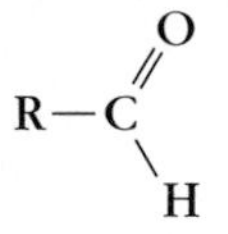

Acid	Base	pK_a
$R-\overset{\overset{+}{O}H}{\underset{H}{\overset{\|}{C}}}$	$R-\overset{\overset{O}{\|}}{\underset{H}{C}}$	≈ -10
HCl	Cl$^-$	≈ -7
$R-\underset{\underset{+OH}{\|}}{C}-R$	$R-\underset{\underset{O}{\|}}{C}-R$	≈ -7
 $\overset{+}{R}\underset{H}{O}R$	ROR	≈ -4
$RCH_2OH_2{}^+$	RCH_2OH	≈ -2
H_3O^+	H_2O	-1.7

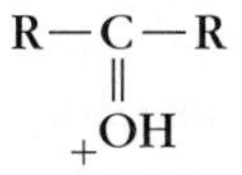

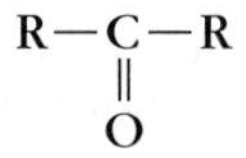

*HCl and H_3O^+ are included in the table as a reference.

14.8   Nucleophilic Additions to the Carbonyl Group Are Important Reactions of Aldehydes and Ketones

Nucleophiles add to the electrophilic carbonyl carbon atom of aldehydes and ketones in a reaction called a **nucleophilic addition reaction.** The overall reaction is the addition of a reagent Nu-H, in which Nu represents the nucleophile, to the carbonyl group, as shown in the following general reaction.

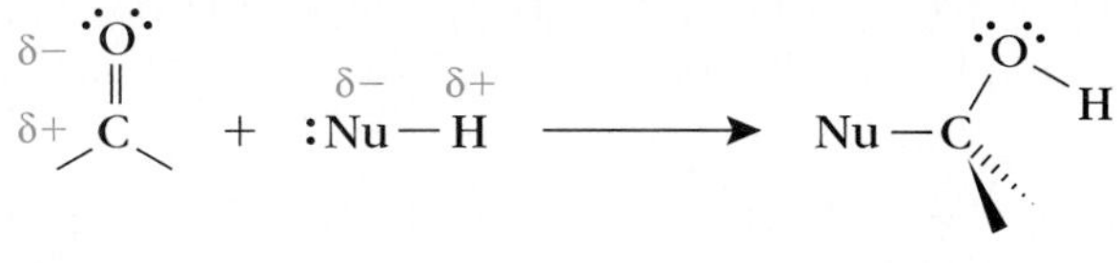

Ketone	Product of
or	nucleophilic
aldehyde	addition

A carbonyl group undergoes this reaction because of the polarization of the carbon-oxygen double bond caused by the difference in electronegativity of the carbonyl carbon and oxygen atoms, as we discussed in Sections 11.6 and 14.2.

We've already encountered two examples of nucleophilic addition reactions of aldehydes and ketones. The first example is the synthesis of alcohols by the reaction of Grignard or organolithium reagents with aldehydes and ketones (Section 11.7). The carbanion-like portions, R$^-$, of organometallic reagents are strong nucleophiles that add to the carbonyl carbon atom to form an alkoxide ion intermediate. Addition of aqueous acid solution converts the alkoxide ion intermediate to an alcohol.

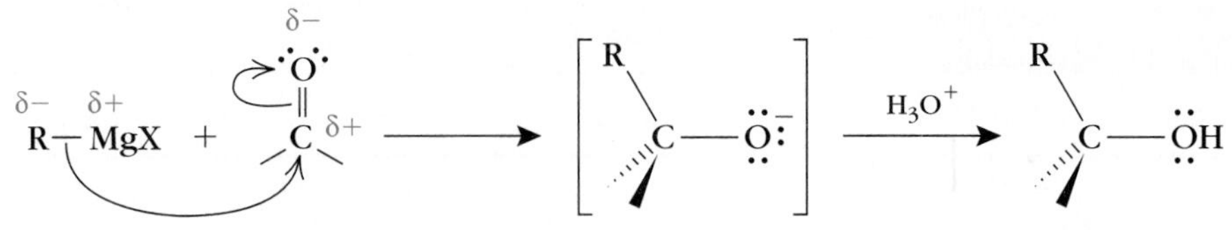

The second example is the synthesis of alcohols by the metal hydride reduction of a ketone or aldehyde (Section 11.10). In this reaction, the nucleophile is a hydride ion ($H:^-$) that adds to the carbonyl carbon atom to form an alkoxide ion. Addition of water or an aqueous acid solution transforms the alkoxide ion to an alcohol.

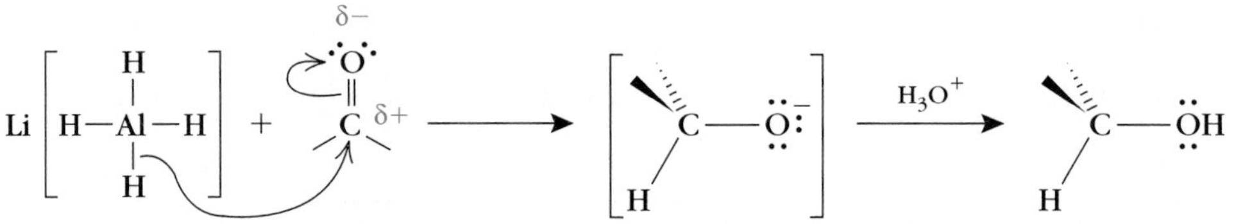

The mechanism of these two reactions is the simplest of the mechanisms of addition of nucleophiles to a carbonyl carbon atom because it consists of a single step, shown in Figure 14.6 for the reaction of formaldehyde and sodium borohydride ($NaBH_4$). The first intermediate formed by the addition of a nucleophile (a hydride ion in Figure 14.6) to a carbonyl carbon atom is called a **tetrahedral intermediate** because the planar trigonal carbonyl carbon atom of the original aldehyde or ketone has been converted into a tetrahedral carbon atom. While the tetrahedral intermediate in Figure 14.6 is represented as an alkoxide ion for simplicity, it is in fact an alkoxyborane, $B(OCH_3)_4^-$. Once the reaction of $NaBH_4$ and formaldehyde is completed, addition of water hydrolyzes the alkoxyborane to the alcohol.

$$B(OCH_3)_4^- + 4\,H_2O \longrightarrow B(OH)_4^- + 4\,CH_3OH$$

Exercise 14.13   Write the structure of the tetrahedral intermediate formed in the mechanism of the reaction of CH_3MgI and CH_3CHO.

Formation of tetrahedral intermediates in the mechanism of the reactions of metal hydrides and organometallic compounds is *irreversible.* Once formed, the tetrahedral intermediates do not undergo any further reaction until water or aqueous acid solution is added to form the desired alcohol. In contrast, addition of other nucleophiles to a carbonyl group is a reversible reaction.

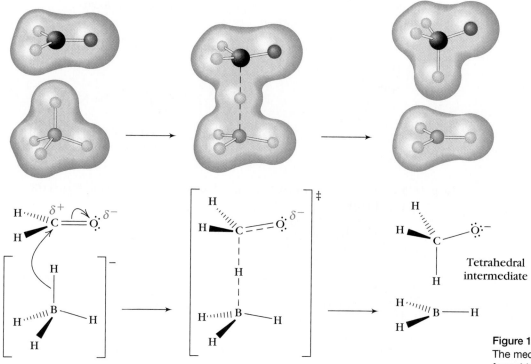

Reactants   Transition state

H
|
H·····C   O:⁻
|
H
Tetrahedral
intermediate

Figure 14.6
The mechanism of the reaction of formaldehyde and $NaBH_4$. Hydride ion (a strong nucleophile) adds to the carbonyl carbon atom to form a tetrahedral intermediate. The tetrahedral intermediate is represented as an alkoxide ion for simplicity.

14.9   Nucleophilic Addition of Cyanide Ion: Cyanohydrin Formation

Aldehydes and ketones react with a mixture of sulfuric acid and an excess of sodium cyanide (NaCN) to form **cyanohydrins.**

Aldehyde
or ketone

A cyanohydrin

Acetaldehyde

Ethanal cyanohydrin
(100% yield)

The net result of this reaction is the reversible addition of HCN to the carbonyl group of aldehydes and ketones.

While there is an equilibrium between cyanide ion and the weak acid HCN (pK_a = 9.1) in the reaction mixture, cyanide ion is the nucleophile. In Step 1 of

■ *Cyanohydrins are named by IUPAC rules as hydroxy derivates of nitriles. Naming of nitriles is not discussed until Chapter 16, so we will name cyanohydrins as derivatives of the aldehyde or ketone from which they are formed.*

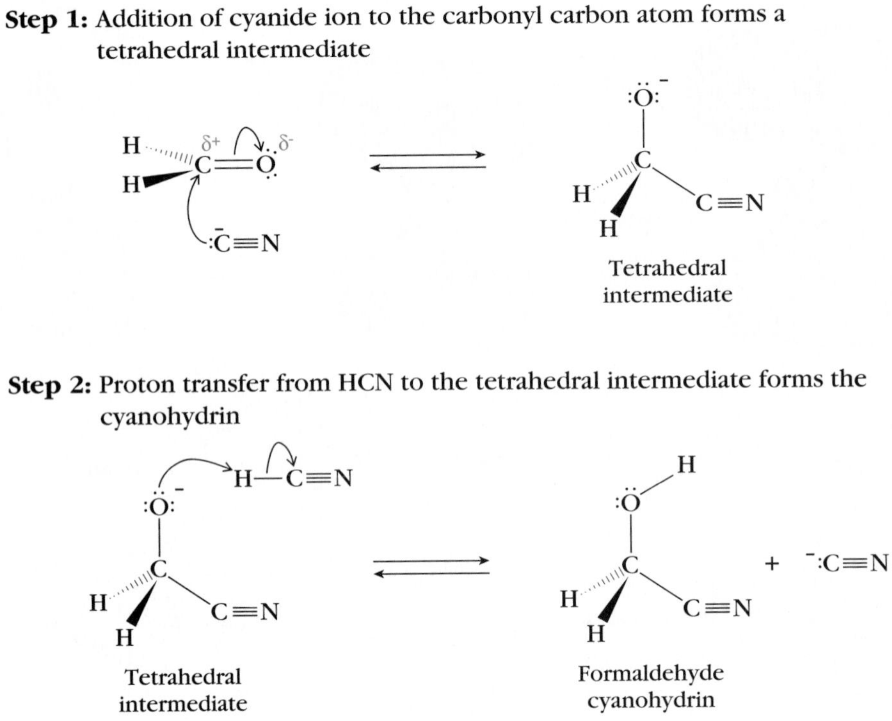

Figure 14.7
The mechanism of the addition of HCN to formaldehyde. Cyanide ion adds to the carbonyl carbon atom in Step 1 to form a tetrahedral intermediate, which rapidly accepts a proton in Step 2 to form the cyanohydrin and more cyanide ion.

Step 1: Addition of cyanide ion to the carbonyl carbon atom forms a tetrahedral intermediate

Tetrahedral
intermediate

Step 2: Proton transfer from HCN to the tetrahedral intermediate forms the cyanohydrin

Tetrahedral
intermediate

Formaldehyde
cyanohydrin

the mechanism, cyanide ion adds to the carbonyl carbon atom to form a tetrahedral intermediate, which rapidly accepts a proton from HCN to form the product in Step 2, as shown in Figure 14.7 for the addition of HCN to formaldehyde.

In both the mechanism of cyanohydrin formation, shown in Figure 14.7, and the mechanism of hydride reduction of aldehydes, shown in Figure 14.6, the slow step is the addition of the nucleophile to the carbonyl carbon atom. Cyanohydrin formation, however, is a reversible reaction, so the tetrahedral intermediate must be formed reversibly. Why is the tetrahedral intermediate formed reversibly by cyanide addition, while hydride addition irreversibly forms a tetrahedral intermediate?

The reversible formation of a tetrahedral intermediate occurs when one of the atoms or groups of atoms bonded to the tetrahedral carbon atom is a good leaving group. The ability of an atom or group of atoms to leave an sp^3 hybrid carbon atom is roughly related to its base strength and ability to stabilize a negative charge, as we learned in Chapter 12. Cyanide ion, for example, is a weaker base and is better able to stabilize a negative charge than either hydride ion or a carbanion, so cyanide ion is the better leaving group. Hydride ion or carbanions are such strong bases that they are almost never leaving groups. Addition of cyanide ion to aldehydes or ketones, therefore, is a reversible reaction because both the forward (addition of cyanide ion to the carbonyl carbon atom) and the reverse reaction (loss of cyanide ion from the tetrahedral intermediate) occur. In contrast, additions of hydride ions and the carbanion-like part of Grignard reagents to carbonyl carbon atoms are irreversible reactions because only the forward reaction (addition of hydride ion) occurs.

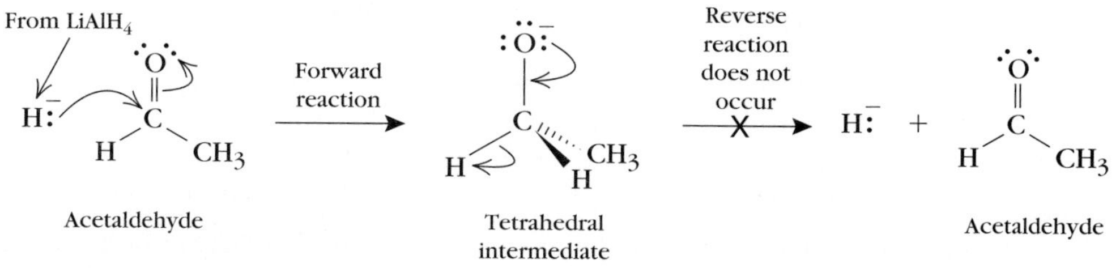

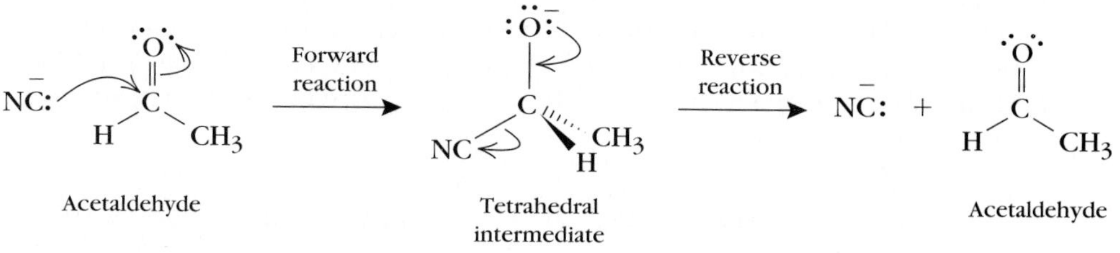

Acetaldehyde   Tetrahedral intermediate   Acetaldehyde

Acetaldehyde   Tetrahedral intermediate   Acetaldehyde

Exercise 14.14   Write the structure of the product of the reaction of cyclopentanecarbaldehyde with each of the following reagents.

(a) $CH_3(CH_2)_3CH_2MgI$ in diethyl ether followed by addition of an aqueous acid solution

(b) $LiBH_4$ in methanol followed by water

(c) Excess NaCN in a sulfuric acid solution

Exercise 14.15   Write a mechanism for the following reaction.

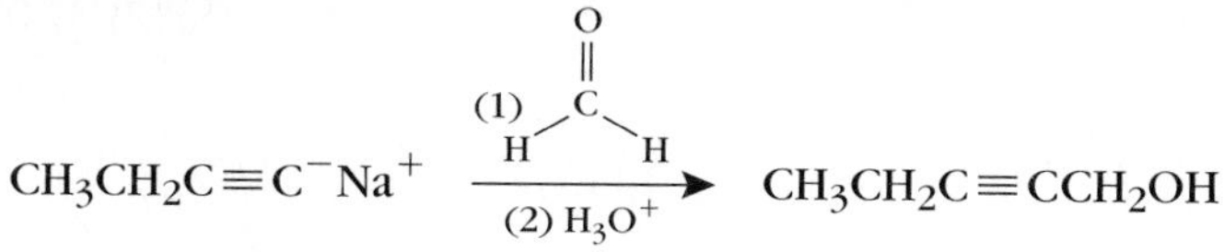

Hydration of aldehydes and ketones is another reversible nucleophilic addition reaction.

14.10 Nucleophilic Addition of Water: Hydration of Aldehydes and Ketones

Hydration of aldehydes and ketones is the addition of water to their carbonyl group to form products called **geminal (gem) diols** or **hydrates.**

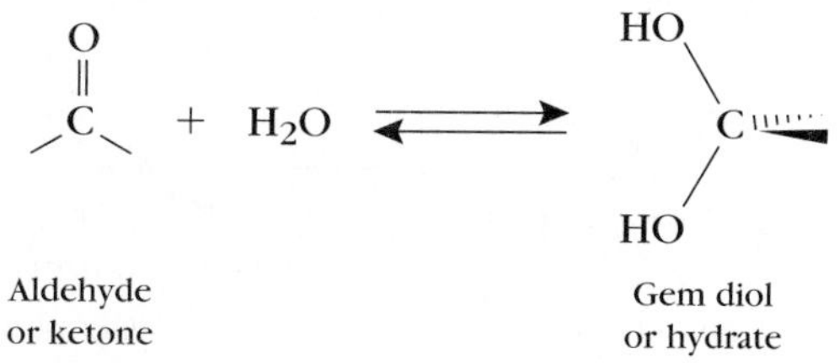

Aldehyde or ketone   Gem diol or hydrate

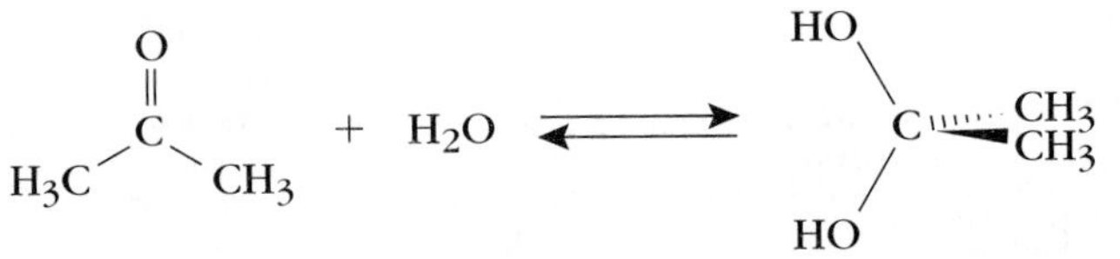

$$
\underset{\text{Acetone}}{\underset{\text{H}_3\text{C}\quad\text{CH}_3}{\overset{\text{O}}{\underset{\|}{\text{C}}}}} \;+\; \text{H}_2\text{O} \;\rightleftharpoons\; \underset{\text{2,2-Propanediol}}{\underset{\text{HO}}{\overset{\text{HO}}{\text{C}}}\overset{\text{CH}_3}{\underset{}{\text{CH}_3}}}
$$

In aqueous solution, an equilibrium exists between an aldehyde or a ketone and its hydrate. Except for low molecular weight aldehydes (formaldehyde, acetaldehyde, and propanal), the position of the equilibrium favors the aldehyde or ketone. An aqueous solution of formaldehyde, for example, contains only about 0.1% formaldehyde and 99.9% methanediol. An aqueous solution of acetone, in contrast, contains only about 0.1% 2,2-propanediol and the rest acetone.

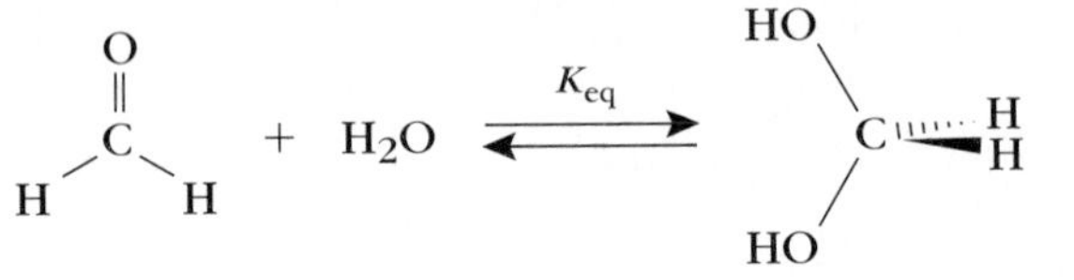

$$
\underset{\text{Formaldehyde}}{\underset{\text{H}\quad\text{H}}{\overset{\text{O}}{\underset{\|}{\text{C}}}}} \;+\; \text{H}_2\text{O} \;\overset{K_{eq}}{\rightleftharpoons}\; \underset{\text{Methanediol}}{\underset{\text{HO}}{\overset{\text{HO}}{\text{C}}}\overset{\text{H}}{\underset{}{\text{H}}}} \qquad K_{eq} = 2.3 \times 10^3
$$

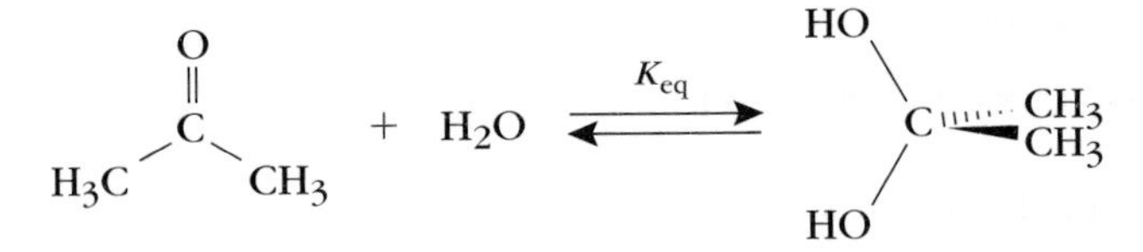

$$
\underset{\text{Acetone}}{\underset{\text{H}_3\text{C}\quad\text{CH}_3}{\overset{\text{O}}{\underset{\|}{\text{C}}}}} \;+\; \text{H}_2\text{O} \;\overset{K_{eq}}{\rightleftharpoons}\; \underset{\text{2,2-Propanediol}}{\underset{\text{HO}}{\overset{\text{HO}}{\text{C}}}\overset{\text{CH}_3}{\underset{}{\text{CH}_3}}} \qquad K_{eq} = 1.4 \times 10^{-3}
$$

Hydration of aldehydes and ketones is slow in pure water but is catalyzed by hydroxide ion. The mechanism of hydroxide ion–catalyzed hydration, shown in Figure 14.8, is similar to the mechanisms of the reaction of aldehydes and ketones with cyanide ion (Section 14.9), metal hydrides (Section 14.8), and Grig-

Figure 14.8
Mechanism of hydroxide ion-catalyzed hydration of a ketone or aldehyde. Hydroxide is a catalyst because it is a more reactive nucleophile than a neutral water molecule.

Step 1: Addition of hydroxide ion to the carbonyl carbon atom forms a tetrahedral intermediate

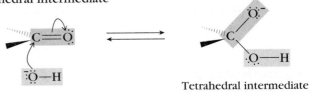

Tetrahedral intermediate

Step 2: Rapid proton transfer from water to the tetrahedral intermediate forms a gem diol

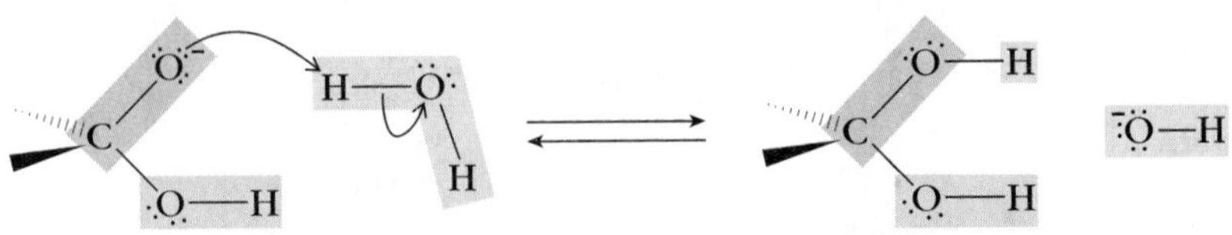

Gem diol

nard reagents (Section 14.8). Step 1 of the mechanism is the reversible addition of hydroxide ion to the carbonyl carbon atom to form a tetrahedral intermediate. Rapid proton transfer from a water molecule to the tetrahedral intermediate in Step 2 forms the gem diol and regenerates the hydroxide ion catalyst.

Hydration of aldehydes and ketones is also catalyzed by acids. The mechanism of acid-catalyzed hydration is slightly different than the mechanisms of nucleophilic additions that we've discussed so far because it contains an additional step. Acid-catalyzed hydrations occur by first forming the conjugate acid of the aldehyde or ketone by protonating the carbonyl oxygen atom, as shown in Figure 14.9. Addition of water to the carbonyl carbon atom of the conjugate acid of the aldehyde or ketone occurs in Step 2 to form a tetrahedral intermediate. Transfer of a proton from the tetrahedral intermediate to a water molecule in Step 3 forms the gem diol and regenerates the catalyst.

Notice the differences in the mechanisms of hydration in a neutral water solution, in a basic solution (hydroxide ion-catalyzed hydration), and in an acid solution (acid-catalyzed hydration). In a neutral solution, the nucleophile is a water molecule. Reaction occurs faster in basic solution than in water because water is converted into hydroxide ion, which is a better nucleophile than water. Reaction

Step 1: Protonation of the carbonyl oxygen atom forms the conjugate acid of an aldehyde or ketone

Aldehyde
or
ketone

Conjugate acid of aldehyde or ketone

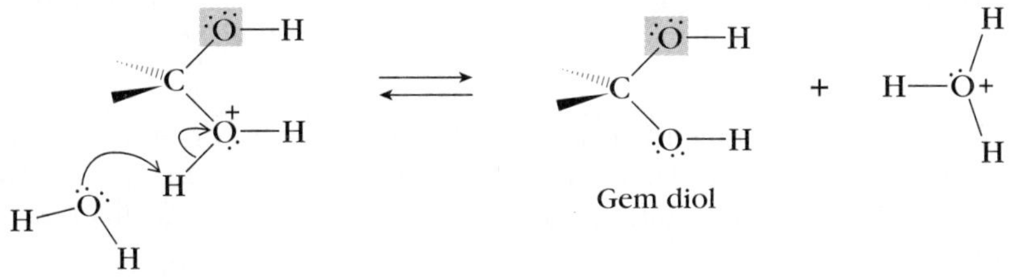

Figure 14.9
Mechanism of acid-catalyzed hydration of aldehydes and ketones. The acid catalyzes the reaction by forming the conjugate acid of the aldehyde or ketone, which makes the carbonyl carbon atom more electrophilic and more reactive.

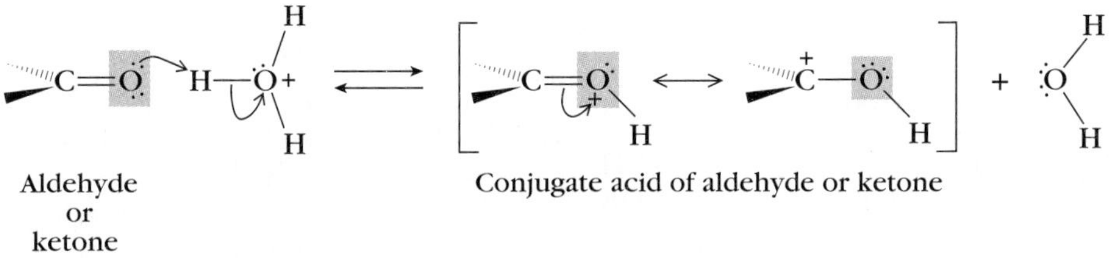

Step 2: Addition of a water molecule to the carbonyl carbon atom of the conjugate acid of an aldehyde or ketone forms a tetrahedral intermediate

Tetrahedral
intermediate

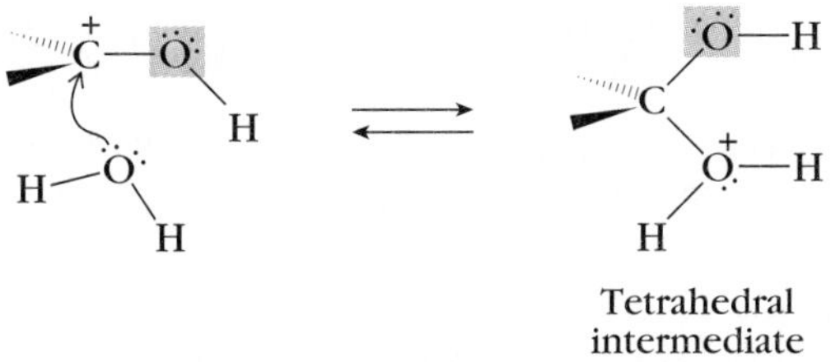

Step 3: Rapid proton transfer from the tetrahedral intermediate to water forms the gem diol

Gem diol

also occurs faster in an acid solution than in water because the carbonyl carbon atom is made more electrophilic by the formation of the conjugate acid of the aldehyde or ketone.

Exercise 14.16 Write the mechanism of the hydration of formaldehyde in neutral solution.

We've seen in this section that water is a nucleophile that adds to the carbonyl group of aldehydes and ketones. Alcohols are nucleophiles, too, and they also add to the carbonyl group of aldehydes and ketones.

14.11 Nucleophilic Addition of Alcohols: Acetal Formation

- *The prefix* hemi- *means half, so a hemiacetal is half an acetal.*

The initially formed product of many of the nucleophilic addition reactions of carbonyl-containing compounds reacts further to form other products under the reaction conditions. An example is the acid-catalyzed addition of alcohols to aldehydes and ketones. The initial product of addition of one equivalent of alcohol to the carbonyl group of aldehydes and ketones is called a **hemiacetal.** In acid solution, the hemiacetal reacts further to form an **acetal.**

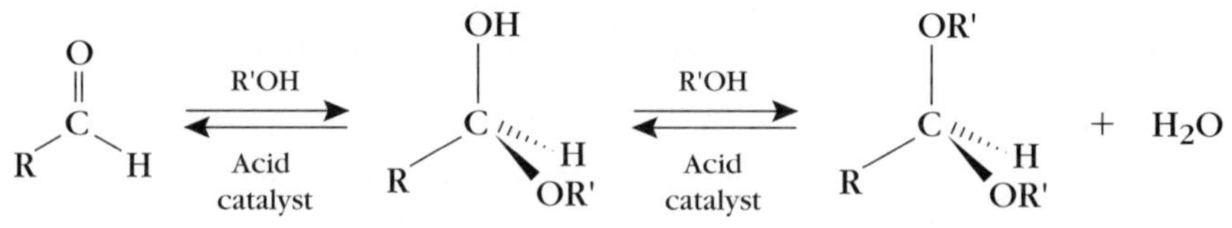

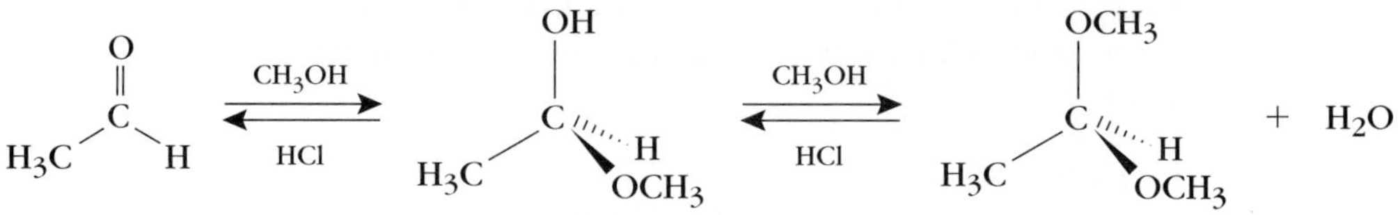

- *Acetals formed from ketones were originally called* ketals. *The name* ketal *was dropped recently from the IUPAC system of naming compounds, so the products of addition of two equivalents of alcohol to both aldehydes and ketones are called* acetals.

Exercise 14.17 Write the structure of the hemiacetal and acetal formed by the acid-catalyzed reaction of each of the following aldehydes or ketones with ethanol.

(a) Propanal **(b)** Ethyl methyl ketone **(c)** Benzaldehyde

---

The mechanism of acetal formation is a combination of two mechanisms that we have already learned. First, the intermediate hemiacetal is formed by an acid-catalyzed nucleophilic addition mechanism analogous to the hydration of aldehydes and ketones.

Second, the acetal is formed from the hemiacetal by a mechanism that is analogous to the S_N1 mechanism of the reaction of a tertiary alcohol with HCl (Section 12.9). Step 1 of the mechanism, shown in Figure 14.10, is protonation of the
hydroxy group of the hemiacetal to form its conjugate acid. Loss of water from
the hemiacetal forms an **oxonium ion** in Step 2. Oxonium ions are relatively sta-

Step 1: Protonation of the hydroxy group forms the conjugate acid of the
hemiacetal

Figure 14.10
Mechanism of acid-catalyzed formation of an acetal from a hemiacetal.
The strongest acid in an acidic
alcohol solution is $R\overset{+}{O}H_2$, so this
acid protonates the hemiacetal in
Step 1.

Hemiacetal   Conjugate acid
                  of hemiacetal

Step 2: Conjugate acid of hemiacetal loses water to form a resonance stabilized
oxonium ion

Conjugate acid   Resonance stabilized oxonium ion
of hemiacetal

Step 3: Reaction of oxonium ion with alcohol forms the conjugate acid of the
acetal

One resonance   Conjugate acid
form of the   of acetal
oxonium ion

Step 4: Transfer of a proton from the conjugate acid of the acetal to a base forms
the acetal

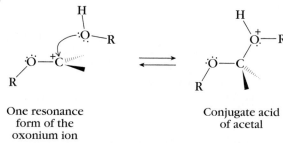

Conjugate acid   Acetal
of acetal

■ *The effect of substituents on the stability of carbocations is discussed in Section 7.15.*

ble carbocations because the strong electron-donating alkoxy group stabilizes the adjacent electron-deficient carbon atom. Reaction of an alcohol with the oxonium ion occurs in Step 3 to form the conjugate acid of the acetal, which transfers a proton to a base in Step 4 to form the acetal.

The acid-catalyzed formation of acetals is reversible. Consequently, acetals in the presence of an acid catalyst and excess water are converted rapidly into the corresponding aldehyde or ketone and an alcohol. This reaction is called **acetal hydrolysis.** Because acetal hydrolysis is the exact reverse of acetal formation, the mechanism of one is the exact reverse of the other.

Exercise 14.18	Write the three steps involved in the mechanism of the formation of a hemiacetal by the acid-catalyzed reaction of acetaldehyde and methanol.

Exercise 14.19	Write the mechanism of the acid-catalyzed hydrolysis of 1,1-diethoxypropane.

The equilibrium constant for the formation of acetals from aldehydes is generally unfavorable. Carrying out the reaction in the alcohol as solvent, however, increases the concentration of acetal at equilibrium. The equilibrium constant for the formation of acetals from ketones is even more unfavorable. To obtain acetals in good yields from ketones, special techniques must be used that remove water from the reaction mixture as it forms.

Cyclic acetals are formed by the reaction of either aldehydes or ketones with one equivalent of 1,2- or 1,3-diols. For example, the reaction of butanal and 1,2-ethanediol in benzene with *p*-toluenesulfonic acid as the acid catalyst forms 2-propyl-1,3-dioxolane, a cyclic acetal.

| Butanal | 1,2-Ethanediol (Ethylene glycol) | | 2-Propyl-1,3-dioxolane | |

Cyclic acetal formation is a reversible reaction. At equilibrium, formation of the aldehyde or ketone is favored over formation of the acetal. As in the case of other acetals, however, removing water from the reaction mixture as it forms shifts the position of the equilibrium in favor of the acetal, making it possible to obtain excellent yields of cyclic acetals.

The mechanism of cyclic acetal formation from one equivalent of ethylene glycol is the same as that presented earlier for the formation of acetals from two equivalents of a monoalcohol. The only difference is that both oxygen atoms of the acetal come from the same molecule rather than from different molecules.

Compounds that contain cyclic hemiacetals and acetals are important to living organisms. Monosaccharides, for example, are cyclic hemiacetals that combine to form sugars and other carbohydrates, as we will learn in Chapter 25.

Exercise 14.20   Write the structure of the products formed in each of the following reactions.

(a) CH_3CHO + $HOCH_2CH_2OH$ $\underset{\text{catalyst}}{\overset{\text{Acid}}{\rightleftarrows}}$

(b) 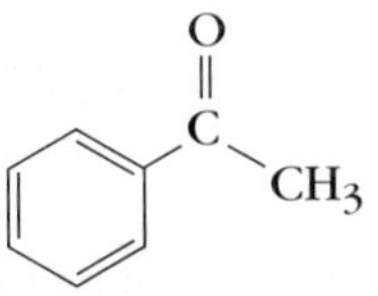 [phenyl]$\underset{\underset{CH_3}{|}}{\overset{\overset{O}{\|}}{C}}$ + $HOCH_2CH_2OH$ $\underset{\text{catalyst}}{\overset{\text{Acid}}{\rightleftarrows}}$

(c) $CH_3CH_2CH_2CHO$ + $HOCH_2CH_2CH_2OH$ $\underset{\text{catalyst}}{\overset{\text{Acid}}{\rightleftarrows}}$

Exercise 14.21   Write the mechanism of the acid-catalyzed formation of an acetal by the reaction of acetaldehyde with 1,2-ethanediol.

Notice that an acetal is a diether in which both ether oxygen atoms are attached to the same carbon atom. Consequently, acetals, like other ethers, are stable to basic conditions. This makes acetals useful as protecting groups.

14.12 Acetals Are Used to Protect Carbonyl Groups of Aldehydes and Ketones in Organic Synthesis

The compounds in most of the reactions we have studied so far have contained only a single functional group. Reactions of compounds that contain more than one functional group are inherently more complex and will be discussed in detail in Chapter 24. For now, however, the chemistry of acetals provides an opportunity to introduce the use of a protecting group, which is an important technique frequently used in synthesis with compounds that contain two or more functional groups.

Consider a compound that contains two functional groups, X and Y. If both X and Y are known to react with a particular reagent, how do we react only X with this reagent and leave Y unaffected? The way to protect group Y is to transform it into another functional group that does not react with the reagent. After X reacts, the protection is removed from Y. In this way, a net reaction occurs only with group X. The preparation of 4-octynal from 3-iodopropanal illustrates this technique.

$$CH_3(CH_2)_2C\equiv CCH_2CH_2CHO \quad \text{from} \quad ICH_2CH_2CHO$$

4-Octynal   3-Iodopropanal

At first glance the reaction of 3-iodopropanal with 1-pentynyllithium seems like a good synthesis. The idea is to replace the iodine atom by a 1-pentynyl group. Recall from Section 9.11 that methyl and primary haloalkanes alkylate acetylide ions. However, we also know that an acetylide ion can add to the carbonyl group of an aldehyde to form an alcohol, after hydrolysis. As a result, exclusive replacement of the iodine atom by a 1-pentynyl group is impossible.

$$CH_3(CH_2)_2C\equiv CCH_2CH_2CHO$$

$$CH_3(CH_2)_2C\equiv CLi \quad + \quad ICH_2CH_2CHO$$

$$\overset{OLi}{\underset{H}{ICH_2CH_2C-C\equiv C(CH_2)_2CH_3}}$$

Suppose we make the acetal of 3-iodopropanal. The acetal is an ether and does not react with the basic acetylide ion. In this way, a molecule with two reactive sites, the carbonyl group and the carbon-iodine bond, is transformed into one with only a single reactive site. After the reaction, the aldehyde functional group is regenerated by mild acid-catalyzed hydrolysis of the acetal. Under the reaction conditions used to hydrolyze the acetal, water does not add to the carbon-carbon triple bond. Using a large excess of water ensures that the aldehyde and the diol predominate at equilibrium. The following sequence of reactions illustrates the complete synthesis of 4-octynal.

Reaction 1: Protection of the carbonyl group by formation of a cyclic acetal

$$ICH_2CH_2CHO \quad + \quad HOCH_2CH_2OH \quad \xrightarrow[\text{catalyst}]{\text{Acid}} \quad \overset{O\quad O}{\underset{ICH_2CH_2\quad H}{\diagdown\diagup}} \quad + \quad H_2O$$

Reaction 2: Reaction of the cyclic acetal with the acetylide ion

$$CH_3(CH_2)_2C\equiv C\overset{-}{:} \quad \overset{O\quad O}{\underset{\underset{I}{CH_2CH_2\quad H}}{\diagdown\diagup}} \quad \longrightarrow \quad CH_3(CH_2)_2C\equiv C-CH_2CH_2\overset{O\quad O}{\diagdown\diagup}\ H$$

Reaction 3: Acid-catalyzed hydrolysis regenerates the aldehyde group

$$CH_3(CH_2)_2C\equiv C-CH_2CH_2\overset{O\quad O}{\diagdown\diagup}\ H \quad \xrightarrow{H_3O^+} \quad CH_3(CH_2)_2C\equiv CCH_2CH_2CHO$$
$$+$$
$$HOCH_2CH_2OH$$

While preparing a cyclic acetal and then hydrolyzing it adds two steps to the synthesis, both steps are essential to its success. Remember that acetals are hydrolyzed by acids, so they cannot be used to protect carbonyl groups in reactions carried out under acid conditions.

Exercise 14.22   Propose a synthesis of the following compound from 5-bromo-2-pentanone.

$$\overset{OH}{\underset{}{CH_3CHCH_2CH_2CH_2CH_2CH_2OH}}$$

Another important class of compounds that adds to carbonyl groups of aldehydes and ketones is nitrogen-containing nucleophiles.

14.13 Nucleophilic Addition of Primary Amines: Formation of Imines

Alkyl primary amines (RNH_2) and aryl primary amines ($ArNH_2$) add to the carbonyl carbon atom of aldehydes and ketones to initially form a carbinolamine. Under the reaction conditions, the carbinolamine loses water to form an **N-substituted imine** as the final product.

> ■ *The German chemist H. Schiff first described the formation of N-substituted imines, so they are often called Schiff bases.*

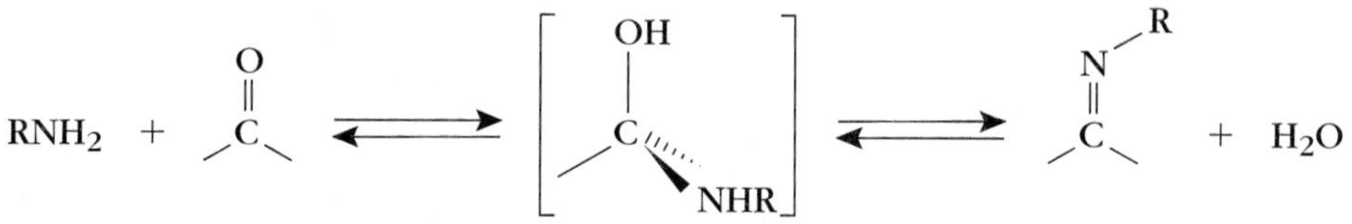

| Primary amine | Aldehyde or ketone | Carbinolamine | N-Substituted imine or Schiff base | Water |

As in the case of acid-catalyzed acetal formation (Section 14.11), the product of nucleophilic addition of ammonia or primary amines to aldehydes or ketones reacts further to form the observed product under the reaction conditions.

The reaction to form imines is reversible. In most cases, the equilibrium favors the aldehyde or ketone, so the water must be removed as it is formed to ensure a good yield of imine. The following examples illustrate the formation of imines.

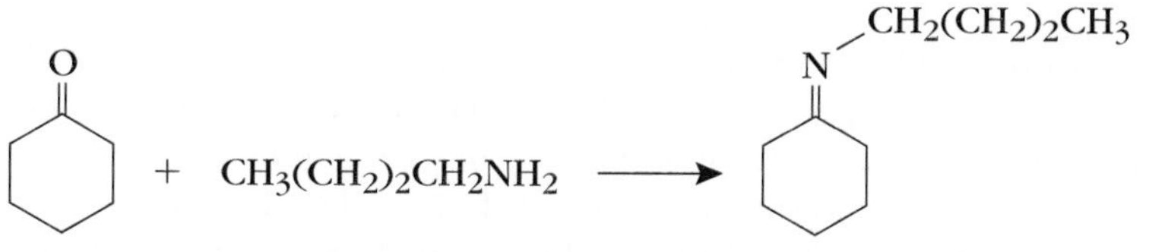

Cyclohexanone + 1-Butanamine → N-Cyclohexylidene-1-butanamine (80% yield)

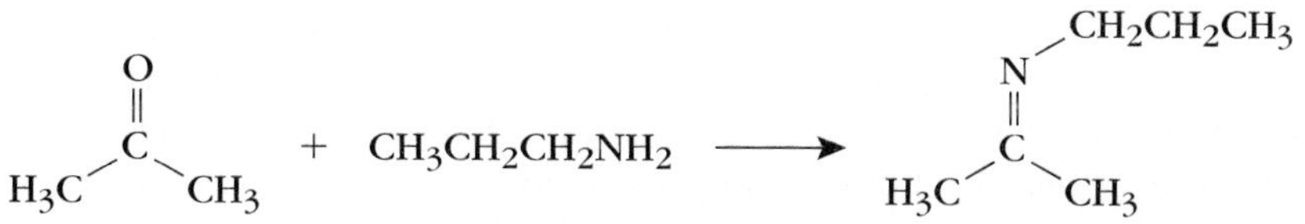

Acetone + 1-Propanamine → N-Isopropylidene-1-propanamine (65% yield)

Step 1 of the mechanism of imine formation is the addition of the nucleophile, in this case a nitrogen nucleophile, to the carbonyl carbon atom of the aldehyde or ketone to form a tetrahedral intermediate. Proton transfer in Step 2 forms a **carbinolamine,** as shown in Figure 14.11. Notice that the carbinolamine is the nitrogen analog of a hemiacetal and that the mechanisms of their formation are very similar.

Figure 14.11

The mechanism of carbinolamine formation. Proton transfers such as the one in Step 2 are believed to occur in several steps. One possible series of steps that can be visualized is a proton transfer from the −[NH₂R]+ group of the tetrahedral intermediate to a base. In a separate step a proton is transferred from an acid to the negatively charged oxygen atom to form the neutral carbinolamine. For simplicity, such individual proton transfer steps are combined in a single step that we will call proton transfer reactions.

Step 1: Nucleophilic addition of amine to the carbonyl carbon atom of an aldehyde or ketone forms a tetrahedral intermediate

Tetrahedral
intermediate

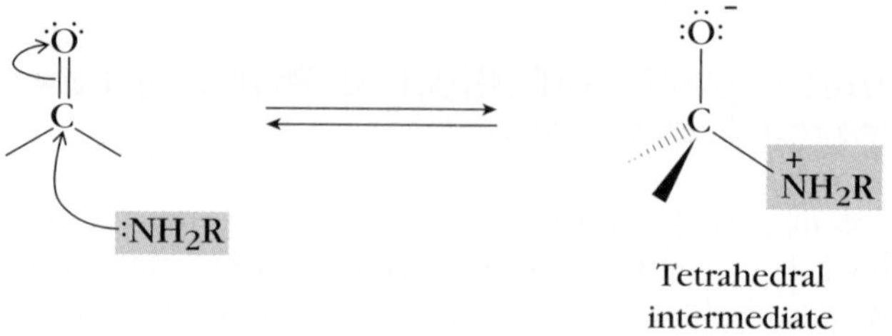

Step 2: Proton transfer forms the neutral carbinolamine

Proton transfer

Carbinolamine

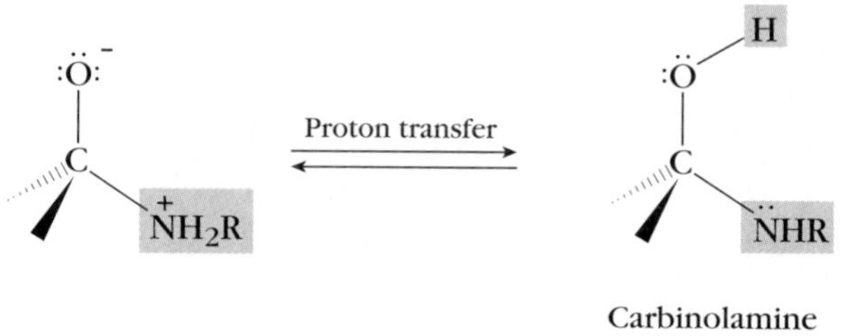

Exercise 14.23   Write the mechanism for the formation of the hemiacetal of acetaldehyde and methanol in neutral solution and compare it with the mechanism for the formation of its carbinolamine by reaction with methanamine (CH_3NH_2). Note carefully any similarities and differences.

Carbinolamines lose water to form imines in the presence of acids. The mechanism outlined in Figure 14.12 is similar to that of the acid-catalyzed dehydration of a tertiary alcohol, which occurs by an E1 mechanism (Section 12.16). Protonation of the hydroxy group occurs in Step 1, followed by loss of a water molecule in Step 2. Transfer of a proton from the intermediate iminium ion forms the imine in Step 3.

Careful control of the pH of the reaction mixture is essential to the successful formation of imines by the reaction of aldehydes or ketones and amines. The dehydration of the carbinolamine is acid catalyzed. Just as in the mechanism of the formation of acetals from hemiacetals, an acid is needed to protonate the hydroxy group to make it a good leaving group. This step of the mechanism will not occur unless there is enough acid present. If too much acid is present, however, the nitrogen atom of the nucleophile will be protonated, too, which converts the primary amine into an *N*-substituted ammonium ion. Because the ammonium ion lacks a pair of electrons, it is nonnucleophilic and the first step of the mechanism, nucleophilic addition, will not occur or will occur extremely slowly.

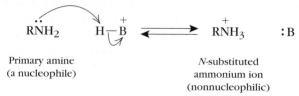

Primary amine
(a nucleophile)

N-substituted
ammonium ion
(nonnucleophilic)

Step 1: Protonation of the carbinolamine

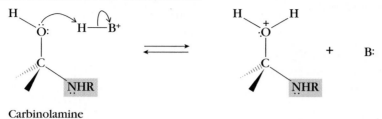

Carbinolamine

Figure 14.12
Mechanism of imine formation by loss of water from the intermediate carbinolamine.

Step 2: Loss of water forms an iminium ion

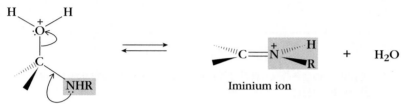

Iminium ion

Step 3: Proton transfer from the iminium ion to base forms an imine

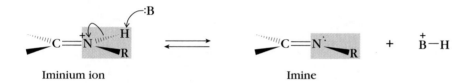

Iminium ion   Imine

To overcome this problem, the reaction is carried out in a weakly acidic buffer solution (pH = 4 to 5). This is a compromise between the need for enough acid to catalyze the dehydration of the carbinolamine and the need for enough nucleophile present for the first step of the mechanism to occur.

Exercise 14.24   Write the structure of the product of each of the following reactions.

(a) [structure: acetophenone] + CH₃NH₂ → (Acidic buffer)

(b) CH₃(CH₂)₅CHO + [cyclohexylamine, NH₂] → (Acidic buffer)

(c) [benzyl amine, CH₂NH₂] + [cyclopentane-CHO] → (Acidic buffer)

Exercise 14.25   Write the equation for the preparation of each of the following imines by the reaction of an aldehyde or ketone and an amine.

(a) $CH_3CH_2CH{=}NCH_3$
(b) $(CH_3)_2C{=}NCH(CH_3)_2$
(c)

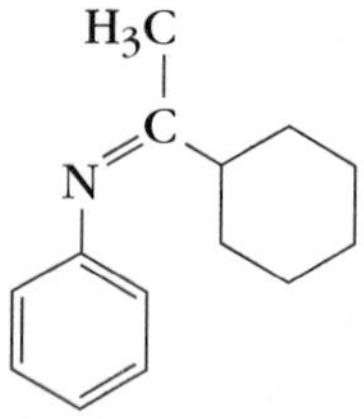

14.14 Nucleophilic Additions of Derivatives of Ammonia

Several other derivatives of ammonia react with aldehydes and ketones to form products that are analogous to imines.

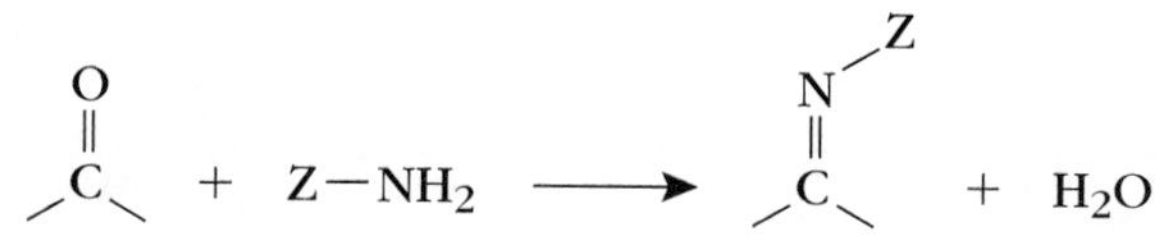

| Aldehyde or ketone | Derivative of ammonia | Imine-like product | Water |

In this reaction, the carbonyl group of the aldehyde or ketone is converted to C=N—Z and a molecule of water is formed. Aldehydes and ketones, for example, react with 2,4-dinitrophenylhydrazine to form **2,4-dinitrophenyl hydrazones.**

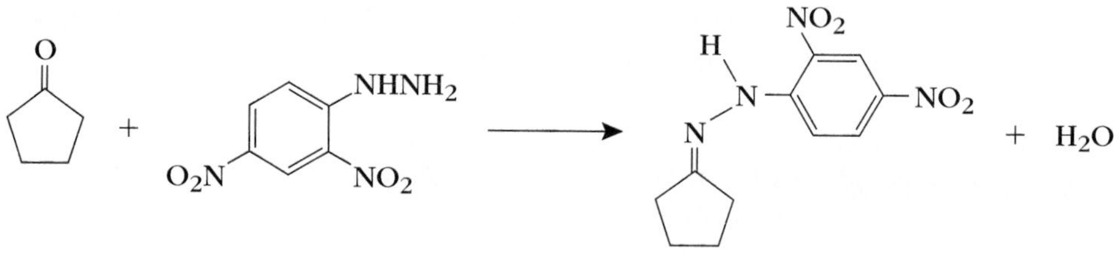

Cyclopentanone   2,4-Dinitrophenylhydrazine   Cyclopentanone 2,4-dinitrophenylhydrazone

Table 14.4 lists several other reagents of the type Z—NH_2 that react with aldehydes and ketones and the structures of the products. Unlike the reaction with primary amines, the equilibrium constants for the reactions in Table 14.4 favor formation of the product.

Oximes, hydrazones, 2,4-dinitrophenylhydrazones, and semicarbazones are solid derivatives of aldehydes and ketones. At one time, the preparation of such solid derivatives was a very important means of identifying liquid aldehydes and ketones. Imagine, for example, that you have isolated a liquid ketone that might be either 3-methyl- or 4-methylcyclohexanone. Both compounds boil at 169 °C, so you can't use boiling points to distinguish between them. However, preparing

Table 14.4	Summary of the Products of the Nucleophilic Addition Reactions of Some Derivatives of Ammonia to Aldehydes and Ketones

GENERAL REACTION

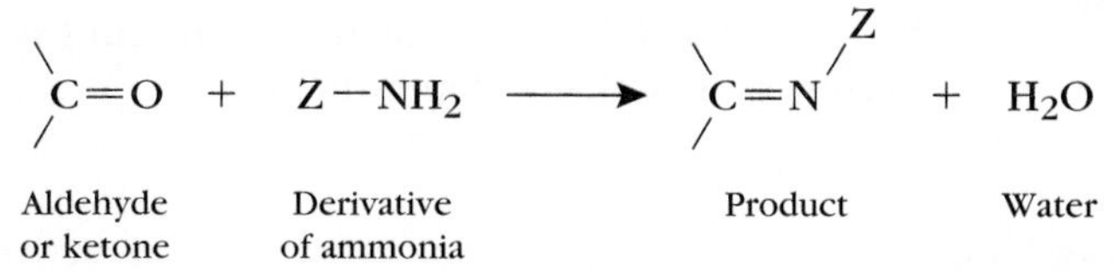

$$\underset{\substack{\text{Aldehyde} \\ \text{or ketone}}}{\backslash C=O} \;+\; \underset{\substack{\text{Derivative} \\ \text{of ammonia}}}{Z-NH_2} \;\longrightarrow\; \underset{\text{Product}}{\backslash C=N^{Z}} \;+\; \underset{\text{Water}}{H_2O}$$

Reagent	Z in Z—NH₂	Product
HO—NH₂ Hydroxylamine	—OH	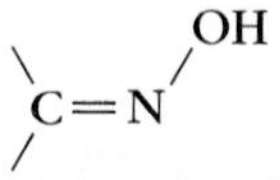 $\backslash C=N^{OH}$ Oxime
H₂N—NH₂ Hydrazine	—NH₂	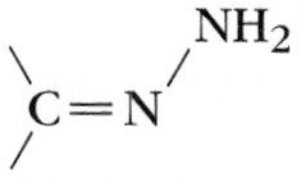 $\backslash C=N^{NH_2}$ Hydrazone
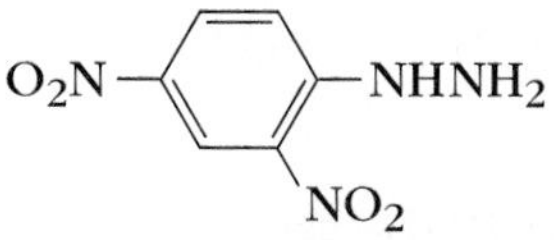 $O_2N{-}\text{(ring)}{-}NHNH_2$, NO₂ 2,4-Dinitrophenylhydrazine	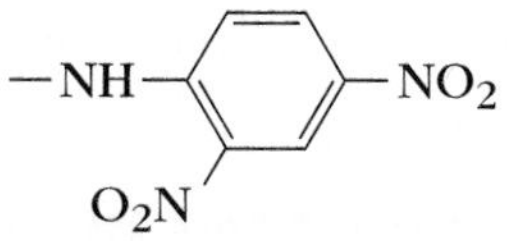 $-NH{-}\text{(ring)}{-}NO_2$, O₂N	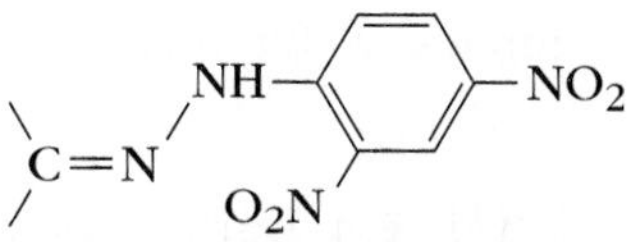 $\backslash C=N{-}NH{-}\text{(ring)}{-}NO_2$, O₂N 2,4-Dinitrophenylhydrazone
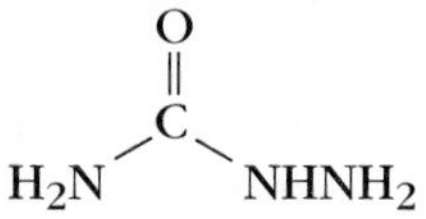 H₂N—C(=O)—NHNH₂ Semicarbazide	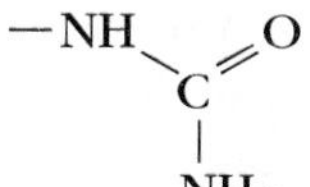 $-NH{-}C(=O){-}NH_2$	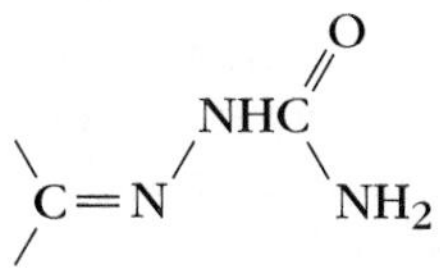 $\backslash C=N{-}NHC(=O){-}NH_2$ Semicarbazone

the 2,4-dinitrophenylhydrazone of the liquid ketone and determining its melting point quickly establishes its identity. If the solid 2,4-dinitrophenylhydrazone of the ketone melts at 130 °C, the ketone must be 4-methylcyclohexanone. If it melts at 155 °C, the ketone must be 3-methylcyclohexanone.

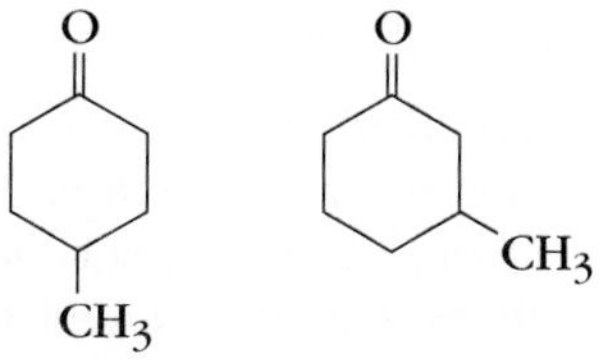

	4-methylcyclohexanone	3-methylcyclohexanone
Boiling point of ketone	169 °C	169 °C
Melting point of 2,4-dinitrophenylhydrazone	130 °C	155 °C

The widespread use of NMR to identify compounds has made the use of such solid derivatives less important as a means of identification.

Exercise 14.26   Write the equations for the preparation of the oxime, hydrazone, 2,4-dinitrophenylhydrazone, and semicarbazone of 4-methylcyclohexanone.

Exercise 14.27   What differences in the ^{1}H and ^{13}C NMR spectra of 3-methyl- and 4-methylcyclohexanone would distinguish each isomer?

Exercise 14.28   Write the structure of the product in each of the following reactions.

(a) 2-Butanone   + NH_2OH ⟶

(b) Benzaldehyde   +   2,4-Dinitrophenylhydrazine   ⟶

(c) Cyclopentanone   + $CH_3CH_2NH_2$ $\xrightarrow[\text{Acidic buffer}]{}$

Imines are intermediates in the degradation of amino acids in mammals.

14.15 An Imine Is an Important Intermediate in Enzyme-Catalyzed Transamination Reactions

Imines or Schiff bases play an important role in the chemical reactions that degrade α-amino acids in mammals. α-Amino acids are compounds that contain both a carboxylate group ($-COO^-$) and an ammonium group ($-NH_3{^+}$) on the same carbon atom. Glycine, alanine, and phenylalanine are examples of α-amino acids.

■ *In the physiological pH range (near pH = 7), the amino group of an α-amino acid exists as an ammonium ion ($-\overset{+}{N}H_3$), and the carboxylic acid group exists as a carboxylate anion ($-COO^-$).*

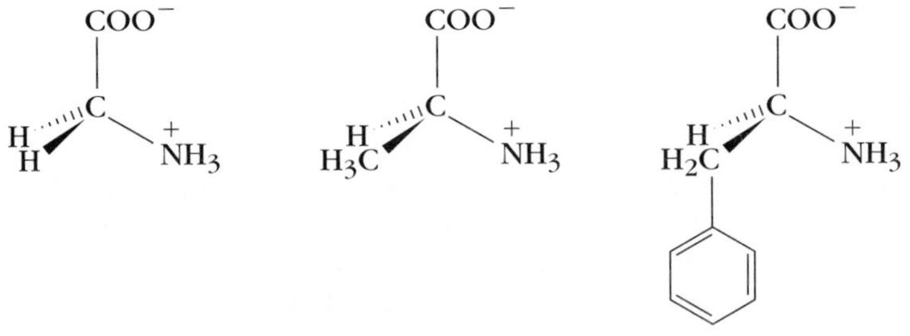

Glycine   Alanine   Phenylalanine

Proteins are synthesized from α-amino acids, as we will learn in Chapter 26. α-Amino acids must be obtained from food or synthesized by the organism. Mammals synthesize some amino acids, called nonessential amino acids, but must obtain others, called essential amino acids, in their diet. Excess dietary amino acids cannot be stored in the body nor can they be excreted, so the amino group of surplus amino acids is removed in a reaction called deamination. The amino group of the amino acid that is removed by deamination is eventually converted to ammonia, and the carbon skeleton of the amino acid is used in the synthesis of carbohydrates and fatty acids.

The deamination of α-amino acids is a **transamination reaction** in which the amino groups of amino acids are transferred to α-ketoglutarate to yield the α-keto acid of the original amino acid and glutamate.

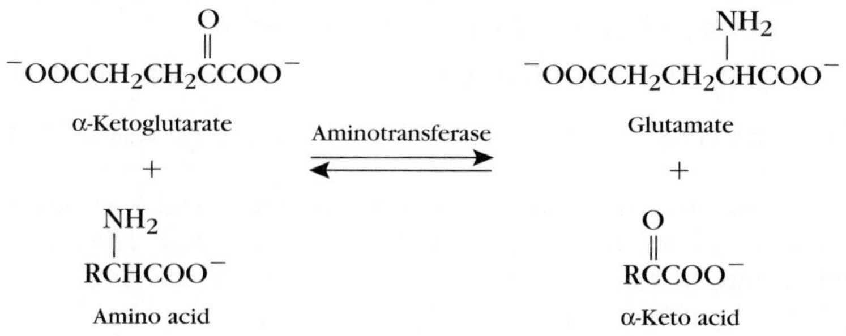

$$^-OOCCH_2CH_2\overset{\overset{\textstyle O}{\|}}{C}COO^-$$

α-Ketoglutarate

$$^-OOCCH_2CH_2\overset{\overset{\textstyle NH_2}{|}}{C}HCOO^-$$

Glutamate

Aminotransferase

+

$$\overset{\overset{\textstyle NH_2}{|}}{R}CHCOO^-$$

Amino acid

+

$$R\overset{\overset{\textstyle O}{\|}}{C}COO^-$$

α-Keto acid

Transamination reactions are catalyzed by a class of enzymes called **aminotransferases.** Aminotransferases need the help of a coenzyme to catalyze the transfer of amino groups. The coenzyme in this reaction is a compound called pyridoxal phosphate, PLP, which is obtained from vitamin B_6 (pyridoxal).

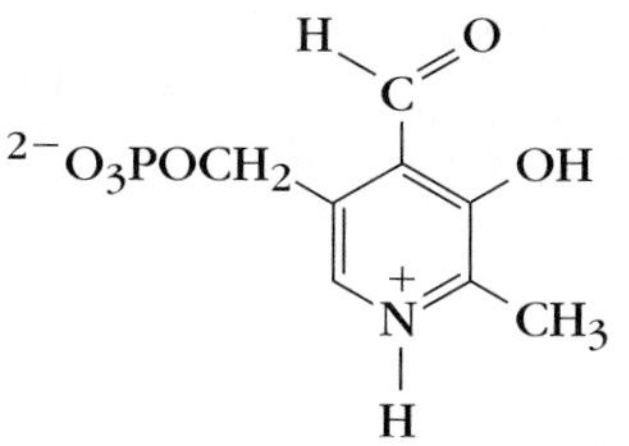

Pyridoxal phosphate
(PLP)

PLP is covalently attached to the enzyme by means of an imine (Schiff base) bond, which is formed between the aldehyde group of PLP and an amino group of the enzyme.

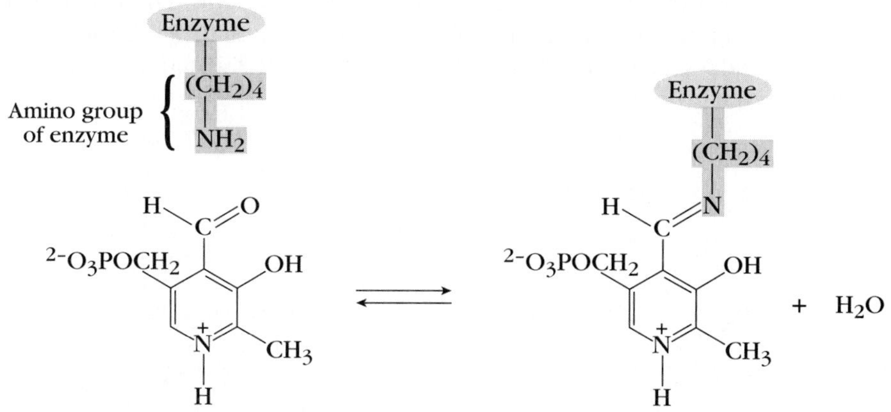

Pyridoxal phosphate
(PLP)

Schiff base of PLP
and enzyme

This imine bond is the focus of the catalytic activity of the coenzyme.

Aminotransferase-catalyzed transamination reactions occur in two stages:

Stage 1. The amino group of an amino acid is transferred to the enzyme with the formation of the corresponding α-keto acid.

Enzyme   +   α-Amino Acid   ⇌   NH$_2$-Enzyme   +   α-Keto acid

Stage 2. The amino group of the enzyme is transferred to α-ketoglutarate to form glutamate and regenerate the enzyme.

NH$_2$-Enzyme   +   α-Ketoglutarate   ⇌   Enzyme   +   Glutamate

These two stages consist of three steps each. Let's begin by examining the three steps of Stage 1.

Step 1 of Stage 1 of the transamination of an amino acid is its reaction with the enzyme-bound PLP. The reaction forms a new Schiff base called an aldimine, as shown in Figure 14.13.

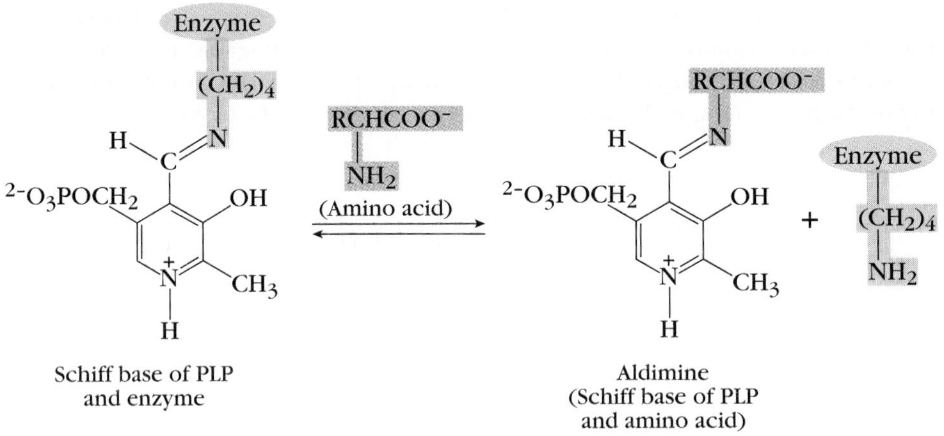

Figure 14.13
The first step in the degradation of amino acids by the enzyme amino-transferase is the formation of an aldimine, a Schiff base, between PLP and an amino acid. This reaction is called a transimination reaction because one imine (the Schiff base of PLP and the enzyme) is converted to another (the aldimine).

Schiff base of PLP
and enzyme

Aldimine
(Schiff base of PLP
and amino acid)

Step 2 of Stage 1 is a proton transfer that converts the aldimine to a ketimine. This is believed to occur by the two reactions shown in Figure 14.14. First, removal of a proton from the aldimine forms a quinoid intermediate. Then addition of a proton to the quinoid intermediate forms a ketimine. The overall reaction shown in Figure 14.14 is the establishment of an equilibrium between two Schiff bases. One is an aldimine formed by the reaction of PLP and the amino acid substrate, and the other is its isomer, the ketimine.

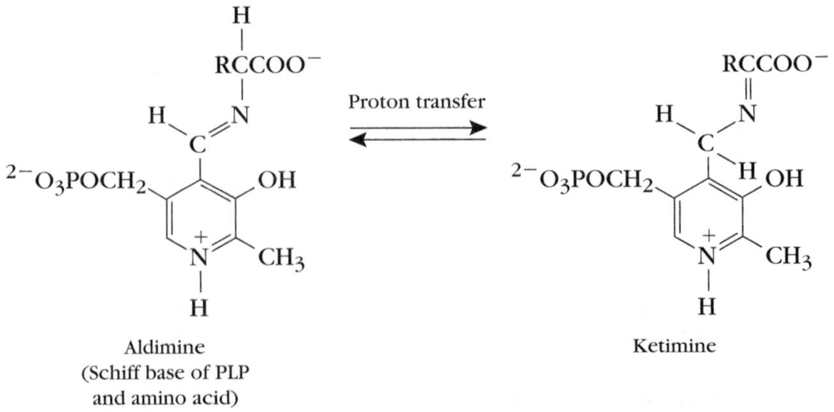

Aldimine
(Schiff base of PLP
and amino acid)

Ketimine

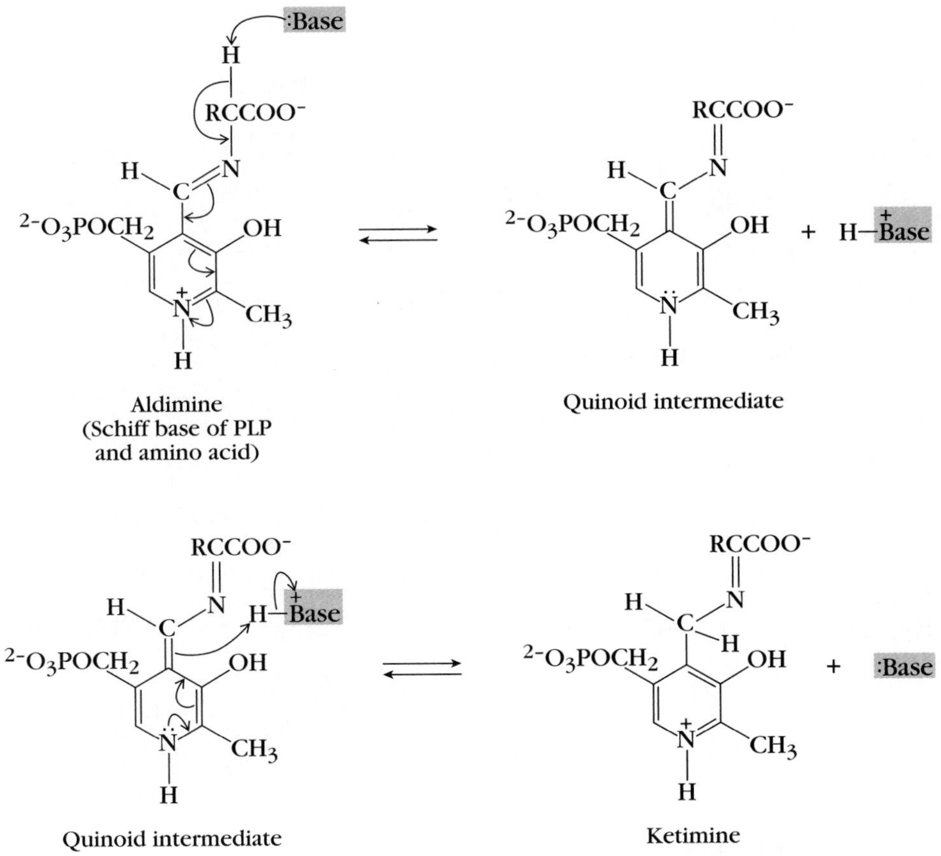

Figure 14.14
The mechanism by which proton transfer converts an aldimine to its isomeric ketimine.

Step 3 of Stage 1 is the hydrolysis of the ketimine to form pyridoxamine phosphate (PMP) and an α-keto acid.

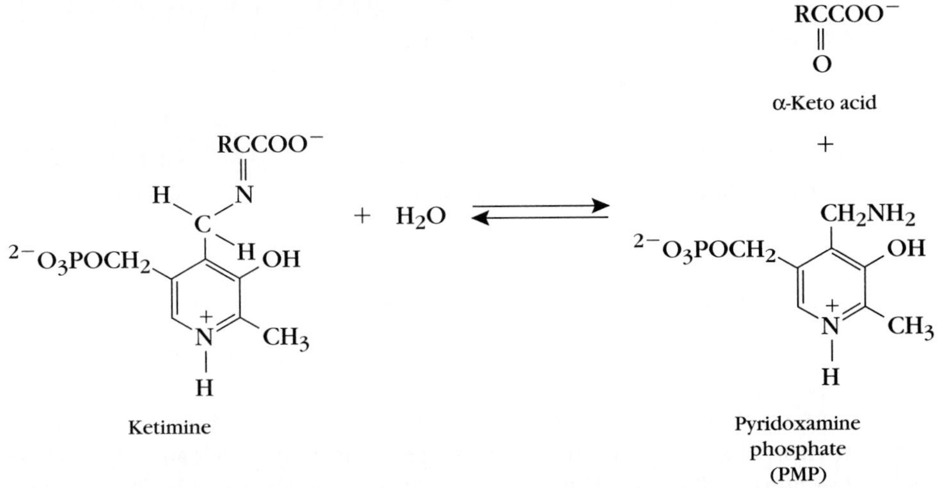

Stage 2 of the aminotransferase-catalyzed transamination reaction converts PMP back to the Schiff base of PLP and the enzyme. This involves the same three steps as in Stage 1 but in reverse order. Step 4 is the reaction of PMP with an α-keto acid to form a ketimine (reverse of Step 3, Stage 1). Step 5 converts the

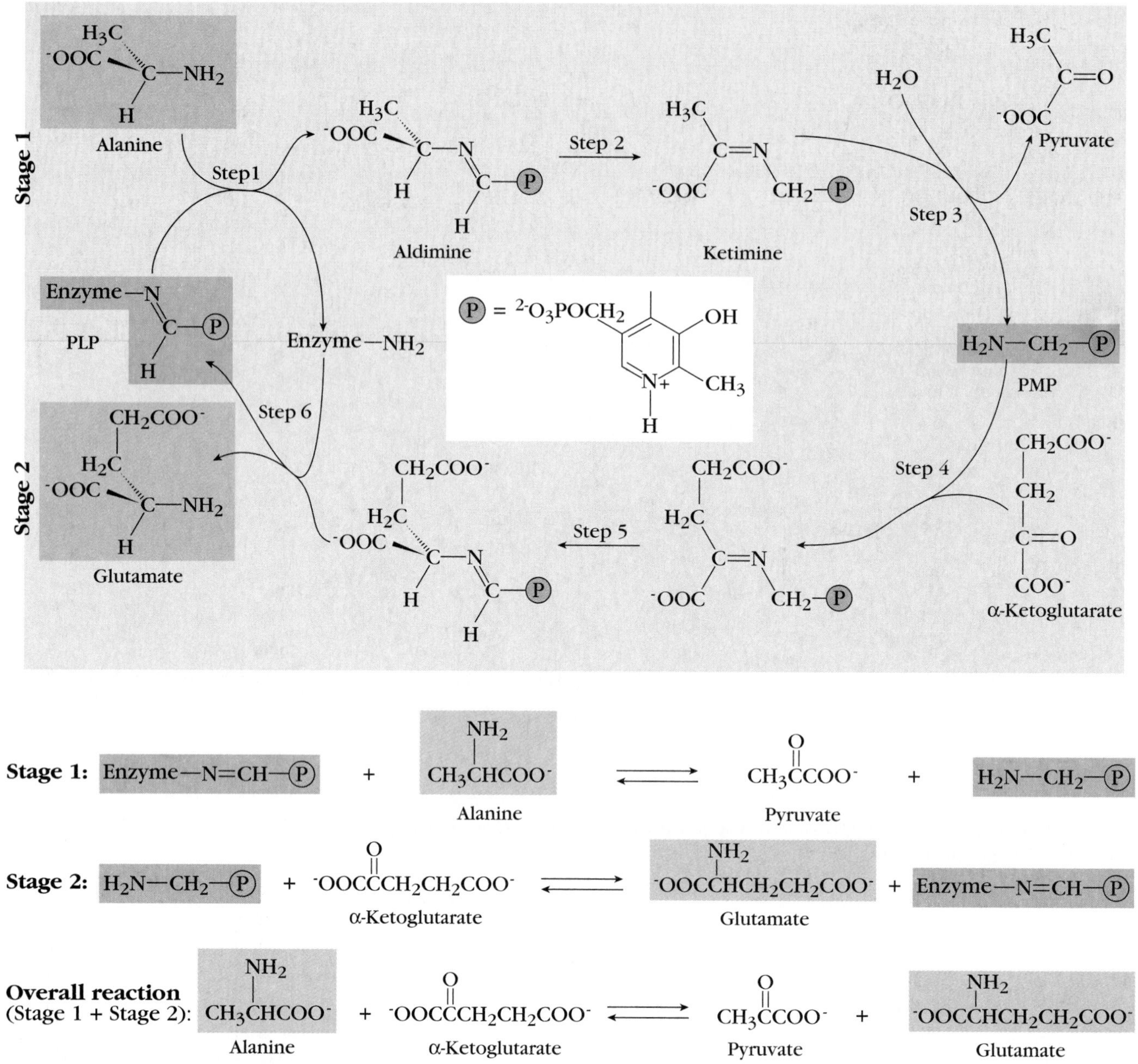

Figure 14.15
The six steps of the PLP-dependent transaminase-catalyzed transamination of alanine. The first stage, which consists of three steps (1, 2, and 3), transfers the α-amino group of alanine to PLP, forming PMP and pyruvate. Stage two, which also consists of three steps (4,5, and 6), transfers the amino group of PMP to α-ketoglutarate to yield glutamate and the Schiff base of PLP and the enzyme.

ketimine to an aldimine by proton transfer (reverse of Step 2, Stage 1). Step 6 forms the new amino acid and regenerates the Schiff base of PLP and the enzyme (reverse of Step 1, Stage 1). The cyclic nature of these six steps is illustrated by the deamination of alanine shown in Figure 14.15.

Imines are the key intermediates in the transaminase-catalyzed transamination of an α-amino acid. Their formation and subsequent hydrolysis are the key steps in the transfer of an amino group from one amino acid to another.

14.16 Converting Aldehydes and Ketones to Alkenes by the Wittig Reaction

Another reaction whose first step involves nucleophilic addition to the carbonyl carbon atom of aldehydes and ketones is the Wittig reaction. The Wittig reaction is named after its discoverer, the German chemist George Wittig, who shared the Nobel prize in chemistry for it in 1979. The Wittig reaction uses phosphorus-stabilized carbanions, called **ylides,** as the nucleophile to convert aldehydes and ketones to alkenes.

■ *The word* ylide *is pronounced "ill' id."*

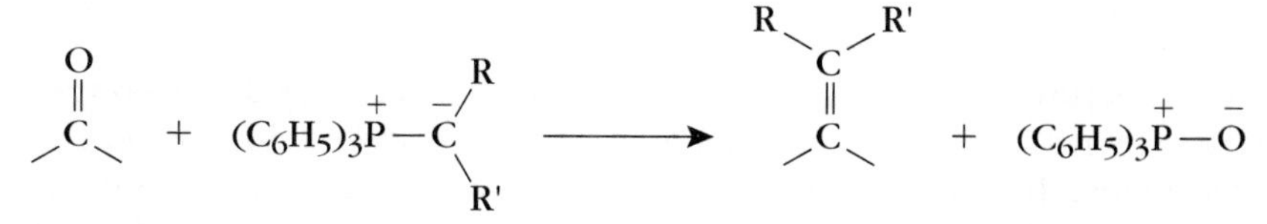

| Aldehyde or ketone | Triphenyl-phosphonium ylide | Alkene | Triphenylphosphine oxide |

| Cyclohexanone | Methylenetriphenyl-phosphorane | Methylene-cyclohexane (85% yield) | Triphenyl-phosphine oxide |

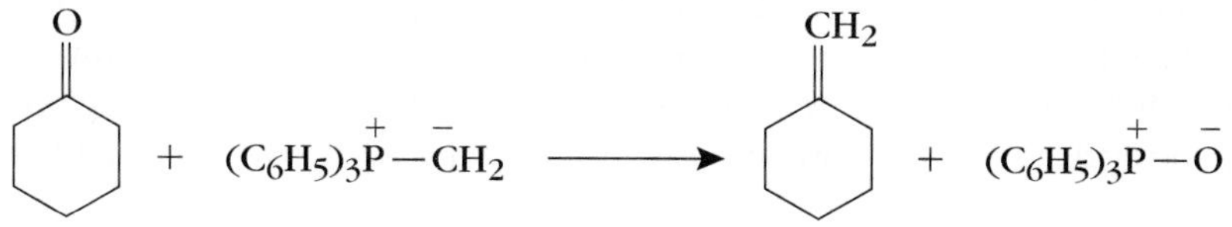

■ *Recall that* C_6H_5— *represents a phenyl group*

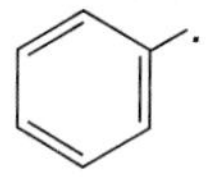

Ylides are compounds that have a structure with opposite charges on adjacent atoms, each of which has an octet of electrons. In the Wittig reaction, the ylide is a **phosphorus ylide,** which is prepared in a sequence of two reactions. Formation of a tetraorganophosphonium salt by a nucleophilic substitution reaction of triphenylphosphine with a methyl, primary, or secondary haloalkane is the first reaction.

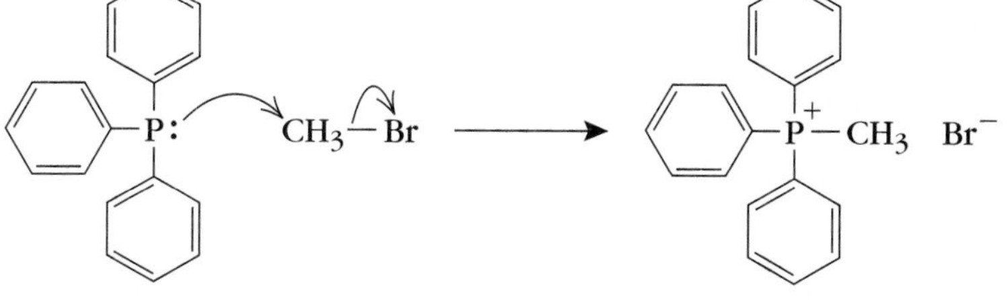

| Triphenylphosphine | Methyltriphenyl-phosphonium bromide |

Triphenylphosphine is a very good nucleophile but only a weak base, so it reacts readily with methyl, primary, and secondary haloalkanes.

The hydrogen on the carbon atom adjacent to the positively charged phosphorus atom is slightly acidic and is removed by a strong base such as butyllithium in the second reaction to form the phosphorus ylide.

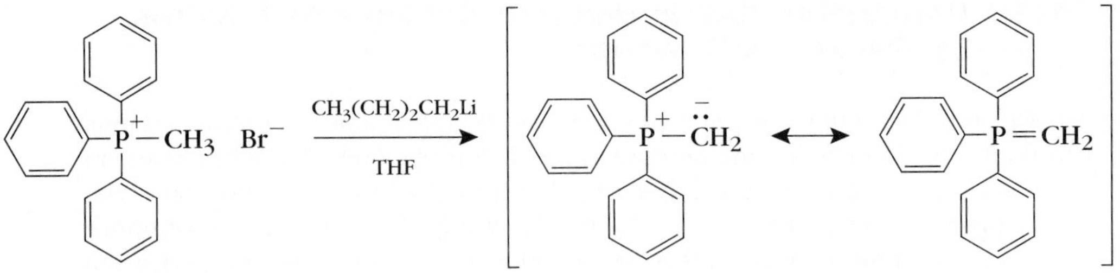

Methyltriphenyl-
phosphonium bromide

Methylenetriphenylphosphorane
(A phosphorus ylide)

The structure of a phosphorus ylide is a resonance hybrid of two resonance structures. One structure contains a double bond between the carbon and phosphorus atoms; the other has a negative charge on the carbon atom and a positive charge on the phosphorus atom. The structure with a double bond requires 10 electrons about the phosphorus atom. The only way 10 electrons can be placed about a phosphorus atom is by the use of phosphorus $3d$ atomic orbitals. Because of their different sizes, overlap is poor between a phosphorus $3d$ orbital and a carbon p orbital. As a result, the carbon-phosphorus double bond is weak and the structure with the separation of charges is the major contributor to the resonance hybrid. To emphasize this, we will use the structure with separation of charges to represent phosphorus ylides, as shown in the following examples.

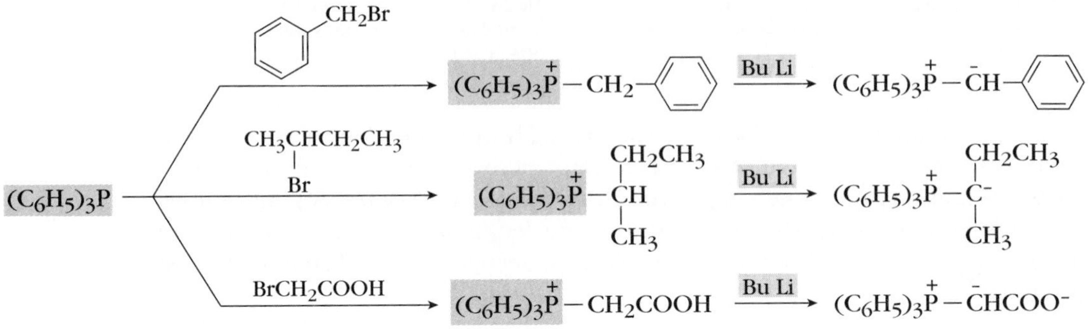

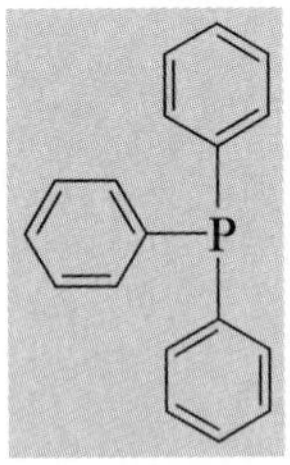

Phosphorus ylides with a wide variety of structures can be prepared. The only limitation is that the structure of the haloalkane used in the formation of the phosphonium salt must allow it to undergo a substitution reaction by an S_N2 mechanism.

Step 1 of the mechanism of the Wittig reaction, like the other reactions of aldehydes and ketones that we have studied in this chapter, is the addition of the nucleophile to the carbonyl carbon atom, as shown in Figure 14.16. The ylide carbon atom is the nucleophile in Step 1 because of its carbanion character. It adds to the carbonyl carbon atom of an aldehyde or ketone to form a dipolar intermediate called a **betaine.** Betaines are a class of compounds that have opposite charges on nonadjacent atoms.

In Step 2, the anionic oxygen atom of the betaine reacts with the positively charged phosphorus atom to form a cyclic intermediate called an **oxaphosphetane,** which spontaneously breaks apart to an alkene and triphenylphosphine oxide in Step 3.

Step 1: Nucleophilic addition of ylide carbon atom to the carbonyl carbon atom of an aldehyde or ketone forms a betaine

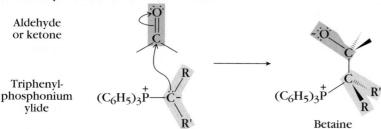

Aldehyde or ketone

Triphenyl-phosphonium ylide

Betaine

Figure 14.16
The mechanism of the Wittig reaction. Step 1 is the nucleophilic addition of the ylide to the carbonyl carbon atom of an aldehyde or ketone to form a betaine. The betaine closes to form a four-membered ring oxaphosphetane in Step 2, which breaks apart to an alkene and triphenylphosphine oxide in Step 3.

Step 2: Fast formation of four-membered oxaphosphetane ring

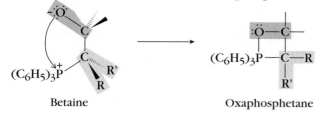

Betaine   Oxaphosphetane

Step 3: The oxaphosphetane breaks apart to form an alkene and triphenylphosphine oxide

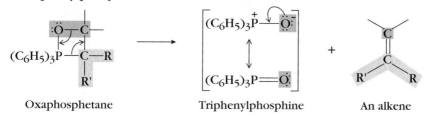

Oxaphosphetane   Triphenylphosphine   An alkene

Exercise 14.29   Write the reaction for the formation of each of the following phosphorus ylides starting with triphenylphosphine and a halogenated compound.

(a) $(C_6H_5)_3\overset{+}{P} - \overset{-}{C}HCH_2CH_2CH_3$ **(c)** $(C_6H_5)_3\overset{+}{P} - \overset{-}{C}(C_6H_5)_2$

(b) $(C_6H_5)_3\overset{+}{P} - \overset{-}{C}HC \equiv CCH_2CH_3$ **(d)** $(C_6H_5)_3\overset{+}{P} - \overset{-}{C}HCH_2 -$⬡

The Wittig reaction is an important synthetic method for the preparation of mono-, di-, and trisubstituted alkenes because the position of the double bond is never in doubt. It is always exactly where the carbonyl group was in the starting aldehyde or ketone. This makes it possible to synthesize alkenes (such as 2-hexene) that are difficult to prepare by elimination reactions. Reaction of 2-chlorohexane with potassium methoxide in methanol forms a mixture of 1-hexene (33%) and 2-hexene (67%). In contrast, reaction of butanal with the ylide formed from ethyltriphenylphosphonium bromide forms exclusively 2-hexene.

SYNTHESIS OF 2-HEXENE
By Elimination Reaction

$$\underset{\substack{\text{2-Chlorohexane}}}{CH_3(CH_2)_3\overset{\overset{\displaystyle Cl}{|}}{C}HCH_3} \xrightarrow[\substack{CH_3OH}]{K^{+\,-}OCH_3} \underset{\substack{\text{1-Hexene}\\(33\% \text{ yield})}}{CH_3(CH_2)_3CH{=}CH_2} + \underset{\substack{\text{2-Hexene}\\(67\% \text{ yield})}}{CH_3(CH_2)_2CH{=}CHCH_3}$$

(Mixture of *E*- and *Z*- isomers)

By Wittig Reaction

$$(C_6H_5)_3P \xrightarrow[\substack{(2)\ C_4H_9Li}]{(1)\ CH_3CH_2Br} (C_6H_5)_3\overset{+}{P}\overset{-}{C}HCH_3 \xrightarrow{CH_3CH_2CH_2CHO} CH_3(CH_2)_2CH{=}CHCH_3$$

2-Hexene
(Mixture of *E*- and *Z*- isomers)

Exercise 14.30   Write the structure of the product of each of the following reactions.

(a) 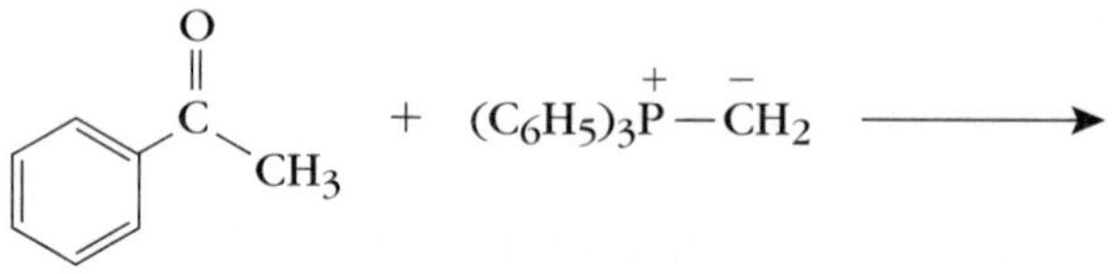 $+\ \ (C_6H_5)_3\overset{+}{P}{-}\overset{-}{C}H_2 \longrightarrow$

(b) $CH_3CH_2CH_2CHO\ +\ \ (C_6H_5)_3\overset{+}{P}{-}\overset{-}{C}(CH_3)_2 \longrightarrow$

(c) Cyclopentanone $+\ \ (C_6H_5)_3\overset{+}{P}{-}\overset{-}{C}H\overset{\overset{\displaystyle O}{\|}}{C}\overset{\diagdown}{}_{OCH_3} \longrightarrow$

(d) $(CH_3)_2C{=}O\ +\ \ (C_6H_5)_3\overset{+}{P}{-}\overset{-}{C}HCH{=}CH_2 \longrightarrow$

The Wittig reaction has two limitations. The first is that a mixture of *E*- and *Z*-isomers is formed in the synthesis of internal alkenes. Advanced techniques have been developed to overcome this limitation, but these techniques are beyond the scope of this text. The second is that tetrasubstituted alkenes cannot be synthesized by this method, probably because steric hindrance prevents formation of the betaine intermediate.

Despite these two limitations, the Wittig reaction is an excellent method of preparing alkenes. Let us learn how to plan the Wittig synthesis of an alkene.

14.17 Planning the Synthesis of an Alkene by the Wittig Reaction

We learned in Section 9.12 that a logical way to plan a synthesis is to start with the desired product and work backwards until we arrive at the starting materials. The same method can be used to plan the synthesis of an alkene by the Wittig reaction. Sample Problem 14.1 illustrates this method with the synthesis of 2-methyl-2-butene.

Sample Problem 14.1   Propose a synthesis of 2-methyl-2-butene that uses the Wittig reaction.

Solution:

In the Wittig reaction, one carbon atom of the double bond comes from the haloalkane used to prepare the ylide, while the other comes from the carbonyl carbon atom of the aldehyde or ketone. Either carbon atom of the double bond can come from the haloalkane. In principle, therefore, two different Wittig reactions can be used to synthesize a particular alkene.

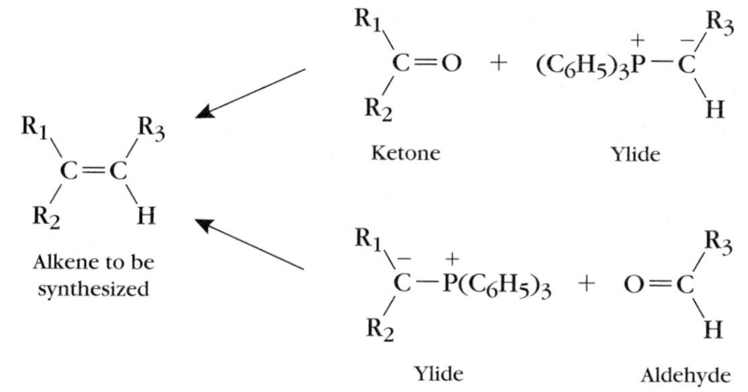

In the case of the synthesis of 2-methyl-2-butene, both Wittig reactions will form the desired product.

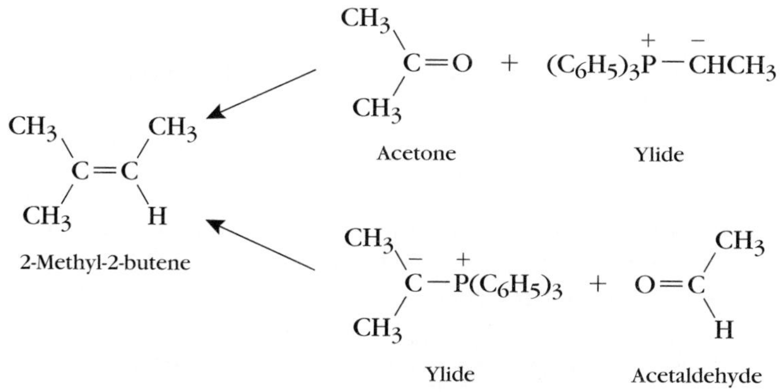

The choice of which reaction to use depends on the availability of the various starting materials.

Exercise 14.31	Starting with triphenylphosphine, cyclohexanone, any organic compound containing four or fewer carbon atoms, any inorganic reagent, and any solvent, propose a synthesis of each of the following alkenes.

(a) $\underset{\underset{\text{CH}_3}{|}}{\overset{\overset{\text{CH}_3\text{CH}_2}{\diagdown}}{\text{C}}}=\text{CH}_2$ (b) [cyclohexane ring]=CHCH$_3$ (c) $\text{CH}_3\text{CH}_2\text{CH}=\text{C(CH}_3)_2$

Aldehydes and ketones do not always react similarly. One major difference is their relative ease of oxidation.

14.18 Oxidation of Aldehydes

Aldehydes are much more easily oxidized than ketones. Recall from Section 11.18 that aldehydes, formed by the oxidation of primary alcohols, easily undergo further oxidation to carboxylic acids. Aldehydes are oxidized so easily, in fact, that special oxidation reagents had to be developed to stop the oxidation of primary alcohols at the aldehyde stage (Section 11.18).

In addition to Cr(VI)-containing compounds in acid solution, aldehydes are easily oxidized by Tollen's reagent and Fehling's solution. Only a few functional groups, including aldehydes but not ketones (except α-hydroxy ketones), are oxidized by Tollen's reagent and Fehling's solution. **Tollen's reagent,** a convenient qualitative test for aldehydes, is based on the ability of silver(I) salts to oxidize aldehydes but not ketones. The oxidizing agent in Tollen's reagent is diamminosilver(I) ion [$\text{Ag(NH}_3)_2^+$]. Aldehydes and α-hydroxy ketones reduce silver(I) to silver metal, which is deposited as a silver mirror on the sides of the test tube. If the compound is a ketone, no silver mirror is formed (no reaction has occurred).

$$\underset{\text{Aldehyde}}{\text{RCHO}} + \underset{\underset{\text{reagent}}{\text{Tollen's}}}{\text{Ag(NH}_3)_2^+} \longrightarrow \text{RCOO}^- + \underset{\underset{\text{mirror}}{\text{Silver}}}{\text{Ag}}$$

$$\underset{\text{Ketone}}{\overset{\overset{\displaystyle\text{O}}{\|}}{\underset{\text{R}\quad\text{R'}}{\text{C}}}} + \underset{\underset{\text{reagent}}{\text{Tollen's}}}{\text{Ag(NH}_3)_2^+} \longrightarrow \text{No Reaction}$$

Fehling's solution is an alkaline solution of Cu(II) in a tartarate buffer. Aldehydes but not ketones (except α-hydroxy ketones) reduce Cu(II) to cuprous oxide, which precipitates as a brick-red solid.

$$\text{RCHO} + 2\text{Cu}^{2+} + 5\text{OH}^- \xrightarrow[\text{buffer}]{\text{Tartarate}} \text{RCOO}^- + 3\text{H}_2\text{O} + \underset{\underset{\text{(Brick-red precipitate)}}{\text{Cuprous oxide}}}{\text{Cu}_2\text{O}}$$

The ease of identifying aldehydes by ^{1}H NMR spectroscopy (Section 14.4) has diminished the importance of the Tollen's and Fehling's tests for aldehydes.

14.19 Summary

The functional group of both aldehydes and ketones is a **carbonyl group,** which is a carbon-oxygen double bond. Recall from Section 11.18 that aldehydes are conveniently prepared by the oxidation of primary alcohols with pyridinium chlorochromate (PCC), while ketones are obtained by the oxidation of secondary alcohols.

An important reaction of aldehydes and ketones is **nucleophilic addition** to the carbonyl group. The wide variety of nucleophiles that add to the carbonyl group of aldehydes and ketones can be classified according to their nucleophilic atom. **Grignard** and **organolithium reagents, cyanide ion,** and **Wittig reagents** contain a nucleophilic carbon atom. Grignard and organolithium reagents add to aldehydes and ketones to form alcohols (primary and secondary alcohols from aldehydes and tertiary alcohols from ketones); cyanide ion adds to form cyanohydrins; and Wittig reagents add to form alkenes. **Primary amines** contain a nucleophilic nitrogen atom that reacts to form **imines.** Imines are a class of compounds that contain a carbon-nitrogen double bond. Imines are intermediates in the **transamination reaction.** Transamination is an enzyme-catalyzed reaction that is important in the degradation of amino acids in which an amino group is transferred from an amino acid to α-ketoglutarate. Acid-catalyzed addition of **alcohols,** which contain a nucleophilic oxygen atom, forms acetals. Acetals are used in synthesis to protect the carbonyl group of aldehydes and ketones from unwanted reactions.

14.20 Summary of Preparation of Aldehydes and Ketones

1. **Preparation of Aldehydes**

   a. **Oxidation of primary alcohols**   SECTION 11.18

   Oxidation of primary alcohols can be stopped at the aldehyde stage only by using special reagents like pyridinium chlorochromate (PCC).

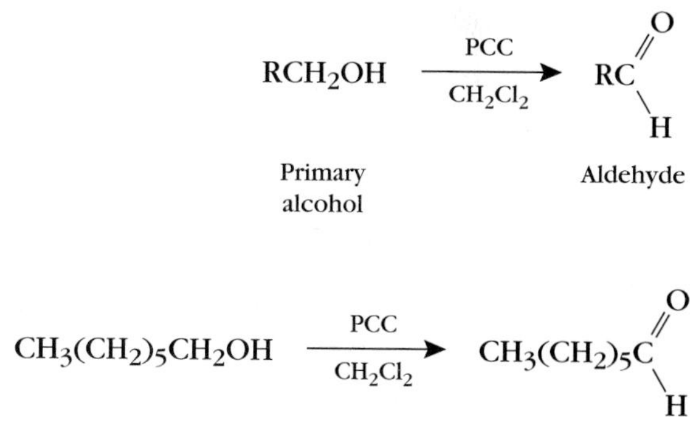

   b. **Ozonolysis of alkenes**   SECTION 14.5

   Synthetically useful only with terminal alkenes, symmetrically disubstituted alkenes, and cyclic alkenes.

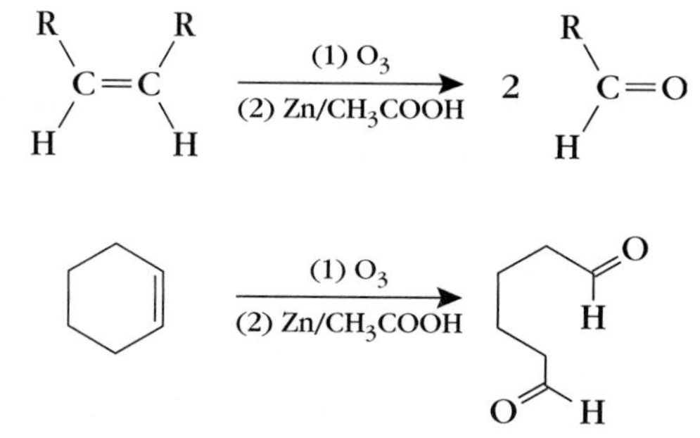

c. **Partial reduction of acyl chlorides**   SECTION 16.14, A

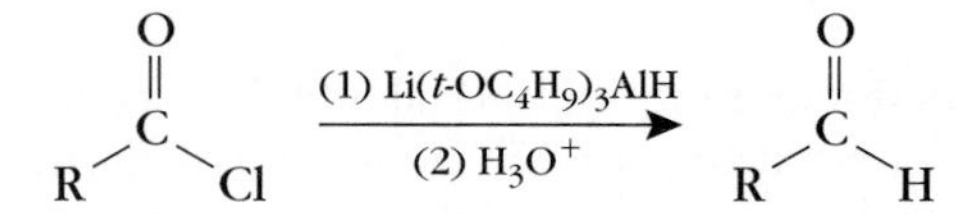

Acyl chloride → Aldehyde

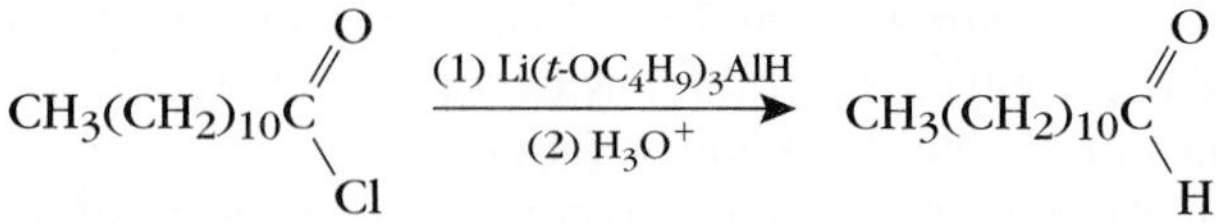

$$\underset{\text{Acyl chloride}}{\overset{O}{\underset{R}{\overset{\|}{C}}}-Cl} \xrightarrow[\text{(2) } H_3O^+]{\text{(1) Li(t-OC}_4H_9)_3\text{AlH}} \underset{\text{Aldehyde}}{\overset{O}{\underset{R}{\overset{\|}{C}}}-H}$$

$$CH_3(CH_2)_{10}\overset{O}{\overset{\|}{C}}-Cl \xrightarrow[\text{(2) } H_3O^+]{\text{(1) Li(t-OC}_4H_9)_3\text{AlH}} CH_3(CH_2)_{10}\overset{O}{\overset{\|}{C}}-H$$

d. **Hydroboration-oxidation of terminal alkynes**   SECTION 9.7, B

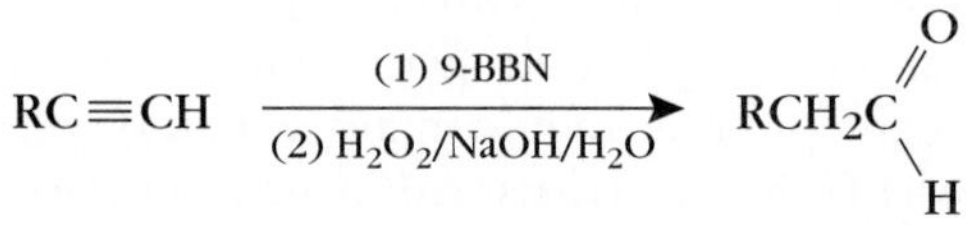

$$\underset{\substack{\text{Terminal}\\\text{alkyne}}}{RC\equiv CH} \xrightarrow[\text{(2) } H_2O_2/NaOH/H_2O]{\text{(1) 9-BBN}} \underset{\text{Aldehyde}}{RCH_2\overset{O}{\overset{\|}{C}}-H}$$

$$CH_3CH_2C\equiv CH \xrightarrow[\text{(2) } H_2O_2/NaOH/H_2O]{\text{(1) 9-BBN}} CH_3CH_2CH_2\overset{O}{\overset{\|}{C}}-H$$

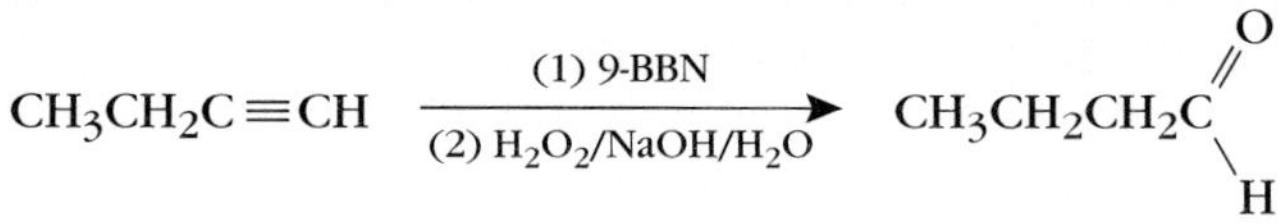

2. **Preparation of Ketones**

a. **Oxidation of secondary alcohols**   SECTION 11.18

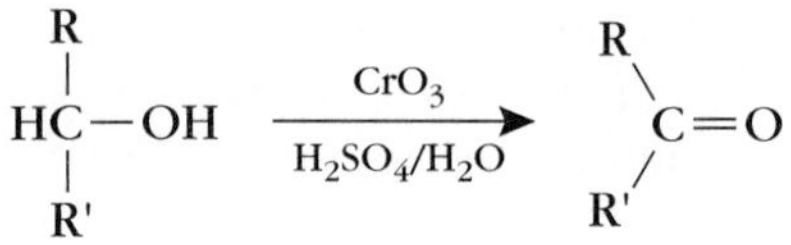

$$\underset{\substack{\text{Secondary}\\\text{alcohol}}}{\overset{R}{\underset{R'}{\overset{|}{HC}-OH}}} \xrightarrow[H_2SO_4/H_2O]{CrO_3} \underset{\text{Ketone}}{\overset{R}{\underset{R'}{\overset{|}{C}=O}}}$$

$$\overset{OH}{\bigcirc} \xrightarrow[H_2SO_4/H_2O]{CrO_3} \overset{O}{\bigcirc}$$

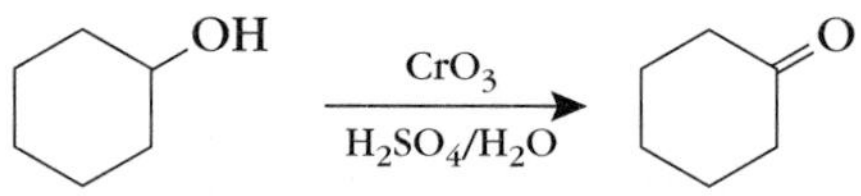

b. **Ozonolysis of alkenes**   SECTION 14.6

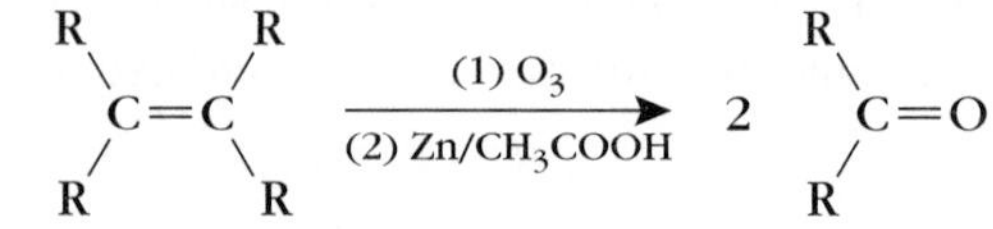

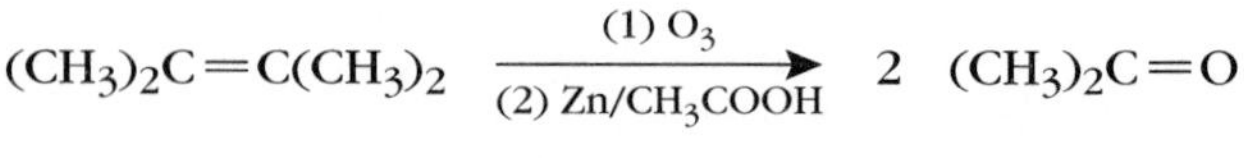

$$\underset{R}{\overset{R}{C}}=\underset{R}{\overset{R}{C}} \xrightarrow[\text{(2) Zn/CH}_3COOH]{\text{(1) O}_3} 2 \underset{R}{\overset{R}{C}}=O$$

$$(CH_3)_2C=C(CH_3)_2 \xrightarrow[\text{(2) Zn/CH}_3COOH]{\text{(1) O}_3} 2 \; (CH_3)_2C=O$$

c. Friedel-Crafts acylation — SECTION 21.9

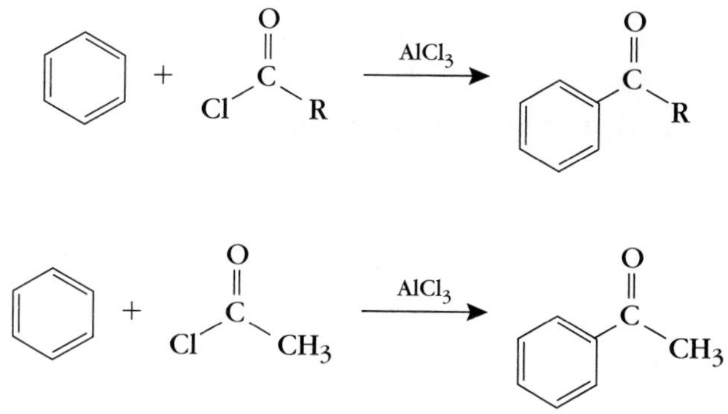

d. Hydration of alkynes — SECTION 9.7, A

$$RC{\equiv}CH \xrightarrow[HgSO_4]{H_3O^+} R\overset{O}{\overset{\|}{C}}CH_3$$

$$CH_3(CH_2)_3C{\equiv}CH \xrightarrow[HgSO_4]{H_3O^+} CH_3(CH_2)_3\overset{O}{\overset{\|}{C}}CH_3$$

e. Organocopper reagent and acyl chlorides — SECTION 16.14, B

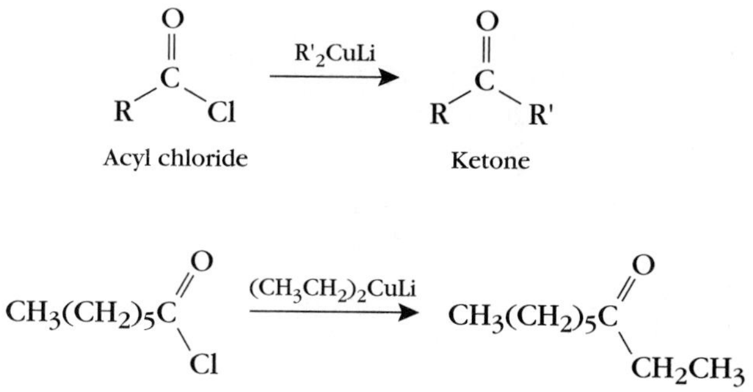

f. Hydroboration-oxidation of a symmetrically substituted and terminal alkynes — SECTION 9.7, B

$$RC{\equiv}CR \xrightarrow[(2)\ H_2O_2/NaOH/H_2O]{(1)\ 9\text{-BBN}} R\overset{O}{\overset{\|}{C}}CH_2R$$

Symmetrically substituted alkyne — Ketone

14.21 Summary of Reactions of Aldehydes and Ketones

1. **Nucleophilic Addition Reactions of Aldehydes and Ketones**

   a. Reduction   SECTION 11.10

   (i) Hydrogen and metal catalyst

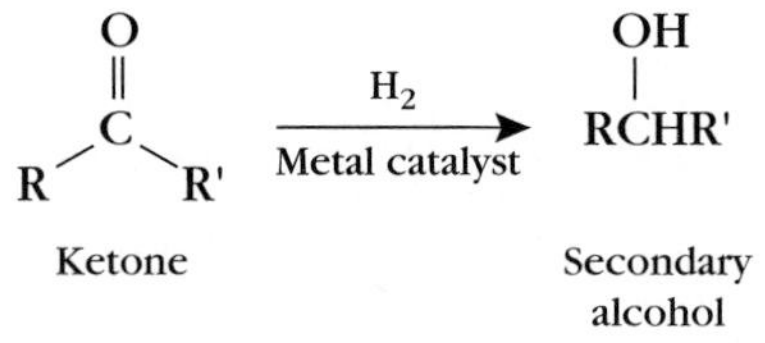

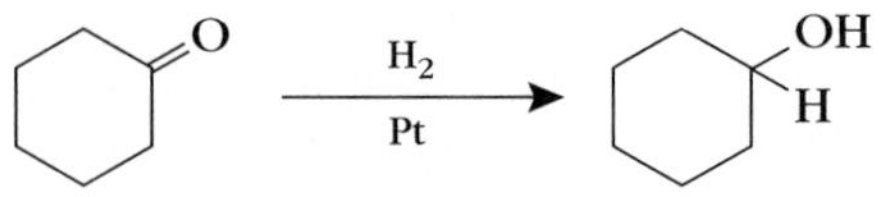

   (ii) Metal hydride reduction

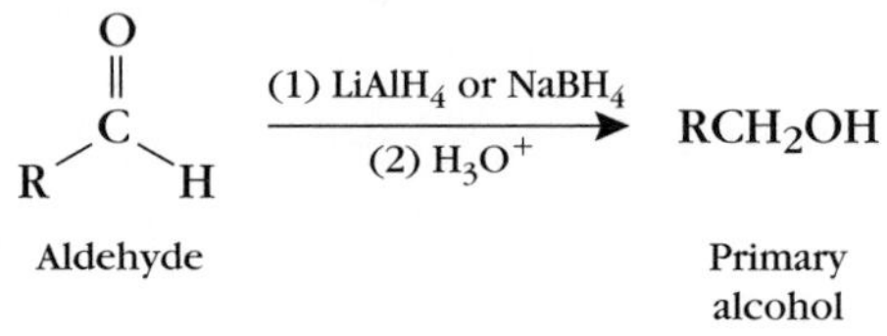

$$CH_3(CH_2)_4CHO \xrightarrow[\text{(2) } H_3O^+]{\text{(1) LiAlH}_4} CH_3(CH_2)_4CH_2OH$$

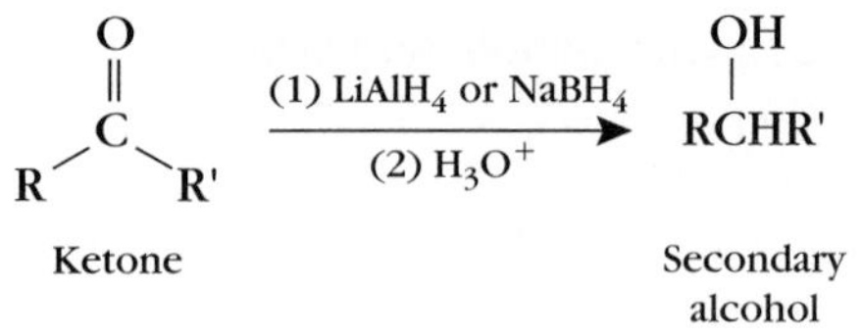

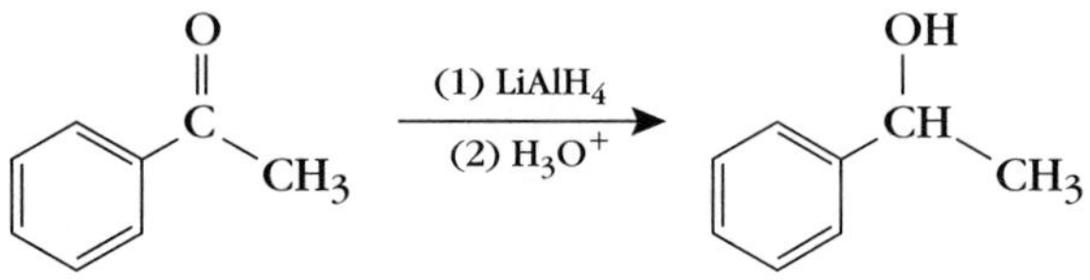

   b. Grignard and organolithium reagents SECTION 11.7

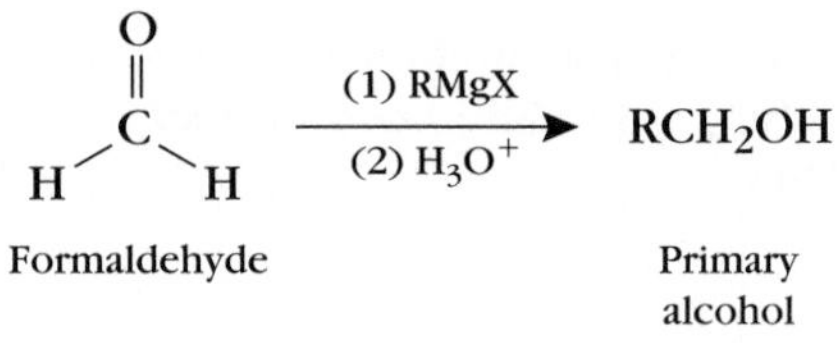

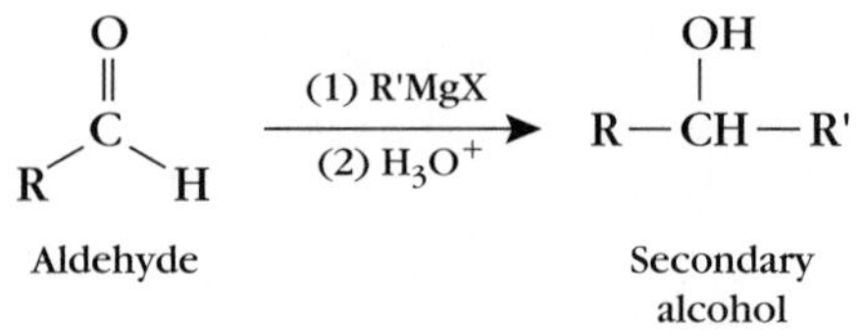

$$\underset{\underset{\text{Ketone}}{R}}{\overset{\overset{\displaystyle O}{\|}}{C}}\overset{}{\underset{}{R'}} \xrightarrow[\text{(2) } H_3O^+]{\text{(1) } R''MgX} \underset{\underset{\underset{\text{Tertiary}}{R''}}{|}}{R-\overset{\overset{\displaystyle OH}{|}}{C}-R'}$$

Ketone   Tertiary alcohol

$$CH_3(CH_2)_3CHO \xrightarrow[\text{(2) } H_3O^+]{\text{(1) } CH_3MgI} CH_3(CH_2)_3\overset{\overset{\displaystyle OH}{|}}{C}HCH_3$$

$$\bigcirc=O \xrightarrow[\text{(2) } H_3O^+]{\text{(1) } C_2H_5MgI} \bigcirc\overset{OH}{\underset{C_2H_5}{}}$$

c. HCN   SECTION 14.9

$$\underset{\underset{\substack{\text{Aldehyde} \\ \text{or ketone}}}{R}}{\overset{\overset{\displaystyle O}{\|}}{C}}\overset{}{\underset{}{H}} \underset{H_2SO_4}{\overset{NaCN}{\rightleftharpoons}} \underset{\underset{\text{Cyanohydrin}}{}}{R\overset{\overset{\displaystyle OH}{|}}{C}HCN}$$

Aldehyde or ketone   Cyanohydrin

$$CH_3CHO \underset{H_2SO_4}{\overset{NaCN}{\rightleftharpoons}} CH_3\overset{\overset{\displaystyle OH}{|}}{C}HCN$$

d. Water   SECTION 14.10

Hydration of aldehydes and ketones can be acid or base catalyzed.

$$\underset{\underset{\substack{\text{Aldehyde} \\ \text{or ketone}}}{R}}{\overset{\overset{\displaystyle O}{\|}}{C}}\overset{}{\underset{}{H}} \overset{H_2O}{\rightleftharpoons} \underset{\underset{OH}{|}}{R-\overset{\overset{\displaystyle OH}{|}}{C}-H}$$

Aldehyde or ketone

$$CH_3CHO \underset{H_3O^+}{\rightleftharpoons} CH_3-\overset{\overset{\displaystyle OH}{|}}{\underset{\underset{OH}{|}}{C}}-H$$

e. Alcohols   SECTION 14.11

The first formed product is a hemiacetal that reacts with alcohol in an acid solution to form an acetal.

$$\underset{\underset{\substack{\text{Aldehyde} \\ \text{or ketone}}}{}}{\overset{\overset{\displaystyle O}{\|}}{C}} \underset{H_2SO_4}{\overset{2R'OH}{\rightleftharpoons}} \underset{\underset{\underset{\text{Acetal}}{OR'}}{|}}{-\overset{\overset{\displaystyle OR'}{|}}{C}-}$$

Aldehyde or ketone   Acetal

$$CH_3CHO \underset{H_2SO_4}{\overset{2CH_3OH}{\rightleftarrows}} \begin{array}{c} OCH_3 \\ | \\ CH_3CH \\ | \\ OCH_3 \end{array}$$

f. Primary amines   SECTION 14.13

Careful pH control of the reaction is needed to form an imine.

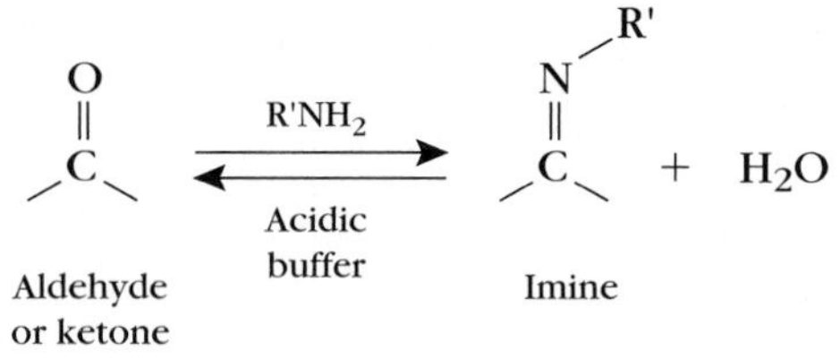

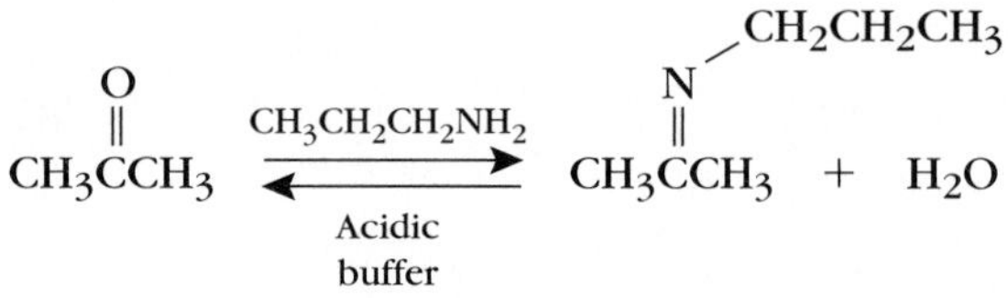

$$\begin{array}{c} O \\ \| \\ \diagdown C \diagup \end{array} \underset{\text{Acidic buffer}}{\overset{R'NH_2}{\rightleftarrows}} \begin{array}{c} N^{\diagup R'} \\ \| \\ \diagdown C \diagup \end{array} + H_2O$$

Aldehyde   Imine
or ketone

$$\begin{array}{c} O \\ \| \\ CH_3CCH_3 \end{array} \underset{\text{Acidic buffer}}{\overset{CH_3CH_2CH_2NH_2}{\rightleftarrows}} \begin{array}{c} N^{\diagup CH_2CH_2CH_3} \\ \| \\ CH_3CCH_3 \end{array} + H_2O$$

g. Wittig reagents   SECTION 14.16

Excellent synthetic method of alkenes except for tetra-substituted alkenes.

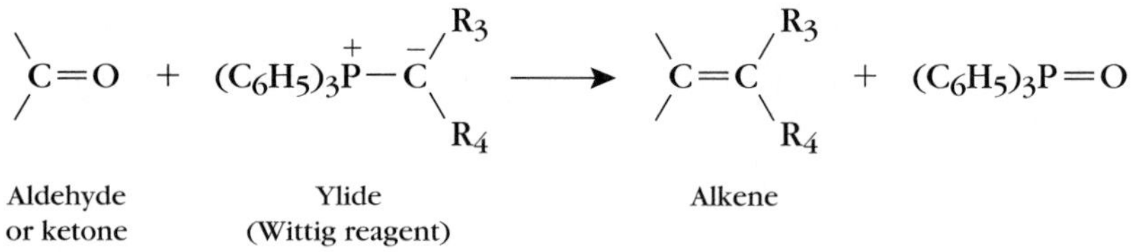

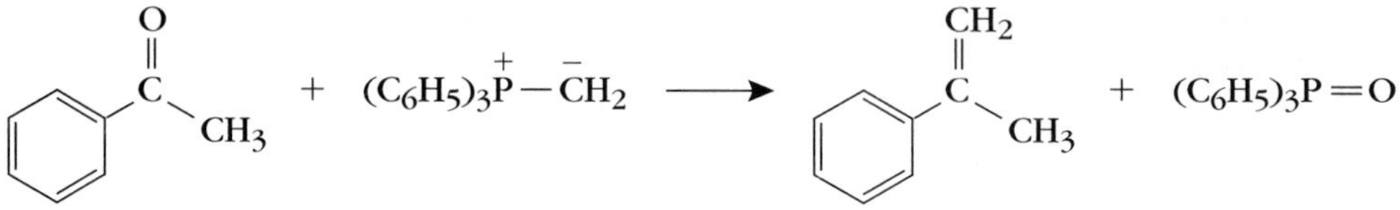

$$\begin{array}{c} \diagdown \\ C=O \\ \diagup \end{array} + (C_6H_5)_3\overset{+}{P} - \overset{-}{C} \diagup_{R_4}^{R_3} \longrightarrow \begin{array}{c} \diagdown \\ C=C \\ \diagup \end{array}_{R_4}^{R_3} + (C_6H_5)_3P=O$$

Aldehyde   Ylide   Alkene
or ketone   (Wittig reagent)

$$\begin{array}{c} O \\ \| \\ C \diagdown CH_3 \end{array} + (C_6H_5)_3\overset{+}{P} - \overset{-}{C}H_2 \longrightarrow \begin{array}{c} CH_2 \\ \| \\ C \diagdown CH_3 \end{array} + (C_6H_5)_3P=O$$

2. Oxidation of Aldehydes by Tollen's Reagent and Fehling's Solution   SECTION 14.18

Only aldehydes and α-hydroxy ketones are oxidized.

Tollen's reagent

$$RCHO + Ag(NH_3)_2^+ \overset{NH_3/H_2O}{\longrightarrow} RCOO^- + Ag$$

Fehling's solution

$$RCHO + 2Cu^{2+} + 5OH^- \overset{\text{Tartarate buffer}}{\longrightarrow} RCOO^- + 3H_2O + Cu_2O$$

14.22 Summary of Mechanisms of Nucleophilic Addition Reactions of Aldehydes and Ketones

The mechanisms of all nucleophilic addition reactions of aldehydes and ketones share a common step. This common step is the addition of a nucleophile to the carbonyl carbon atom to form a tetrahedral intermediate. In neutral or basic solutions, this is the first step of the mechanism.

The common first step in the neutral or basic solutions is nucleophilic addition to the carbonyl carbon atom to form a tetrahedral intermediate.

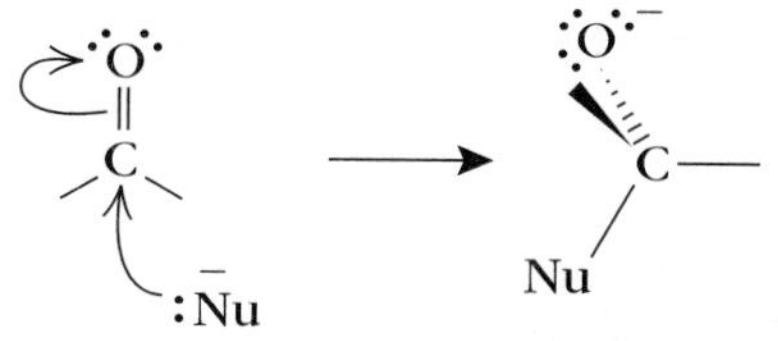

Aldehyde   Tetrahedral
or ketone   intermediate

In acid solution, protonation of the carbonyl oxygen atom is the first step, followed by addition of the nucleophile to the carbonyl carbon atom of the conjugate acid.

In acid solutions, nucleophilic addition to the conjugate acid of aldehyde or ketone is usually the second step.

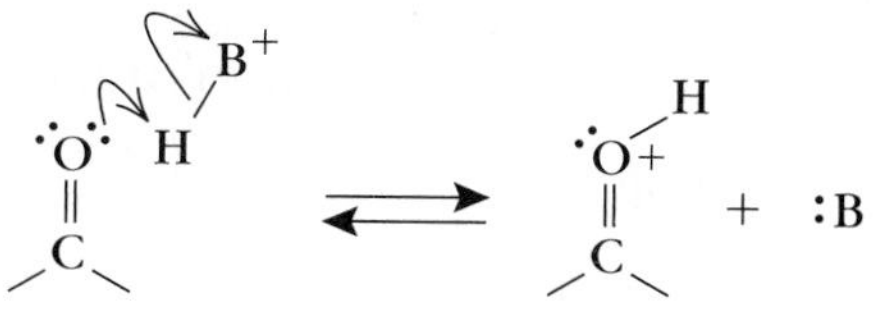

Aldehyde   Conjugate acid of
or ketone   aldehyde or ketone

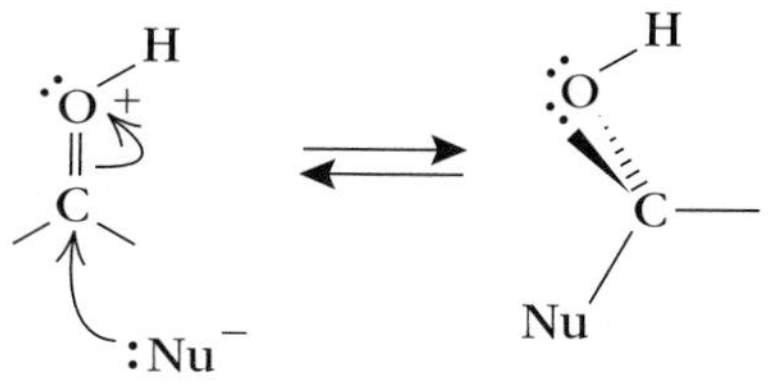

Tetrahedral
intermediate

The product of nucleophilic addition reactions of aldehydes and ketones depends on what happens to the tetrahedral intermediate. Reactions in which the product is formed directly by proton transfer to or from the tetrahedral intermediate are listed in Table 14.5.

In certain acid-catalyzed nucleophilic addition reactions, the first formed product reacts further. These reactions are listed in Table 14.6.

| Table 14.5 | | Addition Reactions to Aldehydes and Ketones Where Proton Transfer to or from the First Formed Tetrahedral Intermediate Leads Directly to the Product | |

Reagent	Structure of Tetrahedral Intermediate	Nature and Source of Acid or Base in Second Step	Structure of Final Product
$NaBH_4$ or $LiAlH_4$	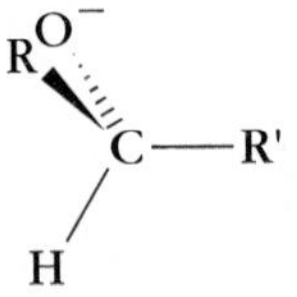 R–C(O^-)(H)–R'	H_3O^+ added to reaction mixture once reaction is completed	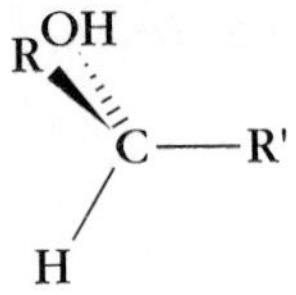 R–C(OH)(H)–R'
R''MgX or LiR''	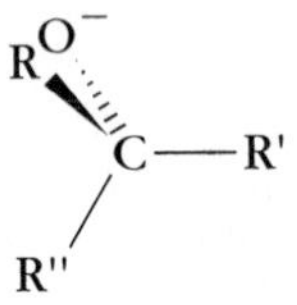 R–C(O^-)(R'')–R'		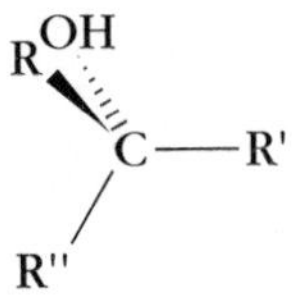 R–C(OH)(R'')–R'
$NaCN/H_2SO_4$	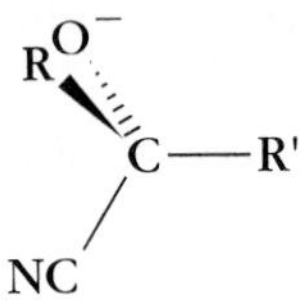 R–C(O^-)(NC)–R'	HCN in reaction mixture	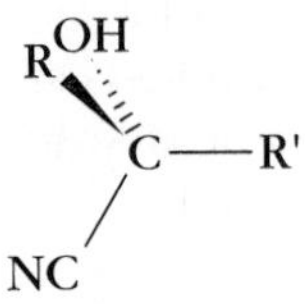 R–C(OH)(NC)–R'
H_2O/OH^-	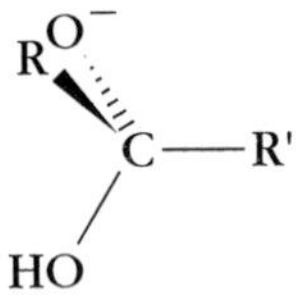 R–C(O^-)(HO)–R'	Solvent (H_2O)	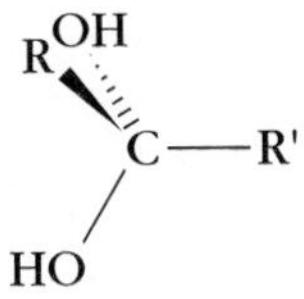 R–C(OH)(HO)–R'
H_3O^+	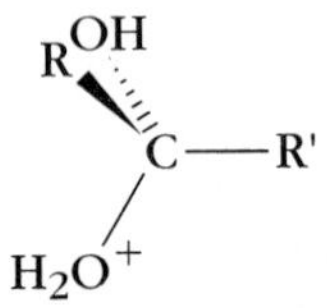 R–C(OH)(H_2O^+)–R'	Solvent (H_2O)	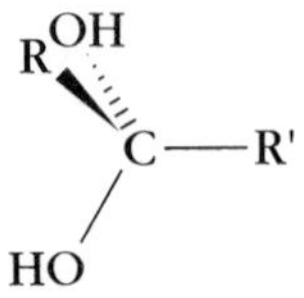 R–C(OH)(HO)–R'
H_2O	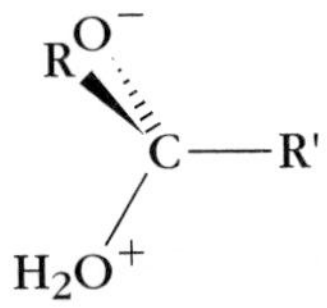 R–C(O^-)(H_2O^+)–R'	Solvent (H_2O)	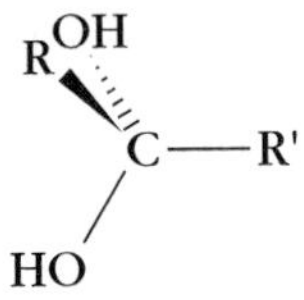 R–C(OH)(HO)–R'
Alcohol (ROH)/RO^-	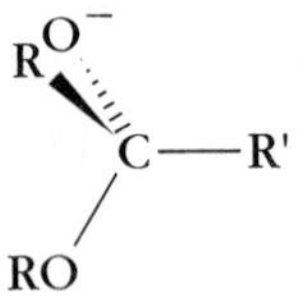 R–C(O^-)(RO)–R'	Solvent (ROH)	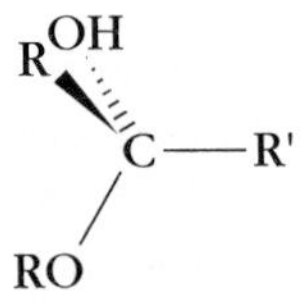 R–C(OH)(RO)–R'

Table 14.6		Addition Reactions of Aldehydes and Ketones in Which the Initial Product Reacts Further to Form the Observed Product			
Reagent	Initial Product		Second Formed Intermediate	Final Product	Reference
RNH$_2$/Acidic buffer	:OH C NHR Carbinolamine		C=N—H R Iminium ion	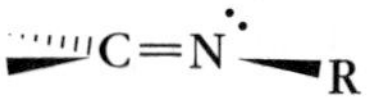 C=N—R Imine	14.13
ROH/Acid solution	:OH R'—C—R O: R" Hemiacetal		[R'—C—R O: R" ↔ R'—C—R O: R"] Oxonium ion	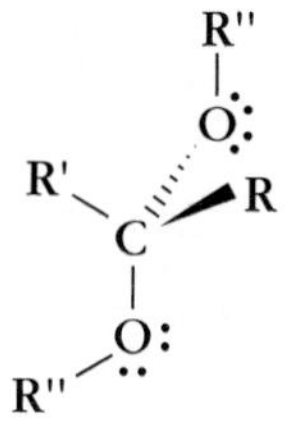 R" O: R'—C—R O: R" Acetal	14.11

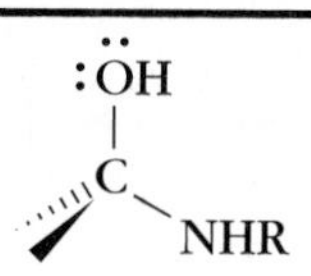

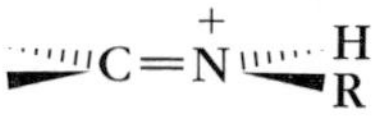

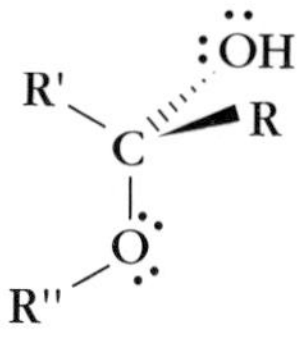

Additional Exercises

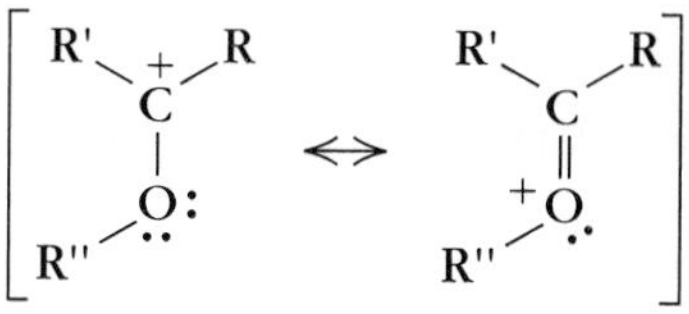

14.32 Define each of the following terms in words, by a structure, or by means of a chemical equation.

(a) Hemiacetal
(b) Nucleophilic addition reaction of a ketone
(c) Imine
(d) Acetal
(e) Semicarbazone

(f) Tetrahedral intermediate
(g) Oxime
(h) Cyanohydrin
(i) 2,4-Dinitrophenylhydrazone

14.33 Give the IUPAC name of each of the following compounds.

(a)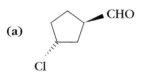

(b) 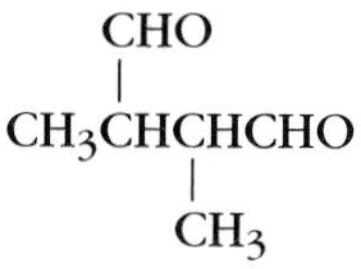 CH$_3$CHCHCHO with CHO and CH$_3$ substituents

(c) 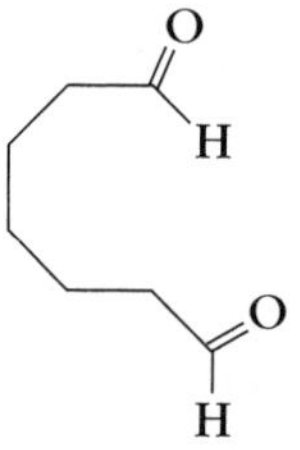 structure with two CHO groups

(d) 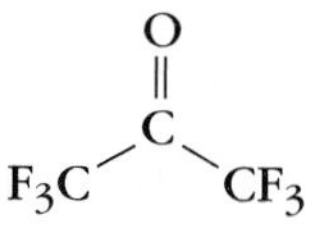 F$_3$C—C(=O)—CF$_3$

(e) cyclohexanone structure with O, O and CH$_3$

(f) Cl and CH$_3$, H and CHO on C=C

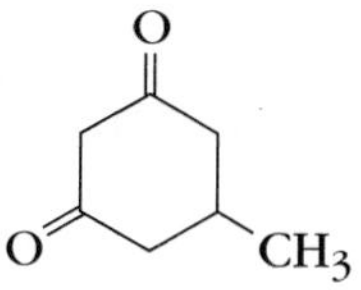

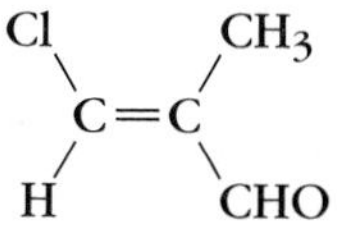

14.34 Write the structure of the compound corresponding to each of the following names.

(a) 1,3-Cyclopentanedione **(c)** 2-Cyclopentenone **(e)** (*Z*)-2-Octenal
(b) Diisopropyl ketone **(d)** 2,3-Dichloro-4-ethylheptanedial **(f)** 3-Hydroxy-4-pentenal

14.35 Replace the question marks by the proper starting materials, reagents, or products in each of the following equations.

(a) ?

(b)

(c) ? $\xrightarrow[\text{(2) } H_3O^+]{\text{(1) } Li(t\text{-}OC_4H_9)_3AlH}$ $CH_3CH_2CH_2CHO$

(d) ?

(e) Cyclopropanecarbaldehyde $\xrightarrow[\text{(2) } H_3O^+]{\text{(1) } CH_3MgI}$?

(f)

(g)

(h) Acetone $\underset{H_2SO_4}{\overset{H_2O}{\rightleftharpoons}}$?

(i) ? $\xrightarrow{Ag(NH_3)_2^+}$ $CH_3CH_2CH_2COO^-$

(j)

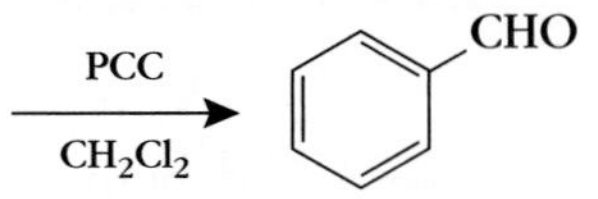

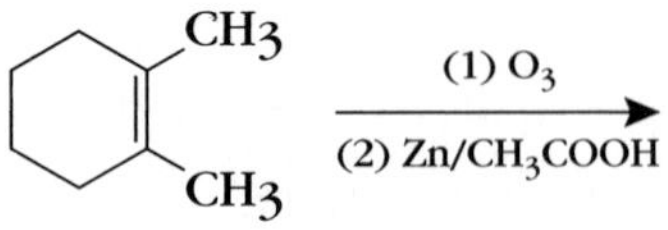

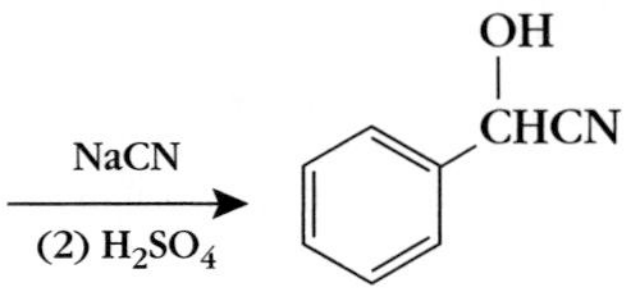

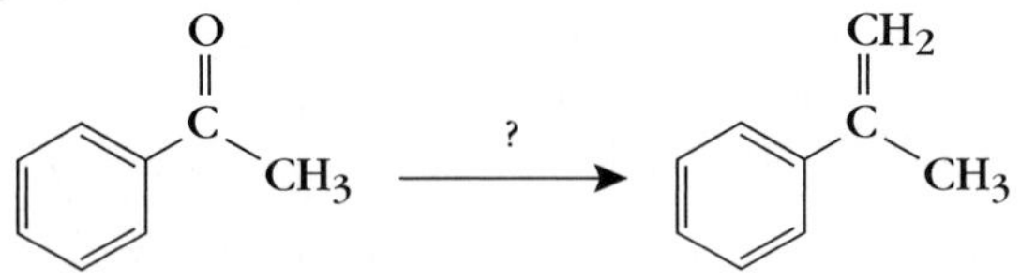

14.36 Write the equation and the structure of the product (if any) formed by the reaction of pentanal with each of the following reagents.

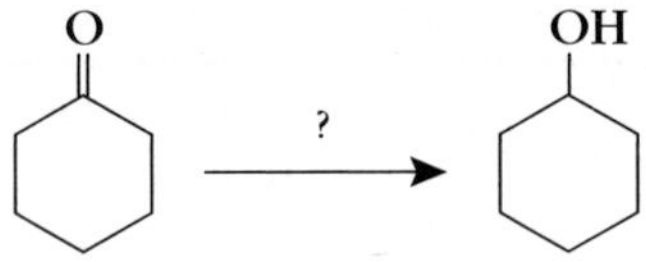

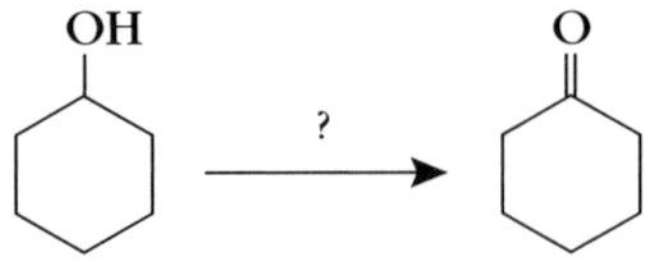

(a) Tollen's reagent
(b) Ethanol, dry HCl
(c) NaCN in dilute aqueous HCl
(d) Hydroxylamine in a weakly acidic buffer solution
(e) $LiAlH_4$ followed by H_3O^+
(f) Methylmagnesium iodide followed by H_3O^+
(g) $CrO_3/H_2SO_4/H_2O$
(h) $HC{\equiv}C^-Na^+$ followed by H_3O^+
(i) 2,4-Dinitrophenylhydrazine in a weakly acidic buffer solution
(j) $(C_6H_5)_3\overset{+}{P}{-}\overset{-}{C}HCH_3$

14.37 Repeat exercise 14.36 using 3-pentanone instead of pentanal.

14.38 Specify the reagents needed to accomplish the following sequence of reactions.

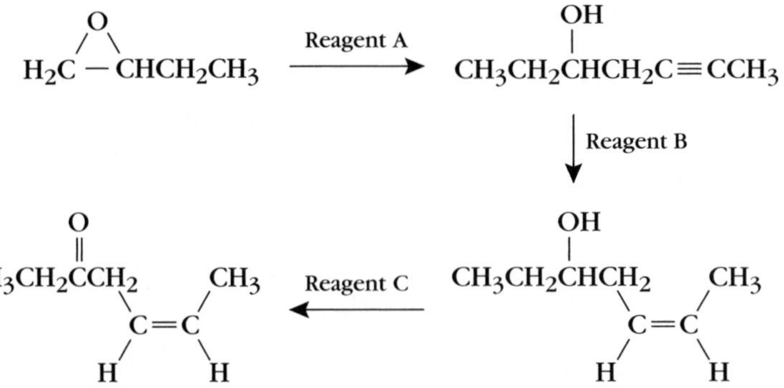

14.39 Write the structure of the product formed in each of the following reactions.

(a) 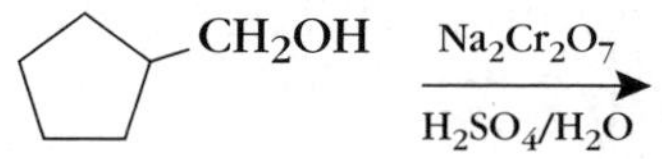 cyclopentyl-CH$_2$OH $\xrightarrow[\text{H}_2\text{SO}_4/\text{H}_2\text{O}]{\text{Na}_2\text{Cr}_2\text{O}_7}$

(b) 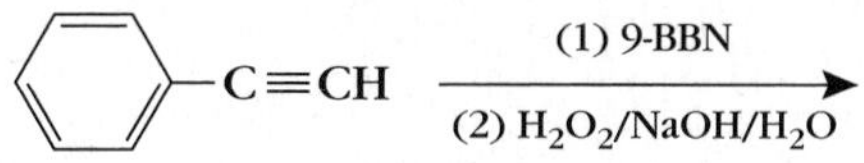 phenyl-C≡CH $\xrightarrow[\text{(2) H}_2\text{O}_2/\text{NaOH/H}_2\text{O}]{\text{(1) 9-BBN}}$

(c) 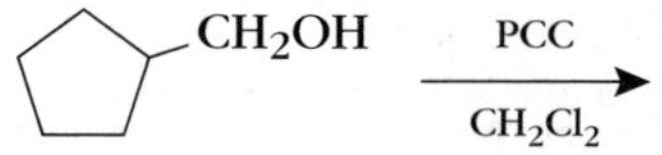 cyclopentyl-CH$_2$OH $\xrightarrow[\text{CH}_2\text{Cl}_2]{\text{PCC}}$

(d) 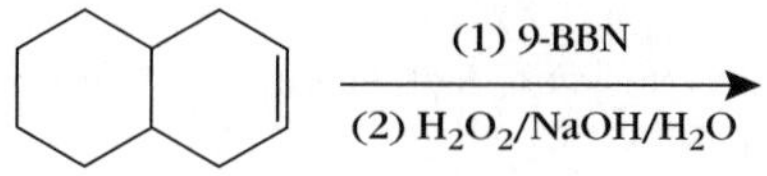 decalin-ene $\xrightarrow[\text{(2) H}_2\text{O}_2/\text{NaOH/H}_2\text{O}]{\text{(1) 9-BBN}}$

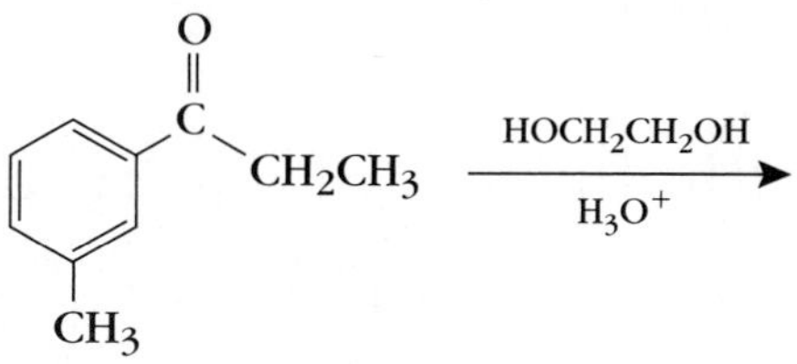

(e) 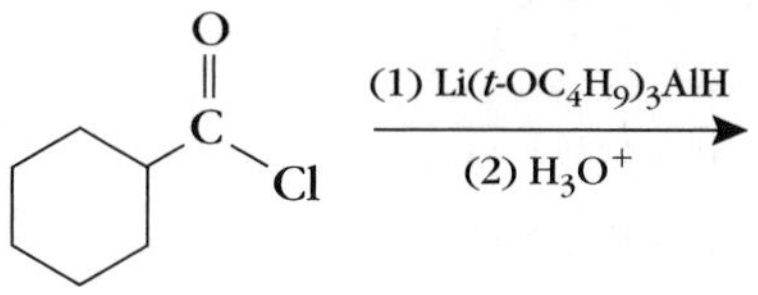 cyclohexyl-C(=O)-Cl $\xrightarrow[\text{(2) H}_3\text{O}^+]{\text{(1) Li(}t\text{-OC}_4\text{H}_9\text{)}_3\text{AlH}}$

(f) 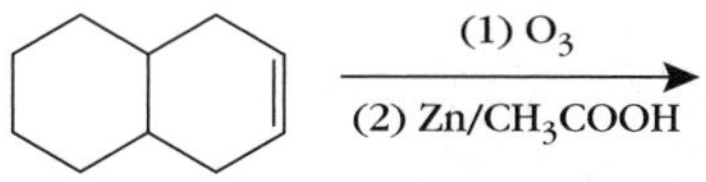 decalin-ene $\xrightarrow[\text{(2) Zn/CH}_3\text{COOH}]{\text{(1) O}_3}$

(g) aryl ketone with CH$_2$CH$_3$ and CH$_3$ $\xrightarrow[\text{H}_3\text{O}^+]{\text{HOCH}_2\text{CH}_2\text{OH}}$

(h) CH$_3$CH$_2$-C(=O)-Cl $\xrightarrow{\text{(CH}_3\text{)}_2\text{CuLi}}$

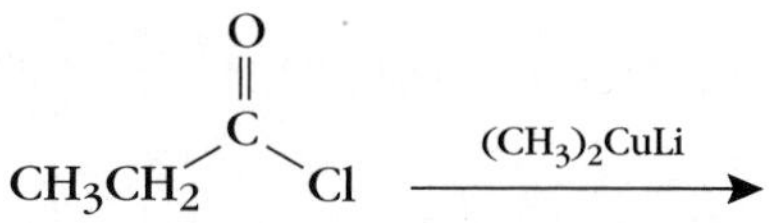

(i) 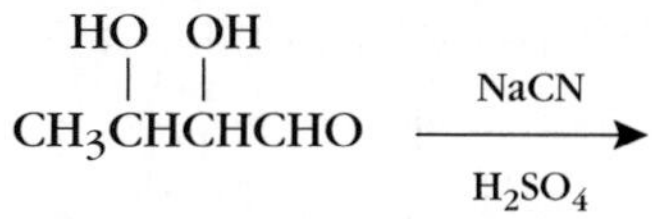 CH$_3$CHCHCHCHO with HO OH $\xrightarrow[\text{H}_2\text{SO}_4]{\text{NaCN}}$

(j) 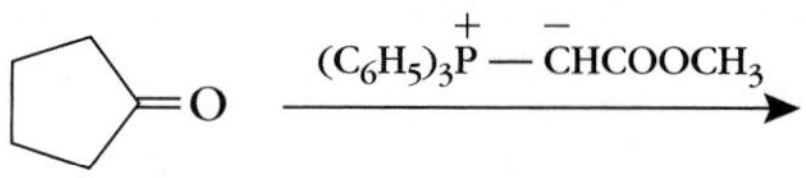 cyclopentylidene=O $\xrightarrow{\text{(C}_6\text{H}_5\text{)}_3\overset{+}{\text{P}} - \overset{-}{\text{C}}\text{HCOOCH}_3}$

14.40 Write the structures of compounds A, B, and C.

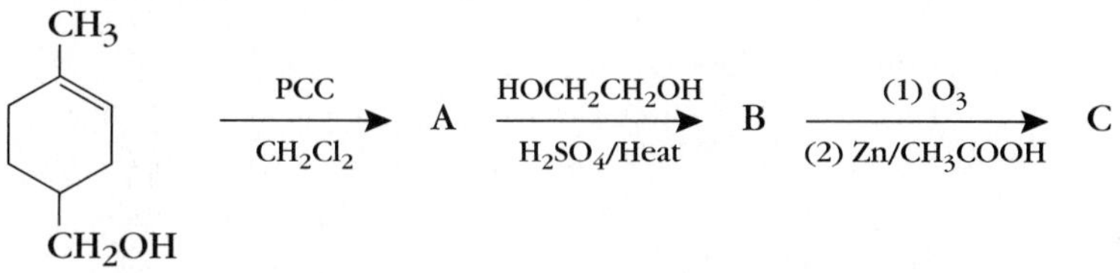

cyclohexene with CH$_3$ and CH$_2$OH $\xrightarrow[\text{CH}_2\text{Cl}_2]{\text{PCC}}$ A $\xrightarrow[\text{H}_2\text{SO}_4/\text{Heat}]{\text{HOCH}_2\text{CH}_2\text{OH}}$ B $\xrightarrow[\text{(2) Zn/CH}_3\text{COOH}]{\text{(1) O}_3}$ C

14.41 Propose a series of reactions to carry out the following synthesis.

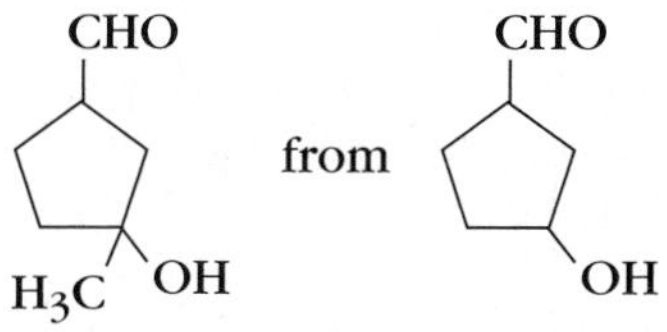

cyclopentane with CHO, H$_3$C, OH **from** cyclopentane with CHO, OH

14.42 Propose a synthesis for each of the following compounds, starting with butanal and any other organic compound and inorganic reagent.

(a) 1,3-Heptadiene
(b) 3-Hexanol
(c) 1-Pentanol
(d) CH$_3$CH$_2$CH$_2$COOH
(e) CH$_3$CH$_2$CH$_2$CH=CHCOOC$_2$H$_5$

(f) 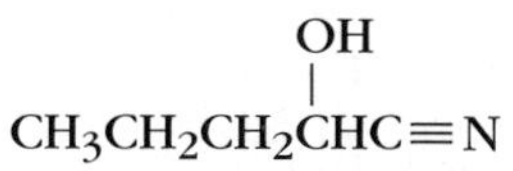 CH$_3$CH$_2$CH$_2$CHC≡N with OH

(g) 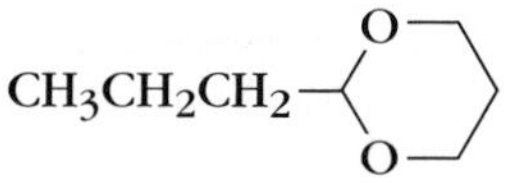 CH$_3$CH$_2$CH$_2$- dioxane ring

14.43   Methanethiol (CH_3SH) is the sulfur analog of methanol. Propanal reacts with methanethiol in the presence of a Lewis acid such as BF_3 to form thioacetals in high yield.

$$CH_3CH_2CHO \quad + \quad 2\ CH_3SH \quad \xrightarrow{\ BF_3\ } \quad CH_3CH_2\overset{\displaystyle SCH_3}{\underset{\displaystyle SCH_3}{\overset{|}{\underset{|}{CH}}}} \quad + \quad H_2O$$

A thioacetal

Based on your knowledge of the acid-catalyzed mechanism of the formation of acetals, propose a mechanism for this reaction.

14.44   What spectroscopic technique or techniques would you use to distinguish between each of the following pairs of compounds? Indicate clearly the differences you expect to find.

(a)   [structure of a ketone]   and   [structure of a ketone]   **(c)** $CH_3(CH_2)_3COOH$ and $CH_3\overset{\displaystyle OH}{\overset{|}{CH}}CH_2CH_2CHO$

(b)   [cyclohexanone structure]   and   [HO-substituted cyclohexene structure]

14.45   Propose a mechanism for the formation of an oxime by the reaction of acetone and hydroxylamine hydrochloride in a weakly acidic buffer solution.

14.46   When acetaldehyde is dissolved in neutral ethanol, the only product formed is the hemiacetal. No acetal is formed. Propose a mechanism for the formation of the hemiacetal and explain why the acetal is not formed.

14.47   When placed in a large excess of acetone, hydrazine (H_2NNH_2) slowly reacts to form the following azine.

$$(CH_3)_2C{=}N{-}N{=}C(CH_3)_2$$

Azine

Propose a mechanism for the formation of this product.

14.48   Write the mechanism for the acid-catalyzed hydrolysis of the imine formed by the reaction of acetaldehyde and ethanamine ($CH_3CH_2NH_2$).

14.49   The natural abundance of the nonradioactive oxygen isotope ^{18}O in acetone and water is very small. When acetone is dissolved in water that contains 80% ^{18}O ($H_2{}^{18}O$), it slowly begins to incorporate ^{18}O in its carbonyl oxygen atom. The incorporation of ^{18}O is catalyzed by strong acids and by strong bases. Write a mechanism for the incorporation of ^{18}O in neutral aqueous solution, in acidic solution, and in basic solution.

14.50   In aqueous acid solution, vinyl methyl ether is hydrolyzed to acetaldehyde and methanol. When vinyl methyl ether containing an excess of ^{18}O is hydrolyzed in an acid solution, the excess ^{18}O is found exclusively in methanol.

$$CH_2{=}CH^{18}OCH_3 \quad \xrightarrow{\ H_3O^+\ } \quad CH_3CHO \quad + \quad CH_3{}^{18}OH$$

Propose a mechanism that accounts for this observation.

14.51 Compound H, C_9H_{10}, is prepared by a Wittig reaction. Write the structure of compound H from its ^{1}H NMR spectra given below, and write the equation for the preparation of compound H by a Wittig reaction.

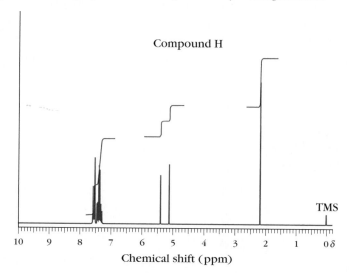

14.52 Write the structural formula of each of the following compounds based on their molecular formulas and the spectral information provided.

  (a) C_4H_7ClO IR 1722 cm^{-1}. ^{13}C: δ 203.2; δ 59.1; δ 25.7; δ 20.1. ^{1}H NMR: δ 1.58 d (3H); δ 2.30 s (3H); δ 4.28 q (1H).
  (b) $C_5H_8O_2$ IR 1715 cm^{-1}. ^{13}C: δ 199.8; δ 197.5; δ 29.2; δ 23.7; δ 6.9. ^{1}H NMR: δ 1.10 t (3H); δ 2.23 s (3H); δ 2.72 q (2H).
  (c) $C_3H_6O_2$ IR 1720 cm^{-1}; 2980 cm^{-1}. ^{13}C: δ 161; δ 59.9; δ 14.2. ^{1}H NMR: δ 1.30 t (3H); δ 4.22 q (2H); δ 8.04 s (1H).
  (d) $C_6H_{12}O_2$ IR 1710 cm^{-1}; 3250 cm^{-1}. ^{13}C: δ 210.7; δ 69.5; δ 53.9; δ 31.8; δ 29.3. ^{1}H NMR: δ 1.25 s (6H); δ 2.18 s (3H); δ 2.60 s (2H).

14.53 NaBH$_4$ reduction of a racemic sample of 3-phenyl-2-butanone forms a mixture of four stereoisomeric alcohols.

3-Phenyl-2-butanone

  (a) Write unambiguous stereochemical representations of the two stereoisomers formed by addition to the Re face of the C=O bond and give each its correct IUPAC name.
  (b) Write unambiguous stereochemical representations of the two stereoisomers formed by addition to the Si face of the C=O bond and give each its correct IUPAC name.
  (c) Which pair of the four stereoisomeric alcohols MUST be formed in equal amounts in this reduction?
  (d) Which pair of the four stereoisomeric alcohols MAY be formed in *different amounts* in this reduction?

14.54 Compounds I, J, K, and L are isomers of molecular formula $C_6H_{12}O$. The IR spectra of all three compounds have an absorption band in the region of 1700 to 1730 cm^{-1}. The ^{1}H NMR spectra of the four isomers are given below. Assign a structure to each of the compounds and indicate which hydrogen atoms give rise to which peaks in each ^{1}H NMR spectrum.

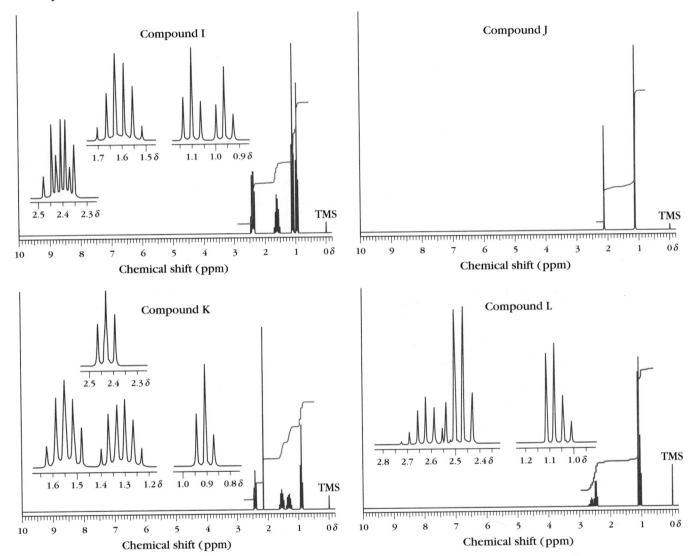

14.55 Write the structural formula of each of the following compounds whose molecular formulas, ^{1}H NMR spectra, and other spectral data are given below.

(a) C_4H_8O; IR: 1710 cm^{-1}

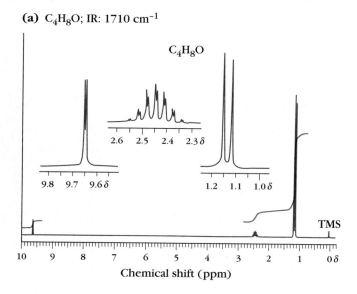

(b) $C_5H_8O_2$

(c) $C_4H_{10}O_2$

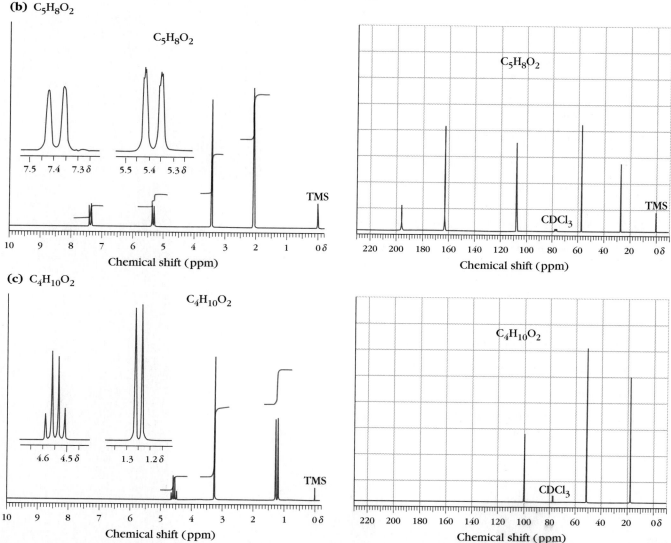

CHAPTER

CARBOXYLIC ACIDS, their SALTS, and their ESTERS

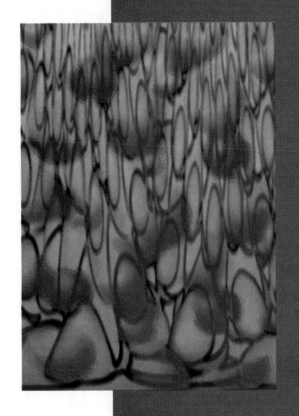

Carboxylic acids, their salts, and their esters are a group of compounds whose chemistry is so closely linked in nature that it is logical to study them together.

Recall that in Section 2.14, G, we defined compounds that contain one or more **carboxyl groups** as **carboxylic acids.**

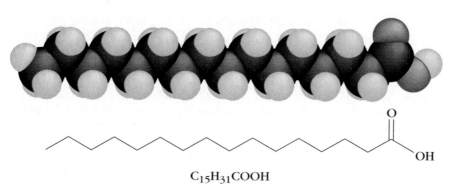

A carboxyl group   Condensed structural   Two equivalent ways of
                 formula of the   writing the structure of
                 carboxyl group   formic acid

> ■ The carboxyl group is sometimes abbreviated as —CO_2H, although this abbreviation is not recommended by IUPAC.

Many simple carboxylic acids are found in nature. Formic acid, HCOOH, for instance, was first isolated from ants and is partly responsible for the irritating sting of red ants. Acetic acid, which makes up 4% to 5% of vinegar, gives vinegar its tangy taste. Butanoic acid, $CH_3CH_2CH_2COOH$, was first isolated from rancid butter and is responsible for its sharp odor. Hexanoic acid, $CH_3(CH_2)_4COOH$, is responsible for the strong odors of goats and dirty socks.

Carboxylic acids with long, unbranched chains of carbon atoms are called **fatty acids** because they are formed from fats by a hydrolysis reaction called **saponification.** Palmitic acid is an example of a fatty acid.

> ■ We will examine saponification reactions in detail in Section 15.12.

$C_{15}H_{31}COOH$

Common name: Palmitic acid (because it can be obtained from palm oil)
IUPAC name: Hexadecanoic acid

The functional group of a carboxylic acid salt is the carboxylate anion.

Carboxylate   Condensed structural   Two equivalent ways of
anion   formula for   writing the structure of
               the carboxylate anion   sodium acetate

Sodium acetate, which is formed by the reaction of acetic acid and sodium hydroxide, is a simple carboxylic acid salt.

Carboxylic acid esters are functional group derivatives of carboxylic acids because they are hydrolyzed to carboxylic acids under acidic or basic conditions, as we will discuss in Section 15.12. In esters, the hydroxy of the carboxyl group is replaced by an alkoxy group.

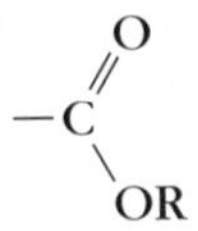

$$
\begin{array}{ccc}
\underset{\text{An ester}}{\underset{\text{group}}{-\overset{\displaystyle O}{\overset{\|}{C}}-OR}} & \underset{\substack{\text{Condensed structural}\\\text{formula of}\\\text{the ester group}}}{-COOR} & \underset{\substack{\text{Two equivalent ways of writing}\\\text{the structure of ethyl acetate}}}{H_3C\overset{\displaystyle O}{\overset{\|}{C}}OCH_2CH_3 \qquad CH_3COOCH_2CH_3}
\end{array}
$$

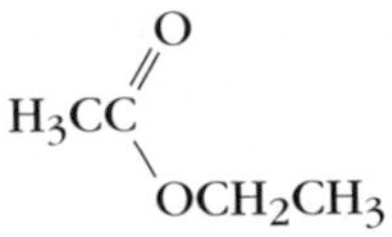

The distinctive aromas of many fruits and flowers are due to the presence of simple, volatile esters as well as small amounts of alcohols, aldehydes, and ketones. Many pure esters have the characteristic aroma of a particular fruit or flower. The following are a few examples:

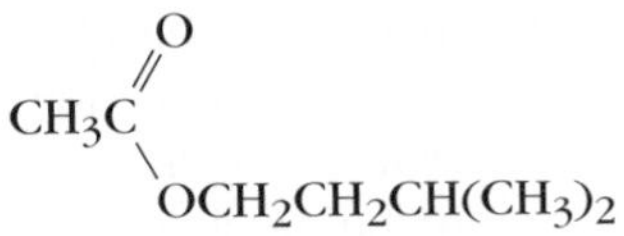

i-Amyl acetate
Banana-like odor

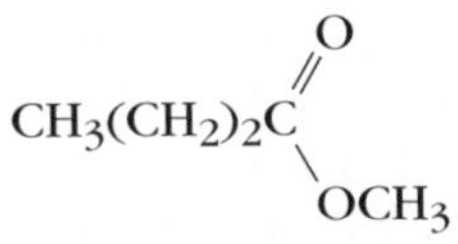

Methyl butyrate
Apple-like odor

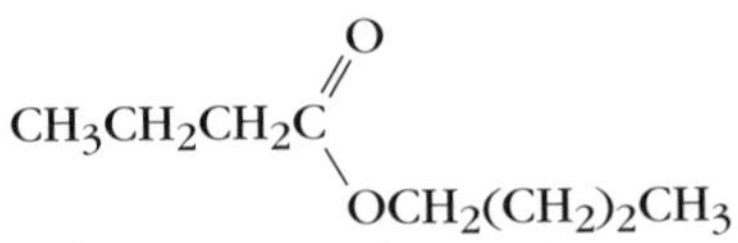

Butyl butyrate
Pineapple-like odor

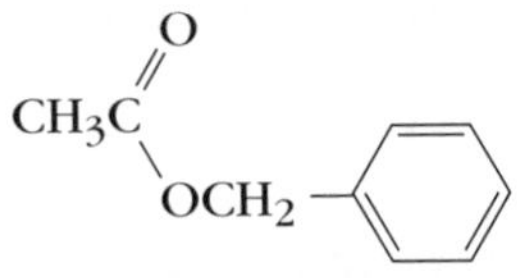

Benzyl acetate
Jasmine-like odor

We will see later in this chapter that many esters in nature are fats and oils.

Let's begin our study of carboxylic acids, their salts, and esters by learning how to name them.

15.1   Naming Carboxylic Acids, Salts, and Esters

Many of the simpler carboxylic acids were isolated and characterized long before the invention of the IUPAC system of naming organic compounds. As a result, they have well-entrenched common names. These names are so well entrenched, in fact, that IUPAC rules allow the use of many of them, giving rise to two systems for naming carboxylic acids.

Carboxylic acids are systematically named by identifying the longest chain of carbon atoms that contains the carboxyl group. This establishes the parent alkane name. The terminal -*e* of the parent alkane name is then replaced by the suffix -*oic acid* to identify the compound as a carboxylic acid. Carboxyl groups, like aldehyde groups (Section 14.1), are always at the end of a chain, so the carbonyl carbon atom is always assigned number 1.

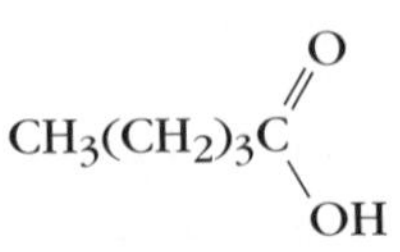

Pentanoic acid

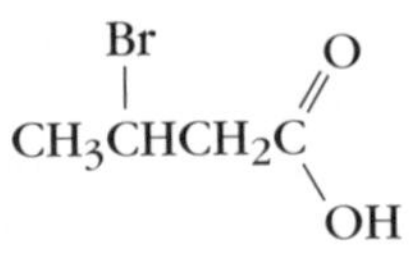

3-Bromobutanoic acid

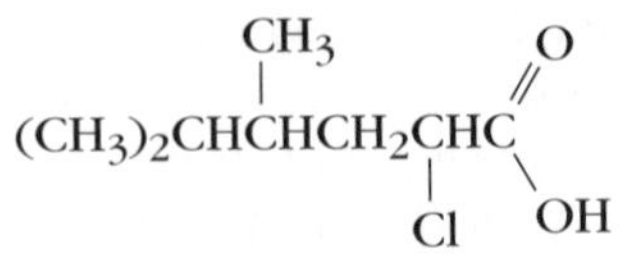

2-Chloro-4,5-dimethyl-
hexanoic acid

When naming dicarboxylic acids, the final -*e* of the parent name is retained.

$$HOOCCH_2CH_2CH_2CH_2COOH$$

1,6-Hexanedioic acid

Carboxylic acids whose carboxyl group is attached to a ring are named by adding the suffix -*carboxylic acid* to the name of the ring. This is similar to the method used in Section 14.1 to name aldehydes whose carbonyl carbon atom is attached to a ring.

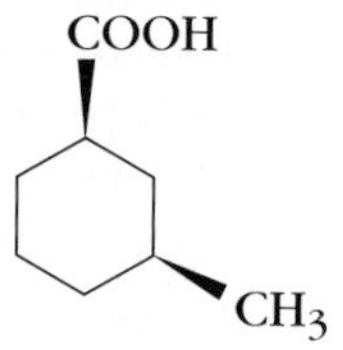

cis-3-Methylcyclohexanecarboxylic acid

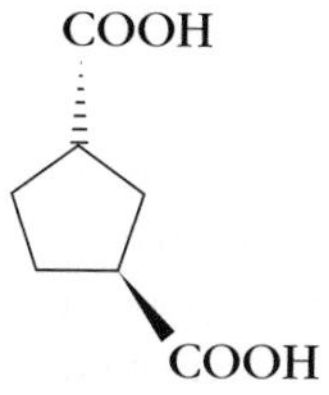

trans-1,3-Cyclopentanedicarboxylic acid

The common and systematic names of a number of the simpler mono- and dicarboxylic acids are listed in Table 15.1. We will use systematic names of carboxylic acids in this book, with the exception of the common names printed in bold type in Table 15.1.

The common name of 2-bromobutanoic acid is α-bromobutyric acid, not 2-bromobutyric acid, because the positions of substituents are designated in common names by Greek letters, not numbers.

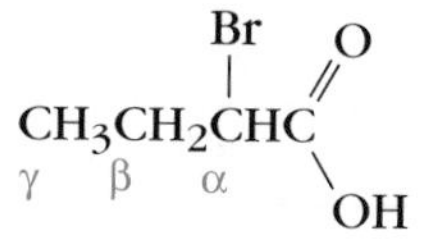

IUPAC name: 2-Bromobutanoic acid
Common name: α-Bromobutyric acid (not 2-bromobutyric acid)

A similar method was used in the common names of substituted aldehydes and ketones (Section 14.1).

Exercise 15.1 Give a systematic name to each of the following compounds.

(a) $(CH_3)_3CCOOH$

$$CH_3$$
(b) 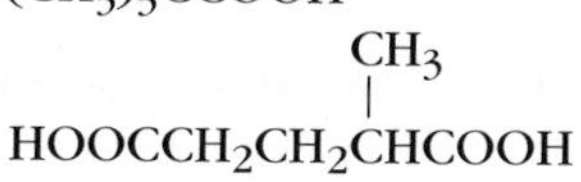 $HOOCCH_2CH_2CHCOOH$

(c)

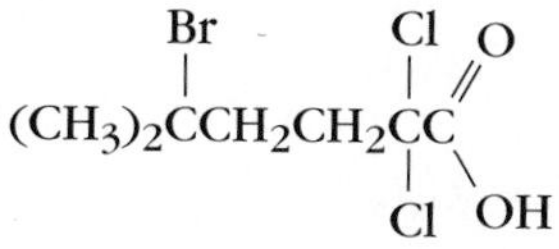

(d)

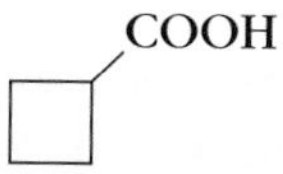

(e) $(CH_3)_2C{=}CHCOOH$

(f) $CH_3C{\equiv}CCOOH$

Table 15.1	Names and Structures of Simple Carboxylic Acids	
Structure	**Common Name**	**Systematic Name**
HCOOH	**Formic acid**	Methanoic acid
CH₃COOH	**Acetic acid**	Ethanoic acid
CH₃CH₂COOH	Propionic acid	Propanoic acid
CH₃CH₂CH₂COOH	Butyric acid	Butanoic acid
(CH₃)₂CHCOOH	Isobutyric acid	2-Methylpropanoic acid
CH₃(CH₂)₃COOH	Valeric acid	Pentanoic acid
(CH₃)₂CHCH₂COOH	Isovaleric acid	3-Methylbutanoic acid
(CH₃)₃CCOOH	Pivalic acid	2,2-Dimethylpropanoic acid
CH₂=CHCOOH	**Acrylic acid**	2-Propenoic acid
CH₃CH=CHCOOH	**Crotonic acid**	2-Butenoic acid
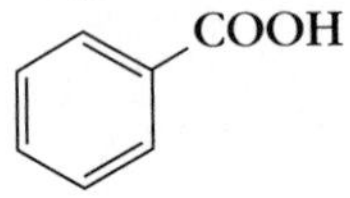 ⬡—COOH	**Benzoic acid**	Benzoic acid
HOOCCOOH	**Oxalic acid**	Ethanedioic acid
HOOCCH₂COOH	**Malonic acid**	Propanedioic acid
HOOCCH₂CH₂COOH	**Succinic acid**	Butanedioic acid
HOOCCH₂CH₂CH₂COOH	**Glutaric acid**	Pentanedioic acid
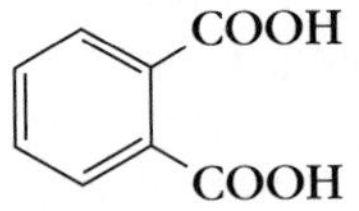 ⬡(COOH)(COOH)	**Phthalic acid**	1,2-Benzenedicarboxylic acid
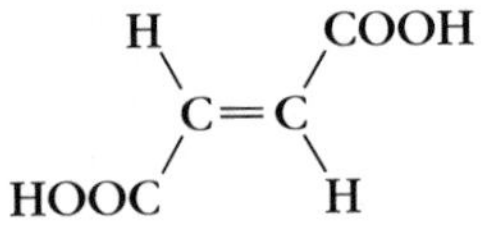 (H)(COOH)C=C(HOOC)(H)	**Fumaric acid**	(E)-2-Butenedioic acid
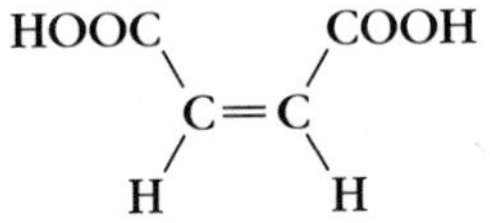 (HOOC)(COOH)C=C(H)(H)	**Maleic acid**	(Z)-2-Butenedioic acid

■ *The structure of acetic acid is sometimes abbreviated as HOAc.*

Exercise 15.2	Write the structure of each of the following compounds.

(a) α-Bromoglutaric acid
(b) *cis*-1,3-Cyclohexanedicarboxylic acid
(c) 4-Methylhexanoic acid
(d) (Z)-4-Chloro-2-butenoic acid
(e) α,α-Diiodoacetic acid

The names of salts and esters of carboxylic acids consist of two words, such as sodium acetate (a salt) or ethyl acetate (an ester).

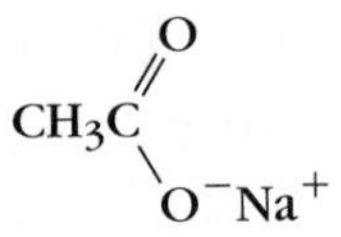

$$CH_3C \overset{O}{\underset{O^- Na^+}{\big\|}} \qquad CH_3C \overset{O}{\underset{OCH_2CH_3}{\big\|}}$$

Sodium acetate 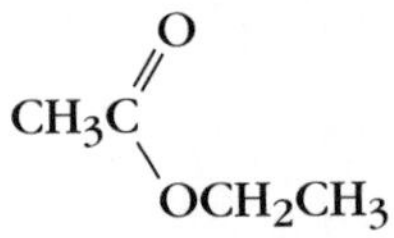   Ethyl acetate

The first word identifies the name of the cation in salts or the alkyl or aryl group bonded to the carboxyl oxygen atom in esters. The second word identifies the carboxylic acid from which the carboxylate group is derived by replacing the *-ic acid* ending of its name with *-ate*.

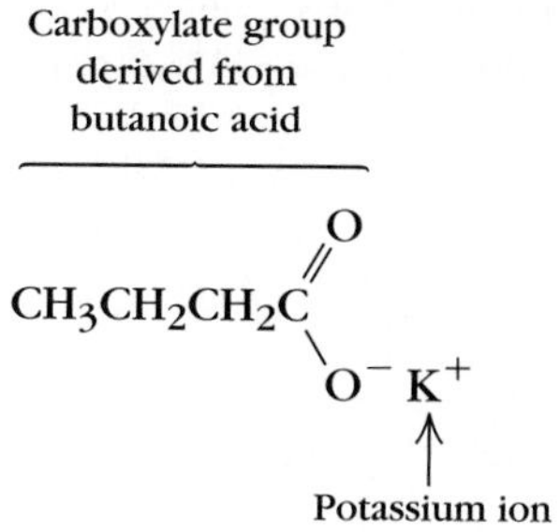

Carboxylate group derived from butanoic acid

Carboxylate group derived from propanoic acid

$$CH_3CH_2CH_2C \overset{\overset{\displaystyle O}{\|}}{\underset{\underset{\displaystyle \uparrow}{O^- K^+}}{}}$$

Potassium ion

Potassium butanoate

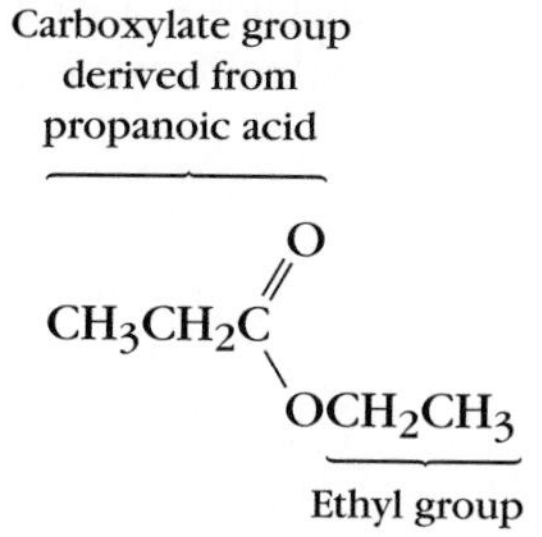

$$CH_3CH_2C \overset{\overset{\displaystyle O}{\|}}{\underset{\underset{\displaystyle \text{Ethyl group}}{OCH_2CH_3}}{}}$$

Ethyl propanoate

The following examples illustrate the names of several other salts and esters.

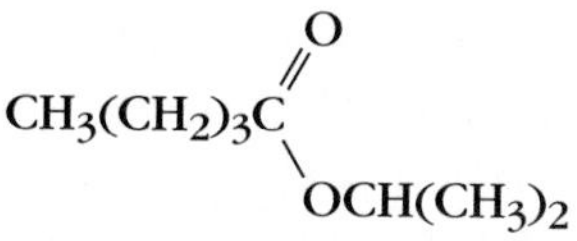

$$CH_3(CH_2)_3C \overset{\overset{\displaystyle O}{\|}}{\underset{OCH(CH_3)_2}{}}$$

Isopropyl pentanoate

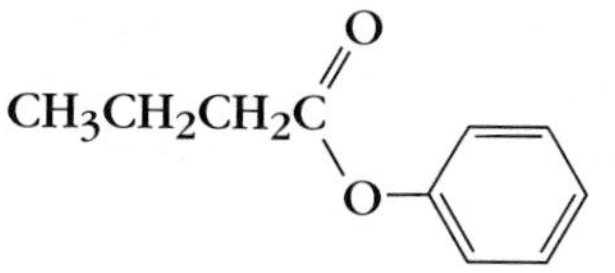

$$CH_3CH_2CH_2C \overset{\overset{\displaystyle O}{\|}}{\underset{O}{}}\bigcirc$$

Phenyl butanoate

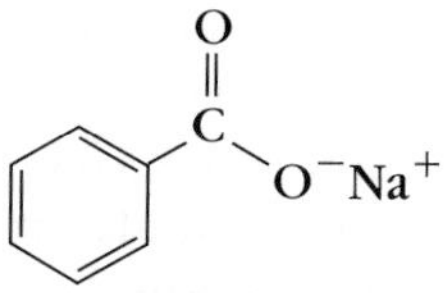

$$\bigcirc \overset{\overset{\displaystyle O}{\|}}{C} O^- Na^+$$

Sodium benzoate

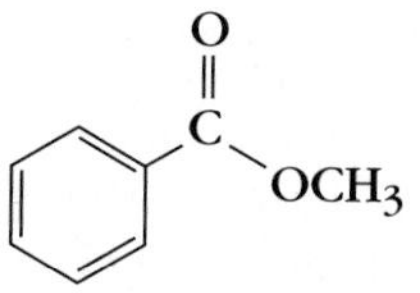

$$\bigcirc \overset{\overset{\displaystyle O}{\|}}{C} OCH_3$$

Methyl benzoate

$$C_2H_5O \overset{\overset{\displaystyle O}{\|}}{C}CH_2CH_2\overset{\overset{\displaystyle O}{\|}}{C} OC_2H_5$$

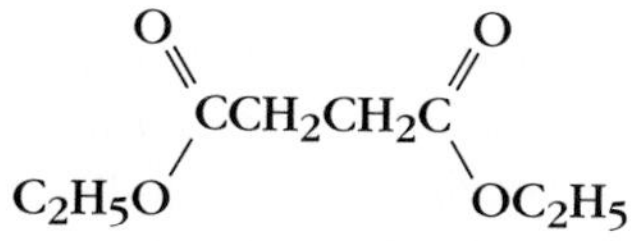

Diethyl succinate

$$CH_3CH_2COO^- K^+$$

Potassium propanoate

Exercise 15.3 Give a systematic name to each of the following compounds.

(a) $CH_3CH_2CH_2COOCH(CH_3)_2$

(b) $(CH_3)_2CHCHCOOCH_2(CH_2)_4CH_3$ with Cl on the CH

(c) 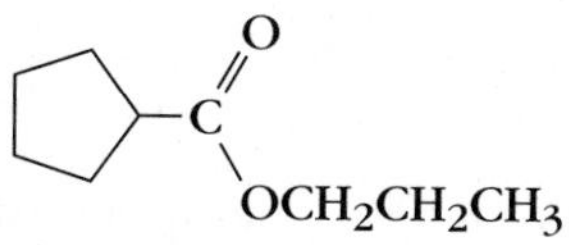 cyclopentane $-C \overset{\overset{\displaystyle O}{\|}}{} OCH_2CH_2CH_3$

(d) $(CH_3)_2CHCOO^- Na^+$

(e) $CH_2COOCH(CH_3)_2$
    |
    $CH_2COOCH(CH_3)_2$

(f) $CH_2{=}CHCOO^- K^+$

Exercise 15.4 Write the structure of each of the following compounds.

(a) Methyl (*Z*)-3-heptenoate
(b) Isobutyl cyclobutanecarboxylate
(c) Diethyl maleate

(d) Butyl 2-phenylpropanoate
(e) Calcium propanoate

15.2   Structure of Carboxylic Acids and Their Esters

The structures of carboxylic acids and esters resemble the structures of aldehydes and ketones because the carbonyl group and the atoms attached to it all lie in a plane. The data in Figure 15.1 demonstrate, moreover, that bond angles and bond lengths are remarkably constant throughout a series of analogous carbonyl compounds.

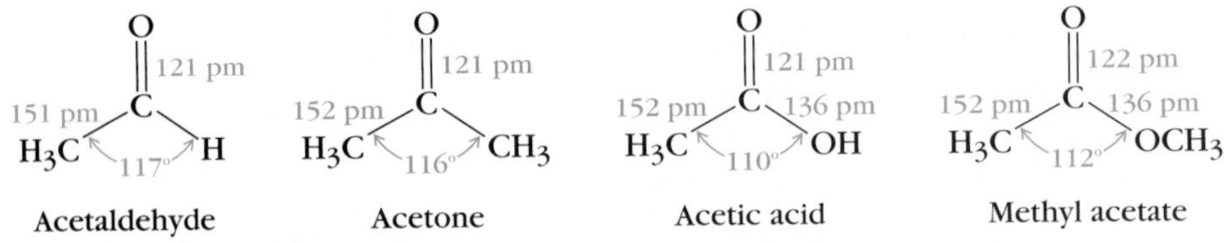

Figure 15.1
Comparison of the bond lengths and bond angles in analogous aldehydes, ketones, carboxylic acids, and esters.

Acetaldehyde   Acetone   Acetic acid   Methyl acetate

The π electron structure of the carboxyl group can be described by the simplified LCAO method used to describe the π bonds of the carbonyl group of aldehydes and ketones (Section 14.2). Recall that the σ-bond framework of a carbonyl group is described by localized valence bond orbitals, and the LCAO method is used only to describe the π bonds. We use an sp^2 hybrid carbon atom for the carboxyl carbon atom because its planar trigonal geometry most closely resembles the geometry in simple carboxylic acids. The atomic orbitals used in the LCAO method and the shapes of the resulting π molecular orbitals for formic acid are shown in Figure 15.2.

The lowest energy molecular orbital of the carboxyl group in Figure 15.2 shows that there is some π bonding between the carboxyl carbon atom and the —OH oxygen atom. This delocalization of electrons from the p orbital of the oxygen atom of the —OH group to the π bond of the carbonyl group reduces the electron deficiency on the carboxyl carbon atom, compared to the carbon atom of the carbonyl group of aldehydes and ketones. As a result, nucleophilic additions to the carboxyl carbon atom are more difficult than additions to the carbonyl carbon atom of aldehydes and ketones, as we will discuss in more detail in Section 16.7.

A similar picture of the π electron structure of the formic acid carboxyl group is obtained by representing it as a resonance hybrid of the three following resonance structures:

■ *The simplified LCAO method was first described for ethylene in Section 1.13.*

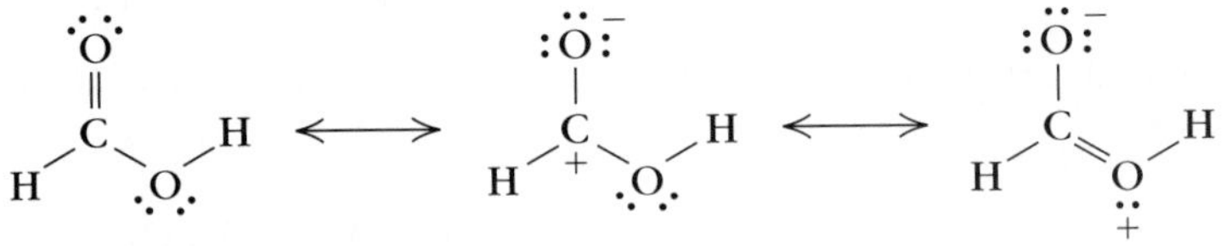

The delocalized π electron structure of carboxylate ions can be represented as either a resonance hybrid or by molecular orbitals constructed by the simplified LCAO method, as we discussed in Sections 1.12 and 1.13. Recall that the resonance hybrid structure of the formate ion is represented by the following resonance structures:

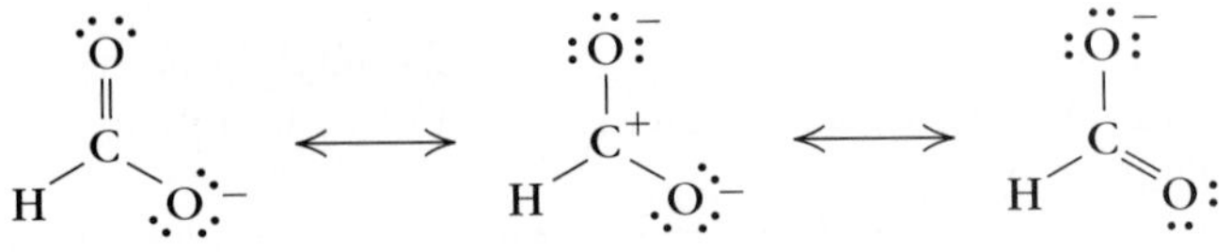

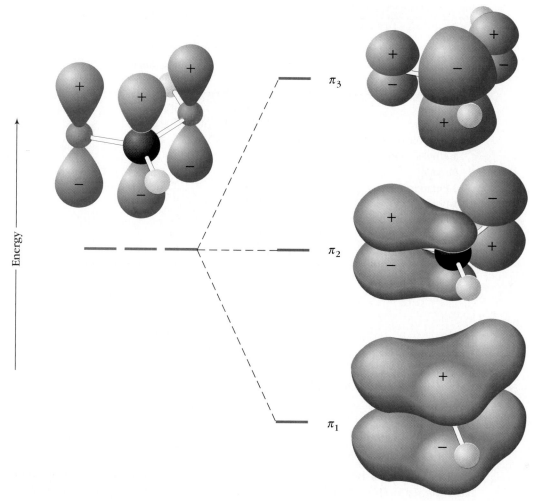

Figure 15.2
The π molecular orbitals of formic acid. The molecular orbitals are constructed by combining the *p* orbital of the *sp²* hybrid carboxyl carbon atom, the *p* orbital of the *sp²* hybrid carboxyl oxygen atom, and a *p* orbital of the oxygen atom of the —OH group.

Both the resonance and LCAO descriptions of the formate ion (Figure 1.25, p. 35) show very well the delocalization of π electrons in the carboxylate ion. For simplicity, we will use only one of the resonance structures to represent the delocalized carboxylate ion in this text.

| Exercise 15.5 | Show how to use orbital overlap to construct the localized valence bond orbitals of the σ-bond framework of formic acid. |

15.3 Physical Properties

The data in Table 15.2 show that the boiling points of carboxylic acids are much higher than those of other organic compounds of similar molecular weights. The boiling points of carboxylic acids are unusually high because carboxylic acids exist largely in a dimeric form. Carboxylic acids dimerize because hydrogen bonds can form between the —OH hydrogen atom of one molecule and the carbonyl oxygen atom of another.

| | Table 15.2 | Comparison of the Boiling Points of Organic Compounds of Similar Molecular Weights | | |

Name	Structure	Molecular Weight	Boiling Point (°C)
Pentane	$CH_3CH_2CH_2CH_2CH_3$	72	36
1-Pentene	$CH_3CH_2CH_2CH{=}CH_2$	70	30
Propanoic acid	CH_3CH_2COOH	74	141
Methyl acetate	CH_3COOCH_3	74	58
Butanal	$CH_3CH_2CH_2CHO$	72	76
2-Butanone	$CH_3C(O)CH_2CH_3$	72	80
1-Butanol	$CH_3CH_2CH_2CH_2OH$	74	118
2-Butanol	$CH_3CH(OH)CH_2CH_3$	74	99
1-Chloropropane	$CH_3CH_2CH_2Cl$	78	47
Ethoxyethane	$CH_3CH_2OCH_2CH_3$	74	35

Acetic acid   Acetic acid dimer

More energy must be added to break these hydrogen bonds, so carboxylic acids have higher boiling points than other organic compounds of similar molecular weight that do not form dimers by hydrogen bonding.

Esters cannot dimerize because the hydrogen atom of the hydroxy group has been replaced by an organic group that cannot form hydrogen bonds with the carbonyl oxygen atom of another ester molecule. As a result, the boiling point of an ester is much lower than that of a carboxylic acid of the same molecular weight. In Table 15.2, for instance, the boiling point of methyl acetate (MW = 74) is 58 °C, whereas the boiling point of propanoic acid (MW = 74) is 141 °C.

The melting and boiling points of carboxylic acid salts are very high because they are ionic compounds. Low molecular weight carboxylic acid salts, such as sodium acetate, are very soluble in water, too. As the size of the hydrocarbon portion of carboxylic acid salts increases, however, their water solubility decreases and they form micelles. Micelles will be discussed in Section 15.14.

Low molecular weight carboxylic acids are very soluble in water because they can form hydrogen bonds with water. As the size of the hydrocarbon portion increases, the solubility of carboxylic acids in water decreases. Acetic acid, for example, is miscible with water, while only 0.003 g of decanoic acid, $C_9H_{19}COOH$, dissolves in 100 mL of water.

Hydrogen bonding between esters and water is poorer than between carboxylic acids and water, so esters are generally less soluble in water than carboxylic acids. Ethyl hexanoate, for example, is insoluble in water, while 1.1 g of hexanoic acid dissolves in 100 mL of water.

15.4 Spectroscopic Properties of Carboxylic Acids and Their Esters

Carboxylic acids have two highly characteristic absorption bands in their infrared spectra, as we discussed in Section 5.5, F. Recall that one absorption band, due to the O—H bond stretching, usually overlaps with the C—H stretching frequency to form a broad, relatively intense absorption band that extends from 3400 to 2400 cm^{-1}. The other is the carbonyl bond stretching frequency, which is found between 1720 and 1680 cm^{-1}.

Esters are easy to distinguish from carboxylic acids because esters lack the O—H bond stretching absorption band. Esters have two characteristic absorption bands. One relatively intense absorption band is due to the carbonyl stretching and is found in the region 1750 to 1730 cm^{-1}. The other is due to the C—O bond stretch, and it is found in the region 1300 to 1050 cm^{-1}. Both of these bands are visible in the IR spectrum of ethyl acetate in Figure 15.3.

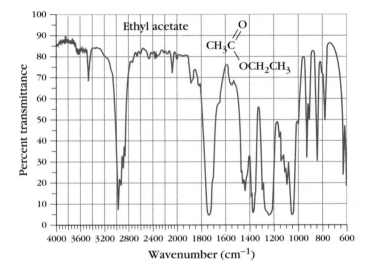

Figure 15.3
The IR spectrum of ethyl acetate. The absorption band due to the C=O stretching is found at about 1750 cm^{-1}. The absorption band due to the C—O bond stretching is found at about 1250 cm^{-1}.

The ^{1}H NMR chemical shifts of α-hydrogens in carboxylic acids and their esters are all found near δ 2.1. This is the same chemical shift region where the α-hydrogens of aldehydes and ketones are found. Thus it is difficult to distinguish among aldehydes, ketones, carboxylic acids, and carboxylic esters by using the chemical shifts of these hydrogens. Sometimes, it is possible to use the chemical shift of the hydrogen of the carboxyl group, —COOH, to identify carboxylic acids. The chemical shift of this hydrogen atom is found in the region δ 10 to 13. The signal is always a singlet, and it can be broad or sharp, depending on the solvent and concentration of the carboxylic acid. As in the case of alcohols, the hydrogen atom of a carboxyl group can be easily exchanged by deuterium when D$_2$O is added to the sample tube, causing the —COOH peak to disappear from the spectrum.

Table 15.3	Comparison of the ^{13}C Chemical Shifts of Carbonyl Carbon Atoms of Representative Aldehydes, Ketones, Carboxylic Acids, and Esters	
Structure	Name	δ
CH_3CHO	Acetaldehyde	200.5
(acetone structure)	Acetone	206.7
CH_3COOH	Acetic acid	176.9
$CH_3COOC_2H_5$	Ethyl acetate	170.7

The ^{13}C NMR chemical shifts of the carboxyl carbon atoms of carboxylic acids and their esters are found in the range δ 165 to 180, as shown in Table 15.3. The data in Table 15.3 show that the ^{13}C chemical shifts of aldehyde and ketone carbonyl carbon atoms are found in the δ 190 to 210 range, which is downfield of the carboxylic acids and esters. Generally speaking, it is easier to distinguish carboxylic acids and their esters from aldehydes and ketones by the ^{13}C chemical shift of their carbonyl carbon atoms than it is to distinguish between specific carboxylic acids and their specific esters. Once a compound has been shown by ^{13}C NMR spectroscopy to be a carboxylic acid or an ester, other techniques are usually required to distinguish between the two.

Exercise 15.6	What spectral technique would you use to distinguish between the following pairs of compounds? Indicate clearly the difference you expect to find.

(a) 2-Butanone and acetone   (d) Benzaldehyde and benzoic acid
(b) Acetone and acetaldehyde   (e) Ethyl acetate and propanal
(c) Acetic acid and acetone   (f) Methyl acetate and acetic acid

Figure 15.4
^1H NMR spectrum for Exercise 15.7.

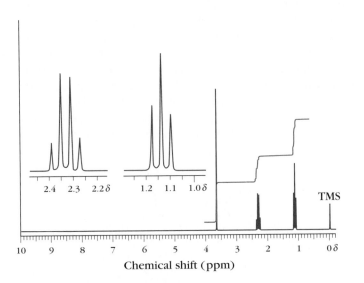

Chemical shift (ppm)

Exercise 15.7   Determine if the compound whose ^{1}H NMR spectrum is given in Figure 15.4 is

$$\underset{H_3C}{\overset{\displaystyle O}{\underset{\displaystyle \Vert}{C}}}\underset{O}{\diagup}CH_2CH_3 \quad \text{or} \quad \underset{CH_3CH_2}{\overset{\displaystyle O}{\underset{\displaystyle \Vert}{C}}}\underset{O}{\diagup}CH_3$$

15.5   Synthesis of Carboxylic Acids by Oxidation Reactions

Oxidation of primary alcohols with a strong oxidizing agent such as any Cr(VI) salt in aqueous H_2SO_4 forms carboxylic acids, as we discussed in Section 11.18.

$$RCH_2OH \xrightarrow{\text{Strong oxidizing agent}} RCOOH$$

Primary alcohol   Carboxylic acid

$$CH_3CH_2CH_2\overset{\displaystyle CH_3}{\overset{\displaystyle |}{C}}HCH_2CH_2CH_2OH \xrightarrow[H_2SO_4/H_2O]{Na_2Cr_2O_7} CH_3CH_2CH_2\overset{\displaystyle CH_3}{\overset{\displaystyle |}{C}}HCH_2CH_2COOH$$

4-Methyl-1-heptanol   4-Methylheptanoic acid
                                                       (78% yield)

Aldehydes, too, are oxidized to carboxylic acids by Cr(VI) salts in H_2SO_4. Oxidation of aldehydes, therefore, is also a method of preparing carboxylic acids.

■ *Recall that an aldehyde is an intermediate product in the oxidation of a primary alcohol to a carboxylic acid.*

$$\underset{H}{\overset{\displaystyle O}{\underset{\displaystyle \diagup}{RC}}} \xrightarrow{\text{Oxidizing agent}} \underset{OH}{\overset{\displaystyle O}{\underset{\displaystyle \diagup}{RC}}}$$

Aldehyde   Carboxylic acid

$$\underset{H}{\overset{\displaystyle O}{\underset{\displaystyle \diagup}{CH_3CH_2CH_2CH_2C}}} \xrightarrow[H_2SO_4/H_2O]{CrO_3} \underset{OH}{\overset{\displaystyle O}{\underset{\displaystyle \diagup}{CH_3CH_2CH_2CH_2C}}}$$

Pentanal   Pentanoic acid
                                        (88% yield)

Oxidation of alkenes is useful synthetically only with either a symmetrical disubstituted alkene or a terminal alkene.

$$RCH{=}CHR \xrightarrow{\text{Strong oxidizing agent}} 2\ RCOOH$$

Symmetrically   Carboxylic acid
disubstituted
alkene

$$CH_3CH_2CH{=}CHCH_2CH_3 \xrightarrow[H_2SO_4/H_2O]{CrO_3} 2\ CH_3CH_2COOH$$

(*E*)- or (*Z*)-3-Hexene   Propanoic acid
                                        (68% yield)

$$RCH{=}CH_2 \xrightarrow{\text{Strong oxidizing agent}} RCOOH \;+\; CO_2 \;+\; H_2O$$

Terminal alkene   Carboxylic
                                                             acid

Strong oxidizing agents oxidize terminal alkenes first to a carboxylic acid and formic acid ($HCOOH$), which is further oxidized to CO_2 and H_2O.

$$CH_3(CH_2)_4CH{=}CH_2 \xrightarrow[\text{H}_2\text{SO}_4/\text{H}_2\text{O}]{\text{K}_2\text{Cr}_2\text{O}_7} CH_3(CH_2)_4COOH \;+\; CO_2 \;+\; H_2O$$

1-Heptene   Hexanoic acid
                                              (82% yield)

Oxidation of unsymmetrical alkenes is less useful synthetically because a mixture of two carboxylic acids is usually formed. Oxidation of alkenes is particularly useful for the synthesis of dicarboxylic acids from cycloalkenes.

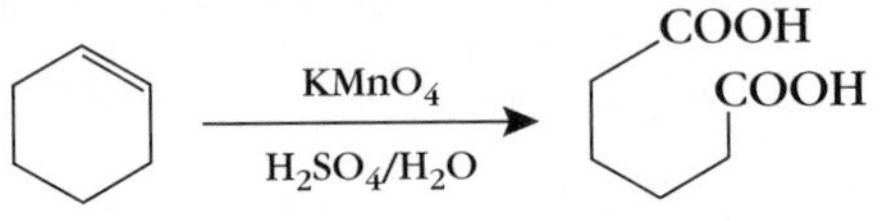

Cyclohexene   Hexanedioic acid
                                        (76% yield)

Oxidation of an alkyl side chain of a benzene ring forms substituted benzoic acids.

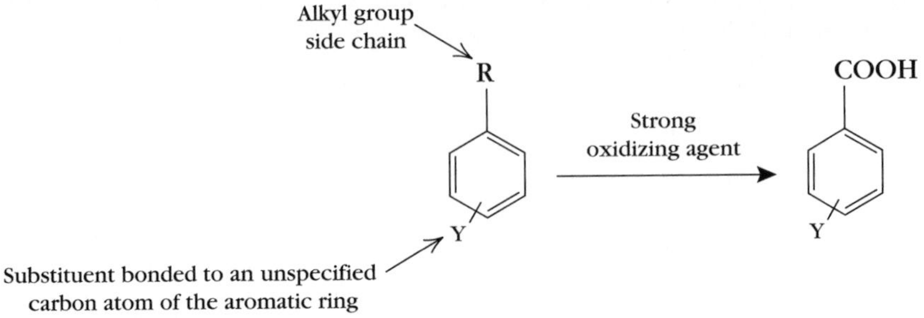

Substituted
alkyl benzene

Substituted
benzoic acid

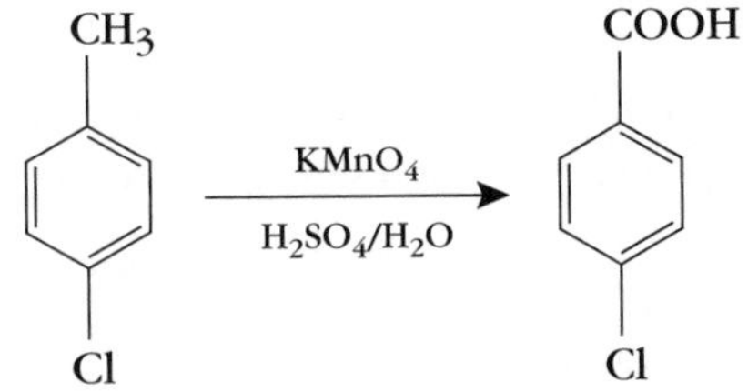

4-Chlorotoluene   4-Chlorobenzoic
                                       acid
                                   (85% yield)

Primary and secondary alkyl groups are oxidized in this reaction, but tertiary alkyl groups are not. We will discuss this reaction in more detail in Chapter 20.

Exercise 15.8 Write the equations for the reactions that would prepare the following carboxylic acids from the starting materials indicated.

(a) 3-Methylbutanoic acid from an alcohol
(b) 3-Methylbutanoic acid from an alkene
(c) Pentanedioic acid from an alkene
(d) from a derivative of benzene

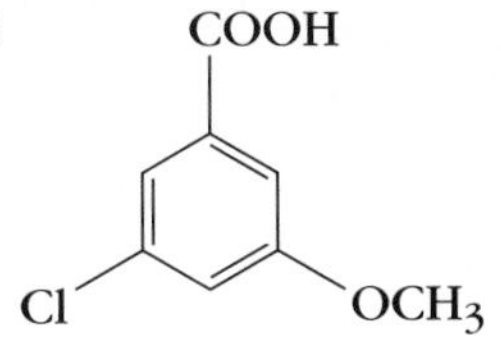

15.6 Synthesis of Carboxylic Acids by Carboxylation of Grignard Reagents

Grignard reagents react with carbon dioxide to form carboxylate salts, which form carboxylic acids on treatment with a strong aqueous acid solution. This reaction is called a **carboxylation reaction** because it adds a carboxylic acid functional group to an organic compound. Carboxylation reactions of Grignard reagents are usually carried out in the laboratory by adding the Grignard reagent to dry ice (solid CO_2).

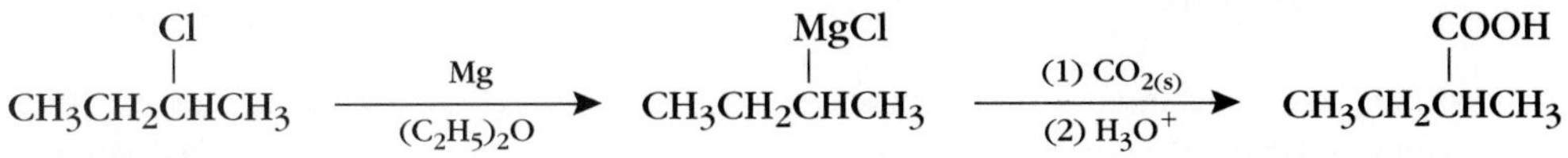

2-Chlorobutane

2-Methylbutanoic acid (87% yield)

A similar reaction occurs between organolithium reagents and carbon dioxide.

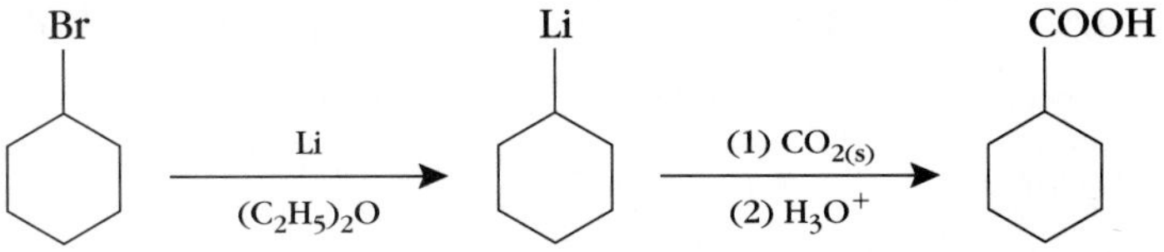

Bromocyclohexane

Cyclohexanecarboxylic acid (82% yield)

Carboxylation of Grignard and organolithium reagents forms carboxylic acids that have one more carbon atom than the starting haloalkane. The only limitation to this synthetic method is that the haloalkane must be capable of forming a Grignard or organolithium reagent.

The mechanism of the reaction of Grignard reagents and carbon dioxide is another example of nucleophilic addition to the carbonyl carbon atom of a carbonyl group. In Section 11.7, it was the carbonyl carbon atom of aldehydes and ketones; this time, it's the carbon atom of carbon dioxide.

Step 1. Nucleophilic addition of the carbon portion of the Grignard reagent to the carbon atom of CO_2.

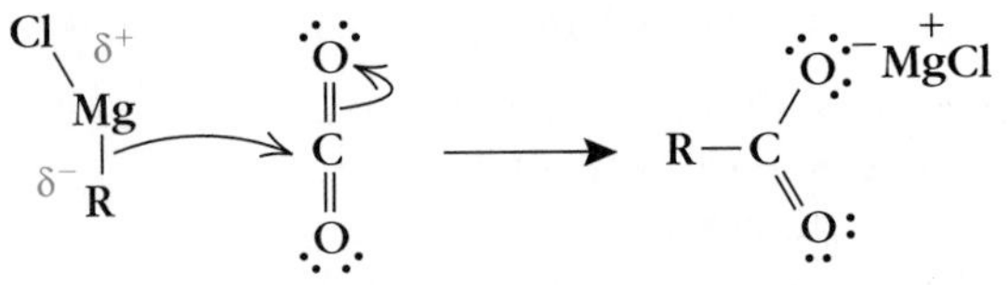

Step 2. Addition of aqueous acid forms a carboxylic acid.

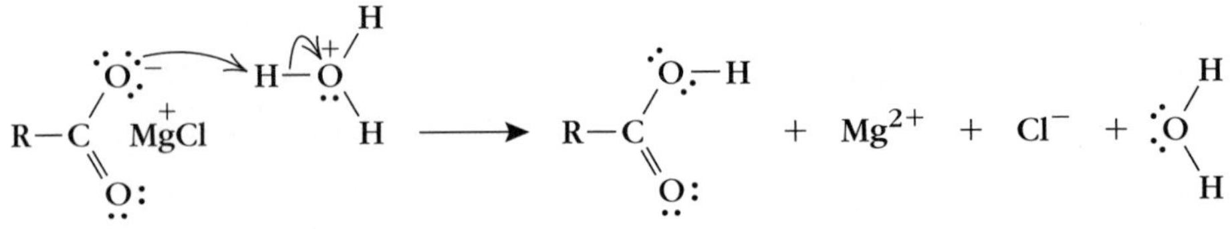

Exercise 15.9	Propose a series of reactions to form each of the following compounds from the designated starting material.

(a) Cyclopentanecarboxylic acid from bromocyclopentane
(b) 1-Ethylcyclohexanecarboxylic acid from cyclohexanone
(c) Hexanoic acid from 1-bromobutane
(d) Pentanoic acid from 1-butene

15.7   Synthesis of Carboxylic Acids by Hydrolysis of Nitriles

Primary and secondary haloalkanes can be converted to carboxylic acids by a sequence of two reactions. The first reaction is the conversion of the haloalkane to a nitrile, $R—C\equiv N$, and its hydrolysis to a carboxylic acid is the second reaction.

$$RX \xrightarrow{\text{NaC}\equiv\text{N}} RC\equiv N \xrightarrow{H_3O^+} RCOOH + NH_4^+$$

Primary or secondary haloalkane   Nitrile   Carboxylic acid

Notice that the carboxylic acid produced in this two-step reaction has one more carbon atom than the starting haloalkane.

Recall from Section 12.18 that nitriles are prepared by nucleophilic substitution reactions of haloalkanes that occur by an S_N2 mechanism. The nitrile is usually isolated and then hydrolyzed by heating in an aqueous solution of a strong acid to form the carboxylic acid.

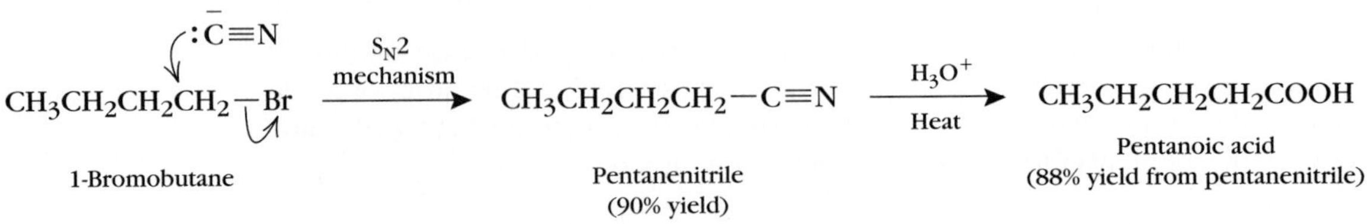

$$CH_3CH_2CH_2CH_2—Br \xrightarrow[\text{mechanism}]{S_N2} CH_3CH_2CH_2CH_2—C\equiv N \xrightarrow[\text{Heat}]{H_3O^+} CH_3CH_2CH_2CH_2COOH$$

1-Bromobutane   Pentanenitrile (90% yield)   Pentanoic acid (88% yield from pentanenitrile)

The mechanism of the hydrolysis of a nitrile is discussed in detail in Section 16.15, C.

Exercise 15.10	Each of the following series of reactions will not occur as written. Explain why not.

(a) $HOCH_2CH_2CH_2Br$ $\xrightarrow[\substack{(2)\ CO_2 \\ (3)\ H_3O^+}]{(1)\ Mg/\ (C_2H_5)_2O}$ $HOCH_2CH_2CH_2COOH$

(b) $(CH_3)_3CCl$ $\xrightarrow{NaCN}$ $(CH_3)_3CCN$ $\xrightarrow[Heat]{H_3O^+}$ $(CH_3)_3CCOOH$

15.8   Acidity of Aliphatic Carboxylic Acids

The pK_a values of unsubstituted aliphatic carboxylic acids are in the range of 4 to 5, as we saw in Section 3.4. Except for formic acid, their pK_a values do not change very much as the number of carbon atoms in the alkyl group increases, as shown by the data in Table 15.4. A pK_a of 4 to 5 makes carboxylic acids stronger acids than alcohols ($pK_a \approx 16$) and strong enough to give an acid reaction with litmus paper but makes them weaker acids than mineral acids such as HCl ($pK_a \approx -7$) or H_2SO_4 ($pK_a \approx -9$).

In practical terms, a pK_a of 4 to 5 means that only about 1% of the carboxylic acid molecules are ionized in a 0.1 M aqueous solution of acetic acid, compared to 100% for aqueous acid solutions of HCl or H_2SO_4. In aqueous solution, therefore, acetic acid exists predominantly as the carboxylic acid.

In aqueous solutions in mammals whose pH is about 7, in contrast, carboxylic acids exist predominantly as the carboxylate ion. This can be verified by a simple calculation using the Henderson-Hasselbalch equation. If we take a value of 4.7 as the pK_a of a typical water soluble aliphatic carboxylic acid and place it in a solution whose pH is 7, the ratio of carboxylate ion to carboxylic acid in the solution can be calculated as follows:

$$pH = pK_a + \log \frac{[RCOO^-]}{[RCOOH]} \quad \text{(Henderson-Hasselbalch equation)}$$

$$\log \frac{[RCOO^-]}{[RCOOH]} = pH - pK_a = 7.0 - 4.7 = 2.3$$

$$\frac{[RCOO^-]}{[RCOOH]} = 10^{+2.3} = 2.0 \times 10^2$$

Table 15.4	**Acidity of Selected Aliphatic Carboxylic Acids**		
Name	Structure	K_a	pK_a
Formic acid	HCOOH	17.7×10^{-5}	3.77
Acetic acid	CH_3COOH	1.76×10^{-5}	4.74
Propanoic acid	CH_3CH_2COOH	1.34×10^{-5}	4.87
Butanoic acid	$CH_3CH_2CH_2COOH$	1.54×10^{-5}	4.82
Hexanoic acid	$CH_3(CH_2)_3CH_2COOH$	1.31×10^{-5}	4.88

Figure 15.5
Plot of the concentrations of a typical carboxylic acid, RCOOH, and its conjugate base, RCOO$^-$, as a function of pH. The pK_a of the carboxylic acid is arbitrarily chosen as 5. RCOOH is the predominant form of the acid in the pH = 0–4 range, RCOO$^-$ is the predominant form in the pH = 6–14 range, and both are present in the pH = 4–6 range.

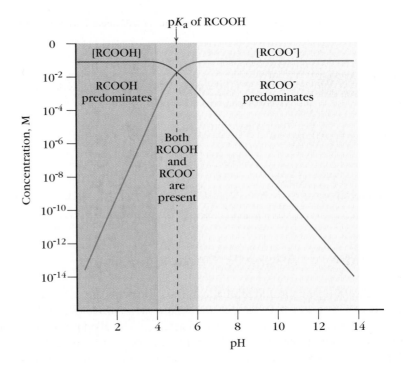

From this calculation, we conclude that the concentration of carboxylate ion is 200 times that of the carboxylic acid in body fluids of pH near 7.0. In these body fluids, therefore, carboxylic acids exist as their conjugate base, the carboxylate ion, and not as the undissociated carboxylic acid. As a result, carboxylic acids in body fluids are not important sources of protons because their concentration is so low.

In general, if the pH of a solution is within about one unit of the pK_a of an acid, appreciable concentrations of both the acid and its conjugate base are present in solution. Outside of this narrow range, only one of the species, either the acid *or* its conjugate base, is present. These three regions are depicted in Figure 15.5, where the relative concentrations of a typical carboxylic acid and its conjugate base are plotted as a function of pH.

Exercise 15.11	An acid whose pK_a is 8.5 is dissolved in each of the following buffered solutions whose pH values are as indicated. What is the predominant species in each solution?

(a) pH = 7.0   **(b)** pH = 9.0   **(c)** pH = 11.0   **(d)** pH = 3.0

The pK_a values of saturated aliphatic carboxylic acids can be greatly increased by placing electronegative substituents on the α-carbon atom. The data in Table 15.5 show, for example, that the acidity of acetic acid is greatly increased by replacing one or more α-hydrogen atoms by a nitrile group or halogen atom. Chloroacetic acid (pK_a = 2.86), for instance, is a much stronger acid than acetic acid (pK_a = 4.74). Replacing a single α-hydrogen atom of acetic acid by an electron-withdrawing atom or group of atoms generally makes the substituted acetic acid about 100 times stronger than acetic acid. The greater the number of electron-withdrawing atoms or groups of atoms on the α-carbon atom of acetic acid, the stronger the acid. Trihaloacetic acids, for example, are about 1000 times stronger than acetic acid. This makes trihaloacetic acids almost as strong as nitric acid ($pK_a = -1.44$).

Table 15.5	Strengths of Some Substituted Aliphatic Carboxylic Acids[*]		
Name	Structure	K_a	pK_a
Ethanol	CH_3CH_2OH	$\approx 10^{-16}$	≈ 16
Acetic acid	CH_3COOH	1.8×10^{-5}	4.74
Fluoroacetic acid	FCH_2COOH	2.6×10^{-3}	2.59
Chloroacetic acid	$ClCH_2COOH$	1.4×10^{-3}	2.86
Cyanoacetic acid	$N{\equiv}CCH_2COOH$	3.4×10^{-3}	2.46
Dichloroacetic acid	$Cl_2CHCOOH$	5.5×10^{-2}	1.26
Trichloroacetic acid	Cl_3CCOOH	0.23	0.64
Trifluoroacetic acid	F_3CCOOH	0.59	0.23
Hydronium ion	H_3O^+	55.5	-1.7
Butanoic acid	$CH_3CH_2CH_2COOH$	1.5×10^{-5}	4.82
2-Chlorobutanoic acid	$CH_3CH_2CHClCOOH$	139×10^{-5}	2.86
3-Chlorobutanoic acid	$CH_3CHClCH_2COOH$	8.9×10^{-5}	4.05
4-Chlorobutanoic acid	$CH_2ClCH_2CH_2COOH$	3.0×10^{-5}	4.52

[*]Acetic acid, butanoic acid, ethanol, and the hydronium ion are included for comparison.

The effect of substituents on the acidity of carboxylic acids decreases as the substituent is moved farther from the carboxyl group. The acidities of the chlorobutanoic acids given in Table 15.5 demonstrate this effect. The chlorine atom of 2-chlorobutanoic acid is bonded to the α-carbon atom, so this makes it a stronger acid ($pK_a = 2.86$) than butanoic acid ($pK_a = 4.82$). 3-Chlorobutanoic acid ($pK_a = 4.05$) is still a stronger acid than butanoic acid, but it is a weaker acid than 2-chlorobutanoic acid because the effect of the chlorine is diminished by its distance from the carboxyl group. The effect of the chlorine atom on the acidity of 4-chlorobutanoic acid ($pK_a = 4.52$) has decreased so much that it is about as strong an acid as butanoic acid ($pK_a = 4.82$).

In summary, the pK_a of unsubstituted aliphatic carboxylic acids is 4 to 5, which makes them the most acidic family of organic compounds. Placing electronegative atoms or groups of atoms on the α-carbon atom of a carboxylic acid increases its acidity.

15.9 Why Are Carboxylic Acids so Acidic?

We can explain why carboxylic acids are so acidic by comparing the ionization of a carboxylic acid, such as formic acid, with the ionization of water.

Ionization of formic acid

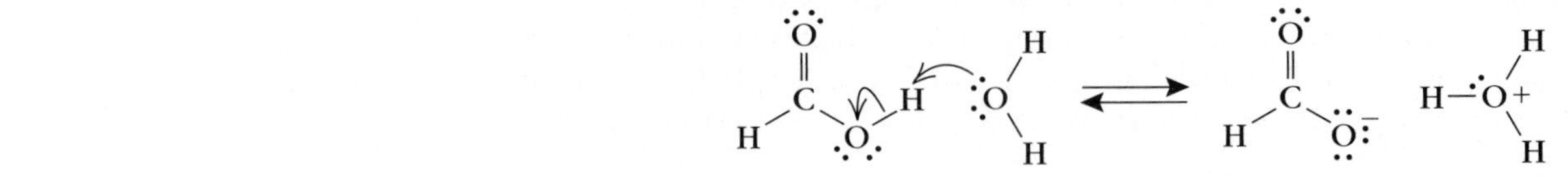

$pK_a = 3.77$

Ionization of water

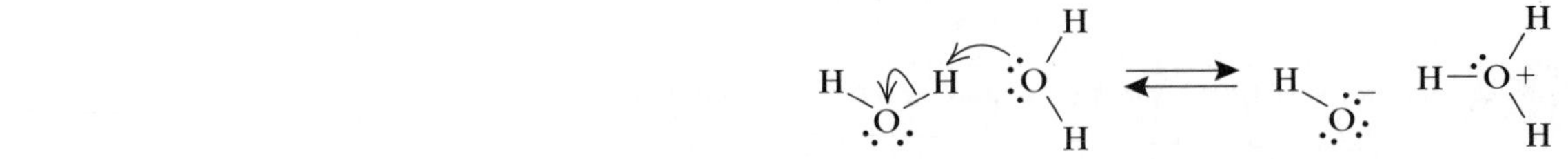

$pK_a = 15.7$

Formic acid can be viewed as being formed from water by replacing one hydrogen atom of water by a $H-C\overset{O}{\underset{\diagdown}{\parallel}}$ group. The effect of such a change in structure is to place a group more electron withdrawing than a hydrogen atom next to the oxygen atom of the O—H bond. The adjacent strong electron-withdrawing $H-C\overset{O}{\underset{\diagdown}{\parallel}}$ group makes the oxygen atom of the O—H bond of formic acid more electron deficient than the oxygen atom of water. As a result, a proton is held more tightly to the oxygen atom of water than it is to the oxygen atom of the O—H bond of formic acid, which makes formic acid the stronger acid.

In general, therefore, the greater the electron-withdrawing ability of the substituent Z in an acid of the type Z—OH, the stronger the acid. The following acids and their pK_a's illustrate this trend:

Z	Z—OH	pK_a
O_2N-	O_2N-OH	-1.44
$H-C\overset{O}{\underset{\diagdown}{\parallel}}$	$H-C\overset{O}{\underset{\diagdown}{\parallel}}OH$	3.75
$Cl-$	$Cl-OH$	7.54
$H-$	$H-OH$	15.7

(left axis label: Electron withdrawing ability of Z, increasing)

We can now explain why α-chlorocarboxylic acids are more acidic than their unsubstituted counterparts. Chlorine is an electronegative element that exerts an electron-withdrawing inductive effect on its neighboring atoms. The inductive effect of an α-chlorine atom makes the electron-withdrawing ability of the $Cl-\overset{\mid}{\underset{\mid}{C}}-C\overset{O}{\underset{\diagdown}{\parallel}}$ group greater than the $H-\overset{\mid}{\underset{\mid}{C}}-C\overset{O}{\underset{\diagdown}{\parallel}}$ group, so the chlorine-substituted carboxylic acid is the stronger acid.

The relative acidities of water and formic acid are also influenced by the relative stabilities of their conjugate bases. The conjugate base of formic acid (formate ion) is resonance stabilized (Section 1.12). Hydroxide ion, the conjugate base of water, in contrast, cannot be stabilized by resonance. The tendency to form the more stable conjugate base also makes formic acid more acidic than water.

In summary, both the formation of a resonance-stabilized carboxylate ion and the electron-withdrawing effect of the $R-C\overset{O}{\underset{\diagdown}{\parallel}}$ group in a carboxylic acid contribute to make carboxylic acids more acidic than water.

Exercise 15.12   In each of the following pairs of acids, identify the stronger one.

(a) CH_3COOH and CH_3OCH_2COOH
(b) FCH_2COOH and F_3CCOOH
(c) O_2NCH_2COOH and CH_3CH_2COOH
(d) $HCOOH$ and CH_3COOH

15.10 Interconverting Carboxylic Acids and Their Salts

An aqueous solution of a strong base such as sodium hydroxide converts a carboxylic acid into the sodium salt of its carboxylate ion.

$$R-C\overset{O}{\underset{O-H}{\parallel}} \quad + \quad NaOH \quad \longrightarrow \quad R-C\overset{O}{\underset{O^-Na^+}{\parallel}} \quad + \quad H_2O \qquad (15.1)$$

Carboxylic acid   Aqueous   Sodium salt of the   Water
                     solution   carboxylic acid
                     of NaOH

The carboxylic acid can be regenerated by reacting its carboxylate salt with an aqueous solution of a strong acid, such as HCl.

$$R-C\overset{O}{\underset{O^-Na^+}{\parallel}} \quad + \quad HCl \quad \longrightarrow \quad R-C\overset{O}{\underset{O-H}{\parallel}} \quad + \quad NaCl \qquad (15.2)$$

Sodium salt of the   Aqueous   Carboxylic acid   Sodium
carboxylic acid   solution   chloride
                       of HCl

Chemists make use of this interconversion of carboxylic acids and their carboxylate salts to separate carboxylic acids from nonacidic compounds. This is possible because most carboxylic acids and their salts have different solubilities in water.

Like most salts, carboxylic acid salts are crystalline solids with relatively high melting points. The sodium, potassium, and ammonium salts of low molecular weight carboxylic acids are soluble in water but insoluble in nonpolar solvents such as diethyl ether, benzene, or dichloromethane. This contrasts with the solubilities of carboxylic acids, which are soluble in nonpolar solvents but insoluble in water (except for those containing four or fewer carbon atoms).

A carboxylic acid can be separated from nonacidic compounds by washing

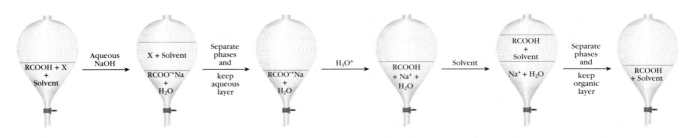

RCOOH = Carboxylic acid
        X = Non–acid compound
Solvent = Organic solvent

Figure 15.6
The sequence of steps used to separate a mixture of a carboxylic acid and a nonacidic organic compound.

the organic solvent containing the carboxylic acid and the nonacidic compound with an aqueous solution of sodium hydroxide, as shown in Figure 15.6. The sodium hydroxide converts the carboxylic acid to its salt (according to equation 15.1, p. 669), which dissolves in the aqueous layer. The nonacidic compound remains in the organic solvent. The two layers are separated, and addition of aqueous HCl to the aqueous layer regenerates the carboxylic acid (according to equation 15.2, p. 669). Adding fresh organic solvent redissolves the carboxylic acid in the organic layer, leaving NaCl in the aqueous layer.

This technique can also be used when the nonacidic compound is the desired product and the acid is the impurity. For example, an aldehyde prepared by the oxidation of a primary alcohol is sometimes contaminated by the presence of the carboxylic acid formed by further oxidation. The carboxylic acid impurity can be removed simply by washing the product with dilute aqueous sodium hydroxide. This forms the sodium salt of the carboxylic acid, which is soluble in the aqueous phase, leaving the desired aldehyde in the organic layer.

Exercise 15.13	We saw in Section 15.6 that the reaction of a Grignard reagent with dry ice (solid CO_2) forms carboxylic acids. When this reaction is carried out in the laboratory, the dry ice is frequently coated with a thin layer of water, particularly on a hot, humid day. It is difficult to remove all the water from the surface of the dry ice, so some of the Grignard reagent will be hydrolyzed to a hydrocarbon. Draw a sequence of steps to show how you would separate the carboxylic acid from any hydrocarbon formed as an impurity.

15.11 Synthesis of Esters: The Fischer Esterification

The formation of a carboxylic acid ester (usually called simply an ester) by the acid-catalyzed reaction of a carboxylic acid and an alcohol is called the **Fischer esterification** reaction.

$$R-C{\overset{\displaystyle O}{\underset{\displaystyle OH}{\big\|}}} \quad + \quad R'OH \;\underset{\text{catalyst}}{\overset{\text{Strong acid}}{\rightleftarrows}}\; R-C{\overset{\displaystyle O}{\underset{\displaystyle OR'}{\big\|}}} \quad + \quad H_2O$$

Carboxylic acid   Alcohol   Carboxylic   Water
                                                                  acid ester

Fischer esterification is a reversible reaction, and generally appreciable concentrations of both the carboxylic acid and the ester are present at equilibrium. Therefore, to obtain a good yield of the ester, it is necessary to force the reaction to completion either by removing the water as it is formed or by using an excess of one of the reactants. The equilibrium is frequently shifted to products by using the alcohol as both a solvent and a reactant. For example, a good yield of methyl acetate can be obtained by the acid-catalyzed reaction of acetic acid in methanol as solvent.

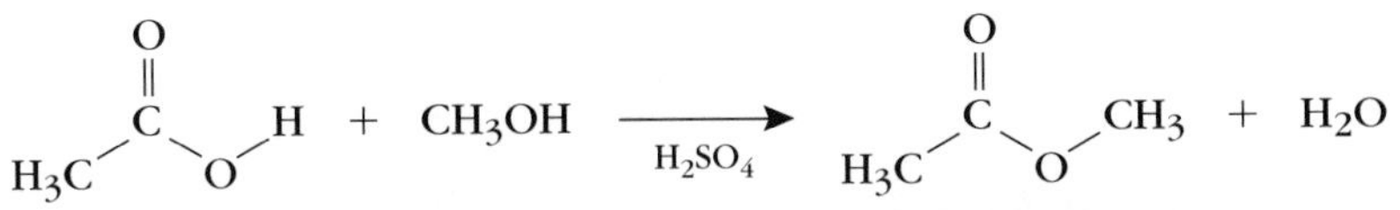

Acetic acid   Methanol (solvent)   Methyl acetate (85% yield)

Before a mechanism for the Fischer esterification can be proposed, we must know if the oxygen atom of the alkoxy group comes from the alcohol or the carboxylic acid. For example, does the oxygen atom of the methoxy group of methyl benzoate come from methanol or does it come from benzoic acid?

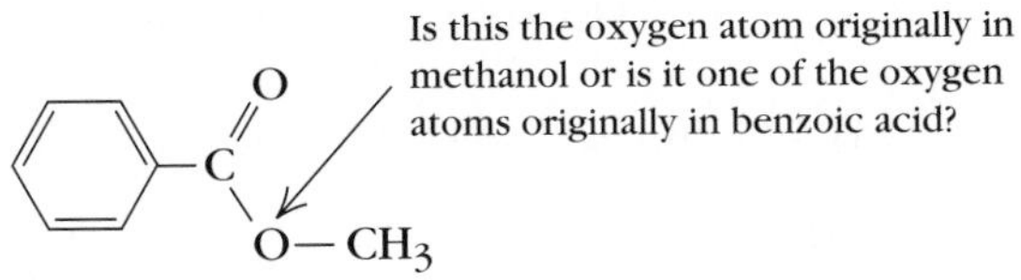

Is this the oxygen atom originally in methanol or is it one of the oxygen atoms originally in benzoic acid?

Methyl benzoate

The answer to this question was obtained by the isotopic tracer experiments of the American chemists Irving Roberts and Harold Urey. Benzoic acid containing only its natural abundance of ^{18}O (oxygen-18) was reacted with methanol enriched with ^{18}O in the presence of an acid catalyst. The water formed in the reaction was isolated and shown to contain no more ^{18}O than its natural abundance. This result means that the C—O bond of the alcohol is *not* broken during the reaction, so the methoxy oxygen atom must come from the oxygen of the alcohol.

■ *Recall that oxygen-18 is a nonradioactive isotope of oxygen whose natural abundance is 0.204%.*

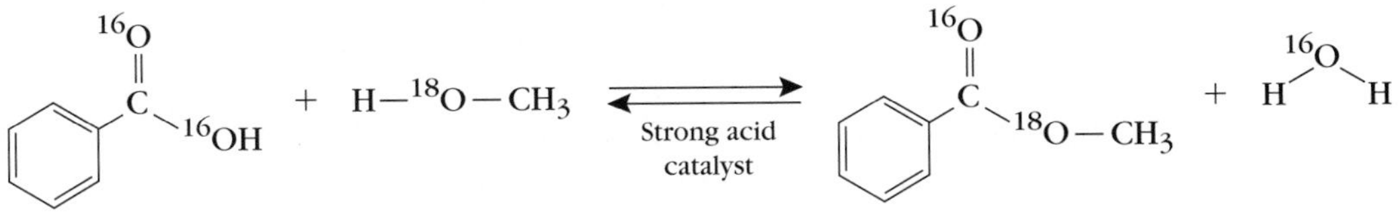

From this result, the Fischer esterification is not a nucleophilic displacement reaction that occurs by an S_N2 mechanism. If it did occur by an S_N2 mechanism, the water should be enriched in ^{18}O, contrary to the experimental results.

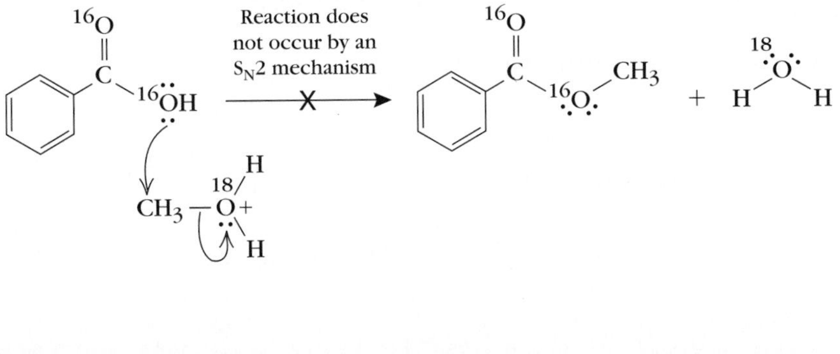

Based on this and many other experiments, the accepted mechanism of the Fischer esterification reaction of benzoic acid and methanol is illustrated in Figure 15.7.

Step 1 is protonation of the carboxylic acid. A carboxyl group is not reactive enough to add a nucleophile (for reasons that will be discussed in Chapter 16), so a strong acid catalyst is needed to enhance the reactivity of the carboxyl group. Protonation activates the carboxyl group in the first step by increasing the positive character of the carboxyl carbon atom. Protonation of the carboxyl group

Figure 15.7
The mechanism of the formation of methyl benzoate by the Fischer esterification reaction. Protonation of the carboxyl group in Step 1 activates it to *addition* of methanol in Step 2. Step 3 is a proton transfer reaction that probably requires several steps. Step 4 is *elimination* of a water molecule to form the conjugate acid of the ester, which loses a proton in Step 5 to form the ester.

Step 1: Protonation of the carboxyl group

Resonance stabilized cation

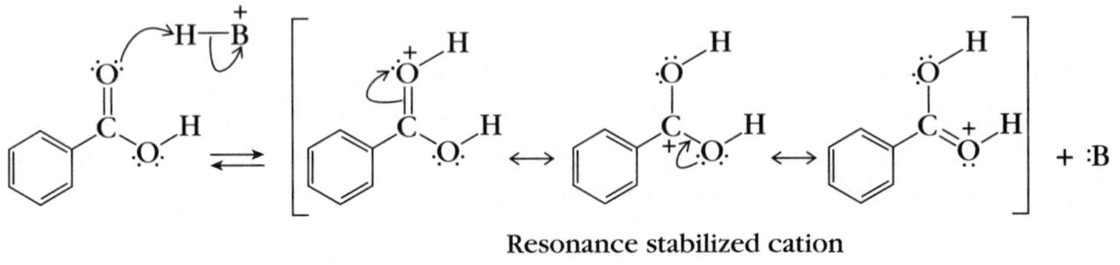

Step 2: Addition of methanol to the carboxyl carbon atom forms a tetrahedral intermediate

Tetrahedral intermediate

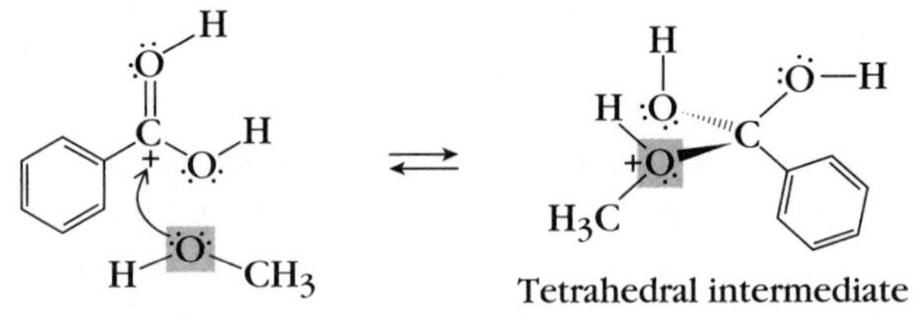

Step 3: Proton transfer

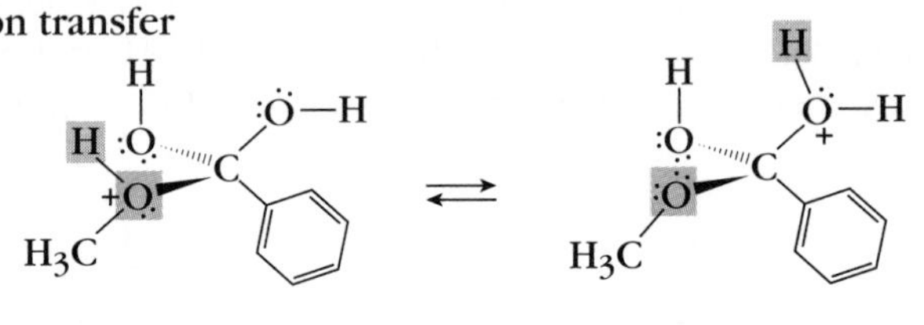

Step 4: Elimination of water from the tetrahedral intermediate

Conjugate acid of ester

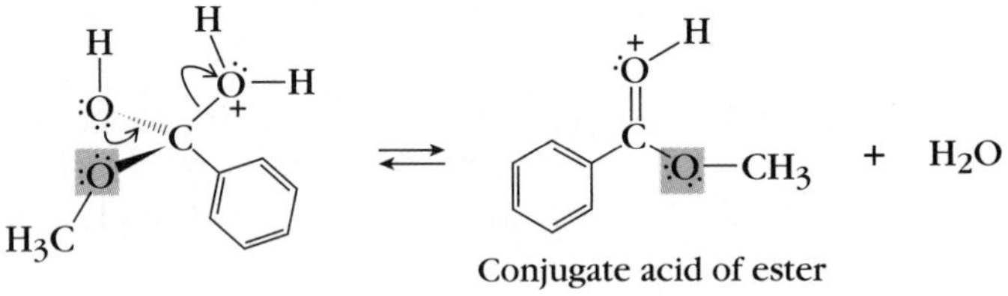

Step 5: Proton transfer forms methyl benzoate

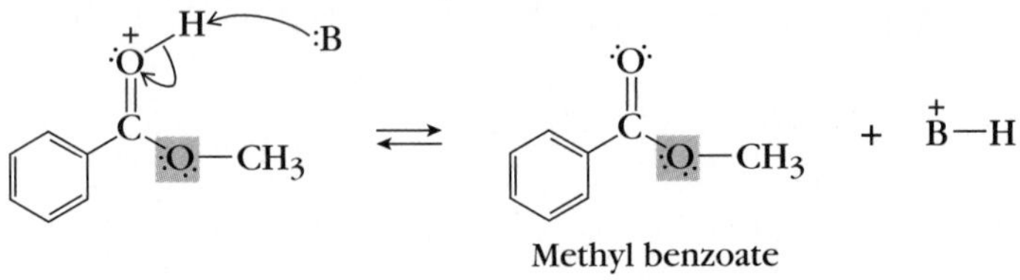

Methyl benzoate

occurs on the oxygen of the C=O bond, not on the oxygen of the —OH group. Protonation could occur on the O—H group, but the resulting conjugate acid is less stable because it is not resonance stabilized.

Nucleophilic addition of methanol to the carboxyl carbon atom occurs in Step 2 to form a tetrahedral intermediate. This step is analogous to Step 2 of the acid-catalyzed mechanism of nucleophilic addition of an alcohol to aldehydes and ketones to form a hemiacetal (Section 14.11).

Step 3 represents a series of proton transfer reactions in which a proton is

$$\overset{\overset{\displaystyle H}{|}}{\underset{+}{}}$$

removed from the —OCH$_3$ group and added to the oxygen atom of one of the —OH groups. This proton transfer converts the —OH group into a good leaving group. We've seen this method of converting a hydroxy group into a good leaving group in the mechanism of acid-catalyzed dehydration of tertiary alcohols (Section 7.18) and the mechanism of the formation of tertiary haloalkanes by the reaction of tertiary alcohols and hydrogen halides (Section 11.17).

Step 4 of the mechanism is the elimination of a water molecule to form the conjugate acid of the ester. Finally, in Step 5, transfer of a proton from the conjugate acid of the ester to a base forms the ester.

The five steps of this mechanism represent three types of reactions. The first type is addition of a nucleophile to the carboxyl carbon atom to form a tetrahedral intermediate (Step 2). The second type is the elimination of a good leaving group (Step 4) from the tetrahedral intermediate. The third type is the proton transfer reactions. In Step 1, proton transfer activates the carboxyl group. In Step 3, proton transfer occurs to form a good leaving group. In Step 5, loss of a proton forms the product. In Chapter 16, we will learn that these three types of reactions are part of the mechanisms of *all* acid-catalyzed reactions of carboxylic acid derivatives.

The five steps in the mechanism of the esterification reaction are all reversible because the reverse reaction, the acid-catalyzed hydrolysis of an ester to form a carboxylic acid and an alcohol, can also occur. The mechanism of acid-catalyzed ester hydrolysis is the reverse of the esterification mechanism given in Figure 15.7, so the same three types of reactions are involved. First, the ester functional group is activated by protonation of the oxygen atom of the C=O bond. Second, water adds to form a tetrahedral intermediate, which loses an alcohol molecule to form the carboxylic acid.

■ *In the mechanism shown in Figure 15.7, the letter B is used to represent a base because we cannot specify the nature of the base any more precisely. The base could be an alcohol molecule, a water molecule, or the conjugate base of the acid catalyst.*

Exercise 15.14   Write the mechanism for the acid-catalyzed hydrolysis of methyl acetate.

While esters can be hydrolyzed with aqueous acid, it is easier to hydrolyze them under aqueous basic conditions.

15.12 Saponification of Esters

Esters react with aqueous base to form a carboxylate salt and an alcohol. The carboxylic acid is obtained by adding a mineral acid to the reaction mixture.

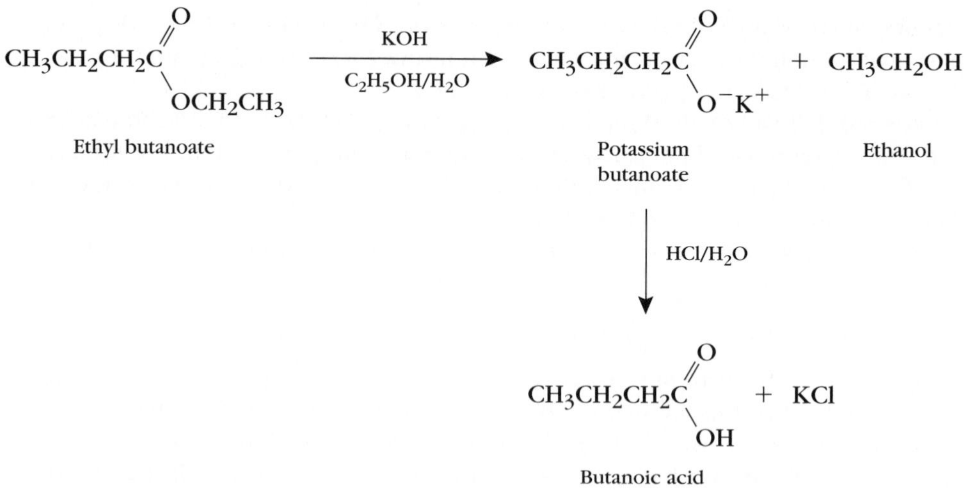

$$\text{CH}_3\text{CH}_2\text{CH}_2\overset{\displaystyle O}{\overset{\|}{\text{C}}}\text{--OCH}_2\text{CH}_3 \quad\xrightarrow[\text{C}_2\text{H}_5\text{OH/H}_2\text{O}]{\text{KOH}}\quad \text{CH}_3\text{CH}_2\text{CH}_2\overset{\displaystyle O}{\overset{\|}{\text{C}}}\text{--O}^-\text{K}^+ \quad+\quad \text{CH}_3\text{CH}_2\text{OH}$$

Ethyl butanoate   Potassium   Ethanol
                                               butanoate

$$\downarrow \text{HCl/H}_2\text{O}$$

$$\text{CH}_3\text{CH}_2\text{CH}_2\overset{\displaystyle O}{\overset{\|}{\text{C}}}\text{--OH} \quad+\quad \text{KCl}$$

Butanoic acid

Hydroxide ion is actually a reagent instead of a catalyst because it is consumed in the reaction.

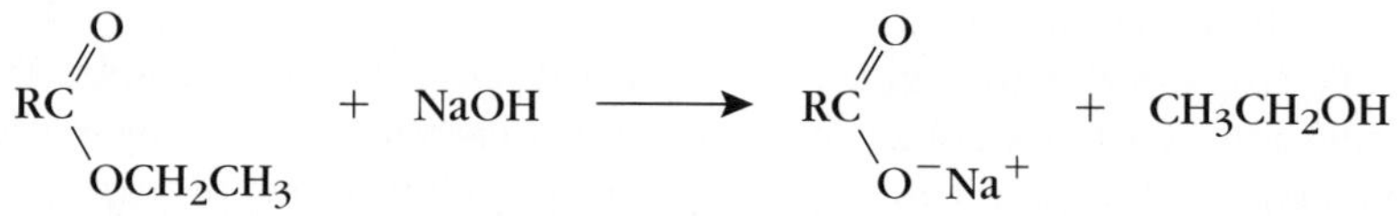

$$\text{RC}\overset{\displaystyle O}{\overset{\|}{}}\text{--OCH}_2\text{CH}_3 \quad+\quad \text{NaOH} \quad\longrightarrow\quad \text{RC}\overset{\displaystyle O}{\overset{\|}{}}\text{--O}^-\text{Na}^+ \quad+\quad \text{CH}_3\text{CH}_2\text{OH}$$

Hydrolysis of esters in an aqueous basic solution is called **saponification** (from the Latin *sapo* meaning *soap*) because this reaction has been used since antiquity to prepare soaps from fats. Despite its connection with soap making, the term *saponification* is still used today to indicate the hydrolysis of esters in aqueous basic solution.

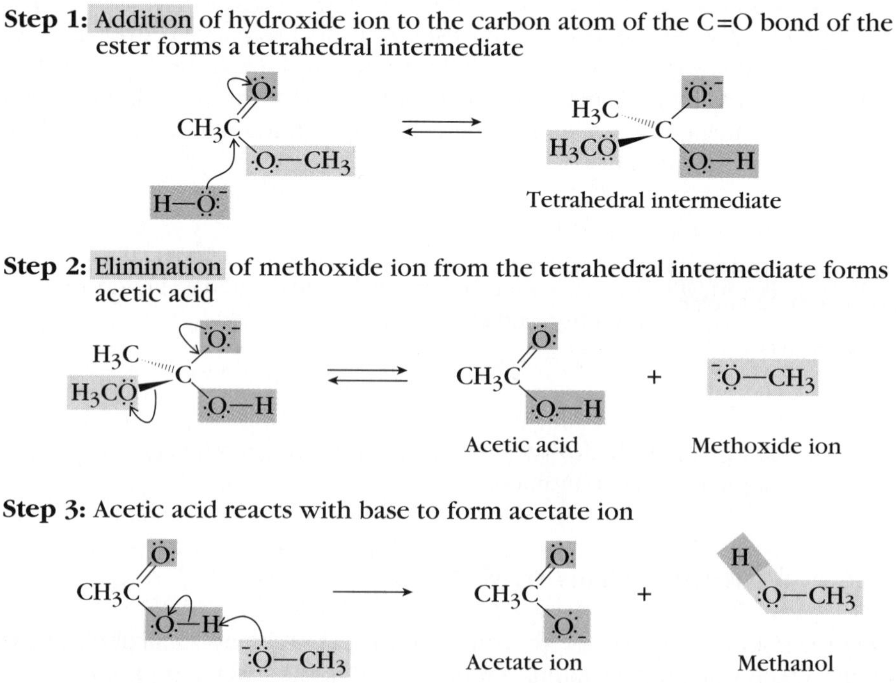

Figure 15.8
Mechanism of the saponification of methyl acetate.

Step 1: Addition of hydroxide ion to the carbon atom of the C=O bond of the ester forms a tetrahedral intermediate

Tetrahedral intermediate

Step 2: Elimination of methoxide ion from the tetrahedral intermediate forms acetic acid

Acetic acid   Methoxide ion

Step 3: Acetic acid reacts with base to form acetate ion

Acetate ion   Methanol

The proposed mechanism of saponification of an ester is shown in Figure 15.8. Step 1 is addition of a nucleophile to the carbon atom of the C=O bond of the ester to form a tetrahedral intermediate. There are two nucleophiles in aqueous base, hydroxide ion and water. Hydroxide ion is the better nucleophile, however, so it adds faster than water.

Step 2 of the mechanism is the elimination of an ion from the tetrahedral intermediate. Only two ions can be eliminated, the hydroxide ion that was just added or methoxide ion, CH_3O^- ($^-CH_3$ is not a leaving group). Elimination of the hydroxide ion, however, is unproductive because it leads to the starting material, while elimination of the methoxide ion leads to the product, acetic acid.

In Step 3, acetic acid rapidly reacts with a strong base such as methoxide ion or hydroxide ion to form the carboxylate ion and an alcohol. Because of the large difference in acidity between a carboxylic acid ($pK_a \approx 5$) and an alcohol ($pK_a \approx 16$), Step 3 is irreversible. This makes saponification an irreversible reaction.

Saponification is the preferred method of hydrolysis of esters because it is an irreversible reaction and so excess water is not needed to ensure completion of the reaction as in the acid-catalyzed hydrolysis of esters. Actually only one equivalent of water is needed in the saponification reaction, so saponification is particularly useful for the hydrolysis of esters that are insoluble in water. Saponification, therefore, can be carried out in solvents made up of one equivalent of water and another liquid (such as an alcohol) that will dissolve the ester in the reaction mixture.

| **Exercise 15.15** | In which product would you expect to find the ^{18}O after the saponification of the following ester? |

$$CH_3CH_2 - {}^{18}O$$
$$C=O$$
$$CH_3CH_2$$

Esters, particularly the high molecular weight esters that make up many fats, oils, and waxes, are found widely in nature.

15.13 Waxes, Fatty Acids, and Fats

Many of the commercial products that we use to protect and beautify cars, floors, and furniture are **waxes.** In nature, waxes are found on the surfaces of stems and leaves of plants, where they protect the plant from loss of moisture and attack by harmful insects.

Waxes are mixtures of esters made up of long-chain unbranched alcohols and carboxylic acids that contain an even number of carbon atoms. Waxes generally contain a total of between 40 and 72 carbon atoms. Carnuba wax, which is widely used to protect floors and automobiles, is a complex mixture of several esters, hydrocarbons, and alcohols, but its major component is myricyl cerotate.

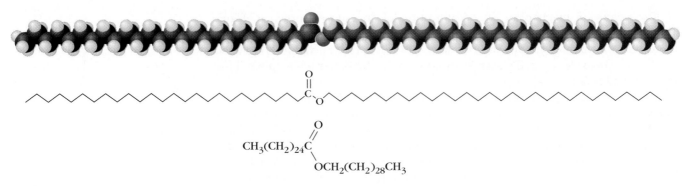

$$CH_3(CH_2)_{24}C \overset{O}{\underset{OCH_2(CH_2)_{28}CH_3}{\parallel}}$$

Myricyl cerotate (a major component of carnuba wax)

Animal fats, such as butter and lard, tend to be solids, while vegetable oils, such as corn, peanut, and olive oils, tend to be liquids. Despite their physical differences, however, fats and oils have similar chemical structures. Both are triesters of the alcohol glycerol (1,2,3-propanetriol) and three long-chain carboxylic acids (fatty acids). These triesters, which are called **triacylglycerols,** are the major reservoir of fatty acids in mammals.

> ■ *The distinction between fats and oils is based on their melting points. Oils melt below 25 °C, and fats melt above 25 °C.*

$$\begin{array}{c} H_2COH \\ | \\ HCOH \\ | \\ H_2COH \end{array} \qquad \begin{array}{c} H_2CO-\overset{O}{\overset{\parallel}{C}}R \\ | \\ HCO-\overset{O}{\overset{\parallel}{C}}R' \\ | \\ H_2CO-\overset{O}{\overset{\parallel}{C}}R'' \end{array}$$

Glycerol   A triacylglycerol (the general structure of a fat or oil)   Portion derived from long-chain carboxylic acids

Fatty acids have three functions in mammals. First, they are important components of biological membranes; second, they serve as hormones and intracellular messengers; and third, they are a major source of energy. Fatty acids are made available to organisms by the enzyme-catalyzed hydrolysis of triacylglycerols.

$$\begin{array}{c} \overset{O}{\overset{\parallel}{R'C}}-O-\overset{CH_2O-\overset{O}{\overset{\parallel}{C}}R}{\underset{CH_2O-\overset{\parallel}{C}R''}{\overset{|}{C}-H}} \\ \overset{\parallel}{O} \end{array} \quad + \quad 3H_2O \quad \xrightarrow{\text{Enzyme}} \quad \begin{array}{c} CH_2OH \\ | \\ CHOH \\ | \\ CH_2OH \end{array} \quad + \quad \begin{array}{c} \overset{O}{\overset{\parallel}{RC}}{\diagdown}_{O^-} \\ \overset{O}{\overset{\parallel}{R'C}}{\diagdown}_{O^-} \\ \overset{O}{\overset{\parallel}{R''C}}{\diagdown}_{O^-} \end{array}$$

Triacylglycerol   Glycerol   Fatty acids

Fatty acids, like low molecular weight carboxylic acids, exist as their carboxylate ions at the pH (pH ≈ 7) of living systems, as we saw in Section 15.8.

Fatty acids obtained by hydrolysis in biological systems usually contain an

Table 15.6		Common Fatty Acids Found in Naturally Occurring Fats and Oils	
Name	Total number of carbon atoms	Structure	Melting Point (°C)
Myristic acid	14	$CH_3(CH_2)_{12}COOH$	58
Palmitic acid	16	$CH_3(CH_2)_{14}COOH$	63
Stearic acid	18	$CH_3(CH_2)_{16}COOH$	70
Arachidic	20	$CH_3(CH_2)_{18}COOH$	75
Oleic	18	$(Z)\text{-}CH_3(CH_2)_7CH{=}CH(CH_2)_7COOH$	4
Linoleic	18	$(Z,Z)\text{-}CH_3(CH_2)_4CH{=}CHCH_2CH{=}CH(CH_2)_7COOH$	−5
Linolenic acid	18	$(Z,Z,Z)\text{-}CH_3CH_2(CH{=}CHCH_2)_2CH{=}CH(CH_2)_7COOH$	−11

even number of carbon atoms. Several of the typical fatty acids that contain between 14 and 20 carbon atoms are listed in Table 15.6. The most common saturated fatty acids are palmitic acid (16 carbon atoms) and stearic acid (18 carbon atoms). The alkyl chain is generally unbranched in fatty acids, such as stearic acid obtained from beef fat.

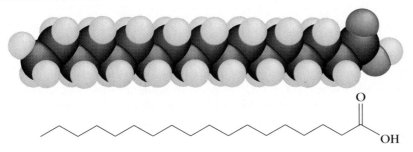

$C_{17}H_{35}COOH$

Stearic acid

The alkyl chain may also contain one or more double bonds, which are usually in the (Z) configuration. The most common monounsaturated fatty acid is oleic acid, which contains 18 carbon atoms.

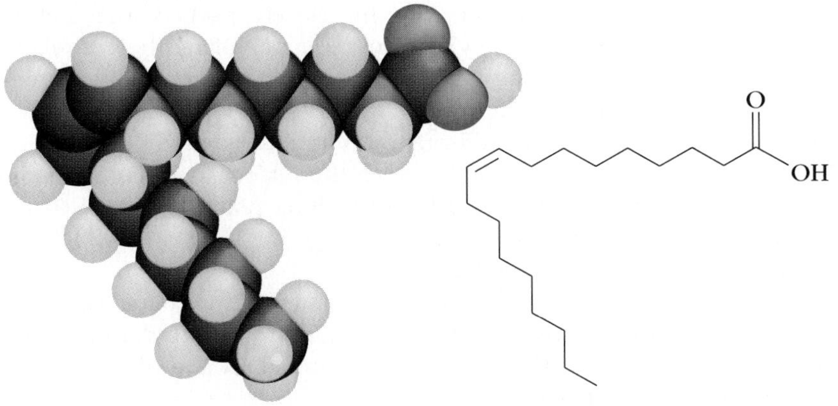

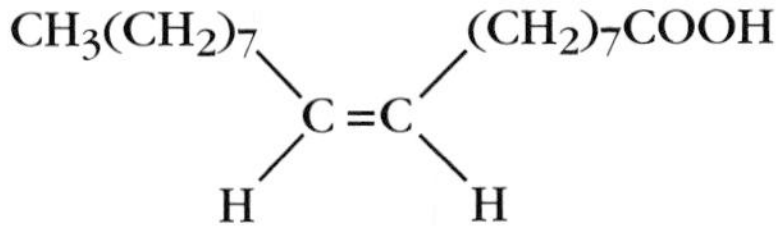

$$CH_3(CH_2)_7 \quad\quad (CH_2)_7COOH$$
$$C{=}C$$
$$H \quad\quad\quad H$$

Oleic acid

Polyunsaturated fatty acids contain two or more double bonds that are separated by at least one methylene group. The most common polyunsaturated fatty acids are linoleic acid and linolenic acid, both of which contain 18 carbon atoms.

Natural fats and oils vary widely in composition. The three fatty acids that form part of a particular triacylglycerol may be the same or different. While the proportions of various fatty acids vary from fat to fat, each fat has its characteristic composition, which is fairly constant from sample to sample. For example, all samples of olive oil contain between 80% and 95% unsaturated fatty acids (mostly oleic and smaller quantities of linoleic acids), while butter contains a lower percentage (about 35%) of unsaturated fatty acids (again mostly oleic and a smaller amount of linoleic acids). Generally the triacylglycerols of vegetable oils contain a higher percentage of unsaturated fatty acids than do animal fats. As a result, vegetable oils are said to be high in unsaturated fats, while animal fats are said to be high in saturated fats. Unsaturated fats are generally regarded as being more healthful for humans because excessive consumption of saturated fats has been linked to heart disease.

Exercise 15.16   Saponification of a triacylglycerol forms only sodium palmitate and glycerol as products. Write the structure of the triacylglycerol.

Notice in Table 15.6 that the melting points of unsaturated fatty acids are lower than the melting points of saturated ones. Notice as well that the melting points of unsaturated fatty acids decrease as the number of double bonds increases. The melting point of oleic acid (18 carbon atoms, 1 double bond), for instance, is 4 °C, whereas the melting point of linolenic acid (18 carbon atoms, 3 double bonds) is −11 °C. This trend in melting points is also reflected in their esters. The higher the percentage of unsaturated alkyl chains in the fatty acid portion of triacylglycerols, the lower their melting points. Vegetable oils, therefore, which are highly unsaturated, are liquids, while animal fats, which are more highly saturated, are solids.

Why is there this relationship between the composition of a triacylglycerol and its melting point? To answer this question, let us look at the shapes of saturated and unsaturated triacylglycerols. The carbon chains of saturated triacylglycerols are fully extended in order to minimize repulsions between neighboring methylene groups. Consequently, their shapes are fairly uniform and they can fit closely together in a solid. Because they are closely packed, the attraction between molecules of saturated triacylglycerols is large, giving them a relatively high melting point. In unsaturated triacylglycerols, on the other hand, the (Z) configuration of the double bond introduces bends and kinks into the hydrocarbon chain that make it more difficult for them to fit closely together in a solid. As a result, the van der Waals interaction between the molecules of unsaturated triacylglycerols is reduced, giving them a lower melting point. Figure 15.9 shows how the difference in shape between saturated and unsaturated triacylglycerols affects their ability to fit closely together.

Humans need fats in order to live. Some we manufacture ourselves; others, obtained in the food we eat, are synthesized by plants and animals. Fats are also important industrially because they are the raw material used to produce most of the soaps and detergents we use.

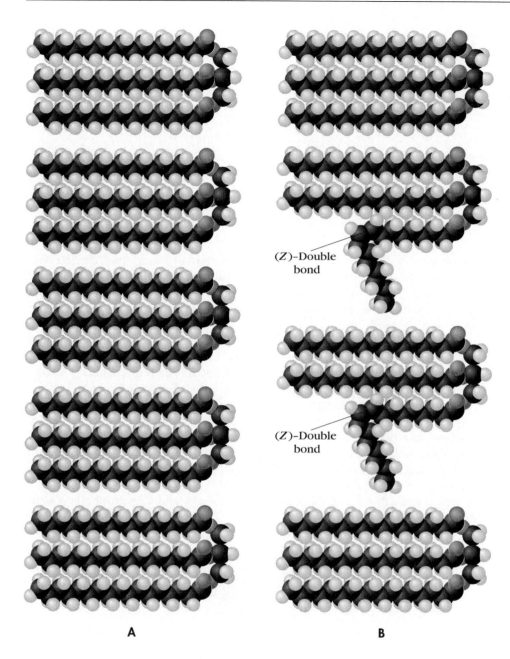

A   **B**

(Z)-Double
bond

(Z)-Double
bond

Figure 15.9
A, The carbon chains of saturated triacylglycerols are fully extended so they fit tightly together in the solid. **B,** The (Z)- double bond of unsaturated triacylglycerols interferes with their ability to fit tightly together. As a result, saturated triacylglycerols tend to be solids and unsaturated triacylglycerols tend to be liquids.

15.14 Soaps and Detergents

For centuries people have made soap by boiling animal fat with the aqueous extract of the ashes of a fire (a basic solution). Today a variation of this method is used. Instead of an aqueous extract of ashes, however, an aqueous solution of sodium hydroxide is used to hydrolyze (saponify) triacylglycerols to the sodium salts of fatty acids (which we call soap) and glycerol.

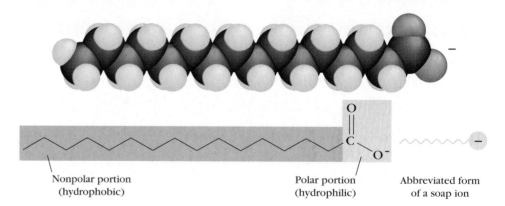

Soaps vary in their composition and their method of processing. The fats used to prepare soaps are a mixture of different triacylglycerols, so saponifying them forms the sodium salts of a mixture of fatty acids. This does not create a problem because a mixture works as well for washing as the salt of a single pure carboxylic acid.

Crude soaps are nevertheless purified by boiling the crude soap with a large amount of water and adding sodium chloride to precipitate (salt out) the soap. Soaps can be further processed by adding a variety of different chemicals for specific effects or uses, such as dyes to give them color, perfumes for aroma, germicides to make them antiseptic, and sand or pumice to make them abrasive enough for scouring. Soaps can be made softer by precipitating them as the potassium salts of fatty acids instead of the sodium salt; bubbling air into the molten soap will make a soap that floats. Regardless of the extra processing (and the extra cost), most soaps are chemically about the same and work in much the same way.

Soaps contain both a hydrophilic (water loving) and a hydrophobic (water hating) portion.

The hydrophilic part is the polar carboxylate group, while the long alkyl chains are the nonpolar hydrophobic (or lipophilic) part.

When soap is added to water, a "soap solution" is formed. This is not a true solution, however, in which the individual molecules or ions are separate and solvated. Soaps exist in water as **micelles,** which are spherical clusters of hundreds of carboxylate ions dispersed throughout the water phase. Figure 15.10 shows a schematic drawing of two micelles in aqueous solution.

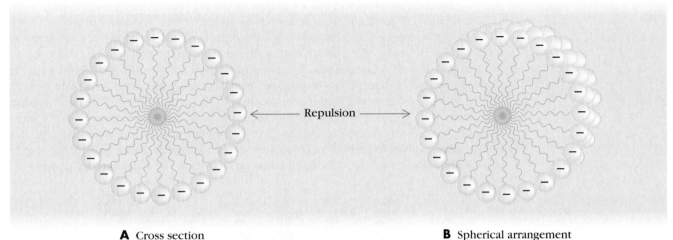

A Cross section   **B** Spherical arrangement

Figure 15.10
A, The cross section of a soap micelle. **B,** Spherical arrangement of the soap molecules in a micelle. The nonpolar alkyl chains of the soap molecules cluster in the center of the spherical micelle, while the carboxylate ions are located at the surface in contact with water. The two micelles repel each other because the surface of each is negatively charged.

Soaps form micelles because their hydrophobic portion seeks a nonpolar environment and their hydrophilic portion seeks an aqueous environment. In water, a nonpolar environment can be achieved if the alkyl chains of a number of molecules cluster together to form a sphere (a micelle) in which the alkyl chains point to the center of the sphere. The carboxylate ions, the hydrophilic portion, are on the surface of the sphere projecting out into the polar aqueous environment, as shown in Figure 15.10. Aligning the carboxylate groups puts a negative charge on the surface of the micelle that repels any similarly charged micelle. As a result, micelles are dispersed throughout the aqueous phase.

Water alone cannot dissolve the hydrophobic greases, oils, and fats that soil fabrics and surfaces. When soap is added to the water, however, the hydrophobic part of the micelles dissolves the grease. This, in turn, removes the grease from the surface and transfers it to a micelle. Repulsion between micelles prevents the oil from coalescing, so that a stable emulsion of water and oil is formed. The net result is solubilized grease that can be easily rinsed away with water. This process is shown schematically in Figure 15.11.

Figure 15.11
Schematic representation of the cleaning action of a soap.

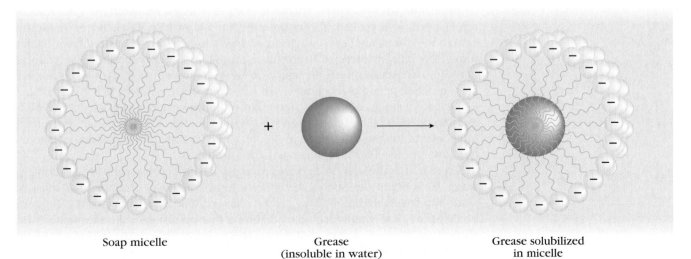

Soap micelle   Grease   Grease solubilized
                             (insoluble in water)   in micelle

Soaps work better in theory than they do in practice. For instance, hard water contains calcium and magnesium ions that convert sodium carboxylates into insoluble calcium and magnesium carboxylates. These compounds are responsible for the ring of so-called soap scum in many bathtubs or sinks and the grey tinge to some white clothes. Chemists have tried to solve this problem by preparing synthetic detergents that do not form insoluble salts with ions such as Ca^{2+} and Mg^{2+}. There is considerable variation in the chemical structures of these synthetic detergents, but they all have the same basic structure as soaps. Both have a large nonpolar hydrocarbon tail that is hydrophobic and a polar head that is water soluble. Examples include the sodium alkyl sulfates (Section 11.16) such as sodium lauryl sulfate, which is a common ingredient in many laundry detergents.

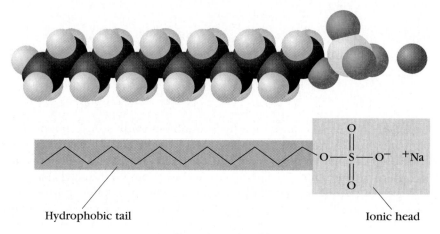

Hydrophobic tail   Ionic head

Sodium lauryl sulfate
(Sodium dodecanyl sulfate)
$C_{12}H_{25}OSO_3^-Na^+$

Certain compounds in humans act as detergents or emulsifying agents to assist in the digestion of fats.

15.15 Biological Detergents

The first step in the digestion of the fats that humans eat is enzyme-catalyzed hydrolysis or saponification. This does not occur in the stomach but rather in the upper portion of the small intestine with the aid of **bile salts.** Bile salts are the salts of the conjugate bases of bile acids. Bile salts are biosynthesized from cholesterol. They are synthesized in the liver, stored and concentrated in the gallbladder, and then released into the small intestine. The two major bile salts in humans are taurocholate and glycocholate, whose structures are shown in Figure 15.12.

Glycocholate is an effective detergent or emulsive agent because it, like soaps and detergents, has polar regions and nonpolar regions. The polar part of the molecule, highlighted in green in Figure 15.12, is hydrophilic, while the remainder of the molecule is nonpolar and hydrophobic.

Fats resist hydrolysis by enzymes in the stomach because they are insoluble in water. As a result, fats enter the small intestine as small globules that must be made water soluble before they can be digested. Bile salts carry out this function by dispersing the water-insoluble fats into a fine emulsion, which facilitates their absorption by the intestine and makes them susceptible to the action of the fat-hydrolyzing enzymes.

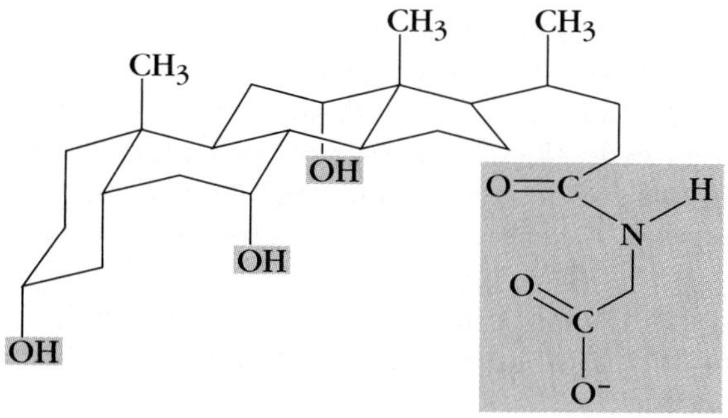

Glycocholate

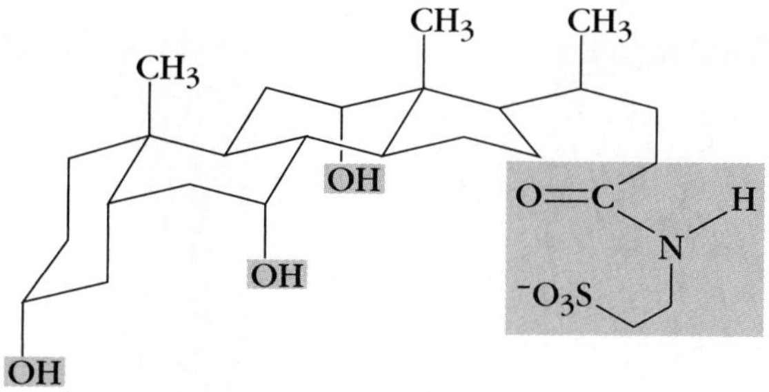

Taurocholate

Figure 15.12
Structures of cholesterol, glyco-cholate, and taurocholate. The polar portion of the ions, highlighted in green, is hydrophilic, while the remainder of the molecule is hydrophobic.

In the intestine, enzyme-catalyzed hydrolysis of fats and other esters occurs by a mechanism very similar to the one that we learned in Section 15.11 for the acid-catalyzed hydrolysis of esters in the laboratory.

15.16 Enzyme-Catalyzed Hydrolysis of Esters

The enzyme-catalyzed hydrolysis of fats into their constituent fatty acids and glycerol is an example of the hydrolysis of esters in living systems. Although the result is the same as the acid-catalyzed hydrolysis of esters, the conditions under which the two reactions occur are very different. In the laboratory, hydrolysis occurs rapidly only in acidic or basic solutions. In living systems, in contrast, the solution pH of cells is near 7.0, so the acid and base concentrations are too low to speed up hydrolysis. In living systems, therefore, the rapid hydrolysis of esters is not due to the presence of high concentrations of hydroxide ions or protons in the cell solution. How then do enzymes catalyze reactions in living systems?

The answer is found in the structure of the enzyme. The hydrolysis of esters occurs at the active site of the enzyme. Acidic or basic functional groups found at specific locations in the active site catalyze the reaction by donating or accepting protons during the course of the reaction. Conjugate acids of amines, for example, are proton donors, while amines and carboxylate ions are proton acceptors. By transferring protons to and from these groups to the substrate, an enzyme carries out acid- and base-catalyzed reactions within the active site. We can demonstrate these properties of enzymes by examining the mechanism of the enzyme-catalyzed hydrolysis of esters.

Esterases are a family of enzymes that catalyze the hydrolysis of esters. The mechanism of this enzyme-catalyzed reaction has been extensively studied by chemical and x-ray techniques. Two functional groups that are part of the active site have been identified as being particularly important to the catalytic process. One is a hydroxy group that acts as a nucleophile, and the other is the nitrogen atom of an amine, which accepts a proton and then gives it back during the reaction.

The mechanism of esterase-catalyzed hydrolysis consists of two stages, each of which consists of two steps. Step 1 of Stage 1 is nucleophilic addition to form a tetrahedral intermediate:

Step 1, Stage 1:

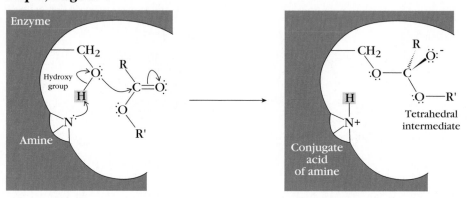

The nucleophile is the oxygen atom of an —OH group on the enzyme. It adds to the carbon atom of the C=O bond of the ester. Notice that as this is happening, the hydroxy hydrogen atom is transferred to the nitrogen atom of an amine located nearby.

In Step 2 of Stage 1, a proton is transferred from the conjugate acid of the amine to the alkyl oxygen atom of the substrate at the same time as an alcohol molecule is lost from the tetrahedral intermediate to form an acyl enzyme intermediate:

Step 2, Stage 1:

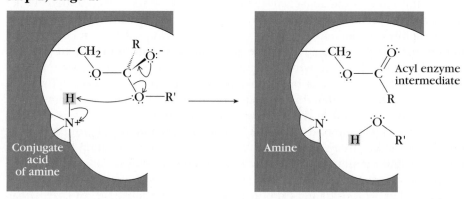

This ends the first stage of the mechanism. The second stage is the hydrolysis of the acyl enzyme intermediate, which also consists of two steps.

In Step 1 of Stage 2, the oxygen atom of a water molecule adds to the carbon atom of the C=O bond of the acyl enzyme intermediate with simultaneous transfer of a proton to the amine nitrogen to form a second tetrahedral intermediate.

Step 1, Stage 2:

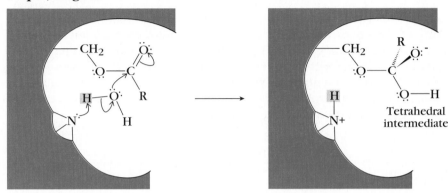

In Step 2 of Stage 2, the enzyme oxygen atom of the tetrahedral intermediate is eliminated at the same time as a proton is transferred from the conjugate acid of the amine, forming the carboxylic acid.

Step 2, Stage 2:

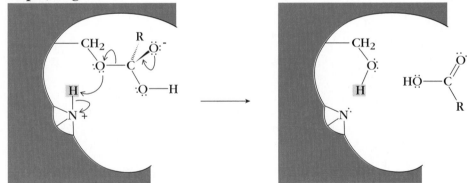

The carboxylic acid subsequently loses a proton to a base in solution to form a carboxylate ion. Notice that the enzyme is restored in Step 2 of Stage 2 so it can immediately catalyze the hydrolysis of another molecule of ester.

The biggest difference between the mechanisms of enzyme-catalyzed and nonenzymatic hydrolysis of esters is the formation of an acyl enzyme intermediate in the enzyme-catalyzed reaction. Note, however, that this intermediate, which contains a covalent bond between the enzyme and acyl part of the ester, is formed and hydrolyzed by a familiar mechanism. That mechanism is the addition of a nucleophile to the carbon atom of a C=O bond to form a tetrahedral intermediate, which eliminates a group to form the product. Proton transfer occurs in both mechanisms, too. The presence of neighboring acidic and basic functional groups in the active site, however, makes proton transfer much more rapid in enzyme-catalyzed hydrolysis.

Despite the differences between the two, both reactions occur by an addition-elimination mechanism. Thus the mechanism that describes the enzyme catalysis of esters is fundamentally the same as that which describes the hydrolysis of esters in the laboratory.

15.17 Reduction of Carboxylic Acids and Their Esters

Carboxylic acids can be reduced to primary alcohols with reducing agents, such as $LiAlH_4$ ($NaBH_4$ does not reduce carboxylic acids to alcohols) and H_2, and a metal catalyst. Reduction with $LiAlH_4$ is more convenient and is the method usually used in the laboratory. Carboxylic acids react with $LiAlH_4$ to form first the carboxylate ion. The lithium carboxylate is then reduced further to the salt of the corresponding primary alcohol. Addition of an aqueous acid solution to the reaction mixture converts the alkoxide ion to an alcohol.

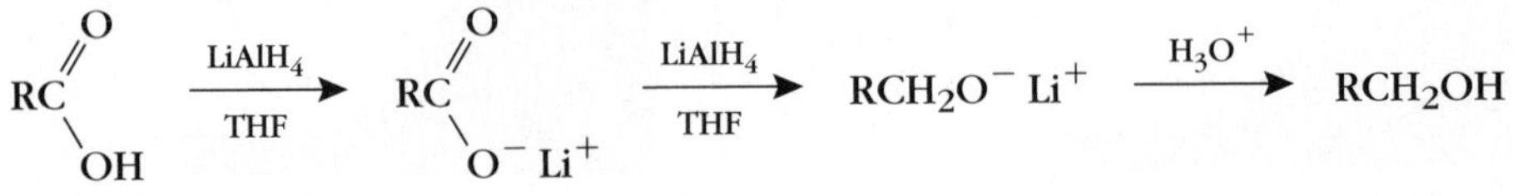

Carboxylic acid — 1° Alcohol

Carboxylic acid esters are reduced by $LiAlH_4$ to two alcohols on reduction. One is a primary alcohol corresponding to the acid portion of the ester. The other alcohol may be a primary, secondary, or tertiary alcohol, depending on the nature of the alkoxy part of the ester.

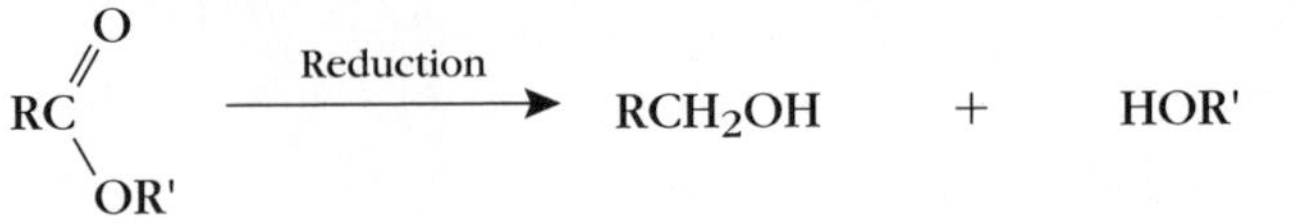

Carboxylic acid ester — 1° Alcohol from carboxylic acid part of ester — Alcohol from alkoxy part of ester

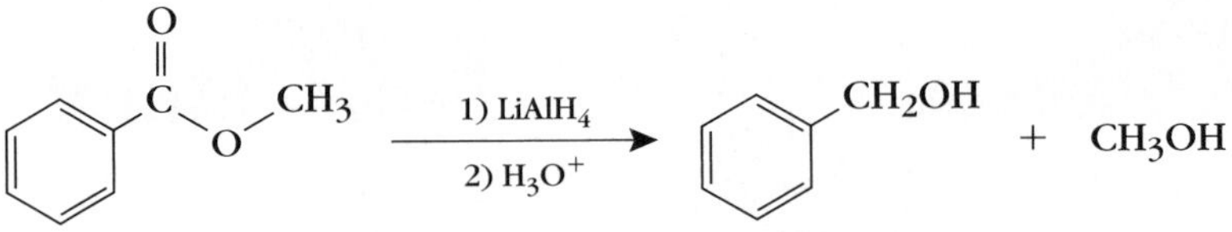

Methyl benzoate — Benzyl alcohol (90% yield) — Methanol

The mechanism of these reductions is discussed in Section 16.12.

Industrially catalytic hydrogenation of carboxylic acids and their esters is an important source of primary alcohols, particularly the long straight-chain primary alcohols that are obtained by the reduction of fatty acids (either directly or as their esters). Many detergents are made commercially from the alcohols obtained by the reduction of these fats.

$$\text{Fats} \xrightarrow[\text{Pd}]{H_2} RCH_2OH \xrightarrow{H_2SO_4} RCH_2OSO_3H \xrightarrow{\text{NaOH}} RCH_2OSO_3^{-}Na^{+}$$

Mixture of long-chain 1° alcohols — Mixture of alkyl hydrogen sulfates — Mixture of alkyl sulfates (a detergent)

Exercise 15.17 Propose a synthesis of each of the following compounds using decanoic acid, $C_9H_{19}COOH$, as the starting material.

(a) 1-Bromodecane
(b) Undecanoic acid, $CH_3(CH_2)_9COOH$
(c) Dodecanoic acid, $CH_3(CH_2)_{10}COOH$
(d) Decane

(e) 1-Decene
(f) 1-Decyne
(g) 2-Dodecanone
(h) $CH_3(CH_2)_9C{\equiv}CCH_2CH_3$

15.18 Summary

Carboxylic acids are compounds that contain a carboxyl group, —COOH. They are relatively polar compounds and exist in solution as dimers because hydrogen bonds can form between the hydrogen of the —OH group of one molecule and the oxygen atom of the C=O bond of another. As a result, low molecular weight carboxylic acids have relatively high boiling points and water solubilities.

Aliphatic carboxylic acids are weak acids, with pK_a values between 4 and 5. At pH values of 7 and above, they exist in solution exclusively as their carboxylate ions, $RCOO^-$. The acidity of carboxylic acids is increased by the presence of electron-withdrawing substituents on the α-carbon atom. This increase in acidity can be substantial. Trichloroacetic acid, for example, is almost as strong an acid as nitric acid.

Carboxylic acids are prepared by (1) oxidation of aldehydes and primary alcohols, (2) oxidation of alkylbenzenes, (3) oxidation of alkenes, (4) carboxylation of Grignard and organolithium reagents, and (5) hydrolysis of nitriles. The latter two methods are particularly useful because they allow haloalkanes to be converted into carboxylic acids that contain one carbon atom more than the starting haloalkane.

Esters are compounds that contain the —COOR functional group. They are prepared by acid-catalyzed **esterification** of carboxylic acids and alcohols. Esters are hydrolyzed under acidic or basic conditions to a carboxylic acid (or its conjugate base in basic solution) and an alcohol. In basic solution the reaction is called **saponification.**

Fats and **oils** are called **triacylglycerols** because they are the triesters of long-chain fatty acids and glycerol (1,2,3-propanetriol). Fatty acids are carboxylic acids that contain an even number of carbon atoms arranged in a straight chain. **Soaps,** the salts of fatty acids, are formed commercially by the saponification of fats. Soaps have both a hydrophobic and a hydrophilic region. When placed in water, soaps form clusters of several hundred molecules called **micelles,** in which their hydrophobic parts bunch together on the inside of the micelle and the ionic or hydrophilic parts lie on the outside where they are solvated by water. Micelles disperse hydrophobic substances in water. In addition to soaps, commercial detergents and biological detergents such as the bile salts also have hydrophobic and hydrophilic regions and they act like soaps. Reduction of fats and fatty acids is an important commercial method of preparing long-chain primary alcohols.

The mechanisms of the acid-catalyzed preparation and hydrolysis of esters in both the laboratory and living systems have several features in common. Protonation of the oxygen atom of the C=O bond occurs in both cases before addition of a nucleophile can occur to form a tetrahedral intermediate. Elimination of a group from the tetrahedral intermediate reforms the C=O bond in the product. A major difference is the source of the acid catalyst. In the laboratory, the acid is added to the solvent, while in living systems the acid comes from acidic groups adjacent to the reactive site of the enzyme.

15.19 Summary of Preparations and Reactions of Carboxylic Acids and Esters

1. **Preparation of Carboxylic Acids**

   a. **Oxidation reactions**   SECTION 15.5

   (i) **Primary alcohols**

   The most commonly used oxidizing agents are $Na_2Cr_2O_7/H_2SO_4/H_2O$; $CrO_3/H_2SO_4/H_2O$; and $KMnO_4/H_2SO_4/H_2O$.

   $$RCH_2OH \xrightarrow{[O]} RCOOH$$

   1° Alcohol   Carboxylic acid

   $$(CH_3)_2CHCH_2OH \xrightarrow[H_2SO_4/H_2O]{KMnO_4} (CH_3)_2CHCOOH$$

   (ii) **Aldehydes**

   Aldehydes are easily oxidized to carboxylic acids. See Sections 11.18 and 14.18.

   $$RCHO \xrightarrow{[O]} RCOOH$$

   Aldehyde   Carboxylic acid

   $$(CH_3)_3CCHO \xrightarrow[\text{reagent}]{\text{Tollen's}} (CH_3)_3CCOOH$$

   (iii) **Alkenes**

   Synthetically useful only with terminal alkenes or symmetrically disubstituted alkenes.

   $$RCH=CHR \xrightarrow{[O]} 2 \ RCOOH$$

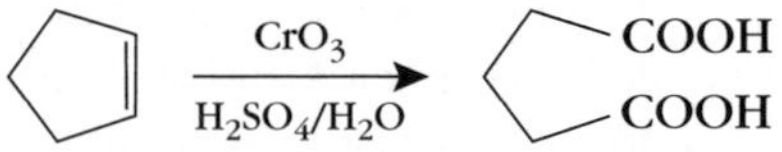

   (iv) **Alkylbenzene derivatives**

   Benzene ring is not oxidized by most common oxidizing agents.

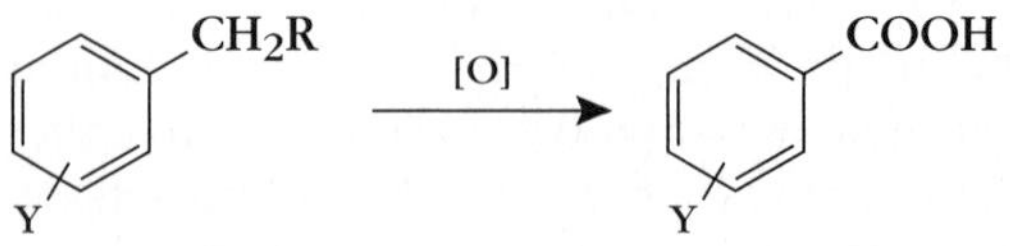

   Alkylbenzene   Benzoic acid

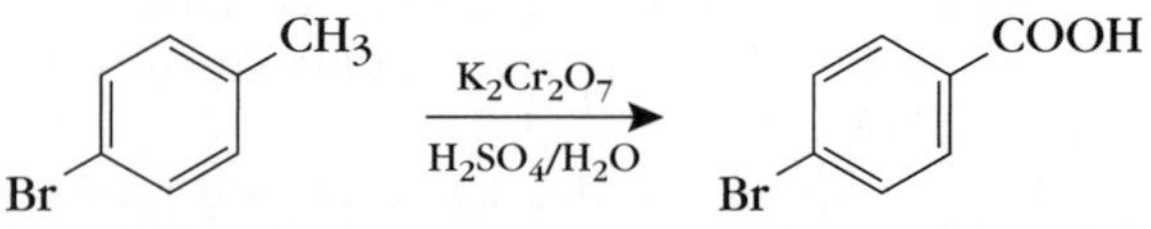

b. Carboxylation of Grignard reagents   SECTION 15.6

A similar reaction occurs with organo-lithium reagents.

$$RMgX \xrightarrow[\text{(2) } H_3O^+]{\text{(1) } CO_2} RCOOH$$

Grignard reagent   Carboxylic acid

$$(CH_3)_2CHCH_2CH_2MgBr \xrightarrow[\text{(2) } H_3O^+]{\text{(1) } CO_2} (CH_3)_2CHCH_2CH_2COOH$$

c. Hydrolysis of nitriles   SECTION 15.7

For synthesis of nitriles, see Section 12.18.

$$RC\equiv N \xrightarrow[\text{Heat}]{H_3O^+} RCOOH$$

Nitrile   Carboxylic acid

$$CH_3(CH_2)_3CH_2CN \xrightarrow[\text{Heat}]{H_3O^+} CH_3(CH_2)_3CH_2COOH$$

2. Reactions of Carboxylic Acids

a. Acidity   SECTION 15.8

pK_a of unsubstituted aliphatic carboxylic acids is 4 to 5.

$$RCOOH + H_2O \rightleftharpoons RCOO^- + H_3O^+$$

Carboxylic acid   Carboxylate ion

$$CH_3COOH + H_2O \rightleftharpoons CH_3COO^- + H_3O^+ \quad pK_a = 4.74$$

b. Fischer esterification   SECTION 15.11

To obtain a good yield of ester, water must be removed as it forms, or an excess of one of the reactants must be used.

$$RCOOH + R'OH \underset{\text{Acid catalyst}}{\overset{K}{\rightleftharpoons}} RCOOR' + H_2O$$

Carboxylic acid   Alcohol   Ester   Water

$$(CH_3)_2CHCOOH + CH_3OH \underset{\text{Acid catalyst}}{\rightleftharpoons} (CH_3)_2CHCOOCH_3 + H_2O$$

c. Reduction   SECTION 15.17

$NaBH_4$ does not reduce carboxylic acids, but both $LiAlH_4$ and catalytic hydrogenation do.

$$RCOOH \xrightarrow[\text{(2) } H_3O^+]{\text{(1) } LiAlH_4} RCH_2OH$$

Carboxylic acid   1° Alcohol

$$CH_3(CH_2)_4CH_2COOH \xrightarrow[\text{(2) } H_3O^+]{\text{(1) } LiAlH_4} CH_3(CH_2)_4CH_2CH_2OH$$

3. **Preparation of Carboxylic Acid Esters**

a. **Fischer esterification** SECTION 15.11

To obtain a good yield of ester, water must be removed as it forms, or an excess of one of the reactants must be used.

$$RCOOH + R'OH \underset{\text{Acid catalyst}}{\overset{K}{\rightleftarrows}} RCOOR' + H_2O$$

Carboxylic acid   Alcohol   Ester   Water

$$CH_3COOH + CH_3OH \underset{\text{Acid catalyst}}{\rightleftarrows} CH_3COOCH_3 + H_2O$$

b. **Nucleophilic substitution reaction** SECTION 12.18

Reaction occurs by an S_N2 mechanism.

$$RCOO^- + R'X \longrightarrow RCOOR' + X^-$$

Carboxylate ion   Haloalkane

$$CH_3COO^- + CH_3(CH_2)_4CH_2Br \longrightarrow CH_3COOCH_2(CH_2)_4CH_3 + Br^-$$

4. **Reactions of Carboxylic Acid Esters**

a. **Acid-catalyzed hydrolysis** SECTION 15.11

Reverse of Fischer esterification; use excess water to drive hydrolysis to completion.

$$RCOOR' + H_2O \underset{\text{Acid catalyst}}{\rightleftarrows} RCOOH + R'OH$$

Ester   Water   Carboxylic acid   Alcohol

$$CH_3COOCH(CH_3)_2 + H_2O \underset{\text{Acid catalyst}}{\rightleftarrows} CH_3COOH + (CH_3)_2CHOH$$

b. **Saponification** SECTION 15.12

Irreversible reaction.

$$RCOOR' + OH^- \xrightarrow{H_2O} RCOO^- + R'OH$$

Ester   Hydroxide ion   Carboxylate ion   Alcohol

$$(CH_3)_2CHCOOCH_3 + OH^- \xrightarrow{H_2O} (CH_3)_2CHCOO^- + CH_3OH$$

c. **Reduction** SECTION 15.17

Two alcohols are formed as product.

$$RCOOR' \xrightarrow[\text{(2) } H_3O^+]{\text{(1) LiAlH}_4} RCH_2OH + R'OH$$

Ester   Alcohol   Alcohol

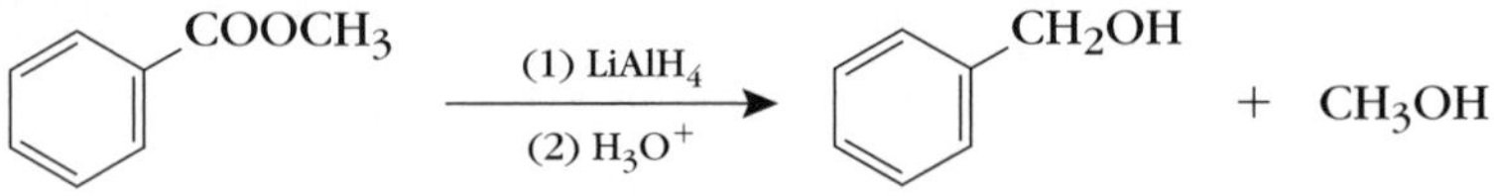

Additional Exercises

15.18 Define each of the following terms in words, by a structure, or by a chemical equation.

    (a) Carboxylic acid   **(f)** Saponification of an ester   **(j)** Fat

    (b) Carboxylate ion   **(g)** Acid-catalyzed esterification   **(k)** Detergent

    (c) Fatty acid   **(h)** Salt of a carboxylic acid   **(l)** Saturated fat

    (d) Unsaturated fat   **(i)** Soap   **(m)** Fischer esterification

    (e) Ester

15.19 Give a systematic name to each of the following compounds:

    (a) 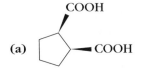   **(d)**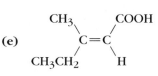

    (b) 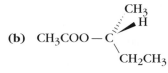   **(e)**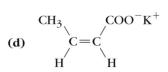

    (c) 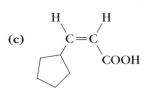   **(f)**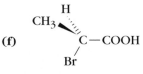

15.20 Write the structural formula for each of the following names:

    (a) Phenylacetic acid

    (b) Magnesium formate

    (c) 2-Methylbutanoic acid

    (d) Sodium benzoate (used as a food preservative)

    (e) (R)-2-Chloro-3-butenoic acid

    (f) (E)-3-Pentenoic acid

    (g) Hexyl acetate

    (h) *trans*-2-Nitrocyclohexanecarboxylic acid

    (i) Isopropyl 3-methylbutanoate

    (j) 2,2-Dimethylpropyl *cis*-2-methylcyclopropanecarboxylate

15.21 Write the structures and give names to nine esters of molecular formula $C_5H_{10}O_2$.

15.22 Arrange each of the following groups of compounds in order of increasing acid strength.

    (a) Acetic acid, ethanol, formic acid

    (b) Butanoic acid, 2-chlorobutanoic acid, 3-chlorobutanoic acid

    (c) Sulfuric acid, 1-butyne, butanoic acid

    (d) 1-Butyne, 1-butanol, water

15.23 In each of the following pairs of compounds, identify the stronger base.

    (a) Sodium acetate, sodium chloroacetate   **(d)** Sodium acetate, sodium hydroxide

    (b) Sodium acetate, sodium acetylide   **(e)** Sodium acetate, sodium chloride

    (c) Sodium acetate, sodium ethoxide

15.24 Write the acid-base reactions between the following pairs of reactants and predict if the products or reactants are favored at equilibrium.

    (a) Butanoic acid + ammonia   **(c)** Benzoic acid + sodium ethoxide

    (b) Acetic acid + sodium trichloroacetate

15.25 The pK_a of dodecanoic acid, $CH_3(CH_2)_{10}COOH$, is 4.89. Calculate the equilibrium constant for the following reaction (you may wish to review Section 3.5).

$$CH_3(CH_2)_{10}C\overset{O}{\underset{OH}{}} + OH^- \overset{K}{\rightleftharpoons} CH_3(CH_2)_{10}C\overset{O}{\underset{O^-}{}} + H_2O$$

15.26 The following compounds all have about the same molecular weight. Arrange them in order of increasing boiling point.

     (a) Butanoic acid   **(b)** Ethyl acetate   **(c)** 1-Chlorobutane   **(d)** 2-Pentanone

15.27 Explain how the following mixtures of compounds can be separated by chemical means to obtain each component in pure form.

     (a) Hexanoic acid and ethyl hexanoate   **(c)** Hexanoic acid and 1-hexanol
     (b) Dipropyl ether and propanoic acid

15.28 Write equations for the reaction (if any) of propanoic acid with each of the following reagents:

     (a) KOH/H_2O **(c)** 1-Propanol/acid catalyst   **(e)** $KMnO_4/H_2SO_4/H_2O$
     (b) $LiAlH_4$ followed by H_3O^+ **(d)** $NaBH_4$ followed by H_3O^+

15.29 Write equations for the reaction (if any) of ethyl propanoate with the reagents listed in Exercise 15.28.

15.30 Write the structure of the major organic product in each of the following reactions.

     (a) Potassium butanoate $\xrightarrow[H_2O]{HCl}$ **(f)** $(CH_3)_2CHCOOH \xrightarrow[H_2O]{KOH}$

     (b) 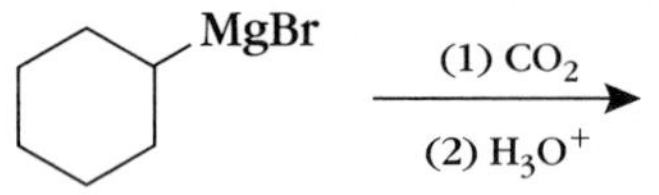 [cyclohexyl]–MgBr $\xrightarrow[(2) H_3O^+]{(1) CO_2}$ **(g)** Oleic acid $\xrightarrow[CCl_4]{I_2}$

     (c) 1-Bromopentane $\xrightarrow{KCN}$ **(h)** Linolenic acid $\xrightarrow[Ni]{Excess H_2}$

     (d) Product of **(c)** $\xrightarrow[Heat]{H_2SO_4/H_2O}$

     (i)
$$\begin{array}{l} CH_2OOC(CH_2)_{16}CH_3 \\ | \\ CHOOC(CH_2)_{14}CH_3 \\ | \\ CH_2OOC(CH_2)_7CH{=}CH(CH_2)_7CH_3 \end{array} \xrightarrow[(2) HCl/H_2O]{(1) NaOH/H_2O}$$

     (e) Oleic acid $\xrightarrow[H_2SO_4/Heat]{KMnO_4}$ **(j)** 3-Chlorohexanoic acid + 2-butanol $\xrightarrow{H_2SO_4}$

15.31 Write the structure of the product or products formed in each of the following reactions.

     (a) 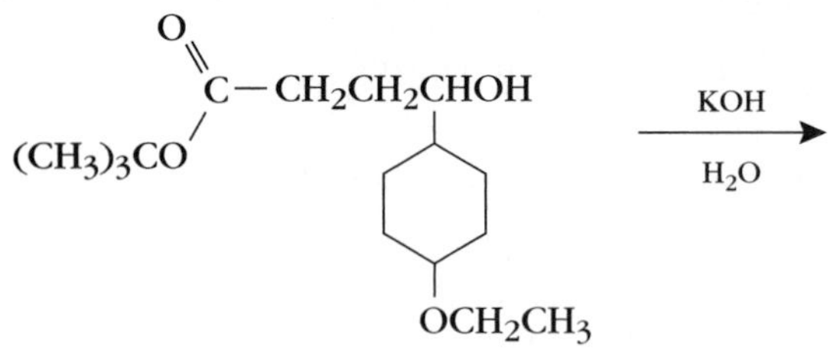
$$(CH_3)_3CO\overset{O}{{-}}C{-}CH_2CH_2CHOH\text{[cyclohexyl with }OCH_2CH_3] \xrightarrow[H_2O]{KOH}$$

     (b) 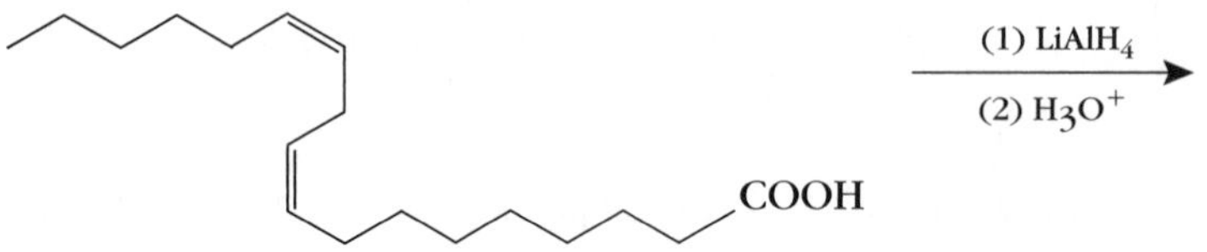 [long-chain diene] COOH $\xrightarrow[(2) H_3O^+]{(1) LiAlH_4}$

(c)

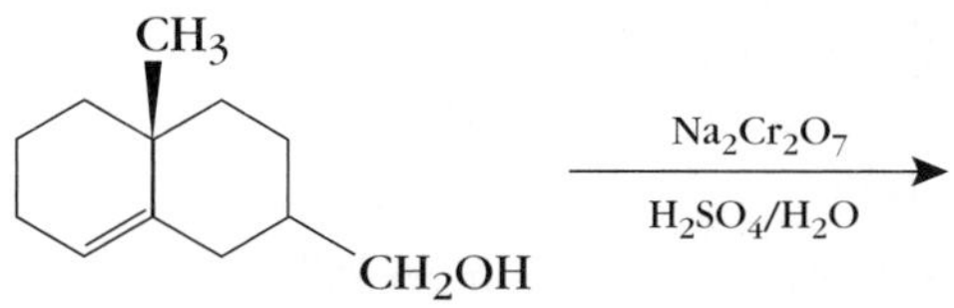

$$\xrightarrow[\text{H}_2\text{SO}_4/\text{H}_2\text{O}]{\text{Na}_2\text{Cr}_2\text{O}_7}$$

(d)

$$\xrightarrow[\text{PtO}_2]{\underset{\text{H}_2}{\text{1 Equivalent}}}$$

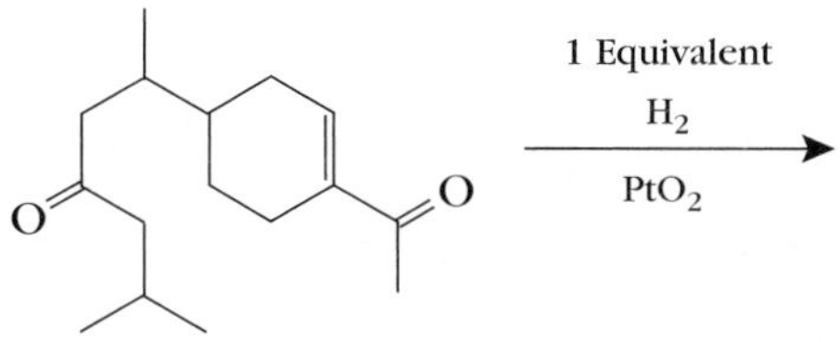

15.32 Write the equation for the reaction or reactions needed to carry out the following transformations.

(a)

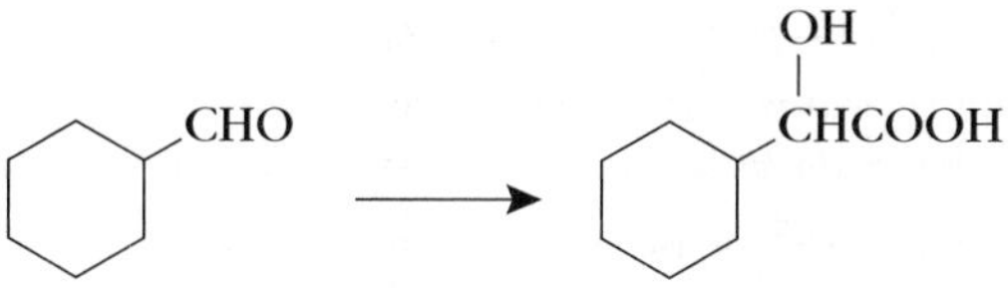

(b) $(CH_3)_2CHBr \longrightarrow (CH_3)_2CHCOOH$

(c) Bromocyclopentane $\longrightarrow$ CH_2COOH

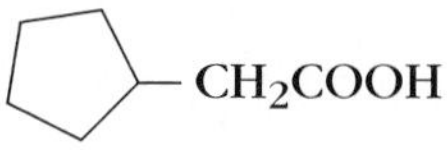

(d) $CH_3(CH_2)_3C{\equiv}CH \longrightarrow CH_3(CH_2)_6COOCH_3$

(e) 1-Butene $\longrightarrow$ 1-bromo-2-methylbutane

15.33 Propose a synthesis for each of the following compounds from the indicated starting material.

    (a) 3,4-Dibromobutanoic acid from 3-bromopropene
    (b) Pentanoic acid from 1-bromopropane
    (c) Methyl (Z)-2-butenoate from propyne
    (d) 1-Chloropentane from pentanoic acid
    (e) 1,5-Pentanedioic acid from 1,3-propanediol

15.34 Using $^{13}CO_2$, $^{13}CH_3OH$, and $H_2{}^{18}O$ as the sources of carbon-13 and oxygen-18, propose a synthesis for each of the following compounds.

(a) 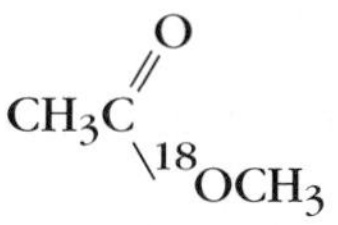   **(c)** 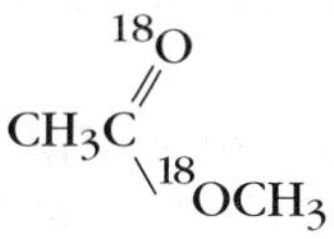   **(e)**

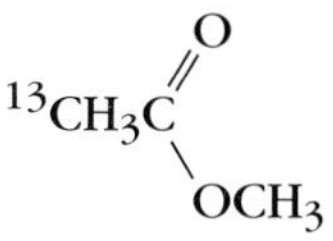

(b) 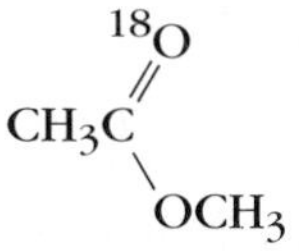   **(d)**

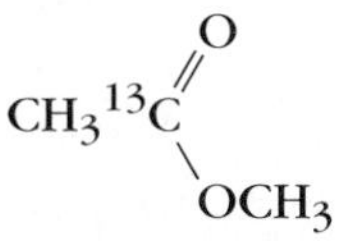

15.35 What spectral technique(s) would distinguish between the following pairs of compounds?

    (a) Ethyl acetate and methyl acetate
    (b) Chloroacetic acid and 2-chloropropanoic acid
    (c) Acrylic acid and propanoic acid

    (d) 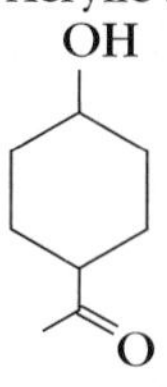 and

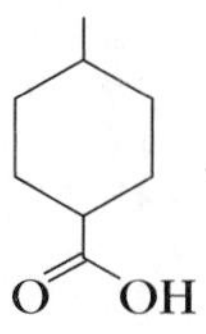

    (e) 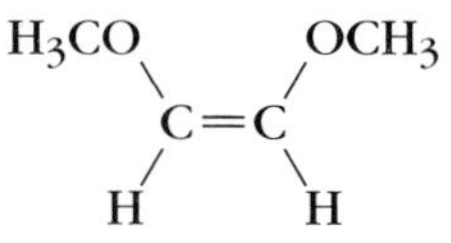 and $CH_3CH_2{-}C$

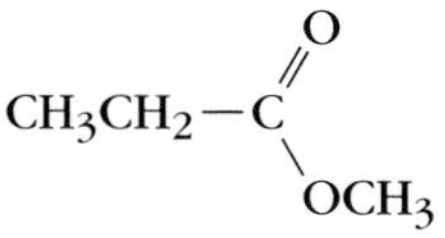

15.36 On the basis of the information given, deduce the structures of Compounds A to E.

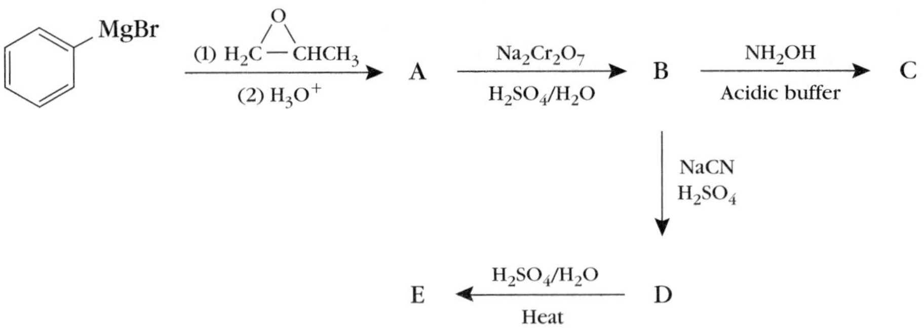

15.37 Write the structure of a triacylglycerol whose saponification forms one equivalent of glycerol, one equivalent of sodium stearate, and two equivalents of sodium oleate.

15.38 Heating acetic acid in an aqueous solution of oxygen-18 enriched water slowly forms acetic acid, in which oxygen-18 is found in the oxygen atoms of the carboxyl groups. Write a mechanism for this reaction that includes all intermediates.

15.39 Carboxylic acids add to alkenes in the presence of an acid catalyst, as shown by the following reaction.

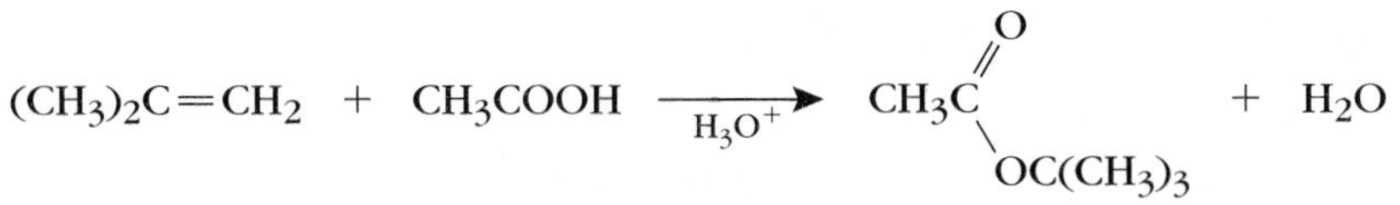

$$(CH_3)_2C\!=\!CH_2 \ + \ CH_3COOH \ \xrightarrow[H_3O^+]{} \ CH_3C\underset{OC(CH_3)_3}{\overset{O}{\Vert}} \ + \ H_2O$$

Write a mechanism for this reaction that includes all intermediates.

15.40 On the basis of the following information, deduce the structures of compounds **F** to **K**.

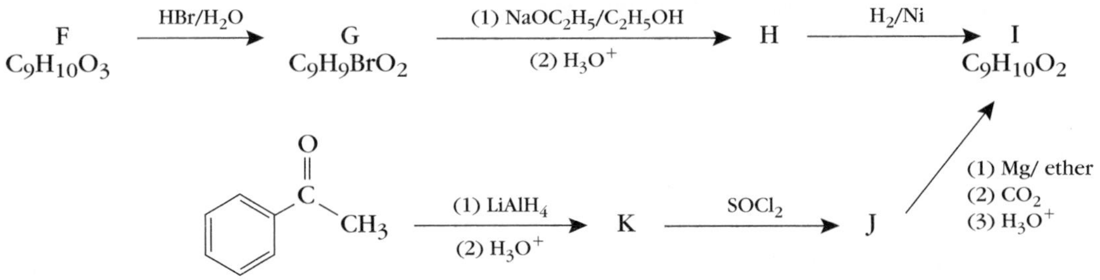

15.41 Write the structural formula of each of the following compounds based on their molecular formula and the spectral data provided.

(a) $C_5H_9ClO_2$ IR: 3400-2600 cm^{-1}; 1705 cm^{-1}. ^{13}C NMR: δ 182.1; δ 51.5; δ 44.5; δ 23.1. ^{1}H NMR: δ 1.33 s (3H); δ 3.62 s (1H).

(b) $C_5H_9ClO_2$ IR: 1740 cm^{-1}. ^{13}C NMR: δ 170.9; δ 61.2; δ 41.2; δ 31.6; δ 20.9. ^{1}H NMR: δ 2.07 s (3H); δ 2.10 quintet (2H); δ 3.62 t (2H); δ 4.22 t (2H).

(c) $C_4H_8O_3$ IR: 3400-2600 cm^{-1}; 1730 cm^{-1}. ^{13}C NMR: δ 175.7; δ 67.5; δ 67.4; δ 14.9. ^{1}H NMR: δ 1.26 t (3H); δ 3.63 q (2H); δ 4.14 s (2H).

(d) $C_5H_8O_4$ IR: 3300-2900 cm^{-1}; 1710 cm^{-1}. ^{13}C NMR: δ 170.9; δ 53.3; δ 21.9; δ 11.8. ^{1}H NMR: δ 0.94 t (3H); δ 1.80 quintet (2H); δ 3.10 t (1H).

15.42 The ^{1}H NMR spectra of two isomeric compounds of molecular formula $C_3H_5ClO_2$ are designated Spectrum A and B below. The IR spectra of both compounds have intense absorption bands in the regions 3400 to 2400 cm^{-1} and 1720 to 1680 cm^{-1}. Write the structural formula of each isomer and indicate which hydrogen atoms give rise to which peaks in each ^{1}H NMR spectrum.

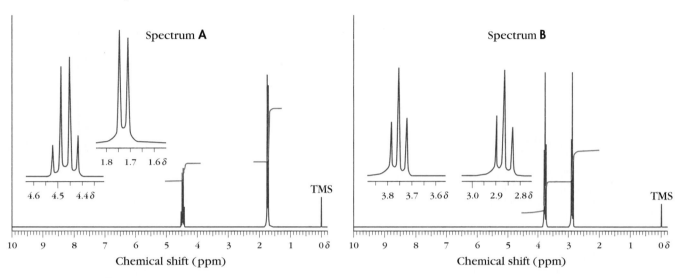

15.43 Sketch the ^{1}H NMR spectrum of each of the following compounds, giving the approximate chemical shifts of equivalent groups of hydrogen atoms and their spin-spin coupling patterns.

   (a) $(CH_3)_2CHCOOH$ **(b)** $BrCH_2COOCH_2CH_3$

15.44 Write the structural formula of each of the following compounds based on their molecular formula, IR, and ^{1}H NMR spectra.

   (a) $C_5H_{10}O_2$ **(b)** $C_5H_8O_2$ **(c)** $C_4H_6O_2$

(a) $C_5H_{10}O_2$

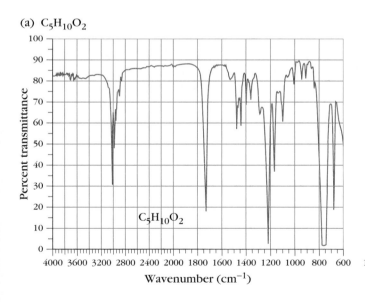

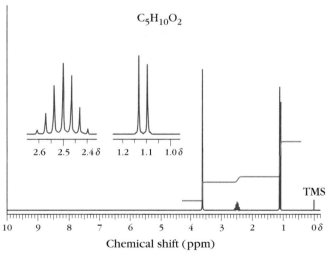

(b) $C_5H_8O_2$

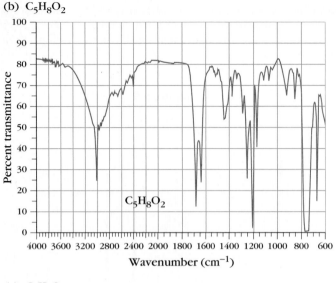

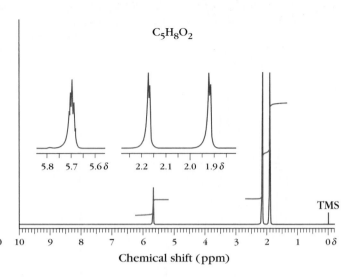

(c) $C_4H_6O_2$

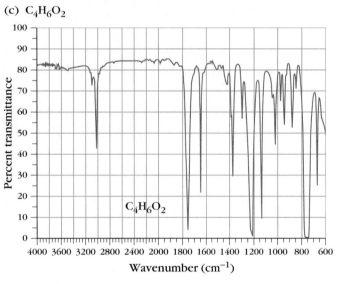

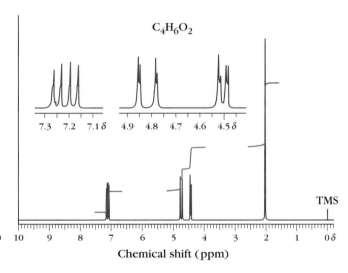

CHAPTER 16

ACYL TRANSFER REACTIONS
Interconversion of Carboxylic Acid Derivatives

Carboxylic acid esters, which are discussed in Chapter 15, are one example of a class of compounds called **carboxylic acid derivatives.** They are classed as derivatives because they form a carboxylic acid on hydrolysis. These compounds are also called **acyl derivatives** because they contain an **acyl group,** $R-C\overset{\displaystyle O}{\underset{\diagdown}{}}$.

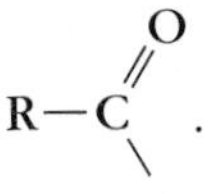

The structure of carboxylic acid derivatives or acyl derivatives consists of two parts, an acyl group and an electronegative substituent, which we designate by the letter Y.

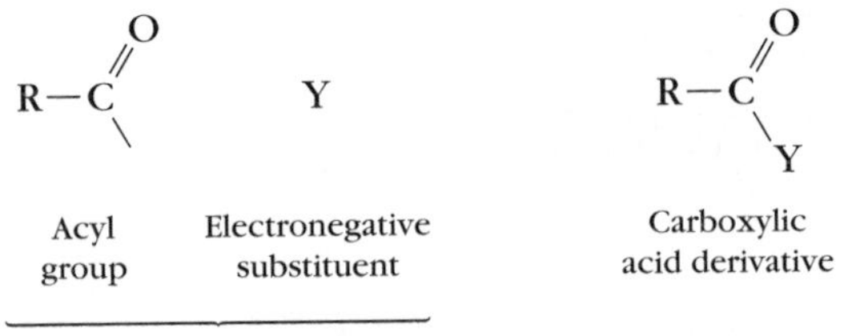

| Acyl group | Electronegative substituent | Carboxylic acid derivative |

Two parts of carboxylic acid derivatives

In this chapter the study of the chemistry of acyl derivatives, which began in Chapter 15 with carboxylic acids, their esters, and salts, is extended to carboxylic acid halides (simply referred to as acyl halides), carboxylic acid anhydrides (also referred to as acid anhydrides), thiocarboxylic acid esters (also referred to as thioesters), and amides. We also describe the reactions of nitriles, which are a class of compounds distantly related to carboxylic acids. The general structural formulas of these six carboxylic acid derivatives are compared to the corresponding carboxylic acid in Figure 16.1.

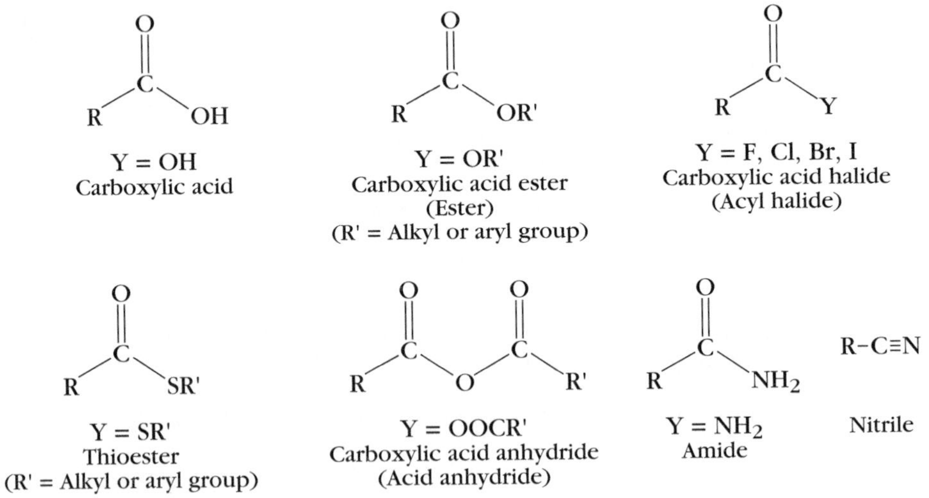

Figure 16.1
General structural formulas of carboxylic acids and their derivatives. All of the classes of compounds, except nitriles, contain an acyl group, $R-C\overset{\displaystyle O}{\underset{\diagdown}{}}$.

Y = OH
Carboxylic acid

Y = OR'
Carboxylic acid ester
(Ester)
(R' = Alkyl or aryl group)

Y = F, Cl, Br, I
Carboxylic acid halide
(Acyl halide)

Y = SR'
Thioester
(R' = Alkyl or aryl group)

Y = OOCR'
Carboxylic acid anhydride
(Acid anhydride)

Y = NH₂
Amide

Nitrile

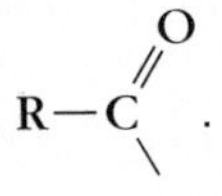

The predominant reaction of all acyl derivatives is an **acyl transfer reaction** in which the acyl group is transferred to an atom of another molecule. In the following hydrolysis reaction, for example, the acyl group of the acyl derivative is transferred from Y to the oxygen atom of a water molecule.

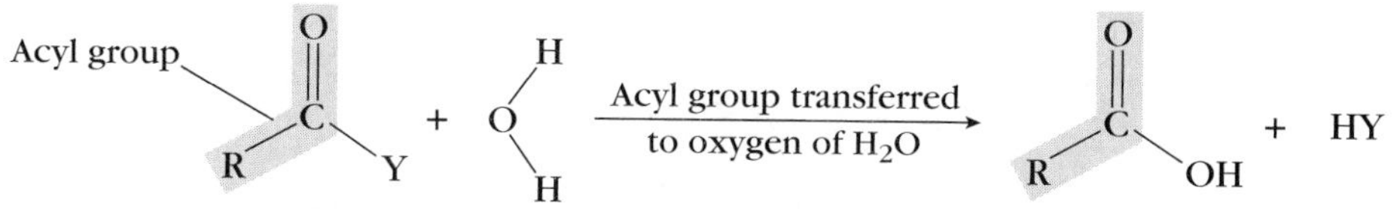

Acyl transfer reactions are also called **nucleophilic acyl substitution reactions** because this name more accurately describes the mechanism of such reactions, as we will learn shortly. Let's begin our study of the chemistry of carboxylic acid derivatives by learning how to name them.

16.1 Naming Carboxylic Acid Derivatives

Because the common feature of carboxylic acid derivatives is an acyl group, we begin by discussing how to name such groups.

Acyl groups are named by replacing the *-ic* acid ending of the corresponding carboxylic acid with *-yl.*

Acyl group

Formyl group | Acetyl group | Benzoyl group | 3-Butenoyl group

Corresponding carboxylic acid

Formic acid | Acetic acid | Benzoic acid | 3-Butenoic acid

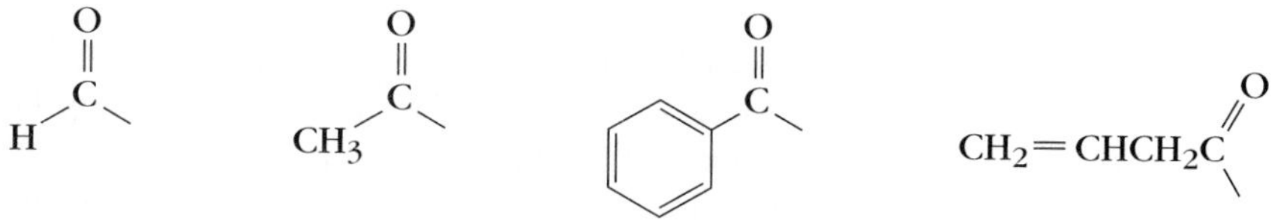

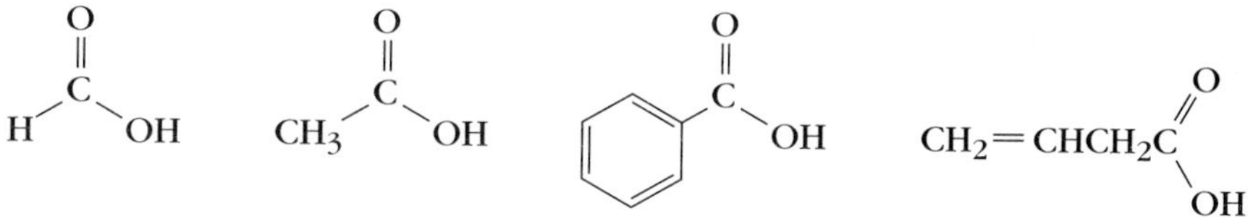

Carboxylic acid halides are named by adding the name of the appropriate halide after the name of the acyl group.

Acetyl chloride | Benzoyl bromide | 3-Butenoyl chloride

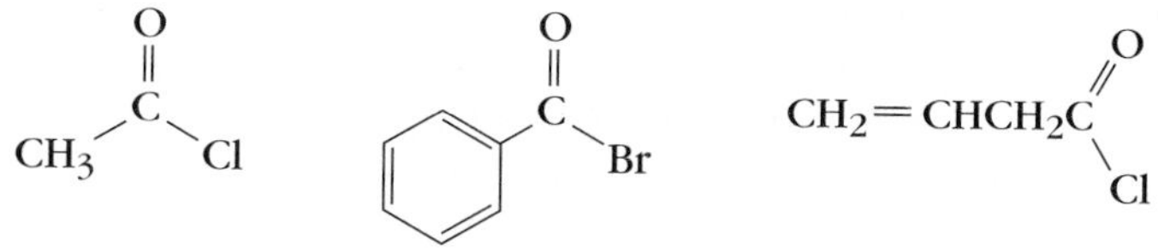

Carboxylic acid anhydrides contain two acyl groups. When the two acyl groups are the same, the name of the anhydride is obtained from the name of the corresponding carboxylic acid by replacing the word *acid* with *anhydride*. When the acyl groups are different, the anhydride is called a *mixed anhydride* and it is named by listing alphabetically the names of the two corresponding carboxylic acids and again replacing the word *acid* with *anhydride*. For example:

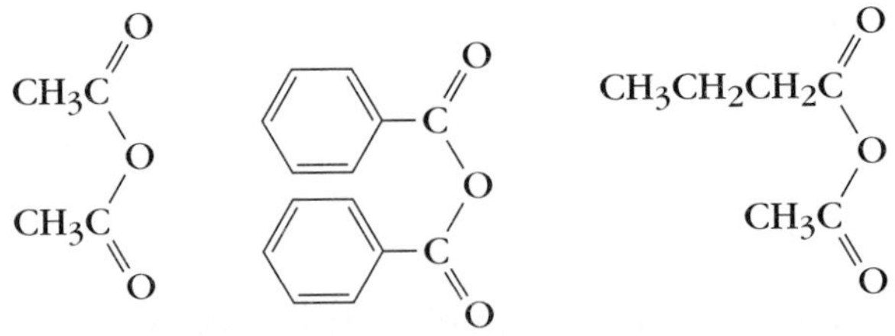

Acetic anhydride   Benzoic anhydride   Butanoic ethanoic anhydride

Cyclic anhydrides formed from dicarboxylic acids are named in an analogous fashion.

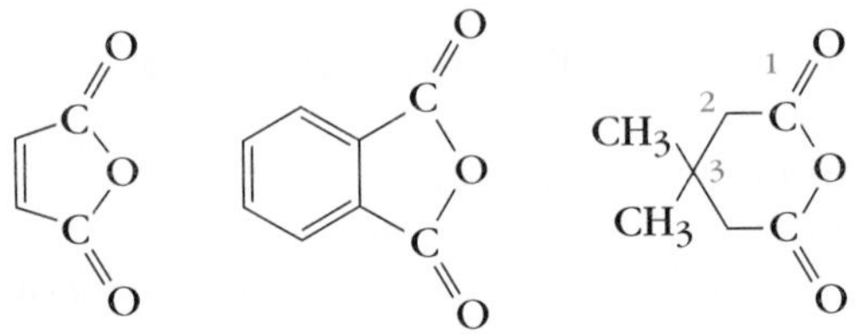

| Maleic anhydride | Phthalic anhydride | 3,3-Dimethylglutaric anhydride |

The names of **thiocarboxylic acid esters,** like carboxylic acid esters, consist of two words. The first word identifies the alkyl or aryl group bonded to the carboxyl sulfur atom. A prefix *S-* is added to the first word to specify that the group is bonded to a sulfur atom. The second word identifies the acyl group. It is made up by adding *thioate* to the end of the name of the alkane corresponding to the longest chain of carbon atoms of the acyl group. The word *thioate* indicates that the sulfur atom in the molecule is bonded to the carbon atom of the C=O bond. For example:

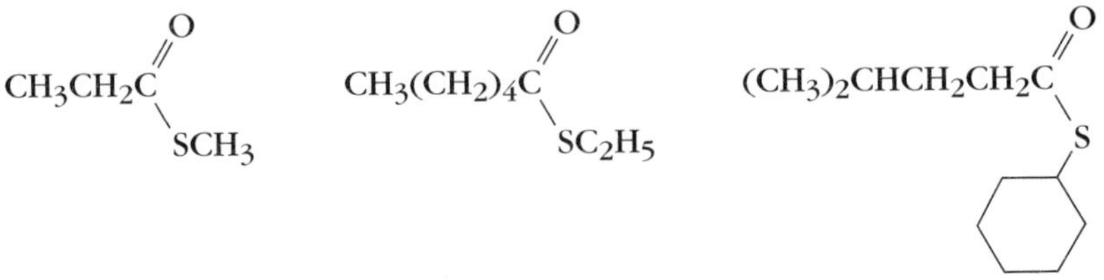

| *S*-Methyl propanethioate | *S*-Ethyl hexanethioate | *S*-Cyclohexyl 4-methyl-pentanethioate |

The names of unsubstituted **amides** are derived from the names of the corresponding carboxylic acids by replacing the ending *-oic acid* (in its systematic name) or *-ic acid* (in its common name) by *amide* or by replacing the *-carboxylic acid* ending with *-carboxamide.*

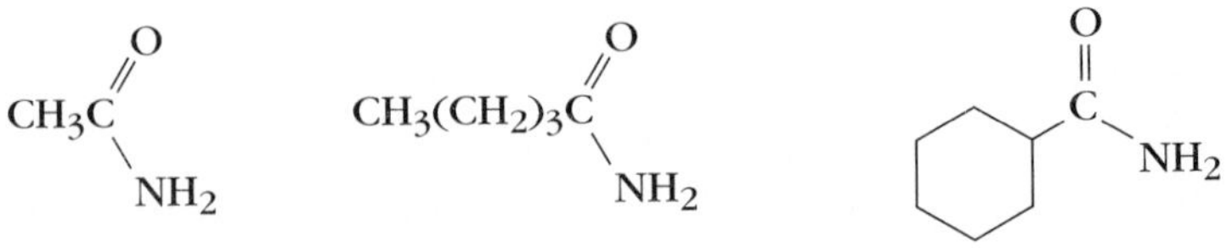

| Acetamide (from acetic acid) | Pentanamide (from pentanoic acid) | Cyclohexanecarboxamide (from cyclohexanecarboxylic acid) |

Amides that contain additional substituents bonded to the nitrogen atom are named as derivatives of the parent amide. The name of the single substituent with a prefix *N-* or the names of two substituents with the prefix *N,N-* are placed before the name of the parent amide. If the two substituents on the nitrogen atom are different, their names are listed in alphabetical order.

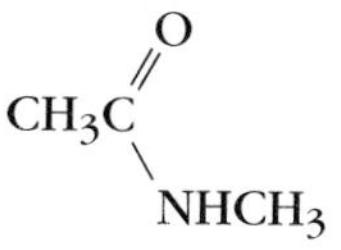

$$CH_3C \begin{matrix} O \\ \| \\ \\ NHCH_3 \end{matrix}$$

$$CH_3(CH_2)_3C \begin{matrix} O \\ \| \\ \\ N(C_2H_5)_2 \end{matrix}$$

N-Methylacetamide

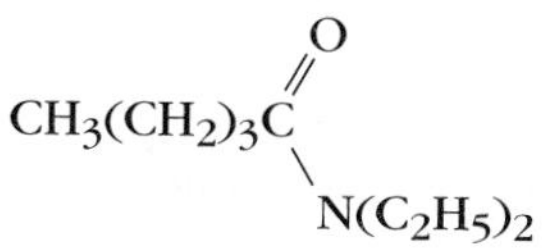

N, N-Diethylpentanamide

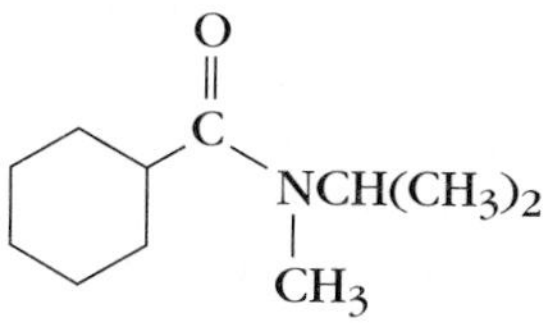

N-Isopropyl-*N*-methyl-cyclohexanecarboxamide

Nitriles are named by adding the ending *-nitrile* to the name of the alkane corresponding to the longest continuous chain of carbon atoms that *includes the carbon atom of the CN group.* The carbon of the nitrile group is designated as carbon 1. Nitriles are also named by replacing the *-oic acid* or *-ic acid* ending of the name of the corresponding carboxylic acid by *-onitrile* or replacing the *-carboxylic acid* ending with *-carbonitrile:*

$$CH_3C \equiv N \qquad CH_3(CH_2)_3C \equiv N \qquad C \equiv N$$

Acetonitrile

Pentanenitrile

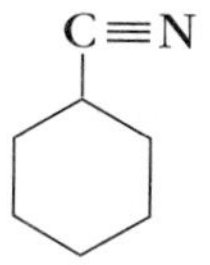

Cyclohexane-carbonitrile

Exercise 16.1   Give the IUPAC name for each of the following compounds.

(a) 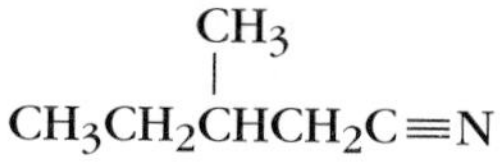 CH₃CH₂CHCH₂C≡N
       |
      CH₃

(b)

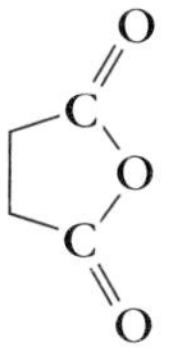

(c) 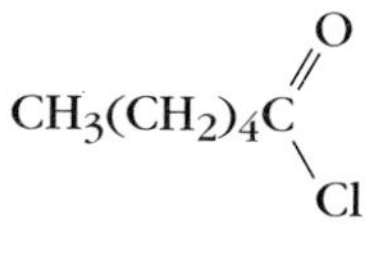 $CH_3(CH_2)_4C \begin{matrix} O \\ \| \\ \\ Cl \end{matrix}$

(d) 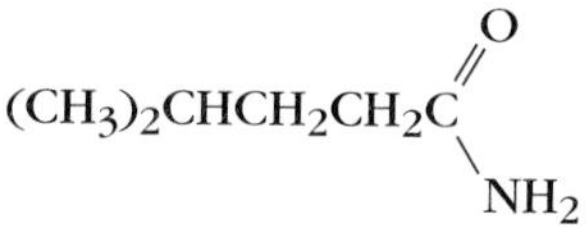 $(CH_3)_2CHCH_2CH_2C \begin{matrix} O \\ \| \\ \\ NH_2 \end{matrix}$

(e) 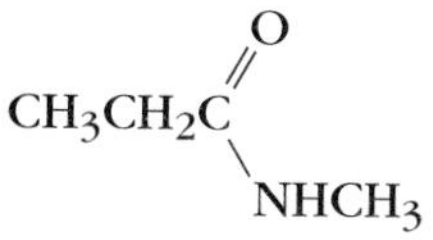 $CH_3CH_2C \begin{matrix} O \\ \| \\ \\ NHCH_3 \end{matrix}$

(f)

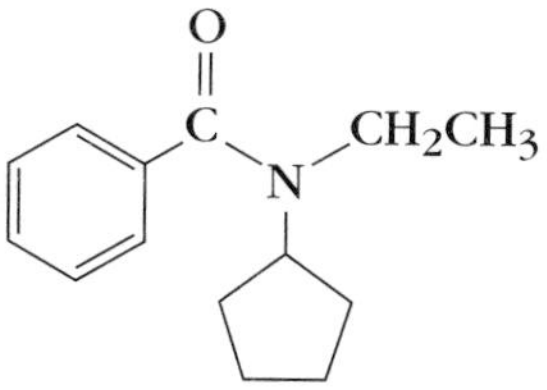

Exercise 16.2   Write the structure of each of the following compounds.

(a) Butyramide
(b) *N*-Propylpropenamide
(c) 3,3-Dichloropentanenitrile
(d) Propanoyl chloride
(e) *S*-Isopropyl butanethioate

(f) Heptanoic anhydride
(g) Cyclohexylacetamide
(h) *S*-Methyl-2,3-dimethyl-2-butenethioate
(i) Cyclopentyl 2-methylpropanoate
(j) *N*-Ethyl-*N*-isopropyl-4-pentenamide

16.2   Structure of Carboxylic Acid Derivatives

The structures of carboxylic acid derivatives resemble the structures of carboxylic acids, esters, aldehydes, and ketones because the C=O bond and the atoms bonded to it all lie in a plane. The data in Figure 16.2 demonstrate, as well, that the bond angles and bond lengths are similar for ketones, carboxylic acids, esters, acyl chlorides, and amides.

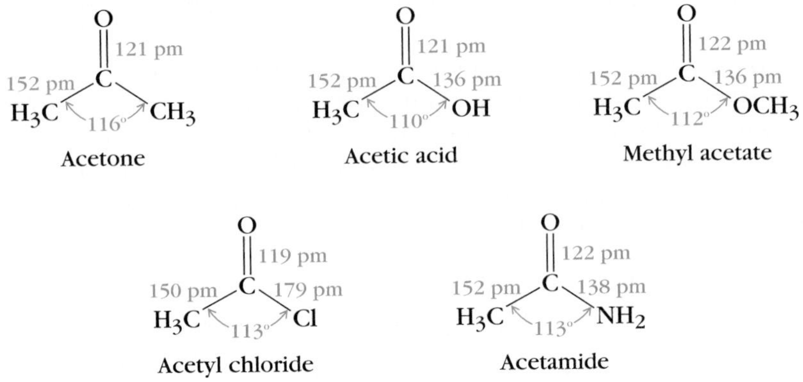

Figure 16.2
Comparison of bond lengths and bond angles of acetone, acetic acid, methyl acetate, acetyl chloride, and acetamide. In these typical examples of these classes of compounds, the C=O bond and the atoms bonded to it all lie in a plane.

The substituent Y bonded to the carbon atom of the C=O bond in acyl chlorides, anhydrides, thioesters, esters, and amides has at least one pair of nonbonding electrons that can interact with the π bond of the C=O group to form a delocalized π electron system. Figure 16.3 shows the atomic orbitals that overlap to form a delocalized π system that encompasses both the C=O bond and the substituent Y.

Figure 16.3
The *p* atomic orbital of the carbon atom of the C=O bond, the *p* atomic orbital of the oxygen atom of the C=O bond, and an atomic orbital of atom Y overlap to form a delocalized π system that encompasses both the C=O and Y in acyl chloride, anhydride, amide, ester, and thioester functional groups.

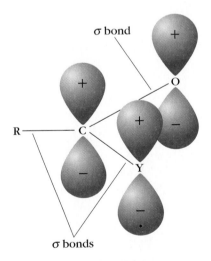

The delocalized π electron structure of carboxylic acid derivatives can be represented as resonance hybrids of the following resonance structures.

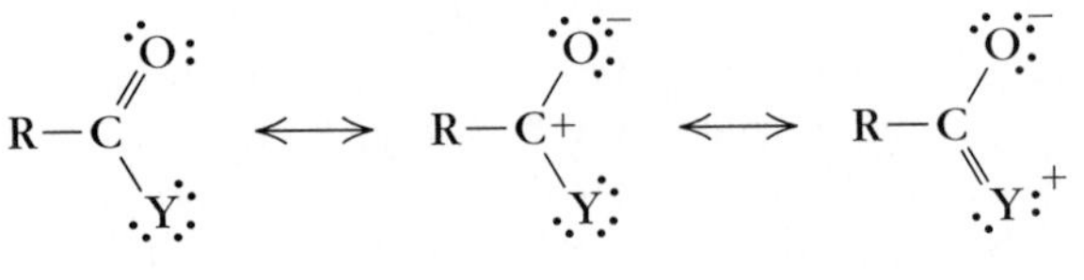

The extent of electron delocalization from the substituent Y to the π bond of the C=O group depends on the electron-donating properties of Y. In acyl chlorides and thioesters, the nonbonding pair of electrons of chlorine and sulfur are in 3p orbitals. Overlap of 3p atomic orbitals with the 2p atomic orbitals of the carbon atom of the C=O bond is poor because of the difference in their energies. As a result, there is little delocalization of the nonbonding electron pair from the chlorine or sulfur atom into the π bond of the C=O group. This means that the resonance structure with a positive charge on Y contributes little to the overall resonance hybrid of acyl chlorides and thioesters.

Delocalization of π electrons is better in acid anhydrides and esters than in acyl chlorides or thioesters because the lone pair of electrons on oxygen (which are in a 2p atomic orbital) overlap better with the 2p orbital of the carbon atom of the C=O bond. Delocalization of π electrons in acid anhydrides involves both C=O bonds and is represented by a resonance hybrid of the following resonance structures.

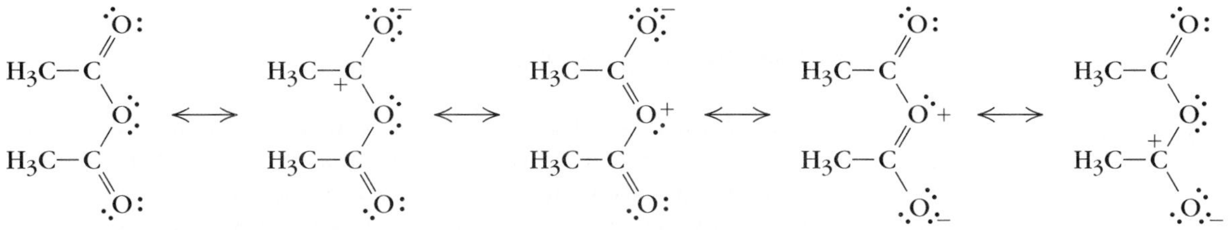

Nitrogen is better able to donate electrons to the π bond of the C=O group than oxygen is because nitrogen is less electronegative. The structure of amides is affected by the delocalization of the electrons of the nitrogen atom. Unlike the pyramidal structure of ammonia, the bonds to nitrogen in amides lie in the same plane. For example, microwave measurements of the structure of formamide indicate that the entire molecule is planar, as shown in Figure 16.4.

The two N—H bonds in formamide have different lengths, 101.3 pm and 100.2 pm. These two hydrogen atoms are different, therefore, and this difference can be observed in the ^{1}H NMR spectrum of formamide. One hydrogen atom is found at δ 7.7, the other at δ 7.9. Because these two hydrogen atoms are distinguishable at room temperature, there must be a barrier to rotation about the C—N bond. In formamide, this barrier is 18 kcal/mol (75 kJ/mol). In general, the barrier to rotation about the C—N bond in all amides is somewhere between 16 and 20 kcal/mol (70 and 85 kJ/mol). This is a high barrier to rotation, so the structures of amides are not only planar but also very rigid. The relatively high barrier to rotation about the C—N bond is consistent with the amide group as a resonance hybrid of the following three resonance structures.

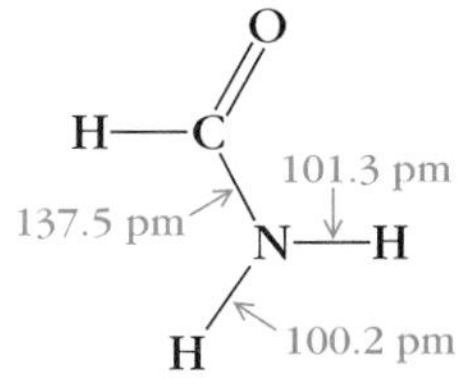

Figure 16.4
Planar structure of formamide. The molecular plane is the plane of the page.

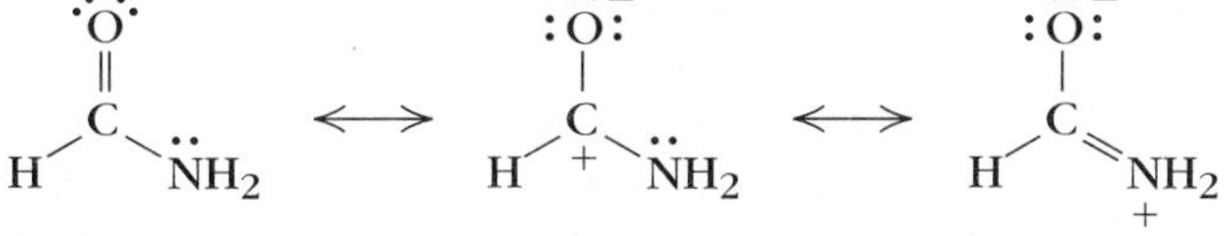

The resonance structure on the right shows a double bond between the carbon and the nitrogen atom. The C—N bond, therefore, has double bond character, and this double bond character accounts for the barrier to rotation.

The C—N bond length of amides is usually about 138 pm, whereas in amines it is usually about 146 pm. The shorter C—N bond observed in amides is consist-

■ The rigidity of the amide functional group is important to the structure of amino acids and peptides as we will learn in Chapter 26.

ent with electron delocalization of the lone pair of electrons on the nitrogen atom to the C=O bond, which results in partial double bond character in the C—N bond.

In summary, electron delocalization of a nonbonding pair of electrons from chlorine or sulfur to the C—O bond is relatively unimportant to the π electron structure of acyl chlorides and thioesters. In contrast, electron delocalization from oxygen and nitrogen atoms to the C=O bond of anhydrides, esters, and amides is important and contributes greatly to their π electron structure. Electron delocalization in the amide and ester functional groups is not as great, however, as it is in carboxylate ions. In Section 16.7, we will discuss how the extent of π electron delocalization in a particular carboxylic acid derivative affects its reactivity.

The nature of the electronegative atom Y bonded to the carbon atom of the C=O bond of carboxylic acid derivatives also greatly affects their physical properties.

16.3   Physical Properties

In Chapter 14, we learned that the polarity of a carbonyl-containing compound and its ability to form hydrogen bonds are the two major factors that determine its boiling point and water solubility. The same is true for carboxylic acid derivatives. All derivatives of carboxylic acids are more or less polar compounds, but they differ greatly in their ability to form or accept hydrogen bonds. Carboxylic acids, as we discussed in Chapter 15, are polar compounds that are good hydrogen bond donors and acceptors. They are so good that they form dimers even in aqueous solution. Consequently, the low molecular weight carboxylic acids such as acetic and butanoic acids are water soluble and have high boiling points (118 °C and 164 °C, respectively).

Table 16.1	Boiling Points and Melting Points of Compounds with Similar Molecular Weights but Different Functional Groups		
Name	Structure	Molecular Weight	Boiling Point (°C)
Butanamide	$CH_3(CH_2)_2CONH_2$	87	Solid (mp 116)
N-Methylpropanamide	$CH_3CH_2CONHCH_3$	87	Solid (mp 35)
N,N-Dimethylacetamide	$CH_3CON(CH_3)_2$	87	165
Butanoic acid	$CH_3(CH_2)_2COOH$	88	164
Pentanenitrile	$CH_3(CH_2)_3C{\equiv}N$	83	145
1-Butanol	$CH_3(CH_2)_2CH_2OH$	88	118
3-Pentanone	$CH_3CH_2COCH_2CH_3$	86	101
Ethyl acetate	$CH_3COOCH_2CH_3$	88	77
S-Methyl ethanethioate	CH_3COSCH_3	90	55
1-Chlorobutane	$CH_3(CH_2)_2CH_2Cl$	92	78
Butanal	$CH_3(CH_2)_2CHO$	86	76
Hexane	$CH_3(CH_2)_4CH_3$	86	69
Acetyl chloride	CH_3COCl	79	51

Unsubstituted amides are another example of polar compounds that are good hydrogen bond acceptors and donors. The hydrogen atom bonded to nitrogen of one amide molecule forms a hydrogen bond with the oxygen atom of the C=O bond of a neighboring amide molecule or a water molecule.

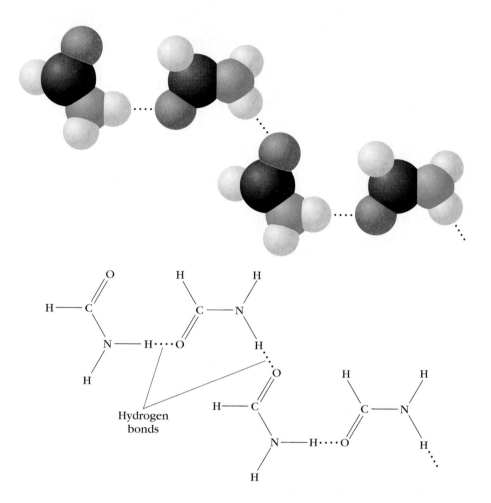

■ *Representation of the hydrogen bonds between molecules of formamide.*

As a result, low molecular weight amides have such unusually high boiling points that most are solids at room temperature. The importance of this hydrogen bonding in amides is shown in Table 16.1, where the boiling points of butanamide, *N*-methylpropanamide, and *N,N*-dimethylacetamide are compared. All three of these compounds have the same molecular weight, but butanamide and *N*-methylpropanamide, the unsubstituted and *N*-monosubstituted amides, are solids whereas *N,N*-dimethylacetamide, the *N,N*-disubstituted amide, is a liquid. Unsubstituted and *N*-monosubstituted amides are solids because they can form hydrogen bonds, whereas *N,N*-disubstituted amides cannot form hydrogen bonds. Note also that the melting point of butanamide (116 °C) is significantly higher than the melting point of *N*-methylpropanamide (35 °C). The difference probably arises because butanamide has one more hydrogen atom available for hydrogen bonding than *N*-methylpropanamide. In general, amides, substituted or unsubstituted, have higher boiling points than compounds with other functional groups of about the same molecular weight, as shown by the data in Table 16.1.

Acyl chlorides and acid anhydrides are not very polar molecules, so their boiling points are not unusually high. The boiling point of acetyl chloride, for example, is about the same as other relatively nonpolar compounds of similar molecular weight, as shown in Table 16.1.

Nitriles are among the most polar organic compounds. As a result, the boiling points of nitriles are unusually high for compounds that are not hydrogen bond donors and are poor hydrogen bond acceptors. The boiling point of pentanenitrile (145 °C), for example, is nearly as high as the boiling point of butanoic acid (164 °C).

Differences in boiling points and water solubilities are clearly evident in low molecular weight compounds with different functional groups. Keep in mind, however, that these differences slowly disappear as the size of the hydrocarbon portion of these molecules increases.

Exercise 16.3   Which one of each of the following pairs of compounds has the higher boiling point?

(a) $CH_3(CH_2)_3COOH$ and 1-chloropentane
(b) Pentanamide and ethyl butanoate
(c) and hexanenitrile
(d) Butanoyl chloride and 1-hexanol

16.4   Spectroscopic Properties

All compounds that contain a carbonyl group show a strong absorption in their IR spectrum in the region 1650 to 1850 cm^{-1}. This absorption is due to the stretching of the C=O bond, and its exact position depends on the kind of carbonyl-containing functional group present in the compound. Infrared spectroscopy, therefore, provides valuable information to distinguish between different carboxylic acid derivatives.

The IR data in Table 16.2 show, for example, that acid anhydrides can be identified because they show two absorptions in the carbonyl region; one is near 1760 cm^{-1} and the other is near 1820 cm^{-1}. The carbonyl absorption of esters, as mentioned in Chapter 15, is found near 1740 cm^{-1}, which is at higher wavenumbers than that of aldehydes or ketones. The carbonyl absorption of amides is usually found near 1670 cm^{-1}, which is the lower end of the carbonyl region. In addition to a carbonyl absorption, amides that contain an N—H bond show an N—H stretching absorption band in the 3050 to 3550 cm^{-1} region. Finally thioesters show a carbonyl absorption near 1690 cm^{-1}.

The type of functional group in a compound often cannot be identified from the C=O stretching absorption bands alone because there are several carbonyl-containing functional groups that absorb within a given wavenumber range. Other regions of its IR spectrum as well as its ^{13}C and ^{1}H NMR spectra must be consulted to determine the structure of the compound. A single absorption band in the region 1700 to 1750 cm^{-1} of an IR spectrum, for example, could indicate that a compound of unknown structure is an ester, an aldehyde, or a ketone. The presence of two absorption bands near 2700 and 2800 cm^{-1} in its IR spectrum would identify the compound as an aldehyde, as would a peak near δ 9 in its

Table 16.2	Absorption Bands due to the Carbonyl Stretching of Carboxylic Acid Derivatives*	

The spectral properties of nitriles are discussed in Section 16.15, A.

Carbonyl-Containing Functional Group	Structure	Infrared Absorption (cm^{-1})
Acyl chlorides	$\overset{\displaystyle O}{\underset{\displaystyle Cl}{-C}}$	1815–1790
Anhydrides	structure	1790–1740 1850–1800
Esters	$\overset{\displaystyle O}{\underset{\displaystyle OR}{-C}}$	1750–1730
Amides	$\overset{\displaystyle O}{\underset{\displaystyle NH_2}{-C}}$	1690–1650
N-Substituted amide	$\overset{\displaystyle O}{\underset{\displaystyle NHR}{-C}}$	1685–1670
N,N-Disubstituted amide	$\overset{\displaystyle O}{\underset{\displaystyle NR_2}{-C}}$	1670–1650
Aldehydes	$\overset{\displaystyle O}{\underset{\displaystyle H}{-C}}$	1730–1720
Ketones	$\overset{\displaystyle O}{\underset{\displaystyle R}{-C}}$	1715–1700
Carboxylic acids	$\overset{\displaystyle O}{\underset{\displaystyle OH}{-C}}$	1720–1680
Thioesters	$\overset{\displaystyle O}{\underset{\displaystyle SR}{-C}}$	1695–1690

*For comparison, the C=O stretching absorption bands of ketones, aldehydes, esters, and carboxylic acids are included in this table.

Table 16.3	^{13}C NMR Chemical Shifts of the Carbonyl Carbon Atoms of Carboxylic Acid Derivatives	
Structure	Name	δ for Carbon Atom of C=O Bond
CH_3COOH	Acetic acid	176.9
$CH_3C(O)OCH_3$	Methyl acetate	170.6
CH_3COCl	Acetyl chloride	170.1
CH_3CONH_2	Acetamide	171.8
CH_3COSCH_3	S-Methyl ethanethioate	195.4
$(CH_3CO)_2O$	Acetic anhydride	166.6

^{1}H NMR spectrum. The presence of a strong broad band near 1200 cm[−1] would suggest that the compound is an ester.

The carbon atoms of the C=O bond of carboxylic acid derivatives are found in their ^{13}C NMR spectra in the range δ 160 to 220. Specific examples are listed in Table 16.3. It is difficult to use ^{13}C NMR to precisely determine the carbonyl-containing functional group in an unknown because the carbon atoms of the C=O bonds of all the carboxylic acid derivatives except thioesters are found in about the same range. The carbon atoms of the C=O bond of thioesters are found in their ^{13}C NMR spectra near δ 200, which is the same region of the spectrum where the carbonyl carbon atoms of aldehydes and ketones are found (Section 14.4). The identity of the functional group is difficult to determine precisely from ^{1}H NMR spectra, too, because the α-hydrogens of all carboxylic acid derivatives are found near δ 2.0.

Exercise 16.4	How could IR or NMR spectroscopy be used to distinguish between the following pairs of compounds?

(a) Acetamide and acetyl chloride
(b) Acetic anhydride and CH_3COOH
(c) Acetone and acetic acid
(d) S-Methyl ethanethioate and CH_3COOCH_3

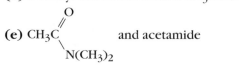

(e) and acetamide

Let's begin our study of the chemistry of carboxylic acid derivatives by discussing how to prepare acyl chlorides because all other carboxylic acid derivatives can be prepared from them.

16.5 Acyl Chlorides Are Important Intermediates in the Preparation of Carboxylic Acid Derivatives

Most carboxylic acid derivatives are prepared in the laboratory from acyl chlorides. Acyl chlorides, in turn, are prepared from carboxylic acids. Thus most carboxylic acid derivatives can be prepared from carboxylic acids in a sequence of two reactions. Because acyl chlorides are important intermediates in the synthesis of carboxylic acid derivatives, we begin by explaining how acyl chlorides are prepared and then we will discuss the reactions that they undergo.

Carboxylic acids react with thionyl chloride ($SOCl_2$), phosphorus trichloride (PCl_3), or phosphorus pentachloride (PCl_5) to form acyl chlorides.

| Pentanoic acid | Thionyl chloride | Pentanoyl chloride (95% yield) |

| 3 Benzoic acid | Phosphorus trichloride | 3 Benzoyl chloride (92% yield) |

Thionyl chloride is a particularly convenient reagent for this reaction because the side products, SO_2 and HCl, are gases, so they are easily separated from the acyl chloride. Any excess thionyl chloride can be easily removed by distillation because its boiling point (78 °C) is lower than that of most acyl chlorides.

Acyl chlorides react with a variety of nucleophiles to form most other carboxylic acid derivatives. In this reaction the nucleophile replaces the chlorine atom of the acyl chloride. In effect, the acyl group is transferred from chlorine to an oxygen, nitrogen, or sulfur atom of the nucleophile.

| Acyl chloride | Carboxylic acid derivative |

If the nucleophile is an alcohol, an ester is formed. If the nucleophile is an amine, an amide is formed. Thiols react with acyl chlorides to form S-thioesters, and carboxylate ions react to form acid anhydrides. If water is the nucleophile, the corresponding carboxylic acid is formed in the reverse of the reaction that formed the acyl chloride in the first place.

Esters are formed rapidly by the reaction of acyl chlorides with alcohols. The reaction is typically carried out in the presence of the weak base pyridine to neutralize the HCl that also forms.

The lone pair of electrons on its nitrogen atom makes pyridine a base, so it reacts with HCl and other strong acids to form its conjugate acid (the pyridinium ion) in an acid-base reaction.

| Pyridine | Pyridinium ion |

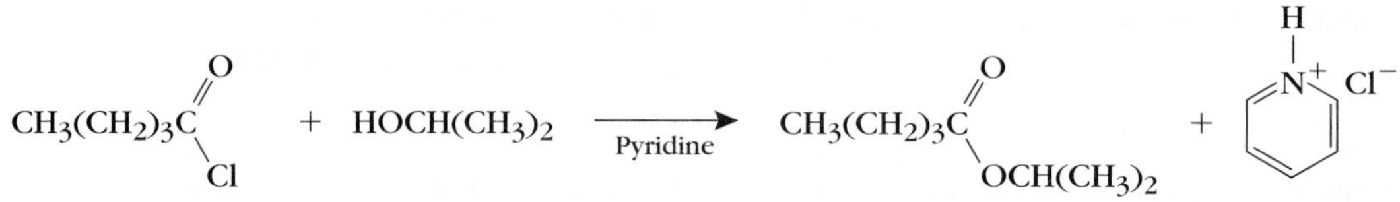

Pentanoyl
chloride

2-Propanol

Isopropyl pentanoate
(94% yield)

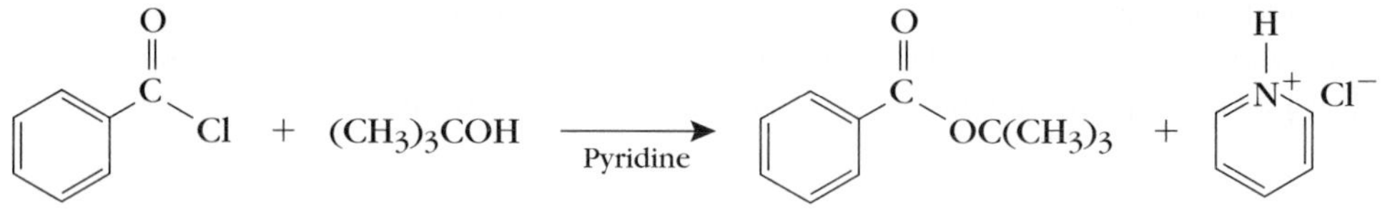

Benzoyl chloride

2-Methyl-2-
propanol

tert-Butyl benzoate
(76% yield)

The reaction of acyl chlorides with alcohols is similar to the reaction of the acid chlorides of arenesulfonic acid (arenesulfonyl chlorides, Section 11.16) with alcohols. Both reactions form esters; acyl chlorides form carboxylic acid esters and arenesulfonyl chlorides form arenesulfonic esters.

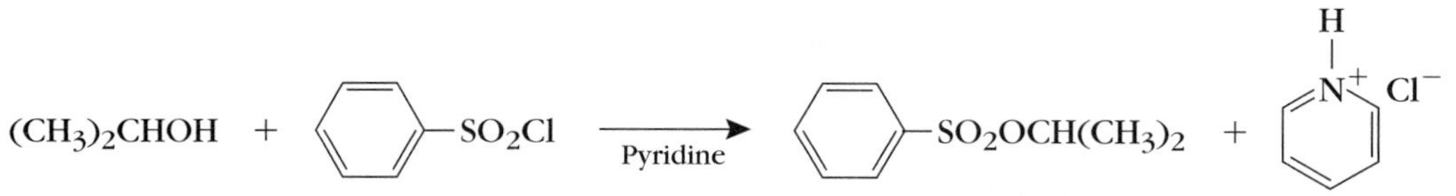

2-Propanol

Benzenesulfonyl
chloride

Isopropyl benzenesulfonate
(78% yield)
(ester of benzenesulfonic acid)

Acyl chlorides react with ammonia to form unsubstituted amides, with primary amines to form *N*-substituted amides, and with secondary amines to form *N,N*-disubstituted amides.

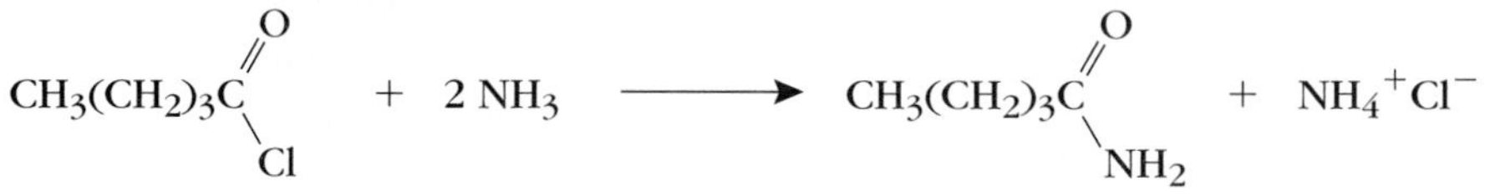

Pentanoyl
chloride

Ammonia

Pentanamide
(83% yield)

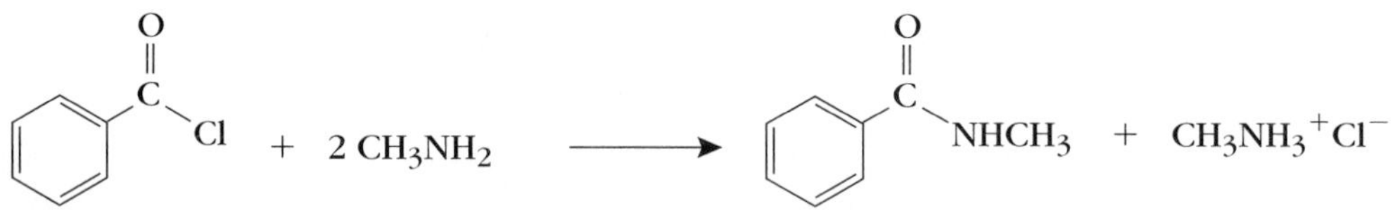

Benzoyl chloride

Methylamine
(a primary amine)

N-Methyl benzamide
(92% yield)

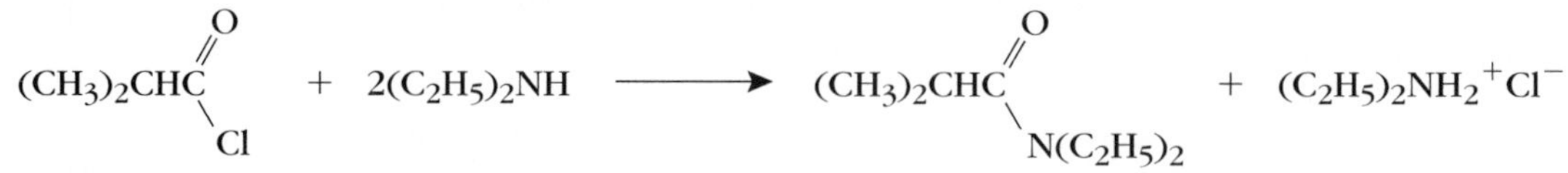

2-Methylpropanoyl
chloride

Diethylamine
(a secondary
amine)

N,N-Diethyl 2-methyl-
propanamide
(80% yield)

Notice that two equivalents of amine are used for each equivalent of acyl chloride in these reactions because HCl is a reaction product, too, and it reacts with any base in solution. If the amine is the only base present, it acts as both a nucleophile and a base. Consequently, for each equivalent of amide formed, an equivalent of amine is protonated. If the amine is too expensive to be used in excess, a tertiary amine such as pyridine is used as the base instead.

Tertiary amines, R_3N, react with acyl chlorides to form acyl ammonium salts. Pyridine, for example, reacts with benzoyl chloride to form benzoylpyridinium chloride.

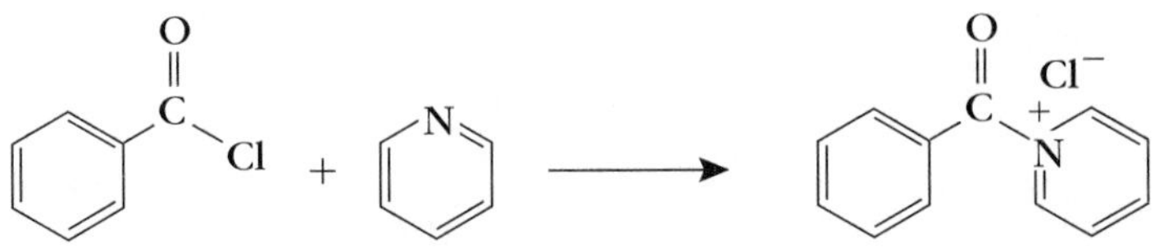

Benzoyl chloride   Pyridine   Benzoylpyridinium chloride

Acyl ammonium salts are very reactive, so benzoylpyridinium chloride reacts quickly with any nucleophile that is present. If the nucleophile is an alcohol, the product is an ester.

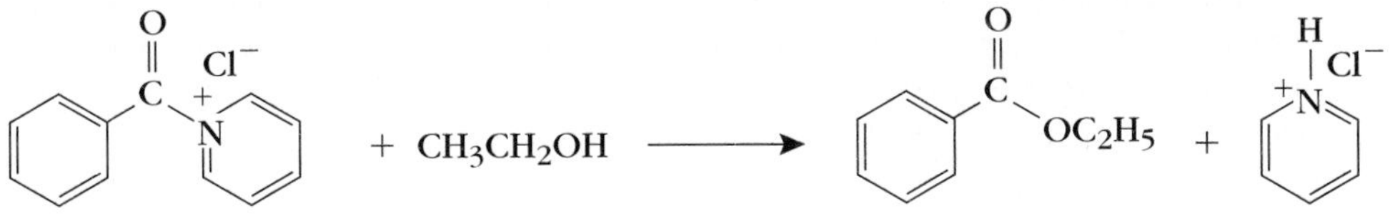

Benzoylpyridinium chloride   Ethanol   Ethyl benzoate

Tertiary amines cannot form stable amides, so tertiary amines can be used to neutralize the HCl that forms in the reactions of acyl chlorides. Even though acylpyridinium salts are formed, they don't interfere with the formation of esters or amides because they rapidly react with the nucleophile to form the desired product.

Acyl chlorides react with thiols to form S-thioesters. This reaction is analogous to the formation of esters from the reaction of acyl chlorides and alcohols. HCl is a product in both reactions, so a base must be used to neutralize it as it forms.

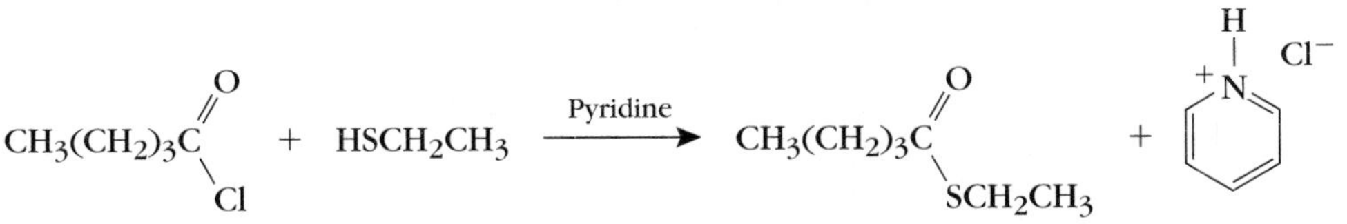

Pentanoyl chloride   Ethanethiol   S-Ethyl pentanethioate

Acid anhydrides, including mixed anhydrides, are formed by the reaction of carboxylate ions with acyl chlorides.

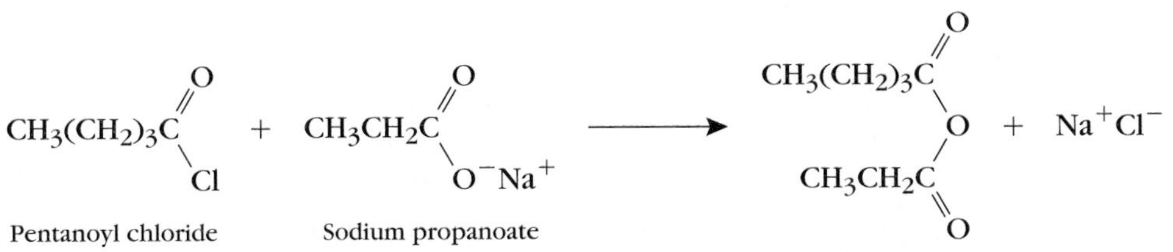

Pentanoyl chloride   Sodium propanoate

Pentanoic propanoic
anhydride

■ *The reactions of acyl chlorides to form esters, amides, S-thioesters, and acid anhydrides must be carried out in the absence of water, because any water present will react with the acyl chloride to form the corresponding carboxylic acid as an unwanted side product.*

Finally, acyl chlorides react with water to form carboxylic acids. In an aqueous basic solution, the salt of the carboxylate ion is the product. This reaction is not synthetically useful, however, because acyl chlorides are prepared from carboxylic acids.

$$CH_3(CH_2)_3C\!\!\!\overset{O}{\underset{Cl}{\Big\langle}} \;+\; H_2O \;\longrightarrow\; CH_3(CH_2)_3C\!\!\!\overset{O}{\underset{OH}{\Big\langle}} \;+\; HCl$$

Pentanoyl chloride   Pentanoic acid

Benzoyl chloride + 2 NaOH $\xrightarrow{\;H_2O\;}$ Sodium benzoate + Na^+Cl^- + H_2O

Benzoyl chloride   Sodium benzoate

The preparation of carboxylic acid derivatives from acyl chlorides is summarized in Figure 16.5.

The reactions shown in Figure 16.5 are only a few examples of acyl transfer reactions that interconvert carboxylic acid derivatives. The mechanism of acyl transfer reactions is similar in many ways to the mechanism of hydrolysis of esters described in Chapter 15.

Figure 16.5
Synthesis of carboxylic acid derivatives from acyl chlorides.

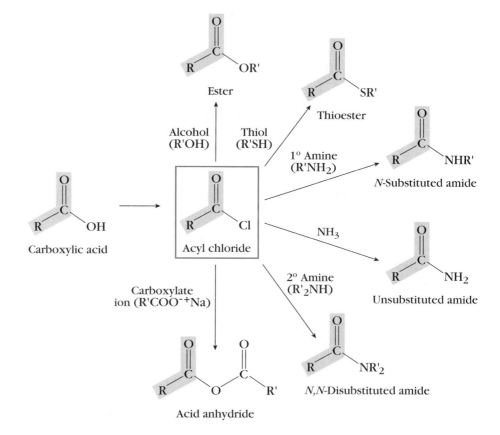

Exercise 16.5	Write the name and the structure of the product of the reaction of propanoyl chloride with each of the following reagents:

(a) Ethanol/pyridine
(b) Ethylamine ($C_2H_5NH_2$)
(c) Propanethiol/pyridine
(d) Aqueous sodium hydroxide

(e) Sodium benzoate
(f) Ammonia
(g) Diethylamine [(C_2H_5)$_2$NH]
(h) Cyclohexanol/pyridine

16.6   Acyl Transfer Reactions Occur by an Addition-Elimination Mechanism

The mechanism of acyl transfer reactions is similar in many respects to the mechanism of the saponification of esters. Step 1 in both is the addition of a nucleophile to the carbon atom of the C=O bond to form a tetrahedral intermediate, and Step 2 is the elimination of a good leaving group. Proton transfer in Step 3 forms the product. These three steps are illustrated in Figure 16.6 for the reaction of ammonia with acetyl chloride to form acetamide.

Step 1: Addition of a nucleophile to the carbon atom of the C=O bond forms a tetrahedral intermediate

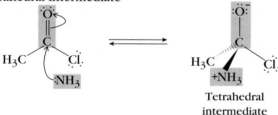

Tetrahedral intermediate

Step 2: Elimination of the chloride ion from the tetrahedral intermediate forms the conjugate acid of the amide

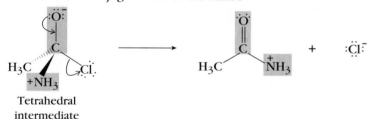

Tetrahedral intermediate

Step 3: Proton transfer forms an amide

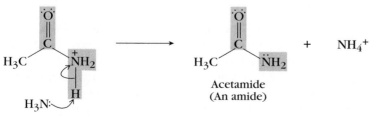

Acetamide (An amide)

Figure 16.6
Three-step addition-elimination mechanism of the reaction of acetyl chloride and ammonia to form acetamide.

Ammonia, acting as a nucleophile, adds to the carbon atom of the C=O bond of acetyl chloride to form a tetrahedral intermediate in Step 1 of the mechanism. This step is analogous to Step 1 of the mechanism for the reaction of ammonia and primary amines with aldehydes and ketones discussed in Chapter 14.

Step 2 of the mechanism is an elimination reaction. This step is analogous to Step 2 in the mechanism of the saponification of esters discussed in Chapter 15. If the $-NH_3^+$ group is eliminated from the tetrahedral intermediate, the starting reagents, acetyl chloride and ammonia, are regenerated. Elimination of chloride ion, which forms the conjugate acid of the amide, is irreversible because chloride ion is too weak a nucleophile to add to the conjugate acid of acetamide. Finally, acetamide is formed in Step 3 by the transfer of a proton from the protonated amide to ammonia, the strongest base in the reaction mixture.

Exercise 16.6   Using arrows to indicate electron movement, write the detailed mechanisms for the reaction of acetyl chloride with each of the following reagents:

(a) Aqueous NaOH solution   (c) Sodium acetate
(b) Methanol   (d) Methanethiol (CH_3SH)

Exercise 16.7   Propose a method for the synthesis of each of the following compounds using an acyl chloride as one of the starting reagents.

(a) Cyclopentyl 2-methylpropanoate   (c) Diethyl succinate
                                             ($C_2H_5OOCCH_2CH_2COOC_2H_5$)

(b)

(d) S-Ethyl butanethioate

Exercise 16.8   Both methyl benzenesulfonate and methyl acetate react with an aqueous solution of sodium hydroxide to form methanol according to the following equations:

Methyl benzenesulfonate

$$CH_3COOCH_3 \xrightarrow[\text{H}_2\text{O}]{\text{NaOH}} CH_3OH + CH_3COO^-$$

Methyl acetate

If the reaction is carried out in water enriched with oxygen-18, the methanol obtained from the reaction with methyl benzenesulfonate is found to be enriched with oxygen-18, while methanol from the reaction with methyl acetate is not enriched with oxygen-18. Based on the mechanisms of these two reactions, explain these observations.

Although carboxylic acid derivatives all undergo acyl transfer reactions, their relative reactivities vary greatly.

16.7 Relative Rates of Acyl Transfer Reactions

Carboxylic acid derivatives are hydrolyzed to carboxylic acids, but the rates at which the different derivatives are hydrolyzed vary greatly. The reaction conditions necessary for hydrolysis to occur are one way to express this difference in rate. Thus unreactive derivatives such as amides must be heated in the presence of acid or base in order to hydrolyze them. More reactive derivatives such as acyl chlorides, on the other hand, hydrolyze rapidly at room temperature in the absence of acid or base. Acid anhydrides, esters, and thioesters hydrolyze slowly at room temperature, but their rates can be increased in either acidic or basic solutions. The relative reactivities of carboxylic acids and their derivatives in acyl transfer reactions are as follows:

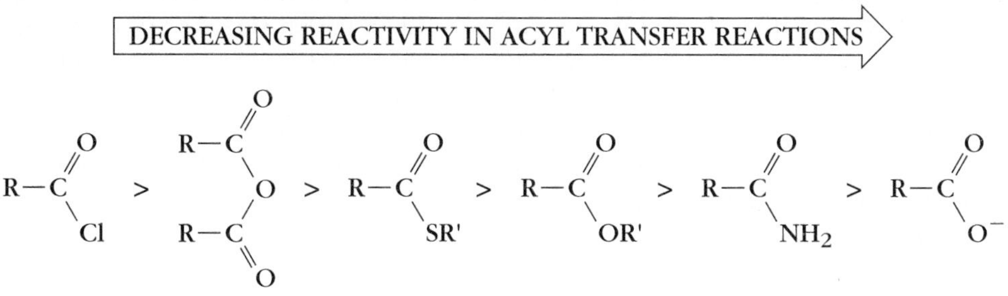

Acid chloride · Acid anhydride · Thioester · Ester · Amide · Carboxylate ion

Why is there such a difference in reactivity between the various carboxylic acid derivatives? In order to answer this question, we must look more closely at the mechanism of acyl transfer reactions.

Although both the addition and the elimination steps can affect the overall rate of acyl transfer reactions, it is usually the addition step that determines reactivity. Consequently, the free energy of activation, $\Delta G^{\ddagger}$, of this step determines the rate. Anything that stabilizes the carboxylic acid derivative or destabilizes the transition state, whose structure and free energy resemble the tetrahedral intermediate, will increase $\Delta G^{\ddagger}$ and decrease the rate of reaction.

Recall that according to the Hammond Postulate, discussed in Section 7.13, the free energy and structure of the transition state leading to a reactive intermediate is similar to the structure and free energy of that intermediate.

What are the factors that stabilize or destabilize the various carboxylic acid derivatives? There are two, the inductive effect of the substituent Y and the ability of Y to donate a nonbonding pair of electrons to the π bond of the C=O group. We can analyze these effects by examining how they affect the relative stabilities of the three resonance structures that contribute to the resonance hybrid structure of carboxylic acid derivatives.

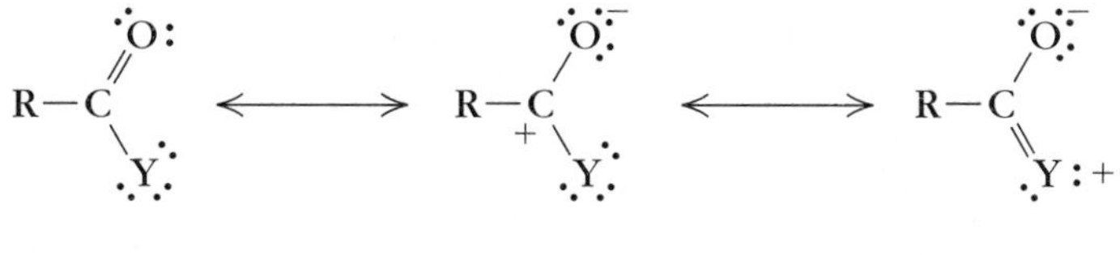

Electronegative atoms such as chlorine, oxygen, sulfur, and nitrogen are inductively electron withdrawing. Attaching an electron-withdrawing group to the partially positive carbon atom of the C=O bond is energetically unfavorable, which destabilizes the carboxylic acid derivative. The more electron withdrawing the substituent Y, the less stable is the carboxylic acid derivative. In terms of its resonance hybrid structure, inductive electron withdrawal by Y decreases the contribution of the two dipolar resonance structures (the two structures on the right) to the overall hybrid structure.

The abilities of the substituents Y to donate a pair of nonbonding electrons were discussed in Section 16.2. The greater the ability of Y to donate a pair of electrons, the more stable is the carboxylic acid derivative. Electron donation from Y to the C=O bond stabilizes the carboxylic acid derivative by increasing the importance of the dipolar resonance structure, which has a positive charge on Y.

The inductive and electron-donating effects of Y oppose each other: one destabilizes, while the other stabilizes, the carboxylic acid derivative. Which is the more important? The answer depends on the exact nature of the substituent Y. In acyl chlorides and thioesters (Y = Cl and S), the inductive electron-withdrawing effects of chlorine and sulfur are more important than their electron-donating abilities because overlap is poor between the $3p$ orbitals on sulfur or chlorine and the $2p$ orbitals on carbon (Section 16.2). In esters and amides (Y = O and N), the opposite is true; the electron-donating abilities of oxygen and nitrogen atoms are more important than their inductive electron-withdrawing ability. In acid anhydrides, neither factor is dominant. The balance of the two is such that acid anhydrides are more stable than acyl chlorides and slightly less stable than thioesters. By this analysis, the effect of Y on the stability of carboxylic acid derivatives is as follows:

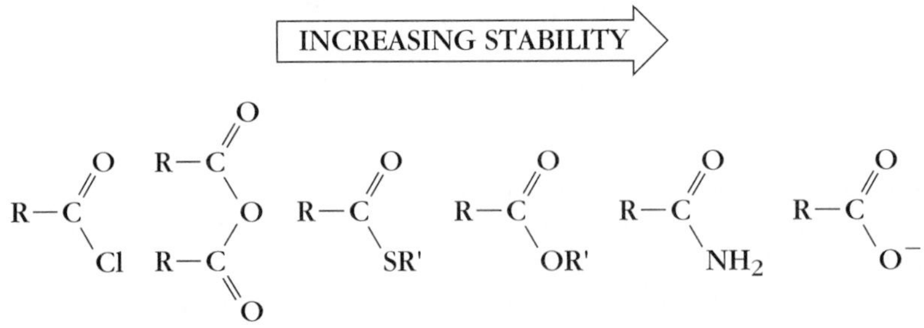

Before concluding that this is the major effect on the relative reactivity, we must examine how Y affects the stability of the transition states, since it is $\Delta G^{\ddagger}$ that determines the rate, not just the stability of the carboxylic acid derivatives.

The transition states of acyl transfer reactions resemble the tetrahedral intermediate. Because there is less positive charge on the carbon atom of the C—Y bond in the tetrahedral intermediate or transition state than in the carboxylic acid derivative, the effect of Y is diminished in the transition state. Thus Y affects mainly the free energy of the reactants, not the transition state. The effect of Y on the free energy of the carboxylic acid derivatives and their transition states for acyl transfer reactions is shown by the free energy-reaction path diagram in Figure 16.7.

Recall from Section 16.5 that a practical consequence of the relative reactivities of carboxylic acid derivatives is that acyl chlorides, the most reactive of all carboxylic acid derivatives, can be readily converted into all of the other carboxylic acid derivatives. In general, interconversion of carboxylic acid derivatives occurs from the most reactive to the least reactive. These interconversions are summarized in Figure 16.8 (p. 718). Thus one carboxylic acid derivative is converted to another if the reaction leads to the more stable carboxylic acid derivative. An ester, for example, can be easily formed in one step from the more reactive acyl chloride, but an ester cannot be easily converted in one step to an acyl chloride.

■ Remembering the order of reactivities of carboxylic acid derivatives is a convenient way of keeping track of the large number of reactions by which they are interconverted.

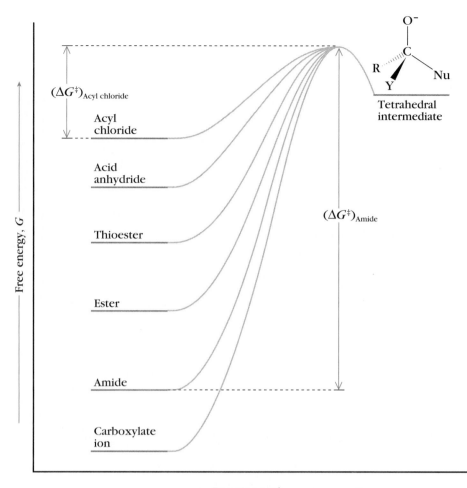

$(\Delta G^{\ddagger})_{\text{Amide}} > (\Delta G^{\ddagger})_{\text{Acyl chloride}}$ so acyl chlorides are more reactive than amides

Figure 16.7
Free energy-reaction path diagram for acyl transfer reactions of carboxylic acid derivatives. The free energies of the transition states are not identical but differ much less than the free energies of the corresponding carboxylic acid derivatives. Consequently, they are placed at the same energy level for purposes of comparison.

Exercise 16.9 Which one of each of the following pairs of compounds will react faster with water at room temperature?

(a) CH_3C(=O)OCH_2CH_3 or CH_3C(=O)Cl

(b) *S*-Methyl propanethioate or methyl butanoate

(c) Acetic anhydride or acetamide

(d) or

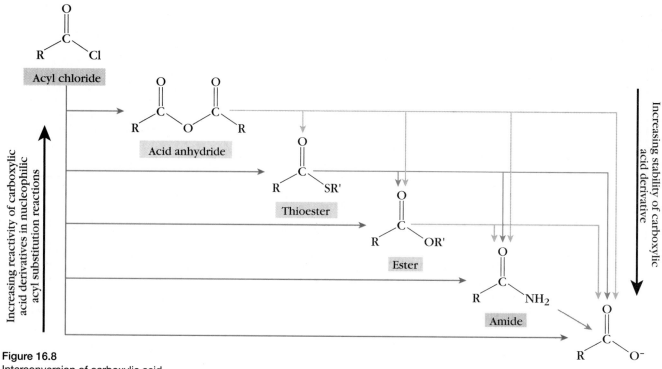

Figure 16.8
Interconversion of carboxylic acid derivatives occurs from the more reactive to the less reactive derivative.

We will examine several of these reactions in detail beginning with the hydrolysis of amides.

16.8   Hydrolysis of Amides

Hydrolysis is an important nucleophilic acyl substitution reaction of amides. Amides are very unreactive so their hydrolysis occurs only under strongly acidic or basic conditions. Amides are hydrolyzed in a strong aqueous acid solution to form a carboxylic acid and the conjugate acid of the amine, an ammonium ion.

| Amide | Hydronium ion | | Carboxylic acid | Ammonium ion |

Hydrolysis of amides in basic solution forms a carboxylate ion and an amine.

| Amide | Hydroxide ion | | Carboxylate ion | Amine |

Although the conditions needed to hydrolyze amides are more severe than those required for the hydrolysis of esters, the mechanisms of both reactions are similar. The mechanism of the hydrolysis of amides in acid solution is shown in Figure 16.9. Step 1 of the mechanism is activation of the amide functional group by protonation of the oxygen atom of the C=O bond to form the resonance-stabilized conjugate acid of the amide. Step 2 is nucleophilic addition of water to

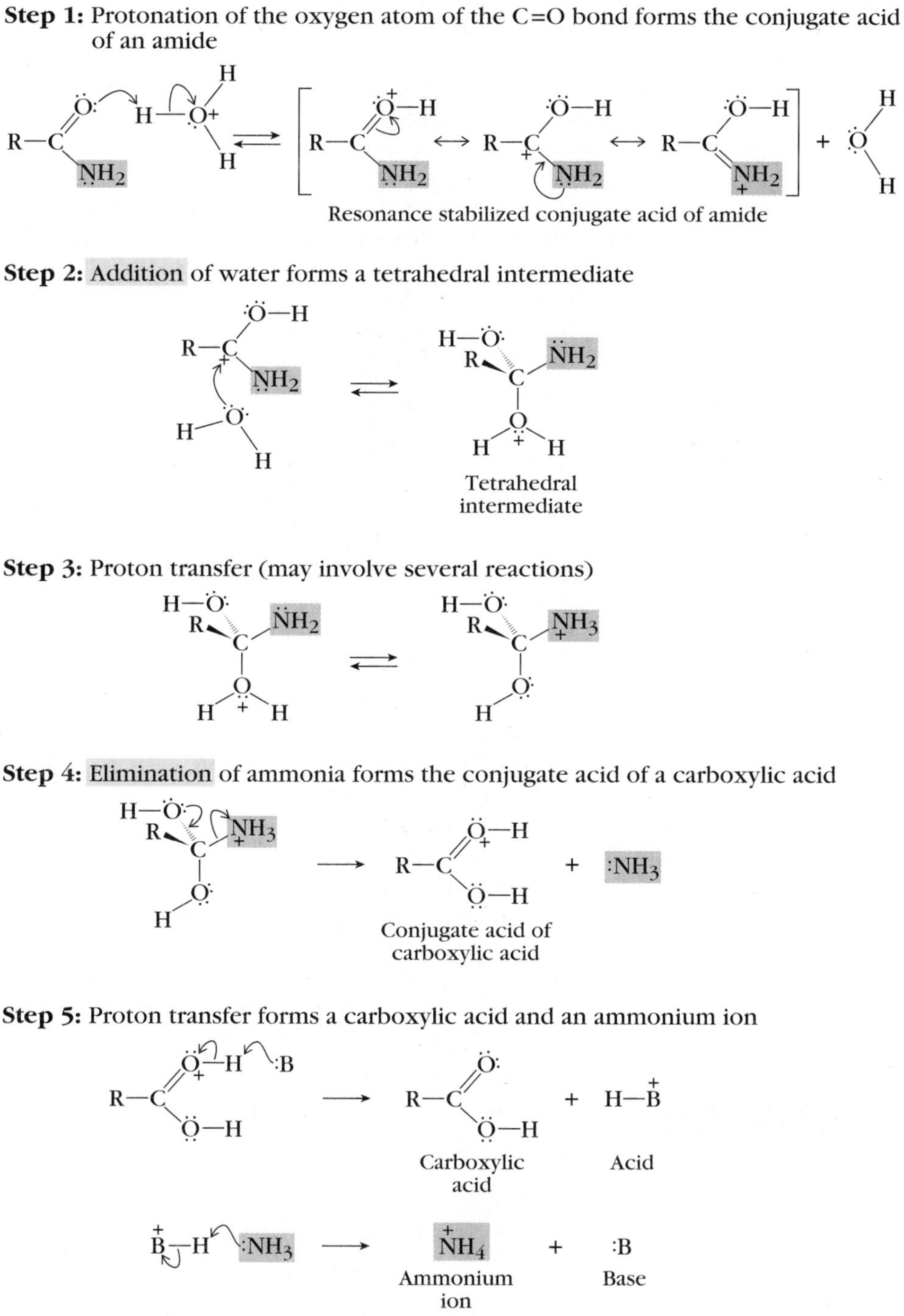

Step 1: Protonation of the oxygen atom of the C=O bond forms the conjugate acid of an amide

Resonance stabilized conjugate acid of amide

Step 2: Addition of water forms a tetrahedral intermediate

Tetrahedral intermediate

Step 3: Proton transfer (may involve several reactions)

Step 4: Elimination of ammonia forms the conjugate acid of a carboxylic acid

Conjugate acid of carboxylic acid

Step 5: Proton transfer forms a carboxylic acid and an ammonium ion

Carboxylic acid

Acid

Ammonium ion

Base

Figure 16.9
The addition-elimination mechanism of the hydrolysis of an amide in acid solution. Formation of the tetrahedral intermediate in Step 2 is the addition step, while the elimination step is loss of ammonia in Step 4. Steps 3 and 5 are proton transfers that are believed to occur by more than one reaction. For simplicity they are collected together here as a single step. The letter B refers to any base in the reaction mixture, such as water or ammonia.

form a tetrahedral intermediate. The proton transfer in Step 3 is not a single reaction but a number of reactions that transfer a proton from an oxygen atom to a nitrogen atom of the tetrahedral intermediate. For simplicity, we combine these reactions in a single step. Elimination of ammonia in Step 4 forms the conjugate acid of the carboxylic acid and ammonia. The carboxylic acid and the ammonium ion are formed in Step 5 by a number of proton transfer reactions.

This mechanism, while specific for the acid-catalyzed hydrolysis of amides, illustrates two reasons why acids accelerate acyl transfer reactions of carboxylic acid derivatives. Acids accelerate acyl transfer reactions first by protonating the oxygen atom of the C=O bond to form the conjugate acid of the carboxylic acid derivative. While the conjugate acid is present in solution in only small amounts, the carbon atom of its protonated C=O bond is very electrophilic. Nucleophiles, therefore, add to the conjugate acid of the carboxylic acid derivative faster than they do to the neutral carboxylic acid derivative.

Protonation of the tetrahedral intermediate is the second reason why acids accelerate acyl transfer reactions. Protonation converts one of the groups into a better leaving group. Protonation of the —NH$_2$ group of the tetrahedral intermediate in Step 3 in Figure 16.9 allows it to leave as a neutral ammonia molecule (NH$_3$), which is a less basic leaving group than an amide anion (—NH$_2$). Protonation of the oxygen atom of the C=O bond and conversion of one of the groups of the tetrahedral intermediate into a better leaving group are two common steps of the mechanisms of all acid-catalyzed acyl transfer reactions.

| **Exercise 16.10** | Write all the steps in the hydrolysis of *N*-ethyl acetamide in acid solution. |

| **Exercise 16.11** | Propose a mechanism for the hydrolysis of *N*-ethyl acetamide in basic solution. |

The mechanism of the hydrolysis of amides in basic solution, shown in Figure 16.10, is similar to the saponification of esters (Section 15.12). Step 1 is the nucleophilic addition of hydroxide ion to the carbon atom of the C=O bond of the amide to form a tetrahedral intermediate. A number of proton transfer reactions convert the tetrahedral intermediate into an ammonium ion in Step 2, which eliminates ammonia to give the conjugate acid of a carboxylic acid in Step 3. Proton transfer from the conjugate acid to hydroxide ion in Step 4 forms the carboxylate ion. In aqueous basic solution, the carboxylate ion is the weaker base and water is the weaker acid (Section 3.3), so the equilibrium constant for the formation of a carboxylate ion is so large that Step 4 is irreversible.

Step 1: Addition of a hydroxide ion to the carbon atom of the C=O bond forms a tetrahedral intermediate

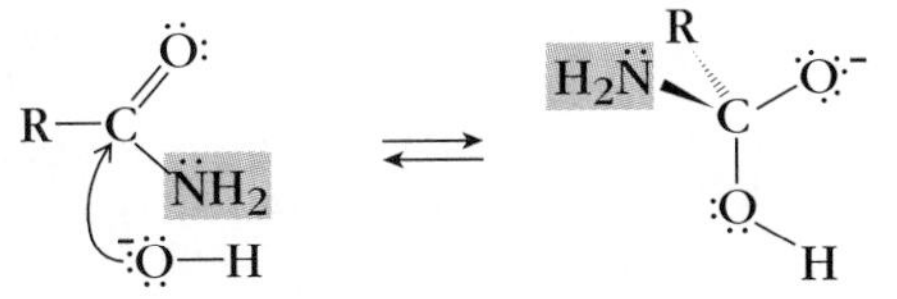

Tetrahedral
intermediate

Step 2: Proton transfer (may involve several reactions)

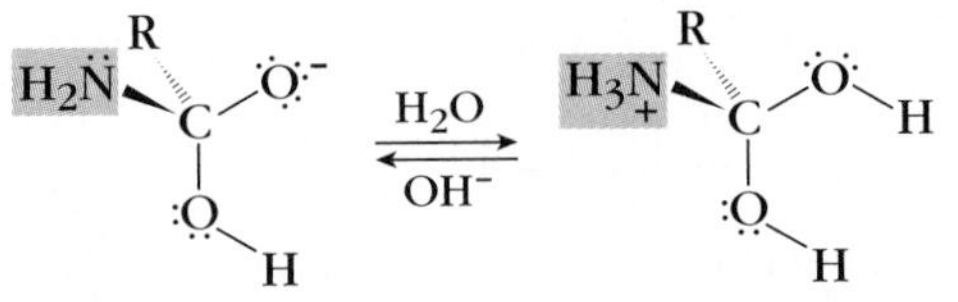

Step 3: Elimination of ammonia forms the conjugate acid of a carboxylic acid

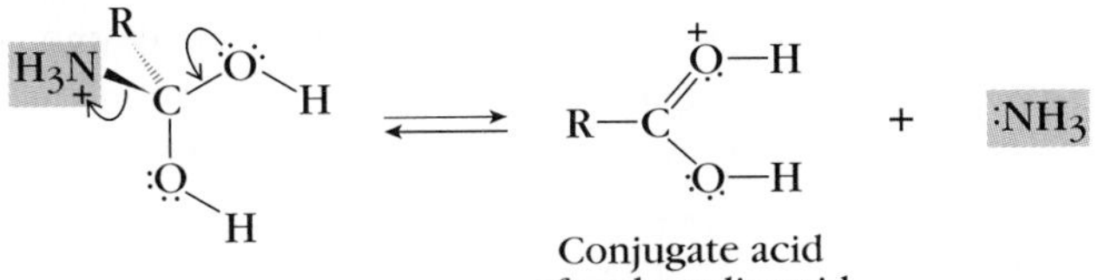

Conjugate acid
of carboxylic acid

Step 4: Proton transfer reactions form a carboxylate anion and water

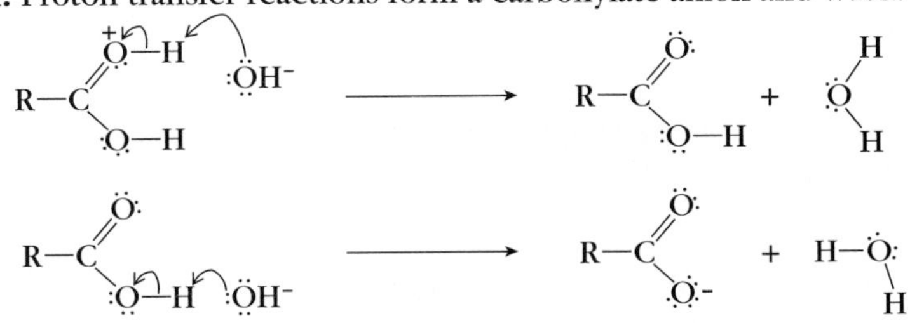

Carboxylate
anion

Figure 16.10
The addition-elimination mechanism of the hydrolysis of amides in basic solution. Addition is Step 1 in which a hydroxide ion adds to the carbon atom of the C=O bond. Elimination of ammonia in Step 3 forms the conjugate acid of a carboxylic acid, which is transformed into a carboxylate ion by proton transfer in Step 4.

16.9   Acyl Transfer Reactions of Anhydrides, Thioesters, and Esters

Acyl transfer reactions to oxygen and nitrogen atoms are the most common reactions of carboxylic acid anhydrides, thioesters, and esters. These reactions are summarized in Figure 16.11.

While carboxylic acid anhydrides undergo the same type of acylation reactions as thioesters and esters, only one of the acyl groups of the anhydride is transferred to the nucleophile in an acylation reaction. The other one becomes the acyl group of the carboxylic acid or carboxylate ion.

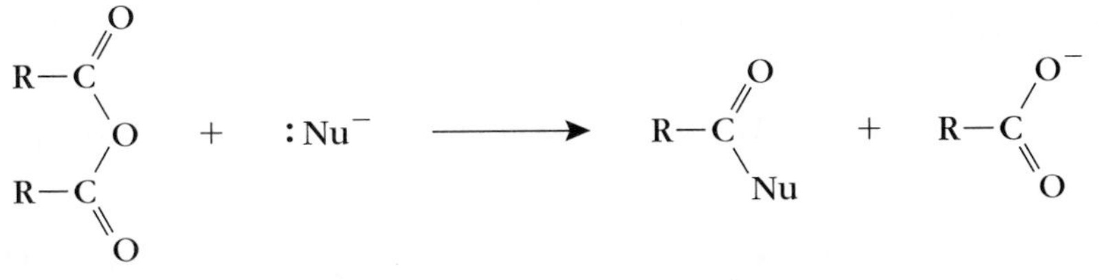

Carboxylic acid   Nucleophile   Acylated   Carboxylate
anhydride   nucleophile   ion

Figure 16.11
Summary of the most common acyl transfer reactions of carboxylic acid anhydrides, thioesters, and esters. Acyl transfer reactions of anhydrides also form one equivalent of carboxylic acid, RCOOH, or the carboxylate ion, RCOO⁻, depending on the pH of the reaction mixture.

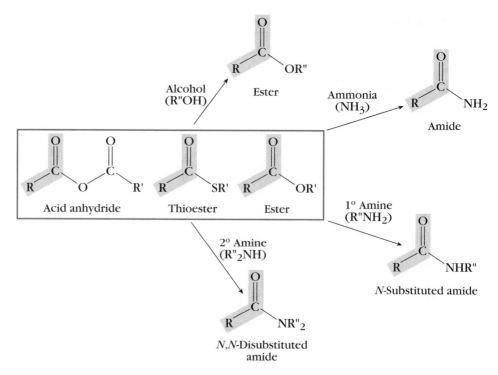

The following are specific examples of acyl transfer reactions of anhydrides, thioesters, and esters.

Esters react with alcohols in the presence of an acid catalyst to form a new ester.

$$\underset{\text{Ethyl acetate}}{\underset{|}{\underset{\text{OCH}_2\text{CH}_3}{CH_3C}}\overset{\overset{O}{\|}}{}} \quad + \quad \underset{\text{Excess 1-pentanol}}{HO(CH_2)_4CH_3} \quad \xrightarrow[\text{catalyst}]{\text{Acid}} \quad \underset{\substack{\text{Pentyl acetate}\\(90\%\ \text{yield})}}{\underset{|}{\underset{\text{O(CH}_2)_4\text{CH}_3}{CH_3C}}\overset{\overset{O}{\|}}{}} \quad + \quad \underset{\text{Ethanol}}{CH_3CH_2OH}$$

This reaction, which is called **transesterification,** is a reversible reaction in which neither ester is favored at equilibrium. Either ester can be made the product by using an excess of the appropriate alcohol. Use of excess 1-pentanol, for instance, converts ethyl acetate into pentyl acetate.

These acyl transfer reactions occur by the same addition-elimination mechanism discussed in Section 16.6.

Exercise 16.12 Write the structure of the product or products of the reaction of each of the following compounds with methanol and any other reagent or catalyst, if needed.

(a) Propanoyl chloride

(b)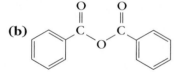

(c)

(d) Isopropyl pentanoate
(e) *N,N*-Dimethyl butanamide

Exercise 16.13 Write the structure of the product or products of the reaction of each of the following compounds with ammonia.

(a) Pentanoyl chloride
(b) Acetic anhydride

(c) *S*-Methyl butanethioate
(d) Isopropyl 4-methylpentanoate

Exercise 16.14 Write the structure of the product(s) of the reaction of each of the compounds in Exercise 16.13 with methylamine (CH$_3$NH$_2$).

Exercise 16.15 Write the structure of the product(s) of the reaction of each of the compounds in Exercise 16.13 with dimethylamine [(CH$_3$)$_2$NH].

| **Exercise 16.16** | Write the mechanism of the following acid-catalyzed transesterification reaction. |

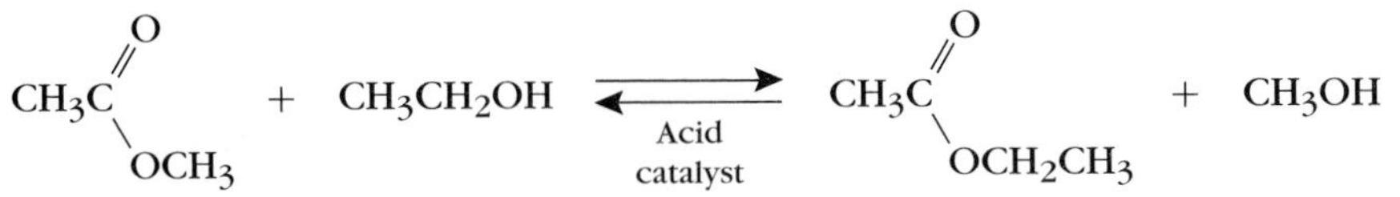

$$
CH_3C\underset{OCH_3}{\overset{O}{\diagup}} + CH_3CH_2OH \underset{\underset{catalyst}{Acid}}{\overset{}{\rightleftarrows}} CH_3C\underset{OCH_2CH_3}{\overset{O}{\diagup}} + CH_3OH
$$

16.10 Acyl Transfer Reactions in Living Systems

Acyl transfer reactions occur in living systems just as they do in the laboratory. In living systems, thioesters are used as the acyl transfer reagent rather than other carboxylic acid derivatives for two reasons. First, hydrolysis of thioesters does not form acidic products the way the hydrolysis of acyl chlorides and acid anhydrides does. Acidic hydrolysis products are unwanted because they upset the pH balance of living systems. Second, the hydrolysis of thioesters is thermodynamically more favorable than the hydrolysis of esters.

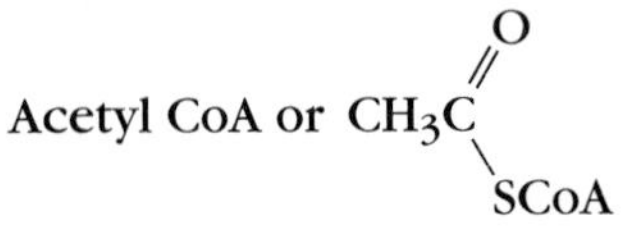

The thioester used in living systems for acetyl transfer reactions is **acetyl coenzyme A.**

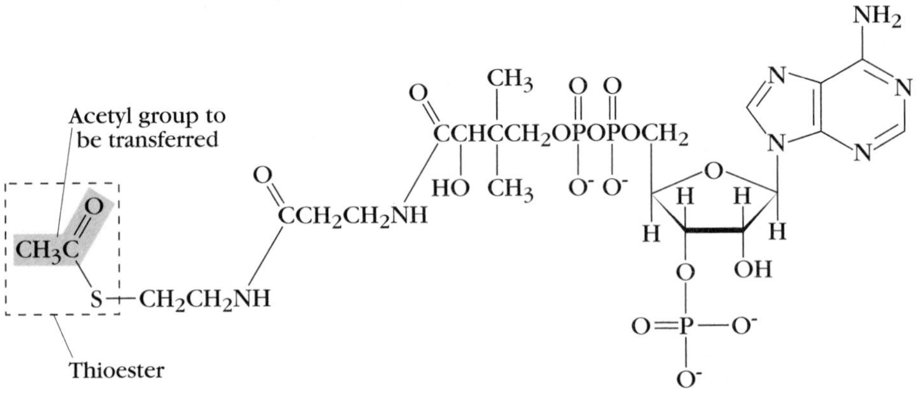

■ *Acetyl coenzyme A is usually abbreviated as*

$$
Acetyl\ CoA\ or\ CH_3C\underset{SCoA}{\overset{O}{\diagup}}
$$

Acetyl coenzyme A

Acetyl CoA is a much more complicated molecule than acetyl chloride or acetic anhydride, yet it functions in exactly the same way. It transfers an acetyl group to an oxygen or nitrogen atom of another molecule.

Acetyl CoA transfers an acetyl group to the nitrogen atom of galactosamine to form *N*-acetylgalactosamine, an important component of fibrous connective tissue.

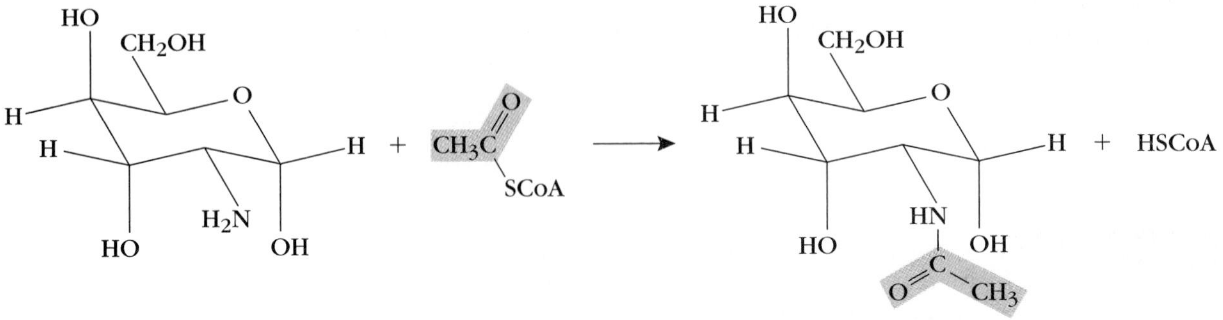

Galactosamine   Acetyl CoA   *N*-Acetyl galactosamine   Coenzyme A

Acetyl CoA also transfers an acetyl group to the oxygen atom of choline to form acetylcholine in a reaction that is catalyzed by the enzyme choline acetyltransferase.

$$HOCH_2CH_2\overset{+}{N}(CH_3)_3 \;+\; CH_3\overset{\overset{O}{\|}}{C}{\diagdown}_{SCoA} \;\xrightarrow[\text{acetyltransferase}]{\text{Choline}}\; CH_3\overset{\overset{O}{\|}}{C}{\diagdown}_{OCH_2CH_2\overset{+}{N}(CH_3)_3} \;+\; HSCoA$$

Choline   Acetyl CoA   Acetylcholine   Coenzyme A

Acetylation of choline and its reverse, hydrolysis of acetylcholine, play very important roles in the transmission of nerve impulses between nerve cells in many living organisms. Nerve impulses are transmitted from one nerve cell to another at junctions called synapses. Because the synapse constitutes a space or gap between adjacent nerve cells, the transmission of nerve impulses occurs when small diffusible molecules move across the gap. These molecules are called chemical transmitters, and acetylcholine is one of them.

The arrival of a nerve impulse triggers the release of acetylcholine from the nerve cell into the synapse gap. Acetylcholine diffuses through the gap to the membrane of the nerve cell on the other side of the synapse and combines with specific receptor molecules. The binding of acetylcholine to these receptors causes a chemical reaction that transmits the nerve impulse to the other nerve cell. After the nerve impulse has been received and propagated, the nerve cell must return to its original state in order to receive the transmission of the next nerve impulse. This is accomplished by the hydrolysis of acetylcholine in a reaction catalyzed by the enzyme acetylcholinesterase.

The mechanism of the acetylcholinesterase-catalyzed hydrolysis of acetylcholine is similar to the mechanism of the enzyme-catalyzed hydrolysis of esters discussed in Section 15.16. In the first step, a hydroxy group located on the active site of the enzyme adds to the carbon atom of the C=O bond of acetylcholine to form a tetrahedral intermediate that eliminates choline to form an acetyl-enzyme intermediate.

$$Enzyme\text{-}OH \;+\; H_3C{-}\overset{\overset{O}{\|}}{C}{\diagdown}_{OCH_2CH_2\overset{+}{N}(CH_3)_3} \;\longrightarrow\; Enzyme\text{-}O{-}\overset{\overset{O}{\|}}{C}{\diagdown}_{CH_3} \;+\; HOCH_2CH_2\overset{+}{N}(CH_3)_3$$

Acetylcholine   Acetyl-enzyme intermediate   Choline

Subsequent hydrolysis of the acetyl-enzyme intermediate forms acetate ion and regenerates the enzyme.

$$Enzyme\text{-}O{-}\overset{\overset{O}{\|}}{C}{\diagdown}_{CH_3} \;\xrightarrow{H_2O}\; Enzyme\text{-}OH \;+\; {}^{-}O{-}\overset{\overset{O}{\|}}{C}{\diagdown}_{CH_3}$$

Acetyl-enzyme intermediate   Acetate ion

Exercise 16.17   Write the mechanism of the reaction of enzyme-OH and acetylcholine to form the acetyl-enzyme intermediate.

Anything that interferes with acetylcholinesterase-catalyzed hydrolysis disrupts the transmission of nerve impulses in living organisms. Compounds that cause these kinds of disruptions include insecticides and nerve gases.

16.11 Nerve Gases and Insecticides Inhibit the Action of Acetylcholinesterase

Compounds that prevent the normal function of an enzyme, either temporarily or permanently, are called **enzyme inhibitors.** Many compounds are known to inhibit acetylcholinesterase. Because some are toxic, the mechanism of their action is of considerable practical importance.

World War I was the first time that chemical weapons were ever used. It was the beginning of one of the most horrible aspects of modern warfare. The first chemical weapon was chlorine gas. Much later, biological weapons such as nerve gases were developed. Most nerve gases are organophosphate compounds that work by inhibiting acetylcholinesterase. One of these nerve gases is diisopropylfluorophosphate (DIFP).

$$(CH_3)_2CHO-\overset{\overset{\displaystyle O}{\|}}{\underset{\underset{\displaystyle F}{|}}{P}}-OCH(CH_3)_2$$

Diisopropylfluorophosphate
(DIFP)

Substitution of the fluorine atom of DIFP by the hydroxy group on the active site of acetylcholinesterase forms a very stable phosphorylated enzyme intermediate.

$$\text{Enzyme-OH} \;+\; F-\overset{\overset{\displaystyle OCH(CH_3)_2}{|}}{\underset{\underset{\displaystyle OCH(CH_3)_2}{|}}{P}}=O \;\longrightarrow\; \text{Enzyme-O}-\overset{\overset{\displaystyle OCH(CH_3)_2}{|}}{\underset{\underset{\displaystyle OCH(CH_3)_2}{|}}{P}}=O$$

Diisopropyl-
fluorophosphate   Phosphorylated
enzyme intermediate

This phosphorylated enzyme intermediate is formed by the reaction of the same hydroxy group on the enzyme that adds to the carbon atom of the C=O bond of acetylcholine. This effectively prevents acetylcholine from occupying the active site on acetylcholinesterase. As a result, the hydrolysis of acetylcholine cannot occur. DIFP inhibition is long term, furthermore, because the phosphorylated enzyme intermediate is resistant to hydrolysis. Long-term inhibition causes respiratory paralysis, which is fatal to the organism.

Other organic phosphate compounds that have been synthesized for use as agricultural insecticides or nerve gases include sarin and parathion.

$$C_2H_5O-\overset{\overset{\displaystyle O}{\|}}{\underset{\underset{\displaystyle O}{|}}{P}}-OC_2H_5$$

Parathion

$$(CH_3)_2CHO-\overset{\overset{\displaystyle O}{\|}}{\underset{\underset{\displaystyle F}{|}}{P}}-OCH_3$$

Sarin

The mechanism of their action is the same as DIFP. The atom or group substituted by the enzyme-bound hydroxy group is highlighted in blue.

To counteract the effects of the nerve gases used in warfare, antidotes were soon developed. One of the most successful is based on hydroxylamine (NH_2OH), which was known to remove the phosphoryl group and reactivate acetylcholinesterase. Hydroxylamine is toxic, however, so it has no value as an antidote. After a great deal of research, pyridine aldoxime methiodide (PAM), a nontoxic derivative of hydroxylamine, was developed as an antidote to organophosphate poisoning. Its hydroxy group attacks the phosphoryl group of the inhibited enzyme and forms a phosphorylated PAM and the reactivated acetylcholinesterase.

Inhibited enzyme   Pyridine aldoxime   Reactivated   Phosphorylated PAM
                          methiodide (PAM)   acetylcholinesterase

PAM is a very effective antidote to organophosphate poisoning because it is nontoxic and because it is about a million times more effective than hydroxylamine in reactivating acetylcholinesterase.

The discussion in this and the preceding section of acyl transfer reactions in nature illustrates a very important point. No matter where an acyl transfer reaction occurs, either in the laboratory or in living organisms, its mechanism is essentially the same. All acyl transfer reactions occur by an addition-elimination mechanism, in which addition of a nucleophile to the carbon atom of the C=O bond occurs to form a tetrahedral intermediate. This is the first step of the mechanism in basic solution and the second step in acidic solution. Subsequent elimination of a group forms another acyl compound. While this mechanism was first proposed from results obtained in the laboratory, its application to living organisms has contributed to our understanding of the mechanisms of such diverse processes as the hydrolysis of fats and proteins in the intestines, transmission of nerve impulses, and the action of nerve gases and insecticides. Clearly, the application to living organisms of organic chemistry principles gained in the laboratory is a major reason understanding organic chemistry is an important part of understanding the life sciences.

Let's return now to the laboratory to learn about the reduction of carboxylic acid derivatives.

16.12 Reduction of Carboxylic Acid Derivatives

In Chapter 15, we discussed how $LiAlH_4$ reduces carboxylic acids and esters to form primary alcohols. $LiAlH_4$ reduces other carboxylic acid derivatives, too, as shown in the following examples.

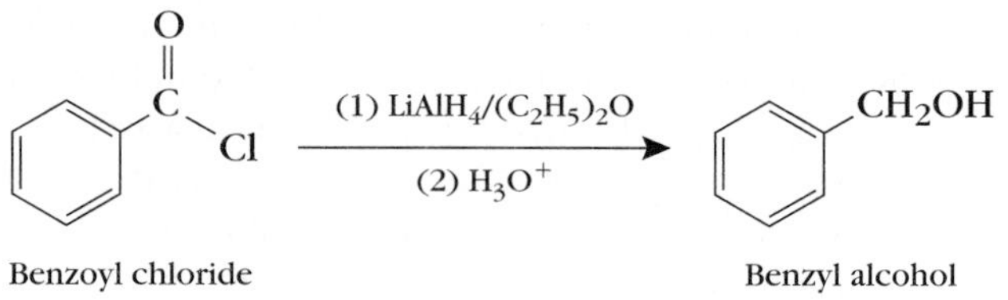

Benzoyl chloride → Benzyl alcohol

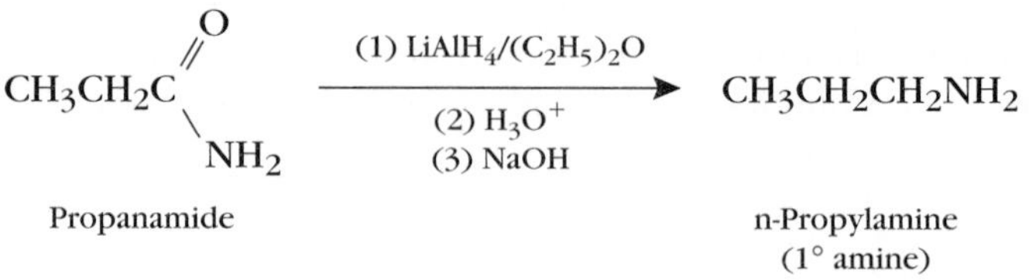

(1) LiAlH$_4$/(C$_2$H$_5$)$_2$O
(2) H$_3$O$^+$

Propanamide → n-Propylamine (1° amine)

(1) LiAlH$_4$/(C$_2$H$_5$)$_2$O
(2) H$_3$O$^+$
(3) NaOH

NaBH$_4$ in methanol, which is a much milder reducing agent than LiAlH$_4$, reduces the carbonyl group of aldehydes and ketones rapidly at room temperature (Section 14.8) but is essentially inert to other carbonyl-containing functional groups. The data in Table 16.4 illustrate the difference in reactivity between LiAlH$_4$ and NaBH$_4$.

The differences in reactivity between LiAlH$_4$ and NaBH$_4$ make it possible to reduce one carbonyl-containing functional group in the presence of another. For example, the carbonyl group of an aldehyde can be reduced without reducing an amide group by using NaBH$_4$ as the reducing agent.

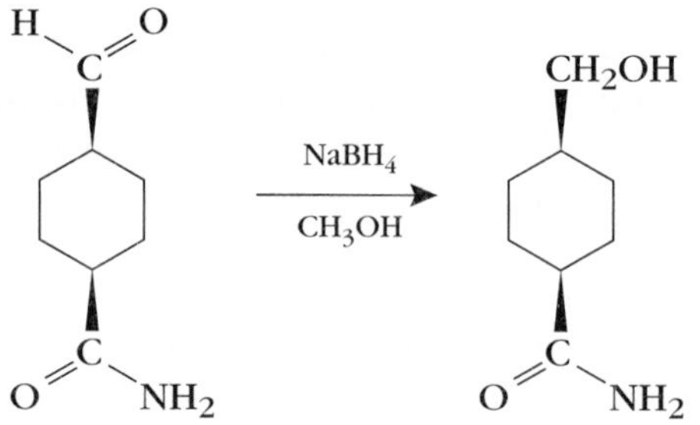

NaBH$_4$
CH$_3$OH

LiAlH$_4$ would reduce the aldehyde group *and* the amide group.

In addition to differences in reactivity, the data in Table 16.4 show that a primary alcohol is the product of the reaction of LiAlH$_4$ with all carboxylic acid derivatives except amides; amides are reduced to amines. Why are amides different from the others? We can answer this question by comparing the tetrahedral intermediates formed by the addition-elimination mechanisms of the reduction of a typical amide, acetamide, and a typical ester, ethyl acetate.

Recall from Section 14.8 that the hydride ion, H$^-$, is the nucleophile in LiAlH$_4$-reduction reactions. Hydride ion adds to the carbonyl carbon atom of the ester to form a tetrahedral intermediate in the first step of the mechanism.

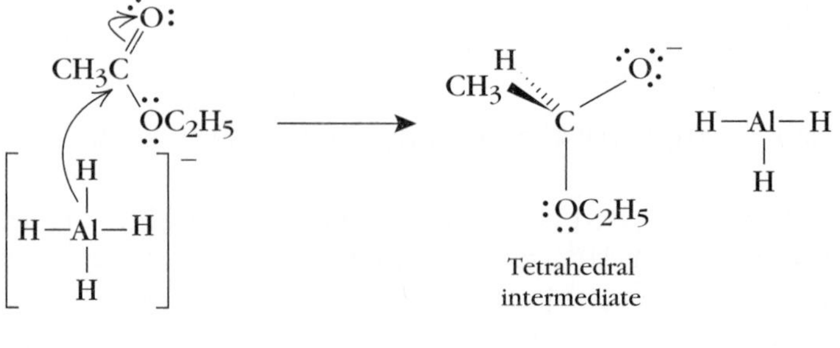

Tetrahedral intermediate

Table 16.4	Comparison of the Ability of $LiAlH_4$ and $NaBH_4$ to Reduce Various Compounds*		
		Reaction Products	
Compound	Structure	$LiAlH_4$ in Diethyl Ether	$NaBH_4$ in Methanol
Aldehyde	RCHO	1° Alcohol	1° Alcohol
Ketone	$R_2C{=}O$	2° Alcohol	2° Alcohol
Acyl chloride	RCOCl	1° Alcohol	Acyl chloride reacts with solvent
Thioester	RCOSR′	1° Alcohol	Slow reduction
Ester	RCOOR′	1° Alcohol	Slow reduction
Carboxylic acid	RCOOH	1° Alcohol	$NaBH_4$ reacts with acidic hydrogen, no reduction
Carboxylate ion	RCOO⁻	1° Alcohol	No reaction
Amide	$RCONH_2$	1° Amine	No reaction
Nitrile	RC≡N	1° Amine	No reaction
Alkene	RCH=CHR	No reaction	No reaction

*Compounds are arranged in order of decreasing reactivity with the metal hydrides.

Step 2 of the mechanism is elimination of the best leaving group (weakest base) from the tetrahedral intermediate. The ethoxide group is the best leaving group in the tetrahedral intermediate formed by addition of $LiAlH_4$ to ethyl acetate, so its elimination forms an aldehyde.

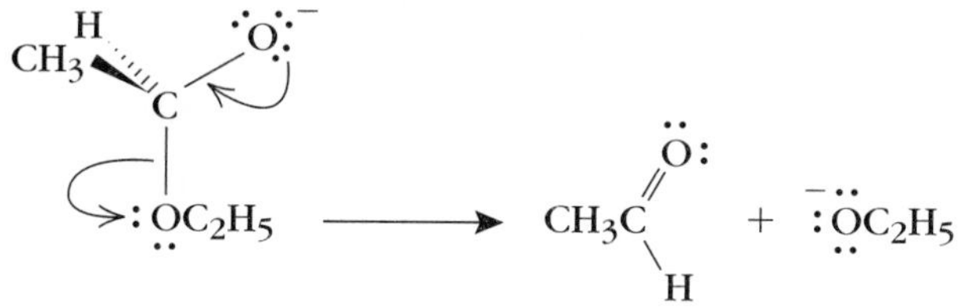

Acetaldehyde

Unlike acyl transfer reactions, the reaction does not stop at this point because the aldehyde readily reacts with the reducing agent to form an alcohol after aqueous acid addition.

■ *Recall that $LiAlH_4$ reduction of esters forms two molecules of alcohol, one from the RO group of the ester and the other from the RC=O group.*

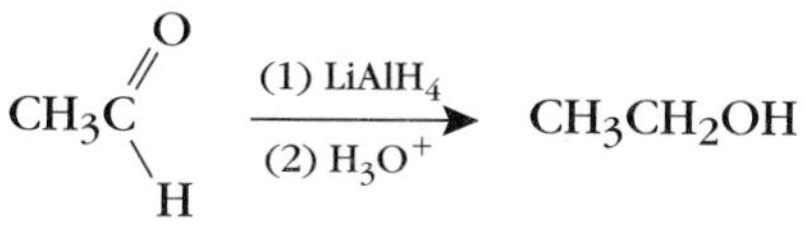

When an amide is added to a solution of $LiAlH_4$ in diethyl ether, an acid-base reaction first occurs between the strongly basic $LiAlH_4$ and the weakly acidic N—H bond of the amide to form the lithium salt of the amide.

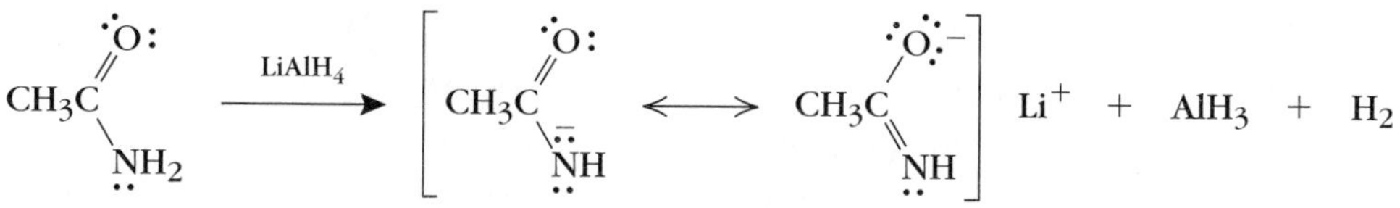

A tetrahedral intermediate is formed by addition of a hydride ion from either LiAlH$_4$ or AlH$_3$ to the carbon atom of the C=O bond. At the same time, a bond is formed between the aluminum atom of the AlH$_2$ group and the carbonyl oxygen atom.

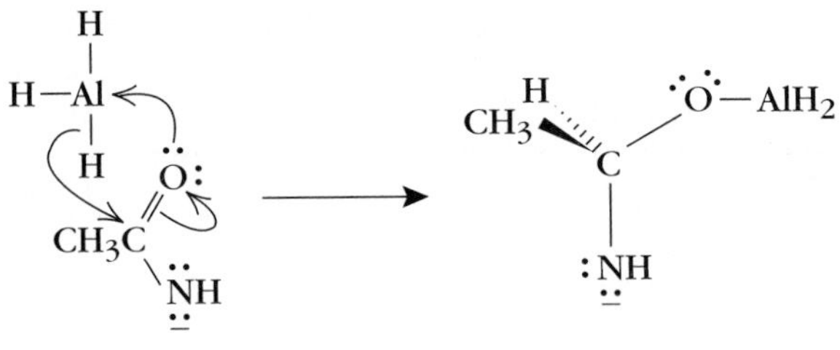

Tetrahedral intermediate

Elimination of the best leaving group from the tetrahedral intermediate occurs in the next step of the mechanism. The —OAlH$_2$ group is the best leaving group (weakest base), so it is eliminated to form an imine.

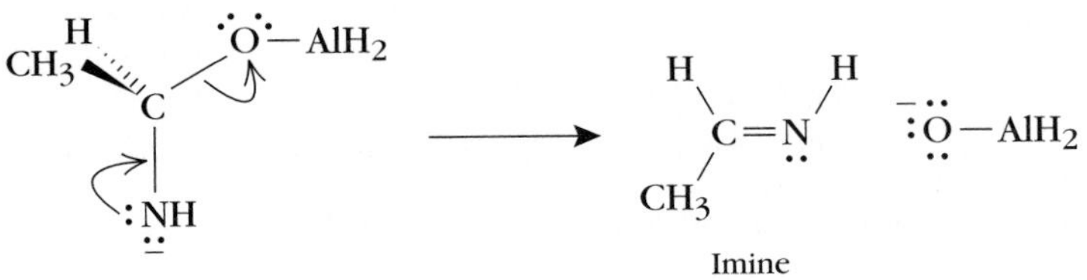

Imine

A hydride ion adds to the imine to form the lithium salt of the amine. Addition of aqueous acid followed by base forms the amine.

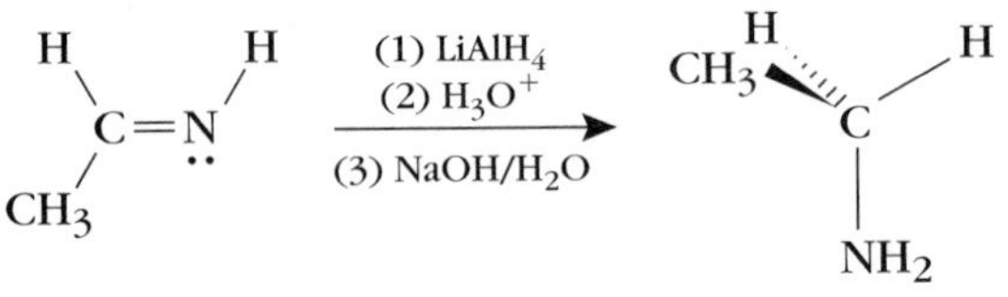

Imine   An amine

Exercise 16.18   Write the structure of the product of the reaction of each of the following compounds with LiAlH$_4$ followed by reaction with an aqueous acid solution.

(a) 3-Methylbutanoyl chloride

(b)

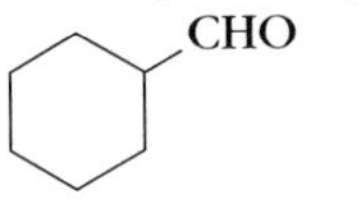

(c) 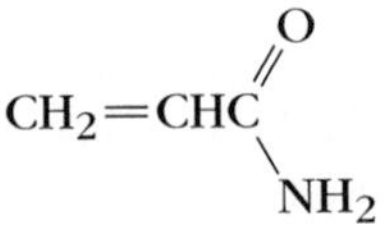 CH$_2$=CHC(=O)NH$_2$

(d) 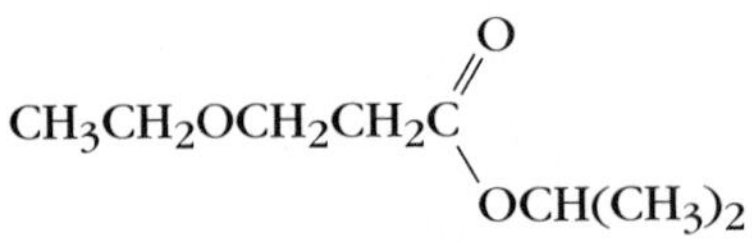 CH$_3$CH$_2$OCH$_2$CH$_2$C(=O)OCH(CH$_3$)$_2$

(e) *N,N*-Dimethyl pentanamide

(f)

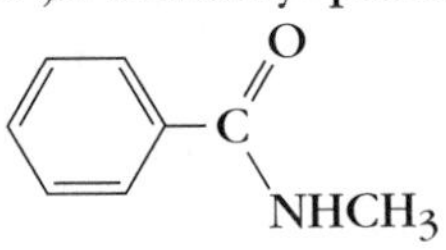

(g) Propanoic anhydride

(h) 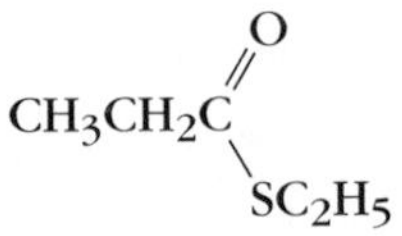 CH$_3$CH$_2$C(=O)SC$_2$H$_5$

In summary, reduction of amides forms amines, rather than alcohols, because the oxygen atom of the C=O bond of an amide is converted into a better leaving group than the nitrogen atom of the $-NH_2$ group in the tetrahedral intermediate.

Exercise 16.19   Write the structure of the product, if any, of the reaction of $NaBH_4$ in methanol with each of the compounds in Exercise 16.18.

The reactions of carboxylic acid derivatives with organometallic reagents are similar in many respects to their reactions with metal hydrides. Both reactions yield alcohols, and both occur by similar mechanisms.

16.13   Reactions of Grignard Reagents with Carboxylic Acid Derivatives

All carboxylic acid derivatives except carboxylic acids and amides react with Grignard reagents. While acyl chlorides, acid anhydrides, thioesters, and esters all react with two equivalents of Grignard reagent to form alcohols after aqueous acid addition, the reaction with esters is the most useful synthetically because esters are easier to prepare and handle in the laboratory. Ethyl 3-methylbutanoate reacts with two equivalents of methylmagnesium iodide, for example, to form 2,4-dimethyl-2-pentanol.

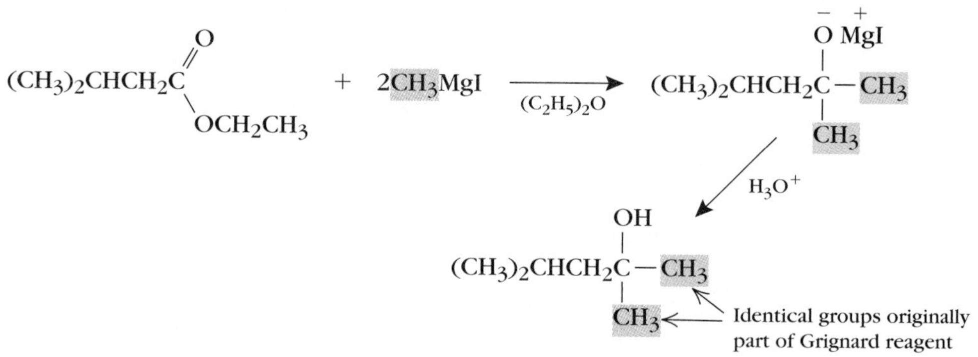

2,4-Dimethyl-2-pentanol
(82% yield)

The reaction of esters with Grignard reagents is particularly useful for the preparation of tertiary alcohols in which two identical groups and a hydroxy group are bonded to the same carbon atom. The groups are identical because they both were the alkyl group of the Grignard reagent. Formyl esters are the only exception. They react with Grignard or organolithium reagents to form secondary alcohols.

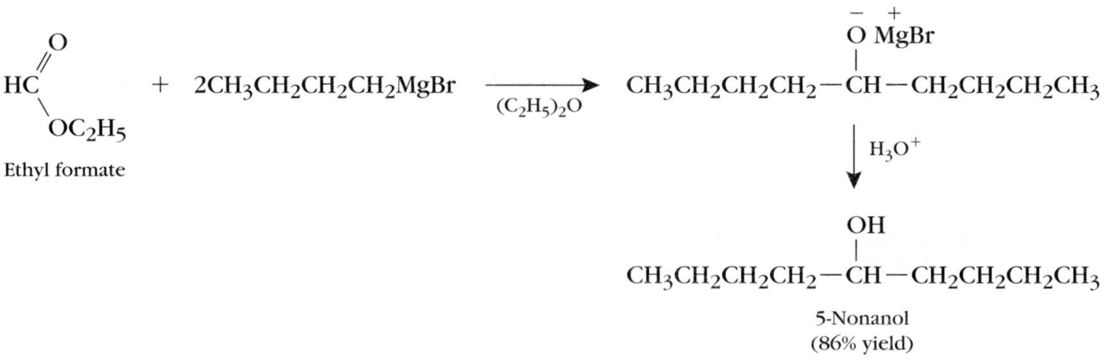

$$\text{HC} \overset{\displaystyle O}{\underset{\displaystyle OC_2H_5}{\Big\langle}} \; + \; 2CH_3CH_2CH_2CH_2MgBr \xrightarrow[\;(C_2H_5)_2O\;]{} CH_3CH_2CH_2CH_2 - \overset{\overset{\displaystyle \overset{-}{O} \; \overset{+}{MgBr}}{|}}{CH} - CH_2CH_2CH_2CH_3$$

Ethyl formate

$$\Big\downarrow H_3O^+$$

$$CH_3CH_2CH_2CH_2 - \overset{\overset{\displaystyle OH}{|}}{CH} - CH_2CH_2CH_2CH_3$$

5-Nonanol
(86% yield)

The reaction of esters and Grignard reagents occurs by the addition-elimination mechanism shown in Figure 16.12. Step 1 of the mechanism is addition of the carbon part of the Grignard reagent to the carbon atom of the C=O bond to form a tetrahedral intermediate, which is a magnesium salt of a hemiacetal. Elimination of an alkoxide group in Step 2 forms a ketone, or an aldehyde in the case of formate esters. The ketone or aldehyde is not isolated because aldehydes and ketones are more susceptible to nucleophilic additions than esters. Consequently, the aldehyde or ketone reacts with a second equivalent of Grignard reagent to form the observed alcohol after addition of an aqueous acid solution.

Grignard reagents do not react with carboxylic acids and amides because they remove a proton from each one to form resonance-stabilized anions. This resonance stabilization reduces the reactivity of the carbonyl carbon atoms to nucleophilic addition. Grignard reagents are not nucleophilic enough to react with either of these anions, even in the presence of excess Grignard reagent.

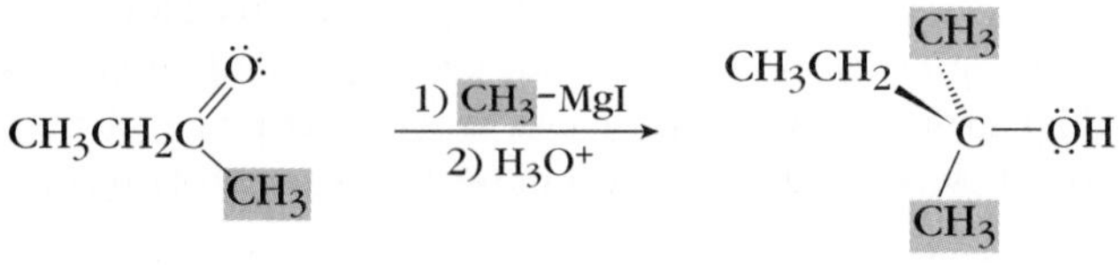

Figure 16.12
First two steps of the addition-elimination mechanism of the reaction of an ester and a Grignard reagent to form an alcohol. The mechanism of the reaction of a ketone and a Grignard reagent is not included in this Figure but can be found in Section 14.8.

Step 1: Addition of a Grignard reagent to the carbon atom of the C=O bond forms a tetrahedral intermediate

Tetrahedral intermediate

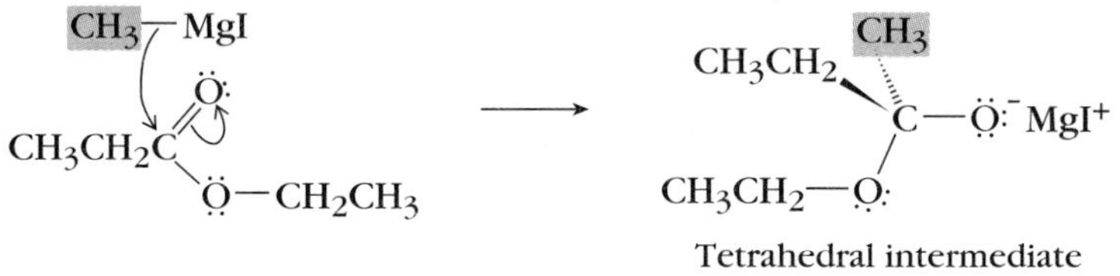

Step 2: Elimination of an ethoxide ion forms a ketone

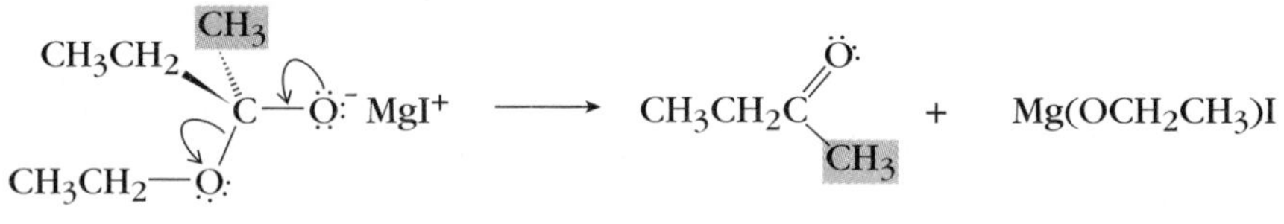

Reaction of the ketone and the Grignard reagent forms an alcohol after addition of aqueous acid solution

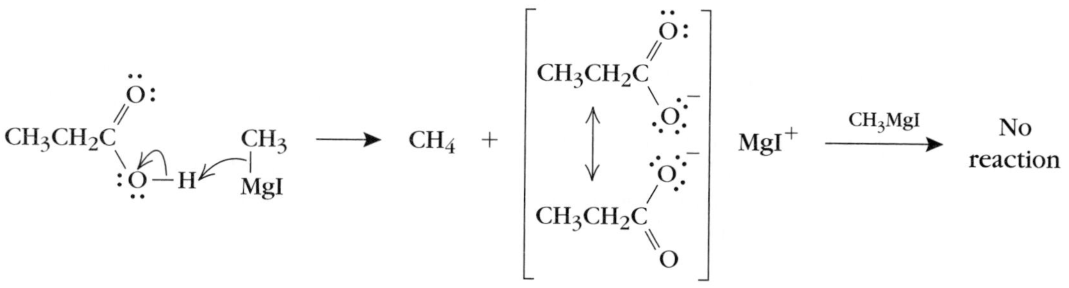

Resonance-stabilized ion

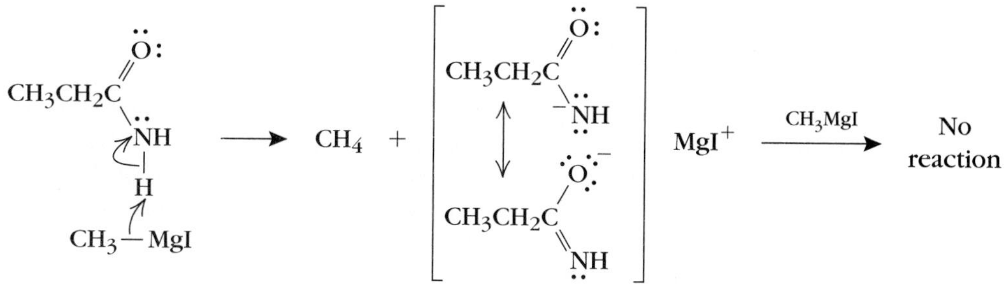

Resonance-stabilized ion

Exercise 16.20   Write the equation for the reaction of each of the following compounds with ethylmagnesium iodide followed by reaction with an aqueous acid solution.

(a) Ethyl butanoate

(b) Acetic anhydride

(c) S-Ethyl propanethioate

(d) $CH_3CH_2CH_2CH_2CH_2COOH$

(e) N-Butyl propanamide

Aldehydes are intermediates in the $LiAlH_4$ or $NaBH_4$ reduction of carboxylic acid derivatives, while ketones are intermediates in their reactions with Grignard reagents. Under certain conditions, these addition reactions can be stopped at the intermediate stage to form the aldehyde or ketone as the final product.

16.14 Preparation of Aldehydes and Ketones from Carboxylic Acid Derivatives

In Section 16.12 we discussed how aldehydes are intermediates in the reduction of some carboxylic acid derivatives. Aldehydes are not the products of reduction, however, because they react further with the reducing agent to form alcohols. Chemists have discovered a number of reagents that stop the reduction at the aldehyde stage, as well as other reagents that convert carboxylic acids to ketones. These reagents provide another synthetic route to aldehydes and ketones. We begin with the synthesis of aldehydes from carboxylic acid derivatives.

A   Aldehydes

Acyl chlorides are more reactive than aldehydes in nucleophilic addition reactions. Therefore it should be possible to find a reducing agent that would reduce acyl chlorides but not aldehydes. The reagent that chemists have developed for this purpose is called lithium aluminum tri(t-butoxy) hydride, Li(t-OC$_4$H$_9$)$_3$AlH. One equivalent of this reagent, which is prepared by treating LiAlH$_4$ with three equivalents of t-butyl alcohol, selectively reduces acyl chlorides to aldehydes.

$$
\text{(1) Li(t-OC}_4\text{H}_9\text{)}_3\text{AlH} \quad \text{(2) H}_3\text{O}^+
$$

This reduction reaction occurs by the familiar addition-elimination mechanism. Step 1 is addition of a hydride ion to the carbon atom of the C=O bond to form a tetrahedral intermediate. Step 2 is elimination of a chloride ion to form an aldehyde. Reduction of acyl chlorides to aldehydes is in effect an acyl transfer reaction to hydrogen.

B   Ketones

Ketones are an intermediate in the reaction of a Grignard reagent with acyl chlorides or esters. It is often possible to stop the reaction at the ketone stage by reversing the order of mixing the reagents. Usually the acyl chloride or ester is added to the Grignard reagent. However, adding *one equivalent* of the Grignard reagent to the carboxylic acid derivative, in what is called an inverse addition, makes it possible to isolate the ketone because the small amount of Grignard reagent in the reaction mixture reacts only with the more reactive acyl chloride.

$$
\text{CH}_3\text{CH}_2\text{CH}_2\text{CH}_2\text{C} \xrightarrow[\text{(2) H}_3\text{O}^+]{\text{(1) CH}_3\text{(CH}_2\text{)}_3\text{CH}_2\text{MgBr}} \text{CH}_3\text{(CH}_2\text{)}_3\text{C(CH}_2\text{)}_4\text{CH}_3
$$

The problem of further addition of an organometallic reagent to the ketone to form an alcohol can be avoided by using less reactive organometallic reagents such as lithium dialkylcuprates. These reagents are prepared by the reaction of two equivalents of an organolithium reagent with one equivalent of copper(I) chloride.

$$
2 \text{ R-Li } + \text{ CuCl } \longrightarrow \text{ R}_2\text{CuLi } + \text{ LiCl}
$$

Lithium dialkylcuprates react rapidly with acyl chlorides and aldehydes, slowly with ketones, and not at all with esters and amides. Because of this range of reactivities, the reaction of lithium dialkylcuprates with acyl chlorides is an excellent method of preparing ketones.

$$
2 \text{ (CH}_3\text{)}_2\text{CHCH}_2\text{C} + \text{ (CH}_3\text{CH}_2\text{)}_2\text{CuLi} \longrightarrow 2 \text{ (CH}_3\text{)}_2\text{CHCH}_2\text{CCH}_2\text{CH}_3
$$

Exercise 16.21   Show how you would prepare each of the following compounds from 2-methylpropanoyl chloride.

(a) (CH₃)₂CHCOOH
(b) *N,N*-Diethyl 2-methylpropanamide
(c) 2,3-Dimethyl-2-butanol

(d) (CH₃)₂CHC⟍(=O)SCH₂CH₂CH₃

(e) 2-Methyl-3-pentanone
(f) (CH₃)₂CHCHO
(g) 2-Methylpropanol
(h) 2-Methylpropanoic propanoic anhydride
(i) Isopropyl isobutyrate

Nitriles and carboxylic acid derivatives are readily interconverted, so nitriles can be thought of as another carboxylic acid derivative.

16.15 Nitriles Are Distant Relatives of Carboxylic Acids

Nitriles can be treated as carboxylic acid derivatives because they form carboxylic acids upon hydrolysis. The structure of the nitrile functional group, however, is very different from that of other carboxylic acid derivatives. Consequently, we begin the study of nitriles by describing their structure, bonding, and spectral properties. Then we learn how to prepare them, and finally we study some of their chemical reactions.

A   Structure, Bonding, and Spectral Properties of the Nitrile Group Resemble Those of Acetylene

The nitrile group contains a carbon-nitrogen triple bond. The N—C—C bond angle in nitriles is 180°, as shown by the structure of acetonitrile in Figure 16.13. Like a C≡C bond, a C≡N bond consists of one σ bond and two π bonds. The localized valence bond orbital description of both triple bonds is very similar. The carbon-nitrogen sigma bond is formed by combining an *sp* hybrid orbital from carbon and an *sp* hybrid orbital from nitrogen. The two π bonds are formed by combining the unhybridized *p* orbitals on each atom. The nonbonding electron pair on the nitrogen atom occupies the remaining *sp* hybrid atomic orbital. The π electron structure of the nitrile group in acetonitrile is shown in Figure 16.14.

The difference in electronegativity between carbon and nitrogen and the presence of a nonbonding electron pair at the nitrogen end of the molecule make nitriles very polar molecules. The dipole moments of nitriles, for example, are much greater than the dipole moments of alkynes.

$$\text{CH}_3\text{C}\equiv\text{N:} \qquad \text{CH}_3\text{CH}_2\text{C}\equiv\text{N:} \qquad \text{CH}_3\text{CH}_2\text{C}\equiv\text{CH}$$
$$\mu = 3.30\,\text{D} \qquad\quad \mu = 3.39\,\text{D} \qquad\qquad \mu = 0.80\,\text{D}$$

Nitriles can be readily identified by their C≡N stretching vibration that appears at about 2255 cm⁻¹. The only other functional group that has an absorption in this region is the C≡C bond.

The ¹³C NMR chemical shift of the carbon atom of a nitrile group is found in the region δ 117 to 125. This is considerably downfield from that of the carbon atoms of a C≡C bond (δ 65 to 85) and is due to the greater electronegativity of nitrogen compared to carbon. Thus the combination of IR and ¹³C NMR spectroscopy can usually be used to distinguish nitriles from alkynes.

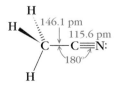

Figure 16.13
The molecular structure of acetonitrile (ethanenitrile).

A **σ Bond formation in acetonitrile, CH₃CN:**

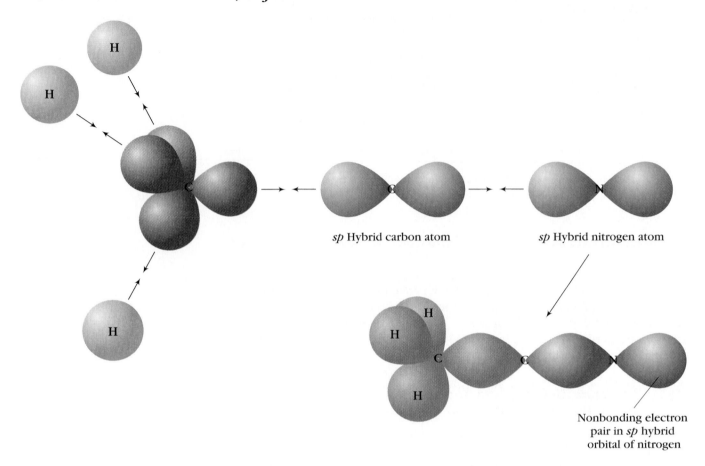

B **π Bond formation in acetonitrile, CH₃CN:**

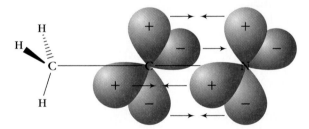

C **Space-filling model of acetonitrile, CH₃CN:**

Figure 16.14
σ and π Electron structure of the nitrile group of acetonitrile.

The ^{1}H NMR chemical shifts of hydrogen atoms adjacent to the nitrile group are found near δ 2. This is about the same position as α-hydrogens in other carboxylic acid derivatives.

B   Preparation of Nitriles

The simplest method of preparing alkanenitriles is by a nucleophilic substitution reaction between cyanide ion and a haloalkane. This reaction, which was introduced in Chapter 12, occurs by an S_N2 mechanism. Reactions that occur by this mechanism are limited to methyl, primary, and sometimes secondary haloalkanes.

$$CH_3CH_2CH_2CH_2Br \ + \ CN^- \ \xrightarrow[\text{mechanism}]{S_N2} \ CH_3CH_2CH_2CH_2CN \ + \ Br^-$$

Dehydration of unsubstituted amides is another method of preparing nitriles. A number of dehydrating agents, such as thionyl chloride ($SOCl_2$), P_4O_{10}, $POCl_3$, and acetic anhydride, have been used to carry out this reaction. Alkanenitriles that cannot be prepared by a nucleophilic substitution reaction can be synthesized by this method.

2-Bromo-2-methylpropane

2,2-Dimethylpropanamide

2,2-Dimethylpropanenitrile

The two most important reactions of nitriles are hydrolysis and reduction.

C   Hydrolysis of Nitriles Forms Carboxylic Acids

The polarity of a nitrile group makes its carbon atom partially positive and susceptible to nucleophilic addition, which is the typical reaction of carboxylic acid derivatives. One reaction of this kind is hydrolysis to form carboxylic acids under conditions similar to those used for the hydrolysis of amides, namely heating in aqueous acidic or basic solution for several hours.

Hydrolysis of both amides and nitriles is an irreversible reaction in the presence of acid or base. Under acidic conditions the products are a carboxylic acid and ammonium ion, while under basic conditions the products are ammonia and a carboxylate ion. In order to obtain the carboxylic acid, subsequent acidification of the basic solution is necessary.

$$CH_3CH_2CH_2CH_2CN \ \xrightarrow[\text{Heat}]{H_2SO_4/H_2O} \ CH_3CH_2CH_2CH_2COOH \ + \ NH_4^+$$

The mechanism of the hydrolysis of nitriles in acid solution is illustrated in Figure 16.15. Nitriles contain a nonbonding pair of electrons on their nitrogen atom, so they are weak bases (the pK_a of conjugate acids of nitriles ≈ -10). Step 1 of the mechanism is protonation of the nitrogen atom of the nitrile group to form its resonance-stabilized conjugate acid. Step 2 is nucleophilic addition of water to the carbon atom of the conjugate acid of the nitrile. Notice that the product of this step is not a tetrahedral intermediate. A tetrahedral intermediate is not formed because addition of water converts the *sp* hybrid carbon atom of the con-

Figure 16.15
The first three steps in the mechanism of the hydrolysis of nitriles in acid solution. The mechanism of the hydrolysis of amides is not included in this Figure but is given in Figure 16.9 (p. 719).

Step 1: Protonation of the nitrogen atom of the nitrile group forms the resonance-stabilized conjugate acid of the nitrile

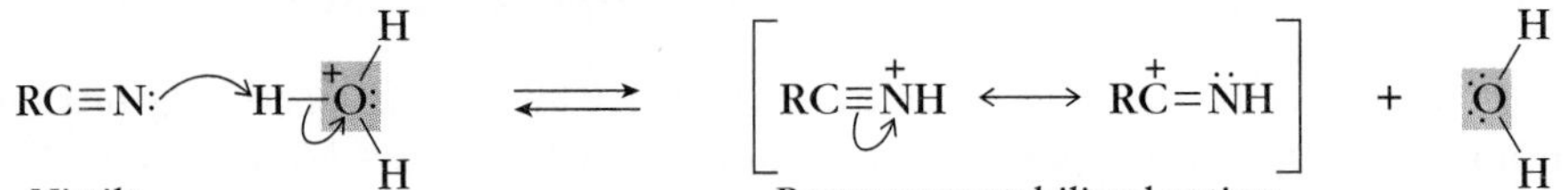

Step 2: Nucleophilic addition of water to the carbon atom of the conjugate acid of the nitrile

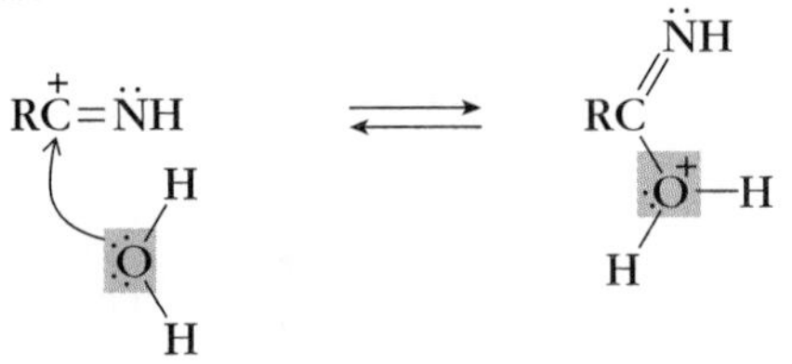

Step 3: A sequence of proton transfer reactions and electron movements forms an amide

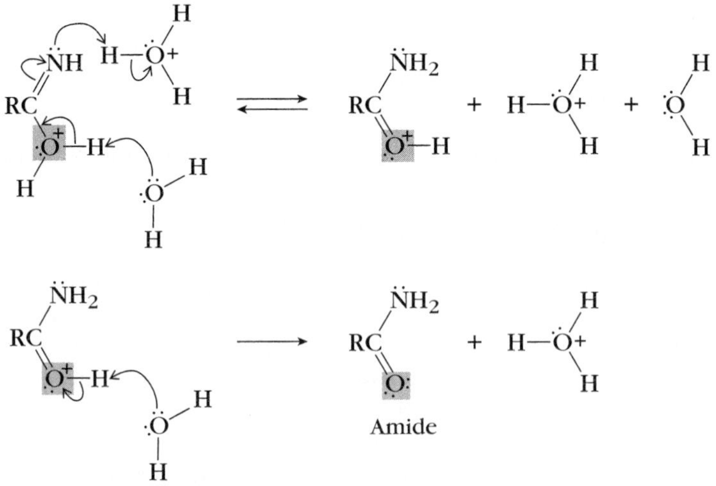

jugate acid of the nitrile into an sp^2 hybrid carbon atom of the product. Step 3 is a series of proton transfer reactions with accompanying electron movement to form an amide. Hydrolysis of the amide by the mechanism discussed in Section 16.8 forms the carboxylic acid and an ammonium ion.

Because an amide is an intermediate in the hydrolysis of a nitrile, the reaction conditions needed to hydrolyze both nitriles and amides are similar.

D   Reduction of Nitriles

The reduction of nitriles with $LiAlH_4$ forms primary amines.

$$CH_3CH_2CH_2CH_2CN \xrightarrow[\text{(2) } H_2O]{\text{(1) } LiAlH_4/(C_2H_5)_2O} CH_3CH_2CH_2CH_2CH_2NH_2$$

Pentanenitrile   Pentanamine
                                                      (62% yield)

The mechanism of this reaction is nucleophilic addition of a hydride ion to the carbon atom of the nitrile to form an imine anion intermediate.

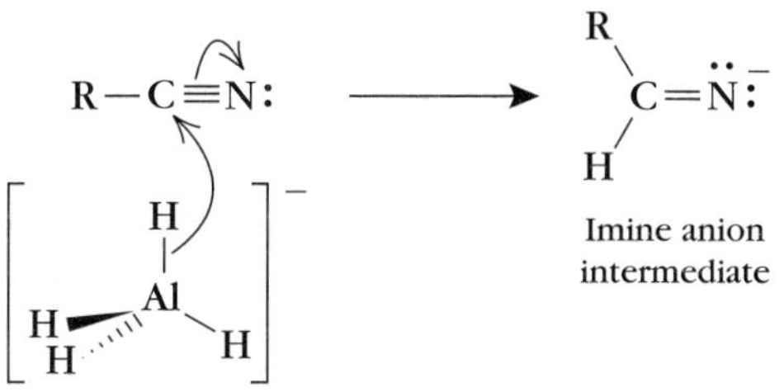

Imine anion
intermediate

When LiAlH$_4$ is the reducing agent, the imine anion intermediate adds a second equivalent of hydride to form an amine as the final product. If a less powerful reducing agent is used, however, such as diisobutylaluminum hydride or lithium aluminum tri(t-butoxy)hydride, a second addition does not occur and the imine anion can be hydrolyzed to yield an aldehyde.

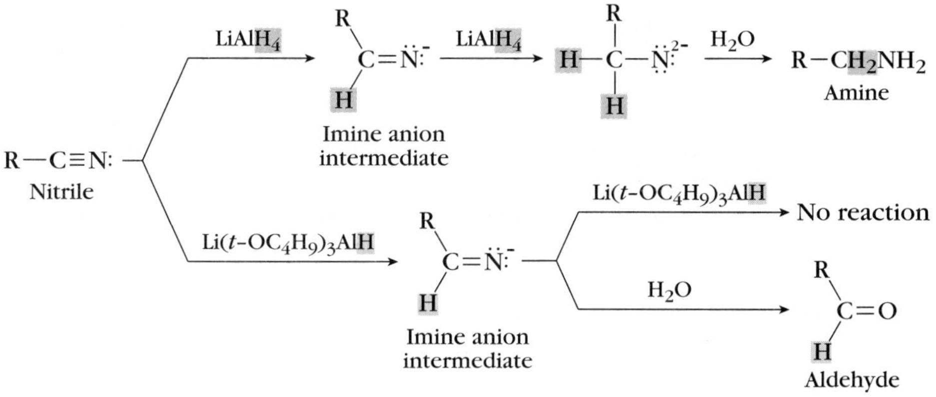

E   Reaction of Nitriles with Grignard Reagents

Grignard reagents add to nitriles to form ketones after hydrolysis of the intermediate magnesium salt.

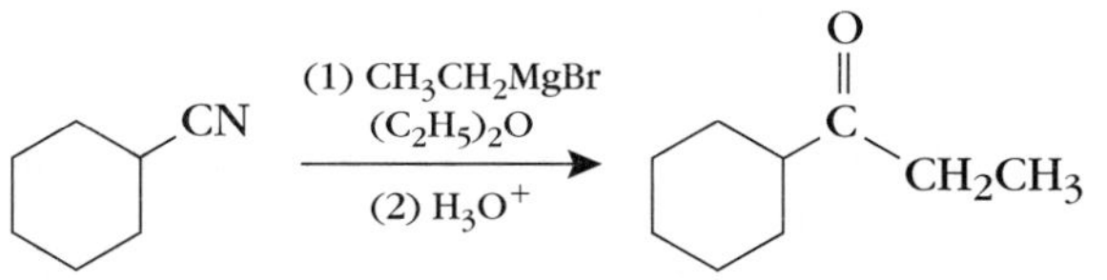

The mechanism of this reaction is analogous to the reduction of nitriles by diisobutylaluminum hydride or lithium aluminum tri(t-butoxy)hydride. The carbanion portion of the Grignard reagent adds to the carbon atom of the nitrile to form an intermediate imine anion. The Grignard reagent is too weak a nucleophile to add again, so the product is formed by the hydrolysis of the imine anion.

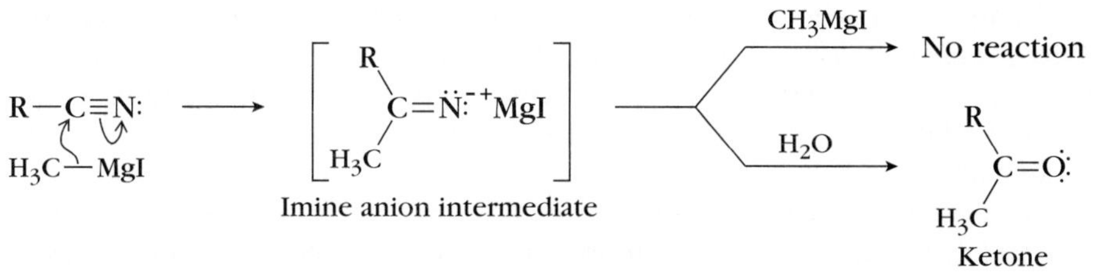

Exercise 16.22   Write equations for the preparation of each of the following compounds from a nitrile.

(a) 3-Pentanone
(b) CH$_3$CH=CHCH$_2$CH$_2$NH$_2$
(c) 5-Methylhexanoic acid
(d) CH$_3$CH$_2$CH$_2$CHO
(e) 2-Methyl-3-hexanone
(f) Sodium benzoate

16.16 Summary

Carboxylic acid derivatives are compounds that form a carboxylic acid upon hydrolysis. The most important such derivatives are **acyl chlorides, acid anhydrides, thioesters, esters,** and **amides.** Organic compounds that contain a carbon-nitrogen triple bond are called **nitriles.** They, too, are classified as carboxylic acid derivatives because they form carboxylic acids upon hydrolysis.

Acyl transfer reactions are important reactions of carboxylic acid derivatives. These reactions all occur by the same **addition-elimination mechanism.** The addition step is addition of a nucleophile to the carbonyl carbon atom to form a tetrahedral intermediate. Elimination of one of the groups of the tetrahedral intermediate is the elimination step. The group lost is the best leaving group, which is usually the weakest base.

The reactivity of carboxylic acid derivatives depends on the electron nature of the group bonded to the acyl group. The observed order of reactivity is:

DECREASING REACTIVITY IN ACYL TRANSFER REACTIONS

| Acid chloride | Acid anhydride | Thioester | Ester | Amide | Carboxylate ion |

The most important reactions of carboxylic acid derivatives are hydrolysis, the acyl transfer to the oxygen atom of water, acyl transfer to the oxygen atom of an alcohol, and acyl transfer to the nitrogen atom of an amine. Grignard reagents react with carboxylic acid derivatives to form alcohols. Reduction of amides forms primary amines, while reduction of all other acid derivatives forms primary alcohols. Aldehydes and ketones can also be prepared from acyl chlorides and esters by use of less reactive reducing agents or less reactive organometallic reagents.

Acyl transfer reactions occur in nature using complex thioesters as the acyl transfer agent. A particularly important acyl transfer agent is **acetyl coenzyme A,** which acetylates alcohols and amines. Acetyl coenzyme A, for example, forms acetylcholine by acetylating choline. Formation of acetylcholine and its enzyme-catalyzed hydrolysis are important reactions in the transmission of nerve impulses in mammals.

The most important reactions of nitriles are their hydrolysis to carboxylic acids, their reduction to primary amines, their reaction with Grignard reagents to form ketones, and their partial reduction to aldehydes. These reactions make nitriles useful starting materials for the preparation of primary amines, carboxylic acids, ketones, and aldehydes.

16.17 Summary of Preparations of Carboxylic Acid Derivatives

1. Preparation of Acyl Chlorides   SECTION 16.5

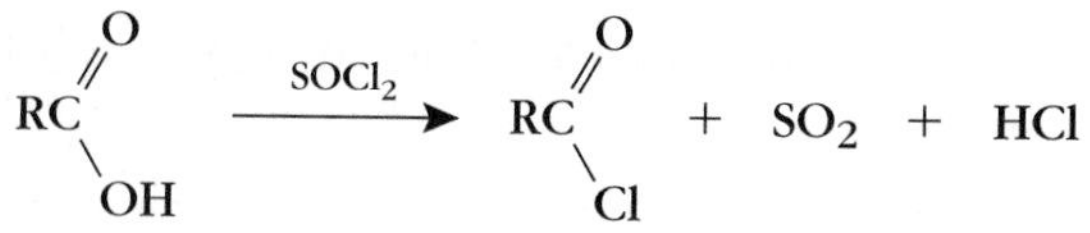

2. Preparation of Acid Anhydrides   SECTION 16.5

Useful synthetic method to prepare symmetrical and unsymmetrical anhydrides.

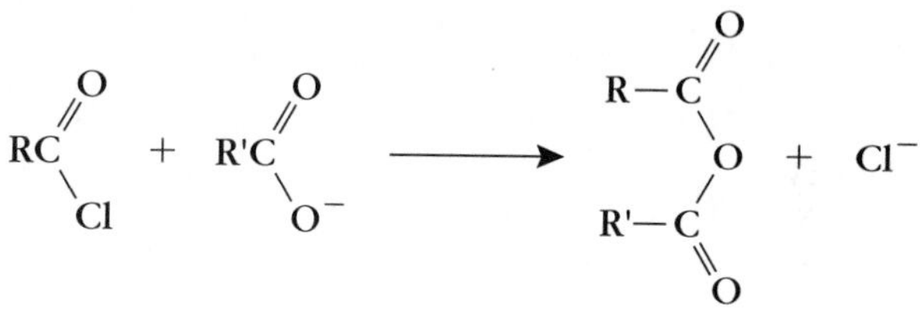

3. Preparation of Thioesters   SECTION 16.5

Reaction is usually carried out in pyridine, which acts as a base to neutralize HCl as it forms.

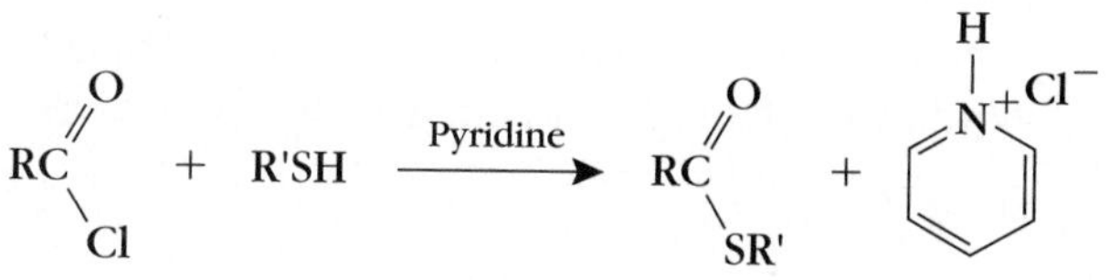

4. Preparation of Esters

a. Fischer esterification   SECTION 15.11

To obtain a good yield of ester, water must be removed as it forms, or an excess of one reactant must be used.

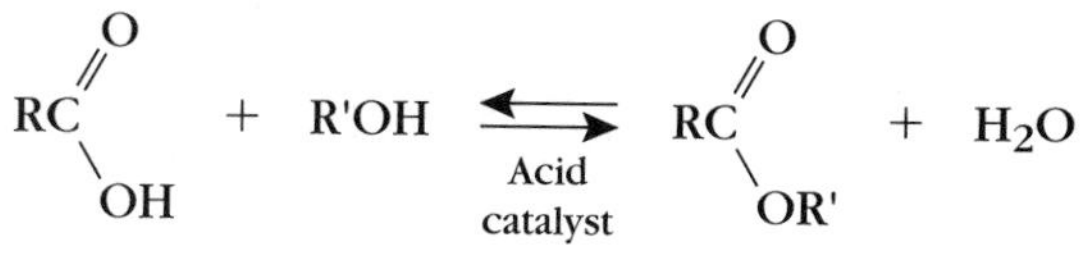

b. Acyl chloride and alcohol   SECTION 16.5

Reaction is usually carried out in pyridine, which acts as a base to neutralize HCl as it forms.

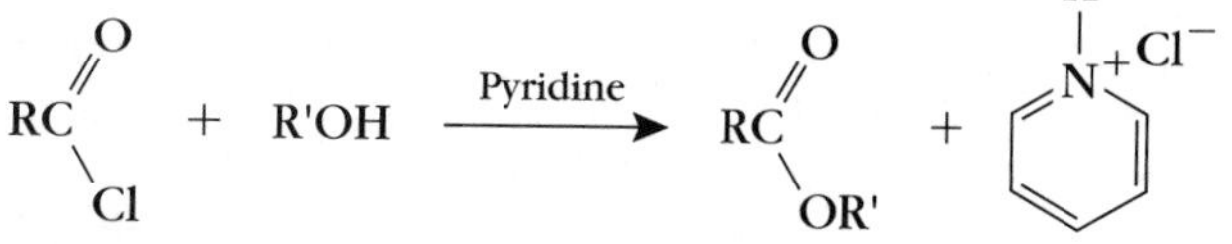

c. Acid anhydride and alcohol   SECTION 16.9

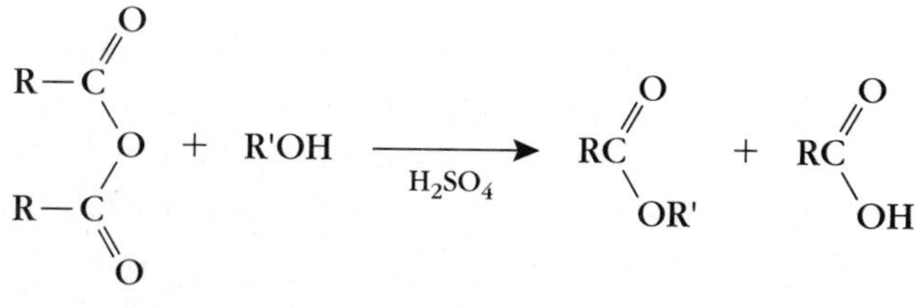

5. **Preparation of Amides**   SECTION 16.5

Acyl chlorides react with NH_3 to form unsubstituted
amides, with primary amines to form *N*-substituted
amides, and with secondary amines to form
N,N-disubstituted amides.

$$RC\overset{O}{\underset{Cl}{\big|}} \;+\; 2\,NHR_2 \longrightarrow RC\overset{O}{\underset{NR_2}{\big|}} \;+\; \overset{+}{N}H_2R_2\;Cl^-$$

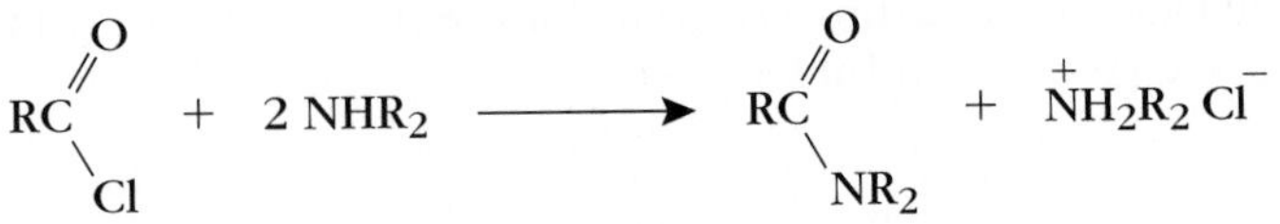

R = H, alkyl group, or aryl group

6. **Preparation of Nitriles**
   a. **Nucleophilic substitution reaction**   SECTION 12.18

   $$RBr \;+\; {}^-CN \xrightarrow[\text{mechanism}]{S_N2} RCN \;+\; Br^-$$

   b. **Dehydration of amides**   SECTION 16.15, B

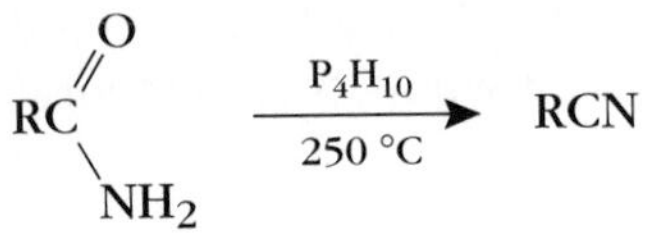

   $$RC\overset{O}{\underset{NH_2}{\big|}} \xrightarrow[250\,°C]{P_4H_{10}} RCN$$

16.18 Summary of Reactions of Carboxylic Acid Derivatives

1. **Acyl Chlorides**
   a. **Reaction with alcohols**   SECTION 16.5

   Pyridine is a base that neutralizes HCl as it forms.

   $$RC\overset{O}{\underset{Cl}{\big|}} \;+\; R'OH \xrightarrow{\text{Pyridine}} RC\overset{O}{\underset{OR'}{\big|}} \;+\; \text{[pyridinium }Cl^-]$$

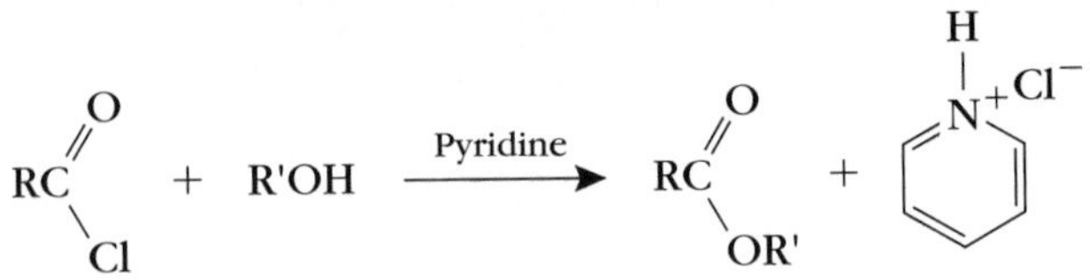

   b. **Reaction with thiols**   SECTION 16.5

   Pyridine is a base that neutralizes HCl as it forms.

   $$RC\overset{O}{\underset{Cl}{\big|}} \;+\; R'SH \xrightarrow{\text{Pyridine}} RC\overset{O}{\underset{SR'}{\big|}} \;+\; \text{[pyridinium }Cl^-]$$

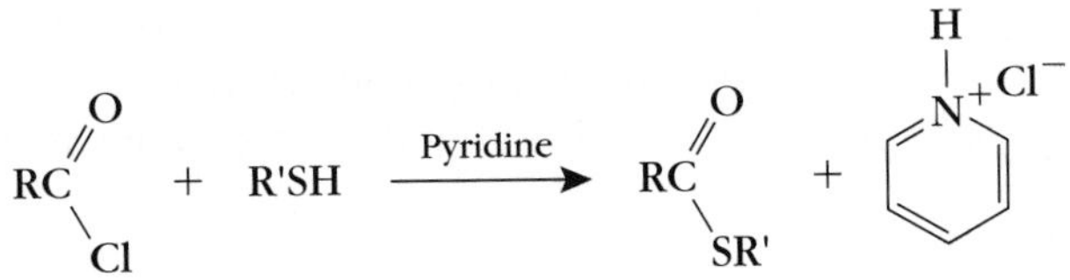

   c. **Reaction with carboxylate anions**   SECTION 16.5

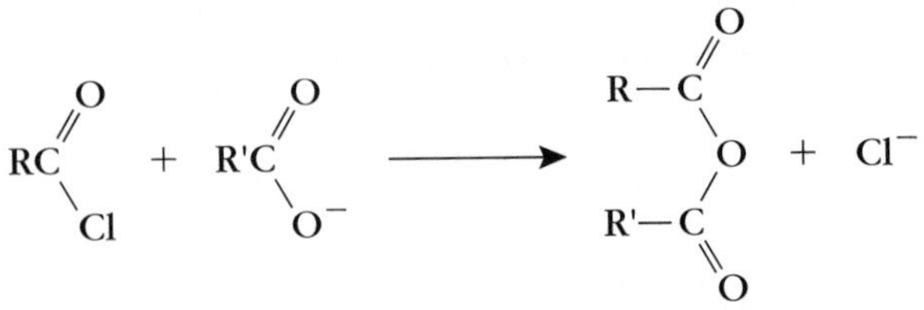

   $$RC\overset{O}{\underset{Cl}{\big|}} \;+\; R'C\overset{O}{\underset{O^-}{\big|}} \longrightarrow \begin{matrix} R-C\overset{O}{\big|} \\ O \\ R'-C\underset{O}{\big|} \end{matrix} \;+\; Cl^-$$

   d. **Reaction with NH_3 and amines**   SECTION 16.5

   Acyl chlorides react with NH_3 to form unsubstituted amides, with primary amines to form
   N-substituted amides, and secondary amines to
   form *N,N*-disubstituted amides.

   $$RC\overset{O}{\underset{Cl}{\big|}} \;+\; 2\,NHR'_2 \longrightarrow RC\overset{O}{\underset{NR'_2}{\big|}} \;+\; R'_2\overset{+}{N}H_2\;Cl^-$$

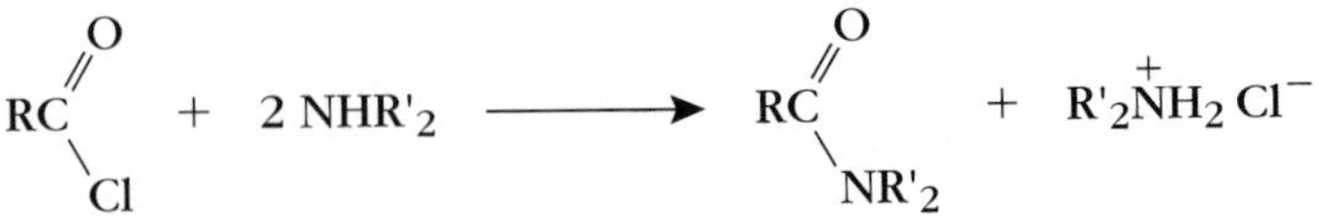

e. Reduction to primary alcohols   SECTION 16.12

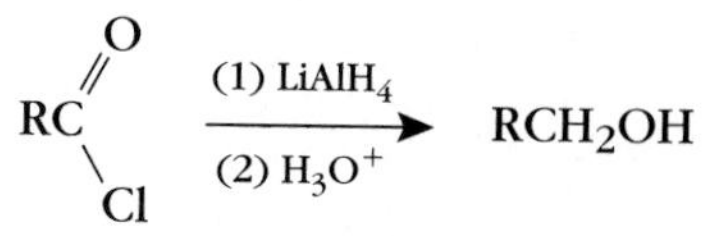

$$RC\underset{Cl}{\overset{O}{\|}} \xrightarrow[\text{(2) } H_3O^+]{\text{(1) } LiAlH_4} RCH_2OH$$

f. Reaction with Grignard reagents   SECTION 16.13

Product is a tertiary alcohol in which two identical groups and a hydroxy group are bonded to a carbon atom.

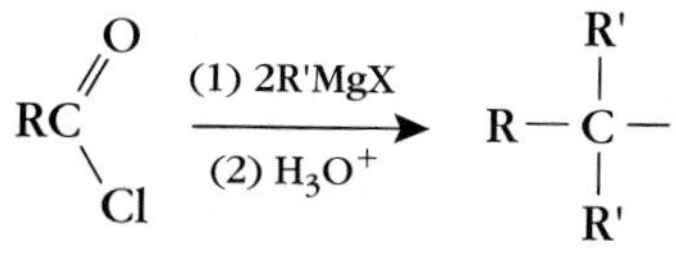

$$RC\underset{Cl}{\overset{O}{\|}} \xrightarrow[\text{(2) } H_3O^+]{\text{(1) } 2R'MgX} R-\underset{R'}{\overset{R'}{\underset{|}{\overset{|}{C}}}}-OH$$

g. Reduction to aldehydes   SECTION 16.14, A

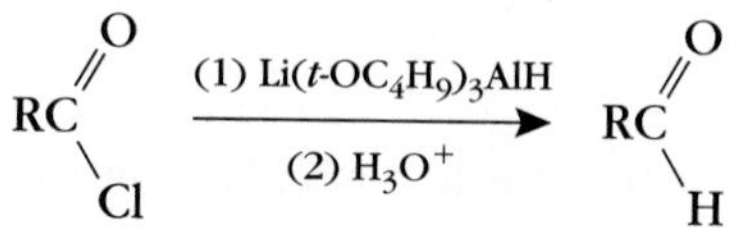

$$RC\underset{Cl}{\overset{O}{\|}} \xrightarrow[\text{(2) } H_3O^+]{\text{(1) } Li(t\text{-}OC_4H_9)_3AlH} RC\underset{H}{\overset{O}{\|}}$$

h. Reaction with lithium dialkylcuprates   SECTION 16.14, B

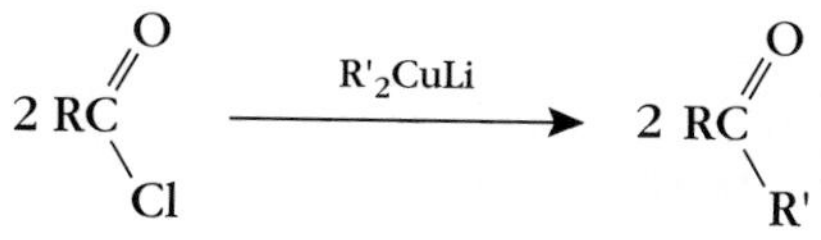

$$2\ RC\underset{Cl}{\overset{O}{\|}} \xrightarrow{R'_2CuLi} 2\ RC\underset{R'}{\overset{O}{\|}}$$

i. Hydrolysis   SECTION 16.5

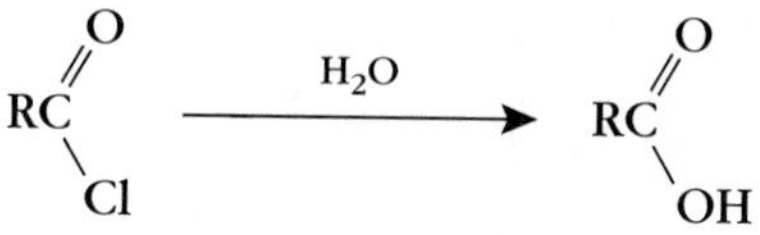

$$RC\underset{Cl}{\overset{O}{\|}} \xrightarrow{H_2O} RC\underset{OH}{\overset{O}{\|}}$$

2. Acid Anhydrides

a. Reaction with alcohols   SECTION 16.9

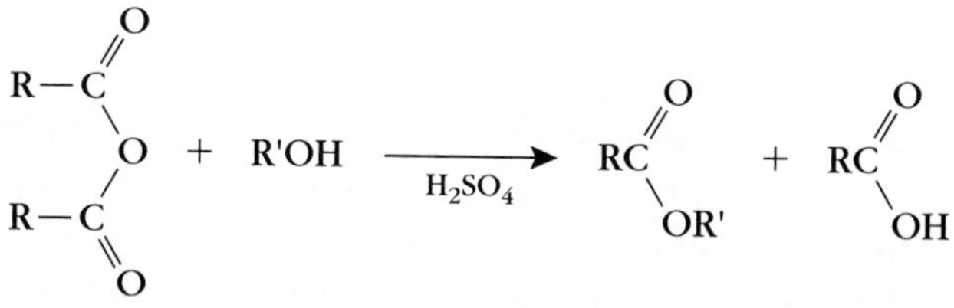

$$\begin{matrix} R-C\overset{O}{\|} \\ \quad\ \searrow \\ \qquad O \\ \quad\ \nearrow \\ R-C\underset{O}{\|} \end{matrix} + R'OH \xrightarrow{H_2SO_4} RC\underset{OR'}{\overset{O}{\|}} + RC\underset{OH}{\overset{O}{\|}}$$

b. Reaction with NH₃ and amines   SECTION 16.9

Reaction with NH_3 forms unsubstituted amides, with primary amines forms N-substituted amides, with secondary amines forms N,N-disubstituted amides.

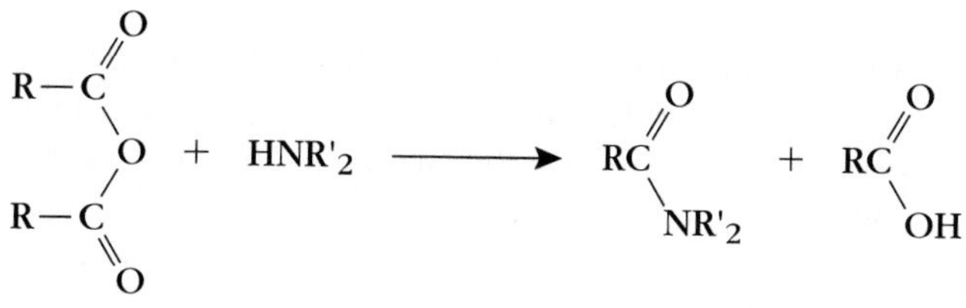

$$\begin{matrix} R-C\overset{O}{\|} \\ \quad\ \searrow \\ \qquad O \\ \quad\ \nearrow \\ R-C\underset{O}{\|} \end{matrix} + HNR'_2 \longrightarrow RC\underset{NR'_2}{\overset{O}{\|}} + RC\underset{OH}{\overset{O}{\|}}$$

c. Hydrolysis   SECTION 16.9

$$R-\overset{\overset{\displaystyle O}{\|}}{C}-O \quad + \quad H_2O \quad \longrightarrow \quad 2\ R\overset{\overset{\displaystyle O}{\|}}{C}-OH$$

3. Thioesters

a. Reaction with alcohols   SECTION 16.9

$$RC\overset{\overset{\displaystyle O}{\|}}{\underset{SR'}{}} \quad + \quad R''OH \quad \longrightarrow \quad RC\overset{\overset{\displaystyle O}{\|}}{\underset{OR''}{}} \quad + \quad R'SH$$

b. Reaction with NH$_3$ and amines   SECTION 16.9

Reaction with NH$_3$ forms unsubstituted amides, with primary amines forms *N*-substituted amides, and with secondary amines forms *N,N*-disubstituted amides.

$$RC\overset{\overset{\displaystyle O}{\|}}{\underset{SR'}{}} \quad + \quad HNR''_2 \quad \longrightarrow \quad RC\overset{\overset{\displaystyle O}{\|}}{\underset{NR''_2}{}} \quad + \quad R'SH$$

c. Hydrolysis   SECTION 16.9

$$RC\overset{\overset{\displaystyle O}{\|}}{\underset{SR'}{}} \quad + \quad H_2O \quad \longrightarrow \quad RC\overset{\overset{\displaystyle O}{\|}}{\underset{OH}{}} \quad + \quad R'SH$$

4. Esters

a. Reaction with alcohols   SECTION 16.9

Transesterification is an equilibrium reaction. Either ester can be made the product by using an excess of the appropriate alcohol.

$$RC\overset{\overset{\displaystyle O}{\|}}{\underset{OR'}{}} \quad + \quad R''OH \quad \underset{\substack{\text{Acid} \\ \text{catalyst}}}{\rightleftharpoons} \quad RC\overset{\overset{\displaystyle O}{\|}}{\underset{OR''}{}} \quad + \quad R'OH$$

b. Reaction with NH$_3$ and amines   SECTION 16.9

Reaction with NH$_3$ forms unsubstituted amides, with primary amines forms *N*-substituted amides, and with secondary amines forms *N,N*-disubstituted amides.

$$RC\overset{\overset{\displaystyle O}{\|}}{\underset{OR'}{}} \quad + \quad HNR''_2 \quad \longrightarrow \quad RC\overset{\overset{\displaystyle O}{\|}}{\underset{NR''_2}{}} \quad + \quad R'OH$$

c. Reduction   SECTION 16.12

Two alcohols are formed as product. One is a primary alcohol obtained from the acyl part of the ester. The other is obtained from the alkoxy (—OR′) part of the ester.

$$RC\overset{\overset{\displaystyle O}{\|}}{\underset{OR'}{}} \quad \xrightarrow[\text{(2) } H_3O^+]{\text{(1) LiAlH}_4} \quad RCH_2OH \quad + \quad R'OH$$

d. Reaction with Grignard reagents   SECTION 16.13

Product is a tertiary alcohol in which two identical groups and a hydroxy group are bonded to a carbon atom.

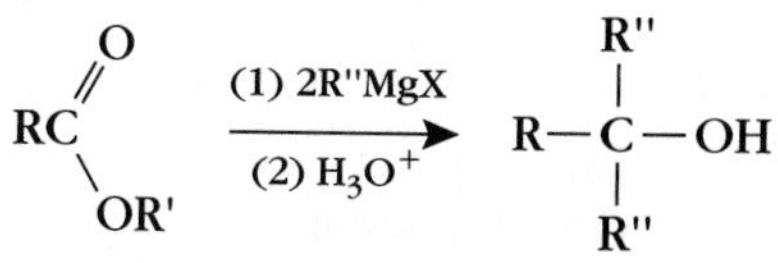

$$RC\overset{O}{\underset{OR'}{\big\backslash}} \xrightarrow[\text{(2) }H_3O^+]{\text{(1) }2R''MgX} R-\overset{R''}{\underset{R''}{\overset{|}{\underset{|}{C}}}}-OH$$

5. Amides

a. Hydrolysis   SECTION 16.8

Hydrolysis occurs readily only under strongly basic or acidic conditions.

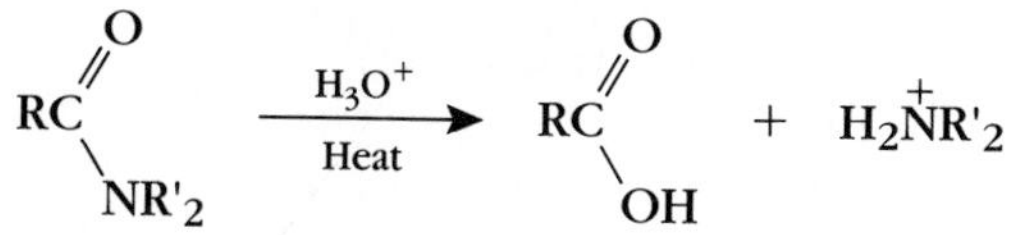

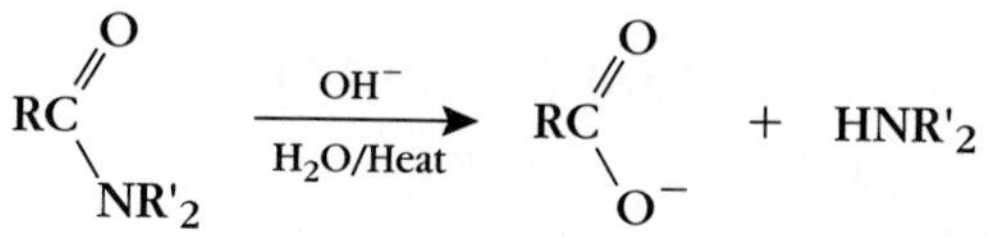

$$RC\overset{O}{\underset{NR'_2}{\big\backslash}} \xrightarrow[\text{Heat}]{H_3O^+} RC\overset{O}{\underset{OH}{\big\backslash}} + H_2\overset{+}{N}R'_2$$

$$RC\overset{O}{\underset{NR'_2}{\big\backslash}} \xrightarrow[\text{H}_2\text{O/Heat}]{OH^-} RC\overset{O}{\underset{O^-}{\big\backslash}} + HNR'_2$$

b. Reduction   SECTION 16.12

Unsubstituted amides are reduced to primary amines, N-substituted amides are reduced to secondary amines, and N,N-disubstituted amides are reduced to tertiary amines.

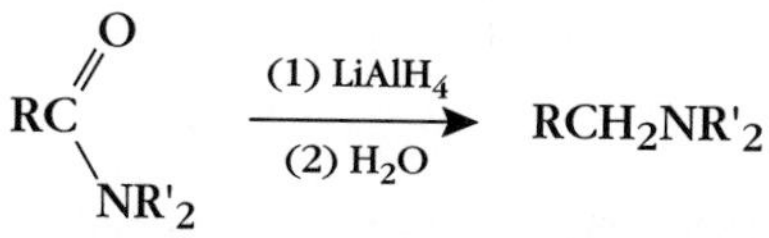

$$RC\overset{O}{\underset{NR'_2}{\big\backslash}} \xrightarrow[\text{(2) }H_2O]{\text{(1) }LiAlH_4} RCH_2NR'_2$$

6. Nitriles

a. Hydrolysis   SECTION 16.15, C

Hydrolysis occurs readily only under strongly basic or acidic conditions.

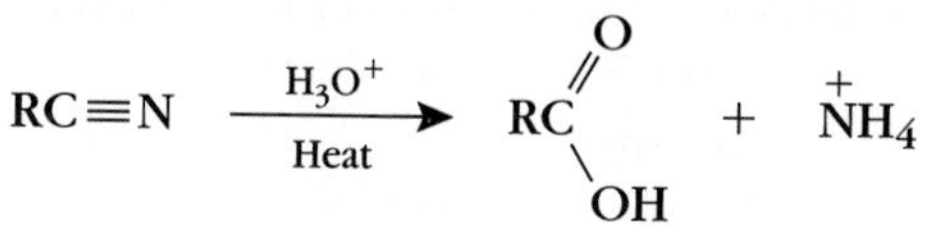

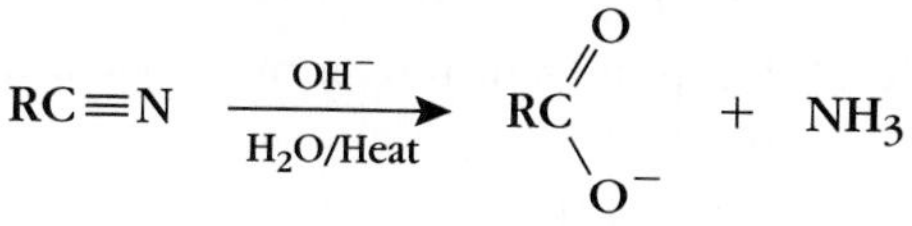

$$RC\equiv N \xrightarrow[\text{Heat}]{H_3O^+} RC\overset{O}{\underset{OH}{\big\backslash}} + \overset{+}{N}H_4$$

$$RC\equiv N \xrightarrow[\text{H}_2\text{O/Heat}]{OH^-} RC\overset{O}{\underset{O^-}{\big\backslash}} + NH_3$$

b. Reduction   SECTION 16.15, D

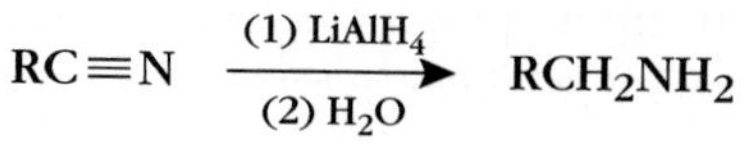

$$RC\equiv N \xrightarrow[\text{(2) }H_2O]{\text{(1) }LiAlH_4} RCH_2NH_2$$

c. Reaction with Grignard reagents   SECTION 16.15, E

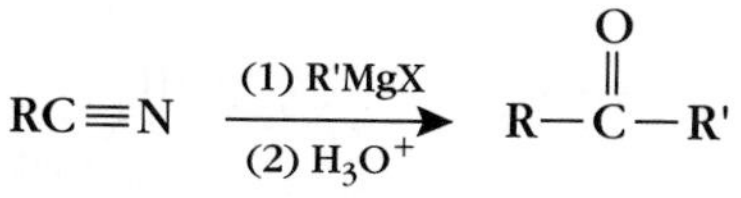

$$RC\equiv N \xrightarrow[\text{(2) }H_3O^+]{\text{(1) }R'MgX} R-\overset{O}{\overset{\|}{C}}-R'$$

Additional Exercises

16.23 Define each of the following terms in words, by a structure, or by a chemical equation.

(a) Acyl chloride   (d) Addition-elimination mechanism   (g) Thioester

(b) Acyl group   (e) Acid anhydride   (h) Enzyme inhibitors

(c) *N*-Substituted amide   (f) Nitrile   (i) Acyl transfer reaction

16.24 Name each of the following compounds.

(a)

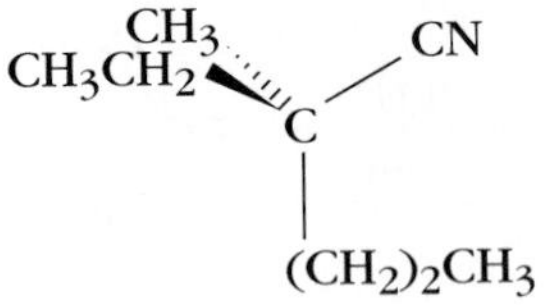

(b)

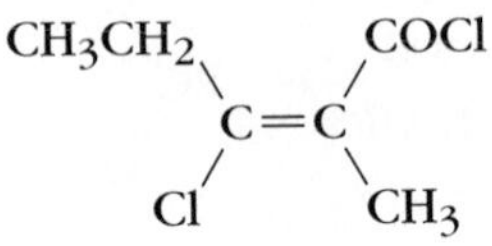

(c)

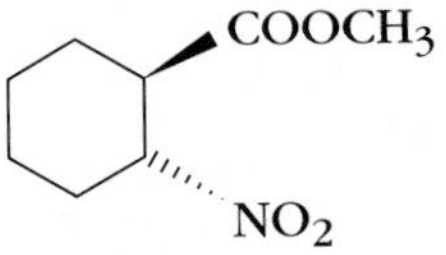

(d)

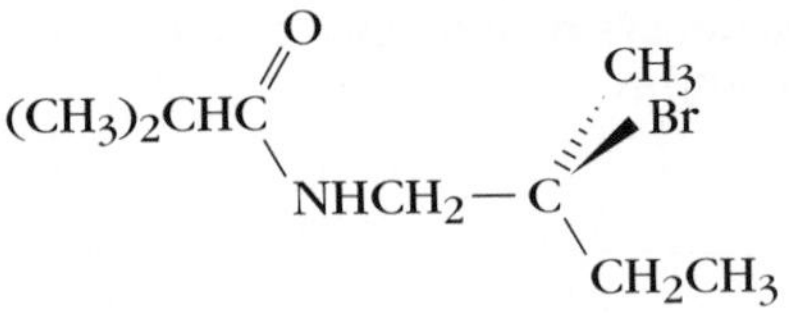

(e)

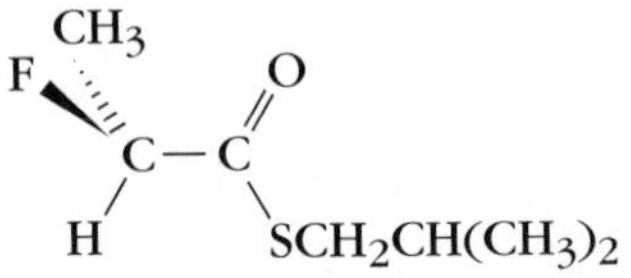

16.25 Write the structural formula for each of the following names.

(a) *S*-2-Butyl (*Z*)-2-pentenethiolate   (d) Pentanoyl chloride

(b) 3-Chloropentanenitrile   (e) *cis*-2-Chlorocyclopentanecarbonyl chloride

(c) *N*-Ethyl-*N*-propylheptanamide   (f) 3-Methyl-1-butyl 2-chloro-5-methylheptanoate

16.26 Write the structure of the product or products, if any, of the reaction of acetic anhydride with each of the following reagents.

(a) NH_3 (e) Excess $LiAlH_4$ followed by H_3O^+

(b) Cyclohexanol   (f) $NaOH/H_2O$

(c) Butanethiol   (g) $NaBH_4/CH_3OH$

(d) Excess CH_3MgI followed by H_3O^+

16.27 Write the structure of the product or products, if any, of the reaction of methyl hexanoate with each of the following reagents.

(a) $NaOH/H_2O$ (d) Large excess of ethanol with an acid catalyst

(b) $NaBH_4/CH_3OH$ (e) $(CH_3)_2CHNH_2$

(c) $LiAlH_4$, followed by H_3O^+

16.28 Write the structure of the product or products, if any, of the following reactions. Include the stereochemistry of the product if appropriate.

(a) $NCCH_2CH_2CN \ + \ H_2O \xrightarrow[\text{Heat}]{\text{HCl}}$

(b) 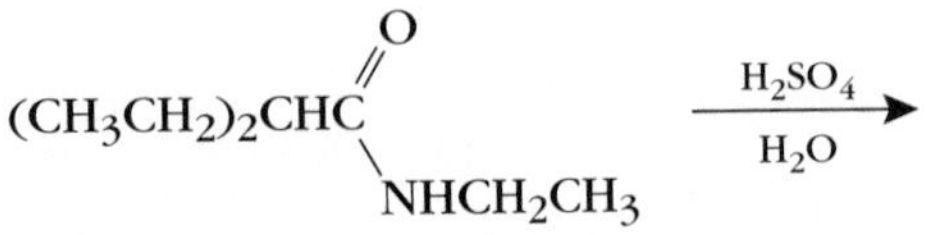 $(CH_3CH_2)_2CHC(=O)NHCH_2CH_3 \xrightarrow[\text{H}_2\text{O}]{\text{H}_2\text{SO}_4}$

(c) $(CH_3)_3COH \ + \ $ Acetic anhydride $\xrightarrow{\text{H}_2\text{SO}_4}$

(d) 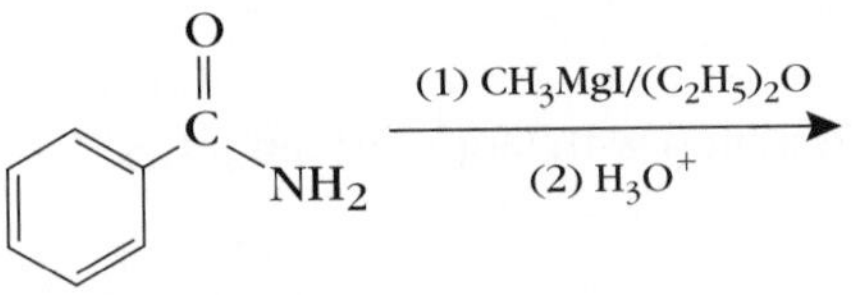 C₆H₅C(=O)NH₂ $\xrightarrow[\text{(2) H}_3\text{O}^+]{\text{(1) CH}_3\text{MgI/(C}_2\text{H}_5)_2\text{O}}$

(e) 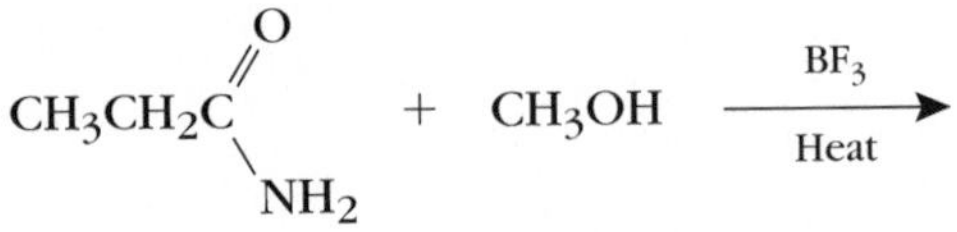 $CH_3CH_2C(=O)NH_2 \ + \ CH_3OH \xrightarrow[\text{Heat}]{\text{BF}_3}$

(f) 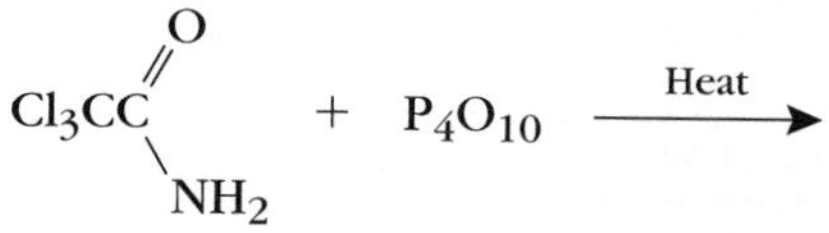 $Cl_3CC(=O)NH_2$ + P_4O_{10} $\xrightarrow{\text{Heat}}$

(g) 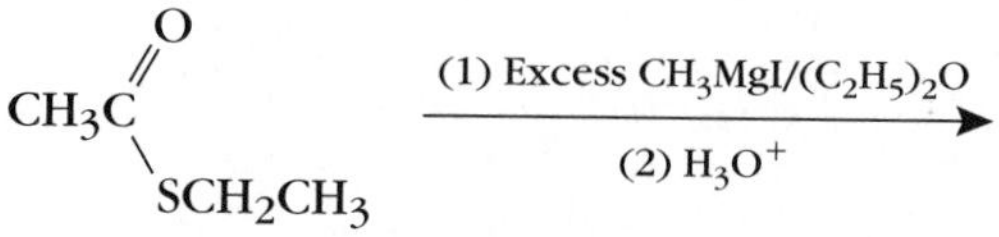 $CH_3C(=O)SCH_2CH_3$ $\xrightarrow[\text{(2) H}_3\text{O}^+]{\text{(1) Excess CH}_3\text{MgI/(C}_2\text{H}_5)_2\text{O}}$

(h) $C_6H_5C(=O)SCH_2CH_3$ + $CH_3CH_2NH_2$ $\longrightarrow$

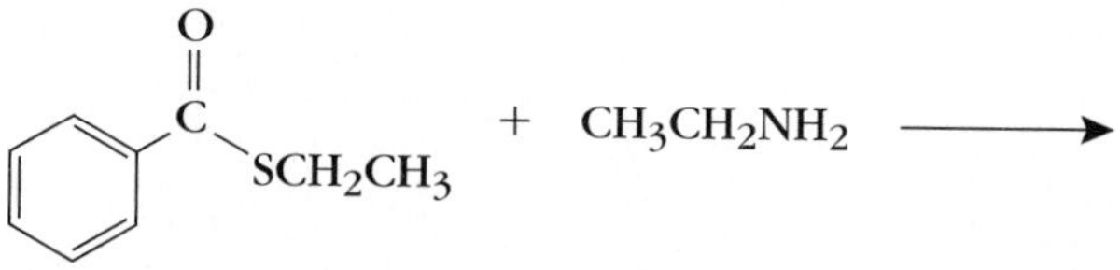

(i) Benzoic anhydride + CH_3CH_2SH $\longrightarrow$

(j) 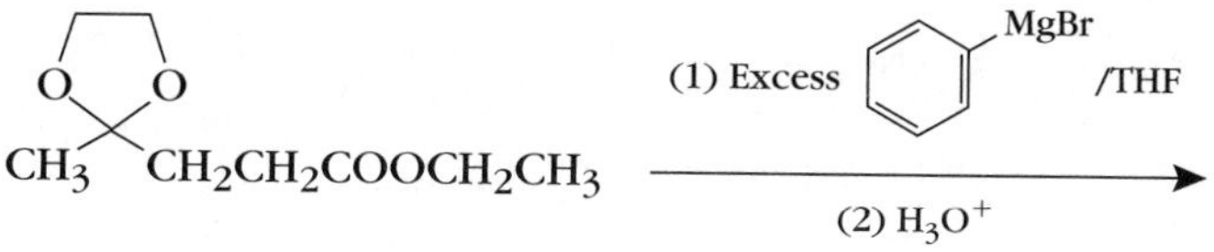 $CH_3C(OCH_2CH_2O)CH_2CH_2COOCH_2CH_3$ $\xrightarrow[\text{(2) H}_3\text{O}^+]{\text{(1) Excess } C_6H_5\text{MgBr /THF}}$

(k) 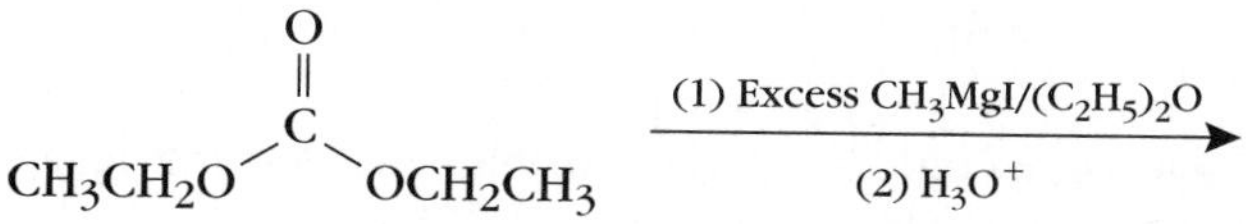 $CH_3CH_2O-C(=O)-OCH_2CH_3$ $\xrightarrow[\text{(2) H}_3\text{O}^+]{\text{(1) Excess CH}_3\text{MgI/(C}_2\text{H}_5)_2\text{O}}$

16.29 Write the structure of the missing reactant, product, or reagent in each of the following reactions.

(a) CH_3COOH + ? $\underset{\longleftarrow}{\overset{?}{\longrightarrow}}$ $CH_3COOCH_2CH(CH_3)_2$ + H_2O

(b) 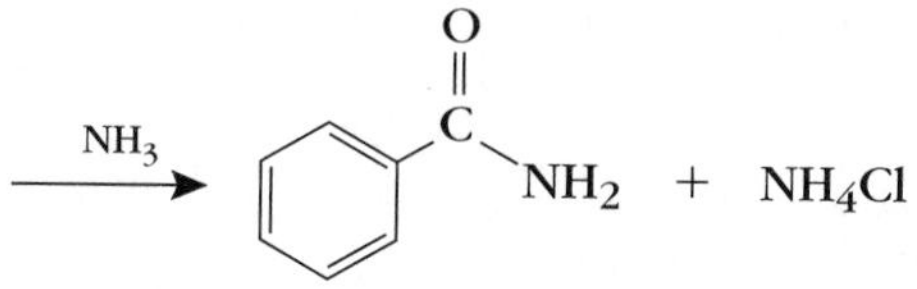 ? $\xrightarrow{NH_3}$ $C_6H_5C(=O)NH_2$ + NH_4Cl

(c) cyclopentanone =O + ? $\longrightarrow$ (cyclopentylidene)$=NNH-C_6H_3(NO_2)_2$ with NO_2 and O_2N

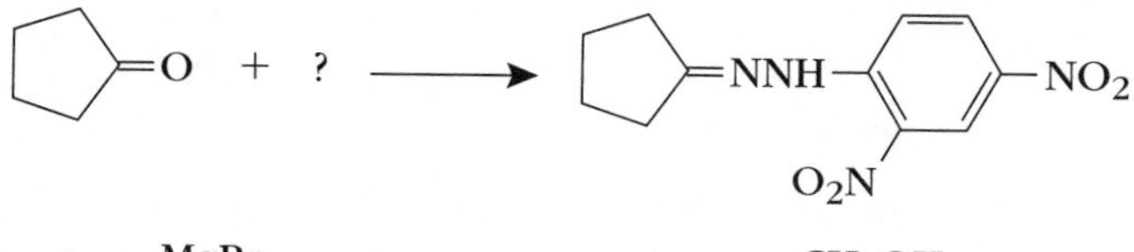

(d) 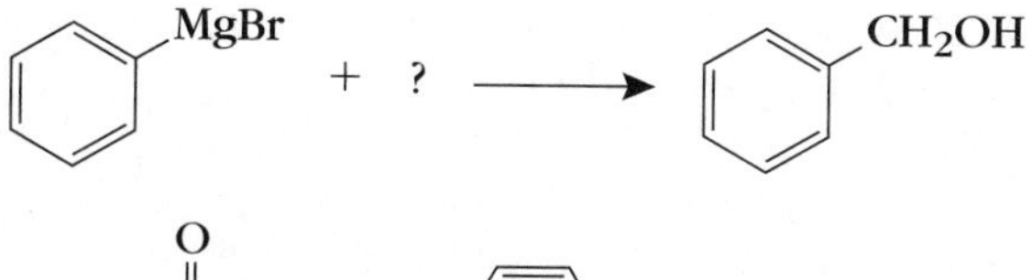 C_6H_5MgBr + ? $\longrightarrow$ $C_6H_5CH_2OH$

(e) 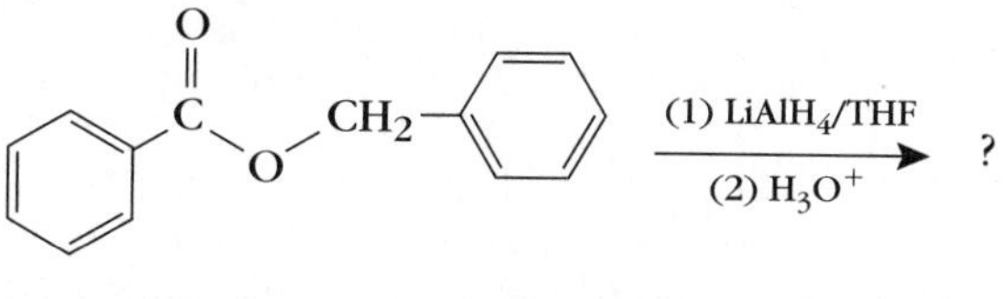 $C_6H_5C(=O)O-CH_2-C_6H_5$ $\xrightarrow[\text{(2) H}_3\text{O}^+]{\text{(1) LiAlH}_4\text{/THF}}$?

(f) $CH_3CH_2COOCH_2CH_2CH_3$ $\xrightarrow{HN(CH_2CH_3)_2}$?

(g) 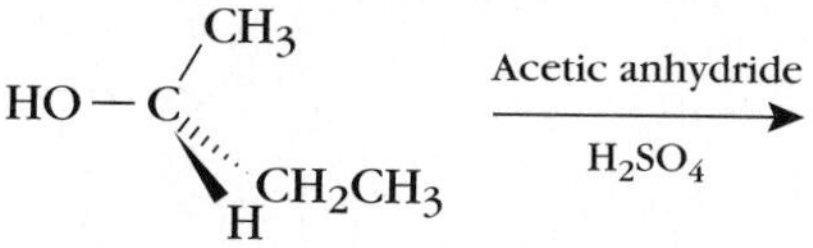 $HO-C(CH_3)(CH_2CH_3)(H)$ $\xrightarrow[\text{H}_2\text{SO}_4]{\text{Acetic anhydride}}$?

16.30 Write the structure of each compound designated by a letter.

(a)  $(CH_3)_3CLi$ + $CH_3CH\overset{O}{\overbrace{-}}CH_2$ ⟶ A $\xrightarrow{\text{Acetic anhydride}}$ B

(b) 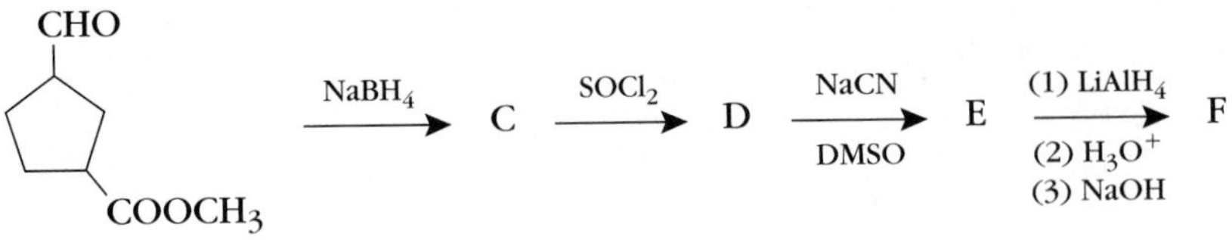 $\xrightarrow{\text{NaBH}_4}$ C $\xrightarrow{\text{SOCl}_2}$ D $\xrightarrow[\text{DMSO}]{\text{NaCN}}$ E $\xrightarrow[\substack{(2)\ H_3O^+ \\ (3)\ NaOH}]{(1)\ LiAlH_4}$ F

(c) $\xrightarrow[\text{Remove water}]{\substack{\text{HOCH}_2\text{CH}_2\text{OH} \\ \text{Acid catalyst}}}$ G $\xrightarrow[(2)\ H_3O^+]{(1)\ LiAlH_4}$ H

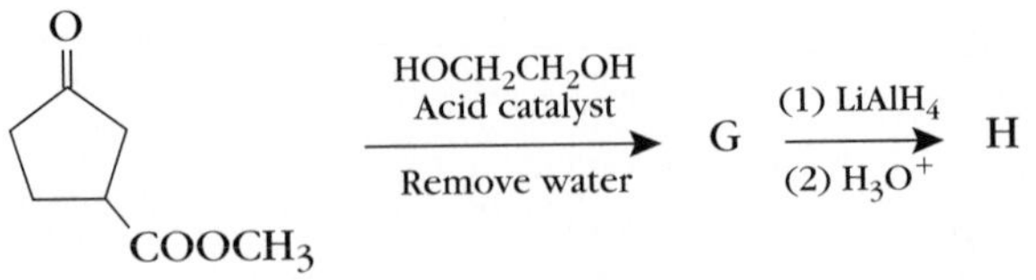

(d) $CH_3C\equiv CH$ $\xrightarrow[\text{NH}_3]{\text{NaNH}_2}$ I $\xrightarrow{\text{Br(CH}_2)_3\text{Cl}}$ J $\xrightarrow[\substack{(2)\ CO_2 \\ (3)\ H_3O^+}]{(1)\ Mg}$ K $\xrightarrow[\text{NH}_3]{\text{Li}}$ L

L $\xrightarrow[\text{Pyridine}]{\text{SOCl}_2}$ M $\xrightarrow[(2)\ H_3O^+]{(1)\ \text{Li}(t\text{-OC}_4H_9)_3AlH}$ N

16.31 Starting with propanoyl chloride, any other organic compound, and any inorganic reagent, write equations for the reactions necessary to synthesize each of the following compounds.

(a) Propanal   **(d)** 2-Butanone   **(f)** *S*-Propyl propanethiolate
(b) Propanamide   **(e)** 1-Propanol   **(g)** Benzoic propanoic anhydride
(c) Ethyl propanoate

16.32 Write the structure of the product or products, if any, of the following reactions.

(a) 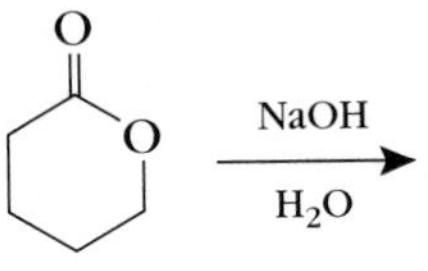 $\xrightarrow[\text{H}_2O]{\text{NaOH}}$ **(c)** 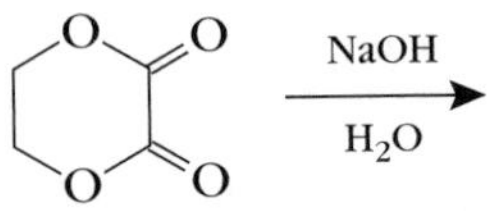 $\xrightarrow[\text{H}_2O]{\text{NaOH}}$

(b) 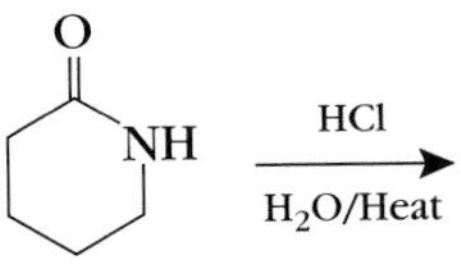 $\xrightarrow[\text{H}_2O/\text{Heat}]{\text{HCl}}$ **(d)** 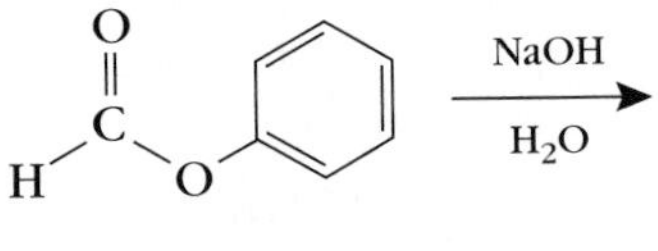 $\xrightarrow[\text{H}_2O]{\text{NaOH}}$

(e)

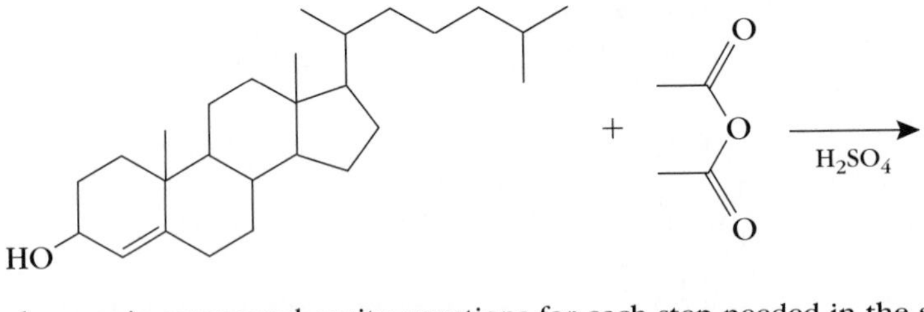

16.33 Starting with ethanol as the only organic compound, write equations for each step needed in the synthesis of each of the following compounds. More than one step may be needed.

(a) Acetic acid   **(c)** 3-Methyl-3-pentanol   **(e)** Propanenitrile
(b) Ethyl acetate   **(d)** Sodium acetate

16.34 Using arrows to indicate electron movement, propose a mechanism for the following reaction.

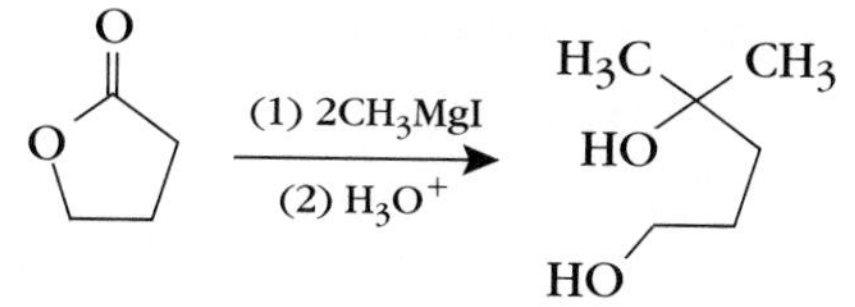

16.35 Starting with propanoic acid, any other organic compound, and any inorganic reagent, write equations for each step in the preparation of the following compounds. More than one step may be needed.

(a) 2-Methyl-2-butanol
(b) 1-Propene
(c) 1-Propanol
(d) Butanenitrile
(e) 1-Butanol
(f) Propanamide
(g) $CH_3CH_2CH_2NH_2$
(h) 1-Pentanol

16.36 Using arrows to indicate electron movement, propose a mechanism for the following reaction.

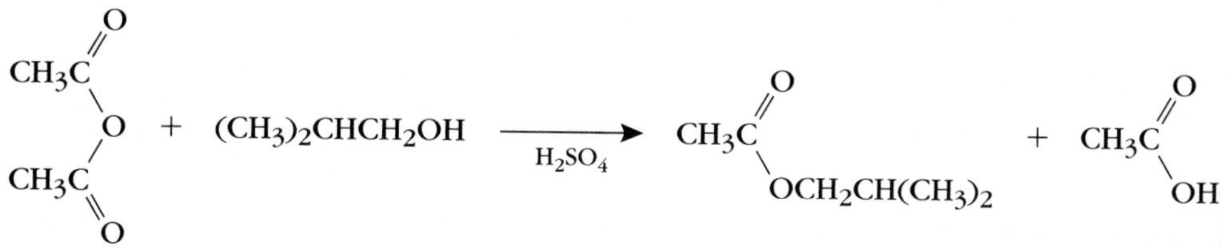

| Acetic anhydride | 2-Methyl-1-propanol | | 2-Methyl-1-propyl ethanoate (isobutyl acetate) | Acetic acid |

16.37 List the following compounds in order of increasing reactivity with water.

(a) CH_3COOCH_3
(b) CH_3COSCH_3
(c) $(CH_3CO)_2O$
(d) CH_3COCl
(e) CH_3CONH_2

16.38 Acyl hydrazines are prepared by the reaction of hydrazine and an acyl chloride according to the following equation.

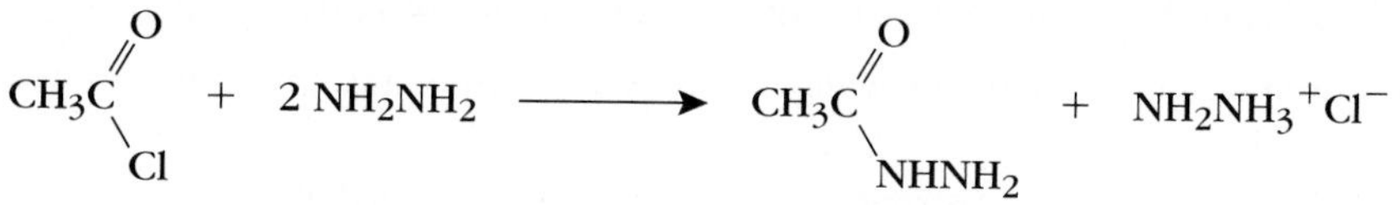

Hydrazine   Acyl hydrazine

Using arrows to indicate electron movement, propose a mechanism for this acyl transfer reaction.

16.39 Acid-catalyzed hydrolysis of an ester of molecular formula $C_8H_{16}O_2$ forms a carboxylic acid, compound A, and an alcohol, compound B. Reaction of compound B with acidic $KMnO_4$ forms compound A. Write the structure of the original ester.

16.40 Phosgene, $Cl_2C{=}O$, was used in World War I as a poisonous gas. Today it is used as a starting material for the synthesis of organic compounds. It undergoes the usual reactions of acyl chlorides, except that it reacts with two equivalents of nucleophiles. Write the equation for the reaction of phosgene with an excess of each of the following reagents.

(a) Ethanol
(b) Water
(c) CH_3CH_2MgI
(d) Ammonia

16.41 Using arrows to indicate electron movement, write a mechanism for the reaction of phosgene and excess ammonia.

16.42 Propose a synthesis of each of the following compounds from the indicated starting material.

(a) $OHC(CH_2)_6CH_2COOH$ from

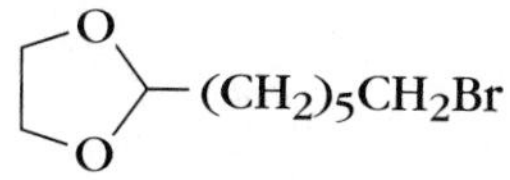

(b) $CH_3C{\equiv}CCHO$ from $CH_3C{\equiv}CCOOCH_3$

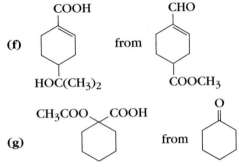

(c) $CH_3(CH_2)_5\overset{\overset{\displaystyle OH}{|}}{C}HC\equiv CCOOCH_3$ from $HC\equiv CCOOCH_3$

(d) (cyclohexene with CH_2CD_2OH) from (cyclohexene with CH_2COOCH_3)

(e)   Hexanoic acid from 1-pentene

(f) (cyclohexene with COOH and $HOC(CH_3)_2$) from (cyclohexene with CHO and $COOCH_3$)

(g) (cyclohexane ring with CH_3COO and COOH) from (cyclohexanone)

16.43   In water containing more than the natural abundance of ^{18}O, acid-catalyzed hydrolysis of methyl acetate forms methyl alcohol that contains no more than the natural abundance of ^{18}O and ^{18}O labelled acetic acid. In the same ^{18}O enriched water, however, acid-catalyzed hydrolysis of *t*-butyl acetate forms ^{18}O labelled *t*-butyl alcohol and acetic acid that contains no more than the natural abundance of ^{18}O. Propose a mechanism to account for this observation.

16.44   Write the structural formula of each of the following compounds based on their molecular formula and the spectral data provided.

  (a) $C_6H_{10}O_3$ IR: 1820 cm^{-1}; 1750 cm^{-1}. ^1H NMR: δ 1.19 t (3H); 2.49 q (2H)
  (b) C_3H_7NO IR: 1680 cm^{-1}. ^{13}C NMR: δ 162.4; δ 36.4; δ 31.4. ^1H NMR: δ 2.88 s (3H); δ 2.98 s (3H); δ 8.00 s (1H)
  (c) $C_5H_6Cl_2O_2$ IR: 1795 cm^{-1}. ^{13}C NMR: δ 173.1; δ 45.0; δ 20.3. ^1H NMR: δ 2.08 quintet (2H); δ 3.03 t (4H)
  (d) $C_5H_9ClO_2$ IR: 1735 cm^{-1}. ^1H NMR: δ 1.21 t (3H); δ 2.82 t (2H); δ 3.80 t (2H); δ 4.20 q (2H)

16.45   Write the structural formula of each of the following compounds based on their molecular formula, IR, and ^1H NMR spectra.

  (a) $C_5H_{10}O_2$ (b) $C_6H_{13}NO$ (c) C_4H_6BrN (d) $C_4H_6O_2$

(a) $C_5H_{10}O_2$

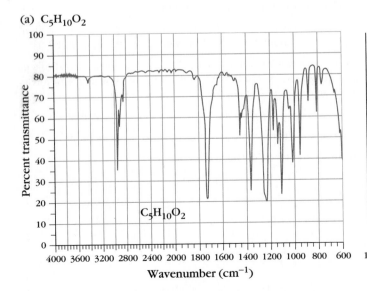

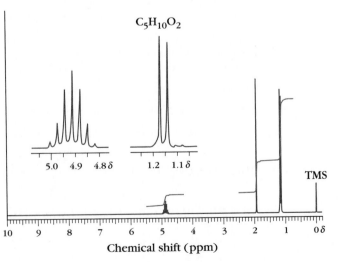

(b) C$_6$H$_{13}$NO

(c) C$_4$H$_6$BrN

(d) C$_4$H$_6$O$_2$

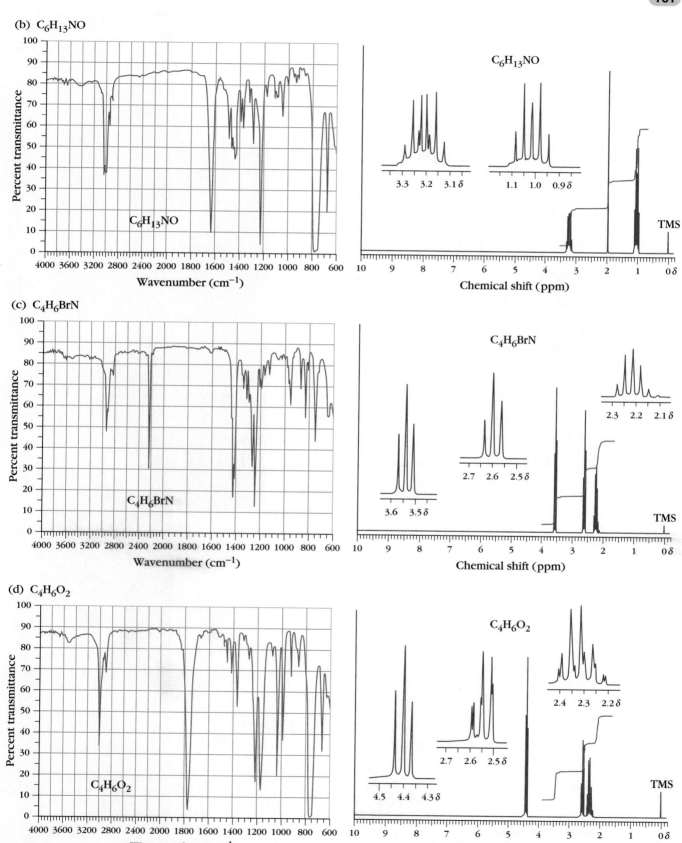

CHAPTER 17

ENOLS and ENOLATE ANIONS

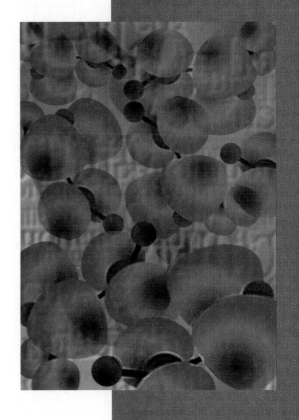

17.1 Keto and Enol Tautomers

In Chapters 14, 15, and 16, we studied the chemistry of carbonyl-containing compounds, with particular emphasis on reactions at the carbon atom of the C=O bond. In this chapter, we begin to describe the reactions at their α-carbon atoms. Let's start by reexamining the structure of 2,4-pentanedione.

$$\overset{\text{O}\quad\text{O}}{\underset{\text{2,4-Pentanedione}}{\text{CH}_3\text{CCH}_2\text{CCH}_3}}$$

The ^{1}H NMR spectrum of 2,4-pentanedione, shown in Figure 17.1, is not in accord with this structure. Its spectrum consists of four singlets at δ 2.05, δ 2.22, δ 3.61, and δ 5.52. We would expect the ^{1}H NMR spectrum of 2,4-pentanedione to consist of two singlets, one at about δ 2.1 due to the methyl hydrogen atoms, and the other at about δ 3.5 due to the methylene hydrogen atoms.

Why are there additional signals in the ^{1}H NMR spectrum of 2,4-pentanedione? The answer is that 2,4-pentanedione exists in solution as an equilibrium mixture of two isomers, a **keto** isomer and an **enol** isomer.

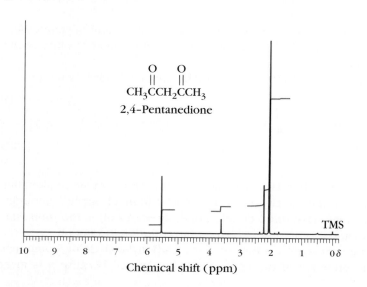

Keto form   Enol form

These keto and enol forms are constitutional isomers. They differ in the location of a hydrogen atom and the double bond. The interconversion of keto and enol isomers, which occurs rapidly in acidic or basic solution, is called **keto-enol tautomerization.** The individual isomers are called **tautomers.**

All carbonyl-containing compounds that contain at least one α-hydrogen atom exist as an equilibrium mixture of tautomers. The amount of enol present at equi-

$$\overset{\text{O}\quad\text{O}}{\underset{\text{2,4-Pentanedione}}{\text{CH}_3\text{CCH}_2\text{CCH}_3}}$$

TMS

10   9   8   7   6   5   4   3   2   1   0 δ

Chemical shift (ppm)

Figure 17.1
^{1}H NMR spectrum of 2,4-pentane-dione.

Table 17.1	Keto-Enol Equilibrium Constants

Compound	K_E
Keto $\underset{\longleftarrow}{\overset{K_E}{\rightleftharpoons}}$ Enol	$K_E = \dfrac{[\text{Enol}]}{[\text{Keto}]}$
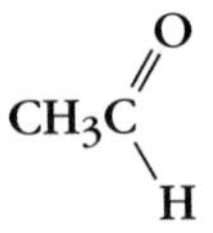 $CH_3C\overset{\displaystyle O}{\underset{\displaystyle H}{\big\backslash}}$	5.9×10^{-7}
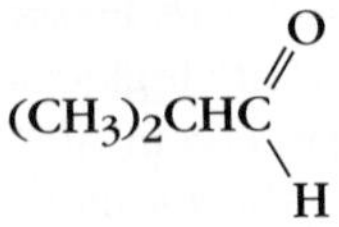 $(CH_3)_2CHC\overset{\displaystyle O}{\underset{\displaystyle H}{\big\backslash}}$	1.3×10^{-4}
$CH_3\overset{\displaystyle O}{\underset{}{\overset{\|}{C}}}CH_3$	4.7×10^{-9}
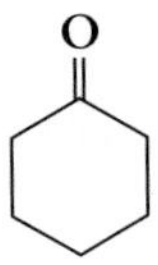 (cyclohexanone)	4.1×10^{-7}

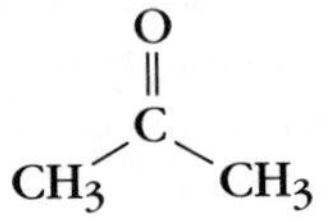

librium, called the **enol content,** depends on the structure of the carbonyl-containing compound as well as the solvent. The enol content in water of a number of carbonyl-containing compounds, expressed as a keto-enol equilibrium constant (K_E), is given in Table 17.1.

Exercise 17.1	Write the structure of the enol of each of the compounds in Table 17.1.

From the data in Table 17.1, we see that the enol content of simple aldehydes and ketones is very small. This means that the keto isomer is more stable than the enol isomer.

An unsymmetrical ketone such as 2-butanone can form two enols.

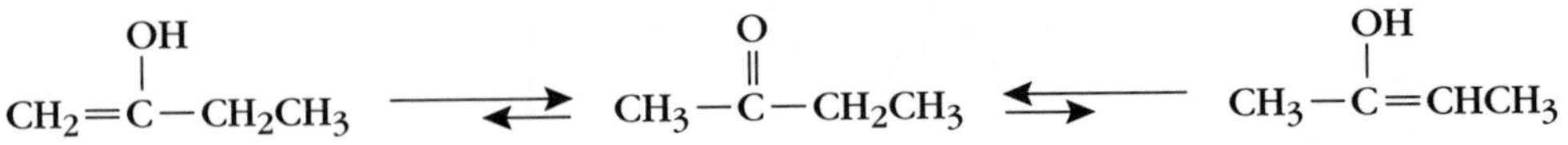

$$\underset{\substack{\text{1-Buten-2-ol}\\\text{enol}}}{\overset{\overset{\displaystyle OH}{|}}{CH_2{=}C{-}CH_2CH_3}} \;\rightleftharpoons\; \underset{\substack{\text{2-Butanone}\\\text{keto}}}{\overset{\overset{\displaystyle O}{\|}}{CH_3{-}C{-}CH_2CH_3}} \;\rightleftharpoons\; \underset{\substack{\text{2-Buten-2-ol}\\\text{enol}}}{\overset{\overset{\displaystyle OH}{|}}{CH_3{-}C{=}CHCH_3}}$$

As expected from the data in Table 17.1, 2-butanone exists predominantly as the keto tautomer and the two enols are present in much smaller amounts. The enol with the more substituted double bond, 2-buten-2-ol, is the more stable of the two enols, and it is present in higher concentration in solution.

Although the enol content of many carbonyl-containing compounds is very small, the presence of the enol cannot be ignored because it is more reactive than the keto tautomer in many reactions. Only the enol will react, moreover, if the keto tautomer can be rapidly converted to the enol to maintain the small enol concentration. This can be done with an acid, a base, or an enzyme as a catalyst. Let's first examine the use of an acid to catalyze enolization.

17.2   Acid-Catalyzed Enolization

Enolization is the process of converting a keto tautomer to its enol. The mechanism of acid-catalyzed enolization involves a series of proton transfer reactions between the solvent and the keto-enol tautomers, as shown in Figure 17.2.

Step 1: Proton transfer from a hydronium ion to the oxygen atom of the C=O bond forms the conjugate acid of the keto tautomer

Figure 17.2
The mechanism of acid-catalyzed enolization in aqueous solution.

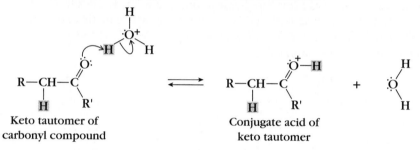

Keto tautomer of carbonyl compound

Conjugate acid of keto tautomer

Step 2: Removal of an α proton by a water molecule forms the enol tautomer

Enol tautomer

Step 1 of the mechanism is a proton transfer from a hydronium ion to the oxygen atom of the C=O bond to form the conjugate acid of the keto form. A water molecule acts as a Brønsted base to remove a proton from the α-carbon atom to form the enol in Step 2. Each step of the mechanism is reversible. Consequently, the mechanism of the enolization of a carbonyl compound is the exact reverse of the mechanism by which an enol is converted to its keto tautomer. Recall from Chapter 9 that we encountered the conversion of an enol to its keto form in the hydration of alkynes.

17.3   Enzyme-Catalyzed Enolization

Enolization of carbonyl compounds is an important step in many biological reactions. The conversion of dihydroxyacetone phosphate to glyceraldehyde-3-phosphate, which is catalyzed by the enzyme triose phosphate isomerase, is one example.

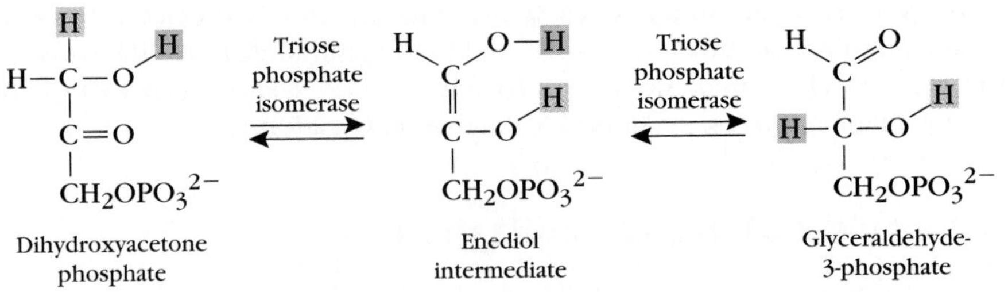

Dihydroxyacetone phosphate ⇌ (Triose phosphate isomerase) Enediol intermediate ⇌ (Triose phosphate isomerase) Glyceraldehyde-3-phosphate

The conversion of dihydroxyacetone phosphate to glyceraldehyde-3-phosphate is a reversible reaction. The first step is the enolization of dihydroxyacetone phosphate to an enediol intermediate. The enediol intermediate then isomerizes to the keto form of glyceraldehyde-3-phosphate.

The conversion of dihydroxyacetone phosphate to glyceraldehyde-3-phosphate is a key reaction in the biochemical process that converts glucose into energy, CO_2, and H_2O. This biochemical process is known as glycolysis. In glycolysis, a molecule of fructose 1,6-bisphosphate, which contains six carbon atoms, is split into dihydroxyacetone phosphate and glyceraldehyde-3-phosphate, which contain three carbon atoms each.

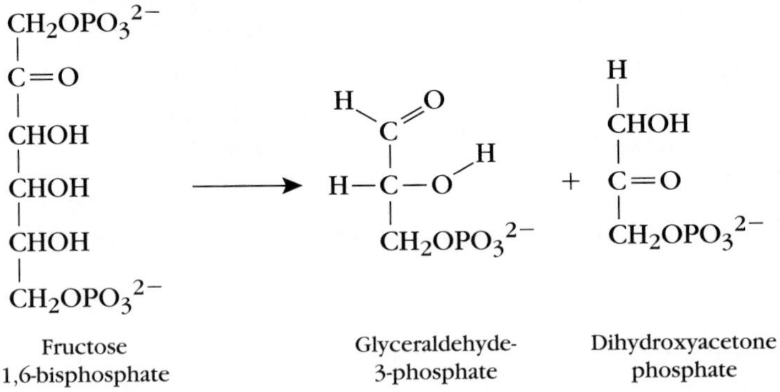

Fructose 1,6-bisphosphate → Glyceraldehyde-3-phosphate + Dihydroxyacetone phosphate

Only glyceraldehyde-3-phosphate can be utilized further in glycolysis. Thus, without an enzyme to convert dihydroxyacetone phosphate into glyceraldehyde-3-phosphate, half the energy stored in glucose would be lost and glycolysis would be a much less efficient process.

Let's now consider the chemical reactions of enols. Because enols contain a carbon-carbon double bond, they react like alkenes.

17.4 Halogenation of Enols

Enols react with electrophiles in much the same way as alkenes because both contain a carbon-carbon double bond. The enol double bond is more reactive, however, because electron donation from the hydroxy group puts a partial negative charge on one of the carbon atoms of the double bond. This makes the double bond of an enol more nucleophilic than the double bond of an alkene. The following resonance structures illustrate how the hydroxy group donates electrons to the double bond.

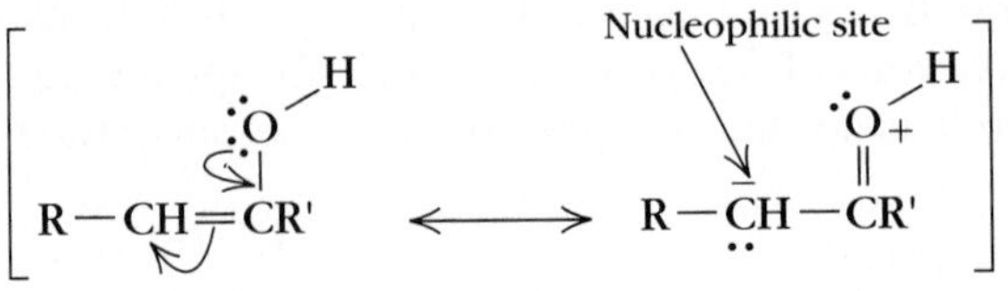

Electron donation makes the original α-carbon atom of the carbonyl compound a nucleophilic site. Reaction of an enol with an electrophile at this carbon atom results in the formation of an α-substituted carbonyl compound. These reactions, therefore, are called **α-substitution reactions.** The halogenation of aldehydes and ketones is an example of an α-substitution reaction.

When aldehydes or ketones react with one equivalent of chlorine, bromine, or iodine in acidic solution, only one α-hydrogen atom is substituted by a halogen atom.

(61%–66% yield)

(69%-72% yield)

| **Exercise 17.2** | Write the structure of the product formed by the reaction of each of the following compounds with one equivalent of bromine in acidic solution. |

(a) Acetone

(b) Cyclopentanone

(c)

(d) Acetaldehyde

(e) $(CH_3)_2CHCHO$

The mechanism of the halogenation of ketones is illustrated in Figure 17.3, using the reaction of acetone and bromine as an example. Step 1 is an acid-catalyzed enolization of acetone. Once formed, the enol reacts rapidly with bromine in Step 2 to form a resonance-stabilized cationic intermediate. Loss of a proton from the cationic intermediate forms the product, bromoacetone, in the final step of the mechanism.

From the rate law, we can establish which of the three steps in the mechanism is rate-determining. Lapworth and Hann in 1920 reported that the rate law of the reaction of all three halogens with acetone is second order overall; it is first order in acid concentration and first order in acetone concentration.

$$\text{Rate of halogenation} = k \ [H_3O^+][\text{Acetone}]$$

Varying the concentration of the halogen has no effect on the rate of halogenation of acetone. The rate law, therefore, is independent of the concentration of the halogen.

In mechanistic terms, the rate law for the halogenation of acetone means that the halogen does not participate in the reaction until *after* the rate-determining step. Consequently, the first step of the mechanism, enolization of the carbonyl compound, must be the rate-determining step in the halogenation of aldehydes and ketones in acid solution.

Figure 17.3
Mechanism of the acid-catalyzed bromination of acetone. Once the enol forms in Step 1, it is quickly brominated in Step 2. The resulting resonance-stabilized intermediate is then deprotonated in Step 3.

Step 1: Rate-determining acid-catalyzed enolization

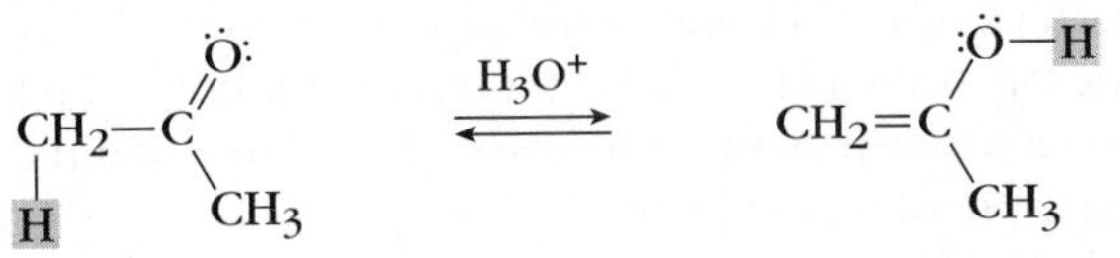

Keto form of acetone   Enol form of acetone

Step 2: Fast electrophilic addition of bromine to the enol

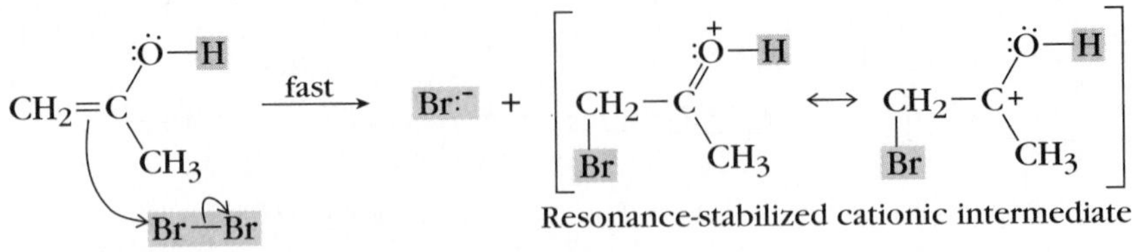

Resonance-stabilized cationic intermediate

Step 3: Deprotonation of the cationic intermediate

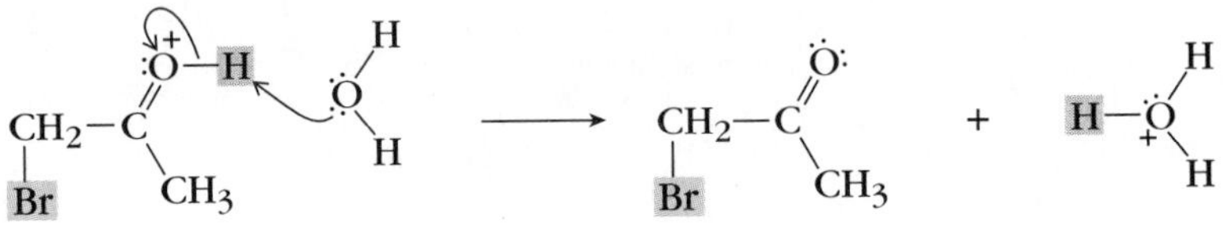

Bromoacetone

α-Halogenated aldehydes and ketones usually form enols more slowly than the corresponding unsubstituted aldehyde or ketone. Bromoacetone, for example, forms an enol more slowly than acetone. As a result, the substitution of each successive α-hydrogen atom by a halogen atom becomes progressively more difficult so it is possible to introduce one, two, or more halogen atoms to a ketone or aldehyde in a controlled manner by simply limiting the amount of halogen added.

The racemization of optically active aldehydes and ketones in acid solution also occurs by a mechanism that involves the rate-determining formation of an enol. For example, a solution of (R)-(+)-sec-butyl phenyl ketone in acetic acid containing nitric acid as the acid catalyst gradually loses its optical activity. The ketone that can be isolated after all optical activity has disappeared is a racemic mixture.

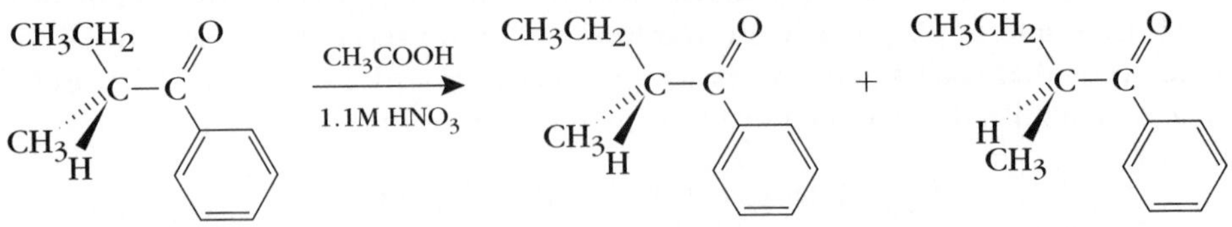

(R)-(+)-sec-Butyl
phenyl ketone

(±)-sec-Butyl
phenyl ketone

The rate of racemization is exactly equal to the rate of iodination of (+)-sec-butyl phenyl ketone under the same experimental conditions. Thus the rate-determining step in both iodination and racemization must be enolization of the ketone.

Racemization occurs because the ketone is slowly and reversibly enolized to an achiral enol in the presence of acid. Addition of a proton to this achiral enol occurs at the same rate to either side of the double bond, so equal amounts of the two enantiomers are formed.

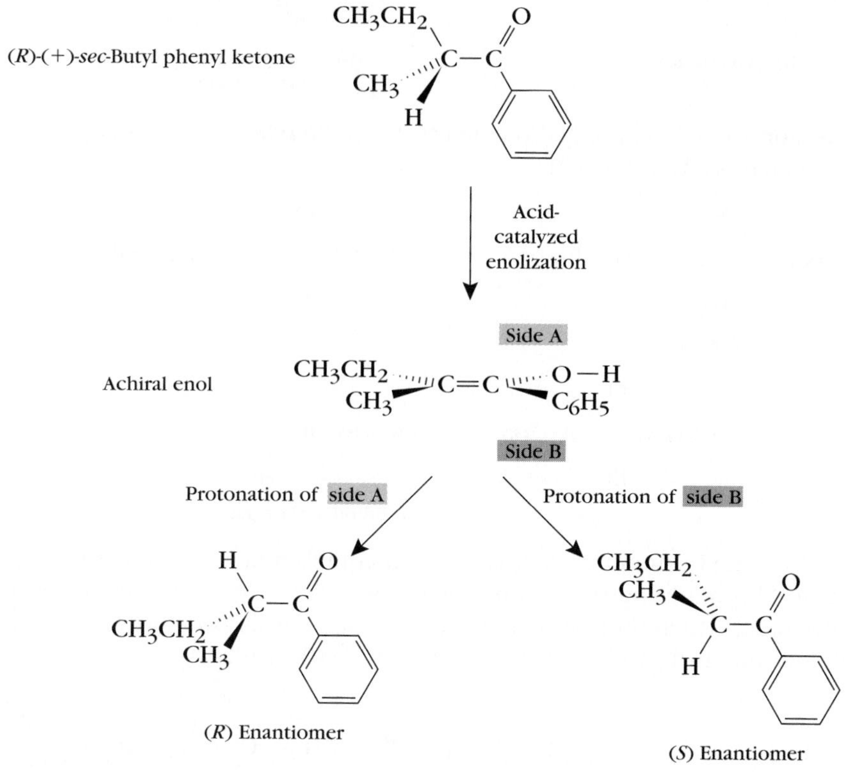

(R)-(+)-sec-Butyl phenyl ketone

Acid-catalyzed enolization

Side A

Achiral enol

Side B

Protonation of side A

Protonation of side B

(R) Enantiomer

(S) Enantiomer

Exercise 17.3

Explain why the rates of chlorination, bromination, and iodination of (R)-(+)-sec-butyl phenyl ketone in acetic acid solution containing nitric acid are found to be the same within experimental error.

Exercise 17.4

Explain why placing (S)-2,2,4-trimethyl-3-heptanone in an acid solution causes it to slowly lose its optical activity, while (S)-2,2,5-trimethyl-3-heptanone retains its optical activity under the same conditions.

α-Bromination of a carboxylic acid can also be accomplished in the presence of a small amount of PBr$_3$. This reaction, which is called the **Hell-Volhard-Zelinsky reaction,** can be used to prepare α-bromocarboxylic acids. The following are examples of this method of preparing α-bromocarboxylic acids.

$$CH_3CH_2CH_2COOH \ + \ Br_2 \ \xrightarrow{\text{PBr}_3} \ CH_3CH_2\overset{\overset{\displaystyle Br}{|}}{C}HCOOH \ + \ HBr$$

Butanoic acid

2-Bromobutanoic acid
(85% yield)

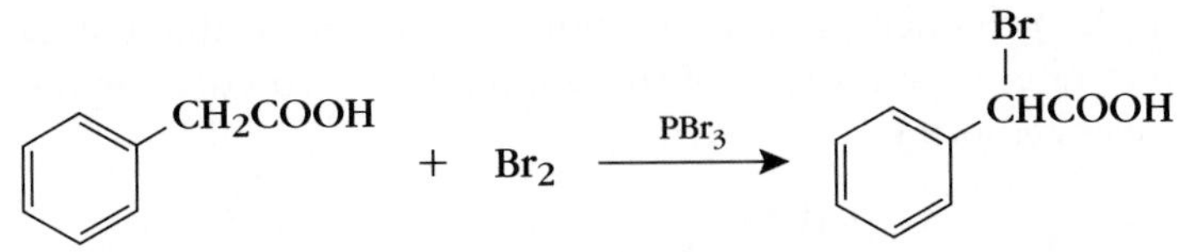

Phenylacetic acid   α-Bromophenylacetic acid
                                            (90% yield)

The purpose of the PBr$_3$ is to convert the carboxylic acid to its acyl bromide, which undergoes bromination.

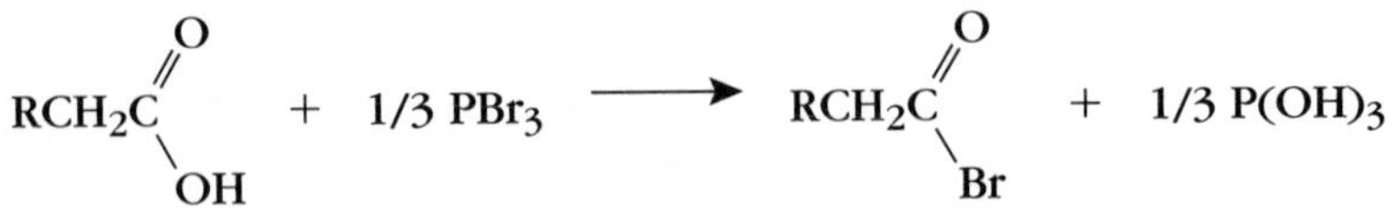

Carboxylic acid   Acyl bromide

α-Bromoacyl bromide

When a small amount of PBr$_3$ is used, an equilibrium is set up between the α-bromoacyl bromide and the starting carboxylic acid. This equilibrium forms more acyl bromide, which then starts the cycle of reactions all over again. Under these conditions the final product is an α-bromocarboxylic acid.

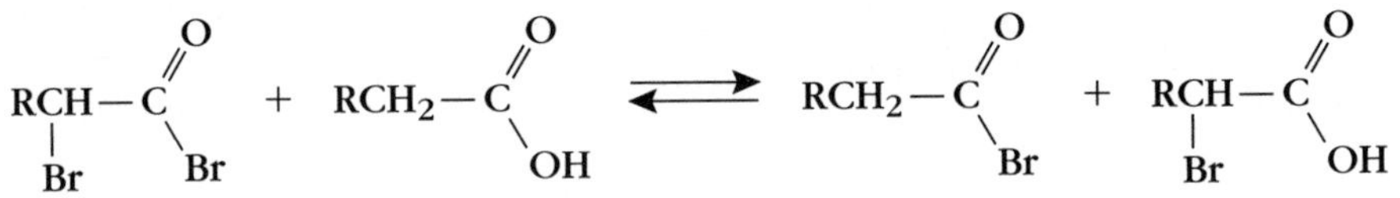

α-Bromoacyl bromide   Carboxylic acid   Acyl bromide   α-Bromocarboxylic
                                                                        acid

---

Exercise 17.5   Write the structure of the product formed in each of the following reactions.

(a) Cyclohexanone + 1 equivalent Br$_2$ in acetic acid ⟶
(b) 3-Pentanone + 2 equivalents Cl$_2$ in aqueous HCl ⟶
(c) 2-Methylbutanoic acid + 1 equivalent Br$_2$/PBr$_3$ ⟶

α-Halogenated aldehydes, ketones, and carboxylic acids are very useful synthetic intermediates because they can undergo substitution and elimination reactions just like haloalkanes. For example, reaction of an α-haloketone or carboxylic acid with base eliminates HX and forms an α,β-unsaturated carbonyl compound.

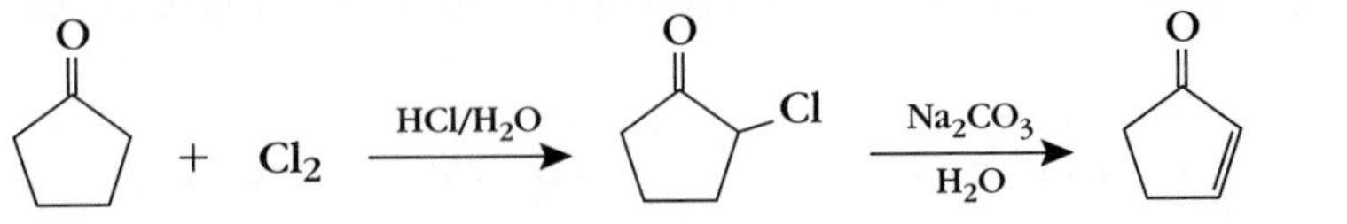

Cyclopentanone   2-Chlorocyclopentanone   2-Cyclopentenone
                                                               (72% yield)

$$\underset{\substack{\text{2-Bromobutanoic acid}}}{CH_3CH_2\overset{\displaystyle Br}{\overset{|}{C}}HCOOH} \xrightarrow[\text{Ethanol}]{\text{KOH}} \underset{\text{Potassium 2-butenoate}}{CH_3CH{=}CHCOO^-K^+} \xrightarrow{H_3O^+} \underset{\substack{\text{2-Butenoic acid}\\(65\%\ \text{yield})}}{CH_3CH{=}CHCOOH}$$

In this way, a double bond can be introduced adjacent to a carbonyl group.

α-Halogenated carbonyl compounds also undergo nucleophilic substitution reactions that occur by an S_N2 mechanism.

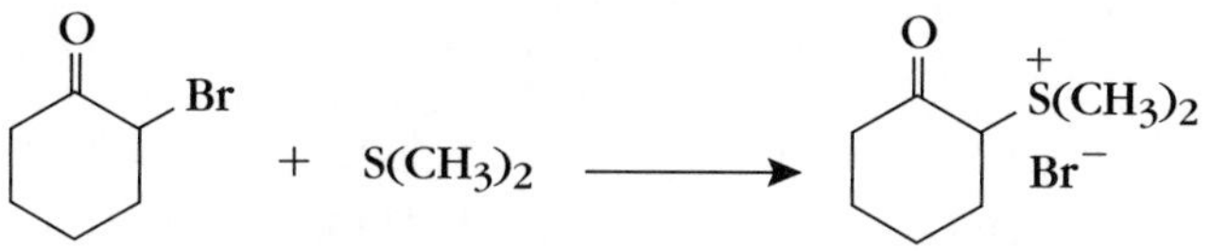

$$\underset{\substack{\text{}}}{CH_3CH_2\overset{\displaystyle Br}{\overset{|}{C}}HCOOH} \xrightarrow[\text{H}_2\text{O}]{\text{KCN}} CH_3CH_2\overset{\displaystyle CN}{\overset{|}{C}}HCOOH$$

Exercise 17.6   Write the structure of the product formed in each of the following reactions.

(a) $ClCH_2COOH$ + $NaOH/H_2O$ $\longrightarrow$

(b) Phenylacetic acid + 1 equivalent Br_2/PBr_3

(c) 2-Chlorocyclopentanone + $S(CH_3)_2$ $\longrightarrow$

When aldehydes and ketones are placed in a basic solution, an enolate ion is formed.

17.5 Enolate Ions Are Formed From Keto and Enol Tautomers in Basic Solution

When placed in an aqueous sodium hydroxide solution, both the keto and enol forms of acetone react with hydroxide ion to form an enolate ion according to the following acid-base reaction.

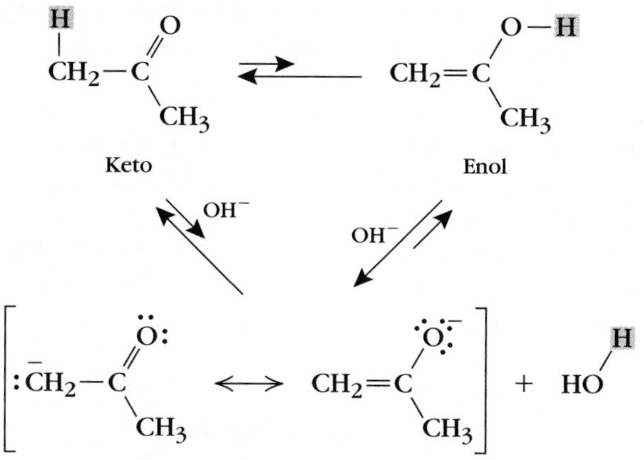

Table 17.2	Values of pK_a of Selected Compounds			

Compound	Structure of Acid	Structure of Conjugate Base	pK_a	K_a
Acetic acid	CH_3COOH	CH_3COO^-	4.75	1.8×10^{-5}
Enol of acetaldehyde	$\overset{\overset{\displaystyle OH}{\displaystyle \vert}}{CH_2{=}CH}$	$\overset{\overset{\displaystyle O^-}{\displaystyle \vert}}{CH_2{=}CH}$	10.5	3.2×10^{-11}
Enol of acetone	$\overset{\overset{\displaystyle OH}{\displaystyle \vert}}{CH_2{=}CCH_3}$	$\overset{\overset{\displaystyle O^-}{\displaystyle \vert}}{CH_2{=}CCH_3}$	10.9	1.3×10^{-11}
Water	H_2O	HO^-	15.7	2×10^{-16}
Ethanol	CH_3CH_2OH	$CH_3CH_2O^-$	16.0	10^{-16}
Acetaldehyde	CH_3CHO	$^-CH_2CHO$	16.7	2×10^{-17}
Acetone	$\overset{\overset{\displaystyle O}{\displaystyle \|}}{CH_3CCH_3}$	$\overset{\overset{\displaystyle O}{\displaystyle \|}}{CH_3CCH_2^-}$	19.3	5×10^{-20}
Ethyl acetate	$CH_3COOCH_2CH_3$	$^-CH_2COOCH_2CH_3$	24.5	3.2×10^{-25}
Acetonitrile	$CH_3C{\equiv}N$	$^-CH_2C{\equiv}N$	25	10^{-25}
Diisopropylamine	$[(CH_3)_2CH]_2NH$	$[(CH_3)_2CH]_2N^-$	38	10^{-38}
1-Propene	$CH_3CH{=}CH_2$	$^-CH_2CH{=}CH_2$	42	10^{-42}
Ethane	CH_3CH_3	$CH_3CH_2^-$	50	10^{-50}

Acetone reacts with hydroxide ion because it is weakly acidic. In general, carbonyl compounds that contain α-hydrogens are more acidic than most organic compounds, as shown by a comparison of the pK_a values of a number of compounds listed in Table 17.2.

Compared to ethane ($pK_a \approx 50$), aldehydes, ketones, carboxylic acid esters, and nitriles ($pK_a \approx 17$–25) are acids. They are not as acidic, however, as carboxylic acids ($pK_a \approx 5$). Based on the data in Table 17.2, the presence of a carbonyl or a nitrile group increases the acidity of an α-hydrogen by about 10^{25} over the acidity of an alkane hydrogen.

Why are α-hydrogen atoms of carbonyl-containing compounds acidic? Carbonyl-containing compounds are acidic because their conjugate bases, enolate ions, are stabilized by π electron delocalization. The simplified LCAO method can be used to describe the π electron delocalization of an enolate ion, as shown in Figure 17.4. The π molecular orbitals of the enolate ion are formed by combining the p atomic orbital of the α-carbon atom of the enolate ion with the p atomic orbitals of the sp^2 hybrid carbonyl carbon and carbonyl oxygen atoms. The π molecular orbitals of the enolate ion are shown in Figure 17.4, B. The lowest energy π molecular orbital is delocalized over all three atoms of the enolate ion.

The delocalization of π electrons in the enolate ion can also be represented as a resonance hybrid of the following two resonance structures.

$$\left[\quad \overset{..}{\underset{..}{R\ddot{C}H}}{-}\overset{\overset{\textstyle R'}{\big|}}{\underset{\underset{\displaystyle \ddot{O}{:}}{\|}}{C}} \quad \longleftrightarrow \quad RCH{=}\overset{\overset{\textstyle R'}{\big|}}{\underset{\underset{\displaystyle {:}\ddot{O}{:}^{-}}{}}{C}} \quad \right]$$

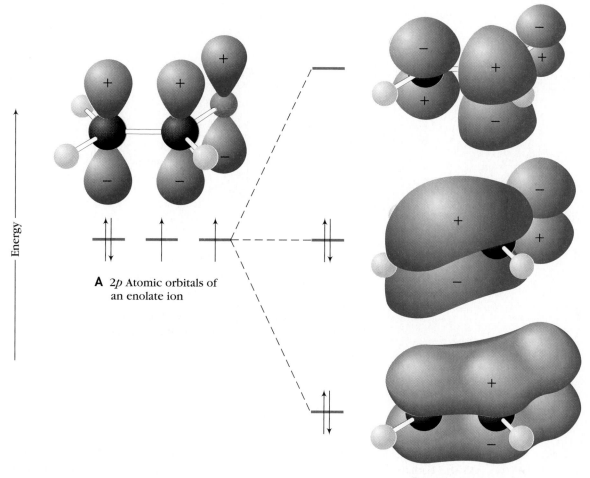

Energy

A 2*p* Atomic orbitals of an enolate ion

B π Molecular orbitals of an enolate ion

Figure 17.4
A, Atomic orbitals that combine to form the molecular orbitals of an enolate ion. **B,** The π molecular orbitals formed by overlap of the three *p* orbitals in **A.**

Both the resonance hybrid and the LCAO method accurately describe the enolate ion as a π electron delocalized ion.

The π electron delocalization in the enolate ion is an important factor in the greater acidity of the keto and enol forms of a carbonyl-containing compound compared to the acidity of an alkane. The negative charge on the carbon atom of the conjugate base of ethane cannot be delocalized. The negative charge on the α-carbon of carbonyl compounds, on the other hand, can be delocalized into the carbonyl group. The result is that the enolate ion is stabilized relative to the conjugate base of an alkane, as shown in Figure 17.5, and consequently, the carbonyl compound is the stronger acid.

Although the α-hydrogen atoms of aldehydes, ketones, esters, and nitriles are acidic, they are still very weak acids. This can be demonstrated by calculating the equilibrium constant for the reaction of acetone in an aqueous solution of sodium hydroxide. In order to do this, we need the pK_a values of the keto and enol forms of acetone and the pK_a value of water (the conjugate acid of hydroxide ion), all of which can be found in Table 17.2.

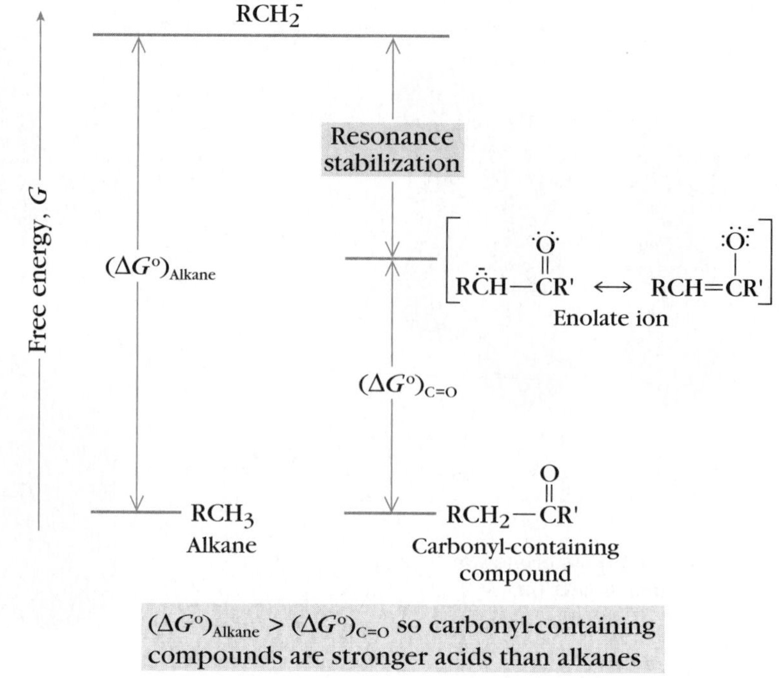

Figure 17.5
Resonance stabilization of the enolate ion makes carbonyl-containing compounds more acidic than alkanes. The free energies of the alkane and the carbonyl-containing compound are arbitrarily set equal.

$(\Delta G^\circ)_{\text{Alkane}} > (\Delta G^\circ)_{\text{C=O}}$ so carbonyl-containing compounds are stronger acids than alkanes

The strongest acid in solution is the enol form of acetone ($pK_a = 10.9$). It is a much stronger acid than water ($pK_a = 15.7$), so the enol is completely converted to the enolate ion by hydroxide ion. Consequently, the only two species related to acetone present in an aqueous sodium hydroxide solution of acetone are the enolate ion and the keto tautomer of acetone.

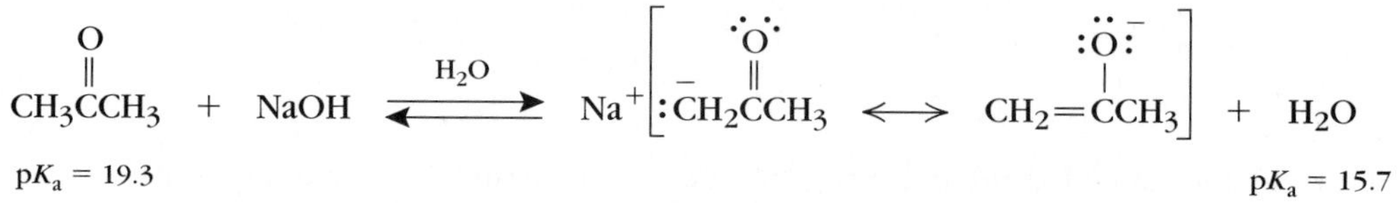

$pK_a = 19.3$ $pK_a = 15.7$

Therefore, we need only consider these two species when we estimate the equilibrium constant for the reaction of acetone in aqueous NaOH in the following way:

$$K = \frac{[\text{Enolate}][H_2O]}{[\text{Acetone}][OH^-]} = \frac{K_a\,(\text{Acetone})}{K_a\,(\text{Water})} = \frac{10^{-19.3}}{10^{-15.7}} \approx 10^{-3}$$

This calculation shows that the amount of enolate ion present is small in aqueous solutions where the strongest base is hydroxide ion.

| **Exercise 17.7** | Using the data in Table 3.1 (p. 90) and Table 17.2 (p. 762), estimate the equilibrium constant for the reaction of acetone with each of the following bases. |

(a) NaNH$_2$/NH$_3$ **(b)** NaH/THF   **(c)** NH$_3$/H$_2$O   **(d)** NaOCH$_3$/CH$_3$OH

| **Exercise 17.8** | Write the resonance structures of the resonance hybrid of each of the following ions. |

(a) $^-CH_2COOCH_3$

(b) $^-CH_2CHO$

(c) $CH_2{=}\overset{..}{\underset{|}{C}}H$ with $:\overset{..}{O}:^-$ above

(d) $^-CH_2CH{=}CH_2$

(e) $^-CH_2C{\equiv}N$

Why are we so concerned about the amount of enolate ion in solution? The answer is that the products of the reaction of enolate ions depend greatly on the relative amounts of enolate ion and keto forms present in solution. Let us begin by learning what happens in a solution in which both keto and enolate forms are present.

17.6 Reactions of Enolate Ions: Aldol Condensations of Aldehydes and Ketones

When acetaldehyde is placed in an aqueous solution of sodium hydroxide, 3-hydroxybutanal is formed. 3-Hydroxybutanal was given the trivial name **aldol** many years ago because it contains both an **ald**ehyde and an alco**hol** functional group.

$$CH_3CHO \ + \ CH_3CHO \ \underset{5\,°C}{\overset{NaOH/H_2O}{\rightleftarrows}} \ CH_3\overset{\overset{\displaystyle OH}{|}}{C}HCH_2CHO$$

Acetaldehyde   Acetaldehyde   3-Hydroxybutanal
(aldol)

The mechanism of the reaction of acetaldehyde and aqueous sodium hydrox-

Step 1: Formation of enolate ion

Step 2: Nucleophilic addition to carbonyl carbon atom

Tetrahedral intermediate

Step 3: Proton transfer to tetrahedral intermediate

3-Hydroxybutanal
(aldol)

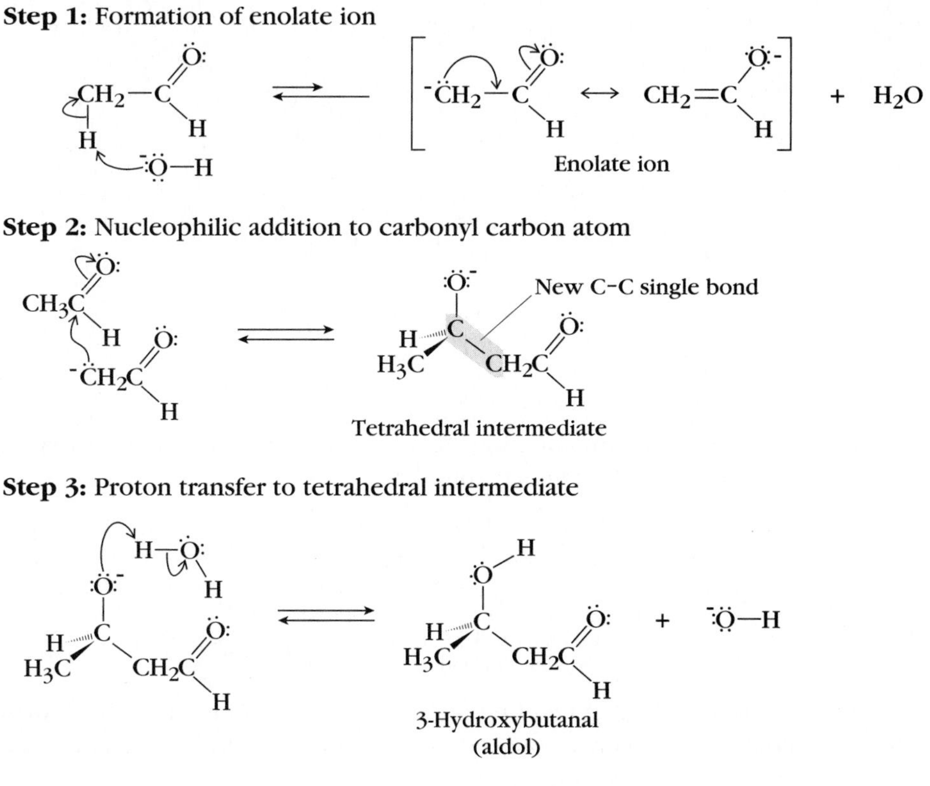

Figure 17.6
The mechanism of the reaction of acetaldehyde and aqueous sodium hydroxide. Step 1 is formation of the enolate ion, which adds to the carbonyl carbon atom of acetaldehyde to form a tetrahedral intermediate in Step 2. Proton transfer to the tetrahedral intermediate in Step 3 forms aldol.

ide occurs by the sequence of three steps given in Figure 17.6. Step 1 is the formation of the enolate ion discussed in Section 17.5. The enolate ion is a nucleophile because both its carbon and oxygen atoms carry a partial negative charge. In Step 2, the enolate ion generated from one molecule of acetaldehyde adds to the carbonyl carbon atom of another molecule of acetaldehyde to form a tetrahedral intermediate. This is a typical nucleophilic addition reaction of aldehydes, as we discussed in Chapter 14. A new carbon-carbon single bond is formed in this reaction between the carbon atom of the enolate ion of one molecule and the carbonyl carbon atom of the other. Finally, proton transfer from the solvent to the tetrahedral intermediate yields aldol in Step 3. This series of three reversible reactions is known as an **aldol condensation,** and it is a general reaction in a basic solution of aldehydes and ketones that have α-hydrogen atoms.

The mechanism of the aldol condensation is another example of nucleophilic addition to the carbonyl carbon atoms of aldehydes and ketones. While an aldol condensation may seem more complicated, it is no different in concept from any other carbonyl addition reaction such as the addition of a Grignard reagent.

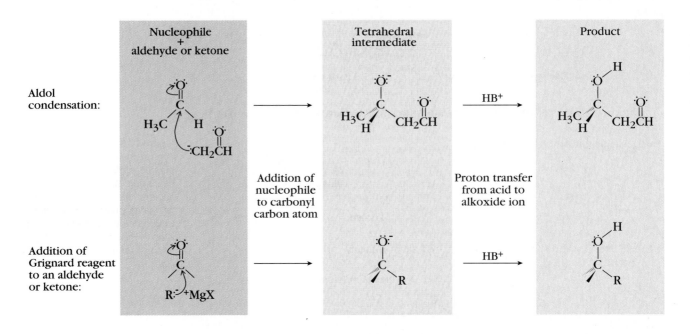

The products of aldol condensations all have a common structural feature, a hydroxy group β to a keto functional group.

Common structural feature of
the products of aldol condensations

This structural feature is obtained by bonding the α-carbon atom of the enolate ion of one aldehyde molecule to the carbonyl carbon atom of the other aldehyde molecule. The enolate ion is called the donor because it donates a pair of electrons to the carbonyl carbon atom of the aldehyde molecule. The aldehyde molecule, in turn, is called the acceptor.

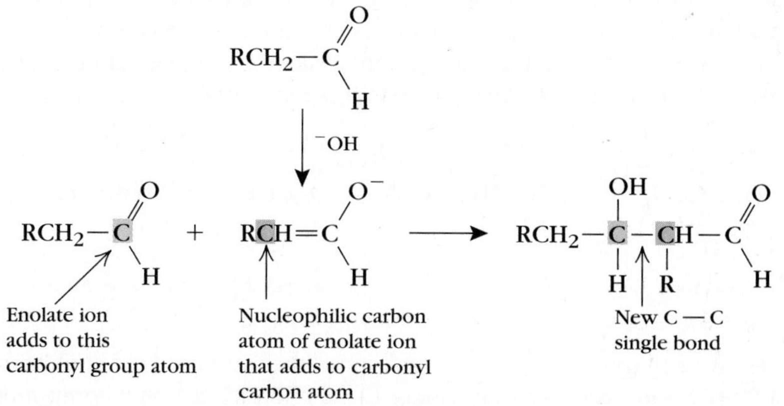

The net effect of an aldol condensation is an α-substitution reaction. The α-hydrogen atom of one aldehyde molecule is replaced by a group derived from another aldehyde molecule.

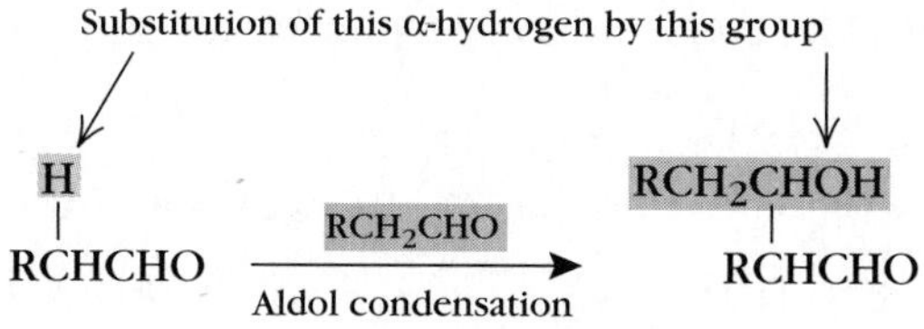

Exercise 17.9   Write the structure of the product formed in the reaction of each of the following aldehydes with a solution of aqueous sodium hydroxide.

(a) Propanal   **(b)** 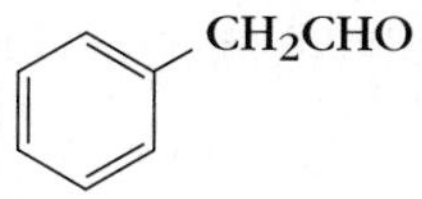   **(c)** 2-Methylbutanal

---

Exercise 17.10   Write the equation for the preparation of each of the following compounds from the starting material given using an aldol condensation. More than one step may be needed.

(a) Butanol from acetaldehyde
(b) 2-Ethyl-1-hexanol from butanal

(c) 1,3-Diphenylbut-2-ene-1-one from

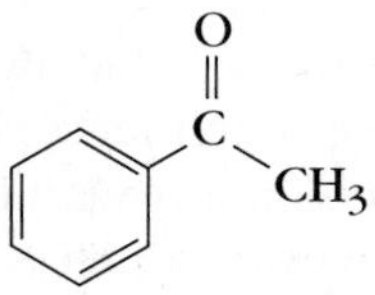

The yield of product formed in an aldol condensation normally depends on the equilibrium constant of Step 2, nucleophilic addition to the carbonyl group. Like other reversible carbonyl addition reactions discussed in Chapter 14, the equilibrium constant is less favorable for nucleophilic additions to ketones than it is for aldehydes. As a result, the aldol condensation of ketones gives much lower yields of product. The aldol condensation of acetone, for example, forms only about 5% of the product, 4-hydroxy-4-methyl-2-pentanone.

$$
\underset{\text{Acetone}}{CH_3\overset{\overset{\displaystyle O}{\|}}{C}CH_3} \;+\; \underset{\text{Acetone}}{CH_3\overset{\overset{\displaystyle O}{\|}}{C}CH_3} \;\;\underset{}{\overset{NaOH/H_2O}{\rightleftharpoons}}\;\; \underset{\substack{\text{4-Hydroxy-4-methyl-2-pentanone}\\ \text{(5\% yield)}}}{CH_3-\underset{\underset{\displaystyle CH_3}{|}}{\overset{\overset{\displaystyle OH}{|}}{C}}-CH_2\overset{\overset{\displaystyle O}{\|}}{C}CH_3}
$$

There are a number of ways to overcome the unfavorable equilibrium constant of Step 2. One way is to dehydrate the β-hydroxycarbonyl compound to an α,β-unsaturated aldehyde or ketone (which is also called an enone). Dehydration can be accomplished either by heating the basic reaction mixture or by first acidifying the reaction mixture, then heating. Dehydration is usually energetically favorable, so all equilibria shift in the direction of product. The result is to increase the yield of aldol condensation product by forming an enone.

$$
\underset{\substack{\beta\text{-Hydroxyaldehyde}\\ \text{(aldol condensation product)}}}{RCH_2\underset{\underset{\displaystyle R}{|}}{\overset{\overset{\displaystyle OH}{|}}{CH}}-CH-\overset{\displaystyle O}{\underset{\displaystyle H}{C}}} \;\;\underset{\substack{H_3O^+ \text{ or}\\ H_2O/OH^-}}{\overset{Heat}{\rightleftharpoons}}\;\; \underset{\substack{\alpha,\beta\text{-Unsaturated aldehyde}\\ \text{(an enone)}}}{RCH_2CH=\underset{\underset{\displaystyle R}{|}}{C}-\overset{\displaystyle O}{\underset{\displaystyle H}{C}}} \;+\; H_2O
$$

Normally if the α,β-unsaturated aldehyde or ketone is the desired product, the condensation reaction is carried out at higher temperatures. Under these conditions the first formed aldol is dehydrated to the enone.

The mechanism of the dehydration of β-hydroxyaldehydes or ketones in aqueous base involves formation of an enolate ion, which then loses a hydroxide ion to form the enone.

We have repeatedly stressed that hydroxide ion is a poor leaving group in either elimination or substitution reactions. Why then, do we propose a dehydration mechanism of an alcohol in which hydroxide ion is the leaving group? Aldol is a special alcohol, however, because the hydroxy group is β- to a carbonyl group. The presence of the carbonyl group allows formation of an enolate ion, which makes it easier for a hydroxide ion to be eliminated. Also, the product is an α,β-unsaturated carbonyl compound, which is particularly stable. This too makes it easier for hydroxide ion to be eliminated. Even so, elimination of hydroxide ion from β-hydroxycarbonyl compounds is not an easy reaction because vigorous reaction conditions—heat and a concentrated base solution—are needed.

Exercise 17.11   Write the structure of the product formed in the reaction of each of the following compounds with a hot aqueous solution of sodium hydroxide.

(a) Butanal   (b) 3-Pentanone   (c) 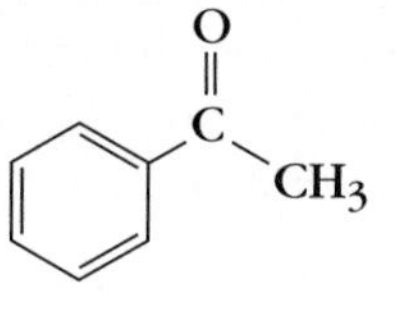   (d)

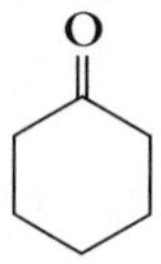

Carboxylic acid esters also form enolate ions in basic solution. These enolate ions also undergo condensation reactions.

17.7   Reactions of Enolate Ions: Claisen Condensations

Carboxylic acid esters also undergo condensation reactions. For example, the reaction of ethyl acetate with sodium ethoxide in ethanol forms ethyl acetoacetate (also called acetoacetic ester).

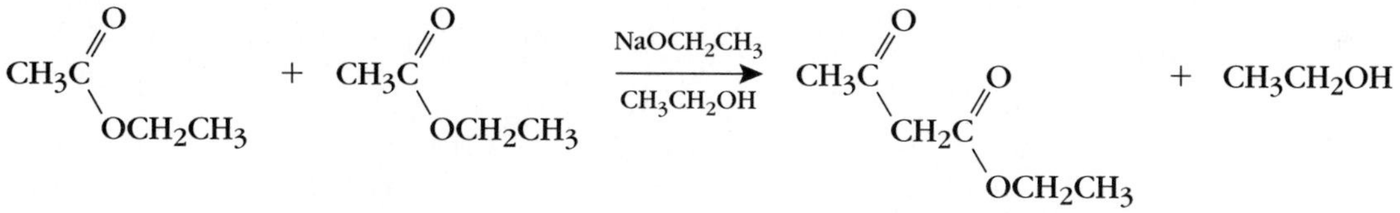

Ethyl acetate   Ethyl acetate   Ethyl acetoacetate
(acetoacetic ester)
a β-keto ester

Ethyl acetoacetate is an example of a class of compounds called **β-keto esters.** β-Keto esters are compounds with a ketone carbonyl group β- to an ester group.

The formation of ethyl acetoacetate is an example of a **Claisen condensation** reaction, which is a general reaction of esters that contain at least two hydrogens α- to the ester functional group. A Claisen condensation is much like an aldol condensation reaction because both are α-substitution reactions. A Claisen condensation can also be viewed as an acyl transfer reaction because the acyl group of one ester molecule is transferred to the α-carbon atom of another ester molecule.

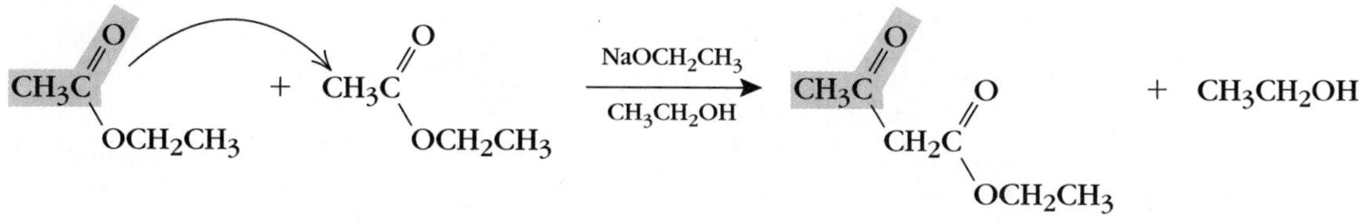

Viewed in this way, the mechanism of the Claisen condensation reaction, shown in Figure 17.7, is similar to the addition-elimination mechanism that we discussed in Chapter 16. Step 1 of the mechanism of the Claisen condensation is the reversible formation of a small amount of the nucleophile (the enolate ion). In Step 2, the enolate ion formed from one ester molecule adds to the carbon atom of the C=O bond of another to form a tetrahedral intermediate. This is the familiar addition step of the mechanisms of acyl transfer or nucleophilic substitution reactions of carboxylic acid derivatives. In this reaction a carbon-carbon single bond is formed. The tetrahedral intermediate eliminates an ethoxide ion to form the product in Step 3.

The formation of ethyl acetoacetate would seem to be the end of the mechanism because we have formed the product. Unfortunately, however, the equilibrium constants of Steps 1 to 3 favor the starting material. The desired product is formed in a Claisen condensation because the equilibrium constant of Step 4, the

Figure 17.7
The mechanism of the Claisen condensation reaction.

Step 1: Formation of an enolate ion

Step 2: Addition of the enolate ion to the carbon atom of the C=O bond

Step 3: Elimination of the ethoxide ion from the tetrahedral intermediate

Step 4: Deprotonation of ethyl acetoacetate

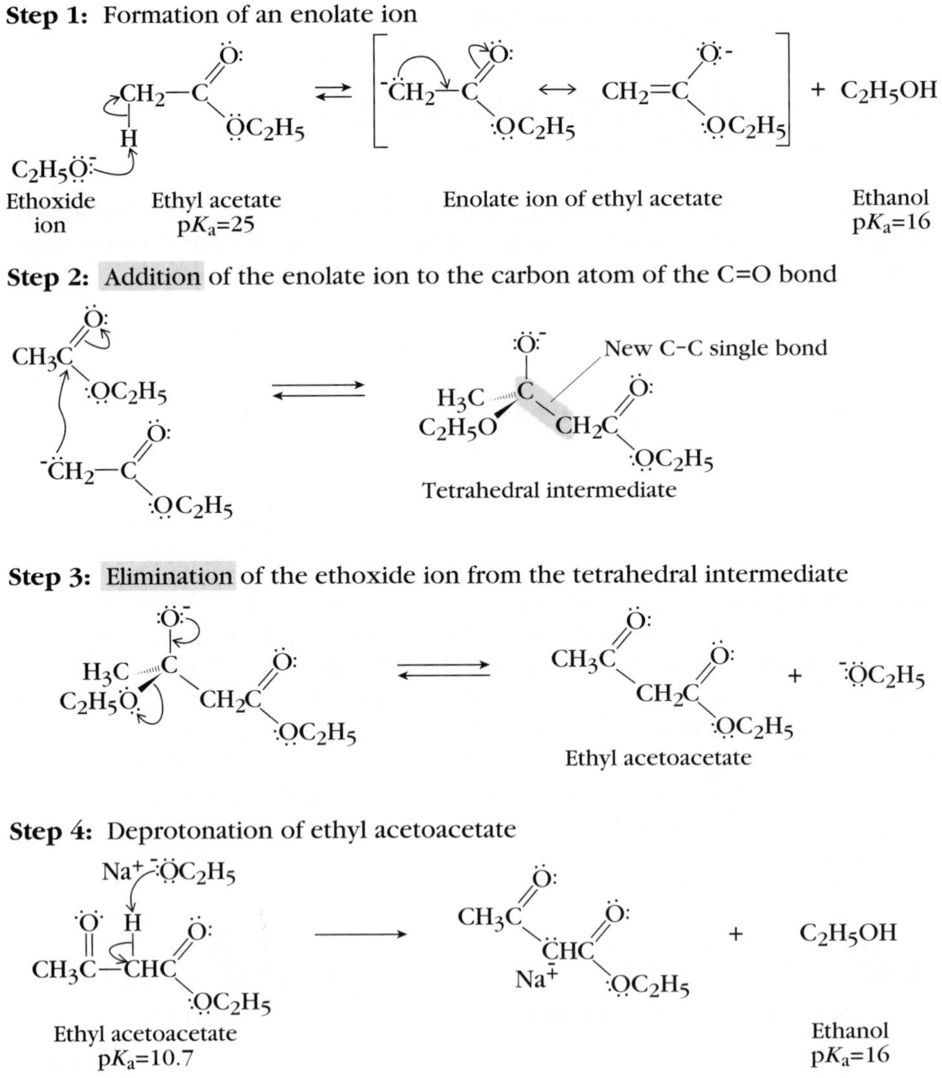

deprotonation of ethyl acetoacetate by ethoxide ion, favors the products. The products are favored because they are the weakest base and acid in solution.

We know that the formation of the enolate ion of the β-keto ester is the key step to product formation because esters that have only one α-hydrogen do not undergo the Claisen condensation. For example, no condensation products are obtained when ethyl 2-methylpropanoate is added to a solution of sodium ethoxide in ethanol, the usual conditions of Claisen condensations.

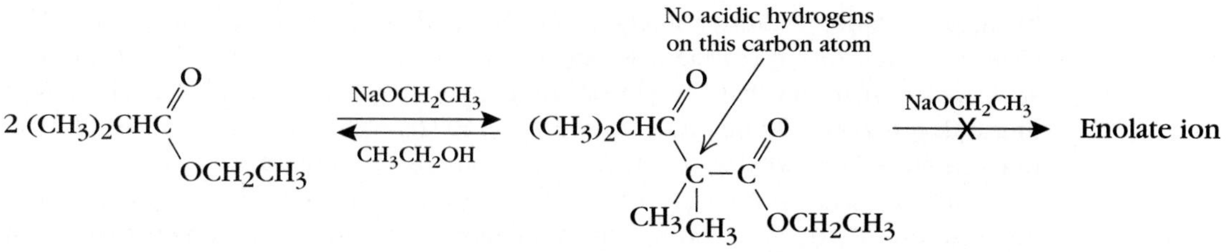

In summary, to ensure that a Claisen condensation goes to completion, enough base must be added to completely convert the β-keto ester formed in Step 3 to its enolate ion. This is done in the laboratory by carrying out the reaction in two distinct steps. First, one equivalent of base is reacted with two equivalents of ester to form the enolate ion of the β-keto ester.

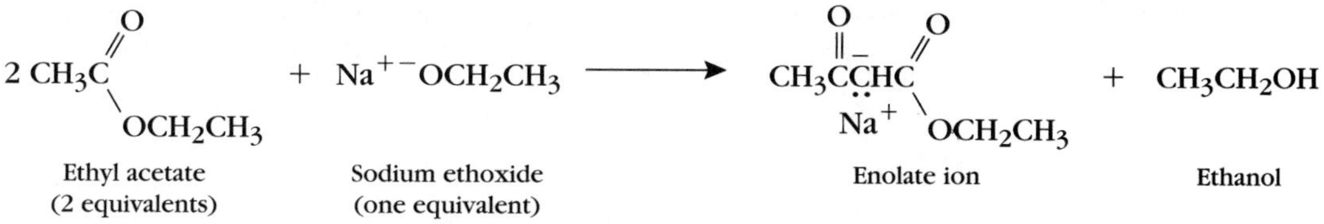

Second, the β-keto ester is isolated by acidification of the reaction mixture.

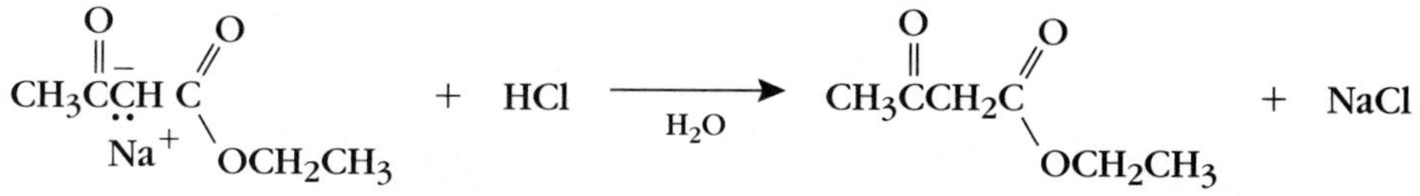

Exercise 17.12   Write the product of the Claisen condensation reaction of each of the following esters.

(a) Ethyl propanoate   **(b)** Ethyl phenylacetate   **(c)** Ethyl pentanoate

---

Exercise 17.13   Write the mechanism of the Claisen condensation reaction of ethyl butanoate.

Exercise 17.14   Explain why the two following esters do not undergo the Claisen condensation reaction.

(a) Ethyl benzoate   **(b)** Ethyl 2-methylbutanoate

Living systems use a Claisen condensation reaction to form carbon-carbon bonds.

17.8 Claisen Condensation Reactions in Living Systems

We learned in Section 16.10 that acetyl transfer reactions in living systems occur by means of the thioester acetyl CoA rather than carboxylic acid esters. The Claisen condensation reaction, which was introduced in Section 17.7, is another acyl transfer reaction that acetyl CoA undergoes. A specific example can be found in the degradation of fatty acids to acetyl CoA. This is an important series of reactions because it generates energy from fats and other foodstuffs.

Under certain conditions, the concentration of acetyl CoA formed by the degradation of fatty acids builds up in the system because it is formed faster than it can be used in other biological processes. To reduce the excess amount of acetyl CoA, it is converted into acetoacetyl CoA by an enzyme called thiolase.

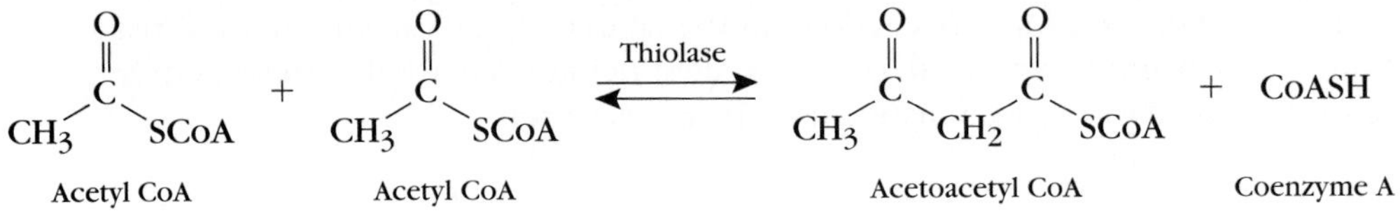

Figure 17.8
Mechanism of thiolase-catalyzed condensation of acetyl CoA.

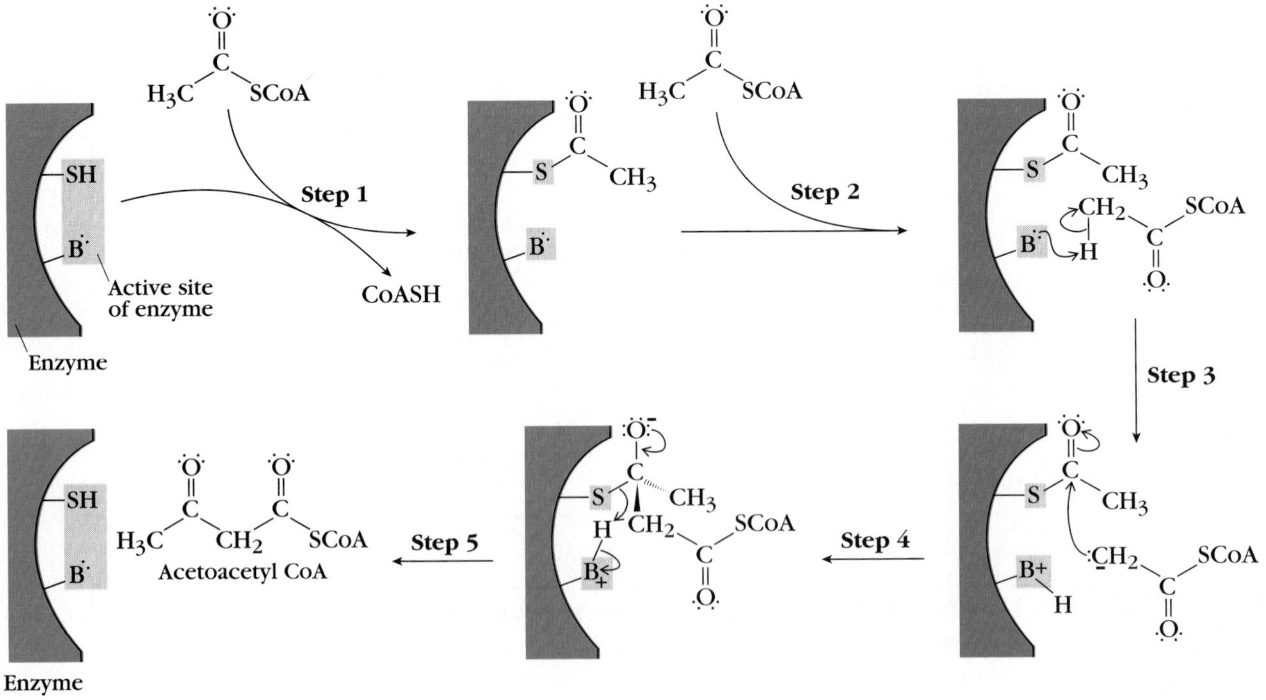

This reaction is analogous to the Claisen condensation of ethyl acetate discussed in Section 17.7. The major difference is that acetoacetyl CoA is a β-keto thioester of coenzyme A, not a β-keto ester. The mechanism of the thiolase-catalyzed condensation reaction of acetyl CoA is shown in Figure 17.8.

Step 1 is the transfer of an acetyl group from acetyl CoA to the thiol group (SH) at the active site of the enzyme. Coenzyme A (CoASH) is released in the process. Steps 2 and 3 can be visualized as the removal of an α-hydrogen atom from acetyl CoA to form an enolate ion. The enolate ion then adds to the carbonyl carbon atom of the enzyme-bound acetyl group to form a tetrahedral intermediate in Step 4. Finally, elimination of the enzymatic thiol group in Step 5 regenerates the enzyme and yields acetoacetyl CoA as the final condensation product.

As we have learned from previous examples of enzyme-catalyzed reactions, their mechanisms are very similar to the same reaction carried out in nonliving systems. Thus the mechanism of the enzyme-catalyzed Claisen condensation reaction outlined in Figure 17.8 occurs by the same addition-elimination mechanism that is observed in the reaction of ethyl acetate in an alcohol solution of sodium ethoxide. The major difference is the initial formation of an acetyl-enzyme intermediate. As we learned in Chapter 16, even this intermediate is formed by a mechanism familiar to us. Again we find that the mechanistic information obtained from the study of simple organic reactions can be directly applied to their more complex biological analogs.

| Exercise 17.15 | Write the structure of the product formed by the thiolase-catalyzed condensation reaction of propionyl CoA. |

So far, the condensation reactions that we have studied involve only one compound, which serves as both the donor and the acceptor. Sometimes it is possible to choose conditions so that condensation reactions can occur between two different compounds. One compound acts as the donor, the other as the acceptor.

17.9   Mixed Condensation Reactions

The condensation reactions presented so far are **self-condensation reactions** because one compound serves as both the donor and the acceptor. Chemists have attempted to broaden the synthetic utility of these reactions by carrying out condensation reactions between two different compounds. The reactions are called **mixed** or **crossed condensation reactions.** The problem with mixed condensation reactions is that up to four products can be formed. For example, when two aldehydes, such as acetaldehyde and propanal, are placed in an aqueous basic solution, two products form by the mixed condensation reactions of acetaldehyde and propanal, while two more form by self-condensation reactions.

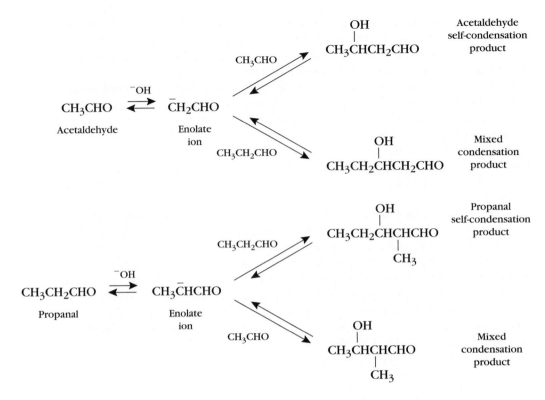

In order to make this reaction synthetically useful, we must do something to reduce the number of products. One way is to use an acceptor that does not have any α-hydrogen atoms. Compounds without α-hydrogen atoms cannot form enolate ions, so they can function only as acceptors. This immediately limits the number of possible products to two. For example, the following are the only two possible products formed by the mixed condensation reaction of formaldehyde and acetaldehyde.

$$CH_2CO \xrightarrow{\ ^-OH\ } \text{No α-hydrogens so no enolate ion can form}$$

Formaldehyde

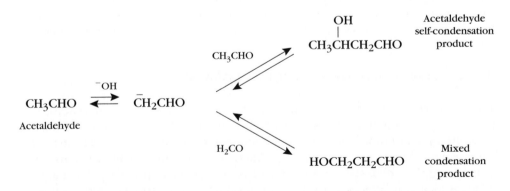

The mixed condensation product is the only product if the compound that forms the enolate ion is either a ketone or an ester. Only 4-phenyl-3-buten-2-one is formed, for example, when equivalent amounts of benzaldehyde and acetone are mixed in an aqueous basic solution.

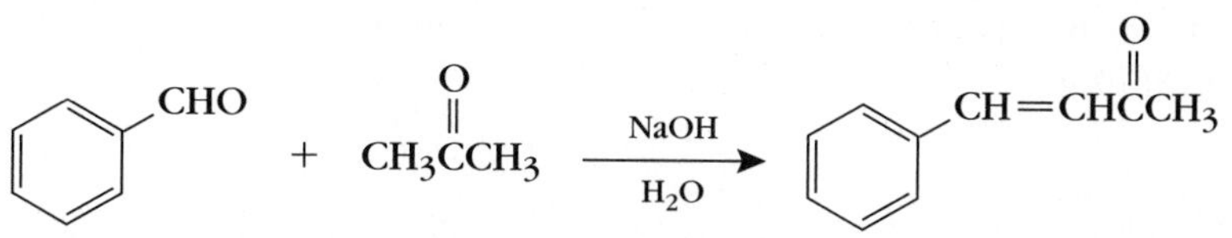

Benzaldehyde   Acetone   4-Phenyl-3-buten-2-one
(no α-hydrogen atoms)   (65%-78% yield)

The self-condensation product of acetone is not formed for two reasons. First, addition of the enolate ion occurs faster to the carbonyl carbon atom of an aldehyde than a ketone. (Remember that the relative ease of addition to the carbonyl carbon atom is aldehyde > ketone > carboxylic acid ester.) Second, even if addition of the enolate ion to another molecule of acetone occurs, the equilibrium constant for self-condensation of ketones is less favorable than addition to aldehydes. Consequently, both rate and equilibrium for addition of the enolate ion to benzaldehyde are more favorable than for addition to acetone. The end result, therefore, is the formation of a single product.

The same technique can be applied to Claisen condensation reactions. Again, the most successful mixed Claisen condensation reactions occur when one ester has no α-hydrogen atoms such as the following:

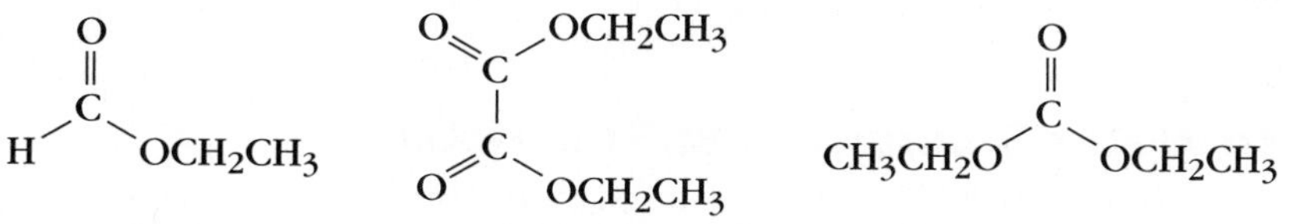

Ethyl formate   Diethyl oxalate   Diethyl carbonate

The reaction of ethyl propanoate and an excess of diethyl oxalate with one equivalent of sodium ethoxide in ethanol is a mixed Claisen condensation. The

---

Exercise 17.16   Write the structure of the product or products in each of the following reactions.

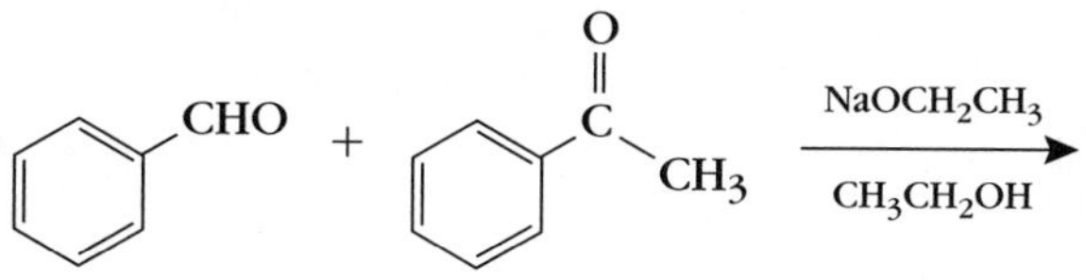

(b) Methyl benzoate + Methyl propanoate $\xrightarrow[\text{(2) H}_3\text{O}^+]{\substack{\text{(1) 1 equivalent}\\\text{NaOCH}_3}}$

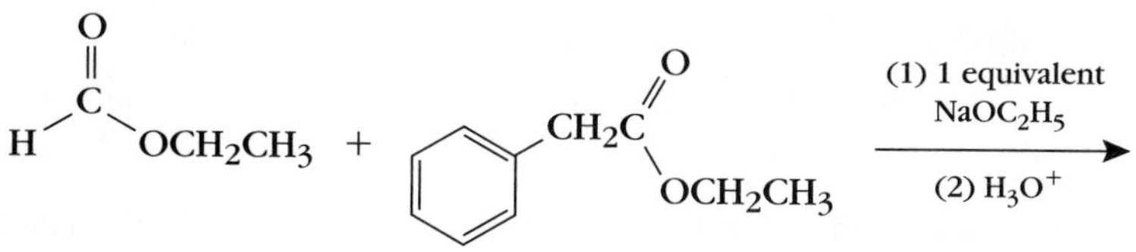

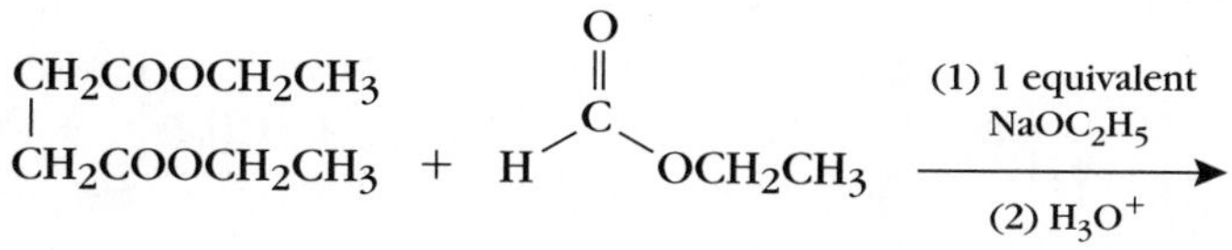

reaction mixture can be acidified to isolate the product as in Claisen self-condensation reactions.

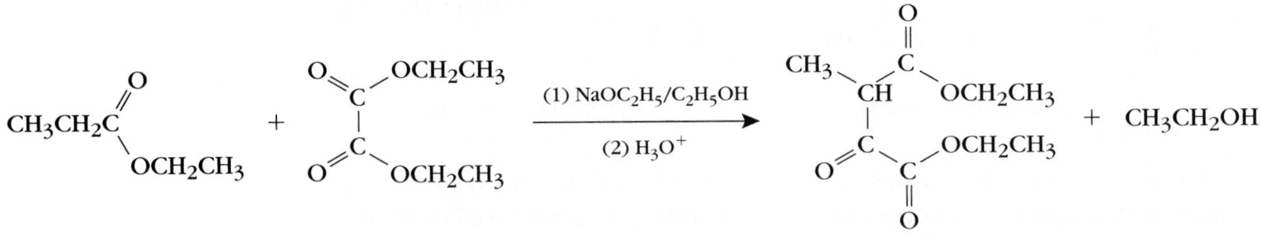

Ethyl propanoate   Diethyl oxalate   (70% yield)

In summary, mixed aldol and Claisen condensations are most successful when one of the reactants has no α-hydrogens and has a particularly reactive carbonyl group. In these cases, the major product forms by addition of the enolate ion to the carbonyl carbon atom of the compound without α-hydrogen atoms.

The major emphasis so far has been on the use of condensation reactions to form carbon-carbon single bonds. Because condensation reactions are reversible, the reverse reaction can be used to break a carbon-carbon single bond. This is an important reaction in the degradation of fatty acids and carbohydrates in living systems.

17.10 Retro-Condensation Reactions

We learned in Section 17.6 that aldol condensations are reversible reactions. So far we have concentrated on the use of aldol and Claisen condensations to form

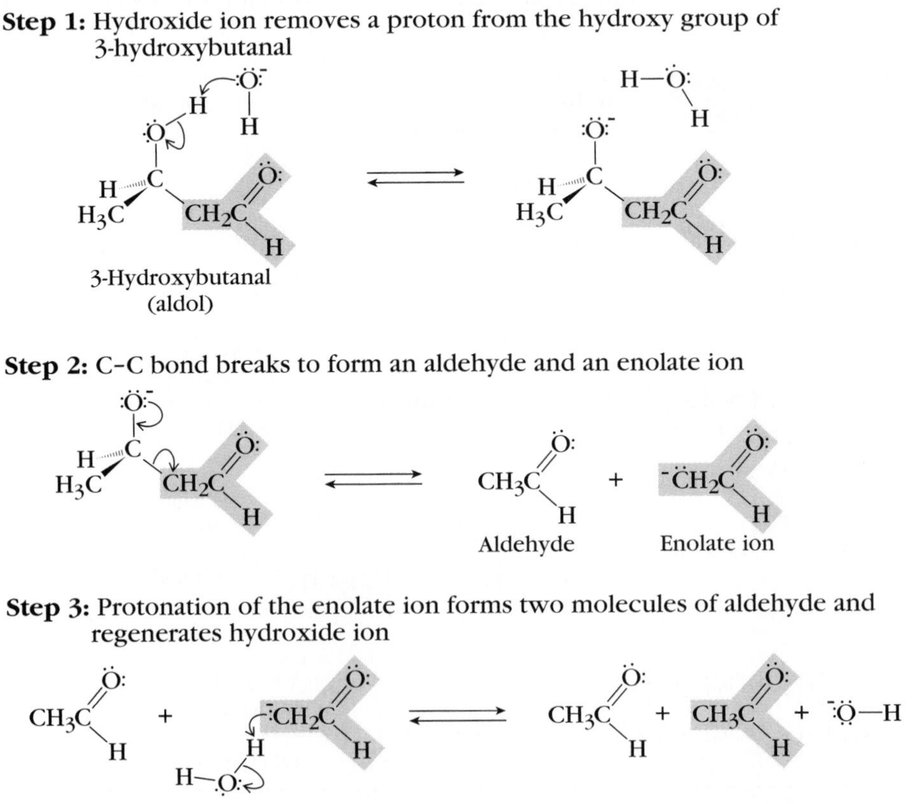

Figure 17.9
The mechanism of the retro-aldol condensation reaction of 3-hydroxy-butanal is the reverse of the mechanism of the aldol condensation of acetaldehyde, as shown in Figure 17.6 (p. 765).

Step 1: Hydroxide ion removes a proton from the hydroxy group of 3-hydroxybutanal

3-Hydroxybutanal
(aldol)

Step 2: C–C bond breaks to form an aldehyde and an enolate ion

Aldehyde   Enolate ion

Step 3: Protonation of the enolate ion forms two molecules of aldehyde and regenerates hydroxide ion

carbon-carbon single bonds. Their reverse reactions, however, which are called **retro-condensation reactions,** are also important. The mechanism of a retro-aldol condensation is simply the reverse of the forward mechanism, as shown in Figure 17.9 for 3-hydroxybutanal.

While retro-aldol condensations are of limited synthetic value in the laboratory, they are very important in living systems. Fructose 1,6-bisphosphate, for example, is a six-carbon carbohydrate that is cleaved in the aldose-catalyzed retro-aldol reaction to form dihydroxyacetone phosphate and glyceraldehyde 3-phosphate, two molecules containing three carbon atoms each. This reaction is part of a series of enzyme-catalyzed reactions that degrades glucose in living systems. We can propose a mechanism for this reaction, shown in Figure 17.10, that is analogous to that of the retro-aldol condensation reaction of 3-hydroxybutanal, except that a base at the active site of the enzyme, rather than a hydroxide ion, removes a proton from the β-hydroxy group.

Step 1: Removal of the proton from the hydroxy group by a base on the enzyme

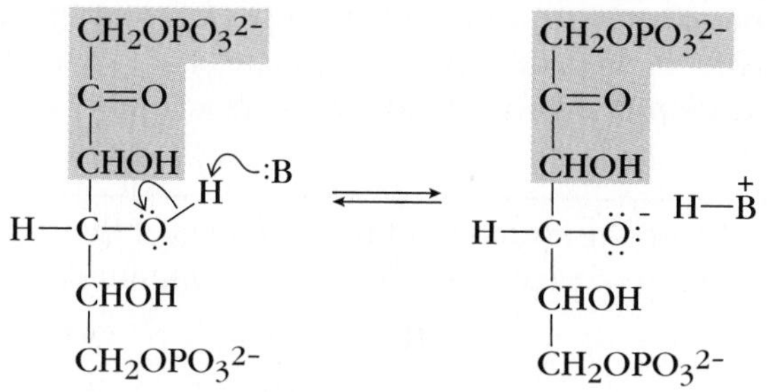

Fructose 1,6-bisphosphate

Step 2: Breaking of the C–C bond

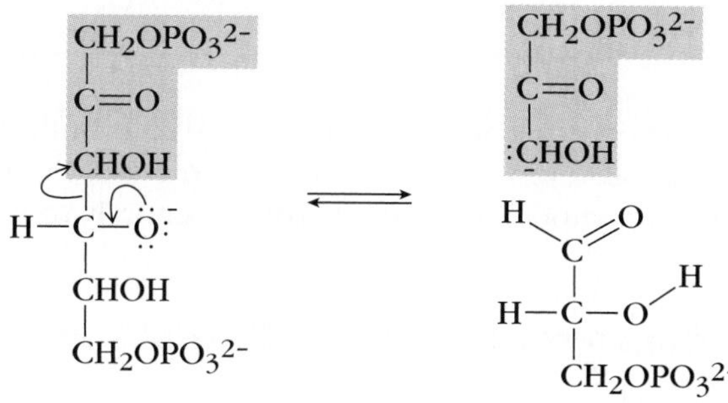

Glyceraldehyde 3-phosphate

Step 3: Protonation of the enolate ion

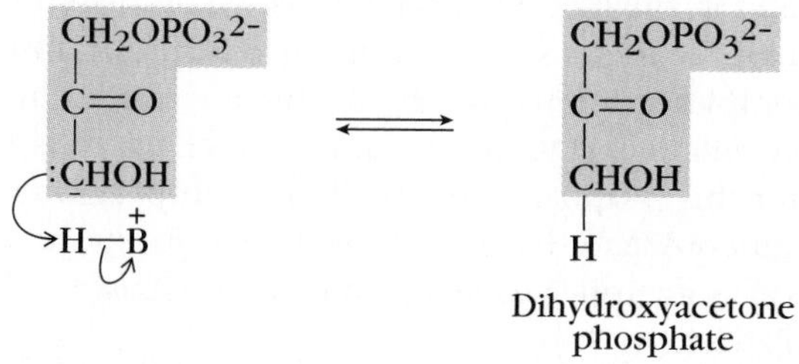

Dihydroxyacetone
phosphate

Figure 17.10
A proposed mechanism for the retro-aldol reaction of fructose 1,6-bisphosphate to form glyceraldehyde 3-phosphate and dihydroxyacetone phosphate that resembles the mechanism of the retro-aldol reaction of 3-hydroxybutanal shown in Figure 17.9.

This mechanism is an oversimplification, as we will learn in Section 17.15, but it does illustrate again how applying mechanistic principles of organic chemistry to biological systems allows us to understand their chemical reactions.

An example of a retro-Claisen condensation reaction occurs in the degradation of fatty acids. Fatty acids are degraded by the sequential removal of two-carbon units, starting from the carboxyl end of the fatty acid molecule. The actual carbon-carbon cleavage reaction is catalyzed by thiolase, and its mechanism is the reverse of that given in Figure 17.8 (p. 772).

β-Keto fatty acyl CoA
R = Alkyl chain

Fatty acid shortened
by two carbon atoms

Acetyl CoA

The two-carbon unit removed each time is acetyl CoA. This reaction is repeated until the fatty acid is completely converted to acetyl CoA, which is eventually oxidized to CO_2 and H_2O.

In summary, retro-condensation reactions are possible in basic solutions for β-hydroxy aldehydes, β-hydroxy ketones, and β-keto esters. These reactions are of limited value in synthesis because we usually want to *form* carbon-carbon single bonds rather than *cleave* them. Their biological analogs, however, are important reactions in the degradation of carbohydrates and fatty acids.

Exercise 17.17   Write the structure of the products of the retro-condensation of each of the following compounds.

Condensation reactions occur under conditions in which both the keto form of a compound and its enolate ion exist in solution at the same time. Under certain conditions a carbonyl compound can be converted entirely to its enolate ion.

17.11 Quantitative Preparation of Enolate Ions

The condensation reactions that we have discussed so far are carried out by bases such as hydroxide ion in water or ethoxide ion in ethanol. Because aldehydes, ketones, and esters are much less acidic than the solvent in these cases, enolate ions form reversibly and in small concentrations, as we discussed in Section 17.5. If a strong enough base is used, such as lithium diisopropylamide, a carbonyl compound can be completely converted to its enolate ion. Lithium diisopropylamide, which is abbreviated as LDA, has a number of characteristics that make it particularly suitable for this purpose. First of all, diisopropylamine is a very weak acid ($pK_a \approx 38$), so its conjugate base, LDA, is a very strong base. Its ability to convert a ketone (such as acetone) to its enolate can be illustrated by the following equilibrium calculation.

| Acetone $pK_a = 19.3$ | Lithium diisopropylamide (LDA) | | Enolate ion | Diisopropylamine $pK_a = 38$ |

$$K = \frac{[\text{Enolate}][\text{HN}[\text{CH}(\text{CH}_3)_2]_2]}{[\text{Acetone}][\text{Li}^{+-}\text{N}[\text{CH}(\text{CH}_3)_2]_2]} = \frac{K_a \text{ of acetone}}{K_a \text{ of diisopropylamine}} = \frac{10^{-19.3}}{10^{-38}} \approx 10^{+20}$$

An equilibrium constant of approximately 10^{20} means that the concentrations of acetone and LDA are negligible at equilibrium. Because acetone is completely and irreversibly converted to its enolate ion, its self-condensation reaction cannot occur.

LDA is a useful reagent because it is easy to prepare by the reaction of an organolithium compound, such as methyllithium or butyllithium, with diisopropylamine in 1,2-dimethoxyethane (DME) or tetrahydrofuran (THF) as solvent.

| Butyllithium | Diisopropylamine | Lithium diisopropylamide (LDA) | Butane |

LDA is soluble in organic solvents, moreover, so it can be used as a base with any compound soluble in organic solvents like THF and DME. Bases are also nucleophiles, so they can add to the carbonyl carbon atom in what is an unwanted reaction in this case. LDA, however, is too bulky to undergo this reaction. Finally LDA reacts rapidly with a carbonyl-containing compound to form an enolate ion. For these reasons, LDA is frequently employed when a very strong base is needed in a polar aprotic solvent.

The quantitative formation of an enolate ion in this way can be applied to ketones, carboxylic acid esters, carboxylic acids, and *N,N*-dialkylamides. It cannot be applied to aldehydes, however, because the self-condensation of most aldehydes is so rapid that it competes successfully with the deprotonation of the α-carbon atoms by LDA. The following are examples of enolate ions that can be formed by the reaction of carbonyl compounds with LDA.

Cyclohexanone

Ethyl propanoate

■ *Do not confuse the word* amide *in the name* lithium diisopropylamide *(LDA) with the same word used to identify an* amide *functional*

group NH_2. *The* NH_2^- *ion is called an* amide *ion so its salts are named by placing the name of the cation before the word* amide:

$$\text{Na}^{+-}\text{NH}_2 \qquad \text{Li}^{+-}\text{NH}_2$$
Sodium amide   Lithium amide

$$\text{Li}^{+-}\text{N}[(\text{CH}(\text{CH}_3)_2]_2$$
Lithium diisopropylamide

LDA is named, therefore, as a derivative of lithium amide, LiNH_2.

$$(CH_3)_2CHCOOH \xrightarrow[-78\,°C]{2\ LDA/THF} (CH_3)_2C=C\begin{matrix} O^-Li^+ \\ \\ O^-Li^+ \end{matrix} + 2\ HN[CH(CH_3)_2]_2$$

Methylpropanoic acid

$$CH_3CH_2C\overset{O}{\underset{N(CH_3)_2}{\diagdown\!\!\!\!\parallel}} \xrightarrow[-78\,°C]{LDA/THF} CH_3CH=C\begin{matrix} O^-Li^+ \\ \\ N(CH_3)_2 \end{matrix} + HN[CH(CH_3)_2]_2$$

N,N-Dimethyl-
propanamide

Amides and *N*-substituted amides do not form enolate ions because the N—H hydrogen atom has a pK_a of about 15, while the α-hydrogen atom has a pK_a of about 30. Therefore α-hydrogen atoms are removed only from *N,N*-disubstituted amides, where there are no N—H bonds.

Nitriles do not form enolate ions, but they do react with LDA to lose a proton to form a resonance-stabilized ion. These ions and enolate ions undergo many of the same reactions.

$$CH_3CH_2CH_2C\equiv N \xrightarrow[\substack{\text{Diethyl ether} \\ 0\,°C}]{LDA} \left[\begin{matrix} CH_3CH_2\overset{\cdot\cdot}{C}H-C\equiv N: \\ \updownarrow \\ CH_3CH_2CH=C=N\overset{\cdot\cdot}{\underset{\cdot}{:}}{}^{-} \end{matrix} \right] + HN[CH(CH_3)_2]_2$$

Butanenitrile   Resonance-
stabilized ion

■ *The preparation and reactions of acetylide ions are discussed in Section 9.11.*

Enolate anions are very useful in synthesis because they react as nucleophiles with many compounds to form carbon-carbon single bonds. In this way enolate ions resemble acetylide ions. Both undergo two kinds of reactions that are synthetically very important, namely nucleophilic additions to carbonyl carbon atoms and nucleophilic substitution reactions at saturated carbon atoms.

Carbon nucleophiles react with haloalkanes in a nucleophilic substitution reaction that occurs by an S_N2 mechanism. The product is the alkylation of the nitrile or carbonyl compound at its α-position.

$$\text{(cyclohexane)}\!-\!COOC_2H_5 \xrightarrow[\text{(2) } CH_3(CH_2)_6I]{\text{(1) LDA/THF/}-78\,°C} \text{(cyclohexane)}\!\begin{matrix} COOC_2H_5 \\ (CH_2)_6CH_3 \end{matrix}$$

(90% yield)

$$CH_3CH_2CH_2C\equiv N \xrightarrow[\text{(2) } CH_3CH_2Br]{\text{(1) LDA/THF/0\,°C}} CH_3CH_2\underset{\underset{CH_2CH_3}{|}}{C}HC\equiv N$$

(81% yield)

Elimination reactions always compete with nucleophilic substitution, particularly in cases such as these where the anion is a strong base as well as a nucleophile. Consequently, alkylation of enolate ions or the anion derived from nitriles is successful only when primary haloalkanes or their derivatives are used. Secondary or tertiary haloalkanes and their derivatives form mostly products of elimination.

Exercise 17.18 Propose a synthesis of each of the following compounds by a reaction of an appropriate haloalkane and the starting material given.

(a) 2-Ethylcyclohexanone from cyclohexanone
(b) Ethyl pentanoate from ethyl acetate
(c) 3-Phenylpropanenitrile from acetonitrile

Exercise 17.19 Write the mechanism for the following reaction sequence.

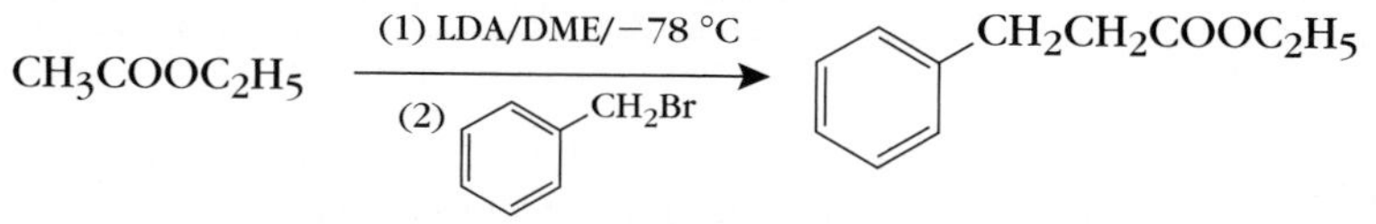

$$CH_3COOC_2H_5 \xrightarrow[\text{(2)}]{\text{(1) LDA/DME}/-78\ °C} \quad \text{[phenyl]}\ CH_2CH_2COOC_2H_5$$

(2) [phenyl]CH$_2$Br

Enolate ions also add to the carbonyl carbon atom of aldehydes, ketones, acyl chlorides, and esters.

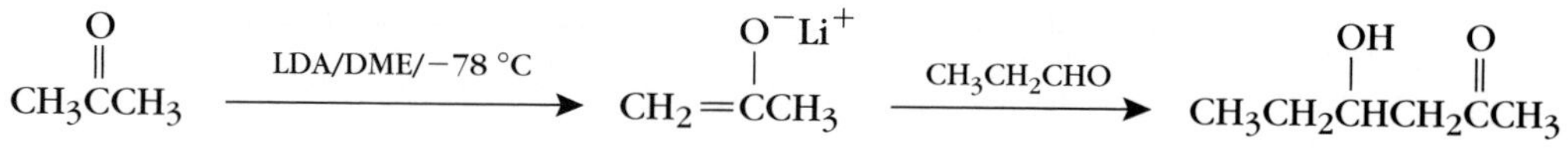

$$\underset{CH_3CCH_3}{\overset{O}{\|}} \xrightarrow{\text{LDA/DME}/-78\ °C} \underset{CH_2=CCH_3}{\overset{O^-Li^+}{|}} \xrightarrow{CH_3CH_2CHO} \underset{CH_3CH_2CHCH_2CCH_3}{\overset{OH \quad O}{|\quad\quad\|}}$$

(86% yield)

$$\underset{CH_3C}{\overset{O}{\diagdown}}_{OCH_2CH_3} \xrightarrow{\text{LDA/THF}/-78\ °C} \underset{CH_2=C}{\overset{O^-Li^+}{\diagup}}_{OCH_2CH_3} \xrightarrow{(CH_3)_3CC\overset{O}{\diagup}_{Cl}} (CH_3)_3CCCH_2C\overset{O}{\diagup}_{OCH_2CH_3}$$

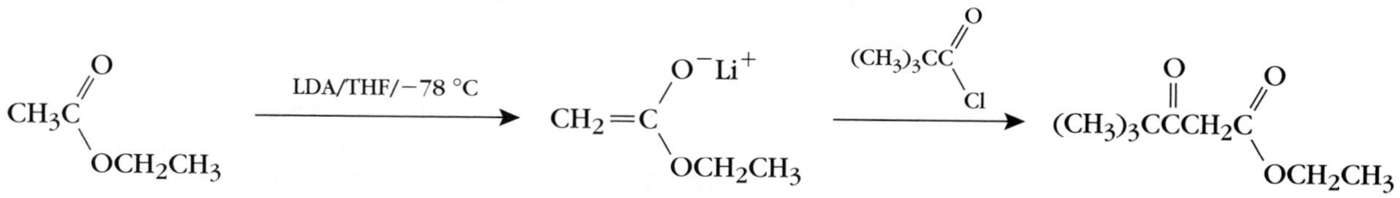

(70% yield)

These reactions are carried out by first forming the enolate ion, then slowly adding the carbonyl-containing compound to the reaction mixture. The product is isolated after an aqueous acid addition.

To summarize, strong bases such as lithium diisopropylamide completely convert ketones, esters, carboxylic acids, *N,N*-disubstituted amides, and nitriles to resonance-stabilized ions. Slowly adding a primary haloalkane or a carbonyl-containing compound to the preformed ion results in an α-substitution reaction in which a carbon-carbon single bond is formed.

There are still a few things about enolate ions that need to be explained. We know, for example, that the charge on an enolate ion is shared between the oxygen and carbon atoms, yet the reactions of enolate ions presented so far have occurred only at the carbon atom. What factors determine whether reaction occurs at the carbon or oxygen atom?

Exercise 17.20   Write the structure of the product of each of the following reactions.

(a) Ethyl acetate $\xrightarrow[\substack{(2)\ \text{Cyclohexanone} \\ (3)\ H_3O^+}]{(1)\ \text{LDA/DME}/-78\ °C}$

(b) Methylpropanoic acid $\xrightarrow[\substack{(2)\ \text{3-Pentanone} \\ (3)\ H_3O^+}]{(1)\ 2\ \text{LDA/DME}/-78\ °C}$

(c) Acetonitrile $\xrightarrow[\substack{(2)\ \text{3-Pentanone} \\ (3)\ H_3O^+}]{(1)\ \text{LDA/THF}/-70\ °C}$

(d) *N,N*-Dimethylacetamide $\xrightarrow[\substack{(2)\ \text{Cyclohexanone} \\ (3)\ H_3O^+}]{(1)\ \text{LDA/DME}/-78\ °C}$

Exercise 17.21   Write the mechanism for the following reaction sequence.

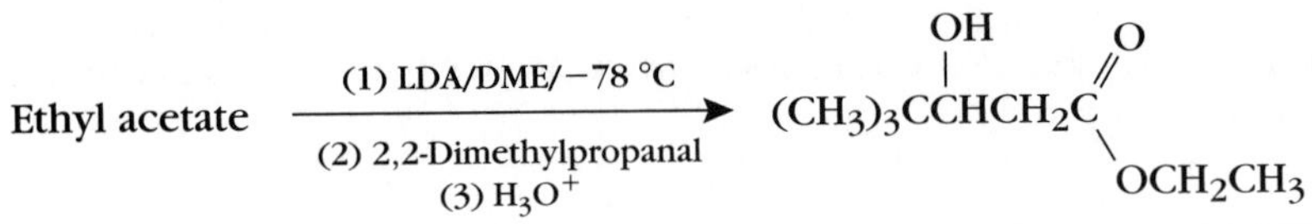

Ethyl acetate $\xrightarrow[\substack{(2)\ \text{2,2-Dimethylpropanal} \\ (3)\ H_3O^+}]{(1)\ \text{LDA/DME}/-78\ °C}$ $\underset{OCH_2CH_3}{\overset{\displaystyle OH \quad\quad O}{(CH_3)_3CCHCH_2C}}$

17.12 Why Do Enolate Ions React Principally at Carbon?

We know from the structure of an enolate ion that it has electron density at both its carbon and oxygen atoms. Consequently, it should react at both of these sites to form an enol ether by alkylation at the oxygen atom and an α-alkylated ketone by alkylation at the carbon atom.

O-Alkylation

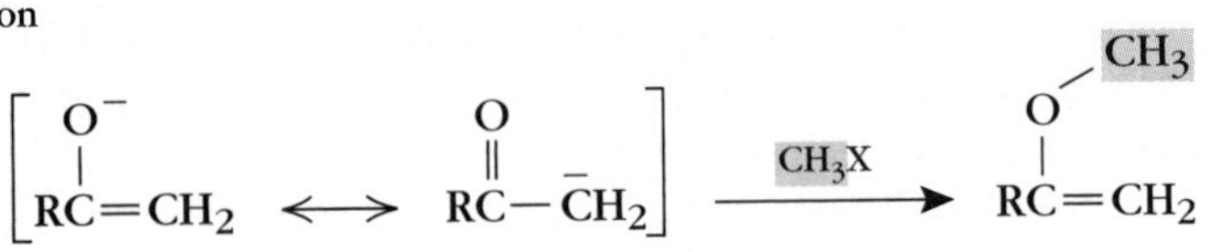

$$\left[\overset{\displaystyle O^-}{\underset{}{RC}}=CH_2 \quad\longleftrightarrow\quad \overset{\displaystyle O}{\underset{}{RC}}-\overset{-}{C}H_2 \right] \xrightarrow{CH_3X} \overset{\displaystyle O}{\underset{}{RC}}=CH_2 \overset{\displaystyle \nearrow CH_3}{}$$

Enol ether

C-Alkylation

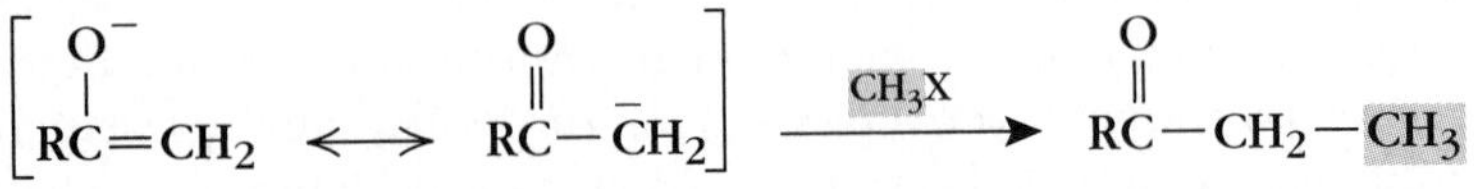

$$\left[\overset{\displaystyle O^-}{\underset{}{RC}}=CH_2 \quad\longleftrightarrow\quad \overset{\displaystyle O}{\underset{}{RC}}-\overset{-}{C}H_2 \right] \xrightarrow{CH_3X} \overset{\displaystyle O}{\underset{}{RC}}-CH_2-CH_3$$

Ketone

An anion that reacts at either of two positions is called an **ambident anion** or **ambident nucleophile.**

Reactions of enolate ions formed from simple ketones and esters occur almost exclusively at the carbon atom because the oxygen-alkylated products are less stable than the carbon-alkylated products. An enol ether, like the enol form of a ketone, is less stable than its isomeric ketone. If this stability is reflected in the transition states for formation of the two products, then the free energy of activation for formation of the carbon-alkylated product will be lower than that for formation of the enol ether. This favors carbon alkylation.

A second factor that favors reaction at the carbon atom is the nature of the cation associated with the enolate ion. Certain metal cations, such as lithium, calcium, or magnesium, are known to be tightly coordinated with the oxygen atom of the enolate ion. Tight coordination interferes with reaction at the oxygen atom, thus making the carbon atom the preferred site of alkylation.

The nature of the electrophile is also important. As we have noted, enolates usually undergo nucleophilic substitution reactions at the carbon atom. However, some electrophiles, such as silyl chlorides, react predominantly at the oxygen atom.

The greater strength of the Si—O bond, 108 kcal/mol (451 kJ/mol), compared to 72 kcal/mol (301 kJ/mol) for the Si—C bond is believed to be the dominant factor in this reaction. It makes the compound with the Si—O bond the more stable product. Again, the stability of this product is reflected in the transition state, which favors reaction at oxygen.

To summarize, enolate ions can react at the carbon or oxygen atom, but reactions at the carbon atom are more useful synthetically. Consequently, organic chemists have developed experimental techniques that make reaction at the carbon atom predominate. For the chemistry in this text, therefore, enolate ions react under conditions that form predominantly or exclusively products of reaction at the carbon atom.

So far we have discussed the formation of enolate ions from compounds that have only one kind of α-hydrogen atom. What happens when a compound with nonequivalent α-hydrogens is treated with a strong base?

17.13 Isomeric Enolate Ions

Alkylation of an unsymmetrical ketone under the usual conditions forms two isomeric products, as shown in the following example.

If we could control the direction of alkylation so that only a single product is formed, then alkylation of unsymmetrical ketones would be a valuable synthetic reaction. In order to do this, we must first understand what determines the relative concentrations of the two enolate ions formed by the reaction of a ketone with base.

The reaction of an unsymmetrical ketone with base forms two isomeric enolate ions. The relative concentrations of the two enolate ions depend on the reaction conditions under which they are formed. For example, the reaction of 2-methylcyclohexanone with a relatively weak base, such as triethylamine in dimethylformamide (DMF), forms an equilibrium mixture of its two isomeric enolate ions.

(78% yield)
More stable enolate
ion has more substituted
double bond

(22% yield)
Less stable enolate
ion has less substituted
double bond

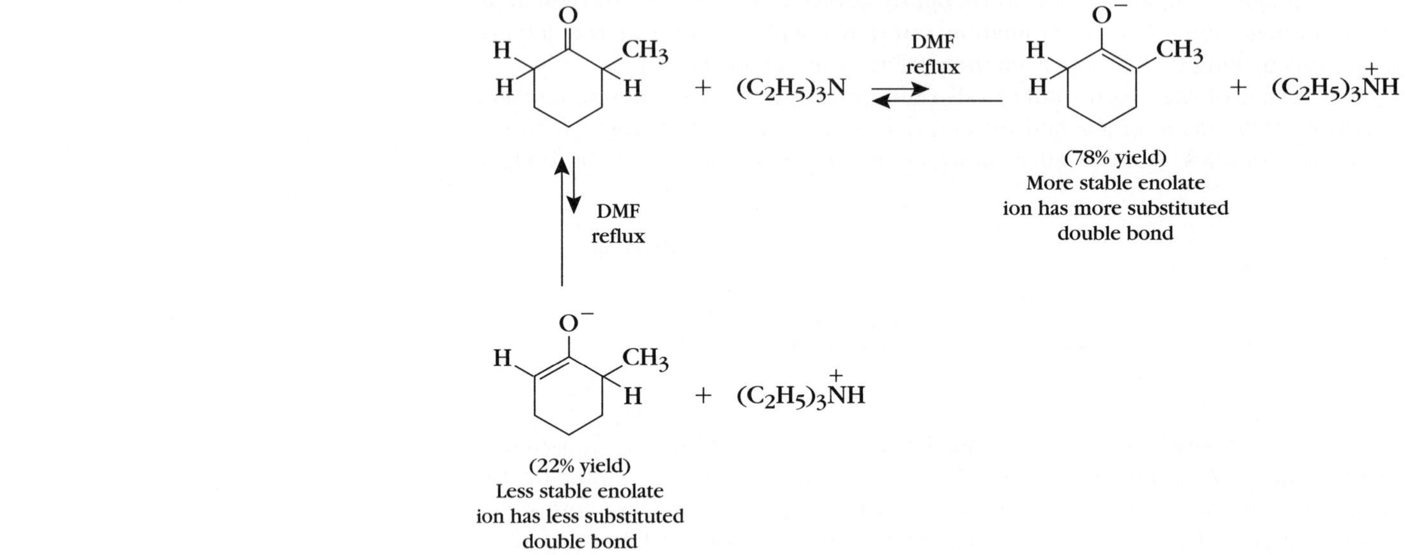

Triethylamine is not a strong enough base to convert the ketone completely to the enolate ions, so equilibria are established. The equilibrium concentrations of products represent their relative thermodynamic stabilities. Products formed under equilibrium conditions are said to be under **thermodynamic control.** Because the enolate ion with the most substituents predominates at equilibrium, it is the more stable enolate ion. Thus the more stable enolate ion is formed under thermodynamic control.

If the ketone is added slowly to a solution of a strong base, such as LDA in 1,2-dimethoxyethane at $-78\ °C$, a different mixture of enolate ions is produced.

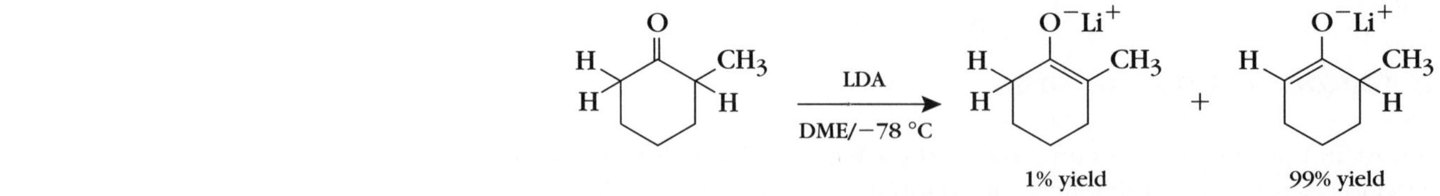

1% yield   99% yield

LDA converts the ketone completely to its enolate ions. Under these conditions, the predominant enolate is the one that is formed the fastest. This product is not necessarily the most stable one. In fact, the least stable enolate ion is formed the fastest in the reaction of 2-methylcyclohexanone with LDA. The product that forms the fastest in a chemical reaction is called the product of **kinetic control.** The enolate ion that is usually formed the fastest is the isomer with the least number of substituents.

The ability to prepare predominantly one or the other enolate ion by the proper choice of experimental conditions makes it possible to selectively alkylate an unsymmetrical ketone. Thus conditions of thermodynamic control (a relatively weak base, high temperatures, and excess ketone) form predominantly the more stable enolate ion (the one with the most substituents). Under these conditions alkylation of the enolate occurs at the most-substituted α-carbon atom.

Kinetic control (a strong base, low temperature, and excess base), on the other hand, forms predominantly the isomeric enolate ion with the least number of substituents. Under kinetic control, alkylation occurs predominantly at the least-substituted α-carbon atom. The following examples illustrate the differences between kinetic and thermodynamic control of the alkylation of enolate ions.

Kinetic control

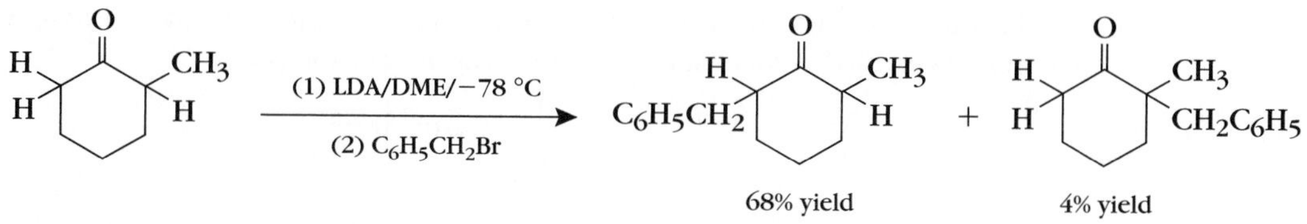

Thermodynamic control

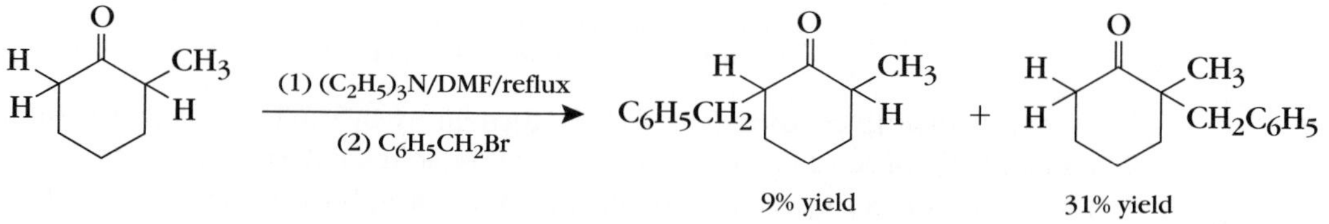

This technique is not perfect. Mixtures are still obtained, but clearly a different isomer predominates in each case.

Exercise 17.22   Propose a method of preparing each of the following compounds, starting with 2-methylcyclohexanone.

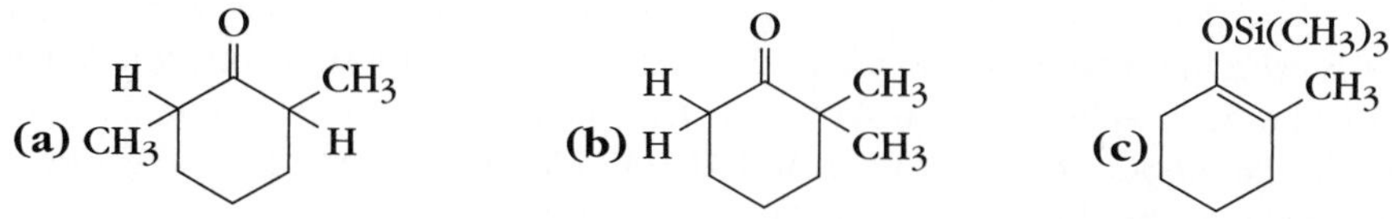

The alkylation of an intermediate enamine is another technique developed for the selective alkylation of a ketone.

17.14 Enamines Are Enolate Equivalents

Enamines are compounds that contain an amino group bonded to a carbon-carbon double bond. Enamines are the nitrogen analogs of enols.

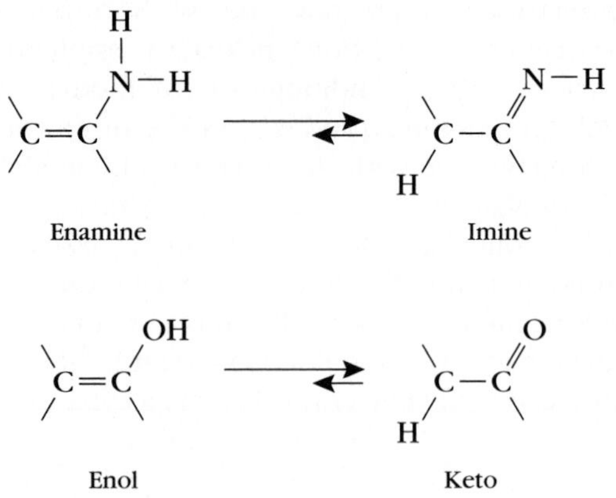

Enamine   Imine

Enol   Keto

Enamines, like enols, are generally unstable, so they quickly convert to their isomeric imines. When the nitrogen atom of the enamine is tertiary, however, the enamine is stable because such tautomerization cannot occur.

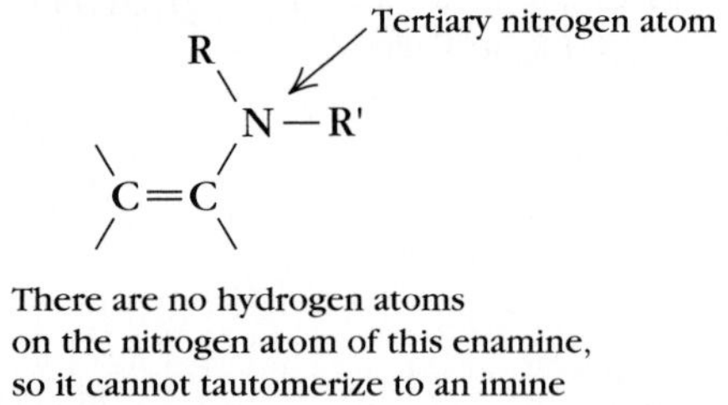

Tertiary nitrogen atom

There are no hydrogen atoms
on the nitrogen atom of this enamine,
so it cannot tautomerize to an imine

Stable enamines, therefore, can be formed by heating a secondary amine with an aldehyde or ketone in the presence of an acid catalyst. This reaction is similar to the one used to prepare imines discussed in Section 14.13. The reaction is an equilibrium that can be shifted to form the enamine by removing the water as it forms.

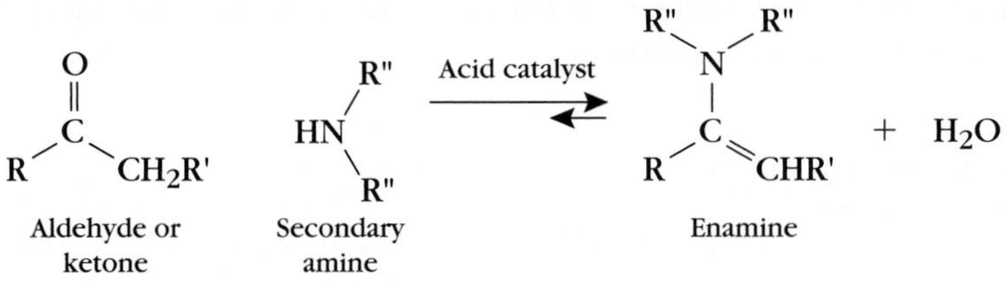

Aldehyde or   Secondary   Enamine
ketone   amine

The principal reason for forming enamines is to use them as intermediates in the alkylation of aldehydes and ketones. The β-carbon atom in an enamine bears a substantial amount of negative charge due to electron release from the nitrogen atom, as shown by the following resonance structures.

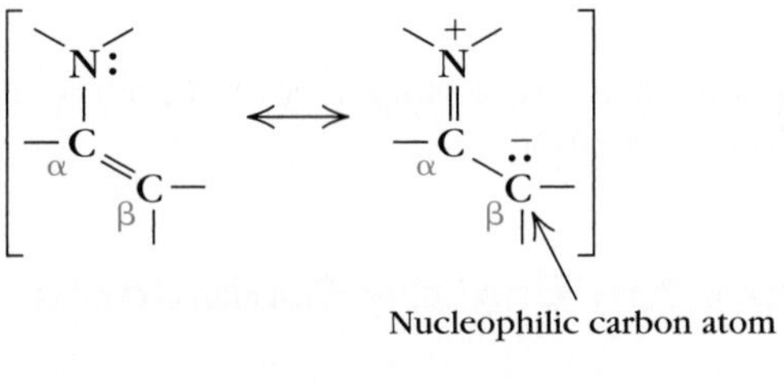

Nucleophilic carbon atom

Electron donation from the nitrogen atom is analogous to the electron release by the oxygen atom of an enolate ion. Thus the electron structure of an enamine resembles that of an enolate ion. Enamines and enolate ions, therefore, undergo similar reactions. For example, both undergo alkylation reactions with primary haloalkanes.

Iminium ion

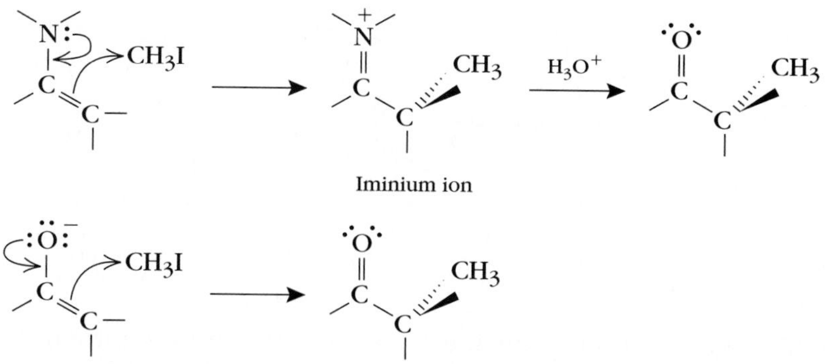

Notice that hydrolysis of the iminium ion formed by alkylation of the enamine forms the same product as the alkylation of an enolate ion. This means that the same product can be prepared by either alkylation and hydrolysis of an enamine or alkylation of the structurally similar enolate ion. As a result, enamines are frequently referred to as enolate ion equivalents. The one difference between enamines and enolate ions, however, is that enamines are alkylated under milder conditions.

Enamines prepared by the reaction of pyrrolidine with aldehydes and ketones are widely used in synthesis. These enamines react rapidly with haloalkanes in a nucleophilic substitution reaction that occurs by an S_N2 mechanism. As a result, the best yields of the iminium salt are obtained with primary haloalkanes.

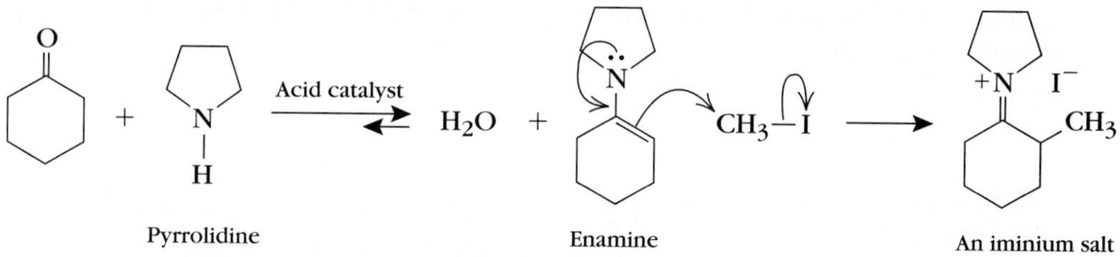

Pyrrolidine   Enamine   An iminium salt

Hydrolysis of the iminium salt forms the alkylated ketone and pyrrolidine.

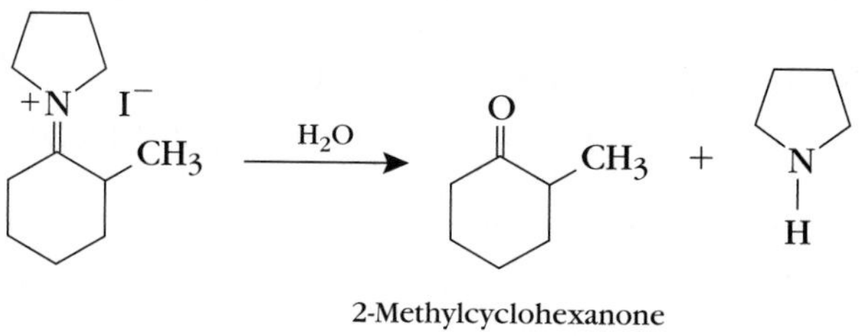

2-Methylcyclohexanone

Exercise 17.23   Starting with cyclohexanone and pyrrolidine, propose a synthesis for each of the following compounds.

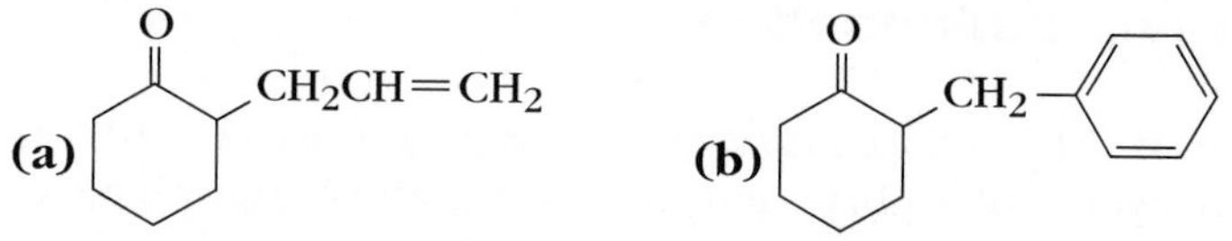

(a)   (b)

The enamine method can be used to selectively alkylate unsymmetrical ketones, particularly unsymmetrically substituted cyclohexanones. In general, the predominant enamine formed is the one with the least-substituted double bond.

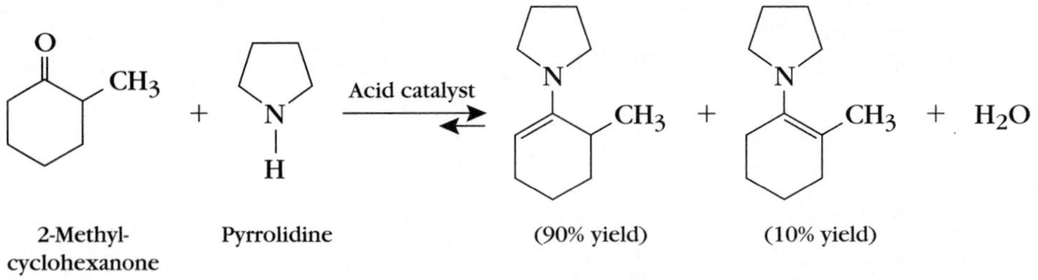

2-Methyl-   Pyrrolidine   (90% yield)   (10% yield)
cyclohexanone

As a result, alkylation occurs primarily at the less-substituted α-carbon atom. We can take advantage of this selectivity to prepare 2,6-disubstituted cyclohexanones.

(66% yield)

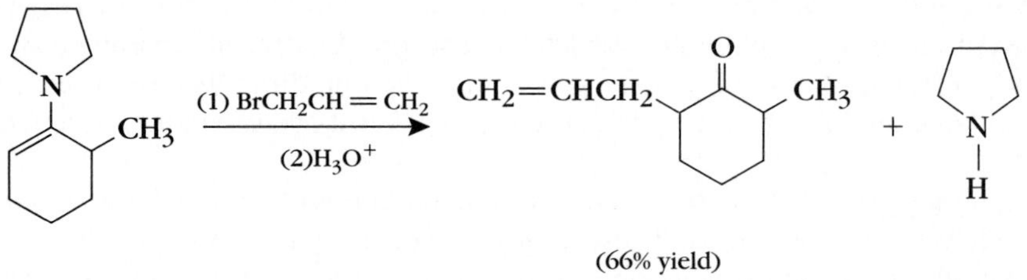

This sequence of reactions, namely enamine formation followed by alkylation and finally hydrolysis, complements the preparation of alkylated ketones by alkylating their enolate ions.

Exercise 17.24   Prepare each of the following compounds by the enamine method from pyrrolidine, the appropriate haloalkane, and any ketone or aldehyde.

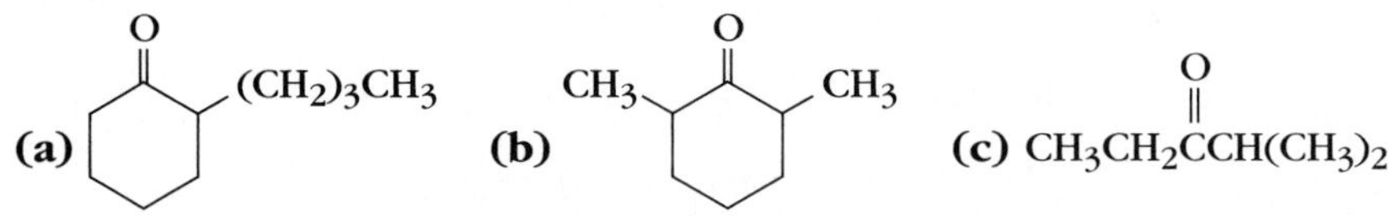

(a)   **(b)**   (c) $CH_3CH_2CCH(CH_3)_2$

Enamines are also intermediates in certain enzyme-catalyzed reactions.

17.15 Enamines Are Intermediates in Aldolase-Catalyzed Condensations

In Section 17.10, we cited the aldolase-catalyzed cleavage of fructose 1,6-bisphosphate into glyceraldehyde 3-phosphate and dihydroxyacetone phosphate as an example of a reversible aldol condensation in living systems.

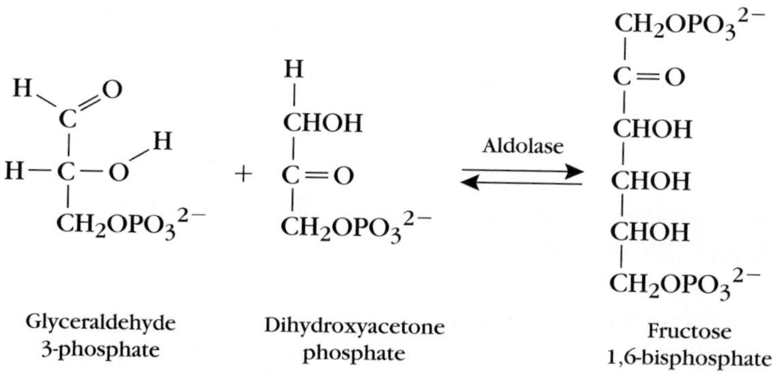

Glyceraldehyde 3-phosphate + Dihydroxyacetone phosphate → (Aldolase) → Fructose 1,6-bisphosphate

Unlike reversible aldol condensations in basic solution, the mechanism of this reaction does not involve an enolate ion intermediate. Instead, an enamine (an enolate ion equivalent) is formed, which is more stable and easier to form than an enolate ion. The mechanism of the aldolase-catalyzed reaction is shown in Figure 17.11.

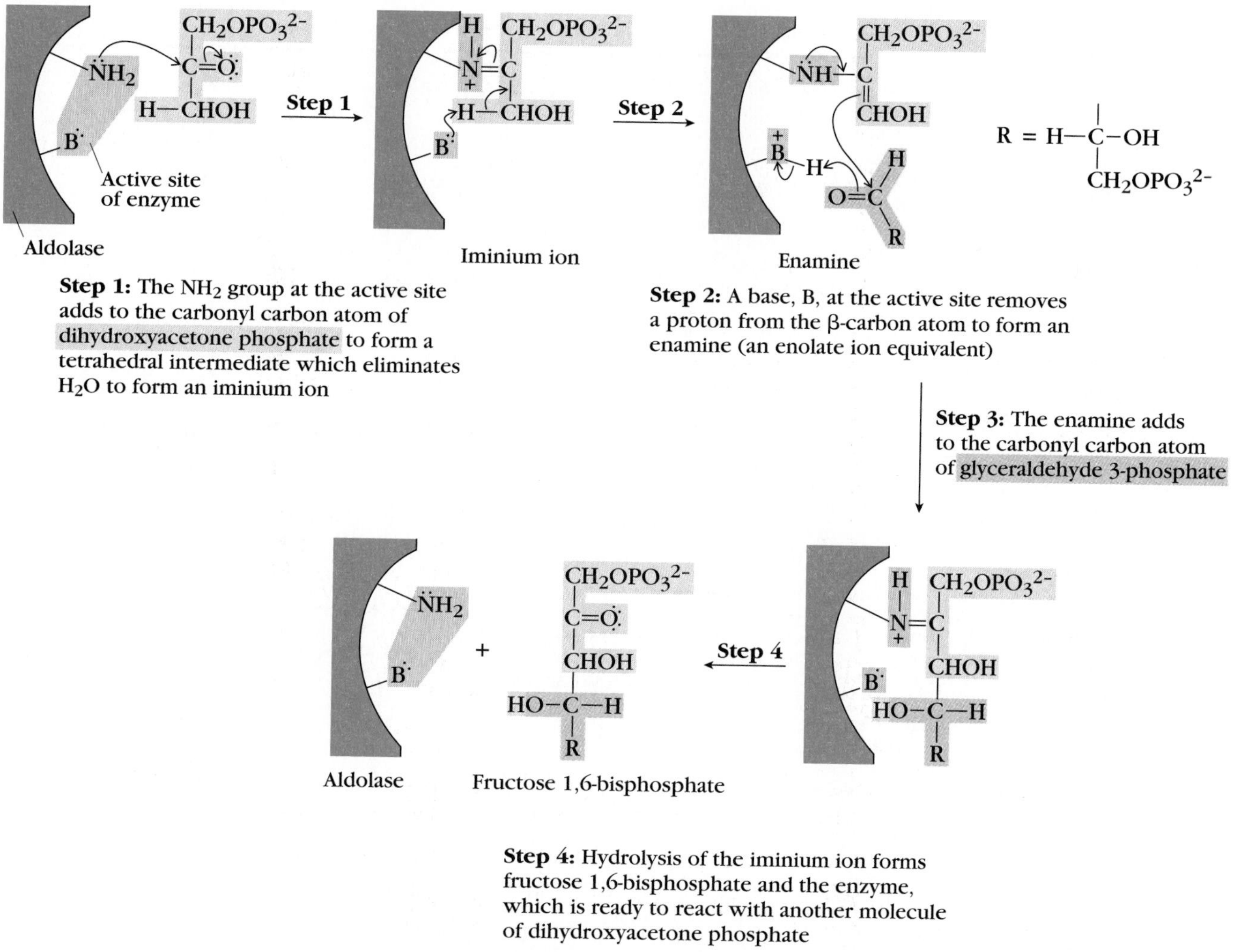

Figure 17.11
Mechanism of the aldolase-catalyzed condensation of dihydroxyacetone phosphate and glyceraldehyde 3-phosphate to form fructose 1,6-bisphosphate. Although other acidic and basic groups on the enzyme also participate in each step of the mechanism, they have been omitted for simplicity.

Step 1: The NH_2 group at the active site adds to the carbonyl carbon atom of dihydroxyacetone phosphate to form a tetrahedral intermediate which eliminates H_2O to form an iminium ion

Step 2: A base, B, at the active site removes a proton from the β-carbon atom to form an enamine (an enolate ion equivalent)

Step 3: The enamine adds to the carbonyl carbon atom of glyceraldehyde 3-phosphate

$$R = H-\overset{|}{\underset{|}{C}}-OH$$
$$CH_2OPO_3^{2-}$$

Step 4: Hydrolysis of the iminium ion forms fructose 1,6-bisphosphate and the enzyme, which is ready to react with another molecule of dihydroxyacetone phosphate

The mechanism outlined in Figure 17.11 demonstrates that carbon-carbon bonds are formed in nature by means of condensation reactions just as in nonliving systems. Because the formation of carbon-carbon single bonds is universally important, the methods of forming them are summarized in Section 17.16.

17.16 A Summary of Reactions That Form Carbon-Carbon Bonds

The two most common reactions that form carbon-carbon bonds are nucleophilic substitution reactions and carbonyl addition reactions. The details of nucleophilic substitution reactions that form carbon-carbon bonds are summarized in Table 17.3, while the details of carbonyl addition reactions are summarized in Table 17.4.

Table 17.3	Nucleophilic Substitution Reactions That Form Carbon-Carbon Bonds			
Nucleophile	**Structure**	**Type of Haloalkane**	**Structure of Product**	**Reference**
Cyanide ion	CN^-	1° (RCH$_2$X)* or 2° (R$_2$CHX)	RCH$_2$—C≡N or R$_2$CH—C≡N	12.18 12.18
Acetylide ion	R'C≡C$^-$	1° (RCH$_2$X)	R'C≡C—CH$_2$R	9.11
Enolate ion†	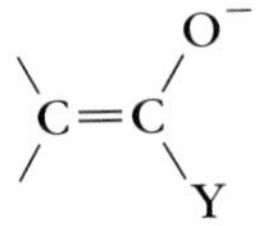 (C=C with O⁻ and Y)	1° (RCH$_2$X)	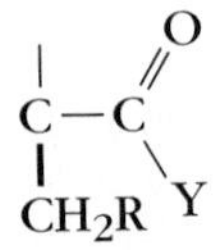 (C—C with O, CH$_2$R, Y)	17.11
Enamine	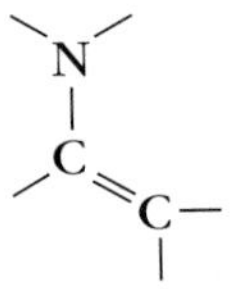 (C=C with N)	1° (RCH$_2$X)	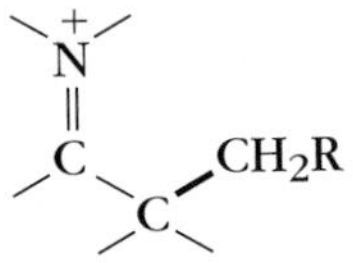 (C with N⁺, CH$_2$R)	17.14
Conjugate base of alkylnitrile	R'CHC≡N	1° (RCH$_2$X)	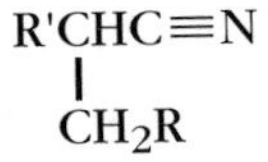 R'CHC≡N CH$_2$R	17.11

*X = halide or other good leaving group.

† The enolate ion can be formed from any one of the following: a ketone, a carboxylic acid ester, a carboxylic acid, or an N,N-dialkylamide.

| Table 17.4 | | Addition Reactions to Carbonyl Groups That Form Carbon-Carbon Bonds | | |

Nucleophile or Donor	Structure	Carbonyl Compound or Acceptor	Structure of Product	Reference
Cyanide ion	CN^-	Aldehyde RCHO	$\overset{\displaystyle OH}{\underset{\displaystyle \vert}{RCH}}$—C≡N	14.9
Cyanide ion	CN^-	Ketone $R_2C=O$	$\underset{\displaystyle \underset{\vert}{R}}{\overset{\displaystyle \overset{OH}{\vert}}{RC}}$—C≡N	14.9
Grignard or organolithium reagent	R′MgX or R′Li	Aldehyde RCHO	2° alcohol $\overset{\displaystyle R'}{\underset{\displaystyle \vert}{RCHOH}}$	11.7
Grignard or organolithium reagent	R′MgX or R′Li	Ketone $R_2C=O$	3° alcohol $\overset{\displaystyle R'}{\underset{\displaystyle \vert}{R_2COH}}$	11.7
Grignard or organolithium reagent	R′MgX or R′Li	Ester RCOOR	3° alcohol $\overset{\displaystyle R'\ \ R'}{\underset{\displaystyle \diagdown\diagup}{RCOH}}$	16.13
Grignard or organolithium reagent	R′MgX or R′Li	Formaldehyde $H_2C=O$	1° alcohol R'—CH_2OH	11.7
Grignard or organolithium reagent	R′MgX or R′Li	Acyl chloride RCOCl	3° alcohol $\overset{\displaystyle R'\ \ R'}{\underset{\displaystyle \diagdown\diagup}{RCOH}}$	16.13
Grignard or organolithium reagent	R′MgX or R′Li	Nitrile RC≡N	Ketone $\underset{\displaystyle R'\diagup \ \diagdown R}{\overset{\displaystyle \overset{O}{\overset{\|}{C}}}{}}$	16.15, E
Acetylide ion	$R'C≡C^-$	Aldehyde RCHO	$R'C≡C$—$\overset{\displaystyle OH}{\underset{\displaystyle \vert}{CHR}}$	16.13
Acetylide ion	$R'C≡C^-$	Ketone $R_2C=O$	$R'C≡C$—$\overset{\displaystyle OH}{\underset{\displaystyle \vert}{CR_2}}$	16.13
Organocopper reagent	R'_2CuLi	Acyl chloride RCOCl	Ketone $\underset{\displaystyle R'\diagup \ \diagdown R}{\overset{\displaystyle \overset{O}{\overset{\|}{C}}}{}}$	16.14, B

Continued.

Table 17.4—cont'd Addition Reactions to Carbonyl Groups That Form Carbon-Carbon Bonds

Nucleophile or Donor	Structure	Carbonyl Compound or Acceptor	Structure of Product	Reference
Enolate ion*	$\diagdown$C=C$\diagup$ with O^- and Y	Aldehyde RCHO	RCH(OH)—C—C($=O$)Y	17.11
Enolate ion*	$\diagdown$C=C$\diagup$ with O^- and Y	Ketone $R_2C{=}O$	R_2C(OH)—C—C($=O$)Y	17.11
Enolate ion*	$\diagdown$C=C$\diagup$ with O^- and Y	Ester RCOOR	RC($=O$)—C—C($=O$)Y	17.11
Enolate ion*	$\diagdown$C=C$\diagup$ with O^- and Y	Acyl chloride RCOCl	RC($=O$)—C—C($=O$)Y	17.11
Wittig reagent	$(C_6H_5)_3P{=}CHR'$	Aldehyde RCHO	Alkene $RCH{=}CHR'$	14.16
Wittig reagent	$(C_6H_5)_3P{=}CHR'$	Ketone $R_2C{=}O$	Alkene $R_2C{=}CHR'$	14.16

*The enolate ion can be formed from any of the following: a ketone, a carboxylic acid ester, a carboxylic acid, or an N,N-dialkylamide.

17.17 Summary

Aldehydes, ketones, and carboxylic acid derivatives exist as equilibrium mixtures of **keto** and **enol tautomers.** For most simple carbonyl-containing compounds, the enol content is very small. The interconversion of keto and enol tautomers, which can be catalyzed by acids, bases, or enzymes, is called **tautomerism.**

Although the concentration of the enol tautomer is low, its presence is still very important because the enol tautomer is more reactive than the keto. For example, the enol form of aldehydes or ketones is rapidly halogenated to form α-haloaldehydes or α-haloketones. In order to brominate carboxylic acids in the α-position, a small amount of PBr_3 must be added. The reaction of carboxylic acids with Br_2/PBr_3 is called the **Hell-Volhard-Zelinsky reaction.** α-Halogenated carbonyl-containing compounds undergo the typical nucleophilic substitution and elimination reactions of haloalkanes.

In basic solution, enolate ions are formed from keto and enol tautomers. Enolate ions are nucleophiles in which their negative charge is shared (or delocalized) between the carbon and oxygen atoms. Enolate ions react with electrophilic reagents predominantly at the carbon atom. The concentration of enolate ion in solution depends on the strength of the base used. With relatively weak bases, both the keto and enolate ion exist in solution. With strong bases such as **lithium diisopropylamide (LDA),** the keto form is converted completely to its enolate ion. The reactions of enolate ions depend on whether or not any carbonyl-containing compound is present in solution.

If the keto and enolate ion are both present in solution, a **carbonyl condensation reaction** occurs. In condensation reactions, the nucleophilic enolate ion (the donor) adds to the electrophilic carbonyl carbon atom (the acceptor) of the keto form to make a new carbon-carbon single bond. Condensation reactions are a particularly useful way of forming carbon-carbon bonds in nature as well as the laboratory.

A reaction in which one molecule of a carbonyl compound is the donor and another molecule of the same carbonyl compound is the acceptor is called a **self-condensation.** The **aldol condensation** and the **Claisen condensation** are examples of self-condensation reactions. **Mixed condensations** are condensations between two different carbonyl compounds. Mixed condensation reactions are useful synthetically only if one of the compounds has no α-hydrogens and has a particularly reactive carbonyl group.

An enolate ion formed by the reaction of LDA reacts as a nucleophile in nucleophilic substitution reactions with haloalkanes to form α-alkylated carbonyl compounds. **Enamines** prepared from aldehydes and ketones also react with haloalkanes to form **iminium ions.** Hydrolysis of the iminium ion forms α-alkylated aldehydes or ketones. The enamine synthesis complements the method that uses LDA as a base. Enolate ions formed by the reaction of LDA also add to a variety of carbonyl compounds to form β-hydroxy and β-keto carbonyl compounds.

17.18 Summary of the Formation and Reactions of Enols and Enolates

1. Formation of Enols   SECTION 17.1

Enolization catalyzed by acids and bases.

Keto
tautomer

Enol
tautomer

2. Halogenation of Enols   SECTION 17.4

a. Aldehydes and ketones

One, two, or more α-H can be replaced by a halogen atom by limiting the amount of halogen used.

$$CH_3CH_2CCH_2CH_3 \xrightarrow{Br_2} CH_3CHCCH_2CH_3$$

b. Carboxylic acids (Hell-Volhard-Zelinsky reaction)

$$CH_3CH_2CH_2COOH \xrightarrow[PBr_3]{Br_2} CH_3CH_2CHCOOH$$

3. Formation of Enolate Ions

a. Reversible formation — SECTION 17.5

Reversible formation of enolate ion occurs by reaction of aldehydes and ketones with a base whose conjugate acid has a $pK_a < 17$.

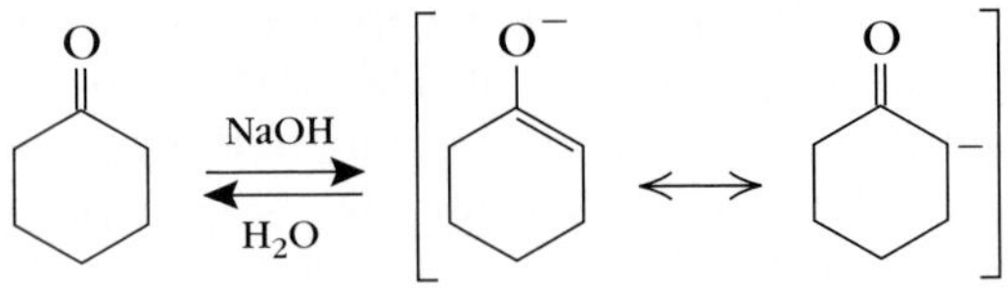
NaOH
H₂O

Enolate ion

b. Irreversible formation — SECTION 17.11

Irreversible formation of enolate ion occurs by reaction of ketones and aldehydes with a base whose conjugate acid has a $pK_a > 20$.

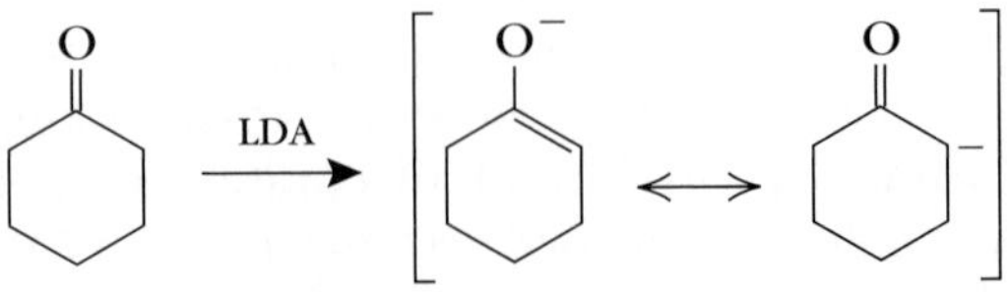
LDA

LDA = $LiN[CH(CH_3)_2]_2$

4. Reactions of Enolate Ions

a. Aldol condensation — SECTION 17.6

Reaction occurs when both enolate ion and aldehyde are present in the reaction mixture.

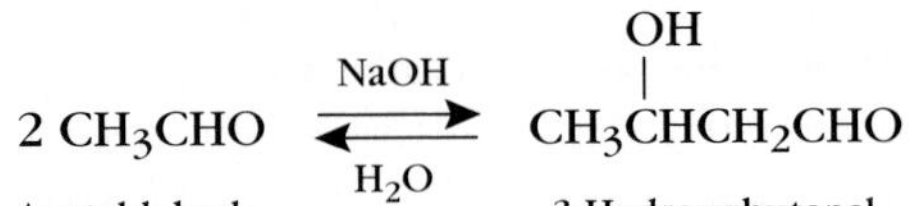
2 CH3CHO
NaOH
H2O
OH
CH3CHCH2CHO

Acetaldehyde — 3-Hydroxybutanal (aldol)

b. Claisen condensation — SECTION 17.7

Reaction occurs when both enolate ion and ester are present in the reaction mixture.

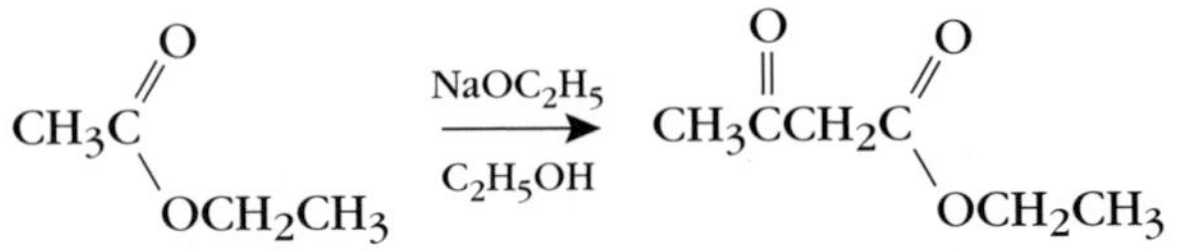
CH3C
OCH2CH3
NaOC2H5
C2H5OH
CH3CCH2C
OCH2CH3

Ethyl acetate — Ethyl acetoacetate

c. Reaction as nucleophiles — SECTION 17.11

(i) Nucleophilic substitution reaction that occurs by an S_N2 mechanism.

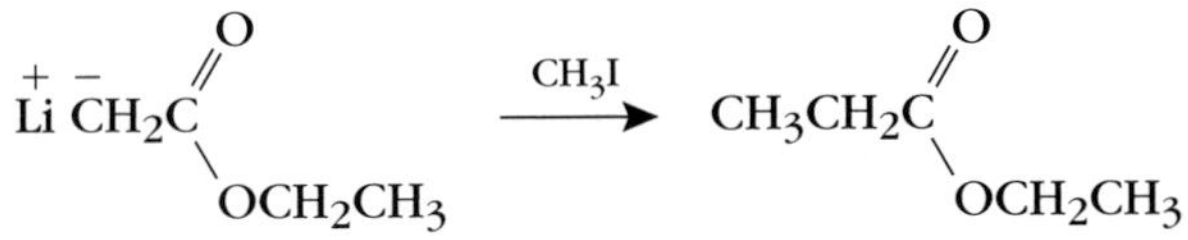
Li CH2C
OCH2CH3
CH3I
CH3CH2C
OCH2CH3

(ii) Carbonyl addition reactions

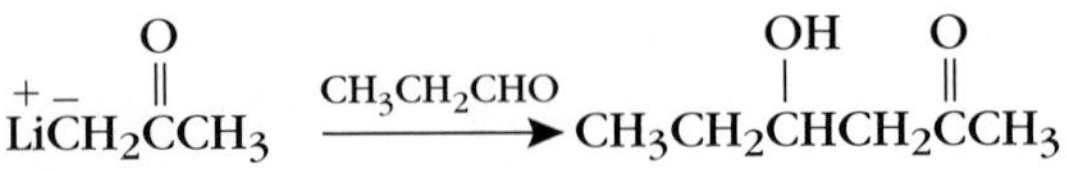
LiCH2CCH3
CH3CH2CHO
OH
CH3CH2CHCH2CCH3

5. Formation of Enamines   SECTION 17.14

Water must be removed as it forms to obtain enamine.

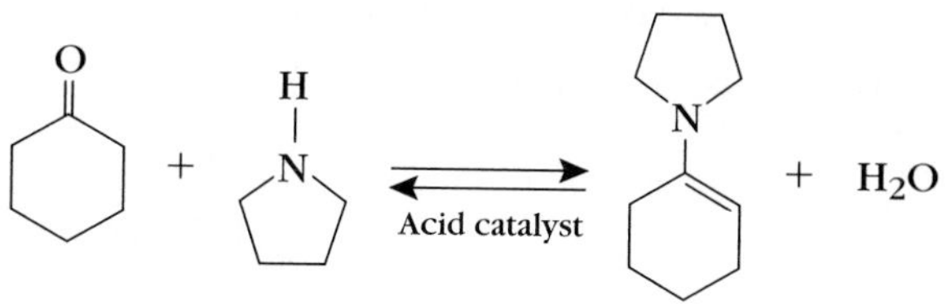

Enamine

6. Alkylation of Enamines   SECTION 17.14

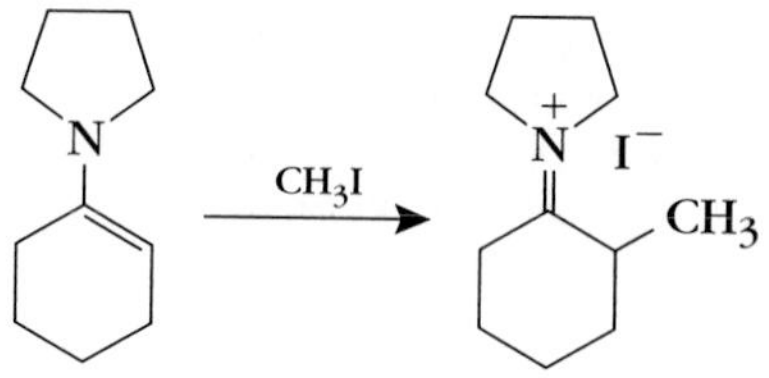

Additional Exercises

17.25 Define each of the following terms in words, by a structure, or by a chemical equation.

(a) Enol tautomer
(b) Claisen condensation
(c) Keto tautomer
(d) Tautomerism

(e) Hell-Volhard-Zelinsky reaction
(f) Aldol condensation
(g) Mixed Claisen condensation

(h) Enamine
(i) Enolization
(j) Mixed aldol condensation

17.26 Tautomerism is not limited to the keto and enol tautomers of carbonyl compounds. Each of the following compounds exists in solution as a pair of tautomers. Write the structure of the missing tautomer in each case.

(a) $(CH_3)_2CHN=O$ ⇄

(b)

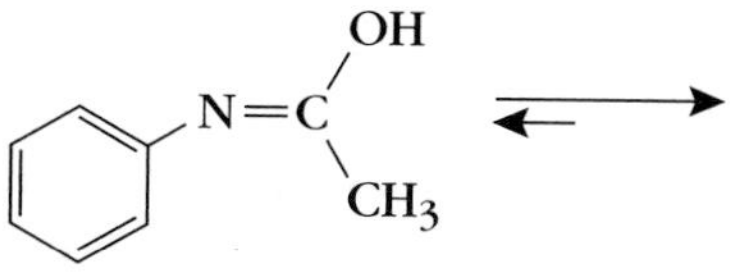

(c)

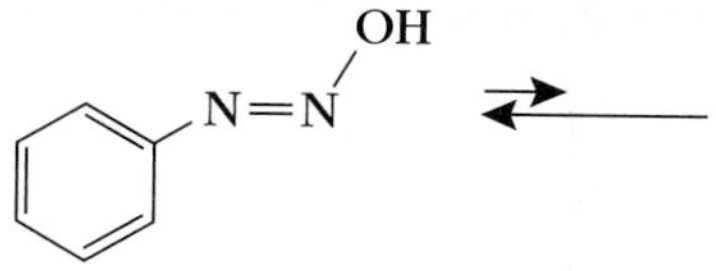

17.27 Write the structure of each of the following compounds and indicate which are acids whose $pK_a < 30$. Circle the hydrogen atom or atoms responsible for the acidity of the compound.

(a) 2-Pentanone
(b) Propanoic anhydride

(c) Butanoic acid
(d) 3-Methyl-2-butenal

(e) Methyl 2-hydroxypropanoic acid

17.28 Nitromethane reacts with bases to form a resonance-stabilized anion. Write the structure of its two contributing resonance structures.

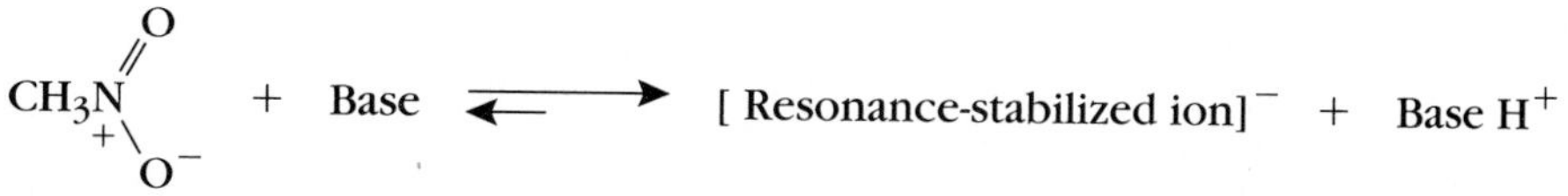

17.29 Arrange the following compounds in order of increasing acidity.

(a) 1-Butyne
(b) Ethyl butyrate
(c) Butanoic acid
(d) 2-Butanone
(e) 1-Butene

17.30 Enols of simple aldehydes and ketones have recently been prepared in aqueous solution and they are stable in the absence of acid or base for as long as several hours before they tautomerize. However, the slightest trace of acid or base rapidly converts them to their keto tautomer. Using the enol of methylpropanal as an example, write the mechanism of both acid- and base-catalyzed keto-enol tautomerism.

17.31 In acidic deuterium oxide (D_2O), acetone undergoes the following hydrogen-deuterium exchange reaction.

$$CH_3\overset{\overset{\displaystyle O}{\|}}{C}CH_3 \ + \ D_2O \ \xrightarrow{D_3O^+} \ CH_3\overset{\overset{\displaystyle O}{\|}}{C}CH_2D$$

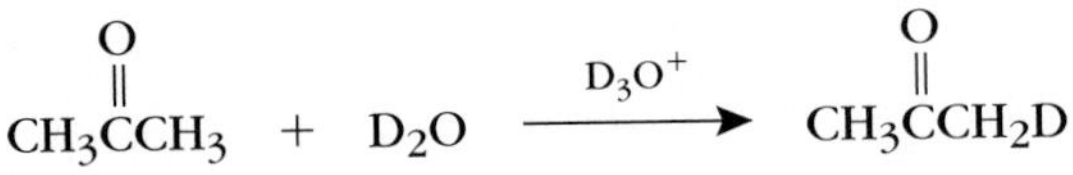

Under identical reaction conditions, the rate constant of bromination and hydrogen-deuterium exchange are the same. Propose a mechanism for the hydrogen-deuterium exchange reaction that is in accord with this observation.

17.32 Write the structure of the product of the reaction of cyclopentanone with each of the following reagents.

    (a)   1 equivalent of Br_2 in acetic acid   **(c)** $(C_6H_5)_3P{=}CHCH{=}CH_2$

    (b)   NaOD in D_2O **(d)**   LDA in THF followed by $BrCH_2CH_2CH_3$

17.33 Explain the following observations. Compound A isomerizes to compound B in basic aqueous solution. Under the same reaction conditions, however, compound C *does not* isomerize to compound D.

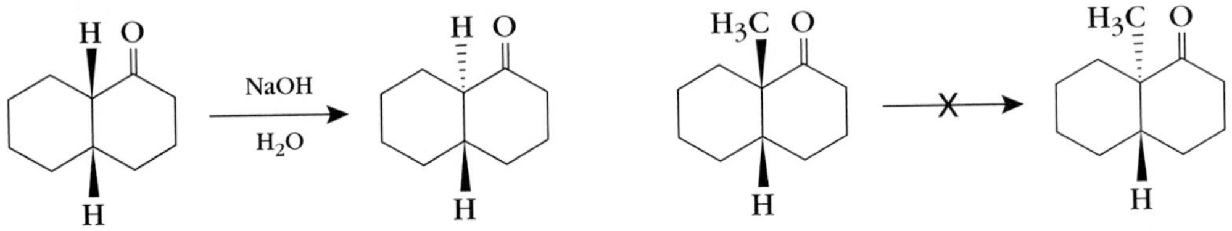

17.34 Beginning with cyclohexanone, propose a synthesis for each of the following compounds. More than one step may be required.

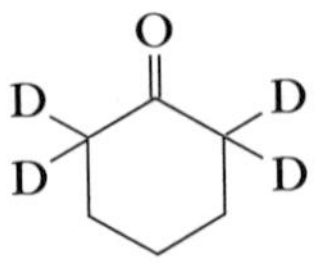

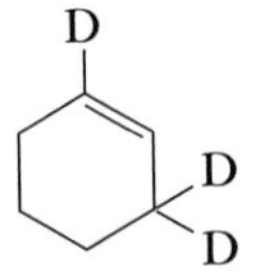

17.35 Write the structure of the product or products in each of the following reactions.

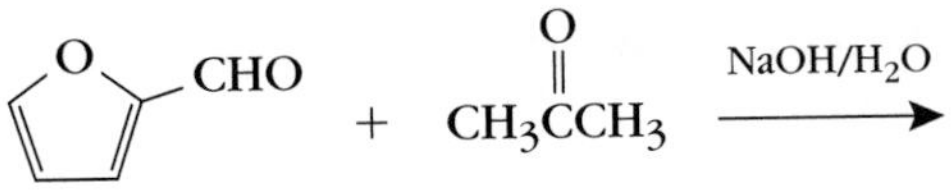

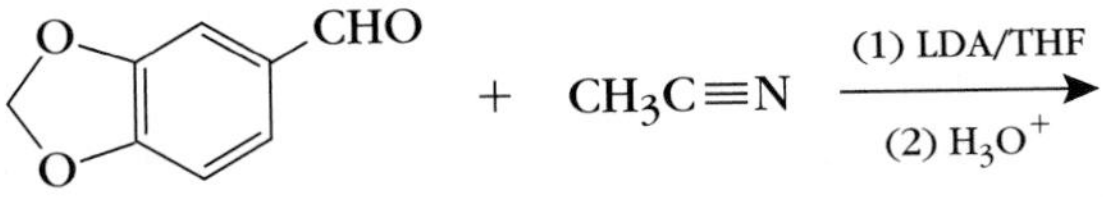

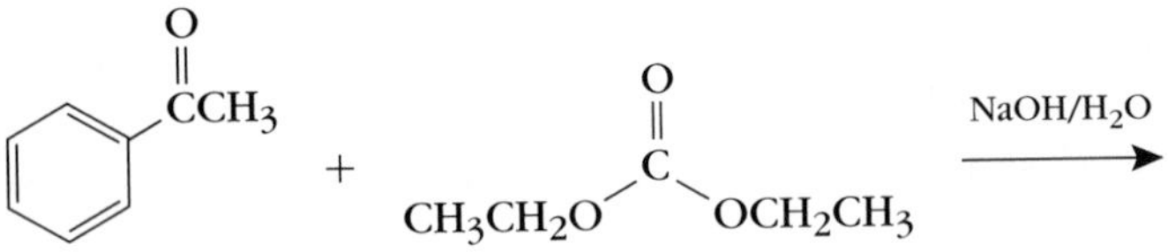

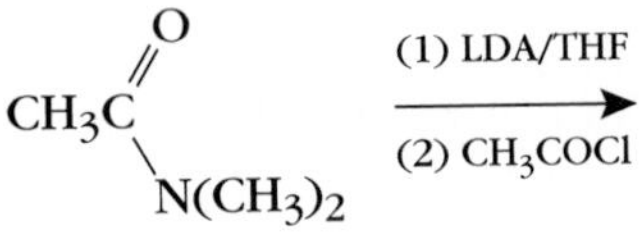

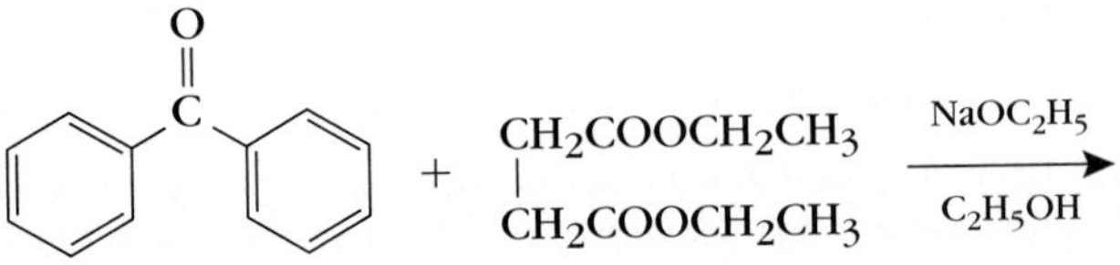

(f) 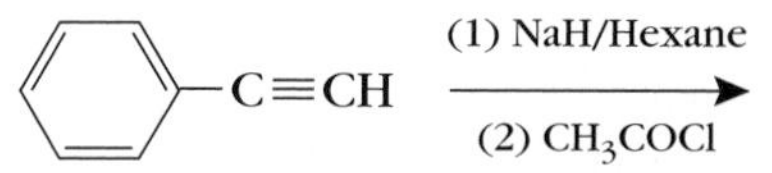 —C≡CH $\xrightarrow[\text{(2) } CH_3COCl]{\text{(1) NaH/Hexane}}$

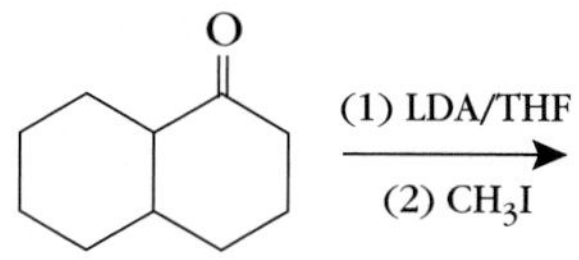

(g) 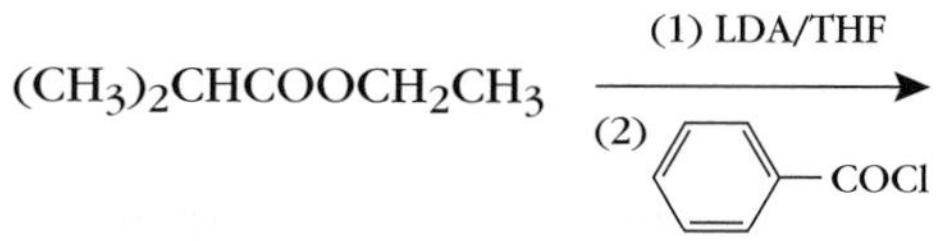 $(CH_3)_2CHCOOCH_2CH_3$ $\xrightarrow[\text{(2)}]{\text{(1) LDA/THF}}$ —COCl

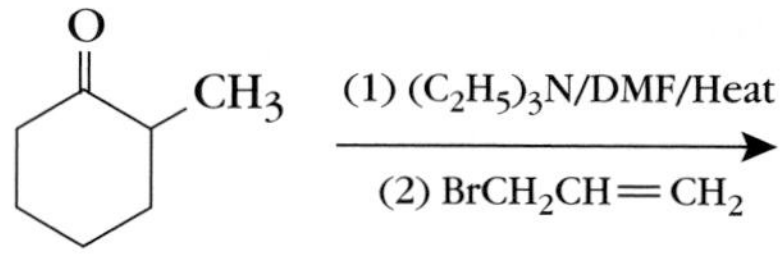

(h) $CH_3\overset{\overset{\displaystyle O}{\|}}{C}CH_3$ $\xrightarrow[\text{(2) } CH_3CH_2COOCH_2CH_3]{\text{(1) LDA/DME}}$

(i) $\xrightarrow[\text{(2) } CH_3I]{\text{(1) LDA/THF}}$

(j) $\xrightarrow[\text{(2) } BrCH_2CH=CH_2]{\text{(1) } (C_2H_5)_3N/DMF/Heat}$

17.36 Propose a mechanism for the following acid-catalyzed isomerization:

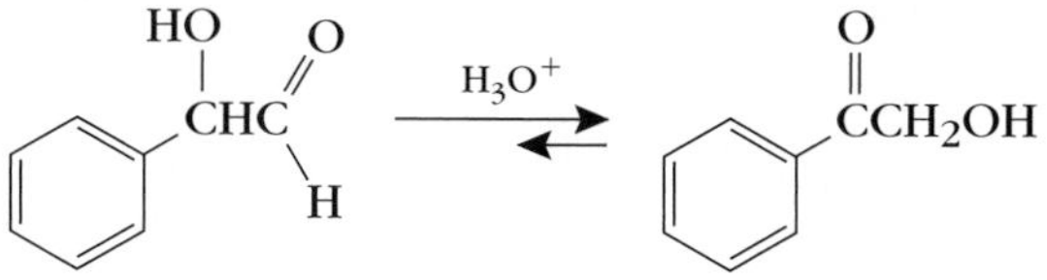

17.37 Write the structure of the missing reagents or products in each of the following reactions.

(a) 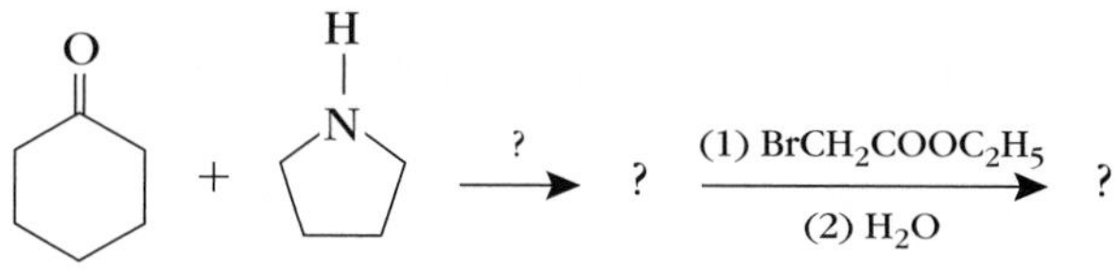 + $\xrightarrow{?}$? $\xrightarrow[\text{(2) } H_2O]{\text{(1) } BrCH_2COOC_2H_5}$?

(b) 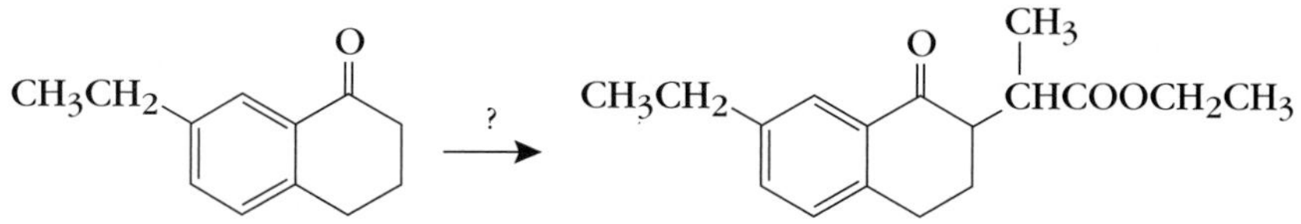 $\xrightarrow{?}$

(c) $\xrightarrow[\text{(2) } CH_3(CH_2)_6I]{\text{(1) LDA/THF/}-78\,°C}$ 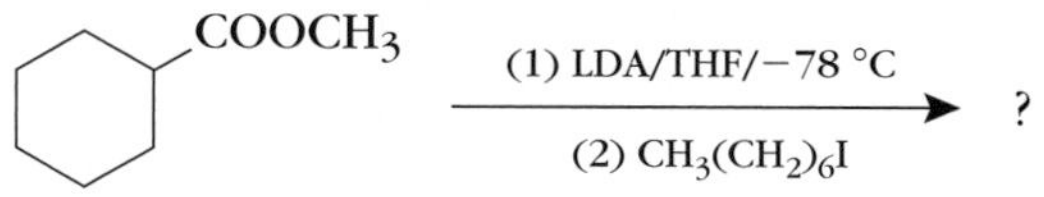 ?

(d) $(CH_3)_2CHCOOH$ $\xrightarrow{?}$ $CH_3(CH_2)_2\overset{\overset{\displaystyle CH_3}{|}}{\underset{\underset{\displaystyle CH_3}{|}}{C}}COOH$

(e) 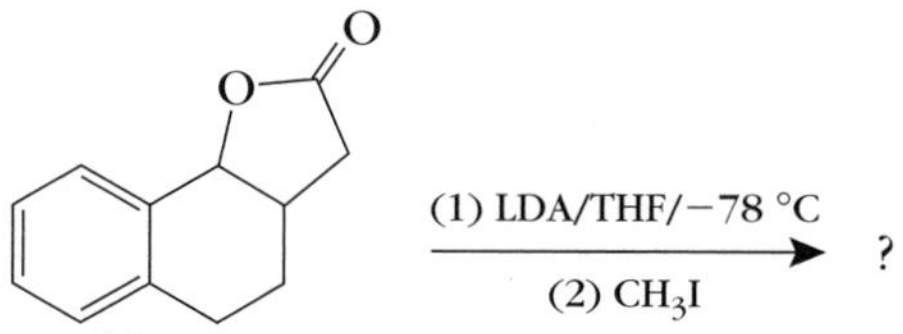 $\xrightarrow[\text{(2) } CH_3I]{\text{(1) LDA/THF/}-78\,°C}$?

17.38 Write equations for all reactions in the synthesis of each of the following compounds. Start with butanal and utilize any other reagents needed. More than one step may be necessary.

(a) 2-Bromobutanal
(b) 2-Ethyl-3-hydroxyhexanal
(c) Butanoic acid
(d) 2-Ethyl-4-hexenoic acid
(e) 2-Butenal
(f) 3-Ethyl-1,5-hexadiene

17.39 The following is an example of a Perkin condensation reaction. Propose a mechanism for this reaction.

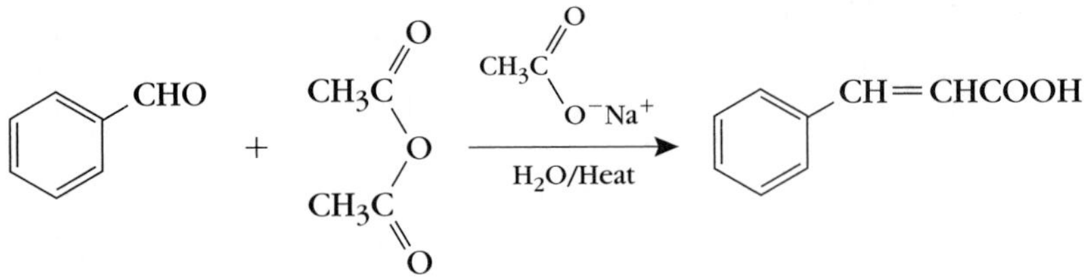

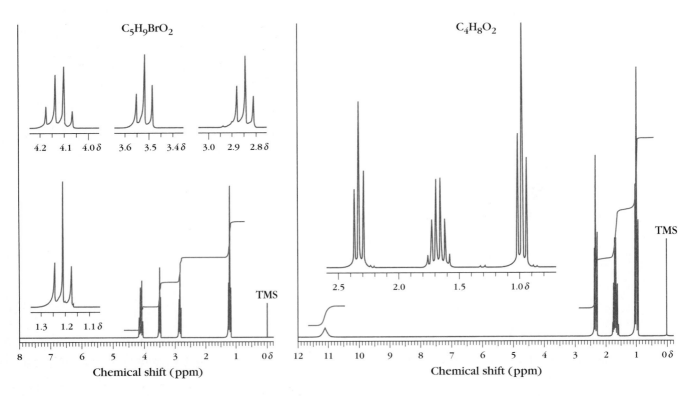

17.40 Write equations for all reactions in the synthesis of each of the following compounds. Begin with 3-pentanone and utilize any other reagents needed. More than one step may be necessary.

  (a)   2-Ethyl-1-butene   (c)   3-Methyl-3-chloropentane   (e)   2-Methyl-3-pentanol

  (b)   2-Ethyl-1-bromobutane   (d)   3-Methyl-3-pentanol   (f)   3-Ethyl-2-pentene

17.41 The following reaction is an example of the acylation of an enamine.

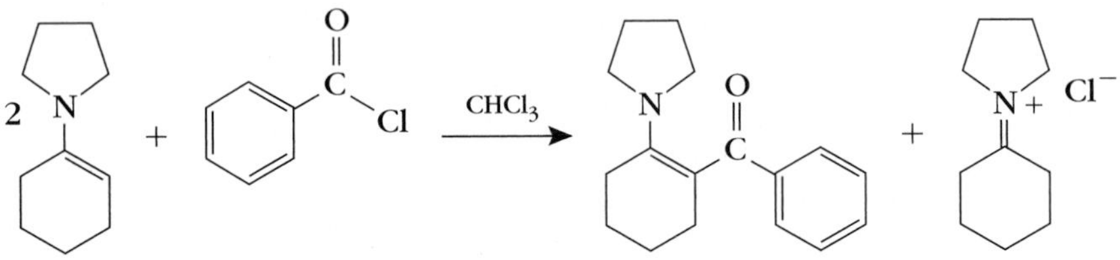

Propose a mechanism for this reaction. Compare your mechanism with the mechanism of the reaction of acyl chlorides and ammonia and note any similarities and differences.

17.42 Write the structural formula of each of the following compounds whose molecular formula and ^{1}H NMR spectra are given below.

CHAPTER

FREE RADICAL REACTIONS

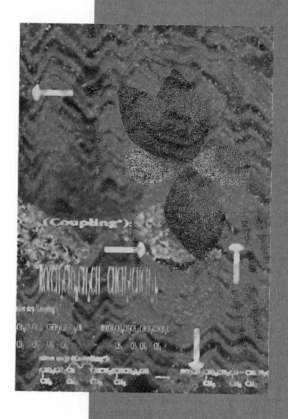

The reactions whose mechanisms we've discussed so far have been ionic reactions. Bond breaking in ionic reactions occurs so that one of the atoms of the covalent bond leaves with the two electrons. Bond breaking to form ions is called **heterolytic bond breaking.**

Heterolytic bond breaking:

$$M\!-\!L \longrightarrow M^+ + :L^-$$

For example:

$$\overset{H}{\underset{H}{\diagdown}}\ddot{O}: \quad H\!-\!\ddot{C}l: \longrightarrow \overset{H}{\underset{H}{\diagdown}}\overset{+}{\ddot{O}}\!-\!H + :\ddot{C}l:^-$$

A bond can also break so that each atom of the covalent bond leaves with an electron. This process is called **homolytic bond breaking.**

Homolytic bond breaking:

$$M\!-\!L \longrightarrow M\cdot + \cdot L$$

For example:

$$H\ddot{O}\!-\!\ddot{O}H \longrightarrow H\ddot{O}\cdot + \cdot\ddot{O}H$$

Unpaired
electron

- We designate the movement of a single electron in homolytic bond breaking by using a single barbed, curved arrow: ⌢ .

When bonds break homolytically, two species are formed that have unpaired electrons. Species with unpaired electrons are called **free radicals** or usually simply radicals. While a free radical usually has no charge, it is still very reactive because it lacks an octet of electrons. Because of this electron deficiency, free radicals are strongly electrophilic.

It takes energy to break a covalent bond into two free radicals. Energy is usually added by heating or irradiating with light. Compounds called dialkyl peroxides, for example, contain a relatively weak oxygen-oxygen bond. Heating dialkyl peroxides causes homolytic breaking of the oxygen-oxygen bond to form two alkoxy radicals.

$$R\ddot{O}\!-\!\ddot{O}R \xrightarrow{\text{Heat}} R\ddot{O}\cdot + \cdot\ddot{O}R$$

Dialkyl peroxide   Alkoxy radicals

The bond in halogen molecules is also relatively weak. Heating or irradiating iodine, for example, causes homolytic bond breaking to form iodine atoms.

$$:\ddot{I}\!-\!\ddot{I}: \xrightarrow[\text{or heat}]{h\nu} :\ddot{I}\cdot + \cdot\ddot{I}:$$

Iodine   Iodine atoms
(free radicals)

- The symbol hν placed above or below the arrow means that energy is added to the reaction by irradiating with light.

An iodine atom is a free radical because it contains an unpaired electron.

The Lewis structures of free radicals are usually simplified by omitting all electrons except the unpaired electron. Figure 18.1 shows the Lewis structures of several free radicals and their condensed structures.

					H
Lewis structure	$: \ddot{I} \cdot$	$: \ddot{Cl} \cdot$	$H - \ddot{O} \cdot$	$H - C \cdot$	
					H
Condensed structure	$I \cdot$	$Cl \cdot$	$HO \cdot$	$CH_3 \cdot$	
	Iodine atom	Chlorine atom	Hydroxy radical	Methyl radical	

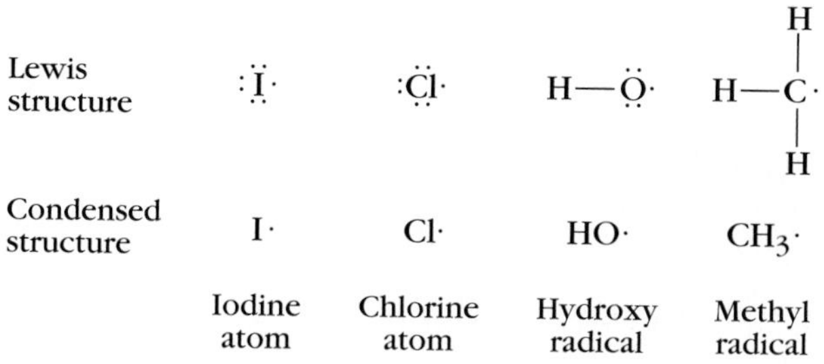

Figure 18.1
The Lewis structures of several free radicals and their condensed structures.

Exercise 18.1 By the use of single barbed, curved arrows, show the movement of electrons in each of the following reactions.

(a) $Cl - Cl \longrightarrow 2\ Cl \cdot$

(b) $CH_3O - OCH_3 \longrightarrow 2 \cdot OCH_3$

(c) $(CH_3)_2\overset{\overset{\displaystyle CN}{|}}{C} - N = N - \overset{\overset{\displaystyle CN}{|}}{C}(CH_3)_2 \longrightarrow 2(CH_3)_2\overset{\overset{\displaystyle CN}{|}}{C} \cdot\ +\ N \equiv N$

Exercise 18.2 In each of the following reactions, indicate whether it occurs by homolytic or heterolytic bond breaking. Use the correct type of arrow to indicate electron movement for each reaction.

(a) $HBr \longrightarrow H \cdot\ +\ Br \cdot$

(b) $H_2O\ +\ H_2O \longrightarrow H_3O^+\ +\ OH^-$

(c) $CH_2 = CH_2\ +\ H - Cl \longrightarrow \overset{+}{C}H_2 - \overset{\overset{\displaystyle |}{|}}{C}H_2\ +\ Cl^-$
with H below

(d) $CH_2 = CH_2\ +\ \cdot SCH_3 \longrightarrow \cdot CH_2 - CH_2$ with SCH_3 below

(e) $NH_3\ +\ HCl \longrightarrow NH_4Cl$

(f) $\cdot CCl_3\ +\ CH_2 = CH_2 \longrightarrow \cdot CH_2CH_2CCl_3$

In this chapter, we will examine the reactions of several functional groups that occur by a free radical mechanism. We begin with the chlorination of methane.

18.1 Chlorination of Methane

Chlorination is one of the few reactions that alkanes undergo. Recall from Section 2.9 that heating or irradiating chlorine and methane forms a mixture of chlorinated products.

$$CH_4$$

$$\downarrow \text{Cl}_2 \quad \begin{array}{c}\text{Heat or}\\ h\nu\end{array}$$

$$CH_3Cl \;+\; HCl$$

$$\xrightarrow[\begin{array}{c}\text{Heat or}\\ h\nu\end{array}]{Cl_2} \; CH_2Cl_2 \;+\; HCl$$

$$\xrightarrow[\begin{array}{c}\text{Heat or}\\ h\nu\end{array}]{Cl_2} \; CHCl_3 \;+\; HCl$$

$$\xrightarrow[\begin{array}{c}\text{Heat or}\\ h\nu\end{array}]{Cl_2} \; CCl_4 \;+\; HCl$$

The exact product composition depends on the relative concentrations of the two reactants and the time allowed for the reaction.

The free radical mechanism of the chlorination of methane consists of three steps, **initiation, propagation,** and **termination.** Each step may include one or more individual reactions. Let's examine each of these steps in detail.

Initiation Step   The initiation step of a free radical mechanism starts the reaction by producing free radicals. The initiation step in the chlorination of methane is the homolytic breaking of the chlorine-chlorine bond to produce two chlorine radicals. Irradiating chlorine with light or heating it to a high temperature provides the energy to break the chlorine-chlorine bond.

$$Cl_2 \xrightarrow[\substack{h\nu \text{ or} \\ \text{heat}}]{} Cl\cdot \;+\; Cl\cdot$$

Chlorine   Chlorine atoms
(free radicals)

Propagation Step   Once a chlorine radical is formed, it begins a cycle of two reactions called a chain reaction. In the first propagation reaction, a chlorine atom abstracts a hydrogen atom from methane to form a new radical, a methyl radical, and HCl, one of the products of the reaction.

$$Cl\cdot \quad H-CH_3 \longrightarrow Cl-H \;+\; \cdot CH_3$$

Chlorine   Methane   Hydrogen   Methyl
atom   chloride   radical

The reactive methyl radical continues the propagation step in a second reaction by abstracting a chlorine atom from molecular chlorine to form chloromethane, the other product of the reaction, and regenerate a chlorine atom.

$$Cl-Cl \quad \cdot CH_3 \longrightarrow Cl\cdot \; + \; Cl-CH_3$$

Chlorine   Methyl   Chlorine   Chloromethane
           radical   atom

The chlorine radical formed as a product in the second propagation step is a reactant of the first propagation step. As a result, a chlorine atom formed as a product in the second propagation step can start the first propagation step again. This sequence of two reactions constitutes a cycle that forms one molecule of product. The sum of the two propagation reactions is equal to the overall chemical equation of the chlorination of methane.

First reaction of propagation step:

$$Cl\cdot \; + \; CH_4 \longrightarrow HCl \; + \; \cdot CH_3$$

Second reaction of propagation step:

$$\cdot CH_3 \; + \; Cl_2 \longrightarrow CH_3Cl \; + \; Cl\cdot$$

Sum of two propagation reactions is the overall chemical equation for chlorination of methane:

$$CH_4 \; + \; Cl_2 \longrightarrow CH_3Cl \; + \; HCl$$

The sequence of two propagation reactions continues until one of the reagents is exhausted or some other reaction consumes the radical intermediates.

Termination Step   A termination step breaks the propagation cycle and slows or stops the reaction. Any reaction that destroys the free radical intermediates may terminate the propagation cycle. The most common termination steps are reactions that consume free radicals without forming new ones. The following are some of the possible termination reactions in the chlorination of methane.

$$Cl\cdot \; + \; Cl\cdot \longrightarrow Cl_2$$

$$Cl\cdot \; + \; \cdot CH_3 \longrightarrow ClCH_3$$

$$\cdot CH_3 \; + \; \cdot CH_3 \longrightarrow CH_3CH_3$$

During the chlorination of methane, the concentration of radicals is very low. The chance that two radicals will encounter each other and react to terminate the reaction is much smaller than the chance that they will encounter a molecule of reactant and continue the propagation step. As a result, termination reactions rarely occur compared to propagation reactions, so most of the product of chlorination of methane arises from propagation reactions, not termination reactions.

The free radical chain mechanism of the chlorination of methane is summarized in Figure 18.2.

Figure 18.2
The free radical chain mechanism of the chlorination of methane. The initiation step increases the concentration of free radicals by forming them, the propagation step maintains the concentration of free radicals because one radical is formed for every one consumed, and the termination step decreases the concentration of radicals by converting them to stable products.

Initiation step: Starts the reaction by forming free radicals

$$Cl_2 \xrightarrow[\text{or } h\nu]{\text{Heat}} Cl\cdot \ + \ Cl\cdot$$

Propagation step: Cyclic chain reaction that maintains the number of free radicals in the reaction mixture

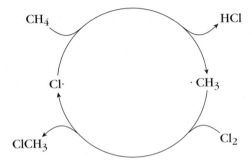

Termination steps: Any reaction that forms a species that cannot continue the propagation steps because it consumes free radicals

$$R\cdot \ + \ R\cdot \longrightarrow R{-}R$$

Free radicals   A stable species
that can continue   that cannot continue
propagation steps   propagation steps

Exercise 18.3   Write the propagation steps in the free radical mechanism of the reaction of chlorine and chloromethane to form dichloromethane.

Exercise 18.4   The number of products formed by chlorination of methane can be controlled to a certain extent by varying the concentrations of chlorine and methane. Which reagent would you use in excess to form principally chloromethane as product? Which one would you use in excess to form tetrachloromethane?

18.2   Monochlorination of Alkanes That Contain More Than Two Carbon Atoms

Free radical chlorination of alkanes that contain different types of hydrogen atoms forms mixtures of isomeric monochlorinated products. The free radical chlorination of propane, for example, forms two monochlorinated products. The majority of the monochlorinated product (55%) results from substitution of a secondary (2°) hydrogen atom by a chlorine atom to form 2-chloropropane. The remainder of the monochlorinated product (45%) results from substitution of a primary (1°) hydrogen atom by a chlorine atom to form 1-chloropropane.

$$CH_3CH_2CH_3 \xrightarrow[h\nu]{Cl_2} CH_3CH_2CH_2Cl + CH_3\overset{\displaystyle Cl}{\underset{|}{C}HCH_3}$$

Propane   1-Chloropropane   2-Chloropropane
            (45% yield)   (55% yield)

The ratio of 1-chloropropane to 2-chloropropane shows that the substitution of the 1° and 2° hydrogen atoms of propane by chlorine atoms is not a random process. Propane has six 1° hydrogen atoms and two 2° hydrogen atoms, a ratio of 1° to 2° hydrogen atoms of 6/2 or 3/1. If the reaction were random, we would expect three times more 1-chloropropane than 2-chloropropane. This is not observed, so substitution of the 2° hydrogen atoms of propane by chlorine atoms must occur faster than substitution of the 1° hydrogen atoms.

A similar analysis of the products of chlorination of many other alkanes has shown that 3° hydrogen atoms are abstracted more readily by chlorine atoms than 2° hydrogen atoms. As a result, the following order of the ease of abstracting various types of hydrogen atoms by chlorine atoms has been established.

$H-CH_3$	$H-CH_2CH_3$	$H-CH(CH_3)_2$	$H-C(CH_3)_3$
Type of hydrogen atom	Methyl	1°	2°	3°

INCREASING EASE OF ABSTRACTION BY CHLORINE ATOMS $\Rightarrow$

Exercise 18.5 Indicate which hydrogen or hydrogens in each of the following compounds would be the most easily replaced by a chlorine atom.

(a) $CH_3CH_2CH_2CH_3$
(b) 2-Methylbutane
(c) [structure of methylcyclohexane with CH_3]
(d) 2,2-Dimethylbutane
(e) $(CH_3CH_2)_4C$

To explain the difference in the ease of abstraction of various hydrogen atoms, we must examine the stabilities of alkyl free radicals.

18.3   Free Radical Stabilities and Product Composition of the Chlorination of Alkanes

The relative stabilities of free radicals can be determined by comparing their homolytic bond dissociation energies. Propane, for example, requires 98 kcal/mol (410 kJ/mol) to break a C—H bond to form a primary propyl radical, whereas only 95 kcal/mol (397 kJ/mol) are needed to form a secondary propyl radical.

$$CH_3CH_2CH_3 \longrightarrow CH_3CH_2CH_2\cdot + \cdot H \quad \Delta H° = 98 \text{ kcal/mol}$$

Propane   1° Propyl radical   (410 kJ/mol)

$$CH_3CH_2CH_3 \longrightarrow CH_3\overset{\displaystyle \cdot}{C}HCH_3 + \cdot H \quad \Delta H° = 95 \text{ kcal/mol}$$

Propane   2° Propyl radical   (397 kJ/mol)

Figure 18.3
Less energy is needed to form a
secondary propyl radical from
propane than a primary propyl radical, so the secondary radical is the
more stable.

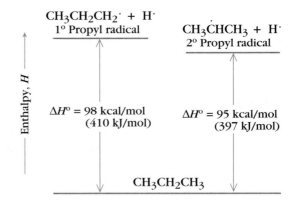

The more stable a free radical, the less energy is needed to form it, so a secondary propyl radical is more stable than a primary propyl radical, as shown in
Figure 18.3.

By comparing the bond dissociation energies of the C—H bonds of other
alkanes, the following order of free radical stabilities can be obtained.

$$\cdot CH_3 \qquad CH_3CH_2\overset{\centerdot}{C}H_2 \qquad (CH_3)_2\overset{\centerdot}{C}H \qquad (CH_3)_3C\cdot$$

Methyl   1° Radical   2° Radical   3° Radical
radical

$$\boxed{\text{INCREASING STABILITY} \Longrightarrow}$$

The order of stabilities of free radicals is the same as the order of stabilities of carbocations, namely stability increases from methyl to tertiary. Methyl and other
alkyl groups, therefore, stabilize both carbocations and free radicals by electron
donation to an electron-deficient carbon atom.

How do relative free radical stabilities affect the product composition of the
chlorination of alkanes? To answer this question, we must reexamine the free radical chain mechanism of the chlorination of alkanes. Recall from Section 18.1 that
almost all of the product of reactions that occur by a free radical chain mechanism is formed in the propagation step. The first reaction of the propagation step
is abstraction of a hydrogen atom to form an alkyl radical. The structure of the
radical formed in this step determines the structure of the product. In the chlorination of propane, for instance, formation of the primary radical leads to the formation of 1-chloropropane as product, whereas formation of the secondary
propyl radical leads to the formation of 2-chloropropane as product.

$$CH_3CH_2CH_3$$
Propane

$CH_3CH_2CH_2\cdot \xrightarrow{Cl_2} CH_3CH_2CH_2Cl + Cl\cdot$
1° Propyl
radical   1-Chloropropane

$CH_3\overset{\centerdot}{C}HCH_3 \xrightarrow{Cl_2} CH_3\overset{\overset{\displaystyle Cl}{|}}{C}HCH_3 + Cl\cdot$
2° Propyl
radical   2-Chloropropane

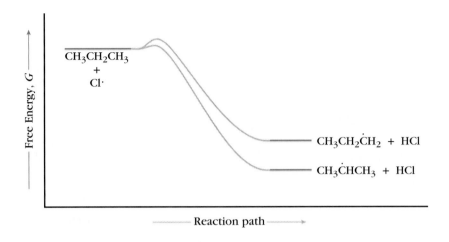

Figure 18.4
The free energy–reaction path diagram for the first reaction in the propagation cycle of the chlorination of propane. The relative stabilities of the two radicals are reflected in the transition states leading to their formation. The starting material is the same, so the energies of the transition states are directly proportional to the free energies of activation of formation of the two radicals. The free energy of activation is less for formation of the secondary radical than for the primary radical, so the secondary radical is formed in preference to the primary one.

More 2-chloropropane is formed than 1-chloropropane in the chlorination of propane because the secondary free radical is more stable than the primary free radical. As shown in Figure 18.4, the difference in stability of the two radicals is reflected in the transition states leading to their formation. As a result, the free energy of activation for formation of the secondary propyl radical is less than the free energy of activation for formation of the primary radical, so the secondary radical is formed in preference to the primary one.

Exercise 18.6   Place the following free radicals in order of increasing stability.

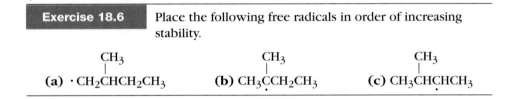

Exercise 18.7   Write the free radical chain mechanism for the chlorination of methylpropane. Sketch the free energy-reaction path diagram for the first step of the propagation cycle.

Free radicals resemble carbocations in their relative stabilities. The structures of free radicals and carbocations are also similar.

18.4   Stereochemistry of Free Radical Chlorination of Alkanes

One of the products of the free radical chlorination of butane is 2-chlorobutane. A sample of 2-chlorobutane, separated from the other chlorinated products by gas chromatography, is found to be optically inactive. An explanation of this observation is that the structure of the 2-butyl radical is planar, and optically inactive 2-chlorobutane is formed by reaction of a chlorine molecule with both sides of the planar radical to form an equal mixture of (*R*)- and (*S*)-2-chlorobutane.

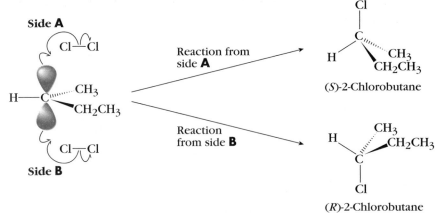

(S)-2-Chlorobutane

(R)-2-Chlorobutane

Compounds that contain a stereocenter formed by a free radical halogenation reaction, therefore, must be optically inactive.

A planar structure requires that the alkyl groups or hydrogen atoms bonded to the electron-deficient carbon atom of an alkyl free radical all lie in a plane with bond angles of about 120°, as shown in Figure 18.5.

To construct the valence bond representation of the electron structure of an alkyl free radical with this geometry, we must represent the electron-deficient carbon atom as an sp^2 hybrid. Its three sp^2 hybrid orbitals lie in a plane with the lobes of the p orbital above and below the plane, as shown in Figure 18.6. The unpaired electron of the radical is in the p orbital. The structure of an alkyl free radical is very similar to the structure of an alkyl carbocation, except that the p orbital of the free radical contains a single electron.

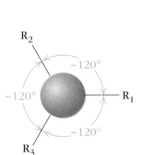

A Side view

B Top view

Figure 18.5
The planar geometry of an alkyl free radical. The R—C—R bond angles are all about 120°.

| **Exercise 18.8** | One of the products of irradiating a mixture of chlorine and (R)-3-methylhexane with light is 3-chloro-3-methylhexane. Write the mechanism for the formation of 3-chloro-3-methylhexane showing clearly the configuration at the stereocenter. |

Some addition reactions of alkenes are known to occur by a free radical mechanism.

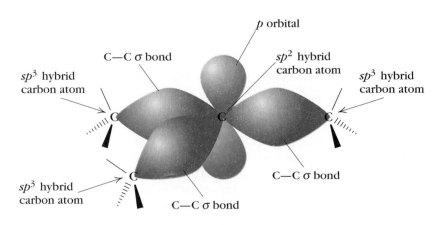

Figure 18.6
Valence bond representation of the electron structure of a tertiary alkyl free radical.

18.5 Free Radical Addition of HBr to Carbon-Carbon Double Bonds

Before 1933, the regiochemistry of HBr addition to alkenes was the source of much confusion. Conflicting results were reported in the chemical literature for additions of HBr to the same alkene under what seemed to be identical experimental conditions. The addition of HBr to 3-bromopropene, for example, was reported to form both 1,2-dibromopropane and 1,3-dibromopropane in a number of laboratories but in widely different ratios.

$$CH_2=CHCH_2Br \xrightarrow{\text{HBr}} BrCH_2-CHCH_2Br \; + \; HCH_2-CHCH_2Br$$

$$\underset{\text{H}}{\mid} \qquad\qquad \underset{\text{Br}}{\mid}$$

| 3-Bromopropene | 1,3-Dibromopropane (Anti-Markownikoff product) | 1,2-Dibromopropane (Markownikoff product) |

> ■ Recall that the product formed by addition to an alkene contrary to the Markownikoff rule has the anti-Markownikoff orientation and is usually called the **anti-Markownikoff product.**

The conflict was resolved when the American chemists M. S. Kharasch and F. R. Mayo found that products obtained in the presence of free radical initiators in the reaction mixture were different than the products obtained in the absence of free radical initiators. A **free radical initiator** is any compound that can initiate a free radical reaction.

The presence of a free radical initiator in the reaction mixture caused the addition of HBr to occur by a free radical mechanism, which formed products with anti-Markownikoff orientation. In the absence of a free radical initiator or in the dark, HBr addition occurs by an ionic mechanism to form products with Markownikoff orientation.

> ■ Additions to alkenes that occur according to the Markownikoff rule are discussed in Section 7.11.

$$CH_3CH=CH_2 \; + \; HBr \xrightarrow[\text{Heat}]{\text{Free radical initiators}} CH_3CH-CH_2$$

$$\underset{\text{H}}{\mid}\;\underset{\text{Br}}{\mid}$$

Propene   1-Bromopropane (Anti-Markownikoff product)

$$CH_3CH=CH_2 \; + \; HBr \xrightarrow[\substack{\text{Reaction carried} \\ \text{out in the dark}}]{\text{No free radical initiators}} CH_3CH-CH_2$$

$$\underset{\text{Br}}{\mid}\;\underset{\text{H}}{\mid}$$

Propene   2-Bromopropane (Markownikoff product)

To ensure that HBr addition occurs by a free radical mechanism to form the product of anti-Markownikoff orientation, we can purposely add a free radical initiator to the reaction mixture. Diacyl and dialkyl peroxides, for example, are frequently used as free radical initiators because they are stable compounds at room temperature but decompose to free radicals when heated.

$$RO-OR \xrightarrow{\text{Heat}} 2\,RO\cdot \qquad R-\overset{\overset{\text{O}}{\|}}{C}-O-O-\overset{\overset{\text{O}}{\|}}{C}-R \xrightarrow{\text{Heat}} 2\,R-\overset{\overset{\text{O}}{\|}}{C}-O\cdot$$

Dialkyl peroxide   Diacyl peroxide

If the product of Markownikoff orientation is desired, on the other hand, we must ensure that the free radical reaction does not interfere with the ionic addition of HBr. To avoid addition by a free radical mechanism, either we carry out the reaction in the dark (to prevent light-induced formation of free radicals) or we add a free radical inhibitor. **Free radical inhibitors,** or simply inhibitors, are compounds that prevent the free radical chain mechanism from occurring. Inhibitors react with free radicals formed in the initiation or propagation step to form compounds or radicals that are not initiators of new chains. The inhibitor 2,6-di-*t*-butyl-4-methylphenol, often referred to as BHT, reacts with free radicals to form a phenoxy radical.

■ *The abbreviation BHT comes from **b**utylated **h**ydroxy **t**oluene, a common name for 2,6-di-t-butyl-4-methylphenol.*

2,6-Di-*tert*-butyl-4-methylphenol
(BHT)

"A phenoxy radical"

The phenoxy radical is so unreactive that it does not initiate any new chains. As a result, the free radical mechanism is slowed or stopped.

We can obtain either regioisomer of the addition of HBr to an alkene by choosing the proper reaction conditions. Conditions that favor an ionic mechanism (presence of an inhibitor or in the dark) form products with Markownikoff orientation, whereas conditions that favor a free radical mechanism (presence of an initiator) form products with anti-Markownikoff orientation.

Exercise 18.9	Write the equation for the addition of HBr to methylpropene **(a)** in the presence of a free radical initiator and **(b)** in the dark in the presence of a free radical inhibitor.

18.6   Free Radical Mechanism of HBr Addition to Alkenes

Kharasch and Mayo proposed that HBr can add to an alkene by the free radical chain mechanism shown in Figure 18.7.

Recall from Section 18.1 that a free radical mechanism starts by an initiation step. Free radical addition of HBr to an alkene starts when a source of free radicals is present in the reaction mixture. This can occur accidently when an impurity decomposes on heating or in the presence of light to form free radicals, or deliberately by the addition of an initiator such as a dialkyl peroxide. When heated, the oxygen-oxygen bond of a peroxide breaks to form two alkoxy radicals. The alkoxy radicals then abstract the hydrogen atom from a molecule of HBr to form an alcohol and a bromine atom.

The bromine atom, formed in the initiation step, adds to the π bond of the alkene to form an alkyl radical in the first reaction of the propagation step. In the

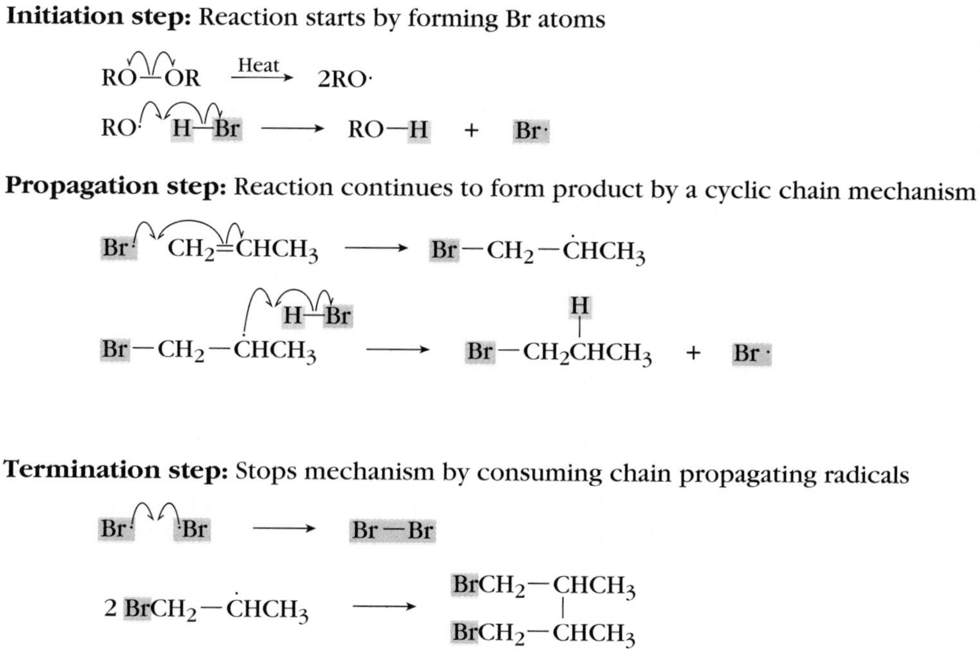

Initiation step: Reaction starts by forming Br atoms

$$RO\!-\!OR \xrightarrow{\text{Heat}} 2RO\cdot$$

$$RO\cdot \quad H\!-\!Br \longrightarrow RO\!-\!H \ + \ Br\cdot$$

Figure 18.7
The free radical chain mechanism of
the addition of HBr to propene. The
mechanism contains three steps,
which are called initiation, propagation, and termination.

Propagation step: Reaction continues to form product by a cyclic chain mechanism

$$Br\cdot \quad CH_2\!=\!CHCH_3 \longrightarrow Br\!-\!CH_2\!-\!\dot{C}HCH_3$$

$$Br\!-\!CH_2\!-\!\dot{C}HCH_3 \quad H\!-\!Br \longrightarrow \overset{H}{\underset{|}{Br\!-\!CH_2CHCH_3}} \ + \ Br\cdot$$

Termination step: Stops mechanism by consuming chain propagating radicals

$$Br\cdot \quad Br \longrightarrow Br\!-\!Br$$

$$2\,BrCH_2\!-\!\dot{C}HCH_3 \longrightarrow \begin{array}{c} BrCH_2\!-\!CHCH_3 \\ | \\ BrCH_2\!-\!CHCH_3 \end{array}$$

second reaction of the propagation step, the alkyl radical removes a hydrogen atom from HBr to form the product and regenerate a bromine radical.

As in the free radical mechanism of the chlorination of alkanes, each cycle of the propagation step forms one molecule of product. As well, the sum of the two propagation steps is equal to the overall chemical equation.

Propagation step:

$$CH_3CH\!=\!CH_2 \ + \ Br\cdot \longrightarrow CH_3\dot{C}H\!-\!CH_2\!-\!Br$$

$$CH_3\dot{C}H\!-\!CH_2\!-\!Br \ + \ HBr \longrightarrow \overset{H}{\underset{|}{CH_3CHCH_2}}\!-\!Br \ + \ Br\cdot$$

Net overall reaction:

$$CH_3CH\!=\!CH_2 \ + \ HBr \longrightarrow \overset{H}{\underset{|}{CH_3CHCH_2}}\!-\!Br$$

Finally free radicals are consumed in the termination step, which slows or stops the propagation cycle.

Why does free radical addition of HBr form the anti-Markownikoff product? Recall from Section 18.2 that the structure of the product of a reaction that occurs by a free radical mechanism is determined by the structure of the radical formed in the first reaction of the propagation step. A bromine radical adds to the alkene in the first reaction of the propagation step to form the more stable radical. While addition of a bromine atom to propene, for example, can occur to form two radicals, addition occurs only to form the more stable secondary radical.

$$CH_3CH = CH_2 \quad + \quad \overset{\cdot}{Br} \longrightarrow CH_3\overset{\cdot}{CH} - CH_2 - Br \overset{HBr}{\longrightarrow} \overset{\overset{H}{|}}{CH_3CH} - CH_2 - Br \quad + \quad \overset{\cdot}{Br}$$

Secondary free radical

Observed product
with anti-Markownikoff
orientation

$$\overset{X}{\longrightarrow} CH_3CH - \overset{\cdot}{CH}_2$$
$$\underset{Br}{|}$$

Primary free radical

The secondary free radical reacts with HBr to form the observed product with anti-Markownikoff orientation.

Notice a common feature in the mechanisms of acid additions and free radical additions to alkenes. *A free radical or a proton adds to the double bond to form the more stable intermediate, either an alkyl radical or a carbocation.* We can illustrate this similarity by comparing the free radical addition of HBr with the ionic addition of HBr to propene.

Free radical addition of HBr to propene:

$$CH_3CH = CH_2 \overset{\overset{\cdot}{Br}}{\longrightarrow} CH_3\overset{\cdot}{CH} - CH_2 - Br \overset{HBr}{\longrightarrow} \overset{\overset{H}{|}}{CH_3CH} - CH_2 - Br \quad + \quad \overset{\cdot}{Br}$$

Secondary free radical

1-Bromopropane
(anti-Markownikoff product)

Ionic addition of HBr to propene:

$$CH_3CH = CH_2 \overset{H_3O^+}{\longrightarrow} CH_3\overset{+}{CH} - CH_2 - H \overset{Br^-}{\longrightarrow} \overset{\overset{Br}{|}}{CH_3CH} - CH_2 - H$$

Secondary carbocation

2-Bromopropane
(Markownikoff product)

A bromine radical adds first to the double bond of propene to form the more stable secondary radical, which abstracts a hydrogen atom in the next step to form the anti-Markownikoff product. In contrast, the proton adds first to form the more stable secondary carbocation in the ionic addition of HBr to propene. The carbocation subsequently reacts with bromide ion to form the Markownikoff product. The important point to remember is that in both mechanisms, *addition occurs to form the more stable intermediate.*

Exercise 18.10   Write the equation and the mechanism for the reaction that occurs when a mixture of HBr, 1-methylcyclohex-ene, and di-*tert*-butyl peroxide, $(CH_3)_3COOC(CH_3)_3$, is heated to 60 °C.

18.7 Autooxidation of Organic Compounds

Autooxidation is the slow oxidation reaction that almost all organic compounds undergo with atmospheric oxygen. Autooxidation occurs when compounds are exposed to air and light.

The first product of many autooxidation reactions is a hydroperoxide formed by the conversion of a C—H bond to a C—O—O—H group.

■ *Slow oxidation means that oxidation of the organic compound occurs without combustion.*

The hydroperoxides are rarely isolated because they react further to form a variety of products. Autooxidation is responsible for the deterioration of foods, lubricating oils, rubber, and many other organic substances in nature.

Autooxidation occurs because of the electron structure of oxygen. In the ground state, molecular oxygen has two unpaired electrons, so an oxygen molecule is said to be a **diradical,** which can be represented as $\cdot O{-}O\cdot$.

Autooxidation occurs by a free radical mechanism. Oxygen is not reactive enough to initiate a free radical chain mechanism by itself. Molecular oxygen initiates the chain by reacting with any free radical present in the surroundings to form a peroxy radical.

■ *The highest occupied molecular orbital of molecular oxygen is the degenerate pair of π antibonding molecular orbitals, each of which contains one electron.*

Initiation step:

$$M{-}M \xrightarrow[\text{or heat}]{\text{Light}} 2M\cdot$$

Any compound   Free
capable of forming   radicals
free radicals

$$M\cdot + \cdot O{-}O\cdot \longrightarrow M{-}O{-}O\cdot$$

Peroxy radical

■ *The action of sunlight on many substances in the environment, both organic and inorganic, forms free radicals. Consequently, free radicals are continually being generated and consumed all around us.*

The peroxy radical abstracts a hydrogen atom in the first reaction of the three reactions that make up the propagation step.

Propagation step:

$$M{-}O{-}O\cdot + H{-}R \longrightarrow M{-}O{-}O{-}H + R\cdot$$

Peroxy radical

$$R\cdot + \cdot O{-}O\cdot \longrightarrow R{-}O{-}O\cdot$$

$$R{-}O{-}O\cdot + H{-}R \longrightarrow R{-}O{-}O{-}H + R\cdot$$

The termination steps of autooxidation reactions are so complicated that we will not discuss them in detail. In most of the termination reactions, the oxygen-

oxygen bonds of the radicals formed in the propagation step are broken to eventually form a mixture of alcohols, ketones, and carboxylic acids.

Autooxidation reactions, like any reaction that occurs by a free radical mechanism, can be slowed by free radical inhibitors. BHT, for example, is added to food to retard spoilage by autooxidation. Recall from Section 18.5 that BHT is a free radical inhibitor that slows the free radical mechanism by trapping free radicals that propagate the chain.

Certain natural products, called **antioxidants,** are free radical inhibitors that protect living cells from the destructive effects of free radicals. Vitamin C (ascorbic acid) and bioflavonoids, such as quercetin and Vitamin E (α-tocopherol), are all known to have antioxidant properties.

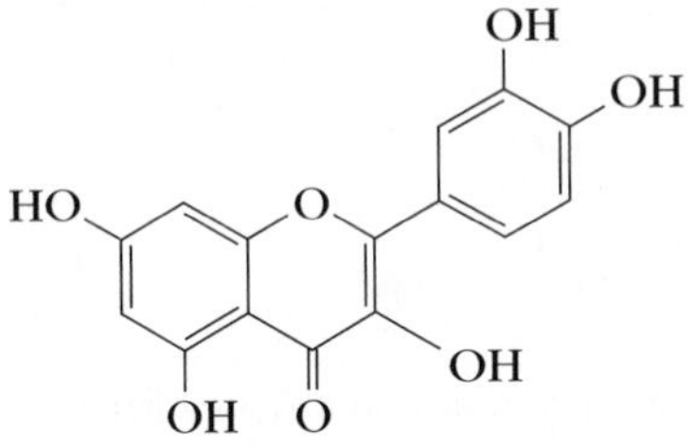

Quercetin

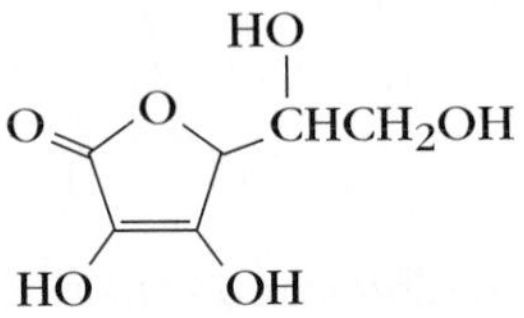

Vitamin C
(ascorbic acid)

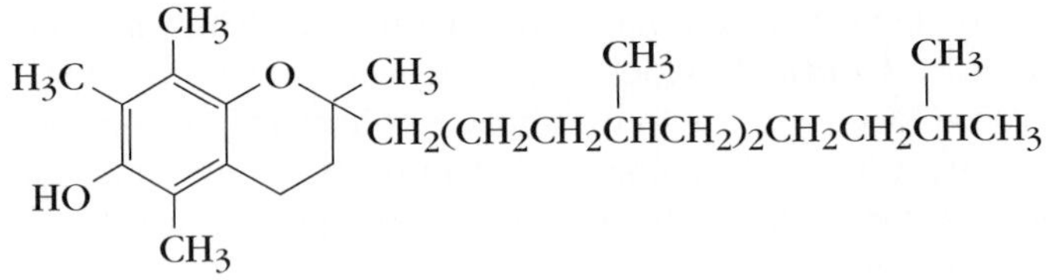

Vitamin E
(α-tocopherol)

Free radicals react with quercetin, Vitamin E, and Vitamin C by abstracting a hydrogen atom from an —OH group to form free radicals that are so unreactive that they do not initiate a propagation step.

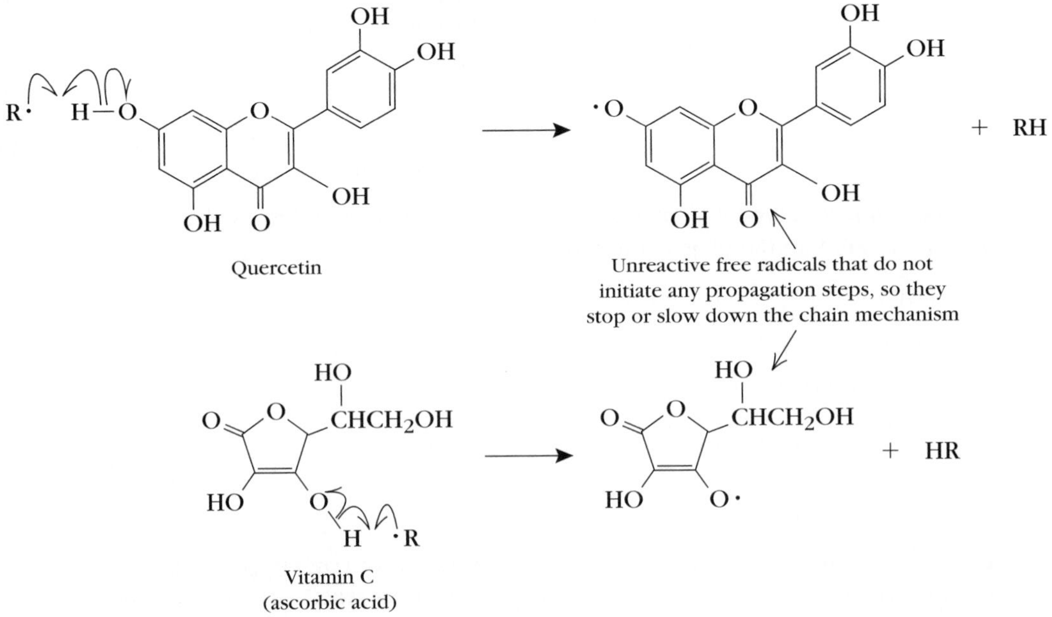

Quercetin

Unreactive free radicals that do not initiate any propagation steps, so they stop or slow down the chain mechanism

Vitamin C
(ascorbic acid)

As a result, any free radicals that damage the living cell are destroyed, and any chain mechanisms that they initiate are stopped or slowed down.

| Exercise 18.11 | Write the reaction by which Vitamin E (α-tocopherol) destroys the hydroxy radical, HO ·. |

Many important plastics are produced by free radical reactions.

18.8   Addition Polymers

Polymers, or **macromolecules,** are high molecular weight compounds that are made up of repeating low molecular weight units called **monomers.**

Polysaccharides (Chapter 25), proteins (Chapter 26), and nucleic acids (Chapter 27) are **natural polymers** or **biopolymers** because they are formed by plants and animals. Since the 1930s, **synthetic polymers** have been produced industrially for use in the manufacture of a wide variety of consumer products.

Polymerization is the term used for chemical reactions that form polymers. The first reaction in polymerization is the joining of two monomers to form a **dimer** (two parts). Further polymerization forms **trimers, tetramers,** and finally the polymer molecules. The polymers that we discuss in this section are called **addition polymers** because they are formed by the addition of monomers to each other without the formation of any other product.

Polymers are named by adding the prefix poly- to the name of the monomer. Ethylene, for example, forms a polymer called polyethylene.

> ■ *The words* polymer *and* monomer *come from the Greek.* Poly- *and* meros *mean* many parts, *while* mono- *and* meros *mean* one part.

Repeating units of the monomer

$$CH_2{=}CH_2 \longrightarrow {-}CH_2{-}CH_2{-}CH_2{-}CH_2{-}CH_2{-}CH_2{-}$$

Ethylene
(monomer)

Polyethylene
(polymer)

The equations of polymerization reactions are conveniently represented in the following way where n is any large integer.

$$n\ CH_2{=}CH_2 \longrightarrow {\left[CH_2{-}CH_2\right]}_n$$

Ethylene   Polyethylene

Substituted ethylenes can also form addition polymers. Polymerization of substituted ethylenes forms polymers that contain a substituent at regular intervals along the carbon skeleton of the polymer.

$$H_2C{=}CH\underset{R}{|} \longrightarrow {-}CH_2CH\underset{R}{|}CH_2CH\underset{R}{|}CH_2CH\underset{R}{|}CH_2CH\underset{R}{|}{-}$$

Vinyl monomer   Polymer

These substituted ethylenes are called **vinyl monomers,** and their polymerization is known as **vinyl polymerization.** A number of vinyl monomers that are used to produce commercially valuable polymers are listed in Table 18.1.

Table 18.1	Examples of Addition Polymers and Some of Their Uses	
Trade Name	**Uses**	**Monomer**
Polyethylene	Bottles, tubing, and other molded objects	$CH_2{=}CH_2$ Ethylene
Polypropylene	Indoor-outdoor carpets, rope	$CH_2{=}CHCH_3$ Propene
Poly(vinylchloride) PVC	Protective films, pipes for household and industrial plumbing	$CH_2{=}CHCl$ Vinyl chloride
Teflon	Nonstick pan linings, valves, and chemical-resistant films	$F_2C{=}CF_2$ Tetrafluoroethylene
Polystyrene	Rigid foam, packaging filler, and molded household utensils	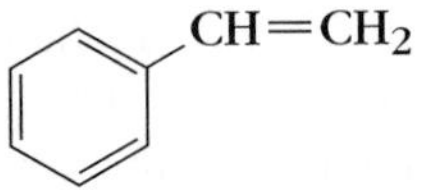 $CH{=}CH_2$ attached to benzene ring Styrene
Poly(vinyl acetate)	Adhesives and latex paints	$CH_2{=}CHOCOCH_3$ Vinyl acetate
Orlon	Textile fibers	$CH_2{=}CHC{\equiv}N$ Acrylonitrile
Lucite, Plexiglass	Transparent, nonshattering material	$\overset{\displaystyle CH_3}{\underset{\displaystyle }{CH_2{=}CCOOCH_3}}$ Methyl methacrylate

Polymerization of ethylene and other vinyl monomers can be carried out under free radical conditions.

18.9　Mechanism of Free Radical Polymerization

The free radical polymerization of ethylene is carried out industrially at high temperatures (150 to 250 °C) and high pressures (1000 to 3000 atm) in the presence of a compound that is a source of free radicals such as a dialkyl peroxide. The mechanism of free radical polymerization has many of the same features as the free radical chain mechanism of the chlorination of methane described in Section 18.1. The initiation, propagation, and termination steps of the free radical polymerization of propene are shown in Figure 18.8.

The first reaction of the initiation step is the thermal decomposition of the dialkyl peroxide to form two alkoxy radicals. One of the alkoxy radicals formed in the first reaction adds to propene in the second reaction of the initiation step to form a carbon radical.

The first reaction of the propagation step is the addition of the carbon radical

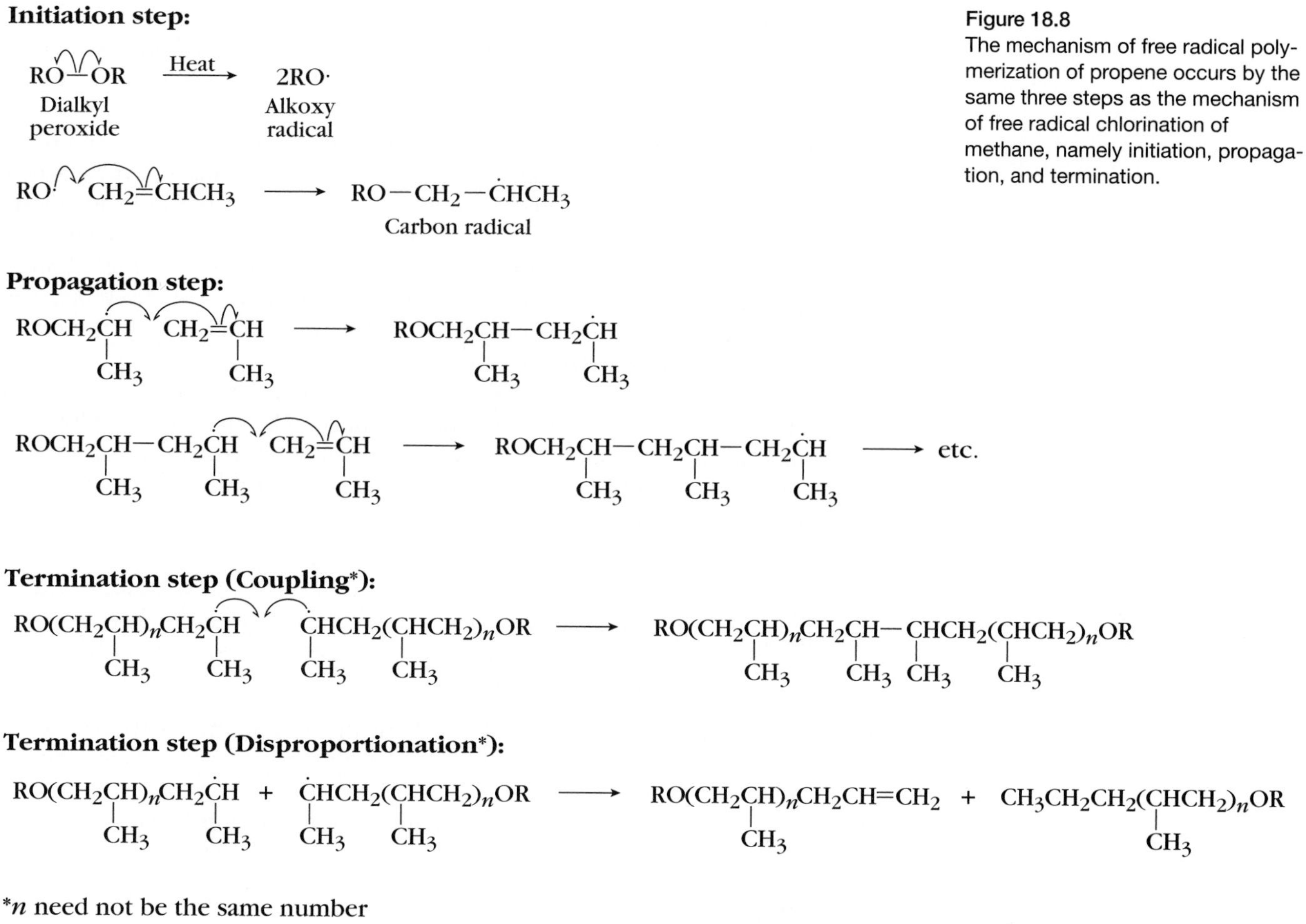

Initiation step:

RO—OR $\xrightarrow{\text{Heat}}$ 2RO·
Dialkyl   Alkoxy
peroxide   radical

RO· CH$_2$=CHCH$_3$ $\longrightarrow$ RO—CH$_2$—ĊHCH$_3$
                                        Carbon radical

Figure 18.8
The mechanism of free radical polymerization of propene occurs by the
same three steps as the mechanism
of free radical chlorination of
methane, namely initiation, propagation, and termination.

Propagation step:

ROCH$_2$ĊH   CH$_2$=ĊH $\longrightarrow$ ROCH$_2$CH—CH$_2$ĊH
  | | | |
  CH$_3$ CH$_3$ CH$_3$ CH$_3$

ROCH$_2$CH—CH$_2$ĊH   CH$_2$=ĊH $\longrightarrow$ ROCH$_2$CH—CH$_2$CH—CH$_2$ĊH $\longrightarrow$ etc.
  | | | | | |
  CH$_3$ CH$_3$ CH$_3$ CH$_3$ CH$_3$ CH$_3$

Termination step (Coupling*):

RO(CH$_2$CH)$_n$CH$_2$ĊH   ĊHCH$_2$(CHCH$_2$)$_n$OR $\longrightarrow$ RO(CH$_2$CH)$_n$CH$_2$CH—CHCH$_2$(CHCH$_2$)$_n$OR
   | | | | | | | |
   CH$_3$ CH$_3$ CH$_3$ CH$_3$ CH$_3$ CH$_3$ CH$_3$ CH$_3$

Termination step (Disproportionation*):

RO(CH$_2$CH)$_n$CH$_2$ĊH + ĊHCH$_2$(CHCH$_2$)$_n$OR $\longrightarrow$ RO(CH$_2$CH)$_n$CH$_2$CH=CH$_2$ + CH$_3$CH$_2$CH$_2$(CHCH$_2$)$_n$OR
   | | | | | |
   CH$_3$ CH$_3$ CH$_3$ CH$_3$ CH$_3$ CH$_3$

*n need not be the same number

to another propene molecule. Addition occurs to form the more stable carbon free radical.

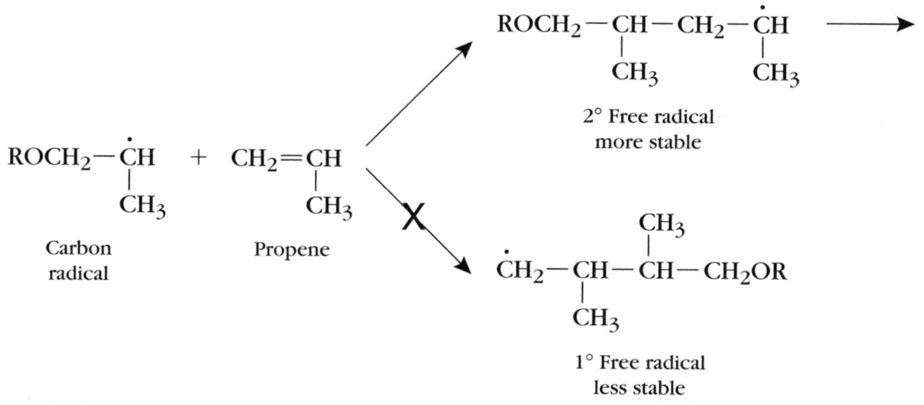

ROCH$_2$—ĊH + CH$_2$=CH
  | |
  CH$_3$ CH$_3$
Carbon   Propene
radical

$\nearrow$ ROCH$_2$—CH—CH$_2$—ĊH $\longrightarrow$
              | |
              CH$_3$ CH$_3$
           2° Free radical
           more stable

$\searrow$ X ĊH$_2$—CH—CH—CH$_2$OR
                  | |
                  CH$_3$ CH$_3$
           1° Free radical
           less stable

Repeating the addition of a carbon radical to a propene molecule continues the propagation reaction. Each new addition forms a chain extended free radical.

Finally, the reaction is terminated by reactions that consume but do not form free radicals. Combination and disproportionation reactions are possible termination reactions. In a combination reaction, two free radicals, either of the same or different lengths, combine to form a stable molecule. In a disproportionation reaction, one radical is oxidized to an alkene, while another is reduced to an alkane.

The number of monomer units in a polymer depends on the method of preparation and can vary greatly. Commercial synthetic polymers usually have between 100 and 1500 monomer units. In a sample of a polymer, all the polymer molecules are not the same size because it is impossible to ensure that all termination reactions occur after exactly the same number of propagation reactions. Unlike ordinary small molecules, therefore, a polymer does not have an exact molecular weight. Instead, an average molecular weight is determined, which is in the range 10^5 to 10^6 g/mol for many synthetic polymers.

The polymers produced by the mechanism described in Figure 18.8 have end groups that are obtained from the free radical initiator. The end groups, while different from the repeating monomer units in the rest of the molecule, do not affect the chemical and physical properties of the polymer because they are only a small part of the polymer.

Exercise 18.12 Write the structure of a short segment (four units) of the polymer formed by polymerization of each of the following vinyl monomers.

(a) Styrene   **(b)** Vinyl acetate   **(c)** Acrylonitrile   **(d)** Methyl methacrylate

Polymers can also be formed by the polymerization of different monomers.

18.10 Copolymers

The polymers that have been discussed so far are made up of only one monomer, so they are called **homopolymers.** Polymers can also be formed from two or more monomers, however, and these are called **copolymers.** For example, polymerization of a mixture of styrene and acrylonitrile in the same flask forms a copolymer that contains units derived from both monomers.

Many commercially important polymers are copolymers. A number of examples are shown in Table 18.2.

The sequence of monomer units in a copolymer chain can vary depending on the method and mechanism of synthesis. A copolymer can be random, regular, or block.

When no discernible pattern of monomers exists in a copolymer, it is said to be a **random copolymer.** Thus the following copolymer of the monomers A and B would be a random copolymer.

Trade Name	Uses	Monomers	
Kel-F	Molded articles, O-rings, and rocket motors	$ClFC{=}CF_2$ Chlorotrifluoro-ethylene	$+$ $CH_2{=}CF_2$ Vinylidene fluoride
Saran	Food wrapping	$CH_2{=}CHCl$ Vinyl chloride	$+$ $CH_2{=}CCl_2$ Vinylidene chloride
Teflon FEP	Wire insulation, tubing, and printed circuits	$CF_2{=}CF_2$ Tetrafluoro-ethylene	$+$ $CF_3CF{=}CF_2$ Hexafluoro-propene
Dynel	Texile fibers	$CH_2{=}CHC{\equiv}N$ Acrylonitrile	$+$ $CH_2{=}CHCl$ Vinyl chloride
Poly(ethylene-vinyl acetate) copolymer	Medical tubing, syringes, toys, and cable insulation	$CH_2{=}CH_2$ Ethylene	$+$ $CH_2{=}CHOCOCH_3$ Vinyl acetate

Table 18.2   **Examples of Copolymers and Their Uses**

$$-B-A-A-B-B-B-A-B-A-A-A-B-$$

Random copolymers are often formed by free radical copolymerization of vinyl monomers. Their properties are usually quite different from those of the homopolymers obtained from either of the two monomers.

A **regular copolymer** contains a regular alternating sequence of the two monomer units.

$$-B-A-B-A-B-A\left[B-A\right]_n$$

The properties of regular copolymers are usually very different from those of the homopolymers obtained from either monomer.

A **block polymer** contains a large number (a block) of one kind of monomer bonded to a block of another kind:

$$-A\left[A\right]_n A-B\left[B\right]_m A\left[A\right]_x B\left[B\right]_y$$

Unlike regular and random copolymers, block copolymers retain many of the physical properties of the homopolymers obtained from either monomer.

The properties of the monomer or monomers that make up a polymer determine the properties of the polymer. Two other factors, the linearity and stereoregularity of the polymer chain, are also important.

18.11 Formation of Branched Polymers

The polymerization reactions discussed so far are idealized. That is, we have assumed that when polymer chains are formed on initiation, they propagate lin-

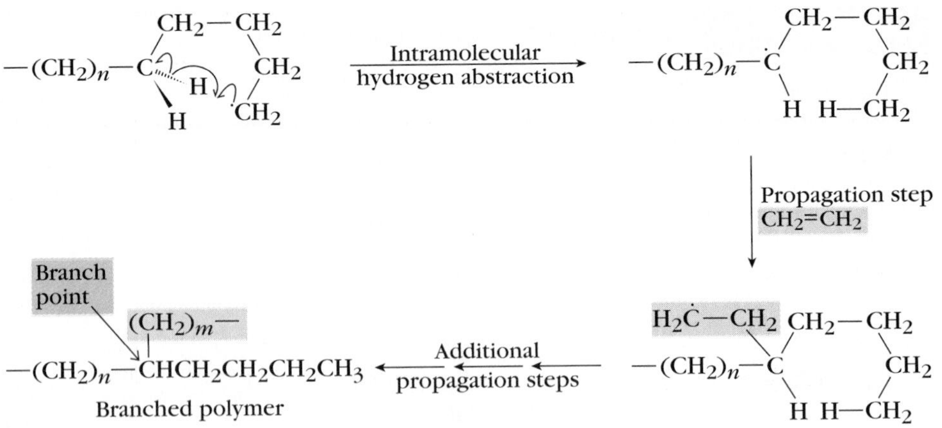

Figure 18.9
Intramolecular hydrogen abstraction
to form a branched polymer.

early, and are finally terminated to form a linear polymer. Such an idealized poly-merization rarely occurs in free radical polymerization.

Deviation from ideality occurs when a polymer radical reacts not by addition to a molecule of monomer but by abstraction of a hydrogen atom. The abstracted hydrogen can be from the same radical chain (intramolecular hydrogen abstrac-tion) or from another molecule (intermolecular hydrogen abstraction).

Intramolecular hydrogen abstraction occurs when a radical chain curls back on itself and abstracts a hydrogen atom from its own chain (it bites its own tail). Most of the hydrogen atoms abstracted are located four or five carbon atoms from the radical end of the polymer. In order to remove these hydrogen atoms, a five- or six-atom cyclic transition state must be formed. Five- or six-atom cyclic transi-tion states, like five- and six-membered rings, are highly favored. Intramolecular hydrogen abstraction reactions form a secondary radical that, on further propaga-tion, forms a branched polymer. This intramolecular hydrogen abstraction is illus-trated in Figure 18.9 for the free radical polymerization of ethylene.

Intermolecular hydrogen abstraction between the radical end of one chain and a hydrogen atom located in the middle of another polymer chain can also occur. This also leads to branched polymers. Intermolecular hydrogen abstraction is illustrated in Figure 18.10 for the free radical polymerization of ethylene.

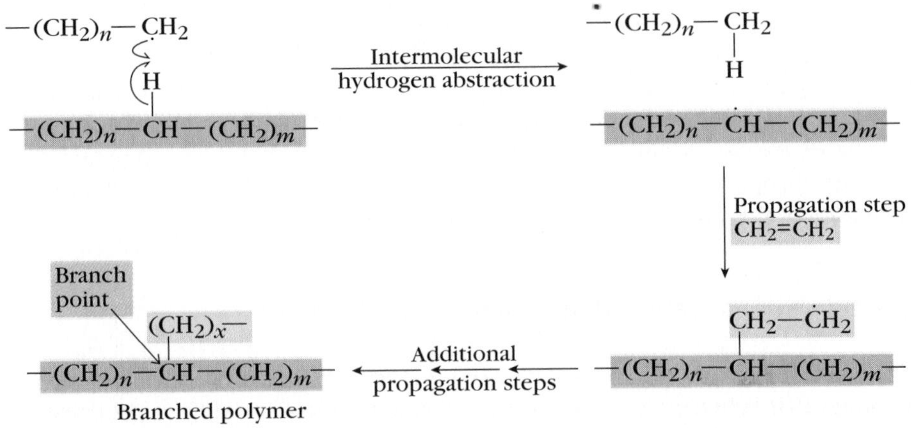

Figure 18.10
Intermolecular hydrogen abstraction
to form a branched polymer.

Formation of chain branched polymers is not limited to polyethylene. Almost all free radical vinyl polymerizations form branched polymers. Thus polystyrene and polypropylene both contain branched chains. Controlling the amount of branching during polymerization is very important because the amount of branching affects the properties of a polymer. It was discovered in the 1950s that polymerization catalyzed by certain transition metal salts, called Ziegler-Natta catalysts, yields unbranched polymers.

18.12 Use of Ziegler-Natta Catalysts to Form Unbranched and Stereoregular Polymers

Until 1953, almost all polymers of commercial importance were formed by free radical polymerization. These polymers, as we discussed in Section 18.11, have highly branched chains. Highly branched chains fit together poorly in the solid state, so these polymers are said to lack crystallinity. Polymers with low crystallinity have low melting points and are mechanically weak. Both of these properties decrease the commercial value of a polymer.

In 1953, two chemists, Karl Ziegler in Germany and Giulio Natta in Italy, developed catalysts that make it possible to prepare linear polymer molecules and control the stereochemistry of the polymer.

The Ziegler-Natta catalysts, as they are now called, consist of a transition metal salt, generally titanium chloride ($TiCl_4$ or $TiCl_3$) and an alkyl metal compound, such as triethylaluminum [$(CH_3CH_2)_3Al$] or diethyl zinc [$(CH_3CH_2)_2Zn$]. Polyethylene produced with a Ziegler-Natta catalyst is linear. These unbranched chains fit together well in the solid state, so the polyethylene has a high crystallinity. Polyethylene produced in this way is called high-density polyethylene because it has a higher density than polyethylene formed by free radical polymerization (known as low-density polyethylene). High-density polyethylene also has a higher melting point and is mechanically much stronger, both of which are commercially valuable properties.

Besides forming linear polymers, the stereochemistry of many polymers can be controlled by the choice of Ziegler-Natta catalyst. Many polymers are formed from vinyl monomers that contain a prochiral center. Polymerization of these alkenes forms polymers with stereocenters along the linear chains. Polymerization of propene, for example, forms a polymer in which every other carbon atom is a stereocenter.

$$CH_2{=}CHCH_3 \xrightarrow{\text{Polymerization}} -CH_2\overset{*}{C}HCH_2\overset{*}{C}HCH_2\overset{*}{C}HCH_2\overset{*}{C}H-$$

Three sequences of (R,S)- configurations are possible for the stereocenters. These sequences are called atactic, isotactic, and syndiotactic.

A polymer is said to be **atactic** if the sequence of (R,S)- configurations is random. The structure of polypropylene shown in Figure 18.11 is atactic. In atactic polypropylene, the methyl groups are randomly arranged above and below the plane of the carbon skeleton of the polymer.

An atactic polymer can also be defined by assigning the (R) or (S) configuration to each stereocenter in the polymer. We do this by arbitrarily assigning a higher priority to one end of the polymer chain than the other. If the sequence of (R,S)- designations along the chain is random, the polymer is defined as atactic.

Figure 18.11
Atactic polypropylene. In this representation of polypropylene, the carbon chain is in the plane of the page, and the methyl groups are randomly located above and below the plane of the page. Thus the sequence of (*R,S*)-designations of the stereocenters along the chain is random.

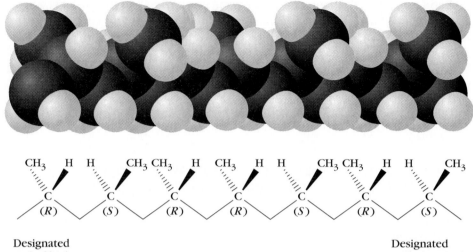

Designated
end of highest
priority

Designated
end of second
highest priority

A polymer is said to be **isotactic** if the substituents, which are methyl groups in polypropylene, are all located on the same side of the plane of its carbon chain. This means that the sequence of (*R,S*)- configurations of the stereocenters is all (*R*) or all (*S*), depending on which side of the polymer is assigned the higher priority. Isotactic polypropylene is shown in Figure 18.12.

A polymer is said to be **syndiotactic** if the substituents alternate from one side of the plane of its carbon chain to the other. This means that the configurations of stereocenters alternate (*R*), (*S*), (*R*), (*S*), (*R*), (*S*), and so on. Syndiotactic polypropylene is shown in Figure 18.13.

Figure 18.12
Isotactic polypropylene. All the methyl groups of isotactic polypropylene are located on the same side of the plane of its carbon chain. As a result, the configurations of the stereocenters are all identical.

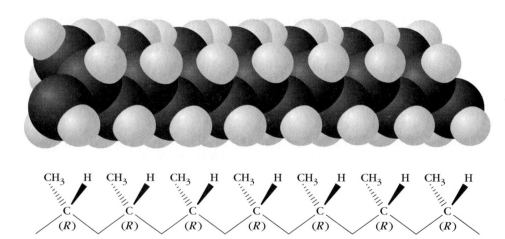

Designated
end of highest
priority

Designated
end of second
highest priority

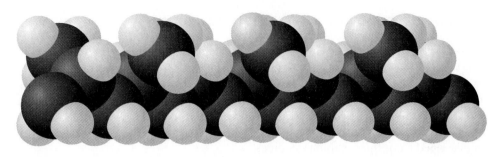

Figure 18.13
Syndiotactic polypropylene. The methyl groups of syndiotactic polypropylene alternate from one side of the plane of its carbon chain to the other. As a result, the configurations of the stereocenters alternate, too.

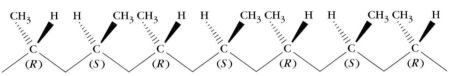

Designated
end of highest
priority

Designated
end of second
highest priority

With the proper choice of catalyst, all three different stereoisomers of polypropylene have been prepared. Atactic polypropylene, formed by high-temperature and high-pressure free radical polymerization of propylene, is a soft and elastic material. Both syndiotactic and isotactic polypropylene, which are prepared with Ziegler-Natta catalysts, are highly crystalline because of the regularity of their structures. The regularity of their structures allows their chains to fit tightly together in the solid. Isotactic polypropylene is one of the major synthetic fibers produced commercially today. Several billion pounds are produced annually to be molded or extruded into sheets and fibers.

| Exercise 18.13 | By the proper choice of catalysts, all three stereoisomers of polystyrene have been prepared. Write the structures of atactic, isotactic, and syndiotactic polystyrene. |

| Exercise 18.14 | Atactic, isotactic, and syndiotactic polystyrenes each contain a large number of stereocenters. Are samples of any of these polymers optically active? Explain. |

| Exercise 18.15 | Is it possible to form syndiotactic and isotactic polyvinylidine by the polymerization of vinylidene chloride ($CH_2=CCl_2$)? Explain. |

Free radicals are also one of the products formed when high-energy electrons collide with organic compounds.

18.13 Introduction to Mass Spectrometry

Mass spectrometry (MS) is a technique of structure determination that is different from IR spectroscopy or NMR spectroscopy, which we studied in Chapters 5 and 10, because it does not depend on the selective absorption of particular frequencies of electromagnetic radiation. Mass spectrometry, instead, analyzes the results of bombarding a molecule with high-energy electrons.

When a high-energy electron collides with an organic compound, an electron is dislodged from the organic molecule:

$$RZ \quad + \quad e^- \quad \longrightarrow \quad RZ\overset{+}{\cdot} \quad + \quad 2e^-$$

Organic   High-   Molecular
molecule   energy   ion
           electron

Z=H or any functional group

We say that in this process, the molecule has been ionized by electron impact. The ionized species is called a **molecular ion.** The molecular ion is both a radical because it contains an unpaired electron and a cation because it has lost an electron, so a molecular ion is also called a **radical-cation.**

The electron dislodged from a molecule on electron impact is usually one from a high-energy molecular orbital. In general, the order of decreasing energy of molecular orbitals is nonbonding molecular orbitals > π bonding molecular orbitals > σ bonding molecular orbital. If a compound contains a pair of electrons in a nonbonding molecular orbital, a nonbonding electron is lost on electron impact. If a compound lacks nonbonding electrons but contains a pair of electrons in a π bonding molecular orbital of a carbon-carbon double bond or triple bond, an electron is lost from one of these bonds. If neither nonbonding electrons nor unsaturated bonds are present, then an electron is lost from a σ bonding molecular orbital.

In the collision between a high-energy electron and an organic compound, much of the energy of the electron is transferred to the organic molecule. Only a small part of the energy is needed to dislodge an electron from the organic molecule. The rest of the energy causes the molecular ion to dissociate into smaller fragments by bond breaking. The process by which a molecular ion breaks apart into many smaller molecular fragments is called **fragmentation.** Fragmentation of a molecular ion forms a neutral fragment and a positively charged fragment. The positively charged specie often undergoes further fragmentation. The molecular ion of methane, for example, fragments to form a carbocation and a neutral hydrogen atom. The carbocation then fragments to form a radical-cation and a hydrogen atom.

Methane   Molecular ion   Carbocation   Hydrogen atom
                                           (positively charged species)   (neutral species)

Radical-cation   Hydrogen atom
(positively charged species)   (neutral species)

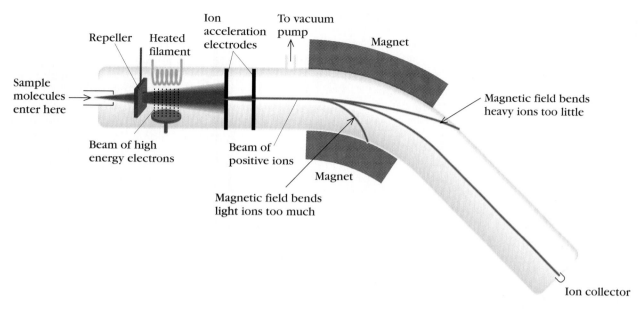

Figure 18.14
Schematic diagram of the parts of a single-focusing, magnetic sector mass spectrometer. A beam of high energy electrons, produced by a heated filament, interacts with a stream of vaporized sample molecules to form positive ions and other products. The positive ions are pushed out of the ionizing region by the repeller, and then they are accelerated into a magnetic field. As the magnetic field is increased, ions of increasing m/z ratio are focused on the mass collector.

■ In the ratio m/z, m is the mass of the positively charged ion and z is its number of charges.

Electron bombardment of organic molecules is carried out in an instrument called a **mass spectrometer** whose design is shown schematically in Figure 18.14. In a mass spectrometer, a small sample of an organic compound is vaporized in a high vacuum. The vapor is passed into the ionization chamber, where the ionization of the organic compound occurs. The positive ions formed by electron impact (the molecular ion and any positively charged ions formed by fragmentation of the molecular ion) are accelerated and pass into a magnetic field. The magnetic field causes the ions to deflect from their original path and adopt a circular path. The radius of the circular path depends on their **mass to charge ratio (m/z).** Ions of small values of m/z are deflected more than those with large values of m/z. Neutral fragments are not deflected by the magnetic field and are lost by collision with the walls of the instrument. By varying the magnetic field, the ions of a particular value of m/z can be focused on the mass collector where they are recorded. Scanning all the values of m/z and recording them on a chart gives the distribution of positive ions, which is called a mass spectrum.

18.14 The Mass Spectrum

A **mass spectrum** of a compound is a plot or tabulation of the relative abundance of each ion versus its m/z value. An example is the mass spectrum of methane, shown in Figure 18.15. The m/z values of the ions are plotted on the horizontal axis, while the relative numbers of ions of that value of m/z (relative abundance) are plotted on the vertical scale.

Figure 18.15
Mass spectrum of methane presented as a bar graph and in tabular form.

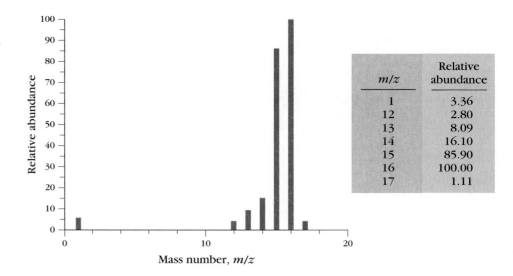

m/z	Relative abundance
1	3.36
12	2.80
13	8.09
14	16.10
15	85.90
16	100.00
17	1.11

The most intense peak in Figure 18.15 is the one at $m/z = 16$. This peak is called the base peak, and it is assigned the arbitrary intensity value of 100. The other peaks in the spectrum are scaled relative to the base peak. The base peak, $m/z = 16$, in the mass spectrum of methane (Figure 18.15), corresponds to the molecular ion $(M^{+\bullet})$. The peak of next highest intensity, $m/z = 15$, corresponds to the carbocation $(CH_3{}^+)$.

Exercise 18.16   Write the Lewis structure of the molecular ion formed by the reaction of high-energy electrons with each of the following compounds.

(a) Methanol   **(b)** Formaldehyde (H_2CO) **(c)** Methyl amine (CH_3NH_2)

If we look carefully at the mass spectrum of methane, we find that there is a peak at m/z value one mass unit greater than the molecular ion peak. This peak is due to the presence of isotopes of the atoms in the molecule.

18.15 Isotopes in Mass Spectra

The small peak one mass unit higher than the molecular ion peak, $M^{+\bullet}$, in Figure 18.15 corresponds to molecules of methane that contain ^{13}C in place of ^{12}C. This peak is called the $(M+1)^{+\bullet}$ peak. A mass spectrometer is capable of distinguishing between the ions of methane that contain the ^{13}C isotope of carbon and the ions that contain only ^{12}C because they have different masses.

$$\left[\begin{array}{c} {}^{1}H \\ | \\ {}^{1}H-{}^{13}C-{}^{1}H \\ | \\ {}^{1}H \end{array}\right]^{+} \qquad \left[\begin{array}{c} {}^{1}H \\ | \\ {}^{1}H-{}^{12}C-{}^{1}H \\ | \\ {}^{1}H \end{array}\right]^{+}$$

$$m/z = 17 \qquad\qquad m/z = 16$$

Table 18.3		Natural Abundance and Exact Masses of Important Isotopes	
Element	Isotope	Relative Abundance (%)	Exact Mass
Hydrogen	^1H	99.985	1.007825
Carbon	^{12}C	98.89	12.00000
	^{13}C	1.11	13.00335
Nitrogen	^{14}N	99.634	14.00307
	^{15}N	0.366	15.00011
Oxygen	^{16}O	99.759	15.99491
	^{18}O	0.204	17.99992
Fluorine	^{19}F	100	18.99984
Phosphorus	^{31}P	100	30.99376
Sulfur	^{32}S	95.00	31.97207
	^{33}S	0.76	32.97146
	^{34}S	4.22	33.96786
Chlorine	^{35}Cl	75.53	34.96885
	^{37}Cl	24.47	36.9659
Bromine	^{79}Br	50.54	78.9183
	^{81}Br	49.46	80.9163
Iodine	^{127}I	100	126.90004

As a result, all of the peaks except the $m/z = 1$ peak of methane and all organic compounds are accompanied by a smaller peak one mass unit higher.

Other elements such as nitrogen, oxygen, sulfur, bromine, and chlorine all have isotopes of higher mass that give rise to additional mass peaks in the mass spectra of compounds that contain these elements. The important isotopes of these elements are given in Table 18.3.

Deuterium, ^2H, the isotope of hydrogen of higher mass, is not included in this list because its natural abundance is so low (0.015%) that it does not contribute to the peaks in the mass spectrum. Consequently, the natural abundance of deuterium in a molecule is usually ignored in analyzing mass spectra.

Several of the isotopes listed in Table 18.3 result in the appearance of an isotope peak two mass units higher than the molecular ion peak, so it is called an $(M+2)^{+\bullet}$ peak. Thus organic compounds that contain oxygen, sulfur, chlorine, or bromine all have $(M+2)^{+\bullet}$ peaks in their mass spectra due to the presence of the isotopes ^{18}O, ^{34}S, ^{37}Cl, and ^{81}Br, respectively. The mass spectrum of chlorobenzene, for example, has two intense peaks at $m/z = 112$ and $m/z = 114$. The peak at $m/z = 112$ corresponds to $C_6H_5{}^{35}Cl$, whereas the peak at $m/z = 114$ corresponds to $C_6H_5{}^{37}Cl$, as shown by the mass spectrum on p. 828.

The intensities of the peaks $(M+1)^{+\bullet}$ and $(M+2)^{+\bullet}$ relative to the $M^{+\bullet}$ peak depend on the number of atoms of the element present in the compound and the abundance of its isotope. For example, in methane, 1.1% of all carbon atoms are ^{13}C. Therefore the $(M+1)^{+\bullet}$ peak is 1.1% of the abundance of the $M^{+\bullet}$ peak. This relative abundance is usually expressed as the ratio $(M+1)^{+\bullet}/M^{+\bullet}$; for methane, $(M+1)^{+\bullet}/M^{+\bullet} = 1.1\%$. In chlorobenzene, about 25% of the chlorine atoms are ^{37}Cl. Therefore the $M^{+\bullet}$ peak is about three times as intense as the $(M+2)^{+\bullet}$ peak.

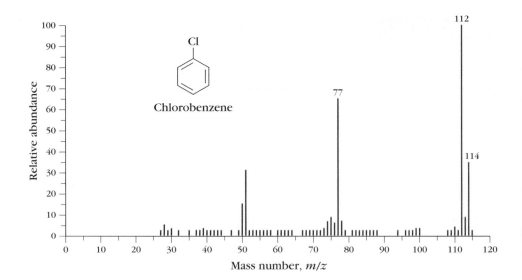

Exercise 18.17   Calculate the value of m/z for the molecular ion in each of the following compounds.

(a) Ethane   **(b)** 2-Chloropropane   **(c)** Methanol

Exercise 18.18   Which of the compounds in Exercise 18.17 would have a peak at $(M+1)^{+\bullet}$? Which would have a peak at $(M+2)^{+\bullet}$?

18.16 Structural Information From Fragment Ions

Additional information about an organic compound can be obtained from the ions formed by fragmentation of the molecular ion. The relative energies of the various fragmentation pathways determine which ions will be formed. Fragmentation occurs generally at the weakest bond and/or fragmentation occurs to form the most stable fragment in the greatest abundance. Fragmentation is not a random process because it occurs by well-defined pathways. We can make use of fragmentation reactions to gain information about structures of organic compounds in two ways.

First, the pattern of ions formed by fragmentation is unique for each organic compound. The mass spectrum of a compound, therefore, is another fingerprint; this time it is a molecular one. The mass spectrum, like an IR spectrum, can be used to identify an unknown compound by the technique of matching spectra. If the mass spectrum of an unknown exactly matches that of a known compound, the two are identical. Such identifications have been facilitated in recent years by the advent of computers. Computer matching of the spectrum of an unknown with spectra of over 130,000 compounds stored in the memory of a computer can be done in a few minutes.

The second way to obtain structural information about a compound is to examine the fragments in its mass spectrum. The presence of certain fragments gives information about the functional groups present in the molecule. The fragments produced in a mass spectrometer are small parts of the whole molecule. Thus to obtain information about the structure of the molecule we must work backwards. We try to fit the fragments together, like a jigsaw puzzle, to finally get the overall structure. However, we do have one advantage. Tabulation of hundreds of thousands of mass spectra has revealed that each functional group has a tendency to fragment in certain characteristic ways. While we cannot go into great detail in this introduction to mass spectrometry, we can mention a few of the more important fragmentation reactions of each functional group. Let us begin with the alkanes.

A   Alkanes

The important fragmentation reactions of alkanes are illustrated by comparing the mass spectra of pentane, 2-methylbutane, and 2,2-dimethylpropane, shown in Figure 18.16.

The only similarity in the three spectra of the isomers of molecular formula C_5H_{12} is that the molecular ion peak ($m/z = 72$) is very small in each because fragmentation occurs extensively and rapidly. The peak at $m/z = 43$ is the base peak in the mass spectrum of pentane. Since it corresponds to the loss of a C_2H_5 unit, it is called the $(M-C_2H_5)^+$ peak and it is formed in the following way.

Loss of sigma electron

$$CH_3CH_2CH_2 \!:\! CH_2CH_3 \quad \xrightarrow{\;e^-\;} \quad CH_3CH_2CH_2 \overset{+}{\cdot} CH_2CH_3$$

Fragmentation

$$CH_3CH_2CH_2 \overset{+}{\underset{\frown}{\cdot}} CH_2CH_3 \quad \longrightarrow \quad \underset{m/z\,=\,43}{CH_3CH_2CH_2{}^+} \quad + \quad \underset{\substack{\text{Not observed}\\ \text{in mass spectrum}}}{\cdot CH_2CH_3}$$

Other peaks are found at $m/z = 29$ $(C_2H_5)^+$ and $m/z = 57$ $(M-CH_3)^+$

The fragmentation patterns of the mass spectra of methylbutane and pentane are similar. However, the relative abundance of the peaks differs. The peak at $m/z = 43$ is still the base peak, but abundance of the peak at $m/z = 57$ is greatly increased. This difference is due to the branching effect in methylbutane. In general, branching in a hydrocarbon leads to fragmentation at the branch because of the relative stability of the cations formed. Thus the major fragmentation of methylbutane occurs as follows.

$$\underset{\underset{H}{|}}{\overset{\overset{CH_3}{|}}{CH_3 \overset{+}{\cdot} C - CH_2CH_3}} \quad \longrightarrow \quad CH_3 \cdot \quad + \quad \underset{\underset{H}{|}}{\overset{\overset{CH_3}{|}}{{}_+C - CH_2CH_3}}$$
$$\hspace{8cm} m/z = 57$$

$$\underset{\underset{H}{|}}{\overset{\overset{CH_3}{|}}{CH_3 - C \overset{+}{\cdot} CH_2CH_3}} \quad \longrightarrow \quad \underset{\underset{H}{|}}{\overset{\overset{CH_3}{|}}{CH_3 - C_+}} \quad + \quad \cdot CH_2CH_3$$
$$\hspace{4cm} m/z = 43$$

Figure 18.16
Mass spectra of pentane, 2-methylbutane, and 2,2-dimethyl-propane.

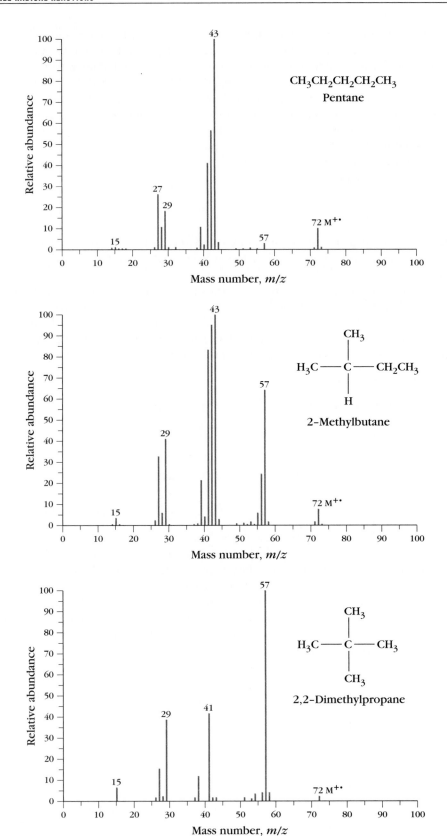

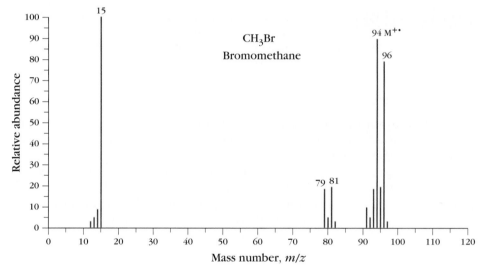

$$CH_3-\underset{\underset{\overset{|}{\cdot+}}{\overset{CH_3}{|}}{C}}-CH_2CH_3 \longrightarrow CH_3-\underset{\underset{+}{\overset{CH_3}{|}}}{C}-CH_2CH_3 \ + \ \cdot H$$

$m/z = 71$

The branching effect is even more pronounced in the mass spectrum of 2,2-dimethylpropane. Loss of a methyl radical from the molecular ion produces a *tert*-butyl cation as the base peaks at $m/z = 57$.

Peaks at $m/z = 41$ and 29 are found in the mass spectra of all the C_5H_{12} isomers, even though their molecular ions cannot directly form fragments corresponding to these values of m/z. These fragments are the result of complex structural reorganization of intermediate cations and radicals.

B   Chloro- and Bromoalkanes

The mass spectra of chloro- and bromoalkanes are characterized by two peaks: one at the molecular ion and the other at $(M+2)^{+\bullet}$, as illustrated by the spectra of bromomethane and chloromethane, shown in Figure 18.17.

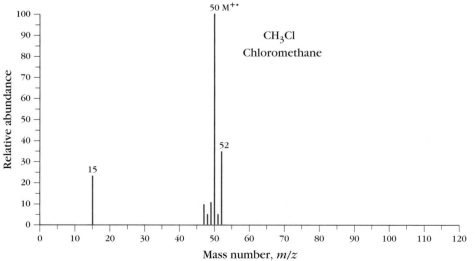

Figure 18.17
Mass spectra of bromomethane and chloromethane.

The peak for bromomethane that appears at $m/z = 94$ contains ^{79}Br and is the molecular ion peak, while the other peak at $m/z = 96$, the $(M+2)^{+\bullet}$ peak, contains ^{81}Br. The relative abundance of the two peaks is about equal in accord with the natural abundance of the two isotopes of bromine (50.5% ^{79}Br and 49.5% ^{81}Br). Similarly the mass spectrum of chloromethane has a molecular ion peak at $m/z = 50$ corresponding to $[CH_3{}^{35}Cl]^{+\bullet}$, while the peak at $(M+2)^{+\bullet}$ corresponds to $[CH_3{}^{37}Cl]^{+\bullet}$. The appearance of $M^{+\bullet}$ and $(M+2)^{+\bullet}$ peaks is characteristic of bromoalkanes and chloroalkanes, as is the appearance of a $(M-Halogen)^+$ peak.

Exercise 18.19   In the mass spectrum of bromomethane (Figure 18.17), there are small peaks at $m/z = 79$, $m/z = 81$, and a large peak at $m/z = 15$. Write the equation for the fragmentation of the molecular ion of bromomethane that accounts for these peaks.

Exercise 18.20   What should be the relative abundance of the $M^{+\bullet}$ and $(M+2)^{+\bullet}$ peaks in the mass spectrum of chloromethane (Figure 18.17)?

C   Alcohols

The molecular ion peak is usually weak or nonexistent in the mass spectra of alcohols because of the ease of fragmentation, as shown by the mass spectrum of 1-butanol in Figure 18.18.

One important fragmentation pathway of the molecular ions of alcohols is the loss of water. For 1-butanol this results in a peak at $m/z = 56$. The $(M-H_2O)^{+\bullet}$ peak at $m/z = 56$ is also the base peak. The second most abundant peak is found at $m/z = 31$, which is due to a second major fragmentation process.

Figure 18.18
The mass spectrum of 1-butanol shows a weak molecular ion peak at $m/z = 74$ and a base peak at $m/z = 56$, which is due to loss of water. The peak at $m/z = 31$ is due to $H_2C=OH^+$.

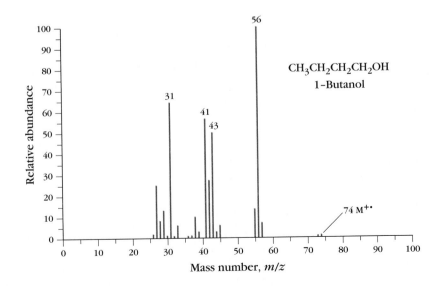

$$CH_3CH_2-\overset{\overset{\displaystyle H}{|}}{\underset{\overset{\displaystyle |}{H}}{C}}-\overset{\overset{\displaystyle H}{|}}{\underset{\overset{\displaystyle |}{H}}{C}}\overset{\cdot\,+}{\underset{\cdot\cdot}{O}H} \longrightarrow CH_3CH_2-\overset{\overset{\displaystyle H}{|}}{\underset{\overset{\displaystyle |}{H}}{C}}\cdot \quad + \quad \overset{H}{\underset{H}{}}C=\overset{+}{\underset{\cdot\cdot}{O}}H$$

Molecular ion
$m/z = 74$

$m/z = 31$

D   Aldehydes and Ketones

The molecular ions of aldehydes and ketones commonly undergo cleavage of the bond between the α-carbon atom and carbonyl carbon atom (a so-called α-cleavage) to form a neutral alkyl group and an acylium ion.

$$\left[\overset{O}{\underset{R}{\overset{||}{\underset{}{C}}}}\diagdown_{CH_2R'}\right]^{\overset{+}{\cdot}} \longrightarrow R-\overset{+}{C}=O \quad + \quad R'CH_2^{\cdot}$$

Molecular ion   Acylium ion

Aldehydes and ketones that have a hydrogen atom on their γ-carbon atoms undergo a characteristic fragmentation reaction called the **McLafferty rearrangement.** The McLafferty rearrangement occurs by an intramolecular transfer of a hydrogen atom on the γ-carbon atom to the carbonyl oxygen atom with the breaking of the bond between the α- and β-carbon atoms to form a neutral alkene fragment.

$$\underset{H_2C\diagdown_{\underset{H_2}{C}}\diagup^{C}\diagdown_{R}}{\overset{R}{\underset{\gamma}{\diagdown}}CH}\overset{H}{\diagup}\overset{O}{||} \longrightarrow \underset{CH_2}{\overset{R}{\underset{\gamma}{\diagdown}}\overset{||}{C}}\overset{H}{\diagup} \left[\underset{H_2C\diagup^{\underset{}{C}}\diagdown_{R}}{\overset{H}{\diagdown}\overset{O}{\diagup}}\right]^{\overset{+}{\cdot}}$$

Fragment ions from both α-cleavage and the McLafferty rearrangement can be seen in the mass spectrum of 2-pentanone, shown in Figure 18.19. Alpha cleavage of 2-pentanone occurs mostly at the more substituted side of the carbonyl

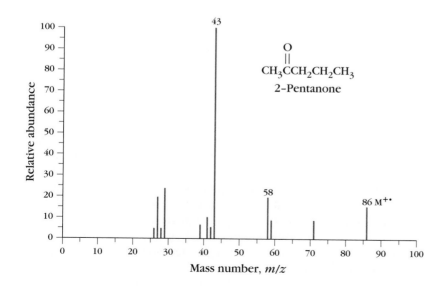

O
||
$CH_3CCH_2CH_2CH_3$
2-Pentanone

Figure 18.19
The mass spectrum of 2-pentanone shows a weak molecular ion peak at m/z = 86 and the base peak at m/z = 43 that is due to an α-cleavage. The peak at m/z = 58 is due to McLafferty rearrangement.

group because a methyl radical is less stable than a propyl radical, leading to a $[CH_3C≡O]^+$ fragment with $m/z = 43$. The fragment with $m/z = 58$ is due to McLafferty rearrangement and loss of ethylene.

While these are general observations of the major fragments observed in the mass spectra of various types of organic compounds, we must remember that fragmentation patterns of compounds with the same functional groups are still very dependent on their structure. In this short introduction to mass spectrometry, we cannot go into any more detail about fragmentation patterns. However, the principles presented in this section along with infrared, ^1H NMR, and ^{13}C NMR spectroscopy provide us with very powerful tools to determine the structure of many organic compounds.

Exercise 18.21   Suggest structures and fragmentation patterns to account for the peaks observed in the mass spectrum of each of the following compounds.

(a) 1-Propanol; $m/z = 31$; $m/z = 60$ **(c)** Butane; $m/z = 58$; $m/z = 43$
(b) 2-Propanol; $m/z = 44$ **(d)** 2-Methylpropane; $m/z = 43$

Exercise 18.22   Write the equation for the formation of the molecular ion; show how each of the indicated fragments is formed; and propose a structure for the fragment for each of the following compounds.

(a) Heptane; $m/z = 100$; $m/z = 43$
(b) 2,3-Dimethylpentane; $m/z = 57$; $m/z = 43$; $m/z = 41$
(c) 1,2-Dichloroethane; $m/z = 62$; $m/z = 27$
(d) 2-Methyl-1-butanol; $m/z = 57$
(e) 2-Methyl-2-propanol; $m/z = 59$

18.17 Summary

Free radicals are species with unpaired electrons that are formed by the **homolytic breaking** of a covalent bond. The energy needed to homolytically break a bond to form free radicals can be obtained by heating or irradiating a compound with light.

The light-induced chlorination of alkanes occurs by a free radical mechanism. A free radical mechanism consists of three steps, which are called initiation, propagation, and termination. The **initiation step** starts the reaction by forming free radicals. The **propagation step** maintains the concentration of free radicals in the solution because one radical is formed for every one consumed. The **termination step** stops or slows the propagation step because it consists of reactions that consume free radicals without forming new ones. Given a choice of hydrogen atoms to abstract, a free radical preferentially abstracts the hydrogen atom that leads to the most stable free radical.

$\cdot CH_3$ $CH_3CH_2\overset{\displaystyle\cdot}{C}H_2$ $(CH_3)_2\overset{\displaystyle\cdot}{C}H$ $(CH_3)_3C\cdot$

Methyl radical   1° Radical   2° Radical   3° Radical

INCREASING STABILITY ⟩

Free radical initiators are compounds that decompose to free radicals and can initiate a free radical chain mechanism. A **free radical inhibitor,** on the other hand, reacts with the free radicals formed in the propagation step to form compounds or other free radicals that are incapable of continuing a chain reaction. In the presence of free radical initiators, HBr can add to alkenes by a free radical mechanism to form products of anti-Markownikoff orientation.

Autooxidation is the slow oxidation that organic compounds undergo with atmospheric oxygen. Molecular oxygen is a diradical, and it promotes the oxidation of compounds by free radical mechanisms.

Polymers are high molecular weight compounds that consist of repeating low molecular weight units called **monomers.** Many commercial polymers are produced by the free radical vinyl polymerization of substituted alkenes. Free radical vinyl polymerization produces polymers that are highly branched. Linear polymers are formed by vinyl polymerization in the presence of alkyl transition metal compounds called **Ziegler-Natta catalysts.**

Mass spectrometry is a technique of structure determination that analyzes the results of bombarding a molecule with high-energy electrons. Part of the energy transferred to the molecule dislodges an electron to form a **molecular ion.** The rest of the energy causes the molecular ion to break apart into smaller molecules in a process called **fragmentation.** The fragments that have a positive charge are separated according to their m/z values in a **mass spectrometer** and recorded as a **mass spectrum.** Structural information about the compound is obtained from the m/z value of its molecular ion and the ions formed by fragmentation.

Additional Exercises

18.23 Define each of the following terms in words, by a structure, or by means of a chemical equation.

(a) Free radical	**(h)** Initiation step	**(n)** Syndiotactic polymer
(b) Homolytic bond breaking	**(i)** Atactic polymer	**(o)** Free radical initiator
(c) Copolymer	**(j)** Termination step	**(p)** Free radical chain mechanism
(d) Molecular ion	**(k)** Isotactic polymer	**(q)** Free radical inhibitor
(e) Addition polymer	**(l)** Heterolytic bond breaking	**(r)** Mass spectrum
(f) Monomer	**(m)** Propagation step	**(s)** Ziegler-Natta catalyst
(g) Free radical polymerization		

18.24 Identify each of the following reactions as occurring in initiation, propagation, or termination steps.

(a) $HO-OH \longrightarrow 2\,HO\cdot$

(b) $2\,(CH_3)_2CH\overset{\displaystyle\cdot}{C}H_2 \longrightarrow (CH_3)_2CH=CH_2 + (CH_3)_2CHCH_3$

(c) $Cl\cdot + CH_2{=}CH_2 \longrightarrow ClCH_2\overset{\displaystyle\cdot}{C}H_2$

(d) $2\,RCH_2\overset{\displaystyle\cdot}{C}H_2 \longrightarrow RCH_2CH_2CH_2CH_2R$

(e) $(CH_3)_3CO\cdot + HBr \longrightarrow (CH_3)_3COH + Br\cdot$

(f) $(CH_3)_3COCl \longrightarrow (CH_3)_3CO\cdot + Cl\cdot$

18.25 The free radical chlorination of a hydrocarbon that has a molecular weight of 72 forms only a single monochlorinated product. What is the structure of the hydrocarbon and its monochlorinated derivative?

18.26 Write the structure of the major product formed by the reaction of 1-methylcyclohexene and each of the following reagents.

(a) HBr in diethyl ether and a dialkyl peroxide   (d) 9-BBN followed by H_2O_2/NaOH
(b) HBr in acetic acid   (e) Dilute aqueous solution of H_2SO_4
(c) Br_2 in CCl_4

18.27 Write the mechanism of the light-initiated reaction of chlorine and cyclopentane to form chlorocyclopentane.

18.28 Write the structures of the monomers that polymerize to form each of the following polymers.

(a) $\left[CF_2CF_2CF_2CF_2CF_2CF_2 \right]_n$

(e) $\left[CH_2CHClCH_2CHClCH_2CHCl \right]_n$

(b) $\left[\underset{\underset{CN}{|}}{CH_2CH}CH_2\underset{\underset{CN}{|}}{CH}CH_2\underset{\underset{CN}{|}}{CH} \right]_n$

(f) $\left[CH_2CH_2CH_2\underset{\underset{OOCCH_3}{|}}{CH}CH_2CH_2\underset{\underset{OOCCH_3}{|}}{CH} \right]_n$

(c) $\left[\underset{\underset{NO_2}{|}}{CH_2CH}CH_2\underset{\underset{NO_2}{|}}{CH}CH_2\underset{\underset{NO_2}{|}}{CH} \right]_n$

(g)

(d) $\left[\overset{\overset{OOCCH_3}{|}}{CH_2CH}CH_2\overset{\overset{OOCCH_3}{|}}{CH}CH_2\underset{\underset{OOCCH_3}{|}}{CH} \right]_n$

(h) $\left[\underset{\underset{CH_3}{|}}{\overset{\overset{CH_3}{|}}{C}}CH_2\underset{\underset{CH_3}{|}}{\overset{\overset{CH_3}{|}}{C}}CH_2\underset{\underset{CH_3}{|}}{\overset{\overset{CH_3}{|}}{C}}CH_2 \right]_n$

18.29 Heating a benzene solution of propene, chloroform, and a small amount of dialkyl peroxide forms 1,1,1-trichlorobutane. Write the mechanism of this reaction.

$$CH_3CH{=}CH_2 \ + \ HCCl_3 \ \xrightarrow{\overset{\text{Dialkyl}}{\underset{}{\text{peroxide}}}} \ CH_3CH_2CH_2CCl_3$$

18.30 Write the structures and name all the dichlorinated products formed by the light-induced reaction of chlorine and (S)-2-chlorobutane. Which of the dichlorinated butanes are optically active? Explain.

18.31 Write a three-dimensional representation of a four-unit segment of each of the following polymers.

(a) Atactic polyacrylonitrile   (b) Isotactic polyvinyl acetate   (c) Syndiotactic polystyrene

18.32 While neutral molecules are not recorded in the mass spectrum of a compound, they are, nevertheless, frequently lost in the fragmentation of molecular ions in a mass spectrometer. (1) Suggest reasonable structures for each of the following neutral molecules that are lost on fragmentation. (2) How would you know that the molecular ion had lost such a neutral fragment?

(a) Mass of 18 units and contains no carbon atoms.
(b) Mass of 28 units and contains neither carbon nor hydrogen atoms.
(c) Mass of 28 units and contains only carbon and hydrogen atoms.
(d) Mass of 36 units and contains no carbon atoms.
(e) Mass of 44 units and contains no hydrogen atoms.

18.33 Explain what differences you would expect to see in the mass spectrum of CH_3CH_2OH containing the normal abundance of isotopes and the mass spectra of the following isotopically labelled samples of ethanol.

(a) $CH_3{}^{13}CH_2OH$ (b) $CH_3CH_2{}^{18}OH$ (c) ${}^{13}CH_3CH_2OH$

18.34 Write a mechanism for the reaction that forms the first three units of the polymer formed by the polymerization of methyl methacrylate in the presence of benzoyl peroxide.

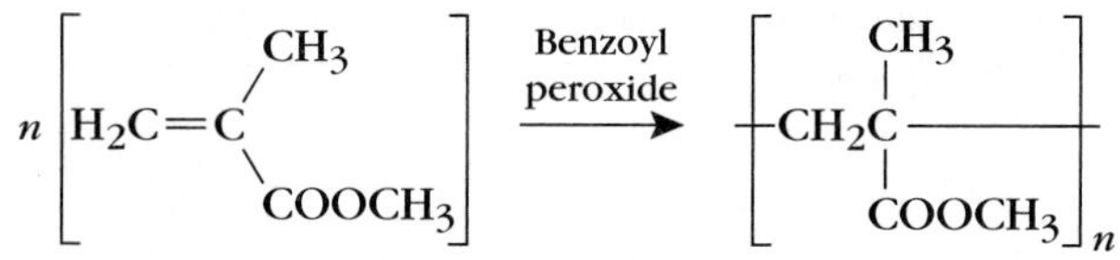

$$n \begin{bmatrix} H_2C=C \begin{smallmatrix} CH_3 \\ \\ COOCH_3 \end{smallmatrix} \end{bmatrix} \xrightarrow[\text{peroxide}]{\text{Benzoyl}} \begin{bmatrix} CH_3 \\ | \\ CH_2C \\ | \\ COOCH_3 \end{bmatrix}_n$$

18.35 The light-initiated bromination of propene forms 3-bromopropene, 1,2,3-tribromopropane, and 1,2-dibromopropane. Write a mechanism to account for this observation.

18.36 Suggest structures and fragmentation patterns to account for the peaks observed in the mass spectrum of each of the following compounds.

  (a) 2-Butanone; $m/z = 72$; $m/z = 43$
  (b) Butanal; $m/z = 72$; $m/z = 57$; $m/z = 44$; $m/z = 29$
  (c) 5-Methyl-2-hexanone; $m/z = 114$; $m/z = 58$; $m/z = 43$
  (d) 3-Hexanol; $m/z = 73$; $m/z = 59$

18.37 What peaks in the mass spectra of the following compounds would result from the McLafferty rearrangement of the molecular ion?

  (a) 2-Octanone   **(b)** 3-Methylpentanal   **(c)** 5-Nonanone

18.38 How would you use mass spectrometry to distinguish between the following pairs of compounds?

  (a) Butanal and 2-butanone   **(c)** 2-Pentanone and 3-pentanone
  (b) 2-Methylbutanal and 3-methylbutanal

18.39 The three ^{1}H NMR spectra that are numbered 1 to 3 correspond to three of the following compounds. Match the compound to its correct spectrum.

  (a) $CH_2=CHC\equiv N$

  (b) $CH_2=CHBr$

  (c) 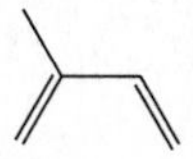 (structure)

  (d) 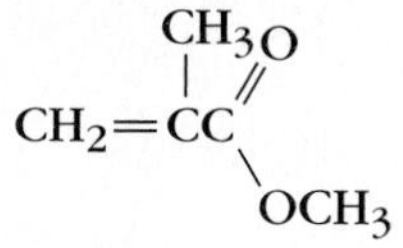 $CH_2=CC \begin{smallmatrix} CH_3O \\ | \;\; \| \\ \\ OCH_3 \end{smallmatrix}$

  (e) 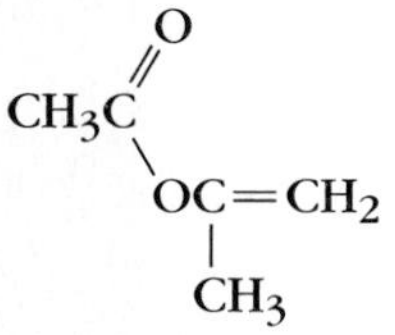 $CH_3C \begin{smallmatrix} O \\ \| \\ \\ OC=CH_2 \\ | \\ CH_3 \end{smallmatrix}$

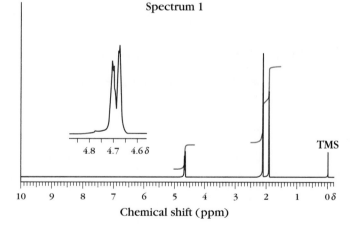

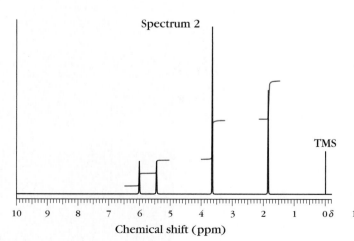

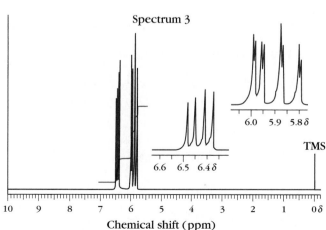

CHAPTER

π ELECTRON DELOCALIZATION in ACYCLIC COMPOUNDS and INTERMEDIATES

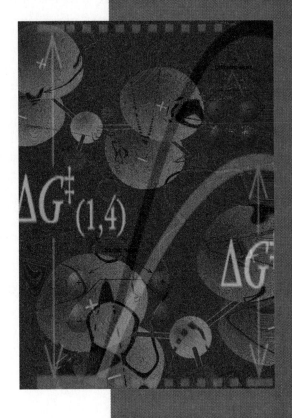

Alkenes are organic compounds that contain a single functional group, the carbon-carbon double bond. Additions to the carbon-carbon double bond, as we learned in Chapters 7 and 8, are their principal reactions. However, alkenes also undergo reactions at carbon atoms *adjacent* to the double bond. The three-carbon unit that contains a carbon atom bonded to a carbon-carbon double bond is commonly called an **allyl group** (IUPAC name 2-propenyl). The carbon atom adjacent to the double bond is called an allylic carbon atom, and hydrogen atoms bonded to it are called allylic hydrogen atoms.

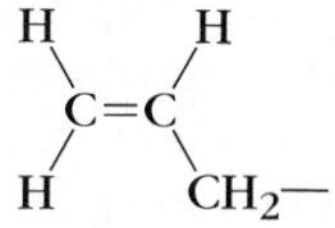

Allyl group
(2-Propenyl group)

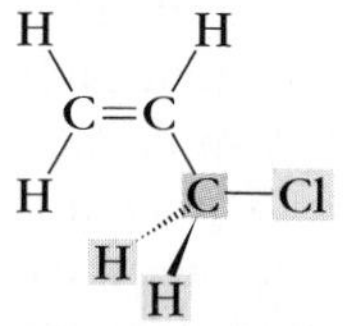

Allylic hydrogen atoms
Allylic carbon atom
Allylic chlorine atom

Substituents bonded to an allylic carbon atom are more reactive than those bonded to the corresponding saturated carbon atom. The solvolysis of 3-chloro-3-methyl-1-butene in 50% aqueous ethanol, for example, is 260 times faster than it is for 2-chloro-2-methylpropane, even though both compounds react by an S_N1 mechanism.

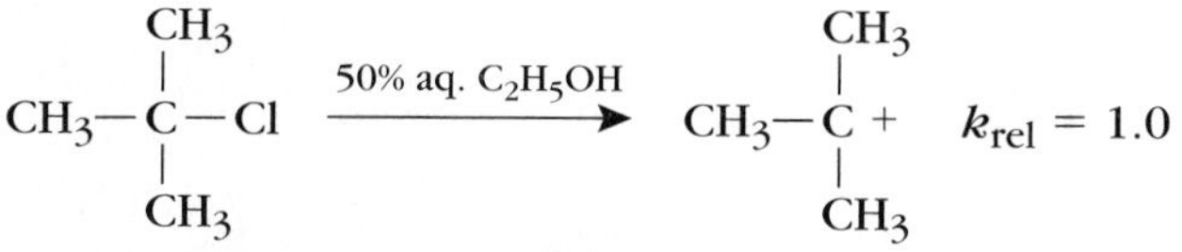

2-Chloro-2-methylpropane

tert-Butyl cation

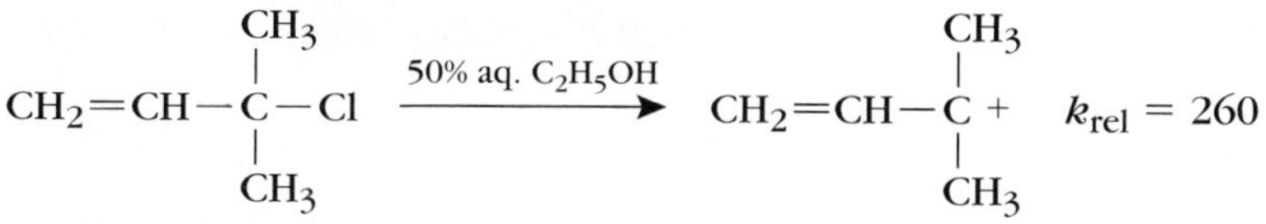

3-Chloro-3-methyl-1-butene

2-Methyl-3-buten-2-yl cation

An allyl group with a positive charge is called an **allyl cation,** while more highly substituted analogs, such as the 2-methyl-3-buten-2-yl cation, are called **allylic cations.**

Exercise 19.1   In each of the following compounds, identify all allylic hydrogen atoms.

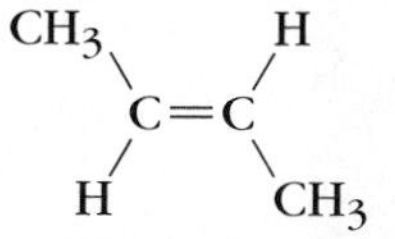

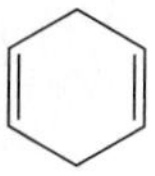

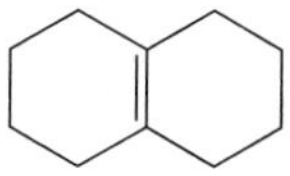

Why are substituents bonded to an allylic carbon atom more reactive than those bonded to the corresponding saturated carbon atom? To answer this question, we must examine the electron structure of the allyl cation.

19.1   The Electron Structure of the Allyl Cation

The electron structure of an allyl cation can be described by the simplified LCAO method, which we introduced in Section 1.13. Recall that this method uses localized valence bond orbitals to describe the σ bonds, and the π molecular orbitals are constructed by the LCAO method. An allyl system consists of three sp^2 hybrid carbon atoms that all lie in a plane. The molecular orbitals of the allyl system are constructed from the three p orbitals (one from each carbon atom) that lie perpendicular to the plane of the carbon atoms. Mathematically combining these three atomic orbitals produces the three molecular orbitals shown in Figure 19.1.

Of the three new π molecular orbitals, the one of lowest energy (π_1) is bonding and has no nodes. The next lowest (π_2) is nonbonding and has one node. It is called nonbonding because it has the same energy as the p atomic orbitals combined to form the molecular orbitals. Finally the one of highest energy (π_3) is antibonding and has two nodes.

We obtain the electron structure of the allyl cation by adding π electrons to the molecular orbitals of the allyl system according to the aufbau principle. The results are shown in Figure 19.2. To obtain the electron structure of the allyl cation, we place its two π electrons in π_1, the MO of lowest energy. Since this MO extends over all three atoms, its two electrons are delocalized over these three atoms.

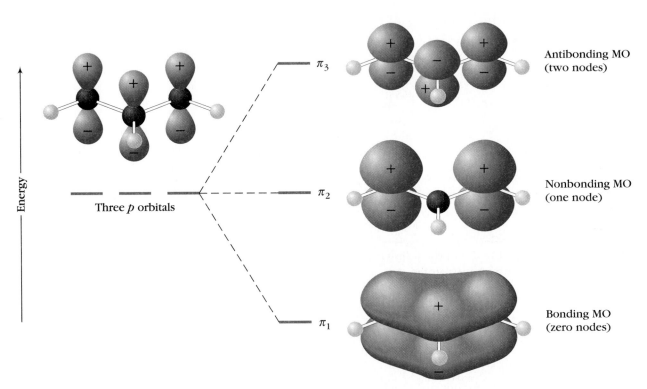

Figure 19.1
The π molecular orbitals of the allyl system.

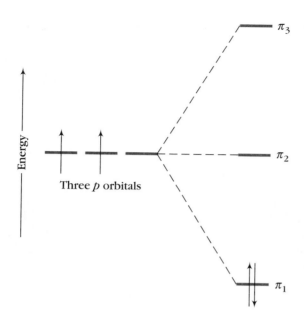

Figure 19.2
Two electrons are added to the π molecular orbitals of the allyl system according to the aufbau principle to obtain the electron structure of an allyl cation.

Electron delocalization makes allyl cations more stable than their saturated counterparts. We can understand why π electron delocalization stabilizes the allyl system by comparing the energy levels of the allyl molecular orbitals with those of a molecule that contains a double bond isolated from an sp^2 hybrid carbon atom. To do this let us use the 4-penten-1-yl intermediate as our model of a system where no delocalization of π electrons can occur because the double bond is isolated from the sp^2 hybrid carbon by two CH_2 groups.

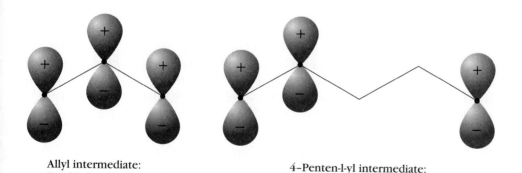

Allyl intermediate:
Double bond adjacent
to a p atomic orbital

4–Penten-l-yl intermediate:
Double bond isolated
from a p atomic orbital

The three π molecular orbitals for the 4-penten-1-yl intermediate are shown on the left in Figure 19.3. The molecular orbital of lowest energy, π_1, is identical to the bonding molecular orbital of ethylene, which is described in Section 1.13. The next lowest orbital is a nonbonding orbital, which is actually an isolated p atomic orbital. The highest energy molecular orbital, π_3, is identical to the antibonding molecular orbital of ethylene.

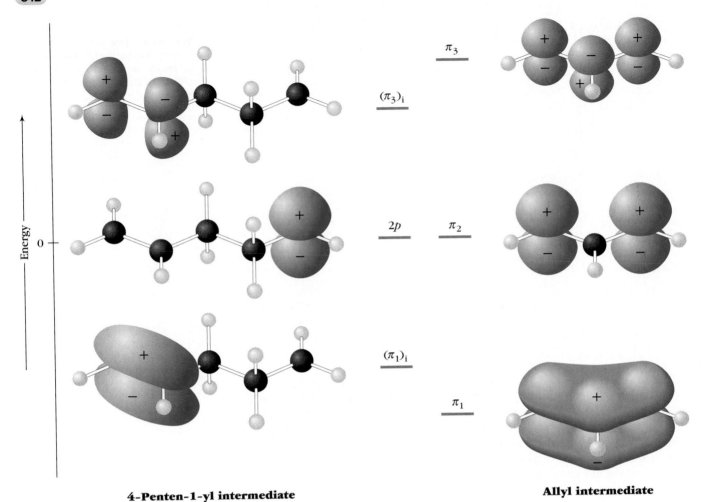

4-Penten-1-yl intermediate

Allyl intermediate

Figure 19.3
A comparison of the energy levels of the molecular orbitals of the 4-penten-1-yl and the allyl systems. The energy of the π_2 molecular orbital of the allyl intermediate is equal to that of an isolated p atomic orbital, which is defined as zero. The π_1 molecular orbital, however, is lower in energy than $(\pi_1)_i$, whereas π_3 is higher in energy than $(\pi_3)_i$.

According to molecular orbital theory, the energy of any chemical species depends on the sum of the energies of all occupied molecular orbitals. Applying this idea to both the allyl and the 4-penten-1-yl cations, we can calculate their energies as follows:

$$\text{Total energy of the allyl cation} = E\pi_1$$
$$\text{Total energy of the pentenyl cation} = E(\pi_1)_i$$

Figure 19.3 shows that π_1 is lower in energy than $(\pi_1)_i$, so it follows that $E\pi_1 < E(\pi_1)_i$ and the allyl cation must be more stable. Generally molecules or ions whose π electrons are delocalized are more stable than similar molecules or ions whose π electrons *are not* delocalized.

The delocalized electron structure of an allyl cation can be equally well represented as a resonance hybrid of the following resonance structures.

Exercise 19.2	Write a second resonance structure for each of the following allyl cations.

(a) [cyclopentenyl cation structure]   (b) $\overset{+}{C}H_2CH=C(CH_3)_2$ (c) [cyclohexenyl =CH_2 cation structure]

■ Recall from Section 7.13 that generally the more stable carbocation is formed faster than its less stable isomers because the stabilities of the carbocations are reflected in their transition states. The lower the energy of the transition state leading to a carbocation, the lower its free energy of activation and the faster the carbocation is formed. This is not always true, as we will discuss in Section 19.9.

The π electron structure of the allyl cation is described by resonance and molecular orbital (LCAO) methods as being delocalized over the three carbon atoms of the allyl system, with the positive charge being shared between the two terminal carbon atoms. These two methods, therefore, are different but complementary ways of visualizing the electron structure of an allyl cation. Because they lead to the same conclusion in this case, we will use only one of the resonance structures to describe the π electron structure of allyl cations. We must remember, however, that the π electrons of allyl cations are delocalized.

The consequences of the enhanced stability of these intermediates are evident in their chemical reactions. We have already seen that allylic cations are formed more easily than their saturated analogs. Another difference is that nucleophilic substitution reactions of allylic compounds form rearranged products.

19.2   Nucleophilic Substitution Reactions of Allylic Derivatives Form Rearranged Products

Most allylic compounds readily undergo nucleophilic substitution reactions to form a mixture of products. An example is the facile hydrolysis of 3-chloro-3-methyl-1-butene at room temperature to form a mixture of alcohols consisting of 15% 3-methyl-2-buten-1-ol and 85% 2-methyl-3-buten-2-ol.

$$(CH_3)_2CCH=CH_2 \xrightarrow[\substack{Na_2CO_3 \\ 25\,°C}]{H_2O} (CH_3)_2CCH=CH_2 \;+\; (CH_3)_2C=CHCH_2OH$$
$$\overset{|}{Cl} \qquad\qquad\qquad\qquad \overset{|}{OH}$$

3-Chloro-3-methyl-1-butene   2-Methyl-3-buten-2-ol   3-Methyl-2-buten-1-ol
                                         (85% yield)   (15% yield)

This distribution of products is consistent with an S_N1 mechanism that involves an intermediate allylic cation.

$$(CH_3)_2CCH=CH_2 \longrightarrow \left[(CH_3)_2\overset{+}{C}-CH=CH_2 \longleftrightarrow (CH_3)_2C=CH-\overset{+}{C}H_2 \right]$$
$$\overset{|}{Cl}$$

$$\Big\downarrow H_2O$$

$$(CH_3)_2CCH=CH_2 \;+\; (CH_3)_2C=CHCH_2OH_2{}^+$$
$$\overset{|}{O}H_2{}^+$$

$$\Big\downarrow \text{loss of a proton}$$

$$(CH_3)_2CCH=CH_2 \;+\; (CH_3)_2C=CHCH_2OH$$
$$\overset{|}{OH}$$

The allylic cation reacts in the second step with the nucleophile (H_2O) at either positive center to give a mixture of products. One product has the same structure as the original allylic chloride, while the other has a rearranged structure. The formation of mixtures of isomeric products in the reaction of allylic compounds is often referred to as an **allylic rearrangement** and is a common feature of reactions that involve allylic cations.

Exercise 19.3 Write the structures of all the isomeric products expected for the reaction of 2-buten-1-ol with hydrogen bromide at 0 °C.

In some cases, the same allylic cation is formed by the reaction of two isomers. Thus hydrolysis of not only 3-chloro-3-methyl-1-butene but also 1-chloro-3-methyl-2-butene forms 2-methyl-3-buten-2-ol (85%) and 3-methyl-2-buten-1-ol (15%) because both proceed by a mechanism involving the same allylic cation.

$$(CH_3)_2CCH\!=\!CH_2 \qquad\qquad (CH_3)_2C\!=\!CHCH_2Cl$$
$$\underset{\textstyle Cl}{|}$$

3-Chloro-3-methyl-1-butene   1-Chloro-3-methyl-2-butene

$$\underset{H_2O}{\overset{Na_2CO_3}{\underset{25\ °C}{\searrow}}} \qquad \underset{H_2O}{\overset{}{\swarrow}}\ \underset{25\ °C}{\overset{Na_2CO_3}{}}$$

$$\left[(CH_3)_2\overset{+}{C}\!-\!CH\!=\!CH_2 \ \longleftrightarrow \ (CH_3)_2C\!=\!CH\!-\!\overset{+}{C}H_2\right]$$

$$\Big\downarrow H_2O$$

$$(CH_3)_2CCH\!=\!CH_2 \ + \ (CH_3)_2C\!=\!CHCH_2OH_2{}^{+}$$
$$\underset{\textstyle OH_2{}^{+}}{|}$$

$$\Big\downarrow \text{loss of a proton}$$

$$(CH_3)_2CCH\!=\!CH_2 \ + \ (CH_3)_2C\!=\!CHCH_2OH$$
$$\underset{\textstyle OH}{|}$$

(85% yield)   (15% yield)

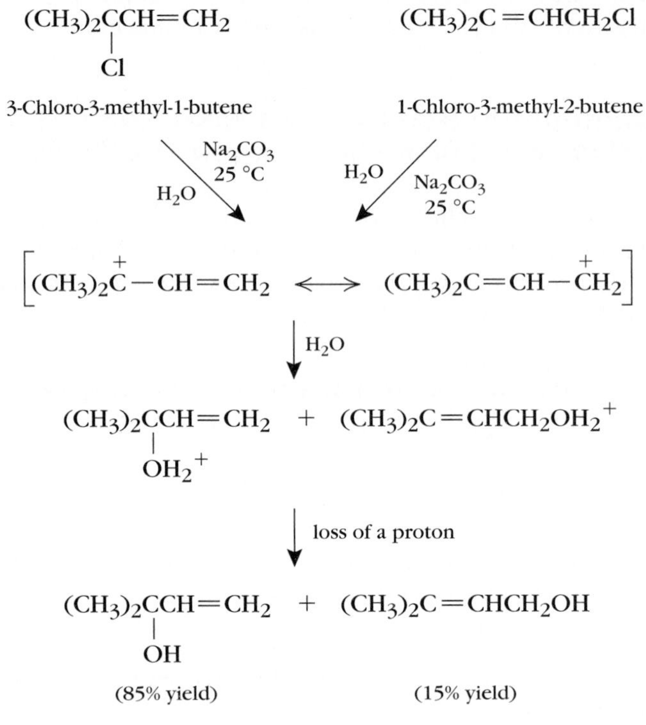

Exercise 19.4 Each of the following compounds forms two products on hydrolysis. Write the structures of the two products formed; write the structure of another compound that forms these two products on hydrolysis; and write the resonance structures that contribute to the resonance hybrid of the intermediate allylic cation.

(a) 3-Bromo-1-butene   **(b)** 3-Chloro-3-methylcyclohexene

Exercise 19.5 A chemist wants to determine the specific rotation of a sample of (2S)-(E)-3-penten-2-ol known to be optically active. The sample, dissolved in ethanol, is placed in a polarimeter tube that, unknown to the chemist, contains a trace of HCl. Much to the surprise of the chemist, the sample is found to be optically inactive, yet on isolation the product is still (E)-3-penten-2-ol. Explain this result.

Notice that hydrolysis of 1-chloro-3-methyl-2-butene, although a primary alkyl halide, occurs by an S_N1 mechanism. Most allylic derivatives undergo nucleophilic substitution reactions in very polar solvents by this mechanism. The major exception is 3-chloropropene, which seems to react predominantly by an S_N2 mechanism, as shown by the data in Table 19.1.

The data in Table 19.1 show that the ratios of relative rates of 3-chloropropene to 1-chloropropane are small and do not change very much despite changing the reaction conditions from one favoring an S_N2 mechanism (ethoxide ion in ethanol) to one favoring an S_N1 mechanism (the very polar and nonnucleophilic formic acid). In formic acid, for example, 3-chloropropene reacts 26 times faster than 1-chloropropane, while in the reaction with ethoxide ion in ethanol, it reacts at about the same rate, 37 times faster than 1-chloropropane. This result suggests that both 1-chloropropane and 3-chloropropene react by the same mechanism under all three conditions.

In contrast, the rate of reaction of 1-chloro-2-butene with ethoxide ion in ethanol is almost 100 times faster than that of 1-chloropropane. The ratio of the two rates increases dramatically, moreover, as the solvent becomes more polar and less nucleophilic. In formic acid, the reaction of 1-chloro-2-butene is 54,210 times faster than 1-chloropropane. This suggests that 1-chloro-2-butene reacts by an S_N1 mechanism, even with ethoxide ion in ethanol.

The fact that most primary allylic halides undergo nucleophilic substitution reactions by an S_N1 mechanism is evidence of the greater stability of allylic cations compared to alkyl carbocations. How stable are allylic cations compared to alkyl carbocations? The results of solvolysis reactions suggest that a 2-propenyl cation is about as stable as a secondary alkyl carbocation, a secondary allylic cation is about as stable as a tertiary alkyl carbocation, and a tertiary allylic cation is more stable than a tertiary alkyl carbocation.

Table 19.1 Relative Reactivities of Allylic Compounds

	Compound		
Nucleophile	$CH_3CH_2CH_2Cl$ 1-Chloropropane	$CH_2{=}CHCH_2Cl$ 3-Chloropropene (allyl chloride)	$CH_3CH{=}CHCH_2Cl$ 1-Chloro-2-butene
Ethoxide ion in ethanol, 45 °C	1.0	37	97
50% Aqueous ethanol, 45 °C	1.0	14	1300
Formic acid, 100 °C	1.0	26	54210

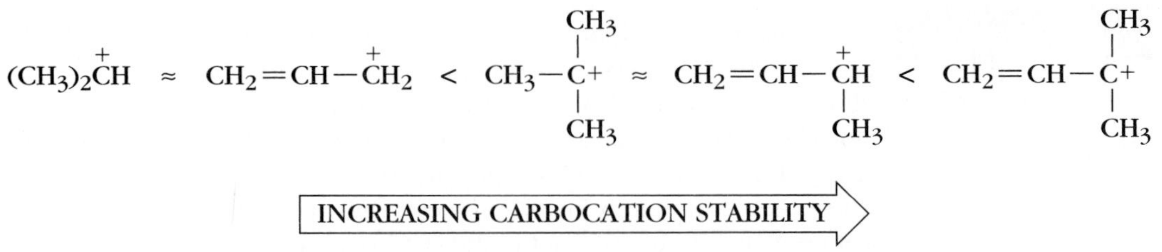

$$(CH_3)_2\overset{+}{C}H \approx CH_2{=}CH{-}\overset{+}{C}H_2 < CH_3{-}\underset{CH_3}{\overset{CH_3}{\underset{|}{\overset{|}{C^+}}}} \approx CH_2{=}CH{-}\underset{CH_3}{\overset{|}{\overset{+}{C}H}} < CH_2{=}CH{-}\underset{CH_3}{\overset{CH_3}{\underset{|}{\overset{|}{C^+}}}}$$

INCREASING CARBOCATION STABILITY ⟩

In summary, most allylic compounds undergo nucleophilic substitution reactions via an S_N1 mechanism involving an allylic cation. The major exception is 3-chloropropene, which undergoes nucleophilic substitution reactions predominantly via an S_N2 mechanism. Allylic cations react with nucleophiles at the two electropositive centers to form a mixture of products.

19.3 Allylic Radicals Are Intermediates in the Allylic Bromination of Alkenes

When bromination of an alkene, such as propene, is carried out at very low concentrations of bromine in a nonpolar solvent such as tetrachloromethane, addition to the double bond (Section 8.8) does not occur. A different reaction occurs, called **allylic bromination,** in which the bromine atom replaces an allylic hydrogen atom. The allylic bromination of propene by a low concentration of bromine in tetrachloromethane is one example.

$$CH_2{=}CHCH_3 \ + \ \underset{\substack{\text{Low}\\\text{concentration}}}{Br_2} \ \xrightarrow{\ CCl_4\ } \ CH_2{=}CHCH_2Br \ + \ HBr$$

Maintaining a low concentration of bromine is difficult to do, so it is more convenient to carry out allylic bromination in the laboratory with the commercially available reagent *N*-bromosuccinimide (NBS).

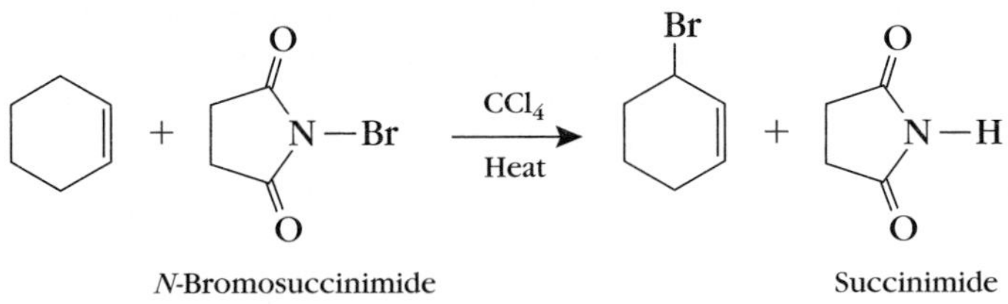

N-Bromosuccinimide   Succinimide
(NBS)

The actual brominating agent in this reaction is bromine, which is formed in low concentrations by the reaction of *N*-bromosuccinimide with traces of hydrogen bromide.

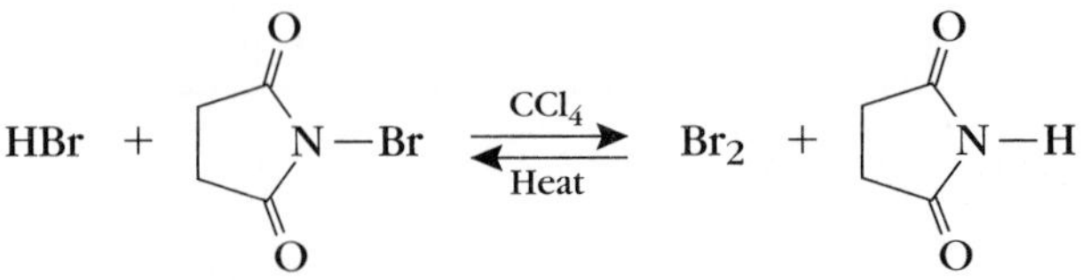

N-Bromosuccinimide   Succinimide
(NBS)

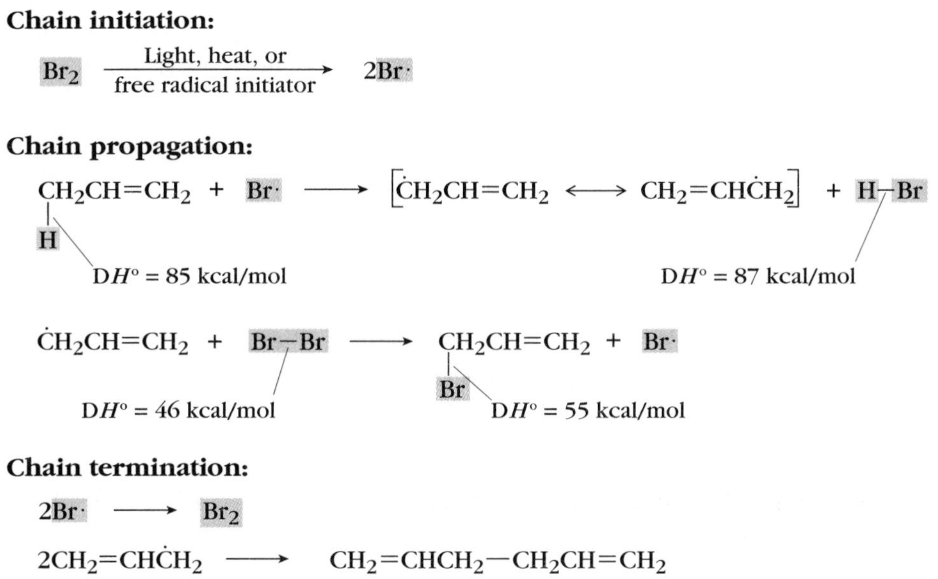

Chain initiation:

$$Br_2 \xrightarrow{\text{Light, heat, or}\atop\text{free radical initiator}} 2Br\cdot$$

Figure 19.4
The free radical chain mechanism of
the bromination of propene by NBS.
The weakest C—H bond of propene
is the allylic C—H bond, so abstraction of the allylic hydrogen atom
occurs in the chain propagation
step.

Chain propagation:

$$\underset{\overset{|}{H}}{CH_2CH{=}CH_2} + Br\cdot \longrightarrow \left[\dot{C}H_2CH{=}CH_2 \longleftrightarrow CH_2{=}CH\dot{C}H_2\right] + H{-}Br$$

$DH° = 85$ kcal/mol $DH° = 87$ kcal/mol

$$\dot{C}H_2CH{=}CH_2 + Br{-}Br \longrightarrow \underset{\overset{|}{Br}}{CH_2CH{=}CH_2} + Br\cdot$$

$DH° = 46$ kcal/mol $DH° = 55$ kcal/mol

Chain termination:

$$2Br\cdot \longrightarrow Br_2$$

$$2CH_2{=}CH\dot{C}H_2 \longrightarrow CH_2{=}CHCH_2{-}CH_2CH{=}CH_2$$

Allylic bromination is believed to occur by a free radical chain mechanism (Section 18.1). Recall from Chapter 18 that a free radical chain mechanism occurs by three steps, which are initiation, propagation, and termination. These three steps are shown for the NBS bromination of propene in Figure 19.4.

In free radical reactions, the hydrogen atom abstracted is the one whose C—H bond is the weakest. In this case, the allylic C—H bond is the weakest. Consequently, bromine atoms, formed in the initiation step, abstract allylic hydrogen atoms in the first propagation reaction. This reaction is almost thermoneutral, however, because the energy absorbed by breaking an allylic C—H bond (85 kcal/mol or 355 kJ/mol) is almost equal to the energy released by forming an H—Br bond (87 kcal/mol or 364 kJ/mol). In the second propagation reaction, the allylic radical reacts with more bromine to form 3-bromopropene and a bromine atom, which continues the chain reaction. This reaction is exothermic, because the overall propagation step is exothermic. The HBr formed in the first reaction reacts with NBS to generate more bromine to continue the reaction.

The π electron structure of the allyl radical can be described as a resonance hybrid of the following two resonance structures.

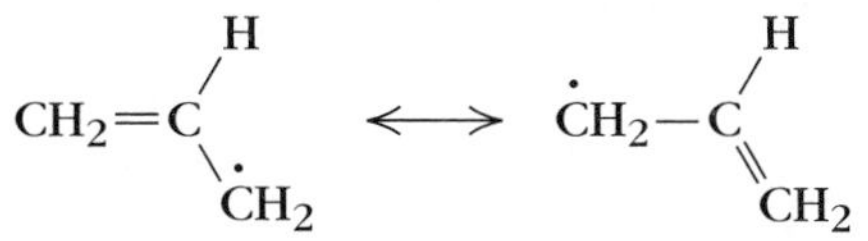

$$CH_2{=}\overset{\overset{\displaystyle H}{|}}{\underset{\underset{\displaystyle CH_2}{\cdot}}{C}} \longleftrightarrow \dot{C}H_2{-}\overset{\overset{\displaystyle H}{|}}{\underset{\underset{\displaystyle CH_2}{\parallel}}{C}}$$

Consequently, the allyl radical has a delocalized π electron structure similar to the allyl cation.

| **Exercise 19.6** | Describe the π electron structure of the allyl radical by adding electrons according to the aufbau principle to the π molecular orbitals of the allyl system, shown in Figure 19.2 (p. 841). |

Best results in the allylic bromination of alkenes are obtained with symmetrical alkenes. When an alkene is unsymmetrical, the corresponding delocalized allyl radical intermediate is also unsymmetrical, so a mixture of products is usually formed.

$$CH_3(CH_2)_4CH_2CH{=}CH_2$$

1-Octene

$$\frac{CCl_4}{Heat}\Big|\ NBS$$

$$\Big[CH_3(CH_2)_4CH{=}CH\dot{C}H_2 \longleftrightarrow CH_3(CH_2)_4\dot{C}H{-}CH{=}CH_2\Big]$$

Unsymmetrical electron delocalized allyl radical

$$CH_3(CH_2)_4CH{=}CHCH_2 \quad + \quad CH_3(CH_2)_4CHCH{=}CH_2$$
$$\underset{Br}{|} \qquad\qquad\qquad \underset{Br}{|}$$

1-Bromo-2-octene   3-Bromo-1-octene
(54% yield)   (17% yield)

If bromine is the actual bromination agent in the NBS reaction, why doesn't it add to the double bond as well? Under the conditions of low bromine concentration and nonpolar solvent, the rate law of electrophilic addition of bromine is second order in bromine, so it is very slow. It is so slow, in fact, that allylic bromination is the only reaction that occurs to any noticeable extent.

Exercise 19.7 Write the structures of the products formed by the reaction of one equivalent of NBS in CCl_4 with each of the following alkenes.

(a) 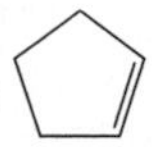 (pentagon with double bond)   (b) $(CH_3)_2C{=}C(CH_3)_2$ (c) 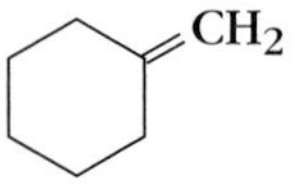 (cyclohexane with $=CH_2$)

Exercise 19.8 Starting with propene, NBS, any solvent, and any inorganic reagent, propose a synthesis of each of the following compounds.

(a) $CH_2{=}CHCH_2CN$ (c) $CH_2{=}CHCH_2SH$
(b) $CH_2{=}CHCH_2I$ (d) $CH_2{=}CHCH_2OOCCH_3$

Allylic intermediates are also formed by addition reactions of conjugated dienes and polyenes.

19.4 Dienes and Polyenes

A **diene,** as we learned in Chapter 7, is a hydrocarbon that contains two (di) carbon-carbon double bonds (ene). Dienes are classified according to the relative locations of their two double bonds. When the two double bonds are separated by one single bond, the diene is called a **conjugated diene.** These double bonds are said to be **conjugated double bonds.**

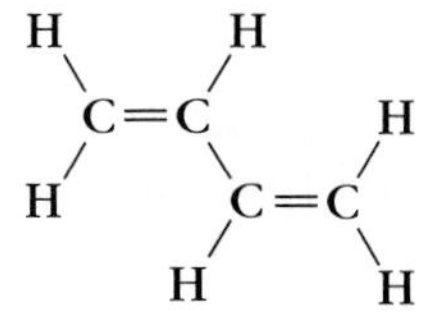

1,3-Butadiene
A conjugated diene

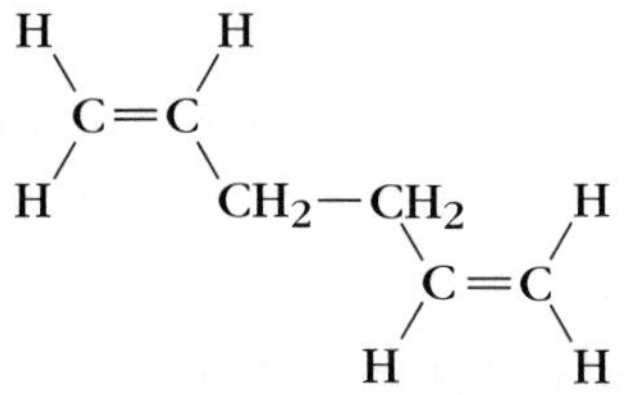

1,5-Hexadiene
A nonconjugated diene

Dienes in which the double bonds are separated by two or more single bonds are called **nonconjugated dienes,** and the double bonds are said to be isolated. The two isolated double bonds of a nonconjugated diene react independently of each other.

Cumulenes are dienes in which one carbon atom is part of two carbon-carbon double bonds. The double bonds in these compounds are called **cumulative double bonds.** The simplest cumulene is 1,2-propadiene, which is better known by its common name, allene. The name allene is not only the name of a single compound, but is often used to denote the class of compounds that contains two cumulative double bonds.

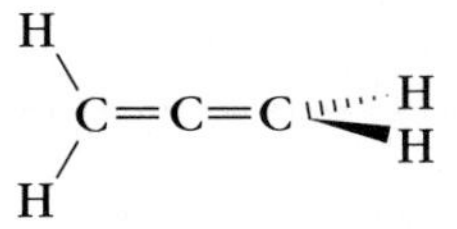

1,2-Propadiene (allene)
The simplest example of
a cumulene

Compounds with conjugated carbon-carbon double bonds are common in nature. β-Carotene, which is responsible for the color of carrots, is one example.

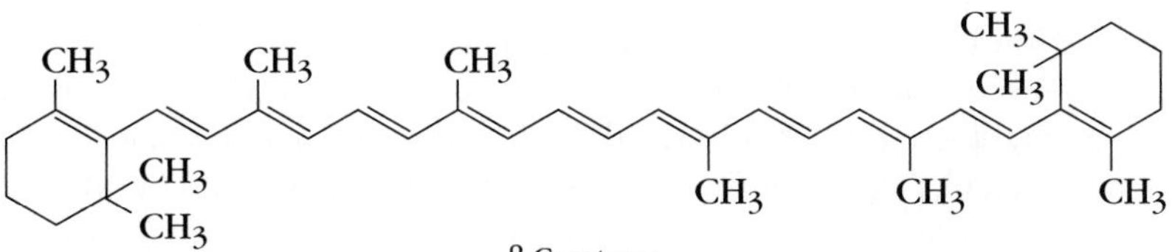

β-Carotene

Hydrocarbons that contain multiple double bonds, such as β-carotene, are called **polyenes.** β-Carotene is a conjugated polyene.

■ *Recall that the prefix* poly- *means* many.

Exercise 19.9 Indicate which of the following hydrocarbons are conjugated dienes and circle the conjugated double bonds.

(a)

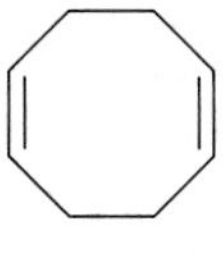

(c)

(b)

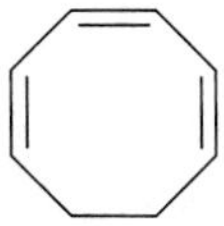

(d) 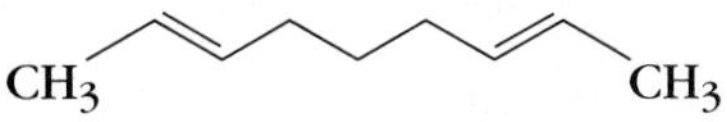 CH$_3$ CH$_3$

| Exercise 19.10 | What is the IUPAC name of each of the compounds in Exercise 19.9? |

Conjugated dienes are similar to isolated alkenes in many of their reactions. There are several differences, however, one of which is their stability.

19.5   Stability of Conjugated Dienes

The stability of allylic intermediates was attributed to electron delocalization in Section 19.1. The structures of allylic intermediates and conjugated dienes both have several sp^2 hybrid carbon atoms adjacent to each other. Thus we might expect that the π electrons of conjugated dienes would be delocalized, making them more stable than dienes with two isolated double bonds. An examination of the heats of hydrogenation of the alkenes and alkadienes given in Table 19.2 suggests that this is true.

Table 19.2	Heats of Hydrogenation	
Compound	Heats of Hydrogenation $\Delta H°$ kcal/ mol (kJ/mol)
$CH_3CH_2CH{=}CH_2$	−30.3 (−127)
$CH_3CH_2CH_2CH{=}CH_2$	−29.8 (−125)
$CH_3CH_2CH_2CH_2CH{=}CH_2$	−30.0 (−126)
$CH_2{=}CHCH_2CH{=}CH_2$	−60.8 (−254)
$CH_2{=}CHCH_2CH_2CH{=}CH_2$	−60.5 (−253)
$CH_2{=}CHCH{=}CH_2$	−57.1 (−239)

Hydrogenation of an alkene containing a terminal double bond, such as 1-butene, 1-pentene, or 1-hexene, releases about 30 kcal/mol (126 kJ/mol) of energy. We might expect, therefore, that the heat of hydrogenation of a compound that contains two noninteracting isolated terminal double bonds would be about twice this value. The heats of hydrogenation of 1,4-pentadiene and 1,5-hexadiene, which are about 60 kcal/mol (251 kJ/mol), are consistent with this hypothesis. The heat of hydrogenation of 1,3-butadiene, on the other hand, is only 57.1 kcal/mol (239 kJ/mol), which is about 3 kcal/mol (13 kJ/mol) less than that of 1,4-pentadiene or 1,5-hexadiene. The double bonds in 1,3-butadiene are conjugated, which suggests that 1,3-butadiene is stabilized by about 3 kcal/mol compared to nonconjugated dienes. This stabilizing energy due to the presence of conjugated double bonds in a molecule is called the **delocalization energy** of a conjugated diene.

| Exercise 19.11 | Hydrogenation of both 2-methyl-1-butene and 2-methyl-1,3-butadiene forms 2-methylbutane. On the basis of their heats of hydrogenation, calculate the delocalization energy of 2-methyl-1,3-butadiene. |

$\Delta H°$ (2-methyl-1-butene) = −26.9 kcal/mol (−113 kJ/mol)
$\Delta H°$ (2-methyl-1,3-butadiene) = −53.4 kcal/mol (−223 kJ/mol)

Electron delocalization over two conjugated double bonds can best be described by molecular orbital theory.

19.6 Molecular Orbital Description of 1,3-Butadiene

The molecular orbitals of 1,3-butadiene are constructed according to the LCAO method by mathematically combining four p orbitals (one from each carbon atom). The result is the four molecular orbitals shown qualitatively in Figure 19.5. The lowest molecular orbital, π_1, is a bonding orbital and contains two π electrons. Next higher in energy is molecular orbital π_2, which is also a bonding orbital. It contains the other two π electrons. Finally, the antibonding molecular orbitals, π_3 and π_4, complete the molecular orbital description of 1,3-butadiene.

Quantitative molecular orbital calculations show that the sum of the energies of the two bonding molecular orbitals of 1,3-butadiene is slightly less than the sum of the energies of the bonding orbitals of two isolated double bonds in 1,4-pentadiene. This means that placing four electrons in the two bonding molecular orbitals of 1,3-butadiene is a more stable arrangement than placing four electrons in the bonding molecular orbitals of two isolated double bonds of 1,4-pentadiene. Thus 1,3-butadiene is another example where electron delocalization leads to a system that is more stable than one in which no delocalization of electrons is possible.

Figure 19.5
The four molecular orbitals of 1,3-butadiene. Its four π electrons are placed in the two π molecular orbitals of lowest energy, π_1 and π_2. Both of them are bonding. Molecular orbitals π_3 and π_4 are antibonding.

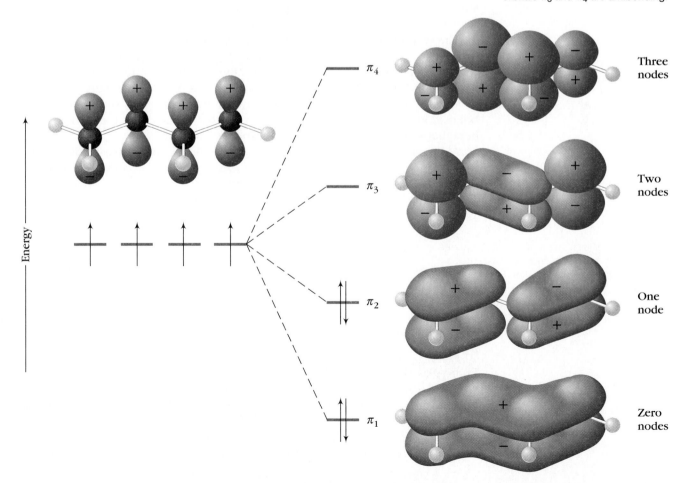

Three nodes

Two nodes

One node

Zero nodes

Several features of the structure of 1,3-butadiene are believed to be the direct result of electron delocalization.

19.7   The Structure of 1,3-Butadiene

The lengths of the carbon-carbon single bonds in ethane, propene, and 1,3-butadiene decrease in the order ethane (154 pm) > propene (151 pm) > 1,3-butadiene (146 pm).

sp^3-sp^3 Single bond	sp^3-sp^2 Single bond	sp^2-sp^2 Single bond
(154 pm)	(151 pm)	(146 pm)

The single bond between C2 and C3 of 1,3-butadiene is particularly short. Two explanations have been proposed to account for this fact. The first is based on the hybridization of the carbon atoms that form the single bonds. In ethane, the single bond forms between two sp^3 hybrid carbon atoms. In 1,3-butadiene, however, the single bond forms between two sp^2 hybrid carbon atoms. This single bond has more s character than single bonds in alkanes, so it is shorter and stronger. The single bond in propene is intermediate in length because it forms between an sp^2 and an sp^3 hybrid carbon atom.

The second explanation is that the C2—C3 bond in 1,3-butadiene is shortened because of π electron delocalization in the bonding molecular orbitals. According to this explanation, π electron delocalization results in the formation of a partial double bond between carbon atoms 2 and 3. As a result, the length of this bond should be between that of a single bond in an alkane (154 pm) and a double bond in an alkene (133 pm). According to this explanation, the partial double bond between carbon atoms 2 and 3 stabilizes the diene.

Although electron delocalization may stabilize 1,3-butadiene, it does not prevent rotation about the C2—C3 bond. Rotation makes two conformations possible, the s-*cis* and the s-*trans* conformations.

■ *The prefix s in s-cis and s-trans refers to conformations about a single bond; in this case the single bond is the one between carbons 2 and 3.*

$E_a = 3.9$ kcal/mol

$\Delta H° = -2.8$ kcal/mol

s-*cis* Conformation of 1,3-butadiene

s-*trans* Conformation of 1,3-butadiene

In the s-*cis* conformation, two hydrogen atoms on C1 and C4 come in close contact. The resulting steric interference, which does not occur in the s-*trans* conformation, makes the s-*cis* 2.8 kcal/mol (11.7 kJ/mol) less stable than the s-*trans* conformation. Interconversion of the s-*cis* and s-*trans* conformations of 1,3-butadiene requires only 3.9 kcal/mol (16.3 kJ/mol), which is about the same as the energy needed to interconvert the conformations of butane. As a result, the two conformations of 1,3-butadiene easily interconvert by rotation about the C2—C3 bond.

Recall from Chapters 7 and 8 that electrophiles add to alkenes. Electrophiles add to conjugated dienes as well.

19.8   Electrophilic Additions to Conjugated Dienes

Addition of one equivalent of HCl to 1,5-hexadiene forms the expected product, 5-chloro-1-hexene, with Markownikoff orientation. Addition of a second equivalent of HCl forms 2,5-dichlorohexane. Typical of nonconjugated dienes, the two double bonds react as if they were in different molecules.

$$CH_2{=}CHCH_2CH_2CH{=}CH_2 \xrightarrow{\;HCl\;} \underset{\underset{H}{|}}{CH_2}{-}\underset{\underset{Cl}{|}}{CH}CH_2CH_2CH{=}CH_2$$

1,5-Hexadiene   5-Chloro-1-hexene
(a nonconjugated diene)

2nd Equivalent
HCl

$$\underset{\underset{H}{|}}{CH_2}{-}\underset{\underset{Cl}{|}}{CH}CH_2CH_2\underset{\underset{Cl}{|}}{CH}{-}\underset{\underset{H}{|}}{CH_2}$$

2,5-Dichlorohexane

Conjugated dienes, on the other hand, react with one equivalent of HCl to form two products.

$$CH_2{=}CHCH{=}CH_2 \xrightarrow{\;HCl\;} \underset{\underset{H}{|}}{CH_2}{-}\underset{\underset{Cl}{|}}{CH}{-}CH{=}CH_2 \;+\; \underset{\underset{H}{|}}{CH_2}{-}CH{=}CH{-}\underset{\underset{Cl}{|}}{CH_2}$$

1,3-Butadiene   3-Chloro-1-butene   1-Chloro-2-butene
                             (80% yield)   (20% yield)
                             1,2-addition product   1,4-addition product

The major product, 3-chloro-1-butene, is called the **1,2-addition product** because it is formed by addition of HCl to adjacent carbon atoms of the diene. The other product, 1-chloro-2-butene, is called the **1,4-addition product** because HCl adds to carbon atoms 1 and 4 of the conjugated system.

Many other electrophiles add to conjugated dienes to form both 1,2- and 1,4-addition products. For example:

$$CH_2{=}CHCH{=}CH_2 \xrightarrow[-15\,°C]{\;Br_2\;} \underset{\underset{Br}{|}}{CH_2}{-}\underset{\underset{Br}{|}}{CH}{-}CH{=}CH_2 \;+\; \underset{\underset{Br}{|}}{CH_2}{-}CH{=}CH{-}\underset{\underset{Br}{|}}{CH_2}$$

1,3-Butadiene   3,4-Dibromo-1-butene   1,4-Dibromo-2-butene
                             (46% yield)   (54% yield)
                             1,2-addition product   1,4-addition product

How does the 1,4-addition product form when HCl is added to 1,3-butadiene? The 1,4-addition product forms because an allylic cation is formed in the mechanism of addition of HCl, as shown in Figure 19.6.

Step 1: Electrophilic addition of a proton to the diene forms the more stable allylic cation

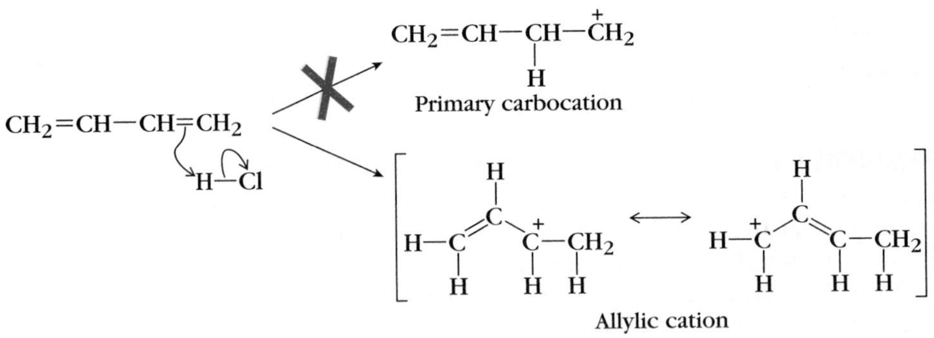

Step 2: Chloride ion reacts with both electron-deficient carbon atoms of the allylic cation to form a mixture of two products

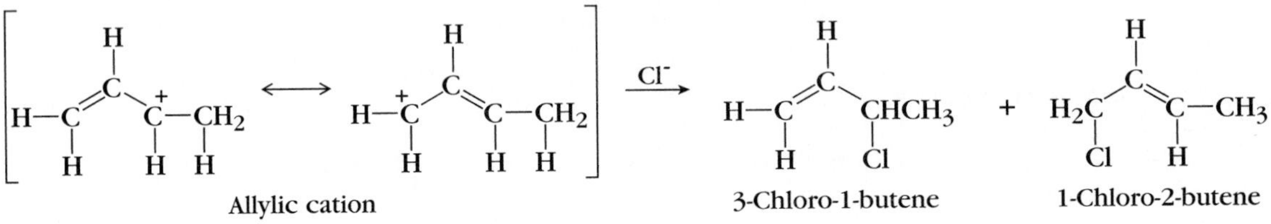

Figure 19.6
The two-step mechanism for the addition of HCl to 1,3-butadiene occurs via an allylic cation, so 1,2- and 1,4-addition products are formed.

Step 1 of the mechanism is the protonation of the conjugated diene. This is the same as Step 1 in the mechanism of the protonation of alkenes. The proton does not add to carbon 2 because this would form the less stable primary cation. Instead, protonation occurs on carbon 1 to form the more stable allylic cation. In Step 2, the nucleophile (Cl^-) reacts at the two electron-deficient sites (electrophilic sites) of the allylic cation to form a mixture of products.

In summary, the intermediate in electrophilic addition reactions of conjugated alkenes is a delocalized allylic cation that reacts with nucleophiles to give a mixture of products. An interesting feature of these reactions is that the product composition changes with time or with an increase in reaction temperature. The reason for this behavior is discussed next.

| **Exercise 19.12** | Reaction of HBr and 1,3,5-hexatriene forms three addition products. Write their structures and propose a mechanism that explains their formation. |

19.9 Kinetic vs. Thermodynamic Control of Product Composition

In Section 19.8, we learned that many reactions of conjugated dienes give mixtures of 1,2- and 1,4-addition products. When such reactions are carried out at temperatures below room temperature, the 1,2-addition product is formed in greater amounts. The addition of HCl to 1,3-butadiene at $-15\,°C$ is one example.

$$CH_2{=}CHCH{=}CH_2 \xrightarrow[-15\,°C]{HCl}$$

CH₂—CH—CH=CH₂   +   CH₂—CH=CH—CH₂
 | | | |
 H   Cl   H   Cl

1,3-Butadiene   3-Chloro-1-butene   1-Chloro-2-butene
                              (82% yield)   (18% yield)

This mixture of products is not at equilibrium because the product composition changes on standing for a long time, or after the mixture is heated. It turns out, in fact, that the 1,4-addition product is the major product at equilibrium.

CH₂—CH—CH=CH₂ $\xrightarrow[]{\text{Heat}}\rightleftharpoons$ CH₂—CH=CH—CH₂
 | | | |
 H   Cl   H   Cl

    (25% at equilibrium)   (75% at equilibrium)

The more stable product is always favored at equilibrium, so the 1,4-addition product is the more stable of the two isomers.

The fact that more 1,2- than 1,4-addition product forms at $-15\,°C$ means that it forms faster than the 1,4-addition product, not that it is necessarily the more stable isomer. In this case, in fact, the reaction that forms the less stable isomer is faster than the one that forms the more stable product.

CH₂=CHCH=CH₂   +   HCl

— Fast →
CH₂—CH—CH=CH₂
 | |
 H   Cl
(82% yield)
1,2-Addition product
(less stable isomer)

+

— Slow →
CH₂—CH=CH—CH₂
 | |
 H   Cl
(18% yield)
1,4-Addition product
(more stable isomer)

Because the product distribution of the addition of HCl to 1,3-butadiene at low temperature is determined by the relative *rates* of formation of the two products, the reaction is said to be under **kinetic control.** At higher temperatures, however, the product distribution of the addition of HCl to 1,3-butadiene is determined by the relative *stabilities* of the two products and consequently the reaction is said to be under **thermodynamic control.** The free energy–reaction path diagram in Figure 19.7 illustrates how this can arise.

The composition of the products from the reaction of HCl and 1,3-butadiene is determined by Step 2 of the mechanism (reaction of the allylic cation with Cl⁻), so we need consider only that step to explain the effect of temperature on the product composition. The 1,2-addition product forms the fastest because its free energy of activation, $\Delta G^{\ddagger}_{(1,2)}$, is less than that for the 1,4-addition product, $\Delta G^{\ddagger}_{(1,4)}$. To understand how this diagram explains kinetic and thermodynamic control, consider what happens at two different temperatures.

At high temperatures, there is enough energy present in the system for both the forward and reverse reactions to occur rapidly. The molecules, therefore, have enough energy to overcome the energy barriers in both the forward and reverse directions, so Step 2 of the mechanism is reversible. When this is true,

Figure 19.7
The free energy-reaction path diagram for Step 2 of the mechanism of the addition of HCl to 1,3-butadiene. The less stable 1,2-addition product forms first because $\Delta G^{\ddagger}_{(1,2)} < \Delta G^{\ddagger}_{(1,4)}$.

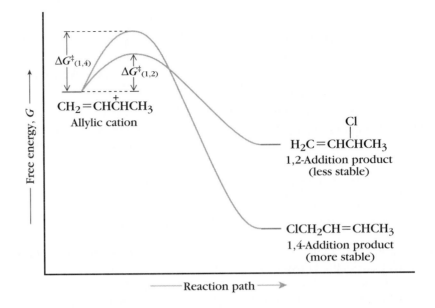

the 1,2- and 1,4-addition products can interconvert rapidly by an allylic rearrangement. Under conditions where 1,2- and 1,4-addition products rapidly equilibrate, the thermodynamic stability of the two products is more important than their rates of formation. Consequently, while the 1,2-addition product forms more rapidly, it is readily reconverted to the allylic cation, which can then form the 1,4-addition product. In other words, there is a path for the 1,2-addition product to be easily converted to the more stable 1,4-addition product. The reverse path can convert the 1,4-addition product to the 1,2-addition product. The final result, then, is a product mixture determined by the relative *stabilities* of the products, not the rates at which they form.

At lower temperatures, it takes longer for the reaction to reach equilibrium, and if the reaction is stopped before equilibrium is attained, the product composition is determined by the free energies of activation, $\Delta G^{\ddagger}_{(1,2)}$ and $\Delta G^{\ddagger}_{(1,4)}$. The result is a product mixture determined by the relative *rates* of formation of the products. These conditions favor the 1,2-addition product because $\Delta G^{\ddagger}_{(1,2)}$ is less than $\Delta G^{\ddagger}_{(1,4)}$.

The product compositions of many organic reactions are formed under conditions of kinetic control. It is important to keep this fact in mind because it means that the more stable product is not always formed in the greatest amounts in a chemical reaction. While the relative thermodynamic stabilities of compounds can be estimated fairly well, there is still no way to predict if the products of a reaction are formed under conditions of kinetic or thermodynamic control because there is no way to predict the energies of activation of chemical reactions. Thus we may know the relative stabilities of compounds, but we cannot predict how fast they will be formed.

19.10 Electrocyclic Reactions of Polyenes

Electrocyclic reactions are concerted interconversions of conjugated polyenes and cyclic compounds. The interconversion of conjugated dienes and cyclobutenes is an example of an electrocyclic reaction.

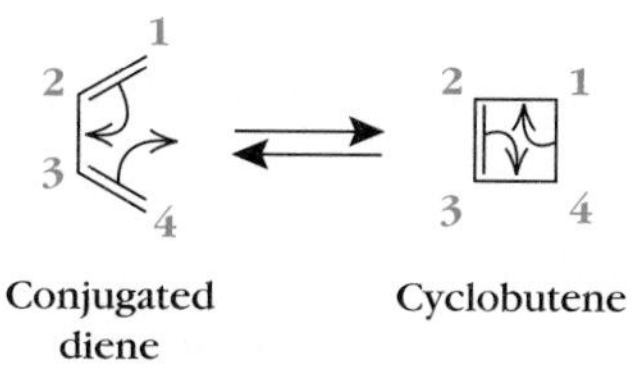

Conjugated   Cyclobutene
diene

The position of the equilibrium favors the conjugated diene.

Electrocyclic reactions are stereospecific. When *cis*-3,4-dimethylcyclobutene, for example, is heated to 175 °C, *only* (2E,4Z)-hexadiene is formed:

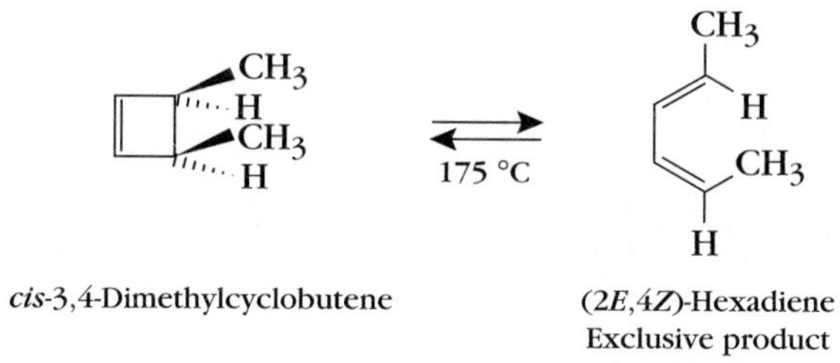

cis-3,4-Dimethylcyclobutene   (2E,4Z)-Hexadiene
                                       Exclusive product

The stereochemistry of the product of an electrocyclic reaction depends on whether the reaction occurs by heating (thermal reaction) or by irradiation with ultraviolet light (photochemical reaction). The photochemical reaction of *cis*-3,4-dimethylcyclobutene, for example, forms *exclusively* (2E,4E)-hexadiene, *not* (2E,4Z)-hexadiene, which is the product of the thermal reaction of *cis*-3,4-dimethylcyclobutene.

cis-3,4-Dimethylcyclobutene   (2E,4E)-Hexadiene
                                       Exclusive product

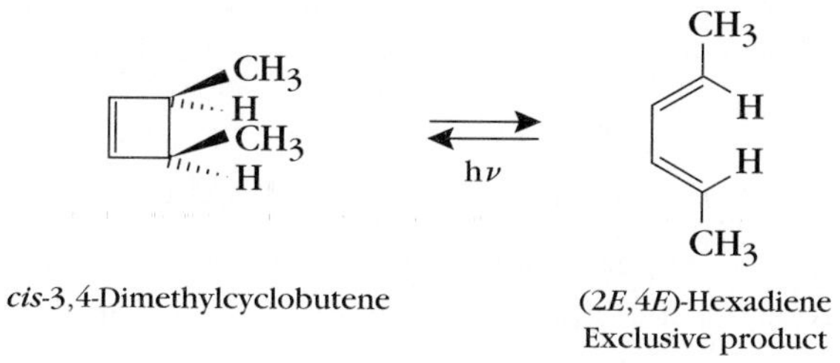

Thus the products of thermal and photochemical electrocyclic reactions of 3,4-dimethylcyclobutene have opposite stereochemistry.

To explain this remarkable stereospecificity, let's look more closely at the movement of the atoms during the reactions. As the electrocyclic reactions of 3,4-dimethylcyclobutene occur, the σ bond between carbon atoms 3 and 4 must break and the bonds between carbon atoms 2 and 3 and 1 and 4 must rotate.

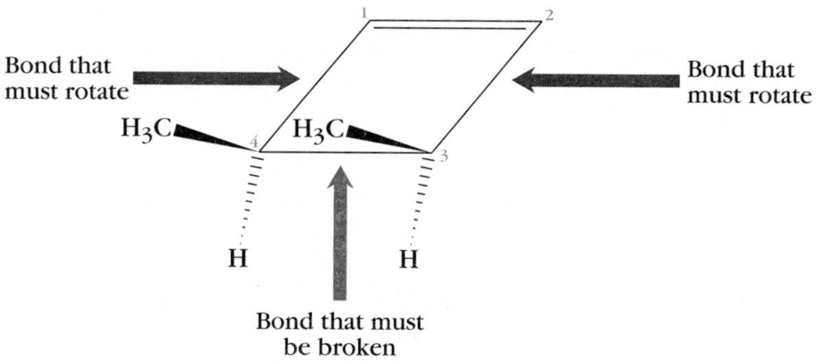

Bond that must rotate   Bond that must rotate

Bond that must be broken

The rotation can occur so that both methyl groups move in the same direction, either both clockwise or both counterclockwise. This rotation is called **conrota-**

tory. Rotation can also occur so that one methyl group moves in a counterclockwise direction and the other moves in a clockwise direction. This rotation is called **disrotatory.**

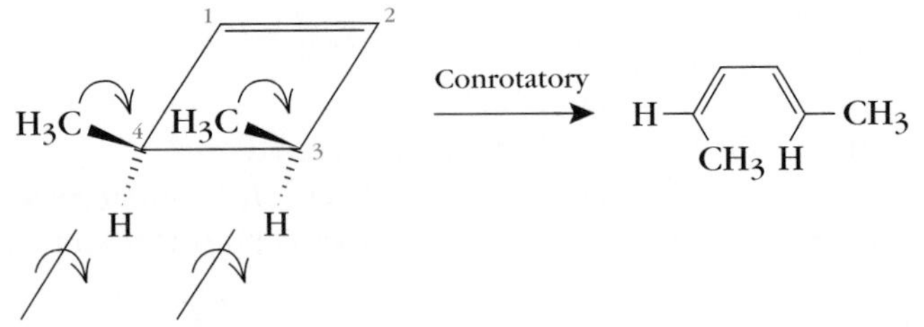

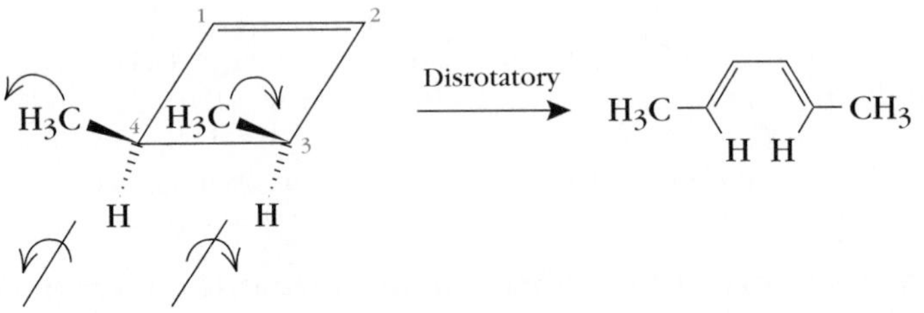

The thermal reaction of *cis*-3,4-dimethylcyclobutene must occur by conrotatory ring opening to form (2*E*,4*Z*)-hexadiene.

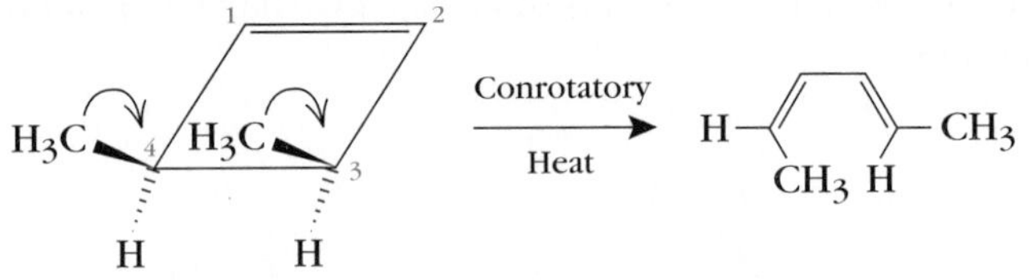

cis-3,4-Dimethylcyclobutene   (2*E*,4*Z*)-Hexadiene

In order to form (2*E*,4*E*)-hexadiene, the photochemical reaction of *cis*-3,4-dimethylcyclobutene, on the other hand, must occur by disrotatory ring opening.

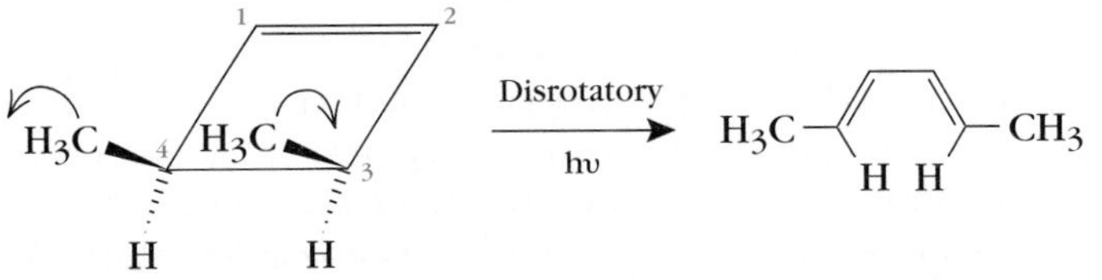

cis-3,4-Dimethylcyclobutene   (2*E*,4*E*)-Hexadiene

The reversible electrocyclic reactions that interconvert cyclobutenes and conjugated dienes are stereospecific because only conrotatory rotation can occur in the thermal reaction, whereas only disrotatory rotation can occur in the photochemical reaction.

Exercise 19.13   Write the structure of the product formed by heating *trans*-3,4-dimethylcyclobutene to 175 °C.

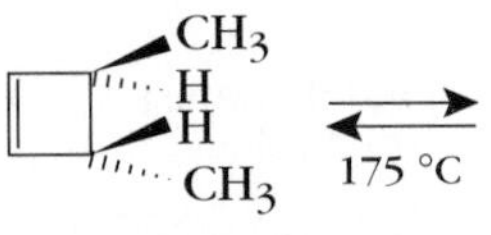

trans-3,4-Dimethylcyclobutene

Before we explain why only certain rotations are allowed, let's look at another example.

Conjugated trienes also undergo concerted electrocyclic reactions to form cyclohexadienes.

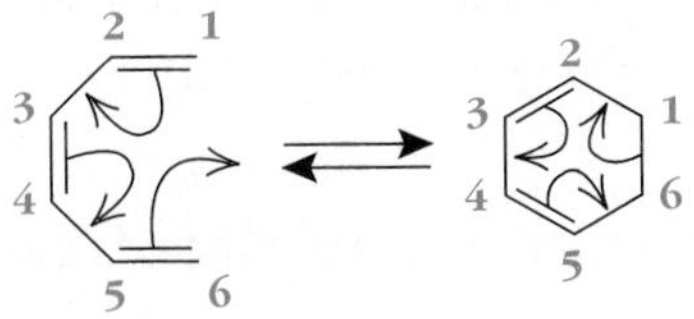

The interconversion of 1,3,5-hexatrienes and 1,3-cyclohexadienes is also reversible, and the position of the equilibrium favors the cyclohexadiene.

The interconversion of 1,3,5-hexatrienes and 1,3-cyclohexadienes is also stereospecific, and the stereochemistry of the products also depends on the experimental conditions. The thermal reaction of (2E,4Z,6E)-octatriene, for example, forms *cis*-5,6-dimethyl-1,3-cyclohexadiene.

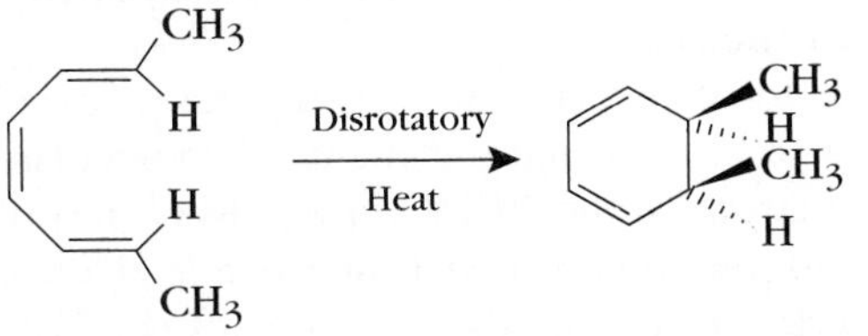

(2E,4Z,6E)-Octatriene   *cis*-5,6-Dimethyl-
                                          1,3-cyclohexadiene

The thermal electrocyclic reactions of conjugated trienes occur by disrotatory ring closure, unlike the thermal electrocyclic reactions of conjugated hexadienes, which occur by conrotatory ring closure.

The stereochemistry of the products of electrocyclic reactions of conjugated trienes is reversed when the reaction occurs by ultraviolet irradiation. Thus the photochemical reaction of (2E,4Z,6E)-octatriene forms *trans*-5,6-dimethyl-1,3-cyclohexadiene.

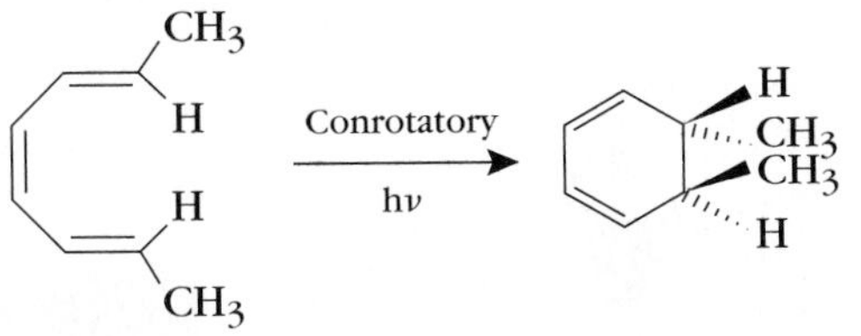

(2E,4Z,6E)-Octatriene   *trans*-5,6-Dimethyl-
                                          1,3-cyclohexadiene

Exercise 19.14   Write the structure of the product in each of the following electrocyclic reactions.

(a) 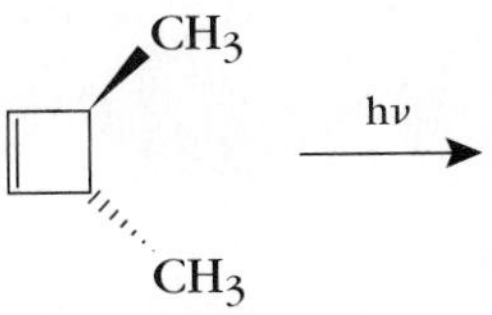 [structure with CH₃ groups, hν]

(c) 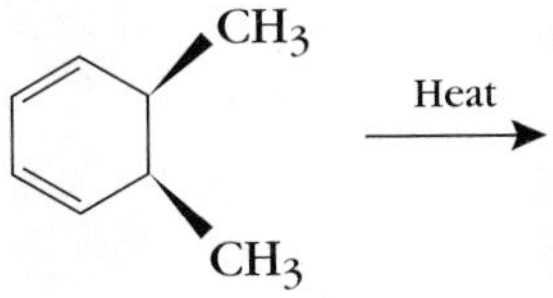 [structure with CH₃ groups, Heat]

(b) (2E,4Z,6Z)-Octatriene $\xrightarrow{\text{Heat}}$

(d) (2E,4E)-Hexadiene $\xrightarrow{h\nu}$

Unlike photochemical electrocyclic reactions of conjugated hexadienes, which occur by a disrotatory ring closure, photochemical electrocyclic reactions of conjugated trienes occur by conrotatory ring closure.

Why is there a difference between the stereochemistry of the products of electrocyclic reactions of dienes and trienes under the same experimental conditions? Examination of the molecular orbitals of polyenes provides the answer.

19.11 Molecular Orbitals and Electrocyclic Reactions

An explanation of the electrocyclic reactions discussed in Section 19.10 was provided by the American chemists Robert B. Woodward and Roald Hoffmann in 1965. They formulated the set of rules for electrocyclic reactions by examining the symmetry properties of molecular orbitals. A complete analysis of electrocyclic reactions using the method of Woodward and Hoffmann is beyond the scope of this text. A simplified approach, however, can be used that is easy to visualize and is accurate in most instances. This simplified approach is based on the **frontier orbital theory.**

According to the frontier orbital theory, the reactions of a polyene are determined by the symmetry of its highest occupied molecular orbital (HOMO). We can regard the π electrons of the HOMO of a polyene just like we regard valence electrons of atoms. In the same way that the valence electrons are responsible for the chemical properties of atoms, the electrons of the HOMO are the ones that govern the course of reactions of polyenes.

To apply the frontier orbital theory to electrocyclic reactions, we must first examine how the orbitals interact during the reaction. The two orbitals at the ends of the π system of a polyene must rotate in a concerted fashion so that they can overlap (and rehybridize) and form the σ bond of the ring. For a σ bond to form, favorable overlap must occur; that is, a positive lobe must overlap a positive lobe or a negative lobe must overlap a negative lobe. When the two lobes of the same sign are on the same side of the polyene, disrotatory rotation leads to σ bond formation by overlap of lobes of the same sign. When the two lobes of the same sign are on opposite sides of the polyene, conrotatory rotation leads to σ bond formation:

- *Concerted electrocyclic reactions are reversible, so the path for the forward reaction is the same as that for the reverse reaction. It is generally easier to analyze these reactions by following the cyclization reaction of the polyene.*

- *The HOMO of a polyene is simplified by drawing only that part of the HOMO at the end carbon atoms of the π-system and connecting them with a line that represents the rest of the HOMO.*

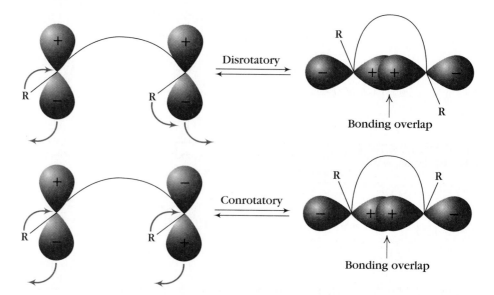

Let's now apply the frontier orbital theory to the thermal electrocyclic interconversion of (2*E*,4*Z*)-hexadiene and 3,4-dimethylcyclobutene. The symmetry of the HOMO of 1,3-butadiene (π_2 in Figure 19.5, p. 851) requires a conrotatory ring closure in order for bonding to occur.

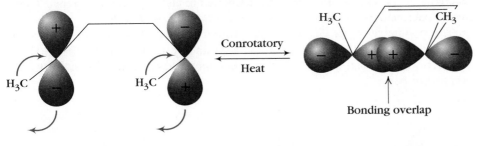

HOMO (π_2) of
(2*E*,4*Z*)-Hexadiene

cis-3,4–Dimethylcyclobutene

> ■ In the analysis of the electro-
> cyclic reaction of 2,4-hexadi-
> ene, we make the approxi-
> mation that the molecular
> orbitals of 2,4-hexadiene are
> identical to those of 1,3-
> butadiene. That is, we
> assume that the methyl
> groups at the ends of the
> polyene can be ignored
> when we construct the
> molecular orbitals of 2,4-
> hexadiene.

Disrotatory ring closure, in contrast, does not lead to σ bond formation since orbitals of opposite sign overlap, which leads to an antibonding (higher energy) state.

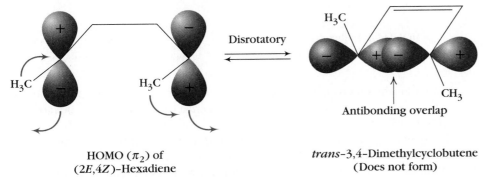

HOMO (π_2) of
(2*E*,4*Z*)-Hexadiene

trans-3,4–Dimethylcyclobutene
(Does not form)

Thermal electrocyclic reactions of (2*E*,4*Z*)-hexadiene occur only by a conrotatory rotation because the symmetry of its HOMO requires a conrotatory rotation to form the σ bond of *cis*-3,4-dimethylcyclobutene.

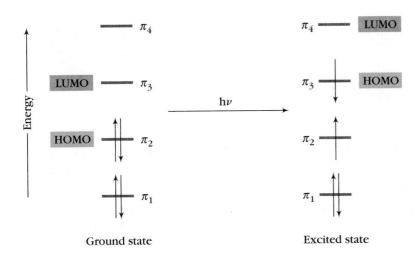

Ground state

Excited state

Figure 19.8
Ground state and excited state electron configurations of 1,3-butadiene. Absorption of energy promotes an electron from the HOMO of the ground state (π_2) to the LUMO of the ground state (π_3). As a result, the HOMO of excited 1,3-butadiene is π_3.

In a photochemical reaction, energy from the absorbed light promotes an electron from the HOMO of 1,3-butadiene to its lowest unoccupied molecular orbital (LUMO, orbital π_3 in Figure 19.5, p. 851), as shown in Figure 19.8. 1,3-Butadiene with one of its four electrons in its LUMO is in an excited state. The reactions of the excited state of 1,3-butadiene are also determined by its HOMO, but for excited 1,3-butadiene, *molecular orbital π_3 is its HOMO.*

The symmetry of the HOMO of the ground state is different than the symmetry of the HOMO of the excited state. The symmetry of the HOMO of excited 1,3-butadiene (π_3 in Figure 19.5) requires a disrotatory ring closure in order for bonding to occur.

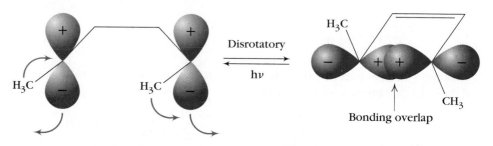

HOMO (π_3) of excited
(*2E,4Z*)–Hexadiene

trans–3,4–Dimethylcyclobutene

To analyze the electrocyclic interconversions of trienes and cyclohexadienes, we must examine the symmetry of the molecular orbitals of 1,3,5-hexatriene, shown in Figure 19.9.

The symmetry of the HOMO of the ground state electron structure determines the stereochemistry of the product formed by a thermal electrocyclic reaction. Disrotatory rotation must occur to form a bond between the terminal carbon atoms of a conjugated triene. The thermal electrocyclic reaction of (*2E,4Z,6E*)-octatriene, therefore, occurs by disrotatory ring closure to form *cis*-5,6-dimethyl-1,3-cyclohexadiene.

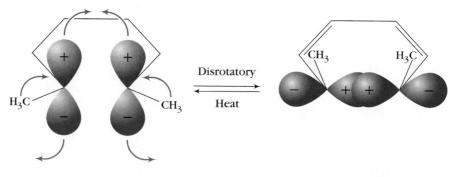

(*2E,4Z,6E*)–Octatriene

cis–5,6–Dimethyl–1,3–cyclohexadiene

Notice that ring-opening and ring-forming thermal reactions of conjugated dienes occur by a conrotatory rotation, whereas conjugated trienes undergo the same electrocyclic reactions by a disrotatory rotation. This difference is due to the different symmetries of the HOMO of the ground states of dienes and trienes.

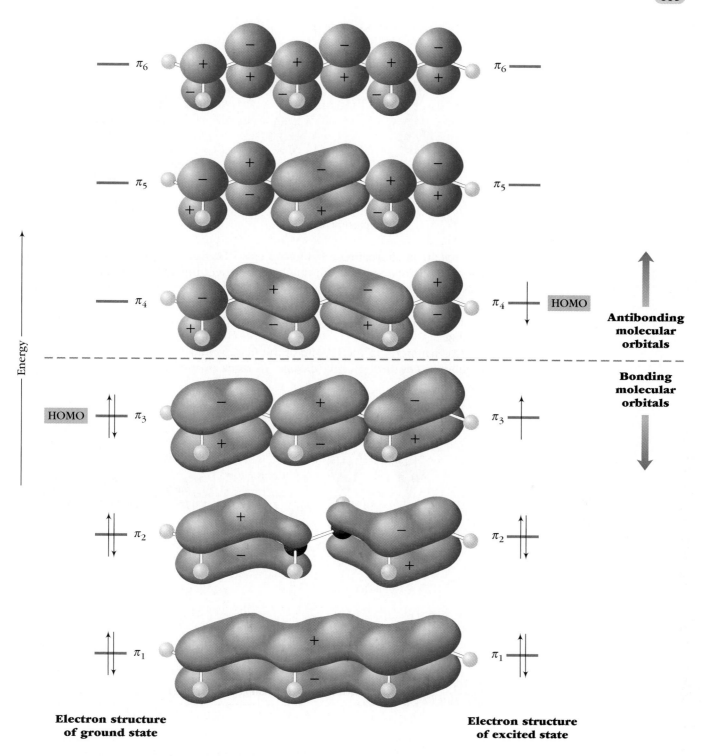

Figure 19.9
The six molecular orbitals of 1,3,5-hexatriene. In the HOMO of the ground state electron structure, the lobes of the same sign at the terminal carbon atoms are on the same side of the triene,
whereas in the HOMO of the first excited state, the lobes of the same sign at the terminal carbon
atoms are on different sides of the triene.

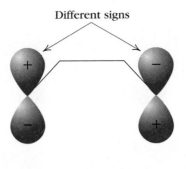

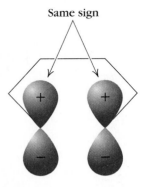

HOMO of ground state
of conjugated diene

HOMO of ground state
of conjugated triene

As a result, an alternating relationship exists between the type of rotation, conrotatory or disrotatory, and the number of π electrons in the conjugated polyene. Polyenes with $4n$ π electrons (where n is an integer) undergo thermal electrocyclic reactions by conrotatory ring opening or closing, whereas polyenes with $4n + 2$ π electrons undergo the same reactions by disrotatory ring opening or closing.

The symmetry of the HOMO of the first excited state of (2E,4Z,6E)-octatriene requires a conrotatory ring closure to form *trans*-5,6-dimethyl-1,3-cyclohexadiene.

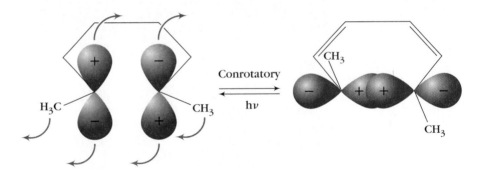

(2E,4Z,6E)-Octatriene

trans-5,6-Dimethyl-1,3-cyclohexadiene

Table 19.3	Woodward-Hoffmann Rules for Electrocyclic Reactions	
Number of π Electrons in the Polyene	Reaction Conditions	Type of Allowed Rotation
$4n$	Thermal reaction	Conrotatory
$4n$	Photochemical reaction	Disrotatory
$4n + 2$	Thermal reaction	Disrotatory
$4n + 2$	Photochemical reaction	Conrotatory

Notice that ring-opening and ring-forming photochemical reactions of conjugated dienes occur by a disrotatory rotation, whereas conjugated trienes undergo the same electrocyclic reactions by a conrotatory rotation. This difference is again due to the different symmetries of the HOMO of the first excited states of dienes and trienes.

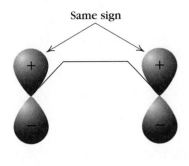

Same sign   Different signs

HOMO of first excited   HOMO of first excited
state of conjugated diene   state of conjugated triene

As a result, polyenes with $4n$ π electrons undergo photochemical electrocyclic reactions by disrotatory ring opening or closing, whereas polyenes with $4n + 2$ π electrons undergo the same reactions by conrotatory ring opening or closing.

In summary, two factors determine the stereochemistry of the products of an electrocyclic reaction, the number of π electrons in the polyene and the reaction conditions (thermal or photochemical). These factors have been incorporated by Woodward and Hoffmann into a set of rules given in Table 19.3. Use of these simple rules makes it possible to predict the stereochemistry of electrocyclic reactions.

Exercise 19.15   Write the structure and the stereochemistry of the product in each of the following reactions.

(a) 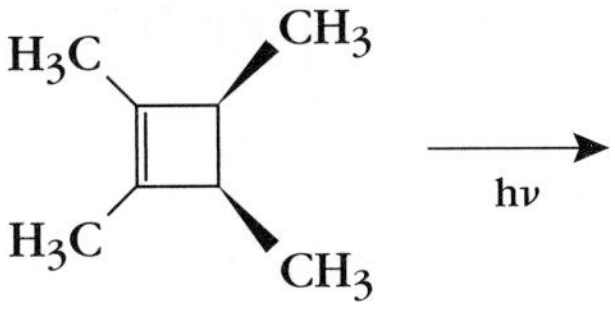 [structure] $\xrightarrow{h\nu}$

(b) (2E,4Z,6Z,8E)-decatetraene $\xrightarrow{\text{Heat}}$

(c) (2E,4Z,6Z,8E)-decatetraene $\xrightarrow{h\nu}$

(d) 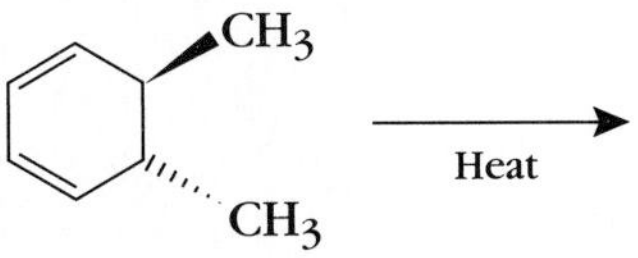 [structure] $\xrightarrow{\text{Heat}}$

19.12 The Diels-Alder Reaction

Otto Diels and Kurt Alder discovered in 1928 that conjugated dienes react with certain alkenes to form derivatives of cyclohexene. A typical example of this reaction, which is called the **Diels-Alder reaction** after its discoverers, is the formation of cyclohexene-4-carboxaldehyde from 1,3-butadiene and propenal.

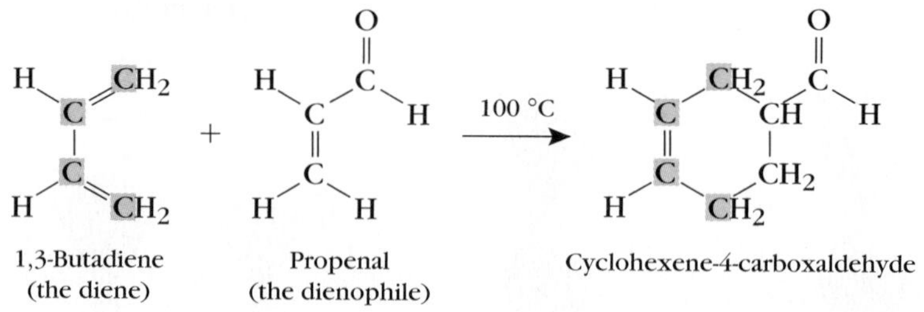

| 1,3-Butadiene (the diene) | Propenal (the dienophile) | Cyclohexene-4-carboxaldehyde |

Propenal is called the **dienophile** (a diene-loving reagent) in this reaction because it reacts with 1,3-butadiene, a conjugated diene. The Diels-Alder reaction results in a 1,4-addition because the carbon atoms of the carbon-carbon double bond of the dienophile add to the 1,4-carbon atoms of the conjugated diene to form a compound that contains a six-membered ring. Both new carbon-carbon single bonds and the new double bond are formed as the three double bonds in the reactants break, so the reaction is concerted.

The Diels-Alder reaction is very important synthetically because it forms a six-membered ring that contains two new carbon-carbon single bonds in one step. Diels-Alder reactions are an example of cycloaddition reactions. **Cycloaddition reactions** are intermolecular reactions in which two unsaturated molecules add to each other to form a cyclic product. Diels and Alder shared the Nobel prize in chemistry in 1950 for their extensive work on this reaction.

A Diels-Alder reaction occurs readily when the dienophile contains a strong electron-withdrawing group bonded to its carbon-carbon double bond. The most commonly used dienophiles are α,β-unsaturated carbonyl and related compounds. Without a strong electron-withdrawing group bonded to the double bond of the dienophile, the reaction occurs only slowly or not at all. The Diels-Alder reaction between the simplest dienophile, ethylene, and 1,3-butadiene, for example, occurs only very slowly at elevated temperature and pressure. The

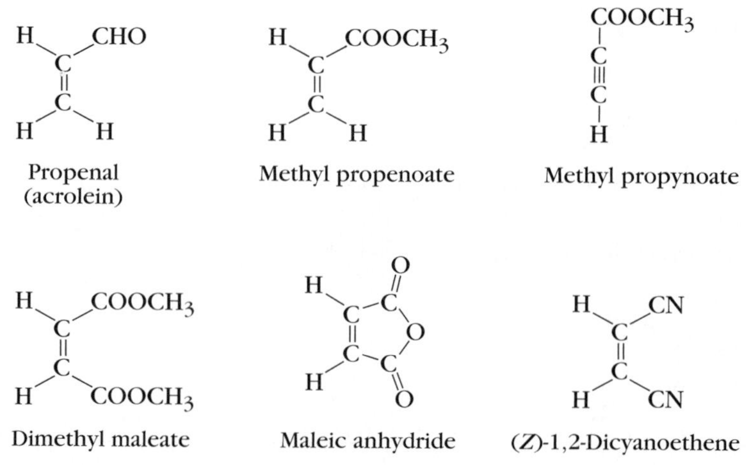

Figure 19.10
The double bond of good dienophiles is bonded to the carbon atom of the carbonyl group of aldehydes, ketones, esters, or anhydrides. Compounds with a carbon-carbon double bond adjacent to a nitrile group are also good dienophiles as are compounds containing a triple bond adjacent to the carbon atom of a carbonyl group.

| Propenal (acrolein) | Methyl propenoate | Methyl propynoate |

| Dimethyl maleate | Maleic anhydride | (Z)-1,2-Dicyanoethene |

Exercise 19.16   Which of the following compounds would you expect to be good dienophiles in the Diels-Alder reaction?

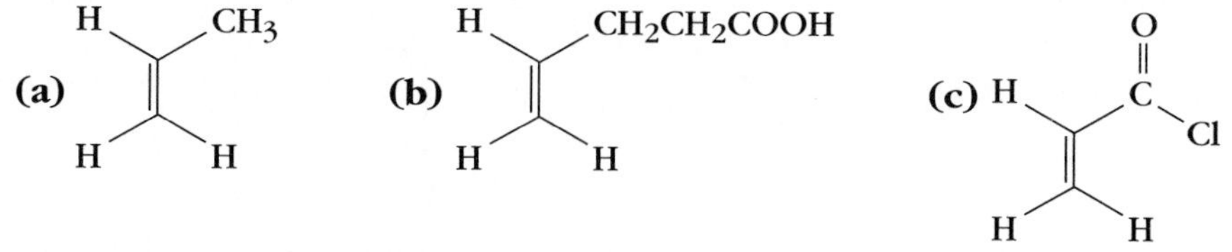

structures of a number of good dienophiles, including one with a carbon-carbon triple bond, are given in Figure 19.10.

One of the consequences of the concerted nature of the Diels-Alder reaction is that the stereochemistry at the double bond of the dienophile is retained in the product. For example, 1,3-butadiene reacts with dimethyl maleate (which contains a *cis-* double bond) to form dimethyl *cis*-4-cyclohexene-1,2-dicarboxylate, while it reacts with dimethyl fumarate (the *trans-* isomer) to form dimethyl *trans*-4-cyclohexene-1,2-dicarboxylate.

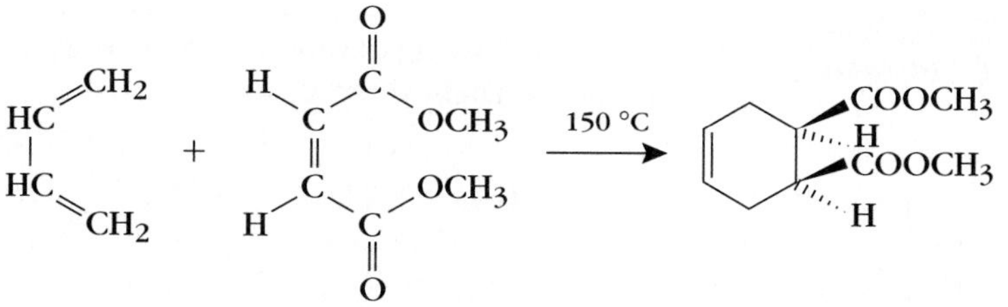

Dimethyl maleate
(*cis-* starting material)

Dimethyl *cis*-4-cyclohexene-1,2-dicarboxylate
(*cis-* product)

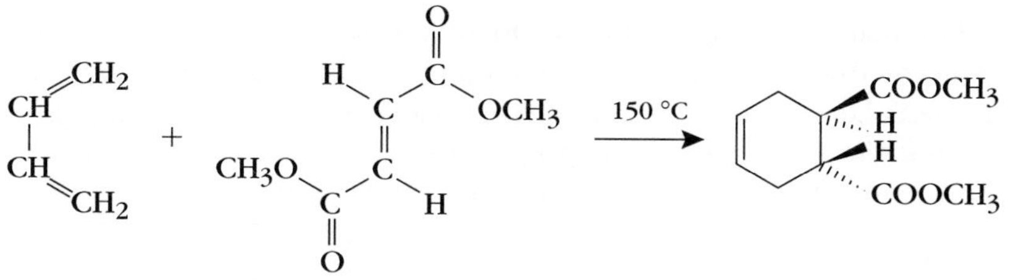

Dimethyl fumarate

Dimethyl *trans*-4-cyclohexene-1,2-dicarboxylate
(*trans-* product)

Dienes that undergo the Diels-Alder reaction must be in the s-*cis* conformation, because conjugated dienes that cannot adopt the s-*cis* conformation do not undergo the Diels-Alder reaction. 3-Methylenecyclohexene, for example, is a diene that is locked into a rigid s-*trans* conformation.

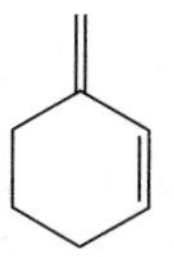

3-Methylenecyclohexene

It cannot undergo a Diels-Alder reaction. Conjugated dienes that are locked into a *cis-* conformation do undergo Diels-Alder reactions. Furan, cyclopentadiene, and 1,3-cyclohexadiene are examples of rigid s-*cis* dienes.

Furan   Cyclopentadiene   1,3-Cyclohexadiene

Because the Diels-Alder reaction is concerted, carbon atoms 1 and 4 of the conjugated diene must be close enough to both carbon atoms of the dienophile in order to react.

s-*cis* Orientation:
Favorable for concerted
formation of two bonds
between diene and dienophile

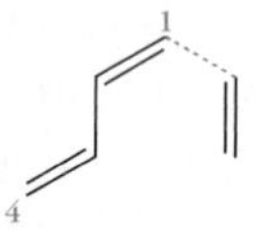

s-*trans* Orientation:
Unfavorable for concerted
bond formation

Exercise 19.17 Which of the following compounds would be suitable as dienes in the Diels-Alder reaction?

(a)   **(b)**   **(c)**   **(d)**

Exercise 19.18 Write the structures of the products of the following Diels-Alder reactions.

(a) 1,3-Butadiene + methyl propenoate ⟶

(b) 2,3-Dimethyl-1,3-butadiene + methyl propynoate ⟶

(c) 1,3-Butadiene + (Z)-1,2-dicyanoethene ⟶

(d) 1,3-Butadiene + (E)-1,2-dicyanoethene ⟶

(e) 1,3-Butadiene + 1,1-dicyanoethene ⟶

Exercise 19.19 Write the structures of the dienes and dienophiles that you would use to prepare each of the following compounds by the Diels-Alder reaction.

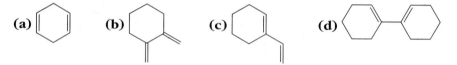

(a) 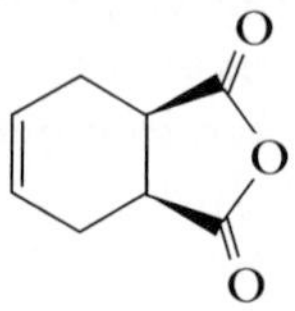   **(b)** 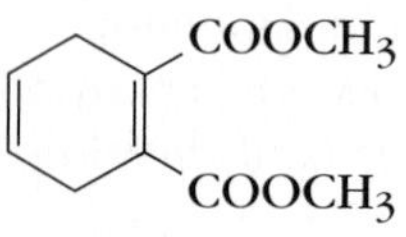 $COOCH_3$ / $COOCH_3$ **(c)** 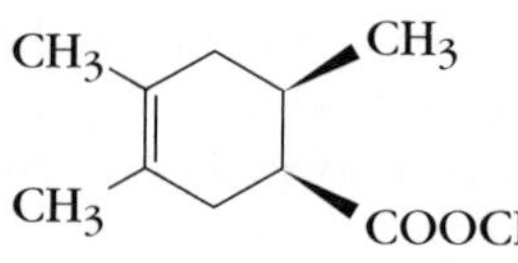 CH_3, CH_3 / CH_3, $COOCH_3$

An explanation of the mechanism of cycloaddition reactions in general and the Diels-Alder reaction in particular was provided by Woodward and Hoffmann based on the symmetry properties of the molecular orbitals involved in the reaction. As in the case of electrocyclic reactions, we can use the frontier orbital theory instead of the detailed analysis of Woodward and Hoffmann to visualize the Diels-Alder reaction.

According to the frontier orbital theory, a cycloaddition reaction occurs when the HOMO (highest occupied molecular orbital) of one reactant forms bonds by overlapping with the LUMO (lowest unoccupied molecular orbital) of the other reactant. Let's select the HOMO of the diene and the LUMO of the dienophile (we could equally well use the LUMO of the diene and the HOMO of the dienophile). Bonding occurs between the terminal carbon atoms of the diene and the dienophile because the signs of the orbitals permit overlap that results in formation of a bond.

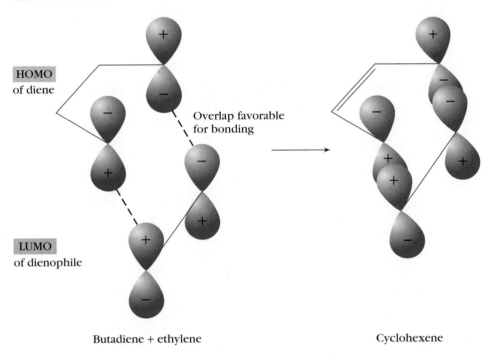

HOMO
of diene

Overlap favorable
for bonding

LUMO
of dienophile

Butadiene + ethylene   Cyclohexene

Exercise 19.20   Show that use of the LUMO of 1,3-butadiene and the HOMO of ethylene also leads to formation of a cyclohexene ring.

In summary, the Diels-Alder reaction is important in organic synthesis because it is one of the easiest ways to form compounds containing cyclohexane rings, including bicyclic compounds, by the concerted formation of two carbon-carbon single bonds. Furthermore, the reaction is stereoselective, and the stereochemistry of the dienophile is retained in the product.

19.13 Natural and Synthetic Rubber by Polymerization of Conjugated Dienes

Conjugated dienes, such as 1,3-butadiene, polymerize by 1,4-addition to form polymers that still contain a carbon-carbon double bond. The configuration of these double bonds can be either *cis*- or *trans*-.

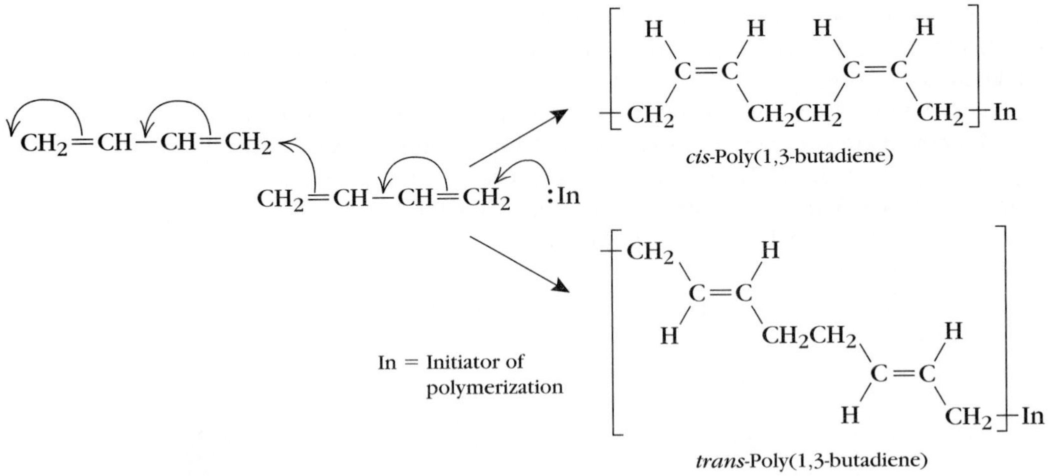

cis-Poly(1,3-butadiene)

In = Initiator of
polymerization

trans-Poly(1,3-butadiene)

Natural rubber is a homopolymer of 2-methyl-1,3-butadiene whose common name is isoprene. Natural rubber is mostly polyisoprene made up of 5,000 to 12,000 isoprene units. The double bonds in the isoprene units are *cis*-.

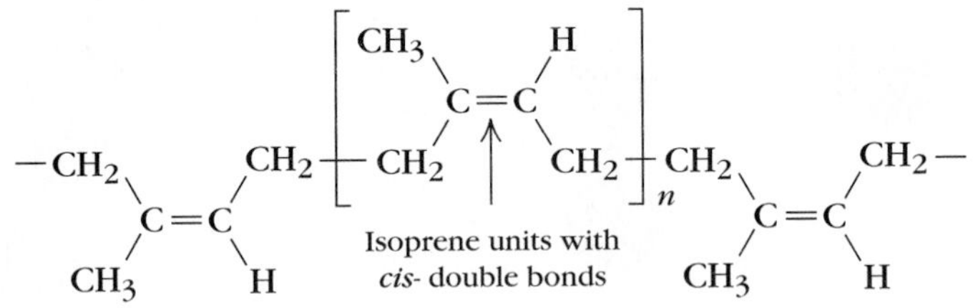

Isoprene units with
cis- double bonds

Natural rubber

Natural rubber is obtained from the liquid, called latex, that is exuded by a number of tropical plants. Chief among these tropical plants is the rubber tree, *Hevea brasiliensis*. Other trees produce an isomeric polyisoprene called **gutta-percha,** which consists of 100 to 150 isoprene units. In gutta-percha, the double bonds in the isoprene units are *trans*-.

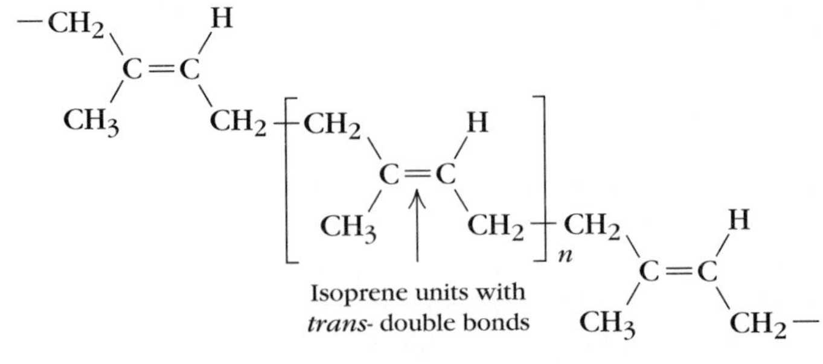

Isoprene units with
trans- double bonds

Gutta-percha

Rubber and gutta-percha have different properties because the configurations about their double bonds are different. The *trans*- double bonds of gutta-percha allow its chains to pack more closely in the solid state. As a result, gutta-percha is a more crystalline solid and is harder, less flexible, and more brittle than natural rubber. Gutta-percha is used commercially as a cover for golf balls and electrical cables.

Raw natural rubber is not a particularly useful material. It becomes soft and sticky on hot days and hard on cold days. It reacts readily with light, oxygen, and ozone, and it tends to swell and dissolve in organic hydrocarbons such as oil and gasoline. To improve its properties, raw rubber is heated with a few percent by weight of sulfur. This reaction is called **vulcanization** and was discovered by Charles Goodyear in 1839. Sulfur adds to the double bonds of the polymer chains of raw rubber to form cross-links between them.

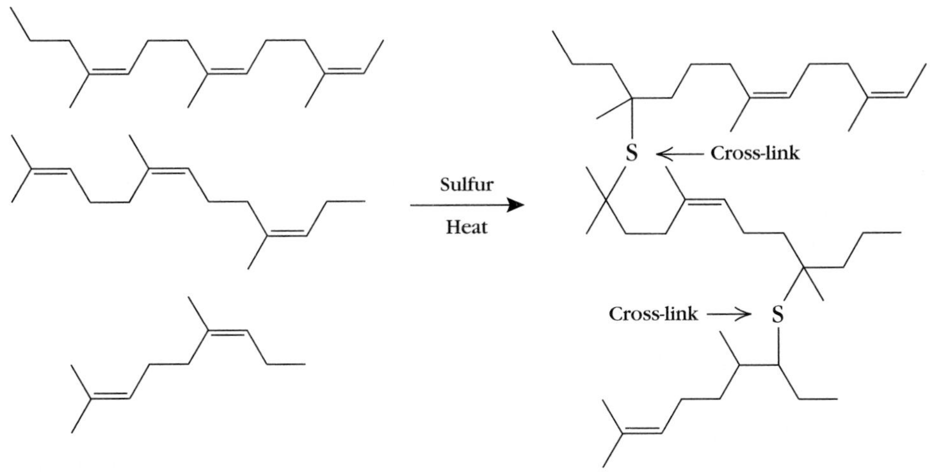

Polymer chains of raw rubber

Cross-linked polymer chains of vulcanized rubber

Vulcanizing rubber makes it less sticky, stronger, more elastic, less soluble in hydrocarbon solvents, and more flexible at low temperatures than raw rubber.

Polyisoprene with the *cis-* or *trans-* configuration about the double bond can also be prepared by choosing the proper Ziegler-Natta catalyst (Section 18.12). These synthetic polymers have properties like natural rubber and gutta-percha, respectively.

Several different synthetic rubbers are produced commercially by the polymerization of conjugated dienes. Examples include neoprene, which is produced commercially by the polymerization of chloroprene (2-chloro-1,3-butadiene), and butadiene rubber, which is produced by the polymerization of 1,3-butadiene.

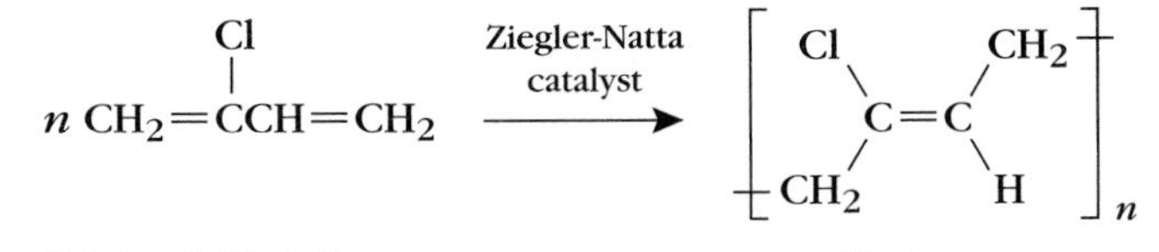

2-Chloro-1,3-butadiene

Neoprene

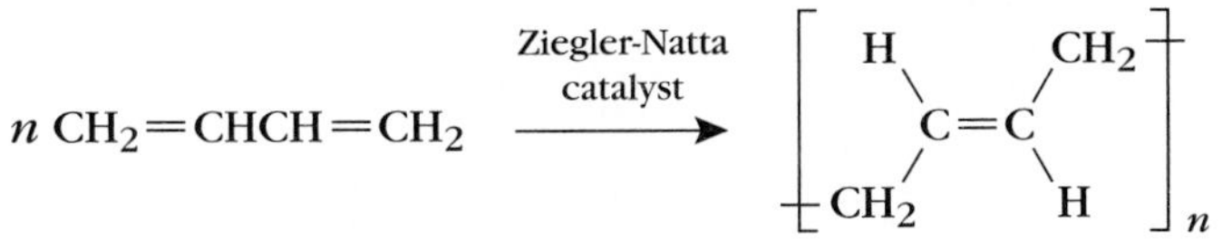

1,3-Butadiene

Butadiene rubber

Most synthetic rubbers are now copolymers prepared by polymerization of a diene and an alkene. A number of examples are given in Table 19.4.

Table 19.4	Synthetic Rubbers Prepared by Copolymerization of a Diene and an Alkene	
Trade Name	Uses	Monomers
Nitrile rubber	Gaskets, oil seals, and oil tank linings	$CH_2{=}CHC{\equiv}N$ + $CH_2{=}CHCH{=}CH_2$ Acrylonitrile 1,3-Butadiene
Styrene-butadiene rubber (SBR)	Tire treads, footwear	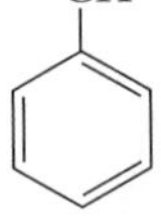 Styrene + $CH_2{=}CHCH{=}CH_2$ Styrene 1,3-Butadiene
Butyl rubber	Raincoats, inner tubes for tires, and caulking compounds	$(CH_3)_2C{=}CH_2$ + $CH_2{=}C(CH_3)CH{=}CH_2$ Isobutylene Isoprene

19.14 Terpenes

Terpenes are a large class of natural products in which the total number of carbon atoms in their carbon skeletons is a multiple of five. Terpenes are further subdivided, depending on the total number of carbon atoms in the molecule. **Monoterpenes** are compounds with 10 carbon atoms, **sesquiterpenes** contain 15 carbon atoms, and **diterpenes** contain 20 carbon atoms. **Steroids** are derived from **triterpenes,** which contain 30 carbon atoms, while the **carotenoids,** compounds that give plants color, are **tetraterpenes,** because they contain 40 carbon atoms. The following are examples of the subclasses of terpenes.

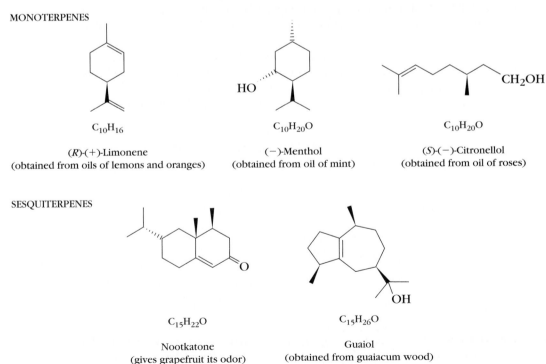

MONOTERPENES

$C_{10}H_{16}$

(*R*)-(+)-Limonene
(obtained from oils of lemons and oranges)

$C_{10}H_{20}O$

(−)-Menthol
(obtained from oil of mint)

$C_{10}H_{20}O$

(*S*)-(−)-Citronellol
(obtained from oil of roses)

SESQUITERPENES

$C_{15}H_{22}O$

Nootkatone
(gives grapefruit its odor)

$C_{15}H_{26}O$

Guaiol
(obtained from guaiacum wood)

DITERPENES

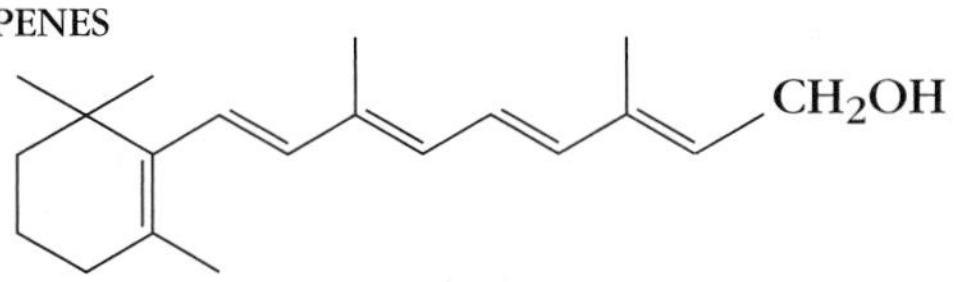

$C_{20}H_{30}O$

Vitamin A

TRITERPENES

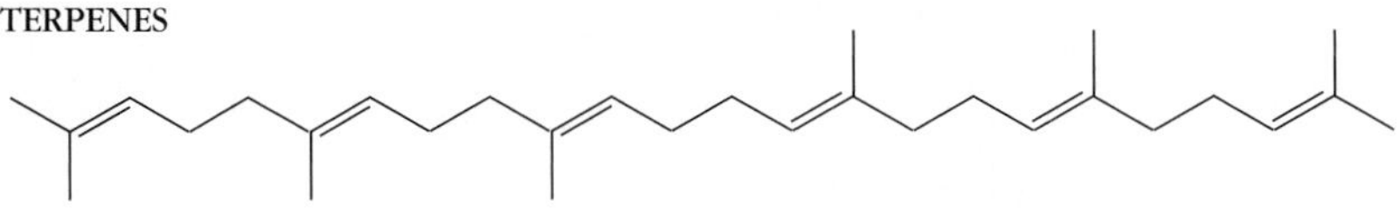

$C_{30}H_{50}$

Squalene

In nature, terpenes are synthesized from dimethylallyl pyrophosphate and isopentenyl pyrophosphate, the biological equivalents of isoprene units.

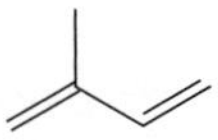

| 2-Methyl 1,3-butadiene (isoprene) | Dimethylallyl pyrophosphate | Isopentenyl pyrophosphate |

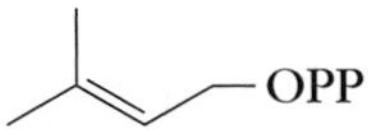

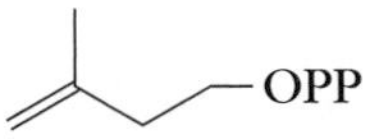

OPP = pyrophosphate group =

An enzyme-catalyzed reaction of dimethylallyl pyrophosphate and isopentenyl pyrophosphate forms the 10-carbon compound geranyl pyrophosphate.

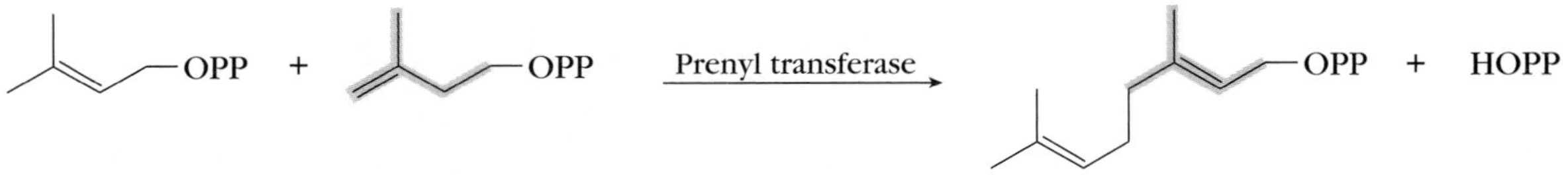

| Dimethylallyl pyrophosphate | Isopentenyl pyrophosphate | Geranyl pyrophosphate |

The mechanism of the enzyme-catalyzed reaction of dimethylallyl pyrophosphate and isopentenyl pyrophosphate is believed to occur by the S_N1 mechanism shown in Figure 19.11. In Step 1 of the mechanism, an allylic cation forms by elimination of the pyrophosphate group. Addition of the allylic cation to the double bond of isopentenyl pyrophosphate occurs in Step 2, followed by loss of a proton to form geranyl pyrophosphate in Step 3.

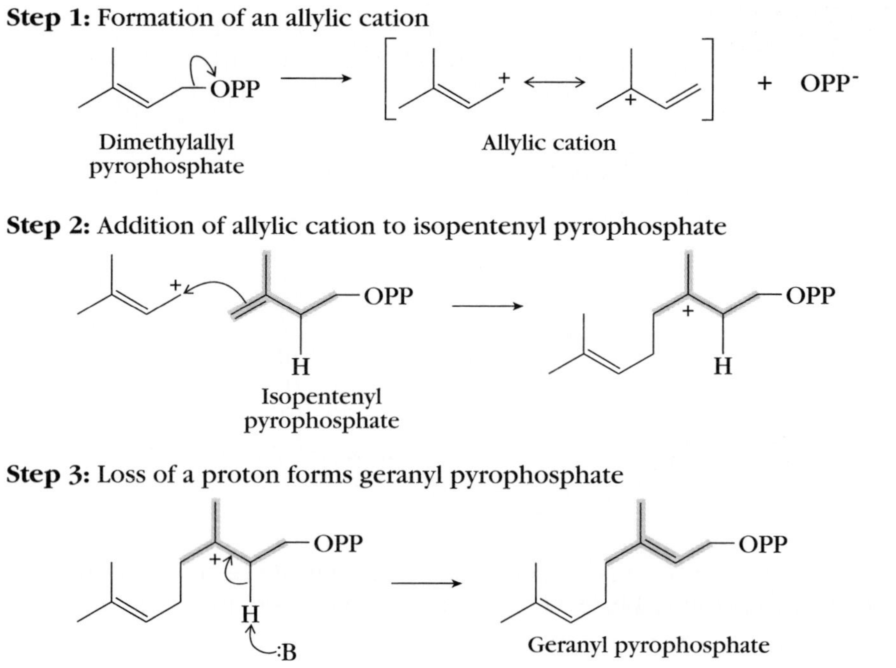

Step 1: Formation of an allylic cation

Dimethylallyl pyrophosphate   Allylic cation

Step 2: Addition of allylic cation to isopentenyl pyrophosphate

Isopentenyl pyrophosphate

Step 3: Loss of a proton forms geranyl pyrophosphate

Geranyl pyrophosphate

Triterpenes are synthesized in nature by joining two farnesyl pyrophosphate molecules to form squalene, a precursor to the steroids.

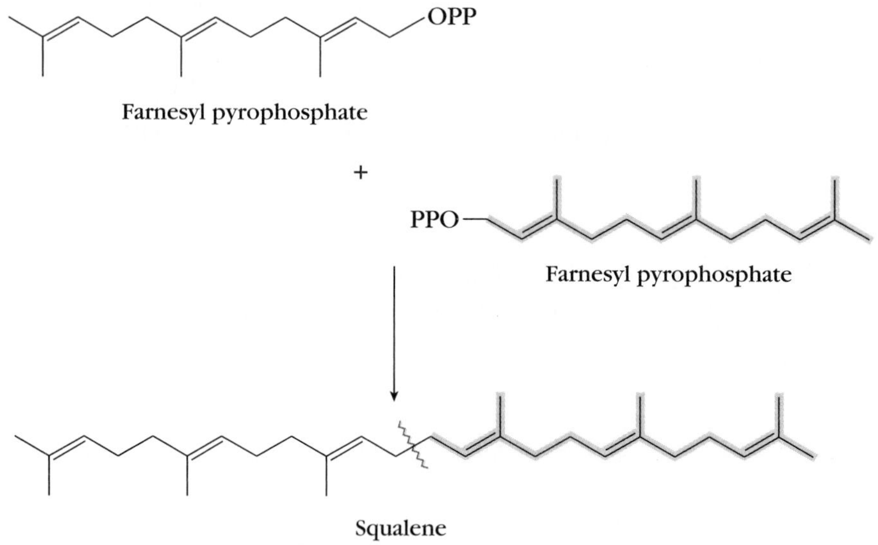

Farnesyl pyrophosphate

+

Farnesyl pyrophosphate

Squalene

In summary, the basic carbon skeleton of most terpenes, steroids, and carotenoids is assembled by a chemical reaction that is familiar to us, namely a nucleophilic substitution reaction, catalyzed by an enzyme, that occurs by an S_N1 mechanism. While many more reactions are necessary to introduce the functional groups and the precise stereochemistry of individual compounds, the general reaction by which large terpenes, steroids, and carotenoids are formed from smaller alkenes or polyenes is well established.

Many of the terpenes, steroids, and carotenoids synthesized in nature are easily identified by ultraviolet and visible spectroscopy.

| Exercise 19.21 | Myrcene and ocimene are obtained in nature by the elimination of HOPP from geranyl pyrophosphate. Propose a mechanism for this reaction. |

Geranyl pyrophosphate   Myrcene   Ocimene

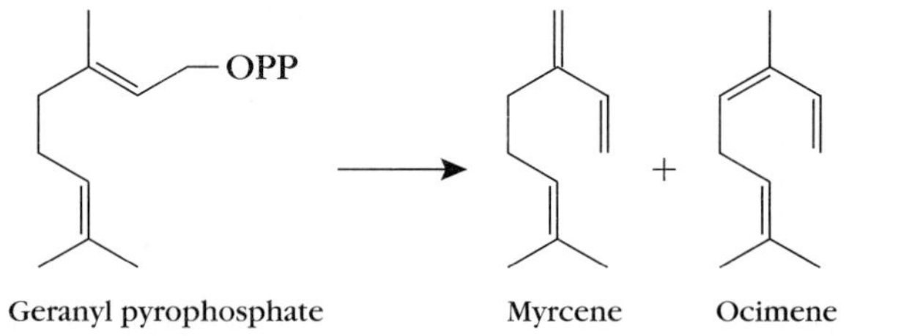

| Exercise 19.22 | In the laboratory, acid-catalyzed dehydration of geraniol also forms myrcene and ocimene. Propose a mechanism for this reaction. |

Geraniol   Myrcene   Ocimene

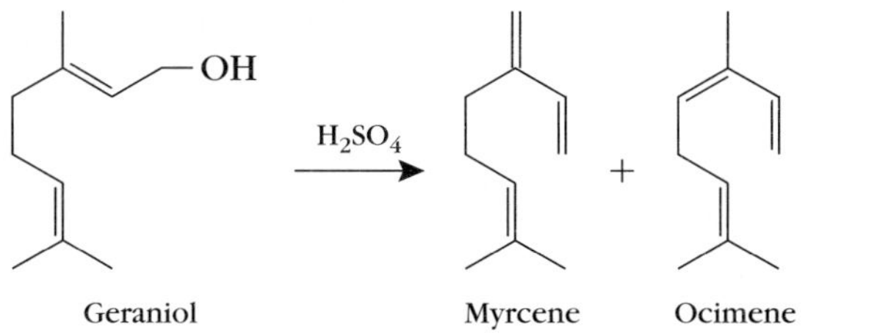

19.15 Ultraviolet and Visible Spectroscopy

We learned in Chapters 5 and 10 that measuring the electromagnetic radiation absorbed by organic compounds helps determine their structures. Infrared spectroscopy, for instance, identifies most of the functional groups present, ^{13}C NMR spectroscopy tells us the number and kinds of carbon atoms of the carbon skeleton, and ^{1}H NMR spectroscopy tells us the number and kinds of hydrogens attached to the carbon skeleton.

Ultraviolet (UV) and visible spectroscopy is another technique used by chemists to determine the structure of a compound. Molecules that contain π electrons usually absorb electromagnetic radiation in the ultraviolet, 200 to 400 nm, and visible, 400 to 800 nm, regions of the spectrum.

A compound is prepared for UV and visible absorption spectroscopy by dissolving a sample of it in a solvent that does not absorb in the UV or visible regions of the spectrum. Solvents commonly used for this purpose are methanol, 95% ethanol, and cyclohexane because they do not have π electrons. The solution is placed in a spectrometer that is similar in construction to an IR spectrometer (Chapter 5), except that the light source provides radiation in the 200 to 800 nm region of the electromagnetic spectrum.

The graph of the UV or visible light absorbed by the compound versus its wavelength is called its UV or visible spectrum; the UV spectrum of 2-methyl-1,3-butadiene is shown in Figure 19.12. Notice in Figure 19.12 that the wavelength of the radiation (in nanometers) is plotted on the horizontal axis, whereas absorbance (the amount of UV/visible light absorbed by the sample) is plotted on the vertical axis.

Figure 19.12
UV spectrum of
2-methyl-1,3-butadiene.

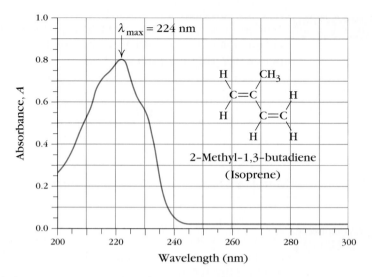

If the light entering the sample has an intensity I_0, and the light emerging has an intensity I, absorbance, A, is defined as the logarithm of the ratio I/I_0.

$$A = \log I/I_0$$

The absorbance at any wavelength depends on the number of molecules in the light path. As a result, the absorbance, A, is proportional to the concentration of the sample, c, in moles per liter, and the length of the light path through the sample, l, in cm.

$$A \propto lc$$

The proportionality constant that links A to l and c is called the molar extinction coefficient or simply the *extinction coefficient, ϵ.*

$$A = \epsilon lc$$

$$\epsilon = A/lc$$

The *maximum* of every absorption peak in a UV/visible spectrum of a compound has a characteristic extinction coefficient, designated ϵ_{max}. Values of ϵ_{max} also depend on solvent and temperature, and they vary from below 100 to several hundred thousand. The larger the value of ϵ_{max}, the greater the amount of light the compound absorbs at that particular wavelength.

UV/visible spectra can be abbreviated by listing two values. The first is the wavelength at the maximum of the absorption peak, which is called λ_{max} (pronounced "lambda max"), and the second is ϵ_{max} (pronounced "e max"). The position and size of the absorption peak in the spectrum of 2-methyl-1,3-butadiene in pentane, shown in Figure 19.12, can summarized as follows:

$$\lambda_{max} \text{ (pentane)} = 224 \text{ nm } (\epsilon_{max} = 10{,}500)$$

Exercise 19.23	Ten milligrams of a compound of molecular weight 240 are dissolved in 10 mL of 95% ethanol. The solution is placed in a 1 cm UV cell, and the UV spectrum is recorded. The spectrum contains an absorption at λ_{max} = 327 nm, with a maximum absorbance of 0.65. Calculate the value of ϵ_{max}.

Table 19.5	The λ_{max} and ϵ_{max} of the Longest Wavelength UV/Visible Absorptions for Alkenes and Conjugated Polyenes		
Name	Alkene Structure	λ_{max} nm	ϵ_{max}
Ethene		171	15,000
1-Octene		177	12,000
1,3-Butadiene		217	21,000
2-Methyl-1,3-butadiene		224	10,800
(E)-1,3,5-Hexatriene		268	36,300
(3E,5E)-1,3,5,7-Octatetraene		302	—
(3E,5E,7E)-1,3,5,7,9-Decapentaene		334	125,000

What can UV/visible spectroscopy tell us about the electron structure of an alkene? We can answer this question by examining the values of λ_{max} and ϵ_{max} of the alkenes and conjugated polyenes listed in Table 19.5. Notice that the values of λ_{max} increase as the number of conjugated double bonds in the molecule increases. For example, λ_{max} for 1,3-butadiene (two double bonds) is at 217 nm, while λ_{max} for (E)-1,3,5-hexatriene (three double bonds) is at 268 nm. Compounds with isolated double bonds, in contrast, such as ethylene and 1-octene have λ_{max} values that are less than 200 nm.

Why does an increase in the number of conjugated double bonds affect the position of λ_{max}? To answer this question, we must remember that the absorption of electromagnetic energy causes a molecule to be excited from one energy state, usually the ground or unexcited state, to one of higher energy. This excitation can occur only when the energy of the incident radiation corresponds exactly to the difference between the unexcited and excited states of the molecule. In the case of alkenes and polyenes, energy absorbed in the 200 to 800 nm

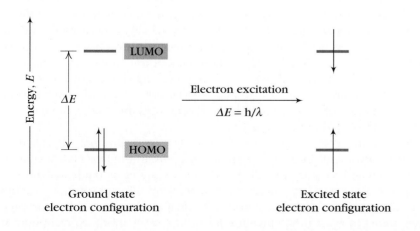

Ground state
electron configuration

Excited state
electron configuration

Figure 19.13
UV light can excite an electron from the HOMO of ethylene to the LUMO.

Figure 19.14

The π molecular orbital diagram of ethylene, 1,3-butadiene, and 1,3,5-hexatriene. The distance and consequently the energy between the HOMO and LUMO decrease as the number of conjugated double bonds increases. As a result λ_{max} for ethylene, 1,3-butadiene, and 1,3,5-hexatriene shifts progressively to longer wavelengths.

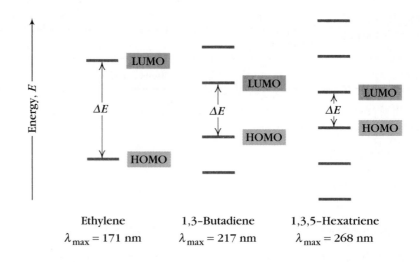

region of the electromagnetic spectrum excites the π electrons from the highest occupied molecular orbital (HOMO) to the lowest unoccupied molecular orbital (LUMO), as shown in Figure 19.13.

As the energy difference between the HOMO and LUMO increases, the value of λ_{max} shifts to shorter wavelengths. Conversely, the value of λ_{max} shifts to longer wavelengths as the HOMO-LUMO energy difference decreases. As the number of conjugated double bonds increases in a molecule, the energy difference between the HOMO and the LUMO decreases, so the values of λ_{max} shift to longer wavelengths. This is illustrated in Figure 19.14 where the relative energies of the HOMO and LUMO of ethylene, 1,3-butadiene, and 1,3,5-hexatriene are plotted on the same scale.

As we can see from Figure 19.14, it takes light of higher energy (shorter wavelength) to excite an electron from the HOMO to the LUMO of ethylene than in either 1,3-butadiene or 1,3,5-hexatriene. The amount of energy needed to accomplish these electron excitations decreases with increasing numbers of conjugated double bonds. Consequently, λ_{max} is found at longer and longer wavelengths as the number of conjugated double bonds in the alkenes increases.

19.16 Summary

The stability of the allyl (2-propenyl) system, which is due to electron delocalization, is the dominant factor in the chemistry of allylic derivatives and dienes. According to molecular orbital theory, the π electron structure of the allyl system is described by three π molecular orbitals. One of the molecular orbitals is bonding, one is nonbonding, and one is antibonding. Any charges or odd electrons in the allyl system are equally distributed between the two end carbons. Quantitative calculations show that an allyl system is more stable than a double bond isolated from an sp^2 hybrid carbon atom.

The reaction that replaces an allylic hydrogen by bromine is called **allylic bromination.** An allylic radical is an intermediate in the free radical chain mechanism of allylic bromination. Allylic cations are stable enough that most allylic derivatives, even primary ones, undergo nucleophilic substitution reactions in polar solvents by an S_N1 mechanism. The major exception is 3-chloropropene, which seems to react predominantly by an S_N2 mechanism. Nucleophilic substitution reactions of allylic compounds produce a mixture of products formed by reaction of the nucleophile with the two electropositive centers of the allylic cation. The formation of a rearranged product is called an **allylic rearrangement.**

A **conjugated diene** is a compound, such as 1,3-butadiene, that contains two double bonds separated by a single bond. 1,5-Hexadiene is an example of a **nonconjugated diene** because it contains a diene with two **isolated double bonds.** The double bonds of nonconjugated dienes are separated by two or more carbon atoms. Conjugated dienes are slightly more stable than nonconjugated dienes.

Conjugated dienes react with electrophiles to form products of **1,4-addition** as well as the usual 1,2-addition. Both products are formed from an allylic cation intermediate, which is formed by electrophilic addition to a conjugated diene. The ratio of 1,2- and 1,4-addition products depends on the reaction conditions. Under conditions of **kinetic control,** usually low temperatures, the 1,2-addition product is formed in greater amounts. Under conditions of **thermodynamic control,** the products are at equilibrium and the more stable 1,4-addition product is formed in greater amounts.

Thermal or photochemical interconversions of conjugated polyenes and cyclic compounds are called **electrocyclic reactions.** The stereochemistry of the products formed by electrocyclic reactions depends on the number of π electrons in the polyene and whether the reaction occurs thermally or photochemically.

The concerted reaction of conjugated dienes and dienophiles to form cyclohexene or cyclohexadiene derivatives by 1,4-addition is called a **Diels-Alder reaction. Dienophiles** are alkenes that have one or more electron-withdrawing groups bonded to the carbon atoms of their double bond. The Diels-Alder reaction is stereospecific, so the stereochemistry at the double bond of the dienophile is retained in the product.

Terpenes are a class of naturally occurring compounds whose total number of carbon atoms is a multiple of five. The biological equivalents of isoprene are **isopentenyl pyrophosphate** and **dimethylallyl pyrophosphate.** They react in an enzyme-catalyzed nucleophilic substitution reaction to form the 10-carbon compound geranyl pyrophosphate from which all other terpenes are formed.

Ultraviolet/visible spectroscopy is a useful technique for determining the identity of a compound that contains π electrons. The UV/visible spectrum can be abbreviated as the values of the wavelength at the maximum of the absorption peak, λ_{max}, and the extinction coefficient, ϵ_{max}, which is a measure of the intensity of absorption. These two quantities are characteristic of a compound. In general, the values of λ_{max} shift to longer wavelengths as the number of conjugated double bonds increases in a molecule.

19.17 Summary of Reactions

1. **Nucleophilic Substitution of Allylic Derivatives** Section 19.2

Nucleophilic substitution generally forms a mixture of products. The reaction occurs by an S_N1 mechanism involving an allylic cation as an intermediate. The exceptions are the allyl halides, which react by an S_N2 mechanism.

$$RCH{=}CHCH_2X \xrightarrow{\ :Nu\ } RCH{=}CHCH_2Nu \ + \ \overset{\displaystyle Nu}{\overset{|}{R}CHCH{=}CH_2}$$

X = halogen   :Nu = nucleophile

2. Allylic Bromination   SECTION 19.3

Reaction occurs by a free radical chain mechanism. An allylic radical is an intermediate.

$$RCH_2CH\!=\!CHCH_2R \xrightarrow[CCl_4]{NBS} RCH_2CH\!=\!CHCHR$$
(with Br substituent above the final CHR)

3. Reactions of Conjugated Dienes

a. Electrophilic additions   SECTION 19.8

1,2-Addition product is formed predominantly under kinetic control, whereas 1,4-addition product is formed under thermodynamic control.

$$CH_2\!=\!CHCH\!=\!CH_2 \xrightarrow{HCl} CH_2\!=\!CHCHCH_3 \;+\; CH_2CH\!=\!CHCH_2$$

(with Cl on second product; H and Cl on third product)

     1,2-Addition   1,4-Addition
       product   product

b. Electrocyclic reactions   SECTION 19.10

Stereochemistry of the ring-opening and ring-closing reactions depends on the number of π electrons in the polyene and whether the reaction is photochemical or thermal.

c. Diels-Alder reaction   SECTION 19.12

Excellent method of preparing compounds that contain a six-membered ring.

     Diene   Dienophile

Additional Exercises

19.24 Define each of the following terms in words, by a structure, or by means of a chemical equation.

(a) Conrotatory rotation	**(g)** Concerted reaction	**(m)** s-*cis* Conformation
(b) Conjugated diene	**(h)** HOMO	**(n)** Products of thermodynamic control
(c) Dienophile	**(i)** Allyl cation	**(o)** 1,4-Addition
(d) Diels-Alder reaction	**(j)** λ_{max}	**(p)** Disrotatory rotation
(e) 1,2-Addition	**(k)** Allyl radical	**(q)** Conjugated triene
(f) Nonbonding MO	**(l)** Products of kinetic control	**(r)** Cumulative double bonds

19.25 Give the IUPAC name for each of the following compounds.

(a) (cyclopentadiene structure)

(c) $CH_2\!=\!C\!=\!CHCH(CH_3)_2$

(d)

(b) $CH_2\!=\!CH(CH_2)_4CH\!=\!C\!=\!CH_2$

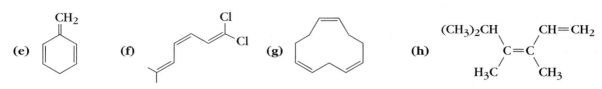

(e) **(f)** **(g)** **(h)**

19.26 Write structural formulas corresponding to each of the following names.

(a) 2,3-Nonadiene
(b) (2E,5E)-Nonadiene
(c) 6-Allyl-1,4-cyclohexadiene
(d) 2,4-Dichloro-1,3-pentadiene
(e) *trans*-1,3-Divinylcyclohexane
(f) (E)-1,4-Dimethoxy-2-butene

19.27 Classify each of the following as polyenes that contain isolated double bonds, cumulative double bonds, or conjugated double bonds.

(a) 1,5-Octadiene
(b) 1,3-Octadiene
(c) 1,3,5-Octatriene
(d) 1,2-Heptadiene
(e) 1,3,5-Cyclooctatriene

19.28 Write the resonance structures for each of the following species.

 (a)

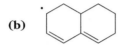 (b)

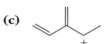 (c)

 (d)

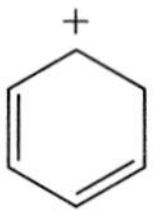

19.29 Write the equations for the reaction of 1,3-cyclopentadiene with each of the following reagents.

(a) 1 Equivalent Cl_2 in CCl_4
(b) NBS in CCl_4
(c) H_2/Pt
(d) O_3 followed by Zn/H_3O^+
(e) 1 Eq HCl in diethyl ether

19.30 Suggest the appropriate reagents to carry out each of the following reactions.

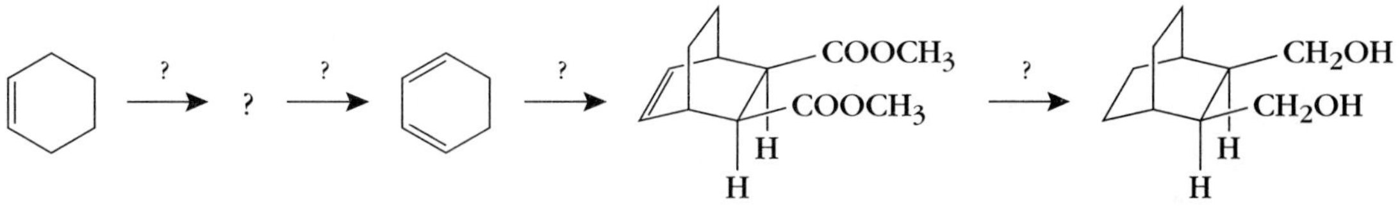

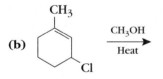

19.31 Write the structure and the stereochemistry, if appropriate, of the product or products in each of the following reactions.

(a) Cyclopentene $\xrightarrow[\text{CCl}_4]{\text{NBS}}$

(b) $\xrightarrow[\text{Heat}]{\text{CH}_3\text{OH}}$

(c) (3E,5E)-Octadiene $\xrightarrow{h\nu}$

(d) Cyclopentadiene + CH_2=CHCOOCH₃ $\xrightarrow{\text{Heat}}$

(e) Cyclopentadiene $\xrightarrow{\text{Heat}}$

(f) $\xrightarrow{\text{Heat}}$

(g) 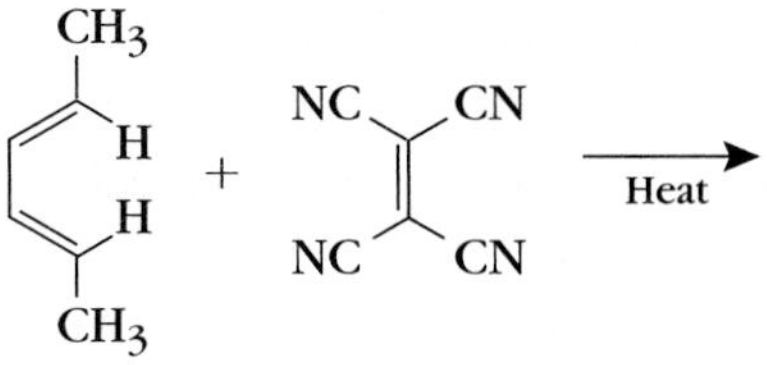 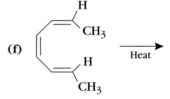 $\xrightarrow{\text{Heat}}$

19.32 Arrange the following compounds in order of increasing wavelength of their λ_{max}.

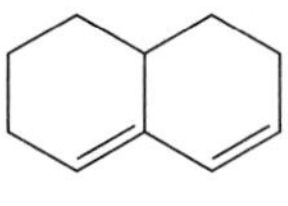

A 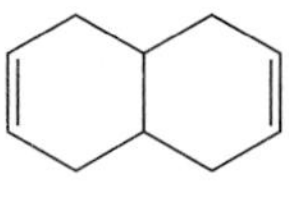 **B** 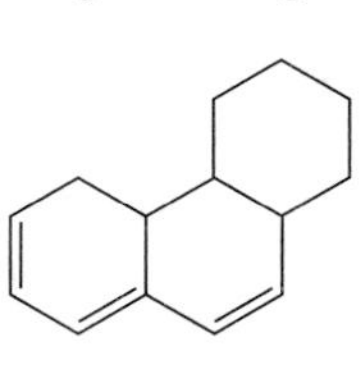 **C**

19.33 The reaction of HCl and 3-buten-2-ol forms two products, 3-chloro-1-butene and 1-chloro-2-butene. Write a mechanism to explain this observation.

19.34 For each of the following electrocyclic reactions, decide if the reaction occurs photochemically or thermally.

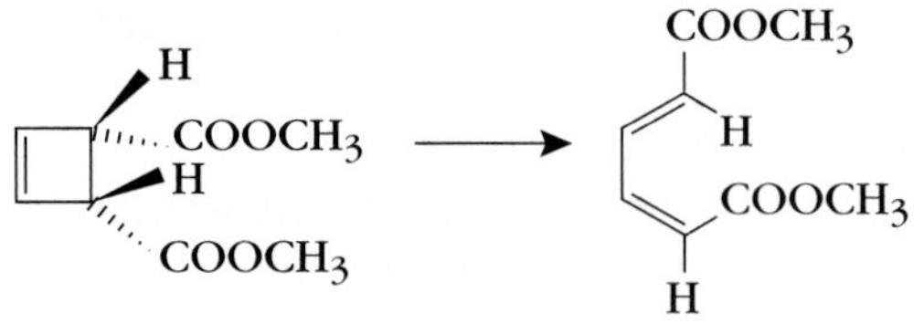

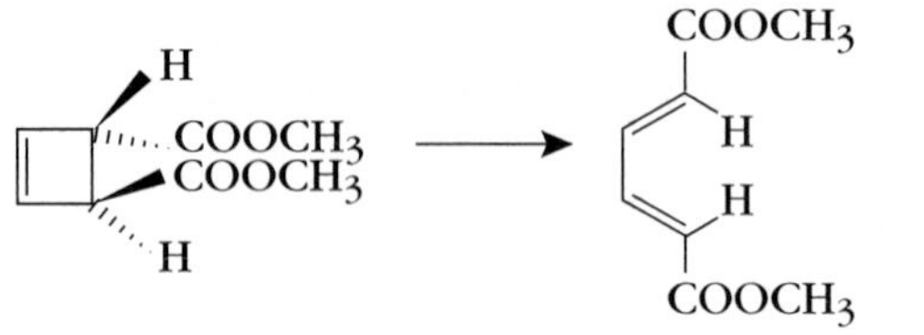

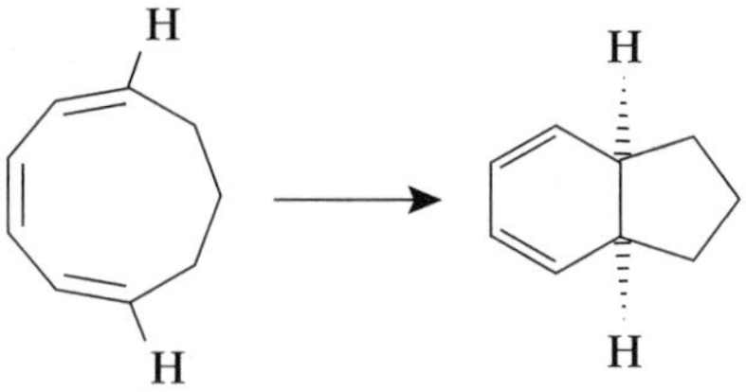

19.35 2,3-Pentadiene is a chiral molecule. Write three-dimensional representations of the two stereoisomers of 2,3-pentadiene.

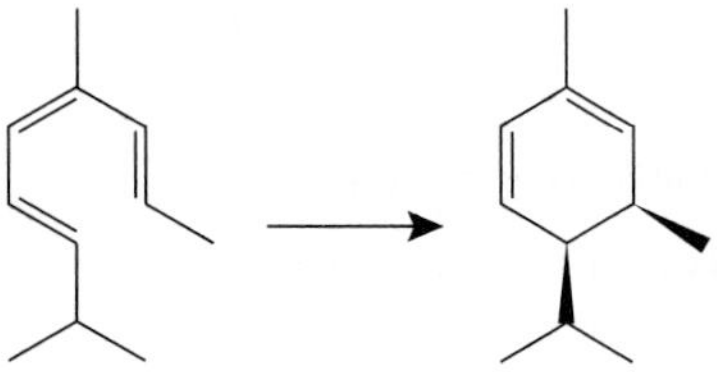

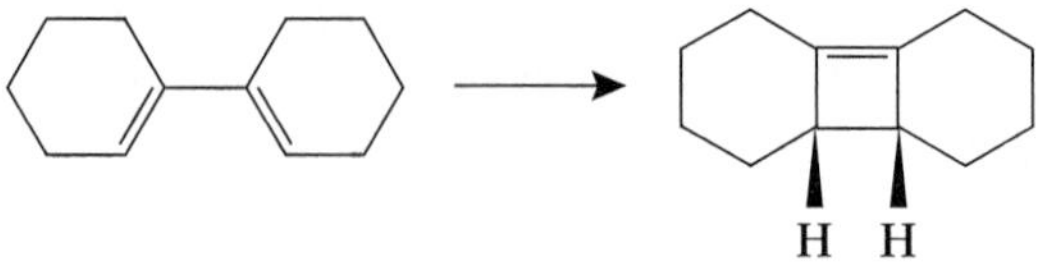

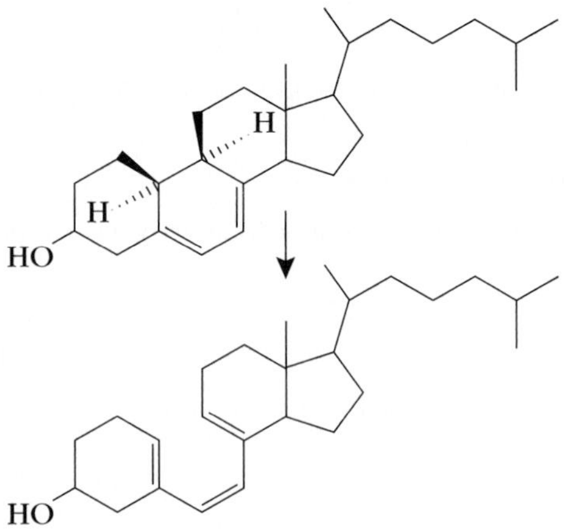

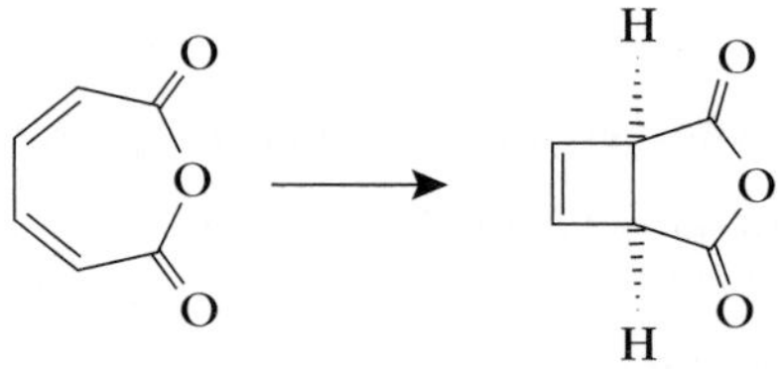

19.36 The addition of HCl to isoprene (2-methyl-1,3-butadiene) in diethyl ether at −15 °C forms a mixture of 3-chloro-3-methyl-1-butene (73%) and 1-chloro-3-methyl-2-butene (27%). When the reaction is carried out at room temperature, an equilibrium mixture is formed that consists of 30% 3-chloro-3-methyl-1-butene and 70% 1-chloro-3-methyl-2-butene. Which isomer is the product of kinetic control? Which isomer is the product of thermodynamic control? Sketch a free energy-reaction path diagram that explains these observations.

19.37 Thermal ring opening of *trans*-3,4-dimethylcyclobutene can occur by two conrotatory rotations in which the two methyl groups both rotate clockwise or counterclockwise. Write the structure of the two products formed by these rotations. In fact only one of the products is formed. Which one would you expect to be formed? Explain your answer.

19.38 Write the structure of the product in each of the following Diels-Alder reactions.

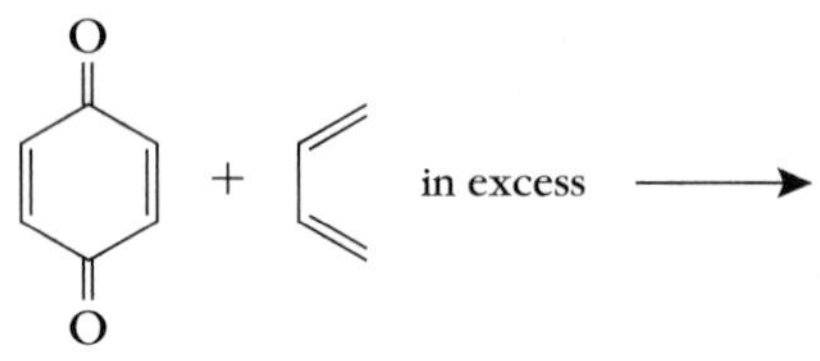

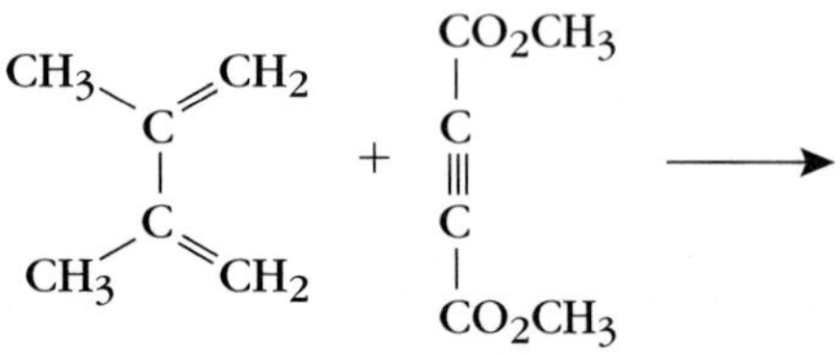

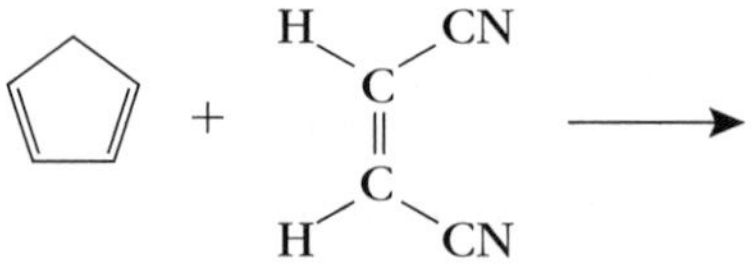

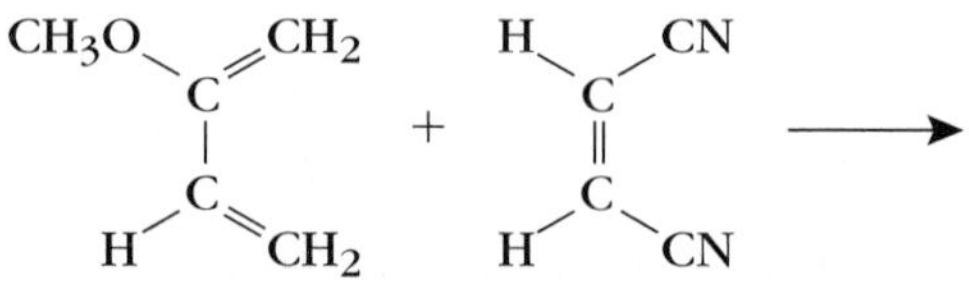

19.39 Write the structures of a diene and a dienophile that would combine in a Diels-Alder reaction to form each of the following compounds.

(a)

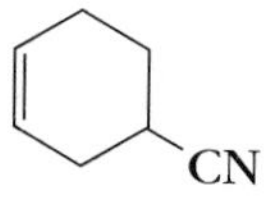

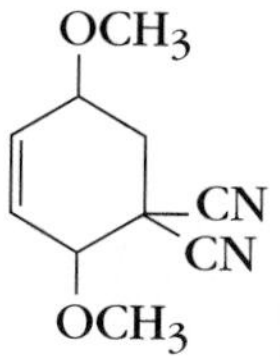

(d)

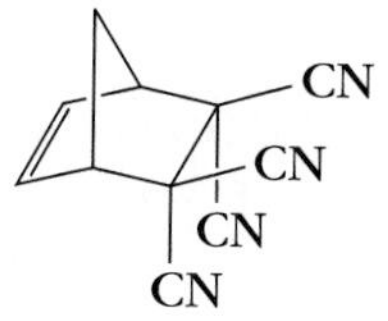

(b)

(e)

(c)

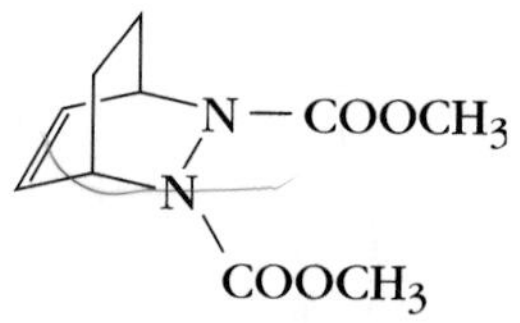

19.40 The reaction of chlorine and 3,4-dimethylcyclobutene in CCl_4 at room temperature forms compound I, $C_6H_{10}Cl_2$. If the solution of 3,4-dimethylcyclobutene in CCl_4 is first heated and then chlorine is added, compounds II and III, isomers of compound I, are formed. Write the structures of the products and write the equations to show how they are formed.

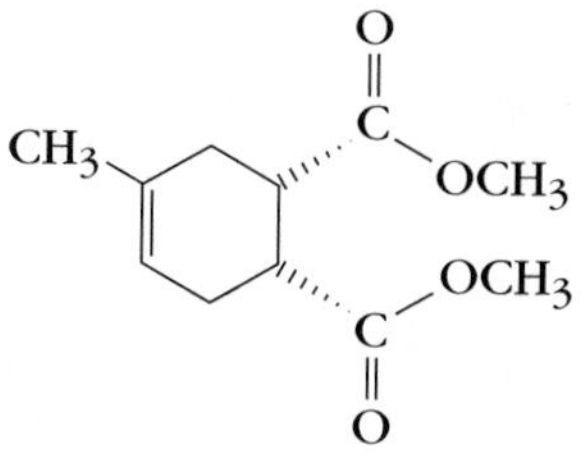

19.41 In certain plants, geranyl pyrophosphate cyclizes to form α-terpineol, a monocyclic terpene, and α-pinene and β-pinene, bicyclic terpenes. Propose a mechanism for these cyclization reactions in which a carbocation is formed as an intermediate.

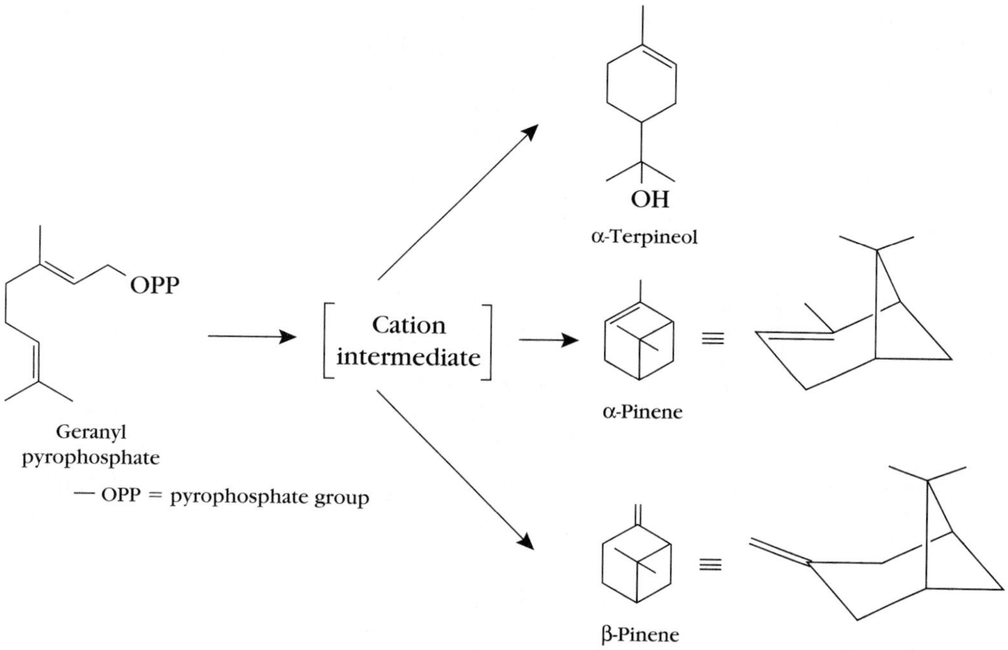

Geranyl
pyrophosphate

— OPP = pyrophosphate group

α-Terpineol

α-Pinene

β-Pinene

19.42 Write the structural formula of each of the following compounds whose molecular formula, ^{1}H NMR spectra, and other spectral data are given below.

   (a) $C_8H_{16}O$ IR: 1710 cm^{-1}
   (b) $C_8H_{12}O_4$ IR: 1720 cm^{-1}
   (c) $C_3H_5BrO_2$ IR: 3400-2400 cm^{-1}; 1700 cm^{-1}
   (d) $C_6H_{12}O_2$ IR: 3400-2400 cm^{-1}; 1700 cm^{-1}

(a) $C_8H_{16}O$; IR: 1710 cm^{-1}

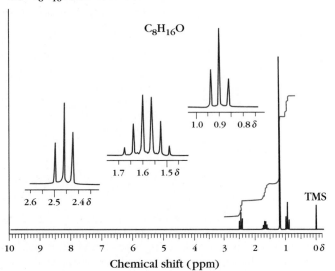

(b) $C_8H_{12}O_4$; IR: 1720 cm^{-1}

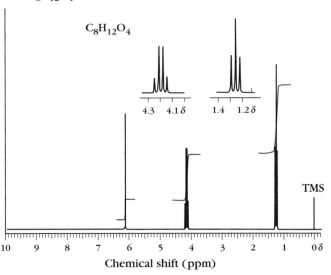

(c) $C_3H_5BrO_2$; IR: 3400–2400 cm^{-1}, 1700 cm^{-1}

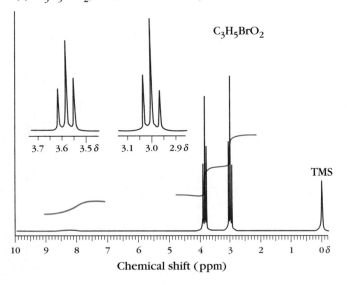

(d) $C_6H_{12}O_2$; IR: 3400–2400 cm^{-1}, 1700 cm^{-1}

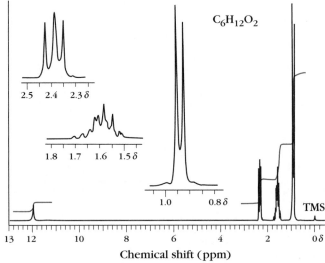

CHAPTER 20

AROMATICITY
π Electron Delocalization in Cyclic Compounds

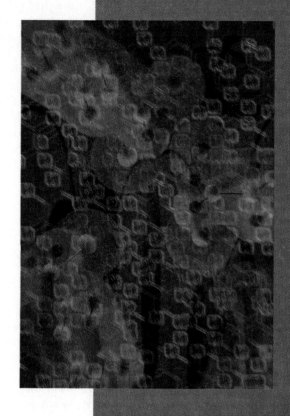

20.1   Discovery of Benzene and Its Unusual Properties

During the latter half of the 18th century, the gas used to light homes and city streets in England was manufactured by the thermal decomposition of whale oil and other liquid fats. When this gas was compressed in tanks for distribution, a liquid separated. In 1825, Michael Faraday, the same person known for his pioneering work on the laws governing electrolysis of aqueous solutions, isolated from this liquid a compound with the molecular formula C_6H_6.

This compound, which is now called *benzene,* is unusual because it has a high carbon-to-hydrogen ratio. Benzene, therefore, must be highly unsaturated. Benzene, however, does not react as if it were a polyene. Unlike alkenes or polyenes, benzene is inert to the usual laboratory oxidizing agents such as $KMnO_4$ or $K_2Cr_2O_7$. Furthermore it does not easily react with chlorine and bromine. A catalyst is needed for the halogenation of benzene, and the reaction turns out to be a substitution instead of an addition. Only one monosubstituted bromination product of benzene has ever been obtained, moreover, which implies that the six hydrogen atoms of benzene are equivalent.

$$C_6H_6 \ + \ KMnO_4 \ \xrightarrow[\ H_2O\ ]{\ H_2SO_4\ } \ \text{No reaction}$$

$$C_6H_6 \ + \ Br_2 \ \begin{array}{c} \xrightarrow{\ CCl_4\ } \ \text{No addition reaction} \\[2ex] \xrightarrow[\text{catalyst}]{} \ C_6H_5Br \ + \ HBr \end{array}$$

During the late 19th century, scientists isolated many compounds from natural sources that have a similar high carbon-to-hydrogen ratio. Many of these compounds have pleasant aromas, and so they were called aromatic compounds. Examples include the essential components of the volatile oils of cloves, cinnamon, bitter almonds, and wintergreen. The Austrian chemist Joseph Loschmidt was the first to suggest a connection between these compounds and benzene when he proposed that aromatic compounds were derivatives of the hydrocarbon benzene. Today we classify benzene and its derivatives as aromatic compounds. Many derivatives of benzene do not have particularly pleasant aromas, however, so the association between aroma and aromatic compounds has long been lost.

Chemists in the latter part of the 19th century were faced with a problem, namely how to describe the structure of benzene, a highly unsaturated yet relatively unreactive hydrocarbon. A satisfactory solution to this problem was not found until the development of resonance and molecular orbital theories in the 20th century. Let us begin by examining the resonance hybrid description of bonding in benzene.

20.2   Bonding in Benzene: A Resonance Hybrid Description

What do we know about the structure of benzene? From experimental data, we know that benzene is a planar molecule in which the six carbon atoms of its skeleton form a regular hexagon. All carbon-carbon bond lengths are 139 pm. All carbon-hydrogen bonds are equivalent, too, because only one compound has ever been found for each monosubstituted benzene, C_6H_5X. From these observations and the requirement that each carbon atom must have four bonds, two equivalent resonance structures can be written for benzene.

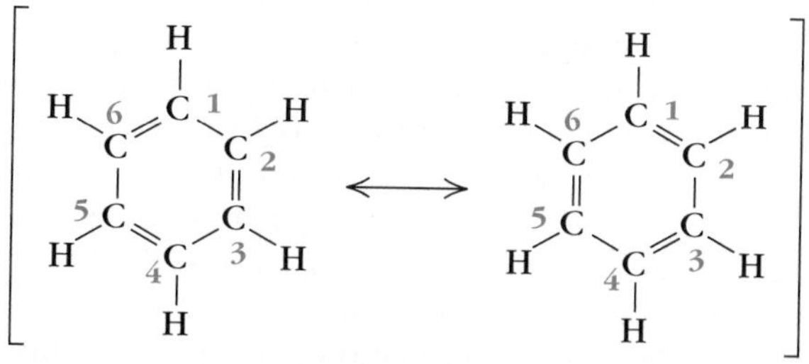

These two resonance structures are called *Kekulé structures* in honor of Fredrich Kekulé, who first suggested a hexagonal structure for benzene.

Compounds or ions for which we can write several resonance structures have an electron structure that is a resonance hybrid of all the resonance structures, as we discussed in Section 1.12. Thus the electron structure of benzene is a resonance hybrid of the two Kekulé structures shown above. This means that the carbon-carbon bonds of benzene are neither double nor single bonds. This is evident, moreover, from the experimentally measured carbon-carbon bond lengths of benzene, which are all the same (139 pm) and which are midway between the value of the carbon-carbon single bond length in butadiene (146 pm) and its carbon-carbon double bond length (134 pm).

The resonance hybrid description of the π electron structure of benzene represents the π electrons of benzene as being delocalized around the ring. Chemists sometimes represent this π electron delocalization by writing the structure of benzene as a circle inscribed inside a hexagon.

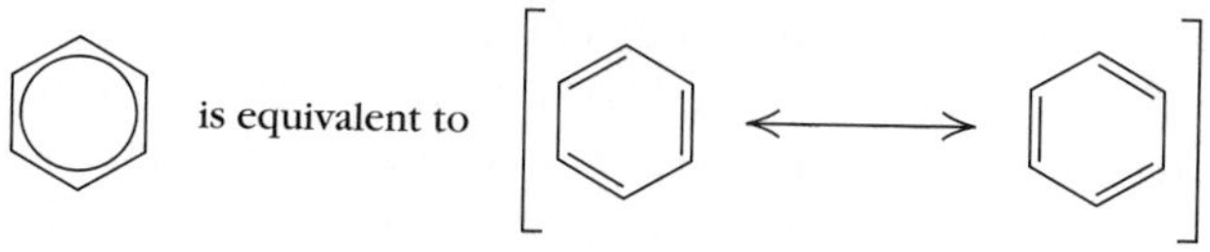

While this representation is convenient and reminds us of the delocalization of π electrons in benzene, it is not very useful when we must keep track of the electrons in the chemical reactions of benzene. In this text, therefore, we use a single Kekulé structure to represent benzene. Keep in mind, however, that the real structure of benzene is actually a resonance hybrid.

Increased stability is another feature of compounds or ions that are represented as resonance hybrids. This increased stability can be estimated from heats of hydrogenation data.

20.3   An Estimate of the Stability of Benzene from Heats of Hydrogenation

In the presence of the usual metal catalysts, benzene reacts with three equivalents of hydrogen gas at high temperature and pressure to form cyclohexane.

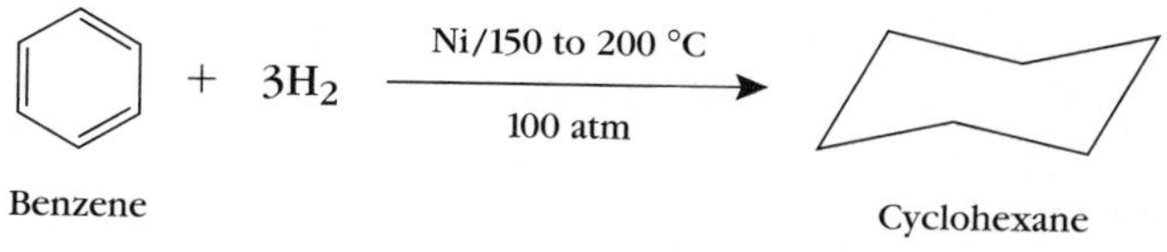

$\Delta H° = -49.3$ kcal/mol
$(-206.3$ kJ/mol$)$

The heat of hydrogenation for this reaction, -49.3 kcal/mol (-206.3 kJ/mol), can be used to estimate the stability of benzene. In previous chapters, the relative stabilities of various compounds were determined from their experimental heats of hydrogenation. In Section 8.12, for example, heats of hydrogenation were used to show that monosubstituted alkenes are less stable than disubstituted ones. In order to determine the relative stabilities of these compounds, all we had to do was compare their experimentally determined heats of hydrogenation. In the case of benzene, however, we want to determine how π electron delocalization affects the stability of benzene; that is, we want to compare the heat of hydrogenation of benzene *with* π electron delocalization to the heat of hydrogenation of benzene *without* π electron delocalization.

Benzene without π electron delocalization does not exist, however. How then can the heat of hydrogenation of a real benzene molecule be compared with that of an imaginary benzene molecule without π electron delocalization? The answer is to try to calculate the heat of hydrogenation of this imaginary molecule called "cyclohexatriene." One approach is to assume that its heat of hydrogenation is 85.8 kcal/mol (360 kJ/mol), which is three times the heat of hydrogenation of cyclohexene. The experimental value of the heat of hydrogenation of benzene is only 49.3 kcal/mol (206 kJ/mol). This is 36.5 kcal/mol (152 kJ/mol) less than the calculated amount for cyclohexatriene. All of these values are illustrated in Figure 20.1.

According to Figure 20.1, catalytic hydrogenation of benzene evolves 36.5 kcal/mol (152 kJ/mol) less energy than predicted for the imaginary cyclohexatriene. This means that delocalization of π electrons stabilizes benzene by about 36 kcal/mol (152 kJ/mol). This stabilization is called the **resonance energy** of benzene. Other methods of estimating the stability of benzene have arrived at values of its resonance energy of anywhere from 30 to 80 kcal/mol (125 to 335 kJ/mol). The actual value of the resonance energy is not our major concern. The important point is that π electron delocalization is the major factor that contributes to the stability of benzene and its derivatives. This greater stability places benzene and its derivatives in a separate category of compounds that have unique chemical reactions, as we will learn in detail in Chapter 21.

Figure 20.1
Estimating the resonance energy of benzene from the heats of hydrogenation of benzene and cyclohexene. The observed heat of hydrogenation of cyclohexene is 28.6 kcal/mol (120 kJ/mol), so we estimate that the heat of hydrogenation of "cyclohexatriene" is 3 × 28.6 kcal/mol or 85.8 kcal/mol (360 kJ/mol). The observed heat of hydrogenation of benzene is 49.3 kcal/mol (206 kJ/mol). The difference between the estimated heat of hydrogenation of "cyclohexatriene," 85.8 kcal/mol (360 kJ/mol), and the heat of hydrogenation of benzene, 49.3 kcal/mol (206 kJ/mol), is 36.5 kcal/mol (152 kJ/mol)—the estimated resonance energy of benzene.

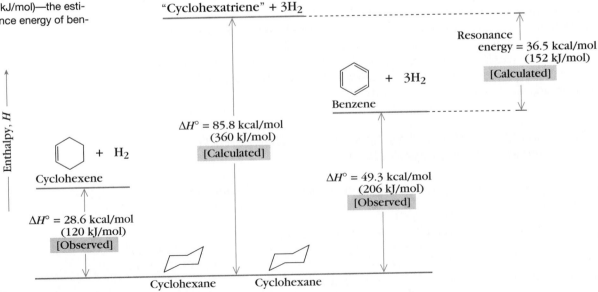

Exercise 20.1	Hydrogenations of 1,3,5-hexatriene and benzene do not form the same product, yet a comparison of their heats of hydrogenation has been used to estimate the resonance energy of benzene. Write the equations for the hydrogenations of 1,3,5-hexatriene and benzene. If the heat of hydrogenation of (Z)-1,3,5-hexatriene is -80.5 kcal/mol (-336.8 kJ/mol), what value of the resonance energy of benzene is obtained by this comparison?

Cyclobutadiene and cyclooctatetraene are two hydrocarbons that can also be represented by more than one resonance structure. Are they also aromatic compounds?

20.4 Cyclobutadiene and Cyclooctatetraene

Cyclobutadiene and cyclooctatetraene are cyclic compounds whose structures can be written as two equivalent resonance structures with alternating single and double bonds.

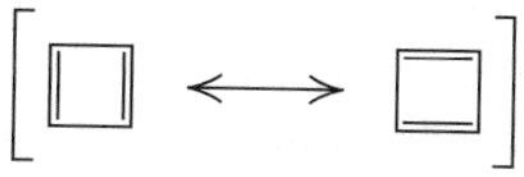

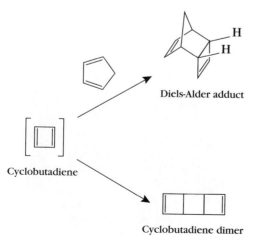

Cyclobutadiene 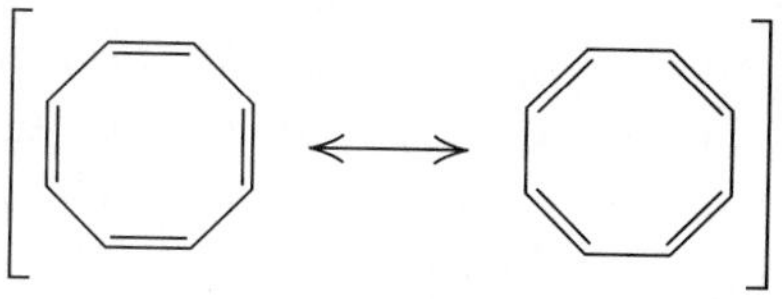   Cyclooctatetraene

Because of this similarity to benzene, chemists expected these compounds and their derivatives to be stabilized by π electron delocalization, too. Despite repeated attempts, however, cyclobutadiene and cyclooctatetraene were not synthesized until the 20th century. Evidence for the successful synthesis of cyclobutadiene was first reported by the American chemist R. Pettit in the 1960s. It was found that cyclobutadiene is so reactive that it cannot be isolated except at low temperatures, below 35 K, where it can be trapped in a frozen gas. At higher temperatures, it reacts rapidly with itself to form a dimer. In the presence of a diene such as cyclopentadiene, it reacts to form a Diels-Alder adduct.

Diels-Alder adduct

Cyclobutadiene

Cyclobutadiene dimer

Figure 20.2
The bond lengths in cyclobutadiene, 1, 3-butadiene, benzene, and cyclooctatetraene.

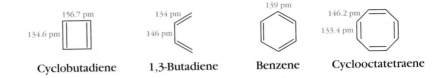

Cyclobutadiene   1,3-Butadiene   Benzene   Cyclooctatetraene

The bond angles and bond lengths of cyclobutadiene, obtained from cyclobutadiene frozen at low temperatures, show that its structure is not a square but a rectangle with alternating single and double bonds. The bond lengths in cyclobutadiene are compared with those of 1,3-butadiene, benzene, and cyclooctatetraene in Figure 20.2.

Cyclooctatetraene was first synthesized by the German chemist Richard Willstatter in 1911. The carbon-carbon bond lengths alternate around the ring as expected for a polyene, as shown in Figure 20.2. Heats of hydrogenation demonstrate no special stability due to π electron delocalization, and cyclooctatetraene undergoes the normal electrophilic addition reactions of alkenes and polyenes.

Clearly electron delocalization is absent in both cyclobutadiene and cyclooctatetraene, so they differ from benzene in stability and chemical reactivity. Why is there no π electron delocalization in cyclobutadiene and cyclooctatetraene? The structures of both compounds can be written as resonance hybrids just like benzene. We can only conclude that their resonance hybrid structures do not adequately represent their real structures. This is worrisome because the organization of organic chemistry is based on the principle that compounds with similar structural features, such as functional groups, have similar chemical and physical properties. Benzene, cyclobutadiene, and cyclooctatetraene seem to violate this principle because they all have the same structural features, yet they differ in reactivity. The German chemist Erich Hückel provided a solution to this problem.

20.5 Hückel's Rule of Aromaticity

In 1937, Erich Hückel used molecular orbital theory to define aromaticity. His approach is an extension of the simplified LCAO method introduced in Chapter 1 to obtain the π molecular orbitals of ethylene, and in Chapter 15 to obtain the π molecular orbitals of the allyl systems and 1,3-butadiene. We can illustrate Hückel's method first with benzene and then apply it to other molecules.

Benzene is a planar molecule containing six sp^2 hybrid carbon atoms bonded

Figure 20.3
The six *p* atomic orbitals of benzene are parallel to each other and perpendicular to the benzene ring. Each *p* atomic orbital contains one electron.

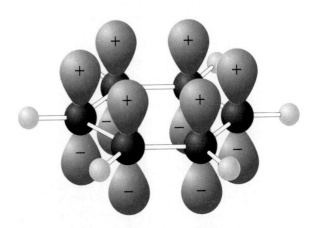

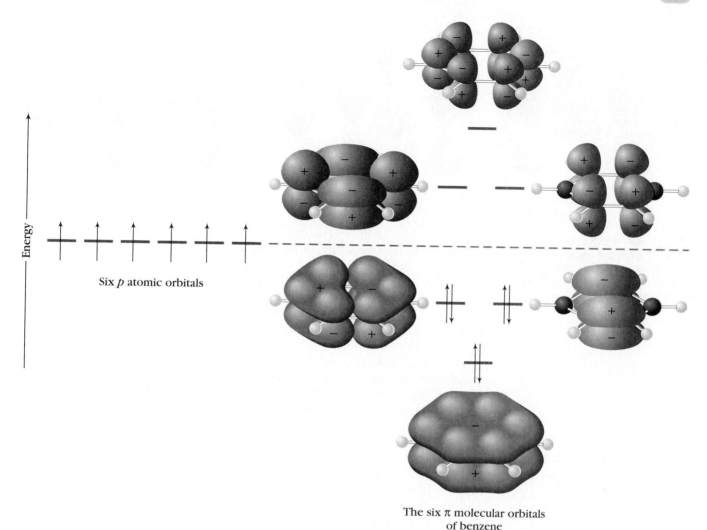

The six π molecular orbitals
of benzene

Figure 20.4
The energy levels of the six π molecular orbitals of benzene.

to six hydrogen atoms. Benzene, therefore, has six σ C—H bonds made by combining an sp^2 hybrid orbital on each carbon atom with an s orbital of a hydrogen atom. The framework of C—C bonds is made by combining two sp^2 hybrid orbitals on each carbon atom with the sp^2 hybrid orbitals of two adjacent neighboring carbon atoms. The remaining six p atomic orbitals, one on each carbon atom, contain one electron each and lie parallel to each other, perpendicular to the plane of the ring, as shown in Figure 20.3.

The π molecular orbitals of benzene are obtained by mathematically combining the six p atomic orbitals to form six π molecular orbitals, three of which are bonding molecular orbitals and three of which are antibonding molecular orbitals. The six π molecular orbitals and their relative energies are shown in Figure 20.4.

Notice that two of the bonding π molecular orbitals of benzene have the same energy. Orbitals with the same energy are called **degenerate orbitals.** Two of the higher energy antibonding π molecular orbitals of benzene are also degenerate.

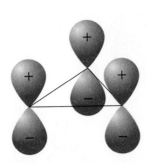

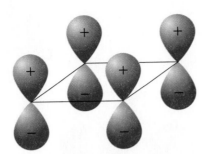

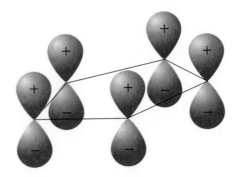

Three sp^2 hybrid carbon
atoms in a planar ring

Four sp^2 hybrid carbon
atoms in a planar ring

Five sp^2 hybrid carbon
atoms in a planar ring

Figure 20.5
Three-, four-, and five-membered
rings of sp^2 hybrid carbon atoms.
Their p atomic orbitals, containing
one electron each, are used to construct their molecular orbitals.

The π electron structure of benzene can be obtained by placing the six π
electrons in the π molecular orbitals according to the aufbau principle. When
this is done, as shown in Figure 20.4, the three bonding molecular orbitals contain two electrons each. Notice that the six π electrons exactly fill the three
bonding π molecular orbitals but none of the antibonding molecular orbitals.
This must be a very stable π electron structure to account for the observed stability of benzene.

How does this electron structure account for the lack of aromaticity of
cyclobutadiene and cyclooctatetraene? Hückel provided the answer to this question by carrying out molecular orbital calculations of cyclic compounds containing sp^2 hybridized carbon atoms joined together in a planar ring. The rings containing the carbon atoms are polygons, namely, a triangle, a square, a pentagon, a
hexagon, and so forth. The first three members of the series, which contain
three, four, and five sp^2 hybrid carbon atoms in a ring, are shown in Figure 20.5.

The π molecular orbitals of these molecules are constructed by mathematically combining their p atomic orbitals. The energies of the molecular orbitals
obtained from these combinations for rings of three to nine carbon atoms are
compared in Figure 20.6.

We know that cyclobutadiene is not an aromatic compound because its properties are very different from those of benzene. The energy level diagrams in Figure 20.6 demonstrate, moreover, that cyclobutadiene and benzene also differ in
their π electron structures. The π electrons of cyclobutadiene, unlike those of
benzene, are not all located in bonding molecular orbitals. Two π electrons are in
the bonding molecular orbital, but the other two are in nonbonding molecular
orbitals. The two π electrons in the nonbonding molecular orbitals are unpaired
because the nonbonding molecular orbitals are degenerate.

The difference between the π electron structures of benzene and cyclobutadiene is the basis for the difference in their properties. Compounds that contain p
atomic orbitals on adjacent atoms in a planar ring are particularly stable if all their
π electrons completely fill *only* the bonding molecular orbitals. Benzene is particularly stable, therefore, because its six π electrons completely fill its three bonding molecular orbitals. Cyclobutadiene, on the other hand, is not as stable as benzene because the four π electrons of cyclobutadiene fill not only its bonding
molecular orbital but also half fill each of its two nonbonding molecular orbitals.

A pattern emerges from the energy level diagrams shown in Figure 20.6. The
numbers of bonding molecular orbitals for the compounds shown in Figure 20.6
are one, three, or five. Therefore the numbers of π electrons that completely fill
only the bonding orbitals are two (one M.O.), six (three M.O.s), or ten (five M.

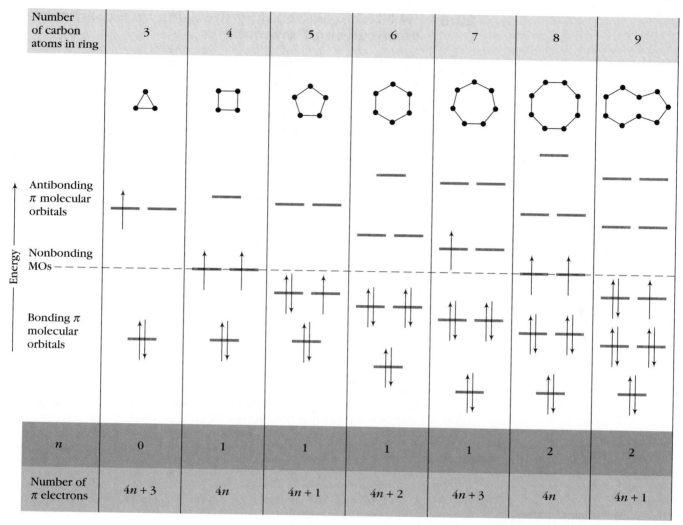

Figure 20.6
The energy levels of the π molecular orbitals obtained by combining the *p* atomic orbitals of cyclic molecules containing three to nine sp^2 hybrid carbon atoms.

O.s). The numbers two, six, and ten represent a series given by the general formula $4n + 2$, where n is any integer (0, 1, 2, etc.). As a result, any compound that contains *p* atomic orbitals on adjacent atoms in a planar ring is aromatic if it contains $4n+2$ π electrons. This is called Hückel's rule of aromaticity.

To summarize, a compound is aromatic if it conforms to each of the following conditions:

•It must be cyclic.
•It must have a *p* atomic orbital on each atom of the ring.
•It must be planar, so that overlap can occur between the *p* orbitals on adjacent atoms.
•It must have $4n+2$ electrons in its *p* orbitals, where n is any integer (0, 1, 2, etc.). Aromatic compounds have, therefore, 2, 6, 10, 14, etc., π electrons.

So far benzene is the only aromatic compound that has been presented. In order to find other aromatic compounds, we need to know what to look for. Experimental criteria for determining aromaticity include increased stability based on heats of hydrogenation, reactivity unlike alkenes or acyclic polyenes, and spectral properties. The best of these criteria, however, is ^{1}H NMR spectroscopy.

20.6 ^{1}H NMR Spectroscopy Provides an Experimental Verification of Aromaticity

The ^{1}H NMR spectrum of benzene consists of a singlet in the region δ 7.24 to 7.27, which is far downfield from the usual chemical shifts of vinyl hydrogens of alkenes. Hydrogen atoms bonded to benzene rings in substituted benzenes have chemical shifts in the region δ 6.5 to 8.0. No other hydrogens are commonly found in this region, so ^{1}H NMR spectroscopy can serve as a criterion for the identification of substituted benzenes. Can deshielding of benzene hydrogen atoms be used as a *general* criterion of aromaticity? In order to answer this question, we must first learn why hydrogen atoms bonded to a benzene ring are deshielded compared to other hydrogens.

Electron delocalization of the π electrons in benzene's aromatic ring is the reason the hydrogen atoms bonded to the carbon atoms of benzene are so deshielded. The π electron density in benzene is found in two doughnut-shaped regions above and below the plane of the ring. When an aromatic ring is oriented perpendicular to the applied magnetic field, as shown in Figure 20.7, its π electrons generate an electric current called a **ring current.**

The ring current induces a small local magnetic field, which opposes the applied magnetic field in the center of the ring but reinforces the applied field outside the ring where the aromatic hydrogens are located. As a result, the aromatic hydrogens experience a magnetic field greater than the applied magnetic field and therefore come into resonance at a lower applied field. This is observed experimentally as a large downfield chemical shift. The deshielding effect of the π electrons of a benzene ring is similar to the deshielding effect of the π electrons of an alkene, except that it is much stronger.

The model illustrated in Figure 20.7 predicts that hydrogen atoms located inside the ring should be shielded. These hydrogen atoms should be found at

Figure 20.7
The hydrogen atoms of benzene are deshielded because the π electrons generate a ring current and the ring current, in turn, induces a magnetic field, which deshields the hydrogen atoms.

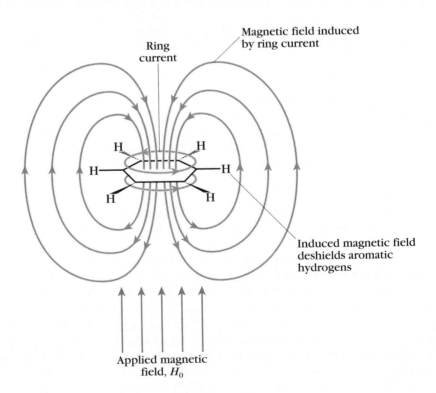

chemical shifts higher than normal. This has been found experimentally, as is discussed in Section 20.7.

| Exercise 20.2 | Carbon atoms of double bonds in alkenes have similar ^{13}C NMR chemical shifts to the aromatic carbon atoms in benzene and its derivatives. Use the model illustrated in Figure 20.7 to suggest an explanation. |

Both aromaticity and ring current are the end results of π electron delocalization in molecular orbitals formed by combining the p orbitals of a continuous cyclic array. We conclude, therefore, that all aromatic compounds must have a ring current that influences the chemical shift of the hydrogen atoms bonded to the aromatic ring. The chemical shift of these hydrogens depends on their position relative to the ring. Those outside the ring are found downfield, usually in the δ 6.5 to 8.5 region. Those inside the ring, on the other hand, are found upfield and often they are found upfield of TMS. This provides a powerful yet simple experimental method of determining if a compound is aromatic. We now apply this criterion to test the validity of Hückel's $4n+2$ rule.

20.7   Testing the 4*n*+2 Rule

A great deal of research followed the formulation of Hückel's $4n+2$ rule of aromaticity. This research focused on the synthesis of various compounds to test the validity of Hückel's prediction. These new compounds, as well as many others already known, support Hückel's $4n+2$ rule. We can demonstrate the validity of the $4n+2$ rule by applying it to four classes of compounds, namely annulenes, polycyclic benzenoid compounds, heterocyclic aromatic compounds, and charged aromatic compounds. We begin with the annulenes.

A   Annulenes

The term **annulene** refers to completely conjugated monocyclic polyenes. They are all given the name annulene with the number of carbon atoms in the ring designated by a numerical prefix. Thus cyclobutadiene is [4]annulene, benzene is [6]annulene, and cyclooctatetraene is [8]annulene.

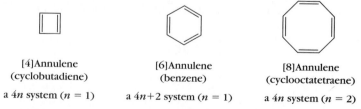

[4]Annulene	[6]Annulene	[8]Annulene
(cyclobutadiene)	(benzene)	(cyclooctatetraene)
a $4n$ system ($n = 1$)	a $4n+2$ system ($n = 1$)	a $4n$ system ($n = 2$)

Recall from Section 20.4 that [4]annulene and [8]annulene react differently than [6]annulene. On this basis, neither [4]annulene nor [8]annulene are aromatic. Their structures also indicate a lack of aromaticity. They are not regular polygons because their bond lengths alternate in length consistent with a polyene rather than an aromatic structure. Cyclooctatetraene, moreover, is not even planar; it is tub-shaped, as shown in Figure 20.8. Cyclooctatetraene, therefore, is

Figure 20.8
The tub conformation of cyclooctatetraene. The planar conformation is estimated to be 10.9 to 12.1 kcal/mol (45.6 to 50.6 kJ/mol) less stable than the tub conformation.

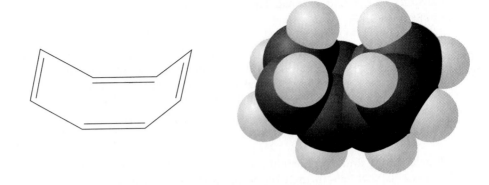

not a system to which Hückel's rule applies. This is further confirmed by the ^{1}H NMR spectrum of cyclooctatetraene, which shows a singlet at δ 5.69, the usual chemical shift region for vinyl hydrogen atoms, not aromatic hydrogen atoms.

According to Hückel's rule, the next larger aromatic monocyclic planar ring should be [10]annulene ($4n+2$; $n=2$). In larger monocyclic rings, however, *trans* double bonds can be incorporated, giving rise to a number of geometric isomers. In [10]annulene, serious steric strain in all of its isomers prevents them from becoming planar. The *Z,E,Z,Z,E*-isomer, for example, has minimal bond-angle strain but suffers a severe nonbonded repulsion between the two hydrogen atoms inside the ring.

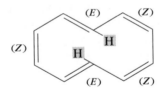

The steric problem of the *Z,E,Z,Z,E*-isomer was overcome by preparing compound **20.1,** in which the π system is planar. The ^{1}H NMR spectrum of **20.1** consists of two sets of signals; a singlet of relative area 1 is found at δ −0.50, and a multiplet of relative area 4 is found at δ 6.9 to 7.3. The eight hydrogen atoms outside the ring are found in the deshielded aromatic region of the ^{1}H NMR spectrum, while the other two hydrogen atoms are so strongly shielded that there must be a ring current. The ^{1}H NMR spectrum of **20.1** indicates, therefore, that it is aromatic.

A negative chemical shift, such as δ −0.50, means that the peak appears to the right (upfield) of the tetramethylsilane (TMS) peak in the ^{1}H NMR spectrum of the compound.

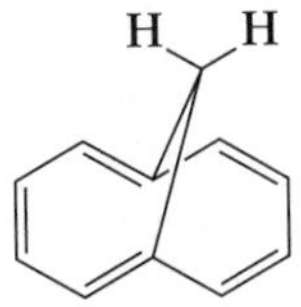

20.1

Steric hindrance between the hydrogen atoms inside the ring of [14]annulene prevents its ring from being planar. One way of eliminating this hydrogen interference is to introduce one or more triple bonds into the ring, as in the case of dehydro[14]annulene.

Triple bond
contributes
2 of the 14
π electrons

Dehydro[14]annulene

■ *The prefix* dehydro- *in a name means that the compound has two fewer hydrogen atoms in its structure than the parent compound. Dehydro[14]annulene, $C_{14}H_{12}$, for example, has two fewer hydrogen atoms that [14]annulene, $C_{14}H_{14}$.*

The two π electrons of the second π bond of the triple bond do not form part of the delocalized π electron system. Instead they simply exist as a localized π bond in the plane of the ring. The ten outside ring hydrogens are found at δ 7.3 to 8.5, while the two inner hydrogen atoms are found at δ −0.70 of the ^{1}H NMR spectrum of dehydro[14]annulene.

Cyclic conjugated rings that contain up to 24 π electrons have been synthesized either as annulenes or dehydroannulenes. The ^{1}H NMR spectra of these compounds suggest that [14], [18], and [22]annulenes or dehydroannulenes are aromatic compounds (all contain $4n+2$ π electrons), while the [12], [16], [20], and [24]annulenes or dehydroannulenes are not aromatic (all contain $4n$ π electrons). An aromatic [18]annulene and a nonaromatic large conjugated ring compound, 1,5,9-tridehydro[12]annulene, are shown in Figure 20.9.

Exercise 20.3   Label each of the following compounds as aromatic or nonaromatic.

(a)    (b) 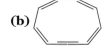   (c)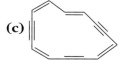

The synthesis of annulenes and dehydroannulenes validates Hückel's rule for neutral monocyclic compounds. A number of polycyclic compounds also have aromatic properties.

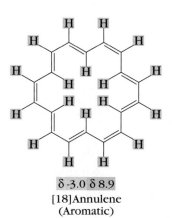

δ -3.0 δ 8.9
[18]Annulene
(Aromatic)

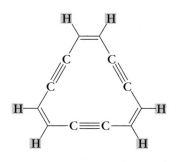

δ 4.42
1,5,9-Tridehydro[12]annulene
(Nonaromatic)

Figure 20.9
Examples of aromatic and nonaromatic large conjugated ring compounds.

B   **Polycyclic Benzenoid Compounds**

A number of polycyclic compounds are known that contain two or more benzene rings fused or joined at adjacent carbon atoms in such a way that each ring shares two carbon atoms. These compounds are called **polycyclic benzenoid compounds.** The simplest polycyclic benzenoid compound is naphthalene. It consists of two fused benzene rings. Other examples are the tricyclic compounds anthracene and phenanthracene.

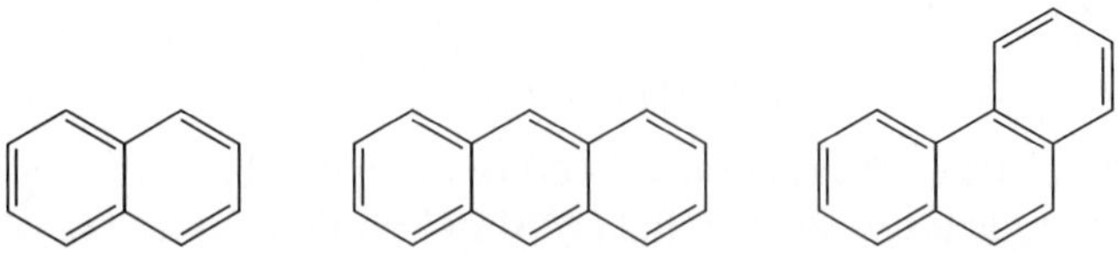

Naphthalene   Anthracene   Phenanthracene

Hückel's rule of aromaticity does not strictly apply to polycyclic benzenoid compounds because they are not monocyclic. They are classified as aromatic, however, because they undergo substitution reactions like benzene, and their ring hydrogen atoms are found in the aromatic region of their ^{1}H NMR spectra.

Compounds such as naphthalene, anthracene, and phenanthracene are essentially assemblies of benzene units, so like benzene they have substantial resonance energies. The resonance energies of naphthalene, anthracene, and phenanthracene are estimated from heats of hydrogenation to be about 60 kcal/mol (251 kJ/mol), 80 kcal/mol (334 kJ/mol), and 90 kcal/mol (376 kJ/mol), respectively.

Polycyclic benzenoid compounds such as benzo[a]pyrene and dibenz[a,h]anthracene are some of the most potent carcinogenic (cancer-causing) chemicals known. These compounds are found in the products of combustion, such as soot and tobacco smoke.

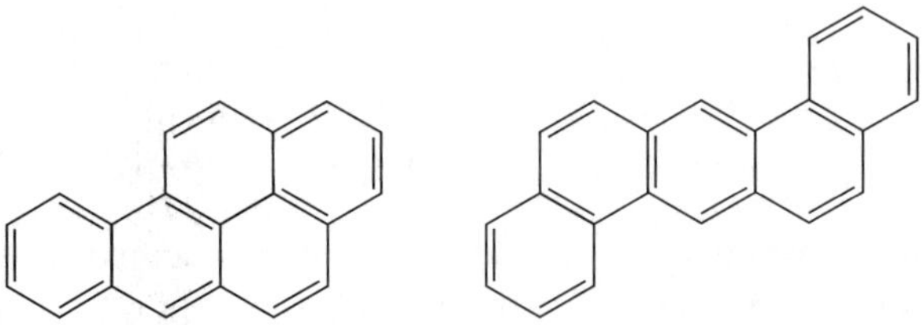

Benzo[a]pyrene   Dibenz[a,h]anthracene

The actual carcinogenic compounds are not benzo[a]pyrene and dibenz[a,h]anthracene, but metabolic products formed in the cell. Benzo[a]pyrene, for example, undergoes an enzyme-catalyzed oxidation to a diol epoxide after it enters the cell.

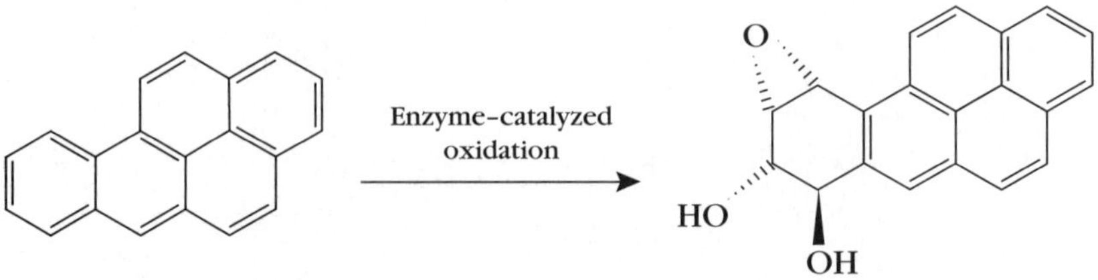

Benzo[a]pyrene   Diol epoxide

The purpose of this oxidation is to render the compound more water soluble so that it can be eliminated from the cell. Unfortunately, the diol epoxide is carcinogenic; it causes mutations that eventually cause the cell to lose its ability to control replication, thus leading to uncontrolled growth.

Aromaticity is not restricted to hydrocarbons. Replacing a carbon atom of an aromatic ring with a heteroatom such as oxygen, nitrogen, or sulfur forms compounds that satisfy Hückel's rule, too.

C   Heterocyclic Aromatic Compounds

Cyclic compounds that contain one or more ring heteroatoms (atoms other than carbon), are called **heterocyclic compounds.** If the ring is planar and contains $4n+2$ π electrons, it is a **heterocyclic aromatic compound.** The most common heterocyclic aromatic compounds contain either nitrogen atoms or oxygen atoms as the heteroatom. Four important heterocyclic aromatic compounds that contain nitrogen or oxygen atoms are shown in Figure 20.10. The compounds in Figure 20.10, like benzene, undergo chemical reactions typical of aromatic compounds. That is, they undergo substitution rather than addition reactions and their ^{1}H NMR spectra show the effect of a ring current.

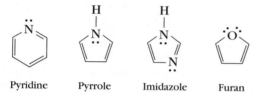

Pyridine   Pyrrole   Imidazole   Furan

Figure 20.10
Examples of nitrogen- and oxygen-containing heterocyclic aromatic compounds.

Each of the compounds in Figure 20.10 contains six π electrons in a ring, so each obeys Hückel's rule. The heteroatom may contribute one or two electrons to the π system. This can be seen by comparing the atomic orbitals used to construct the π molecular orbitals of pyridine and pyrrole.

The p atomic orbitals of pyridine are shown in Figure 20.11. Pyridine is a planar molecule containing five sp^2 hybrid carbon atoms and one sp^2 hybrid nitrogen atom. Each ring atom forms two sp^2–sp^2 σ bonds with its neighboring ring atoms, and each carbon atom forms one sp^2–s σ bond to a hydrogen atom. The

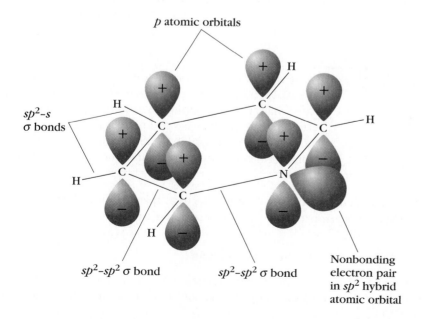

p atomic orbitals

sp^2–s σ bonds

sp^2–sp^2 σ bond sp^2–sp^2 σ bond

Nonbonding electron pair in sp^2 hybrid atomic orbital

Figure 20.11
The six *p* atomic orbitals that combine to form six π molecular orbitals of pyridine.

Figure 20.12
The five *p* atomic orbitals containing six electrons that combine to form the six π molecular orbitals of pyrrole.

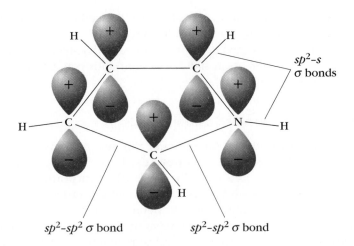

remaining nitrogen sp^2 atomic orbital contains a pair of electrons (the nonbonding electron pair). The *p* atomic orbitals of each ring atom combine to form six molecular orbitals (three bonding and three antibonding). The six *p* electrons, one from each carbon atom and one from nitrogen, fill the three bonding molecular orbitals, thus giving pyridine its aromatic character.

The *p* atomic orbitals used to construct the π molecular orbitals of pyrrole are shown in Figure 20.12. Pyrrole is a planar molecule, too, containing four sp^2 hybrid carbon atoms and one sp^2 hybrid nitrogen atom. Each ring atom forms two $sp^2—sp^2$ σ bonds with its neighboring ring atoms, and each forms one $sp^2—s$ σ bond to a hydrogen atom. The *p* atomic orbitals of each ring atom combine to form five molecular orbitals (three bonding and two antibonding). The six *p* electrons, one from each carbon atom and *two* from nitrogen, fill the three bonding molecular orbitals, thus giving pyrrole its aromatic character.

The nitrogen atoms in pyridine and pyrrole are different. In pyridine, the nitrogen atom has a lone pair of electrons, which is readily available for reaction with a proton. In pyrrole, on the other hand, the lone pair of electrons on the nitrogen atom is part of the aromatic sextet, so it is not available for reaction with a proton. Pyridine, therefore, is a stronger base than pyrrole. The pK_a of the conjugate acid of pyrrole is -4.4, while the pK_a of the conjugate acid of pyridine is $+5.0$.

Exercise 20.4   Draw an atomic orbital structure of furan similar to Figure 20.12. Indicate the hybridization of each atom; indicate what orbitals combine to form each σ bond; indicate what atomic orbitals combine to form the π molecular orbitals; and indicate the number of electrons contributed by each atom to the aromatic sextet.

So far, only neutral aromatic compounds have been discussed. However, rings containing positive or negative charges can also be aromatic.

D   Charged Aromatic Compounds

Hückel's rule of aromaticity is not restricted to neutral compounds. One of Hückel's rules for aromaticity requires that all the π electrons completely fill *only*

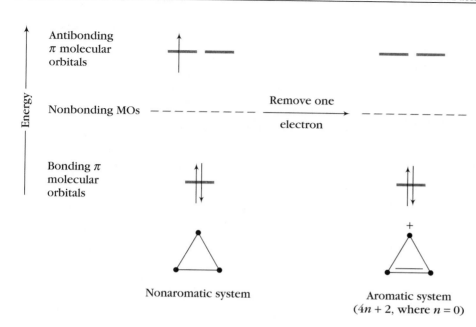

Figure 20.13
The formation of an aromatic system by removing one electron from the ring containing three sp^2 hybrid carbon atoms.

the bonding π molecular orbitals. Several structures in Figure 20.6 (p. 893), which consists of cyclic arrays of p atomic orbitals, contain either too many or too few electrons to exactly fill only the π bonding molecular orbitals. In these cases, it is possible to form an aromatic system by removing or by adding one or more electrons until all bonding π molecular orbitals are just filled.

The system containing three sp^2 carbon atoms in a ring, for example, has too many electrons to be aromatic. The π molecular orbital energy diagram for this system is reproduced in Figure 20.13. Two of its π electrons occupy the bonding molecular orbital, while the third electron occupies an antibonding molecular orbital. If we remove the electron from the antibonding molecular orbital, we obtain a system that has only two electrons in the bonding molecular orbital. Because an electron was removed, this system has a positive charge. This charged, two-electron, three-member ring is called a cyclopropenyl carbocation. It is aromatic because it is planar and contains $4n+2$ π electrons ($n = 0$).

Aromaticity makes the cyclopropenyl carbocation stable enough that it can be prepared and isolated in the laboratory. When 3-chlorocyclopropene reacts with antimony pentachloride ($SbCl_5$), a strong Lewis acid, cyclopropenium hexachloroantimonate is formed.

Cl

$\triangle$ + $SbCl_5$ $\longrightarrow$ $\overset{+}{\triangle}$ $SbCl_6^{-}$

3-Chlorocyclopropene   Cyclopropenium hexachloroantimonate

The product of this reaction is stable indefinitely at room temperature, provided moisture is excluded. The hydrogens of the cyclopropenyl carbocation are found further downfield at δ 11.10 in its ^1H NMR spectrum because of its aromatic character and the positive charge on the ring.

The cyclopentadienyl anion and the cycloheptatrienyl carbocation are two more charged aromatic compounds. Their electron distributions are shown in Figure 20.14. Both these ions contain six π electrons in three π bonding molecular orbitals.

Figure 20.14

π Electron structure of the cyclopentadienyl anion and the cycloheptatrienyl carbocation. Both ions are aromatic because they have six π electrons in three π bonding molecular orbitals.

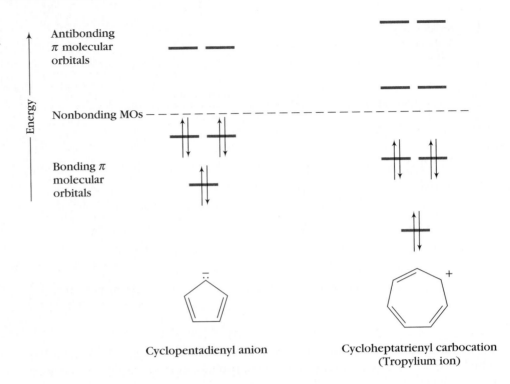

Cyclopentadienyl anion

Cycloheptatrienyl carbocation
(Tropylium ion)

Aromaticity greatly enhances the stability of the cyclopentadienyl anion. Its conjugate acid, for example, is one of the most acidic hydrocarbons known. While the pK_a values of most hydrocarbons are greater than 45, the pK_a of cyclopentadiene is 16, which makes it comparable in acidity to water and the alcohols.

$:H^-$ +   [Cyclopentadiene] ⟶ [Cyclopentadienyl anion]   + H_2

Hydride ion   Cyclopentadiene $pK_a = 16$ Cyclopentadienyl anion   Hydrogen $pK_a \approx 36$

Cycloheptatrienyl carbocation is easily formed by the reaction of bromine and cycloheptatriene. In this reaction, bromine is reduced to bromide ion and cycloheptatriene is oxidized to the carbocation. Cycloheptatrienylium bromide is stable at room temperature.

[Cycloheptatriene] + Br_2 ⟶ [Cycloheptatrienylium bromide] Br^- + HBr

Cycloheptatriene   Cycloheptatrienylium bromide

There is experimental evidence that charged ring compounds with $4n$ π electrons are not stabilized by electron delocalization. For example, cycloheptatrienyl anion has been prepared in solution and its acidity is about the same as a typical hydrocarbon ($pK_a \approx 46$).

+

Cycloheptatriene	Cycloheptatrienyl	Butane
$pK_a \approx 46$	anion ($4n$ π electrons)	$pK_a \approx 46$

Cyclopentadiene, in contrast, is strongly acidic, thus indicating that ions that contain $4n+2$ are more stable than those with $4n$ π electrons.

The four π electron cyclopentadienyl cation shows no evidence of enhanced stability. Solvolysis of cyclopentadienyl bromide is extremely slow, even though the compound is doubly allylic. The heterolytic bond dissociation energy to form the cation from cyclopentadiene is 258 kcal/mol (1078 kJ/mol), which is substantially more than the value for formation of an allylic cation and only slightly more than the value for formation of a secondary carbocation.

These examples clearly illustrate that Hückel's rule of aromaticity provides a simple method of distinguishing between charged ring compounds that are stabilized by π electron delocalization and those that are not.

Exercise 20.5 Indicate which of the following ions are aromatic.

(a) (b) (c) (d)

Exercise 20.6 How many electrons would you need to add or remove in order to make aromatic a cyclic molecule containing nine sp^2 hybrid carbon atoms? (Hint: Use the molecular orbital diagram in Figure 20.6, p. 893.)

20.8 Aromaticity: A Summary

Hückel's $4n+2$ rule of aromaticity provides us with a powerful yet simple method of recognizing compounds and ions that are stabilized by π electron delocalization. It is a great improvement over the resonance hybrid description of aromaticity. While resonance hybrids can explain the aromaticity of benzene, they fail to identify other aromatic systems. This is dramatically illustrated by the following ions:

The π electron structures of these four ions can be described as resonance hybrids consisting of a number of equivalent resonance structures for each, so all four ions are predicted to be highly stabilized, contrary to the experimental results. Hückel's rule of aromaticity, on the other hand, predicts that only the ions with six $(4n+2)$ π electrons will be aromatic. The other two ions should be nonaromatic because they have only four $(4n)$ π electrons despite their conjugated π electron systems. The prediction based on Hückel's rule agrees with the experimental results.

| **Exercise 20.7** | Write all resonance structures of the cyclopentadienyl anion, the cyclopentadienyl cation, the cycloheptatrienyl anion, and the cycloheptatrienyl cation. |

| **Exercise 20.8** | How many signals would you expect to find in the ^{1}H and ^{13}C NMR spectra of the cyclopentadienyl anion and the cycloheptatrienyl cation? |

The study of organic chemistry in this text is organized according to structure because compounds that contain the same functional group generally have similar chemical and physical properties. Knowledge of the reactions of a few alcohols, for example, can be used to predict the reactions of almost all other alcohols. In this chapter, however, we have seen that a knowledge of the reactions of a few molecules that contain conjugated π electrons in a cyclic molecule or ion is insufficient to predict the stability and reactivity of all members of that class of compound. Thus knowing the reactions of benzene does not mean we can predict the reactions of cyclobutadiene and cyclooctatetraene. Does this mean that the structural theory of organic chemistry is breaking down? No, it means that we must apply Hückel's rule to the structure of these molecules and ions because it is not simply the presence of π electrons but their total number, either $4n+2$ or $4n$, that determines the stability and reactivity of these compounds. Therefore, Hückel's rule of aromaticity becomes a part of the structural theory of organic chemistry.

| **Exercise 20.9** | The ^{1}H NMR spectrum of cyclopropenone shows a singlet at δ 9.08. In addition cyclopropenone has a high dipole moment. These observations seem inconsistent with the following structure of cyclopropenone. Explain. |

Cyclopropenone

Delocalization of π electrons can also extend outside of the aromatic ring.

20.9   π Electron Delocalization Outside the Aromatic Ring: Benzyl Radicals

The idea that an atom adjacent to a functional group is more reactive than it would be in the absence of the functional group was first introduced in Chapter

2. In subsequent chapters, a number of examples where this is true were presented. Thus the presence of a carbonyl group makes its α-hydrogen atoms more acidic than those of an alkane where no carbonyl group is present. Similarly the enhanced reactivity of the allylic position of an alkene compared to its saturated analog is due to the presence of the carbon-carbon double bond.

An aromatic ring also influences the reactivity of adjacent atoms and groups, as we will demonstrate with the reactions of alkylbenzenes. **Alkylbenzenes** are a class of compounds in which one or more alkyl groups is attached to a benzene ring.

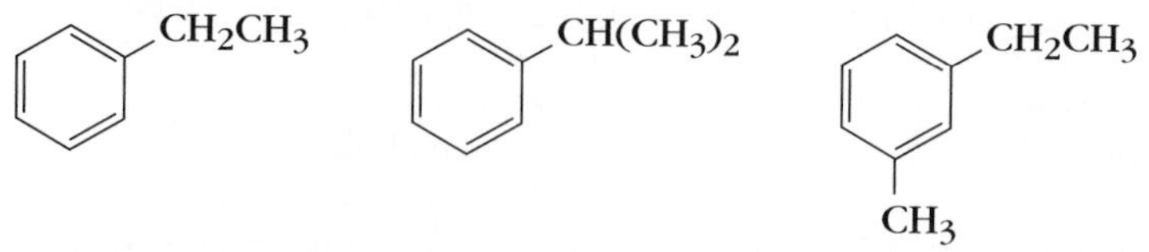

The aliphatic portion of the compound is commonly called the **side chain.** The carbon atom directly attached to the benzene ring in alkylbenzenes is called a **benzylic carbon atom.** The name **benzylic** is also used to identify any atoms bonded to this carbon atom, while **benzyl** is the common name given to the $C_6H_5CH_2-$ group.

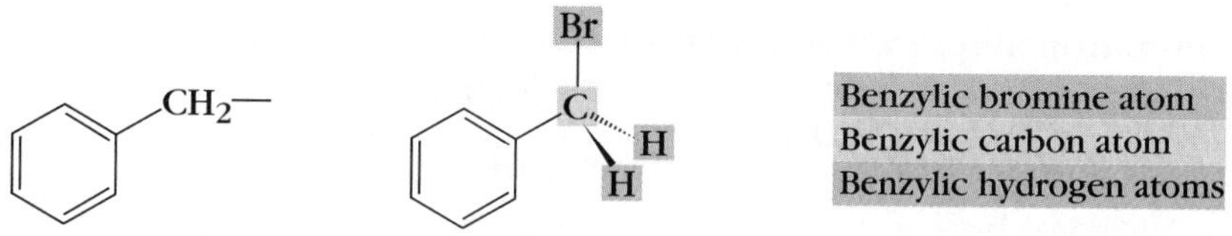

A functional group often enhances the reactivity of an adjacent site by stabilizing the intermediate. It stabilizes the intermediate and the transition state leading to that intermediate in the reaction by electron delocalization. For example, hydrogens on the α-carbon of aldehydes and ketones are more acidic than those on alkanes because of the delocalized electron structure of the enolate anion. Similarly a benzene ring stabilizes an adjacent radical, anion, or carbocation by electron delocalization. The first example of this effect that we will examine is the benzylic radical formed during the free radical bromination and oxidation of alkylbenzenes.

A   Free Radical Bromination of Alkylbenzenes

In the presence of light or heat, alkylbenzenes react with bromine exclusively at the benzylic carbon atom.

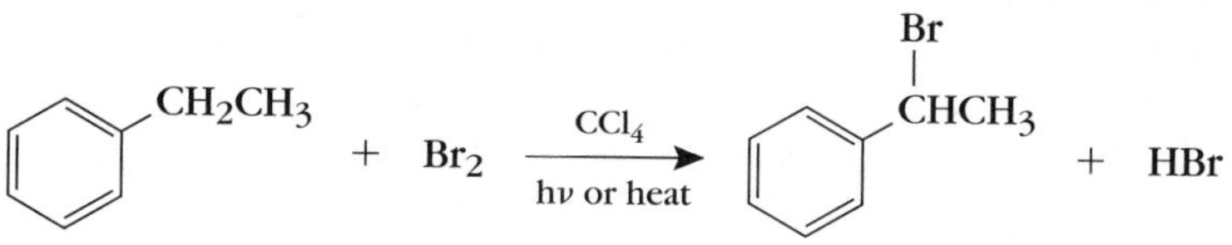

This regioselective substitution reaction occurs by the free radical chain mechanism, shown in Figure 20.15. The mechanism includes initiation, propagation, and termination steps analogous to the free radical chain mechanism, discussed in Chapter 18.

Figure 20.15
Chain mechanism for the free radical bromination of ethylbenzene.

Initiation step: Light-induced breaking of the bromine–bromine bond to form bromine radicals

$$Br_2 \; \underset{\longleftarrow}{\overset{Light}{\longrightarrow}} \; 2Br\cdot$$

Propagation steps: Bromine radical abstracts benzylic hydrogen atom to form benzylic radical

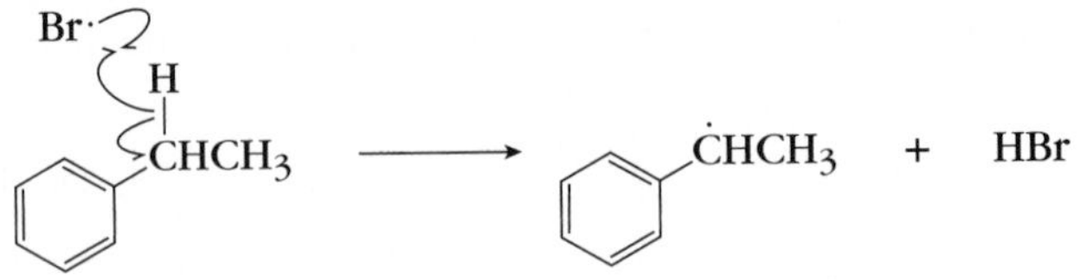

Benzylic radical reacts with a molecule of bromine to form product and a bromine radical, which continues the chain reaction

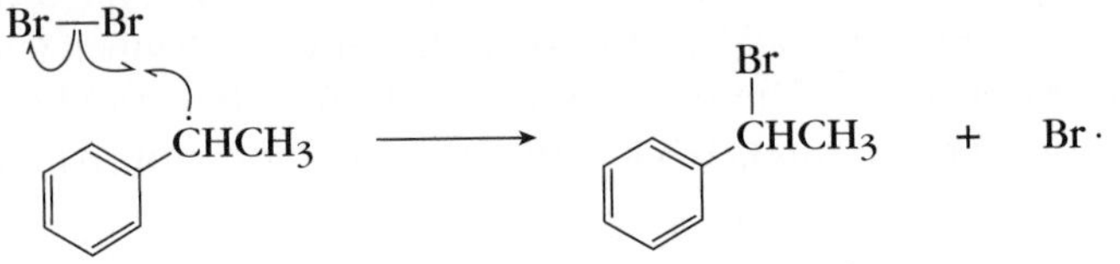

Termination step: Two radicals combine to form a stable product

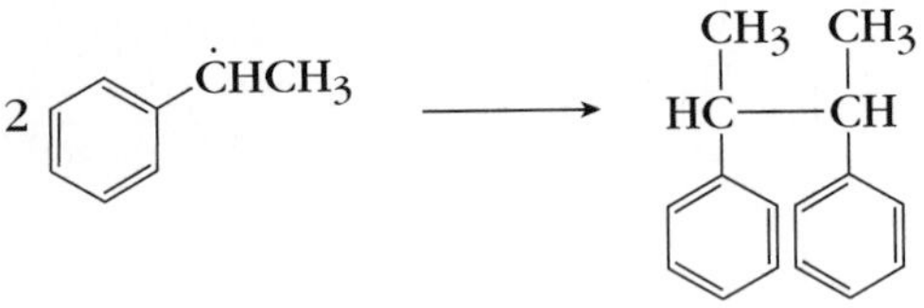

$$2Br\cdot \; \underset{\longleftarrow}{\longrightarrow} \; Br_2$$

Only the benzylic hydrogen atom is abstracted in the propagation step of this mechanism because the benzylic C—H bond is relatively weak for a C—H bond. For instance, the dissociation energy of the benzylic C—H bond is only 85 kcal/mol (355 kJ/mol), compared to a dissociation energy of 95 kcal/mol (397 kJ/mol) for a typical secondary C—H bond of an alkane. The benzylic C—H bond is weak because after it breaks, the lone electron on the benzylic carbon atom can be delocalized into the adjacent benzene π electron system. Thus a benzylic radical, just like the allylic radical, is stabilized by electron delocalization into a π electron system.

Delocalization into the benzene π electron system can be depicted by a resonance hybrid of the following resonance structures:

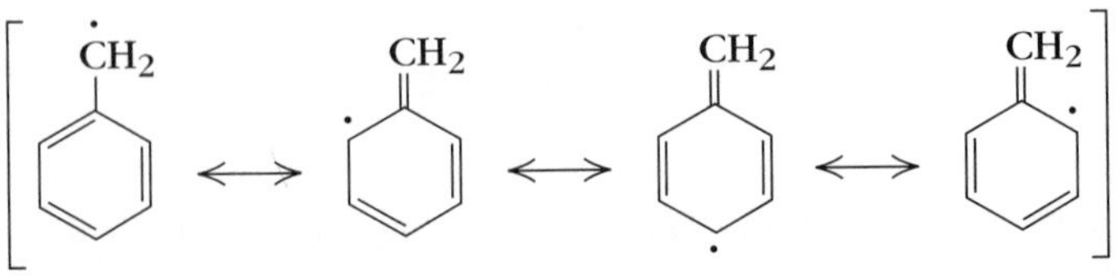

■ *Delocalization of π electrons in the allylic system is discussed in Section 19.1.*

The resonance hybrid representation of the benzylic radical places an odd electron on certain carbon atoms of the benzene ring. By analogy with the allylic system, we might expect products of the reaction of a bromine molecule with these ring carbon atoms, as well as reaction with the benzylic carbon atom.

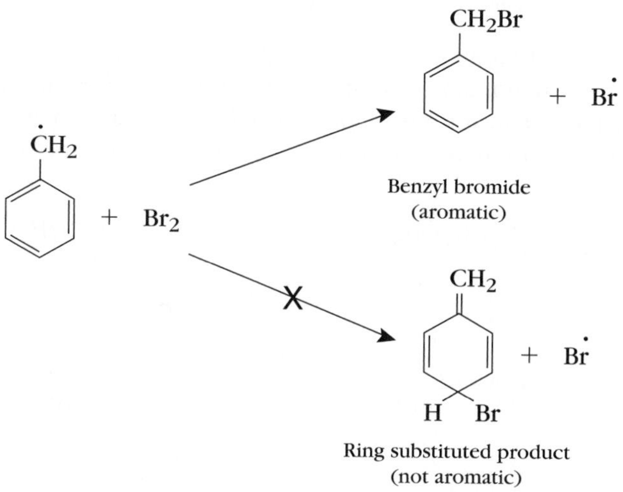

This is not observed, however, because reaction at any carbon but the benzylic carbon atom destroys the aromaticity of the benzene ring. Because benzyl bromide contains a benzene ring, it is about 35 kcal/mol (146 kJ/mol) more stable than the ring-substituted product. The energy difference between the products is reflected in the transition states for their formation, so reaction at the benzylic carbon atom is energetically much more favorable than reaction at the ring carbon atoms.

In Chapter 19, we discussed how allylic bromination can be accomplished with N-bromosuccinimide (NBS) as the brominating agent. Benzylic brominations can also be performed with this reagent in CCl_4 as solvent in the presence of a free radical initiator such as a peroxide. As the following example shows, the NBS reaction is another way to exclusively substitute a benzylic hydrogen atom with a bromine atom.

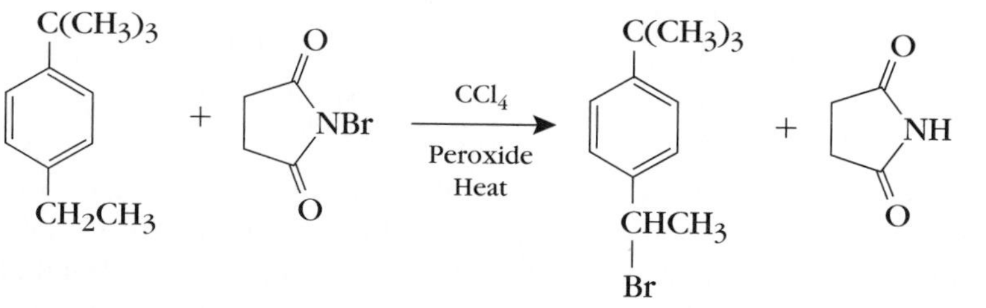

An aromatic ring also affects the reactivity of benzylic positions in the oxidation of alkylbenzenes.

B   Oxidation of Alkylbenzenes

Both alkanes and benzene are inert to most of the oxidizing agents that are used in organic chemistry.

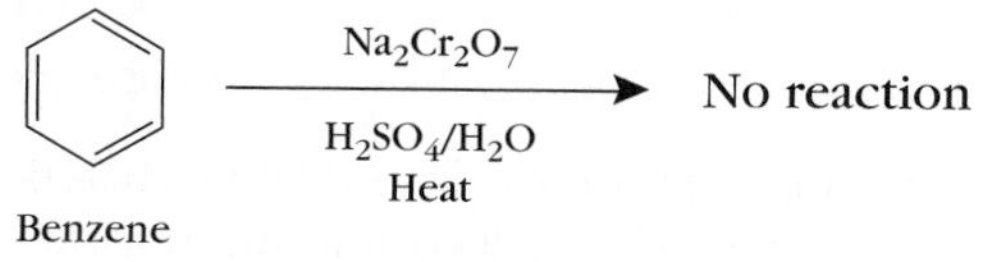

$$CH_3(CH_2)_nCH_3 \xrightarrow[\substack{H_2SO_4/H_2O \\ Heat}]{Na_2Cr_2O_7} \text{No reaction}$$

Alkane

The alkyl group of an alkylbenzene, on the other hand, is oxidized to a carboxylic acid group by hot acidic solutions of either sodium dichromate or potassium permanganate. The net effect of this oxidation reaction is to convert an alkylbenzene into an aromatic carboxylic acid, as shown by the following examples.

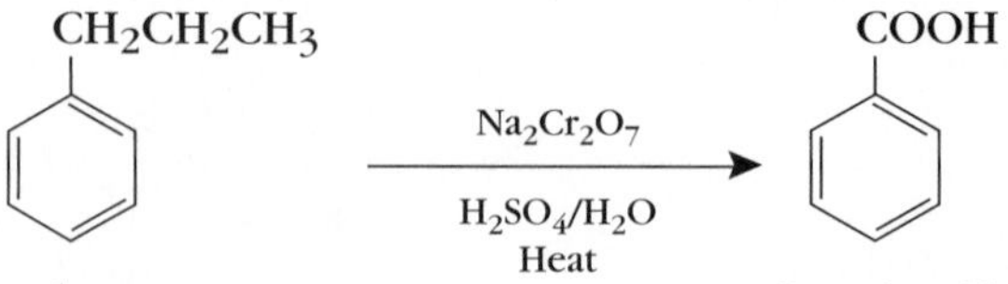

1-Phenylpropane → Benzoic acid (85% yield)

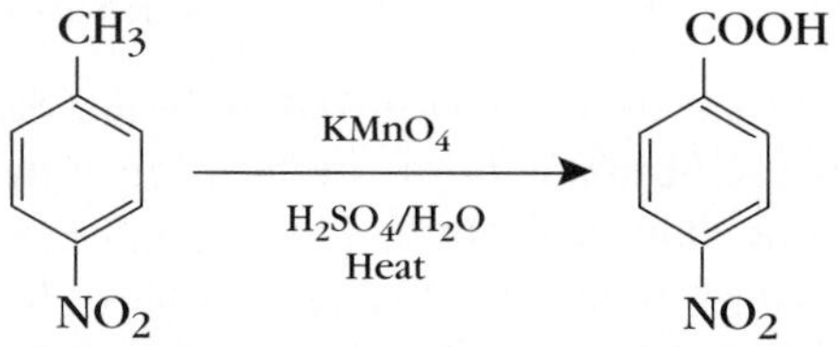

4-Nitrotoluene → 4-Nitrobenzoic acid (82% yield)

If two or more alkyl groups that have at least one benzylic hydrogen atom each are bonded to an aromatic ring, all are oxidized to a carboxyl group.

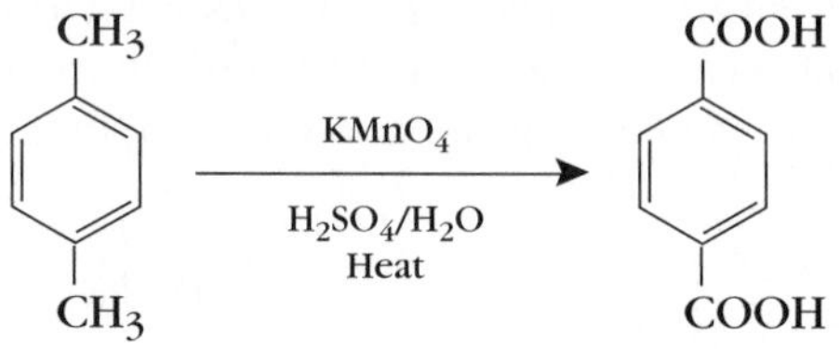

p-Xylene → Terephthalic acid (72 % yield)

The preceding examples demonstrate that even if the alkyl group or groups on the aromatic ring have more than one carbon atom, vigorous oxidation still forms an aromatic carboxylic acid in which the carboxyl group is bonded directly to the aromatic ring. The one exception is a tertiary alkyl substituent. Because they do not contain benzylic hydrogen atoms, tertiary alkyl groups are not oxidized under these conditions.

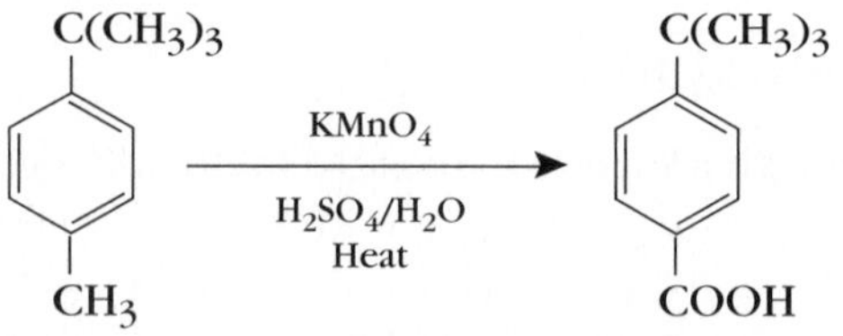

4-*t*-Butyltoluene → 4-*t*-Butylbenzoic acid

As a result, the mechanisms of these oxidation reactions are all believed to involve abstraction of a benzylic hydrogen atom to form a benzylic radical. From

20.10   π ELECTRON DELOCALIZATION OUTSIDE THE AROMATIC RING II: NUCLEOPHILIC SUBSTITUTION REACTIONS OF BENZYLIC DERIVATIVES

909

that point on, the mechanism depends on the exact nature of the oxidizing agent.

Exercise 20.10   Write the structure of the product formed in each of the following reactions.

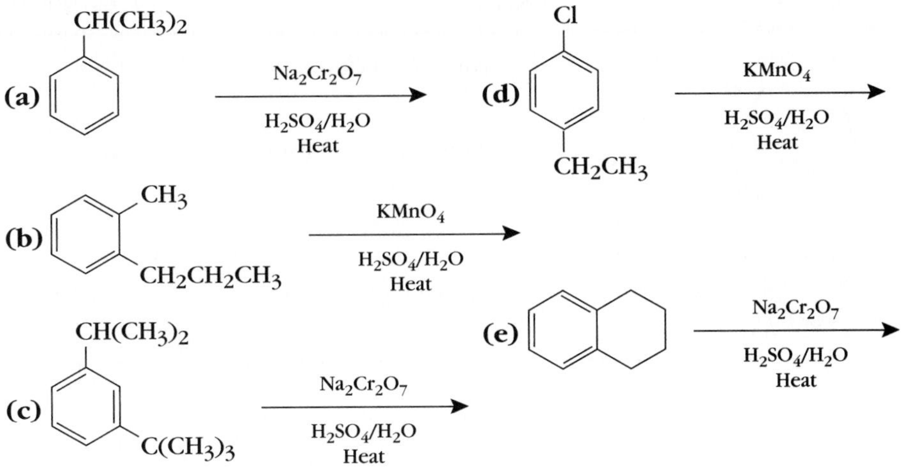

20.10 π Electron Delocalization Outside the Aromatic Ring II: Nucleophilic Substitution Reactions of Benzylic Derivatives

Benzylic halides or arenesulfonates undergo nucleophilic substitution reactions by either an S_N1 or S_N2 mechanism, depending on whether the benzylic derivative is primary, secondary, or tertiary.

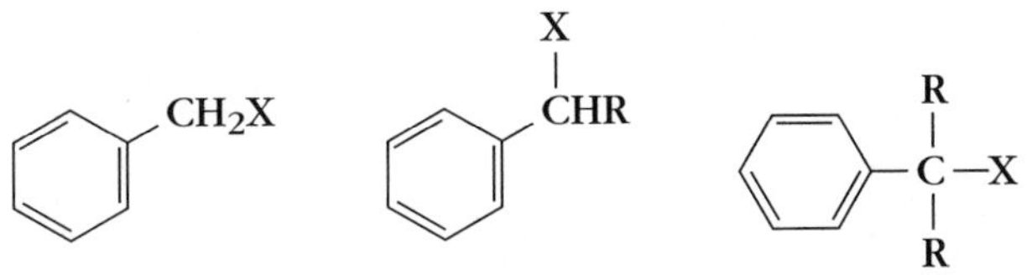

1° Benzylic derivative   2° Benzylic derivative   3° Benzylic derivative

R = alkyl or aryl group

Tertiary alkyl halides and arenesulfonates undergo nucleophilic substitution reactions only by an S_N1 mechanism, as we learned in Chapter 12. All tertiary derivatives (benzylic and nonbenzylic) react by an S_N1 mechanism, but tertiary benzylic derivatives react much faster than their nonbenzylic analogs. Some representative examples are listed in Table 20.1. Why are tertiary benzylic derivatives so much more reactive than their nonbenzylic analogs? The answer is that a benzylic carbocation is stabilized by electron delocalization, whereas a nonbenzylic carbocation is not. For example, the structure of a tertiary butyl cation is adequately represented by a single structure.

Table 20.1	Comparison of the Rates of Hydrolysis of Tertiary Benzylic and Nonbenzylic Alkyl Chlorides	

$$H_2O \ + \ RCl \ \xrightarrow[\text{25 °C}]{\text{90\% Aqueous acetone}} \ ROH \ + \ HCl$$

Compound	Name	Relative Rate
$(CH_3)_3C-Cl$	2-Chloro-2-methylpropane	1
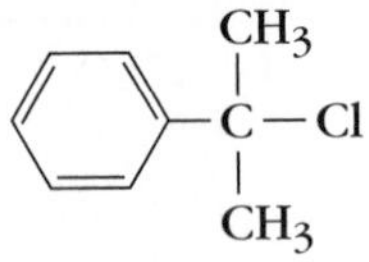 [phenyl]-C(CH$_3$)$_2$-Cl	2-Chloro-2-phenylpropane	620
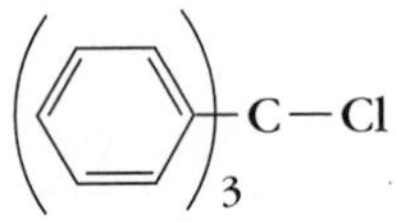 (phenyl)$_3$-C-Cl	Chlorotriphenylmethane	≈650,000

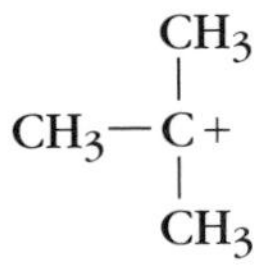

The carbocation formed in the first step of the hydrolysis of 2-chloro-2-phenyl-propane, on the other hand, is stabilized by electron donation from the benzene ring to the positively charged benzylic carbon atom. This delocalization is adequately described as a resonance hybrid of the following resonance structures.

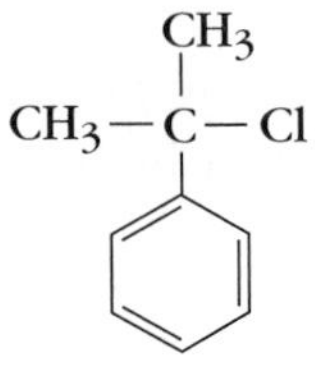

2-Chloro-2-phenylpropane

+

Cl$^-$

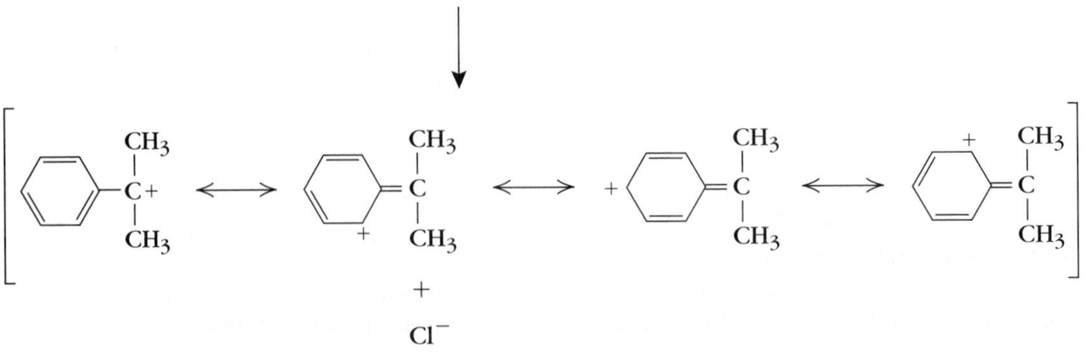

π Electron delocalization makes the tertiary benzylic carbocation more stable than a tertiary alkyl carbocation. According to the Hammond Postulate, the difference in stability is reflected in the transition states for formation of the two carbo-

20.10   π ELECTRON DELOCALIZATION OUTSIDE THE AROMATIC RING II:
NUCLEOPHILIC SUBSTITUTION REACTIONS OF BENZYLIC DERIVATIVES

911

cations. Consequently the free energy of activation for formation of a tertiary benzylic carbocation is lower and it forms faster than a nonbenzylic tertiary carbocation. As a result, the hydrolysis of 2-chloro-2-phenylpropane is faster than the hydrolysis of 2-chloro-2-methylpropane.

Placing additional phenyl groups adjacent to a carbocation stabilizes the carbocation even more. In the hydrolysis of chlorotriphenylmethane [$(C_6H_5)_3CCl$], for example, the positive charge of the carbocation is stabilized by electron delocalization over all three phenyl rings. The stability of this ion is the reason chlorotriphenylmethane is hydrolyzed more than 1000 times faster than 2-chloro-2-phenylpropane.

Exercise 20.11	Write all ten resonance structures of the carbocation formed as an intermediate in the hydrolysis of chloro-triphenylmethane.

Primary benzyl halides react with nucleophiles to form only products of substitution because the competing elimination reaction cannot occur.

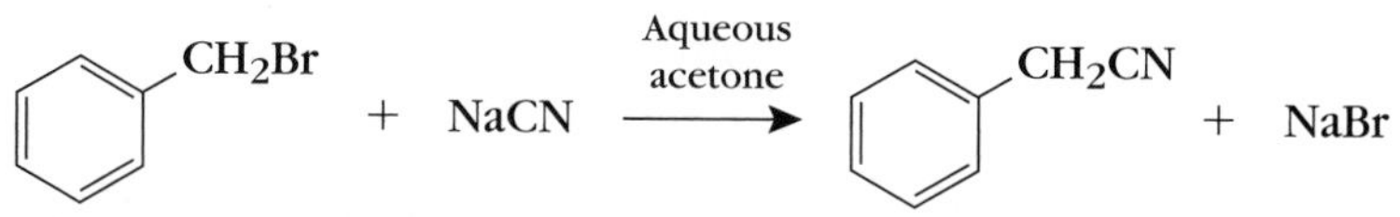

Benzyl bromide

Phenylacetonitrile
(80% yield)

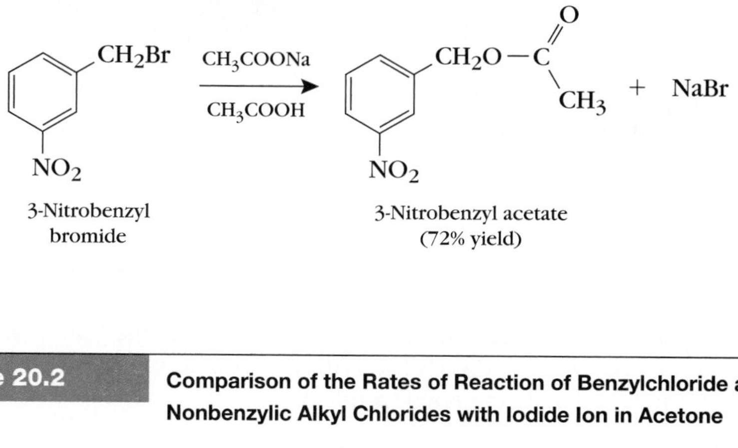

3-Nitrobenzyl bromide

3-Nitrobenzyl acetate
(72% yield)

Table 20.2	Comparison of the Rates of Reaction of Benzylchloride and Nonbenzylic Alkyl Chlorides with Iodide Ion in Acetone

$$RCl + I^- \xrightarrow{\text{Acetone}} RI + Cl^-$$

Compound	Name	Relative Rate
$(CH_3)_2CHCH_2Cl$	1-Chloro-2-methylpropane	1
CH_3Cl	Chloromethane	3,000,000
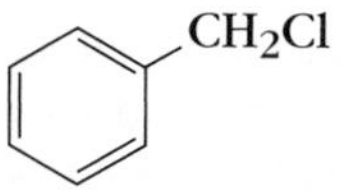 CH_2Cl	Benzyl chloride	10,000,000

The nucleophilic substitution reactions of primary benzyl halides typically occur by an S_N2 mechanism. As in the case of tertiary benzylic derivatives, the rate of nucleophilic substitution of benzyl chlorides is faster than their nonbenzylic analogs. Several examples are listed in Table 20.2 for the reaction of iodide ion in acetone (conditions that favor an S_N2 mechanism).

Reactions of benzyl compounds that occur by an S_N2 mechanism are accelerated because their transition states are stabilized by the adjacent aromatic ring, as shown in Figure 20.16. Recall from Section 12.2 that in the structure of the transition state of the S_N2 mechanism, the incoming nucleophile and the leaving group are partially bonded to the p orbital of the sp^2 hybrid carbon atom at the reactive site. In the transition state of the reaction of iodide ion with benzyl chloride, the p orbital of the benzylic sp^2 hybrid carbon atom can combine with the π electrons of the ring to form an extended delocalized electron system, which stabilizes its transition state. No such extended delocalization can occur in nonbenzylic compounds. As a result, benzylic compounds react faster than structurally similar nonbenzylic compounds in nucleophilic substitution reactions that occur by an S_N2 mechanism.

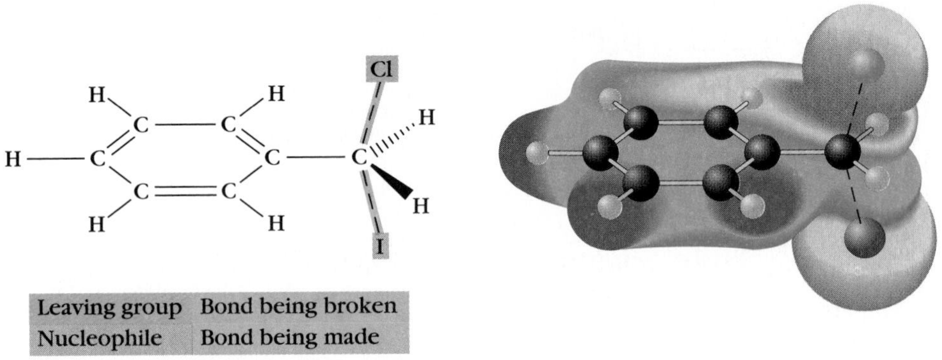

Figure 20.16
Transition state of a substitution reaction that occurs by an S_N2 mechanism at a benzylic carbon atom. This transition state is stabilized compared with that of an S_N2 mechanism at a nonbenzylic carbon atom by delocalization of the π ring electrons with the p orbital of the reactive site.

| Leaving group | Bond being broken |
| Nucleophile | Bond being made |

Secondary and tertiary benzylic halides and arenesulfonates undergo elimination reactions under the same conditions and by the same mechanisms as their nonbenzylic analogs. If isomeric alkenes can be formed in the elimination reactions of benzylic compounds, the alkene with the double bond conjugated with the aromatic ring is generally the major product.

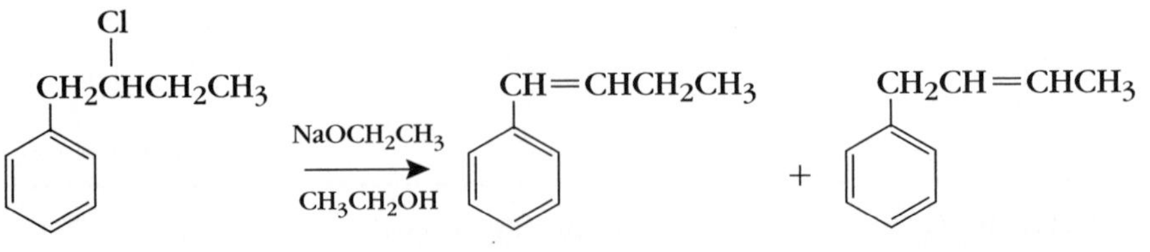

2-Chloro-1-phenyl butane   1-Phenyl-1-butene (major product)   1-Phenyl-2-butene (minor product)

■ *Elimination reactions of alkyl compounds are discussed in Sections 12.10 to 12.15.*

1-Phenyl-1-butene, whose double bond is conjugated with the benzene ring, gains extra stability from the extended conjugation, so it is more stable than 1-phenyl-2-butene, whose double bond is not conjugated with the benzene ring. As a result, this elimination reaction follows the Saytzeff Rule (Section 12.14).

| Exercise 20.12 | Write the structure of the product of the reaction of benzyl chloride with each of the following reagents. |

(a) NaI in acetone
(b) NaCN in ethanol
(c) Potassium *t*-butoxide in *t*-butyl alcohol
(d) Sodium methoxide in methanol

| Exercise 20.13 | Write the structure of the major product of the reaction of 2-chloro-2-phenylpropane with each of the reagents given in Exercise 20.12. |

The reactions of a double bond are also affected by the presence of an adjacent aromatic ring.

20.11 π Electron Delocalization Outside the Aromatic Ring III: Additions to Alkenylbenzenes

An alkenylbenzene is an aromatic compound whose side chain contains one or more double bonds. The double bond can be either conjugated or nonconjugated with the aromatic ring.

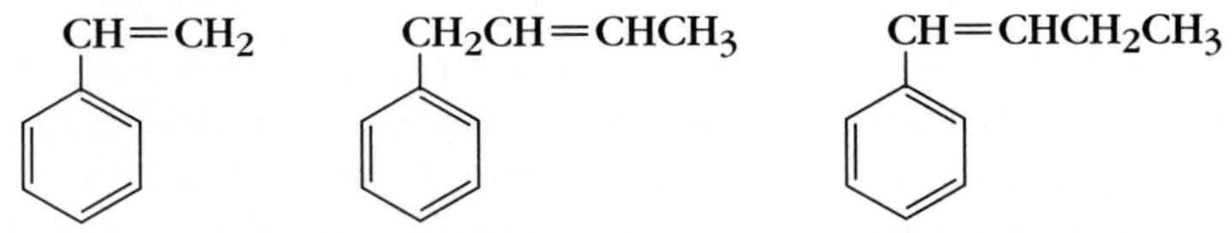

$$CH=CH_2 \qquad CH_2CH=CHCH_3 \qquad CH=CHCH_2CH_3$$

| Styrene | 1-Phenyl-2-butene | 1-Phenyl-1-butene |
| (conjugated) | (nonconjugated) | (conjugated) |

Conjugated alkenylbenzenes undergo the usual addition reactions of alkenes. For example, addition of hydrogen halides and acid-catalyzed hydration of 1-phenylpropene, either the (*E*) or (*Z*) isomer, forms only the product in which the chlorine and hydroxy group are bonded to the benzylic carbon atom in each case.

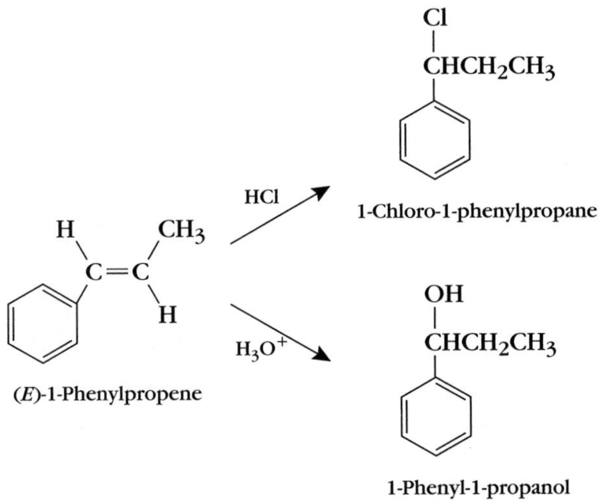

$$\overset{Cl}{\underset{}{|}}$$
$$CHCH_2CH_3$$

1-Chloro-1-phenylpropane

(*E*)-1-Phenylpropene

$$\overset{OH}{\underset{}{|}}$$
$$CHCH_2CH_3$$

1-Phenyl-1-propanol

Recall from Section 7.13 that the more stable carbocation forms as the intermediate in the mechanisms of additions of acids to alkenes. Additions to conjugated alkenylbenzenes are no exception. Of the two possible carbocations that can be formed by addition of a proton to 1-phenylpropene, the benzylic carbocation is the more stable. As a result, the free energy of activation for formation of the benzylic cation is lower than the free energy of activation for formation of the 1-phenyl-2-propyl cation. Consequently, addition of acids occurs regiospecifically to form the benzylic cation in the first step. In the second step, the benzylic cation reacts with a nucleophile to form the product.

Exercise 20.14 Write the mechanism of the ionic addition of HCl to styrene.

Based on a comparison of the reactions of benzylic and alkyl derivatives, the stability of benzylic carbocations relative to other carbocations can be summarized as follows:

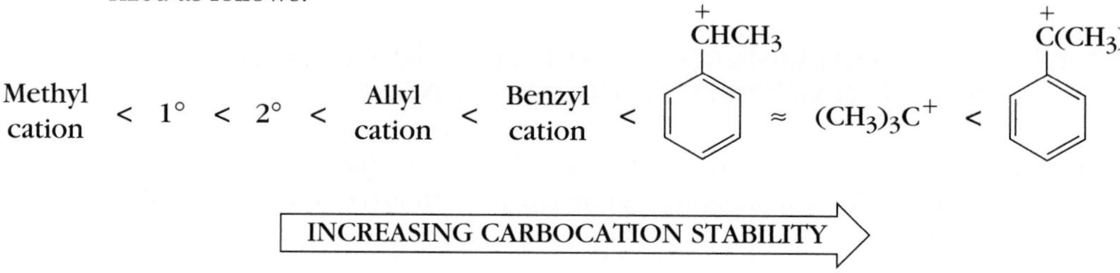

Methyl cation $<$ 1° $<$ 2° $<$ Allyl cation $<$ Benzyl cation $<$ [phenyl-$\overset{+}{C}HCH_3$] $\approx$ $(CH_3)_3C^+$ $<$ [phenyl-$\overset{+}{C}(CH_3)_2$]

INCREASING CARBOCATION STABILITY $\Longrightarrow$

A secondary benzylic cation is about as stable as the tertiary butyl cation but more stable than a benzyl cation. This is in keeping with the stabilizing effect of a double bond or benzene ring adjacent to a positively charged carbon atom.

Additions of other electrophiles to conjugated alkenylbenzenes generally occur in the same way as additions to nonconjugated or simple alkenes, as shown by the following examples:

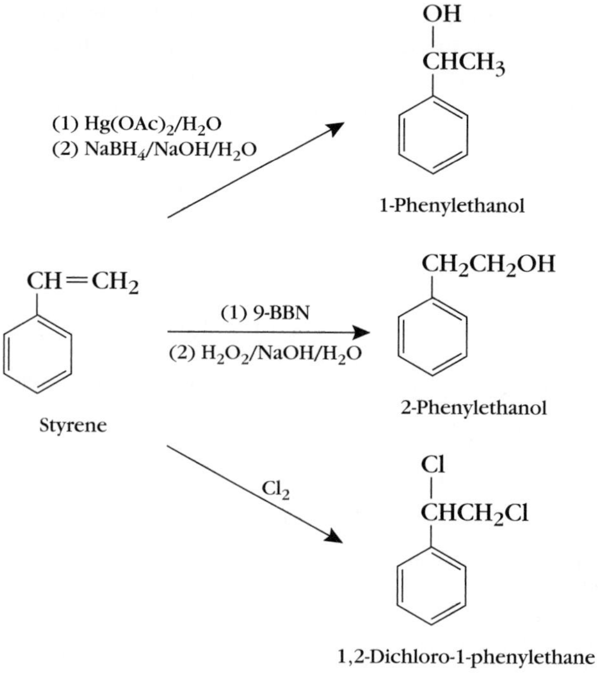

Styrene, $C_6H_5CH=CH_2$

(1) $Hg(OAc)_2/H_2O$
(2) $NaBH_4/NaOH/H_2O$ → 1-Phenylethanol, $C_6H_5CHOH\cdot CH_3$

(1) 9-BBN
(2) $H_2O_2/NaOH/H_2O$ → 2-Phenylethanol, $C_6H_5CH_2CH_2OH$

Cl_2 → 1,2-Dichloro-1-phenylethane, $C_6H_5CHCl\cdot CH_2Cl$

Hydroxylation of the double bond to form a glycol (Section 8.5) can be accomplished with OsO_4/H_2O_2. A stronger oxidizing agent, such as an acidic solution of $K_2Cr_2O_7$, oxidizes the side chain to an aromatic carboxylic acid.

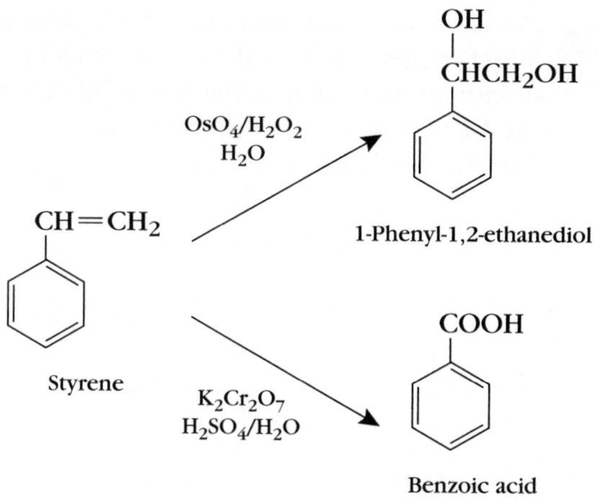

1-Phenyl-1,2-ethanediol

Benzoic acid

The regiochemistry of free radical additions to conjugated alkenylbenzenes is opposite to the regiochemistry of electrophilic additions.

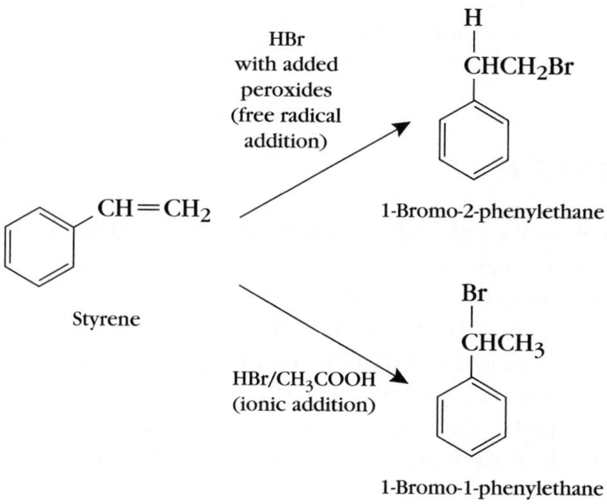

1-Bromo-2-phenylethane

1-Bromo-1-phenylethane

In both reactions, addition occurs to form the more stable intermediate. In ionic additions to styrene, the benzylic cation is formed, as we have previously discussed. In free radical additions to styrene, the benzylic radical (the 2-bromo-1-phenylethyl radical) is more stable than the alkyl radical (the 2-bromo-2-phenylethyl radical). As a result, the free energy of activation for formation of the benzylic radical is lower than the free energy of activation for formation of the alkyl radical. Consequently, free radical additions occur regiospecifically to form the benzylic radical in the first step. In the second step, the benzylic radical abstracts a hydrogen atom to form the product.

Exercise 20.15 Write the mechanism of the free radical addition of HBr to (E)-1-phenylpropene.

In summary, the major effect of an aromatic ring on the reactions of an adjacent double bond is to direct the addition so that a positive charge or free radical forms preferentially at the benzylic position.

Exercise 20.16 Write the structure of the major product of the reaction of indene with each of the following reagents. Indicate the stereochemistry of the product where appropriate.

Indene

(a) $HCl/(C_2H_5)_2O$
(b) $OsO_4/H_2O_2/H_2O$
(c) H_3O^+
(d) Br_2/H_2O
(e) $Hg(OAc)_2/NaBH_4/NaOH/H_2O$
(f) $9\text{-BBN}/H_2O_2/NaOH/H_2O$
(g) $9\text{-BBN}/CH_3CH_2COOH$
(h) $KMnO_4/H_2SO_4/H_2O$
(i) HBr with peroxides
(j) Peroxyacetic acid in acetic acid

20.12 Summary

According to **Hückel's rule of aromaticity,** any compound that contains p atomic orbitals on adjacent atoms of a planar ring will be aromatic if it contains $4n+2$ π electrons. Benzene is the most common example of an aromatic compound. Other examples include the **annulenes, heterocyclic aromatic compounds** (such as pyridine and pyrrole), and **charged ring compounds** (such as the cyclopropenyl cation and the cyclopentenyl anion). Heterocyclic aromatic compounds have atoms other than carbon in the ring (nitrogen and oxygen are the most common). **Polycyclic benzenoid compounds** are not strictly aromatic systems according to Hückel's rule because they are not monocyclic. They are classified as aromatic, however, because their physical and chemical properties are similar to those of benzene.

When an aromatic compound is placed in a magnetic field, its π electrons generate an electric current called a **ring current.** The ring current, in turn, induces a magnetic field, which influences the chemical shift of hydrogen atoms bonded to the carbon atoms of the aromatic ring. Hydrogen atoms outside the ring are deshielded, so they are found downfield in the δ 6.5 to 8.5 region. Hydrogen atoms inside the ring, on the other hand, are shielded so they are found upfield. Often they can be upfield of TMS. In this way ^{1}H NMR provides a general criterion of aromaticity.

A site adjacent to an aromatic ring is more reactive than it would be in the absence of the ring. As a result, alkylbenzenes undergo free radical bromination exclusively at the benzylic position, while vigorous oxidation converts alkylbenzenes to an aromatic carboxylic acid in which the carboxyl group is directly bonded to the aromatic ring. Such regiospecificity is due to the stability of the benzylic radical compared to any isomeric nonbenzylic radical that may be formed in the reaction. The benzylic radical is stabilized by the delocalization of its benzylic electron into the adjacent benzene π electron system. The increased stability of benzylic radicals is reflected in the transition states for their formation. Consequently, the free energies of activation for formation of benzylic radicals are lower and they form faster than any isomeric nonbenzylic radical.

An adjacent benzene ring also enhances nucleophilic substitution reactions of benzylic derivatives. In this case the adjacent ring stabilizes both the carbocation intermediate in reactions that occur by an S_N1 mechanism and the transition states of reactions that occur by an S_N2 mechanism.

Addition reactions to conjugated alkenylbenzenes occur so that the more stable benzylic intermediate, either a radical or a carbocation, is formed.

20.13 Summary of Reactions of Alkyl and Alkenyl Side Chains of Benzene

1. Free Radical Bromination   SECTION 20.9, A

Substitution occurs exclusively at benzylic carbon atom.

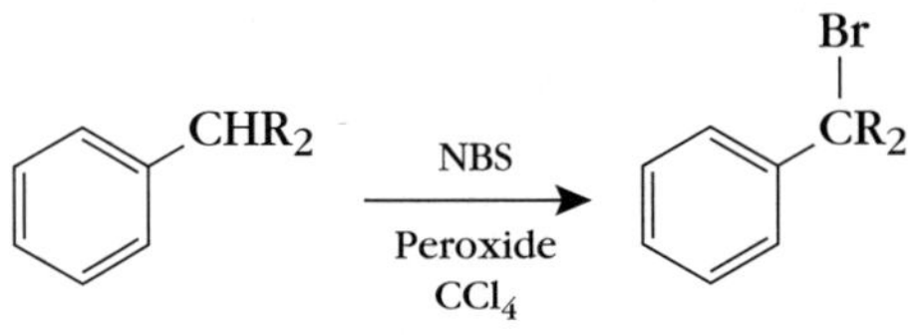

2. Oxidation   SECTION 20.9, B

Vigorous oxidation occurs at benzylic carbon atoms containing at least one benzylic hydrogen atom to form benzoic acid.

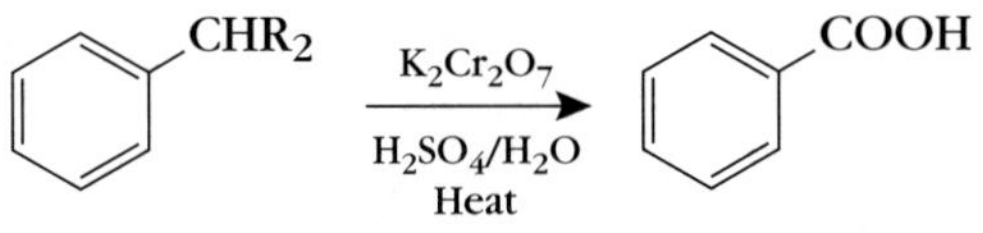

3. Nucleophilic Substitution   SECTION 20.10

1° Benzylic derivatives react by an S_N2 mechanism; 2° and 3° benzylic derivatives react by an S_N1 mechanism.

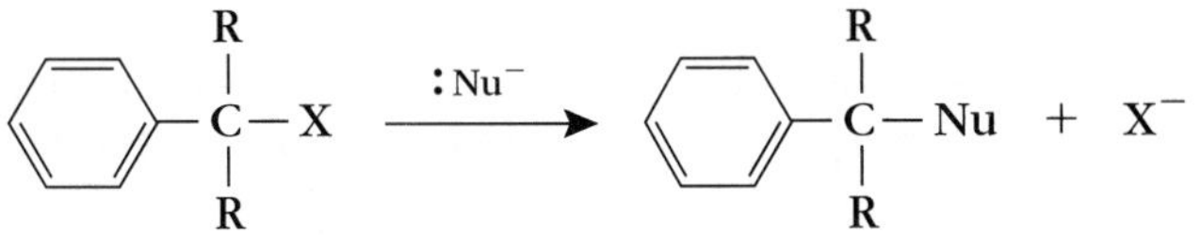

R = alkyl or aryl group

4. Electrophilic Addition   SECTION 20.11

Addition occurs to form a benzylic cation as an intermediate to give products with Markownikoff orientation.

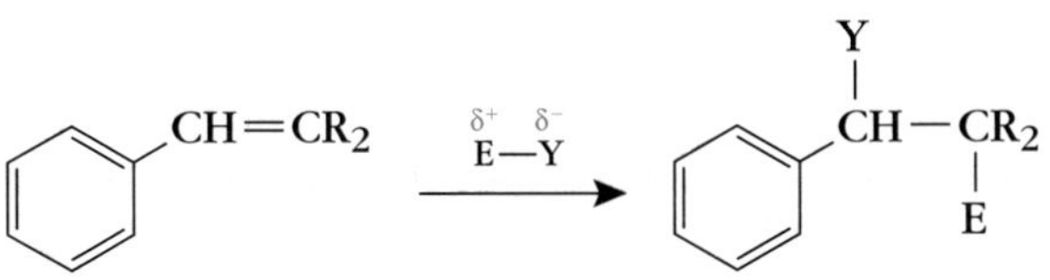

5. Free Radical Addition   SECTION 20.11

Addition occurs to form a benzyl radical as an intermediate to give products with anti-Markownikoff orientation.

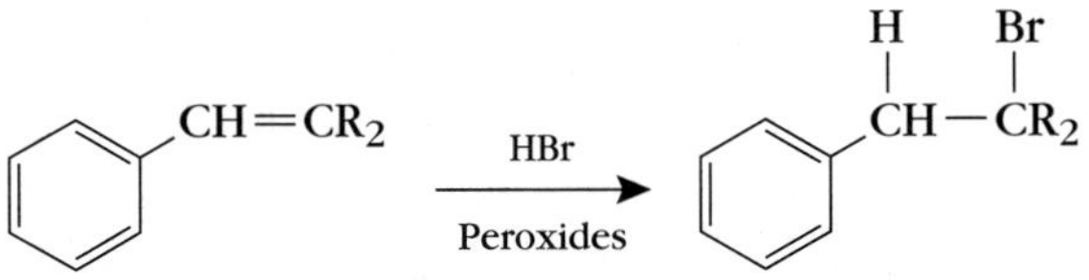

Additional Exercises

20.17 Define each of the following terms and give an example.

    (a) Resonance energy   **(e)** Ring current
    (b) Hückel's rule of aromaticity   **(f)** A heterocyclic aromatic compound
    (c) An annulene   **(g)** A charged aromatic compound
    (d) A polycyclic aromatic hydrocarbon

20.18 Apply Hückel's rule of aromaticity to each of the following compounds and decide which are aromatic.

(a)    **(b)** 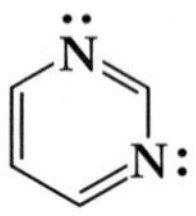   **(c)** 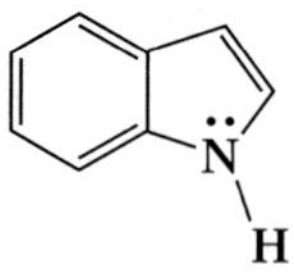   **(d)** 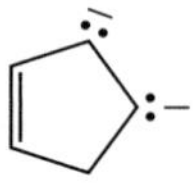   **(e)** 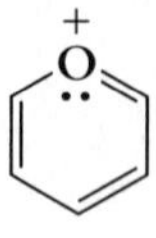   **(f)** 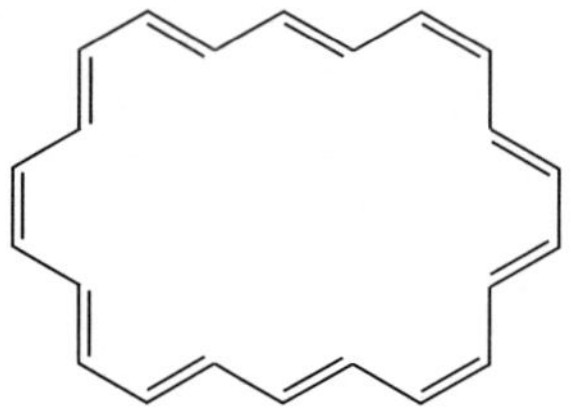   **(g)** 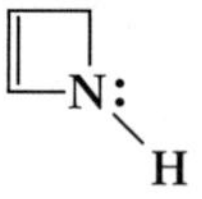   **(h)** 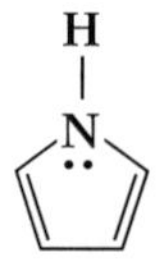   **(i)** 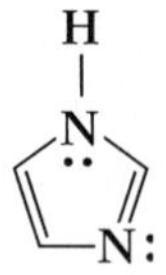   **(j)**    **(k)**

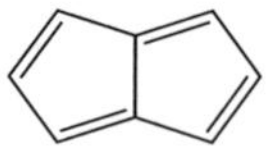

20.19 Explain why each of the following compounds has a large dipole moment. Remember that a dipole moment indicates that the centers of positive and negative charge in a molecule do not coincide.

(a)   **(b)** 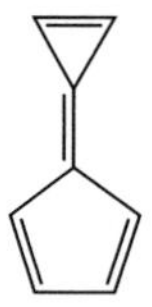   **(c)**

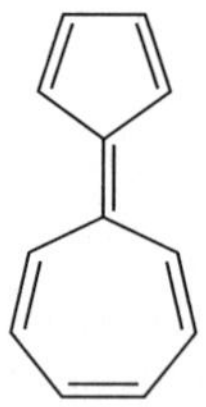

20.20 Each of the following reactions occurs very easily. Write the structure of the product and explain why the reaction occurs so rapidly.

(a) 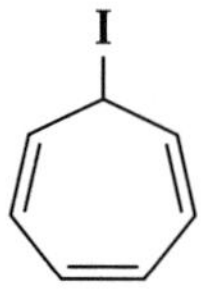 + AgNO₃ $\xrightarrow{\text{H}_2\text{O}}$

(c) 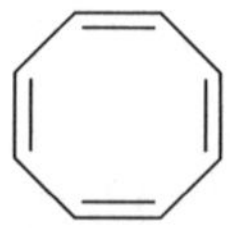 + 2K $\xrightarrow{\text{Heptane}}$

(b) 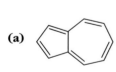 $\xrightarrow[\text{(2) CH}_3\text{I}]{\text{(1) NaNH}_2}$

(d)  + Excess HCl $\longrightarrow$

20.21 Write the structure of the product in each of the following reactions. Explain why the reactions occur with great difficulty.

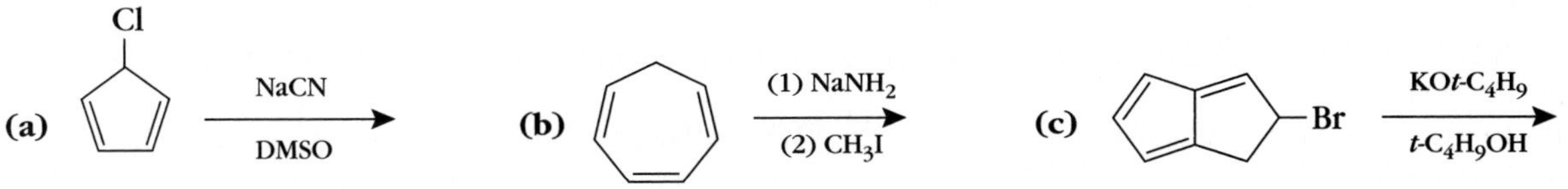

(a) [structure: cyclopentadiene with Cl] $\xrightarrow[\text{DMSO}]{\text{NaCN}}$ **(b)** [cycloheptatriene structure] $\xrightarrow[\text{(2) CH}_3\text{I}]{\text{(1) NaNH}_2}$ **(c)** [bicyclic structure]—Br $\xrightarrow[t\text{-C}_4\text{H}_9\text{OH}]{\text{KO}t\text{-C}_4\text{H}_9}$

20.22 Explain why tropolone is a stronger acid than most alcohols.

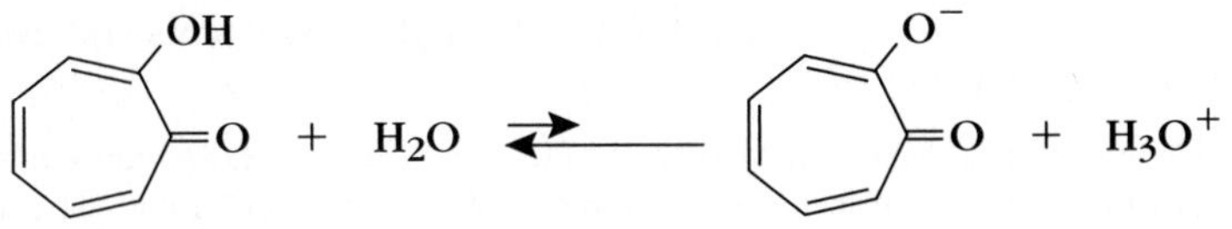

[structure: tropolone with OH] $+ \; H_2O \; \rightleftarrows \;$ [structure with O$^-$] $+ \; H_3O^+$

Tropolone

20.23 Which one of the following pairs of compounds would you expect to be the stronger acid? Explain.

(a) [cycloheptatriene structure] or [cyclopentadiene structure] **(b)** [azetidinium structure with N$^\pm$—H] or [pyridinium structure with H—N$^+$]

20.24 Explain why protonation of 4-pyrone occurs on the carbonyl oxygen atom rather than the ring oxygen atom.

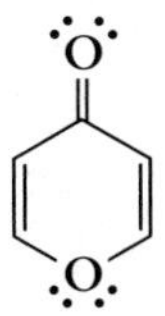

[structure of 4-pyrone]

4-Pyrone

20.25 Explain why cyclopentadienone is unknown, despite many attempts to prepare it, yet cycloheptatrienone is easily prepared.

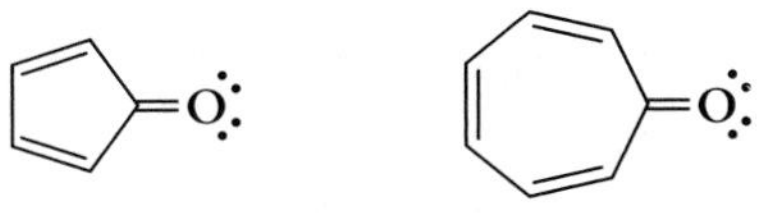

[structure of cyclopentadienone] [structure of cycloheptatrienone]

Cyclopentadienone   Cycloheptatrienone

20.26 Write the structure of the major organic product formed in each of the following reactions. Indicate the stereochemistry of the product where appropriate.

(a) (Z)-1-Phenylpropene + m-chloroperoxybenzoic acid
(b) Benzyl bromide + sodium phenolate (NaOC$_6$H$_5$)
(c) 1-Phenyl-1-bromoethane + NaOC$_2$H$_5$ in C$_2$H$_5$OH
(d) (Z)-1-Phenylpropene + aqueous H$_2$SO$_4$
(e) (Z)-1-Phenylpropene + hot acidic solution of K$_2$Cr$_2$O$_7$
(f) (Z)-1-Phenylpropene + OsO$_4$/H$_2$O$_2$/H$_2$O

20.27 Starting with styrene, C$_6$H$_5$CH=CH$_2$, any other organic compound, and any inorganic reagents, propose a synthesis for each of the following compounds.

(a) C$_6$H$_5$CHBrCH$_2$Br
(b) C$_6$H$_5$CHBrCH$_3$
(c) C$_6$H$_5$C≡CH
(d) C$_6$H$_5$(CH$_2$)$_3$CH$_3$
(e) C$_6$H$_5$CH$_2$CH$_2$C≡CH
(f) C$_6$H$_5$CH$_2$CH$_2$CH(OH)CH$_2$Br

20.28 Write the structural formula of each of the following compounds based on their molecular formula and the spectral data provided.

(a) C_9H_{12} ^{1}H NMR: δ 1.25 d (6H); δ 2.89 sept (1H); δ 7.26 broad multiplet (5H). ^{13}C NMR: δ 24.0; δ 34.3; δ 125.7; δ 126.3; δ 128.2; δ 148.7

(b) C_4H_4O ^{1}H NMR: δ 6.30 d (2H); δ 7.38 d (2H). ^{13}C NMR: δ 109.9; δ 143.0

(c) $C_4H_4N_2$ ^{1}H NMR: δ 8.63 s. ^{13}C NMR: δ 144.9

(d) $C_4H_4N_2$ ^{1}H NMR: δ 7.55 d (2H); δ 9.24 d (2H). ^{13}C NMR: δ 126.5; δ 151.4

20.29 2-Chloro-3-phenylbutane eliminates HCl on reaction with $NaOC_2H_5$ in ethanol. Write the structure of the elimination product. There are four stereoisomers of 2-chloro-3-phenylbutane. Write their structures and decide which give the (Z)-alkene and which give its isomer on elimination of HCl.

20.30 Addition of chlorine to (Z)-2-pentene forms only two of the four possible stereoisomers as products. In contrast, addition of chlorine to (Z)-1-phenylpropene forms all four stereoisomers. Write the mechanism of the addition of chlorine to (Z)-2-pentene. How must this mechanism be modified for the addition of chlorine to (Z)-1-phenylpropene to account for the stereochemistry of the product?

CHAPTER 21

CHEMISTRY of BENZENE and its DERIVATIVES

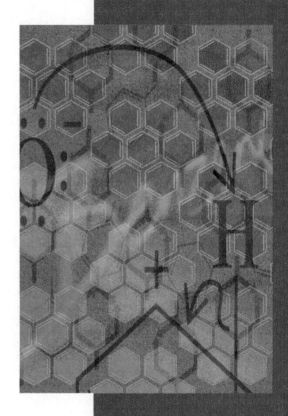

Derivatives of benzene are numerous and serve many functions in plants and animals. Cinnamaldehyde, for example, is found in cinnamon, acetylsalicylic acid (ASA or aspirin) is an analgesic, and epinephrine is a hormone produced by the adrenal medulla.

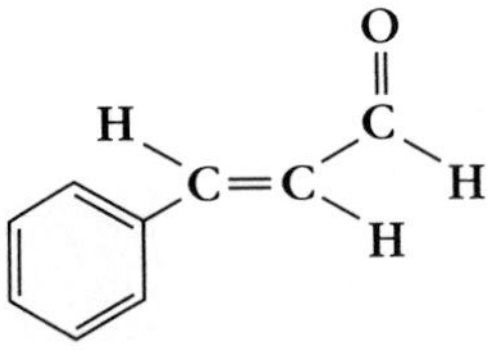

Cinnamaldehyde
A constituent of
natural cinnamon oils

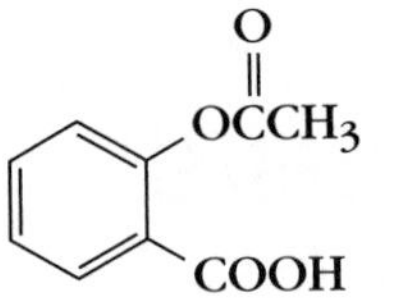

Aspirin
Acetylsalicylic acid (ASA)

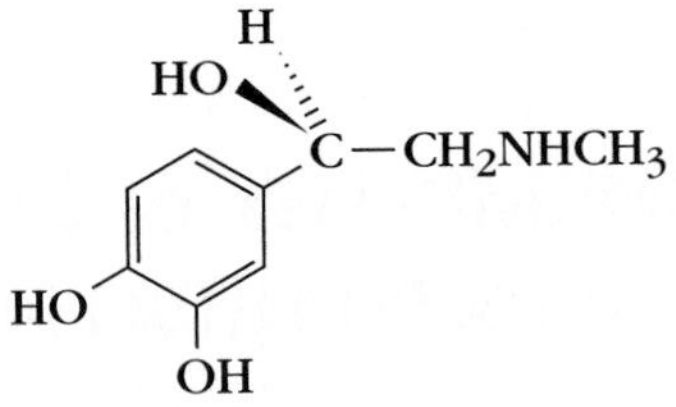

Epinephrine
A hormone produced by the
adrenal medulla

While the structures of benzene derivatives vary widely, they all undergo electrophilic aromatic substitution reactions. In this chapter, the principles of the mechanism of these substitution reactions will be established. We begin, however, by discussing how to name the derivatives of benzene.

21.1   Naming Derivatives of Benzene

Many derivatives of benzene are named by using the name of the substituent as a prefix to the word benzene. These names are written as one word. All the hydrogens on benzene are equivalent, so only one monosubstituted benzene exists for each substituent. As a result, no numbers are needed to locate a single substituent on the ring.

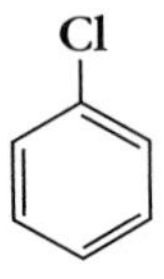

Chlorobenzene 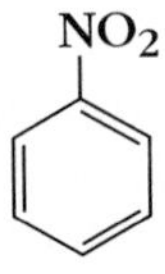   Nitrobenzene 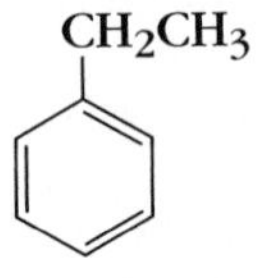   Ethylbenzene

Many simple monosubstituted derivatives of benzene were given common names long before the introduction of the IUPAC system, and many of these have been retained. The following are compounds whose common names are acceptable in the IUPAC system.

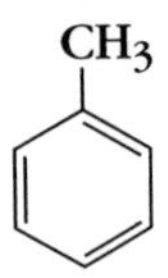

Toluene 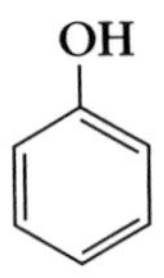   Phenol 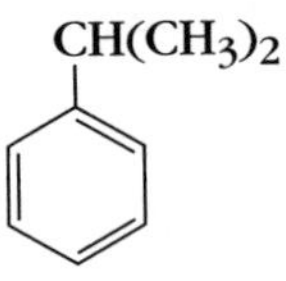   Cumene 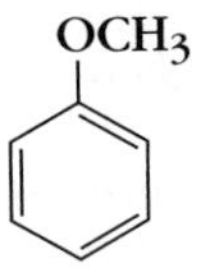   Anisole 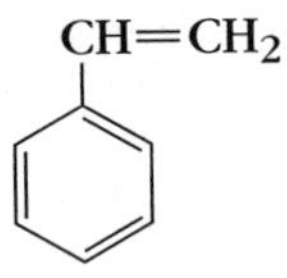   Styrene

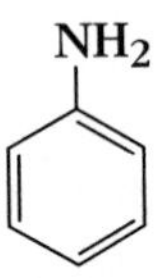

Aniline 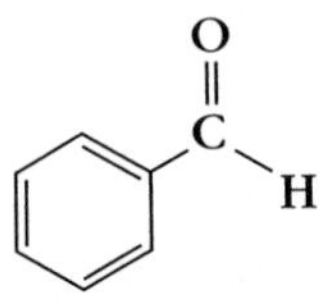   Benzaldehyde 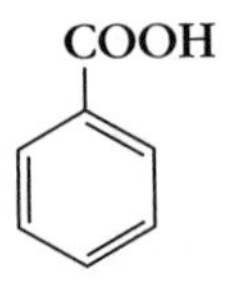   Benzoic acid 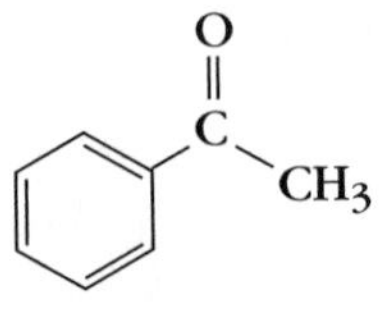   Acetophenone 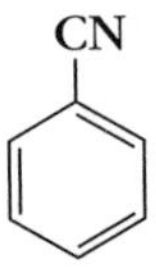   Benzonitrile

Disubstituted benzenes are named by using the prefixes *ortho-*, *meta-*, and *para-* (usually abbreviated as *o-*, *m-*, *p-*). In an *ortho-*disubstituted benzene, the two substituents are bonded to adjacent ring carbon atoms so they are in the 1,2-positions. The two substituents are in the 1,3-positions in a *meta-*disubstituted benzene and in the 1,4-positions in a *para-*disubstituted benzene.

o-Dibromobenzene	*m*-Dibromobenzene	*p*-Dibromobenzene
1,2-disubstituted	1,3-disubstituted	1,4-disubstituted

When the two substituents are different, the prefixes are placed in alphabetical order.

o-Bromochlorobenzene	*m*-Fluoronitrobenzene	*p*-Bromoethylbenzene

When one of the substituents corresponds to one of the monosubstituted benzenes that has a special name, the disubstituted benzene is named as a derivative of that compound.

o-Chlorotoluene	*m*-Nitrobenzoic acid	*p*-Bromoaniline

For more than two substituents, their positions are indicated by numbers, which are assigned to give them the lowest possible numbers. The substituents are arranged alphabetically when writing the name.

2-Chloro-4-nitrotoluene	1,4-Dichloro-	2,4,6-Tribromophenol
not	2-nitrobenzene	
6-chloro-4-nitrotoluene		

2-Chloro-4-nitrotoluene and 2,4,6-tribromophenol are named as derivatives of toluene and phenol, respectively, rather than benzene because both compounds contain a substituent that defines a common name when it is bonded to a benzene ring. This name becomes the parent name of the polysubstituted benzene.

Moreover, the group that defines the parent name (methyl for toluene and hydroxy for phenol) is assigned the number 1 position of the benzene ring.

Derivatives of benzene are usually called **arenes,** and the symbol Ar is used to signify an aryl group just as R is used to represent an alkyl group. We are already familiar with the use of the word phenyl to signify that a benzene ring is a substituent.

Exercise 21.1   Give the IUPAC name of each of the following aromatic compounds.

(a)

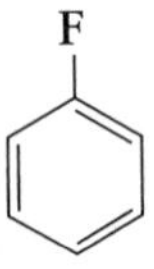

(b)

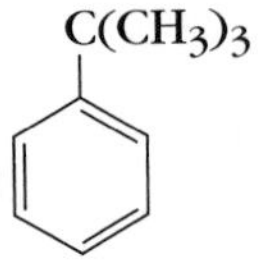

(c)

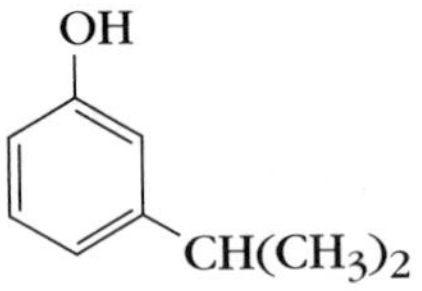

(d)

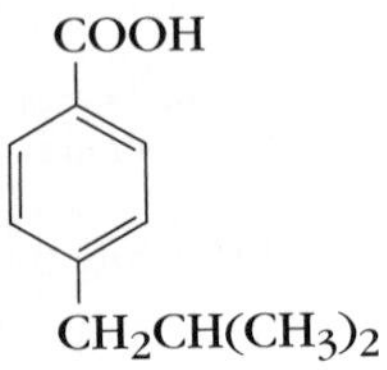

(e)

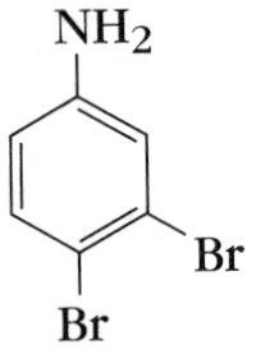

(f)

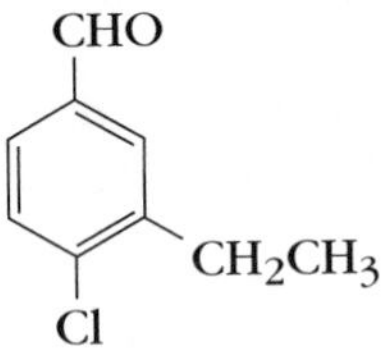

(g)

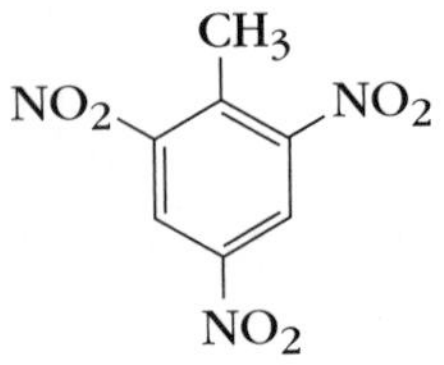

(h)

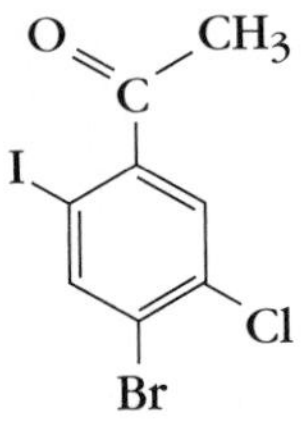

Exercise 21.2   Write the structure of the compound corresponding to each of the following names.

(a) Iodobenzene
(b) o-Fluorotoluene
(c) p-Propylbenzoic acid
(d) 3-Nitroaniline

(e) m-t-Butylnitrobenzene
(f) p-Fluorophenol
(g) 2,4,6,-Trinitrofluorobenzene
(h) 2,4-Dinitroanisole

21.2   Spectroscopic Properties of Benzene Derivatives

Recall from Chapter 20 that the π electrons of aromatic rings generate a ring current when placed in a magnetic field, and that the ring current, in turn, generates a new local magnetic field. This results in the deshielding of hydrogen atoms located outside the aromatic ring and the shielding of nuclei inside or above the ring. The ^{1}H NMR spectrum of benzene, for example, shows a sharp singlet for the six equivalent hydrogen atoms in the region δ 7.24 to 7.27.

While benzene has a single peak in its ^{1}H NMR spectrum, substituted benzenes may have more complicated spectra. For example, the hydrogen atoms in the o-, m-, and p- positions of a monosubstituted benzene are nonequivalent and

Figure 21.1
Aromatic region of the ^1H NMR spectra of **(A)** chlorobenzene and **(B)** toluene.

■ *Recall from Section 10.14 that first order spectra arise only if the chemical shift difference (Δ δ) between nonequivalent hydrogen atoms, expressed in Hz, is about 10 times larger than their coupling constants, J. That is, Δ δ > 10J.*

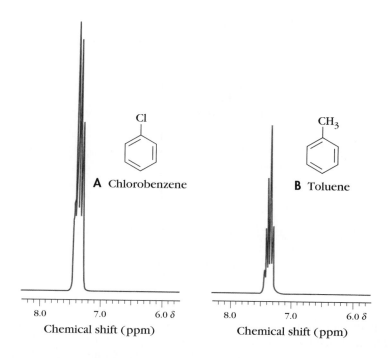

may have different chemical shifts. Depending on the differences in the chemical shifts and the coupling constants of the various ring hydrogen atoms, the ^1H NMR spectra of substituted benzenes may or may not approximate first order spectra.

The aromatic regions of the ^1H NMR spectra of monosubstituted benzenes whose substituent is neither strongly electron withdrawing nor strongly electron donating are nearly first order. The ^1H NMR spectra of chlorobenzene and toluene, shown in Figure 21.1, are two examples of such monosubstituted benzenes. The aromatic regions of their ^1H NMR spectra consist of a broad band between δ 7.2 and 7.5 that shows further small splitting.

The chemical shifts of the *ortho-*, *meta-*, and *para-*hydrogen atoms in monosubstituted benzenes containing strongly electron donating or strongly electron withdrawing groups, on the other hand, are usually sufficiently different that they are separated from one another in the aromatic region of the ^1H NMR spectrum. For example, the ^1H NMR spectrum of phenol, shown in Figure 21.2, while not first order, consists of a complex multiplet at δ 6.8 to 7.0 of relative area three separated from another complex multiplet of relative area two at δ 7.2 to 7.4. Depending on the substituent, the complex multiplet of area two may be upfield of the other multiplet of area three. Both patterns are characteristic of a single strong electron withdrawing or electron donating substituent bonded to a benzene ring.

Figure 21.2
Aromatic region of the ^1H NMR spectrum of phenol.

While the aromatic region of the ^1H NMR spectra of most disubstituted benzenes is complex, certain patterns are characteristic of the nature and the position of the substituents. The aromatic region of the ^1H NMR spectra of *p*-disubstituted benzenes, for example, in which the two substituents have opposing electronic effects, is approximately first order. An example is the ^1H NMR spectrum of 4-bromoanisole, which is shown in Figure 21.3.

The spectrum of 4-bromoanisole consists of a set of two doublets in the aromatic region, as well as a singlet at δ 3.75 due to the methoxy methyl group. The

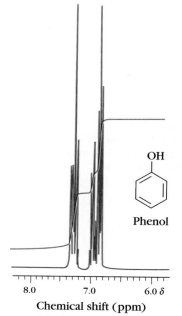

Figure 21.3
^{1}H NMR spectrum of 4-bromoanisole.

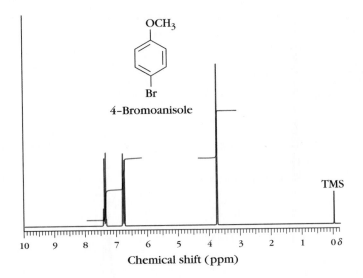

pattern in the aromatic region arises because there are two sets of chemically equivalent hydrogen atoms. One set is *ortho-* to the bromine atom, while the other is *ortho-* to the methoxy group. Because the chemical shift of these two sets of hydrogen atoms is sufficiently large compared to their coupling constant, its coupling pattern approximates a first order spectrum. Thus the hydrogen atoms *ortho* to both Br and OCH$_3$ are split into a doublet by the adjacent hydrogen atom. This pattern is characteristic of *p*-disubstituted benzenes in which one substituent donates electrons to the ring, while the other withdraws electrons.

It is more difficult to identify *o*- and *m*-disubstituted benzenes. The aromatic region of the ^{1}H NMR spectrum of 2-bromoanisole, shown in Figure 21.4, *A*, consists of three complex multiplets of relative areas 1:1:2. This pattern is frequently observed in the aromatic region of the ^{1}H NMR spectrum of *o*-disubstituted benzenes in which one substituent donates electrons to the ring, while the other withdraws electrons. The aromatic region of the ^{1}H NMR spectra of 3-bromoanisole, shown in Figure 21.4, *B*, is the most complex of the three isomers of bromoanisole. Generally, the aromatic region of the ^{1}H NMR of the *meta-*

Figure 21.4
Aromatic region of the ^{1}H NMR spectra of **(A)** 2-bromoanisole and **(B)** 3-bromoanisole.

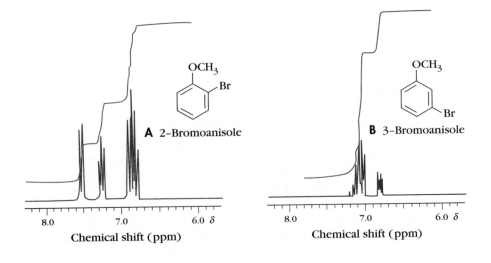

isomer is the most complex and least symmetrical of all of the isomeric di-substituted benzenes.

The chemical shifts of benzylic hydrogen atoms in ^{1}H NMR spectra are shifted downfield compared to those found in alkanes. The chemical shifts of the benzylic hydrogens of toluene (δ 2.30) and ethylbenzene (δ 2.54) are typical.

In contrast to ^{1}H NMR, the ring current of aromatic compounds does *not* affect the chemical shifts of ^{13}C nuclei. There are two reasons for this observation. First, the chemical shifts of ^{13}C nuclei are largely determined by charge and hybridization, which are unaffected by the ring current. Second, the carbon atoms of the aromatic ring lie *between* the deshielding and shielding areas of the aromatic ring. As a result, the ^{13}C chemical shifts of the ring and benzylic carbon atoms of benzene derivatives are similar to those of a comparable alkene.

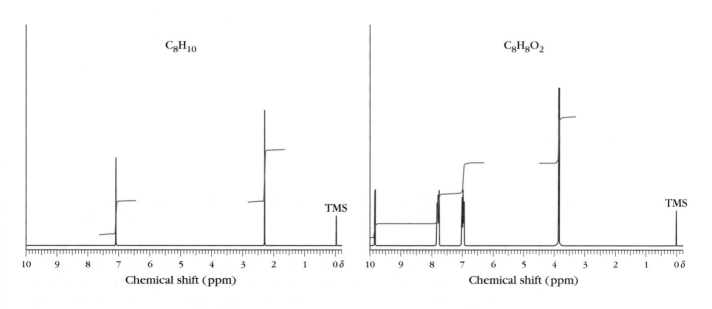

The number of lines in the ^{13}C NMR spectrum of a polysubstituted benzene is characteristic of the substitution pattern. For example, the ^{13}C NMR spectrum of 1,3,5-trichlorobenzene has two lines, its 1,2,3-isomer has four lines, and its 1,2,4-isomer has six lines.

Exercise 21.3   Write the structural formula of each of the following compounds whose molecular formulas and ^{1}H NMR spectra are given below.

(a) C_8H_{10} **(b)** $C_8H_8O_2$

| Exercise 21.4 | How many lines would you expect to find in the ^{13}C NMR spectrum of each of the following compounds? |

(a) CH₃, CH₃ (ortho-dimethylbenzene)

(b) Cl, Cl (para-dichlorobenzene)

(c) Cl, NO₂ (para, with NO₂)

(d) Cl, NO₂ (meta)

(e) Cl, NO₂ (ortho)

The presence of a benzene ring in a compound can also be established by ultraviolet spectroscopy. The delocalized π electron system in simple alkylbenzenes gives rise to a relatively strong band near 205 nm and much weaker ones in the 255 to 275 nm range. The spectrum of toluene in hexane as solvent, shown in Figure 21.5, is typical: λ_{max} = 208 nm (ϵ = 7900); 262 nm (ϵ = 230). The band at 208 nm is not visible because it is obscured by solvent absorption.

Figure 21.5
The ultraviolet spectrum of toluene.

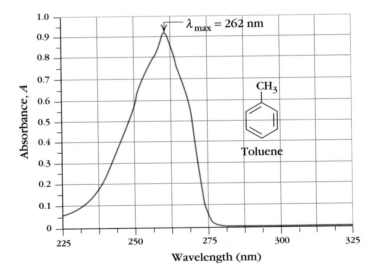

λ_{max} = 262 nm

Toluene

21.3   Typical Electrophilic Aromatic Substitution Reactions

Benzene and its derivatives are nucleophiles just like alkenes because they all have π electrons. As a result, many of the same electrophiles that add to alkenes also react with benzene and its derivatives. Alkenes, however, form products of addition, whereas most aromatic compounds react with electrophiles to form products of substitution. Thus the principal reaction of aromatic compounds is

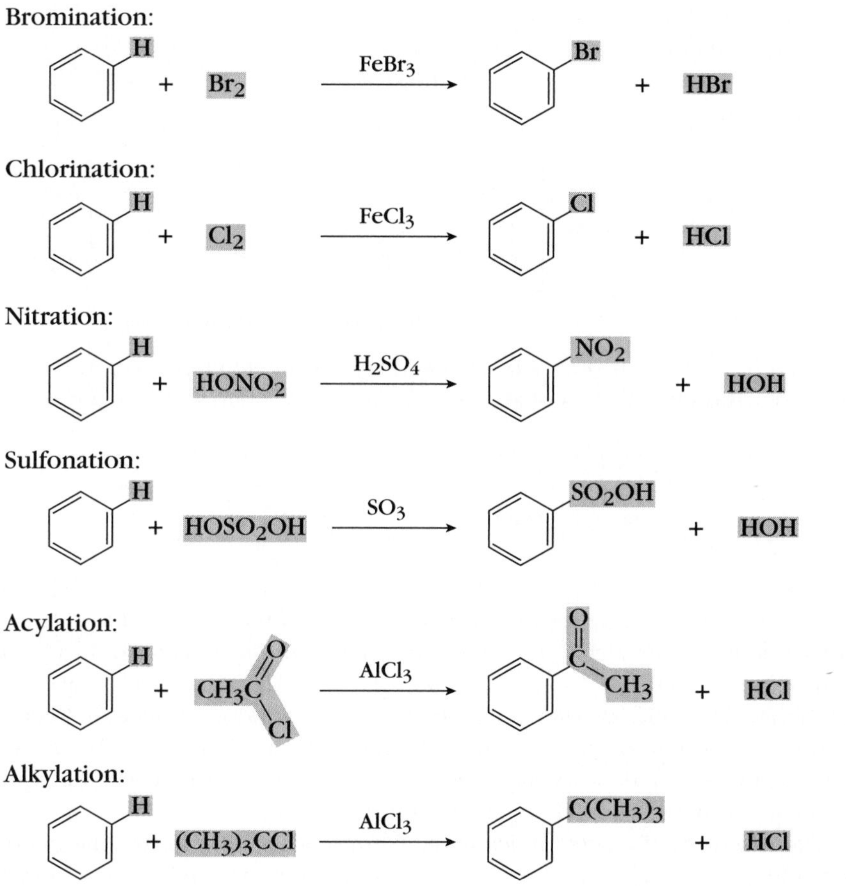

Bromination:

Chlorination:

Nitration:

Sulfonation:

Acylation:

Alkylation:

Figure 21.6
A number of typical electrophilic aromatic substitution reactions of benzene.

electrophilic aromatic substitution. A number of typical electrophilic substitution reactions are given in Figure 21.6.

To understand why aromatic compounds react with electrophiles by substitution rather than addition reactions, we must understand the general mechanism of electrophilic aromatic substitution reactions.

21.4   A General Mechanism of Electrophilic Aromatic Substitution

The general mechanism of most electrophilic aromatic substitution reactions has three steps. Step 1 is formation of the electrophile. In Step 2, the electrophile adds to the benzene ring, and in Step 3 a proton is removed to regenerate the aromatic ring. Let's examine each of these steps in detail.

The actual electrophile in aromatic substitution reactions is not necessarily the reagent that is added to the reaction mixture. In most cases, the function of the catalyst is to form the electrophile from one of the reagents. In the mechanism of each aromatic substitution reaction, therefore, we must pay particular attention to the nature of the electrophile and how it is formed in Step 1.

Step 2 of this mechanism is generally the rate-determining step. It is the addition of the electrophile to benzene to form a carbocation intermediate, which no longer has an aromatic structure. Instead, it is a carbocation with four π electrons delocalized over five carbon atoms. The sixth carbon atom is sp^3 hybridized. This intermediate carbocation can be pictured as a resonance hybrid of the following resonance structures.

Resonance-stabilized intermediate

In Step 3, the final step, the aromatic ring is regenerated from the intermediate carbocation by the loss of a proton from the sp^3 hybrid carbon atom.

Reactions of electrophiles with benzene and with alkenes differ in two important ways. First, benzene is much *less* reactive than most alkenes. As a result, the usual reagents that add to alkenes (Br_2, Cl_2, HCl, HBr, H_3O^+, etc.) are not sufficiently electrophilic to react with benzene, so a catalyst is needed. Second, the products are different; alkenes form addition products, whereas benzene forms substitution products. Why are there such differences? To answer this question, we must compare the mechanism of both reactions.

The reactions of electrophiles with alkenes and with benzene have a common rate-determining step, which is the addition of an electrophile to a π electron system to form a carbocation intermediate. The relative rates of reactions of elec-

Figure 21.7
Comparison of the relative free energies of activation of the rate-determining step in the mechanism of the reaction of an electrophile (E^+) with an alkene and with benzene. The free energy of activation of the reaction of benzene and an electrophile is greater than the free energy of activation of the reaction of an alkene, so benzene reacts more slowly with the electrophile.

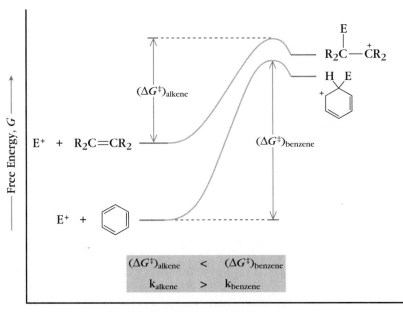

trophiles with benzene and with alkenes depend on the relative free energies of activation of these two addition steps. We can estimate these free energies of activation by comparing the relative free energies of the two starting materials and their carbocation intermediates.

We learned in Chapter 20 that aromaticity makes benzene much more stable than any alkene. However, addition of an electrophile to benzene forms a carbocation intermediate that has lost its aromaticity. As a result, the free energy difference between benzene and its intermediate carbocation is greater than the free energy difference between an alkene and its intermediate. The reaction of an electrophile with benzene, therefore, has a larger free energy of activation, so benzene reacts more slowly than alkenes with electrophiles. The free energy-reaction path diagram in Figure 21.7 illustrates the difference between $\Delta G^{\ddagger}$ for the reaction of benzene and $\Delta G^{\ddagger}$ for the reaction of an alkene with electrophiles.

We must look at Step 2 of the mechanism in order to understand why the carbocation intermediate formed from benzene loses a proton to regenerate an aromatic ring. If the carbocation reacts with an anion, the product is not aromatic. Consequently, a nonaromatic product of addition would be some 36 kcal/mol (150 kJ/mol) less stable than the aromatic product formed by loss of a proton. The free energy-reaction path diagram in Figure 21.8 shows, moreover, that the transition state for formation of the aromatic product is energetically more favorable than the transition state for the nonaromatic product.

In summary, stability of the benzene ring due to its aromaticity is the chief factor responsible for the differences between the rates of reactions of alkenes and benzene with electrophiles. The addition of an electrophile to benzene disrupts its aromaticity, so it reacts more slowly than an alkene. Benzene can restore its aromaticity by the loss of a proton from its carbocation intermediate, however, so benzene undergoes substitution rather than addition reactions.

We can now apply this general mechanism to specific examples of aromatic substitution reactions of benzene.

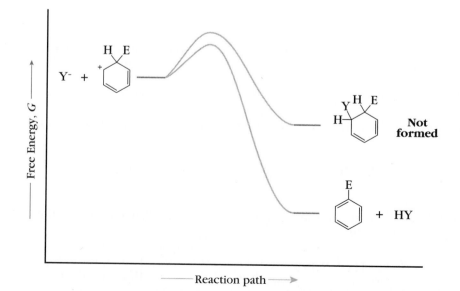

Figure 21.8
Free energy–reaction path diagram for the product-forming step in an electrophilic aromatic substitution reaction. The free energy of activation for formation of the aromatic product is smaller than the free energy of activation for the nonaromatic product. As a result, formation of the substitution product is favored.

21.5 Halogenation of Benzene

Nowhere is the difference between alkenes and benzene more pronounced than in their reactions with bromine. Bromine in a CCl_4 solution reacts instantly at room temperature with most alkenes but does not react with benzene. Benzene only reacts with bromine or chlorine in the presence of a catalyst. Typical catalysts are Lewis acids such as $FeCl_3$, $AlCl_3$, or $ZnCl_2$ for chlorine, and $FeBr_3$ for bromine. These Lewis acids catalyze the halogenation of benzene because they polarize the halogen-halogen bond by forming a complex. Complex formation makes the halogen molecule more electrophilic by producing an electron-deficient halogen atom. Complex formation to create a more electrophilic chlorine or bromine is Step 1 in the mechanism of chlorination and bromination of benzene, as shown in Figure 21.9.

In Step 2 of the bromination mechanism, benzene reacts with the electron-deficient bromine or chlorine of the complex to form a intermediate carbocation. Finally, in Step 3, $FeBr_4^-$ (or $FeCl_4^-$ in the case of chlorination) acts as a base to remove the ring proton. In this step, bromobenzene and hydrogen bromide are formed and the catalyst is regenerated.

The order of reactivity of halogens with benzene is $F_2 > Cl_2 > Br_2 > I_2$. As in the case of alkenes, fluorine is too reactive to be used to prepare aromatic fluorine compounds. Iodine, on the other hand, is unreactive toward most aromatic rings. Iodination can be accomplished, however, by making molecular iodine more electrophilic. Nitric acid, for example, oxidizes molecular iodine into the more electrophilic iodonium ion (I^+), which reacts with benzene to form iodobenzene:

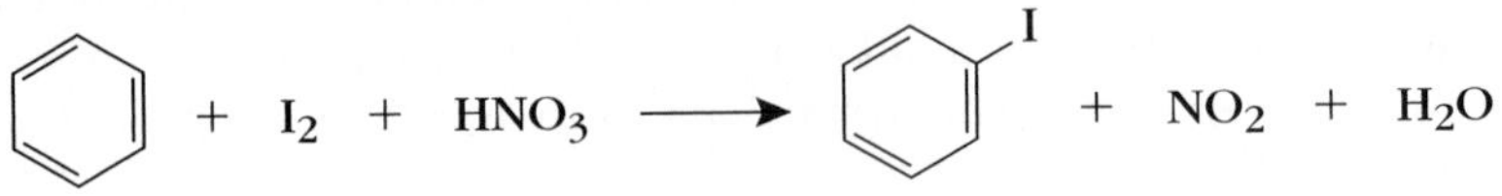

Nitric acid is *not* a catalyst in this reaction because it is consumed in the preparation of the electrophilic iodonium ion.

While we have concentrated on halogenation of the aromatic ring, remember that if a benzene derivative contains a functional group on a side chain that can

Figure 21.9
The three steps of the mechanism of the bromination of benzene. Step 1 is the formation of the electrophile, which adds to benzene in Step 2. Rearomatization of the benzene ring occurs in Step 3.

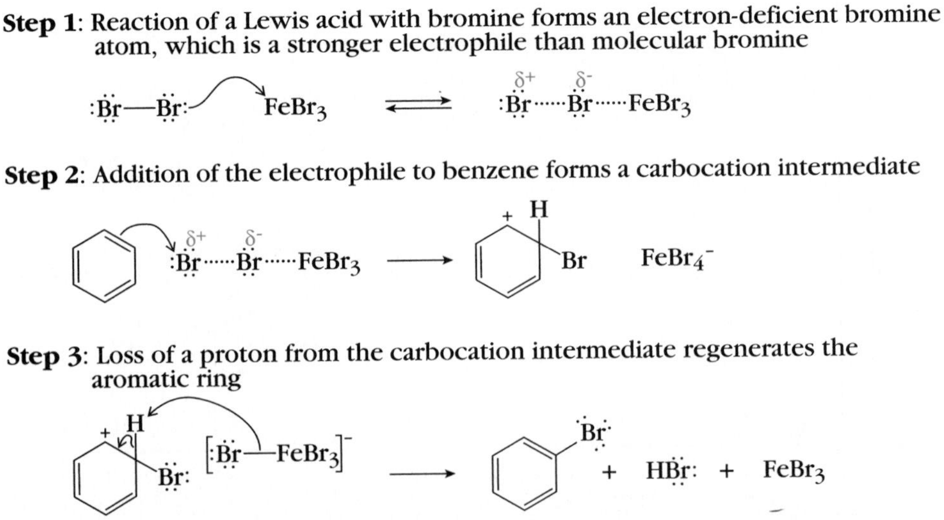

Step 1: Reaction of a Lewis acid with bromine forms an electron-deficient bromine atom, which is a stronger electrophile than molecular bromine

Step 2: Addition of the electrophile to benzene forms a carbocation intermediate

Step 3: Loss of a proton from the carbocation intermediate regenerates the aromatic ring

react with chlorine and bromine, reaction also occurs at that site. Functional groups on a side chain that will react with bromine and chlorine (which we have mentioned in previous chapters) include the double bond of alkenes (electrophilic addition reactions) and the benzylic carbon atom (free radical reactions). In fact, benzene rings are so much less reactive that uncatalyzed reactions usually occur only on the side chain. Styrene, for example, reacts readily with bromine in the absence of a catalyst to form only the product of addition to the side chain double bond.

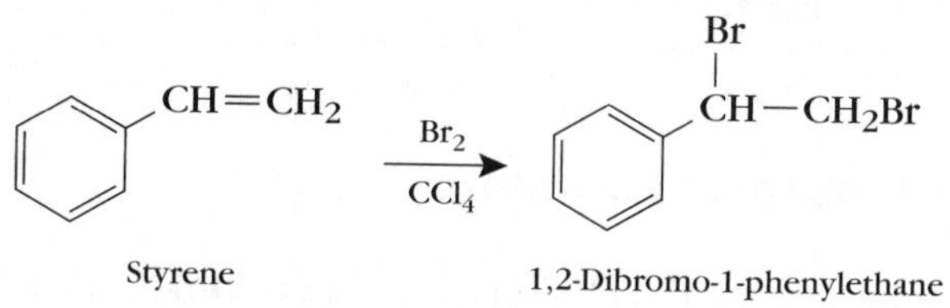

$$\underset{\text{Styrene}}{\text{C}_6\text{H}_5\text{CH}=\text{CH}_2} \xrightarrow[\text{CCl}_4]{\text{Br}_2} \underset{\text{1,2-Dibromo-1-phenylethane}}{\text{C}_6\text{H}_5\text{CHBr}-\text{CH}_2\text{Br}}$$

It is often possible, therefore, to control the site of bromination and chlorination by the proper choice of reaction conditions.

| **Exercise 21.5** | Write the mechanism of the reaction of chlorine and benzene in the presence of $FeCl_3$. |

21.6 Nitration of Benzene

Benzene reacts with a mixture of nitric and sulfuric acids to form nitrobenzene. The actual electrophile in this nitration reaction is the nitronium ion, NO_2^+, formed in Step 1 of the mechanism. It is formed in a sequence of two reactions,

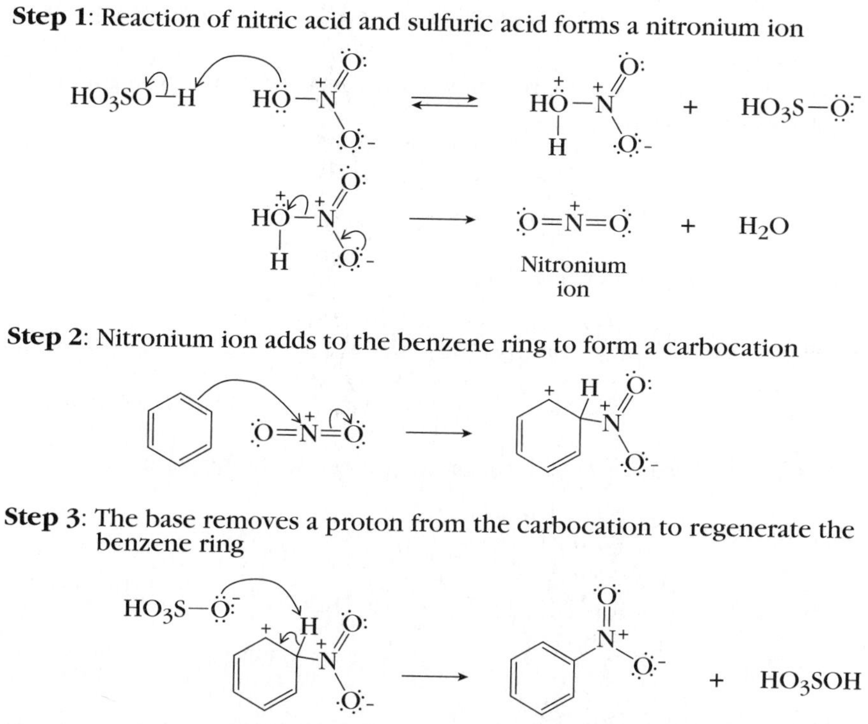

Step 1: Reaction of nitric acid and sulfuric acid forms a nitronium ion

Step 2: Nitronium ion adds to the benzene ring to form a carbocation

Step 3: The base removes a proton from the carbocation to regenerate the benzene ring

Figure 21.10
The mechanism of the nitration of benzene. The electrophile, a nitronium ion, is formed in Step 1. It reacts with the benzene ring in Step 2, and loss of a proton in Step 3 forms nitrobenzene.

as shown in Figure 21.10. The first is an acid-base reaction in which sulfuric acid protonates nitric acid, followed by dehydration of the protonated nitric acid to form the nitronium ion. In Step 2, the nitronium ion adds to benzene to form a carbocation intermediate, which loses a proton in Step 3 to form nitrobenzene and regenerate the acid catalyst.

Exercise 21.6	Why does sulfuric acid protonate nitric acid and not the reverse?

21.7 Sulfonation of Benzene

■ *Fuming sulfuric acid is made by dissolving 8% SO_3 in concentrated sulfuric acid.*

When benzene is heated with a solution of fuming sulfuric acid, a hydrogen atom on the benzene ring is substituted by a sulfonic acid group ($-SO_3H$). The electrophile in fuming sulfuric acid is believed to be the SO_3 molecule. Despite being a neutral molecule, SO_3 is a powerful electrophile because its sulfur atom is electron deficient.

Exercise 21.7	The following species have been proposed as sulfonating electrophiles in the sulfonation of benzene. Write an equation for the formation of each from H_2SO_4.

(a) $H_3SO_4^+$ **(b)** SO_3 **(c)** HSO_3^+ **(d)** $H_2S_2O_7$

In fuming sulfuric acid, Step 1 of the general mechanism is unnecessary because the solution contains dissolved SO_3, which reacts directly with benzene, as shown in Figure 21.11.

The two steps of the mechanism for the sulfonation of benzene and its derivatives are the usual ones, namely addition to the benzene ring followed by elimination of a proton.

Because of their structural resemblance to sulfuric acid, aromatic sulfonic acids are also strong acids ($pK_a \approx -6$).

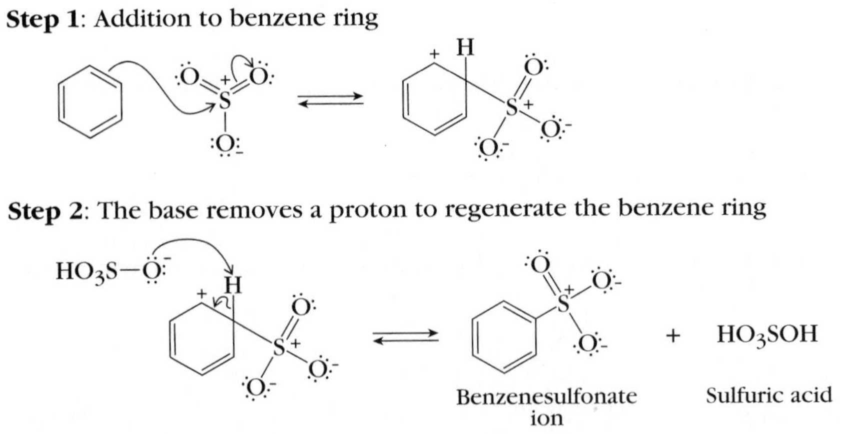

Figure 21.11
Mechanism of the sulfonation of benzene consists of only two steps because SO_3, the electrophile, is already present in fuming sulfuric acid. Addition of SO_3 to benzene and removal of a proton to form benzenesulfonate ion are the two steps of the mechanism.

Step 1: Addition to benzene ring

Step 2: The base removes a proton to regenerate the benzene ring

Benzenesulfonate ion   Sulfuric acid

21.8   Friedel-Crafts Alkylation of Benzene

Any reaction that forms a carbon-carbon bond is particularly valuable in organic synthesis. The French chemist Charles Friedel and his American coworker James Crafts found in 1877 that haloalkanes, but *not* haloaromatic or vinyl halides, react with benzene in the presence of Lewis acids such as $FeBr_3$ or $AlCl_3$ to form alkylbenzenes. This reaction, which forms a carbon-carbon single bond, is now called the **Friedel-Crafts alkylation reaction.**

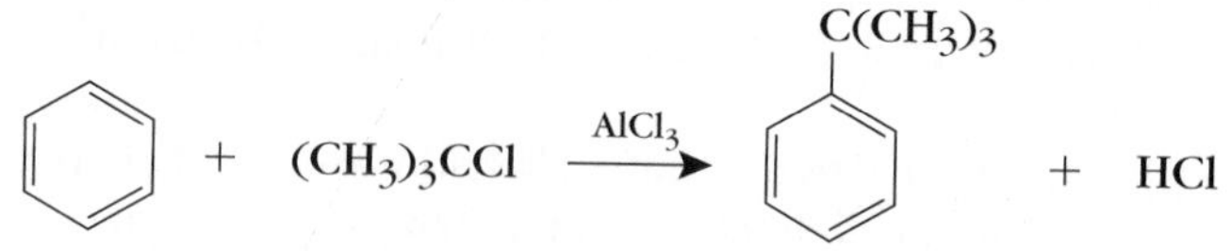

| Benzene | 2-Chloro-2-methylpropane | *t*-Butylbenzene (65% yield) |

Step 1 of the mechanism of this reaction, shown in Figure 21.12, is the generation of the electrophile by the reaction of a Lewis acid, such as $AlCl_3$, with a haloalkane. The Lewis acid reacts with the haloalkane to form a complex in which the carbon-halogen bond is highly polarized.

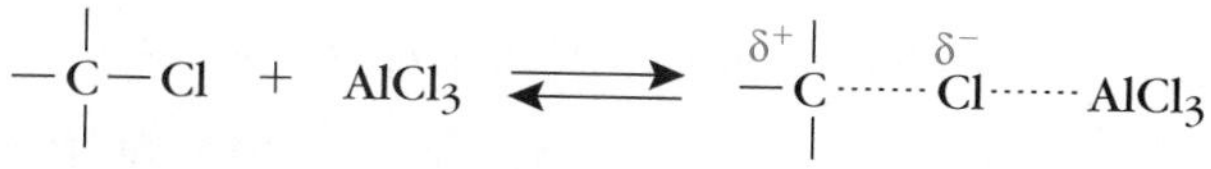

The Lewis acid acts as a catalyst in this reaction in the same way as $FeBr_3$ does in the bromination of benzene.

While the carbon portion of the Lewis acid complex probably is not a free carbocation (particularly in the case of a 1° haloalkane), it reacts very much like one. Thus Step 2 of the mechanism is the addition of the electrophilic carbon portion of the Lewis acid complex to a benzene ring. Loss of a proton in Step 3 completes the reaction.

Friedel-Crafts alkylation reactions are of limited synthetic utility because they have two major limitations. The first is that the reaction is often difficult to stop after the introduction of one alkyl group because alkylbenzenes are more reactive in electrophilic aromatic substitution reactions than benzene. Thus the Friedel-

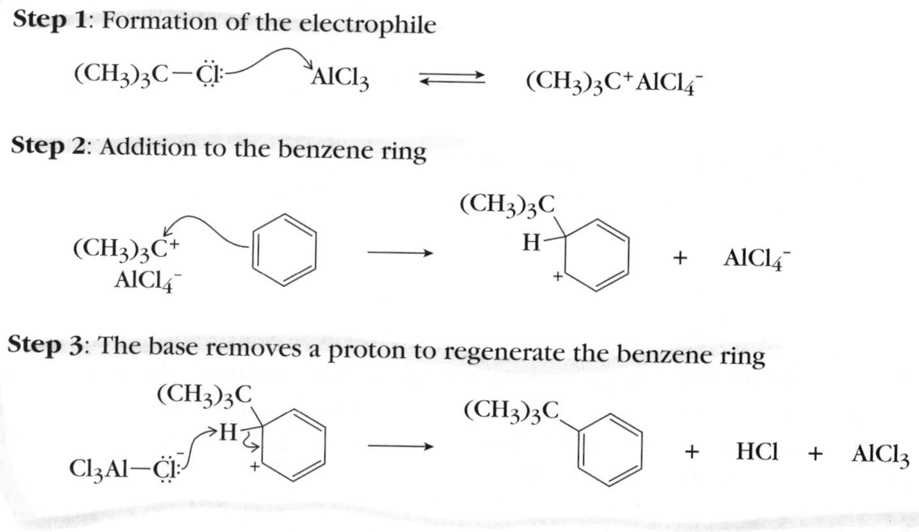

Step 1: Formation of the electrophile

Step 2: Addition to the benzene ring

Step 3: The base removes a proton to regenerate the benzene ring

Figure 21.12
The mechanism of the Friedel-Crafts alkylation reaction.

■ *The effects of substituents on further electrophilic aromatic substitution reactions are discussed in more detail in Section 21.11.*

Crafts alkylation of benzene frequently forms polyalkylated products. For example, the reaction of equal amounts of chloromethane and benzene with AlCl₃ forms a complex mixture of tri-, tetra-, and pentamethylbenzenes as product.

$$C_6H_6 \xrightarrow[CH_3Cl]{AlCl_3} C_6H_5CH_3 \xrightarrow[\underset{faster}{CH_3Cl}]{AlCl_3} C_6H_4(CH_3)_2 \xrightarrow[\underset{faster}{CH_3Cl}]{AlCl_3} \text{etc.}$$

The only practical way of overcoming this limitation is to keep benzene in large excess. While this solves the problem for benzene, which is a relatively inexpensive compound, it is impractical with most derivatives of benzene, which are generally more expensive.

The second limitation of Friedel-Crafts alkylation reactions is the tendency of the electrophile to undergo skeletal rearrangement. For example, the reaction of 1-chloropropane and benzene with AlCl₃ forms isopropylbenzene as the major product.

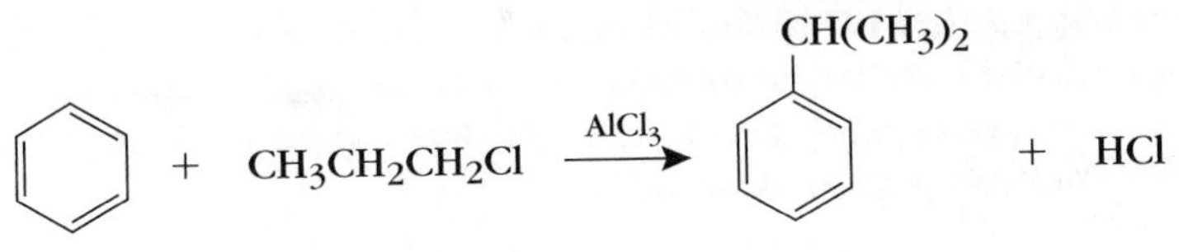

Benzene   1-Chloropropane   Isopropylbenzene

In the presence of Lewis acids, the starting chloroalkane rearranges to the thermodynamically more stable isopropyl cation.

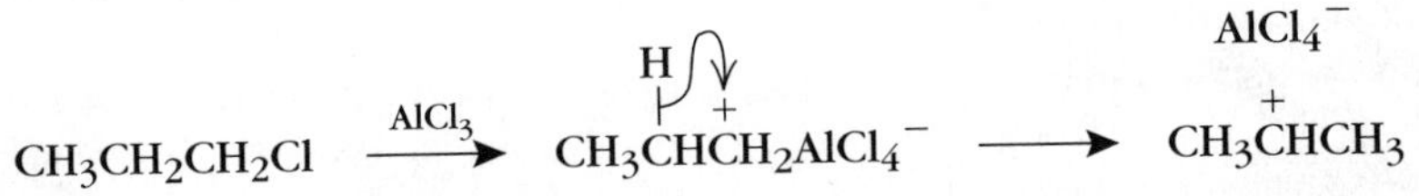

Addition of the isopropyl cation to the benzene ring, followed by loss of a proton forms the observed product.

Exercise 21.8   The reaction of 1-chloro-2-methylpropane and benzene with AlCl₃ forms *t*-butylbenzene as product. Write the mechanism for this reaction.

Exercise 21.9   Industrially *t*-butylbenzene is prepared by the reaction of methylpropene and benzene with HF/BF₃ as catalyst. Write a mechanism for this reaction.

Although rearrangements of the haloalkane severely limit the synthetic utility of the Friedel-Crafts alkylation reaction, there is a solution. There are electrophilic carbon species that can add to a benzene ring, do not rearrange, and deactivate the ring to further substitution. These carbon electrophiles are formed in Friedel-Crafts acylation reactions.

21.9   Friedel-Crafts Acylation of Benzene

Benzene reacts with an acyl chloride in the presence of aluminum chloride ($AlCl_3$) catalyst to form an acylated benzene derivative.

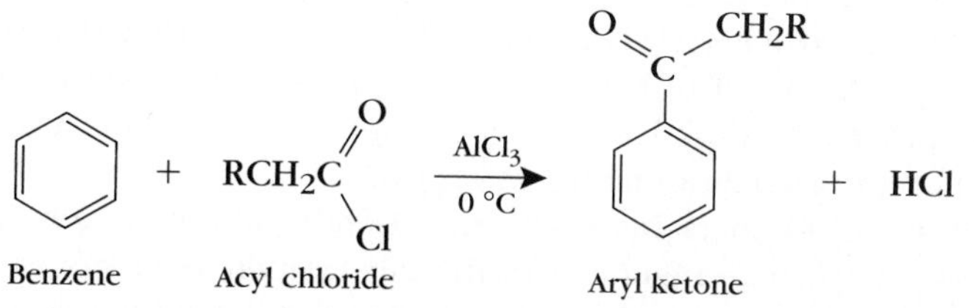

The acylation of benzene is an acyl transfer reaction because an acyl group is transferred from a chlorine atom to a ring carbon atom. This general reaction, which is called a **Friedel-Crafts acylation reaction,** is an excellent way to synthesize ketones in which the carbon of the carbonyl group is bonded to an aromatic ring. This class of ketones is called **aryl ketones.**

Exercise 21.10   Use a Friedel-Crafts acylation reaction to synthesize each of the following aryl ketones.

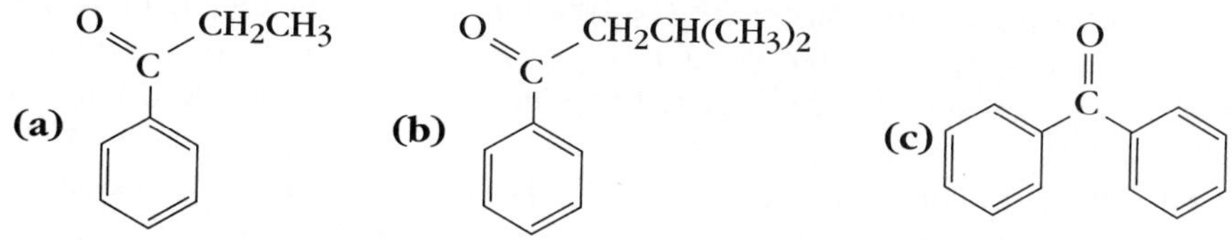

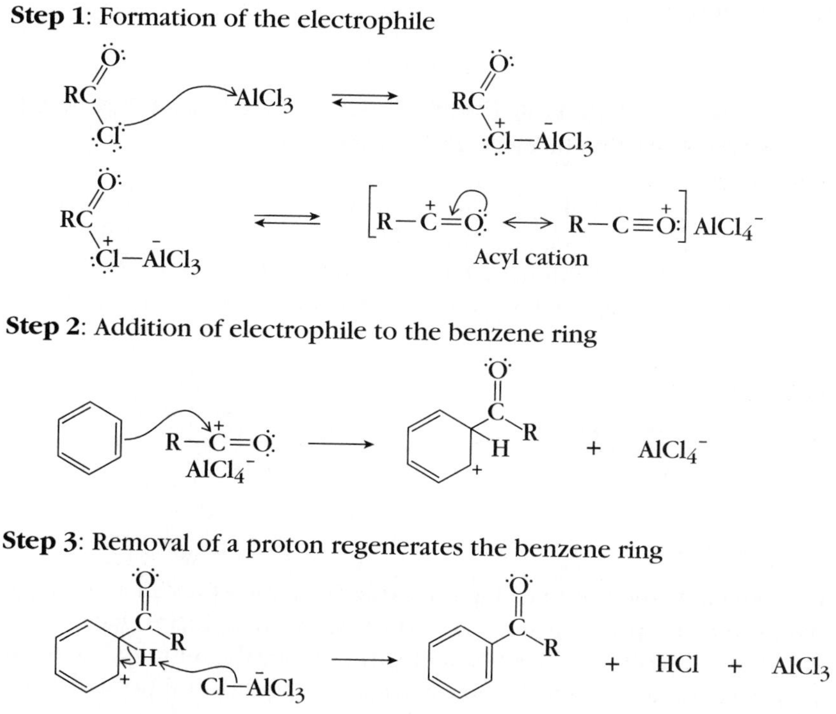

Step 1: Formation of the electrophile

Figure 21.13
The mechanism of the Friedel-Crafts acylation reaction.

Step 2: Addition of electrophile to the benzene ring

Step 3: Removal of a proton regenerates the benzene ring

The mechanism of the Friedel-Crafts acylation reaction is shown in Figure 21.13. Step 1 is the reaction of an acyl chloride with AlCl$_3$ to form a complex. The complex dissociates to yield a small equilibrium concentration of the electrophile, which is an acyl cation.

The structure of the acyl cation can be written as a resonance hybrid of the two resonance structures shown in Figure 21.13. The carbonyl carbon atom of the acyl cation is the electrophilic site, and it adds to benzene in Step 2 in the same way as other electrophiles. Step 3 is removal of a proton from the carbocation by a base to regenerate the benzene ring.

There are two differences between Friedel-Crafts alkylation and acylation that are noteworthy. First, rearrangement of the electrophile does not occur in acylation reactions. Consequently, the acyl cation is transferred to the benzene ring intact.

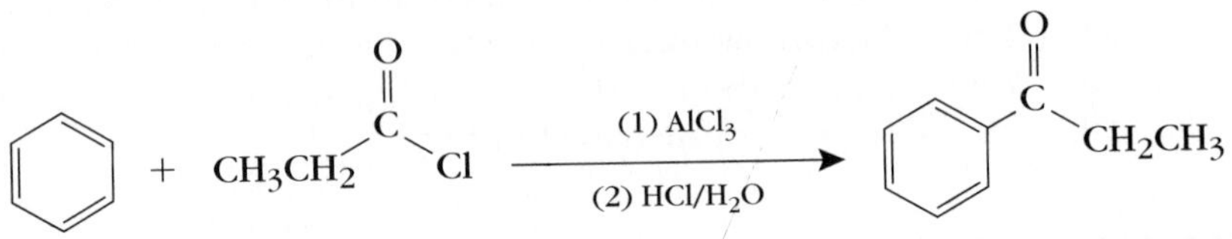

1-Phenyl-1-propanone
(85% yield)

Second, multiple substitutions of the benzene ring do not occur in acylation reactions because the acylated product is always less reactive than the starting material. The reasons why the acylated product is less reactive are discussed in Section 21.11.

Acylation is the more synthetically useful Friedel-Crafts reaction because one substitution product is formed without rearrangement. Moreover, the carbonyl group can be converted in one or more steps to a variety of other functional groups. Consequently, Friedel-Crafts acylation reactions can be used to prepare derivatives of aromatic compounds that cannot be prepared by Friedel-Crafts alkylation reactions.

21.10 Acylation Is the First Step in the Synthesis of Many Derivatives of Benzene

Problems of polysubstitution and rearrangement in the synthesis of alkylbenzenes by Friedel-Crafts alkylation can be avoided by using a sequence of two reactions. First, a Friedel-Crafts acylation reaction is carried out to form an aryl ketone. Second, the carbonyl group of the aryl ketone is reduced to a methylene group (—CH$_2$—).

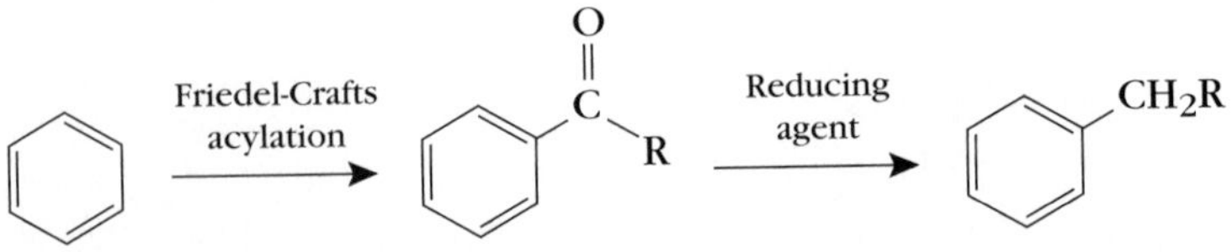

Benzene   Aryl ketone   Alkylbenzene

One general method of reducing the carbonyl group of ketones and aldehydes to a methylene group uses hydrazine (NH$_2$NH$_2$) as the reducing agent. This reaction is called the **Wolff-Kishner reduction.** It is carried out in the laboratory by heating a mixture of hydrazine, the carbonyl compound, and potassium hydroxide in a high boiling solvent such as diethylene glycol ($HOCH_2CH_2OCH_2CH_2OH$, bp 245 °C) or triethylene glycol ($HOCH_2CH_2OCH_2CH_2OCH_2CH_2OH$, bp 287 °C). A recent modification of this procedure is to use dimethyl sulfoxide [DMSO, $(CH_3)_2SO$] as the solvent and potassium *tert*-butoxide as the base. Under these conditions, the reaction occurs at much lower temperatures (between 25 and 100 °C).

■ *Hydrazine is available commercially as an 85% aqueous solution called hydrazine hydrate.*

Ethyl phenyl ketone	Hydrazine (85% aqueous solution)		1-Phenylpropane

Alternatively, a carbonyl group of an aldehyde or a ketone can be reduced to a methylene group using the **Clemmensen reduction.** In the Clemmensen reaction, the carbonyl-containing compound is treated with concentrated hydrochloric acid and a catalyst of zinc amalgam, Zn(Hg).

■ *Zinc amalgam, Zn(Hg), is an alloy of zinc dissolved in mercury.*

1-Phenyl-1-hexanone   1-Phenylhexane

The reduction of a **thioacetal,** the sulfur analog of an acetal (Section 14.11), is a third method of reducing a carbonyl group to a methylene group. Thioacetals are formed by the reaction of thiols, the sulfur analogs of alcohols, with aldehydes and ketones. The formation of the thioacetal is acid catalyzed, but a Lewis acid such as BF_3 in diethyl ether is used instead of a proton catalyst. In practice, a dithiol such as 1,2-ethanedithiol is usually used to form a cyclic thioacetal.

Aldehyde or ketone	1,2-Ethane-dithiol	Cyclic thioacetal	Reduced product

Once the thioacetal is formed, a specially prepared metal catalyst called **Raney nickel** (Raney Ni) is used to replace the sulfur atoms of the thioacetal with hydrogen atoms. This reaction is called **desulfurization.** The hydrogens in the desulfurization reaction come from hydrogen gas, which is absorbed on the surface of the Raney nickel catalyst during its preparation.

The three methods for reducing carbonyl groups of aldehydes and ketones to methylene groups complement each other. Their major differences are the conditions under which the reactions occur. Wolff-Kishner reduction occurs in strongly basic solution, Clemmensen reduction occurs in strongly acid solution, and the desulfurization of thioacetals occurs in neutral solution. The ability to carry out this reduction reaction at any pH level makes the two-step sequence of acylation-reduction a particularly valuable method of preparing alkylbenzenes.

Exercise 21.11	By analogy with the mechanism of the formation of acetals by the acid-catalyzed reaction of alcohols and aldehydes and ketones, write the mechanism of the formation of a thioacetal by the Lewis acid-catalyzed reaction of methanethiol (CH_3SH) and acetophenone.

The products of Friedel-Crafts acylation reactions are useful synthetic intermediates because their carbonyl groups can be converted into a wide variety of other functional groups. Aryl ketones, for example, undergo the usual addition reactions of ketones discussed in Chapter 14, as well as condensation reactions discussed in Chapter 17. The Friedel-Crafts acylation reaction, therefore, is the first step in the synthesis of benzene rings with side chains that contain a wide variety of functional groups.

Exercise 21.12	Starting with acetophenone and any other organic compound, inorganic reagent, and solvent, propose a synthesis for each of the following compounds.

(a) $C_6H_5CH_2CH_3$

(b) $C_6H_5\overset{\overset{\displaystyle OH}{|}}{C}HCH_3$

(c) $C_6H_5\overset{\overset{\displaystyle OH}{|}}{\underset{\underset{\displaystyle C_2H_5}{|}}{C}}CH_3$

(d) $C_6H_5\overset{\overset{\displaystyle Br}{|}}{C}HCH_3$ (2 methods)

(e) $C_6H_5\overset{\overset{\displaystyle O}{\|}}{C}CH_2CH_2OH$

We have learned that acylation in the Friedel-Crafts reaction never occurs more than once on a benzene ring. The reason for this is discussed next.

21.11 Effects of Substituents on Further Electrophilic Substitution

Electrophilic substitution reactions on a benzene ring that already contains a substituent can form *ortho-, meta-,* and *para-* disubstituted products. If substitution were random, the three products would form in the ratio 2:2:1. Experimentally, it is found that substitution of a second group is not a random process; instead, the substituent already on the ring affects the product distribution.

Some monosubstituted benzenes react to form mainly *ortho-* and *para-* disubstituted products. Examples include the products formed by nitration of toluene and bromobenzene.

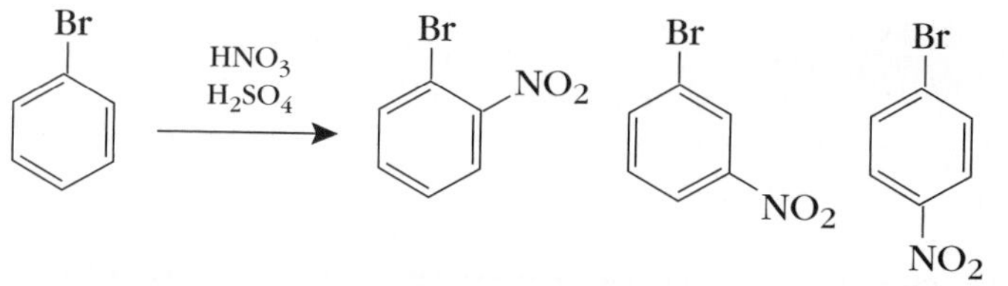

(43% yield)   (1% yield)   (56% yield)

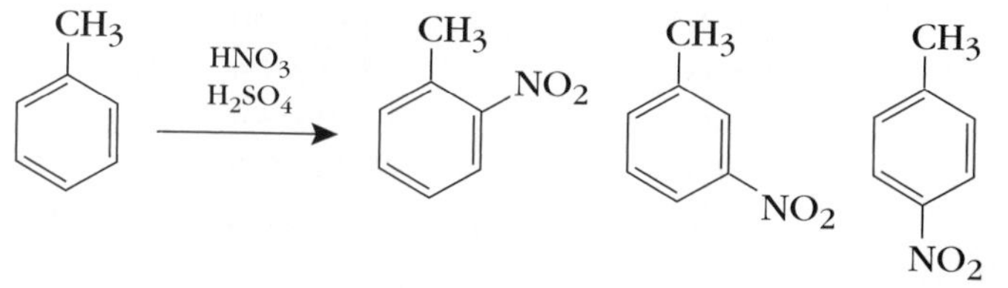

(62% yield)   (5% yield)   (33% yield)

When a monosubstituted benzene undergoes further electrophilic substitution predominantly at the *ortho-* and *para-* positions, the original substituent is called an ***ortho, para*-directing group.** Thus methyl groups and bromine atoms are *ortho, para*-directors.

Other monosubstituted benzenes undergo electrophilic substitution reactions to form mostly *meta-* disubstituted products with little or no *ortho-* and *para-* products.

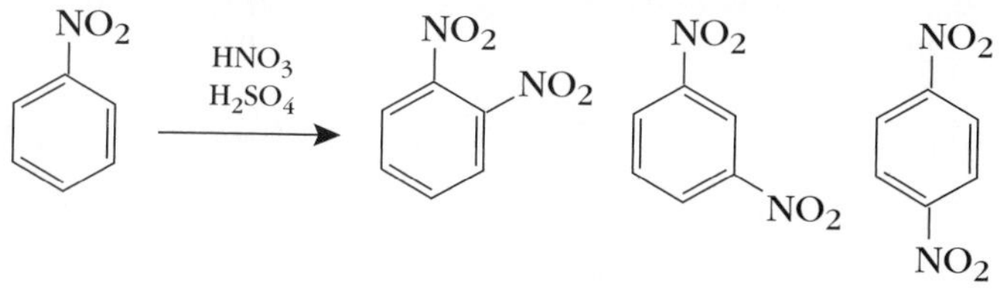

(7% yield)   (91% yield)   (2% yield)

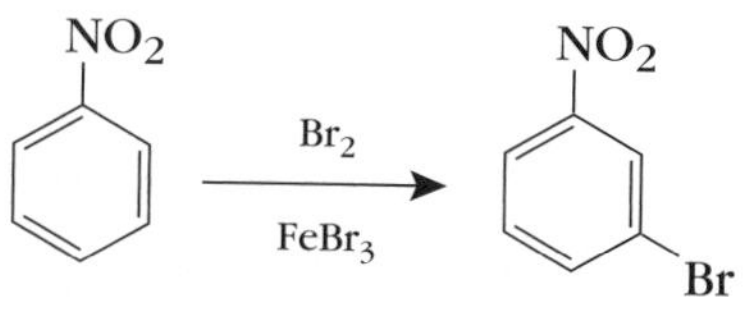

Only product

A substituent on a monosubstituted benzene that directs an electrophile to a position *meta* to itself is called a ***meta*-directing group.** The nitro group is a *meta*-directing group.

All substituents can be categorized as either *ortho, para*-directing or *meta*-directing. No substituent is solely *ortho*-directing; *para*-directing; *ortho, meta*-directing; or *meta, para*-directing.

Another way to characterize substituent groups is by their effect on the rate of further substitution reactions. If a monosubstituted benzene reacts more slowly than benzene itself, its substituent is said to be a **deactivating group.** Nitro groups are deactivating because nitrobenzene reacts slower than benzene in all electrophilic substitution reactions. The bromination of nitrobenzene, for example, is almost 100,000 times slower than the bromination of benzene.

Substituents that contain a carbonyl group are also deactivating groups. Thus

| Table 21.1 | Summary of Effects of the Most Common Substituents on Reactivity and Product Composition* | | |

Substituent	Name	Directing Effect	Effect on Rate
$-NH_2$	Amino	*ortho, para*	Very strongly activating
$-NHR$	Alkylamino	*ortho, para*	
$-NR_2$	Dialkylamino	*ortho, para*	
$-OH$	Hydroxy	*ortho, para*	
$-OR$	Alkoxy	*ortho, para*	Strongly activating
$-NHC\overset{O}{\underset{R}{\|}}$	Acylamino	*ortho, para*	
$-OC\overset{O}{\underset{R}{\|}}$	Acyloxy	*ortho, para*	
$-R$	Alkyl	*ortho, para*	Activating
$-Ar$	Aryl	*ortho, para*	
$-H$	Hydrogen		Standard
$-F, -Cl, -Br, -I$	Halogens	*ortho, para*	Deactivating
$-C\overset{O}{\underset{R}{\|}}$	Acyl	*meta*	Strongly deactivating
$-C\overset{O}{\underset{OH}{\|}}$	Carboxy	*meta*	
$-C\overset{O}{\underset{NH_2}{\|}}$	Carboxamido	*meta*	
$-C\overset{O}{\underset{OR}{\|}}$	Ester	*meta*	
$-SO_3H$	Sulfonic acid	*meta*	
$-CN$	Cyano	*meta*	
$-CF_3$	Trifluoromethyl	*meta*	
$-NO_2$	Nitro	*meta*	Very strongly deactivating

*Groups are arranged in decreasing order of activation.

the products of Friedel-Crafts acylation reactions are less reactive than benzene to further substitution. This is why it is easy to stop the acylation reaction after formation of the monosubstituted product. In fact, aromatic rings that contain a strongly deactivating group do not undergo Friedel-Crafts acylation reactions. The nitro group is so deactivating, for example, that nitrobenzene is often used as a solvent in Friedel-Crafts reactions.

If a monosubstituted benzene reacts faster than benzene in an electrophilic aromatic substitution reaction, its substituent is called an **activating group.** Methyl groups are activating groups. Nitration of toluene, for example, is 45 times faster than benzene under the same experimental conditions. Toluene reacts faster than benzene in all other electrophilic substitution reactions, too.

All alkyl groups are activating groups, so monoalkylbenzenes are more reactive than benzene in electrophilic substitution reactions. As a result, the products of Friedel-Crafts alkylation reactions react faster than benzene to form polysubstituted products. Alkyl groups are sufficiently activating that it is difficult to avoid polysubstituted products in Friedel-Crafts alkylation reactions.

Table 21.1 summarizes the directing effects of the most common substituents and whether they are activating or deactivating relative to benzene. The substituents in Table 21.1 may be placed into three groups:

• Activating and *ortho, para*-directing groups. Functional groups in this category include the first nine entries in Table 21.1.
• Deactivating and *ortho, para*-directing groups. The halogens are the only functional groups in this category given in Table 21.1.
• Deactivating and *meta*-directing groups. The last eight entries in Table 21.1 are in this category.

The data in Table 21.1 show that all activating groups are *ortho, para*-directing and all *meta*-directing groups are deactivating. These generalizations come from a great deal of experimental data and form a set of empirical rules that can be used to design organic syntheses.

Exercise 21.13   Write the structure of the major organic product or products, if any, in each of the following reactions. Indicate whether the reaction occurs faster or slower than with benzene.

(a) Toluene + fuming sulfuric acid
(b) Fluorobenzene + HNO_3/H_2SO_4
(c) Benzaldehyde + acetyl chloride/$AlCl_3$
(d) Acetophenone + $Br_2/FeBr_3$
(e) Benzonitrile + $Cl_2/FeCl_3$
(f) Benzoic acid + HNO_3/H_2SO_4
(g) Ethyl benzoate + $Cl_2/FeCl_3$
(h) Anisole + acetyl chloride/$AlCl_3$

Why do these substituents have such effects on electrophilic aromatic substitution reactions?

21.12 Electron Theory of Substituent Effects

To understand the effect of substituents, we must examine Step 2 of the mechanism of electrophilic aromatic substitution reactions. Recall from Section 21.4 that Step 2, which is rate determining, is the addition of an electrophile to the π electron system of a benzene ring to form an intermediate carbocation. The rate-determining step is an endergonic reaction because the aromaticity of the benzene ring is lost on forming the carbocation.

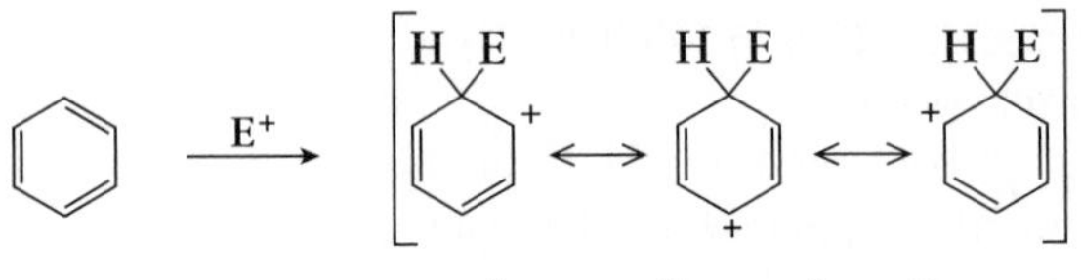

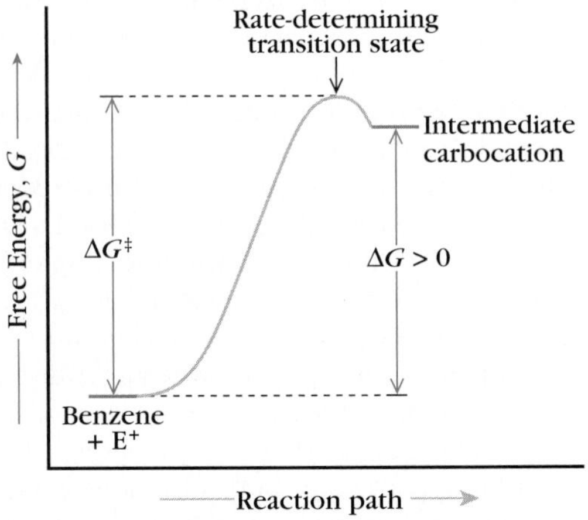

The relative rates of reaction at the *ortho, meta,* and *para* positions of a monosubstituted benzene are determined by their relative free energies of activation. For a substitution reaction of a monosubstituted benzene, the differences in the free energies of activation are determined only by the free energies of the cation-like transition states because the same monosubstituted benzene is the starting material for reaction at all three positions. According to the Hammond Postulate (Section 7.11), the structure and free energy of the transition state in the endergonic Step 2 of the mechanism is similar to that of the carbocation intermediate. Factors that affect the relative free energy of this intermediate also affect the free energy of the transition state leading to that intermediate. We can understand the role of substituents, therefore, by assessing their effects on the relative free energies of the carbocation intermediates formed by reaction at each of the three positions. We begin by examining the effect of substituents that are activating and *ortho, para*-directing.

A   Activating and *ortho, para*-Directing Substituents

The methyl group of toluene is an activating and *ortho, para*-directing group. The structure of the intermediate formed by the addition of any electrophile, E^+, to the *ortho* position of toluene can be described as a resonance hybrid of the following resonance structures.

ortho Addition:

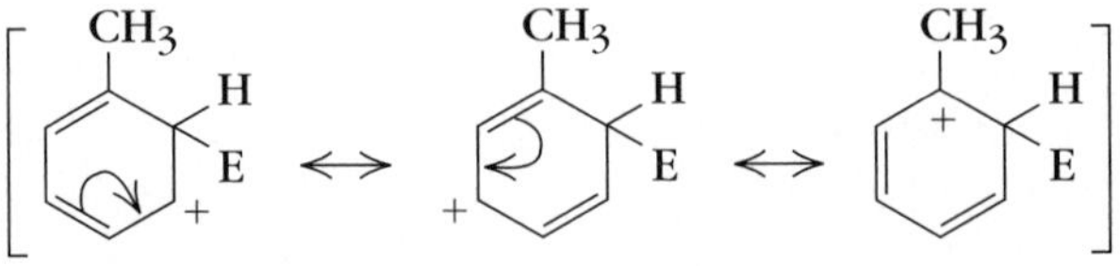

In a similar way we can describe the structures of the intermediates formed by addition of a nucleophile E^+ to the *para* and *meta* positions of toluene as resonance hybrids.

para Addition:

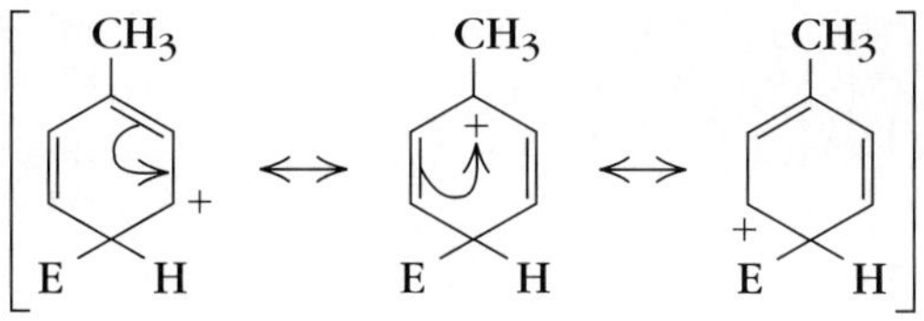

meta Addition:

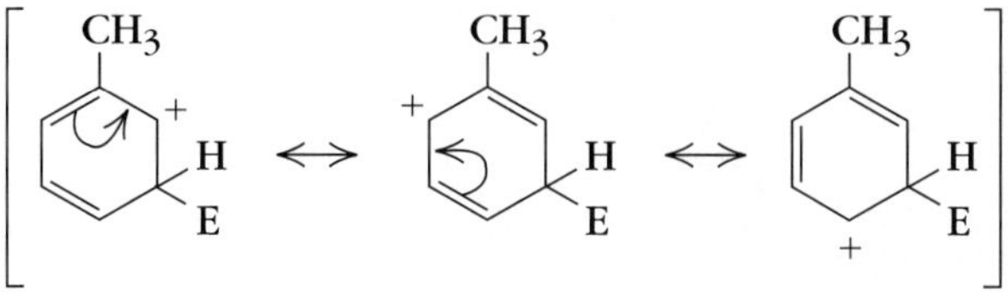

The resonance hybrid structures show that three of the six carbon atoms of the carbocation intermediates are electron deficient. In the carbocations formed by addition to the *ortho* and *para* positions of toluene, the methyl group is bonded to one of the electron-deficient carbon atoms. In the carbocation formed by addition to the meta position of toluene, in contrast, the methyl group is *not* bonded to one of the electron-deficient carbon atoms. An electron-donating methyl group stabilizes an adjacent electron-deficient carbon atom more than an electron-deficient carbon atom that is one carbon atom removed.

■ *Recall from Section 7.10 that a methyl group stabilizes an adjacent electron-deficient carbon atom.*

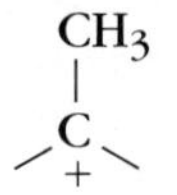

Stabilization of adjacent electron-deficient carbon atom by a methyl group

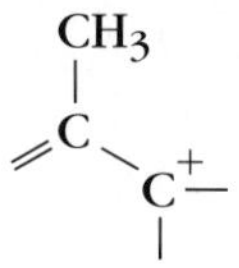

Less stabilization of electron-deficient carbon atom by a methyl group

As a result, the intermediate carbocations formed by addition of an electrophile to the *ortho* and *para* positions are stabilized more by the methyl group than is the intermediate formed by *meta* addition. Consequently, the intermediate carbocations for addition *ortho* and *para* and their transition states are more stable and form faster than the corresponding intermediate and transition state leading to *meta* substitution. Therefore the *ortho* and *para* positions of toluene are more reactive to electrophilic substitution than its *meta* position.

In Section 7.10, we discussed how a methyl group stabilizes a carbocation. In electrophilic aromatic substitution reactions, a methyl group is an activating substituent because it stabilizes all the carbocation intermediates more than a hydrogen atom stabilizes the corresponding carbocation intermediate formed in the addition to benzene. As a result the intermediates formed in addition to toluene (and hence their transition states) are more stable and react more quickly than the corresponding intermediate and transition state for benzene. The difference in free energies of activation for electrophilic substitution of toluene and benzene is illustrated in Figure 21.14.

Figure 21.14
Free energy-reaction path diagram for the rate-determining steps in the mechanism of electrophilic substitution to benzene and toluene. The free energies of activation for addition of an electrophile to any position of toluene are less than the free energy of activation for addition to benzene. Thus electrophiles react faster with toluene than benzene. The free energies of activation for addition to the *ortho* and *para* positions are less than the free energy of activation for addition to the *meta* position, so substitution occurs preferentially at the *ortho, para* positions. For comparison purposes, the energies of the starting material, benzene, toluene, and the electrophile, are placed at the same free energy level.

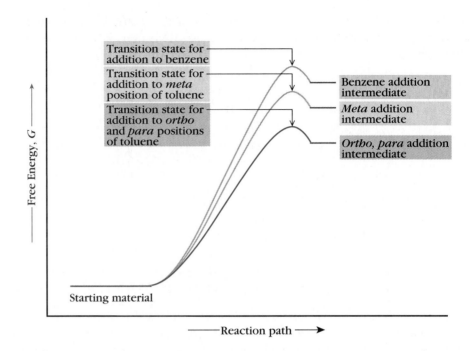

These principles apply equally well to other types of activating and *ortho, para*-directing substituents. For example, the electron structure of the carbocation intermediates formed by addition of an electrophile to the *ortho* and *para* positions of anisole can be represented as a resonance hybrid of the following resonance structures.

ortho Addition:

Oxonium ion

para Addition:

Oxonium ion

The electron structure of the carbocation intermediate formed by addition to the *meta* position of anisole can be described as a hybrid of the following resonance structures.

meta Addition:

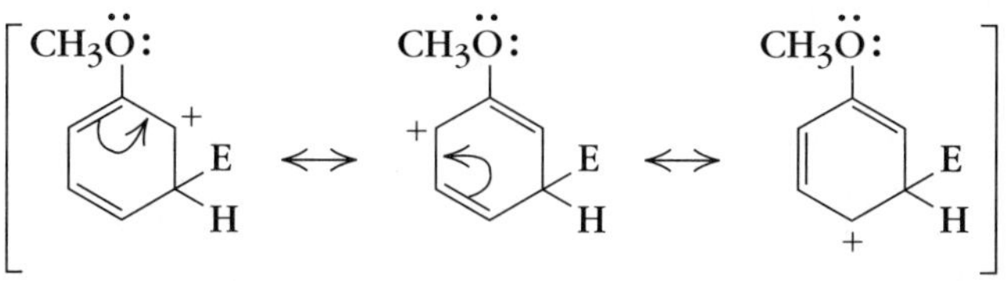

In the carbocations formed by addition to the *ortho* or *para* positions of anisole, the methoxy group is bonded to one of the electron-deficient carbon atoms. In the carbocation formed by addition to the *meta* position of anisole, in contrast, the methoxy group is *not* bonded to one of the electron-deficient carbon atoms.

A methoxy group stabilizes an adjacent electron-deficient carbon atom by donating a pair of electrons from its oxygen atom. This electron donation is represented by the oxonium ion resonance structure in the hybrid structure of additions to the *ortho* and *para* positions. No such electron donation by the methoxy group can occur in the intermediate for addition to the *meta* position. As a result, the methoxy group stabilizes the intermediate formed by addition to the *ortho* and *para* positions more than the intermediate formed by *meta* addition. Consequently, the intermediate carbocations for addition *ortho* and *para* and their transition states are more stable and form faster than the corresponding intermediate and transition state leading to *meta* substitution. Therefore, the *ortho* and *para* positions of anisole are more reactive to electrophilic substitution than is its *meta* position.

B   Deactivating and *meta*-Directing Substituents

Deactivating and *meta*-directing substituents all have one thing in common. The atom of the substituent bonded to the aromatic ring is electron deficient. This feature is illustrated in Figure 21.15, which shows the resonance structures that contribute to the electron structures of deactivating *meta*-directing substituents.

The *meta*-directing effect of the groups shown in Figure 21.15 can be explained by an argument similar to that used in the case of *ortho, para*-directing

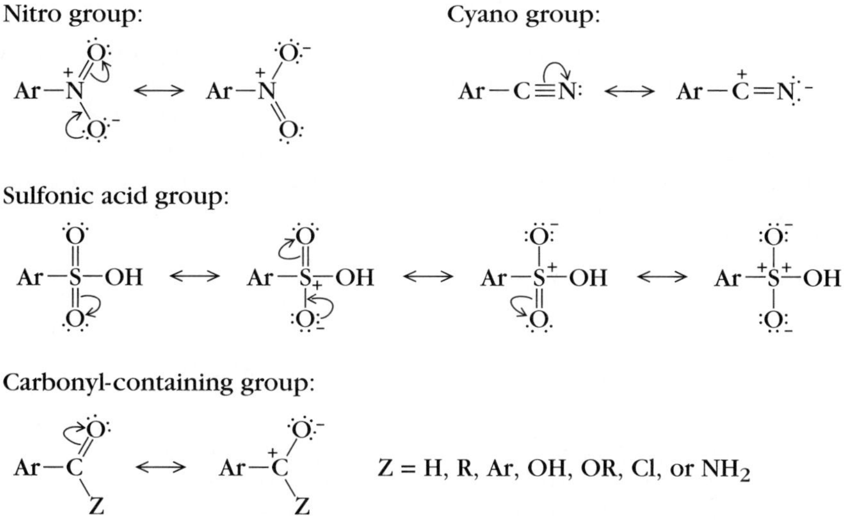

Figure 21.15
Contributing structures to the resonance hybrids of deactivating, *meta*-directing substituents. Each of the substituents has an electron-deficient atom bonded to the aromatic ring.

and activating substituents. We begin by examining the electron structures of the three carbocation intermediates formed in Step 2 of the mechanism of electrophilic substitution reactions of nitrobenzene.

ortho Addition:

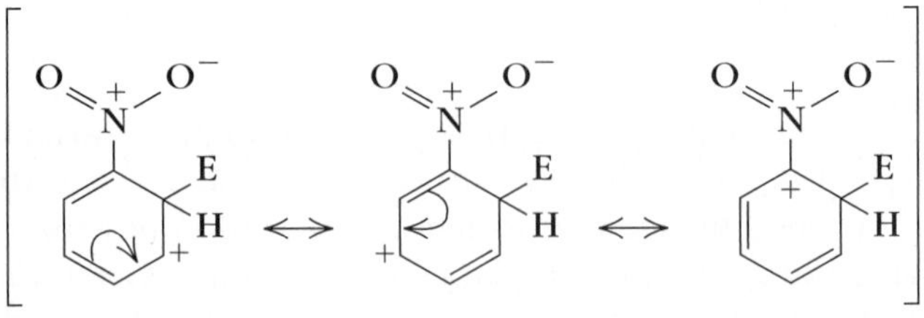

para Addition:

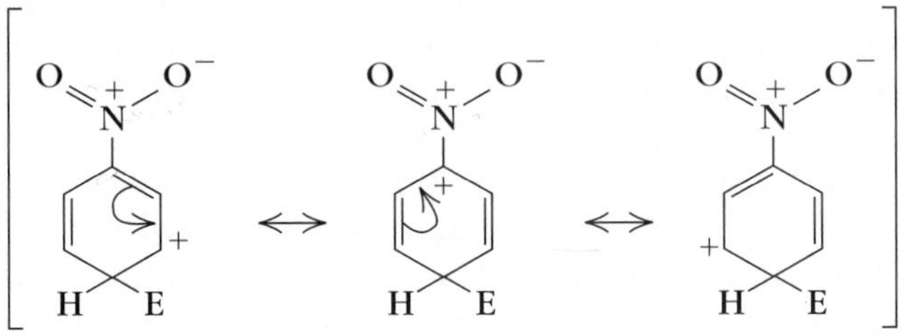

meta Addition:

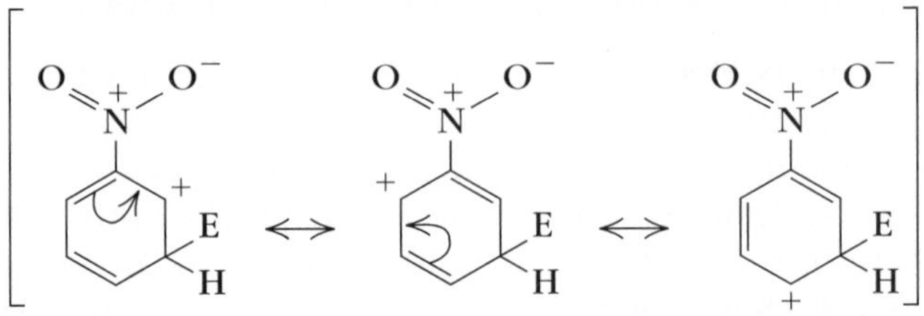

Just like in the additions to toluene and anisole, the substituent is bonded to an electron-deficient carbon atom in the intermediate formed by electrophilic additions to the *ortho* and *para* positions. Also, the substituent is not bonded to an electron-deficient carbon atom in the intermediate formed by addition to the *meta* position. Replacing a hydrogen atom of a carbocation by an electron-withdrawing nitro group destabilizes the carbocation. The carbocation is destabilized because the positive charge of the nitrogen atom of the nitro group is placed adjacent to the electron-deficient carbon atom. A nitro group destabilizes an adjacent electron-deficient carbon atom more than an electron-deficient carbon atom that is one carbon atom removed.

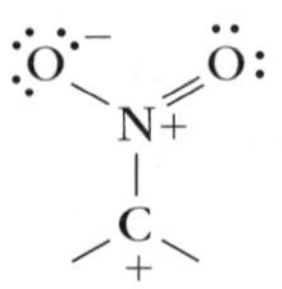

Destabilization of adjacent
electron-deficient carbon
atom by a nitro group

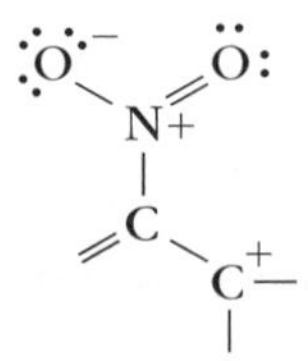

Less destabilization of
electron-deficient carbon
atom by a nitro group

As a result, the intermediate carbocations formed by addition to the *ortho* and *para* positions are destabilized more by a nitro group than is the intermediate formed by *meta* addition. Consequently, the intermediate carbocations for addition *ortho* and *para* and their transition states are less stable and form slower than the corresponding intermediate and transition state leading to *meta* substitution. Therefore, the *meta* position of nitrobenzene is more reactive to electrophilic substitution than its *ortho* and *para* positions.

A nitro group, and the other substituents in Figure 21.15 as well, deactivate a benzene ring because their adjacent full or partial positive charges make *all* the carbocation intermediates and the transition states leading to them less stable than the carbocation intermediate formed by addition to benzene. Thus deactivating substituents deactivate all positions on a ring compared to an unsubstituted benzene ring. The *meta* position, however, is deactivated less than either the *ortho* or *para* positions, as shown in the free energy–reaction path diagram in Figure 21.16.

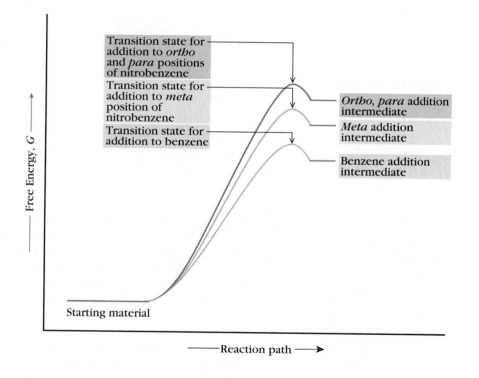

Transition state for addition to *ortho* and *para* positions of nitrobenzene

Transition state for addition to *meta* position of nitrobenzene

Transition state for addition to benzene

Ortho, para addition intermediate

Meta addition intermediate

Benzene addition intermediate

Free Energy, *G*

Starting material

Reaction path

Figure 21.16

Free energy–reaction path diagram for the rate-determining steps in the mechanism of electrophilic substitution to benzene and nitrobenzene. The free energy of activation for addition to the *meta* position is smaller than the free energies of activation for addition to the *ortho* and *para* positions, so electrophilic addition occurs preferentially at the *meta* position of nitrobenzene. The free energy of activation for addition of an electrophile to any position of nitrobenzene is greater than the free energy of activation for addition to benzene. Thus electrophiles react slower with nitrobenzene than benzene. For comparison purposes the energy of the starting material, benzene, nitrobenzene, and the electrophile, are placed at the same energy level.

C   The Halogens: *ortho, para*-Directing, and Deactivating

Halogens differ from other substituents because they are deactivating and *ortho, para*-directing substituents. How can we explain this? To answer this question we must again examine the electron structures of the intermediate carbocations formed by addition of an electrophile at the *ortho, para,* and *meta* positions of chlorobenzene.

ortho Addition:

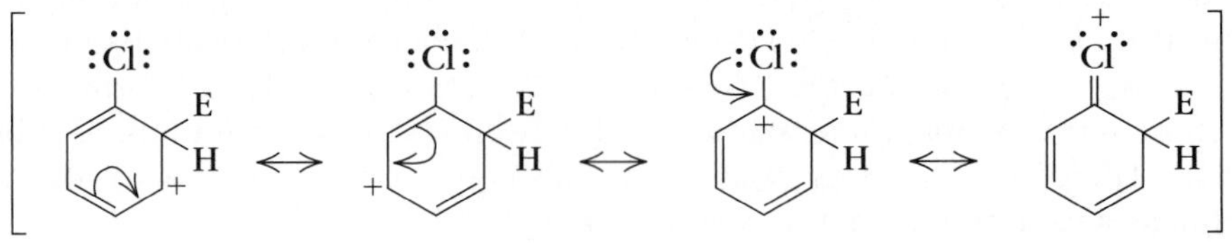

Delocalization
of charge onto
chlorine atom

para Addition:

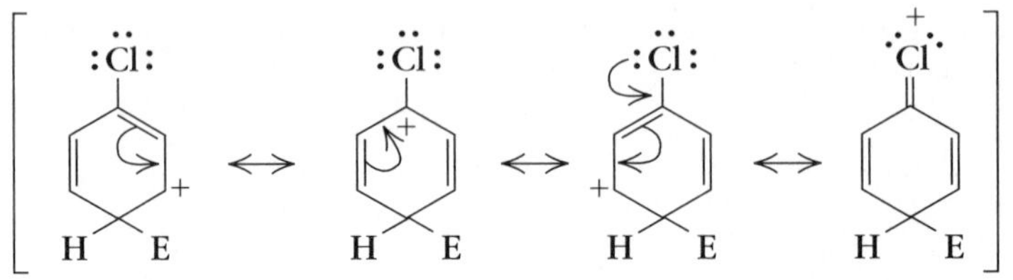

Delocalization
of charge onto
chlorine atom

meta Addition:

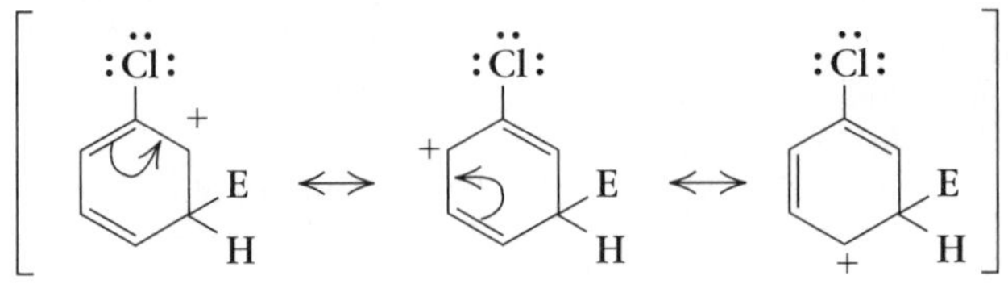

Halogens, like the oxygen atoms of a methoxy group, have unshared electrons that can be donated to an adjacent positively charged carbon atom. Donating electrons in this way stabilizes the intermediate formed by addition to the *ortho* and *para* positions. No comparable structures can be written for the intermediate formed by addition to the *meta* position, so halobenzenes react faster with electrophiles at the *ortho* and *para* positions. Halogens, therefore, are *ortho, para*-directing substituents.

Halogens are electronegative elements so they inductively withdraw electrons away from the ring carbon atom to which they are attached. The electron-withdrawing ability of a halogen destabilizes all the carbocation intermediates and the transition states leading to them relative to the transition state for formation of the carbocation in the reaction with benzene. As a result, halobenzenes are less reactive than benzene in electrophilic substitution reactions. Halogens, however, are not as good as oxygen or nitrogen atoms at delocalizing the positive charge on an adjacent electron-deficient carbon atom, as we learned in Section 16.7. While electron donation from the halogen atom makes the *ortho* and *para* positions more reactive than the *meta* position, it does not compensate for the deactivating effect caused by the inductive electron-withdrawing ability of the halogens. The net effect is that halogens are both deactivating and *ortho, para*-directing substituents.

■ *The ability of halogens to both stabilize an adjacent electron-deficient carbon atom by donating a pair of nonbonding electrons and destabilize it by an electron-withdrawing effect was first encountered in Section 16.7.*

| **Exercise 21.14** | Write the structures that contribute to the resonance hybrid electron structure of the carbocation intermediates formed by addition of an electrophile, E^+, to one ring of biphenyl, [biphenyl structure]. Explain why a phenyl group is an activating and *ortho, para*-directing group. |

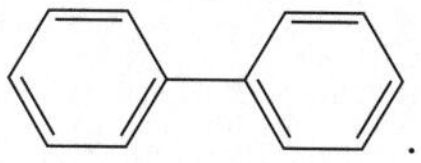

21.13 Synthetic Uses of Electrophilic Aromatic Substitution Reactions

Most disubstituted benzene derivatives are synthesized from benzene by a sequence of two electrophilic substitution reactions. The relative positions of the two substituents in the final product are determined by the substituent introduced in the first reaction. For example, in order to prepare *p*-bromonitrobenzene from benzene, we need to carry out a nitration and a bromination. Which one of these reactions should we carry out first?

To answer this question we must work backwards from the product. The two substituents in the product are located *para* to each other. Therefore the first substituent introduced must be an *ortho, para*-director. Because a nitro group is a *meta*-director while bromine is an *ortho, para*-director, bromination should be carried out first.

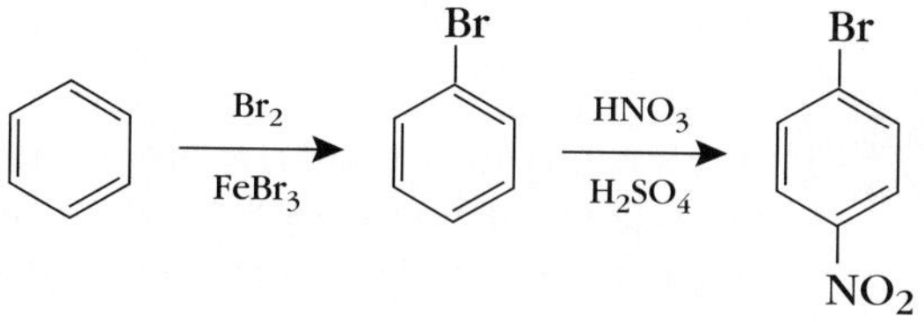

| **Exercise 21.15** | What product is formed when benzene undergoes nitration followed by bromination? |

Next consider the synthesis of *m*-nitroacetophenone from benzene. The two substituents in the product are located *meta* to each other. Because both a nitro group and an acetyl group are *meta*-directing, it may seem that either nitration or a Friedel-Crafts acylation can be carried out in the first reaction. Nitro groups are deactivating, however, and recall from Section 21.9 that Friedel-Crafts reactions do not occur on highly deactivated rings. Therefore, the Friedel-Crafts reaction must be carried out first.

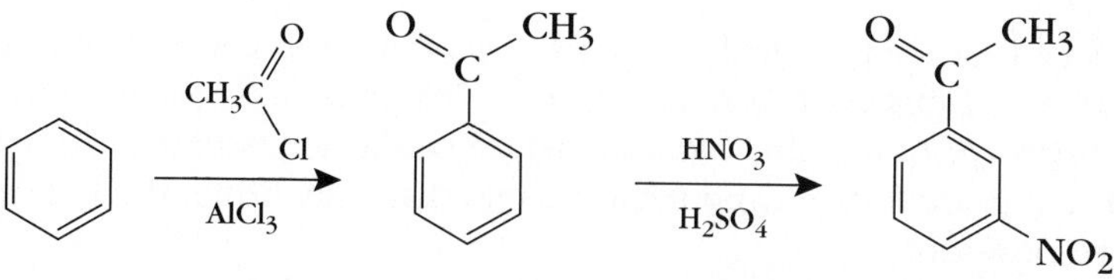

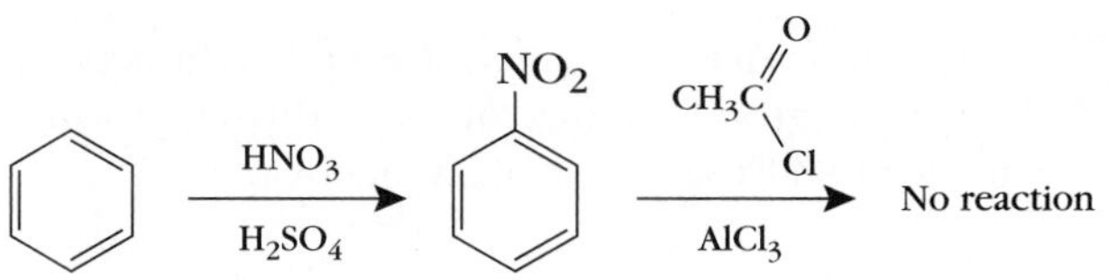

Sometimes a group is introduced in the first reaction only because of its directing ability in the second reaction. After it has done its job, the group is then chemically transformed into the desired substituent. In the synthesis of *m*-bromopropylbenzene from benzene, for example, both bromine and the propyl group are *ortho*, *para*-directors. Therefore neither bromination of propylbenzene nor alkylation of bromobenzene will form a product with substituents *meta* to each other. An alkyl group, however, can be prepared by a combination of Friedel-Crafts acylation followed by reduction, and acyl groups are *meta*-directors.

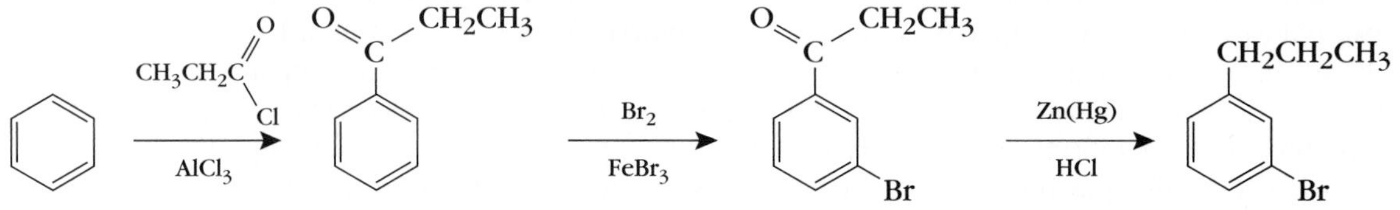

Thus introducing the appropriate group in the first reaction makes it possible to brominate the *meta* position in the second reaction. Once the acyl group has done its job in the second reaction, it is reduced to an alkyl group by the Clemmensen reaction.

Exercise 21.16	Starting with benzene and any other organic and inorganic reagents, propose a synthesis of each of the following compounds. More than two steps may be required in some cases.

(a) *m*-Dinitrobenzene
(b) *m*-Nitroethylbenzene
(c) *p*-Nitroethylbenzene
(d) *m*-Chlorobenzenesulfonic acid
(e) *m*-Nitrobenzoic acid
(f) 1-(3-bromophenyl)ethanol

What if we want to carry out an electrophilic substitution reaction on a compound whose benzene ring has two substituents? Can we predict where the reaction will occur? The answer is surprisingly simple.

21.14 Orientation in Disubstituted Benzenes

Further substitution in a disubstituted benzene is governed by the same substituent effects discussed in Section 21.11. With two substituents, however, the directive effects of both substituents must be taken into account. Sometimes the two substituents are located about the ring so that they direct the entering group to the same position.

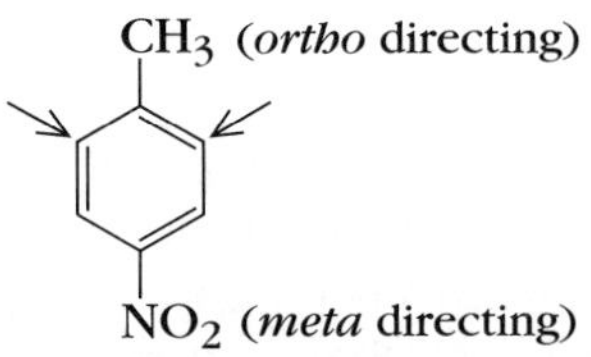

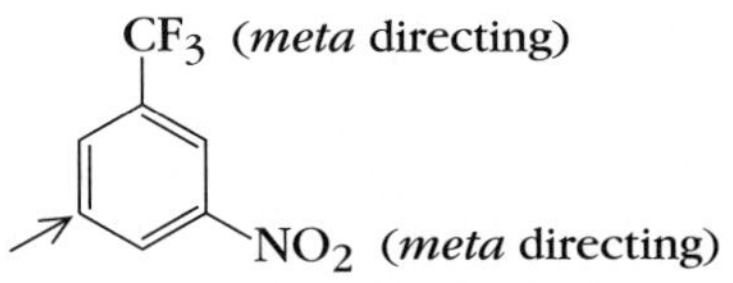

CH$_3$ (*ortho* directing)

NO$_2$ (*meta* directing)

CF$_3$ (*meta* directing)

NO$_2$ (*meta* directing)

If the two substituents are located so that their directing effects oppose each other, the more powerful activating group generally determines the position of substitution. For example, nitration of *o*-fluoroanisole occurs primarily at the positions *ortho* and *para* to the methoxy group because alkoxy groups are stronger activating groups than are halogens.

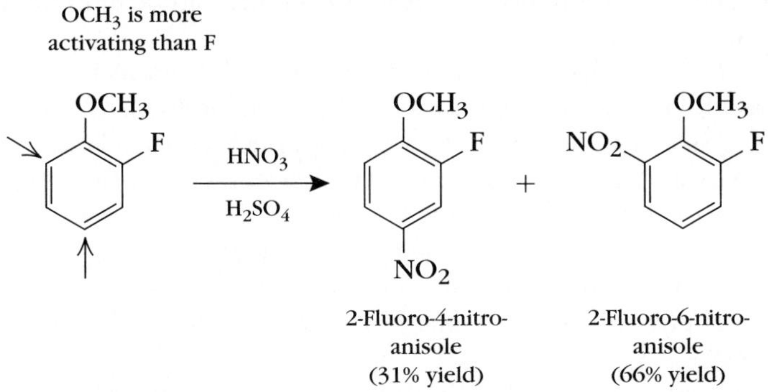

OCH$_3$ is more
activating than F

OCH$_3$ OCH$_3$ OCH$_3$

F $\xrightarrow[\text{H}_2\text{SO}_4]{\text{HNO}_3}$ F   +   NO$_2$ F

NO$_2$

2-Fluoro-4-nitro-   2-Fluoro-6-nitro-
anisole   anisole
(31% yield)   (66% yield)

ortho, para-Directing groups determine the position of further substitution because they are activating, while *meta*-directing groups are deactivating.

If both substituents have about the same activating effect, then mixtures of all possible products are usually formed, as shown in the following example.

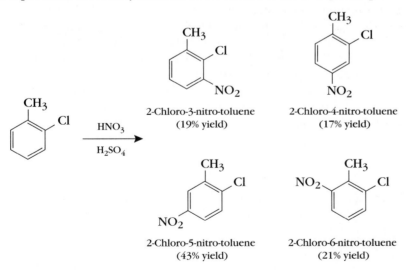

CH$_3$

Cl $\xrightarrow[\text{H}_2\text{SO}_4]{\text{HNO}_3}$

CH$_3$
Cl
NO$_2$

2-Chloro-3-nitro-toluene
(19% yield)

CH$_3$
Cl

NO$_2$

2-Chloro-4-nitro-toluene
(17% yield)

CH$_3$
Cl

NO$_2$

2-Chloro-5-nitro-toluene
(43% yield)

CH$_3$
NO$_2$ Cl

2-Chloro-6-nitro-toluene
(21% yield)

Exercise 21.17   Predict the product or products of nitration of each of the following compounds.

(a) *m*-Nitrobenzoic acid   (d) *p*-Methoxybenzonitrile
(b) *o*-Chloronitrobenzene   (e) *m*-Nitrotoluene
(c) *p*-Methylphenol

21.15 Summary

The principal reactions of aromatic compounds are **electrophilic aromatic substitution** reactions. In these reactions, a hydrogen atom on the aromatic ring is substituted by another atom or group. The hydrogen is replaced by a nitro group ($-NO_2$) in **nitration,** a sulfonic acid group ($-SO_3H$) in **sulfonation,** a halogen in **halogenation,** an acyl group in **Friedel-Crafts acylation,** and an alkyl group in **Friedel-Crafts alkylation.**

All of these reactions occur by a similar three-step mechanism. Step 1 is the formation of the electrophile. In Step 2, which is rate-determining, the electrophile adds to the π electrons of the aromatic ring to form a carbocation intermediate. In Step 3, a base removes a proton from the carbocation intermediate to regenerate the aromatic ring.

Friedel-Crafts reactions are particularly important reactions because they form carbon-carbon bonds. They suffer, however, from the following limitations:

- Strongly deactivated aromatic rings do not undergo Friedel-Crafts reactions.
- Haloalkanes and acyl chlorides react with arenes in Friedel-Crafts reactions, but haloaromatics and vinyl halides do not.
- Friedel-Crafts alkylation frequently forms products of polysubstitution.
- Rearrangement of alkyl groups can occur in Friedel-Crafts alkylation reactions, particularly with primary halides of the type RCH_2CH_2X or R_2CHCH_2X.

Substituents on the ring affect both the reactivity of the ring toward further substitution and the position of substitution. All substituents fall into one of the following three categories:

- Activating and *ortho, para*-directing.
- Deactivating and *ortho, para*-directing.
- Deactivating and *meta*-directing.

When disubstituted benzenes undergo electrophilic substitution reactions, the two groups already present exert their directive effects independently. Sometimes, the two substituents are located so that they both direct the incoming group to the same position. If their directing effects oppose each other, however, the more activating substituent determines the site of substitution. If both substituents have about the same activating effect, then a mixture of all possible products is usually formed.

21.16 Summary of Electrophilic Aromatic Substitution Reactions

1. **Halogenation**

   a. **Chlorination**   SECTION 21.5

   b. **Bromination**

c. Iodination

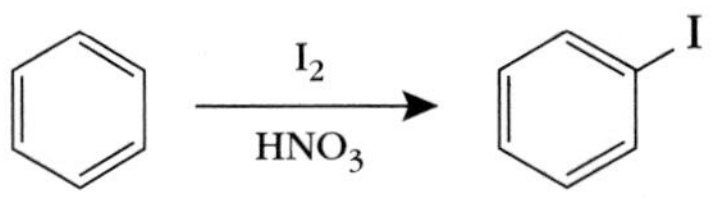

2. Nitration   SECTION 21.6

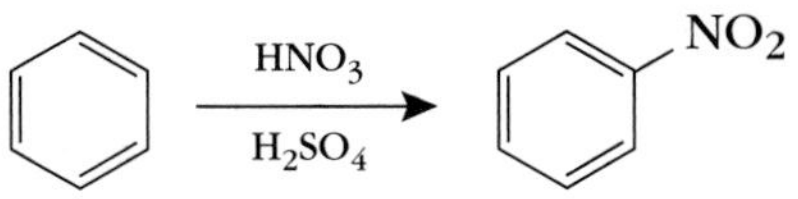

3. Sulfonation   SECTION 21.7

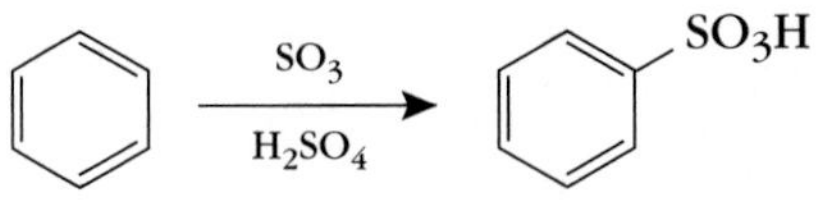

4. Friedel-Crafts Reaction
a. Alkylation   SECTION 21.8
Polyalkylated products are often formed. Rearrangement of the alkyl halide during the reaction can occur.

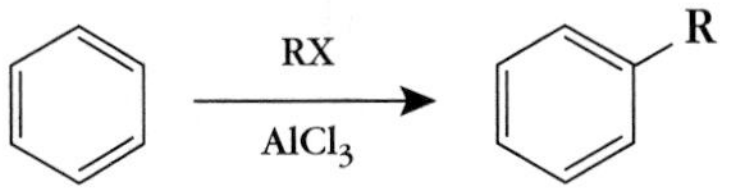

R = alkyl group

b. Acylation   SECTION 21.9

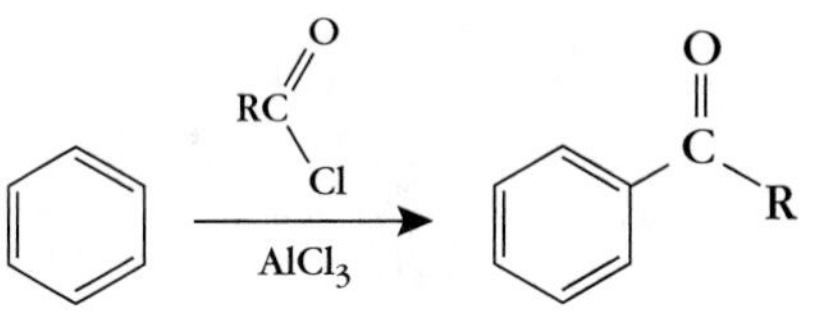

Additional Exercises

21.18 Define each of the following terms in words, by a structure, or by means of a chemical equation.

(a) An *ortho, para*-directing substituent	**(f)** Friedel-Crafts acylation reaction
(b) Friedel-Crafts alkylation reaction	**(g)** A *meta*-directing substituent
(c) An activating group	**(h)** Wolff-Kishner reduction
(d) Clemmensen reduction	**(i)** A thioacetal
(e) A deactivating group	

21.19 Write the structures and give the names of all possible tetramethylbenzenes.

21.20 There are three possible isomeric dibromobenzenes. Isomer A reacts with $Br_2/FeBr_3$ to form tribromobenzenes D, E, and F. Under the same reaction conditions, isomer B forms only tribromobenzenes D and E, while isomer C forms only tribromobenzene E. Write the structures of compounds A, B, C, D, E, and F.

21.21 Write the structures of the major organic product or products formed by the bromination, nitration, and sulfonation of each of the following compounds. Which of them reacts faster than benzene?

(a) Toluene	**(c)** Anisole	**(e)** (Trifluoromethyl)benzene, $C_6H_5CF_3$
(b) Nitrobenzene	**(d)** Chlorobenzene	

21.22 Explain why the nitroso group (—N=O) is an *ortho, para*-directing substituent.

21.23 Write the structure of the products in each of the following reactions.

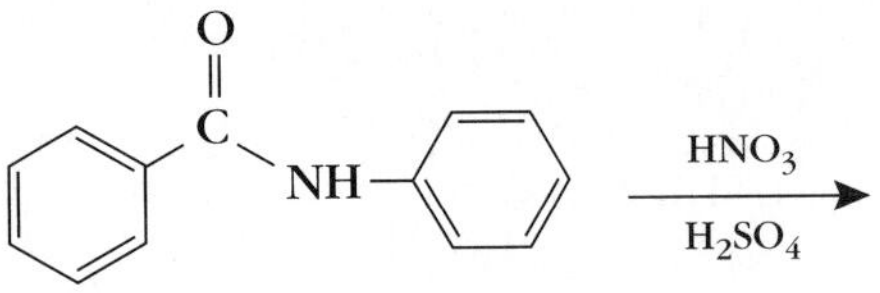

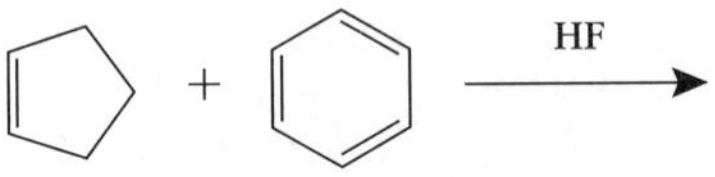

21.24 Propose a synthesis for each of the following compounds starting with benzene, any other organic compound, and any inorganic reagent. More than one step may be necessary.

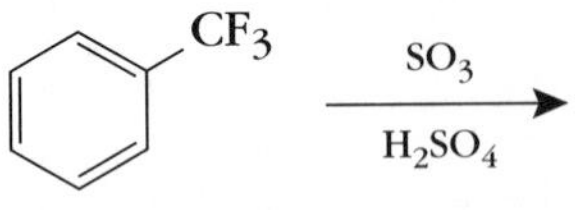

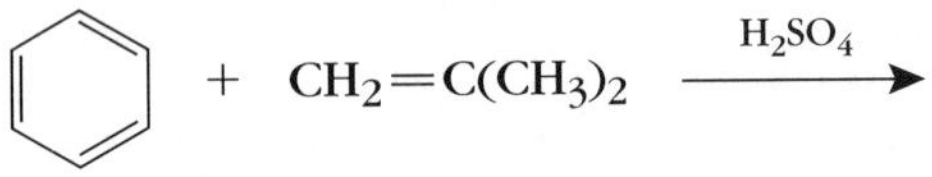

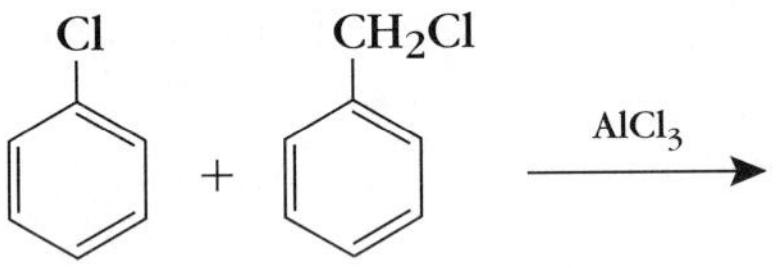

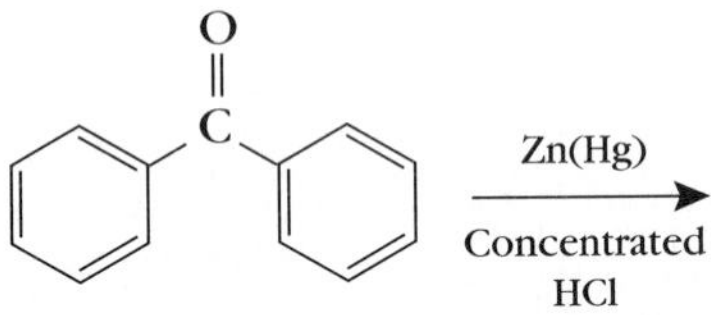

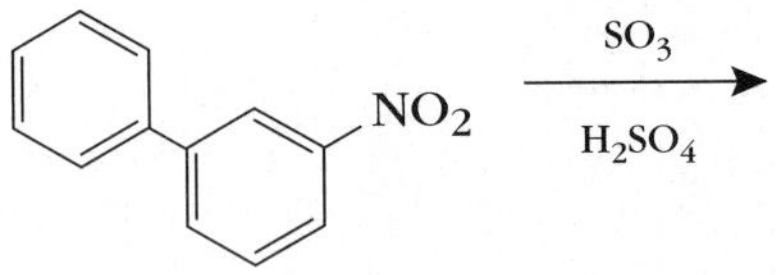

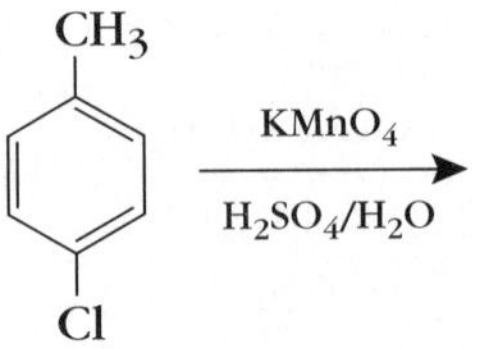

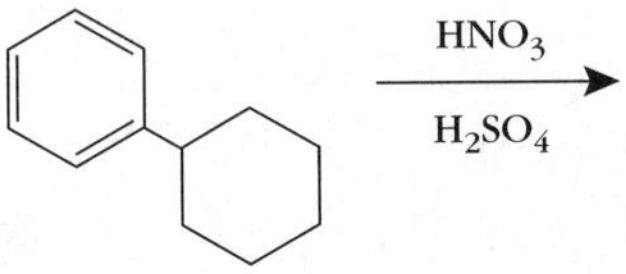

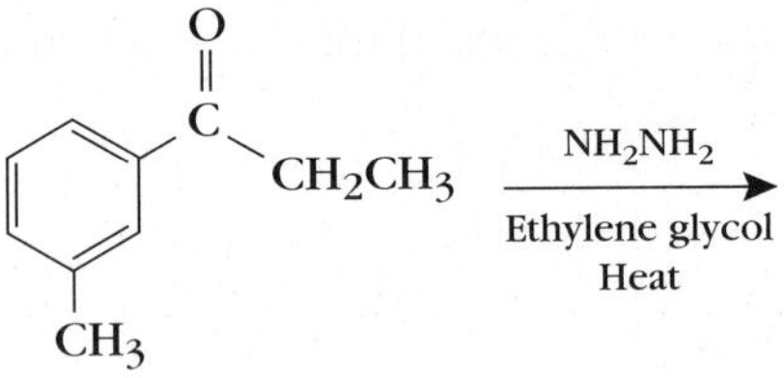

(a) Ethylbenzene
(b) *p*-Bromotoluene
(c) Styrene
(d) Benzonitrile
(e) 1-Chloro-2-methyl-2-phenylpropane
(f) 1,1-Diphenylethane
(g) *p*-Chloroacetophenone
(h) 2,4-Dinitrochlorobenzene

21.25 Arrange each of the following sets of compounds in increasing order of reactivity with HNO_3 in H_2SO_4.

(a) Nitrobenzene, benzene, chlorobenzene
(b) Phenol, *t*-butylbenzene, toluene
(c) Benzenesulfonic acid, anisole, styrene
(d) Benzoic acid, benzonitrile, acetophenone
(e) Ethyl benzoate, phenyl acetate, benzamide

21.26 Nitronium tetrafluoroborate, $NO_2^+BF_4^-$, is another nitrating agent. Write a two-step mechanism for the reaction of $NO_2^+BF_4^-$ with benzene. Write structures for all intermediates and use curved arrows to indicate electron movement.

21.27 Anisole reacts with iodine monochloride to form *p*-iodoanisole and HCl as products. Propose a mechanism for this reaction. Write structures for all intermediates and use curved arrows to indicate electron movement.

21.28 Nucleophilic substitution reactions of 2-chloro-2-phenylpropane occur by an S_N1 mechanism. From your knowledge of the effects of various substituents on aromatic substitution reactions, predict which of the following two compounds will react faster in a nucleophilic substitution reaction. Hint: Use resonance hybrid structures of the intermediate carbocation to explain your answer.

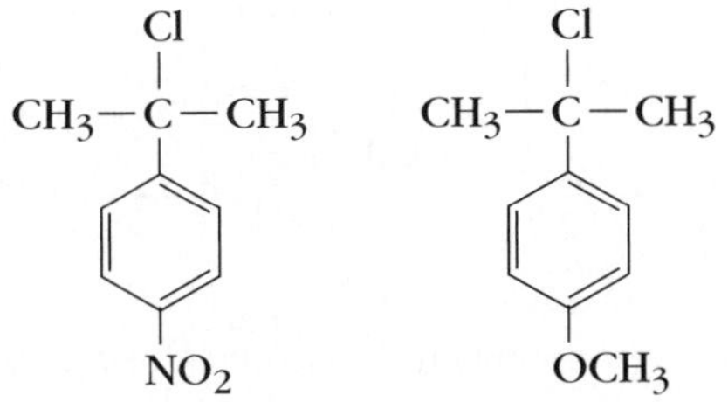

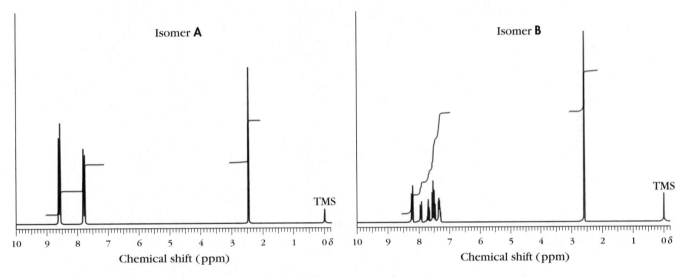

21.29 Nitration of toluene forms two products, *o*- and *p*-nitrotoluene. The ^{1}H NMR spectra of the two isomers are shown in the following figure. Assign each spectrum to the correct isomer and explain your assignment.

21.30 Write the structures of the products A to J in the following series of reactions.

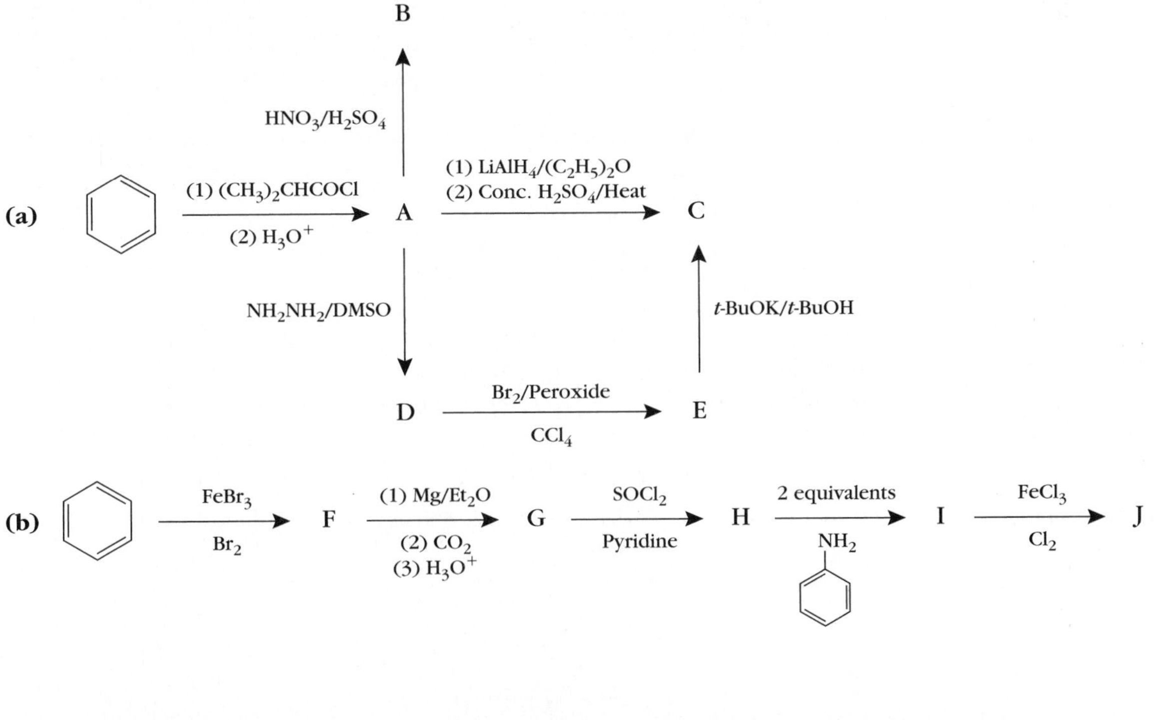

21.31 For each of the following compounds, sketch its ^{1}H NMR spectrum. Indicate the approximate chemical shift of each set of nonequivalent hydrogen atoms and their splitting patterns.

(a) 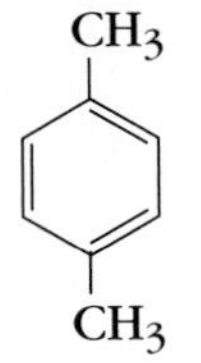
(b) 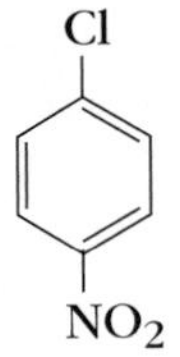
(c) 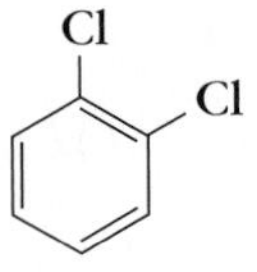
(d) 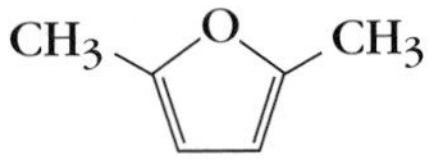

21.32 Propose a one-step synthesis of each of the following compounds using a Friedel-Crafts reaction starting with any two organic compounds.

(a) 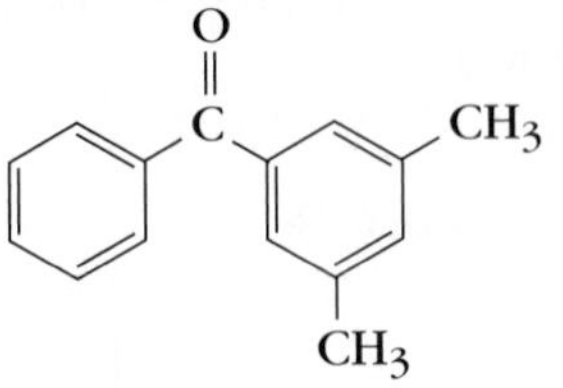
(b) 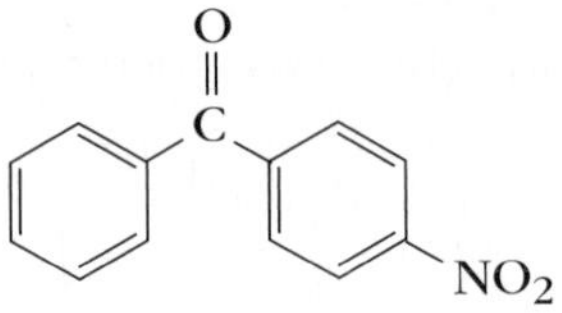
(c) 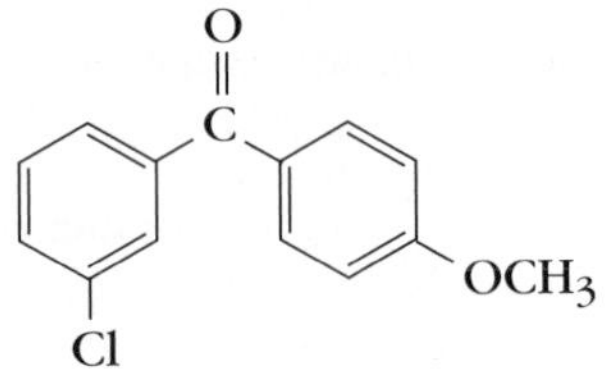

21.33 1,3,5-Trimethylbenzene reacts with a solution of formaldehyde and HCl to form 2,4,6-trimethylbenzyl chloride. Propose a mechanism for this reaction. Write structures for all intermediates and use curved arrows to indicate electron movement.

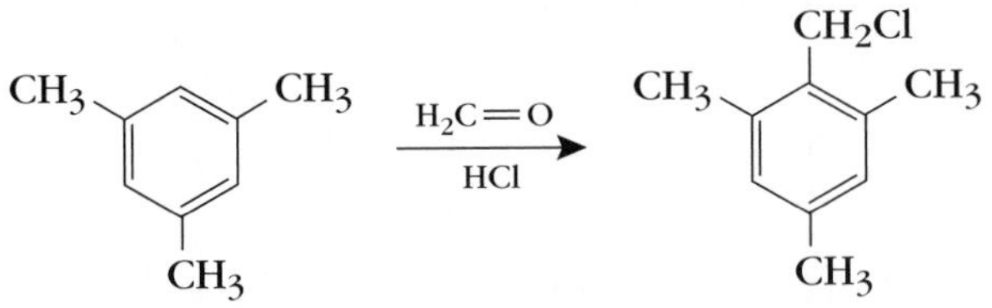

21.34 Propose a mechanism for the following reaction. Write structures for all intermediates and use curved arrows to indicate electron movement.

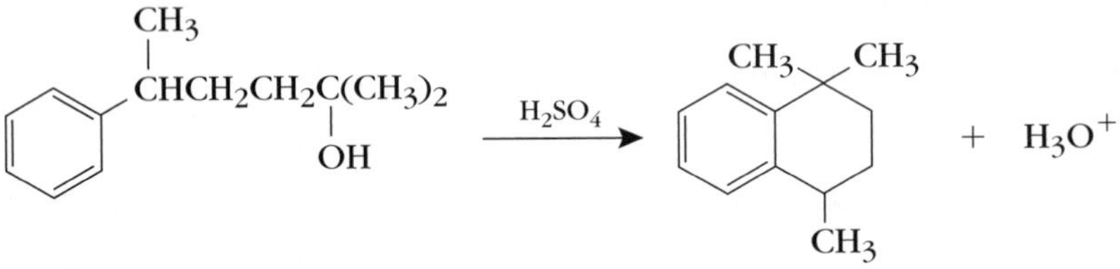

21.35 (*E*)-1,2-Diphenylethene (stilbene) reacts with a diethyl ether solution of HCl to form 1-chloro-1,2-diphenylethane.

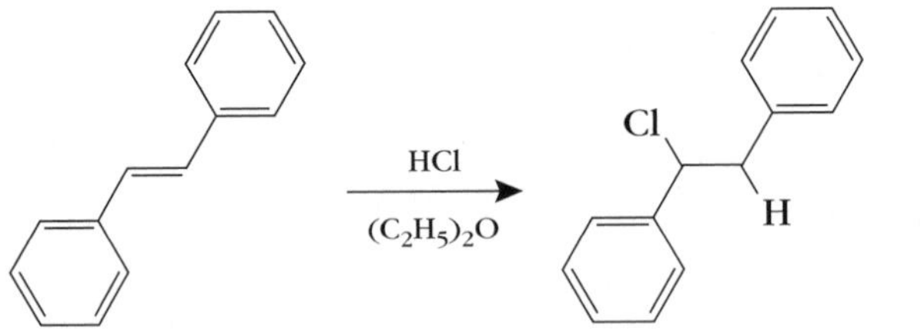

(*E*)-1,2-Diphenylethene (stilbene) 1-Chloro-1,2-diphenylethane

Based on your knowledge of the mechanism of the addition of HCl to alkenes and the effect of substituents in electrophilic aromatic substitution reactions, predict the product of addition of HCl to the following derivative of stilbene.

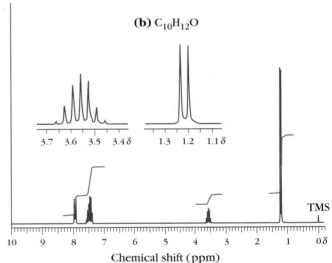

21.36 Write the structural formula of each of the following compounds whose molecular formula, ^1H NMR spectra, and other spectral data are given.

(a) C_8H_9Cl ^{13}C NMR: δ 138.0; δ 128.7; δ 128.5; δ 126.8; δ 44.9; δ 39.1.
(b) $C_{10}H_{12}O$ IR: 1685 cm^{-1}
(c) C_8H_8O IR: 1700 cm^{-1}
(d) C_9H_8O IR: 1675 cm^{-1} ^{13}C NMR: δ 193.4; δ 152.5; δ 134.0; δ 131.2; δ 129.0; δ 128.6; δ 128.4.
(e) $C_7H_6BrNO_2$ ^{13}C NMR: δ 148.4; δ 139.6; δ 135.0; δ 129.8; δ 123.8; δ 123.2; δ 31.1.
(f) C_7H_7Br ^{13}C NMR: δ 136.7; δ 131.2; δ 130.7; δ 119.0; δ 20.9.
(g) $C_9H_{10}O_2$ IR: 1680 cm^{-1}
(h) $C_{10}H_{11}BrO$ IR: 1685 cm^{-1}

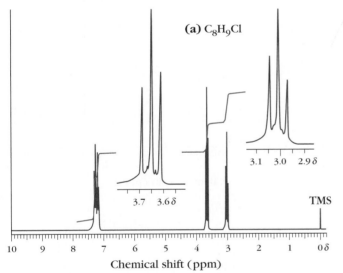

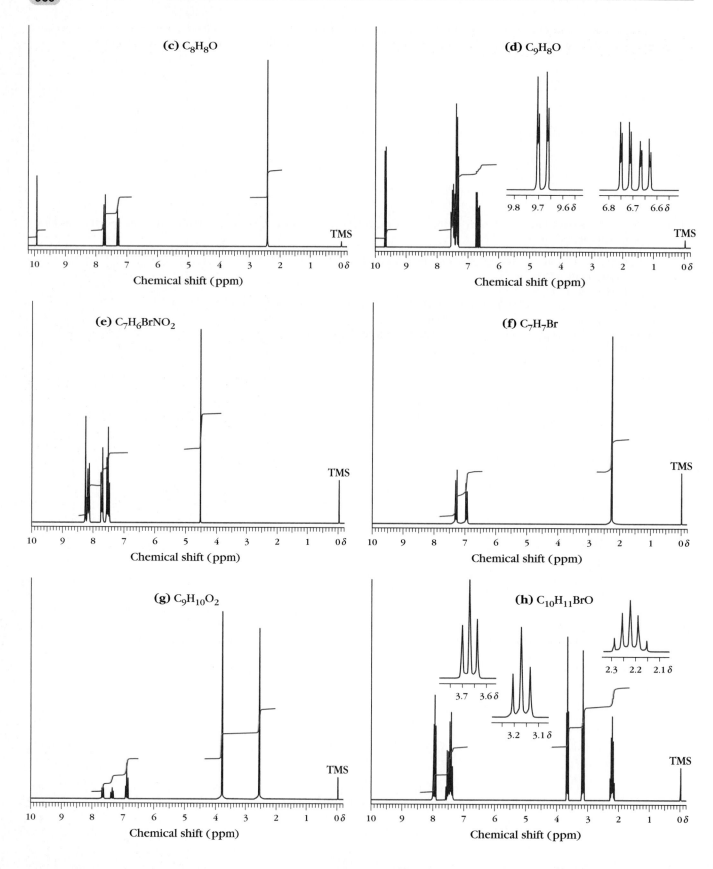

CHAPTER

22

AMINES

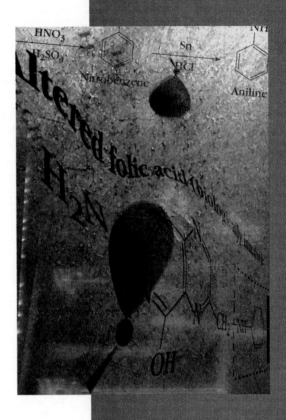

Amines are organic derivatives of ammonia in which one, two, or all three hydrogens of ammonia are replaced by alkyl or aryl groups. Recall from Section 2.14, E, that amines are classified as primary (1°), secondary (2°), or tertiary (3°). **Primary amines** have one organic group bonded to nitrogen, **secondary amines** have two organic groups, while **tertiary amines** have three.

$$
\begin{array}{ccc}
\text{H} & \text{R} & \text{R} \\
| & | & | \\
\text{R—N:} & \text{R—N:} & \text{R—N:} \qquad\qquad \text{R = alkyl or aryl group}\\
| & | & | \\
\text{H} & \text{H} & \text{R}
\end{array}
$$

Primary amine   Secondary amine   Tertiary amine

The organic groups bonded to the nitrogen atom may be any combination of alkyl or aryl groups, and the nitrogen atom may be part of a ring.

The terms primary, secondary, and tertiary used to classify amines differ slightly from the way they are used to classify alcohols and haloalkanes. Applied to alcohols or haloalkanes, these terms refer to the number of carbon atoms on the carbon atom bonded to the hydroxy group or halogen atom. For amines, these terms refer to the number of organic groups on the nitrogen atom.

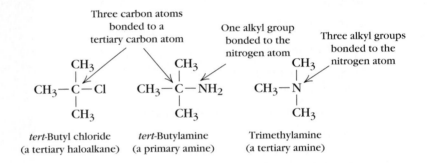

Three carbon atoms bonded to a tertiary carbon atom

One alkyl group bonded to the nitrogen atom

Three alkyl groups bonded to the nitrogen atom

tert-Butyl chloride (a tertiary haloalkane)   *tert*-Butylamine (a primary amine)   Trimethylamine (a tertiary amine)

Ammonium ion (NH_4^+), which is the conjugate acid of ammonia, also forms organic derivatives. One, two, three, or all four hydrogen atoms of the ammonium ion can be replaced by any combination of alkyl or aryl groups. Compounds in which all four hydrogen atoms of the ammonium ion are replaced with organic groups are called **quaternary ammonium salts.** As in the ammonium ion, the nitrogen atom of all substituted ammonium ions carries a positive charge.

$$
\begin{array}{cc}
\text{H} & \text{H} \\
|+ & |+ \\
\text{R—N—H} \;\; \text{X}^- & \text{R—N—H} \;\; \text{X}^- \\
| & | \\
\text{H} & \text{R}
\end{array}
$$

Ammonium salt of a primary amine   Ammonium salt of a secondary amine

$$
\begin{array}{cc}
\text{R} & \text{R} \\
|+ & |+ \\
\text{R—N—R} \;\; \text{X}^- & \text{R—N—R} \;\; \text{X}^- \\
| & | \\
\text{H} & \text{R}
\end{array}
$$

Ammonium salt of a tertiary amine   Quaternary ammonium salt

Exercise 22.1 Classify each of the following amines as either primary, secondary, or tertiary.

(a) $CH_3(CH_2)_2NHCH_2CH_3$

(b) 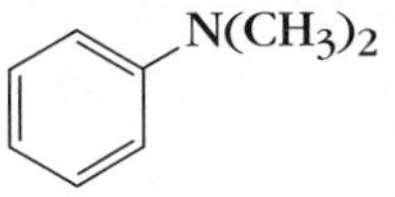 [structure: benzene ring with $N(CH_3)_2$ substituent]

(c) 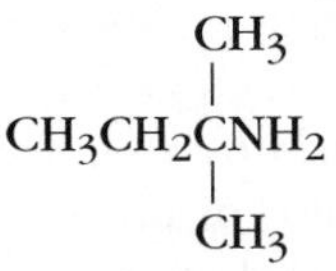 $CH_3CH_2\overset{\overset{\displaystyle CH_3}{|}}{\underset{\underset{\displaystyle CH_3}{|}}{C}}NH_2$

(d) 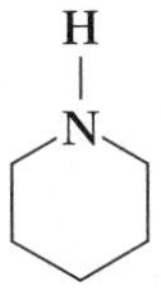 [structure: piperidine ring with N–H at top]

(e) 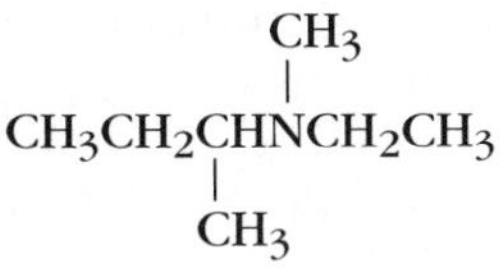 $CH_3CH_2\overset{\overset{\displaystyle CH_3}{|}}{C}HN\underset{\underset{\displaystyle CH_3}{|}}{C}H_2CH_3$

Exercise 22.2 Classify each of the following as a quaternary ammonium salt or as a salt of a primary, secondary, or tertiary amine.

(a) $CH_3CH_2CH_2NH_3{}^+Cl^-$

(b) $(CH_3)_4N^+Br^-$

(c) 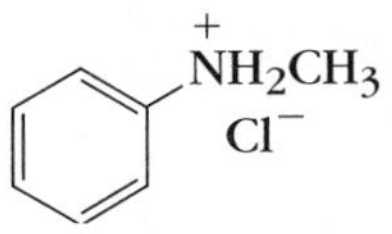 [structure: benzene ring with $\overset{+}{N}H_2CH_3$ substituent and Cl^-]

(d) 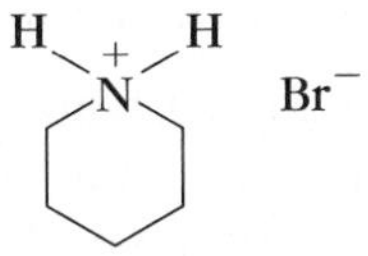 [structure: piperidine ring with $\overset{+}{N}$ bearing two H atoms and Br^-]

(e) 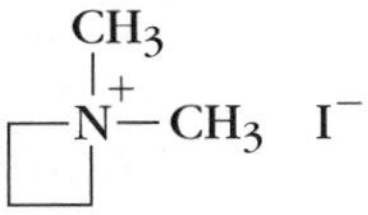 [structure: cyclobutane ring attached to $\overset{\overset{\displaystyle CH_3}{|}}{\underset{}{N}}{}^+{-}CH_3$ with I^-]

22.1 Naming Amines

Simple primary amines are named by adding the suffix *amine* to the name of the alkyl group bonded to the nitrogen atom of the amine. The name of the amine is written as one word.

$(CH_3)_2CHNH_2$ $CH_3CH_2\underset{\underset{\displaystyle NH_2}{|}}{C}HCH_3$ [cyclohexane ring with NH_2]

Isopropylamine   *sec*-Butylamine 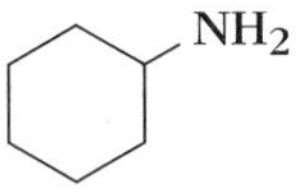   Cyclohexylamine

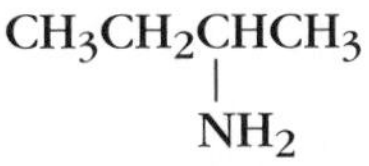

Alternatively, primary amines can be named by replacing the final *e* of the IUPAC name of the parent alkane with *amine*.

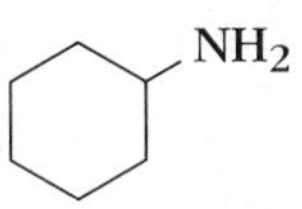

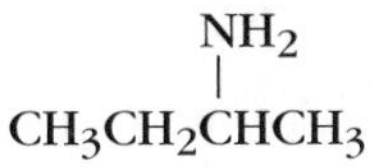

[cyclohexane ring with NH_2] $CH_3CH_2\overset{\overset{\displaystyle NH_2}{|}}{C}HCH_3$ $(CH_3)_3CNH_2$

Cyclohexanamine   2-Butanamine   2-Methyl-2-propanamine

Symmetrical secondary and tertiary amines are named by adding the prefix *di-* or *tri-* to the name of the alkyl group.

$$(CH_3CH_2)_2NH$$

Diethylamine   Triphenylamine

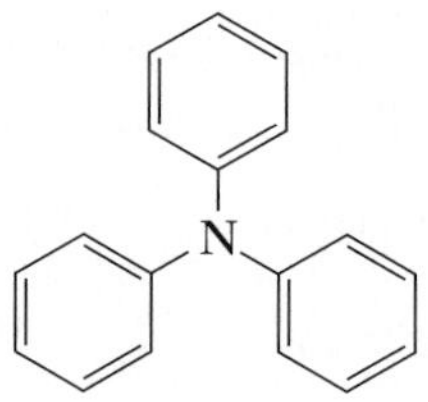

Unsymmetrically substituted secondary or tertiary amines are named as *N*-substituted primary amines. The largest of the alkyl substituents is chosen as the parent chain. The use of the letter *N* indicates that the alkyl groups are attached to the nitrogen atom and not to a carbon atom of the parent alkyl chain. The names of the alkyl groups bonded to the nitrogen atom are listed alphabetically.

$$CH_3CH_2CH_2NHCH_3$$

N-Methyl-1-propanamine
(not *N*-propylmethanamine)

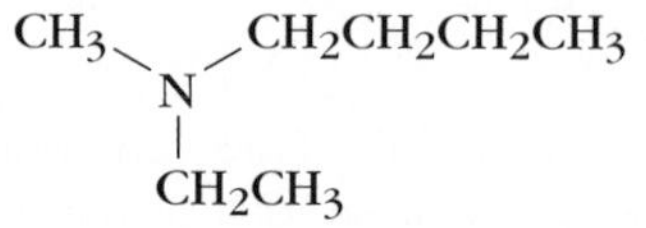

N-Ethyl-*N*-methyl-1-butanamine
(not *N*-methyl-*N*-ethyl-1-butanamine)

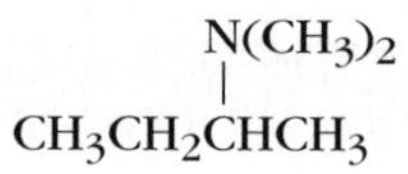

$$N(CH_3)_2$$
$$CH_3CH_2CHCH_3$$

N,N-Dimethyl-2-butanamine

The —NH_2 group of amines whose structures are more complicated is called an **amino** group. The position of the amino group on the chain is indicated by a number just like any other substituent.

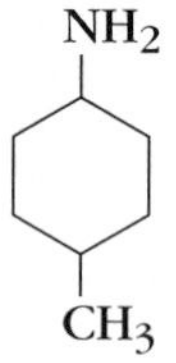

4-Methylcyclohexanamine

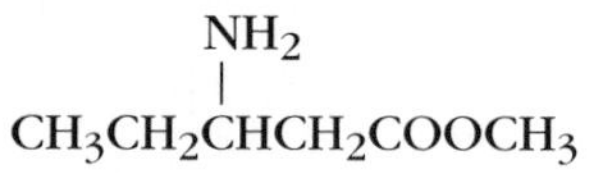

$$CH_3CH_2CHCH_2COOCH_3$$

Methyl 3-aminopentanoate

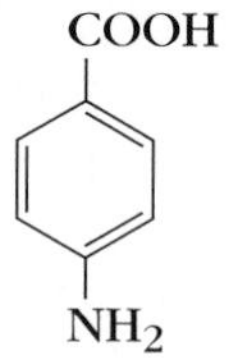

4-Aminobenzoic acid

Arylamines are compounds with an amino group bonded directly to an aryl group. Arylamines in which the aryl group is a benzene ring are usually named as derivatives of aniline.

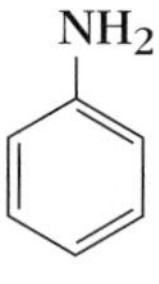

Aniline 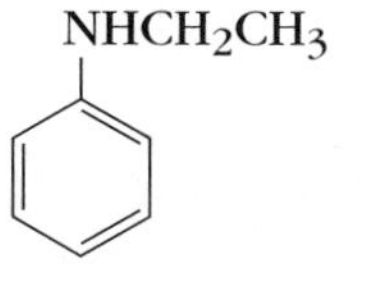   *N*-Ethylaniline 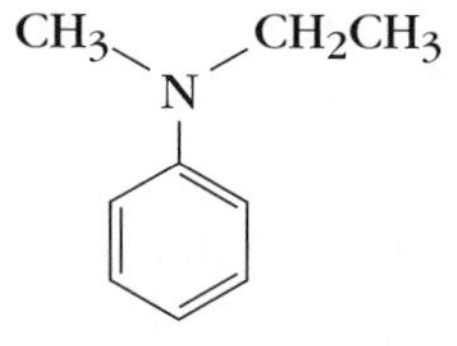   *N*-Ethyl-*N*-methylaniline

Most arylamines are named in this way, but others are better known by their common names, which have been accepted as part of the IUPAC system of naming amines. The following are some of the more important examples:

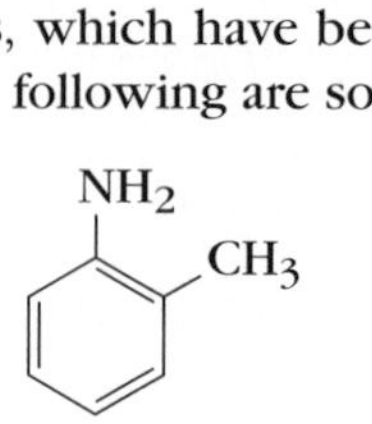

o-Toluidine
(*o*-Aminotoluene)

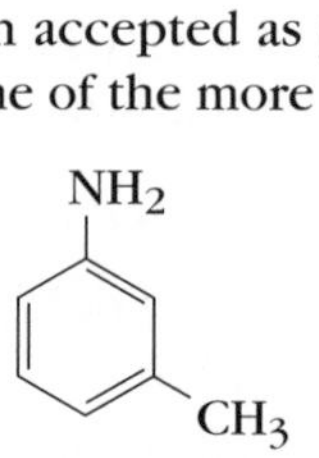

m-Toluidine
(*m*-Aminotoluene)

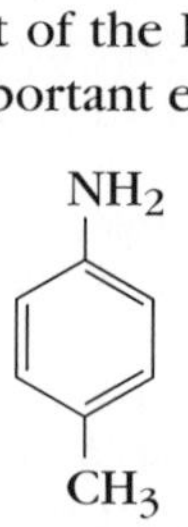

p-Toluidine
(*p*-Aminotoluene)

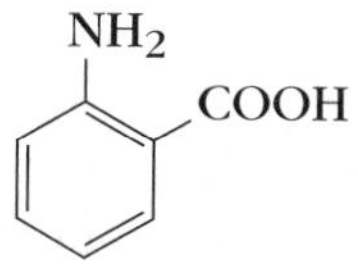

Anthranilic acid
(*o*-aminobenzoic acid)

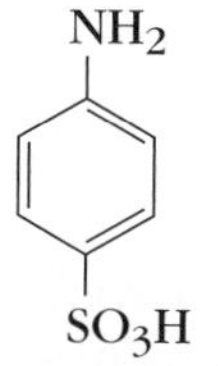

Sulfanilic acid
(*p*-aminobenzenesulfonic acid)

A similar set of rules is used to name ammonium salts, except the suffix *ammonium* replaces *amine* and the identity of the anion must be included as part of the name.

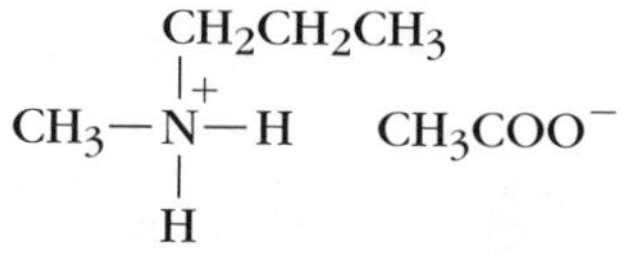

N-Methylpropylammonium acetate

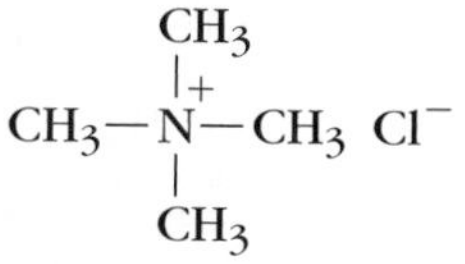

Tetramethylammonium chloride

Compounds with two amino groups are named by adding the suffix *diamine* to the name of the corresponding alkane or arene.

1,3-Propanediamine

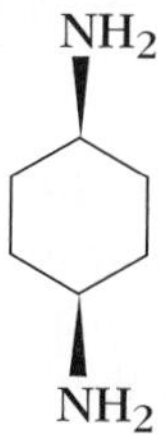

cis-1,4-Cyclohexane-diamine

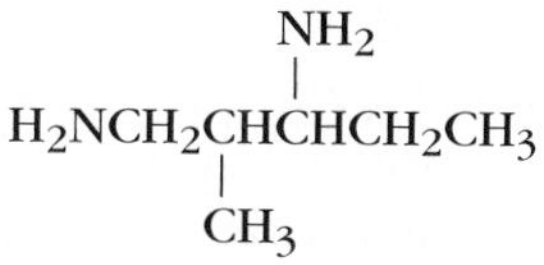

2-Methyl-1,3-pentanediamine

Notice that the final *e* of the name of the parent chain is retained when naming diamines.

Exercise 22.3 Write the structure of each of the following compounds.

(a) 1-Butanamine
(b) *N*-Ethyl-3-heptanamine
(c) Diisobutylamine
(d) *N,N*-Diethylaniline

(e) *trans*-1,3-Cyclopentanediamine
(f) Butylammonium chloride
(g) Trimethyldecylammonium acetate
(h) 3-Chlorocyclopentanamine

Exercise 22.4 Give a name to each of the following compounds.

(a) 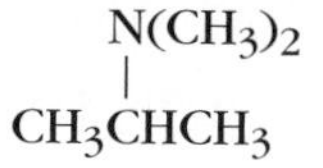 $CH_3\overset{\displaystyle N(CH_3)_2}{\underset{\displaystyle |}{C}}HCH_3$

(b) ◁—NHCH₃

(c) N(CH₃)₃ I⁻

(d) (benzene with NH₂ groups at 1,3 positions)

(e) (cyclohexane with N(CH₃)(CH₂CH₃))

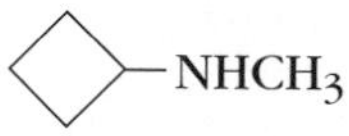

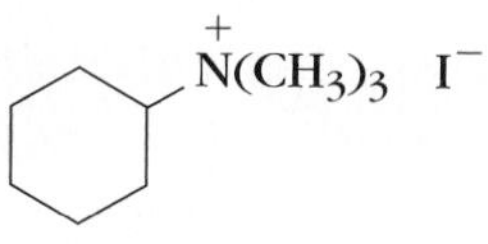

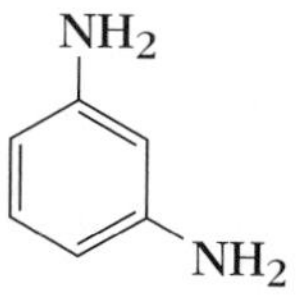

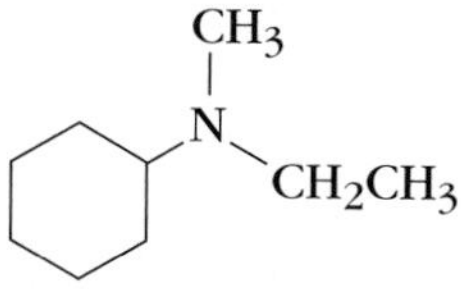

(f) [(CH₃)₃C]₂NH

22.2   Structure of Amines

Amines are pyramidal in shape just like ammonia. The valence bond description of amines is similar to ammonia. We use an sp^3 hybrid nitrogen atom because it gives the correct geometry about the nitrogen atom of amines and ammonia. The hydrogen atoms or the carbon atoms of the substituents form bonds with three of the sp^3 hybrid orbitals of the nitrogen atom, while the lone pair occupies the remaining sp^3 orbital.

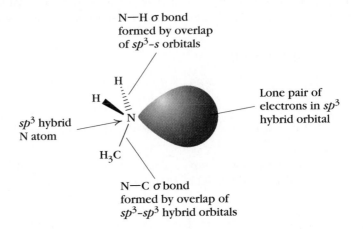

We learned in Chapter 6 that an sp^3 hybrid carbon atom bonded to four different substituents is a stereocenter. The nitrogen atom of an amine resembles a carbon stereocenter if we regard the lone pair of electrons as a fourth substituent. Picturing an amine in this way means that a nitrogen atom with three different substituents bonded to it should also be a stereocenter and should exist as a pair of enantiomers. Amines such as ethylmethylamine are indeed chiral compounds, but the enantiomers cannot be isolated because they rapidly interconvert by a **pyramidal inversion.** A pyramidal inversion is similar to the inversion of the carbon atom at the reactive site in the S_N2 mechanism. Pyramidal inversion occurs through a planar transition state in which the nitrogen is sp^2 hybridized.

The activation energy for pyramidal inversion of a number of acyclic amines has been determined experimentally to be in the range of 6 to 10 kcal/mol (25 to 45 kJ/mol). These values are sufficiently low (only about twice as large as the barrier to rotation about a carbon-carbon single bond) that pyramidal inversion occurs so rapidly at room temperature that the two enantiomers of most acyclic amines cannot be isolated.

Aziridine is a cyclic amine in which the nitrogen atom is incorporated into a three-membered ring.

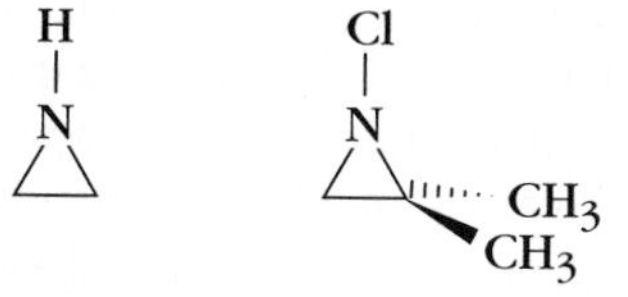

Aziridine   1-Chloro-2,2-dimethylaziridine

The activation energy for pyramidal inversion in aziridines is much higher than the barrier to inversion of acyclic amines because of additional strain in the planar transition state, caused by the three-membered ring. The barrier to pyramidal inversion for 1-chloro-2,2-dimethylaziridine, for example, is sufficiently high that its two enantiomers can be isolated.

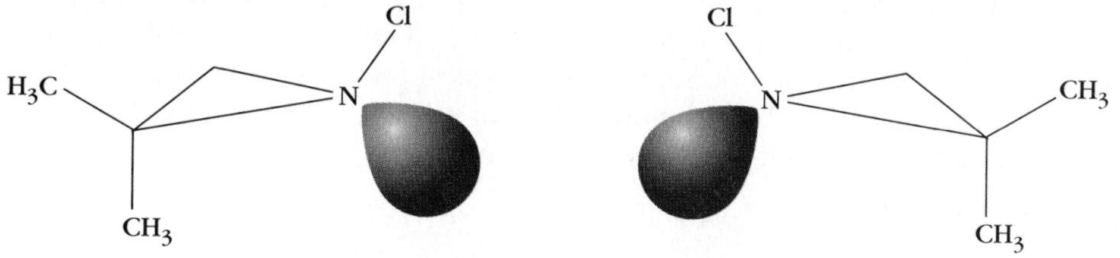

(*R*)-1-Chloro-2,2-dimethylaziridine   (*S*)-1-Chloro-2,2-dimethylaziridine

Quaternary ammonium salts that contain four different groups bonded to the nitrogen atom are chiral, and they can be resolved because they lack a pair of nonbonding electrons to undergo pyramidal inversion. The salts of *N*-ethyl-*N*-isopropyl-*N*-methylanilinium ion, for example, can be resolved into enantiomers.

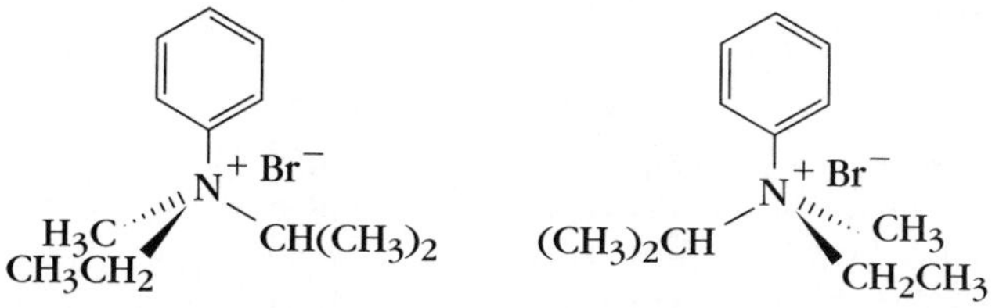

(*R*)-*N*-Ethyl-*N*-isopropyl-   (*S*)-*N*-Ethyl-*N*-isopropyl-
N-methylanilinium bromide   *N*-methylanilinium bromide

| **Exercise 22.5** | Draw a three-dimensional representation of the two enantiomers of each of the following compounds. Indicate which enantiomers would undergo rapid pyramidal inversion at room temperature. |

(a) *N*-Methylaniline
(b) *N*-Ethyl-*N*-isopropyl-*N*-methylammonium chloride
(c) *N*-Methyl-*trans*-2,3-dimethylaziridine

22.3   Physical Properties

The boiling points and melting points of alkylamines are compared with those of other types of organic compounds of similar molecular weight in Table 22.1.

The data in Table 22.1 illustrate two important trends. First, the boiling points of primary amines are higher than those of alkanes, ethers, and haloalkanes of similar molecular weight but lower than those of alcohols and carboxylic acids. Second, the boiling points of amines of the same molecular weight decrease in the order 1° > 2° > 3°.

The ability of low molecular weight primary and secondary amines to form hydrogen bonds, similar to alcohols, is the accepted explanation for these trends. Nitrogen is not as electronegative an element as oxygen, however, so the intermolecular hydrogen bonds in alcohols are stronger than in amines. Thus alcohols have higher boiling points than amines. Primary and secondary amines have higher boiling points than tertiary amines because primary and secondary amines can participate in intermolecular hydrogen bonding while tertiary amines cannot.

The solubilities of amines are similar to those of alcohols. Amines that contain six or fewer carbon atoms are very soluble in water. As the size of the alkyl portion of the amine increases, however, water solubility decreases. This effect of increasing size of the alkyl portion on the physical properties of many functional groups has been encountered before. Therefore the trends shown by the data in Table 22.1 are evident only in amines of low molecular weight.

| **Table 22.1** | Comparison of Melting and Boiling Points of Alkylamines and Other Organic Compounds of Similar Molecular Weight | | | |

Name	Structure	M.W.	M.P. (°C)	B.P. (°C)
1-Propanamine	$CH_3CH_2CH_2NH_2$	59	−83	50
N-Methylethanamine	$CH_3NHCH_2CH_3$	59	−50	34
Trimethylamine	$(CH_3)_3N$	59	−117	3
1-Propanol	$CH_3CH_2CH_2OH$	60	−127	97
Ethyl methyl ether	$CH_3OCH_2CH_3$	60	—	10.8
Propanal	CH_3CH_2CHO	58	−81	50
Acetic acid	CH_3COOH	60	17	118
1-Chloroethane	CH_3CH_2Cl	64	−136	12
Butane	$CH_3CH_2CH_2CH_3$	58	−139	0

Amines are unpleasant-smelling compounds. Low molecular weight amines smell much like ammonia, while higher molecular weight amines have a distinct fishlike aroma. It turns out that the presence of volatile amines in their tissues gives fish their characteristic odors. Two diamines that are released by decaying flesh have putrid odors and have been given common names to indicate both their odor and their source.

$$H_2NCH_2CH_2CH_2CH_2NH_2 \qquad H_2NCH_2CH_2CH_2CH_2CH_2NH_2$$

Putrescine
(1,4-butanediamine)

Cadaverine
(1,5-pentanediamine)

22.4 Spectroscopic Identification of Amines

Recall from Section 5.5, D that primary and secondary amines have characteristic N—H stretching absorptions in the 3300 to 3500 cm^{-1} region of their infrared spectra. Alcohols also absorb in this region, but the absorption bands of amines are usually both sharper and less intense than the absorption bands due to the O—H group.

Primary amines have two sharp absorption bands and secondary amines have a single sharp absorption band in this region. Tertiary amines show no absorption in this region because they lack an N—H bond. These differences are illustrated by the IR spectra in Figure 5.11 (p. 184).

The ^1H NMR chemical shifts of hydrogen atoms bonded to carbon near the nitrogen atom of amines are deshielded compared to alkanes because of the electronegativity of the nitrogen. However, the downfield shifts are not as great as in alcohols. The data in Table 22.2 compare the chemical shifts of CH_3, CH_2, and CH hydrogens in alkanes, amines, and alcohols.

The ^1H NMR chemical shifts of the hydrogen atoms bonded to the nitrogen atom in primary and secondary amines are variable and depend on a number of factors such as the purity of the sample, the solvent in which it is dissolved, the concentration of the sample, and the temperature at which the measurement is made. The peak is usually broad, and coupling to neighboring hydrogens is not observed due to exchange with the solvent. Thus hydrogen atoms bonded to a nitrogen atom of amines, like the hydrogen atoms of the hydroxy group of alcohols, are best identified by adding a small amount of D_2O to the sample tube, which makes them disappear.

The ^{13}C NMR chemical shifts of carbon atoms adjacent to the nitrogen atom

Table 22.2		Comparison of the ^1H NMR Chemical Shifts of Methyl, Methylene, and Methine Hydrogen Atoms in Alkanes, Amines, and Alcohols			
Compound	CH_3 δ	Compound	CH_2 δ	Compound	CH δ
CH_3CH_3	0.86	$CH_3CH_2CH_3$	1.33	$(CH_3)_3CH$	1.50
CH_3NH_2	2.47	$CH_3CH_2NH_2$	2.74	$(CH_3)_2CHNH_2$	3.07
CH_3OH	3.39	CH_3CH_2OH	3.59	$(CH_3)_2CHOH$	3.94

Table 22.3	Comparison of the ^{13}C NMR Chemical Shifts of Methyl, Methylene, and Methine Hydrogen Atoms in Alkanes, Amines, and Alcohols				
Compound	CH_3 δ	Compound	CH_2 δ	Compound	CH δ
CH_3CH_3	7.3	$CH_3CH_2CH_3$	15.9	$(CH_3)_3CH$	25.0
CH_3NH_2	28.3	$CH_3CH_2NH_2$	36.9	$(CH_3)_2CHNH_2$	43.0
CH_3OH	50.2	CH_3CH_2OH	57.8	$(CH_3)_2CHOH$	64.0

are also shifted downfield relative to alkanes. Again the shift is not as large as in the case of alcohols, as shown by the data in Table 22.3.

Amines are easily identified by mass spectrometry due to a fragmentation pattern characteristic of α-cleavage.

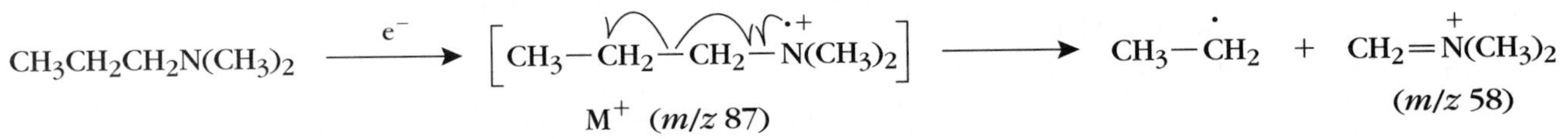

$$CH_3CH_2CH_2N(CH_3)_2 \xrightarrow{e^-} \left[CH_3-CH_2-CH_2-\overset{\cdot+}{N}(CH_3)_2 \right] \longrightarrow CH_3-\overset{\cdot}{C}H_2 \ + \ CH_2=\overset{+}{N}(CH_3)_2$$

$$M^+ \ (m/z \ 87) \qquad\qquad\qquad (m/z \ 58)$$

Amines readily undergo α-cleavage because the carbocation formed is stabilized by the adjacent nitrogen atom.

Exercise 22.6	The only significant peaks in the mass spectrum of the compound whose ^1H NMR spectrum is shown below are found at $m/z = 59$ and $m/z = 30$. Write the structure of the compound and the structures of the species responsible for the two peaks in its mass spectrum.

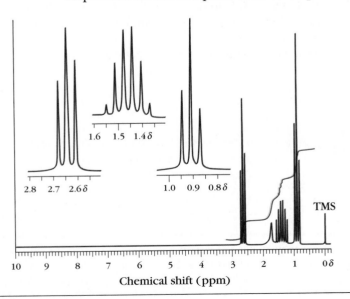

22.5   Amines Are More Basic Than Most Other Organic Compounds

Amines, like ammonia, are bases because they have a nonbonding pair of electrons on nitrogen. One of their principal reactions is protonation by Brønsted-Lowry acids such as HCl.

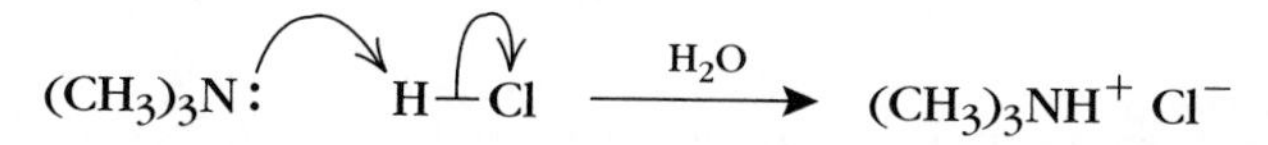

$$(CH_3)_3N: \quad H-Cl \xrightarrow{H_2O} (CH_3)_3NH^+ \ Cl^-$$

Trimethylamine   Trimethylammonium chloride

The basicity of amines is usually expressed as the dissociation constant (K_a) of their conjugate acids (Section 3.3). Thus for the equilibrium

$$RNH_3^+ \ + \ H_2O \ \rightleftharpoons \ RNH_2 \ + \ H_3O^+$$

K_a is defined by the following equation:

$$K_a = \frac{[RNH_2][H_3O^+]}{[RNH_3^+]}$$

The values of K_a of amines are often quoted as pK_a, which is -log K_a. Remember that the more positive the pK_a value, the weaker the acid. Also, the weaker the alkylammonium ion, the more basic is the alkylamine, its conjugate base. Values of pK_a for typical alkylammonium ions are given in Table 22.4. The pK_a of ammonium ion is provided for comparison.

Two important conclusions can be reached from the data in Table 22.4. First, amines are weak bases, but they are among the most basic of all simple neutral

Table 22.4	Acidity of Typical Alkylammonium Ions in Water		
Compound	Structure	Conjugate Acid	pK_a of the Conjugate Acid
Ammonia	NH_3	NH_4^+	9.25
PRIMARY AMINES			
Methylamine	CH_3NH_2	$CH_3NH_3^+$	10.63
Ethylamine	$CH_3CH_2NH_2$	$CH_3CH_2NH_3^+$	10.75
t-Butylamine	$(CH_3)_3CNH_2$	$(CH_3)_3CNH_3^+$	10.68
Cyclohexylamine	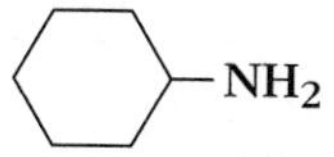 ⬡—NH_2	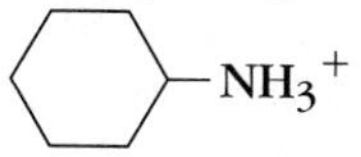 ⬡—NH_3^+	10.64
SECONDARY AMINES			
Dimethylamine	$(CH_3)_2NH$	$(CH_3)_2NH_2^+$	10.73
Diethylamine	$(CH_3CH_2)_2NH$	$(CH_3CH_2)_2NH_2^+$	10.94
Piperidine	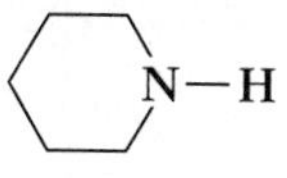 ⬡N—H	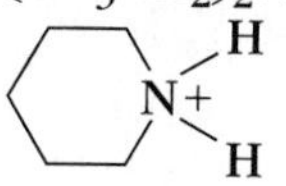 ⬡N$^+$(H)(H)	11.20
TERTIARY AMINES			
Trimethylamine	$(CH_3)_3N$	$(CH_3)_3NH^+$	9.79
Triethylamine	$(CH_3CH_2)_3N$	$(CH_3CH_2)_3NH^+$	10.75

organic compounds. This conclusion can be verified by comparing the pK_a values listed in Table 22.4 with those in Table 3.1 (p. 90). Second, the values of the pK_a of all simple alkylammonium ions are in the range 10 to 11, so alkylamines are all slightly more basic than ammonia. In neutral aqueous solution, therefore, alkylamines exist as their conjugate acids. For instance, in blood, whose pH is about 7, ammonia and alkylamines exist in their protonated forms.

Exercise 22.7   Using the Henderson-Hasselbalch equation (Section 15.8, p. 665), calculate the ratio [butanamine]/[butylammonium ion] in aqueous solution at the following pH.

(a) pH 5.0   (b) pH 7.4   (c) pH 10.5   (d) pH 12.0

While it is convenient to express the basicity of amines as the pK_a of their conjugate acids, care must be taken to distinguish between the pK_a for the dissociation of an alkylammonium ion and the pK_a for the dissociation of an alkylamine. These are two different quantities. The pK_a of an alkylammonium ion refers to the reaction in which an alkylammonium ion acts as an acid, whereas the pK_a of an alkylamine refers to the reaction in which the alkylamine acts as an acid. This difference is illustrated for the case of diisopropylamine and its conjugate acid by the two following equations:

Diisopropylammonium ion as an acid

$$[(CH_3)_2CH]_2NH_2^+ \ + \ H_2O \ \underset{K_a}{\xrightarrow{\hspace{1cm}}} \ [(CH_3)_2CH]_2NH \ + \ H_3O^+$$

Diisopropylamine as an acid

$$[(CH_3)_2CH]_2NH \ + \ C_4H_9Li \ \xrightarrow{\hspace{1cm}} \ [(CH_3)_2CH]_2N^-Li^+ \ + \ C_4H_{10}$$

Lithium diisopropylamide
(LDA)

Diisopropylamine ($pK_a \approx 38$) is a much weaker acid than diisopropylammonium ion ($pK_a = 10.9$). It is important to remember, therefore, that the pK_a of an amine is not the same as the pK_a of its conjugate acid.

Exercise 22.8   Using the data in Tables 22.4 and 3.1 (p. 90), calculate the equilibrium constant for each of the following reactions.

(a) $CH_3NH_3^+ \ + \ CH_3OH \ \rightleftharpoons \ CH_3NH_2 \ + \ CH_3OH_2^+$
(b) $CH_3CH_2NH_2 \ + \ H_2O \ \rightleftharpoons \ CH_3CH_2NH_3^+ \ + \ OH^-$
(c) $(CH_3CH_2)_2NH \ + \ CH_3COOH \ \rightleftharpoons \ (CH_3CH_2)_2NH_2^+ \ + \ ^-OOCCH_3$

Arylamines are generally less basic than alkylamines in aqueous solution. For example, the pK_a value of anilinium ion, the conjugate acid of aniline, is lower than the pK_a value of a typical primary alkylammonium ion such as cyclohexylammonium ion.

Anilinium ion   Aniline $pK_a = 4.60$

Cyclohexyl-   Cyclohexanamine $pK_a = 10.64$
ammonium ion

Substituted anilines may be more or less basic than aniline. The data in Table 22.5 show, for example, that electron-donating substituents, such as $-OCH_3$ and $-CH_3$ groups, increase the basicity of arylamines. Electron-withdrawing substituents, such as $-NO_2$ and $-CN$ groups, decrease the basicity of arylamines.

Table 22.5	pK_a Values of p-Substituted Anilinium Ions in Water at 25 °C						
Substituent	OCH_3	CH_3	H	Cl	Br	CN	NO_2
pK_a	5.34	5.10	4.60	3.98	3.86	1.74	1.00

Exercise 22.9	Using the Henderson-Hasselbalch equation (Section 15.8, p. 665), calculate the ratio [aniline]/[anilinium ion] in aqueous solution at the following pH.

(a) pH 5.0   **(b)** pH 7.4   **(c)** pH 10.5   **(d)** pH 12.0

To explain the effect of substituents on the basicity of amines, let us consider the two following acid-base reactions:

$$NH_4^+ \; + \; H_2O \; \rightleftharpoons \; NH_3 \; + \; H_3O^+$$

$$YNH_3^+ \; + \; H_2O \; \rightleftharpoons \; YNH_2 \; + \; H_3O^+$$

Recall from Section 3.10 that the free energy difference between products and reactants, ΔG, determines the equilibrium constant, K, of a reaction. By determining how replacement of a hydrogen atom on NH_4^+ and NH_3 by a group Y affects the relative stabilities of the reactants and products, we can estimate the relative values of K_a for the two reactions.

If Y is an atom or group that donates electrons by an inductive effect, it stabilizes the positive charge on the nitrogen atom of YNH_3^+ relative to the hydrogen atom of NH_4^+, so YNH_3^+ is more stable than NH_4^+. Inductive electron donation, however, destabilizes YNH_2 relative to NH_3, so NH_3 is more stable than YNH_2, which makes ΔG for the dissociation of YNH_3^+ more positive than ΔG for NH_4^+. As a result, YNH_3^+ is the weaker acid, so its conjugate base is a stronger base than NH_3, as shown in Figure 22.1. Replacing hydrogen atoms by atoms or groups that donate electrons by an inductive effect, such as methyl groups, therefore, increases the basicity of amines.

Figure 22.1
Free energy diagram comparing the dissociation of YNH_3^+ and NH_4^+. Because $\Delta G°$ is larger for the dissociation of YNH_3^+ than $\Delta G°$ for the dissociation of NH_4^+, YNH_3^+ is a weaker acid than NH_4^+. As a result, YNH_2 is a stronger base than NH_3.

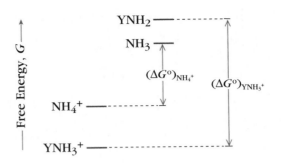

When Y is an atom or a group that can delocalize the unshared electron pair on the nitrogen atom by either an inductive effect or a resonance effect, it stabilizes YNH_2 compared to NH_3. Electron-withdrawing groups generally destabilize YNH_3^+ compared to NH_4^+, so YNH_3^+ is a stronger acid than NH_4^+, which makes YNH_2 a weaker base than ammonia.

Exercise 22.10   Construct a free energy diagram similar to Figure 22.1 comparing the ionization of YNH_3^+ and NH_4^+ when Y is an electron-withdrawing substituent.

Exercise 22.11   Aniline is stabilized by delocalization of its lone pair of electrons into the benzene ring. Electron delocalization can be described in terms of a resonance hybrid structure. Write the resonance structures of this hybrid.

Amides are derived from ammonia or amines by replacing a hydrogen atom bonded to the nitrogen atom by an acyl group. This simple change in structure greatly changes the basicity of amides relative to amines.

22.6   Amides Are Much Less Basic Than Amines

Amides are such weak bases that their conjugate acids have pK_a values close to zero. Amides, therefore, are about 10 powers of ten ($\approx 10^{10}$) less basic than alkyl-amines or ammonia. In practical terms, this means that amides, in contrast to amines or ammonia, do not form salts with aqueous solutions of mineral acids such as HCl.

Exercise 22.12   Verify that amines react with aqueous acids, while amides do not, by calculating the equilibrium constants for the following reactions.

(a) Acetamide with aqueous HCl ($pK_a = -2.2$)
(b) Ethylamine with aqueous HCl

Amides can be regarded as ammonia or amines in which one hydrogen atom is replaced by an acyl group. Compared to a hydrogen atom an acyl group is electron withdrawing, as we learned in Section 21.11. Amides are weaker bases than ammonia and amines, therefore, because of the electron-withdrawing effect of the acyl group.

The acyl group withdraws electrons by delocalization of the lone pair of electrons on the amide nitrogen into the adjacent carbonyl group. As we discussed in Section 16.2, this electron delocalization can be represented as a hybrid of the following resonance structures:

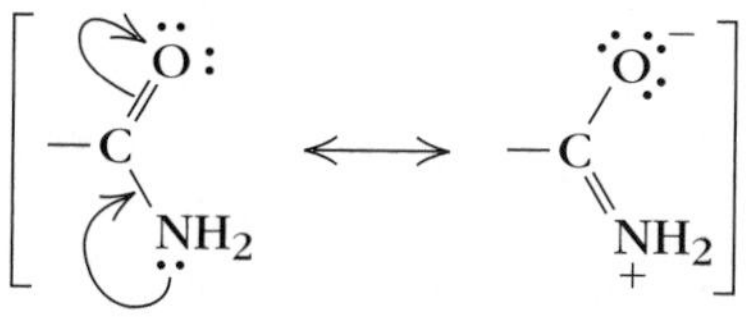

Exercise 22.13   Arrange each of the following sets of compounds from least basic to most basic.

(a) Aniline, acetamide, diethylamine
(b) Acetic acid, aniline, *p*-nitroaniline
(c) Acetic acid, ethylamine, aniline

22.7 Amines Are Widely Distributed in Plants and Animals

Because of their basicity, amines can be easily separated from other plant material by extraction with aqueous acid solutions. Because they are easily regenerated by treatment with aqueous base, amines were some of the first organic compounds isolated. Naturally occurring amines obtained from plants in this way are called **alkaloids.**

The chemistry of alkaloids encompasses thousands of compounds of many structural types. Alkaloids are important compounds because of their physiological activity. Some have beneficial medicinal effects, while others are highly toxic. In this section, representative examples of each are presented.

Many years ago, crude extracts of the poppy, *Papaver somniferum,* were found to relieve pain. This observation led eventually to the discovery of morphine and codeine and their use as analgesics.

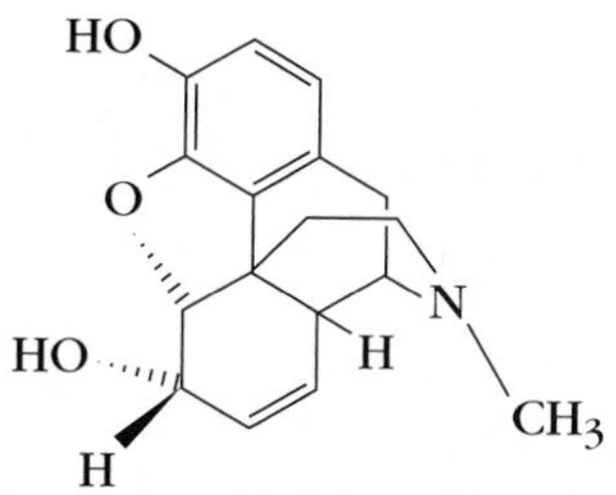

Morphine 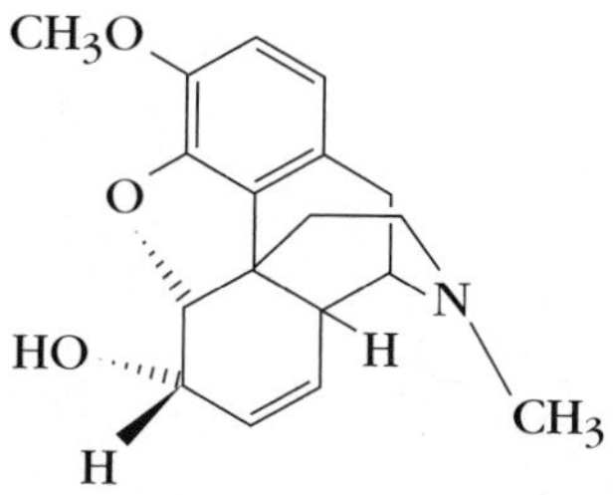   Codeine

Although morphine and codeine are effective analgesics, they pose enormous social problems because they are addictive.

The antimalaria drug quinine is another medicinally useful alkaloid. It was first obtained by extraction of the bark of the South American *Cinchona* tree.

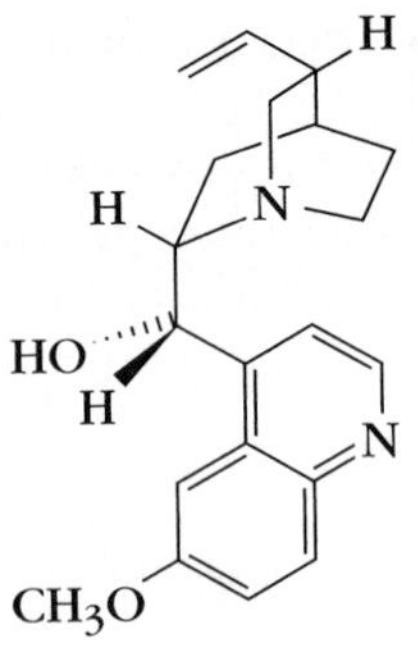

Quinine

Other alkaloids are less benign to humans. Coniine, for example, which is present in the extract of the water hemlock, is toxic even at low concentrations. Others, such as cocaine and nicotine, can be ingested at low concentrations but are poisonous at high concentrations.

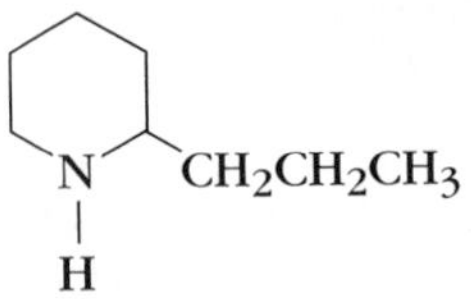

Coniine 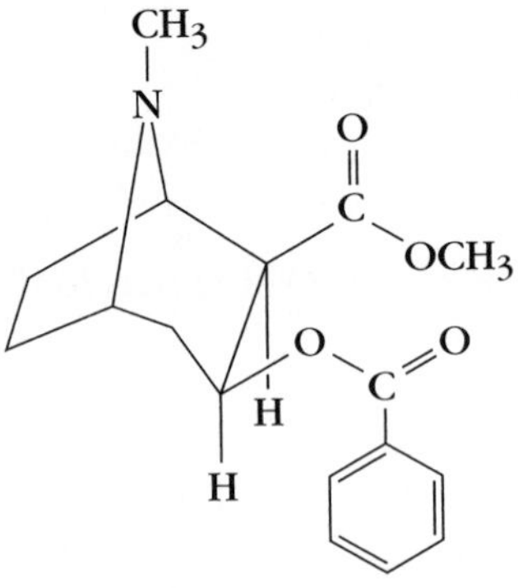   Cocaine 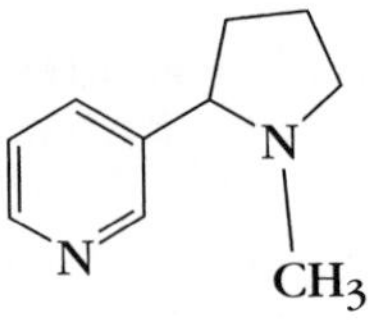   Nicotine

Other physiologically active amines can be found in humans and other animals. These compounds are not classified as alkaloids because they are not isolated from plants. Muscle activity leads to release of (-)-epinephrine by the adrenal medulla. Serotonin stimulates the contraction of smooth muscles and causes vasoconstriction. Histamine is released from body tissue in allergic reactions.

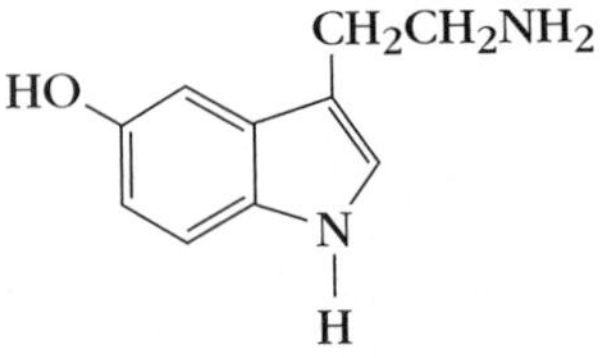

Serotonin 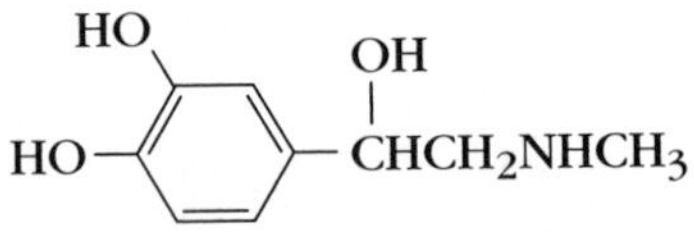   Epinephrine 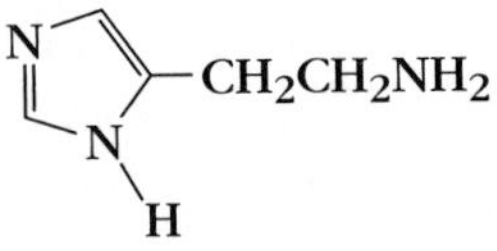   Histamine

Although the structures of most alkaloids are complex, all of them are primary, secondary, or tertiary amines just like the simpler amines described earlier in this chapter.

22.8   Synthesis of Amines: A Review

In previous chapters, we discussed a number of reactions that form amines as products. These reactions are reviewed in Table 22.6. The reductions of nitriles and amides are both good methods of preparing primary amines. Secondary and tertiary amines can also be prepared by the reduction of suitable N-substituted amides. The reactions of ammonia and amines with α-bromocarboxylic acids and epoxides are nucleophilic substitution reactions that produce amines with a second functional group.

The last method of preparing amines in Table 22.6, which is called the alkylation of amines, is not as simple as it looks. Generally a mixture of mono-, di-, and trialkylamines is formed by alkylation of amines because of the following reactions.

Iodomethane and ammonia first react to form methylammonium iodide:

$$CH_3I \ + \ NH_3 \ \longrightarrow \ CH_3NH_3{}^+I^-$$

As soon as it is formed, an acid-base equilibrium is set up between methylammonium ion and ammonia.

$$CH_3NH_3{}^+ \ + \ NH_3 \ \rightleftharpoons \ CH_3NH_2 \ + \ NH_4{}^+$$

Table 22.6	Methods of Preparing Amines Discussed in Previous Chapters	
1. Reduction		
a. Nitriles	Section 16.15, D	$RC{\equiv}N \xrightarrow[\text{(2) NaOH/H}_2\text{O}]{\text{(1) LiAlH}_4} RCH_2NH_2$
b. Amides	Section 16.12	
Unsubstituted amides are reduced to 1° amines, N-substituted amides are reduced to 2° amines, and N,N-disubstituted amides are reduced to 3° amines.		$\underset{NR'_2}{\overset{O}{RC}} \xrightarrow[\text{(2) H}_2\text{O}]{\text{(1) LiAlH}_4} RCH_2NR'_2$
2. Reaction of α-Bromocarboxylic Acids with Ammonia or Amines	Section 17.4	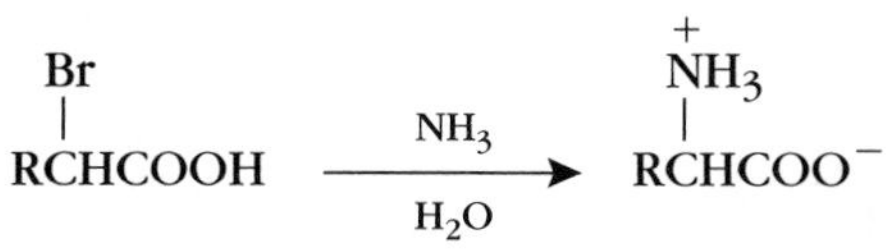 $\underset{RCHCOOH}{\overset{Br}{\mid}} \xrightarrow[\text{H}_2\text{O}]{\text{NH}_3} \underset{RCHCOO^-}{\overset{\overset{+}{NH_3}}{\mid}}$
3. Reaction of Epoxides with Ammonia or Amines	Section 13.12, A	
Ammonia and amines react at the least-substituted carbon atom of epoxide ring.		epoxide $\xrightarrow[\text{H}_2\text{O}]{\text{NH}_3}$ amino alcohol
4. Nucleophilic Substitution Reaction	Section 12.18	$RCH_2X \xrightarrow{\text{Excess NH}_3} RCH_2NH_2$
Reaction occurs via an S_N2 mechanism.		(Major product)

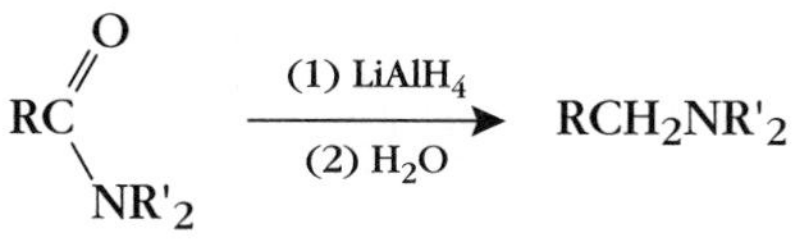

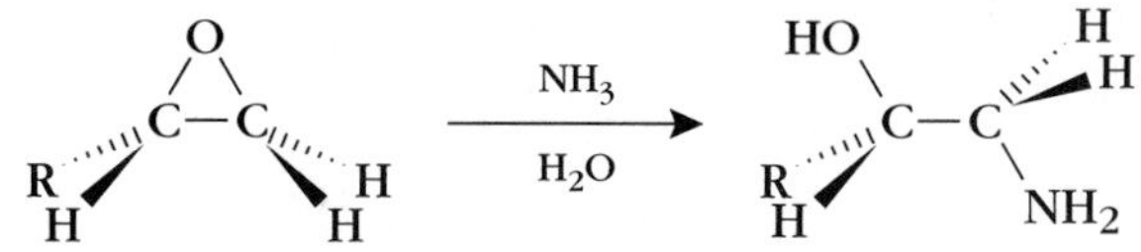

As a result, another nucleophile (methylamine) is introduced into the solution. Methylamine reacts with iodomethane to form dimethylammonium iodide:

$$CH_3NH_2 \ + \ CH_3I \ \longrightarrow \ (CH_3)_2NH_2{}^+I^-$$

Dimethylammonium ion transfers a proton to ammonia and another nucleophile, dimethylamine, is formed. These reactions continue until either all the ammonia or all the haloalkane is consumed.

| **Exercise 22.14** | Calculate K_{eq} for the following equilibrium. |

$$CH_3NH_3{}^+ \ + \ NH_3 \ \rightleftharpoons \ CH_3NH_2 \ + \ NH_4{}^+$$

A primary amine is rarely formed as the sole product in the alkylation reaction, even with an excess of ammonia. Generally, with an excess of ammonia, the primary amine is the major product. While alkylation of ammonia is rarely used in the laboratory to prepare amines, it is used successfully in industry to prepare large quantities of amines. The mixtures of amines are separated by distillation.

A better way to prepare primary amines from haloalkanes is by the Gabriel amine synthesis.

22.9   Gabriel Amine Synthesis

Preparing pure primary amines by the alkylation of amines is difficult to do because complex mixtures of amines are frequently formed. This problem can be overcome by using a nitrogen nucleophile that can be alkylated only once. One such nitrogen nucleophile is the potassium salt of phthalimide. Phthalimide is acidic enough ($pK_a = 8.3$) that it reacts with potassium hydroxide to form potassium phthalimide.

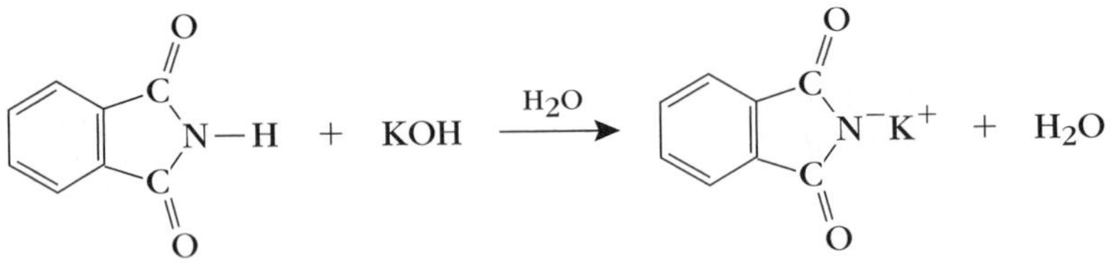

Phthalimide   Potassium phthalimide

The nitrogen atom of potassium phthalimide has a considerable negative charge, so it acts as a nucleophile in nucleophilic substitution reactions. The reaction of potassium phthalimide with primary haloalkanes occurs by an S_N2 mechanism to form an *N*-alkyl imide (a diacyl derivative of an amine). Hydrolysis of the *N*-alkyl imide in either acid or base solution forms the desired primary amine in high yields. Benzylamine, for example, can be prepared from benzyl chloride by the following sequence of reactions.

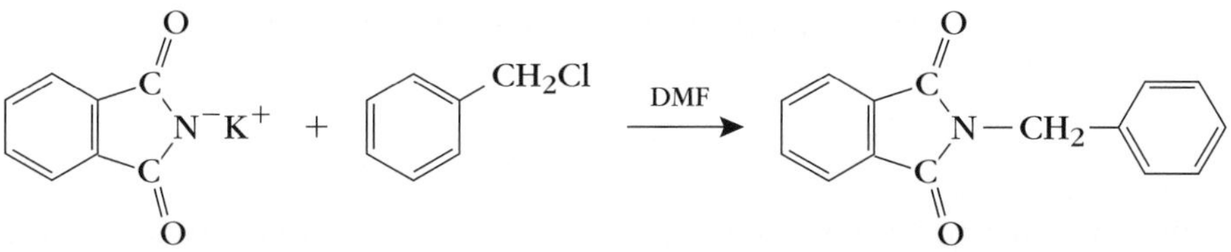

Potassium phthalimide   Benzyl chloride   *N*-Benzylphthalimide

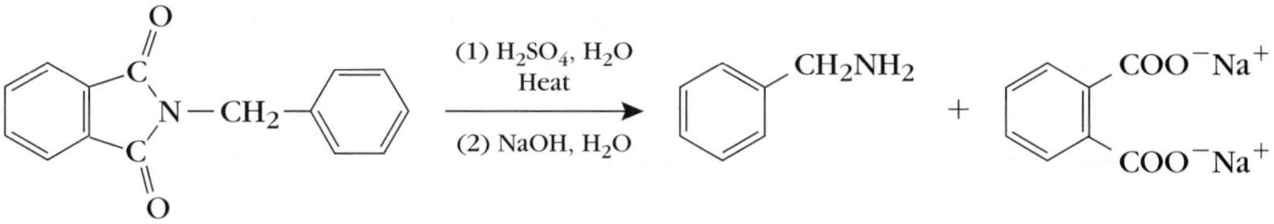

N-Benzylphthalimide   Benzylamine

Amines cannot be formed from vinyl halides or aryl halides by the Gabriel amine synthesis because these halides cannot undergo nucleophilic substitution with potassium phthalimide in the first step of the synthesis. In addition to haloalkanes, primary alkyl tosylates can also be used as starting material because they undergo nucleophilic substitution reactions by an S_N2 mechanism.

Potassium phthalimide can only be alkylated once, so no secondary or tertiary amines can be formed in this reaction. The Gabriel amine synthesis, therefore, is a particularly valuable way of preparing primary amines.

Exercise 22.15	Write the equations for the preparation of each of the following amines by a Gabriel amine synthesis.

(a) 1-Octanamine   **(b)** Isobutylamine   **(c)** α-Aminoacetic acid

Exercise 22.16	A student tries to prepare *t*-butylamine by the Gabriel synthesis using *t*-butyl chloride and potassium phthalimide as starting materials. After hydrolysis of the imide and work up, the only product isolated was phthalimide. The perplexed student comes to you for help. What went wrong? What happened to the original *t*-butyl chloride?

Reduction of nitrogen-containing compounds is another general method of the synthesis of amines.

22.10 Synthesis of Amines by the Reduction of Nitrogen-Containing Compounds

Most nitrogen-containing compounds form amines when reduced. Two examples, the reduction of nitriles and amides, have already been discussed in earlier chapters (see Table 22.6, p. 977).

Reduction of a nitro group is the most common method of synthesizing primary arylamines.

$$ArNO_2 \xrightarrow{\text{Reducing agent}} ArNH_2$$

Reduction of aromatic nitro compounds to primary arylamines can be accomplished by hydrogenation with Pt, PtO$_2$, Pd, or Ni as a catalyst. Other common reducing agents are metals, such as Fe, Zn, or Sn, in hydrochloric acid. The wide variety of aromatic nitro compounds available from the nitration of aromatic compounds makes reduction a general route to arylamines. Thus a two-step sequence of nitration of an aromatic compound followed by reduction is a standard synthesis of arylamines.

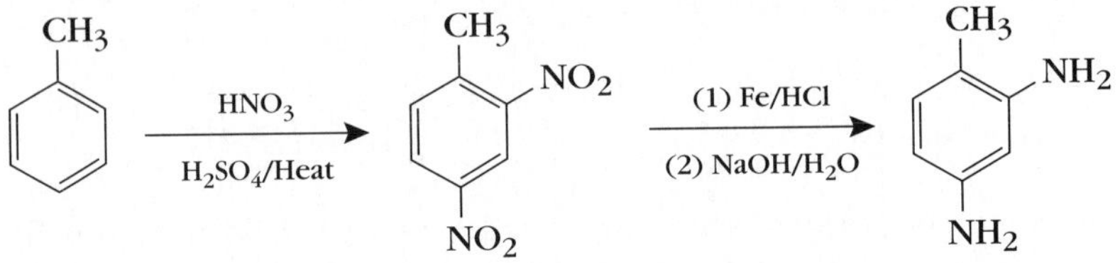

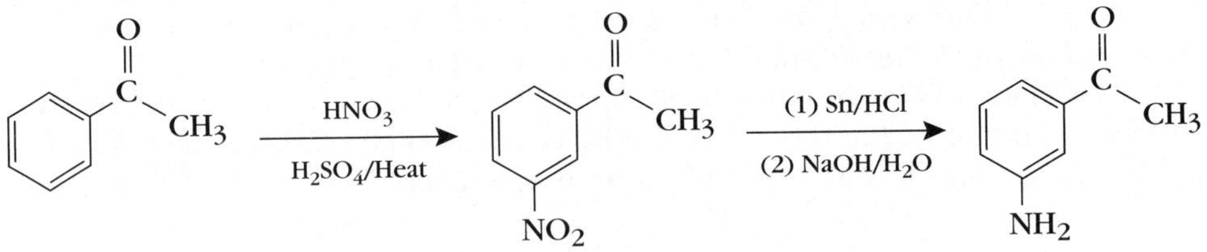

Reduction of imines, formed by the reaction of ammonia with aldehydes and ketones, is another way to synthesize primary amines. The imine prepared from ammonia is not usually isolated. Instead, a solution of excess ammonia and an aldehyde or ketone is reduced to a primary amine by hydrogen gas in the presence of a catalyst.

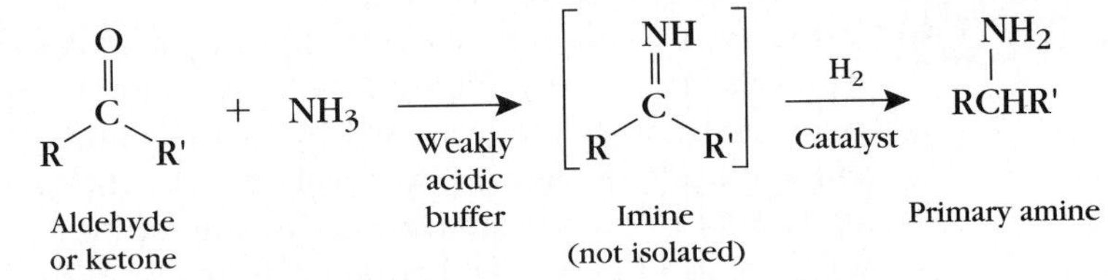

- *Recall from Section 14.13 that the reaction of aldehydes and ketones with ammonia or primary amines is carried out in a weakly acidic buffer solution (pH = 4 to 5).*

This reaction, which is called **reductive amination,** is particularly useful because it converts aldehydes and ketones to primary, secondary, and tertiary amines. The reaction can be done in a single step, moreover, because the intermediate need not be isolated. Originally hydrogen gas and a catalyst were used as the reducing agent. A more convenient variation of this method uses sodium or lithium cyanoborohydride (NaBH$_3$CN or LiBH$_3$CN) in place of hydrogen and a catalyst. These reducing agents are similar to NaBH$_4$ except that they are stable to the moderately acidic reaction conditions needed for reductive amination.

Secondary amines can be prepared by reduction of a mixture of an aldehyde or ketone and a primary amine. The *N*-substituted imine, or Schiff base, is believed to be the species that undergoes reduction.

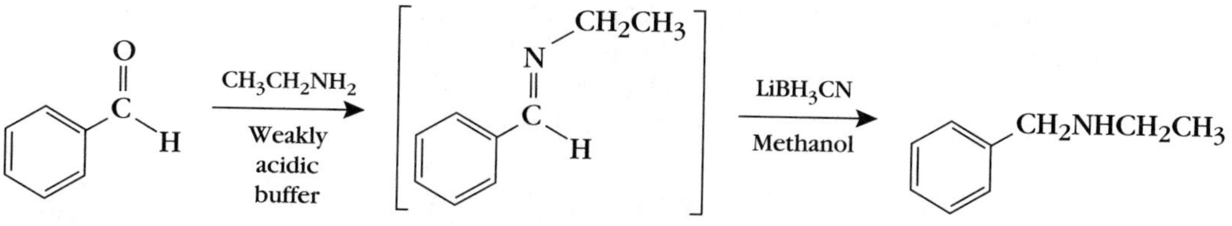

Benzaldehyde — $CH_3CH_2NH_2$ / Weakly acidic buffer → Intermediate imine is not isolated — $LiBH_3CN$ / Methanol → $CH_2NHCH_2CH_3$ Benzylethylamine

Tertiary amines can be prepared by a similar reaction starting with a secondary amine and an aldehyde or ketone.

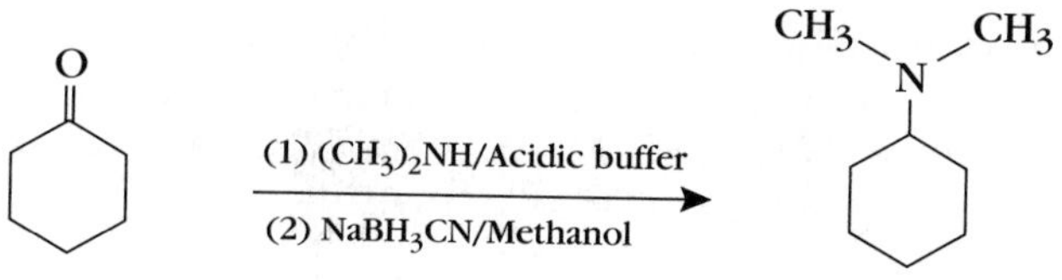

Cyclohexanone

(1) $(CH_3)_2NH$/Acidic buffer
(2) $NaBH_3CN$/Methanol

N,N-Dimethyl-cyclohexanamine

An imine cannot form in the reaction of a secondary amine with aldehydes and ketones, so a carbinolamine or an iminium ion is believed to be the species that is reduced in this reaction.

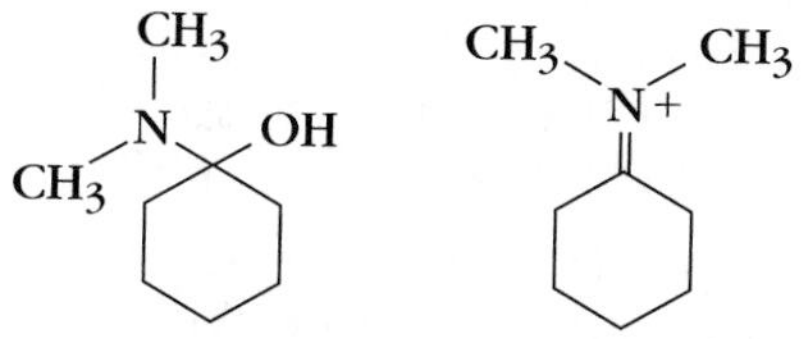

Carbinolamine

Iminium ion

Alkylamines can also be prepared by the reduction of alkyl azides.

$$CH_3(CH_2)_6CH_2N_3 \xrightarrow[\text{(2) NaOH/H}_2\text{O}]{\text{(1) LiAlH}_4} CH_3(CH_2)_6CH_2NH_2$$

1-Azidooctane

1-Octanamine
(70% yield)

Alkyl azides are readily prepared by the nucleophilic substitution reaction of sodium azide and a haloalkane (Section 12.18).

$$CH_3(CH_2)_6CH_2I \xrightarrow{\text{NaN}_3} CH_3(CH_2)_6CH_2N_3$$

1-Iodooctane

1-Azidooctane

Exercise 22.17 Which of the following aromatic amines can be prepared by the sequence of nitration and reduction of an aromatic compound? Which cannot be prepared in this way?

(a) m-Aminobenzoic acid   **(b)** m-Methoxyaniline   **(c)** p-Chloroaniline

| Exercise 22.18 | Write equations for the preparation of each of the following amines by reductive amination. |

(a) 1-Hexanamine   **(c)** *N,N*-Diethylbenzylamine

(b) *N*-Isopropylaniline   **(d)** *N*-Cyclohexylpiperidine,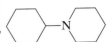

22.11 Synthesis of Primary Amines from Alkenes

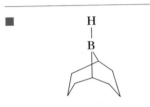

9-Borabicyclo[3.3.1.]nonane
(or 9-BBN) is frequently
abbreviated as

Alkenes react with borane and 9-BBN to form organoboranes. The boron atom of these organoboranes can be replaced by a number of different atoms and groups. Recall from Section 8.4 that alcohols can be formed by the reaction of organoboranes with hydrogen peroxide.

$$3RCH{=}CH_2 \xrightarrow[\text{THF}]{BH_3} [RCH_2CH_2]_3B \xrightarrow[\text{NaOH}]{H_2O_2} 3RCH_2CH_2OH$$

Alkene   Trialkylborane   Alcohol

Primary amines can be formed by the reaction of organoboranes with chloramine (H_2NCl) or hydroxylamine-O-sulfuric acid (NH_2OSO_3H).

$$[RCH_2CH_2]_3B \xrightarrow{H_2NCl} 3RCH_2CH_2NH_2$$

Trialkylborane   Primary amine

The overall result of this sequence of two reactions is formation of a primary amine by the anti-Markownikoff addition of NH_3 to the carbon-carbon double bond of an alkene. As in the case of the synthesis of alcohols, the product is formed by syn addition.

CH₃   H
        (1) BH₃•THF   CH₃
     ──────────────▶   H
        (2) H₂NCl
                                NH₂

1-Methylcyclopentene   *trans*-2-Methyl-
                             cyclopentanamine
                             (85% yield)

Enzyme-catalyzed additions of ammonia to the double bond of an alkene are also known. For example, the enzyme aspartase catalyzes the addition of ammonia to fumaric acid.

H   COOH   H₂N   COOH
  C   NH₃   C
  ‖   ──────────▶   H
  C   Aspartase
HOOC   H   CH₂COOH

Fumaric acid   *L*-Aspartic acid

This reaction, which is used commercially to produce several hundred tons of *L*-aspartic acid every year, differs in one important respect from the hydroboration-amination reaction. The product of the enzyme-catalyzed reaction is formed by anti addition to the double bond.

| Exercise 22.19 | Write the structure of the major product formed in the reaction of each of the following alkenes with borane in THF (BH_3/THF), followed by hydroxylamine-O-sulfuric acid and then aqueous sodium hydroxide workup. |

(a) 1-Octene   **(b)** 2-Methyl-1-pentene   **(c)** 1-Methylcycloheptene

In nature, mammals cannot incorporate nitrogen directly into organic compounds. Many plants and microorganisms can, however, and those that can use ammonium ions as the source of nitrogen atoms to synthesize amines by a process similar to reductive amination.

22.12 Microorganisms and Plants Use Ammonium Ions as the Source of Nitrogen in the Biosynthesis of Amines

Plants, animals, and microorganisms need a source of nitrogen atoms for the biosynthesis of any nitrogen-containing compound. Ammonia, in the form of ammonium ion, is the major inorganic source of nitrogen atoms used by microorganisms and plants for the biosynthesis of nitrogen-containing organic compounds. Some microorganisms use ammonium ions found in their growth medium, while others have the enzymes needed to produce NH_4^+ by the reduction of atmospheric nitrogen. Still others have enzymes that reduce nitrate (NO_3^-) and nitrite ions (NO_2^-) to NH_4^+.

Besides using ammonium ions, all microorganisms can obtain nitrogen by the degradation of amino acids. Humans and most other animals cannot use NH_4^+ directly in biosynthesis, so they must get their nitrogen atoms from the degradation of proteins and other nitrogen-containing compounds in food.

Ammonium ion is incorporated into plants and microorganisms by means of a reaction that converts it into glutamate. The enzyme glutamate dehydrogenase catalyzes the synthesis of glutamate from NH_4^+ and α-ketoglutarate. A biological reducing agent, NADPH, is needed in this reaction.

■ *The structure of NADPH is shown in Section 11.12 (Figure 11.8, p. 476).*

$$
\begin{array}{l}
COO^- \\
| \\
C{=}O \\
| \\
CH_2 \\
| \\
CH_2 \\
| \\
COO^-
\end{array}
\; + \; NH_4^+ \; + \; NADPH \;
\xrightarrow[\text{dehydrogenase}]{\text{Glutamate}} \;
\begin{array}{l}
COO^- \\
| \\
CHNH_3^+ \\
| \\
CH_2 \\
| \\
CH_2 \\
| \\
COO^-
\end{array}
\; + \; NADP^+ \; + \; H_2O
$$

α-Ketoglutarate   Glutamate

The formation of glutamate from NH_4^+ and α-ketoglutarate is the biological equivalent of the laboratory synthesis of amines by reductive amination (Section 22.10). In nature, NADPH instead of a metal hydride or catalytic hydrogenation reduces the intermediate imine to an amine.

The reaction of ammonium ions with glutamate to form glutamine is another pathway for the biosynthesis of nitrogen-containing organic compounds. The enzyme glutamine synthetase catalyzes this reaction, and energy is provided by the hydrolysis of ATP.

$$
\begin{array}{c}
\text{COO}^- \\
|\\
\text{CHNH}_3{}^+ \\
|\\
\text{CH}_2 \\
|\\
\text{CH}_2 \\
|\\
\text{COO}^-
\end{array}
+ \text{NH}_4{}^+ + \text{ATP}
\xrightarrow[\text{synthetase}]{\text{Glutamine}}
\begin{array}{c}
\text{COO}^- \\
|\\
\text{CHNH}_3{}^+ \\
|\\
\text{CH}_2 \\
|\\
\text{CH}_2 \\
|\\
\underset{\text{O}}{\overset{}{\text{C}}}\!\!-\!\text{NH}_2
\end{array}
+ \text{ADP} + \text{PO}_4{}^{3-}
$$

Glutamate   Glutamine

The formation of glutamine from glutamate and $NH_4{}^+$ is an enzyme-catalyzed formation of an amide from a carboxylic acid. The analogous laboratory reaction is discussed in Section 16.5.

Glutamate and glutamine are pivotal molecules in the biosynthesis of other nitrogen-containing organic compounds in nature. A few of the many pathways leading from glutamate and glutamine are shown in Figure 22.2.

Glutamate is a key intermediate in the formation of several nonessential amino acids in animals. These reactions are called transaminations because the —NH₂ group of glutamate is transferred to the carbonyl group of an α-ketocarboxylate. In the cell, α-ketocarboxylates are obtained by the biodegradation of carbohydrates.

Recall from Section 22.5 that —NH₂ groups exist as their conjugate acid, an —NH₃⁺ group, at the pH level of living systems (pH ≈ 7).

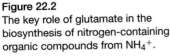

$$
\begin{array}{c}
\text{O}\\
\|\\
\text{R}\!-\!\text{C}\!-\!\text{C}\!-\!\text{O}^-\\
\|\\
\text{O}
\end{array}
+
\begin{array}{c}
\text{COO}^-\\
|\\
\text{CHNH}_3{}^+\\
|\\
\text{CH}_2\\
|\\
\text{CH}_2\\
|\\
\text{COO}^-
\end{array}
\rightleftharpoons
\begin{array}{c}
\text{NH}_3{}^+\\
|\\
\text{R}\!-\!\text{CH}\!-\!\text{COO}^-
\end{array}
+
\begin{array}{c}
\text{COO}^-\\
|\\
\text{C}=\text{O}\\
|\\
\text{CH}_2\\
|\\
\text{CH}_2\\
|\\
\text{COO}^-
\end{array}
$$

α-Ketocarboxylate   Glutamate   Amino acid   α-Ketoglutarate

These reactions are catalyzed by enzymes called transaminases (aminotransferases), which use pyridoxal phosphate as a coenzyme. The mechanism of transamination reactions is discussed in Section 14.15.

Amino acids in living organisms are continually being synthesized and degraded. The first step in the biodegradation of many amino acids in animals is the reverse of the reactions just mentioned; that is, the amino acids lose their

Figure 22.2
The key role of glutamate in the biosynthesis of nitrogen-containing organic compounds from $NH_4{}^+$.

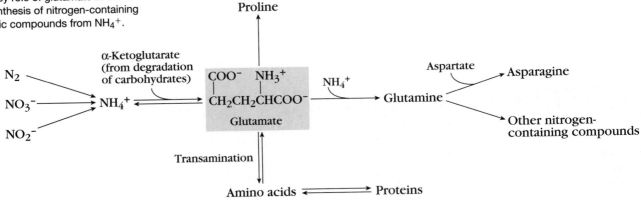

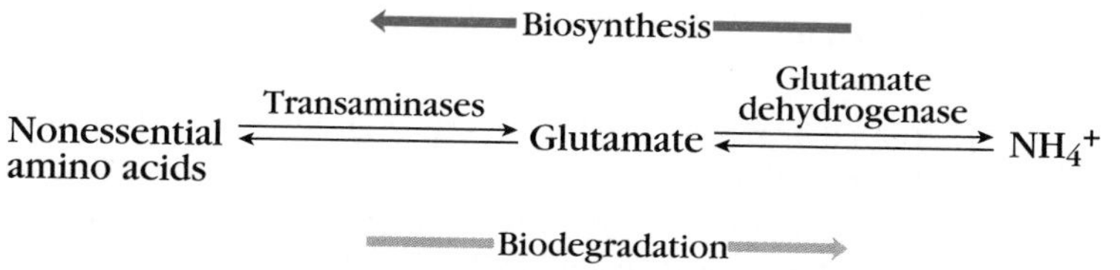

$$\text{Nonessential amino acids} \underset{\text{Transaminases}}{\rightleftarrows} \text{Glutamate} \underset{\text{Glutamate dehydrogenase}}{\rightleftarrows} NH_4^+$$

← Biosynthesis

Biodegradation →

Figure 22.3
The reversible reactions involved in the biosynthesis and biodegradation of nonessential amino acids in animals.

amino group in a transamination reaction to form glutamate. The amino group of glutamate is then converted into NH_4^+ in the glutamate dehydrogenase–catalyzed reaction (this is the reverse of the reaction that incorporates ammonium ion into organic compounds). The steps in the biosynthesis and biodegradation of most nonessential amino acids are summarized in Figure 22.3.

The fate of NH_4^+ produced in the biodegradation of amino acids depends on the requirements of the organism. If the organism needs the nitrogen of the ammonia, then the excess ammonia is quickly recycled by reforming glutamate and glutamine. If the organism does not need the nitrogen of the ammonia, then the excess ammonia must be excreted because ammonia and ammonium ion are toxic in high concentrations. Fishes excrete ammonia directly, while animals convert excess ammonia or ammonium ions into urea in the urea cycle, then excrete the urea. Birds and reptiles convert excess ammonia into uric acid, which is then excreted.

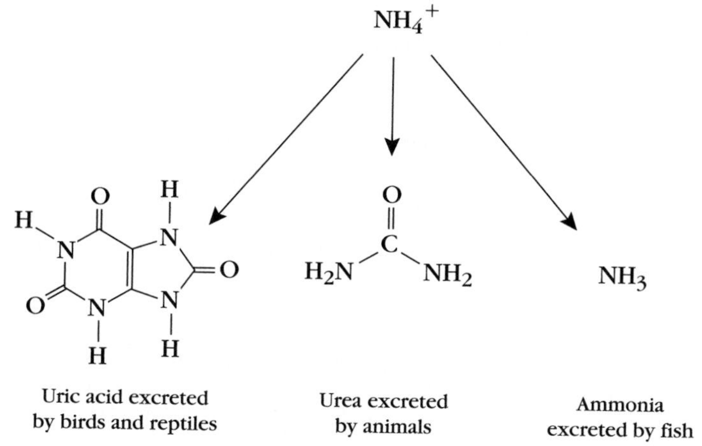

Uric acid excreted by birds and reptiles

Urea excreted by animals

Ammonia excreted by fish

In summary, glutamate and glutamine are pivotal compounds for both the incorporation of nitrogen atoms into organic molecules in living systems and the removal of nitrogen from living systems. Glutamate is formed by a reaction that is the biological equivalent of reductive amination. Glutamine is formed by the reaction of ammonium ion and glutamate. Both are soluble carriers of ammonia and are intermediates in the formation of a wide variety of nitrogen-containing organic compounds.

| Exercise 22.20 | Write the structure of the product of each of the following transamination reactions. |

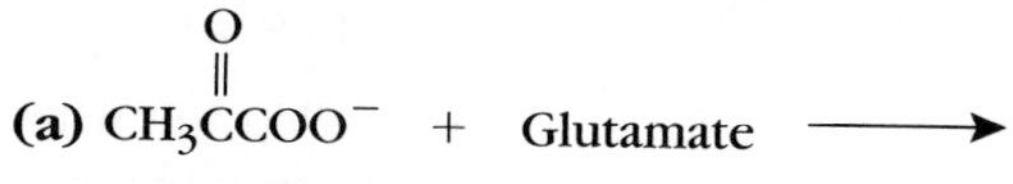

(a) $CH_3\overset{O}{\overset{\|}{C}}COO^-$ + Glutamate ⟶

Pyruvate

(b) $^-OOC\overset{O}{\overset{\|}{C}}CH_2OPO_3^{2-}$ + Glutamate ⟶

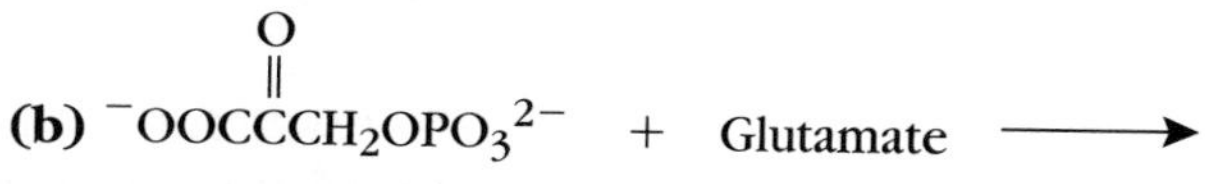

3-Phosphoglycerate

22.13 Reactions of Amines: A Review

The lone pair of electrons on the nitrogen atom of amines is responsible for both their basicity and nucleophilicity. The basicity of amines was discussed in Section 22.5, while several reactions in which amines act as nucleophiles were discussed in previous chapters and are summarized in Table 22.7.

Table 22.7	Reactions of Amines Discussed in Previous Chapters

1. Reaction with Aldehydes and Ketones

a. Ammonia and 1° amines

Nucleophilic addition of amine to carbonyl carbon atom forms a carbinolamine intermediate. Loss of water from the carbinolamine intermediate forms the imine.

Section 14.13

$$\underset{\substack{\text{Aldehyde} \\ \text{or ketone}}}{\overset{\displaystyle O}{\underset{R}{\overset{\|}{C}}\underset{R'}{}}} + \underset{\substack{\text{Ammonia} \\ \text{or 1° amine}}}{H_2NR''} \xrightarrow[\substack{\text{Weakly} \\ \text{acidic} \\ \text{buffer}}]{} \underset{\text{Imine}}{\overset{\displaystyle NR''}{\underset{R}{\overset{\|}{C}}\underset{R'}{}}}$$

b. 2° Amines

Nucleophilic addition of 2° amine to carbonyl carbon atom forms a carbinolamine intermediate. Loss of water from the carbinolamine intermediate forms the enamine.

Section 17.14

$$\underset{\substack{\text{Aldehyde} \\ \text{or ketone}}}{\overset{\displaystyle O}{\underset{R}{\overset{\|}{C}}\underset{CH_2R'}{}}} + \underset{\text{2° amine}}{HNR''_2} \xrightarrow[\substack{\text{Weakly} \\ \text{acidic} \\ \text{buffer}}]{} \underset{\text{Enamine}}{\overset{\displaystyle R''\;\;\;\;R''}{\underset{\underset{H}{\overset{|}{C}-R'}}{\overset{\displaystyle N}{\underset{R}{C}}}}}$$

2. Reaction of Ammonia and 1° or 2° Amine with Carboxylic Acid Derivatives

Nucleophilic addition to carbonyl carbon atom forms a tetrahedral intermediate. Loss of group Y from tetrahedral intermediate forms an amide.

Section 16.5 and 16.9

$$\underset{\substack{\text{Carboxylic} \\ \text{acid} \\ \text{derivative}}}{\overset{\displaystyle O}{\underset{R}{\overset{\|}{C}}\underset{Y}{}}} + \underset{\substack{\text{Ammonia,} \\ \text{1°, or 2°} \\ \text{amine}}}{HNR''_2} \longrightarrow \underset{\text{Amide}}{\overset{\displaystyle O}{\underset{R}{\overset{\|}{C}}\underset{NR''_2}{}}}$$

Y = Cl, OOCR, SR', OR'

3. Nucleophilic Substitution Reaction

Section 12.18

$$RCH_2X + \underset{\substack{\text{Ammonia, 1°, 2°,} \\ \text{or 3° amine}}}{NR'_3} \longrightarrow RCH_2\overset{+}{N}R'_3\;X^-$$

Because of their nonbonding pair of electrons, amines undergo oxidation reactions. An example is their reaction with nitrous acid.

22.14 Nitrosation of Alkylamines

The reaction of amines with nitrous acid (H—O—N=O) is called the **nitrosation of amines.** Nitrous acid is unstable, so it is prepared in the reaction mixture (*in situ*) by mixing sodium nitrite ($NaNO_2$) with cold dilute hydrochloric acid.

The species in solution that reacts with the amine is believed to be the nitrosyl cation formed in the following sequence of reactions:

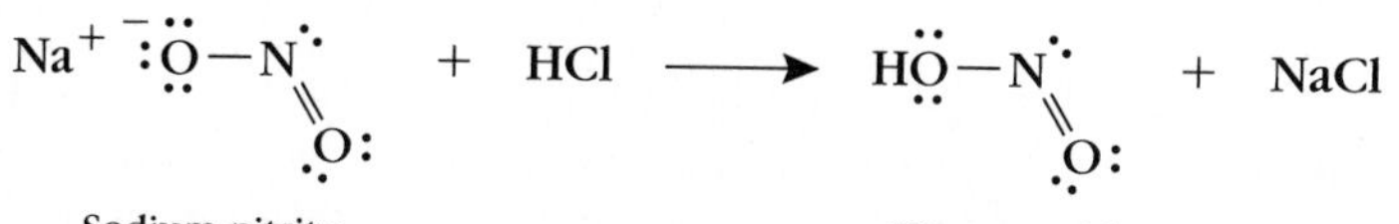

Sodium nitrite   Nitrous acid

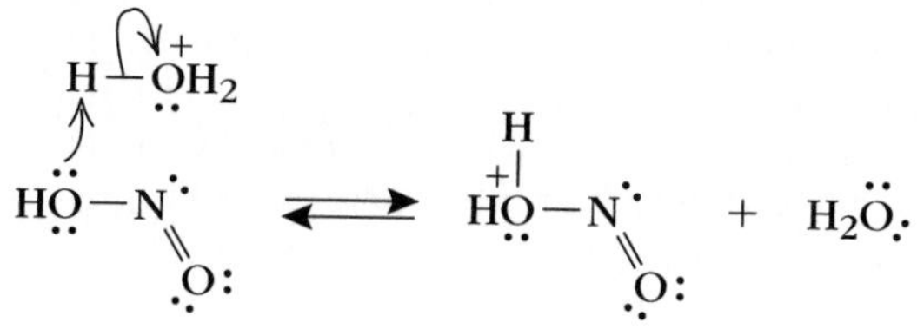

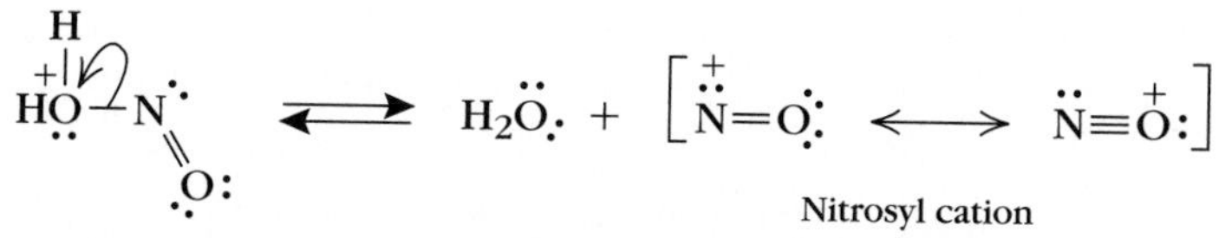

Nitrosyl cation

The nitrosyl cation is electrophilic, so it reacts with the lone pair of electrons on the nitrogen atom of amines to form an *N*-nitrosoammonium compound.

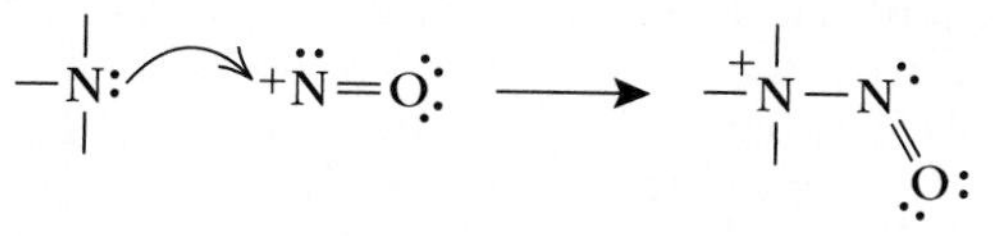

Amine   Nitrosyl   *N*-Nitrosoammonium
               cation   compound

What happens next depends on the structure of the amine. The product of nitrosation of secondary alkyl or aryl amines is an *N*-nitrosodialkylamine, also called a **nitrosamine,** that is formed by the loss of a proton from the *N*-nitrosoammonium compound.

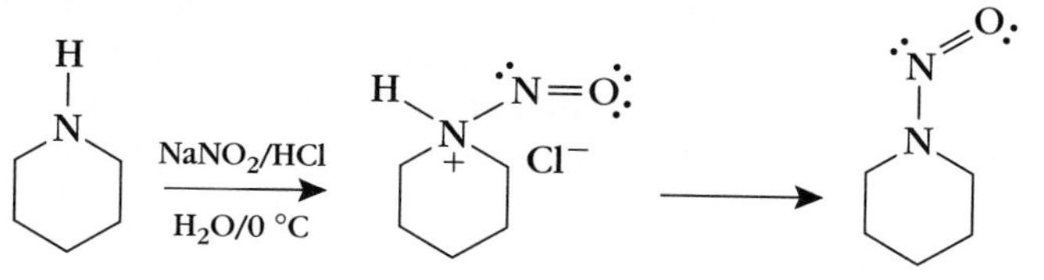

Piperidine   *N*-Nitrosopiperidine

Nitrosamines are potent carcinogens in a variety of animals and are suspected of causing cancer in humans (see Box 22.1).

The nitrosation of primary alkylamines forms alkanediazonium ions, $R-\overset{+}{N}\equiv N:$, by the mechanism shown in Figure 22.4 (p. 989). Step 1 of the mechanism of formation of an alkanediazonium ion from a primary amine is a Lewis acid-base reaction followed by removal of a proton to form an *N*-nitrosoamine in Step 2. Step 3 is an acid-catalyzed tautomerization reaction in which the tautomer with a nitrogen-oxygen double bond is converted to the tautomer with a nitrogen-nitrogen double bond. This tautomerization reaction is similar to the keto-enol tautomerization discussed in Section 17.1. Protonation of the hydroxy group in Step 4 followed by loss of water in Step 5 forms the alkanediazonium ion.

Box 22.1   Nitrosamines and Cancer

The curing of meat with aqueous solutions of sodium nitrite and sodium nitrate is a long-established practice in the meat-packing industry. Sodium nitrite apparently plays an important role in the curing process and also prevents the growth of the bacterium *Clostridium botulinum*, which causes botulism, a deadly form of food poisoning.

A number of years ago, concern was raised that nitrosamines, which were known to be carcinogens in animals, could be formed when humans ate meats cured with sodium nitrite. Nitrosamines were detected, in fact, in a variety of cured meats. For example, *N*-nitrosodimethylamine is found in hot dogs and *N*-nitrosopyrrolidine is formed when bacon cured with sodium nitrite is fried.

$$CH_3 \quad \overset{\cdot\cdot}{N}-\overset{\cdot\cdot}{N} \overset{O}{=} \qquad \qquad$$

N-Nitrosodimethylamine   *N*-Nitrosopyrrolidine

It was then discovered that nitrosamines are produced in humans by compounds that are all obtained naturally. Spinach and beets and other vegetables, for example, contain relatively large amounts of nitrate, which is enzymatically reduced to nitrite in the body. Many constituents of the food we eat are secondary amines, so they react with nitrite in the acidic environment of the stomach to form nitrosamines.

How can we determine if nitrites in processed food are responsible for causing cancer in humans? The answer is we can't. In fact, it's still uncertain whether naturally produced nitrosamines or nitrites added to our diet play a role in causing human cancers.

Alkanediazonium ions rapidly lose nitrogen at room temperature to form carbocations, which undergo their usual reactions such as elimination of a proton, reaction with a nucleophile, and rearrangement. So many products are formed by the loss of nitrogen from an alkanediazonium ion that it is of little practical use.

$$RCH=CH_2 \qquad \qquad RCH_2CH_2OH$$

$$RCH_2CH_2NH_2 \xrightarrow[\substack{H_2SO_4 \\ H_2O}]{NaNO_2} \left[RCH_2CH_2\overset{+}{N}\equiv N \right] \longrightarrow RCH_2\overset{+}{C}H_2 \; + \; N_2$$

Alkanediazonium ion   Carbocation

More products of loss $\longleftarrow$ $RCH\overset{+}{C}H_3$
of proton and reactions
with nucleophiles   Rearranged
                                         carbocation

Step 1: Lewis acid-base reaction of the nitrosyl cation and a 1° amine

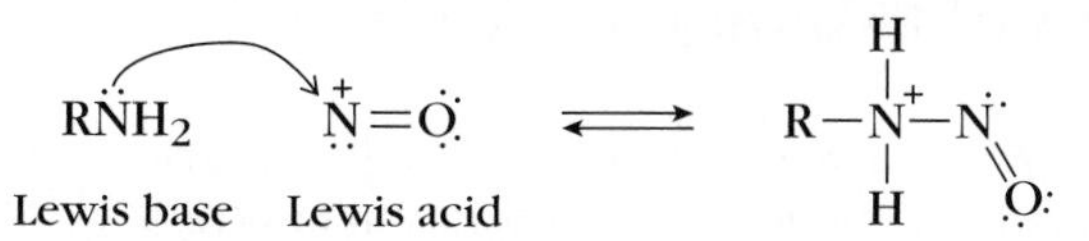

$$R\ddot{N}H_2 \quad \overset{+}{\ddot{N}}=\ddot{O} \quad \rightleftharpoons \quad R-\overset{H}{\underset{H}{\overset{|}{\underset{|}{N}}}}-\ddot{N}\overset{..}{\underset{\ddot{O}:}{}}$$

Lewis base   Lewis acid

Figure 22.4
The mechanism of nitrosation of primary amines to form an alkanediazonium ion. The letter *B* represents any base in solution such as a water molecule or a molecule of amine.

Step 2: Removal of a proton to form an *N*-nitrosoamine

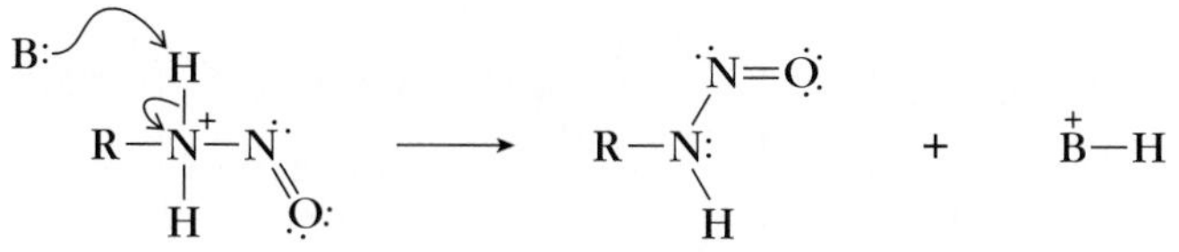

$$B:\quad R-\overset{H}{\underset{H}{\overset{|}{\underset{|}{\overset{+}{N}}}}}-\ddot{N}\overset{..}{\underset{\ddot{O}:}{}} \quad \longrightarrow \quad R-\underset{H}{\overset{|}{N}}:\,\overset{\ddot{N}=\ddot{O}}{} \quad + \quad \overset{+}{B}-H$$

N-Nitrosoamine

Step 3: Acid-catalyzed tautomerization

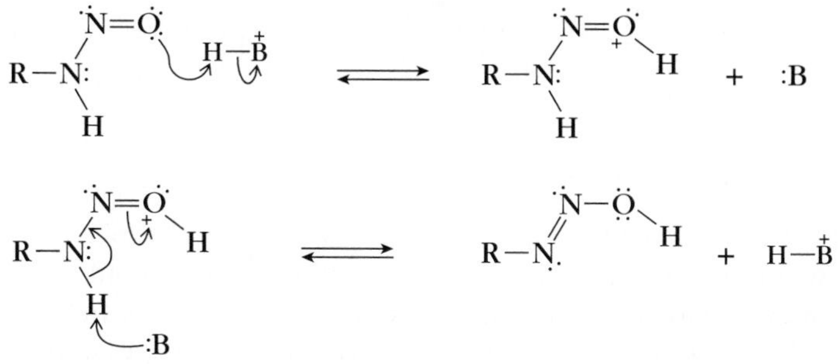

$$R-\underset{H}{\overset{|}{N}}:\,\overset{\ddot{N}=\ddot{O}}{}\quad H-\overset{+}{B} \quad \rightleftharpoons \quad R-\underset{H}{\overset{|}{N}}:\,\overset{\ddot{N}=\ddot{O}}{}\,H \quad + \quad :B$$

$$R-\underset{H}{\overset{|}{N}}:\,\overset{\ddot{N}=\ddot{O}}{}\,H \quad :B \quad \rightleftharpoons \quad R-\ddot{N}.\,\overset{\ddot{N}-\ddot{O}}{}\,H \quad + \quad H-\overset{+}{B}$$

Step 4: Protonation of the OH group

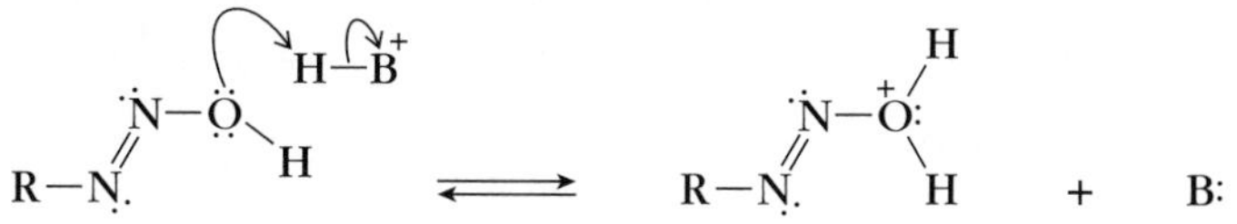

$$R-\ddot{N}.\,\overset{\ddot{N}-\ddot{O}}{}\,H \quad H-\overset{+}{B} \quad \rightleftharpoons \quad R-\ddot{N}.\,\overset{\ddot{N}-\overset{+}{O}:}{}\overset{H}{\underset{H}{}} \quad + \quad B:$$

Step 5: Loss of water to form the diazonium ion

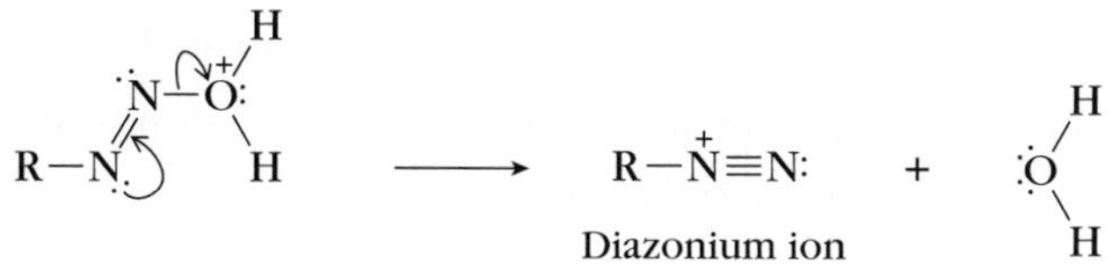

$$R-\ddot{N}.\,\overset{\ddot{N}-\overset{+}{O}:}{}\overset{H}{\underset{H}{}} \quad \longrightarrow \quad R-\overset{+}{N}\equiv N: \quad + \quad :\overset{H}{\underset{H}{\ddot{O}}}$$

Diazonium ion

The nitrosation of tertiary amines is very complicated and will not be discussed because no useful chemistry is associated with these reactions. Nitrosation of primary arylamines, on the other hand, is a very useful reaction that we will discuss in detail in Section 22.15.

Exercise 22.21   Write the structure of the product of nitrosation of each of the following amines.

(a) *N*-Methylaniline   **(b)** Diisopropylamine   **(c)** *N*-Ethyl-2-butanamine

22.15 Preparation of Arenediazonium Ions by the Nitrosation of Primary Arylamines

Primary arylamines react with an acidic solution of sodium nitrite to form **arene-diazonium ions,** which are stable in aqueous solution at 0 to 5 °C. The process of converting an arylamine into an arenediazonium ion is called **diazotization** and occurs with a wide variety of aniline derivatives.

$$ArNH_2 \xrightarrow[\substack{\text{Aqueous } H_2SO_4 \\ 5\,°C}]{NaNO_2} Ar\overset{+}{N}{\equiv}N + HSO_4^-$$

Substituted aniline   Arendiazonium ion

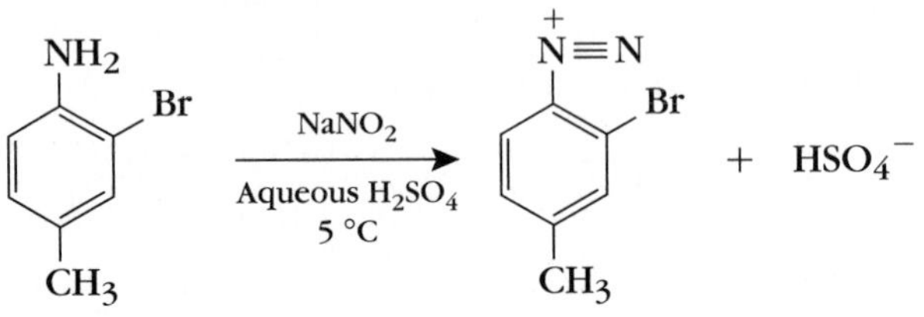

2-Bromo-4-methylaniline

Arenediazonium salts are not usually isolated because they tend to decompose violently when in the solid state. Solutions of arenediazonium salts are relatively stable, however, and they react with a number of reagents to form a variety of substituted benzenes plus nitrogen gas, as we will learn in Section 22.16.

Only primary arylamines form diazonium salts. Secondary arylamines, just like secondary alkylamines, form *N*-nitrosoamines under the same reaction conditions.

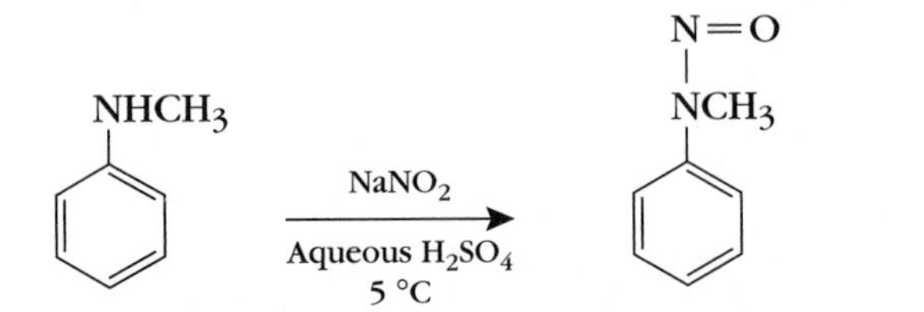

N-Methylaniline   *N*-Methyl-*N*-nitrosoaniline

Nitrosation of tertiary arylamines results in an electrophilic substitution reaction in which the nitrosyl cation is the electrophile.

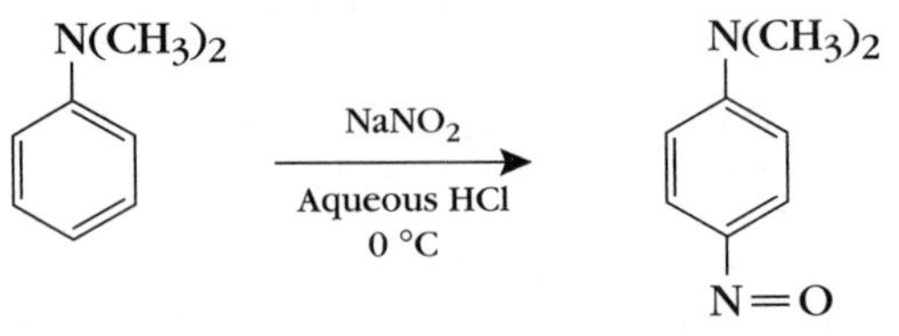

N,N-Dimethylaniline   *p*-Nitroso-*N,N*-dimethylaniline

Exercise 22.22   Write the structure of the product of the reaction of an aqueous acidic solution of sodium nitrite with each of the following compounds.

(a) *N*-Methyl-2-butanamine   **(b)** 2,4,6-Trichloroaniline   **(c)** Diphenylamine

Arenediazonium ions are important in organic synthesis because they undergo a number of reactions in which the diazonium group (N_2^+) is replaced by other atoms or groups.

22.16 Replacement of the Diazonium Group in Arenediazonium Ions

Many substituents can be introduced into a benzene ring by electrophilic substitution reactions (Section 21.3). Many substituents can also be introduced into a benzene ring by replacing the diazonium group of arenediazonium ions with another substituent. In this way, a primary amino group on a benzene ring can be transformed into a wide variety of substituents. The types of substitution reactions that arenediazonium ions undergo are summarized in Figure 22.5.

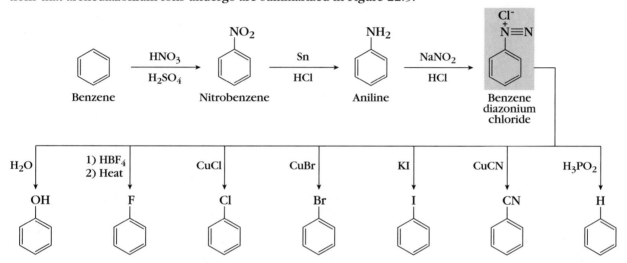

Figure 22.5
Summary of the preparation of substituted aromatic compounds by the replacement of the diazonium group of an arenediazonium ion by other atoms or groups.

Certain substituted aromatic compounds that cannot be prepared directly by electrophilic aromatic substitution can be synthesized easily by the reaction of arenediazonium ions. The cyano group, for example, cannot be introduced directly into an aromatic ring either by electrophilic aromatic substitution or by direct substitution because most aromatic halides do not undergo nucleophilic substitution reactions. However, a cyano group can be introduced by the reaction of an arenediazonium ion with copper(I)cyanide. The replacement of the diazonium group by the cyano group is regiospecific, as illustrated by the conversion of *p*-toluidine to *p*-toluonitrile.

Aromatic nitriles prepared in this way can be converted to many other compounds, including a carboxylic acid by hydrolysis, a primary amine by reduction, and a ketone by reaction with a Grignard reagent.

Aromatic chlorides and bromides can also be prepared by the reaction of arenediazonium salts with Cu(I)Cl and Cu(I)Br.

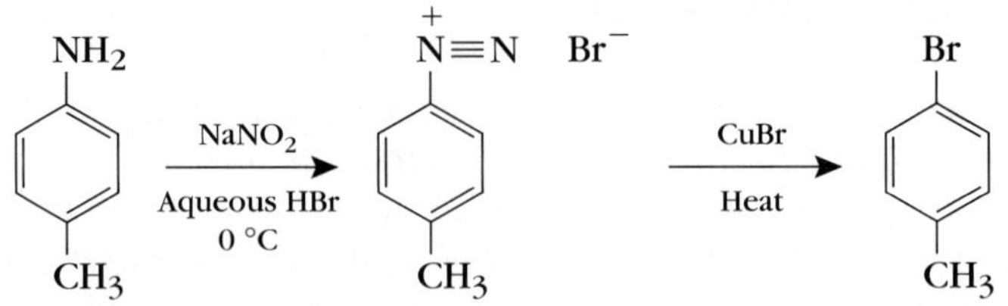

p-Toluidine   *p*-Toluenediazonium bromide   *p*-Bromotoluene

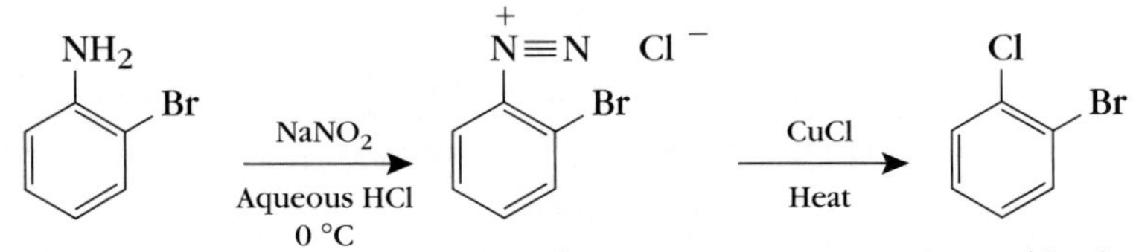

o-Bromoaniline   *o*-Bromobenzene-   *o*-Bromochlorobenzene
                         diazonium chloride

These reactions, which require copper(I) salts, are called **Sandmeyer reactions.**

The principal method of preparing aromatic fluorine compounds is by means of arenediazonium tetrafluoroborate salts. Most solid arenediazonium salts are unstable at room temperature, but their tetrafluoroborate salts are an exception. Arenediazonium tetrafluoroborate salts are prepared by diazotization with sodium nitrite and tetrafluoroboric acid (HBF_4). Heating either the dry salt or the salt in an inert solvent such as THF forms the desired aryl fluoride.

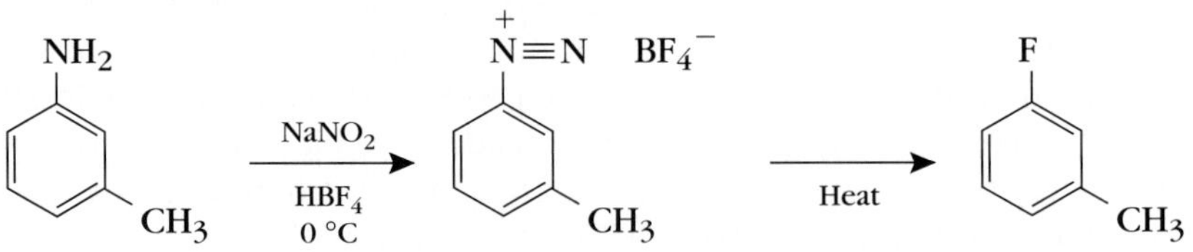

m-Toluidine   *m*-Toluenediazonium   *m*-Fluorotoluene
                       tetrafluoroborate

Aryl iodides can be prepared from arenediazonium ions by simply adding potassium iodide to a cold aqueous solution of the arenediazonium salt.

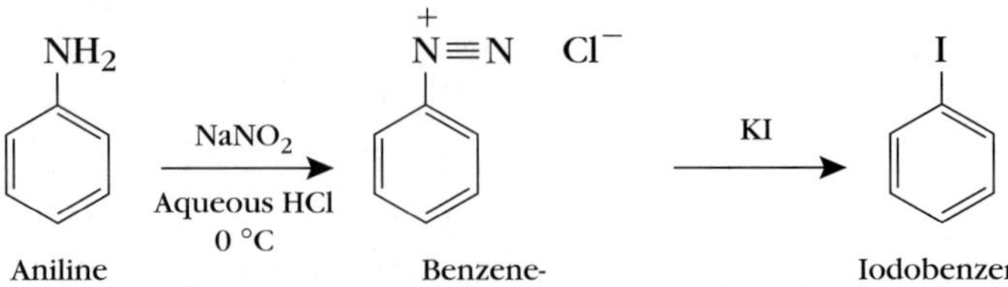

Aniline   Benzene-   Iodobenzene
                 diazonium
                 chloride

In all the reactions of arenediazonium salts discussed so far, small amounts of phenols are formed by means of a side reaction in which water reacts with the diazonium salt. The replacement of the diazonium group by a hydroxy group becomes the major reaction when the bisulfate salt of an arenediazonium is heated. Experimentally, this is accomplished by using aqueous sulfuric acid and sodium nitrite as the diazotization medium. Simply heating the reaction mixture forms a phenol. This reaction is useful because there are relatively few methods of introducing a hydroxy group into an aromatic ring.

m-Nitroaniline → m-Nitrobenzene-diazonium hydrogen sulfate → m-Nitrophenol

Finally, arenes are formed by the reduction of arenediazonium salts by hypophosphorus acid (H_3PO_2). The overall effect is to replace the diazonium group with a hydrogen atom.

$$ArNH_2 \xrightarrow[\text{Aqueous } H_2SO_4 \\ 5\,°C]{NaNO_2} Ar\overset{+}{N}\equiv N \ HSO_4^- \xrightarrow{H_3PO_2} ArH$$

Substituted aniline → Arenediazonium hydrogen sulfate → Arene

It may seem strange to remove a substituent after so much time and effort has been spent introducing it into a benzene ring in the first place. Sometimes it is synthetically useful, however, to introduce a group, take advantage of its directing effect in electrophilic substitution reactions, and then remove it. For example, the preparation of 1,3,5-tribromobenzene from aniline utilizes this synthetic strategy.

Aniline → 2,4,6-Tribromoaniline → 2,4,6-Tribromobenzene-diazonium chloride → 1,3,5-Tribromobenzene

An amino group is so strongly activating that bromine reacts with aniline without a catalyst to form 2,4,6-tribromoaniline. Diazotization followed by reaction with hypophosphorus acid removes the amino group and forms 1,3,5-tribromobenzene. This sequence of three reactions is necessary because 1,3,5-tribromobenzene can't be formed by direct bromination of benzene.

In summary, formation of arenediazonium compounds by the diazotization of arylamines is a useful synthetic reaction because the diazonium group can be replaced by a number of other groups and atoms in a subsequent reaction. These reactions make it possible to prepare aromatic compounds that cannot be synthesized by aromatic substitution reactions. In this way, the diazotization of arylamines complements the introduction of substituents by electrophilic aromatic substitution.

| **Exercise 22.23** | Starting with either benzene or toluene, propose a synthesis for each of the following compounds. |

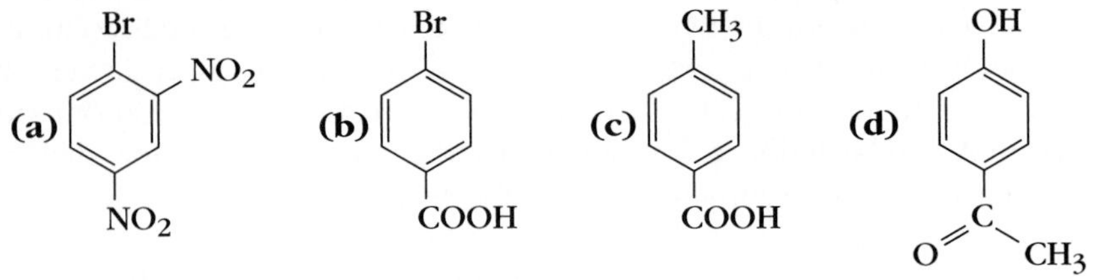

(a)   (b)   (c)   (d)

Not all reactions of arenediazonium ions involve loss of nitrogen. An important group of reactions of arenediazonium ions is the aromatic substitution of the diazonium ion to form azo compounds.

22.17 Diazo Coupling of Arenediazonium Ions

Arenediazonium ions are very weak electrophiles that react with highly activated aromatic rings to form **azo compounds** in a reaction called **diazo coupling.**

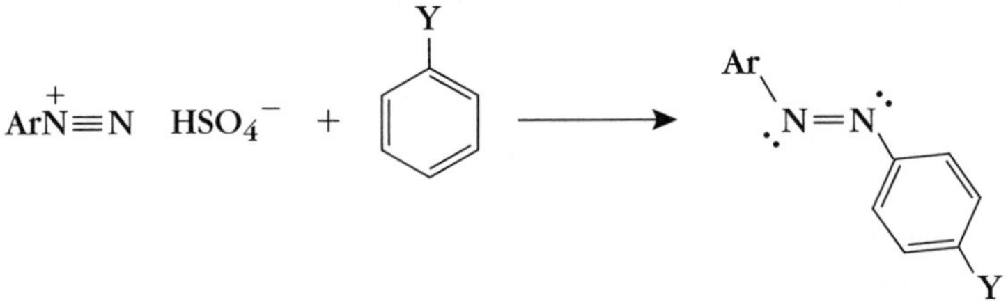

Arendiazonium salt   Y = OH or NR_2 An azo compound

Diazo coupling reactions are electrophilic aromatic substitutions in which the diazonium cation is the electrophile. A diazonium cation is such a weak electrophile that it reacts only with highly activated aromatic rings. In practice this means that diazo coupling reactions occur only with phenols and arylamines. Reaction occurs usually at the *para* position. If the *para* position is blocked, reaction takes place at the *ortho* position.

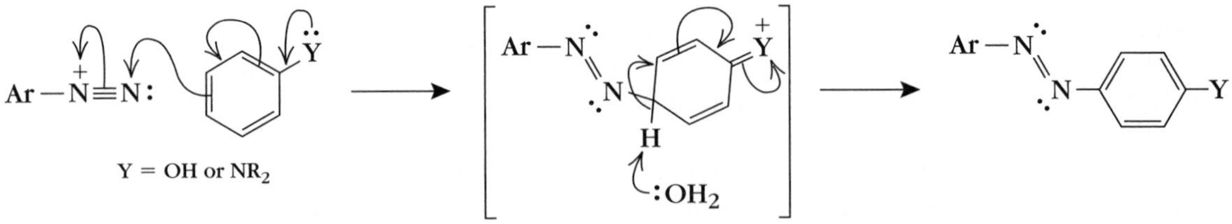

The azo group (—N=N—) brings the two aromatic rings into conjugation and extends or delocalizes their π electron system. As a result, these compounds absorb in the visible region of the electromagnetic spectrum. Because of their intense color and their easy, low-cost preparation, azo compounds are used widely as fabric dyes and as food coloring additives. The following are examples:

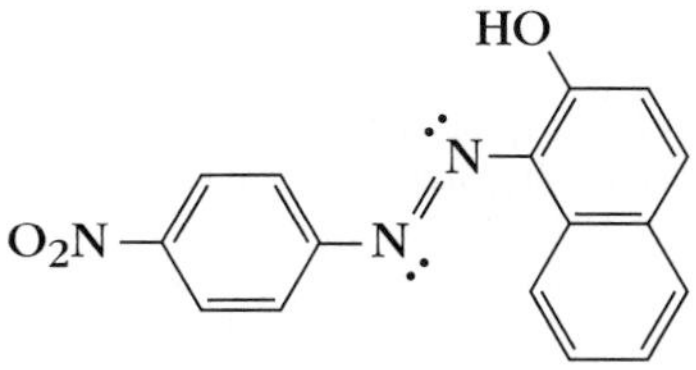

Para red
(a cotton dye)

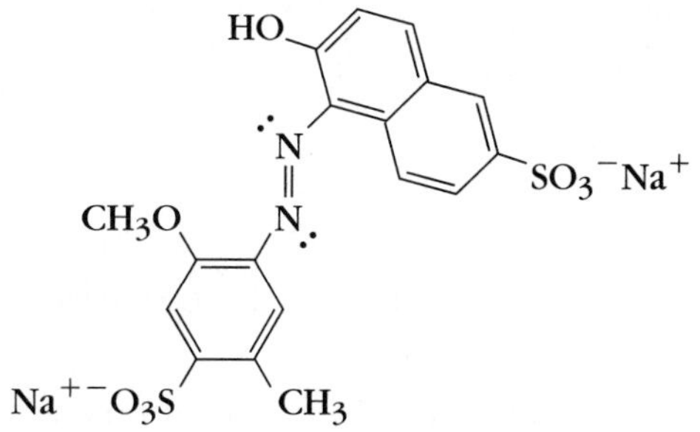

Allura red
(food color additive)

Exercise 22.24   The compound *p*-(dimethylamino)azobenzene, known commercially as butter yellow, was once used as a food coloring for margarine. It's no longer used for this purpose, however, because it is a known carcinogen. Propose a synthesis for butter yellow from benzene and *N,N*-dimethylaniline.

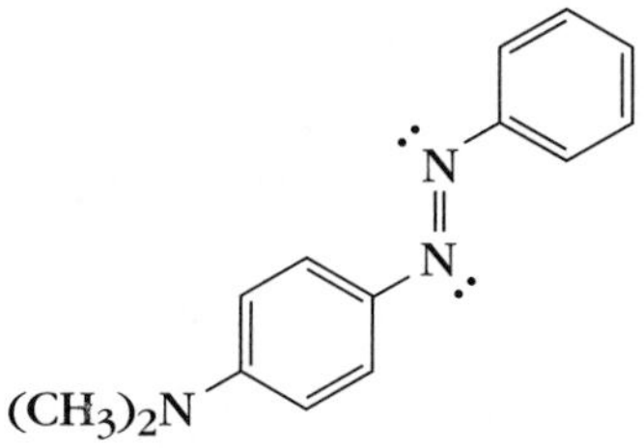

p-(Dimethylamino)azobenzene
(butter yellow)

Aniline is an important compound because many valuable compounds can be obtained from it; besides dyes and food coloring additives, sulfa drugs are another valuable class of compound that is derived from aniline.

22.18 Sulfa Drugs Are Synthesized by Reaction of Amines with Arenesulfonyl Chlorides

The reaction of amines with arenesulfonyl chlorides is similar to the reaction of amines with carboxylic acid chlorides (Section 16.5). Primary alkyl and arylamines react with arenesulfonyl chlorides to form *N*-substituted arenesulfonamides, while secondary amines form *N,N*-disubstituted arenesulfonamides. Neither acyl chlorides nor arenesulfonyl chlorides react with tertiary amines to form stable products.

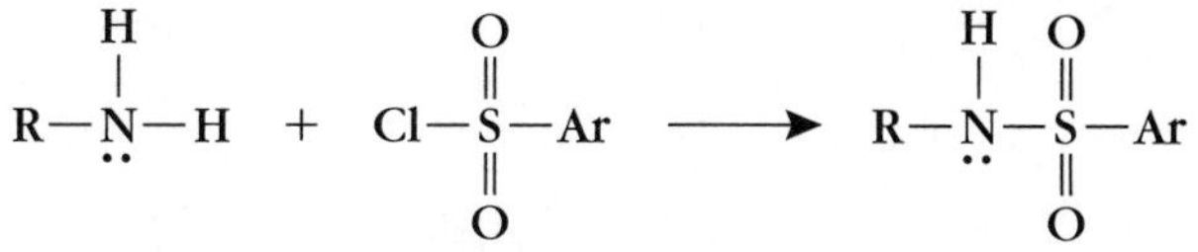

| 1° Amine | Arenesulfonyl chloride | *N*-Substituted arenesulfonamide |

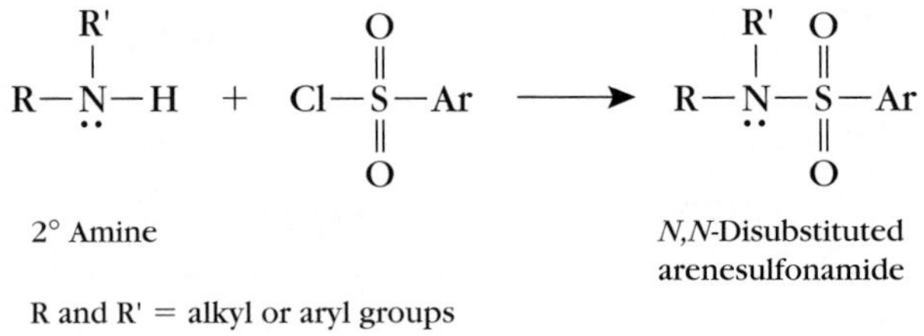

2° Amine   *N,N*-Disubstituted
arenesulfonamide

R and R' = alkyl or aryl groups

Arenesulfonamides, like carboxylic acid amides, are hydrolyzed to the amine salt in aqueous acid. The hydrolysis of arenesulfonamides, however, is much slower than the hydrolysis of carboxylic acid amides.

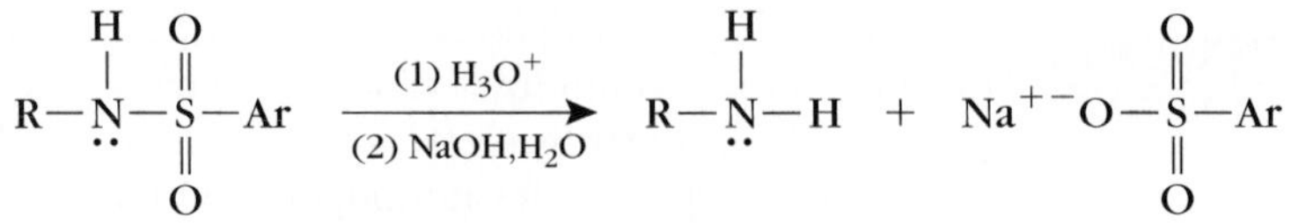

The reaction of amines and arenesulfonyl chloride is a key step in the synthesis of sulfanilimide, one of the sulfa drugs, from aniline.

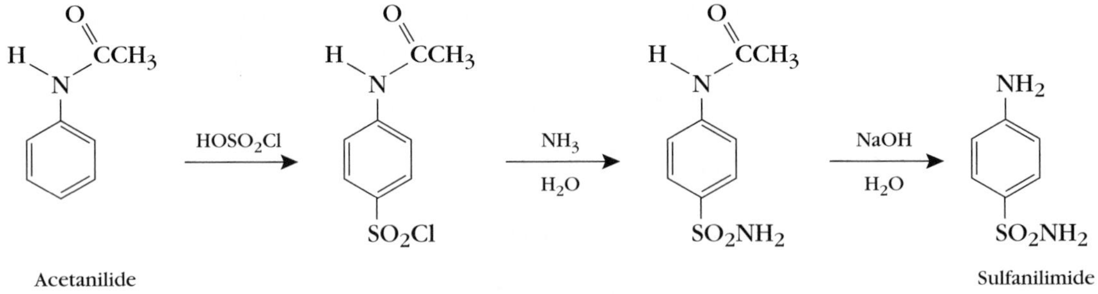

Acetanilide   Sulfanilimide

The first reaction in this sequence is the protection of the amino group of aniline by acetylation to form acetanilide. Acetanilide then undergoes an electrophilic aromatic substitution reaction with chlorosulfonic acid (HOSO$_2$Cl) followed by reaction with ammonia or another primary amine to form a diamide (an amide of both a carboxylic acid and sulfonic acid). The final reaction is hydrolysis of the carboxylic acid amide to form sulfanilimide.

Analogous sequences of reactions have been used to synthesize many other sulfa drugs, including the following:

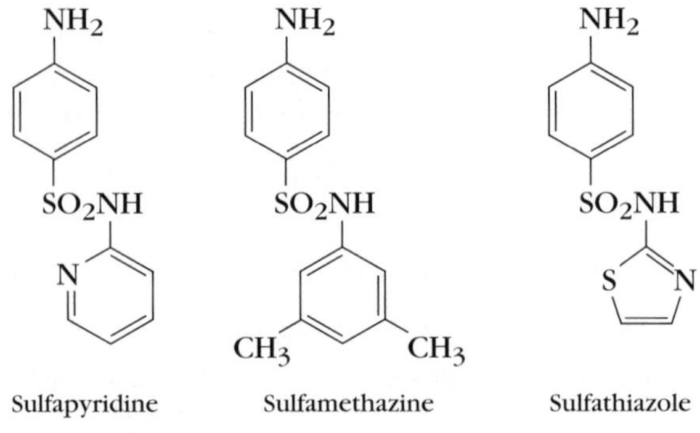

Sulfapyridine   Sulfamethazine   Sulfathiazole

Exercise 22.25	By analogy with the synthesis of sulfanilimide from aniline shown on p. 996, propose a synthesis for sulfapyridine, sulfamethazine, and sulfathiazole.

Sulfa drugs are effective in combating infectious diseases because their structures resemble the structure of 4-aminobenzoic acid. Many bacteria need 4-aminobenzoic acid to synthesize the coenzyme folic acid. In the presence of sulfa drugs, the bacteria use the sulfa drug instead of 4-aminobenzoic acid in the synthesis of folic acid to form an altered folic acid. Part of the structure of altered folic acid is obtained from the sulfa drug instead of 4-aminobenzoic acid, as shown in Figure 22.6. The altered folic acid, however, is ineffective as a coenzyme in the bacteria, so the infectious organism dies. Humans are not affected by sulfa drugs because we cannot synthesize folic acid (we obtain folic acid from food).

Folic acid (a coenzyme):

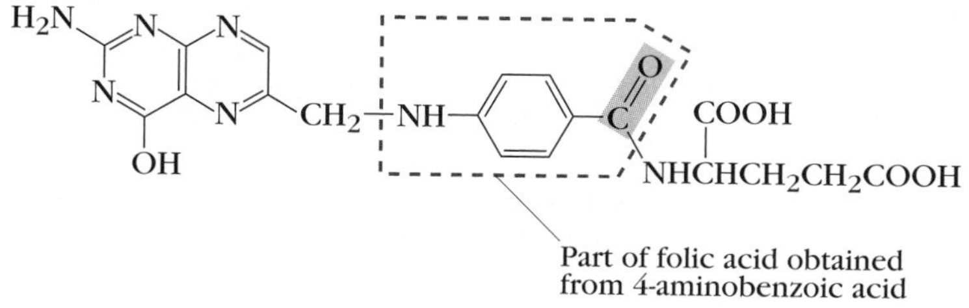

Part of folic acid obtained from 4-aminobenzoic acid

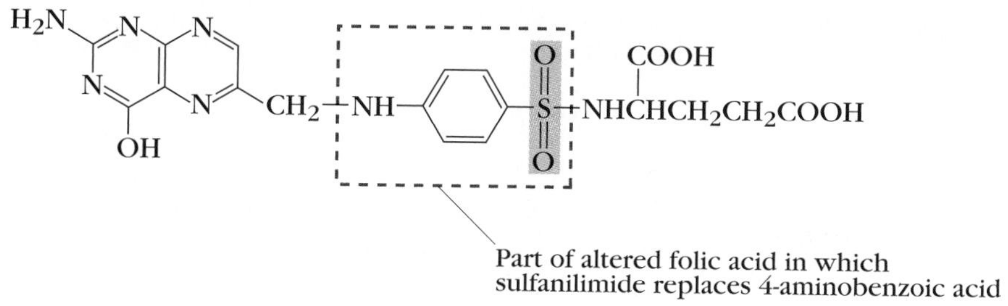

Figure 22.6
The structure of folic acid and altered folic acid.

Altered folic acid (biologically inactive):

Part of altered folic acid in which sulfanilimide replaces 4-aminobenzoic acid

22.19 Summary

Amines are derivatives of ammonia in which one or more hydrogen atoms has been replaced by alkyl or aryl groups. Alkylamines, like ammonia, are pyramidal in shape. Their nonbonding electron pair is localized in one of the four orbitals of the sp^3 hybrid nitrogen atom. In arylamines, the nonbonding electron pair is delocalized into the π electron system of the ring, so arylamines are less basic than alkylamines.

The nonbonding pair of electrons on nitrogen makes amines both bases and nucleophiles. The basicity of amines is usually expressed as the pK_a of their conjugate acid (alkylammonium or anilinium ions). The pK_a values of alkylammonium ions are in the range 9 to 11. Anilinium ions are more acidic, with pK_a values in the range 4 to 6. Groups that deactivate the benzene ring in electrophilic substitution reactions increase the acidity of anilinium ions (make arylamines weaker bases), while groups that activate the ring to electrophilic substitution decrease the acidity of anilinium ions (make arylamines stronger bases).

22.20 Summary of Preparations of Amines

In this section, only the preparations of amines discussed in this chapter are summarized. For other methods, see Table 22.6 (p. 977).

1. **Reduction**

   a. **Nitro groups**   SECTION 22.10

   Reducing agent can be H_2/metal catalyst or metals, such as Fe, Zn, or Sn in HCl.

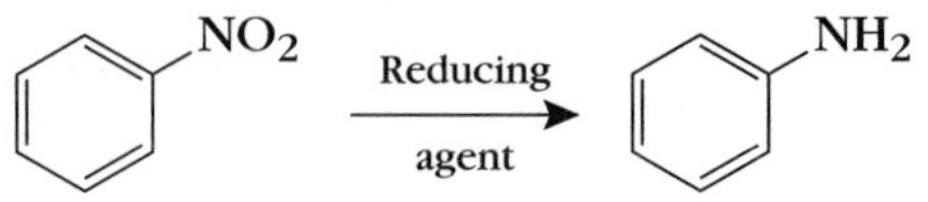

   b. **Reductive Amination**

   Reaction with ammonia forms 1° amines; reaction with 1° amines forms 2° amines; reaction with 2° amines forms 3° amines.

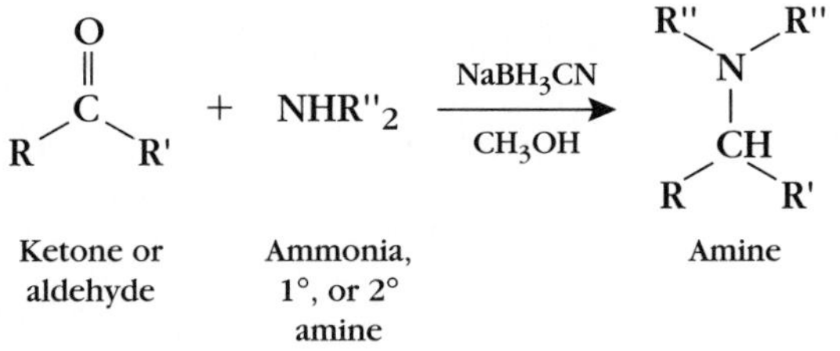

   c. **Azides**

   The nucleophilic substitution reaction of sodium azide (NaN_3) and a haloalkane occurs by an S_N2 mechanism.

   $$RCH_2Br \xrightarrow{NaN_3} RCH_2N_3 \xrightarrow{LiAlH_4} RCH_2NH_2$$

2. **Hydroboration-Amination**   SECTION 22.11

   Syn anti-Markownikoff addition of NH_3 to a carbon-carbon double bond.

   $$RCH{=}CH_2 \xrightarrow[\text{2) } H_2NCl]{\text{1) 9-BBN}} RCH_2CH_2NH_2$$

3. **Gabriel Synthesis**   SECTION 22.9

   The reaction of potassium phthalimide and a haloalkane occurs by an S_N2 mechanism.

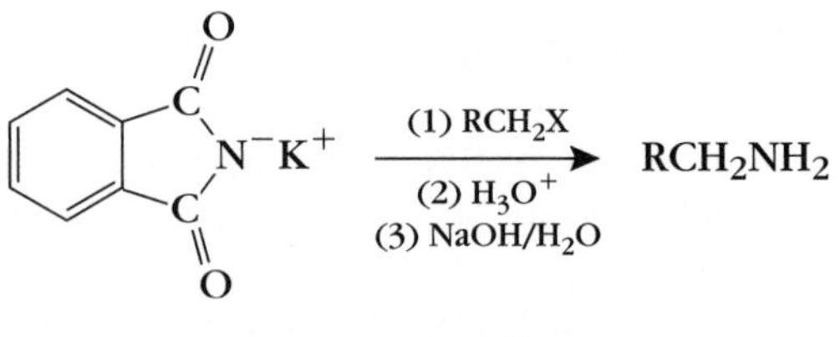

   X = Halogen or arenesulfonate

22.21 Summary of Reactions of Amines

In this section, only the reactions of amines discussed in this chapter are summarized. For other reactions, see Table 22.7 (p. 986).

1. Basicity   Section 22.5

Arylamines are generally weaker bases than alkylamines.

$$R_3N: \; + \; HB^+ \; \rightleftharpoons \; R_3\overset{+}{N}H \; + \; :B$$

2. Nitrosation   Section 22.14

   a. 2° Amines

   R = alkyl or aryl group

$$R_2\overset{..}{N}H \; \xrightarrow[\text{H}_2\text{O/0 °C}]{\text{NaNO}_2/\text{HCl}} \; R_2\overset{..}{N}-\overset{..}{N}=O$$

N-Nitrosamine

   b. 1° Amines

   R = alkyl or aryl group

$$R\overset{..}{N}H_2 \; \xrightarrow[\text{H}_2\text{O/0 °C}]{\text{NaNO}_2/\text{HCl}} \; RN_2{}^+$$

Alkanediazonium ion

3. Reaction with Arenesulfonyl Chlorides   Section 22.18

$$ArSO_2Cl \; + \; R_2\overset{..}{N}H \; \longrightarrow \; ArSO_2NR_2$$

Arenesulfonyl   Ammonia,   Arenesulfonamide
chloride   1°, or 2°
         alkylamine or
         arylamine

22.22 Summary of Reactions of Arenediazonium Ions

1. Replacement of Diazonium Group   Section 22.16

   a. By Cl

$$Ar\overset{+}{N}_2 \; \xrightarrow{\text{CuCl}} \; ArCl \; + \; N_2$$

   b. By Br

$$Ar\overset{+}{N}_2 \; \xrightarrow{\text{CuBr}} \; ArBr \; + \; N_2$$

   c. By I

$$Ar\overset{+}{N}_2 \; \xrightarrow{\text{CuI}} \; ArI \; + \; N_2$$

   d. By F

$$Ar\overset{+}{N}_2 \; \xrightarrow[\text{Heat}]{\text{HBF}_4} \; ArF \; + \; N_2$$

   e. By OH

$$Ar\overset{+}{N}_2 \; \xrightarrow{\text{H}_2\text{O}} \; ArOH \; + \; N_2$$

f. By CN

$$\overset{+}{ArN_2} \xrightarrow{CuCN} ArCN + N_2$$

g. By H

$$\overset{+}{ArN_2} \xrightarrow{H_3PO_2} ArH + N_2$$

2. Diazo Coupling   SECTION 22.17

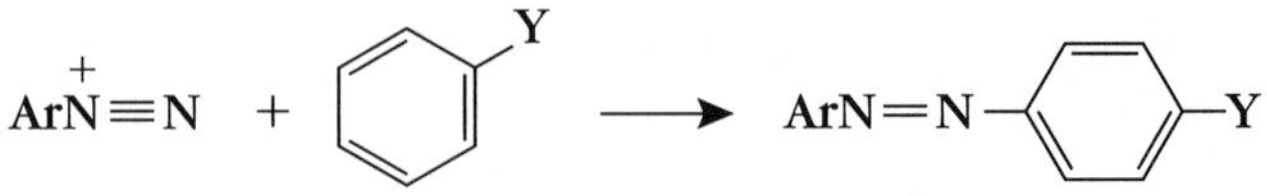

$$\overset{+}{ArN}\equiv N + \underset{}{\bigcirc}-Y \longrightarrow ArN=N-\bigcirc-Y$$

$$Y = -OH \text{ or } -NR_2$$

Additional Exercises

22.26 Define each of the following terms in words, by a structure, or by means of a chemical equation.

(a) Gabriel amine synthesis	**(e)** Arenediazonium ion	**(i)** Quaternary ammonium salt
(b) A sulfa drug	**(f)** *N*-Nitrosamine	**(j)** Reductive amination
(c) A 1° amine	**(g)** Diazotization	**(k)** A 3° amine
(d) Nitrosation of amines	**(h)** Sandmeyer reaction	**(l)** Arenesulfonamide

22.27 Write the structure of each of the following compounds.

(a) Isobutylamine	**(d)** 4-Chloro-*N,N*-dimethylaniline	**(g)** 5-Bromo-1,3-pentanediamine
(b) *N*-Benzylcyclopropanamine	**(e)** 4-Bromo-2-pentanamine	**(h)** *cis*-4-Aminocyclohexanol
(c) Tetraallylammonium acetate	**(f)** *N*–Benzyl-*N*-ethyl-1-octanamine	

22.28 Give a name to each of the following structures.

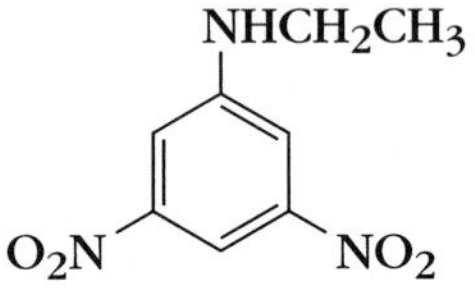

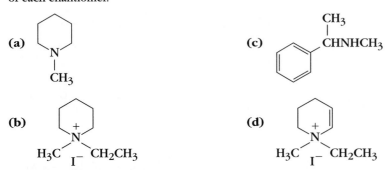

(b) $CH_3CHCH_2CH_2CHCH_3$ with NH_2 and NH_2

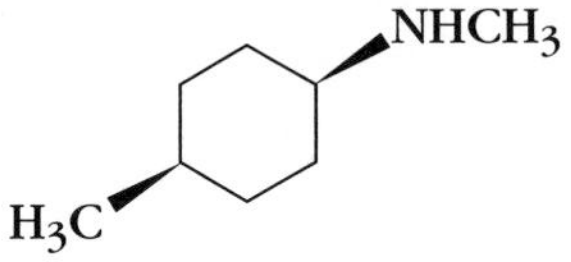

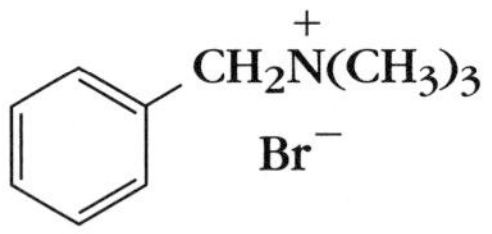

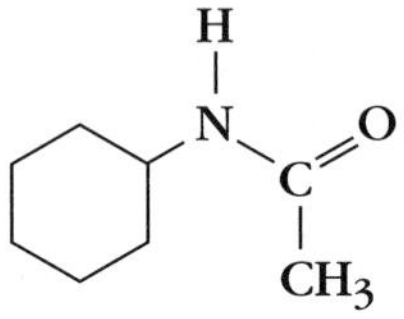

(f) $(C_6H_5)_3N$

(g) $CH_2=CHCH_2CH_2NH_2$

22.29 Classify each of the compounds in Exercise 22.28 as a 1° amine, a 2° amine, a 3° amine, or a quaternary ammonium ion.

22.30 Decide which of the following compounds can be resolved into enantiomers, and draw a three-dimensional representation of each enantiomer.

22.31 List each set of compounds in order of increasing basicity.

(a) Aniline; ammonia; methylamine

(c) H_3C-C (=O, NH_2) ; NH_3; [structure: 4-chloroaniline with NH_2 and Cl]

(b) Aniline; methylamine; 4-nitroaniline

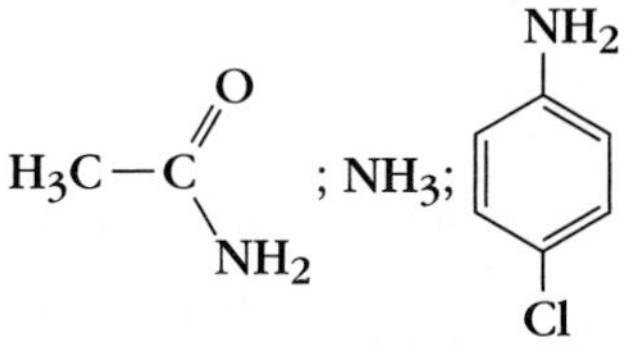

(d) CH_3COOH; CH_3NH_2; $C_6H_5NH_2$

22.32 Complete the following acid-base reactions. Decide if the products or reactants are favored at equilibrium.

(a) CH_3COOH + $CH_3CH_2NH_2$ $\rightleftharpoons$

(b) 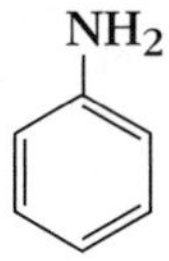 [benzene ring with NH_2] + CH_3COOH $\rightleftharpoons$

(c) 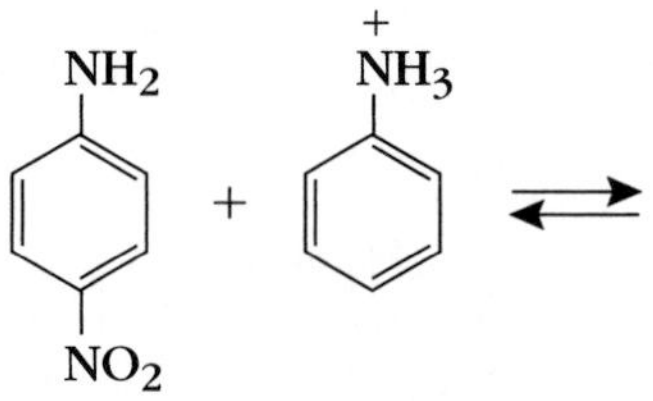 [benzene ring with NH_2 and NO_2] + [benzene ring with $\overset{+}{NH_3}$] $\rightleftharpoons$

22.33 By means of a free diagram similar to the one in Figure 22.1 (p. 974), explain why 4-nitroaniline is a weaker base than aniline.

22.34 Write the equations for the reaction or reactions that convert aniline into each of the following compounds.

(a) Acetanilide
(b) Iodobenzene
(c) Phenol
(d) 4-Nitroaniline

(e) Benzoic acid
(f) *N,N,N*-Trimethylanilinium iodide
(g) Benzene

22.35 Write the structure of the product or products in each of the following reactions.

(a) 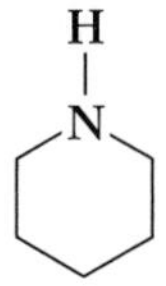 [piperidine with N-H] $\xrightarrow{\text{(CH}_3\text{CO)}_2\text{O} \\ \text{Acetic anhydride}}$

(b) Acetone $\xrightarrow[\substack{\text{NaBH}_3\text{CN} \\ \text{Weakly acidic buffer}}]{\text{NH}_3}$

(c) *N*-Ethylacetamide $\xrightarrow[\text{(2) NaOH/H}_2\text{O}]{\text{(1) LiAlH}_4}$

(d) Nitrobenzene $\xrightarrow[\text{(2) NaOH/H}_2\text{O}]{\text{(1) Sn/HCl}}$

(e) Butanenitrile $\xrightarrow[\text{(2) NaOH/H}_2\text{O}]{\text{(1) LiAlH}_4}$

(f) 1-Methylcyclohexene $\xrightarrow[\text{(2) H}_2\text{NCl}]{\text{(1) 9-BBN}}$

(g) Aniline $\xrightarrow{\text{NaNO}_2\text{/HCl}}$

(h) Product of (g) 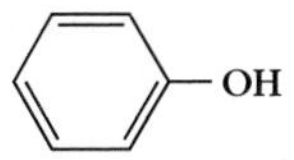 $\xrightarrow{\text{[benzene ring with OH]}}$

(i) Product of (g) $\xrightarrow{\text{CuCN}}$

(j) Product of (g) $\xrightarrow[\text{(2) Heat}]{\text{(1) HBF}_4}$

22.36 Write equations for the preparation of each of the following compounds from 1-pentanol and any other organic and inorganic reagents needed.

(a) 1-Pentanamine
(b) *N,N*-Dimethyl-1-pentanamine

(c) Hexanenitrile
(d) Pentanal

(e) Pentanoic acid
(f) *N*-Pentylacetamide

22.37 How would you prepare 1-pentanamine from each of the following starting materials and any other organic and inorganic reagents needed?

   (a)   1-Bromopentane   **(c)**   Pentanamide   **(e)**   Pentanoic acid

   (b)   1-Butanol   **(d)**   1-Pentene   **(f)**   Pentanal

22.38 Write the structure of the product (or products) formed in each of the following reactions.

(a) $CH_3CH_2NH_2$ $\xrightarrow[\text{Pyridine}]{\text{Acetyl chloride}}$

(b) 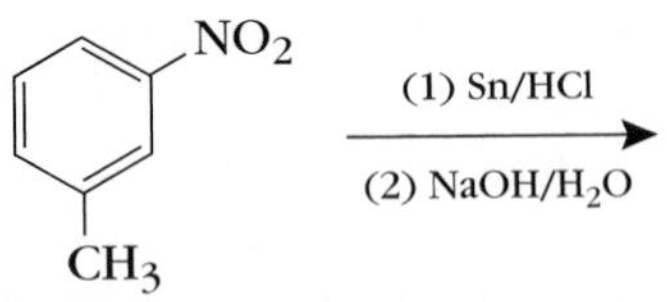   [structure: 3-nitrotoluene] $\xrightarrow[\text{(2) NaOH/H}_2\text{O}]{\text{(1) Sn/HCl}}$

(c) 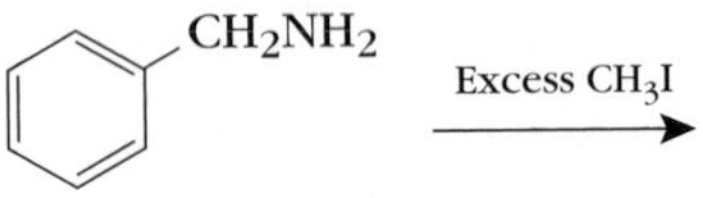   [structure: benzyl amine, CH_2NH_2] $\xrightarrow{\text{Excess CH}_3\text{I}}$

(d)   *N*-Methylbenzamide $\xrightarrow[\text{(2) NaOH/H}_2\text{O}]{\text{(1) LiAlH}_4}$

(e) $(CH_3)_2CHCN$ $\xrightarrow[\text{(2) NaOH/H}_2\text{O}]{\text{(1) LiAlH}_4}$

(f) $(CH_3)_2C{=}CH_2$ $\xrightarrow[\text{(2) ClNH}_2]{\text{(1) 9-BBN}}$

(g) 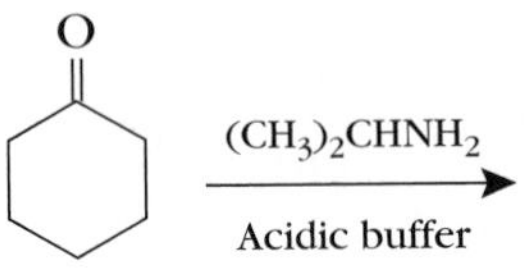   [structure: cyclohexanone] $\xrightarrow[\text{Acidic buffer}]{(CH_3)_2CHNH_2}$

(h) 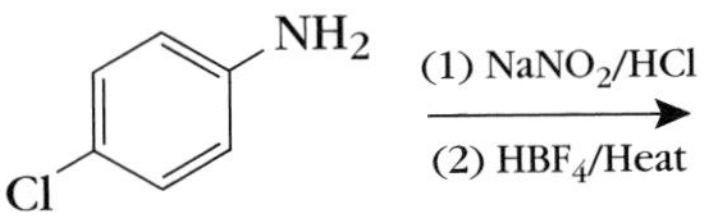   [structure: 4-chloroaniline] $\xrightarrow[\text{(2) HBF}_4/\text{Heat}]{\text{(1) NaNO}_2/\text{HCl}}$

(i) 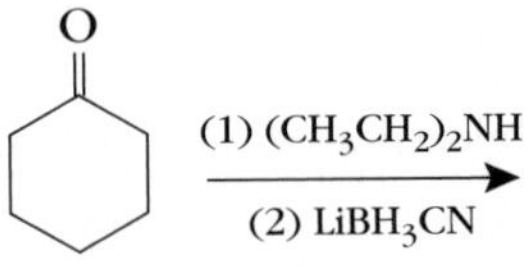   [structure: cyclohexanone] $\xrightarrow[\text{(2) LiBH}_3\text{CN}]{\text{(1) (CH}_3\text{CH}_2)_2\text{NH}}$

(j) 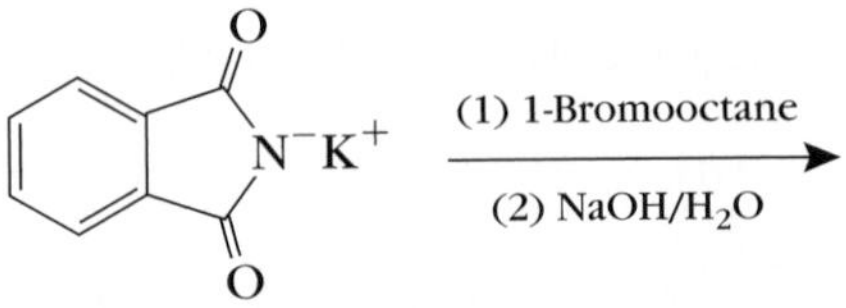   [structure: potassium phthalimide] $\xrightarrow[\text{(2) NaOH/H}_2\text{O}]{\text{(1) 1-Bromooctane}}$

(k)   3-Bromoheptane $\xrightarrow[\text{CH}_3\text{CH}_2\text{OH}]{\text{NaCN}}$

(l) 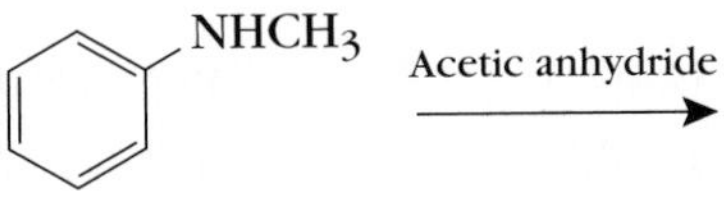   [structure: *N*-methylaniline, $NHCH_3$] $\xrightarrow{\text{Acetic anhydride}}$

(m)    [structure: 3-nitrophenylacetic acid, CH_2COOH / NO_2] $\xrightarrow[\text{FeBr}_3]{\text{Br}_2}$

(n) 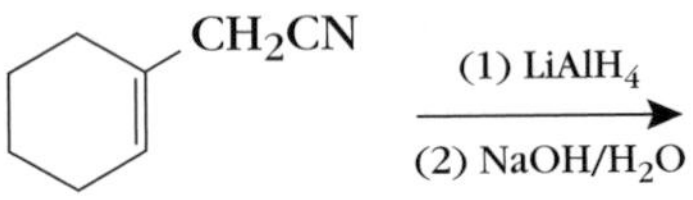   [structure: 1-cyclohexenylacetonitrile, CH_2CN] $\xrightarrow[\text{(2) NaOH/H}_2\text{O}]{\text{(1) LiAlH}_4}$

(o)   [structure: acetanilide, $NHCCH_3$] $\xrightarrow[\text{H}_2\text{SO}_4]{\text{HNO}_3}$

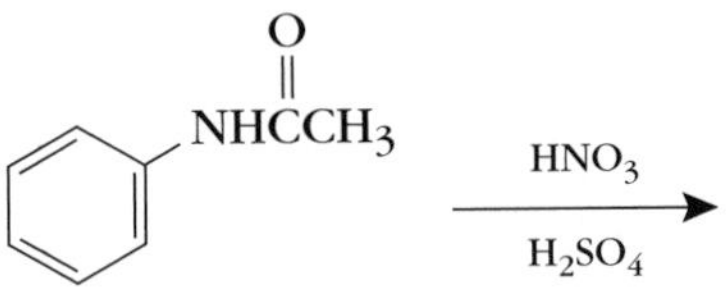

22.39 Write the structure of the missing reactant, reagent, or product in each of the following reactions.

(a)   ? $\xrightarrow{\text{Isopropylamine}}$ $CH_3CH_2C(=O)NHCH(CH_3)_2$

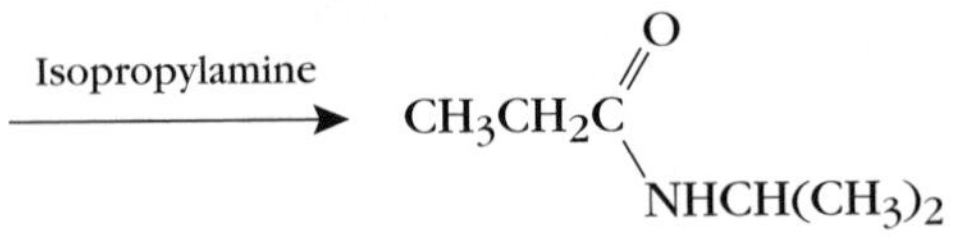

(b)   ? $\xrightarrow[\text{(2) ?}]{\text{(1) NaNO}_2/\text{HCl}}$ [structure: 3-iodotoluene, I / CH$_3$]

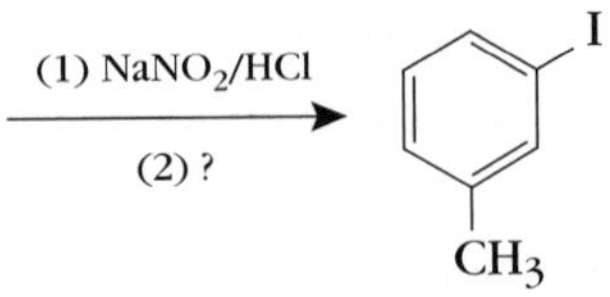

(c)   Hexanal $\xrightarrow{?}$ $CH_3(CH_2)_5NH_2$

(d) Benzylamine $\xrightarrow{\text{?}}$ *N,N,N*-Triethylbenzyl ammonium iodide

(e) ? $\xrightarrow[\text{(2) NaOH/H}_2\text{O}]{\text{(1) Zn/HCl}}$ 4-Chloroaniline

(f) 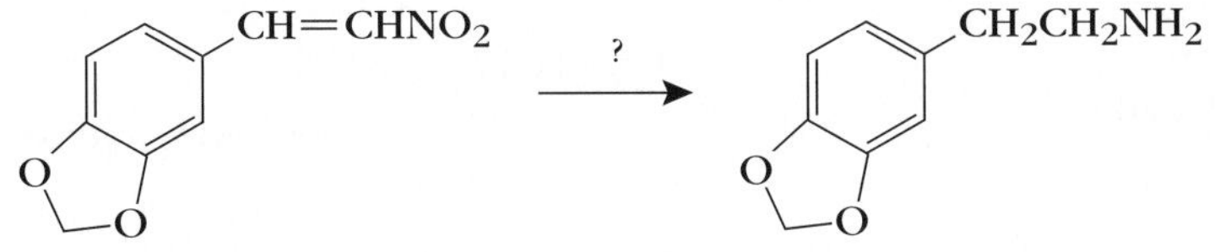 $\xrightarrow{\text{?}}$

(g) *N*-Methylaniline $\xrightarrow{\text{NaNO}_2\text{/HCl}}$?

(h) Cyclohexylamine $\xrightarrow{\text{Benzenesulfonyl chloride}}$?

(i) Benzenediazonium chloride $\xrightarrow{\text{}}$?

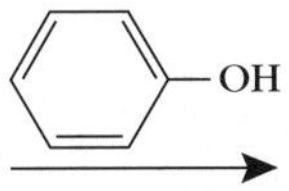

22.40 When the reaction of *N*-ethyl propanamide with an aqueous HCl solution is heated for only a short time, the product that results is a mixture of ethyl ammonium chloride, propanoic acid, and unreacted *N*-ethyl propanamide. By means of a flow diagram, show how you would separate the mixture into its pure components.

22.41 The reaction of 1-butanamine with a cold aqueous solution of NaNO$_2$ and HCl forms a mixture containing 1-butanol (20%), 2-butanol (12%), 1-chlorobutane (5%), 2-chlorobutane (4%), 1-butene (26%), *Z*-2-butene (3%), and *E*-2-butene (7%). Write a mechanism that explains the formation of all these products.

22.42 Propose a synthesis of the following compounds starting from the compounds given. More than one step may be necessary.

(a) Cyclohexanone $\longrightarrow$

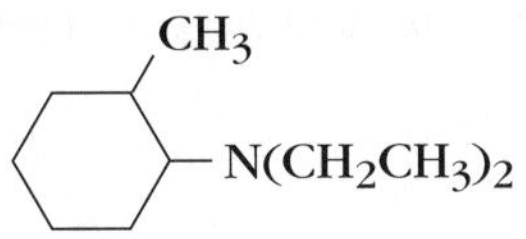

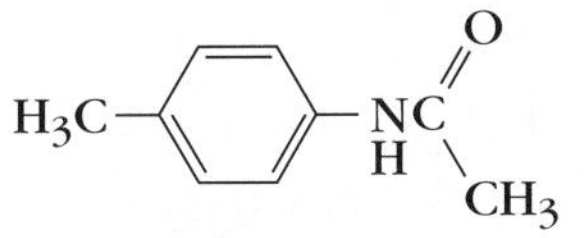

(c)  $\longrightarrow$ 1,4-Butanediamine

(b) Toluene $\longrightarrow$

(d) 1-Bromobutane $\longrightarrow$ 3-Hexanamine

22.43 The most intense peak in the mass spectrum of *N,N*-dimethyl-1-butanamine occurs at $m/z = 58$. Propose a fragmentation of the molecular ion to account for this peak.

22.44 Propose synthetic schemes to convert 3-chloroaniline into each of the following compounds. Use any other organic or inorganic reagent as needed.

(a)

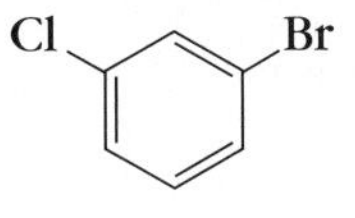

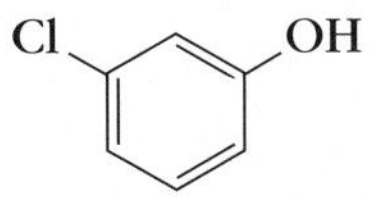

(d)

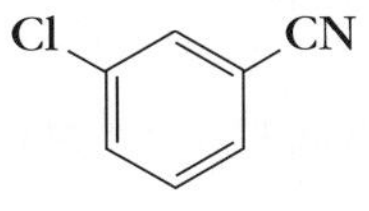

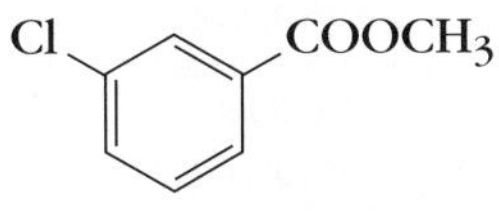

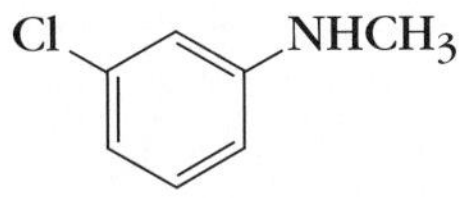

(g)

(b)

(e)

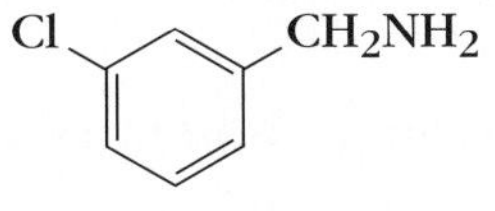

(c)

(f)

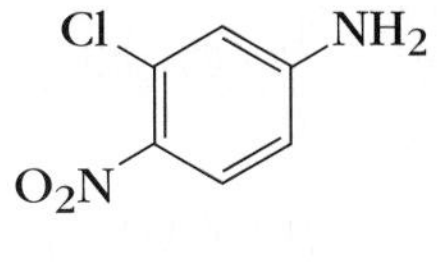

(h)

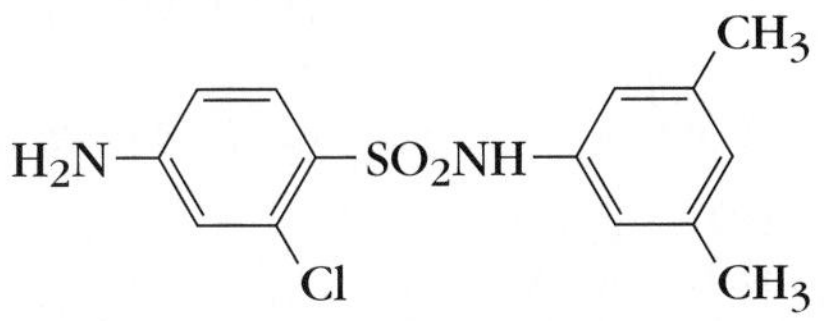

22.45 Propose a synthesis of each of the following compounds from the designated starting material. Use any other organic or inorganic reagent as needed.

(a) 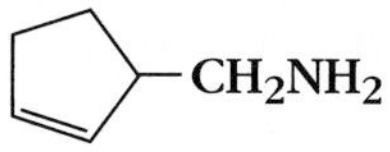 ⟨cyclopentene⟩—CH_2NH_2 from bromocyclopentane

(b) 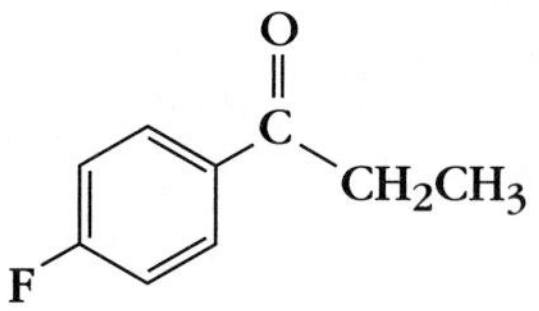 ⟨4-fluorophenyl ketone with CH_2CH_3⟩   from benzene

(c) 2-Bromotoluene from 4-nitrotoluene

(d) 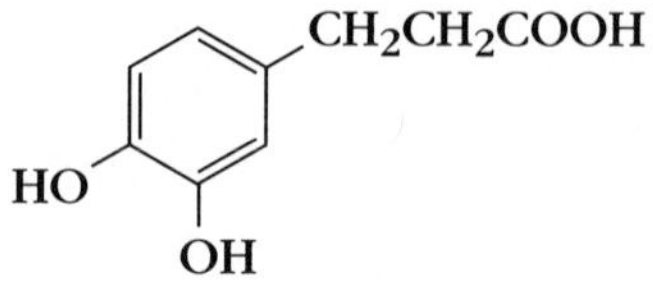 ⟨3,4-dihydroxyphenyl with CH_2CH_2COOH⟩ from ⟨methylenedioxybenzaldehyde, CHO⟩

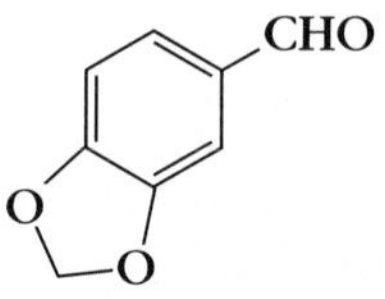

(e) 3,5-Dichloronitrobenzene from 4-nitroaniline

(f) 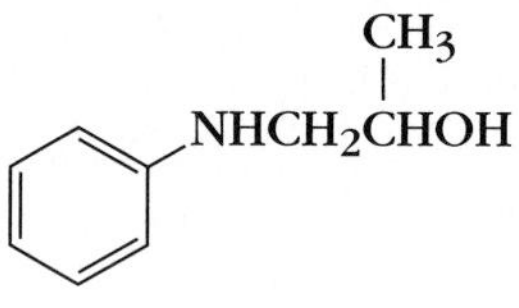 ⟨phenyl $NHCH_2CHOH$ with CH_3⟩ from aniline

(g)  ⟨phenyl⟩—$CH_2NHCH_2CH(CH_3)_2$ from benzene

22.46 Write the structures, including stereochemistry, of compounds A to E.

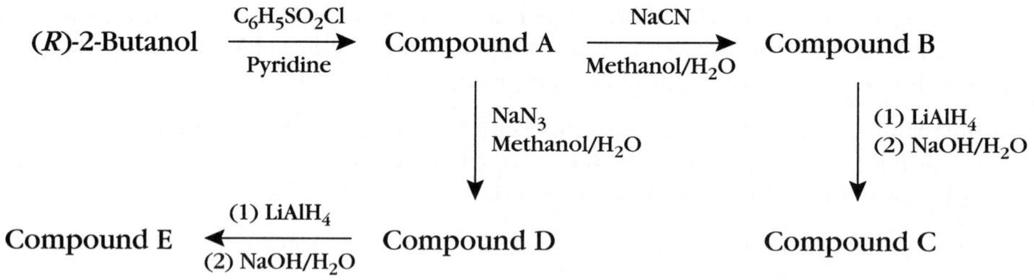

(R)-2-Butanol $\xrightarrow[\text{Pyridine}]{C_6H_5SO_2Cl}$ Compound A $\xrightarrow[\text{Methanol/H}_2\text{O}]{\text{NaCN}}$ Compound B

Compound A $\xrightarrow[\text{Methanol/H}_2\text{O}]{\text{NaN}_3}$ Compound D

Compound B $\xrightarrow[\text{(2) NaOH/H}_2\text{O}]{\text{(1) LiAlH}_4}$ Compound C

Compound E $\xleftarrow[\text{(2) NaOH/H}_2\text{O}]{\text{(1) LiAlH}_4}$ Compound D

22.47 Starting with aniline and any other organic or inorganic reagent needed, propose a synthesis of the acid-base indicator methyl orange.

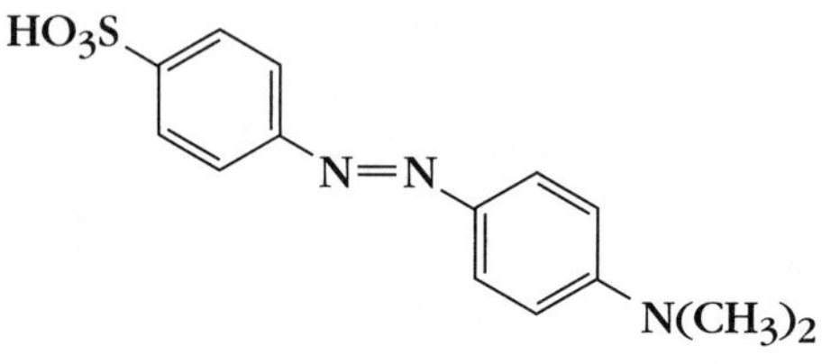

HO_3S—⟨benzene ring⟩—$N{=}N$—⟨benzene ring⟩—$N(CH_3)_2$

Methyl orange

22.48 Reductive amination forms N,N-dimethylamines when an excess of formaldehyde is used, as shown by the following example. Propose a mechanism for this reaction.

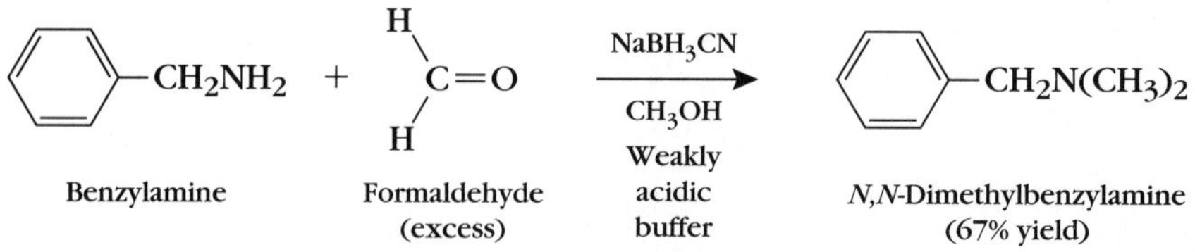

⟨phenyl⟩—CH_2NH_2 + $\underset{H}{\overset{H}{C}}{=}O$ $\xrightarrow[\substack{\text{CH}_3\text{OH} \\ \text{Weakly} \\ \text{acidic} \\ \text{buffer}}]{\text{NaBH}_3\text{CN}}$ ⟨phenyl⟩—$CH_2N(CH_3)_2$

Benzylamine   Formaldehyde N,N-Dimethylbenzylamine
               (excess)   (67% yield)

22.49 Write the structural formula of each of the following compounds based on its molecular formula, IR, and NMR spectra.

    (a) $C_4H_{11}N$
    (b) $C_8H_9NO_2$
    (c) C_8H_9NO
    (d) $C_9H_{11}NO$

(a) $C_4H_{11}N$

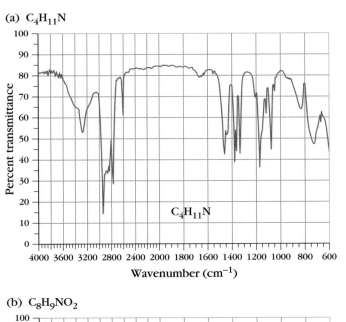

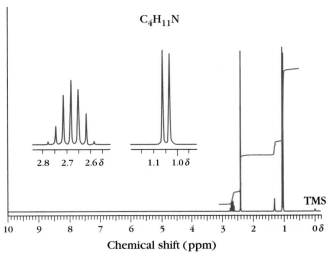

(b) $C_8H_9NO_2$

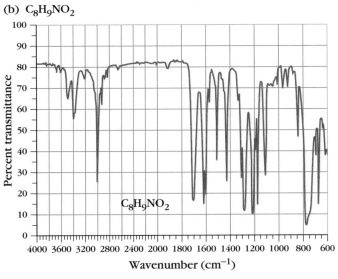

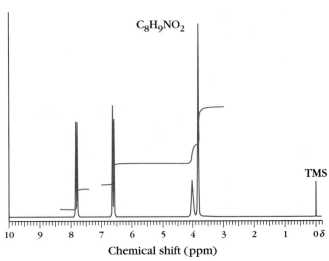

(c) C_8H_9NO

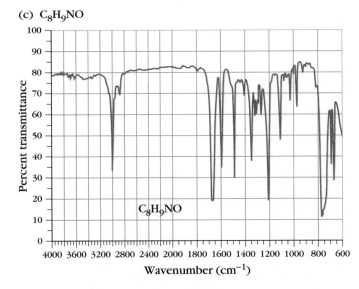

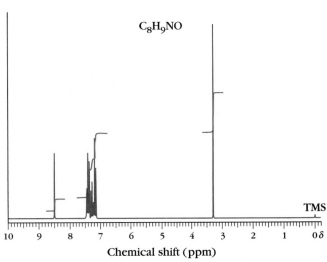

(d) $C_9H_{11}NO$

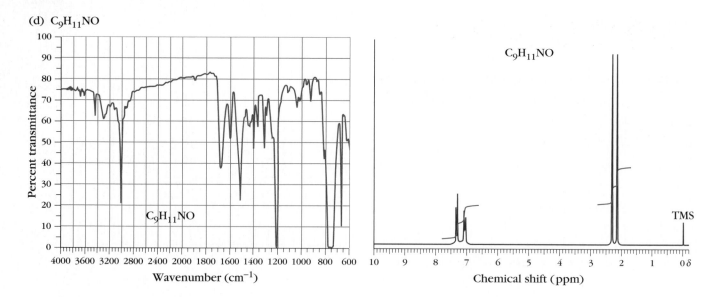

22.50 Write the structural formula of each of the following compounds based on their molecular formula and the spectral data provided.

(a) C_3H_9N IR: 3300 cm^{-1}. ^{1}H NMR: δ 1.05 t (3H); δ 2.54 q (2H); δ 2.29 s (3H); δ 4.83 s (1H).

(b) $C_6H_{16}N_2$ ^{1}H NMR: δ 2.24 s (3H); δ 2.39 s (1H).

(c) $C_8H_{11}N$ IR: 3320 cm^{-1}; 3410 cm^{-1}. ^{1}H NMR: δ 1.18 t (3H); δ 2.53 q (2H); δ 3.50 broad (2H); δ 6.58 d (2H); δ 6.96 d (2H).

(d) C_4H_9NO IR: 3330 cm^{-1}; 3180 cm^{-1}; 1640 cm^{-1}. ^{13}C NMR: δ 179.1; δ 34.0; δ 31.4. ^{1}H NMR: δ 1.06 d (6H); δ 2.40 heptet (1H); δ 6.60 broad (1H); δ 7.10 broad (1H).

CHAPTER

HALOBENZENES, PHENOLS, and QUINONES

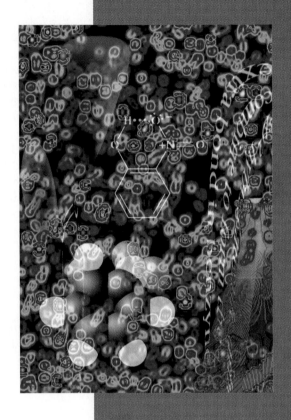

Recall from Chapter 21 that substituents on a benzene ring affect its reactivity. The benzene ring in turn affects the chemistry of the groups or atoms bonded directly to it. In Section 20.9, for example, we discussed how the reactivity of substituents on carbon atoms α to a benzene ring is enhanced. In this chapter, the effect of a benzene ring on the chemistry of a hydroxy group and halogen atoms will be examined. We begin with the chemistry of halobenzenes.

23.1   Halobenzenes

Halobenzenes are compounds that have a halogen atom bonded directly to a benzene ring. These compounds can be divided into two classes according to their reactivity with nucleophiles.

In one class are the halobenzenes that contain only halogen atoms or halogen atoms and substituents that are electron releasing. These halobenzenes undergo **nucleophilic aromatic substitution reactions** in which the halogen atom is substituted by a nucleophile only with difficulty. Chlorobenzene, for example, reacts with hydroxide ion to form phenol only at high temperatures (370 °C) and high pressure (300 atm).

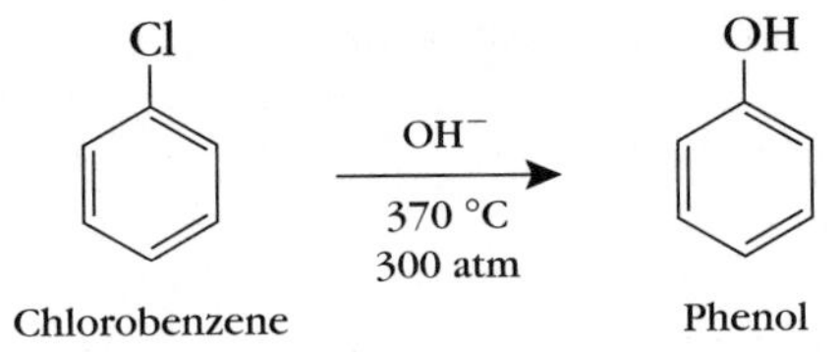

Chlorobenzene $\xrightarrow[\text{300 atm}]{\substack{\text{OH}^- \\ \text{370 °C}}}$ Phenol

The following are other examples of halobenzenes that react with nucleophiles only with difficulty:

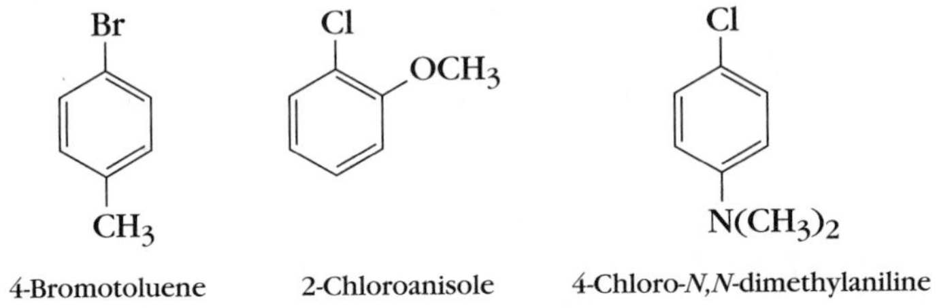

4-Bromotoluene   2-Chloroanisole   4-Chloro-*N,N*-dimethylaniline

The other class contains halobenzenes in which the halogen atom is *ortho*- or *para*- to a strongly electron-withdrawing, *meta*-directing substituent. These compounds readily undergo nucleophilic aromatic substitution reactions. Thus the electron-withdrawing substituents activate the ring of these compounds to nucleophilic aromatic substitution reactions. For example, 2,4-dinitrochlorobenzene reacts with aqueous NaOH at room temperature to form 2,4-dinitrophenol.

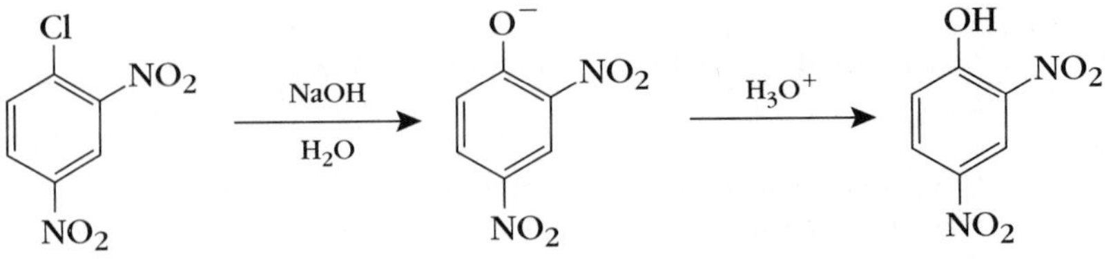

The following are other examples of halobenzenes that readily undergo nucleophilic aromatic substitution reactions:

2-Bromonitrobenzene   4-Chloronitrobenzene   2,4-Dinitrofluorobenzene

---

Exercise 23.1   Write the structure of the product, if any, of the reaction of each of the following compounds with sodium ethoxide in ethanol.

(a) 2,4-Dinitrofluorobenzene
(b) 4-Chloro-*N,N,N*-trimethylanilinium chloride
(c) 4-Bromotoluene
(d) 2,4,6-Trinitrobromobenzene

Activated and relatively unreactive halobenzenes undergo nucleophilic aromatic substitution reactions by two different mechanisms. We begin by discussing the mechanism of nucleophilic aromatic substitution of activated halobenzenes.

23.2   Addition-Elimination Mechanism of Nucleophilic Aromatic Substitution

The general features of the reaction of 2,4-dinitrochlorobenzene with aqueous NaOH resemble the nucleophilic substitution reactions of haloalkanes that occur by an S_N2 mechanism. The nucleophiles are the same (e.g., OH^-, CH_3O^-, RNH_2), the reaction is second order overall (first order in nucleophile and first order in halide), and, for a particular halobenzene, the more nucleophilic the reagent the faster the reaction occurs. Nucleophilic substitution of halobenzenes can't occur by an S_N2 mechanism, however, because back side attack of the halogen-bearing carbon atom requires an approach through the benzene ring. This reaction path is physically impossible.

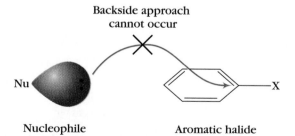

Nucleophile   Aromatic halide

The mechanism of nucleophilic aromatic substitution reactions of activated halobenzenes occurs by the two steps shown in Figure 23.1. In Step 1, the nucleophile adds to the halogen-bearing carbon atom of the halobenzene to form an intermediate anion. In Step 2, elimination of the halide ion from the intermediate anion regenerates the aromatic ring in the product. Because addition precedes elimination, the mechanism is called an **addition-elimination mechanism.**

Figure 23.1
Addition-elimination mechanism for nucleophilic aromatic substitution reactions.

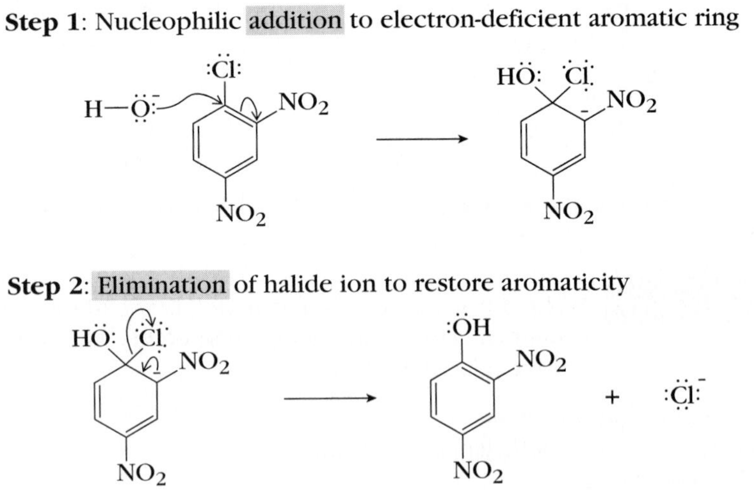

Step 1: Nucleophilic addition to electron-deficient aromatic ring

Step 2: Elimination of halide ion to restore aromaticity

In order to understand why this mechanism is energetically favorable only for activated halobenzenes, we must examine the electron structure of the anionic intermediates. As an example, consider the substitution of the fluorine atom by methoxide ion in the reaction of 4-fluoronitrobenzene.

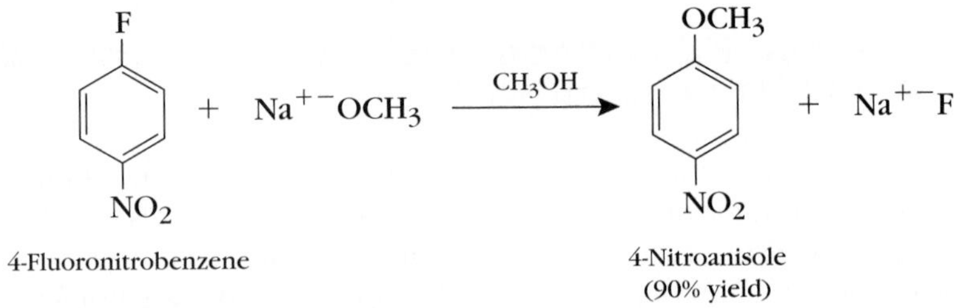

4-Fluoronitrobenzene   4-Nitroanisole
                                         (90% yield)

The electron structure of the anionic intermediate formed by addition of methoxide ion to the aromatic ring can be described as a resonance hybrid of the following resonance structures:

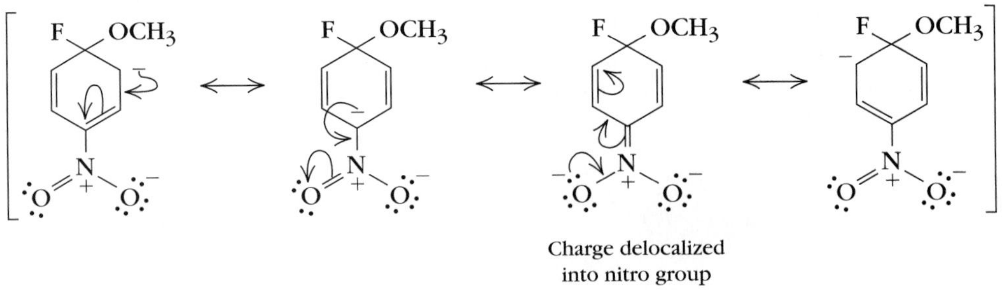

Charge delocalized
into nitro group

In one resonance structure, the negative charge is delocalized into the nitro group. Similar resonance structures illustrating electron delocalization into the nitro group can be written for a nitro group *ortho-* to the fluorine atom.

When substituents are in the *meta-* position, on the other hand, it is impossible to delocalize the negative charge into the nitro group. The following resonance hybrid shows that the negative charge cannot be delocalized into the nitro group:

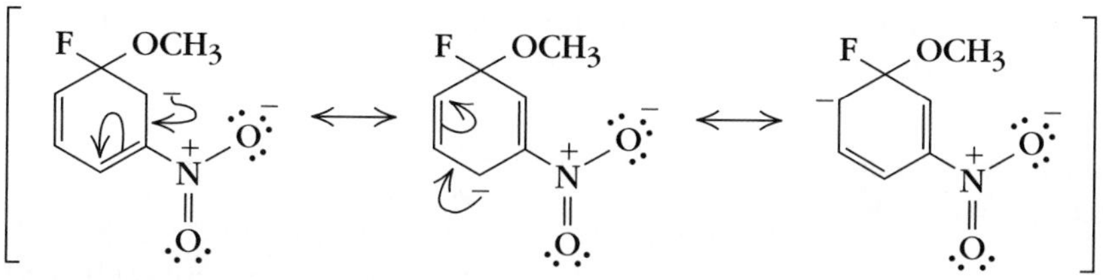

Thus only when an anionic intermediate is stabilized by a nitro group or other strong electron-withdrawing group located *ortho-* or *para-* to the leaving group will its energy and the transition state leading to it be low enough for reaction to occur by this mechanism. As a result, *o-* or *p-*fluoronitrobenzene reacts almost 10^5 times faster with methoxide ion than *m-*fluoronitrobenzene.

In a few cases, stable compounds that resemble the proposed anionic intermediates have been isolated. For example, sodium methoxide reacts with 2,4,6-trinitrophenetole to form the same product as the reaction of sodium ethoxide with 2,4,6-trinitroanisole.

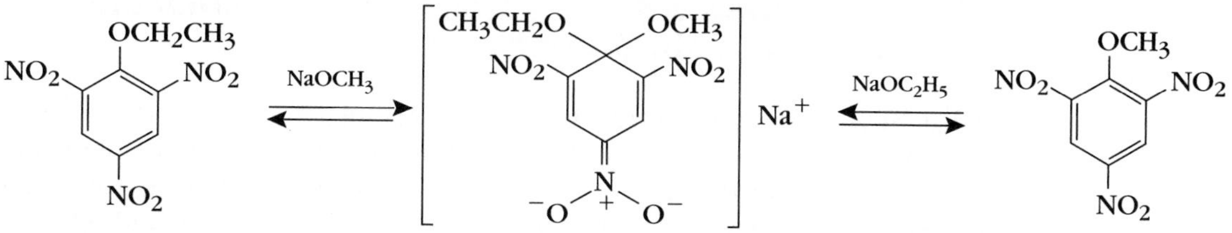

2,4,6-Trinitrophenetole   2,4,6-Trinitroanisole

This result is evidence in support of the addition-elimination mechanism of nucleophilic aromatic substitution.

A comparison of the effects of substituents on electrophilic and nucleophilic aromatic substitution reactions is given in Table 23.1. The important thing to conclude from the data in Table 23.1 is that the effect of substituents differs in the two reactions. Electron-withdrawing groups deactivate aromatic rings toward electrophilic substitution but activate them for nucleophilic substitution. This

Table 23.1	Effect of Substituents on Electrophilic and Nucleophilic Aromatic Substitution Reactions		
Type of Substituent	Example	Effect on Nucleophilic Substitution	Effect on Electrophilic Substitution
Electron-withdrawing	Nitro, nitroso, cyano, halogen, carboxylate, trialkylammonium, and carbonyl-containing groups	**Activating** when located *o-* or *p-* to the leaving group. Less effect when *m-* to leaving group	**Deactivating** and *m-* directing except for halogen, which is deactivating and *o-*, *p-* directing
Electron-donating	Alkoxy, alkyl, dialkylamino, acylamino, acyloxy	**Deactivating** when located *o-* or *p-* to the leaving group. Less effect when *m-* to leaving group	**Activating** and *o-*, *p-* directing

suggests that describing substituents as activating or deactivating is meaningless without also specifying the reaction under consideration.

Exercise 23.2	Write the resonance structures that contribute to the electron structure of the anionic intermediate formed in the reaction of o-fluoronitrobenzene with methoxide ion.

Under certain conditions, nonactivated chlorobenzene can be made to undergo nucleophilic aromatic substitution reactions. However, this reaction proceeds by an entirely different mechanism.

23.3   Elimination-Addition Mechanism of Nucleophilic Aromatic Substitution

Most halobenzenes without electron-withdrawing substituents can be made to undergo nucleophilic aromatic substitution reactions. The reaction of chlorobenzene with OH^- to form phenol, for instance, requires high temperatures and pressures, as we discussed in Section 23.1. Other nucleophilic aromatic substitution reactions occur by using a very strong base. For example, the strong bases sodium amide ($NaNH_2$) or potassium amide (KNH_2) can be used to convert aromatic chlorides, bromides, and iodides to aniline in liquid ammonia at $-33\ °C$.

Bromo-
benzene

Sodium
amide

Aniline

These nucleophilic substitution reactions are more complicated than they appear. The reaction of amide ion and chlorobenzene, for example, was carried out with chlorobenzene in which the carbon atom bonded to chlorine was labeled with ^{14}C. In about half the aniline isolated as product, the amino group was bonded to a ^{14}C atom. In the rest, the ^{14}C was in a position *ortho-* to the amino group.

* = location of ^{14}C (48% yield)   (52% yield)

The mechanism proposed to explain the position of the ^{14}C in aniline involves the three steps shown in Figure 23.2. Step 1 is an elimination of hydrogen halide to form a highly reactive intermediate called **benzyne.** In Step 2, benzyne reacts rapidly with the solvent (ammonia). Proton transfer in Step 3 completes the reaction. Because elimination of HX occurs *before* addition of the nucleophile, this is called an **elimination-addition mechanism.**

Step 1: Elimination of HCl by amide ion, a strong base

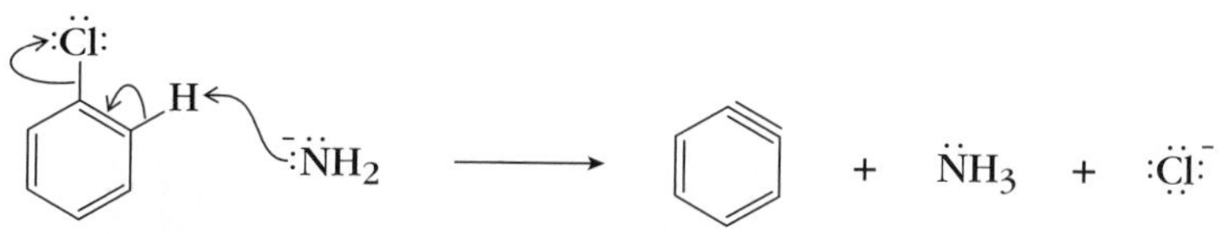

Benzyne

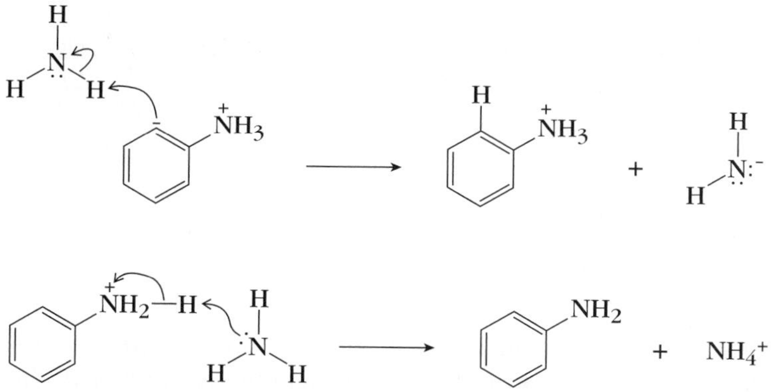

Figure 23.2
Elimination-addition mechanism of nucleophilic aromatic substitution. Step 1 forms a benzyne intermediate, which reacts with ammonia in Step 2. Proton transfer in Step 3 forms aniline.

Step 2: Addition of ammonia to one of the carbons of the triple bond

Benzyne

Step 3: Proton transfer to complete the reaction

The electron structure of benzyne consists of an aromatic π system and an extra bond made by combining two sp^2 hybrid atomic orbitals from two adjacent carbon atoms.

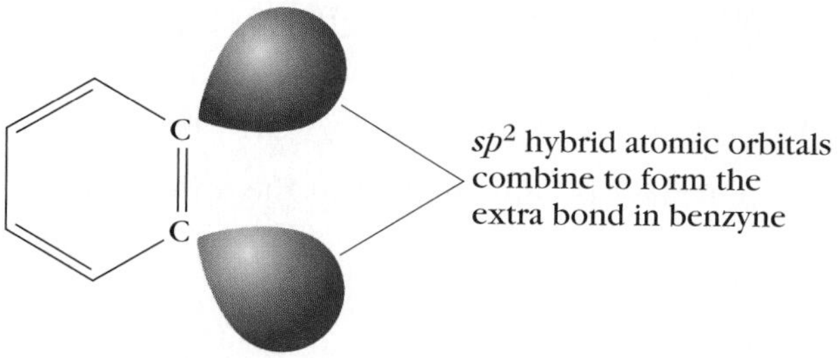

sp^2 hybrid atomic orbitals combine to form the extra bond in benzyne

Benzyne

These sp^2 hybrid atomic orbitals lie in the plane of the benzene ring, so they do not interact with the aromatic π system. While these two sp^2 atomic orbitals are in the same plane, they are not properly oriented for best combination. As a result, this bond is very weak, which makes benzyne a very reactive intermediate.

The elimination-addition mechanism explains the results of the ^{14}C labeling experiment. In ^{14}C labeled benzyne, one carbon atom of the "triple bond" is labeled, the other is not. Addition of amide ion occurs to either carbon atom. As a result, the amino group in about half the aniline molecules is bonded to a ^{14}C atom, while the rest is not.

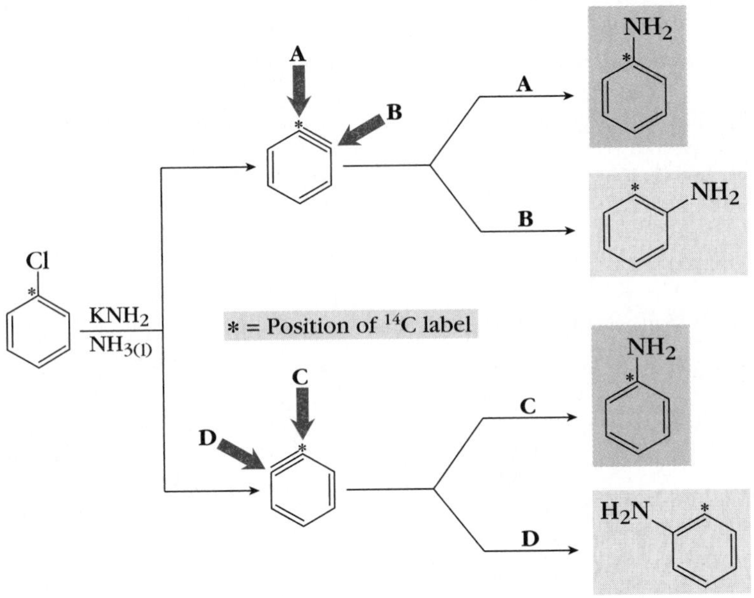

* = Position of ^{14}C label

The strong bases used to produce benzyne intermediates in these reactions are all derived from the nucleophile that adds in the addition step. However, benzyne can also be formed without the use of strong bases. For example, the diazotization of anthranilic acid forms a diazonium salt that decomposes on heating to form benzyne.

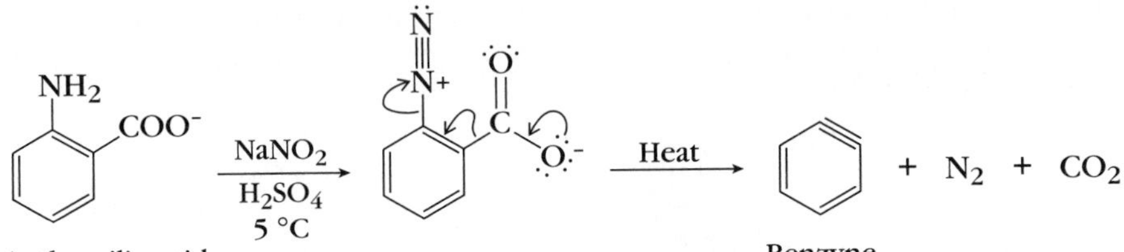

Anthranilic acid

Benzyne

When benzyne is generated by this method in the absence of a nucleophile, it dimerizes.

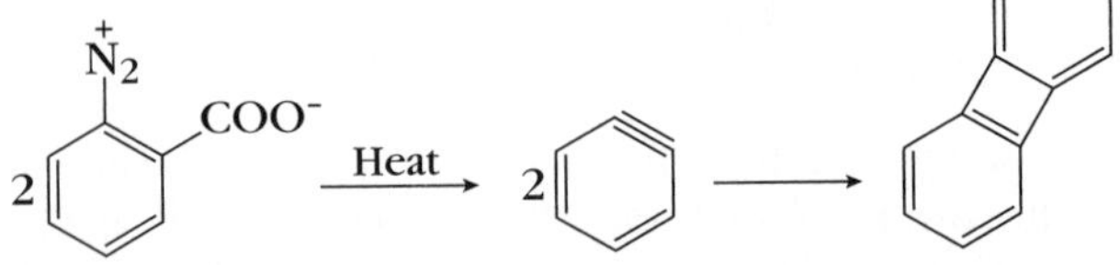

Benzyne dimer

When the intermediate arenediazonium ion is heated in the presence of added nucleophiles, however, the benzyne intermediate reacts as fast as it is formed with the nucleophile. Benzyne formed in this way is a particularly good dieneophile, and it readily undergoes Diels-Alder reactions.

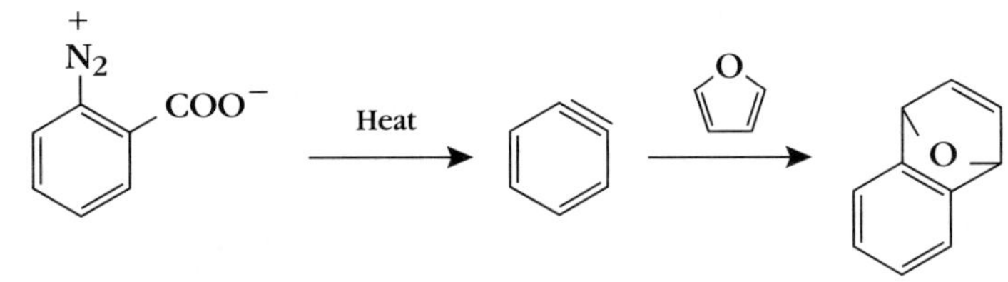

Exercise 23.3   Write the structures of the products obtained by the reaction of 4-bromotoluene and potassium amide in liquid ammonia at $-33\ ^\circ C$.

Exercise 23.4   Halobenzenes, like haloalkanes, form Grignard and lithium reagents (Section 11.7). Write equations for the preparation of each of the following compounds by the reaction of a Grignard or lithium reagent obtained from a halobenzene and the appropriate organic compound.

(a) 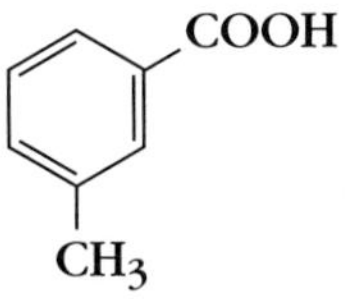   (b) 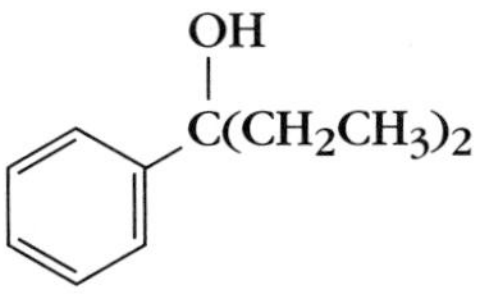   (c)

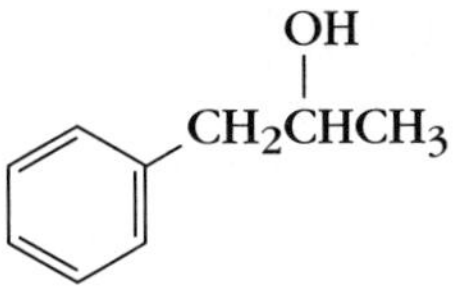

The reactions of phenols, like halobenzenes, differ in many ways from their alkyl counterparts.

23.4   Phenols

Phenols are compounds that contain a hydroxy group bonded directly to a benzene ring. The word phenol refers to both a specific compound, which is also called hydroxybenzene, as well as the general name of this class of compounds. Most phenols are named according to the rules given in Section 21.1. When naming phenols that contain carboxy or carbonyl functional groups, these groups take precedence over the hydroxy group. The following examples illustrate these rules.

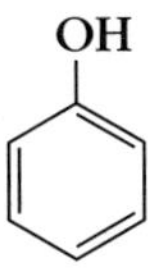

Phenol   4-Fluorophenol 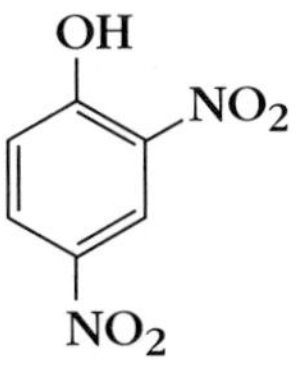   2,4-Dinitrophenol 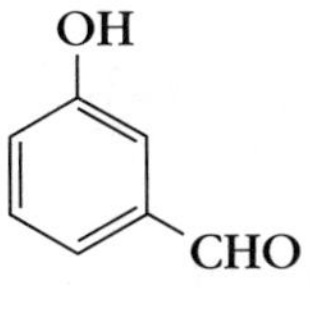   3-Hydroxybenzaldehyde 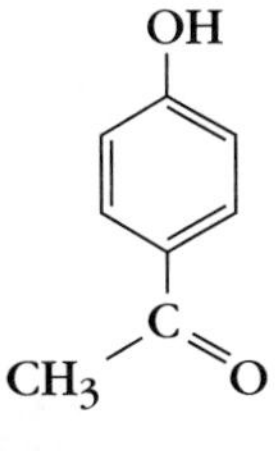   4-Hydroxyacetophenone
(hydroxybenzene)

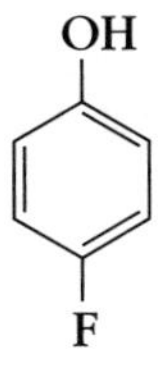

The following phenols are better known by their common names.

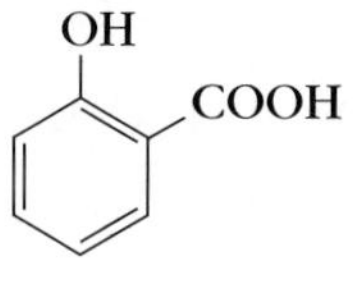

Salicylic acid
(o-hydroxybenzoic acid)   o-Cresol
(o-methylphenol)   m-Cresol
(m-methylphenol)   p-Cresol
(p-methylphenol)

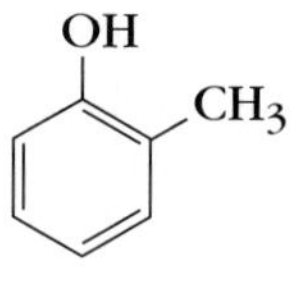

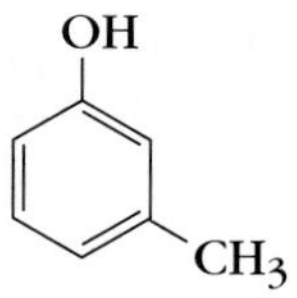

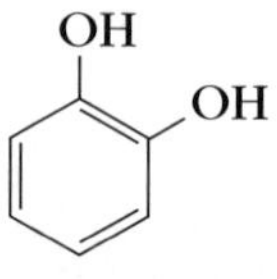

Catechol
(1,2-benzenediol)   Resorcinol
(1,3-benzenediol)   Hydroquinone
(1,4-benzenediol)

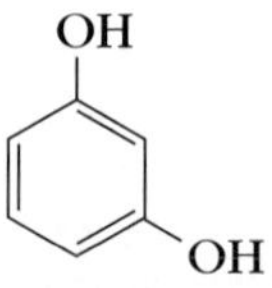

Alkyl phenyl ethers are named as alkoxybenzenes, while the $C_6H_5O—$ group is called phenoxy.

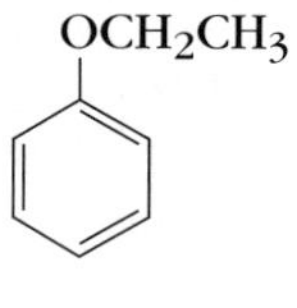

Ethoxybenzene
(Phenetole)   4-Chloro-3-phenoxyheptane

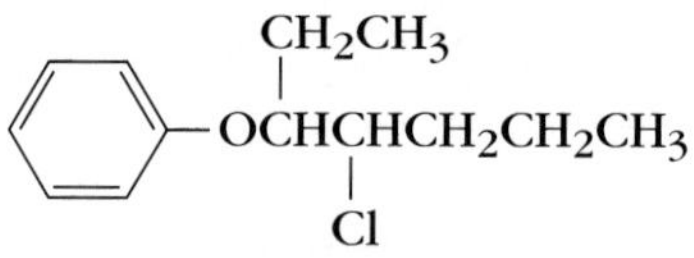

The hydroxy derivatives of naphthalene, whose chemical reactions are similar to those of phenol, are called 1-naphthol and 2-naphthol; both are acceptable IUPAC names.

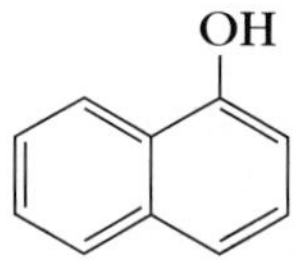

1-Naphthol 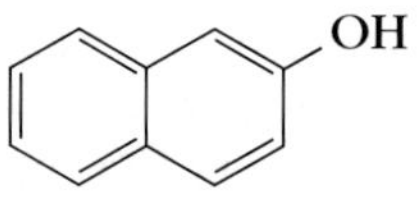   2-Naphthol

---

Exercise 23.5   Write the structure for each of the following compounds.

(a) 3-Nitrophenol
(b) 4-t-Butylphenol
(c) 2-Fluoro-4-propylphenol
(d) 3-Isopropoxyphenol
(e) 3-Phenoxypentane

Exercise 23.6	Write the IUPAC name of each of the following compounds.

(a)

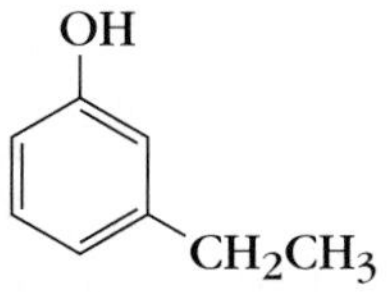

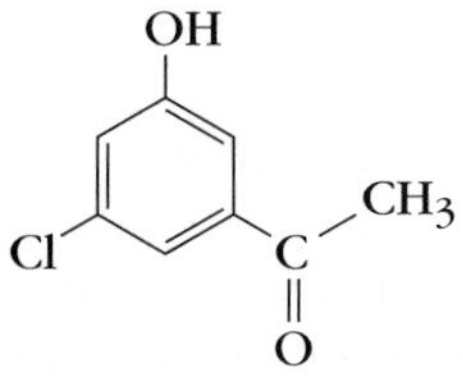

(b)

(c)

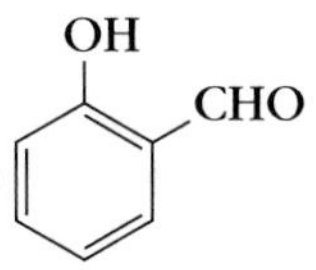

(d)

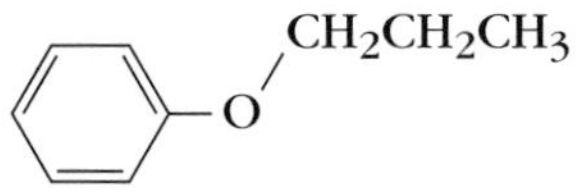

23.5 Commercially Important Phenols

Phenols occur widely in nature. They are particularly prevalent in plants, where they perform a variety of functions. Some are found as sap-soluble pigments of flowers and fruits, while others contribute to their aroma. Since antiquity, many of these substances have been isolated from plants for use as condiments and coloring materials. With the growth of the chemical industry during the last 75 years, many of these compounds are now made industrially, including the following:

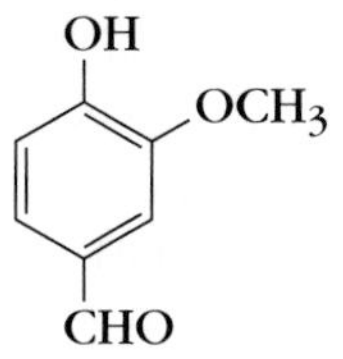

Vanillin
(found in vanilla bean)

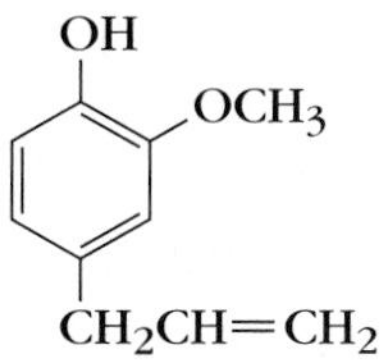

Eugenol
(found in clove oil)

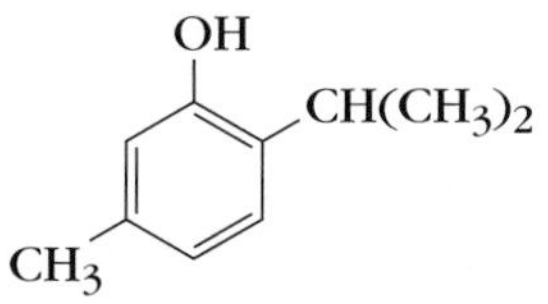

Thymol
(found in oil of thyme)

It has been known for a long time that an aqueous solution of carbolic acid (another common name for phenol) is a disinfectant. Other phenols, as a result, have been synthesized commercially for their germicidal properties. Hexylresorcinol, for example, is the active ingredient in throat lozenges and several mouthwashes.

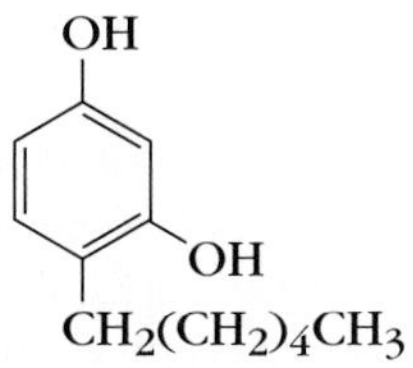

Hexylresorcinol

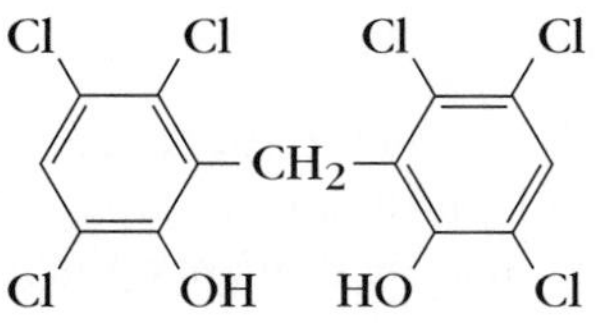

Hexachlorophene

Hexachlorophene, moreover, was used in germicidal soaps, some toothpastes, and deodorants. Its use has been discontinued, however, because of compelling evidence that it is absorbed through the skin in sufficient quantities to be dangerous to infants.

Because of their commercial value, several methods have been devised to synthesize phenols.

23.6 Synthesis of Phenols

The methods of preparing phenols vary depending on whether the synthesis occurs in the laboratory, in nature, or in industry. We begin by describing the relatively small scale preparation of phenols in the laboratory.

A Laboratory Synthesis

■ *Diazotization of the derivatives of aniline is discussed in Section 22.15.*

The most widely used method for synthesizing phenols is the hydrolysis of arenediazonium salts. This is a very versatile synthesis because most other substituents on the ring are not affected by the mild reaction conditions. Yields of products are generally good, too.

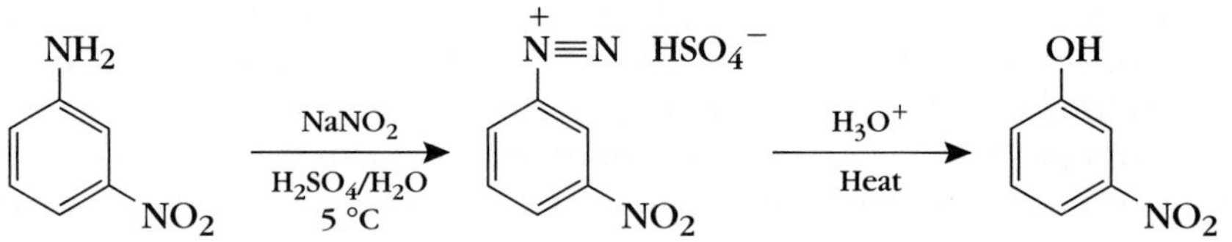
NH2
NO2
NaNO2
H2SO4/H2O
5 °C
N≡N HSO4
NO2
H3O+
Heat
OH
NO2

m-Nitroaniline

m-Nitrophenol (74%–80% yield)

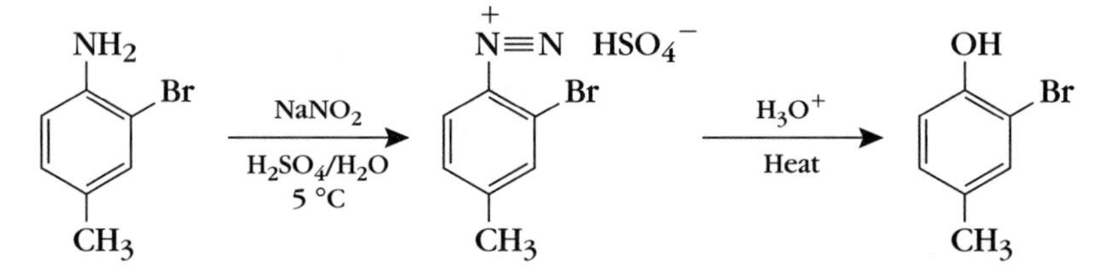
NH2
Br
CH3
NaNO2
H2SO4/H2O
5 °C
N≡N HSO4
Br
CH3
H3O+
Heat
OH
Br
CH3

2-Bromo-4-methylaniline

2-Bromo-4-methylphenol

B Synthesis in Nature

Phenols and their ethers are the largest group of aromatic compounds in nature. The biosynthesis of aromatic compounds in general and phenols in particular occurs almost exclusively in plants, fungi, and microorganisms. Relatively few are produced in animals, and those that are are usually formed by modification of aromatic compounds obtained in food.

■ *Claisen and aldol condensation reactions are discussed in Sections 17.7 and 17.6, respectively.*

The biosynthesis of most naturally occurring phenols and other aromatic compounds occurs by two pathways. One sequence begins with acetyl coenzyme A. Four molecules of acetyl CoA undergo a succession of Claisen condensations to form a β-polyketone intermediate.

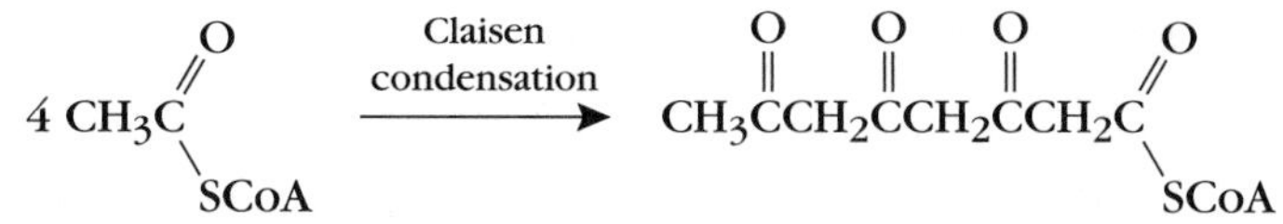

4 CH₃C—SCoA (Acetyl coenzyme A) → Claisen condensation → CH₃CCH₂CCH₂CCH₂C—SCoA (β-Polyketone intermediate)

Exercise 23.7   Write a mechanism for the four successive Claisen condensations by which four molecules of acetyl CoA form the β-polyketone intermediate.

The β-polyketone intermediate can cyclize to form phenols in two ways. One way is by an internal Claisen-like condensation reaction to form a cyclohexane derivative. The cyclohexane derivative then enolizes to form 2,4,6-trihydroxyacetophenone.

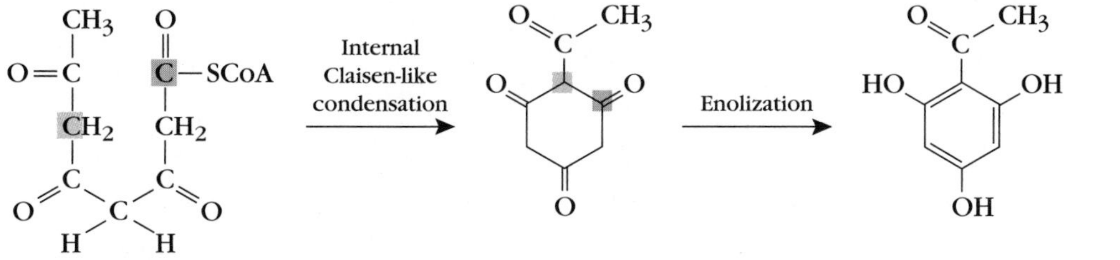

β-Polyketone → Internal Claisen-like condensation → Cyclohexane derivative → Enolization → 2,4,6-Trihydroxyacetophenone

The β-polyketone intermediate can also cyclize by an internal aldol-like condensation. In this case, it forms a different cyclohexane derivative, which then forms orsellinic acid.

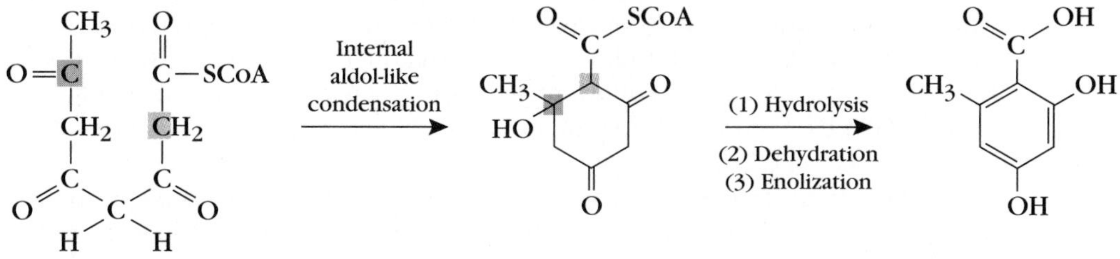

β-Polyketone → Internal aldol-like condensation → (1) Hydrolysis (2) Dehydration (3) Enolization → Orsellinic acid

In each of these biosynthetic pathways, internal condensation reactions are the key steps used in nature to form a carbon-carbon bond of a cyclohexane ring, which eventually becomes aromatized.

Exercise 23.8   The β-polyketone intermediate, whose structure is given above, undergoes both a Claisen-like and an aldol-like condensation reaction. Propose a mechanism for each of these reactions.

C   Industrial Synthesis

Phenol can be prepared industrially by heating chlorobenzene and a concentrated solution of sodium hydroxide under pressure.

This method is called the **Dow process.** It is a nucleophilic aromatic substitution reaction that is believed to occur by the elimination-addition mechanism involving a benzyne intermediate, as discussed in Section 23.3.

Most of the phenol in the world is produced today by a sequence of reactions that begins with benzene and propene and ends with the formation of phenol and acetone.

The first reaction in the sequence is a Friedel-Crafts alkylation of benzene to form cumene. Autooxidation of cumene forms cumene hydroperoxide, which forms a mixture of phenol and acetone on treatment with dilute sulfuric acid. This process is particularly important because it starts with two relatively inexpensive chemicals and transforms them into two more valuable ones, phenol and acetone.

The three reactions of this process occur by different mechanisms. In the first reaction, the Friedel-Crafts alkylation occurs by the mechanism discussed in Section 21.8. In the second reaction, autooxidation of cumene occurs by the radical chain mechanism shown in Figure 23.3. The mechanism involves formation of a cumyl radical, which reacts with oxygen to form a cumylperoxy radical. The cumylperoxy radical then abstracts a hydrogen atom from cumene to form a cumyl radical and cumene hydroperoxide. The chain continues because a cumyl radical is both a product and a reactant in different reactions of the propagation step. Termination of the chain occurs by a number of reactions of which only the disproportionation reaction is shown in Figure 23.3.

The final reaction in the formation of phenol from cumene is a rearrangement that occurs by the mechanism shown in Figure 23.4 (p. 1022). Step 1 is the protonation of cumene hydroperoxide followed by migration of the phenyl group from the carbon atom to the oxygen atom with loss of water to form a carbocation in Step 2. The reaction in Step 2 of the mechanism is similar to carbocation rearrangements (Section 7.17) except that the group migrates from a cabon atom to an oxygen atom rather than to another carbon atom. Reaction of the carbocation with water forms the conjugate acid of a hemiacetal in Step 3, which decomposes after a proton transfer in Step 4 to form acetone and phenol in Step 5.

■ *The radical chain mechanism of autooxidation is discussed in Section 18.7.*

Propagation step:

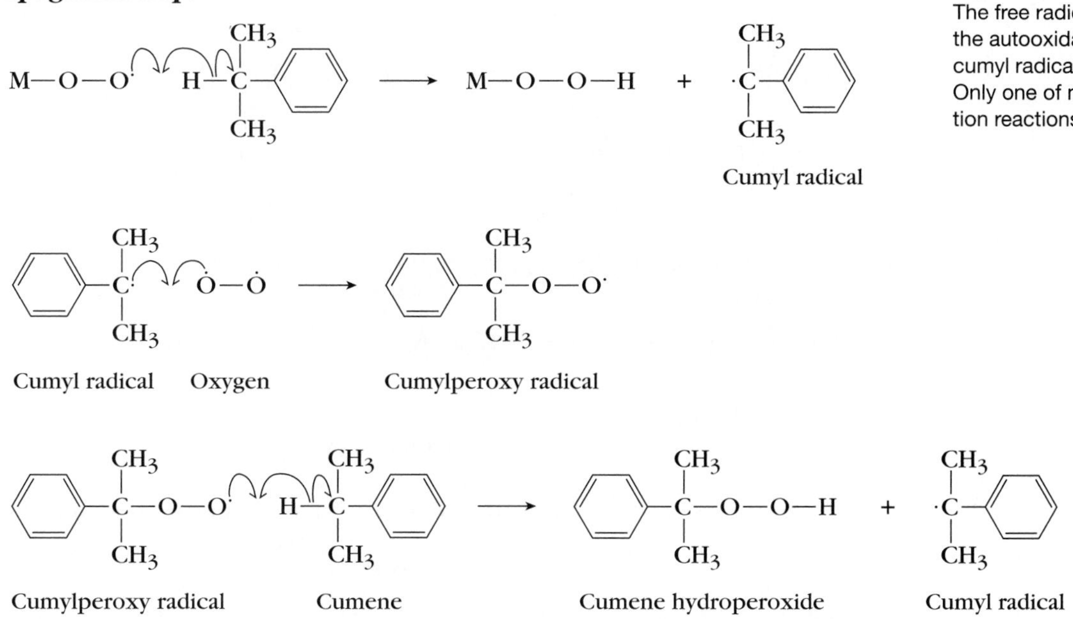

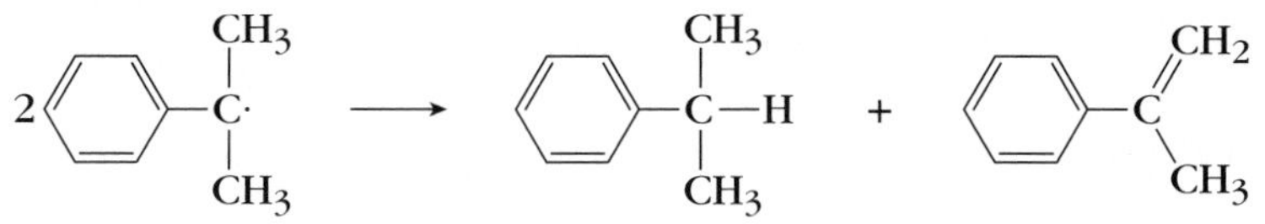

Figure 23.3
The free radical chain mechanism of the autooxidation of cumene. The cumyl radical propagates the chain. Only one of many possible termination reactions is shown.

Cumyl radical

Cumyl radical   Oxygen   Cumylperoxy radical

Cumylperoxy radical   Cumene   Cumene hydroperoxide   Cumyl radical

Termination step:

Exercise 23.9   Write the mechanism for the preparation of cumene by the reaction of benzene and propene in phosphoric acid (H_3PO_4).

One of the major differences between phenols and alcohols is their acidity.

23.7 Phenols Are Generally Stronger Acids Than Alcohols

While both phenols and alcohols are acids, their acidities differ greatly. The pK_a of phenol is 10, for example, while that of most alcohols falls in the range 15 to 18. Phenol, therefore, is about 10^6 times more acidic than a typical alcohol.

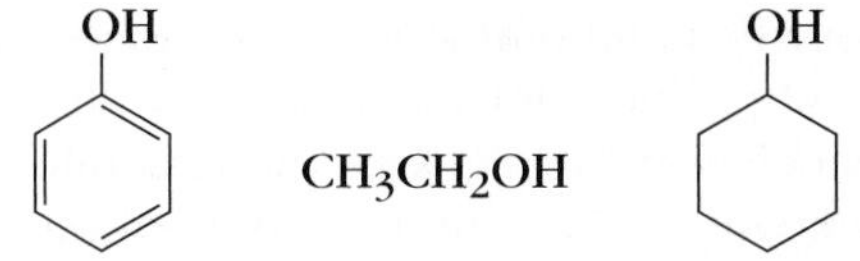

Phenol   Ethanol   Cyclohexanol
$pK_a = 10.0$ $pK_a = 15.5$ $pK_a = 17.1$

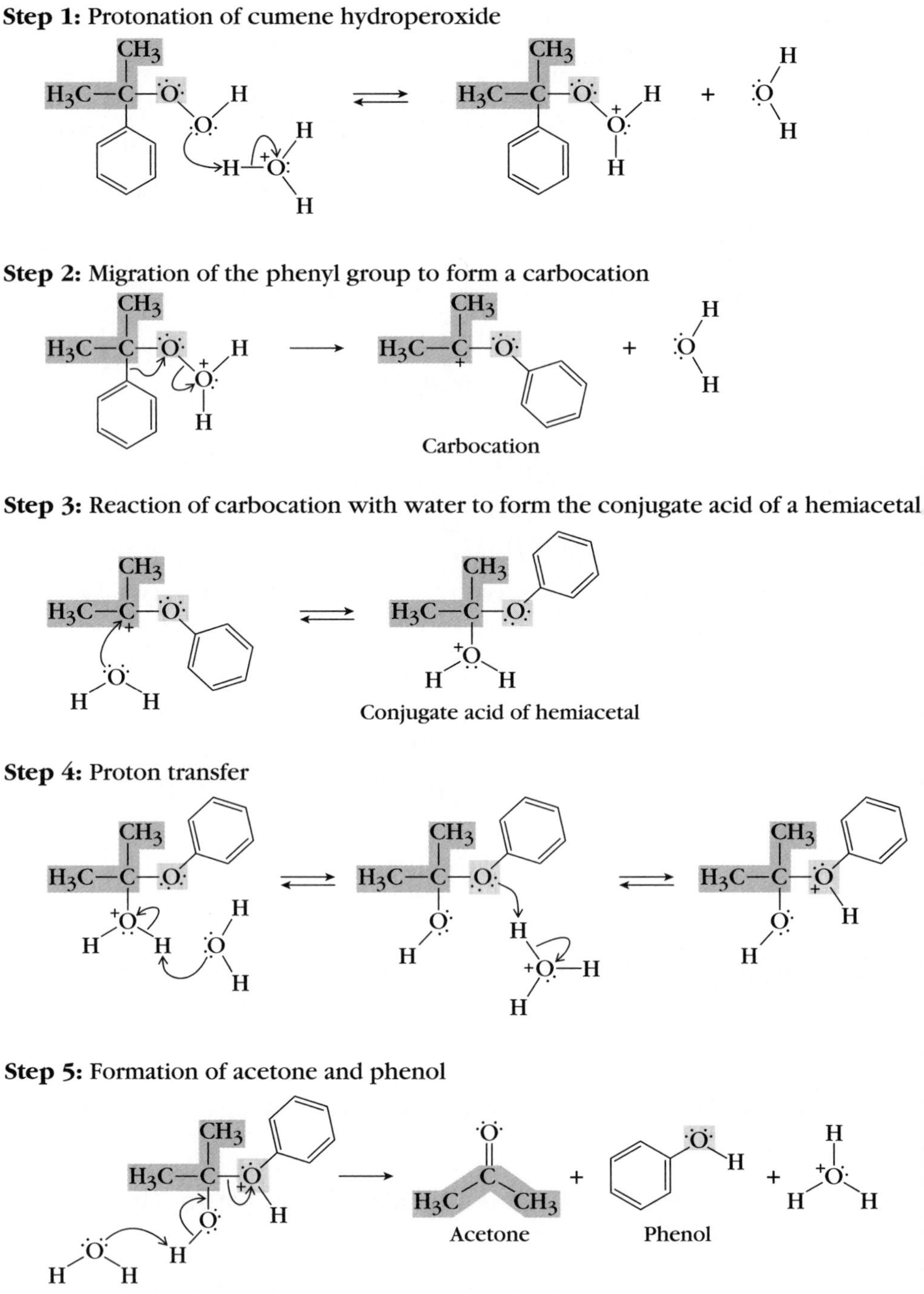

Step 1: Protonation of cumene hydroperoxide

Step 2: Migration of the phenyl group to form a carbocation

Carbocation

Step 3: Reaction of carbocation with water to form the conjugate acid of a hemiacetal

Conjugate acid of hemiacetal

Step 4: Proton transfer

Figure 23.4
The mechanism of the decomposition of cumene hydroperoxide. Step 1 is the protonation of cumene hydroperoxide followed by migration of the phenyl group in Step 2 to form a carbocation. Reaction of the carbocation with water in Step 3 forms the conjugate acid of a hemiacetal, which, after a proton transfer in Step 4, decomposes to acetone and phenol in Step 5.

Step 5: Formation of acetone and phenol

Acetone   Phenol

As in the case of aromatic anilinium ions, substituents on the benzene ring of phenols affect their acidity. The data in Table 23.2 show, however, that substituent effects are generally very small, with the exception of the nitrophenols. For example, o- and p-nitrophenol are several hundred times stronger acids than phenol. Increasing the number of nitro groups on the ring greatly increases the acidity of nitrophenol. Thus the acidity of 2,4-dinitrophenol ($pK_a = 4.09$) is comparable to most carboxylic acids (pK_a = 4–5), while 2,4,6-trinitrophenol (pK_a = 0.25) is almost as strong an acid as nitric acid (pK_a = -1.44).

23.8 Practical Uses of the Acidity of Phenols

Although phenols are acidic, they tend to be less acidic than carboxylic acids, as shown by the data in Table 23.3.

Hydroxide ion is the strongest base listed in Table 23.3, so it reacts with both carboxylic acids and phenol to form their conjugate bases. Bicarbonate ion, on the other hand, is a strong enough base to react with carboxylic acids but not with phenol.

Table 23.2	
Acidities of Phenols	
Phenol	pK_a
Phenol	10.0
o-Cresol	10.29
m-Cresol	10.09
p-Cresol	10.26
o-Methoxyphenol	9.98
m-Methoxyphenol	9.65
p-Methoxyphenol	10.21
o-Nitrophenol	7.22
m-Nitrophenol	8.39
p-Nitrophenol	7.15
2,4-Dinitrophenol	4.09
2,4,6-Trinitrophenol	0.25

Phenol	Hydroxide ion	Phenoxide ion	Water
(stronger acid)	(stronger base)	(weaker base)	(weaker acid)

Phenol	Bicarbonate ion	Phenoxide ion	Carbonic acid
(weaker acid)	(weaker base)	(stronger base)	(stronger acid)

The difference between the reaction of phenol with hydroxide ion and bicarbonate ion is the basis of a simple chemical test that can be used to distinguish between carboxylic acids and phenols. The procedure is to shake the unknown compound with an aqueous solution of sodium bicarbonate. Carboxylic acids react with bicarbonate and liberate CO_2, thus causing the solution to effervesce. Most phenols do not react with bicarbonate, so no CO_2 is evolved.

The same principle can be used to separate phenols from carboxylic acids. The procedure is to dissolve the water insoluble phenol and carboxylic acid in a solvent that is immiscible with water, such as diethyl ether. A solution of sodium bicarbonate is added to the diethyl ether solution and the two layers are shaken. The carboxylic acid reacts with $NaHCO_3$ to form the water soluble sodium salt of

Table 23.3	**pK_a Values of Selected Acids**		
Acid	Name	Conjugate Base	pK_a
CH_3COOH	Acetic acid	CH_3COO^-	4.74
H_2CO_3 ($H_2O + CO_2$)	Carbonic acid	HCO_3^-	6.35
C_6H_5OH	Phenol	$C_6H_5O^-$	10.0
HCO_3^-	Bicarbonate ion	CO_3^{2-}	10.33
C_2H_5OH	Ethanol	$C_2H_5O^-$	15.5
H_2O	Water	OH^-	15.74

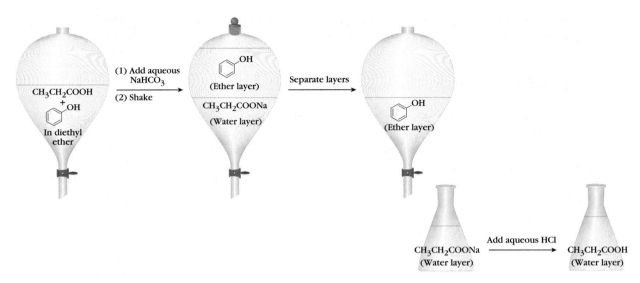

Figure 23.5
The procedure used to separate a mixture of phenol and propanoic acid dissolved in diethyl ether.

its conjugate base. The phenol remains in the diethyl ether layer. Separating the layers and adding HCl to the aqueous layer reforms the carboxylic acid. This technique of separating phenols and carboxylic acids is summarized in Figure 23.5.

Exercise 23.10   Using the data in Tables 23.2 and 23.3 (p. 1023), calculate the equilibrium constant for each of the following reactions.

(a) Phenol + HCO_3^- ⇌ Phenoxide ion + H_2O + CO_2

(b) Acetic acid + HCO_3^- ⇌ Acetate ion + H_2O + CO_2

(c) Phenol + CO_3^{2-} ⇌ Phenoxide ion + HCO_3^-

(d) 2,4,6-Trinitrophenol + HCO_3^- ⇌ 2,4,6-Trinitrophenoxide ion + H_2O + CO_2

Exercise 23.11   Arrange the following sets of compounds in decreasing order of acidity.

(a) Phenol, acetic acid, ethanol   **(c)** p-Nitrophenol, $NaHCO_3$, phenol
(b) p-Nitrophenol, p-methoxyphenol, phenol   **(d)** HCl, 2,4,6-trinitrophenol, $NaHCO_3$

Phenols and alcohols transfer a proton to a base by breaking an O—H bond. This process is common to phenols and alcohols in a number of other reactions.

23.9   Reactions That Are Common to Alcohols and Phenols

Reactions that break the O—H bond of the hydroxy group are common reactions of phenols and alcohols. One such reaction, the acid-base reaction of phenols and alcohols, is discussed in Section 23.7. Another example is the formation of esters by the reaction of phenols and alcohols with acyl chlorides and carboxylic acid anhydrides (Chapter 16). The reactions of acyl chlorides with phenols, like those of alcohols (Section 16.5), are carried out in the presence of pyridine to neutralize the HCl that also forms.

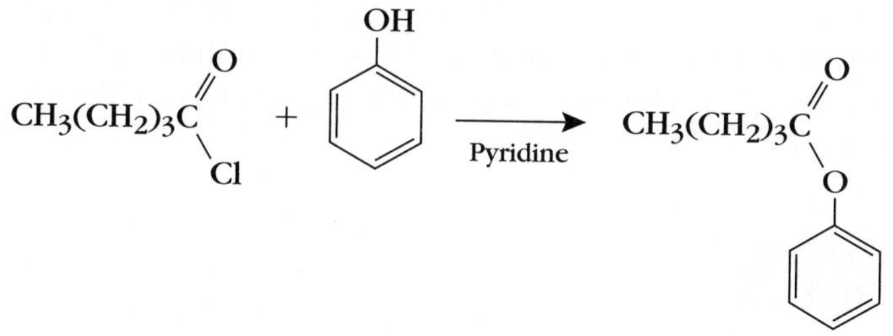

Pentanoyl chloride   Phenol   Phenyl pentanoate
                                                            (90%–96% yield)

An acid catalyst is usually needed when a carboxylic acid anhydride is used as the acylating agent. An acid catalyst, such as a few drops of concentrated sulfuric or phosphoric acid, acts by increasing the electrophilic character of the carbonyl carbon atom of the acid anhydride.

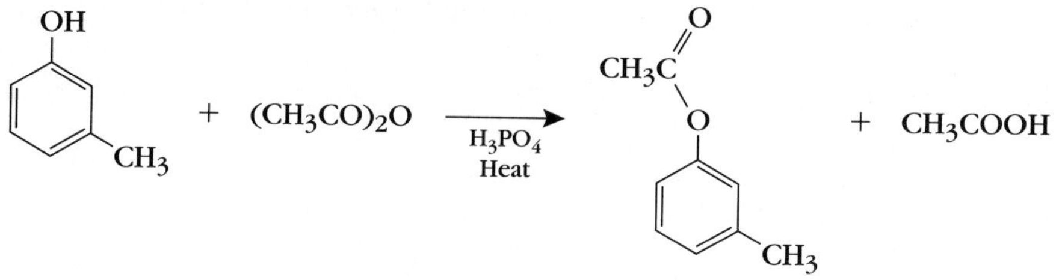

m-Cresol   Acetic   *m*-Tolylacetate   Acetic
                    anhydride   (88%–92% yield)   acid

A base such as NaOH can be used in the acylation of phenols by acid anhydrides. The NaOH increases the nucleophilicity of the phenols by converting them to their phenoxide ions.

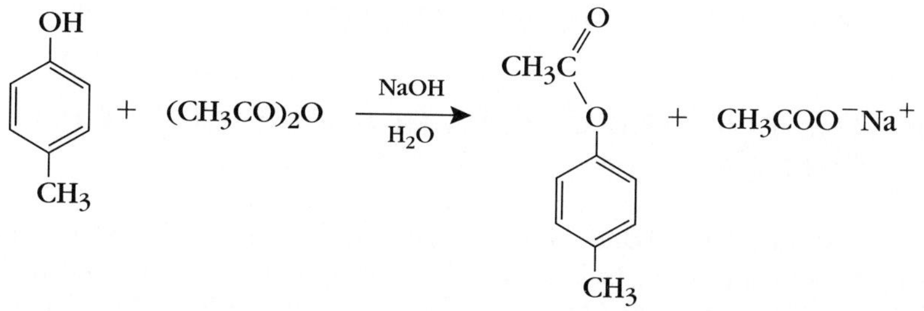

p-Cresol   Acetic   *p*-Tolylacetate   Sodium
                    anhydride   (85%–92% yield)   acetate

Another reaction common to alcohols and phenols is the formation of ethers by the Williamson ether synthesis (Section 13.8, C). Aryl ethers are best prepared by the reaction of a phenoxide ion with a haloalkane. The reaction is usually carried out by heating a mixture of the phenol, haloalkane, and potassium carbonate (K_2CO_3) in a suitable solvent.

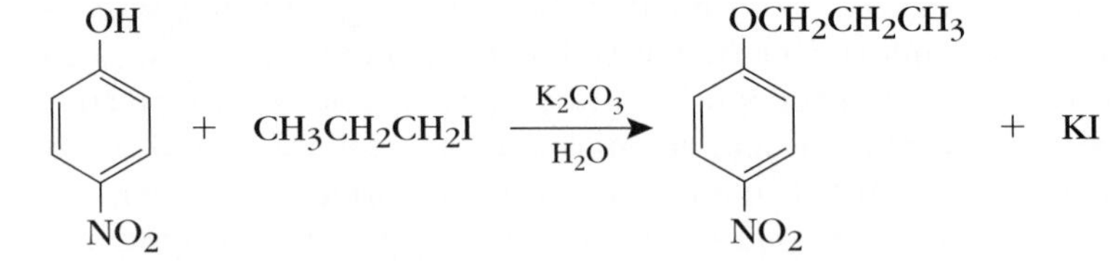

| p-Nitrophenol | 1-Iodopropane | p-Nitrophenyl propyl ether (80%–85% yield) |

Because the Williamson ether synthesis is a nucleophilic substitution reaction that occurs by an S_N2 mechanism, best results are obtained with primary and secondary haloalkanes.

Exercise 23.12   Propose a mechanism for the acid-catalyzed reaction of acetic anhydride and phenol to form phenyl acetate.

Exercise 23.13   Heroin, a derivative of morphine, does not occur naturally but is synthesized by reacting morphine with two equivalents of acetic anhydride. Based on the structure of morphine given below, write the equation for the preparation of heroin and write its structure.

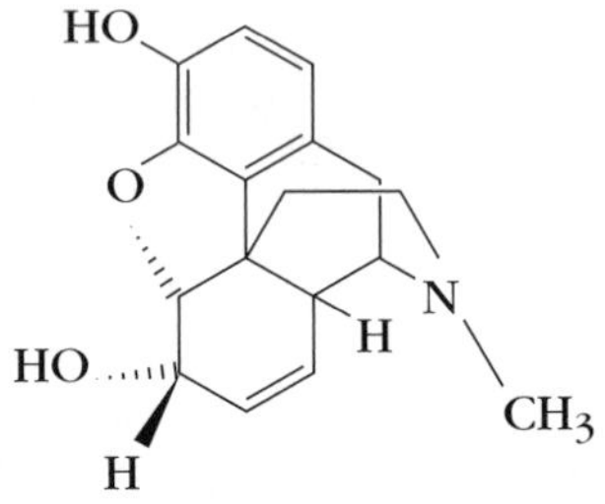

Morphine

Phenols rarely undergo reactions in which the aromatic C—O bond is broken. This is a major difference between the reactions of alcohols and phenols.

23.10 The Aromatic C—O Bond Is Difficult to Break in Phenols

A typical reaction of an alcohol is the replacement of its —OH group by a halide. This can be accomplished by a variety of reagents, as we discussed in Section 11.17. *Phenols,* on the other hand, *do not react with any of these reagents to form halobenzenes.* Phenols do not react with these reagents because nucleophilic aromatic substitution at an unactivated benzene carbon atom is extremely difficult (see Section 23.1). Thus strong acids such as HBr can protonate phenol, but no further reaction occurs.

The lack of reactivity of the aromatic C—O bond of phenols accounts for the products of the acid cleavage of alkyl aryl ethers. Cleavage forms a phenol and a haloalkane. No halobenzene is formed.

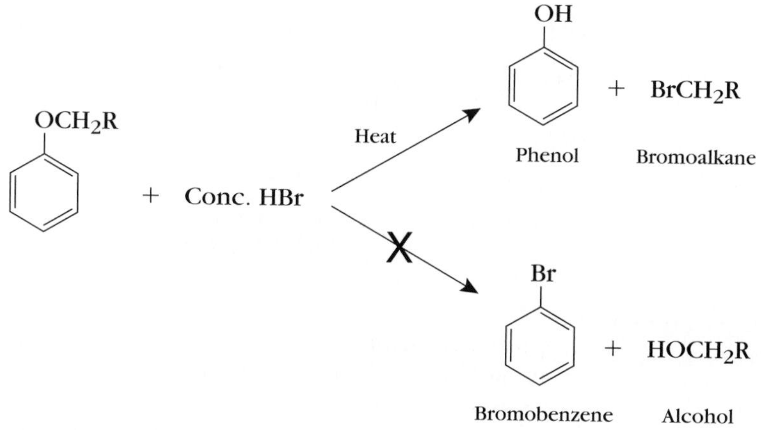

The mechanism of this reaction is similar to the acid cleavage of dialkyl ethers (Section 13.9) and is shown in Figure 23.6.

Exercise 23.14   Anisole reacts with a concentrated aqueous solution of HI at 100 °C to form phenol and iodomethane. In contrast, diphenyl ether $(C_6H_5)_2O$ is recovered unchanged after reaction with a concentrated aqueous solution of HI even at 200 °C. Explain this difference in reactivity.

A phenol is converted into a halobenzene in compounds where the halogen-bearing carbon atom of the aromatic ring is activated by nitro groups *ortho-* or *para-* to it. In these cases, the hydroxy group is first transformed into a good leav-

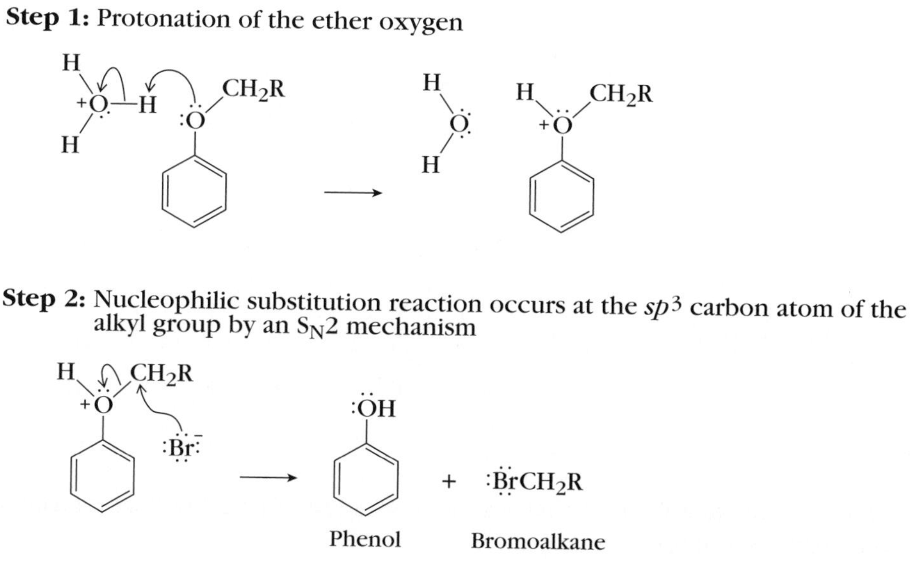

Step 1: Protonation of the ether oxygen

Figure 23.6
The mechanism of the acid cleavage of alkyl aryl ethers. The products are phenol and a haloalkane, not a halobenzene and an alcohol.

Step 2: Nucleophilic substitution reaction occurs at the sp^3 carbon atom of the alkyl group by an S_N2 mechanism

ing group. It then undergoes a nucleophilic aromatic substitution reaction that occurs by an addition-elimination mechanism. Thus 2,4-dinitrochlorobenzene is formed by the reaction of 2,4-dinitrophenol and PCl_5.

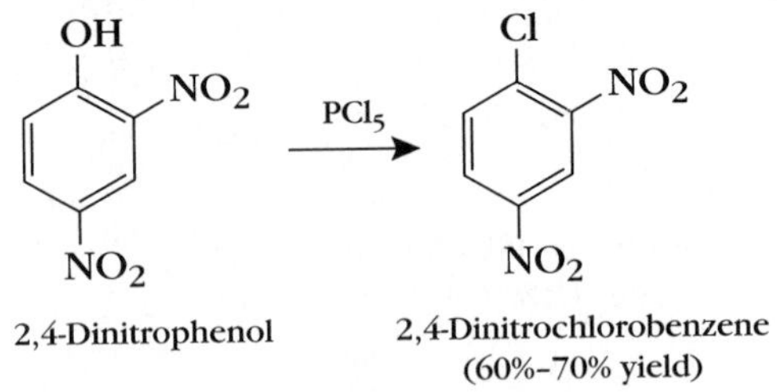

2,4-Dinitrophenol   2,4-Dinitrochlorobenzene
(60%–70% yield)

Exercise 23.15   Propose a mechanism for the formation of 2,4-dinitrochlorobenzene by the reaction of PCl_5 and 2,4-dinitrophenol.

23.11 Electrophilic Aromatic Substitution Reactions of Phenols and Phenoxide Ions

■ *The mechanism of electrophilic aromatic substitution is discussed in Section 21.4.*

The —OH and —O⁻ groups of phenols and phenoxide ions, respectively, like the —NH_2 group of aniline, strongly activate aromatic rings to electrophilic aromatic substitution reactions. They activate aromatic rings because they stabilize the intermediate by donating electrons into the benzene ring. This delocalizes the positive charge of the intermediate carbocation, as illustrated by the following resonance hybrid structures.

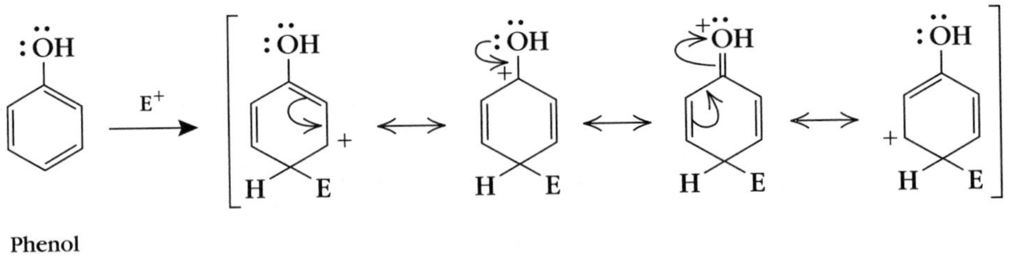

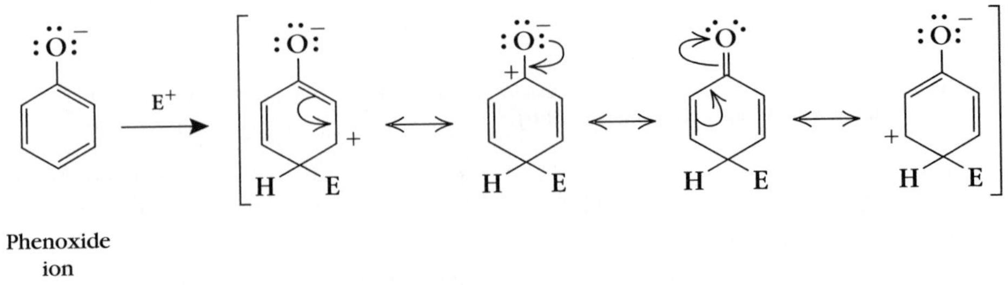

Phenol

Phenoxide
ion

Phenol, for example, reacts with a dilute solution of bromine in acetic acid to form mostly 4-bromophenol.

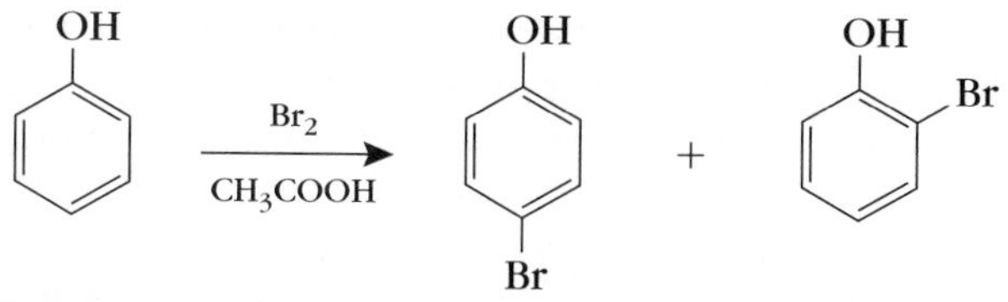

Phenol   4-Bromophenol (89% yield)   2-Bromophenol (11% yield)

If phenol is brominated in aqueous solution, the product is 2,4,6-tribromophenol.

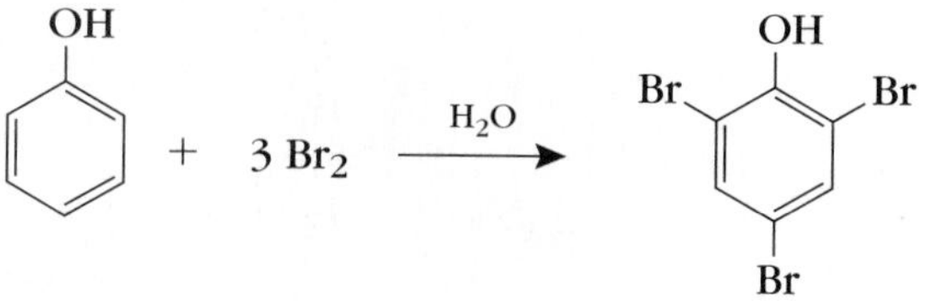

Phenol   2,4,6-Tribromophenol

The bromination of phenol yields different products in acetic acid as solvent than in water as solvent because the reactive species is different in the two solvents. In an acidic solvent such as acetic acid or in nonpolar solvents such as CCl_4, phenol does not ionize and bromination occurs at the *para*-position of phenol. In polar solvents like water, phenol and phenoxide ion are in equilibrium and it is the phenoxide ion that undergoes electrophilic aromatic substitution. Even though there is less phenoxide ion in solution, it is more reactive than phenol. Once the phenoxide ion reacts, the phenol-phenoxide equilibrium is restored by more phenol converting to phenoxide. Eventually all of the phenol is converted to phenoxide in this way.

Polysubstitution occurs because the first substitution product, a monobromophenol, is a stronger acid than phenol and a greater fraction of it is in the form of its phenoxide ion. Consequently, monobromophenoxide ion reacts more rapidly than phenoxide ion to form a dibromophenol, which is still more acidic. This process continues until all the phenol is consumed.

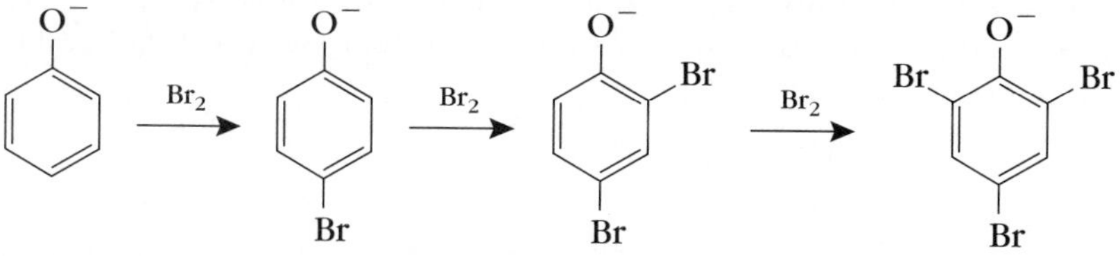

Nitration of phenol is possible, but care must be exercised to avoid nitric acid oxidation of phenol to quinone, a reaction discussed in Section 23.14. Phenol is nitrated in dilute nitric acid to form both the *ortho*- and *para*- isomers.

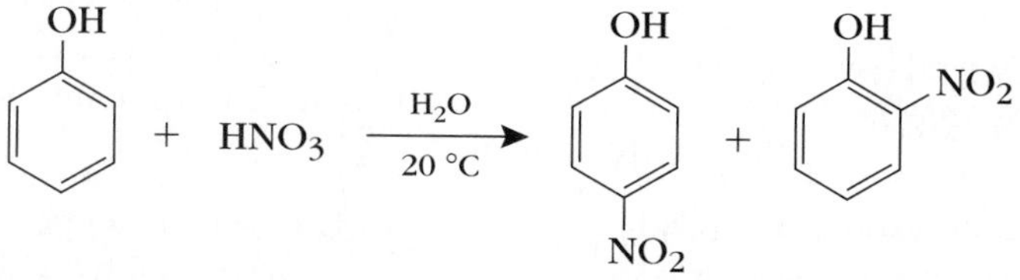

Phenol   4-Nitrophenol (60% yield)   2-Nitrophenol (40% yield)

The *ortho-* and *para-* isomers can be separated by steam distillation. The *ortho-* isomer is very volatile because it forms strong intramolecular hydrogen bonds. The *para-* isomer is less volatile because it forms intermolecular hydrogen bonds.

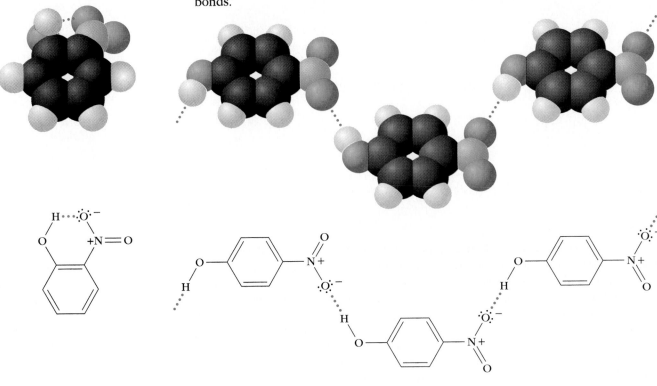

ortho isomer
(Intramolecular hydrogen bonding)

para isomer
(Intermolecular hydrogen bonding)

Phenols undergo a number of other electrophilic substitution reactions that are possible only because of their high reactivity. For example, they react with the weakly electrophilic diazonium ion to form azo compounds. This reaction was introduced in Section 22.17. Another weak electrophile, the nitrosyl cation (NO^+), formed by the reaction of sodium nitrite and sulfuric acid, reacts with phenols to form nitrosophenols.

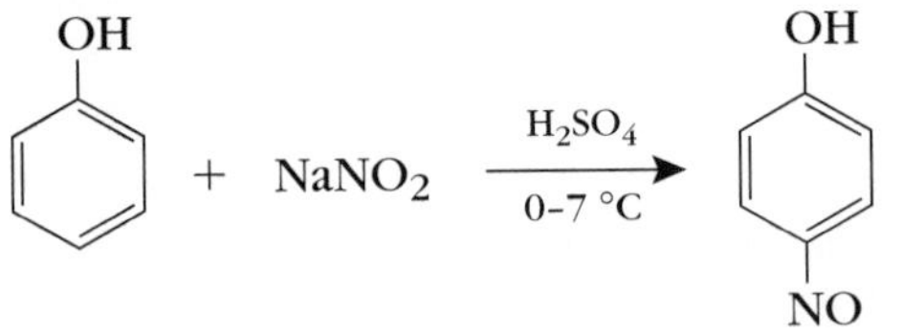

$$\text{OH} + \text{NaNO}_2 \xrightarrow[\text{0-7 °C}]{\text{H}_2\text{SO}_4} \text{OH} \quad \text{NO}$$

(78% yield)

Exercise 23.16   Write the equation for the reaction of phenol with each of the following reagents.

(a) Benzenediazonium chloride
(b) Concentrated sulfuric acid
(c) An aqueous solution of chlorine
(d) Chlorine in acetic acid
(e) Benzenesulfonyl chloride

Exercise 23.17   Propose a mechanism for the following reaction.

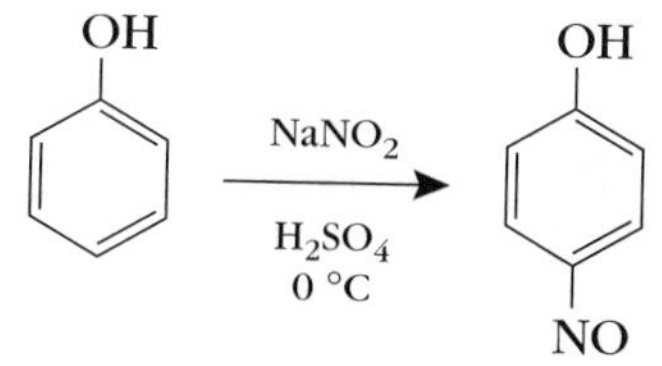

Several commercially important products are formed by the electrophilic aromatic substitution of phenoxide ions with carbon electrophiles.

23.12 Reactions at the Ring Carbon Atoms of Phenoxide Ions

Phenoxide ions, just like enolate ions, can be alkylated at either their oxygen or ring carbon atoms. The reaction of phenoxide ion and CO_2 at 125 °C and 100 atm pressure to form sodium salicylate is an example of an alkylation reaction that occurs at one of the ring carbon atoms of the phenoxide ion.

■ Recall from Section 17.12 that enolate anions are ambident nucleophiles. That is, enolate ions can be alkylated at either their carbon or oxygen atoms.

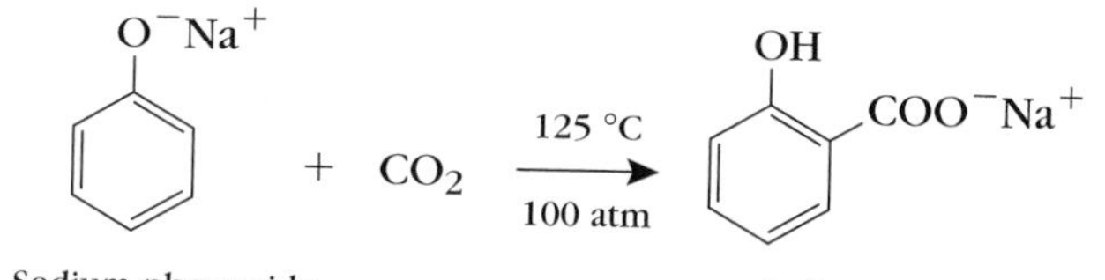

This reaction, discovered over 100 years ago by the German chemist Hermann Kolbe, is called the **Kolbe reaction.**

The mechanism of the Kolbe reaction may be visualized as a nucleophilic addition of the phenoxide ion to the carbon atom of CO_2, as shown in Figure 23.7. The Kolbe reaction is an equilibrium reaction. The salicylate ion is favored at equilibrium because it is a weaker base than the phenoxide ion. The salicylate ion is also a weaker base than its isomer, 4-hydroxybenzoate, so the salicylate ion, formed by reaction at the *ortho-* position of the phenoxide ion, is the preferred product at equilibrium.

The Kolbe reaction is the key step in the industrial synthesis of acetylsalicylic acid (ASA or aspirin). Once sodium salicylate is formed, it is acidified to form salicylic acid. Salicylic acid is then treated with acetic anhydride to form aspirin.

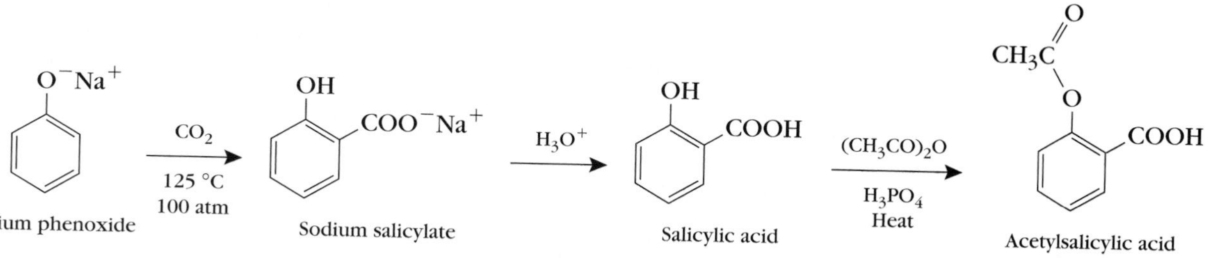

Figure 23.7
The mechanism of the Kolbe reaction. Step 1 resembles the mechanism of the addition of a Grignard reagent to CO_2. In the reaction of CO_2 with both a Grignard reagent and a phenoxide ion, a carbon-carbon single bond is formed.

Step 1: Nucleophilic addition to the carbon atom of CO_2

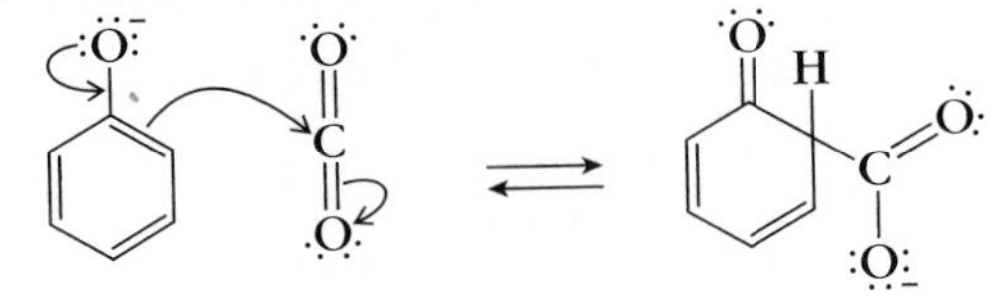

Phenoxide
ion

Step 2: Loss of proton regenerates benzene ring

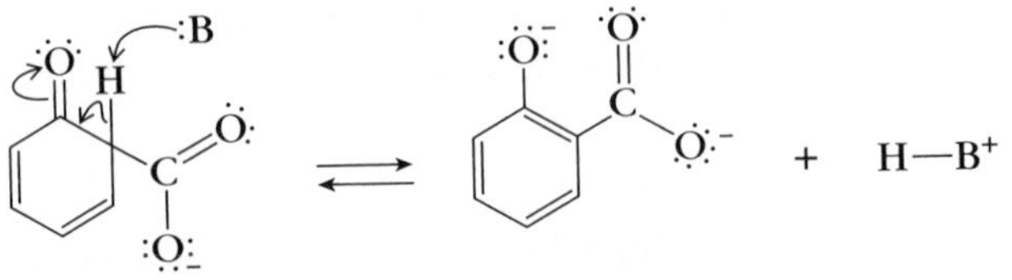

$+$ H—B^+

Step 3: Proton transfer reaction

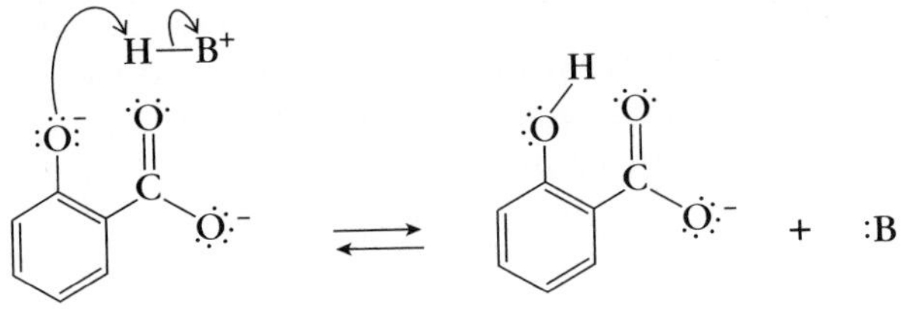

$+$:B

Salicylate ion

Exercise 23.18   Propose a mechanism for the following reaction of phenol with formaldehyde in an aqueous basic solution.

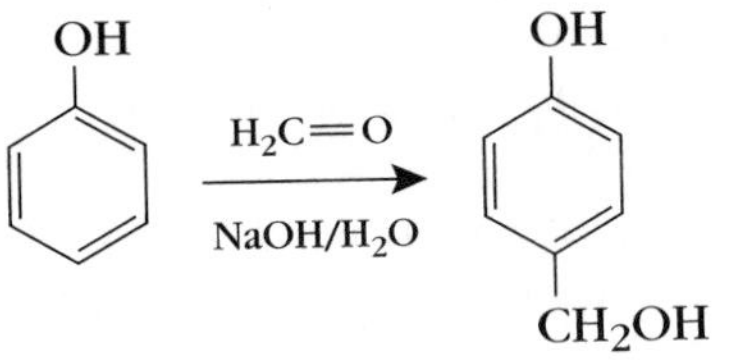

23.13 Claisen Rearrangements of Allyl Aryl Ethers

When an allyl aryl ether is heated to 200 °C, its allyl group migrates from the oxygen atom to the ring carbon atom *ortho-* to it. This reaction is called the **Claisen rearrangement.** Heating allyl phenyl ether, for example, forms 2-allylphenol by a Claisen rearrangement.

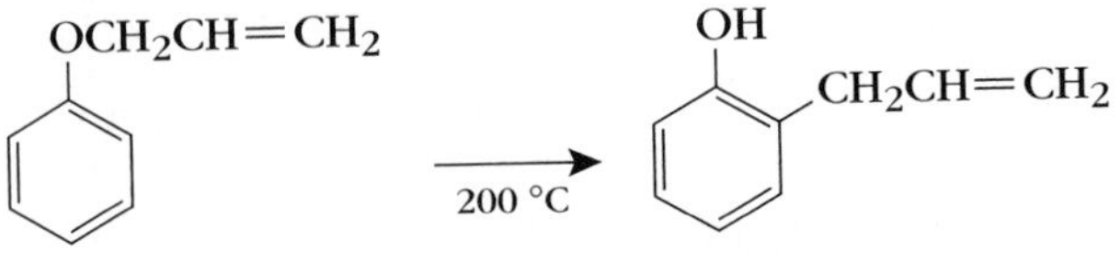

Allyl phenyl ether

2-Allylphenol
(70%–75% yield)

Claisen rearrangement of allyl phenyl ether containing a ^{14}C label at the terminal carbon atom of the allyl group forms 2-allylphenol in which the ^{14}C label is located on the carbon atom of the allyl group bonded to the benzene ring. This observation can be explained by a cyclic mechanism in which a concerted reorganization of bonding electrons occurs in the transition state. High temperatures are needed for the Claisen rearrangements because the benzene ring of the reactant is converted into a nonaromatic cyclohexadienone intermediate. The intermediate then enolizes in a second step to reform the aromatic ring.

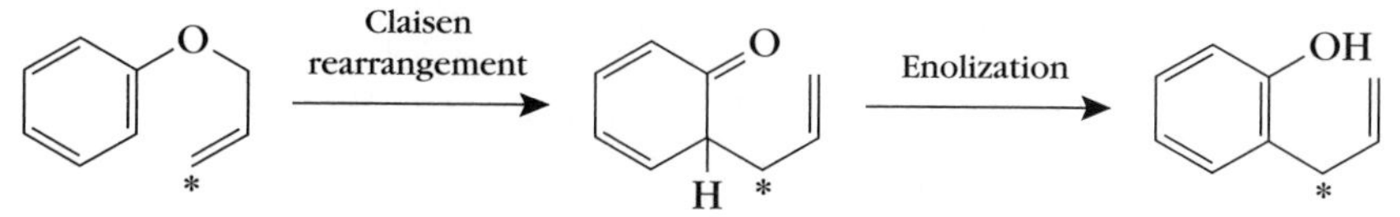

| Allyl phenyl ether | Cyclohexadienone intermediate | 2-Allylphenol |

* = location of ^{14}C

Exercise 23.19   Write the structure of the product formed by the Claisen rearrangement of 2-butenyl phenyl ether.

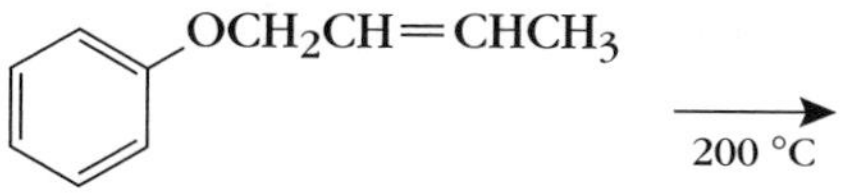

2-Butenyl phenyl ether

Because phenols are electron-rich compounds, they are very susceptible to oxidation.

23.14 Formation of Quinones by Oxidation of Phenols

Phenols, like alcohols, can be oxidized to carbonyl-containing compounds. The oxidation of phenol by chromic acid, for example, forms a **quinone.**

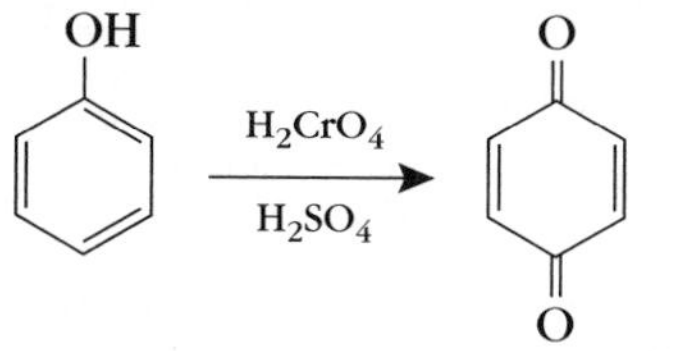

Phenol   A quinone

Quinones are named as derivatives of the aromatic compound from which they are obtained on oxidation. The simplest quinones are 1,2-benzoquinone and 1,4-benzoquinone, both of which are derived from benzene.

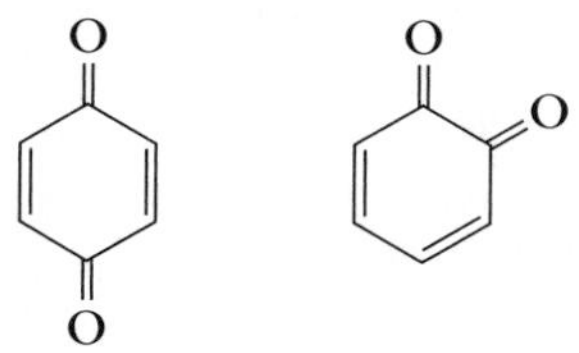

p-Benzoquinone   *o*-Benzoquinone
1,4-benzoquinone   1,2-benzoquinone
(quinone)

The word quinone refers to this class of compounds, but it is also the common name of 1,4-benzoquinone.

Many quinones are known that contain fused benzene rings. Examples include naphthoquinone, whose name is derived from napthalene, and anthroquinone, whose name is derived from anthracene.

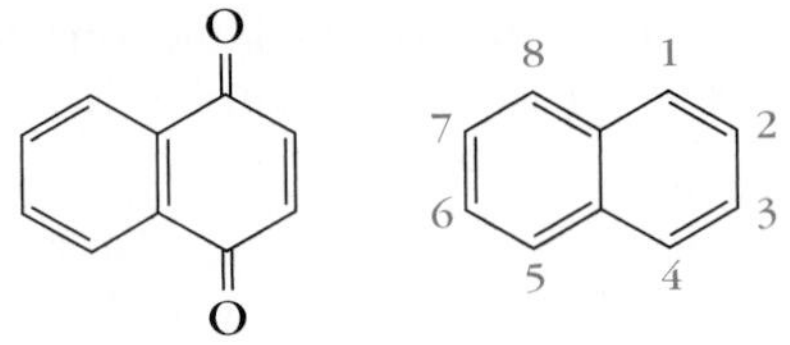

1,4-Naphthoquinone   Naphthalene

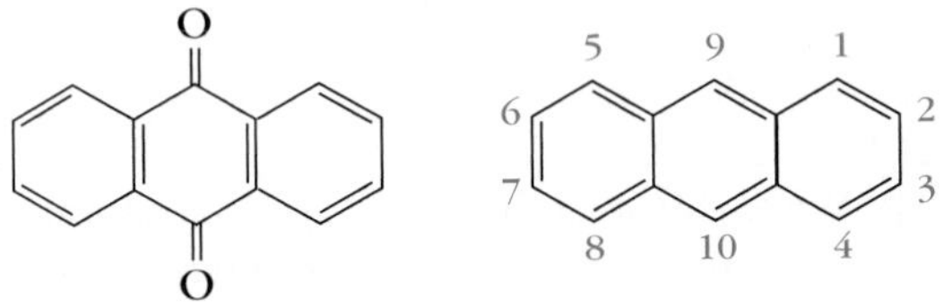

9,10-Anthroquinone   Anthracene

Exercise 23.20 Write the structures of each of the following compounds.

(a) 2,3-Dichloro-1,4-benzoquinone   **(c)** 4-Methyl-1,2-benzoquinone
(b) 1,4-Dimethyl-9,10-anthroquinone

■ *The word* hydroquinone *is both the common name of 1,4-dihydroxybenzene and the name of a class of compounds that contain two —OH groups on benzene ortho or para to each other.*

Mono- and polycyclic aromatic hydrocarbons that contain two hydroxy groups can also be oxidized to quinones. Examples include 1,2-dihydroxybenzene (catechol) and 1,4-dihydroxybenzene (hydroquinone).

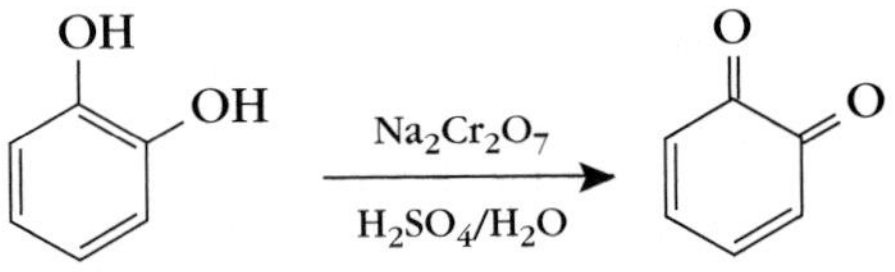

1,2-Dihydroxybenzene   1,2-Benzoquinone
(catechol)

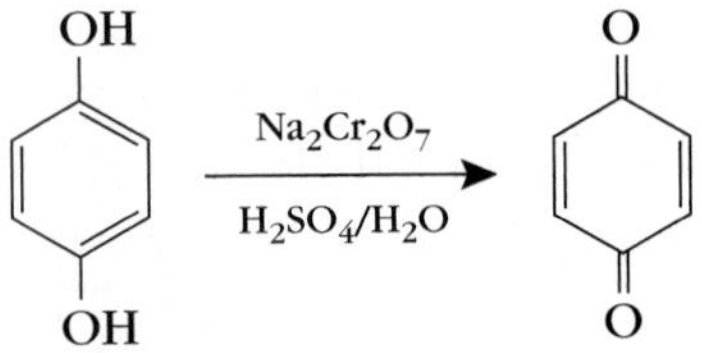

1,4-Dihydroxybenzene   1,4-Benzoquinone
(hydroquinone)   (quinone)

Aniline and its derivatives are electron-rich aromatic compounds that can be oxidized to quinones.

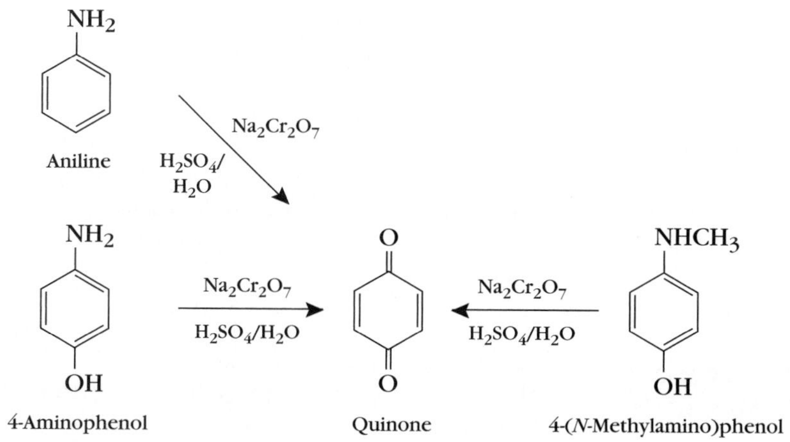

Aniline

4-Aminophenol   Quinone   4-(N-Methylamino)phenol

Exercise 23.21   Write the structure of the product formed by the reaction of each of the following compounds with an aqueous sulfuric acid solution of $Na_2Cr_2O_7$.

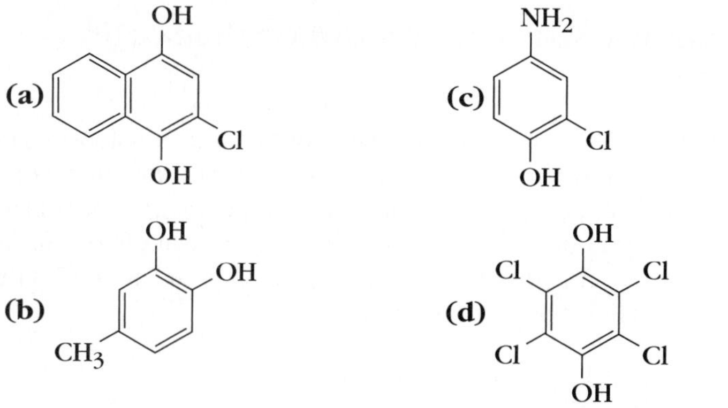

(a)

(c)

(b)

(d)

23.15 Addition Reactions of Quinones

Quinones are not aromatic compounds. Instead, they are α,β-unsaturated ketones, so they undergo 1,4-addition reactions. The 1,4-addition of HCl to 1,4-benzoquinone, for example, forms a product that enolizes to form 2-chloro-1,4-dihydroxybenzene.

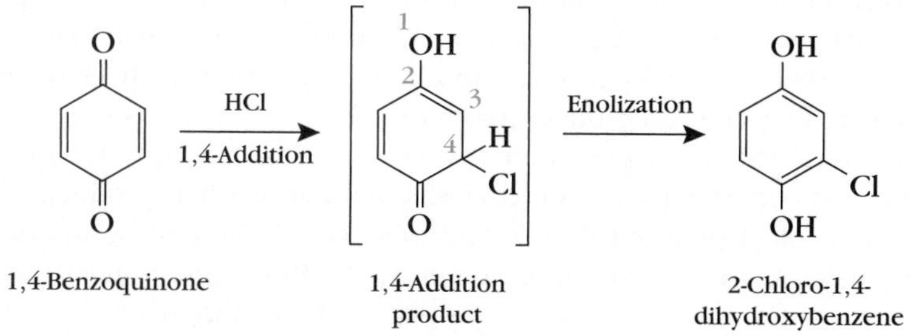

1,4-Benzoquinone   1,4-Addition   2-Chloro-1,4-
product   dihydroxybenzene

Quinones also readily undergo Diels-Alder reactions.

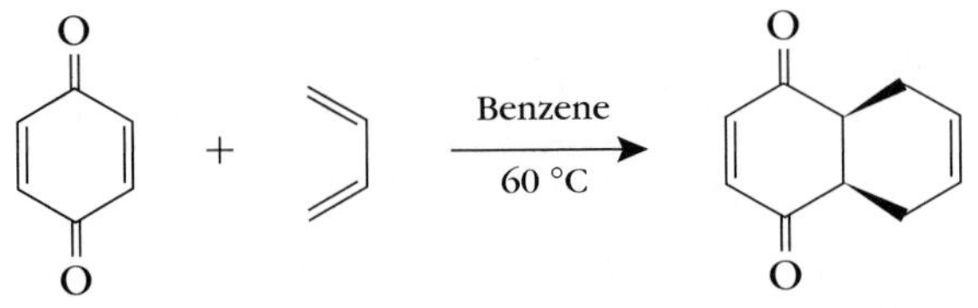

1,4-Benzo-   1,3-Butadiene   5,8,9,10-Tetrahydro-
quinone   1,4-napthoquinone

Exercise 23.22   Write the structure of the major product in each of the following reactions.

(a) 1,4-Naphthoquinone + 1,3-Butadiene ⟶

(b) 1,4-Benzoquinone + Cyclopentadiene ⟶

23.16 Hydroquinone-Quinone Oxidation-Reduction Equilibria

Phenols resemble alcohols when it comes to oxidation-reduction reactions. Phenols, for example, are oxidized to carbonyl-containing compounds, just as alcohols are. Furthermore, reduction of the carbonyl-containing compound regenerates an alcohol or a hydroquinone (a dihydroxybenzene). An example of this kind of oxidation-reduction reaction is the interconversion of hydroquinone (1,4-dihydroxybenzene) and quinone (1,4-benzoquinone).

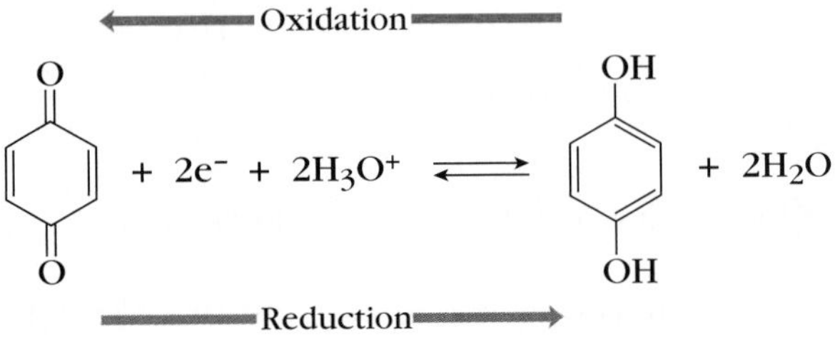

Quinone   Hydroquinone
(1,4-Benzoquinone)   (1,4-Dihydroxybenzene)

Interconversion of hydroquinone and quinone can be accomplished with the usual chemical oxidizing or reducing agents. Thus strong oxidizing agents such as chromic acid oxidize hydroquinone, while reducing agents such as metals in acid and catalytic hydrogenation reduce quinones.

The hydroquinone-quinone interconversion is unusual among organic reactions because it occurs rapidly and reversibly with an easily reproduced electrode potential in an electrochemical cell. The position of the hydroquinone-quinone equilibrium is proportional to the square of the hydrogen ion concentration. Consequently, the electrode potential is sensitive to the acidity of the solution. Prior to the development of the glass electrode, the half-cell potential of this equilibrium reaction was used to determine the pH level of aqueous solutions.

Exercise 23.23 Write the structure of the dihydroxybenzene formed by the reduction of each of the following quinones.

(a)

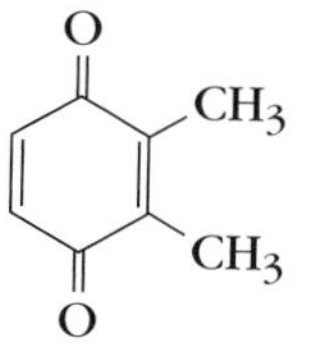

(b)

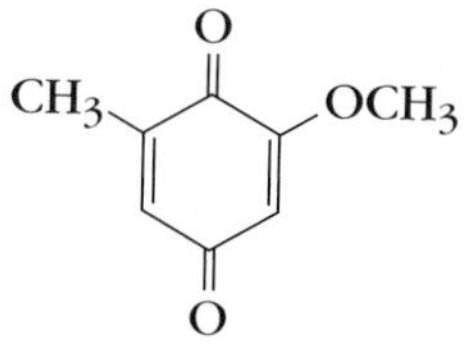

(c)

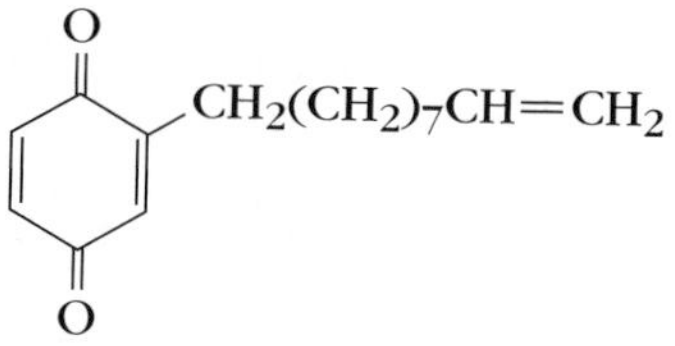

Humans and other organisms use reversible hydroquinone-quinone redox reactions to obtain energy from food.

23.17 Quinones Are Important in Obtaining Free Energy from Food

Living organisms need a continuous supply of free energy. Without it, the organism is not able to perform mechanical work such as muscle contraction, transport of molecules and ions between its various parts, or synthesis of molecules unavailable from food.

Humans and other mammals obtain this free energy by oxidizing food. Part of this free energy is transformed into the compound adenosine triphosphate (ATP), which is the most important chemical form of free energy in cells. Formation of most ATP is coupled to a process called **oxidative phosphorylation.** The overall process is the formation of NAD^+ by the oxidation of NADH. Molecular oxygen (O_2) is the oxidizing agent, and it is reduced to water in the process.

> ■ *The structure of NAD^+, nicotinamide adenine dinucleotide, is given in Figure 11.9 (p. 477).*

Oxidation

$$\tfrac{1}{2}O_2 + NADH + H_3O^+ \rightleftharpoons 2H_2O + NAD^+$$

Reduction

This redox reaction does not occur in a single step. Instead it occurs via a series of electron-transport complexes made up of proteins and cofactors that catalyze different steps in the overall oxidation of NADH by O_2.

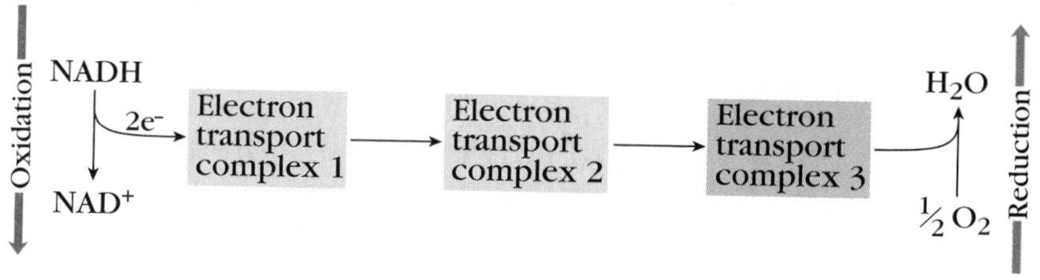

Oxidation → NADH $\xrightarrow{2e^-}$ NAD$^+$ → Electron transport complex 1 → Electron transport complex 2 → Electron transport complex 3 → $\tfrac{1}{2}O_2$ → H$_2$O ← Reduction

ATP is formed in each of these complexes by the reaction of ADP and PO_4^{3-} (inorganic phosphate).

A key electron carrier in several of these complexes is ubiquinone, which is also known as coenzyme Q. Ubiquinone is a quinone that contains a number of substituents, including a long chain made up of isoprene units. The length of this chain depends on the organism; in humans and most other mammals it is 10 units long.

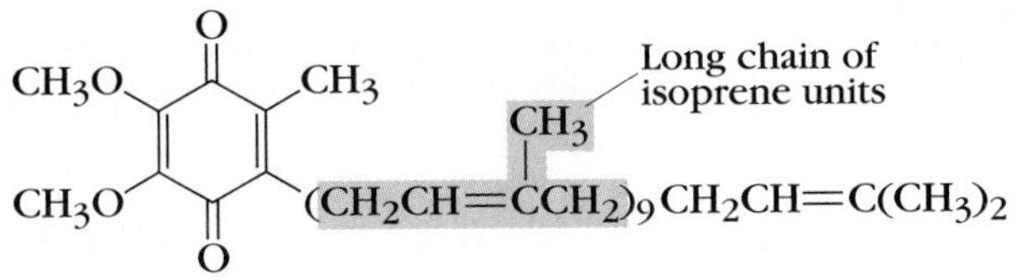

Ubiquinone
(Coenzyme Q)

Ubiquinone carries electrons in a sequence of two reactions. In the first reaction, a reducing agent that is part of the complex transfers two electrons to ubiquinone (Q). This reaction forms ubiquinol (QH_2), the reduced form of ubiquinone.

Oxidized form of ubiquinone (Q)

+

Z—H

Reducing agent transfers
two electrons and two protons to Q

Reduced form of ubiquinone (QH_2)

+

Z

Oxidized form of
the reducing agent

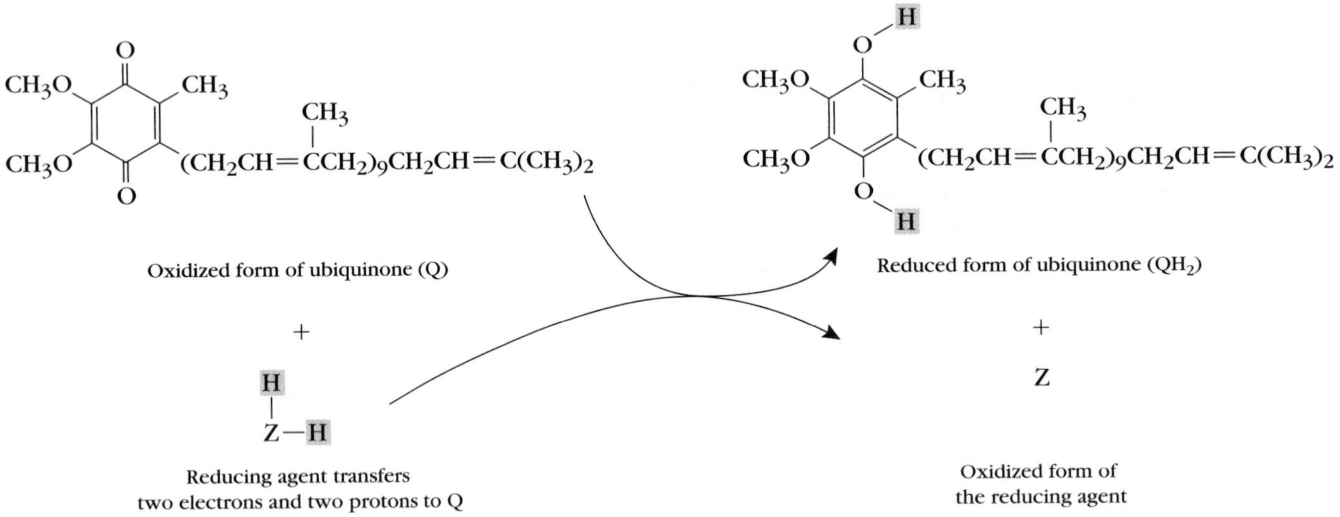

QH_2 is mobile, so it can travel to another complex where it reacts with an oxidizing agent attached to the complex to form Q.

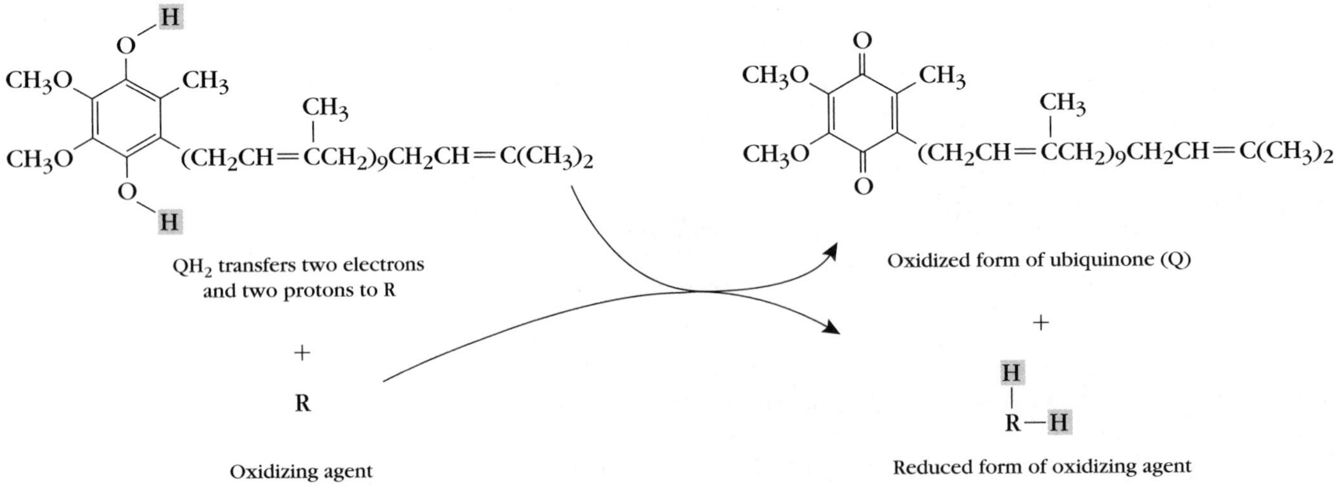

QH_2 transfers two electrons
and two protons to R

+

R

Oxidizing agent

Oxidized form of ubiquinone (Q)

+

R—H

Reduced form of oxidizing agent

By means of these two reactions, Q and QH_2 shuttle electrons from complex 1 to complex 2. In complex 1, Q accepts two electrons and two protons and is reduced to QH_2. QH_2 then moves to complex 2, where it delivers two electrons and two protons and is oxidized to Q again. Q then moves back to complex 1 to restart the cycle. This cyclic redox process can be summarized as follows:

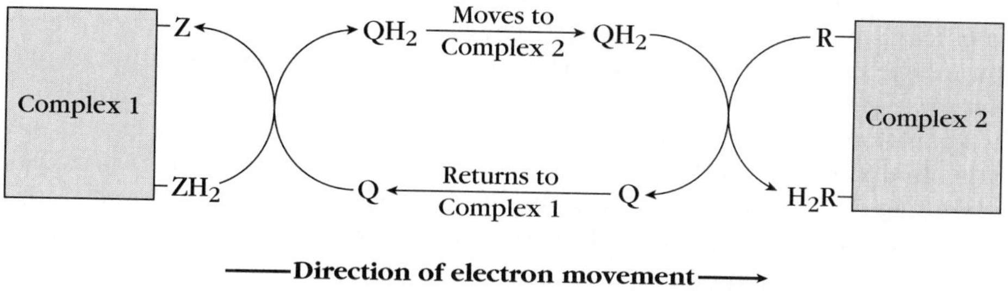

Plastoquinone (PQ) and its reduced form (PQH$_2$) are another quinone-hydroquinone electron-transporting pair. They shuttle electrons associated with photosynthesis in plants.

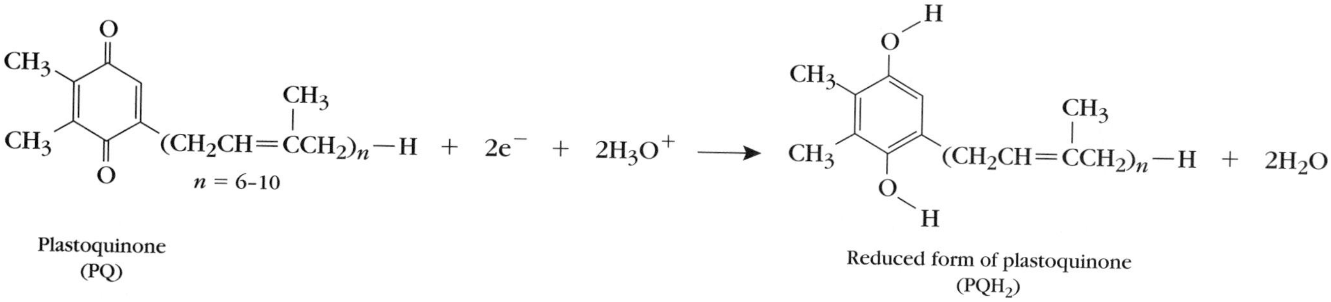

Plastoquinone
(PQ)

Reduced form of plastoquinone
(PQH$_2$)

In summary, redox reactions of the quinone-hydroquinone type are extremely important in living organisms. They serve as key shuttles in electron transfer during the oxidation of food. As a result, they are key players in the processes that provide the free energy needed for organisms to live.

23.18 Summary

Halobenzenes are compounds that have a halogen atom bonded directly to a benzene ring. They undergo **nucleophilic aromatic substitution reactions** by two mechanisms. Halobenzenes that contain strongly electron-withdrawing groups in the *ortho-* and/or *para-* positions undergo substitution by a mechanism that involves addition of the nucleophile to the ring to form an anionic intermediate, which then eliminates a halide ion to form the product. Unsubstituted halobenzenes or those not activated by electron-withdrawing substituents react by a mechanism that involves first elimination of HX followed by addition of the nucleophile to the intermediate **benzyne**.

Phenols are compounds that contain a hydroxy group bonded directly to a benzene ring. They are prepared in the laboratory by hydrolysis of arenediazonium salts, which are formed by the diazotization of derivatives of aniline. Biosynthesis of phenols occurs almost exclusively in plants, fungi, and microorganisms by two complex pathways. One is an enzyme-catalyzed Claisen-like condensation of acetyl CoA to form a β-polyketone, which cyclizes to eventually form phenols. Industrially, phenol is prepared by heating chlorobenzene and a concentrated solution of NaOH under pressure (the **Dow process**), and by air oxidation of cumene, which is prepared by the Friedel-Crafts reaction of benzene and propene.

The most important characteristic of phenols is their acidity; they are some 10^6 times more acidic than typical alcohols. Phenols undergo many of the same reactions as alcohols. Thus the hydroxy group can be converted into an ester or an ether group. The hydroxy group strongly activates the benzene ring of phenol toward the usual electrophilic aromatic substitution reactions. Oxidation of phenols form **quinones.**

Reduction of quinones either by chemical reagents or in electrochemical cells forms **hydroquinones.** Hydroquinones and quinones can be interconverted easily and rapidly via a reversible oxidation reduction reaction. In nature, a similar set of redox reactions involving ubiquinone transfers electrons in a process called **oxidative phosphorylation.**

23.19 Summary of Reactions

1. **Nucleophilic Substitution Reactions of Halobenzenes**   SECTION 23.1

   Reaction occurs much more easily when strongly electron-withdrawing groups are located on the ring *o-* and *p-* to the halogen atom.

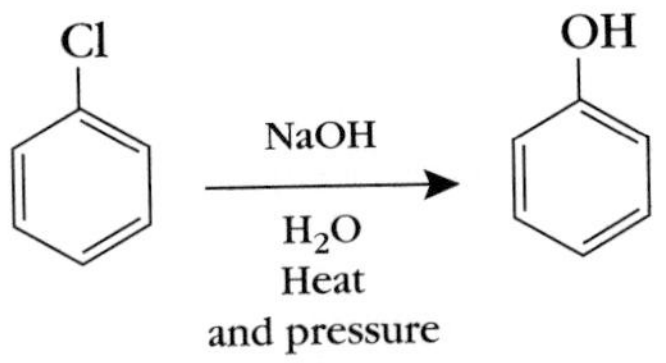

2. **Preparation of Phenols by Hydrolysis of Diazonium Salts**   SECTION 23.6

   Good general synthesis of substituted phenols because substituents on ring are not affected by mild reaction conditions.

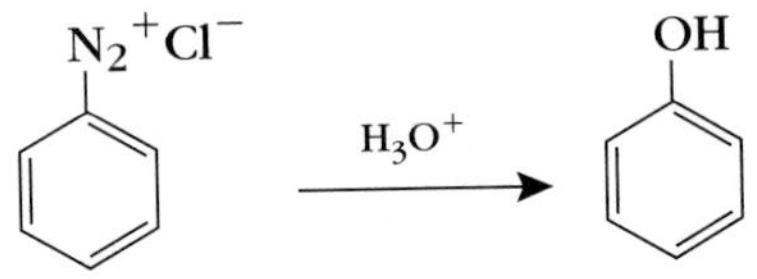

3. **Reactions of Phenols**

   a. **Acidity**   SECTION 23.7

   pK_a of phenols is about 10 except for nitrophenols, which are much stronger acids.

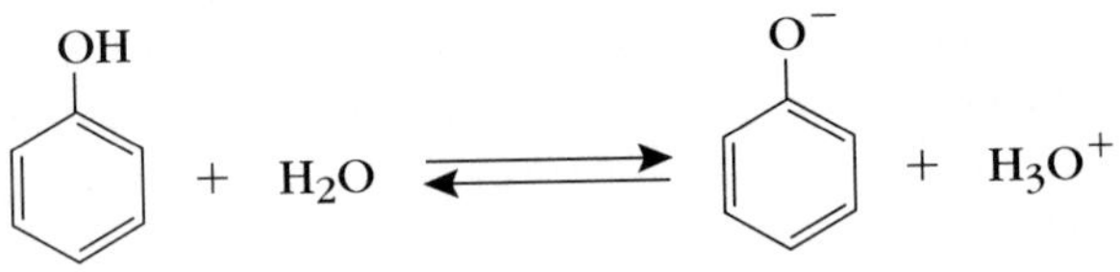

   b. **Ester formation**   SECTION 23.9

   Phenols also react with both acyl chlorides and carboxylic acid anhydrides to form esters.

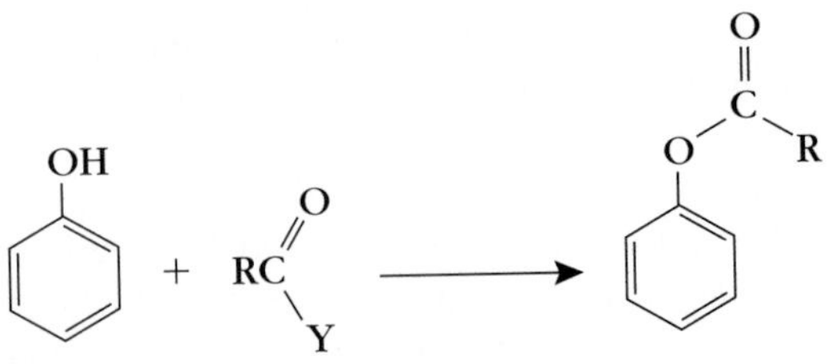

Y = Cl or OOCR

c. Williamson ether synthesis   SECTION 23.9

Reaction occurs by an S_N2 mechanism, so best results are obtained with primary and secondary alkyl derivatives.

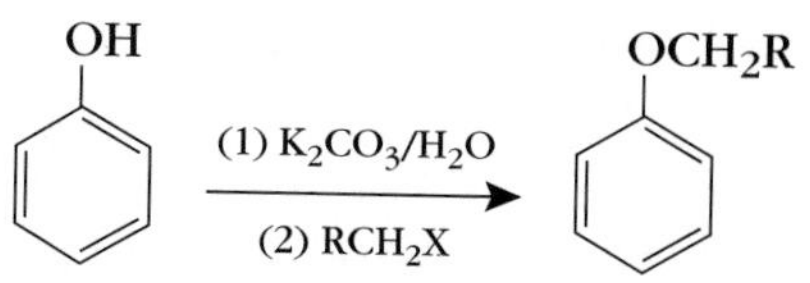

d. Electrophilic substitution reactions   SECTIONS 23.11 and 21.11

The —OH group is strongly activating and *o-*, *p-* directing.

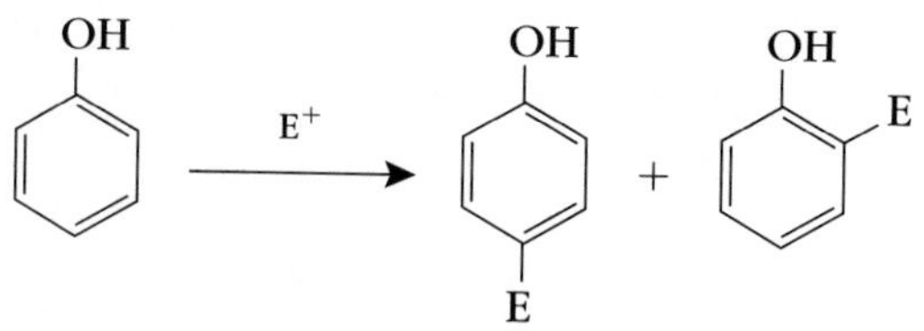

E = Electrophilic reagent

e. Kolbe reaction   SECTION 23.12

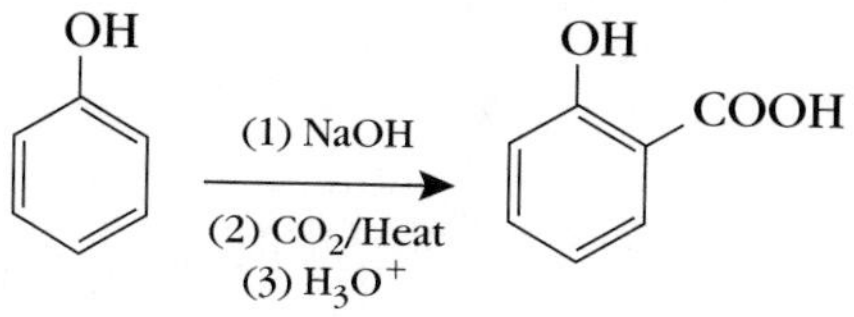

f. Claisen rearrangement   SECTION 23.13

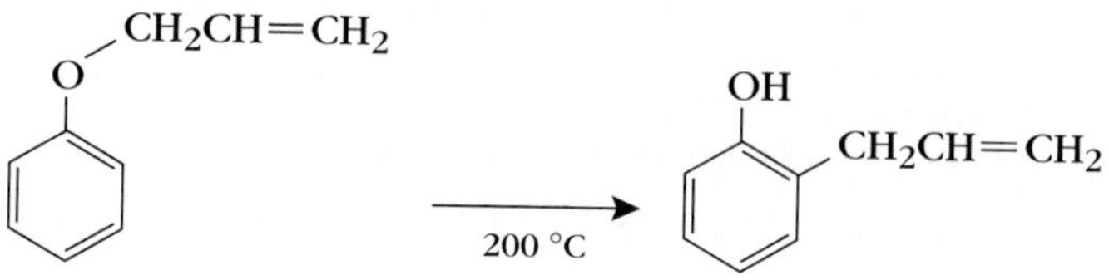

4. Formation of Quinones by Oxidation of Dihydroxybenzenes   SECTION 23.14

Oxidation of phenol also forms 1,4-benzoquinone.

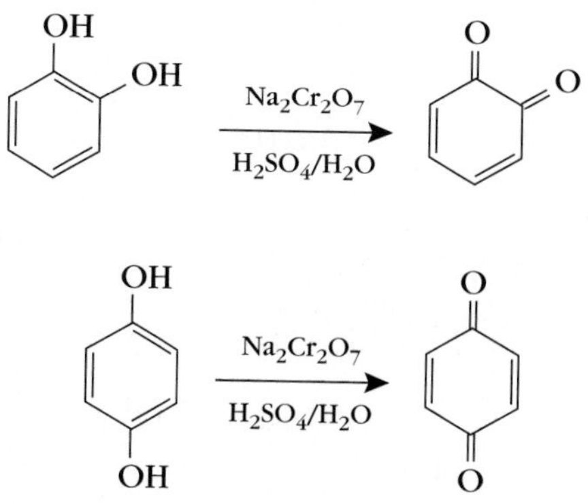

5. Reactions of Quinones
    a. Diels-Alder reaction   SECTION 23.15

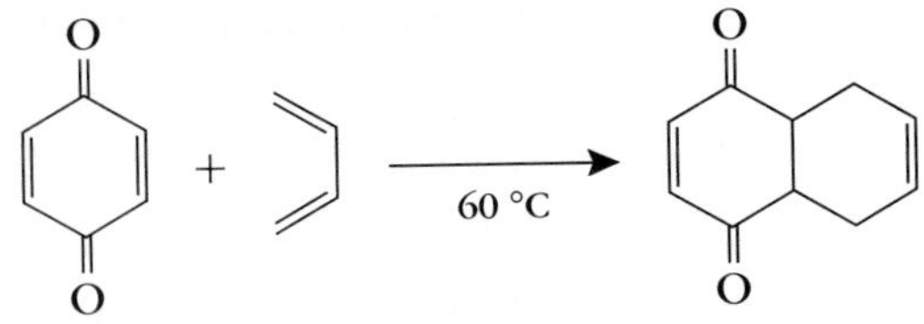

    b. Quinone-hydroquinone
        oxidation-reduction equilibria   SECTION 23.16

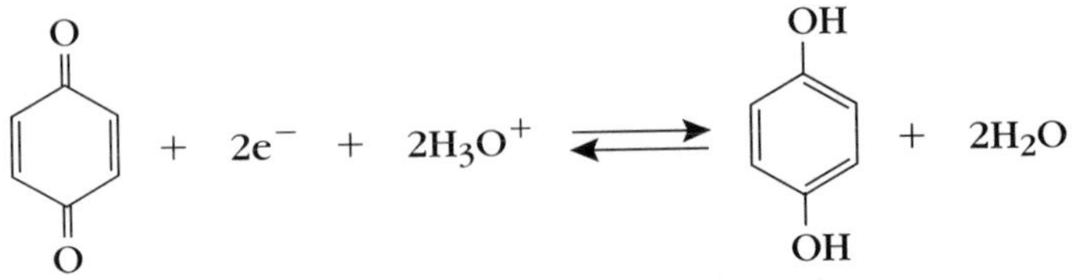

Additional Exercises

23.24 Define each of the following terms in words, by a structure, or by a chemical equation.

(a) A phenol
(b) Nucleophilic aromatic substitution reaction
(c) A hydroquinone
(d) Kolbe reaction
(e) A halobenzene

(f) Claisen rearrangement
(g) Addition-elimination mechanism
(h) Benzyne
(i) Elimination-addition mechanism

23.25 Name each of the following compounds.

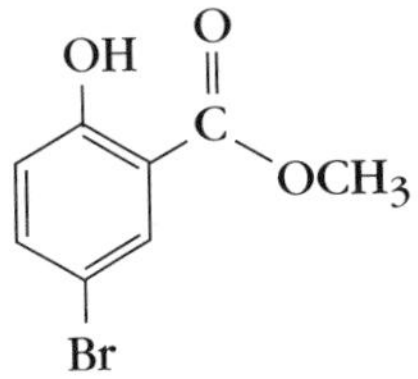

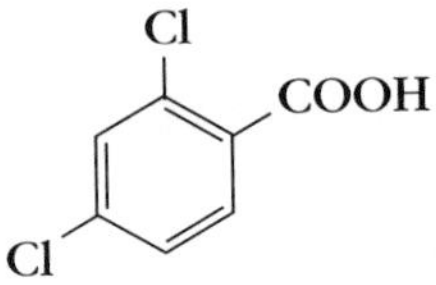

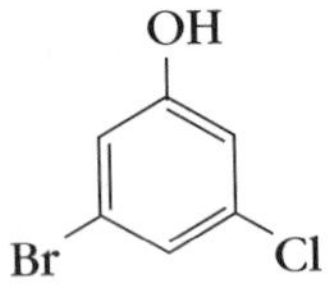

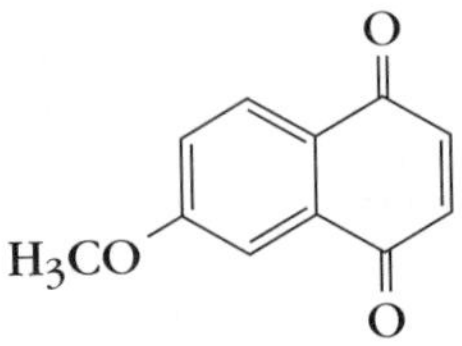

23.26 Write the structural formula for each of the following compounds.

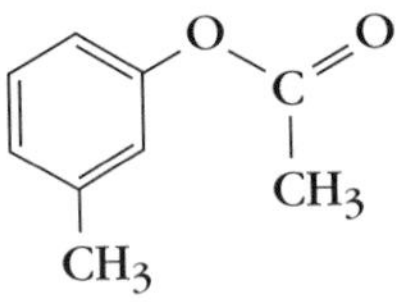

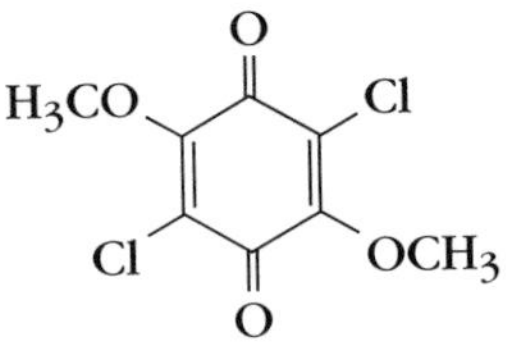

(a) 4-(2-Butyl)phenol
(b) 3-Bromo-1,2-benzoquinone
(c) 2-Ethoxy-1,4-naphthoquinone

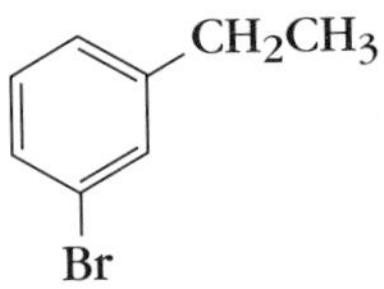

(d) 2-Methoxyphenol
(e) 4-Bromobenzyl bromide
(f) 2-Fluorostyrene

(g) 1,3,5-Trihydroxybenzene
(h) 3,4-Dihydroxytoluene
(i) Acetylsalicylic acid

23.27 Write the structure of the product or products of the reaction of 4-methylphenol with each of the following reagents.

(a) $NaOH/CH_3I$
(b) Br_2/CH_3COOH
(c) $(CH_3CO)_2O/CH_3COOH$
(d) Br_2/H_2O
(e) $C_6H_5N_2^+ Cl^-/NaOH/H_2O$
(f) $NaOH/CO_2$/heat
(g) $NaNO_2/HCl$
(h) H_2SO_4

23.28 Write the structure of the product or products of the reaction of 4-chloroanisole with each of the following reagents.

(a) HBr/H_2O heat
(b) $NaNH_2$/liquid NH_3
(c) HNO_3/H_2SO_4
(d) Mg followed by acetone then H_3O^+
(e) $NaOH/H_2O$/heat/pressure.
(f) $Br_2/FeBr_3$

23.29 Would you expect 2-hydroxyacetophenone or 4-hydroxyacetophenone to be the more volatile? Explain.

23.30 Write the structure of the product or products of each of the following reactions. If isomers are formed in appreciable amounts, write the structures of all of them.

(a) 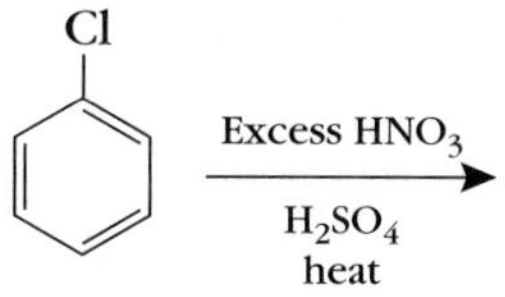

(b) Product of (a) $\xrightarrow{NH_3}$

(c) 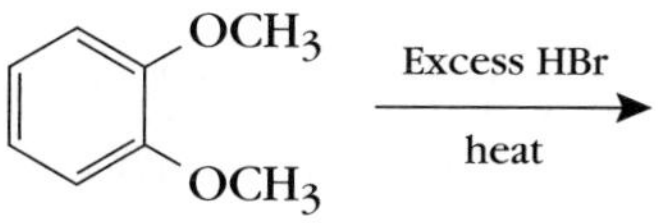

(d) Aromatic product of (c) $\xrightarrow[H_2SO_4]{Na_2Cr_2O_7}$

(e) 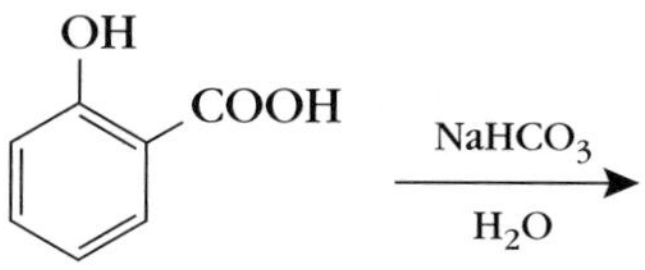

(f) 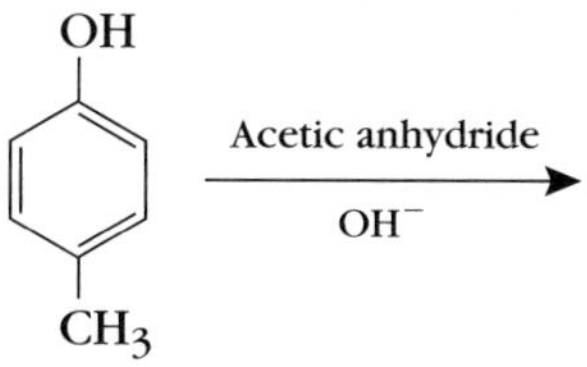

(g) 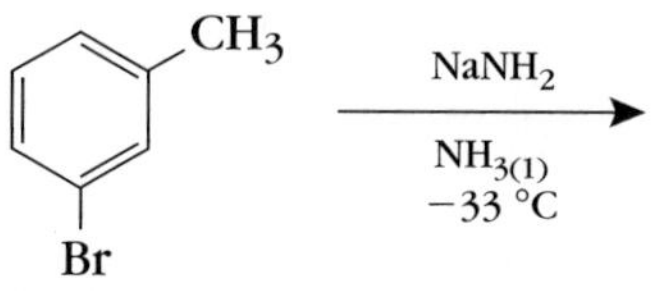

(h) 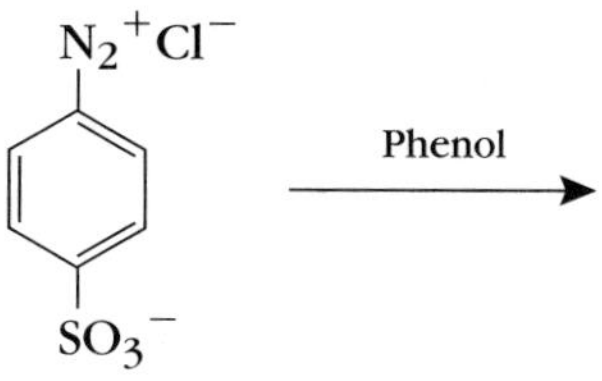

(i) 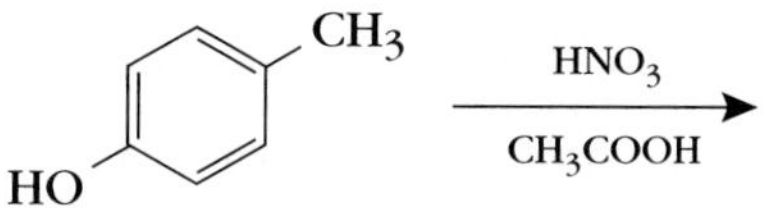

(j) 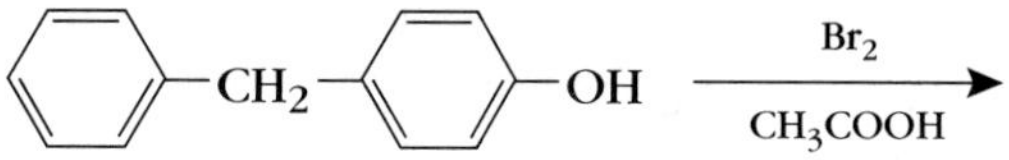

(k) 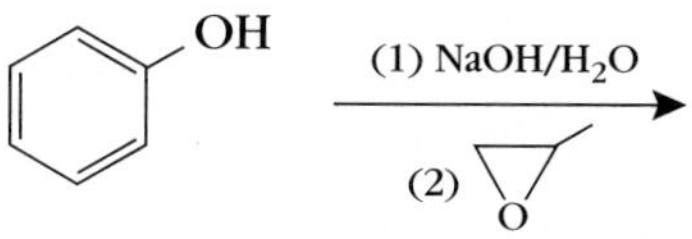

(l) 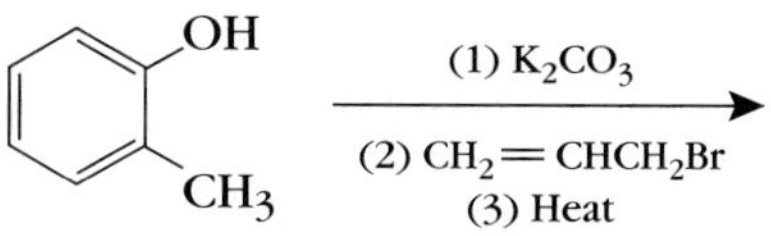

23.31 Propose a mechanism for the following reaction.

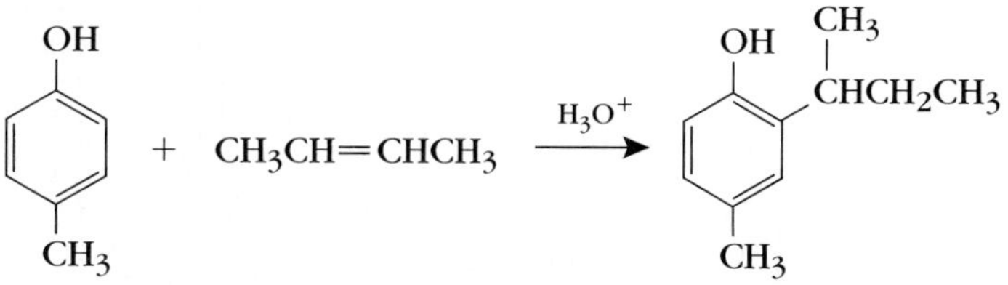

23.32 Propose a mechanism for the following reaction.

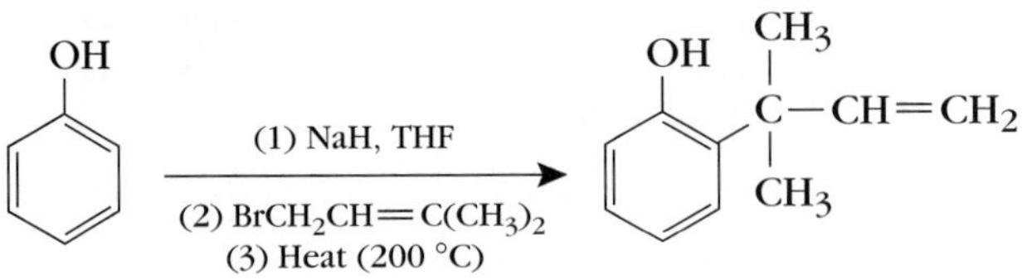

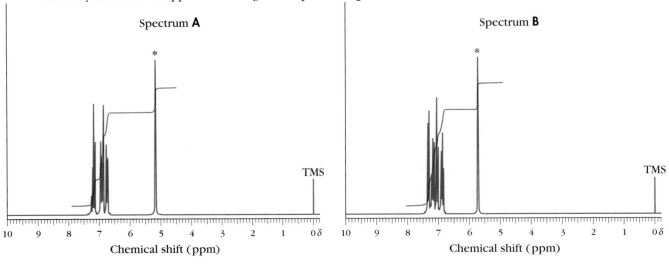

23.33 When the product of the Diels-Alder reaction of 1,4-benzoquinone and 1,3-butadiene is treated with acid, the following isomerization reaction occurs. Propose a mechanism for this reaction.

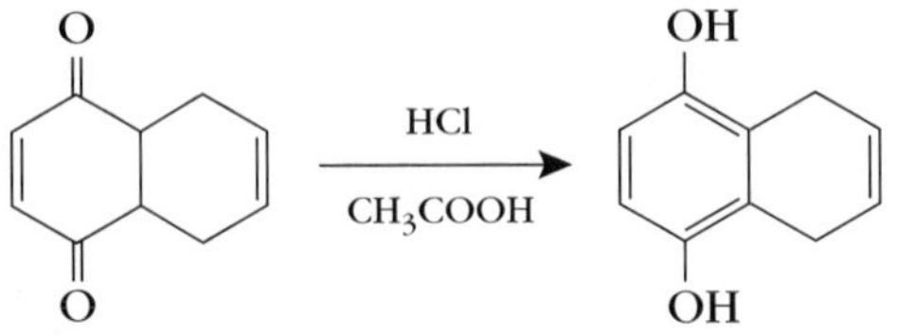

23.34 The herbicide 2,4,5-trichlorophenoxyacetic acid (2,4,5-T) is prepared from 1,2,4,5-tetrachlorobenzene by the following sequence of two reactions. Propose a mechanism for both reactions of the synthesis of 2,4,5-T.

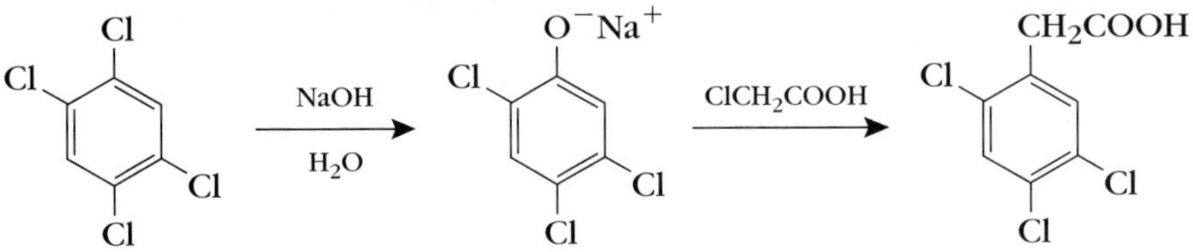

23.35 By means of a flow diagram, show how you would separate a mixture of benzene, 2,4-dinitrophenol, and phenol.

23.36 The following is an example of the Hoesch reaction. Propose a mechanism for this reaction and suggest a structure for compound A.

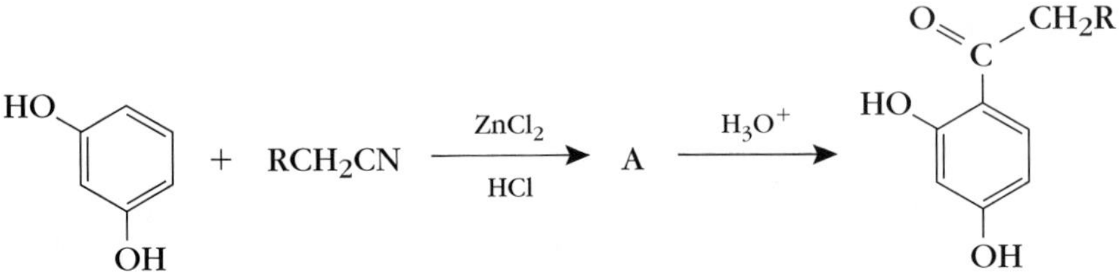

23.37 The ^{1}H NMR spectra of two isomeric compounds of molecular formula C_6H_5ClO are designated A and B below. The peaks marked by an asterisk disappear on shaking the sample with D_2O. Write the structural formulas of the two isomers.

23.38 The ^{1}H NMR spectra of two isomeric compounds of molecular formula C_7H_8O are designated C and D below. The peaks marked by an asterisk disappear on shaking the sample with D_2O. Write the structural formulas of the two isomers.

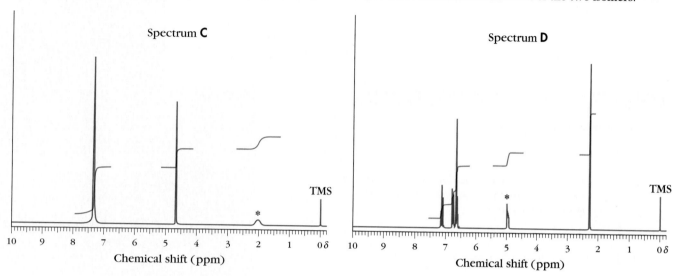

CHEMISTRY of DIFUNCTIONAL COMPOUNDS

In the preceding chapters, we have learned the chemical and physical properties of a number of important functional groups. Generally, a single functional group is the focus of each chapter. However, most compounds in nature contain more than one functional group. In this chapter, we consider how the presence of more than one functional group affects the reactions of di- or polyfunctional compounds.

24.1   The Reactions of Difunctional Compounds Are Determined by the Relative Positions of Their Functional Groups

The factor that most affects the reactivity of a di- or polyfunctional compound is the relative locations of the functional groups. If the functional groups in a molecule are sufficiently far apart, they react independently of each other. Lanosterol, for example, contains two carbon-carbon double bonds and a hydroxy group. These functional groups are far enough apart, however, that they undergo the usual reactions of alkenes and alcohols independently of each other.

Lanosterol

If two functional groups are close together, one may affect either the rate or equilibrium constant of reactions of the other functional group. Recall from Chapter 15, for example, that electronegative substituents bonded to the α-carbon atom of carboxylic acids increase their acidity. Notice that the second functional group does not change the fundamental reaction of the carboxylic acid. It is still an acid, but the substituent affects its acid dissociation constant.

Sometimes adjacent functional groups interact so strongly that they form a new functional group. A single ring with conjugated double bonds that have $4n + 2$ π electrons, for example, is an aromatic compound. These conjugated double bonds interact to form a new functional group whose chemical and physical properties differ greatly from those of typical alkenes or polyenes.

Finally, the two functional groups in a molecule may react with each other. A carboxylic acid and an alcohol react in the presence of an acid catalyst to form an ester. Depending on their relative positions in the molecule, ester formation may occur intramolecularly to form a cyclic ester or intermolecularly to form a polyester.

In this chapter, we will examine in detail two complex functional groups. The first example is 1,3-dicarbonyl compounds, which contain two carbonyl groups bonded to the same carbon atom. The second example is α,β-unsaturated carbonyl-containing compounds, which contain a carbon-carbon double bond in conjugation with a carbonyl group. Then we will examine the factors that determine whether two functional groups in the same molecule undergo intramolecular reaction to form a ring or intermolecular reaction to form dimers, trimers, and larger molecules. We begin by examining the chemistry of 1,3-dicarbonyl compounds.

24.2 1,3-Dicarbonyl Compounds

Compounds that contain two carbonyl groups bonded to the same carbon atom are called 1,3-dicarbonyl compounds or β-dicarbonyl compounds.

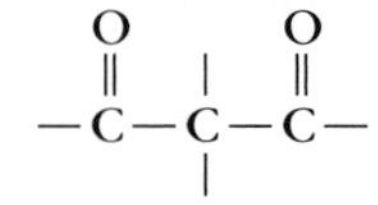

$$-\overset{\overset{\displaystyle O}{\|}}{C}-\overset{|}{\underset{|}{C}}-\overset{\overset{\displaystyle O}{\|}}{C}-$$

1,3-Dicarbonyl group
(β-dicarbonyl group)

The most important types of compounds that contain the 1,3-dicarbonyl group are 1,3-diketones (β-diketones), 3-oxoaldehydes (β-ketoaldehydes), 3-oxo-carboxylic acids (β-ketocarboxylic acids), 3-oxoesters (β-ketoesters), and 1,3-diesters (β-diesters).

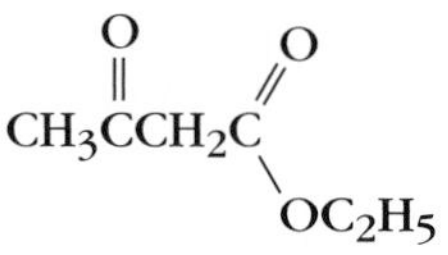

Ethyl acetoacetate
(acetoacetic ester)
A β-ketoester

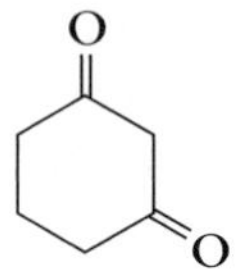

1,3-Cyclohexanedione
A β-diketone

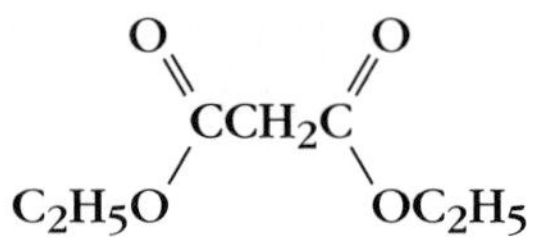

Diethyl malonate
(malonic ester)
A β-diester

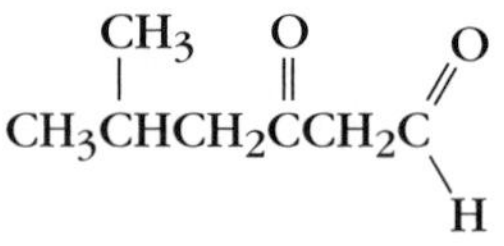

5-Methyl-3-oxohexanal
A β-ketoaldehyde

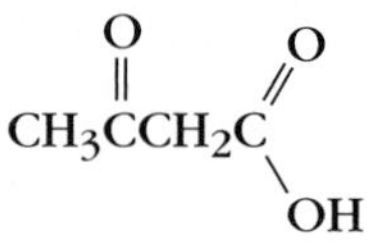

3-Oxobutanoic acid
A β-ketocarboxylic acid

β-Ketoesters, β-ketocarboxylic acids, and β-ketoaldehydes are named according to the IUPAC rules by using the prefix *oxo*-, along with a number, to indicate the position of the carbonyl group.

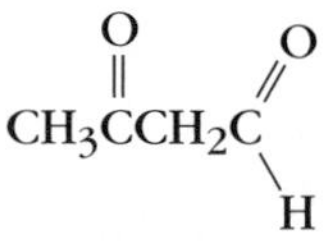

3-Oxobutanal

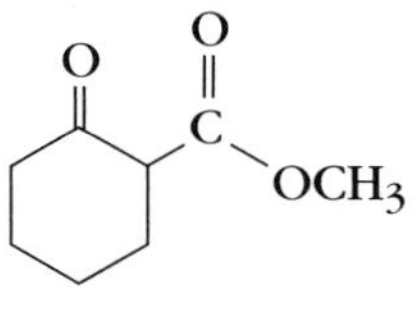

Methyl (2-oxocyclohexane)-carboxylate

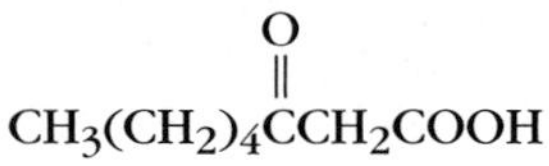

3-Oxooctanoic acid

Exercise 24.1 Give the IUPAC name of each of the following compounds.

(a) 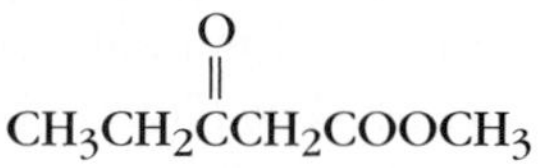 $CH_3CH_2CCH_2COOCH_3$

(b) 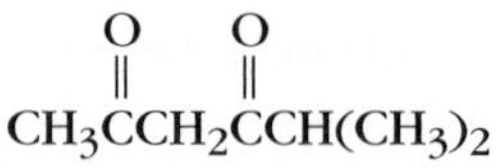 $CH_3CCH_2CCH(CH_3)_2$

(c) 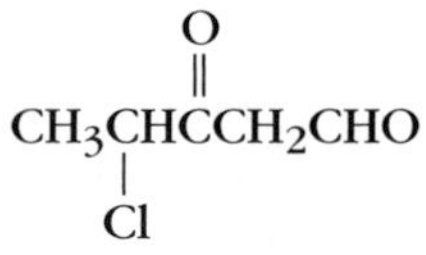 CH_3CHCCH_2CHO, Cl

(d) 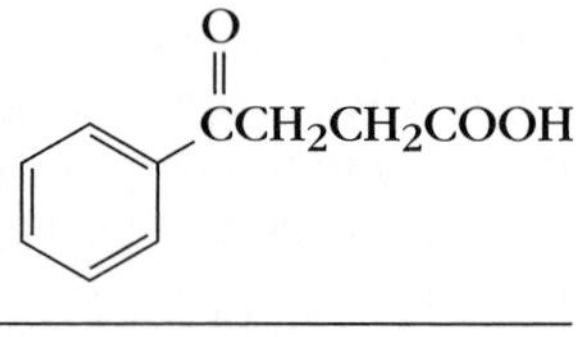 CCH_2CH_2COOH

24.3 Synthesis of 1,3-Dicarbonyl Compounds

Generally 1,3-dicarbonyl compounds are prepared by a Claisen or mixed Claisen condensation reaction. For example, the reaction of ethyl acetate with sodium ethoxide in ethanol forms ethyl acetoacetate.

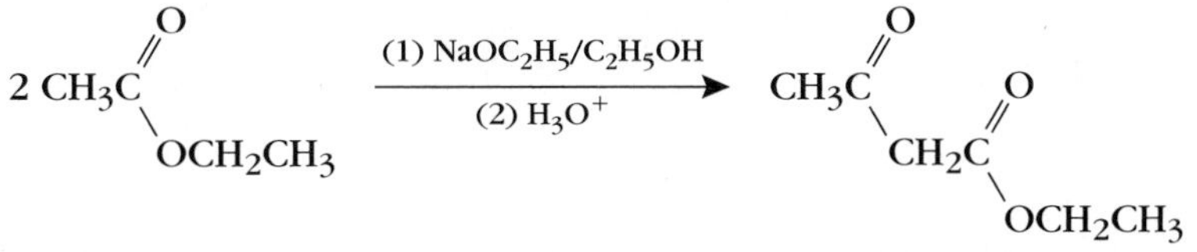

Ethyl acetate   Ethyl acetoacetate
                                 (72%–78% yield)

Mixed Claisen condensation reactions between ketones and esters are a general synthesis of β-diketones and β-ketoaldehydes.

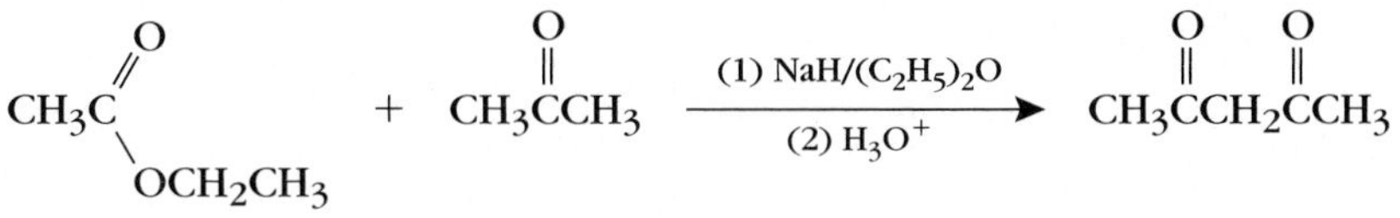

Ethyl acetate   Acetone   2,4-Pentanedione
                                                   (72%–75% yield)

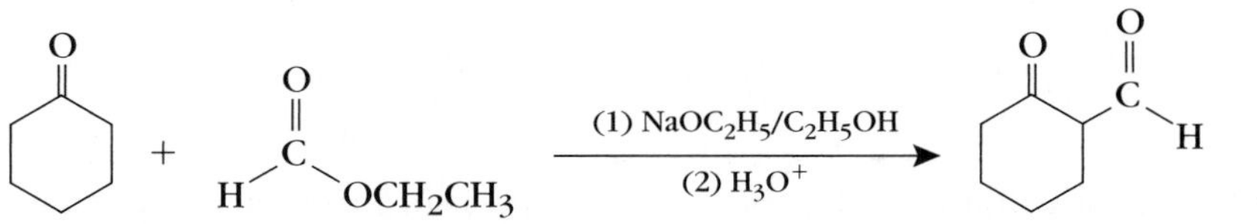

Cyclohexanone   Ethyl formate   2-Formylcyclohexanone
                                                (72%–75% yield)

The success of mixed condensation reactions is due to the difference in acidities of ketones and esters. The pK_a values of acetone and ethyl acetate are 19.3 and 24.5, respectively. Thus in a basic medium such as sodium ethoxide in ethanol or sodium hydride in diethyl ether, deprotonation generally occurs to a larger extent in acetone (and other ketones) than in esters.

Exercise 24.2   Calculate K_{eq} for the following equilibria.

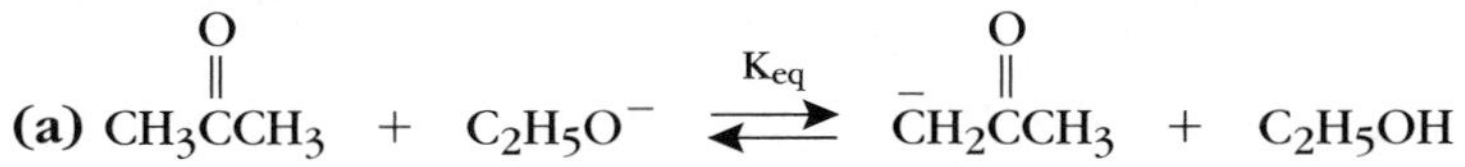

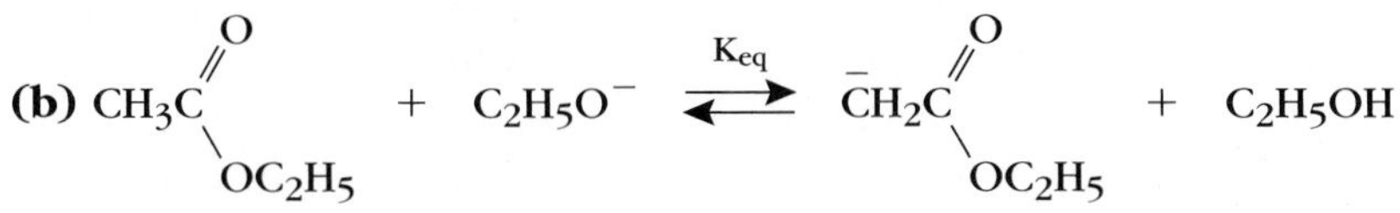

The pK_a of ethanol is 16.0.

Another factor that favors mixed Claisen condensation reactions is that aldol condensation reactions of ketones are reversible and their equilibrium constants are unfavorable. The last step of a mixed Claisen condensation reaction, on the other hand, is the irreversible formation of the enolate ion of the β-keto compound (Section 17.7). This drives a mixed Claisen condensation reaction to form the β-keto compound in good yield.

Exercise 24.3	Write all the reactions that occur in a diethyl ether solution of ethyl acetate, acetone, and sodium hydride. Identify the reactions that are involved in the formation of the product, 2,4-pentanedione.

Like simple aldehydes, ketones, and esters, 1,3-dicarbonyl compounds also exist as keto-enol equilibrium mixtures.

24.4   Keto-Enol Equilibria in 1,3-Dicarbonyl Compounds

> ■ *The keto-enol equilibria of simple aldehydes and ketones are discussed in Section 17.1.*

Simple aldehydes, ketones, and esters exist predominantly in the keto form. 1,3-Dicarbonyl compounds, in contrast, contain a large amount of enol in equilibrium with the dione form. 2,4-Pentanedione, for example, exists in aqueous solution as a mixture of 16% enol form and 84% dione form.

$$
\underset{(84\%)}{CH_3CCH_2CCH_3} \;\;\rightleftharpoons\;\; \underset{(16\%)}{CH_3C{=}CHCCH_3}
$$

Under similar conditions, the concentration of enol of acetone is about 0.0001%.

Two structural features are responsible for stabilizing the enol form of 1,3-dicarbonyl compounds. The first is the conjugation of the carbon-carbon double bond with the remaining carbonyl group. The second is the formation of an intramolecular hydrogen bond between the enolic hydroxy group and the oxygen atom of the carbonyl group.

The importance of this intramolecular hydrogen bond is evident from the fact that the enol content of 1,3-dicarbonyl compounds is higher in nonpolar solvents than it is in water. The enol content of 2,4-pentanedione, for instance, increases from 16% in water to 92% in hexane. In water or other protic solvents, both the dicarbonyl compound and the enol can form hydrogen bonds with the solvent. Consequently, the intramolecular hydrogen bond affords little additional stability. In nonpolar solvents, however, the intramolecular hydrogen bond greatly stabilizes the enol.

The presence of a second carbonyl group not only increases the enol content of 1,3-dicarbonyl compounds compared to simple aldehydes and ketones, it also increases the acidity of the methylene hydrogens between the two carbonyl groups.

24.5 1,3-Dicarbonyl Compounds Are Strong Carbon Acids

We discussed in Chapter 17 that the carbonyl groups of aldehydes, ketones, and esters are responsible for the acidity of their α-hydrogen atoms. The data in Table 24.1 show that 1,3-dicarbonyl compounds with hydrogen atoms alpha to two carbonyl groups are even more acidic than simple aldehydes and ketones.

The pK_a values given in Table 24.1 show that 1,3-dicarbonyl compounds are 10^7 to 10^{10} times more acidic than simple aldehydes, ketones, and esters. 1,3-Dicarbonyl compounds are so acidic because their enolate anions are stabilized by delocalization of the negative charge into both carbonyl groups. This electron

Table 24.1	pK_a Values of Representative 1,3-Dicarbonyl and Related Compounds*		
Acid		**Conjugate Base**	**pK_a**
			9
			9
			11
$N\equiv CCH_2C\equiv N$		$N\equiv C\ddot{C}HC\equiv N$	11
			13
			16.7
			19.3
			24.5

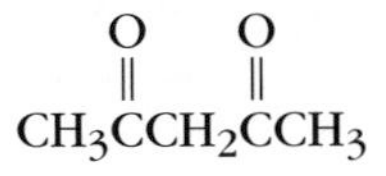

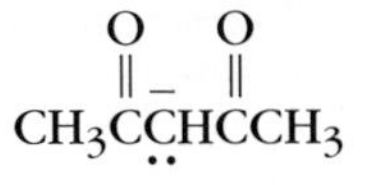

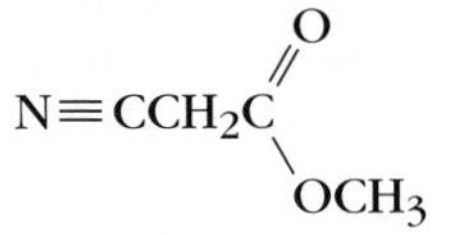

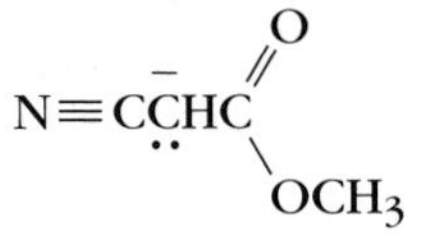

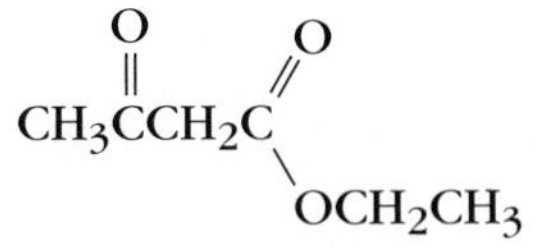

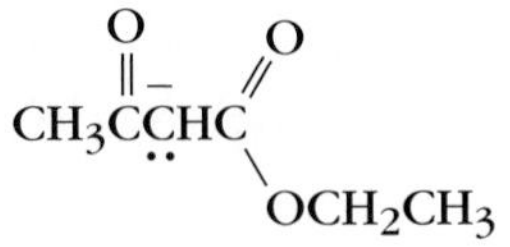

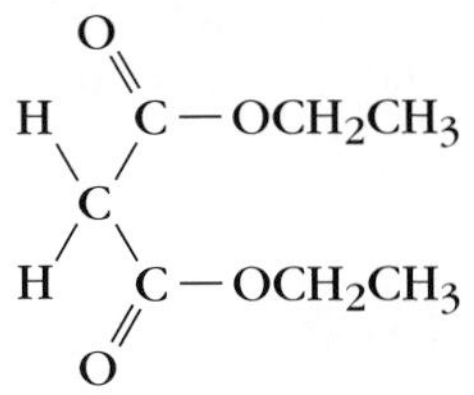

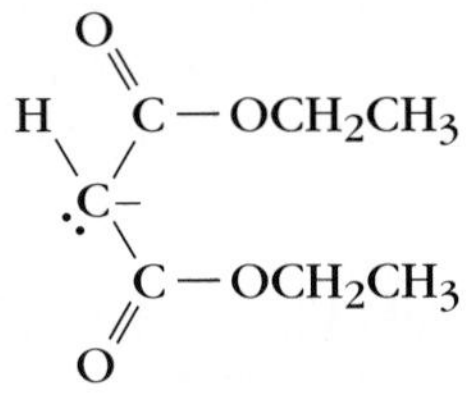

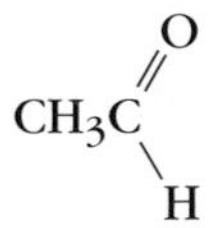

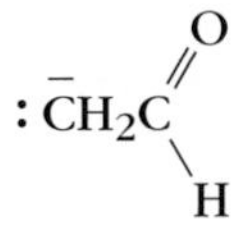

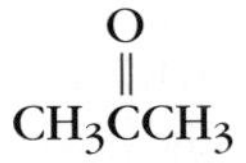

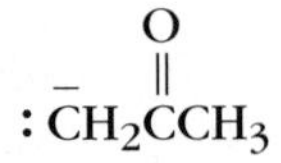

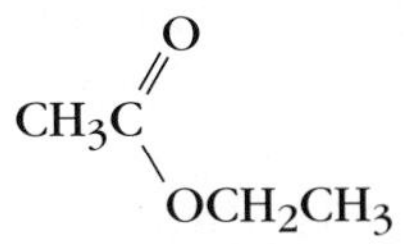

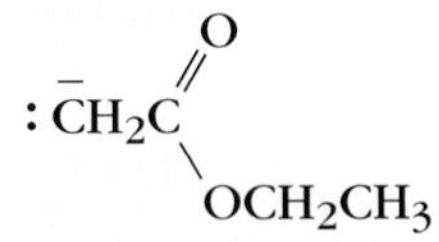

*Acetaldehyde, acetone, and ethyl acetate are included for comparison.

delocalization is represented in resonance theory as a resonance hybrid of the following resonance structures.

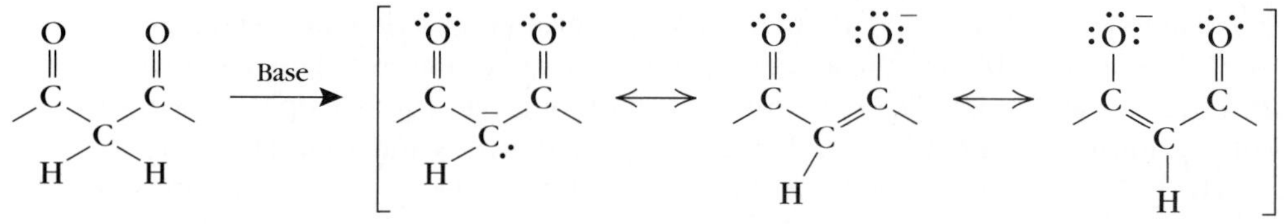

1,3-Dicarbonyl compound

Enolate anion

The cyano compounds listed in Table 24.1 are also relatively acidic because the negative charge of their anions is stabilized by delocalization into the cyano group.

Exercise 24.4 Write the resonance structures that represent the electron delocalization of the negative charge from the carbon atom into the nitrile groups in the anion formed from 1,3-propanedinitrile ($N{\equiv}CCH_2C{\equiv}N$).

The acidities of 1,3-dicarbonyl compounds are sufficiently high that they are converted completely to their enolate anions by alkoxide ions in alcohol solvents. In Section 24.7, we will see how these easily prepared enolate anions are useful synthetic intermediates.

$$\underset{pK_a = 11}{CH_3\overset{O}{\overset{\|}{C}}CH_2COOC_2H_5} + Na^{+\,-}OC_2H_5 \xrightarrow{C_2H_5OH} CH_3\overset{O}{\overset{\|}{C}}\underset{Na^+}{\overset{-}{C}}HCOOC_2H_5 + \underset{pK_a = 16}{C_2H_5OH}$$

Another difference between 1,3-dicarbonyl compounds and simple carbonyl compounds is the ease of decarboxylation of β-ketocarboxylic acids and 1,3-dicarboxylic acids compared to monocarboxylic acids.

24.6 Decarboxylation of β-Ketocarboxylic and 1,3-Dicarboxylic Acids

In a **decarboxylation reaction,** a molecule of carbon dioxide is lost from a carboxylic acid.

$$RCOOH \longrightarrow RH + CO_2$$

Simple monocarboxylic acids undergo decarboxylation only very slowly, and the reaction is rarely observed. Both β-ketocarboxylic acids and 1,3-dicarboxylic acids, in contrast, easily undergo decarboxylation when heated to 100 to 150 °C.

$$\underset{\text{3-Oxobutanoic acid}}{CH_3\overset{O}{\overset{\|}{C}}CH_2COOH} \xrightarrow{100\ ^\circ C} \underset{\text{Acetone}}{CH_3\overset{O}{\overset{\|}{C}}CH_3} + CO_2$$

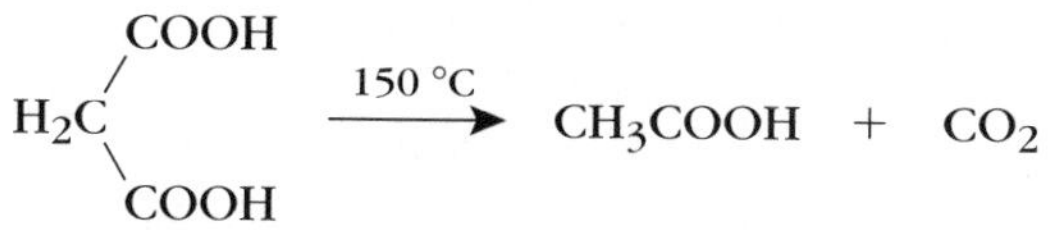

$$
\underset{\text{Malonic acid}}{\text{H}_2\text{C}\begin{array}{c}\text{COOH}\\[-2pt]\\[-2pt]\text{COOH}\end{array}} \xrightarrow{150\,°\text{C}} \underset{\text{Acetic acid}}{\text{CH}_3\text{COOH}} + \text{CO}_2
$$

Decarboxylation occurs so readily because the β-carbonyl group takes part in the mechanism of the reaction. The oxygen atom of the β-carbonyl group accepts a proton from the carboxy group with concerted loss of carbon dioxide to generate an enol. Rapid conversion of the enol to the keto form completes the decarboxylation reaction.

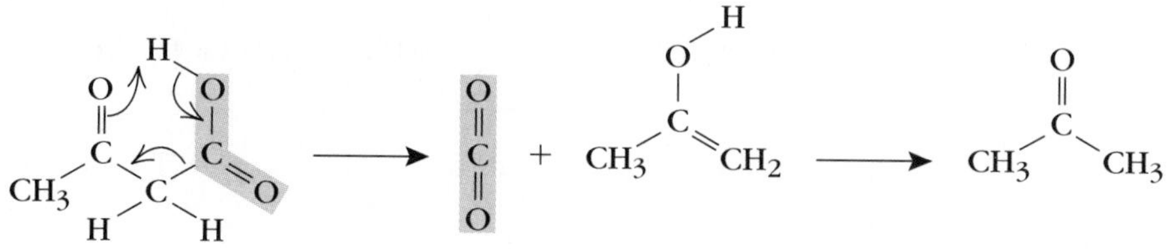

| Enol form of acetone | Keto form of acetone |

Exercise 24.5   Write a mechanism for the decarboxylation of malonic acid ($\text{HOOCCH}_2\text{COOH}$) to form acetic acid and CO_2.

A sequence of two reactions, alkylation or acylation followed by decarboxylation, makes 1,3-dicarbonyl compounds useful starting materials for the synthesis of ketones and carboxylic acids.

24.7 Alkylation of Enolate Anions of 1,3-Dicarbonyl Compounds: the Malonic Ester Synthesis of Carboxylic Acids

The **malonic ester synthesis** is a sequence of two reactions in which diethyl propanedioate, commonly called diethyl malonate or malonic ester, reacts with haloalkanes to form a carboxylic acid that contains two more carbon atoms than the starting haloalkane.

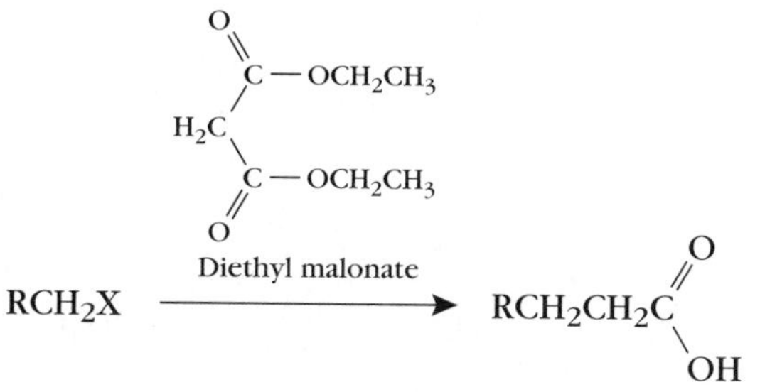

Diethyl malonate

$$
\underset{\substack{\text{Haloalkane}\\(\text{contains } n \text{ carbon atoms})}}{\text{RCH}_2\text{X}} \longrightarrow \underset{\substack{\text{Carboxylic acid}\\(\text{contains } n+2 \text{ carbon atoms})}}{\text{RCH}_2\text{CH}_2\text{C}\overset{\text{O}}{\underset{\text{OH}}{\big\|}}}
$$

The first step of this sequence is the quantitative formation of the enolate anion of diethyl malonate.

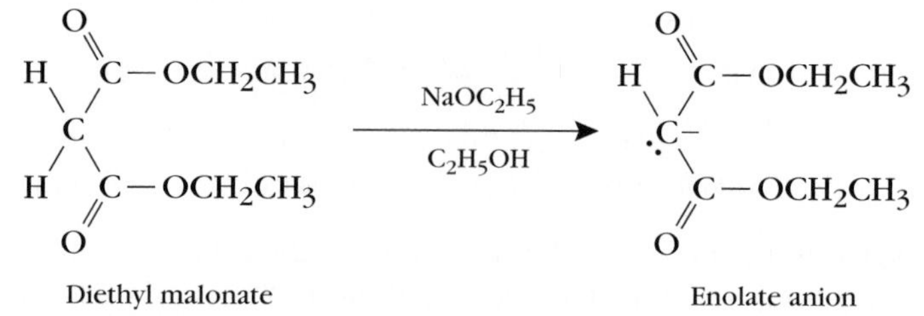

Diethyl malonate   Enolate anion

The enolate anion is a good nucleophile, so it reacts with haloalkanes by an S_N2 mechanism to form α-substituted malonic esters.

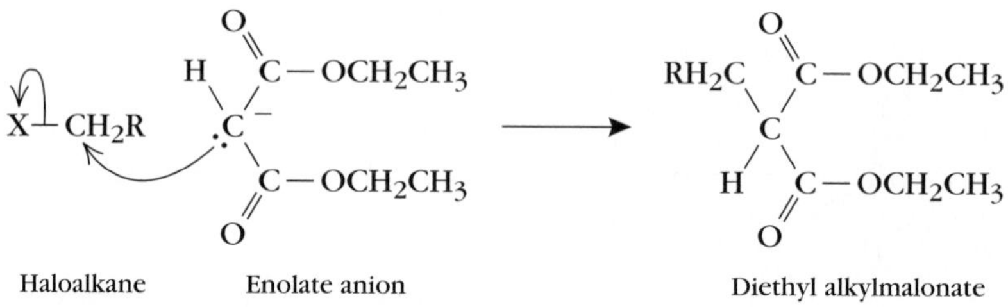

Haloalkane   Enolate anion   Diethyl alkylmalonate

The diethyl alkylmalonate still has one acidic α-hydrogen atom, so the alkylation reaction can be repeated a second time to form a diethyl dialkylmalonate.

Diethyl alkylmalonate   Diethyl dialkylmalonate

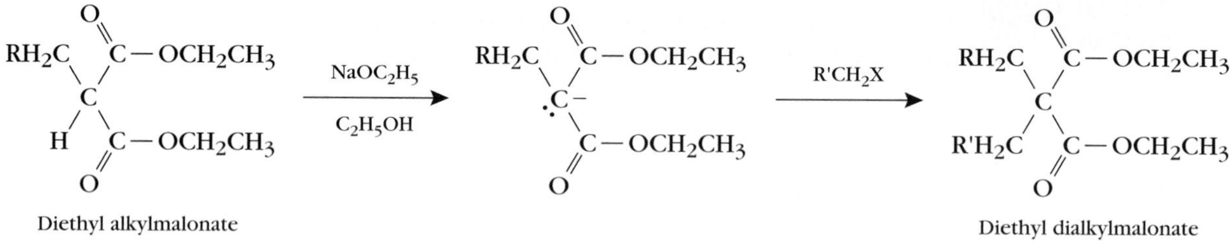

Either the mono- or dialkyl malonic ester can then be hydrolyzed and decarboxylated to form a carboxylic acid. Ester hydrolysis can be carried out in either acidic or basic solution. If basic conditions are used, the solution must be acidified in order to form a substituted malonic acid, which decarboxylates on heating (Section 24.6).

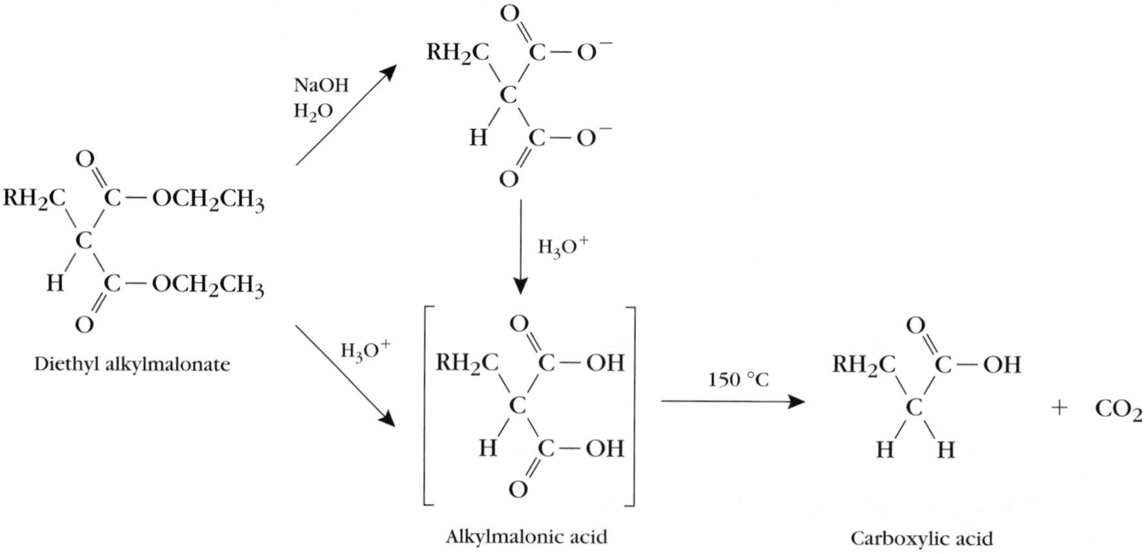

Diethyl alkylmalonate   Alkylmalonic acid   Carboxylic acid

While this sequence of reactions seems complicated, it is experimentally very easy because the intermediate products need not be isolated. Thus enolate anion formation and addition of the haloalkane (alkylation) can be carried out successively in the same flask. Hydrolysis and decarboxylation are usually carried out in a single step without isolating the alkyl malonic acid. The following is a specific example of the malonic ester synthesis.

Diethyl malonate

Enolate anion

Hexanoic acid
(70%–75% overall yield)

Diethyl butylmalonate

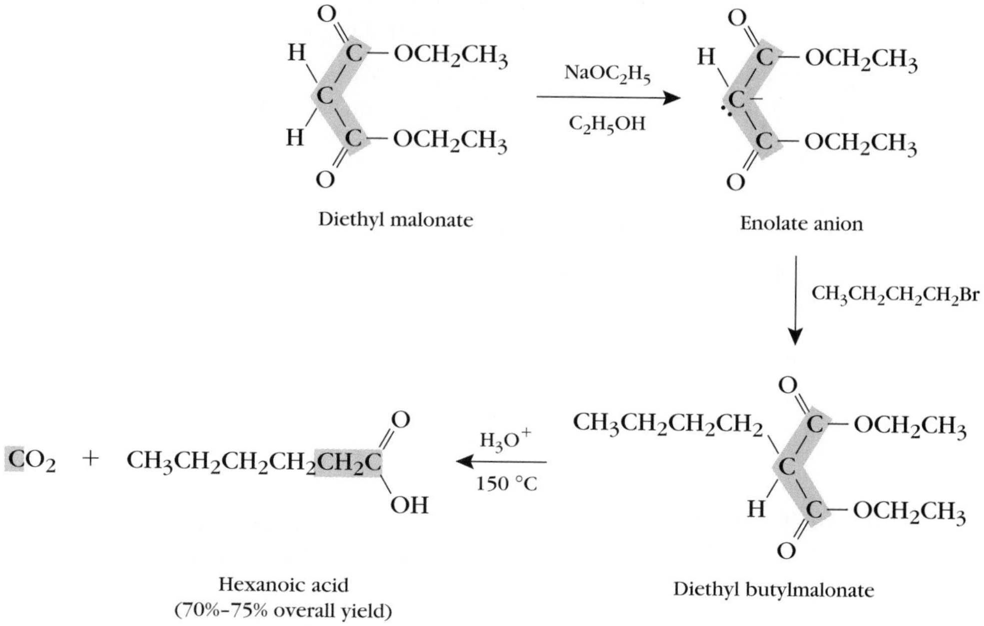

The carboxylic acid formed in the malonic ester synthesis contains the following two-carbon unit, which was originally part of diethyl malonate.

$$\text{CHCOOH}$$

The alkyl group or groups bonded to the α-carbon atom of the carboxylic acid were originally the alkyl groups of the haloalkane. Thus the origins of the various parts of a carboxylic acid prepared by a malonic ester synthesis are as follows:

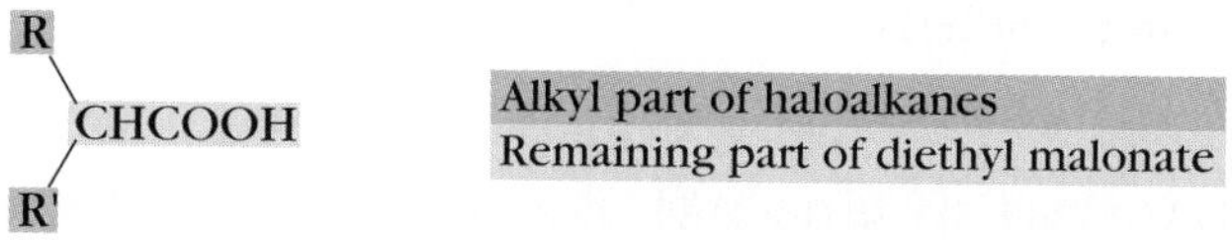

Alkyl part of haloalkanes
Remaining part of diethyl malonate

Viewing the product in this way allows us to plan the synthesis of a carboxylic acid by a malonic ester synthesis. The method is illustrated in Sample Problem 24.1.

Sample Problem 24.1

Prepare 4-methylpentanoic acid using the malonic ester synthesis.

Solution:

First, examine the structure of the carboxylic acid to determine which groups will come from diethyl malonate and which groups will come from the haloalkane. By mentally removing the alkyl group or groups bonded to the α-carbon atom of the carboxylic acid, we see that there is one alkyl group and that it is an isobutyl (2-methyl-1-propyl) group.

$$(CH_3)_2CHCH_2CH_2COOH$$

Part of carboxylic acid obtained from haloalkane   Part of carboxylic acid obtained from diethyl malonate

Therefore alkylation of diethyl malonate with 1-bromo-2-methylpropane followed by hydrolysis and decarboxylation forms 4-methylpentanoic acid, the desired product.

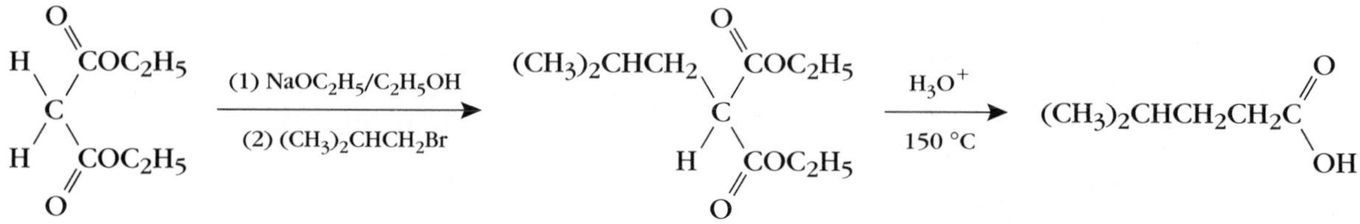

Diethyl malonate   (1) NaOC$_2$H$_5$/C$_2$H$_5$OH   (2) (CH$_3$)$_2$CHCH$_2$Br   Diethyl 2-methyl-1-propylmalonate   H$_3$O$^+$ 150 °C   4-Methylpentanoic acid

---

Exercise 24.6   Starting with diethyl malonate and any haloalkane, propose a synthesis for each of the following carboxylic acids.

(a) 3-Phenylpropanoic acid   **(b)** Octanoic acid   **(c)** 2-Methylhexanoic acid

---

Exercise 24.7   Explain why the following carboxylic acids cannot be prepared by means of the malonic ester synthesis.

(a) 2-Phenylacetic acid   **(b)** 2,2-Dimethylpropanoic acid

The **acetoacetic ester synthesis** is a way of preparing ketones that is similar to the malonic ester synthesis.

24.8   Alkylation of Enolate Anions of 1,3-Dicarbonyl Compounds: Acetoacetic Ester Synthesis of Methyl Ketones

The **acetoacetic ester synthesis** is a general method of preparing methyl ketones from ethyl 3-oxobutanoate, which is commonly called ethyl acetoacetate or acetoacetic ester.

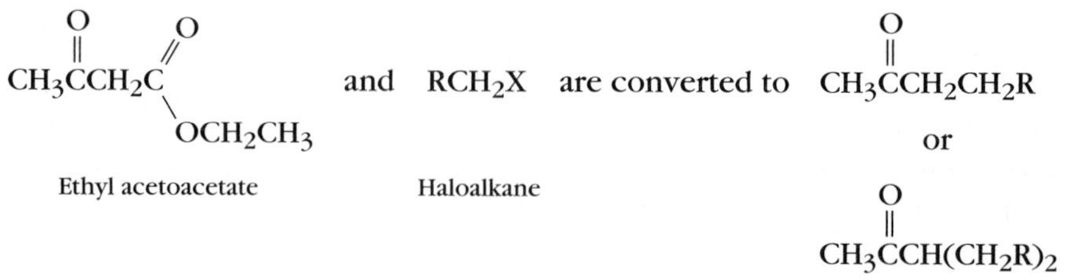

$$CH_3CCH_2C \underset{OCH_2CH_3}{\overset{O \quad O}{\|\quad\|}}$$

and   RCH$_2$X   are converted to $CH_3CCH_2CH_2R$

Ethyl acetoacetate   Haloalkane   or

$$CH_3CCH(CH_2R)_2$$

Ethyl acetoacetate, like diethyl malonate, is a relatively strong acid that is converted completely to its enolate anion by reaction with sodium ethoxide in ethanol. Alkylation of the anion followed by hydrolysis in dilute aqueous base (or acid) forms the corresponding β-ketocarboxylic acid. The β-ketocarboxylic acid decarboxylates on heating to yield a methyl ketone, as shown in the following reaction sequence.

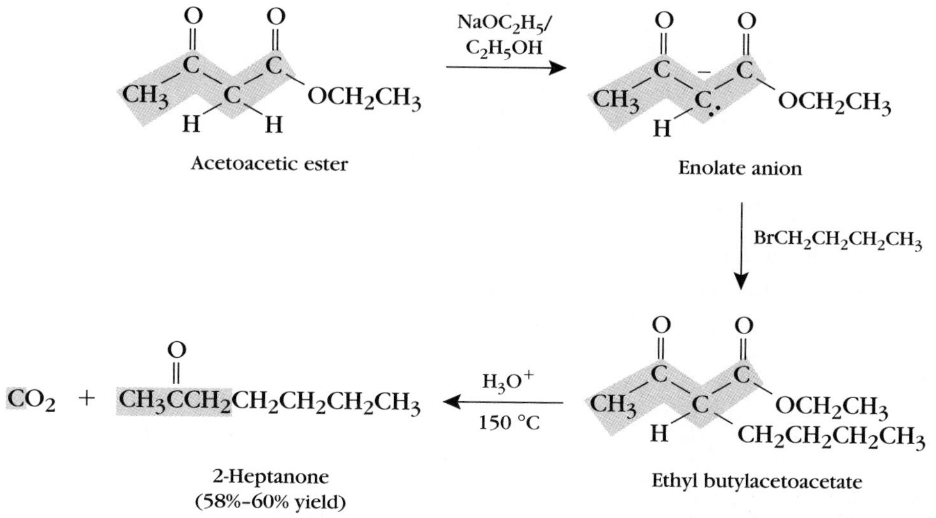

Acetoacetic ester, like diethyl malonate, can be alkylated once or twice. In the previous example, hydrolysis of the ethyl alkylacetoacetate forms a ketone with one alkyl substituent on the α-carbon atom. Dialkylation of ethyl acetoacetate followed by hydrolysis and decarboxylation forms a product with two alkyl substituents on the α-carbon atom.

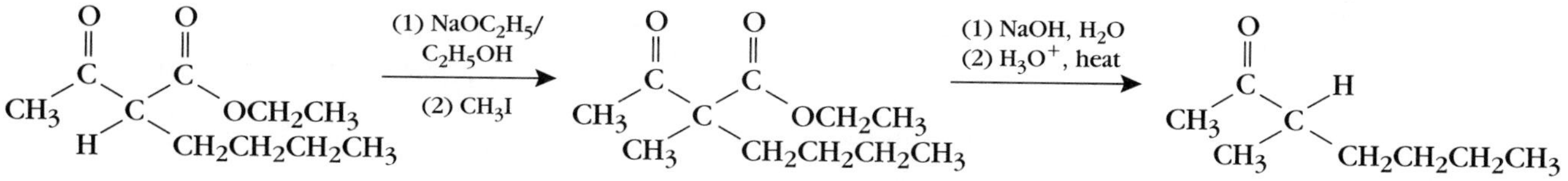

3-Methyl-2-heptanone
(60%–65% yield)

Ethyl acetoacetate provides the following three-carbon unit to the ketone:

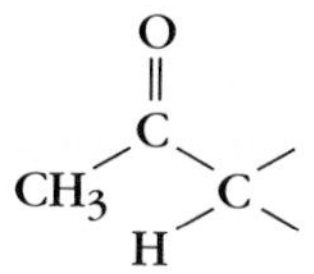

The alkyl group or groups bonded to the α-carbon atom are obtained from the haloalkane. Viewing the product in this way simplifies the preparation of a methyl ketone by the acetoacetic ester synthesis. Sample Problem 24.2 illustrates how to plan the synthesis of a methyl ketone using the acetoacetic ester synthesis.

| **Sample Problem 24.2** | Propose a synthesis of 3-methyl-2-hexanone using the acetoacetic ester synthesis. |

Solution:

First, examine the structure of the ketone to identify which parts can be derived from ethyl acetoacetate and which must be obtained by alkylation.

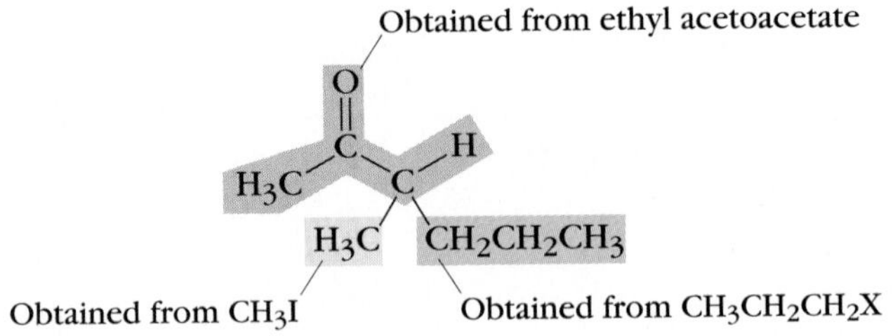

Obtained from ethyl acetoacetate

$$H_3C - \overset{\overset{\displaystyle O}{\|}}{C} - \overset{\overset{\displaystyle H}{|}}{\underset{\underset{\displaystyle H_3C \quad CH_2CH_2CH_3}{}}{C}}$$

Obtained from CH_3I Obtained from $CH_3CH_2CH_2X$

Two alkylations of ethyl acetoacetate are needed to form 3-methyl-2-hexanone. The first is with 1-bromopropane, and the second is with iodomethane. Hydrolysis and decarboxylation of ethyl dialkylacetoacetate forms the desired product.

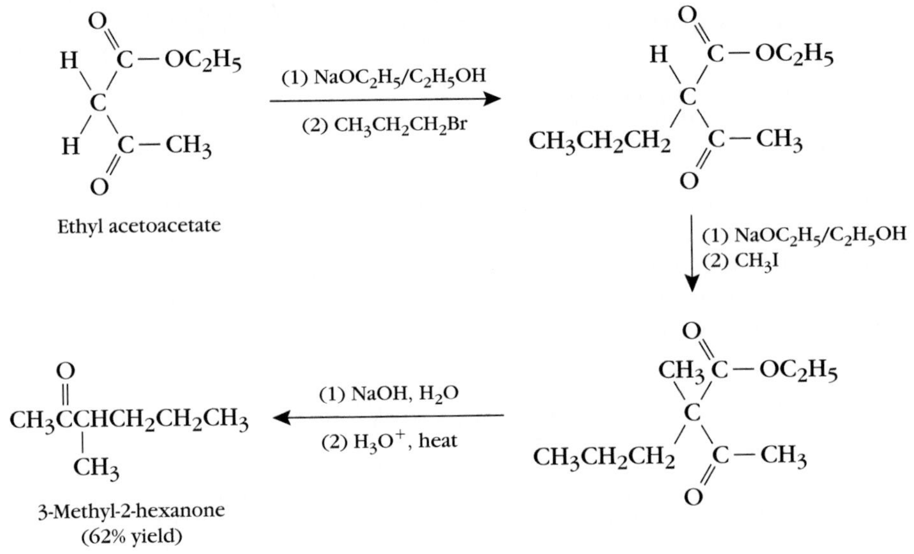

$$\underset{\text{Ethyl acetoacetate}}{\overset{\overset{\displaystyle O}{\|}}{H - C - OC_2H_5} \atop \underset{\underset{\displaystyle O}{\|}}{H - C - CH_3}}$$

$\xrightarrow[\text{(2) } CH_3CH_2CH_2Br]{\text{(1) } NaOC_2H_5/C_2H_5OH}$

$$\overset{\overset{\displaystyle O}{\|}}{H - C - OC_2H_5} \atop CH_3CH_2CH_2 - \underset{\underset{\displaystyle O}{\|}}{C - CH_3}$$

$\downarrow \begin{array}{l}\text{(1) } NaOC_2H_5/C_2H_5OH\\ \text{(2) } CH_3I\end{array}$

$$\overset{\overset{\displaystyle O}{\|}}{CH_3 C - OC_2H_5} \atop CH_3CH_2CH_2 - \underset{\underset{\displaystyle O}{\|}}{C - CH_3}$$

$$\underset{\underset{\displaystyle CH_3}{|}}{CH_3\overset{\overset{\displaystyle O}{\|}}{C}CHCH_2CH_2CH_3}$$

$\xleftarrow[\text{(2) } H_3O^+, \text{ heat}]{\text{(1) } NaOH, H_2O}$

3-Methyl-2-hexanone
(62% yield)

The acetoacetic ester synthesis complements the direct alkylation of the enolate anions of ketones (Section 17.11) as a method of preparing ketones. The main advantage of the acetoacetic ester synthesis is that its delocalized enolate anion is far less basic than the enolates of ketones. As a result, there is less chance of elimination products being formed when the less basic anions react with haloalkanes.

| **Exercise 24.8** | Using ethyl acetoacetate and any other inorganic or organic reagents needed, propose a synthesis for each of the following compounds. |

(a) 5-Methyl-2-hexanone   **(b)** 4-Phenyl-2-butanone   **(c)** 5-Hexen-2-one

In summary, success of the malonic ester and acetoacetic ester syntheses is due to the presence of two carbonyl groups bonded to a single carbon atom. Their methylene hydrogen atoms are relatively acidic, which makes it easy to form their corresponding enolate anions. These delocalized anions are good nucleophiles, and they react with haloalkanes by an S_N2 mechanism. On hydrolysis, the alkylated product forms a β-ketocarboxylic acid, which readily undergoes decarboxylation on heating.

Diethyl malonate and ethyl acetoacetate can also react with carbonyl-containing compounds.

24.9   Other Condensation Reactions of 1,3-Dicarbonyl Compounds

The aldol (Section 17.6) and Claisen condensation (Section 17.7) are reactions in which compounds with acidic α-hydrogen atoms react with a second carbonyl compound. There are a number of similar reactions that involve 1,3-dicarbonyl compounds. The **Knoevenagel reaction,** for example, is the reaction of aldehydes and unhindered ketones with 1,3-dicarbonyl compounds to form α,β-unsaturated carbonyl compounds.

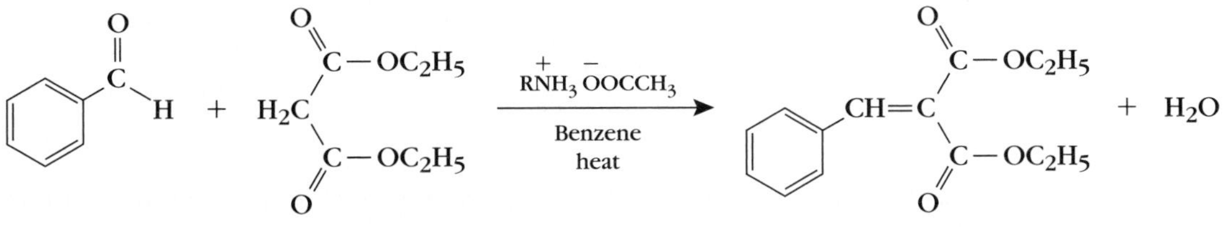

Benzaldehyde   Diethyl malonate   (78%–82% yield)

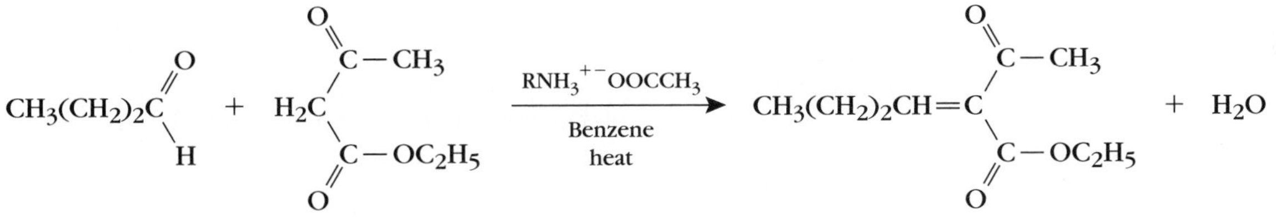

Butanal   Ethyl acetocetate   (72%–76% yield)

The Knoevenagel reaction is carried out in the presence of an alkyl ammonium salt as catalyst in refluxing benzene as solvent. The reaction is forced to completion by continuously removing water as it forms. The initial products of the Knoevenagel reaction can be hydrolyzed and decarboxylated to form α,β-unsaturated carboxylic acids or ketones.

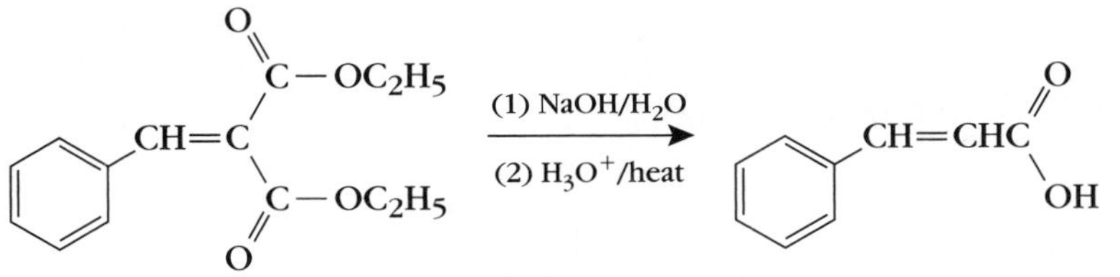

(88%–92% yield)

> **Exercise 24.9**   Using the Knoevenagel condensation reaction, propose a synthesis for each of the following compounds.

(a) 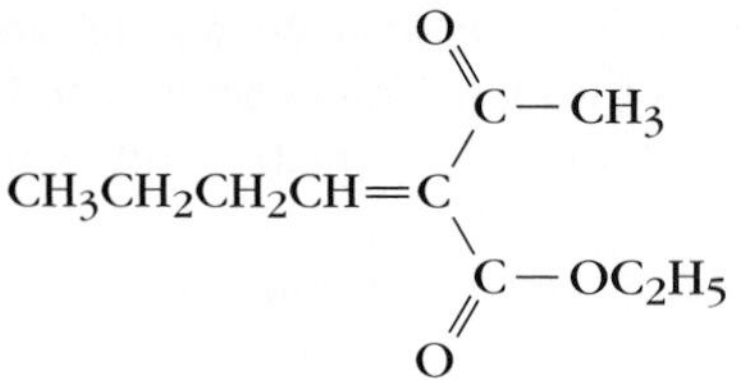 CH$_3$CH$_2$CH$_2$CH=C $\begin{array}{l} \overset{O}{\overset{\|}{C}}-CH_3 \\ \overset{\|}{\underset{O}{C}}-OC_2H_5 \end{array}$

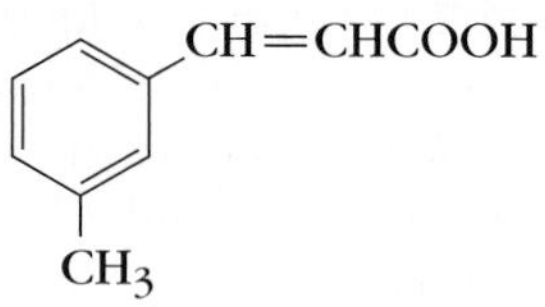

(c) [aromatic ring with CH=CHCOOH and CH$_3$ substituents]

(b) CH$_3$CH=CHCOOH

(d) (CH$_3$)$_2$C=CHCCH$_3$ (with O above the carbonyl C)

Enolate anions of 1,3-dicarbonyl compounds also react with acyl chlorides or carboxylic acid anhydrides.

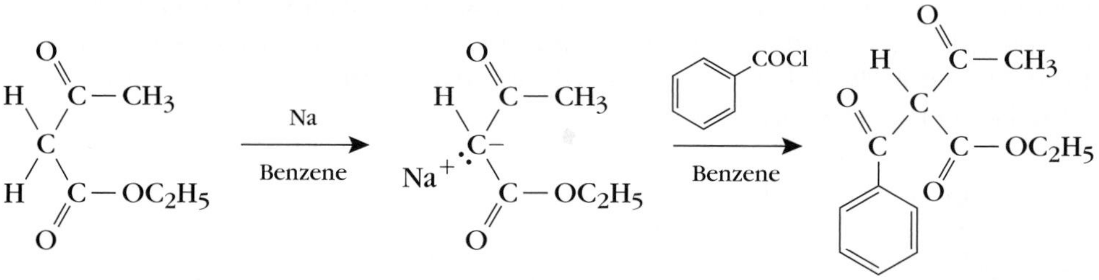

Ethyl acetoacetate   (63%–75% yield)

This reaction is an acyl transfer reaction whose mechanism we discussed in Section 16.6. In the reaction of ethyl acetoacetate and benzoyl chloride, sodium metal is used to form the enolate anion and benzene is used as the solvent rather than ethanol because ethanol reacts with an acyl chloride.

> **Exercise 24.10**   Write a mechanism for the reaction of the enolate anion of ethyl acetoacetate and benzoyl chloride in benzene.

1,3-Dicarbonyl compounds appear as intermediates in synthesis in nature, too.

24.10 1,3-Dicarbonyl Compounds Are also Used in Nature as Synthetic Intermediates

We learned in previous chapters that plants and animals use condensation and acylation reactions of acetyl CoA to form carbon-carbon bonds in the synthesis of complex compounds. Sometimes a more powerful carbanion equivalent than acetyl CoA is needed. In those cases, acetyl CoA is converted into a β-dicarbonyl compound called malonyl coenzyme A.

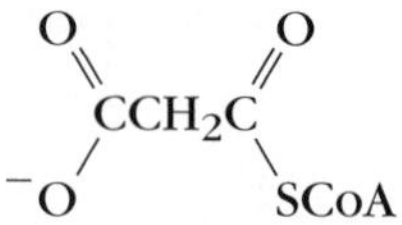

■ *The structure of acetyl CoA is given in Section 16.10. For convenience, we use the following abbreviation for acetyl CoA:*

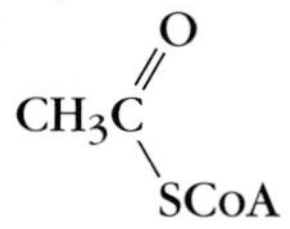

CH$_3$C (with O above) —SCoA

Acetyl CoA

[structure: malonyl coenzyme A]
$^-$O—CCH$_2$C—SCoA (with O above each carbonyl)

Malonyl coenzyme A

Animals use malonyl coenzyme A to synthesize fatty acids. Fatty acids are synthesized by successive cycles of seven enzyme-catalyzed reactions. Each cycle extends the fatty acid chain by two carbon atoms. Acyl transfer to malonyl CoA is the key reaction that forms the carbon-carbon bond in each cycle.

The first reaction in the first cycle is the acetyl CoA carboxylase-catalyzed conversion of acetyl CoA to malonyl CoA. The hydrolysis of ATP to ADP provides the energy for this reaction.

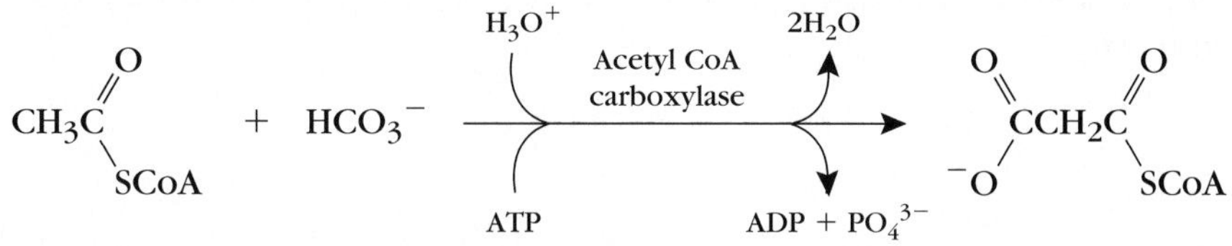

Acetyl coenzyme A   Malonyl coenzyme A

The enzyme-catalyzed conversion of acetyl CoA to malonyl CoA is related to the Kolbe reaction (Section 23.12). In both cases the overall result is the formation of a C—C bond by addition of CO_2 to an enol or enolate.

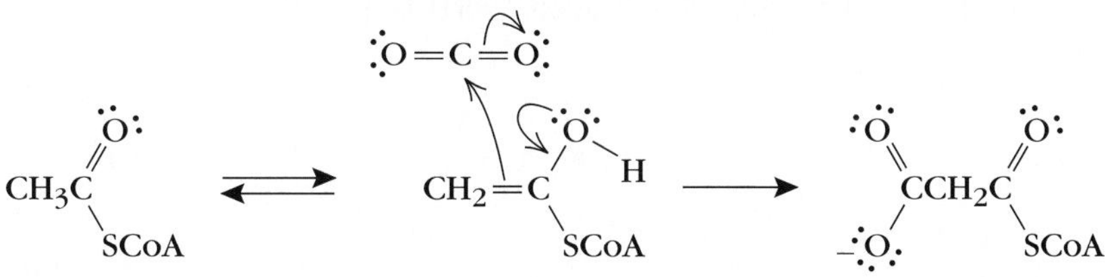

Keto form of   Enol form of   Malonyl coenzyme A
acetyl coenzyme A   acetyl coenzyme A

In the next two steps, the malonyl and acetyl groups are transferred to the thiol group of a protein known as acyl carrier protein (ACP).

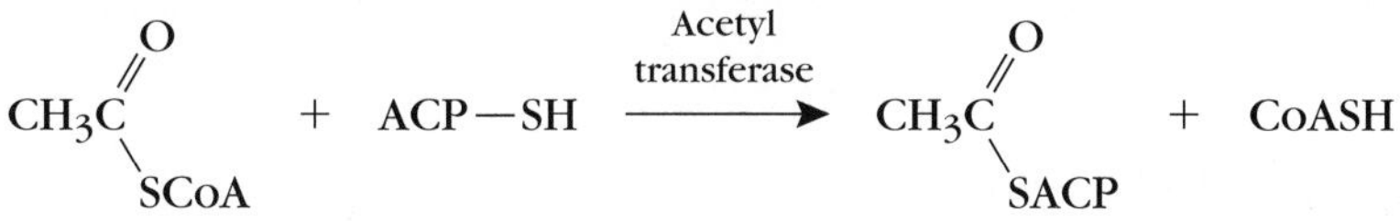

Acetyl coenzyme A   Acetyl ACP

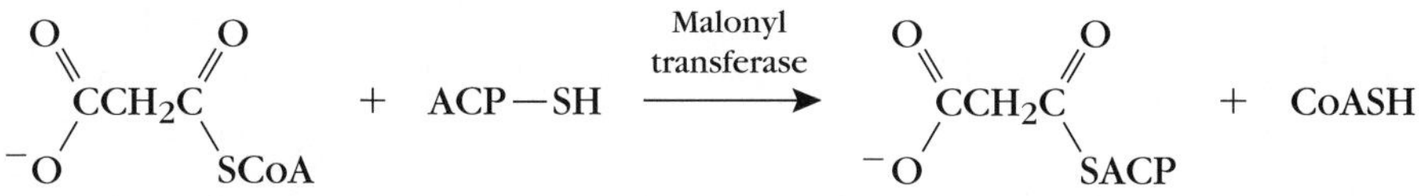

Malonyl coenzyme A   Malonyl ACP

The elongation step is an acyl transfer reaction between acetyl ACP and malonyl ACP and is catalyzed by the acyl malonyl ACP condensing enzyme.

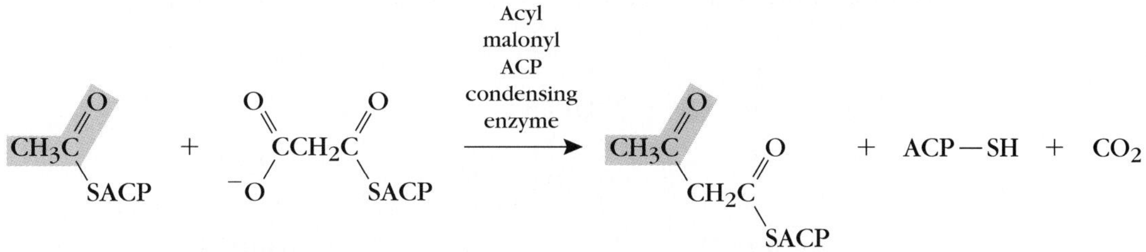

Acetyl ACP   Malonyl ACP   Acetoacetyl ACP

This reaction is a typical acyl transfer reaction in which a new carbon-carbon single bond is formed between the methylene carbon atom of malonyl ACP and the carbonyl carbon atom of acetyl ACP. This reaction extends the carbon chain by two carbon atoms.

To carry out an acetylation reaction in the laboratory, the enolate anion is formed by reaction with a base. In nature, however, the enzyme catalyzes the decarboxylation of malonyl ACP to form a carbanion equivalent. The carbanion equivalent adds to the carbonyl carbon atom of acetyl ACP to form a tetrahedral intermediate.

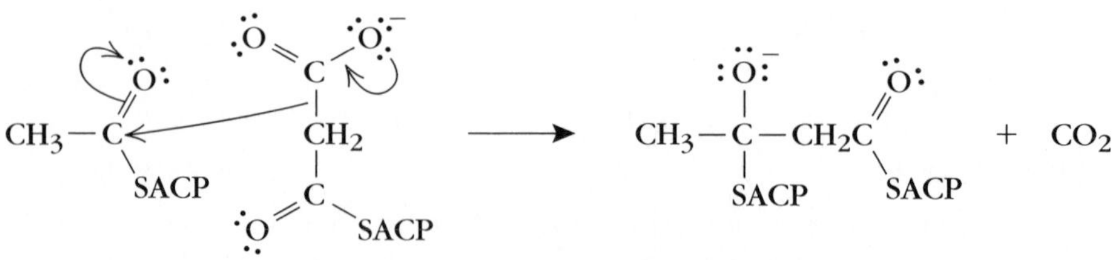

Acetyl ACP   Malonyl ACP   Tetrahedral intermediate

Elimination of the SACP group forms acetoacetyl ACP.

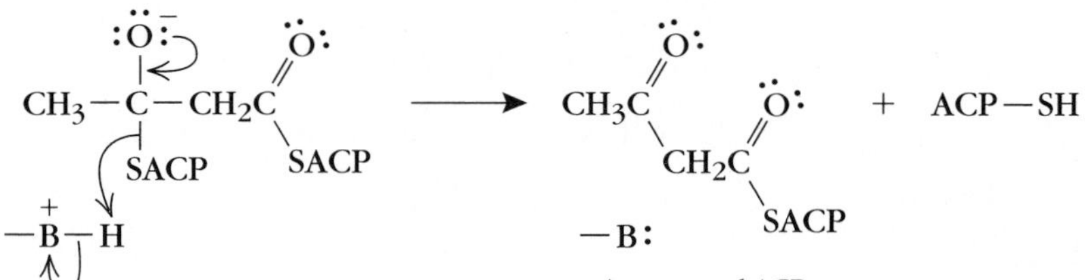

Acetoacetyl ACP

The mechanism of this acyl transfer reaction is very similar to the addition-elimination mechanism of the acyl transfer reactions discussed in Section 16.6.

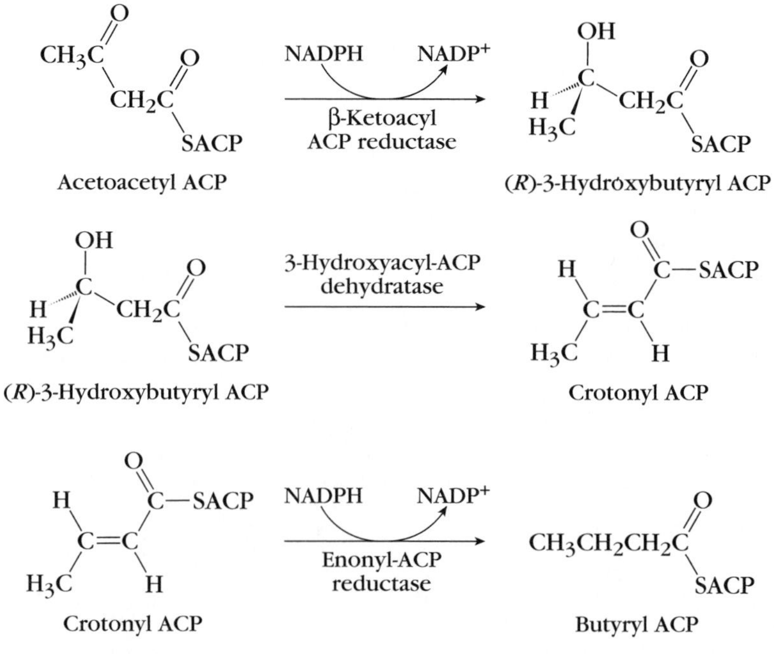

Figure 24.1
Reduction of acetoacetyl ACP to butyryl ACP in fatty acid biosynthesis. The enzyme β-ketoacyl ACP reductase catalyzes the reduction of acetoacetyl ACP to (*R*)-3-hydroxybutyryl ACP. The enzyme 3-hydroxyacyl-ACP dehydratase then catalyzes the dehydration of (*R*)-3-hydroxybutyryl ACP to crotonyl ACP. Finally, the enzyme enonyl-ACP reductase catalyzes the reduction of crotonyl ACP to butyryl ACP.

In the final three reactions, which are shown in Figure 24.1, acetoacetyl ACP is reduced to butyryl ACP. This completes the cycle that adds a two-carbon unit to acetyl ACP. The product of the first cycle, butyryl ACP, undergoes a second cycle to add another two carbon atoms. These cycles are repeated until the required number of carbon atoms is added, two at a time, to form the desired fatty acid.

The cycle of reactions in the biosynthesis of fatty acids demonstrates how similar types of reactions occur in both nature and the chemical laboratory. In Table 24.2, the reactions in the biosynthesis of fatty acids are compared with their analogous reactions in the laboratory.

While these reactions are similar, there are several major differences. First, enzyme-catalyzed reactions are particularly selective. Reactions usually occur at only one of several functional groups in the molecule. An example is the enzyme-catalyzed reduction of the carbonyl group but not the thiolester group in acetoacetyl ACP. Enzyme-catalyzed reactions are so selective because enzymes bind the substrates in such a way that only one functional group is involved in the reactive site of the enzyme.

Second, the reactions in living organisms must occur within a limited pH range. This means that strong acids and bases cannot catalyze reactions in nature. Enzymes compensate for their limited pH range by having functional groups in and around the reactive site that can donate or accept protons. Thus enzyme-bound amino or carboxylate groups accept protons from the substrate, while carboxylic acid and ammonium ions donate protons to the substrate.

While these differences alter the mechanism in subtle ways, the basic mechanistic principles learned from reactions studied in the laboratory still apply.

24.11 α,β-Unsaturated Carbonyl Compounds

A carbon-carbon double bond in conjugation with a carbonyl group forms a new functional group called an **α,β-unsaturated carbonyl group.**

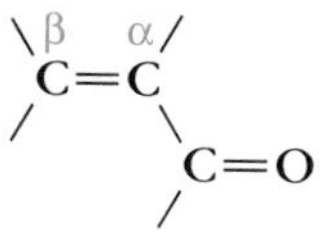

$$\underset{\beta}{C}=\underset{\alpha}{C}\quad C=O$$

α,β-Unsaturated carbonyl group

Compounds that contain an α,β-unsaturated carbonyl group are called α,β-unsaturated carbonyl compounds. The most common α,β-unsaturated carbonyl compounds are α,β-unsaturated ketones, aldehydes, esters, and carboxylic acids. The structures of representative α,β-unsaturated carbonyl compounds, their IUPAC names, and their common names are given in Figure 24.2.

Figure 24.2
Examples of α,β-unsaturated carbonyl compounds.

$CH_2{=}CHCHO$	$(CH_3)_2C{=}CHCCH_3$	$CH_2{=}CCOOH$		
Propenol (acrolein)	4-Methyl-3-penten-2-one (mesityl oxide)	2-Methylpropenoic acid (methacrylic acid)	Ethyl E-3-phenylpropenoate (ethyl *trans*-cinnamoate)	Dimethyl E-butenedioate (dimethyl fumarate)

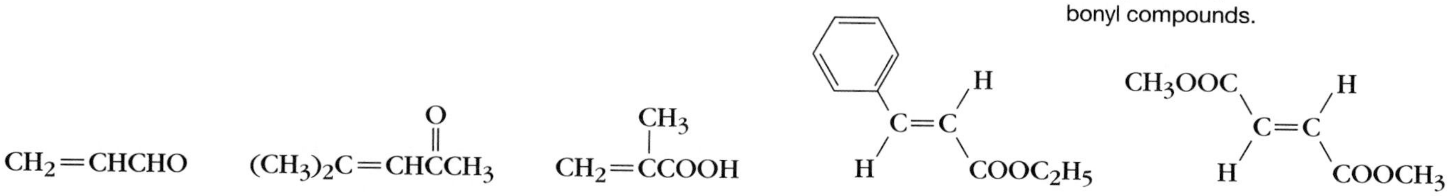

Table 24.2	Reactions that Occur in both the Laboratory and in the Biosynthesis of Fatty Acids

Reaction Type	Enzyme-Catalyzed Reaction in Fatty Acid Biosynthesis
Carboxylation (addition to carbonyl carbon atoms)	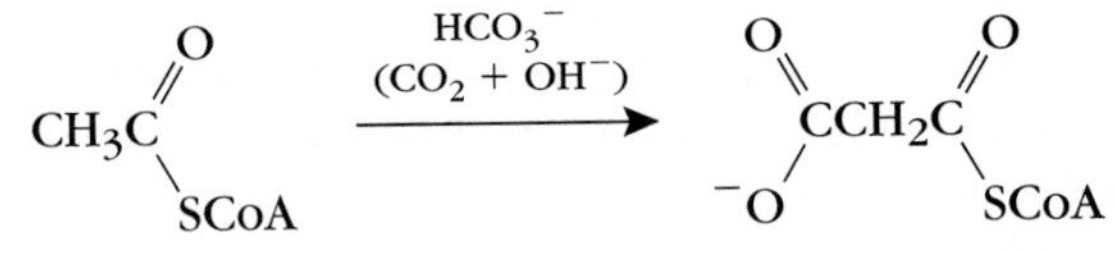
Acyl transfer to sulfur atom	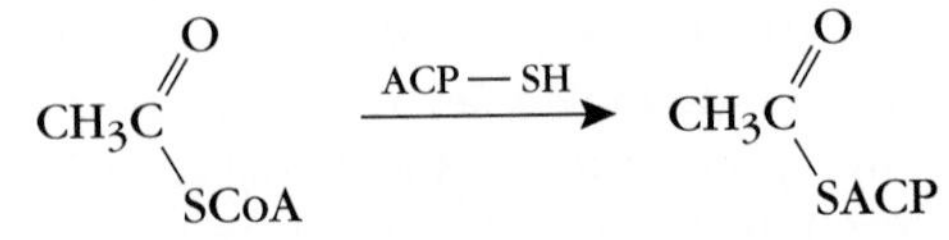
Acyl transfer to carbon atoms (Claisen-like condensation reaction)	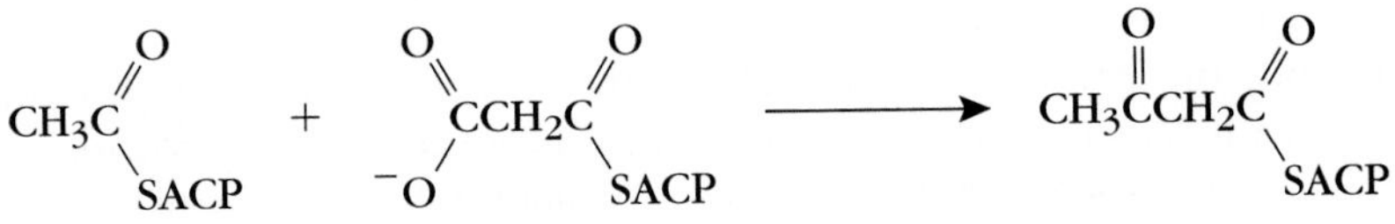
Reduction of a carbonyl group to an alcohol	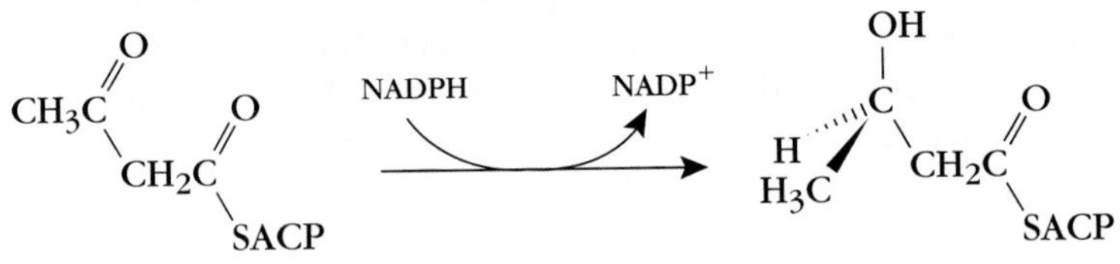
Dehydration of an alcohol	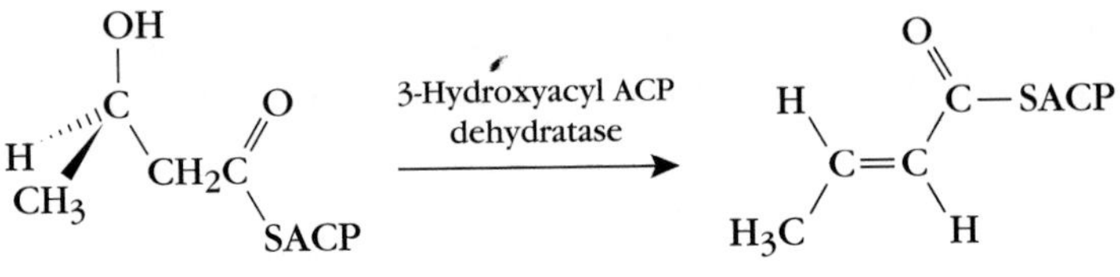
Reduction of the double bond of an α,β-unsaturated carbonyl compound	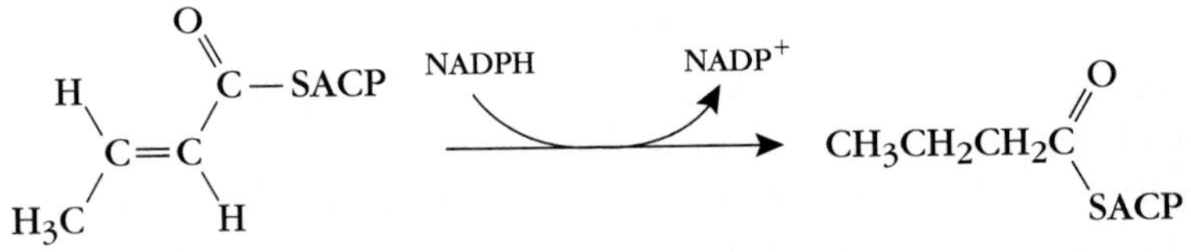

Analogous Laboratory
Reaction

$$RMgX \xrightarrow[\text{(2) } H_3O^+]{\text{(1) } CO_2} RCOOH$$

Grignard reaction

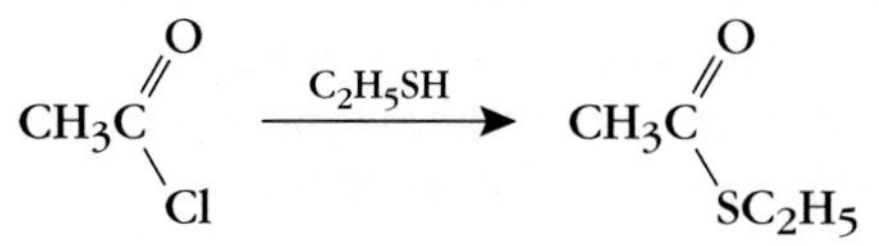

Acetyl chloride → S-Ethyl ethanethiolate

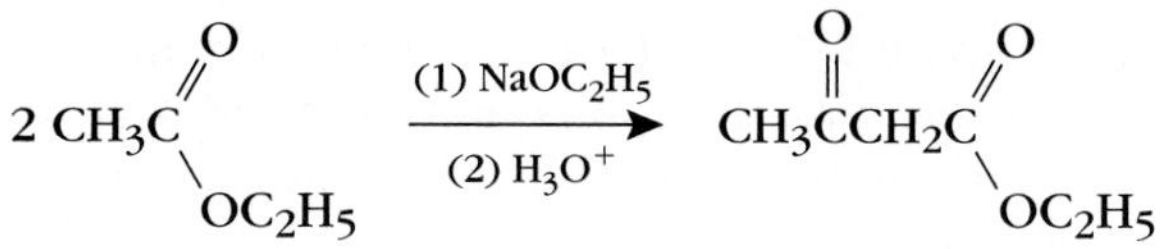

Ethyl acetate → Acetoacetic ester

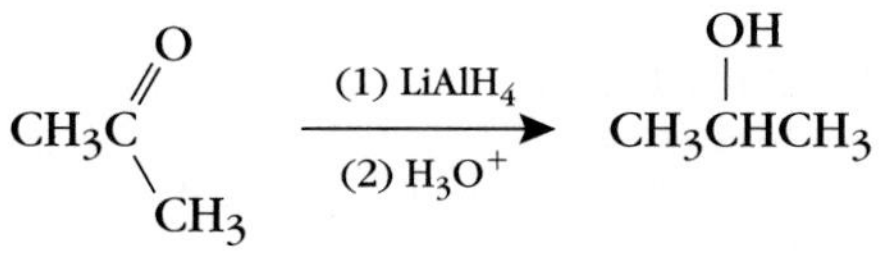

Acetone → 2-Propanol

$$(CH_3)_3COH \xrightarrow[\text{H}_2\text{SO}_4]{\text{Conc}} CH_2{=}C(CH_3)_2$$

2-Methyl-2-propanol → Methylpropene

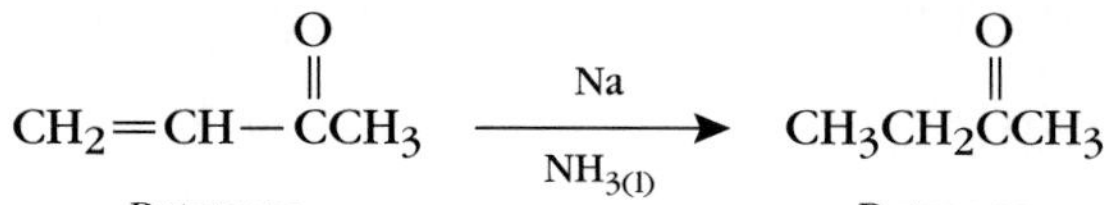

$$CH_2{=}CH{-}\overset{\overset{\displaystyle O}{\|}}{C}CH_3 \xrightarrow[\text{NH}_{3(l)}]{\text{Na}} CH_3CH_2\overset{\overset{\displaystyle O}{\|}}{C}CH_3$$

Butenone → Butanone

α,β-Unsaturated nitriles and compounds with a carbon-carbon triple bond conjugated to a carbonyl group are also included in this class of compounds because they undergo many of the same reactions.

$$CH_2=CHC\equiv N \qquad\qquad H-C\equiv CCOOCH_3$$

Propenenitrile   Methyl propynoate
(acrylonitrile)

α,β-Unsaturated compounds are prepared in general by a number of reactions that we have already discussed. α,β-Unsaturated aldehydes and ketones, for example, can be prepared by the aldol condensation (Section 17.6), α,β-unsaturated carboxylic acids can be prepared by dehydrohalogenation of α-haloacids (Section 17.4), and the products of the Knoevenagel reaction can be hydrolyzed and decarboxylated to form α,β-unsaturated carboxylic acids or ketones.

Exercise 24.11   Propose a synthesis for each of the following compounds, starting with the compound indicated. Use any other organic compound and any inorganic reagent or solvent needed.

(a) Ethyl 3-phenylpropenoate from benzaldehyde
(b) 4-Phenyl-3-buten-2-one from acetone
(c) Propenoic acid from ethanol
(d) 3-Phenylpropenoic acid from acetic anhydride

24.12 Nucleophilic Addition Reactions of α,β-Unsaturated Carbonyl Compounds

Certain nucleophiles react with α,β-unsaturated carbonyl compounds to form products of 1,4-addition. 2-Butenal, for example, reacts with potassium methoxide in methanol to form 3-methoxybutanal by 1,4-addition.

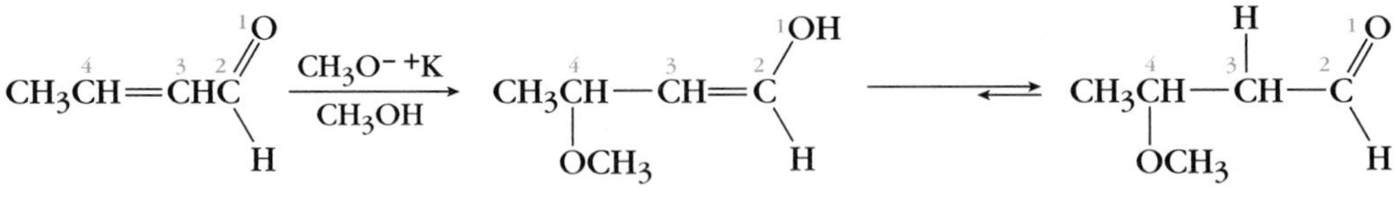

2-Butenal   Enol form   Keto form
3-Methoxybutanal

■ *Recall from Chapter 19 that the terms 1,2-addition and 1,4-addition mean that the addition is to atoms 1 and 2 or 1 and 4 of a conjugated system. The numbers are not the same as the numbers given to the carbon atoms in the IUPAC name of the compound that contains the conjugated system.*

Other nucleophiles react with α,β-unsaturated carbonyl compounds to form products of 1,2-addition. Methyllithium, for example, reacts with 4-methyl-3-penten-2-one to form 2,4-dimethyl-3-penten-2-ol.

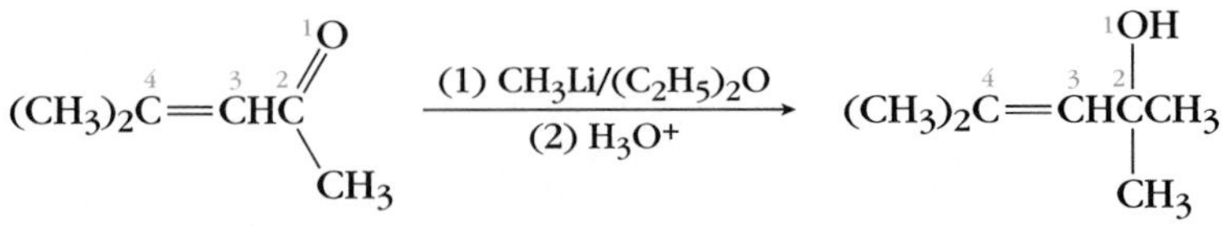

4-Methyl-3-penten-2-one   2,4-Dimethyl-3-penten-2-ol

Nucleophilic additions to α,β-unsaturated carbonyl compounds form products of 1,2- or 1,4-addition because the carbonyl carbon atom and the β-carbon atom of the α,β-unsaturated carbonyl group are both electrophilic sites. We can understand why an α,β-unsaturated carbonyl group has two electrophilic sites by examining its π electron structure. The π electron structure of an α,β-unsaturated carbonyl group can be described as a resonance hybrid of the following three resonance structures:

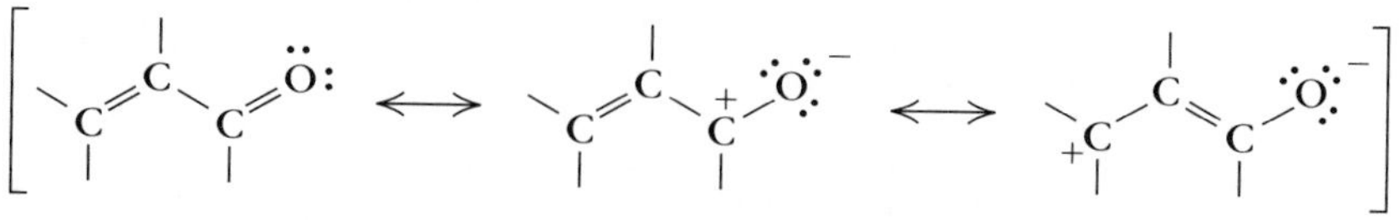

The electronegative oxygen atom polarizes the π electrons of the conjugated system so that both the carbonyl carbon atom and the β-carbon atom are electron deficient. This polarization of the conjugated system diminishes the π electron density of the α,β-double bond compared to the π electron density of an alkene. As a result, an electrophilic reagent (Chapter 8) adds more slowly to the α,β-double bond of an α,β-unsaturated carbonyl compound than it does to a simple alkene.

The polarization of the π electron structure makes the double bond of α,β-unsaturated carbonyl compounds much more reactive toward nucleophilic reagents than the double bond of simple alkenes.

Nucleophiles react with α,β-unsaturated carbonyl compounds to form 1,2-addition products by a mechanism similar to the mechanism of nucleophilic addition to a carbonyl group, which we discussed in Section 14.8. The mechanism of the reaction of methyllithium and 4-methyl-3-penten-2-one occurs by addition of the nucleophile to the carbonyl carbon atom to form a tetrahedral intermediate. Protonation of the tetrahedral intermediate forms the product.

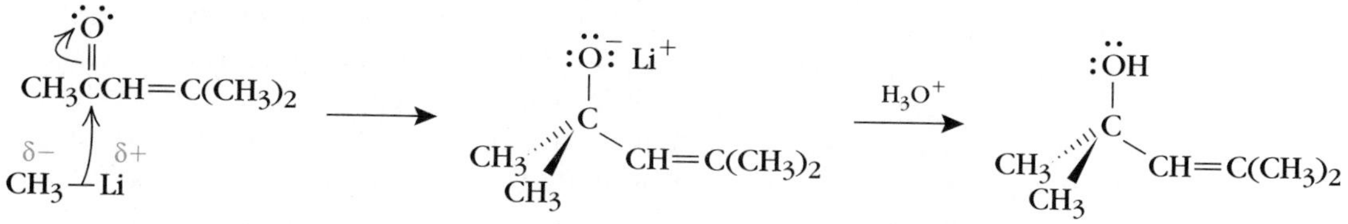

Tetrahedral intermediate

In the mechanism of 1,4-addition, the nucleophile adds to the β-carbon atom to form an enolate anion intermediate. Protonation leads to formation of the carbonyl compound.

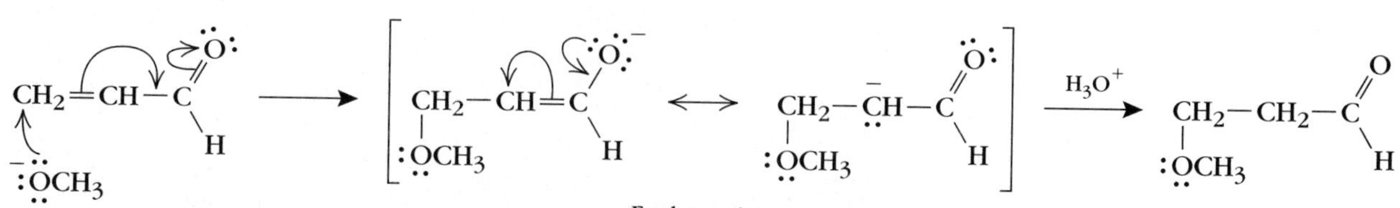

Enolate anion

In the examples presented so far, addition has been either 1,2- or 1,4. This is not always the case. Usually mixtures of 1,2- and 1,4-addition products are formed in which one or the other product predominates. Predominant or exclusive 1,2-addition usually occurs in the reaction of highly basic nucleophiles such as organolithium reagents and $LiAlH_4$. Weaker bases such as CN^- or amines usually form products of 1,4-addition.

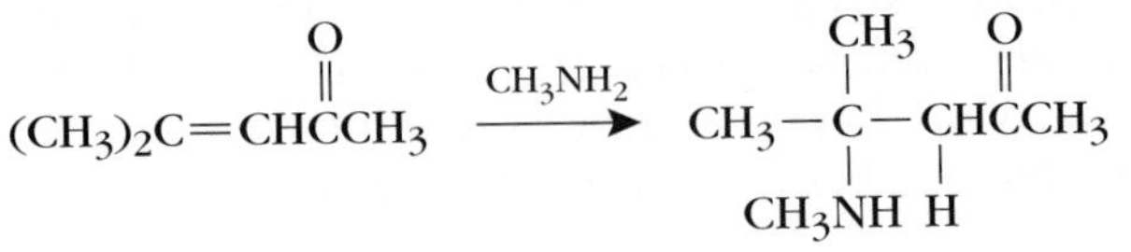

(68%–72% yield)

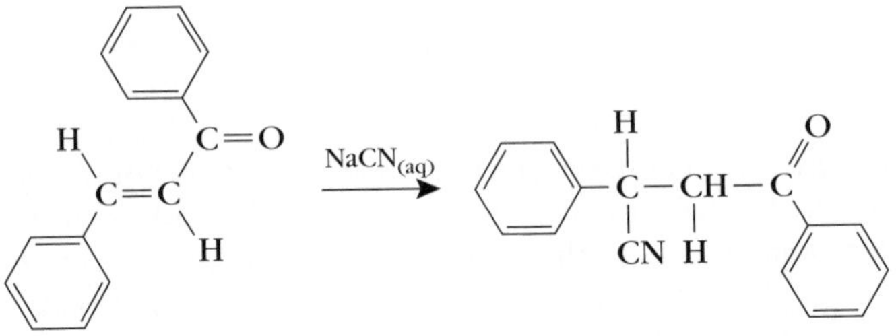

(93%–96% yield)

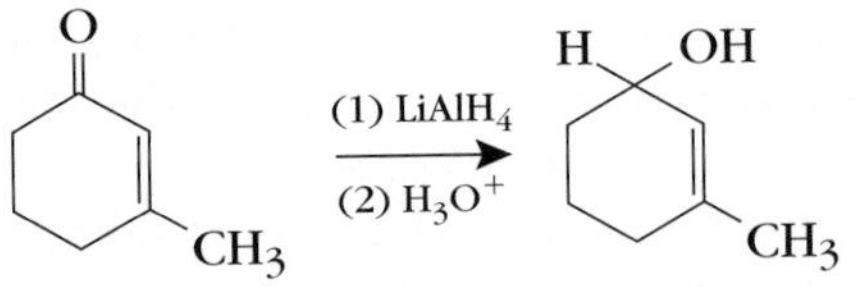

(80%–84% yield)

Grignard reagents react with α,β-unsaturated carbonyl compounds to form products of either 1,2- or 1,4-addition, or both 1,2- and 1,4-additions. Grignard reagents react at the least hindered position, so the steric hindrance at the electrophilic site affects the course of the reaction. If steric hindrance about the β-carbon atom is greater than steric hindrance about the carbonyl carbon atom, formation of 1,2-addition products is favored. If there is more steric hindrance about the carbonyl carbon atom, on the other hand, formation of the 1,4- product is favored. The effect of steric hindrance on product composition is illustrated by the reaction of phenylmagnesium bromide (C_6H_5MgBr) with the following α,β-unsaturated ketone.

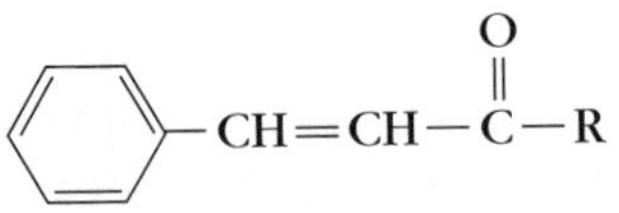

The addition, 1,2- or 1,4-, depends on the nature of the group R. If R is hydrogen, only 1,2-addition product is formed. However, if R is *t*-butyl, only 1,4-addition product is formed. Other alkyl groups (methyl, ethyl, and isopropyl) give

---

Exercise 24.12   Write the structure of the organic product formed by the reaction of 3-penten-2-one with each of the following reagents.

(a) Aqueous NaCN   **(c)** CH_3Li followed by acid hydrolysis
(b) $LiAlH_4$ followed by acid hydrolysis   **(d)** Aqueous NH_3

Table 24.3	Products of Addition of Phenylmagnesium Bromide to

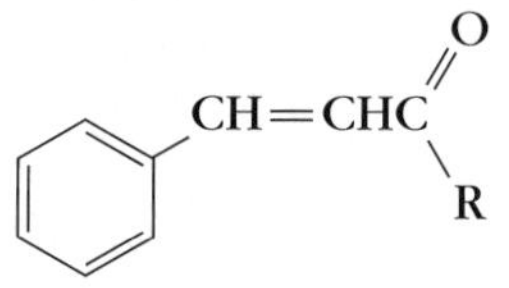

$$CH{=}CHC\overset{\displaystyle O}{\underset{\displaystyle R}{\diagdown}}$$

R	1,4-Addition product (%)	1,2-Addition product (%)
H	0	100
CH_3	12	88
CH_2CH_3	40	60
$CH(CH_3)_2$	88	12
$C(CH_3)_3$	100	0

mixtures as shown by the data in Table 24.3. Thus as the size of R increases, steric hindrance to 1,2-addition increases. As a result, less 1,2-addition product is formed as steric hindrance about the carbonyl group increases.

Enolate anions are nucleophiles that can also add to α,β-unsaturated carbonyl compounds.

24.13 The Michael Addition Reaction

The 1,4-addition of an enolate anion to an α,β-unsaturated carbonyl compound is called a **Michael addition reaction.** The enolate anion obtained from diethyl malonate, for example, adds to 3-buten-2-one to form a 1,4-addition product that yields 5-oxohexanoic acid on acid hydrolysis and decarboxylation.

$$CH_2(COOC_2H_5)_2 \xrightarrow[C_2H_5OH]{Na^{+-}OC_2H_5} {}^-CH(COOC_2H_5)_2 \xrightarrow[(2)\ H_3O^+/heat]{(1)\ CH_3CCH=CH_2} CH_3CCH_2CH_2{-}CH_2COOH$$

Diethyl malonate

5-Oxohexanoic acid
(86%–92% yield)

The nucleophilic enolate anion is called a **Michael donor** because it donates a pair of electrons in the reaction. The compound in the Michael addition reaction that contains a polarized double bond is called a **Michael acceptor** because it accepts the pair of electrons.

Exercise 24.13	Write the structure of the product of the reaction of each of the following Michael acceptors with the enolate anion of ethyl acetoacetate followed by heating with aqueous acid.

(a) Propenal   **(b)** Ethyl 2-butenoate   **(c)** Propenenitrile   **(d)**

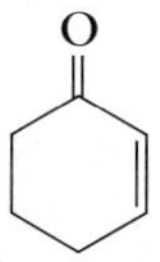

Table 24.4	Some Common Michael Donors and Acceptors		
Michael Donors		**Michael Acceptors**	
Structure of Enolate Anion	Compound from Which Enolate Anion Is Obtained	Structure	Type of Compound
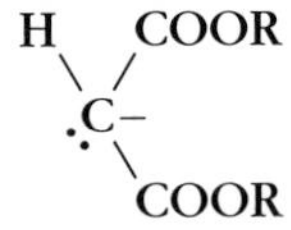 H–C(COOR)(COOR) ·⁻	Malonic ester such as diethyl malonate	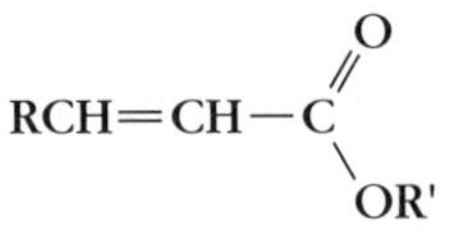 RCH=CH–C(=O)OR'	α,β-Unsaturated ester
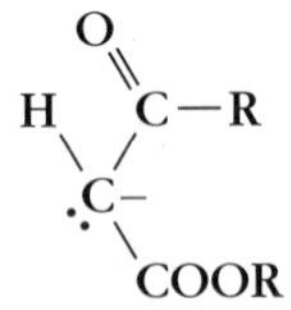 (O=)C–R, H, ·C⁻–COOR	β-Ketoester such as ethyl acetoacetate	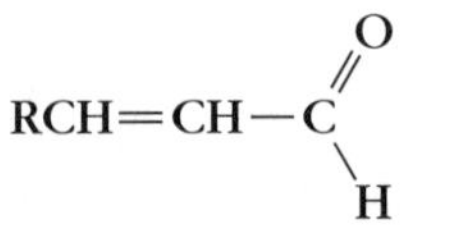 RCH=CH–C(=O)H	α,β-Unsaturated aldehyde
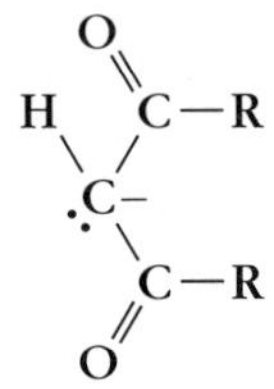 (O=)C–R, H, ·C⁻–C(=O)R	1,3-Diketones such as 2,4-pentanedione	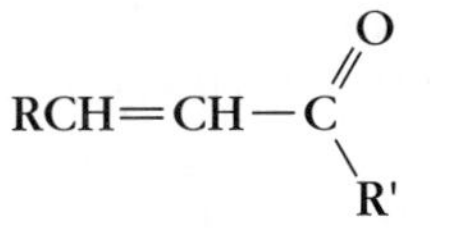 RCH=CH–C(=O)R'	α,β-Unsaturated ketone
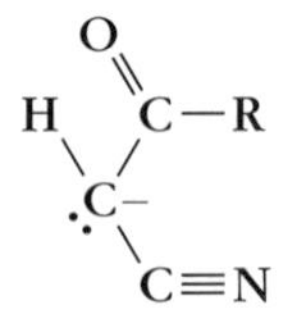 (O=)C–R, H, ·C⁻–C≡N	β-Keto nitriles	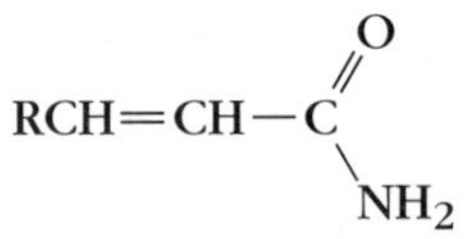 RCH=CH–C(=O)NH₂	α,β-Unsaturated amide
$\overline{R}CHNO_2$	Nitro compound	RCH=CH–C≡N	α,β-Unsaturated nitrile

Best results in the Michael reaction are obtained when particularly stable enolate anions are used, such as those obtained from 1,3-dicarbonyl compounds discussed in Section 24.5. The Michael addition reaction is useful synthetically because of the wide variations possible in the structures of the Michael donors and acceptors. Table 24.4 lists some common Michael donors and Michael acceptors.

Exercise 24.14   Write the structure of the product of the reaction of the enolate anions of each of the following Michael donors with 3-penten-2-one followed by reaction with aqueous acid.

(a) Diethyl malonate   (c) Ethyl acetoacetate
(b) 3,5-Heptanedione   (d) 2-Nitropropane

The Michael addition reaction occurs by the addition of the enolate anion to the β-carbon atom of the Michael acceptor, according to the mechanism shown in Figure 24.3.

Step 1: Formation of enolate anion

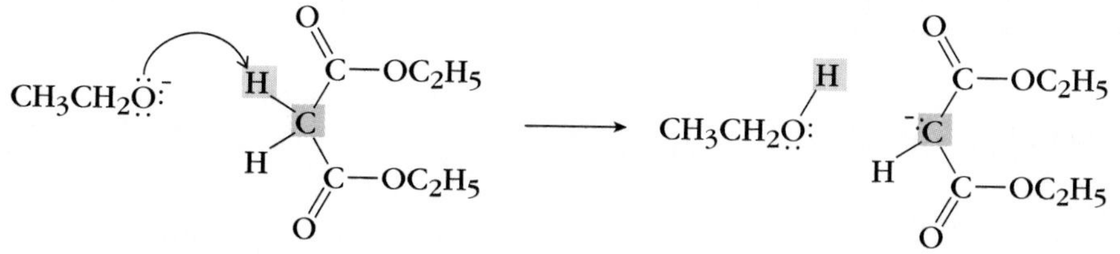

Diethyl malonate — Enolate anion

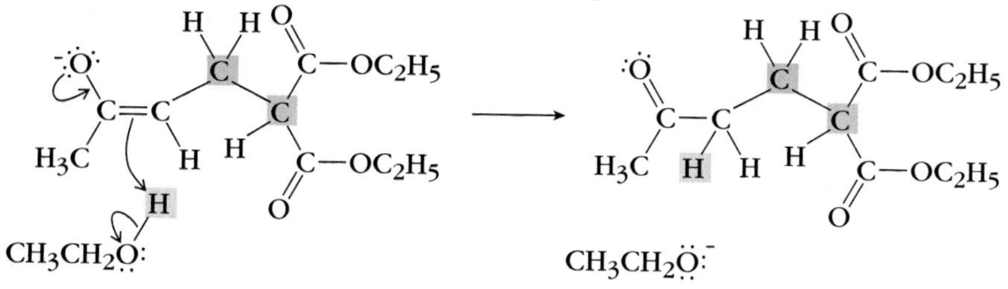

Figure 24.3
The mechanism of the Michael addition reaction. The enolate anion formed in Step 1 adds to the β-carbon atom of the α,β-unsaturated carbonyl compound in Step 2 to form another enolate anion. Protonation of the second enolate anion in Step 3 forms the 1,4-addition product.

Step 2: Enolate anion adds to the β-carbon atom of the α,β-unsaturated ketone

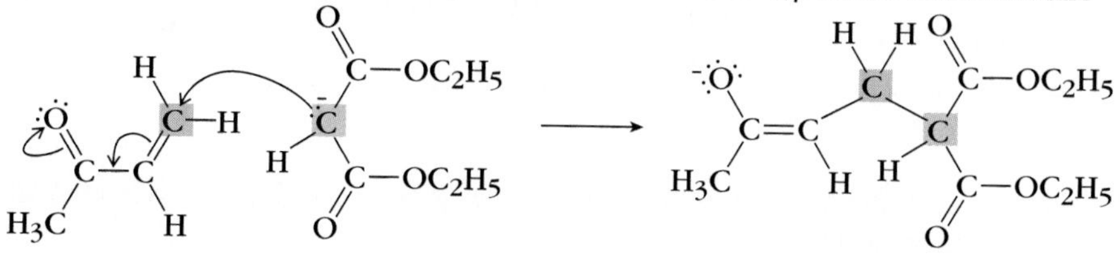

Step 3: Proton transfer to form final addition product

The product of the Michael addition reaction of the enolate anion of diethyl malonate and ethyl 3-phenylpropenoate forms 3-phenylpentanedioic acid on acid-catalyzed hydrolysis and decarboxylation.

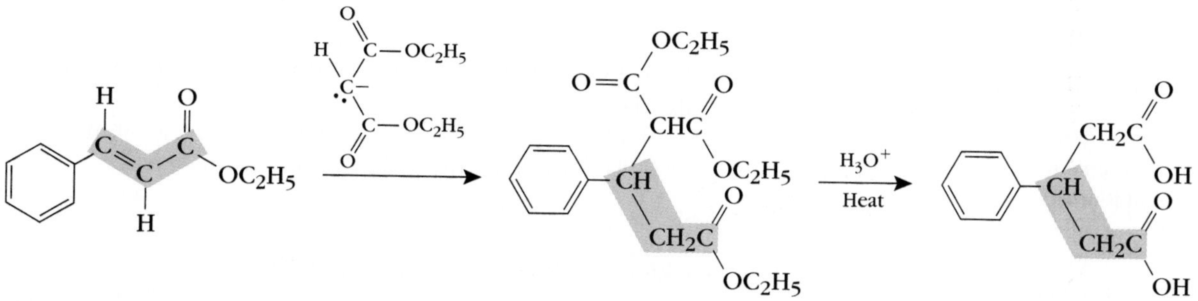

Ethyl (E)-3-phenylpropenoate

3-Phenylpentanedioic acid

Exercise 24.15   Write the equations for the preparation of 2,6-heptanedione, starting with the Michael addition reaction of ethyl acetoacetate and 3-buten-2-one.

Exercise 24.16 Write the structure of the missing products or reactants in each of the following reactions.

(a) $CH_2(COOC_2H_5)_2$ $\xrightarrow[\text{C}_2\text{H}_5\text{OH}]{\text{NaOC}_2\text{H}_5}$? $\xrightarrow{CH_3CH=CHCOOC_2H_5}$? $\xrightarrow[\text{Heat}]{\text{H}_3\text{O}^+}$

(b) $H_2C\begin{subarray}{l} \diagup COOC_2H_5 \\ \diagdown C\equiv N \end{subarray}$ $\xrightarrow[\text{C}_2\text{H}_5\text{OH}]{\text{NaOC}_2\text{H}_5}$? $\xrightarrow{CH_2=CHCHO}$? $\xrightarrow[\text{Heat}]{\text{H}_3\text{O}^+}$

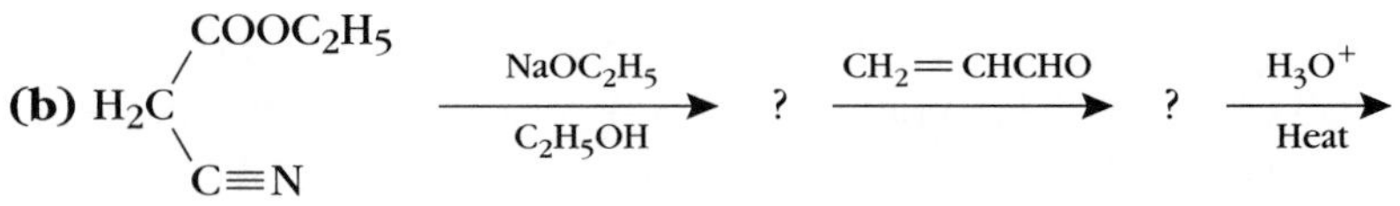

(c) $NCCH_2CN$ $\xrightarrow[\text{C}_2\text{H}_5\text{OH}]{\text{NaOC}_2\text{H}_5}$? $\xrightarrow{CH_2=CHCN}$? $\xrightarrow[\text{Heat}]{\text{H}_3\text{O}^+}$

24.14 Two Substituents Properly Placed in a Molecule React to Form a Ring

Most reactions that form new bonds between functional groups in separate molecules can also form rings if the reactive functional groups are both present in the same molecule. For example, the intermolecular nucleophilic substitution reaction of excess ammonia and methyl bromide forms methyl ammonium bromide.

$$H_3N: \quad CH_3-Br \longrightarrow \overset{+}{H_3N}-CH_3 \quad Br^-$$

Excess

When a bromine atom and a primary amino group are both present in the same molecule, an intramolecular cyclization reaction can occur to form a cyclic ammonium bromide.

$$Br(CH_2)_xNH_2 \xrightarrow{\substack{\text{Cyclization} \\ \text{reaction}}} \underset{(CH_2)_x}{\overset{H\diagdown \overset{+}{N}\diagup H}{\bigcirc}} \quad Br^-$$

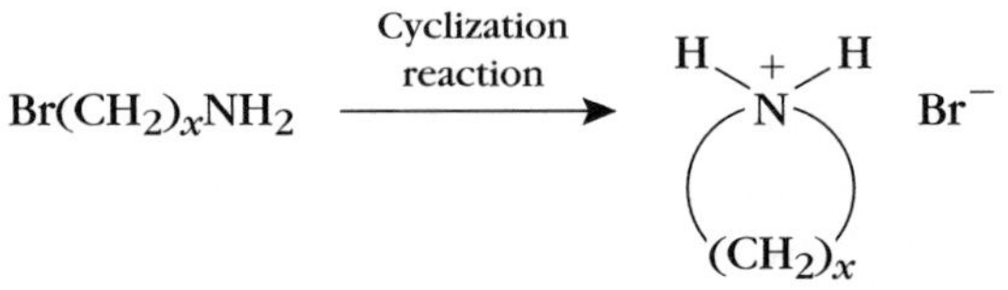

Acyclic bromoamine   Cyclic ammonium bromide

The rate of ring formation depends greatly on the size of the ring to be formed, as shown by the data in Table 24.5.

Based on the data in Table 24.5, the rate of formation of a cyclic ammonium salt depends on the size of the ring to be formed in the following way:

Ring Size $5 > 6 \gg 3 > 4 \approx 7 > 8$ or more atoms

$\boxed{\text{DECREASING RATE OF RING FORMATION}} \Rightarrow$

A different order of the rates of cyclization is obtained from a study of the following reaction:

$$Br(CH_2)_n CH \begin{subarray}{l} \diagup \overset{O}{\overset{\|}{C}}-OC_2H_5 \\ \diagdown \underset{\overset{\|}{O}}{C}-OC_2H_5 \end{subarray} \xrightarrow[\text{t-C}_4\text{H}_9\text{OH}]{\text{KO } t\text{-C}_4\text{H}_9} (CH_2)_n C \begin{subarray}{l} \diagup \overset{O}{\overset{\|}{C}}-OC_2H_5 \\ \diagdown C-OC_2H_5 \\ \overset{\|}{O} \end{subarray}$$

Table 24.5

Relative Rates of Ring Closure to Form a Cyclic Ammonium Salt as a Function of Ring Size

Size of Ring to Be Formed	Relative Rate
3	.07
4	0.001
5	100
6	1
7	0.002
8	0.00004

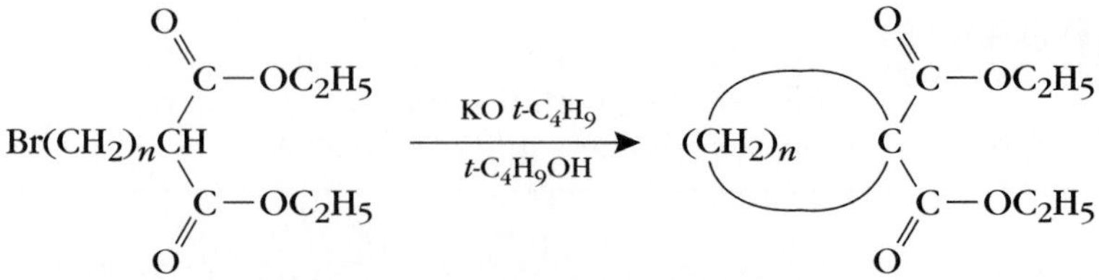

The rate of ring formation of this internal alkylation reaction depends on the size of the ring to be formed in the following way:

Ring Size   3 > 5 > 6 > 7 > 8 > 4 or more atoms

> DECREASING RATE OF RING FORMATION

From these and other observations, it has been found that a three-membered ring is formed faster than a five- or six-membered ring in some reactions but slower in others. How do we explain these observations? To answer this question, we must examine the two factors that determine whether or not a cyclization reaction readily occurs:

1. How easy is it (what is the probability) for the two functional groups to come into contact? In an acyclic molecule, it becomes more difficult for them to come into contact and ring formation becomes less likely the farther apart the two functional groups.
2. How much must the normal tetrahedral bond angles (109.5°) deform to form the ring? The greater the deformation, the more strained the ring and the less likely is ring formation. The strain in a ring is also increased if hydrogens on adjacent carbon atoms are not in the staggered conformation.

These two factors affect rates of ring formation in the following ways:

Rings of five and six members form rapidly because the two functional groups are sufficiently close that they easily come into contact and the bond angles between the atoms forming the ring are deformed only slightly from the normal angles (109.5°) in acyclic compounds.

Formation of three-membered rings is favored because the two functional groups are very close to each other, so they easily react. Ring formation is retarded, on the other hand, because of the great distortion in bond angles required to form a three-membered ring from an acyclic compound. For example, the 109.5° bond angle between atoms in a chain of an acyclic compound must be deformed to about 60° for a three-membered ring to form. Depending on the functional groups involved in the cyclization reaction, one or the other of these two factors will dominate, so in some reactions a three-membered ring will form faster, while in other reactions it will form slower than a five- or six-membered ring.

Four-membered rings usually form less rapidly than three-, five-, and six-membered rings. The two functional groups are farther apart than for the formation of a three-membered ring, and formation of four-membered rings still requires a great distortion of the normal tetrahedral bond angle of the acyclic precursor.

Rings containing seven or more atoms form relatively slowly, however, because the probability of the two functional groups finding each other is very low.

When two functional groups in an acyclic molecule are far enough apart, they may react with a functional group on another molecule rather than the same functional group in the same molecule. When this happens, dimers, trimers, and polymers are formed. The formation of dimers, trimers, and polymers by intermolecular reactions is discussed in Section 24.17.

Most of the bond formation reactions that we have studied in previous chapters can occur intramolecularly. A few examples are shown in Figure 24.4.

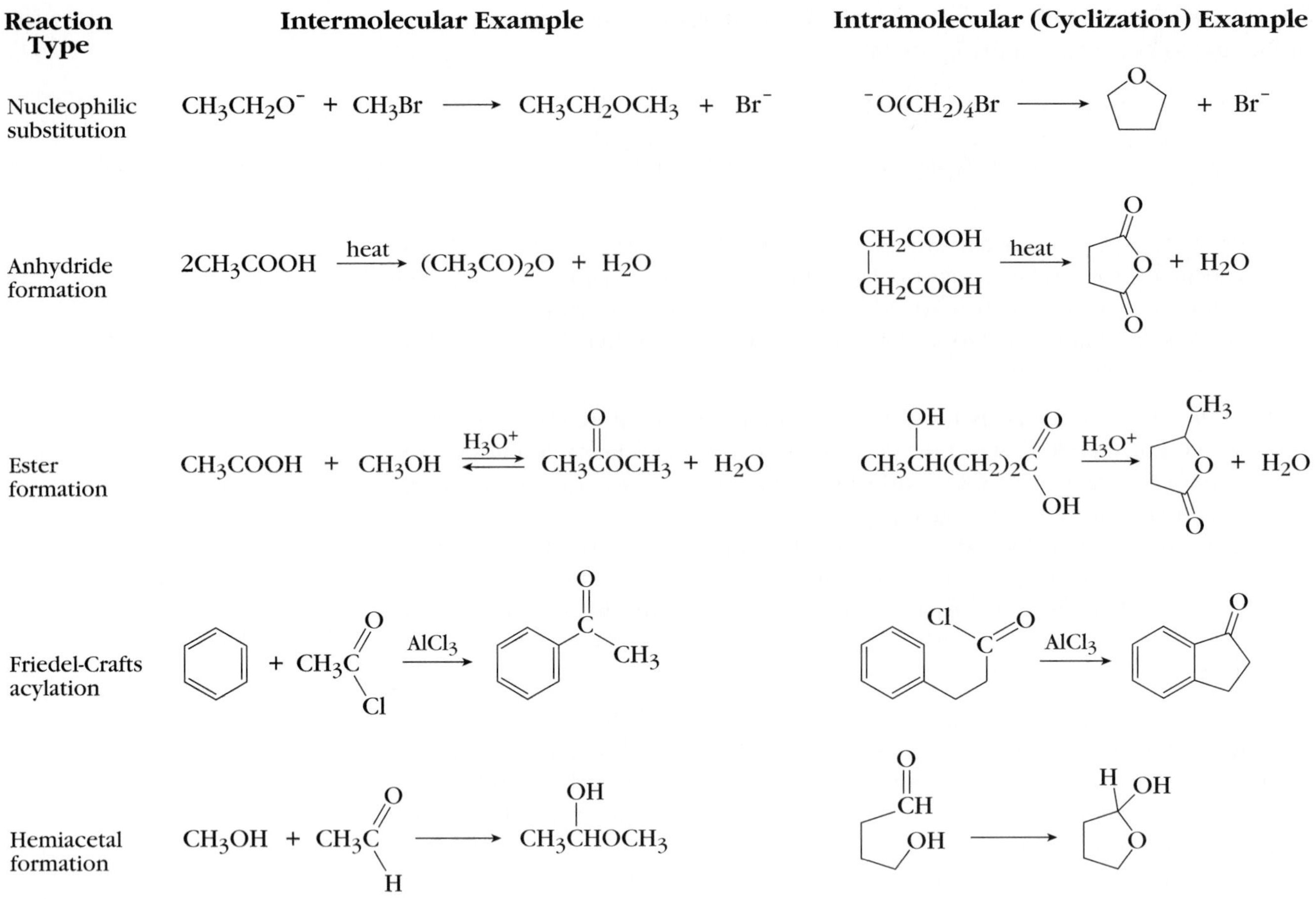

Reaction Type	Intermolecular Example	Intramolecular (Cyclization) Example
Nucleophilic substitution	$CH_3CH_2O^- + CH_3Br \longrightarrow CH_3CH_2OCH_3 + Br^-$	$^-O(CH_2)_4Br \longrightarrow$ [tetrahydrofuran ring] $+ Br^-$
Anhydride formation	$2CH_3COOH \xrightarrow{heat} (CH_3CO)_2O + H_2O$	$\begin{array}{l}CH_2COOH\\ \mid \\ CH_2COOH\end{array} \xrightarrow{heat}$ [succinic anhydride] $+ H_2O$
Ester formation	$CH_3COOH + CH_3OH \rightleftharpoons \overset{O}{\overset{\|}{CH_3COCH_3}} + H_2O$ $\xrightarrow{H_3O^+}$	$CH_3CH(CH_2)_2\overset{O}{\overset{\|}{C}}\underset{OH}{\overset{OH}{\mid}} \xrightarrow{H_3O^+}$ [γ-methyl lactone] $+ H_2O$
Friedel-Crafts acylation	[benzene] $+ CH_3\overset{O}{\overset{\|}{C}}Cl \xrightarrow{AlCl_3}$ [acetophenone]	[3-phenylpropanoyl chloride] $\xrightarrow{AlCl_3}$ [indanone]
Hemiacetal formation	$CH_3OH + CH_3\overset{O}{\overset{\|}{C}}H \longrightarrow \overset{OH}{\overset{\mid}{CH_3CHOCH_3}}$	[4-hydroxybutanal] $\longrightarrow$ [cyclic hemiacetal]

Figure 24.4
Examples of typical bond-forming reactions that occur intermolecularly and intramolecularly.

Exercise 24.17   Write the structure of the product of each of the following reactions.

(a) $\overset{O}{\overset{\|}{CH_3CCH_2CH_2CHOH}}$ with $\underset{CH_3}{\mid}$ $\xrightarrow{H_3O^+}$

(b) $HOCH_2(CH_2)_3CH_2Br \xrightarrow[H_2O]{NaOH}$

(c) $HOCH_2CH_2CH_2COOC_2H_5 \xrightarrow{H_3O^+}$

Intramolecular condensation reactions can be used to prepare substituted cyclopentane and cyclohexane rings.

24.15 Ring-Forming Condensation Reactions

Carbonyl condensation reactions are among the most valuable methods of synthesizing compounds with rings containing five and six carbon atoms. Each of the condensation reactions that we have studied in previous chapters has an intramolecular analog.

A Intramolecular Aldol, Claisen, and Mixed Condensation Reactions

When correctly placed in a molecule, an enolate anion reacts with a carbonyl group of an aldehyde or ketone to form a ring by an intramolecular aldol condensation.

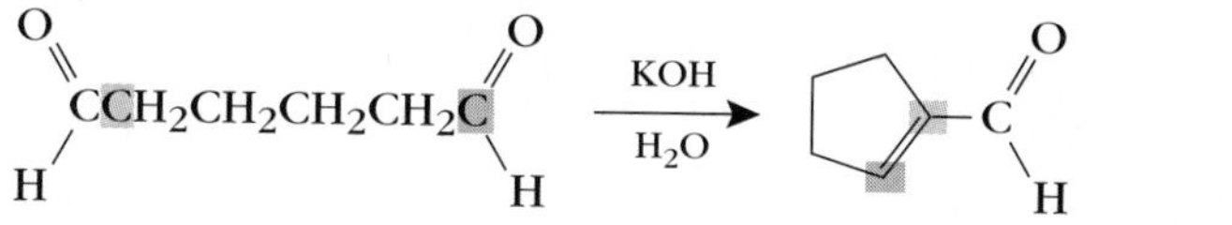

Hexanedial   1-Cyclopentenecarboxaldehyde

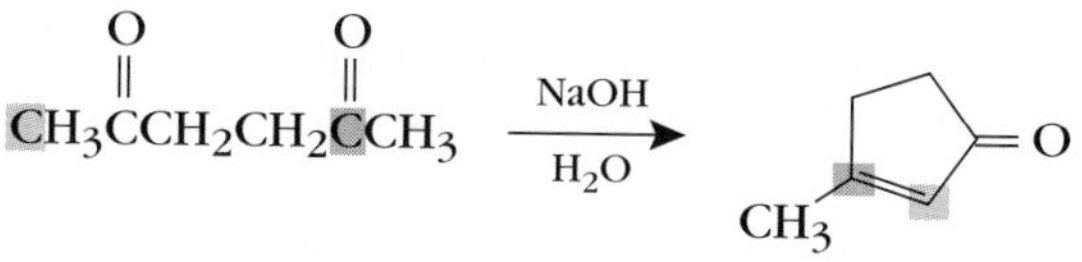

2,5-Hexanedione   3-Methyl-2-cyclopentenone

An intramolecular aldol condensation reaction can, in theory, give several products with different ring sizes. Aldol condensation reactions are reversible (Section 17.6), so the more stable product is formed. The more stable product in an intramolecular aldol condensation generally contains a five- or six-membered ring. The aldol condensation of 2,7-octanedione, for example, forms a five-membered ring, not a seven-membered ring.

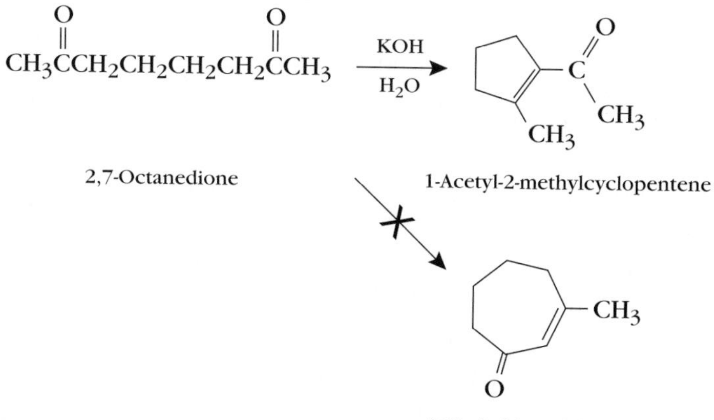

2,7-Octanedione   1-Acetyl-2-methylcyclopentene

3-Methyl-2-cycloheptenone

An intramolecular Claisen condensation reaction is called a **Dieckmann condensation.** The products of a Dieckmann condensation reaction are β-keto esters.

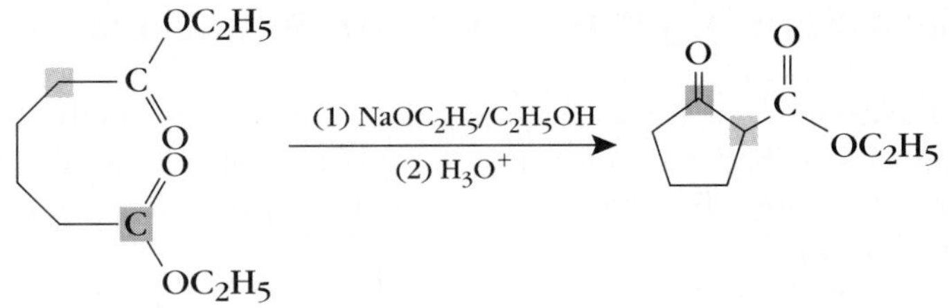

Diethyl hexanedioate   2-Carbethoxycyclopentanone

Mixed condensation reactions can also be used to form ring compounds, as shown by the following example:

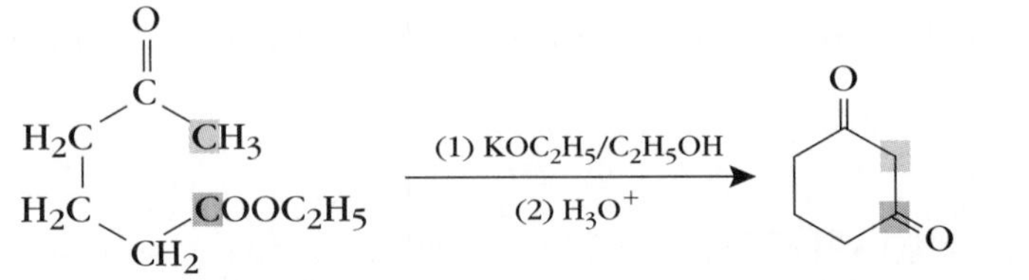

Ethyl 5-oxohexanoate   1,3-Cyclohexanedione

The formation of 1,3-cyclohexanedione from ethyl 5-oxohexanoate is another example of a mixed condensation reaction between a ketone and an ester. The more acidic hydrogens adjacent to the carbonyl group of the ketone functional group are removed to form an enolate anion, which adds to the carbonyl carbon atom of the ester group to form a six-membered ring.

Exercise 24.18   Write the structure of the products of each of the following reactions.

(a) 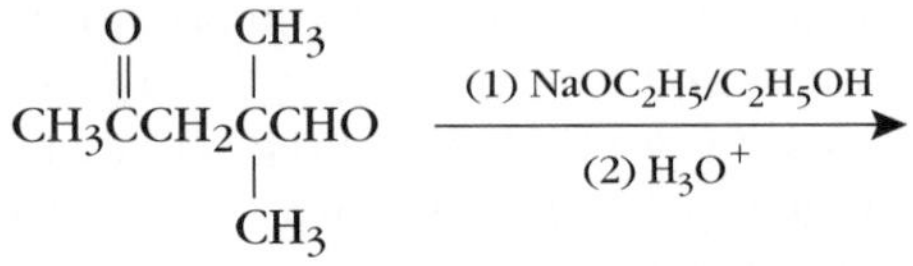 CH_3CCH_2CCHO $\xrightarrow{\text{(1) NaOC}_2\text{H}_5/\text{C}_2\text{H}_5\text{OH}}_{\text{(2) H}_3\text{O}^+}$

(b) $NCCH_2CH_2CH_2CH_2CN$ $\xrightarrow{\text{(1) NaOC}_2\text{H}_5/\text{C}_2\text{H}_5\text{OH}}_{\text{(2) H}_2\text{SO}_4/\text{H}_2\text{O/heat}}$

(c) 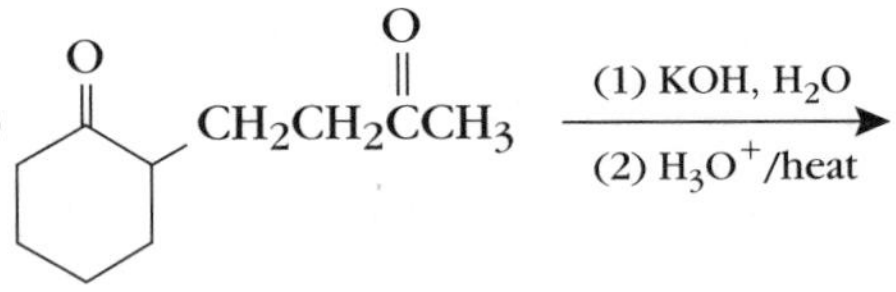 $\xrightarrow{\text{(1) KOH, H}_2\text{O}}_{\text{(2) H}_3\text{O}^+/\text{heat}}$

A Michael addition reaction followed by a carbonyl condensation reaction is a useful method of preparing multiring compounds.

B   The Robinson Annulation Reaction

The word annulation *is derived from the Latin word* annulus, *meaning a ring.*

A very useful synthetic technique for the preparation of complex molecules containing several rings is to carry out a Michael addition reaction (Section 24.13) followed by an aldol condensation reaction. This sequence of reactions is called the **Robinson annulation reaction.**

First Reaction: Michael addition

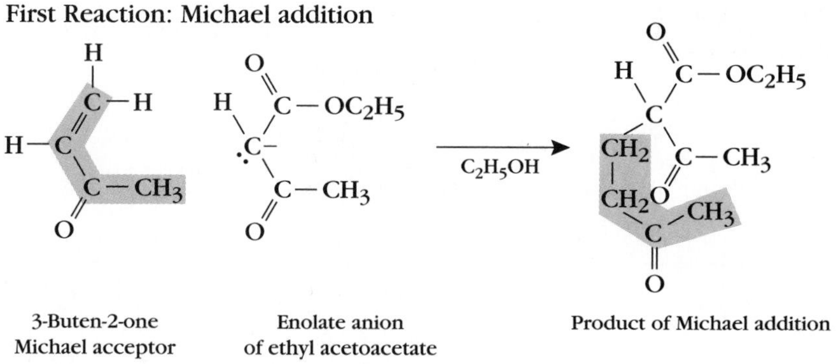

3-Buten-2-one   Enolate anion   Product of Michael addition
Michael acceptor   of ethyl acetoacetate

Second Reaction: Intramolecular aldol condensation

Product of annulation

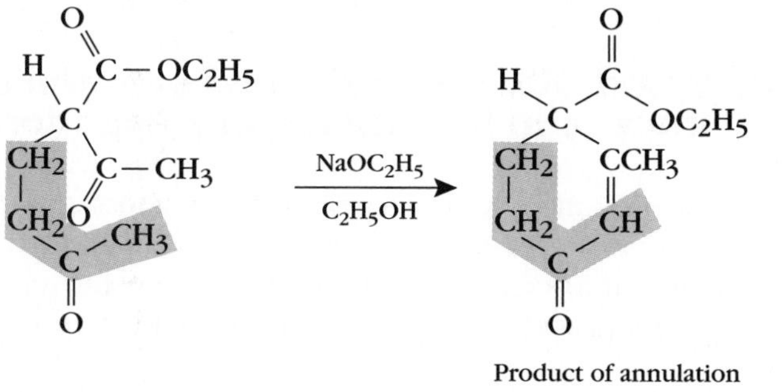

The product of the Michael addition is a 1,5-diketone, which undergoes an intramolecular aldol condensation reaction (Section 24.15, A) to create a new ring.

Exercise 24.19 Write the mechanism for both reactions in the Robinson annulation reaction of 3-buten-2-one and ethyl acetoacetate given previously.

Polycyclic compounds can be formed by starting with a cyclic Michael donor such as cyclohexanone.

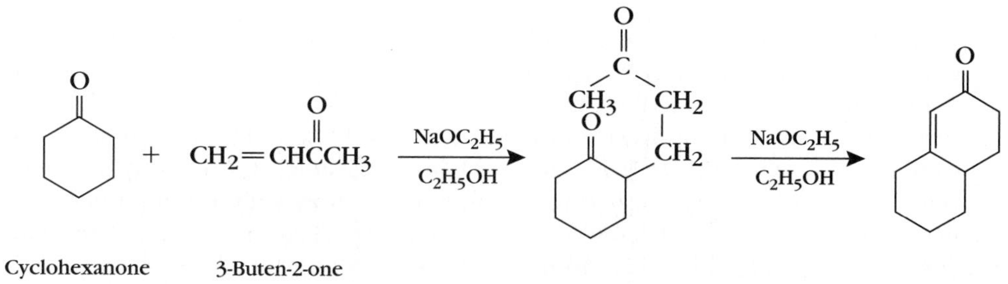

Cyclohexanone   3-Buten-2-one

| **Exercise 24.20** | Write the structure of the product in the following Robinson annulation reaction. |

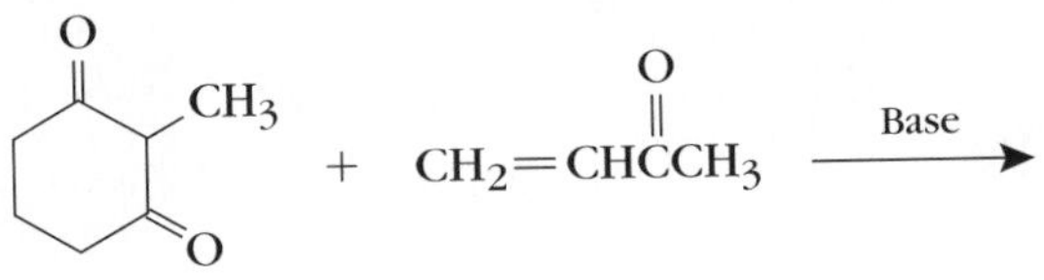

Molecules containing two functional groups that react with each other but that are improperly located for intramolecular reaction form polymers by intermolecular reactions instead.

24.16 Some Difunctional Molecules Undergo Both Intra- and Intermolecular Reactions

The most favored intramolecular reactions are those that form rings containing five or six atoms. Reactions that form rings of other sizes not only occur more slowly but also must compete with intermolecular side reactions. For example, 7-hydroxyheptanoic acid can form products of both intra- and intermolecular reactions.

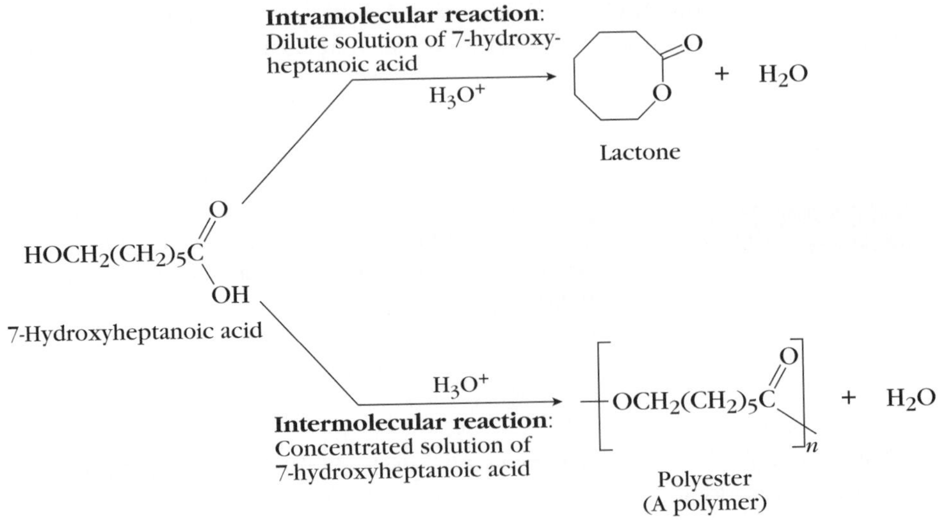

Cyclization forms a lactone, while the intermolecular reactions form a **polyester.** This polyester is formed by joining together many molecules of 7-hydroxyheptanoic acid into a polymer. Polymers are usually formed in a sequential manner. In the case of the polyester formed from 7-hydroxyheptanoic acid, polymerization starts when the carboxy group of one molecule of 7-hydroxyheptanoic acid reacts with the hydroxy group of another to form a dimer. The dimer still has a carboxy group and a hydroxy group that can react intermolecularly with more 7-hydroxyheptanoic acid to form a trimer. The trimer can react with more monomer to form a tetramer and so on until a polymer is formed.

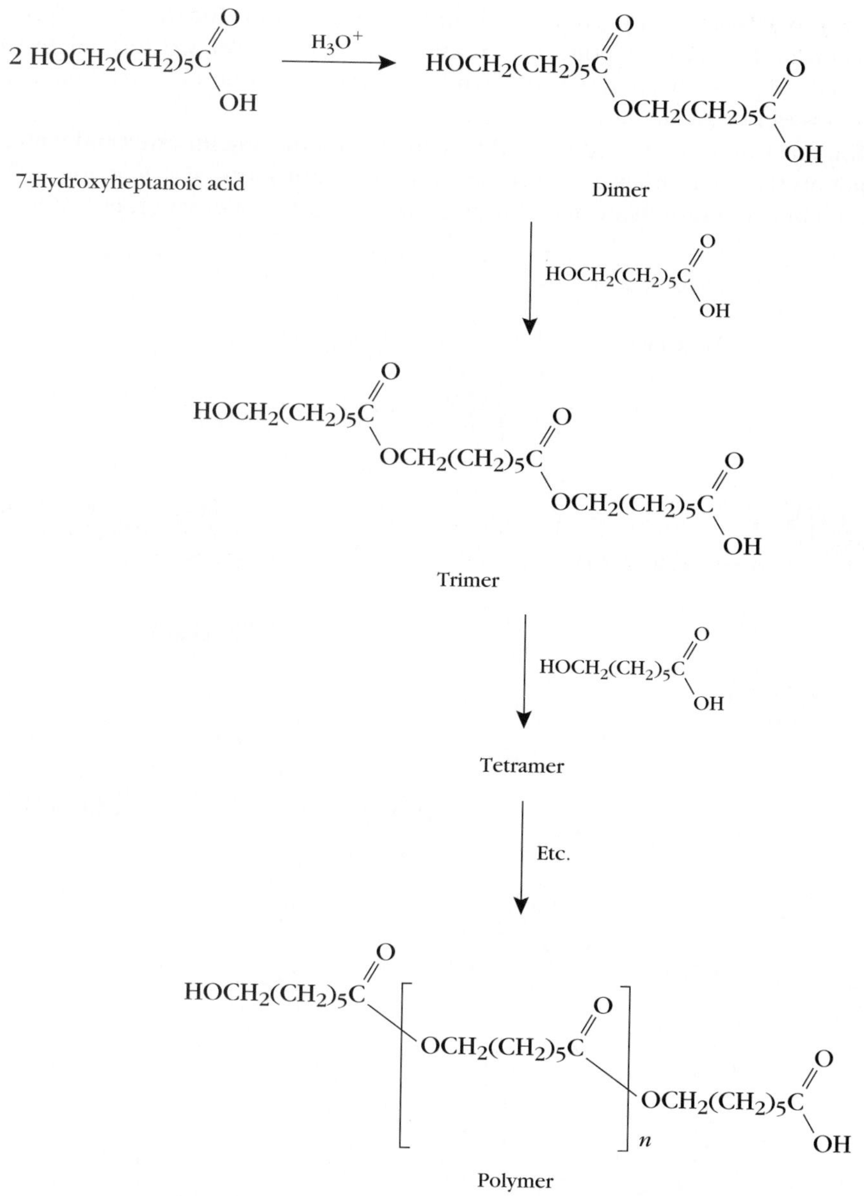

$$2 \ HOCH_2(CH_2)_5C\overset{O}{\underset{OH}{\big\|}} \xrightarrow{H_3O^+} HOCH_2(CH_2)_5C\overset{O}{\big\|}OCH_2(CH_2)_5C\overset{O}{\underset{OH}{\big\|}}$$

7-Hydroxyheptanoic acid   Dimer

$HOCH_2(CH_2)_5C\overset{O}{\underset{OH}{\big\|}}$

$HOCH_2(CH_2)_5C\overset{O}{\big\|}OCH_2(CH_2)_5C\overset{O}{\big\|}OCH_2(CH_2)_5C\overset{O}{\underset{OH}{\big\|}}$

Trimer

$HOCH_2(CH_2)_5C\overset{O}{\underset{OH}{\big\|}}$

Tetramer

Etc.

$HOCH_2(CH_2)_5C\overset{O}{\big\|}\left[OCH_2(CH_2)_5C\overset{O}{\big\|} \right]_n OCH_2(CH_2)_5C\overset{O}{\underset{OH}{\big\|}}$

Polymer

For compounds that undergo both intra- and intermolecular reactions, one reaction can be favored over the other if the reaction conditions are chosen properly. Formation of cyclic products is usually favored by carrying out the reaction at low concentrations of monomer. This works because the rate law of the cyclization reaction is first order in monomer, while the formation of polyester is second order. At low concentrations of monomer, the first order reaction (cyclization) occurs more quickly than the second order reaction (polymerization).

In most cases, however, polymerization is the desired reaction.

24.17 Condensation Polymers

Synthetic polymers are categorized according to the type of polymerization reaction used. We discussed the formation of addition polymers by the polymerization of ethylene and vinyl compounds in Chapter 18. **Condensation polymers** are another category of polymers. Condensation polymers are formed by the polymerization of monomers accompanied by the formation of a small molecule such as water or an alcohol.

The reaction of a carboxylic acid and an alcohol to form an ester and water is a reaction that can be used to form condensation polymers. The reaction of 1,2-ethanediol and terephthalic acid, for example, forms a condensation polymer.

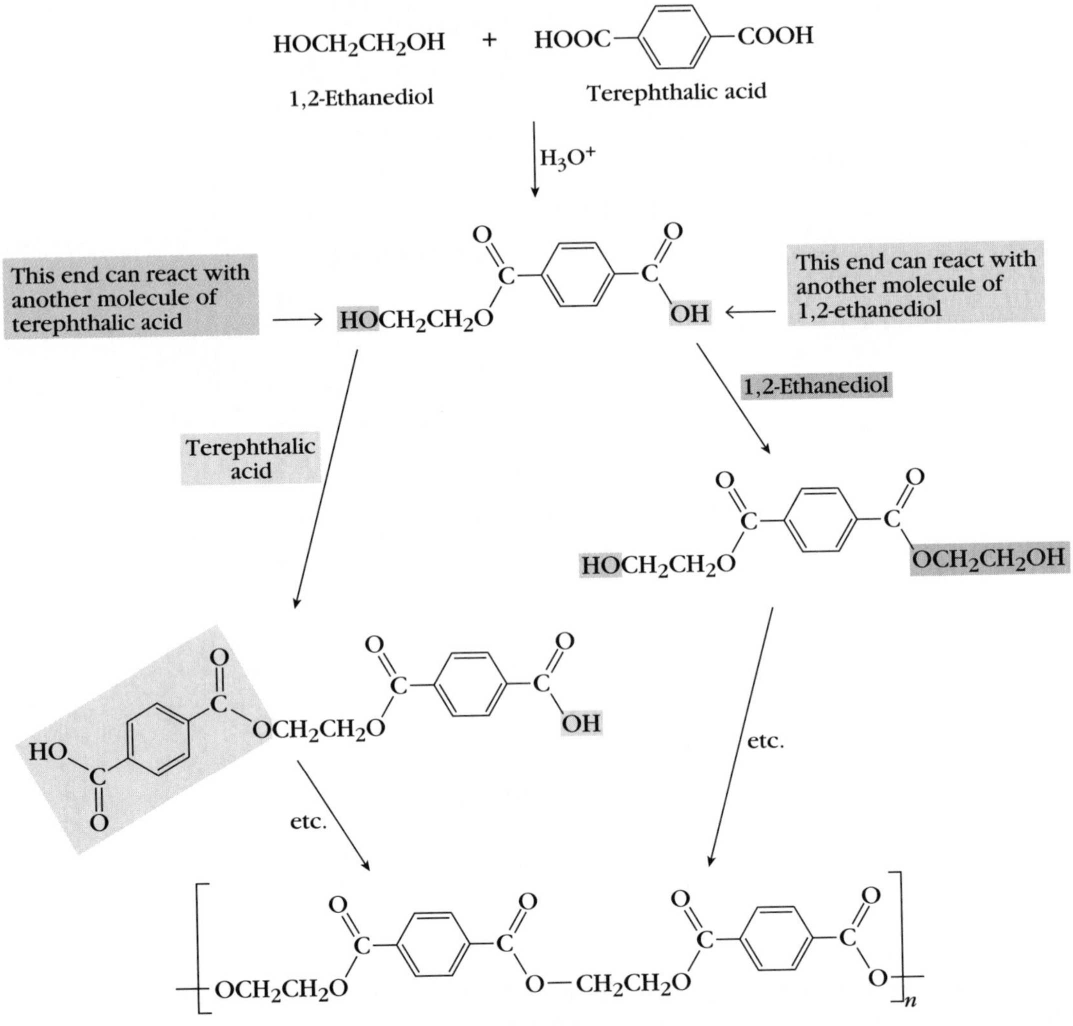

$HOCH_2CH_2OH$ + $HOOC$—⬡—$COOH$

1,2-Ethanediol   Terephthalic acid

H_3O^+

This end can react with another molecule of terephthalic acid

This end can react with another molecule of 1,2-ethanediol

$HOCH_2CH_2O$ ... OH

Terephthalic acid

1,2-Ethanediol

$HOCH_2CH_2O$ ... OCH_2CH_2OH

HO ... OCH_2CH_2O ... OH

etc.

etc.

$[OCH_2CH_2O \cdots O{-}CH_2CH_2O \cdots O]_n$

Polymer

The functional groups of terephthalic acid and 1,2-ethanediol are located so that intramolecular ester formation is impossible. Consequently, the only possible reaction of terephthalic acid and 1,2-ethanediol is polymerization. The condensation polymer grows by a series of reactions. In each reaction, an ester functional group is formed.

Heating a carboxylic acid and an amine forms an amide and water.

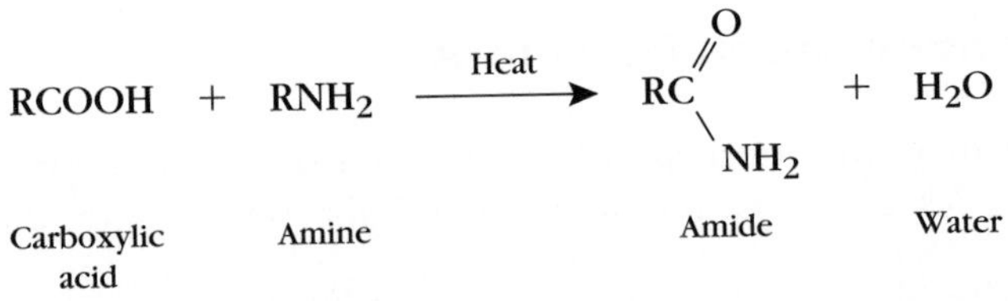

$RCOOH$ + RNH_2 $\xrightarrow{\text{Heat}}$ $RC(=O)NH_2$ + H_2O

Carboxylic   Amine   Amide   Water
acid

Table 24.6	Examples of Condensation Polymers and Some of Their Uses		
Functional Group	Trade or Common Name of Polymer	Uses	Reactants
Polyester	Dacron, Mylar, Fortrel	As a fiber, used in tire cords, yacht sails, and electrical insulation. As a film, used as base for photographic films and recording tapes.	$HOCH_2CH_2OH$ 1,2-Ethanediol (ethylene glycol)  + HOOC—⟨benzene⟩—COOH Terephthalic acid
Polyamide	Capran (Nylon-6)	As a fiber, used in textiles and tire cords.	$H_2N(CH_2)_5COOH$ 6-Aminohexanoic acid
Polyamide	Nylon (Nylon-66)	As a fiber, used in textiles, rope, tire cords, thread, and cloth. Polymer used in molded objects that are subject to high impact.	$HOOC(CH_2)_4COOH$ Adipic acid + $H_2N(CH_2)_6NH_2$ Hexamethylene diamine
Polycarbonate	Lexan	High strength, optically transparent polymer. Used as safety glass, in skylights, lighting fixtures, food containers, and as the face shield of astronauts' space suits.	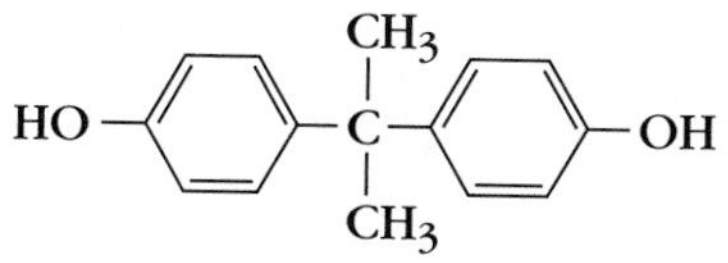 $\overset{O}{\overset{\|}{ClCCl}}$ Phosgene + HO—⟨benzene⟩—$\overset{CH_3}{\underset{CH_3}{C}}$—⟨benzene⟩—OH Bisphenol A

Amides are not usually prepared in the laboratory by this reaction, but it is used as an industrial method of forming polyamides. Reaction of carboxylic acids and amines forms a salt, which on heating forms an amide and water. The polyamide nylon-66, for example, is formed by heating 1,6-diaminohexane, also called hexamethylene diamine, and hexanedioic acid, also called adipic acid.

$$H_2N(CH_2)_6NH_2 \ + \ HOOC(CH_2)_4COOH \ \longrightarrow \ H_2N(CH_2)_6NH_3^{+\ -}OOC(CH_2)_4COOH$$

1,6-Diaminohexane   Hexanedioic acid

↓ Heat

$$\left[\begin{array}{c} \\ -HN(CH_2)_6NH \end{array} \overset{\displaystyle O}{\overset{\|}{C}}(CH_2)_4\overset{\displaystyle O}{\overset{\|}{C}} \ HN(CH_2)_6NH \ \overset{\displaystyle O}{\overset{\|}{C}}(CH_2)_4\overset{\displaystyle O}{\overset{\|}{C}} \right]_n$$

Nylon-66

These and other condensation polymers and their uses are listed in Table 24.6.

Exercise 24.21   Write the structure of the repeating unit of Lexan (Table 24.6) formed by the polymerization of phosgene and bisphenol A.

Exercise 24.22   The reaction of 1,4-butanediol and terephthalic acid forms a tough polymer that is used in automobile ignition systems because of its resistance to water absorption. Write the structure of the repeating unit in this polymer.

24.18 Summary

Many organic compounds in the world contain more than one functional group. The reactions that such compounds undergo depend on the relative locations of their functional groups. If the functional groups are sufficiently far apart, each functional group undergoes its normal reactions unaffected by the other functional groups.

If two functional groups are bonded to the same carbon atom, one may affect the rate or equilibrium constant of reactions of the other. Two carbonyl-containing functional groups on one carbon atom is an example. Such compounds are called **1,3- or β-dicarbonyl compounds.** A second carbonyl group enhances the enol content of the compound; makes the methylene hydrogens more acidic than hydrogens α- to a single carbonyl group; and, in the case of β-keto acids and β-dicarboxylic acids, facilitates decarboxylation. Generally the second functional group does not change the fundamental reactions of the first.

1,3-Dicarbonyl compounds are widely used as starting materials for the synthesis of ketones and carboxylic acids. The key step in all these methods is the quantitative formation of an enolate anion by reaction of the 1,3-dicarbonyl compound with base. The enolate anion is a nucleophile, and it reacts with haloalkanes and the carbonyl carbon atoms of most carbonyl-containing compounds to form a new carbon-carbon bond.

The reaction of the enolate anion of diethyl malonate with haloalkanes followed by acid hydrolysis and decarboxylation to form a carboxylic acid is called the **malonic ester synthesis.** The formation of ketones by a similar sequence of reactions with the enolate anion of ethyl acetoacetate is called the **acetoacetic**

ester synthesis. Addition of an enolate anion to the carbonyl carbon atom of aldehydes or ketones to form an α,β-unsaturated ketone, aldehyde, or carboxylic acid is called a **Knoevenagel condensation reaction.**

Nature also makes use of a 1,3-dicarbonyl compound, malonyl CoA, in the biosynthesis of a variety of compounds. The biosynthesis of fatty acids is an example of how nature uses malonyl CoA. Fatty acids are synthesized by successive cycles of seven enzyme-catalyzed reactions. Each cycle extends the fatty acid chain by two carbon atoms. Acyl transfer to malonyl CoA is the key reaction that forms the carbon-carbon bond in each cycle.

Nucleophiles add to α,β-unsaturated carbonyl compounds to form 1,2- and 1,4-addition products. Predominant or exclusive 1,2-addition usually occurs in the reaction of strongly basic nucleophiles such as organolithium reagents and LiAlH₄. Weaker bases such as CN⁻ or amines usually form products of 1,4-addition. The 1,4-addition of enolate anions of 1,3-dicarbonyl compounds to α,β-unsaturated carbonyl compounds is called a **Michael addition reaction.**

If two functional groups that can react with each other are present in the same molecule, reaction may occur either intra- or intermolecularly. When the two functional groups are placed so that a five- or six-membered ring can be formed, intramolecular ring formation or **cyclization** is favored.

If the two reacting functional groups are located so that ring formation is difficult or impossible, intermolecular reaction occurs to form a dimer, a trimer, and finally a polymer. **Condensation polymers** are formed by means of condensation reactions. Two examples are **polyamides** (nylons), formed by reaction of a diacid and a diamine, and **polyesters,** formed by reaction of a diacid and a diol.

24.19 Summary of Reactions

1. **Reaction of 1,3-Dicarbonyl Compounds with Base Forms an Enolate Anion**   SECTION 24.5

2. **Reactions of Enolate Anions Derived from 1,3-Dicarbonyl Compounds**

    a. Alkylation   SECTIONS 24.7 AND 24.8

    Enolate anion is the nucleophile in a nucleophilic substitution reaction that occurs by an S_N2 mechanism.

b. Knoevenagel Condensation Reaction   SECTION 24.9

Reaction of an aldehyde or ketone and a 1,3-dicarbonyl compound forms an α,β-unsaturated carbonyl-containing compound.

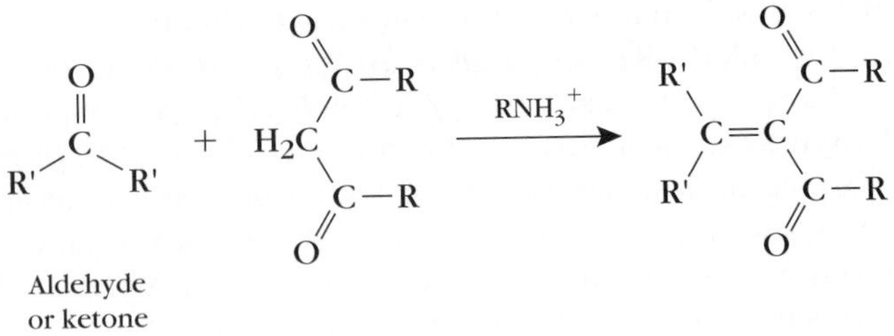

Aldehyde
or ketone

c. Michael Addition Reaction   SECTION 24.13

1,4-Addition of an enolate anion to an α,β-unsaturated carbonyl compound.

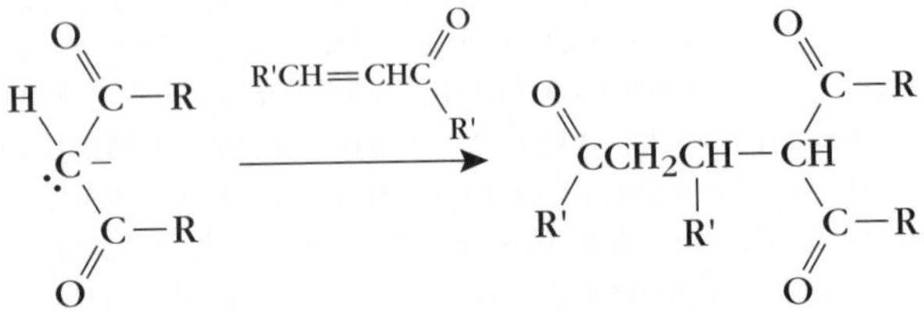

3. Nucleophilic Addition Reactions of α,β-Unsaturated Carbonyl Compounds   SECTION 24.12

Weak basic nucleophiles form mostly products of 1,4-addition. Strongly basic nucleophiles form mostly products of 1,2-addition.

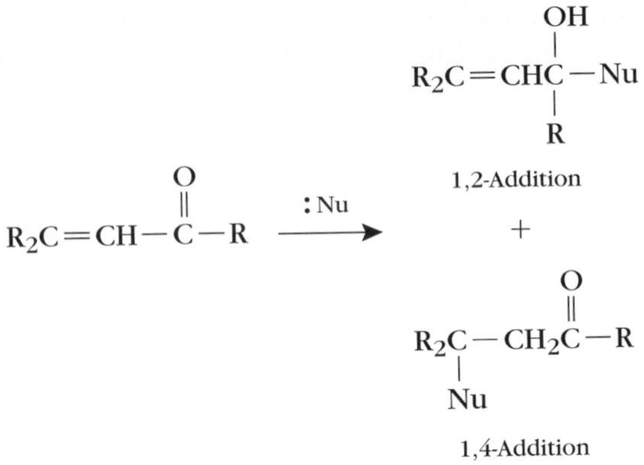

4. Ring-Forming Reactions   SECTION 24.14

X and Y are any two functional groups that can react to form a single bond. Reaction to form a three-, five- or six-membered ring ($n = 1, 3,$ or 4) is favored over any other size ring.

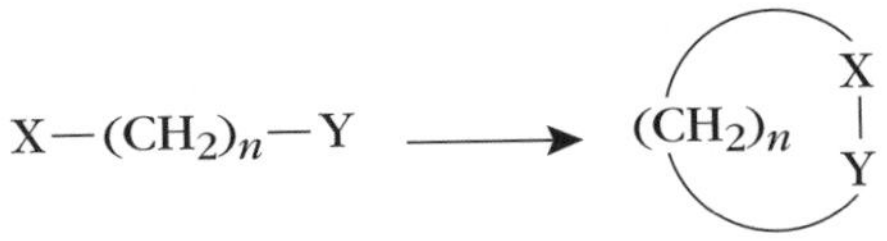

$$X-(CH_2)_n-Y \longrightarrow (CH_2)_n \begin{smallmatrix} X \\ | \\ Y \end{smallmatrix}$$

5. Ring-Forming Condensation Reactions

a. Intramolecular aldol, Claisen, and mixed condensation reactions   SECTION 24.15, A

Suitably located carbonyl group and nucleophilic carbon atom in a molecule react to form a product containing a five- or six-membered ring.

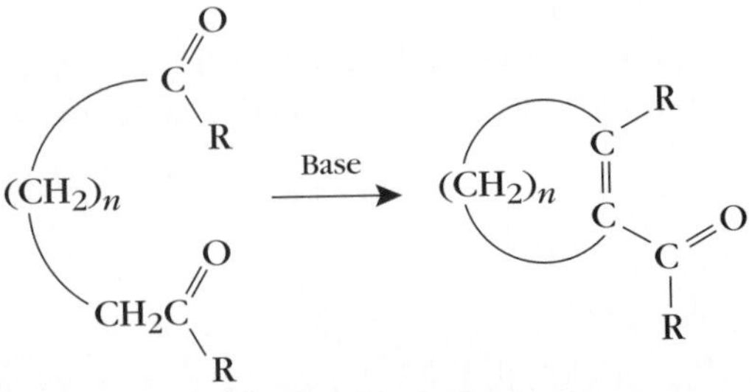

b. Robinson Annulation Reaction   SECTION 24.15, B

A sequence of a Michael addition of an enolate anion to an α,β-unsaturated carbonyl compound followed by an intramolecular aldol condensation reaction.

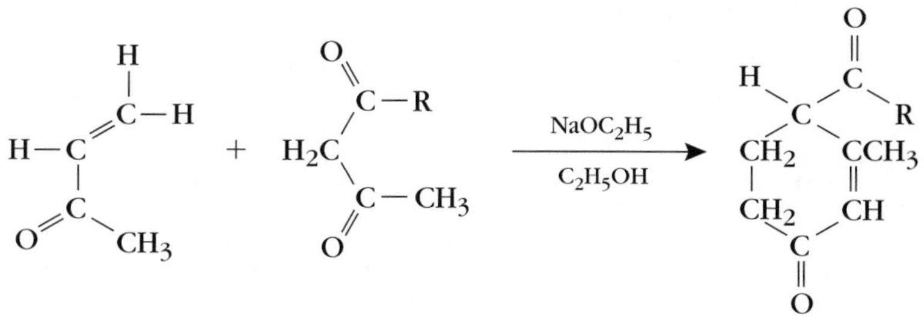

Additional Exercises

24.23 Define each of the following terms in words, by a structure, or by a chemical equation.

(a) 1,3-Dicarbonyl compound
(b) Michael addition reaction
(c) α,β-Unsaturated carbonyl compound
(d) Michael acceptor
(e) Condensation polymer
(f) Michael donor

(g) Robinson annulation reaction
(h) Knoevenagel reaction
(i) Intramolecular reaction
(j) Malonic ester synthesis
(k) Acetoacetic ester synthesis
(l) Intermolecular reaction

24.24 Name each of the following compounds.

(a)

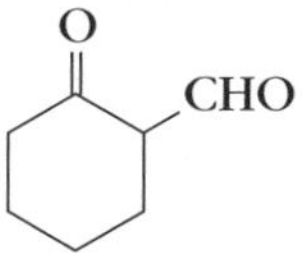

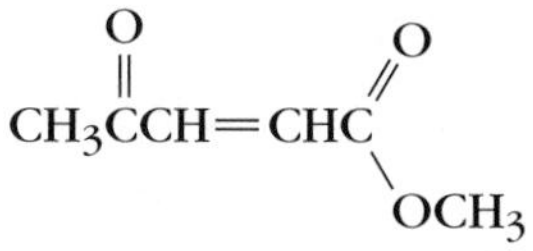

(b)

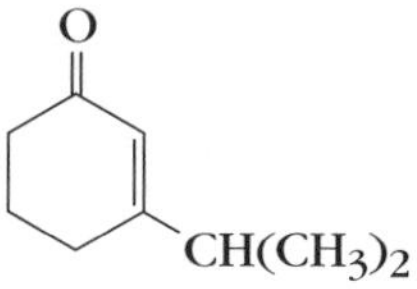

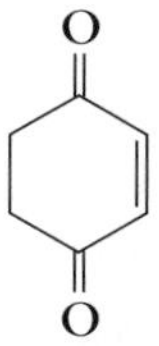

(c)

(d) $CH_3CCH\!=\!CHC$

(e)

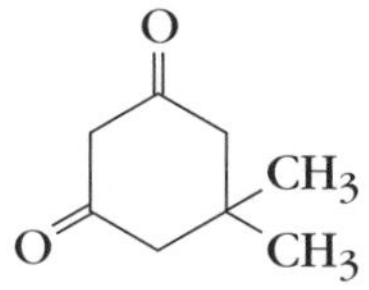

(f)

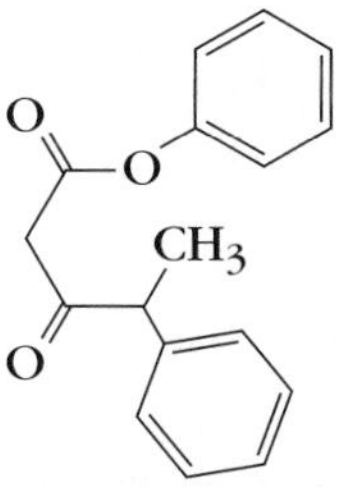

24.25 Write the structural formula for each of the following compounds.

(a) 4-Methyl-3-penten-2-one
(b) 2,2-Dimethylcyclohexanedione
(c) 4-Oxo-2,4-diphenylbutanenitrile

(d) Ethyl (2-oxocyclopentane)carboxylate
(e) Ethyl 4-ethyl-2-methyl-3-oxooctanoate

24.26 Write the structure of the product of each of the following reactions.

(a)

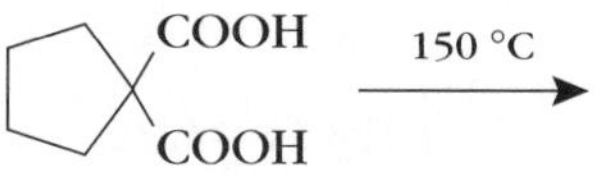

(b) Diethyl malonate $\xrightarrow[\substack{(2)\ BrCH_2(CH_2)_6CH_3 \\ (3)\ H_3O^+/heat}]{(1)\ NaOC_2H_5/C_2H_5OH}$

(c) $(CH_3)_2C{=}CHCCH_3$ (with C=O) $\xrightarrow[(2)\ H_3O^+]{(1)\ CH_3MgI}$

(d) $HO(CH_2)_4CHO \xrightarrow[H_3O^+]{}$

(e) $(CH_3)_2C{=}CHCHO \xrightarrow{NaCN/CH_3OH}$

(f) (acetophenone) + (benzaldehyde) $\xrightarrow[(2)\ H_3O^+/heat]{(1)\ NaOC_2H_5/C_2H_5OH}$

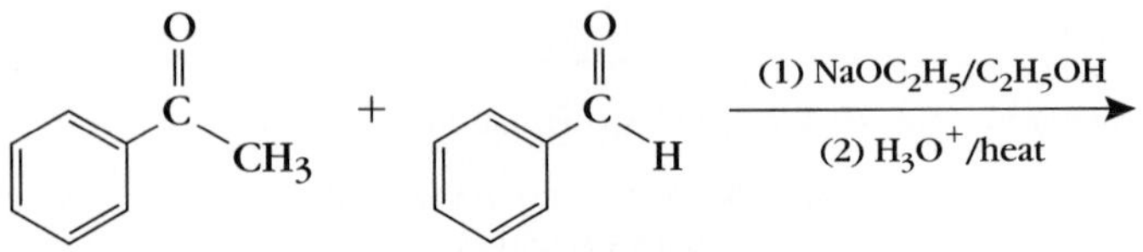

(g) 2,5-Hexanedione $\xrightarrow[(2)\ H_3O^+/heat]{(1)\ NaOC_2H_5/C_2H_5OH}$

(h) $HOCH_2(CH_2)_2CH_2Br \xrightarrow[Hexane]{NaH}$

(i) $CH_3\overset{O}{\underset{SCoA}{C}}$ + $CO_2 \xrightarrow{\substack{Acetyl\ CoA \\ carboxylase}}$

(j) $(CH_3)_2C{=}CHCCH_3$ (with C=O) $\xrightarrow{CH_3NH_2}$

(k) (C₆H₅)CH=CHCHO + $H_2C\overset{\overset{O}{\|}{C-CH_3}}{\underset{\underset{O}{\|}{C-OC_2H_5}}{}}$ $\xrightarrow{Piperidine}$

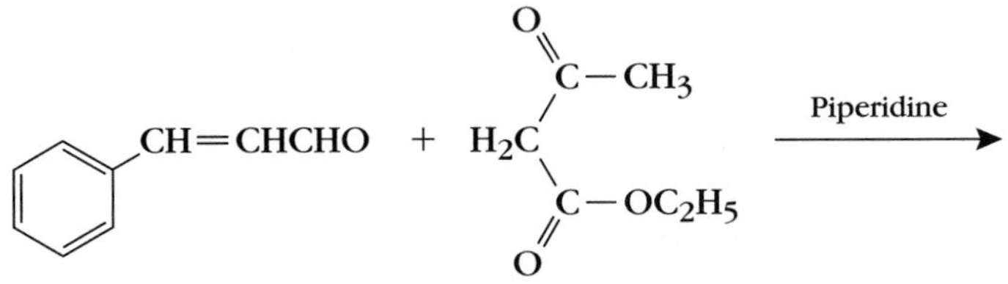

(l) Ethyl acetoacetate $\xrightarrow[\substack{(2)\ 3\text{-Bromocyclohexene} \\ (3)\ H_3O^+/heat}]{(1)\ NaOC_2H_5/C_2H_5OH}$

24.27 Supply the missing reagents A through E for each of the following reactions.

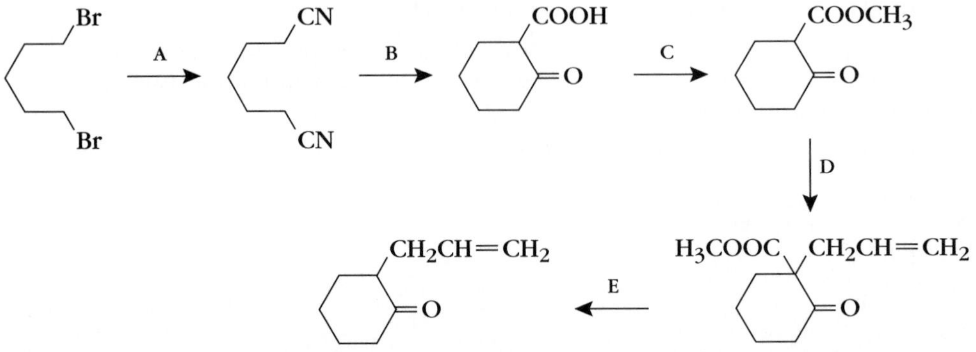

24.28 Starting with diethyl malonate and any other organic or inorganic compounds, propose a synthesis for each of the following compounds.

(a) 3-Methylbutanoic acid

(c) 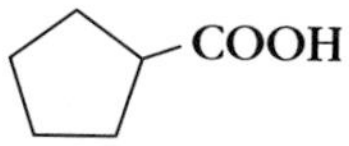 —COOH

(b) 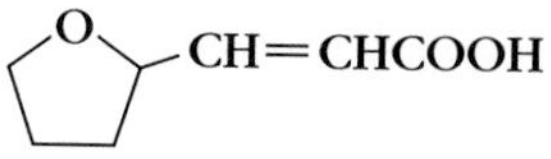 —CH=CHCOOH

(d)

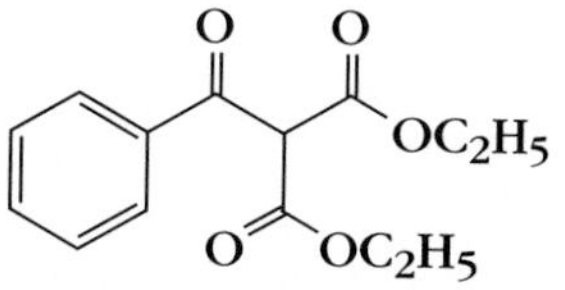

24.29 Starting with ethyl acetoacetate and any other organic or inorganic compounds, propose a synthesis for each of the following compounds.

(a) 3-Methyl-2-pentanone

(c) 2,4-Hexanedione

(b) 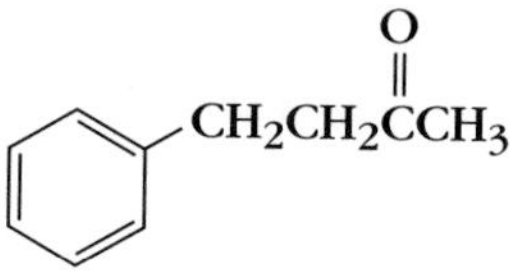 $CH_2CH_2\overset{O}{\overset{||}{C}}CH_3$

(d) 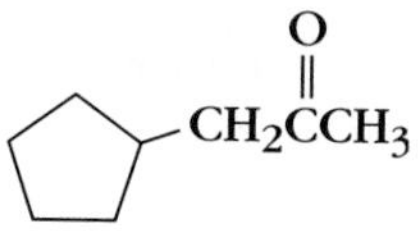 $CH_2\overset{O}{\overset{||}{C}}CH_3$

24.30 Write a mechanism for the following cyclization reaction.

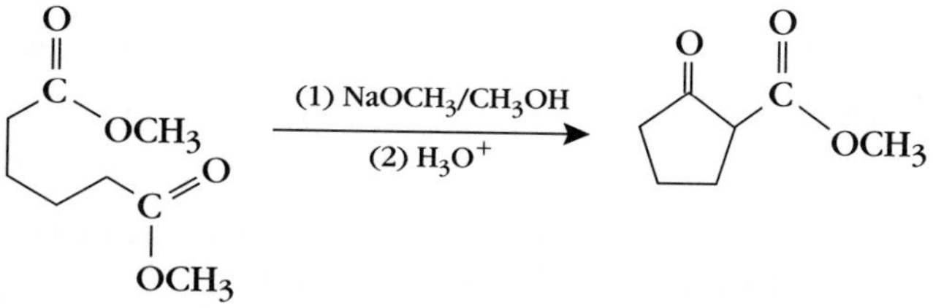

24.31 Starting with either diethyl malonate or ethyl acetoacetate and any other organic or inorganic compound, propose a synthesis for each of the following compounds.

(a) 5-Oxohexanoic acid

(c) 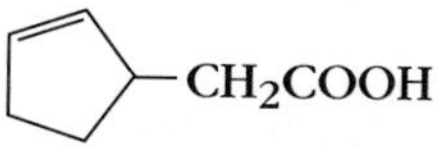 —CH$_2$COOH

(b) $CH_3CH_2CH_2CH=CH\overset{O}{\overset{||}{C}}CH_3$

(d) Ethyl 2-methylbutanoate

24.32 Write the structure of the product in each of the following reactions.

(a) Benzaldehyde + Diethyl malonate $\xrightarrow[\text{(2) } H_3O^+/\text{heat}]{\substack{\text{(1) NaOC}_2H_5 \\ C_2H_5OH}}$

(b) $HOCH_2CH_2CH_2COOH \xrightarrow{H_3O^+}$

(c) 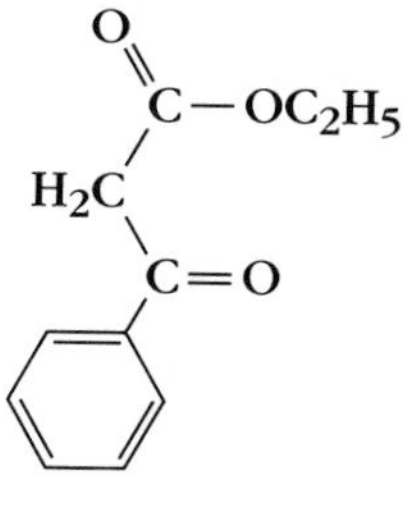 $\xrightarrow[\substack{\text{(2) 4-Methyl-1-bromohexane} \\ \text{(3) } H_3O^+/\text{heat}}]{\substack{\text{(1) NaOC}_2H_5 \\ C_2H_5OH}}$

(d)   Ethyl benzoate   +   ethyl acetate $\xrightarrow[\text{(2) } H_3O^+]{\text{(1) } NaOC_2H_5/C_2H_5OH}$

(e)   Diethyl pentanedioate $\xrightarrow[\text{Benzene}]{\text{Na}}$

(f)   Diethyl malonate $\xrightarrow[\substack{\text{(2) Benzoyl chloride}\\ \text{(3) } H_3O^+/\text{heat}}]{\text{(1) } NaOC_2H_5/C_2H_5OH}$

24.33 The barbiturate phenobarbital can be prepared by the following sequence of three reactions. Propose a mechanism for each of these three reactions.

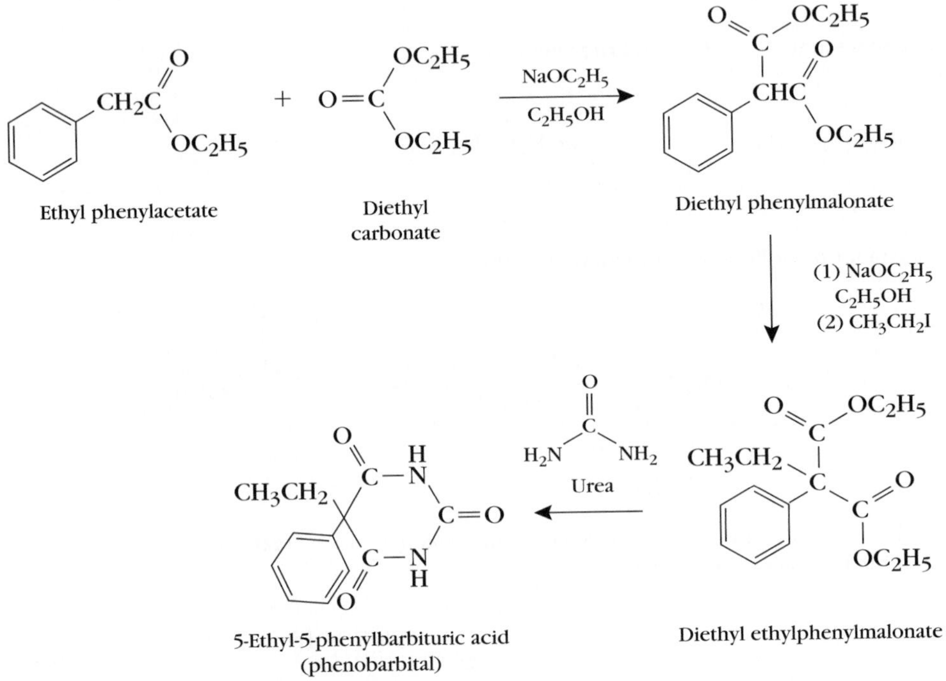

Ethyl phenylacetate   Diethyl carbonate   Diethyl phenylmalonate

(1) $NaOC_2H_5$ C_2H_5OH
(2) CH_3CH_2I

5-Ethyl-5-phenylbarbituric acid (phenobarbital)   Diethyl ethylphenylmalonate

24.34 Write the structure (including the stereochemistry) of compounds A to C.

(R)-2-Butanol $\xrightarrow[\text{Pyridine}]{CH_3C_6H_4SO_2Cl}$ A $\xrightarrow[C_2H_5OH]{Na^{+-}CH(COOC_2H_5)_2}$ B $\xrightarrow{H_3O^+}$ C

24.35 Show how to carry out each of the following conversions. More than one reaction may be necessary.

(a) $CH_3\overset{\displaystyle O}{\overset{\|}{C}}CH_2CH_2COOH$ $\longrightarrow$

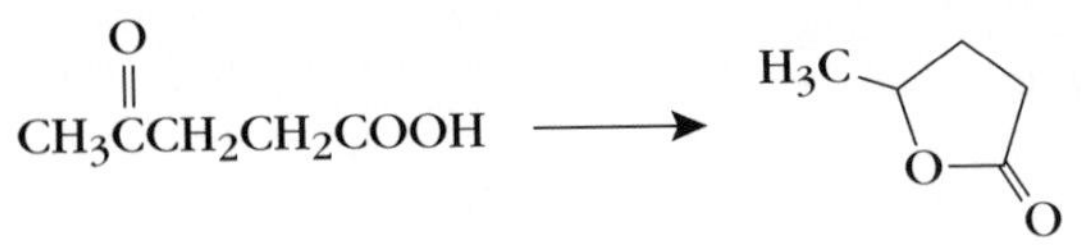

(b)

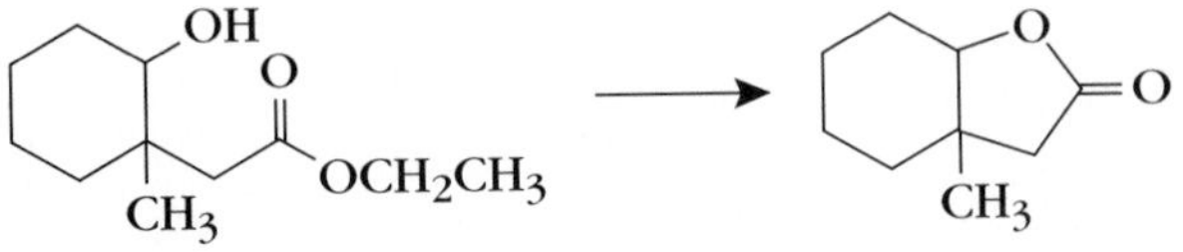

(c)

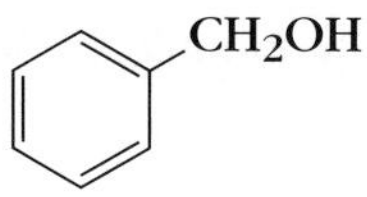

⟶ 3-Methyl-4-phenyl-2-butanone

(d)

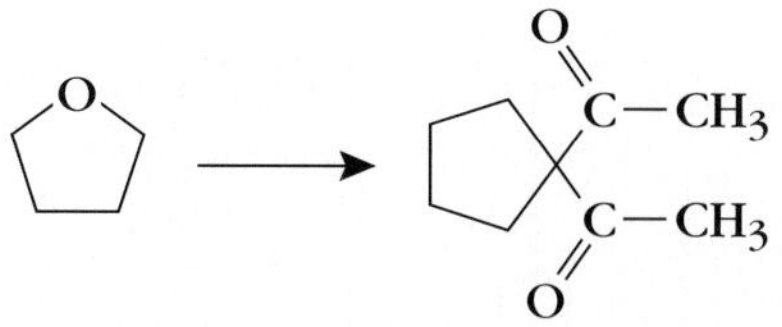

(e)

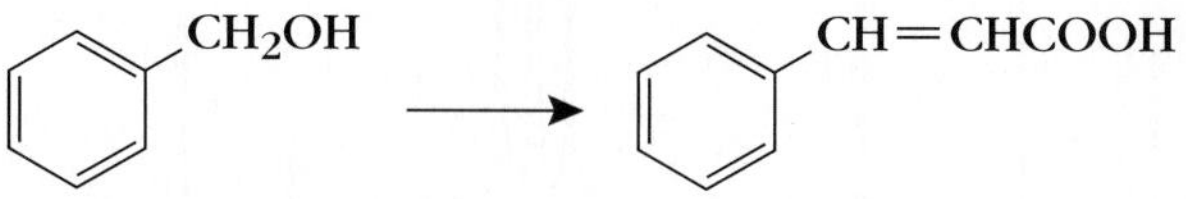

24.36 Propose a mechanism for the following reaction.

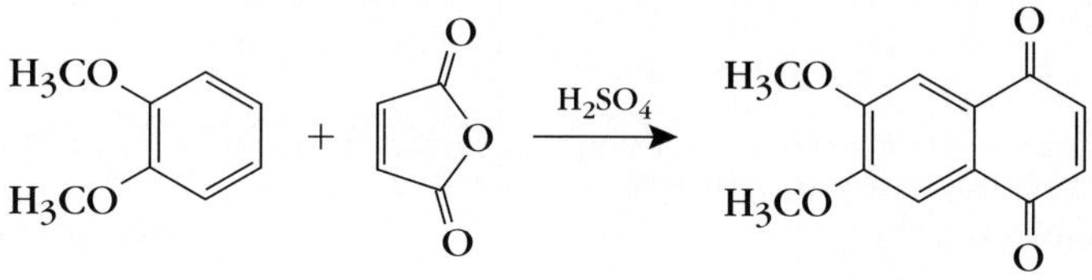

24.37 Using the Robinson annulation reaction and starting with the indicated Michael donor, propose a synthesis for each of the following compounds.

(a)

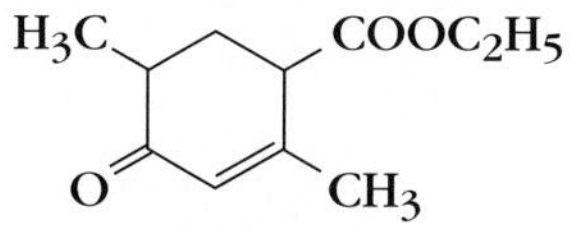

Using ethyl
acetoacetate as the
Michael donor

(b)

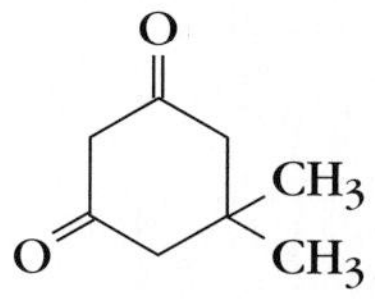

Using diethyl
malonate as the
Michael donor

(c)

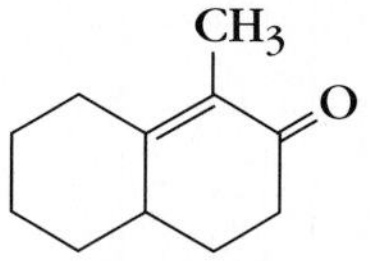

Using cyclo-
hexanone
as the Michael
donor

24.38 HCl adds slowly to the double bond of 3-buten-2-one. Write the structure of the product and propose a mechanism for the reaction.

24.39 Propose a mechanism for the following reaction.

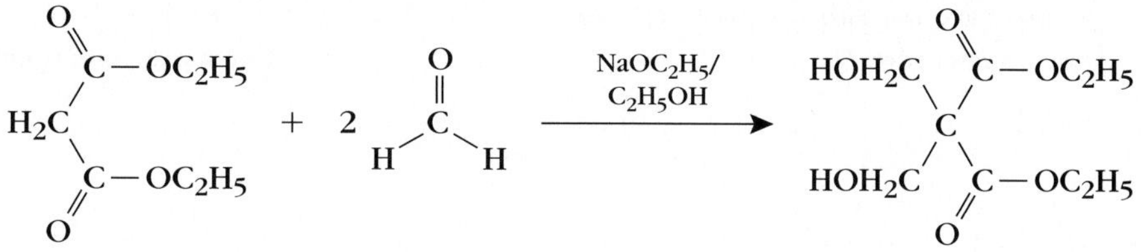

24.40 Write the structural formula of each of the following compounds based on their molecular formula, IR, and ^{1}H NMR spectra.

(a) $C_7H_{12}O_4$

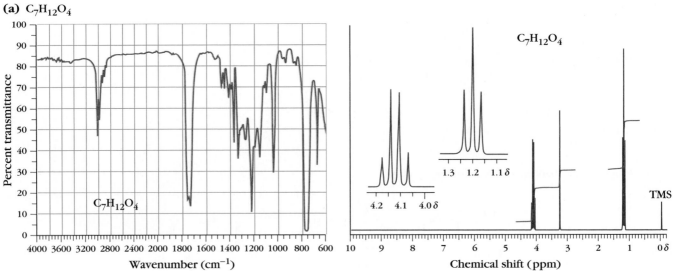

(b) $C_8H_{14}O$

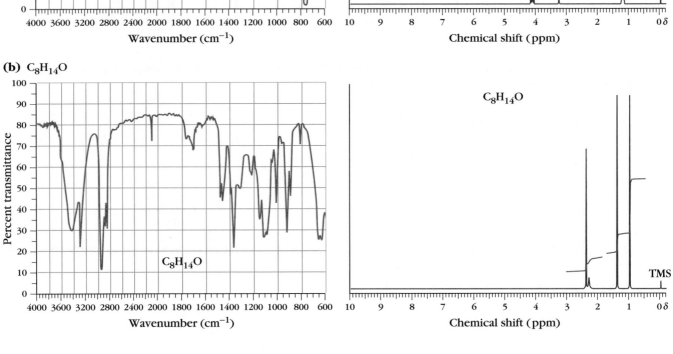

(c) $C_6H_{10}O_3$

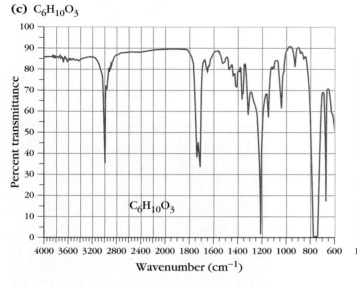
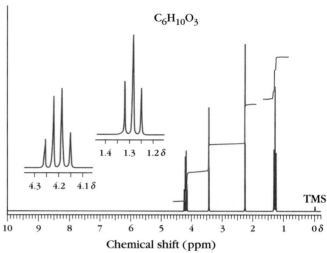

(d) $C_{11}H_{12}O_2$

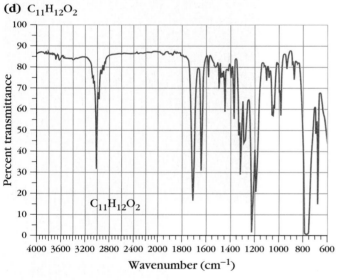
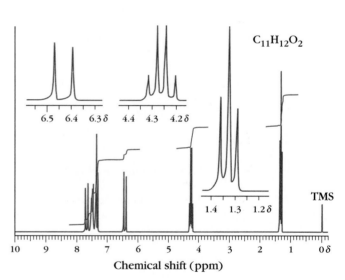

(e) $C_{11}H_{20}O_4$

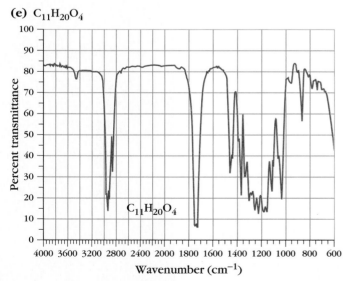
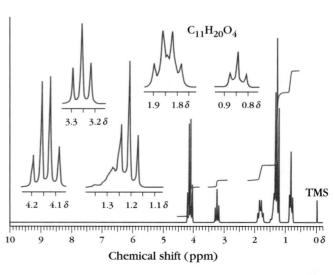

CARBOHYDRATES

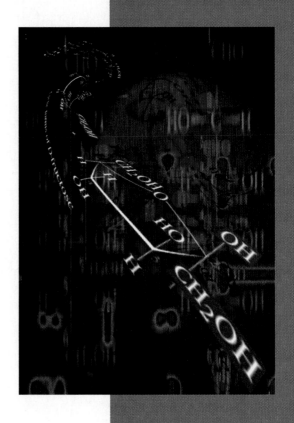

People consume carbohydrates every day, either in the form of sugar or as one of the principal nutrients in their food. Carbohydrates are essential to life because they provide the energy needed for humans and other animals to function. Moreover, carbohydrates are the structural components of the walls of plant cells and the wood of trees, and some carbohydrate derivatives form part of the coating around cell membranes.

25.1   Classification

The first sugars that were purified and analyzed had molecular formulas that correspond to $C_n(H_2O)_n$, so they were considered to be hydrates of carbon or carbohydrates. While this view was soon discarded, the name carbohydrates remains. Today carbohydrates are defined as polyhydroxyaldehydes and polyhydroxyketones, or substances that yield such compounds on acid hydrolysis.

Carbohydrates are classified on the basis of their acid-catalyzed hydrolysis products. **Monosaccharides,** or simple sugars, are carbohydrates that cannot be hydrolyzed into smaller compounds. The simplest monosaccharides are glyceraldehyde (2,3-dihydroxypropanal) and dihydroxyacetone (1,3-dihydroxypropanone).

$$\underset{\substack{\text{Glyceraldehyde}\\ \text{(2,3-dihydroxypropanal)}}}{\overset{\overset{\displaystyle OH}{|}}{HOCH_2CHCHO}} \qquad \underset{\substack{\text{Dihydroxyacetone}\\ \text{(1,3-dihydroxypropanone)}}}{\overset{\overset{\displaystyle O}{\parallel}}{HOCH_2CCH_2OH}}$$

A **disaccharide** can be hydrolyzed into two monosaccharides, which may be the same or different. Sucrose (table sugar) is a disaccharide that forms one molecule of glucose and one molecule of fructose on hydrolysis.

$$\underset{\text{Disaccharide}}{\text{Sucrose}} \xrightarrow{H_3O^+} \underset{\text{Monosaccharide}}{\text{Glucose}} + \underset{\text{Monosaccharide}}{\text{Fructose}}$$

Oligosaccharide is a general term given to carbohydrates that yield a few (3 to 10) monosaccharides on hydrolysis. Finally, hydrolysis of **polysaccharides** yields several thousand monosaccharides. Cellulose is a polysaccharide that forms several thousand molecules of glucose when completely hydrolyzed.

$$\underset{\text{Polysaccharide}}{\text{Cellulose}} \xrightarrow{H_3O^+} \text{Several thousand glucose molecules}$$

The suffix *-ose* is used to designate a carbohydrate. Monosaccharides are further classified according to the number of carbon atoms they contain by adding the prefixes *tri-, tetr-, pent-, hex-,* and so forth. For example, a four-carbon monosaccharide is called a tetrose, a five-carbon monosaccharide is called a pentose, and so forth.

The kind of carbonyl group in a monosaccharide can be specified by adding the prefix *keto-* for a ketone and *aldo-* for an aldehyde Thus an aldohexose is a six-carbon monosaccharide that contains an aldehyde group and a ketopentose is a five-carbon monosaccharide that contains a ketone group. The classification of monosaccharides is summarized in Table 25.1.

Table 25.1	Classification of Monosaccharides by the Number of Carbon Atoms and Type of Carbonyl Group	
Number of Carbon Atoms	Aldehyde-Containing Monosaccharides	Ketone-containing Monosaccharides
3	Aldotriose	Ketotriose
4	Aldotetrose	Ketotetrose
5	Aldopentose	Ketopentose
6	Aldohexose	Ketohexose

One of the most convenient ways of writing the structure of monosaccharides is by means of Fischer projections.

25.2 Fischer Projection Formulas of Monosaccharides

Fischer projection formulas (Section 6.15) were designed in the early part of the 20th century as a standard way of drawing the structures of carbohydrates. They are particularly useful for representing the configuration of the many stereocenters in monosaccharides.

Recall that a carbon atom is located where the lines cross in a Fischer projection. By convention, vertical lines represent bonds that are directed below the page, while horizontal lines represent bonds that are directed above the page. The Fischer projection of one enantiomer of the simplest monosaccharide, (R)-glyceraldehyde (2,3-dihydroxypropanal), is shown as follows:

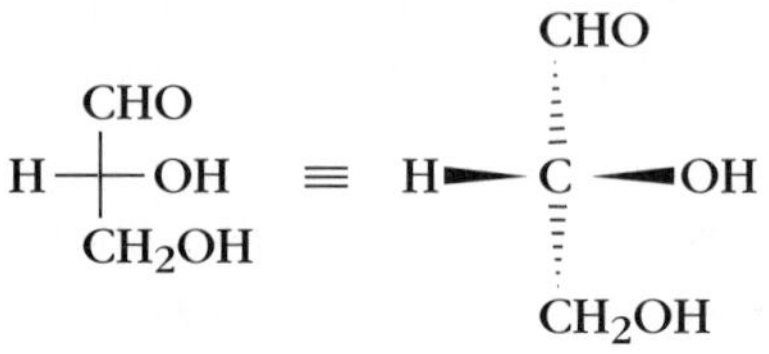

Fischer projection
of (R)-glyceraldehyde

Fischer projections of monosaccharides with more than one stereocenter are obtained by simply placing the stereocenters one above the other. Again, the intersection of vertical and horizontal lines represents a carbon stereocenter. By convention, the aldehyde group of aldoses is always placed at the top and the CH_2OH group at the bottom. The relationship between the Fischer projection

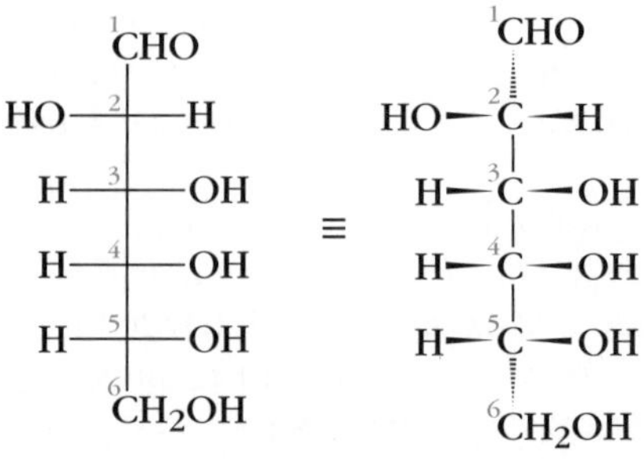

Figure 25.1

The relationship between the Fischer projection of an aldohexose and its three-dimensional representation.

Exercise 25.1	Write the Fischer projection formulas for each of the following three-dimensional representations.

(a) 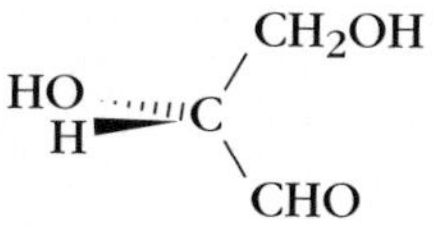
$$\overset{\displaystyle CH_2OH}{\underset{\displaystyle CHO}{\overset{HO}{\underset{H}{\rule{0pt}{1em}}}\!C}}$$

(c)
$$\underset{\displaystyle H}{\overset{\displaystyle HO}{C}}\!\!\begin{array}{l}CH_2OH\\CHO\end{array}$$

(b) 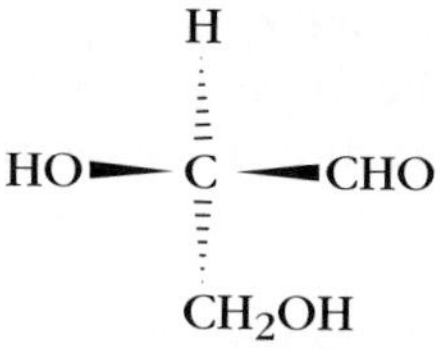 HO ━ C ━ CHO
    with H on top, CH$_2$OH on bottom

(d) 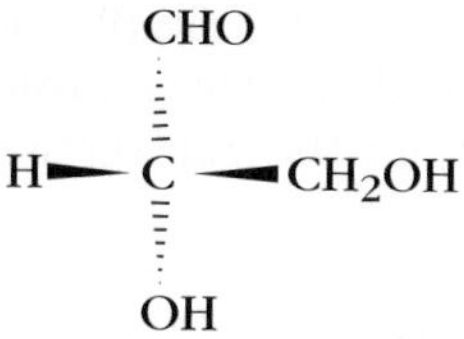 H ━ C ━ CH$_2$OH
    with CHO on top, OH on bottom

formula of an aldohexose and its three-dimensional representation is shown in Figure 25.1.

Carbohydrates undergo the usual reactions of aldehydes, ketones, and alcohols.

25.3   Reactions of Monosaccharides

Monosaccharides undergo the usual reactions of aldehydes and ketones. Since monosaccharides are polyfunctional compounds, some of their reactions are more complicated than those of simple aldehydes and ketones. In this section, we review the typical aldehyde and ketone reactions of monosaccharides, then discuss more complicated reactions that are characteristic of monosaccharides.

A   Epimerization

Hydrogen atoms α- to the carbonyl group of a monosaccharide, like those in simple aldehydes and ketones, are relatively acidic. As a result, monosaccharides form enolate anions in basic solution. Because the carbon atom next to the carbonyl group in most monosaccharides is a stereocenter, its configuration is lost by forming an enolate anion. Thus reprotonation of this enolate anion can occur from either face to form a mixture of the original monosaccharide and its diastereomer. The reaction of aldotetrose **25.1** with base, for example, forms a mixture of **25.1** and its diastereomer **25.2.**

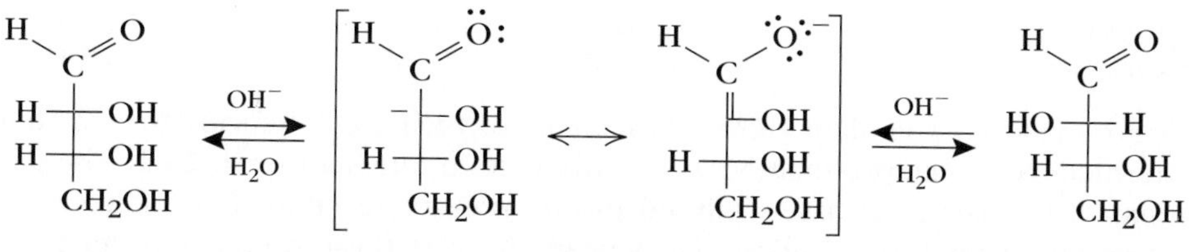

25.1   Enolate anion   **25.2**

Diastereomers such as **25.1** and **25.2**, which differ in the configuration of *only one* of their stereocenters, are called **epimers.** This conversion of pure **25.1** (or pure **25.2**) to a mixture of **25.1** and **25.2** is called **epimerization.** To avoid unwanted epimerization, reactions of monosaccharides are carried out in neutral or acidic solution.

Enolate anion formation is the first step in a more complicated series of reactions that converts an aldose into a ketose.

B   Enediol Rearrangement

An enolate anion formed from a simple aldehyde or ketone can be protonated either on its oxygen or carbon atoms. Protonation on the oxygen atom forms an enol, while protonation on the carbon atom forms a carbonyl-containing compound.

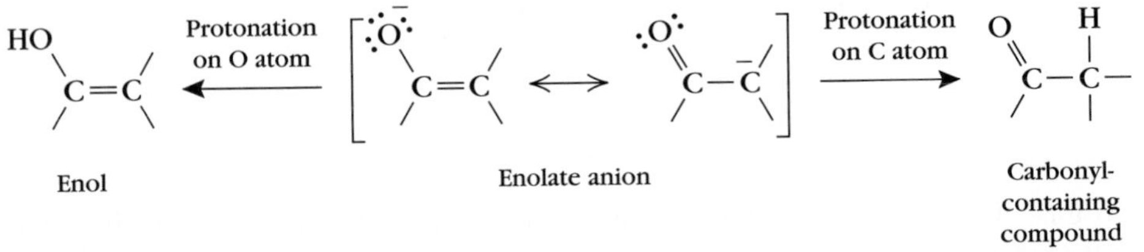

Similarly, the very small amount of enolate anion formed from an aldose can be protonated on either its carbon or oxygen atoms. Protonation at the carbon atom results in epimerization, while protonation on the oxygen atom forms an **enediol intermediate.**

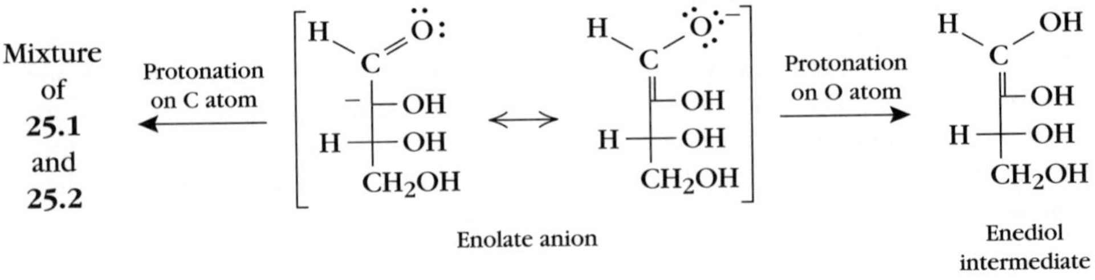

In a basic solution, loss of a proton from the hydroxy group on carbon atom 2 of the enediol intermediate forms an enolate anion **25.3**, which is an isomer of the one originally formed from the aldotetrose **25.1.**

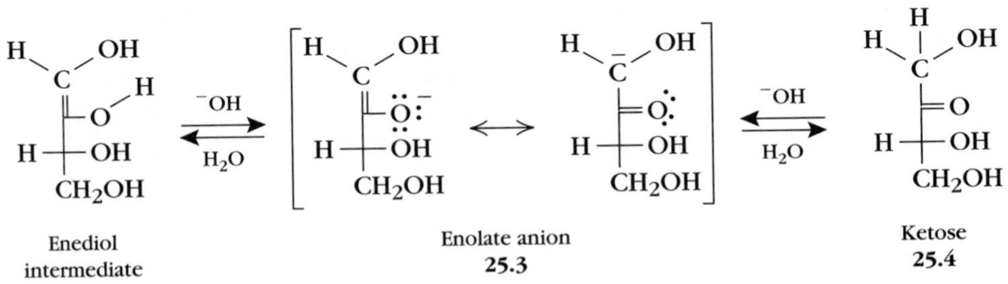

Protonation of enolate anion **25.3** on its oxygen atom reforms the enediol intermediate, while protonation at its carbon atom forms a ketose **25.4.** The net result of this series of base-catalyzed proton transfer reactions is to convert an aldose into a mixture of aldose and ketose. Because these proton transfer reactions are all reversible, the same mixture of aldose and ketose can be obtained by treating the ketotetrose **25.4** with base.

Base-catalyzed interconversion of aldoses and ketoses occurs with all monosaccharides and is summarized in Figure 25.2. These isomerizations can be

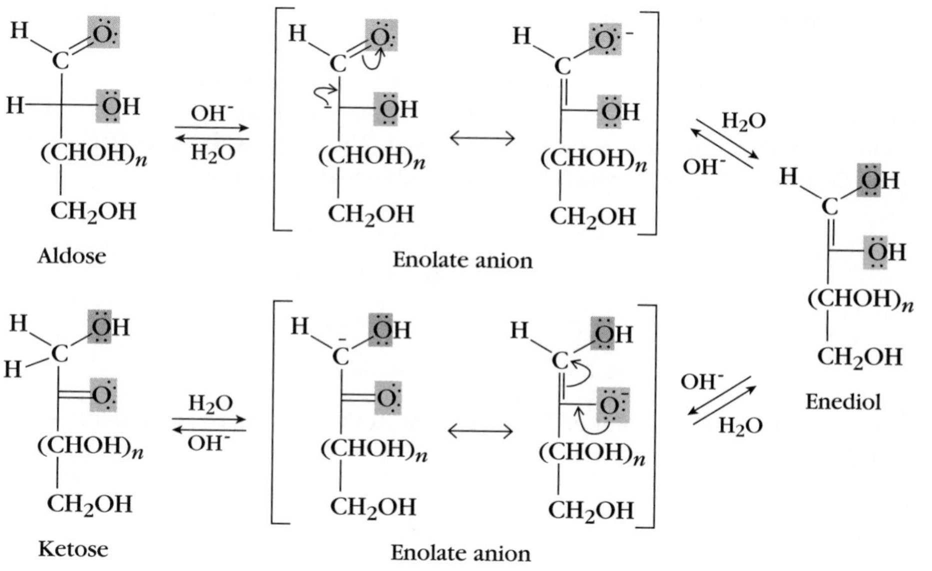

Figure 25.2
Base-catalyzed interconversion of aldoses and ketoses.

avoided by carrying out the reactions of monosaccharides in neutral or acidic solution.

C Oxidation

Aldoses, like simple aldehydes, are oxidized by Fehling's solution and Tollen's reagent. Sugars that reduce Fehling's solution and Tollen's reagent are called **reducing sugars.** Aldoses are also oxidized by bromine water and nitric acid, which are two oxidizing agents widely used in carbohydrate chemistry.

Fehling's solution and Tollen's reagent are often used to distinguish between simple aldehydes and ketones. In the case of monosaccharides, however, these two reagents give a positive test with both aldoses and ketoses. Positive tests are obtained with both reagents because they are carried out in alkaline solutions. In alkaline solution, an enediol rearrangement converts ketoses to aldoses, which are then oxidized. Thus oxidation by Fehling's solution or Tollen's reagent does not distinguish between aldoses and ketoses. As well, oxidation by Fehling's solution or Tollen's reagent is not a good method of preparing carboxylic acids from aldoses. The basic reaction mixture forms both aldoses and ketoses, which on oxidation gives a complex mixture of products.

A better method of preparing carboxylic acids from aldoses is with bromine water. Bromine water is an acidic reagent that does not isomerize the starting monosaccharide. Furthermore, it does not oxidize the hydroxy groups of either aldoses or ketoses. As a result, bromine water can be used to distinguish between aldoses and ketoses. The carboxylic acids produced by bromine water oxidation of aldoses are called **aldonic acids.**

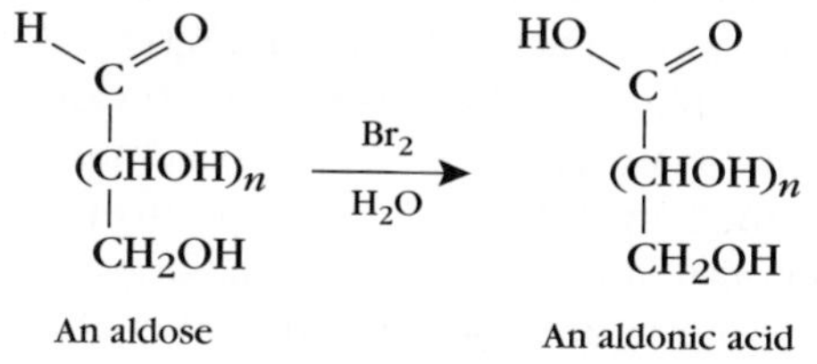

> ■ *Oxidation of aldehydes by Tollen's reagent and Fehling's solution is discussed in Section 14.18.*

Specific aldonic acids are named by adding the suffix -*onic* to the root name of the monosaccharide.

Nitric acid is a stronger oxidizing agent than bromine water or Fehling's solution and Tollen's reagent. It oxidizes both the aldehyde group and the CH_2OH group of an aldose to form a dicarboxylic acid. The name **aldaric acids** is given to the dicarboxylic acids formed by nitric acid oxidation of aldoses.

$$
\begin{array}{c}
CHO \\
| \\
(CHOH)_n \\
| \\
CH_2OH
\end{array}
\xrightarrow{HNO_3}
\begin{array}{c}
COOH \\
| \\
(CHOH)_n \\
| \\
COOH
\end{array}
$$

An aldose   An aldaric acid

Specific aldaric acids are named by adding the suffix -*aric* to the root name of the monosaccharide.

D Reduction

Like other aldehydes and ketones, aldoses and ketoses can be reduced to the corresponding alcohols, which are called **alditols.** The most commonly used reagents are sodium borohydride and catalytic hydrogenation on a nickel catalyst.

$$
\begin{array}{c}
CHO \\
| \\
(CHOH)_n \\
| \\
CH_2OH
\end{array}
\xrightarrow{NaBH_4}
\begin{array}{c}
CH_2OH \\
| \\
(CHOH)_n \\
| \\
CH_2OH
\end{array}
$$

An aldose   An alditol

Reduction of ketoses (except dihydroxyacetone) generates a new stereocenter, thus creating two possible configurations. Reduction can occur at either face of the carbonyl group, so two epimers are formed.

$$
\begin{array}{c}
CH_2OH \\
| \\
C=O \\
| \\
(CHOH)_n \\
| \\
CH_2OH
\end{array}
\xrightarrow{NaBH_4}
\begin{array}{c}
CH_2OH \\
H-\!\!-OH \\
(CHOH)_n \\
CH_2OH
\end{array}
+
\begin{array}{c}
CH_2OH \\
HO-\!\!-H \\
(CHOH)_n \\
CH_2OH
\end{array}
$$

A ketose

E The Kiliani-Fischer Synthesis: Lengthening the Carbon Chain of Aldoses

The Kiliani-Fischer synthesis is a sequence of reactions that lengthens an aldose chain by one carbon atom. An aldopentose, for example, is converted to an aldohexose by this method.

The German chemist Heinrich Kiliani reported in 1886 that aldoses react with HCN to form cyanohydrins. Furthermore, he reported that hydrolysis of a cyanohydrin forms an aldonic acid that has one more carbon atom than the starting aldose. Several years later, another German chemist, Emil Fischer, reported that aldonic acids formed by hydrolysis of cyanohydrins form aldonolactones when heated. Fischer also found that aldonolactones are reduced by sodium amalgam at pH 3.0 to 3.5 to a new aldose. This series of reactions, which is now called the **Kiliani-Fischer synthesis,** is outlined in Figure 25.3.

■ *The reaction of HCN and aldehydes to form cyanohydrins is discussed in Section 14.9.*

■ *The factors that affect ring closure are discussed in Section 24.14.*

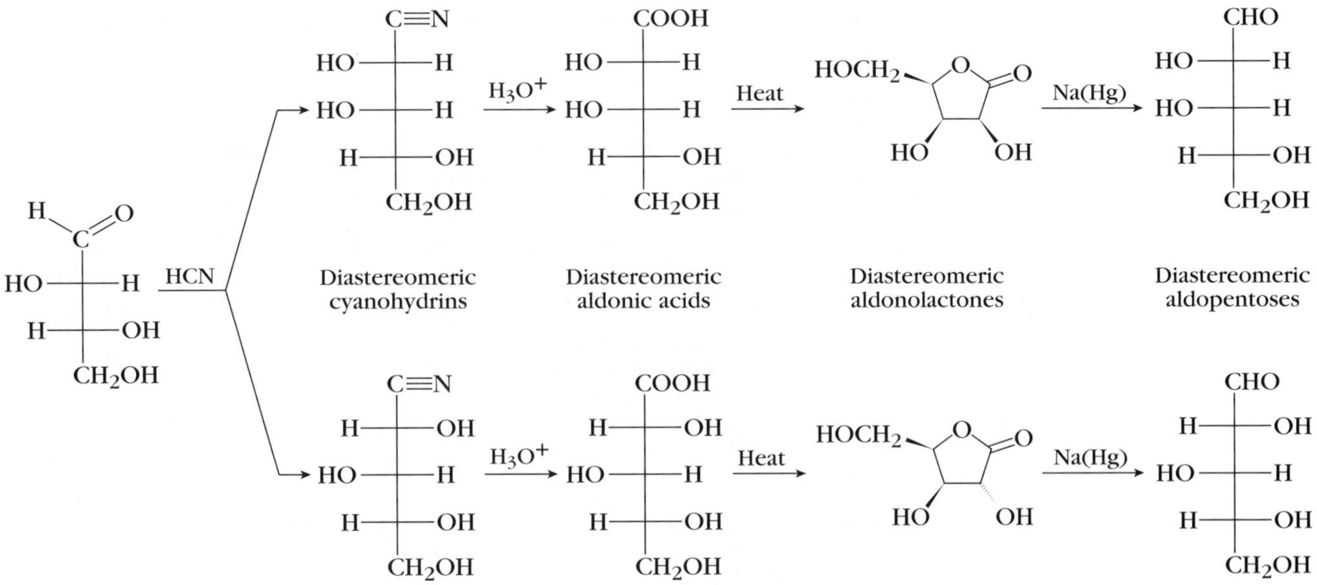

Figure 25.3
The Kiliani-Fischer synthesis converts an aldotetrose into diastereomeric aldopentoses.

Addition of cyanide ion to the carbonyl group of an aldose generates a new stereocenter, about which there are two possible configurations. Consequently, two diastereomeric cyanohydrins are formed, which yield two aldonic acids and finally two diastereomeric aldoses.

Diastereomers formed in the Kiliani-Fischer synthesis, like all diastereomers, differ in physical properties, so they can be separated. They are usually separated at the aldonic acid stage because carboxylic acids form easily purified crystalline salts. The diastereomerically pure aldonic acids can then be converted to diastereomerically pure aldoses.

F The Ruff Degradation for Shortening the Carbon Chain of Aldoses

In 1896, the German chemist Otto Ruff reported that calcium salts of aldonic acids can be oxidized by hydrogen peroxide in the presence of an Fe(III) salt catalyst. The product is an aldose with one fewer carbon atom in its chain than the original aldonic acid.

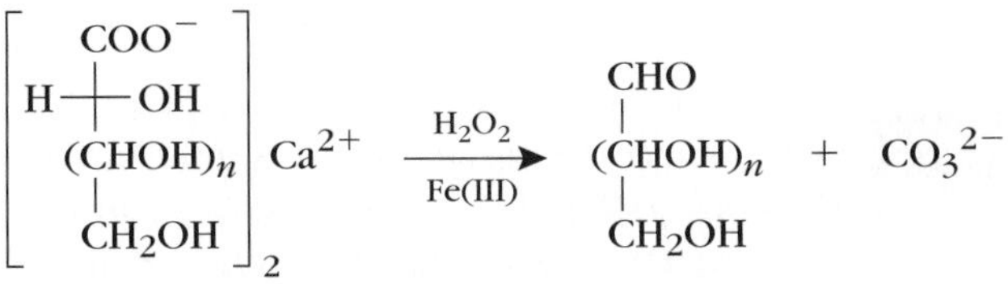

Calcium salt of aldonic acid containing $n + 3$ carbon atoms

Aldose containing $n + 2$ carbon atoms

Aldonic acids are readily formed by the bromine water oxidation of aldoses, so this two-step sequence of reactions, which is called the **Ruff degradation,** is a way of converting an aldose into an aldose containing one fewer carbon atom. Ruff degradation destroys the configuration at carbon atom 2, however, so two aldoses that differ only in configuration at carbon atom 2 form the same aldose with one less carbon atom.

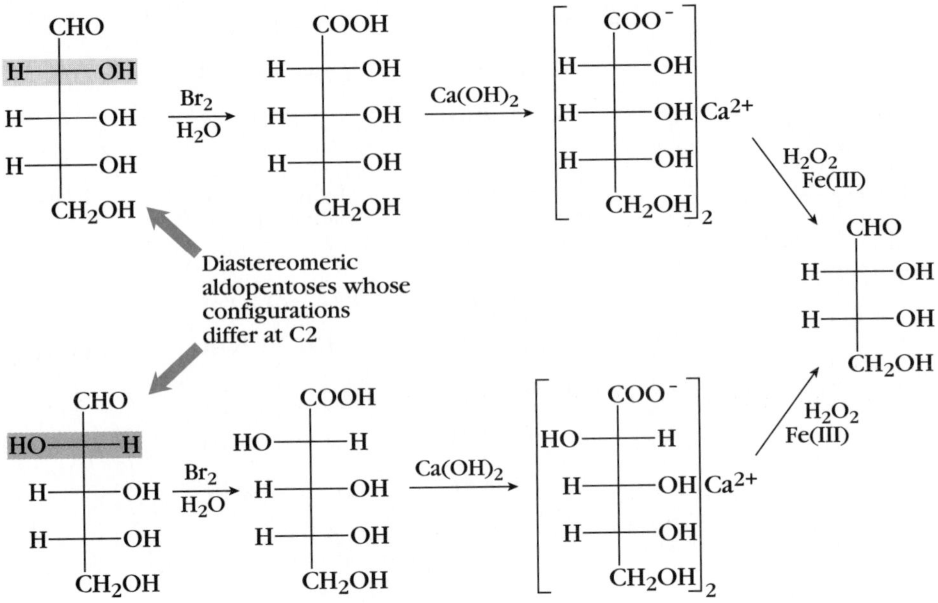

Exercise 25.2

Write the structure of the product of the reaction of (−)-threose with each of the following reagents.

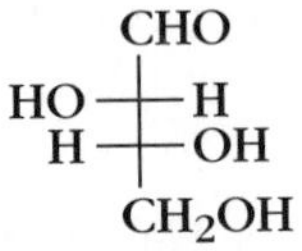

$$
\begin{array}{c}
\text{CHO} \\
\text{HO} \!\!-\!\! \text{H} \\
\text{H} \!\!-\!\! \text{OH} \\
\text{CH}_2\text{OH}
\end{array}
$$

(−)-Threose

(a) Tollen's reagent   **(b)** NaOH/H_2O **(c)** NaBH$_4$ **(d)** Br$_2$/H_2O **(e)** HNO$_3$

Exercise 25.3

Write the equations and the intermediate products in the conversion of (−)-arabinose to a pair of aldohexose diastereomers by the Kiliani-Fischer synthesis.

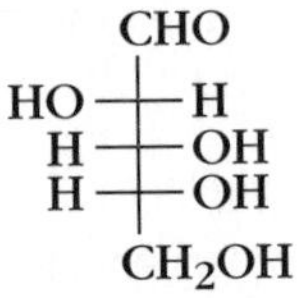

$$
\begin{array}{c}
\text{CHO} \\
\text{HO} \!\!-\!\! \text{H} \\
\text{H} \!\!-\!\! \text{OH} \\
\text{H} \!\!-\!\! \text{OH} \\
\text{CH}_2\text{OH}
\end{array}
$$

(−)-Arabinose

Exercise 25.4	Write the equations and the intermediate products in the conversion of (−)-ribose into (−)-erythrose by the Ruff degradation.

$$
\begin{array}{c}
\text{CHO} \\
\text{H} \!-\! \text{OH} \\
\text{H} \!-\! \text{OH} \\
\text{H} \!-\! \text{OH} \\
\text{CH}_2\text{OH}
\end{array}
\quad\longrightarrow\quad
\begin{array}{c}
\text{CHO} \\
\text{H} \!-\! \text{OH} \\
\text{H} \!-\! \text{OH} \\
\text{CH}_2\text{OH}
\end{array}
$$

(−)-Ribose   (−)-Erythrose

25.4 Relating Configurations of Aldoses

Organic chemists of the late 19th century isolated and purified three carbohydrates, (+)-glucose, (+)-mannose, and (+)-galactose. All three were found to be aldohexoses. The concept of a tetrahedral carbon atom helped explain the existence of these stereoisomers, but a challenging question still remained. An aldohexose has four stereocenters, so it has 16 possible stereoisomers ($2^4 = 16$). Which of the 16 possible stereoisomers of aldohexose corresponds to (+)-glucose, or (+)-mannose, or (+)-galactose?

Organic chemists in the late 19th century did not have X-ray crystallography to determine the absolute configuration of a compound. Nevertheless, they managed to establish the relative configurations of the various stereocenters of many monosaccharides.

The idea behind their method of relating configurations of carbohydrates was to either increase or decrease the length of the carbon chain of a monosaccharide by chemical reactions that do not affect the other stereocenters. We can illustrate this method by relating the configuration of the aldotetroses, (−)-erythrose and (−)-threose, to the aldotriose (+)-glyceraldehyde. The configurations of these three aldoses, all of which are available from natural sources, will be related by preparing the two aldotetroses from (+)-glyceraldehyde by the Kiliani-Fischer synthesis.

Before starting, the configuration of (+)-glyceraldehyde must be established. Because organic chemists had no way of determining its absolute configuration, the Fischer projection formula of the enantiomer in which the —OH group points to the right was arbitrarily given the configuration of (+)-glyceraldehyde. Later results proved that this choice was the correct one.

$$
\begin{array}{c}
\text{H} \diagdown \quad \diagup \text{O} \\
\text{C} \\
\text{H} \!-\! \text{OH} \\
\text{CH}_2\text{OH}
\end{array}
$$

(+)-Glyceraldehyde

The synthesis of two aldotetroses from (+)-glyceraldehyde by the Kiliani-Fischer synthesis is shown in Figure 25.4.

The result of this Kiliani-Fischer synthesis is the creation of a new stereocenter at C2 of the aldotetrose by adding a carbon atom to the carbonyl end of (+)-glyceraldehyde. The stereocenter of (+)-glyceraldehyde is not disturbed by

■ Recall from Section 6.9 that a compound with n stereocenters has a maximum of 2^n stereoisomers.

Figure 25.4
Using the Kiliani-Fischer synthesis to establish the relative configurations at carbon 3 of two aldotetroses to that of (+)-glyceraldehyde.

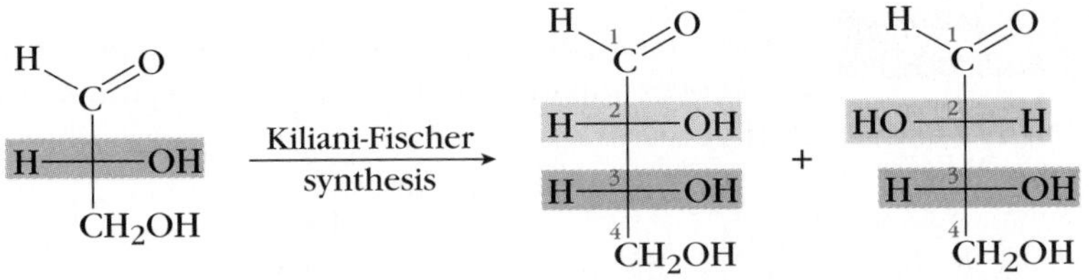

(+)-Glyceraldehyde   (−)-Erythrose   (−)-Threose

Both aldotetroses have the same configuration at carbon 3 as (+)-glyceraldehyde

the reactions that convert it into a pair of aldotetroses. As a result, the configuration about C3 in both aldotetroses is the same as that in (+)-glyceraldehyde. The effect of this sequence of reactions is to relate the configuration of one stereocenter, the one next to the —CH$_2$OH group of the aldotetroses, to (+)-glyceraldehyde. However, we still don't know which aldotetrose is (−)-erythrose. To identify the two aldotetroses, the configuration at the other stereocenter must be determined. Its configuration can be determined from the experimental data shown in Figure 25.5.

Nitric acid oxidation of a pure sample of (−)-erythrose forms optically inactive *meso*-tartaric acid. Of the structures of the two aldotetroses, only structure **25.5** yields optically inactive tartaric acid. Therefore **25.5** is the structure of (−)-erythrose. Nitric acid oxidation of pure (−)-threose forms optically active tartaric acid. Thus **25.6** is the structure of (−)-threose because nitric oxidation of **25.6**, not **25.5**, forms optically active tartaric acid.

The series of reactions illustrated in Figures 25.4 and 25.5 determine the relative configurations of four compounds, (+)-glyceraldehyde, (−)-erythrose, (−)-threose, and (−)-tartaric acid. All four have the same configuration at the carbon stereocenter farthest from the carbonyl group.

Figure 25.5
Determining the relative configurations of (−)-erythrose and (−)-threose.

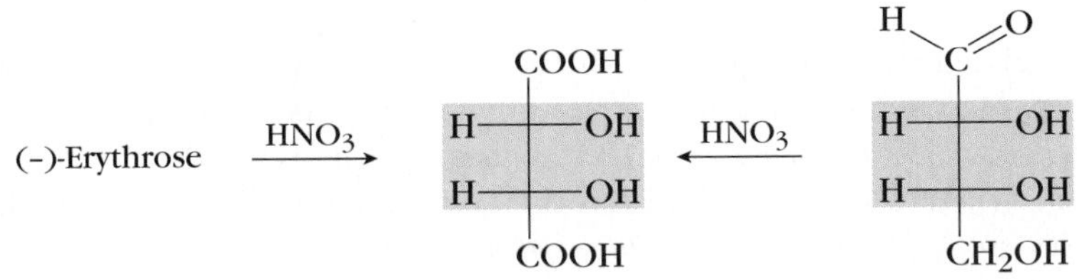

(−)-Erythrose

meso-Tartaric acid
(Optically inactive)

25.5

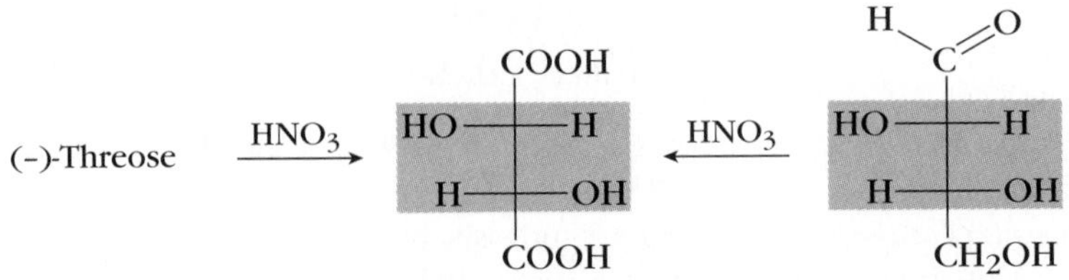

(−)-Threose

(−)-Tartaric acid

25.6

| D-(+)-Glyceraldehyde | D-(−)-Erythrose | D-(−)-Threose | D-(−)-Tartaric acid |

Chemists in the early part of the 20th century added the capital letter D to the names of monosaccharides that have the same configuration as (+)-glyceraldehyde at the stereocenter the farthest from the aldehyde group. Thus the names of (−)-erythrose, (−)-threose, and (−)-tartaric acid become D-(−)-erythrose, D-(−)-threose, and D-(−)-tartaric acid, respectively, to indicate that the configuration at their stereocenter farthest from the carbonyl group is the same as that of D-(+)-glyceraldehyde. In the $(R),(S)$ convention, D-(+)-glyceraldehyde has the (R) configuration.

Exercise 25.5   On the basis of the following observations, write the Fischer projection formulas of the aldopentose, (+)-xylose.

$$(+)\text{-Xylose} \xrightarrow[\text{degradation}]{\text{Ruff}} (-)\text{-Threose} \xrightarrow{\text{HNO}_3} (-)\text{-Tartaric acid}$$

$$(+)\text{-Xylose} \xrightarrow{\text{HNO}_3} \text{An optically inactive dicarboxylic acid, } C_5H_8O_7$$

We have seen how it is possible to relate the configuration of two aldotetroses to (+)-glyceraldehyde. But what about monosaccharides that contain five and six carbon atoms? Determining which of the 16 stereoisomers of aldohexoses is glucose seems like an overwhelming task. In 1891, however, Emil Fischer succeeded in establishing the correct structure of glucose. For this work he received the Nobel prize in chemistry in 1902. His elegant series of logical deductions has become known as the Fischer proof.

25.5   The Fischer Proof of the Configuration of (+)-Glucose

The Fischer proof of (+)-glucose is based on an elegant series of logical deductions from the data obtained in many careful experiments. The following experimental data and the deductions that lead to the structure of (+)-glucose are somewhat modified but are essentially the same ones that Fischer used.

Fischer realized that the 16 stereoisomers of aldohexoses consist of eight pairs of enantiomers. To simplify the problem, Fischer assumed that the OH on carbon 5 of (+)-glucose pointed to the right in its Fischer projection.

H—C=O (C1)
|
(2)
|
(3)
|
(4)
|
H——OH (5)
|
CH₂OH (6)

Fischer assigned this configuration to C5 of (+)-glucose

This reduces the number of possible stereoisomers to eight. Fischer had a 50% chance of being correct, but this does not affect the proof because the other stereocenters are determined relative to carbon atom 5. It turned out many years later that Fischer's assumption of the configuration about carbon atom 5 of (+)-glucose was correct.

Let's examine the experimental evidence and line of reasoning that establishes the relative configuration of (+)-glucose.

Experiment 1. Kiliani-Fischer chain extension of (−)-arabinose, a naturally occurring aldopentose, yields both (+)-glucose and (+)-mannose.

Conclusion. Configurations at carbon atoms 2, 3, and 4 of (−)-arabinose are the same as carbon atoms 3, 4, and 5 of (+)-glucose and (+)-mannose. This also means that (+)-glucose and (+)-mannose differ only by their configuration at carbon atom 2.

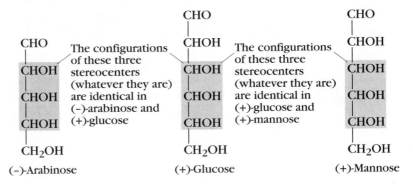

CHO
|
CHOH
| The configurations of these three stereocenters (whatever they are) are identical in
CHOH (−)-arabinose and (+)-glucose
|
CHOH
|
CH₂OH

(−)-Arabinose

CHO
|
CHOH
|
CHOH
| The configurations of these three stereocenters (whatever they are) are identical in
CHOH (+)-glucose and (+)-mannose
|
CHOH
|
CH₂OH

(+)-Glucose

CHO
|
CHOH
|
CHOH
|
CHOH
|
CHOH
|
CH₂OH

(+)-Mannose

Experiment 2. Nitric acid oxidation of (−)-arabinose yields an optically active dicarboxylic acid.

Conclusion. Because the —OH group on the stereocenter farthest from the CHO group is assigned to the right, the —OH group on the stereocenter next to the CHO group of (−)-arabinose must be to the left. If it were to the right, an optically inactive dicarboxylic acid would result.

CHO
|
HO——H
|
CHOH
|
H——OH
|
CH₂OH

One possible structure of (−)-arabinose

$\xrightarrow{\text{HNO}_3}$

COOH
|
HO——H
|
CHOH
|
H——OH
|
COOH

Optically active dicarboxylic acid

CHO
|
H——OH
|
CHOH
|
H——OH
|
CH₂OH

The other possible structure of (−)-arabinose

$\xrightarrow{\text{HNO}_3}$

COOH
|
H——OH
|
- - - CHOH - - - - - Plane of symmetry
|
H——OH
|
COOH

Optically inactive dicarboxylic acid

At this point, the following partial structure of (+)-glucose can be written:

```
        CHO
         |
HO ——————|
         |
         |—— OH
        CH2OH
```

Partial structure
of (+)-glucose

Experiment 3.   Nitric acid oxidation of both (+)-glucose and (+)-mannose forms optically active dicarboxylic acids.

Conclusion.   The —OH group on C4 of (+)-glucose and (+)-mannose must be on the right. If it were on the left, one of the dicarboxylic acids would be an optically inactive *meso* form.

Possible structures of (+)-glucose and (+)-mannose with OH on carbon atom 4 on the right:

```
      CHO                CHO
H ———— OH          HO ———— H
HO ———— H          HO ———— H
H ———— OH          H ———— OH
H ———— OH          H ———— OH
     CH2OH              CH2OH
       |HNO3              |HNO3
       ▼                  ▼
     COOH               COOH
H ———— OH          HO ———— H
HO ———— H          HO ———— H
H ———— OH          H ———— OH
H ———— OH          H ———— OH
     COOH               COOH
Optically active   Optically active
dicarboxylic acid  dicarboxylic acid
```

Possible structures of (+)-glucose and (+)-mannose with OH on carbon atom 4 on the left:

```
      CHO                CHO
H ———— OH          HO ———— H
HO ———— H          HO ———— H
HO ———— H          HO ———— H
H ———— OH          H ———— OH
     CH2OH              CH2OH
       |HNO3              |HNO3
       ▼                  ▼
     COOH               COOH
H ———— OH          HO ———— H
HO ———— H          HO ———— H
HO ———— H          HO ———— H
H ———— OH          H ———— OH
     COOH               COOH
Optically inactive  Optically active
dicarboxylic acid   dicarboxylic acid
```

Plane of symmetry (shown as dashed line across the third structure)

With this information, we can add the configuration of C4 to our partial structure of (+)-glucose. We can also now write the complete structure of (−)-arabinose because the configurations of carbon atoms 3, 4, and 5 of (+)-glucose are the same as carbon atoms 2, 3, and 4 of (−)-arabinose.

Experiment 4.   Fischer developed a series of reactions that interchange the —CH2OH and —CHO groups at the ends of aldoses. This method is complicated, so we will not consider the details here. Instead we will focus on the results of

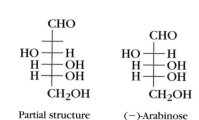

Partial structure
of (+)-glucose

(−)-Arabinose

this series of reactions on (+)-glucose. Chemically interchanging the —CHO and —CH₂OH groups on (+)-glucose forms a new aldohexose, called (+)-gulose.

Conclusion. The —OH group of C2 of (+)-glucose must be on the right because only this structure can produce a different aldohexose. If the —OH group is on the left, the same aldohexose is formed by the interchange of the —CHO and —CH₂OH groups.

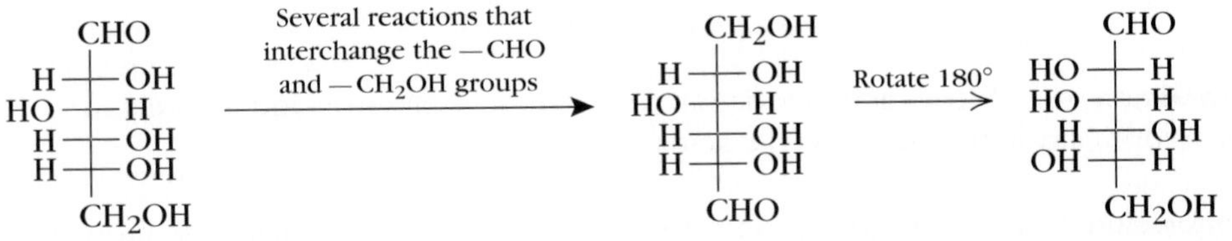

Different structure than starting aldohexose

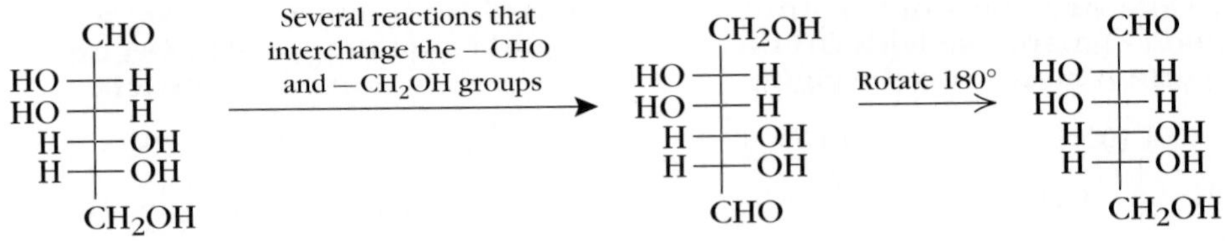

Identical structure to starting aldohexose

We now have the relative configuration of all the stereocenters of (+)-glucose, so we can write its Fischer projection as follows:

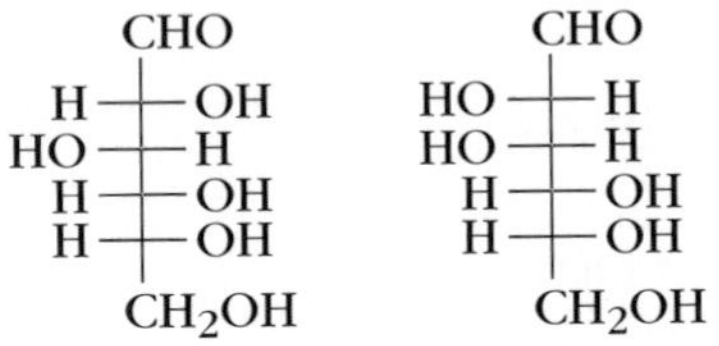

(+)-Glucose   (+)-Mannose

Because (+)-glucose and (+)-mannose differ only by the configuration at C2 (Experiment 1), we can also write the Fischer projection of (+)-mannose.

By reasoning from carefully obtained experimental results, Fischer and others were able to establish the relative configurations of all aldotetroses, aldopentoses, and aldohexoses. Once their stereochemistry was established, it was possible to group them into families.

25.6 The *D*- and *L*-Families of Monosaccharides

All monosaccharides belong to either the *D* or the *L* family. The difference between the two families is the configuration of the stereocenter farthest removed from the carbonyl group. In members of the *D*- family, this stereocenter has the same configuration as *D*-(+)-glyceraldehyde. In members of the *L*- family, this stereocenter has the same configuration as *L*-(−)-glyceraldehyde.

The Fischer proof of the configuration of (+)-glucose does not allow us to assign it to either the *D*- or *L*- family because we do not know which of the two enantiomeric structures, **25.7** or **25.8**, represents its absolute configuration. Fischer arbitrarily chose structure **25.7**, in which the —OH group of the stereocenter farthest from the carbonyl group points to the right.

25.7
D-(+)-glucose

25.8
L-(−)-glucose

The proof that (+)-glucose is related to *D*-(+)-glyceraldehyde was established by a number of reaction sequences. Based on these reactions, structure **25.7** represents *D*-(+)-glucose and structure **25.8** represents *L*-(−)-glucose.

Given the Fischer projection of any monosaccharide, it is possible to determine whether it belongs to the *D*- or *L*- family by examining the position of the —OH group on the stereocenter farthest from the carbonyl group. If the —OH group points to the right, it belongs to the *D*- family. If it points to the left, the compound belongs to the *L*- family. Consider (−)-ribose. The configuration of the stereocenter farthest from the carbonyl group is the same as that of *D*-(+)-glyceraldehyde, so (−)-ribose belongs to the *D*- family. It is called *D*-(−)-ribose.

D-(-)-Ribose

These carbon atoms have the same configuration

D-(+)-Glyceraldehyde

The Fischer projections of the entire *D*- family of aldoses up to aldohexoses are given in Figure 25.6. Notice that there is no relationship between the direction in which the monosaccharides in Figure 25.6 rotate polarized light and whether they are in the *D*- family or the *L*- family. For example, (−)-arabinose and (+)-mannose are both members of the *D*- family but they rotate plane-polarized light in opposite directions.

All of the members of the *D*- family can be obtained either from natural sources or by synthesis in the laboratory. The most common naturally occurring pentoses are *D*-(−)-ribose, *D*-(−)-arabinose, and *D*-(+)-xylose. *D*-(−)-Ribose is part of numerous biologically important substances such as nucleic acids (Chapter 27), while *D*-(−)-arabinose and *D*-(+)-xylose are obtained by hydrolysis of the polysaccharides present in plants and wood, respectively.

The most abundant, best-known, and most important aldohexose is *D*-(+)-glucose. It is formed in plants by photosynthesis from carbon dioxide, water, and light. It is also an important source of energy for all forms of animal life.

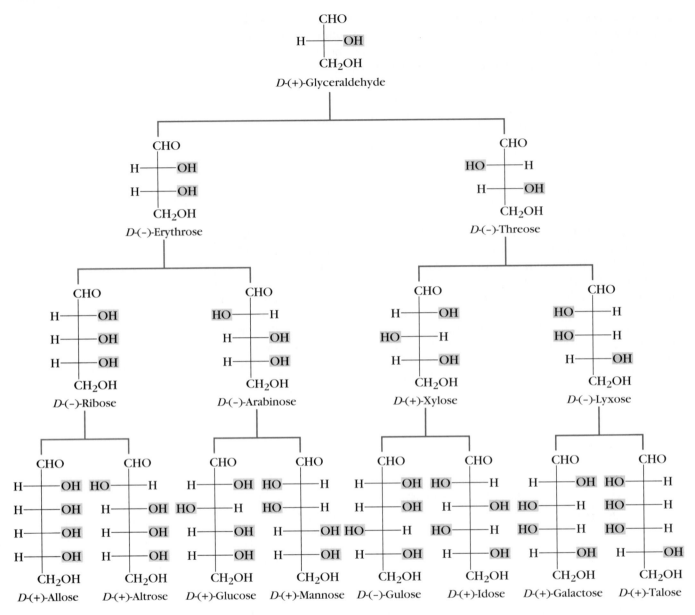

Figure 25.6
The Fischer projections of the configurations of the *D*- family of aldoses up to aldohexoses.

Exercise 25.6   Classify each of the following as a *D*-monosaccharide or an *L*-monosaccharide.

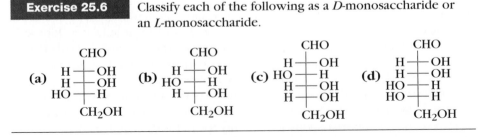

Exercise 25.7	Write the Fischer projection formula of each of the following monosaccharides.

(a) *L*-Erythrose   **(b)** *L*-Ribose   **(c)** *L*-Arabinose   **(d)** *L*-Galactose

Until now we have written the structures of aldoses and ketoses as open chain compounds. Actually, one hydroxy and the carbonyl group in a monosaccharide react to form a cyclic hemiacetal.

25.7   Cyclic Forms of Monosaccharides

In Chapter 24 we discussed that two functional groups suitably located in a molecule may undergo intramolecular reactions to form a ring. Monosaccharides are no exception. Intramolecular nucleophilic addition of one —OH group to the carbonyl carbon atom of a monosaccharide forms cyclic hemiacetals. The aldotetrose *D*-(−)-threose, for example, exists in aqueous solution primarily as a five-membered **furanose** ring formed by intramolecular nucleophilic addition of the hydroxy group on C4 to the carbonyl carbon atom.

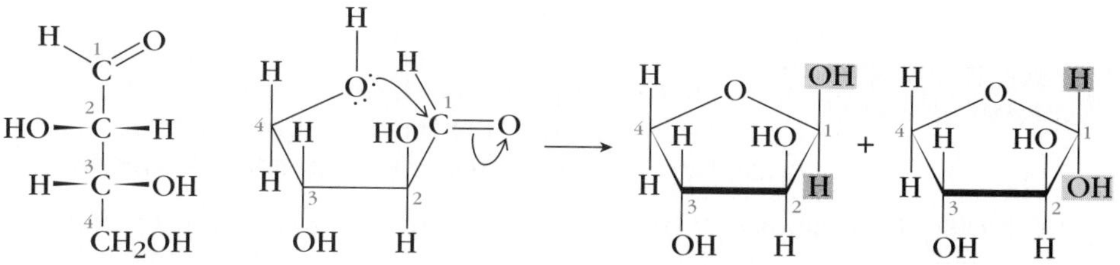

Open-chain form
of *D*-threose

Cyclic forms of *D*-threose

The cyclic forms of *D*-threose are written as **Haworth formulas** rather than Fischer projections. In a Haworth formula, the hemiacetal ring is written as if it were flat and viewed from one edge with the ether oxygen atom at the rear. The —OH groups that project to the right in Fischer projections project down in the Haworth formulas.

Cyclization of *D*-(−)-threose to form a furanose results in a new stereocenter at the hemiacetal carbon atom, which was originally the carbonyl carbon atom of the open chain form. The two diastereomers formed by cyclization are called **anomers,** and the hemiacetal carbon atom is called an **anomeric center.** The anomers are distinguished by the Greek letters α and β. In the α-anomer, the —OH group on carbon atom 1 is down in the Haworth formula, while the —OH group is up in the β-anomer.

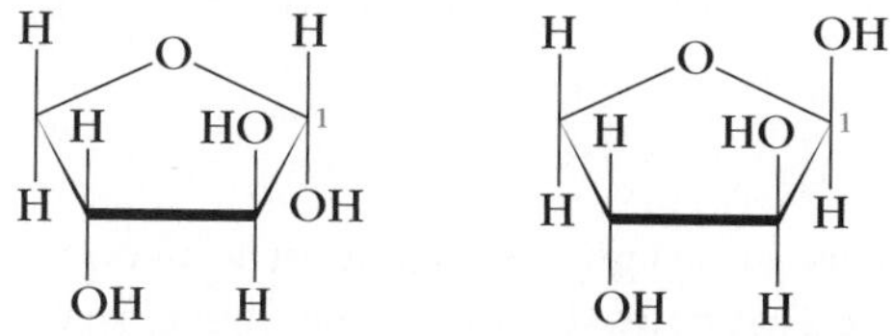

α-*D*-(−)-Threofuranose   β-*D*-(−)-Threofuranose

The cyclic forms of monosaccharides are named to convey the following five pieces of information: the configuration at the anomeric center, the family (*D*- or *L*-), the direction in which plane-polarized light is rotated (frequently omitted), the name of the monosaccharide from which the cyclic form is derived, and the size of the ring.

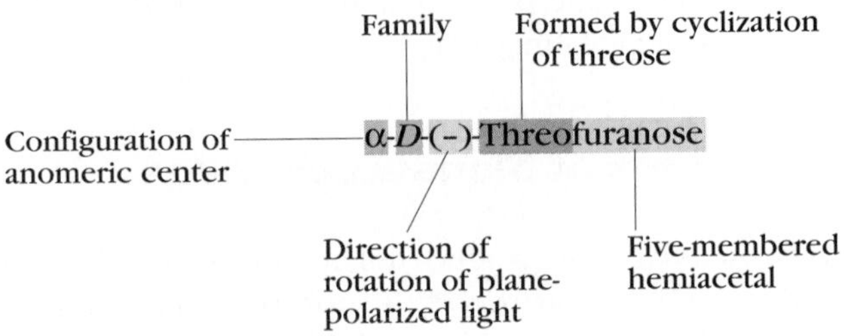

In α-*D*-(−)-threofuranose, the name furanose is derived from furan, which is the name of the simplest five-membered ring ether. The two anomeric furanose forms of *D*-threose are named α-*D*-(−)-threofuranose and β-*D*-(−)-threofuranose.

■ *The names of cyclic ethers are given in Section 13.1.*

Exercise 25.8   Using the Fischer projection formulas in Figure 25.6 (p. 1108) as a guide, write Haworth formulas of the cyclic forms of each of the following monosaccharides.

(a) β-*D*-Erythrofuranose   **(b)** α-*D*-Xylofuranose   **(c)** β-*D*-Lyxofuranose

---

About 20% of *D*-fructose, a ketohexose, exists as a five-membered furanose ring. The furanose ring is formed by the reaction of the −OH group on C5 with the carbonyl group on C2.

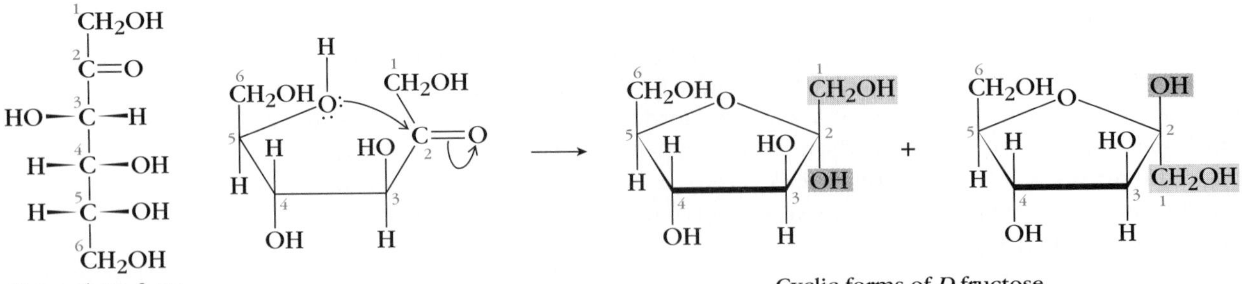

Open-chain form of *D*-fructose   Cyclic forms of *D*-fructose

Aldohexoses such as *D*-(+)-glucose exist in aqueous solution almost completely as six-membered **pyranose** rings. The name pyranose is derived from pyran, which is the name of the simplest six-member cyclic ether.

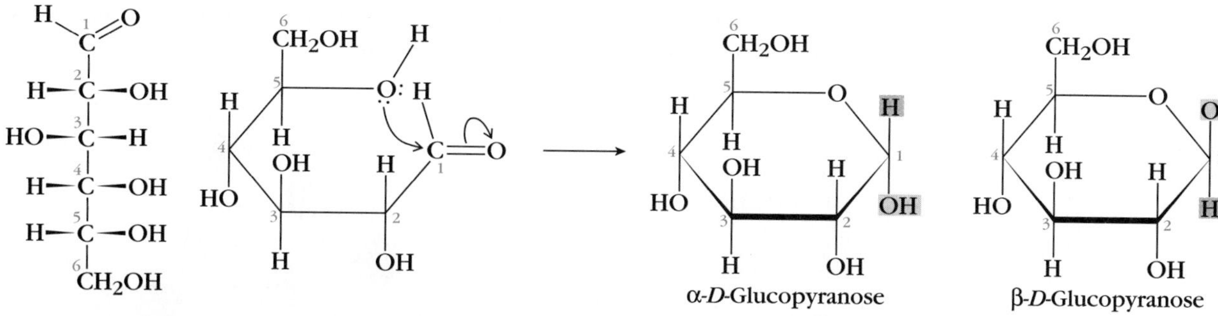

Open-chain form
of D-glucose

α-D-Glucopyranose   β-D-Glucopyranose

Cyclic forms of D-glucose

As in the case of threose, the anomeric forms of D-(+)-glucopyranose are identified by the letters α and β. In their Haworth formulas, the —OH group on the anomeric carbon atom (C1) of α-D-(+)-glucopyranose is *trans* to the —CH₂OH group at the stereocenter that designates the *D* configuration (C5 in hexoses). The anomer with these two groups *cis* is called β-D-(+)-glucopyranose.

So far, the cyclic structures of monosaccharides have been represented by Haworth formulas. These formulas do not accurately represent the conformation of the pyranose ring, however, so we represent the pyranose ring in its chair conformation. When the pyranose form of an aldohexose such as D-(+)-glucopyranose is written in its chair conformation, the —OH group on C1 and the —CH₂OH group on C5 are *trans* in the α-anomer and *cis* in the β-anomer.

■ The *cis-* and *trans-relationship* between two substituents on a pyranose ring are the same as those on a disubstituted cyclohexane ring given in Table 4.4, p. 161.

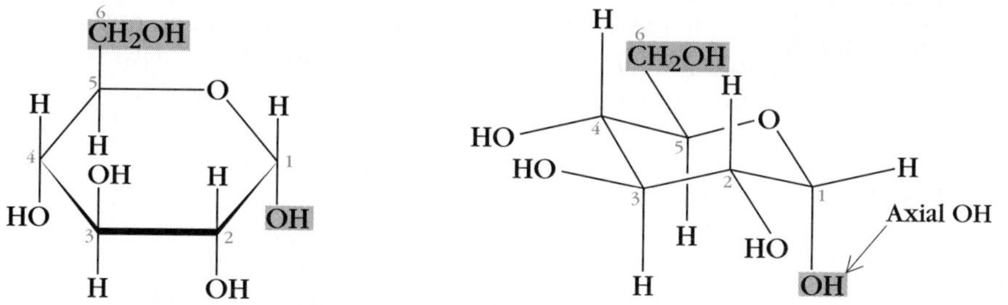

α-**D-Glucopyranose**
CH₂OH and OH are *trans*

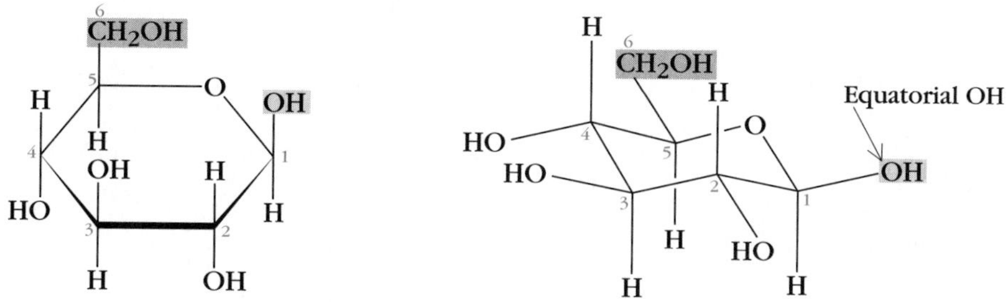

β-**D-Glucopyranose**
CH₂OH and OH are *cis*

When representing the structure of the cyclic hemiacetal forms of D-(+)-glucose with the chair form of pyranose, all the substituents on the ring of β-D-(+)-

glucopyranose are equatorial. As a result, β-*D*-(+)-glucopyranose is the least sterically crowded and the most stable of the *D*- aldohexoses.

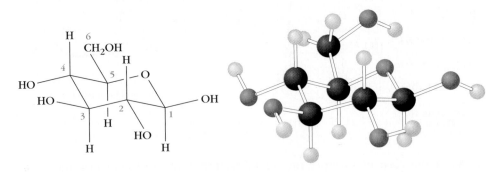

β–*D*–(+)–Glucopyranose

Exercise 25.9   Using the Fischer projection formulas given in Figure 25.6 (p. 1108) as a guide, write a chair representation of the cyclic form of each of the following monosaccharides.

(a) α-*D*-Glucopyranose   **(b)** β-*D*-Mannopyranose   **(c)** β-*D*-Galactopyranose

---

Exercise 25.10   Circle the hemiacetal functional group in each of the following monosaccharides. Indicate whether the anomer shown is the α- form or the β- form.

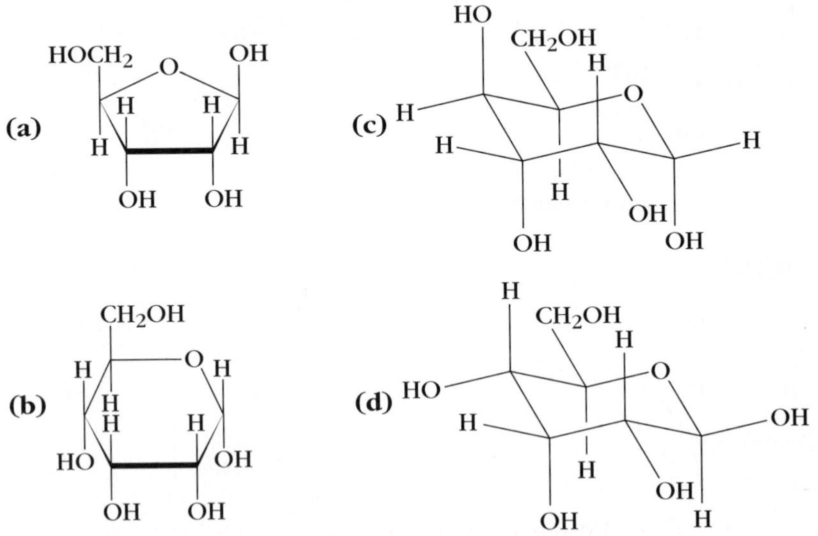

In solution, monosaccharides exist as an equilibrium mixture of the open chain and the cyclic forms of both anomers.

25.8   Mutarotation

Both α- and β-D-(+)-glucopyranose can be isolated as pure, stable solids. When either of these anomers is dissolved in water, the optical rotation of the solution slowly changes. A freshly prepared aqueous solution of α-D-(+)-glucopyranose has a specific rotation of $+112°$. The rotation of this solution gradually changes with time until a constant value of $+53°$ is reached. Starting with pure β-D-(+)-glucopyranose, the initial optical rotation $(+19°)$ gradually changes to the same constant value of $+53°$. This process is called **mutarotation.**

Each solution initially contains a single anomer, so what causes mutarotation? The answer is that each solution undergoes equilibration through an open chain intermediate to form the same mixture of α- and β-D-(+)-glucopyranose. At equilibrium, D-(+)-glucopyranose consists of 36% α- anomer, 64% β- anomer, and 0.002% open chain form.

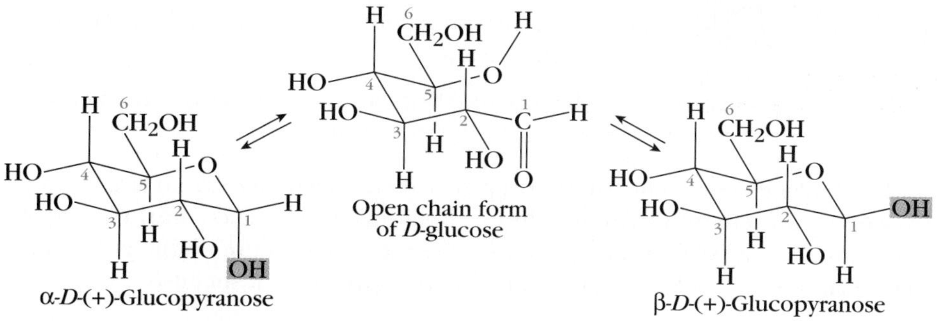

α-D-(+)-Glucopyranose   Open chain form
of D-glucose   β-D-(+)-Glucopyranose

The interconversion between the α-anomer, the β-anomer, and the open chain form is so facile in water that it is usually impossible to specify the configuration at the anomeric carbon atom. Consequently, the bond between the anomeric carbon atom and the —OH group is often represented by a squiggle, which means that the configuration of the anomeric carbon atom can be α- or β- or a mixture.

This squiggle means that the configuration of the anomeric carbon atom can be α- or β- or a mixture

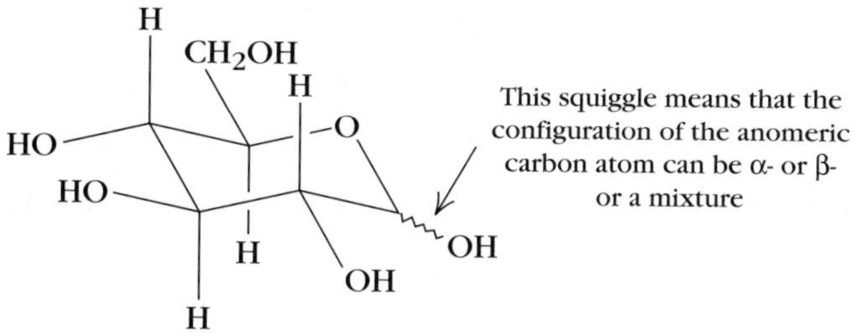

While the majority of carbohydrates correspond to the general formula $C_n(H_2O)_n$, a few have minor structural variations.

25.9   Structural Variations in Carbohydrates

There are many polyhydroxyaldehydes or ketones whose structures are similar but not identical to those of monosaccharides. These compounds, whose molecular formulas are not $C_n(H_2O)_n$, are still classified as carbohydrates because of the similarity of their reactions to those of monosaccharides. The most common variation in monosaccharide structure is the replacement of one or more —OH groups by another atom or group.

Deoxy sugars resemble monosaccharides, except that one of their —OH groups is replaced by a hydrogen atom. 2-Deoxy-D-ribose, for example, is a deoxysugar that is part of DNA (deoxyribonucleic acid). It is similar to D-ribose, except that the —OH group at C2 of D-ribose is replaced by a hydrogen atom.

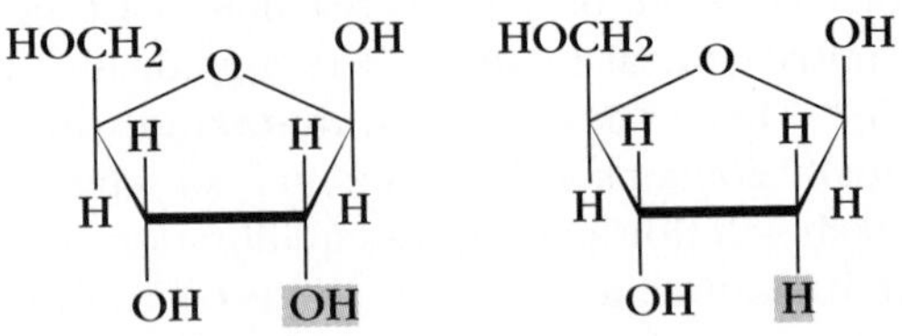

β-D-Ribose   2-Deoxy-β-D-ribose

Exercise 25.11   Write the Fischer projection formula for each of the following deoxy sugars.

(a) 2-Deoxy-D-ribose   (b) 3-Deoxy-L-ribose   (c) 3-Deoxy-D-glucose

---

The replacement of an —OH group of a monosaccharide by an amino group is another structural variation. These compounds are called **amino sugars.** 2-Acetamido-2-deoxy-β-D-glucopyranose, the repeating unit of chitin, is an amino sugar. Chitin is the polysaccharide found in the hard shell of shellfish.

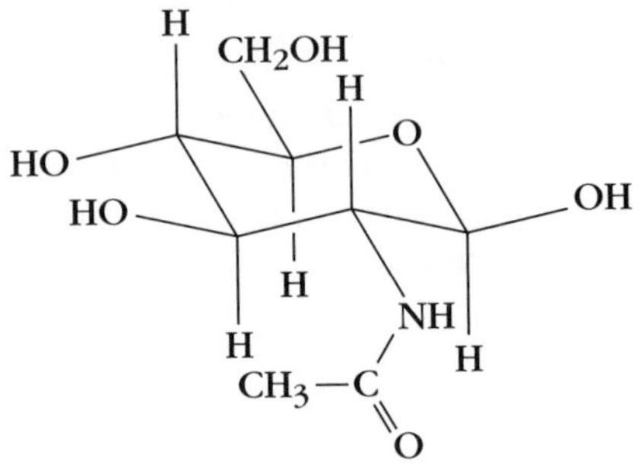

2-Acetamido-2-deoxy-β-D-glucopyranose

Carbohydrates contain many —OH groups that undergo reactions typical of alcohols.

25.10 Esters and Ethers of Monosaccharides

So far, we have concentrated on the reactions of the carbonyl group of monosaccharides. In this section, we discuss the reactions of the —OH groups of monosaccharides. They react like simple alcohols to form esters and ethers.

Esterification can usually be accomplished by reacting the carbohydrate with acetic anhydride below 0 °C in the presence of sodium acetate or pyridine as catalyst. At low temperatures, the interconversion of anomers is slower than acetylation, so the product is the pentaacetate with the same configuration as the original carbohydrate. For example, β-D-glucopyranose reacts with acetic anhydride and sodium acetate at low temperatures to form penta-O-acetyl-β-D-glucopyranose. At higher temperatures, a mixture of α- and β-isomers of penta-O-acetyl-D-glucopyranose is formed.

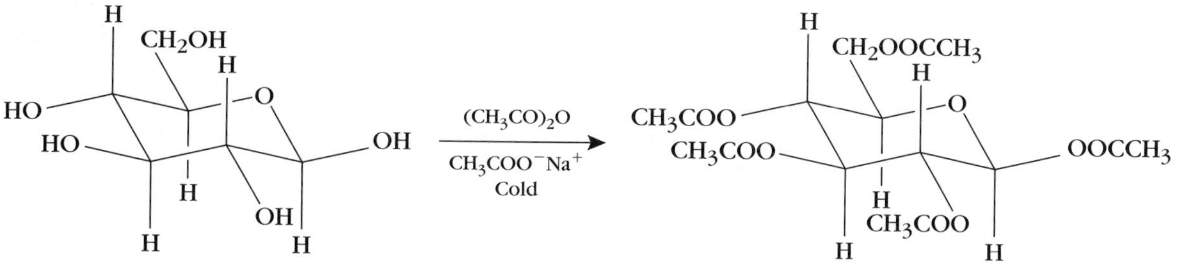

β-D-Glucopyranose   Penta-O-acetyl-β-D-glucopyranose

When a monosaccharide such as β-D-glucopyranose is treated with excess dimethyl sulfate [$(CH_3O)_2SO_2$] in aqueous sodium hydroxide, all the —OH groups are converted to methoxy groups.

■ *The O in the name penta-O-acetyl-β-D-glucopyranose means that the acetyl groups are bonded to oxygen atoms.*

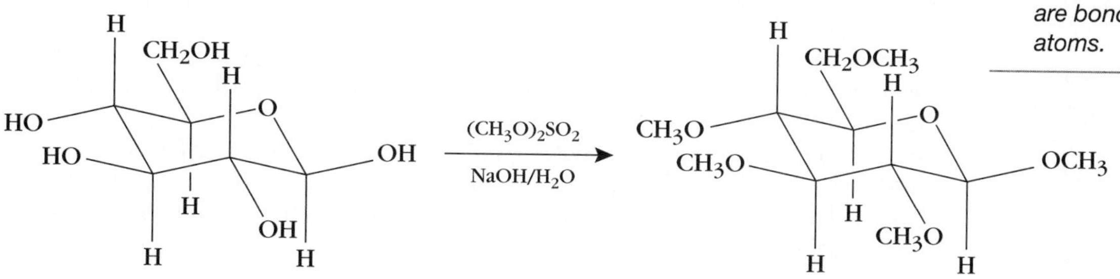

β-D-Glucopyranose   Methyl tetra-O-methyl-β-D-glucopyranoside

Monosaccharides also react with haloalkanes in aqueous sodium hydroxide to form ethers. The conversion of β-D-glucopyranose to methyl tetra-O-methyl-β-D-glucopyranoside is a multiple Williamson ether synthesis with a minor difference. Usually, a stronger base than OH^- must be used to form the alkoxide ion in the Williamson ether synthesis. The —OH groups of monosaccharides are stronger acids than simple alcohols, however, because of the electron-withdrawing inductive effect of the many adjacent electronegative oxygen atoms. As a result, NaOH is a sufficiently strong base to convert the hydroxy groups to alkoxide ions, which then react with dimethyl sulfate or haloalkanes by means of an S_N2 mechanism.

■ *The Williamson ether synthesis is discussed in Section 13.8, C.*

Exercise 25.12   Write the structure of the product, if any, of the reaction of α-D-galactopyranose with each of the following reagents.

(a) Acetic anhydride/sodium acetate
(b) $(CH_3O)_2SO_2$/NaOH
(c) H_3O^+

(d) Tollen's reagent
(e) Br_2/H_2O

One of the five hydroxy groups of α- or β-D-glucopyranose is part of a hemiacetal. Consequently, it can be selectively converted into an acetal. In carbohydrate chemistry, an acetal is called a glycoside.

25.11 Glycoside Formation

In Chapter 14, we learned that hemiacetals react with alcohols in the presence of an acid catalyst to form acetals. The same reaction occurs with the cyclic hemiacetal form of monosaccharides to yield an acetal in which the anomeric —OH group is replaced by an alkoxy group. The alkoxy group can occupy either the α- or β- position. For example, the reaction of either α- or β-D-glucopyranose with an acidic solution of methanol forms a mixture of two methyl acetals.

α- or β-D-glucopyranose

Methyl
β-D-glucopyranoside

Methyl
α-D-glucopyranoside

■ *For clarity, the hydrogen atoms on the ring carbon atoms will no longer be included in the structures of pyranose rings. Only the hydrogen atoms on the anomeric carbon atom will be included to indicate stereochemistry.*

Carbohydrate acetals are called **glycosides.** Specific glycosides are named in the following way. The first word indicates the alkyl group of the alkoxide, and the second word is obtained by replacing the suffix *-ose* of the specific cyclic form of the monosaccharide with *-oside*. The methyl of methyl β-D-glucopyranoside, for example, indicates that a methyl group is bonded to the oxygen atom of the acetal. The word β-D-glucopyranoside is obtained by replacing the suffix *-ose* of β-D-glucopyranose by *-oside*. β-D-Glucopyranose is the name of the monosaccharide from which the glycoside was formed. The following are the names of other glycosides:

Ethyl
β-D-mannopyranoside

Methyl
α-D-galactopyranoside

The proposed mechanism for formation of glycosides from monosaccharides is shown in Figure 25.7. Step 1 is the protonation of the —OH group of the anomeric carbon atom, then water is lost by an S_N1-like mechanism in Step 2 to form a stable oxonium ion. The oxonium ion reacts with a molecule of an alcohol (in this example, methanol) in Step 3 to form the conjugate acid of a glycoside, which loses a proton in Step 4 to form the glycoside. Addition of methanol occurs to either face of the oxonium ion in Step 3, so a mixture of the two anomeric methyl glycosides is formed.

Glycosides are acetals, so they are stable in neutral and basic solutions but can be hydrolyzed in acid solution. Glycosides do not undergo mutarotation in the absence of acid, and they are unaffected by reagents that react with carbonyl groups because their anomeric carbon atom is blocked. As a result, glycosides give a negative test with Fehling's and Tollen's reagent. Glycosides, therefore, are **nonreducing sugars.**

Certain enzymes also hydrolyze glycosides with remarkable structural speci-

Step 1: Protonation of the OH group on the anomeric carbon atom

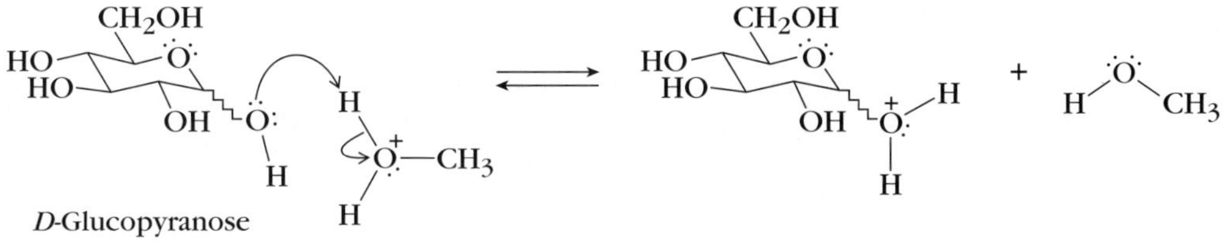

D-Glucopyranose

Step 2: Loss of water forms an oxonium ion

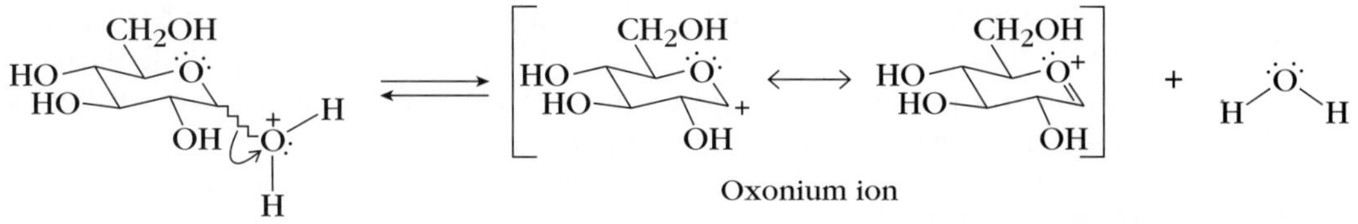

Oxonium ion

Step 3: Reaction of methanol and oxonium ion

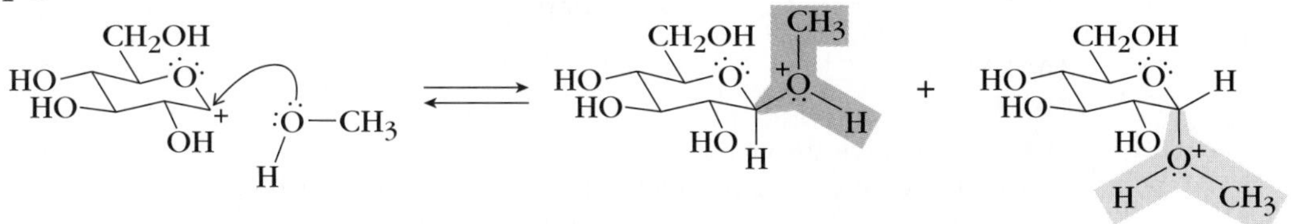

Step 4: Loss of proton forms a mixture of isomeric glycosides

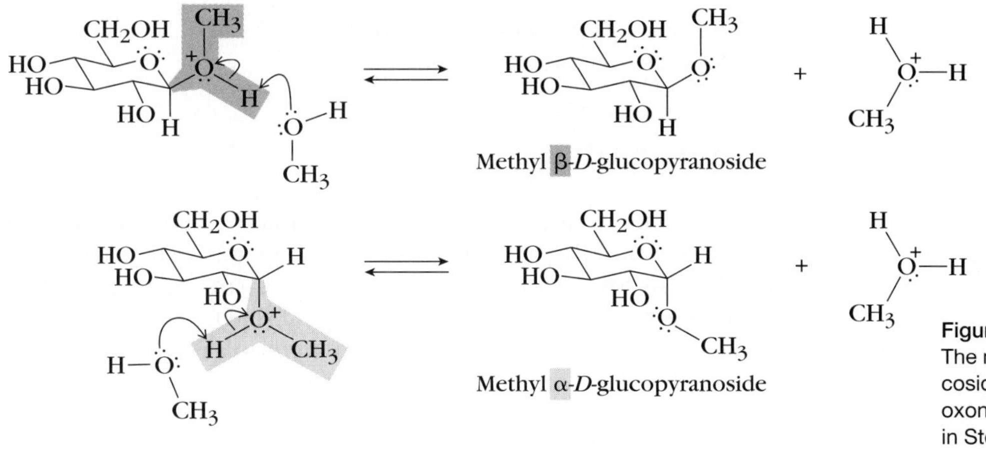

Methyl β-*D*-glucopyranoside

Methyl α-*D*-glucopyranoside

Figure 25.7
The mechanism for formation of glycosides from monosaccharides. An oxonium ion intermediate is formed in Step 2 by an S_N1 mechanism. Methanol adds to both sides of the oxonium ion in Step 3 to form both anomers.

ficity. α-Galactosidase, for example, catalyzes the hydrolysis of only α-galactoside bonds. Thus it catalyzes the hydrolysis of methyl α-*D*-galactopyranoside but not the hydrolysis of methyl β-*D*-galactopyranoside.

Methyl α-*D*-galactopyranoside   α-*D*-Galactopyranose

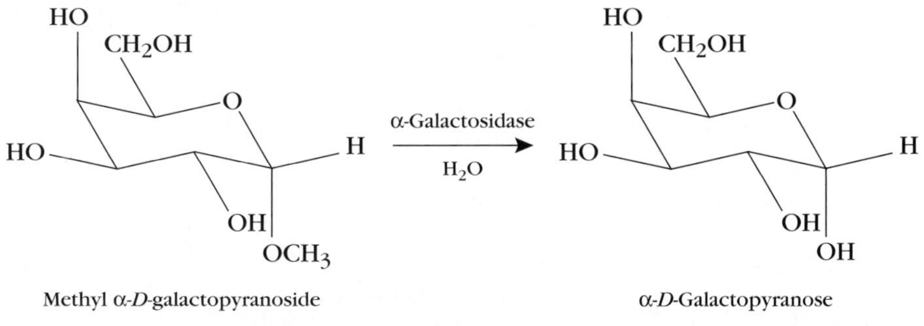

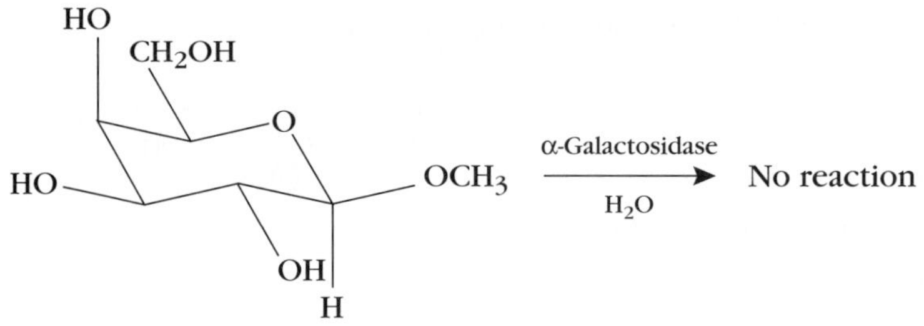

Methyl β-*D*-galactopyranoside

β-Galactosidase, on the other hand, catalyzes only the hydrolysis of β-galacto-sides. β-Galactosidase, therefore, catalyzes the hydrolysis of methyl β-*D*-galactopy-ranoside but not methyl α-*D*-galactopyranoside. These reactions are further exam-ples of the remarkable specificity of enzymes.

Exercise 25.13   Write the structure of the product, if any, of the reac-tion of methyl α-*D*-mannopyranoside with each of the following reagents.

(a) HCl/H_2O
(b) Br_2/H_2O
(c) Fehling's solution
(d) An enzyme that catalyzes the hydrolysis of α-glycosides
(e) An enzyme that catalyzes the hydrolysis of β-glycosides

If the —OH group of the alcohol used to form a glycoside is part of another monosaccharide, a disaccharide is formed.

25.12 Disaccharides

Disaccharides are dimers made up of two monosaccharide molecules. The mono-saccharides may be either the same or different. The monosaccharides are joined by a glycoside linkage between the anomeric carbon of one monosaccharide and an —OH group of the other. The most common glycoside linkage is between the 4-hydroxy group of one monosaccharide and the α- or β- position of the anomeric carbon. This kind of a link is called an α-1,4- or β-1,4-glycoside linkage (or bond), depending on the configuration at the anomeric carbon atom.

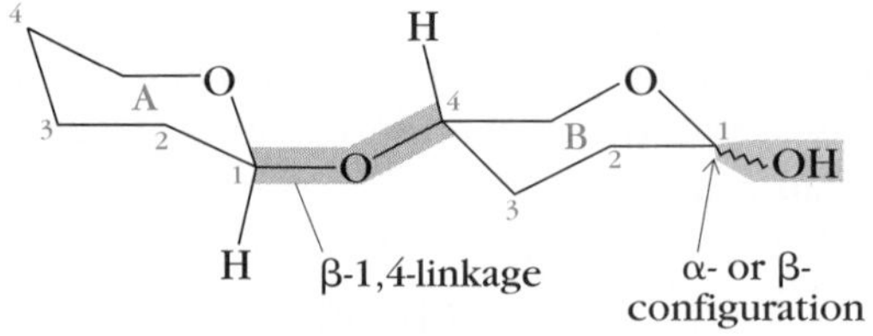

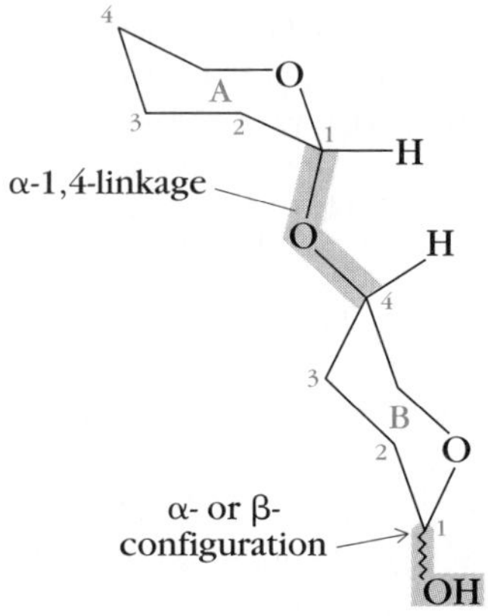

Rotating ring B 180° relative to ring A gives a second representation of a β-1,4-linkage:

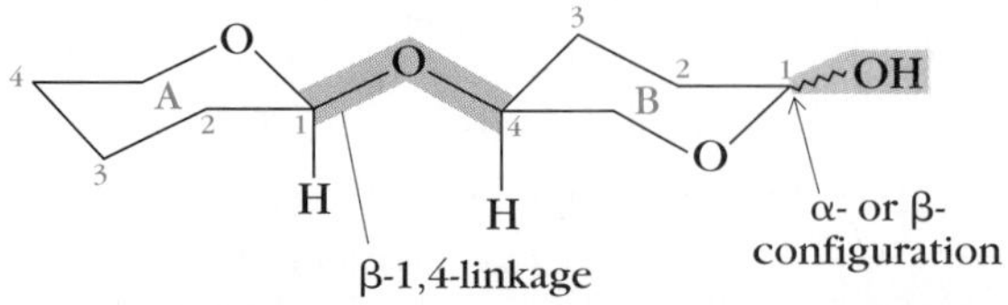

Maltose, cellobiose, and lactose are disaccharides with α-1,4- or β-1,4-glycoside linkages.

Maltose is the principal product of the acid- or enzyme-catalyzed hydrolysis of starch. It contains two D-glucose units, both in the pyranose form, joined by an α-1,4-linkage. Maltose exists in both the α- and β- anomeric forms.

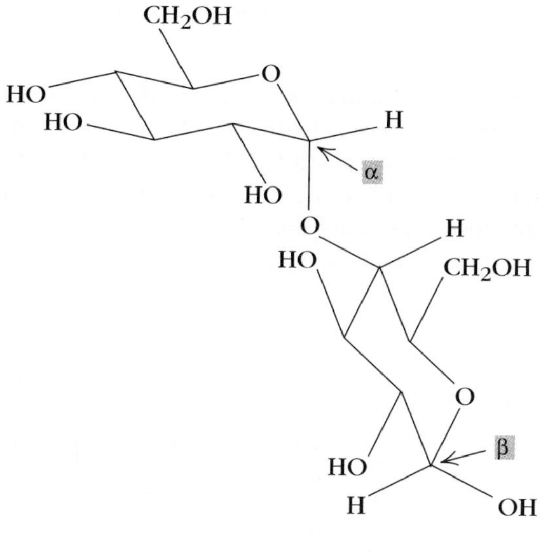

β–Maltose
4-O-(α-D-glucopyranosyl)-β-D-glucopyranoside

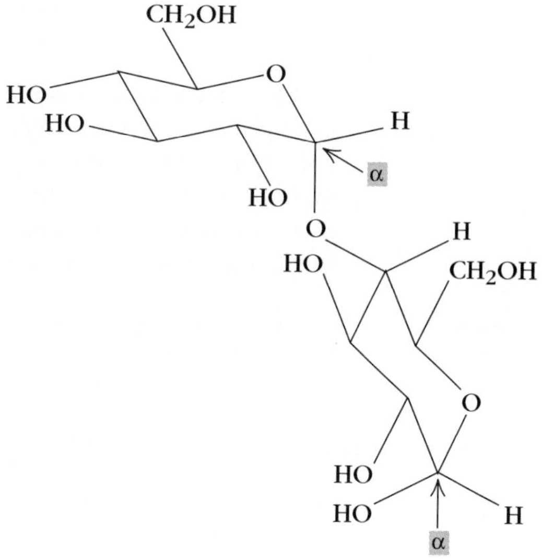

α–Maltose
4-O-(α-D-glucopyranosyl)-α-D-glucopyranoside

One ring of maltose contains a carbonyl group that is still in a hemiacetal form. This carbonyl group can undergo the normal reactions of a carbonyl group of an aldose. Maltose, therefore, reduces Fehling's solution and Tollen's reagent, so it is a reducing sugar. It undergoes mutarotation and can be oxidized to a carboxylic acid by bromine water.

Exercise 25.14   Write the structure of the product of the reaction of β-maltose with each of the following reagents.

(a) H_3O^+ (b) Br_2/H_2O (c) Tollen's reagent   (d) $(CH_3O)_2SO_2/NaOH/H_2O$

Cellobiose is a disaccharide that is obtained by the partial hydrolysis of cellulose. It contains two D-glucose units joined by a β-1,4-glycoside bond, so it is isomeric with maltose.

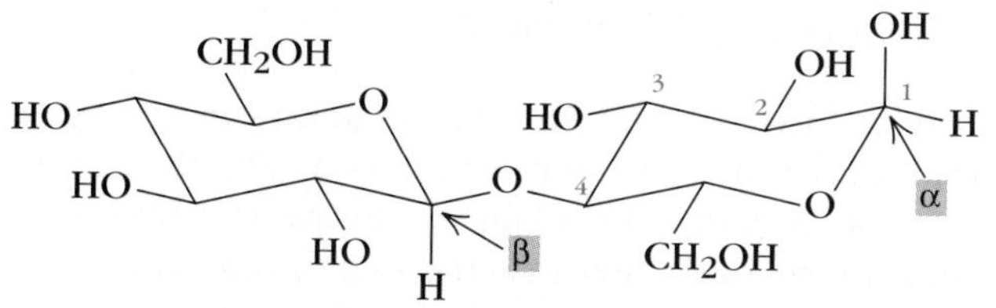

α-Cellobiose
4-O-(β-D-Glucopyranosyl)-α-D-glucopyranoside

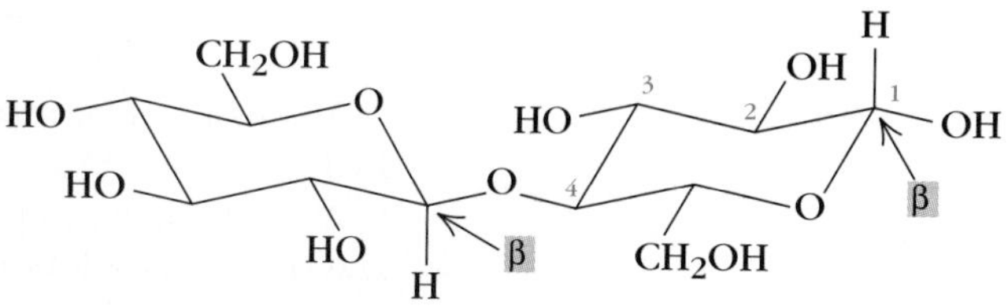

β-Cellobiose
4-O-(β-D-Glucopyranosyl)-β-D-glucopyranoside

Like maltose, cellobiose exists as two anomers, is a reducing sugar, undergoes mutarotation, and is oxidized to a carboxylic acid by bromine water.

Exercise 25.15   Write the structure of the product, if any, of the reaction of α-cellobiose with each of the following reagents.

(a) H_3O^+

(b) Br_2/H_2O

(c) $(CH_3CO)_2O/Na^+ \ ^-OOCCH_3$

(d) $NaBH_4$

(e) β-Glucosidase

(f) Fehling's solution

Lactose, a disaccharide that occurs in both human and cow's milk, contains a β-1,4-glycoside bond. Unlike maltose and cellobiose, hydrolysis of lactose forms one equivalent of glucose and one equivalent of galactose. The two are both in their pyranose forms, and the β-glycoside bond joins the anomeric carbon of galactose with C4 of glucose. Like maltose and cellobiose, lactose is a reducing sugar, undergoes mutarotation, and exists as two anomers.

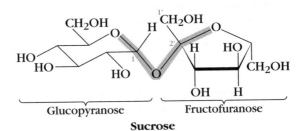

α-Lactose
4-O-(β-D-galactopyranosyl)-α-D-glucopyranoside

β-Lactose
4-O-(β-D-galactopyranosyl)-β-D-glucopyranoside

Sucrose (ordinary table sugar) is one of the most widespread carbohydrates in nature. It is obtained commercially from sugar beets and sugar cane. Hydrolysis of sucrose forms one equivalent of D-glucose and one equivalent of D-fructose. Sucrose is not a reducing sugar, so it cannot have a hemiacetal group. This means, in turn, that the pyranose form of glucose and the furanose form of fructose are joined by an acetal link between the two anomeric carbons.

Sucrose
2-O-(α-D-Glucopyranosyl)-β-D-fructofuranoside

Sucrose is dextrorotatory with $[\alpha]_D = +66°$. Either acid- or enzyme-catalyzed hydrolysis of (+)-sucrose produces an interesting phenomenon. The dextrorotatory solution of pure (+)-sucrose becomes levorotatory during the hydrolysis reaction. This solution is called "invert sugar" because the sign of optical rotation inverts during its preparation. The reason the sugar inverts becomes apparent when we look at the optical rotations of glucose and fructose. The equilibrium mixture of fructose isomers has a negative optical rotation, $[\alpha]_D = -92°$, while the equilibrium mixture of glucose isomers has a positive optical rotation, $[\alpha]_D = +52°$. The optical rotation of fructose is so highly negative that the 1:1 mixture of fructose and glucose formed on complete hydrolysis of sucrose is also levorotatory; $[\alpha]_D = -20°$. Based on the sign of their optical rotations, glucose is sometimes referred to as "dextrose" and fructose as "levulose." Honeybees have enzymes called invertases that catalyze the hydrolysis of sucrose to fructose and glucose. Honey is largely a mixture of *D*-glucose, *D*-fructose, and sucrose.

25.13 Polysaccharides

Most carbohydrates in nature exist as high–molecular-weight polymers called polysaccharides. Polysaccharides are formed by joining hundreds and thousands of monosaccharides by means of glycoside bonds. The most common monosaccharide present in polysaccharides is *D*-glucose.

Polysaccharides have two quite different roles in nature. Some polysaccharides serve as a means of storing chemical energy, while others serve as structural units. The remarkable fact is that the polysaccharides that function in these two ways both consist of 1,4-linked polyglucose. Their different roles are due to the fact that two *D*-glucopyranose molecules can be joined by either β-1,4- or α-1,4-glycoside bonds. Let's first learn how a β-1,4-glycoside bond affects the properties of a 1,4-linked polyglucose.

Straight chains are the most stable arrangement of *D*-glucopyranose molecules that are joined together by β-1,4-glycoside bonds, as shown in Figure 25.8. Notice that every pyran ring in the straight chain is rotated by 180° relative to the previous one. This results in an almost fully extended polysaccharide chain. These extended chains pack tightly together because many hydrogen bonds form between the chains. As a result, these chains are strong and chemically relatively inert.

Figure 25.8
A, The most stable arrangement of *D*-glucopyranose molecules joined by β-1,4 bonds is a straight chain. Notice that every pyran ring in the chain is rotated 180° relative to the previous one. **B,** The chains formed by *D*-glucopyranose molecules joined by β-1,4 bonds pack tightly together because many hydrogen bonds form between adjacent chains.

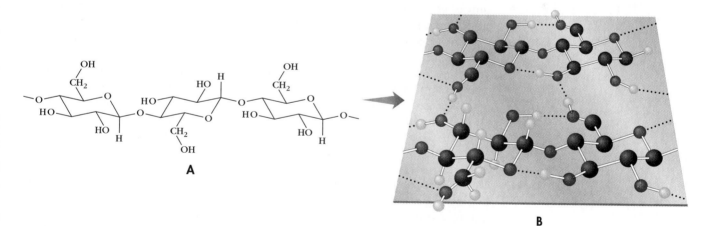

A

B

Cellulose is an example of a polysaccharide consisting of tightly packed extended chains of *D*-glucopyranose molecules joined by β-1,4 glycoside bonds. Cellulose is the most abundant of all polysaccharides. It is the major constituent of cell membranes and other structural features of plants. Cotton fiber and filter paper are almost entirely cellulose, while wood is about 50% cellulose.

Instead of linear chains, the chains of *D*-glucopyranose molecules joined by α-1,4 glycoside bonds adopt a helical configuration, as shown in Figure 25.9.

Figure 25.9
A, Two *D*-glucopyranose units joined by an α-1,4-linkage. **B,** A chain of *D*-(+)-glucopyranose molecules joined by α-1,4 glycoside bonds adopts a helical configuration.

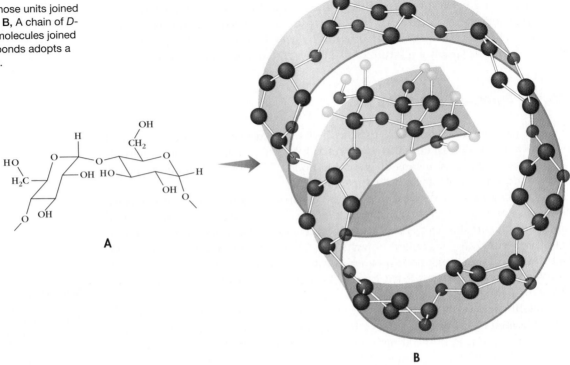

A

B

Starch is an example of a polysaccharide consisting of coiled chains of *D*-glucopyranose molecules joined by α-1,4 glycoside bonds. Starch is the major source of carbohydrates in the human diet. It is a mixture of about 20% amylose and 80% amylopectin. Amylose is a straight-chain polysaccharide made up entirely of *D*-glucopyranose molecules joined by α-1,4-glycoside bonds.

Amylopectin differs from amylose in that an occasional glucopyranose molecule in the polymer forms an additional α-glycoside bond at C6. Amylopectin, therefore, has a branched chain structure. It is estimated that branching occurs at one in every 20 to 25 glucopyranose molecules in the amylopectin chain.

Glycogen is another example of a polysaccharide consisting of coiled chains of *D*-glucopyranose molecules joined by α-1,4-glycoside bonds. Glycogen is used to store carbohydrates in animals. Like amylopectin, glycogen is a branched polymer of glucopyranose molecules joined by α-1,4- and α-1,6-glycoside bonds. It is formed from *D*-(+)-glucose obtained by hydrolysis of starch in the food we eat. The amount of glucose obtained in this way is often more that the body needs, so the excess is converted to glycogen and stored in the body. Between meals and during fasting, the body hydrolyzes glycogen to glucose as it is needed.

The extended linear chain of *D*-glucopyranose molecules joined by β-1,4-glycoside bonds is used in nature for cell membranes and the structural features of plants and trees. The coiled chain of *D*-glucopyranose molecules joined by α-1,4-glycoside bonds, on the other hand, is used in nature as an energy-storing polymer. Correlated with this difference in their functions is the widespread distribution of enzymes capable of hydrolyzing starch and glycogen and the very limited distribution of enzymes that hydrolyze cellulose. Cellulose is hydrolyzed in the gastrointestinal tract of herbivores, such as the cow, or in insects, such as termites, by a protozoan that synthesizes the enzyme cellulase, which hydrolyzes

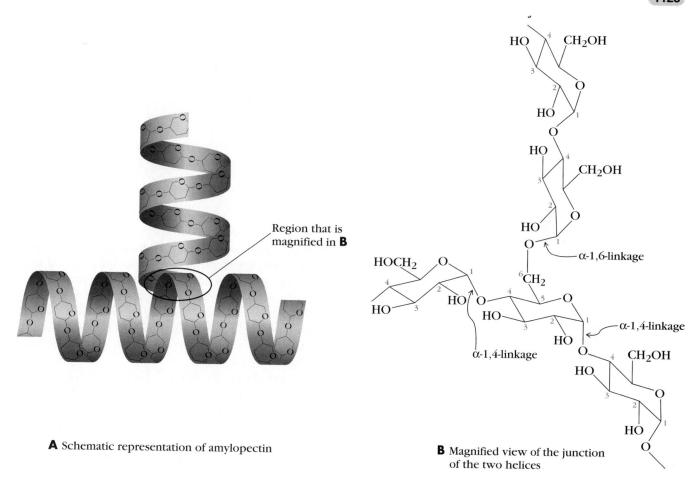

A Schematic representation of amylopectin

B Magnified view of the junction of the two helices

the β-1,4-glycoside bonds of cellulose. Humans do not have this enzyme, so we cannot use cellulose as a source of *D*-glucose.

25.14 The Fate of Carbohydrates in Humans

Enzyme-catalyzed hydrolysis of carbohydrates in the body produces monosaccharides that are used as a major source of energy. This process occurs by means of several sequences of chemical reactions. These sequences are known as pathways, and the principal ones are shown in Figure 25.10. We begin by examining the overall picture before studying the chemical reactions involved in several pathways in more detail.

Some of the carbohydrates needed by adults are obtained from lactose (milk sugar), sucrose (table sugar), and starch found in potatoes and cereals. Their digestion starts in the mouth. Saliva contains the enzyme α-amylase, which randomly hydrolyzes α-1,4-bonds on either side of a branch point of the polysaccharide. The hydrolysis of α-1,6-bonds (the glycoside bonds found at branch points) is not catalyzed by α-amylase. As a result, the polysaccharide remaining after hydrolysis by α-amylase is highly branched. The branches are removed by the debranching enzyme amylo-α-1,6-glucosidase. The straight chain polysaccharides that result from debranching can then be further hydrolyzed by α-amylase.

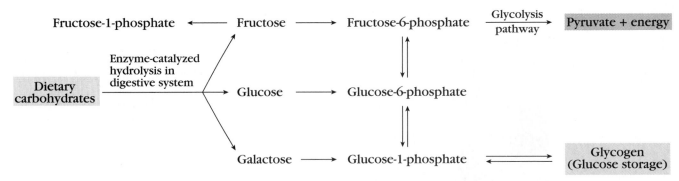

Figure 25.10
Hydrolysis of dietary carbohydrates forms monosaccharides, which are either stored or converted to energy and smaller molecules.

Three hexoses are obtained by digestion of dietary carbohydrates. Glucose is obtained from the hydrolysis of starch, glycogen, and maltose; glucose and fructose are obtained from the hydrolysis of sucrose; and glucose and galactose are obtained from the hydrolysis of lactose. After hydrolysis, glucose, fructose, and galactose enter the bloodstream and are carried to tissues, especially the liver. In the liver, galactose is phosphorylated and epimerized to glucose-1-phosphate, fructose is converted into either fructose-1-phosphate or fructose-6-phosphate, and glucose is converted into glucose-6-phosphate.

Glucose is the major source of energy for the brain and is an important source of energy for other organs such as the heart, kidneys, and muscles. Glucose is carried to these organs in the bloodstream, so adequate blood glucose levels must be maintained. The liver regulates blood glucose levels with chemical reactions that remove glucose from the blood and either store it as glycogen or convert it into carbon dioxide, water, and energy. Other chemical reactions in the liver hydrolyze glycogen stored there to provide glucose for the blood.

These sequences of reactions all start from a pool of three readily interconvertible compounds: fructose-6-phosphate, glucose-1-phosphate, and glucose-6-phosphate. At equilibrium, the pool contains 3% glucose-1-phosphate, 65% glucose-6-phosphate, and 32% fructose-6-phosphate. These hexose phosphates are continually consumed and replenished according to the needs of the body. Glucose, in excess of that needed for immediate use, is stored in the body as glycogen, which is formed from glucose-1-phosphate. Glucose-1-phosphate is also the first product obtained by hydrolysis of glycogen when the body needs to obtain energy from its stored glycogen. Glucose-6-phosphate is the first hexose phosphate formed in the pathway that converts free glucose to pyruvate and energy. Fructose-6-phosphate is the first hexose phosphate formed by gluconeogenesis, which is a series of reactions that synthesizes glucose from lactate. Fructose-6-phosphate is also the entry point for glycolysis that converts glucose to pyruvate and energy.

Let's now examine in detail the chemical reactions involved in several of these pathways. We begin with glycogen formation.

25.15 Biosynthesis of Glycogen

Biosynthesis of glycogen in the body occurs by an enzyme-catalyzed process in which glucopyranose units are added stepwise to a growing glycogen polymer. In order to react with the polymer, the glucopyranose must be activated by forming an α-*D*-glucopyranosyl nucleotide diphosphate. The most common example of an activated glucopyranose is uridine diphosphate glucose (UDP-glucose).

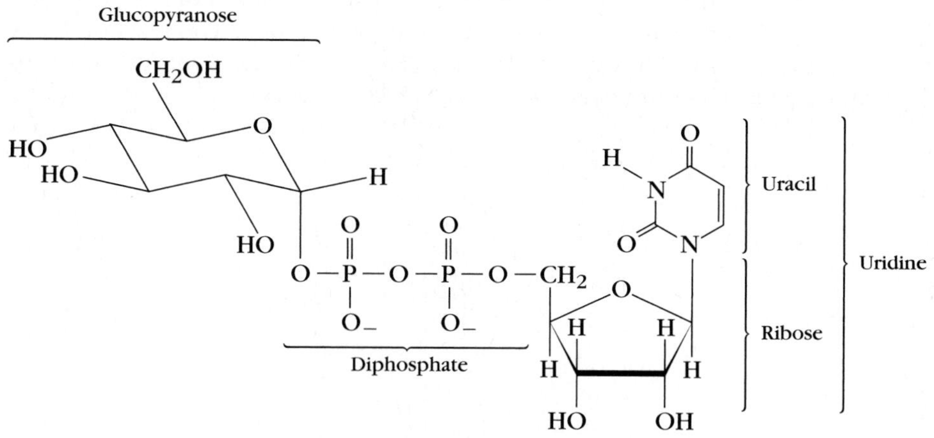

Uridine diphosphate glucose
(UDP-glucose)

UDP-glucose is synthesized from glucose-1-phosphate and uridine triphosphate (UTP) in a reaction catalyzed by UDP-glucose pyrophosphorylase. The pyrophosphate liberated in the reaction comes from the outer two phosphoryl groups of UTP.

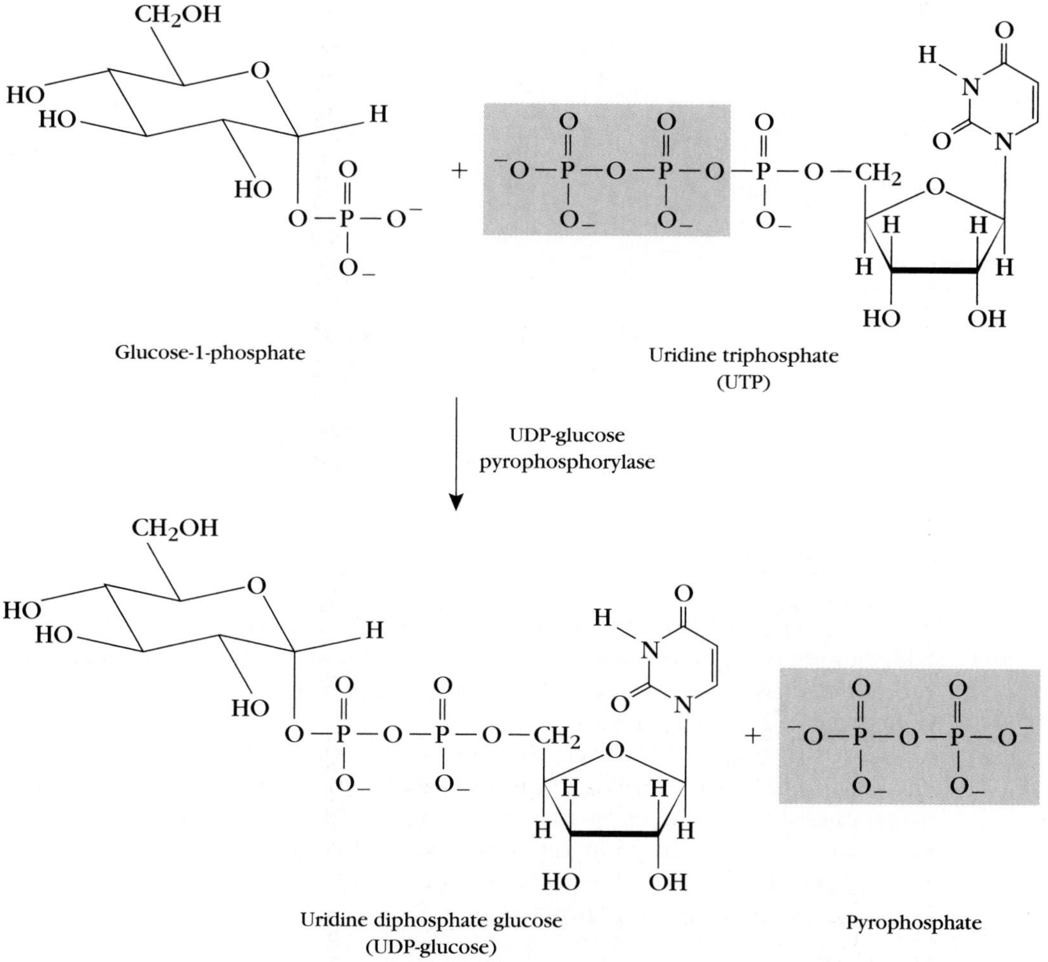

Glucose-1-phosphate

Uridine triphosphate
(UTP)

UDP-glucose
pyrophosphorylase

Uridine diphosphate glucose
(UDP-glucose)

Pyrophosphate

The glucopyranose of UDP-glucose is added to glycogen by a nucleophilic substitution reaction (see Chapter 12). In UDP-glucose, the uridine diphosphate ion is the leaving group and the nucleophile is the oxygen atom of the —OH group on C4 of the terminal glucopyranose unit of glycogen. The reaction is catalyzed by the enzyme glycogen synthetase. The overall reaction adds one glucopyranose unit to the nonreducing end of a glycogen chain.

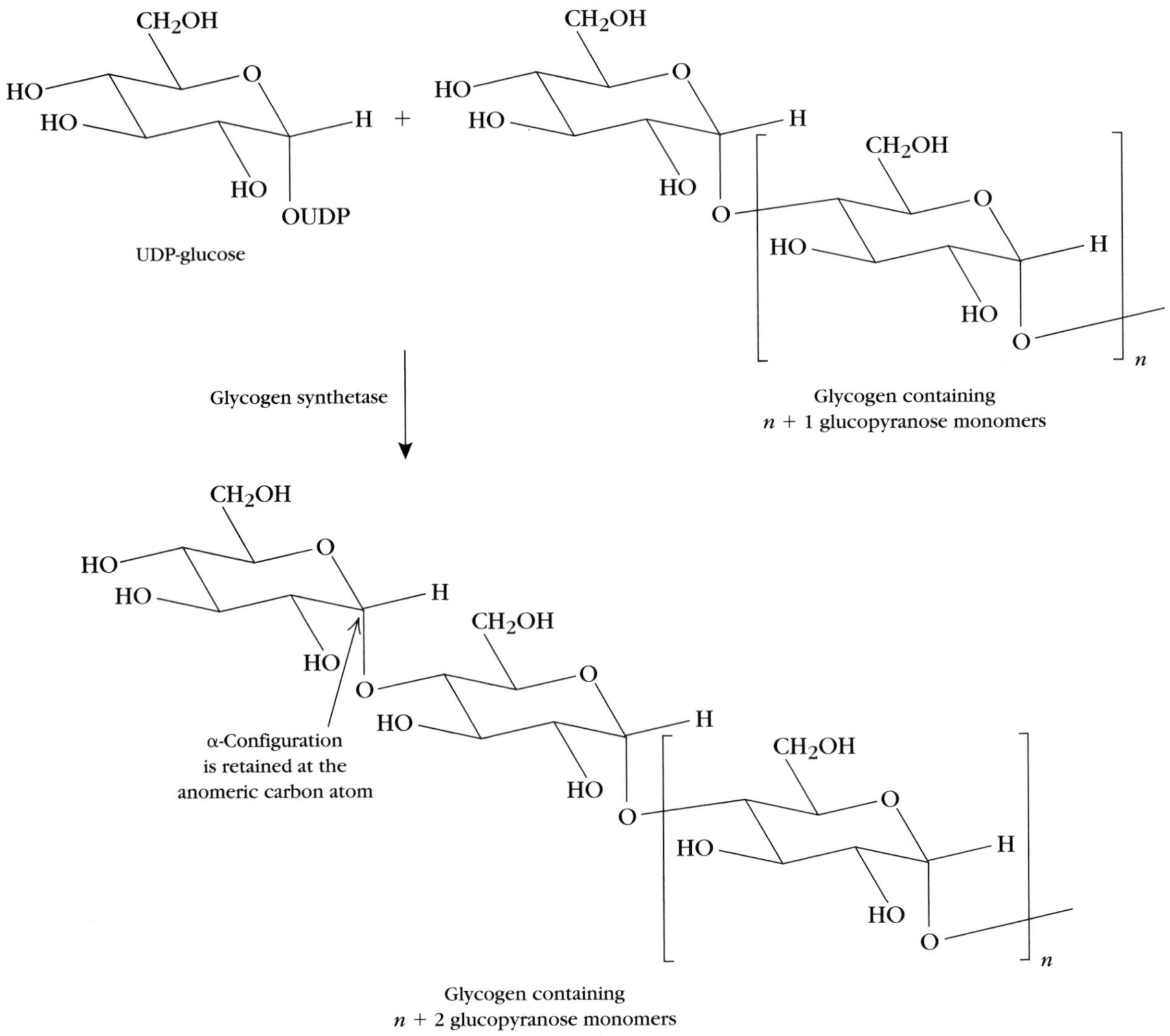

UDP-glucose

Glycogen containing
$n + 1$ glucopyranose monomers

Glycogen synthetase

α-Configuration
is retained at the
anomeric carbon atom

Glycogen containing
$n + 2$ glucopyranose monomers

The configuration at the anomeric carbon atom of the glucopyranose unit added to the chain is retained. It is not yet known whether this enzyme-catalyzed reaction occurs by a two-step S_N1 mechanism or two consecutive reactions that both occur by S_N2 mechanisms.

In an S_N1 mechanism, shown in Figure 25.11, a carboxyl group at the active site may protonate the leaving group to form a carbocation intermediate, which is stabilized by a neighboring carboxylate anion. The cation reacts in a second step with the —OH group at C4 of the glucopyranose ring at the end of the glycogen chain. The enzyme shields one face of the carbocation so that the reaction occurs with retention of configuration.

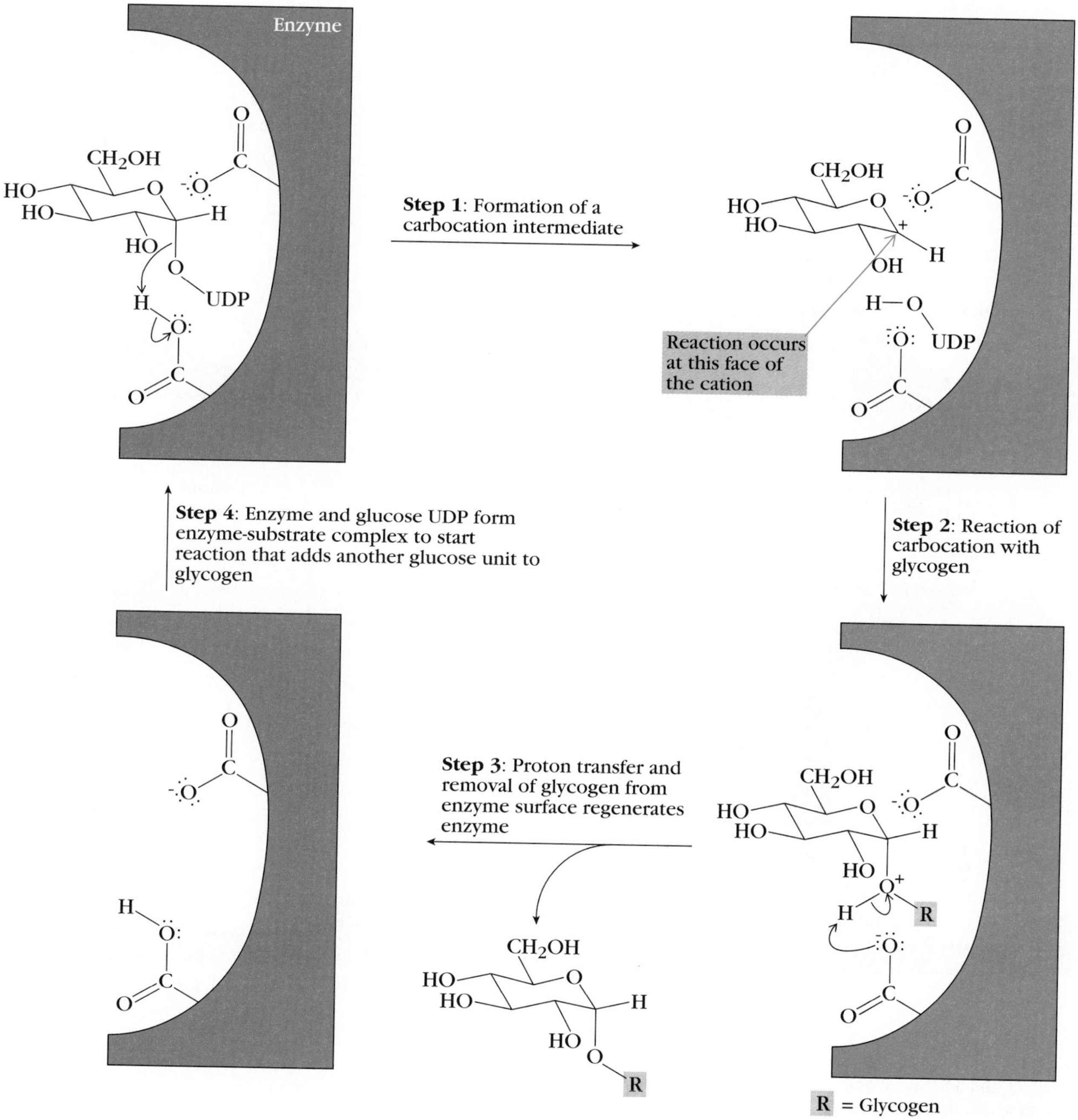

R = Glycogen

Figure 25.11
Enzyme-catalyzed addition of one glucose unit to glycogen by an S_N1 mechanism. One side of the carbocation formed in Step 1 of the mechanism is shielded so reaction occurs only at the opposite side to form a product with retention of configuration at the anomeric center.

In an alternative mechanism, shown in Figure 25.12, a carboxylate anion may serve as the nucleophile, which displaces the UDP group by an S_N2 mechanism to form a covalent glycosyl-enzyme intermediate. The reaction of this intermediate with the —OH group at the end of the glycogen chain also occurs by an S_N2 mechanism to add a glucopyranose molecule to the glycogen chain. Two nucleophilic displacement reactions that both occur by S_N2 mechanisms form a product with retention of configuration. The first reaction inverts the configuration at the reaction site, then the second reaction inverts it back to the original configuration.

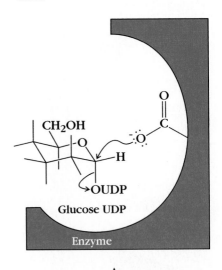

Step 1: Nucleophilic displacement by an S$_N$2 mechanism

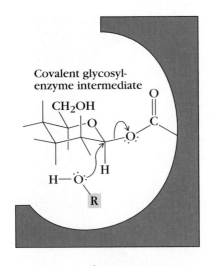

Step 4: Enzyme and glucose UDP form enzyme-substrate complex to start reaction that adds another glucose unit to glycogen

Step 2: Second displacement reaction occurs by an S$_N$2 mechanism

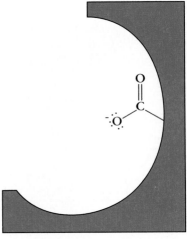

Step 3: Regeneration of enzyme catalyst

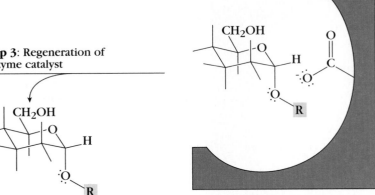

Glycogen with one glucose unit added

R = Glycogen

Figure 25.12
Enzyme-catalyzed addition of one glucose unit to glycogen that occurs by formation of a glycosyl-enzyme intermediate. The formation of the glycosyl-enzyme intermediate in Step 1 occurs by an S$_N$2 mechanism. The reaction of the glycosyl-enyme intermediate with glycogen in Step 2 also occurs by an S$_N$2 mechanism. In Step 1, the configuration at the anomeric carbon atom is inverted, then in Step 2 the configuration of the anomeric carbon atom is inverted back to its original configuration. The result is a product with retention of configuration.

Exercise 25.16   Biosynthesis of several important disaccharides involves the transfer of a glycosyl group from UDP-glucose to a hydroxy group of a monosaccharide. Write the equation for the formation of sucrose phosphate by the enzyme-catalyzed transfer of a glycosyl group from UDP-glucose to the hydroxy group on C2 of the cyclic hemiacetal form of fructose-6-phosphate.

Let's now examine the process by which glycogen is degraded to glucose in the liver.

25.16 Conversion of Glycogen to Glucose

Glycogen degradation occurs when the level of glucose in the blood is low. The enzyme glycogen phosphorylase catalyzes the cleavage of glycogen to form α-*D*-glucose-1-phosphate. This reaction takes place at the nonreducing ends of the polymeric glycogen chains and proceeds in a stepwise manner. One glucose molecule is removed at a time.

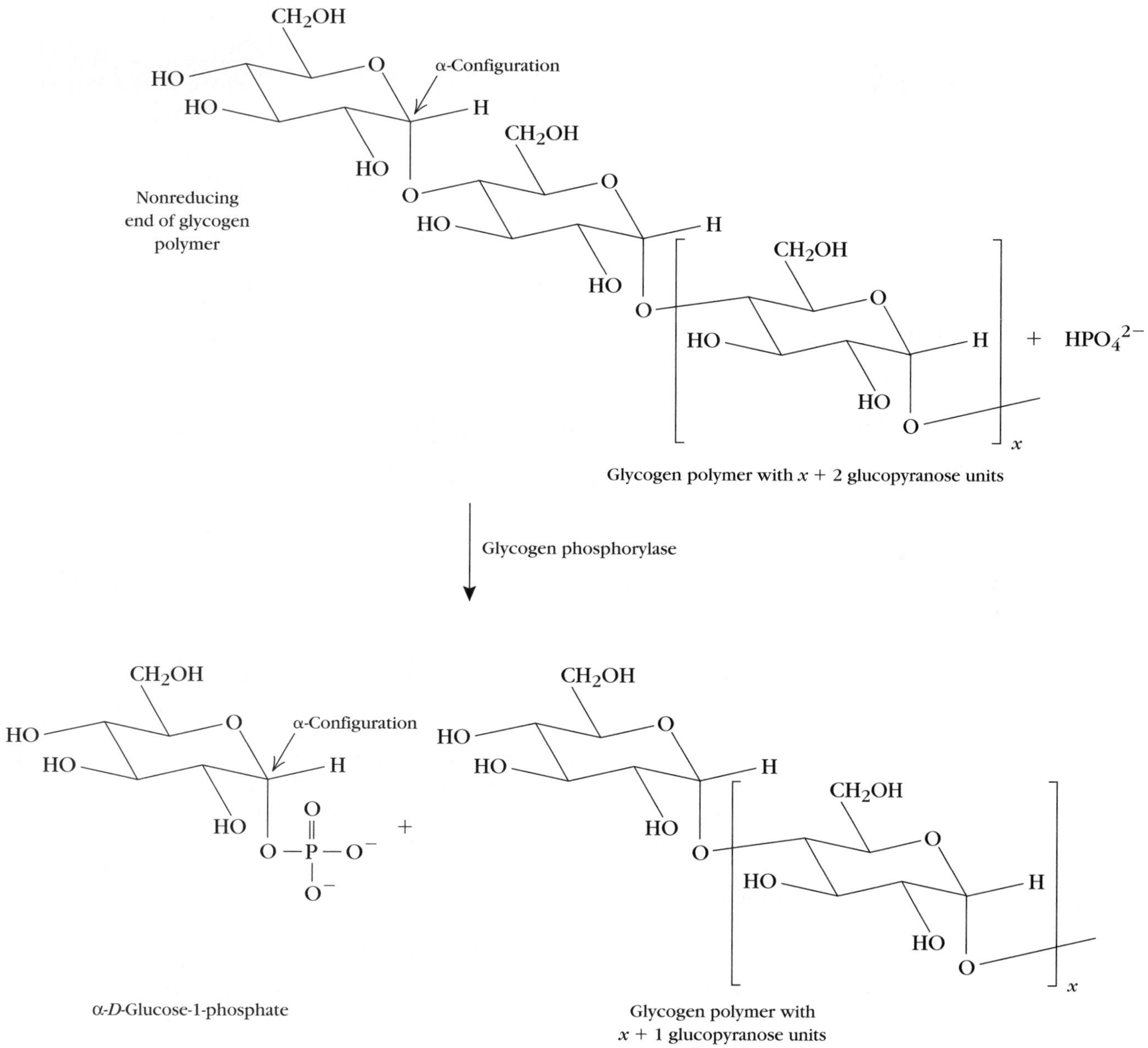

Nonreducing end of glycogen polymer

α-Configuration

$+ \ HPO_4^{2-}$

Glycogen polymer with $x + 2$ glucopyranose units

Glycogen phosphorylase

α-*D*-Glucose-1-phosphate

Glycogen polymer with $x + 1$ glucopyranose units

As in the biosynthesis of glycogen, the cleavage of glycogen to form α-*D*-glucose-1-phosphate occurs with retention of configuration. The most likely mechanism involves the carbocation intermediate shown in Figure 25.13.

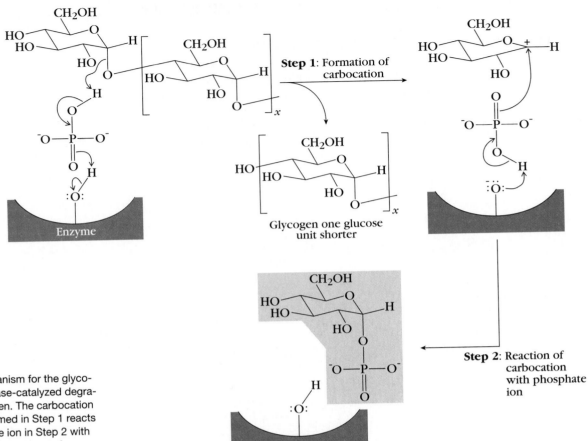

Figure 25.13
Proposed mechanism for the glycogen phosphorylase-catalyzed degradation of glycogen. The carbocation intermediate formed in Step 1 reacts with a phosphate ion in Step 2 with retention of configuration at the anomeric carbon atom to form glucose-1-phosphate.

25.17 Glycolysis

The sequence of 10 enzyme-catalyzed reactions that converts glucose into pyruvate and ATP is called **glycolysis.** These reactions are summarized in Figure 25.14.

The first reaction of glycolysis is the hexokinase-catalyzed transfer of a phosphoryl group from ATP to the hydroxy group on C6 of glucopyranose to form glucose-6-phosphate. This reaction is a nucleophilic substitution reaction at a phosphorus atom of ATP, in which the —OH group on C6 of glucopyranose is the nucleophile and ADP is the leaving group.

Glucopyranose   Adenosine triphosphate
(ATP)

Hexokinase

Glucose-6-phosphate

Adenosine diphosphate
(ADP)

The next reaction is the isomerization of glucose-6-phosphate to fructose-6-phosphate, which is catalyzed by phosphoglucose isomerase. The isomerization of glucose-6-phosphate to fructose-6-phosphate is the enzyme-catalyzed equivalent of the base-catalyzed enediol rearrangement discussed in Section 25.3, B. The proton transfer occurs between acidic and basic groups at the active site of the enzyme, as shown in Figure 25.15.

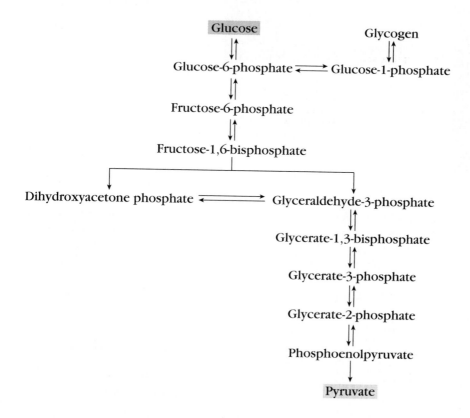

Figure 25.14
The major chemical reactions of glycolysis, which converts glucose into pyruvate and ATP.

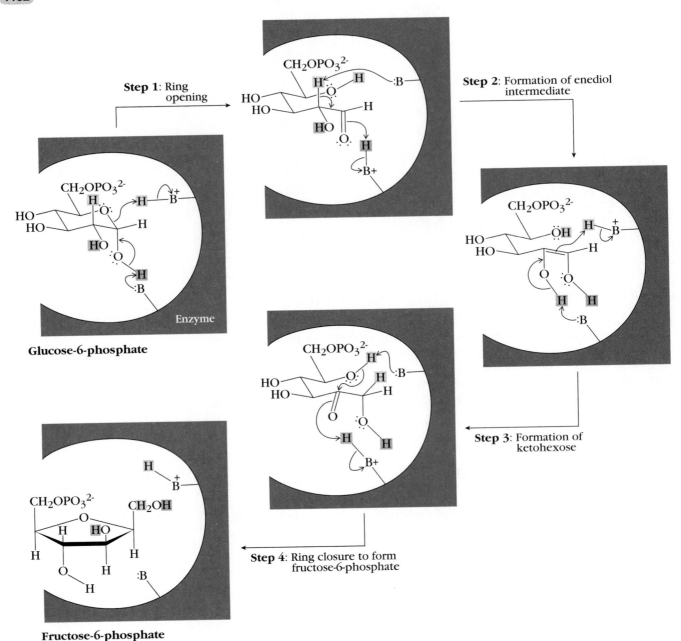

Figure 25.15
Mechanism of the glucose-6-phosphate isomerase-catalyzed isomerization of glucose-6-phosphate to fructose-6-phosphate.

Isomerization of glucose-6-phosphate to fructose-6-phosphate is followed by a second phosphorylation reaction, which forms fructose-1,6-bisphosphate. As in the formation of glucose-6-phosphate, a phosphoryl group is transferred from ATP to fructose-6-phosphate.

Fructose-6-phosphate $\xrightarrow[\text{ATP} \quad \text{ADP} + H_3O^+]{\text{Phosphofructokinase}}$ Fructose-1,6-bisphosphate

In the fourth reaction, fructose-1,6-bisphosphate (a hexose) is split into two trioses, dihydroxyacetone and glyceraldehyde-3-phosphate. The enzyme aldolase catalyzes this retroaldol condensation.

Fructose 1,6-bisphosphate   Dihydroxyacetone phosphate   Glyceraldehyde 3-phosphate

The mechanism of the retroaldol condensation of fructose-1,6-bisphosphate is discussed in Section 17.10.

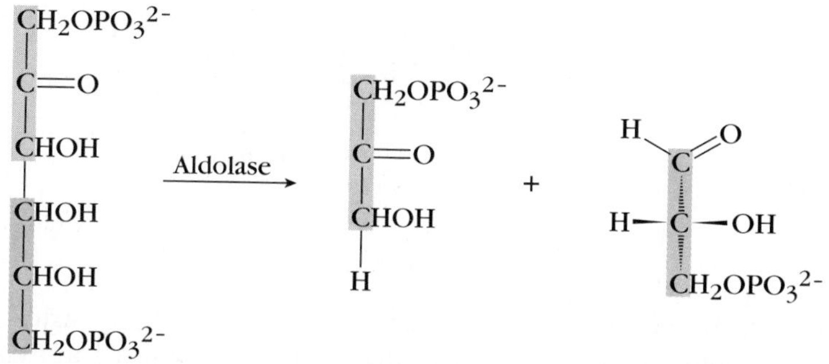

Of these two trioses, only glyceraldehyde-3-phosphate is used in glycolysis. While dihydroxyacetone phosphate is the three-carbon unit used for the biosynthesis of the glycerol part of fatty acids, most of it is converted to glyceraldehyde-3-phosphate by triose phosphate isomerase. In this way, each hexose is converted into two glyceraldehyde-3-phosphate molecules. In practical terms, this means that all the carbon atoms of hexoses are available for the production of energy by glycolysis.

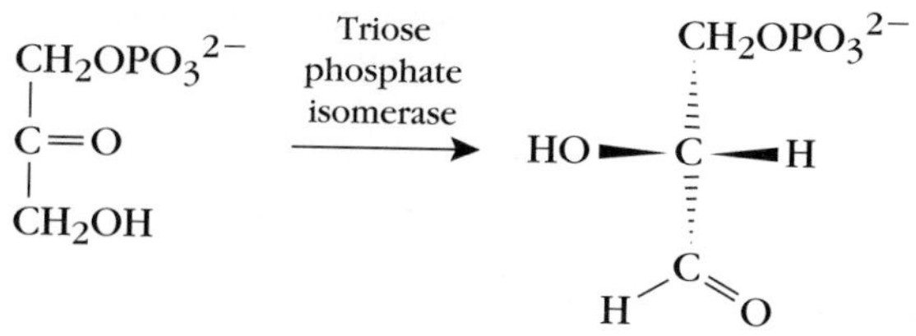

Dihydroxyacetone phosphate   Glyceraldehyde-3-phosphate

Glyceraldehyde-3-phosphate is an aldotriose, and dihydroxyacetone phosphate is its isomeric ketotriose. They have the same structural relationship as glucose-6-phosphate (an aldohexose) and fructose-6-phosphate (its isomeric ketohexose). As a result, both pairs undergo aldose-ketose isomerization reactions whose mechanisms involve an enediol intermediate. An enediol intermediate is a common feature of the mechanism of aldose-ketose isomerization reactions.

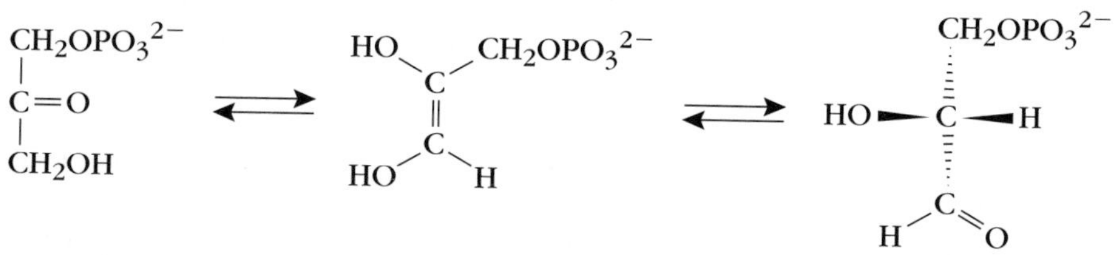

Dihydroxyacetone phosphate   Enediol intermediate   Glyceraldehyde-3-phosphate

The first oxidation reaction in the glycolysis pathway is the conversion of glyceraldehyde-3-phosphate to glycerate-1,3-bisphosphate. The reaction is catalyzed by glyceraldehyde-3-phosphate dehydrogenase and requires the presence of NAD$^+$. Glycerate-1,3-bisphosphate is an acyl phosphate, which is a mixed anhydride of phosphoric acid and a carboxylic acid.

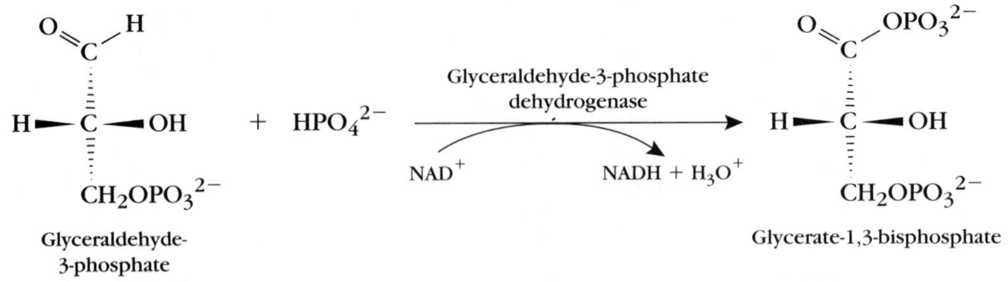

Glyceraldehyde-3-phosphate   Glycerate-1,3-bisphosphate

■ *The structures of NAD$^+$ and NADH are given in Section 11.12.*

The mechanism of this reaction is shown in Figure 25.16. After forming an enzyme-substrate complex in Step 1, glyceraldehyde-3-phosphate reacts with an enzyme-bound thiol group to form a thiohemiacetal intermediate in Step 2 (Section 14.11). The thiohemiacetal is oxidized in Step 3 by NAD$^+$ to form a thioester. Addition of phosphate ion to the carbonyl carbon atom of the thioester in Step 4 forms a tetrahedral intermediate, which breaks apart in Step 5 to form glycerate-1,3-bisphosphate. Finally, the NADH on the enzyme is oxidized to NAD$^+$, which regenerates the enzyme.

Acyl phosphates such as glycerate-1,3-bisphosphate are good acylating or phosphorylating agents in much the same way as acyl halides and carboxylic acid anhydrides are useful acylating reagents in organic synthesis. Thus in the presence of the enzyme 3-phosphoglycerate kinase, glycerate-1,3-bisphosphate transfers a phosphoryl group to ADP and regenerates ATP. In this way, part of the energy stored in hexoses is transferred to ATP, the major carrier of energy in living organisms.

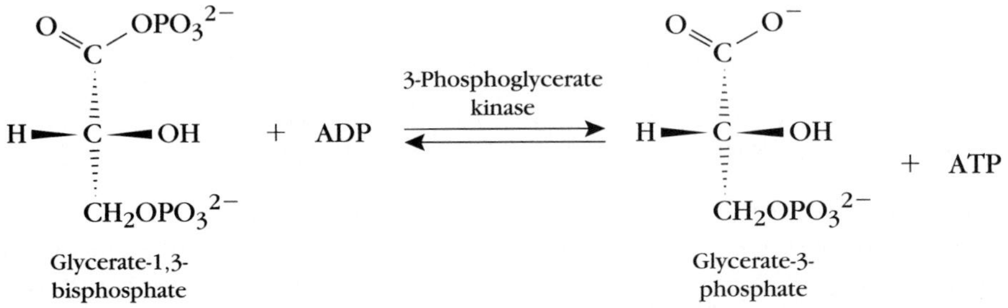

Glycerate-1,3-bisphosphate   Glycerate-3-phosphate

The conversion of glycerate-3-phosphate to glycerate-2-phosphate is not an intramolecular migration of a phosphoryl group from C3 to C2. Instead, the reaction occurs by the intermediate formation of glycerate-2,3-bisphosphate.

Glycerate-3-phosphate   Glycerate-2,3-bisphosphate   Glycerate-2-phosphate

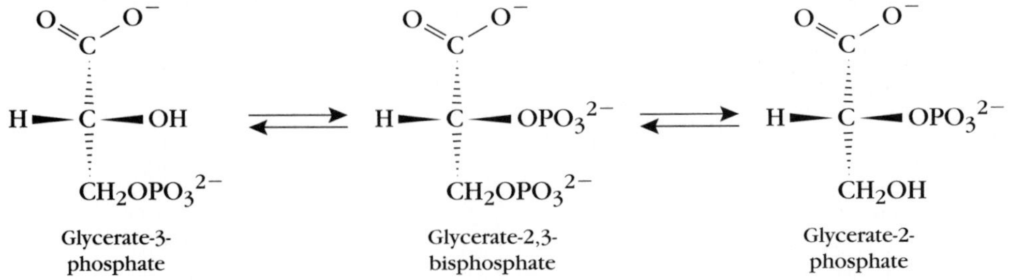

A nitrogen base in the active site of the enzyme acts as a phosphoryl group donor or acceptor in the mechanism of the conversion of glycerate-3-phosphate to glycerate-2-phosphate, as shown in Figure 25.17. The nitrogen base donates a

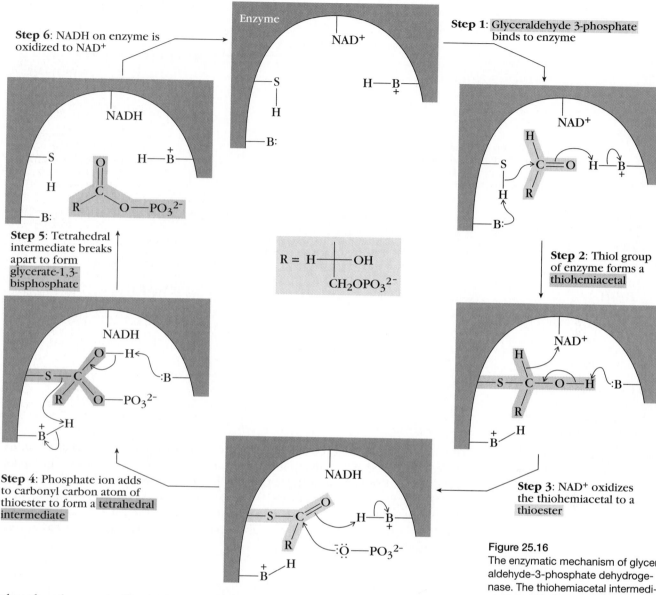

Step 6: NADH on enzyme is oxidized to NAD⁺

Step 1: Glyceraldehyde 3-phosphate binds to enzyme

Step 5: Tetrahedral intermediate breaks apart to form glycerate-1,3-bisphosphate

$$R = H-\!\!\!\!\begin{array}{c} | \\ | \end{array}\!\!\!\!-OH$$
$$CH_2OPO_3{}^{2-}$$

Step 2: Thiol group of enzyme forms a thiohemiacetal

Step 4: Phosphate ion adds to carbonyl carbon atom of thioester to form a tetrahedral intermediate

Step 3: NAD⁺ oxidizes the thiohemiacetal to a thioester

Figure 25.16
The enzymatic mechanism of glyceraldehyde-3-phosphate dehydrogenase. The thiohemiacetal intermediate formed in Step 2 is oxidized in Step 3 by NAD⁺ to form a thioester intermediate. The thioester intermediate reacts with phosphate ion in Step 4 to form a tetrahedral intermediate, which breaks apart to form glycerate-1,3-bisphosphate in Step 5. The enzyme is regenerated by the oxidation of NADH to NAD⁺.

phosphoryl group to the oxygen atom on C2 in Step 2 of the mechanism to form glycerate-2,3-bisphosphate as an intermediate. In Step 3, the nitrogen base accepts a phosphoryl group from C3 to form glycerate-2-phosphate. In this way, a phosphoryl group is rapidly transferred between glycerate-3-phosphate and glycerate-2-phosphate.

The next step is the enolase-catalyzed elimination of water from glycerate-2-phosphate to form phosphoenolpyruvate.

Glycerate-2-phosphate   Enolase   Phosphoenolpyruvate

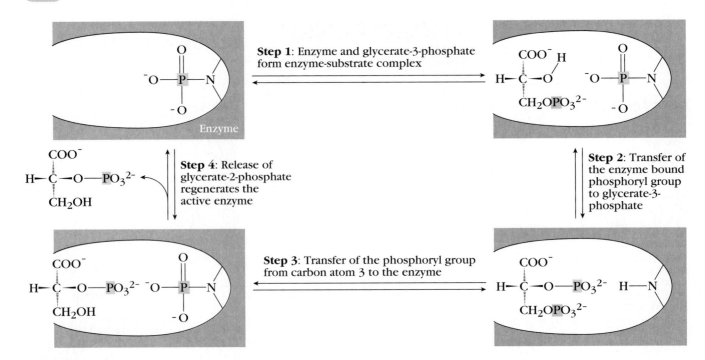

Figure 25.17
Mechanism for the conversion of glycerate-3-phosphate to glycerate-2-phosphate. An enzyme-bound phosphoryl group is transferred to glycerate-3-phosphate in Step 2 to form an intermediate glycerate-2,3-bisphosphate. Transfer of the phosphoryl group on carbon atom 3 to the enzyme in Step 3 forms glycerate-2-phosphate.

The mechanism is believed to occur by the rapid formation of a carbanion intermediate followed by elimination of the OH group on C3, as shown in Figure 25.18. An acid and a base at the active site facilitate the reaction.

The last reaction of glycolysis is the pyruvate kinase-catalyzed transfer of a phosphoryl group from phosphoenolpyruvate to ADP to form ATP and enolpyruvate. This is the second reaction of glycolysis in which part of the energy stored in hexoses is transferred to ATP.

The mechanisms of several of the reactions discussed in this section are similar to those we have learned in earlier chapters. The isomerizations of glucose-6-phosphate to fructose-6-phosphate and dihydroxyacetone phosphate to glyceraldehyde-3-phosphate, for example, are the enzyme-catalyzed equivalent of the enolization reactions that we first encountered in Chapter 17. The reaction that cleaves fructose-1,6-bisphosphate into dihydroxyacetone phosphate and glyceraldehyde-3-phosphate is the enzyme-catalyzed equivalent of a retroaldol reaction, another reaction first introduced in Chapter 17. These examples illustrate once again that learning the mechanistic principles of organic chemistry is an important part of understanding biologically important reactions.

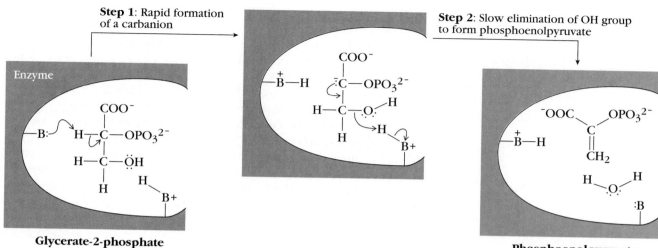

Step 1: Rapid formation of a carbanion

Step 2: Slow elimination of OH group to form phosphoenolpyruvate

Enzyme

Glycerate-2-phosphate

Phosphoenolpyruvate

Figure 25.18
The mechanism of the enolase-catalyzed elimination of water from glycerate-2-phosphate to form phosphoenolpyruvate. The carbanion intermediate, formed in Step 1 of the mechanism, eliminates a hydroxide ion in Step 2 to form phosphoenolpyruvate. An acid and a base at the active site facilitate the reaction.

25.18 Summary

Carbohydrates are polyhydroxy aldehydes and ketones or substances that yield such compounds on acid-catalyzed hydrolysis. Carbohydrates are further classified according to their hydrolysis products. **Monosaccharides** or simple sugars are carbohydrates that cannot be hydrolyzed into smaller compounds. **Disaccharides** can be hydrolyzed into two monosaccharides, and hydrolysis of **polysaccharides** yields several thousand monosaccharides.

Monosaccharides can be classified according to the number of carbon atoms and the kind of carbonyl group they contain. A five-carbon monosaccharide that contains a keto group, for example, is classified as a ketopentose. Glucose, which contains six carbon atoms and an aldehyde group, is classified as an aldohexose.

Monosaccharides undergo the usual reactions of aldehydes, ketones, and alcohols. The carbonyl group of aldoses can be oxidized by Tollen's reagent or Fehling's solution and by bromine water to form **aldonic acids;** warm nitric acid oxidizes aldoses and hexoses to **aldaric acids;** and $NaBH_4$ reduces the carbonyl groups of aldoses and ketoses to **alditols.** The **Kiliani-Fischer synthesis** increases the number of carbon atoms in the parent chain of a monosaccharide by one, while the **Ruff degradation** shortens the parent chain by one carbon atom.

Monosaccharides exist primarily as **cyclic hemiacetals** rather than open chain aldehydes or ketones. A **furanose** is a five-membered ring hemiacetal, and a **pyranose** is a six-membered ring hemiacetal. A new stereocenter, called an **anomeric center,** is formed on cyclization. As a result, carbohydrates can exist as two diastereomeric hemiacetals, which are called alpha (α-) and beta (β-) anomers. Equilibration of the two anomers occurs in aqueous solution through the open chain form as an intermediate by a process called **mutarotation.**

The stereochemical relationships among monosaccharides can be shown by **Fischer projection formulas** in which the stereocenters are represented by pairs of crossed lines. Five- and six-membered cyclic hemiacetals are often portrayed by **Haworth projections.** The chair form of the pyranose ring most accurately represents the stereochemistry of six-membered cyclic acetals.

All monosaccharides belong to either the *D*- or *L*- family. In members of the *D*-family, the —OH group on the stereocenter farthest removed from the carbonyl group points to the right in their Fischer projection formulas.

Disaccharides contain two monosaccharides joined by a **glycoside bond.** The two monosaccharides can be the same (as in maltose) or different (as in sucrose). The most common glycoside bonds are 1,4-bonds, which join the anomeric carbon atom of one monosaccharide with the —OH group on C4 of the other monosaccharide. The glycoside bond can be either alpha (α-), as in maltose, or beta (β-), as in cellobiose.

Carbohydrates are a major source of energy for the human body. The energy is obtained from carbohydrates by a long sequence of chemical reactions that begins by converting carbohydrates to *D*-(+)-glucose. A sequence of 10 enzyme-catalyzed reactions called **glycolysis** then converts *D*-(+)-glucose to pyruvate and energy (as ATP). Many of the reactions of glycolysis illustrate the similarities of the mechanisms of reactions that we have described previously in this text and their analogs in the body.

25.19 Summary of Reactions

1. **Reactions of Carbonyl Group**

   a. **Enediol rearrangement**   SECTIONS 25.3, A and 25.3, B

   Aldoses or ketoses both isomerize to a mixture of the aldose and ketose by means of an enediol intermediate when placed in a basic solution. Configuration at C2 of the aldose is lost by epimerization.

   $$\text{Aldose} \underset{H_2O}{\overset{^-OH}{\rightleftharpoons}} \text{Enediol} \underset{H_2O}{\overset{^-OH}{\rightleftharpoons}} \text{Ketose}$$

   Aldose:
   CHO
   H—OH
   (CHOH)$_n$
   CH$_2$OH

   Enediol:
   H—C—OH
   ‖—OH
   (CHOH)$_n$
   CH$_2$OH

   Ketose:
   CH$_2$OH
   ‖O
   (CHOH)$_n$
   CH$_2$OH

   b. **Oxidation**   SECTION 25.3, C

   Bromine water and Tollen's reagent or Fehling's solution oxidize aldoses to an aldonic acid. Aldonic acids are best prepared by bromine oxidation.

   $$\text{Aldose} \xrightarrow{\text{Oxidation}} \text{Aldonic acid}$$

   Aldose:
   CHO
   H—OH
   (CHOH)$_n$
   CH$_2$OH

   Aldonic acid:
   COOH
   H—OH
   (CHOH)$_n$
   CH$_2$OH

   c. **Reduction**   SECTION 25.3, D

   NaBH$_4$ and catalytic hydrogenation are the usual methods of reducing the carbonyl groups of carbohydrates.

   $$\text{An aldose} \xrightarrow{\text{Reduction}} \text{An alditol}$$

   An aldose:
   CHO
   H—OH
   (CHOH)$_n$
   CH$_2$OH

   An alditol:
   CH$_2$OH
   H—OH
   (CHOH)$_n$
   CH$_2$OH

d. Kiliani-Fischer chain extension   SECTION 25.3, E

Nucleophilic addition of HCN to the carbonyl group followed by hydrolysis of the nitrile forms an aldonolactone when heated. Reduction of the aldonolactone by sodium amalgam forms a mixture of the stereoisomers of an aldose containing one more carbon atom than the starting aldose.

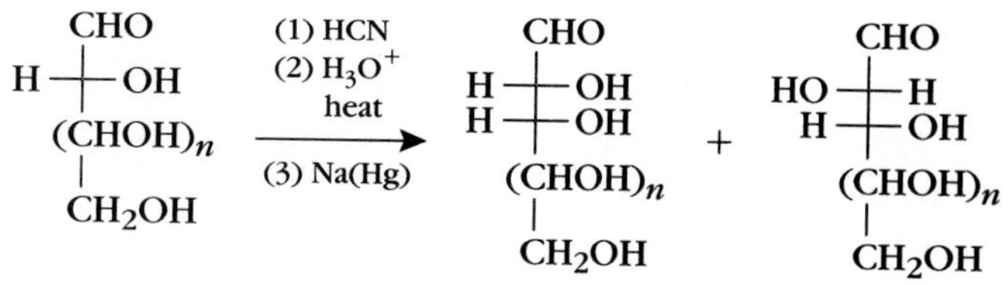

e. Ruff degradation chain shortening   SECTION 25.3, F

A method of shortening the carbon chain of an aldose by one carbon atom.

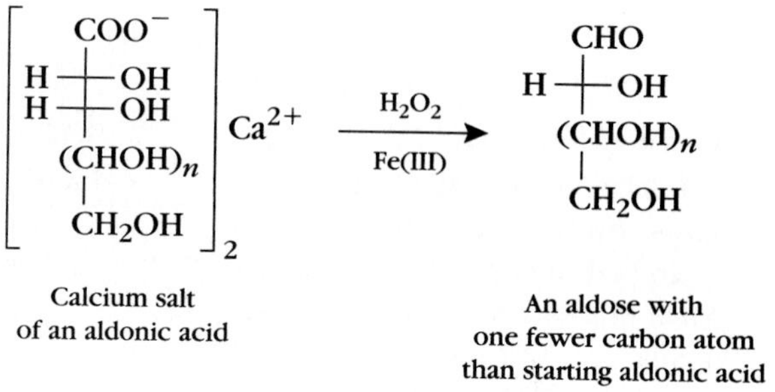

Calcium salt
of an aldonic acid

An aldose with
one fewer carbon atom
than starting aldonic acid

f. Mutarotation   SECTION 25.8

When a pure sample of either α- or β-D-(+)-glucopyranose is placed in water, an equilibrium mixture of the α-anomer, the β-anomer, and a very small amount of the open chain form of glucose is formed.

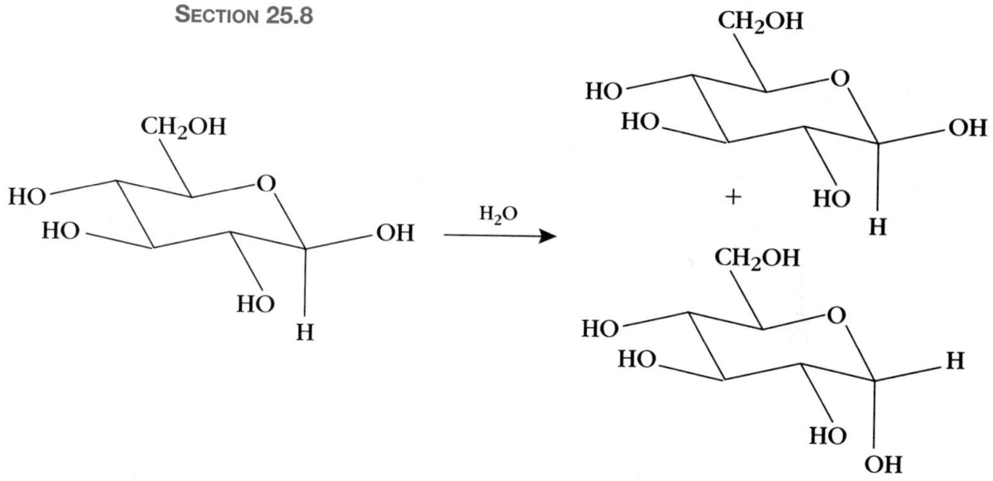

g. Glycoside formation   SECTION 25.11

The product is a mixture of the α- and β-anomers.

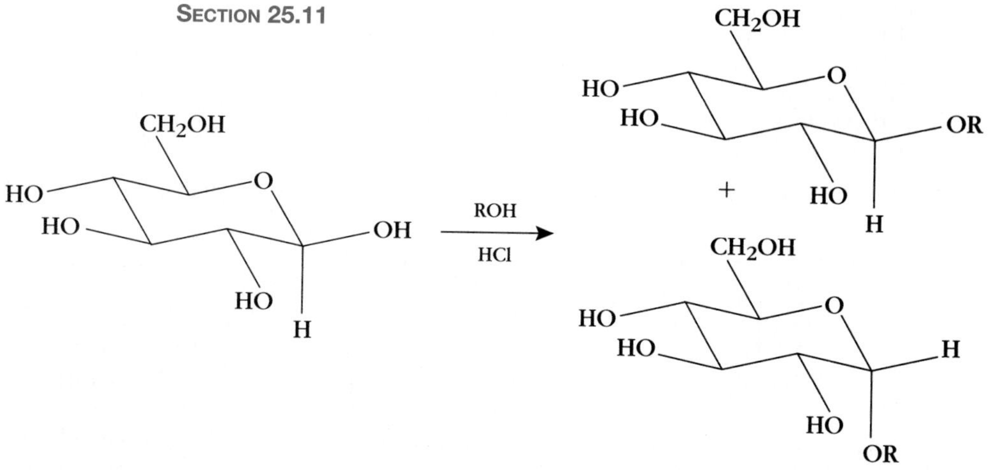

2. Reactions of Hydroxy Groups

a. Ester formation

SECTION 25.10

Reaction of carbohydrates with acylating agents forms esters with all available hydroxy groups.

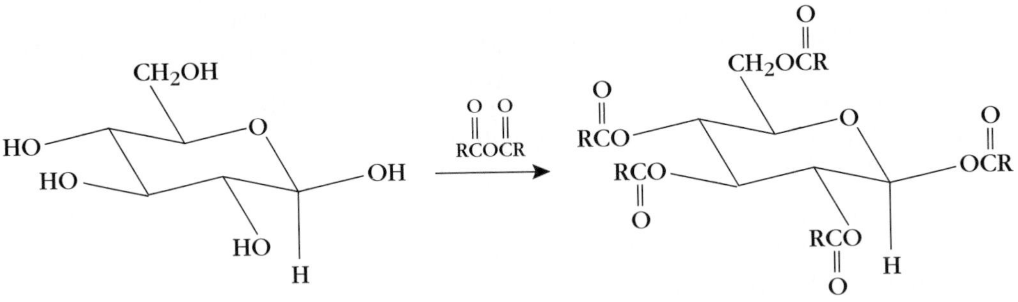

b. Ether formation

SECTION 25.10

Haloalkanes or dialkyl sulfates and NaOH react with carbohydrates to form ethers with all the available —OH groups. This is an application of the Williamson ether synthesis to carbohydrates.

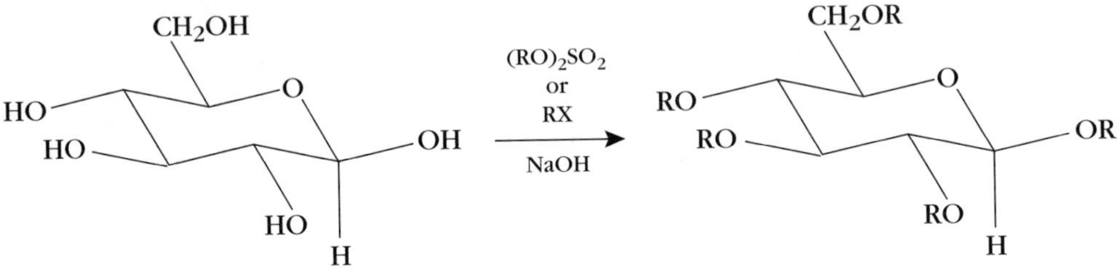

c. HNO₃ oxidation

SECTION 25.3, C

Nitric acid oxidizes both the aldehyde and the —OH group of the —CH_2OH group to form an aldaric acid.

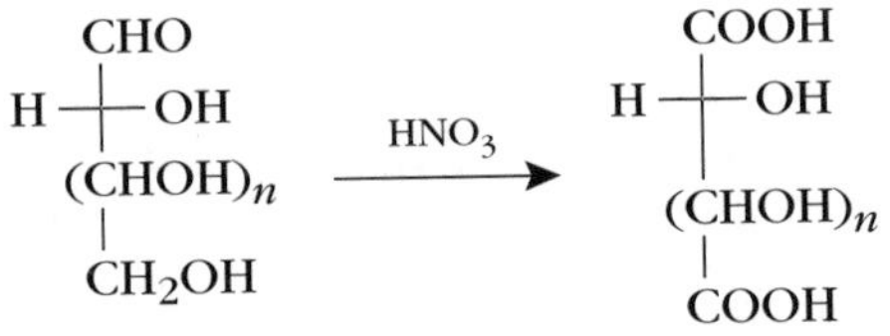

An aldose   An aldaric acid

Additional Exercises

25.17 Define each of the following terms in words, by a structure, or by a chemical equation.

- (a) Monosaccharide
- (b) D-(+)-glucose
- (c) Polysaccharide
- (d) Epimers
- (e) An aldonic acid
- (f) Mutarotation
- (g) Kiliani-Fischer synthesis
- (h) Ruff degradation
- (i) Furanose ring
- (j) Anomer
- (k) Pyranose ring
- (l) Deoxy sugar
- (m) Amino sugar
- (n) A glycoside
- (o) Glycogen
- (p) Disaccharide
- (q) Glycolysis
- (r) Enediol intermediate

25.18 Classify each of the following monosaccharides according to the number of carbon atoms and the type of carbonyl group it contains.

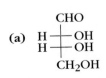

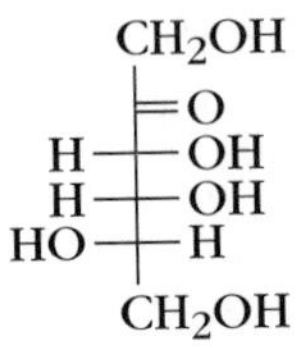

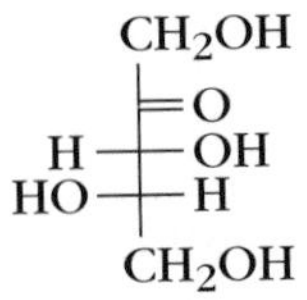

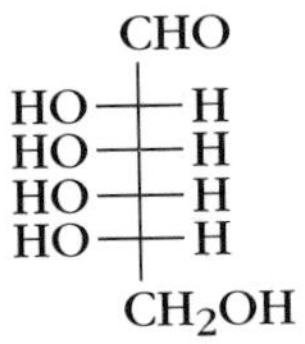

25.19 Label the stereocenters in each of the monosaccharides in Exercise 25.18 by an asterisk and determine the maximum number of stereoisomers of each.

25.20 Assign each of the monosaccharides in Exercise 25.18 to either the *D*- or *L*- family.

25.21 Write the cyclic hemiacetal structures for each of the monosaccharides in Exercise 25.18. Indicate which is the α-anomer and which is the β-anomer.

25.22 Write the Fischer projection formula for each of the following.

    (a) A ketopentose   **(b)** A ketotetrose   **(c)** An aldoheptose   **(d)** A ketohexose

25.23 Write the Fischer projection formula for each of the following cyclic monosaccharides.

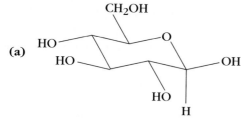

25.24 Write the Fischer projection formulas for the products of $NaBH_4$ reduction of each of the following monosaccharides. Which of the products are optically active?

    (a) *D*-(−)-Ribose   **(b)** *D*-(−)-Arabinose   **(c)** *L*-(+)-Xylose   **(d)** *D*-(−)-Lyxose

25.25 Write the Fischer projection formulas for the products of the reaction of HNO_3 with each of the monosaccharides in Exercise 25.24. Which of the products are optically active?

25.26 Write the structure of the products, if any, of the reaction of α-*D*-galactopyranose with each of the following reagents.

    (a) Acetic anhydride/sodium acetate   **(c)** HCl/H_2O **(e)** Br_2/H_2O
    (b) $(CH_3O)_2SO_2/NaOH/H_2O$ **(d)** Fehling's solution

25.27 Write the structure of the products, if any, of the reaction of Br_2/H_2O with each of the following compounds.

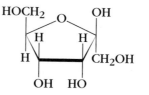
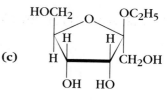

25.28 Which pentose gives the same aldaric acid on oxidation as does the oxidation of *D*-(-)-arabinose?

25.29 Write the structures of the aldonic acids and aldaric acids obtained by oxidation of each of the following monosaccharides.

25.30 Does ascorbic acid belong to the *D*- or the *L*- family?

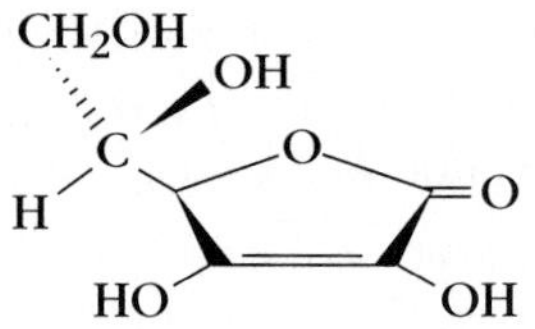

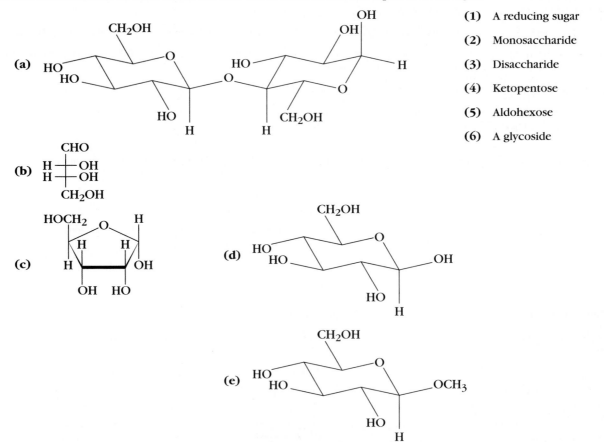

25.31 Write the equations that illustrate the mutarotation of β-maltose (the structure of β-maltose is given on p. 1119).

25.32 Recall from Chapter 14 that 1,2- and 1,3-diols react with aldehydes and ketones to form cyclic acetals. Certain monosaccharides with hydroxy groups in the proper location react with acetone to form a bicyclic acetal. Write a mechanism for this reaction.

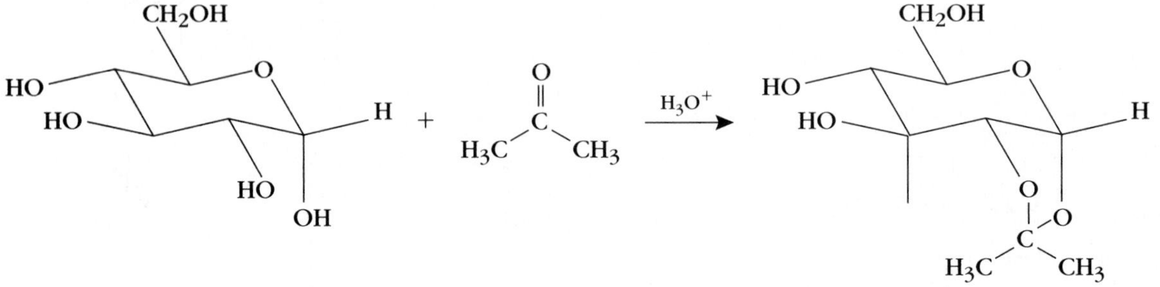

25.33 Match each of the structures on the left with one or more correct descriptions on the right.

(1)   A reducing sugar

(2)   Monosaccharide

(3)   Disaccharide

(4)   Ketopentose

(5)   Aldohexose

(6)   A glycoside

25.34 Write the mechanism of the acid-catalyzed hydrolysis of methyl α-*D*-glucopyranoside. How does your mechanism explain that both α- and β-*D*-glucopyranosides give the same mixture of α- and β-*D*-glucopyranose?

25.35 *D*-Fructose exists in aqueous solution as a mixture of β-*D*-fructofuranose, α-*D*-fructofuranose, β-*D*-fructopyranose, α-*D*-fructopyranose, and the open chain form. Write stereochemical representations of all five forms of *D*-fructose.

25.36 Talose has the following cyclic structure:

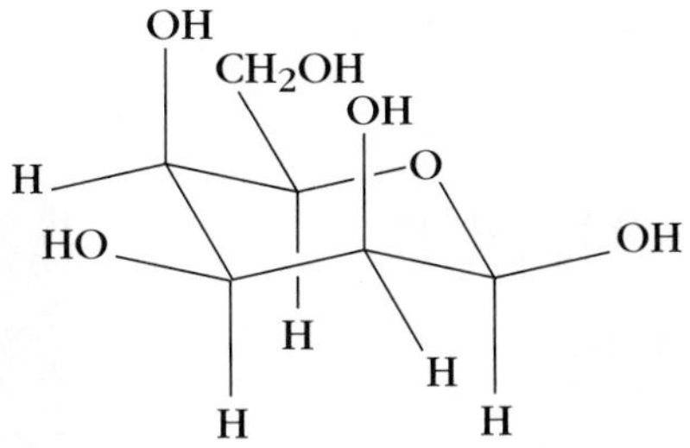

Talose

(a) Is this the furanose or pyranose form of talose?
(b) Is this the structure of the α- or β- anomer?
(c) Is this the structure of *D*-talose or *L*-talose?
(d) Is talose a reducing sugar?
(e) Write the Fischer projection formula of talose.

25.37 When treated with acid, *D*-glucaric acid undergoes the following reaction:

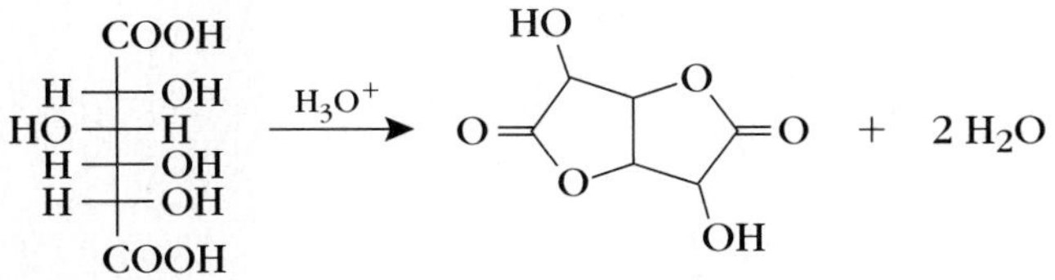

D-Glucaric acid

The bicyclic product is formed by two successive cyclization reactions. Propose a mechanism for this reaction.

25.38 On the basis of the following observations, deduce the structure and stereochemistry of trehalose.

(a) The molecular formula of trehalose is $C_{12}H_{22}O_{11}$.
(b) Trehalose does not reduce Tollen's reagent or Fehling's solution.
(c) Acid-catalyzed hydrolysis of one equivalent of trehalose yields two equivalents of *D*-glucose.
(d) Methylation of one equivalent of trehalose with $(CH_3O)_2SO$/NaOH followed by acid-catalyzed hydrolysis yields two equivalents of the following compound:

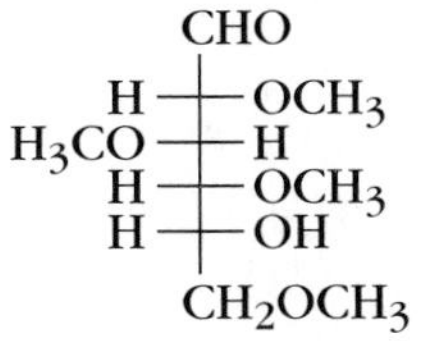

(e) Trehalose is hydrolyzed by enzymes that catalyze the hydrolysis of α-glycosides but not by enzymes that catalyze the hydrolysis of β-glycosides.

25.39 When placed in a basic aqueous solution *D*-(+)-glucose isomerizes to a mixture of *D*-(+)-glucose, *D*-(+)-mannose, and (*D*)-(−)-fructose. Using Fischer projection formulas, write a mechanism, including any intermediates, for the base-catalyzed isomerization of *D*-(+)-glucose.

25.40 Determine which carbon atom of pyruvate would contain ^{14}C if glucose containing ^{14}C at C1 is metabolized by the glycolysis pathway.

25.41 Write the structure of four monosaccharides joined together so that they represent part of the structure of amylose. Clearly indicate the stereochemistry of each glycoside linkage.

25.42 Using the answer to Exercise 25.41 as the starting material, write a mechanism for the acid-catalyzed hydrolysis of amylose to form four monosaccharides.

CHAPTER

AMINO ACIDS, POLYPEPTIDES, PROTEINS, and ENZYMES

Proteins are natural polymers made up of α-amino acid monomers. Proteins are the major components of muscles, skin, and bones; they transport small molecules and ions within systems; and other proteins known as antibodies recognize and neutralize invading foreign substances such as viruses and bacteria.

Despite the variety of their roles, their differences in molecular weights (which range from several thousand to several million amu), and the differences in their physical properties (silk, skin, and bone are all protein), proteins are sufficiently similar in their molecular structure that we can classify them as a single family of compounds. In fact, proteins are another example of the enormous variety in chemical and physical properties that can be obtained by varying a simple structural theme.

We begin the study of proteins by examining the 20 α-amino acids that join together to form proteins.

■ *The word* protein *is derived from the Greek* proteios, *meaning of first importance.*

26.1   Structure, Physical Properties, and Stereochemistry of α-Amino Acids

All proteins are composed of the 20 common α-amino acids shown in Table 26.1. All α-amino acids, with the exception of proline, have a hydrogen atom, a primary amino group, and a carboxyl group bonded to a tetrahedral carbon atom (the α-carbon atom). The 20 α-amino acids differ in the structure of the fourth group bonded to the α-carbon atom. This group is called the **amino acid side chain,** and it is designated by the letter *R*.

α–Carbon atom
Amino acid side chain

Proline is an exception to this general structure because it has a secondary amino group bonded to the α-carbon atom. As a result, proline is actually an α-imino acid, but it is nevertheless commonly referred to as an amino acid.

In neutral solution (pH ≈ 7), the amino group of an amino acid exists as an ammonium ion and the carboxylic acid group exists as a carboxylate ion.

■ *Recall that in neutral aqueous solutions, carboxylic acids exist predominantly as their carboxylate ions (Section 15.8), while amines exist predominantly as their substituted ammonium ions (Section 22.5).*

Because of this separation of charges, amino acids are a kind of internal salt, so their physical properties are characteristic of ionic compounds. Most α-amino acids, therefore, have large dipole moments, have melting points near 300 °C, and are very soluble in water but largely insoluble in nonpolar solvents.

The 20 α-amino acids listed in Table 26.1 are commonly classified according to the polarities of their side chains (*R* group). The first category contains nine α-amino acids with relatively nonpolar side chains; the second category contains six α-amino acids with uncharged polar side chains; and the third category con-

Table 26.1	The Structures and Names of the 20 Common α-Amino Acids

Amino acids with nonpolar side chains:

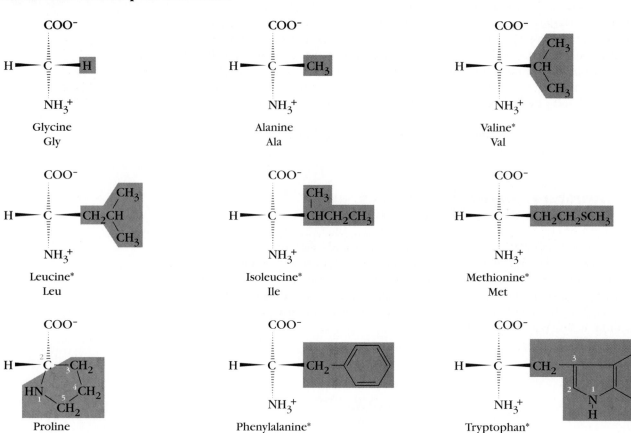

Amino acids with uncharged polar side chains:

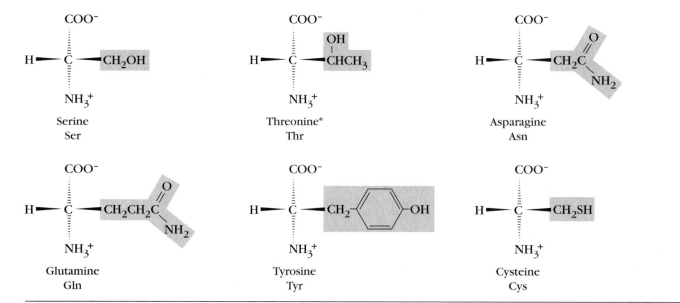

Table 26.1—cont'd	The Structures and Names of the 20 Common α-Amino Acids

Amino acids with charged polar side chains:

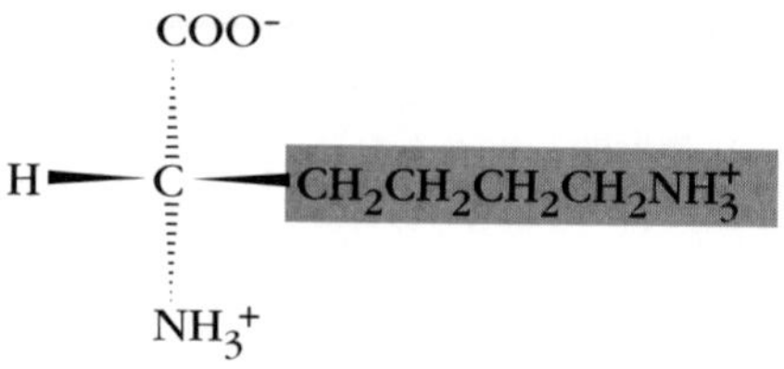

Lysine*
Lys

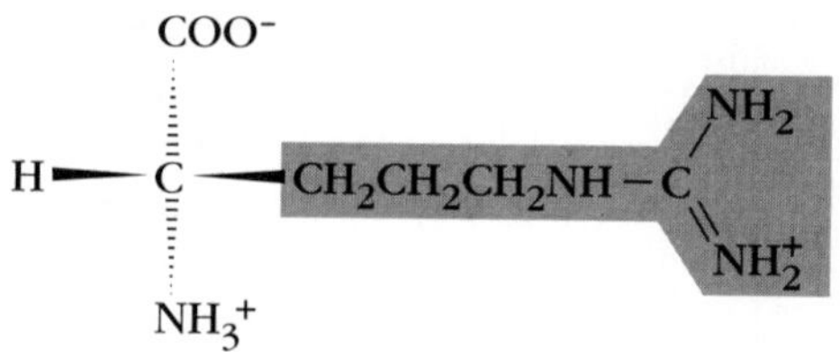

Arginine*
Arg

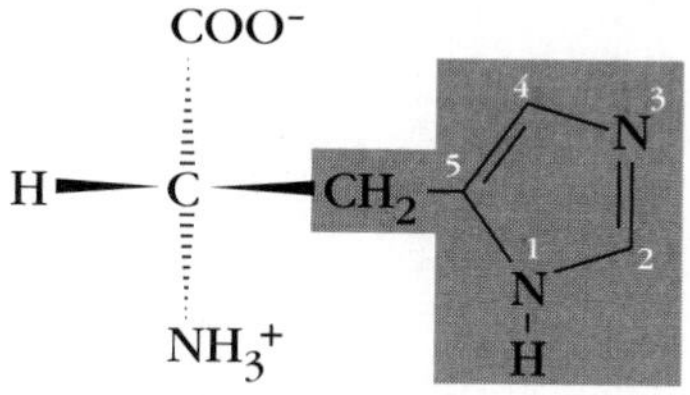

Histidine*
His

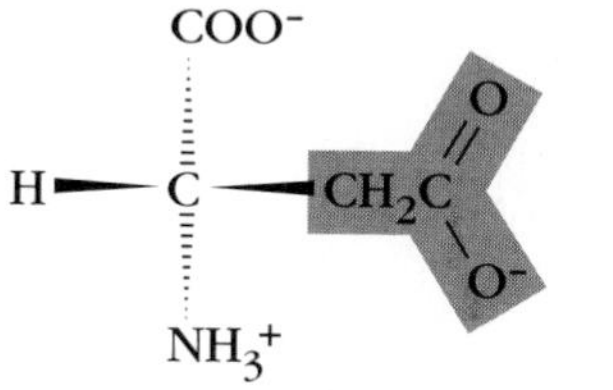

Aspartic acid
Asp

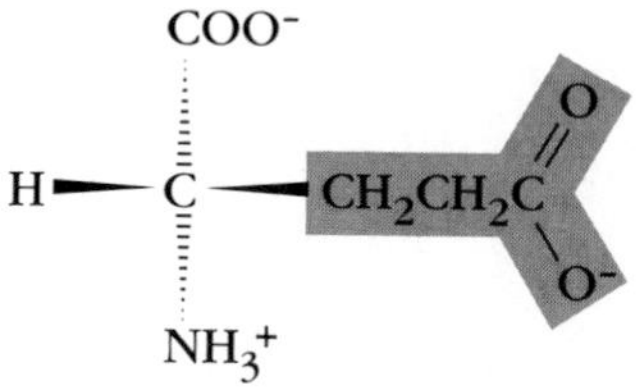

Glutamic acid
Glu

* = Essential amino acids

tains five α-amino acids with charged polar side chains. The charged polar side chains can be either positively or negatively charged.

Half of the amino acids listed in Table 26.1 must be obtained from the food we eat because they cannot be synthesized by the human body. These 10 α-amino acids are called **essential amino acids,** and they are indicated by an asterisk in Table 26.1.

The α-carbon atoms of all α-amino acids obtained from plants and animals (except glycine, whose α-carbon atom is not a stereocenter) have the *S* configuration. The *S* configuration of α-amino acids has been related to the configuration of *L*-glyceraldehyde, so natural α-amino acids belong to the *L* family. The stereochemical relationship between *L*-glyceraldehyde and the *L* family of α-amino acids can be represented by Fischer projection formulas in which the carboxyl, amino, hydrogen, and *R* group of the *L*-α-amino acid correspond with the aldehyde, hydroxy, hydrogen, and hydroxymethyl group of *L*-glyceraldehyde, respectively.

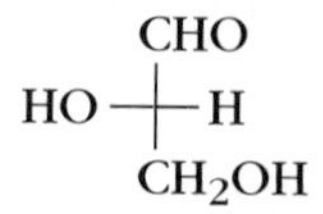

L-Glyceraldehyde

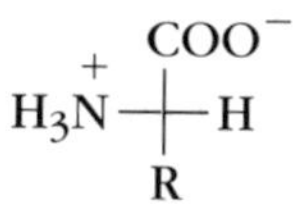

L-α-Amino acid

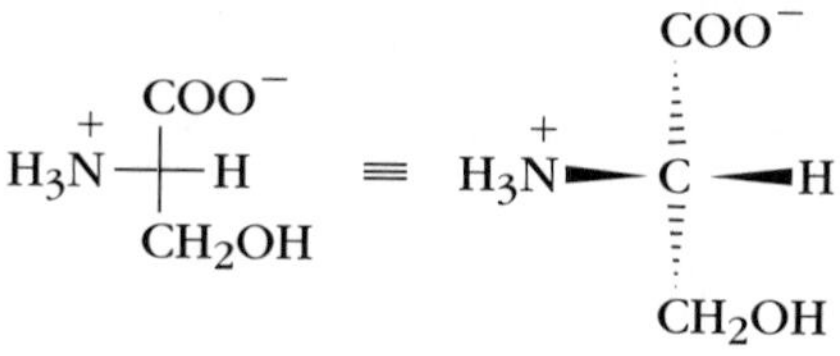

(*S*)-Serine
L-Serine

Table 26.2	Abbreviations for the α-Amino Acids	
Amino acid	Three-Letter Abbreviation	One-Letter Symbol
Alanine	Ala	A
Arginine	Arg	R
Asparagine	Asn	N
Aspartic acid	Asp	D
Cysteine	Cys	C
Glutamic acid	Glu	E
Glutamine	Gln	Q
Glycine	Gly	G
Histidine	His	H
Isoleucine	Ile	I
Leucine	Leu	L
Lysine	Lys	K
Methionine	Met	M
Phenylalanine	Phe	F
Proline	Pro	P
Serine	Ser	S
Threonine	Thr	T
Tryptophan	Trp	W
Tyrosine	Tyr	Y
Valine	Val	V

The 20 α-amino acids are often designated by either the three-letter abbreviations or the one-letter symbols given in Table 26.2. The three-letter abbreviations are the first three letters of the name of the amino acid, except for asparagine (Asn), glutamine (Gln), isoleucine (Ile), and tryptophan (Trp). The one-letter symbols have been agreed on by convention.

Exercise 26.1	Write the Fischer projection formula of each of the following amino acids.

(a) *L*-Alanine   **(b)** *L*-Phenylalanine   **(c)** (*S*)-Histidine   **(d)** *L*-Threonine

26.2   Acid-Base Properties of Amino Acids

Alanine and other α-amino acids with nonionizable side chains are amphoteric because they contain both an acidic functional group (the ammonium ion, NH_3^+) and a basic functional group (the carboxylate ion, $-COO^-$). As a result, alanine can exist as a cationic form, a zwitterion, or an anionic form depending on the pH of the solution.

Cationic form of alanine	Zwitterion form of alanine	Anionic form of alanine

A **zwitterion** is a neutral dipolar molecule in which the positive and negative charges are not adjacent to each other.

The cationic form of alanine and its zwitterion are a conjugate acid-base pair. The equilibrium between these two forms is the first acid dissociation constant of alanine $(K_a)_1$.

$$\overset{+}{N}H_3CHCOOH \quad + \quad H_2O \quad \xrightleftharpoons{(K_a)_1} \quad \overset{+}{N}H_3CHCOO^- \quad + \quad H_3O^+$$
$$\overset{|}{CH_3} \hspace{9cm} \overset{|}{CH_3}$$

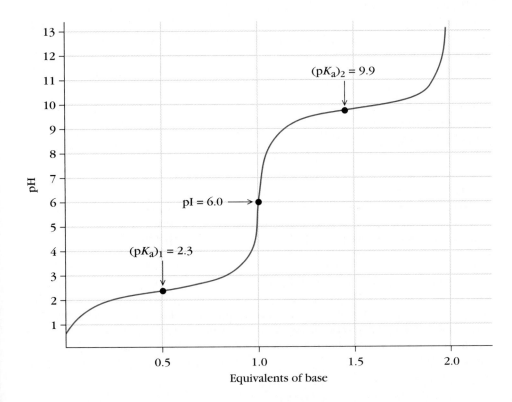 *Conjugate acid and base pairs are discussed in Section 3.2.*

The zwitterion of alanine and its anionic form are another conjugate acid-base pair. The equilibrium between these two forms is the second acid dissociation constant of alanine, called $(K_a)_2$.

$$\overset{+}{N}H_3CHCOO^- \quad + \quad H_2O \quad \xrightleftharpoons{(K_a)_2} \quad NH_2CHCOO^- \quad + \quad H_3O^+$$
$$\overset{|}{CH_3} \hspace{9cm} \overset{|}{CH_3}$$

The values of $(K_a)_1$ and $(K_a)_2$ for a particular amino acid are determined from its titration curve. We can use the titration curve of alanine, shown in Figure 26.1, to determine its two acid dissociation constants. The first inflection point of the titration curve is reached when one half equivalent of hydroxide ion is added to one equivalent of alanine. At this inflection point, the concentrations of the zwitterion and the cationic form are equal, so pH = $p(K_a)_1$ according to the Henderson-Hasselbalch equation. Thus $p(K_a)_1$ of alanine is 2.3.

The pH of the solution when one equivalent of hydroxide ion has been added to one equivalent of an amino acid with a nonionizable side chain is called the

Recall that the Henderson-Hasselbalch equation is:

$$pH = pK + \log \frac{[A^-]}{[HA]}$$

Figure 26.1
The titration curve of alanine. Other amino acids with nondissociating side chains have similar titration curves. For alanine, $p(K_a)_1$ = 2.3 and $p(K_a)_2$ = 9.9.

Figure 26.2
The concentrations of the cationic, zwitterion, and anionic forms of alanine as a function of pH. At pH < 2.0, alanine exists in the cationic form; at pH > 11.0, it exists in the anionic form, while at pH = 4 to 8 it exists predominantly in the zwitterion form.

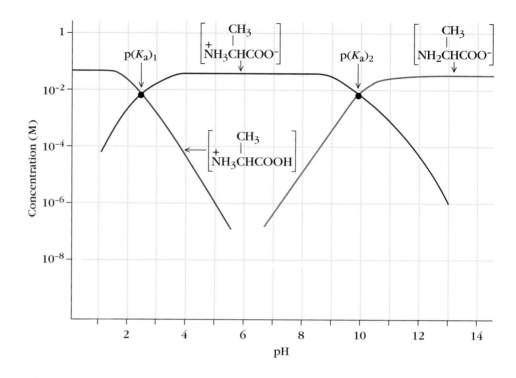

isoelectric pH or **isoelectric point (pI).** At the isoelectric point, the amino acid carries no net electric charge. For alanine, pI is 6.0. A second inflection point at pH = 9.9 is reached when 1.5 equivalents of hydroxide ion have been added. At this point, the pH is equal to $p(K_a)_2$, so $p(K_a)_2$ of alanine is 9.9.

The two pK values of alanine are sufficiently different so that the Henderson-Hasselbalch equation can be used to determine the concentration of the various species of alanine as a function of pH. The results in Figure 26.2 show that alanine exists almost exclusively in the anionic form in strong base (pH > 11); almost exclusively in the cationic form in strong acid (pH < 2); and as the zwitterion in neutral solution (pH = 4 to 8).

Exercise 26.2	Using the Henderson-Hasselbalch equation, calculate the following ratios for alanine at the pH values indicated.

(a) $[^+H_3NCH(CH_3)COOH]/^+H_3NCH(CH_3)COO^-]$ at pH = 1.5
(b) $[^+H_3NCH(CH_3)COOH]/[^+H_3NCH(CH_3)COO^-]$ at pH = 4.0
(c) $[^+H_3NCH(CH_3)COO^-]/[H_2NCH(CH_3)COO^-]$ at pH = 8.2
(d) $[^+H_3NCH(CH_3)COO^-]/[H_2NCH(CH_3)COO^-]$ at pH = 12.0

α-Amino acids with an ionizable side chain have a pK_a value in addition to $p(K_a)_1$ (for the —COOH group) and $p(K_a)_2$ (for the —NH$_3^+$ group). The third acid dissociation constant is due to the ionizable side chain, and it is called pK_R. The values of pK_R vary much more than the values of $p(K_a)_1$ and $p(K_a)_2$ for α-amino acids, as shown in Table 26.3. The values of $p(K_a)_1$ are close to 2, and the values of $p(K_a)_2$ are close to 9.5 for all the 20 α-amino acids. The values of pK_R, in contrast, vary from 3.9 to 12.5.

Table 26.3	pK values for the 20 common amino acids			
α-Amino Acid	p(K_a)$_1$ (α-COOH Group)	p(K_a)$_2$ (α-NH$_3^+$ Group)	pK_R (Side Chain Group)	pI
Alanine	2.3	9.9	—	6.02
Arginine	1.8	9.0	12.5	10.8
Asparagine	2.1	8.8	—	5.4
Aspartic Acid	2.0	9.9	3.9	2.8
Cysteine	1.9	10.8	8.3	5.1
Glutamic Acid	2.1	9.5	4.1	3.2
Glutamine	2.2	9.1	—	5.7
Glycine	2.3	9.8	—	5.97
Histidine	1.8	9.3	6.0	7.6
Isoleucine	2.3	9.8	—	6.0
Leucine	2.3	9.7	—	5.98
Lysine	2.2	9.2	10.8	9.74
Methionine	2.1	9.3	—	5.7
Phenylalanine	2.2	9.2	—	5.5
Proline	3.0	10.6	—	6.3
Serine	2.2	9.2	—	5.7
Threonine	2.1	9.1	—	5.6
Tryptophan	2.4	9.4	—	5.9
Tyrosine	2.2	9.1	10.1	5.7
Valine	2.3	9.7	—	5.96

Exercise 26.3   Write the structure of the predominant form of each of the following amino acids at the pH of blood, 7.4.

(a) Valine   (b) Glutamic acid   (c) Lysine   (d) Glycine

Exercise 26.4   Write the structure of the predominant form of lysine in each solution of the following pH:

(a) pH = 0.2   (b) pH = 9.8   (c) pH = 13.0   (d) pH = 5.0

26.3   Proteins Are Polymers of α-Amino Acids Joined by Peptide Bonds

The formation of proteins from α-amino acids can be viewed as the joining of the α-carboxyl group of one amino acid to the amino group of another α-amino acid with the elimination of water.

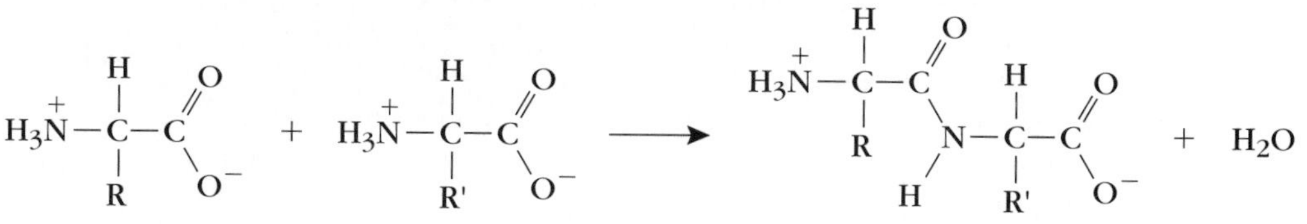

The bond between amino acids is called a **peptide bond** or **peptide linkage.** A peptide bond or peptide linkage is an amide bond.

Recall from Section 16.2 that rotation is restricted about the C—N bond of an amide group because of its delocalized electron structure. As a result, peptide bonds are rigid and planar.

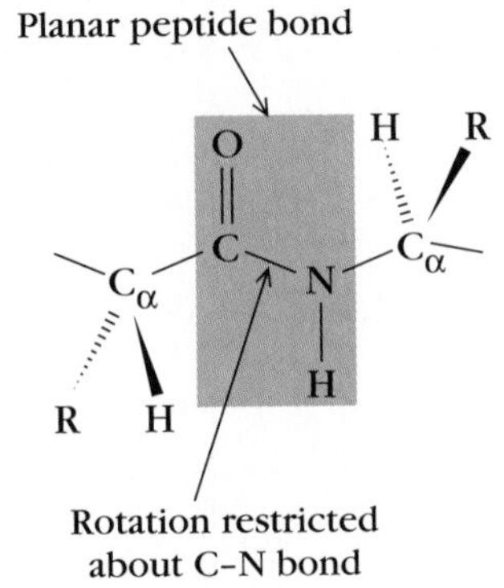

In most peptide bonds, the hydrogen atom of the substituted amino group is *trans* to the carbonyl oxygen atom. The *trans*-conformation is the more stable because steric interference is less in the *trans*- than the *cis*-conformation.

Exercise 26.5	Write a three-dimensional structure of the *cis*-conformation of a peptide bond. What steric interactions are present in the *cis*-conformation that are absent in the *trans*-conformation?

A polymer formed by joining two α-amino acids by a single peptide bond is called a **dipeptide;** a polymer composed of three amino acids is called a **tripeptide;** and a polymer composed of many amino acids is called a **polypeptide.** **Proteins** are molecules that consist of one or more polypeptide chains. An amino acid unit in a polypeptide or protein is called an **amino acid residue.** Polypeptides are linear polymers because the carbonyl group of one amino acid

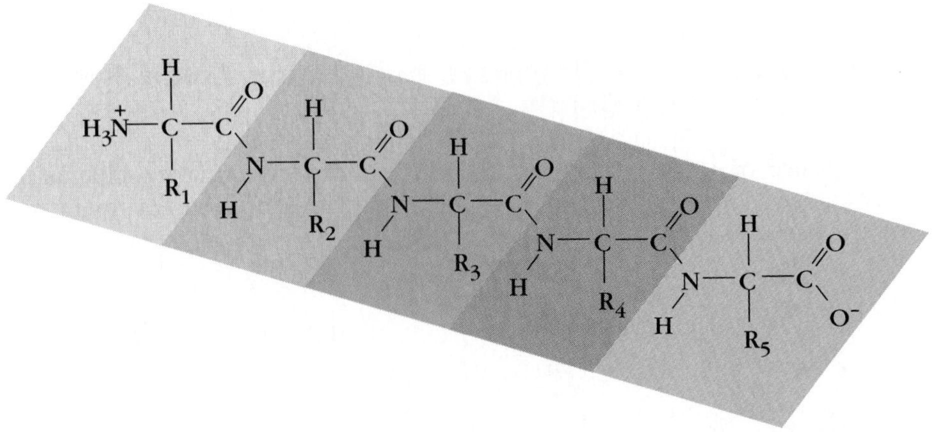

Figure 26.3
A linear pentapeptide. The five constituent amino acid residues are outlined. A peptide bond joins the carbonyl group of one amino acid residue with the α-amino group of an adjacent amino acid residue.

residue is linked to the α-amino group of the next amino acid residue, as shown in Figure 26.3.

A polypeptide chain has direction because its building blocks have different ends. One end has a carboxyl group, and the other end has an amino group. By convention, the structures of polypeptides are written so that the free ammonium ion (NH_3^+) is to the left and the free carboxylate ion (COO^-) is on the right. Written in this way, the amino acid at the left end of the chain is called the **N-terminal residue** and the amino acid at the other end is called the **C-terminal residue.** For example, alanine is the N-terminal residue and serine is the C-terminal residue in the following tripeptide:

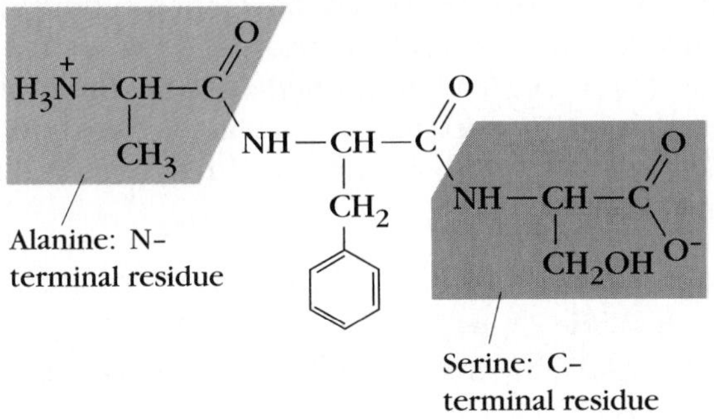

A more convenient way of writing the structures of polypeptides is to use the three-letter or one-letter abbreviations of the amino acids given in Table 26.2.

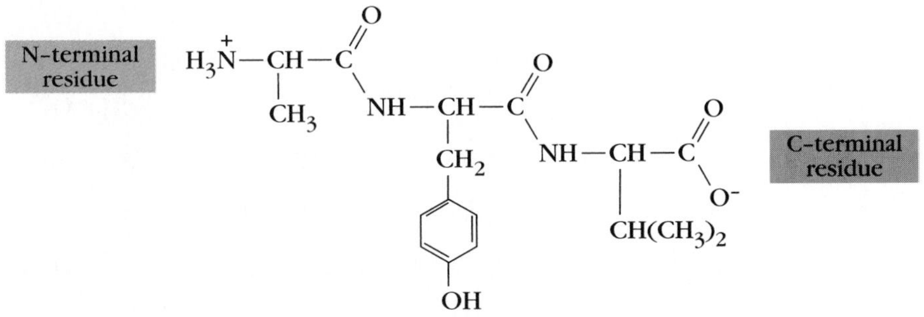

Constituent amino acids:	Alanine	Tyrosine	Valine
Three letter abbreviation:	Ala ————	Tyr ————	Val
One letter symbol abbreviation:	A	Y	V

A polypeptide chain consists of two parts. The first part is the main chain or backbone, which consists of repeating NH—CHC=O units. The second part consists of the side chains. The main chain is the same in most all polypeptides. The side chains, on the other hand, vary from one polypeptide to another. The characteristic properties of a particular polypeptide, in fact, are determined by the number and types of its side chains, as we will learn in Section 26.11.

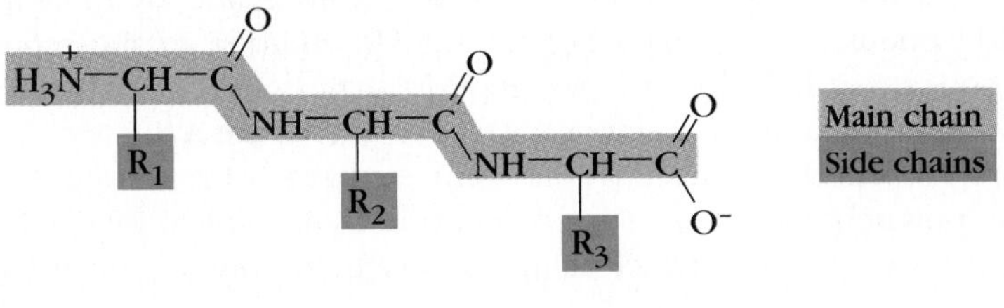

Exercise 26.6	Write the complete structure of each of the following polypeptides.

(a) Val-Glu **(b)** Arg-Gln-Phe **(c)** Ala-Tyr-Asp-Gly

Many polypeptides contain another covalent bond called a disulfide bond.

26.4 Disulfide Bonds

Covalent peptide bonds form the backbone of polypeptides. Disulfide bonds (also called disulfide linkages or disulfide bridges) are a second kind of covalent bond present in many proteins and polypeptides. Recall from Section 11.19 that a disulfide bond is formed by the mild oxidation of two thiol-containing compounds. The reaction is reversible because mild reduction reduces the disulfide to two molecules of thiol.

$$RSH \quad + \quad HSR \quad \underset{\text{[Reduction]}}{\overset{\text{[Oxidation]}}{\rightleftarrows}} \quad RS-SR$$

Thiol   Thiol   Disulfide

Cysteine is the amino acid that contains a thiol group in its side chain. Air oxidation of the thiol groups of two cysteine molecules forms cystine, which is the disulfide-linked dimer of cysteine.

Cysteine   Cystine

The oxidation of two cysteine residues in a single polypeptide chain forms a disulfide bond that is part of a ring. The nonapeptide oxytocin, for example, is a hormone secreted by the pituitary gland; its structure is shown in Figure 26.4. Oxytocin contains a ring formed by a disulfide bond linking the N-terminal cysteine residue with a cysteine residue in the middle of the chain.

Polypeptide chains of proteins are frequently linked by disulfide bonds. Bovine insulin, for example, consists of two polypeptide chains, one containing 21 amino acid residues and the other containing 30 amino acid residues, linked by disulfide bonds, as shown in Figure 26.5. In addition to the intermolecular disulfide bonds that link the two polypeptide chains, an intramolecular disulfide bond between two cysteine residues forms a ring in Chain A.

The first step in the determination of the structure of a protein is the separation of the polypeptide chains. For polypeptide chains linked by disulfide bonds, this separation is achieved by cleaving the disulfide bonds. Two methods com-

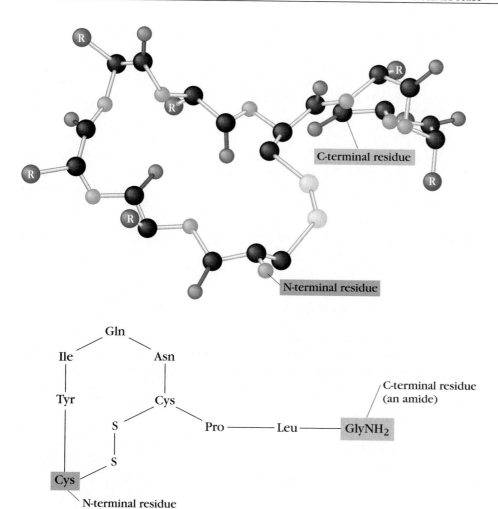

Figure 26.4
The structure of oxytocin. The disulfide bond between two cysteine residues forms part of a large ring. Notice that the C-terminal residue of oxytocin is not a free carboxylate ion but an amide.

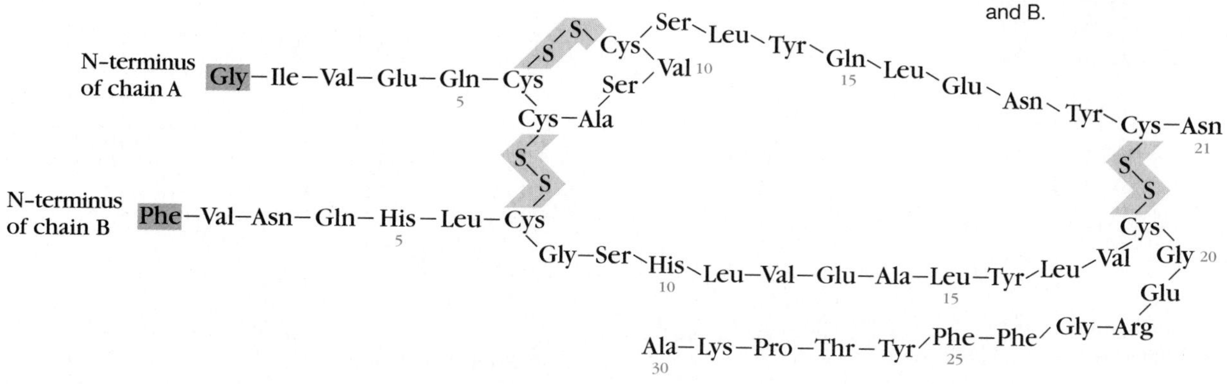

Figure 26.5
The structure of bovine insulin. Bovine insulin has an intramolecular disulfide bond that forms a ring in Chain A and two intermolecular disulfide bonds that link Chains A and B.

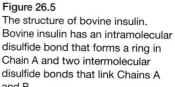

monly used to cleave disulfide bonds are reductive cleavage and oxidative cleavage.

Reductive cleavage can be accomplished by reaction of β-mercaptoethanol and the disulfide bonds in polypeptides and proteins, which converts the cystine functional groups into cysteine residues.

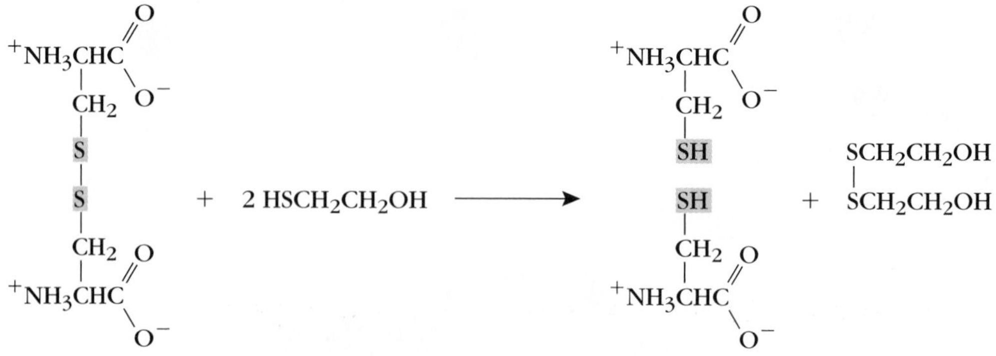

Cystine   β-Mercaptoethanol   Cysteine

The free thiol groups must be alkylated to prevent the cysteine units from reforming disulfide linkages by air oxidation. Reaction of the free thiol groups with iodoacetate forms an S-alkylated cysteine residue that resists air oxidation.

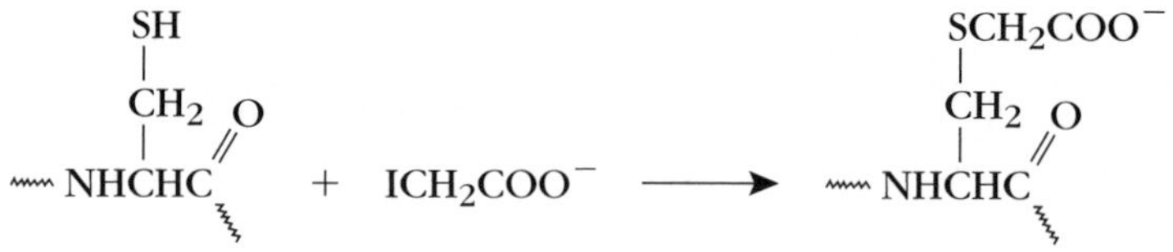

Cysteine residue   Iodoacetate   S-Alkylated cysteine
of polypeptide chain   residue is not oxidized
is easily oxidized   by oxygen in the air
to cystine

Oxidative cleavage can be accomplished by reaction of performic acid with disulfide bonds. Performic acid oxidizes the sulfur atoms of the disulfide bonds to $-SO_3^-$ groups.

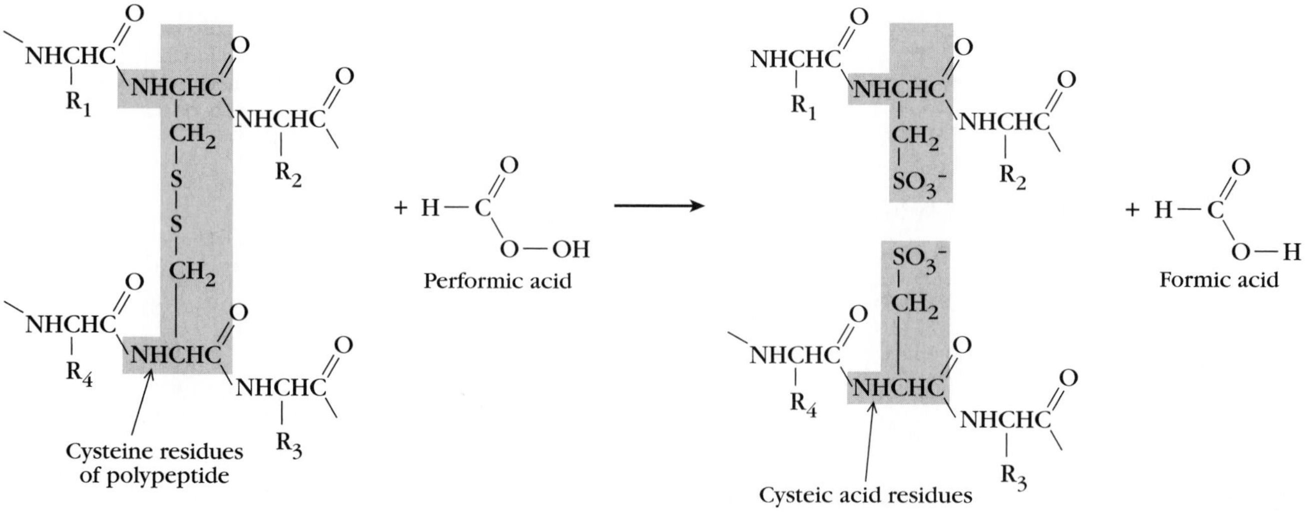

Cysteine residues   Performic acid   Cysteic acid residues   Formic acid
of polypeptide

Cysteic acid is not further oxidized and is stable in both acid and base solutions.

Exercise 26.7 Write the structure of the product of the reaction of oxytocin, whose structure is given in Figure 26.4 (p. 1155), with each of the following reagents:

(a) β-Mercaptoethanol followed by sodium iodoacetate
(b) Performic acid

Amino acid composition is the next step in the structure determination of a polypeptide.

26.5 Amino Acid Composition of Polypeptides

The **amino acid composition** is defined as the number of each type of amino acid residue present in the protein or polypeptide. The amino acid composition of a protein is determined by cleaving the polypeptide chain into its constituent amino acids, separating the free amino acids, and measuring the quantities of each.

Peptide bonds are usually cleaved by dissolving the protein in 6 M HCl in an evacuated tube (to prevent air oxidation of the sulfur-containing amino acid residues) and heating the solution for 10 to 100 hours at 100 to 120 °C.

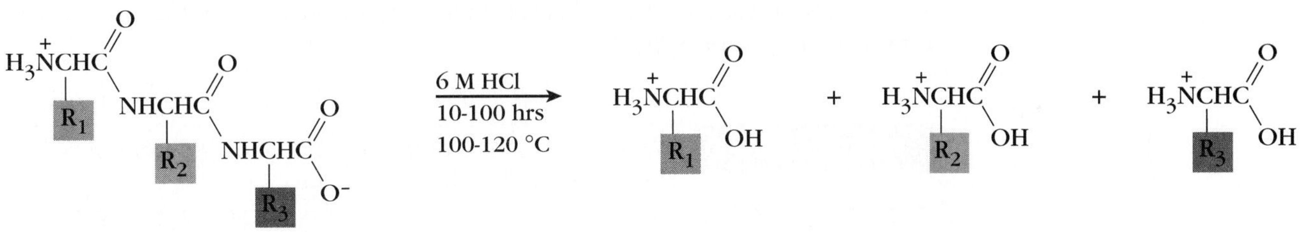

The vigorous conditions of acid hydrolysis of proteins cause partial destruction of the tryptophan residues. Furthermore, the amide functional groups on the side chains of glutamine and asparagine are hydrolyzed to form glutamic and aspartic acids, respectively.

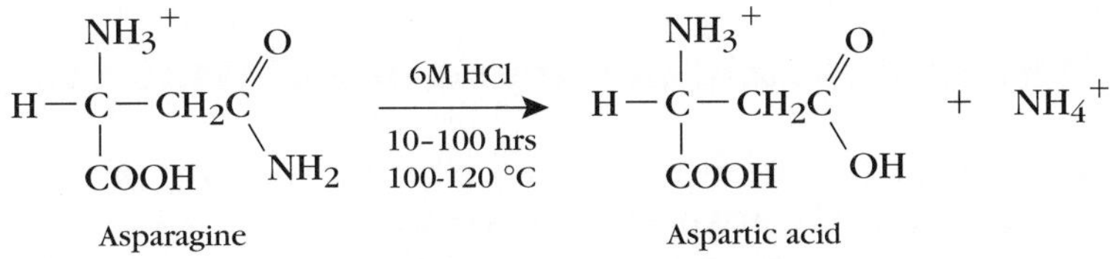

Consequently, the measured amount of glutamic acid obtained by acid hydrolysis of proteins includes the concentrations of both glutamine and glutamic acid residues. Similarly the measured amount of aspartic acid includes both asparagine and aspartic acid.

An automated amino acid analyzer is used to quantitatively determine the amino acid content of the product of hydrolysis of a protein. An amino acid analyzer separates amino acids by ion-exchange chromatography. Recall from Section

5.1 that chromatography separates mixtures because the intermolecular attractions differ between the various compounds of the mixture and the stationary phase. The stationary phase in ion-exchange chromatography, which is called an ion-exchange resin, is a relatively inert polymer that has groups that are positively charged at pH 7. Amino acids such as lysine that are positively charged at pH 7 are not attracted by a positively charged stationary phase. Aspartic acid, on the other hand, is negatively charged at pH 7 so it is attracted to a positively charged stationary phase. Passing a mixture of aspartic acid and lysine through a column containing a positively charged stationary phase would separate the two amino acids. The progress of aspartic acid through the column would be retarded compared to the progress of lysine because of the greater affinity of the aspartic acid for the positively charged stationary phase.

The time required for a given amino acid to elute from the ion-exchange column is reproducible, so the identity of the amino acids in a polypeptide of unknown composition can be determined from their elution times.

Each amino acid as it elutes from the end of the chromatography column is mixed with a solution of ninhydrin. Ninhydrin reacts with α-amino acids to form a colored reaction product.

An α-amino acid   Ninhydrin   Purple color

The amount of each amino acid is determined by the intensity of the color, which is measured by a spectrometer.

| **Exercise 26.8** | Write the structure of the product obtained from the reaction of ninhydrin and leucine. |

The amino acid analyzer determines the amino acid composition of proteins and polypeptides, but it does not reveal their sequence (the order in which they are linked together). Special techniques have been developed to determine the amino acid sequence of polypeptides.

26.6   The Amino Acid Sequence of Polypeptides

The sequence of amino acids is called the **primary structure** of polypeptides. The primary structure of a polypeptide is determined by removing and identifying successive amino acid residues from either the N-terminal or the C-terminal end of the polypeptide chain.

The **Edman degradation** is a common method of N-terminal residue identification. In reaction 1 of the Edman degradation procedure, shown in Figure 26.6, phenyl isothiocyanate reacts with the free amino group of the N-terminal residue to form a phenyl thiocarbamyl (PTC) derivative. The reaction of the PTC derivative with anhydrous trifluoroacetic acid (TFA) in reaction 2 cleaves the N-terminal residue as its thiazolinone derivative but does not hydrolyze any other peptide

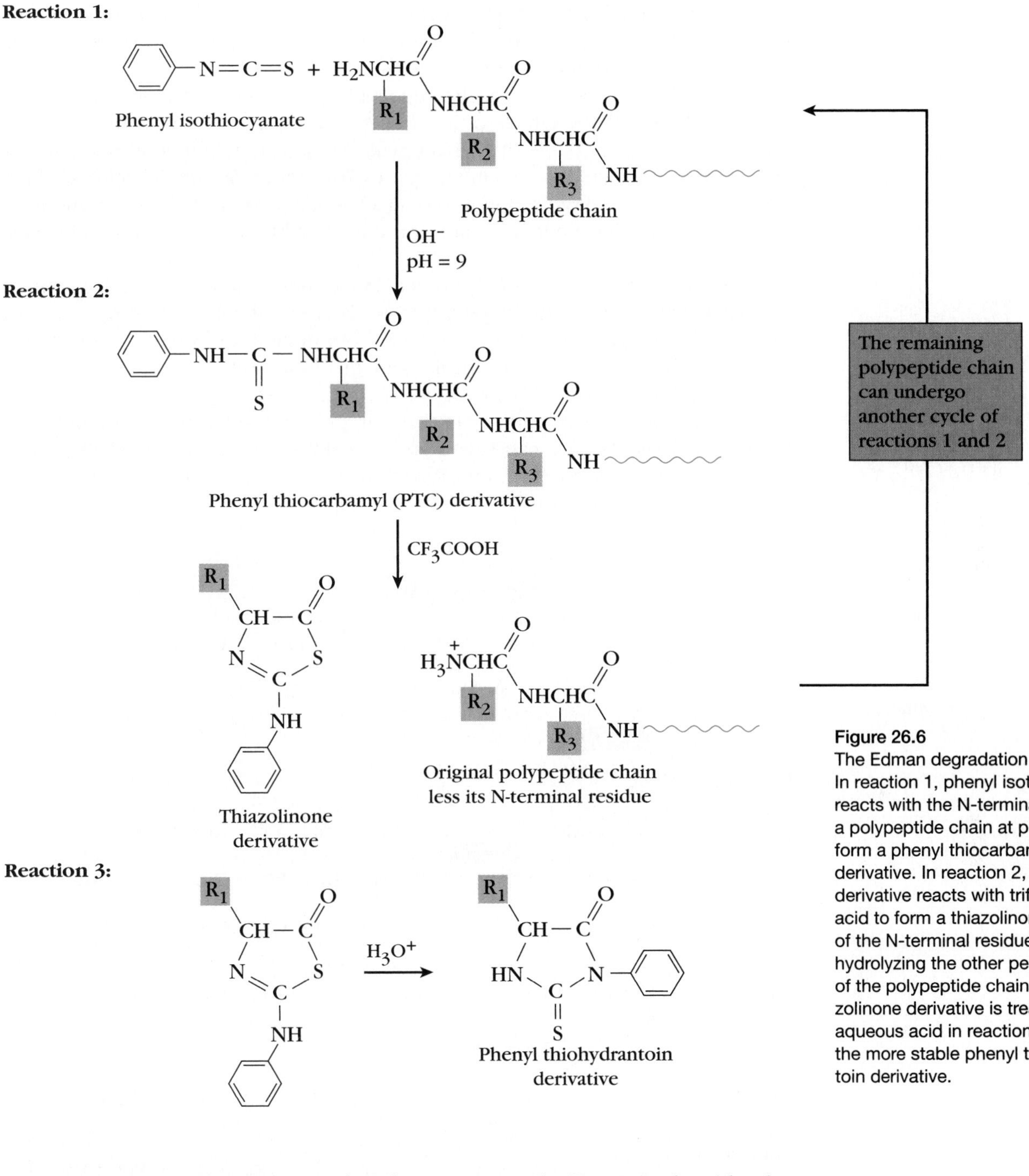

Reaction 1:

Reaction 2:

Phenyl thiocarbamyl (PTC) derivative

The remaining polypeptide chain can undergo another cycle of reactions 1 and 2

Thiazolinone derivative

Original polypeptide chain less its N-terminal residue

Reaction 3:

Phenyl thiohydrantoin derivative

Figure 26.6
The Edman degradation procedure. In reaction 1, phenyl isothiocyanate reacts with the N-terminal residue of a polypeptide chain at pH 9.0 to form a phenyl thiocarbamyl (PTC) derivative. In reaction 2, the PTC derivative reacts with trifluoroacetic acid to form a thiazolinone derivative of the N-terminal residue without hydrolyzing the other peptide bonds of the polypeptide chain. The thiazolinone derivative is treated with aqueous acid in reaction 3 to form the more stable phenyl thiohydrantoin derivative.

bonds. *The Edman degradation,* therefore, *removes the N-terminal residue but leaves intact the rest of the polypeptide chain.* In reaction 3, the thiazolinone derivative is extracted into an organic solvent and is converted into the more stable phenyl thiohydrantoin (PTH) derivative by treatment with aqueous acid. The PTH derivatives of the 20 amino acids can be identified by means of chromatography.

The Edman degradation procedure is a valuable technique because the amino acid sequence of a polypeptide chain can be determined from the N-terminal end by subjecting the polypeptide chain to repeated Edman degradations and identifying the newly released PTH derivative after every cycle. Repeated Edman degradations have been automated by means of devices called sequenators, resulting in great savings of time and material.

The Edman degradation can be repeated up to about 60 times on a single polypeptide before the effects of incomplete reactions, side reactions, and accumulated loss of the polypeptide chain make the amino acid identification unreliable. Polypeptide chains containing more than about 60 amino acid residues, therefore, must be broken down into smaller polypeptide chains that are short enough for the Edman degradation to produce reliable results of their amino acid sequence.

Partial hydrolysis of a polypeptide chain can be carried out chemically or by the use of enzymes. Acid-catalyzed hydrolysis is unselective and leads to a mixture of short polypeptide chains resulting from random cleavage. Cyanogen bromide (N≡CBr), however, specifically and quantitatively cleaves the C-terminal end of methionine residues in the polypeptide. The enzyme trypsin catalyzes hydrolysis of the polypeptide chain only at the C-terminal end of arginine and lysine, and chymotrypsin catalyzes hydrolysis at the C-terminal end of the amino acids phenylalanine, tyrosine, and tryptophan. These sites of cleavage are illustrated by the hydrolysis of the following polypeptide by N≡CBr, trypsin, and chymotrypsin.

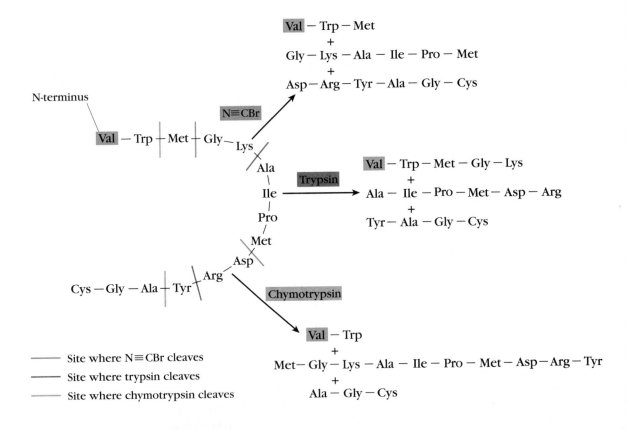

Exercise 26.9   Indicate where N≡CBr, trypsin, and chymotrypsin will cleave the following polypeptide chain.

Ala-Val-Lys-Met-Ile-Pro-Tyr-Thr-Arg-Ser-Met-Leu-His-Gln

The amino acid sequence of the smaller polypeptide fragments obtained by hydrolysis is determined by Edman degradation. The sequence of amino acids in the original polypeptide is then determined by comparing the amino acid sequences of one set of polypeptide fragments with those of a second set whose specific cleavage sites overlap those of the first set. This method is illustrated in Sample Problem 26.1.

Sample Problem 26.1   A polypeptide is subjected to the following reactions. From the results of these reactions, determine its amino acid sequence.

Reaction 1   Amino acid analysis of the polypeptide reveals the presence of 12 different amino acids. One equivalent of the polypeptide yields one equivalent of Phe, Trp, Lys, Val, Pro, Asp, Arg, Cys, and Gln, and two equivalents of Met, Gly, and Ala.

Reaction 2   An N-terminal residue analysis by the Edman degradation reveals that the polypeptide has a phenylalanine residue at the N-terminus.

Reaction 3   Trypsin-catalyzed partial hydrolysis yields the following polypeptides, whose amino acid sequences were determined by the Edman degradation:

   a. Phe-Trp-Met-Gly-Ala-Lys

   b. Val-Pro-Met-Asp-Gly-Arg

   c. Cys-Ala-Gln

Reaction 4   Reaction of the polypeptide with BrCN yields the following polypeptides, whose amino acid sequences were determined by the Edman degradation:

   a. Phe-Trp-Met

   b. Gly-Ala-Lys-Val-Pro-Met

   c. Asp-Gly-Arg-Cys-Ala-Gln

Solution:

Phe is the N-terminal residue, so we start by writing the longest peptide chain that contains Phe as the N-terminal residue:

Phe-Trp-Met-Gly-Ala-Lys

Then we match the overlapping regions of the various polypeptides, which provides the full amino acid sequence of the polypeptide.

Phe-Trp-Met-Gly-Ala-Lys
Phe-Trp-Met
            Gly-Ala-Lys-Val-Pro-Met-
                  Val-Pro-Met-Asp-Gly-Arg
                          Asp-Gly-Arg-Cys-Ala-Gln
                                    Cys-Ala-Gln

Phe-Trp-Met-Gly-Ala-Lys-Val-Pro-Met-Asp-Gly-Arg-Cys-Ala-Gln

Finally, check to see if the amino acid composition of the polypeptide agrees with the amino acid analysis (Reaction 1). The polypeptide contains 12 different amino acids, one each of Phe, Trp, Lys, Val, Pro, Asp, Arg, Cys, and Gln, and two of Met, Gly, and Ala. This agrees with the amino acid analysis.

This polypeptide is sufficiently small that its amino acid sequence could have been determined by a sequenator. The techniques and logic used in this example, however, are the same as those used to determine the amino acid sequence in single polypeptide chains containing more than 400 amino acids.

Exercise 26.10   Determine the amino acid sequence of the following polypeptides, whose partial hydrolysis gives the peptide fragments indicated.

(a) A hexapeptide that gives the following fragments on partial hydrolysis: Leu-Tyr-Ile, Asp-Arg-Leu, and Tyr-Ile-Phe
(b) A nonapeptide that gives the following fragments on partial hydrolysis: Ile-Gln-Asn-Cys, Cys-Tyr-Ile-Gln-Asn, Gln-Asn-Cys-Pro, Pro-Leu-Gly, and Cys-Pro-Leu-Gly

---

Once scientists were able to determine the amino acid sequence of polypeptides, they developed techniques for their synthesis.

26.7 Chemical Synthesis of Polypeptides

Polypeptides are chemically synthesized by forming peptide (or amide) bonds between different amino acids. Forming amide bonds between different amino acids is a more complex process than forming amide bonds between simple amines and carboxylic acid derivatives. Amino acids are difunctional compounds, so the reaction between different amino acids can form several different peptide links. Imagine, for instance, that we wish to synthesize a dipeptide by reacting two different α-amino acids with a reagent that will form amide bonds. All four possible dipeptides will be formed in this reaction because each amino acid has both an amino group and a carboxyl group. We can force the reaction to form only one product by *protecting* all the amino and carboxyl groups, except for those that we want to react. One dipeptide will be formed, therefore, only when an amino group of one amino acid and a carboxyl group of the other amino acid are available for reaction. The use of protective groups permits the synthesis of polypeptides in which amide bonds link their constituent amino acids in their *correct sequence*.

■ *The synthesis of amide bonds between simple amines and carboxylic acid derivatives is discussed in Sections 16.5 and 16.9.*

Exercise 26.11   Write the structure of the four possible dipeptides formed by reacting alanine and valine with a reagent that forms amide bonds.

Many different amino- and carboxyl-protecting groups have been devised. A *t*-butyloxycarbonyl (*t*-Boc) group is frequently used as a protecting group for the amino groups of α-amino acids. A *tert*-butyloxycarbonyl group is introduced by the acyl transfer reaction of an amino acid, such as alanine, with di-*t*-butyl dicarbonate.

■ *Acyl transfer reactions are discussed in Chapter 16.*

$$(CH_3)_3COC \overset{O}{\underset{}{\big\|}} \underset{(CH_3)_3COC}{\overset{O}{\underset{\big\|}{}}} O \quad + \quad H_2NCHCOOH \overset{CH_3}{\underset{}{\big|}} \longrightarrow (CH_3)_3COC \overset{O}{\underset{}{\big\|}} \underset{NHCHCOOH}{\overset{CH_3}{\big|}} \quad + \quad (CH_3)_3COH \quad + \quad CO_2$$

Di-*t*-butyl dicarbonate   Alanine   *t*-Boc alanine   *t*-Butyl alcohol

A *t*-butyloxycarbonyl protecting group is easily removed by brief treatment of the *t*-Boc amino acid with trifluoroacetic acid.

$$(CH_3)_3COC \overset{O}{\underset{}{\big\|}} \underset{NHCHCOOH}{\overset{CH_3}{\big|}} \xrightarrow{CF_3COOH} H_3\overset{+}{N}CHCOOH \overset{CH_3}{\underset{}{\big|}} \quad + \quad CH_2{=}C(CH_3)_2 \quad + \quad CO_2$$

t-Boc alanine   Alanine   Methylpropene

Exercise 26.12   Write an acceptable mechanism for the reaction of alanine and di-*t*-butyl dicarbonate to form *t*-Boc-alanine.

Carboxyl groups are often protected by converting them to their methyl or benzyl esters by standard methods of ester formation. The ester functional group is easily removed by hydrolysis with aqueous sodium hydroxide.

■ *Synthesis and hydrolysis of esters are discussed in Chapter 15.*

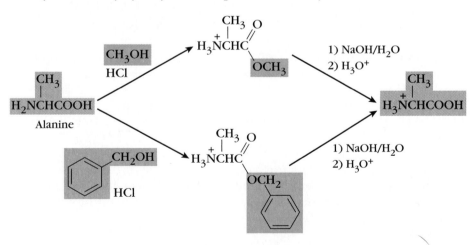

In order to form a peptide bond, the free carboxyl group of the amino acid must be activated. In the laboratory, the more reactive acyl chloride is used to react with amine to form simple amides. In the chemical synthesis of polypeptides, an amide bond is usually formed by reacting an equimolar mixture of *N,N'*-dicyclohexylcarbodiimide (abbreviated DCC), a carboxylate ion, and an amine.

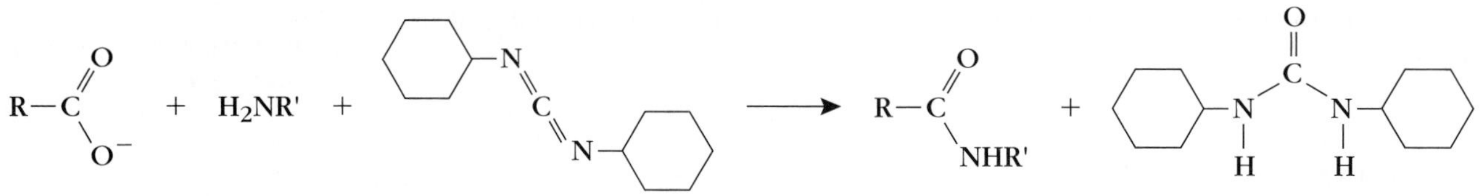

| Carboxylate ion | Amine | N,N'-Dicyclohexylcarbodiimide (DCC) | | Amide | N,N'-Dicyclohexyl urea (DCU) |

A mechanism for the reaction of DCC, a carboxylate ion, and an amine similar to the mechanism of acyl transfer reactions is shown in Figure 26.7. Step 1 is the addition of the carboxylate ion to the strongly electrophilic carbon atom of DCC to form an activated acylating reagent, whose reactivity with nucleophiles is similar to the reactivities of acyl halides and carboxylic anhydrides. Nucleophilic addition of the amine to the acylating reagent in Step 2 forms a tetrahedral intermediate, which loses dicyclohexylurea in Step 3 to form the conjugate acid of an amide that loses a proton to form an amide.

How we use protecting groups in the synthesis of polypeptides can be illustrated by the preparation of the dipeptide Ala-Val. We begin this synthesis with the amino acid that is the N-terminal residue of the peptide and add an amino acid to its C-terminus:

■ *The mechanism of acyl transfer reactions is discussed in Section 16.6.*

Reaction 1. Alanine is the N-terminal residue, so we begin by protecting the amino group of alanine by forming its t-Boc derivative:

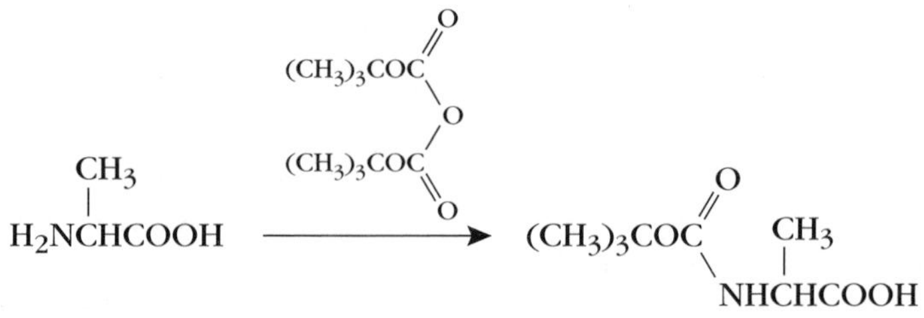

Alanine　　　　　　　　　　　　　　　t-Boc-Ala

Reaction 2. We want to link valine to alanine, so the carboxyl group of valine must be protected by forming its methyl ester:

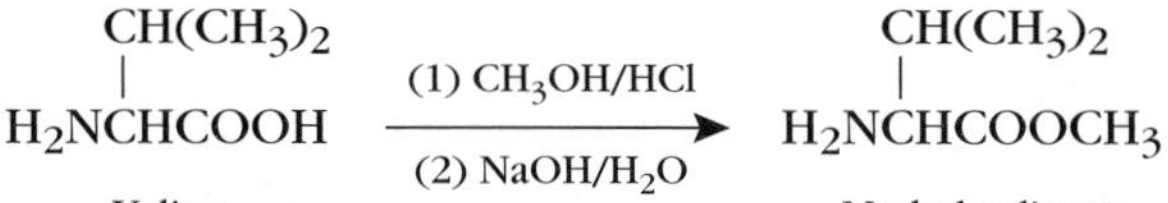

Valine　　　　　　　　　　　　　　Methyl valinate

Reaction 3. The free carboxyl group of t-Boc-Ala and the free amino group of methyl valinate form an amide bond by reacting them with DCC:

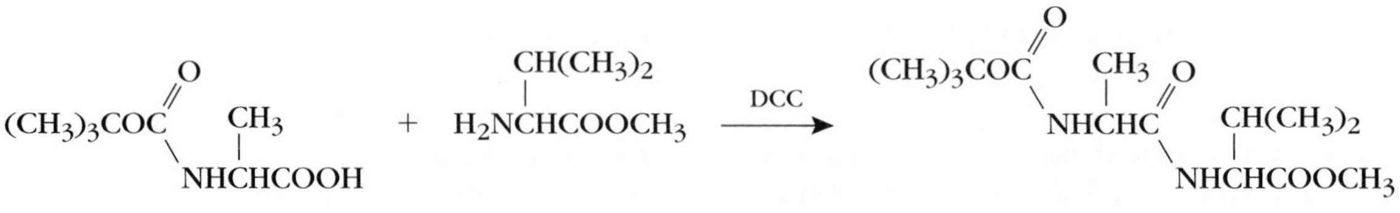

t-Boc-Ala　　　　　　　Methyl valinate

Step 1: Addition of carboxylate ion to DCC

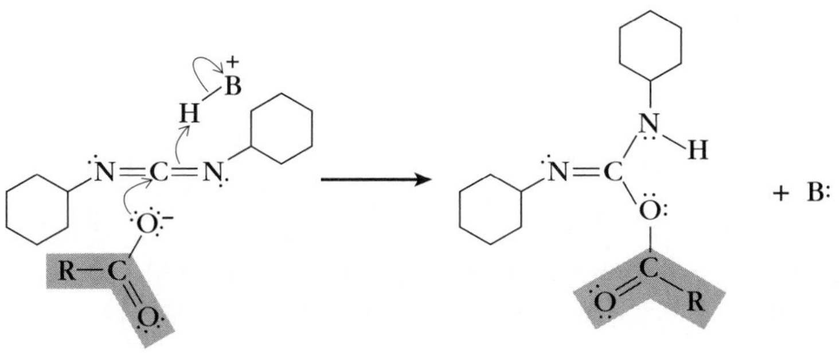

Step 2: Addition of amine to the carbonyl carbon atom froms a tetrahedral intermediate

Tetrahedral intermediate

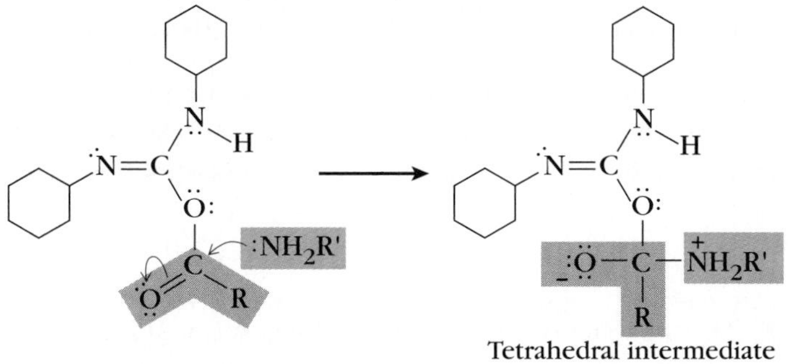

Step 3: Elimination of DCU from the tetrahedral intermediate forms the conjugate acid of an amide, which loses a proton to form an amide

N,N'-Dicyclohexylurea (DCU)

Conjugate acid of amide

Amide

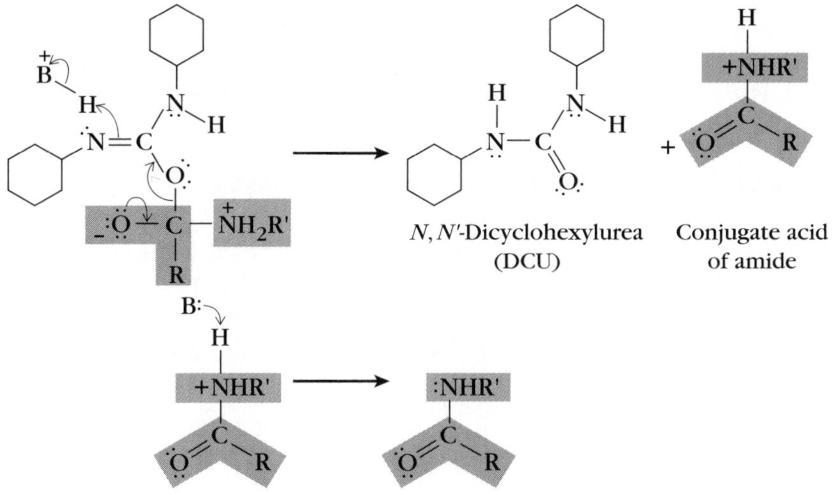

Figure 26.7

Mechanism of the formation of an amide by the reaction of DCC, a carboxylate ion, and an amine. Addition of the carboxylate ion to DCC in Step 1 forms a reactive acylating reagent. Nucleophilic addition of the amine to the acylating reagent forms a tetrahedral intermediate in Step 2. In Step 3, the tetrahedral intermediate loses dicyclohexylurea to form DCU and the conjugate acid of an amide, which loses a proton to form an amide.

Reaction 4. The *t*-Boc protecting group is removed by reaction with trifluoroacetic acid:

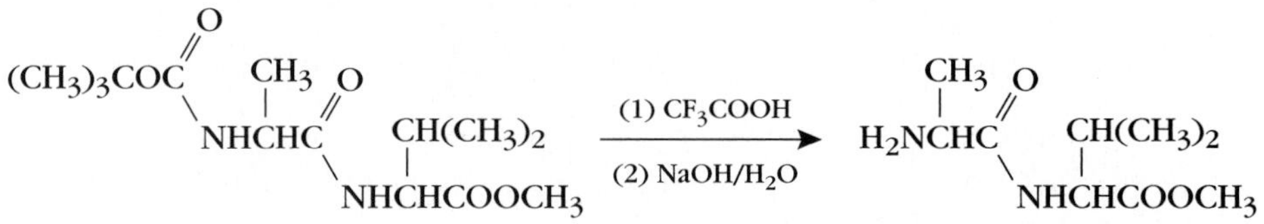

Reaction 5. Hydrolysis of the methyl ester by reaction with aqueous sodium hydroxide yields the desired dipeptide, Ala-Val:

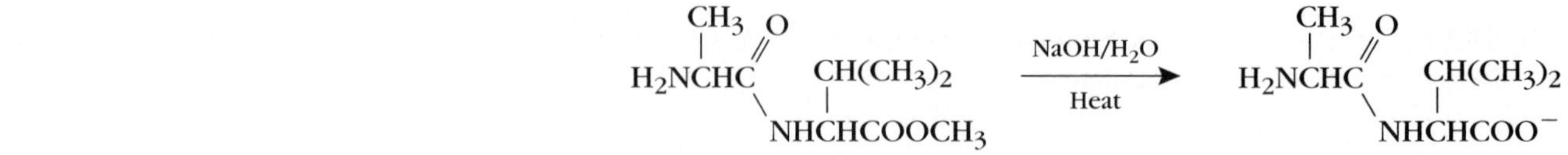

These five steps can be repeated to add amino acids, one at a time, to a growing peptide chain. Using this method, Frederick Sanger synthesized human insulin, which is composed of two polypeptide chains containing a total of 51 amino acids.

Exercise 26.13	List the reactions needed to synthesize Leu-Ala-Val from leucine, alanine, valine, DCC, and di-*t*-butyl dicarbonate.

The synthesis of polypeptides by the sequential addition of one amino acid at a time to a growing chain is time-consuming. This problem can be overcome by using a solid-phase polypeptide synthesis that can be automated.

26.8   Solid-Phase Polypeptide Synthesis

In 1962, Robert B. Merrifield developed a technique called the solid-phase synthesis that greatly simplifies the chemical synthesis of polypeptides. In the Merrifield solid-phase method, polypeptide synthesis occurs on the surface of a solid support. The solid support consists of beads of a copolymer prepared by copolymerizing styrene with a few percent of 4-(chloromethyl)styrene.

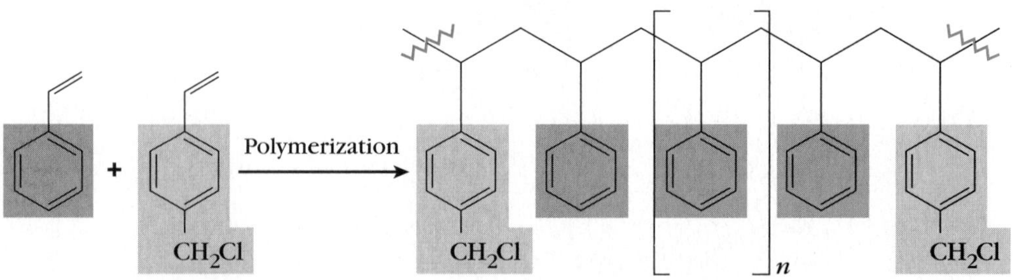

Styrene   4-(chloromethyl)-   Copolymer of styrene and
            styrene   4-(chloromethyl)styrene

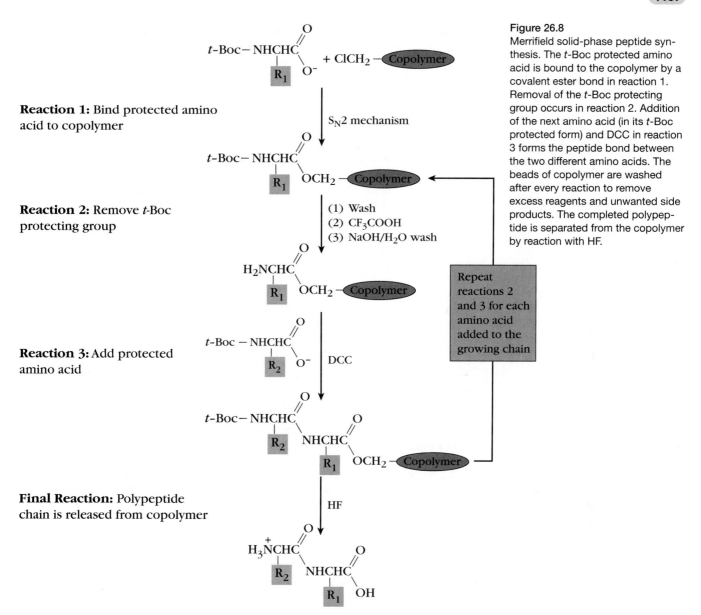

Reaction 1: Bind protected amino acid to copolymer

Reaction 2: Remove *t*-Boc protecting group

Reaction 3: Add protected amino acid

Final Reaction: Polypeptide chain is released from copolymer

Figure 26.8
Merrifield solid-phase peptide synthesis. The *t*-Boc protected amino acid is bound to the copolymer by a covalent ester bond in reaction 1. Removal of the *t*-Boc protecting group occurs in reaction 2. Addition of the next amino acid (in its *t*-Boc protected form) and DCC in reaction 3 forms the peptide bond between the two different amino acids. The beads of copolymer are washed after every reaction to remove excess reagents and unwanted side products. The completed polypeptide is separated from the copolymer by reaction with HF.

In the copolymer, about one of every 100 or so benzene rings contains a chloromethyl group. The four reactions of the Merrifield solid-phase synthesis, shown in Figure 26.8, occur at exposed $ClCH_2$-groups on the copolymer.

Reaction 1. A covalent ester bond is formed by the nucleophilic substitution reaction of the carboxyl group of a *t*-Boc protected amino acid and the chloromethyl groups of the copolymer. This attaches the protected amino acid to the copolymer.

Reaction 2. After formation of the ester bond is complete, the copolymer with the attached amino acid is washed to remove any excess reagents. Reaction of the copolymer-bound amino acid with trifluoroacetic acid removes the *t*-Boc protecting group. This reaction frees the amino group of the bound amino acid, so it can form a peptide bond.

Reaction 3. A second *t*-Boc protected amino acid is added to the copolymer along with DCC to form a peptide bond. The amino group of the second amino acid is protected, so the only peptide bond that can form is between the two different amino acids. After the peptide bond is formed, the copolymer and the attached protected dipeptide are again washed to remove any excess reagents.

If we wish to add additional amino acids to form a longer peptide, we must repeat the sequence of reaction 2 and reaction 3 for every amino acid added to the growing peptide chain.

Reaction 4. After the required number of amino acids have been linked together and the desired peptide has been synthesized, the peptide is released from the copolymer by adding anhydrous HF, which cleaves the ester bond without disrupting any of the peptide bonds. HF also removes the *t*-Boc protecting group.

Computer-controlled peptide synthesizers using the Merrifield solid-phase synthesis technique are available that automatically repeat the removal of the protecting group (reaction 2) and formation of the peptide bond (reaction 3) with different amino acids to form polypeptides containing as many as 150 amino acid residues. In fact, Merrifield has synthesized bovine pancreatic ribonuclease containing 124 amino acid residues.

Exercise 26.14	Show the steps needed to prepare the tripeptide Ala-Val-Phe by the solid-phase peptide synthesis starting with alanine, valine, phenylalanine, and a styrene-4-(chloromethyl)-styrene copolymer.

In their natural state, polypeptides and proteins exist in a number of preferred conformations called their secondary structure.

26.9   The Secondary Structure of Polypeptides

Rotation about the bond between the nitrogen atom and the carbonyl carbon atom of an amide functional group is restricted, so a peptide bond is rigid. Rotation can occur, however, about the two single bonds that join the nitrogen atom and the carbonyl carbon atom of the amide group to adjacent α-carbon atoms.

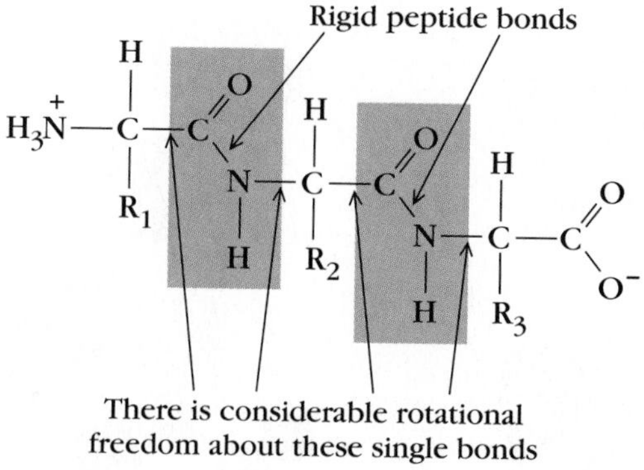

Rigid peptide bonds

There is considerable rotational freedom about these single bonds

Rotation about these bonds as well as the bonds in amino acid side chains defines the conformation of a polypeptide chain. In theory, the backbone of a polypeptide chain can adopt many different conformations by rotation about these bonds. In fact, only a limited number of conformations are possible because of steric hindrance between various groups of atoms. A polypeptide adopts the most stable conformation (the one of lowest energy). The conformations of a polypeptide that are repeated regularly along its main chain are called its **secondary structure.** We will consider in detail three types of secondary structures, the α-helix, the β-pleated sheet, and the β-bend.

A The α-Helix

The **α-helix,** shown in Figure 26.9, is the most commonly observed protein secondary structure. An α-helix can be right- or left-handed. A right-handed α-helix turns clockwise as we proceed toward the N-terminal end of the main chain, while a left-handed α-helix turns counterclockwise. A right-handed α-helix composed of *L*-amino acids is more stable than a left-handed α-helix composed of *L*-amino acids because of the steric hindrance between the carbonyl group and the side chains in the left-handed α-helix. Only the right-handed α-helix has been found in naturally occurring polypeptides.

An α-helix has a rodlike structure. The polypeptide backbone forms the inner part of the rod, and the side chains extend outward. A polypeptide in an α-helix conformation is stabilized by hydrogen bonds between the carbonyl oxygen atom of every amino acid residue and the amide hydrogen of the residue four amino acids away.

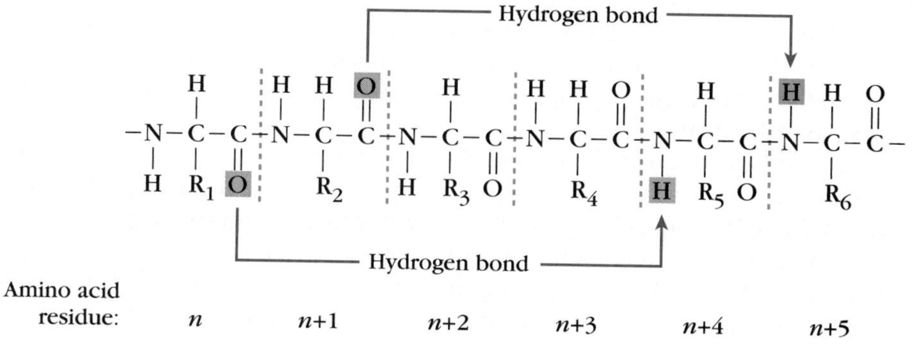

Hydrogen bond

Hydrogen bond

Amino acid residue: n $n+1$ $n+2$ $n+3$ $n+4$ $n+5$

Figure 26.9
A, The carbon and nitrogen atoms that form the backbone of the polypeptide chain arranged in a right-handed α-helix. A right-handed α-helix turns clockwise as we proceed toward the N-terminal end of the polypeptide chain. **B,** A ball-and-stick model of an α-helix. The dots represent hydrogen bonds.

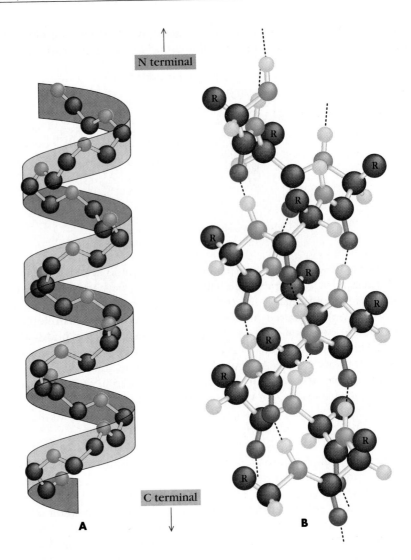

N terminal

C terminal

A   B

In an α-helix, therefore, almost all the carbonyl oxygen atoms and amide hydrogen atoms of the main chain are hydrogen bonded.

An α-helix has a regular structure because all the residues have nearly the same conformation. In the regular structure of an α-helix, there are approximately 3.6 amino acid residues per turn of helix. Amino acids that are three or four residues apart in the linear representation of a polypeptide, therefore, are quite close to each other in an α-helix. Amino acids that are only two residues apart, on the other hand, are located on opposite sides of an α-helix so they are unlikely to make contact.

Some amino acids are more common in an α-helix conformation than in other conformations. Generally, amino acids with small or uncharged side chains, such as alanine, leucine, and phenylalanine, are often found in α-helices. The polar side chains of amino acids, such as arginine, glutamine, and serine, tend to repel each other and destabilize the helix, so they are less frequently found in α-helices. Proline is never found in an α-helix because its five-membered ring cannot assume the required conformation.

The α-helix content varies greatly in proteins whose three-dimensional structures are known. In myoglobin, for instance, the α-helix is the major structural feature. In contrast, the α-helix is only a small part of the structure of the enzyme chymotrypsin.

B   The β-Pleated Sheet

The **β-pleated sheet** is a second commonly occurring protein secondary structure. A β-pleated sheet is formed when two or more almost fully extended polypeptide chains are placed side by side so that hydrogen bonds can form between the carbonyl oxygen atoms of an amino acid in one chain and the amide hydrogen atoms of an amino acid in the other sheet.

Adjacent chains in a β-pleated sheet are either parallel or antiparallel, as shown in Figure 26.10. In a parallel β-pleated sheet, the polypeptide chains run in the same direction. In an antiparallel β-pleated sheet, in contrast, adjacent polypeptide chains run in opposite directions.

The pleated structure is seen more clearly in the representation of the antiparallel β-sheet shown in Figure 26.11. In both the parallel and antiparallel β-pleated sheets, the side chains of the polypeptide chains extend alternatively above and below the plane of the sheet so that side chains on alternate amino acid residues are on opposite sides of the sheet.

The backbone of each polypeptide chain has its amide hydrogen atoms and carbonyl oxygen atoms *trans* to each other, so it is possible to extend a β-pleated sheet into a multistrand structure by simply adding successive chains to the sheet. The β-pleated sheets in many proteins contain anywhere from 2 to 15 parallel or antiparallel polypeptide chains.

Amino acids with small, nonpolar side chains such as methionine, valine, and isoleucine are often found in β-pleated sheets. Amino acids with bulky and

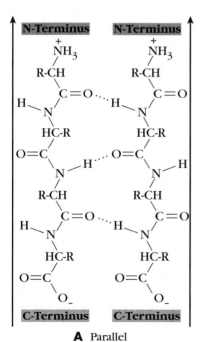

A Parallel   **B** Antiparallel

Figure 26.10
The two forms of the β-pleated sheet. **A**, Adjacent polypeptide chains run in the same direction in a parallel β-sheet. **B**, Adjacent polypeptide chains run in opposite directions in an antiparallel β-sheet. Dotted lines indicate hydrogen bonds between the chains.

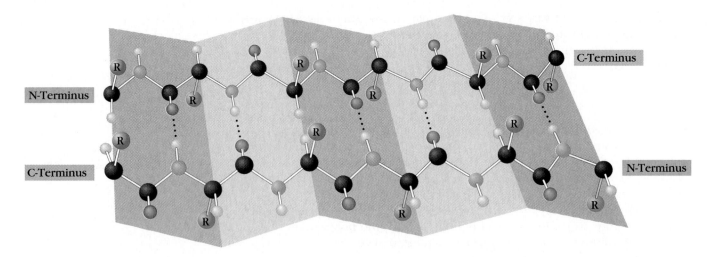

Figure 26.11
A two-strand ball-and-stick representation of an antiparallel β-pleated sheet drawn to emphasize its pleated structure. Notice that the side chains on adjacent residues, indicated by *R,* are on opposite sides of the sheet.

charged side chains are less frequently found in β-pleated sheets, probably due to unfavorable interactions of the side chains. Proline, which never occurs in α-helices, sometimes occurs in β-pleated sheets, although it tends to produce kinks in the structure.

C   The β-Bend

Polypeptide chains change direction in proteins by means of a **reverse turn** or **β-bend.** A reverse turn forms a tight loop in which the carbonyl oxygen of one residue forms a hydrogen bond with the amine hydrogen atom of the residue three positions farther along the polypeptide chain. This characteristic structure of a β-bend is shown in Figure 26.12. Glycine, proline, and one or more hydrophilic amino acids such as asparagine or serine are frequently found as the amino acid residues in β-bends.

26.10 Tertiary and Quaternary Structures of Globular Proteins

Globular proteins are water soluble proteins that adopt a spherical shape. The **tertiary structure** of a globular protein is its biologically active or native confor-

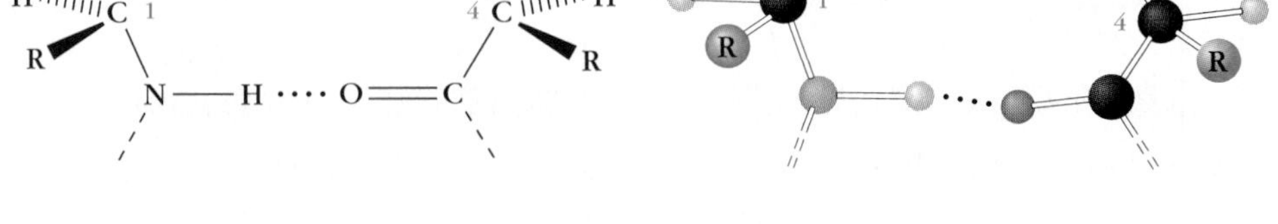

Figure 26.12
The structure of a reverse turn or β-bend in polypeptides. The carbonyl group of amino acid residue 1 shown here is hydrogen bonded to the hydrogen atom of the NH group of residue 4. Hydrogen bonds are represented by dotted lines.

N–Terminus

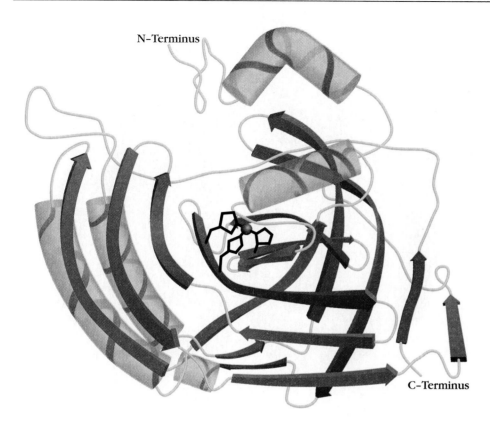

C–Terminus

Figure 26.13
The tertiary structure of the enzyme carbonic anhydrase. The regions of the protein that exist as α-helices are represented as cylinders, the regions that exist as β-pleated sheets are represented by flat arrows pointing towards the N-terminus of the polypeptide, and the polypeptide chains linking the secondary structures are represented as a thin ribbon. The gray ball in the middle represents a Zn^{2+} ion that is coordinated to the three histidine residues.

mation. In other words, the tertiary structure is the folding of the secondary structures of a globular protein together with the spatial arrangement of the side chains of its amino acid residues to obtain the lowest energy conformation that places most of the hydrophilic groups on the exterior and most of the hydrophobic groups on the interior of the protein. Only about half of a globular protein exists as α-helices or β-pleated sheets (on average about 27% α-helices and 23% β-pleated sheets). The rest of the globular protein consists of β-bends and unstructured segments of polypeptide chains.

The tertiary structure of the enzyme carbonic anhydrase, a globular protein, is illustrated in Figure 26.13. In Figure 26.13, the sections of the protein that exist as β-pleated sheets are represented as flat arrows pointing toward the N-terminus of the polypeptide. The sections that exist as α-helices are represented as cylinders. The polypeptide chains linking various secondary structures are represented as a thin ribbon. The gray ball represents a Zn^{2+} ion that is coordinated to three histidine residues.

Folding in the tertiary structures of proteins is important because it brings amino acid residues that are very far apart in the primary structure closer together. The three histidine residues that complex with the Zn^{2+}, for example, are located far apart in the primary structure of the enzyme carbonic anhydrase.

The **quaternary structure** of a protein is the assembly of two or more individual polypeptide chains held together by noncovalent forces or covalent bonds. The assembly is often called an *oligomer,* and the constituent polypeptide chains are called *subunits.* The subunits of an oligomeric protein can be identical or very different in their primary, secondary, and tertiary structures.

Figure 26.14
Four folded polypeptide chains, two α subunits and two β subunits, assemble to form the quaternary structure of hemoglobin.

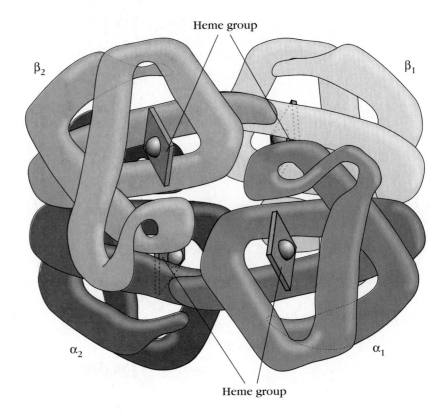

Heme group

β_2 β_1

α_2 α_1

Heme group

Figure 26.15
The levels of protein structure. **A**, The linear sequence of amino acid residues defines the primary structure. **B**, The most common conformations that determine the secondary structure of polypeptides are α-helices and β-pleated sheets. **C**, The tertiary structure is the three-dimensional conformation of the protein. **D**, The assembly of two or more fully-folded polypeptide chains forms the quaternary structure. The oligomeric protein in this example consists of two subunits.

The quaternary structure of hemoglobin, whose function is to deliver oxygen from the lungs to other tissues, is shown in Figure 26.14. Hemoglobin consists of two α subunits, each with 141 amino acid residues, and two β subunits, each with 146 amino acid residues. Each subunit contains one heme group, which is responsible for the characteristic red color of blood and is the site that binds one molecule of oxygen.

The levels of protein structure are summarized in Figure 26.15.

Ala—Val—Thr—Asp—Pro—Leu—Gly

A Primary structure

α-Helix

β-Pleated sheet

B Secondary structure

C Tertiary structure

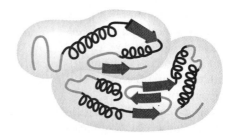

D Quaternary structure

26.11 The Information for Folding Is in the Primary Structure of Polypeptides

A series of experiments carried out in the early 1960s demonstrates the importance of the amino acid sequence to the properties of polypeptides. In one of the experiments, described schematically in Figure 26.16, the biologically active enzyme ribonuclease, a globular protein, was placed in a solution containing 8 M urea and β-mercaptoethanol. When placed in this solution, the enzyme is made biochemically inactive, a process called **denaturation.** Denaturation of enzymes is frequently irreversible. In the case of the denatured enzyme ribonuclease, however, when it is removed from the urea and β-mercaptoethanol solution and exposed to air oxidation, the enzyme is spontaneously restored to its original biochemically active form. Analysis by several methods showed that the renatured product was indistinguishable from the native ribonuclease.

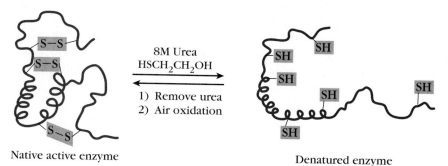

Native active enzyme

8M Urea
HSCH₂CH₂OH

1) Remove urea
2) Air oxidation

Denatured enzyme

Figure 26.16
The denaturation and spontaneous renaturation of the enzyme ribonuclease. Denaturation of the enzyme occurs by placing it into a solution of 8 M urea and β-mercaptoethanol. The denatured enzyme is spontaneously converted to its native active form when it is removed from urea and β-mercaptoethanol and exposed to air oxidation.

Folding is spontaneous because it is an energetically favorable process. That is, in going from the unfolded to the folded state, the free energy change must be negative.

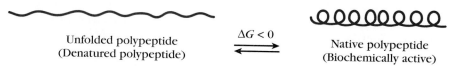

Unfolded polypeptide
(Denatured polypeptide)

$\Delta G < 0$

Native polypeptide
(Biochemically active)

What does this mean at the molecular level? Polypeptides and proteins form folded or native structures because of a fine balance among various noncovalent forces to which they are subjected. These noncovalent forces are ionic and dipolar interactions, hydrogen bonding, and hydrophobic interactions. Hydrophobic interactions is a name given to those influences that cause nonpolar substances to minimize their contacts with water. These noncovalent forces all contribute to the stability of the folded polypeptide chain relative to the unfolded polypeptide.

Polypeptide chains of globular proteins fold so that their nonpolar side chains are out of contact with water to maximize hydrophobic attractions and minimize destabilizing interactions between nonpolar groups and water. This folding also places the polar or charged side chains on the surface, so that the stabilizing effects of ionic and dipolar interactions and hydrogen bonding with water are at a maximum. *The polypeptide chains of globular proteins, therefore, fold spontaneously so that the hydrophobic side chains are placed in the interior away from water and the polar and charged side chains are on the surface in contact with water.*

The primary structure of a polypeptide specifies the location of side chains

■ *Noncovalent forces are discussed in Section 1.16.*

and consequently determines how the polypeptide will fold to form a hydrophobic interior and a hydrophilic surface. Although the information needed to fold a polypeptide is contained in its primary structure, anything that alters the balance of noncovalent forces that maintain its native structure causes denaturation of the polypeptide. Heating a polypeptide in solution causes a change in its native structure. Varying the pH of the aqueous environment of a polypeptide changes its charge distribution and hydrogen bonding requirements. High concentrations of water soluble alcohols such as ethanol interfere with the hydrophobic forces that stabilize the polypeptide. Thus heating, changing the pH, or adding alcohols all denature polypeptides by causing its natural structure to unfold into a random structure that is biologically inactive.

26.12 The Structure of the Enzyme Lysozyme

Enzymes are proteins that catalyze biochemical reactions. Lysozyme, for example, is a protein that catalyzes the hydrolysis of the polysaccharide components of bacterial cell walls. The cell wall polysaccharide consists of alternating molecules of N-acetylglucosamine (NAG) and N-acetylmuramic acid (NAM). NAG and NAM are derivatives of glucosamine in which the amino group is acetylated, as shown

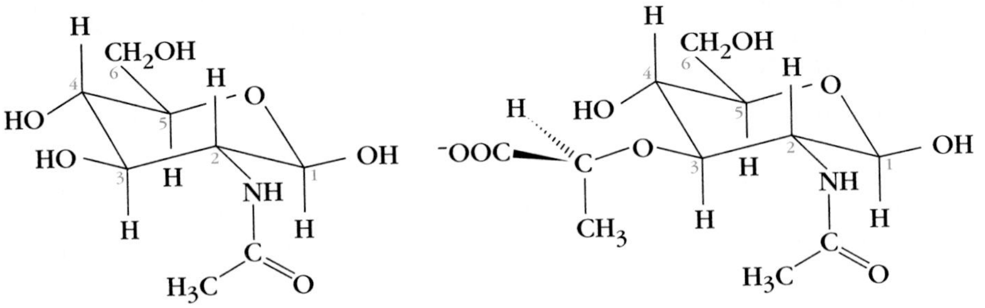

Figure 26.17
The structures of N-acetylglucosamine (NAG) and N-acetylmuramic acid (NAM). NAG and NAM are derivatives of glucosamine in which the amino group is acetylated. NAM also contains a lactyl group,

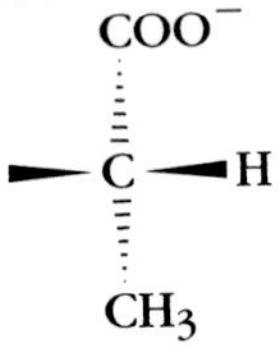

$$COO^-$$

—C—H, attached to carbon

$$CH_3$$

atom 3 of the ring by an ether bond.

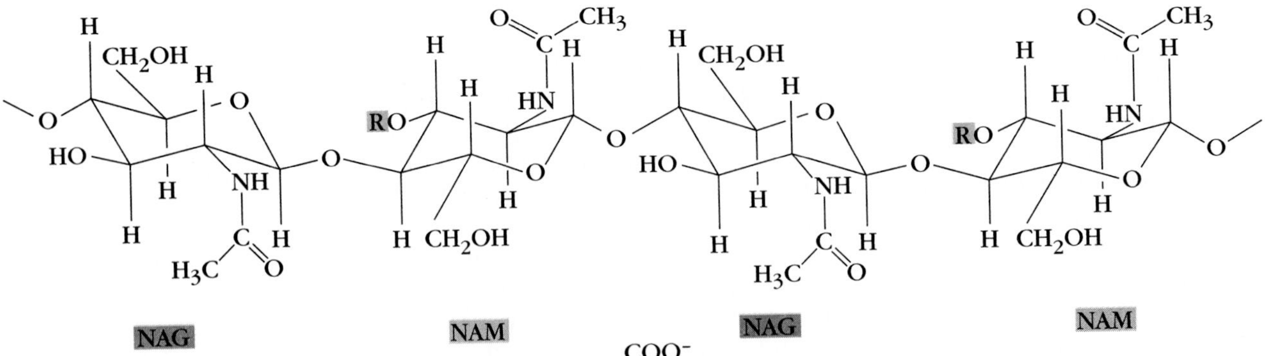

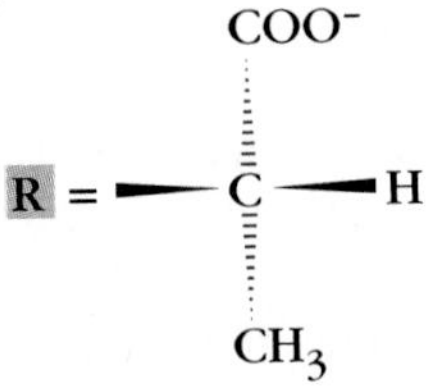

N–Acetylglucosamine (NAG)

N–Acetylmuramic acid (NAM)

NAG   NAM   NAG   NAM

Figure 26.18
A short segment of the polysaccharide portion of bacterial cell walls. N-Acetylglucosamine (NAG) is linked to N-acetylmuramic acid (NAM) by β-1,4-glycoside bonds. Notice that the structures of NAM are flipped 180° relative to the structures of NAG.

$$COO^-$$

$$R = —C—H$$

$$CH_3$$

in Figure 26.17. NAM also contains a lactyl side chain attached to carbon atom 3 of the ring by an ether bond. The alternating NAG and NAM molecules in the polysaccharide portion of bacterial cell walls are joined by β-1,4-glycoside bonds, as shown in Figure 26.18.

Lysozyme-catalyzed hydrolysis of the polysaccharide components of cell walls in water enriched with ^{18}O yields a product in which ^{18}O is found in the hemiacetal functional group of NAM. This result indicates that lysozyme hydrolyzes the glycosidic bond between NAM and NAG. The hydroxy group of the terminal NAM unit is in the β-configuration, so hydrolysis occurs with retention of configuration.

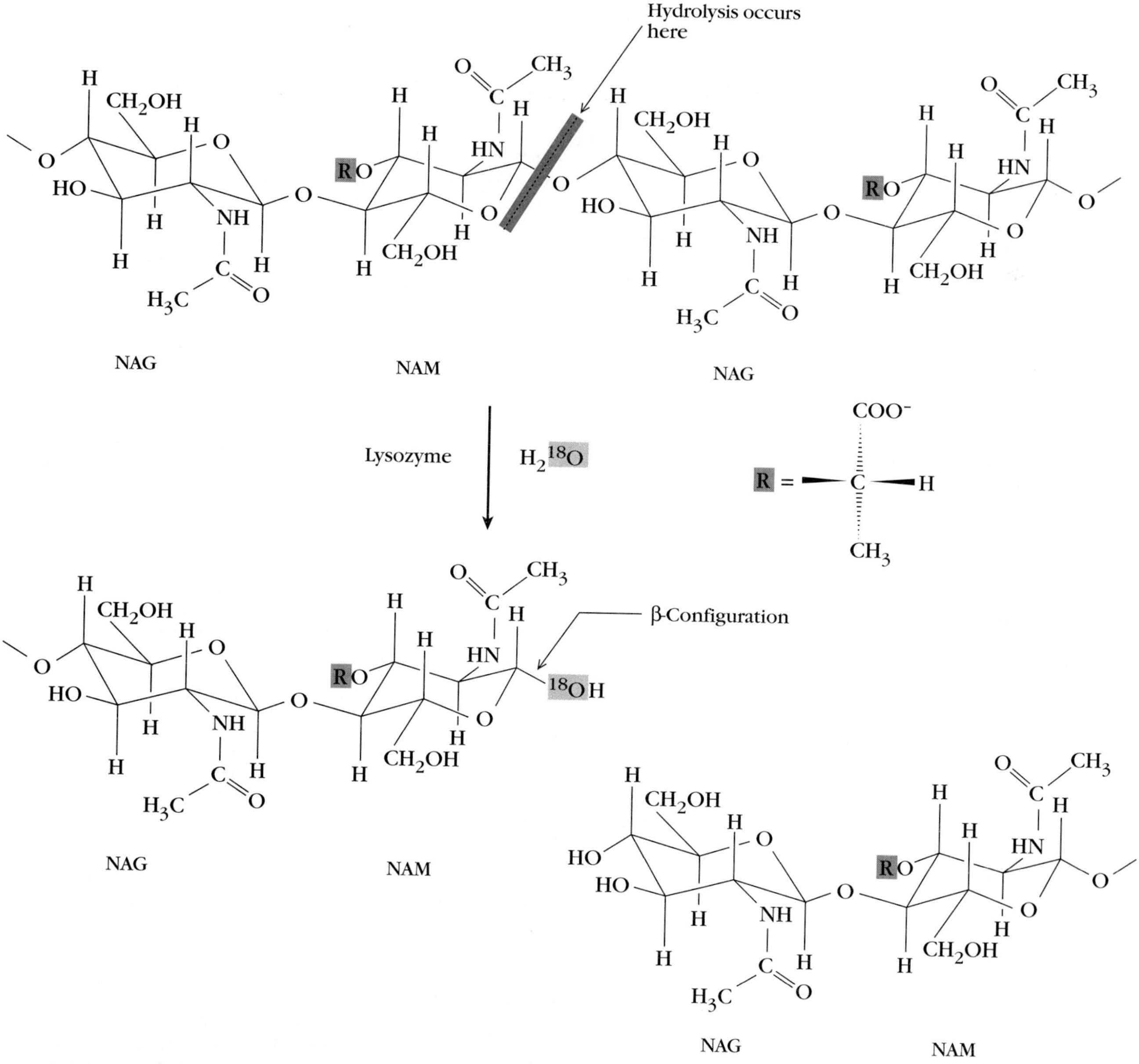

Hen egg white (HEW) lysozyme is the most widely studied species of lysozyme. HEW lysozyme is a globular protein of molecular weight 14,600 amu, whose single polypeptide chain contains 129 amino acid residues and four disulfide bonds. The primary structure of HEW lysozyme is shown in Figure 26.19.

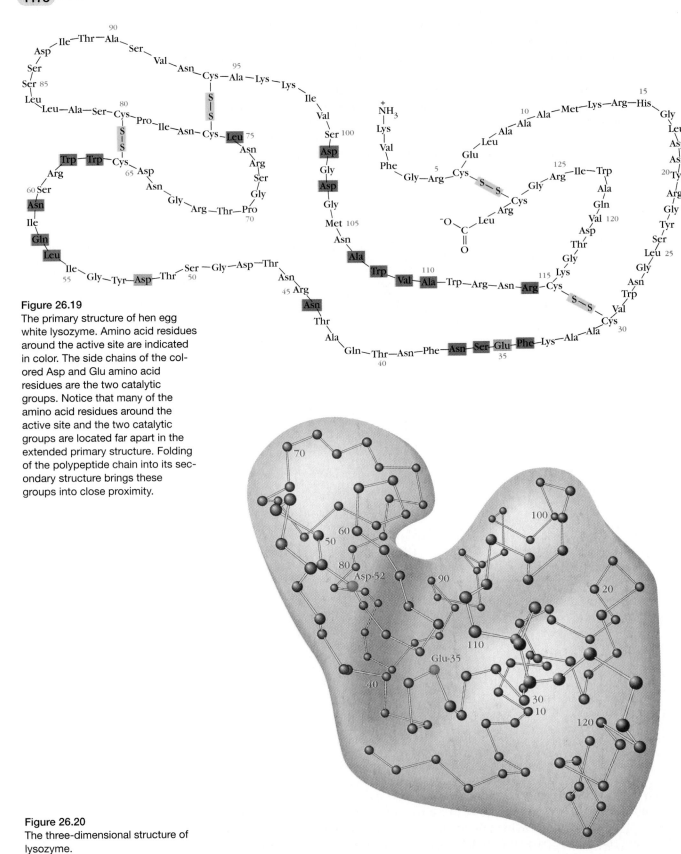

Figure 26.19
The primary structure of hen egg white lysozyme. Amino acid residues around the active site are indicated in color. The side chains of the colored Asp and Glu amino acid residues are the two catalytic groups. Notice that many of the amino acid residues around the active site and the two catalytic groups are located far apart in the extended primary structure. Folding of the polypeptide chain into its secondary structure brings these groups into close proximity.

Figure 26.20
The three-dimensional structure of lysozyme.

David Phillips and colleagues determined the three-dimensional structure of lysozyme by x-ray diffraction in 1965. A molecule of lysozyme is roughly ellipsoidal in shape. The most striking feature of the molecule is a prominent cleft that cuts across one face of the molecule, as shown in Figure 26.20. Most of the nonpolar side chains are located in the interior of the molecule, shielded from contact with aqueous solvent.

The catalytic mechanism of lysozyme is not immediately revealed by a detailed knowledge of its three-dimensional structure because the location of the active site is not obvious. Location of the active site, therefore, is the next step in determining the catalytic mechanism of lysozyme.

26.13 The Active Site in Lysozyme

Recall that the mechanism of an enzyme-catalyzed reaction occurs by first forming an enzyme-substrate complex (ES). Reaction occurs in the enzyme-substrate complex to form the product and regenerate the enzyme.

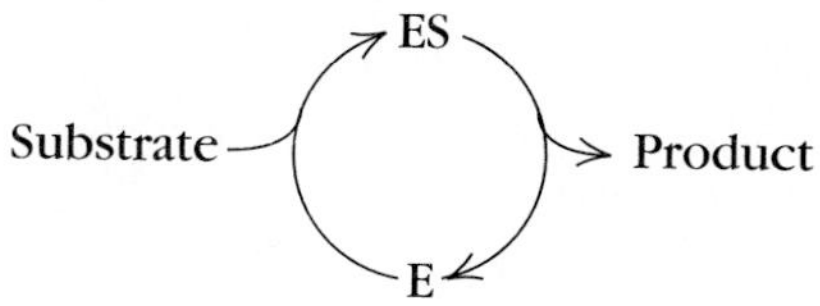

The enzyme-substrate complex is formed by binding the substrate to the active site of the enzyme.

We cannot use x-ray methods to determine the structure of an enzyme-substrate complex undergoing reaction because the conversion of substrate into product is so rapid that the enzyme-substrate cannot be observed. Information about the active site of an enzyme can be obtained by using x-ray methods to determine the three-dimensional structure of a complex of the enzyme with an inhibitor. The trimer of N-acetylglutamine, (NAG)$_3$, which is made in the laboratory and not found in nature, is an inhibitor of lysozyme.

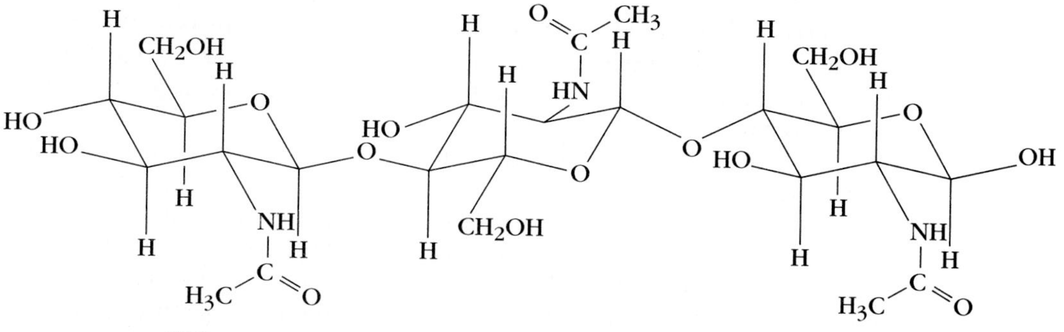

X-ray studies of the (NAG)$_3$-lysozyme complex show that the trimer binds to the right half of the enzyme cleft. The three pyranose rings of the trimer are designated as A, B, and C in Figure 26.21.

By the use of molecular models of lysozyme, it was determined that there is space for three additional NAG molecules in the cleft. These three additional units are designated D, E, and F in Figure 26.21. While NAG molecules E and F fit

Figure 26.21

The structure of the complex formed between lysozyme and a hexamer of *N*-acetylglucosamine. The six pyranose rings of the substrate are designated A through F. Pyranose ring F contains the reducing end of the substrate. The positions of the rings A, B, and C were determined by x-ray studies of the complex of (NAG)$_3$ and lysozyme. The positions of rings D, E, and F were inferred from molecular models. Rings A to F occupy sites A to F, respectively, on the active site.

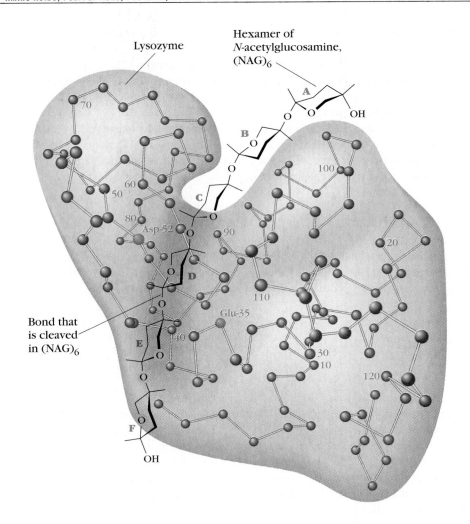

easily in the cleft, D does not fit unless its ring is distorted from the usual chair conformation to that of a half chair. Molecular models were used to determine that six NAG molecules can fit into the cleft. This was confirmed by the observation that (NAG)$_6$ is rapidly hydrolyzed by lysozyme. The lysozyme-catalyzed hydrolysis of (NAG)$_6$ occurs at a glycoside bond between the second and third molecule from its reducing end.

These observations and a molecular model of lysozyme were used to determine the structure of the complex between the polysaccharide of cell walls and lysozyme. Recall that the polysaccharides of cell walls are oligomers of alternating molecules of *N*-acetylglucosamine (NAG) and *N*-acetylmuramic acid (NAM). Molecular models showed that a NAM molecule is too large to fit into site C because of its lactyl side chain. The NAM residues, therefore, cannot bind to site C and must bind to sites B, D, and F.

The observation that lysozyme hydrolyzes the β-1,4-glycoside bonds between NAM and NAG (Section 26.12) implies that the glycoside bond cleavage occurs between rings B and C or between rings D and E (Section 26.12). Because (NAG)$_3$ binds to the active site but is not cleaved by the enzyme while occupying sites B and C, the probable cleavage site is the glycoside bond between rings D and E. Thus the active site of lysozyme consists of six pyranose binding sites, and hydrolysis of the glycoside bond occurs between rings D and E of the bound substrate, as shown in Figure 26.21.

26.14 Mechanism of Lysozyme Catalysis

The proposal that the hydrolysis of (NAG)$_6$ occurs between rings D and E led to a search for possible catalytic groups near the glycoside bond that is cleaved. Proton transfer is a critical step in the mechanism of most hydrolysis reactions, so the most likely catalytic groups, therefore, are groups that can serve as proton donors or acceptors. The only two groups near the glycoside bond cleaved by lysozyme that can serve as proton donors or acceptors are Asp 52 and Glu 35. The aspartic acid residue is on one side of the glycoside bond, and the glutamic acid residue is on the other side. The acidic side chains of Asp 52 and Glu 35 are in different environments. Asp 52 is in a hydrophilic environment, so it should exist as a carboxylate anion throughout the pH range of 3 to 8 in which lysozyme is biologically active. Glu 35, on the other hand, is in a hydrophobic environment, so it is likely to remain protonated throughout the pH range of 3 to 8.

The mechanism proposed by Phillips for the lysozyme-catalyzed hydrolysis of (NAG)$_6$ is shown in Figure 26.22. Step 1 of the mechanism is proton transfer from Glu 35 to the glycoside oxygen atom between rings D and E. Glycoside bond cleavage occurs in Step 2 to form an oxonium ion on ring D. Recall that in the enzyme-substrate complex ring D is distorted into a half-chair conformation. Formation of an oxonium ion is facilitated because the half-chair conformation of ring D resembles the structure of the oxonium ion and the transition state leading to the oxonium ion. Oxonium ion formation is also facilitated by charge-charge interactions between the oxonium ion and the carboxylate ion of Asp 52, which stabilize the transition state and the oxonium ion. At this point the dimer (NAG)$_2$, which consists of rings E and F, is released from the enzyme surface. Asp 52 also shields the oxonium ion so that reaction with water occurs from the opposite side in Step 3 to form a product with retention of configuration. Finally proton transfer to Glu 35 regenerates the biologically active enzyme.

■ *The designation Asp 52 means that the 52nd amino acid residue from the N-terminus of lysozyme is an aspartic acid residue.*

Exercise 26.15	Write the mechanism of the acid-catalyzed hydrolysis of methyl β-*D*-glucopyranoside. List the similarities and differences between the acid-catalyzed hydrolysis of methyl β-*D*-glucopyranoside and the lysozyme-catalyzed hydrolysis of (NAG)$_6$.

The formation and reaction of the (NAG)$_6$-lysozyme complex illustrate many of the principles of organic chemistry discussed in previous chapters. Noncovalent attractive and repulsive forces are important in the formation of the enzyme-substrate complex. Proton transfers are as important in the mechanisms of enzyme-catalyzed reactions as in the mechanisms of nonbiological reactions in the laboratory. The major difference is the nature of the proton transfer reagents. Strong acids such as sulfuric acid are commonly used as proton donors in the laboratory. Enzymes are usually destroyed by strong mineral acids, so they use acidic groups on the side chains of amino acid residues at the active site as proton transfer groups. The most common groups are ammonium ions and carboxylic acids. Hydroxide ion is commonly used as the proton acceptor in nonbiological reactions, while enzymes use carboxylate ions and amino groups as proton acceptors.

The similarity between the mechanisms of enzyme-catalyzed reactions and the mechanisms of the analogous reactions in a nonbiological environment means that the basic principles of organic chemistry are directly applicable to biochemical reactions such as the catalytic role of enzymes.

Step 1:
Proton transfer
from Glu 35 to the
glycoside oxygen
atom between
rings D and E

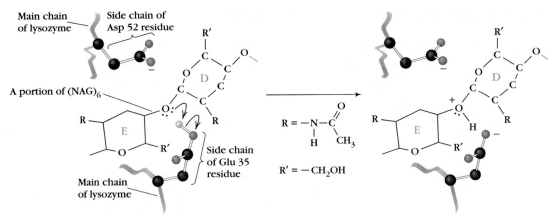

Step 2:
Cleavage of the glycoside
bond forms an oxonium ion

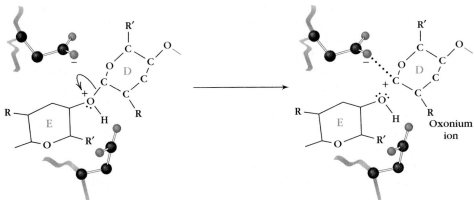

Step 3:
Reaction of oxonium ion
with water

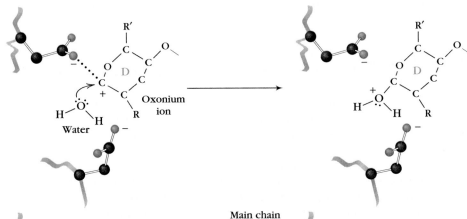

Step 4:
Proton transfer regenerates
active form of lysozyme

Figure 26.22
The Phillips mechanism of the
lysozyme-catalyzed hydrolysis of
(NAG)₆.

26.15 Summary

Polypeptides are linear biopolymers that consist of **α-amino acid residues** joined together by amide or peptide bonds. **Proteins** are molecules that consist of one or more polypeptide chains. Twenty α-amino acids are commonly found in proteins, and except for glycine they all contain a stereocenter. The stereocenter has been related to *L*-glyceraldehyde, so the α-amino acids in proteins belong to the *L*-family. According to the Cahn-Ingold-Prelog Rules, the stereocenters of the α-amino acids in proteins have the *S*-configuration.

Determining the structure of polypeptides and proteins requires several steps. The first step is the cleavage of any covalent disulfide bonds. Acid-catalyzed hydrolysis of the polypeptide into its constituent amino acids is the next step. The constituent amino acids are separated and identified by chromatography, and their amount is determined by spectroscopy. Next the sequence of amino acids of the polypeptide, called the **primary structure**, is determined. One common method is the **Edman degradation**, in which treatment of the polypeptide with phenyl isothiocyanate removes the N-terminal amino acid residue as a phenyl thiocarbamyl derivative but leaves intact the rest of the polypeptide chain. Repeated Edman degradations allow the sequence of amino acids to be determined in polypeptides containing up to about 60 amino acid residues.

Chemical synthesis of polypeptides is accomplished by adding amino acids one at a time to the lengthening polypeptide chain by the use of **protecting groups.** A successful synthesis requires protecting all the reactive groups, except the one carboxyl group and the one amino group that will react to form the desired peptide bond. Amines are commonly protected as their *tert*-butoxycarbonyl (*t*-Boc) derivatives, and carboxylic acids are protected as their esters. Once the protecting groups are in place, dicyclohexylcarbodiimide (DCC) is used to form a peptide bond between the unprotected amino group and the unprotected carboxyl group. Once the peptide bond is formed, the protecting groups are removed and the sequence of two reactions is repeated. This chemical synthesis has been greatly simplified by the **Merrifield solid-phase method,** in which the growing polypeptide chain is anchored to an insoluble polymer support.

The conformations of a polypeptide that are repeated regularly along its main chain are called its **secondary structure.** Three common secondary structures are the **α-helix,** the **β-pleated sheet,** and the **β-bend.** The **tertiary structure** is the folding of the structures of proteins together with the spatial arrangements of the side chains of its amino acid residues. The **quaternary structure** of a protein is the assembly of two or more individual polypeptide chains held together by noncovalent forces or covalent bonds.

The information needed to fold a polypeptide is contained in its primary structure. In general, a polypeptide chain folds spontaneously so that the hydrophobic side chains are placed in the interior away from water and the polar and charged side chains are on the surface in contact with aqueous solutions.

Lysozyme is a protein that catalyzes the hydrolysis of the polysaccharide components of bacterial cell walls. The mechanism of lysozyme-catalyzed hydrolysis of polysaccharides is similar in many ways to the acid-catalyzed hydrolysis of acetals. This similarity illustrates how the basic principles of organic chemistry are directly applicable to biochemical reactions such as the catalytic role of enzymes.

Additional Exercises

26.16 Define each of the following terms in words, by a structure, or by a chemical equation.

(a) Zwitterion
(b) α-Amino acid
(c) Disulfide bond
(d) Primary structure of a polypeptide
(e) Peptide bond

(f) Essential amino acid
(g) Secondary structure of a polypeptide
(h) Solid-phase peptide synthesis
(i) Amphoteric compound
(j) Isoelectric point

26.17 Write three-dimensional structures for each of the following amino acids.

(a) *L*-Valine   (b) (*S*)-Cysteine   (c) (*R*)-Glutamine   (d) (*S*)-Proline

26.18 Write Fischer projection formulas for each of the amino acids in Exercise 26.17.

26.19 Write three-dimensional structures for each of the following:

(a) Phe   (b) Trp   (c) Ala-Ser   (d) Gly-Val-Glu   (e) Cys-Lys-Arg-His

26.20 Write the structure of each of the following amino acids in solution at pH = 3, pH = 7, and pH = 11.

(a) Leu   (b) Met   (c) Asp   (d) Lys

26.21 The titration curve of an unknown amino acid is shown below.

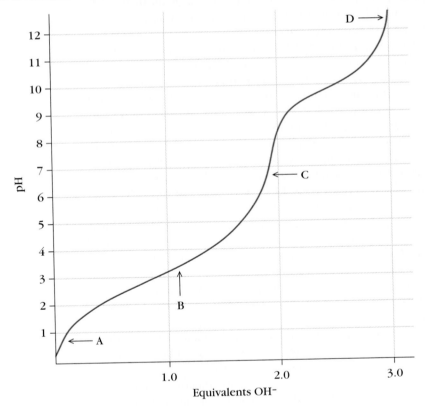

(a) Estimate the pK_a values of the ionizable groups.
(b) Determine the identity of the amino acid by comparing the pK_a values obtained in (a) with those in Table 26.3 (p. 1151).
(c) Write the structure of the amino acid, charged or uncharged, at points *A* to *D* on the titration curve.

26.22 Lys has three acid dissociation constants whose values are $p(K_a)_1 = 2.2$; $pK_R = 10.8$; and $p(K_a)_2 = 9.2$. Write the equation for the proton transfer reaction involved in each acid ionization.

26.23 Explain why there is a difference of 2.4 units between the $p(K_a)_1$ of alanine (2.3) and the pK_a of acetic acid (4.7).

26.24 Which of the side chains of the 20 amino acids listed in Table 26.1 are charged at pH = 7?

26.25 Write the structure of the intermediate products A to C in the following synthesis of methionine.

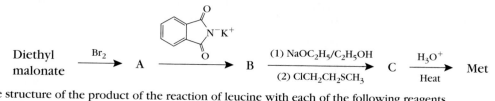

26.26 Write the structure of the product of the reaction of leucine with each of the following reagents.

(a) CH_3OH/HCl

(b) Basic aqueous solution of benzoyl chloride

(c) Di-*tert*-butyl dicarbonate

(d) Acetic anhydride

26.27 Write the structure of the products formed in each of the following reactions.

(a) Asn $\xrightarrow[\text{Heat}]{\text{NaOH/H}_2\text{O}}$

(b) Val-Gln-Lys $\xrightarrow[\text{Heat}]{\text{HCl/H}_2\text{O}}$

(c) *t*-Boc-alanine + Methyl valinate $\xrightarrow{\text{DCC}}$

(d) Asp $\xrightarrow[\text{CH}_3\text{OH}]{\text{HCl}}$

(e) 2 Ninhydrin + Phe $\longrightarrow$

26.28 Propose a one-step synthesis of racemic alanine, starting with 2-bromopropanoic acid and any other reagents.

26.29 Show where trypsin, chymotrypsin, and BrCN would cleave the following polypeptides.

(a) Gly-Met-Gly-Phe-Ala-Val-Arg-Met-Leu-Tyr-Lys-Gly-Ala

(b) Phe-Ile-Met-Lys-Tyr-Asp-Gly-Arg-Ala-Val-Leu-Pro-Cys

(c) Val-Cys-Gly-Glu-Met-Pro-Leu-Arg-Ala-Ile-Tyr-Gly-Ala

26.30 Write the structures of the phenyl thiohydrantoin (PTH) derivatives formed by the Edman degradation of each of the following polypeptides.

(a) Ala-Leu-Tyr

(b) Asn-Gly-Glu-Thr

(c) Val-Pro-Ile-His

26.31 The sweetener Nutrasweet® is the methyl ester of a dipeptide consisting of aspartic acid and phenylalanine. How would you determine whether the amino acid sequence in Nutrasweet® is Asp-Phe or Phe-Asp?

26.32 Due to a malfunction of the sequenator, the pH of reaction 1 of the Edman degradation was set at pH 3.0 instead of pH 9.0. What effect will this have on the Edman degradation?

26.33 Repeated Edman degradations are carried out on a hexapeptide. The results are as follows:

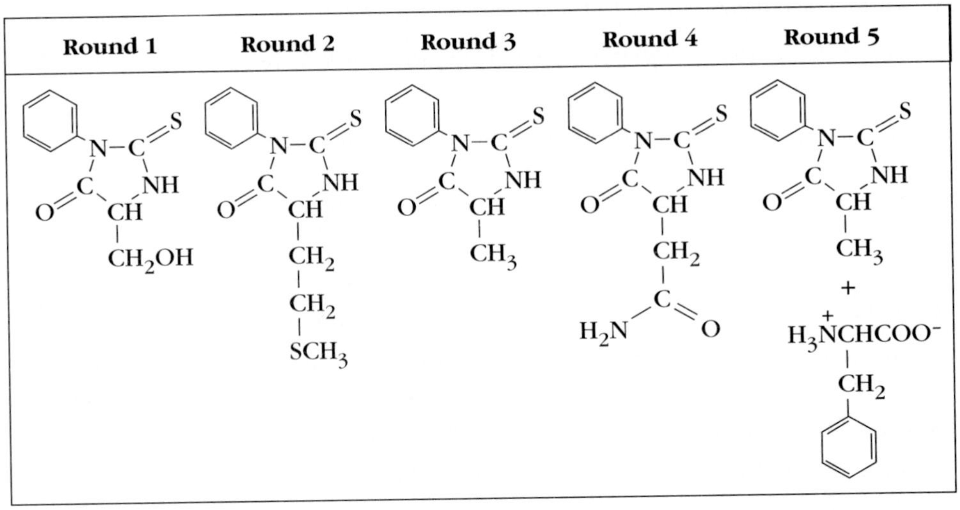
Round 1
Round 2
Round 3
Round 4
Round 5

(a) What is the amino acid composition of the hexapeptide?
(b) What is its amino acid sequence?

26.34 The primary structure of lysozyme is shown in Figure 26.19 (p. 1178). The disulfide bonds are oxidized by performic acid to form a linear polypeptide chain. Write the structure of the products of the reaction of this linear polypeptide with each of the following reagents:

(a) N≡CBr **(b)** Trypsin **(c)** Chymotrypsin

26.35 2,4-Dinitrofluorobenzene is another reagent that can be used to determine the N-terminal amino acid residue of a polypeptide. The product of the reaction of the polypeptide and 2,4-dinitrofluorobenzene is hydrolyzed and the N-terminal amino acid is identified because only it can form an N-2,4-dinitrophenyl derivative. Propose a mechanism for the reaction of 2,4-dinitrofluorobenzene with the N-terminal amino acid of a polypeptide chain.

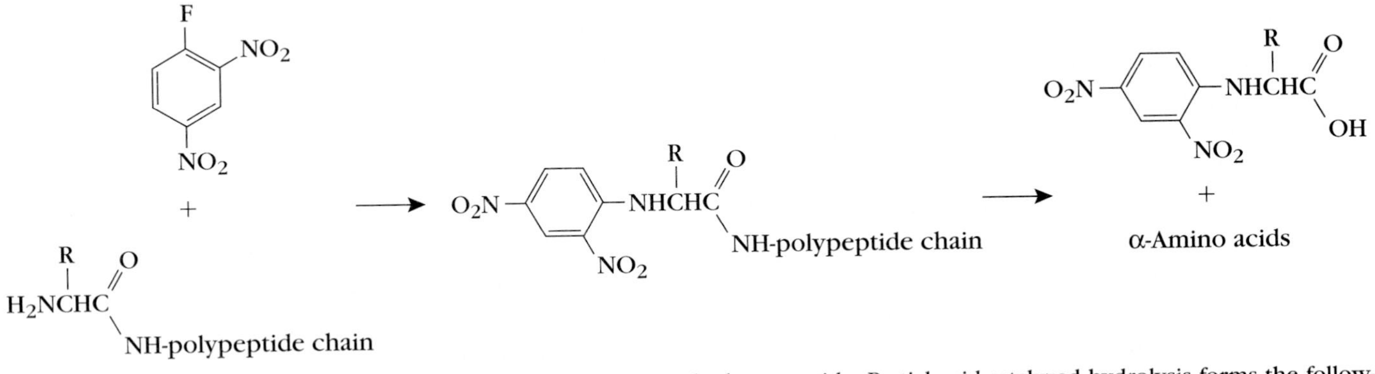
NH-polypeptide chain
α-Amino acids

26.36 Trypsin reacts with a heptapeptide to form arginine and a hexapeptide. Partial acid-catalyzed hydrolysis forms the following fragments:

Arg-Gly-Pro
Pro-Leu-Phe-Ile-Val
Gly-Pro-Leu

Assign the primary structure of the heptapeptide.

26.37 Complete hydrolysis of a polypeptide gives the following amino acid composition:

Val Trp (Met)$_2$ (Gly)$_2$ Lys (Ala)$_2$ Ile Pro Asp Arg Tyr Cys

Trypsin-catalyzed hydrolysis of the polypeptide formed the following fragments:

Val-Trp-Met-Gly-Lys
Ala-Ile-Pro-Met-Asp-Arg
Tyr-Ala-Gly-Cys

Chymotrypsin-catalyzed hydrolysis of the polypeptide formed the following fragments:

Ala-Gly-Cys
Met-Gly-Lys-Ala-Ile-Pro-Met-Asp-Arg-Tyr
Val-Trp

Deduce the primary structure of this polypeptide.

26.38 Write the reactions needed to synthesize Ala-Gly-Leu-Val-Phe, starting with copolymer-bound *t*-Boc-Leu-Val-Phe.

26.39 Write the reactions needed for the solid-phase synthesis of Gly-Val-Leu.

CHAPTER

NUCLEIC ACIDS

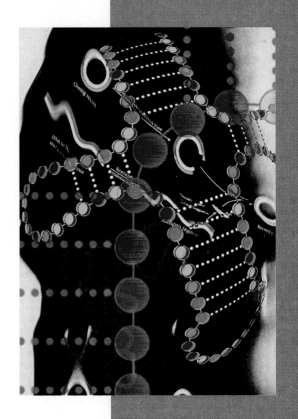

The essential property of all living systems is the ability to reproduce or replicate themselves. For **replication,** an organism must possess a complete description of itself. This description is essentially a set of instructions that specifies every step that is required for a cell to construct an exact replica of itself. The information required for replication of an organism is contained in molecules called nucleic acids.

The two classes of nucleic acids are **deoxyribonucleic acid (DNA)** and **ribonucleic acid (RNA).** Deoxyribonucleic acid is the master repository of genetic information in cells. DNA directs its own replication during cell division. The genetic information in DNA is copied into RNA by a process called **transcription.** Subsequent **translation** of the information in RNA forms proteins. The genetic information that is stored in DNA, therefore, is passed from DNA to RNA to proteins. We begin our study of nucleic acids by examining the chemical composition of nucleic acids.

27.1   Composition of Nucleic Acids

Nucleic acids, just like proteins, are polymers made of individual units that are linked to form long chains. In proteins, the units are α-amino acids. In nucleic acids, the units are called **nucleotides.** Enzyme-catalyzed cleavage of a nucleotide forms a **nucleoside** and a phosphate ion. Each nucleoside, in turn, can be hydrolyzed to form a pentose and heterocyclic base.

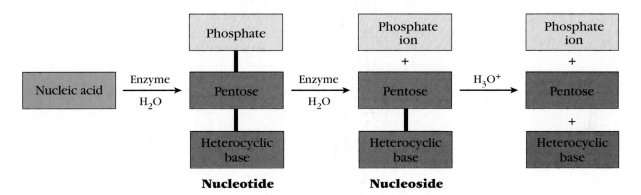

In DNA, the pentose portion is 2-deoxy-β-*D*-ribose, while in RNA the pentose portion is β-*D*-ribose. Ribose and 2-deoxyribose exist as five-membered furanose rings in both DNA and RNA.

2-Deoxy-β-*D*-ribose   β-*D*-Ribose

When 2-deoxy-β-*D*-ribose and β-*D*-ribose are part of the structure of nucleosides and nucleotides, their carbon atoms are identified by primed numbers.

The heterocyclic bases in DNA and RNA are derivatives of purine and pyrimidine.

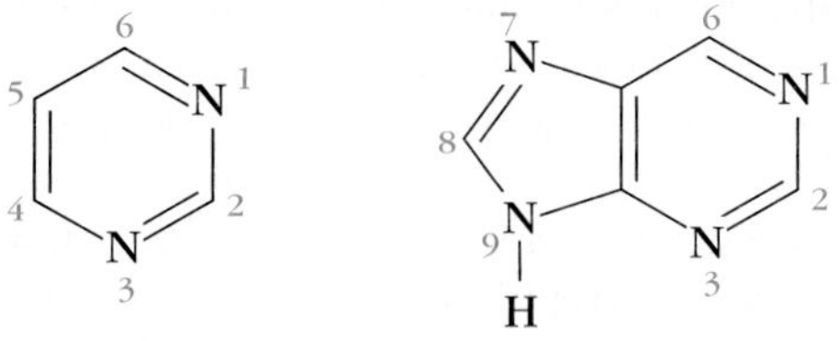

Pyrimidine   Purine

There are four different heterocyclic bases in DNA. Adenine and guanine are substituted purines, and cytosine and thymine are substituted pyrimidines.

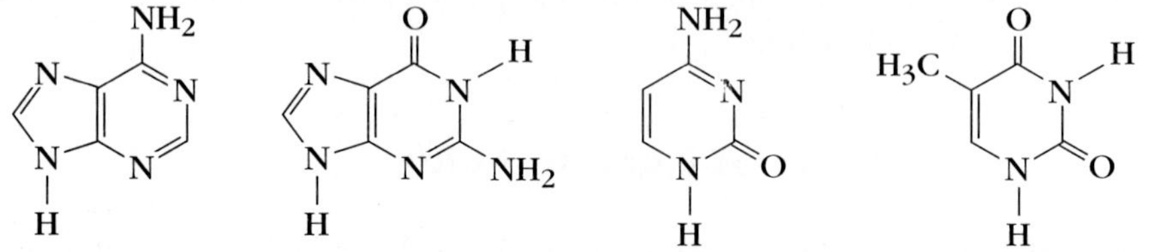

Adenine   Guanine   Cytosine   Thymine
(A)   (G)   (C)   (T)

Derivatives of purine   Derivatives of pyrimidine
found in DNA   found in DNA

The first letter of their names is used as an abbreviation of these heterocyclic bases.

RNA also contains four different heterocyclic bases, but thymine is replaced by uracil, a derivative of pyrimidine, so the four bases in RNA are adenine, guanine, cystosine, and uracil.

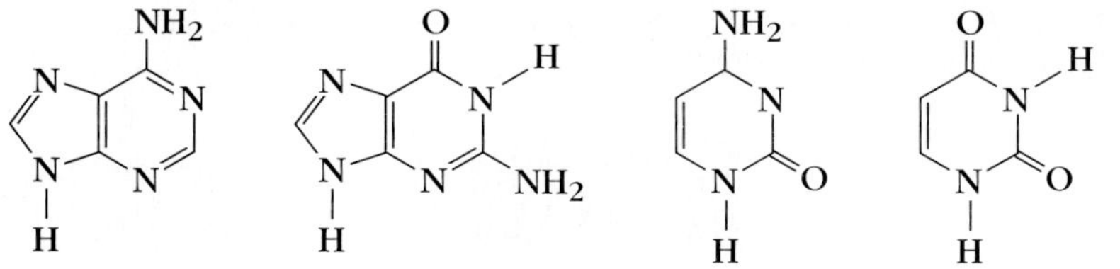

Adenine   Guanine   Cytosine   Uracil
(A)   (G)   (C)   (U)

Derivatives of purine   Derivatives of pyrimidine
found in RNA   found in RNA

Table 27.1	The Composition of DNA and RNA	
	Nucleic Acid	
Component	DNA	RNA
Purine derivatives	Adenine	Adenine
	Guanine	Guanine
Pyrimidine derivatives	Cytosine	Cytosine
	Thymine	Uracil
Pentose sugar	2-Deoxy-β-D-ribose	β-D-ribose
Phospate linkage	Phosphate group	Phosphate group

The composition of DNA and RNA is summarized in Table 27.1. While adenine, guanine, cytosine, uracil, and thymine can exist as tautomers, they exist in the form shown above when present in nucleic acids.

Exercise 27.1	Write the structures of the two tautomeric forms of adenine, guanine, cytosine, uracil, and thymine.

27.2 The Structure of Nucleosides and Nucleotides

The structures of the nucleosides of both DNA and RNA consist of the heterocyclic base bonded to C1′ of the pentose molecule by a *N*-β-glycosidic bond. The general structure of nucleosides and the structure of 2′-deoxyadenosine and 2′-deoxycytidine, nucleosides found in DNA, are shown in Figure 27.1.

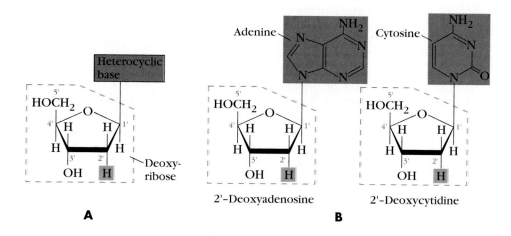

Figure 27.1
A, The general structure of a nucleoside found in DNA. **B,** The structures of 2′-deoxyadenosine and 2′-deoxycytidine. The heterocyclic bases are attached by an *N*-β-glycoside bond to the C1′ of 2′-deoxyribose. Numbers with a prime refer to positions on the furanose ring of the pentose.

The nucleosides found in RNA have the same general structure, except that the pentose is ribose instead of deoxyribose. The structures of uridine and guanosine, both found in RNA, are shown in Figure 27.2. The name for each pentose-heterocyclic base combination in RNA and DNA is given in Table 27.2.

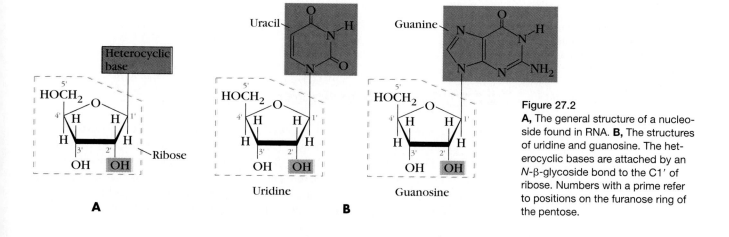

Figure 27.2
A, The general structure of a nucleoside found in RNA. **B,** The structures of uridine and guanosine. The heterocyclic bases are attached by an *N*-β-glycoside bond to the C1′ of ribose. Numbers with a prime refer to positions on the furanose ring of the pentose.

Table 27.2	Names of Nucleosides	
Base	Name When Combined with Ribose	Name When Combined with 2-Deoxyribose
Cytosine	Cytidine	2'-Deoxycytidine
Adenine	Adenosine	2'-Deoxyadenosine
Guanine	Guanosine	2'-Deoxyguanosine
Uracil	Uridine	—
Thymine	—	2'-Deoxythymidine

Exercise 27.2   Write the structure of the nucleoside formed by combining each of the following pairs of heterocyclic bases and pentoses.

(a) Ribose and guanine   **(c)** Uracil and 2-deoxyribose
(b) Thymine and 2-deoxyribose

The general structures of the nucleotides of DNA and RNA are shown in Figure 27.3. Both contain a phosphate group bonded to the C5′ position of the 2′-deoxyribose unit in DNA and the C5′ position of the ribose unit in RNA. The complete structures of the four deoxyribonucleotides and the four ribonucleotides are shown in Figure 27.4.

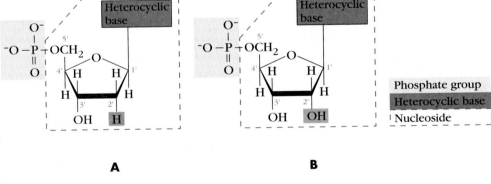

Figure 27.3
A, The general structure of a nucleotide found in DNA. The phosphate group is bonded to C5′ of the 2′-deoxyribose unit of the nucleoside. **B,** The general structure of a nucleotide found in RNA. The phosphate group is bonded to C5′ of the ribose unit of the nucleoside.

27.3   Structure of Nucleic Acids

Nucleotides in DNA and RNA are linked by phosphate ester bonds between C5′ of the pentose of one nucleotide and C3′ of the pentose of another nucleotide, as shown in Figure 27.5. The end of the nucleic acid that has a free phosphate group at C5′ is called the 5′-end, and the end that has a free hydroxy group at C3′ is called the 3′-end. The sequence of nucleotides in a nucleic acid is written by convention so that the 5′-end is placed at the left end of the structure.

A The four nucleotides found in DNA:

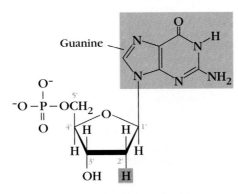

Adenine

2'-Deoxyadenosine 5'-monophosphate

Guanine

2'-Deoxyguanosine 5'-monophosphate

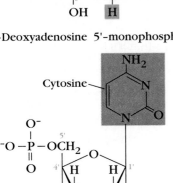

Cytosine

2'-Deoxycytidine 5'-monophosphate

Thymine

2'-Deoxythymidine 5'-monophosphate

B The four nucleotides found in RNA:

Adenine

Adenosine 5'-monophosphate

Guanine

Guanosine 5'-monophosphate

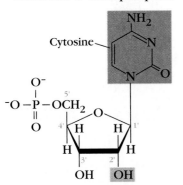

Cytosine

Cytidine 5'-monophosphate

Uracil

Uridine 5'-monophosphate

Figure 27.4
A, Structures and names of the four nucleotides found in DNA. **B,** Structures and names of the four nucleotides found in RNA.

Figure 27.5
General structure of DNA and RNA. Individual nucleotides are linked by phosphate ester bonds between C5′ of the pentose of one nucleotide and C3′ of the pentose of another nucleotide.

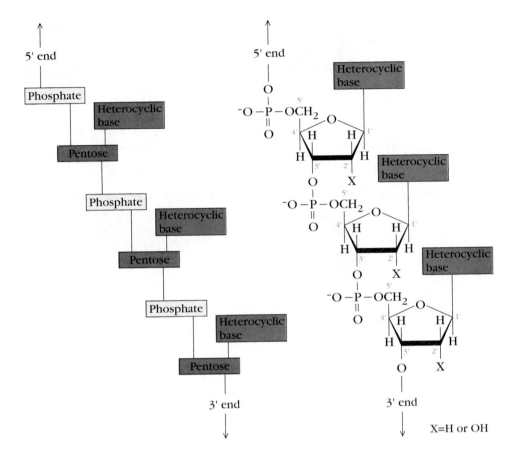

The structure of nucleic acids resembles the structure of proteins in many ways. A nucleic acid, just like a protein, has two parts. The pentose and phosphate groups form the backbone to which are attached heterocyclic bases at regular intervals.

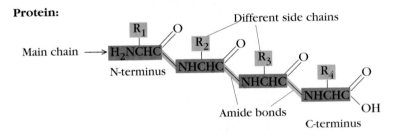

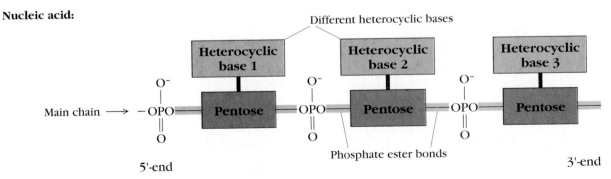

Although chemically and structurally similar, RNA and DNA are very different in size. DNA molecules are much larger than RNA molecules. DNA molecules contain as many as 46 million bases and have molecular weights up to 150 billion amu. RNA molecules, in contrast, contain only about 50,000 bases and have molecular weights of about 150 million amu.

27.4 The Three-Dimensional Structure of DNA Is a Double Helix

Chemical analysis of samples of DNA isolated from different species shows that the mole percent of each of the four heterocyclic bases varies widely. These data from the DNA of several species are given in Table 27.3. The data in Table 27.3 show that every DNA sample has a molar ratio of G/C and A/T of about 1. The molar ratio G/T, in contrast, varies from 0.56 to 1. In every DNA sample, therefore, the number of moles of guanine equals the number of moles of cystosine and the number of moles of adenine equals the number of moles of thymine. This means that the heterocyclic bases occur in pairs; guanine and cystosine are paired, and adenine and thymine are paired.

Table 27.3	Base composition and base pair ratios of DNA of various species						
	Heterocyclic Base Composition (Mol %)				Base Pair Ratios		
Source of DNA	A*	G*	C*	T*	G/T†	G/C†	A/T†
Escherichia coli	26.0	24.9	25.2	23.9	1.04	0.988	1.09
Yeast	31.7	18.3	17.4	32.6	0.56	1.05	0.972
Ox	29.0	21.2	21.2	28.7	0.74	1.00	1.01
Human	30.4	19.9	19.9	30.1	0.66	1.00	1.01

*A = adenine, G = guanine, C = cytosine, T = thymine.
†Variations from 1 are due to experimental uncertainties.

In 1953, James Watson and Francis Crick proposed a model for the three-dimensional structure of DNA that explains the pairing of these bases. The Watson-Crick model consists of two nucleic acid chains that wind about a common axis with a right-handed twist to form a double helix. This double helix is shown in Figure 27.6. The two pentose-phosphate backbones run in opposite directions and wind about the outside of the double helix like a staircase. As a result, the polar groups of the pentose and phosphate group are exposed to aqueous solution. The heterocyclic bases occupy the center of the double helix, and their planes are nearly perpendicular to the helix axis.

Only two types of heterocyclic base pairs can be accomodated by this double helix structure. Each adenine base in one chain must be hydrogen bonded to a thymine base on the other chain and vice versa. As well, each guanine base on one chain must be hydrogen bonded to a cytosine base on the other chain and vice versa. These specific base pairs mean that the two chains of DNA are com-

Figure 27.6
The structure of DNA. **A,** A ball-and-stick model of DNA. The pentose-phosphate backbones that wind about the periphery of the molecule are colored blue, and the bases that occupy the interior are colored red. **B,** The space-filling model of DNA. Notice that the two nucleic acid chains run in opposite directions. In this model, C atoms are blue, N atoms are dark blue, H atoms are white, O atoms are red, and P atoms are yellow.

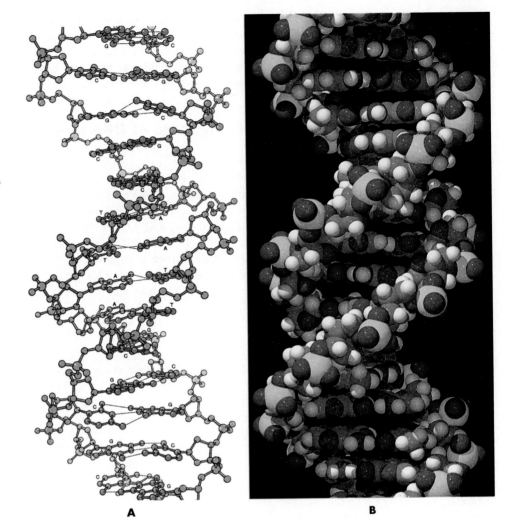

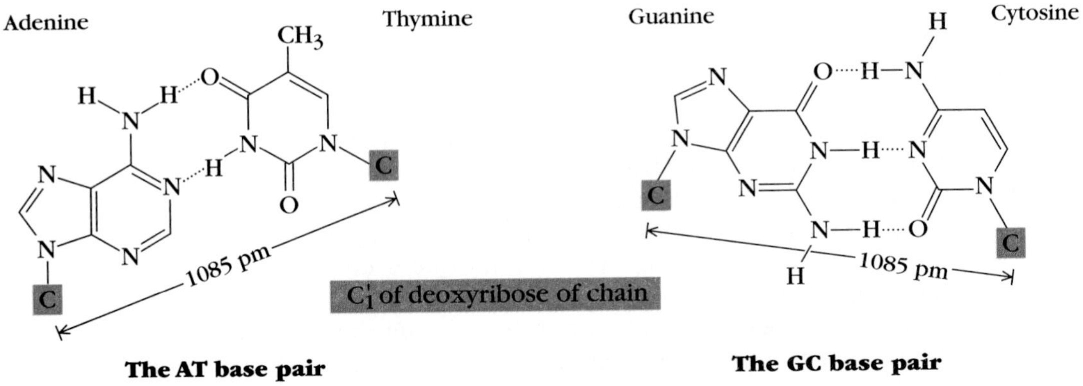

A

B

Figure 27.7
The complementary base pairs of DNA. The dimensions of an AT base pair and a GC base pair are sufficiently similar that they both fit well within the center of the DNA double helix formed by antiparallel nucleic acid chains. Two hydrogen bonds form between A and T, and three hydrogen bonds form between G and C.

Adenine   Thymine

Guanine   Cytosine

1085 pm

C_1' of deoxyribose of chain

1085 pm

The AT base pair

The GC base pair

plementary. Whenever thymine appears in one chain, adenine must appear oppo-site it on the other chain. This complementary base pairing is shown in Figure 27.7.

The sizes of the base pairs, AT and GC, are sufficiently similar that they can be interchanged without disturbing the structure of the double helix. Any AT base pair in the double helix, therefore, can be replaced by a GC pair. Likewise, the structure of the double helix is unaffected by changing a CG pair to GC or an AT pair to TA. Any other combination of bases, such as AC or GT, requires consider-able reorganization of the pentose-phosphate backbone.

The double helix structure of DNA can accomodate any sequence of bases on one nucleic acid chain if the opposite chain has the complementary base sequence. This accounts for the observation shown in Table 27.3 that the ratios of A/T and G/C are equal to 1. Furthermore, the double helix structure suggests that hereditary information is encoded in the sequence of bases on either chain.

The double helix structure of DNA explains more than just base pairing. It provides a model of how DNA produces an exact replica of itself.

27.5 Replication of DNA

The process by which DNA molecules reproduce themselves in the nucleus of cells is called **replication.** Replication of DNA is an enzyme-catalyzed process, which begins by partially unwinding the double helix. The chains progressively separate and the separated portions act as templates on which enzyme-catalyzed synthesis of a new complementary chain occurs. The nucleotides that form the new chain are chosen by their ability to form complementary base pairs with the template chain so that the newly synthesized chain forms a double helix with the template chain. This process is illustrated in Figure 27.8. In each new mole-cule of DNA, therefore, one chain of the double helix comes from the original DNA molecule and the other chain is newly synthesized.

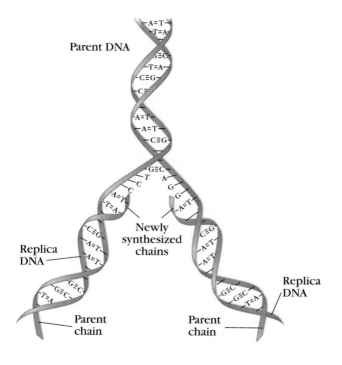

Figure 27.8
Replication of DNA. Both replica DNA molecules contain one newly synthesized chain and one from the parent DNA molecule.

The sequence of heterocyclic bases of DNA also serves as the template for the synthesis of RNA.

27.6 Synthesis of RNA

Synthesis of RNA occurs by a RNA polymerase-catalyzed process called **transcription,** which is shown schematically in Figure 27.9. The process starts by unwinding a section of the double helix of DNA to expose a short segment of chain. The sequence of bases on one of the two exposed chains, called the template strand, is the code for the synthesis of RNA. The ribonucleotides assemble along the template strand by pairing with the bases of DNA. The pairings are the same as for DNA, except that uracil replaces thymine, so RNA is the complement of that portion of the DNA chain that synthesized it.

Figure 27.9
The RNA polymerase–catalyzed synthesis of RNA. The DNA is unwound at one end of the RNA polymerase and rewound at the other end. The template strand, shown in green, pairs with the correct ribonucleotides from the solution to form the nascent RNA shown in blue.

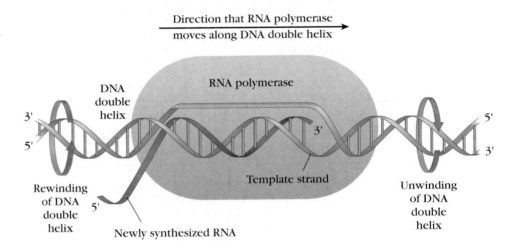

While both DNA and RNA are polymers of nucleotides, the pentose in RNA is ribose instead of 2-deoxyribose in DNA and RNA contains uracil instead of thymine found in DNA. Most molecules of RNA consist of a single nucleotide chain rather than a double helix. The three major classes of RNA are messenger RNA (mRNA), ribosomal RNA (rRNA), and transfer RNA (tRNA).

Messenger RNA (mRNA) carries the genetic information from DNA to the ribosomes, where they direct protein synthesis. Only a small percentage of the RNA of cells is messenger RNA.

Ribosomal RNA (rRNA) combines with proteins to form the intracellular substructures called ribosomes, where proteins are synthesized. Ribosomes, are composed of about 60% rRNA and 40% protein. Although ribosomes are the site of protein synthesis, rRNA does not direct protein synthesis.

Transfer RNA (tRNA) transports specific amino acids to ribosomes, where they are joined together to form proteins. Each of the 20 α-amino acids that combine to form proteins has at least one tRNA molecule that carries it to the ribosome. All tRNA molecules have similar physical and chemical properties. Each tRNA molecule consists of a single chain containing a relatively small number of nucleotide units (between 70 and 90 units) whose shape can be represented in two dimensions as the cloverleaf shown in Figure 27.10. The two important parts of tRNA are the amino acid joined to the 3′-end and the three-base sequence

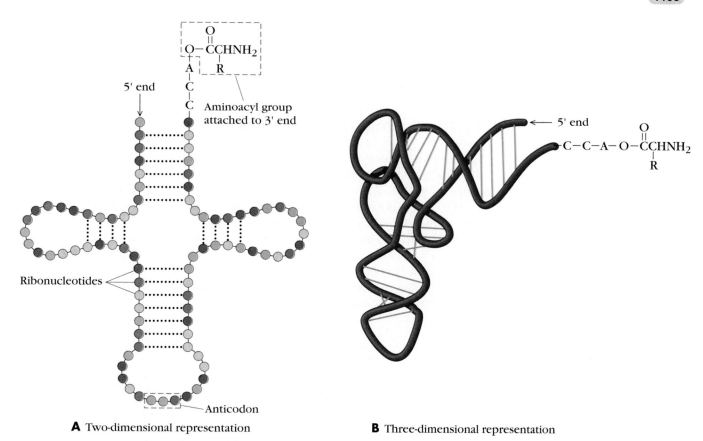

A Two-dimensional representation

B Three-dimensional representation

Figure 27.10
Representations of tRNA. **A,** A two-dimensional structure showing its cloverleaf shape. Each circle represents a ribonucleotide. **B,** A three-dimensional representation. The outlines emphasize the folded L shape.

located at the bottom of the molecule as drawn in Figure 27.10. The three-base sequence, which is a specific code for the amino acid carried by the tRNA, is called the **anticodon** of the tRNA. We will discuss the importance of anticodons in Section 27.8.

The three-dimensional structure of tRNA resembles an L-shaped structure, with the 3′- and the 5′-ends at one end of the L and the anticodon at the other end of the L, as shown in Figure 27.10, *B*. The 3′-end of all tRNA molecules terminates with the sequence cytosine-cytosine-adenine (C-C-A). The specific α-amino acid is joined to this end of tRNA by an ester linkage to the 3′-hydroxy group of the ribose unit of the terminal ribonucleotide.

Exercise 27.3	The 3′-end of tRNA that transports leucine can be abbreviated as —A-ribose-Leu. Write the complete structure of this portion of the tRNA molecule.

27.7 Reading the Codons of mRNA

Protein biosynthesis occurs by the integrated actions of mRNA, many types of tRNA (which deliver the required α-amino acids to mRNA), ribosomes, and many enzymes. mRNA directs protein biosynthesis by a process called **translation.**

Translation occurs by reading the sequence of nucleotide bases, three at a time, along the chain of mRNA from its 5′-end to its 3′-end. Each set of three bases is called a **codon.** Each codon directs the incorporation of a specific amino acid into a growing polypeptide chain. The codons for the amino acids, which are the same for all known life forms, have been determined and are listed in Table 27.4. The codon GUU, for example, specifies the incorporation of a valine residue into the growing polypeptide chain.

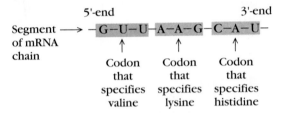

The sequence of bases is read three at a time
from the 5′-end to the 3′-end of mRNA

More than one codon in Table 27.4 can specify the incorporation of a particular α-amino acid. The codons UCG, UCA, UCC, UCU, AGU, and AGC are all signals that a serine residue is to be incorporated into a protein. Certain codons also have special functions. The codon AUG, for example, serves as both a signal to incorporate methionine into the growing polypeptide chain and as a signal to initiate the start of a polypeptide synthesis. The codons UAA, UAG, and UGA, on the other hand, signal the stop of polypeptide synthesis.

Exercise 27.4	Find the six codons in Table 27.4 that are a signal to incorporate arginine into a growing polypeptide.

Exercise 27.5	Using Table 27.4, name the amino acid specified by each of the following codons.

(a) AUU   **(b)** GAU   **(c)** GCU   **(d)** GGG   **(e)** CGA   **(f)** CAU

Exercise 27.6	Write the sequence of amino acids in the tetrapeptide specified by the following sequence of codons:

UCAUUGACCGAG

Table 27.4	Codons for the α-Amino Acids				

First Base (5'-End)	Center Base				Last Base (3'-End)
	U	C	A	G	
U	Phe	Ser	Tyr	Cys	U
U	Phe	Ser	Tyr	Cys	C
U	Leu	Ser	End	End	A
U	Leu	Ser	End	Trp	G
C	Leu	Pro	His	Arg	U
C	Leu	Pro	His	Arg	C
C	Leu	Pro	Gln	Arg	A
C	Leu	Pro	Gln	Arg	G
A	Ile	Thr	Asn	Ser	U
A	Ile	Thr	Asn	Ser	C
A	Ile	Thr	Lys	Arg	A
A	Met*	Thr	Lys	Arg	G
G	Val	Ala	Asp	Gly	U
G	Val	Ala	Asp	Gly	C
G	Val	Ala	Glu	Gly	A
G	Val	Ala	Glu	Gly	G

*The codon AUG serves both as a code for methionine and as a code for starting protein synthesis.

Let's now examine how the codons on mRNA actually direct the synthesis of polypeptides.

27.8   Protein Biosynthesis

The key step in protein biosynthesis occurs when the codons on an mRNA molecule are read by tRNA molecules in a process called **translation.** By reading, we mean that the anticodons of tRNA recognize and combine with the complementary bases of the codons on mRNA. The codon C-U-C present on an mRNA molecule, for instance, is recognized by a leucine-containing molecule of tRNA that has the complementary anticodon base sequence G-A-G.

The step-by-step growth of a polypeptide chain, illustrated in Figure 27.11, starts by an initiation step in which the codon AUG specifies the start of the synthesis by placing an N-formylmethionine residue as the first amino acid of the polypeptide chain to be synthesized. The codon AUG specifies both the incorporation of methionine in the interior of the polypeptide and the start of the peptide synthesis with N-formylmethionine (fMet). The tRNA molecule that brings N-formylmethionine to start the polypeptide chain, however, is slightly different than the tRNA that brings methionine, so mRNA can distinguish between the two. Before the synthesis is completed, the formyl group or even the methionine residue may be removed by hydrolysis.

Polypeptide chain elongation occurs by placing a tRNA with the correct anticodon sequence at the next codon. The next codon on the mRNA chain shown

Figure 27.11
Protein biosynthesis.

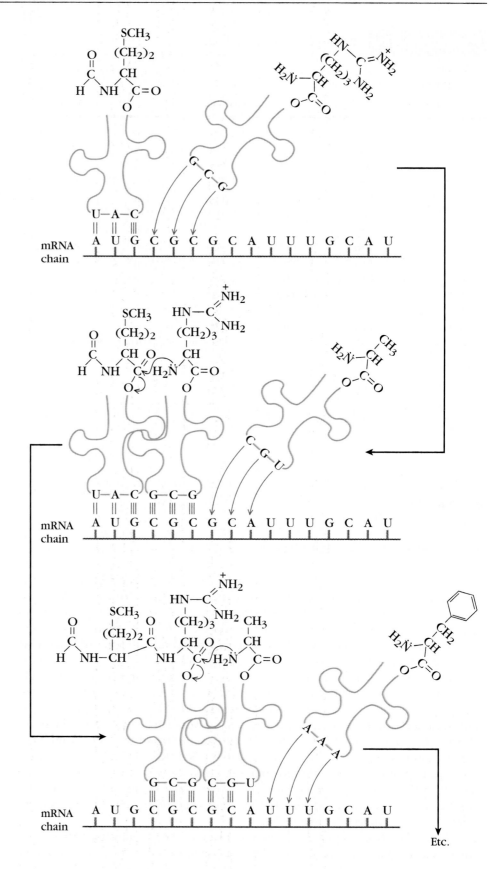

in Figure 27.11 is CGC, which is the codon that specifies arginine. An arginine-containing tRNA with the complementary anticodon GCG attaches itself to this site. The two amino acids are properly located so that enzyme-catalyzed peptide bond formation occurs. After bond formation, the ribosome moves along the mRNA chain to the next codon. In Figure 27.11, the next codon is GCA, which specifies alanine, so an alanine-containing tRNA attaches itself to this site. Enzyme-catalyzed peptide bond formation again occurs to attach the alanine to the growing chain. This process is repeated again and again until the ribosome encounters one of the codons that signifies STOP. Finally protein-releasing enzymes catalyze hydrolysis reactions that separate the polypeptide, mRNA, and ribosome.

The synthesis of a specific polypeptide is directed by a single mRNA. A single mRNA molecule can synthesize more than one molecule of a polypeptide at a time because more than one ribosome at a time can read the sequence of codons on a single mRNA. Several polypeptide molecules, therefore, can be synthesized in assembly line fashion along the mRNA molecule.

The relationship between DNA and RNA in protein synthesis is summarized in Figure 27.12. DNA contains the instructions for its replication and the synthesis of RNA in its sequence of bases. Instructions for protein synthesis are transcribed from DNA to mRNA. Ribosome-bound mRNA translates these instructions to tRNA, which results in the synthesis of proteins.

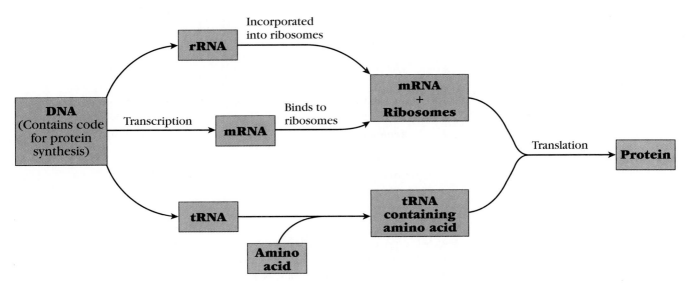

Figure 27.12
The relationship between DNA and RNA in protein synthesis.

Exercise 27.7   Write the structure of the portion of a polypeptide that would be formed from the following sequence of bases in mRNA.

(a) 5′-end A-C-A-U-C-A-U-U-U-G-C-U-C-A-U 3′-end
(b) 5′-end A-U-U-G-G-U-G-U-G-U-A-C-C-A-G-G-G-G-U-A-G 3′-end

Occasionally something happens to one or more of the bases in DNA, which leads to mutations.

27.9   Mutations

Mutations are any change that alters the sequence of bases in DNA molecules. Anything that causes mutation is called a **mutagen.** One kind of mutation that involves an error in a single base of DNA is called a point mutation. Point mutations cause DNA to direct the synthesis of proteins with incorrect primary structures. Some point mutations may occur spontaneously, while others may result from radiation or a variety of chemical compounds.

One point mutation, called **insertion,** is the addition or insertion of one extra nucleotide into one of the chains of DNA. **Deletion,** which is the opposite of insertion, is the deletion of one nucleotide from one of the chains of DNA. The mutation in DNA caused by insertion or deletion is passed on to mRNA in transcription, which results in a shift in the base sequences, as shown in Figure 27.13. This shift in base sequence in mRNA causes a change in the codons, which during translation results in the formation of a protein with an incorrect primary structure. As a result of its incorrect primary structure, the mutant protein cannot carry out its normal biological functions.

Figure 27.13
The effect of insertion point mutations of DNA on protein synthesis. An extra 2-deoxythymidine 5'-monophosphate is inserted in the template strand of DNA. This defect is passed along to mRNA during transcription, which results in the formation of a polypeptide with the incorrect primary structure.

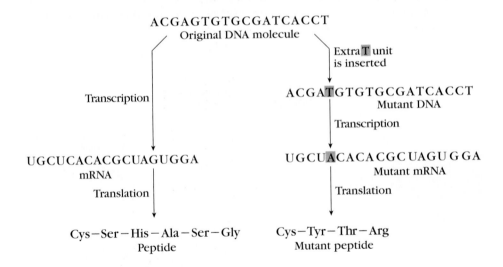

Another type of point mutation occurs when one nucleotide substitutes for another. Replacement of a 2'-deoxyadenosine 5'-monophosphate by a 2'-deoxyguanosine 5'monophosphate, for example, causes the replacement of the original A-T base pair by a G-C base pair. This mutation, too, can result in the formation of mutant proteins.

Exercise 27.8   The base sequences of the template strand of a DNA molecule and its mutant formed by deleting one base are given below. Write the abbreviated structure of the protein formed from both the original DNA and its mutant.

Base sequence in original DNA: A-G-T-A-C-G-C-G-A-A-C-T-G-C-A-T-G-C
Base sequence in mutant DNA: A-G-A-C-G-C-G-A-A-C-T-G-C-A-T-G-C

27.10 Recombinant DNA Technology

The technique of recombinant DNA technology, which is also called genetic engineering, consists of two stages. In the first stage, a DNA molecule of interest is cleaved to form an insert DNA molecule, which is then joined to another DNA molecule, called the vector DNA, to form a recombinant DNA molecule. The vector DNA is also called the cloning vector or cloning vehicle because it carries the insert DNA molecule into a host cell where it can be replicated.

■ *A clone is a collection of identical organisms that are derived from a single ancestor.*

The recombinant DNA, which contains both the insert DNA and vector DNA joined together, is introduced into a compatible host in the second stage. The host cell containing a single recombinant DNA molecule divides to form a colony of cells derived from the original host cell. Host cells that contain recombinant DNA molecules can be distinguished from cells lacking recombinant DNA molecules by the presence of an identifiable gene on the cloning vehicle.

Plasmids are one type of cloning vector. Plasmids are naturally occurring circular DNA molecules that are found in bacteria. Plasmids have the required ability to replicate autonomously, and they usually carry genes conferring resistance to antibiotics. This resistance to antibiotics is a way to distinguish between host cells that contain the cloning vector and other cells that do not.

DNA molecules are cleaved by restriction enzymes such as type II restriction endonucleases. Most restriction endonucleases recognize a specific base sequence of four to eight bases in the double helix of DNA and cut both chains of the double helix. The most useful restriction endonucleases do not cut the chains evenly. One chain, either the 3′- or 5′-end, is cut so that it is up to four nucleotides longer than the other chain. These single chain ends are called sticky ends.

When the same restriction enzyme is used to cut both the DNA molecule to be cloned and the cloning vector, the sticky ends of the insert DNA and the vector DNA are complementary. As a result, the insert DNA pairs to the complementary bases of the termini on the cut vector DNA. The insert DNA and the vector DNA are then covalently bonded (spliced) by the action of an enzyme called DNA ligase. This process is shown in Figure 27.14.

27.11 Summary

The two classes of nucleic acids are **deoxyribonucleic acids (DNA),** which are the master repository of genetic information in cells, and **ribonucleic acids**

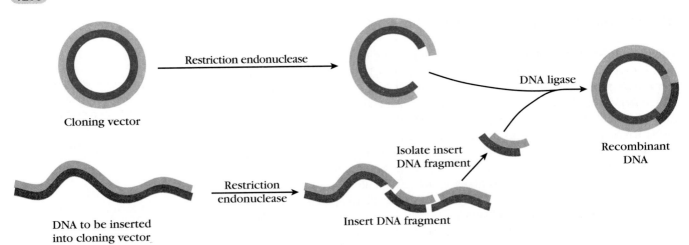

Restriction endonuclease

Cloning vector

DNA ligase

Isolate insert
DNA fragment

Recombinant
DNA

Restriction
endonuclease

DNA to be inserted
into cloning vector

Insert DNA fragment

Figure 27.14
The formation of a recombinant DNA
molecule by placing an insert DNA
molecule into the cut of a vector
DNA molecule. The same restriction
enzyme is used to cut both the DNA
molecule to be cloned and the vector DNA so the ends of each fragment are complementary.

(RNA), which implement the genetic information in the cell. Both DNA and RNA
are polymers consisting of repeating nucleotide units. Nucleotides in DNA and
RNA are linked by phosphate ester bonds between C5′ of the pentose of one
nucleotide and C3′ of the pentose of another nucleotide.

A **nucleotide** contains a phosphate group, a pentose, and a heterocyclic base.
In DNA, the pentose is 2-deoxy-β-*D*-ribose and the heterocyclic bases are adenine
(A), guanine (G), thymine (T), and cystosine (C). In RNA, the pentose is β-*D*-
ribose and uracil (U) replaces thymine as one of the four heterocyclic bases.

Molecules of DNA consist of two nucleic acid chains that wind about a common axis with a right-handed twist to form a **double helix.** The two chains are
held together by hydrogen bonds between heterocyclic bases on the different
chains. Adenine forms hydrogen bonds with thymine; cytosine forms bonds with
guanine. The deoxyribose and phosphate groups are located on the outside of
the double helix, and the bases are located in the center. The sequence of bases
in DNA contains the genetic information of a particular species. DNA forms exact
copies of itself by a process called **replication.**

DNA is also the template for the synthesis of the three kinds of RNA, each of
which has a specific role. **Ribosomal RNA (rRNA)** combines with proteins to
form ribosomes, which are the site of protein synthesis. **Messenger RNA
(mRNA)** carries the genetic information from DNA to the ribosomes, where they
direct protein synthesis. **Transfer RNA (tRNA)** transports specific amino acids
to ribosomes, where they are joined together to form proteins.

Successive sequences of three bases along the chain of mRNA are called
codons. Each codon designates a particular amino acid to be incorporated into a
growing polypeptide chain. The base sequence in a codon is complementary to
the sequences of three bases in tRNA called **anticodons.** The key step in protein
biosynthesis occurs when the codons on an mRNA molecule are recognized by
the anticodons of tRNA molecules in a process called **translation.**

Additional Exercises

27.9 Define each of the following terms.

    (a) DNA replication   **(e)** Translation   **(i)** Anticodon
    (b) Codon   **(f)** tRNA   **(j)** mRNA
    (c) Transcription   **(g)** Nucleotide   **(k)** Nucleoside
    (d) Mutagen   **(h)** Clone

27.10 Nucleic acids, proteins, and carbohydrates are all biopolymers. Write the structures of the repeating monomers in each of these polymers.

27.11 Give the name and write the structure of the pentose in:

    (a) DNA   **(b)** RNA

27.12 How do DNA and RNA differ in their chemical composition?

27.13 How do DNA and RNA differ in their biochemical functions?

27.14 Write the structure of each of the following:

    (a) A purine derivative found in DNA   **(c)** A pyrimidine derivative not found in DNA
    (b) A pyrimidine derivative found in RNA

27.15 Give the names and write the structures of the complementary base pairs in:

    (a) DNA   **(b)** RNA

27.16 Write the structure and give the name of the nucleosides formed by combining the following pairs of bases and pentoses.

    (a) Adenine and β-*D*-ribose   **(b)** Uracil and β-*D*-ribose   **(c)** Cytosine and 2-deoxy-β-*D*-ribose

27.17 Write the structure and give the name of the nucleotide formed by combining a phosphate group and the following pairs of bases and pentoses.

    (a) Cytosine and β-*D*-ribose   **(b)** Adenine and 2-deoxy-β-*D*-ribose   **(c)** Thymine and β-*D*-ribose

27.18 ATP is the abbreviation of adenosine triphosphate. Based on the structure of adenosine 5'-monophosphate, propose a structure for ATP.

27.19 A portion of one chain of a DNA molecule has the following base sequence: 5'-end AGGCTATTCGT 3'-end

Write the sequence of bases in the complementary chain of this portion of the DNA molecule.

27.20 Recall from Section 27.4 that the molar ratios G/C and A/T are 1 in DNA. Explain why there is no such relationship between the molar quantities of the heterocyclic bases A, U, G, and C in most RNA molecules.

27.21 Find the codons in Table 27.4 (p. 1201) that are a signal to incorporate each of the following amino acids into a growing polypeptide chain.

    (a) Leu   **(b)** Val   **(c)** Glu   **(d)** Ile   **(e)** Pro   **(f)** His

27.22 Write the sequence of codons on a section of mRNA that directs the synthesis of each of the following polypeptides.

    (a) Arg-Phe-Leu-Thr-Met-Ser   **(b)** Gly-Pro-Ile-Cys-Asp-Tyr-Gln-Glu

27.23 Write the DNA base sequence that transcribes each of the following codons.

    (a) U-U-A   **(b)** C-A-G   **(c)** A-U-A   **(d)** G-G-G   **(e)** A-A-C

27.24 How are codons and anticodons related?

27.25 Write the base sequence in the anticodons of tRNA that transports each of these amino acids to mRNA molecules.

    (a) Val   **(b)** Ala   **(c)** Leu   **(d)** His   **(e)** Met   **(f)** Ile

27.26 Write all possible codons on mRNA that can produce the following dipeptides.

    (a) Arg-Glu   **(b)** Val-Ser   **(c)** Ala-Tyr   **(d)** Glu-Arg

27.27 Write the structure of the polypeptide formed by the codons on each of the following mRNA molecules.

   (a) 5'-end C-U-A-A-U-C-G-U-U 3'-end   **(c)** 5'-end C-C-C-A-U-C-U-A-A-A-A-A-G-C-U-U-G-U-C-C-C 3'-end

   (b) 5'-end U-U-U-A-C-A-G-C-U-A-U-U-U-U-A 3'-end

27.28 The codon U-G-G of mRNA signals the incorporation of Trp into a growing polypeptide chain.

   (a)   Write the sequence of bases in DNA that produces these codons on transcription.

   (b)   Write the sequence of anticodons in tRNA that transports Trp.

27.29 A portion of a DNA chain consists of the following sequence of bases: A-A-G-C-C-G-U-A-C-G-A-G

   (a)   Write the order of bases in the mRNA chain formed by transcription of this part of the DNA.

   (b)   Write the structure of the amino acids, using three-letter abbreviations, coded by this part of the DNA chain.

27.30 Mutation of the portion of the DNA chain given in Exercise 27.29 changes its sequence of bases to the following: A-A-G-C-C-G-U-A-G-G-A-G. Write the structure of the amino acids coded by this mutant DNA molecule.

27.31 A molecule of DNA carries the instructions for the synthesis of the following polypeptide: Cys-Ser-His-Ala-Ser-Gly. After mutation, the following polypeptide is formed instead: Cys-Tyr-Thr-Arg.

Write the sequence of bases on **(a)** the original DNA molecule and **(b)** the mutant DNA molecule. Where is the difference in base sequence between the original and mutant DNA and why is the mutant polypeptide shorter than the original polypeptide?

ANSWERS TO SELECTED EXERCISES

This appendix gives brief answers to selected Exercises. The *Study Guide and Solutions Manual* contains answers to all Exercises and Additional Exercises and explains in detail how answers are obtained.

CHAPTER 1

1.3 **(a)** 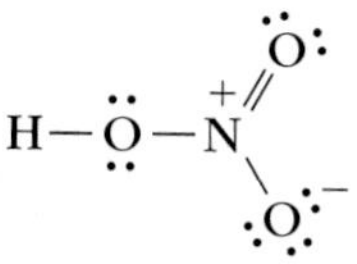 [structure: H—Ö—N with +/ and O above, O⁻ below]   **(b)** 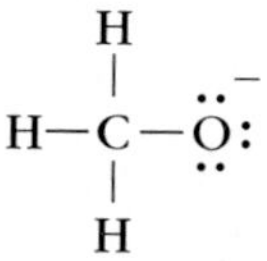 [structure: H—C—Ö⁻ with H above and below]   **(e)** No formal charges on any atom

1.5 **(a)** Tetrahedral   **(b)** Linear   **(d)** Trigonal pyramid

1.11 **(a)** 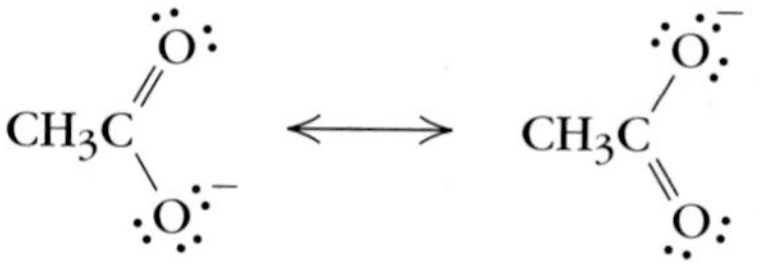 CH_3C [resonance structures with O] ⟷ CH_3C [resonance structures]   **(c)** $CH_2\!=\!CHCH_2^+$ ⟷ $\overset{+}{C}H_2\!-\!CH\!=\!CH_2$

1.14 HCl, ICl, and BrCl have polar covalent bonds

1.16 **(a)** Hydrogen bonds   **(b)** London forces   **(c)** Dipole-dipole interaction

CHAPTER 2

2.2 C_5H_{12} : [structures] ; [structure] ; [structure]

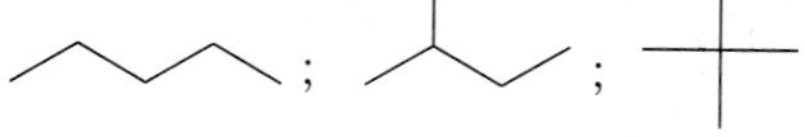

C_6H_{14} : [structures] ; [structure] ; [structure] ; [structure]

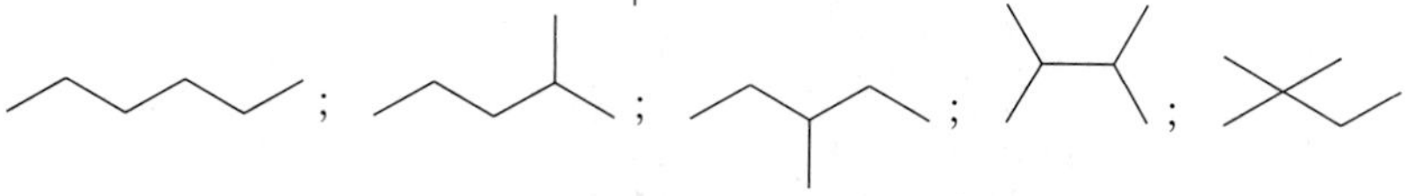

2.3 **(a)** $CH_3CH_2CH_2CH_2-$ **(c)** $CH_3(CH_2)_9CH_2-$ **(e)** $CH_3(CH_2)_{11}CH_2-$

2.4 **(a)** 3-Methylpentane   **(f)** 3-Ethyl-2,4-dimethylheptane

2.5 **(b)** $CH_3CH_2CHCH_2CH_2CH_2CH_3$ with CH_2CH_3 **(d)** $(CH_3)_3CCH_2CH_2CH_2CHCH_2CH_3$ with CH_2CH_3

2.7 **(b)** 4-Methyl-5-(1-methylethyl)decane

2.10 **(e)** 1-Methyl-3-(1-methylethyl)cyclodecane

2.11 (c) 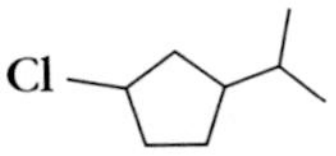 Cl—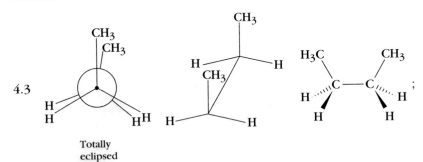（cyclopentane structure with isopropyl group）

2.14 CH_4

2.15 (b) 1-Bromo-2-chloro-3-fluorocyclopropane

2.16 (b)

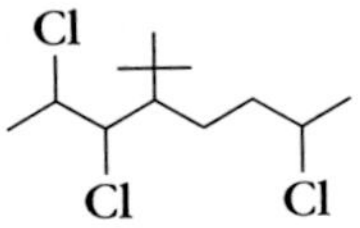

2.17 (a) $(CH_3)_3CCH_2OH$

2.18 (a) Double bond   **(f)** Alcohol
    (d) Carbonyl group of a ketone   **(i)** Aldehyde

2.19 (a) 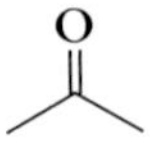   **(b)**

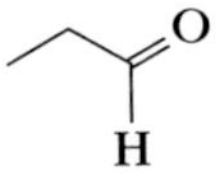

2.20 (a) 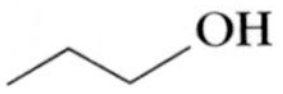 ⟶OH ; ⟶OH   **(b)** 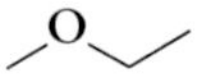 ⟶O⟶

CHAPTER 3

3.1 (a) HOH   **(e)** NH_4^+

3.2 (a) ClO_4^- **(f)** $H_2PO_4^-$

3.3 (a) HBr/Br^- and H_2O/H_3O^+

3.6 (a) Methanesulfonic acid   **(c)** Glucose

3.7 (a) HCO_3^- **(c)** CN^-

3.8 (b) 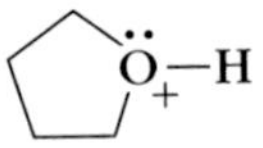   **(e)** 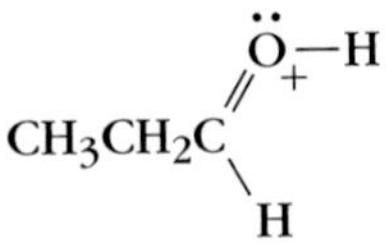 CH_3CH_2C

3.9 (c) 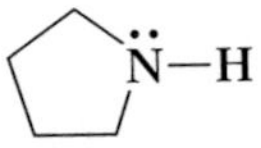   **(d)** 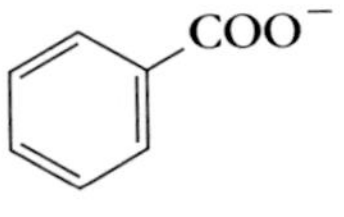   **(f)**

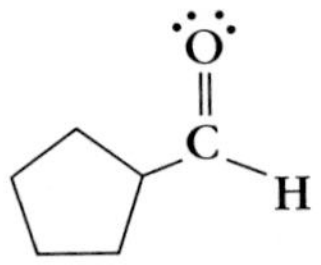

3.10 (a) Yes   **(c)** No   **(e)** No

3.11 (a) No   **(d)** Yes   **(e)** Yes

3.12 (a) Right   **(b)** Left

3.13 (b) $K_{eq} = 5.0 \times 10^{-3}$

3.19 (a) $K_{eq} = 3.58 \times 10^{15}$

3.20 (a) $\Delta G° = 6.47$ kcal/mol (27.06 kJ/mol)

CHAPTER 4

4.3

Totally
eclipsed
$(\theta = 0°)$

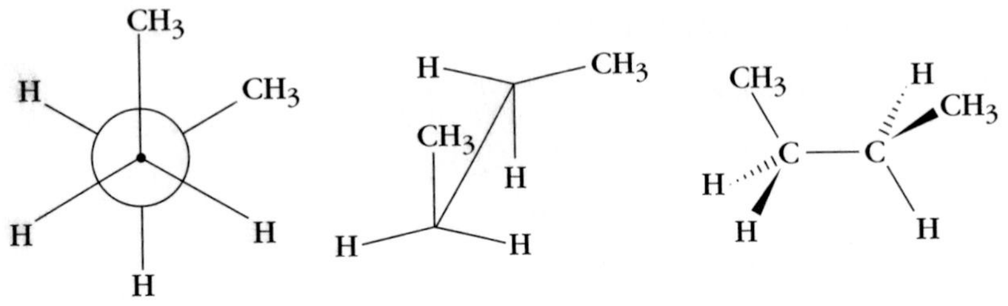

Gauche
$(\theta = 60°)$

4.8 **(a)** Equatorial

4.9 **(b)** *cis*-1-Bromo-2-chlorocyclopentane

4.11 1.8 kcal/mol (7.6 kJ/mol)

4.13 **(a)** *trans*　　**(c)** *cis*　　**(d)** *trans*　　**(f)** *trans*

CHAPTER 5

5.1 **(b)** 2.92 m

5.2 **(c)** 1.03×10^{-21} kcal/mol (4.31×10^{-21} kJ/mol)

5.3 **(e)** $\lambda = 10^{-3}$ m $< \lambda = 10^{-9}$ m $< \nu = 10^{26}$ Hz ($\lambda = 3 \times 10^{-18}$ m)

5.5 **(b)** Hydroxy group; strong absorption band centered near 3300 cm^{-1}
　　(g) Carboxyl group of a ketone; strong absorption band in the region 1715 to 1700 cm^{-1}

5.6 Spectrum A: Carboxylic Acid　　Spectrum C: Alcohol　　Spectrum D: Alkane

5.8 **(a)** $CH_2=CH(CH_2)_3CH_3$　　**(b)** $(CH_3)_2CHCH_2COOH$　　**(c)** $CH_3CH_2CH_2CH_2OH$

5.9 **(b)** 23.9

5.10 **(a)** 2　　**(c)** 7　　**(e)** 2

5.13 **(a)** 1　　**(b)** 4　　**(d)** 1

5.14 2 Degrees of unsaturation; $HC\equiv CCH_2CH_3$; $CH_2=CHCH=CH_2$; $\triangleright\!\!=\!CH_2$; $\square$

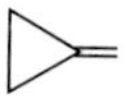

5.16 **(a)** CH_3CH_2OH　　**(b)** $CH_3CH_2\overset{\overset{O}{\|}}{C}CH_2CH_3$　　**(c)** $(CH_3)_3CCOOH$

CHAPTER 6

6.1 **(a)** Achiral　　**(d)** Chiral　　**(f)** Chiral

6.2 **(a)** Achiral　　**(d)** Chiral　　**(f)** Achiral

6.5 **(b)** Chiral　　**(d)** Chiral

6.7 **(a)** Enantiomers
　　(b) Superposable
　　(c) Enantiomers

6.8 **(b)** $-CH_3 < -CH_2CH_3 < -CH(CH_3)_2 < -C(CH_3)_3$

6.9 **(a)** (*S*)-2-Chloro-3-methylbutane

6.12 **(a)** (*S*)　　**(b)** (*S*)　　**(c)** (*S*)

6.13 $+163.2°$

6.14 $+66.6°$

6.15 0.0808 g/mL

6.18 (*S*)

6.24 **(b)** CH_3CH_2COOH　　**(c)**

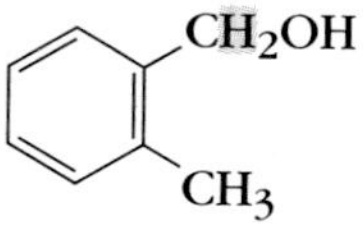

6.26 **(a)** No　　**(b)** No

CHAPTER 7

7.1 Only (a) and (b) can exist as diastereomers

7.2 **(a)** 4-Bromo-2,5-dimethyl-2-heptene

7.3 **(a)** 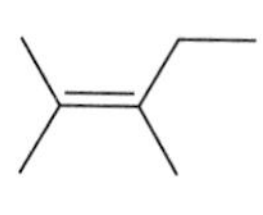　　**(c)**

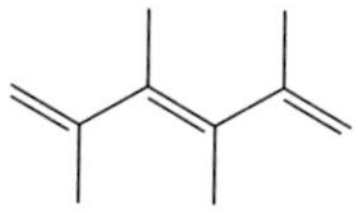

7.4 **(c)** 1,2-Dimethylcyclobutene

7.5 **(a)**

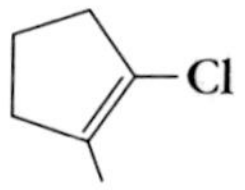

7.8 **(d)** (*E*)-1-Chloro-2-methyl-2-pentene

7.9 **(d)**  Cl⎯⎯⎯⎯Br

7.13

7.17 Step 1 $CH_2{=}CH_2$ ⇌ $\overset{+}{C}H_2{-}CH_2{-}H$ +

Step 2 $CH_3\overset{+}{C}H_2$ ⇌ $CH_3CH_2\overset{..}{\underset{..}{O}}{-}\overset{O}{\overset{\|}{C}}CF_3$

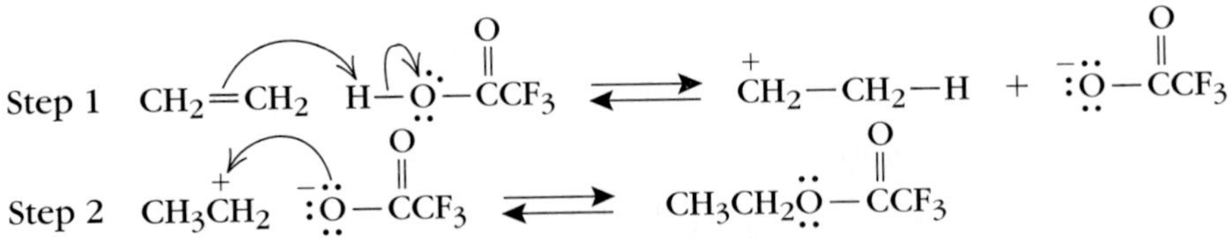

7.18 **(a)** 2°   **(b)** 1°   **(c)** 3°   **(d)** 3°

7.19 (c) < (b) < (a) < (d)

7.20 **(b)** 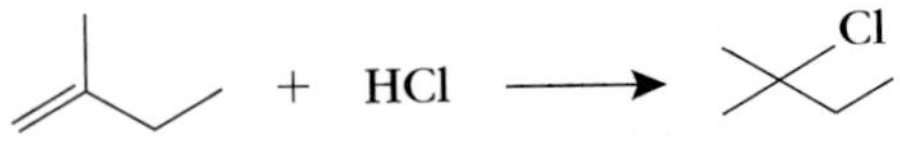   +   HCl   ⟶

7.24

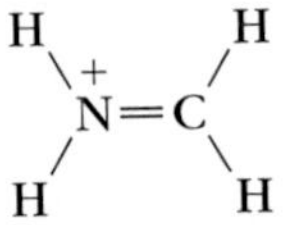

CHAPTER 8

8.4 **(b)** 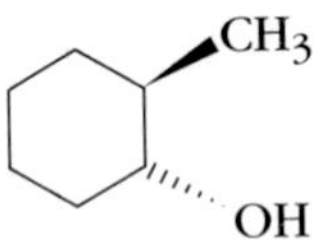   **(d)**

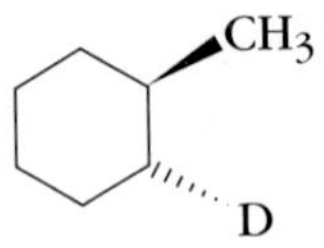

8.5 **(b)**

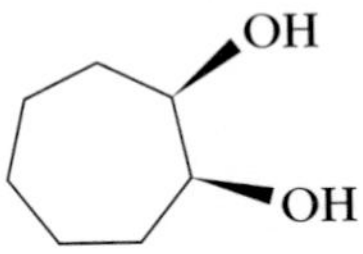

8.7 **(b)** 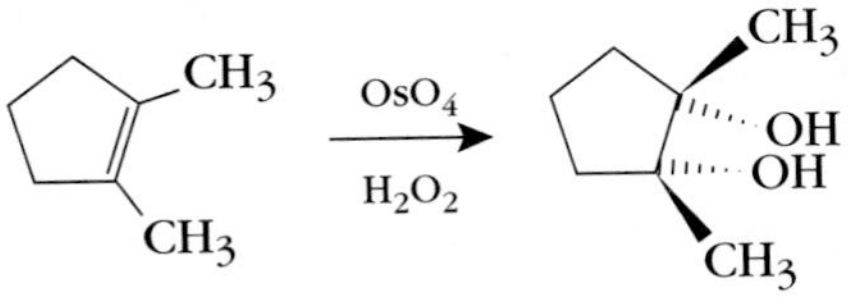 $\xrightarrow[\text{H}_2\text{O}_2]{\text{OsO}_4}$

8.8 **(d)**

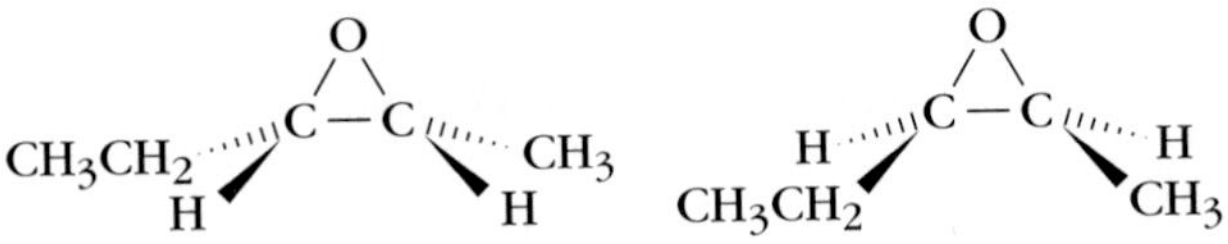

8.9 **(b)**

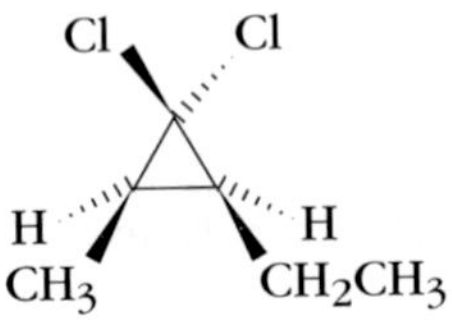

8.10 **(a)**

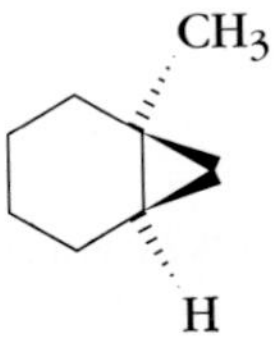

8.11 (c) 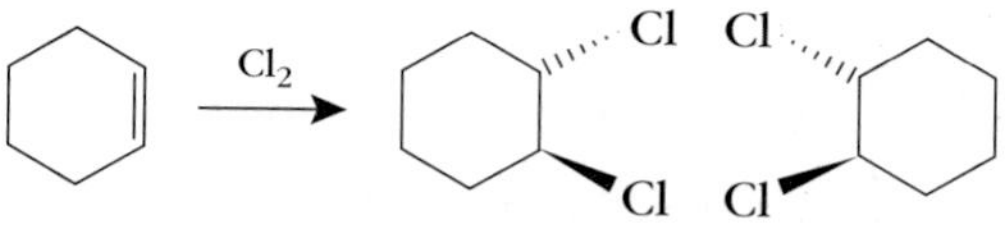 $\bigcirc\!\!\!=$ $\xrightarrow{Cl_2}$ [cyclohexane with Cl, Cl trans/cis isomers]

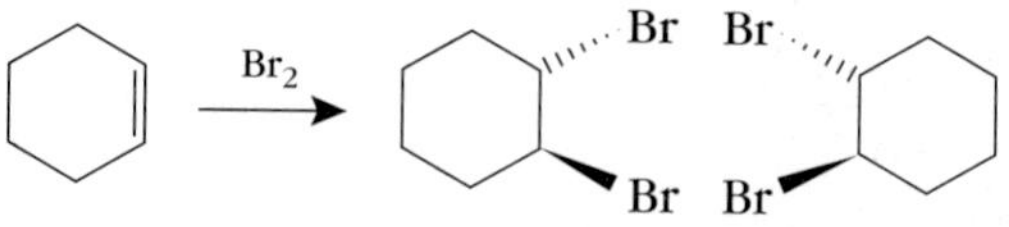

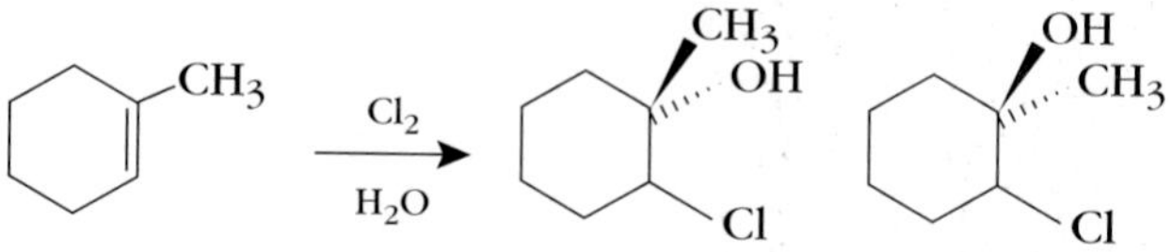

8.20 (b) 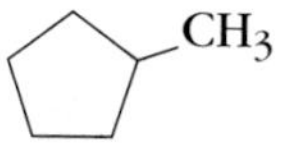 [methylcyclopentane]

8.22 $\xrightarrow[\text{increasing stability}]{E < A < B < D < C}$

$\bigcirc\!\!\!=$ $\xrightarrow{Br_2}$ [cyclohexane with Br, Br isomers]

8.15 (c) [1-methylcyclohexene] $\xrightarrow[H_2O]{Cl_2}$ [two chlorohydrin products with CH$_3$, OH, Cl]

CHAPTER 9

9.1 (b) 6-Bromo-3,8-dimethyl-4-nonyne

9.2 (d) 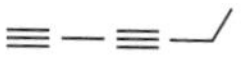 $\equiv\!-\!\equiv\!\diagup$ **(h)** 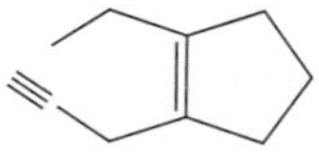 [bicyclic structure]

9.5 (b) $HC\equiv C(CH_2)_3CH_3 \xrightarrow{H_3O^+} CH_3\overset{\overset{O}{\|}}{C}(CH_2)_3CH_3$

9.6 (b) $CH_3CH_2\overset{\overset{O}{\|}}{C}CH_3$

9.7 (b) 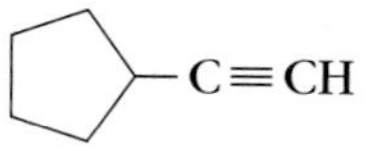 [cyclopentyl]$-C\equiv CH$

9.13 No

9.15 $HC\equiv C(CH_2)_3CH_3 + H_2 + NaOH$

9.16 (a) $CH_3C\equiv CCH_2CH_2CH_3$
(b) $CH_3CH_2C\equiv CCH_2CH_2CH_2CH_2C\equiv CCH_2CH_3$

9.18 (a) $CH_3CH_2\overset{\overset{Br}{|}}{C}=CH_2 \xleftarrow{HBr} CH_3CH_2C\equiv CH \xleftarrow{CH_3CH_2Br} HC\equiv CNa \xleftarrow[NH_{3(l)}]{NaNH_2} HC\equiv CH$

(c) $CH_3CH_2CH_2CH_2OH \xleftarrow[(2)\ H_2O_2/NaOH/H_2O]{(1)\ 9\text{-}BBN} CH_3CH_2CH=CH_2 \xleftarrow[\substack{\text{Lindlar's}\\\text{catalyst}}]{H_2} CH_3CH_2C\equiv CH$ (Prepared in 9.18a)

9.19 93%

CHAPTER 10

10.1 Increasing shielding $\longrightarrow$
(a) δ 9.56, δ 5.34, δ 2.74
(c) 750 Hz, 450 Hz, 150 Hz

10.3 8 Hz

10.6 (a) Actual ratio 6/2; Integration ratio 3/1

10.9 (a) 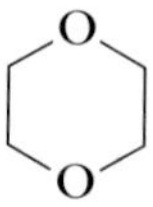 [1,4-dioxane] **(b)**  CH_3-[benzene]$-CH_3$ **(c)** $(CH_3)_3C\overset{\overset{O}{\|}}{C}CH_3$

10.10 Spectrum A: $CH_3C\begin{smallmatrix}O\\\\OCH_2CH_3\end{smallmatrix}$ Spectrum B: Cl_2CHCH_3 10.13 $BrCH_2CH_2CH_2Br$

10.20 **(b)** 300 Hz

10.11 Spectrum A is 1-Bromopropane

CHAPTER 11

11.1 **(a)** 2°   **(c)** 3°   **(e)** 1°

11.2 **(b)** *n*-Pentyl alcohol

11.3 **(b)** 1-Pentanol

11.4 **(d)** 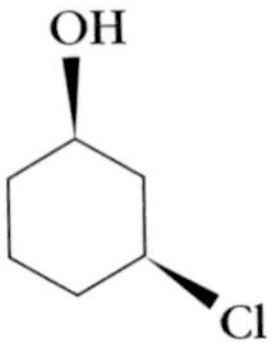   **(h)**

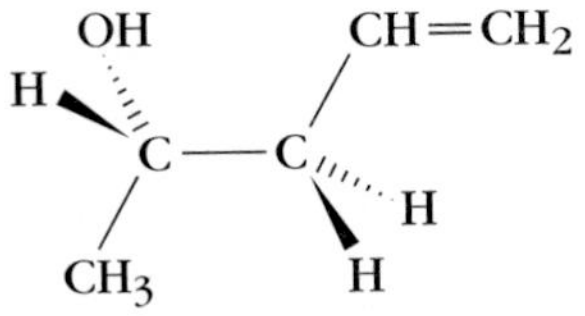

11.5 **(c)** 5,5-Dibromo-2-methyl-1-hexanol   **(e)** (*R*)-3-Ethyl-5-iodo-3-hexanol

11.6 A: $CH_3CH_2\overset{\overset{\displaystyle CH_3}{|}}{\underset{\underset{\displaystyle CH_3}{|}}{C}}OH$ B: $(CH_3)_2CHCH_2OH$

11.10 **(b)**

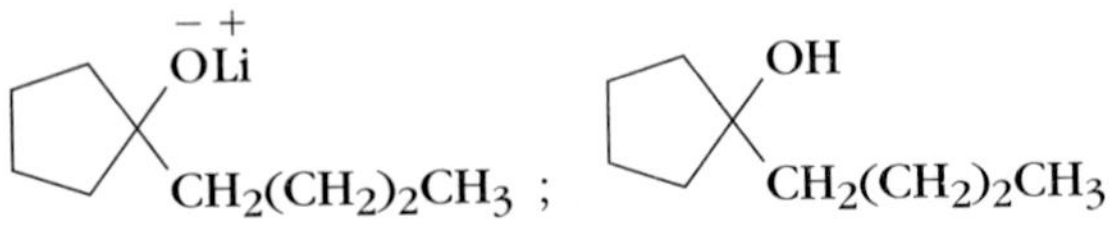

11.11 **(c)** CH_3MgI

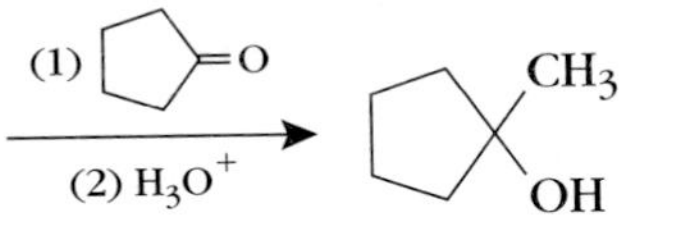

11.12 **(c)** $CH_2=\overset{-}{C}H\overset{+}{Li}$ $\xrightarrow[(2)\ H_3O^+]{(1)\ CH_3CHO}$ $CH_3\overset{\overset{\displaystyle OH}{|}}{C}HCH=CH_2$

11.14 **(b)** Neither   **(c)** Reduction   **(d)** Oxidation

11.15 **(a)** Reducing agent   **(c)** Oxidizing agent

11.16 **(b)** $(CH_3)_2CHCH_2\overset{\overset{\displaystyle OH}{|}}{C}HCH_3$

11.17 **(b)**

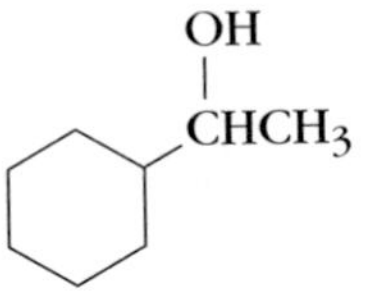

11.20 **(a)** Re   **(b)** Si

11.23 **(c)**

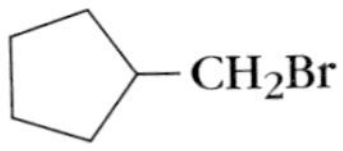

11.24 **(a)** $KMnO_4$ or $K_2Cr_2O_7/H_2SO_4/H_2O$
      (b) Collins' reagent or PCC in CH_2Cl_2

CHAPTER 12

12.2 **(b)** $CH_3CH_2CH_2Cl$

12.4 **(a)** $ClCH_2(CH_2)_4CH_2C\equiv CCH_2CH_3$

12.5 $CH_3OTs > CH_3I > CH_3Br > CH_3O\overset{\overset{\displaystyle O}{||}}{C}CH_3$

12.8 **(a)** S_N2 **(b)** S_N1

12.14   I **(b)** $CH_3CH_2CH_2CH_2OCH_3$ I **(d)** $CH_3CH_2CH=CH_2$

      II **(b)** $CH_3CHCH_2CH_3$ II **(d)** $CH_3CH_2CH=CH_2$

              $\overset{\displaystyle |}{SCH_3}$

      III **(a)** $(CH_3)_2CCH_2CH_3$ III **(b)** $(CH_3)_2CCH_2CH_3$ + $(CH_3)_2C=CHCH_3$

              $\overset{\displaystyle |}{OH}$ $\overset{\displaystyle |}{OCH_2CH_3}$

                                     (major)   (minor)

CHAPTER 13

13.1 **(a)** 2-Ethoxypropane   13.3 **(a)** (2*R*,3*S*)-3-Methoxy-2-methyloxane

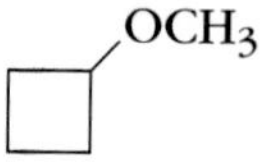

13.2 **(a)**

13.4 **(a)**

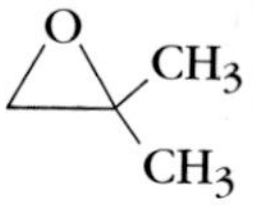

13.9 **(a)**

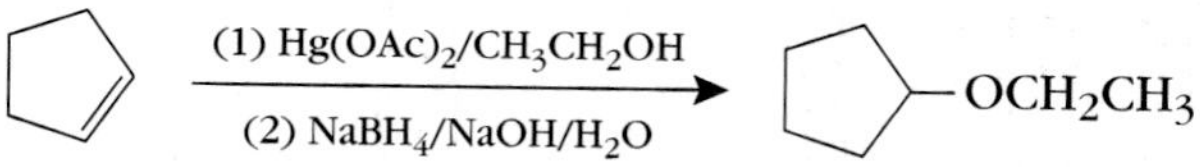

13.11 **(d)**

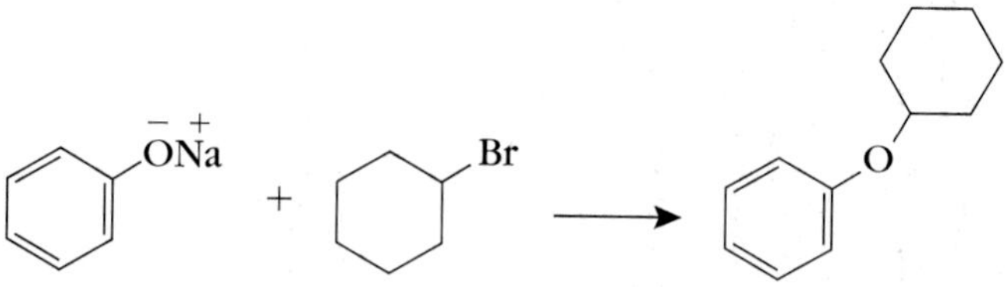

13.13 **(a)** $ICH_2CH_2CH_2CH_2CH_2I$ **(c)**

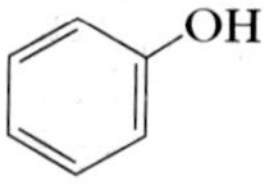

13.19 **(b)** $HOCH_2CH_2CN$ **(d)** $CH_3CH_2CH_2CH_2OH$

13.22 **(a)** $HOCH_2\overset{\displaystyle \overset{Br}{|}}{C}(CH_3)_2$

13.23 **(a)** $\overset{O}{\overset{/\backslash}{CH_2CH_2}}$ + $Na\overset{+}{C}\overset{-}{\equiv}CCH_2CH_3$ ⟶ $HOCH_2CH_2C\equiv CCH_2CH_3$

13.24 **(b)**

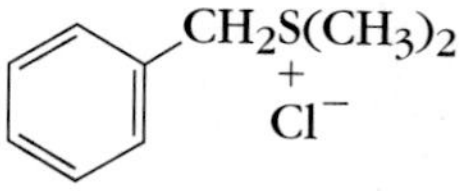

           (c)

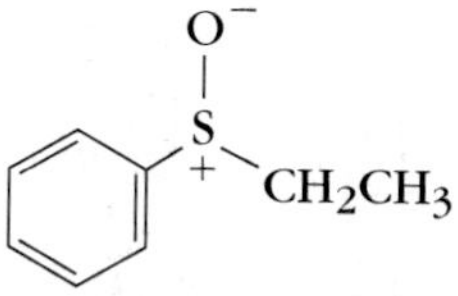

CHAPTER 14

14.1 **(a)** Hexanal   **(c)** *cis*-2-Methylcyclobutanecarbaldehyde

14.2 **(c)** $\overset{\textstyle CHO}{\underset{\textstyle CHO}{|}}$

14.3 **(e)** 2,4-Cyclopentadienone

14.4 (b) $CH_3CH_2\overset{\overset{\displaystyle O}{\|}}{C}CH{=}CHCH_2CH_3$ **(f)** $CH_3\overset{\overset{\displaystyle O}{\|}}{C}CH_2\underset{\underset{\displaystyle OH}{|}}{C}HCH_3$

14.9 Compound I 3-pentanone   Compound II Pentanal   Compound III 2,2-Dimethylpropanal

14.10 (b)

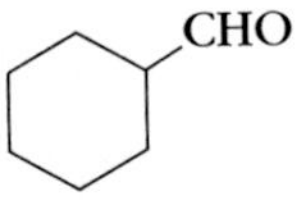

   (c) $(CH_3)_2CHCHO$

14.11 (a) PCC/CH_2Cl_2 **(c)** $LiAl(Ot\text{-}Bu)_3H/H_3O^+$

14.12 (b) 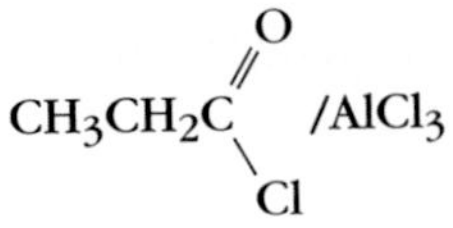 $CH_3CH_2\overset{\overset{\displaystyle O}{\|}}{\underset{\underset{\displaystyle Cl}{|}}{C}}$ /$AlCl_3$

14.14 (b)

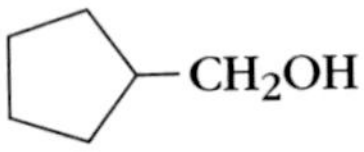

$-CH_2OH$

14.17 (b) Hemiacetal $\underset{\underset{\displaystyle CH_3CH_2O \quad OH}{}}{CH_3\overset{|}{\underset{|}{C}}CH_2CH_3}$

   Acetal $\underset{\underset{\displaystyle CH_3CH_2O \quad OCH_2CH_3}{}}{CH_3\overset{|}{\underset{|}{C}}CH_2CH_3}$

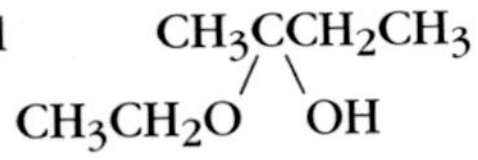

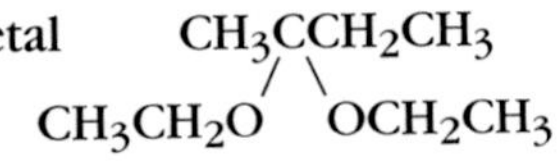

14.20 (a) 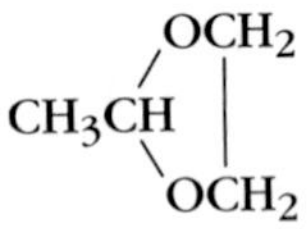 $CH_3\underset{\underset{\displaystyle OCH_2}{|}}{\overset{\overset{\displaystyle OCH_2}{|}}{CH}}$ **(c)** 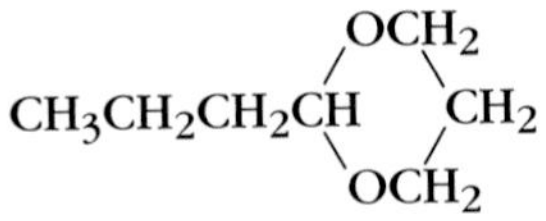 $CH_3CH_2CH_2CH\underset{OCH_2}{\overset{OCH_2}{\diagdown\,\diagup}}CH_2$

14.24 (a)

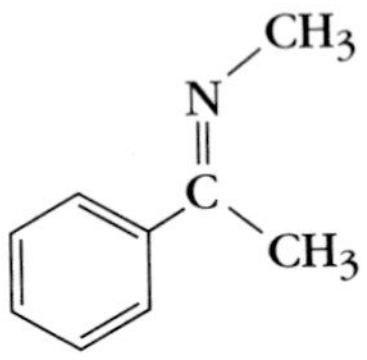

14.25 (b) $(CH_3)_2C{=}O$ + $(CH_3)_2CHNH_2$ $\xrightarrow[\text{pH 4-5}]{\text{Buffer}}$ $(CH_3)_2C{=}N{-}CH(CH_3)_2$

14.28 (a) 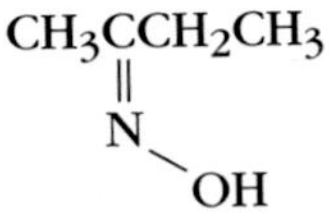 $\underset{\underset{\displaystyle N}{\|}}{CH_3\overset{}{C}CH_2CH_3}$
        $\underset{\displaystyle OH}{\diagdown}$

14.29 (b) $(C_6H_5)_3P$ $\xrightarrow[\text{(2) Bu Li / THF}]{\text{(1) } CH_3CH_2C{\equiv}CCH_2Br}$ $(C_6H_5)_3\overset{+}{P}CH\overset{-}{C}{\equiv}CCH_2CH_3$

14.30 (c)

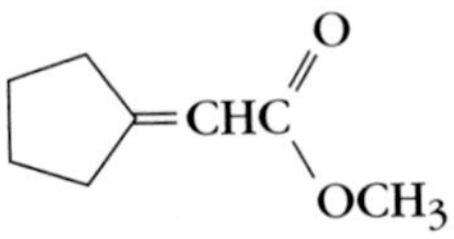

${=}CHC\overset{\overset{\displaystyle O}{\|}}{}_{\diagdown OCH_3}$

14.31 (a) $CH_3\overset{\overset{\displaystyle CH_2}{\|}}{C}CH_2CH_3$ $\longleftarrow$ $CH_3\overset{\overset{\displaystyle O}{\|}}{C}CH_2CH_3$ + $(C_6H_5)_3\overset{+}{P}\overset{-}{C}H_2$ $\xleftarrow{\text{Bu Li/THF}}$ $(C_6H_5)_3\overset{+}{P}CH_3$ $\xleftarrow{CH_3I}$ $(C_6H_5)_3P$

CHAPTER 15

15.1 (d) Cyclobutanecarboxylic acid   **(e)** 3-Methyl-2-butenoic acid

15.2 (d)

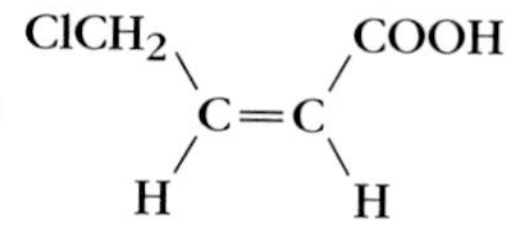

$$ClCH_2\text{ }\diagdown\hspace{-0.2em}C\!=\!C\diagup\text{ }COOH$$
with H below each carbon

15.7 CH_3CH_2C with =O and OCH_3

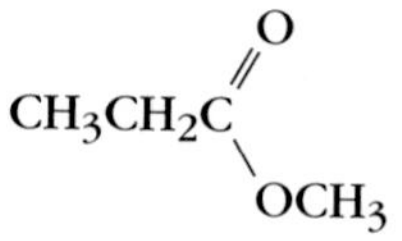

15.3 (f) Potassium propenoate

15.4 (c)

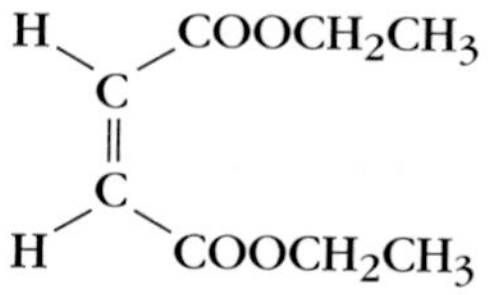

H, COOCH$_2$CH$_3$ on top carbon; H, COOCH$_2$CH$_3$ on bottom carbon of C=C

15.8 (c) 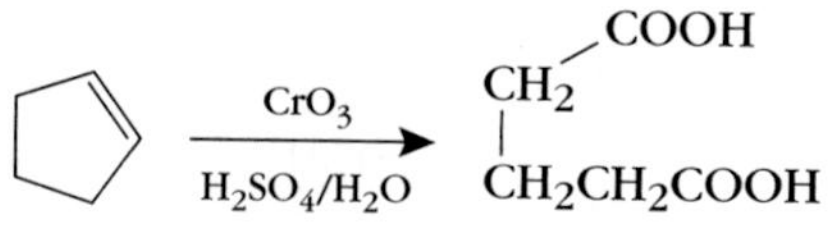 cyclopentene $\xrightarrow[H_2SO_4/H_2O]{CrO_3}$ CH_2COOH / CH_2CH_2COOH

15.9 (c) 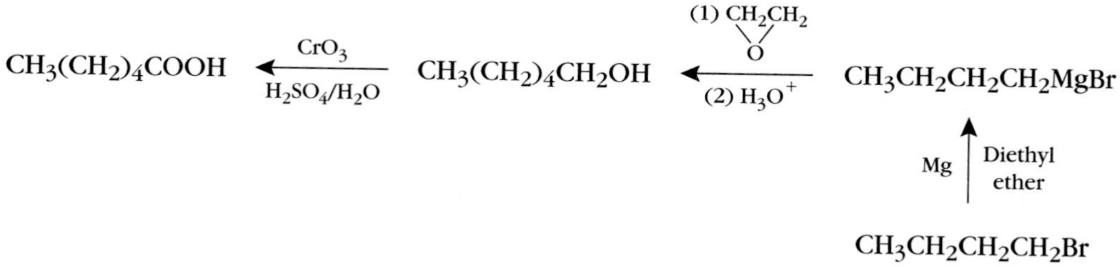 $CH_3(CH_2)_4COOH$ $\xleftarrow[H_2SO_4/H_2O]{CrO_3}$ $CH_3(CH_2)_4CH_2OH$ $\xleftarrow[(2)\ H_3O^+]{(1)\ CH_2CH_2\ O}$ $CH_3CH_2CH_2CH_2MgBr$

$\uparrow$ Mg | Diethyl ether

$CH_3CH_2CH_2CH_2Br$

15.11 (a) Carboxylic acid   **(c)** Carboxylate ion

15.12 (c) O_2NCH_2COOH

15.16 $CH_2OOC(CH_2)_{14}CH_3$
$CHOOC(CH_2)_{14}CH_3$
$CH_2OOC(CH_2)_{14}CH_3$

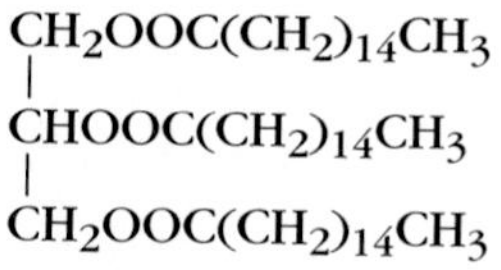

15.17 (b) 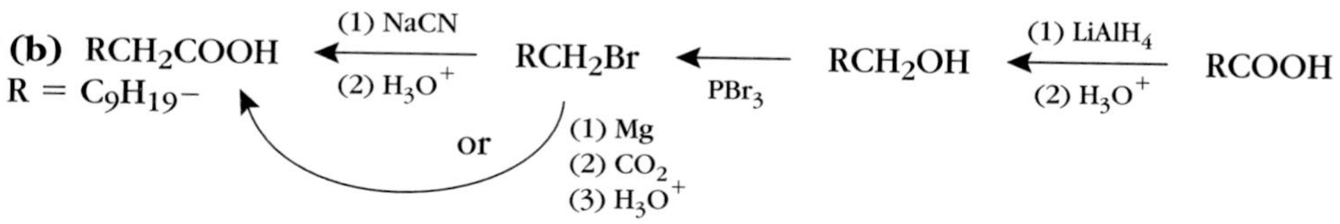 RCH_2COOH $\xleftarrow[(2)\ H_3O^+]{(1)\ NaCN}$ RCH_2Br $\xleftarrow[PBr_3]{}$ RCH_2OH $\xleftarrow[(2)\ H_3O^+]{(1)\ LiAlH_4}$ $RCOOH$
$R = C_9H_{19}-$
or $\begin{array}{l}(1)\ Mg \\ (2)\ CO_2 \\ (3)\ H_3O^+\end{array}$

CHAPTER 16

16.1 (e) *N*-Methylpropanamide

16.2 (g) 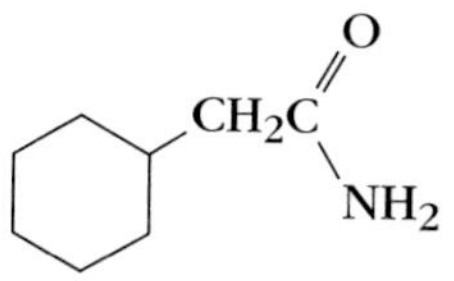 cyclohexane–CH_2C with =O and NH_2 **(i)** 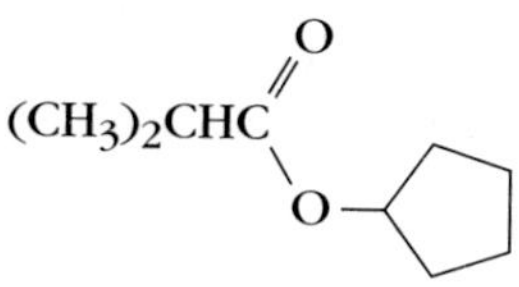 $(CH_3)_2CHC$ with =O and O–cyclopentane

16.5 (a) 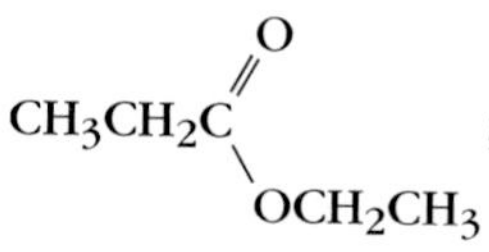 CH_3CH_2C with =O and OCH_2CH_3 ; Ethyl propanoate   **(g)** 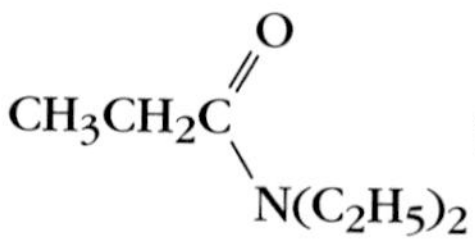 CH_3CH_2C with =O and $N(C_2H_5)_2$; *N, N*-Diethylpropanamide

16.7 (a) 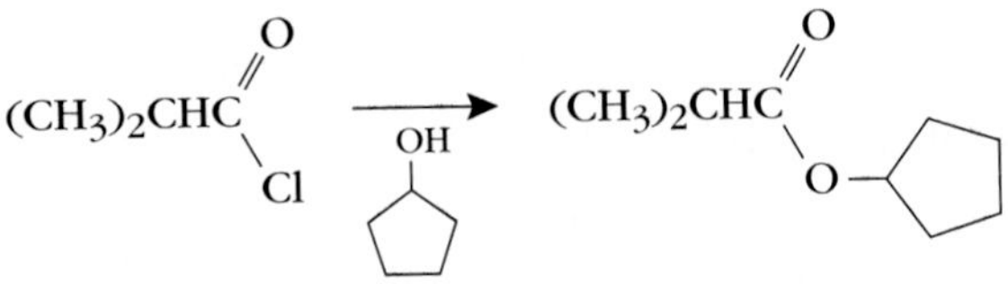 $(CH_3)_2CHC(=O)Cl \xrightarrow{\text{(cyclopentanol) OH}} (CH_3)_2CHC(=O)O\text{–(cyclopentyl)}$

16.9 (c) Acetic anhydride

16.12 (d) 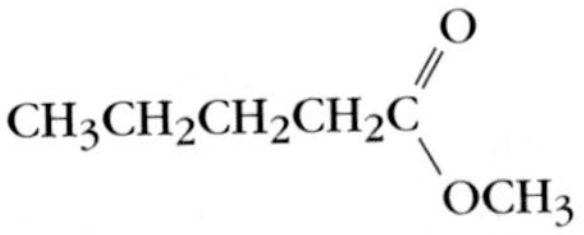 $CH_3CH_2CH_2CH_2C(=O)OCH_3$

16.13 (b) 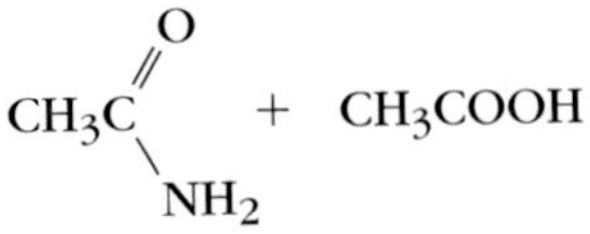 $CH_3C(=O)NH_2$ + CH_3COOH

16.14 (b) 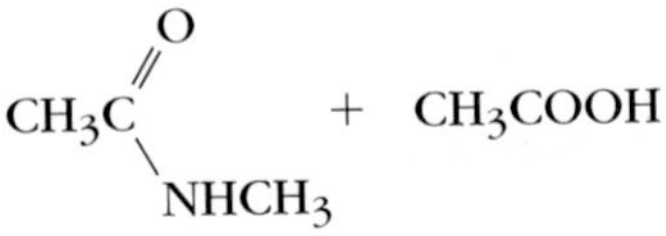 $CH_3C(=O)NHCH_3$ + CH_3COOH

16.15 (b) 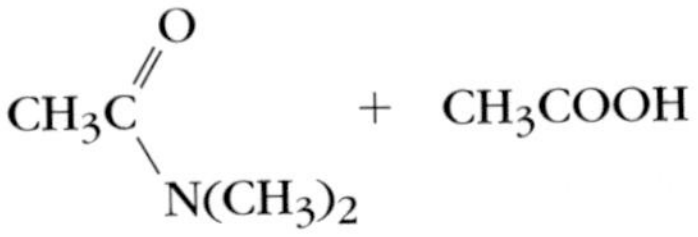 $CH_3C(=O)N(CH_3)_2$ + CH_3COOH

16.18 (b) 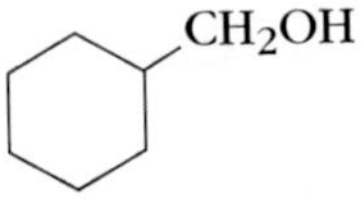 (cyclohexyl)CH$_2$OH **(c)** $CH_2{=}CHCH_2NH_2$ **(f)** 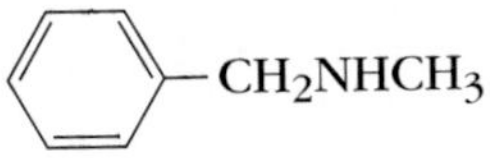 (phenyl)$-CH_2NHCH_3$

16.20 (a) 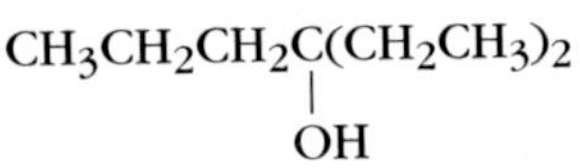 $CH_3CH_2CH_2C(CH_2CH_3)_2$ with OH   **(d)** $CH_3CH_2CH_2CH_2CH_2COO^-$
(OH on central carbon)

16.21 (c) 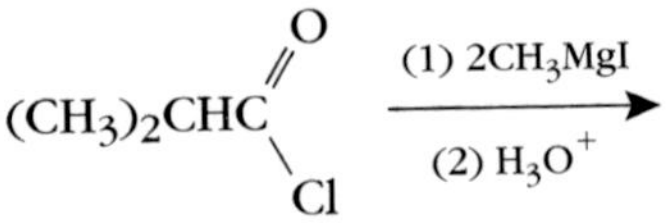 $(CH_3)_2CHC(=O)Cl \xrightarrow[\text{(2) } H_3O^+]{\text{(1) } 2CH_3MgI}$

16.22 (b) $CH_3CH{=}CHCH_2CN \xrightarrow[\text{(2) } H_3O^+]{\text{(1) } LiAlH_4}$ **(f)** 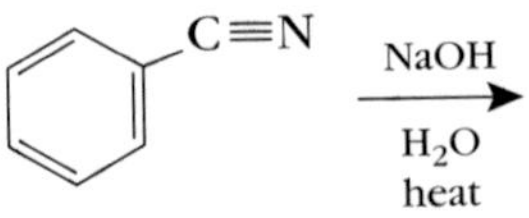 (phenyl)$C{\equiv}N \xrightarrow[\substack{H_2O \\ heat}]{NaOH}$

CHAPTER 17

17.1 $(CH_3)_2CHC(=O)H \rightleftharpoons (CH_3)_2C{=}C(OH)H$

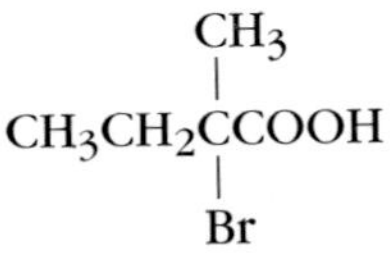

17.2 (c) 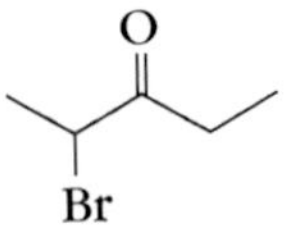 (CH$_3$–CH(Br)–C(=O)–CH$_2$CH$_3$)

17.5 (c) $CH_3CH_2C(CH_3)COOH$ with Br

17.6 (c) 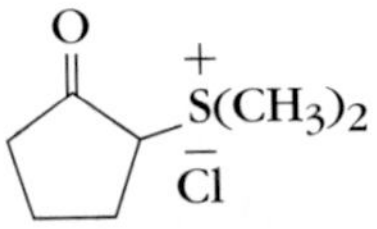 (cyclopentanone with $\overset{+}{S}(CH_3)_2$ and Cl^-)

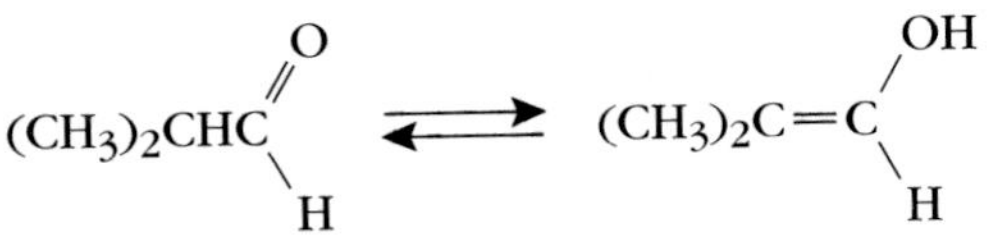

17.8 (a) 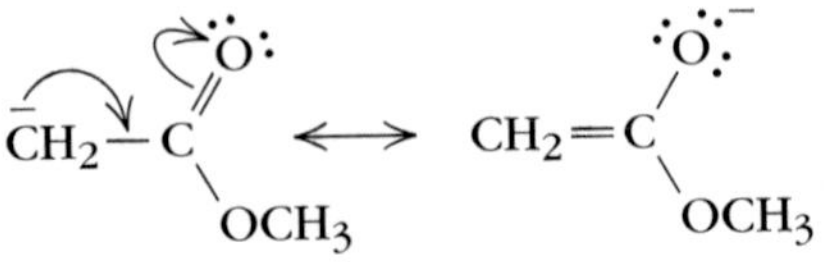 $^-CH_2{-}C(\cdot\ddot{O}\cdot)(OCH_3) \longleftrightarrow CH_2{=}C(\cdot\ddot{O}\cdot^-)(OCH_3)$ **(e)**  $^-CH_2{-}C{\equiv}N: \longleftrightarrow CH_2{=}C{=}\ddot{N}:^-$

17.9 (b) 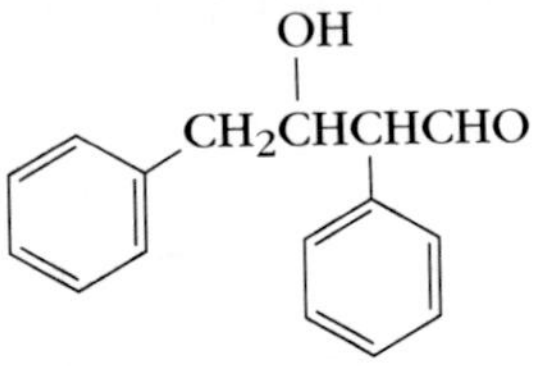 (phenyl)$CH_2CH(OH)CH(phenyl)CHO$

17.10 (a) $CH_3CH_2CH_2CH_2OH$ $\xleftarrow[PtO_2]{H_2}$ $CH_3CH=CHCHO$ $\xleftarrow[heat]{}$ $\underset{\underset{CH_3CHCH_2CHO}{|}}{\overset{HO}{\overset{|}{}}}$ $\xleftarrow[H_2O]{NaOH}$ $2CH_3CHO$

17.11 (a) $CH_3CH_2CH_2CH=\underset{\underset{CH_2CH_3}{|}}{C}CHO$

(c)

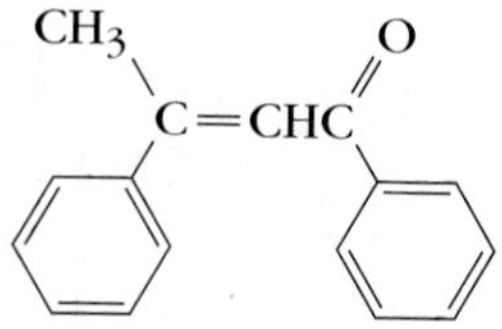

17.12 (a) 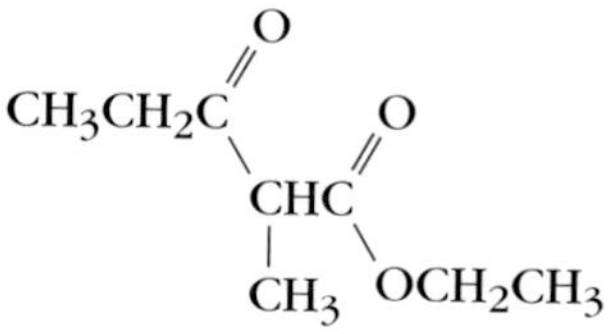 $CH_3CH_2\overset{\overset{O}{\|}}{C}$ $\underset{\underset{CH_3}{|}}{CH}\overset{\overset{O}{\|}}{C}OCH_2CH_3$

17.16 (a)

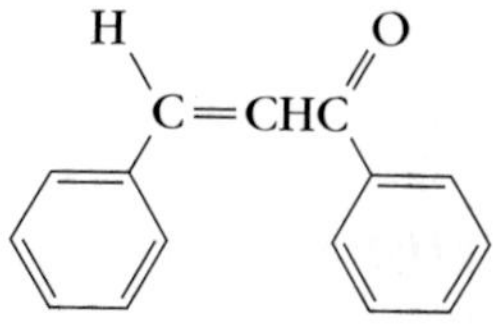

17.17 (a) $2CH_3CH_2CHO$

(c)

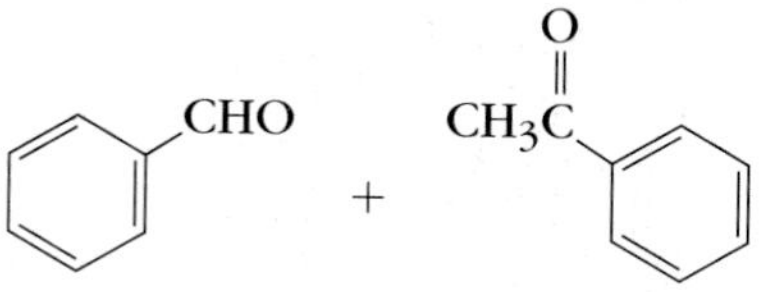

$+$

CHAPTER 18

18.2 (b) Heterolytic   (d) Homolytic

18.5 (a) $CH_3CH_2CH_2CH_3$ (c)

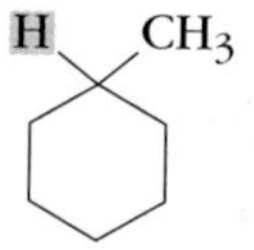

18.6   Increasing stability
        $\longrightarrow$
        (a) < (c) < (b)

18.12 (a) 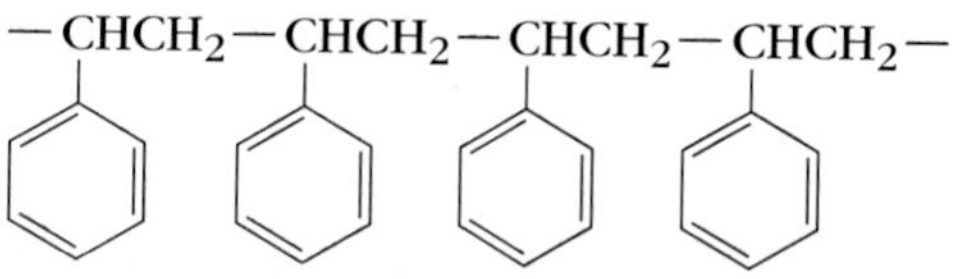 $-CHCH_2-CHCH_2-CHCH_2-CHCH_2-$

18.17 (a) $m/z = 30$ (c) $m/z = 32$

18.20 $M^{+\cdot}/(M+2)^{+\cdot} \cong 3$

CHAPTER 19

19.2 (b) $CH_2=CH-\overset{+}{C}(CH_3)_2$

19.3

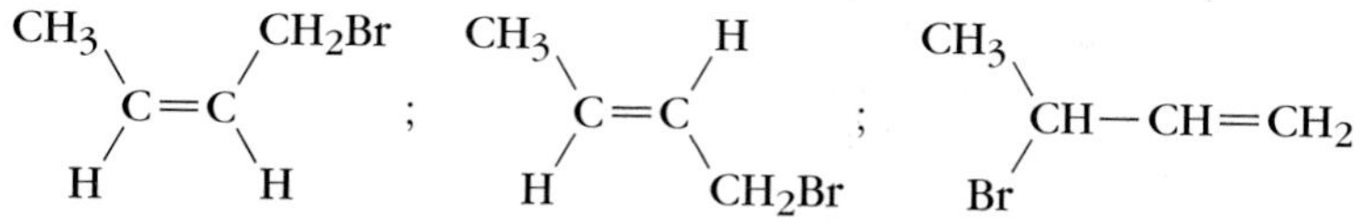

19.7 (a)

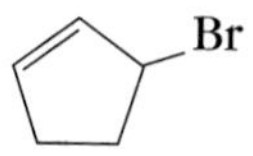

19.8 (a) $CH_2=CHCH_2CN$ $\xleftarrow[C_2H_5OH]{NaCN}$ $CH_2=CHCH_2Br$ $\xleftarrow[CCl_{4,}\ heat]{NBS}$ $CH_2=CHCH_3$

19.9 (b) and (c) are conjugated dienes

19.13 CH$_3$—═—CH$_3$ (with H, H below)

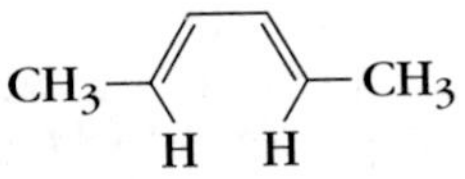

19.14 (b) 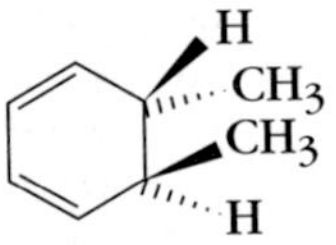 (structure with H, CH$_3$, CH$_3$, H)

19.17 (b), (c), and **(d)** are suitable

19.18 (c) 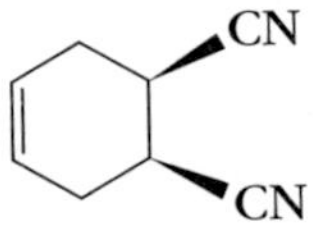 (cyclohexene with two CN groups)   **(e)** 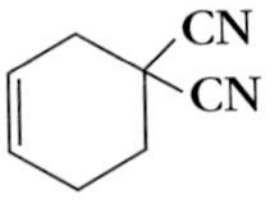 (cyclohexene with two CN groups)

19.19 (c) 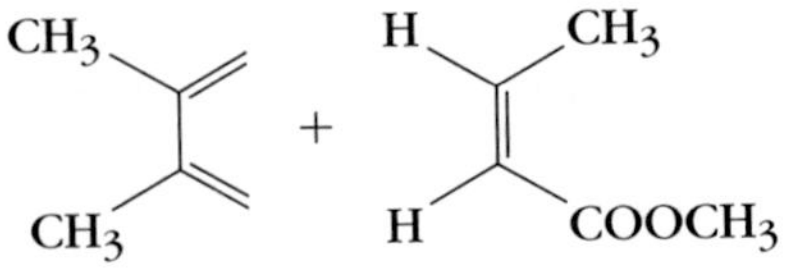 (diene structure with CH$_3$ groups) + (structure with H, CH$_3$, H, COOCH$_3$)

CHAPTER 20

20.3 (a) Nonaromatic   **(b)** Aromatic

20.5 (b), (c), and **(d)** are aromatic

20.8 One

20.10 (b) 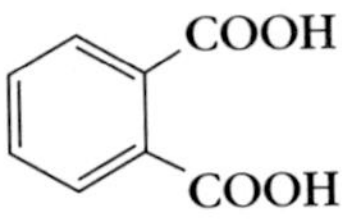 (benzene with two COOH groups, ortho)   **(c)** 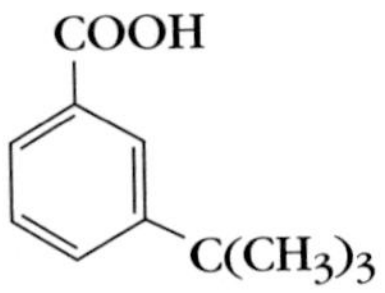 (benzene with COOH and C(CH$_3$)$_3$, meta)

20.12 (b) 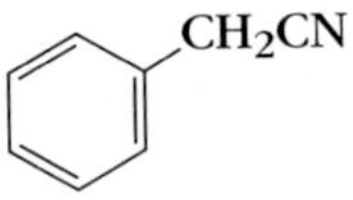 (benzene with CH$_2$CN)   **(d)** 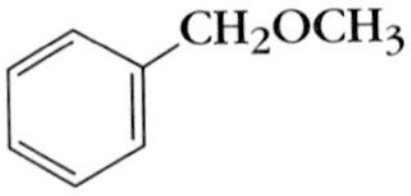 (benzene with CH$_2$OCH$_3$)

20.16 (c) 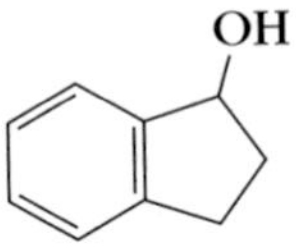 (indane with OH)   **(h)** 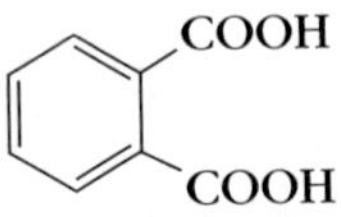 (benzene with two COOH groups, ortho)   **(j)** 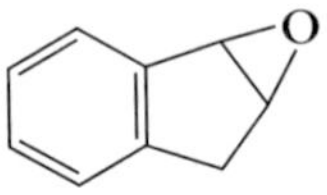 (indane with epoxide O)

CHAPTER 21

21.1 (c) 3-Isopropylphenol   **(g)** 2,4,6-Trinitrotoluene

21.2 (d) 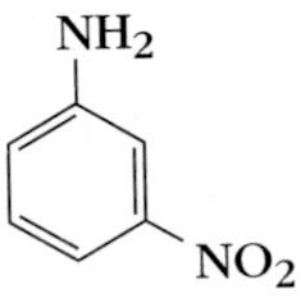 (benzene with NH$_2$ and NO$_2$, meta)   **(h)** 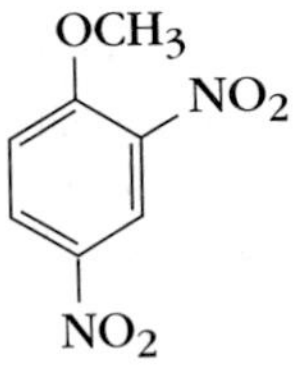 (benzene with OCH$_3$, NO$_2$, NO$_2$)

21.3 (a)  (benzene with CH$_3$ and CH$_3$, para)   **(b)** 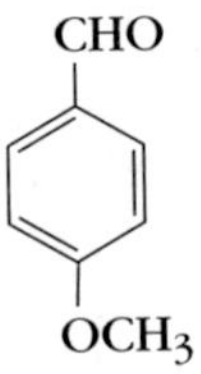 (benzene with CHO and OCH$_3$, para)

21.10 (a) 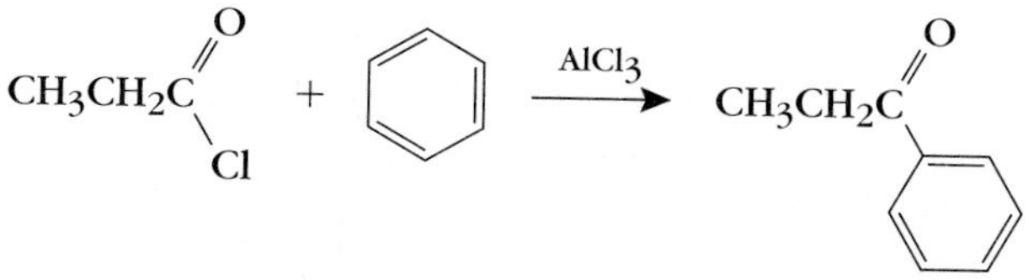 CH₃CH₂C(=O)Cl + ⬡ →[AlCl₃] CH₃CH₂C(=O)— (phenyl ketone)

21.12 (c) 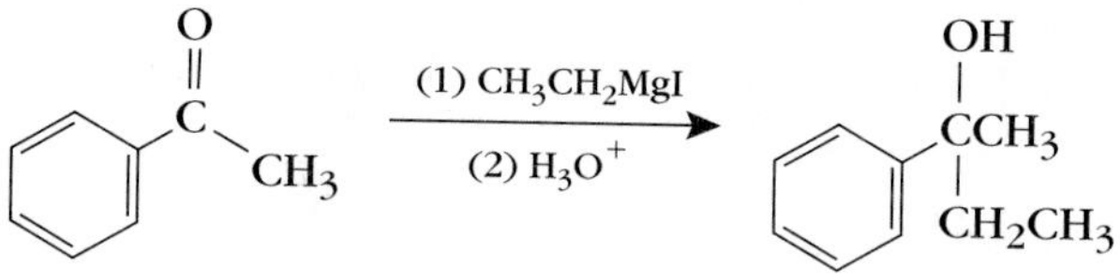 acetophenone (phenyl C(=O)CH₃) →[(1) CH₃CH₂MgI] [(2) H₃O⁺] phenyl—C(OH)(CH₃)CH₂CH₃

21.15 3-nitro-bromobenzene (NO₂ and Br on benzene)

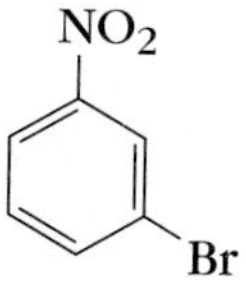

21.17 (d) 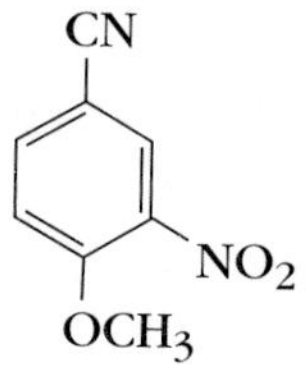 benzene ring with CN, NO₂, OCH₃ substituents

CHAPTER 22

22.3 (b) 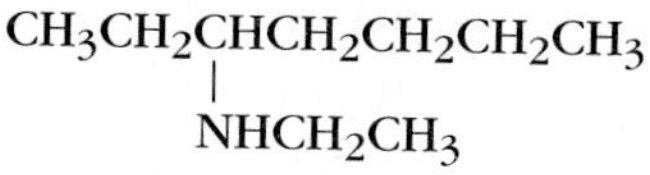 CH₃CH₂CHCH₂CH₂CH₂CH₃ with NHCH₂CH₃ substituent

22.6 CH₃CH₂CH₂NH₂

22.15 (b) 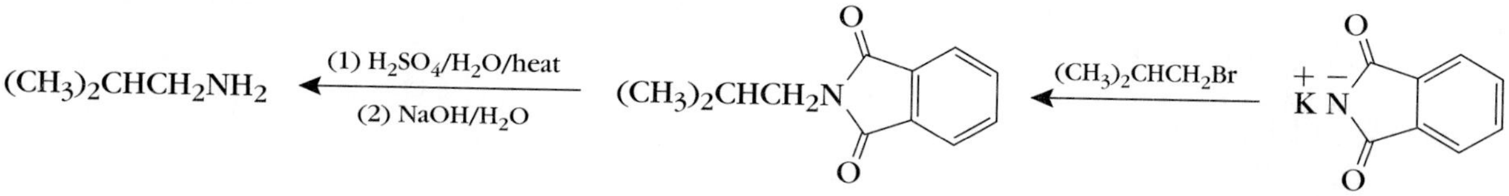 (CH₃)₂CHCH₂NH₂ ←[(1) H₂SO₄/H₂O/heat] [(2) NaOH/H₂O] (CH₃)₂CHCH₂N(phthalimide) ←[(CH₃)₂CHCH₂Br] K⁺ ⁻N(phthalimide)

22.18 (b) 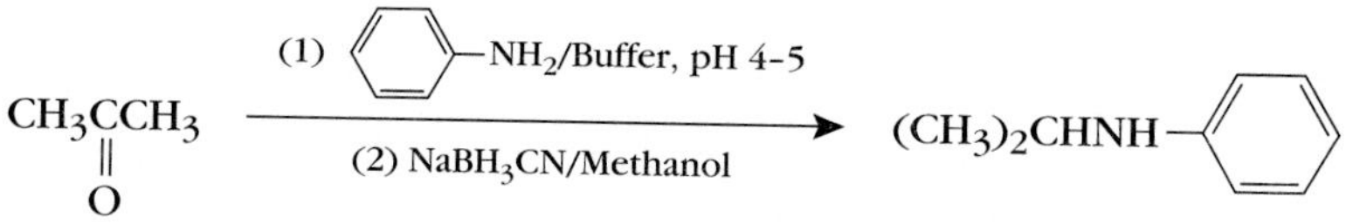 CH₃CCH₃ (with O) →[(1) phenyl—NH₂/Buffer, pH 4-5] [(2) NaBH₃CN/Methanol] (CH₃)₂CHNH—phenyl

22.21 (a) 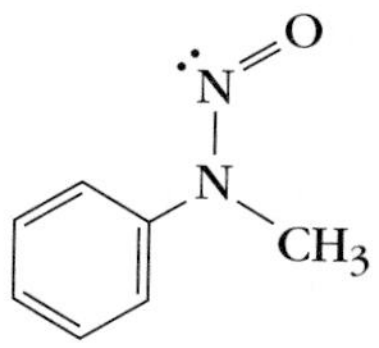 N-nitroso-N-methylaniline (phenyl—N(CH₃)—N=O)

22.22 (b) 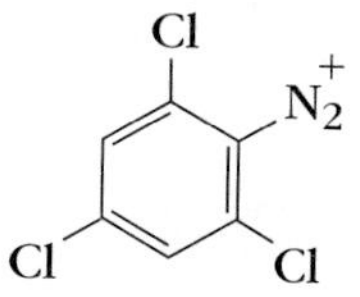 2,4,6-trichlorobenzenediazonium (benzene ring with N₂⁺ and three Cl)

22.23 (c) 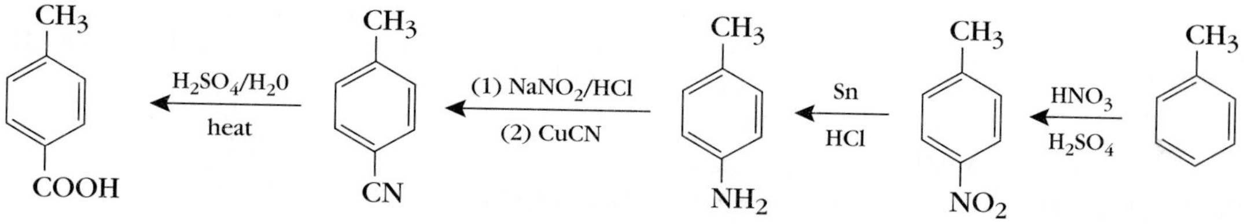 toluene (CH₃—phenyl) →[HNO₃ / H₂SO₄] CH₃—phenyl—NO₂ →[Sn / HCl] CH₃—phenyl—NH₂ →[(1) NaNO₂/HCl] [(2) CuCN] CH₃—phenyl—CN →[H₂SO₄/H₂O / heat] CH₃—phenyl—COOH

CHAPTER 23

23.1 (a)

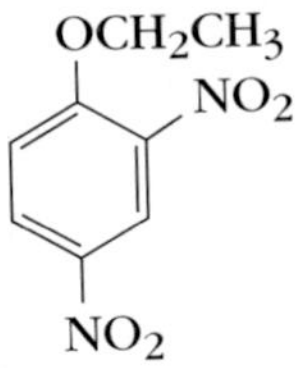

23.3

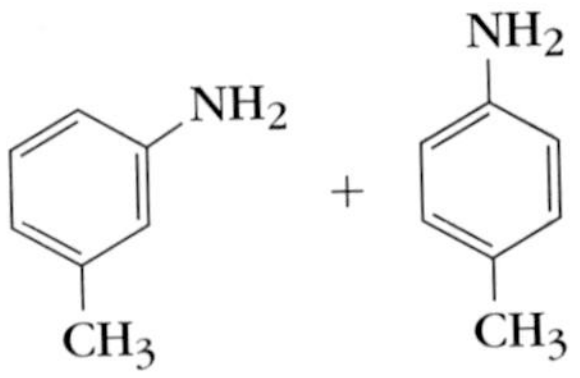

23.4 (a)

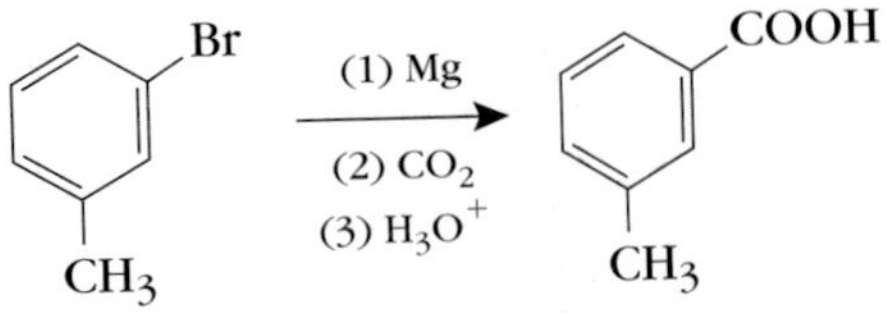

23.11 (a) Acetic acid > phenol > ethanol

23.16 (a)

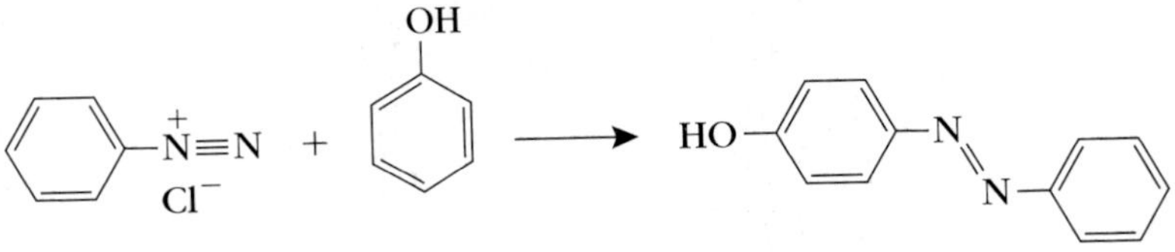

23.19

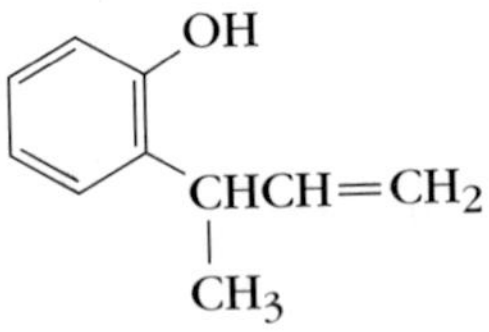

23.20 (a)

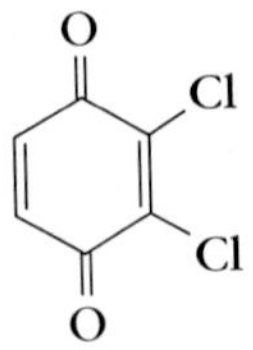

23.21 (a)

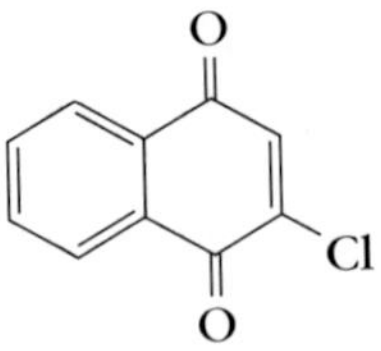

23.22 (a)

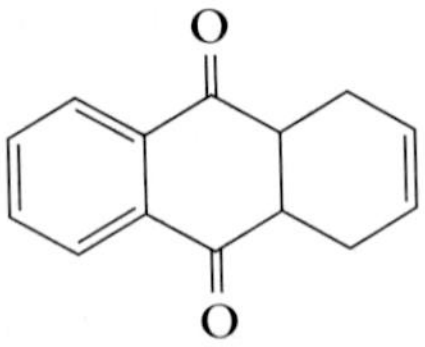

CHAPTER 24

24.1 (a) Methyl 3-oxopentanoate

24.4

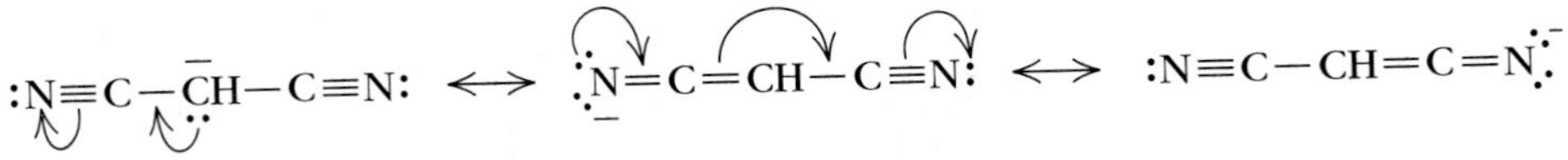

24.6 (a)

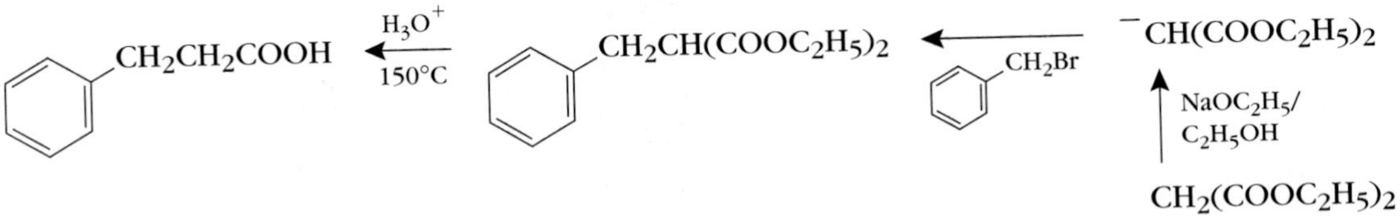

24.9 (a)

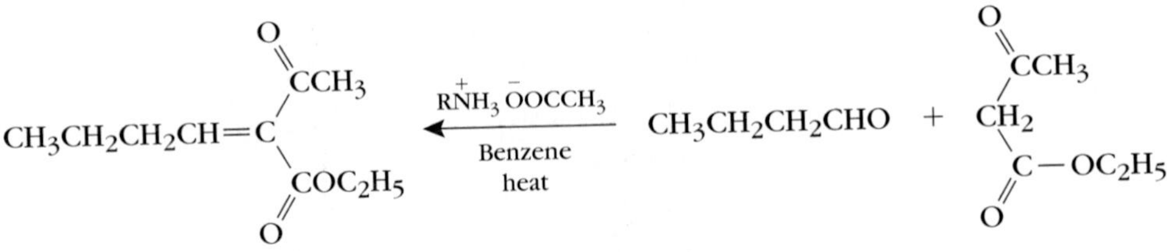

24.12 (a) 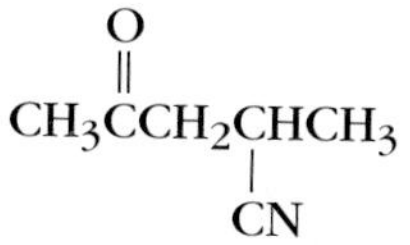 $CH_3\overset{\overset{\displaystyle O}{\|}}{C}CH_2CHCH_3$
$\quad\quad\quad\quad\quad\quad |$
$\quad\quad\quad\quad\quad\quad CN$

24.13 (a) 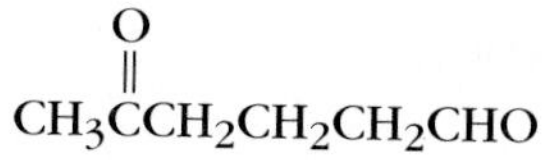 $CH_3\overset{\overset{\displaystyle O}{\|}}{C}CH_2CH_2CH_2CHO$

24.14 (b) 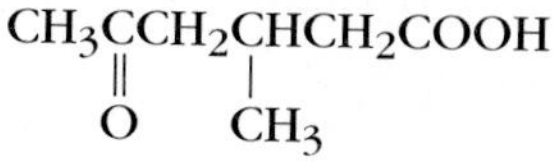 $CH_3\overset{}{C}CH_2\overset{}{C}HCH_2COOH$
$\quad\quad\quad\quad\quad \| \quad\quad |$
$\quad\quad\quad\quad\quad O \quad\quad CH_3$

24.17 (a)

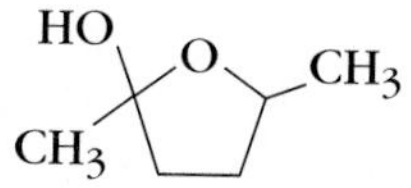

24.18 (a)

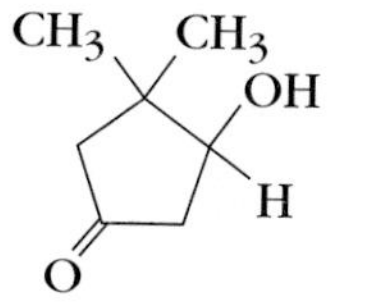

24.21

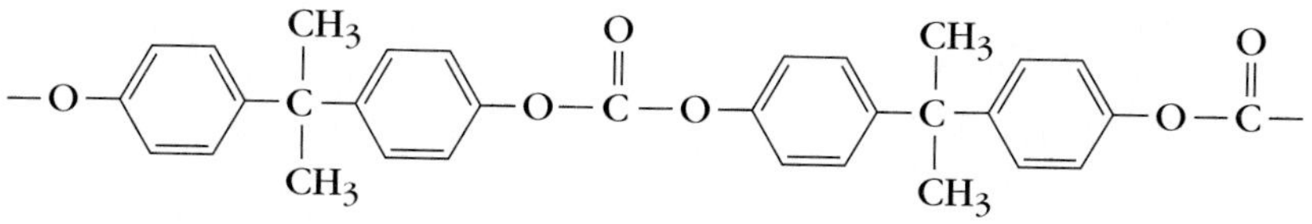

CHAPTER 25

25.1 (a) 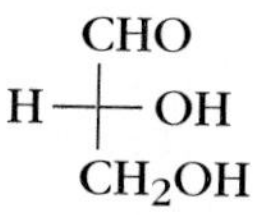
$$\begin{array}{c} CHO \\ H\!-\!\!\!-\!OH \\ CH_2OH \end{array}$$

25.2 (c) 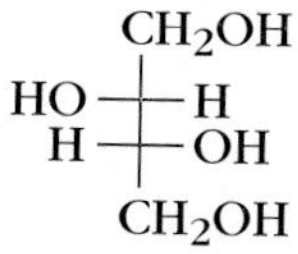
$$\begin{array}{c} CH_2OH \\ HO\!-\!\!\!-\!H \\ H\!-\!\!\!-\!OH \\ CH_2OH \end{array}$$

25.5
$$\begin{array}{c} CHO \\ H\!-\!\!\!-\!OH \\ HO\!-\!\!\!-\!H \\ H\!-\!\!\!-\!OH \\ CH_2OH \end{array}$$

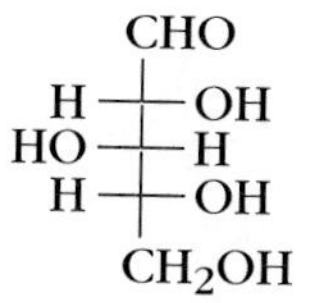

25.6 (a) L (c) D

25.7 (a) 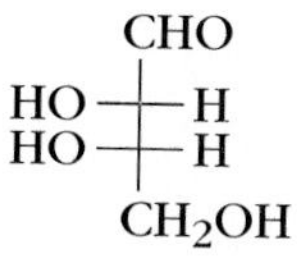
$$\begin{array}{c} CHO \\ HO\!-\!\!\!-\!H \\ HO\!-\!\!\!-\!H \\ CH_2OH \end{array}$$

25.12 (a)

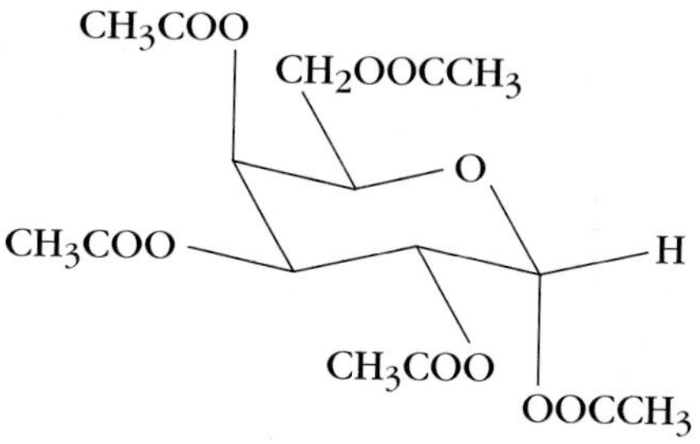

25.14 (a)

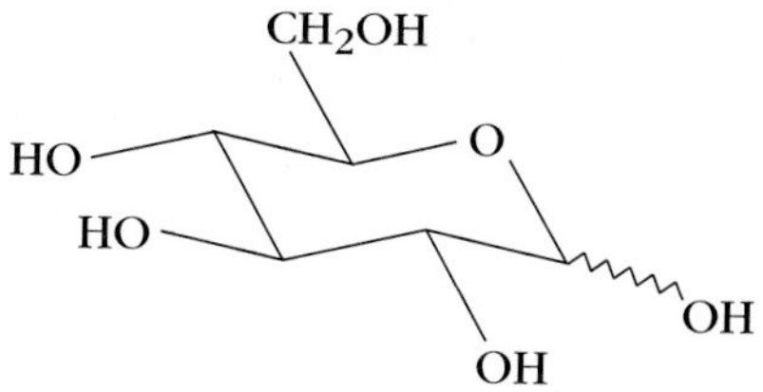

Mixture of α and β anomers

CHAPTER 26

26.1 **(a)** 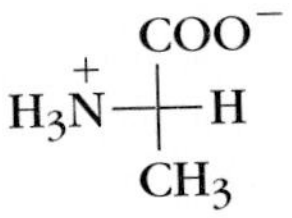 $H_3\overset{+}{N}-\overset{\underset{\displaystyle CH_3}{\big|}}{\underset{}{\overset{\displaystyle COO^-}{\big|}}}H$

26.3 **(a)** 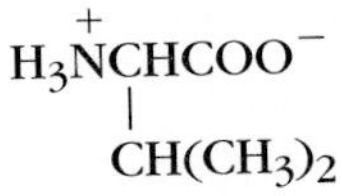 $H_3\overset{+}{N}CHCOO^-$
$\quad\quad\quad\quad\overset{\big|}{CH(CH_3)_2}$

26.4 **(a)** 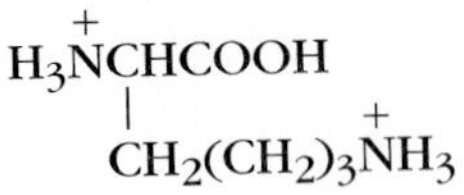 $H_3\overset{+}{N}CHCOOH$
$\quad\quad\quad\overset{\big|}{CH_2(CH_2)_3\overset{+}{N}H_3}$

26.6 **(a)** 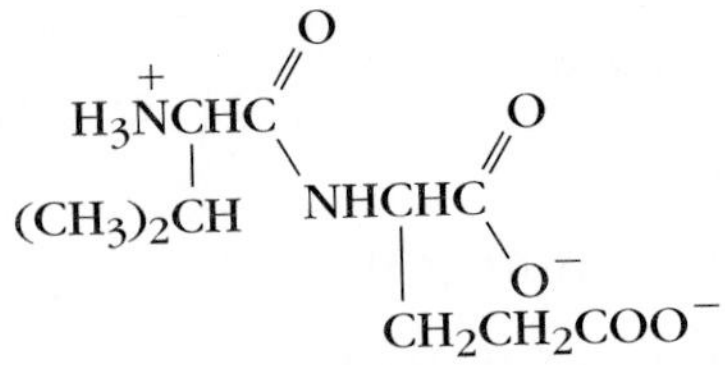 $H_3\overset{+}{N}CHC$ ... $NHCHC$
$\quad\quad\quad\quad(CH_3)_2CH \quad\quad\quad CH_2CH_2COO^-$

26.10 **(a)** Asp-Arg-Leu-Tyr-Ile-Phe

CHAPTER 27

27.1

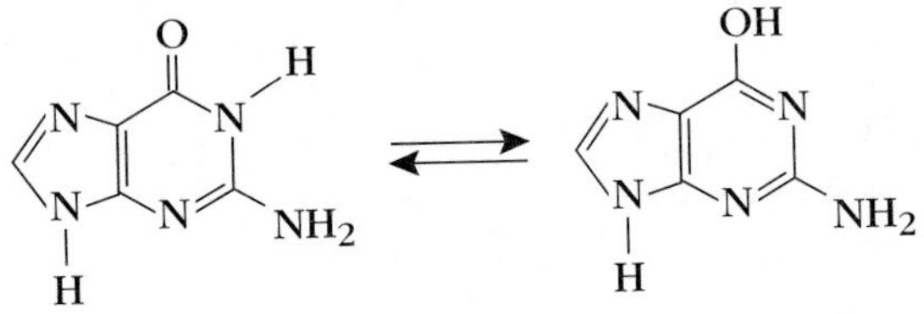

Guanine

27.4 CGC; CGU; CGA; CGG; AGA; AGG

27.5 **(a)** Ile

GLOSSARY

Absolute configuration *(Section 6.8)* The actual orientation in space of the four different substituents bonded to a stereocenter.

Acetal *(Section 14.11)* A functional group containing two alkoxy groups bonded to the same carbon atom $R_2C(OR')_2$.

Achiral *(Section 6.1)* A term that refers to objects that are superposable on their mirror images.

Activating group *(Section 21.11)* A substituent which, when present in place of a hydrogen atom, increases the rate of a particular reaction. Notice that the designation *activating* for a particular substituent depends on the reaction. (See Section 23.2.)

Acyclic *(Section 2.1)* Molecular structure that does not contain rings of atoms.

Acyl group *(Section 16.1)* The group $R\overset{O}{\underset{}{\overset{\|}{C}}}$ where R may be alkyl or aryl.

Acylation *(Section 16.1)* The transfer of an acyl group from one molecule to another; also called *acyl transfer.*

1,2-Addition *(Section 19.8)* The addition of an unsymmetrical reagent to both carbon atoms of one double bond of a diene or polyene.

1,4-Addition *(Section 19.8)* The addition of an unsymmetrical reagent to the terminal carbon atoms of a diene.

Alcohol dehydrogenase *(Section 11.13)* An enzyme that catalyzes the oxidation of alcohols to aldehydes and ketones.

Aldaric acid *(Section 25.3, C)* The dicarboxylic acid formed by the oxidation of an aldose.

Alditol *(Section 25.3, D)* The polyalcohol formed by the reduction of the carbonyl group of a carbohydrate.

Aldonic acid *(Section 25.3, C)* The monocarboxylic acid formed by the mild oxidation of an aldose.

Alkaloids *(Section 22.7)* Naturally occurring amines obtained from plants.

Alkyl halide *(Section 2.9)* Alkane in which one or more hydrogen atoms is replaced by halogen; also called *haloalkane.*

Alkyl shift *(Section 7.17)* The intramolecular migration of an alkyl group with a pair of electrons from one carbon atom to another.

Alkylation *(Section 9.11)* A reaction that introduces an alkyl group into a molecule. Examples include alkylation of enolate anions (Section 24.7) and alkylation of acetylide ions (Section 9.11).

Allylic *(Introduction, Chapter 19)* A term that designates a position on a carbon atom adjacent to a carbon-carbon double bond.

Ambident nucleophile *(Section 17.12)* A nucleophile that reacts at either of two positions.

Amino sugar *(Section 25.9)* Monosaccharide in which an —OH group is replaced by an amino group.

Amino acid residue *(Section 26.3)* An amino acid unit in a polypeptide or protein.

Angle strain *(Section 4.7)* The increase in energy in a molecule caused by distorting a bond angle from its normal value. Angle strain is severe in cyclopropane, for example, because the normal carbon bond angles of 109.5° must be compressed to 60°.

Annulenes *(Section 20.7)* Completely conjugated monocyclic polyenes.

Anomeric carbon atom *(Section 25.7)* The hemiacetal carbon atom formed by the cyclization of an aldose or ketose.

Anti addition *(Section 8.8)* Addition of two atoms or groups to the two carbon atoms of a double bond from opposite faces or sides of the plane of a double bond.

Anti conformation *(Section 4.5)* The conformation about a single bond when the two largest substituents are 180° apart as viewed in a Newman projection formula.

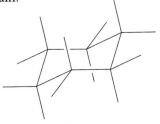

Anti conformation of butane

Anticodon *(Section 27.6)* A three-base sequence in a molecule of tRNA that is complementary to the codon of mRNA for a particular amino acid.

Aprotic solvents *(Section 3.13)* Solvents that do not contain easily exchangeable protons. Chloroform, acetone, and benzene are examples of aprotic solvents.

Aromaticity *(Section 20.8)* A term that summarizes the special properties associated with aromatic compounds. These properties include unusual stability, characteristic ^{1}H NMR chemical shifts caused by a ring current, and substitution rather than addition reactions.

Atactic polymer *(Section 18.12)* A polymer whose substituents are randomly located on either side of the plane of its carbon atom chain.

Atomic number *(Section 1.2)* The number of protons in the nucleus of a particular atom.

Autooxidation *(Section 18.7)* Slow oxidation of organic compounds that occurs without combustion.

Axial bonds *(Section 4.10)* The six C—H bonds of cyclohexane that lie along the axis perpendicular to the general plane of the cyclohexane ring.

Benzylic *(Section 20.9)* The term used to designate a carbon atom (and any atoms attached to this carbon atom) directly bonded to a benzene ring.

Benzyne *(Section 23.3)* The reactive intermediate involved in the reaction of halobenzenes and strong bases such as sodium amide.

Betaine *(Section 14.16)* Compound that has opposite charges on nonadjacent atoms.

Boat conformation *(Section 4.8)* The conformation of cyclohexane that roughly resembles a boat.

The boat conformation has no angle strain but has torsional strain so it is less stable than the chair conformation.

Branch point *(Section 2.1)* The carbon atom of a chain of carbon atoms that is bonded to more than two other carbon atoms.

Branched chain alkanes *(Section 2.1)* Alkanes that contain at least one carbon atom bonded to more than two other carbon atoms.

Brønsted-Lowry acid *(Section 3.2)* Any compound or ion that can donate a proton (a hydrogen ion, H^+).

Brønsted-Lowry base *(Section 3.2)* Any compound or ion that can accept a proton (a hydrogen ion, H^+).

Carbene *(Section 8.7)* A neutral reactive intermediate that contains a divalent carbon atom with only six valence electrons ($R_2C:$).

Carbocation *(Section 7.9)* A positive ion in which the positive charge resides on a carbon atom.

Chair conformation *(Section 4.8)* The conformation of cyclohexane that roughly resembles a chair. The chair conformation is the most stable conformation of cyclohexane because it has neither angle strain nor torsional strain.

Chemical shift *(Sections 5.9 and 10.2)* The position on an NMR spectrum where a nucleus (either ^{13}C or ^{1}H) absorbs energy. The chemical shift of tetramethylsilane (TMS) is taken by convention as zero and the chemical shifts of all other peaks are given relative to it.

Chiral *(Section 6.1)* A term that refers to objects that are nonsuperposable on their mirror images.

Chromatography *(Section 5.1)* A method of separating mixtures into their pure components. Chromatography separates mixtures because the intermolecular attractions between the various compounds of the mixture and the stationary phase differ.

cis- *(Section 4.12)* A prefix indicating that two substituents are located on the same side of a ring or a double bond. (See *trans-*.)

Clone *(Section 27.10)* Collection of identical organisms derived from a single ancestor.

Codon *(Section 27.7)* Set of three bases on the chain of mRNA. Each codon directs the incorporation of a specific amino acid into a growing peptide chain.

Coenzymes *(Section 11.12)* Molecules that act with enzymes to carry out biological reactions.

Concerted reaction *(Section 19.10)* A reaction that occurs in a single step without intermediates.

Condensation reaction *(Section 17.6)* A reaction in which two molecules react to form a product by the elimination of some small stable molecule such as H_2O.

Conformations *(Section 4.3)* Different arrangements of atoms in a molecule that are converted into another by rotation about a single bond.

Conjugate acid *(Section 3.2)* The species formed when a Brønsted-Lowry base accepts a proton.

Conjugate base *(Section 3.2)* The species that remains after a Brønsted-Lowry acid has donated a proton.

Conrotatory *(Section 19.10)* The rotation of p orbitals in the same direction during an electrocyclic reaction.

Constitutional isomers *(Section 2.3)* Isomers that differ in the way their atoms are connected.

Copolymers *(Section 18.10)* Polymers that are formed by combining two or more different monomers.

Coupling constant *(Section 10.8)* The distance between adjacent peaks in a multiplet in an NMR spectrum.

Covalent bonds *(Section 1.14)* Bonds formed by the sharing of electrons between two nuclei. Equal sharing of electrons between identical nuclei results in a nonpolar covalent bond. Unequal sharing of electrons between two nuclei results in a polar covalent bond.

Crown ethers *(Section 13.6)* Macrocyclic polyethers that contain four or more ether functional groups in a ring of twelve or more atoms.

Cumulative double bond *(Section 19.4)* A functional group in which one carbon atom is part of two double bonds: $H_2C=C=CH_2$.

Cyanohydrin *(Section 14.9)* A functional group that contains a hydroxy and a nitrile on the same carbon atom.

Cycloaddition reaction *(Section 19.12)* A reaction in which two unsaturated compounds add together in a single step to form a cyclic compound. The Diels-Alder reaction is one of the best known examples of a cycloaddition reaction.

Deactivating group *(Section 21.11)* A substituent which, when present in place of a hydrogen atom, decreases the rate of a particular reaction. Notice that the designation *deactivating* for a particular substituent depends on the reaction. (See Section 23.2.)

Decarboxylation *(Section 24.6)* The loss of a molecule of carbon dioxide from a carboxylic acid.

Degree of unsaturation *(Section 5.11)* The number of pairs of hydrogen atoms that must be added to the molecular formula of a compound of unknown structure to obtain the molecular formula of the alkane with the same number of carbon atoms.

Dehydrohalogenation *(Section 12.10)* A reaction in which HX is lost from the starting material. An example is the reaction of a haloalkane with base to form an alkene by the loss of a proton from one carbon atom and a halogen atom from the adjacent one.

Denaturation *(Section 26.11)* Any process that renders an enzyme biologically inactive.

Diastereomers *(Section 6.9)* Stereoisomers that are not mirror images.

1,3-Diaxial interaction *(Section 4.11)* Steric strain between substituents on the same side of a cyclohexane ring.

Diazotization *(Section 22.15)* The reaction of a primary amine with nitrous acid to form a diazonium salt, $RN_2^+X^-$.

Dienophile *(Section 19.12)* A double bond-containing compound that reacts with a diene in a Diels-Alder reaction.

Dipole moment *(Section 1.15)* A measure of the charge separation within a molecule, which is defined as the product of the magnitude of the charges forming the dipoles and the distance between them.

Dipole-dipole interaction *(Section 1.16)* The weak interaction between the negative end of the dipole of one polar molecule and the positive end of the dipole of an adjacent polar molecule.

Disrotatory *(Section 19.10)* The rotation of p orbitals in opposite directions during an electrocyclic reaction.

DNA, deoxyribonucleic acid *(Section 27.1)* A biopolymer consisting of deoxyribonucleotide units.

Double bond *(Section 1.5)* Bond formed by the sharing of two pairs of electrons between adjacent atoms.

E-Z Configuration of alkenes *(Section 7.3)* A system for designating the configuration about a carbon-carbon double bond. Each group on the two carbon atoms of the double bond is assigned a priority according to the Cahn-Ingold-Prelog Rules and the priorities are compared. The isomer in which the group of higher priority on each carbon atom is on the same side of the double bond is designated (*Z*) and the other is designated (*E*).

Eclipsed conformations *(Section 4.3)* Conformations in which bonds to substituents on adjacent carbon atoms are aligned. The C—H bonds on one carbon atom of ethane are aligned with C—H bonds on the adjacent carbon atom in the eclipsed conformation as shown by the following Newman projection formula:

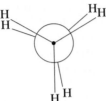

Eclipsed conformation
of ethane

Electrocyclic reaction *(Section 19.10)* The use of light or heat to interconvert a cyclic compound and a conjugated polyene by means of a concerted reorganization of electrons.

Electromagnetic spectrum *(Section 5.2)* The range of electromagnetic radiation from about 10^6 m to 10^{-14} m.

Electron wave function *(Section 1.3)* A mathematical expression that defines the energy of an electron in an atom or a molecule. The square of the wave function is the probability function that defines the shape of orbitals.

Electron-donating group *(Section 7.10)* An atom or group of atoms whose ability to donate electrons to a nucleus is greater than a hydrogen atom.

Electron-withdrawing group *(Section 7.10)* An atom or group of atoms whose ability to withdraw electrons from a nucleus is greater than a hydrogen atom.

Electronegativity *(Section 1.14)* The intrinsic ability of an element to attract electrons to itself. Electronegativity generally increases going from left to right across a period and decreases going down a group in the periodic table so fluorine is the most electronegative element.

Electrophile *(Section 7.9)* Electron-loving or electron-seeking reagent. Electrophiles are also Lewis acids because they react by accepting electrons.

Elementary reaction *(Section 3.13)* A reaction in which only a single molecular event takes place between reactants and products.

1,2-Elimination *(Section 12.10)* Reaction in which a double or triple bond is formed by loss of atoms on adjacent carbon atoms.

Enantiomeric excess *(Section 6.6)* The fractional excess of one enantiomer over the other. A racemic mixture has zero enantiomeric excess while a pure enantiomer has 100% enantiomeric excess.

Enantiomers *(Section 6.1)* Stereoisomers that are nonsuperposable mirror images.

Enantiotopic *(Section 6.14)* Two identical atoms or groups that upon replacement of first one then the other by a different group can give rise to enantiomers.

Endergonic reaction *(Section 3.9)* A chemical reaction with $\Delta G > 0$. An endergonic reaction will not take place spontaneously as written but the reverse reaction is spontaneous.

Endothermic reaction *(Section 3.9)* A chemical reaction that absorbs heat when reactants are converted into products.

Energy of activation $\Delta G^{\ddagger}$ *(Section 3.13)* The difference between the free energy of the reactants and the free energy of the rate-determining transition state. The free energy of activation determines the rate of reaction.

Enol form *(Section 17.1)* Compounds that contain the structure $R_2C{=}CR$ with an OH group on the CR.

Enolate ion *(Section 17.5)* The conjugate base of the keto and the enol form of a carbonyl-containing compound.

Enolization *(Section 17.2)* Process by which enol and keto tautomers are interconverted.

Enthalpy change ΔH *(Section 3.9)* The change of heat content in a process. The enthalpy change in a chemical reaction is the heat lost or gained when reactants are transformed into products.

Entropy change ΔS *(Section 3.9)* The change in amount of molecular disorder in a process. The entropy change in a chemical reaction is a measure of the difference in molecular disorder between the products and reactants.

Enzymes *(Section 3.17)* Giant protein molecules that function as catalysts for biological reactions.

Epimers *(Section 25.3)* Diastereomers that differ in the configuration of only one of their stereocenters.

Epoxidation *(Section 8.6)* The reaction of an alkene with a peroxy acid to form an epoxide (oxirane).

Equatorial bonds *(Section 4.10)* The six C—H bonds of cyclohexane that are roughly parallel to the general plane of the cyclohexane ring.

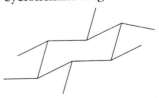

Essential amino acids *(Section 26.1)* The α-amino acids that must be obtained from the food we eat because they cannot be synthesized by the human body.

Excited state *(Section 19.11)* Any electron structure of higher energy than the ground state electron structure of a molecule.

Exergonic reaction *(Section 3.9)* A chemical reaction with $\Delta G < 0$. An exergonic reaction is spontaneous in the direction written.

Exothermic reaction *(Section 3.9)* A term used to describe a chemical reaction that evolves heat when reactants are converted into products.

Fat *(Section 15.3)* A solid triacylglycerol obtained from animal sources.

Fatty acids *(Section 15.3)* The long-chain carboxylic acids obtained along with glycerol by the hydrolysis of animal fats and vegetable oils.

Fingerprint region *(Section 5.5)* The 1400 to 625 cm^{-1} region of an infrared spectrum that is unique for every compound.

First order NMR spectra *(Section 10.7)* Spectra that conform to the $n + 1$ rule of spin-spin coupling.

Fischer projection formula *(Section 6.15)* A projection of a stereocenter onto a flat surface. The four bonds to the stereocenter are represented by a cross. The horizontal lines represent bonds projecting toward the viewer and the vertical lines represent bonds projecting away from the viewer.

Formal charge *(Section 1.6)* The difference in the number of electrons owned by an atom in a molecule and by the same atom in its elemental state. The formal charge on an atom can be calculated by the formula:

Formal Charge on an Atom in a Lewis Structure	=	Number of valence electrons in the isolated atom	−	Number of valence electrons assigned in the Lewis structure

Fragmentation *(Section 18.13)* The process by which a molecular ion breaks apart into many smaller molecular fragments.

Free energy–reaction path diagram *(Section 3.13)* A plot of the conversion of the reactants into products in which the free energy change is plotted as a function of the progress of the reaction along the reaction path.

Frequency *(Section 5.2)* The number of wave crests or troughs that pass a given point in one second.

Functional groups *(Section 2.12)* The groups of atoms and their associated bonds that define the structure and chemical properties of a particular family of compounds.

Gauche conformation *(Section 4.5)* Conformation about a single bond where the largest substituents are 60° apart when viewed in a Newman projection formula.

Gauche conformation of butane

Geminal *(Section 10.7)* Two atoms or substituents that are located on the same atom.

Globular proteins *(Section 26.10)* Water soluble proteins that adopt a spherical shape.

Glycolysis *(Section 25.17)* A sequence of 10 enzyme-catalyzed reactions that converts glucose into pyruvate and ATP.

Glycoside *(Section 25.11)* A carbohydrate acetal formed by the reaction of a carbohydrate cyclic hemiacetal with an alcohol.

Ground state *(Section 19.11)* The lowest energy, most stable electron structure of a molecule.

Haloalkane *(Section 2.9)* Alkane in which one or more hydrogen atoms is replaced by halogen; also called *alkyl halide*.

Halogenation reaction *(Section 2.9)* Chemical reaction of a halogen with an organic compound. The reaction may result in the substitution of a hydrogen atom by a halogen atom or the addition of halogen to the molecule.

Halohydrin *(Section 8.9)* A vicinal halo alcohol.

Halonium ion *(Section 8.8)* A reactive intermediate that contains a positively charged divalent halogen and two carbon atoms in a three-membered ring.

Heterocyclic compounds *(Section 20.7, C)* Cyclic compounds that contain one or more atoms other than carbon atoms.

Heterolytic bond breaking *(Introduction, Chapter 18)* Bond breaking so that one of the atoms of the covalent bond leaves with both electrons. Heterolytic bond breaking forms ions.

HOMO *(Section 19.11)* An acronym for *highest occupied molecular orbital.* The HOMO is the highest energy molecular orbital that contains at least one electron.

Homologous series *(Section 2.5)* A group of structurally related compounds in which any one differs from its neighbors by only one $-CH_2-$ group.

Homolytic bond breaking *(Introduction, Chapter 18)* Bond breaking so that each of the original atoms of the covalent bond leaves with an electron. Homolytic bond breaking forms radicals.

Hückel's rule *(Section 20.5)* Monocyclic conjugated planar cyclic compounds containing $4n + 2$ π electrons show unusual properties associated with aromaticity.

Hund's rule *(Section 1.4)* When orbitals of identical energy are available, electrons occupy these orbitals singly.

Hybrid atomic orbitals *(Section 1.10)* Atomic orbitals obtained by various mathematical combinations of the *s, p,* and *d* ground state atomic orbitals of an atom. (See *sp, sp²,* and *sp³* hybrid atomic orbitals.)

Hydration *(Section 11.5)* Addition of water. In organic chemistry this usually means addition of the elements of water (H, OH) to a multiple bond.

Hydride shift *(Section 7.17)* The intramolecular migration of a hydrogen atom with a pair of electrons from one carbon atom to another.

Hydrocarbons *(Section 2.1)* Compounds that contain only the elements carbon and hydrogen.

Hydrogen bond *(Section 1.16)* The attraction between a hydrogen atom bonded to oxygen or nitrogen atoms in one molecule and the nonbonding electron pair on oxygen or nitrogen atoms in another molecule.

Hydrogenation *(Section 8.11)* The addition of hydrogen to a carbon-carbon double or triple bond to form the saturated product.

Hydrophilic *(Section 11.3)* A term given to substances that are soluble in water (water-loving) but insoluble in nonpolar hydrocarbon-like solvents.

Hydrophobic *(Section 11.3)* A term given to substances that are insoluble in water (water-hating) but soluble in nonpolar hydrocarbon-like solvents.

Hydroxylation *(Section 8.5)* The reaction in which a hydroxy group is added to each of the two carbon atoms of a double bond.

Inductive effect *(Section 7.10)* The electron-donating or electron-withdrawing effect of atoms or groups that is transmitted by polarization of sigma bonds.

Infrared spectroscopy *(Section 5.3)* A spectroscopic technique that uses infrared radiation to detect the functional groups in molecules.

Infrared spectrum *(Section 5.4)* The record of the absorption of light in the infrared region by a compound.

Ion-dipole interactions *(Section 1.16)* Weak attraction between an ion and the oppositely charged end of a polar molecule.

Ionic bonds *(Section 1.14)* A bond formed by the electrostatic attraction between ions of different charge.

Ionophores *(Section 13.7)* Compounds that increase the permeability of cell membranes to sodium, potassium, and other ions.

Isoelectric point *(Section 26.2)* The pH at which an α-amino acid carries no electric charge.

Isomers *(Section 2.3)* Different compounds that have the same molecular formula but different structures.

Isotactic polymer *(Section 18.12)* A polymer whose substituents are all located on the same side of the plane of its carbon atom chain.

Kinetic control of a reaction *(Section 17.14)* Reactions whose major product is the one that is formed the fastest.

Leaving group *(Section 12.1)* The group replaced in a substitution reaction.

Lewis acid *(Section 3.7)* Molecule or ion that reacts by accepting a pair of electrons.

Lewis base *(Section 3.7)* Molecule or ion that reacts by donating a pair of electrons.

Lewis structure *(Section 1.5)* A representation of the structure of a molecule showing covalent bonds as a pair of electron dots between atoms. Two dots (or a line) represent a single bond; four dots (or two lines) represent a double bond; and six dots (or three lines) represent a triple bond.

Line or skeletal structures *(Section 2.2)* A way of writing structural formulas in which neither carbon nor hydrogen atoms are shown explicitly. A carbon atom is assumed to be at the end of each line and at the intersection of any two, three, or four lines. The correct number of hydrogen atoms is assumed to be bonded to each carbon atom to complete the four bonds.

Linear combination of atomic orbitals (LCAO) *(Section 1.13)* A method of obtaining molecular orbitals by the linear combination of all orbitals of the atoms that make up the molecule.

Lipophilic *(Section 11.3)* A term applied to substances that are insoluble in water but soluble in nonpolar hydrocarbon-like solvents (fat-loving).

Lipophobic *(Section 11.3)* A term applied to substances that are soluble in water but insoluble in nonpolar hydrocarbon-like solvents (fat-hating).

Localized valence bond orbital model *(Section 1.9)* The formation of a bond between two atoms by superimposing the wave functions of the valence electrons of the two atoms.

London Forces *(Section 1.16)* Instantaneous molecular dipole moments caused by momentary unsymmetrical electron distribution in a molecule.

LUMO *(Section 19.11)* An acronym for *lowest unoccupied molecular orbital.* The LUMO is the lowest energy molecular orbital that contains no electrons.

Markownikoff rule *(Section 7.13)* Addition of an unsymmetrical reagent to an unsymmetrical double bond occurs so that the positive part of the reagent adds to the carbon atom of the double bond that has the greater number of hydrogen atoms. Another way of stating the rule is that an electrophile reacts with a carbon-carbon double bond to form the more stable carbocation.

Mass number *(Section 1.2)* The total number of protons and neutrons in the nucleus of a particular atom.

Mass spectrometry *(Section 18.13)* A technique that ionizes a molecule and examines the various ions formed by its fragmentation to obtain information about its structure.

Mass spectrum *(Section 18.14)* A plot or tabulation of the relative abundance of each ion versus its mass/charge *(m/z)* ratio.

***meso* compounds** *(Section 6.10)* Molecules that have stereocenters but are achiral because they have a plane of symmetry.

Messenger RNA, mRNA *(Section 27.6)* The RNA that carries the genetic information from DNA to the ribosomes where protein synthesis occurs.

meta- *(Section 21.1)* Substituents *meta-* on a benzene ring are situated in a 1,3-position.

***meta-* directing group** *(Section 21.11)* A substituent on a benzene ring that directs an incoming group to positions *meta-* to itself.

Methylene group *(Section 2.5)* The —CH_2— group.

Micelle *(Section 15.14)* Clusters of molecules that have long hydrophobic tails (organic portion) and hydrophilic heads (ionic portion). The ionic heads of the molecules lie on the outside of the micelle where they interact with water while the hydrophobic tails remain inside the micelle away from water.

Mobile phase *(Section 5.1)* The solvent, either gas or liquid, used in chromatography to transport the mixture through the stationary phase.

Molecular ion *(Section 18.13)* The radical-cation formed by the loss of an electron from the parent molecule.

Molecular orbitals *(Section 1.13)* Orbitals that can encompass more than just two atoms. Molecular orbitals are either bonding, nonbonding, or antibond-

ing. Bonding molecular orbitals are lower in energy than the atomic orbitals used to construct them, nonbonding molecular orbitals are equal in energy, and antibonding molecular orbitals are higher in energy.

Monomers *(Section 18.8)* The stable low molecular weight molecules that combine to form a polymer.

Mutagen *(Section 27.9)* Anything that causes mutation.

Mutarotation *(Section 25.8)* The spontaneous change in the optical rotation observed when a pure anomer is dissolved in water.

Mutation *(Section 27.9)* Any change that alters the sequence of bases in DNA molecules.

Newman projection formula *(Section 4.2)* A method of depicting conformations by means of a projection of the view along the single bond connecting two carbon atoms onto the page. The front carbon atom is represented by a dot and the rear carbon atom by a circle. The remaining bonds are drawn symmetrically around each carbon atom.

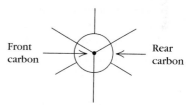

Nonpolar molecule *(Section 1.15)* A molecule that has no dipole moment.

Nuclear magnetic resonance (NMR) spectroscopy *(Section 5.9 and Chapter 10)* A spectroscopic technique that provides information about the carbon skeleton (^{13}C NMR) and the attached hydrogen atoms (^{1}H NMR) of a compound.

Nucleophile *(Section 7.9)* Nucleus-loving or nucleus-seeking reagents. Nucleophiles are also Lewis bases because they react by donating electrons.

Nucleoside *(Section 27.1)* The portion of a nucleic acid that contains a sugar (either ribose or deoxyribose) bonded to a purine or pyrimidine base.

Nucleotide *(Section 27.1)* The monomer unit of DNA and RNA that consists of a sugar (either ribose or deoxyribose) joined to a purine or pyrimidine base and to a phosphate group.

Optical activity *(Section 6.3)* The ability of a compound or substance to rotate plane-polarized light. A compound must be chiral to be optically active.

Optical purity *(Section 6.6)* The ratio of the specific rotation of an optically active sample to the specific rotation of one of its pure enantiomers.

Orbital *(Section 1.3)* A region near an atom or molecule where there is a high probability of finding an electron.

Organometallic compound *(Section 11.7)* Compound that contains a carbon-metal bond.

ortho- *(Section 21.1)* Substituents *ortho-* on a benzene ring are situated in a 1,2-position.

ortho-para directing group *(Section 21.11)* A substituent on a benzene ring that directs an incoming group to positions *ortho-* and *para-* to itself.

Oxonium ion *(Section 14.11)* A general name for a carbocation in which an alkoxy group is bonded to the electron-deficient carbon atom.

para- *(Section 21.1)* Substituents *para-* on a benzene ring are situated in a 1,4-position.

Paraffins *(Section 2.9)* A trivial name for alkanes and cycloalkanes.

Pauli exclusion principle *(Section 1.4)* Only two electrons may exist in the same orbital and they must have opposite spins.

Peptide *(Section 26.3)* A polymer formed by joining two or more α-amino acids by peptide bonds.

Plane of symmetry *(Section 6.1)* An imaginary plane that cuts a molecule (or object) into equal halves that are mirror images of each other. Molecules with a plane of symmetry are achiral.

Polar molecule *(Section 1.15)* A molecule that has a dipole moment.

Polyenes *(Section 19.4)* Hydrocarbons that contain multiple double bonds.

Polymerization *(Section 18.8)* The term used for the chemical reactions that form polymers from monomers.

Polymers *(Section 18.8)* High molecular weight compounds that are made up of repeating low molecular weight units.

Polypeptide *(Section 26.3)* A linear biopolymer composed of many amino acids joined by amide (peptide) bonds.

Primary carbon atom *(Section 2.7)* A carbon atom that is bonded to only one other carbon atom.

Primary structure *(Section 26.6)* The sequence of amino acids in a polypeptide or protein.

Prochiral atom *(Section 6.14)* An atom which upon substitution by a different atom or group creates a stereocenter.

Protecting group *(Section 14.12)* A group that is introduced to temporarily alter a functional group so that it is made inert to the conditions under which a reaction occurs somewhere else in the molecule.

Proteins *(Section 26.3)* Molecules that consist of one or more polypeptide chains.

Protic solvent *(Section 3.15)* Solvents that contain easily exchangeable protons. Water and alcohols that contain —OH groups are examples of protic solvents.

Quaternary carbon atom *(Section 2.7)* A carbon atom that is bonded to four other carbon atoms.

Quaternary structure *(Section 26.10)* The assembly of two or more individual polypeptide chains of a protein held together by noncovalent forces or covalent bonds.

Racemic mixture *(Section 6.5)* An equal mixture of two enantiomers. The racemic mixture is optically inactive because it contains equal concentrations of the two enantiomers.

Radical *(Introduction, Chapter 18)* Species with unpaired electrons.

Radical inhibitor *(Section 18.5)* Compound that prevents a free radical mechanism from occurring by reacting with radicals formed in the propagation step.

Radical initiator *(Section 18.5)* Compound that decomposes to form radicals that can initiate a radical mechanism.

Radical-cation *(Section 18.13)* A species that is both a radical because it contains an unpaired electron and a cation because it has lost an electron.

Rate law *(Section 3.12)* A mathematical expression that relates the rate of a reaction to the concentrations of its reactants raised to various powers.

Rate-determining step *(Section 7.9)* The slowest step in a multistep mechanism. No reaction can occur faster than its rate-determining step.

Reaction mechanism *(Section 3.13)* The path that describes in detail the sequence of steps by which reactants are converted to products.

Rearrangement *(Section 7.17)* The intramolecular migration of an atom or group from one atom to another in a reaction.

Reducing sugar *(Section 25.3, C)* Sugar that reduces Fehling's solution and Tollen's reagent.

Regiospecific reactions *(Section 7.11)* Reactions that can potentially form two or more constitutional isomers but actually form only one.

Relative configuration *(Section 6.8)* The orientation in space of the four different substituents bonded to a stereocenter relative to the configuration of another stereocenter in the same or different molecules.

Replication *(Section 27.5)* The process by which DNA molecules reproduce themselves in the nuclei of cells.

Resolution *(Section 6.13)* The separation of a racemic mixture into its enantiomers.

Resonance effect *(Section 7.10)* The electron-donating or electron-withdrawing effect of atoms or groups that is transmitted by orbital overlap with neighboring π bonds.

Resonance energy *(Section 20.3)* Extent to which a compound is stabilized by electron delocalization.

Resonance hybrid *(Section 1.12)* A molecule whose electron structure cannot be represented by a single Lewis structure. The true electron structure of the molecule is regarded as a hybrid of the various Lewis structures that can be written for the molecule. The various Lewis structures differ only in the positions of the electrons, not their nuclei.

Ribosomal RNA, rRNA *(Section 27.6)* The RNA that combines with proteins to form the intracellular substructures called *ribosomes* where protein synthesis occurs.

Ring current *(Section 20.6)* The electric current generated by the π electrons of an aromatic ring when placed in a magnetic field.

Ring inversion or **ring flip** *(Section 4.10)* The molecular motion that converts one chair conformation of cyclohexane into another chair conformation. Ring inversion converts the axial bonds of one conformation into the equatorial bonds of the other conformation.

RNA, ribonucleic acid *(Section 27.1)* A biopolymer consisting of ribonucleotide units.

Saccharides *(Section 25.1)* Sugars.

Saponification *(Section 15.12)* Hydrolysis of carboxylic acid ester in basic solution to form an alcohol and a carboxylate salt.

Saturated hydrocarbons *(Section 2.1)* Hydrocarbons that contain the maximum number of hydrogen atoms per carbon atom. Another term for *alkanes.*

Secondary carbon atom *(Section 2.7)* A carbon atom that is bonded to two other carbon atoms.

Secondary structure *(Section 26.9)* The conformations of a polypeptide or protein that are repeated regularly along its main chain. The α-helix, the β-pleated sheet and the β-bend are three examples of secondary structures of proteins.

Single bond *(Section 1.5)* Bond formed by the sharing of a single pair of electrons between adjacent atoms.

Soap *(Section 15.14)* Sodium or potassium salts of straight chain carboxylic acids obtained by the hydrolysis of fats in basic solution.

sp **Hybrid atomic orbitals** *(Section 1.10)* Atomic orbitals obtained by combining an *s* and a *p* atomic orbital of about the same energy.

sp^2 **Hybrid atomic orbitals** *(Section 1.10)* Atomic orbitals obtained by combining one *s* and two *p* atomic orbitals of about the same energy.

sp^3 **Hybrid atomic orbitals** *(Section 1.10)* Atomic orbitals obtained by combining one *s* and three *p* atomic orbitals of about the same energy.

Specific rotation [α] *(Section 6.4)* The specific rotation of a chiral compound is a physical constant that is defined as follows:

$$[\alpha] = \frac{[\text{observed rotation in degrees}]}{[\text{path length, l, in dm}]\,[\text{conc., C, in g/mL}]}$$
$$= \frac{\alpha}{l \cdot C}$$

Spin-spin coupling *(Section 10.7)* The splitting of an NMR signal into a multiplet due to coupling of the nuclear spins of neighboring nuclei.

Staggered conformations *(Section 4.3)* Conformations in which bonds to substituents on adjacent carbon atoms are as far away from one another as possible. In the staggered conformation of ethane, the C—H bonds on one carbon atom exactly bisect the bond angles of the adjacent carbon atom as viewed in a Newman projection formula.

Stationary phase *(Section 5.1)* The solid material packed in a chromatography column. The compounds to be separated adsorb to the stationary phase and are eluted by the mobile phase. The stationary phase is usually silica gel (hydrated SiO_2) or alumina (Al_2O_3).

Stereocenter *(Section 6.1)* An atom bearing several groups of such a nature that an interchange of two groups will produce a stereoisomer. Thus a carbon atom bonded to four different groups or atoms is a stereocenter.

Stereoisomers *(Section 4.12)* Isomers whose atoms are connected in the same way but differ in the arrangement of their atoms in space. The term *stereoisomer* includes enantiomers and diastereomers but not constitutional isomers.

Stereospecific reaction *(Section 8.8)* Reaction in which stereoisomeric reactants form stereoisomeric products.

Steric hindrance or steric strain *(Section 4.5)* A repulsive interaction that occurs when two groups are forced closer to each other than their atomic radii permit.

Straight chain or unbranched alkanes *(Section 2.1)* Alkanes whose carbon atoms are joined together in a straight chain.

Syn addition *(Section 8.2)* Addition of two atoms or groups to the two carbon atoms of a double bond from the same face or side of the plane of a double bond.

Syndiotactic polymer *(Section 18.12)* A polymer whose substituents alternate from one side of the plane of its carbon atom chain to the other.

Tautomerism *(Section 9.7)* The process of interconverting tautomers.

Tautomers *(Section 9.7)* Constitutional isomers that are readily interconvertible by rapid equilibration.

Terminal alkyne *(Section 9.2)* Alkynes of the general structure RC≡CH.

Terpenes *(Section 19.14)* Natural products whose total number of carbon atoms is a multiple of five.

Tertiary carbon atom *(Section 2.7)* A carbon atom that is bonded to three other carbon atoms.

Tertiary structure *(Section 26.10)* The biologically active or native conformation of a protein.

Thermodynamic control of a reaction *(Section 17.14)* Reactions that can form several products but the more stable product is formed in preference because the conditions permit the products to equilibrate.

Torsional strain *(Section 4.3)* Destabilization of a molecule caused by the eclipsing of bonds on adjacent atoms.

trans- *(Section 4.12)* A prefix indicating that two substituents are located on opposite sides of a ring or a double bond. (See *cis-*.)

Transamination *(Section 14.15)* The transfer of the amino group of α-amino acids to α-ketoglutarate.

Transcription *(Section 27.6)* The process by which the genetic information on DNA is used to synthesize RNA.

Transfer RNA, tRNA *(Section 27.6)* The RNA that transports specific amino acids to ribosomes where they are joined together to form proteins.

Transition state *(Section 3.13)* The point of maximum energy between reactants and products in an elementary reaction.

Translation *(Section 27.7)* The process by which the information transcribed from DNA to mRNA is used by tRNA to direct protein biosynthesis.

Tree diagram *(Section 10.13)* A diagram that shows the spin-spin couplings responsible for the splitting of NMR signals.

Triple bond *(Section 1.5)* Bond formed by the sharing of three pairs of electrons between adjacent atoms.

Ultraviolet-visible spectroscopy *(Section 19.15)* Spectroscopic technique using ultraviolet and visible radiation to provide information about the extent of π-electron conjugation in organic molecules.

Unsaturated hydrocarbons *(Section 2.1)* Compounds such as alkenes, alkynes, and aromatic compounds that do not contain the maximum number of hydrogen atoms per carbon atom.

Valence shell electron pair repulsion model (VSEPR) *(Section 1.7)* The positions of atoms or groups about a central atom are determined principally by minimizing electron pair repulsion. Four electron pairs arrange themselves into a tetrahedral geometry, three electron pairs adopt a trigonal planar geometry, and two electron pairs assume a linear arrangement.

Vicinal *(Section 8.5)* Two atoms or substituents that are located on adjacent atoms.

Vinylic *(Sections 7.2 and 9.6)* A term that designates the position of a substituent bonded to one of the carbon atoms of a carbon-carbon double bond.

Wavelength *(Section 5.2)* The distance between successive crests or troughs of a wave.

Wavenumber *(Section 5.4)* The conventional unit of infrared spectroscopy defined as the number of wavelengths in one cm.

Ylide *(Section 14.16)* Compound that has a structure with opposite charges on adjacent atoms.

Zwitterion *(Section 26.2)* A neutral dipolar molecule in which the two unlike charges are not adjacent to each other. The zwitterion form of an α-amino acid, for example, contains an ammonium group and a carboxylate group.

INDEX

Characteristic IR Absorption Bands

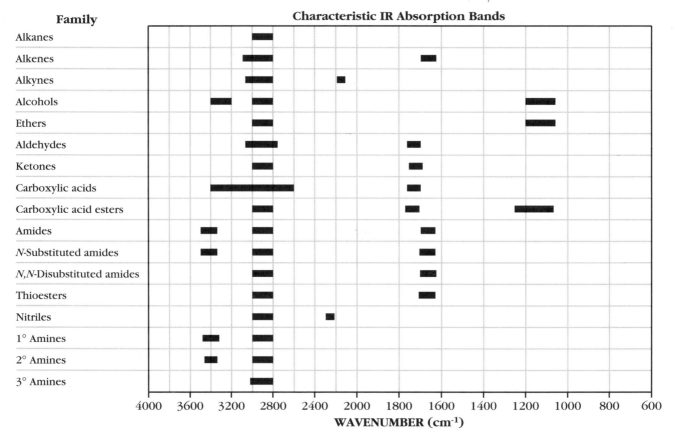

Family
Alkanes
Alkenes
Alkynes
Alcohols
Ethers
Aldehydes
Ketones
Carboxylic acids
Carboxylic acid esters
Amides
N-Substituted amides
N,N-Disubstituted amides
Thioesters
Nitriles
1° Amines
2° Amines
3° Amines

WAVENUMBER (cm^{-1})

Approximate ^{13}C NMR Chemical Shifts

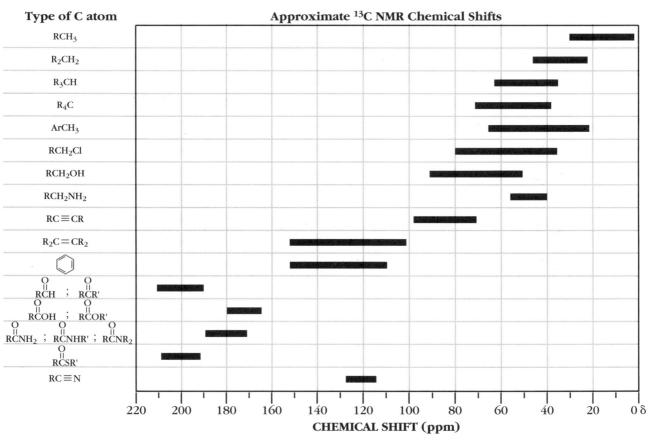

CHEMICAL SHIFT (ppm)